FACILITIES ENGINEERING AND MANAGEMENT HANDBOOK

FACILITIES ENGINEERING AND MANAGEMENT HANDBOOK

Commercial, Industrial, and Institutional Buildings

Paul R. Smith, P.E., P.M.P., M.B.A., M.S.M.E.   Editor in Chief

Principal, Peak Leadership Group, Division of Paul R. Smith & Associates, Boston, Massachusetts

Anand K. Seth, P.E., C.E.M., C.P.E., M.S.M.E.   Editor

Director of Utilities and Engineering, Partners HealthCare System, Inc., Boston, Massachusetts

Roger P. Wessel, P.E.   Editor

Principal, RPW Technologies, Inc., West Newton, Massachusetts

David L. Stymiest, P.E., SASHE, C.E.M., M.E.E.P.E.   Editor

Senior Consultant, Smith Seckman Reid, Inc., New Orleans, Louisiana

William L. Porter, FAIA, Ph.D., M.Arch.   Editor

Professor of Architecture and Planning; Formerly Dean, Massachusetts Institute of Technology School of Architecture and Planning, Cambridge, Massachusetts

Mark W. Neitlich, B.Ch.E., M.B.A.   Editor

Owner, CEO, and Chief Engineer of Chemical Manufacturer (Retd.), New Haven, Connecticut

McGRAW-HILL

New York   San Francisco   Washington, D.C.   Auckland   Bogotá
Caracas   Lisbon   London   Madrid   Mexico City   Milan
Montreal   New Delhi   San Juan   Singapore
Sydney   Tokyo   Toronto

Library of Congress Cataloging-in-Publication Data

Facilities engineering and management handbook : commercial, industrial, and
institutional buildings : Paul R. Smith, editor-in-chief.
    p.   cm.
  ISBN 0-07-059323-X
    1. Plant engineering—Handbooks, manuals, etc. 2. Facility management—
Handbooks, manuals, etc. I. Smith, Paul R.
  TS184 .F35 2000
  658.2—dc21
                                                            00-062452

McGraw-Hill

A Division of The **McGraw·Hill** *Companies*

P/N 0-07-137347-0
Part of
ISBN 0-07-059323-X

*The sponsoring editor for this book was Linda Ludewig and the production supervisor
was Sherri Souffrance. It was set in Times Roman by North Market Street Graphics.*

Printed and bound by R. R. Donnelley & Sons Company.

CONTENTS

Part 2   Facilities Engineering

Chapter 4. Planning and Programming Process

Chapter 5. Engineering and Design Process

Chapter 6. Construction, Modifications/Renovation, and Demolition/Site Restoration   6.1

Chapter 7. Facility Operations, Maintenance, and Assessment

Chapter 8. Codes and Standards

Part 3   Facilities: Buildings and Complexes

Chapter 9. Health Care Facilities

Chapter 10. Laboratories   10.1

Chapter 11. Industrial and Manufacturing Facilities   11.1

Chapter 12. College and University Facilities   12.1

Chapter 13. Airports, Government Installations, and Prisons   13.1

Chapter 14. Data Centers   14.1

STEERING COMMITTEE

Cindy Aiguier *International Facility Management Association Committee, IOI, Inc., Boston, Massachusetts*

Joseph J. Albanese *Vice President, Shawmut Design and Construction, Boston, Massachusetts*

Leonard D. Albano, Ph.D., P.E. *Worcester Polytechnic Institute, Worcester, Massachusetts*

Eugene M. Bard, P.E. *President, BR+A Consulting Engineers, Inc., Boston, Massachusetts*

Bart D. Bauer P.E. *Senior Project Manager/Account Executive, Edwards and Kelcey, Inc., Boston, Massachusetts*

John E. Bender *Manager of Engineering Training, FM Global, Norwell, Massachusetts*

Stephen Biello *Senior Vice President of Facilities, Raytheon Technical Services Company, Burlington, Massachusetts*

Dan P. Bittel, C.P.E. *Association for Facilities Engineering (AFE), Principal, Bittel Enterprises, Inc., Merrimack, New Hampshire*

Michael L. Brainerd, P.E. *Principal, Simpson Gumpertz & Heger, Inc., Arlington, Massachusetts*

Brian Brenner, P.E. *Parsons Brinckerhoff, Boston, Massachusetts*

Harvey J. Bryan, Ph.D., F.A.I.A. *Arizona State University, Tempe, Arizona*

Charles T. Buuck *Vice President and General Manager, Turner Construction Company, Boston, Massachusetts*

Fred Clark, A.I.A. *Architect, Shepley, Bulfinch, Richardson, and Abbott, Boston, Massachusetts*

Donald L. Colavecchio *President and CEO, George B. H. Macomber Company, Boston, Massachusetts*

Douglas Christensen *Brigham Young University, Provo, Utah*

Thomas K. Davies, A.I.A. *Executive Vice President, Vanderweil Facility Advisors, Boston, Massachusetts*

Kathleen A. DeMello *International Facility Management Association Committee, Manager Space Planning and Design, John Hancock, Boston, Massachusetts*

Jim Downey *International Facility Management Association Committee, Stride Rite, Lexington, Massachusetts*

Norman E. Faucher, C.P.E. *New England Regional Vice President and Director (volunteer position), Association for Facilities Engineering (AFE), President, CAC Industries, Leominster, Massachusetts*

Robert W. Fitzgerald *Worcester Polytechnic Institute, Worcester, Massachusetts*

Joseph R. Flocco Jr., P.E. *Sen. Mem. IEEE, Account Manager and Consulting Engineer, Siemens Energy & Automation, Inc., Alpharetta, Georgia*

Bruce Kenneth Forbes *President and CEO, ARCHIBUS, Inc., Boston, Massachusetts*

David C. Foster, Ph.D. P.E. *Director of Engineering, Stone & Webster Engineering Corporation, Boston, Massachusetts*

Steve Glazner *The Association of Higher Education Facilities Officers (APPA), Alexandria, Virginia*

Joseph Griffin, M.S. *Director, Harvard University Environmental Health & Safety, Cambridge, Massachusetts*

Robert A. Gracilleri, P.E.   *President, Shooshanian Engineering Associates, Inc., Boston, Massachusetts*

Larry Healy   *Vice President of Facilities, Blue Cross and Blue Shield of Massachusetts, Boston, Massachusetts*

Walter E. Henry, P.E.   *Vice President, Syska & Hennessy New England, Cambridge, Massachusetts*

Joel Higginson   *President, Advantage Engineering, Wakefield, Massachusetts*

Graham K. Hill, P.E., B.Sc.(Eng.)   *President and CEO, Tomlin, Inc., Falmouth, Massachusetts*

Kristin R. Hill, A.I.A.   *International Facility Management Association Committee Chair, President of IFMA/Boston Chapter, Principal, Design Management Corporation, Natick, Massachusetts*

Kerrie L. Julian   *International Facility Management Association Committee, Project Designer, Design Management Corporation, Natick, Massachusetts*

John F. King, Jr., L.C.S.   *Principal, CID Associates Inc., Boston, Massachusetts*

Harvey A. Kirk, A.I.A.   *Associate, The Stubbins Associates Inc., Cambridge, Massachusetts*

Richard P. Leber, P.E.   *Partner, Cosentini Associates, Cambridge, Massachusetts*

Bruce M. McGregor, P.E.   *Shooshanian Engineering Associates, Inc., Boston, Massachusetts*

Heather G. Merrill, Esq.   *Hill & Barlow, Boston, Massachusetts*

Joseph Milando   *FIT Technologies, Newton Upper Falls, Massachusetts*

Stephen P. Miscowski   *Manager Electrical Operations, Massachusetts Institute of Technology, Cambridge, Massachusetts*

Mark W. Neitlich, M.B.A.   *Owner, CEO, and Chief Engineer of Chemical Manufacturer (Retd.), New Haven, Connecticut*

Christopher L. Noble, Esq.   *Hill & Barlow, Boston, Massachusetts*

William L. Porter, Ph.D., F.A.I.A.   *Professor of Architecture and Planning; formerly Dean, Massachusetts Institute of Technology School of Architecture and Planning, Cambridge, Massachusetts*

Michael K. Powers, P.E.   *President and CEO, Symmes Maini & McKee Associates, Cambridge, Massachusetts*

Cornelius Regan, P.E., CLEP   *Principal, C. Regan Associates, Rockland, Massachusetts; Project Electrical Engineer (retired), Stone & Webster Engineering Corporation, Boston, Massachusetts*

William Ronco, Ph.D.   *President, Gathering Pace, Inc., Bedford, Massachusetts*

Gregory A. Schmellick   *Senior Associate, Symmes Maini & McKee Associates, Cambridge, Massachusetts*

Bonnie Seaberg   *International Facility Management Association Committee, Wolfers Lighting Associates, Allston, Massachusetts*

Anand K. Seth, P.E., C.E.M., C.P.E.   *Director of Utilities and Engineering, Partners HealthCare System, Inc., Boston, Massachusetts*

Kevin T. Sheehan   *Principal, BR+A Consulting Engineers, Inc., Boston, Massachusetts*

Sheila Sheridan, C.F.M., C.P.M.   *International Facility Management Association Committee, Director Facilities and Services, Harvard University John F. Kennedy School of Government, Cambridge, Massachusetts*

Scott Simpson, F.A.I.A.   *Principal, The Stubbins Associates, Cambridge, Massachusetts*

Paul R. Smith, P.E., M.B.A., P.M.P.   *Principal, Peak Leadership Group, Division of Paul R. Smith & Associates, Boston, Massachusetts*

Dennis G. Stafford   *Account Manager Drives, Siemens Energy & Automation, Inc., Alpharetta, Georgia*

Gary Stirgwolt   *Turner Construction Company, Boston, Massachusetts*

Kenneth H. Stowe, P.E.   *Director of Project Services, George B. H. Macomber Company, Boston, Massachusetts*

David L. Stymiest, P.E., SASHE, C.E.M.   *Senior Consultant, Smith Seckman Reid, Inc., New Orleans, Louisiana (was Senior Electrical Engineer at Partners HealthCare System, Inc., Boston, Massachusetts, for much of the preparation of this book).*

A. Eugene Sullivan, P.E.   *President, R. W. Sullivan Company, Inc., Boston, Massachusetts*

Scott S. Tibbo   *International Facility Management Association Committee, Director Real Estate Advisory Services, Expense Management Solutions, Inc., Southboro, Massachusetts*

Gary Vanderweil, P.E.   *President, R. G. Vanderweil Engineers, Inc., Boston, Massachusetts*

Roger P. Wessel, P.E.   *Principal, RPW Technologies, Inc., West Newton, Massachusetts*

William P. Wohlfarth, P.E.   *Formerly President, Mass. Society of Professional Engineers, Senior Electrical Engineer, Massachusetts Institute of Technology, Cambridge, Massachusetts*

Cynthia M. Zawadski   *Director of Marketing Communications, ARCHIBUS, Inc., Boston, Massachusetts*

CONTRIBUTORS

Cindy Aiguier   *International Facility Management Association Committee, IOI, Inc., Boston, Massachusetts*

Matthew C. Adams, P.E.   *APPA, Adams Consulting Group, Wellesley, Massachusetts*

Joseph J. Albanese   *Vice President, Shawmut Design and Construction, Boston, Massachusetts*

Frank Addeman   *Walt Disney Imagineering, Los Angeles, California*

Leonard D. Albano, Ph.D., P.E.   *Worcester Polytechnic Institute, Worcester, Massachusetts*

William P. Bahnfleth, Ph.D., P.E.   *Department of Architectural Engineering, The Pennsylvania State University, State College, Pennsylvania*

Guillermo J. Banchiere   *Director Environmental Services, Massachusetts General Hospital, Boston, Massachusetts*

Eugene M. Bard, P.E.   *President, BR+A Consulting Engineers, Inc., Boston, Massachusetts*

John W. Basch   *Vice President, Cini*Little Schachinger, LLC, Dagsboro, Delaware*

Bart D. Bauer, P.E.   *Senior Project Manager/Account Executive, Edwards and Kelcey, Inc., Boston, Massachusetts*

Steve Bauer   *Assistant Vice President, CGU, Foxborough, Massachusetts*

Richard T. Battles, P.E.   *President and Treasurer, Thompson Consultants, Inc., Marion, Massachusetts*

Janet Baum, A.I.A.   *Principal, HERA, Inc., St. Louis, Missouri*

David Beard   *RTKL Associates, Baltimore, Maryland*

Franklin Becker, Ph.D.   *Professor, Cornell University, Ithaca, New York*

John E. Bender   *Manager of Engineering Training, FM Global, Norwell, Massachusetts*

Robert S. Berens, INCE   *Acentech, Inc., Cambridge, Massachusetts*

Steven L. Bernstein, P.E.   *Daylor Consulting Group, Inc., Braintree, Massachusetts*

Joseph Best   *CEO, Project Management Services, Inc., Atlanta, Georgia*

Stephen Biello   *Senior Vice President of Facilities, Raytheon Technical Services Company, Burlington, Massachusetts*

Richard P. Bingham   *Manager, Products/Technology, Dranetz-BMI, Edison, New Jersey*

David Blanchard, M.B.A.   *Blanchard Management Associates, Inc., Hingham, Massachusetts*

Dan P. Bittel, C.P.E.   *Association for Facilities Engineering (AFE), Principal, Bittel Enterprises, Inc., Merrimack, New Hampshire*

Michael L. Brainerd, P.E.   *Principal, Simpson Gumpertz & Heger, Inc., Arlington, Massachusetts*

Brian Brenner, P.E.   *Parsons Brinckerhoff, Boston, Massachusetts*

Robert A. Broder, A.I.A.   *Monacelli Associates, Inc., Cambridge, Massachusetts*

Richard G. Brunner, P.E.   *Principal Engineer, Stone & Webster Engineering Corporation, Boston, Massachusetts*

Harvey J. Bryan, Ph.D., F.A.I.A.   *Arizona State University, Tempe, Arizona*

Charles T. Buuck   *Vice President and General Manager, Turner Construction Company, Boston, Massachusetts*

Donna K. Burnette   *Senior Vice President, Paradigm Learning, Tampa, Florida*

Mary Cancian   *Taylor and Partners, Inc., Boston, Massachusetts*

Robert S. Capaccio, P.E.   *President, Capaccio Environmental Co., Inc., Sudbury, Massachusetts*

Daniel J. Caufield   *Chief Electrician, Massachusetts General Hospital, Boston, Massachusetts*

Robert D. Celmer, Ph.D., P.E.   *University of Hartford, West Hartford, Connecticut*

Kenneth S. Charest, CIPE, CET   *Associate, Robert W. Sullivan Company, Inc., Boston, Massachusetts*

Dan Chisholm   *Publisher, "Healthcare Circuit News"; President, Motor and Generator Institute, Winter Park, Florida*

Douglas Christensen   *Brigham Young University, Provo, Utah*

Michael Clancy   *IOI, Inc., Boston, Massachusetts*

Margaret Dempsey Clapp, M.S.   *Director of Pharmacy Operations, Massachusetts General Hospital, Boston, Massachusetts*

Fred Clark, A.I.A.   *Architect, Shepley, Bulfinch, Richardson, and Abbott, Boston, Massachusetts*

Robert Clark   *Vice President, BCM Control Systems, Woburn, Massachusetts*

Donald L. Colavecchio   *President and CEO, George B. H. Macomber Company, Boston, Massachusetts*

Michael Colburn   *SES Company, Inc. (SESCO), Hingham, Massachusetts*

Steve Collins   *R. G. Vanderweil Engineers, Inc., Boston, Massachusetts*

Peter Coxe, A.I.A.   *Peter Coxe Associates, Marblehead, Massachusetts*

George D. Coombs, P.E.   *Symmes Maini & McKee Associates, Cambridge, Massachusetts*

Denis R. Cormier   *Facilities Management Solutions, Inc., San Diego, California*

Thomas K. Davies, A.I.A.   *Executive Vice President, Vanderweil Facility Advisors, Boston, Massachusetts*

Robert F. Daylor, P.E.   *Daylor Consulting Group, Inc., Braintree, Massachusetts*

Jack W. Dean, P.E.   *Electrical Department Head, Novare Engineers, Inc., Providence, Rhode Island*

Robert V. DeBonis   *Associate, Robert W. Sullivan Company, Inc., Boston, Massachusetts*

Kathleen A. DeMello   *International Facility Management Association Committee, Manager Space Planning and Design, John Hancock, Boston, Massachusetts*

Harold J. De Monaco, M.S.   *Director of Drug Therapy Management, Massachusetts General Hospital, Massachusetts General Physicians Organization, Boston, Massachusetts*

Stephen Denker   *GTE Internetworking*

Richard de Neufville, Ph.D.   *MIT Technical and Policy Program, Massachusetts Institute of Technology, Cambridge, Massachusetts*

Charles Dever   *SES Company, Inc. (SESCO), Hingham, Massachusetts*

Louis DiBerardinis   *Industrial Hygiene Office, Massachusetts Institute of Technology, Cambridge, Massachusetts*

Roderick A. Dike   *Honeywell, Boston, Massachusetts*

Anthony T. DiStefano, P.E., CIPE   *Principal, Robert W. Sullivan Company, Inc., Boston, Massachusetts*

David A. Dow   *Director Facility Engineering, Walt Disney Imagineering, Anaheim, California*

Jim Downey   *International Facility Management Association Committee, Stride Rite, Lexington, Massachusetts*

John J. Downing, P.M.P.   *Project Manager Network and Systems Integration Services, Compaq Computer Corporation, Lexington, Massachusetts*

Gregory O. Doyle   *Director Buildings and Grounds Department, Massachusetts General Hospital, Boston, Massachusetts*

James L. Drinkard, P.E.   *HNTB Corporation, Atlanta, Georgia*

John Driscoll, CHPA   *Assistant Director Department of Police, Security and Parking, Massachusetts General Hospital, Boston, Massachusetts*

Sean Duffy, B.S.   *Robert W. Sullivan, Inc., Boston, Massachusetts*

William Duncan, P.M.P.   *Project Management Partners, Lexington, Massachusetts*

Paul J. DuPont, P.E.   *President, DuPont Dobbs & Kearns Engineers, LLC, Portland, Oregon*

Michael Dykens   *Environmental Health and Engineering, Inc., Newton, Massachusetts*

Robert Edwards   *Industrial Hygiene Office, Massachusetts Institute of Technology, Cambridge, Massachusetts*

David M. Elovitz, P.E.   *President, Energy Economics, Inc., Natick, Massachusetts*

Jorge A. Emmanuel, Ph.D., P.E., CHMM   *The Environmental and Engineering Research Group; Consultant, EPRI Healthcare Initiative, Palo Alto, California*

Douglas Erickson, FASHE   *President, DSE Consulting LLC, Christiansted, Virgin Islands*

Norman E. Faucher, C.P.E.   *New England Regional Vice President and Director (Volunteer Position), Association for Facilities Engineering (AFE), President, CAC Industries, Leominster, Massachusetts*

Harold Feinleib   *Aperture Technologies, Inc., Stamford, Connecticut*

Kimball Ferguson, P.E.   *Engineering Manager, Duke Medical Center, Durham, North Carolina*

Friedrich W. Finger III, P.E.   *The Finger Company, Inc., Duxbury, Massachusetts*

Richard Fink   *Biosafety Office, Massachusetts Institute of Technology, Cambridge, Massachusetts*

Robert W. Fitzgerald   *Worcester Polytechnic Institute, Worcester, Massachusetts*

Joseph R. Flocco Jr., P.E.   *Sen. Mem. IEEE, Account Manager and Consulting Engineer, Siemens Energy & Automation, Inc., Alpharetta, Georgia*

Robert Foley, P.E.   *Consultant, Duxbury, Massachusetts*

Allan R. Forbes   *ARCHIBUS, Inc., Charlotte, North Carolina*

Bruce Kenneth Forbes   *President and CEO, ARCHIBUS, Inc., Boston, Massachusetts*

Janice E. Forbes   *Global Practices Group, IBM, Charlotte, North Carolina*

Julia M. Forbes   *ARCHIBUS, Inc., Boston, Massachusetts*

Richard E. Forbes   *Retired IBM, Concord, North Carolina*

Edwin J. Fortini   *Senior Electrical Engineer, Siemens-Westinghouse Technical Services, Inc., Needham, Massachusetts*

James W. Fortune, M.B.A.   *Chairman and Chief Executive Officer, Lerch Bates & Associates, Inc., Littleton, Colorado*

David C. Foster, Ph.D., P.E.   *Director of Engineering, Stone & Webster Engineering Corporation, Boston, Massachusetts*

Richard A. Fricklas   *Roofing Industry Educational Institute, Englewood, Colorado*

Darryl E. Galletti, P.E.   *Senior Associate, R. G. Vanderweil Engineers, Inc., Boston, Massachusetts*

Mitchell Galanek   *Radiation Safety Office, Massachusetts Institute of Technology, Cambridge, Massachusetts*

Daniel Gainsboro   *President, Genesis Planning and Delphi Construction, Waltham, Massachusetts*

Frederick M. Gibson, A.I.A.   *Partner, Taylor and Partners, Inc., Boston, Massachusetts*

Joseph Gibson   *Manager, Duke Hospitals, Durham, North Carolina*

Steve Glazner   *The Association of Higher Education Facilities Officers (APPA), Alexandria, Virginia*

Sheridan A. Glen   *Principal, Thermetric, Inc., Madison, Wisconsin*

Peter S. Glick   *Symmes Maini & McKee Associates, Cambridge, Massachusetts*

Leslie A. Glynn, A.I.A.   *Symmes Maini & McKee Associates, Cambridge, Massachusetts*

William J. Goode, P.E., C.P.M.   *Principal and President, W. J. Goode Corporation, Walpole, Massachusetts*

Harold L. Gordon, P.E.   *Principal Engineer, Stone & Webster Engineering Corporation, Boston, Massachusetts*

Glen J. Goss, B.S.E.E.   *GJ Associates, Stow, Massachusetts*

Dana L. Green   *Substation Test Company, Forestville, Maryland*

Joseph Griffin, M.S.   *Director, Harvard University Environmental Health & Safety, Cambridge, Massachusetts*

Robert A. Gracilieri, P.E.   *President, Shooshanian Engineering Associates, Inc., Boston, Massachusetts*

Thomas W. Grottke, C.P.A.   *KPMG, Hartford, Connecticut*

Brian Hagopian   *Fluid Solutions, Inc., Boston, Massachusetts*

Patrick E. Halm, P.E.   *Symmes Maini & McKee Associates, Cambridge, Massachusetts*

David J. Hanitchak, R.A.   *Director Planning and Construction, Massachusetts General Hospital, Partners HealthCare System, Inc., Boston, Massachusetts*

Charles R. Harrell   *ProModel Corporation, Oren, Utah*

Mark Hasso, Ph.D., P.E.   *Wentworth Institute of Technology, Boston, Massachusetts*

Kevin S. Hastings   *Sullivan Code Group, Boston, Massachusetts*

David N. Hayes, P.E.   *BSC Group, Boston, Massachusetts*

Larry Healy   *Vice President of Facilities, Blue Cross and Blue Shield of Massachusetts, Boston, Massachusetts*

Penny Henderson-Maher   *Lightolier, Div. of Genlyte Thomas, Garland, Texas*

Walter E. Henry, P.E.   *Vice President, Syska & Hennessy New England, Cambridge, Massachusetts*

Brian T. Herteen, M.B.A.   *Johnson Controls, Inc., Lynnfield, Massachusetts*

Michael Hickey, P.E.   *Partner, Thompson Consultants, Inc., Marion, Massachusetts*

Joel Higginson   *President, Advantage Engineering, Wakefield, Massachusetts*

Graham K. Hill, P.E., B.Sc.(Eng.)   *President and CEO, Tomlin, Inc., Falmouth, Massachusetts*

Kristin R. Hill, A.I.A.   *International Facility Management Association Committee Chair, President of IFMA/Boston Chapter, Principal, Design Management Corporation, Natick, Massachusetts*

Yolanda Hill   *Vice President of Development, Facility Information Systems, Inc., Camarillo, California*

Jack Hug   *APPA, University of California at San Diego, San Diego, California*

William Hughes   *Johnson Controls, Inc., Lynnfield, Massachusetts*

David Hutchens   *Consultant, Paradigm Learning, Tampa, Florida*

Vincent Iannucci   *Aperture Technologies, Inc., Stamford, Connecticut*

Janice F. Jonas   *ARCHIBUS, Inc., Boston, Massachusetts*

Dennis R. Julian, P.E., RCDD   *Communications Group Leader, Carter and Burgess, Inc., Boston, Massachusetts*

Kerrie L. Julian   *International Facility Management Association Committee, Project Designer, Design Management Corporation, Natick, Massachusetts*

Harvey H. Kaiser, M.Arch, R.A., Ph.D.   *President, HHK and Associates, Inc., Syracuse, New York*

Charles Kalasinsky, P.E.   *R. G. Vanderweil Engineers, Inc., Boston, Massachusetts*

Michael J. Kearns, C.C.S., R.C.I.   *Senior Project Manager, CID Associates Inc., Boston, Massachusetts*

James King   *Senior Fire Protection Specialist, Office of State Fire Marshal, Delaware Fire Service Center, State of Delaware, Dover, Delaware*

John F. King, Jr., L.C.S.   *Principal, CID Associates Inc., Boston, Massachusetts*

Eugene B. Kingman, CIPE   *Principal, Robert W. Sullivan, Inc., Boston, Massachusetts*

Harvey A. Kirk, A.I.A.   *Associate, The Stubbins Associates Inc., Cambridge, Massachusetts*

Fred Klammt   *Aptek Associates, Plymouth, California*

Paul Konz, P.E.   *Electrical Department Head, The Ritchie Organization, Newton, Massachusetts*

Edwin A. Kotak, Jr., P.E., C.E.T., C.I.P.E.   *Senior Associate, Robert W. Sullivan, Inc., Boston, Massachusetts*

Lawrence P. Lammers, P.E.   *Lammers & Associates, Reston, Virginia*

Brian W. Lawlor, P.E.   *Symmes Maini & McKee Associates, Cambridge, Massachusetts*

Dawn LeBaron, CHPA   *Director Facilities Services, Newton-Wellesley Hospital, Newton, Massachusetts*

Richard P. Leber, P.E.   *Partner, Cosentini Associates, Cambridge, Massachusetts*

Olga Leon, P.E.   *President, Diversified Concepts, Sharon, Massachusetts*

Bernard T. Lewis, P.E., C.P.E., D.B.A.   *Consultant, Potomac, Maryland*

Joan Licitra, Esq.   *The Law Office of Joan Licitra, Boston, Massachusetts*

Marshall Long, Ph.D., P.E.   *Marshall Long Acoustics, Sherman Oaks, California*

Vaughn L. Lotspeich, P.E.   *Project Engineering Department Manager, INEEL-Bechtel BWXT Idaho LLC, Idaho Falls, Idaho*

Stephen K. Lowe   *Environmental Science Faculty, Art Center College of Design, Pasadena, California; Creative Director, Facility Information Systems, Inc., Camarillo, California*

John Ludwig   *Environmental Health and Engineering, Newton, Massachusetts*

Scott Maddern   *Vice President of Sales and Marketing, Integris, Billerica, Massachusetts*

Richard Marchi   *President Technical Affairs, Airports Council International, Washington, DC*

Frank Massaro   *Director of Pharmacy, North Shore Medical Center, Salem, Massachusetts*

Harrison McCampbell, A.I.A.   *Principal, McCampbell & Associates, Brentwood, Tennessee*

John F. McCarthy, Sc.D., C.I.H.   *Environmental Health and Engineering, Inc., Newton, Massachusetts*

Heather C. McCormack   *Symmes Maini & McKee Associates, Cambridge, Massachusetts*

Bruce M. McGregor, P.E.   *Shooshanian Engineering Associates, Inc., Boston, Massachusetts*

Howard McKew, P.E., C.P.E.   *Principal, Sebesta Blomberg and McKew, Topsfield, Massachusetts*

Thomas F. McNamara, R.C.D.D.   *Manager of Telecommunication/Data Engineering Services, BR+A Consulting Engineers, Inc., Boston, Massachusetts*

Thomas J. McNicholas   *Consultant; Commissioner of Inspectional Services (Retired), City of Boston, Massachusetts, Charlestown, Massachusetts*

Heather G. Merrill, Esq.   *Hill & Barlow, Boston, Massachusetts*

Bonnie Michelman, C.P.P., C.H.P.A.   *Director Department of Police, Security and Parking, Massachusetts General Hospital, Boston, Massachusetts*

Joseph Milando   *FIT Technologies, Newton Upper Falls, Massachusetts*

Kurt Milligan, P.E., INCE   *Acentech, Inc., Cambridge, Massachusetts*

Stephen P. Miscowski   *Manager Electrical Operations, Massachusetts Institute of Technology, Cambridge, Massachusetts*

Arthur Mombourquette   *Director Environmental Services, Brigham & Womens Hospital, Boston, Massachusetts*

Diane L. Morgan, M.P.A.   *Senior Telecommunications Analyst, BR+A Consulting Engineers, Boston, Massachusetts*

John Morganti   *International Facility Management Association, Boston, Massachusetts*

Robert Morin, P.E.   *Pizzagalli Construction Company, South Burlington, Vermont*

Nanette E. Moss, M.S., C.I.H.   *Environmental Health and Engineering, Inc., Newton, Massachusetts*

Mark W. Neitlich, M.B.A.   *Owner, CEO, and Chief Engineer of Chemical Manufacturer (Retd.), New Haven, Connecticut*

Christopher L. Noble, Esq.   *Hill & Barlow, Boston, Massachusetts*

John M. Nevison   *Oak Associates, Inc., Maynard, Massachusetts*

Ronald P. O'Brien, P.E.   *Daylor Consulting Group, Inc., Braintree, Massachusetts*

H. Thomas O'Hara, Ph.D.   *Suffolk University, Boston, Massachusetts*

Roland Ouellette   *President, REB Training International, Stoddard, New Hampshire*

Joe Akinori Ouye, Ph.D.   *Director of Strategic Planning, Gensler, San Francisco, California*

James C. Parker, P.E.   *Simpson Gumpertz & Heger, Inc., Arlington, Massachusetts*

Arlene Parquette   *Fluid Solutions, Inc., Boston, Massachusetts*

Roy A. Pedersen, R.A.   *The Stubbins Associates, Inc., Cambridge, Massachusetts*

Syed M. Peeran, Ph.D., P.E.   *Sen. Mem. IEEE, Stone & Webster Engineering Corporation, Boston, Massachusetts*

George Player   *Director of Engineering, Brigham and Women's Hospital, Boston, Massachusetts*

William L. Porter, Ph.D., F.A.I.A.   *Professor of Architecture and Planning; Formerly Dean, Massachusetts Institute of Technology School of Architecture and Planning, Cambridge, Massachusetts*

Michael K. Powers, P.E.   *President and CEO, Symmes Maini & McKee Associates, Cambridge, Massachusetts*

Arjun B. Rao, P.E.   *Vice President, BR+A Consulting Engineers, Boston, Massachusetts*

Cornelius Regan, P.E., CLEP   *Principal, C. Regan Associates, Rockland, Massachusetts; Project Electrical Engineer (retired), Stone & Webster Engineering Corporation, Boston, Massachusetts*

Michael J. Reilly   *Symmes Maini & McKee Associates, Cambridge, Massachusetts*

Douglas T. Reindl, Ph.D., P.E.   *HVAC&R Center & Thermal Storage Applications Research Center, University of Wisconsin, Madison, Wisconsin*

James B. Rice, Jr., M.B.A.   *Massachusetts Institute of Technology, Cambridge, Massachusetts*

David P. Richards, C.Eng.   *Ove Arup & Partners, New York, New York*

Lawrence F. Richmond   *Vice President, Bennett Electrical, Inc., Quincy, Massachusetts*

Paul J. Rinaldi   *Director Space Management Department, Boston University, Boston, Massachusetts*

A. Todd Rocco, P.E. *Principal, Diversified Consulting Engineers, Quincy, Massachusetts*

Cathleen M. Ronan, A.I.A. *Symmes Maini & McKee Associates, Cambridge, Massachusetts*

William Ronco, Ph.D. *President, Gathering Pace, Inc., Bedford, Massachusetts*

Jack Rose *Robert W. Sullivan Company, Inc., Boston, Massachusetts*

Carl J. Rosenberg, A.I.A. *Acentech, Inc., Cambridge, Massachusetts*

J. Lyndon Rosenblad, Ph.D., P.E. *Consulting Engineer, Westwood, Massachusetts*

Christopher Russo, P.E. *Partner, Newcomb & Boyd, Atlanta, Georgia*

John Saad *Senior Vice President, R. G. Vanderweil Engineers, Inc., Boston, Massachusetts*

Kurt Samuelson *Beacon-Skanska USA, Boston, Massachusetts*

Nancy Johnson Sanquist, C.F.M., IFMA Fellow, Associate A.I.A. *Director of Strategic Initiatives, Peregrine Systems, Inc., San Diego, California*

William J. Schlageter, P.E. *Principal, Novare Engineers, Inc., Providence, Rhode Island*

Gregory A. Schmellick *Senior Associate, Symmes Maini & McKee Associates, Cambridge, Massachusetts*

Bonnie Seaberg *International Facility Management Association Committee, Wolfers Lighting Associates, Allston, Massachusetts*

Stephen M. Sessler, P.E. *Newcomb and Boyd, Atlanta, Georgia*

Anand K. Seth, P.E., C.E.M., C.P.E. *Director of Utilities and Engineering, Partners HealthCare System, Inc., Boston, Massachusetts*

Kevin T. Sheehan *Principal, BR+A Consulting Engineers, Inc., Boston, Massachusetts*

Sheila Sheridan, C.F.M., C.P.M. *International Facility Management Association Committee, Director Facilities and Services, Harvard University John F. Kennedy School of Government*

Scott Simpson, F.A.I.A. *Principal, The Stubbins Associates, Cambridge, Massachusetts*

William Sims, Ph.D., C.F.M., IFMA Fellow *Professor, Cornell University, Ithaca, New York*

Jason Slibeck, M.Eng. *Massachusetts Institute of Technology, Organic, Inc., Cambridge, Massachusetts*

Paul R. Smith, P.E., M.B.A., P.M.P. *Principal, Peak Leadership Group, Division of Paul R. Smith & Associates, Boston, Massachusetts*

Teerachai Srisirikul, P.E., C.E.M., C.D.S.M. *Senior Mechanical Engineer, Partners Healthcare System, Inc., Boston, Massachusetts*

Dennis G. Stafford *Account Manager Drives, Siemens Energy & Automation, Inc., Alpharetta, Georgia*

Charles Stein, P.E. *R. G. Vanderweil Engineers, Inc., Boston, Massachusetts*

John A. Stevermer, A.I.A. *Symmes Maini & McKee Associates, Cambridge, Massachusetts*

Blake Stewart *Johnson Controls, Inc., Lynnfield, Massachusetts*

Gary Stirgwolt *Turner Construction Company, Boston, Massachusetts*

Kenneth H. Stowe, P.E. *Director of Project Services, George B. H. Macomber Company, Boston, Massachusetts*

Michael Stowe *The Boeing Corporation, Seattle, Washington*

Thomas Stratford *McClier Group, Chicago, Illinois*

Douglas H. Sturz, INCE *Acentech, Inc., Cambridge, Massachusetts*

David L. Stymiest, P.E., SASHE, C.E.M. *Senior Consultant, Smith Seckman Reid, Inc., New Orleans, Louisiana (was Senior Electrical Engineer at Partners HealthCare System, Inc., Boston, Massachusetts for much of the preparation of this book.)*

Eugene Sullivan, P.E., F.A.S.C., F.A.C.E.C.   *Chairman, Robert W. Sullivan, Inc., Boston, Massachusetts*

Mark J. Sullivan, C.E.T., C.I.P.E., C.D.A.   *Treasurer, Robert W. Sullivan, Inc., Boston, Massachusetts*

Paul D. Sullivan, P.E.   *President, Sullivan Code Group, Robert W. Sullivan, Inc., Boston, Massachusetts*

Michael J. Sweeney, P.E.   *Symmes Maini & McKee Associates, Cambridge, Massachusetts*

J. Richard Swistock   *The Association of Higher Education Facilities Officers (APPA)*

Peter K. Taylor, P.E.   *Geotechnical Consultant, Weston, Massachusetts*

Raymond J. Taylor   *Director, Housekeeping Management Services, Inc., South Hamilton, Massachusetts*

Eric Teicholz   *President, Graphic Systems, Inc., Cambridge, Massachusetts*

Scott S. Tibbo   *International Facility Management Association Committee, Director Real Estate Advisory Services, Expense Management Solutions, Inc., Southboro, Massachusetts*

Nicholas N. Timpko, A.I.A., C.S.I.   *Anderson-Nichols/Goodkind & O'Dea, Boston, Massachusetts*

Robert Trenck   *Aperture Technologies, Inc., Stamford, Connecticut*

Kevin Tunsley   *Symmes Maini & McKee Associates, Cambridge, Massachusetts*

James O. Turner, P.E., C.E.M.   *Utilities Manager, Partners HealthCare System, Inc., Boston, Massachusetts*

Gary Vanderweil, P.E.   *President, R. G. Vanderweil Engineers, Inc., Boston, Massachusetts*

William VanSchalkwyk   *Safety Office, Massachusetts Institute of Technology, Cambridge, Massachusetts*

Dennis L. Wagner, P.M.P.   *formerly Vice President of Operations, Stone & Webster Communications Services Group, Boston, Massachusetts*

Kenneth S. Weinberg, M.Sc., Ph.D.   *Director of Safety, Massachusetts General Hospital, Boston, Massachusetts*

Roger P. Wessel, P.E.   *Principal, RPW Technologies, Inc., West Newton, Massachusetts*

Christopher Wheeler   *Johnson Controls, Inc., Lynnfield, Massachusetts*

John P. White   *Department of Police, Security and Parking, Massachusetts General Hospital, Boston, Massachusetts*

William P. Wohlfarth, P.E.   *formerly President, Mass. Society of Professional Engineers, Senior Electrical Engineer, Massachusetts Institute of Technology, Cambridge, Massachusetts*

A. Vernon Woodworth, A.I.A., C.B.O.   *Sullivan Code Group, Boston, Massachusetts*

Cynthia M. Zawadski   *Director of Marketing Communications, ARCHIBUS, Inc., Boston, Massachusetts*

FOREWORD

David A. Dow, Director Facility Engineering
Walt Disney Imagineering, Anaheim, California

The Walt Disney organization recognizes the importance of facilities engineering and management. As the company becomes global, more theme parks are being designed and built to keep up with the demand for the Disney product. It has become a monumental task to dream up, design, construct, operate, and maintain the facility assets of the company. Throughout the process, from the early conceptual plan through to the maintenance of the entire physical plant, the goal of every "cast" member is to cater to "guest" satisfaction.

The facilities manager has evolved from being a mechanic in the early 1900s to a highly skilled, multidisciplined individual. Through all of the phases mentioned, the facilities manager must have the same knowledge as the mechanic and must also be computer literate, Internet comfortable, a planner, a scheduler, a motivator of people, an integrator, a problem solver, and probably the biggest change of all, a great manager. To wear all of the hats required, the facilities managers must have the management skills to lead a team of staff that will pull together all of the required facets of their businesses. Now, they probably sit at the boardroom table with the other department heads that make company decisions. The budget for facility management is probably looked at in the most detail and needs to be justified by payback analysis, comparable numbers, and generally, the amount of money every dollar spent will save. This can be difficult in maintenance functions where not performing the tasks can mostly be measured only in problems that may show up in the years ahead.

The Disney experience and approach exemplify the thrust of this book, which is that one must always give more consideration to the individual attributes of modern facilities to succeed in our present competitive business environment. Facilities engineering and facilities management now interrelate as never before. Please come along with us for a moment as we explore the Disney experience. Consider how the Disney experience relates to your own facility experience by substituting "customer" for "guest," "employee" for "cast member," and "process, industry, product, or business" for "show, magic, or attraction." We believe that you will be able to draw parallels to your own situation quite easily.

In its way, the Disney environment sets a standard that any organization must meet. Just as the guests in any Disney environment must be transported into their world of fantasy, employees and customers in any work or commercial environment must be well served to achieve their highest aspirations and to fulfill the missions of the host organizations. Walt Disney Imagineering is the design arm of the Disney organization. Within its hallways, creative departments coexist with the engineering groups, and themed designers and architects mingle with the structural engineers. A synergy exists on campus at the heart of the operation that takes creative ideas and produces new attractions from simple brick and mortar, that immerses the guest into a fantasy location where, for a short period of time, the rigors of ordinary life are left behind. Facility engineers are just one member of this team that is under constant pressure to provide the expected product, also to keep costs down, and to use the most efficient method and technology throughout the process.

Once an attraction is open and the guests are enjoying the magic, the monumental task of operations and maintenance takes over. All attractions must maintain a very high operational rate, similar to many other modern facilities. Attractions cannot be allowed to be inoperative. Guests expect to see certain attractions when they enter a park and any downtime can lead to guest dissatisfaction. Preventive maintenance tasks must be scheduled and cycled periodically, depending on the type of ride or show. All scenes in a show must be fully operational or the show is considered down. (The storyline must be maintained.) When a show or attraction does go 101 (inoperative), crews must be available to repair the problem and bring the facility back on line as soon as practically possible. The readiness of the parks and facilities must all be maintained, as mentioned before, despite the added complications of holding down utility costs, minimizing maintenance crews, and obtaining the best possible replacement parts for the least amount of money.

In today's economy, the shareholders expect growth—growth coupled with saving costs wherever possible. This, of course, cannot in any way compromise the safety and enjoyment of the guests. Growth is not possible without repeat visits from our guests.

Each one of the areas discussed—design, operation, and maintenance—now takes place for the Disney organization on a global scale. Design for each facility may be happening at any one time in almost anywhere in the world. The architect may reside in one place in the United States, and the engineers may be located either in one office or in several offices, in another area of the country. In the meantime, coordination is needed with a ride vendor in Europe and possibly other special effects and show-related items in another part of the world. As the Internet becomes a factor in design coordination and collaboration, a whole new business of e-commerce has developed that provides the project team with methods for pulling the entire package together.

This book pulls together the different knowledge bases of facilities engineering and management and offers a valuable tool to anyone involved in the fast-growing and evolving profession of facility management.

FACILITIES ENGINEERING AND MANAGEMENT HANDBOOK

P · A · R · T · 1

FACILITIES MANAGEMENT

Paul R. Smith, P.M.P., P.E., M.B.A., Chapter Editor
Peak Leadership Group, Boston, Massachusetts

Anand K. Seth, P.E., C.E.M., Chapter Editor
Partners HealthCare System, Inc., Boston, Massachusetts

Roger P. Wessel, P.E., Chapter Editor
RPW Technologies, Inc., West Newton, Massachusetts

David L. Stymiest, P.E., SASHE, C.E.M., Chapter Editor
Smith Seckman Reid, Inc., New Orleans, Louisiana

William L. Porter, Ph.D., F.A.I.A., Chapter Editor
Massachusetts Institute of Technology, Cambridge, Massachusetts

Mark W. Neitlich, M.B.A., Owner, CEO, and Chief Engineer of Chemical Manufacturing (Retired), Chapter Editor
New Haven, Connecticut

Part 1 deals with facilities management. We begin in Chap. 1 by discussing contemporary pressures on organizations and then setting out the challenges for facilities management. In Chap. 1, we also suggest how facilities management can be related to corporate strategy and indicate the need to see facilities management as part of a broader picture of corporate infrastructure. We then look at financial analysis and reporting in Chap. 2, presenting necessary background and concepts, as well as methods of accounting, budgeting, and evaluating projects. We also lay out the new roles of facilities managers in their organizations, indicating how they can create value for their organizations and affect corporate earnings and valuations. Through this, the new facilities manager can become part of the corporation's senior management team. Chapter 3 then gives many new tools to facilities mangers. In Chap. 3, we look at the facility over its entire life cycle, beginning with planning and budgeting and then dealing with the management of design and construction over the entire life cycle of a project. We treat crosscutting facilities management functions like automation and supply chain management. We close Chap. 3 with some important new developments in corporate operations management that will have increasingly strong effects on facilities management.

CHAPTER 1
PERSPECTIVES

Paul R. Smith, P.M.P., P.E., M.B.A., Chapter Editor
Peak Leadership Group, Boston, Massachusetts

William L. Porter, Ph.D, F.A.I.A., Chapter Editor
*Massachusetts Institute of Technology,
Cambridge, Massachusetts*

INTRODUCTION TO THE FACILITIES ENGINEERING AND MANAGEMENT HANDBOOK

To all facilities engineers and managers:

You spend untold hours compiling information from brochures, professional publications, books, and other sources on how to manage your facilities. This handbook will simplify your job. This handbook is intended to simplify facilities professionals' jobs—and lives—by helping them to:

- Run their facilities better
- Satisfy their customers better
- Incorporate good management practices into their facilities
- Relate effectively to the goals and strategies of their organization
- Coordinate facilities with the management of information technology (IT) and human resources (HR)
- Incorporate computer technologies into facilities management
- Improve their career skills for the next millennium

These are just a few reasons why you will want to own and use this book.

ORGANIZATION OF THE HANDBOOK

This handbook is organized in three parts.
   Part I is "Facilities Management." It sets the stage for the handbook, providing the needed background and rationale. This chapter discusses contemporary pressures on organizations and sets out the challenges for facilities management. It suggests how facilities management can be related to corporate strategy and indicates the need to see it as part of a broader picture of corporate infrastructure.

In Chapter 2, we look at financial analysis and reporting, presenting necessary background and concepts, as well as methods of accounting, budgeting, and evaluation of projects. It also lays out the role of the facilities managers (FMs) in their organizations and indicates how they can affect corporate earnings and valuations.

Finally, Chapter 3 looks at the facility over its entire life cycle beginning with planning and budgeting and then dealing with the management of design and construction over the entire life cycle of a project. It treats crosscutting facilities management functions like automation and supply chain management. The chapter closes with some important new developments in corporate operations management that will have increasingly strong effects on facilities management.

Part II is entitled "Facilities Engineering." It includes Chap. 4, "Planning and Programming Process" for buildings and facilities, both new and old. It provides a step-by-step process to enable the FM to capture the owner's requirements and translate them into process and technical baseline requirements for the facility's design and maintenance.

Chapter 5, the largest of the handbook, continues with a discussion of the major systems that support buildings and facilities. Using a systems engineering approach, the major systems engineering topics for facilities are presented, including issues for planning and programming processes; engineering and design processes; construction, modification, and renovation processes; and project management processes. Separate chapters address the following topics:

- Architectural and structural systems
- Electrical systems
- Lighting systems
- Mechanical systems
- Instrumentation and control systems
- Construction processes for both new and renovated facilities
- Maintenance, operations, and assessment of facilities and systems
- Building codes, regulations, and standards including performance-based codes

Part III addresses all types of facilities including buildings and complexes. It includes health care, laboratories, and industrial and manufacturing facilities, as well as colleges, malls, military bases, airports, and high-technology facilities. For each, it highlights the most important systems, the associated management operations, and the problems distinctively associated with that type of facility. The section also stresses the development and management of the infrastructure, and includes other issues particular to multiple buildings in extended environments. Part III does not repeat the work of Part II. Instead, it is devoted to problems that are specific to these kinds of facilities.

Figure 1-1 shows the organization of the handbook. Parts I and II deal with management and engineering issues and are on the vertical axis of the diagram. Part III, on the horizontal axis, deals with facilities. The proportions of the diagram indicate roughly the number of pages in each chapter and part.

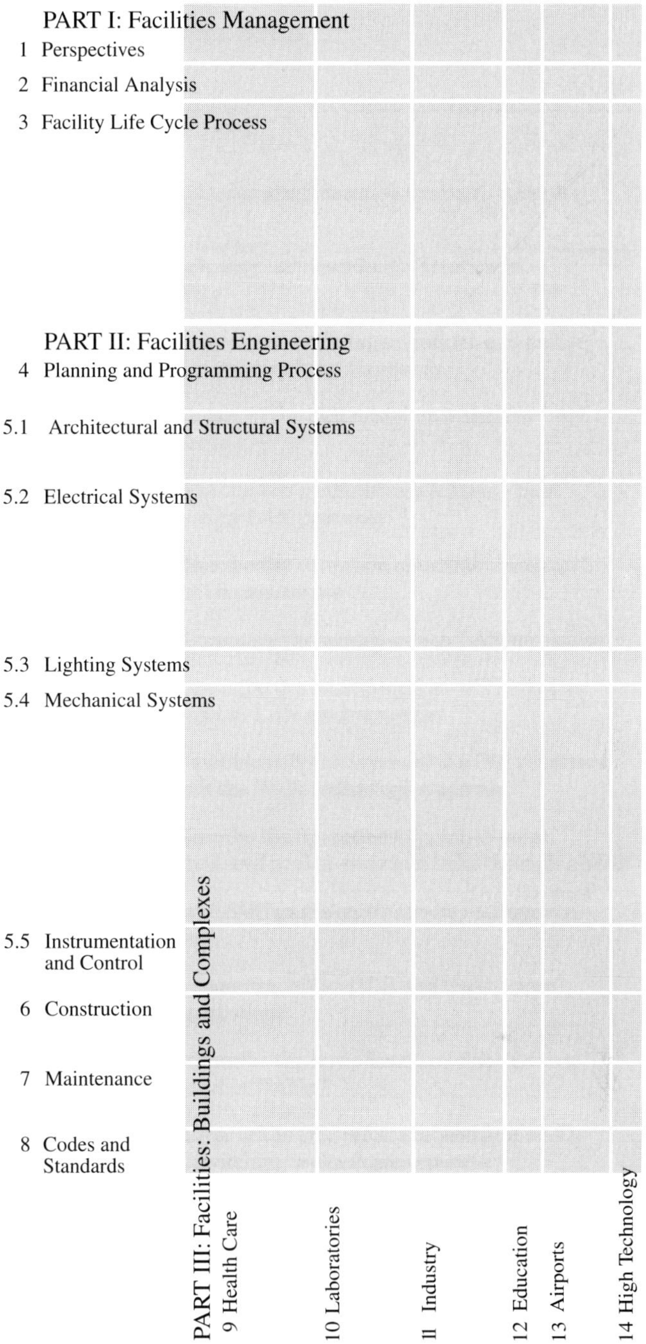

PART I: Facilities Management

1 Perspectives

2 Financial Analysis

3 Facility Life Cycle Process

PART II: Facilities Engineering

4 Planning and Programming Process

5.1 Architectural and Structural Systems

5.2 Electrical Systems

5.3 Lighting Systems

5.4 Mechanical Systems

5.5 Instrumentation and Control

6 Construction

7 Maintenance

8 Codes and Standards

PART III: Facilities: Buildings and Complexes

9 Health Care
10 Laboratories
11 Industry
12 Education
13 Airports
14 High Technology

FIGURE 1-1 Organization of the Handbook. Parts I and II deal with facilities management and engineering issues that cut across all facilities and complexes. Part III deals with facilities and includes buildings and complexes. It does not repeat earlier presentation of issues but supplements and refers to them as needed. The distances along each axis in the figure are a rough guide to the number of pages in each section. (*Source: Figure courtesy of Paul R. Smith, Peak Leadership Group, and W. L. Porter, Massachusetts Institute of Technology.*)

SECTION 1.1
INTRODUCTION AND HISTORICAL PERSPECTIVE TO FACILITIES ENGINEERING AND MANAGEMENT

Jack Hug
University of California at San Diego, San Diego, California

Bruce K. Forbes
ARCHIBUS, Boston, Massachusetts

The facilities management profession has changed substantially over the years. As with any successful profession, its vitality has come from dedicated practitioners who have succeeded in gaining positive public support and recognition for the value of professionally managed facilities. All sectors of the economy have benefited from the work of many dedicated and service-minded facilities managers (FMs).

The demands placed on FMs have increased. These new demands are the result of two relentless drivers: (1) increasing complexity and (2) rapid change. The evidence of this manifests itself daily and adds to the challenges that are faced by every FM. Complexity and change will continue to dominate the agenda and propel the profession well into the twenty-first century.

The fact that life has become more complex for the FM is, in itself, no great revelation. The content of this book will help us understand in substantial detail what is at the root of this complexity, and it will fuel the incentive to continuously learn and master the change.

Although the importance of facilities has been acknowledged throughout history, the facilities management profession has not enjoyed the same level of recognition. The FMs, in the early years of modern America and other parts of the world, were clearly working behind the scenes. Throughout history, however, the facilities professional has consistently focused on the management of an asset's life cycle (Fig. 1.1-1).

FIGURE 1.1-1 Facilities managers as stewards of corporate assets. (*Source: Courtesy of Bruce Forbes, ARCHIBUS.*)

To understand today's facilities management profession, it is useful to understand the role played by its predecessors. These earlier FMs were, in every sense, leaders on the cutting edge of an emerging and developing new profession. History has shown that every profession has its own way of doing things and the management of facilities is no exception.

Honoring the past practices of the profession by looking at the way things used to be can provide important clues to significant events that have influenced how things are done today. One is cautioned, however, when looking in the rearview mirror, not to get trapped into thinking that past practices—a return to the "good old days"—are a solution for today's challenges. Honoring the present emphasizes the continuous need to be able to bridge one's actions—and one's thinking—from a solid foundation of relevant past practices to the new realities of today's professional requirements.

The FMs' predecessors saw themselves as optimists who believed that the focus of responsibility and accountability resided with them. They practiced stewardship of the facilities and managed financial and human resources entrusted to them, as if they were their personal assets. They had pride in ownership and pride in workmanship. The physical plant was their baby; the organization was their family. They led by example and, although quite informally, they taught others by coaching and mentoring. Many of these early FMs came up through the organizational ranks. They had performed successfully all of the jobs required by the facilities organization. They were hands-on managers. Many were perceived as the experts. It was in their expertise that building owners invested, trusted, and relied exclusively to make sure that things worked. These early FMs embodied high standards of performance in their everyday activities. They were disciplined and had a strong sense of honor, duty, and obligation. They had no expectations of entitlement, did not draw attention to themselves, and felt they were part of a noble profession.

In his superb book, *The Greatest Generation,* author Tom Brokaw tells the story of a generation of American citizens who came of age during the Great Depression and World War II, and who went on to build modern America. "The war had taught them what mattered most in their lives"[1]; Brokaw continues, "for they had survived an extraordinary ordeal, but now they were eager to reclaim their ordinary lives of work, family, church and community." This profound experience shared by many FMs served as the cornerstone of the profession's solid character traits and service attributes. Facilities management predecessors, for the most part, welcomed their roles at home and at work. They eagerly pursued the development of competencies in the right combination to ensure success. Some of the core competencies addressed by the facilities management professional are shown in Fig. 1.1-2.

EMERGENCE OF THE FACILITIES ENGINEERING AND MANAGEMENT PROFESSIONS

If facilities engineering has a birthplace, our guess is that it was the boiler room. The boiler room and its required auxiliary equipment—its distribution piping systems, water, oil, gas, sewer, steam, condensate and electrical systems—were the most complex, and potentially treacherous, parts of a facility.

In a large building or in clusters of the buildings built in the 1940s and earlier, the boiler room served as the heart of the building operations. Not only did this space (typically located in the bowels of the building) contain the building's life-support systems, the boiler room also served as the central nervous system and the brain. It was a place of mystery and intrigue. It was the home of the building engineer.

The boiler room was off limits to the building occupants, and most building owners had complete trust and confidence in their engineer. The building engineer had ownership, and he (in the old days, it usually was a *he*) was expected to know everything about the building—its construction, its components, where things were located, how things worked, and how to fix them when they did not work. Generally, the building engineer did not have to concern him-

FIGURE 1.1-2 Facilities management core competencies. (*Source: Courtesy of Bruce Forbes, ARCHIBUS.*)

self with housekeeping, landscape maintenance, or even tenant improvements. The job requirements largely emphasized the more traditional engineering skills and competencies. For some, the engineer may have had responsibilities for building hardware, painting, and carpentry, but these were viewed as secondary to the engineering, power plant, electrical, plumbing, heating, and ventilation system responsibilities.

Inside the boiler room, there was no central control room, as is common today. The boiler operators monitored and controlled the steam cycle manually. The operator would make continuous rounds, checking dials and gauges located on the equipment and recording the conditions observed, such as pressures, temperatures, flow rates, and fluid levels. In more modern buildings, pneumatic control technology (developed in the 1940s) was used.

Somewhere inside the boiler room was a space carved out for the building engineer's office. Sometimes, there was a small workshop for maintenance. Rarely was this space designed into the original construction plans; it was usually an afterthought. It was an icon attesting to the innovation and creativity of the building engineer. Hot, smelly, poorly ventilated, noisy, and cluttered with parts, catalogs, and trade magazines—these accurately describe the conditions of the engineer's workspace.

Whenever there was an emergency, the building engineer would be right in the middle taking a leadership role in resolving the emergency. Whether a power outage, a flood, a major steam leak, or a boiler explosion, the engineer was always expected to be in the thick of things, covering the full spectrum of technical issues that today's modern building requires an army of specialists to address.

Facilities management is generally performed within a broad and diverse context of requirements and circumstances. The forces of change throughout history have required the FM to be constantly vigilant, flexible, and adaptable to changing requirements. This inherent stewardship responsibility has been carried forward. Ensuring that the facility remains useful and functionally adequate continues to be a fundamental part of the job requirement of the contemporary FM.

The fundamental practices of the past have been subjected to continuous change. Practices and procedures that were valid yesterday can easily become invalid and unnecessary in no time at all. For example, many past practices demonstrate that the building engineer was primarily expected to perform in a stable, steady, consistent, according-to-the-book manner, and only occasionally to react to change. Today, however, every successful facilities organization must be designed to respond to change as the norm, and it is even encouraged to create change. Due to the complexities of organizational change, caused in part by the e-commerce phenomenon, the role of the facilities professional will become even more significant. It is expected that real estate, facilities, and infrastructure expenses will consume over 50 percent of an organization's revenue.

During the past several decades, significant events have helped thrust the FM into the middle of things—the energy crisis; population growth; increased environmental concerns; advances in science resulting in new construction materials, methods, and equipment; the recognition of management as a profession; and the impact of technology—all have contributed to the evolution of facilities management as we know it today. Perhaps the most significant changes in the last 5 years have been the e-commerce and business-to-business revolutions. The bricks-and-mortar organizational style that have served us since the industrial revolution is quickly being replaced by clicks-and-mortar and clicks-and-clicks opportunities.

The 1950s and 1960s were significant periods for facilities expansion. The post–World War II boom fueled the need for FMs and engineers. It is difficult to comprehend the explosive growth in all areas of the FM's professional responsibilities. The convergence and interaction of the new forces impinging on facilities management raised the standard by which the FM's performance is now measured (Fig. 1.1-3).

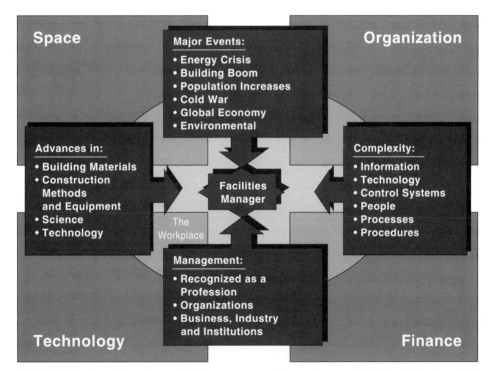

FIGURE 1.1-3   Forces impinging on facilities management. (*Source: Courtesy of Jack Hug, University of California, and Bruce Forbes, ARCHIBUS.*)

Not only has there been enormous growth in all types of new, more complex, and more costly space, but the existing inventory of aging buildings adds a substantial requirement for capital renewal and modernization. Enormous efforts are required to keep the country's institutional, business, and industrial facilities in sound condition.

THE TECHNOLOGY FACTOR

The challenges facing the FM today are awesome. The job has never been easy, but the 1990s have placed more demands than ever before upon the profession. All of the major drivers referred to earlier continue to affect the FM. The most powerful force bearing down on the profession is the quantity, extent, and speed of change in technology.

The facilities management prophet for the year 2000 might proclaim, "When things all around you are changing and you're not, the end is near." The technological shift with its relentless drive during the 1990s is now familiar and we are deeply embedded in the cyber world. This revolution is influencing the type of facility services and the manner in which they are performed. Microchips and computers have been used to create smart buildings, smart systems, smart security, smart access, and smart telephones. The twenty-first century is guaranteed to spawn additional industry-changing innovations. The development of effective infrastructure management enterprises will be critical as facilities professionals help the organizations they serve enter the knowledge age.

Facilities managers are cautioned not to underestimate the impact of this technology. A sudden new world of data communications, local area networks, electronic data interchange, image technology, and the relentless flood of new applications, tools, and technologies are as much a part of the physical plant as are the traditional utility systems. In fact, the data infrastructure not only ranks high in importance with traditional utilities, but also is proving to be one of the most important factors in sustaining business operations, productivity, and profitability. This inexorable fact of life, taking place in multiple facilities organizations around the world, has changed the dynamics of the FM's workplace. A fundamental shift has occurred in the way business is conducted.

STEWARDSHIP, THE FM, AND CYBERSPACE

"Civilization as we know it today owes its existence to the engineers. These are the men and women who down through the centuries have learned to exploit the properties of matter and the sources of power for the benefit of mankind. The story of civilization is in a sense, the story of engineering—the long and arduous struggle to make the forces of nature work for man's good."[2]

The stewardship responsibility that is appropriately assigned to the FM is born out of a long, arduous struggle to make the forces of nature work for the good of humanity. There is a sense of moral responsibility for the care and use of a facility. Today's FM has accepted this responsibility with respect to the principles and needs of those whom the facility is intended to serve. This stewardship perspective has served the profession well, and it will continue to be an important dimension that will help define the future environment of facilities management. Proper maintenance and care of buildings, grounds, and infrastructure demonstrate confidence in the future and recognition of the investment in the physical assets.

We have seen that facilities management is a very broad area of endeavor. It is not regarded in the traditional sense of different branches of engineering such as civil, mechanical, electrical, structural, or chemical engineering. Facilities engineering is a more generalized branch of the engineering discipline. It requires the practitioner to summon many different kinds of engineering expertise in a cohesive, coordinated, and, sometimes, collaborative manner. Today's FMs come from all sorts of formal and informal educational backgrounds. No credible research has been done that would lead to a conclusion that an FM must be an engineer to be successful. This does not imply that facilities management is not a disciplined profession or that a degreed engineer cannot be a successful FM. Indeed, facilities management requires special expertise and clearly carries with it a very special consciousness and a sense of obligation to serve.

Facilities management is a term of broad application used by those who, with proper preparation and training, are occupied with solving the problems of a wide range of disciplines. It is a profession today from whose ranks its members can be called upon to thoroughly advise on questions involving facilities throughout the world.

In the cyber world of the twenty-first century, FMs will have to acquire new competencies, and specialists will emerge from within the ranks to work with the new information technologies. There will be new possibilities for more effective management of facilities, including much better instrumentation and control. This, in turn, will increase the effectiveness of personnel, and it will greatly increase the efficiency of the organization. The facilities management profession will be transformed.

INTEGRATING INFORMATION TECHNOLOGY

The difference that technology makes in the way we organize, structure, and empower our lives and our workplaces is the most significant determinant in changing the way that FMs deliver services.

There is evidence that increasing numbers of FMs no longer have responsibility for information technology systems management. In facilities across the globe, the failure or downtime of data systems, information technology infrastructures, or computer networks is as critical to a business or industry as an electrical power failure. Currently, only a small number of FMs are charged with the responsibility for these critical systems and infrastructures. Interestingly, the same elements of complexity, reliability, capacity, condition, and flexibility that are present in telecom data systems exist in the utility infrastructure.

Unfortunately for the facilities management profession, the computer network, infrastructure, hardware, and software are largely the domain of a new breed: the telecommunications manager, who does not report to the FM. In fact, this position enjoys equal or greater status in the organization than the FM. Yet, the world of data and information, local area networks, the Internet, and the relentless flow of new software applications, tools, and technologies are as much a part of the physical plant as are the traditional utility systems. It is now a fact of life that data and information technology are proven to be important factors in sustaining successful business operations, productivity, and profitability.

In the transition from the information age to the knowledge age, the components of the new workplace (i.e., space, organization, finance, and technology—and the relationships among them) are in a state of flux. The facilities managers are the most appropriate professionals to provide a leadership role in the emergence and implementation of these new opportunities. Although facilities managers have proven to be highly adept at serving the needs of facility occupants, many observers conclude that some facilities management professionals have a long way to go in their understanding and effective management of the requirements of the knowledge age and its associated technologies and best practices.

This critical assessment of current conditions clearly identifies an area of opportunity for the facilities manager. Now is the time to think strategically and to reposition the facilities management profession—to recapture this critical facilities service. There are risks; however, our experience with other challenges has taught us that if we wait until the path is perfectly clear and the risks have disappeared, the opportunities will have passed as well.

SECTION 1.2
FACILITIES ENGINEERING AND MANAGEMENT RECONSIDERED

Paul R. Smith, P.M.P., P.E., M.B.A.
Peak Leadership Group, Boston, Massachusetts

William L. Porter, Ph.D., F.A.I.A.
Massachusetts Institute of Technology, Cambridge, Massachusetts

NEW PERSPECTIVES ON FACILITIES MANAGEMENT

Facilities Management Challenges in the New Millennium

The past century has seen the dawn of several revolutions. The industrial and business revolutions have given way to the information and knowledge revolutions. Shifts in the availability of information and increased processing power virtually eliminate information float. Today, information reaches customers quickly, enabling them to become increasingly informed about available alternatives. This means that customers are likely to be less loyal, but more sophisticated and more demanding than in the past. Meanwhile, competitors are delivering innovative products to the marketplace with greatly reduced lead times. Perhaps most important, products and technologies are changing rapidly, making it absolutely necessary for organizations to shorten their response time to new opportunities.

Moreover, competition is not just from within but also from outside traditional business categories, and it often results from innovative redefinitions of such categories. Globalization is occurring in all organizations—manufacturing, service, and hybrid; governmental and nonprofit—with profound consequences for how and where organizations must operate.

Wider issues are also having a direct effect on organizations. The earth's ecosystem is being threatened, many resources are being depleted faster than they are being replaced, and the degradation of the quality of the environment is now a matter of public concern. As a result, the regulatory environment is shifting and standards are rising for the energy performance of buildings, for the environmental standards for its occupants, and for ecologically responsible ways of handling waste materials. New concepts and technologies are emerging that may form the basis for future standards and will influence what organizations purchase, as well as what they produce.

Agility is the characteristic of organizations that can respond successfully to today's market, economic, and environmental pressures for change. For facilities managers, fostering agility requires a high degree of adaptability to provide and manage facilities that can meet the rapidly evolving needs of the organization.

Organizations are responding to these challenges in a variety of ways, including downsizing, outsourcing, and just-in-time procedures. Thus, the collection of activities that constitutes an organization is distributed geographically, and so, too, is the workforce. At the same time, the formation of high-performance, cross-functional teams is increasing throughout the business sector. Cross-functional teams usually consist of people from several locations who depend heavily on communications and information technology, as well as on travel, for needed face-to-face communication. And these teams may operate at all times of the day and night, depending on where the team members are. They may operate simultaneously or asynchronously as information from one group is passed along to another in a different time zone.

Changes of this scope affect the organization's vision, mission, and core values, its management structure and organizational models, its work systems and processes, its performance expectations, and its behavior at individual and group levels. For example, because of the need to draw on talents across the workforce to meet competitive challenges, reward and recognition systems must place increased emphasis on empowering employees throughout the organization and on sharing the risk of failure and the rewards of success.[1] Thinner profit margins, however, are reducing the flexibility of action and leaving much less room for mistakes.

For organizations to be agile in responding to new conditions, *adaptability* is the watchword for facilities managers. Downsizing in one or more locations causes shifts in the use of facilities, forcing onto the market those that are no longer needed. Outsourcing may remove operations from the specialized facility for which there is no other obvious use, and shifting functions from one region to another may force rapid shifts of buildings to new uses. Finally, the workplace of today's organization is changing from the traditional workplace, associated with a single building or location and with a fixed and regular workday. The new workplace requires both cyber and physical space, and it requires that the two types of space be integrated.

Against this background, the most pressing challenges for the facilities manager are:

- To ensure rapid adaptation of existing facilities to new programmatic demands
- To carry out rapid construction of new facilities
- To improve all services and integrate the new information infrastructure
- To enable all facilities to become integral parts of the new workplace of the contemporary organization
- To make these changes consistent with the core strategy of the organization

Figure 1.2-1 summarizes the points just made in this part of the chapter, showing that organizations need agility to respond to the shifting global context and that facilities management must be highly adaptable to enable the organization to change.

FIGURE 1.2-1   Facilities management challenges. Adaptability in facilities management is necessitated by the organization's need to respond rapidly to the shifting global context. (*Source: Courtesy of Paul R. Smith, Peak Leadership Group, and William Porter, Massachusetts Institute of Technology.*)

Facilities Management and Organizational Change

As most organizations grow, they pass through a series of development phases.[2] Each phase begins with a period of evolution, with steady growth and stability, and ends with a revolutionary period. This is true of all organizations, no matter how fast they grow and regardless of their age. For example, at one phase, centralization practices lead to demands for decentralization. The same organizational practices are not maintained throughout a long life span. Therefore, a company must plan its movement from one phase to the other or its health, growth, and prosperity will be affected. To serve the organization effectively, a facilities manager must be aware of the company's phase of development.

Phases of Growth.   Figure 1.2-2 shows five phases of growth for a company.[2] Every evolutionary period is characterized by a dominant management style that is used to achieve

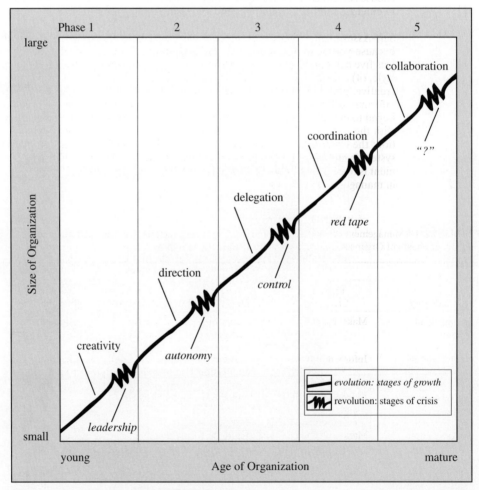

FIGURE 1.2-2   Five phases of growth. The continued successful development of an organization oscillates between evolutionary stages and crises that require new styles of leadership for the next evolutionary stage. (*Source: Courtesy of L. E. Greiner, Harvard Business Review.*)

growth, and it is followed by a revolutionary episode out of which must come a new style of leadership if the company is to continue to grow and prosper. In phase 1, *creativity* is the birth stage of an organization, and the emphasis is on the product and the market. The founders are usually technical or entrepreneurial and tend to disdain management activities. In phase 2, *direction* has a capable business manager installed whose directive leadership channels employees' energy into more efficient growth. During phase 3, *delegation* is provided by installing a decentralized organizational structure that allows the company to grow through heightened motivation of managers at lower levels. In phase 4, *coordination* requires a formal system for achieving greater coordination; top executives take responsibility for initiating and administering the system. Phase 5 calls for strong interpersonal *collaboration* in an attempt to overcome the red-tape crisis.

For each phase, managers are limited in what they can do if growth is to occur. For example, a company that is experiencing an autonomy crisis in phase 2 cannot return to directive management for a solution; it must adopt a new style: delegation. Too often, it is tempting to choose solutions that were tried before but that actually make it impossible for the new phase to emerge.

Self-Assessment. Knowing where you are in the development sequence is important because you need different organizational practices for each of the five phases. Table 1.2-1 lists five categories: (1) management focus, (2) organizational structure, (3) top management style, (4) control system, and (5) management reward emphasis.[2] For example, in the first, or creative, phase, management focus is to make and sell; the organizational structure can be informal; top management style is individualistic and entrepreneurial; the control system is keyed to market results; and the management reward emphasis is on ownership. Contrast this with the last, or collaborative, phase, during which the focus is problem solving and innovation; the organizational structure is a matrix of teams; the style is participative; the control system is mutual goal setting; and the rewards are team bonuses. The task for senior management is to be aware of the phases of growth and to recognize the signs and the time for change in management practices.

TABLE 1.2-1   Management of Growth. Different Management Practices Are Needed during the Different Phases of Organizational Growth and Change.

Category	Phase of Growth				
	Phase 1: Creativity	Phase 2: Direction	Phase 3: Delegation	Phase 4: Coordination	Phase 5: Collaboration
Management focus	Make and sell	Efficiency of operations	Expansion of market	Consolidation of market	Problem solving and innovation
Organizational structure	Informal	Centralized and functional	Decentralized and geographical	Line staff and product groups	Matrix of teams
Top management style	Individualistic and entrepreneurial	Directive	Delegative	Watchdog	Participative
Control system	Market results	Standards and cost centers	Reports and profit centers	Plans and investment centers	Mutual goal setting
Management reward emphasis	Ownership	Salary and merit increases	Individual bonus	Profit sharing and stock options	Team bonus

Source: Courtesy of L. E. Greiner and *Harvard Business Review*

Impediments to Change.   The art of managing change in organizations rests in large part on being able to spot the impediments to change and to deal effectively with them. These impediments include the following:

- Resistance to change by individuals and groups, addressed in part by looking more closely at their needs and responding to them and by creating clear indicators of progress
- Lack of consensus, addressed by creating a common vision statement, assembling a broadly drawn guiding coalition, identifying a champion for the changes, and opening the process to wider participation
- Unrealistic expectations, addressed in part only by putting forth alternatives within a defined budgetary frame
- Lack of cross-functional teams, addressed by requiring such teams from the outset of the project
- Lack of team skills, addressed in part by training accompanying the change processes
- Failure to consider information technologies, addressed by introducing them early in the change process
- Too narrow a project charter, addressed by extending the time of project definition at the outset and by providing for project redefinition of scope and priorities as late as possible in the process

Implications for Facilities Managers.   These are just some of the issues that may accompany change in organizations. Facilities are often an integral part of change within organizations, and today the facilities manager is increasingly a part of the top management team that addresses and manages organizational change. New facilities need to be designed and renovated to serve the particular phase in which the organization finds itself. Therefore, the facilities manager must be aware of how the organization is changing and where it is in the change process. Recognizing the factors that affect organizational change will therefore help him or her to achieve the organization's objectives.

Recognition of these factors defines the objectives of facilities management itself by:

- Clarifying why changes in work practices are necessary
- Clarifying the link between facilities objectives and corporate objectives
- Incorporating successful change management strategies into facilities management procedures themselves

A fundamental shift to a comprehensive integrated approach, indeed a reengineered process, is required to meet these objectives. This integrated approach takes several forms, each making its contribution to facilities managers' contributions to aligning facilities and corporate objectives. The first is integration between life cycle phases (see Chap. 3), which invites the following:

- Strategies for work efficiency
- Design for construction efficiencies
- Design for flexibility
- Materials chosen in the light of long-term maintenance considerations
- Other strategies, sometimes described as *right-to-left* thinking

This will help the facility manager to meet near-term objectives and never lose sight of long-term facility value optimization. The second form of integration is between disciplines. The facility manager is an organizational coach, dealing with people who are expert in finance, law, design, operations, estimating, construction—the list goes on. These professionals, each with their special wisdom and training, can and will make important contributions to the improvement of facilities, given the right environment and motivation. To get these myriad personalities to work together requires careful treatment of the cultural, personal, contractual, and emotional relationships. The final form of the modern integrated approach is the integrated building system. Designing for integrated mechanical, electrical, control, structural, or other systems can improve construction savings and long-term operational economy.

Facilities Management as Part of Corporate Strategy

Success in the global marketplace requires a business model based on new methodologies and techniques that incorporate evolutionary and revolutionary concepts. Typically, these techniques foster problem-solving and innovation skills; they cultivate collaborative strategies and create the appropriate management and individual reward structures; they engage entrepreneurs and inventors with customers; and they shift from centralized control to mutual goal setting. Moreover, they define strategy and continuously improve performance by creating high-performance work teams utilizing your greatest assets: people, processes, technology, and infrastructure. In general, these techniques tend to hasten the organization's maturation in productive ways.

For years, American Big-Five consulting firms like Arthur Andersen and Ernst & Young have been telling their clients they must align their people, processes, technology, and infrastructure to support the business strategy. They remind their clients that a particular strategy will be different depending on their business environment. Therefore, consulting on productivity improvement could begin by helping you analyze why you are in business today, what the products and services are that you sell to an external or internal customer, and how continuous improvement is implemented.

The consequence for facilities managers is that the systems and processes of the organization, including its physical facilities and informational infrastructure, must adapt quickly to these shifting contexts. Facilities managers must be knowledgeable concerning the factors that underlie them, and they must understand and become an integral part of corporate strategy. We suggest that three ongoing activities make it possible for an organization to achieve success in a changing world: (1) direction, (2) alignment, and (3) implementation. *Direction* is the activity that gives rise to the organization's core strategy; *alignment* gives rise to the organizational strategy; and *implementation* results in the production of products and services (see Fig. 1.2-3).

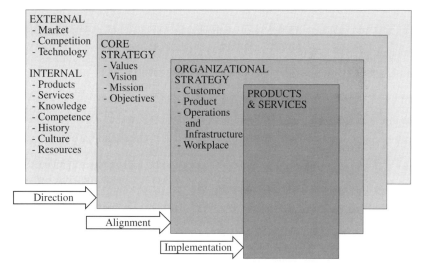

FIGURE 1.2-3   Strategic factors. Direction, alignment, and implementation are the keys to an organization's success. (*Source: Courtesy of Paul R. Smith, Peak Leadership Group, and William Porter, Massachusetts Institute of Technology.*)

Although these three activities are presented sequentially in this text, they actually must occur constantly, each informing the other, to lead to creative transformation of the organization over time.

Direction and Core Strategy.   Direction is the activity that formulates the core strategy out of external and internal factors. External factors include the market into which the organization is trying to penetrate, including the customers in that market, their preferences, and their potential; the competition aiming at the same market; and the technologies of the day that affect design, production, and marketing. Internal factors include the firm's current capabilities to produce products and services, its knowledge and competencies, its history and culture, and the resources on which it can draw. Consideration of these factors should give rise to an integrated core strategy.

A core strategy includes values, vision, mission, and objectives. Guided by its core strategy, a company should be prepared to change everything about itself as it moves through corporate life. Otherwise, the world will pass it by. The core strategy contains the elements of greatest permanence—particularly its core values.

Core values are not to be confused with a specific culture or operating principles; they are not aimed at financial gain or short-term expediency. For example, Wal-Mart's number one value is, "We put the customer ahead of everything else . . . If you're not serving the customer, or supporting the folks who do, then we don't need you." A core value is clear, simple, straightforward, and powerful. Successful companies tend to hold tenaciously to their core values.

A. D. Little studied companies that prospered over the long term, such as Hewlett-Packard, founded in 1938; Motorola, founded in 1928; Merck, founded in 1892; and Procter & Gamble, founded in 1837. These visionary companies attained extraordinary success decade after decade, regardless of the change and difficulties facing them. They achieved this by adhering to the same timeless mission, that today's leaders can use, that they developed as a result of preparing themselves for change, along with growth and transformational strategies.

An organization requires fundamental objectives beyond just making money. A company's core objective need not be unique, but it must guide and inspire. Walt Disney captured his core objective when he said, "Disney will never be completed, as long as there is imagination left in the world." Boeing, likewise, can never stop pushing the envelope in aerospace technology. An organization gives direction by formulating and expressing its core strategy and by letting that be its guiding force.

Alignment.   The activity of alignment brings external and internal factors into a strong and mutually reinforcing relationship that is guided by the core strategy. Alignment results in an organizational strategy consisting of five major parts that address the customer, the product, the operations, the infrastructure, and the workplace.

The Customer.   The customers and their changing buying preferences must be key elements of organizational strategy. Consider how some wholesalers sell used cars in Japan. Until the mid-1980s, vehicles were transported to be sold at live auctions. But only about 45 percent of the products at any given auction were sold, resulting in low efficiency. Fujisaki saw an opportunity to transform the marketplace. He created a proprietary computer and satellite communications system called *AUCNET.* Each week an AUCNET inspector inspects the cars and collects photos of them. The information is digitized, put on laser disc, and shipped to subscribing dealers. AUCNET staffers then moderate an auction that takes place on computer screens all over the country. The purchased cars are then delivered to the appropriate lots.

The AUCNET system has made the physical location of inventory and the actual site of buying cars irrelevant. The traditional face-to-face marketplace interaction between buyer and seller has been eliminated. Now a buyer and seller can meet in marketspace. Consider the implications for space, facilities, and information technology! As indicated earlier, facilities management had to be extremely adaptable to permit the organization to move with agility into its new condition.

The Product.   Some companies change from a product orientation to a service orientation without even knowing it. Consider the near demise of *Encyclopedia Britannica,* one of the best-known brand names in the world. Evans and Wurst report that its sales had plummeted 50 percent since 1990. How is this possible?

The *Encyclopedia Britannica* sells for somewhere in the range of $1500 to $2200. An encyclopedia on CD-ROM, such as Microsoft's *Encarta,* sells for around $50. The cost of producing a set of print encyclopedias—printing, binding, and physical distribution—is about $200 to $300. The cost of producing a CD-ROM encyclopedia is about $1.50.

Imagine what the people at *Britannica* thought was happening. The editors probably viewed CD-ROMs as nothing more than electronic versions of inferior products and toys. When the threat became obvious, *Britannica* did create a CD-ROM version, but to avoid undercutting the sales force, the company included it free with the print version and charged $1000. The best salespeople left, and the company was sold. Under new management, the company is now trying to rebuild the business around the Internet.

Britannica's downfall demonstrates how quickly and dramatically the new economics of information can change the rules of competition, allowing new players and substitute products to render obsolete such traditional sources of competitive advantage as a sales force, a supreme brand, and even the world's best content. Finally, it demonstrates the dangers of not carrying out an effective reassessment of the organization's products, processes, and services.

An effective product strategy requires realignment with the external environment and its windows of opportunity; restriction to a few core competencies in the race to stay ahead of rivals; and alignment of core competencies with the strategy. Alignment may force changes in the core strategy if an organization has developed dysfunctional or outdated approaches to its market. Once again, space, facilities, and information technology had to be highly adaptable in order to support these absolutely fundamental and necessary shifts in organizational strategy.

Operations and Infrastructure.   There is nothing remarkable about an organization's effort to coordinate the deployment of different kinds of resources, of course. What *is* new, however, is that "riding the waves of change" means moving beyond notions of coordination alone to include concepts of integration and coinvention. In other words, *agility seems to imply that how work is done and how work is supported need to be considered as one process.* "Not all the time, not every time, but some times."[3]

Operations strategy guides the way the organization goes about its work. It organizes people, plant, equipment, and other resources to implement the objectives expressed in the core strategy. It creates business processes to carry out the organization's objectives. Typically, infrastructure is thought of as supporting operations,[4] but infrastructure and operations may converge and influence one another. Amazon.com provides a good example of the way the infrastructure that was created to market books opened up other marketing opportunities (e.g., toys) that were not included in the earlier marketing plans.

The Workplace.   The aim is to create a mutually reinforcing relationship between work practice and the workplace. Over the next several years, the nature of work will continue to change, with powerful implications for changes in the place and style of work as well. Work will be where the worker is, rather than where the place of work is. Work at the end of the twentieth century relies increasingly on information and on digital technologies that permit its easy storage and retrieval. In many industries, work is much more than what is done at one's desk at the place of employment. People work in a variety of settings in addition to the places of work owned or leased by the employer. They work while traveling—in transit and at distant locations, while at the client's office, while at home, and wherever they must work or find it convenient to do so. Moreover, even at the place of employment, many employees are mobile but must stay in communication with people for management purposes or for expertise and with sources of information and computational power.

Wherever they are, employees need access to people and information that are not physically present. The nature of their work, therefore, requires support for information and communication every bit as much as it does for facilities. In relation to these new conditions of work, facilities management has two levels of responsibility that it could assume. It could maintain strict boundaries around facilities, dealing only with the workplace as traditionally understood (i.e., mainly the buildings and service infrastructure owned and leased by the corporation). Alternatively, it could extend its reach into the new workspace, including the required infrastructure. Most obviously, this includes information technology, but the infrastructure for work could also be thought of more broadly as the support for all of the management and other services that help create optimal conditions for work.

Of course, for most organizations, the major places of employment will continue to be centrally important in sustaining work culture, and these are the focus of this handbook. These places are evolving in their own right, however, responding increasingly to the local culture of work and to the highly attractive workplaces that other organizations are creating to compete for talented employees.

Implementation and Infrastructure.   The lesson for today's organizations is that infrastructural strategy must be shaped at the highest level of the organization. If it is viewed as a collection of services, typically including space, information technology, human resources, and finance, it must be coordinated and related to operations. It may also be viewed, as in the case of Amazon.com, as creating new organizational capacity, enabling it to open up new markets. Coordination at the highest level is all the more needed.

Figure 1.2-4 is intended to indicate that infrastructural resource strategy should be considered at the same level as all of the other important components of the organization. It should

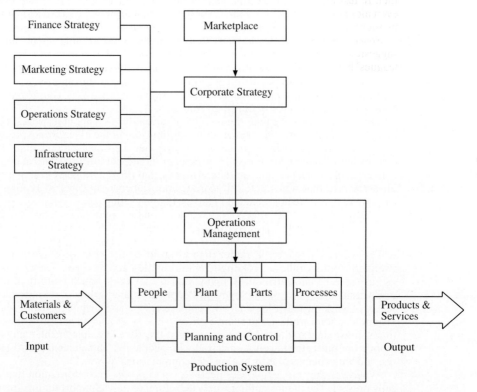

FIGURE 1.2-4   The place of infrastructural resource planning in the organization. Infrastructural resource planning, including space, information technology, human resources, finance, and other areas needs to be carried out at the highest level of the organization to increase its agility in responding to new market pressures and to new ideas. (*Source: Courtesy of Paul R. Smith, Peak Leadership Group, and William L. Porter, Massachusetts Institute of Technology.*)

not be seen just as service. The form this will take for each organization will, of course, be different.

Facilities Management and Workplace Change

Our economy was previously based on economic factors that, in turn, were based on property, plant, and equipment as the primary agents of capital formation. We are entering an economy based on information and knowledge. Therefore, the workforce and all of its understanding, competencies, and sense of social responsibility must become the more important part of the value chain measurement. The implication for facilities managers is to increase the importance of serving the workforce directly as customers in the same way that the corporation must serve its customers if it is to stay competitive. Thus, it is essential that facilities managers see their work as part of the creation of the new workplace, not only the provision of specific services.

The New Workplace.   The components of the new workplace include *organization, finance,* and *technology,* as well as *space.* These four elements interact to provide the environment for any work practice (see Fig. 1.2-5). Their interactions can be mutually beneficial or counterproductive. For example, if the space is set up to reinforce working in teams, but neither the organizational nor the financial incentives encourage teamwork, you have an obvious conflict. If, furthermore, the information technology is not there to support teamwork, failure is even more likely. We'll discuss each element briefly and then return to the opportunities to set up more productive relationships among the elements.

Space.   Several authors have written perceptively about types of workplaces and have suggested how they may change in the future. For example, for offices, Duffy and his colleagues[5] have described four types of offices: (1) the hive, (2) the den, (3) the cell, and (4) the

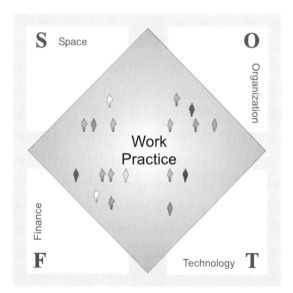

FIGURE 1.2-5 The "SOFT" diagram. Space, organization, finance, and technology must each be designed, and in dynamic relation to one another to create a workplace that supports the core strategy of the organization. (*Source: Courtesy of John Wiley & Sons.*)

club. Repetitive work is associated with the hive, with many workstations, low autonomy, and low interaction among workers. An example might be repetitive backroom work to support banks and other financial institutions. The cell is associated with increased autonomy and might be best illustrated by faculty offices in a college or university. The den is associated with heightened interaction, where most of the work is done in teams—an architectural office, for instance. The club is associated both with increases in autonomy and with increases in interaction; it is exemplified by many research and development groups. Organizations tend to move from the hive to the other types over time, but most medium to large organizations exhibit all four types within their space inventory. In general, organizations have to accommodate an increasing variety of work patterns, including many varieties of telecommuting, and a wide range of types of computing equipment, all of which require more complex and sophisticated information technology support. Space types are increasing in variety as organizations come to understand the great variety required to support service and knowledge work. Many organizations need all four types of space suggested by Duffy: the hive, the cell, the den, and the club.

Two important implications for facilities management flow from this analysis. The first has to do with the type of building that is most supportive of the particular office type, and the second has to do with the heating, ventilating, and air-conditioning (HVAC) systems most appropriate to each. For example, the only office type for which deep buildings are appropriate is the hive, and the only office type for which shallow buildings are appropriate is the club. As the autonomy of the worker and the team increases, and as the needs for interaction increase, HVAC systems have to become increasingly flexible and subject to user control. The raised floor combined with radiative-air systems, increased integration of natural ventilation, mixed-mode systems, and distributed control are all responses to the demands of the new work styles.[6]

Some innovative office buildings are already reflecting these new ideas. The British Airways building at Heathrow, primarily a hive-type office, uses its central atrium to link the control and management functions on the upper floors with the receiving, checking, and briefing of thousands of transient air crew staff each day. In a new factory in the Czech Republic, the offices for management have been placed inside the oval assembly line. Therefore, managers and workers interact on a daily basis, and each can see what the other is doing. Also in that factory, just-in-time deliveries are made from both sides of the building and can be delivered to the assembly line without crossing other operations. The same architect created a so-called main street for the faculty of mechanical engineering complex of Munich University. It contains auditoriums, dining rooms, meeting rooms, and places for casual encounters. These new arrangements result from a deep understanding of the program and a recognition of the need to interact casually, as well as formally, to achieve the mission of the organization. They expand the range of space types and conditions that need to be managed.

Organization.   Organizational innovation takes many forms:

- High-performance, cross-functional teams operating largely autonomously
- Multiple-project teams with many individuals playing roles on several teams
- A mixture of stable laboratories and changing project teams that develop new products
- Collaboration, partnerships, and alliances among parts of geographically distant organizations
- Nonterritorial offices where occupancy is handled by hoteling, in which an individual arranges in advance for space to be used for a stated duration.

These new forms may call for new arrangements of space and technology to support the work. One such arrangement is a set of *town commons,* each of which forms a node around which teams form, but among which there can be technologically sophisticated communications. In some traditional research building layouts, laboratories are located in the center with offices along the perimeter. And in a traditional campus model, buildings are linked by the infrastructure of roads, utilities, and services. Activities tend to be segregated by function. In a recent proposal for a leading research group in a large organization, the focus of the research is located in an innovative node that is networked with related laboratories in other buildings, thus creating a virtual team that is larger than the core team that was spatially collocated. The corridor is transformed into an interactive workplace and news center. Boundaries are blurred between spaces and between functional areas that are traditionally separated; individual offices are replaced by cubicles feeding energy into the Common, fostering collaboration and communication, and encouraging appropriation and use of the shared spaces. The implications for the campus design are profound: multiple linkages, integration of various processes, and sharing of information and services.

Finance.   Financial incentives for work have to reinforce the new work arrangements needed for product and idea development. These include point-of-work accounting that recognizes the work where it is done rather than assuming it is done at the main place of employment. And incentives will be created to reward teams, as well as individuals, for successful results.

Other incentives will have to take account of the location and preferences of the customer by rewarding those individuals and teams best able to respond to new and emerging demands. These new distributed financial incentives may change the demand for facilities and other improvements, making the clients for facilities management more numerous and complex than in the past.

Technology.   If you think the knowledge revolution is not transforming your business, think again. Economic value is being transformed at lightning speed. In some cases, the information about the product or service can have as much effect on bottom-line profits as the product itself. Information-defined transactions—value creation and extraction in the marketspace—are creating new ways of thinking about making money and, thus, are changing the definition of value creation. Until recently, voice-mail services were available only to large corporations supported by private networks. Today, the services are available at your home as sophisticated voice storage and retrieval systems. They may forward calls, store messages, forward messages with notes attached, and broadcast messages to numerous users, all at the same time. Information technology has also allowed the replacement of a physical product (the answering machine) with an information-based service. With voice mail, the phone companies have identified and exploited tremendous marketspace potential. The answering machine is being challenged by a new service that has no products for the customer to buy and no appliance to maintain, and all charges are included in the phone bill—even though the cost of that service may exceed the amortized cost of an answering machine. Some answering-machine companies are losing out because they didn't see the importance of information-defined transactions.

Companies that do not understand the new technologies will miss opportunities, customers, and new ways of conducting business. Often, the business strategies and systems and technology implementation are not coordinated, resulting in a significant gap between the two. Rather than providing synergy, this gap results in the following issues:

- Lack of clear and consistent overall sponsorship and direction
- Lack of integration among applications
- Large technology investments without adequate returns on investments
- Insufficient coordination among organizational units
- Poor connectivity
- Difficulty with data interface
- Disorganized data storage

Workspace Change Processes and New Perspectives.   Because of the rapidly changing market and the increasing need for flexibility and speed in adjusting to new conditions, workplace change processes require a fresh look. Facilities management needs to consider the same set of ideas that corporations are using for product and idea innovation. High-performance teams are necessary to understand the new patterns of work and to suggest how these new patterns are best supported. These teams must be drawn from expertise in facilities management, in information and infrastructural technology, in human resources, and in organizational strategy to guarantee fit with the central mission of the organization.

The traditional methods associated with top-down approaches and rigid schedules simply will not do. Success is more likely found by fostering genuine and informed participation in evaluation, design, and decision making.[7] This approach emphasizes collaborative engagement of all stakeholders. It incorporates the informal and spontaneous activities that usually take place outside the traditional approaches and connects them to the workplace-making process. This is accomplished by the addition of activities that are not typical of traditional processes: increased attention to setting up the process of workplace change, and discovering and defining the problems to be addressed; and a focus on building the client and professional teams and on achieving an agreed-upon project procedure and schedule. As contrasted with traditional processes, these activities are best revisited from time to time as the project proceeds. For greater detail on approaches to programming, an essential part of the newer approaches, see Chap. 4.

Facilities Managers' Responsibilities

Facilities managers have two important ways in which they can respond to contemporary corporate strategy.

The first of these is to take their place in the corporate management team. This implies that facilities must be seen as important resources to be deployed in ways that are most effective for the strategic approach under consideration. This may mean considering ways of utilizing facilities that are radically different from the ways they are used today. Also, it will require being able to demonstrate the value of alternative approaches to the use of facilities within a particular corporate strategy.

The second is for facilities managers to use the corporate strategy approach in their own area of the organization. This implies understanding new directions for facilities management, including some approaches now used by top management.

Facilities management sits typically within the operations management area of an organization. It is usually thought of as the support for the production system that permits an organization to create its products from incoming materials and supplies.

An implementation strategy requires a vision for the operations process that establishes and formulates a set of objectives for decision making. This vision should result from and complement the corporate strategy. The strategy should be the basis for developing a plan that establishes a consistent pattern of decision making, resulting in an operational and competitive advantage for the company. Through an operations strategy, an organization can convert its core strategy into an operational plan.

The Implementation Plan. A comprehensive, measurable, attainable, and understandable implementation plan must be developed and shared throughout the organization. It should include assessment of the firm's strengths and weaknesses, performance measures for evaluating results, and a benchmarking plan aimed at achieving best practices. The plan should be developed collaboratively with those who have a valid stake in the outcome. In addition, it should include specifications for changes in all relevant workspace elements: space, organization, finance, and technology.

Redesign Tools. In recent years, performance measures have evolved as a way of checking the vital signs of corporate health. Performance measures are usually classified under the three general categories of cost, quality, and time. Companies are finding out that improving quality and productivity while reducing cost go hand in hand with enhanced customer satisfaction. At the same time, they have found that the best way to ensure external customer satisfaction is to satisfy every internal customer at each step of the process. This is true for both manufacturing and service operations.[8]

Among redesign tools, benchmarking has been used with great success. Benchmarking is the activity by which an organization measures its performance against standards that it wishes to achieve. There is an accumulating and valuable body of case studies; following is a description of how an organization can provide continuing support. According to Vicki Powers, writing in *APQC,* its objectives should include the following:

- Developing a system that facilitates continuous improvement through regular use
- Identifying areas of excellence and making improvements to reach the level of best practice
- Building a system that can be used in public to demonstrate the value of its services[9]

Powers goes on to distinguish between performance benchmarking and process benchmarking. *Performance benchmarking* is used to analyze relative performance among similar organizations. It typically includes identifying the critical success areas for the organization (or organizational unit), identifying the key performance indicators for each of those areas, identifying the gaps between the organization's performance and the best practices with which to compare performance, and identifying the highest-priority areas for improvement.

Process benchmarking is similar, but it focuses on the organizational means to bring about the needed improvements identified through performance benchmarking. It identifies the organizational processes necessary to bring about the improvements, develops performance indicators, identifies the gaps between these processes and best practices identified as appropriate standards for comparison, and brings about changes in the relevant area.

Both types of benchmarking require the identification of best practices against which to compare the performance of the organization. There is no single best practice, because best for one may not be best for another. Every organization is unique in some way. Corporations have different missions, cultures, environments, and technologies. What are meant by *best* are those practices that have been shown to produce superior results; selected by a systematic process; and judged to be exemplary, good, or successfully demonstrated. Best practices are then adapted to fit a particular organization.[10]

One of the easiest ways to improve your processes and practices is to benchmark yourself against the "best in the class." By finding out what works well somewhere else, you may be able to introduce innovative ideas to improve your organizational unit's processes. For example, facilities managers might focus on the process of space programming, purchasing of furniture and equipment, or contracting for small jobs as the key areas to improve to react more quickly to the formation of high-performance teams, a corporate objective resulting from a performance benchmarking exercise.

One of the most famous benchmarking groups is the International Benchmarking Clearinghouse (IBC) set up by the American Productivity & Quality Center. The blue-chip corporations that make up the IBC pool their information. They provide their corporate vital signs as a basis for comparison. Other corporations may use the IBC information base to compare themselves against the best in the class.

Other redesign tools include reengineering and criteria and procedure lists from many authors.[11] It is essential to choose tools that are convenient to use and easy to reuse over time. Also, it is essential to use commonly understood measures of performance. Any tool, however, must be chosen in the light of the organization's mission and its current circumstances. When an organization is striking out in a new direction, the best practices of others may not be the best standards to use!

Commonsense Rules.   No matter what redesign tool you select, you should balance it against the aims of the implementation plan and against common sense. It should be the framework within which you carry out your redesign procedures and your implementation plan (see Fig. 1.2-6). Here are some commonsense rules that are as applicable to the facilities management area as they are to the corporation as a whole:

- Identify current problems and eliminate or rectify the source of the problem.
- Ensure that information and data are available in the first step and in the right format for all subsequent steps.

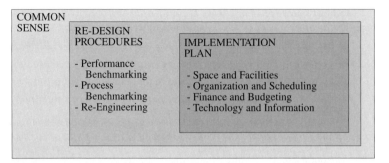

FIGURE 1.2-6 Facilities management's responsibility. Common sense must be the fundamental framework for evaluation and project planning. (*Source: Courtesy of Paul R. Smith, Peak Leadership Group, and William L. Porter, Massachusetts Institute of Technology.*)

- Find the right sequence of activities to occur, and collect information just in time to process it.
- Simplify the process.
- Replace specialist-to-generalist functions by increasing the scope of responsibility of personnel's job descriptions.
- Replace after-the-fact reconciliation with front-end edits, validation, and QA checks.
- Eliminate approval/authorization processes. If still warranted, make them after the fact.
- Make the process logical, rational, and fact-based, and include creative, intuitive, and instinctive elements of strategic insight.
- Implement a clear decision-making process and clarity in the organizational design. The management team must understand the market into which it is selling and create an effective strategy to maximize shareholder value. The corporate strategy is concerned with the intermediate term of a couple years out; the operational management is concerned with the current year.
- Make sure that the strategy has full executive support.
- Expect the development process to make all participants uncomfortable initially, but give them a legitimate stake in the outcome.
- Ensure that the corporation's ambitions and objectives are clearly understood.
- Understand that the corporate strategy development is a learning process that should develop consensus, new insights, and clear communications.

Global and local competition can provide intense pressure to be the best in your industry. To stay the best, you must search out and find the new tools that will keep you ahead of your competition. In making change within facilities management, implement change quickly but "don't outrun your headlights."

NOTES

Section 1.1

1. Tom Brokaw, *The Greatest Generation,* Random House, New York, 1998.
2. L. Sprague DeCamp, *The Ancient Engineer,* Doubleday, Garden City, NY, 1963.

Section 1.2

1. H. S. Resnick, *Business Process Reengineering: An Executive Resource for Implementation,* Work Systems Associates, Marlborough, MA, 1994.

2. L. E. Greiner, "Evolution and Revolution as Organizations Grow," *Harvard Business Review* (May-June 1998).

3. M. L. Joroff and B. Feinberg, "How CIR Accommodates Efficiency and Transformation," IDRC paper (July 1999).

4. These ideas are developed in several publications: M. L. Joroff and B. Feinberg, op. cit.; F. Becker, and F. Steele, *Workplace by Design,* Jossey-Bass, San Francisco, 1997; and in T. Horgen, M. Joroff, W. Porter, and D. Schön, *Excellence by Design: Transforming Workplace and Work Practice,* Wiley, New York, 1999.

5. F. Duffy, with contributions from K. Powell, *The New Office,* Conran Octopus, London, 1997.

6. F. Duffy, D. Jaunzens, A. Laing, and S. Willis, *New Environments for Working: The Re-design of Offices and Environmental Systems for New Ways of Working,* E & F SPON, London, 1998.

7. T. Horgen, M. Joroff, W. Porter, and D. Schön, op. cit.

8. H. James Harrington, *Business Process Improvement: The Breakthrough Strategy for Total Quality Productivity, and Competitiveness,* McGraw Hill, New York, 1991.

9. The American Productivity & Quality Center (APQC) "helps enterprises manage change, improve processes, leverage knowledge, and increase performance by becoming more agile, creative, and competitive. These pursuits require high-quality information, strategies, skills, knowledge, experience, contacts, and best practice. . . . Founded in 1977, APQC is a nonprofit education and research organization supported by more than 550 founders, Clearinghouse designers, and members." From http://www.apqc.org/apqcglan/glance4.htm. See V. J. Powers, "Benchmarking in Hong Kong," *APQC,* no. 11.

10. The APQC created the International Benchmarking Clearinghouse (IBC) as a resource for organizations interested in using benchmarking as a tool for breakthrough improvement. This is based on the premise that all businesses in the world perform essentially the same basic processes (e.g., develop strategies, hire people, purchase resources, make products or deliver services, and bill customers). They all have operational and management and support processes. For further information, see http://www.apqc.org/best/bmk.

11. Another of the strategies for improving organizations is called Business Process Reengineering (BPR). The reengineering trend has had major acceptance in America since Hammer & Champy, the reengineering gurus, issued their book *Reengineering the Corporation: A Manifesto for Business Revolution* (Harperbusiness, 1994). Recently, however, it has also caused a backlash, some of it justified and some guilt by association. Many reengineering efforts have failed in the past. Michael Hammer and James Champy reported a 70 percent failure rate in the efforts they have seen. An added problem in assessing success and failure is that some people use the term *reengineering* for just any downsizing or rightsizing activity that may be a butchering rather than surgical approach for cost cutting. Stowe Boyd, in his article "Business Process and Information Technology," states that the major message to learn from slash-and-burn reengineering is that it is possible, but not necessary, to rapidly inject a process-centered management style into old-style business. And AT&T has a concept called *Integrated Diversity,* which, translated, means plan from the top down for modeling and implement from the bottom up for building. Many authors have listed practical ways of improving management practice. These include Mark Youngblood in *Eating the Chocolate Elephant: Take Charge of Change Through Total Process Management* (Richardson: Micrografx, Inc., 1994), and Steven Rayner in *Recreating the Workplace* (John Wiley & Sons, New York, 1995).

CHAPTER 2
FACILITY FINANCIAL MANAGEMENT

Paul R. Smith, P.M.P., P.E., M.B.A., Chapter Editor
Peak Leadership Group, Boston, Massachusetts

Thomas W. Grottke, C.P.A.
KPMG, Hartford, Connecticut

Douglas Christensen, C.P.A.
Brigham Young University, Salt Lake City, Utah

H. Thomas O'Hara
Suffolk University, Boston, Massachusetts

SECTION 2.1
INTRODUCTION

Why is it that some managers always have their pet projects funded, whereas others do not? Could it be that the winners have learned how to speak the language of business and package their requests in the appropriate business context? We hope that reading this chapter will help you articulate and present your ideas in a manner that gets results and helps you communicate in terms that business managers understand.

Many companies have stories about brilliant strategies that were not implemented. Everyone supported the new strategy, but somehow it was not implemented or was implemented haphazardly. The answer lies with managers themselves—or more specifically, with the way managers direct their energies. Managerial energy is the organization's most important and scarcest resource. When new opportunities arise—such as falling trade barriers, deregulation, emerging market changes, and technological breakthroughs—all of these changes tug on management's attention and business resources. Which opportunity should the corporation pursue and why? This chapter focuses on developing a clear and concise business case to drive supportive business decision-making.

Our goals in this chapter are to provide sufficient knowledge (1) to help the facility manager become part of the senior management team, (2) to establish a standard by which facility managers understand the information needed to have a well-run facility, (3) to combine theory and applications, and (4) to provide a document that facility managers find user-friendly. As with Chapter 1, we did not look at other facility management books and try to improve on them. Instead, we started with a blank page and decided, on the basis of our knowledge and experience with executives and financial officers in business today, what the facility manager should know.

This chapter will identify a process that is critical to the success of the facility manager. Each facility manager has (1) a specific role, given the mission assigned, (2) to plan for and align to the future, (3) to be responsible for the plan, and (4) to report the results, given the resources assigned. This process shows up in many of the things managers try to control. Whether it is a project, a service request, or an annual budget, the manager needs to understand certain key elements within a business context that are part of this process and responsibility.

SECTION 2.2
BACKGROUND INFORMATION

AN OVERVIEW OF FACILITY MANAGEMENT

It took General Dynamics, a major defense contractor, 43 years to increase the value of its stock to $2.7 billion. Netscape Communications accomplished the same feat in about one hour in August 1995, when investors had their first chance to buy the stock on the open market. This fact is amazing, considering that Netscape had never earned a profit at that time and had been giving away its primary product free of charge on the Internet. Why is this example pertinent to facility managers? Understanding the linkage between a company's market value, its publicly reported earnings, and underlying business decisions will help facility managers become key individuals in their organizations.

Facility management has many aspects. Very few industries view facilities management in the same way. Some industries focus on real estate, others focus on property management, and still others focus on asset management in general. A critical part of facility managers' roles is to understand their missions and levels of responsibility. Facility managers are entrusted with a stewardship role over the physical assets and overall security of an enterprise. Generally, their work involves facility planning, use, and design; maintenance and operation of assets; ongoing replacement of assets; and retrofitting and updating of assets as needed over time.

These responsibilities, if central to the day-to-day aspects of facilities management, are sometimes referred to as *core competencies*. Usually there are resources associated with each competency. The role of a facility manager is to plan for and manage these functions and resources.

Effective facility managers should use balanced approaches in dealing with all aspects of their areas of responsibility. Kaplin and Norton established the balanced scorecard approach to ensure that *all* of the resources, including financial resources, are coordinated and focused on supporting the achievement of the organization's overall objectives. Managers tend to focus on issues and easily identifiable work rather than looking at all aspects impacting performance. The balanced scorecard approach directs a manager to consider goals and requirements from the perspectives of the customer, employee, process/technology, and the owner. The internal processes affect how things get done; the customer's perspective helps identify the quality of the work; the innovation and learning perspective focuses on how well the employees learn and advance the mission of the organization; and the owner's perspective typically deals with financial returns. Even though much of the focus is on the financial perspective, ignoring the other perspectives could affect the manager's success. In most cases, accounting for the amount of resources and showing how they were used and managed is a big part of a manager's job.

FACILITY MANAGEMENT FUNCTIONS WITHIN THE ORGANIZATION

Facility management provides an organization with the skills and knowledge to manage its physical assets. This service to an organization impacts a significant portion of any available resources. Therefore, facility managers need to understand the nature of corporate governance, accounting and its principles, and the general principles surrounding the financial obligations associated with the way a company or institution is organized, the type of industry within which it operates, and what considerations may be specific to that industry. Facility managers live within an organization and typically are responsible for managing depreciable assets or significant costs that a company must use efficiently.

SECTION 2.3
OVERVIEW OF FINANCIAL MANAGEMENT

THE NATURE OF ACCOUNTING FOR RESOURCES

Most of the world's work is done through organizations that deploy resources to meet some specific objective. In doing work, organizations use *human* resources, materials, various services, buildings, and equipment. These resources cost the organization money. To work effectively, the people in the organization need information about these resources. Accounting is a key department in the process to provide this information.

Organizations can be broadly classified into two categories: *for-profit* and *nonprofit.* As these names suggest, the dominant reasons for a corporation's existence are to make a profit and increase the shareholder value, whereas nonprofit organizations have some other reason for existence and, at a minimum, must break even. In the United States, approximately two-thirds of employed persons work in for-profit organizations, and the remaining one-third work in governmental and nonprofit organizations. The objectives of accounting are similar in both types of organizations.

Accounting information is traditionally divided into three categories: *financial accounting, operating information,* and *management accounting,* as described in the following sections.

Financial Accounting

This information is intended for company managers and for parties external to the company, including shareholders (trustees in nonprofit organizations), banks and other creditors, governmental agencies, investment advisors, and the general public. Shareholders who invest in a company want to know the financial health and financial performance of the company. Financial accounting information provides this information using generally accepted practices as a basis. This consistency of financial information allows the general public to review the financial condition, income or loss, and cash flows in a consistent and prescribed manner.

Operating Information

Most of what any company does is captured in internal reports. Personnel hired; time consumed during the day; hours of operation; goods or services received, created, stored, transported, and provided to end users (e.g., customers); and other relevant data regarding business activities of a company are captured, stored, modified, and reported in one form or another for supervisors, managers, and others to review.

Although the financial accounting systems of a company are built to capture and reflect all of the business activities of the company, financial accounting ignores nonmonetary activities when capturing and reporting. Thus, operating information exists to provide such nonmonetary data along with financial data.

Management Accounting

Managers may not have time to review the details provided in the operating information, so they turn to summaries provided by management accounting information. These summaries are used to plan, implement, and control departments or projects. Examples include fiscal budget reports, reports summarizing staffing head counts, product line sales reports, and other business unit performance reports.

In this chapter and this handbook, operating information and management accounting information together will be referred to as *internal reporting*. Financial accounting will be referred to as *external reporting*.

Before moving ahead with our discussion of financial management and management decision making, let's review a few key terms and aspects of accounting. These terms represent the language used within a company to summarize the financial activities of the business.

Generally Accepted Accounting Principles (GAAP).   The foundation of accounting rules and guidelines is a set of Generally Accepted Accounting Principles (GAAP). These principles are currently established by the Financial Accounting Standards Board (FASB). Companies are not legally required to adhere to GAAP as established by the FASB unless they are regulated by the Securities and Exchange Commission of the United States. However, as a practical matter, there is strong pressure to do so. When a certified public accountant (CPA) audits a company, it is to render an opinion as to the company's adherence to GAAP.

Financial Statements.   Financial statements are an important management tool for managers. When correctly prepared and interpreted, they contribute to an understanding of a company's current financial health, financial performance, and potential. GAAP require that financial statements include three reports: (1) a balance sheet (sometimes called a Statement of Financial Condition), (2) an income statement (sometimes called a Statement of Operations or Profit and Loss Statement), and (3) a Statement of Cash Flow.

1. *Balance Sheet.*   The balance sheet presents a picture of assets, liabilities, and owner's equity of a company or entity as of a specific date. The balance sheet portrays the entity's financial condition or health. The balance sheet is so named because it represents the following equation:

Assets	=	liabilities	+	owner's equity
(resources of the business)		(amounts owed to others)		(amounts owed to the owners)

This basic equation holds at any time, although the amounts assigned to the individual elements will fluctuate by type of business and relevant industries.

2. *Income Statement.*   The income statement is a summary of the revenues (e.g., sales) and the expenses (e.g., cost of doing business) and clearly states the net income or loss for a specific period of time. It is very important to understand that the income statement represents a period of time, typically a quarter or a year. The income statement portrays the operating results of a company for the applicable period of time.

3. *Statement of Cash Flow.*   The statement of cash flow reports the sources and uses of cash for the same period as the income statement. The cash flow statement consists of three major classifications:

- Cash provided by or used by operations
- ± Cash provided by or applied to investment activities
- ± Cash provided by or applied to financing activities

The sum of these net increases or decreases in cash is shown in this statement.

To be a part of a company's leadership team, a facility manager needs to understand the nature of financial reporting. Understanding how your responsibilities affect the bottom line (e.g., net income) is critical. Most facility management areas represent an expense to a corporation. Knowing the financial impact of resources is important. Accounting is the language businesses use to communicate their performance.

In addition to accounting principles for financial reporting, there are differences stemming from the kind of entity the business is. Financial management is driven by certain accounting principles associated with the ownership structure and the type of industry. Facility managers need to be aware of the way a particular financial management structure can impact the way it functions in the organization.

GENERAL PRINCIPLES OF FINANCIAL MANAGEMENT

Accounting personnel in organizations are not the only individuals concerned with and responsible for the finances and record keeping in a company. Accounting typically is responsible for preparing financial reports using accumulated financial results and activities of the company. The extent and type of financial reporting depends on a number of factors. First, the legal entity typically drives the accounting and financial reporting processes of a business. Second, the type of industry or business has a significant impact on the accounting and financial reporting practices of a company. Third, and finally, a clear distinction must be made and understood regarding the objectives of accounting. The Accounting Department typically is responsible for two basic types of financial reporting, *external* (or *public*) reporting and *internal* (or *management*) reporting. Therefore, financial management is impacted by a company's ownership structure, the type of industry, and key accounting practices.

Ownership Structure Drives Financial Management

The purpose of accounting is to keep in balance the investment (traditionally called *assets* in accounting terms) and financing (traditionally called *liabilities and equity* in accounting terms) activities of a legal entity. What is a legal entity? All businesses that are registered with a governmental body become formal legal entities. Certainly, it is obvious that General Dynamics is a legal entity. But at times, most start-up business ventures are not so clearly "entities," let alone "legal entities."

A legal entity has documents that define its organization, its ownership, its business purpose, its management structure, and its operating policies. These documents are typically the Bylaws, Articles of Incorporation, and Statement of Purpose. The documents create a board of directors that is responsible for managing the business. These documents create the foundation for legal ownership interests and the rights of stockholders. Although not all companies are stock companies (e.g., mutual companies and many not-for-profit organizations), our focus in this book is on companies with stockholder ownership structures.

There are three major types of business organizations: (1) *sole proprietorships,* (2) *partnerships,* and (3) *corporations.* There are also several hybrid forms, such as *limited liability partnerships* (LLP), which are beyond the scope of this book.

Sole Proprietorship.   A sole proprietorship is an unincorporated business owned by an individual. This is the simplest form of organization and can be set up with limited expense. If you use your own name and tax identification number, you may not need to register your company with anyone. However, if you use another name, you should register the name in the form of "doing business as" with the state and federal governments. The sole proprietor receives all the profits or losses generated by the endeavor.

Partnership.   A partnership is formed when two or more individuals decide to conduct business together in an unincorporated form. Partnerships may operate under different degrees of formality, ranging from an informal handshake to a formal agreement drawn up by attorneys and filed with the state where the partnership was formed. The partnership agreement stipulates who owns what and who shares in the profits or losses of the enterprise.

Corporation.   A corporation is a legal entity incorporated by the state in which it was formed. It is a legal entity separate from its owners and managers. Generally, if a corporation has more than 300 shareholders, it must file with the Securities and Exchange Commission of the United States (SEC). This action also allows the corporation to become a "listed" stock or one of the public stock-trading markets like the NYSE. However, it is very important to note that many corporations are "closely held" or subsidiaries of larger companies. These types of corporations typically do not have stock traded on public stock markets.

Sole proprietors and partners have unlimited liability against losses while corporation executives have limited liability.

Type of Industry Drives Financial Management

Have you ever read in newspapers or magazines or heard the news media refer to a company's performance as better or worse than "peers" or "industry norms?" Distinct businesses that manufacture, build, distribute, and/or sell similar products and services are grouped into industries. Obviously, these grouped companies compete with each other at times. But they also pool their collective needs to lobby government actions. However, for accounting and financial management purposes, it is very important to recognize that these industry groups create a natural source of performance comparison. The senior management team of a company typically is acutely aware of how well their financial results compare to those of their peers. Peers are those companies considered similar enough to be grouped in the same industry.

External financial statements are traditionally the source of comparative financial performance. Therefore, a major consideration impacting management decisions, such as a major facility initiative, will be the expected effect on the company's reported financial results compared with those of peers, not just the positive or negative direction of its own earnings.

The potential market value of a company's common stock depends on industry comparisons made by investment banking firms and analysts. The value of a company's common stock is a fundamental component in a free market society that allows the public to invest in companies. The capital markets are the open trading exchanges such as the New York Stock Exchange (NYSE) and the National Association of Securities Dealers (NASDAQ). Today, many small companies are publicly owned (shares of common stock have been sold to the public through the capital markets, and their stocks are traded on one of the exchanges). The external financial statements prepared by a company's Accounting Department provide most of the data used by investment bankers to evaluate and compare financial performance with that of other companies.

The management team must make a range of financial decisions to keep its company growing and prospering. One of the first decisions, the form of the business organization, was made at the company's inception. Other decisions, such as the amount and type of future financing needed, which projects are to be funded in the future, and how to manage cash flow will bring new challenges to the management team of any new or dynamic company. If things go well, at some point earnings will be generated beyond what is needed to keep the organization going, and the company will decide how much of its earnings it will pay out as dividends to shareholders. Most of all, management must be aware of the mission, vision, and goals of the company; how their actions bring them closer to or further from meeting these goals; and how the impact of these actions affects the earnings and therefore the stock price of the company. Performance measures consisting of key financial ratios help the company define and report its goals in measurable terms. Several common financial ratios used to assess a company's health are listed here:

$$\text{Return on equity (ROE)} = \frac{\text{net income}}{\text{average stockholders' equity}}$$

$$\text{Earnings per share (EPS)} = \frac{\text{net income}}{\text{average shares of common stock outstanding}}$$

$$\text{Current ratio} = \frac{\text{current assets}}{\text{current liabilities}}$$

$$\text{Leverage ratio} = \frac{\text{debt}}{\text{shareholder equity}}$$

$$\text{Price to earnings ratio (P/E)} = \frac{\text{market price per share}}{\text{EPS}}$$

Understanding the impact of management decisions on these ratios is very important in gaining the support and confidence of senior management. The financial impact needs to be evaluated as of today, as well as over the future life of the project under consideration. Demonstrating the impact on earnings and financial position from the investment and results of a facility project will go a long way toward gaining acceptance from the executive officers of a company. The following basic example highlights how a modest new facility costing $500,000 to set up can impact reported financial performance:

Example: Purchase of New Facility

New Facility purchase and setup cost $500,000

New facility operating cost $80,000 (annually), excluding depreciation

Current operating results and key balance sheet facts of our example company (before the new facility):

Net Income	$500,000 (annually)
Long-term debt	$2,500,000
Shareholder equity	$1,500,000

Key financial results—impact	Calculation	ROE	Leverage ratio
Before the new facility	500,000/1,500,000	33%	1.67 times
Immediately after buying the new facility		N/A	2.00 times
After the first year of using the new facility	420,000/1,500,000	28%	N/A

This example holds all other variables constant, assumes no incremental growth in revenue, and reflects no change in equity or debt pay down.

As you can see, if only the cost of a project is presented to management against historical operations, the result will be a negative impact. This will not typically get the buy-in of senior officers. Thus, the facility manager needs to demonstrate the benefits in financial terms in addition to the costs.

Key Accounting Practices Drive Financial Management

The financial management practices of an organization are closely related to its business planning processes. In theory, the board of directors and its management team periodically prepare a strategic plan that states the company's mission, vision, values, goals, and objectives. The primary benefit of the strategic plan is to focus the efforts of the organization on achieving its goals and identifying significant investments necessary for the company to achieve its goals.

Upon completing the strategic plan, management prepares its annual technology plan, capital plan, and financial budget. This is the culmination of management's planned initiatives translated into the estimated impact on the company's financial statements. The importance of the technology plan is the consideration of the cost of new equipment, facilities, and software. Each defined department or division prepares its annual plans with the associated income to be generated, expenses to be incurred, cash flow created and used, and capital needs. Table 2.3-1 gives an example of how each department in a company may impact the combined financial results.

We can all understand that income represents the sales revenue of the company, and expenses represent costs incurred. But the impact on a company's cash flow and the capital needs of a department or division may need further discussion. The Accounting Department of a company will attempt to identify, capture, and account for the differences between cash and the accrual basis of transactions and events. Cash basis accounting is as it sounds. The accounting entries are made to reflect the receipt and disbursement of cash. The accrual basis of accounting attempts to reflect the economics of business transactions and activities. What does that mean? Economics? What we are referring to is the fact that a sale or expense has occurred without the actual receipt or disbursement of the company's cash.

For example, if a company had only a single transaction before its December 31 year end, what would the company's financial statements show on December 31?

TABLE 2.3-1   Summary Budget Worksheet Combining Departments

	Projected expenses in dollars (cash outlays)					
Departments	Executive	Accounting	Production	Sales	All other administration areas	Total combined company
Expenses						
Salaries	350,000	400,000	1,250,000	500,000	300,000	2,800,000
Benefits	100,000	125,000	350,000	190,000	114,000	879,000
Supplies	20,000	40,000	20,000	50,000	25,000	155,000
Equipment	0	0	0	0	200,000	200,000
Promotion	0	0	0	250,000	250,000	500,000
Other	25,000	50,000	200,000	50,000	325,000	650,000
Depreciation	395,000	615,000	1,675,000	475,000	3,160,000	6,320,000
Total expenses	890,000	1,230,000	3,495,000	1,515,000	4,374,000	11,504,000
Capital cost						
Furniture and equipment	10,000	25,000	100,000	75,000	10,000	220,000
Facilities	0	0	250,000	0	0	250,000
Computers	10,000	10,000	100,000	15,000	10,000	145,000
Other	0	0	0	0	10,000	10,000
Total	20,000	35,000	450,000	90,000	30,000	625,000

Timing	Cash vs. accrual basis example	Accounting entries		
On Dec. 15	Company sells inventory on credit to a customer for $100		Debit	Credit
		Accounts receivable	$100	
		Sales	$100	
On Jan. 15	Customer pays for inventory (30 days later) it bought with no discounts		Debit	Credit
		Cash:	$100	
		Accounts receivable		$100

In this example, the sale is reflected by the company, as of December 31, because the economic transaction, the sale of the inventory, occurred. However, the company reports that it has receivables of $100 on December 31, not cash, reflecting the credit extended to its customer. This is the accrual basis of accounting for the sale.

Again, understanding the timing of cash flows stemming from a major initiative and recognizing the impact of a facilities initiative on the financial statements of a company is how senior officers assess whether to go ahead with a plan.

Capital needs represent the purchase or development of property, plant, and/or equipment. This is clearly very relevant to facility managers and major facilities initiatives. In accounting terms, the difference between an expense and a capital expenditure is the purpose and expected benefit to be derived from the disbursements. Generally, if the benefits to be derived from the expenditures are expected to last beyond one year, it is classified as a capital expenditure. This is also a very important concept that creates another difference between cash and accrual accounting.

Have you ever heard of the concept of depreciation? Of course, we all know that the day we buy a new car and drive it off the auto dealer's lot, it depreciates in value by a large percentage! But what does that mean in accounting terms?

Except for land, most items of plant and equipment have limited useful lives. They will provide services to the entity for a limited number of future accounting periods (e.g., years). Therefore, a fraction of the cost of the asset is properly chargeable as an expense in each of the accounting periods in which the asset provides services to the facility. The accounting process for this gradual conversion of plant and equipment capital cost into expense is called *depreciation*.

We mentioned that capital needs are for acquiring property, plant, or equipment expected to provide benefit to the company beyond one year. Therefore, accounting attempts to reflect this future benefit to the company by providing an annual depreciation expense in the financial statements. The entire cost of the capital item is depreciated over its estimated useful life, even though the typical expenditure (e.g., cash disbursement) is made for the capital purpose in the year acquired or built. Let's look at another cash versus accrual accounting example as it relates to the purchase of a new car.

Example of Cash Outlay and Depreciation

Purchase price of new item	$25,000
Immediate depreciation	50%

If accounting was to record these events, it would look like this:

	Debit	Credit
Purchase using cash		
Debit automobile	$25,000	
Credit cash		$25,000
Depreciation		
Debit depreciation (expense)	$12,500	
Credit accumulated depreciation		$12,500

The company balance sheet would be impacted as follows:

	Purchase of car	
	Before	After
Assets		
Cash	$25,000	0
Auto	0	25,000
Less accumulative depreciation	0	(12,500)
Total assets	$25,000	$12,500

Wow! What happened to $12,500 of the company's assets? Well, that is what depreciation does to assets. It periodically and consistently reduces the assets for the estimated and expected reduction in value or usefulness. Thus, the company's $25,000 in cash it had previously reported on its balance sheet is now a car worth only $12,500. The company's income statement would have shown an expense for depreciation of $12,500.

This example holds true for the purchase and/or building of any assets of any company. The Accounting Department will use accrual accounting to reflect the depreciation expense of the capital outlay.

A number of business situations will create this type of noncash annual accounting for the economic substance of an event or investment in a capital item such as a plant. In addition to purchasing physical assets, financial statements may include such items as goodwill, which is amortized to expense it over its estimated life. Reserves for bad debt are also established to reflect the fact that certain of the company's accounts receivable may not be fully collected from its customers. Either deterioration or obsolescence limits the useful life of a tangible long-lived asset. Deterioration is the physical process of wearing out. Obsolescence is the loss of usefulness because an improved product or process is developed, styles have changed, or for other reasons unrelated to the physical condition of the asset. The useful life is usually determined as the time it takes the asset to wear out. The time before it becomes obsolete is usually called its *service life*. Other noncash accounting entries include:

- Economic events accrued for in financial statements without a cash outlay or cash receipt
- Sales made on credit terms
- Expected sales warranties (discounts to be provided to customers)
- Expected bad debts (not all sales made on credit will be paid)
- Inventory obsolesce (some inventory will not sell)
- Idle property, plant, and equipment (some assets are not fully used)
- Commitments to buy (contractual liability)
- Excess purchase price versus fair market value (goodwill)

Last, besides external reporting the financial management function of a company includes accounting and reporting internal management information. These reports may (and often do) differ from the external financial statements of the company. Management reports are for internal use and should be confidential. Competitive and other sensitive data about a company's business may be included. These reports need not comply with GAAP (Generally Accepted Accounting Principles). They are developed in a manner that best captures the data critical to management to run its business. Accounting Departments must therefore use the accounting systems to capture and report for internal and external purposes.

Generally, internal management reports are provided to department heads and used in management meetings. However, as we discussed earlier, the impact of a business decision on the external financial statements of a company must not be overlooked.

SECTION 2.4
FACILITY FINANCIAL PLANNING

After the facility manager understands the mission, functions, and resources that have been assigned and has a clear understanding of the type of financial structure and accounting principles needed to be successful in the organization, the next step is to plan and establish financial objectives. It is critical that facility managers understand the importance of planning. To be effective, all levels of the organization need to plan. A summary of planning within facilities management should align with the organization's overall plan. This is especially critical in planning for an organization's use of short-term and long-term financial resources.

Planning is the process of deciding what future actions will or will not be taken. A single plan should be made for the entire organization to show the corporation's health at a glance.

Making the planning process effective and efficient is always a question to management. Does planning pay dividends? Is the effort worth the time and attention? The only way to deal with the changes surrounding business today is to plan. The business professional will always need to plan. A tool to help focus on management's ability to plan is hard to measure, but a qualitative measurement has been suggested that would help measure the effectiveness of planning.

RETURN ON MANAGEMENT

When management's attention becomes diluted or is distracted from looking at too many opportunities, even the best strategy stands little chance of being implemented. So, making sure that management focuses its attention on the best opportunities is the most important thing to look at. Simons and Davila recommend that managers use a new ratio called *return on management* (ROM) that is expressed as follows:[1]

$$ROM = \frac{\text{productive organizational energy released}}{\text{management time and attention invested}}$$

Like its cousins *return on equity* (ROE) and *return on assets* (ROA), ROM measures the payback from the investment of a scarce resource. In this case, ROM measures a manager's time and attention. Because both the numerator and denominator are estimates, the results are more qualitative than the traditional quantitative measures for ROE and ROA. Nevertheless, by computing ROM for various options, specific clear strategic priorities for the return on time and energy may be estimated to select the most promising option.

The allies of high ROM are defined as:

1. Clearly defining which customers, projects, investments, or activities are beyond the organization's boundaries.

2. Critical performance measures are selected for one purpose—to keep everyone looking at the results that count.

3. Managers all know the critical performance measures, of which there are no more than seven at any time.

4. Managerial processes and paperwork exist only when they add value to the bottom line. Employees know on what and where to focus their attention.

In addition to the planning pressures created by a changing world, other management issues impact an organization's financial areas. In this millennium, the scope and business conducted will challenge those that manage assets.

SECTION 2.5
FINANCIAL MANAGEMENT IN THE NEW MILLENNIUM

Two key issues are exerting a pervasive impact on financial and facility management—the globalization of business and the advent of information technology.

GLOBALIZATION OF BUSINESS

The globalization of business will continue because of several major factors. These include (1) improvements in transportation and communication, which lower shipping costs and make more international trade possible; (2) growing numbers of intelligent consumers who desire high-quality, low-cost products; (3) lowering of trade barriers; (4) deregulation of numerous commodities (i.e., telecommunications and utilities); (5) technological advances that have reduced the cost and time to deliver products and services to the marketplace; (6) global acceptance of the Internet for trade; and (7) the world populated with multinational firms that can shift production to places where production cost is the lowest.

Many manufacturing companies have gone offshore to reduce their labor and material costs, while service companies such as banks, insurance, utilities, accounting firms, telecommunication firms, and consultants are seizing global opportunities to expand their customer bases and better serve their global customers.

INFORMATION TECHNOLOGY

The new millennium will see a revolutionary way of making financial decisions by individual businesses. The Accounting Department's role is changing from that of a number cruncher to partnering with other departments to add value to the financial aspects of its services. All personnel in business must become extremely proficient with technology to access and use the wealth of information available.

The World Wide Web or Internet will challenge the way business is conducted, and firms that fully understand the information highway and learn the new rules of engagement will have an advantage. Videoconferencing will allow face-to-face meetings with distant colleagues, and the Internet will allow real-time simultaneous access to data and information by multiple users.

THE FINANCIAL MANAGER'S NEW RESPONSIBILITIES

Financial managers are chartered with the responsibility to maximize shareholder (or the owner's) value. These are some of their specific roles:

Strategic Planning

The Accounting Department develops a financial plan that complements the strategic plans of the company. This plan should be a living document that is updated continually as the marketplace and business change. The impact on projected financial plans is critical to the success of any facilities project. New facilities initiatives have a direct impact on financial plans.

Partnering with Internal Customers

The Accounting Department's role is changing from a command and control position to one where fellow employees are treated as customers. With this new role, the finance person understands what the needs are and helps provide alternative solutions. This is very important. Developing a good relationship with key personnel in accounting will help the facilities manager as well as the financial person.

Managing Financial Investments and Other Financial Decisions

All businesses are changing in some form every year, typically with sales increases or decreases. As sales levels increase or decrease, these changing circumstances must be reflected in the use of assets, including plant, equipment, inventory, people, land, and facilities. The financial manager is responsible for analyzing the projected company growth and determining how to adjust the resources on a macro level to support the business activities profitably. Should the company get the required resources by taking on debt, equity, or some combination of the two? If debt is used, another major decision made by the Accounting Department is whether it should be long-term or short-term.

How to maintain or improve profit is another major responsibility of the Accounting Department. How much the company should provide as a dividend to its shareholders, and whether it should share profits with its employees, are decisions that involve the Accounting Department, working with the Human Resources Department and other departments.

Coordination and Control

The company must have controls set up to ensure that all parts of the organization are within the financial limits established. Providing a spending budget usually does this. The annual budget process is the best way to initiate discussion and analysis, and to coordinate a strategic initiative. Facilities moves, additions, or changes are strategic initiatives for most companies. They impact communities, incur substantial up-front cost, and generally modify existing cost structures. When the Accounting Department is involved in its budget process, it is a great time to gain an understanding of the financial ramification of a facilities initiative. Thus, the Accounting Department must ensure that budgets are made to require that units remain within these financial limits or have the company's performance objectives changed.

Dealing with Financial Organizations

A continuous flow of money and capital is required to keep an organization going. The financial organization must decide where and how the funds are raised, where and when the company's securities are traded, and when profits are reinvested into the company to maximize profitability.

Risk Management

All businesses face risks from disasters such as floods, storms, fires, loss of power, uncertainty in financial markets, interest rate changes, and foreign exchange rate changes. Purchasing insurance or hedging in derivatives markets can reduce many of these risks. The financial managers usually have a responsibility for risk management and will hedge their assets in the most effective and efficient manner.

SUMMARY

In summary, the *facility* and financial managers are integral in decisions regarding which assets a company will acquire, how the assets should be financed, and determining the required returns on investment to drive overall profitability. If these responsibilities are performed optimally, the *facility manager along with the financial manager* has helped to maximize the value of the firm, and this should benefit customers and employees.

SECTION 2.6
MANAGEMENT DECISION MAKING

Because we now have a general understanding of financial and facilities management in business, it is critical to recognize that the financial aspects of managing a business are only part of the story. This section discusses and emphasizes decision-making from a company perspective. These same techniques and methods should be used when making decisions within the facilities organization. Decision making is a critical part of planning for and managing assets. The use of the S.W.O.T. (Strengths, Weaknesses, Opportunities, and Threats) analysis technique in determining strategic direction is a tool used in dealing with stakeholders' needs. Determining strategic needs is the first step in establishing what kind of decisions management needs to make. Reviewing the role of a senior financial officer helps us understand how involved and focused a facility manager needs to be to understand the business impacts surrounding the decision-making process. How to be a good partner in achieving the strategic needs is critical for strategic success.

The senior financial officer of a company oversees the Accounting Department and other areas of a business. In most corporations, the senior finance officer is called the *Chief Financial Officer* (CFO) and works closely with the *Chief Executive Officer* (CEO) and the board of directors. In small businesses, the senior finance manager may be the owner, or a general office manager who assists the owner or president. In all organizations, financial management involves the nearly daily activities of managing the liquid assets such as cash and investment securities against the payment of current obligations. The second major aspect of financial management addresses the longer-term financial considerations of borrowing to finance a company's operations and investment activities. Last, the finance manager is primarily responsible for analyzing profitability. Forecasting profitability and cash flows are integral in determining borrowing or capital needs. Earlier in the chapter, we discussed the importance of cash versus noncash business activities or transactions. Determining the timing of cash inflows and outflows and managing the borrowing and financing needs are key roles of the senior financial manager.

In addition to financial matters, management decision-making considers the strategic business plan, management incentives (e.g., the rewards earned by management stemming from the results of decisions), social responsibilities, and the situation.

STRATEGIC BUSINESS PLANNING

Strategic business planning typically consists of the following eight major activities:

1. Defining the business mission
2. Analyzing external opportunities and threats
3. Analyzing internal strengths and weaknesses
4. Formulating goals
5. Formulating strategy
6. Formulating programs
7. Implementing programs
8. Feedback and control

In discussions about funding specific projects or tasks, the finance officer focuses the attention of the company on determining the value added by these projects to one of these eight specific activities. Let's briefly review key elements of planning as it relates to a desired facilities initiative.

Business Mission

As mentioned earlier, an organization exists to accomplish something: to make cars, lend money, provide a night's lodging, and so forth. Its specific mission or purpose is usually clear once the business starts. Over time, some managers may lose interest in the mission, or the mission may lose relevance because of changes in market conditions. The mission may also become unclear as the corporation adds new products, services, and markets to its portfolio. Generally, any significant investments made by a company should enhance its ability to achieve the organization's mission.

External Environment Analysis (Opportunities and Threats)

Management should be extremely aware of the business environment in which the company operates. Is it static, dynamic, unknown, complex, simple, or confusing? The business environment will create or eliminate opportunities. The business environment can be threatening to an organization. Senior managers are primarily accountable for monitoring the business environment and directing the resources of a company to take advantage of opportunities and address threats to the company's performance.

Internal Environmental Analysis (Strengths and Weaknesses)

It is one thing to discern opportunities in the environment; it is another to have the competence to succeed in making the most of these opportunities. Thus each business must evaluate its internal strengths and weaknesses. Clearly, the business does not have to correct all weaknesses, nor should it gloat about all of its strengths. The big question is whether the business should limit itself to opportunities for which it possesses the required strengths or consider better opportunities where it might have to acquire or develop certain strengths.

Goals

Business decisions should not be made in a vacuum but as part of a unified strategic plan for the company's good health, well-being, and success. The goal of most for-profit companies is to maximize shareholder wealth, which translates into maximizing the firm's common stock price typically driven by earnings or opportunity.

Strategy, Programs, Implementation, and Control

Executing the strategic plan requires programs. Typically, programs are developed annually. The purpose of the annual plan is to determine which projects and expenditures should be funded in the upcoming year. The annual plan is to ensure that the company achieves the sales, profits, and other goals established in the annual plan in a manner well-linked to the company's strategy in plan. The last element of the annual plan is control. A typical approach to control is management by objectives. These four steps are involved in the process:

1. Goal setting: What do we want to achieve?
2. Performance measures: What is happening?
3. Performance diagnosis: Why is it happening?
4. Corrective action: What should we do about it?

Managerial Incentives

Stockholders own the company and elect the board of directors, which in turn hires the management team. Management in turn is supposed to operate in the best interests of the stockholders. However, in many companies, the stock is widely held or closely held and leaves a large amount of autonomy to the management team. The incentives for management should be based upon what it takes to accomplish both the short-term and long-term interests of the shareholders.

Companies are increasingly tying management compensation to the company's profit and stock price. Whether incentive plans include executive stock options, performance shares, or profit-based bonuses, they are supposed to accomplish two major goals: (1) provide executives with an incentive to take actions that will contribute to shareholder wealth, and (2) attract and retain managers possessing enough confidence to stake their financial futures on their abilities and efforts to motivate the company to generate profits or accomplish its strategic goals.

Social Responsibilities

One might question whether return on investment is the only thing that management looks at when deciding to invest in a project. The welfare of the employees, customers, and the community in which the company operates are other considerations. For example, building a new facility in an area frequented by endangered animals may be logical but not socially responsible.

SITUATIONS

Situations are the key to all management decision making. What do we mean by situations? Well, we mean that before putting pencil to paper, you need to do your homework. We recognize that this can be difficult in certain organizations. Politics, culture, fear, and lack of trust can stop the process of gaining an understanding of the situation before you even get going. But the key is perseverance!

The strategic plan, the annual budget and capital plans, public perceptions regarding the company, and recent events such as layoffs, union disputes, management turnover, industry change, and economic pressures can be critical to the timing of a management decision.

Typically, a cost justification is required to gain approval for a major facility change. Such an analysis requires assumptions. These assumptions are always the major sticking points in any management decision. Understanding the situation surrounding the company is crucial to building well-thought-out and logical assumptions. Besides the assumptions, the basic accounting treatment of the business case is also a key area typically leading to a lack of management support. You must not provide the decision-makers with an easy out (e.g., through poor calculations or inaccurate accounting treatment of the initiative). So, understanding the situation and providing an accurate accounting treatment will position your proposal for management approval.

SECTION 2.7
FINANCIAL ANALYSIS

A basic approach to a well-thought-out financial analysis consists of five basic steps: (1) obtaining background and planning, (2) gathering data, (3) analysis, (4) developing conclusions, and (5) making recommendations. The most important part of the planning, managing, and decision-making process is reporting and analyzing how well the plan succeeded. A good information center can certainly make the next projections better and more representative of what is happening. To analyze what is happening, the current status of a number of performance ratios and indicators are used. Common analytical financial tools that help managers evaluate various aspects of the cost/benefit performance of the business include the following:

LIFE CYCLE COST

When a facility manager makes a request to purchase new equipment, the first question might be, "What is the economic basis for the equipment purchase?" Some companies use a simple payback period of two years or less to justify equipment purchases. Others require a life cycle cost analysis with no fuel price inflation considered. Still other companies allow for a complete life cycle cost analysis, including the impact of fuel price inflation and the energy tax credit. These examples are presented to illustrate that when comparing several cost alternatives, you need to ensure that you are comparing apples to apples for an accurate cost assessment.

USING THE PAYBACK METHOD

This method of cost analysis determines the time required for recovering a capital investment from profits or cost savings. The payback method is usually used when funding is limited and it is important to know how fast the investment will be returned to replenish available funds for other projects. The payback period is determined by computing the following ratio:

$$\text{Payback period} = \frac{\text{initial investment}}{\text{after-tax savings}}$$

This method is used to prioritize projects when funding is limited. The advantage of this method is its simplicity. It also emphasizes the cash flow in early years, which is more certain than that of later years. The drawback of the method is that it does not account for the time value of money. This method ignores all savings beyond the payback period, thus penalizing projects that have a long life potential in favor of those that offer high savings in the relative short run. Despite these shortcomings, this method is very helpful as a prescreening technique.

USING LIFE CYCLE COSTING

Life cycle costing is an analytical method of determining the total cost of a system, device, machine, and so forth over its anticipated useful life. Life cycle costing is difficult because one must comprehensively identify all costs associated with the system, not just the initial capital outlay. The costs most commonly included are the negative costs consisting of the initial in-place cost (equipment and installation cost), maintenance cost (personnel and materials), and interest on the investment. The salvage value is an added cost "return" that is used to offset the cost outlay.

Two factors that must be estimated are (1) the expected life of the system or the period after which the system will become obsolete, and, once the life of the system is determined, (2) the effect of interest rates as applied by using one of several expressions for the time value of money. When using alternative methods for particular calculations, the system with the lowest life cycle cost will usually be the first choice. This expression assumes there is little difference in performance among the various alternatives. Other considerations such as installation time and difficulty, pollution effects, aesthetics, delivery lead times, and owner preference may also need to be factored into the decision based on other means/factors. The life cycle cost analysis method still needs judgments about assumptions pertaining to interest rates, useful life, and inflation rates.

NET PRESENT VALUE (NPV) (TIME VALUE OF MONEY)

When a facility purchases a long-lived asset, it makes an investment decision similar to that made by a bank when it lends money. The basic principle is that cash is committed today in the expectation of recovering that cash plus some additional cash in the future. Therefore, when facility managers are deciding whether to purchase long-term assets, they want to know whether the future capital inflows are large enough to make the investment. If we look at a potential investment that projects a cash inflow of $500/year over a three-year period, the result is a total inflow of $1,500. If the investment (outflow) to achieve this inflow is reasonably less than $1,500, the net present value is greater than zero, and therefore it is an acceptable investment. When various project options are compared, the one with the highest NPV is usually selected.

THE TIME VALUE OF MONEY

When making an investment decision, the owners look at their capital and decide what is the best use of it. Should they invest their limited capital in the stock market, buy real estate for resale, or what? Owners know that a dollar received today is worth more than a dollar promised at a later time. They also understand that in weighing options the risk of losing their investment must be considered. People who invested in the overseas stock market quickly found during the late 1990s that they could lose a substantial amount of their portfolio if the market suddenly drops. For these reasons, a time value of money must be placed on all cash flows into and out of the company.

When a choice is made between alternatives that involve different receipts and disbursements, it is essential that interest be considered. Economic studies in facility management generally involve decisions between such alternatives. When a facility manager is evaluating alternative solutions to a problem, the dollar value must be made comparable. The time value of money allows these comparisons.

Monetary financial transactions look at the cash flows to and from the company. The investment decisions take into account alternative investment opportunities and the minimum return on investment. To determine the rate of return on an investment, it is necessary to find the interest rate that equals payment outgoing and incoming, as well as the present and future values of capital. The discounted cash flow method is used to determine the rate of return.

RISK AND RETURN

Risk is defined as the probability of the course of an unfavorable outcome. Risks are generally classified into two categories: (1) *systematic,* which is the risk outcome from the general market conditions resulting from economic, political, or social changes, and (2) *unsystematic,* which is the risk variability of an outcome caused by events unique to an industry, such as labor strikes, management errors, new inventions, advertising campaigns, shifts in consumer taste, and new governmental regulations.

The four major types of systematic risk are as follows:

1. Operating risk caused by variations in operating earnings before interest and taxes
2. Financial risk caused by a variation in earnings per share that is used as leverage in the capital structure
3. Market risk caused by external elements that affect the economy in general and that may affect earnings
4. Purchasing power risk caused mainly by inflation that reduces the purchasing power of savings or invested wealth

Return is defined as the benefit received from incurring a certain cost. Returns that provide a better benefit for their costs are the most attractive as defined by the equation

$$R = \text{rate of return} = \frac{\text{net benefit}}{\text{cost}}$$

BREAK-EVEN ANALYSIS

A major task of facility managers is to choose financial alternatives. Facility managers want to utilize their scarce funds available for getting the job done in the most cost-effective manner. The following are some typical problems:

1. Whether to contract a certain service or to use in-house crews?
2. Whether to buy certain equipment that is reported to save money in maintenance?
3. Which utility type to buy (gas, oil, or coal)?

4. Whether to purchase or lease equipment?

5. How frequently and in what quantity to purchase stock items?

Break-even cost for a volume of products is calculated from the following equation:

$$\text{B-E-V} = \text{break-even volume} = \frac{\text{fixed cost}}{\text{unit cost contribution}}$$

The break-even point is of little interest to most corporations because their major aim is to make a profit. However, when preparing to start a new production process, you need to understand the break-even point as part of your risk and return analysis.

INVESTMENT DECISION MAKING

To make investment decisions, the finance manager usually follows one simple principle: Relate annual cash flows and lump sum deposits to the same account base for comparison. The six following categories used for investment decisions convert cash from one time to another. Because the company has various financial objectives, these factors can be used to solve an investment issue.

Single Payment Compound Amount (F/P)

The *F/P* factor is used to determine the future amount *F* that a present sum *P* will accumulate at *i* percent interest in *n* years. If *P* (present worth) is known and *F* (future worth) is to be determined, then Eq. (2.1) is used:

$$F = P \times (1 + i)^n \tag{2.1}$$

or

$$F/P = (1 + i)^n \tag{2.2}$$

The *F/P* ratio can be computed by an interest formula, but usually its value is found by using the interest tables shown in Table 2.7-1. Interest rates are compounded annually in these tables. Linear interpolation is commonly used for interest rates falling between the numbers shown in these tables.

Single Payment Present Worth (P/F)

The *P/F* factor is used to determine the present worth *P* of a future amount *F* invested at *i* percent interest for *n* years. Thus, if *F* is known and *P* is to be determined, then Eq. (2.3) is used:

$$P = F \times \frac{1}{(1 + i)^n} \tag{2.3}$$

$$P/F = \frac{1}{(1 + i)^n} \tag{2.4}$$

TABLE 2.7-1   Fifteen Percent Interest Factors

Period n	Single-payment compound-amount (F/P) Future value of 1 $(1+i)^n$	Single-payment present-worth (P/F) Present value of 1 $\dfrac{1}{(1+i)^n}$	Uniform-series compound-amount (F/A) Future value of uniform series of 1 $\dfrac{(1+i)^n-1}{i}$	Sinking-fund payment (A/F) Uniform series whose future value is 1 $\dfrac{i}{(1+i)^n-1}$	Capital recovery (A/P) Uniform series with present value of 1 $\dfrac{i(1+i)^n}{(1+i)^n-1}$	Uniform-series present-worth (P/A) Present value of uniform series of 1 $\dfrac{(1+i)^n-1}{i(1+i)^n}$
1	1.150	0.8696	1.000	1.00000	1.15000	0.870
2	1.322	0.7561	2.150	0.46512	0.61512	1.626
3	1.521	0.6575	3.472	0.28798	0.43798	2.283
4	1.749	0.5718	4.993	0.20027	0.35027	2.855
5	2.011	0.4972	6.742	0.14832	0.29832	3.352
6	2.313	0.4323	8.754	0.11424	0.26424	3.784
7	2.660	0.3759	11.067	0.09036	0.24036	4.160
8	3.059	0.3269	13.727	0.07285	0.22285	4.487
9	3.518	0.2843	16.786	0.05957	0.20957	4.772
10	4.046	0.2472	20.304	0.04925	0.19925	5.019
11	4.652	0.2149	24.349	0.04107	0.19107	5.234
12	5.350	0.1869	29.002	0.03448	0.18448	5.421
13	6.153	0.1625	34.352	0.02911	0.17911	5.583
14	7.076	0.1413	40.505	0.02469	0.17469	5.724
15	8.137	0.1229	47.580	0.02102	0.17102	5.847
16	9.358	0.1069	55.717	0.01795	0.16795	5.954
17	10.761	0.0929	65.075	0.01537	0.16537	6.047
18	12.375	0.0808	75.836	0.01319	0.16319	6.128
19	14.232	0.0703	88.212	0.01134	0.16134	6.198
20	16.367	0.0611	102.444	0.00976	0.15976	6.259
21	18.822	0.0531	118.810	0.00842	0.15842	6.312
22	21.645	0.0462	137.632	0.00727	0.15727	6.359
23	24.891	0.0402	159.276	0.00628	0.15628	6.399
24	28.625	0.0349	184.168	0.00543	0.15543	6.434
25	32.919	0.0304	212.793	0.00470	0.15470	6.464
26	37.857	0.0264	245.712	0.00407	0.15407	6.491
27	43.535	0.0230	283.569	0.00353	0.15353	6.514
28	50.066	0.0200	327.104	0.00306	0.15306	6.534
29	57.575	0.0174	377.170	0.00265	0.15265	6.551
30	66.212	0.0151	434.745	0.00230	0.15230	6.566
35	133.176	0.0075	881.170	0.00113	0.15113	6.617
40	267.864	0.0037	1779.090	0.00056	0.15056	6.642
45	538.769	0.0019	3585.128	0.00028	0.15028	6.654
50	1083.657	0.0009	7217.716	0.00014	0.15014	6.661
55	2179.622	0.0005	14524.148	0.00007	0.15007	6.664
60	4383.999	0.0002	29219.992	0.00003	0.15003	6.665
65	8817.787	0.0001	58778.583	0.00002	0.15002	6.666

Uniform Series Compound Amount (F/A)

The *F/A* factor is used to determine the amount *F* to which an equal annual payment *A* will accumulate in *n* years at *i* percent interest. If *A* (uniform annual payment) is known and *F* (the future worth of these payments) is required, then Eq. (2.5) is used:

$$F = A \times \frac{[(1 + i)^n - 1]}{i} \tag{2.5}$$

$$F/A = \frac{[(1 + i)^n - 1]}{i} \tag{2.6}$$

Sinking Fund Payment (P/A)

The *P/A* factor is used to determine the present amount *P* that can be paid in equal payments of *A* (uniform annual payment) at *i* percent interest for *n* years. If *A* is known and *P* is required, then Eq. (2.7) is used:

$$P = A \times \frac{(i)}{(1 + i)^n} \tag{2.7}$$

$$P/A = \frac{(i)}{[(1 + i)^n - 1]} \tag{2.8}$$

Capital Recovery (A/P)

The capital recovery factor is used to determine the annual payment *A* required to pay off a present amount *P* at *i* percent interest for *n* years. If the present sum of money *P* spent today is unknown and the uniform payment *A* needed to pay back *P* over a stated period of time is required, then Eq. (2.9) is used:

$$A = P \times \frac{i(1 + I)^n}{[(1 + i)^n - 1]} \tag{2.9}$$

$$A/P = \frac{i(1 + I)^n}{[(1 + i)^n - 1]} \tag{2.10}$$

Sinking Fund Payment

The *A/F* factor is used to determine the equal annual amount *R* that must be invested for *n* years at *i* percent interest to accumulate a specific future amount. If *F* (the future worth of a series of annual payments) is known and *A* (the value of those annual payments) is required, then Eq. (2.11) is used:

$$A = \frac{F[(1 + i)^n - 1]}{i(1 + i)} \tag{2.11}$$

$$A/F = \frac{[(1+i)^n - 1]}{i(1+i)^n} \qquad (2.12)$$

Internal Rate of Return

The internal rate of return (IRR) method computes the rate of return that equals the present value of the cash outflow with the present value investment. Thus if we use any of the previously cited methods to determine the present value of the future years' outlay and the expression IRR = total investment − total inflows, any rate that makes the IRR equal to zero is acceptable, as long as it exceeds the cost of funds for its risk class.

Marginal Cost

Marginal cost refers to variable cost. Variable costs are those that vary proportionately with change in the volume of output. The marginal cost of a product is the cost of producing one additional unit of that product.

It is important to note that simplicity is best with management decision making. It is fine and even desirable to prepare very detailed studies, calculations, and analyses for the justification. *But do not present these detailed computations.* Summarize the financial results and impacts, and provide clear depictions of the impact of the initiative on the key financial performance measures today and over the course of the project.

The enemies of high ROM are:

1. A company has a "sky-is-the-limit" strategy driven by vague or overly broad mission statements.
2. "Politically correct" performance measures are in place that are selected so as to not exclude or offend any constituency in the organization.
3. People are not sure what they are accountable for, or they face so many measures that they are overwhelmed.
4. Planning, budgeting, and controlling systems have lives of their own.
5. Employees have little or no awareness of senior management priorities and performance measures.

BENCHMARKING

Today financial and facility managers need to understand and practice *benchmarking*. Benchmarking is a measurement used to determine the "best in class." Benchmarking is used in two different ways. Equipment like a computer benchmarks the time it takes to complete a certain process. This benchmark measurement, usually in time, is then compared to the performance of similar equipment given the same process. Another approach is to compare the processes used by different industries to get the best results. An example would be the case of an organization that wants to improve pizza delivery comparing the process used by a parcel post service organization to deliver packages all over the world, to determine if the process used by the parcel post organization could help deliver pizzas better. The benchmarking of that process could be viewed as a "best-in-class" process.

Benchmarking in today's facility and financial management profession can be a key to the future. The search to find better ways to carry out business processes and add value is a goal toward which we all must continue striving.

One of the most famous benchmarking groups is the International Benchmarking Clearinghouse (IBC). The IBC founders include AT&T, Arthur Andersen, DRI/McGraw-Hill, General Motors Corporation, IBM Corporation, PriceWaterhouse, and Xerox. These corporations pooled their information to share it among themselves. They provide the vital signs of their corporations as a basis for comparison. In 1991, the American Productivity & Quality Center (APQC) and 86 companies designed the APQC International Benchmarking Clearinghouse membership to help managers find and adapt best practices. Clearinghouse members include hundreds of companies, government agencies, health care providers, and educational institutions. With the clearinghouse's assistance, organizations discover best practices through many forms of benchmarking and learn from one another through network benchmarking studies, systematic knowledge transfer, and sharing outstanding practices.

Additional corporations can and do provide information to populate the information base. Corporations may also use the IBC information base to compare themselves against the best in the class. The IBC basic principle is that all businesses in the world perform essentially the same basic processes (e.g., develop strategies, hire people, purchase resources, make products or deliver services, and bill customers). They all have (1) *operational* and (2) *management and support* processes. The IBC operational processes include eight processes:

1. Understand market and customers
2. Develop vision and strategy
3. Design products and services
4. Design and construct network
5. Market and sell
6. Produce and deliver for manufacturing operations
7. Produce and deliver for services organization
8. Invoice and service customers

The IBC management and support processes are:

1. Develop and manage human resources
2. Manage information
3. Manage financial and physical resources
4. Execute environmental program
5. Manage external relationships
6. Manage improvement and change

These 14 IBC key processes provide the foundation to compare and contrast your process against the best in class. By classifying the operating and management and the support processes in the same way, a standard process for all corporations is created to provide information. Once information is collected consistently for standardization; information within its database can be catalogued, stored, searched, and sorted. Typical questions include information on the following topics against which to benchmark your company:

Best practices: Organizational practices that are noted for innovation, productivity, or effectiveness

Benchmarking studies/interest registry: Registry of business practices from member organizations that they have benchmarked, are currently benchmarking, or want to benchmark

Case studies and articles: Documents including surveys, studies, and potential partners

IW's interactive benchmarking database: This powerful interactive CD-RO database contains performance metrics on the manufacturing practices—and performance results—from 2,800 manufacturing facilities. Use it to view performance metrics based on company SIC, product type, plant size, union/nonunion status, and companies' use of advanced manufacturing practices and technologies such as:

- Empowered work teams
- Supplier rationalization
- Quick changeover techniques
- Just-in-time (JIT)/continuous-flow production
- Cellular manufacturing
- Advanced MRP II
- Strategic outsourcing
- Agile manufacturing strategies
- Total quality management
- Pay for performance
- ISO 9000

SECTION 2.8
IMPLEMENTATION PLAN

A comprehensive, measurable, attainable, and understandable implementation plan must be developed and shared throughout the organization. It should consist of a rigorous assessment of the firm's strengths and weaknesses, performance measures for evaluating results, and a benchmarking plan aimed at achieving best practices.

There is no single "best practice," because best is not best for everyone. Every organization is different from every other organization in one or more ways—different missions, cultures, environments, and technologies. What is meant by "best" are those practices that have been shown to produce superior results; are selected by a systematic process; and are judged exemplary, good, or successfully demonstrated. Best practices are then adapted to fit a particular organization.

In recent years, performance measures have evolved as a way of checking the vital signs of corporate health. Performance measures are classified under the three general categories of cost, quality, and service time. Companies are learning that improving quality and productivity go hand in hand with cost. At the same time, they have found out that the best way to ensure external customer satisfaction is to satisfy every internal customer at each step of the process. This condition is true for both manufacturing and service operations.

Once you have selected the desired criteria, view the performance of plants that match those criteria. Then, analyze how much your facility could benefit in terms of higher productivity, greater efficiency, and decreased costs and product defects.

SECTION 2.9
REENGINEERING

Now that your corporation has decided upon a new strategy, you must decide how to adapt the new strategy to the existing organization, a process sometimes labeled Business Process Reengineering (BPR). The reengineering trend has had major acceptance in America (since Hammer & Champy, the reengineering, gurus, issued their book *Reengineering the Corporation*). Recently, however, it has also caused a backlash, some of it justified and some guilt by association. Many reengineering efforts have failed in the past. Michael Hammer and James Champy reported a 70 percent failure rate of the efforts they have seen. An added problem in assessing success and failure is that some people use the term *reengineering* for just any downsizing or rightsizing activity, which may be butchery rather than a surgical approach to cost cutting. Stowe Boyd, in his article, "Business Process and Information Technology," states the major message to learn from slash-and-burn reengineering is that it is possible (but not necessary) to rapidly inject a process-centered management style into old-style business.

Usually a top-down process model is created of the area under study. This high-level process identification allows you to look at your important processes that are done to accomplish the mission, goals, and objectives of the company. This high-level model should not have any organizational or technology constraint. By definition a process as a collection of sequential and logical activities which when performed add value to the product or service. A process produces distinct deliverables whose quality can be measured. It has a beginning and an end, and is usually repetitive. Processes usually coexist in companies, with two types being most common. One type of process is organized along functional lines, in which the input has its value added to from an output within a single organization. These vertically organized processes are easier to improve because there is no major departmental interface (e.g., word processing) during the process.

When processes span one or more vertical (function such as information systems management) departments, they are called *cross-functional* processes. Usually, with cross-functional processes, no single organization or person is in charge of the entire process, thereby making the process complex. Most large companies have hundreds of processes and thousands of subprocesses.

A critical step is to identify your core business processes. Most companies have defined their core processes. This step is key to beginning a reengineering effort because it helps management to focus on the important areas where work gets done. Some organizations find it difficult to determine their core processes because they tend to think in terms of departments or functions, not in terms of the work they actually do and the manner in which this work is performed. Without a core process, the corporation would cease to function properly. Core business processes should always consider the core competencies of an organization. Process analysis should look at the following considerations: Is the process effective? Does the process output meet customers' and users' needs (products and services)?

The product should be efficient [cost, time frame, and full-time equivalents (FTEs) included]. When one decides to do process REENGINEERING, it forces a fundamental examination of the core purpose of the key processes to be reengineered and may require a major adjustment of many of the ways in which the organization conducts its business.

Considerable effort has recently been devoted to studying manufacturing processes and has resulted in improvements such as just-in-time manufacturing, continuous-flow manufacturing, computer-integrated manufacturing, total quality management, materials requirements, and supply-chain management. Many of the techniques for analyzing manufacturing companies can also be used to analyze service companies, although you may need a different set of tools for the analysis. In manufacturing, however, the essential ingredient is the machine and its efficiency, whereas in service companies the key ingredient is people.

Service companies deal with providing services and products in a non-mass-production (nonmanufacturing) environment. The increasing global competition will require that service businesses minimize the "time" it takes to service customers with quality, so as to increase profits. Service companies should look at the human rather than the nonhuman aspects of organizations, in the form of Human Performance Technology (HPT).

Some reengineering projects do not result in downsizing. Instead, they free resources from wasteful or obsolete processes and redeploy them for more important or effective processes. For example, the traditional hospital pharmacy dispensing process of counting pills and filling vials is facing obsolescence from automation. Automated point-of-use dispensing machines and unit-of-use packaging eliminate most of the dispensing work, freeing as much as $200,000 per year in pharmacy resources for clinical roles such as medication management. These new programs, in turn, generate additional savings from improved medication usage and improved patient care outcomes.

Mark Youngblood, in his book *Eating the Chocolate Elephant,* lists 32 ways to improve business processes. Most of these principles have been used by industrial engineers and applied to production problems for decades. However, it is important to list some of them because they are areas where you can get quick results in process improvement:

- Organize multifunctional teams.
- Simplify processes.
- Eliminate duplicate activities.
- Combine related activities.
- Outsource inefficient activities.
- Implement demand-pull.
- Eliminate movement of work
- Eliminate multiple reviews and approvals

The following are some guiding principles aimed at reducing cost and enhancing effectiveness:

- Compress time.
- Align toward customers.
- Organize around outcomes.
- Provide end-to-end solutions.
- Empower people to make recommendations for change.
- Quality is number 1.

SUMMARY

This chapter has explained the financial role of an organization through the eyes of a financial manager. The lesson gleaned from this view should make every facility manager aware of what takes place in a professional business, and more important, what it will take to become part of senior management. This standard of information will assist any facility manager in realizing how roles are shared in an effective organization. There is a melding and sharing of responsibilities between the facility manager and the financial manager. To be successful in the current business climate, each has to understand the other.

REFERENCE

1. R. Simons and A. Davila, "How High Is Your Return on Management?" *Harvard Business Review* (January–February 1998).

CHAPTER 3
FACILITY LIFE-CYCLE PROCESS

Paul R. Smith, P.M.P., P.E., M.B.A., Chapter Editor
Peak Leadership Group, Boston, Massachusetts

William L. Porter, Ph.D., F.A.I.A., Chapter Editor
Massachusetts Institute of Technology, Cambridge, Massachusetts

In Chap. 1, we describe the evolution of the facilities manager (FM) from the boiler room to the board room! Traditionally focused on the management of facilities over their life cycle, the FM today must rise to contemporary challenges and participate in decisions critical to the future of the organization. Today the components of the workplace include not just space, but also technology, finance, organization, and the relationships between them. And these are each in a state of flux that must be recognized and mobilized to support the mission of the organization. Facilities managers have a leadership role to play in realizing these new opportunities. But to play a leadership role requires that FMs learn more about effective management practices and about the requirements of the knowledge age and its associated technologies and best practices.

Chapter 2 provides sufficient information to (1) help the FM become part of the senior management team, (2) establish standards so that the FM understands the information needed for a well-run facility, (3) combine the theory and the applications of facilities management, and (4) provide a chapter that FMs will find user-friendly.

Chapter 3 identifies processes that are critical to the success of the facility and the FM's role in managing the facility throughout its entire life cycle. And it identifies issues in corporate operations management that facilities managers must increasingly take into account. For each facility, there must be management of planning and budgeting of the design and construction process and then of the facility, once it is in operation. We give special emphasis to the role and to the potential of information technology. We also underline the importance of corporate operations management by discussing some new developments in that field.

Some FMs would question why operations management, facilities management automation systems, and infrastructure management topics are discussed in this book. The answer is that the FM must pay attention to the way the corporation creates value in both the physical and virtual world. Whether it is a project, a service request, or an annual budget, the manager needs to understand certain key elements within the business context that are part of this process and are the FM's responsibility. The FMs of the future must be able to communicate with their peers on computer networks, infrastructure, hardware, and software. They can no longer ignore their information technology and their sisters and brothers down the hall. Rather they must get very close to them, learn their language, and become trained in their technological skill sets and toolkits. They must help them design the new processes, practices, and technological solutions that will help us navigate this complex economic environment and bring economic value to all of our organizations, whether industrial, institutional, or commercial buildings. And these solutions are all built on the fuel for this new economy—knowledge.

Now is the time to think strategically and to reposition the facilities management profession. First, it must be closely connected to the mission and management of an organization. And second, it must be closely coordinated with the management of technology, human resources, and finance. Third, it must be prepared to grow and to change, depending on the demands of its particular organization, sometimes incorporating untraditional functions and sometimes forming new management teams and partnerships. These are dynamic times, and there are risks associated with them. However, our experience with other challenges has taught us that if we wait until the path is perfectly clear and the risks have disappeared, the opportunities will also have passed.

SECTION 3.0
OVERVIEW

Kristin Hill, A.I.A.; Cindy Aiguier, John Morganti, and Bonnie Seaberg
International Facility Management Association (IFMA), Boston, Massachusetts

Norman Faucher
Association for Facilities Engineering (AFE), CAC Industries, Inc., Leominster, Massachusetts

Chapter 3 is written to equip facility managers with the tools to succeed in the new millennium. Here you will encounter advanced thoughts and ideas combined with solid industry knowledge. The authors have unleashed information that represents the best of practices so as to empower and excite aspiring facility managers and facility professionals who wish to refine their performance capability.

The chapter is formatted to take the reader through the facility management process as it logically unfolds around a facility project. Section 3.1, for instance, contains information having to do with planning and budgeting, the first steps to getting a facility project underway. Section 3.2 offers insight into the next phase, design and construction. Following these are new developments in knowledge management, in Sec. 3.3, and new developments in operations management, in Sec. 3.4. These last two sections explore the many aspects of facility management encountered after a "project" has been completed and "facility management" has begun.

Chapter 3 does not attempt to provide a step-by-step guide to facilities management. We believe a step-by-step guide would be impossible due to the nature of facilities management. What this chapter seeks to do is to impart the wisdom that facilities management is evolutionary and revolutionary (as discussed in Chap. 1) and is responding to the new technological advances and influences in business today. "Tried and true" can mean outdated and debilitating. Just as it is inefficient today to use an abacus to solve mathematical problems, so too is following a stale formula for facilities management. Traditional methods of approach and operation become the burdens of yesterday.

And so this chapter is also about connection—the connection between technology and change and the way we use each to affect the other. The technological advances in industry, communication, improved methodologies in manufacturing, financial reporting, and other systems used to conduct business become the catalysts for change. In response to these influences, facilities themselves will undergo change to adapt to the advances. But it does not stop there. The new technologies that are driving the changes in facilities are also providing facility managers and their teams with new tools for planning and budgeting, designing and constructing, managing the change, the knowledge, and the operations. Success for facility managers lies in recognizing these technological forces and employing them to their best effect.

The tidings in this chapter are about a different way to approach a facilities project. Understanding that it is vital to maintain a dynamic program by seeing in new ways, the facility manager today must develop a vision for new possibilities and new methods. Don't make the mistake of believing that this will be easy. It is never easy to sell new ideas. It is human nature to rally against anything unusual or deviant, and change almost always meets resistance. The professional gain that will be achieved by successfully championing this effort will, however, be immeasurable.

The basic tools for accomplishing the facility manager's goals are not radical. Common sense, an unclouded vision of both the present and the future, and commitment to innovation in problem solving are all it takes to recognize needs and set new directions. To facilitate a change in old methods of approaching projects, one must set the course, weigh the potential gain or loss, and see the change through. So read on. This is a brave chapter designed to encourage you to be a changemaker, and to reach for the leading edge. In the new millennium, the facility managers who survive and prosper will be those who foresee, seek, and strive for change.

SECTION 3.1
PLANNING AND BUDGETING

Kristin Hill, A.I.A.; Cindy Aiguier, John Morganti, and Bonnie Seaberg
International Facility Management Association (IFMA),
Boston, Massachusetts

Norman Faucher
Association for Facilities Engineering (AFE), CAC Industries, Inc.,
Leominster, Massachusetts

Planning and budgeting are initial steps in every project, whether sketched on a cocktail napkin or compiled into a multiple-volume presentation. One must begin with the end in mind. Plan and budget for the FUTURE, not to catch up with the past. It is essential to spend the time to produce a solid plan to avoid unanticipated costs and project overruns that will destroy the budget and ultimately threaten the plan. Goals must be clearly identified.

To have a project funded, it is important to know the owner and the leadership. This is the place to identify the principal players and their concerns while keeping in mind the goals of the proposed project. Learning and speaking the language can ensure that you understand the goals, cost, and schedule while moving them toward a successful outcome. The approval process in financing a project can result in both positive and negative outcomes—it is important to know both! Planning and budgeting defines the level of risk in a potential project. The information in these articles is essential for understanding the project process and the stakeholders' goals.

In these articles you will learn to identify all of the project stakeholders and the impact they will have on the project. Their concerns for the project may range from the contribution it will make to the success of the business, the design and functionality of the space, and the return on capital investment, to the ability to adapt the space for future uses. Each of these concerns carries with it a risk and consequence of failure that must be addressed during the project planning and budgeting phases. One must be prepared to show how a particular project will benefit the company. It is important that the project is aligned with the company's mission and objective statements.

Every stakeholder in the project could be considered a speaker of a dialect of the common project language. The financial and approval process represents a common understanding of the goals, parameters, and structure set for the project during the planning and budgeting phase. During this process, risks are evaluated, negative impacts are mitigated, and positive outcomes are strengthened.

The facility manager's role in the company structure is to provide plans, action, evaluation, correction, and results. Armed with the knowledge of this role and the reality of available resources, the facility manager is prepared to develop a plan and begin to view it in many different ways. Learn how to propose projects successfully and get them approved, so you can move forward with confidence.

ARTICLE 3.1.1
OWNER REQUIREMENTS

Matthew C. Adams, P.E.
Association of Higher Education Facilities Officers (APPA),
Adams Consulting Group, Wellesley, Massachusetts

Facility owners are becoming increasingly sophisticated in their expectations that the "highest and best" methods of management will be applied in their operations. Certain self-improvement criteria now affect departments in all operating areas. There are seven key criteria for empowering a governing or management organization to achieve high performance standards. These interlocking criteria for success substantially increase the ability to achieve operational superiority in all areas affected by managerial actions.

ADHERE TO THE MISSION

The driving purpose of management within any type of organization must always be to direct operations toward attaining its mission. The highest level of accountability to which management can be held is the degree to which executive decisions advance the cause of the organization as a whole. Managers do not operate in a vacuum; their role is to manage operations within their responsibility to advance the highest and best interests of the organization.

It is essential that the organization have a clearly understandable mission statement. Without such a guiding plan, there is no foundation to provide direction to management groups within the organization.

An interesting exercise for departmental managers is to assess the mission-directedness of the management team members by asking them what they perceive as the mission. This can take the form of a written comparison, in which each team member is asked to write down the organization's mission. The responses are reviewed to determine the cohesiveness of the team and the degree to which members are working toward achieving a common mission. A strong team will display significant areas of overlap, if not complete consistency; both with one another and with the organization's formal mission statement. However, if responses vary significantly or if some members are unable to formulate a coherent mission statement, it is time to undertake serious internal work to ensure that the management team members get on track so that they see themselves as accountable for achieving the organization's essential mission.

FOCUS ON HOLISTIC THINKING

The old system, by which management made decisions through a process of representational governance, is being replaced by a trend toward holistic or systemic thinking. Each member of management is expected to look at the operation as a whole and to work toward achieving the mission of the greater entity. When each management team member focuses on the best interests of the entire system, the competition between formerly competing elements will be eliminated, and decisions that better benefit the organization as a whole will be facilitated. Previous procedures by which many executives were expected to "bring home the bacon" for their perceived constituency are being made obsolete by the expectation that management will work for the greater good, as a cohesive team without hidden agendas.

Through holistic thinking, the organization's strategic goals are advanced. Executives will be empowered through their unity of purpose rather than divided by conflicting special interests. This new emphasis on holistic thinking will certainly affect the ways in which the management team is selected—and thereby also affect the structure of the team. In addition, members of the management team will be expected to focus on a mission that will entail a willingness to forego the idea that they have been placed on the team to represent a particular constituency. The resulting management emphasis on holistic thinking will affect all management decisions from budgeting to operational methodology.

In the long run, the particular interests of component groups or constituencies within the organization will be best served by the holistic approach. As the greater interests of the organization are advanced, all of the various components will find themselves in stronger positions.

IMPLEMENT STANDARDS OF MANAGEMENT ACCOUNTABILITY

An effective management team achieves success by evaluating the performance of its members in terms of certain established criteria. These will vary from organization to organization, but include structural measures, such as regularity of meeting attendance and completion of continuing education goals, and such qualitative measures as the ability to contribute to the decision-making process, preparedness for decision-making sessions, and the overall level of support for management policies and activities.

Failure to upgrade the performance of weak managers or to replace them with strong managers can result in a negative dynamic that undermines managerial effectiveness and drives away stronger managers who are eager to be on a winning team.

The composition of the management team must be periodically updated to reflect changing requirements. The need to infuse new skills and new areas of expertise grows stronger as the pace of change increases.

CREATE THE FUTURE

This same trend to ever-faster rates of change means that a successful management team must be prepared to implement forward-thinking techniques and to actively plan and initiate action to cope with all types of progress. A streamlined decision-making process will prove more adaptable to changing requirements than approaches that resist change.

Fear of change lies at the heart of institutional inertia. It is not a new problem but it is one that becomes more critical as the pace of change accelerates. Rather than simply monitoring the past, effective managers anticipate the future. They use resources of all kinds, including valuable managerial time, to create the future, develop new ways of doing things, and envision new paradigms. By utilizing planning, policymaking, and the decision-making process, effective managers concentrate on future requirements.

DEVELOP STRATEGIC INFORMATION

This emphasis on the future requires that effective managers ensure that they receive the kinds of information needed for making decisions that affect the future. Reviewing past performance should require no more than 25 percent of the total information intake. The rest of the information on which managers rely will reflect current and predicted conditions of the operations for which they are responsible. By defining long-range operational policies, it will be possible to develop the kinds of data that will most significantly impact decision-making processes. By controlling the flow of information and emphasizing information with long-range value, leaders will be in a better position to formulate policies for the future success of their organizations.

LET THE MISSION CONTROL THE STRUCTURE

The ways in which authority is divided can work either for or against an organization. Outmoded management structures that have numerous subsidiary organizations or too many members can impede progress because they are not designed to meet changing needs. An obsolete management structure is another example of institutional inertia.

Ineffective management structures can result in either gridlock or conflict as executives strive to reach meaningful decisions. If the existing system tends to drag out decision making because of multiple committees or unclear lines of authority, decision making will be untimely and unresponsive.

Highly effective management teams continually assess and fine-tune their structures to facilitate their ability to accomplish goals. Ideally, management structures reflect functions, areas of responsibility, and strategy.

CLARIFY JOB DESCRIPTIONS

As management structure changes to reflect current conditions, individual jobs within the management team will undergo corresponding transformations. To avoid confusion about who is supposed to do what and how the various responsibilities interrelate, it is necessary to update job descriptions for individual roles.

All members of the management team should have a shared understanding of the group purpose. In addition, each should have an understanding of the particular roles of the individual team members. Responsibilities and relationships with other leaders should be defined to increase the streamlining process and to achieve clarity of understanding. Job descriptions for each management role enable members to assess their performance and to improve it as appropriate, based on a clear vision of the standards. In particular, all must understand that the role of the head of the team includes the responsibility for teaching holistic thinking, changing team selection procedures, and spearheading the other operations of an effective managerial team.

ARTICLE 3.1.2
FINANCIAL MODELING

Thomas K. Davies, A.I.A.
Vanderweil Facility Advisors, Boston, Massachusetts

The financial methods, techniques, and tools that organizations employ to manage their large building portfolios are directly correlated with the state of the facilities now and the dollars needed to keep their condition acceptable well into the future. Research exists to support such an opinion, and failure to recognize this concept often leads organizations to defer needed maintenance year after year, eventually paying much higher "balloon payments" down the road to fix building failures, working reactively instead of proactively.* Considering the consequences of doing nothing or little in the way of managing their facilities, it is no wonder that several organizations have elected to go down the path of proactive facilities management to maintain the substantial value of their investments in facilities. A cornerstone of this proactive approach is the financial model.

WHAT IS FINANCIAL MODELING?

Financial modeling is an integral part of proactive facilities management. The process involves a parametric analysis of several variables and is ultimately intended to forecast future costs of maintaining real plant and equipment, to provide a vehicle for understanding the magnitude of the problems at hand, and to provide a foundation of realistic costs and assumptions for building a workable long-term business plan for capital investment in the facilities. The financial model utilizes "baseline" information on the condition of the facilities as one of the parameters. Another parameter utilizes commonly accepted cost models for certain types of buildings. Still other parameters take into account the desired time frame of the analysis, the current age of building components, and the life expectancy of building components.

Financial scenario analysis enables flexibility through dynamically altering variables and estimating the resulting changes in the financial model. Improved financial decisions are facilitated through analysis of accurate baseline facilities information, projections based upon proven models, and established benchmarks.

---

* Many of the concepts presented in this article represent joint efforts and projects including, but not limited to, the Society of College and University Planners (SCUP), the National Association of College and University Planners (NACUBO), the Association of Higher Education Facilities Officers (APPA), and Coopers & Lybrand. Specific credit is noted to *Managing the Facilities Portfolio,* a publication prepared by the foregoing with contributions by others.

WHICH TOOLS ARE USED IN FINANCIAL MODELING?

Several tools are used in the financial modeling process for facilities. First and foremost is the raw data collected during the Facility Condition Assessment (FCA). Such data provide a fundamental snapshot of the building's condition. The Facility Condition Index (FCI) is the ratio of the cost of remedying facility deficiencies to current replacement value (CRV).

$$\text{Facility Condition Index (FCI)} = \frac{\text{deficiencies}}{\text{current replacement value}}$$

The FCI is a recurring theme in financial modeling. It is the basis of understanding multiple funding scenarios to improve the condition of the facilities and reduce the FCI.

One such tool is the FCI versus Annual Funding graph, an example of which is shown in Fig. 3.1.2-1. For each funding option, this graph illustrates the annual dollar amounts that would be funded under the option and the forecasted FCI that results from that option. Building replacement costs are adjusted annually by the inflation rate identified, 4.7 percent in this case. If the organization anticipates accelerated growth, this rate is reflected in both the replacement cost and the overall FCI, assuming that growth is via new construction. Anticipated new physical plant deficiencies are computed as the sum of two factors. The first is the further deterioration of current deficiencies not corrected, at the backlog deterioration rate identified as 2 percent. The second is a forecast of annual renewal requirements, using a best-fit trend line. The vertical bars indicate annual dollars invested in the building, as marked on the left axis, and the graph lines indicate the effect of this spending on the FCI, as marked on the right axis. This particular graph represents a 22-year forecast, but longer or shorter forecasts can be generated.

Option 1 explores the effects of minimal investment in the building, reflected by 0.5 percent of the facility replacement cost. This minimal funding option is included for comparison only. This funding strategy is unrealistic because it projects an unacceptably high FCI. Options Two and Three take the opposite approach—these establish the FCI desired and determine the funding required to achieve that FCI.

Option 2 determines the funding requirement to maintain the FCI at its current level.

Option 3 explores the improvement of the portfolio, in which significant building improvements are distributed over a 10-year period, bringing the FCI down to an efficient 0.05 level, and maintaining it at this level thereafter.

Another useful tool is the Facility Renewal Forecast graph (see Fig. 3.1.2-2) that illustrates the amount of dollars that will need to be spent to renew the facilities over time. The annual forecast of renewal costs, which is based on the accumulation of the data on the current condition and costs of the building components, is useful for projecting capital funding required to maintain status quo. The moving average of projected annual expenditures and the straight-line trended regression should form one basis of funding justification for ongoing renewal. Finally, long-term average funding requirements provide a rational baseline for average ongoing funding and expenditure. Lower average spending levels translate to increased deferred renewal and increased costs to maintain steady state of facility conditions and functionality.

Yet another tool, the Investment Gains/Losses by Year graph, illustrates the effect of the three funding options shown in Fig. 1 on the value of the physical plant over time, as shown in Fig. 3.1.2-3. Plant value is considered the current replacement value minus the aggregate costs to correct deferred maintenance and deferred renewal.

The comparison of Total Funding and Change in Plant Value graph (see Fig. 3.1.2-4) illustrates a tool for comparing each funding option, the cumulative renewal expenditures to the cumulative change in the value of the physical plant, and permits comparing such dollar amounts across the various funding options.

The Return on Investment graph (see Fig. 3.1.2-5) is based on the preceding graph and shows a return on investment analysis by computing the ratio of (1) the "profit" (i.e., the

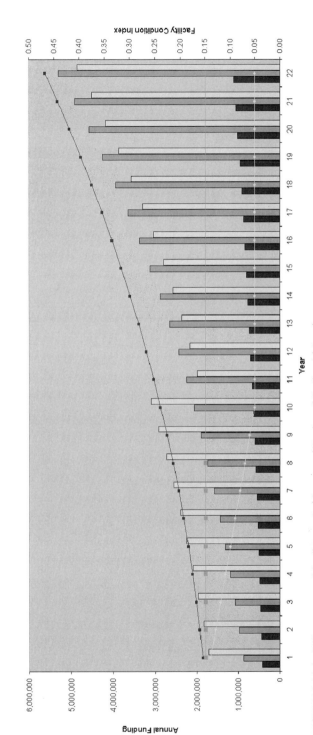

FIGURE 3.1.2-1 FCI versus annual funding graph. (*Courtesy of Vanderweil Facility Advisors.*)

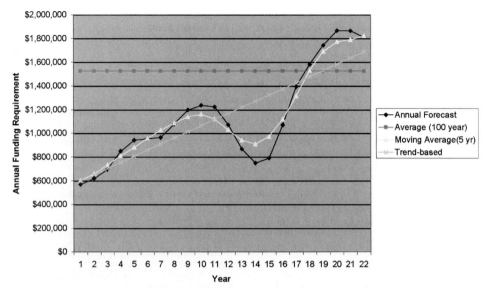

FIGURE 3.1.2-2   Facility renewal forecast. (*Courtesy of Vanderweil Facility Advisors.*)

cumulative increase in value less the cumulative renewal expenditures) to (2) the cumulative renewal expenditures.

These illustrative graphs, coupled with sophisticated facilities management software programs, are the major tools employed in state-of-the-art financial modeling.

WHY IS FINANCIAL MODELING IMPORTANT?

Organizations are increasingly establishing mission statements and business plans for their facilities-related businesses. This is a critically important process, as any business manager knows. Financial modeling of facilities is used to support key decisions regarding the direction of the organization's business plan or mission. The business plan is both rudder and filter, establishing defined criteria against which potential actions can be measured.

The quality of the business plan is often based on key assumptions and factual cost data that have been loaded into the parametric analysis of the financial model. If the financial model uses incomplete or erroneous methodologies or relies on unsubstantiated cost data, the resulting data may negatively influence the organization's business plan. The result could be that the business plan steers the organization in the wrong direction. Consider a typical research university where this "wrong direction" could be made manifest. Say, for example, that the university's financial model fails to accurately depict the true costs for maintenance and renewal. Data from the financial model find their way into the business plan, leading to insufficient funding of campus facilities. The subsequent inability to update 1950s-era laboratories results in degraded appearance and functionality. This in turn leads to declining enrollment—and ultimately the organization's mission of being a premier research university is short-circuited. A sound financial model could have prevented the organization from going in

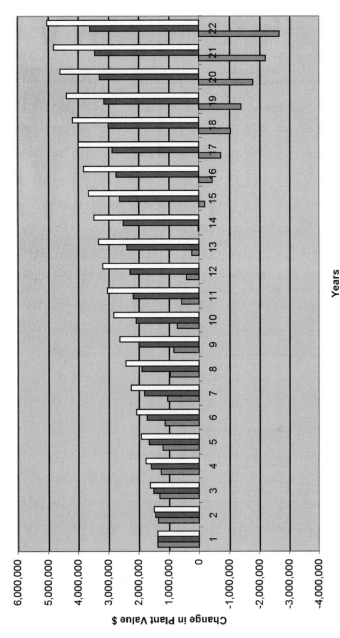

FIGURE 3.1.2-3 Investment gains/losses by year. (*Courtesy of Vanderweil Facility Advisors.*)

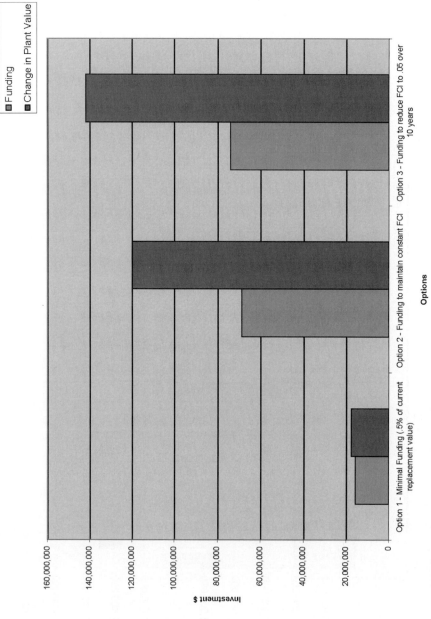

FIGURE 3.1.2-4 Comparison of total funding and change in plant value. (*Courtesy of Vanderweil Facility Advisors.*)

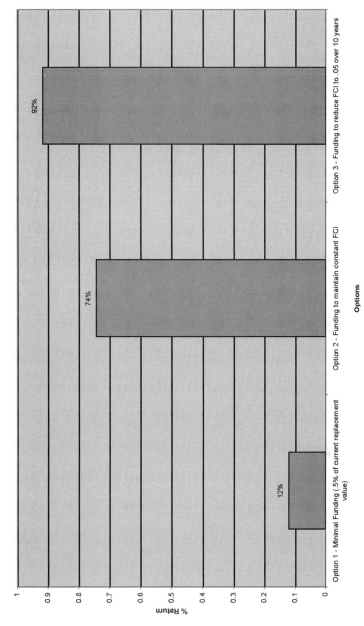

FIGURE 3.1.2-5 Return on investment. (*Courtesy of Vanderweil Facility Advisors.*)

the wrong direction and kept it focused on its true mission and business plan. Financial modeling is at the core of any sound long-term plan.

WHO IS INVOLVED IN FINANCIAL MODELING?

Business and technology shifts are altering the landscape of facilities management. Strategic facilities management is primarily a communication-driven process. Decision support must involve input, planning, and management from several areas, including facilities directors, operational management, and planners. Moreover, the owners of this process must engage in roll-up reporting to and executive direction from CFOs and CEOs. As such, technology plays a critical role.

Web and related advanced technologies have been developed and are being deployed to link strategic facilities management with the organizational mission of the enterprise. Technology is increasingly changing the process of baseline data gathering, capital requirements planning, capital project planning, and project execution. Redistribution of information, capability, and assets is an ongoing process. Software and technology systems that allow several decision makers to be involved simultaneously in cutting-edge financial modeling remain the key to sound planning.

WHAT ARE THE RESULTS OF FINANCIAL MODELING?

Financial modeling allows an organization to see into the future and take action. If properly designed and methodically executed, financial modeling systems can greatly improve an organization's ability to understand, address, and solve complex capital spending plans that are a necessary component of long-term capital planning. Consider the fact that organizations often have a large list of building-related problems to be fixed, but usually only a small list of available funds with which to do the work. The value of financial modeling becomes clear when an organization needs to squeeze the most value out of every facilities dollar. Financial modeling allows the organization to establish funding priorities with scarce resources and to stay on track with the organization's long-term vision.

ARTICLE 3.1.3
FACILITIES MANAGEMENT SYSTEMS

J. Richard Swistock
The Association of Higher Education Facilities Officers (APPA),
Alexandria, Virginia

Facilities management systems and processes must be integrated with and support the institution's or company's broad goals. Public or private facilities rarely exist in their own right, and their existence can be justified only by their support of the institution's and company's mission and goals. The exception to this is historic structures and facilities that can properly exist in their own right. The existence of all other facilities can be justified only by their contribution to the larger institutional mission. Preserving tradition and history is an important element in the broad goal of institutions such as colleges, universities, churches, museums, and so forth.

The justification for facilities should frequently be tested against the broader institutional or company goals, and just because a facility exists is not adequate justification to continue owning and maintaining it. Private companies are more likely than public institutions to frequently test the needs for facilities against corporate goals and objectives.

Facilities should be managed as assets, and, as with all other assets, require attention and should provide a positive return on investment. Ideally this return can be quantified in dollars and reflected in financial statements. This may not be reasonable for many public facilities. However, the return on investment in facilities should be expressed in some manner such as more effective use of research resources, recruitment and retention of faculty, or some other identified return.

Many institutions such as public and private colleges and universities do not treat facilities as assets but rather only as expenses, and therefore frequently find themselves owning and maintaining facilities that are liabilities which do not contribute to the institutions' missions and goals.

An integrated facilities system will support the goals of the institution and will consist of master planning; space planning and utilization; facilities construction, renovation, and maintenance; and utilities and building services. Good process management of these systems will assure good customer service to those who use the facilities and good cost-effectiveness in their acquisition, renovation, maintenance, and operation, and in the procurement of utilities and related services.

This article discusses these elements of facilities systems and the related process management for each part of the system required to provide and maintain a productive environment as cost-effectively as possible for users of the facilities.

DEFINITION OF FACILITIES MASTER PLANNING

Facilities master planning is a process that defines the physical requirements needed to achieve the institution's vision, mission, and goals. When the master planning process is done effectively, it leads to cost-effective decisions for facilities development.

The facilities planning process is an extension of the institutional planning process and must acknowledge the needs of institutional constituents and the community in which it functions. This process results in the *facilities master plan,* which defines current facility requirements and makes provisions for future facilities that are likely to be required to support future institutional goals. The facilities master plan usually is a snapshot in time that defines the facilities planning process. The facilities master plan usually has a useful life of two to five years. The process should be continuous, and published plans should be frequently updated.

The facilities master plan contains the following four major elements:

1. Inventory of current facilities, including buildings, grounds, and infrastructures.
2. Currently needed additions, deletions, or major changes to the inventory that are required to support the institution's current goals and objectives.
3. Likely future changes, additions, or deletions to the inventory that may be required as institutional goals and objectives change.
4. Design guidelines that describe the quality of the environment desired, including landscaping, architectural styles and limitations, and specific infrastructural and utility requirements. These guidelines should ensure consistency across facilities while permitting latitude for creative architectural and engineering designs. They can be included in the master plan or clearly referenced if they are consolidated into another comprehensive document.

GOALS AND OBJECTIVES

The goal of the master planning process is to produce a master plan. This master plan, supported by a financial plan, is the road map for institutional facilities development for the next two to five or more years. The master plan document is a vehicle for communication that clearly shows the long-term facilities objectives and the near-term development plans of the institution to all interested parties.

A well-developed master plan is a powerful tool in obtaining community, constituent, and legislative support of management goals and objectives. For this reason, the document must be easily readable, as well as technically complete.

The plan is a starting point for all future development and construction.

Techniques and Procedures

Successful master planning of facilities includes input from major constituents and facilities users, as well as senior management. The process of getting this input is important to ensure future support of the plan. In large institutions with diverse constituencies and facility users, formal meetings or hearings with these groups may be appropriate. This is especially important when public or quasi-public facilities are involved.

Managers and planners must control the input to avoid unrealistic expectations from constituents and users. Boundaries and guidelines for the input procedure must be presented to and understood by interested groups before input from them is received. The boundaries and guidelines are the anticipated financial limits of the future facilities development, the corporate vision and goals for the future facilities, and the legal and physical boundaries to facilities development. After corporate objectives, boundaries, and constituents are identified, user group inputs should be obtained.

Before the final plan is prepared, a conceptual plan that meets corporate objectives and constituent and user needs within the planning constraints should be developed and shared with all those who provided input. At this point, plan developers should seek buy-in concurrence from major users. The greater the buy-in and support at this stage, the better the chances are that the plan will ultimately be fulfilled.

Facilities Inventory

To properly manage facilities as assets, an accurate facilities inventory in a useful, conveniently available form is essential. Inventory methods include facilities maps, building plans, and documentation of infrastructure and building systems and equipment.

Facilities maps locate buildings, roads, parking facilities, and all other ground improvements and physical facilities. Maps should show topographical features and document significant trees, flower beds, historical structures, and other significant items. The mapping system should also locate and document all underground facilities and utilities, as well as property boundaries, easements, and right-of-ways.

Accurate building plans are essential for space management and compliance with fire and other safety regulations. Effective maintenance plans require a complete inventory and location of all building electrical, mechanical, and structural systems and a database for developing equipment histories. The Americans with Disabilities Act (ADA) requires that public spaces are accessible or a plan is developed to provide equal access to all programs and activities. Accurate building plans are essential for developing a comprehensive ADA compliance plan.

Geographic information systems (GIS) are computer-based systems that contain the complete facilities inventory and allow retrieval and presenting of information in several useful formats. These systems are built in layers starting with base maps and building plans, with consecutive layers that display additional information. Because these systems contain digitized information and are stored in computer files, they can easily be updated and kept current. This is a powerful tool that allows exploring various alternatives during the planning and development process. A comprehensive GIS can combine all facilities inventory—including utility systems, property and real estate maps, underground infrastructures, ground topographical features and improvements, and building plans—into a single database.

GIS is a rapidly developing technology that should be considered for all facilities inventory and master planning.

Facilities Needs

Facilities master plans must be carefully linked to corporate and institutional planning. The facilities master plan should clearly indicate the facilities required to support overall corporate goals and objectives. These facility requirements should include prudent deletions from the facilities inventory, as well as additions and new construction.

The facilities master plan must be supported by the corporate business plan to achieve the institutional goals and objectives. The corporate business plan for a university may, for example, envision the expansion of medical research activities and the recruitment of a prestigious medical research faculty. In this case, the facilities master plan must address these facilities needs for expanded medical research facilities.

The facilities changes needed to support corporate goals, objectives, and strategies must be coordinated with the financial plan for developing these facilities. Although facilities master planning is frequently initiated independently of the corporate financial plan, the facilities master plan process will not be fully effective until it is supported by a capital financing strategy and plan.

The fully integrated facilities master plan must support the corporate mission and goals and should be closely coordinated with the corporate business and financial plans.

Design Guidelines

Design guidelines are not necessarily required for effective master planning, but they are essential for effective development and operation of facilities. Including them usually by reference in the master plan helps ensure their integration into the construction that is likely to result from the master plan. Design guidelines ensure consistency with institutional architectural concepts and compatibility with all building and infrastructural systems. Design guidelines document the architectural themes in various zones of the campus or grounds, ensuring that new facilities are consistent with the desired architectural themes. Standard elements such as streetlights, paving types, building finishes, and hardware should be detailed in these guidelines. Landscape details such as species of trees desired on campus or sections of campus should also be specified.

The design guidelines detail building systems that should be specified in all facilities. Such things as campuswide building automation systems, security, and lock and key systems should also be specified. Utility system details (especially local systems such as electricity, steam, or chilled water distribution parameters) should be clearly spelled out in the design guidelines. Central building automation and fire alarm and reporting systems must be clearly identified.

Building systems preferences such as desired roofing systems should be indicated in the guidelines. Proprietary systems in use, such as telephones and master keys, should be specified. Special performance requirements, such as high-reliability utility distribution system designs, should be included.

Many institutions have developed information systems requirements such as size and electrical requirements in telephone and computer/information systems rooms, closets, raceways, and outlets that should be included in the design guidelines.

A well-organized comprehensive design guidelines manual will eliminate many costly redesigns of new facilities and help ensure that user needs are met.

Need for Space Planning and Administration

Acquiring and owning or leasing facilities can be justified only by their value in helping meet the institution's missions and goals. Effective space planning and administration are essential if facilities are to optimally support the institution's mission and goals. Facilities are assets and like all assets must be carefully managed to achieve the desired return on the investment made in them. Space planning and administration are important in achieving this return on investment.

Space Utilization and Cost

Every organization has various space needs, which must be quantified as the first step in effective utilization. Administrative, classroom, laboratory, industrial, showroom/retail, and storage are some common categories of building space. Grounds areas can also be categorized by use. Parking, recreational, residential, storage, access, and "green" are some common categories.

All spaces have acquisition, operating, and maintenance costs associated with them, which should be quantified. Usually the owning and operating costs of buildings are expressed in dollars per square foot. When building space is categorized by use (administrative, research, production, etc.), facilities costs to support these functions can be quantified.

Development, Process, and Standards

To manage and administer building space effectively, an accurate inventory of space is required. The GIS and CAD systems discussed in master planning are effective systems for storing, organizing, and managing building space inventories. Space utilization guidelines allow space managers to make effective decisions for space allocation. These guidelines usually take the form of square feet per unit. For administrative space in categories such as clerical, supervisory, managerial, and senior managerial, standards are frequently expressed as square feet per person. These standards include circulation and support space for the individual person or workstation.

Space guidelines for common office support such as mail and reproduction rooms, conference rooms, and lunchrooms are not usually included in the guidelines for square footage per person. Separate guidelines for such facilities are frequently developed on the basis of the number of people served by the spaces.

All buildings have several categories of space, which should be identified and quantified for effective space management and administration. The following are the major categories of building space:

- *Gross square feet:* The total square footage of a building, including walls (interior and exterior), basements, mechanical, electrical, and other building support spaces, as well as lobbies, stairwells, and hallways.

- *Net square feet:* Usually gross square feet less exterior walls, mechanical, and other building spaces normally not accessible to building occupants.

- *Assignable square feet:* Usually considered net square footage less common areas such as lobbies, hallways, public restrooms, and so forth. This is sometimes referred to as *rentable square footage.*

Space planning and administration is primarily associated with management of assignable building spaces.

Space utilization guidelines should be developed for all categories of building space. The guidelines should be realistic and shared with all space users. Table 3.1.3-1, an example of space guidelines, is published by the Commonwealth of Virginia.

TABLE 3.1.3-1   Space Guidelines for State Agencies Leasing Real Property

Space type		Maximum area (square feet)	
Office space	Pay grade	Private	Open
Offices (excludes receptionists)			
Department or agency head	18 and above	256	
Assistant department or agency head	15–17	192	—
1st level administrator	13–14	168	96
2nd level administrator	11–12	144	96
Other	10 and below	120	64

Note: Field office personnel who are routinely out of the office 50 percent or more of the normal work week shall be restricted to 64 sf per person.

Reception areas (including receptionist)			
1–4 visitors (peak)			240
Over 4 visitors add (per additional visitor, peak)			20
Conference rooms (per chair, peak)			
First 10		30	
All over 10		20	

Note: Conference rooms shall be shared among work units where possible to avoid excessive space requirements.

Interview areas		80	80
Testing, training, or hearing rooms (per chair, peak)			
Seminar seating		15	
Auditorium seating		10	
Lounge/break rooms (requires prior approval of the Division of Engineering and Buildings)			
Equipment		60	
Plus (per chair peak)		20	
Furniture/equipment (except personal offices)			
Copier (freestanding)			25
Copy room (including copier)			50
Plan/flat file			25
Lateral file			10
Vertical file (letter)			10
Vertical file (legal)			12

Circulation (to compute total usable space)
  If the total office area is less than 50% open office space, add 25% to the office area.
  If the total office area is 50% or more open office space, add 30% to the office area.

Note: the aggregate usable area shall not exceed 250 square feet per person without the prior approval of the Division of Engineering and Buildings.

Application of Guidelines

After realistic space utilization guidelines have been developed, all inventoried space should be tested against the guidelines. Space utilization guidelines are guidelines, not standards, and considerable judgment is needed when applying guidelines to existing spaces. The guidelines, however, can be used as standards when new space is being planned, constructed, or leased.

Successful application of the guidelines requires good communication with all levels of management and with the individuals who occupy or will occupy the spaces in question.

When space utilization guidelines are used with design guidelines, they provide the means to quantify facility requirements during the master planning process. The result is an integrated approach to facilities planning and development that stands the test of meeting the facilities requirements to support the institution's mission and goals.

With accurate and readily available building and space utilization information in the facilities files, managers can build effective maintenance and energy management programs to support the facilities operations in the most effective manner possible.

Documentation and Updating Process

Automated files and GIS inventories make it feasible to update information systems continuously so that current inventory and facility utilization information is always easily available. A formal feedback process to the files is necessary to ensure that an accurate database is maintained. The feedback information is from the following three sources:

1. New construction and renovation information
2. Maintenance files
3. Space utilization and requirement surveys

All new construction and renovation "as built" information must be recorded in the GIS. This is easily done by requiring that construction and renovation designers and project managers provide digitized information as part of the project management and close-out procedures. Final payment to designers and contractors should not be made until such information is provided in the desired format.

As the preventive and corrective maintenance programs are executed, information about any equipment changed and/or replaced should automatically be provided to the facilities database managers. It is important that such information as roof, HVAC system, or electrical system component replacement be documented in the facilities files.

Space usage frequently changes as a result of staffing changes, production and office equipment upgrades, and changes in the work assignments of building occupants. Although most of these individual changes have minimal impact on facilities planning and management, the impact can be significant in aggregate and over time. For effective space utilization and management, these changes should be periodically quantified and compared to space utilization guidelines and building systems capabilities. This information will enable corporate managers to make space reassignments when appropriate and will be important for the continuing master planning process. A convenient method for gathering this information is an annual survey of space utilization that will allow space managers to document any changes in the preceding 12 months.

FACILITIES OPERATIONS AND SERVICES

Objectives

Facilities should exist only to support the institution's missions and goals. The mission of facilities management and operation is to maximize the benefit of all corporate facilities and ensure their continued benefit in the future. Facilities support the institution's goals by providing the most effective environment possible for the people, equipment, and processes that contribute to achieving the institution's mission and goals. Except for those exceptional facilities that have recognized historical or esthetic significance, facilities should be managed as assets. Investments in these assets should yield quantified returns measured by increased productivity and positive effects on people who use the facility.

The public image of an institution is frequently reflected in its facilities. This is especially true of colleges and universities and other public institutions. Therefore, this image must be reflected in facilities planning, designing, and construction and also in maintenance and operations throughout the facilities' lives.

Because facilities exist to help meet corporate goals primarily through people who use the facilities, effective facilities managers view these people as valuable customers. Routine maintenance activity is more than facilities maintenance; it is customer service. Facilities operations and services include the following:

1. Preventive and corrective maintenance of all buildings and structural, electrical, mechanical, and communication systems.

2. Building services such as housekeeping and in some cases moving, package delivery, and other services.

3. Utilities such as electricity, water, steam, gas, chilled and hot water, and, where appropriate, compressed air and other piped gases. Grounds, parking lots, and all other property should be similarly managed.

Facilities maintenance and building and grounds services are usually overhead costs to the institution and subject to appropriate scrutiny at the corporate level. Therefore, the objective for facilities managers is to provide high-quality facilities services in the most cost-effective manner possible.

Customer Communications

Building occupants and users (customers) ultimately determine the effectiveness of facilities management and operations. Good customer communication with the facilities organizations is critical to good customer service. There are several levels of communication between facilities operations and customers. The communications pyramid (Fig. 3.1.3-1) shows the various levels of these communications. At the top of the pyramid are formal policies, directives, and instructions. This is the most formal level of communications, and communication with customers becomes less formal through the lower levels of the pyramid. However, the amount, intensity, and influence of communications on the effectiveness (or the perceived effectiveness) of the facilities organization increases as communications pass down through the pyramid. The most important communications frequently occur at the base of the pyramid. These communications between facilities management staff who directly provide services (housekeepers, air conditioning technicians, work reception clerks, etc.) and the customers who personally and directly receive these services have the most effect on daily operations and on the effectiveness of the facilities operation.

First-line supervisors and service providers are empowered individuals. They frequently work independently, not under direct supervision, in the presence of customers, and frequently perform tasks that require making judgments and decisions on the spot. First-line workers frequently deal with the concerns of frustrated customers, and in a very real sense they are the facilities management organization in the eyes of the customer at that time. Plumbers or multicraft maintenance persons responding to a call concerning a leaking sink may find anything from a dripping faucet to a flooded restroom. They are empowered to speak for the facilities organization, commit resources, and accomplish the task at hand either well or poorly. Empowerment is by default or through positive means.

Empowerment of first-line workers by default occurs when

1. They have not been given clear guidelines within which to work and make field decisions

2. Management does not keep first-line workers informed of organizational mid- and long-term plans and strategies.

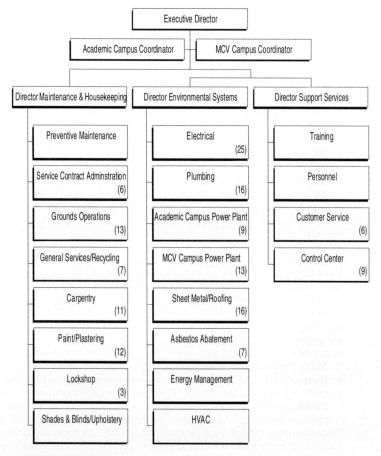

FIGURE 3.1.3-1 Traditional organization.

3. First-line workers have not received adequate technical training to deal with field situations.

4. An atmosphere of distrust exists between management, supervisors, and first-line workers.

When any of these conditions exists, the most likely results will be poor completion of tasks or at least the perception by the customer that the work is poor. Frequently the perception (and sometimes the reality) are that costs were too high. Empowerment by default can exist in any facilities maintenance service organization. Services provided by contract, as well as those provided by in-house forces, are equally subject to empowerment by default.

Because first-line workers are empowered when they are in the field, they must be *positively* empowered if the mission of facilities organization is to be achieved. Positive empowerment increases the probability that the field task will be completed correctly in a timely way and that customers will be satisfied with the results. Positive empowerment of first-line workers occurs when

1. They have a clear understanding of their responsibility and authority.

2. They understand the organization's goals and objectives, and especially the goals and objectives of their unit.

3. Workers are technically competent, have confidence in their ability, and have the proper tools and equipment to do the job.

4. They have confidence and trust in their supervisors and managers.

Training is essential for all facilities management personnel. First-line workers must stay current with developments in their profession, and in many areas such as plumbing, electrical, and air conditioning, they must keep licenses and certifications current. First- and second-line supervisors usually receive a considerable amount of training in managing the workforce and dealing with problem employees, OSHA, and safety requirements. These are the key people who interface daily with customers, and they should also receive training to develop and sharpen their skills in customer relations. From telephone etiquette to empathizing with customer concerns, these people have important interactions with customers and should receive formal customer service training.

Appropriate individuals should be encouraged to continue individual education and earn certificates and even advanced degrees that are work-related. Some tasks involving refrigerants and asbestos require certification before an individual can work with the products in the field. Electricians and plumbers should be required to take tests for and obtain journeyman and master licenses and certifications. Multicraft technicians should be encouraged to obtain all certifications applicable. Although certification is not required for many maintenance tasks, it gives a facilities worker credibility in the field and leads to a more competent workforce that is more likely to respond positively to field situations. When this is recognized, technical competence is complemented with skills in customer relations, customer satisfaction increases significantly.

Skills in customer relations are developed through formal training classes, teaching, and coaching by supervisors and managers. Telephone and other communication skills, appropriate personal dress and grooming, and trust and pride in the organization are all influenced for better or worse by the attitude and behavior of the first-line supervisor.

Many first-line supervisors and lower level managers were field workers before advancing to their current positions. Management must ensure that these individuals are properly trained to lead and empower their subordinates positively.

Middle managers frequently are required to deal with customer frustrations and receive compliments from customers. How well they deal with these situations will significantly affect the perceived quality of customer service in the organization. These individuals should also receive training in customer relations.

Control and Allocation of Resources

Capital construction is almost always specifically funded through direct appropriations, bond issuance, or some other quantified means. Obviously, adequate resources are identified, and fund-raising is specifically planned, or the project will not be completed.

Once construction is complete and construction contracts closed out, resources are required to operate and maintain the facility over its life. Depending on the life planned for the facility, it is not unusual for operating and maintenance costs to exceed construction costs. Some common methods of providing these resources include space rent, continuing governmental appropriations, facility user fees, increased corporate earnings, and benefactor endowments. Some of these, such as rent or lease revenue, are used frequently to service capital debt for the project and to provide resources for operations and maintenance. Predicting these costs can be challenging, especially when future periodic capital renewals and component replacement are considered. Superimposed on this demand for operating and maintenance resources are the possible additional requirements that may occur during the life of the facility. Ten or 20 years ago, few would have anticipated the cost of dealing with chlorofluorocarbons (CFCs) and ADA requirements. Because of these factors, many facilities managers do not have what they consider to be adequate resources to maintain and operate their facilities. As a result, decisions that affect allocations of operating and maintenance resources are continually being made, and frequently result in deferring maintenance items. It is estimated that in the United States the backlog of unfunded facilities maintenance at colleges and universities is $26 billion.

On a daily recurring basis, facilities managers are allocating resources by setting priorities for work to be done that affect the level of customer service or perceived service. Three broad categories of maintenance and operations compete for available resources:

1. Those that are immediately essential

2. Those required to protect the asset for its planned life and ensure continuous satisfactory facilities performance

3. Those required to maintain the desired level of customer service and satisfaction

Examples of the first category are utility services, elevator certifications, maintenance of building structural integrity, and operation of HVAC systems. Examples of the second category are preventive maintenance, basic housekeeping services, major component replacements such as roofs, air conditioning systems, and compliance issues such as NFPA and building code requirements. The third category may include painting, carpet replacement, higher level housekeeping services, flower bed maintenance, etc. The first category of expenses must occur if the facility is to open its doors and meet its basic function. Very few of these expenditures are considered at all discretionary.

The second category of expenditures allows some discretionary spending by postponing building component replacement or delaying compliance with some regulatory requirements. Generally these expenses can be delayed but not eliminated. Most of the maintenance backlog at colleges and universities falls into this category.

The third category of expenses is usually the most visible to building occupants and is the greatest source of expressed customer dissatisfaction. Allocation of resources between and within the second and third categories is one of the most difficult and at the same time among the most important decisions facilities managers make.

Facilities must be functional and attractive to the individuals who visit and work in the buildings and grounds, and adequate resources must be dedicated to this end. However, basic preventive maintenance and corrective maintenance, major repairs, and component replacement must all be accomplished when required to ensure that the facility is functional during its programmed life.

FACILITIES MAINTENANCE AND SERVICE ORGANIZATIONS

The traditional building maintenance organization at large institutions is made up of applicable trade shops. These shops include carpentry; painting; electrical; plumbing; and heating, ventilating, and air conditioning (HVAC). Additionally, specialty shops such as a locksmith, plasterer, roofer, signage maker, motor rewinding, electronics, and asbestos abatement may be included in the facilities organization. Housekeeping services, ground maintenance, and moving/general labor are frequently part of the facilities operations and service organization.

Many of these shops' functions, especially the specialty shop services, are obtained through contract services rather than from in-house shops. Whether contract specific services or to provide them using in-house staff is one of the most complex decisions made by facilities managers.

In general, consideration should be given to contract services if any of the following circumstances exists:

1. The specialty shop staff is not fully employed on a year-round basis doing specialty work. When specialty personnel perform a significant amount of work outside their specialty areas, a close examination of shop operating costs will frequently reveal that the actual specialty work could be done more effectively by contract. Examples of this are motor repair and rewinding, furniture repair and refinishing, asbestos abatement, and in some cases, air conditioning and refrigeration repair.

2. Rapidly changing technologies require a continuous reinvestment in specialty training and diagnostic equipment. Examples of this are building automation and HVAC controls, elevator repair, and sign making.

3. Low technology, labor-intensive operations are being performed by people with significant longevity. Some examples of this are housekeeping/janitorial services and routine grounds care. In many of these cases costs will be much higher than with vendor-provided services.

Although many companies provide full-service facilities management services, ultimate management responsibility cannot be contracted out and must rest with the institutional administration. Large institutions with diverse and complex facilities often require a stable, dedicated workforce with intimate knowledge of the facilities to provide the high level of dependable customer service and responsiveness expected. It is more difficult to obtain this level of service from vendors. For these reasons the best mix of vendor-provided and in-house-provided maintenance and building services is difficult to determine.

Zone maintenance is an organizational concept that focuses on the whole building or facilities maintenance and operation rather than on specific systems and components (electrical, plumbing, etc.). This organizational concept is gaining wider acceptance in facilities maintenance and building services. The zone maintenance organization consists of multicraft, skilled facilities maintenance teams responsible for all aspects of facilities and maintenance in a relatively small number of buildings. The team is staffed with multicraft technicians to perform routine preventive and correct maintenance. The team leader is responsible for all facilities maintenance in the zone and has access to specialty shops or vendors, as required. The team leader is responsible for routine maintenance budgets and expenditures. The zone organization replaces the institution's carpentry, electrical, plumbing, paint, and HVAC shops. Specialty services such as roofing, locksmithing, and elevator maintenance are procured as needed by the zone maintenance team leader. These services may be procured from the in-house specialty shops or from vendors. This concept employs traditional competitive market forces to assure high-quality, cost-effective services.

Figure 3.1.3-1 (shown previously) is a typical facilities maintenance organization for institutions with 150 buildings and 4 million gross square feet of space. Figure 3.1.3-2 is an example of a zone maintenance organization for the same size facilities.

Facilities rarely exist in their own right. Most exist to provide an environment in which people pursue their organization's mission and goals. Therefore, facilities managers must understand and respond to the needs of building users in the most cost-effective way possible. In manufacturing and other productive facilities, it is relatively straightforward to define the facilities environment that enhances the production mission and the facilities' unit cost of output can usually be reasonably quantified. In an office, academic, or service environment where the organizational missions and goals are pursued by individuals with fewer measurable output units, it is more difficult for facilities managers to identify and quantify cost-effective facilities management strategies. In this people service environment, facilities managers must establish good communications with facilities users to understand their facilities needs and educate them regarding building and facility capabilities and limitations, as well as the cost of providing the facilities environment in which they function. Because people service organizations do not frequently measure unit output and even less frequently the cost associated with this output, it is difficult to quantify facilities cost per unit of output meaningfully.

In large institutions with many buildings spread over a campus, customer coordinators are frequently employed to optimize the effectiveness of the facilities organization. Customer coordinators are members of the facilities management organization whose task is to help building occupants and users obtain the most effective service possible. Coordinators cham-

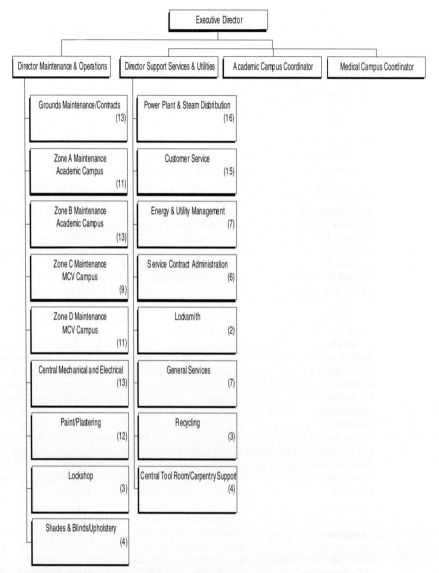

FIGURE 3.1.3-2 Zone maintenance organization.

pion the building users' needs in the facilities management organization and ensure that the individual requirements of building customers are addressed in the facilities management organization. The coordinators educate building users concerning reasonable expectations, limits, and capabilities of the building systems and facilities services organization. Effective customer coordinators maximize the effectiveness of the facilities maintenance and service organization in meeting individual customer requirements.

In a multiple-building campus environment, building managers should be established to work with customer coordinators and expedite communications with the facilities management organization. The duties of building managers are collateral duties assigned to an administrator or a business manager of the principal department or unit that occupies the building. The duties of the building manager include the following:

1. Being the principal contact for reporting building systems problems to the facilities management organization
2. Being the principal contact for facilities management to disseminate information to the building occupants concerning such things as utility shutdowns and renovations and design work schedules that may affect building occupants

Building managers and customer coordinators must develop a continuous dialogue that greatly helps in the process of delivering effective facilities maintenance and building services.

Budget and cost management is critical for facilities managers in most effectively providing the best physical environment for building users. After capital investments and construction are completed, realistic budgets must be developed to preserve the facility for its programmed life and to provide the maintenance building services to optimize customer satisfaction in the building. Most facilities are maintained and operated on the basis of an annual budget. Funds for this budget may be appropriated from central sources or may be from rental or lease charges or other sources. Budgets should be identified in each of the three categories that compete for resources as previously discussed. Deviations from this budget plan should be minimized and made very thoughtfully. Such deviations usually result in diversion of resources from basic maintenance to customer desire, and discretionary projects increase the deferred maintenance backlog of the facility. Budget and cost accounting systems should clearly document such diversions, and senior management must address the issue to prevent large increases in deferred maintenance.

Work order systems that are integrated with cost accounting systems will automatically accumulate cost against specific budget line items. Standing and open work orders should not be used. All maintenance and service work should be on specific work orders and should clearly identify the work to be accomplished. Even routine work such as housekeeping services should be accumulated specifically by building, and for large, complex buildings, by subcategories of the building. First-line productive workers should account for all of their time every day against specific work orders. Overhead functions and activities such as training time, allowed time, leave, and tool costs should be accumulated into overhead accounts. Contractual service costs should be accumulated on specific work orders in the same way. Such a system allows management to review organizational cost-effectiveness and to compare it to benchmark costs. Commonly used benchmarks are building maintenance costs per square foot for different categories of space, housekeeping cost per square foot, and overhead cost as a percentage of direct cost.

Facilities Maintenance

Refer to Chap. 7 for several detailed facilities maintenance articles that will complement the following overview. Preventive and predictive maintenance are central to any effective maintenance program. *Preventive* maintenance includes any scheduled maintenance effort that improves the serviceability and dependability of building systems. Preventive maintenance includes periodic maintenance work recommended by equipment manufacturers that ranges from changing filters to periodic equipment overhauls. It also includes exercising and operating equipment such as emergency generators and switchgear on a scheduled basis. *Predictive* maintenance includes examining and analyzing equipment and components to predict specific maintenance or component replacement to reduce the probability of failures. Such activities range from basic tasks such as visual inspection of drive belts and couplings to more sophisticated testing such as current analysis of heat exchangers in boilers, chiller tubes, and drums. Predictive maintenance is included with preventive maintenance when air filters are changed based on measured air-pressure drops across filters and when experienced mechanics decide to adjust the tension on drive belts based on visual and audio observation of drive systems while they are operating. Therefore, preventive maintenance usually includes elements of predictive and preventive maintenance.

A basic automated preventive maintenance (PM) system schedules this work, estimates the task time required, produces detailed work orders, tracks costs, and produces backlog reports of completed PM work. More sophisticated systems will schedule PM work, taking into account manpower availability and production or building schedules; will develop equipment maintenance cost analysis; and will manage inventories of spare and replacement parts. These systems will manage all preventive and predictive maintenance activities from daily visual inspections to multiple-year equipment overhauls. Data from these systems provide a quantified basis for capital replacement programs and predict manpower and equipment needed to maintain equipment most effectively.

All dynamic equipment from sump pumps to elevators and emergency generators should be maintained on a PM system. Equipment such as door closures, piping and plumbing valves and devices, switchgear, panel boards, transformers, and other building system components should be on a PM system to assure maximum system dependability. Roofing systems should be included in the PM system to ensure that rainwater drainage systems are clean and functioning properly and that water is not penetrating roof membranes.

Corrective maintenance is activity to correct malfunctioning or broken system components. This work is frequently identified by building occupants when an obvious failure occurs. Too-hot and too-cold calls, information regarding plumbing or roof leaks, and complaints regarding burned-out lights and doors not closing properly are examples where building occupants identify the need for corrective maintenance. An effective PM program will minimize the need for corrective maintenance, but a responsive service call system is required to meet customer requirements for corrective maintenance.

Scheduled maintenance includes planned work that is usually included in periodic work plans and budgets. This includes scheduled painting, major roof repair or replacement, and replacement or overhaul of other major building system components. When scheduled maintenance and preventive maintenance are not completed as planned, the need for corrective maintenance increases, and eventually a backlog of maintenance requirements accumulates. This deferred maintenance will eventually require major building reinvestments and replacement of building systems. Deferring maintenance when building components are scheduled to be upgraded or replaced in the near future because of service code or other external requirements is the only legitimate reason for allowing deferred maintenance to accumulate. Examples of this include HVAC systems that have exceeded planned life and are scheduled to be replaced with more energy-efficient systems using non-CFC refrigerants, and planned replacement of roofing systems with better insulated systems and modern roof membranes. Except in the case of such planned replacement, deferred maintenance should be minimized.

The Facilities Condition Index (FCI), discussed in detail in Art. 3.1.2, is a measure of the general maintenance condition of a building. It is expressed as a percentage calculated by dividing the total estimated cost of deferred maintenance by the replacement value of the building. An FCI of 5 percent or less is an indication of a properly maintained building. An FCI of 5 to 10 percent indicates that additional maintenance and repair effort are needed, and an FCI of more than 10 percent indicates that the facility is generally poorly maintained and requires a significant reinvestment.

Building Services.   From the perspective of a customer and building occupant, housekeeping services are part of facilities services. To be most effective, housekeeping services should be integrated with building maintenance systems. Housekeeping services are frequently provided after normal working hours and provide an excellent opportunity for spot relamping and monitoring of building and security lighting systems. The housekeeping staff can very effectively document and report building system malfunctions ranging from improperly operating door closers to restroom plumbing leaks and malfunctions. When housekeeping operations are integrated into the building maintenance effort, total building services are improved. Refer to Art. 7.1.4, "Environmental Services," for a more detailed discussion of these issues.

Solid waste management and recycling operations are most effective when they are part of the integrated facilities services system. The housekeeping staff can perform initial separation of recyclable material at the source. This is especially true for office paper and discarded cardboard. By making best use of freight elevators, loading docks, and spaces normally managed by facilities maintenance operations, solid waste and recycling effectiveness can frequently be significantly improved.

Relocation, moving, and storage services within a building are frequently part of the integrated facilities service system. Most administrative classroom laboratory and medical facilities require frequent relocation of furniture and equipment. To be most effective, these services require close coordination with facilities management, and many building services such as minor painting, electrical outlet relocation, and so forth are part of the complete moving and relocation service for the facility's users.

Minor space renovations are frequently associated with moving and relocation. These renovations should be managed by the facilities maintenance organization to ensure continued compliance with building and fire codes. To provide good, cost-effective facilities service to building occupants, such minor renovations should be integrated into the facilities management systems.

Modern building security involves locking, lighting, and monitoring systems, as well as physical security. Electronic locking systems using various sensing technologies provide a high level of access control and a wealth of security and access control information. To function most effectively, these systems require that doors and door hardware operate properly and that the building emergency power supply is dependable. Good maintenance of the associated building systems is essential to the best operation of these security systems.

Lighting systems, security cameras, and other security devices require dependable building support systems. Therefore, to be most effective, these systems should be part of the integrated facilities management systems.

Utilities.   Management of building utilities is an important part of the integrated facilities management process. Utilities often represent a large part of the cost of building operations and maintenance or housekeeping. Modern lighting systems and related controls, automated building management systems, and efficient, optimally managed and operated heating and cooling systems can reduce utility costs significantly. Refer to Sec. 5.5, "Instrumentation and Control Systems," for a detailed treatment of these instrumentation and controls systems.

Facility managers must understand utility and energy markets in order to procure electricity, natural gas, and fuel oil at the lowest possible cost. Refer to Art. 7.1.6, "Fuel and Energy Procurement," for a detailed discussion of this topic. The electric power industry restructuring will offer challenges and opportunities for facilities managers to manage energy procurement more cost-effectively. Natural gas markets have been deregulated in most areas for many years, and facilities managers have considerable flexibility in purchasing natural gas. Where possible, the use of natural gas and fuel oil should be integrated to obtain the lowest BTU cost of energy while ensuring a dependable supply of fuel. Facilities managers would have access to contractor and other physical plant technical support on a reimbursable basis, as required.

- A central electrical mechanical team would manage and provide maintenance for large chilled-water generation systems, large electrical switchgear, fire pumps, emergency generators, and elevators.

- A central support shop with lock shop, roofing, asbestos abatement, and other trade and craft capabilities would be available to support the zones on reimbursable basis, as needed.

Lighting technologies offer many opportunities to reduce lighting costs while improving the quality of light for building users. Some frequently used technologies include converting fluorescent lighting systems from T-12 tubes with magnetic ballast to T-8 tubes with electronic ballast; the use of compact fluorescent bulbs in lieu of incandescent lamps; the use of metal highlight lights and geometric reflector systems; the use of motion and ambient light sensors to switch and dim lights; and integration of lighting systems with building automation systems. Section 5.3, "Lighting Systems," offers a detailed presentation of these issues.

HVAC technologies include high-efficiency chillers and cooling towers, high-efficiency dual-fuel boilers, thermal storage, variable-volume air flows, variable-frequency drives for fan and pump motors, the use of premium efficiency electric motors, and sophisticated building automation systems. Sections 5.4, "Mechanical Systems," and 5.5, "Instrumentation and Control Systems," provide further information on these subjects.

The U.S. Department of Energy sponsors energy star and green light programs to encourage conversion to and use of energy-efficient building components and systems. These programs provide a clearinghouse for information and efficient energy technologies, offer training and education programs about these technologies, and publicize companies and institutions that reduce and manage energy usage.

As the electric power industry moves toward deregulation and competition, facilities managers must position themselves and their companies and institutions to take advantage of the deregulated environment. The pace and details of deregulation vary from state to state, and facility managers should understand how deregulation will affect the electric power industry in their states.

Facilities managers should be familiar with the various rate schedules available to them from their local electric power companies and ensure that electrical energy is purchased at the most favorable rate or rates available.

Natural gas markets have been deregulated for several years, and larger users of gas usually have several options for purchasing natural gas. Dual-fuel boilers using fuel oil and natural gas give the facility manager considerable leverage in negotiating natural gas prices with gas producers, interstate pipeline companies, and local distribution companies (LDC). Some large users can effectively purchase natural gas at the wellhead and pay interstate pipeline companies and LDCs to transport natural gas to the boiler considerably more economically than by direct purchase of natural gas from the LDC. Having the dual fuel capability may also allow negotiation of gas prices directly with the LDC. When dual-fuel boilers are used, it allows facility managers to purchase interruptible gas service at lower rates when they agree to switch to fuel oil at times of extremely high regional or local natural gas demand. When world oil markets are depressed, these depressed prices give additional leverage when negotiating with gas suppliers and LDCs.

Natural-gas brokers are available to help users purchase natural gas at the most favorable prices. When the electric power industry is deregulated, similar opportunities are likely to exist in the purchase of electrical energy.

In some regions and localities, high water and wastewater rates provide an incentive for reducing water consumption and justify investment in water-saving devices. Some localities require water use reduction or other restrictions, adding further justification for controlling water usage and consumption.

Reduced-flow toilets, faucets, and showerheads are available and should always be considered when replacing existing devices. Separate metering of cooling tower makeup water and irrigation water should be considered when wastewater charges are based on water usage. The responsibility for water reduction strategies varies widely with local conditions and regulations but should be considered in integrated management strategies.

Dependability and the need for alternate and backup utilities should be considered in integrated facilities strategies. The use of emergency generation, dual-fuel boilers, alternate electric feeders, and water storage or wells are common strategies to ensure dependable utility services for critical buildings.

SUMMARY

Facilities should exist only to support an institution's or company's mission and goals. A few facilities, such as historic buildings, may exist in their own right. But even these facilities should be integrated into the institution's mission and goals.

Master planning is the process of identifying and quantifying the facilities required and outlining a strategy to acquire and maintain these facilities. Space planning and administration is the process of effectively and efficiently utilizing the facility assets of an institution. Facilities operation is the process of maintaining and operating the facility to the maximum continuing benefit of the institution and its constituents.

Facilities operation is the process that includes facilities maintenance; building services such as housekeeping, moving services, and renovations; and dependable utility services. It is the responsibility of facility managers to provide these services to building users in the most cost-effective manner possible.

Investment in facility assets can be justified only by the return on investment. The return is in the form of a physical environment that facilitates the accomplishment of the institutional mission. Tools to measure this return include facilities investment, operating cost per unit of output, and user satisfaction with facilities services.

ARTICLE 3.1.4
BUILDING LIFE CYCLE— AN INTEGRATED APPROACH

Bart D. Bauer, P.E.
CID Associates, Boston, Massachusetts

Kenneth H. Stowe, P.E.
George B.H. Macomber Company, Boston, Massachusetts

A NEW WAY OF THINKING

Facilities professionals have the opportunity to make profound contributions to their companies when they find ways to contribute *at every phase* of their buildings' life cycles. This can be accomplished by taking an *integrated approach* to the building life cycle—making decisions and taking actions in one phase to improve overall performance in many others. These facility professionals are promoting and defending a philosophy—that every facility investment should produce maximum value. This article focuses on the vital need to consider all phases of a life cycle when investment decisions are made. Some call it "right-to-left thinking."

The need of forward-thinking corporations in the recent past and in the far-reaching future has been and will be to minimize the effects of facility costs on the bottom line. This is accomplished by striving for optimal value from every facility. The well-informed facility manager has the opportunity to participate in value-related discussions at the boardroom level and provide the understanding and information to maximize the facility's value. Maximizing the long-term service of a facility and minimizing its long-term costs generate optimal facility value. The service of a facility is expressed in intangible forms as

- Operational productivity
- Aesthetic value or public image
- Comfort
- Flexibility

The costs show up in these tangible forms:

- Construction capital cost
- Energy costs
- Maintenance and repair

WHAT IS AN INTEGRATED LIFE-CYCLE APPROACH?

The integrated life-cycle approach has one fundamental tenet—that experience at every phase is valuable and that the wisdom gained from many professionals over many phases must be considered.

Reengineered Systems

The enlightened facilities professional solicits input from many experts involved in all aspects of the facility. For example, the design team should receive consistent and mature input from operations, maintenance, construction, and procurement. Input from these areas of expertise ensures that design decisions are not made in a vacuum—achieving short-term goals but sacrificing efficiency in other aspects. Collecting, processing, manipulating, organizing, and presenting the data that are levered into position at the boardroom table are the facility manager's responsibility. The pieces of facility data may have been collected individually for years; the power that the new facility manager has is in combining those individual pieces into a cradle-to-grave facility picture that is all-encompassing.

Similarly, the modern builder is collecting vital electronic as-built computer-aided design (CAD) information, photographs, and testing and manufacturing data during the construction phase for efficient use during operation and maintenance. Changes to the facility during operation are documented in up-to-date CAD libraries to facilitate the design of a future renovation.

Other examples of life-cycle approaches include:

- Integrating the CAD approach
- Investing for flexibility—structure
- Investing for low maintenance and operating costs
- Designing for low maintenance and operating costs
- Designing for the owner's philosophy and risk preferences
- Involving participants from other aspects for decision making

There is additional power in leveraging available technology. During the last 15 years, technology has been used to collect and analyze individual pieces of data, and the technologically literate facility manager could use individual computer applications to prepare presentations based on those individual pieces of data. The additional power of this new integrated approach is found in leveraging the ever-growing body of computer applications, to combine and synchronize data from one application to the next to create models that work across all phases of the building life cycle. An example of this approach is the three-dimensional computer-aided design and drafting (CADD) model that has color references to tie into a schedule, that the architect uses for presentations to the client, that the estimator uses for true quantity takeoffs, that the facility manager uses as living as-builts, and that allows all of the design engineers and construction trades to prepare true, coordinated drawings highlighting physical and scheduling conflicts long before the materials are on-site for installation. Space planners and marketing staffs can use this three-dimensional model to sell the space in the building and an architect retained to perform tenant fit out as the space turns over throughout the life of the building can use it. Although this example uses CADD as the leveraged tool, the application of technology is not limited to a single computer application or program. Rather it is the combination of the programs and the use of the programs output at all levels and phases that more closely defines the use of technology in this new leveraged way of thinking.

The nature of the facilities professional's challenge is that there will be conflict when decisions are made, and decisions about life-cycle value are no exception. This article is designed to equip you with an approach to facilitate the decision-making process, so that your personnel can reach agreement and communicate decisions in a compelling way.

The Phases of a Facility Life Cycle

The phases of a facility life cycle (see Fig. 3.1.4-1) are as follows:

- *Definition of need*—documenting specifically how an existing facility no longer serves as an optimally functional space for the intended use, and understanding the reasons (e.g., shortcomings or functional failures). Additionally understanding the requirements of a new way of doing business and what those requirements will be into the future use of the building. (This phase is often combined with the program phase into one large phase.)
- *Program*—A set of owner/occupant-defined guidelines or specifications that reflect how the needs defined in the previous phase can be met by a facility. A designer uses these guidelines to set goals with which to guide the process of designing a facility.
- *Design*—The collaborative, creative process that produces a set of documented instructions, both written and diagrammed, that address the program requirements and are used to build a new facility or renovate an existing one.
- *Build*—Procuring, expediting, and constructing the building components to produce the physical entity that was specified in the design phase.
- *Operate/maintain*—The process and time period when the owner-occupant is using the building for its intended use and is doing the upkeep.
- *Decision phase*—The analysis of options leading to a decision to renovate a facility or choose the "end-of-useful life" option. This phase is reached when the overall cost of using a building equals or exceeds the value received from the occupancy or the value of building a new facility in its place. This decision requires reentry into the facility life cycle either through a total renovation of the existing facility or construction of a new one.

These phases and their associated costs make up the life cycle of a component of a specific building, a group of components of a building system, a facility as a whole, a group of facilities that comprise a campus, or a group of facilities scattered on sites across the globe that is owned and operated by a single organization.

Communication

The key to the new integrated approach is achieving a global view. The global view encompasses all stakeholders—all critical players at all phases of the facility life cycle. The methodology to realize this and to achieve success is communication among all the stakeholders. All stakeholders must be involved early in the process and remain engaged throughout the entire life cycle. Lessons learned and life experiences are the conduit for the value realized from the integrated approach.

The stakeholders include the owner, bankers, limited partners, architect and engineers, contractor, estimators, production management personnel, and subcontractors. Critical to the success is the involvement of the tenant and the building maintenance team. All of these

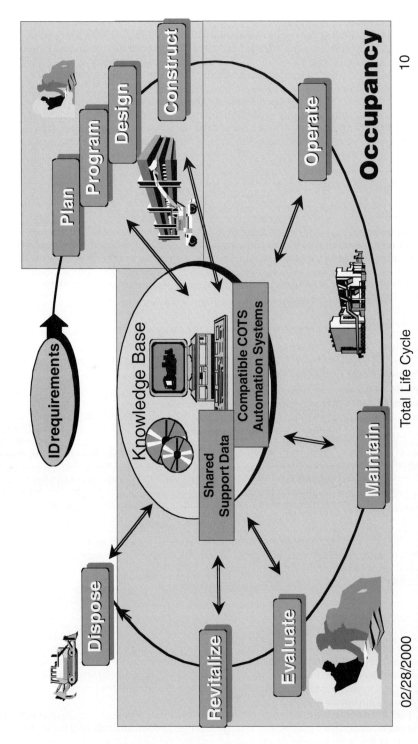

FIGURE 3.1.4-1 A facility life-cycle model. Information collected by many people can be useful during many phases. (*Courtesy of Bart D. Bauer, P.E., CID Associates, and Kenneth H. Stowe, P.E., George B.H. Macomber Company.*)

stakeholders must be involved in the process at the earliest possible moment. Why? Let us revisit the one tenet of the integrated approach—"That experience at every phase is valuable and the wisdom gained from many professionals over many phases must be considered". The combined value of all of the wisdom and knowledge will create the closest-to-perfect facility imaginable. This wisdom appears in many forms when the opportunity is presented to all of the stakeholders:

- The contractor shares in the planning phases with realistic estimates and schedules that come from real-world experience.
- The maintenance team ensures that the products and materials are compatible with the staff's knowledge and experience and are durable and maintainable with the user group in mind.
- The architect is a working member of the team, rather than an adversary, and participates with the contractor and subcontractors to find and resolve conflicts preemptively.
- The owner presents a fully coordinated team and project to investors, and the project team can respond to the owner's concerns about time, scope, or money because all of the team members are aware of the project restraints.

The earlier in the process this wisdom is shared, the more value the wisdom has. The more value the wisdom, the less likely that the project team will spend time on costly tangents with no real chance of success. The less time wasted, the faster and more efficiently the facility is produced, reducing financing costs to the owner, general condition costs to the contractor, and allowing the design team to move on to the next project. So, there is a direct financial benefit to all the stakeholders who participate in an integrated approach.

The integrated approach is not something that requires a Ph.D. in computer technology. Any project team in any corner of the world can adopt advanced communication strategies. Project teams can be formed, and work together around a conference table or a pile of sand and share wisdom. As project teams become broader and more disciplines and areas of expertise are introduced, they no longer center geographically around the client's home office, and as the team members are selected for their expertise, not their geographic office locations, computers can be used to make these distances and logistical problems disappear. Transferring CADD files over the Internet, videoconferencing, redlining drawings from remote facilities, e-rooms, and bulletin boards are all tools that facilitate an easier implementation of the new integrated approach. Additionally, the power and speed of computer technology allows for fast and accurate generation of different scenarios such as different phasing plans, floor plan layouts, and schedule priorities. The power of the technology is also used by combining applications, such as linking the schedule, the budget, and CADD drawings, so that each is updated by the mouse stroke in the other application.

Life Cycle Costs—The Story

For any facility produced in the modern age, the overall financial picture of life-cycle costs throughout its life can generally be grouped in the following six broad categories:

- Design, or *soft* costs
- Initial construction, or *hard* costs
- Indirect expenses
- Operating and maintenance expenses
- Capital replacement costs for replacing building components
- The cost of money, or the interest paid on the monies expended

According to Alphonse Dell'Isola, author of *Value Engineering: Practical Applications*, during the life of a building, these costs break down into the percentages shown in Fig. 3.1. 4-2 (see Ref. 1).

Every facility and every facility operational system is comprised of individual components. All of these individual components have varied but fixed life expectancies, usually referred to as *expected useful life* (EUL). A well-designed project will have facility and system components whose life expectancies complement other related building components or the owner's maintenance philosophy. An example of failing to integrate the components of a building system would be the use of polyvinyl chloride (PVC) masonry flashing that has a short (5 to 10 year) life expectancy in a brick wall that has a 75 to 100+-year life expectancy. An example of building components that can be integrated with the philosophy of an owner who is a long-range thinker and wants to lower the cost of life cycle maintenance would be the use of masonry for exterior sheathing. If, on the other hand, the owner has a short-range financial focus, then maybe an exterior insulation finish system (EIFS) that has a 15-year life expectancy would be used.

During the course of a system component's useful life, there are costs associated with it. These costs are generally divided into initial purchase costs (which is a component of the overall facility construction budget), operating and maintenance costs, and finally capital replacement costs (see Fig. 3.1.4-3).

The operating and maintenance costs of building components can be generally divided into routine and nonroutine maintenance and capital expenditures. As an example, the routine maintenance costs for a HVAC system may include preventive maintenance such as replacing filters. The nonroutine costs include replacing a belt when one breaks, or recharging the refrigerant. Capital replacement may be the replacement of the compressor. In a facility design that is well integrated, both routine and nonroutine maintenance can be predicted and will be evenly spread out over the expected useful life of the building.

At each capital expenditure, a decision process is undertaken. The facility manager questions whether there is still value in renovating the unit by replacing the belt. In this case, the belt is definitely a small cost compared to the cost of replacing the entire machine. When a compressor breaks down and needs to be replaced, the decision process is again visited: Is the

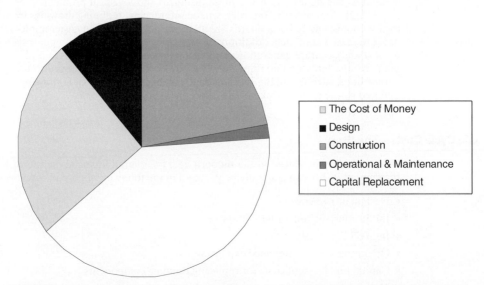

FIGURE 3.1.4-2   Distribution of expenses. Cost breakdown for a facility's life cycle. (*Courtesy of Bart D. Bauer, P.E., CID Associates, and Kenneth H. Stowe, P.E., George B.H. Macomber Company.*)

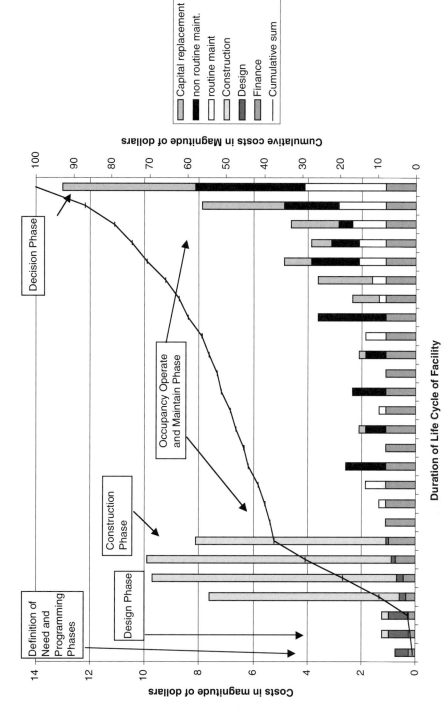

FIGURE 3.1.4-3 Distribution of costs during a facility's life cycle. (*Courtesy of Bart D. Bauer, P.E. CID Associates, and Kenneth H. Stowe, P.E., George B.H. Macomber Company.*)

cost of the compressor equal to the replacement or the value received from the entire unit? No. Therefore, the unit is renovated again. When the compressor breaks down again, and the belt is worn and frayed, the case in which the unit is housed is rusted and can no longer keep the weather out, the fan blades are rusted, the bearings need to be replaced, and the motor needs to be rewound, then the final decision phase is reached: Should the unit be replaced or totally rebuilt? This is the end-of-life decision that is the last phase of the life cycle.

No matter, however, whether the focus is short-range or long-range or whether the facility is well integrated, it eventually reaches a point of diminishing financial returns. Then it becomes worthwhile to evaluate the alternatives of building a new facility. This decision is based upon the cost effectiveness or the return on an investment to replace major pieces of capital equipment versus investing those monies in a new facility. This is potentially the final stage of a facility's life cycle.

In our HVAC example, all of the previous costs have covered work and parts within the confines of the original box and cover. No matter how many times those internal components are replaced, there is a point at which the basic structure of the HVAC unit decays and must be replaced. This is a major capital outlay. Before this time, an analysis may have determined the cost-effectiveness of repair or replacement, but once this major structure has deteriorated, a major capital outlay is the only option. The facility manager can choose to buy a new unit or to totally rebuild the current unit.

No matter what the life expectancy of a facility, whether it is a masonry building with a 100+-year life expectancy or a building with shorter life-span components, there is a time at which that facility reaches the end of its useful life. This does not mean that the facility is standing one day and collapses the next. Rather this is the point at which the majority of the major systems require considerable capital maintenance; routine maintenance has grown in scope and costs; and nonroutine maintenance has occurred so frequently that it is bordering on routine. This point in a well-integrated facility project can be predicted on the basis of an understanding of escalating maintenance costs, end of useful life (EUL) of the major components, and the usage and related physical depreciation of the facility. That escalating expense is shown in graphical form in Fig. 3.1.4-3. A conscious decision must be made at this end-of-life point.

Based upon philosophy, facility performance, capital funding, and risk and benefit analysis, the options presented are to renovate the existing facility or move and build a new facility. There are benefits and consequences to each decision, as shown in Table 3.1.4-1.

If the decision is made to renovate the existing facility, then the facility manager and owner essentially have the opportunity to reenter and start the life cycle over again. Needs are

TABLE 3.1.4-1 The Decision to Renovate or Build New

Decision	Possible benefits	Possible negative consequences
Renovate	These are discussed in chapter 6, and include the following: Working within the existing shell, even in a total rehabilitation, is often less expensive than new construction. When working with the existing facility, generally there are no new property acquisition costs. The existing location is a known entity to employees and clients alike.	It may be more difficult to institute new technology and industrial methodology in an existing facility. Even in a total rehabilitation, some existing building fabric remains, and therefore all capital planning cannot assume time zero as an age as in new construction.
Build new	All components of the facility fabric are new, and life-cycle capital planning can start at time zero. New locations, new images, and building around new and current technology are possible.	Additional costs are incurred for land acquisition, moving, and employee and client orientation.

assessed, and a program is established to understand the scope of work that is required to rejuvenate the building.

There are other ways for a facility manager to reenter the life cycle that are unrelated to the end of the facility's useful life. If the current technology has changed so dramatically that the product can no longer be produced in a facility of the current design, the functionality of the building has become obsolete despite the fact that the building components have not exceeded their EUL. Or if the user has added so many employees or developed new business units, it is possible to exceed the facility's capacity, and therefore the usefulness of the facility must be reevaluated again based on functionality, rather than on the basis of EUL.

CONCLUSION

An integrated approach to the life-cycle performance of a facility can make a huge difference in the net value of a facility engineer/manager.

REFERENCE

1. Alphonse Dell'Isola, P. E., *Value Engineering: Practical Applications,* Page xviii, Construction Publishers and Consultants, 1997.

ARTICLE 3.1.5

CORPORATE LEASING, DESIGN, AND CONSTRUCTION: A PROJECT MANAGEMENT APPROACH FROM THE TENANT'S PERSPECTIVE

Scott S. Tibbo
International Facility Management Association (IFMA),
Boston Massachusetts, and Expense Management Solutions, Inc.,
Southboro, Massachusetts

AN OVERVIEW OF FACILITIES PROJECT MANAGEMENT FROM THE TENANT'S PERSPECTIVE

In today's dynamic business environment, a company's need for office space can change rapidly, and it is essential that the corporate real estate department be prepared to respond to the business challenge of effectively managing the company's real estate transactions while still meeting the corporate goal of controlling occupancy costs. Because most real estate portfolios today, for both small and large corporations, tend to be comprised of leased office space rather than owned building assets, it is critical that the team that represents the tenant use an effective transaction process. This article provides an overview of the major steps in a typical real estate transaction, and provides insight from a tenant's perspective into proven project management methodologies that engender up-front planning, organized communication, and informed decision making. Table 3.1.5-1 summarizes the major phases and tasks within the process in a matrix format.

Phase I: Project Initiation/Needs Assessment

Initiate the Project. When the initial request is made to the corporate real estate (CRE) department by an end user group to obtain more space or to modify an existing office environment, it is essential that CRE obtain clear direction as to the size, length of term, and business need for this new requirement. It can be very beneficial to establish a standard project initiation process that involves a formal sign-off on the scope and intent of the new project by the requesting department's senior management. Without this end user management control in place, the CRE department could invest significant effort and cost in pursuing a new space requirement, only to find out later that the end user's senior management did not support or agree with the original direction. A simple, one-page project initiation form that is approved by the requester and their senior management can help minimize these occurrences.

Analyze the Current Environment. Before proceeding to discuss the new space requirement, it be very beneficial first to evaluate and confirm the conditions and costs of the end user's current environment. In addition to providing CRE insight as to how the existing space has been performing, it will establish an internal benchmark that both CRE and the end user can use to compare the quality and cost of any new space alternatives. This assessment of the current con-

TABLE 3.1.5-1 Summary of Transaction Process

Real estate transaction process				
Phase I Needs assessment	Phase II Analysis of options	Phase III Negotiation and approvals	Phase IV Design and construction	Phase V Project close-out
• Project initiation • Analyze current environment • Review real estate market • Define business needs & strategy • Develop project schedule • Define project team • Update client on schedule and project team	• Confirm staffing and space needs • Detailed market search • Complete property tours • Issue RFPs • Analyze RFPs • Test-fit drawings • Due diligence • Develop/ compare occupancy costs between options	• Issue counterproposals • Summarize proposals • Detailed financial analysis • Issue recommendation memorandum • Obtain approval • Negotiate lease • Confirm design and construction terms	• Design development • Construction documents • Integrate I/T • Incorporate furniture and equipment specs • GC prequal • Bidding & award • Construction phase • Install furniture and cabling • Move-in activities	• Punch list • Reconcile accounting • Issue close-out report • Complete project evaluation • Finalize project files • Enter data into real estate management system • Transition files to lease administration • Monitor occupancy issues

ditions should confirm how the space is being utilized, identify any environmental or property management concerns, and clearly document the full occupancy costs of this current space.

Review the Current Real Estate Market Conditions.   Based on the general information provided in the project initiation form, CRE can research the current market rental rates and space availability within the desired geographic area. This market data, in addition to the information gathered on the current conditions, can be very instrumental in conducting strategic discussions with the client regarding specific long-term space needs.

Define the Business Needs and Confirm the Real Estate Strategy.   After the information previously mentioned has been gathered and evaluated, it is recommended that CRE share and review that data with the end user. This would be followed up by a meeting or conference call with the requester and a senior management representative to discuss the specific business and real estate needs for this new space in greater detail. Specific direction regarding site location, image, access, visibility, employee demographics, and so forth is critically important so that the CRE team clearly understands the situation before it begins a detailed space/real estate search.

Develop a Preliminary Project Schedule and Establish the Project Team.   Once CRE has obtained this baseline information and project direction from the client, there should be sufficient information to develop a project time line (schedule) that outlines the subsequent phases of the project and the anticipated amount of time needed to complete each major phase. It is helpful if CRE can establish and publish schedules in a standard format for all clients and members of the project team. At this point in the project, it is recommended that CRE establish the expanded core team that will be responsible for supporting the execution of this transaction (project manager, architect, engineers, leasing agent, information systems coordinator, etc.). It is also recommended that CRE issue a contact list that includes all of the core team member names, addresses, phone numbers, e-mail addresses, and so forth.

Phase II: Analysis of Options

Confirm Staffing and Space Needs.   Before conducting any detailed market search, it is essential that CRE first confirm their long-range staffing and space needs with the end-user group(s). This information should be clearly detailed and summarized within a standard programming document. Quite often, this information is gathered and evaluated by a space planner or architect who has met with the key client contacts, on-site, to evaluate their current work environment and discuss long-range space needs. Depending on the size of the organization, the department's senior management may want to approve this long-range staffing and space need information.

Conduct a Detailed Market Survey.   Once the project team has a clear understanding of the space needs, along with the specific real estate criteria already gathered, it can commence a detailed market search to identify all available property locations within the defined search area. Using a standard set of selection criteria, such as building type, access, visibility, rental rate, ownership, floor-plate design, and so forth, the team can establish a "long list" of properties available that may meet the requirements and qualify for a property tour.

Complete Property/Site Tours.   At this point, representatives from the core team (e.g., project manager, tenant rep, architect) should conduct facility tours of all long-list properties with the building owner's representative to evaluate how these facilities meet the defined needs. Based on the team's assessment of the property tours, a "short list" of potential buildings should be confirmed for further evaluation.

Issue RFPs to Selected Sites and Landlords.   Detailed requests for proposals (RFP) should be sent to the owner's representatives from each short-listed building. The RFP should request information on critical lease issues such as space availability, term, rent, operating expenses, tenant fit-up allowances, building systems, ownership, other tenants in the building, and so forth.

Analyze the Requests for Proposals.   Once all of the RFPs are completed by the building owners/representatives, it is important that all members of the core team have an opportunity to review the detailed responses. Quite often, the tenant representative will compile a matrix document that summarizes the major deal points and building features to help compare, on a line-by-line basis, the qualitative and financial differences between the various building candidates. If critical pieces of information are missing or appear inconsistent, invest the time to obtain clarification at this point in the process. Based on the detailed review, the team should have sufficient information to reduce the list of buildings under consideration to three primary choices. It is very important to keep at least two to three options open and in play until a final recommendation has been accepted by the client and all critical lease terms are negotiated and approved by both parties.

Complete Test Fits and Building Due Diligence.   In addition to considering the business and financial terms of the various lease proposals, it is valuable if the tenant can develop some preliminary "test-fit" drawings to define more accurately how much space will be required in each prospective facility. The project designer or architect can often quickly develop these test-fit drawings from the approved staffing and space program documents already gathered. In addition to indicating how much usable square footage will be needed in each location, it will help define the total rentable square footage per location, which can vary dramatically, depending on the building's configuration. These test-fit drawings are also helpful in determining how much new construction, or use of existing architecture, will be required in each of the facilities. This information is critically important in developing probable cost estimates for tenant improvement expenses and for evaluating how well each building fits the space requirements. It is also highly advisable at this point in the process to engage a team of engineering consultants on the tenant's behalf to fully evaluate the quality and performance of the base building infrastructure for each of the buildings under consideration. These studies may include examination of all major mechanical, structural, electrical, and fire protection systems, including life safety and Americans With Disabilities Act (ADA) accommodations, the telecommunications infrastructure, and the amount of lighting in the designated parking areas. It is critically important to recognize any major building issues that may exist along with their associated costs and risk factors early in evaluating prospective buildings—before short-listing or any lease negotiations.

Develop Full Occupancy Cost Models.   It is essential that a comprehensive occupancy cost model is completed by the core team before any recommendations are made to the end-user group. These financial models should take into account the proposed rent and operating and common area maintenance costs over the term, including escalations and real estate taxes.

The models also need to include all soft costs related to design and engineering fees, tenant construction costs (minus any tenant improvement allowances provided by the landlord), plus any additional costs for signage, security, moving, telecommunications infrastructure, or any landlord management fees that may be required. Determine whether your Finance Department requires that you include any "cost of capital" or internal financing charges related to capital outlay. Also factor in any holdover penalties or duplicate rent from your current location and any landlord management fees, if involved. These financial models should determine for you a total annual occupancy cost and the total commitment over the lease term. It is also helpful to compare the average "occupancy cost per person" among the various options. Comparing the true total occupancy costs among the building alternatives is the best approach. The decision on which building to lease should never be made on basic rent alone.

Phase III: Internal Approvals and Lease Negotiation

Issue Counterproposals.   Based on the qualitative and financial comparisons that have been completed at this point and the needs of the business unit, the project team should be able to define the top two locations that best meet the overall requirements. It is advisable to keep at least two options "in play" and to continue to refine and negotiate business/lease terms in the event that one of the space options is leased to another tenant or if a deal falls through. This is also a good time to research and leverage any incentives that the landlord, community, or the state/regional economic development sector may have to offer your company (e.g., tax relief, parking concessions, energy discounts).

Complete Financial Analysis.   Once the final evaluation for each building has been completed and cash flows have been calculated for each option, summarize the results in a matrix format to be used as a financial exhibit with the recommendation package that the project team will issue to the internal customer.

Issue a Recommendation Memo.   Present the results of the process in a standard memorandum format for the client that summarizes the defined primary business objectives, the research completed, the building options considered, and a recommendation as to which location best meets the defined criteria, including the cost considerations. The recommendation memo should include the detailed financial model along with descriptions of the building, location maps, and copies of the test-fit drawings. If there are multiple business units involved in the deal, it is helpful to break out each unit's pro rata share of the total project costs.

Obtain Senior Management Approval.   Issue the recommendation memo to all necessary levels of management within CRE, finance, and the business unit(s) that requested the project and clearly communicate which level of management needs to approve the transaction and in what time frame. It is very helpful if the organization can have preestablished approval authority levels for new transactions based on the size of the financial commitment.

Review, Negotiate, and Execute the Lease Document.   Once internal approval has been confirmed for the preferred location, the project team, along with legal counsel, should finalize all financial and business terms related to the lease document. Make sure that the team members who represent the design, engineering, and construction aspects of the project have the opportunity to review and approve any lease language or exhibits that relate to design, construction management, schedule, access to the site, landlord allowances, square foot calculations, and so forth, before executing the lease.

Phase IV: Design, Construction, and Occupancy

Complete Design Development.   Although information regarding staffing and space needs were gathered earlier in the process and preliminary layouts were generated during the test-fit stage, it is advisable at this point to have the design team reconfirm the initial requirements with the end users to make sure there have not been any major changes in their business plan that would impact space needs. This is also the point in the process when the specific details of the space plan should be finalized with the user group. It is imperative that the design team understands how each specific unit functions, what its ideal work flow arrangement is, and that it also understands how the various departments function as an organization. Those work-flow needs along with their unique conference, filing, and equipment requirements need to be effectively incorporated within the space plan design. The final design development documents need to effectively represent an office environment that enhances the functional work-flow needs of the user group; otherwise the result will be an ineffective, unproductive workspace.

Finalize the Construction Documents.   Once the design development phase is complete, the expanded core team can be engaged to coordinate a full set of working drawings and specifications that will be used for bidding, permitting, and construction. It is essential that the architectural and engineering disciplines have thoroughly examined the existing conditions of the proposed space and understand how the building systems are intended to operate, so that their modifications and improvements represent logical, constructable, and efficient designs. Any building system modifications should also be reviewed with the landlord's building management and approved by an authorized representative of the owner. For example, if there are postoccupancy mechanical problems in cooling the tenant space, it is very advantageous if the tenant can establish that all HVAC modifications to the space were coordinated and previously approved by the owner, who is ultimately responsible for operating the base building systems in accordance with the heating/cooling/fresh air requirements referenced in the lease document.

Integrate All I/S Telecom Requirements.   Make sure that all IS/telecommunications infrastructure requirements are incorporated into the plans and specifications. Incorporate any special I/S requirements related to building access (T1 line, fiber, dual feeds), conduit for riser cabling between floors, uninterruptible power supply (UPS) needs for the computer/server rooms, and generator backup requirements for critical equipment.

Complete Furniture and Equipment Plans and Specifications.   Coordinate all electrical and/or mechanical requirements for tenant furniture and equipment with the architects and engineers. Make sure that power supply, circuitry, and installation methods for the furniture system are clearly defined in the drawings. Also make sure that all ancillary equipment such as fax machines, printers, and copiers are specifically located on the floor plan and that all special electrical or HVAC requirements are engineered and included in the original bid set. This will help avoid last-minute scrambles during the weekend of the move and will eliminate costly change orders.

Bidding and Award of Construction Contract.   Depending on the location, size, complexity, and schedule for the project, the tenant may elect to contract the actual work in one of several different acceptable methods. Stipulated sum, cost plus fee, guaranteed maximum price, and design build are a few of the more common approaches, and in some cases the tenant will elect to have the landlord manage the entire bidding, award, and construction phase. In any case, it is advisable that both the tenant and the landlord mutually agree on the short list of contractors who are proposed to bid the work.

Regardless of which method is used, the following are a few key factors for success:

1. Make sure that the construction documents are clear, accurate, and as thorough as possible. Weak drawings can result in change orders, delays, and performance issues.
2. Take the time to prequalify the contractors who are proposed to bid the work. Consider each firm's financial resources, annual volume, experience with similar type work, qualifications of the proposed project manager and superintendent, and the company's overall reputation for completing work on time, within budget, and in a professional manner.
3. When evaluating the final bids, carefully review the major cost components among the bidders and note major variances by trade or any exclusions that may indicate a problem area within the drawings. Make sure that the project schedule is clear and demonstrates that the contractor can accomplish the work within your required time frame.

Construction Phase.   Once the bid has been successfully awarded and all permits and contracts have been approved, it is very beneficial to conduct a construction kickoff meeting where the client (whether it be the tenant or the landlord) meets with the primary contacts from the construction firm (GC) to discuss schedule, process, and project administration. Quite often the GC will include representatives from the major trades including mechanical,

HVAC, electrical, and plumbing. If possible, it is also advantageous to include representatives regarding furniture installation, floor covering, security hardware, and voice/data wiring to incorporate their particular coordination issues into the overall master schedule and to agree on sequencing and the way these various trades will integrate on the job site.

At the kickoff meeting, it is also important to communicate to the contractor who is authorized (by the tenant's organization) to direct changes in scope (i.e., the project manager from CRE, not the end user) and to establish a clear process for identifying, pricing, and approving change orders. During the construction phase, it is essential that the tenant has some representation at the weekly project meeting with the contractor, whether it be an internal CRE project manager or a representative from the tenant's architectural (or construction consulting) firm. Close monitoring and documentation of construction progress including schedule, budget, outstanding issues, and request for information is essential. With today's technology, it is also possible to require that the contractor digitally photograph construction progress and/or site issues and forward these images electronically through a project site on the Web. At the conclusion of the construction phase, it is also very important to have the contractor produce a project manual that includes a contact list for all major trades for follow-up work/warranty issues, copies of the certificate of occupancy, any operational and maintenance guides, HVAC balancing reports, and important information that the owner, end-user contact, or CRE may need to refer to in the future.

Complete Furniture and Cabling Installation.   Toward the end of the construction phase, it is important to confirm that the installation of systems furniture and voice/data wiring are being installed according to schedule, leaving sufficient time for punch-listing the furniture installation and testing all voice/data wiring connections prior to move-in activities. Make sure that base plate power is in place for the ancillary printers and fax machines and that all outlets for large copier machines have the appropriate plug configurations installed.

Move-In Activities.   At this point, if all of the subsequent phases of activity have been completed properly, the site should be readied for move-in activities that typically involve relocation of boxes, computers, equipment, filing cabinets, and assorted support furniture. In addition to communicating all moving requirements to the employees involved, it is important to make sure that all move-in activities are coordinated carefully with building management regarding loading dock and elevator access, security, trash removal, and cleaning. It is also advantageous to have an electrician and cabling installer on site or on call in the event there are some critical last-minute electrical or telecommunication wiring needs that must be completed before the start of business the next day.

Phase V: Project Closeout

Complete Punch Lists.   In the event all punch list items have not been completed by the contractor, furniture installer, or cabling installer, these remaining activities should be documented, prioritized, and scheduled for immediate completion. Holding back on the final payment is often the best leverage for getting these completed in a timely manner. Before making final payments, also make sure to obtain all lien releases from the general contractor to ensure that all subcontractors have been paid in full and do not have any outstanding claims that could cause them to place a lien against the building, which in turn could place your lease in default.

Reconcile all Project Accounting.   As all of the final payments to contractors, suppliers, and consultants are being finalized, be sure to reconcile final costs against your contracts with each party, challenge and reconcile any outstanding change orders, and collect any landlord allowances due. Compare your final project costs with the budgeted amounts, identify any variances, and determine why each variance occurred.

Issue Project Closeout Report. You should generate a final project close-out document that summarizes for CRE management, as well as for the end user/business unit management, all final costs against the budget numbers presented when the project was approved. Summarize the activity completed, the basic terms of the lease (abstract), critical notification dates (notice to terminate, expansion options, etc.), final project costs, and explanations of variances.

Complete a Project Evaluation. It can be very helpful if the project team conducts a brief conference call with the entire core team when the project is completed, to discuss the degree of success of the project and to identify what worked well and what did not. This information should be shared with other members of the CRE department to help share "lessons learned" with the goal of continuous improvement. It is also recommended that someone from CRE obtain input from the end users regarding how they felt the project management met their needs and expected levels of service.

Finalize the Project File. It is essential that all critical documentation relating to the project, including contracts, financial data, specifications, drawing file references, customer approvals regarding changes in scope, and copies of payment applications are organized in a project file that will be maintained in a central filing area within Corporate Real Estate.

Enter Critical Data into CRE Facilities Information Management System. Once the project closeout is finalized, all critical legal or financial data regarding this project should be entered into a central (CIFM) database where this information can be accessed and reported on to assist with internal portfolio/lease management responsibilities. This information will be essential for CRE in tracking and strategically managing critical lease dates, rental rates, escalation schedules, square footage under lease, excess space, opportunities for consolidation, and so forth.

Postoccupancy Evaluation. Approximately six months after the users have occupied the new space, it can be very beneficial to follow up with the end-user group to seek input as to how well the overall space is meeting their needs. A standard survey can be utilized that seeks feedback regarding satisfaction with individual workstations, ergonomics, the adequacy of common support areas, any positive or negative impacts on work productivity, and general questions regarding how well the building is being maintained and the responsiveness of the property manager.

ARTICLE 3.1.6
PROPERTY LOSS PREVENTION

John Bender
FM Global, Johnston, Rhode Island

Though designated corporate individuals deal with providing a business with property and liability insurance, no one is better positioned to maintain business continuity than the facility manager. And these days, if a business of any kind suffers an unanticipated interruption, the likelihood of regaining its former status is in jeopardy because of competition being what it is. It is, therefore, incumbent on the facility manager to understand the threats against the business and to take appropriate action to prevent, or at least minimize, the possibility of property loss.

VULNERABILITIES TO PROPERTY DAMAGE

Fire and Other Perils

When facility managers stand on the hill overlooking their property and ponder all the threats to their business, one that may come to mind is fire. This is because fire happens so often and gets so much press. And, despite the relatively recent rash of natural hazards that have doomed so many businesses, property insurers still find that fire causes the most losses. Of course, to have a fire, there have to be the elements of the fire triangle: ever-present oxygen, fuel, and the ignition source. So, as easy as it sounds, to avoid a business-threatening fire loss, one has to manage combustibles and control their ignition sources. The property insurer can assist you with both measures.

To manage combustibles, one has to know what and where they are, plan carefully for their safest storage configuration, and protect them adequately. Control of ignition sources is much harder. Table 3.1.6-1 shows ignition sources in order of severity. Note that *electrical* has been the leading ignition source for years. How do you control that? It boils down to installing adequate overcurrent protection, testing the protection, and maintaining the electrical system. Hot work, formerly known by the more restrictive term *cutting and welding,* is number two on the list. Did you know that over one-half of the fires that were started as a result of hot work were the responsibility of contractors? Therefore, to control this, you have to manage not only the in-house operations, but watch over the contracted operations as well.

Another in the top 10 of ignition sources is arson. Arson includes not only the usually thought-of "burning for profit," but also intentionally set fires for other reasons like the disgruntled employee who just got laid off or the pyromaniac who loves to watch things burn. How can that ignition source be managed? Well, it comes down to security. Did you ever ask yourself how easy would it be to get into your facility unnoticed? Would your employees recognize and challenge nonemployees? Do you leave piles of combustibles like pallets next to the building for an arsonist to light? This ignition source *is* manageable.

Now, how about the other perils that were alluded to. Over the last decade, there have been tremendous property losses due to such things as winds, earthquakes, freeze-ups, floods, and building collapses due to snow accumulation or rain ponding due to poor drainage. Why are so many of these natural hazards happening lately? Well, weather patterns have changed. And businesses have changed too, choosing to locate in storm-prone

TABLE 3.1.6-1   1989–1998 Fires and Explosions by Probable Cause*

Probable cause	No. of losses	Gross loss ($U.S.)
Electrical	1,497	$787,354,883
Hot work	565	696,302,342
Hot surface/radiant heat	597	482,883,379
Arson/incendiary	759	362,930,132
Spontaneous ignition	313	285,533,985
Exposure	232	225,647,028
Friction	289	223,008,409
Burner flame	238	176,255,610
Overheating	487	140,149,881
Smoking	280	122,107,182
Overpressure	64	116,851,760
Combustion sparks	136	70,116,310
Static electricity	77	46,141,023
Lightning	69	23,080,895

* Losses reported to Factory Mutual Insurance Company.

areas. Construction has changed with the advent of synthetics and a preference to install pre-engineered metal buildings.

Hurricane Andrew taught about weak roof construction and failure to shutter or otherwise protect glazing. The early 1990s floods in the Midwest found unnoticed vulnerabilities as many businesses were quite unprepared for the magnitude of water damage that resulted. Unpredictable earthquakes took their toll in this decade, and we learned about the devastating aftereffects when unleashed gas caught fire while protection systems lay hopelessly damaged and unable to protect. Unusual weather patterns brought freezing temperatures to areas not built to withstand them and the result was devastating. Industries quickly learned to avoid winter shutdowns and have a protection plan ready at all times. The heavy snows of the early 1990s collapsed many weak buildings, causing reevaluation of the capability of roofs to hold the weight. And plans were developed to keep the excess cleared off.

There is another loss potential to think about. Is there any possible exposure from the property adjacent to yours? How vulnerable are neighboring facilities to property loss, and if there is a problem, will it affect you? If there is a threat, you need to plan for it. Suppose, for instance, that the company has a building very close to yours. You know there are hazardous processes within and there isn't any automatic sprinkler protection either. The side of your building that faces theirs has lots of windows. You might consider eliminating the windows on that side of your building, knowing there is a high probability of uncontrolled fire from next door. That would be good planning.

Equipment

Once you've inventoried the threats to your property from fire and natural hazard perils, consider what can happen to your business if key equipment ceases to function: Loss of steam due to boiler malfunction; loss of cooling due to a compressor, chiller, or cooling tower failure; loss of electrical power because of a weak cable or failed circuit breaker; a production line down because a gear set failed and there is no spare; a press part cracked that should have had nondestructive examination but didn't, or. . . . How vulnerable is your equipment? Is the protection the best it can be, and is it regularly tested and maintained? Are operators formally trained and empowered to take quick action to minimize an impending failure?

A facility manager has to evaluate all of these business threats, and others. Then, perhaps with the help of the property insurer, the threats must be prioritized by likelihood and potential impact. Each needs to be evaluated further to determine if protective measures are in place to prevent and/or control losses. Finally, employees need to be made aware and trained to do their part—all with the backing of the highest levels of company management.

Management sets the tone and imbeds the loss prevention culture through its policies and procedures, and through its personal follow-up by everyone on the management team. Employees, once educated, play key roles in emergency response, protection system maintenance, and generally watching over the facility for threats of any kind.

PROTECTION PROVIDED

Prevention and Control

At this point, you should be either elated or concerned, depending on how many of the vulnerabilities you've considered before. Providing a suitable degree of protection against the threats we've just discussed will require some professional help. Perhaps the best organization to turn to is your property insurer whose trained loss prevention engineers are in a good position to help you evaluate your needs.

There are two elements of protection that you must consider—prevention and control. Prevention measures should stop a property loss from happening in the first place, but there also has to be a control function to take over quickly if a loss does happen and minimize the effects.

Effective protection depends on both the peril (fire, flood, etc.) and the contributing factors that are unique to the business being protected. For instance, let's say you are in the paper manufacturing business and looking at fire prevention and control. Fire in the wood-yard is prevented by the arrangement of the raw stock and taking measures to prevent smoking and other ignition sources from starting a fire. The control aspect is perhaps some monitoring towers ready to deluge a fire if it starts. However, the monitoring towers are useless without the proper water supply. The dry ends of the paper machines require some sort of automatic sprinkler protection that is adaptable to the higher temperatures, but the prevention aspect is how clean the area is kept. The warehouse has its own automatic sprinkler needs based on the type of paper, how it is stored, and how much there is. The prevention in this case is the proper storage, perhaps using electric instead of internal combustion fork trucks, and a robust antismoking policy.

On the other hand, if you are managing a public building, chances are that the sprinkler requirements will be less robust, but there are pressure needs (adequate water on the 60th floor) and standpipe locations to consider. If you have to retrofit an existing building, there are aesthetics to deal with. For prevention, there are such things as need of a smoking policy, daily refuse removal, strict policy on space heater use (and other appliances), infrared scans of the electrical system, and a good electrical testing and maintenance program.

Now you have a prioritized list of perils that are likely to affect your facility at one time or another. You should take each of the perils, make two columns labeled "Prevention" and "Control," and begin to fill them in with necessary measures. Keep in mind that both prevention and control contain human factors, too.

During the Planning Stage

The best time to consider loss prevention and control is when a new facility is in the early planning stage. Considerations include where best to site the structure, what construction methods and materials are best given the location and function, what installed fire protection will best protect the facility based on the processes involved, and many other related concerns (e.g., insurance company plan review, protection during construction, building code compliance, using materials and methods approved by standard setting institutions, etc.). Protection can be very affordable as a percent of the building cost when considered during the planning stage. This is the time when your property insurer should be working with you on the project.

Change as Circumstances Change

If the facility and/or equipment is not new, you are in a maintenance mode. However, you should view each change that takes place in your building use or the processes contained within as a threat to property loss prevention or control. For example, to enhance productivity, you took a two-week shutdown to exchange a production and a storage area. However, the automatic sprinkler protection in each area was never considered in the changeout. You may have just compromised protection for your finished goods because the automatic sprinkler system, designed for its former use, may not have the proper strength for the type of material now being stored in the new location.

Another example—after a recent merger, the entire workforce underwent a tremendous change and ultimate downsizing. However, you never checked the adequacy of your emergency response organization after the changes. Only after a devastating fire might you discover that only one-half of the positions were covered, and by ill-trained people at that. Property loss prevention and control must be a consideration after every change, no matter how small. Failure to do so will result in a false sense of security and a more severe loss if and when a loss occurs.

THE HUMAN FACTOR

The term *human factor* encompasses a vast area, starting at the very top of the organization and involving every person. Policies and procedures related to property loss prevention and control are a human factor. Communication up, down, and sideways is a human factor. Learning from incidents so they never happen again is a human factor. Other human factors include educating all levels as to perils and protection, empowering employees to take action to protect assets, designing and constructing proper buildings and equipment, and managing contractors. Human factors need to be periodically assessed and corrective action planned when deficiencies arise. On average, 70 percent of property losses have a contributing factor that is related to human factors. This is simply stated as *actions or in-actions of people on all levels that either cause a property loss or contribute to its severity.*

Emergency Response

One of the most important parts of human factor planning is emergency response, or how employees are organized and trained to combat an emergency situation. No matter how hard you try to manage prevention, there will be inevitable incidents. How quickly and properly these are dealt with will determine the severity of the loss. To get there, one has to revisit the original prioritized peril list and think about how to respond to the most important or most likely ones. For instance, a planned response to a fire will surely involve someone designated to call the fire department. It should also include someone to make sure the automatic sprinkler control valves are open, that the fire pump (if installed) is running, and maybe even tackling the fire (but read your OSHA regulations first to make sure you are aware of the requirements).

If your facility is vulnerable to flood or surface water, do you have people trained to respond? Do they have the proper supplies they need? Do they know how to use them? How about an approaching hurricane if you are on the East Coast or an earthquake if you are on the West Coast? It is up to you to determine to what degree you plan to, but there is no substitute for a trained and organized response. And it needs to be kept up to date. Data on previous losses clearly shows that an inadequate or ineffective response can make a loss greater than 10 times more severe than the situation where the response is both adequate and effective.

Maintenance of Protection

All of this may be for naught, however, if there is a failure to maintain what was put into place. Why do things deteriorate? Protection systems are, for the most part, passive. And they do not directly contribute to the end product or bottom line. What happens, therefore, is that dwindling resources get devoted to producing the end product, often at the expense of the needs of the protection system maintenance. It takes belief and determination to keep a reasonable level of maintenance on both the hard systems, like automatic sprinkler protection system components, and people programs, like emergency response, that keep the facilities loss-free. It usually takes higher management backing to prevent protective system maintenance from being the backlog. An ill-maintained system or human element program cannot be counted on to work when needed.

Contingency Planning

Contingency planning is evaluating possible scenarios and planning for those that are most likely to occur. If you are sited next to a river that has overflowed its banks twice in the last 50 years, a contingency plan for flood is very much in order. The plan would allow for emergency response but would go beyond that. Basically, a contingency plan is designed to maintain business continuity in the face of a probable outage. Your contingency plan may include moving equipment and supplies or using a competitor's facility or obtaining raw product from alternative suppliers. Contingency planning can, and should, be used on a smaller scale, too. For instance, if a spare part for a key machine is not kept in stock, a contingency plan would list where to get the part, how to transport the part, and how long the suspected outage would be.

PROPERTY INSURANCE

Now that the vulnerabilities to your business have been discussed in terms of property loss causes and their prevention and control, you should be aware that more than likely, your company has a degree of property insurance. Having property insurance does not relieve the facility engineer from pursuing a realistic level of loss prevention and control. However, insurance is limited and can never bring back customers lost because of a business outage caused by a property loss. And there may be a sizable deductible that will surely impact the bottom line. Here is a brief primer on property insurance.

An insurance policy is a contract in which one party (the *insurer*) agrees for a price (the *premium*) to reimburse another party (the *insured,* or *policyholder*) for loss that is caused by insured events. Insurance policies usually specify items such as, but not limited to, the following:

- The location (where the insured property is located)
- The item(s) to be insured and for what occurrence
- The amount of insurance purchased
- The premium
- The *deductible,* if applied (the specified amount of an otherwise insured loss that must be sustained by the insured)
- The length of time, or *term,* for which the policy is issued
- The exclusions and restrictions (no policy covers all property for all possible occurrences)

Insurance coverage may be described in terms of the following:

- *Perils.* Specific events that may cause loss (e.g., fire, explosion, wind, hail, lightning, or equipment breakdown).
- *Exposure.* The vulnerability to damage or destruction by some peril.
- *Hazard.* A condition or activity that increases the probable frequency or severity of loss.
- *Endorsement.* A document attached to a policy that modifies its terms.
- *Exclusions.* Policies or endorsements may contain these, which limit the scope of coverage.

The facility manager should be aware of the property insurance coverage for their facility(ies) so they can better plan their loss prevention and control strategy. What specifically is covered? More important, what is not covered? What is the deductible, or, in other words, how much of the risk has the business chosen to assume itself? This could be significant. Often, properties are underwritten as *highly protected risk* (HPR). This means that the business being insured has exhibited genuine interest in loss prevention by management and employees, is appropriately protected, manages its human element programs, and maintains a high level of interest in property protection in general. All businesses should strive to be classified as an HPR.

You should work as closely as possible with your property insurer. Pay attention to their findings made during evaluation visits. Seek their advice any time a process, machine, or building undergoes change, no matter how insignificant it may seem. When new construction is planned, the insurer should be notified at the earliest opportunity so that they can advise on appropriate siting, construction, and protection. The facility manager is in a key position to ensure that the business is not interrupted by an unanticipated property loss.

ARTICLE 3.1.7
VALUE IN RISK MANAGEMENT

Robert W. Fitzgerald and Leonard D. Albano, Ph.D., P.E.
Worcester Polytechnic Institute, Worcester, Massachusetts

As values in buildings increase, risk management is becoming an increasingly important element in a comprehensive facilities management program. Value includes the quality and performance of a building in terms of user satisfaction and changes in user needs, life-cycle costs, and delivery time. Value is in the systems that comprise physical facilities, in the information base housed within the facility, and in the impact of facility downtime on the well-being of a company and its employees. Risk may arise from natural forces such as earthquakes, tornadoes, hurricanes, and floods. Risk may also be caused by man-made influences such as fire, explosion, sabotage, and terrorism. This article describes a process for developing a risk management program. Although the general process may be applied to all natural and man-made hazards, fire hazard is used here as the model for study.

ELEMENTS OF THE PROCESS

A program to improve fire prevention or fire protection in a building commonly originates from a fire department inspection that identifies conditions that do not meet regulations. Alternatively, the program may be a response to an insurance carrier's evaluation of the status of the building's operations and the existing loss exposure. Often, the corrective action recommended by an insurance carrier is related to changes in premiums, so a cost analysis may be incorporated relatively easily. This type of analysis merely relates the cost of corrective action to the change in the insurance premium. Rarely does the analysis incorporate the change in risk as a result of the physical changes in the building. The fire risk process outlined in this article relates an engineering analysis of risk to decision making and value. The process enables management to evaluate the mix of alternative options in accepting the risk, transferring the risk through insurance coverage, and reducing the risk through changes to the building or the facility operations.

The complete process is described by the following seven elements:

1. Understand the problem.
2. Define the building.
3. Analyze incident scenarios to understand the performance of the building.
4. Characterize the risk to people, property, and operational continuity. The risk to the environment and to neighbors can also be incorporated into this evaluation.

5. Evaluate prevention effectiveness.

6. Structure a decision analysis.

7. Evaluate alternatives, and decide the best course of action.

Understand the Problem

The most important part of a risk management program is to understand what is at risk from the consequences of the hazard. For fire risk, one first identifies the value contained within the building. A convenient technique is to use a space utilization plan and to identify the cost of a complete loss of each room or designated space with regard to people, property, and mission or operational continuity. For example, one would value the replacement cost of the contents of a room or space. The replacement costs include both the contents and the room (building) structure itself. For operational continuity, the value of the loss of the facility in terms of production downtime, loss in market share, business interruption, or other factors that are sensitive to the operational continuity of the facility would be shown on the space plan. The risk to humans is more difficult to quantify in this manner. Therefore, this identification typically describes the number of individuals and their physical characteristics when in the space.

Second, when the value of the contents and operations of the facilities have been identified, their sensitivities to the products of the disaster are described. For fire, one would describe the sensitivity of the contents and operations to flame–heat, smoke and toxic gases, and water. For example, some materials may be very sensitive to only a small rise in temperature or to the corrosive effects of the gases that are released by a fire. Computer software and data storage may be particularly sensitive to the products of combustion. In any event, the sensitivity of the contents and operations to the natural products of the fire must be identified.

A third important part of identifying and understanding the problem is a description of the type of loss that management is willing to accept. For example, is the loss of the complete contents of Rooms A and B acceptable but the loss of only accounts receivable or data storage in Room C unacceptable? Is a two-month downtime in the operations of Space D acceptable but a one-day downtime of Space E unacceptable? Again, an analysis of the space utilization plans will provide a cost-effective and efficient way to understand the potential value of losses and the special sensitivity that some spaces will require. Because it is possible to *harden* selected spaces against hazards without necessarily providing the same protection to the remainder of the building, the building can be adapted to accommodate selected protection, if one understands the nature of the problem.

Define the Building

Defining the building means identifying the physical structure and all of the components that influence the hazard. For example, for a fire risk, building definition includes the type and location of detection and notification instruments, the coverage provided by automatic suppression systems, the ease or difficulty of fire department access, and the location of barriers to the passage of flame–heat or smoke and toxic gases. An existing building is usually easier to describe in terms of the existing active and passive fire defenses. The responsible designers are expected to provide this information for a new building being designed. Although this description is not difficult, it is essential for the subsequent evaluation of building performance.

Evaluate Performance

Predicting the performance of a building for specific fire scenarios is a necessary part of the risk assessment and management program. A series of fire scenarios is identified to understand the building and its performance. An important aspect of this analysis is to recognize that prevention measures may fail. Do not incorporate prevention effectiveness into this analysis. The goal is to identify the likely outcome from an ignition and established burning. Corporate management must understand what can happen when prevention fails. This puts prevention and what happens after prevention fails into perspective.

Fire performance may be organized into three major components: (1) flame–heat movement, (2) smoke–gas movement, and (3) structural frame performance. Each of these components may be evaluated independently. The results of each analysis are then incorporated into the risk characterization described next. The evaluation of each component is discussed.

Flame-Heat Movement. The evaluation of flame-heat movement involves the following factors: fire growth potential, automatic sprinkler system, fire department extinguishment, and barriers.

Fire Growth Potential. Evaluation of flame-heat movement starts with an estimate of the relative potential for fire growth in the space. Some spaces make it very difficult for a fire to grow, whereas other spaces have conditions in which the fire growth potential is very great. A general classification that gives an indication of the fire growth potential posed by the interior design of the space is a first step in describing the relative hazard posed by the combustible contents and their arrangement. The fire growth potential also gives an indication of the type of fire that the active and passive fire defenses are expected to resist.

Automatic Sprinkler System. The most common and effective fire defense for many types of building operations is the automatic sprinkler system. Refer to Art. 5.4.7, "Sprinklers and Fire Protection," for a more complete treatment of this important subject. No other fire defense can demonstrate the success of automatic sprinkler systems in controlling unwanted hostile fires, but these systems require care in installation and maintenance. The quality of an automatic sprinkler system can be evaluated to a certain extent by assessing its reliability and design effectiveness. The reliability of a sprinkler system depends on ensuring that the water control valves are open when a fire occurs and that water will reach the fused sprinklers in a timely manner. Valve supervision and company attitude are the major factors that influence the control valve condition. The reliability of the exhausters or accelerators of dry pipe systems, as well as pump design and continuity of power supply, are the main influences on water movement to the fused sprinklers. Design effectiveness involves several considerations. One is the temperature sensitivity of the sprinkler to the fire growth. In addition, the water supply must be sufficient to maintain control and limit fire spread until manual suppression can complete extinguishment. The influence of building elements (e.g., partitions, furniture, air conditioning ducts, or beams and joists of floor systems) on water discharge is another major consideration. A well-designed sprinkler system can be rendered ineffective by occupants or contractors who are not sensitive to the importance of allowing the sprinkler spray pattern to control the fire.

Fire Department Extinguishment. Manual suppression by a community fire department is another way by which a fire can be extinguished. The analysis of fire department extinguishment involves comparing two time lines. One is the expected time of fire growth, that is, what is the expected time for fire propagation? The second is the accumulated duration for the following sequence of events: fire detection, notification of the fire department, fire department arrival at the site, and application of first water to the fire.

The time for detection depends on the size of the fire, the flow path from the fire to the detection location, and the sensitivity of the detector. Detection may involve human sensitivity, instrument detection, or both. The selection of the detection system should be consistent with the needs of the building, and is discussed more completely in Art. 5.5.4, "Fire Alarm Systems." For example, smoke detectors are sensitive to the airborne particles released during combustion. Heat detectors may either respond to a fixed temperature felt by the detector or to a rate of temperature rise over a relatively short period of time. A variety of specialized types of detectors are available to sense fires in industrial occupancies.

The delay between detecting and notifying the fire department often is a major determinant of the success or failure of manual suppression to limit the fire to small areas of a building. Delayed notification will enable the fire to grow large, requiring the local fire department to change from an aggressive attack to a defensive mode of fire fighting. Thus, the time for suppression and the amount of damage are directly related to the time between detection and notification. Detection and notification may be greatly influenced by the availability of humans in the process. Consequently, two different fire department suppression analyses are normally done. One is for a building that is occupied and the other is for the same building unoccupied. The time for detection and notification and the amount of damage that can be expected can differ significantly, depending upon the occupancy condition.

After the fire department arrives at the site, the time to first water application depends on how much time the firefighting forces need to complete the following sequence of activities: size up the situation, determine the best strategy for allocating resources, develop the water supply, lay the attack lines through building obstacles, and apply water to the fire. In addition to the challenges imposed by reduced visibility and smoke, the building and site design and the community's firefighting resources affect the time to first water application.

The time to first water application will give a good indication of the amount of damage that can be expected, and this damage estimate is the final part of a fire department analysis. Damage is influenced predominantly by the fire size at first water application. Larger fire sizes require more hose lines and a reliable water supply. The water flow provided by the local community will depend upon the staffing and equipment that is available at the fire scene, as well as the source of supply water. Supply water may be provided by underground piping and hydrants or may be drafted from unfrozen ponds or streams. If neither of these sources is available, water must be transported to the fire by tankers.

Barriers.   The remaining part of a flame-heat analysis involves the effectiveness of barriers to delay or prevent fire propagation. Note that the term used is *barrier effectiveness,* not fire resistance. During the early stages of a fire, the fire resistance of barriers is not as significant as their status in protecting openings. Open doors allow fire to propagate into the next space regardless of the fire resistance of the barrier. Consequently, openings for doors, windows, and ducts are important factors in evaluating the effectiveness of a barrier to prevent or delay fire propagation into an adjoining space. Barriers are also important in assisting sprinkler and fire department suppression. Containing the hot gases of a fire will enable the sprinkler system to control the fire more effectively. Barriers are also an important ancillary feature in manual fire suppression. They prevent rapid fire propagation from space to space, and they enable the conversion of water to steam to become an effective heat absorption factor in extinguishing flames more quickly.

Smoke-Gas Movement.   The airborne products of combustion require an analysis separate from the flame-heat movement because the speed and ease of spread are so different. Most individuals who have not been involved with a large fire do not realize the speed with which a large volume of smoke is produced. Visibility is reduced or completely obscured very quickly. It is not uncommon for individuals to comment about how dark the space becomes and how disoriented a person can become in these situations.

It is possible to estimate the volume of smoke that is produced. The architecture of the building, obstructions to airflow, and air movement from air-handling systems influence the movement of smoke in a fire. Beyond the room of origin, smoke flows to other spaces along the paths of least resistance. The objective of a smoke movement analysis is to estimate the time in which target spaces become untenable. Therefore, one must identify the target space and the concept of an untenable condition.

A target space may be an exit access or an exit if one were evaluating the time for egress. The target space may also be a room with equipment that is sensitive to the corrosive gases of a fire. In either case, the target space beyond the space of origin should be identified to enable an estimate of the time before the space becomes untenable.

The definition of an untenable condition can be more difficult. The most convenient definition is expressed in terms of visibility, that is, what is the maximum distance that one should be able to see in a smoky condition? Ten feet? Twenty feet? Across a room to the next door? The definition of an untenable condition should be specific and based on the activities in the building. An untenable condition for sensitive equipment may be defined in terms of the corrosive products that are produced in a fire. This type of definition is more difficult to apply. However, recognizing that some equipment is sensitive to the corrosive products of combustion is often sufficient to make the building operators aware that a fire in remote spaces may cause unanticipated consequences to operational continuity and property protection. The engineering and design aspects of systems intended to mitigate the effects of smoke are presented in Art. 5.5.5, "Smoke Control Systems."

Structural Frame Performance. The third major part of a building performance analysis involves excessive deflection or collapse of the structural frame. A fire that continues to burn will cause deterioration of the structural frame. A structural analysis can provide an estimate of the likelihood that the structure will remain in place for the duration of the fire.

Risk Characterization

After an estimate of the building performance is made, the risk to people, property, and operational continuity may be characterized. The risk to the environment and to neighbors can also be incorporated into this evaluation if desired. A risk characterization estimates the likelihood that people, contents, and operations will be threatened by a fire. The process must recognize when people and too many products of combustion (i.e., an untenable condition) are present in the same space at the same time. Similar evaluations of combustible products are used to characterize the risk to equipment, computer software, data, and other storage.

Evaluate Prevention Effectiveness

Fire prevention was not incorporated into the process at the intuitively logical position before the building performance evaluation. This omission is intentional because management must recognize the outcomes to the building, its occupants, and its operations when prevention fails. Often, when the consequences of an unwanted ignition are recognized, the importance of fire prevention takes on a completely different value.

Fire prevention consists of two major parts: preventing ignition and preventing established burning. Ignition prevention is the traditional way of viewing fire prevention. However, from an engineering viewpoint, the initial part of a fire from ignition to a flame about 10 inches high (which is called *established burning*) is very fragile, and the behavior of the fire during this stage is difficult to predict. In addition, most small fires are extinguished by occupants if they happen to be in the vicinity of the fire. Room fire analysis starts at established burning and estimates future outcomes because these are much easier to predict after the flame reaches about 10 inches in height.

Fire prevention also incorporates the occupant in an evaluation. This has an advantage because building performance can be separated from occupant characteristics. This separation simplifies the analytical process and provides a better understanding of building performance. For example, any initial suppressive actions by an occupant can be clearly established and evaluated. A value analysis of prevention and initial occupant extinguishment activities can be assessed more easily.

Structure a Decision Analysis

A risk management program includes a decision analysis to organize the study so that it provides value. A decision analysis includes three parts: (1) identifying the alternatives, (2) selecting and structuring the type of analysis that is most appropriate for the needs of management, and (3) evaluating the alternatives. Risk management generally considers three alternatives for addressing risk. The first is to accept any losses that may occur. This is normally the approach for the military and other governmental properties. Business decisions also incorporate this alternative in the form of insurance deductibles. A business may be able to write an insurance policy with a deductible of $5,000, $50,000, $500,000, $5,000,000, or more. Each of these approaches has implications as both a business loss and an expense (in the form of insurance premiums paid to an insurance carrier).

A second alternative involves transferring risk by purchasing insurance. As noted in the previous paragraph, premium expense is related to the deductible of the policy. In addition, the influence of the loss of contents and the cost of business interruption can be more clearly defined because of the information that was obtained during the identification of the problem (i.e., the first part of the process).

The third alternative involves changing the risk by changing the building through the incorporation of different active or passive fire defenses. The cost of each of these alternatives can be identified clearly. The effectiveness would be understood by the engineering evaluation, which is the third part of the process.

A decision analysis would identify the mix of these three alternatives, along with the cost, expense of losses, and the probability of performance that relates to any particular alternative under consideration. Business management practices of decision analysis offer techniques for structuring and identifying values for the possible outcomes. These can be expressed in terms of expected monetary value or utility value.

Decide the Best Action

The final step in the process is the shortest because it is based on understanding all that unfolded during the previous parts of the process. Management can decide to accept the building as it is or to change the building by incorporating different fire defenses. When management opts for building changes, an additional analysis may be useful to identify the changes and their cost-effectiveness. Chapter 2 presents useful guidance for conducting financial life cycle cost analyses. Alternatively, management may decide to accept losses or to buy different insurance coverage. Regardless of the decision, it will have been made with a better understanding of the risk and more appropriate management of that risk.

Closing

The design of any facility involves trade-offs among a variety of factors, including life-cycle cost, delivery time, operations, architectural image, and risk due to natural and man-made forces. In many instances, levels of risk can be traded against capital investments, insurance premiums, and the expenses that result from loss of functions. To make informed decisions, facility owners and managers must understand each source of risk, such as fire, earthquakes, hurricanes, explosions, and terrorism; the consequential risks to human safety and the value of the facility; and the economic impact of alternative solutions for managing risk. In addition, the design team must communicate the results of their engineering analyses and use a systems-based approach so that interfaces and interactions between building subsystems are included in the decision process. For example, significant drift of the structural frame during an earthquake may be tolerable from the standpoint of life safety, but the resulting impact on the building's nonstructural systems may produce life-threatening fires and unanticipated losses of function. The process described in this article enables an enterprise to understand the risks inherent in the design, construction, and operation of facilities and to make more informed decisions for managing that risk. The understanding that unfolds during the study enables the consideration of a wide variety of alternatives for changing the risk and the inclusion of the facility's value in the decision process.

ARTICLE 3.1.8
THE NEW FACE OF THE PROJECT
TEAM MEMBER

Donna K. Burnette and David Hutchens
Paradigm Learning, Tampa, Florida

Read the following project management case study carefully. A quiz will follow! Ready? Here's the scenario: In the Chevron Corporation, a project team was commissioned to manage the design, funding, and building of a large plant to manufacture a product that was in great demand and offered significant margins. After many months of work, the project team finally completed the project—ahead of schedule and under budget. The plant was opened to great fanfare, and the project team celebrated its achievement.

Got it? Here's your one-question quiz: Was the Chevron project a success? (Take your time answering. No one's looking over your shoulder.)

By now, you may have suspected that the question contains some hidden trick. You're right. The fact is that you have not been given enough information to evaluate the success or failure of the Chevron project. A correct answer in this case would have been to ask for more information about the larger organizational context or to inquire about what happened next. Okay, so the above exercise was a little manipulative. But it illustrates a common trend in project management: the tendency to evaluate project success too quickly, too narrowly, or to use short-sighted criteria. The traditional measurement criteria of schedule, budget, and quality, on their own, may no longer be sufficient.

Case in point: Let us return to Chevron for the rest of the story. Very soon after its glorious opening, the Chevron manufacturing plant failed and was sold. As it turned out, the market for the high-demand, high-market product was volatile and prone to change quickly. And change it did. Had the project team considered the possibility of this market change, they surely would have explored the option of shelving the project altogether.

Now, which criteria would you use to determine success? If you were to consider only the traditional criteria of budget and schedule, you would have concluded that the project was a smashing success. But if the criterion value is added to the business, the project failed abysmally. And here we can observe the need for project team members to begin thinking about their work in different ways.

But isn't such market forecasting the responsibility of leadership? Isn't it unrealistic to expect that kind of strategic thinking from our already-stressed, neophyte, project team members? Perhaps that was once true. But the world has changed.

A NEW WAY TO THINK ABOUT PROJECTS

In the new world of work, the familiar project has become a very different entity. Carl Pritchard, PMP and Principal of Pritchard Management Associates, has lectured and written extensively on the subject of project management. He sees many new trends in project management, not the least of which is the changing profile of the people who are being called upon to lead projects. "What I see in the classroom is interesting," he says. "It's an amalgam of old-timer project managers who have been doing it for years without any formal processes. They're in there with a lot of new folks who are just entering the practice. In many instances, these new people are joining project management on the heels of success they had in some other part of the organization. They were pulled off some organizational effort, and management is now looking for a new home for them. Poof! They're expected to be project managers."

What's behind this new trend of putting more and more people on a rapidly growing number of projects? Pritchard suggests that one major cause is the nearly obsessive concentration on the customer that has become the focal point of organizing for so many businesses. This intense customer focus will require a special touch in fostering new processes and relationships. And that touch is going to come from individuals who can marshal and control resources and who can recognize when a project is faring well—and can also tell when it needs more intense time and attention.

One would be hard-pressed to complain about a customer service orientation. But the new sense of urgency is thrusting a lot of people onto project teams who are not fully equipped. Many still are learning "the hard way"—let 'em learn it on the fly and hope they don't foul it up. At the very least, many are being trained in the basic principles and tool of PMI's Project Management Body of Knowledge (PMBOK).

But how many are being trained to ask the kinds of questions that could have prevented the painful Chevron fiasco?

> "We should be training our project teams to be more. They need to be entrepreneurs. Consultants. Even agents of change."
>
> *—Carl Pritchard, PMP*

If you're thinking PMBOK is enough grounding, think again. Those are implementation techniques—a great place to start, but hardly a full toolbox. More is needed. Pritchard agrees. "We should be training our project teams to be more," he says. "They need to be entrepreneurs. Consultants. Even agents of change. We need to build the confidence of these individuals to serve the organization with a willingness to affect change without apology and to implement change with an understanding of all those it will affect."

The profile of successful project teams is changing. Now, success will require the people on those teams to take on new skills and responsibilities that include the following:

- Strategic, big-picture thinking
- Continual reassessment of the risk and opportunities throughout the project
- Sensitivity to all critical stakeholders and sponsors
- An environment of openness and trust

Strategic, Big-Picture Thinking

As the Chevron project team members learned painfully, projects do not exist in a vacuum. In the end, the project teams that deliver value to the organization will be those who are keenly attuned to the organization's strategies and objectives. Lynne Hambleton, Manager of Learning Systems at Xerox, concurs. "I'm aware that many organizations measure project success using the traditional criteria—like time, cost, scope," she says. "But that just doesn't fly in our culture, where quality is important and our Malcolm Baldrige award continues to set a legacy for excellence. Those traditional criteria just don't get us there. That's why at Xerox, our projects must always speak to value provided whether that's value to Xerox or to the client. We make it clear that project teams must offer solutions that support strategy. That's a lot more complex than the old way of thinking."

Continual Reassessment of Risk and Opportunities

The project team may be doing things the right way—but are they still doing the right thing? That is the kind of question that characterizes today's new breed of project managers.

At Xerox, project team members are trained to assess both risks and opportunities continually, from the inception to the completion of the project. "Just about everyone involved in the project is asking questions like what's going on in the larger context? What could derail our efforts?" says Lynne. "Project team members certainly ask those questions. And the operations side of the team is definitely asking those questions a lot because they'll be the ones running the effort after the project is done. They are constantly held accountable because they're going to be living with it."

Sensitivity to Critical Stakeholders and Sponsors

Nothing scintillating here; it is in all the basic texts. Yet the critical process of managing stakeholders continues to be ignored. It is the oldest story in the books: The project was delivered skillfully, but a key stakeholder was left out of the process . . . and the project choked. It is sobering to realize that these stakeholder interests are almost never hidden. In fact, they are often quite overt. But if the project team members do not ask, their interests will not be spelled out in the statement of work or the contract. The make-it-or-break-it criteria never even show up on the project team's radar screen.

Pritchard offers this example:

> I met with my contractor just this morning to talk about the addition we're adding to our home. He hammered me with all the right questions. "Who will I talk to?" "Who might I talk to?" "Who will sign the changes?" "Who will be around the day they place the Porta-John?" They were great questions! The barrage was almost too much for me. I started thinking about the many project managers who would rather put off those difficult discussions until later, after the relationship develops. But my contractor is building our relationship now. He's not wasting time. As a result, he's going to have a very clear understanding of my expectations—just as I will have a clear understanding of his.

Trust and Openness

Perhaps it is the many technical tools of project management—the risk assessment grids, the Gantt charts, the budget sheets—that lull well-intentioned project teams into the mistaken belief that their role is itself technical. But success is not built exclusively on deadlines, budget, and quality. Remember, there are people involved here. And the wheels of relationships are greased by those old nonquantifiables of trust and openness.

Pritchard agrees. "Project management has long been a somewhat furtive practice where we [the practitioners] take it upon ourselves to 'hide' extra money and time without letting the customer know we're doing it and why," he says. "Honest project management is going to be the successful project management over the long term. Some organizations cannot support overt communication about risk, time and cost slippage, and team relationships. Those organizations should be out of the project management business in the not too distant future."

> The "skill-building" approach to project management is giving way to a new paradigm: the culture-creation approach.

Ripples of Change

The "skill-building" approach to project management is giving way to a new paradigm: the culture-creation approach, in which project management is a holistic process, nurtured by the organization's shared beliefs, attitudes, and infrastructure.

But cultures can be slippery. How does one create a culture where project team members enthusiastically embrace the new attitudes, beliefs, and skills that lead to success? At Xerox, it is a matter of education. And not just the dry flip-chart-and-lecture kind of education; Hambleton immerses people in the subject in the most compelling way of all—experientially. Using a learning simulation called Countdown®, a strategy game for project teams developed by Tampa-based Paradigm Learning, Lynne equips project team members in a low-risk, high-involvement learning environment. "Countdown has proven to be a marvelous tool," Lynne confirms. "It introduces people at Xerox not just to the skills and tools of project management but also to the culture for creating value-added solutions. It is teaching project team members at Xerox what it means to be a consultant and a strategic partner."

As more and more companies begin to embrace this strategic partner approach to project management, success stories are emerging and offering a tantalizing glimpse at the possibilities. At AT&T, project team members from many divisions are brought in at the earliest conceptual planning stages of an effort. At Bell Atlantic, formal reporting processes have been established to ensure that project teams have the ongoing involvement of critical stakeholders. And at NCR, team members begin assessing opportunity and risk on day one . . . and continue until completion.

To be sure, at each of these organizations, the traditional tools of PMBOK are still firmly in place. So what distinguishes them from others? Look a little deeper, and you will find some new assumptions at work: Project work teams are most enabled when members see themselves as leaders, strategic partners, and entrepreneurs. And leaders, partners, and entrepreneurs are developed only when organizations embrace employees as whole people.

As is often the case with success stories, the victory arises not in the tools but from human beings. Now *there* is an assumption worth embracing.

The Learning Connection

So how does one communicate new principles about project management in a way that leads to a culture shift? Based on her own experience with Paradigm Learning's Countdown® simulation, Xerox's Lynne Hambleton offers some suggestions:

- Make it active. There is still a place for flip charts and PowerPoint presentations in classrooms of corporate America. But adult learning theory confirms time and again that people learn by doing. At some point, put away the flip chart and get people on their feet.

- Make it fun. "Let's face it. Project management is a pretty dry subject," confides Lynne. "I have the most success engaging learners when they're having fun. That was one big reason the Countdown simulation worked so well at Xerox. It was fun."

- Make it practical. After engaging people in the fictional world of the learning simulation, Lynne's work is only half done. The next critical piece is connecting the experience back to the reality of daily work. How can we actually practice what we've learned? How do we actually do this in our work?

If you do not tie it back, people will not ever own it.

SECTION 3.2
DESIGN AND CONSTRUCTION PROCESS MANAGEMENT FUNCTIONS

Kristin R. Hill, A.I.A.; Cindy Aiguier, John Morganti, and Bonnie Seaberg
International Facility Management Association (IFMA),
Boston, Massachusetts

Norman E. Faucher, C.P.E.
Association for Facilities Engineering (AFE),
CAC Industries, Inc., Leominster, Massachusetts

The largest investment of capital, personnel, and time culminates in the design construction process. In turn, this is where the greatest problems can occur. Facilities management departments can be negatively impacted when expenses get out of control, because a large portion of the costs managed by them are tenant and capital improvement expenses. Adequate procedures, standards, systems, and controls need to be addressed by a solid effective team that can deal with the many complicated issues that will arise. The design and construction process is the most volatile of all the processes and could have the greatest impact on the financial stability of the facility.

Knowing what constitutes the process and service of a facility keeps you on track for initiating the design of the facility; function follows form. Knowledge of objectives, requirements, and the specifics of different types of facilities is your guide to that design. Realize that the driving forces for these goals are the advances in technology, communication, and financial reporting. This paves the way for the tools necessary to succeed in the new millennium. The goals and objectives should focus on the true life cycle of the facility. Maintaining control of the goals and objectives as processes aids the definition of the investment and payback of the facility's life cycle. Ultimately, you can control that which you can define or measure.

Equally important are the roles of your design team. How do they accomplish these roles and responsibilities? Which type of contract is most appropriate for the project? A contract can take on many forms; knowing which one or which part best serves the project goals and objectives will enhance your position.

Learn to assemble a great team with knowledge of the advantages and disadvantages of the players and where and how they interact. What are the tools necessary to manage these players, architects, engineers, construction managers, and general contractors? Keep in mind that the success of the project depends on the players as they relate to a unified team.

Change is inevitable within the design process, but managing it brings about a controlled change. It is important to be able to identify the project baseline and learn how to facilitate changes. A controlled change can lead to a cost-effective design. The entire project team must understand the change process.

Know what documents are required to start up the facility; what records, policies, and requirements are needed to maintain the facility. Keep accurate records on the procedures that have been formulated by the design team. This enables a smoother transition from construction to commissioning to occupying the facility. Remember, design and construction, like all projects, have an end—not an end in itself but a signal of closure that sets the agenda for the informational management and operational management of the facility.

ARTICLE 3.2.1
THE PROJECT MANAGEMENT PROCESS

Paul R. Smith, P.M.P., P.E. M.B.A.
Peak Leadership Group, Boston, Massachusetts

Kenneth H. Stowe, P.E.
George B. H. Macomber Co., Boston, Massachusetts

William Duncan, P.M.P.
Project Management Partners, Lexington, Massachusetts

Successful project management depends on clear communication. All participants must share a common understanding of the terminology they use if problems are to be avoided. Yet project management's diverse roots have led to many different "dialects." Consequently, project-related exchanges are foggy at best and dangerous at worst if we do not appreciate that different people may use the same term to convey a different meaning. This article is written as an overall description of project management. Most facility managers have projects ongoing throughout their facility's life cycle. These projects may be subcontracted to third parties. However, once completed, the revised facility must be taken back and operated by the facility manager. Therefore, there is a continuous flow of information between the project manager and the facility manager.

For the facility manager who faces a renovation, restoration, or new construction, balancing competing goals relating to scope, cost, and schedule is part art and part science. Project management is a vital contribution to achieving the highest value from facilities investments. Project management involves a group of activities that mesh together; action in one activity affects others. These interactions often require trade-offs among project objectives, goals, and missions. In this article, the reader will be presented with all of the knowledge areas that must be integrated to manage a project. We begin by explaining the basic behavioral aspects of a project and later give detailed information about how to run a project.

BEHAVIORAL ASPECTS OF A PROJECT

What Is a Project?

Projects are the way persons or organizations define related jobs that perform work to create deliverable(s). Projects can be used to do work in any of the life cycles of a facility, from conception to closeout. Usually, they are temporary undertakings to create unique products or services. By being temporary, projects have definite beginnings and endings. By being unique, there is some distinguishing way in which the product or service differs from similar products or services.

Projects are performed at all levels of the organization. They may involve a single person or many thousands of people. Projects may require less than 100 hours to complete or more than 10 million hours to complete. Projects are often critical components of the performing organization's business strategy. In facility engineering and management, a project can be, for example:

- Preparing a plan for the feasibility of developing a product or service
- Creating a cross-functional team to determine the conceptual design of a facility
- Preparing the design basis, drawing, specifications, and contracts to construct or modify a facility

Because the project is unique, we will spend the majority of our discussion on defining and managing a project.

What Does "Temporary" Mean?

"Temporary" means that there is a definite beginning and a planned ending. The beginning of a project is usually after a contract is authorized or when your supervisor gives you the authority to spend money on a task or group of tasks. A project ends when you complete the customer's requirements to their satisfaction, or it is completed when you determine that the project must stop because the final results cannot provide the scope of the work within the constraints of the costs and the schedule. The results of projects need not be temporary. The Egyptian pyramids are an example of projects that have lasted for millenniums.

Products or Services

Projects provide deliverables that are measurable, attainable, and unique. Even though something was done before, it still can be unique. A second pyramid built by the Egyptians could have been unique and a project, even though one had already been built. It contained a process including the initiation, planning, execution, control, and closeout of the project.

What is Project Management?

Project management describes an organizational approach to the management of ongoing operations. Project management is the application of knowledge, skills, tools, and techniques to project activities to meet or exceed stakeholder needs and expectations from a project. Project management is necessary throughout the entire life cycle of a project.

What is a Project Manager?

A project manager (PM) is an individual responsible for managing a project. The PM is an organizer of the project. A PM is a single person who can work through others to accomplish the objectives of the project. The PM must coordinate and motivate people who sometimes owe their allegiance to other managers in their functional areas. In addition, the people who work on a project frequently possess specialized knowledge and skills that the PM lacks. The PM must often function in an environment that has many uncertainties. Even so, budgets, schedules, and resource constraints are usually imposed that create pressure on both the PM, who has limited authority, and on the project personnel whom the PM supervises.

Areas of Authority for Project Managers

The role and responsibility of a project manager depends on the type of organizational structure adopted by the corporation. The following are some typical areas of authority for project managers:

- Customer contact
- Project scope definition
- Project management plan (PMP) preparation
- Change control
- Resource allocation
- Staffing assignments and selection

Project Sponsor

Any successful project has a senior manager who champions the project. As a minimum, this sponsor must provide the following support:

- Believe in the cause of the project
- Knock down all roadblocks to the project's success
- Appoint and empower the project manager
- Define the project scope and boundaries
- Ensure project buy-in at management levels
- Ensure that resources are provided in a timely manner
- Resolve major policy issues
- Police sustained adherence to constraints
- Provide guidance and direction for key business strategies
- Support the project team in the resolution of cross-functional issues
- Stay informed about the project's status
- Deliver regular positive feedback to the team on performance versus expectations.

Tools for Successful Project Managers

There are many tools for gaining and maintaining control over a project that a PM can use, including the following:

- Having a project (senior level) sponsor
- A written statement of authority
- Sufficient rank to deal with the project personnel and stakeholders
- Access to decision makers on the topic of the project
- Communication skills and use
- Progress reports
- Knowing hidden agenda (if applicable)

Why Care About Project Management?

For the business community to succeed in a world that demands lightning-fast response to rapidly changing markets and economic conditions, we need to become project experts. As consumer markets undergo radical political and social change more frequently, more and more of the work we do is done in a project mode.

Project management is the only tested, cohesive management approach that allows you to pull a team of experts together rapidly, focus them on a specific job, disband them when they are finished, and start all over again as soon as a new need arises. An effective project manager system enables project teams to deliver the highest-quality products and services possible in the shortest time and at the lowest cost.

FORMAL PROJECT MANAGEMENT

Project management was a successful methodology in the construction and engineering world long before the business world discovered it. For decades, the U.S. Navy, Department of Defense (DOD), and Department of Energy (DOE) have actively contributed to the tools and techniques required for good project management. Their work has standardized the behavior of project managers throughout the United States and also in technology transfer throughout the world. American industries in aerospace, shipbuilding, construction, and telecommunications have benefited from this technology.

The Project Management Institute (PMI) is an international organization that is concerned with developing and disseminating good project management practices. PMI issues its *Project Management Body of Knowledge* (PMBOK) to assist users within the project management profession. The PMBOK includes knowledge of proven and widely applied traditional practices, as well as knowledge of innovative and advanced practices that have seen more limited use.

PMI's guide defines five major process groups in project management, as shown in Fig. 3.2.1-1 and defined as follows:

- *Initiating process.* Based upon the fact that projects have a beginning phase, there must be well-defined project goals, objectives, deliverables, and team spelled out in a project charter.
- *Planning process.* Devises and maintains a workable scheme to accomplish the work and the business needs for the project by defining the major activities and producing a project plan.
- *Executing process.* Coordinates the resources to carry out the plan and manage the constraints of deliverables, schedule, and budget.
- *Controlling process.* Ensures that the project objectives are meet and that the constraints are not exceeded before corrective actions or change controls are initiated.
- *Closing process.* Formal acceptance of the project deliverables brings the project to an orderly end. Documents are archived, and retrospective reviews of lessons learned are conducted.

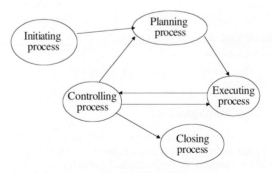

FIGURE 3.2.1-1 Links among PM process groups. [*Courtesy of the Project Management Institute (PMI).*]

The process groups are linked by the results they produce such as the product or service of one process to feed another process. The arrowheads shown in Fig. 3.2.1-1 illustrate the direction of information flow among the process groups. The processes overlap each other and are not sequential, as is shown in Fig. 3.2.1-2.

These five process groups are not one-time events that stand alone without any interaction. The levels of interaction and activity overlie and vary with the processes, as shown in Fig. 3.2.1-2. The end of one process group may provide input to another process group that requires deliverables to be accepted (such as a project charter approved) before the planning process can begin. The five processes repeat themselves in each phase of a project, as shown in Fig. 3.2.1-3.

The PMBOK also has nine knowledge areas. Each knowledge area includes the activities and processes required to ensure the following:

- *Project integration.* This knowledge area ensures that the various other knowledge areas of the project are properly coordinated.
- *Scope management.* Includes the processes required to ensure that the project includes all work required to complete the project successfully.
- *Time management.* Includes the processes required to ensure timely completion of the project.
- *Cost management.* Includes the processes required to ensure that the project is completed within the approved budget.
- *Quality management.* Includes the processes required to ensure that the project will satisfy the needs for which it was undertaken.
- *Human resources management.* Includes the processes required to make the most effective use of the people involved in the project.
- *Communication management.* Includes the processes required to ensure timely and appropriate generation, collection, dissemination, storage, and ultimately disposal of project information.
- *Risk management.* Includes the processes for identifying, analyzing, and responding to project risk.
- *Procurement management.* Includes the processes required to acquire goods and services outside of the organization.

Figure 3.2.1-4 is a cross-reference among the five processes of a project and the nine areas defined in the PMBOK. The project management outputs/deliverables of each process are

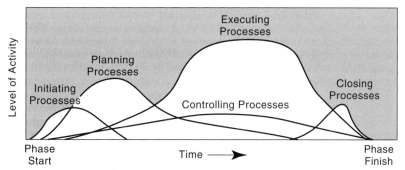

FIGURE 3.2.1-2 Overlap of process groups. [*Courtesy of the Project Management Institute (PMI).*]

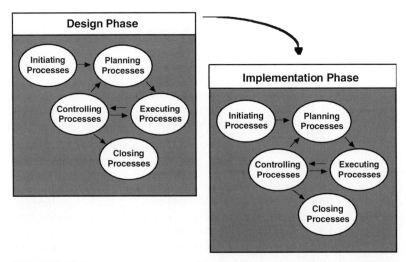

FIGURE 3.2.1-3   Process groups repeat. [*Courtesy of the Project Management Institute (PMI). Chart adapted from* A Guide to the PMBOK. *Used with permission. Copyright 1998, Project Management Partners.*]

shown as a separate box after the knowledge areas at the bottom of each process. For example, the deliverable of the initiating process is a project charter. The five process processes of a project just defined are further described in the following paragraphs:

Facility Phases

Article 3.1.4 discusses six major facility phases, or life cycles, which include the following:

- Definition of need
- Planning and programming
- Design
- Construction (build)
- Operate/maintenance
- Decision for next use

Each of these phases has five process groups called:

- Initiating
- Planning
- Execution
- Controlling
- Closing

These five groups are discussed in this article. The reader should also refer to Art. 3.1.4 for a discussion on the six facility phases.

Project Management

Knowledge Areas	Initiating	Planning	Executing	Controlling	Closing
Integration Management	Plan Development	Plan Development	Plan Execution	Overall Change Control	Overall Project Documentation
Scope Management	Initiation	Scope Planning Scope Definition	Scope Verification	Scope Change Control	Scope Documentation
Time Management	High-Level Time Management	Activity Definition Activity Sequence Activity Definition Estimating Schedule Estimate	Schedule Development	Schedule Control	Schedule Documentation
Cost Management	High-Level Cost Estimating	Resource Planning Cost Estimating Cost Budgeting	Cost Budgeting	Cost Control	Cost Documentation
Quality Management	Define Quality Approach	Quality Management	Quality Assurance	Quality Control	Quality Documentation
Human Resource Management	Define Human Resource Approach	Organizational Planning	Staff Acquisition Team Development	Team Development	Team Development
Communication Management	Communications Planning	Communications Planning	Information Distribution Performance Reporting	Performance Reporting	Results Development
Risk Management	Risk Identification	Risk Identification Risk Assessment Risk Response Development	Risk Identification Risk Assessment Risk Response Development	Risk Identification Risk Assessment Risk Response Control	Risk Documentation
Procurement Management	Define Procurement Approach	Procurement Planning Solicitation Planning Solicitation	Solicitation Source Selection Contract Administration	Contract Administration	Contract Close-Out

PM Outputs Deliverables	Initiating	Planning	Executing	Controlling	Closing
	Project Charter High-Level Work Breakdown Structure (WBS) Initial Staffing Resource Plan Project Central Repositions	Project Plan Communication Plan Issues Management Plan Change Management Plan Risk Management Plan	Project Progress Reports Updated Project Plan Issues Log Update Risk Plan Update Change Request/Control Log	Performance Measurements Updates Project Plan Risk Plan to date Change Control Update	Project Documents Lessons Learned Report User Acceptance Signoff Business Value Realization Plan

FIGURE 3.2.1-4 Cross-references between process groups and knowledge areas. (*Courtesy of Peak Leadership Group and Project Manager Partners.*)

Project Initiating Process

The initiating process identifies the project assumptions and the project objectives, products, and services to be delivered, and produces a project charter. The inputs to this process phase are the project business case, corporate policies and procedures, and lessons learned from previous projects. The process reviews and analyzes the inputs and identifies, in the project charter, the major actions to produce the project deliverables such as a definition of the business sponsor, project title, project objectives, statement of value, scope of the work, order of magnitude estimates, key deliverables, assumptions, major risks and constraints, desired time frame and sponsor approval.

Stakeholders and Their Potential Conflicts

During the project initiating process phase, all stakeholders should be identified and made part of the project team. The project team gets its lead from the financial sponsors (executive sponsor) of the project. They are likely to be looking for a specific financial performance from the facility, often described formally in a pro forma financial statement that describes the financial performance of the investment. There are the users, who will be dedicated to the functional and the aesthetic performance of the facility. These two groups can often be at odds—users demanding "bigger, better," and sponsors demanding "less, leaner." There is the operations group, which will be looking for operational and maintenance efficiency, perhaps many years, even decades, hence. There is the environmental/building codes "audience," whose demands can conflict with both sponsors and users. Finally, there is the project team— charged with developing consensus among all of the stakeholders on what to achieve, when, how, and with what resources, and also charged, of course, with executing the plan. Other potential conflicts include

- Speed versus waste
- Speed versus ongoing operations
- Cost versus quality
- Capital cost now versus operational efficiency later
- Fiscal phasing versus cash flow
- Aesthetics versus cost and/or schedule
- Environment or code upgrades versus user needs

The list of potential conflict can go on indefinitely. The project manager must work with all stakeholders to resolve or minimize conflicts and keep the project moving forward. Methods appear later in this article for refining the plan so as many as possible of the stakeholders benefit and so that the conflicts are successfully resolved.

Project Planning Process

The planning process produces a comprehensive project plan that identifies the activities necessary to produce the required deliverables of products and/or services to the customer. The plan identifies and manages all knowledge areas of the project, including integration, scope, time, cost, quality, human resources, communications, risk, and procurement, as shown in Fig. 3.2.1-4. The project charter defines the level of detail with which each of these knowledge areas will be addressed in the project. Because changes happen throughout a project, a requirement in the planning process is to monitor the execution process to ensure that the project stays on track with the project charter. Any proposed changes must be compared with the project plan to ensure that scope creep (work done outside the project charter) is not allowed without client approval of the changes.

To define and present the project scope, time, and cost, the project manager must structure the work that is going to be performed. There are numerous methods that can be used to further define the statement of work and to understand the customers' requirements for the scope of work. One method used by many professionals is to create a work breakdown structure.

Work Breakdown Structure.   Many projects involve a large number of activities. Therefore, planners require a tool to determine exactly what work has to be done, so that they can estimate how long it will take to accomplish the various elements of the project. A work breakdown structure (WBS) is a hierarchical listing of what must be done during the project. A WBS may be in the form of project life-cycle processes (design, construct, commission, closeout), functions (engineering, information systems, manufacturing, construction), or deliverables (main building, annex, infrastructure, landscaping), or a combination of these types. The WBS preparation shown in Fig. 3.2.1-5 is a major first step in planning the project. The WBS preparation is often developed to establish a common understanding of the project scope between the PM, the stakeholders, and the customer. The WBS usually includes the necessary information to define major work deliverables. The WBS approach can provide definition and structure to the work that the project team will provide and helps prevent misunderstandings. In some cases, for project definition clarity, the WBS defines work not in the scope of work or follow-up work.

A WBS is developed in flow chart form as in Fig. 3.2.1-5. A WBS is a method of creating a project plan that subdivides work into hierarchical units of tasks or deliverables, work packages, and units of work. Each element within the WBS serves as an item or subtotal in the budget and as an activity or summary bar in the schedule. There are numerous levels in the WBS, and each one gives more detail to support the higher-level elements of the project. Each item in the WBS is given a unique identifier; collectively, these are called the *code of accounts.* The items at the lowest level of the WBS are often referred to as *work packages,* as shown here:

Level 10—Company

Level 20—Authorization

Level 30—Control

Level 40—Work Package

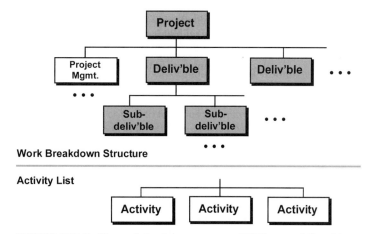

Work Breakdown Structure

Activity List

FIGURE 3.2.1-5 The work breakdown structure (WBS) types deliverable or activity. [*Courtesy of the Project Management Institute (PMI). Copyright 1998, Project Management Team.*]

The WBS should be prepared in conjunction with the scope of work; should include the entire scope of work; and should be developed to the level of detail where responsibility for performance is assigned. Responsibility for each element of a WBS should be established.

A WBS dictionary uses the terms identified in the WBS so that a common language is spoken on the project. Once the language is standardized, information for follow-up activities such as work package development and descriptions, resource loading, and other planning information regarding schedule dates, cost budgets, and staff assignments is consistent.

Work from a WBS can be displayed in a pictorial form, with color schemes to show information such as the actual 3-D work, processes of the project, or even schedules for the project, as shown in Figs. 3.2.1-6 and 3.2.1-7.

Defining Work Scope.   Many projects are unsuccessful because they fail to define structure and do not communicate what the customer really is willing to pay for. When a project is initiated, the following information should be addressed in a project overview:

Project overview:

- Current status, described today
- End status: a description of the process after the requested project is completed

What:

- Customer requirements, statement of work
- Project scope: identify all of the tasks and deliverables included in the project
- *Project objectives.* Overall goals of the project
- *Project outcome.* Target deliverables/results of project
- *Project constraints.* Cost, schedule, and so forth; limits of the current project
- *Key assumptions.* Information that defines the project scope

Why:

- *Business case.* Business objectives of the project
- *Business benefits.* Summary of the business benefits of the project, including financial and tangible benefits

When:

- *Start date.* Project initiation target date
- *Start date drivers.* Factors impacting the start date, such as key resources
- *Key milestone location.* Certain date or time when the status of the project or deliverable is required
- *End date.* Project completion target date
- *End date drivers.* Factors that impact the completion date, such as Year 2000 and government deadlines

Who:

- *Project sponsor(s).* Senior management overseeing the project
- *Process owners.* Process users impacted by the project
- *Lead users.* Lead people who will participate in the project and provide functional expertise
- *Key users.* Other user participants

How:

- *Alternatives.* A summary of the different methods by which the project could be implemented
- *Recommended approach.* Proposed solution to the project

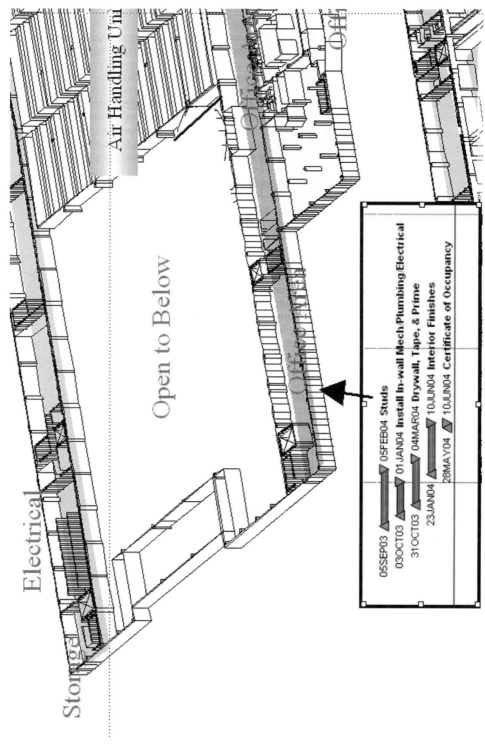

FIGURE 3.2.1-6 Pictorial way to illustrate work. (*Courtesy of the George B. Macomber Company.*)

Electrical

Storage

Open to Below

Air Handling Unit

Office Finish

Office

Office

05SEP03 Studs
03OCT03
31OCT03
23JAN04
28MAY04

05FEB04 Studs
01JAN04 Install In-wall Mech/Plumbing/Electrical
04MAR04 Drywall, Tape, & Prime
10JUN04 Interior Finishes
10JUN04 Certificate of Occupancy

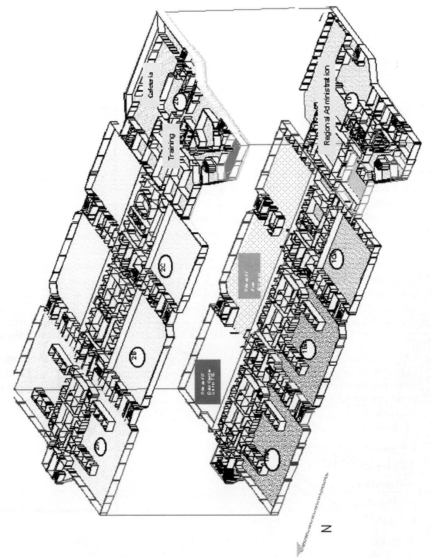

FIGURE 3.2.1-7 Pictorial work scheduling technique. (*Courtesy of the George B. Macomber Company.*)

- *Project work plan.* A high level project work plan, including primary project tasks and milestone dates
- *Deliverables.* Description of project end results (reports, programs, documents, products, and services)
- *Risk management.* The risk management strategy to be used during the project
- *Change request.* The method for gathering issues and change requests

Early communication with the customer is important for each work activity to ensure that both sides understand the work scope sell. People can have different interpretations of descriptions and varying levels of technical expertise, so it is important that you communicate up front what you understand the customer requires and what the deliverables of the project mean to the project team, and that you get the customer's responses.

Work Package Development.   Work is usually defined in a work package that organizes the work task or groups of tasks. Each work package includes detailed schedules, resources, and assumptions. When applicable, standard work packages planning procedures with documentation forms should be developed companywide.

Detailed work package planning involves clearly defining and planning the detailed scope, schedules, and costs for specific elements of the work. In addition, detailed resources planning and budgeting occurs at the work package level. (See the previous discussion on WBS for work package definition.)

It is considered a best practice to use work package planning to establish a basis for scope statements and a basis for the detailed cost and schedule estimates. Commitments from the performing organization can be easily documented on the work package activity assignment section.

Project Schedule, and Cost.   All projects have goals and limited time, space, money, equipment, and intellectual resources with which to achieve them: Rigorous project management, often described as part art and part science, is a vital process for managing unique jobs or activities that have clearly defined scopes, costs, and schedules. The project scope is the most important of the three factors to define. The cost and schedule are the constraints under which the scope must be defined and managed, and sometimes they require readdressing the scope.

Organizational Structure.   All of the resources required to accomplish the work described in the WBS must be clearly identified and scheduled. Often, when the lines of authority need to be clarified, the team should make an *organizational breakdown structure* (OBS). Detailed resources requirements and availability of those resources should be identified before schedules and budgets are formally established and commitments are made. The following are examples of resources:

- Labor (by discipline, skill, organization, number)
- Computer (including special software)
- Materials and supplies
- Physical facilities
- Support services (recruiting, training, and administration)

Resource planning and optimization should be based on detailed cost estimates coordinated with performing organizations. Resource planning such as that shown in Fig. 3.2.1-8, provides the basis for work task managers, as well as for functional organization managers, to plan their resource needs for the project's duration.

Preliminary resource planning can be critical. For instance, functional managers or suppliers may need to be contacted early to ensure availability of personnel, equipment, or supplies.

Enterprise Resource Planning.   Project resource demands often compete for resources required elsewhere in the company. Doctors are needed to attend hospital project planning

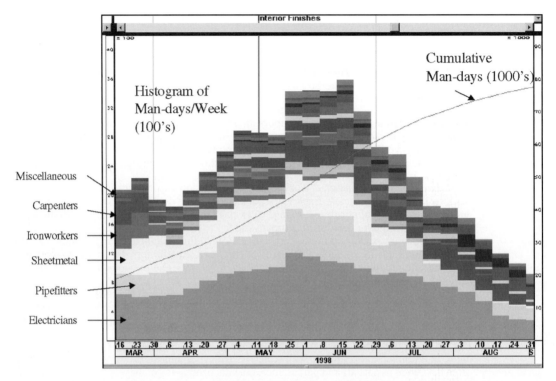

Resource Forecasting
Helps to Select Qualified Bidders and Estimate Labor Premiums

FIGURE 3.2.1-8   Resource forecast loading. (*Courtesy of the George B. Macomber Company.*)

meetings. Production managers are needed to attend design and construction meetings. These conflicts give rise to a need to carefully design the project organization and clarify how much time the project team can expect from these resources. Decisions need to be timely and often require understanding complex inputs. There is no substitute for taking the time and having the skills and experience to digest the matters involved. The skills of the project team must be addressed in all processes of the project to ensure that the right persons are on the job at the right time.

Schedule and Time Techniques.   There are two principal methods for estimating the time that project activities will take. The *backscheduling* technique sets the date of completion and then forces the project to fit the available time. Backscheduling is often dictated by legal, regulatory, or market factors. However, some deadlines are selected arbitrarily without taking cost and quality implications into account. This unstudied development of milestones can be very costly to organizations and demoralizing for the project team. With the backscheduling method, managing project resources is critical, and controlling time rigorously is advisable. There are times when backscheduling may be the only option. Unfortunately, accepting a time period because it is "mathematically necessary" can produce unrealistic project completion dates and in reality can be too costly to meet. Also, this method may cause undue stress within the organization to meet the deadline.

Forward scheduling estimates the time for each activity and estimates a project completion date from these estimates. This approach usually produces more realistic project completion dates. However, there is always the possibility that excessive time will be padded into the schedule.

Time Contingency.   One method of dealing with tight schedules is to acknowledge the risk of missing any one of many critical dates in a project schedule and to create a time contingency. Many professional planners consider planning with time contingency (at the end of the schedule) just like a budget contingency to allow for unknowns.

Tools and Techniques for Documenting Activity Sequencing.   Activity sequencing looks at the proposed tasks and can use two types of diagrams, namely, *activity on node* and *activity on arrow*.

Series Versus Parallel Processing Activities.   When two or more activities can take place in the same time period, the process is called a *parallel process*. When only one activity occurs at a time, it is called *series processing*. Parallel processing saves time, but there are times when a series process cannot be avoided. However, changing a work method may create a parallel process from a series process. For example, traditional home construction methodology dictates that you cannot build a roof until you put in the foundation. However, if you use a prefabricated home kit, the roof can be assembled and ready to install before the foundation is laid.

PERT.   The Program Evaluation and Review Technique (PERT) was originally used by the U.S. Department of Defense (DOD) Research and Deployment Team for the Polaris submarine project.

PERT Estimates.   PERT uses three estimates of duration—optimistic, pessimistic, and that most likely to increase the accuracy of the estimate. In estimating durations, many projects suffer from one or another party's subjective interest in the project. Some estimates are unrealistically low (the head-in-the-sand estimate)—often a result of someone who wants the project to move forward or who wants (too much) to win a bid. Some estimates are grossly overestimated (the sky-could-fall estimate)—the result of a risk-averse person or party. The best estimate is one that is factual, accurate, and objective.

CPM.   Critical path scheduling consists of analytical techniques to plan and control a project. The critical path method (CPM) is quite similar to the PERT method. The basic forms of PERT and CPM focus on finding the longest time-consuming path through a network of tasks as a basis for planning and controlling the project. A good CPM planner creates and/or facilitates the creation of a network of related activities. The activities are related to predecessor and successor relationships with *lag*. Lag, the term to define whether the start or finish of one activity is related to the start or finish of another, may be zero. Firms that offer project planning and control software often have excellent documentation and training programs that introduce CPM.

A good CPM network is a model that forecasts the way the project will perform and as such has the following characteristics:

- The scope is well defined.
- Tasks are discrete and are related accurately.
- Resources are modeled that may impact project performance.
- The model has a database structure that honors the project work breakdown structure so that activities can be sorted, filtered, and colored for effective presentations and summaries.

Slack Time.   Total slack is working time that an activity can slip without affecting the end date or any critical milestone. Slack time is a valuable piece of information. If the project is moving forward according to schedule, activities that have slack time can be postponed. Then, effort can be directed toward other more critical activities. There are two kinds of slack time, *total* and *free*. Total slack is the total nonscheduled time available for the activity. Free slack is the amount of working time an activity can slip without affecting any other activity.

To calculate a slack time, one must calculate four time-values for each activity as follows:

- *Early start time (ES)*. The earliest possible time the activity can start
- *Early finish time (EF)*. The early start time plus the time to conduct the activity
- *Late finish time (LF)*. The latest time an activity can finish without delaying the project
- *Late start time (LS)*. The latest time an activity can start without delaying the project

The total slack of an activity is the lesser of (LS – ES) and (LF – EF).

DESIGN—AN INTEGRATED APPROACH TO REFINING THE SCOPE DOCUMENTS, BUDGET, AND SCHEDULE

Schedule Baseline and Optimization

The most effective way to build a schedule so that it is realistic and so that the project team will perform at a high level is to create a "first-pass" baseline that is a solid network and has the entire team's buy-in as a starting point. In this schedule, all activity durations are acceptable to the team, and activities have realistic relationships. Milestones are established in this baseline schedule as a definitive way to break down the project and allow "at-a-glance" comprehension of the schedule. This baseline schedule (and its implicitly and explicitly implied means and methods) is the basis for a comprehensive baseline estimate of project costs. Attention is drawn to time-variable costs (e.g., staffing, office trailer rental), and seasonal costs (cold weather protection, erosion control). These costs bear watching as the team analyzes options.

Then, optimization analysis follows (usually, but not always accelerated) to achieve the earliest economic completion. The critical activities are established and their durations and relationships are tightened. Each "element of acceleration" is analyzed for its impact on cost, quality, and safety. A strategic team is also engaged to consider the value or "earning power" of the completed project and may be empowered to increase the budget if the increase can show a favorable return on investment.

Resource Loading and Leveling

The project team's job is to anticipate when resources may not be available to accomplish activities. When resources are potentially scarce enough to constrain the pace of a project, resource loading of the activities on the schedule is vital. These constraining resources may be a selected trade (like pipefitters), a material (such as concrete from a batch plant), a piece of equipment (such as a crane), or cash.

Resource Leveling. When resource demand may exceed resource availability, activities often need to be postponed to reduce or flatten the demand. Professional project planning software will allow resource loading and automatic leveling and will postpone activities with lower criticality or priority to reduce demand for critical resources during peaks. If resource-leveling algorithms are employed, they should be used with a measure of caution because they only "reason" as well as the model has integrity. For example, if an activity does not have the correct successors and as such has an inaccurate amount of slack, the algorithm may postpone an activity that may greatly and adversely effect the completion date.

A good method of resolving disagreements between stakeholders about how much to accelerate a schedule is to drive all inputs to a net present value and optimize. The chart in Fig. 3.2.1-9 is an illustration of a project analysis to determine whether the August 1 project deadline is justified when all time-variable costs are considered: the expediting, the general conditions, and the revenue and compromises for each completion date.

Gantt Charts.   Gantt or bar charts are graphical displays of schedule-related information. In this type of chart, activities or project elements are listed down the left side of the chart, dates are shown across the top, and activity durations are shown as date-placed horizontal bars. Gantt charts are to scale, and therefore unlike PERT charts, the actual duration can be determined by glancing at the timetable. Activity sequencing and whether activities run in series and parallel are easily recognized at a glance, as shown in Fig. 3.2.1-10.

Presenting Complex Data.   Often, bar charts and Gantt and PERT diagrams are inadequate communication vehicles. The audience may be unskilled, uninterested, or unwilling to take the time to digest a complex schedule. Then a visual image can be worth a thousand words. Here, a single graphic conveys to the user community where the processes begin and end, when they begin and end, and what the status of the adjacent spaces will be when the user takes occupancy. The result can often be that the users understand more of what to expect and what their obligations are.

Project Execution and Control—Measuring Progress as the Plan Is Executed

The project plan is never static. Every day brings reports of progress, questions, delays, new information, and so forth. The project progress is measured against the plan on a formal and regular basis. Milestones (discrete, measurable, and undebatable progress points) should be built into the schedule to make the comparison of plan versus actual progress very easy. If these milestones represent design releases, they should be defined with a clear set of deliverables. These points can be used to reward the project personnel formally or informally. The frequency of reporting progress should be at the project team's discretion but in any case should be at least monthly. When projects are very dynamic and very tightly scheduled, the frequency may be weekly or even daily. More frequent updates are often easier to accomplish because the information is fresh.

Updating the schedule for the greatest accuracy and making progress payments sometimes requires measuring *earned value*. Earned value is used to protect the owner's interests and to counter any confusion that may enter the discussion around "percentage completion." An activity may be 50% complete in terms of duration, because mobilization and setup time can be extensive, but may be 10% complete in terms of the quantity of material in place. For payment and productivity calculations, the earned value, defined as the completed quantity in place divided by the total quantity multiplied by the value of the activity, may represent an important difference.

Project Status Meetings

Weekly status meetings provide a forum in which participating project team members can communicate with one another. Status meetings allow for discussion of essential issues relating to the project and provide an opportunity for project team members to communicate with each other to ensure understanding of status, issues, and upcoming plans. In addition, the status meetings allow for issues with cross-project area categories to be discussed by parties involved within the project area, allowing for common understanding and joint definition of resolution.

Project meetings are usually conducted on either a Friday afternoon or Monday morning, depending on the number of active projects and the availability of the meeting location. The list of invitees should include the following:

Cost Benefit Analysis of Alternative Project Completion Dates

Affected Budget Line Item	For the Schedule Completing:				
	01- Jan	01- Mar	01- May	01- Jul	
Construction Expediting Contingency	$ 902,000	$ 740,000	$ 520,000	$ 400,000	
Design Expediting Contingency	$ 120,000	$ 90,000	$ 50,000	$ 25,000	
General Conditions-variance from baseline	$ 25,000	$ -	$ 40,000	$ 80,000	
Fees on Acceleration Costs(variance from baseline)	$ 6,510	$ -	$ (6,600)	$ (9,750)	
Net Present Value of Revenue Impact-(variance from baseline)	$ (230,000)	$ -	$ 230,000	$ 460,000	
Other Time/Cost Variables	$ 22,000	$ -	$ (36,000)	$ (11,000)	
Total	**$ 845,510**	**$ 830,000**	**$ 797,400**	**$ 944,250**	
	$ 230,000	estimated net value of operating month at completion			

FIGURE 3.2.1-9 Cost analysis of alternative project completion dates. (*Courtesy of the George B. Macomber Company.*)

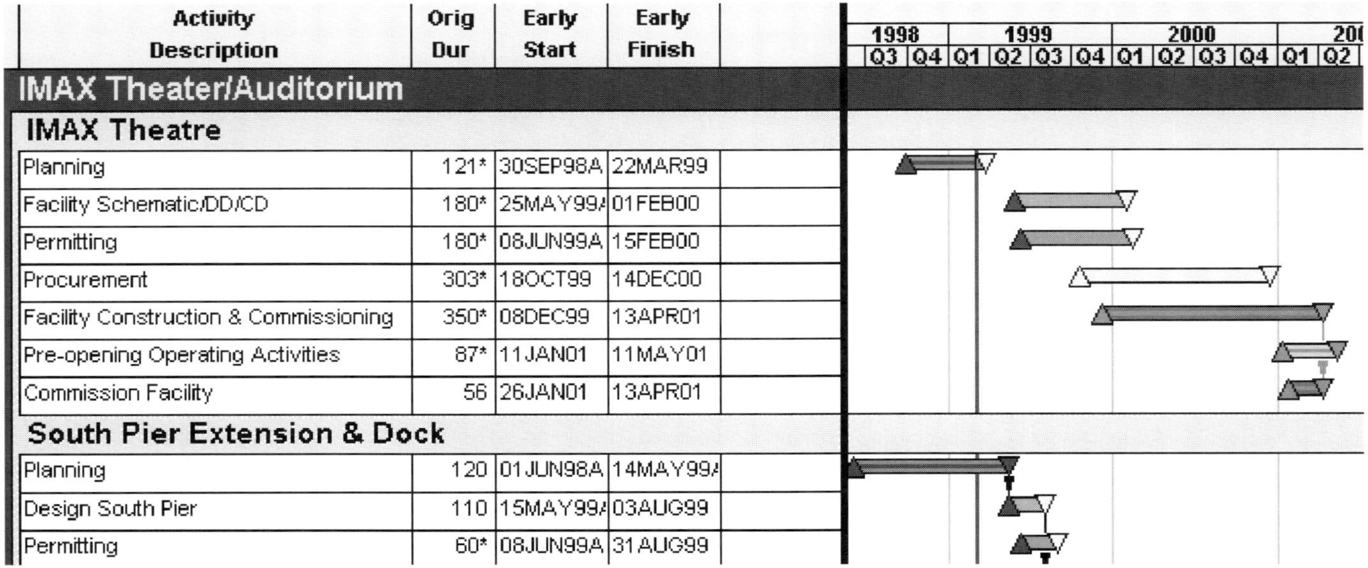

Activity Description	Orig Dur	Early Start	Early Finish
IMAX Theater/Auditorium			
IMAX Theatre			
Planning	121*	30SEP98A	22MAR99
Facility Schematic/DD/CD	180*	25MAY99A	01FEB00
Permitting	180*	08JUN99A	15FEB00
Procurement	303*	18OCT99	14DEC00
Facility Construction & Commissioning	350*	08DEC99	13APR01
Pre-opening Operating Activities	87*	11JAN01	11MAY01
Commission Facility	56	26JAN01	13APR01
South Pier Extension & Dock			
Planning	120	01JUN98A	14MAY99A
Design South Pier	110	15MAY99A	03AUG99
Permitting	60*	08JUN99A	31AUG99

FIGURE 3.2.1-10 Project activity controlling. (*Courtesy of the George B. Macomber Company.*)

- The project management team
- The functional lead for the project area
- At least one representative from each outside party associated with the project
- Any project area participants whose areas are currently active and may have input into the status meeting
- A representative of the project management office, if applicable

Each of these should be invited to the weekly status meeting, but the status group should be kept as small as possible to minimize the impact of the meeting on the participants' progress. The status meeting should cover the following topics:

- A walk-through of the status report to identify accomplishments since the prior week and to highlight deliverables missed and their possible impact
- Identification of issues related to project areas and assignment of responsibility for resolving them
- Identification of cross-project area issues and assignment of responsibility for bringing these issues to the PM
- Discussion of project logistics among team members
- Plans for the upcoming week and arrangement of upcoming meetings

Status meeting minutes should be taken and issued within one working day. The minutes should include the following items:

- A list of attendees and those invited but not in attendance
- A distribution list for meeting minutes
- A summary of key points discussed
- Highlights of accomplishments since the previous week
- A summary of tasks for the coming week, and assignment of responsibility for each task
- Identification of issues and follow-up items discussed but not resolved during the meeting, and assignment of a person to resolve and report the findings
- Modification to key project milestones or the project schedule from the prior week or resulting from the status metering
- A summary of overall project goals for the coming week

Project Management Status Meetings

Project Management Status review meetings bring the entire team up to speed on the progress since the last status meeting. The management review meeting is usually conducted on a regularly scheduled basis (i.e., monthly basis) using the information from the project weekly status reports. The meeting should have an established agenda that includes the following topics:

- Project objectives
- Scheduled deliverables
- Original project timeline
- Original project budget
- The team: resource utilization
- Current project status: includes work completed, deliverables provided, and client perceptions

- Status against budget and timeline
- Key milestones going forward
- Lessons learned
- Open issues

Version Control

As the schedule and scope options are embraced and each is analyzed for the impacts on cost, multiple versions will be published and can confuse the audience and the team members. Version control is a simple naming of versions to eliminate confusion among various published schedules. One easy to follow method is to name the first acceptable baseline "Version 1.0," and label each succeeding version with small changes 1.1, 1.2, and so forth. When a delay, a substantial change, or a schedule is published to a wide or high-level audience, the schedule is named "Version 2.0." Version control gives rise to an important type of report called the *variance report.* It defines the actual or "current forecast" dates for each activity, the "as planned" dates of the baseline or "target," and the variance or difference between the two.

Issue Resolution Strategy

One of the primary roles of the project area managers is identifying, tracking, and resolving project issues, including issues that may change the project scope. A key factor for success for the effort is the development and use of a standard issue tracking and resolution strategy (see Fig. 3.2.1-11). Issues should be identified and communicated through the use of issue communication strategy.

The project issues should be maintained in an issue tracking system. The issue tracking system, which may be custom-developed or purchased, will allow storing issue information in a database that has multiuser entry and retrieval capabilities. Project lead persons should be able to enter resolutions or actions against the issues and to change issue status.

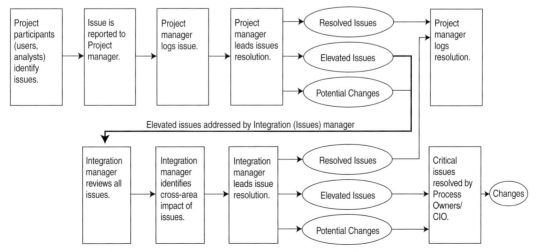

FIGURE 3.2.1-11   Issue resolution process. (*Courtesy of Peak Leadership Group.*)

Change Management

In some cases, issues and their resolution, including changes in cost and effort, will impact the project scope and timeline. These types of issues can be identified as potential changes during the issue resolution process. The change management process is similar to the project initiation process. When the resolution of an issue requires a change in the project, the change management process is initiated. Based upon the analysis of the pros and cons of implementing the change, it may be approved, revised, or the issue resolution may be modified so that the change may not be required. For more details, see Art. 3.2.2, "Change-Control Process."

Project Closeout

The closing process documents the outcome of the project. Products and services are delivered, actual project results are reported, project contracts are closed out, personnel are reassigned, and project lessons learned are documented.

The input to this process includes the final (last) approved charter, final project plan, and project results. The results usually contain the final project status report and final earned value analysis (EVA). A checklist of final actions should include the following items:

- Close out and settle contracts
- Resolve any remaining action items
- Provide final project accounting information, including invoice details and billable expenditures
- Conduct a lessons-learned meeting with the project team
- Document successes
- Document opportunities for improvement
- Communicate results to all stakeholders
- Update the lessons-learned database
- Generate and archive final project documentation
- Produce final scope, cost, quality, and risk documentation
- Produce final as-built documents
- Provide final performance information
- Provide performance feedback to all members
- Provide final documentation to project sponsor, owners, and management for approval

Occupancy is an important phase of a successful facility project. It involves planning, resource analysis, milestone definition (what is the condition of the facility that is acceptable for the owner's equipment, furniture, personnel, etc.), packing, move planning, and control. It often involves weekend and off-shift work. It is the phase that, if unsuccessful, can soil the reputation of a solid project team because it is the most visible phase to many users. If it is treated as an integral part of the project planning and execution, the likelihood of success can be greatly enhanced.

For a discussion of construction and commissioning, see Chap. 6.

CONTRIBUTORS

John J. Downing, P.M.P., Compaq Computer Corporation, Lexington, Massachusetts

John M. Nevision, Oak Associates, Inc., Maynard, Massachusetts

Stephen Bauer, C.G.U., Foxborough, Massachusetts

ARTICLE 3.2.2
CHANGE-CONTROL PROCESS

Paul R. Smith, P.M.P., P.E., M.B.A.
Peak Leadership Group, Boston, Massachusetts

Robert Morin, P.E.
Pizzagalli Construction Co., South Burlington, Vermont

Bill Duncan, P.M.P.
Project Management Partners, Lexington, Massachusetts

In some cases, issues and their resolutions, including their cost and effort, impact a project's scope and time line. These issues are identified as potential changes during the issue tracking effort (see Art. 3.2.1, "The Project Management Process").

A change-control process ensures that facility changes are identified, controlled, and integrated into the facility's system. The steps for this process are evaluating, coordinating, implementing, and documenting the changes. This process maintains consistency among the design requirements, the physical facility's configuration, and its documentation requirements. Change control is the most important process in ensuring that the facility has accurate documentation; as such, change control is usually the first process to be worked on when a facility wants to regain configuration control.

When establishing change control, the facility manager should have established, up-to-date, facility baseline documents. Examples of baseline documents include the master equipment list, as-built master facility drawings, a master facility drawing list, change-control documents, as-built specifications, approved vendor data, operating procedures, and operating and maintenance manuals.

PROCESS STEPS

The first step in the change-control process is to identify, as soon as possible, the change that needs to be tracked, so as to assess the impact of the change on risk, cost, coordination, and time. Those changes that are critical to the facility, costly to purchase or maintain, and/or are complex to manage should be the top priorities. The larger the facility complex or the more numerous the locations, then the more items that should go through this process. In this process, the benefit of tracking the change should outweigh the efforts to produce the change. Keep necessary documentation of the change to a minimum to avoid excessive and unnecessary paperwork.

The second step is maintaining the change-control process. The facility manager is responsible for maintaining this process and should use a system for logging and tracking the steps of the change. This can be a facility change form, a document revision request, a drawing change form, or a construction interface document. These forms can provide a means of formal review, evaluation, and approval of the proposed change.

The third step is configuration management control. This step assures that with the change, the physical facility maintains its conformance with the design and that both the physical facility and the design requirements are accurately reflected in the documents. Configuration management control provides supervised change and integrates information about the facility into a structured, usable format (see Fig. 3.2.2-1, Steps 3.1 through 3.8, for more detail).

The fourth step is implementing the approved change.

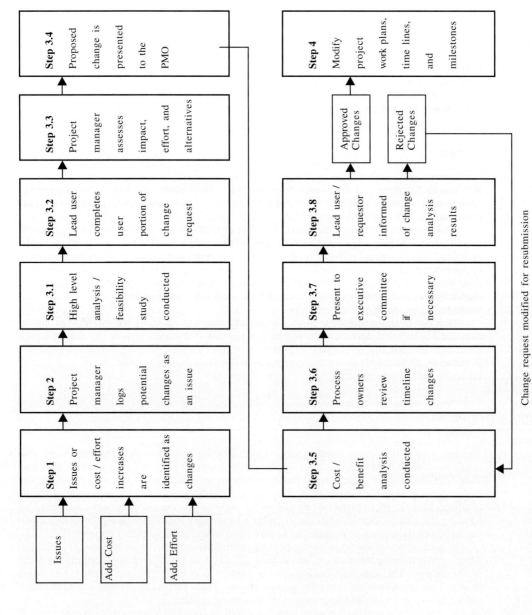

FIGURE 3.2.2-1 The change-control process. (*Courtesy of Paul Smith, P.M.D., P.E., M.B.A., Peak Leadership Group.*)

TYPES OF PROJECT CHANGES

Figure 3.2.2-1 is a high-level diagram that illustrates the control of project changes. The following are some common types of changes and the reasons for tracking them.

Facility Modifications

Facility modifications that alter the facility's equipment or structure need to be consistent with the design, need to list acceptable substitutes, and need to consider maintenance and repair requirements. A list of structures, systems, and equipment critical to the facility operations should be identified.

Maintenance Changes

Maintenance changes alter the facility's maintenance process. Maintenance personnel are the best source for maintenance changes and should be encouraged to submit changes via this change-control process. The facility manager should be cautious and make sure that plant personnel clearly understand the difference between a "change" and "routine maintenance." Maintenance activities that change the facility need to be evaluated for consistency with facility requirements by ensuring that safety and reliability have not been downgraded or compromised and that the change conforms to the design, construction, and operational requirements of the facility. Routine maintenance activities that involve routine repair or restoration of the facility (like-for-like replacement) without altering the physical or functional characteristics need not be evaluated for consistency with design. Any deviation, such as manufacturer equivalent substitutions, changing torque values on a mechanical seal, changing O-ring material, changing glue type on a control component, and so forth, should be evaluated as a maintenance change.

Procedural Changes

Procedural changes (such as in operating procedures) result from a modification and need to be evaluated to ensure that they do not violate the design and safety limits established for the facility.

Document Only Changes

These changes to specifications, vendor data, manuals, and so forth do not affect hardware but alter the accuracy, intent, or context of the facility's documentation. Changes to these documents should be reviewed to ensure that they do not alter the design, regulatory, or operational limits of the facility. The level of approval of the change in these documents should be consistent with the level of the original document approval, unless the change is nontechnical (such as correcting misspelled words). Take care not to alter trade-specific terminology when attempting to correct misspelled words.

Software Changes

Changes to software include the basis for safety analysis, design information, facility configuration, supports of the facility process system, or changes to software used for technical or safety decisions. This software requires a formal evaluation of the basis and an evaluation of its impact on performance, customer requirements, and safety before implementing the change.

Operational Changes

These are changes that deviate from the procedure established for start-up, normal operations, or the shutdown phases of facility operations. An example of an operational change is adjustment in instruments and equipment set points that have secondary effects on engineering considerations such as reductions in safety margins and response time. Additionally, lifting leads, jumpers, and other temporary modifications need careful attention. When established operational limits are exceeded, this could also damage or reduce the qualified life of equipment/components or cause potential accidents.

Safety Questions

These are changes that did not review safety issues. An example is a change made to a facility that reduces the margin of safety, increases the probability of a serious accident, or introduces a possibility of a type of accident not previously identified.

Training Changes

These are changes in hardware, maintenance practices, procedures, documents, software, or operational techniques that require updated training. The form of training can be formal classroom introduction, required reading of a text or manual, on-the-job training, prejob briefings, computer- or video-prompted introduction, self-paced workbooks, and so forth. A written record should be established and maintained by the facility training coordinator to ensure that the required training update has been completed. Dates should be established for completion of training to ensure that the training is done in a timely fashion. The required training for each job title should be in written form. Each employee's personnel file should document completion of this required training.

Minor Changes

These changes include spelling, grammar, punctuation, and inconsequential textual content, changes to personnel positions or procedural title changes, and updating the facility management system document matrix. These changes can be exempt from the same level of review and approval as the original document. The facility manager can authorize these minor changes. The revised minor changes should be distributed to the same personnel as the original document.

Technical Review

These are changes to an approved baseline item that must be reviewed and approved by the "design authority" for that system or process to ensure consistency with the design basis. Technical reviews may be performed by organizations other than the engineering unit, such as maintenance or operating personnel, if they have access to the design basis via a "design envelope" that established preapproval limits or constraints within which changes can be made without exceeding design requirements. If a proposed change is not consistent with the design requirements, then the engineering unit must be consulted. Technical reviews and approvals of changes to the documents previously mentioned should be the same as required for the original issuance of the documents. The changes should be implemented upon receipt of the required approval.

MANAGEMENT REVIEW OF CHANGE

Facility management should review proposed changes to the scope, cost, and schedule of the project to verify that adequate technical reviews have been performed and documented. Management can prevent changes to the project baseline, except as directed and when changes to the project effort are agreed to. All changes to the project scope, cost, and schedule shall be documented and approved in accordance with the requirements defined in the operations program management plan. Management should approve changes before implementation. Changes made before the required approvals are given should be prohibited for safety reasons. The only exception is for minor corrections.

DOCUMENTATION OF CHANGES

Once a document change has been approved, all relevant documents affected by this change shall be identified, modified, and updated in a timely manner. Documents such as those required to operate the facility shall be updated before returning the facility to operation. Facility personnel should document and review the approved changes to the project baseline documents as soon as practical during the normal course of design, construction, and operation.

ADDITIONAL POINTS OF CONTROL

The following areas must be attended to in a successful change-control process.

Technical Vendor Control

Procedures should be established to ensure that important vendor activities and information are consistent with the facility's requirements. The control management program requires reviewing and approving vendor procedures before work commences or imposes the use of facility procedures for work conducted at the vendor's facility. The facility manager is to define when vendor work has been completed satisfactorily and is ready for information turnover with established acceptance criteria to the facility personnel. After turnover, vendor information can be kept current and available by including it in the document control center. Vendor data on systems, equipment, components, or items associated with the facility's operation are available from the manufacturer. Document Control shall receive, catalog, and store all important vendor data for retrieval. Vendor data consists of: shop and fabrication drawings, operating and maintenance instructions, installation procedures, installation instructions, user manuals, training materials, warranties, spare parts lists, parts availability, and operating life expectancies.

Information Equipment Databases

Document Control should maintain information on all systems, equipment, and components that are critical to the facility's operation in one central configuration management database, which may be a hard-copy file. The database will provide updated and consistent information on systems, equipment status, and maintenance to facility management and the project personnel. The database can be based on the facility's comprehensive equipment list, master equipment list, vendor data, and key drawings.

Computer Software Configuration Management

Software configuration management applies to any computer programs, procedures, rules, and associated documentation and data that are essential and critical to the design, analysis, evaluation, assessment, operations, and process control of the facility. Software users must manage and control computer software during its entire life cycle. This is particularly significant in software used for technical decisions.

Configurational Management Procedures and Training

When new projects or tasks are initiated, the facility manager should identify appropriate procedures and engineering standard practices that must be complied with by the project.

Records Management and Document Control

Record management ensures that activities affecting the life cycles of records are planned and managed. These actions ensure compliance with applicable federal, state, and company policies and requirements. Records are broadly defined as papers or other documentary materials. Regardless of their physical form, records are worth preserving temporarily or permanently. The records provide proof of the facility's performance.

Quality Program Plan (QPP)

The facility's Quality Program Plan identifies program requirements, personnel responsibilities, and authorities to provide required quality assurance for operating the facility. The QPP can be organized around the American Society of Mechanical Engineers (ASME) procedure NQA-1 18-point criteria.

Environmental, Safety, & Health (ES&H) Requirements

ES&H design requirements are necessary to (1) prevent or mitigate the consequences of hazards (e.g., toxic or chemical hazards) to personnel off-site, on-site, and local to the facility; (2) provide worker health or industrial safety; and (3) protect the environment. ES&H design requirements are established to comply with safety regulatory requirements imposed on facilities by agencies such as the DOE, EPA, OSHA, and other federal, state, and local agencies that regulate environmental protection, worker safety, and the health and safety of the public.

Mission Requirements

Design requirements critical to the mission are necessary to prevent or mitigate substantial interruptions in facility operation or severe cost impact, or are necessary to satisfy other facility mission considerations. All operations will be conducted in a manner that minimize the impact on the environment.

Design Basis

The principal design requirements for the facility should be established and documented. The following are some documents that can be used to create the design basis:

- Functional and operational requirements
- Design reports
- Engineering design files
- "As-built" drawings
- System descriptions
- Component specifications (when available)
- Master equipment list
- Safety analysis reports/technical safety requirements
- Environmental permits
- Operating and maintenance procedures
- Emergency plan/RCRA contingency plan
- Vendor data

ARTICLE 3.2.3
DIFFERENT TYPES OF DESIGN AND CONSTRUCTION AGREEMENTS

Christopher L. Noble, Esq., and Heather G. Merrill, Esq.
Hill and Barlow, Boston, Massachusetts

There are many ways of structuring the design and construction team for a renovation or new construction project. Various options, including construction management and design/build, are described in Art. 6.2.1, "Construction Documents." In this article, we focus on the traditional system, under which the owner contracts first with an architect or engineer (the "A/E") to design the project and then contracts with a contractor to build it. We focus on the various types of business arrangements that the owner can make with these key members of the team.

A/Es should always be engaged under written contracts, which can take the form of simple purchase orders or more complicated design agreements, depending on the scope of the project and the A/E's services. The most basic form of business arrangement with an A/E involves payment by the hour. If either the project or the scope of the A/E's services is undefined, this may be the only appropriate method of compensation. The hourly charges may be based on flat rates for various categories of personnel or may be expressed as a multiple of salary or "direct personnel expense" (which includes mandatory and customary contributions and benefits). The multiples used by different A/E firms range from two to three times direct personnel expense; two and a half times is the most common.

An hourly fee permits owners to purchase and pay for only the services they desire. However, if the owner is not able to maintain control over the amount of time that is spent by the A/E, it is possible for costs to get out of hand. Conversely, the owner may be tempted to exclude the A/E from some meetings and tasks to save money, when the A/E's participation would actually be useful and efficient in the long run.

When the project and scope of work are sufficiently defined, it may be appropriate to add an upset limit to the A/E's hourly compensation. This system is not preferred by A/Es because it imposes upon them the risk of cost overruns without giving them the opportunity to benefit from efficiencies. Although the owner may believe that it is advantageous to retain the benefit of "savings," this benefit may be illusory because the A/E has no incentive to spend less time than is reflected in the fee limit. Shared savings clauses that are common in construction contracts are virtually unknown in A/E contracts.

A lump-sum fee is often more satisfactory to both the owner and the A/E if the project and the scope of services permit. A lump-sum contract is easier to administer because it does not involve detailed time sheets and justification of hours spent. Progress payments are based on a percentage completion of documents or services. Because of the simplicity of the lump-sum fee arrangement and the relative absence of minor disputes if the fee is appropriately matched with well-defined services, the professional relationship between the owner and A/E is enhanced.

A percentage fee represents a compromise between open-ended hourly charges and a fixed lump sum. Under this arrangement, the A/E is paid on the basis of a percentage of estimated or actual construction cost. Some owners believe that a percentage fee arrangement gives A/Es an incentive to load a project with expensive components, resulting in an increase in the A/E's compensation. In most cases, however, a competent cost estimating and control process minimizes the risk of self-serving manipulation by the A/E.

Methods of compensation are limited only by the creativity and imagination of the parties. Sometimes, for instance, the major variable will be the size of the project, in which case the A/E can be paid a certain amount per square foot. Other contracts employ "blended" methods, such as hourly rates for initial planning services, and full design services are provided for a percentage fee that is transformed into a lump sum when a firm construction cost estimate or guaranteed maximum price is established.

A typical construction contract consists of several parts, including an agreement, general conditions, drawings, and specifications. The basic business terms are usually defined in the agreement, which will state the basis of compensation, the time and method of payment, the time for completion, and such items as special incentives or liquidated damages. As with A/E agreements, construction contracts can be based on several alternative methods of compensating the contractor.

The lump-sum (stipulated sum) agreement is nearly always used when the contractor is to be chosen by competitive bidding. It has the advantage of relative certainty as to price, provided that the drawings, specifications, and other contract documents contain sufficient detail to define fully what the contractor must do and provide. Administration is also relatively easy because payments to the contractor are typically based on the percentage completion of the various components of the work, as shown in a schedule of values agreed upon at the beginning of construction, instead of on the basis of the contractor's actual verified costs. Under this type of agreement, the contractor bears the risk of cost escalation, inefficient work, and miscalculation of cost, but also stands to profit from his or her own efficiency and skillful purchase of materials and choice of subcontractors.

The cost-plus fee agreement is often used when the contract results from negotiation between the owner and the contractor either before or after the completion of detailed drawings and specifications. Under this system, the owner reimburses the contractor for the contractor's actual cost of performing the work and pays him or her either a fixed or percentage fee to cover overhead and profit. Although drawings and specifications are not necessarily well defined under this type of agreement (and it is often used when the exact scope of construction work is not known in advance), defining in detail the types of expenses that are reimbursable as allowable costs and those that must be absorbed by the contractor as part of the fee is very important. Items in question might include certain field office costs, the salaries of certain supervisory personnel, and the cost of correcting damaged or negligently performed construction work. A cost-plus fee agreement may be more difficult for an owner to monitor and administer than a lump sum agreement because of the need to verify hundreds of claimed costs, not just the percentage of completion of the project.

In a typical construction project, it is common for contractors to subcontract a large percentage of the work. These subcontracts are typically bid on a lump-sum basis. Payments to subcontractors under these lump-sum subcontracts constitute reimbursable costs to the general contractor under his or her cost-plus agreement. Therefore, it is not uncommon for a purportedly cost-plus contract actually to consist of several lump-sum components. Indeed, fixing these lump sum subcontract costs by bidding or negotiation provides a significant means of protecting the owner against open-ended cost overruns in a pure cost-plus contract.

When the scope of the project is sufficiently defined, an owner may want a cost-plus contractor to agree that the cost and fee together will not exceed a guaranteed maximum price (GMP). Establishing a GMP creates risks for the contractor similar to those encountered under a lump-sum agreement, without the related opportunity to make additional profits if additional costs are lower than expected. To provide an incentive to the contractor to run the job efficiently and to keep costs down, a *shared savings clause* is often added to a GMP agreement to provide a bonus to the contractor that is equal to some percentage of the amount by which the contractor's actual costs and fee are less than a prenegotiated GMP.

The unit-price agreement is appropriate in specialized situations where the type of work is defined but the quantity of work is not. Contracts for excavation and site work may be let on a unit-price basis because the exact quantity of ledge, for instance, is not known in advance. It is also common to provide that some allowances in an otherwise lump sum contract will be made on a unit-price basis if the work that is subject to the allowances can be easily measured or quantified.

With these basic alternatives in mind, the facilities manager can develop the most appropriate business terms to procure design and construction services. There is no "right" way to structure these arrangements. Any one of them may be best for a specific project.

CONTRIBUTOR

Special thanks to Bart D. Bauer, P.E., of CID Associates, Inc., Boston, Massachusetts for his assistance.

ARTICLE 3.2.4
IMPROVING PRODUCTIVITY THROUGH INTEGRATED WORKPLACE PLANNING[1]

Joe Akinori Ouye, Ph.D.
Gensler, San Francisco, California

Members of the R&D Workplace Performance Consortium*

Many planners who deal with aspects of the workplace, whether they are technologists, organizational change planners, or facility planners, realize that the workplace has to be planned comprehensively. They have come to this realization because they are motivated or driven to make significant changes in group productivity, workplace-related costs, or both, and these changes can no longer be realized by changing only a single aspect of the workplace. Improv-

* Members include 3Com, Adobe Systems, Bay Networks, Cadence Design Systems, Cisco Systems, Hewlett-Packard, Netscape Communications, Octel Communications, Oracle, Sun Microsystems, Tandem Computers.[3]

ing productivity is especially problematic because it is the resultant of many variables, just one of which is the physical workplace. To improve organizational productivity, it is necessary to plan multiple aspects of the organization. This realization has all the symptoms of a classic paradigm shift: The basic definition of the problem has changed, and we are faced with a lack of solutions, techniques, processes, and even expertise to address them.[2]

This article describes an approach to planning workplaces in an integrated fashion by defining, analyzing, and improving key aspects of an organization to improve its performance. It also addresses the issues that characterize this type of problem:

- How do you define better productivity or performance?
- At which level should you attack the problem?
- Who should be involved in defining and solving it?
- What is the role of leadership?
- How do you measure success?
- What are the obstacles to implementation?

This article is based on the efforts and experiences of the members of the R&D Workplace Performance Consortium, comprised of a select group of high-technology firms centered in the Silicon Valley, that have been wrestling with and in some cases dealing successfully with these issues. Although the consortium members share many issues with other "non-high-tech" companies, they are at the forefront in dealing with performance. Because of the competitiveness of their businesses, they are willing to embrace new approaches to the workplace to support or enhance performance or to achieve other strategic advantages.

WHAT IS PRODUCTIVITY?

If we are trying to improve knowledge-worker performance, it is necessary to define it. The definition by the American Productivity and Quality Center is "the relationship between what is put into a piece of work and what is yielded (output)."[4] The application of this broad definition presents a challenge because everyone defines output differently, especially the "knowledge worker" whose work is characterized by intangible, ill-defined, and uncountable outputs, processes, linkage to the company's strategic objectives, performance criteria, and high independence."[5] Finally, the quest for some kind of generic measurement(s) of group performance is a further challenge.

As we reviewed the organizational performance literature, it soon became obvious that each situation is unique and that performance must be defined in terms of the goals and objectives of the specific group versus the individual. In corporations today, generally the group (whether it be called a "department," "group," or "team") is the basic work unit. Management goals and objectives are defined by groups, and because of the interest in quality management techniques such as process improvement, performance is increasingly measured by groups. The American Performance and Quality Center recommends that performance be defined in terms that are most relevant to the groups' outputs and that a "family of measures" be used. "To determine how well an organization is functioning, its leaders must not restrict their focus to just one indicator—one individual, one department, one product, one process, one expenditure, or one measure of success. They must examine an entire family of measures."[6] Using this approach, performance includes a wide group of indicators that may include the excellence of the product or service, alignment with the budget, keeping on schedule or achieving rapid time-to-market, and customer satisfaction. For example, the performance of a software engineering group may be based on the perceived quality of the software, the group's ability to stay within the budget, timeliness (keeping on schedule), and customer satisfaction.

WHAT AFFECTS PERFORMANCE?[7]

The following elements directly or indirectly affect individual performance, and by extension, affect team performance ranging across the spectrum of the workers' physical and social environment:

- Personal factors
- Organizational/management factors
- Process-related factors
- Technology factors
- Physical environment

Personal Factors

- Technical competence in performing the job
- Motivation to work, especially because "the work of knowledge professionals happens inside their heads"
- Compensation, recognition, leadership, physical environment, and just about anything else that affects mood
- Work Strategies: taking initiative, networking, self-management, teamwork effectiveness, leadership, fellowship, perspective, show-and-tell, and organizational savvy

Organizational/Management Factors

- Participation in determining how employees are managed and the way their workspaces are arranged
- Independence and initiative in doing work
- The lack of obstacles to effective work, for example, adequate resources and tools, a clear mandate, and lack of interference
- Clear performance expectations and feedback to keep workers on track and from going on unproductive tangents
- Few and focused meetings and the lack of interruptions to concentrated, focused work
- Compensation and incentives

Process-Related Factors

- More effective and efficient processes, supported by the right kinds of technologies

Technology Factors

- Production tools: computers, appropriate software, printers, scanners, and copiers
- Communication tools: telephone, fax, modem, networks, and videoconferencing

- Automation: computers and other technology to duplicate and even enhance processes performed by workers; for example, workflow software that automatically pulls and sends information as necessary in a work process

Physical Environment

- Spatial comfort: the most important physical factor that affects performance (includes the amount of workspace, adequacy of storage, furniture, and equipment arrangement, furniture comfort, and ergonomics)
- Control of general office noise levels, distracting conversations, equipment noise, and other audible distractions
- Privacy, including phone privacy, visual privacy, and freedom from interruptions
- Air quality: air movement, air freshness, ventilation, odors, humidity, and warmth
- Lighting: brightness, absence of glare, colors, and daylighting
- Support space: availability and adequacy of quiet rooms, large and small meeting rooms, resource centers, and lounge areas

INTEGRATED WORKPLACE PLANNING APPROACH

Because performance is affected by many factors in the workers' environment, it is necessary to design the workplace comprehensively by considering the various factors as a system (see Fig. 3.2.4-1). This is not a new idea. More than 35 years ago, Eric Trist and Fred Emory coined the *sociotechnical systems* (STS) approach, which "considers every organization to be made up of people (the social system) using tools, techniques, and knowledge (the technical system) to produce goods or services valued by customers (who are part of the organization's external environment). How well social and technical systems are designed with respect to one another and with respect to the demands of the external environment determines to a large extent how effective the organization will be. Thus every organization is a socio-technical system, but not every organization is designed using the principles and techniques that have come to be a part of the socio-technical systems approach."[8] Much of the efforts of STS proponents has been to improve work-group performance in manufacturing settings and has not reached out extensively into knowledge-worker performance.

The STS perspective has and is being used by researchers such as Professor Wellford Wilms of the UCLA Graduate School of Education and Information Systems to understand how teams perform effectively in settings ranging from the new United Motors plant in Fremont, California, to Hewlett-Packard's R&D laboratories.[9] Wilms used ethnographic research methods, integrated his team into various organizations, worked alongside the employees at their jobs, and made detailed observations of what people said and did. He found that although some teams manage to work in spite of the context, whether it be the management culture or tools and processes, teams that thrive do so because their social and physical environments provide nurturing conditions for them.

Professor Franklin Becker at Cornell University describes a similar approach as the "Integrated Workplace Strategy," a way of "rethinking how work was being done and creating organizational, technological, and workplace solutions that supported the way work could be done effectively."[10] This approach views the workplace as an "organizational ecology" that "considers these elements of a workplace system as part of an integrated workplace strategy that defines a total workplace in two distinct ways: through the scope of the physical settings considered and through the social process used to plan and manage—and link—the physical settings through time."[11]

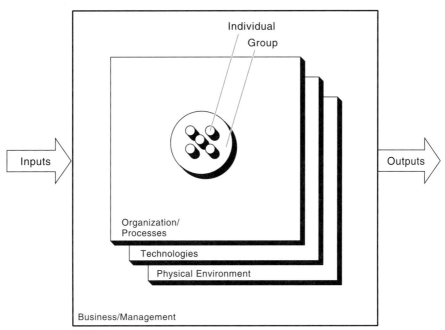

FIGURE 3.2.4-1   Workplace as a system. (*Courtesy of* The Competitive Workplace, *by Joe Ouye and Jean Bellas.*)

And so, settings (and work settings in particular) have been long viewed as systems of people, things, and their relationships that need to be planned as systems to achieve the goals of the group—although successful applications of this approach have been "spotty and results mixed."[12] But recently, as corporations have continued to push for greater performance and more value from their assets, there has been a surge of interest in considering the place of work, the real estate and facilities location and characteristics, as an important part of the ecology or system. This approach is especially relevant as corporations explore new workplace approaches by rethinking how, when, and where people work.

NEW INTEGRATED WORKPLACES

Planning an integrated workplace can result in solutions that change the physical work setting and also parameters such as behaviors, processes, technologies, and other resources to achieve the goals of the planning effort. Each of the resulting solutions is idiosyncratic. There is no single approach or ideal integrated workplace or "office of the future." Any such endeavors are doomed to fail because approaches must vary to suit different corporate intents and different group goals, objectives, and constraints. But these solutions do share common themes:

The Personal Workplace

Specific characteristics of the personal workspace are driven by the functional needs of the occupant for high performance and by the informal organizational culture. This is contrasted to "space standards," which define personal workplaces by organizational or group status. In addition, the change of behavior is considered a "design variable," for example, promoting behaviors for an open office environment: lower voice levels, no use of speakerphones, using other support areas for confidential conversations or focused work, or redefining what in fact is really "confidential" or "private."

Multiple Work Settings

More focus is placed on supporting the productivity of individuals and groups in areas other than the personal workplace. Information technologies continue to support workers with information, whether data or voice, whenever and wherever the worker is working.

Informal Interaction.   Promoting informal interaction recognizes that productivity results not only from individual work and formal interactions but also that really important gains often result from spontaneous, informal interchange.

Teaming and Remote Teaming.   Groups are changed from hierarchical or individually based organizations to team-based organizations, and team rooms, project rooms, and more meeting areas of varying levels of formality or informality are provided.

Alternative Workplace Strategies

Although telecommuting, hoteling, office sharing, and virtual officing are appropriate mostly for workers such as consultants, auditors, and salespersons who spend a great deal of time out of the office, most corporations are exploring or piloting these radical strategies.

Whole Life Needs

Supporting high performers means making their lives simpler and better. Free soda, abundant lunches, espresso and cappuccino bars—the attention paid by these companies to make their workers happy sometimes seems to be at the point of spoiling them.

THE INTEGRATED WORKPLACE PLANNING PROCESS

Planning the workplace for performance and dealing with all of the disparate elements as a system is a radical departure from simply "planning a space" or dealing with each element separately. Going through a process that systematically considers the business context and workplace elements (as outlined in the following paragraphs) is highly recommended. The process necessarily includes tools and techniques from diverse fields, such as organizational analysis, sociotechnical systems analysis, strategic planning, and facility planning.

1. Establish Goals and an Approach to the Project and the Project Team

Meet with the overall leadership and, second, meet with the specific group to define the overall direction and goals, the planning approach, the schedule, and to identify the appropriate team members. This step is important and difficult because of the scope of possible changes and the elusive level of the problem.

The composition of the planning team is often the most important early decision. It must include individuals whose scope of responsibilities encompasses the range of possible workplace changes. Its membership should involve leaders who have the authority to change business organization and processes, information technologies, human resources, and real estate and facilities. Often, additional expertise such as corporate risk, taxation, and purchasing should be engaged.

2. Define Desired Organizational Goals and Outputs

This may be based on an organizational business plan, interviews of key managers, or a workshop, and should focus on the performance goals and objectives of the organization. It is not unusual for the organizational business plan to be nonexistent, far too sketchy to be of much use, or simply outdated, and it must be defined in this process. For many fast-moving high-tech companies, the rule "If it's written down, it's out-of-date," usually holds. But at the lower organizational levels, such plans are increasingly outcomes of continual quality management or other similar processes, and this process can capitalize on them. Existing organizational goals and objectives have the advantages of (probably!) prior buy-in and historical metrics, which can be the basis of measuring organizational performance.

3. Understand and Assess the Existing Organization and Workplace

The existing organization can be understood by reviewing its structure (the organizational chart), its business processes, and its job designs.[13] Process mapping of groups, tasks, information flows, skills, and resources is a way to gain this understanding. A tour of the existing space, a review of floor plans and technology plans, and interviews with key individuals will provide information about the workplace and other softer aspects such as communication patterns, morale and motivation, and management style. A group can be understood by using an organizational profile map that maps the characteristics of a group against a range of organizational, technological, and work-setting variables. The resulting group profile can be readily compared against other profiles within the organization.

4. Identify Obstacles to Improved Performance

Critique the existing organization and workplace by identifying obstacles to performance goals and objectives. Obstacles may be caused by any elements of the organization or workplace. Again, the proper level of the problem has to be respected. It may be true that the president of the company is an "obstacle," but changing that condition is probably outside of the scope of your mandate! Most obstacles may also be perceived as opportunities. For example, the obstacle "There is a lack of cross-group interaction" suggests the opportunity "Improve cross-group interaction."

5. Develop, Evaluate, and Select Workplace Options

Create workplace options that capitalize on the opportunities identified. Some workplace options may resolve multiple obstacles and capitalize on multiple opportunities. For example, a central meeting hub could support more serendipitous interactions, informal group meetings, and meetings of the entire group. The options should be evaluated using the criteria previously developed. Cost analysis should include harder costs such as technology facilities/equipment and services, and also softer costs such as increased work hours due to process improvements. These latter costs usually have to be estimated, and if so, should be discounted very conservatively in anticipation of "doubting Thomases." Other costs, such as tax implications, training costs, and insurance/corporate risks that are not normally considered may have to be included.

6. Implement the Planned Changes

The implementation plan should identify key management and organizational support, training, and personnel/human resource issues. The implementation plan establishes the appropriate steps, specific tasks, resources, and timing. The changes may be piloted with a smaller sub-group to evaluate its effectiveness before rolling it out to the entire group, or in some cases, it may be a full implementation program. A communications plan to inform and involve workers in the change is critical. The failure of many efforts in workplace change is the failure to communicate, rather than the failure to develop a good solution.

7. Evaluate the Results

Monitor and evaluate the results of the workplace changes. The evaluation should consider the impacts of the changes on the group's performance, the effectiveness of the workplace in supporting performance, and satisfaction with the planning process itself. This step is expanded in, "Measuring Performance."

OTHER CONSIDERATIONS

The integrated planning process has some unique characteristics that make the planning processes very "wicked"[14] and fraught with difficulties:

Redefining Performance

The parameters for defining "better performance" vary from company to company and from group to group. As John Igoe, Director of Real Estate and Property Development of Octel notes, "Octel is more interested in performance than sheer square footage reduction. . . . The goal is not to cut space usage; the goal is to make the employee more productive." In Octel's case, the driving force is the need to be first in the market, and the time to achieve this has decreased dramatically in recent years. As Igoe relates, "The life cycle of a typical product in the '50s was 10 years, in the '80s 5 years, now it's 6 to 8 months." Frank Robinson, Vice President of Corporate Real Estate and Services for Tandem Computers, has similar concerns. "We're faced with the challenge of reducing time-to-market for our products. The time frame that used to be 24 to 36 months has been pushed back to 18, 12, sometimes 9 months. Unless the time taken to assemble the group in their workplace shrinks too, we (i.e., Corporate Real Estate and Facilities) become a stumbling block."

In the case of Tandem Computers' sales group, better performance equates to more time spent with the customers. "The key for our sales and service people [formerly] was to bring a customer back to our computer showroom to sell hardware," says Tandem's Frank Robinson. "Today the customer doesn't care what the box looks like. He has a problem. He needs a solution to the problem. Our people are spending far more time out with the customers understanding what the problem is. The offices have changed not because we dreamed up a way to change their world, but from our saying 'Let's take a look at how they're doing their jobs today.' "

Debra Engel, Vice President for Human Resources and Corporate Services at 3Com, is redefining and extending performance beyond the normal realm of specific group outputs to include more sophisticated meanings. An example is the concept of *organizational capability,* which is the ability of the organization to incorporate and instill superior values, superior experience and knowledge, superior learning, and superior processes and systems.

Organizational capability is another edge to differentiate companies from their competition. It is an aspect of visionary companies that "build an organization that fervently preserves its core ideology in specific, concrete ways. . . . Visionary companies translate their ideologies into tangible mechanisms aligned to send a consistent set of reinforcing signals. They impose tightness of fit, [and] create a sense of belonging to something special through practical, concrete items."[15]

Defining the Level of the Problem

Integrated workplace planning can include high business and organizational planning and change efforts, such as reengineering and business process improvement. Although these possibilities should not be automatically excluded, the danger is that they may be beyond the authority of the client group or project mandate and the capabilities of the planning team and may diffuse the efforts of the planning team. But ongoing efforts in business planning and change can be leveraged with great effectiveness, especially because the physical aspects of these changes are very often overlooked.

Setting the Goal of the Process

Keep the goal of the workplace planning in mind, which is, most of all, to support the group's mission or business plan. It is not to cut costs or even to develop an alternative workplace strategy, although they may result! As Igoe of Octel observes, "Octel is more interested in performance than sheer square footage reduction. The goal is not to cut space usage, the goal is to make the employees more productive." The planner should not have any preconceptions about the solution, because the solution may be to do nothing or to improve the social aspects of the group, without doing much to technologies or the workplace. Garcia of Adobe concurs, "Facilities people sometimes want to be evaluated on the basis of costs. A lot of facilities people don't spend enough time with the rest of the organization, besides simply saving money on real estate or operating the facilities because they're not participants in some of the other kinds of decisions. If you spend a large portion of your time thinking about increasing performance, at the end of the day people will look at you as having a contribution to solving these problems."

Planning for Change and Unpredictability

Most high-tech companies (maybe all!) are faced with increasing rates of change and turbulence in their business environments. They in turn are forced to react with corresponding speed and unpredictability. Groups and teams are organized and reorganized; groups grow at high but unpredictable rates; and workers are moved constantly (as much as one and a half or more times per year, based on the total number of moves divided by the total number of employees—what is known as the "churn rate"). Abe Darwish of 3Com, a company that has been growing at 30 to 40 percent per year recently and experiencing 110 percent churn, recommends, "[You have to] anticipate the future and prepare for a high degree of churn. Design your facility for change. You can't rule out change; you have to be prepared, from operations to design; you have to be ready from the beginning." All this is expensive and unproductive. It is expensive to change furniture, technologies, and carpets. In some companies, it may cost as much as $2,000 to $3,000 per employee per move! Because it may take from one to two days to accomplish moves, the costs of unproductive time can be even higher, not to mention the time it takes for the workers to settle into their new spaces and become comfortable and effective again. All this, notwithstanding impacting group performance in time-to-market and scheduling. Hewlett-Packard, San Diego, has reduced the average time per move from 11 hours to just about 3 hours at one of its new R&D buildings. The productivity saving is far more because the disruption to the relocated engineer is much greater—in the range of 40 to 60 hours per move! Given that the churn rate is about 300 percent per year and that there are 400 workers, the saving of productive time is about 48,000 to 72,000 hours per year. This was achieved by "universal planning" (i.e., planning and designing space for changes by anticipating them). Workstations, furniture and equipment, technologies, and other resources are designed to accommodate most future possibilities and are standardized (or "universal"). Therefore, moves are reduced to "box" moves rather than reconfiguring the workplace and all of its components.[16]

Participation

No one likes to be "planned at" or even "planned for," especially when the plans involve changing important aspects of their lives. IBM summarily notifying some of their workers that they no longer had personal space and had to work out of their homes or share hoteling space is an example. Although successful from the viewpoint of reducing real estate costs, this action had far more damaging costs in terms of human relationships. Most significant workplace changes involve meaningful changes in a worker's life—management styles, evaluative methods, communication patterns, work location, information technologies, hardware/software, and personal "territory." Darwish of 3Com is convinced of the importance of employee participation: "You enhance your performance if the business unit fully participates in the solution. You can't use one solution and apply it universally. Part of performance enhancement is getting the user involved in space, information systems, and human resources planning. What that means is that we, as facilities solution providers, can't expect to provide a solution and enforce it universally. You have to customize it. You have to have buy-in." Although businesses may not subscribe to the ethical view that it is the users' "right" to participate in decisions that affect their lives, those businesses should heed the advice that employee participation informs the process for a better solution and also makes it far more likely that the participants will be motivated to make the necessary changes for a successful outcome.

Thoroughness

When you start changing parts of a complex workplace ecology or sociotechnical system, many adjustments may have to be made to stabilize the new system. It is important to thoroughly consider all of the possible factors that may be upset by the changes. For example, in the case of a telecommuting program at a major interstate bank, the payroll tax consequences had to be seriously considered because many of the telecommuters would be working at their homes, which are located in a state other than where the corporation pays payroll taxes.

Cultural Change

The fact that people do not like to change should not be underestimated. Change is painful to most, even if they know it might be better for them! As mentioned before, participation in the planning of the change makes a big difference. Full participation in the change process is ideal, but it is usually not possible for all affected workers to participate. And so constant communication to all involved is necessary so that they can at least understand why the workplace is being changed, how it is being changed, and the extent and timing of the changes. As Tandem's Robinson cautions, "If you view it as a real estate project, you're only going to get so far. You have to view it as a workplace solution project. Real estate is the easy part, the technology is more difficult, and the human behavior side is the hardest. Never take for granted [the resistance to] behavior changes."

Leadership

Leadership from the top can make all the difference in changing the workplace. Leadership is not telling workers what to do but rather motivating and showing them what to do. In other words, to "walk the talk." When Pacific Bell wanted its executives to give up their private offices in order to increase the density of the headquarters, there was little enthusiasm until the CFO decided to "walk the talk" and move out to a cube. After his move, it was a race between the other executives as to who would be the next out on the floor. Needless to say, this was an important symbolic gesture because spaces of other workers were being changed to shared, teamed, or hoteled spaces. "Leaders are no more or less than a manager of transitions."[17]

Measuring Performance

The measurement of performance and the contribution of the workplace to performance have been perceived as major stumbling blocks by the R&D Workplace Performance Consortium. If you can't measure the improvements, it is very difficult to convince management to invest monies in the change of the workplace when those funds could be spent for more marketing or R&D. As Ann Bamesberger, manager of planning and research for Sun Microsystems, observes,

> Where [achieving buy-in] gets tricky is proving that these changes have actual value. Gaining credibility. Nobody expects direct causal linkages. We should expect some intervention that we are responsible for, [which] is at least intuitively linked to some enhanced business outcome.

It would be ideal to have measures of group performance that can be taken from readily available information (i.e., information that is collected in the course of doing business anyway). As an example, Eric Richert, Director of Design and Construction of Sun Microsystems, uses the proposed "Information Productivity Index" (IPI), which is proposed as an index of the value added of information systems to a company. The IPI is defined as follows:

$$\text{Information Productivity Index} = \frac{\text{net income} - \text{costs of equity}}{\text{total costs of sales, general operating and administration and R\&D}}$$

Unfortunately, we remain far from agreement on any similar measure for productivity and workplace effectiveness.

In the absence of such an ideal measure, we have been exploring a technique developed by the American Productivity and Quality Center (APQC), the family of measures approach described earlier. These measures can be defined quantitatively or qualitatively. One is not necessarily better than the other. Logically, all quantitative measurements ultimately have to translate to a qualitative scale to have meaning. For example, how much "better" is beating the budget by 10 percent versus 15 percent? This can be determined only by assigning a value scale to being 10 percent and 15 percent under the budget.

Performance can also be defined in terms of direct and indirect measurements. Some desired characteristics of products or outputs of a group work process can be measured directly, or characteristics of the group, which correlate to performance, could be measured to determine performance. For example, the lack of interruptions and distractions may be an indicator of the performance of knowledge workers who require "flow."

Correlating improvements of organizational performance to specific changes of the workplace is another difficulty. It is clear that because workplace performance may be the result of many kinds of changes (new management, processes, technologies, desks, chairs, etc.), as we described, correlating specific workplace changes to performance is impossible because there are too many intervening variables. But because we are assuming that performance results from the interplay of many factors, perhaps the point is to see how and if performance was improved as a result of applying all these factors, instead of trying to understand the impact of an isolated factor.

Performance and workplace improvements can be visualized as lying along two axes. The first axis measures the change in performance, using the APQC approach. The other axis quantifies the group's assessment of the performance of discrete characteristics of the workplace. If group performance and perceived performance of the workplace both improve, we can be fairly confident that the changes as a whole contributed to improve performance. For example, if too frequent interruptions were identified as a barrier to better performance, and if changes were made in the workplace to deal with it, and the group perceived that there were indeed fewer interruptions, then those changes probably contributed to improved performance.

Actually measuring performance and workplace improvements resulting from workplace changes is a rarity. This is not because of lack of interest but rather due to the inexperience or lack of expertise in performance and workplace measurement techniques of those dealing with workplace changes. It is still regarded by some noted researchers in the field of workplace planning as "not relevant" because of the measurement difficulties and the improbability of correlating performance to specific workplace changes.

These researchers' views notwithstanding, a few companies have had success dealing with this difficult area. At Hewlett-Packard, as part of their Headquarters Renewal Project, the performance and satisfaction with changes in the workplace were evaluated using pre- and postmove surveys of 122 workers.[18] The following changes were included:

- Increasing the overall density and team spaces by sacrificing independent, personal space
- Updating and thoroughly utilizing integrated information technology and taking advantage of state-of-the-art adaptable workstations

The results are convincing that both performance and satisfaction with the workplace environment were significantly improved.

Performance was measured using three indicators (ranging from "excellent" to "poor") that rated the group's performance in the following areas:

- Producing quality services and products
- Producing services and products in a timely manner
- Producing services and products within budget

The perceived performance in all three parameters increased in the range of 30 to 32 percent.

Satisfaction with the workplace was surveyed in the categories of personal workspace, meeting spaces, and technology. The perceived change in satisfaction with the workspace varied from 0.25 on a scale of 3 to 5, but respondents were uniformly more satisfied with the changed workplace. The greatest perceived change was with the meeting spaces (39 percent), followed by personal workspace (38 percent), and then technology (31 percent). Among the specific indicators, improved ergonomics and work surfaces were rated significantly better than the others, whereas the adequacy of the workspace was last.

These results are not surprising, because the main business change driving the workplace changes was the emphasis on facilitating teamwork and cross-group synergies. Increasing teaming and meeting areas were emphasized at the expense of individual workspace areas. The quality of the furnishings, equipment, and technologies of the individual workstations were significantly improved.

We can conclude that performance was significantly improved, along with satisfaction with the workplace. Although we cannot say certainly that the workplace changes led to the improvement in performance, we are fairly confident that they did because these changes were specifically addressed toward eliminating barriers to better performance.

Challenges to Implementation

Improving workplace performance in this integrated fashion is difficult to achieve in the corporate environment. First, organizational issues are not usually tackled in an integrated fashion. They are not defined as problems in that way, nor are integrated solutions sought. Second, an integrated approach to organizational performance does not typically have a corporate "home" or champion. For example, the information technology group will address it as an information problem, human resources will address it as a personnel issue, and real estate/facilities will address it as a physical issue. Third, we do not have a good way of defining problems and solutions in a way that cuts across disciplinary boundaries. These problems can be solved to some extent by working with cross-functional project teams to integrate and leverage corporate-wide initiatives, for example, linking technology initiatives with initiatives to reduce real estate occupancy costs. But a hopeful sign is that some of the leaders in these disciplines are beginning to understand, appreciate, and even advocate the necessity for an integrated workplace approach.

Case Studies

The following paragraphs summarize three case studies from a forthcoming book by Joe A. Ouye and Jean Bellas, entitled *Competitive Workplace*. The book describes how to transform the workplace into a competitive tool to lower costs, accelerate the time-to-market of products and services, infuse companies with creativity, and increase the flexibility and productivity of the organization:

AT&T Global Services.   Vice President of AT&T Global Services Richard Miller had a vision of transforming his sales organization by streamlining his account management processes. The critical needs were to increase the exposure of the sales team to both existing and potential customers and to increase the synergy between his team members.

This vision was achieved by changing team organization and processes and supporting them with new technologies and physical workplaces. Teams were consolidated from multiple locations, so that they were no longer separated by account and function. There were two major profiles of workers: those comprising the administrative team, who worked from the office 100 percent of the time and represented 30 percent of the team, and those making up the sales and technical teams, who worked from the office for 40 to 75 percent of a typical workweek. The administrative team led the other teams, energized the sales team, and set quality standards for the marketing responses. The sales and technical Team sold products and services primarily by "face time" with the customers. The team members were out of the office for 25 to 60 percent of the time. Between client meetings, they needed a place to "touch down" to generate new appointments, follow up on old appointments, and get recharged with ideas and energy.

The space of the new sales centers was reduced by 50 percent by reducing office sizes by 30 percent and by having the sales team share offices at a ratio of one workstation to every three salespersons. A portion of the new space was dedicated to encouraging interaction among staff members—a "Café Break Out" functions as the "kitchen," where staff members can informally meet for coffee and casual conversation. The space is peppered with more than 40 percent more small- to medium-sized meeting rooms (see Fig. 3.2.4-2). All meeting areas, including the Café Break Out, are equipped with network connectivity for group work and with overflow space for individual work.

The prototype is being replicated through the Creative Workplace Strategy initiative, consolidating almost 200 sales centers throughout the country. Since consolidating, the Rick Miller group produced the highest revenue in AT&T in both 1997 and 1998. The group credits this success directly to meeting the needs of the sales team and the new processes, technology systems, and physical workplaces that responded to those needs.

Lucent Technologies' Project Atlas.   Lucent Technologies' Project Atlas is an ambitious effort to "transform their R&D culture and workplace." Launched shortly after Lucent Technologies split from AT&T, Project Atlas (see Fig. 3.2.4-3) was to enhance the productivity of the R&D team and, at the same time, reduce real estate costs of 12.5 million square feet of R&D space to fund the improvements.

The project assembled the following information:

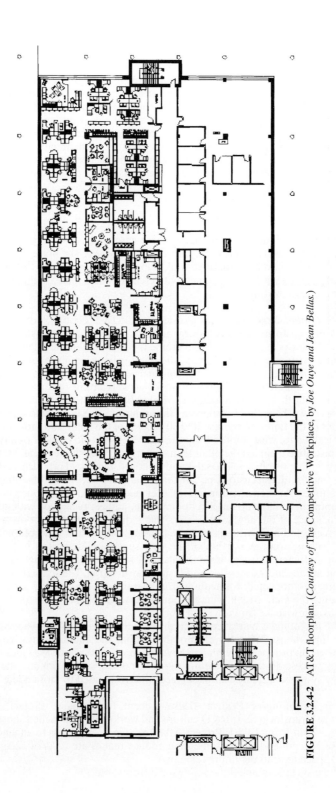

FIGURE 3.2.4-2 AT&T floorplan. (*Courtesy of* The Competitive Workplace, *by Joe Ouye and Jean Bellas.*)

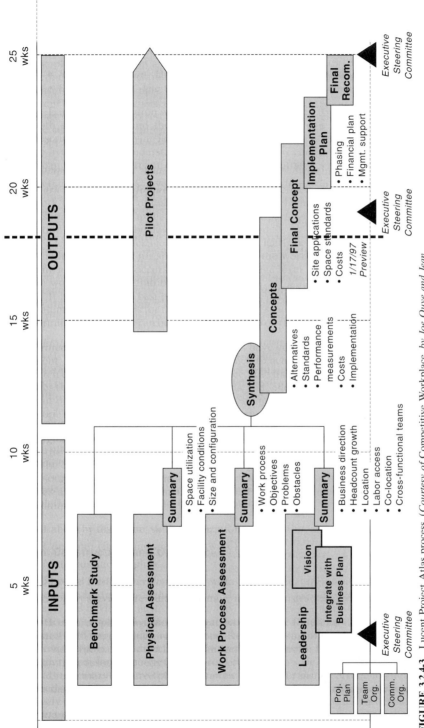

FIGURE 3.2.4-3 Lucent Project Atlas process. (*Courtesy of Competitive Workplace, by Joe Ouye and Jean Bellas.*)

- A benchmark of Lucent competitors' use of space
- An assessment of existing conditions
- Discovery of user needs by analyzing business drivers and work processes
- Development of planning and design concepts and evaluation of their costs and benefits
- Field testing in multiple cities
- A full-scale, phased-implementation program

Workplace planning was conducted concurrently with the planning of new supporting information technologies comprising the "Communication, Computing and Collaboration" project.

The solution supports both individual and group productivity because it was found through focus groups of R&D engineers that projects typically go through cycles of both intensely individual and group collaborative work. R&D engineers are housed in shared offices (see Fig. 3.2.4-4) that can be also used as collaborative team areas. The shared offices are enhanced with liberal amounts of team areas and conference rooms. Workers from multiple groups can gather and meet casually at strategically placed "Café Lucents."

The new workplace optimizes companywide resources by using advanced communication, computing, and collaborative technology. As a side benefit, the new workplace showcases Lucent products, performance, and company image.

The new workplace dramatically reduced operating expenses. The projected cost saving on a 10-year net present value basis is $200 million. This consists mostly of reducing the cost of relocating team members by 90 percent by standardizing workstation technology and furniture, and of 20 percent savings in space costs.

Prudential Customer Services. Faced with dramatic changes of its business environment, Prudential was forced to rethink its way of doing business and radically change the delivery of customer services. The solution was to make fundamental changes in the customer services processes, supporting technologies, and the workplace.

The customer services strategy consolidated more than 25 existing service centers into 5 major sites to reduce redundancies among the service centers and to improve service delivery. Local presence was no longer essential, and many customer service tasks were routine and could be automated. An investment in technology allowed the company to consolidate services, greatly expedite service delivery, and significantly reduce the costs of the customer service function. In the new model, customer service agents have higher levels of technical skill, education, and training so that they can deliver increased levels of information to the customers, including future product offerings beyond claims processes, such as various loan programs, investment consulting, and portfolio management services.

A real estate strategy was coupled with the customer service strategy to reduce overall occupancy costs. The strategy consolidates existing service centers into five existing locations and disposes of the excess properties or gets out of the excess leases. The new workplaces (see Fig. 3.2.4-5) are highly interactive, extremely efficient, and supportive. The new centers include six Imaging Centers to receive and digitize all incoming mail, creating a complete electronic database of customer communication. The data is then accessible from the centralized database by the customer service representatives while they are on the phone with customers.

Ultimately, the project involved more than 30 phases, each averaging about 250,000 square feet. More than 1.5 million square feet were renovated within the first 14 months. Seven thousand workers were relocated into their new homes.

As a result of this workplace transformation, the performance of the workers has improved 40 percent in the daily number of claims processed. Prudential is also continuing to grow into a financial services organization with marketing continuing to feed new services into the Prudential system. The initiative was extensive and expensive, but it positioned Prudential to absorb growth and to lead the financial services industry by positioning its products, the quality of its service, and a streamlined delivery system.

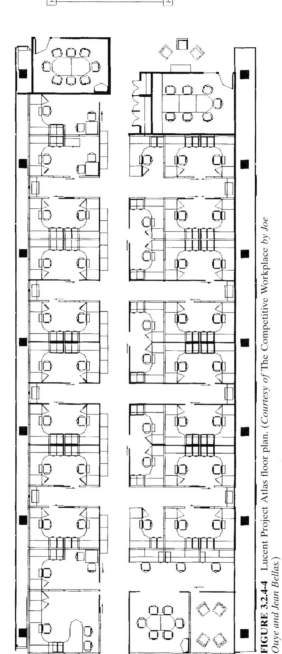

FIGURE 3.2.4-4 Lucent Project Atlas floor plan. (*Courtesy of* The Competitive Workplace *by Joe Ouye and Jean Bellas.*)

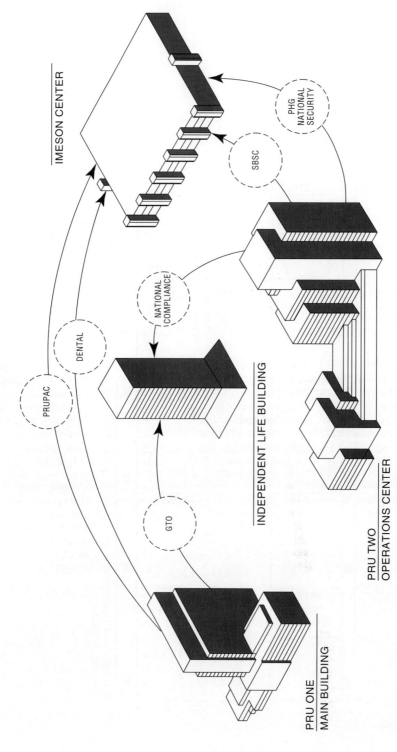

FIGURE 3.2.4-5 Prudential business unit relocations and restacking. (*Courtesy of The Competitive Workplace, by Joe Ouye and Jean Bellas.*)

The R&D Workplace Performance Consortium

The R&D Workplace Performance Consortium pools the resources and knowledge of managers and researchers to investigate the role of the workplace in supporting knowledge-worker performance. The consortium is comprised of leading edge, high-technology companies in the Silicon Valley and San Francisco Bay Area.

NOTES

1. This article is based on Working Paper No. 3 of the R+DWPC, "The Workplace as a Performance Tool," Summer 1996.

2. T. Kuhn, *The Structure of Scientific Revolutions* describes these characteristics of paradigms, University of Chicago Press, Chicago, IL, 3rd Edition, 1996.

3. Members of the Consortium were interviewed by Rob Wittig and include Abe Darwish, Director of Facilities and Site Services of 3Com; Angelo Garcia, Director of Real Estate and Facilities of Adobe Systems; John Igoe, Director of Real Estate and Property Development of Octel Communications; Ann Bamesberger, Manager, Planning and Research, and Eric Richert, Director of Corporate Design and Construction of Sun Microsystems; and Frank Robinson, Vice President Corporate Real Estate and Services of Tandem Computers. Jean Bellas, President, and Paul Heath, Sr. Associate of SpAce SYSTEMs, contributed reviews and helpful comments.

4. APQC. *The Master Measurement Model of Employee Excellence.* American Productivity and Quality Center, Houston, TX, 1990.

5. J. Belcher, Jr. *Productivity Plus,* pp. 155–156, Gulf Publishing Company, Houston, TX, 1987.

6. C. G. Thor, *National Productivity Review/Summer,* p. 112, John Wiley & Sons, Inc., New York, 1995.

7. Much of this section is from *Summary of Productivity Literature,* Working Paper No. 1, RDWPC, March 10, 1994.

8. W. C. Hauck, and V. R. Dingus, *Achieving High Commitment Work Systems: A Practitioners Guide to Sociotechnical System Implementation,* p. 4, Industrial Engineering and Management Press, Norcross, GA, 1993.

9. W. Wilms, Professor of Education and Information Science, UCLA, presentation at the RDWPC meeting of September 8, 1994. Also will be subject of soon-to-be-published, *Restoring Prosperity,* by Random House.

10. F. Becker, *Workplace by Design,* p. 141, Jossey-Bass Publishers, San Francisco, CA, 1995.

11. Ibid., p. 14.

12. W. A. Pasmore, *Designing Effective Organizations: The Sociotechnical Systems Perspective,* p. 155. John Wiley & Sons, New York, 1988.

13. G. Rummler, and A. Brache, *Improving Performance: How to Manage the White Space on the Organization Chart.* Jossey-Bass Publishers, 2nd Edition, San Francisco, CA, 1995. This book is an excellent guide to a structured technique for understanding organizational processes.

14. "Wickedness" is used in the sense of Prof. Horst Rittel's "wicked problems," which are ill-defined and ill-behaved compared to "tame problems." See H. Rittel, "On the Planning Crisis: Systems Analysis of the First and Second Generations." *Bedrifts Okonomen* **8**: October (1972).

15. J. Collins, and J. Porras, *Built to Last: Successful Habits of Visionary Companies.* pp. 135–136, Harper Business, New York, 1997.

16. R. Jakubowski, Project Manager, Hewlett-Packard; and J. Bellas, President, SpAce SYSTEMS, presentation to the R+DWPC, March 16, 1995.

17. Pasmore, op. cit., p. 148.

18. T. Mogi, "Impact on Productivity and the Workplace," Hewlett-Packard Headquarters Renewal, internal white paper report, Facility Technics, Inc., Oakland, CA, May 30, 1996.

ARTICLE 3.2.5

FACILITY LIFE-CYCLE MANAGEMENT TOOLS

Bruce Kenneth Forbes
ARCHIBUS, Inc., Boston, Massachusetts

INTRODUCTION

Facility life-cycle management is not a new application area of facilities management. In fact, many consider it one of the core competency areas in which all facilities professionals should be knowledgeable. Driven by downsizing, outsourcing, reengineering, mergers, acquisitions, e-commerce, and the virtual office, effective facility life-cycle management is discussed in corporate boardrooms around the world. Directors are asking facility professionals to educate them about the costs and opportunities of effective facility life-cycle management and the total costs of ownership (TCO) of corporate assets. Life-cycle costs are constantly evaluated and presented regularly to directors worldwide. Directors are realizing that a good understanding of the management of facility life cycles can mean the difference between a profitable quarter and a very profitable quarter. Thus, the facility professional's access to directors and corporate boardrooms is becoming more and more prevalent.

The depths of facility life-cycle requirements vary, but most boards have directed the facility professional to deal with such areas as capital consumption, energy use, technology deployment and access, asset use and renovation, capital projects, infrastructure management, systems and material replacement programs, and financing.

In an integrated facility life-cycle management environment, FM life-cycle benchmarking and FM life-cycle best practices analyses address numerous areas of concern. These are 10 of the more common areas explored:

- *Capital investment costs.* Initial construction costs, cost of money, cost of delays, opportunity costs
- *Repair and replacement costs.* Costs of repair and restoration activities (people, time, and materials)
- *Maintenance costs.* Predictive, preventive, demand, proactive (people, time, and materials)
- *Renovation, alteration, and improvement costs.* Costs of changes required to meet the changing needs of a facility's users and its customers
- *Lost revenue.* Costs of not undertaking appropriate life-cycle activities
- *Basic operating costs.* Costs of the day-to-day facility activities (heating, ventilation, air-conditioning, electricity, telecommunications, IT infrastructure, and utilities)
- *Occupancy and use costs.* Costs of people, materials, and supplies required to perform and/or enable normal facility management and occupancy functions
- *Financing costs and project liabilities.* Costs of equity and costs of capital (bridge financing, short-term and long-term credit expenses and opportunities)
- *Disaster containment and contingency planning.* Costs of disaster management and containment, as well as contingency planning (backup facilities and alternative delivery systems)

- *Technology realization costs.* Costs of ensuring competitive technological access and empowerment for the users and customers of a facility (how to enable access of the "latest and greatest" technological innovations)

Strategic capital planning and management are other key elements of effective facility life-cycle management. Facility life-cycle management optimizes the performance of spatial and nonspatial assets by deploying condition assessment surveys, optimized maintenance management, regulatory compliance, risk management, project bundling, and effective capital program justification and allocation. In their striving to create customer value, facilities professionals who participate in life-cycle management programs address such areas as life safety, risk mitigation, downtime minimization, and effective asset selection, procurement, redeployment, use, and disposal.

Condition assessment is common to all life-cycle management programs. FM automation and visual infrastructure management solutions enable FM professionals to manage an asset's life cycle more efficiently by allowing rapid access to alphanumeric reports, graphics, and floor plans (see Fig. 3.2.5-1).

The following are the top 15 areas reviewed in a facility's life cycle with visual infrastructure management tools:

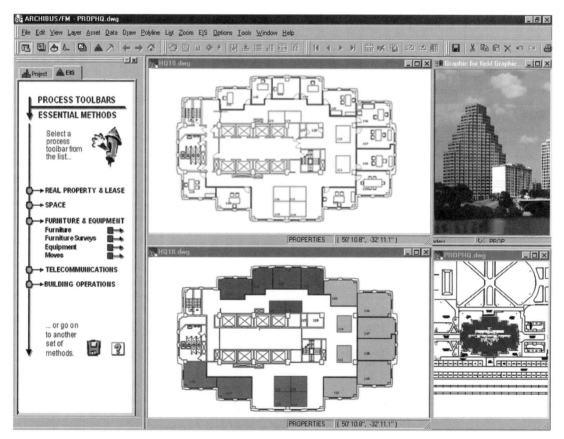

FIGURE 3.2.5-1 Life-cycle management is readily facilitated with FM automation and infrastructure management solutions. (*Courtesy of Bruce Kenneth Forbes, ARCHIBUS, Inc.*).

- Foundations
- Substructure
- Superstructure (floor and roof construction, stairs)
- Exterior closure (exterior walls and fenestration)
- Roofing
- Interior construction (partitions, interior finishes, specialties)
- Conveying systems (elevators, escalators, conveyors)
- Mechanical (plumbing, HVAC, fire protection, special mechanical systems)
- Electrical (service and distribution, lighting and power, special electrical systems)
- Equipment (fixed equipment, furnishings, special construction)
- Site work (site preparation, utilities, improvements, and off-site work)
- Special infrastructures (telecommunications, cables, and wireless systems)
- Special activity and functions (move management, hoteling, room booking, etc.)
- Soft infrastructure (space, organization, technology, and finance management)
- Asset exposure, special needs, and disaster management (direct, indirect, third party)

THE TECHNOLOGICALLY-DRIVEN LIFE CYCLE OF FACILITIES

To understand and manage a facility's life cycle, a facilities professional must have a vision and an understanding of the technological changes that affect that facility. Keeping up with the exponential rate of technological change will be one of the greatest challenges faced by facilities professionals in the new millennium. New technology is driving obsolescence, and obsolescence is considered a primary cause of facility failure and subsequent business failure. Bearing in mind that there is an opportunity with every business failure, it is important to identify both the opportunity of facility-related failure and the opportunity for facility-related success. Any discussion of technology-driven facility lifecycles is most likely to be obsolete in two or three years. At best, we can only identify trends, opportunities, and pitfalls to avoid.

During the next decade, facilities must become enablers of technologically changed corporations and institutions. As such, the facility manager must be able to rapidly adopt and employ technologies and realize the opportunity inherent in technology-driven facility life cycles. An individual facility's life cycle will be controlled by four business criteria (as illus-

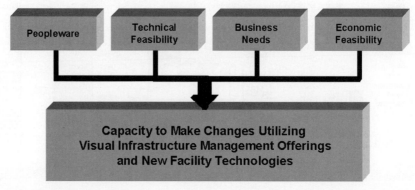

FIGURE 3.2.5-2 Four business criteria affect business life cycles. (*Courtesy of Bruce Kenneth Forbes, ARCHIBUS, Inc.*)

FIGURE 3.2.5-3   Seven generations of automated facility life-cycle integration and management. (*Courtesy of Bruce Kenneth Forbes, ARCHIBUS, Inc.*)

trated in Fig. 3.2.5-2) and by its capacity and/or ability to be reshaped using FM automation and infrastructure management offerings and new facility technological enablers.

Figure 3.2.5-3 lists seven levels of technological integration that have proven beneficial in enabling facility professionals to deliver superior facility life-cycle management services.

CONTRIBUTORS

Julia M. Forbes, Janice F. Jonas, and Cynthia M. Zawadski, ARCHIBUS, Inc., Boston, Massachusetts

ARTICLE 3.2.6

HOW FIT ARE YOUR FACILITIES? PERFORMANCE MEASUREMENTS AND FACILITIES INFORMATION TECHNOLOGY

Joseph Milando

FIT Technologies, Newton Upper Falls, Massachusetts

Measure twice, cut once, is the well-known carpenter's adage. We intuitively grasp the wisdom that it is wasteful, time-consuming, and costly to take action without proper measurements. Trying to squeeze savings from bricks and mortar without proper measurements is like building a house without a tape measure.

How do you measure the performance of your real estate and facilities? What are your key financial indicators? How do you measure maintenance performance or space utilization? The only way to measure is with timely, reliable information. Timely and reliable information sounds so simple—but in reality it takes planning. Like building a building (or renovating an existing structure), you need to gather the requirements, conceive the solution, and communicate the idea before starting construction.

Most of us could benefit from a little improvement in our personal fitness routine. Similarly, most organizations can benefit from a facility fitness program. Many companies vastly undervalue the financial impact of proactive investment in facility information systems. Facility information is the lifeblood of a facility fitness program. You cannot measure fitness or progress without information.

This article covers conceptualizing, prioritizing, and designing a sound information system to support your facility fitness program and demonstrate the recurring value of facility management (see Fig. 3.2.6-1).

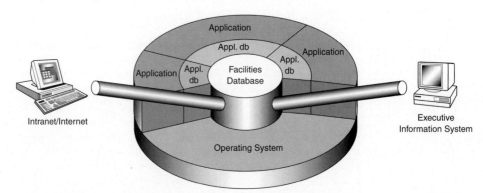

FIGURE 3.2.6-1 Who needs information? What information do they need? How do they need it delivered? Where does/should the information reside? Who maintains each category of data? (*Courtesy of Joseph Milando, FIT Technologies.*)

WHERE DO I START?—GOALS

Start at the end and work your way to the beginning (called *right-to-left thinking*). Since knowledge is the key to intelligent decisions, ask yourself what knowledge you need to make sound decisions. Who else needs information, and how does it need to be delivered to them? By working with the end result in mind, you will stay focused.

Define Your Goals

In facilities management as in fitness training, you will find that a vaguely worded goal (like "become more fit") is not going to drive you toward any worthwhile result. You need to have specific goals, and most likely there will be multiple goals and they will be interrelated. A set of exercise goals might be to lose 15 pounds in the next three months, to be able to run 5 miles in less than 40 minutes, and to achieve a resting heart rate of 64. Notice a few things about these goals:

1. They all rely on specific measurements comparing "before" and "after." (One needs to establish baseline measurements.)
2. The goals are based upon knowledge of what constitutes fitness for that person. Appropriate weight, heart rate, and so forth (industry standard performance metrics).
3. They take into account a vision of realistic progress for that person (assessment of the level of progress that is achievable).

To achieve sustainable fitness, one needs to establish complementary goals in various categories; dietary, scheduling, and goals for each specific exercise. Just as fad diets don't last over the long term, quick fix information projects don't stand up to time either.

Set three goals in each major information category. For illustrative purposes (see Fig. 3.2. 6-2), I have divided facility information goals into segments where commercial applications exist. These may not be the best categories for your needs. The choice of categories is unique to each company, because they are related to many variables: the size of the organization (people, revenues, square feet), the number of buildings, the use of space, centralized versus decentralized control, owned versus leased properties, the level of technical sophistication, the core business of the organization, existing information, and other considerations. The figure may help you identify your key categories.

Triage

People generally have short memories and little patience for long time horizons. The best path to wide-ranging success is usually through demonstration of incremental smaller successes. By demonstrating results, you solidify your reputation and the requisite support to continue your plan. You need to rank your goals based on value, risk, and political considerations. These are three primary considerations:

1. What is the value of the goal in both financial and political terms? Who will make or save money? When will gains be recognized? Who will benefit politically? Who would oppose this goal? Would the organization or personnel assignments change?
2. How much will it cost over what period of time? What is the initial investment? Will recurring investment be needed? What level of staffing is needed?
3. Your intuitive assessment of the likelihood of success must consider the people required to cooperate, the availability of data, and the simplicity or complexity of the project.

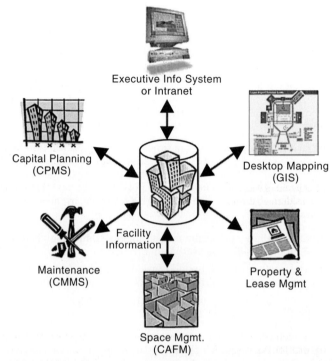

FIGURE 3.2.6-2   Facility information goals segments. (*Courtesy of Joseph Milando, FIT Technologies.*)

To assess these considerations you may wish to use a matrix. At this point, remember that you are just trying to prioritize goals, so don't get caught up in too much detail. (See Table 3.2.6-1.)

Other items to consider for your matrix include

- Dependency information—goals that need to be accomplished first
- Who needs the information?
- How often is it needed?
- How accurate is the source data?
- Who is responsible for maintaining the source data?
- Where are the data? How do you collect the data?
- What technologies support the collection and distribution of the information?

FUNDAMENTAL CONSIDERATIONS

By now, you may be anxious to start your project. But before you do, a little caution is necessary. Here are two fundamental rules that many overlook in their desire to achieve success.

TABLE 3.2.6-1 Ways to Identify Your Goals

Item	Space options	Maintenance options
Goal	Reduce usable square feet per person 5% over 18 months	Increase ratio of preventive maintenance to reactive maintenance by 15% in 18 months
Benefit/value in both financial and political terms	Financial—cost reduction of X, avoid cost of move, avoid X capital expenditures Political—enhance new COO's stature, avoid move	Financial—business production without mechanical failures, increase revenue 1%, reduce operating costs 5% Political—improve image of FM dept., increase employee satisfaction
Budget (cost) over time	Capital investment of X, with X per month to monitor and maintain	Capital investment of X with X per month to monitor and maintain
Likelihood of success considering people, availability of data, and the simplicity of the project	Very high—considering new corporate directives, staffing, and widespread recognition of the issue	Medium—cultural change is required, need buy-in from union, fits with FM department looking to change image

1. The Mousetrap

Achieving a realistic balance between efficiency and effectiveness is the golden rule (see Fig. 3.2.6-3). Just because a system can track very detailed data doesn't mean you should do so. I have seen organizations itemize more than 20 departments in order to split the cost of an 8-square-foot closet rather than simply sharing it among the 22 departments on the floor. The annual cost of the closet was less than $250. Did it matter whether it was split among 20—or 22—departments? I call tracking more information than is worthwhile the "mousetrap." Set practical guidelines to avoid this trap.

Conversely, one can err on the side of shortsightedness, not capturing the extra 5 percent of work that will provide whole new functionality (e.g., while performing a survey to update architectural plans, missing the opportunity to note vacant spaces for planning purposes).

2. The Value Chain

Performance measures rely on information. Reliable information is the result of a continual process of renewing the data. Therefore, unless everyone who needs to participate in the

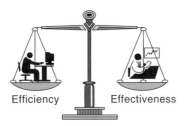

FIGURE 3.2.6-3 Keep efficiency versus effectiveness in balance. (*Courtesy of Joseph Milando, FIT Technologies.*)

information chain gains some benefit from their participation, the data will become unreliable over time. Let's face it, everyone has more work on their plate than they can perform. Shortcuts will be taken wherever possible. However, people will usually not shortchange tasks that benefit them.

DEFINE YOUR SPECIFIC REQUIREMENTS

By stepping through the process described, we have already done a great deal of our work and have certainly thought through the beginning of a plan of attack. But, in the analogy of the carpenter's rule, we are still not ready to cut the wood. We know we are going to build a house, and we know how big it will be and how many rooms it will have, but we still need the design and to communicate that design on paper.

Start again with the matrix, but make sure you consider all of the items (including those listed after the sample matrix). This time list every piece of information required. Consider who is responsible for maintaining each data component and how often it needs to be refreshed. Wherever possible, refresh as infrequently as practical. You can always increase the frequency as the kinks are worked out. At the end of this process, you will understand each piece of required data and how the information needs to flow.

Data Definitions

This step is critical to the long-term integrity of your data. Simply list each allowable name within each data category. This serves three functions: it tests your logic, it communicates a standard definition that everyone will use, and it controls data creep. I have seen organizations that did not tightly control and define their room types. During a period of years, new room types were created to suit personal tastes. Over time, more than 430 room types existed. Obviously, no one could get any real information from such an expansive and widely interpreted list. Keep in mind that you may have to adjust your list upon system selection, but you should perform this exercise before system selection if for no other reason than that it helps you to understand the right system for your needs.

System Selection

Now you are in a good position to evaluate systems and the technology available on the market. Comparing and selecting systems is beyond the scope of this document. But it is usually a good idea to get your information systems group involved along with someone who has set up these types of systems before.

Project Plan

Because each system differs in its setup and operation; it is best to select your system and technology platform before detailing your implementation plan. Your project plan transforms the requirement specifications into the tasks and milestones needed to get to your goal. Devising a good project plan is also beyond the scope of this document, but a few reminders are warranted.

Break the project into small, discrete, manageable sections. Establish milestones and deliverables. As you reach each milestone, communicate your success. The idea is to break the project into pieces so that you can consistently achieve small successes. It is easier to monitor progress and forecast schedules that way, and most important, momentum is maintained and sometimes accelerated.

SUMMARY

Applications of facilities information technology have traditionally been introduced through individual grassroots initiatives, often doing nothing more than expediting manual tasks. Each application is for the most part its own island, often duplicating data from other systems. The result is multiple systems, disparate databases, duplicated and erroneous data, and most important, the inability of information to flow throughout the organization.

The Possibility

An ongoing convergence of technologies (e.g., Internet, Intranet, databases, open architecture) is transforming every aspect of our lives—including the way facilities information systems transform our work. The ability to build and maintain immensely valuable facilities information systems is becoming easier every day. Before starting or revising your system, step back and try to envision the entire need for information within real estate and facilities. If you start with the end in mind and prepare a plan before purchasing and implementing a desired change, you will shift the burden of much data maintenance back to where it belongs. The overall effectiveness of the entire organization will be increased. And most important, you will have reliable information to monitor facility fitness and make strategic decisions.

ARTICLE 3.2.7
MANAGING THE MAGIC

Frank Addeman
Walt Disney Imagineering, Los Angeles, CA

> Whatever we accomplish is due to the combined effort. The organization must be with you or you don't get it done . . . In my organization, there is respect for every individual, and we all have a keen respect for the public.
>
> —*Walter Elias Disney*

IT TAKES MORE THAN "PIXIE DUST" FOR A PROJECT MANAGER AT WALT DISNEY IMAGINEERING TO MANAGE THE DESIGN AND DELIVERY OF A MAJOR ATTRACTION

Walt Disney Imagineering (WDI) has always been the Walt Disney Company's best-kept secret. The reality of a project at WDI is that managing and delivering the magic requires all of the fundamental project control tools and processes typically found in other industries to define and organize the project scope and deliver a quality project within budget and on schedule. The budgets are tight, the schedules are aggressive, Disney standards for design quality and guest experience are high, and the project teams are spartan. Although no ambitious organization can bat 1.000, WDI has an extraordinary track record for delivering projects on budget and on schedule. Creating the magic involves many components, from blue-sky conception to design and implementation, and finally opening the attraction. Don't be fooled. The work doesn't stop there; it just begins in another arena, presenting these new projects to Disney audiences around the world.

So how does a WDI project manager manage the design and delivery of a major attraction? It takes more than "pixie dust" to focus the wide spectrum of Imagineering *dreamers* and *doers* and turn an unproven concept that may initially defy the laws of physics into a cost-effective and reliable attraction that meets guest expectations on Opening Day. In addition to the project challenges found in other industries, creating a major attraction at WDI involves two further unique complexities that must be fully understood and integrated by the project team for a successful delivery.

THE WDI PROJECT ORGANIZATION

The first unique difference is that WDI is in the entertainment industry and its project organization is similar to that for film production, with both a producer and a director who share project responsibilities. Each project team has a show producer who is responsible for the creative vision and a project manager who is responsible for project delivery. Second, WDI projects mobilize a wide range of specialized talent—residing in over 100 specialized work groups in multiple locations—that must be fully integrated and sequenced during the project's life cycle.

These Imagineering specialists are assigned to functional divisions and support each project when their specific skills are required at various points in the project life cycle. Some of the specialized talents required to deliver a major WDI attraction include writers, artists, model makers, sculptors, set designers, artificial foliage designers, lighting designers, special effects designers, audio/video designers, show/ride mechanical engineers, show/ride electronic software engineers, show production and tooling specialists, show animators and programmers, rockwork designers, graphic designers, facility engineers for each discipline, architects, landscape specialists, film production specialists, interior designers, themed carvers, and themed painters. Each of these specialty groups must be engaged at the appropriate time and must provide a timely deliverable to another group to keep the project on track. For example, creating a complex AUDIO ANIMATRONICS figure involves more than two dozen different specialists residing in different groups that must work in sequence from the initial figure movement sketch through a fully functional figure in an Opening Day scene.

Walt Disney Imagineering uses a decentralized project matrix organizational structure to provide the diversity of talent to achieve a large portion of the project scope. It is neither practical nor cost-effective to attempt to dedicate this talent only to one project. A large portion of the work is performed by Imagineers working in more than two dozen separate locations. The WDI functional divisions include Creative Development, Architecture and Facilities Engineering, Show/Ride Design and Production, Theatrical Design and Production, and Finance and Project Controls.

The projects require timely and accurate reporting to maintain cost/schedule visibility in a rapidly evolving project/division environment, as the project matures through its life cycle. A project team is assembled and empowered to be responsible and accountable for achieving the objectives for each individual project. Division work groups are authorized to perform defined work scopes for each project.

One of the biggest challenges for the project team is coordinating the many specialized performing work groups, which adds a very complex dimension to project coordination. To monitor work performance, each performing work group must have consistent and clearly stated objectives and a well-defined and measurable scope of work assigned. By establishing what is to be "accomplished" or "delivered" and who does it, the project manager can maintain the schedule and budget and identify deviation from the original plan.

Each specialized WDI work group can perform work at multiple times on a project during the design or implementation phase of the life cycle. In turn, each division work group must support multiple projects at any time. Most work is performed in a location different from that of the team. Each functional work group within each division is responsible and accountable for achieving the correct creative and technical solutions within its area of responsibility and expertise for each project. Each division provides the appropriate support personnel, including managers and technical specialists, to support the project team.

The most challenging aspects of the project and division management matrix are (1) tracking budgets and (2) scheduling performance for each work group from project inception through each phase of the life cycle to completion. Project scope and continuity are controlled using a standard WDI project work breakdown structure (WBS). The WBS identifies the theme park, land, attraction, major components, and divisions of work. This structure provides the project team with a basis for evaluating where the project is today and, most important, what remains to be done.

An overview of the WBS Level 5, Project division of work (DOW) is shown in Fig. 3.2.7-1. The show and ride portion of the DOW is unique to WDI and is the common denominator for cost and schedule control and communication between the projects and divisions. A division management summary report could be for one DOW, such as DOW 41, Animated Figures, for all projects currently authorized. A project summary could be for all show components of the project for the full project life cycle, including DOW 41. The DOWs are assigned to three major work categories: facility, show, and ride. Facility design and construction scope are organized by the CSI standard for consistency within the construction industry. Show and ride components are assigned to DOWs that were created for WDI. Each show component is assigned to scenes within each attraction to coordinate production and installation sequences with the building construction.

The Project Team Challenge

If you were a project manager for an automotive manufacturer in Detroit, and you were asked to design and build a new high-performance car that looks great, exceeds all of the current safety standards, and is less expensive than last year's model, and were asked to find a way to deliver it one year early with an 800,000-mile warranty, would you accept the challenge?

Many of the Imagineering project teams are required to meet similar objectives when new show or ride technology is required for their attraction. The project financial pro forma is tight. The schedule is usually aggressive, supporting an Opening Day requirement. The attraction needs to receive a high guest satisfaction rating. The show components and ride need to last 20 to 30 years and operate up to 20 hours per day, seven days a week.

So how does a project team meet all of these objectives? It takes a blend of "pixie dust"; an experienced core team that knows where the risks are; world-class talent in each specialty area; proactive communication among the project team, divisions, and contractors; and lots of sweat. Compared to other industries, the project controls staff that provides cost and schedule control is spartan. Realizing that the effort to control cost and schedule increases with the level of reporting detail required, the project manager must limit the level of control detail to the high-risk areas and use high-level reports for the lower risk scope.

To stay on budget and schedule, the team needs to avoid design reiterations by identifying timely design deliverables when they are required for efficient design. For example, the heat load for a pyrotechnic special effect must be finalized before the HVAC can be designed. The size and movement of a large AUDIO ANIMATRONICS dinosaur must be finalized before the building structural, mechanical, and electrical loads can be designed. The energy profile of a ride must be finalized before the show set and facility can be designed. A complex attraction could easily have more than 1000 design deliverables to maintain the design schedule.

Casting the Project Team

Before joining WDI, I worked for a large design and construction firm. Our project teams, typical of the heavy construction industry, consisted of a project manager, project engineer, construction manager, and a project controls manager. Even though the projects were all

LEVEL 4

WORK CATEGORY

Facility Design/Construction
(16 codes)

Show Design/Production/Installation
(23 codes)

> Schedule activities and logic ties are required for each component within each Division of Work as they are sequenced between 6 to 24 different WDI work groups.

Ride Design/Production/Installation
(13 codes)

LEVEL 5

DIVISIONS OF WORK

- 02 Site Work
- 03 Concrete
- ---
- 16 Electrical

- 41 Animated Figures
- ---
- 44 Special Effects
- 45 Projection Hardware
- 48 Audio Software
- ---
- 51 Show Lighting
- 55 Rockwork/Shotcrete
- ---
- 64 Theatrical Rigging

- 71 Load/Unload Conveyance
- ---
- 74 Track/Flume
- 75 Vehicle
- ---
- 79 Ride Control Software

FIGURE 3.2.7-1 At Walt Disney Imagineering, the work breakdown structure (WBS) division of work (DOW) is the common denominator for cost and schedule control and communication between projects and divisions. The DOWs are assigned to three major work categories: facility, show, and ride. Facility design and construction scope is organized by the Construction Specifications Institute (CSI) standard for consistency within the construction industry. Show and ride components are assigned to DOWs that were created for WDI (© *Copyright Walt Disney*).

much bigger and required more resources, the number of specialty groups required was less than one dozen. The WDI project team is organized to support the special demands placed on the project manager to provide an entertainment project and manage the diversity of talent required. The WDI project team has a project manager, a show producer, project core team members, and project control team members.

Project Manager. The project manager (PM) is the focal point for the project team and has full accountability for project results. The PM is the project facilitator and must have the ability and visibility to deal with problems and obstacles from a single, coordinated viewpoint over the total project. A PM must have sufficient influence to ensure that the division priorities support the project requirements. The PM assigns work to the divisions and must resolve priorities with each division manager to support the project schedule.

Show Producer. The WDI show producer is responsible for directing the creative vision throughout the project's life cycle. The show producer represents the Creative Division's executive management on the project team. The show producer manages the intent of the show design from conceptual design through project completion and reconciles operational and budget trade-offs, when required, to meet the project's creative objectives.

Core Team Members. The core team for a major attraction typically consists of an architectural and engineering manager, a ride project engineer, a show program manager, and a construction manager. The project core team members are the communication conduits between the project and their respective divisions. They are responsible for project leadership, decision making, and delivery for their respective scopes. They have dual reporting relationships. They report to the project manager in their roles for accomplishing the project objectives and to their respective functional managers for professional execution of their creative or technical responsibilities in support of the project. They are accountable for project-related work performed by their respective divisions or outside vendors and contractors. They report to both their project manager and their respective director/manager regarding cost, schedule, and technical status.

The project core team members are also responsible for their respective divisions' scope/strategy, plans, schedules, budgets, and design intent. They work with the project team and their division directors to identify their respective division's scope and deliverable requirements for each funding phase and to ensure that the creative vision is achieved.

For successful project delivery, core team members must ensure that the project goals are understood and committed to by their respective division managers and performing work groups and that each work group's schedule is integrated into the project master schedule.

Project Control Specialists. The project team also has project control specialists who report to the project manager for project direction and to their division director for functional direction. The project control specialists include a project estimator, project planner/scheduler, project financial analyst, and a contract manager. They support both the project manager and each one of the project core team members with project control services for both division authorized work and outside contractors and vendors.

PROJECT DELIVERY

Each new attraction evolves through five phases, from a potential new idea through Opening Day. The project life cycle establishes the duration of each phase and the framework for each schedule. An overview of the project life cycle and key milestones is shown in Fig. 3.2.7-2. A unique challenge for a WDI project team is blending creative and technological innovations with Disney storytelling. "Concept" is the evolutionary phase that is used to ensure that an attraction can be designed to integrate the story line into three dimensions through feasible

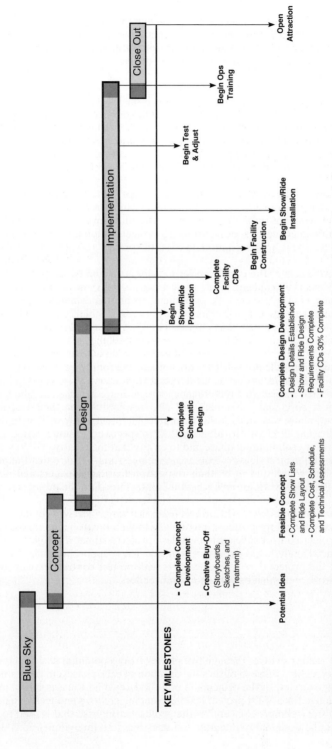

FIGURE 3.2.7-2 Each new attraction evolves through five phases, progressing from a potential "blue-sky" idea through Opening Day. The project life cycle establishes the duration for each phase and the framework for each schedule. *Concept* is the evolutionary phase that is used to ensure that an attraction can be designed to integrate the story line into three dimensions through feasible technology. (*Courtesy of Walt Disney Imagineering. © Copyright Walt Disney.*)

technology. This is the time required to develop a recommended unproven concept into a feasible concept that supports the budget pro forma.

The design phase includes the *schematic design* and *design development* stages. The schematic stage is used to finalize the building and show/ride layouts and to select the appropriate show/ride systems. The design development stage must lock down all of the design details including equipment location and functions.

The implementation phase includes the final design of facility contract documents and show/ride production documents. During this phase, all of the design scope must be completed and appropriately defined in the contract bid documents. All of the work unique to WDI, such as themed construction finishes and ride technical performance requirements, must be well-defined. The contract drawings, specifications, samples, and special conditions must clearly define the work scope for each potential bidder, how it needs to be accomplished, and when it needs to be performed. This is a key transition point for the team. This phase also includes show and ride production, facility construction, show and ride installation, and test and adjust.

The concept and design phases are the most challenging phases for the project team to manage because design is not locked down until implementation. These phases place special demands on the project manager to manage the diversity of specialized talent required, to maintain the creative intent, and to avoid negative impact on the project from design reiteration. The project team must complete these phases on schedule to optimize its implementation budget and schedules. The scope and strategy for these phases must be clearly defined by the project team, and the performing divisions must understand the key project milestones and critical schedule interfaces. WDI deliverable guidelines are used to help each project team establish the appropriate deliverables to achieve the objectives for each stage. They provide a consistent framework between each project and performing division. Producing too many consecutive stage deliverables too early may result in major rework; underachievement of the required stage deliverables may result in errors or omissions. The team needs to avoid both of these scenarios to optimize its budget and schedule.

PROJECT PLANNING AND SCHEDULING CONTROL

The project team must develop a fully integrated project master schedule to successfully deliver the project. The project master schedule must be fully integrated into all of the division work group schedules to support a coordinated work effort and ensure that performing organization resources can support the project's objectives. To do this, the project master schedule must be realistic and sufficiently defined to support communication among the project, the division's performing work groups, and each contractor. It must also provide a quantifiable base for performance measurement. As the project evolves, the scope, strategy, and schedule are expanded.

Project/division integration is achieved by developing sufficient DOW detail to provide a common denominator for both the projects and division work groups to plan and show the status of the work scope progress. The project master schedule must identify the appropriate phase deliverables and objectives to avoid design reiterations and to minimize design errors and omissions. Show/ride design must be planned and sequenced to ensure that it does not impact facility design. All project and division schedule activities must contain a consistent coding structure to support project and division management sorting and selection requirements.

The project manager must establish the schedule goals, conditions, and requirements in harmony with the project team members, division management, and contractors. The master schedule is developed within the framework of the project's life cycle and is expanded during each phase by the project planner/scheduler and the project team, using the most current project scope documents. Working with the project team, the project planner/scheduler integrates all project-related schedules into one overall project master schedule to ensure that all work scope is addressed at the appropriate time.

The project master schedule is a critical path method (CPM) logic schedule that identifies key project milestones and critical interfacial points. It provides the project team, division management, vendors, and contractors with an intermediate level of detail, required times of performance, interrelationships, and project schedule objectives. Key project schedule milestones are identified in the project master schedule to ensure that senior project and division management mutually understand schedule goals.

The project team members and the project planner/scheduler must coordinate the schedule development with the responsible work groups to ensure that detailed work plans are properly integrated and that the appropriate need dates are verified. This helps the team identify schedule conflicts among the various work groups. The project manager must ensure that all detail division schedules are integrated within the framework of the project master schedule. The project planner/scheduler identifies the key interfacial points among the performing organizations and determines the project's critical path and work priorities.

Conflict Resolution

Conflicts are inevitable within any matrix organization. The diversity of disciplines required for delivering an attraction increases the magnitude of interfaces and potential scope, schedule, and financial conflict. Most WDI project conflicts are resolved through timely communications between the project core team members and divisional work groups.

The project manager must resolve conflicts between the project team and divisions or among the core team members before they impact project delivery. Critical unresolved conflicts must be elevated to the portfolio project vice president and divisional vice president for appropriate executive management action. Projects for each geographical portfolio are reviewed monthly by senior management of WDI. Any unresolved conflicts that may impact Opening Day are addressed for his resolution as part of the project update process.

SO, HOW DOES A PM LEAD THE TEAM TO SUCCESS?

At the end of the day, the project manager has to successfully deliver a major attraction and also has to examine the principles it took to get there. No one will ever explain this, nor will it ever be put in any type of manual or job description; however, the project manager's key ingredient to success is to remember that he or she is not only managing the project but the human aspect as well. The other elements that fit into this recipe would be to:

- Influence the team through open communication
- Anticipate risks and have a backup plan ready
- Proactively interface with the division managers and work groups to ensure they share the project's goals
- Work closely with the project team to resolve conflicts
- Proactively maintain a healthy balance between budget, schedule, and quality
- Lead the team with passion, conviction, and courage

I believe it was best said by one of the best project managers at the Walt Disney Company:

Courage is the main quality of leadership, in my opinion, no matter where it is exercised. Usually it implies some risk—especially in new undertakings. Courage to initiate something and to keep it going—pioneering and adventurous spirit to blaze new ways, often, in our land of opportunity.

—*Walter Elias Disney*

RETHINKING THE ARCHITECT OR CONTRACTOR SELECTION PROCESS

William C. Ronco, Ph.D.
Gathering Pace, Inc., Bedford, Massachusetts

Daniel Gainsboro, President
Genesis Planning and Delphi Construction

Facility managers readily acknowledge that the main problem in any construction project is communications. However, when they interview architects, contractors, and engineers, facility managers usually neglect communications issues or even ignore them completely. They examine scale models and drawings, review slides and graphs, and study reports and columns of numbers. They fail to explore the candidates' track records in the communications issues that determine project success.

The roots of many of the problems facility managers have with design and construction begin in the interview process. Too often, traditional methods for interviewing and selecting architects and contractors portray little about the interviewees' skills and track record in management and project communications. Candidates who have weak or difficult histories in communications slip through the interview process unnoticed while selection committees study their slides.

The selection process for architects and contractors naturally focuses facility managers on the building. Have you ever built this kind of building? What did it look like? What did it cost? Can you make it look a little more like this? Could you accommodate that? What would it cost? Following this natural line of questioning provides only a small part of the information clients need to know. Questions about experience with buildings provide the baseline, but they do not touch the problem areas.

Selecting an architect based on the look of the buildings that his or her firm designed is like buying a car based on a picture in a magazine, not considering its reliability or its operation. Your buildings committee can devise an excellent list of questions that explore the building and also the process of communications required to get it built. How will the architects communicate with you? How will they involve your input? How will they work with a committee that has internal disagreements, a client base that does not communicate well? How will they manage their own internal turnover?

Following the tenets of "behavioral interviewing," it is also important to ask questions following particular guidelines. Ask no hypothetical or "what if" questions such as:

- What would you do if you disagreed with a client?
- What do you think is an ideal building?
- Of which jobs are you proudest?
- Do you have any weaknesses?

These classic interview questions provide interesting, often fascinating responses. Perhaps that is why they are classics. However, the responses to hypothetical or "what if" questions in no way predict what the respondent will actually do in the situation. Instead, ask questions based on real events.

Effective questions do a better job in predicting what the respondent will do because they explore what the respondent has already done. Effective questions follow this format (also see Table 3.2.8-1).

TABLE 3.2.8-1 Effective Questions for Selection Interviews

Don't ask hypothetical questions:	Do ask questions about real events. Use this format:
What would you do if . . . ?	Tell me about a time when you _____.
Don't ask speculative questions:	What did you do?
What is your view of an ideal building?	What were the results?
These questions yield responses that are interesting but provide no prediction of behavior.	Based on past behaviors, these questions better predict future actions.

- Tell me about a time when you ____.
- What did you do?
- What were the results of what you did?

Here are 10 questions that facility managers find reveal useful information about candidates in the design and construction process. After each question, the interviewee asks, "What are some actions you took?" and "What were the results of those actions?"

The statements all begin with the phrase, "Tell me about a time when you . . .":

1. Went over budget or beyond schedule
2. Disappointed a customer
3. Disagreed with a customer
4. Disagreed with other members of the project team
5. Had to work with an organization such as ours that does not communicate effectively with its members?
6. Had to work with an organization such as ours that includes some departments that do not get along with each other
7. Had to move key staff off a project
8. Had a client who would not get involved enough in the project
9. Had a client who got too involved in the project
10. Had to work with a project team that had one weak player

We find that few architects and contractors come to selection interviews prepared to answer these questions thoughtfully. Actually, we have seen some architects and contractors who took offense at facility managers who even raised these questions. Needless to say, those are candidates to be concerned about. Many candidates find that they have to return to their offices to gather information to address these questions.

Some architects and contractors cannot answer these questions because they have not really thought about them.

This approach to selection has strong potential for facility managers in design and construction and also in the following areas in which they must select a candidate or a partner:

- Job interviews
- Outsourcing service providers
- Preferred vendors
- Contractors of any kind

In all these cases, it is clear that facility managers must ask these questions and also ensure that candidates provide excellent answers. As long as communications problems impact design and construction projects, it is important for facility managers to understand and manage them in the same way that they understand and manage technical issues.

ARTICLE 3.2.9
ENVIRONMENTAL COMPLIANCE AND MANAGEMENT SYSTEMS

Joel Higginson
Advantage Engineering, Wakefield, Massachusetts

OVERVIEW OF ENVIRONMENTAL COMPLIANCE MANAGEMENT IN MODERN FACILITIES

Since the authorization of the USEPA in 1970, the burden on designers, managers, and operators to build and continuously run their facilities within the framework of local, state, and federal environmental regulations has become increasingly challenging and complex. The Federal Environmental Code, Section 40 of the Code of Federal Regulations, has grown from just over 10,000 pages in 1990 to over 17,000 pages in 2000, and updates and proposed changes to the Federal Code are printed daily in the Federal Register. It should also be noted that the federal regulations are the minimum standard to which our facilities must be managed. State and city codes must be at least as stringent as the federal regulations, and they are often more restrictive.

This article provides a general strategy for managing environmental compliance at your facility. The approach that is described herein is broad and generic. Because of differences from facility to facility, and especially between different types of institutions and industries, not all sections of this article will apply to all facilities. Beyond the scope of this article, it is the responsibility of facility management to identify the specific regulatory requirements that apply to their facility.

All of the regulations that comprise a facility's federal environmental compliance responsibility can be found within the following legislation (Acts):

- *Clean Water Act (CWA).*   Regulates chemical, biological, and particulate pollutants in sewage, industrial effluent, navigable waters, and stormwater runoff.
- *Clean Air Act (CAA).*   Regulates pollutants from combustion sources, manufacturing processes, and storage systems.
- *Toxic Substances Control Act (TSCA).*   Regulates chemical raw materials (primarily imported and new materials) and certain existing materials which represent an "unreasonable hazard."
- *Superfund Amendments and Reauthorization Act (SARA).*   Enacted to minimize the potential of a disaster similar to the chemical release disaster in Bhopal, India, SARA is the umbrella act which covered EPCRA and which modernized and strengthened CERCLA.
- *Emergency Planning and Community Right to Know Act (EPCRA).*   Regulates facilities that store and use greater than threshold amounts of hazardous materials; provides emergency planning information to communities; requires annual usage and release reporting.
- *Comprehensive Environmental Response, Compensation, and Liability Act (CERCLA).*   Regulates processes for identification and cleanup of contaminated (hazardous waste) sites and for responses to spills of hazardous materials.
- *Resource Conservation and Recovery Act (RCRA).*   Regulates hazardous waste generation, storage, transportation, treatment, and ultimate disposal.

In addition to the regulations that enacted under each of the preceding Acts, the federal government and state agencies frequently issue "policies," "guidance documents," and "guidelines" that augment, reinforce, or supplement their legislation. Complying on a day-to-day basis with the "letter of the law" is a legal requirement, yet it is also a formidable task. This article presents a strategy for implementing and maintaining compliance and discusses the issues of importance to facility manager over the life cycle of the facility.

DEVELOPING A STRATEGY FOR ENVIRONMENTAL MANAGEMENT

A successful environmental management system (or program) depends on two key things—*awareness* and *ownership*.

It is generally very easy to tell within minutes of walking through the "nonpublic" (service, maintenance, waste collection and disposal, or utilities) areas of a facility how strong its "culture" (with respect to environmental compliance) is. Companies and facilities that practice reactive environmental management are certain to have more frequent and more serious noncompliance issues than those that are proactive in style.

Awareness of the regulatory responsibilities, requirements, and limits under which a facility must operate in order to remain in compliance can only be attained via continuous education and training. Sources of educational and training materials for facility management include federal, state, and local agencies, independently operated agency hotlines, trade associations, industrial and institutional journals and trade publications, business and industry associations, equipment vendors, suppliers, and consultants. Awareness training should be "pushed" by facility management to each appropriate organizational level of the workforce.

Ownership of compliance responsibilities must be clearly delegated via job descriptions and enforced via their presence in the individuals' performance review program. A compliance program that is designed to establish responsibility for certain environmental activities at the *lowest* levels of the organization will generally create the appropriate culture for successful environmental management.

GETTING STARTED WITH OR RESTRUCTURING AN ENVIRONMENTAL MANAGEMENT PROGRAM

The easiest way to get started is to select a team of people from a geographic *and* functional cross-section of the facility to brainstorm all of the "things" that affect the facility's ability to be "in compliance." The number of "things" may vary in number from 12 or 13 for a simple office building to hundreds for a complex manufacturing facility or institution campus (see Table 3.2.9-1).

Grouping the "Things" into Appropriate Compliance "Modules"

The foundation for effective environmental compliance management begins with a system with concise, coherent structure, which takes into account the following elements:

- Personnel who manage the process that has the potential to contribute to pollution
- Personnel who use the process
- Personnel who maintain or supply the process
- The form of the potential pollution and the way in which it enters the environment

TABLE 3.2.9-1   Compliance Issues for an Office Building

Past history of site
    Site contamination
    Leaking underground storage tanks

Air emissions
    Fossil-fuel-fired boilers, incinerators, emergency generators

Wastewater
    Sewer connection permit

Hazardous waste and hazardous materials handling
    Light bulbs, batteries
    Fluorescent light ballasts
    Cooling tower and boiler water treatment chemicals

Solid waste management
    Waste management and reduction programs

Fuel storage
    Underground storage tanks

Refrigeration management
    Chlorofluorocarbons
    Ammonia

Spill prevention, control, and countermeasures
    Fuel oil storage
    Hydraulic oil (elevators)
    Electrical transformers (PCBs)

The "modular" approach results in a program structure that localizes compliance issues and breaks them down into easily understandable and manageable activities. Shifting gears to a manufacturing environment, an example of modules for a light manufacturing facility may consist of the following:

- *Water pollution control module.*   Takes into account discharges of industrial wastewater that may include suspended solids, toxic contaminants such as heavy metals and organic toxics, BOD (biochemical oxygen demand), FOG (fats, oils, and greases), temperature, and pH

- *Air pollution control module.*   Takes into account emissions from fuel-fired devices such as boilers, heaters, incinerators, dryers, and ovens, and from manufacturing processes such as coating and painting systems, reactors, baking ovens, and so forth

- *Hazardous waste and hazardous materials handling module.*   Relates to the proper disposal of hazardous waste materials and the proper storage, handling, and reporting of hazardous chemicals that are used on the manufacturing site

- *Solid waste management module.*   Deals with the management of solid waste streams that leave the facility, and includes waste minimization, substitution, elimination, and recycling

- *Spill prevention, control and countermeasures (oil spills), and stormwater pollution prevention module.*   Takes into account storage systems, stored materials, and raw materials deliveries that could result in a spill of oily material into the environment (e.g., storm sewers, soil, roadways)

- *Underground storage tanks.*   Relates to the potential for spills from underground storage tanks and piping systems.

A Specific Module Example for a Manufacturing Facility—Identifying Compliance Activities

The module teams should meet to define the specific activities that should be managed in order to ensure compliance. Each module can be effectively managed by implementing 6 to 12 simple management activities. For example, most water pollution control modules can be managed by producing and maintaining the following eight module "tools":

- *Site map with discharge points.* The module team should locate and identify on a site plan the sources and discharge points for process wastewater as well as sanitary wastewater. Care should be taken during the site review to identify for elimination any cross-connections which may allow sanitary wastewater to enter a process wastewater pretreatment system; sources of noncontact "once-through" cooling water (which unnecessarily increases the hydraulic loading on downstream treatment processes); drains from "clean" roofing (which should discharge into stormwater systems); and floor drains in manufacturing areas which bypass wastewater pretreatment systems.

- *Permit or registration summary.* Identify by discharge point whether the discharge requires a permit or a registration. Local regulations, ordinances, and sewer authority guidelines should be reviewed, as they may contain more stringent requirements than federal and state regulations.

- *Summary of permit obligations.* The module team should review the facility wastewater discharge permit and prepare, for management review, a two-page summary of permit obligations. Once a permit is obtained, *most* permit holders quickly lose track of all but the most significant of their compliance obligations. The permit summary should include a listing of what pollutants are controlled (BOD, TSS, FOG, pH, etc.); a summary of the "controlled" parameters (daily wastewater flow, process throughput, hours of operation, pretreatment system functions, and so forth; a summary of record-keeping and reporting requirements; operator licensing requirements; noncompliance notification procedures, phone numbers, and timing; and permit renewal requirements.

- *Compliance and operations data filing system.* The module team should ensure that records and reports are being properly completed, and that they are maintained in an active file for the mandatory records retention period. Records should include discharge monitoring reports; analyses performed by outside laboratories; qualifications documentation on outside laboratories; manufacturing production records (if permits restrict operations); and preventive maintenance and repair records for equipment and control systems.

- *Site accountability listings.* The module team should decide in advance and make generally available to employees on each shift the appropriate chain of command for managing and reporting noncompliance issues on a shift-by-shift basis.

- *Incident reporting action plan.* Facility management and the module team should establish a decision tree or flow chart to identify proper incident-reporting protocols for notification of regulatory agencies as well as corporate headquarters. The plan must include 24-hour emergency notification numbers. One of the most horrifying experiences that corporate senior management can have is to turn on the television to a "live report" from outside one of their facilities.

- *Licensing and certification requirements.* If operator licensing is required, the module team should put in place a system to ensure that annual training requirements are planned for and achieved and that recertification examinations are scheduled and successfully completed.

Record Keeping

The single most important compliance management activity is record keeping. Each facility is obligated to prepare and maintain records that demonstrate its compliance with environmental regulations. Records are the principal tools that regulatory agencies use to determine whether a facility is meeting its obligations. By simply reviewing the filing system, a good field inspector can tell within 15 minutes whether a facility understands and is managing its compliance obligations. A good record keeping system should include:

- *File system index.*   A single point of information that describes the filing system.
- *File location summary.*   A "file locator" which identifies where each type of information is filed. As a general rule, it is recommended that a facility's active environmental files be maintained in a single location, and that limited access be provided to personnel on each shift. For example, current (monthly) boiler or fuel documentation may be maintained in the boiler room, then subsequently filed in the main file location.
- *Separate location for archived files.*   Files that are maintained beyond the minimum statutory records retention requirements (in some cases a corporate requirement) should be maintained in a separate locked location. These files are not part of the facility's regulatory compliance requirements, and they should therefore not be treated as such.

Program Review and Evaluation

Facility environmental compliance programs must be periodically monitored by or on behalf of senior management to ensure that they are being maintained at an acceptable level.

ENVIRONMENTAL COMPLIANCE OVER THE LIFE CYCLE OF A FACILITY

Environmental management is a process that must be implemented and maintained throughout the life cycle of a facility. Facility owners and operators are responsible "from the cradle to the grave" for the emissions, discharges, and wastes that are generated at those facilities. Therefore, compliance issues must be planned for and managed from facility planning through facility closure. Permitting processes often lie on the critical path of facility construction or modification plans. This section addresses some of the issues that must be addressed at each stage of the facility's life cycle.

Site Selection Regulatory Issues

Site selection issues include principally *site condition assessment* and *environmental impact assessment.* Site assessment is a site condition and data review process designed to identify historic uses, review reported soil and groundwater contamination in the region, and evaluate the likelihood of contamination originating onsite. Environmental impact assessment ensures that the intended development and use of the site will not create unacceptable impact on the local environment.

Site Assessment (Phase I/Phase II).   "Site Assessment" is a very specific process for identifying and evaluating risks at a particular location. Site assessments are frequently required by lending institutions on mortgage acquisition or refinance and by insurers on new coverage or increased limits coverage requests. Although the American Society for Testing Materials (ASTM) has developed site assessment specifications, there is great variability in the level of information that the lending institutions and insurers require.[1,2] A Phase I site assessment addresses observed and recorded site contamination issues. A Phase I assessment is therefore a screening process. The minimum Phase I site assessment requirements generally consist of three elements:

- *Environmental database review.*   Several vendors offer environmental database review and report services. The database reviews identify contaminated sites; leaking underground storage tanks; major generators of hazardous waste; hazardous waste treatment, storage, and disposal facilities; landfills and incinerators; and sites of major spills in the vicinity of the subject site.

- *Local department records review.*   A local review consists of reviews of the planning board (historic site plan approvals); building department (ownership, historic use, significant construction and demolition projects); and fire department (underground tank and hazardous materials storage registration) records.

- *Site visit.*   The facility walk-through identifies visible signs of contamination and environmental risk. Key objectives of the site visit are to identify signs of significant spills and leaks; presence of asbestos, lead paint, and PCB's; and evidence of inadequate compliance management.

A Phase II site assessment[3] is generally ordered if there is a significant potential of site contamination and if additional testing and analyses are required to determine the extent of the contamination. Significant contamination can range from issues such as lead paint chips on the surface of the ground (identified and characterized via a soils analysis) to serious soil and groundwater contamination (often from the disposal of industrial chemicals into old septic systems or from a previous history of leaking drums stored onsite).

Environmental Impact Assessment.   The "environmental impact assessment" process originated with the National Environmental Policy Act (NEPA), Article 22-1. This act required federal agencies, departments, and installations to prepare environmental impact statements on proposals for legislative and other major federal activities that have the potential to significantly affect the quality of the natural environment.[4] Many states have adopted the environmental impact assessment process (or various parts or reconfigurations of it) as a means of creating a forum for public input and comment on projects that have the potential of causing significant impact on the region or community. Typically the process, which is triggered by a myriad of potential impacts (including deforestation and loss of natural habitat, wetlands and drainage basin modification, traffic, noise, lighting, impact on community services and infrastructure, air quality, and others), requires the applicant to evaluate the impacts for the preferred project configuration as well as for less intrusive or less impacting alternatives. This process can have a tremendous impact on the site selection process, as it may drive a proposed project to a different location, or it may result in the rejection of the project in its entirety.

SITE PLANNING REGULATORY ISSUES

Specific site-related projects (whether they be new "greenfields" projects, conversion of existing property to an alternative use, or addition to or modification of an existing property) are regulated under state, county, and local programs. Site planning activities, which generally come under the purview of elected and/or appointed boards, encompass things such as zoning (the establishment of specific land use patterns, criteria, and requirements), wetlands protection, community impacts (including traffic, noise, lighting, air emissions), and the availability of utilities and services.

Zoning

Zoning is the process whereby county or local officials regulate land use via a predetermined definition of permissible uses and land use restrictions within the community. Zoning is intended to be protective of the interests of the community as a whole. The zoning process typically addresses the following issues:

- Land uses (which are typically listed as permitted or not permitted within certain sections of the community)
- Density of development (by regulation of minimum lot size; establishment of requirements for the minimum setback distance of facilities, structures, and paving from streets and property lines; and by potentially regulating the specific location of a building within the property boundaries)
- Community impact (by addressing issues such as traffic, air emissions, noise, and lighting)
- Infrastructure impact (including electric power, potable water, and natural gas availability; impact on community safety and educational services; site access; impact on existing roadways; and wastewater disposal issues and alternatives)

Applicants who initiate the zoning process should be aware of the fact that this process will result in the discussion and review of alternatives to the proposed plan, and applicants should prepare to address justification to support the preferred project configuration during the zoning review process.

Requests for exemptions, exceptions, or waivers to specific requirements within the zoning approval process are generally handled by a separate site plan approval authority, or by an appeals board process.

Wetlands Protection

Wetlands issues are regulated at the local level, frequently with oversight and even active involvement at the state level. Wetlands are important to the ecology for several reasons. They provide accumulation volume and settling capacity to prevent downstream damage and pollution, they provide habitat for wildlife, and wetlands plants perform treatment of wastes.

During the initial stages of the project development process, the owner and the site engineer must make a determination of the location and boundaries of wetlands on the project site (and perhaps on adjacent sites if the project has the potential of impacting them). Although wetlands determination requirements vary from state to state, wetlands determinations are generally based on plant species, hydrology, and soil type. Wetlands impacts originate from two sources—the impact on wetlands that must be moved, reconstructed, or replicated due to the construction of the facility and its access roadways; and the impact of stormwater runoff storage requirements on existing wetlands and storm drainage systems. Most states now limit the project's postconstruction runoff to less than or equal to the preconstruction runoff. This means that the additional runoff that is caused by impervious structures such as building roofs and paved areas must be stored onsite and released to existing offsite drainage systems in a controlled manner.

The board, commission, or agency that is responsible for managing the wetlands permitting process will frequently require the applicant to submit alternative analyses to the baseline design in order to ensure that the permitted project will minimize impact on the wetlands.

Community Impacts

Air emissions, noise, and lighting issues are administered at the county or local level via "nuisance ordinances." Nuisance ordinances are often based on subjective evaluation of the impact of the site and its operations on abutters and other nearby residents. Some of the considerations that an owner may make when evaluating a particular site are:

- Direction of prevailing wind and impact on facility odors that may be carried to surrounding properties

- Facility operating schedule and placement of noisy or brightly lit operations on the side of the facility that has the least impact on neighbors
- Site topography, vegetative screening, and access roadway design to minimize impact on neighbors

Availability of Utilities and Services

Potable water may be extracted from the ground by the facility owner, or it may be delivered via a municipal supply system. In either case, the facility's initial as well as proposed water requirements (to accommodate growth) may have a significant bearing on site selection. Aquifer studies and pumping tests can be performed to determine the ultimate capability of the aquifer to support the desired extraction quantities and rates. Owners who are dependent upon extracting potable water from their own site should evaluate surrounding operations to rule out or minimize the potential risk of groundwater contamination due to leaks, spills, or catastrophic events at nearby properties. Owners who plan to rely on municipal potable water supplies should inquire during the project planning activity about the community's ability to meet initial as well as projected water needs.

Of equal importance is the ability of the community to provide treatment of the wastewater from the facility. Nonmanufacturing facilities may estimate the amount of sewerage leaving the facility based on the number of employees occupying the facility and on additional activities such as operation of a commissary or laboratories. Manufacturing facilities may estimate and characterize their wastewater based on similar operations and on the proposed manufacturing system's material balance. The U.S. Environmental Protection Agency (USEPA) has developed categorical wastewater pretreatment standards for some specific industries that generate high-strength wastewaters. The categorical standards define the amount of treatment required at the manufacturing site prior to discharge of the wastewater to a publicly owned treatment works (POTW). The cost of wastewater treatment can be a significant line item of a facility's annual operating budget. In fact, wastewater disposal fees and surcharges may represent a significant negotiation opportunity during a new facility siting process.

REGULATORY ISSUES DURING FACILITY PLANNING AND CONSTRUCTION

During the preconstruction and construction phases of a project, the owner or the owner's representative must be cognizant of the types of proposed activities that may require permits. Some of the permit requirements may have a significant impact on facility design, and others may fall along the critical path of the construction and startup schedule. The types of permits that must be considered include stormwater discharge associated with construction activity (in the case of disturbance of five acres or more of land with the potential to discharge soils and contaminants into stormwater systems and waterways); flammables storage licenses and permits; treated wastewater discharge; stormwater discharge associated with industrial activities; hazardous waste storage; and combustion systems and manufacturing processes operating permits.

Stormwater Discharge (Construction)

Any construction project that will disturb over five acres of land, and that will discharge runoff into an ocean, lake, river, stream, or wetland must obtain an NPDES (National Pollutant Discharge Elimination System) permit. (Discharge into a storm sewer system that subsequently discharges into a waterway or a wetland is considered to be the same as directly discharging into the waterway or wetland.) The NPDES permit will require two things—a pollution prevention plan and installed mitigation measures. Typical mitigation measures include controls to stop the runoff of fine soils and silt (erosion and sedimentation controls) and discharge flow controls (detention basins) which control the volumetric flow of stormwater from the site. The application process for an NPDES permit for stormwater discharge associated with construction is a simple process, and does not generally impact the project schedule. Most facilities choose coverage under existing general permits.

Flammables Storage Permits and Licenses

Permits and licenses to store flammable materials are managed at the state or local level, and permit requirements vary widely. For this reason it is important to determine prior to beginning facility construction what the local authorities are willing to permit. For example, a community with a high groundwater problem may prohibit the installation of underground storage tanks, and the facility plan will have to account for aboveground fuel storage in a room or vault. This may have an impact on the footprint of the proposed building.

Treated Wastewater Discharge

If the proposed facility will be responsible for pretreating the wastewater or for completely treating it to discharge standards, the type of treatment system and its real estate requirements for the proposed initial operation as well as for planned expansion must be taken into account when planning the site layout. Planning should take into account strategies for holding or recycling wastewater so that operations will not have to be shut down in the event of an upset in the treatment system. If the facility proposes to discharge treated wastewater directly to a waterway, the permit application must be submitted at least 180 days prior to initial wastewater discharge.

Stormwater Discharge (Industrial)

A facility that discharges stormwater that has had the potential of contacting raw materials, work in process, finished products, wastes, or equipment and machinery which may contaminate that water, must obtain a NPDES permit for discharge of the stormwater. As with the wastewater treatment NPDES permit application, this permit application must be submitted 180 days prior to startup of the facility. This requirement encompasses facilities that discharge via any "conveyance" (swale, berm, pipe, channel, etc.) into a waterway or wetland or into a storm sewer system which subsequently discharges to a waterway or wetland. Facilities that unload raw materials in unsheltered areas, discharge exhaust from dust collection devices to the roof, or discharge compressor condensate with trace amounts of oil are considered to "have the potential" of contaminating the stormwater. The facility owner may wish to design the facility to perform all activities and to release all discharges within protected (roofed and drained) areas in order to eliminate the need for participation in this permit program.

Hazardous Waste Handling and Storage

Facilities that generate hazardous waste (as defined by the Resource Conservation and Recovery Act—RCRA—40 CFR Part 270) must apply for an EPA ID number and must store and dispose of the waste material in accordance with the regulations. Facilities that propose to treat, store, and dispose of their hazardous wastes (TSDF facilities) must submit a two-part RCRA permit application 180 days prior to *commencing construction* of the treatment, storage, and disposal systems.

Combustion Systems and Manufacturing Process Permitting

Three types of air emissions permits must be considered during facility design and construction, and must be implemented prior to constructing a new facility or modifying an existing facility. Each of these permits is dependent upon the size and scope of operation of the proposed facility as well as upon the air quality status of the region in which the facility is to be located. Under the 1990 Clean Air Act Amendments (CAAA), nearly all air emissions are subject to permit requirements. The individual states have been delegated the authority to issue and administer air emissions permits on behalf of the EPA. Therefore, although each state's program includes as a minimum all of the federal program requirements, individual states may enact more restrictive regulations.

The federal Clean Air Act Amendments regulate a variety of emissions that are categorized as "air toxics" as well as a set of six "criteria pollutants." The criteria pollutants include nitrogen oxides, carbon monoxide, sulfur dioxide, particulate matter, lead, and ozone. The standard for ozone emissions (which is a major component in photochemical smog) is actually enforced by monitoring volatile organic chemicals and chlorofluorocarbons.

The National Ambient Air Quality Standards (NAAQS) have been established to define maximum levels in the environment of the criteria pollutants above which human health is at significantly increased risk. Through a system of data collection, monitoring, and analysis, EPA has defined regions of the country as being either *nonattainment* (meaning that the region has not reduced air pollution below the maximum program levels), or *attainment*. As one might expect, the permit process in nonattainment areas is more restrictive than that in attainment areas.

The Clean Air Act Amendments are also written around two tiers of facilities—*potential major sources* and *minor sources*. Potential major sources are facilities which "have the potential to emit" greater than "threshold amounts" of pollutants. The potential to emit is defined as how much of any given criteria pollutant the facility would emit if it operated at full capacity for 24 hours per day, 365 days per year. The threshold amounts of pollutants differ depending upon whether the region in which the facility is to be located is in an attainment area, a moderate nonattainment area, or a severe nonattainment area.

The New Source Review (NSR) program mandates two types of permits for "potential major" sources—*preconstruction review* permits and *prevention of significant deterioration* (PSD) permits. Any new "major source" or any major source that is proposed to undergo a significant modification in an area which is nonattainment for one or more of the criteria pollutants must apply for a preconstruction review permit. Preconstruction review permits are administered by the state environmental agencies, and they may take up to a year or longer to obtain. Clearly, such a permit will fall along the project's critical path, and it must be considered and initiated at the very beginning stages of the project. Proposed new facilities or existing facilities which are undertaking significant modifications and which are located in attainment areas, must file for prevention of significant deterioration permits. The objective of PSD permits is to ensure that the region in which the facility is located remains an attainment area after startup of the facility. PSD permits can take up to a year to secure also.

A third type of permit, the *operating permit,* also called the "Title V" Permit, is an umbrella permit that has been issued to all facilities which are potential major sources of air toxics or of any one of the criteria pollutants. Operating permits establish facility-wide emissions limits, define criteria for limiting operations of the process or facility, establish monitoring and recordkeeping requirements, and specify criteria for reporting exceptions or exceedances.

In addition to the permits just specified, state environmental agencies and local air quality control boards may require *plan approval applications* for systems or processes that are above a certain size. Plan approval applications are a mechanism by which the permitting authority can regulate nonmajor processes. Plan approval applications typically have a 30 to 90 day approval window.

REGULATORY ISSUES DURING FACILITY OPERATION

Each of the environmental permitting requirements described here carry through into the operation of the facility. The permit restrictions, reporting, and record keeping that were established prior to construction must be implemented and followed throughout the course of facility operation. Permit renewals must be expeditiously filed. For example, most NPDES Permit renewals require that an application be filed 180 days prior to expiration of the current permit.

The facility manager must be aware of the requirements of the initial facility permits. In fact, it is recommended that facilities prepare a two- or three-page summary of the key points and obligations of each permit, and periodically review both the summary and the permit.

The following issues will trigger new permitting or regulatory compliance activities after initial startup and operation of the facility.

Process Changes or Expansion

A change in fuels or a significant change in raw materials used, process conditions and capacity, or operation of manufacturing processes must be evaluated for its impact on licenses and permits. Typical areas to be reviewed include wastewater (NPDES permits or sewer connection permits), air emissions (operating permits or plan approvals), and flammables storage licenses.

Flammables and Hazardous Materials Storage

Once a facility is in operation, it must comply with the requirements of the Community Right-to-Know Act of 1986, also known as Title III of the Superfund Amendments and Reauthorization Act (SARA). This program includes three key elements—*emergency planning and release notification, community right-to-know reporting,* and *toxic chemical release reporting.* Under Emergency Planning and Release Notification, any facility that stores more than the "threshold planning quantity" of substances that are listed as "extremely hazardous substances" must prepare a process hazards analysis. This plan requires the evaluation of potential effects of a release and the development of a plan for notification of response agencies and sensitive receptors. EPA's *risk management plan rule* requires facilities that handle, store, or use any one of 139 hazardous chemicals at or above the listed threshold quantities to prepare a detailed hazard assessment, an integrated prevention program, an emergency response program, and an overall management system. Furthermore, the facility must coordinate its risk management plan (RMP) with the local emergency planning committee (which is comprised of local first-response agencies).

Community Right-To-Know Reporting requires any facility that stores over 10,000 pounds of any material for which it is required to keep a material safety data sheet (MSDS) or over 500 pounds or the threshold planning quantity (whichever is lower) of an "extremely hazardous substance" to report information on how and where the material is stored to the state emergency response commission (SERC), the local emergency planning council (LEPC), and the local fire department. This reporting must be renewed annually prior to March 1. Toxic Chemical Release Reporting (TRI) requires any facility that uses directly in its manufacturing process over 25,000 pounds (or which uses in "ancillary processes" over 10,000 pounds) of "listed chemicals" to report annually to the EPA the amounts used and the amounts released to the environment. TRI reporting must be made annually prior to July 1.

Hazardous Waste Management

Hazardous waste regulations are complex. The key elements to proper hazardous waste management are to properly define the facility's generator category (conditionally exempt small quantity generator; small quantity generator; or large quantity generator), to properly store the waste (in strong containers made of compatible materials, segregating reactive materials, with secure access and proper signage, having adequate ventilation, and with a means of contacting and communicating with personnel outside of the storage area), and to properly manage the waste disposal paperwork. The generator of a hazardous waste is obligated to receive and maintain copies of the shipping manifest from the transporter and from the final treatment or disposal site, or to report absence of the documents within a statutory time frame. Improper management of this paperwork is one of the most frequently cited violations during a regulatory inspection.

REGULATORY ISSUES ASSOCIATED WITH CLOSING A FACILITY

The responsibility for environmental compliance extends through the entire life cycle of a facility—from initial planning and site selection through decommissioning. The owner/operator of a site or a facility must therefore be concerned with the condition of the site upon departure from the site, as well as with its subsequent deterioration. Responsible facility management would suggest that adequate documentation of the condition of the facility and site upon closure is appropriate. The following are among the types of issues with which facility management should concern themselves:

- *Asbestos and lead paint.* As long as the facility remains under the control of the owner/operator, that entity is responsible for exposure of people to these contaminants. Therefore it is encumbent upon the owner to secure the facility against unauthorized access to areas in which these contaminants may be present and to make available any knowledge of these contaminants to construction or demolition crews.

- *PCBs.* Older transformers, switches and capacitors, and even some hydraulic systems may contain cooling oils which are contaminated with polychlorinated biphenyls. When contained within the devices, this material presents minimal risk to humans or to the environment. However, their release constitutes both a potential health risk and a potentially significant cleanup expense. Therefore, owners/operators should consider removing contaminated systems from the facility upon decommissioning or, as a minimum, securing these systems from unauthorized access and vandalism.

- *Oils and maintenance chemicals.* Fuel oils, lubricating oils, coolants, and other maintenance chemicals above and beyond any quantity which is necessary to maintain the integrity of the decommissioned facility should be removed. If the facility is to be demolished, underground tanks should be closed and tank closure documents should be secured.

- *Wells.* If the facility has operated an onsite water well, the well should be tested to document its condition upon closure of the facility, and appropriate measures should be taken to prevent unauthorized access to the well.

In summary, common sense should prevail with respect to closing a facility. Facility management personnel are encouraged to perform a risk analysis ("what-if" assessment) on facility closure issues.

REFERENCES

1. ASTM E 1527, Standard Practice for Environmental Site Assessments: Phase I Environmental Site Assessment Process, American Society for Testing and Materials, W. Conshohocken, PA.
2. ASTM E 1528, Standard Practice for Environmental Site Assessments: Transaction Screen Process, American Society for Testing and Materials, W. Conshohocken, PA.

3. ASTM Standards Related to the Phase II Environmental Site Assessment Process, American Society for Testing and Materials, W. Conshohocken, PA.
4. William T. Ingram, *Standard Handbook for Civil Engineers,* McGraw-Hill, Inc., New York, 1983.

ARTICLE 3.2.10
RELOCATION MANAGEMENT

Cindy Aiguier
IOI, Inc., Boston, Massachusetts

OVERVIEW

In today's working environment, change is constant. As changes occur, they invariably affect the physical office. The need to accommodate company growth, improve business processes, enhance the company image, or even reduce space due to company downsizing are reasons for change—change usually manifested as an office reconfiguration or relocation.

There are many complexities to bringing such facility projects from concept to reality. A number of people and a great many details must be coordinated. Landlords, architects, contractors, furniture vendors, security, telephone and data system installers, and movers all play a part in a facilities change or relocation.

This article explores the process of facility change and/or relocation and provides the facility manager with a checklist that will be useful for ensuring that no aspect of a project is overlooked.

HIRING AN ARCHITECT

The first step in making a facility change is identifying the need for it. Once this is done, the business of making the change typically involves the services of an architect. Depending on the size or scope of the project, the architect may be hired through a bid process.

Programming

Once hired, the architect will work with the owner and other designated parties to establish the project requirements, timetable, and budget.

Next, the specific client needs are assessed. Through a series of interviews with appropriate staff members, an analysis of activities is conducted to establish the client's operating procedures, space requirements, communications relationships, and present and future functional needs.

If the project involves relocation to another site, the following predesign activities ensue:

- Site evaluations for relocation and comparisons of prospective sites
- Lease negotiations (typically do not involve the architect) (see Art 3.1.5, Corporate Leasing Design, and Construction)
- Feasibility studies
- Preliminary block plans

If the project is a renovation of an existing site, then the architect will set about inventorying the existing site conditions. As part of a feasibility study, the architect also reviews code compliance to determine building limitations and design parameters.

To complete the programming phase of the project, the architect will prepare a summary analysis of the programming findings and issue this to the owner for sign-off approval.

Schematic Design

The architect then enters into the schematic design phase of the project, whereby schematic layouts are produced to communicate preliminary concepts for group clustering, furniture layout, and circulation. Color, finish, and materials selections are explored in this phase, and a preliminary cost estimate is also established. Several owner reviews and architectural revisions can be expected before arriving at an approved layout and finish selections.

Design Development

With approvals in place, the architect moves on to the design development phase. Drawings and documents are developed to include more detail with respect to the size and character of the interior design and proposed work. A more complete project budget is formulated as the plans illustrate actual counts of items for specification.

Contract Documents

The contract document phase of the project involves the development of working drawings, necessary documentation for the build-out of the new or renovated space. Along with the drawings, schedules and specifications are produced that detail particular items in the plans such as doors, hardware, finishes, equipment, and so forth. Upon completion, contract documents are sent out to bid to secure a general contractor and/or obtain pricing for finishes, furnishings, equipment, and custom items. These documents ultimately become the "contract" for the awarded general contractor, as well as the instrument used for securing building permits.

Contract Administration

The final phase of the project for the architect is contract administration—a required duty of the architect as mandated by most building codes. Here the progress of construction is monitored to assure compliance with the architect's drawings. Contractor submittals in the form of shop drawings, product data, and samples are reviewed and approved. Change orders are also prepared as necessary. When construction is complete, the architect walks through with the contractor to note any deficiencies in construction. This "punch list" is followed until the architect and the owner are satisfied with all resolutions.

Other Project Aspects

The facility change or relocation project does not begin and end with the architect alone. As previously mentioned, there are many aspects to these projects, all of which must be coordinated carefully and at appropriate intervals to assure project success.

Furniture.   When a facility change involves new furnishings for which corporate standards do not already exist, research and evaluation of prospective furniture lines must be conducted. Aesthetics, ergonomics, and price are only some of the points for consideration in selecting furniture. (See Art. 3.2.11, "Furniture Specification," for more information.) The process, in fact, can become fairly lengthy and involved. Because furniture lead times can sometimes run up to 16 weeks, it becomes necessary to begin the research and evaluation as early in the project as possible. The process will include the following elements:

- Development of workstation "typicals" and furniture standards
- Tour of furniture dealer or manufacturer showrooms (as well as local installations) to view furniture being considered
- Preparation and evaluation of bid package for furnishings
- Institution of product mock-ups for in-house evaluation
- Unbiased evaluation of potential furniture purchases, comparing performance criteria, cost, availability, flexibility, long-term maintenance, and residual value

The following are other furniture issues that must be coordinated during the various phases of the project process:

- Physical inventory of existing on-site furnishings to determine whether any existing pieces are appropriate for reuse in the new plan. The size, finish, and condition of each piece must be carefully evaluated. When reuse is contingent on refurbishment, a financial analysis must be performed to evaluate the cost-effectiveness, contrasted with the likely diminishing life expectancy of the older pieces.
- Furniture disposition. When reuse is not an option, resale opportunities should be explored. Used furniture has become a big industry over the years, and competitive brokerage networks exist that can quickly bring products to the market, sometimes using the Internet and e-mail to broadcast digital images of their inventory. However, should supply and demand determine that old, unused pieces have no residual value, the disposal of these items may include donation or simple discarding.
- Reuse of furniture from inventory when applicable. This process is sometimes facilitated if and when companies maintain an accounting of their warehoused inventory using a bar code system. Bar coding allows the owner to know quickly what is in inventory, its source, location, condition, and cost center.

Once it has been determined if and where existing furnishings will be reused, the information must be incorporated into a plan. From there, the following activities must take place to complete the furniture aspect of the project:

- Submit the approved layout to the furniture vendor for quotation.
- Ensure the review of furniture vendor drawings and specifications for accuracy before order entry.
- Track the purchase orders for furniture.
- Coordinate and track the schedule for delivery and installation.
- Oversee the delivery and installation to ensure that items are damage-free.
- Oversee the punch listing of furniture and expedite the resolution of all items, including missing, damaged, or incorrect products that cannot be installed per the approved plan (see Fig. 3.2.10-1).

Security.   Very often, facility change or relocation projects involve security considerations. If the company has an in-house security department, a representative will meet with the users to identify security requirements, risk levels, internal security needs, security devices (e.g., card

Item	Station/Room	Description	Notes	Status
1	1	Tighten 42″w panel to worksurface (bracket)		Completed 5/3
2	5	Missing 2-way connector base cap		
3	5	Tighten seam between 36″w & 48″w worksurfaces		Completed 5/3
4	5	Missing finished end base cap		
5	7	Alternate recep locations-"X" to be under corner w.s.		Completed 5/3
6	8	Secure finished end		Completed 5/3
7	12	Tighten 42″w panel to worksurface (bracket)		Completed 5/3
8	15	Missing finished end (include top cap)		
9	26	Straighten run at 48″w & 30″w panel		Completed 5/3
10	26	Adjust base at 2-way connector		Completed 5/3
11	30	Replace torn finished end w/left-over in better condition		Completed 5/3
12	31	Tighten seam between 36″w & 48″w worksurfaces		Completed 5/3
13	1217	Missing Allsteel 2 high 36″w lateral file	Shipped damaged— Candy DeLuca making freight claim Replacement to be located *either* between stations 5 & 7 *or* 56 & 58	
14	58	Broken key in pedestal cylinder	Cylinder to be replaced	
15	1208	Missing business accessories visual board	Reference purchase order 409-25	

FIGURE 3.2.10-1   An example of a punchlist that details deficiencies in a furniture installation. (*Reprinted with the permission of Artesania, Boston, MA.*)

reader locations, CCTV camera locations), staff access issues, and operational issues to develop a security plan. This information must be forwarded to the project engineers and architect before initiating construction drawings. Refer to Art. 5.5.3, "Security Systems," for a detailed discussion of this topic.

Telephone/Data.   User needs involving equipment locations, as well as the origins and destinations for telephone and network lines, must be coordinated with the company's telephone and data vendors. The vendors in turn coordinate with the architect to ensure that the requirements are incorporated into the new design. Refer to Art. 5.5.6, "Telecommunications and Data Distribution Systems," for a detailed discussion of this topic.

Project Phasing

Project phasing and the temporary fitting out of "swing space" often becomes a planning consideration in a facility change. To allow a company to continue to conduct business throughout a renovation, swing space is located or created to accommodate staff who will be displaced from the area or phase to undergo change. Much planning must go into the temporary relocation of these individuals. For example:

- Existing furniture configurations must be modified to provide new work areas for the displaced individuals. This often requires the services of the furniture vendor, electrician, and furniture installers.
- Existing telephone and data systems must be modified to provide continuous service to the swing space.
- Common files, equipment, and individuals' personal belongings must be packed, tagged, and moved.
- Furnishings remaining in the vacated area must be torn down and put into storage or otherwise disposed of before the renovations commence.
- Any artwork or signage must be removed and stored.
- Plants must be relocated out of the area of renovation and into new, compatible locations.

All of these efforts must be coordinated and project strategies must be implemented to accomplish the temporary move within a time frame that works within the overall project schedule.

Premove. As the project schedule for construction nears a close, a number of activities should be performed in advance of moving in to ensure that both the move and initial occupancy transpire as smoothly as possible:

- Notify plant vendors of new location.
- Coordinate with building management for building signage and/or directory update.
- Coordinate with computer manufacturers, or the owner's information systems department, to move all computers, printers, scanners, and plotters.
- Coordinate with copy machine company to move all copiers.
- Arrange for final cleaning before moving in.
- Coordinate with proper party to have company telephone directory updated with new phone numbers and locations.
- Notify mailroom of location change.
- Ensure that restrooms are clean and stocked with adequate paper supplies.
- Coordinate for security coverage during move-in.
- Ensure that all employees will have access to the new space beginning on the move-in date.

Move. When the design and construction of the new space are complete, the process of moving in can begin. In planning for the physical move, the sequence of activities can be broken down into four phases.

Phase 1.[1] Survey furniture, equipment, and contents to be relocated to determine the following:

- Load factors based on the inventory of relocatables
- Material requirements (i.e., cartons, bins, conveyors, Masonite, and Koroflex)
- Packing and unpacking requirements at origin and destination

- Manpower requirements at the origin and destination
- Vehicle requirements
- Supervision required at origin and destination
- Inventory of data processing equipment to include manufacturer, model, vendor, dimensions, weight, noting whether owned or leased
- Disassembly/assembly of desks, shelving, and library stacks
- Time frames for all relocation activities by location, floor, and department to anticipate and establish the relocation timetable.

Phase 2. Hire a moving company.

- Select qualified moving companies to receive the bid package.
- Prepare bid package for moving services.
- Evaluate bids and contract for moving services. Analysis will include consideration of a mover's ability to perform the services required in accordance with the anticipated schedule, based on their trucks and personnel. A thorough check of references should be conducted, specifically noting the mover's performance on jobs of similar size and scope.
- Negotiate contract with the chosen bidder.

Phase 3. Establish client user group involvement as a system for reporting and communication:

- Produce and distribute a move instruction manual for appointed move coordinators.
- Produce moving instructions for affected staff.
- Assist with "purge campaign" of obsolete documents.
- Develop the scope of work and logistics for electronic data processing (EDP) relocation with the appointed move coordinator for the data processing group. This will include its own timetable to anticipate pre-moves, special packing, preparation, take down, start up and the scheduling required to relocate effectively without disrupting the data stream.

Phase 4. Preparation and start-up of moving activity.

- Conduct move meeting with staff and movers.
- Prepare detailed move plan and schedule, including the names and phone and pager numbers of all key contacts (see Fig. 3.2.10-2). Distribute the completed sets to the move coordinators and the mover's foremen.
- Act as a liaison between the client and the landlord (both present and new landlords, if applicable).
- Ensure that all vendors provide landlord(s) with the proper certificate of insurance.
- Ensure that all vendors understand the guidelines for doing work or making deliveries into the building to be occupied.
- Ensure that all vendors understand the guidelines for doing work in or taking items out of the building to be vacated.
- Resolve in writing the egress and ingress points of the move—lobbies, loading docks, platforms, and so forth—at all origins and destinations.
- Confirm mover reservations for elevator use.
- Schedule building security and/or an elevator operator, when necessary, for after-hours deliveries and installations.
- Walk the finished space with movers and installers to note any existing damage that is the responsibility of the contractor.

Date	Time	Vendor	Floor	Description	Contact	Company/Position	Number
Friday 4/10	reg hours	Property Services/International	12	Doug Stills and Bob McCarthy to coordinate to box contents of (3) Steelcase bookcases & (1) forms storage unit from wall opposite stations 83 & 85	Bob McCarthy Doug Stills	Property Services Green Co International	(617) 123-3008 (617) 123-3047
Monday 4/13	reg hours	Acme Moving Co.	12	Acme to remove & store at warehouse (3) Steelcase bookcases from wall opposite stations 83 & 85—to be done as part of regularly scheduled 'Acme Run'	Mary Partridge Bob McCarthy Mark Marchello	Acme Moving Co. Property Services Green Co Security	(617) 456-1000 (617) 123-3008 (617) 123-3900
Wednesday 4/15	after 6:00 PM	JL West Cleaners	8	Existing carpet to be cleaned	Bob McCarthy Brad Martin	Property Services JL West	(617) 123-3008 (617) 789-1009
Wednesday 4/15	after 6:00 PM	JL West Cleaners	12	Existing carpet to be cleaned	Bob McCarthy Brad Martin	Property Services JL West	(617) 123-3008 (617) 789-1009
Thursday 4/16	7:00 AM	File USA Systems	8	File USA Systems delivery to include: (6) shelves for Storage Room 807A	Malcolm Deary Bob McCarthy Mark Marchello	File USA Systems Property Services Green Co Security	(508) 888-4410 (617) 123-3008 (617) 123-3900
Thursday 4/16	reg hours	File USA Systems	8	File USA Systems to install (6) shelves for Storage Room 807A	Malcolm Deary	File USA Systems	(508) 888-4410
Thursday 4/16	reg hours	Property Services/International	8	Doug Stills and Bob McCarthy to coordinate to box contents of (3) 4 drwr lateral files located in Open Office 831	Bob McCarthy Doug Stills	Property Services Green Co International	(617) 123-3008 (617) 123-3047
Thursday 4/16	after 5:00 PM	Atlantic Office Furn/Gerard Installations	8	AOF/GI delivery from warehouse to include: (1) Knoll work chair & (1) brass lamp for Reception 802; (1) brass lamp for Private Office 803; product that was removed from the executive secretarial stations & common file area (#83 & 84) to accommodate temporary station relocation	Candy DeLuca Frank May Bob McCarthy Mark Marchello	AOF Proj Manager Gerard Installations Property Services Green Co Security	(617) 444-1902 (978) 111-0304 (617) 123-3008 (617) 123-3900
Thursday 4/16	after 5:00 PM	Atlantic Office Furn/Gerard Installations	8	AOF/GI new furniture delivery to include: (2) Nucraft occaisonal tables for Private Office 803	Candy DeLuca Frank May Bob McCarthy Mark Marchello	AOF Proj Manager Gerard Installations Property Services Green Co Security	(617) 444-1902 (978) 111-0304 (617) 123-3008 (617) 123-3900
Thursday 4/16	after 5:00 PM	Atlantic Office Furn/Gerard Installations	12	AOF/GI new furniture delivery to include: (18) Knoll Morrison workstations, (7) Allsteel files, (2) Versteel conference tables, (1) computer cart for Conference Room 1233, & (19) Hag Credo chairs	Candy DeLuca Frank May Bob McCarthy Mark Marchello	AOF Proj Manager Gerard Installations Property Services Green Co Security	(617) 444-1902 (978) 111-0304 (617) 789-3008 (617) 789-3900

FIGURE 3.2.10-2 An example of a move schedule. (*Reprinted with the permission of Artesania.*)

Date	Time	Vendor	Floor	Description	Contact	Company/ Position	Number
Thursday 4/16	after 5:00 PM	Atlantic Office Furn/Gerard Installations	12	AOF/GI delivery from warehouse to include: (1) Hardwood Visuals whiteboard, (5) gray metal shelving units for Storage 1230, (3) Allsteel 3 drwr 30"w & (3) 3 drwr 36"w files for Open Office 1241, (2) lamps for Waiting 1242, (1) custom whiteboard for Conference Room 1238, (2) light blue metal shelving units for Printing 1231, (1) Knoll Hannah workstation for Waiting 1242, & (2) lounge chairs for Waiting 1242	Candy DeLuca Frank May Bob McCarthy Mark Marchello	AOF Proj Manager Gerard Installations Property Services Green Co Security	(617) 444-1902 (978) 111-0304 (617) 789-3008 (617) 789-3900
Friday 4/17	7:00 AM	C. Walker Assoc.	8	C. Walker delivery & installation to include: (1) desk & (1) credenza for Reception 802	Charles Walker Bob McCarthy Mark Marchello	C. Walker Assoc. Property Services Green Co Security	(603) 333-2028 (617) 123-3008 (617) 123-3900
Friday 4/17	8:00 AM	Finkelman Refinishing/ Gerard Installations	8	Finkelman delivery to loading dock to include: (1) occasional table for Reception 802; (1) table for Conference Room 840; GI to pick up from Finkelman at dock	Sam Finkelman Frank May Bob McCarthy Mark Marchello	Finkelman Refinishing Gerard Installations Property Services Green Co Security	(617) 222-3594 (978) 111-0304 (617) 123-3008 (617) 123-3900
Friday 4/17	8:00 AM	Finkelman Refinishing/ Gerard Installations	12	Finkelman delivery to loading dock to include: (2) occasional tables for Waiting 1242 & (1) table for Library 1202; GI to pick up from Finkelman at dock	Sam Finkelman Frank May Bob McCarthy Mark Marchello	Finkelman Refinishing Gerard Installations Property Services Green Co Security	(617) 288-3594 (978) 688-0304 (617) 664-3008 (617) 664-3900
Friday 4/17	reg hours	Atlantic Office Furn/Gerard Installations	12	AOF/GI to begin installing (18) Knoll Morrison workstations; relocate (1) forms storage unit from wall opposite stations 83 & 85 to Printing 1231	Candy DeLuca Frank May	AOF Proj Manager Gerard Installations	(617) 542-1902 (978) 688-0304
Friday 4/17	12:00 PM	Atlantic Office Furn/Gerard Installations	12	General Contractor to coordinate AC/DC Electric to install power in-feeds & begin cabling (18)Knoll Morrison workstations	Dan Smith Billy Kennedy	Samson Constr AC/DC Electric	(617) 664-8708

FIGURE 3.2.10-2 (*Continued*)

- Coordinate the delivery of boxes, labels, bins, equipment, carriers, and so forth, and supervise "materials positioning" by floor, department, and area so as not to disrupt the work environment.
- Survey the items labeled by staff to ascertain that only appropriate items are moved.

Postmove.   The following activities should be conducted immediately after moving in:

- Ensure that all product and moving tags are removed from the furnishings.
- Ensure that all instructional tags are retained and available for user review.
- Coordinate the removal of empty moving boxes and any remaining debris.
- Coordinate the installation of artwork.
- Walk through the vacated area to ensure that nothing is left behind.
- Coordinate the reconnection of fax machines.
- Coordinate the connection of PC and computer equipment; verify network connection.
- Ensure that security equipment is in proper working order.

Closeout.   Project closeout is the final step in completing the facility change or relocation process. Activities to be performed as part of project closeout include the following:

- Schedule product demonstrations between vendors and user groups. In response to the need for offices to be "ergonomically correct," chairs, keyboard trays, work surfaces, and so forth are being designed with a multitude of adjustments to reduce the risk of workplace injury. Users need to be made aware of and properly instructed in the use of these features.
- Administer postoccupancy evaluations. Postoccupancy evaluations (POE) provide a measure of project success. They determine user satisfaction with the overall project, process, furnishings, and team performance. The instrument for evaluation is typically a questionnaire followed by a meeting, both designed objectively to elicit unbiased responses (see Fig. 3.2.10-3). The information gathered offers valuable insight for future projects and possible project process improvements. A POE should be conducted no sooner than six weeks after moving in, to allow users to become adjusted to their new environment and to allow time for issues to surface.

Execution

From beginning to end, the time frame for making a facility change will vary greatly, depending on the scale of the project and the number of affected individuals. A small change involving only furniture could take under a month to plan and execute. A change that includes space renovation usually takes longer, involving more items with lead times and requiring greater coordination of trades. Likely to be longer still is the project that involves relocation to a new site being built-out to suit the client's needs. These projects can run anywhere from 6 to 7 months up to a year or two in duration.

Good project coordination will bring methodology to the project process to assure a logical and systematic approach to the accomplishment of the facility management objective. As the process unfolds, the following activities should take place as part of project coordination:

- Conduct regularly scheduled project meetings with all project stakeholders such as the architect, contractor, client representative, security planner, telephone and data system planner, landlord, and when appropriate, the furniture vendor and mover. Project meetings become a vehicle for monitoring the schedule and discussing relevant issues. Items that require action must be firmly established and monitored to ensure that the responsible team member provides follow-up in a thorough and timely manner.

POSTOCCUPANCY EVALUATION
PROJECT:
LOCATION:
COMPLETION DATE:
PROJECT TEAM:
ATTENDEES:

PROJECT HISTORY:
____ sf
Original program questionnaires completed:
Program approved:
Schematic design approved:
Construction documents issued for bid:
Construction start:
Furniture installation start:
Construction Complete:
Phase 1 move start:
Phase 2 move start:
Occupancy:

I DESIGN PHASE

A. HOW DID THE DESIGN MEET CLIENT NEEDS?

B. DID THE DESIGN PROCESS USED ASSIST THE CLIENT IN UNDERSTANDING THE IMPLEMENTED SPACE AND FURNISHINGS?

C. TEAM OPINION OF PRE-CONSTRUCTION ACTIVITIES.

D. HOW COULD PRE-CONSTRUCTION PROJECT MANAGEMENT ASSIST THE ARCHITECT IN THE FUTURE ON COORDINATION OF THE DRAWING SET?

II CONSTRUCTION PHASE

A. DISCUSS CONSTRUCTION SCHEDULE—IMPACT ON THE ARCHITECT OR CLIENT.

B. DISCUSS COMMUNICATION DURING CONSTRUCTION PHASE.

C. DISCUSS ANY ISSUES ON PRODUCT SPECIFICATION, I.E.: LIGHTING, CEILING, CARPETING, AND FINISHES.

D. COMMENT ON SUBCONTRACTORS/CONSULTANTS ON PROJECT.

E. DESCRIBE ANY COST SAVINGS ACHIEVED.

III FURNISHINGS & VENDORS

A. COMMENT ON MANUFACTURER AND DEALER PERFORMANCE.

B. DESCRIBE ANY AREAS TO ACHIEVE FURTHER COST SAVINGS.

C. DISCUSS ANY ISSUES WITH THE ACTUAL MOVE AND INITIAL OCCUPANCY.

FIGURE 3.2.10-3 Sample questionnaire used to conduct a postoccupancy evaluation.

- Provide a review of the architect's space plan and programming requirements to ensure that client needs are satisfied.
- Monitor to ensure that all coordination and reviews of the construction documents have occurred at appropriate intervals to meet the project schedule.
- Correlate team member schedules. Although each discipline creates its own schedule, it is important that they be correlated to identify potential conflicts. The careful and proper sequencing of activities is essential to avoiding trade disputes—costly propositions in terms of jeopardizing schedules and incurring additional labor charges.
- Conduct site visits to monitor the progress and quality of the interior work.
- Track purchase orders for fixtures, furniture, and equipment (FF&E). Missed lead times can put the project schedule at risk.
- Coordinate and track the schedule for FF&E delivery and installation.
- Oversee FF&E delivery, installation, and placement to ensure that items are damage-free.

CONCLUSION

The success of a facility change and/or relocation project is determined by the ability of the architect's plan to satisfy the users' needs and also by the successful implementation of the project process. Strategic planning and careful project coordination are paramount in ensuring a smooth transition into a new space.

REFERENCE

1. American Management Association Conference, National Relocation Services, Inc., *Facility Move Management Services Sequence of Activities.*

ARTICLE 3.2.11
FURNITURE SPECIFICATION

Cindy Aiguier and Michael Clancy
IOI, Inc., Boston, Massachusetts

OVERVIEW: FURNISHING THE OFFICE FACILITY

Within the last two decades, awareness of the relationship between a well-furnished facility and the enhanced effectiveness of the physical work environment has been increasing. This relationship can be most easily understood by recognizing the negative effect a poorly furnished work environment has on employee morale. This condition is typically manifested in a loss of enthusiasm, motivation, and productivity among employees. It usually results in high employee turnover at great expense to the company.

In today's world, it is necessary for the work environment to meet ongoing requirements for change, growth, and improvement. To do so, facility managers must learn the skills of integrating the facilities and the work process. The selection of appropriate furnishings is an important factor in meeting this challenge.

This article explores the many considerations a facility manager will have in furnishing an office environment. Although the focus here is on an office facility, the general information presented applies to all types of facilities.

IMPORTANT CONSIDERATIONS IN FURNISHING THE OFFICE FACILITY

Company Image

At a glance, the style, color, pattern, and arrangement of furnishings in a company's office facility send a message about its culture. They may project an image that is conservative and traditional, or perhaps one that is young and innovative. Proper furnishings improve a company's image by communicating such things as accomplishment, competence, credibility, and efficiency. Before any new furniture purchases are made, a facility manager should determine whether the selections communicate the desired image.

Work Process

The physical place should facilitate effective operations and for this to happen the furnishings must assist the work process. Therefore, the FM must have a clear understanding of the nature and functions of the company and its employees.

Basic Human Needs

Individuals who occupy the workplace have basic human needs that must be acknowledged when furnishing a facility. These include the need for both physical and psychological comfort. Physical comfort is achieved when the selection of furnishings incorporates anthropomorphic considerations, provides good support to the body, and allows tasks to be conducted with ergonomic correctness. Psychological needs are addressed by providing furnishings that allow for territoriality, personalization, and group interaction.

Budget

The furnishings budget will affect the way the company image is established, the way the work process is supported, and the way basic human needs are met. Although large budgets allow greater choices of aesthetically pleasing and functionally efficient furniture, small budgets can still provide furniture that meets the basic criteria. When limited resources are a determinant in selecting furnishings, it is important to reconcile the value of items that are initially inexpensive because of reduced quality. These items will have a shorter life expectancy and likely a higher long-term cost of ownership.

TYPES OF FACILITIES

The furnishings selections have a direct relationship with the type of facility a company seeks to create. A facility may be comprised primarily of private offices, open plan systems furniture (workstations), or some combination of the two, depending on some or all of the following factors:

- *Profession.* Members of some professions, such as lawyers, accountants, human resource professionals, engineers, project managers, and scientists, require more privacy for confidentiality or simply for concentration. Private offices best support their needs.

- *Corporate style.* Hierarchical organizations may use private offices as a symbol of status or reward, whereas flatter organizations may wish to avoid such symbols.

- *Facility flexibility.* Open plan systems furniture environments are less expensive to build and easier to reconfigure. They do not involve coordination with building systems such as HVAC and fire sprinkler heads, nor do they require architects, building permits, and so forth.

- *Financial implications.* The tax implications of systems furniture differ greatly from those of a private office build-out. For example, a 10% investment tax credit is extended for furniture and equipment purchases in the first year, and, because they are considered personal property, they qualify to be treated taxwise as a depreciating asset. Office construction, by contrast, is taxed on appreciating assessed values. Another financial consideration concerns potential company moves. A private office build-out must remain behind, but systems furniture can move along with the company.

ALTERNATIVE OFFICING STRATEGIES

Work styles or alternative officing strategies (AOS) also impact the way in which a facility is planned and furnished. "Telecommuting," "hoteling," and "teaming" are some of the new strategies that companies are adopting to increase employee performance, improve productivity, decrease the rate of employee turnover, and reduce real estate investment. (See Art. 3.2.4, "Improving Productivity Through Integrated Workplace Planning," for a more detailed discussion.)

Telecommuting

Enabled by today's technology (fax machines, pagers, laptop computers, cellular phones, modems, e-mail, and voice mail), *telecommuting* describes employees who work from home while staying connected to the office. This AOS offers many benefits to a company. By allowing people to work when and where it is more comfortable for them, talented people, who previously may have had to drop out of the workforce because of competing outside commitments (such as child rearing), no longer have to do so. And as the flexibility in their schedule makes employees feel more valued, productivity tends to increase. "Telecommuting" also helps companies comply with the federal Clean Air Act by reducing employees' work-related use of motor vehicles.

Hoteling

The Chicago-based professional services "Big 5" firm Ernst & Young pioneered the alternative officing strategy of *hoteling*. With hoteling, employees' individual desks or workspaces are eliminated, thereby encouraging them to spend more time working out of their clients' offices. When it occasionally becomes necessary for the employees to work on-site, they call ahead to schedule a place to work.

Teaming

The idea behind *teaming* is that by creating a more open, collaborative environment, employees are enabled to exchange ideas and information more freely. With this dynamic, a company can benefit by realizing fewer repeated mistakes and less time wasted, as individuals wait for answers. As a result, a company may experience such positive outcomes as reduced product development time.

OPEN-PLAN LAYOUT/SYSTEMS FURNITURE

Systems furniture was and continues to be designed to integrate furniture and the office facility as a direct response to the work process (see Fig. 3.2.11-1). By providing a high level of efficiency and productivity with systems, companies better manage the dynamic nature of the work process. The communications effectiveness of company staff is improved with the open plan, and work flow between individuals and departments is simultaneously facilitated. Additionally, systems furniture offers these other advantages:

- More flexible and readily accessible energy and communications distribution to better accommodate technology. Some systems offer power and data access above the work surface that is particularly helpful for laptop computer use (see Fig. 3.2.11-2).
- The use of vertical space provides greater storage in a smaller footprint thereby allowing companies to decrease the amount of space required per person (see Fig. 3.2.11-3).
- Systems panels define an employee's territory and create individual work environments that may be personalized (see Fig. 3.2.11-4).

PRIVATE OFFICES

Although systems furniture has revolutionized work environments and brought down a great number of private office walls, private offices have not and are not likely to disappear entirely. There are many reasons for this, not the least of which is that the private office answers two psychological needs: one for territory, the other for status. For some professionals, such as lawyers, accountants, engineers, and program managers, the private office as a status image is also essential to maintain credibility with their clients.

The furnishing choices for the private office are most typically freestanding case goods comprised of a desk (standard, L-shaped, or U-shaped), credenza, and vertical storage unit. A table and chairs for conferencing and/or a soft seating group sometimes accompany them.

Less common but not unusual is the trend to using systems furniture in enclosed offices. Technology has sprouted many young companies (particularly in the computer industry) within which hierarchy and status is intentionally avoided. In these flatter organizations, there tend to be fewer private offices. However, where they do exist for privacy or security, it is likely that the occupants previously worked in systems furniture environments and have come to rely on the support of the furniture to facilitate the work process. A back wall of hanging product provides greater flexibility than case goods and incorporates the principle of verticality to support more work activity.

TASK SEATING

Task seating is perhaps the most important furnishing item within an office facility. It affects nearly every individual within the office and has tremendous impact on those whose jobs have them seated for the greater part of the day. When a chair is uncomfortable, there is inevitably

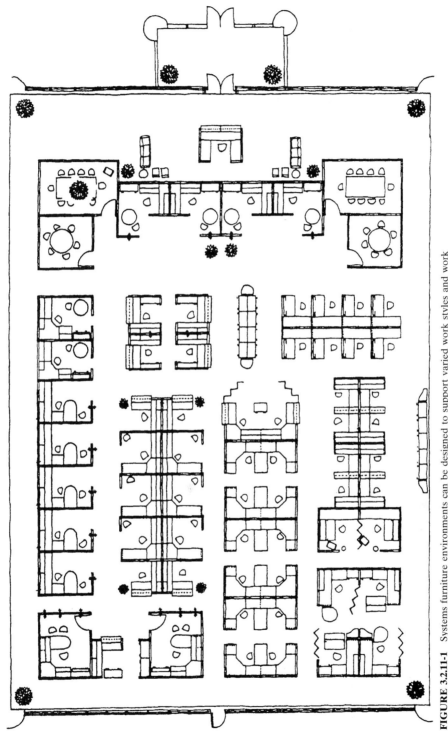

FIGURE 3.2.11-1 Systems furniture environments can be designed to support varied work styles and work processes. (*Reprinted with permission of Herman Miller, Inc., Zeeland, MI.*)

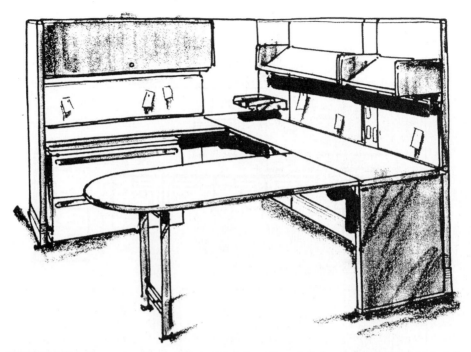

FIGURE 3.2.11-2 Power access is conveniently placed above the work surface. Additional attributes include a peninsula work surface that allows for both a conference area and space to work with an adjacent computer. (*Reprinted with permission of Herman Miller, Inc., Zeeland, MI.*)

a loss in productivity. Individuals find that they must get up and walk about to relieve themselves of the stress and strains caused by their chairs.

Statistics show that bad chairs increase absenteeism and insurance claims, and the latter consequently increase a company's insurance rates. In fact, lower back pain caused by poor seating is the leading cause of absenteeism in the United States and is responsible for $14 billion a year in medical claims, according to Dr. Stover Smith of the Harvard School of Public Health.[1] Additionally, many circulatory problems are attributed to the sometimes hard, flat seats on task chairs. The seat can result in 85 to 100 pounds per square inch of pressure on the thighs. Capillaries begin to restrict blood flow at only 20 pounds per square inch.[2]

In response to these concerns, manufacturers have introduced many chairs labeled "ergonomic." (Ergonomics is the study of the relationship between human physiology and the physical environment.) These chairs are designed to support the body in a variety of positions. All "ergonomic" task chairs are not equal, and a facility manager should carefully consider the following when evaluating seating:

- How well does the chair respond to changing body postures? A chair must respond with the user to changing job functions, social encounters, and intrinsic body shifts.
- How well does the chair provide support to the following body areas: back (including lumbar), seat, thighs, and arms?
- Do the contours of the chair evenly distribute weight loads?
- Does the chair have edges that will not restrict blood flow?

FIGURE 3.2.11-3 The use of vertical space allows for many storage opportunities. Note also the use of a screen to provide selective privacy as well as the two freestanding tables that serve both as auxiliary work surfaces and meeting space. (*Reprinted with permission of Herman Miller, Inc., Zeeland, MI.*)

Passive Versus Active Ergonomics

The industry has developed two types of ergonomic task chairs: *passive* ergonomic and *active* ergonomic. The essential difference between the two is that a passive ergonomic chair responds to changing body postures without much manual adjustment. Conversely, an active ergonomic chair has a number of knobs, paddles, and/or levers to control such things as forward tilt, seat depth, back angle, arm width, and arm height. Both types of chairs are typically offered with either manual or pneumatic height adjustment, and unless cost is the overriding factor, pneumatic height adjustment is recommended for its ease of use.

GENERAL CONCERNS THAT AFFECT FURNISHINGS PLANNING AND SPECIFICATION

Ergonomics

An increasingly important issue, ergonomics goes beyond chair adjustability. A rise in ergonomically based, job-related health problems such as cumulative trauma disorders (CTDs) and vision difficulties has seen a subsequent rise in the number of insurance claims and lawsuits. To alleviate some of the risk involved with intense computer use that leads to these conditions, facility managers should consider the following elements:

FIGURE 3.2.11-4 This semiprivate workstation has varying panel heights that define territory while simultaneously adding visual interest and opening up the office. (*Reprinted with permission of Herman Miller, Inc., Zeeland, MI.*)

- *Keyboard height.* The surface should be between 26 and 28½ inches high and be adjustable.
- *Computer screen glare.* The reduction of glare is typically accommodated in the lighting design or furniture layout and orientation.
- *Angle of video display terminal (VDT).* Such things as monitor lifts or split work surfaces will allow a user to vary the distance or angle of a VDT to provide visual variety. This in turn will promote relief from sustained viewing at a set position.

Fire Code

Many cities and states have specific flammability codes that regulate the specification of furniture. It is essential that furniture be specified to meet all applicable code requirements.

The California Technical Bulletin 133 (CAL TB 133), in effect since October of 1975, is the widely accepted standard by which furniture is tested for approval to meet a minimum fire standard. CAL TB 133 is a full-scale fire test for furniture manufactured for use in public buildings. The standards do not specify how to construct furniture, but only how it should perform when tested by the following procedure: A finished piece of furniture (never single components) is placed in a $10 \times 12 \times 8$ foot chamber and is tested by applying a gas flame for 80 seconds. Thereafter, measurements are taken to determine the rate and amount of heat generated, the temperature of the testing room while the furniture is burning, the smoke opacity, the amount of carbon monoxide generated, and the mass loss of the furniture after one hour or until the fire is completely extinguished.

The Americans With Disabilities Act (ADA)

The ADA was enacted in 1992 to regulate the removal of barriers for the physically disabled. It requires that all employers provide "reasonable accommodations" to individuals with disabilities unless doing so would result in "undue hardship." Although the ADA does not detail particular rules for compliance, it can be expected that over time it will evolve to become more specific and restrictive. This should be a consideration for the layout of the facility and also for the selection of furnishings. Versatility and adjustability become criteria for products that respond well to the needs of disabled individuals. For instance, facility managers should consider the following:

- Desks or work surfaces that are adjustable in height and angle
- Desks or work surfaces that have a minimum knee-well width of 30 inches to ensure accommodating the width of most wheelchairs[3]
- Desks or work surfaces that have a minimum knee-well depth of 19 inches to comfortably accommodate a wheelchair with footrest[4]
- Matte or eggshell work surfaces, as opposed to glossy ones that pose difficulties for individuals with perceptual disabilities
- U-shaped handles that have enough clearance on the inside to permit ease of use by individuals with manual limitations

Furniture Standards Program

Facility managers may wish to consider establishing a furniture standards program. Many companies, particularly larger companies with satellite locations, institute such programs that define the allowable square footage for each level of office or workstation; the manufacturer, model numbers, and finishes of furnishings items; the size and selection of workstation components; and the cost allowances for furnishings items. These programs can be advantageous because they ensure a level of equality for personnel with similar functions or titles in different departments or locations while also providing a consistent corporate image in various office locations. Other merits to standards programs include:

- Predictable cost and square-foot requirements to assist in long-range planning
- Savings through volume purchasing
- Minimal inventory and simplified purchasing procedures for frequent changes in layout
- Lower design fees

There are, however, some disadvantages to standards programs. Consider the following:

- Office standards limit design expression in the facility.
- A poorly conceived standards program can result in the perpetuation of bad design.
- Rigid programs may not address unique individual needs.
- Furniture finish standards may not coordinate with building finishes.

Fabrics and Finishes

In specifying furnishings, fabrics and finishes should be chosen carefully to assure that furnishings last many years without looking dated. A fabric's appropriateness should be evaluated as it relates to the following:

- *Abrasion resistance.* Typically rated with a double rub count known as the Wyzenbeek method. A fabric is considered durable if it withstands 15,000 double rubs.

- *Lightfastness.* Of particular concern if furnishings are located near a window or other natural light source. Fabric testing is performed with a machine called a Fade-O-meter that measures color loss at intervals of 20 hours. A fabric is considered to have a minimum rating at 80 hours' exposure without color loss.

- *Stain resistance.* Teflon- or Scotchgard-coated fabrics have greater stain resistance. Also consider the manufacturer's recommended method of cleaning.

- *Color.* Soiling can be expected to show over time if a fabric choice is too light.

- In considering the use of COMs (customer's own materials) for upholstered items, the advantages and disadvantages should be weighed against one another. For instance, use of a COM allows for unique design expression and perfect color matching or coordination with other elements of the workplace. However, the use of COMs can increase product lead times because furniture production is often delayed pending receipt of the fabric. There is also a greater potential for fabric discontinuance that will affect future furniture orders, and the extra step in the ordering process may create greater complications.

PROCURING OFFICE FURNISHINGS

Method of Procurement

Office furniture is typically priced at "list" and sold through office furniture dealers at a percentage off "list," otherwise known as "net." Many factors affect the discount amount that a dealer will extend on purchases. The dollar volume of the order, the anticipated volume of future orders, the profit margin inherent in the product for sale, and the ability and willingness of the manufacturer to extend a deeper discount to the dealer are all considerations.

In procuring furnishings, a facility manager may issue a request for proposal (RFP) to several dealers for the items that the company wishes to purchase. It is important that this RFP include delivery and installation pricing, because these figures can vary greatly between dealers and will have an overall impact on the purchase price. Alternatively, a facility manager may choose to forgo the RFP process and negotiate the furniture purchase directly with a selected dealer.

Another method of procurement is leasing, which offers the advantages of conserving capital, preserving credit lines, and greater financial flexibility. A lease may or may not be structured to include an option to buy.

MANUFACTURER/DEALER SELECTION

The selection of furnishings is as much about the manufacturer as it is about the product itself. Particularly in the case of systems furniture, the commitment to both the product line and the manufacturer is a long-term one. Therefore, it is important to consider the following elements:

- The manufacturer's proven record in the marketplace
- The manufacturer's financial stability
- Product nonobsolescence policies
- Product warranty
- Point of shipping and associated charges

- Product lead times
- Quick-ship program for projects requiring a rapid turnaround

A manufacturer's furniture line can be either *open* or *closed*. An open line is one that any furniture vendor may represent. A closed line, however, is represented only by select dealers designated by the furniture manufacturer. Because systems furniture lines are generally closed, the selection of a furniture manufacturer will often predetermine the furniture dealer with whom a company will work.

The role of the furniture dealer is to act as a liaison between the purchaser and the furniture manufacturer, facilitating the product order, delivery, and installation process. This role is key to the successful implementation of a furnishings plan, and a dealer's strength and financial stability should be carefully evaluated prior to establishing a relationship. As part of this evaluation, a facility manager should consider a dealer's strength as it relates to level of service and committed follow-up; breadth of service offerings; and ability to control installation, project management, and communications (customer service). For companies that have satellite offices, it will be important to note whether a dealer network exists that may provide service to remote sites.

The relationship between a company and the furniture dealer will extend beyond the initial furnishings purchase. Such things as service issues, the need for additional product, and, in the case of systems furniture, the need for layout reconfigurations will invariably arise. Therefore, a dealer should also be evaluated for its ability to offer the following services:

- A maintenance contract for the installed product
- The ability to refurbish or reupholster the product
- Warehousing and inventory management of customer product
- Trade-in allowances on systems product
- An inventory of recommended touch-up paint, miscellaneous hardware, and replacement parts

CONCLUSION

Furniture must satisfy a wide range of needs, not the least of which is to effectively and efficiently support the company work process. Other functional and practical considerations include comfort, adjustability, durability, aesthetics, and cost. In specifying furniture, it is important for the facility manager to focus on the details of the particular furnishing items and reconcile them with the various selection criteria. Along with changes in culture, technology, and business processes, a well-planned and furnished facility is a contributing factor in helping a company achieve its organizational and business goals.

REFERENCES

1. Herman Miller, Inc., *Seating Brochure,* 1986.
2. Herman Miller, Inc., *Magazine,* 1989.
3. Herman Miller, Inc., "Equal Opportunity Facilities," 1991.
4. J. Mueller, *The Workplace Workbook,* Dole Foundation, Washington, D.C., 1990.

PROJECT COMMUNICATIONS AND PROJECT MODELING

Kenneth H. Stowe, P.E.
George B.H. Macomber Co., Boston, Massachusetts

Bart Bauer, P.E.
Edwards and Kelcey, Boston, Massachusetts

The modern facility manager's ultimate goal is to reduce cost and maximize product life-cycle performance. Can today's high technology enhance a project team's performance? The answer is yes—but not without a thoughtfully reengineered communication process and an integrated philosophy of teamwork across all aspects of the project.

For participants to communicate successfully, the process must be as rapid, effective, efficient, and seamless as possible. The conventional process of hard copies and mail, and of traveling to meetings and coping with unreturned phone calls, is a slow and inefficient way of communicating to all members of a project team, be they owners or accountants, builders or suppliers, designers or users. In this dawn of a new age, it will become increasingly more costly to follow the old linear process of draw, print, package, mail, evaluate, comment, return, revise, analyze, and then finally choose a direction. Technology has provided the opportunity and the power to reengineer the communication process to reduce time and errors and increase value and flexibility. Additionally, this use of technology provides a competitive advantage to all members of the project team who are participating in an integrated philosophical and technological approach.

To understand the reengineered communication model, we must agree on what the components of the communication process really are. We believe that there are three separate elements of the process. The first element is who is communicated with and when; the second, how are they communicated with; and the third, what is communicated.

The value of teamwork has always been appreciated. Traditionally, however, the team was composed of a limited number of members in a finite geographical area and in related disciplines. However, the efforts of building the best project team for any facility should not be limited by geography, disciplines, or an intentional restriction of added value through the sharing of knowledge. An argument can be made to increase the number of members on a project team and, in fact, to adopt a global view—a new philosophy—to involve any individual who is related to the facility from its conception to its ultimate demise, during the formative period of a facility development project. Let us reinforce that here. To communicate efficiently about any item related to any aspect of the project, representatives from all phases of the project life cycle must have the opportunity to participate. Only then is the highest value achieved.

Technology has provided a tool to facilitate these new project teams. The facility manager should be able to build the team for a diversified owner using the best consultants, designers, and contractors. This may draw individuals from tremendous distances and varied collaborations. By using electronic innovations designed for communication and data sharing, the opportunity is available today for the facility manager to reengineer the communication process to involve all the parties.

Once the team has been built, it is critical that its members effectively communicate ideas and concepts, as well as data, throughout the process. Many project teams now elect to use a shared, secure Web site for vital data storage and for advanced communications. The data include master schedules, computer-aided design files, specifications, drawing and submittal logs, site photographs, meeting minutes, requests for information, change orders, daily field reports, and so forth. Sharing eliminates duplication and inaccuracies. Immediate access speeds responses to questions from the site or elsewhere. Elements of the project data that are related can even be "threaded" for better understanding of complex interactions.

Sharing project data has a positive influence on a team. When the entire team observes a request for information being submitted, and when and how it is answered, this sharing increases the sense of unity, accountability, and responsiveness. The team can collaborate with pictures, computer-aided design (CAD), even videos and animation, with fewer time or geographical limits. Other members can comment or intervene if necessary, so that decisions are made quickly, with more participation, and accordingly, with more commitment. Faster, more informed decisions lead to faster time-to-market projects.

INTEGRATING THE MODELS

Creating one secure site for current and archived project information opens the door to computer-integrated construction modeling. This is an integrated approach that streamlines workflow and connects the models that reflect current or explored forecasts of the project plan. An example of this type of integrated models is to create quantity reporting from intelligent CAD models. These data inform the cost model, which generates manpower estimates by trade. These labor projections inform the schedule durations (through crew-size projections). The schedule generates overall project durations and timing, which in turn derive important time-variable costs for the estimate.

Efficiencies of Combining CAD and the Extranet

Sharing project data often results in more accurate models because more sharing means more input into the design effort. This often comes in the form of constructability input, phasing requirements, and cost-saving ideas. Three-dimensional features of the CAD model derive greater value because they reach a greater audience earlier in the process. Nongraphical enhancements to CAD models (e.g., manufacturing data and links to specs) generate efficiencies. The CAD model serves as the trade and vendor coordination vehicle that provides an electronic as-built.

An example of this occurred when the project Web site for one project team enabled CAD to provide efficiencies well beyond design. Centerbrook Architects, working from their Connecticut office on the $20-million phased renovation project at Dartmouth College in Hanover, New Hampshire, posted CAD files to a project Web site. The owner, consultants, and construction manager (CM), located in three other states, had immediate access to them. The construction manager, the George B.H. Macomber Company, downloaded the CAD files, performed some cost and schedule analyses, and posted some very advanced products back to the project site (see Fig. 3.2.12-1).

This collaboration achieved five positive results:

- Faster distribution of the drawings, and savings on printing, packing, and shipping
- Faster and more accurate development of the estimated quantities
- Three-dimensional color output (see Fig. 3.2.12-1) for better understanding of the estimate and the CM's assumptions
- Faster distribution of the estimate and schedule
- Graphical presentation of the phasing and essential dates

Following are more specific examples of the way each segment of project communications can be improved by using the technological tools that are available today.

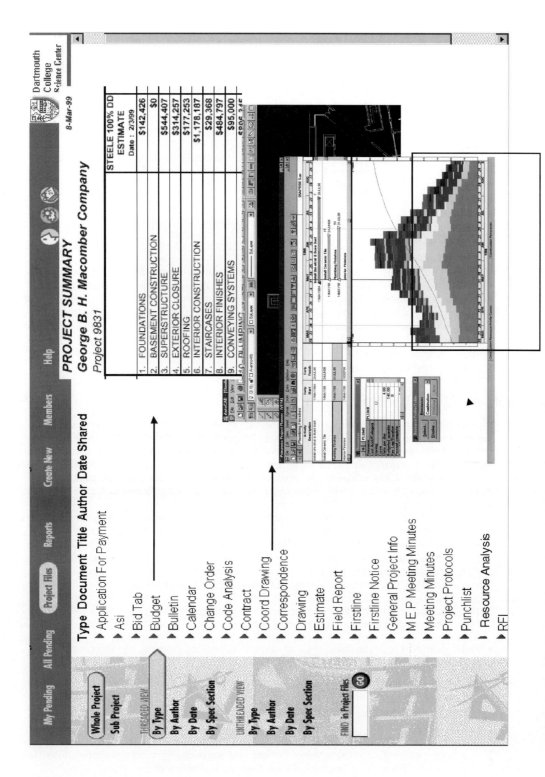

FIGURE 3.2.12-1 Project Web site allows project team members to secure access to shared information. *(Courtesy of Ken Stowe, George B.H. Macomber Company.)*

Sharing CAD Files.   Designers create an electronic design file on CAD that contains an enormous amount of knowledge from a group of design professionals. The construction management team now has the ability to analyze costs and schedules directly from this rich database of design details. CAD files should be sent electronically, and detailed estimates should be taken directly from these electronic files. Schedules can be attached and linked to these estimates, and an initial project plan—received within a few days—will contain a compound document that fully represents the full cost and schedule implications of the project.

Using Advanced Planning Software.   One way to save owners time and money is to develop a comprehensive plan by using advanced critical path method (CPM) scheduling software, such as Primavera Project Planner. To use this software strictly for construction activities wastes its power and potential. Instead, teams should use this technology to create a single database of activities that includes program decisions, design, procurement, construction, commissioning, owners' equipment, and occupancy details. The schedule should be loaded with resources to anticipate which subcontractors have the requisite forces and will be invited to bid.

Due Diligence.   A comprehensive analysis is vital for a company in assessing the purchase of property. For example, a building is selling for $22.5 million, and renovating will require the investment of an additional $50 million and take 12 months to complete. The company is desperate for room to expand and wants to know if it should purchase the property.

This investigation includes an analysis of the roof, HVAC and other skeletal operational systems, the exterior facade, the interior layout, the presence of asbestos and other potential environmental hazards, site evaluation, traffic assessment, insulation, telecommunications data, structural integrity, flexibility, and accessibility compliance.

Consulting experts working individually in each field could require 12 weeks to compile the necessary data by conventional means. Much is at stake in such an analysis. Cost, schedule, risks, and assumptions all need to be factored in. However, the turnaround time of this due diligence effort can be reduced to four weeks. This means that purchase, execution, and occupancy can be achieved earlier.

Expedite Decisions with the Internet.   The Internet is an agent of dramatic change in the world of design and construction. It is the vehicle for improving turnaround time for construction and design projects. Consider these time-savers:

- Web use facilitates communication as a team. Arranging a meeting through e-mail and integrating scheduling options can be accomplished very quickly.
- Most e-mail transmission is instant. Instead of a number of missed telephone messages among five team members and checking everyone's schedule, e-mail allows immediate communication and immediate response. A meeting that formerly took two days to arrange can be scheduled in less than an hour.

On-Line Project Web Sites for Communications.   The Internet allows project meetings to take place on-line. A secure project site allows a team of architects, engineers, designers, and construction consultants to transmit their portions of the design to a site where all other team members can view them and make recommendations.

Consensus can be reached in hours instead of days, and action can be taken immediately. More deadlines can be met through this new technology and more options can be analyzed. And no members of the project team had to leave their respective offices for this meeting.

The ability of a large audience to provide input simultaneously from their desktops dramatically increases the ability to get the right design direction from the start. This increases the reliability of the initial budget and extends the amount of time you will have to look for creative cost saving.

The Competitive Advantage

Ultimately, every day that you reduce in planning and construction is a day of potential revenue from your employees or tenants, once the facility is operational. Shorter schedules also reflect savings in interest on construction and bridge loans and shorten the time to market for your product, the facility. Combining teamwork and technology in your construction project can create a significant competitive advantage.

SECTION 3.3
FACILITY MANAGEMENT FUNCTIONS

Kristin Hill, Cindy Aiguier, John Morganti, and Bonnie Seaberg
International Facility Management Association (IFMA),
Boston, Massachusetts

Norm Faucher
Association for Facilities Engineering (AFE),
Boston, Massachusetts

In the introduction to the design and construction process phase, it was indicated that the completion of the design and construction phase does not signal the end of a project for the facility manager (FM). In fact, it can be considered a beginning—the beginning of the use of the facility. It is here that the knowledge transfer and management functions begin.

All of the requirements and associated documentation need to be assimilated and made available to all facility stakeholders. Moving forward, FMs will have to look at all of the knowledge-related systems in an integrated format. This information needs to be accessible by other departments such as administration, human resources, information systems, and finance. The goal should be to funnel the information from all of these groups together so that each may extract the information needed for decision making. Well-assimilated information will enhance the communication and synergy among all of these groups.

The following articles will show you the tools to assess where you are, to determine where you want to go, and to help you select the tools to aid your passage through the facility's life cycle. Driven by advancements in technology, changes will continue to occur in business products, processes, and procedures that will affect the facility. Facilities management becomes the means for coping with the changes that will occur during the facility's life cycle. It becomes a process for controlling the effects of change. These changes will begin a new cycle of planning, budgeting, design, and construction.

ARTICLE 3.3.1

COMPARISON OF FM AUTOMATION SYSTEMS

Denis R. Cormier

Facilities Management Solutions, Inc., San Diego, California

This article addresses the process and elements for comparison and selection when considering various computer-aided facilities management (CAFM) systems. Before discussing the process and elements of comparison and selection, it may be useful to briefly review some of the more obvious reasons for a business to consider acquiring and implementing a CAFM system:

- To manage and control assets proactively whether they are buildings, building systems, machinery, equipment, computers, or people
- To optimize the business process of managing facilities by measuring and controlling the total life cycle cost of ownership of assets
- To increase productivity and efficient utilization of assets

"Find a way to improve productivity and provide more useful and timely information at less cost. Business as usual just isn't good enough anymore." This message is heard more and more frequently from corporate senior managers as they try to contend with higher real estate and facilities costs. When you consider the total life cycle cost of facilities (buildings and infrastructure) of a corporation, you soon realize that facilities are its second most costly asset. Human resources (employees) are the most costly asset. The extent to which a corporation proactively measures and manages the performance of this dynamic tool, which we call facilities, will to a large extent contribute directly to a corporation's profitability and competitiveness.

In 1995, Arthur Andersen LLP conducted an in-depth study of corporate real estate. One of the significant findings of the study was that corporate senior management generally views real estate (facilities) as a "cost of doing business," rather than a strategic "tool" that can be managed and that can significantly impact how effectively a corporation competes in its respective market.

Historically, corporations have viewed facilities as static shelters to house employees and equipment. In addition, there are usually few processes, if any, in place to measure how effectively they are using their facilities. As a result, corporations often evolve and find occurrences of several areas of clutter filled with unused, abandoned, or obsolete assets (furniture, machines, and equipment) throughout their facilities. In addition to using space ineffectively, the lost revenue from not properly disposing of or redeploying surplus assets is just one of many examples of lost opportunities.

Fortunately, CAFM technology provides a method for dynamically and proactively managing corporate assets. However, the purchase of CAFM software alone will not ensure success in managing corporate facilities. The key to the success of any software solution is proper integration and implementation that closely mirrors well-defined processes. It cannot be stated strongly enough that the processes of your company should dictate the appropriate CAFM solution and not the reverse! For this reason, you must take the time to document your current processes, sometimes referred to as *process mapping*.

One fundamental objective of any process analysis is to provide the ability to create and access information that is commonly understood and shared by everyone who is affected by processes in the organization. To accomplish this analysis, a business process model is created (refer to Art. 3.4.3, "Quality Assurance Practices," as a template). These are the four elements of this process model:

1. "What Is"
2. "Should Be"
3. "What to Do"
4. "How to Do It"

Each of these elements has four critical components:

1. "Inputs"
2. "Outputs"
3. "Controls"
4. "Supports"

In addition, the process map must include all feedback loops that provide updates of a given process.

After completing a process model, you will achieve the following:

1. Clearly understand the existing processes within various departments and organizations.
2. Determine what resources (people, equipment, software) are involved in various processes and thus learn who will be affected by the chosen CAFM system and who will need to be trained in using CAFM.
3. Identify what legacy systems and platforms are used and how CAFM will either extract data from or input data to these legacy systems.

Before investing in CAFM technology, a business must build an intelligent strategy that outlines the performance metrics of the software. In other words, what elements do you want to measure to demonstrate clearly whether the organization is improving? A thorough needs analysis and an assessment of existing processes and procedures will usually enable a business to obtain these elements.

When evaluating a CAFM system, it should be viewed as a decision-support tool. When this tool is properly chosen, integrated, and used in the business process of a company, it can ensure that all of the key elements of your business—people, facilities, and equipment—are working to the best advantage for corporate goals and objectives.

A CAFM system in its simplest form is a huge repository of data that is designed for use as a decision-support tool. Collecting data, whether in a personal computer, mainframe computer, or computer client server, just for the sake of collecting data is of no value to anyone. Only when the data can be easily converted to meaningful reports (graphic or tabular) for use as a decision-support tool has CAFM achieved its goal.

Accurately compiled data can be used to generate reports that clearly illustrate such things as:

- Space planning and forecasting
- Asset management reports
- Changes in space use
- Space chargeback reports
- Change and cost of churn
- Change in personnel requirements relative to work load

An effective CAFM system provides critical decision-support information easily and quickly to anyone in the organization who needs the information in a visual or graphical format or a numerical data format.

Corporations have critical and possibly redundant information residing in various computer systems, sometimes called *islands of information,* that are controlled by various departments. The best CAFM solutions are designed to break down barriers between sources of information. Eliminating islands of information and redundant databases is the hallmark of a well-designed CAFM solution.

To the extent that a company proactively manages its facilities and assets, enabling it to anticipate both internal and external changes, there will be a direct benefit in positioning the company to be more competitive in the marketplace.

METHODOLOGY

The process of evaluating and ultimately selecting an appropriate CAFM system for a corporation can be quite simple or complex, depending on the needs expressed by the various departments that are affected by the CAFM system. If the intent is to have a stand-alone, PC-based CAFM system used primarily within a facilities department for such things as space planning and move management, then the evaluation and selection can be fairly straightforward and can be accomplished by relatively few people—probably members of the facilities department. This is sometimes called a *bottom-up* approach, where just a few modules of a total suite of CAFM software are deemed necessary. However, if a more global enterprise-wide system is warranted, where several departments such as Human Resources, Finance, Real Estate, or Information Technology will be affected, then a more *top-down* approach is probably warranted.

If a top-down approach is viewed as the preferred method, then it is strongly advised that senior management be involved or at least well apprised in the early stages of needs assessment, as well as system definition and specification. Senior management needs to understand clearly the philosophy and intent of any enterprise-wide CAFM system and must support the goals and objectives if any system is to be successful.

The methodology of evaluating and selecting a CAFM system should address as many departmental and organizational process and procedural issues as possible. Try to anticipate and plan for a tool that is functionally flexible and technically robust enough to respond to the specific needs of the facilities department and also to the needs of several other departments, as well as other applicable sites.

In this context of a broad scope for the potential impact of any chosen system, a bottom-up approach can be risky. For example, perceived ease of use for one or two specific departmental processes can lead a team to overvalue system characteristics for a department and undervalue those factors that are critical for enterprise-wide implementation.

The following list outlines the specific methodology steps:

- Project plan
- Functional requirements analysis (needs assessment)
- Technical requirements analysis
- CAFM system recommendations and implementation strategy
- System implementation and testing (pilot application test)
- System-wide transition and training (documentation)
- Evaluation and periodic corporate-wide communication of milestones and achievements

A project plan must be developed to establish the goals and objectives, priorities, and time schedules. The plan should include team members, roles and responsibilities, and milestones.

The next step is to conduct a comprehensive functional requirements analysis using a systematic business process/work flow approach. Sometimes, this is called documenting *what-is* conditions, or documenting *reality*. This is usually done by personally interviewing individuals from various departments who might use a CAFM system or whose business processes could be affected by a CAFM system. Some corporations have well-documented processes and procedures manuals that can be referenced during the interviewing process. Unless a corporation is very diligent in maintaining its procedure manuals, it is quite likely that the actual processes and procedures do not necessarily mirror what is written in the manuals. Quite often, documented processes and procedures do not exist. The resulting report would typically define current business processes in the form of various flowcharts. The report should also point out any inconsistencies between actual procedures and documented procedures. Any functional overlaps, as well as redundant processes, within departments or between departments should also be pointed out.

The technical requirements analysis provides information related to any existing automated tools that are used and what specific data are accessed, how frequently, and by whom. Examples of these are CAD software, personnel databases, fixed asset ledgers, etc. In addition, all legacy computer systems and associated software and platforms should be documented.

The collective knowledge that is gained from both the functional requirements analysis and the technical requirements analysis should be included as part of the evaluative criteria to identify and recommend a specific CAFM suite of tools that are best suited to the needs of the specific corporation.

This is generally achieved by developing a customized matrix that compares the features and benefits of various CAFM software packages in relation to the specific needs that were identified in the functional requirements analysis and technical requirements analysis. However, this "feature comparison" approach focuses attention on a list of features and tends to divert attention from the key, subtle reason for evaluating features in the first place: namely, successful implementation of a CAFM system that improves business results. Products that look equivalent in features may have drastically different records in being used effectively over a period of years to accomplish business process improvements. The only effective means to avoid this mistake in evaluation is to do a thorough job of interviewing several existing user references to the short list of CAFM systems that seem to provide the best job. If possible, contact references in a similar industry and similar facility size.

COMPARISON OF CAFM SYSTEMS

Before discussing the various elements in the comparative tables, note that this is just an illustration of the way a matrix can be used to assist in selecting a CAFM system. The elements and format of any product comparison matrix should be tailored to your company's specific requirements. Furthermore, the data fields are merely depicted as an example and may not represent the latest version of the respective CAFM software provider's product offerings. In addition, the material presented should not be construed as an endorsement of one CAFM solution over another. The information is presented only for illustrative purposes.

Tables 3.3.1-1 through 3.3.1-4 help to categorize and compare the features that are offered in each CAFM software package. The tables have been separated into these four segments:

1. Key features summary
2. Modules and tools summary
3. Technical information summary
4. Cost information summary

TABLE 3.3.1-1 Key Features Summary

Features	SPAN-FM	Aperture	FIS	FM:SPACE	ARCHIBUS/FM Oracle
Maintain drawing accuracy and validity	Space analysis module	Space manager module	Facility inventory manager module or facility drawing coordinator module	Space module	Space management module
Enterprise level functionality	Y	Y	Y	Y	Y
Modular construction	Y	Y	Y	Y	Y
Acceptable RDBMS (Oracle, Sybase, SQL Server)	Y	Y	Y	N	Y
Client–server	Y	Y	Y	N	Y
LAN/WAN	Y/Y	Y/Y	Y/Y	N/N	Y/Y
Ability to import/export AutoCAD (DWG and DWF)	Y	Y	Y	Y	Y

The features that are listed in Table 3.3.1-1 are some of the more important items for consideration, such as drawing accuracy, enterprise level functionality, modular construction, random database management system (RDBMS), client/server capability, local area networking (LAN), wide area networking (WAN), and AutoCAD import/export ability. Additional features could be added to this table in accordance with the needs that are determined to be important to your corporation.

The features listed in Table 3.3.1-2 state the various modules and tools that are offered with the various CAFM software packages. In addition to the basic module that is offered with each package, a suite of modules and tools can be purchased initially or at a later date, depending on the needs of the corporation. If the corporation chooses to implement an enterprise-wide CAFM system, then most if not all of the modules would be purchased. If a more localized solution is preferred, then fewer modules would be purchased. Some of the modules available are lease management, move management, strategic space planning, maintenance management, accounting/chargeback, communication/cable management, and HR personnel management.

The features described in Table 3.3.1-3, "Technical Information Summary," address primarily system-level specifications, such as platform, network access, native database support, database connectivity, user interface, security, reports, file formats, and interoperability.

Finally, Table 3.3.1-4, "Cost Information Summary," lists the retail prices of the various modules of each CAFM system. The prices vary depending on how many concurrent seats or licenses need to be purchased.

Using the information provided in Tables 3.3.1-1 through 3.3.1-4 a thorough evaluation should be made emphasizing scalability of software, cost of software, cost of training, cost and ease of software integration, cost of software maintenance, and after-sale support.

The resulting recommendation should be accompanied by a standard return on investment (ROI) analysis. The ROI should attempt to reflect and quantify quantitative savings, to the greatest extent possible. Some of the quantitative savings to be realized are reduced operational cost due to automating routine manual procedures, reduced occupancy cost due to increased efficiency in space usage, and implementation of a space chargeback system. In addition, reduced maintenance cost and reduced cost associated with churn and relocation should be considered. The recommendation should also include the implementation strategy and associated time schedules.

The system implementation and testing or pilot test should interface various data tables and provide sample interfaces and output formats. Menu-driven windows and floating control boxes should be provided in a *drill-down,* clear and easily understood layout. A report writer and query builder should be provided with plain English terminology. Once the pilot test is

TABLE 3.3.1-2 Modules and Tools Summary

Features	SPAN-FM	Aperture	FIS	FM:SPACE	ARCHIBUS/FM Oracle
Basic Module	SPAN FM 6.3	Aperture Professional	FIS/FM	FM:SPACE 5.1	ARCHIBUS/FM 10
Stacking	WinStack*	Space Manager	Facility Stack & Block	Stacking*	Stacking
Blocking		Space Equip. Mgr.	Facility Stack & Block		Blocking
Lease Management	Lease Management*	Lease Manager	Facility Property Manager	Lease & Property*	Real Property and Lease Management
Property Management	Property Portfolio	Lease Manager	Facility Property Manager	Lease & Property*	Real Property and Lease Management
Move Management	Asset Management*	Personnel Manager	Facility Move Manager	Move Manager	Furniture and Equipment Management
Equipment Management	Asset Management*	Equipment Manager	Facility Inventory Manager*	Asset Management*	Furniture and Equipment Management
Furniture Management	Asset Management*	Furniture Manager	Facility Inventory Manager*	Asset Management*	Furniture and Equipment Management
Strategic Space Planning	Space Analysis*	Space Manager	Facility Master Planner	Strategic Planning*	Space Management
Maintenance Management	Maintenance Management*	Maximo, Datawave		Request Plus	Building Operation Management
Project Budgeting	Project Budgeting*	Facilities Atlas*	Facility Requirements Programmer		
Comm/Cable Management	Cable Management	DocuNet* Data Center Manager	Facility Inventory Manager*		Telecom & Cable Management
Accounting/ Chargeback	Building Occupancy Chargeback	Space Manager	Facility Inventory Manager*	Space Management*	Real Property and Lease Management
Report Writing	Infomaker-Report Writing	Aperture Professional	FIS/FM	FM: SPACE 5.1	ARCHIBUS/FM 10
Web Access	Web Server	Smart Pictures Server	Web Server	FM:WEB	I-Net Tool Kit
HR Personnel Management	Property Portfolio	Personnel Manager	Facility Inventory Manager*	FM:SPACE 5.1	Space Management
Material Handling	Material Handling	Asset Management	Facility Inventory Manager*		

*Included in base price.

implemented and proven, then system-wide transition should proceed along with proper training of all CAFM users as well as process and procedures documentation.

Finally, the CAFM system should be evaluated periodically to ensure that the information and resulting reports provide useful and timely information that supports the corporate strategic decision-making process. As with any change, some will embrace change, and some will resist it. For this reason, the implementation team should publish periodic corporate-wide bulletins stating the various milestones, achievements, and successes of the CAFM system. Use all means of advertisement available (e.g., company newspapers, display exhibits at various visible locations throughout the company, management presentations, and banquets with successful team members). By making heroes of those employees who embrace change, reluctant employees will begin to participate.

TABLE 3.3.1-3 Technical Information Summary

Features	SPAN-FM	Aperture	FIS	FM:SPACE	ARCHIBUS/FM Oracle
Training					
On-Line	N	N	N	N	N
Vendor site	Y	$1980/attendee/4 days	$350/day/attendee	Y	$725/day/attendee
Customer's site	$1000/day up to 3 attendees + expenses; $350/day/ additional attendee	$8500/4 days/up to 10 attendees + expenses	$8400/8 days/3 attendees	$1000/day/ unlimited attendees + expenses	$1500/day/up to 6 attendees + expenses
Platform					
Win 95	Y	Y	Y	Y	Y
Win NT	Y	Y	Y	Y	Y
LAN/WAN	Y/Y	Y/Y	Y/Y	Y/N	Y/Y
Network access					
TCP/IP	Y	Y	Y	Y	Y
SPX/IPX	Y	Y	Y	Y	Y
Native database support					
Aperture DB C++	N	Y	N	N	N
Oracle 7.X	Y	Y via ODBC link	Y	N	Y
DB2	N	Y via ODBC link	N	N	N
Access 7.0	N	Y via ODBC link	N	N	Y
Sybase	Y	Y via ODBC link	N	N	Y
Informix	Y	Y via ODBC link	N	N	N
MS SQL Server 6.0	Y	Y via ODBC link	N	N	Y
Database connectivity					
ODBC	Y	Y	Y	Y	Y
JDBC	N	N	Y	N	N
User interface					
GUI	Y	Y	Y	Y	Y
WEB	Y: work orders and reports	Y: drawings, data, and reports	Y: reports and CAD drawings	Y: work orders and reports	Y: work orders, reports, and queries
Client–server	Y	Y	Y	N	Y
Security					
User	Y	Y	Y	Y	Y
Group	Y	Y	Y	Y	Y
Module	Y	Y	Y	Y	Y
No. of standard reports	300	80	100	N/A	1400
Supports customized reports	Y: with InfoMaker Tool	Y: included w/base package	Y: included w/base package	Y: included w/base package	Y: included w/base package
File formats					
Native CAD drawings	DWG, DWF	Proprietary and DWG, DWF	DWG, DWF	DWG, DWF	DWG, DWF

TABLE 3.3.1-4 Cost Information Summary

	SPAN-FM	Retail price	Aperture	Retail price	FIS	Retail price	FM:SPACE	Retail price	ARCHIBUS/FM Oracle	Retail price
Basic package w/CAD interface	SPAN.FM 6.3 Windows Suite + CAD Integrator Module	$8,500	Aperture Professional Version	$4,855	FIM + FDC	$22,500	FM:SPACE Windows V 5.1	$8,500	ARCHIBUS/FM 10	$8,995
Licensing 5 seats	SPAN.FM 6.3 Windows Suite + CAD Integrator Module	$35,000	Aperture Professional Version	$17,410	FIM + FDC		FM:SPACE Windows V 5.1		ARCHIBUS/FM 10	$8,995
Cost of underlying database 10 seats	Oracle Workgroup	$1,675		included	Oracle Workgroup				Oracle Workgroup	$1,675
	SPAN.FM 6.3 Windows Suite + CAD Integrator Module	$72,250	Aperture Professional Version	$34,820	FIM + FDC		FM:SPACE Windows V 5.1		ARCHIBUS/FM 10	$15,395
Cost of underlying database 60 seats	Oracle Enterprise	$12,900		included	Oracle Enterprise				Oracle Enterprise	$12,900
	SPAN.FM 6.3 Windows Suite + CAD Integrator Module	$408,000	Aperture Professional Version	$208,920	FIM + FDC		FM:SPACE Windows V 5.1		ARCHIBUS/FM 10	$49,140
Additional modules, cost by seat 1 seat	Building Occupancy Chargeback	$3,500	Space Manager	$3,500	FIS Facility Coordinator	N/A	FM: Move Manager	$1,500	Real Property & Lease Mgmt.	$2,334
	Cable Management	$4,500	Personnel Manager	$3,500	FIS Facility Move Manager	N/A	FM: Request Manager Plus	$1,500	Space Management	$2,334
	InfoMaker Report Writing	$250	Equipment Manager	$3,500	FIS Facility Property Mgr.	N/A	FM: CAD	$1,000	ACAD ARX Overlay w/Dsgn. Mgmt.	$1,749
			Furniture Manager	$3,500	FIS Requiremts Programmer	N/A			Furniture & Equip. Mgmt.	$2,334
			Lease Manager	$7,500	FIS Facility Stack & Block	N/A			Telecom & Cable Mgmt.	$2,334
			Data Center Manager	$5,414	FIS Facility Master Plan				Bldg. Operations Mgmt.	$2,334
5 seats	Building Occupancy Chargeback	$15,750	Space Manager	$3,500	FIS Facility Coordinator	$3,750	FM: Move Manager	$7,500	Real Property & Lease Mgmt.	$11,671
	Cable Management	$20,250	Personnel Manager	$3,500	FIS Facility Move Manager	$7,500	FM: Request Manager Plus	$7,500	Space Management	$11,671

TABLE 3.3.1-4 (Continued)

	SPAN-FM	Retail price	Aperture	Retail price	FIS	Retail price	FM:SPACE	Retail price	ARCHIBUS/FM Oracle	Retail price
	InfoMaker Report Writing	$1,125	Equipment Manager	$3,500	FIS Facility Property Mgr.	$7,500	FM: CAD	$5,000	ACAD ARX Overlay w/Dsgn. Mgmt.	$8,746
			Furniture Manager	$3,500	FIS Requiremts Programmer	$7,500			Furniture & Equip. Mgmt.	$11,671
			Lease Manager	$7,500	FIS Facility Stack & Block	$7,500			Telecom & Cable Mgmt.	$11,671
			Data Center Manager	$5,414	FIS Facility Master Plan	$7,500			Bldg. Operations Mgmt.	$11,671
10 seats	Building Occupancy Chargeback	$29,750	Space Manager	$3,500	FIS Facility Coordinator	$7,188	FM: Move Manager	$12,000	Real Property & Lease Mgmt.	$23,342
	Cable Management	$38,250	Personnel Manager	$3,500	FIS Facility Move Manager	$14,375	FM: Request Manager Plus	$12,000	Space Management	$23,342
	InfoMaker Report Writing	$2,125	Equipment Manager	$3,500	FIS Facility Property Mgr.	$14,375	FM: CAD	$12,000	ACAD ARX Overlay w/Dsgn. Mgmt.	$17,492
			Furniture Manager	$3,500	FIS Requiremts Programmer	$14,375			Furniture & Equip. Mgmt.	23,342
			Lease Manager	$7,500	FIS Facility Stack & Block	$14,375			Telecom & Cable Mgmt.	$23,342
			Data Center Manager	$5,414	FIS Facility Master Plan	$14,375			Bldg. Operations Mgmt.	$23,342
60 seats	Building Occupancy Chargeback	$168,000	Space Manager	$3,500	FIS Facility Coordinator	$28,125	FM: Move Manager	$45,000	Real Property & LeaseMgmt.	$140,049
	Cable Management	$216,000	Personnel Manager	$3,500	FIS Facility Move Manager	$56,250	FM: Request Manager Plus	$45,000	Space Management	$140,049
	InfoMaker Report Writing	$12,000	Equipment Manager	$3,500	FIS Facility Property Mgr.	$56,250	FM: CAD	$36,000	ACAD ARX Overlay w/Dsgn. Mgmt.	$104,949
			Furniture Manager	$3,500	FIS Requiremts Programmer	$56,250			Furniture & Equip. Mgmt.	$104,949
			Lease Manager	$7,500	FIS Facility Stack & Block	$56,250			Telecom & Cable Mgmt.	$104,949
			Data Center Manager	$5,414	FIS Facility Master Plan	$56,250			Bldg. Operations Mgmt.	$104,949
Additional tools	Web Server Uniprocessor	$25,000	Aperture SmartPics Server	$9,200					Executive Information Summary	$995
	Web Server Multiprocessor	$59,000							ARCHIBUS/FM 10 Viewer Kit	$2,500
Maintenance cost per seat 1 year	Basic package	$1,530	Basic package	$711	Basic package	$675	Basic package	$800	Basic package	Included

SUMMARY

A well-defined and well-integrated CAFM system can significantly contribute to the goals and profit objectives of a corporation as a strategic decision-support tool. To ensure successful CAFM implementation, certain essential elements must be adhered to. If an enterprise-wide CAFM system is deemed appropriate, the understanding and commitment of top management is extremely important. Also, just as important, there must be a commitment to the adherence to data stewardship and the integrity of data to ensure continuous improvement and efficiencies. If the individuals who use the CAFM system lose confidence in the integrity of the data and the reports generated by the CAFM system, it will eventually be abandoned.

CONTRIBUTORS

Steve Lowe, Facility Information Systems, Inc., Camarillo, California

Joan M. Licitra, Esq., Boston, Massachusetts

GENERAL REFERENCES

Paul R. Smith and Paul J. Fairbourn, *Reengineering the Corporate Structure—Methodology and Technique,* INEL Quality Conference, Idaho Falls, ID, 12 October 1994.

U.S. General Services Administration (GSA); www.gsa.gov.

ARTICLE 3.3.2

THE GREAT GLOBAL INFRASTRUCTURE RESOURCE MANAGEMENT GAME: HOW TECHNOLOGY IS CREATING NEW VALUE PROPOSITIONS IN INFRASTRUCTURE MANAGEMENT FOR THE TWENTY-FIRST CENTURY[1]

Nancy J. Sanquist, CFM, IFMA Fellow, Associate AIA
Peregrine Systems, Inc., San Diego, California

We all thought we knew what value meant. At one time, the ultimate symbol of a business that had the greatest market share in its industry was the large, monolithic corporate headquarters with the most expensive real estate space in a large city. The gigantic high-rise screamed to the world from its lofty perch that it was the most valuable company in the world (think of Philip Johnson's monument to AT&T looming over trendy Madison Avenue in New York City in the 1980s). Following suit, properties all over the globe were gobbled up at incredible costs to the delight of the real estate community.

Something incredible happened a few years later across town, however, on the trading floors of the New York Stock Exchange. Analysts began to award "most valuable player" in the market to companies that no longer cared about making statements through their real estate holdings. Stockholders applauded as the cybermarketspace exploded. Now, a bookstore that has less than $1 million in real assets has a higher market capitalization than its two biggest competitors, which are burdened with a combined real-asset portfolio worth more than $1 billion. Amazon.com's prosperity is reflected in its Web site where shoppers can fill their virtual shopping carts full of the latest best-sellers and click to another Web page to find information about the company and its associative communication links.

That shopper no longer cares what the corporate headquarters looks like and probably does not even know where the building is. The nature of value is shifting before our eyes between physical place and virtual space, and some of us are only beginning to understand what that means for all of our futures in the workplace. It does not mean, at least at this Internet time, that places are going to disappear into the virtual reality. What it does mean is that the value that we have put on places and their relationship to the value of our companies has changed forever.

This change in the creation of value is affecting those of us who were trained to design, manage, and maintain the physical space of an organization, although many are unaware of it. What do we do now that cyberspace is changing that value proposition? How do we understand this new concept of value in today's chaotic and revolutionary economic climate?

We must be aware of some clear patterns emerging that we need to understand in facility management and real estate. We can no longer ignore our information technology (IT) sisters and brothers down the hall. Instead, we must get close to them, learn their language, and train in their technological skill sets and tool kits. We must help them design the new processes, practices, and technological solutions that will help us navigate this complex economic environment and bring value to all our organizations, whether governmental, educational, or business. And these solutions are all built on the fuel for this new economy— information.

Those of us who can create it, use it, optimize it, and synchronize information with other groups, both within and outside of our corporate walls, will be players in the great global infrastructure resource management game. Companies will be attracted to our knowledgeable and networkable skill sets, and we will be rewarded highly and attracted to (and stay with) those organizations that know how to create value in this new world.

WHAT IS VALUE?

The *Wall Street Journal* posed an insightful inquiry in the August 20, 1999, "Money and Investing" section, in an article about the problem of defining value in the investment world: "Does anybody know what *value* is?" The answer is far from simple in this new economic climate. Its simplest definition is the return on investment (ROI) to stakeholders in the organization, stockholders, communities, employees, and suppliers. This ROI must be calculated now in two worlds—the physical and the virtual.

The physical world is composed of the organizations that invest significant capital to acquire the people who inhabit a myriad of places and use lots of things while they create ideas that bring in the capital to fuel corporate growth. And all of these resources are planned for, acquired, deployed, and maintained until the end of their life cycles when they are no longer needed by the organization. We think of infrastructure resource management (IRM) as the mortar that connects the virtual resources with the physical resources to create value (Fig. 3.3.2-1). True IRM unites the unique disciplines of the enterprise service desk, asset management, real estate, facilities, and fleet management through common shared data. It addresses all aspects of organizational infrastructure from IT, including both computers and telecommunications, to the buildings and real estate assets that house the technology and

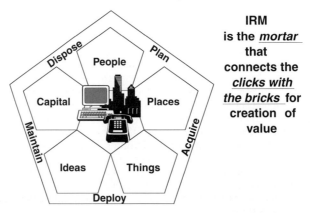

FIGURE 3.3.2-1 Infrastructure resource management (IRM).

people of the organization to the transportation fleets that serve the organization (Fig. 3.3.2-2). The merging of these disciplines results in a more thorough understanding of the impact of events and change upon the investment decisions of an organization.

This is particularly apparent in those organizations where all of these assets have begun to play a greater role in the strategic game plan of their business. Arthur Andersen understands the value of assets in the new business model:

> Businesses are their assets, all of their assets—tangible and intangible, owned and unowned . . . successful business models will combine both old and new economy assets. In fact, it is the combination and interaction of the various assets that will determine economic outcome.[2]

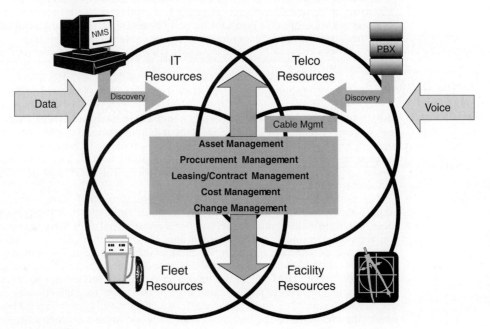

FIGURE 3.3.2-2 Components of IRM.

The e-business tsunami that is impacting the way we review all of our asset portfolios is all about using improved connectivity to optimize the performance of the entire business. And this connectivity is connecting the "clicks and the bricks." Amazon.com is a perfect example of a business that requires a new asset mix. Although it has rejected any off-line retail infrastructure, it has had to invest heavily in the physical world because its business model requires massive warehouse operations like the 800,000-square-foot structure built in Georgia. Location is critical to the immediate distribution of Amazon's products to its customers.

Corporations, universities, hospitals, K–12 institutions, and government agencies are all questioning their existing investments in physical places versus new investments to expand their reach into virtual space. And the plays they make on their particular game boards cannot be made intelligently unless they have the most accurate baseline information on all of their infrastructure. Then, they can benchmark with other organizations, share best practices, and determine what the most critical activities for creating value should be. And the only facilities management and real estate (RE) groups today that are truly playing on this value creation game board are doing it with sophisticated asset repositories and business reporting tools linked to the IT organization.

Boeing

Take the example of Boeing. In the early 1990s, Boeing was under great pressure from its shareholders to respond to the financial problems that were created by deregulation, declining aircraft orders, and divestment in space exploration. By the latter part of the decade, its vertically integrated business model of design, engineering, and manufacturing under hundreds of millions of square feet of roofs had catapulted them into the no-profit zone. At the same time, operating costs were escalating. A new management group was summoned to fix the problems.

Boeing announced its new "flight plan" in the spring of 1999, and it was not about scheduling its fleet of planes. It was a plan to create value for shareholders, customers, and employees. It was a performance-based strategy designed to substantially improve revenue growth, profit margins, and asset use. In doing so, Boeing was announcing a major change in the way the aerospace business measured success. It was turning its focus away from its former concentration on increasing the performance of its major carriers with no consideration of cost in the desperate race to space.

The aerospace giant had finally figured out (like the other industrial manufacturing titans in the automobile, steel, and appliance worlds) that they had to focus on value-based, market-driven cost reduction to compete in the global aerospace game. This was highlighted in an address given by its chief operating officer (COO), who identified cost as the most pressing problem and the "new frontier" in the aerospace business. The COO had now become a change leader for Boeing in its new race to increase value for its stakeholders and decrease costs for its global operations. And it could not begin to make the decisions it had to without state-of-the-art management technology.

New Processes for the Change Leader in Value Creation

Peter Drucker, still our guru of business at 90+ years, has described this new role of the change leader for the twenty-first century, which is to "make the present create the future." The new processes that this leader must adopt include practicing

- Organized abandonment
- Organized improvement
- Exploitation of success

The change leader in infrastructure management must consciously identify, analyze, and abandon any of yesterday's assets (including people, places, and things) that do not produce results. To practice organized abandonment, every asset must be analyzed to determine what it costs the organization. According to Drucker,

> For management purposes there are no "cost-less assets." There are only "sunk costs," the economists' terms for buildings and other fixed investments. The question is never: "*What have they cost?*" The question is "*What will they produce?*" And assets that no longer produce except in accounting terms, that is, assets which produce only because they appear not to "cost" anything, are not assets. There are only sunk costs.[3]

Therefore, as we look at implementing technology, the discussion we should be having should (1) concentrate on cost savings (be they reductions in occupancy costs in FM and RE or reductions in support and procurement costs in the IT world) made by implementing electronic tools and (2) focus on the productivity increases that can be gained when we hang on to the most valuable assets for our strategic missions and sink the nonperformers. When we engage in this type of discussion, we are out of the accounting realm into the economic world of creating value.

Only after abandoning these nonperformers can we seek to make improvements to those that are still valuable to an organization. Drucker discusses drawing from the Japanese *kaizen,* or continuous improvement, at this stage. These improvements to products, processes, services, and technology all depend on accurate information to determine the most significant best practices to apply to the organization.

One interesting example of determining best practices and creating standards is the work of the United Kingdom's Central Computer & Telecommunications Agency (CCTA), the agency that supports the English government in managing information. The CCTA created a methodology to codify best practices for IT services, called the Information Technology Infrastructure Library (ITIL). The ITIL is the only global comprehensive documentation on best practices for IT service management. Now used by hundreds of organizations around the world, it includes best practices for the entire life cycle of the infrastructure, including help desk, problem change and configuration management, contingency planning, and accommodation specification (including environmental and electrical requirements). Infrastructure planning and management tools are now incorporating the ITIL best practices into their product offerings.

Once performance metrics are identified for benchmarking and value scorecards and wins are identified, successes of the group can be exploited within the organization and also in the outside world. Boeing did just that on July 16, 1999, when the results of its own value scorecard for 2Q 1999 were announced to the press. Boeing described four metrics chosen to measure the organization's performance. Two of these metrics involved facilities: facility consolidation and overhead reduction. The Boeing CFO proudly exploited their success in abandoning real estate and consolidating operating groups into 122 million square feet and in reducing overhead by $600 million. Boeing's change leaders presented the results of their value-based performance initiatives to their shareholders, which included reducing overhead and occupancy costs.

The next section addresses specifically how new facility management, real estate, and ITs are allowing our profession to enter a new phase in its evolution as we learn innovative ways to create value in physical place and virtual space for all of our organizations. We can all play in the great global value creation game.

THE FOUR STAGES OF VALUE CREATION

In this section, we explore how the new change leaders in infrastructure management must act more like chief infrastructure officers (the new definition for CIO). They must determine a technology strategy and then the financial justification for these tools that are required to enforce this strategy. Then, they can use the technology to analyze, measure, abandon, and improve the valuable assets in which their organizations have invested to play on the board of the great global value creation game.

To understand how this relates to our world of facility management and real estate, I describe how these processes relate to the introduction and implementation of electronic systems for the following:

- Collecting data (electronic and manual)
- Planning strategy and scenario
- Acquiring
- Tracking
- Monitoring
- Managing
- Maintaining
- Disposing
- Reporting (graphic and alphanumeric)
- Benchmarking, scorecarding, and simulation dashboarding
- Integrating with enterprise resource planning (ERP) systems
- Linking graphics (CAD, GIS, photos, and videos)
- Creating virtual communities

In fact, without installing and implementing the systems just listed, the change leader cannot be successful. Performance cannot be measured, and it would be impossible to determine the value of any changes to business processes made to improve the infrastructure leader's role in supporting the mission and strategies of an organization. To organize improvement, the following four-stage framework has been defined that allows a change leader to move from one stage to another by introducing new management technologies and processes (Fig. 3.3.2-3).

Stage 1. Automation
Stage 2. Utilization

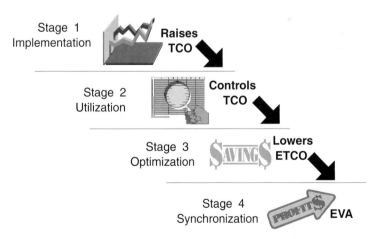

FIGURE 3.3.2-3 Four stages of implementing an IRM system. (*Source: Gartner-Group.*)

Stage 3. Optimization

Stage 4. Synchronization

The first two stages are very reactive environments where IT, FM, and RE manage the infrastructure from their separate organizational silos. In Stage 1, it has already been determined that the infrastructure is business-critical, and there is a need to lower the total cost of ownership (TCO) and occupancy while increasing the quality of service to the customers (the employees of the organization). The organization has determined its automation requirements and justified the investment by using automatic ROI calculators.[4] However, due to the fact that each of the separate groups is separately applying technology without coordination, the initial investment is much larger than attacking the problem in a holistic fashion.

In Stage 2, information is available about the existence and condition of the entire infrastructure resource portfolio—answering the questions, "How much do I own and lease?" and "Where are these assets located?" These are critical baseline data that can serve for internal and external benchmarking. It gives management information as to whether the resources are doing what they should for the maximum success of each of the business units.

After the results of the benchmarking activities have been compiled in Stage 2, Stage 3 is the point where the change leader can move out of a reactive environment into one where strategic planning can optimize the infrastructure resources in each of the respective RE, FM, IT, and telecommunications groups. This is the stage when the TCO finally starts to decrease because of

- Lower real estate costs due to implementation of best practices such as alternative work settings, hoteling, consolidation, and sale and lease back
- Cost recovery due to disposition of sunk assets (e.g., buildings, leases, computers, networks, software, etc.)
- Lower costs of telephony due to accurate call accounting

True IRM occurs in Stage 4 where IT, Finance, HR, RE, FM, Fleet, and Telecommunications are all linked electronically. There are strong linkages particularly with the ERP systems (Fig. 3.3.2-4). Real-time data are collected from one application and fed into another for mon-

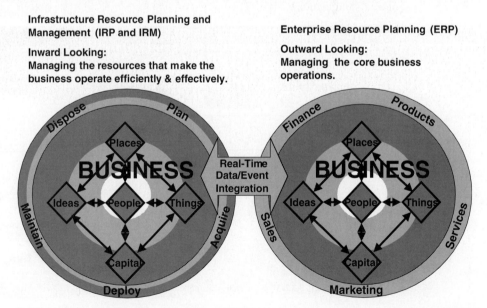

FIGURE 3.3.2-4   Infrastructure resource planning (IRP) integrates with enterprise resource planning (ERP).

itoring the state and condition of the infrastructure from building, data management, and telephony systems. At this stage, true financial benefits can be contributed to the bottom line of the business as measured by net present value analysis or real options pricing. We have moved from an accounting discussion in the first two stages to economic dialog in the last two stages. This is the point where the true value of integration is realized for the stakeholder. These four stages are depicted in Figure 3.3.2-5.

Stage 1: Automation

After you have determined your requirements and justified the investment, you can begin to automate new business processes. The major force behind this stage is cost reduction. Information is captured in one or two repositories from many disparate sources (both graphic and alphanumeric) by each separate silo master of the organization (i.e., FM, RE, and IT). Interfaces and gateways have been established with other groups such as human resources and finance. This short-term view can be accomplished in 1 to 4 months, and assets and resources are still perceived as expenditures and costs to the organization. There is an average of 100 percent churn in this stage and approximately 15 percent of underused because there is no space or asset planning, only a space and asset inventory.

Stage 2: Utilization

After the initial stage, you can move to the second stage where now you can use this information to look at your infrastructure from a global viewpoint (or however large your universe is) and start cutting occupancy costs. This is a space and asset utilization stage where you are look-

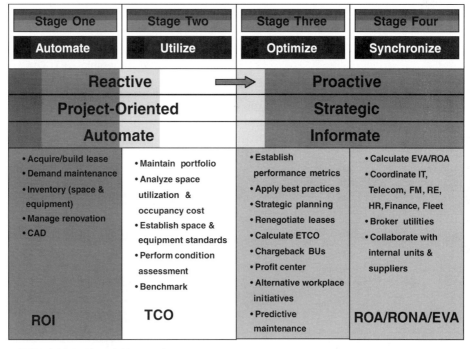

FIGURE 3.3.2-5 Financial impact of IRM.

ing at overall efficiency and effectiveness and establishing policies, procedures, and processes to maximize your portfolio by benchmarking with other groups both within and without the organization. The American Productivity and Quality Center (APQC) has determined that more significant cost savings can occur from internal comparisons than from external benchmarks.

This is the stage for defining how your costs will be reported. It is suggested that you use the Workpoint Accounting System to determine the true costs of the bricks-and-clicks infrastructure provided to support the knowledge worker. This is an accounting standard issued by the Institute of Management Accountants in mid-1997 that includes a chart of accounts and describes categories of IT costs and investments, as well as costs incurred for investment in the physical infrastructure. It was created by the Institute of Management Accountants and is entitled "Practices and Techniques: The Accounting Classification of Workpoint Costs." In this stage, the IT group will use this classification scheme for their TCO, and the FM and RE organization units will account for building and equipment costs (TCO). In the next two stages, they will be brought together for the enterprise total cost of ownership (ETCO), as defined by Mike Bell of the Gartner Group (Fig. 3.3.2-6).[5]

These data are used as accurate baseline data for the third stage, where you can begin strategic planning, benchmarking, and determining what your own value scorecard will look like to measure your performance in optimizing the infrastructure portfolio.

Stage 3: Optimization

In Stage 3, the benchmarking activities that have been performed in Stage 2 are now allowing management to identify the areas where best practices need to be applied. These practices can help to lower the TCO and provide even more superior service for the knowledge workers to help them increase productivity and stay within the corporation. The third stage is the point where you can begin to do what Drucker calls "organized abandonment" and to dispose of "slump" assets. Other more creative activities include examining workplace transformation to enable a true virtual integrated enterprise [VIE (also defined by the Gartner Group)] in the last stage of development.

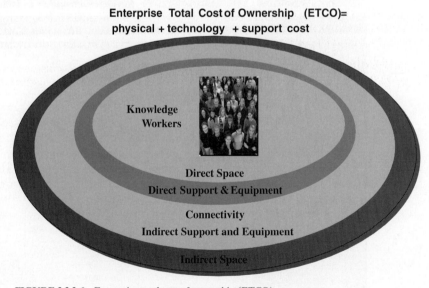

**Enterprise Total Cost of Ownership (ETCO)=
physical + technology + support cost**

Knowledge Workers

Direct Space
Direct Support & Equipment

Connectivity
Indirect Support and Equipment

Indirect Space

FIGURE 3.3.2-6 Enterprise total cost of ownership (ETCO).

Stage 4: Synchronization

As e-business continues to claim resources in every type of organization, this stage is where chief infrastructure officers have now integrated all of their IT applications to synchronize the VIE.[6] The GartnerGroup has defined *VIE* as follows:

> . . . global trends have spawned new business challenges, including enterprise agility, a relentless focus on core competencies and processes, a redefinition of the value chain as flowing backward from customer requirements to the enterprise, instantaneous response and latency (speed rules); the ability to scale resources and infrastructure across geographic boundaries, a "plug and play" IT infrastructure with componentized architectures and an interoperable applications portfolio. In short, as we enter the new millennium, the traditional business model is changing into a new hybrid structure, combining the local dominance of "bricks and mortar" infrastructure with the agility and scalability of digital infrastructure—what we call the VIE.

In this VIE environment, as organizations such as Sun Microsystems and Procter & Gamble combine their IT, FM, and RE functions to decrease the cost of selling and general administration divided by sales revenue (SGA) and to increase the productivity and retention of their workforces globally, they centralize their infrastructure planning and management activities.

The Methodology for Measuring Performance in the Four Stages: The Balanced Scorecard

We are all being held accountable for value creation. But how is this measured? How can we be sure that we have the right road map for applying new technologies and business processes to create the greatest value for our shareholders?

One methodology that has been used successfully since the early 1990s is the Balanced Scorecard.[7] The creators, Kaplan and Norton, designed a management system to link an organization's missions and long-view strategies with short-term action plans. They saw that a company's competitive strength is based "on information, their ability to exploit intangible assets has become far more decisive than their ability to invest in and manage physical assets."[8] Therefore, the Balanced Scorecard measures financial performance, customer satisfaction, internal business processes, and innovation. The Association of Physical Plant Administrators (APPA) has embraced this methodology for physical plant directors of the university community. Jack Hug of the University of California at San Diego has applied the Scorecard to his facility operations:

- *Financial performance.* To ensure financial integrity, to demonstrate stewardship responsibility for financial and capital assets, and to provide services efficiently, productively, and cost-effectively.
- *Customer satisfaction.* Perform "great customer service."
- *Internal business processes.* Implement processes that meet operational requirements and make it easy for customers to receive great service.
- *Innovation and learning.* Create a workplace where people love to come, where they are encouraged to grow and prosper, and where everyone is treated with respect.[9]

Recent research conducted by the APQC, directed by Doug Christianson, relied on the Balanced Scorecard to report on the critical issues underlying space utilization:[10]

> Although financial performance measurements such as efficiency ratios are important, they should not be the only factor affecting space utilization decision. The goals of real estate and facility management include more than just cost cutting. The challenge for corporate real estate and facility management officers is to provide a work environment that increases employee productivity and retention while maximizing the return on the organization's real estate portfolio.

Information technology must be added to this concept because the performance of the increasingly computerized integrated workplace is being measured in its support of the intelligent organizational networks necessary in the new "e" business space.[11]

One can use the Balanced Scorecard as a measurement system in each stage of value creation. To measure specific examples of total life cycle asset management, the work processes can be divided into the following categories:

- General administration and management
- Real estate (acquisition and disposition)
- Planning, design, and construction
- IM maintenance and operations
- Energy management
- IRM asset management
- IRM procurement

As we move from stage to stage in implementing systems in each of these areas, financial discussion will change from an accounting to an economic perspective, improvements in customer satisfaction will move from faster response time to increased productivity, internal business processes will leap from tactical improvements to strategic decision making, and innovation will change from mere automation of processes to new ways of doing business. And these are just some of the ways by which we can navigate this complex value proposition.

CONCLUSION

This article has described how you can navigate from separate distinct silos of activity in the first two stages of systems implementation to integrated asset management that we call *infrastructure resource management* (IRM).

Already, new business processes are being defined by organizations such as KPMG, which has created an infrastructure resource planning (IRP) practice.[12] Just as the Big Five created extremely lucrative practices building on the expertise required for implementing ERP systems, which have changed the physical and virtual supply chain for the core business of an organization, IRP processes will be the next critical movement. These IRP processes will support the changes required to support the complex environment of the technological and physical infrastructure in both old- and new-economy businesses. The question is: Are you creating or destroying value on your own IRM game board?

NOTES

1. A variation of the title of a talk given by Dr. Kjell A. Nordstrom, Assistant Professor, Director Institute of International Business, Stockholm School of Economics, Sweden, at the European IFMA World Workplace in June 1999. The game board concept is attributable to the recent work of Michael Bell at The GartnerGroup. Portions of this article appeared in IFMA's *Facilities Management Journal* (January/February 2000), 22–32.
2. Taken from www.arthurandersen.com Web site excerpted from the April 2000 publication of *Cracking the Value Code: How Successful Businesses Are Creating Wealth in the New Economy,* by Arthur Andersen Partners Richard Boulton, Steve Samek, and Barry Libert.
3. P. Drucker, *Management Challenges for the 21st Century,* HarperBusiness, New York, 1999, p. 75.

4. Peregrine Systems, Inc. has created these electronic calculators, and they are available by contacting the author of this article.

5. As defined by the GartnerGroup in their Workplace Transformation seminar, March 2000, San Antonio, TX.

6. Ibid.

7. See R. S. Kaplan and D. P. Norton, *Harvard Business Review* (January-February 1992), 71–79; (September-October 1993), 134–147; (January-February 1996), 75–85.

8. R. S. Kaplan and D. P. Norton, *Harvard Business Review* 75 (January-February 1996).

9. *A Guide to Leadership and Management Planning for Auxiliary and Plant Services,* University of California, San Diego, 1998.

10. APQC

11. See John Worthington's 1997 *Reinventing the Workplace* for a further discussion of these concepts—particularly Andrew Harrison's article, which predicted that one of the first "casualties" of the moves toward developing intelligent organizational networks would be IT and FM (p. 95).

12. The KPMG pioneers in this arena are Larry Rodda, Jim Mowrer, and Sam Fahr of the Sacramento office.

ARTICLE 3.3.3

LEVERAGING THE FACILITIES MANAGEMENT TECHNOLOGY INFRASTRUCTURE: ALL ASSETS ARE CREATED EQUAL

Eric Teicholz
Graphic Systems, Inc., Cambridge, Massachusetts

It is hardly a surprise that organizations are spending more and more of their resources on their information technology (IT) infrastructure. Not so long ago, this infrastructure encompassed only the hardware and software associated with mainframe computers and the networks that connect terminals to those mainframes. In the 1980s, PCs entered the equation. In the 1990s, the IT infrastructure has expanded to include everything from laptops, palmtops, intranets, extranets, collaborative Web sites, mobile phones, pagers, and personal digital assistants. Given the nature of progress, it is anachronistic to think that facilities managers will continue to control their IT infrastructures independent of other business units.

Indeed, there has been a term coined to describe the growing technological interconnectedness of a company's individual units: *Infrastructure management* (IM) refers to managing the life cycles of all assets within an organization. The goal of IM is to minimize the cost of that asset for the organization as a whole [i.e., the "return on asset"—an increasingly popular facilities management (FM) benchmark]. To accomplish this, IM needs to know about people, locations, and assets—the traditional domain of computer-aided facility management/computer-integrated facility management (CAFM/CIFM). What traditionally has *not* been the domain of CAFM/CIFM systems are calculations for the cost of that asset over time (the life cycle of the asset) and how effectively it is being used to support strategic corporate business objectives. To perform these calculations, new types of interfaces must exist between FM and other mainstream corporate systems, and new business management processes must be established to which all organizational units adhere.

What is key about IM management is that it is done for the organization as a whole—not just the FM department. To optimize return on assets (ROA), the organization must treat all assets alike throughout that organization, and standards must exist for computers and also for data, telecommunications, applications, and even the work processes for managing assets over time.

The implications for FM are profound. It means that FM data must be linked to real estate, human resources, IT, and all aspects of finance and procurement. It means that FM technology systems must function as a part of a whole and that the goals of FM technology must be synchronized with those of the organization as a whole.

THE BEGINNINGS OF IM

Infrastructure management embodies concepts that have arisen as management wakes up to the strategic importance of IT and its influence on all aspects of an organization's business. No business unit, facilities management included, is immune from the centralizing force of IT.

Information technology is already the largest capital expense in many corporations today (it amounts to 7 percent of revenues and 14.2 percent of expenses for service industries), and investment in IT is growing at the rate of 3 percent per year. The corporate/organizational IT infrastructure increasingly reaches across IT to all business units. Because of the potential benefits of IM, however, IT increasingly incorporates these areas:

- PCs and all aspects of computing devices (e.g., laptops, palmtops)
- Communications and telecommunications protocols and vendors
- Vendors and other types of data of corporate-wide interest
- Internal and external Web technology, including the Internet, intranets, and extranets
- Benchmarking and financial metrics for cost-benefit analyses
- Various aspects of mission-critical work processes, such as procurement and financial management

The number of centrally managed technology-based functions is growing, and organizations are increasingly trying to understand and manage the risks and returns of such investments.

THE EFFECT OF IM

The FM cause-and-effect relationships between IM investment, risks, and rewards are just beginning to be understood and quantified. However, enough case studies and feedback are available to reach this conclusion: Organizations that understand such relationships and invest heavily in IM will receive more business value and ROA than those that do not.

For the facilities manager, it means that the role of FM technology will change as IM is better understood and accepted by organizations. It means that an asset (whether a chair, a PC, a paper clip, cabling, a painting, or an entire building) will be processed by the same technology and with the same benchmarks and work processes that are used throughout the rest of the organization. It means that the support structures (e.g., HR, finances, real estate, IT, and purchasing) required for corporate strategic decisions will all use the same technology and business processes and probably even report to the same management. Now, every resource for which an organization is responsible (whether leased or owned) will be part of an integrated strategy that affects the entire life cycle of that asset. In the IM environment, understanding the life cycle of assets from purchase to disposal is critical.

As IM centralizes business functions and processes, it means (at least in larger organizations) that CAFM/CIFM vendors will also be required to integrate with enterprise resource planning (ERP) systems and vendors. Mainstream "mission-critical" systems that control the tasks of purchasing, finances, HR, and IT (typically the domain of ERP vendors) need to share information with FM systems to provide the spatial and financial information required by management. Vendors such as Oracle, PeopleSoft, SAP, and Bain are already working with FM vendors to enable such integration. The line separating FM and ERP technology is still somewhat fuzzy and constantly changing. A number of potential scenarios are possible that would make this line clearer (at least at a particular time):

- ERP vendors might focus on mainstream enterprise-wide business processes, and facilities managers would manage the environment in which these business processes take place.

- ERP vendors might try increasingly to develop FM and real estate functionality within the ERP environment.

- ERP vendors might increasingly partner with FM vendors to provide comprehensive integrated solutions.

Finally, IM will affect how internal and external customers from all business units interact with information. The model of a single point of interface has appeal because it provides a unique location for all organizational data. Thus, the current functionality of help desks that are popular with many existing computerized maintenance management systems (CMMSs) would be expanded to aid the customer with problems associated with all assets. There would perhaps even be a single help desk that would serve all business units within an organization.

Although the benefits of IM for the facilities and real estate manager are just beginning to be understood, some value can be anticipated. Effective IM deployment and management will

- Provide a single repository of information about all occupants, locations, and assets that an organization owns or leases

- Capture, track, and potentially calculate the impact of an asset on an organization (which, in turn, will minimize downtime and expenses) as long as that asset is owned by the organization

- Minimize the cost and time of moving employees, assets, and systems of an organization from one location to another

- Enable deploying new FM and real estate technologies and applications throughout an organization

- Provide much more rapid integration of FM and real estate data with HR, IT, and financial systems

- Leverage the large investment made in ERP systems

- Facilitate defining, collecting, and reporting of financial and operational benchmarks that are consistent throughout the organization

THE FUTURE OF IM

As mentioned, organizations have so far ascertained that investment in IM yields business value and better understanding and control of assets. However, there are trade-offs in just how much centralization should take place in hardware, software, communications, Internet, data, benchmarks, business processes, and so forth. The risk and return of IM are not clearly understood. For example, firms that have more integrated and comprehensive infrastructures can introduce new products faster; however, IT systems and related procedures tend to be complex, and therefore, it takes longer to integrate new technologies and applications into the existing infrastructure. For the facilities manager, technology is constantly changing. Technology vendors are just beginning to move their applications to the Internet on a real-time, transactional basis. New technologies such as workflow and collaborative Web sites have been applied only recently. High-speed Internet, chips embedded in assets, and interoperability are just a few years away. How much will these technologies be influenced, and possibly be slowed, by organizations that have heavily integrated infrastructures? How will different types of investment in IM affect the ROA? No one really knows at this point.

Nevertheless, there is an obvious trend toward increasing investment in an organization's technological infrastructure. It is also clear that the definition of this infrastructure involves more than just hardware, software, and networks. The definition of IM is embracing more functions, including aspects of applications software, Internet and related technologies, databases, benchmarks, profit-and-loss calculations, and even entire business processes. Anything that is required outside of a specific business unit is fair game for IM.

This common organizational integrated technological environment encompasses people, assets, and spatial (locational) data—all of which are critical to the facilities manager. As facilities data are increasingly merged with other corporate data for strategic needs (e.g., occupancy cost calculations, outsourcing analysis, and ROA), the trends toward centralization of IM will continue. Centralized purchasing and e-commerce on the Internet mean that facilities assets will likewise be centrally purchased.

The role of the facilities manager in IM is currently fluid. The role to be played by traditional CAFM vendors vis-à-vis ERP vendors is likewise dynamic. What is clear is that the facilities manager must understand the implications of IM for facilities management and must anticipate the ways in which the traditional functions of space and asset management will be affected by this centralizing force.

ARTICLE 3.3.4
FACILITIES MANAGEMENT AUTOMATION AND INFRASTRUCTURE MANAGEMENT

Bruce Kenneth Forbes
ARCHIBUS, Inc., Boston, Massachusetts

This article is a primer on facilities management (FM) automation and infrastructure management (IM). It bridges the gap between management practices and the application of technology. Issues raised in this article will benefit all facilities professionals and their allied affiliates, no matter what their level of FM automation and IM experience.

Now, FM professionals are using hand-held computers (also known as palmtops), smart chips, cellular connectivity, and bar codes to decrease paperwork, increase quality, and handle larger workloads. Thanks to cutting-edge technology, world-class procedures, benchmarks, best practices, software, and e-commerce, the facilities planning and management field is undergoing a remarkable transformation. Advanced FM automation solutions, IM enterprises, and cyberspace universes for FM have become a reality.

During the last three decades, the FM automation industry has been defined, clarified, realized, and reengineered. Some of the by-products of this process include best-of-class standards and procedures and best practices for the facilities management professional. During this time, the marketplace has been educated about the financial and personal benefits of FM automation. The digital economy of FM automation is evolving rapidly. Comprehensive graphical user interfaces make FM automation solutions easy to understand and use (see Fig. 3.3.4-1). In just 5 minutes, executives can learn how to access information about assets on their sites or anywhere else in the world.

Facilities management automation and IM improve FM knowledge management. The unique needs of dynamic global economies and reengineered business environments are requiring inventive responses. Timely access to corporate information can determine the quality of strategic decision making. Facilities management knowledge management will be used to create unlimited opportunities. The technological aspects of FM knowledge management embrace automation tools, processes, procedures, and best practices for managing corporate assets and will have a direct impact on an organization's effectiveness, efficiency, productivity, and profitability.

Facilities management automation and IM technologies are critical management tools of the virtual networks and enterprises that are vital to the new e-business, e-commerce, and business-to-business (b2b) delivery environments. Yet, for many CEOs and other senior executives, finding easy-to-use resources that help bridge the gap between general management practices and the base FM disciplines and core FM competencies has been a daunting challenge. Among the numerous benefits of deploying FM automation and IM solutions is the ability to use graphical drill-down techniques to parse through billions of pieces of data, information, and knowledge. An executive who has minimal training can use a few clicks of a mouse to locate specific information in a real estate and property abstract, departmental occupancy report, or suite lease expiration report by graphically drilling down and reviewing a desired report (see Fig. 3.3.4-2). The immediate access to strategic information and knowledge is often cited as the executive's top reason for using these systems.

Why You Should Read This Article

This article offers a nontechnical reference to the world of FM automation and IM. By reading it, one can learn how to rapidly deploy FM automation solutions and IM enterprises, implement FM best practices for FM automation, save time and money, and realize greater productivity and profitability.

Facilities professionals are involved with thousands of different activities and would greatly benefit from FM automation and IM if they perform any of the following tasks:

- Real estate acquisition, use, redeployment, maintenance, and disposal
- Long-range facility planning
- Strategic master planning
- Facility financial forecasting and budgeting
- Acquisition, use, redeployment, maintenance, and disposal of assets
- Renovation and/or new construction
- Architecture/engineering planning and design
- Interior design, space planning, workplace specifications, furniture and equipment installation, and space management
- Telecommunications and cable management
- Human resources (HR), security, and general administration
- Predictive, preventive, and demand maintenance
- Mergers and acquisitions

Campus Planning and Management System

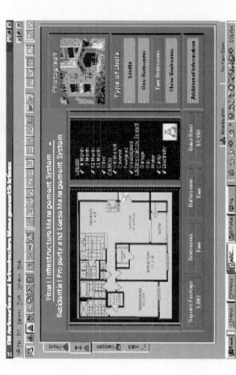

Construction Management System for Campus Planning Office

Corporate Real Estate Portfolio Management System

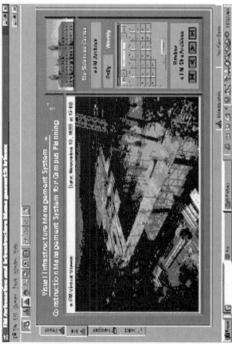

Residential Property and Lease Management System

FIGURE 3.3.4.1 The digital economy of FM is rapidly being shaped by e-FM, e-space, e-commerce, and cyberspace universes for FM. Executives can establish immediate links to their favorite reports, financial analyses, help desk, construction site, work request system, and more. (*Courtesy of Rice University and ARCHIBUS, Inc.*)

Infrastructure Management Enterprises with Graphical Drill Downs *(It is as Easy as 1-2-3...)*

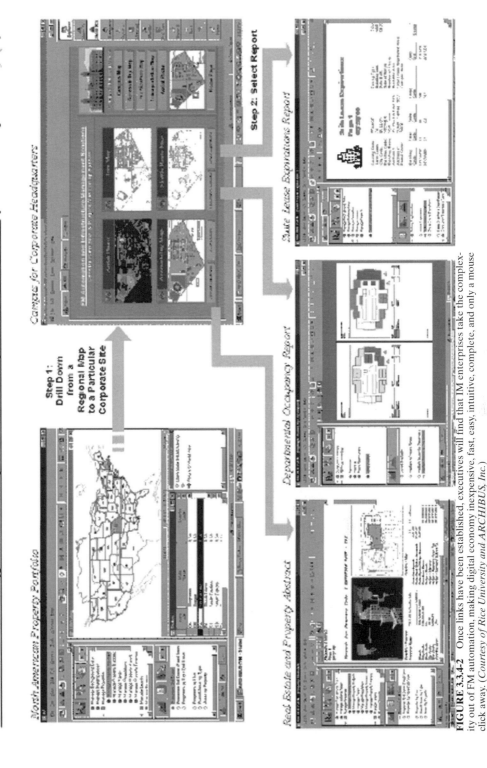

FIGURE 3.3.4-2 Once links have been established, executives will find that IM enterprises take the complexity out of FM automation, making digital economy inexpensive, fast, easy, intuitive, complete, and only a mouse click away. (*Courtesy of Rice University and ARCHIBUS, Inc.*)

THE COMMON ACTIVITIES ADDRESSED BY MOST FM AUTOMATION APPLICATIONS

Facilities management automation deals with the full life cycle of assets. Facilities management automation is currently being used in all of the areas listed previously. The common activities addressed by most FM automation applications include planning, design, acquisition, use, maintenance, redeployment, and disposal. The primary FM automation application areas that are addressed by commercial off-the-shelf solutions include real property and lease management, strategic master planning, space management, CAD overlays with design management, furniture and equipment management, telecommunications and cable management, building operations management, and IM connectivity with Web interfaces and executive information systems (see Fig. 3.3.4-3). When considering where to start to implement FM automation, it is advantageous to review each of the tasks shown in Fig. 3.3.4-3 and determine their relevance to your organization.

BENEFITS OF FM AUTOMATION: TOP REASONS FOR UTILIZING FM AUTOMATION SOLUTIONS

In 1998, a survey of facilities engineering and management professionals revealed what are considered the top benefits of using FM automation (Table 3.3.4-1).

Strategic managers stressed increases in productivity and profitability as the primary drivers for FM automation within their organizations. Business managers cited comprehensive analytical tools for better decision making as a primary benefit. Operations managers use FM automation to reduce costs and get things done more quickly, easily, and efficiently. Administrative and corporate services utilized FM automation to rapidly develop and implement corporate policies, tools, standards, guidelines, FM benchmarks, and FM best practices.

MANAGING TOTAL COST OF OWNERSHIP EFFECTIVELY

Today, an industry-wide dilemma exists: People are looking for panaceas to solve their automation problems. They assume that if they invest in FM automation solutions, all of the problems will be solved. But the reality is that once you make that purchase and commitment to FM automation, you have just begun. To make an FM automation solution work for your organization, you must address a wide variety of issues, and you must make informed decisions that allow your expectations to meet your reality. You must manage the *total cost of ownership* (TCO) effectively.

INFRASTRUCTURE MANAGEMENT ENTERPRISES

In this new millennium, your FM automation solutions must also connect with thousands of information silos and other islands of automation. Infrastructure management enterprises will be able to provide a framework for integration and connectivity. Such integration and connectivity can greatly reduce the TCO. In world-class IM enterprises, the "islands of automation" and "information silos" are assembled, integrated, and accessed via a variety of user interfaces.

An IM enterprise (an example is shown in Fig. 3.3.4-4) is considered a core competency of a virtual-based bricks-and-mortar, clicks-and-mortar, and clicks-and-clicks organization.

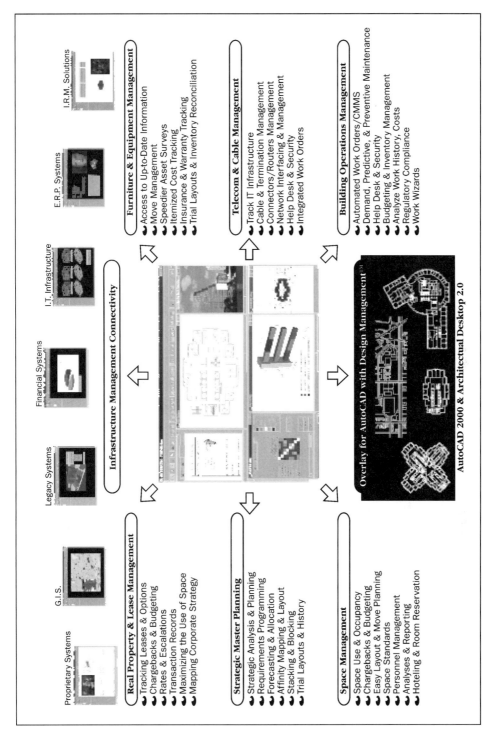

FIGURE 3.3.4-3 Islands of automation and information silos are connected and integrated within FM automation solutions and IM enterprises. The primary application areas are real estate, strategic master planning, space management, CAD overlays, F&E management, telecom, and building operations. (*Courtesy of Bruce Kenneth Forbes, ARCHIBUS, Inc.*)

Proprietary Systems   G.I.S.   Legacy Systems   Financial Systems   I.T. Infrastructure   E.R.P. Systems   I.R.M. Solutions

Infrastructure Management Connectivity

Overlay for AutoCAD with Design Management™

AutoCAD 2000 & Architectural Desktop 2.0

Real Property & Lease Management
- Tracking Leases & Options
- Chargebacks & Budgeting
- Rates & Escalations
- Transaction Records
- Maximizing the Use of Space
- Mapping Corporate Strategy

Strategic Master Planning
- Strategic Analysis & Planning
- Requirements Programming
- Forecasting & Allocation
- Affinity Mapping & Layout
- Stacking & Blocking
- Trial Layouts & History

Space Management
- Space Use & Occupancy
- Chargebacks & Budgeting
- Easy Layout & Move Planning
- Space Standards
- Personnel Management
- Analyses & Reporting
- Hoteling & Room Reservation

Furniture & Equipment Management
- Access to Up-to-Date Information
- Move Management
- Speedier Asset Surveys
- Itemized Cost Tracking
- Insurance & Warranty Tracking
- Trial Layouts & Inventory Reconciliation

Telecom & Cable Management
- Track IT Infrastructure
- Cable & Termination Management
- Connectors/Routers Management
- Network Interfacing & Management
- Help Desk & Security
- Integrated Work Orders

Building Operations Management
- Automated Work Orders/CMMS
- Demand, Predictive, & Preventive Maintenance
- Help Desk & Security
- Budgeting & Inventory Management
- Analyze Work History, Costs
- Regulatory Compliance
- Work Wizards

TABLE 3.3.4-1 Top Reasons for Implementing FM Automation and IM Solutions

FM automation improves	FM automation increases	FM automation reduces
Benchmark data	Accuracy	Asset management costs
Corporate archives	Accountability	Churn rates and vacancies
Corporate knowledge	Asset use and management	Information bottlenecks
Customer satisfaction	Connectivity	Need for new space
Customer value	Control	Portfolio overhead costs
Effectiveness	Efficiency	Redundancy costs
Management	Profitability	Training costs
Responsiveness	Productivity	Training time
Service delivery	Reliability	Vacancy costs
Quality	Responsiveness	Underused assets

Facilities planning and management organizations can successfully create connectivity between their FM automation solutions and the numerous solutions for automation used throughout their organizations. Infrastructure management enables FM professionals to rapidly combine the data, information, and knowledge that has been stored in information silos or that exists in the various islands of automation. Some of the seamlessly integrated applications include the following:

- FM automation solutions
- Enterprise resource planning (ERP) systems
- Building automation systems
- Information technology (IT) infrastructure systems
- Computerized maintenance management systems (CMMS)
- Help desk and telephony systems
- Numerous other automation systems of the new millennium
- Simple return-on-investment (ROI) worksheets for FM automation and IM solutions

There are hundreds of reasons why FM automation has been proven to be a successful tool for facilities professionals. The most-often-cited reason for using FM automation is that organizations often spend up to 50 percent of their revenues in the acquisition, use, maintenance, redeployment, and disposal of assets. These organizations can realize up to 34 percent savings in these costs by effectively implementing and using FM automation.[5] In Table 3.3.4-2, the worksheet following is representative of a simple ROI analysis created for FM automation and IM solutions. This worksheet is based on presentations made at numerous World Workplace conferences, IFMA Best Practices Forums, local IFMA Chapter roundtables, IFMA Computer Applications Council discussions, and the American Institute of Architects—Professional Interest Area for Facilities Management seminars and workshops.

AUTOMATION IN THE TWENTY-FIRST CENTURY

We are rapidly evolving from a marketplace economy to a marketspace economy, courtesy of the opportunities afforded us by e-commerce, e-business, and b2b offerings. Virtual networks will be empowered by IM solutions that

- Are designed for people (providing user interfaces for casual, intermediate, and advanced users)

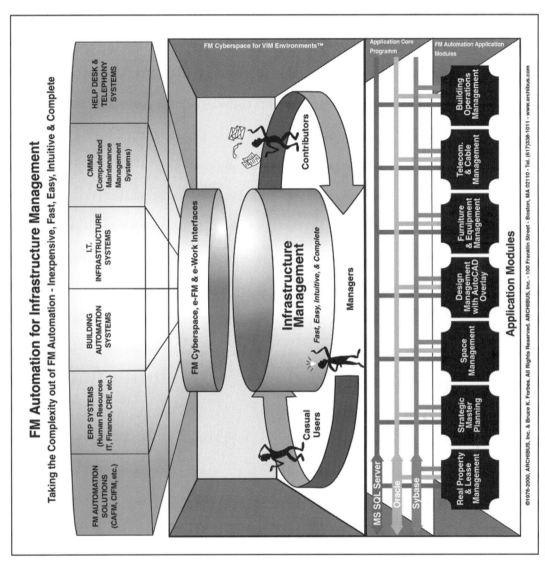

FIGURE 3.3.4-4 Hundreds of systems can be seamlessly brought together in enterprises (of which only a few have been depicted in this diagram). Some of the application categories for the systems more commonly connected include FM automation, ERP, building automation, IT infrastructure, CMMS and BOM, and help desk and telephony. *(Courtesy of Bruce Kenneth Forbes, ARCHIBUS, Inc.)*

TABLE 3.3.4-2 Simple ROI Worksheet to Determine Whether Your Organization Should Consider Using FM Automation and IM Solutions

	Small corporation	Medium corporation	Large corporation	Your organization
AA. Annual revenues of a corporation:	$5,000,000	$50,000,000	$500,000,000	(Estimated revenue)
BB. Facilities and asset-related expenses: Costs associated with the acquisition, use, maintenance, redeployment, and disposal of a corporation's assets (50% of a corporation's revenues)	$2,500,000	$25,000,000	$250,000,000	(Line above times 0.5)
CC. Estimated savings by effectively deploying and using Facilities Management automation and IM solutions (34% of a corporation's facilities-related expenses)	$ 850,000	$ 8,500,000	$ 85,000,000	(Line above times .34)
Estimated automation costs				
Number of square feet managed	25,000	250,000	2,500,000	
A. Total background data capture costs (sq ft × $1.80)	$ 45,000	$ 450,000	$ 4,500,000	
Number of employees and outsourced FTEs* on-site	50	750	10,000	
Number of assets per FTE on-site	12	24	36	
Total number of tracked assets (FTE × number of assets)	600	18,000	360,000	
Asset mgmt. and base maint. costs ("B" × $1.65 × 12 mos)	$ 11,880	$ 365,400	$ 7,171,200	
Recommended FM automation staff				
(1 person for 1,000 FTE on-site @ $60,000/yr)	$12,000	$ 45,000		
(5 persons for 10,000 FTE on-site) @ $300,000/yr)			$ 300,000	
Miscellaneous hardware, software, Internet/ Intranet expenses, customization, training	$ 25,000	$ 325,000	$ 500,000	
Estimated total FM automation expenses (A + C + D + E)	$ 93,880	$1,185,400	$12,471,200	
G. Estimated total ROI (CC–F) (realized over 3 years)	$756,120	$ 7,314,600	$72,528,800	
H. Estimated loss sustained by your organization every 90 days that you do not invest in suitable FM automation and IM solutions (CC ÷ 12 months). Typical payback period is less than 6 months.	Quarterly loss if not automated $ 70,833	Quarterly loss if not automated $ 609,550	Quarterly loss if not automated $ 7,083,333	

The inability to act can cost you millions . . .

* Full time equivalents (employees) (FTE).

- Enable users to save a tremendous amount of time
- Address the TCO issues
- Provide for seamless integration
- Incorporate the latest and greatest technologies and the legacy systems that helped get organizations where they are today

Real-World Savings as Revealed in FM Automation Case Studies

There are hundreds of reasons for creating FM automation initiatives in your organizations. Following are some of the more frequently cited reasons:

- Better management of legal and safety issues
- More rapid completion of projects
- Better maintenance of equipment and assets
- Expedited modernization programs
- Gains in capacity, productivity, and profitability
- High ROI
- Better rent negotiations
- Energy management and savings
- Reduced telecommunications costs
- Reduction of needed space and facilities
- Better procurement methods and pricing
- Savings in labor rates and resource costs
- Increased productivity and profitability
- Reduced churn rates, move, and construction costs
- Reduced architectural, engineering, and interior design costs
- Tax reductions and abatements
- Reduced costs of building operations and maintenance
- Reduced furniture and equipment costs
- Reductions in personnel disruption costs and labor costs
- Incalculable savings of time
- Opportunities for personnel reduction
- Better control of information

Infrastructure management solutions will be accessed via a variety of user interfaces. These interfaces are process oriented and are easy to understand. Using these interfaces, a casual user can be trained in only a few minutes to use cyberspace universes for FM, executive information systems, and FM hot lists.

WHAT IS FACILITIES MANAGEMENT?

Facilities management is the practice of coordinating the physical workplace with the people and the work of the organization; it integrates the principles of business administration, architecture, behavioral, and engineering sciences. Facilities managers oversee assets (e.g., buildings, furniture, equipment, and computers) throughout their life cycles. The life cycle of an asset typically includes its acquisition, use, maintenance, redeployment, and disposal.

WHAT IS THE FACILITIES MANAGEMENT PROFESSION ABOUT?

Simply stated, the primary concern of the facilities management profession is planning and protecting the life cycle of assets and all things relevant to an asset's life cycle. The timely, productive, and profitable acquisition, use, maintenance, redeployment, and disposal of an organization's assets are the prime directives. The relationship of spatial and nonspatial assets changes as an organization evolves from the traditional bricks-and-mortar organization, but the need to manage an asset's life cycle remains consistent.

WHAT IS FACILITIES MANAGEMENT AUTOMATION?

Automated facilities management solutions typically encompass creating, maintaining, and distributing FM data, information, knowledge, and wisdom. These are the primary areas addressed:

- Properties and real estate (parcels, plants, buildings, campuses)
- Physical assets [furniture, fixtures, IT infrastructures, equipment, building infrastructures (including structural, HVAC, mechanical, life safety, emergency, electrical, technical, and plumbing systems), in addition to signage, civil and geosubsystems, and interior and exterior envelopes and walls]
- Spatial assets (financial management analysis and control systems), personnel, and organization (employees, department staffing, vendors, contractors, and workflow processes and procedures)
- FM best practices and procedures
- FM core competencies
- FM benchmarks

State of the Art in Review

COTS Deployment Is "Hip," and Proprietary Solutions Are "Not." Organizations such as the GartnerGroup and Daratech report vast increases in FM automation and IM activities. In the near future, it is estimated that there will be more than 250,000 users of enterprise-based FM automation and IM solutions and 3 million users of cyberspace universes for FM environments. The enterprise solutions will most likely be 95 percent commercial off-the-shelf (COTS) solutions versus 5 percent proprietary and/or heavily customized solutions. The Chief Information Officer (CIO) of the U.S. Navy recently cited numerous benefits of COTS solutions and their deployment in the U.S. Navy. A recent Department of Defense (DoD) mission statement for an enterprise software initiative program reads, "The objectives are to save money and improve information sharing. The initial focus will be on COTS products."

One of the first decisions that an organization pursuing FM automation will have to make is whether its FM organization should buy COTS solutions for FM automation and IM. Historically, most Fortune 1000 FM organizations have opted to create their own FM solutions and invest heavily in customization and proprietary software development. Today, due to staffing, monetary, and time constraints, the same FM organizations are selecting COTS solutions with little or no customization. The ROI in such cases is obtained in less than 1 year, compared with the proprietary systems of the past, which took up to 3 years for an ROI.[6]

ROI Evaluations for FM Automation and IM. A particular organization's ROI for FM automation and IM will vary, depending upon the type of system(s) acquired, level of integration, available financial resources, staffing, and implementation time frames. Now, most corporations are acquiring one or more commercial off-the-shelf (COTS) systems (where the normalized ROI can be less than 1 year compared with a 3-year ROI for proprietary systems). Most organizations use an evaluation matrix when selecting solutions that are appropriate.

The placement of a particular product offering on the "ROI evaluation matrix for FM products" will vary client by client. The ROI evaluation matrix shown in Fig. 3.3.4-5 shows the results of an evaluation of 14 FM automation and IM products, which was done for a product manufacturer that had multiple sites in multiple countries. Criteria for evaluation included examining FM automation costs and the various aspects of FM automation functionality.

FM Automation Costs

- Hardware and software acquisition, implementation, maintenance, and upgrades (the initial software and hardware is only one-seventh of the total cost)
- Training, verification, and support costs
- Personnel costs (direct and indirect)
- In-house and outsourced services for surveys, field verification, data input and verification, customization, legacy system conversion, etc.

FM Automation Functionality

- Extensive graphic user interfaces and ease of use
- Relational access to activities rather than hierarchical
- Integrated report modules, integrated computer-aided design/area network (CAD/AN)
- Environments and robust applications and procedures
- Open-architecture, modular, scalable, enterprise-access-enabled, legacy-system-compatible, and virtual workgroups
- Web access via Internets, intranets, and extranets
- Computer-based training, active agents, and FM objects

Today, it is commonplace for FM organizations to use more than one system. Many will use cyberspace universes for FM with their corporate IT infrastructure to create connective mechanisms for integrating COTS solutions, legacy systems, and proprietary systems. It is estimated that great savings can be realized by effectively deploying and using a COTS solution.

Criteria for Selecting an Appropriate FM Automation Solution

No matter what your business size, an FM automation solution is usually available in a wide variety of configurations and in a range of costs for you and your organization. You can work with a vendor, consultant, IT professional, and/or fellow FM professional to determine which solution would best suit your needs. The criteria examined during the selection of an FM automation solution include but are not limited to the following:

- The size and number of properties and facilities managed
- The types of space and the costs of management (manufacturing, educational, offices, etc.)
- The numbers of people and the annual and quarterly churn rates
- The size and capability of the facilities department and the nature of their business partner relationships (the extent to which outsourcing and outtasking is used)
- The management of facilities-related factors (including people, time, costs, integration, and technology)
- The extents to which standards are enforced and inventory is managed
- The technical and computer competency of staff and outsourcing agents
- The workloads and the management requirements

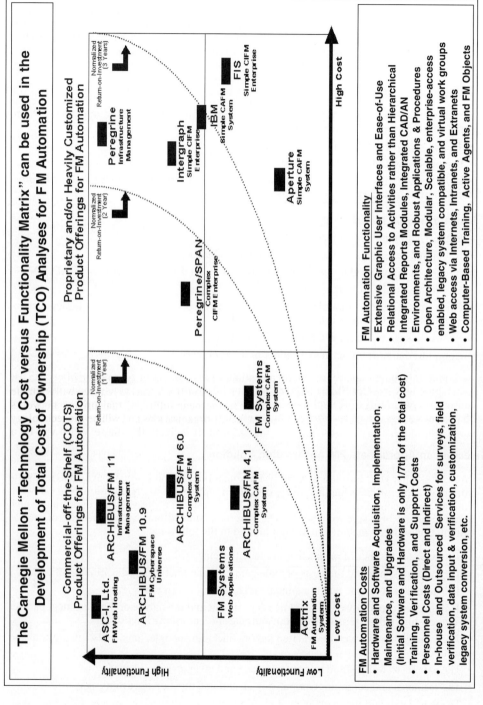

FIGURE 3.3.4-5 The return-on-investment (ROI) and the total-cost-of-ownership (TCO) will vary depending upon the type of systems acquired and implemented. However, it appears most corporations are acquiring one or more commercial off-the-shelf (COTS) systems (where the normalized ROI can be less than 1 year as compared with a 3-year ROI on proprietary systems). The placement of a product on the "FM Product ROI Evaluation Matrix" will vary by client. The sample shown here is an evaluation of 14 products that was done for a product manufacturer with multiple sites in multiple countries. Based on this matrix, a company would evaluate their ROI needs and select one or more of the options available to them. (*Courtesy of Carnegie Mellon University and Facilities Management Techniques, Inc., January 2000.*)

The Carnegie Mellon "Technology Cost versus Functionality Matrix" can be used in the Development of Total Cost of Ownership (TCO) Analyses for FM Automation

Commercial-off-the-Shelf (COTS) Product Offerings for FM Automation

Proprietary and/or Heavily Customized Product Offerings for FM Automation

High Functionality

Low Functionality

Low Cost

High Cost

Normalized Return-on-Investment (1 Year)

Normalized Return-on-Investment (2 Year)

Normalized Return-on-Investment (3 Years)

ASC-I, Ltd.
FM/Web Hosting

ARCHIBUS/FM 11
Infrastructure Management

ARCHIBUS/FM 10.9
FM Cyberspace Universe

ARCHIBUS/FM 6.0
Complex CIFM System

FM Systems
Web Applications

ARCHIBUS/FM 4.1
Complex CAFM System

FM Systems
Complex CAFM System

Actrix
FM Automation System

Peregrine/SPAN
Complex CIFM Enterprise

Peregrine
Infrastructure Management

Intergraph
Simple CIFM Enterprise

IBM
Simple CAFM System

Aperture
Simple CAFM System

FIS
Simple CIFM Enterprise

FM Automation Costs
- Hardware and Software Acquisition, Implementation, Maintenance, and Upgrades (Initial Software and Hardware is only 1/7th of the total cost)
- Training, Verification, and Support Costs
- Personnel Costs (Direct and Indirect)
- In-house and Outsourced Services for surveys, field verification, data input & verification, customization, legacy system conversion, etc.

FM Automation Functionality
- Extensive Graphic User Interfaces and Ease-of-Use
- Relational Access to Activities rather than Hierarchical
- Integrated Reports Modules, Integrated CAD/AN
- Environments, and Robust Applications & Procedures
- Open Architecture, Modular, Scalable, enterprise-access enabled, legacy system compatible, and virtual work groups
- Web access via Internets, Intranets, and Extranets
- Computer-Based Training, Active Agents, and FM Objects

Facilities management automation programs are generally available in these three categories:

1. The Stand Alone or Express Offering (for organizations with small or limited budgets)
2. The Enterprise Offering (for organizations with more substantial budgets and assets)
3. The Cyberspace Universes for FM Offering (for organizations of all sizes and all budgets)

How to Select the Right Tools for the Job.   Selecting appropriate FM automation and IM solutions is a critical activity. Most organizations have eclectic "hybrid" solutions. Solutions exist for organizations of all sizes and capabilities. Consultants and friends who have done this before are great for helping you out in this area. If no one is available, you might begin by answering the three questions that are shown in (see Fig. 3.3.4-6). Having answered these questions, you may be able to determine the appropriate FM automation solution(s) for your organization. This worksheet is based on presentations made at numerous World Workplace conferences and IFMA Best Practices Forums.

Open Architecture, Scalability, and Modularity.   Facilities management automation implementations that are oriented to cyberspace universes for FM and IM are, typically, designed with open architectures, scalability, and modularity. As an organization's FM automation activities expand, applicable legacy systems should be readily integrated with other offerings. Facilities management automation and IM solutions enable the integration of corporate databases readily with express or stand-alone, enterprise, and cyberspace universes for FM offerings (Fig. 3.3.4-7). Typically, corporate databases are used to connect information silos and islands of automation.

The Virtual Opportunity: Extending Classic Wide Area Networks and Enterprise Networks into E-FM and E-Space

As an organization expands its virtual network of employees and business partners, a high degree of feature integration over the wide area network (WAN) can be realized with suitable Intranet, frame relay, and ATM capabilities. In the near future, virtual private networks (VPNs) will become readily available to most facility organizations and will be the technology conduit to and from IM solutions (see Fig. 3.3.4-8). It is likely that these VPNs will be a required part of your corporate strategic IT initiatives.

Infrastructure Management with Cyberspace Universes for FM.   There are, typically, three components of an IM solution: FM automation application programs, FM automation database management engines, and FM legacy systems integration and connectivity.

FM Automation Application Programs.   Facilities management automation application programs usually fall into these seven application areas:

1. Real property and lease management
2. Strategic master planning
3. Space management
4. Design management
5. Furniture and equipment management
6. Telecommunications and cable management
7. Building and operations management

FM Automation Computer-Aided Design Engines.   Facilities management automation products use a variety of computer-aided design (CAD) engines. Most FM automation and IM offerings have external links to one or more CAD products. The more advanced solutions have created seamless CAD interfaces or overlays with their product and/or have OEM CAD solu-

Selecting the Appropriate FM Automation and Infrastructure Management Solutions

What size is your organization?	What do you want to do?	What solution is right for you?
Small business with under 500 people setting up our first FM Automation Solution We do not have an IT Department or IT Specialist Portfolio Size: Less than 100,000 sq. ft.	• Provide basic capabilities • Help business expand • Keep track of critical assets • Outsource FM Web site activities	• Express Products • Stand-Alone Product Offerings • FM Web Central
Medium-size business or department of a larger corporation expanding our FM Automation Activities We have limited access to the IT Department Portfolio Size: Less than 573 Million sq. ft.	• Provide additional capabilities • Help business expand • Keep track of critical assets • Build upon existing data, information, and knowledge • Build FM Web site	• Express Products • Stand-Alone Product Offerings • LAN-based CAFM Systems or Client/Server-based CIFM Enterprises • FM Web Central
Medium to large corporation seeking to achieve strategic competitive advantage using FM Automation We have an IT Specialist and an IT Department Portfolio Size: Unlimited	• Provide advanced capabilities • Help business expand • Keep track of critical assets • Build upon existing data, information, and knowledge • Expand FM Web site • Develop Visual Infrastructure Management Solution	• Express Products • Stand-Alone Product Offerings • LAN-based CAFM Systems or Client/Server-based CIFM Enterprises • FM Web Central

FIGURE 3.3.4-6 Selecting the appropriate FM automation and IM solutions is a critical activity. Most organizations have eclectic "hybrid" solutions. There are solutions for organizations of all sizes and capabilities. (*Courtesy of ARCHIBUS Solution Centers, Ltd. and Facilities Management Techniques, Inc.*)

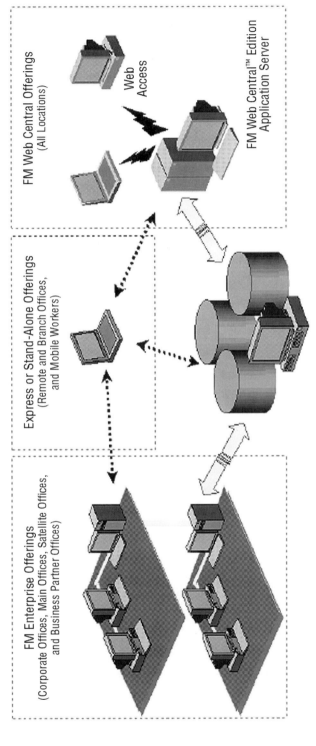

FIGURE 3.3.4-7 Advanced FM automation solutions are modular, scalable, have open architectures, and are readily integrated with existing legacy systems. (*Courtesy of Bruce Kenneth Forbes, ARCHIBUS, Inc.*)

Corporate Infrastructure Management Databases

FM Web Central Offerings
(All Locations)

Web Access

FM Web Central™ Edition Application Server

Express or Stand-Alone Offerings
(Remote and Branch Offices, and Mobile Workers)

FM Enterprise Offerings
(Corporate Offices, Main Offices, Satellite Offices, and Business Partner Offices)

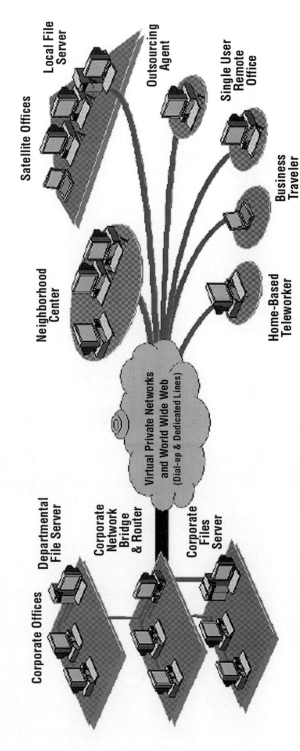

FIGURE 3.3.4-8 Virtual private networks (VPNs) will enable facilities organizations to connect with associates and business partners throughout their buildings and campuses, as well as throughout a city, region, nation, and around the world. (*Courtesy of Bruce Kenneth Forbes, ARCHIBUS, Inc.*)

tions actually embedded into their offerings. In the United States, some of the more frequently used CAD systems are AutoCAD®, Integraph® and Microstation®, Visio®, AutoCAD LT®, AutoCAD Sketch®, and Actrix®. Internet and intranet tools for CAD are being rapidly developed. At this time, some CAD product developers offer free downloads of their CAD-viewing software products. One system that has become very popular is Autodesk's Whip™.

FM Automation Database Management Engines. Facilities management automation products use a variety of database management engines. They range from Microsoft Access® for stand-alone applications to MS SQL Server®, Oracle®, ARCHIBUS/FM®, or Sybase® for the enterprise offerings. Cyberspace universes for FM applications can retrieve and deposit data, information, knowledge, and wisdom to a variety of database management engines.

FM Automation Legacy System Integration and Connectivity. Facilities management automation systems can extend their reach through connectivity with other existing legacy systems and islands of automation. Typically, more than 42 percent of the data and information used in FM automation can come from such existing applications and their associated databases.

The Top Twenty Reasons Why FM Automation Fails

There are many reasons why FM automation has failed within some facility organizations. Some of the top reasons include the following:

- FMA-IM red tape and bureaucracy
- Ineffective FMA-IM management teams
- Lack of proper FMA-IM direction
- Lack of clear FMA-IM goals and objectives
- Lack of tools to achieve goals and objectives
- Lack of suitable FMA-IM value postulates
- Lack of baseline FM data and FM information
- Poor FMA-IM "expectations" management
- Mismanagement by FMA-IM teams
- Anxiety and fear of achieving goals
- Unrealistic FMA-IM goals and expectations
- Poor FMA-IM training and refresher courses
- Lack of FMA-IM audits and reviews
- Poor FMA-IM communications
- Lack of suitable FMA-IM resources and time
- Poor FMA-IM business planning
- Lack of FM best practices and benchmarks
- Lack of FMA-IM knowledge workers
- Poor integration and data mining
- Lack of buy-in for FMA-IM activities

A Brief History of FM Automation

Some of the first uses of computers in the 1960s were for creating data for asset management and financial reporting. Now, we consider that there have been these eight generations of FM automations (see Fig. 3.2.4-9):

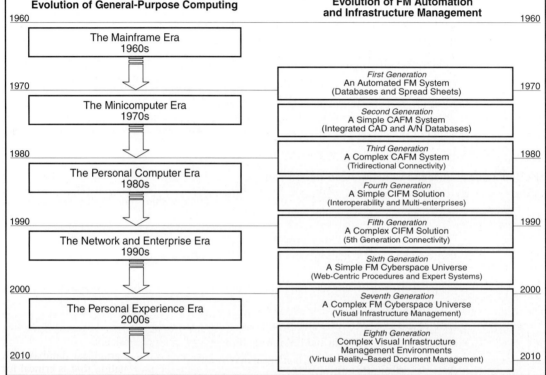

Timeline for the Evolution of FM Automation and Infrastructure Management Solutions

Evolution of General-Purpose Computing	Evolution of FM Automation and Infrastructure Management
1960	1960
The Mainframe Era 1960s	
1970	*First Generation* An Automated FM System (Databases and Spread Sheets) — 1970
The Minicomputer Era 1970s	*Second Generation* A Simple CAFM System (Integrated CAD and A/N Databases)
1980	*Third Generation* A Complex CAFM System (Tridirectional Connectivity) — 1980
The Personal Computer Era 1980s	*Fourth Generation* A Simple CIFM Solution (Interoperability and Multi-enterprises)
1990	*Fifth Generation* A Complex CIFM Solution (5th Generation Connectivity) — 1990
The Network and Enterprise Era 1990s	*Sixth Generation* A Simple FM Cyberspace Universe (Web-Centric Procedures and Expert Systems)
2000	*Seventh Generation* A Complex FM Cyberspace Universe (Visual Infrastructure Management) — 2000
The Personal Experience Era 2000s	*Eighth Generation* Complex Visual Infrastructure Management Environments (Virtual Reality–Based Document Management)
2010	2010

FIGURE 3.3.4-9   The digital economy has become a reality for FM professionals. This figure depicts the evolution of computer technology and FM automation. (*Courtesy of ARCHIBUS Solution Centers, Ltd.*)

- *First generation.* Alphanumeric applications such as spreadsheets and databases (with limited or no CAD system connectivity)
- *Second generation.* Simple CAFM systems [stand-alone or local-area-network (LAN)–based bidirectional connectivity between CAD and A/N databases—a change in a CAD drawing was automatically reflected in the A/N database and vice versa]
- *Third generation.* Complex CAFM systems (stand-alone or LAN based tridirectional connectivity between A/N databases, CAD, and visual display databases such as photographs and video files)
- *Fourth generation.* Simple Computer-Integrated Facilities Management™ (CIFM™) enterprise solutions (stand-alone, LAN, and/or enterprise-based solutions with interoperability and multienterprise connectivity)
- *Fifth generation.* Complex Computer-Integrated Facilities Management Enterprise Solutions (stand-alone, LAN, and/or enterprise-based solutions with interoperability and multienterprise connectivity and/or CIFM-to-CIFM enterprise connectivity at a local, regional, national, and/or global level)

- *Sixth generation.* Simple cyberspace universes for FM (e-commerce enabled Web-centric offerings with expert systems and best-of-class processes and procedures for the FM executive)
- *Seventh generation.* Complex cyberspace universes for FM with connectivity to IM enterprises and cybermedia applications for FM [e-commerce enabled Web-centric offerings with expert systems and best-of-class processes and procedures for the FM executive with limited connectivity to ERP and infrastructure resource planning (IRP) systems]
- *Eighth generation.* Complex IM environments (virtual by design, e-commerce-centric Web offerings with expert systems, artificial intelligence, and best-of-class processes and procedures for the FM executive: The IM environment provides extensive connectivity to legacy systems, IT silos, and data warehouses found in such systems as ERP, help desk, and IT infrastructure systems)

Significant Milestones in the History of FM Automation

Although numerous significant milestones exist in the history of FM automation and IM, it is helpful to examine the generations of FM automation products against similar metrics (see Table 3.3.4-3).

How FM Automation Integration and Utilization Improves
Effectiveness, Productivity, and Profitability

The net result is that FM automation significantly increases productivity and profitability. Now, the three most commonly implemented FM automation systems include computer-aided facilities management (CAFM) systems, computer-integrated facilities management (CIFM) enterprise solutions, cyberspace universes for FM, and IM environments (see Fig. 3.3.4-10). These technologies are changing the way companies do business.

The CAFM systems combine the power of nongraphical and graphical databases. A CAFM system enables you to organize and access data and information that is critical for managing your assets—from office space to furniture and equipment, from real estate to strategic master planning, and from telecommunications and cable management to computerized maintenance management. These CAFM systems are usually configured as stand-alone systems or on LANs.

The CIFM enterprise solutions go a step further and extend the reach of your FM data, information, and knowledge enterprise-wide to provide the information that strategic and business managers within your organization need to make better planning and policy decisions. The CIFM enterprise solutions are usually configured for enterprise-wide access, but they are also available for LANs and stand-alone configurations with enterprise connectivity.

Cyberspace universes for FM combine the power of FM automation technologies such as CAFM systems, CIFM enterprise solutions, information management resource solutions, and ERP solutions with new millennial technologies and best practices into a globally accessible depository of FM data, information, knowledge, and wisdom. The expert systems and knowledgeware cultivated within cyberspace universes for FM enable executives to incorporate relevant best practices and benchmarking applications into the daily activities of their organizations.

Infrastructure management environments seamlessly integrate and combine the power of numerous corporate silos of information management and yield collective FM data, information, knowledge, and wisdom. Typical stand-alone automation silos might include CAFM systems, CIFM enterprises, cyberspace universes for FM, ERP and IM systems, IT infrastructure systems, building automation systems, CMMSs and building operation management systems (BOMSs), help desk and telephony systems, legacy systems, and finance/administration systems.

TABLE 3.3.4-3   A Historical Perspective of FM Automation and IM Solutions

Generation of FM automation products	Stand-alone applications	Stand-alone applications	Integrated CAFM systems	CIFM enterprise solutions	Infrastructure management and cyberspace universes for FM
The five paradigms of computing	Mainframe era Batch processing	Minicomputer era Time-shared environment	Personal computer era PCs and work-stations w/networking	Network and networking with client/servers	Personal-experience era Digital nervous systems
Years	1960s	1970s	1980s	1990s	2000s
Computer location	Computer room "fish bowls"	Terminal rooms	Desktops	Mobile	Virtual
Users	Experts	Specialists	Grouped individuals	Groups	Individuals
Primary directives	Calculate	Calculate and access	Calculate, access, and present	Calculate, access, present, and communicate	Calculate, access, present, communicate, and mind meld
Data	Alphanumeric	A/N, text vectors, and CAD	A/N, CAD, fonts, and bitmaps	A/N, advanced CAD, fonts, and graphics	A/N, advanced CAD and graphics, audiovisual, and virtual simulations
Connectivity	Peripherals	Peripherals and terminals	Peripherals terminals, LANs, and WANs	Peripherals terminals, LANs, WANs, RANs, SANs, GANs, and client/servers	Peripherals, terminals, LANs, WANs, RANs, SANs, GANs, client/servers, and virtual area networks (VANs)
Languages	Fortran, Basic, and COBOL	Fortran, Basic, COBOL, PL/1, and RPG	Fortran, Basic, C, Visual Scripting, and LISP	C, Visual Scripting, LISP, C++, and Java	C, Visual Scripting, LISP, C++, Java, and ActiveX FM Objects™
Applications	Customized	Standardized with moderate customization	Generic applications with moderate customization	Commercial off-the-shelf (COTS) with minimal to moderate customization	COTS with minimal customization, E-FM™, E-Space™, CFM™, and FM Cyberspace™
Application costs	$500,000 to $2,000,000	$50,000 to $200,000	$5,000 to $100,000	$2,500 to $50,000	5 cents per transaction to $25,000

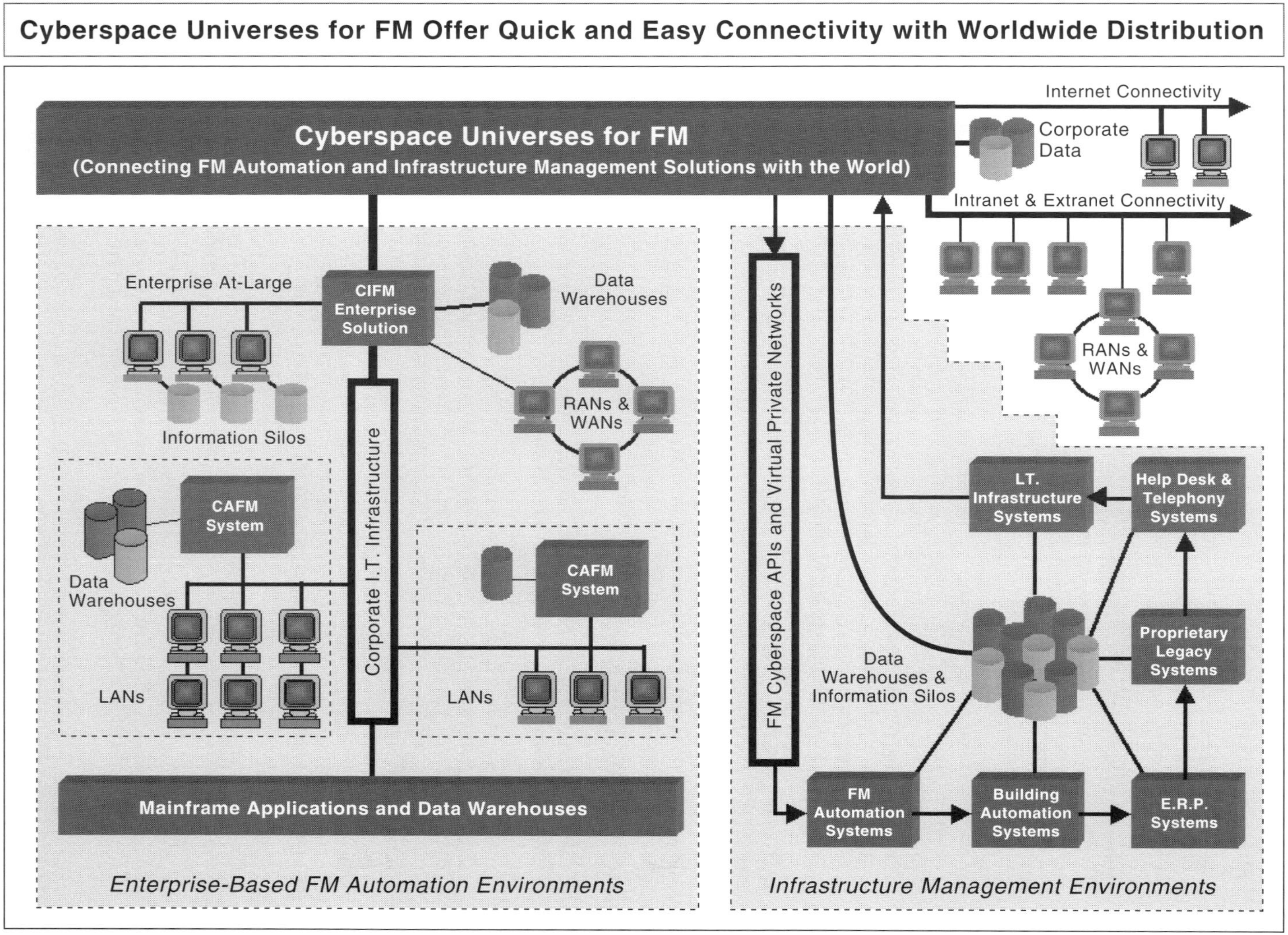

Cyberspace Universes for FM Offer Quick and Easy Connectivity with Worldwide Distribution

FIGURE 3.3.4-10 Cyberspace universes for FM provide seamless connectivity (to FM automation and IM environments) and the TCO is reduced. Such connectivity enables the rapid deployment of automation with full connectivity to exiting legacy systems such as CAFM systems, CIFM enterprises, ERP systems, IT infrastructure systems, and help desk and telephony systems. (*Courtesy of Bruce Kenneth Forbes, ARCHIBUS, Inc.*)

FM Automation—Easy and Intuitive to Use

Familiar Microsoft Windows™ conventions, such as toolbars, pull-down menus, multiple windows, and dialog boxes make working with FM automation easy. Comprehensive and context-sensitive, on-line help systems can support your learning process. Other special task-oriented menus, called FM automation *navigators* and *process toolbars,* provide intuitive interfaces. Just as the names suggest, the navigator and process toolbars guide you step-by-step through complex FM procedures, showing you where to start, what information to enter, and what types of reports you can derive from the information. Many industry-specific FM methodologies and reports are already built into FM automation, so you can put your information to work immediately.

Seamless Integration.   The core application of FM automation software typically links a facilities database with associated CAD drawings, so that changes cascade automatically throughout the system. This activity ensures the accuracy of your data, reduces errors, and eliminates file duplication. The application modules allow you to work simultaneously with multiple facility drawings and data tables. You can even perform all of your FM tasks from within CAD, computer-aided design and drafting (CADD), geographic information systems (GIS), and document management systems.

Dynamic Reports.   Based on user experience, FM automation software providers have developed literally hundreds of preformatted reports that meet the needs of a wide range of users. These reports and database queries come built into many products—simply point and click to select the information you need. You can then easily fine-tune the style and content of your reports using a standard dialog interface.

FM Automation for the World Wide Web

Facilities management automation for the Web is the foundation of cyberspace universes for FM. Best-of-class Web technologies have been expanded into a powerful extension of the CAFM, CIFM, ERP, IRM, and customer relationship management (CRM) environments. Inexpensive and easy to use, cyberspace universes for FM enables you to expand the connectivity of FM automation to anyone, anywhere, who is connected to your corporate intranet or the Internet at large, via the World Wide Web. Anyone who uses your organization's facility information will benefit from the use of FM automation for the Web. Unlike other versions of FM automation, cyberspace universes for FM enables the most casual of users to learn quickly how to interact with the applications. Training time is only a matter of minutes.

THE TOP SEVEN FM AUTOMATION APPLICATION AREAS

Facilities management automation environments are typically built around *application control programs,* or *cores,* with *application modules* that address various FM issues. Facilities management automation enterprises also offer a variety of *enterprise access interfaces* or *user interfaces that are visual* for both the creators and consumers of FM data, information, knowledge, and wisdom. You can put together any combination of these elements to fashion the ideal CAFM-CIFM-cyberspace universes for the environment in your organization.

Real Property and Lease Management Applications

Facilities management automation real property and lease management address the needs of commercial property managers, corporate lease managers, building owners, and tenants, enabling you to record detailed real estate and lease data and to use the information to produce valuable reports and analyses. You can track real-world conditions, such as the impact of lease terms on chargebacks, along with time-dependent elements, including options, escalations, or terminations. You can model complex real estate and lease holdings on a national or even global scale and summarize their revenue potential and performance. A wealth of tools is provided for performing detailed cost tracking, budgeting, and sophisticated financial analyses of your properties.

Tracking Leases and Options.   You can easily track important lease provisions and stay on top of activities that have potentially costly repercussions, such as hazardous material abatement or responding to tax assessment or lease notifications.

Chargebacks.   You can develop chargeback strategies that model the subtleties of real-world conditions and allow you to charge back costs to buildings, leases, or departments. You can designate certain charges to be applied, or not, according to the negotiated terms of each particular lease.

Budgeting.   You can create budgets for properties, buildings, leases, departments, and accounts for categories you define, and you can model your initial budget on your actual costs, on last year's budget, or on costs projected from recurring transactions, such as regular lease revenues or payments.

Maximizing the Use of Space.   By integrating intelligent CAD plans with your real estate database, you can accurately model your holdings and determine how efficiently space is being used. You can create occupancy plans to highlight vacant, rentable space, for example, or reclaim lost space that is currently not included in any negotiated lease area.

Mapping a Corporate Strategy.   With accurate property and lease profiles in hand, corporate executives can monitor the profitability of each property, lease, or building, and make more cost-effective decisions: What did maintenance cost last year, and should we outsource it? What portion of our expenses was due to a high tax rate? We are considering selling off one of our divisions; what is the total asset and liability summary for the property that that division occupies? For answers to these questions and more, decision makers can turn to real property and lease management applications.

Strategic Master Planning Applications

In the power-packed world of mergers and acquisitions, strategic master planning has become a necessity in valuing and repositioning an organization's assets. Strategic master planning applications extend and expand the power of planning from your FM organization to your boardroom. Professionals who manage a single property, a campus, or a diverse portfolio of buildings sprawled across a city or region will gain tremendous benefits by using the techniques and methods incorporated in strategic master planning applications. Value for facilities professionals is created even while dealing with the complexities of downsizing, mergers, acquisitions, and expansions. Strategic master planning divides the tasks of the strategic master planning process into five areas of activity that may be used individually or collectively to optimize business assets and efficiency: (1) requirements programming, (2) forecasting, (3) allocation, (4) layout, and (5) history.

Requirements Programming.   Inventory the data you will need to formulate your organization's strategic planning goals. Data may include departmental affinities, square footage needs, types of space required, duration of need, and so on.

Forecasting. Make informed decisions about future space needs based on those of each department. The data collected during the requirements programming phase of the process allow you to forecast with confidence, using built-in methodologies and user-defined variables.

Allocation. Space is allocated on the basis of your organization's various business goals. One of the most useful automation features available to the facilities manager, the *stack diagram,* is typically available within the allocation activity.

Location. Once you have determined how to best allocate available space, the next step is to design an optimal layout that defines the location of each organizational unit on each floor. One of the most widely used methods for laying out a floor plan is *blocking.* Blocking is based in part on automatically generated stack plans and trial layouts.

History. Maintaining a historical record of the way space was used in the past is very useful in planning for future space requirements. It can provide "snapshots" of your space inventory at various times during your organization's evolution. This activity also enables you to compare trends in the usage of space during different periods.

Space Management Applications

Facilities management automation space management applications enable you to create and maintain an accurate and up-to-date record of all areas in your buildings (a living electronic inventory of your space) and how they are used. By integrating your facilities data with CAD drawings, business graphics, and picture files, FM automation allows you to organize and analyze space information in various ways. Reports can be products that will help your organization make the most of its assets. Users can determine how efficiently space is being used, identify rentable or leasable areas, calculate space costs and chargebacks, and create quick tenant or employee occupancy plans.

Take Control of Your Space. Facilities management automation space management applications make it easy to develop intelligent databases and drawings that track the use of space within your buildings.

Choose the Right Chargeback Method. Once you have developed a space inventory and associated the areas with cost centers, tenants, or other organizations within your company, you can determine how much space each group uses and bill them accordingly.

Analyze Your Data with Predefined Reports. Built upon the experience of thousands of FM operations, the FM automation space management applications typically provide a multitude of preconfigured reports and analyses that immediately allow you to study and interpret your data. You can draw up occupancy plans and determine usable and rentable areas, calculate space efficiency rates or the average area per employee, perform space planning exercises, and more.

FM Automation Overlays for CAD and GIS. Common to most FM automation environments today are supplemental tools, such as FM automation design management overlays that empower CAD- and GIS-based facility drawings and data. Now you can create facility drawings, develop asset symbols, register them in a database, and perform asset symbol edits and database updates—all from within CAD and GIS environments—by using the latest "active objects" drafting tools. These tasks, along with the specialized architectural drafting tools and database tables of the FM automation design management overlays (*overlays*), enable facility managers, architects, and interior designers to create and maintain facility record drawings easily.

Seamless Integration. The features of elegant overlays are indistinguishable from built-in CAD and GIS commands. By adding a series of pull-down menus to a CAD or GIS system's own menu bar, overlays enable you to connect to a project database that contains your facility information and to perform FM tasks. From within a CAD system, for example, you can connect to an FM automation project, create a room polyline, and assign it intelligence, such as a room number and department assignment. This information, along with the polyline's area, is posted directly to the Rooms table of the project database.

Powered by Reactors or Expert Systems. With reactors and expert systems, such as the ARX™ or "AutoCAD Runtime Extension™" for AutoCAD™, the overlays can react directly to notification of events sent by the CAD system. This means that the overlay knows when you load a drawing, save a drawing, or edit, copy, mirror, or erase asset symbols. For instance, if you use the CAD system's copy command to copy an autonumbered furniture item, the overlay automatically creates a record in the database and links it to the new drawing item. Then, if you generate a furniture standards inventory report in the FM automation environment, it will reflect this change.

Asset Symbols Possess Facility Intelligence. Room, group, furniture, and equipment asset symbols know what type of facility object they represent and assume the correct behavior when copied, mirrored, or erased with typical CAD commands. What is more, for asset symbols located within a group, room, or other boundary, you can infer the values of the enclosing boundary. You can also apply values set for the entire drawing, such as building and floor codes, to new asset symbols that you create.

Furniture and Equipment Management Applications

Furniture and equipment assets represent a substantial investment for any organization and can have a direct impact on productivity. This is why the effective management of these assets is vital to your bottom line. Using the simple, but powerful, inventory tracking capabilities of FM automation furniture and equipment (F&E) management applications, you can easily monitor the cost of your assets, calculate depreciation and churn rates, plan moves of furniture and equipment, and more. Bar code support and built-in forms for surveying your assets help you develop and maintain a useful and reliable inventory of information that will enable you to make better planning and policy decisions.

Access Up-to-Date Information. At the core of effective asset management is the need to maintain a historical record of the disposition of your assets—their condition, location, employee, and cost center assignments. By providing easy access to this information in data tables and CAD/GIS drawings, F&E management allows you to define and enforce corporate standards, plan moves, and produce a variety of reports.

Easy Layout and Move Planning. Visualize a variety of new layouts by rearranging or adding asset symbols to your CAD floor plans. Then, create reports that compare the move and acquisition requirements of each proposed layout to determine the most effective plan of action.

Speedier Asset Surveys. Gathering inventory information can be tedious and time-consuming, but F&E management speeds the process by providing built-in survey forms. It also supports bar code technology for faster, easier, and more accurate data entry. Bar coding and chip implants for F&E management are becoming commonplace.

Itemized Cost Tracking. Furniture and equipment management allows you to track financial data throughout the life of an asset: its original value, current value, accumulated depreciation, and, when sold, the gain or loss on the asset. From this data, you can create a variety of summary reports according to cost center, property type, and so on.

Insurance and Warranty Tracking. Locating insurance information for any asset is as simple as entering the item's bar code. Instantly, you have access to its policy value and expiration date. You can also determine appropriate insurance coverage by running a risk management report, comparing an asset's insured value against its original, current, and replacement value.

Track Assets Alphanumerically and/or Graphically.   You can document your F&E assets only in the database or link them to block definitions added to your CAD drawings.

Telecommunications and Cable Management Applications

Voice and data communications are the lifeblood of business today, making the maintenance of this vital infrastructure a mission-critical task. Facilities management automation telecommunications and cable management applications enable you to create a living, electronic inventory of the physical cabling and telecommunications network connections in your facility. With easy access to detailed and up-to-date connectivity information, you can draw network plans, troubleshoot problems, or move employees with minimal disruption.

Track Your Infrastructure.   There is no need to create a complete cable inventory before you begin to reap benefits from a telecommunications and cable management application. Its flexible design allows you to start tracking your network connectivity immediately, at any level of detail, and build gradually on the information. Create an inventory of all of your telecommunications devices and their connections. Identify devices by physical and logical LAN or WAN segments and by their levels in the hierarchy of the overall network.

Speed Cable Connections.   Seamless links with CAD drawings through the FM automation design management overlays enable you to speed your work considerably. You can quickly determine where to lay new cables through a cable distribution network, find open ports on existing equipment, trace the connections from one device to others in its path, highlight devices based on criteria such as type or status, and more. You can also connect devices in your drawings simply by clicking on one and then the other. Interactive queries instantly pull up data for equipment, cables, and their connections. Database queries can graphically trace a particular network device or all devices that would be affected by a given action.

Maintaining Your Infrastructure.   Managing your telecommunications and cables allows you to maintain the most up-to-date databases and drawings of all telecommunications devices and their links for data and voice networks.

Building Operations Management Applications

Facilities management automation building operations management applications are often comprehensive, integrated CMMSs that help keep your facilities running smoothly and minimize your operating costs. With fast and easy access to critical facilities information, you can more efficiently document and monitor maintenance or repair work, schedule employees or outside contractors to perform the work, maintain an accurate inventory of supplies, and budget the costs. A variety of predefined reports allows you to analyze maintenance histories and expenditures to help you prepare for future demands on your facilities.

Automated Work Orders Speed the Process.   Communication is the key to keeping your facilities operating at peak efficiency. By using electronic work requests and work orders to report, track, and analyze problems, you can manage all of the tasks involved in a timely fashion. Anyone in an organization can request work, check on the status of a job, or report on its progress simply by clicking on an icon and entering information in a dialog box. Built-in forms help you process work requests quickly by assigning priority ratings and scheduling resources. Tracking the status of ongoing maintenance work allows you to maximize productivity and resolve backlog issues (see Fig. 3.3.4-11).

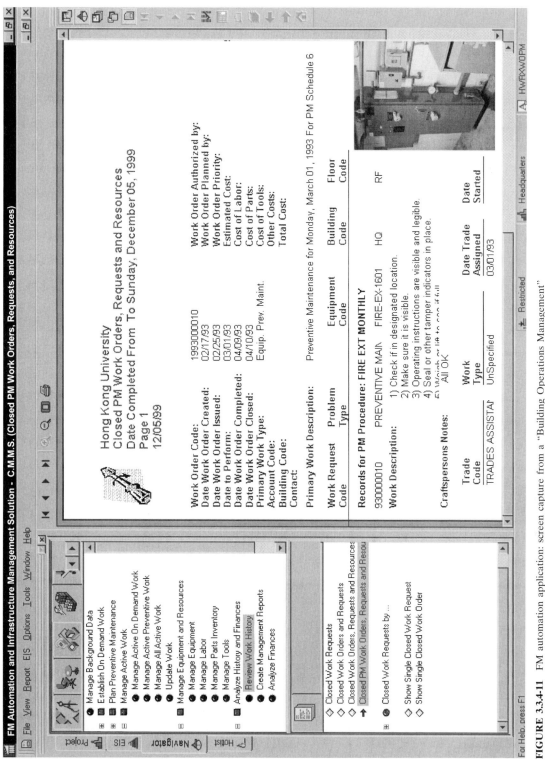

FIGURE 3.3.4-11 FM automation application: screen capture from a "Building Operations Management" application work session (Management Review of Work Requests and Work Order Resolution). (*Courtesy of Bruce Kenneth Forbes, ARCHIBUS, Inc.*)

Maintain Peak Efficiency with Preventive Maintenance.   Automatically scheduling regular preventive maintenance (PM) procedures is the most efficient way to prevent costly breakdowns and downtime. Built-in methodologies can help you schedule regular maintenance or housekeeping tasks at fixed or floating intervals and adjust the schedules to manage conflicting demands on your resources. You can also define complete PM procedures, step-by-step, as well as generate automatic work requests.

Manage Equipment and Resources Effectively.   Cost and availability of equipment and resources can impact many of the operating decisions you make every day. By tracking detailed information on your equipment assets (including location, usage, parts, warranty, and service information), you can generate many ad hoc reports and queries. Tracking workloads and labor availability can help improve the efficiency and performance of your labor force, and maintaining an adequate inventory of spare parts and tools ensures that sufficient resources are available when you need them. You can even use bar code technology for easy, accurate data entry.

Analyze Work History Costs.   Building operations management helps you learn from experience, so that you can forecast future operating costs and requirements and support your long-term facilities plans. Various reports allow you to review and summarize work histories, analyze labor usage, and generate cost expenditure reports by buildings, departments, or cost centers.

Comply with Regulations.   Flexible data collection and customization allows you to maintain data pertinent to EPA and OSHA regulations. Track equipment lockout or tag-out status, mean-time-between-failure or repair (MTBF/MTBR) statistics, or performance histories, and establish PM procedures in compliance with regulations.

Working with Drawings.   If you choose to use CAD drawings with building operations management to show graphical locations of maintenance problems, you have several options for developing your information. Start with existing CAD drawings or create new facility drawings, using the application's simple drawing tools to outline room areas and insert equipment asset symbols, or use facility drawings already developed with the space management or F&E management applications (see Fig. 3.3.4-12).

INFRASTRUCTURE MANAGEMENT STRATEGIES FOR ENTERPRISE WEB CONNECTIVITY

Facilities management applications on the World Wide Web serve as a foundation for cyberspace universes for FM. The world of technology-based best-of-class Web technologies is being expanded into a powerful extension of CAFM and CIFM environments. Inexpensive and easy to use, Internet/intranet applications enable you to expand the connectivity of IM and ERP information to anyone, anywhere, who is connected to your corporate intranet or the Internet at large, via the World Wide Web. In the next few years, Internet/intranet applications will become a critical part of Infrastructure Management™ systems.

In Search of Value for Your Customers

Creating value is increasingly more difficult for facility managers who face corporate and organizational challenges. Their internal customers are more price conscious and knowledgeable about buying FM-related services and products, they face tighter deadlines and need more timely information about their corporate real estate assets, and they are more uncertain about their futures and their facility's needs.

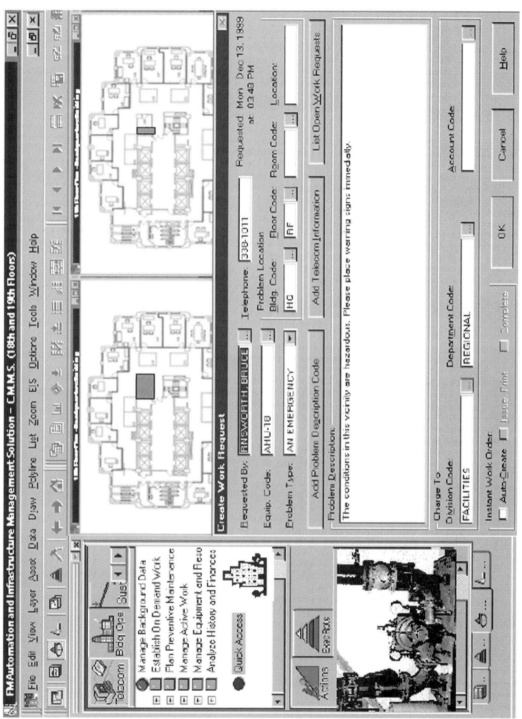

FIGURE 3.3.4-12 FM automation application: screen capture from a "Building Operations Management" application work session (Create a Work Request). (*Courtesy of Bruce Kenneth Forbes, ARCHIBUS, Inc.*)

By Web-enabling their facilities, FM professionals (FMers) can hone their company's competitive edge with business processes unconstrained by time zones or distance. They can communicate their FM vision to the corporation at large, accurately assess conditions throughout their facilities to make the best use of resources, and collect the critical data the company needs to execute its business plan. Internet/intranet applications enable a facility manager to become a key, visible, dependable player in the strategic planning process. Activities previously available only in enterprise solutions are now finding their way to the Web (see Fig. 3.3.4-13).

Forging a New Role with Internet/Intranet Applications

Facilities management cybrarians create value by intelligently using cyberspace universes for FM. They are the "caregivers," facilitators, composers, and orchestra leaders of FM data, information, and knowledge, and they are the creators of FM wisdom. Facilities management cybrarians and their teams work in a place now known as *cyberspace universes for FM*. At the heart of cyberspace universes for FM is a depository and control center of FM automation data, information, knowledge, and wisdom. It is a repository for the accumulated knowledge and wisdom of one's peers and for the experience of one's organization. Instant access to this repository is through a variety of user interfaces. Similar to the user interfaces found in enterprise-based FM automation solutions, which require training, cyberspace universes for FM versions require only a 1- to 5-minute consultation to learn how to use and produce the same result in many cases.

The Key to Success Is Communication

As soon as data are entered into an Internet/intranet application, updated information and active work orders are immediately available to anyone who needs them. Field personnel can call up active work requests in their area on a laptop hooked to a mobile phone. At a meeting in the boardroom, the facility manager can instantly access answers to unexpected questions. Through increased communication, Internet/intranet applications facilitate the creation of value in the enhanced understanding of key issues at headquarters, in the review of branch office requirements, or in the accelerated responsiveness of field personnel. Through Internet/intranet applications, facilities managers can keep their fingers on the pulse of their facilities, while at the same time decreasing the FM department's cost to respond.

Extend Your Organization's Competitive Edge

By connecting your facility to the World Wide Web, Internet/intranet applications will extend your company's competitive edge with business processes not constrained by time zones or distance. It will change the way you do business and give universal access to up-to-the-minute facility information, instantly, anywhere in the world, 24 hours a day, 7 days a week. Internet/intranet applications can serve as a help desk for your customers' self-service information. It will increase the accuracy of your FM information by enabling remote data collection and increase your control of the global enterprise by integrating information from headquarters, regional offices, and branch offices (see Fig. 3.3.4-14). In this example, the FM automation system has combined data and information from an HR system (photographs and personnel profile) with the facilities information (floor plans, F&E databases, telephone and e-mail directories, etc.) to create an "employee—locate and highlight room" report.

Use E-Commerce Tools to Communicate Your Vision

Just as the Web makes your company globally visible, Internet/intranet applications will make your corporate FM efforts globally visible and instantly accessible. Use the Web to promote your FM department's efforts, advertise your capabilities and internal services, and promote

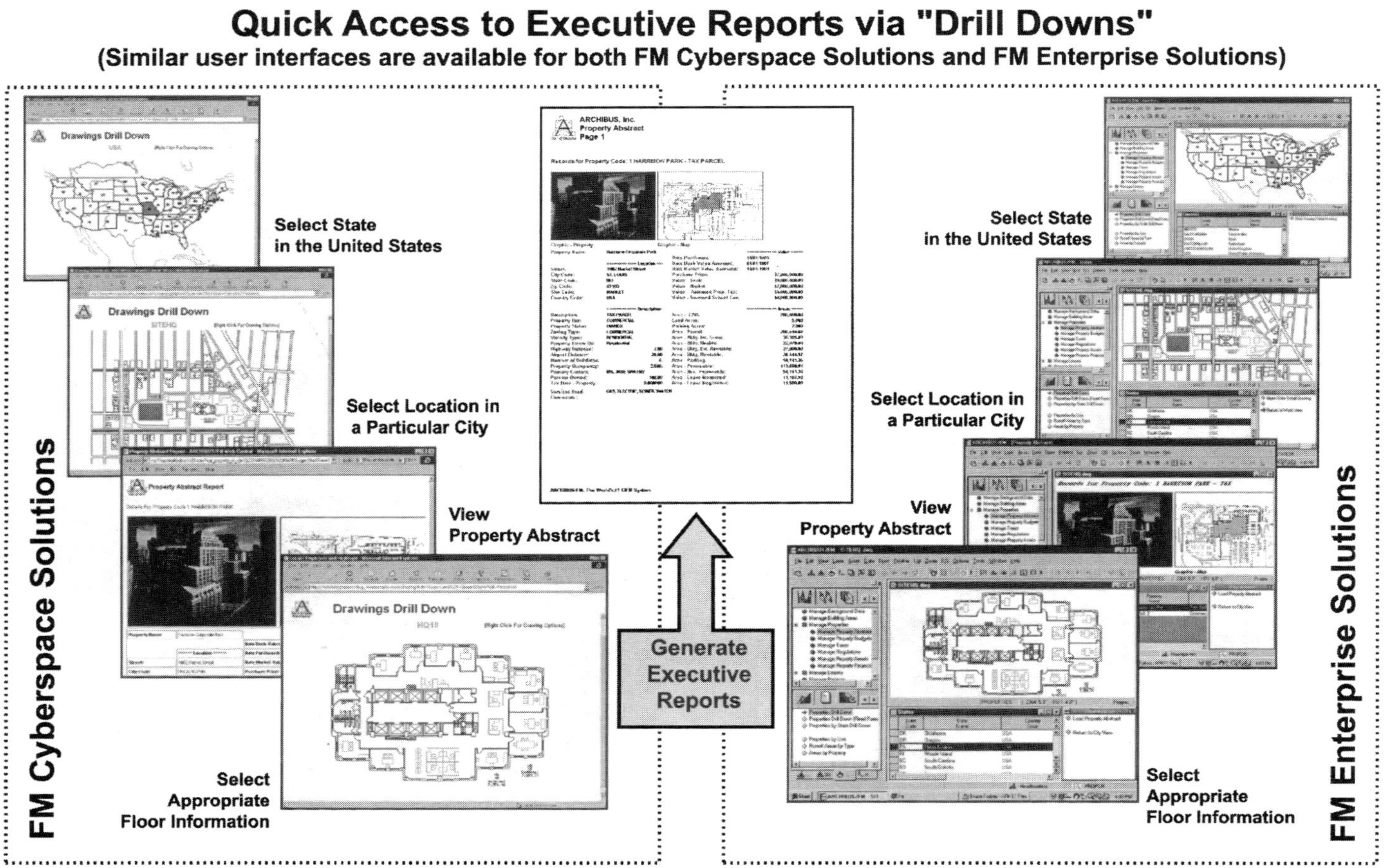
Quick Access to Executive Reports via "Drill Downs"
(Similar user interfaces are available for both FM Cyberspace Solutions and FM Enterprise Solutions)
FM Cyberspace Solutions
Select State in the United States
Select Location in a Particular City
View Property Abstract
Select Appropriate Floor Information
Drawings Drill Down
Property Abstract Report
ARCHIBUS, Inc.
Property Abstract
Page 1
Generate Executive Reports
FM Enterprise Solutions
Select State in the United States
Select Location in a Particular City
View Property Abstract
Select Appropriate Floor Information

FIGURE 3.3.4-13 Infrastructure management environments enable executives to drill down for corporate information using either enterprise-based executive information systems or Web interfaces. Both systems yield fast, intuitive, accurate information in a timely manner. It is as easy as 1-2-3. . . . (*Courtesy of Bruce Kenneth Forbes, ARCHIBUS, Inc.*)

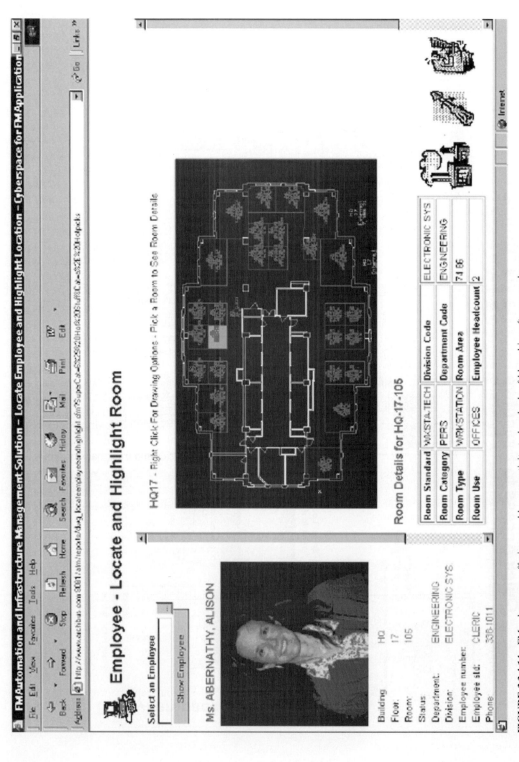

FIGURE 3.3.4-14 FM cyberspace offerings enable connectivity to hundreds of blue-chip software packages via the Internet. In this example, the user can view an AutoCAD drawing showing an office location, SAP HR data, and ARCHIBUS chargeback data simultaneously. (*Courtesy of Bruce Kenneth Forbes, ARCHIBUS, Inc.*)

and enforce your working practices. Use Internet/intranet applications to communicate your FM vision to the corporation at large. Accurately assess conditions within your entire organization to make best use of corporate resources and collect the critical data your corporation needs to execute its business plan. Internet/intranet applications will make you a key, visible, dependable player in the strategic planning process.

Seamless Interface into Your Corporate Web Strategy

Internet/intranet applications are designed to interface into your corporate Web, which has been developed by your corporate MIS department, and leverage their resources. Internet/intranet applications can be entirely open; you can link them to any corporate intranet page; you can use any Internet/intranet application's dynamic form or report on your own Web page. Internet/intranet applications are your productivity powerhouse for dynamic publishing of large numbers of drawings and structured FM data. If there is a report you need that is not provided, you can, in minutes, publish any Internet/intranet application's view to dynamic pages, complete with drill-downs and paging of large results. Internet/intranet applications put all the technological pieces together, so you get the glory (see Fig. 3.3.4-15).

Who Can Benefit from Using Internet/Intranet Applications?   Anyone who uses your organization's facility information will benefit from using Internet/intranet applications. Internet/intranet applications do everything that IM systems can do, from anywhere in the world.

Facility managers can access their facility information anywhere, anytime. They can provide users of facility information with self-service mechanisms that enable them to access information or make work order requests, for example.

Field personnel can easily and efficiently pick up work order requests from remote locations, saving them time and helping them determine what tools are needed for a job before they arrive. Branch office personnel can access home office information in real time.

Strategic planners can immediately provide themselves with the historical information they need to make important planning decisions.

Purchasing managers can verify invoices against inventories and track back orders.

Human resources managers can develop occupancy plans and locate vacant space for new hires.

Building owners can determine rentable areas and bill tenants for the space they occupy, plus their share of common areas.

Corporate real estate managers can track detailed lease characteristics, such as options, payments, and prorated taxes.

Accountants can use space data to charge departments for the areas they occupy, plus their share of common areas.

Risk managers can evaluate assets to obtain adequate insurance coverage.

Tradespeople, maintenance staff, and building tenants can take advantage of remote access to facilities information.

REPORTING CAPABILITIES OF FM AUTOMATION SOLUTIONS

The greatest strength of FM automation lies in its phenomenal reporting capabilities. Based on more than 2 decades of experience, FM automation software providers have developed literally hundreds of formatted reports that have been designed to meet the needs of a wide range of users. To illustrate the value that FM reports can offer your organization, a few key reports from each of the FM automation applications are shown in Table 3.3.4-4.

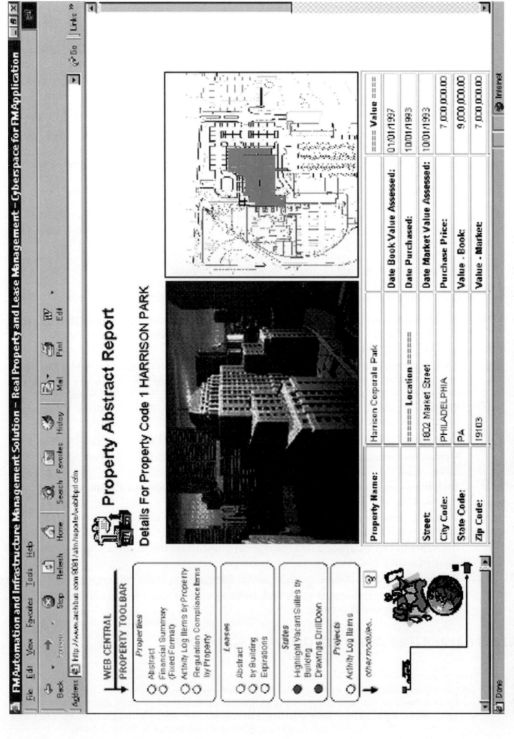

FIGURE 3.3.4-15 Visual user interfaces (such as the FM Web Central Process Toolbar shown) enable rapid training. With just 5 minutes of training/tutoring, corporate real estate executives, CEOs, CFOs, and CIOs can readily access real estate information and hundreds of other reports and analyses. (*Courtesy of Bruce Kenneth Forbes, ARCHIBUS, Inc.*)

TABLE 3.3.4-4 Reports from FM Automation Applications

Real property and lease management	Space management	Telecommunications and cable management
Properties	Gross areas	Data
Property Abstract	Gross Areas by Floor	Data Equipment Plan
Property Abstract Report	Gross Areas by Building	Data Equipment by Room
Properties by Site	Rooms	Voice
Properties by City	Highlight Rooms by Department	Voice Equipment Plan
Properties by State	Highlight Rooms by Standard	Voice Equipment by Room
Properties by Region	Rooms	Jacks
Properties by Country	Rooms by Department	Jack Plan
Properties by All	Rooms by Standard	Jacks by Room
Properties by Primary Contact	Employees	Patch panels
Properties by Account	Occupancy Plan	Patch Panel Plan
Properties by Account by Company	Highlight Vacant Rooms	Patch Panels by Room
Leases	Employees	Software
Lease Abstract Report	Employee Extensions	Client Software Inventory
Lease Abstract	Employees by Room	Server Software Inventory
Lease Expirations	Employees by Department	Software Inventory by Standard
Recurring Base Rents by Leases		
Options by Lease		
Lease Options Exercise Dates		
Leases by Building		
Leases by Account		
Leases by Account by Company		
Lease Measured Areas		
Buildings & floors		
Vertical Penetrations by Building		
Vertical Penetrations Analysis Summary		
Highlight Vertical Penetrations by Standard		
Service Areas by Building		
Highlight Service Areas by Standard		
Building Performance Review		
Suites		
Highlight Vacant Suites		
Highlight Suites by Lease		
Suites by Lease		
Suites by Floor		

F&E management	Building operations management
Furniture	Background data
Furniture Plan	
Furniture	Work orders
Furniture Count by Room	Maintenance Work Order
Furniture Count by Department	Work Order Status
Furniture surveys	Completed Work Orders
Furniture Standards	Cancelled Work Orders
Equipment	Stopped Work Orders
Equipment Plan	Historical Work Orders
Equipment	
Equipment by Room	
Equipment by Department	
Moves	
Unissued Move Orders	
Issued Move Orders	
Cancelled Move Orders	
Closed Move Orders	
Occupancy Plan	

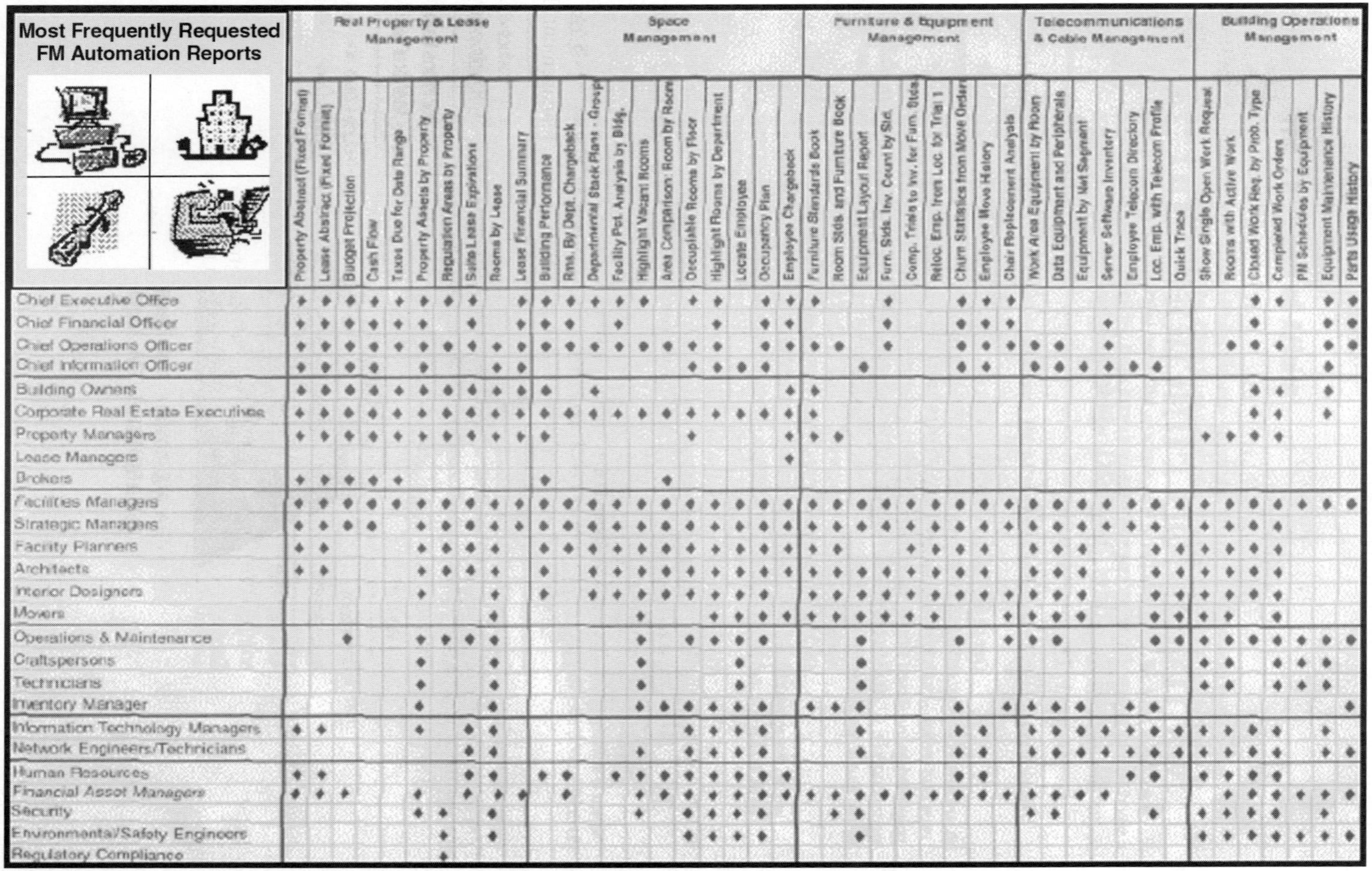

Most Frequently Requested FM Automation Reports

Reports (columns), grouped by category:

Building Operations Management
- Parts Usage History
- Equipment Maintenance History
- PM Schedules by Equipment
- Completed Work Orders
- Closed Work Req. by Prob. Type
- Rooms with Active Work
- Show Single Open Work Request

Telecommunications & Cable Management
- Quick Trace
- Loc. Emp. with Telecom Profile
- Employee Telecom Directory
- Server Software Inventory
- Equipment by Net Segment
- Data Equipment and Peripherals
- Work Area Equipment by Room

Furniture & Equipment Management
- Chair Replacement Analysis
- Employee Move History
- Churn Statistics from Move Orders
- Reloc. Emp. from Loc. for Trial 1
- Comp. Tasks to Inv. for Furn. Stds.
- Furn. Stds. Inv. Count by Std.
- Equipment Layout Report
- Room Stds. and Furniture Book
- Furniture Standards Book

Space Management
- Employee Chargeback
- Occupancy Plan
- Locate Employee
- Highlight Rooms by Department
- Occupiable Rooms by Floor
- Area Comparison: Room by Room
- Highlight Vacant Rooms
- Facility Pct. Analysis by Bldg.
- Departmental Stack Plans - Group
- Rms. By Dept. Chargeback
- Building Performance

Real Property & Lease Management
- Lease Financial Summary
- Rooms by Lease
- Suite Lease Expiration
- Regulation Areas by Property
- Property Assets by Property
- Taxes Due for Date Range
- Cash Flow
- Budget Projection
- Lease Abstract (Fixed Format)
- Property Abstract (Fixed Format)

Individuals (rows):
- Chief Executive Officer
- Chief Financial Officer
- Chief Operations Officer
- Chief Information Officer
- Building Owners
- Corporate Real Estate Executives
- Property Managers
- Lease Managers
- Brokers
- Facilities Managers
- Strategic Managers
- Facility Planners
- Architects
- Interior Designers
- Movers
- Operations & Maintenance
- Craftspersons
- Technicians
- Inventory Manager
- Information Technology Managers
- Network Engineers/Technicians
- Human Resources
- Financial Asset Managers
- Security
- Environmental/Safety Engineers
- Regulatory Compliance

FIGURE 3.3.4-16 Reporting capabilities vary among different FM automation systems. The matrix shown indicates the top 44 FM automation reports requested by 26 individuals within a corporation. (*Courtesy of Bruce Kenneth Forbes, ARCHIBUS, Inc.*)

Facilities management automation reports add a new dimension to your FM data, information, knowledge, and wisdom. The matrix, shown in Fig. 3.3.4-16, indicates the commercial off-the-shelf (COTS) reports that are most frequently used by the various people within an organization. The 26 individuals identified use 44 reports. The most frequently used report, requested by 21 individuals, shows the rooms by lease.

CONTRIBUTORS

Allan R. Forbes, ARCHIBUS, Inc., Charlotte, North Carolina

Julia M. Forbes, ARCHIBUS, Inc. Boston, Massachusetts

Cynthia M. Zawadski, ARCHIBUS, Inc. Boston, Massachusetts

ARTICLE 3.3.5

FACILITIES MANAGEMENT AUTOMATION AND INFRASTRUCTURE MANAGEMENT AUDITS

Bruce Kenneth Forbes

ARCHIBUS, Inc., Boston, Massachusetts

In the area of facilities management (FM) automation and infrastructure management (IM), there are four "audits" that are beneficial to facilities organizations.[1]

1. *The needs assessment and opportunity audit.* This is an examination of an organization to determine the business and operational needs it has for deploying automation. If there is a need, an opportunity audit is performed to determine if the circumstances are conducive for an organization to undertake FM automation and/or IM activities (see later section entitled "Core Competency Areas as They Relate to FM Automation and IM Opportunities").

2. *The technology audit.* A survey and audit of existing/legacy systems and proposed technologies is conducted. The processes for these technologies and the use by the various departments within an organization are evaluated, analyzed, and documented.

3. *The management audit.* An exploration of management groups to determine their knowledge, expectations, and understanding of FM automation and IM.

4. *The implementation audit.* This is an examination of the implementation procedures and practices to determine areas of concern and/or opportunities for improvement.

There are numerous reasons why your organization should conduct FM automation and IM audits regularly. Following are some of the most often stated reasons:

- Increased quality of delivered systems
- Improved development productivity

- Reduced development cycle time
- Reduced development costs
- Increased usability of systems
- Improved alignment of systems to support corporate strategies
- Increased return on investment (ROI)
- Increased human–computer interaction
- Increased partnering results
- Creation of world-class FM value

CONSULTING PROCESS FOR AUDITS: FORMAT AND DEPTH OF THE AUDITS

The format and depth of exploration during an audit will vary from organization to organization, but most audits undertake similar activities. The consulting process for FM automation and IM audits is shown in Fig. 3.3.5-1. The figure highlights the five parts of an audit process mapped against the client's and the consultant's documentation of activity requirements and tasks. Audits should be part of any strategic planning process. The audit is a bridge between management and FM automation and IM leadership teams. The five stages of an audit might be

1. *Definition stage.* Frames early client discussions into articulated sets of expectations and commitments.
2. *Structure stage.* Analyzes the expectations and commitments to structure an approach to engagement.
3. *Gathering stage.* Uses explicit and implicit approaches to gather data, information, and knowledge, and summarizes it into findings.
4. *Synthesis stage.* Synthesizes the results of the data, information, and knowledge gathering into documents, which focus on client opportunities and the ramifications of acting or not acting.
5. *Buyin stage.* Uses the synthesis-stage documents, the consultant's FM knowledge and wisdom, input gained from interactions with client and gathering activities, and an understanding of change management to assist the client in achieving buyin.

Each activity requirement and task can be broken down into a series of subactivities and subtasks. A detailed data gathering activity matrix for FM automation and IM audits is shown in Fig. 3.3.5-2.

FM Automation and IM Needs Assessment and Opportunity Audit

Typical audits include the following roles:[3]

- Identify the needs to be addressed in the audit.
- Document the prescribed process to be undertaken and the expected benefits.
- Structure and schedule activities to be undertaken.
- Select participants and areas to be audited.
- Provide roundtables, interviews, and reviews of legacy information.
- Establish groundwork and foundation activities.

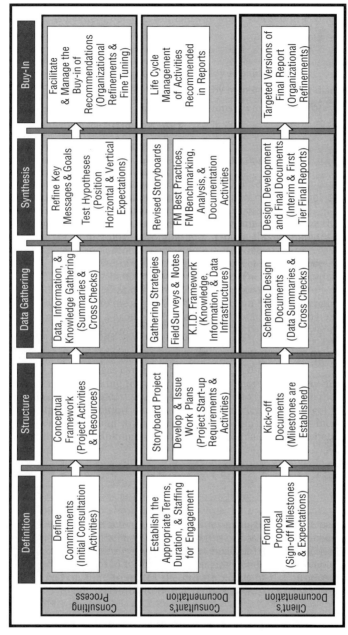

FIGURE 3.3.5-1 The consulting process for FM Automation and Infrastructure Management Related Audits: The Five Primary Stages in an Audit Process as mapped against the client's and the consultant's documentation requirements/tasks. (*Courtesy of ARCHIBUS, Inc.*)

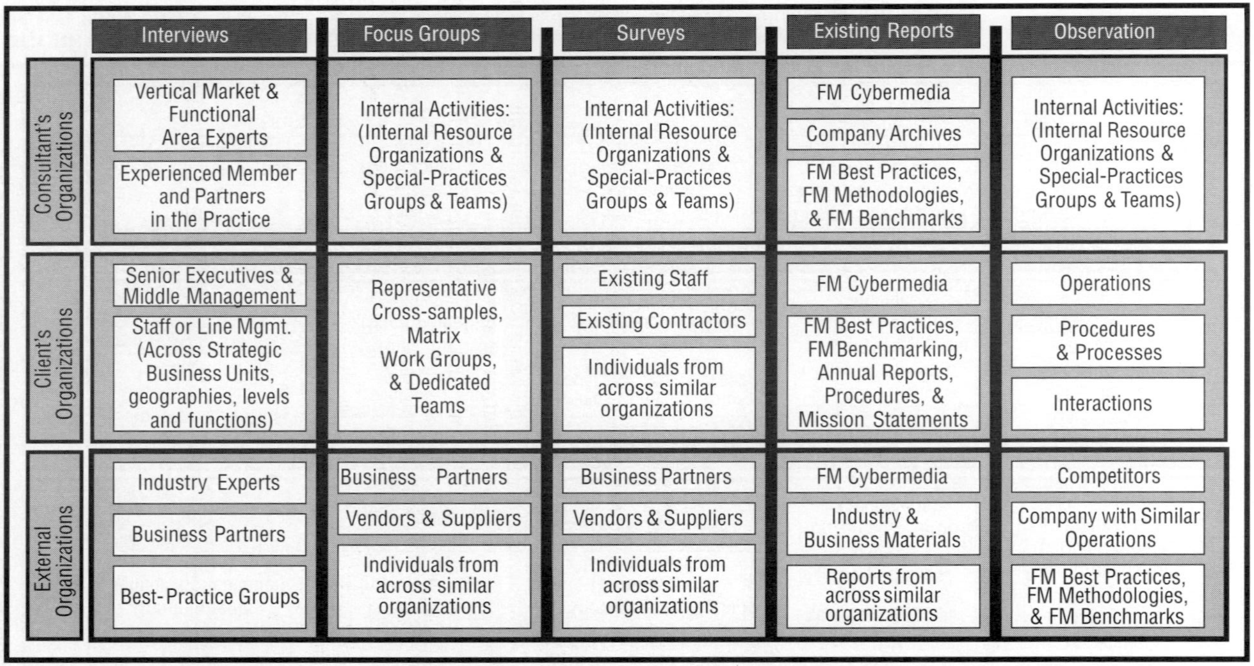

	Interviews	Focus Groups	Surveys	Existing Reports	Observation
Consultant's Organizations	Vertical Market & Functional Area Experts; Experienced Member and Partners in the Practice	Internal Activities: (Internal Resource Organizations & Special-Practices Groups & Teams)	Internal Activities: (Internal Resource Organizations & Special-Practices Groups & Teams)	FM Cybermedia; Company Archives; FM Best Practices, FM Methodologies, & FM Benchmarks	Internal Activities: (Internal Resource Organizations & Special-Practices Groups & Teams)
Client's Organizations	Senior Executives & Middle Management; Staff or Line Mgmt. (Across Strategic Business Units, geographies, levels and functions)	Representative Cross-samples, Matrix Work Groups, & Dedicated Teams	Existing Staff; Existing Contractors; Individuals from across similar organizations	FM Cybermedia; FM Best Practices, FM Benchmarking, Annual Reports, Procedures, & Mission Statements	Operations; Procedures & Processes; Interactions
External Organizations	Industry Experts; Business Partners; Best-Practice Groups	Business Partners; Vendors & Suppliers; Individuals from across similar organizations	Business Partners; Vendors & Suppliers; Individuals from across similar organizations	FM Cybermedia; Industry & Business Materials; Reports from across similar organizations	Competitors; Company with Similar Operations; FM Best Practices, FM Methodologies, & FM Benchmarks

FIGURE 3.3.5-2 Detailed data gathering activities matrix for FM automation and IM audits. (*Courtesy of ARCHIBUS, Inc.*)

- Conduct audit.
- Review data, information, and knowledge.
- Prepare a written document.
- Submit document and publicize applicable findings.
- Implement action items and prepare for next audit cycle.

Having determined the processes and procedures to be undertaken during the audits, your FM automation and IM leadership team is ready to begin.

Needs Assessments and Opportunity Audits Are a Great Place to Start

The process of implementing a computer-aided facilities management (CAFM) system, computer-integrated facilities management (CIFM) solution, cyberspace universes for FM, or IM enterprise poses challenges to even the most well-managed facilities operation. You can easily take advantage of the knowledge and experience of experts in the field. One of the best places to start is an assessment and opportunity audit of FM automation needs.

Assessment of FM Automation Needs

The goal of this procedure is to determine the requirements of an organization for managing its physical assets and facilities processes. An organization should establish milestones to be cyclically reviewed during FM automation and IM audits. These are the six standard milestones for FM automation acquisition:

1. *Awareness of pain.* Begins when you become aware of some level of dissonance between what is and what could be.
2. *Articulation of symptoms.* Requires articulating and defining symptoms, which begin to limit what can constitute a solution or an opportunity.
3. *Envisioning the solution.* Is complete when you have "bought the concept" that a specific service will reduce the dissonance.
4. *Review of available suppliers.* Begins with shopping for the consultant who can best provide the specific service you require.
5. *Definition of technical requirements.* Requires articulating and defining technological requirements, which begin to limit what can constitute a solution or an opportunity.
6. *Selection of business partner.* Is complete when one service provider meets the criteria and demonstrates the ability to optimize your preferences.

Typical Assessment/Audit Activities.   The analysis usually involves 1 or 2 days of an on-site visit followed by 1 or 2 days of analysis and reporting. The following areas will be addressed:

Preliminary assessment. Before the on-site visit, communication is initiated to establish some guidelines and structure for the on-site visit and involves the following:

- Identifying and prioritizing FM needs
- Collecting any relevant materials that the organization can provide to demonstrate current methods, data, reporting, capabilities, controls, processes, and support organizations
- Establishing an initial analysis approach and milestones
- Analyzing corporate and organizational strategies and processes

Analysis of current FM resources and opportunities. The first stage of the on-site visit includes the following key activities:

- Analyzing current FM capabilities and training requirements
- Evaluating processes, procedures, reporting, and presentation requirements
- Reviewing current space and asset standards and FM benchmarks
- Understanding the relevant business issues that need to be addressed

Investigation of data requirements. The next major area of review includes:

- Evaluating current data models
- Understanding the FM organization's current core competency
- Examining current data sources and translation requirements

Analysis of data access and reporting requirements. Maximizing the benefits of the information provided when implementing FM automation and/or IM solutions is analyzed by the following:

- Examining additional reporting benefits
- Analyzing executive information needs
- Identifying key methods and procedures
- Evaluating data flow requirements
- Investigating additional distributed information benefits to other organizational units

Implementation plan and analysis report. The resulting analysis report incorporates the following tasks:

- Translating conceptual requirements into deliverable entities
- Establishing the key system parameters and components
- Identifying key tasks and milestones
- Determining an action plan and timeline alternatives
- Defining ongoing technical needs
- Periodically reviewing core competency areas as they relate to FM automation

Using external consultants to develop and implement regularly scheduled audits and review activities is recommended.

FM Core Competency Areas as They Relate to FM Automation and IM Opportunities

Facilities professionals need competency in numerous areas. The following areas are readily identifiable for FM automation and IM:

- Real estate and property management
- Best practices and environmental management
- Strategic planning and project management
- Benchmarking and facility functions
- Finance and administration
- Quality assessment and operations management
- Marketing and communications
- Operations and management

Each FM organization should develop its own matrix of automation goals and milestones, based on its FM core competency areas. The FM automation and IM applications, which the organization has decided to implement, should be highlighted and monitored (see Table 3.3.5-1).

To Create or Not to Create Your FM Automation Solution

Facilities management organizations can pursue a variety of application development strategies. During the various audits and needs assessments, your organization should determine its strategy for FM automation and IM deployment.

TABLE 3.3.5-1 Each FM Organization Should Develop Its Own Automation Targets Matrix Based on the "The Primary FM Competency Areas" and Those FM Automation Applications That the Organization Has Decided to Implement

Competency areas	Core competencies to be addressed with FM automation	Real estate	Space mgmt.	F&E* mgmt.	Telecom mgmt.	BOM*& CMMS*	FM WEB activity
Real estate and property management	1. Manage and implement the real estate master planning process.	X	X				X
	2. Manage real estate assets.	X	X				X
Best practices and environmental factors	3. Develop and implement practices that promote and protect health, safety, security, and the quality of work life, the environment, and organizational effectiveness.	X	X	X	X	X	X
	4. Develop and manage emergency preparedness procedures.	X	X	X	X	X	X
Strategic planning and project management	5. Develop facility plans.	X	X	X	X	X	X
	6. Plan and manage all phases of projects.	X	X	X	X	X	X
	7. Manage programming and design.	X	X	X	X	X	X
	8. Manage construction and relocation.	X	X	X	X	X	X
Benchmarking and facility functions	9. Plan and organize the facility function.	X	X	X	X	X	X
	10. Manage personnel assigned to the facility.	X	X	X	X	X	X
	11. Administer the facility function.	X	X	X	X	X	X
	12. Manage the delivery of the facility function.	X	X	X	X	X	X
Finance and administration	13. Manage the finances of the facility function.	X	X	X	X	X	X
Quality assessment and innovation management	14. Manage the process of assessing the quality of services and the facility's effectiveness.	X	X	X	X	X	X
	15. Manage the benchmarking process.	X	X	X	X	X	X
	16. Manage audit activities.	X	X	X	X	X	X
	17. Manage development efforts and innovation.	X	X	X	X	X	X
Communications	18. Communicate effectively	X	X	X	X	X	X
Operations and maintenance	19. Oversee building system A-I-O-M.*	X	X	X	X	X	X
	20. Manage the maintenance of interiors.	X	X	X	X	X	X
	21. Oversee F&E A-I-O-M.	X	X	X	X	X	X
	22. Oversee grounds and exterior elements A-I-O-M-D.*	X	X	X	X	X	X

Source: Adapted by ARCHIBUS, Inc. from *Competencies for Facilities Management Professionals*, IFMA.
* AIOM.; AIOMD.; BOM, building operation management; CMMS, computerized maintenance management system; F&E, furniture and equipment.

- *Make.* Engineer a new solution.
- *Update.* Maintain an existing solution.
- *Renovate.* Reengineer an existing solution.
- *Buy.* Purchase a new commercial off-the-shelf (COTS) software solution and customize it.
- *Integrate.* Assemble a new solution from available components.
- *Outsource.* Commission outside development skills with particular responsibilities.
- *Postpone.* Provide other solutions first, based on priorities.
- *Offload.* Enable end users to meet their own computing needs.

Now, existing legacy systems, islands of automation, and information silos have made hybrid solutions very common. Currently, the purchase of COTS software solutions is the most popular alternative in MIS departments. Such groups often cite time and cost savings as primary factors.

FM Automation Requires a Full Range of Support Services

Successful FM automation requires that the facilities professional acquire a full range of support services. Some of the more common services available include:

- *FM automation reviews.* The resources, information, and equipment necessary for FM automation may already exist in your organization. If you have already invested in data collection, software, or hardware, an FM automation consultant or services provider can provide a number of strategies for optimizing these resources.
- *Database creation and verification.* Organizations are generally prepared to maintain their facilities data, but they often need help with the initial process of collecting, entering, and verifying that data. Such intensive data entry projects might support furniture and equipment (F&E) audits or space occupancy inventories.
- *Database creation guidelines.* When different individuals in different departments or outside vendors enter data, it is vital to maintain a consistent format suitable for automatic processing, such as issuing queries. Database guidelines help enforce standards and ensure that everyone enters data in the same way.
- *Drawing translation.* Your FM automation provider can help you extend the usefulness of your existing data by converting paper drawings or drawings from an older CAD system into facility record drawings and integrating them into your FM automation system.
- *System customization.* If your organization has specific reporting needs, your FM automation provider can help you determine which pieces of information are critical and configure your FM automation system accordingly.
- *Design, implementation, and management of enterprise access system.* Increase the value of your facility information by making it more accessible to upper management. Your FM automation provider can help you develop a graphical and alphanumeric model of your facility that provides point-and-click access to strategic information, using any of FM automation's enterprise access solutions.
- *Web-based and professional service organization resources.* There are numerous on-line services that let you connect with other FM automation users, dealers, FM professionals, professional organizations, and technical support personnel over the information superhighway.
- *Telephone technical support.* This is readily available from numerous friendly, knowledgeable, and responsive FM automation customer support teams. A technical support contract can be purchased from the numerous FM automation software and service providers.

- *FM automation users' groups.* Your local FM automation users' group offers the perfect forum for sharing information and experiences with other users in your area. Members of the FM automation community from around the world typically gather annually at International FM Automation Users' Conferences for several days of educational sessions, workshops, and industry forums.
- *Workshops and seminars.* Workshops and seminars that are frequently held in locations across the country and around the world offer free software demonstrations and useful insights designed to help you make the most effective use of FM automation tools.
- *Checklists.* Develop checklists for the numerous audits, reviews, and implementation activities. For examples, see Tables 3.3.5-2 and 3.3.5-3.
- *FM automation work teams.* Perform regular, ongoing FM work using FM automation.

TABLE 3.3.5-2   The IM Master Audit Checklist

✔	Activity	Description	Purpose
	1.0.0	Executive summary of master audit (identify purpose of proposed activity).	Management overview of activity to secure executive management buyin.
	1.1.0	Identify IM activity components.	Software, hardware, technology applications, procedures approaches, methodologies, and previous experience, FM Benchmarks™, FM Best Practices™, and FM Value Matrices™.
	1.2.0	Identify the primary, secondary, and future activity areas.	Integrated support programs, facility management, project management, corporate services, security, finance, accounting/procurement, etc.
	1.3.0	Identify real estate, property, and assets that are to be managed (identify common asset traits and needs).	Distinguish among manufacturing, back-office space, leased versus owned properties, and assets, etc.
	1.4.0	Establish working teams for IM activities.	Review current methods and procedures. Set objectives. Identify base IM needs analysis. Identify and prioritize projects.
	1.5.0	Technology audit.	Review existing systems, IT benchmarks, and their integration.
	1.6.0	Opportunity audit.	Identify goals for IM activities: Improve processes, encourage local/regional/global consistency, match appropriate software with tasks, limit and/or avoid duplication of effort, offer greater flexibility, provide cost-efficient solutions, a variety of user interfaces for easy use, etc.
	1.7.0	Business audit.	Identify best-practice activities and benchmarks for application areas: real estate and lease management, strategic master planning, space management, design management, project management, F&E management, telecommunications and cable management, BOM and computerized maintenance management, finance and administration, ERP, and IM.
	1.8.0	Management audit.	Identify group of IM decision makers. Identify group IM goals. Identify business issue IM drivers. Develop benchmarking techniques. Identify expectations and wish lists. Develop evaluation techniques.
	1.9.0	Executive audit.	Review all of the previous activities.

TABLE 3.3.5-3 The IM Implementation Checklist

✔	Activity	Description	Purpose
	2.0.0	Executive summary of IM implementation activities.	Management overview of activities.
	2.1.0	Identify IM project support business plan.	Needs analysis and business objectives. Budgeting and staffing. Justification of IM (ROI, ROE, ROA, etc.). Risk assessment.
	2.2.0	Identify the primary and secondary implementation activities.	Projects support. Implementation benchmarking. Pilot projects (comprehensive or incremental). Supplemental activities.
	2.3.0	IM technologies design and implementation practices.	Identify COTS, proprietary, customized, and personalized offering requirements with suitable integration.
	2.4.0	Establish working teams for IM activities.	Project sponsors. Business project manager. Business subject matter contributors. IM technical project managers. IM technical experts. IM implementation experts. IM communications liaison. IM systems administrator. IM trainer. IM Web master. IM activity leaders. Casual Infrastructure Users™. Infrastructure Contributors™. Infrastructure Managers™.
	2.5.0	Project management and risk assessment.	Identify open issues and risks. Predict failure rates and identify fixes. Failure mitigation and contingency planning.
	2.6.0	IM requirements and analysis activities.	Business needs and conditions models. Business requirements and expectation models. Project management models. Technical requirements models. Data warehouse and use models. Knowledge and wisdom management models. IM sequencing and sign-off activities.
	2.7.0	Evaluation and reimplementation analyses.	Buy, build, customize, and integrate solutions. Evaluate and select vendors. RFI/RFP models and evaluation. Sign-off activities.
	2.8.0	Derivative IM change management activities.	IM targets. IM-driven organizational changes. IM change agents. IM opportunities.
	2.9.0	Executive implementation audit.	Review all of the previous activities. IM corrective maintenance. IM preventive maintenance. IM predictive maintenance. IM proactive maintenance.

- *FM automation project team.* Knowledge workers drawn from different FM functional units for the purpose of strategic planning, implementing, and managing FM automation and IM solutions.
- *FM automation management team.* Responsible for the overall performance of the auto- mated facilities function and the organizations it serves.
- *FM automation ad hoc network.* Informal collections of individuals, groups, and business part- ners connected by shared interests, purposes, or goals that affect FM automation and IM.
- *Parallel development teams (for CIFM activities).* Special FM-related task forces responsi- ble for coordination, problem solving, and improvement-oriented tasks.

Five Phases of Cyberspace Universes for FM Development and Deployment

The fast-track phases of transforming a business with cyberspace universes for FM solutions, the Internet, and IM offerings was found on the Internet and presented at an FM automation user's conference (see Fig. 3.3.5-3).[1] The five fast-track phases are as follows:

1. The Foundation Phase (Month 1)

    Planning your strategy

    Building an organization and infrastructure

    Stakeholders and Inter/Intra/Extranet Strategy

    Creating an on-line personality and branding strategy

    Publishing your content (includes link strategy)

    Using advanced Internet technologies

2. The Exploitation Phase (Month 2)

    E-mail, fax, and phone communications

    Research the competition

    Research market trends and customer requirements

    Create a free educational resource or educational delivery mechanism

    Establish business relationships worldwide

    Find the lowest-cost item for all purchases

3. The Business Extension Phase (Month 3)

    Interacting with stakeholders

    Using instant and automated interactive strategies

    Extending the reach of your corporate databases

    Extending the reach of your corporate applications

    Linking into back-end work flow processes

    Managing corporate intellectual capital as an asset

4. The Business Innovation Phase (Month 4)

    Embracing/capitalizing on the Internet and IM economy

    Building a community environment

    Proactively reaching out to your stakeholders

    Conducting commercial transactions

    Building Web traffic and activity

    Reengineering your business processes

Activity	1st Month	2nd Month	3rd Month	4th Month
Strategic Planning	- Executive Interviews - Focus Groups - Document Findings - Create Mgmt. T.F.	Present Findings		
FM Automation and Infrastructure Management Opportunity Audit	- Identify Products - Conduct Audits at Existing Sites	- Site visits to Manufacturer and clients where appropriate	- Conduct Pilot/Feasibility Audit - Pilot Legacy System connectivity activities	- Post-Pilot Audit and Evaluation - Select Products, Services, & Suppliers
Design Implementation	- Develop Processes - Meet Op. Mgmt. Team - Develop Electronic Workshop/Backbone	- Collect Data & Info. - Consolidate Findings - Initial Planning	- Review Findings with Op. Mgmt. - Approve Findings - Develop Roll-out	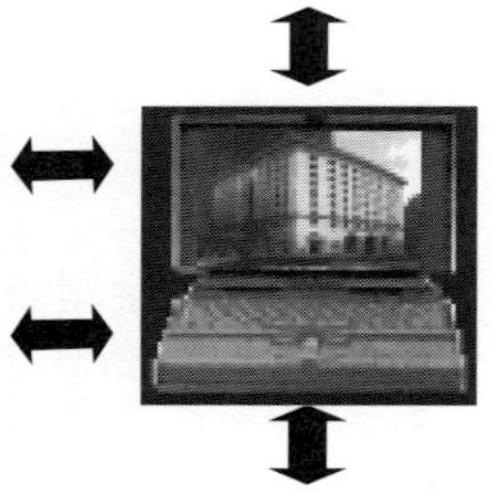
Design Technology	- Design Core Data Architectures - Hire Independent Service Providers	- Build Background Databases - Coordinate all Bldg. & Information Sys.	- Conduct Audits and Accept Submissions	
Orientation/Education	- Establish CIFM™ Orientation and Education Task Forces	- Define Requirement of the Task Forces - Define Goals	- Peopleware Tools - Training - Town Hall Mtgs. - Handbooks	- Implement Programs
Communications/P.R.	- Establish Comm./PR Task Forces - Define Goals, Roles, and Responsibilities	- Design Comm/PR Tools and Formats (Hot Lines, Web-site, newsletters, etc.)	- Implement Comm/PR Activities (Status Reports, Ed., feedback tools, etc.)	- Implement and Evaluate the Communications and P.R. Programs

FIGURE 3.3.5-3 Fast-tracked implementation plans for FM automation and IM solutions require careful planning. (*Courtesy of Bruce Kenneth Forbes, ARCHIBUS, Inc.*)

5. The Strategic Transformation Phase (Month 5)

Meeting the needs of the global community

Targeting micromarket segments

Creating customized content for individual visitors

Exploiting electronic channel strategies

Introducing new channel layers

Aligning with corporate partners

CONTRIBUTORS

Allan R. Forbes, ARCHIBUS, Inc., Charlotte, North Carolina

Janice E. Forbes, Global Practices Group, IBM, Charlotte, North Carolina

Julia M. Forbes, ARCHIBUS, Inc., Boston, Massachusetts

Cynthia M. Zawadski, ARCHIBUS, Inc., Boston, Massachusetts

NOTE

1. B. K. Forbes, D. Packard, and C. M. Zawadski, *Proceedings of the 5th International ARCHIBUS/FM Users' Conference and FM Automation Best Practices Forums,* ARCHIBUS Press, Boston, 2000.

ARTICLE 3.3.6
THE INTERNET, E-COMMERCE, AND FACILITIES MANAGEMENT

Eric Teicholz

Graphic Systems, Inc., Cambridge, Massachusetts

This outernet article is about the Internet. The Internet has already dramatically altered facilities management (FM) and the ways in which FM practitioners interact with their design consultants. This relationship will undergo a dramatic change again because of the rapid deployment of e-commerce.

FM TECHNOLOGICAL EVOLUTION

Computer-aided facilities management (CAFM) goes back almost 30 years when vendors such as ARCHIBUS (www.archibus.com), FM:Systems (www.fmsystems.com), SPAN (www.peregrine.com), Drawbase (www.drawbase.com), and Aperture (www.aperture.com) started linking databases to computer-aided design (CAD) programs to perform facility-related functions such as tracking space availability and physical assets, including furniture and equipment (F&E). Occupancy drawings that depicted the physical location and departmental organization of staff were developed by getting this information from human resources (HR) or information technology (IT) groups and entering it into the CAFM system.

As computers became more powerful and CAFM more mainstream, vendors evolved along two paths: either they developed additional FM applications (work, real estate, cable management, and space planning), or they built links to established third-party software packages that performed these functions. A facility manager might, for example, use a CAFM vendor for CAD drawings and tracking space occupancy while using a second software vendor for space needs forecasting or to generate stacking and blocking diagrams. A design consultant who worked for such a client would have to provide electronic drawings compatible with the CAFM system being used. Because more powerful databases became available in the early 1990s, it became possible to disperse software applications and databases throughout an organization. This, in turn, eliminated much data redundancy, sped up processing time, and made it easier to access other non-CAFM systems and databases. It meant, for example, that CAFM software, operating on a facilities manager's desk, might access HR or financial data from different departments over a network and use this data for occupancy or project management purposes.

THE INTERNET

In the mid-1990s, CAFM vendors initially adopted the Web as a simple reporting mechanism. The CAFM output reports were generated in HyperText Markup Language (HTML) and posted on corporate intranets. These static reports were somewhat clumsy because they had to be posted on the intranet site and then accessed by remote users. The data was only as current as the last posting.

Reports soon became dynamic using technology that enabled generating the report in real time by interacting with the CAFM software directly. In this manner, a user of the corporate intranet would download the latest version of the data when required by a particular application. Again, intranets were deployed to distribute dynamic reports and drawings, if the network speed was sufficient. In short, the intranet was still used as a reporting mechanism. To maintain integrity, the software was still centralized and tightly integrated. The Web's most fundamental advantage, the ability to effortlessly and dynamically link disparate multiple sites together (e.g., clients, consultants, contractors, and subcontractors), was still not being used.

It continues to be difficult and expensive to link different external software programs to CAFM systems on the Internet. Most CAFM vendors try to control the Web environment of their software and do not link very well to other Web-based software. This software model is becoming increasingly unworkable because new vendors are increasingly offering stand-alone FM software applications on the Internet. Stand-alone Internet applications that have appeared recently include vehicle (fleet) management, space planning and forecasting, help desk and maintenance work management, furniture layout, asset management, room scheduling, energy management, and construction project management. These software applications are fragmenting the FM function itself in many organizations because providing these functions is so easy and inexpensive on the Web that much FM functionality is shifting to non-FM professionals. For example, VISIO 2000's (www.visio.com) FM software modules are available for just a few hundred dollars. According to the company, more than 1000 copies of its software are purchased daily on the Web.

This migration of third-party stand-alone Web applications is impacting current CAFM vendors because of overlapping functionality and a very different cost model for Web-based products (people do not seem to want to pay for information provided on the Web). Seeing this trend, some CAFM vendors are trying to incorporate some of the stand-alone applications (i.e., FIS's support and use of VISIO 2000 as a front end to their software). In the long run, however, the Web-based applications are likely to pose a threat to traditional CAFM vendors unless they also change their Internet software architecture. Architectural design consultants (see the case studies following) also need to adapt to this new model.

Project/collaborative Web sites represent another major impact on the way facility managers and their design consultants are doing business and using FM technology. Currently, more than 60 software vendors offer project Web site software—primarily for managing construction projects. These sites traditionally offer document management services and track project information such as drawings, specifications, requests for information (RFIs), various logs, and so forth. The Web sites link the design consultants with the client, engineers, contractors, and subcontractors. These Web sites make data available to corporate users on their intranet as well. For example, HR might want to know what staff is affected by a move; finance might want to track project costs for financial control and capital planning purposes; and business unit managers might want access to floor plans that affect their staff.

Right now, software vendors are charging for project Web site software or are renting software functionality and hosting the software on their computers. However, the valuations of .com companies are based on how many people (called *eyeballs*) visit and use a particular site. This is again changing the financial model in the way vendors charge for software and information. This was dramatically illustrated recently when Autodesk spent an estimated $50 million developing *Buzzsaw.com* that gives away free project Web software simply to get the design industry to use the Autodesk portal. Thus, architects, clients, and contractors all use construction project and document management software provided free of charge by a vendor.

Again, the Internet is playing an increasing role in controlling who accesses, uses, and manages FM data. There is a transition in who is performing FM functions and how that function is being performed. Functions that were traditionally the domain of facility managers are being increasingly performed by both internal and external (e.g., design consultants and service providers) groups.

E-COMMERCE AND FM

To use Internet parlance, we are entering a business-to-business (b2b) application service provider (ASP) world. Financial venture capital funds that fuel Internet start-ups are betting on the b2b, rather than business-to-consumer (b2c), ASP model. The projected numbers are significant: In 1998, b2b revenues were $12 billion, of which services accounted for $1.4 billion [there is no breakdown for architecture/engineering/construction (A/E/C) services]. In 2002, there will be $131.2 billion of b2b commerce, and services will account for $18.4 billion of that figure. This represents more than a 10-fold increase in e-commerce services in 4 years! Project construction Web sites, for example, will share documents; construction materials and services will also most certainly be purchased on-line; and manufacturers will be literally bidding on-line for the business.

To draw a parallel, General Motors and Ford, icons of what the *Wall Street Journal* calls "the Old Economy," announced plans to establish on-line bazaars for all of the goods and services they buy—everything from paper clips to contract manufacturing. The motivation is to cut costs—billions over their current procurement processes. All suppliers will have to go through their sites by the end of 2001. In the A/E/C services world, literally hundreds of companies are being formed that are trying to be the e-marketplace, where it is possible to bid on and procure building supplies from manufacturers worldwide.

The Internet will continue to exert an ever-increasing influence on how we do business. Design consultants need to understand b2b ASP implications and be ready to host design and other services for their clients who are linked to e-commerce sites.

ARTICLE 3.3.7

THE ROLE OF E-COMMERCE AND BUSINESS-TO-BUSINESS ELECTRONIC MARKETPLACES IN FACILITIES MANAGEMENT

Bruce K. Forbes

ARCHIBUS, Inc., Boston, Massachusetts

Traditional procurement organizations and their procedures will be subject to review and reorganization in the new millennium. Current procurement techniques can be labor-intensive and time-intensive, thus driving the costs up and driving down participation by the end user or customer. Once a request is made, the end user has little or no involvement in the procurement process. As such, the end user is not always happy with the buyer's selection of vendors or sellers (see Fig. 3.3.7-1).

In the next few years, thousands of facility professionals will embrace the opportunities afforded them by e-commerce. The collaborative and connective tools of e-commerce will change life, as we know it today, forever. One of the most promising areas of e-commerce is

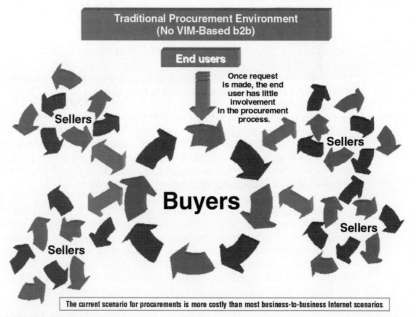

FIGURE 3.3.7-1   Traditional procurement procedures in most companies as they exist today. Buyers must work with a vast array of suppliers to realize great pricing and optimal terms. (*Courtesy of ARCHIBUS, Inc.*)

the business-to-business (b2b) electronic marketplace where end users, buyers, and suppliers have increased access to data, information, and knowledge. Their purchasing processes will be streamlined and they will benefit from leveraged, virtual, on-line sales opportunities. Buyers will benefit from productivity and profitability gains that can be realized by working electronically with public marketplace suppliers and the private network of contracted suppliers (that your organization may have opted to create and use).

INTERNET PROCUREMENT SAVES TIME BY A VARIETY OF METHODS

- Shortened procurement cycles
- Automated purchasing, procurement trails, and routing
- Instant supplier information and communications
- Comprehensive connectivity with legacy systems such as finance and operations
- Comprehensive purchasing activity reports

INTERNET PROCUREMENT SAVES MONEY IN NUMEROUS AREAS

- Through corporate and associative aggregate buying power created throughout your company and your affiliates
- Better prices in competitive bidding scenarios
- Corporate knowledge about suppliers, just-in-time delivery options, etc.

ELECTRONIC MARKETPLACES AND INFRASTRUCTURE MANAGEMENT WILL STREAMLINE PROCESSES

There are three participants in the electronic marketplace—the end user, the buyer, and the seller. The *Internet procurement enterprise* facilitates the electronic collaboration of all interested parties. Participants will use the e-commerce tools of the day to increase productivity and profitability for their respective organizations. In an electronic marketplace where the facilities professional also uses infrastructure management (IM) solutions, the entire process is streamlined from requisition to payment, and on to the various stages of the life cycle of a particular asset. Hundreds of systems can be brought together in IM and Internet procurement enterprises. Buyers benefit from the one-to-many opportunity created by using e-marketplace tools. Better prices and saving time are but a few of the benefits (see Fig. 3.3.7-2).

From the Buyer's Perspective

These are some of the components of an electronic marketplace:

- *E-business sourcing.* Find suppliers by keywords, product searches, product interactions and mapping, and/or SIC codes. Sort via filtering and rating systems.
- *Competitive bidding.* Request bids from the public marketplace suppliers and the private network of contracted suppliers. Send out, receive, and analyze information electronically.

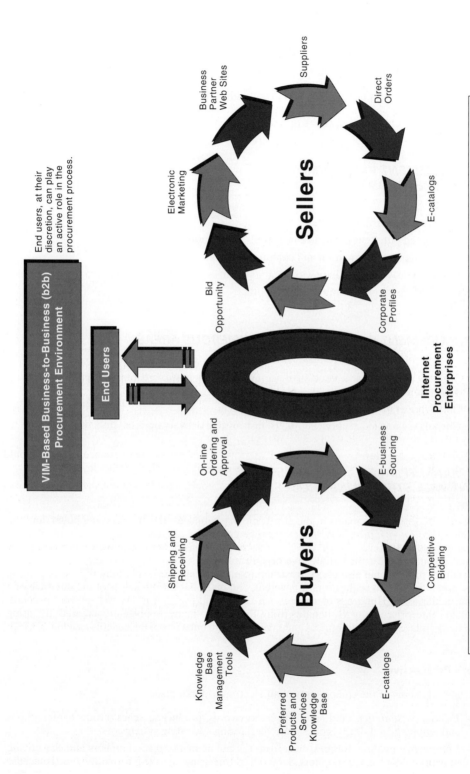

End users, at their discretion, can play an active role in the procurement process.

VIM-Based Business-to-Business (b2b) Procurement Environment

End Users

Sellers

Business Partner Web Sites

Suppliers

Electronic Marketing

Direct Orders

E-catalogs

Bid Opportunity

Corporate Profiles

Internet Procurement Enterprises

On-line Ordering and Approval

E-business Sourcing

Shipping and Receiving

Competitive Bidding

Knowledge Base Management Tools

Preferred Products and Services Knowledge Base

E-catalogs

Buyers

Easy to Use - Great Reporting - End-to-End Messaging - Integration into Legacy Systems - Security - Flexibility - and More

FIGURE 3.3.7-2 Hundreds of systems can be brought together in visual infrastructure management solutions and Internet procurement enterprises. Buyers benefit from the "one-to-many" opportunity created by the utilization of e-marketplace and e-work tools. Better prices and saving time are but a few of the benefits. (*Courtesy of ARCHIBUS. Inc.*)

- *E-catalogs.* Search, review, and analyze electronic catalogs from suppliers. Use "shopping cart" techniques for subsequent acquisition activities.
- *Preferred products and services knowledge base.* Frequently used services or acquired products can be grouped and prioritized. Create "hot pick" list for convenience when ordering.
- *Knowledge base management tools.* A knowledge base of relevant management data, information, and knowledge is collected for creating timely reports and analyses. Archiving of knowledge base materials is automatic and accessible via electronic audit trails and document management systems.
- *Order processing, shipping, and receiving.* All relevant information is passed electronically to the appropriate parties. This includes, typically, the shipping agent coordination and package deployment.
- *On-line ordering and approval.* Create purchase orders automatically for those items put into the shopping cart. "Paper work" is facilitated electronically and via e-mail. Electronic marketplace approval and control systems are used to ensure contract compliance.

From the Seller's Perspective

These are some of the components of an electronic marketplace:

- *Bid opportunity.* Suppliers can work with the buyers and submit bids electronically. Buyers typically feel that electronic submission is usually more responsive and saves time.
- *Electronic marketing.* Electronic marketing can be an effective way to market an organization's products or services. Electronic marketplace suppliers can use Web sites, "hot site links," and E-marketing tools.
- *Business partner Web sites.* Business partners can link their sites to appropriate private network partners and can also be accessed through public vehicles. Web sites can be used for buyer collaboration and information dissemination.
- *Suppliers.* Sellers can also become buyers in the electronic marketplace. Sellers should require their suppliers to be engaged via the electronic b2b tools of the day.
- *Direct orders.* Suppliers can receive, review, and process orders electronically. Payment can also be made electronically.
- *E-catalogs.* Suppliers can provide information electronically for buyers. Purchases can be facilitated via shopping cart techniques.
- *Corporate profiles.* Suppliers can share relevant information with buyers concerning their companies, their products, and their services.

The use of the electronic marketplaces by facilities professionals is still new, but there is little doubt that its presence will be lasting and significant in the years to come. The connectivity provided in cyberspace for facilities management (FM) by IM solutions will enable universal access to new tools offered by the e-marketplace and e-commerce. Buyers, sellers, and end users all benefit from connectivity and information sharing between applications.

RESTRUCTURING FOR E-COMMERCE, FOR FM CYBERSPACE, AND FOR IM

Restructuring for e-commerce, for FM cyberspace, and for IM requires careful examination of the evolving role of the FM cybrarian in organizations that must respond to dramatically changing FM and IM environments. We need to explore how and why FM cybrarians' roles are changing and clarify the impact of that change on managing and administering facilities and infrastructures. The main focus of an FM organization must be on people, processes, technology, time, organization, and, ultimately, the quality of customer care and service. All of this is strongly influenced by the existence of FM cybrarians and their leadership capabilities. Infrastructure management will enable even the most casual of users to rapidly access and use advanced reporting capabilities.

THE "E-OPPORTUNITY"—ARE YOU READY?

E-FM, e-space, and e-business are dramatically and suddenly changing the way the world does business. These automation-based environments will provide new insights into ways business leaders can use process and technology to achieve new efficiencies, create new value, and attain goals previously beyond reach. It is through these e-opportunities that you will be able to examine the competitive threats that come with new business models and to see how new and old competitors will use these environments to outpace their slower-to-react counterparts. As e-pioneers and e-architects, you must explore the wealth of opportunities and threats, as well as the best practices available through Internet technology.

MASTERING THE WEB WITH INFRASTRUCTURE MANAGEMENT

Cyberspace for FM: The Apex of Enterprise Resource Planning and Management in the New Millennium

E-commerce is coming to a company, institution, or organization near you. The results will be all-encompassing. The rapid acceptance, deployment, and use of Internet and intranet technologies, as well as their repercussions on business, is one of the more remarkable innovations and paradigm shifts of the late twentieth century. The consequences to facilities professionals and the organizations and people they serve are also significant.

In the near future, those organizations that are currently transforming themselves and becoming e-commerce-centric, will need as little as 25 percent of the real estate, fixed assets, personnel, and traditional infrastructure that they currently require. By 2005, up to 75 percent of the facilities professionals currently employed by these organizations will have been laid off, forced into early retirement, or, at a bare minimum, retooled. At this time, opportunities within cyberspace for FM are among the best options for facilities professionals seeking alternative careers. Cyberspace for FM combines the power of automation technologies [e.g., computer-aided facilities management (CAFM) systems, computer-integrated facilities management (CIFM) enterprise solutions, and enterprise resource planning (ERP) software] with new millennial technologies and best practices. Such systems are readily transformed into globally accessible depositories of FM data, information, knowledge, and wisdom.

The "tools of the trade" will vary, but all will be enablers of e-commerce and b2b commerce. Facilities management knowledge workers will gain access to information and knowledge through public switched networks (PSNs) that have global access, and application service providers (ASPs) will facilitate virtual private networks (VPNs). The World Wide Web, virtual communities, critical and secured corporate knowledge, E-FM and E-IM universes, and opportunity will all become available via electronic commerce infrastructure for E-FM and E-VIM (see Fig. 3.3.7-3).

A state of co-opetition exists when organizations, which are typically thought to be competitive, establish a domain of cooperation to solve a particular problem or challenge. Cyberspace for FM is becoming the infrastructure of choice, as organizations attempt to deliver co-opetition-based solutions. The individuals who act in FM cyberspace as project managers might also be known as FM cybrarians.

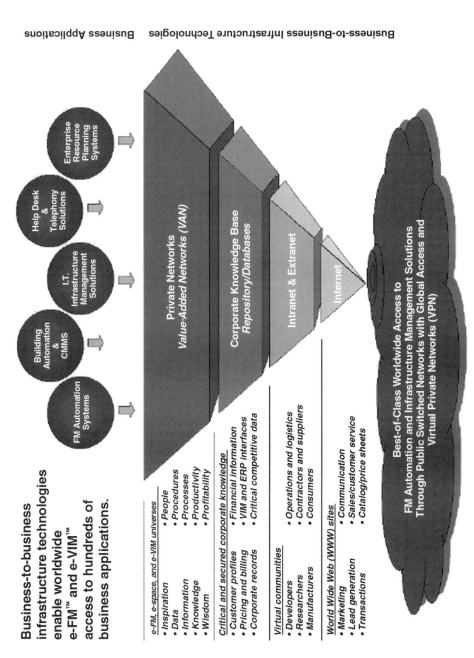

FIGURE 3.3.7-3 Visual User Interfaces™ in FM Cyberspace™ will address an FM organization's portal and content needs. E-commerce infrastructure for e-FM™ and e-VIM™ will enable even the most casual of users to utilize and gain access to FM Data, FM Information, FM Knowledge, and FM Wisdom™. *(Courtesy of ARCHIBUS, Inc.)*

Cyberspace for FM and IM Should Lead Your Organization into the Twenty-First Century

Cyberspace for FM (CFM) (also known as *FM cyberspace*) is being universally accepted as the FM knowledge workers' (also known as *FM cybrarians*) opportunity of a lifetime. In 1999 alone, it was estimated that more than 250,000 new users of CFM technologies would come on line. By 2005, the number of CFM users is expected to exceed 100 million.

CFM enables the rapid assembly, analysis, organization, distribution, and consumption of facilities and operating data, information, knowledge, and wisdom. Although the challenges are great, the costs of establishing such a system are typically one-tenth those of traditional CAFM systems and CIFM enterprise solutions. CFM access will typically cost less than $100 (U.S.) per user per year.

Unique to CFM are user interfaces, which enable even the most casual users to gain rapid access to knowledge. User interfaces can be quickly designed to provide universal access to thousands of applications and hundreds of ERP options. It is expected that CFM and e-space will be accepted and serve as a knowledge infrastructure at all of the successful e-commerce organizations of the new millennium.

Facilities professionals must look for and employ all available technologies that will improve every key corporate function with which they are involved—real estate and lease management, strategic master planning, space management, nonspatial asset management (furniture, equipment, and so forth), telecommunications and cable management, and building operations management (BOM). The help desks and service offerings must be Web-centric. The corporation of tomorrow must successfully intertwine CFM with finance and administrative activities, sales and marketing, human resources (HR), information management, manufacturing, and distribution.

CYBERSPACE FOR FM: THE TECHNOLOGY TOOL OF CHOICE FOR STRATEGIC EXECUTIVES

Computer-aided facilities management and CIFM offerings were representative of the facilities professional's desire to reengineer processes, redeploy resources, and save money by streamlining operations. Cyberspace for FM will be instrumental in an organization's revenue growth, creation of universal awareness, and delivery of informational access to its partners, shareholders, stakeholders, and customers.

Why is CFM so important? It is the executive's user interface to enterprise solutions. Facilities management knowledge workers are enabling the organizations they serve to optimize their basic product and service offerings. Information can be readily communicated and shared in CFM at a fraction of the cost. In the digital economy of tomorrow, knowledge is readily transmitted via enterprise solutions and accessed via a variety of user interfaces for CFM.

Cyberspace for FM command centers will be similar to those of NASA's Mission Control—data are collected and rapidly distributed via user interfaces that have been designed for the casual user, the in-betweener, and the advanced user, alike.

Competition in CFM will be fierce. There are currently many software developers and systems integrators providing software and services for the FM automation, IM, and ERP markets.

The Internet and intranets are fostering a paradigm shift in the way business will operate in the new millennium. Web-centricity is a critical requisite of tomorrow's digital economy.

- *CAFM systems.* Deal with the management and costs of "what has been" and "what is."
- *CIFM enterprise solutions.* Deal with "what has been," "what is," and "what can be."
- *CFM universes.* Deal with "what has been," "what is," "what can be," and also a vast array of expert systems, ActiveX environments, and decision-support systems. These systems can

help strategic business managers determine "what should be done," "with whom," and "what will happen, if . . ."

E-Space: A Critical Part of the E-Commerce Revolution

The one-two combination of CFM and ERP solutions enables broad integration and extensive expansion of FM knowledge works. Innovation, through the e-space sharing of digital economy–based knowledge and wisdom, will be a trademark of the twenty-first century.

BUILDING BLOCKS FOR THE FOUNDATION OF CYBERSPACE FOR FM

The Creation of "CFM Buzz"

Key to creating CFM will be spawning an appropriate "buzz" level—by making it the "talk of the town." Peer-to-peer acknowledgment of the power of CFM will act as a technological enabler. If each FM professional recommended the participation in CFM to five other FM professionals and they, in turn, did the same . . . in no time the vast majority of FM professionals would become catalysts for successful CFM.

Participate in CFM Strategic Master Planning Activities

Audits are critical to the success of CFM and serve as a basis for strategic master planning. Some of the more common CFM audits are management, technology, opportunity, implementation, financial, and return on assets (ROA).

Stratification of CFM

Stratified offerings empower casual users, in-betweeners (facilitators), and advanced users to benefit from CFM. A variety of user interfaces make it easy to become a player in CFM. Horizontal and vertical offerings should complement each other in CFM.

Using CFM to Address Unique Needs

Cyberspace for FM must address both the common and the unique needs of its participants. The expansion of the information highway was the call to action 10 years ago, but now the creation of CFM universes must be the call to action for the new millennium. The unique demands of today's business environments require the rapid creation, deployment, and consumption of FM information and knowledge.

Knowledge About Internetworking

Internetworking, via the Internet, intranets, extranets, and so forth, will become commonplace and critical in the formation of CFM universes. Local area networks (LANs), remote area networks (RANs), global area networks (GANs), cellular area networks (CANs), and satellite area networks (SANs) are but a few that link together PCs, workstations, and personal computing devices so that they can communicate and share information. Some of the technologies deployed in these solutions include adapter cards, hubs, bridges, routers, Ethernet switches, and ATM switches. Facilities management cybrarians should understand basic technologies and be aware of proper "netiquette."

Basic Requirements for Creators of CFM Universes

Various requirements should be imposed upon the creators of your CFM universes. Some of the more common include perpetual increases in the speed of network transmission, advances in the state of networked applications, empowerment of the mobile and virtual workforce, creation of global infrastructures for information distribution and services, demand for open standards, continuous commitment to evolve as the next generation of technology becomes a reality, network management interoperability, and constant undertaking of appropriate CFM audits.

Client/Server Computing in CFM

In client/server networking, personal computers, workstations, and other personal computing devices are linked to communicate and share FM data, information, knowledge, and wisdom. Typically, there is at least one facility-centric server computer connected to the rest. The server houses many of the centralized and scalable applications and databases. Client/server software includes: operating systems; databases; development environments; programs that enable connectivity via the Internet, intranets, and extranets; network management; and program development environments. Client/servers act as part of the foundation for CFM.

FM Cybrarians: Catalyst (Relationship Builder) Between Technologists and Businesspeople

The risks of serious disconnects between technologists and business managers are a primary concern for FM cybrarians. Technologists are often so enmeshed in their technocentricities that they forget that their raison d'être is to serve the needs of the business, the company's shareholders, stakeholders, and customers. Facilities management cybrarians can help both sides understand and function in CFM, thus giving the organization, which they all serve, a competitive advantage.

CFM and the Advantages of Personal Productivity Suites

Personal productivity software should be available to FM professionals today as stand-alone offerings or in suites—collections of programs grouped with seamless interfaces. General office and FM-specific suites enable the rapid deployment of data and information between applications, thus enabling the FM cybrarian to derive FM knowledge and wisdom more quickly.

Multimedia Technologies in CFM

Multimedia technologies in CFM include the integration of alphanumeric data, audio, graphics, animation, video, neural networks, image recognition, and virtual reality data. Advances in CFM require the use of new and improved multimedia technologies in stand-alone, client/server, and web-centric applications. Videoconferencing, chat rooms, enhanced training, design simulations, and next-generation cellular connectivities are but a few of the multimedia applications in CFM.

CFM—Never to Be a Bleeding-Edge Technology

In a day in which fuzzy logic, nanotechnology, neural networking, and virtual reality simulation are quite common, FM cybrarians must develop realistic approaches to CFM to ensure that it never becomes identified with *bleeding edge*. Historically, technologies developed outside the bounds of commercial viability have been created at universities and research labs. Cyberspace for FM was created in commercial space, with the immediate benefit of commercial, market-tested gains in productivity and profitability.

New Millennial Technologies Applied to CFM

There will be millions of new technologies that will impact CFM in the new millennium. Some of the more interesting near-term applications include: hand-held electronic devices; natural language interfaces, like speech and pattern recognition; ActiveX FM objects; high-definition display and communication devices; smart manufacturing; virtual reality–based holograms; e-commerce; e-FM; and e-space.

CFM as a Catalyst for Paradigm Shifts

In the new millennium, CFM will most assuredly serve as a catalyst for paradigm shifts within any organization. Cyberspace for FM will impact an organization's technologies, marketing programs, organizational structures, finances, alliance activities, mergers, acquisitions, and business models.

Characteristics of Highly Effective CFM Management Teams

Organizations that flourish in CFM seem to have CFM management teams that demonstrate numerous technocentric and personal characteristics. These leaders are noted for their vision, charisma, strategic thinking, and ability to execute; they act as enablers for high-caliber recruitment and retention of staff; and they have a solid footing in reality.

Financial Resources Required for CFM

To be successful, the FM cybrarian must ensure that suitable financial resources are established and maintained for the digital universes being created in CFM. Beyond the FM domain, FM cybrarians must be constantly vigilant about their organization's business plan, cash flow, debt burden, revenue and earnings consistency, and the proportion of revenues from recurring sources. In short, to serve the business, the FM cybrarian must know the business.

The Human Being and CFM

Cyberspace for FM cannot exist without human beings. The digital world and its associated nervous circuitry cannot be a substitute for human contact. Cyberspace for FM, however, offers tomorrow's facilities managers the ability to extend the scope, continuity, and even the quality of their interactions with other FM professionals and the customers they serve. Long-term relationships are not usually formed via digital nervous systems but only extended and expanded. Through CFM, the ability to complement, reinforce, and enhance relationships becomes universal. Cyberspace for FM is based on the human being's need for acquisition, analysis, and distribution of FM data, information, knowledge, and wisdom. The "FM net" offers much, yet human beings must determine what kind of effect CFM will have on them, their organizations, their customers, and society at large.

The Digital Nervous System as CFM Matures

Today, it is estimated that 90 percent of all documents in U.S. businesses are created and/or stored electronically. Yet, it is still difficult for people to gain rapid access to them, and printed media are still the "preferred" way by which people communicate with each other. Although e-mail is rapidly changing our mode of interaction, how many e-mails are printed by force of habit each day? The personal computing revolution combined with the various Net offerings will truly enable us to have information at our fingertips and, subsequently, to do something substantial with it. The CFM du jour, where Active-X, FM objects, and infinite bandwidth are connected to every point on the globe and throughout the universe, is only a few years away.

The High-Five Version of CFM Boot Camp

The World Wide Web and its associated nets are still in their infancy. By the year 2000, more than 300 million people will have access worldwide. Facilities management cybrarians must address the so-called High Five questions that are associated with CFM. Perpetual in nature, common in need, the answers to these questions must be constantly reviewed.

1. What can CFM do for me and for the FM organization to which I belong?
2. What can CFM do for my company and the customers I serve?
3. How do I quickly, productively, and cost-effectively continue to be immersed in CFM?
4. How will I do business in and throughout CFM?
5. How do I optimally create, maintain, and use CFM?

INITIATIVES FOR REALIZING CFM SUCCESS AS FM CYBRARIANS

As we enter the new millennium, CFM is a reality! But a decade after its public debut, we are only at the beginning of the journey through the CFM universe. Today, hundreds of thousands of people are already tapping into this universe. By 2005, it is hoped that more than 300 million people will have benefited from the power of CFM. Cyberspace for FM has become a reality because of countless technological innovations and the FM professionals who are willing to step outside the box and become pioneers in the realm of their organization's digital nervous systems via CFM. To date, the following are some key observations about CFM:

- The customer is at the center of the CFM universe.
- Superior products and processes must be at the core of CFM.
- For CFM wisdom to be unique, FM data, information, knowledge, and wisdom must be complete and timeless.
- Cyberspace for FM will be most successful in virtual communities.
- Facilities management cybrarians must perform CFM audits (opportunity, implementation, technology, management, etc.) if CFM is to be successful.
- Cyberspace for FM thrives in strategic master planning environments and requires effective management.

FM BEST PRACTICES AND BENCHMARKS FOR CFM—SOME THOUGHTS ON THE DEVELOPMENT OF A COMMERCIAL OFF-THE-SHELF PRODUCT STRATEGY FOR ENTERPRISE WEB CONNECTIVITY FOR AN FM ORGANIZATION

CFM COTS Products

Web-based commercial off-the-shelf (COTS) products should be the foundation of CFM. Best-of-class Web technologies can be expanded into a powerful extension of the CAFM and CIFM environments. Inexpensive and easy to use, Web-based COTS products enable you to expand the connectivity of your CAFM and CIFM environments to anyone, anywhere, who is connected to your corporate intranet or the Internet at large via the World Wide Web.

Use E-Commerce Tools to Communicate Your Vision

Just as the Web makes your company globally visible, a Web-based COTS product will make your corporate FM efforts globally visible and your information instantly accessible. Use the Web to promote your FM department's efforts, advertise your capabilities and internal services, and promote and enforce your working practices. Use this technology to communicate your FM vision to the corporation at large. Accurately assess conditions within your entire organization to make best use of corporate resources, and collect the critical data that your corporation needs to execute its business plan. A COTS product can make you a key, visible, dependable player in the strategic planning process.

Seamless Interface into Your Corporate Web Strategy

Web-based COTS products are designed to enhance your corporate Web and management information system (MIS) departments and leverage their resources. They should be entirely open, linkable to any corporate intranet page, and allow you to use any dynamic form or report on your own Web page. A Web-based COTS product can be your productivity powerhouse for dynamic publishing of large numbers of drawings and structured FM data. If there is a report you need that is not provided, in minutes you can publish any view to dynamic pages, complete with drill-downs and paging of large results. The product puts all of the technological pieces together, so you get the glory.

Who Can Benefit from Using Web-Based COTS Products?

Anyone who uses your organization's facility information will benefit from the use of Web-based COTS products.

- *Facility managers.* Can access their facility information anywhere, anytime. They can provide users of facility information with self-service mechanisms that enable them to access information or request a work order, for example.
- *Field personnel.* Can easily and efficiently pick up work order requests from remote locations, saving them time and helping them determine what tools are needed for a job before they arrive.
- *Branch office personnel.* Can access home office information in real time.
- *Strategic planners.* Can immediately access the historical information that they need to make important planning decisions.

- *Purchasing managers.* Can verify invoices against inventories and track back orders.
- *Human resources managers.* Can develop occupancy plans and locate vacant space for new hires.
- *Building owners.* Can determine rentable areas and bill tenants for the space they occupy, plus their share of common areas.
- *Corporate real estate managers.* Can track detailed lease characteristics such as options, payments, and prorated taxes.
- *Accountants.* Can use space data to charge departments for the areas they occupy, plus their share of common areas.
- *Risk managers.* Can evaluate assets to obtain adequate insurance coverage.
- *Tradespeople, maintenance staff, and building tenants.* Can take advantage of remote access to facilities information.

For the Facilities/Asset Management and Services Industries

Facilities management cybrarians have come to recognize that the most important part of the job is to process, filter, and prioritize FM data, information, knowledge, and wisdom by using technology that quantifies and focuses it precisely. Systematically networking CFM leads to the timely creation of FM wisdom.

Creating an Information and Knowledge Advantage for Your Customers

Cyberspace for FM solution technology is designed to give companies an unparalleled ability to retrieve the information and knowledge they need, wherever and whenever they need it. This information and knowledge advantage, more than any other, is what will separate the winners from the losers in facilities, asset management, and its allied professions.

The Purpose of CFM

Professionals in FM and the allied professions are constantly looking for an edge in their ultracompetitive markets. The pursuit of normalized FM benchmarks and FM Best Practices™ has led professionals to realize that the old rules no longer apply: Advantage does not lie purely in economies of scale or in cutting-edge product development. Facilities management organizations and the companies they serve separate themselves from their competitors in today's economy on the basis of their ability to swiftly gather, manage, and act on relevant business information and knowledge. Facilities master planning is a critical component, and rapidly collecting and disseminating information through the World Wide Web are the realities of doing business in the new millennium.

Facilities management cybrarians have a common goal: to help clients in their facilities, financial, and professional services organizations become more competitive in automated asset management by using Internet and related technologies to retrieve, filter, sort, and deliver strategic information. To accomplish this mission, FM cybrarians develop and manage strategic intranet portals for senior executives, analysts, and managers who need to have the right information at the right time to increase productivity and profitability.

Facilities management cybrarians must combine technology, information management skills, and a high-level understanding of finance and industry to manage a world of information and knowledge for their clients. They recognize that, although the Internet dramatically changes many business processes, existing processes need to be understood and modified carefully to maintain safeguards, confidentiality, and security. Facilities management cybrarians take great pride in the professional quality of their CFM-based products and services and have an uncompromising determination to achieve excellence in everything they undertake.

Elements of Comprehensive Cyberspace for FM Offerings

Content Management.   The primary function of CFM is the seamless integration of internal resources and external information. Using the flexible Explorer/Navigator (Navigator) platform and an extensive array of vendor-specific APIs and procedures, a CFM solution is uniquely capable of rapidly delivering an intelligent access point for diverse data and information. Such information and knowledge management affords managers, analysts, and executives the simplicity, speed, and accuracy necessary for effective decision making. One example of content management for a real estate portfolio manager might be the access to a building's performance in a single report or "FM snapshot." In such reports, information is gathered from multiple sources. Required information might be gathered from a corporate financial database and a multimedia file of photographs and videotapes. Calculations might be based on data derived from municipal tax ledgers and portfolio cost escalation monitoring systems. Market valuation might be extracted from yesterday's local real estate journal.

Hot Lists.   Facilities management cybrarians work with their clients to identify important content suppliers for inclusion on the customized Navigator (called a *hot list*). Once identified, sources are tagged, prioritized, and integrated into a customized management system that allows single, intelligent searches, and efficient, user-friendly access to all available information. One user of hot lists might be a corporate space planner. By selecting a single action item on the hot list, the corporate space planner can query the corporate database to gather data and establish chargeback information for a particular building. The query results in the generation of alphanumeric and graphical information. A report, such as "Highlighted Rooms by Department by Floor" is rapidly created for each of the floors selected in a particular building.

Custom Interfaces and Personalization Tools.   A CFM solution also offers high-level interactive capabilities. Users have access to interface customization options, enabling individual interfaces and search tools. These tools allow smart and individualized data presentation. In addition, users can create, edit, and distribute documents or objects across the network, allowing the CFM solution knowledge management system to be a dynamic tool that becomes ever more customized in response to the client institution's needs. In this example (see Fig. 3.3.7-4), a "dialog box" for collecting information about the creation of a work request has been created. This custom interface enables the user to extract known information from the corporate building operations systems and denote their perceived sense of urgency. Training to use such an interface is typically only a matter of minutes.

Dynamic Learning.   A CFM solution allows and encourages clients to continuously update and adapt content to meet specific needs. Client-selected Web sites, research information, and other resources are dynamically included in the hot list system. In conjunction with custom client interfaces, the hot list is a constantly growing, flexible system that consistently adapts itself to the needs of each user. A facilities manager who is concerned about completing active work requests can construct an iterative query environment for dynamic learning (see Fig. 3.3.7-5). In this case, a facilities manager can connect and associate information from several systems to assess the status of a particular work request. A report such as "Active Work Requests by Floor Plan" can be created with a single click of a mouse. Through cellular connectivity, the facilities manager could determine if particular staff members had completed activities, their locations, the parts that were consumed, and the ability to be redirected to a project other than those assigned to them at an earlier meeting. Graphical queries are considered extremely helpful in such dynamic learning situations.

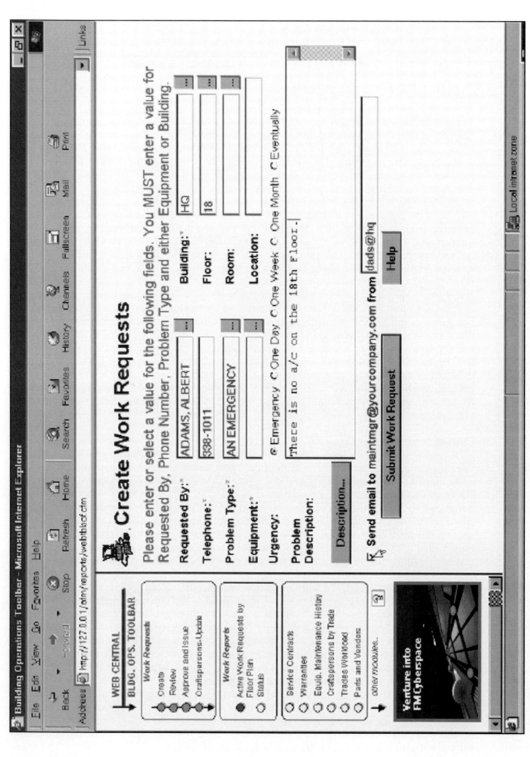

FIGURE 3.3.7-4 The FM Web Central visual user interfaces: FM Crib Notes™ for "Building Operations Management" (Create Work Requests dialog box). (*Courtesy of ARCHIBUS, Inc.*)

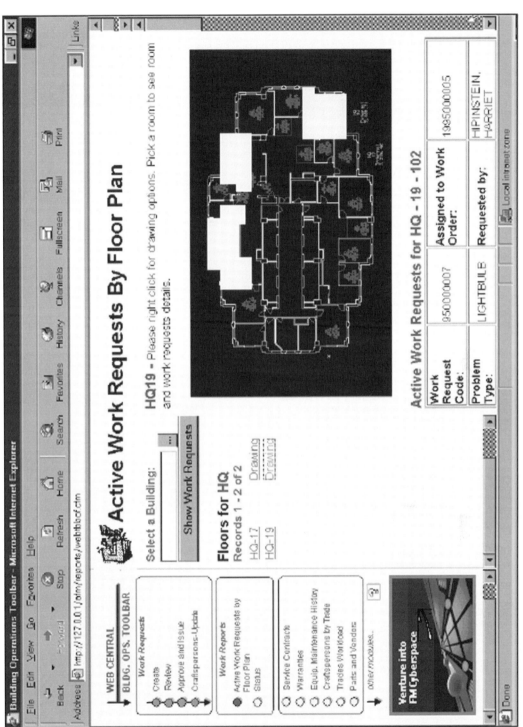

FIGURE 3.3.7-5 The FM Web Central visual user interfaces: FM Crib Notes for "Building Operations Management" (Active Work Requests by Floor Plan). (*Courtesy of ARCHIBUS, Inc.*)

Virtual Libraries.   The World Wide Web has now become the largest repository of information available in the world, and tens of thousands of sites provide valuable, precise, and timely data. Market-moving data, such as U.S. inflation and growth statistics, are now released first on the Internet. Countries around the globe are posting economic information; markets such as NASDAQ are publishing valuable information on listed companies; and markets and companies themselves are posting a plethora of information for investors on their Web sites. The digital economy has become a reality.

OPPORTUNITIES IN THE NEW MILLENNIUM

The ability to execute business strategies based on rapid analysis of available information is vital to any executive or analyst. As a result of the growth of PC and Internet/intranet/extranet usage, the number of electronic news and information retrieval services is exploding. In CFM, FM cybrarians can potentially offer every imaginable type of content and data. They may, unfortunately, toss all of the content haphazardly on the virtual laps of their users. Different approaches to categorization, entitlements, symbology, and ActiveX FM objects force users to waste crucial time locating relevant information. Publishers develop Web-based platforms, and many aggregate a number of different sources, but the problem of information overload for senior executives will ultimately be addressed at the company intranet level and on the user's desktop—not by content providers.

A CFM solution offers a quick and effective solution for managing content and data on the strategic intranets that are being developed for senior executives. It provides a robust, scalable system that is focused on delivering value to the user that can be easily customized and rapidly deployed at a reasonable cost. Figure 3.3.7-6 highlights some of the single-click reports that can be generated by a typical COTS Web offering for corporate executives. In this example, hot list access has been set up to provide immediate and personalized user access to selected reports and analyses, such as sample drill-downs for "Departmental Stack Plan," "Highlight Vacant Rooms by Building," and "Furniture Standards and Allocation Reports."

The Benefits of a CFM Solution

- Seamless integration of internal research, news, and analysis with external feeds from any electronic source.
- Central management and monitoring of external sources of data and information for usage tracking and cost control.
- Ongoing management of content, connections, and technology to ensure ongoing improvements, quality, and freshness of information for satisfied executive intranet users.
- Flexible taxonomy for mapping content in industry, country, company, and event-driven categories to external information services.
- Proprietary symbology mapping to external identifiers for all content sources; API management and contextual delivery for any internal or external information service.
- Intelligent access to valuable information on the World Wide Web developed especially for financial executives and managers including company, industry, and country data.
- Hot lists for the casual user.
- CFM is fast, easy, intuitive, inexpensive, and available.
- CFM enables rapid gains in productivity and profitability.

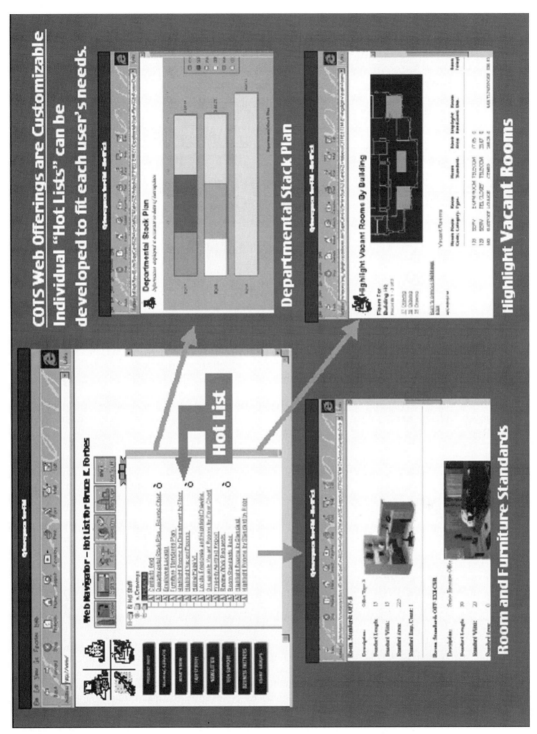

FIGURE 3.3.7-6 COTS Web offerings are customizable. Users can create their own "hot lists" which can provide immediate access to selected reports and analyses. (Sample drill-downs: "Departmental Stack Plan," Highlight Vacant Rooms by Building," and "Furniture Standards and Allocation Reports"). (*Courtesy of ARCHIBUS, Inc.*)

CONTRIBUTORS

Julia M. Forbes, ARCHIBUS, Inc., Boston, Massachusetts

Richard E. Forbes, Retired IBM, Concord, North Carolina

Cynthia M. Zawadski, ARCHIBUS, Inc., Boston, Massachusetts

ARTICLE 3.3.8

HOW TECHNOLOGY IS REENGINEERING FACILITIES MANAGEMENT PROCESSES

Harold Feinleib
Vincent Iannucci
Robert Trenck
Aperture Technologies, Inc., Stamford, Connecticut

Take a close look at your organization. Are your facilities management (FM) processes up to the challenge of global competition? If your FM activities are not running perfectly, then your existing processes may not be working as well as they should. It is likely that they need to be examined and reengineered.

Applying today's technologies properly can transform a process that causes you angst into a process that makes you a hero. Global competition demands near perfection of an organization's products and services. It also demands perfection from its FM operations.

One facility manager recently described his organization's space configuration process as a "nightmare." Another said his company's quarterly chargeback process used to take more than 3 months to prepare. Both used computer and network technologies to reengineer their respective firms' processes. As a result, they have become strategic players within their organizations.

Consider a well-designed automobile assembly line: Everything runs like clockwork along an immaculate, organized path. Parts are delivered at precisely the right moment to precisely the right place, so that a perfect product drops off the end of the assembly line. Not just once, but time after time.

This does not happen by accident. It happens because every aspect of the process has been examined and designed to work flawlessly.

Now, imagine a well-designed FM department where all processes also run like clockwork: Space needs are always forecast accurately for the entire organization. New space is always ready when new hires arrive. Resources are always available just in time to meet business managers' changing demands.

In a smoothly running organization, maintenance and repairs are tackled in a timely manner. Chargeback reports are always accurate and on time, and management information is always available when requested. This is not a dream. The vision of having all facilities resources available at your fingertips can be a reality. Whenever you need a current floor plan or a list of occupants or resources, you can access it instantly and accurately.

Extend the vision one step further by seizing the ability to communicate this pertinent information to business managers in a form that they can grasp and analyze immediately. You do not have to pinch yourself. It can happen at your company. It is not a dream. Can your FM department run like this? If you do not examine and reengineer your FM processes, then the answer will be, "Probably not."

So, how do you know what to do? There is no set formula designed to tell you a system makeover is needed, but you will know it when you see it. Failing processes have these distinct, recognizable symptoms:

- Projects run slowly.
- Things seem to be more difficult than they should be.
- Pulling together information requested by management is a disruption.
- Getting management to make key decisions seems to take forever.
- Too many projects require rework.
- Move-ins are not as smooth as you would like.
- Your customers are impatient with your department.

Although the immediate reaction to process disintegration tends to be, "We need more people," hold off on the hiring, and quickly assess your organization. It is more than likely that the problem can be found in already-established processes.

When you reengineer a process, you are seeking to minimize the number of transactions it takes to get something accomplished. Doing this eliminates unnecessary transactions between people and accomplishes others in just one exchange.

Think about how e-mail has totally reengineered the process of communicating. In the past, a message was handwritten or dictated. Then, it was typed, edited, retyped, copied, addressed, and mailed or delivered to each recipient.

Now, we can write messages in a form that can be easily manipulated. It is just as easy to send the message to a group as it is to an individual, and it can be sent and received instantly. The steps of handwriting, dictating, editing, retyping, copying, mailing, and delivery have been eliminated.

By changing the process, e-mail has made communication dramatically more efficient. The same dramatic changes can be applied to your FM processes.

The value of making these processes work more effectively and more efficiently reaches far beyond the direct cost savings that are likely to occur within the FM department. Now, the parts of the organization impacted by an inefficient process get what they need when they need it, enabling them to fulfill their business missions. This increases the entire organization's competitive ranking in the global marketplace because better decisions are made and things happen faster.

MANAGING SPACE

Of all of the diverse activities that facilities managers oversee each day, space management stands out as one of the most crucial in terms of positive reengineering. Modern organizations run somewhat like factory assembly lines. Space is the raw material that keeps things moving. It must be reconfigured to meet the changing needs of its occupants. Additional space must be delivered on time to match the growth needs of the business. The lack of space should never hold up the execution of business objectives. An excess of space, however, is a waste of money.

Facilities managers walk a fine line. Managing space requires short-, intermediate-, and long-term planning and processes that support efficient decision making and execution. For the short and intermediate term, it is critical to keep space working optimally for its occupants. This includes assigning offices to new employees, moving people, reconfiguring space to accommodate business requirements, producing accurate chargeback reports, and keeping everything in good working order.

The space management process seems straightforward, but it is actually quite complex. You need to assemble a great deal of diverse information and then communicate it and evaluate it. Many decisions must be made. Numerous activities must be coordinated. It is no wonder that this process rarely runs smoothly. Many times, when a space request reaches the FM department, it has already passed the point where FM can satisfy the need in a timely manner. A slow, inefficient space management process will only make matters worse.

Space requests trigger a process that includes a series of planning sessions with the requesting business managers. To prepare for the meeting, the facilities manager will typically carry out these actions:

- Gather information from various sources regarding the people, furniture, network communications, and special requests involved in the move.
- Request or print computer-aided design (CAD) drawings of the affected areas.
- Walk the space to see what actually exists and adjust the CAD drawings.
- Mark information and changes directly onto the CAD drawing (cross out walls and write in people names, department names, and floor areas) to reflect the proposed changes that satisfy the space request.
- Present the marked-up CAD drawings to the business manager who is unable to understand the hand-drawn blueprint presented.
- Spend time going between the business manager and the CAD operations as the drawings get closer and closer to a workable solution.

As this occurs, the requirements usually change, requiring still more interactions. And, as you may have guessed, the process is not problem-free. Stumbling blocks occur when the facilities manager

- Does not know when people are being hired
- Does not know when departments will run out of space
- Cannot determine where space exists to figure out where to put new employees
- Spends too much time gathering data about who is involved in the change, at what levels they are, what furniture and network connections are needed, and whether there are any special requirements
- Relies on CAD-based construction drawings that rarely show the detail required for effective space management, such as actual space use, current occupants, network services present, and security arrangements
- Presents plans to business managers using CAD drawings that are hard or even impossible for the business manager to understand, particularly if there are hand-drawn changes
- Finds difficulty in coordinating the requirements of different department managers to lay out and outfit each space effectively

These, however, are not insurmountable problems. All can be addressed with reengineering and the proper application of technology.

REENGINEERING THE PROCESS

Imagine a space management process that not only minimizes the number of transactions but also one where all the steps happen smoothly! This process would be defined by three major capabilities of the facilities manager:

1. The ability to provide accurate, timely information about all facilities resources
2. The ability to clearly communicate ideas and proposals to business managers
3. The ability to communicate quickly and concisely with service providers who need to set up the space

An efficient process starts with an accurate up-to-date base of information. If the database is not kept current, information will need to be gathered every time a change is considered. This will build up and cause delays and confusion. Eventually, the process will break down.

FM'S COMMITMENT

The information system is the heart of the process, but it is not a silver bullet. The system provides a structure for the facilities manager to succeed, but to achieve success, the FM organization needs to make a commitment to success. First, FM itself must own and operate the system. It is critical that the FM organization controls all aspects of the system. It cannot afford to rely on anyone else to do this for them. Facilities management owns the process, so FM should own the system. For example, a company's chief financial officer (CFO) has announced that a new formula must be used to determine departmental chargebacks. Facilities management must respond to this directive. It can do so quickly when it has its own system and, thus, the ability to make the change in the system.

Facilities management cannot rely on a management information systems (MIS) department, an outside consultant, or even a software vendor to keep the system operational. Facilities management must choose a user-friendly system that its staff can maintain and manage within its own department. It should not be one designed for CAD professionals, architects, engineers, or database administrators. It must specifically take into consideration the needs and functions of the FM organization.

Second, FM must organize and take responsibility for keeping the data current. Processes can be streamlined by tracking space needs and uses with up-to-date space management drawings and data. For example, it is FM's responsibility to ensure that all workers are included in the system. This includes employees (both on-site and off-site), temps, consultants, and contractors. It is not enough to tie the system to a human resources (HR) system because the HR system knows nothing about on-site contractors and consultants.

The FM department cannot rely on the CAD or engineering department or outside consultants to keep this information up to date. If it does, delays and slow turnaround of information will certainly occur when decisions need to be made. Additionally, the information will not be formatted effectively for good management decisions.

This does not preclude CAD or engineering from having an important role in facilities planning. It means that these departments do the work they are best suited to do—creating specification, design, and construction drawings. Computer-aided design and engineering do not know who occupies the space, nor do they understand your customer's needs. You should not expect them to be responsible for maintaining space management drawings or providing information for planning decisions. That is FM's job.

Third, an effective FM must present information in a way that business managers can grasp quickly. Business managers do not have time to worry about space needs. That is FM's job. They expect that FM will clearly communicate space proposals back to them in a way that is easily understood. Business managers do not know how to read or interpret blueprints and CAD drawings. In most cases, it is not necessary, especially if FM is doing its job efficiently.

If information is presented using blueprints or CAD drawings, then the process will require an interpretive step. This wastes time. If the information, instead, is presented in colorful and clearly labeled drawings, business managers can quickly understand the proposal. Then, the business managers and the FM can become a team to solve space requirements, leading to fast customer buyin.

THE NEXT GENERATION: VISUAL INFORMATION SYSTEMS

Attempts have been made in the past 20 years to automate facilities information, but only recently have a number of factors allowed such technology to become successful in practice. These factors include powerful desktop computers with high-resolution graphics; user-friendly, feature-rich software that permits FM administrators to get the job done properly; and a common method (Web browsers) for anyone to access drawings and data using intranets.

In the past, computer-based solutions tried to tie existing CAD products to database systems. Because these products were not FM-driven (nor were they necessarily database-friendly), they often required the support of MIS or an outside consulting organization to customize, administer, and modify software. And these CAD-based products also required a CAD specialist to make all graphic changes. The results usually fell far short of needs and expectations.

The latest improvements in technology have given birth to new software systems—visual information systems—that can be used directly by FM without relying on CAD departments or MIS personnel. Because FM personnel maintain these systems, the time to make changes decreases along with the costs of those changes.

In a visual information system, an easy-to-use drawing system is combined with database management. Objects in a drawing are tightly integrated with the data that describe them. The integration is so tight that variances in data values can change the way the object appears on a drawing. For example, offices can be color-coded based on department assignment or by occupancy. This new approach to space management allows FMs and business managers to visualize new ways to get the most out of existing space using meaningful graphics and data designed to communicate information to executive and department managers. As a result, alternative proposals can be quickly communicated and easily understood by all.

In developing alternatives, facilities managers and business managers can move people and furniture configurations on the drawings. This can even be done during planning sessions to see the effect immediately—leaving little open to interpretation down the road.

Many facilities managers assume that they can accomplish this with CAD drawings. But CAD was designed for people who design things to convey specifications to people who build things. It is great for architects and engineers to create construction drawings for contractors to use in estimating and construction. But CAD falls far short in conveying easily understood, actionable, management information. It is like using a spreadsheet to do the work of a word processor; it can be done, but the results will fall short of expectations.

Computer-aided design floor plans do have an important place in the process: They form the basis of a visual information system. This system adds intelligence to CAD drawings with a database that describes the objects in the drawings. The original CAD base-plan drawings become reference drawings. Any change is reflected immediately—and automatically—in the visual information system.

Surprisingly, a visual information system is also the best way to communicate with architects and outside contractors—the very people who rely so heavily on CAD to perform their duties. The clear visual information conveys the purpose of each and every proposed space, an invaluable tool for professional designers, contractors, furniture dealers, network installers, and electrical technicians. Such a system can help contractors do their work faster, with no misunderstanding or miscommunication.

FACILITIES INFRASTRUCTURE MANAGEMENT

Maintaining a current data and drawing database has many ancillary benefits. For instance, departmental chargeback reports can be produced at any time. Because data are always current and accurate, the streamlined chargeback process includes only the amount of time required to print the report. In fact, browser and intranet technology even eliminate the need to print the report. Business managers can see their current space charges on line in real time any time they wish.

For maintaining space, a visual information system simplifies specification of materials needed for routine repairs. Every object in a drawing has specific data associated with it. Unique information about equipment and systems, such as manufacturer, model number, installation date, and warranty, can help in both repairs and preventive maintenance (PM) because it is easy to see where the maintenance will be performed.

Other departments can also use visual information systems. Data communications, security, electrical, and mechanical, for example, can simply tap into the system and add their area-of-interest objects on new drawing layers. Managers can also track building contents such as PCs, printers, copiers, telephones, and mechanical equipment with ease.

Consequently, organizations can deploy visual information systems as enterprise solutions that coordinate all facilities and technology infrastructure into one system. This enables everyone who maintains part of the infrastructure to see—and better understand—the plans made by other members of the infrastructure team.

THE PROOF IS IN THE PROCESS

At one well-known Fortune 100 corporation, the FM department manages 8 million square feet of space, housed throughout 40 buildings in a campus environment. In the days before installing a visual information system, staffers described their space reconfiguration process as a nightmare. Using hand-rendered changes on CAD drawings, they simply could not communicate reconfiguration plans clearly to business managers. There was little customer agreement because the customer could not understand the impact of the recommended changes. Additionally, the turnaround of CAD updates resulted in too much time between planning meetings.

Installation of a visual information system solved the crisis. In 5 months, the department had more than half the campus up and running on the system. Today, all 8 million square feet are operational in a system that is entirely maintained by FM.

The company's FM process focuses around building managers who oversee space maintenance and reconfiguration for a specific portion of the campus. Managers tap into the visual information system to plan changes and maintain their portions of the system.

When they prepare for planning meetings and other related activities, the building managers distribute colorful drawings that clearly show how each space is to be used. They even make last-minute changes to drawings and data during the planning sessions with business managers. This enables making decisions rapidly and creates strong customer buyin. Thanks to the successful implementation of a visual information system, the FM department's reputation has changed from being an information bottleneck to a strategic department that efficiently meets the challenges facing the business.

REENGINEERING THE CHARGEBACK PROCESS

The real estate department of a large computer manufacturer tracks 300 facilities that comprise 4.5 million square feet. The department is responsible for providing business managers with chargeback information. In the past, the staff used CAD drawings and handwrote pertinent information into each office space in the drawings, including the occupant's name, the department in which the office was located, and the total square footage. These data were entered into the CAD system and then extracted as a file. Then, the business managers reviewed and verified the data and then sent them to IT for validation and compilation. If a name had been misspelled or had been entered incorrectly or a department designation had changed, the data would be rejected.

Completing such a lengthy process took 40 people half of their work time and cost the organization more than $1 million annually to produce. The reporting cycle took more than 3 months to complete, only to have the chargeback information challenged by many business managers who found errors in the data.

To make matters worse, the business managers wanted to receive accurate chargeback information quarterly, rather than annually. Obviously, this was impossible because the reporting process itself took longer than the desired reporting cycle!

They needed a system that would retain one central repository for all facilities information, a system that could be used by everyone in FM. And, most important, one that was easy to learn and use.

A visual information system offered the business and facility managers a workable solution. Seven zone managers use the system to maintain the drawings for all 300 locations, streamlining the process and eliminating administrative clutter.

The system restored the integrity of the reporting process. Because the company now has accurate quarterly reports produced, it has saved upward of $1 million a year in direct costs to produce the reports. By empowering zone managers with a visual information system that they can use to track personnel moves, the way FM is viewed in the corporation has improved dramatically.

RETURN ON INVESTMENT

The success of a visual information system is nearly ensured when you have the backing and commitment of your executive management. Receiving this commitment is contingent on a favorable return on investment (ROI). The ROI for implementing a new process hinges on three questions:

1. Who is impacted?
2. How are they impacted?
3. What is the cost of the impact?

Give the *who* question a lot of thought, and you will soon see how spearheading a reengineered, efficient FM process makes you a financial asset to your organization.

Let us take our first example where space planning was a nightmare. Who was impacted? Certainly, the business manager who made the request. Also, the people directly affected by the request. How are they impacted? The delays may cause them all to spend too much time with the space change, reducing their productivity.

What is the cost of the impact? Multiply the number of days lost by the contribution of each employee multiplied by the number of employees who are moved each year. If the employees such as salespeople produce revenue, the number could be staggering. This could have a large negative impact on the organization's financial results and its ability to compete!

This nightmare process can easily cost the organization millions a year and be the source of lower employee morale and lower employee productivity. Now, you can see how this nightmare process negatively affects earnings and hence the CFO, the COO, and even the CEO!

What about our chargeback example? Business managers need to control their resources, but if they do not receive accurate and timely chargeback reports, then they cannot control one of their largest budget items. They feel out of control and frustrated, which affects their ability to manage their other resources. This leads to a negative effect on corporate operations and on financial results. Again, the CFO, COO, and CEO are affected, not to mention every department and manager between you and the CEO!

FM MORALE

Thanks to this newest form of computer-aided facility management, facilities managers can make a difference in an organization by providing pertinent information quickly and effectively to all customers. Best of all, they finally know that they are making a difference to the organization. This, in turn, raises the esteem of the entire FM department. Moreover, senior executives begin to view the FM organization as more professional because they now see how it contributes to the business mission.

A visual information system can help you to reengineer your FM processes. It places ownership for space planning in your department's hands, helps you reduce your dependency on expensive consultants, streamline the planning process, and communicate more effectively with business managers. This can lead to strong decisions. The direct and indirect cost savings associated with reengineering the space management process can be dramatic. Obviously, cost savings will vary by organization, but the principles reviewed here can be applied to any organization.

Reengineering the FM process can bring dramatic bottom-line results and transform FM into a strategic player. The key is quickly providing executives and department managers with the information they need—in a format they can understand—so that facility decisions can be made faster and more effectively. Armed with the power of this latest technology, members of the FM department will become strategic players, raising their esteem in the eyes of management and peers.

The Characteristics of a Visual Information System

Achieving the unique benefits provided by a visual information system requires a set of important design principles, all of which work together to produce a powerful, cohesive capability. A visual information system designed to be used by FM has these characteristics:

- Does not require professional programmers to customize or modify it. Facilities staff can modify the system to meet the organization's changing needs.
- Is easy to use, so FM can make changes directly in the drawings without requiring CAD operators.
- Allows FM to tie drawings and data together.
- Offers easy access and navigation through drawings and data.
- Features a built-in report writer, so FM can create or modify reports.
- Runs reports against objects in drawings by highlighting objects.
- Runs reports against one or more drawings.
- Moves objects from one space to another, automatically updating the database.
- Can easily access and update data in industry standard databases.
- Accesses drawings and data live over the organization's intranet using a Web browser.
- Accesses drawings and data in a client/server environment over a TCP/IP network.
- Controls access to drawings layer by layer, so different groups can work on the same set of drawings.
- Easily integrates with other applications, such as computerized maintenance management systems (CMMS) or bar code systems.

- Designed to be an information system.
- Has friendly, easy-to-understand graphics.
- Can instantly color-code drawings based on data.
- Can change text on drawings as underlying database information changes.
- Provides information to management in a clear and understandable form.
- Discloses layers of information progressively.
- Locates personnel and equipment easily, no matter in which drawing they are, over the organization's intranet.
- Customizes applications by placing buttons on the drawings that automatically perform predefined functions such as running reports.

ARTICLE 3.3.9

GETTING THE MOST OUT OF INFORMATION TECHNOLOGY PROJECT CONSULTANTS

Joseph Best
Project Management Services Inc., Atlanta, Georgia

Scott Maddern
Integris, Billerica, Massachusetts

You cringe when you think about it—a project that you wholeheartedly sponsored and for which you stuck your neck out to convince other stakeholders that it had "the right stuff" crashes and burns just a few months after its launch. You're in good company.

A recent study by The Standish Group indicates that project failure is a pervasive reality among information technology (IT) executives and reveals that

- Forty percent of IT projects are canceled before completion.
- Thirty-three percent of the remaining projects are challenged by cost and/or time overruns or changes in scope.
- Failed and challenged projects together cost U.S. companies and government agencies an estimated $145 billion per year.

Why do so many projects get off track? There are myriad reasons, but one common factor is mistakes in hiring and managing consultants. Many sponsors unwittingly set their projects up for failure by not establishing clear guidelines up front to determine how to discern when it's the right time to bring in outside help, what to look for in consultants before hiring them, and how to manage consultants effectively once they're on board.

With big money and your management reputation on the line, here's how you can avoid making costly mistakes in working with consultants on your next IT project.

DETERMINE WHEN A PROJECT REQUIRES OUTSIDE HELP

Bringing in consultants before you've fully used your existing resources can be demoralizing to your staff. It conveys to in-house project team members that you think they're not competent enough in their core skills to contribute effectively to the project. Therefore, maintain morale by hiring consultants only when the project's scope makes it necessary. How do you tell? According to Ralf Leszinski, a McKinsey & Co. alumnus and executive vice president of Atlanta-based Project Management Services, Inc., there are five signs:

1. *Your company doesn't have the in-house expertise.* You have no one on your core management team who has the expertise required to manage the project.

2. *Your company lacks the capacity to lead a new project.* You may have the skill set within your company but can't free them up in time to address a critical project.

3. *You need to accelerate the progress.* Suppose you have a new software product ready to go, but the associated systems to control the sales are 3 months behind schedule. If you're at full capacity, bringing in the right consultants will help you expedite the process and diminish loss of potential revenue.

4. *The project's revenues exceed expenditures.* The project will create an additional revenue stream that overcomes its costs.

5. *A project component doesn't fit your core delivery structure.* In other words, gaining knowledge in a particular area is not going to enhance your overall business model. After the project is completed, you can easily part with that knowledge because it's not something you're interested in carrying forward in your company.

KNOW WHAT YOU'RE LOOKING FOR

Once you've determined that it's the right time to bring in consultants, how do you identify who would be a good match for your project? What criteria should you look for? In addition to the standard quantitative issues such as the firm's financial strength and capacity, look for the following qualitative factors as you interview prospective consultants.

What is their track record? Have they done similar work before? Remember to pay for the consultant's knowledge, not training.

Are they open about their failures? There's nobody out there with a perfect track record. Ask about their projects that did not go well. When something went wrong with a project, how did the consultant respond? What steps were taken to salvage the project? What did the consultants learn to prepare them for the next time?

Do they initiate a discussion of your success criteria? Look for consultants who ask you questions like, What is your definition of *success* for the project? If we were to go in and do exactly what you want, what would the postproject environment look like? If the consultant is not clear about your objectives up front, expect to make several time-consuming and money-consuming adjustments down the road to get things back on track. Define success criteria, and keep them in focus throughout the project.

Are they concerned about what drives the profitability of your business? For example, a solution might meet the project's objectives, but if it's too costly, it will have a negative impact on your bottom line. Look for consultants who suggest solutions that meet a particular need and also enhance your company's overall performance.

What is the firm's core competency? Suppose that the project involves several different components: a strategic portion, a tactical portion, specific technical expertise, and so forth. Find out which component the consulting company can handle best, as opposed to looking for a one-stop-shop firm. Why? No firm can be excellent in all facets of a project. Instead, try to match the consultant to the specific project component, or hire a lead consultant as a general contractor who has access to the best coders, network designers, trainers, and so forth.

Whom will you get once the project begins? Beware of a bait-and-switch situation. Many consulting companies show up with their top guns and then supply the client with second stringers after the project starts. Make sure that the contract specifies who will be on your team when the project is launched.

Will the key consultants be around for the long haul? Take steps up front to ensure that the consultants whom you hire will stay with your project until completion (see the sidebar, "Keeping Consultants on the Project").

Before you meet with candidates, design an evaluation matrix by putting four or five (or whatever the number) consultants' names on the top with the previous criteria listed on the left-hand side. During the interview, jot down your observations for each category, and before it's over, ask for a list of three to five references whom you can contact by phone to verify information. After you've completed the interviews and contacted references, look at your notes and incorporate a grading system (e.g., 1–10, 1–5, or whatever works best for you) to codify how comfortable you feel about a consultant in a specific category. Then, add up the scores. In this way, you have the data you need at a glance to compare apples with apples.

INITIATE THE RELATIONSHIP

Once you've decided whom you should contract, how do you ensure that the relationship starts on the right foot? Here are five tips:

1. Notify all team members that you've selected a consultant, and give the reasons behind your decision. Explain to staffers that the consultants will provide necessary support to the team as a whole, while acknowledging the heavy existing workload for employees. In this way, you cause in-house team members to view the idea of bringing in outsiders in a positive light, and you cultivate good morale in the entire team.

2. Identify what facilities, communications, and technical requirements the consultants have, and line them up ahead of time. Consultants may need a laptop, software, access to key information, or even office space. When you address these requirements up front, you help the consultant get off to a good start. Otherwise, you're paying for wasted billable time as you scramble around to equip consultants after the meter has started running.

3. Communicate project expectations, goals, objectives, key milestones, and risks in a formal document (i.e., a project charter) to optimize team performance.

4. Obtain commitments from all project team members before the project begins. Require that your team members stand up in a formal session and commit to their deliverables, as outlined in the project charter. This action solidifies the team members' resolve. Otherwise, you risk project failure because of one or more people who lack genuine commitment to fulfill their project responsibilities.

5. Make sure that consultants understand your standard reporting processes. This is to help you assimilate the data easily and reduce the chances of miscommunication that could hinder the project's progress.

MAXIMIZE THE RELATIONSHIP

Okay, you've done your due diligence by finding the best consultants that the project requires and by getting things started on the right foot. Now, how do you manage the relationship effectively over the long haul to keep the project on track? Here are five guidelines:

1. *Cultivate a project environment in which consultants can report actual status.* Reward honesty when you ask project heads about their status. If the person who is behind is ridiculed, you're not going to get the true status in the next meeting. Instead, provide help, guidance, and additional resources (e.g., equipment, software, human capital, etc.) to get those consultants back up to speed.

2. *Make consultants feel like part of the team.* Consultants work best when they feel a sense of ownership and pride in the project that makes them to want to do whatever it takes to make it successful.

3. *The company should put in the same effort as the consultants.* Part of maximizing the relationship is managing the consultants. If the project is behind, for whatever reason, your resources need to work side by side with the consultants to ensure that the consultants have correct information about your internal structure and to manage the team as a whole to keep the project on track. If the consultants are working on Saturdays to catch up, you and the staffers should be working with them to make the project a success.

4. *Be attentive to the project and project team.* Project status reporting, maintaining schedules, and other supervisory functions are hard work and take you away from core project activities. But if you neglect to review people's status carefully, sooner or later, the consultant will realize it and be more inclined to cut corners or lag behind schedule—and you won't know until it's too late.

5. *Review the project success factors periodically with consultants.* In this way, you and the consultants will stay focused on completing priority tasks, as opposed to being sidetracked by other interesting but less important activities, and will keep the project moving on schedule.

Your decisions in hiring and managing consultants can make or break a project. Therefore, do your homework up front. Make sure that you have a genuine need for consultants, know what to look for, and take steps to maximize the relationship once you've hired them. In this way, you can make informed decisions that boost your project success rates—and your company's bottom line.

KEEPING CONSULTANTS ON THE PROJECT

Although ultimately you can't control people's lives, you can take steps up front to increase the likelihood that the consultant you hire will stay with your project to completion. Here's how:

- When interviewing prospective consultants, ask about their short- and medium-term goals. If the consultant says, "After I get enough knowledge, I'd really like to get out and start my own company," that person might not be around next year.

- If the consultant is under a noncompete agreement, see if you can get a copy of it. The noncompete agreement should include time limits and specify what clients, geographic regions, and services are involved. If it is a tough, enforceable contract, the consultant can't just jump ship to the next employer who offers $5 more.

KEEPING CONSULTANTS ON THE PROJECT (*Continued*)

- Try to secure the person to a project performance bonus that's paid out at the end of the project.
- Require that the consultant sign a nondisclosure agreement (NDA). The NDA should include signatures of the individuals assigned to the project; a clear definition of what constitutes *confidential information*; the rights, restrictions, and obligations of the consultant who receives the information; and the term for which the agreement is in effect.
- Make a provision in the contract that reserves for you the right to review the replacement and specifies that the consulting company will cover the knowledge transfer of the replacement. So if the firm's lead expert leaves, for whatever reason, they do not bill you for time until the replacement is as knowledgeable as the key person who left.

OUTSOURCING: THE KEY TO CONTROL OVER THE ENTERPRISE

Although no one can deny the tremendous benefits of open, client/server computing, few would embrace its corresponding set of complexities. Multivendor madness adds to the complexity of a network, and to IT management's headaches.

According to THG's Information Technology Benchmark, the average company has more than 100 IT hardware, software, and services vendor relationships. With desktop systems, operating systems, and applications frequently supplied by different vendors, the multivendor atmosphere can rapidly grow chaotic. Information technology management holds the responsibility of ensuring harmony across the network. Although the drive toward standards-based open systems has helped, sorting through the complexity is still a major challenge.

Companies that deal with such issues of complexity have found that outsourcing is a sensible remedy.

In addition to complexities, cost is a major challenge. One of the most common problems of client/server technology is the unanticipated expense of deploying such a solution. Rolling out a distributed computing environment brings with it numerous "hidden" costs—maintenance, training, and support—all of which are integral and critical to the ongoing success of the environment.

Outsourcing can deliver a number of cost benefits, as functions such as operations and maintenance become the responsibility of the outsourcing company. For example, implementing an outsourcing strategy allows a business to control its staff costs on a long-term basis.

Personnel Resources and Shortages

It is hardly a surprise to anyone in the business today that, as the demand for and complexity of IT services continue to increase, the experienced labor pool continues to shrink. The nation faces an overall shortage of IT staff skilled in the latest technologies, and the turnover of the existing pool is enormous. Add to this challenge the fact that IT personnel account for about 52 percent of total IT spending, according to Compass America, a leading performance consultancy. The net result? Companies and organizations pay more than one-half of their IT budget for people they have trouble finding in the first place—and holding onto in the second. Then how does this allow the company to remain even moderately ahead of the technology curve?

Add Strategic Repositioning

In addition to complexity, cost, and personnel shortages, today's IT manager also has to confront the Net—most significantly, the growing demands from top management to leverage it, so that the company becomes a competitive e-business. That means that IT is, effectively, no longer the purview of the IT department, but rather a critical, strategic element that affects the entire future of the organization.

There is light at the end of this tunnel, however. Corporations that choose client/server outsourcing over the traditional in-house approach are positioning themselves to achieve considerable benefits. By investing in an external team of IT professionals, a company can achieve new balance in its technology operations. The outsourcing team is more able to deliver higher service levels, and it can also do so while managing corporate costs better. Day-to-day operational needs are met by the outsourcing partner, and internal IT personnel can focus on the requirements driven by the company's core business. Additionally, IT management is free to concentrate on deploying technologies that enhance the company's competitive position and long-term growth.

ARTICLE 3.3.10

IS YOUR CAFM SYSTEM PREPARED FOR TODAY'S NEEDS?

Yolanda Hill

Facility Information Systems, Inc., Camarillo, California

It is most beneficial for facility management (FM) and real estate directors to provide access to enterprise data via a standard intranet Web browser. Enterprise-wide solutions need to provide business intelligence to the staff for day-to-day decisions in multiple locations and often across multiple time zones. Secure intranets allow distributing and receiving mission-critical information anywhere at any time. Before such an implementation, the organization needs to be able to identify computer-aided facility management (CAFM) solutions that are designed to provide the desired information and match the scale of the enterprise database.

Let us examine some of the factors that need to be considered when evaluating and purchasing a CAFM system. First and foremost, a CAFM system must integrate with databases across the entire enterprise to support decision making, problem solving, and planning. Implemented correctly, the system will become a central source of information to authorized participants involved in strategic, tactical, and operational processes during the life cycle of a facility.

The system should also support document and work flow management with technical document management systems (TDMS). Integration with TDMSs ensures that drawings, maps, leases, property documents, and all associated reports are interactively accessible by appropriate participants in every phase of the life cycle.

Many CAFM vendors have rushed their applications to the Web using proprietary and outdated or new and risky technologies in Web deployments. Often these applications do not port to the Web without time-consuming and expensive customizations or ongoing migrations to current technologies and platforms. Many CAFM systems in the marketplace cannot secure the back-end databases nor can they integrate easily with other systems. Often such systems contain features to secure access via forms or screens and leave the back-end database exposed to casual or deliberate tampering.

To determine the readiness of any CAFM application, several critical requirements must be addressed in depth:

- Is the CAFM system built on a current, proven technology?
- Does the database management system (DBMS) scale up to manage very large amounts of data in a secure environment? Can the system be deployed on multiple client/server operating systems? Does the system provide complete functionality and features, including access to computer-aided design (CAD) drawing and geographic information systems (GIS) map data in a Web environment?
- Does the system provide options for secure integration with other corporate or vendor applications?
- Do reporting requirements easily serve the needs of all levels of planning, management, and operation? Corporations that own or lease facilities that have several million square feet of space need the power, integrity, and functions of a system that can answer these questions affirmatively and then provide all of these options.

Is the CAFM System Built on a Current, Proven Technology?

Large FM and real estate organizations need a core system that can be integrated on the Web and in client/server environments throughout a large complex organization. To date, only Oracle development platforms have successfully met this requirement by providing the ability to develop in a variety of languages, including JAVA and XML, as well as its own PL/SQL language. Many DBMS vendors claim the ability to store processing logic in their databases, but Oracle remains the leader in this approach to data management. Only those CAFM vendors that employ 100 percent current technology and methods can ensure delivery of robust, secure, cross-platform solutions.

Does the DBMS Scale Up to Manage Very Large Amounts of Data in a Secure Environment?

In the Fortune 2000 marketspace, a DBMS should scale to multiple terabytes to hold the massive amounts of detailed current and historical data used in daily operations, business-to-business (b2b) transactions, and for global strategic decision support. A CAFM solution must manage a large corporation's complete asset portfolio in a single, secure, well-structured database.

Many CAFM vendors downplay the sheer volume of data that will eventually reside in an enterprise-wide information system and therefore ignore the importance of providing industry standard databases. Too often, important records for valuable assets have been relegated to spreadsheets or file cabinets, or even the heads of key personnel.

All current enterprise data, as well as historical records used for determining trends and forecasts, should reside in a single, secure, well-structured database that can be easily integrated with other corporate, enterprise resource planning (ERP), CAD, and GIS applications. Beware of systems that cannot support all FM and real estate data in a single database.

Does the System Provide Options for Secure Integration with Other Corporate or Vendor Applications?

Inside large organizations, land, buildings, and other structures, plus assets such as furniture and equipment, are all being closely scrutinized for opportunities to increase enterprise profits—especially by using automation to improve decision-making processes. Traditionally, data that are needed to describe and clarify the real estate and FM information are scattered in various manual and automated legacy systems across the corporation.

Today, corporate initiatives require much higher levels of integration and many more data delivery options than ever before, including delivering content to the Web. Work flow and business process automation are standard requirements and features expected in off-the-shelf, enterprise-level applications. E-commerce initiatives and strategies drive organizational goals, including new areas that have not traditionally been considered crucial to a company's success in the marketplace.

The FM and real estate systems that are used to manage and deploy a corporation's physical assets must easily integrate with all corporate, CAD, GIS, ERP, and other vendor systems in real time. They must also support multiple data access options, including serving data to portals and laptops, as well as wireless and hand-held devices—all while providing the appropriate levels of security.

Integration with GIS Products

Geographic information system functionality extends from inside the building to related land and infrastructures outside the building. For example, travel time may be computed starting at any office within a building and terminating inside a building across the street, across town, or across the country. Real estate managers may query the database for available parcels of land based on parameters that specify the desired demographics of the surrounding area (available labor force, distance from major transportation and appropriate housing, the quality of the education system, etc.). Cable lines and connections may be tracked both inside and outside the building. The CAFM vendor must meet the need for large organizations to be able to create, access, and update data securely in real time from anywhere at any time.

Can the System Be Deployed in the Web Environment and on Multiple Client/Server Operating Systems? Is the Entire Set of Features and Functionality Provided in Both Environments?

Too often, excessive manual verification of data for strategic planning causes delays of many months. Delays in getting the data required for making important decisions result in high labor costs, erroneous reporting, and the inability to deploy human, real property, and other assets appropriately and cost-effectively. These delays can cause real harm to an organization.

Oracle remains the leader in scalable, Web-deployable data management. Only those CAFM vendors employing 100 percent current technology and methods can ensure delivery of robust, secure, cross-platform solutions.

Are Reports Provided by the CAFM System Easily Generated? Do They Satisfy the Needs of All Organizational Levels of Planning, Management, and Operation?

Real estate and FM organizations are increasingly the focus of corporate reengineering efforts. Often, data in these areas are not readily available for informed decision making related to process improvements. Outsource vendors often provide deceptively simple, financially attractive alternatives to eliminate in-house FM operations. Many CAFM vendors offer solutions that appear attractive on the surface but that cannot deliver the scalable, corporate-level reporting capabilities that distinguish more robust systems.

A CAFM solution must reliably provide all of the information that executive management needs to evaluate the options and to make fully informed decisions. The information must be distributed to the right people, at the right time, using secure delivery methods.

Checklist of CAFM System Features

A complete best-of-breed CAFM system should provide the following features and functionality:

- Integrate all FM, real estate, and corporate databases with CAD and GIS drawings.
- Capitalize on Oracle 8i object relational database functionality to link database records with the enterprise's tabular data in CAD drawings and GIS maps.
- Work seamlessly with Oracle security, integrity, data manipulation, transaction support, data storage, distributed processing, and performance tuning.
- Provide built-in routines for auditing drawing versus database square footage totals.
- Depict cables and connections in a drawing using GIS network analysis techniques.
- Provide the ability to drill between various levels of regional maps to building and related floor drawings.
- Support relocation planning and move operations.
- Provide parcel maps for properties recorded in the database.
- Support client/server and Web-based publication and distribution of graphic queries of enterprise data on CAD drawings.
- Permit client/server, as well as Web-based, deployment.
- Generate complete floor plan layouts that include organizational adjacency and travel time requirements established by enterprise planners.
- Support the generation and publication of thematic maps used in spatial analysis processes.
- Exploit other GIS functionality for specific niche requirements inside the walls of the buildings.
- Expand the facility adjacency relationships through topology inside the walls of the building.

In summary, the view from the top is that corporate infrastructure resource management without current location data is just a budget total without line items. The Internet changes everything because mission-critical information can be distributed anywhere at any time *and* can be received anywhere at any time.

ARTICLE 3.3.11
HUMAN RESOURCES

Matthew C. Adams, P.E., APPA
Adams Consulting Group, Atlanta, Georgia

The range of functions fulfilled by human resources (HR) professionals is a key component in successfully managing any organization. New concepts are expanding the HR role from that of administration to a source of administrative expertise for the overall functioning of the organization.

Traditional roles met by HR departments include policy and regulatory enforcement, processing hiring and firing procedures, managing the paperwork aspects of benefits programs, and administering compensation decisions. Particularly empowered and proactive departments also may manage recruiting, training, and employee development programs, as well as design workplace diversity initiatives and the like.

Most of these procedures are disconnected from the real work of the organization. However, as the perceived ability of the HR function to positively affect actual operations increases, HR professionals are becoming more involved in processes that affect outcomes, such as creating economic value for owners, creating product or service value for clients, or creating workplace value for employees. New areas for participation include partnering with senior managers and line managers to execute strategy, to organize work more efficiently to reduce costs while maintaining quality, to champion employee concerns and increase employee contribution, and to serve as agents of continuous transformation to improve the organization's ability to change in a rapidly changing world.

In a world that is continually transforming itself at increasing speeds, HR can provide a key competitive advantage by contributing to organizational strength. Qualities such as speed, responsiveness, learning ability, agility, and employee skills will make the difference between success and failure.

STRATEGIC VALUE

By defining organizational architecture, HR can identify the organization's underlying way of doing business, compare it with models of the way things should be accomplished, and work to bridge any gaps that appear. By enabling the organization to clearly see how it functions in terms of strategy, structure, systems, style, skills, and shared values, HR can play a partnership role in ensuring the attainment of strategic goals.

To remedy perceived deficiencies in organizational architecture, HR can propose and create models for best practices to implement programs for culture change that will bring state-of-the-art approaches into play to create more effective structures.

Human resources should be able to assess its traditional activities, including employee initiatives such as pay-for-performance and action-learning experiences, to determine which contribute most to strategic accomplishment. This approach will enable HR to add value to the overall organization by becoming a strategic partner in a business sense.

ADMINISTRATIVE EXPERTISE

Rather than serving merely as rule enforcement police, HR professionals can move from an administrative role to a new capacity as administrative experts in which they help improve the efficiency of both their own departments and that of the entire organization. Processes should be identified that can be accomplished better, faster, and less expensively. Technology can be used in unexpected ways for scanning resumes of applicants and creating databanks of prospective personnel or for developing electronic bulletin boards that enable employees to communicate with senior managers. Human resource professionals throughout the organization can make the same contributions. Systems can be developed to create interdepartmental communication or to enable sharing of services, for example. A center of expertise can coordinate and share important information regarding organizational processes. Such achievements will cut costs and improve efficiency.

EMPLOYEE CHAMPIONS

Traditionally, HR people have attempted to ensure the loyalty and engagement of employees by meeting their social needs and improving family benefit programs. New techniques for creating value for employees include orienting and training staff about the value of high morale, serving as the employees' voice in management discussions, creating opportunities for personal and professional growth, and enabling employees to meet increased demands that may be placed on them.

By proactively identifying causes of low employee morale, for example, HR can help management develop and implement programs to counteract negative trends in productivity. By clarifying goals and providing enabling tools, HR can play a critical role in improving morale problems.

AGENTS OF CHANGE

An organization's ability to adapt, learn, and act quickly can make all the difference in a changing world. Human resources can play a strong role in contributing to an organization's ability to embrace and benefit from change.

By focusing on initiatives such as creating high-performance teams, reducing cycle time for altering procedures, and implementing new techniques, HR can facilitate change. The new HR could also ensure that mission statements are clear and reflect new realities and help to transform those missions into specific workplace behavior that contributes to their attainment.

Human resources can even take the initiative to create a change model, a managerial tool that enables the organization to identify key success factors for change and then assess strengths and weaknesses in meeting each factor. With this information, management can design programs to increase success, which can be monitored by HR to assess their effectiveness.

To facilitate change in culture, HR has four processes at its command that will facilitate success. Initially, the concept of the culture change itself must be defined. Second, the reasons for the necessity for the change must be made clear to all managers and employees. Third, HR should establish ways to assess current and desired cultures to measure the gap. Finally, creative approaches should be developed to close the gap to create the desired culture change.

HUMAN RESOURCES' MANDATE

Increasingly, HR professionals are being required to concentrate on the deliverables rather than on the traditional processes of their work. By focusing on value created, the HR function will develop and implement mechanisms that produce measurable effectiveness in terms of organizational strengths and cultural transformation.

These increased demands can be met if management makes the corresponding investment in the HR function. It must be understood by all that such "soft issues" do matter and that structure and operating paradigms contain value. Without this recognition, strategies become mere hopes rather than realities, and concepts outweigh results.

Management must also hold HR accountable for results in measurable terms. By tracking, measuring, and rewarding performance, management can expect substantially greater value from the HR function.

Investment in innovative HR practices is becoming the norm. Increased communication facilitates awareness of successful practices that can be implemented with appropriate adaptations to an organization's unique structure and mission. The tools, information, and processes are available. They need only be acquired and used.

The professionalism of HR personnel can be upgraded as necessary through continued training and by hiring experts in particular fields. Human resources professionals should understand their organization's purpose, the theory and practice of the HR function, have credibility, and be able to contribute to organizational change. Administrative expertise for facilitating change requires some knowledge of reengineering. With all of these skills, HR professionals will be in a position to effect real and meaningful change that will significantly contribute to an organization's success.

ARTICLE 3.3.12

USING CUSTOMER SURVEYS TO IMPROVE FACILITIES COMMUNICATIONS AND PERFORMANCE

William C. Ronco, Ph.D.
Gathering Pace, Inc., Bedford, Massachusetts

Sheila Sheridan, C.F.M., C.P.M.
Harvard University John F. Kennedy School of Government, Cambridge, Massachusetts

"Why survey customers?" many facility managers challenge.

"We already know what they think," many contend.

"And if we don't, we're not so sure we want to find out," others worry.

Actually, customer satisfaction surveys provide facility managers with an inexpensive, effective tool to improve communications and performance.

However, not all surveys provide reliable data. If amateur survey users are not careful, it is all too possible to invest more meaning into survey results than the tool warrants. Worse, it is easy to misinterpret and misuse survey data. Several key steps ensure that surveys provide facility managers with useful data.

COMMUNICATIONS MISTAKES

Understanding the mistakes facility managers make in communications provides a foundation for understanding the most important communications tools and the need to use them. In addition to the passivity described before, facility managers make three other recurring mistakes when it comes to communications:

1. *Thinking you know what most customers think.* Facility managers explain, "My phone is ringing all the time. Believe me, I know what my customers want." Responding to a ringing phone can tell you what some customers want, but these are usually just a vocal minority, and they often do not represent the sentiments of the quieter majority.

2. *Thinking customers understand what you do.* "How can customers not understand what we do?" facility managers wonder. "We've been here longer than they have." Yet, time after time, facility managers discover that their customers have different expectations, goals, and methods than they do.

3. *Thinking customers understand how best to work with you to add value.* "It's their business," facility managers complain. "How can we tell them how to run it?" Yet, nearly every facility manager knows that customers use the department in reactive, last-minute ways that limit long-term effectiveness and productivity.

These recurring mistakes define the needs that facility communications tools must address. The tools include the following:

- Customer satisfaction surveys
- Internal marketing materials
- Written work plans
- E-mail
- Groupware
- Facility manager

Customer Satisfaction Surveys

Customer satisfaction surveys improve communications coming into the facilities department by providing an accurate picture of the way the overall customer base feels. Survey data are important because they are more accurate, more complete, and more objective than the occasional unsolicited praise or complaint letters that come in. Surveys also provide more specificity, more clarity for action improvements, and clear benchmarks for improvement and charting progress. Figure 3.3.12-1 is an example of a facility management customer survey identifying issues and a means to rate customers' satisfaction.

Not all surveys are useful, however, so it is important to distinguish what constitutes a valid survey:

- *A response of 100 percent or a valid sample.* Basing action conclusions on a 50, 60, or even a 70 percent response rate is misleading. A 100 percent response on a smaller but representative sample provides a more accurate picture of overall sentiment.

- *A tool that distinguishes different levels of sentiment.* "Yes" and "No" responses do not provide the information necessary to take action. The difference between a "Strongly Agree"

The Facility Department:	Strongly Agree	Agree	Slightly Sgree	Slightly Disagree	Disagree	Strongly	Not Applicable	Priority		
								High	Medium	Low
Technical										
1. Has strong technical skills.										
2. Provides innovative solutions.										
3. Solves problems effectively.										
4. Provides a high level of technical service.										
Business value										
5. Tries to understand customer priorities.										
6. Responds to customer business priorities.										
7. Understands what adds value.										
8. Provides solutions that add value.										
Communications										
9. Provides adequate information about what they do and how they work.										
10. Works effectively with customer input.										
11. Provide adequate information while working on projects.										
12. Provides adequate opportunities for customer input during projects.										
13. Listens effectively.										
14. Treats customer interests with respect.										
15. Overall partners effectively.										

16. Things the Facility Department
    does well   Could improve

17. Things I would like to see the Facility Department do:

FIGURE 3.3.12-1   Facility management customer survey—a sample survey draft for comment.

and "Agree" response is a difference that facility managers need to know about and respond to.

- *A one-page form.* Response rates dive after the first page.
- *Both closed- and open-ended questions.* Both kinds of questions provide useful data.

ARTICLE 3.3.13

TOTAL COST OF OWNERSHIP

Stephen K. Lowe
Art Center College of Design, Pasadena, California, and Facility Information Systems, Inc., Camarillo, California

Any facility planner will tell you that the initial design and construction investment in a building is just the tip of the iceberg. During the life cycle of a large enterprise, expenditures for operations, space renovation, occupant relocation, and an abundance of ongoing asset resource budget items quickly eclipse initial construction costs. Facility life cycle economics starts with early planning tasks, continues through all operations, and ultimately ends with disposal issues.

For many of the largest, best-run organizations, facility management (FM) and infrastructure management (IM) are helping to ensure that the capital value of facility information is sustained over time. Moreover, it is critical that enterprise organizations continue to raise the level of understanding and use of financial impact as a key component for infrastructure decision models. Return-on-investment (ROI) evaluation for potential structure, infrastructure, and localized space investments all require real-world tools for measuring cost and estimating value.

The analysis of life cycle cost, which had its roots in the energy crisis of the early 1970s, represents a paradigm shift away from first-costs analysis. For example, before life cycle costing, a facility manager might have considered only the up-front costs of a backup diesel generator, while ignoring the ongoing costs of fuel, maintenance, and other associated infrastructure costs—such as local availability of trained repair personnel.

Today, there is still much discussion about quantifying *all* costs of procurement, operation, maintenance, and even disposal of systems for comparison. Only since the convergence of fast computer hardware and truly scalable database engines have we experienced the horsepower required to calculate the realistic costs of systems ownership for entire enterprises (Figure 3.3.13-1).

- *Distant past.* Workspace was "just another fixed cost."
- *Recent.* Space control reduced costs.
- *Current.* Proactive space management is mission-critical.

The basic principles underlying life cycle cost calculations are the same in all instances, but costs must always be calculated separately for each specific case. The actual comparison values are ideally drawn from current organizational data across the entire enterprise, chosen for relevance to the life cycle of the facility and its subsystems.

In general, costs are calculated at the discounted present value of the cost of components and services, measured in constant dollars for a selected time period, such as 15 to 20 years.

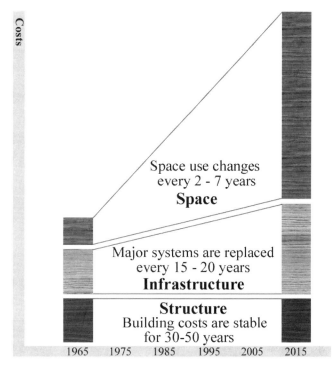

FIGURE 3.3.13-1   The life cycle costs of a facility.

For example, such time horizon and discount rates are commonly used in planning for the supply of electricity.

For simplicity, a linear depreciation schedule is often assumed, so that the residual value at the end of 15, 30, or 50 years is based on the fractional remaining life of the buildings and equipment, and so on. For very large footprint enterprises and federal government organizations, tens to hundreds of millions of dollars are at stake annually and the real cost of asset ownership soars well above initial acquisition costs.

Life cycle facility costing emphasizes ongoing operational and maintenance costs for all organizational systems, purposefully including items such as infrastructure expenditures that may be required for training maintenance and repair crews, and the like, which are often neglected in planning analyses. Costs in this category might be based on the wages and travel costs of contractors or agents who visit the organization, the costs of outside system operators and maintenance personnel, as well as the associated costs of hardware/software upgrades and other equipment. On a 15- to 20-year basis, future costs come into play, and it is often mistakenly assumed that these costs will remain unchanged in constant dollars over time.

Today's highly scalable database engines make the review of the entire facility life cycle—for equipment procurement, operational and maintenance costs, as well as diverse and disparate infrastructural requirements—reasonable now. Everything from installation of special elevators for delivering maintenance equipment, training for operations and maintenance personnel, and much more can now be included in the life cycle analysis for high-level strategic planning to be meaningful to various divisions across the enterprise.

Executives in strategic planning, real estate finance, program and property management, as well as human resources (HR), may gain numerous significant advantages from centralized, scalable software that molds and manages infrastructure delivery.

For large-footprint implementations, software and especially database scaling issues are probably going to be among the top FM technology issues as organizations merge and acquire each other in the years to come. Life cycle cost analyses will tend to require accounting for costs associated with failed pilot FM projects, which are often—especially at very large footprint organizations—directly associated with scaling issues. Facility planners, CIOs, CFOs, VPs of real estate development, and strategic planning executives should embrace scaling as clearly as possible before embarking on pilot implementations. Pilot projects should incorporate real-world growth estimates based on the future expectations of the organization.

Enterprise-wide software solutions may significantly extend traditional areas of FM to include real estate and property management, HR, call center systems, asset management, and computerized maintenance management systems (CMMS). Therefore, the long-term volume of information and ongoing demands on the consolidated data warehouse should be calculated to ensure consistent operation over the full life cycle of the automation system.

The following are examples of scaling-related details to consider for large-footprint FM data consolidation pilot projects:

- How many square feet will be cataloged in the pilot implementation?
- How many square feet are envisioned for the enterprise-wide implementation, assuming a successful pilot project?
- At what rate will additional facility drawings and data be added to the pilot project to complete the enterprise-wide implementation?
- How many people will be cataloged in the pilot versus enterprise-wide implementation?
- Will component technologies or methods of implementation used in the pilot project limit cataloging or differ in any way when cataloging resources are in full-scale implementation?

For planning at the Fortune 500 or federal government levels, planners should assume that virtually unlimited amounts of enterprise component data will be fed into a single database. Measuring costs to maintain, repair, support, upgrade, and manage enterprise properties—and associated buildings, floors, and spaces—is just the start. End uses of the facility by its workers and all ongoing operations to deliver products and services to customers must be considered. For large organizations, this will significantly increase loads on the database engine, and pilot projects should be based on full-load assumptions.

Additionally, at large-footprint organizations, commercial off-the-shelf technology (COTS) that integrates best-of-breed applications and all data sources for shared needs within disparate client organizations is often the specified solution. Life cycle economics requires that we calculate the ongoing costs associated with scaling, maintenance, training, and the like, along with first costs associated with implementation of these systems.

Enterprise resource planning (ERP) is evolving to include the interplay of physical assets, by location, to enable their effective use by the workforce because control of asset deployment to people in workspaces connects planning to productivity. Life cycle costing requires that planners embrace the long view, which will ultimately require unbreakably secure and scalable database engines, fully integrated with industry standard spatial data [computer-aided design (CAD) drawings and geographic information system (GIS) mapping] engines, for client/server, as well as fully Web-enabled implementation.

Facility management executives at many of the largest and best-run organizations are already fully immersed in effective life cycle decision making based on ongoing, long-term strategies for process improvement. Many strategic planners already expect access to mission-critical information at any time and from any location and the ability to view reports the way they want to see them, drawn from common data available across disparate departments, regions, and even time zones.

As the role of facility manager evolves to include meeting the demands of strategic planners adapting to rapid market changes, reliable methodologies for evaluation and review may be the only compass in a sea of organizational turmoil. The ability to view current data in real time—and to display barriers, obstacles, and relationships over time—may be the difference between short-term success and long-term failure of a project, or even an entire organization.

Whether the task is determining the number of light bulbs required to illuminate the executive washroom or analyzing FM process and performance gaps, life cycle cost analysis will be a useful tool to help enhance and integrate business process improvements.

SECTION 3.4
NEW DEVELOPMENTS IN OPERATIONS MANAGEMENT

Kristin Hill, Cindy Aiguier, John Morganti, and Bonnie Seaberg
International Facility Management Association (IFMA),
Boston, Massachusetts

Norm Faucher
Association for Facilities Engineering (AFE), Boston, Massachusetts

A well-managed facility can be likened to a well-oiled machine. If properly maintained, it will provide years of faithful service. Facility management becomes a natural operating mechanism for the facility. However, this is not enough. To simply "maintain" the facility is a stale formula for facilities management. A maintaining mind-set carries the risk of obsolescence. One cannot simply adopt an "if it is not broken, don't fix it," mentality.

The following articles on operations management will help you to understand all of the processes that occur within a facility. We begin by defining the processes that are conducted in the facility and seeing how to improve their effectiveness and efficiency. Next we look at identifying one process that shows that processes begin with understanding customer needs and wants and end with delivering products or services to the customer. The remaining articles identify several ways to improve the operations of the processes within the facility.

The new dictum for facility managers in the twenty-first century is that they must be visionaries. Facility managers must have the ability to foresee what is going to or likely to happen. Without vision, facility managers are living only for today. With vision, they are ready for change and are also actively making and directing it. Therefore, they will avoid having to perform costly damage control by playing catch-up with the future (now the past) after having invested time, money, and resources in maintaining a facility with old practices.

ARTICLE 3.4.1

OVERVIEW OF PROCESS DESIGN IN FACILITIES MANAGEMENT

Charles R. Harrell

ProModel Corporation, Orem, Utah

A facility houses the equipment and personnel that carry out the business processes of an enterprise. It only makes sense that the facility should be designed with the processes in mind. The layout of the facility can have a significant impact on the ability of the enterprise to execute its business processes effectively. This article discusses the role of process design in facilities planning. First, principles and concepts of process design are presented. Second, the relationship between process design and facility design is discussed. Finally, evaluation techniques, particularly computer simulation, are presented that are used to assess the effectiveness of a process design.

WHAT IS A PROCESS?

Before discussing the role of process design in facilities planning, it is helpful to have a clear understanding of the term *process*. A process is defined as a sequence of activities that converts inputs into outputs.[1] A process might be a sequence of production activities that convert raw materials into finished products. In the service sector, a process might be a set of activities required to treat patients in a health care facility.

An integral aspect of a process is the *system* used to carry out the process. A system is defined as a collection of elements that function together to achieve a common goal.[2] Thus, systems encompass processes but also include the resources and controls for carrying out processes. In the initial phase of process design, the focus is often on *what* activities need to be performed. In the final phases of process design, focus is turned toward the system details defining *how, when, where,* and *who.* Ultimately, any implementation of a process is an implementation of the system for carrying out the process. In manufacturing processes, the focus turns toward the machines, operators, and handling equipment used to convert raw parts into finished products. In a service system, such as an emergency room of a hospital, the processing system consists of the nurses, doctors, and equipment used to treat incoming patients.

PROCESS DESIGN

Process design considers the activities that are performed in the facility as well as the system elements that are needed to execute the process. Process design in this larger context is sometimes referred to as *systems engineering,* which can be defined as:

> The effective application of scientific and engineering efforts to transform an operational need into a defined system configuration through the top-down iterative process of requirements definition, functional analysis, synthesis, optimization, design, test and evaluation.[3]

Process design is essentially an iterative process of defining objectives, identifying requirements to meet those objectives, specifying a solution that fulfills the requirements and then evaluating the effectiveness of the solution in meeting overall objectives (see Fig. 3.4.1-1).

In some situations a specific process cannot be completely determined in advance. Job shops, for example, utilize general-purpose equipment that is capable of processing a broad range of products. In this situation, detailed process plans are not generated until an order for a product is actually received. In the case of job shops, the general range of processes anticipated is considered when selecting equipment.

For situations in which processing requirements are known in advance, the basic procedure for coming up with the best specification includes the following steps:

1. Define the overall process.

2. Select applicable processing methods and equipment.

3. Develop a detailed process plan.

4. Evaluate the effectiveness of the plan.

Although these steps are more easily applied when designing a new process for a "green field" facility, they are applicable in principle for improving processes in existing "brown field" facilities. Each of these steps in process design is discussed briefly.

Defining the Overall Process

The first step in process design is to determine the overall process that is best suited for meeting the processing requirements. Processing requirements are driven primarily by (1) the number of different types of products to manufacture or customers to service, (2) product specifications for manufacturing or, in the case of service industries, customer needs, and (3) production or service volumes. In examining the specifications for the finished output, planners should look at the product specifications, standard practices, and any other related documented specifications. Production forecasts showing anticipated production volumes over time also impact the process.

A flowchart is often used in the initial stages of process design to document the required processing steps as dictated by the processing requirements (see Fig. 3.4.1-2). At this stage, the focus is only on *what* activities should be performed and the appropriate sequence.

A flowchart is a visual way to illustrate processing logic and provides a general overview of the process. In its simplest form, it consists of boxes representing the activities and connecting arrows showing their sequential relationship. Symbols can also be used that more specifically characterize the nature of each activity. For example, a set of flowcharting symbols has been defined by the American Society of Mechanical Engineers (ASME) as shown in Table 3.4.1-1

Once a flowchart is constructed, analysis of the process can begin. For simple process improvements, traditional work analysis methods long prescribed by industrial engineering are useful. The most basic approach is to ask the following questions:

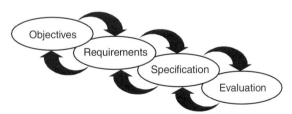

FIGURE 3.4.1-1 Four-phase iterative approach to process design.[4]

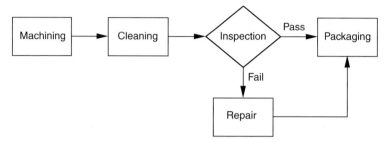

FIGURE 3.4.1-2   Sample process flowchart.

- Can any activity be simplified?
- Can any activity be eliminated?
- Can any activities be combined?

One approach used in process improvement initiatives is called *process value analysis.* This method is basically a systematic way of analyzing the costs and value associated with each step. Process steps or activities that add little or no value to the product (in the eyes of the customer) become candidates for elimination. Process value analysis includes the following steps:

1. Identify possible methods for meeting processing requirements.
2. Evaluate the effectiveness of alternative methods in meeting requirements.
3. Select the methods that meet the requirements at the least cost.

In conducting a process value analysis, the focus should be on essential or value-added activities, those for which the customer is willing to pay. Any activity related to inspection, revision, or rework should be evaluated for possible elimination. It is also useful to look at activities that are optional or occur only occasionally to determine whether there is a way to combine them with other activities so that a common, single flow can occur. The result of doing a process value analysis might be a list of processes needed and perhaps a rough process flow depicting the sequence of activities performed.

To ensure that the resulting process that is defined meets stated objectives and needs, it is recommended that the process flow be reviewed with appropriate managers, stakeholders, customers, and users. Often, more goes on inside a process than a single individual can assess. It always helps to get feedback and suggestions from others.

TABLE 3.4.1-1   Standard Process Flow Symbols

Symbol	Description
○	Operation
▷	Transportation
▽	Storage
D	Delay
☐	Inspection
⬙	Combination operation and inspection

Method and Equipment Selection

Once the overall process flow has been determined, appropriate methods and equipment are selected to carry out the process. This step in process planning requires knowledge of work practices and available relevant technologies. Method and equipment selection should follow sound engineering practices and may even utilize computer-aided process planning (CAPP) software.

Not only should the appropriate resources be selected for performing each job or activity, but the number of resources needs to be estimated. The capacity of a resource is a measure of the number of items or customers it can process in a given period of time, taking into account factors such as availability, efficiency, and so forth. This adjusted capacity is known as the *effective capacity* of the resource. The number of resources required is determined by dividing the required processing volume by the effective capacity of the resource. The required processing volume is typically determined by the *maximum* number of items or customers that must be produced in any given time period.

When only a single type of product is produced or customer serviced, capacity analysis is straightforward. If 20,000 units of a product are required per year and the effective capacity of the resource is 8,000 units, then the number of resources required is 20,000/8000, or 2.5 resources. Because a half resource is usually not possible, three resources are required (unless, of course, overtime is planned to handle excess requirements).

Sometimes capacity planning must be based on a mix of products or customers. Resource sizing in this scenario is based on the combined processing requirements. Suppose, for example, that a machine has the capacity to produce 20,000 units of product A per year and 25,000 units of product B. If the required annual production is 8000 units of product A and 10,000 units of product B, then the load on the machine is (8000/20,000 + 10,000/25,000) × 100 percent, or 80 percent.

When analyzing capacity requirements, it is important to recognize that basing the number of resources on effective capacity does not take into account the interaction of the resource with other resources in the system. This interaction often makes the actual capacity less than the effective capacity. Elements in a system are interdependent, so nothing happens without affecting everything else in the system—like a domino effect. A machine failure at one workstation starves downstream operations and may even create a backup or logjam that halts upstream operations. Another type of interdependency in systems is due to shared resources. Resources are frequently limited and must respond to competing demands for their services. An operator loading one machine in a cell, for example, may be unable to respond immediately to another part waiting to be unloaded from some other machine in the cell.

System interdependency may be tight or loose, depending on how tightly linked the elements are. Elements that are tightly coupled have a greater impact on system operation and performance than elements that are only loosely linked. When an element such as a worker or machine is delayed in a tightly coupled system, the impact is immediately felt by other elements in the system. In a loosely coupled system, any disruption of an activity would have only a minor, delayed impact on other elements in the system. Several methods are used to decouple system elements, including provision for buffer inventories, designing in redundancies, and dedicating resources to single tasks. Unfortunately, these solutions often lead to excessive inventories and underutilized resources. Interdependencies in mixed-product manufacturing are particularly difficult to anticipate because bottlenecks may shift, depending on the current lot being run. For these and other reasons, the need for simulation to predict system performance accurately becomes important.

Developing a Detailed Process Plan

In manufacturing industries, a process plan is often documented in the form of a *route sheet, process chart, operation list, manufacturing data sheet,* and so forth. This process document specifies the sequence of operations for a part, and alternative operations and routings are given wherever feasible. Other processing specifications that can be included on a route sheet are material requirements, machine tolerances, tools, jigs and fixtures required, setup time, and time allowances for each operation.[5]

A detailed process describes how the system will operate (who, what, where, when, and how) in terms of work schedules, processing quantities, resource usage, decision rules, and so forth. The operational specification is best defined from the perspective of the entity flow and should ultimately be extended to include movement in and out of queues, the time and resource requirements for each activity, and so forth. For a complete definition of a process, the following information should be specified for each part or customer at each stop or work-station:

- What activity is performed at the station?
- In what quantities is the activity performed?
- What resources are used to perform the activity?
- Where is the material drawn from and where is it placed afterward?
- To which station is the material sent next?
- In what quantities is the material moved?
- How is the material moved?
- What triggers the move (e.g., schedule, kanban, etc.)?

Process Evaluation

With a complete definition of the way the process is going to be executed, a realistic evaluation can be made of the plan's effectiveness. It isn't enough just to plan the system. What is important to know before the system is implemented is how well the system is going to work. We want to ensure that the system is capable of performing as expected. With the interdependencies and variability inherent in the system, this is not a trivial task. The questions to be answered include the following:

- Where are the bottlenecks?
- What is the expected resource utilization?
- What is the processing capacity of the system?
- How much waiting time can be expected?

Evaluating the effectiveness of a system is where tools such as simulation come into play. Simulation takes into account system interdependencies and variability and therefore accurately predicts how a particular system will operate in actual practice. It can help eliminate problems of overdesign, underdesign, and poor design.

PROCESS DESIGN AND FACILITY LAYOUT

If at all possible, the facility layout should not be done until the process design has been completed. The facility layout should support the flow of material and avoid unnecessary backtracking. The flow of material should be driven by the process design and be a primary consideration in arranging equipment within a facility.

Achieving good material and people flow in a facility is important—yet often neglected. Most facilities are designed with little thought to how objects will move about and whether there is backtracking, congestion points, or even collision points. As organizations and business grow, the impact of poor workflow planning magnifies and, if uncorrected, can become a source of problems in an operation. According to one manufacturing consultant, "Many U.S. plants have horrendous work flow patterns that have grown up over the years, over-laden with a cascade of traditional rules and procedures and systems that create more problems than they solve."[6] This observation is corroborated in an assessment made by a research group from Purdue University that states, "Despite years of effort at both the theoretical and applied levels, we still do not know how to manage the overall flow of work in factories."[7]

In planning the layout of a facility it is important first to have an overall system perspective of material flow. The overall part flow within a manufacturing facility, sometimes referred to as the *material flow system*, begins from the reception of material at the receiving dock and extends to shipping material from the facility (see Fig. 3.4.1-3).

Material flow is largely determined by the layout structure chosen for the system. Manufacturing systems are generally classified as one of the following six types (most of these categories also apply to service systems):

1. *Project shop.* Product produced one at a time remains stationary while equipment is brought to the product. Examples are aircraft manufacturing and patient surgery.

2. *Job shop.* Equipment is arranged by function so that different products produced in low volumes have different routings. Examples include sheet metal fabrication and medical laboratories.

3. *Cellular flow.* Equipment is arranged by process sequence that is common for a family of products produced in small volumes, so that flexible, one-piece flow is supported. Examples are small appliance manufacturing and circuit board assembly.

4. *Batch flow.* Equipment is arranged by process sequence that is common for a set of products produced in medium volumes. Examples include book printing, textile manufacturing, and loan application processing.

5. *Line flow.* Equipment is arranged by process sequence for a product with single-piece flow. Examples are automotive assembly lines and bank drive-thru services.

6. *Continuous flow.* Equipment is arranged by process sequence for continuous product flow. Examples include petrochemical manufacturing and paper mills.

This classification of processing systems does not imply that all systems fall neatly into one of these six categories. On the contrary, many systems have characteristics of more than one type. For example, mixed shops combine both functional and linear equipment structures at different stages in the manufacturing process. The two principal factors that dictate the type of production structure are *product volume* and *product variety*. Figure 3.4.1-4 shows the general applicable ranges of product volume and variety for each structure.

The implementation of the system structure is based largely on material flow. In job shops, machines are arranged according to function (i.e., machines that perform like processes are grouped together) because there is no single flow sequence. Different parts are assigned individual routings that often include backtracking. This type of layout is referred to as a *process layout.* In continuous flow, line flow, and batch flow systems, equipment is arranged according

FIGURE 3.4.1-3   The material flow system.

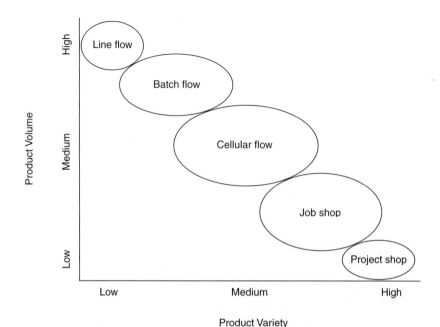

FIGURE 3.4.1-4 Range of application for different types of production systems.

to the processing sequence of a product. Therefore all flow follows essentially the same path. This type of layout is termed a *product layout*. In cellular manufacturing, machines are arranged to process parts of the same family. In cellular manufacturing, a cell usually consisting of less than 10 stations is arranged in the shape of a "U" to provide a common input/output point and enables a single worker to operate the cell. Figure 3.4.1-5 illustrates these three different types of arrangements and their implications for material flow.

PROCESS SIMULATION

Computer simulation is a technique that is frequently used to evaluate the effectiveness of a process design before it is implemented. Simulation itself does not *solve* problems, but it does clearly identify problems and quantitatively evaluate alternative solutions. As a tool for doing what-if analysis, simulation can provide a quantitative assessment of any number of proposed solutions to quickly home in on the best solution. Simulation also provides graphical animation to show the dynamic operation for the system (see Fig. 3.4.1-6).

By using a computer to model a system before it is built or to test operating policies before they are actually implemented, many of the pitfalls that are often encountered in the start-up of a new system can be avoided. Improvements that previously took months and even years of fine-tuning to achieve can be attained through computer simulation in a matter of days or only hours. Modern simulation tools even provide automatic optimization capability that iteratively runs models and changes model parameter settings until the best possible solution is found.

Simulation during the design phase of a system results in cost savings by identifying and eliminating unforeseen problems and inefficiencies. Cost is also reduced by eliminating overdesign and removing excessive safety factors that are added when performance projections are uncertain. It is not uncommon for companies to report hundreds of thousands of dollars in savings on a single project through the use of simulation. One Fortune 500 company

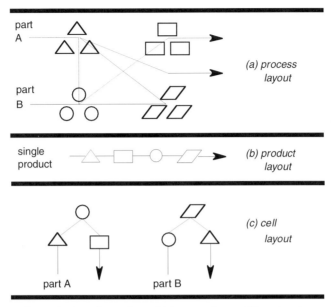

FIGURE 3.4.1-5   Comparison of (*a*) process layout, (*b*) product layout, and (*c*) cell layout.

was designing a facility for producing and storing subassemblies in preparation for the final assembly of a number of different metal products. One of the decisions involved determining the number of containers required for holding the subassemblies. It was initially felt that 3000 containers were needed to handle the activity. However, after a simulation study it was evident that throughput did not significantly change when the number of containers varied between 2250 and 3000. By purchasing 2500 containers instead of 3000, a saving of $528,375 was expected in the first year, with annual savings thereafter of more than $200,000 due to the savings in floor space from having 750 fewer containers.[6]

Even if dramatic savings or improvements are not realized each time a model is built, simulation at least inspires confidence that a particular system design is capable of meeting required performance objectives, and thus minimizes the risk often associated with new start-ups. The economic benefits from gaining others' confidence was evidenced when an entrepreneur, who was attempting to secure bank financing to start a blanket factory, used a simulation model to show the feasibility of the proposed factory. Based on the processing times and equipment lists supplied by industry experts, the model showed that the output projections in the business plan were well within the capability of the proposed facility. Although unfamiliar with the blanket business, bank officials felt more secure in agreeing to support the venture.[7]

Often, simulation can help to achieve productivity improvements without the need to invest heavily in new technologies or facility expansions. By looking at the overall operation of the system in compressed time, long-standing problems such as bottlenecks, redundancies, and inefficiencies that previously went unnoticed start to become more apparent. Consider the following actual examples:

• GE Nuclear Energy was able to increase its output of highly specialized reactor parts by 80 percent. The cycle time required for producing each part was reduced by an average of 50 percent. These results were obtained by running a series of models, with each one solving production problems highlighted by the previous model.[8]

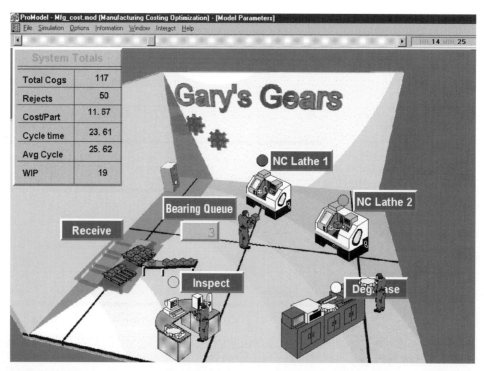

FIGURE 3.4.1-6 Simulation provides animation and performance statistics. (*Courtesy of PROMODEL Corporation.*)

- A large manufacturing company with stamping plants located throughout the world produced stamped aluminum and brass parts on order according to customer specifications. Each plant had from 20 to 50 stamping presses that were utilized at anywhere from 20 to 85 percent of capacity. A simulation study was conducted to experiment with possible ways of increasing capacity utilization. As a result of the study, machine utilization improved from an average of 37 percent to 60 percent.[9]
- A diagnostic radiology department in a community hospital was modeled to evaluate patient and staff scheduling and to assist in expansion planning over the next five years. Analysis using the simulation model enabled improvements in operating procedures that precluded the necessity for any major expansions in department size.[10]

The impact that simulation had on facilities management was significant in each of these examples because it completely eliminated the need for facility expansion by showing how increased capacity could be achieved using the existing facility.

Some of the specific questions that simulation can help address when designing or modifying a facility include the following:

1. What type and quantity of machines or workstations should be used?
2. What type and quantity of auxiliary equipment and operators are needed?
3. How much tooling and fixturing is required?
4. What is the production capability (throughput rate) of a given system?
5. What type and size of material handling system should be used?

6. What are the optimum number and size of storage areas and buffers?
7. What is the best layout of workstations?
8. What is the most effective control logic?
9. What is the optimum unit load size?
10. How effectively are resources being utilized?
11. What effect does a process or method change have on overall production?
12. How balanced is the work flow?
13. Where are the bottlenecks?
14. What is the impact of machine downtime on production (reliability analysis)?
15. What is the effect of setup time on production?
16. What is the effect of centralized versus localized storage?
17. What is the effect of vehicle or conveyor speed on part flow?
18. How many repair personnel are needed?
19. What is the overall effect of automating an operation?

Figure 3.4.1-7 illustrates the use of simulation to compare four conveyor system configurations simultaneously. After running the simulation, performance metrics such as throughput and cycle time were compared to determine the best configuration.

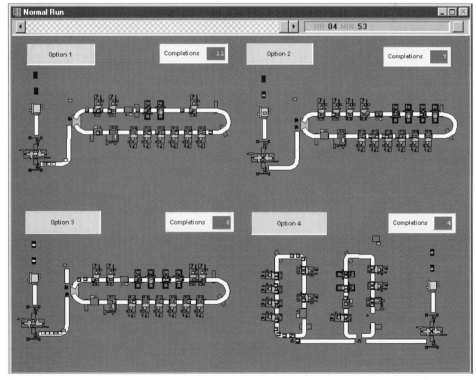

FIGURE 3.4.1-7 Simulating alternative configurations for comparison. (*Courtesy of PROMODEL Corporation.*)

Simulation is finding widespread application in manufacturing and also in service industries. Insurance agencies, health care facilities, airports, restaurants, and numerous other service facilities have all benefited from using dynamic simulation. Figure 3.4.1-8, for example, shows a radiology lab simulation where processing capacity and costs were evaluated to see if facilities were adequate to provide the best level of health care. The visualization of patient flow and staff movement provided an effective way to evaluate the layout for congestion in the waiting area and paths of travel for movement efficiency.

Even before a simulation run is made, the exercise of developing a simulation model alone is beneficial in ensuring that all of the operational issues of the system have been addressed. The philosopher Alfred North Whitehead observed, "We think in generalities, we live detail." This human tendency to overlook detail can lead to poor decisions that go undetected until implementation. Simulation forces one to think through the operational details of a system so they are not left to chance (simulation can work with bad information, but it can't work with insufficient information). The increased discipline that simulation brings to the design process is sometimes a sufficient reason itself for employing simulation. Often improvements present themselves solely as a result of going through the model building exercise—before any simulation run is made.

The benefits of simulation can be summarized as follows:

- Captures system interdependencies
- Accounts for variability in the system
- Is versatile enough to model any system
- Shows performance changes over time

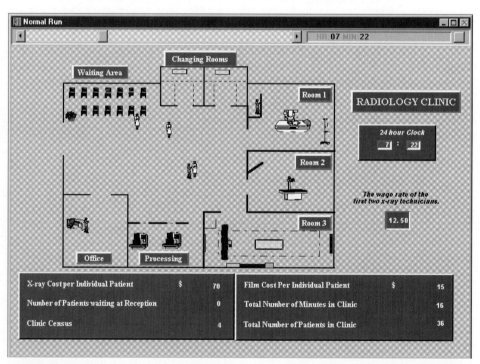

FIGURE 3.4.1-8   Simulating a radiology clinic reduces cost and improves performance. (*Courtesy of PRO-MODEL Corporation.*)

- Is less costly, less time-consuming, and less disruptive than experimenting on the actual system
- Provides information on multiple performance measures
- Is visually appealing and easy to understand
- Is easy to communicate results
- Runs in compressed, real, or even delayed time
- Stimulates interest and participation through animation
- Forces attention to detail in a design

SIMULATION TECHNOLOGY TRENDS

Simulation products have evolved to provide more than simply stand-alone simulation capability. Modern simulation products have open architectures based on component technology and standard data access methods (SQL, etc.) to provide interfacing capability with other applications such as CAD programs and enterprise planning tools. Surveys of the most popular simulation products show that most of them have the following features:

- Input data analysis for distribution fitting
- Point-and-click graphical user interface
- Reusable components and templates
- 2-D or 3-D animation
- Online help and tutorials
- Interactive debugging
- Automatic model generation
- Output analysis tools
- Optimization
- Open architecture and database connectivity

Simulation is a technology that will continue to evolve as related technologies improve and more time is devoted to developing the software. Products will become easier to use, and more intelligence will be incorporated into the software itself. Evidence of this trend can already be seen by optimization and other time-saving utilities that are appearing in simulation products. Animation and other graphical visualization techniques will continue to play an important role in simulation. As 3-D and other graphic technologies advance, these features will also be incorporated into simulation products.

Perhaps the most dramatic change in simulation will be in the area of software interoperability and technology integration. Historically, simulation has been viewed as a stand-alone, project-based technology. Simulation models were built to support an analysis project, to predict the performance of complex systems, and to select the best alternative from a few, well-defined alternatives. Typically these projects were time-consuming and expensive and relied heavily on the expertise of a simulation analyst or consultant. The models produced were generally "single-use" models that were discarded after the project. Now simulation is being integrated with other technologies such as facility layout software.

The trend to integrate simulation as an embedded component in enterprise applications is part of a larger development of software components that can be distributed over the Internet. This movement is being fueled by three emerging information technologies: (1) component technology that delivers true object orientation; (2) the Internet or World Wide Web (WWW), which connects business communities and industries; and (3) distributed computing standards such as Common Object Request Broker Architecture (CORBA) and Distributed Component Object Model (DCOM). These technologies promise to enable parallel and distributed model execution and provide a mechanism for maintaining distributed model repositories that can be shared by many modelers.[11] The interest in Web-based simulation appears likely to continue.

Dedicated simulation products for addressing specific types of industries continue to grow and become easier to use. This is the main factor driving the increase in the use of simulation. PROMODEL Corporation, for example, has products specifically dedicated to modeling manufacturing facilities (ProModel), health care organizations (MedModel), and general service industries such as banking and transporation (ServiceModel).

SUMMARY

Process design in facilities planning is crucial to having a well-running facility. Process design requires a clear understanding of the objectives of the operation in providing goods and services. Process value analysis can help ensure that the best process is designed for delivering value to the customer. This design must then be expanded to specify the methods and technologies that are to be used to execute the process. Finally, consideration can be given to the facility that houses the process. The design of the facility should support the goals of the process. Because of the complexity of processing systems, it becomes challenging to accurately assess the expected performance of a given process design. Process simulation is a technique used to evaluate process designs and improvement ideas. The power of simulation is its ability to account for interdependencies and to show graphical animation of the process or system model. Simulation turns a static facility plan into a dynamic true-to-life model that is a virtual representation of the way the facility will actually function.

REFERENCES

1. M. Hammer and J. Champy, *Reengineering the Corporation,* HarperCollins. New York, 1993.
2. B.S. Banchard, *System Engineering Management,* Wiley, New York, 1991.
3. Ibid.
4. C. Harrell and K. Tumay, *Simulation Made Easy,* IIE Press, Norcross, Georgia, 1994.
5. *APICS Dictionary,* 1980, p. 24.
6. A.M. Law and M.G. McCormas, "How Simulation Pays Off," *Manufacturing Engineering,* February 1988, 37–39.
7. C.R. Harrell, R.E. Bateman, T.J. Gogg, and J.R.A. Mott, *System Improvement Using Simulation,* PROMODEL Corporation, Orem, Utah, 1992.
8. Ibid.
9. W. Hancock, R. Dissen, and A. Merten, "An Example of Simulation to Improve Plant Productivity," *AIIE Transactions,* March, 1977, 2–10.
10. R.F. Perry and R.F. Baum, "Resource Allocation and Scheduling for a Radiology Department." In *Cost Control in Hospital,* Health Administration Press, Ann Arbor, Michigan, 1976.
11. P.A. Fishwick, "Web-Based Simulation," *Proceedings of the 1997 Winter Simulation Conference,* S. Andradottir, K.J. Healy, D.H. Withers, and B.L. Nelson (eds.), 100–109.

ARTICLE 3.4.2
SUPPLY CHAIN MANAGEMENT

James B. Rice, Jr.
Massachusetts Institute of Technology, Cambridge, Massachusetts

Jason Slibeck
Organic Inc., Cambridge, Massachusetts

The purpose of this article is to provide a basic understanding of the key concepts of supply chain management. Supply chain management is a popular topic, and it has a broad and growing impact on business.

This article also identifies relevant issues in applying supply chain management from the perspective of a facilities manager. In particular, the relevant supply chain management issues are primarily those within the four walls of the facility and in the execution of the business, supply chain, and operational strategies.

The parts of the supply chain that are outside the domain of the facility manager are integrating with external partners (customers and suppliers), developing new products, and designing the supply chain. These are important processes and tasks for the supply chain, although outside the scope of the facility manager. The facility manager cannot be entirely insular, however, in that any facility will need to receive materials; manufacturing facilities will also be shipping products. The facility manager will have to coordinate with suppliers and customers, although the company-wide relationship is typically managed by a central entity within the company.

Finally, this article suggests processes and potential solutions for pertinent supply chain issues and in some cases will suggest resources for developing solutions.

SUPPLY CHAIN MANAGEMENT IN DEFINED FACILITY PROCESSES

Supply chain management cuts across most of the nine other facility processes defined in this article and makes the distinction among these critical processes difficult to discern. By its very nature, supply chain management is cross-functional. Contrast this with the traditionally functional nature of business organizations and the overlap becomes clear. "Integrated Workplace Planning" represents concepts that are consistent with forward-thinking supply chain management and takes an enterprise-wide systemic approach to supply chain management to replace the traditional functional optimization that dominates most companies today.

As the reader will see in subsequent sections, our definition of the supply chain includes the manufacturing process. Accordingly, then, this will also include the quality management process. Although this is a critical deliverable of the supply chain, quality management and shop-floor management are best addressed by the organizations traditionally responsible for leading those efforts, the manufacturing organization.

Information management is addressed elsewhere and may have some redundancies with the information technology (IT) issues raised in this section. Although IT issues are addressed, the technical aspects of IT and supply chain management are not detailed.

FOUNDATION OF SUPPLY CHAIN MANAGEMENT AND ELEMENTS

In 1985, John Houlihan introduced the term *supply chain management* in an article in the *International Journal of Physical Distribution & Materials Management.*[1] Supply chain management has been an important concept for nearly 15 years, and the definition and understanding of it have evolved, but a consensus definition is still lacking. The supply chain has evolved from *physical distribution management* to *logistics,* and more recently, *demand chain* and *value chain.* Each has some distinction, but for the purposes of this writing, we focus on *logistics* and *supply chain.*

Definitions

The Council for Logistics Management (CLM) defines *logistics* as follows:

> Logistics is that part of the supply chain process that plans, implements, and controls the efficient, effective flow and storage of goods, services, and related information from the point of origin to the point of consumption in order to meet customers' requirements.[2]

This view of logistics accompanied by an understanding of the manufacturing process leads to an understanding of the broad scope that a supply chain approach can have.

Professor Bernard LaLonde of Ohio State University defines supply chain management as follows:

> The delivery of enhanced customer and economic value through synchronized management of the flow of physical goods and associated information from sourcing to consumption.[3]

Of particular note are the emphasis on enhancing value, synchronized flows, and the broad scope of the supply chain. Other definitions go further by including return processes (disposal, recycling), often called *reverse logistics.*

Using LaLonde's terminology, synchronizing flows creates economic value by coordinating and integrating the separate elements of the supply chain. Traditionally, companies delivered results through their various elements that "focused" on optimizing their respective functions. These are commonly called *silos* or *stovepipes.* Recognizing that this behavior results in suboptimal performance, many companies now place greater emphasis and resources on integrating the separate elements of their supply chain, internally and externally.

Scope of the Supply Chain—The Extended Enterprise™[4]

As suggested in LaLonde's definition, the scope of the supply chain is broad, from "dirt to dirt," as the saying goes. The supply chain scope is best understood by starting with the supply chain internal to the company.

Each company has its own internal supply chain that consists of planning, procurement, receiving, manufacturing, distribution, transportation, and customer service. (Some may also include engineering and product development.) These operations entail contact with suppliers and customers, but the main challenge is coordinating the flows across these different functions to respond to the external customers' demand. How this might look with the various linear flows along the processes is shown in Fig. 3.4.2-1.

Each company also has an external supply chain that consists of its suppliers and customers. In this case, however, the customers and suppliers equate to more than one supply chain. The notion of a single chain is misleading because modern organizations are more

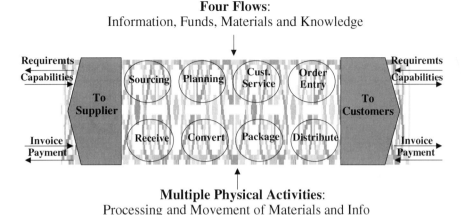

FIGURE 3.4.2-1   Process flow of supply chain management.

often a network of suppliers and customers, or processing nodes; the analogy of a chain directly highlights the dependency of each link on every other link. Companies are reaching out to their supplier's supplier, and to their customer's customer, extending their understanding and interaction further out into the supply chain in search of supply chain performance improvement and elimination of waste. The "supply chain" looks more like a "network" (see external supply chain in Fig. 3.4.2-2). This extended network captures the scope of the supply chain—from the base raw material supplier through all of the converting stages to the end consumer's use and disposal or recycling. Each company in the supply chain has its own internal supply chain not too dissimilar to that shown in the Fig. 3.4.2-1 (internal supply chain).

The definition and understanding of the supply chain will continue to evolve as more companies recognize the need to move beyond functional excellence and toward a process-oriented system-optimized business approach. Collectively, we define the supply chain as the set of processes that plan, source, produce, and deliver value in products or services to customers by managing four flows: physical goods, information, knowledge, and funds. The term

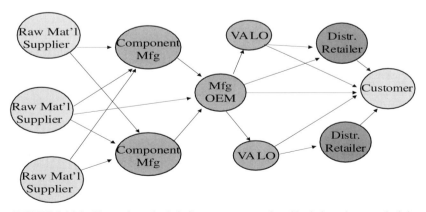

FIGURE 3.4.2-2   External supply chain flows across network and back through reverse logistics.

supply chain is likely to remain as the accepted term, but in the future most companies will be addressing issues of their *network* rather than their singular *chain.*

Background and History—Evolution to Corporate Silos and Independent Systems

Corporate silos and independent systems are the greatest achievements of the industrial revolution. This should not be a surprising statement if one remembers that dinosaurs were ideally adapted to the Jurassic period. In *Wealth of Nations,* Adam Smith analyzed the workings of a pin factory and introduced the world to the principle of division of labor.[5] This division was possible because the pin-making process could be deconstructed into specialized tasks. Advances in industrial technology facilitated labor and allowed one worker to do the work of many. Henry Ford refined the concept even further, and the modern industrial age was nearly complete. However, it was not until Alfred Sloan carried the division of labor concept from the factory floor to management bureaucracy that the subdivision of a process into functionally efficient units was complete. The creation of this industrial management system made the expansion of General Motors possible and ushered in an era of management by financial accounting goals and marketing objectives for functional business units with unique goals and performance measures.[6]

The era of unprecedented growth that fueled this migration to easily scalable hierarchical organizational pyramids was not able to withstand the information age and the three forces of the modern economy: customers, competition, and change.[7] The national mass markets of the 1950s, 1960s, and 1970s have been replaced by an empowered set of business and individual customers. These customers know what they want, what they want to pay, and which service level will accommodate their needs. This strategic shift in power is in part a result of easier access to an enormous amount of information for price comparison, quality judgment, and producer reliability.

Now, the national market forces also face stiff competition, as global trade has become a reality. International differences in factor costs such as labor rates have created enormous sociological changes. Competition has also come from entrepreneurial innovators. Entire sectors of the freight forwarding and distribution industry are now trying to expand the scope of their operations to justify continued existence as more nimble logistics providers force the question, "How do you add value for the customer?"

Finally, change has become a constant state of existence, and it is accelerating. To keep up with nimble global competitors and to react effectively to global market inputs, companies have accelerated the product development life cycle from years down to months. Having only a limited time frame within which to react, demand sensing and highly responsive, flawless execution systems have become critically important. Supply chain integration requires a reexamination of the assumptions of the industrial revolution and a move beyond functional excellence.

Before moving on, it is important to understand that silos have evolved to become detriments to optimal business performance. Internal to the company, the functional "silos" optimize functional performance at the expense of the broader enterprise. External to the company, the "company-focused" behavior of each of the companies in a supply chain results in excessive and redundant inventories carried by each company to protect its respective position, at the expense of higher overall supply chain costs for all.

Relevance of Supply Chain Management

One may ask, "What is the relevance of supply chain management to a facility manager?" Supply chain management has a significant and broad impact on the business. Managing the functions of a business as a customer-serving process—a supply chain—can impact business performance in the following ways:

- Creates a source of competitive advantage
- Satisfies a basic competitive necessity
- Augments financial performance

The importance of supply chain management stems from growing recognition that the supply chain may provide a source of competitive advantage (Dell, Chrysler, and Wal-Mart are good examples). Many others find that supply chain management has become a basic competitive requirement.

Financial Impact of the Supply Chain—Return on Assets

Measuring the financial impact of the supply chain on a business remains a fundamental method of determining the relevance of the supply chain. The financial impact of the supply chain can be understood and illustrated using the Du Pont model devised by F. Donaldson Brown around 1914, at the apex of the industrial revolution.[8] The Du Pont model in Fig. 3.4.2-3 shows the impact of a supply chain order-to-cash process and links with key financial indicators from the corporate balance sheet and income statement.

The model can also be used to show, for example, that a focus on electronic commerce enabled supply chains to impact return on assets (ROA) directly. The focus can be on increasing sales through personalization and customer loyalty, reducing fixed expense and fixed assets by moving from traditional bricks and mortar distribution to electronic storefronts, or reducing inventory and the cost of goods sold by focusing on improved forecast reliability and integrated purchasing systems.

When using the Du Pont model to measure the financial impact of the supply chain, consider the impacts across the scope of the entire supply chain, including all of the operations of the company. Increasingly, companies are finding that product development, process design, and engineering should also be included in the scope of the supply chain because these groups impact the ability of the organization to carry out the fulfillment process.

The use of the Du Pont model is primarily in measuring supply chain impact on costs, not revenues. When the supply chain provides a competitive advantage, it is legitimate to measure the increase in revenues through market share gains or price premiums. This remains an opportunity for many companies, especially when the company is considering making an investment in supply chain capabilities. Most companies must prepare a "business case for supply chain" to win the financial and personal support of the business leadership, yet these analyses often fall short because the true potential has not been adequately captured.

Issues and Solutions

To achieve the potential of the supply chain, several obstacles need to be overcome, including the following:

- Functional silo mentality and measures
- Mismatched incentives
- Requirement for new business capabilities
- Recognition of the requirement for fundamental change

These obstacles can be overcome by emphasizing the following key drivers and enablers of successful supply chain management.

Infrastructure and Organization Design

Too often companies look to the organization structure as the main way to "restructure" the supply chain. This falls far short of the need. According to recent studies, it is the most visible but perhaps not the most important element of organizational design.[9] These elements of organizational infrastructure—performance measures, compensation and incentive systems, communication systems, career development systems, training and education processes, tran-

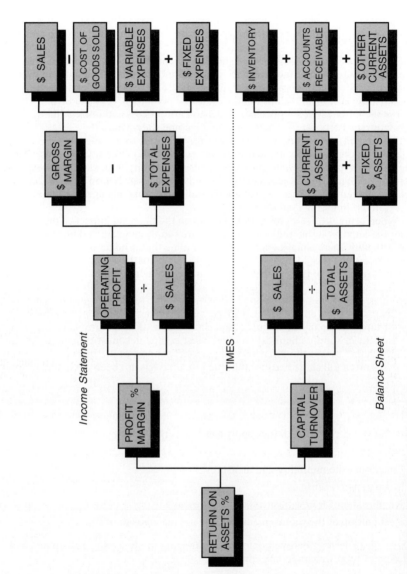

Income Statement

Balance Sheet

FIGURE 3.4.2-3 The Du Pont model.

sition planning, and performance appraisal systems—represent important design choices for delivering high-impact supply chain performance. Each of these gives the organization some leverage and control in establishing the business that has a strong "systems" orientation, instead of the traditional functional orientation. Unfortunately, many companies' performance measures and compensation systems provide incentive for individuals to optimize their function (department) at the expense of the overall business performance. Best approaches include designing each of these elements to support the overall system performance (of the business), not individual functional optimization (of a department).

Organizational structure should not be the primary and driving effort of a transition to process management. Many companies mistakenly depend on organizational structure as the cornerstone for their change efforts. There is a risk that organizational structure will become an obstacle, an end-all, or a focus for the organization, drawing it away from the real focus of the work. "Reorganization is insufficient to produce true integration and should be undertaken only as a natural result of redesigned business processes, and after individuals have internalized the new process objectives."[10]

Importantly, the supply chain infrastructure needs to support the supply chain strategy of the business. In turn, the supply chain strategy must equally align and support the corporate strategy. Corporate and supply chain strategies are not typically within the domain of the facility manager, but it is important to understand them.

Supply Chain Performance Measuring Systems

As the saying goes, "you won't get what you don't measure." To manage the supply chain processes, end-to-end measures must be developed—and actively used by the organization.

Congruent goals are required to align the different elements of the four aforementioned flows or horizontal processes. Congruent goals lead to measures that should be rewarded with incentives when attained. Measurement, goals, and incentives should be integrated when designing new systems. This should include the use of team-based measures for those on cross-functional supply chain teams.

Operational performance measures should change significantly when managing your company as a supply chain. They will shift from purely functional to a mix of functional and system performance that emphasize customer needs. System performance should be optimized, balancing efficient individual operations against system-wide performance measures such as capital deployed and overall service level. Some typical changes are offered as a reference for the facility manager:

- Measures such as "unit cost" and "labor cost per unit" should be augmented with measures such as "delivered on-time" or should be replaced by a balanced set of measures. Using "unit cost" and/or "labor cost per unit" as key measures often results in overproduction of finished product, additional warehousing and logistics costs, and increased product obsolescence. Often the inventory produced bears little or no relation to actual demand.

- The "volume produced, handled, delivered" measure should be replaced with more customer-specific performance measures such as "percent of orders delivered as promised, in proper condition, proper paperwork and as ordered" (perfect order). Using "volume produced" as a key measure often results in the production of products independent of customer needs. Such a production approach results in high markdown costs, high storage costs, and obsolete inventories.

- Equipment utilization measures are balanced against capital and asset productivity. Using equipment utilization as a key measure often results in the production of products and materials without demand or foreseeable demand, incurring additional handling and storage costs and burying quality problems Additionally, atypical performance measures should be considered to capture the impact of the supply chain. These might include inventory productivity, asset productivity, service-delivery performance, product cycle time, cash-to-cash cycle time, customer satisfaction measures, and responsiveness to demand fluctuation (system capability), among others.

Incentive Systems

As noted earlier, incentive systems should be part of an integrated system that involves the goals and measures of the organization. As such, the most important decision regarding incentives is to design them to be consistent with the organization's goals and measures. This includes compensating individuals and teams for team performance, both in functional and process performance, as well as compensating them for optimizing overall system (business) performance at the expense of individual function (department) optimization. This serves to achieve needed balance between the functional and process requirements for employees (at all levels).

Incentives play a key role in aligning the authority and the responsibility for the supply chain. Often, a central staff is assigned the responsibility of the supply chain, yet without any authority to exercise control or change. This "authority/responsibility" predicament is a common obstacle to optimizing business performance for the entire (internal) supply chain. Incentives would be provided for the cross-functional team that has the "shared authority" spanning the horizontal supply chain process.

IMPLEMENTATION/CHANGE MANAGEMENT

Implementing supply chain management takes many forms, from changing organizational infrastructure to implementing software for supply chain optimization. A central staff group often supports and coordinates implementation in a variety of roles, ranging from single-point leader to cross-functional leadership team facilitator.

Many companies use cross-functional teams as a structural element of the supply chain organization. The degree of formality of the team structure, ranging from ad hoc to informal to formal structured leadership team, depends upon the level of coordination and the company situation. No single design works for each situation.[11]

The process of change can take the form of an "evolution" or a "revolution." Evidence exists to support both approaches, although there is a growing belief that change from a revolution will come about only through a revolution due to a crisis. Alternatively, changing through evolution—or over an extended period of time—would require the development of organizational learning capabilities and system thinking.[12]

Information Systems

Information technology has revolutionized the world and the capabilities of the supply chain, but it is not the sole answer to implementing supply chain management. A company needs advanced supply chain management information systems that, preferably, integrate operations with demand and supply in a seamless system. This is a challenge for any business and entails a dedication of significant resources. Enterprise resource planning (ERP) systems promise this capability, although this is a relatively recent development. Individual supply chain software solutions offer best-of-class software capabilities in contrast with a single ERP solution for the entire business. Whichever the choice, these solutions should be selected on the basis of realistic goals and should also be part of an overall supply chain initiative that is integrated with infrastructure changes.

In practice, progressive companies use information systems to facilitate broad information availability and to place more emphasis on real-time information. As with the infrastructure choices that the facility manager makes, information systems should be designed to be consistent with overall desired level of communication and coordination, consistent with the supply chain strategy. Ford CIO James Yost underscores the role of technology in a recent quote: "My view is that technology is not the solution, but an enabler for changing business practices."[13]

SUPPLY CHAIN INTEGRATION

Supply chain integration seeks to achieve optimal performance of the supply chain as a whole by finding the appropriate balance of "focused" excellence and process coordination.[14] Successful supply chain integration requires connecting and coordinating the various flows (information, knowledge, materials, funds) and aligning incentives, measures, organizations, and goals across the internal supply chain and the external network.

Supply chain integration also requires that the coordinating and aligning change as a function of the product life cycle. Depending on the stage of the product life cycle, an integrated supply chain needs to deliver different capabilities to its partners. Early life-cycle supply chains have very uncertain demand profiles and therefore need flexibility (volume). Later life-cycle supply chains have more predictable volume demand but need lower cost and some product mix flexibility that comes with predictable product proliferation at the later stages of the product life cycle.[15]

Again, as suggested previously, to coordinate and align, no single solution works to integrate the supply chain in every case. The challenge is to take learnings from previous successes and reapply them *as appropriate* for your business situation. You still have to do the work.

Outsourcing to Restructuring the Supply Chain

The facility manager will not be tasked with supply chain design choices, but it is likely that the facility manager may wish to outsource some of the facility's needs. Outsourcing falls within the definition of supply chain design, but it is not always a centrally coordinated effort. Outsourcing has become an important aspect for many businesses because, in principle, it enables a company to focus its resources on the limited activities by which the company adds maximum value. It is a redistribution of the value-added tasks to the most efficient and most effective entities. Outsourcing has been widely used by large OEMs such as aircraft manufacturers, automobile manufacturers, and computer manufacturers.

Outsourcing can offer benefits in the following areas, depending on the situation: quality improvements, manufacturing cost reductions, faster cycle times, and rapid access to additional technology and capacity. At the same time, companies choose not to outsource to retain proprietary process and product technology or capabilities or to retain skills and capabilities that differentiate the company and its products in the marketplace.

In assessing whether to outsource tasks, processes, or functions, it is imperative to consider the long-term capabilities and technological impact of those decisions. Professor Charlie Fine of MIT highlighted the potential risk associated with outsourcing with his analysis of IBM's decisions to use outsourcing in the 1980s. While outsourcing operating system development (to Microsoft) and chip development (to Intel) seemed to make sense at time for IBM, it later became apparent that IBM had outsourced its future—resulting in a dramatic loss of market share, profits, and a relinquishing of its market dominance. The operating system and microprocessor later became more important than the hardware and system integration. Applying this learning suggests that it is worthwhile and important to consider the future implications of today's outsourcing decisions.[16]

The increase in the use of outsourcing is making it difficult to distinguish the lines that separate a company from its suppliers and customers, and the lines often change with some frequency. The nature of the buy-sell transaction is evolving from an "exchange" of value between parties into "the blur of fulfillment."[17] Examples of blurred distinction include cases

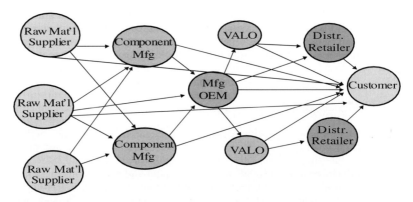

FIGURE 3.4.2-4 Future networked supply chain customer sourcing options increase. More buying options, electronically enabled.

where a manufacturer may have a third-party operator on company premises performing an 'outsourced' task, cases where the third party owns and operates their own equipment on company premises, cases where a third-party operator may take full responsibility for the entire customer service operation in a way that is transparent to the end customer, and cases where an entire facility and assets are outsourced to a third party. These practices are not uncommon even today, where customer service agents for airlines, manufacturers, and retailers are often employees of third-party service providers, unbeknown to the customer.

The capabilities of the Internet, the availability of real-time information, and the increasing potential of information systems integration have extended the potential of the outsourcing concept to actual restructure of the supply chain. These capabilities have enabled Dell Computer to sell directly to its customers, Chrysler to integrate and collaborate with module and component suppliers, and Cisco to integrate both customer and supplier in the product design process. "Disintermediation" (the process of eliminating existing intermediaries in the channel to the customer) represents at its core the redeployment of work, resources, and decision-making along the supply chain towards optimizing the entire extended supply chain "system" rather than the department or even company. In the future, the facility manager can expect to see an increasing application of these concepts, and should consider their potential application (benefits and risks) in day-to-day operations as well as planning for the future of the facility (see Fig. 3.4.2-4). Despite the potential created by technology advances, these are enablers and the success of potential restructuring will depend on careful application of outsourcing and restructuring, and the capabilities of the organization to implement and leverage the potential. This emphasizes all the more the importance of building a systems-oriented organization through organization infrastructure.[18]

REFERENCES

1. J.B. Houlihan, "International supply chain management," *International Journal of Physical Distribution & Materials Management* **17**(**2**): 51–56: (1985).

2. *Council of Logistics Management 1999 Year End Review,* Oakbrook, Illinois, 2000.

3. Bernard J. LaLonde, *Supply Chain Management Review* (Spring 1997).

4. "Extended Enterprise" is a registered trademark of the DaimlerChrysler Corporation.

5. A. Smith, *An Inquiry Into the Nature and Causes of the Wealth of Nations,* 1776, available at *www.geolib.pair.com/essays/smith.adam/woncont.html,* May 1999.

6. A.P. Sloan, Jr., *My Years With General Motors,* Doubleday, Garden City, New York, 1963.

7. M. Hammer, and J. Champy, *Reengineering the Corporation: A Manifesto for Business Revolution,* Harper Collins, New York, 1993, 19.

8. A.D. Chandler, Jr., T. McCraw, and R. Tedlow, *Management Past and Present: A Casebook on the History of American Business,* South-Western College Publishing, Cincinnati, Ohio, 1996, 3–73.

9. *Global Supply Chain Benchmarking and Best Practices Study—Phase I, KPMG Consulting,* New York, 1999.

10. F. Hewitt, *Proceedings of the Council of Logistics Management Annual Conference,* 1992.

11. James B. Rice, Jr., "Spanning the Functional Boundaries Through Horizontal Process Management," *Supply Chain Management Review,* 60–68 (Fall 1997).

12. P. Senge, *The Fifth Discipline,* Currency Doubleday, New York, 1990.

13. *CIO* Magazine, November 15, 1999.

14. M.M. Franciose, "Supply Chain Integration: Analysis Framework and Review of Recent Literature," M.S. Thesis. Massachusetts Institute of Technology, 1995.

15. Marshall L. Fisher, "What is the Right Supply Chain for Your Products?" *Harvard Business Review,* 105–116 (March–April 1997).

16. Charles H. Fine, *Clockspeed,* Perseus Books, Reading, Massachusetts, 1998.

17. S. Davis, and C. Meyer, *Blur: The Speed of Change in the Connected Economy,* Addison-Wesley, Reading, Massachusetts, 1998, 72.

18. For a set of recommended systems-oriented practices, see matrix on page 67 of Rice's "Spanning the Functional Boundaries Through Horizontal Process Management," *Supply Chain Management Review* (Fall 1997).

ARTICLE 3.4.3
QUALITY ASSURANCE PRACTICES

Paul R. Smith, P.E., M.B.A., P.M.P.
Peak Leadership Group, Boston, Massachusetts

To move toward achieving superior quality, it is necessary to have a strong, unified, and consistent application of process control practices. This information is a foundation upon which all process control systems should be built for quality to standardize process control practices in each area.

NATURE OF THE MANUFACTURING PROCESS

Ideally, every manufacturing process attempts to reproduce parts that are identical to each other in every characteristic. However, there are uncontrollable variations in conditions, materials, equipment, operators, and so forth that make each unit slightly different from the design target and from each other. These unit-to-unit differences usually follow a normal distribution, represented by a bell-shaped curve. The standard deviation of the distribution—sigma—is a measuring unit of the amount of variation. The value of sigma decreases as variations are reduced, resulting in a narrower distribution (see Fig. 3.4.3-1).

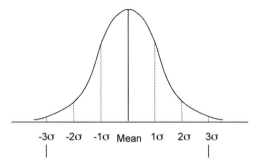

FIGURE 3.4.3-1 Normal distribution of process variation is represented by a bell-shaped curve. (*Courtesy of Peak Leadership Group, Boston, MA.*)

Traditional Process Capability

Traditionally, a process was judged satisfactory with a ±3-sigma capability. This means that if the process specification limits were placed on the process distribution curve, the upper specification limit would be at three sigmas to the right of the process average. The lower specification, likewise, would be at three sigmas to the left of the process average. The area under the curve between the two specification limits–99.73 percent of the total area—represents the products that conform to specifications. The area outside the specification limits—only 0.27 percent of the total area—represents nonconforming or out-of-specification products (see Fig. 3.4.3-2).

At first glance, a three-sigma process may look very good, but it is not good enough by today's standards. The seemingly low defect rate of 0.27 percent means that there will be at least 2,700 defective units for each million units produced. To put this in term's closer to the heart, the three-sigma capability would translate into more than 70,000 incorrect surgical operations each year in the United States. When it comes to health and safety, that rate of defects is obviously unacceptable. Moreover, process shift is an additional factor to be considered. Studied have shown that a process may shift up to 1.5 sigma from the usual process average. When it does, only 93.32 percent of the area under the curve is inside the specification limits. This equals a defect rate of more than 66,000 ppm. Now it is becoming clearer why the traditional three-sigma capability is not acceptable in today's marketplace (see Fig. 3.4.3-3).

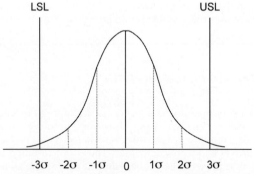

FIGURE 3.4.3-2 Traditional 3-sigma capability. (*Courtesy of Peak Leadership Group, Boston, MA.*)

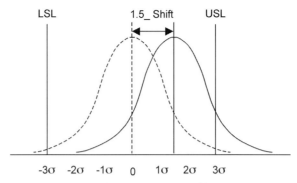

FIGURE 3.4.3-3   Three-sigma capability process with a typical shift. (*Courtesy of Peak Leadership Group, Boston, MA.*)

In reality, very few processes are perfectly centered and remain that way. Process shifts of up to 1.5 sigma are typical in many industries. When factoring in the typical 1.5 sigma shift in a Six-Sigma process, the result would be only 3.4 defects in every million units produced. This is a Six-Sigma goal (see Fig. 3.4.3-4).

AND BEYOND

Measurement Systems Capability

The capability of a measurement system must be ensured before the system is used for process control or for measurements used in capability studies. This prerequisite must be satisfied because the capability of the measurement system contributes significantly to the overall results. Too much variation in a measurement system may mask important variation in the manufacturing process. In addition, the goal of high process capability, as reflected by the C_{pk}

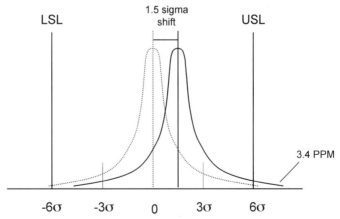

FIGURE 3.4.3-4   A 6-sigma process with a typical 1.5-sigma shift. (*Courtesy of Peak Leadership Group, Boston, MA.*)

index, can never be achieved with a measuring system that has high variability, no matter how much the process is improved.

The capability of a measurement system must be assessed by using appropriate statistical studies of accuracy, repeatability, reproducibility, stability, and linearity. The studies should be repeated periodically. Local organizations should establish time guidelines for repeating these studies.

To understand why a measurement error can make an arbitrary quality goal impossible, imagine a process that has a specification of ±10; the process is centered and has a standard deviation of sigma$_p$. The measurement error of sigma$_m$ is 2.5. A goal of $C_{pk} = 1.5$ is set for the process.

The total observed variation sigma$_t$ is

$$\text{Sigma}_t^2 = \text{sigma}_p^2 + \text{sigma}_m^2$$

Even when the process variation sigma$_p$ is reduced to zero (an impossibility), C_{pk} of this process will never exceed 1.33.

$$\text{Sigma}_t^2 = 0 + \text{sigma}_m^2 = 2.5^2$$

Because the process is centered,

$$C_{pk} = C_p = s\text{pecification width}/6 \text{ sigma}_t$$

$$C_{pk} = 20 = 1.33/6 \text{ sigma}_t$$

The capability goal for this process will never be achieved because the measurement system has low capability. Without that realization, valuable resources might be wasted in trying to improve C_{pk} by reducing the process variation, which is not the root cause in this case.

A measurement system consists of a gauge, the people who use it, and the techniques, procedures, software, and so forth, for measuring. A good measurement system has the following characteristics:

- The system is in statistical control. The variation in the system is due only to "common causes," not special causes.
- The total variability of the measurement system must be small compared with both the manufacturing process variability and the specification limits.
- The scale increments of the measuring device must not be larger than one-tenth of the process variability or the specification limits, whichever is smaller.

Gauge Repeatability and Reproducibility (R&R) Studies.   The most frequent gauge studies performed by engineers and manufacturing personnel are gauge R&R studies. Gauge R&R studies are statistical studies to investigate a gauge's ability to repeat a measurement on the same part (repeatability) and to give the same reading on the same part when used by different people (reproducibility). These studies generally use three operators to make the same measurements on the same set of 10 parts. The result of the study is one single number to represent the combined gauge repeatability and reproducibility errors. This combined gauge error is then compared with the process variability (process spread) to compute the *gauge percentage R&R*. The gauge percentage R&R is the basis for judging whether the gauge is good enough to be used for the process.

Gauge R&R Acceptance Guidelines.   Several factors must be considered when determining the acceptance of gauge repeatability and reproducibility values. The gauge percentage R&R must be judged against the criticality of the measurement being made, the cost of retraining people in using the gauge, the available resources, the customer requirements, and so forth. In most cases the gauge percentage R&R should be compared with the following guidelines to make a decision:

- *Less than 10 percent.*   The present gauge system is acceptable.
- *10 to 30 percent.*   The gauge system may be acceptable depending on the importance of the application and various economic factors. Efforts should be made to improve the gauge capability, if possible. The gauge should be considered unacceptable for critical measurements.
- *More than 30 percent.*   The gauge is not suitable. Efforts must be made to reduce the operator and/or equipment variation. A temporary method for reducing the percentage R&R by using the average of multiple readings is acceptable while permanent improvements are being investigated. In such cases, the required number of multiple readings should be statistically determined.

Traditionally, the gauge error was compared with the part's tolerance to estimate the percentage R&R, resulting in the P/T ratio (precision-to-tolerance ratio). However, this practice is no longer recommended, because our interest is to find out whether a gauge can measure a particular process characteristic that has its own unique variability. Because the process variability is totally independent of the part's tolerance, the gauge's error should be compared with the variability. This concept is very similar to the practice of using statistical control limits (variability), not specification limits (part's tolerance), to control a process.

This concept is especially important to any quality program in companies striving for best practice. With a high-capability process reaching and exceeding the Six-Sigma goal, the process spread is very small compared with the specification tolerance. An otherwise capable gauge, as determined by a P/T ratio (the gauge error is small when compared with the specification limits) may not be suitable for a process with a small process variability (the gauge error is too large compared with the process variability). Such a gauge is not precise enough for effective decision making within the normal process window. Over time, its inability to detect process shifts may mask the deterioration of the process capability and thus inhibit the effectiveness of continuous improvement programs.

To calculate the gauge's percentage R&R, the process variability is defined as 5.15 sigmas (representing 99.7 percent of the area under the normal curve).

Determination of Gauge Calibration Interval.   Traditionally, the calibration intervals of a gauge are established either arbitrarily or loosely based on prior experiences or manufacturer's recommendations. Because no scientific method is used, the calibration intervals may not be frequent enough for the gauge in its real working environment. Then, the risk of using an inaccurate gauge could be significant. On the other hand, the calibration intervals may be too frequent, and resources can be wasted on unnecessary activities.

For these reasons, gauge calibration intervals should be carefully determined. Either an algorithmic method or a statistical method should be employed to establish and adjust the gauge calibration intervals, so as to minimize the risk and optimize resources.

Comparing Two Measuring Instruments.   Occasionally, one measuring instrument must be compared with another similar device to determine whether the two are performing equivalently. In such cases, correlative study is not the best approach. Two measuring instruments often correlate with each other, even if the differences in measuring means are statistically significant.

Instead of correlative study, hypothesis testing should be used, which compares the paired differences in measurements of the same parts on two instruments. These differences should first be tested for normality. If the data are normal, a "t-test" may be used. If the data are not normal, a nonparametric test such as the Wilcoxon matched-pair, signed-rank test should be employed.

Quality Problem-Solving, Diagnostic Tools

Pareto Analysis, Pareto's Law.   Pareto analysis is a very simple way of determining which problem to try to solve first. The problems that seem biggest or most urgent are those that attract our attention. To be an effective problem solver, you must be able to sort out the "vital few" important problems from the "trivial many" other problems. Pareto analysis will help you do this. Pareto analysis was developed by Italian economist Vilfredo Pareto. He determined that approximately 80 percent of the wealth is held by approximately 20 percent of the population. This concept is also known as "Pareto's law." This same principle was adapted by other quality control people and can be used in a manufacturing or administrative process; that is, 80 percent of your problems are due to 20 percent of the causes. Or to put it simply, 80 percent of your headaches are due to only a few problems. So it pays to work on the first few problems and ignore the others until those first few are solved. Problems tend to sort themselves out. When you do some investigating, you usually find that of the many problems that you have, only one or two account for the largest dollar loss, happen most frequently, or account for the most failures.

Procedure. Figure 3.4.3-5 shows a list of failures taken from a soldering operation after one week's production by an automated soldering machine. Later a cumulative or percentage defective curve can be added.

Brainstorming, Fishbone (Ishikawa) Diagrams, Cause-and-Effect Diagrams. You might have heard some or all of these terms used when trying to problem-solve. In the problem-solving process, you need to identify problems, as well as determine their causes. Brainstorming helps you to do both. Kaoru Ishikawa in Japan created the fishbone or Ishikawa diagram in the 1940s. Ishikawa used this method for displaying and organizing the relationship between a problem and the possible root causes of the problem. Brainstorming is an out-pouring of ideas. Have a team facilitator run the brainstorming session; this person will keep the team on track and will be unbiased in viewing the problem. The group should meet in a common meeting room. The room should have the facilities to record all of the ideas stated by the team. Use an overhead projector, chalkboard, computer, or anything else that will enable the group to view the results.

Prerequisites. Choose the subject for the brainstorming. Make sure that everyone has a clear idea of what the real problem is. This is a critical first step and may take some time. Discuss this until everyone agrees.

Form the team. This is also an important step; you must have a group of people who are willing to work together as a team. You should gather together all people from various disciplines of the job at hand. For instance, if this is a manufacturing problem, you may want to include maintenance, manufacturing, engineering, quality assurance, inspection, and so forth. Or for administrative problems, you may want to involve administrative assistance, service centers, the

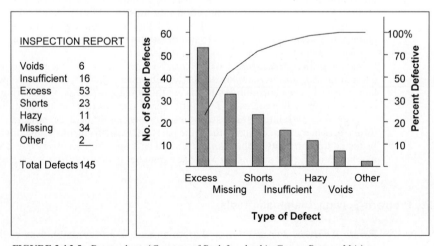

FIGURE 3.4.3-5 Pareto chart. (*Courtesy of Peak Leadership Group, Boston, MA.*)

manager of that operation, and so forth. Involve everyone who is concerned with the problem; this will facilitate more causes, and people will feel more ownership toward solving the problem.

Now you are ready to brainstorm. This can be done in different ways. Some groups find it helpful to go around the room and give each person a turn to express one idea. If people cannot think of ideas, they may "pass" until the next time around. Other groups decide that it is fine to throw out ideas randomly. If you have an idea, just say it, and it will be recorded.

Facilitators are responsible for recording all of the ideas expressed during the brainstorming session. Facilitators should be completely unbiased. Their job is to facilitate the group, to encourage discussion, and to be the timekeeper and mediator.

As a facilitator, your job is to extract *all* of the ideas your group has. The concept that "no idea is a bad idea" should be conveyed to the group. Any and all comments should be tabled until all ideas are exhausted. Do not criticize the ideas of others. If ideas are criticized, this will have the effect of shutting down the ideas of people who are less vocal in the first place. The facilitator also must maintain direction and should write down all ideas stated. Encouraging wild ideas may trigger someone else's thinking.

Procedure.   On your overhead, chalkboard, or computer, draw a line from the left to the right. This will be the main bone of your fishbone diagram. Draw a rectangle to the right of the line, and place the stated "problem" in this area. Draw four additional lines from the main bone. These will act as the general categories of the potential causes of the problem. Typically, we use the four Ms, which are *manpower, machines, methods,* and *materials*. Other general categories may be *Mother Nature* and *measurements*. The team can select whatever categories may be needed for that situation.

Begin your brainstorming. If an idea or possible cause arises that involves a machine, place the idea in the machine category. This can be depicted with the idea and an arrow drawn to the machine category. If ideas or causes of that first idea spawn deeper ideas or causes, place those ideas with arrows pointing toward the original idea. Continue until all ideas are exhausted. This process can last for half an hour to several hours, depending on the complexity of the problem (see Fig. 3.4.3-6).

Process Performance Metrics

Three valuable metrics are used to measure the performance of all significant processes. The potential process capability index (C_p) and the process capability index (C_{pk}) are used to assess process capability. The instability index (S_t) is used to determine the stability of a process.

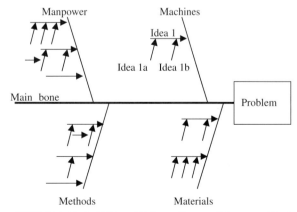

FIGURE 3.4.3-6   Fishbone/Ishikawa diagram. (*Courtesy of Peak Leadership Group, Boston, MA.*)

Potential Process Capability Index (C_p). C_p is defined as the ratio of specification width over the process spread. The specification width is predefined and fixed. The process spread (or process width) is the sole influence on the C_p index. When the spread is wide (more variation), the C_p value is small and indicates low process capability. When the spread is narrow (less variation), the C_p value becomes larger, indicating better process capability (see Fig. 3.4.3-7).

Process Capability Index (C_{pk}). In real life, very few processes are on their desirable target. An off-target process should be "penalized" for shifting from where it should be. C_{pk} is the index to measure this real capability when the off-target penalty is taken into consideration. The penalty, or correction factor, k is defined as

$$k = \frac{\text{target} - \text{process mean}}{\text{½ specification width}}$$

and the process capability index is defined as

$$C_{pk} = C_p (1 - k)$$

When the process is perfectly on target (see Fig. 3.4.3-8),

$$k = 0$$

and

$$C_{pk} = C_p$$

Design of Experiments

Poor quality in any arena can be traced to variation in parts, materials, people, and processes. The larger the variation, the poorer the quality. It has been observed that all work can be bro-

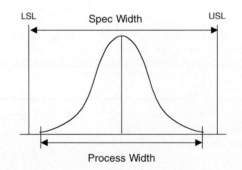

$$C_p = \frac{\text{spec width}}{\text{process width}} = \frac{\text{USL} - \text{LSL}}{6\ \text{Sigma}}$$

Potential Capability Index C_p

FIGURE 3.4.3-7 Process potential. (*Courtesy of Peak Leadership Group, Boston, MA.*)

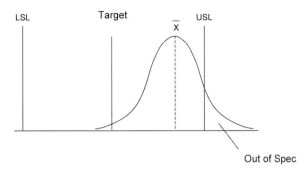

FIGURE 3.4.3-8 Capability index C_{pk}. (*Courtesy of Peak Leader-ship Group, Boston, MA.*)

ken down into processes. All processes, whether manufacturing or administrative, have inherent variability that affect their qualities.

Basic statistical process control (SPC) techniques such as Pareto analysis, cause and effect diagrams, graphical data presentation and analysis, and control charts have been widely incorporated in many businesses. These SPC tools are useful in identifying variation and keeping the process from deteriorating in quality. These tools do little, however, to make dramatic improvements in quality, and they are used primarily for maintenance and monitoring.

Immediate and continuous process improvements require additional tools beyond basic SPC. These techniques are generally classed as *design of experiments* (DOE). DOE encompasses a broad spectrum of approaches and tools. A large number of "quality gurus" have been active in the field of DOE. The most well-known approaches are the "classical," such as full factorials and fractional factorials. Next are the Taguchi and Shainin techniques. The classical approach is the most thorough and the most mathematically complex technique. The Taguchi approach is similarly mathematically complex, but allows shortcuts to the classical approach and thus is simpler and less expensive to perform. Although statistically well-founded, the Shainin approach is the least mathematically complex of these common techniques.

Cross-Functional Process Mapping

Mapping is a tool that can be used in a manufacturing or an administrative area. The concept of mapping out a process or system is simple. A cross-functional team is formed to challenge a problem or business issue. The team concentrates its initial efforts on how they do things presently (for example, how does billing flow from start to finish, or, how does this product flow through my production line?). The group will write down in detail how their process/system currently operates. This will give the team a very good idea of how the process works and what all the current problems/issues may be. A process map results in a list of all of the problems/issues.

The process proceeds to examine the following three steps of the total cycle time implementation process:

1. Identify key customer issues that impact cycle-time reduction through customer and organizational analysis.

2. Apply concepts and tools to reduce cycle time for a clearly defined customer or critical business issue.

3. Develop and implement an action plan to reduce cycle time based on a clearly defined customer or critical business issue.

As a cross-functional group, a team of individuals moves through each step. The team will practice applying the concepts and tools and developing their skills in using organizational and process mapping techniques and mathematical formulas to determine "as-is" and "should-be" cycle times. The APQC best practices discussed in Chap. 2 are excellent sources of potential should-be processes maps. The maps identify a customer issue or critical business issue relevant to their work environment, which is used as the basis for the application activities of the implementation process. After going through this process, a group will be able to describe the total cycle-time implementation process, draw an as-is and should-be process map, establish as-is and should-be cycle times, and develop an action plan to support should-be process and cycle time. The plan usually determines and prioritizes the enhancements and changes needed to transition from the as-is to the should-be process. A gap analysis may help decide what to do and how to do it.

Why Create a Process Map? A process map (see Fig. 3.4.3-9) can be helpful to:

- Identify areas for cycle-time improvement
- Identify bottlenecks, rework, and non-value-added steps for elimination
- Provide documentation of how the process is conducted
- Demonstrate sufficient detail to make improvement recommendations
- Identify opportunities for simplification by comparing with best practices
- Visualize upstream controls

Establishing a Process Control System

An effective process control system requires careful research in the planning phase. The potential failure modes of each process and product characteristic must be explored and understood. The knowledge gained in this investigation is then used to plan the control system, such as determining the important process characteristics to be controlled, how to control them, how to react when the process is unstable, and so forth.

Generally, the *failure mode and effects analysis* (FMEA) for each process should be generated first to assess potential problems. The knowledge gained in developing the process FMEA is invaluable in developing the *control plan,* which is a detailed plan to control all pro-

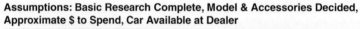

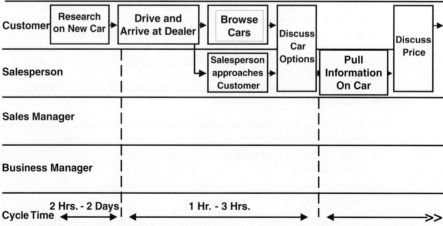

FIGURE 3.4.3-9 Process mapping (purchasing a car). (*Courtesy of Peak Leadership Group, Boston, MA.*)

cesses for a family of products. The process FMEA is equally important in driving the development of the total control methodology, which is designed to give complete and simple instructions to control each process at the production level.

Process Failure Mode and Effects Analysis (FMEA). A process FMEA is an analytical technique, which must be used for all significant processes to ensure that potential failure modes have been considered and addressed. An FMEA should be a summary of an engineer's analysis of what could go wrong with the process, the potential impact of such a failure, and how to deal with or correct the problem. A process FMEA should be developed for all new changed processes. The FMEA should also be reviewed periodically and revised when more knowledge is gained about the process. Normally, an FMEA will serve the following purposes:

- Identifies known or potential process failure modes
- Assesses the potential risk effects of the failures to the customer
- Identifies the potential manufacturing or assembly process causes and control actions
- Identifies significant process variables to focus controls for detecting and reducing similar failures

An FMEA will have a list of potential failure modes ranked according to their effect on the customer, thus establishing a priority system for corrective action. In a process FMEA, a "customer" may also mean a downstream process or operation, as well as the end user. When a FMEA is properly developed, it is a good tool for several purposes. An FMEA can help in developing new machines and processes, as well as troubleshooting old processes that perform below expectation. From a process control viewpoint, an FMEA is also important in the development of several components of the control plan and total control methodology (see Fig. 3.4.3-10).

Control Plan The development of a control plan is a key phase in establishing a process control system. A control plan is a summary description of the procedures for controlling all significant characteristics of the processes used to manufacture a family of similar products. It identifies the important processes used to produce a product, the significant characteristics of each process, and the way to control each of those significant characteristics.

A suitable control plan must be generated to successfully control a manufacturing line. Typical elements of a control plan include the following information:

- *Process.* Identifies the manufacturing process to be controlled.
- *Equipment.* Types of equipment or machines used in the process, either generic or specific.
- *Significant characteristics.* A list of all significant characteristics to be controlled. These should be input variables, but may also be output variables.
- *Measurement techniques and/or equipment used.* For measuring and/or method for obtaining the data.
- *Minimum.* The minimum acceptable number of units or readings for each sample size and monitor and the minimum acceptable frequency of monitoring.
- *Frequency analysis method.* Techniques to analyze the data collected. These can be statistical techniques such as X-bar and R charts, attribute charts, or nonstatistical techniques, such as Positrol logs.

Total Control Methodology. Several separate processes usually exist within each product line. Each one of these processes should have an individual plan to control significant machine, process, and product characteristics. The plan must be complete enough to cover all

Process Name/No. Grilling Hamburgers   Parts Affected   Prepared By: See Attached

Mfg Responsibility   Product Utilizing Part   FMEA Date: (Rev) O

Other Areas Involved   Engineering Release Date: Key Production Date

Process Description / Process Purpose	Potential Failure Mode	Potential Effect(s) of Failure	SEVERITY	Potential Cause(s) of Failure	OCCURRENCE	Current Controls	DETECTION	R.P.N.	Recommended Action(s)	Area/Individual Responsible & Completion Date	Action(s) Taken	SEV	OCC	DET	RPN
Grilling Hamburgers / Enjoy to eat	Overcooked burgers	• Hungry people	4	• Grilled to long • Coals were too hot	6	• Set timer to make sure coals are properly lit • Spray grill with water too cool the coals	1	24	• Raise the grill top • Set automatic timer vs wrist watch • Grill tool test						
	Undercooked burgers	• Sick people • Angry people	9 7	• Not grilled long enough • Coals were too cold	2 7	• Set timer to make sure coals are properly lit • Pierce burgers with a fork to see if juice is too pink	2 2	36 28	• Lower the grill top • Set automatic timer vs wrist watch • Grill tool test						
	Bad fluid tasting burgers	• Hungry people • Angry people	4 7	• Started cooking burgers too soon • Sprayed charcoal fluid on grill top	4 2	• Set timer to make sure coals are properly lit	1 2	16 28	• Set automatic timer • Use pretreated charcoal "all in one" bag						

FIGURE 3.4.3-10 Process FMEA. (*Courtesy of Peak Leadership Group, Boston, MA.*)

periodic process control requirements. But the plan must also be simple and should be considered as the operator's complete and easy instructions and reference for controlling each specific process. Such a system is called a *total control methodology,* or TCM.

A TCM must be developed and implemented at each process within a product line. A cross-functional team approach should be used to develop and implement the TCMs. The design and contents of a TCM should be flexible enough to adapt to different environments, and contain the following major elements:

Positrol plan. A Positrol plan is one element of a TCM, and it is usually the first section of the package. A Positrol plan is a matrix in which different aspects of the control system are characterized.

The *rows* of the Positrol plan matrix define:

1. *What.* The process variable or characteristic to be controlled and related criteria.

2. *How.* How to perform the required control on the characteristic specified. Additional instructions such as procedures for measurements may be needed.

3. *Who.* The person responsible for performing specified actions.

4. *When.* The frequency of monitoring.

5. *Types of control techniques.* X-bar and R charts, P or N_p charts, Positrol plans/logs, and so forth.

The *columns* of the Positrol plan matrix define:

1. *Setup requirements.* Procedures to set up a machine/process and all necessary buy-off requirements before actual production run.

2. *Input characteristics.* All important input variables to be controlled; input variables often associated with the machine/process.

3. *Output characteristics.* All important output variables to be controlled; output variables often associated with the product, such as product measurement.

4. *Preventive maintenance (PM).* All PM requirements done by the operators.

Positrol Log. Control charts may not be necessary for some characteristics. However, these characteristics may still need to be validated periodically to make sure they are within control and/or specification limits., A log sheet known as a *Positrol log* is used to record the reading of those characteristics. A Positrol log generally contains:

- Characteristics to be monitored
- Frequency of monitoring for each characteristic
- Control limits or specification limits used as control limits
- Reactions when reading excess limits
- Other information such as date, time, initials, and so forth

Control Charts. A control chart is a common form of control required by a Positrol plan. The charts must be appropriately selected and used. See the separate section for control chart usage guidelines. The reverse sides of the control charts contain corrective action logs. These logs are required to record actions taken when a control chart shows an out-of-control condition. The action taken should conform to the *out-of-control action plan* (OCAP).

Out-of-Control Action Plan (OCAP). Sometimes called a *decision tree,* an OCAP (see Fig. 3.4.3-11) is a plan to help production personnel react to each specified type of out-of-control situation. It gives instructions for both product and process on what specific actions are needed to dispose of products and restore control when there is an indication of out-of-control conditions. An OCAP is a powerful tool to capture the engineer's knowledge and make it available for the operators. With an OCAP, many correct engineering decisions can

Rev Letter: Rev Date: 4 - Sheet 1 Of 1		**Out of Control Action Plan**	Date Time Operator

To be used in the event of an Out-of-Control Condition
on the Porting Control Chart or the Positrol Plan/Log

___ Yes 1.   Is the calculation and point plotted on the control chart correct? Continue with PCL.

___ No   Correct the calculation and/or point plotted, and have Cell Leader sign CAL and continue to run.

___ Yes 2.   Was The Out-Of-Control Condition; A Single Point Above The UCL; 2 Out Of 3 In Zone A Or Above; 4 Out Of 5 In Zone B Or Above; 8 In Zone C Or Above; Six Points In A Row Steadily Increasing? Adjust Adhesive Reservoir Pressure, down within specified limits. After adjusting the pressure perform a half dozen dispenses in order for the pressure to equalize. Check adhesive weight twice in a row and plot both points on the control chart. Continue to run.

___ No   Continue with PCL.

___ Yes 3.   Was The Out-Of-Control Condition; A Single Point Below The LCL; 2 Out Of 3 In Zone A Or Below; 4 Out Of 5 In Zone B Or Below; 8 In Zone C Or Below; Six Points In A Row Steadily Decreasing? Adjust Adhesive Reservoir Pressure, up within specified limits. After adjusting the pressure perform a half dozen dispenses in order for the pressure to equalize. Check adhesive weight twice in a row and plot both points on the control chart.

___ No   Continue with PCL.

___ Yes 4.   Was the System Air Pressure within specified limits?
              Continue with PCL.

___ No   Adjust System Air Pressure, up or down within specified limits. Check adhesive weight again, and plot on control chart.

___ Yes 5.   Was the Tower Air Pressure within specified limits?
              Continue with PCL.

___ No   Adjust Tower Air Pressure, up or down within specified limits. Check adhesive weight again, and plot on control chart.

___ Yes 6.   Was the Pedestal Rotation Speed within specified limits?
              Continue with PCL.

___ No   Adjust Pedestal Rotation Speed to left or right within specified limits. Check adhesive weight again, and plot on control chart.

___ Yes 7.   Was the Chuck Alignment within specified limits?
              Continue with PCL.

___ No   Adjust Chuck Alignment, align hole on chuck with hole on base of machine. Use small alignment hole for all other ports. Use large alignment hole for axial ports. Check adhesive weight again, and plot on control chart.

___ Yes 8.   Is the Weight Scale (Gauge) in calibration?
              Continue with PCL.

___ No   Check calibration sticker date, if OK Check adhesive weight again, and plot on control chart. If not, call Maintenance.

           9.   If all of the above has been tried and the dispensing is still not correct place "machine down" sign on, and call Maintenance.

FIGURE 3.4.3-11   Out-of-control action plan (OCAP). (*Courtesy of Peak Leadership Group, Boston, MA.*)

be made instantly by the operators. Generally an OCAP may have one of the three following formats:

- Decision tree in flow chart or simple box diagrams, showing diagnostic journey, with remedies.

- Narrative explaining the diagnostic, containment, and corrective actions.

- Interactive computer dialogue that appears when an out-of-control situation occurs, providing the necessary information to lead the operator through the diagnostics and remedies.

- *Setup checklist.*   Summary of all important machine setup items that must be checked on a regular basis to make sure that the machine is operating at the baseline requirements.
- *PM schedule.*   Schedule that specifies when equipment components are to be validated and checked for conformance, maintained, overhauled, or replaced.
- *PM checklist.*   Assists validation that all components due for PM have been completed and are correctly set up to ensure that the machine can perform up to its inherent capability.
- *PM procedures.*   General instructions and criteria detailing how equipment is to be maintained; usually includes brief acceptance/checkout procedures to be performed before release back to production.
- *Reference/backup.*   A TCM should include supporting documents such as detailed PM procedures, machine setup procedures, process specs, and so forth. If the material is too extensive to be included conveniently, the name, document number, and location of the material should be referenced in the list of the TCM.

Quality Function Deployment (QFD)

Objectives:

- Highlight inviolate customer principles
- Translate the "voice of the customer" into product specifications
- Further deploy product specifications into part specifications, part specifications into process specifications, and process specifications into production/quality control
- Outline a variety of techniques that can capture the "voice of the customer":
  1. Value research : simple, effective
  2. Sensitivity analysis : a fine-tuning
  3. Multiattribute evaluation : systematic, effective
  4. Quality function deployment : comprehensive, effective
- Develop QFD as an umbrella technique, covering several disciplines in new product development

Benefits:

- New design in half the time with half the manpower, half the defects, and half the cost of previous designs
- Moving from customer satisfaction to *customer delight*
- Leapfrogging competition
- Quick transfer of knowledge to new engineers
- Making good engineers into excellent engineers

Continuous Improvement.   A well-designed, well-executed process control system will keep a process under control at an existing level. The control charts will identify and even predict when a process is unstable. The OCAP will specify which actions are to be taken under which circumstances to bring an out-of-control process back under control and to dispose of the product. The control plan and TCM will specify in detail how to control a whole product line or any single process.

However, there is one thing a process control system cannot do: It cannot directly improve the capability of a process. A process can be improved only if we understand the significant characteristics that influence the final output, know how to optimize those characteristics, and keep them at the optimum level.

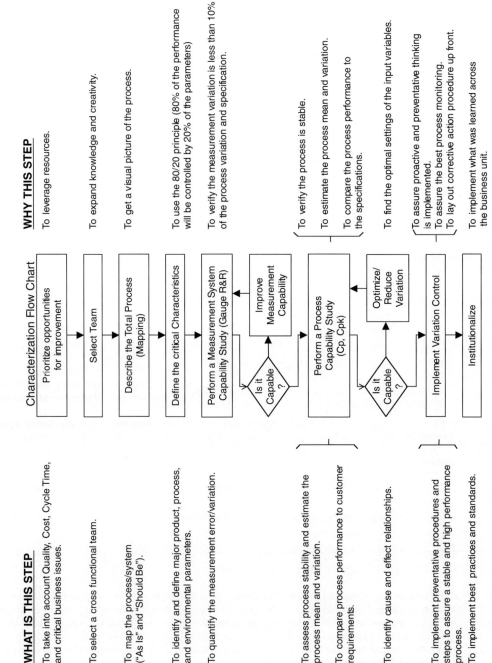

WHAT IS THIS STEP

To take into account Quality, Cost, Cycle Time, and critical business issues.

To select a cross functional team.

To map the process/system ("As Is" and "Should Be").

To identify and define major product, process, and environmental parameters.

To quantify the measurement error/variation.

To assess process stability and estimate the process mean and variation.

To compare process performance to customer requirements.

To identify cause and effect relationships.

To implement preventative procedures and steps to assure a stable and high performance process.

To implement best practices and standards.

WHY THIS STEP

To leverage resources.

To expand knowledge and creativity.

To get a visual picture of the process.

To use the 80/20 principle (80% of the performance will be controlled by 20% of the parameters)

To verify the measurement variation is less than 10% of the process variation and specification.

To verify the process is stable.

To estimate the process mean and variation.

To compare the process performance to the specifications.

To find the optimal settings of the input variables.

To assure proactive and preventative thinking is implemented.
To assure the best process monitoring.
To lay out corrective action procedure up front.

To implement what was learned across the business unit.

Characterization Flow Chart

Prioritize opportunities for improvement

Select Team

Describe the Total Process (Mapping)

Define the critical Characteristics

Perform a Measurement System Capability Study (Gauge R&R)

Is it Capable?

Improve Measurement Capability

Perform a Process Capability Study (Cp, Cpk)

Is it Capable?

Optimize/ Reduce Variation

Implement Variation Control

Institutionalize

FIGURE 3.4.3-12 Characterization flow chart. (*Courtesy of Peak Leadership Group, Boston, MA.*)

In the total picture of continuous improvement, process control is only one component among a number of other equally important components. These continuous-improvement components should be performed in a general sequence. The flow chart shown in Fig. 3.4.3-12 describes the main components and the logical sequence for making continuous improvements in a process. In reality, some flexibility to add other components and/or slight modification of the logical sequence may be needed to adapt to individual situations.

ARTICLE 3.4.4

INFORMATION-DRIVEN PROJECT MANAGEMENT FOR DESIGN AND DEVELOPMENT

Michael Stowe
The Boeing Corporation

Stephen Denker
GTE Internetworking

In today's business environment, many organizations have multiple design projects running simultaneously. Our inability to see the full impact of our own work and that of others or to quickly assess the changes we all can expect during a project is a significant contributor to schedule creep, uncontrolled resource usage, and cost overruns. Complicating matters further, projects as a whole often extend beyond their own business boundaries to external suppliers and customers.

Information-driven project management (IDPM)[1] shows how changes made during a project affect people individually and as a team. Everyone can see clearly how their individual information needs and responsibilities affect the entire project's desired deliverable, cost, and schedule goals. This added situational visibility opens the way to more ideas and innovations.

More than schedule and resource analysis, we need to design our projects. Definition and organization of the project steps to be performed are critical to project success. Designing an outstanding project is itself a creative, iterative project management process. Information-driven project management encourages using behaviors conducive to holistic project optimizations.

DESIGN AND ANALYZE PROJECT PLANS

Projects are sequences of dependent process steps that produce products or services. Over time, project complexity generally increases and changes. To cope, we need increased situational visibility. An integrated set of project visualization techniques and associated software tools, IDPM provides situational visibility by enabling project team members to design and analyze all of the significant information and resource relationships among a project's process steps that are necessary to meet specific goals. This visibility is available to project team members throughout both project planning and execution.

Information-driven project management takes us from a logistics view (managing tasks and deliverables) to a dependency view (managing arrays of dependent entities). Information-driven project management provides project management with a clearer view of all intertask dependencies and an ability to resolve them. Information-driven project management surfaces project information dependencies and project resource constraints. Information-driven project management then deals with them—identifies opportunities. Information-driven project management combines the best of existing project planning methods with two new management capabilities—the Dependency Structure Matrices (DSMs; information dependency structure) and the Critical Chains (resource constraint structures). The DSMs lay out information dependencies in an easily understood spreadsheet-like format and also provide a topology-based foundation to manage dependency structures. Critical Chains differ from Critical Paths by including resource constraints. The Critical Chain, as the critical sequence of resource-constrained project steps, determines when a project is done, that all of its deliverables are delivered. Combined with DSMs, Critical Chains improve project plans by ensuring that they are feasible. The two conditions are interdependent—information dependencies determine the Critical Chain as much as do resource constraints.

IDPM PATHWAY

Within IDPM, projects are viewed as information processing systems. Systems require inputs and produce outputs. Inputs transformed by process steps use resources to produce outputs. We view these inputs and outputs as information. This insight led directly to creation of the IDPM systems approach. The IDPM system has six components just as the cube has six faces. Our overall goal is increased visualization of all of a project's perspectives. Figure 3.4.4-1 represents the complete six-step IDPM pathway.

Although there is only one entry point into the IDPM pathway portrayed in Figure 3.4.4-1, side trails adjacent to the main path permit flexible access to all steps within the IDPM system. The IDPM information processing approach can be repeated as necessary throughout a project's life. The path we actually take depends on how well we can understand and document our work processes and project goals. During the life of a project, process steps or their sequence may need to change because some information or resource is not actually ready when required.

IDPM STEPS

Step 1: Prepare Initial Goals

Right from the start, the better we can describe our project's development and performance goals, the better our chances for creating successful project plans. In many cases, however, the description of a new project and its goals only becomes clearer later, as the project proceeds. This is especially the case where we have little prior experience or timely access to information about similar projects. A project team should start a project goal's document by describing their final goals first and then working backward through each significant project phase to describe their corresponding goals. Be sure you understand all of the project requirements; if they are fuzzy, put in an early step to clarify them with your customer. For every significant project phase, three goals should be documented:

1. Deliverables (for each scope of work)
2. Time
3. Cost

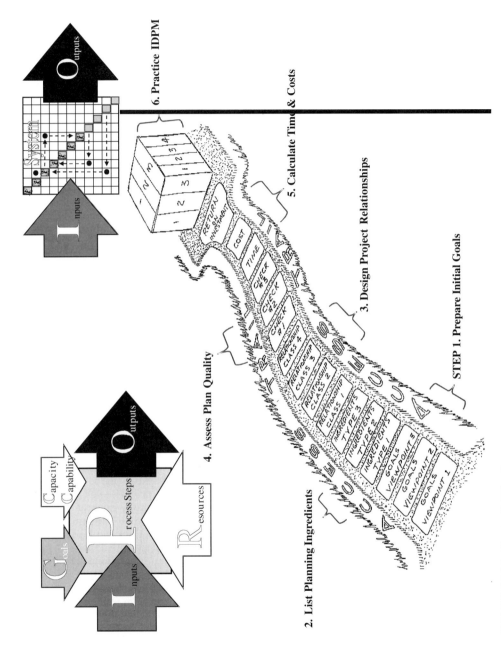

6. Practice IDPM

5. Calculate Time & Costs

3. Design Project Relationships

STEP 1. Prepare Initial Goals

4. Assess Plan Quality

2. List Planning Ingredients

FIGURE 3.4.4-1 Six-step IDPM pathway.

Step 2: List Ingredients

Every project has information dependencies and resource constraints. Interview all team members to obtain this information. Knowing what questions to ask and how to arrange their answers for use is absolutely critical. We list descriptions of the project deliverables, schedules, and cost goals. We also list each team member's individual information dependencies, resource constraints, associated project step lists, and their durations. We update, as necessary, throughout the life of the project. During interviews, managers and project team staff must constantly ask themselves, What is the essential information that I need to make this decision? What do I need to start, update, or complete my work? Project ingredients identified in this step and designed in the next step are analyzed in Step 4 to confirm that the project's goals (specified in Step 1) are met. The three information lists include:

1. Resources
2. Process steps
3. Inputs and outputs

Resources are all those things that energize projects. People, tools, and facilities used within a project system are resources. Each resource has unique capabilities and availability. At this point, simply create lists of resources in any order, along with the associated quantities needed.

Some of the inputs-to and outputs-from process steps are external to the project. Project team members are responsible for bringing external information into the project (external inputs). Some or all of the information they create is considered internal outputs to be shared with project team members. Some of these internal outputs created by the project team are likely to serve as external outputs needed by others outside the project. It is extremely important to be able to identify internal and external inputs and outputs. When interviewing project team members about their respective project roles, we want to identify and list any input and output information both internal and external to the project. Clear definition of the input/output information that each team member provides ensures that the necessary interfaces between their roles match. One technique, a data dictionary, can be used to ensure that different words do not mean the same thing or that the same words do not mean different things.

Step 3: Design Project Relationships

This step makes all of the information dependencies among team members visible. The project team can see how information dependencies might create iterations and work out of sequence. Opportunities for concurrent project steps that increase throughput can be clearly identified. The project team can design a project that achieves its deliverable goals and also identifies key risks and the strategies for overcoming them. The project team uses the structured modeling (DSMs) to represent and optimize complex dependency relationships to arrive at the executable project step sequence before proceeding to resource constraint analysis (Critical Chains).

We build an initial project model based on information collected about person-specific and team-specific process steps. In IDPM, we choose to represent information dependencies using diagrams or matrices. To represent more complex relationships, we could use data flow diagrams. Some IDPM Working Group members use a hierarchical box and arrow language called IDEF0 to considerable advantage. Figure 3.4.4-2 illustrates the IDPM process using the IDEF0 language. Other IDPM Working Group members use DSMs to model their projects directly.

Dependency Structure Matrices offer an effective way to visualize the relationships among activities in a project. The DSMs were developed to solve problems based on the structure of the problem. The technique has been applied to many problem sets, including the design of project plans. The project's process steps have constraints and predecessors. In the design of the project plan, we want to organize process steps to provide the best plan. The DSMs map the dependencies between the project's process step inputs and outputs. We visualize and

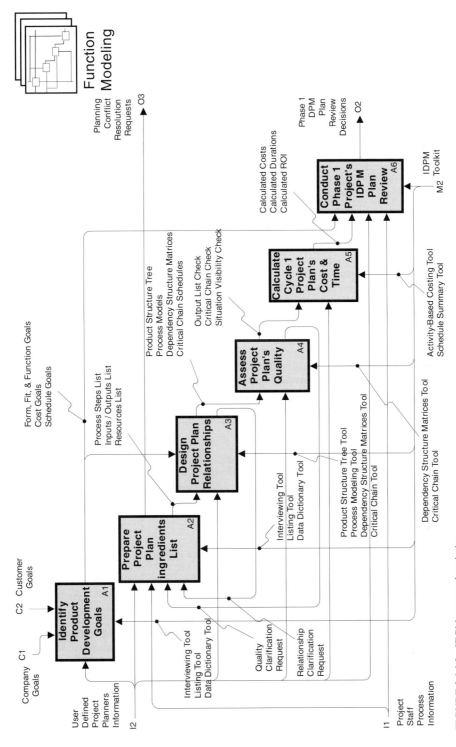

FIGURE 3.4.4-2 IDPM process description.

understand relationships between process steps by tracing information dependencies through a DSM to see all likely consequences. We can then visualize and understand relationships between process steps by tracing dependencies through a DSM to see all likely consequences. The approach presented in this subchapter follows that adapted by Boeing and NASA.

A DSM map can be used to transform the components (inputs, goals, capacity and capability, and resources) that enter process steps into external outputs. It visualizes the internal outputs created when the process steps transform any combination of external and internal inputs. The DSM map uses a square matrix as illustrated in Fig. 3.4.4-3. For simplicity, process step names in the initial matrices could be ordered alphabetically across the rows and columns.

A project plan that has N process steps determines the size of the square $N \times N$ DSM matrix. Two additional matrices that identify external inputs and external outputs are adjacent to the internal process step sequence matrix (the DSM) in the center. Some process steps produce both internal and external outputs. The external input matrix is above the DSM and the external output matrix is to the right of the DSM. These externals could determine the project team's success, and these external outputs are necessary to others outside the project. They depend on them.

A mark in each of the matrix cells in the DSM means that an input or an output relationship exists between the process steps. When the mark is above the diagonal, information is fed forward horizontally until it meets the vertical column of the process step that needs it. When the mark is below the diagonal, then information is fed backward horizontally until it meets the vertical column of the process step that needs it. Any backward flow means that an earlier process step might need to be repeated or that it is out of sequence.

Usually, we initially identify only input and output information exchanges between process steps internal to the project for our internal process step sequence matrix (the DSM). If any resource that performs one of these internal process steps cannot find its inputs in the DSM or if they are not visible, then they must be external inputs. Whatever outputs exit from the DSM to external destinations really determine when a process step is complete and presumably ready for another activity.

This matrix representation allows project team members to visualize a logical order necessary to create and manage its own project and identify all dependencies on external vendors and customers. It also quickly identifies all of the rework cycles that the project team's own internal sequence of process steps may be causing by identifying the backward feeds shown by marks in matrix cells below the diagonal of the center DSM matrix. Some of the backward feeds can be easily removed by simply changing their relative locations in the sequence matrix of internal process steps, so that they feed forward above the diagonal. For example, any necessary rework cycles can be planned now that they have been made visible. Also visible are the process steps that occur concurrently because there is no information dependency among them. This view of concurrence assumes that there are no resource conflicts among those process steps.

Because resource constraints are often significant, IDPM project planning always considers them. The Critical Chain includes the resource constraints that define the overall longest path (constraint) of the project. The method resolves all resource constraints while determining the project Critical Chain. For a project without resource constraints, the Critical Chain will be the same as the Critical Path.

The Critical Chain has the largest impact on any project's completion date and throughput. The Critical Path through a project's process steps represents the path of longest duration from the project's start to finish. Any changes made to the critical process steps have the most significant impact on the project's duration path. Goldratt[2] introduced the powerful viewpoint that a project is a set of chained process steps, one of which is the Critical Chain. Chains exist because there is some dependency between the adjoining chain links. To improve throughput and remove bottlenecks in the chain, special care must be taken when planning the capacity of each of the individual chain links. The DSM represents all of the chains formed by information dependency. The Critical Chain represents opportunities for significant improvement, if the project cannot meet its initial development time or cost goals. We must resolve the combined effects of information dependency and resource conflict within a proj-

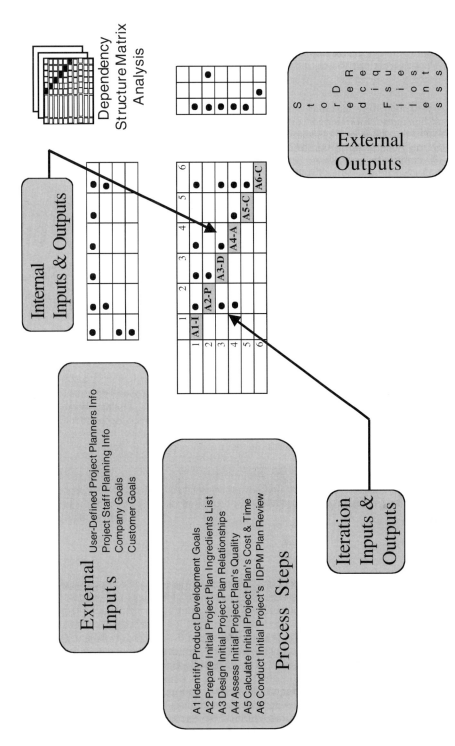

FIGURE 3.4.4-3 Dependency Structure Matrix (DSM) mapping.

ect and across multiple projects. We must manage our reserve of resources. Conflicts in resource usage could modify opportunities for concurrent project steps identified in Step 3. Another very important reason for resource identification on the chain is to prevent multitasking. When people experience resource conflicts due to multitasking on the Critical Chain, projects do not tend to meet their development time and cost goals.

Critical Path project planning often assumes that an acceptable way to account for potential resource constraints on the project is first to identify the critical path and then perform resource leveling. For most, application of resource leveling algorithms lengthens the overall schedule. For this reason, few projects use resource leveling tools. The essential change introduced by Critical Chain Project Management (relative to the current Critical Path practices) is development of the Critical Chain, using both process step information dependency logic and resource constraints.

There are four key IDPM planning relationships. As changes occur while designing the plan, these relationships can be updated in any manner convenient for the project teams. The four relationships include:

1. Product structure
2. Functions
3. Inputs and outputs
4. Resources

The process step functional relationships are a mapping of the inputs, outputs, controls, and resource relationships between process steps. Resource constraints form Critical Chains. Information-driven project management develops a Critical Chain, rather than a Critical Path, as the primary focus of a project manager. Carried out after DSM analysis, the Critical Chain includes both information dependence logic and resource constraints. We must resolve the combined effects of information dependency and resource conflict within a project and across multiple projects. Information-driven project management establishes the Critical Chain after removing resource contentions, rather than before. As for DSM analysis, Critical Chain analysis focuses on the whole project to provide focus and clear decision criteria for the project manager. Information-driven project management matches resource constraints (people, tools, and facilities) to the information-dependent process step sequences established in Step 3. We must manage our reserve of resources. Conflicts in resource usage could modify opportunities for concurrent project steps identified using DSMs.

Step 4: Assess Plan Quality

Now, the three primary quality checks of the project plan can be conducted. They result from the relationships designed in Step 3. They include:

1. Outputs
2. Resource identification
3. Critical Chain multitasking

We will not address the quality of the project produced when your plan is implemented. Good plans can and do produce bad products. Is your list of output names from the process steps complete? This check can be best accomplished by looking at internal and external outputs, represented by the cell marks in the DSM. Some process steps may require external inputs or produce outputs both consumed internally and externally. Project phase deliverables required for exiting that phase could be the creation of a set of external outputs. By reviewing the external outputs' list, we would then satisfy this planning quality check.

Have all the project resources (people, tools, and facilities) and their scheduled durations been identified and reviewed? This quality check ensures that all of the resources (people, tools, and facilities) are available when indicated by the plan. All members of the project team should be able to find their names, associated process steps, inputs and outputs, and durations. All must have situation visibility as to the impact of their activities on the completion of the project. Other resources such as tools and facilities needed by project team members whose project steps are on the Critical Chain of the project must be reserved.

Are any of the resources along the Critical Chain (people, tools, and facilities) multitasking? This quality check reviews the Critical Chain relationship. For the project team resources identified as on the Critical Chain, it is important to ensure they are not scheduled to do any other work during their own Critical Chain process steps. Otherwise, people would be multitasking within or across projects. That would result in lengthening the time of all of the project steps for which they are responsible, and project(s) would be completed later than planned. Multitasking is performing multiple project activities at the same time. Most people think of multitasking project activities as a good way to improve efficiency. It ensures that everyone is busy all of the time. Often, we have to wait for inputs or for someone to call back before we can get on with our work. Multitasking makes good use of this time. But focus on local efficiency could damage the overall performance of a project. If this is a Critical Chain process step, this practice directly extends the duration of the project. Information-driven project management seeks to eliminate this type of multitasking by eliciting 100 percent focus on the process step at hand by considering all resources that support the project. Thus, eliminating "fractional head counts" is a primary Critical Chain consideration in planning a project.

Step 5: Calculate Time and Costs

If the project is interested in making a profit for the business, then the project team could determine a project market price at which to sell the project. The project's profit is equal to the difference between the combined materials and resource usage costs and the price the customer pays for the project. The return-on-investment (ROI) calculation is equal to the profit divided by the investment (total project costs). In Step 1, an initial list of the time and cost goals of the project was collected. The duration and costs of the associated process step can now be calculated. All of the resources (people, tools, and facilities) required by the project, as listed in Step 2, and the durations required by these resources and their usages, as designed in Step 3, form the basis for these calculations. The input and output lists may include some external inputs as bought materials and some external outputs as products sold for a price. We calculate the project's duration by summing up the durations from all of the unbundled process steps along the Critical Chain. The project team resources working these process steps determine how long it takes your entire project team to handle its work. The Critical Chain relationship designed in Step 3 gives the capacity answer. Adding the duration estimates performs the duration calculation for the full project. As soon as a process step is completed, the Critical Chain resource has capacity to handle more work. The calculation of the project cost is simply project step durations multiplied by the cost of using each of the resources required (people, tools, and facilities), added to any cost of any purchased input materials. As a project executes, a graph of the cost as a function of time can be depicted.

Step 6: Practice IDPM

For decades, great athletic teams have harbored one simple secret only a few select business teams have discovered, and it is this: To play and win together, you must practice together. A project team has now completed only Steps 1 to 5 make up the IDPM thinking process. To successfully deploy IDPM, we must motivate all team members to want to apply and practice IDPM as often as possible. So IDPM Step 6 is practice. Before project teams apply IDPM generally to their projects, they must practice together. As experience is gained, the time and cost that are associated with IDPM use diminishes significantly. This occurs for a competitive champion sports team. Practicing and using IDPM often enough should enable everyone to think through effective plans more intuitively.

NOTES

1. *Information Driven Project Management (IDPM) Concept Paper,* IDPM Working Group. Visit our Forum on Information-Driven Project Management at http://www.gte.com/aboutGTE/GTO/bbnt/IDPM.
2. Eliyahu M. Goldratt, *Critical Chain,* North River Press, April 1997.

ARTICLE 3.4.5
ACHIEVING ISO 9000 CERTIFICATION

Graham K. Hill
Tomlin, Inc., Falmouth, Massachusetts

We look at a high-gain, high-speed approach that builds on client relationship, business planning, process analysis, and the acquisition of best practices in project management. For most firms, the step to ISO 9000 registration can be much smaller than they may realize. If the concepts of good business practice as laid out elsewhere in this book (strong relationship with customers, plans for growth, process analysis, continuous improvement, and the best practices in project management and design control) are already in place, then each firm has the necessary ingredients for simple ISO 9000 achievement. The myths of 2-year long ISO 9000 programs at M$1 of cost can be shattered.

This article will explain how existing pieces of a business can simply be brought together for ISO 9000 registration with reduced cost and documentation, and within an accelerated timeline. This approach has already successfully provided for engineering and design consultant firms[1]

- Eighty percent less documentation
- Thirty-five percent less resource demand (cost)
- Six months or less to full compliance (for a practice with up to 250 employees)

We will also show how to easily translate a manufacturing-based standard into a design and construction service firm environment. It will also show how ISO 9000 can directly benefit bottom-line performance, ensure project schedule and budget attainment, raise customer satisfaction, and stimulate growth for the business. The approach is based upon an electronic, paperless system easily compiled into an intranet environment through the use of standard Microsoft® tools.

This article is written for management considering or already committed to ISO 9000 implementation. It does not discuss the justification or merits of ISO 9000 for a design firm nor describe what ISO 9000 is. Resources to guide a firm and provide specialist assistance are listed in the appendix. *Note:* We will focus on successful management of an ISO 9000 implementation project. As such, it will not go into detailed explanation of the content of ISO 9000 and its separate elements, nor touch on the history and justification for ISO 9000. The reader may find numerous alternate reference articles and books on these subjects.

BACKGROUND

Engineering design and consultant firms have demanded a greatly simplified approach to ISO 9000 that acknowledge both their inherent history of good practices and the elevated educational and professional standing of their employees.

Manufacturing industry experiences with ISO 9000, born out of defense standards, regulated industry requirements (nuclear and drug), and a working environment of repetitive and highly prescriptive tasks have little place in a creative, project-centered business such as design and construction. The attempt to directly transfer those experiences into design and construction has justifiably created great skepticism and resistance within the design community. The history of bureaucracy, paperwork, and procedural overload associated with the defense, nuclear, and drug environments is equally out of place within a dynamic engineering practice.

What is repeatedly requested is an approach that:

- Fits the business, naturally
- Acknowledges the skills of personnel readily
- Uses the language of the industry
- Merges with (not replaces) existing processes
- Is attainable with an efficient use of time

CASE STUDY—HOW CAN IMPLEMENTATION BE SIMPLIFIED?

The following example is based on an engineering design firm with five regional offices and approximately 250 employees. They served primarily industrial and commercial clients. Although the company had practiced quality improvement, there was no previous experience with ISO 9000.

Many myths and fallacies around ISO 9000 taken from implementation in manufacturing industries had conspired to make management in the firm believe that ISO 9000 in design firms is unnecessarily difficult. The first step was to discard those ideas and begin afresh. Let's look at the challenges and questions posed by management.

Myth 1: Do We Have to Document Everything?

The standard requires that procedures be written ". . . where their absence would adversely affect quality . . ." Within regulated manufacturing industries, the quality experts insisted on writing each and every instruction. Paralysis by paperwork.

Solution.   With personnel extensively trained in professional techniques, given guidance in project management, and expected to routinely exercise judgment and creativity, there existed a simple need just for compact guidance procedures: documents that indicate major steps—not detailed tasks. Samples and boilerplates from prior projects provided a simple library of good practices and eliminated the need for detailed work instructions. Preparing documents that recognize these factors reduced by 50 percent the level of documentation needed.

Myth 2: Does Implementation Have to Take 15 to 24 Months?

Implementation has traditionally been taken as an opportunity for a firm to discover their business processes, rewrite their operating procedures, and seek improvements and best practices as they go. The debate and wrestling with ISO 9001 interpretation and language extends project duration as personnel, in ignorance, pitch for a safe position of the highest expectations. The habits and practices that are commonplace in a firm are often assumed to be invalid for ISO 900 needs. Only cash-rich firms can afford this cost-loaded approach.

Solution.   Well-led, focused, high-energy compact teams, using simple proven templates that fit their industry and working on high-level management procedures, completed all of the charts and supporting procedures and forms within 3 to 5 months. Some other firms have achieved ISO 9001 compliance within 4 months.

Myth 3: Must Documents Be Written in Governmental, Textual Style?

Examples often contain highly structured formats—Policy, Purpose, Responsibilities, Reference documents, and the like. Engineers, in particular, and the majority of the working population prefer to use diagrams, sketches, graphics, or maps to express ideas. They abhor the use of long, text-based procedures.

Solution.   The use of charts to express ideas greatly simplified information communication and the amount of paperwork required. Charts were found to rapidly communicate each of the major process steps. Ideal charts were based on the business process and ignored any departmental or territorial ownership. As such, they clearly showed how different parts of the business link together. Charts were set at high level for management of the business. Graphics and clip art reduced the need for text definitions. They also aided orientation and training of personnel.

Charts, like those shown later, will meet the criteria of acceptable ISO 9001 procedures and can save over 40 percent in the actual use of paperwork. Figure 3.4.5-1 shows a sample success chart.

Myth 4: Does the Approach Require Many Teams and Resources?

Traditional implementation projects have allocated 20 separate teams—one for each element of ISO 9001, plus teams for ancillary activities. Is this necessary and expensive high demand on resources?

Solution.   A resource-efficient approach was to create just a few, compact teams that reflect how the firm actually does business. It was not necessary to rigorously adhere to the ISO 9000 sequence of 20 elements. The firm chose to rearrange those elements to fit their current practices and habits. The release of first-draft procedures from each team, for immediate use by employees, accelerated acceptance for use within the firm.

Teams were formed around natural business groupings of the firm—not around the 20 ISO 9000 elements. Each process team was resourced by:

- *Process owner.*   This was the team leader. The firm's best mentor for that topic and with ideal qualities in motivating and developing personnel. The process owner was responsible for best practices in a key business area for the firm.
- *Team members.*   Personnel with *direct* content knowledge on that topic. Personnel with indirect knowledge attended team sessions as ad hoc members.

As with any significant company project, management formed part of the leadership group. See the project structure that follows. *Note:* ISO 9001 elements 4.9 Process Control, 4.10

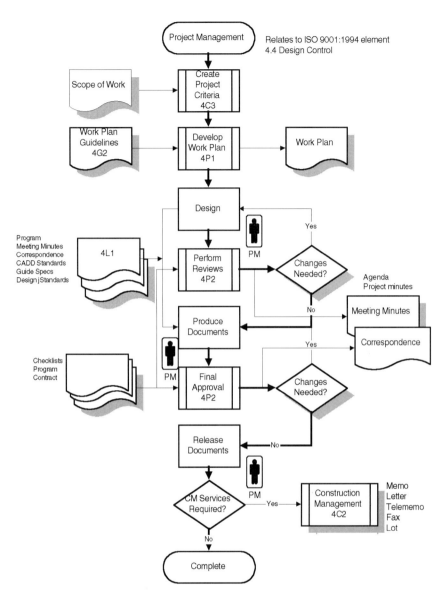

Project Management — Relates to ISO 9001:1994 element 4.4 Design Control

Scope of Work → Create Project Criteria 4C3

Work Plan Guidelines 4G2 → Develop Work Plan 4P1 → Work Plan

Design

PM — Changes Needed? — Yes

Program Meeting Minutes Correspondence CADD Standards Guide Specs Design jStandards | 4L1 → Perform Reviews 4P2

No — Agenda Project minutes — Meeting Minutes

Produce Documents — Yes — Correspondence

Checklists Program Contract | PM → Final Approval 4P2 → Changes Needed?

Release Documents — No

PM — CM Services Required? — Yes → Construction Management 4C2 — Memo Letter Telememo Fax Lot

No → Complete

This document is proprietary and exclusive property of Tomlin Inc. Falmouth MA, USA. This document may not be reproduced in any form whatsoever, without written permission of Tomlin Inc.		
Copies of this document without control stamp are NOT controlled	Stamp	
page 1 of 1	**Design Control** Process Owner: George Chakaris	Doc: MP - 4.4P1 Rev: 1 Date: 02/15/96

FIGURE 3.4.5-1   Sample Success Chart® of an approved ISO 9001 procedure (shown 75 percent of full size).

Inspection & Test, 4.11 Measurement & Test Equipment (Calibration), and 4.12 Inspection & Test Status do not directly apply to an engineering design firm. These can be omitted from the action items in an ISO 9001 project.

> **Team 1: Management.**   4.1 Management Responsibility; 4.2 Quality System; 4.14 Corrective & Preventive Action; 4.17 Internal Quality Audits; 4.19 Statistics
>
> **Team 2: Design Project Control.**   4.3 Contract Review; 4.4 Design Control (Design Project Management); 4.8 Product ID & Traceability (Project Numbers); 4.13 Nonconforming Product (Design Reviews); 4.15 Storage, Labeling, Handling, & Delivery (Drawing, Document Storage, & Delivery)
>
> **Team 3: Documentation / Administration.**   4.5 Document & Data Control; 4.16 Quality Records (Archiving)
>
> **Team 4: Purchasing.**   4.6 Purchasing; 4.7 Customer Supplied Product
>
> **Team 5: Training.**   4.18 Training

It was recommended that internal projects be organized and established in the same way as any client project: scope, criteria defined, Project Manager, teams, budgets, timelines and review, and reporting mechanisms. The project organization structure is shown in Fig. 3.4.5-2.

Myth 5: What Do We Do? Won't This Be Complicated?

Stories indicate a need for external help, major efforts, and complicated tasks, each disrupting the normal flow of work.

Solution.   An ISO 9000 project is very similar to a design project. Clarify scope and elements, gather the best talent in teams, produce documents, have these reviewed and then released for implementation.

Experiences with continuous improvement and process analysis described in this book fitted perfectly with the desired approach to implementation. The four primary steps—plan, do,

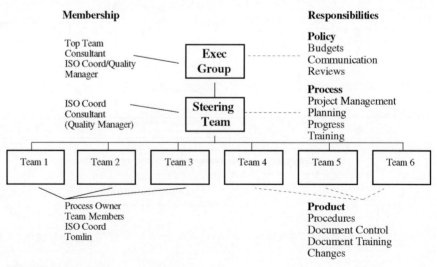

FIGURE 3.4.5-2   Project organization structure chart.

check, act—applied equally well for this firm. Figure 3.4.5-3 shows what a typical implementation project road map can look like.

Myth 6: Won't This Take a Long Time?

Early attempts at understanding and implementing ISO 9000 climbed up a long and slow learning curve. This no longer applies. Well-structured solutions exist for rapid deployment of solutions.

Solution.   Management set aggressive expectations for the project and for their employees. Particular attention was given to the creation of high-energy teams and focused workshops. Teams met monthly for not more than 2 hours and were led by an external facilitator. Deliverables were clearly set for each team. Meeting 1: core processes; meeting 2: inputs and outputs; meeting 3: supporting forms and training of employees, release for use. Six hours of team meetings for standard components. Eight to ten hours for complex components.

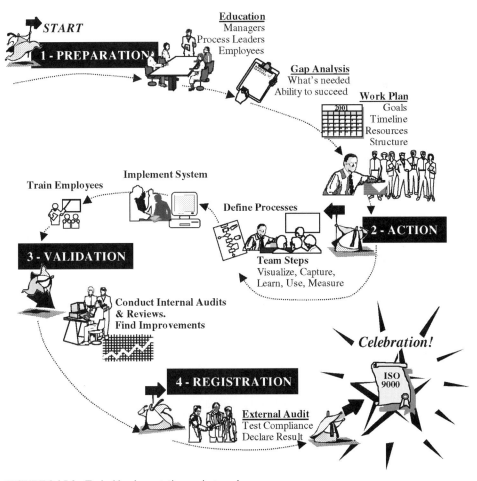

FIGURE 3.4.5-3 Typical implementation project road map.

The timeline ensured rapid orientation and training to the teams during the preparation phase. Go from kickoff to action fast. Each team was scheduled over the next 3 to 4 months for meetings and assignments, and deliverables were locked to dates. Testing effectiveness through audits included opportunities for improvement and fine-tuning. Registrar dates were locked in. Final preparation allowed for history and correction if needed. Figure 3.4.5-4 shows pages 1 (Fig. 3.4.5-4*a*) and 2 (Fig. 3.4.5-4*b*) of the actual timeline plan for the company.

Myth 7: What Are the Key Steps or Actions Once We Decide to Proceed?

Many firms gain an insight into the benefits of ISO 900 or even decide to proceed, but they are not sure what the first or significant steps should be.

Solution.   First things first. The firm had reached consensus among the top management team for ISO 9000. Clarification of scope; setting project criteria; and discussing options, timelines, and budgets followed. The steps included the following preparation decisions:

1. Choosing the appropriate ISO 9000 Standard that fits the business

The proper designation for ISO 9000 is the *ISO 9000 Series*. The family of standards includes:

- *ISO 9000.* The guidelines to help select the appropriate standard
- *ISO 9001.* The standard applicable to design firms
- *ISO 9002.* The standard applicable to manufacturing companies
- *ISO 9003.* The standard applicable to inspection-only firms (e.g., distributors)
- *ISO 9004.* The guidelines for best practice in the application of the standard

As will be seen from this, any engineering design firm will make use of the ISO 9001 Standard. For the rest of this article, ISO 9001 will be used.

There are other associated standards to the ISO 9000 series regarding quality manuals, audits, calibration, and so on that are not needed for this article. These are not discussed.

2. Define scope of coverage

ISO 9001 requires that project-based operations or processes *directly* affecting client deliverables be covered. That is, those activities covered by a contract with a client and payable as direct fee or expense work, are to be included.

Indirect activities not covered in a contract with a client are excused from ISO 9001 coverage. These may include such activities as management accounting, personnel recruitment, pay and benefits, general marketing and promotions activities (not project-specific), some general administrative activities, and some management activities. A considerable amount of unnecessary procedural work can be saved through this segmentation.

A first question for an engineering design firm is to clearly define the nature of the work performed by the business. This will determine the full scope of interface with ISO 9001: Will this be a total or partial overlap? Two assumptions are made here:

1. The firm performs purely a design and/or construction contract administration service (construction is actually performed by a contractor to the owner).
2. The firm performs a design/build service.

In situation 1, by focusing just on design as their core business, the firm has only partial overlap with all 20 of the required elements in ISO 9001. As a consequence, four elements of ISO 9001 (4.10 Process Control, 4.11 Measurement & Test Equipment, 4.12 Inspection & Test Status, and 4.13 Nonconforming Product) do not apply and need not be described in detail. A saving of approximately 20 percent on the extent of paperwork or procedures needed for compliance.

This scenario will apply to the majority of engineering design practices. The role of a firm in administering a construction contract on behalf of an owner places the direct responsibility

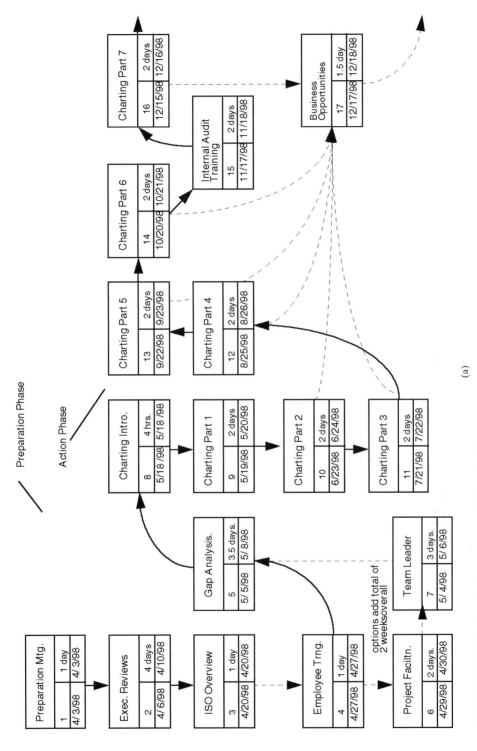

Preparation Phase

Action Phase

Preparation Mtg.		
1	1 day	
4/ 3/98		4/ 3/98

Exec. Reviews		
2	4 days	
4/ 6/98		4/10/98

ISO Overview		
3	1 day	
4/20/98		4/20/98

Employee Trng.		
4	1 day	
4/27/98		4/27/98

Gap Analysis.		
5	3.5 days.	
5/ 5/98		5/ 8/98

Project Faciltn.		
6	2 days.	
4/29/98		4/30/98

options add total of
2 weeksoverall

Team Leader		
7	3 days.	
5/ 4/98		5/ 6/98

Charting Intro.		
8	4 hrs.	
5/18/ 98		5/18 /98

Charting Part 1		
9	2 days	
5/19/98		5/20/98

Charting Part 2		
10	2 days	
6/23/98		6/24/98

Charting Part 3		
11	2 days	
7/21/98		7/22/98

Charting Part 5		
13	2 days	
9/22/98		9/23/98

Charting Part 4		
12	2 days	
8/25/98		8/26/98

Charting Part 6		
14	2 days	
10/20/98		10/21/98

Internal Audit Training		
15	2 days	
11/17/98		11/18/98

Charting Part 7		
16	2 days	
12/15/98		12/16/98

Business Opportunities		
17	1.5 day	
12/17/98		12/18/98

(a)

FIGURE 3.4.5-4 Timeline plan for the company. (*a*) Page 1 of the plan.

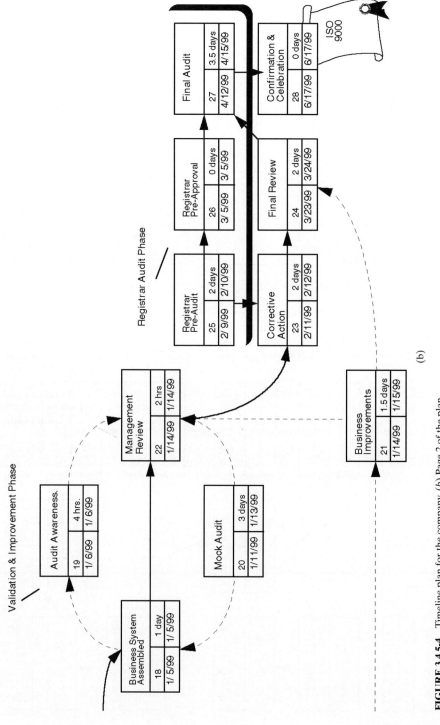

FIGURE 3.4.5-4 Timeline plan for the company. (b) Page 2 of the plan.

Validation & Improvement Phase

Audit Awareness.		
19	4 hrs.	1/ 6/99
	1/ 6/99	

Business System Assembled		
18	1 day	1/ 5/99
	1/ 5/99	

Mock Audit		
20	3 days	1/13/99
	1/11/99	

Management Review		
22	2 hrs	1/14/99
	1/14/99	

Business Improvements		
21	1.5 days	1/15/99
	1/14/99	

Registrar Audit Phase

Registrar Pre-Audit		
25	2 days	2/10/99
	2/ 9/99	

Corrective Action		
23	2 days	2/12/99
	2/11/99	

Registrar Pre-Approval		
26	0 days	3/ 5/99
	3/ 5/99	

Final Review		
24	2 days	3/24/99
	3/23/99	

Final Audit		
27	3.5 days	4/15/99
	4/12/99	

Confirmation & Celebration		
28	0 days	6/17/99
	6/17/99	

ISO 9000

(b)

for construction with that owner—not with the firm. The design firm acts merely as an administrative interface. As such, the firm falls outside of the ISO 9001 requirements for documentation on construction.

In situation 2, if the firm has direct responsibility for design and uses a direct subcontract process to perform the build operation, then the subcontractors will be given the responsibility, by contract, to validate quality. As such, the role of subcontractors is regulated under element 4.6 Purchasing and does not require additional quality control procedures. However, if the firm employs their own direct labor and equipment to perform the build stage, then all 20 elements of ISO 9001 will apply. This firm will need to cover all aspects of ISO 9001 in their procedures.

3. Confirm number of business locations

A firm with a single-site location for its offices will apply ISO 9001 to their direct project activities.

A firm with operations at separate locations has basically two options:

1. Engage in one implementation project involving all locations simultaneously.
2. Have each location implement ISO 9001 independently.

Where a firm presents to its clients a unified position and shares project work among locations, the decision should be to engage in one implementation project across all locations— one firm, one ISO 9001 certificate. The advantage of a unified approach is consistency in practices and lower costs through integrated and shared implementation activities. The potential risk is in ensuring uniformity in pace and quality. Failure at a single location to pass the ISO 9001 certification audit can mean a delay for all locations.

Where a firm offers distinctly separate services at each location, operating autonomously, a decision may be made to engage in independent implementation projects—separate firms, separate ISO 9001 certificates. The advantage of an independent approach is the ability to respond to market needs in a planned and staged way and to customize the certification location by location to match different customer and business focus. Each certification will stand for itself. The potential downside is the cost of certification paid for each separate location ($10,000 to $20,000 each time) and the inability (a regulated restriction) of other locations to claim certification through a sister location. Likewise, the corporation cannot claim total coverage on ISO 9001 until each and every location has its ISO 9001 certificate.

Historically, it was possible for firms to segment out separate portions of their business and thereby gain ISO 9001 more easily and quickly by working on just a portion of their business. This is now frowned on by ISO 9001 regulators and not advised.

4. A/E/C operation and the fit with ISO 9001

Many firms follow the ISO 9001 20-element sequence when writing their quality manual. This is not required. It would be better that the ISO 9001 elements be rearranged to reflect the normal way of doing business. No firm need be "bent out of shape" to suit this external standard.

The different elements (or requirements) of ISO 9001 and their fit with a design practice may be shown through the following diagrams.

THE BASIC WORK PROCESSES   (See Fig. 3.4.5-5.)

The events include the following:

1. Contract stage (proposal and contract setting)
2. Project control (design) activity
3. Transmittal of contract documents (drawings)
4. Construction administration (C/A)

At each stage, A/E/C activities will be indicated first, followed by the relevant ISO 9001 requirement.

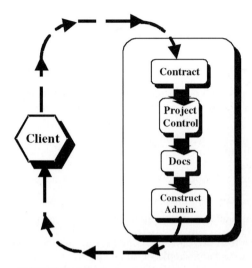

FIGURE 3.4.5-5 Basic work process.

To comply with ISO 9001, each of these requirements *must* be addressed.

Typically, an initial survey of readiness (gap analysis audit) of a firm shows that good activities are performed but often poorly documented.

As you proceed, you may wish to perform a mental assessment of the extent to which your firm already has these requirements in place—and as a result make an early and broad estimate about how much work may be involved in achieving ISO 9001 compliance.

THE 20 ELEMENTS OF ISO 9001

4.1	Management Responsibility	4.12	Inspection and Test Status
4.2	Quality System	4.13	Control of Nonconforming Product
4.3	Contract Review	4.14	Corrective and Preventive Action
4.4	Design Control	4.15	Handling, Storage, Packaging,
4.5	Document and Data Control		Preservation, and Delivery
4.6	Purchasing	4.16	Control of Quality Records
4.7	Customer-Supplied Product	4.17	Internal Quality Audits
4.8	Product ID and Traceability	4.18	Training
4.9	Process Control	4.19	Servicing
4.10	Inspection and Testing	4.20	Statistical Techniques
4.11	Control of Inspection, Measuring, and Test Equipment		

Note: For an A/E/C firm, five elements can be excused from full documentation (a reduction of 20 percent). The elements 4.9, 4.10, 4.12, and 4.13 apply to manufacturing. These can be "folded into" element 4.4 Design Control. Element 4.19 often does not apply and can equally be excused from full documentation.

STAGE **1:** BASIC BUSINESS MODEL   (See Fig. 3.4.5-6.)

A/E/C descriptors are shown in **bold upright font.**

The ISO 9001–associated elements are shown in *italic regular font.*

Each of these needs to be defined in documents (procedures) available to all personnel.

Documents can be in the form of traditional text-based procedures, flow charts, drawings, forms, videotapes, or any combination of the above.

Any media can be used to capture and document the designated process.

Simplification of these procedures, both in terms of content and structure, will greatly reduce complexity for implementation, training, and maintenance.

STAGE **2:** ADMINISTRATIVE PROCEDURES AND SUBCONTRACTING ADDED   (See Fig. 3.4.5-7.)

Clear descriptions are needed of the firm's policies and procedures for:

- Guiding work
- Providing training to employees
- Archiving project records

Additionally, subcontractor quality and how the purchases of materials or goods are handled needs to be described.

There is also a need to show how owner-supplied information (data, prior drawings, etc.) is protected and made secure while in the firm's care.

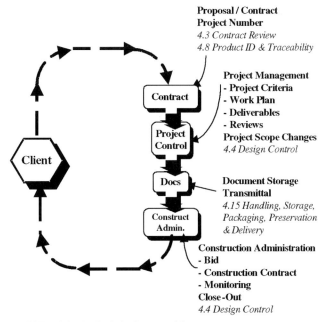

FIGURE 3.4.5-6   Basic business model.

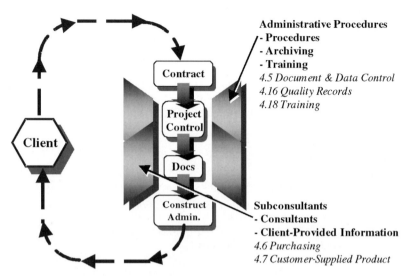

FIGURE 3.4.5-7   Administration and contractors.

STAGE 3: ADDING THE MANAGEMENT COMPONENTS   (See Fig. 3.4.5-8.)

A clear organizational structure, stated company policies with objectives and plans, and someone responsible for quality need to be defined.

The management team should perform routine reviews of the business to test its effectiveness.

Additionally, management should have established measures for the key components of the business and for customer satisfaction.

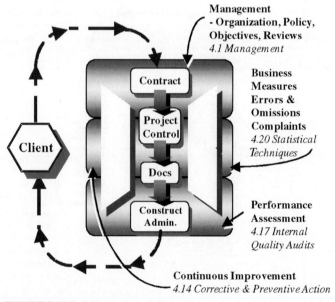

FIGURE 3.4.5-8   Management components.

To collect data for these reviews, there should be planned internal audits of effectiveness.

Finally, a formal process to identify and take action on problems and opportunities for improvement should be in place.

This corrective action and improvement process should be accessible to employees and include subcontractor performance and customer complaints.

ISO 9001 FOR ARCHITECTURE AND DESIGN FIRMS: AN EXPLANATORY SYNOPSIS

What or Who Is ISO?

ISO comes from a Greek word that means *equal*—as in *iso*sceles triangle, *iso*bars, and so on. The word *ISO,* when linked to ISO 9000, is used to signify equal acceptance of standards around the world—the creation of a level playing field.

The organization that publishes these International standards is known as the *Organization for International Standards,* or OIS, a confusing similarity.

The OIS is made up of representatives from each member country (currently over 130 member countries) who attempt to eliminate national quirks in the application of standards by harmonizing understanding and use in a large number of areas, such as safety, measurement, materials, food, testing, and quality. The OIS is a kind of United Nations forum of quality and technical experts from each country. The aim is to permit freer, more cost-effective trade among nations and companies by avoiding disputes of interpretation on technical grounds—an attempt to work at the logical level and avoid the political.

Each member country agrees to apply and uphold the standards it has approved at this international forum—both for their international trade and within their country. The United States has a strong voice within this group and is a principal sponsor (author) of many standards.

Standards are designated identification numbers in the same way that ASME, ASPF, ASTM, BOCA, CABO, or ASHRAE will designate numbers to their codes or standards.

ISO 9000 is the identifier for a series of standards related to quality systems and quality assurance. There are many other standards published by OIS under the ISO designation with different numbers. Many of these have nothing directly to do with quality.

Why was the ISO 9000 Series written? The standards were written because too many businesses did not sufficiently care about *how* they were working. Customers were unhappy about quality and needed a benchmark standard upon which to demand reliability and improvement in service. Hence, the ISO 9000 standard.

Customers also wanted a standard that placed the responsibility for quality directly with the supplier, eliminating the costly need for the customer to check (or recheck) supplied work.

A principal goal of ISO 9000 is to regulate the contractual relationship between the customer and the supplier regarding quality of service. This provides the customer with a process to select or reject potential suppliers based not just upon their technical and professional expertise, but also particularly upon their ability to deliver superb quality.

This results in a standard that requires management to consciously design business systems in a way such that quality, reliability, consistency, and improvement are natural parts of the business, a process that provides assurance around proactive management techniques—not reactive ones. ISO 9000 is about *how* the business is planned and run—not just *what* it does. This condition provides customers with greater confidence in their selection of a particular supplier.

The ISO 9000 standard covers a wide range of business activities affecting the customer relationship—not just the production of the service, and not just quality. The ISO 9000 standard covers the goals, plans, organization, operational performance, administration, employee training, measurement, and corrective and improvement processes of the business.

What Does ISO 9000 Require?

Customers (through the authors of ISO 9000) defined their expectations of a good supplier as being the company or firm where the following requirements are vigorously practiced.

Management.   Objectives, goals, and approaches to quality within the business are clearly expressed. Responsibilities are clear. Company plans are clearly understood at all levels within the firm and are effectively implemented, well tracked, reported, and reviewed. Plans are aligned with customer needs. There is a common structured process for fixing problems and making improvements. The company relies on factual, objective data and trends regarding its business.

Operations.   Best practices and expectations are clearly defined for all personnel—across the complete customer relationship cycle. These include initial contact, proposal, contract stage, project/scope definitions, project control, project reviews and release, delivery, project closeout, support, performance evaluation, customer satisfaction, and so on.

Administration.   Support structures for administrative processes are equally well designed. These cover document and data management, purchasing, archiving/records, and the training/professional development of employees

What Is the Relevance to A&E Design Firms?

The External Pressures.   Industrial customers have developed a strong history of quality. They have learned how to run their own businesses more effectively and what to expect from other businesses that work with them. Their visions of mutual and peer-based relationships have made them highly selective about whom they chose to work and grow with.

Industry has increasing demands about what constitutes good service, based upon their own experience and ability to change the norms of performance.

They want to work and grow only with others who share the same values for constantly learning and being the best. They were the leading edge in ISO 9000 application. They do not want to partner with those closed to the notion of change or improvement. They do not want to partner with those that say ISO 9000 or quality is only a manufacturing issue.

Industrial companies see that now is the time to move their expectations to second-level suppliers: the professional service firms.

State authorities have also seen the benefit of demanding ISO 9000–style performance from supplier firms when it comes to managing tax dollars. Pennsylvania DOT (PennDOT), Maryland DOT, New Jersey DOT, New York DOT, California DOT, and Massachusetts DOT are among those demanding or giving notice of ISO 9000 requirements from their design consultants.

Government departments, such as GSA Building Division, use ISO 9000 themselves and expect similar performance from their design consultants.

Global activity now makes ISO 9000 a minimum requirement for RFP's in Europe, Australia, and the Southeast Asian communities. Canada, Mexico, and other Latin American countries are also leaders in ISO 9000 application for design consultants. The United States lags many of these nations in the use of ISO 9000.

Commercial marketplace customers are becoming savvier about quality and the specific benefits of ISO 9000, but lag behind the industrial marketplace. Few individual commercial clients have embraced ISO 9000 as a model. Their time will come.

Health, education, and local government groups have not, as of yet, adopted ISO 9000 as an operational need—primarily from a lack of resources, rather than a lack of knowledge or desire.

Because of the risk of being accused of trade restriction, no customer will actually *mandate* ISO 9000 as a requirement for continued business. Most requests from customers are expressed as a preference for the supplier to have a quality program or to be ISO 9000–registered. The lack of a clear mandate can be misunderstood as the lack of customer need or interest. However stated, the implication is still there—listen to your customers' needs.

Benefits.   As a strategic goal, the design consultant firm should decide if it wishes to anticipate and support customer requests regarding quality and be among the leading group. Or the design consultant firm can resist until actual client losses and competitor gains become all too clear.

The implementation of ISO 9000 *may* gain a firm some new clients or market activity. However, the lack of ISO 9000 or a strong quality program will eventually lead to the loss of clients and contracts. In this context, ISO 9000 is similar to the implementation of CADD or EDI technologies during the late 1980s and early 1990s.

Liability insurance organizations provide discounts to professional design consulting firms registered to ISO 9000. Some design consultant membership organizations provide funds directly to firms to aid ISO 9000 registration.

The Internal Pressures.   Many design consultant firms face a dilemma. Customers demand higher quality, higher functionality, and faster production and lower cost (fees). They also expect higher levels of service, innovation, and special treatment.

- When it comes to managing their assets, the customer demands discipline, control, and faultless delivery—*no surprises, please!*
- When it comes to solutions, they want imagination, creativity, and verve—*surprise me, please!*

A design consultant firm sells time combined with a unique professional skill. Therefore, the effective use of time and the application of that skill determine success or failure, profit or loss. Anything that robs a project of time or the wise use of skill diminishes the chance to succeed.

Benefits.   ISO 9000 can manage the allocation of time and skill in a planned, structured approach during production and operations. This is how time and budgets are maintained, and it is how efficiency is achieved.

ISO 9000 can deliver significant improvements in the use of time and how work is managed (productivity) through significant changes in habits. This is how effectiveness is achieved.

ISO 9000 does not hamper creativity. It can help to reduce wasteful time. Improvements in the management of time can mean less time on production and more time for creativity.

Systematic methods implemented through ISO 9000 can be used to create larger margins in the cost/profit profile. Consistency and speed are among the keys, plus a "right first time" mentality (accuracy). This is project management par excellence.

Scope drift is an enemy to the design consulting firm. ISO 9000 processes can be used to clarify, define, and control client demands. Nurturing the client includes proactively managing scope definition. ISO 9000 can reduce scope drift through poor management of the client at the start of the project and poor control during project design phases. This is customer service par excellence.

ISO 9000 methods ensure that all employees work from a common model of efficiency and effectiveness in project management. New employees are brought "on stream" more quickly.

Overall benefits to ISO 9000 include the following:

To Marketing

Customer relations improved

Clearer agreements about expectations (fewer changes?)

Increased reliability (in budget, on time)

Recognition (firm "benchmarked")

To Operations

Commonly understood best practices

Focus on project management effectiveness

Reduction in project design rework

Enlarged resource capacity (throughput)

Data based management

Less variability in techniques

Reduced project overruns (time and cost)

ISO 9000 is about

Developing "best practices" in the business

Leadership and energy

Simplicity and guidance

Quality management by everyone

Staying with what you do best

Improving the business

Shaping ISO 9000 to suit your business

ISO 9000 is not about

Controlling innovation or creativity

Abandoning individuality

Regimentation or repetitiveness

Paperwork, paperwork, paperwork

Quality engineering

Changing the business to suit ISO 9000

RESOURCES

American National Standards Institute
11 West 42nd Street
13th Floor
New York, NY 10036

American Society for Quality (ASQ)
300 West Wisconsin Avenue
Milwaukee, WI 53202
Tel: (800) 248-1946
www.asq.org

ISO Central Secretariat
Case postale 56
CH-1211 Genève 20
Switzerland
Tel: 41 22 749 01 11
Fax: 41 22 733 34 30

Internet

International Organization of Standards. All the ISO standards and information from the international body are available from this site. American Society for Quality (ASQC): the largest quality organization in the U.S. with chapters all over the country.

http://mastercontrol.com/links.htm

CFITS Standards Document Library. Standards organized by the different publishing bodies. Updated regularly by SDL-Webmaster for ANSI (American National Standards Institute).

http://www-library.itsi.disa.mil/by_org.html

NOTE

1. Compared with traditional approaches.

SECTION 3.5
CLOSURE

Kristin Hill, Cindy Aiguier, John Morganti, and Bonnie Seaberg
International Facility Management Association (IFMA),
Boston, Massachusetts

Norman E. Faucher
Association for Facilities Engineering (AFE),
Boston, Massachusetts

Is there closure for a facility's life cycle? At any time, it may appear as though the facility's life cycle is ended. But the truth may be that it is beginning another phase of the cycle. The life cycle is a continuum. Even upon demolition of a facility, the land is likely to be reused. Possibly a new facility will be built that meets the needs of another plan. The facility's life cycle, like that of the facility manager's role, is continual and evolutionary. Facility management is the art and science of reinventing facility management as changes take place in business, technology, and processes to achieve the original goals.

It is important along the way to take time to assess the process that the facility has been through. The relationships of those involved from the initial plan through the cycle of management should be evaluated. Question whether the process supported the progress. Has a particular project improved the facility so that the company is better positioned for the future?

A facility and its management team will constantly be presented with new opportunities by the constantly changing business environment. These opportunities may include new tools, technologies, or methods of conducting business. The key is for the initial facility to be adaptable to meet these opportunities. The importance of planning for the future and not to catch up with the past is ever present. The initial goals must include the adaptability of the facility for the unknown future.

P · A · R · T · 2

FACILITIES ENGINEERING

Paul R. Smith, P.M.P., P.E., M.B.A., Chapter Editor
Peak Leadership Group, Boston, Massachusetts

Mark Neitlich, M.B.A., Chapter Editor
Massachusetts Institute of Technology, Cambridge, Massachusetts

William L. Porter, Ph.D., F.A.I.A., Chapter Editor
Massachusetts Institute of Technology, Cambridge, Massachusetts

Anand K. Seth, P.E., C.E.M., Chapter Editor
Partners HealthCare System, Inc., Boston, Massachusetts

David L. Stymiest, P.E., SASHE, C.E.M., Chapter Editor
Smith Seckman Reid, Inc., New Orleans, Louisiana

Roger P. Wessel, P.E., Chapter Editor
RPW Technologies, Inc., West Newton, Massachusetts

The organization of this part was a source of numerous discussions. We wanted to provide a single repository for common information pertaining to the most traditional types of buildings and facilities. Thus, Part 2 of this handbook contains information pertaining to facilities engineering for most types of buildings, including commercial buildings. The unique added engineering features required for special types of buildings such as industrial or institutional buildings are addressed in Part 3. Thus, the facilities engineer for an industrial building should refer to Part 2 for normal engineering topics and to Part 3 for the unique issues of specific building types.

Part 2 introduces the planning and programming process in Chap. 4 as a natural precursor to the engineering and design process in Chap. 5. Chapter 6, following in the most common sequence for creating a facility, covers construction-related issues. Then, Chap. 7 addresses facility operations, maintenance, and assessments. Finally, Chap. 8 rounds out our treatment of facilities engineering issues of common building types with a discussion of codes and standards.

CHAPTER 4

PLANNING AND PROGRAMMING PROCESS

William L. Porter, Ph.D., F.A.I.A., Chapter Editor
Massachusetts Institute of Technology, Cambridge, Massachusetts

A good program may not guarantee a good building, but it is a very good head start. And it can spare the facility manager countless headaches later in the process. As Sims and Becker (the leading experts in the field and the authors of this chapter) point out, without programming "facilities opportunities are missed and the likelihood of mistakes or simply poor value for money is high." "Correcting mistakes, or simply making changes, becomes more expensive at each successive stage of facility development." Therefore, it is the programming phase that offers the least expensive opportunity for satisfying complex needs and for cost savings. Good programs have other extraordinary benefits that include detailed understanding of the requirements that the building must satisfy and the forging of consensus among those who will use the facility and among those whose decisions are critical to its realization. This chapter describes the processes that comprise programming. These include forming the project team, determining the kinds of data that must be collected, choosing methods of data collection and analysis, and developing specific requirements for proposed buildings. It also discusses how programming can contribute to the evaluation of existing buildings. This chapter does not suggest a formula for programming in every circumstance. Instead, it offers resources that are essential for inventing a programming process uniquely suited to each specific situation.

SECTION 4.1

WHAT IS PROGRAMMING?[1]

**William Sims, Ph.D., C.F.M., IMFA Fellow, Professor,
and Franklin D. Becker**
Cornell University, Ithaca, New York

Programming is the front end of the planning/design process. William Peña,[2] recognized widely as the grandfather of architectural programming, defines *programming* as problem seeking and *design* as problem solving. Duerk[3] describes it as "the problem definition phase of the design process. It is done by gathering and analyzing information about the context within which the design must be done and by stating the qualities that the project design must have to be successful." It is a process of collecting, analyzing, and organizing the information needed to guide the search for an architectural solution. In turn, it is useful for evaluating proposed solutions.

INTRODUCTION

Why Is It Important?

All organizations need to prepare a design program at some point. A renovation, a move to a new space, or construction of a new facility all require that the organization determine what it is trying to accomplish with the project before it seeks solutions. Unfortunately, many client organizations and their designers immediately jump to solutions based on assumptions, conventional wisdom, or fashion, be it hoteling, a new headquarters in a greenfield site, or an intelligent building. When this happens, facilities decisions are made in a way that no manager would ever envision when dealing with the other resources of the organization. When done in this way, facilities opportunities are missed and the likelihood of mistakes or simply poor value for money is high. The program focuses the design efforts on the important issues and makes the design process more effective, resourceful, and perhaps most important, accountable. It enables the designer and client to understand why certain issues were selected, the priorities and trade-offs among the project requirements, and the context and constraints within which the project must operate.

A facility program and the process used to produce it are critically important to the designer and the client/user organization of a building project. Buildings are incredibly complex things, and they must satisfy many diverse needs. The needs must be identified, organized, prioritized, and ratified by decision makers in order to design properly. Thus, to design a building responsibly requires analytical tools and the skills to use them—a programming process, a skilled programmer, and a creative designer.

The information required for decisions at various stages in the planning and design process varies in amount, type, and source. But inevitably the information is of an amount and complexity to require some form of systematic process to ensure that all of the relevant issues and needs are uncovered and evaluated and to develop and manage the information during the process. Similarly, a process is needed to ensure that decisions are made by the appropriate individuals and are based on objective and factual evidence to the fullest extent possible.

What Are the Benefits?

A programming process requires the participation of all stakeholders at the critical early stages of decision making when most of the resources are committed to a project and when most of the decisions that affect the cost and direction of the project are made. Most important, it is at this stage in the project that redesigning work processes or organizational structures should be done before the building is designed.

By providing objective information and priorities regarding the goals, constraints, and other factors influencing a project, a programming process can actually increase the designer's ability to solve the problem in a creative and cost-effective way. It ensures that the client's needs are addressed comprehensively, that important issues are not overlooked, and that the design does not become too focused on a few fashionable issues at the expense of others. Having a good program actually decreases the search time (and cost) involved in arriving at a solution by enabling the designer to focus on "problem solving" rather than "problem seeking."

Being able to correct mistakes early in the process is one of the benefits of good programming. Correcting mistakes, or simply making changes, becomes more expensive at each successive stage of facility development, and the ability to impact or reduce the total cost of the project is lessened. (See Fig. 4.1-1.)

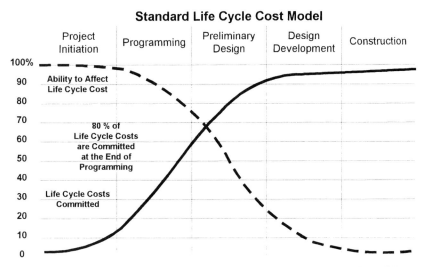

FIGURE 4.1-1 Impact on cost of decisions compared with expenditure of funds at different stages. (*Courtesy of William Sims and Frank Becker.*)

LEVELS OF PLANNING AND DESIGN

Decisions about building projects are made at a variety of levels. Each of these requires some form of problem identification and analysis or programming. Each addresses particular kinds of issues and requires different types of information. The decisions made in subsequent stages build on those of the previous. These levels include:

- Strategic planning
- Master planning
- Site selection and design
- Building design
- Interior design
- Postoccupancy evaluation (POE)

Strategic Planning

Programming can engage the fundamental strategies of the organization. Good programming goes beyond simply collecting information about the current area, equipment, and adjacency requirements. *It can also act as a powerful mechanism for organizational development.* It can frame a better understanding of the way the organization works and lead to insights for better organizing and carrying out its mission.

If employee involvement is included in this process, it can also lead to better understanding of organizational, and improvements in employee, goals and programs. By involving staff in meaningful ways, programming can go beyond generating accurate information toward strengthening staff commitment to the decisions and to the organization generally. These are issues that senior management understands and appreciates. Viewing the programming process as a form of organizational development shifts facility planning and design to concern for the ways in which planning and designing facilities can contribute to employee motivation and commitment, to clarifying and attaining corporate goals and objectives, to rethinking roles and relationships, and to reformulating and clarifying strategy and tactics.

Master Planning

Master planning looks at the long-range needs of the organization. It is based on the current conditions and anticipated future needs. It should provide strategic direction about the moves, construction, and real estate activities of the organization. It provides a context for making coordinated decisions about building projects. It answers questions such as: How much and what types of space will we need in the future? What are our space requirements and our space availability over the next 5 years? Where should departments be located to maximize performance and minimize cost?

The master plan must take into account the strategic business objectives of the organization, real estate costs, markets, taxation, growth trends, regulatory environment, and so on. The time horizon and planning cycle should coincide with those of the business plan. The value of the master plan is not in its precision or accuracy in anticipating future needs, but rather it is in its ability to quickly model or assess alternative futures. "If we grow faster than anticipated, how will we handle it? Is it better to retain some of the additional space freed up after a consolidation or to sublease it?" (See Fig. 4.1-2.)

Site Selection and Design

Feasibility studies are used to evaluate options such as the suitability of a particular building or site for relocating a business unit. Does the zoning permit the proposed use? Is it large

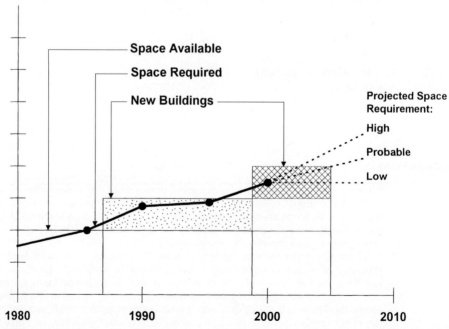

FIGURE 4.1-2   Space projection versus space availability. (*Courtesy of William Sims and Frank Becker.*)

enough? Will it permit future expansion? Will it lay out efficiently? Is accessibility adequate for employees, customers, suppliers, and so forth? (See Fig. 4.1-3.)

Building Design

Many organizations such as government agencies, hospitals, and universities are required to conduct a basic needs assessment study as a part of the initial justification for a project. This involves a careful analysis of needs. For example, a request for an additional building would require a utilization study of the existing facility coupled with projections of future use to document the need for additional space.

Architectural programs are required for renovations or new building projects. They define floor and building sizes; height requirements; structural, mechanical, and other shell, core, and service requirements. This program usually also includes the site requirements as well. Parking, landscaping, pedestrian and service circulation, lighting, security, and signing issues are among the items typically included. The key to architectural programming is identifying the clients' requirements in enough detail to ensure that their needs are understood while generalizing the fixed elements of the building so that it will be flexible enough to adjust to future needs of this client and to other occupants and uses. Buildings tend to last much longer than the initial occupant's requirements. Because buildings typically outlast their occupants, flexibility or adaptability to other occupants and uses should be considered not only as a programmatic requirement, but also as adding to the building's future economic value.

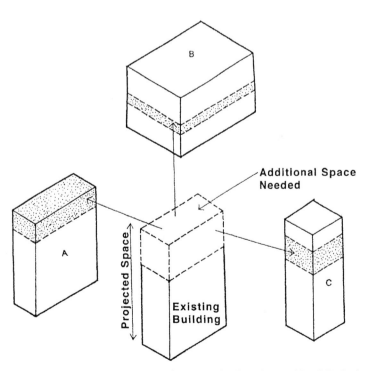

FIGURE 4.1-3   Relocation options. (*Courtesy of William Sims and Frank Becker.*)

Interior Design

Interior Design Programs.   Interior design programs are the most detailed and must deal with all of the elements that make up the interior environment of a building and the relationships among them. This includes rooms, circulation, finishes, furnishings, equipment, signing, acoustics, lighting, adjacencies, image, and the like. For offices, the space and equipment requirements of the different types and numbers of employees (current and projected), along with the necessary auxiliary and support spaces, are the primary drivers of the space requirements. For certain types of facilities such as hospitals, laboratories, or manufacturing, equipment and special spaces may be the primary elements that drive the need for space. Although most interiors last for shorter periods than buildings, the issue of flexibility to deal with future changes needs to be considered.

Postoccupancy Evaluation

Evaluation throughout the construction process and after completion is also very important. A POE should be done to compare the finished project with the objectives agreed to in the Statement of Needs and the Strategic Program. When used by clients, who build serially, such evaluations can help to shape the management, procedures, and content of future construction projects and thus cumulatively improve performance. But even for clients with smaller, one-off projects, a POE can identify changes needed to make the building perform better.

BRITISH PRACTICE

Francis Duffy[4] describes standard programming and design practice in the United Kingdom as one involving regular feedback throughout a project among clients, advisers, and the design or project team. He notes that, because projects involve and integrate so many people and interests, even the simplest requires a series of documents, each taking the preceding one to a greater level of detail.

To get the maximum benefit from a project, a client must regularly and systematically ensure that relevant options for design solutions are evaluated, that appropriate decisions are made, and that there are sufficient resources for programming throughout the project to manage and take ownership of the series of programming documents that will be developed.[5] There are four initial stages:

1. *Statement of Needs.*   A document prepared for board approval at the outset of a project, which defines the objectives and needs of the client organization in relation to a specific property or construction opportunity.

2. *Options Appraisal.*   A formal review of the relative value of the chief options available to the client, including the possible use of existing resources rather than constructing, moving to a new building, or acquiring more space, and the calculation of the benefits, drawbacks, and risks associated with any such option.

3. *Strategic Program.*   The setting out of the broad scope and purpose of the project and its key parameters, including the overall budget and timetable; it should include an outline specification that explains in clear terms what is expected of the project.

4. *Project Execution Plan.*   An explanation of the way the project will be implemented, including details of the procurement system that will be used and the appointments to be made by the project team.

The design and construction phase of a project should develop in three further steps:

5. *Project Program.*   This converts the strategic program into construction terms, establishing initial sizes and quantities for each element of the project and giving each of them an outline budget.

6. *Concept Design.*   Once the project sponsor establishes the project program, the project team can begin to test the design options that will contribute toward the eventual concept design. The tests should include the cost and schedule of the construction program and should also examine whether the concept design is likely to meet the criteria expressed in the client's business case for the project. Duffy warns that it is wise to beware of the bias toward supply-side criteria that is characteristic of the thinking of most in the construction and property industries.

7. *Detailed Design.*   When the concept design is agreed to and signed off, the project team can begin the development of detailed design, specifying the performance requirements for all of the elements of the new project. The detailed design should freeze as much of the design as possible, defining and detailing every component of the construction.

WHERE DOES PROGRAMMING FIT IN THE BUILDING DESIGN PROCESS?

A typical version of the design process consists of a series of steps or phases of activity usually with feedback loops and decision points between each phase. The phases are as follows:

1. Programming
- Intelligence
- Design requirements

2. Design
- Schematic design
- Evaluation of alternative proposals
- Selection of preliminary design
- Design development
- Construction documents

3. Construction
- Construction
- Commissioning

4. Occupancy
- Move-in
- Operation and maintenance
- POE

The feedback loops often relate back to the programming phase as decisions become more specific and as questions arise that require more information. Most variations on this model involve further disaggregation of the phases into more specific steps or collapsing them into more general ones.

Programming Phases

There are two basic phases during programming. The first, the intelligence phase, is concerned with identifying the potential goals for the project, the assumptions or givens upon which it must be based, and the constraints within which it must operate. The second, the requirements phase, is about reaching agreement on the specific requirements, assumptions, givens, or constraints within which the project must operate.

Intelligence Phase: Establishing the Context.   In the intelligence phase for a renovation project, one of the concerns is determining the existing conditions of the facility that may affect the design. These include the conditions of the building and furnishings, personnel and job functions, and such things as ownership versus lease tenure in the building, as well as the conditions in the area surrounding the building such as the other uses, availability of parking, building maintenance, and access to amenities. These may have a bearing on the nature of the design, how long it will remain appropriate, and on the quality and character of the space. It is important to collect information from a variety of sources and confirm it with site visits and interviews. Because the design will be in effect for a number of years, it is important to anticipate the changes that are likely to occur in these existing conditions.

What are the constraints within which the project must operate? Budget and public regulations such as building codes and zoning ordinances significantly determine what can and cannot be done. It is important to know both the formal and informal internal policies of the organization. These conditions are not "cast in stone" but rather are constraints, which must either be accepted or steps must be taken to alter them. It is important to have formal and explicit agreement on them from the relevant decision makers.

Intelligence Phase: Establishing Potential Project Goal.   The second major concern of the intelligence phase is to identify a list of goals that the project might pursue. These potential goals are identified with only cursory attention to resource and time constraints, but all must have implications for design. It is from this list of potential goals that the actual project requirements will be selected. The four basic types of potential goal statements are (1) *problems* posed by the existing or projected conditions that should be eliminated or corrected, (2) *assets* that should be preserved, (3) *opportunities* that should be seized, and (4) *goals* that should be pursued.

Requirements Phase.   In the requirements phase, the concern is with choosing the subset from all of the potential objectives that the design will try to achieve. This is a winnowing activity, and it is critical to remember that this important decision phase must involve those persons empowered to make such choices.

Another key aspect of this phase is operationalizing the selected goals and converting them into measurable objectives and into design or project requirements. Ease of operationalization alone should not determine what is included in the program. Some characteristics (lighting) are easily converted into operational statements because the underpinning science is well developed, whereas others equally important (aesthetics) are not because the knowledge base is less developed. The list also needs to be prioritized. It is important to get a formal sign-off by the relevant decision makers on the program at this stage even though it will change as the process moves forward.

Design

Schematic Design.   The previous phases have been primarily analytical or political. The design phase involves problem solving and is one of synthesis, involving activities of imagining or creating solutions. It is holistic and integrative. As each solution is generated, it is continuously and informally evaluated against the program requirements and assumptions. When a design "feels" as though it satisfies these conditions, it can be put aside for more formal and careful evaluation in the next phase. Other possible solutions should be generated if time and resources permit.

Evaluation and Selection.   When two or three alternative design proposals have been developed that appear to satisfy the requirements, then a more formal and careful process of evaluation is in order. This involves carefully assessing how well each satisfies the program requirements and determining whether there are any other impacts or consequences that need to be considered. Following approval of the recommendation, the selected alternative will be refined.

Design Development.   In this phase, the selected alternative is refined and developed. The program should be used to guide this process. Using a matrix, weak areas can be pinpointed and improved (see Fig. 4.3-3 on page 4.75). Once this process is complete, final approval is sought to proceed with construction documents.

Construction Documents.   Once the decision is made to proceed, it is still necessary to devise the measures that will make a reality from what is still a paper proposal. This involves developing the construction documents—working drawings and specifications. This is often as creative an activity as the initial imagining of the design. The design goes through a lot of change as the difficulties of making it into a reality are encountered.

Construction

Next is the process of selecting a contractor and constructing the building. The processes of value engineering—of seeking more cost-effective ways to achieve the same level of design intent—often occurs during this phase, and the design again goes through considerable change. Finally, change orders inevitably occur as difficulties, errors, omissions, and changes in needs result in yet more adjustments to the design.

Commissioning.   Once construction is completed, the building should be commissioned. During this phase, the facility is accepted and readied for use. The operator's staff is trained, and performance of the building and all its systems is verified under normal and extreme operating conditions.

Occupancy

Once construction is completed and the building is commissioned, then the users move in and the operations and maintenance and general facility management (FM) process begins. Often the building is in need of immediate change because the design must be frozen during the construction process to get it built. Yet the organization has not been frozen during this period and continues to change and evolve. Changes occur that could not have been anticipated during the programming process. Departments may have been merged or eliminated, work processes may have changed, or new equipment purchased.

Postoccupancy Evaluation.   After the occupancy phase has been completed and users have had a chance to settle in and adapt to the new facility, it is a good idea to conduct a POE to see how the design is working. This information can identify needed changes and can feed forward both successes and failures to future projects.

PROGRAMMING MODELS

Relay Model of Programming

There are numerous versions of the planning and design process, but one critical difference centers around the role and position of programming. The first is what we have termed the *Relay Model*. In this case, a staff separate from the designers develops the program, and the program is handed off to the designers who then go about designing the building. Peña[2] argues forcefully for the clear separation of programming from design as a means of maintaining the integrity of each. This is also the policy of Public Works Canada, the Canadian equivalent of the U.S. General Services Administration, the government landlord. In standard American Institute of Architects (AIA) documents, the design process is similarly divided into predesign (programming, feasibility, master planning, prototype development), schematic design, design development, construction documents, construction administration, and postdesign (POE, users' manuals, evaluation research). In the AIA model, the program document becomes the legally binding agreement between the architect and the client as to the scope, focus, and direction of the design project.

The relay model has many problems. First, it assumes that all of the major decisions can be made at the front end of the process and that all of the requirements identified and contained in a document can be handed off to the designers. Most likely this is not the case. If the program is based on one set of assumptions about the problem that must be changed as later information comes to light, then the program document often becomes irrelevant and is ignored.

Rugby Model

The *Rugby Model* envisions that the programmer and designer participate in a cross-functional project team with other expert professionals and user stakeholders who work together from the project's inception to completion. The process is iterative and cyclical. The roles shift with the programmer in the lead during the early stages of problem definition, and with the designer assuming the lead in later stages of problem solving, at each of the levels of the process, such as master planning, architectural design, and interior design. As new possibilities emerge during design, clients and users are critical participants in this process by helping to think through new ways of accomplishing the mission and functions that the building is to accommodate and foster. Designers and programmers can be catalysts in the process by offering insights into organizing the building and the particular process in a new way. Often, this is achieved by showing examples of ways other organizations have solved the same or similar problems.

This cross-functional approach, where designers participate during the problem definition stages, enables them to understand the rich background of information, deliberations, and decisions that went into the definition of the problem, which can never be fully captured in a report. Similarly, it enables the designers to ask questions about assumptions, to offer suggestions and insights based on their experience in solving similar problems, or simply to offer creative insights into new ways to define or solve the problem. The designer can challenge the programmer to be more explicit or to clarify particular requirements.

Similarly, when the designer assumes the lead role during the problem-solving phases, the programmer can be on hand to ask questions. "How does this design solve the problem, that we identified in the program phase, of enabling teams to have rapid and spontaneous collaboration and at the same time offer individuals the opportunity to carry out tasks requiring high levels of concentrated activity?" Also, this enables the programmer to offer insights into ways of fine-tuning the design to better achieve an objective.

This continuous involvement does not mean that there are no decision points, where formal agreement is reached on the program, at key points in the process. It simply recognizes that there will be changes as more is learned about the problem and as alternative approaches to solutions are examined and rejected. In this cross-functional rugby model there is a continuous process of programming and evaluation of proposals based on the program. Similarly, the need for more information to make subsequent decisions requires more programming.

GATHERING INFORMATION

Guiding the search for and evaluating design solutions requires information about the organization, the operating environment, job functions and work styles, space furniture and equipment, building systems and equipment, and the site and surrounding context. Information about the number and type of employees, adjacencies and existing space, furniture and equipment standards—the typical approach to programming—is necessary but not sufficient. Information should be collected about individuals, groups, and departments and their current and projected interaction, current and expected technical requirements, professional identities and work styles, and implicit and explicit organizational goals and values. Good design is based on what works well (and should be preserved) and what needs to be improved (and should be changed). This means understanding the overall operating environment and organizational culture and goals and identifying factors that constrain design options, as well as those that might stimulate change and innovation.

This information must be collected for the current situation and for at least 3 to 5 years into the future. In general, it is sensible to opt to be approximately right rather than precisely wrong. In other words, there is little benefit from spending enormous efforts to be extremely precise about figures that can be only approximations. But for space forecasting purposes, in particular, approximations can be useful in setting basic boundaries and identifying the magnitude of a problem. Things change, and the lesson to be learned is not to abandon efforts to understand future directions but to keep monitoring and updating information and the planning based on it.

Use Multiple Methods

There is no single best data collection method. All have benefits and drawbacks. Every method can be done in an elaborate or large-scale fashion or quite simply, while still retaining the basic characteristics of the approach. It is better to use a variety of techniques in a "quick and dirty" way than it is to use one very sophisticated technique that relies on only one type or source of data—this is called *triangulation.* Data from a variety of sources and users can be obtained without chaos within a brief time frame and with limited resources. These are outlined later. It is usually necessary to derive or extract goals and objectives from the data through a judgmental process of analysis and inference. How elaborate the method is depends on the time frame, budget, and complexity of the problem and the amount of familiarity or existing knowledge. Even the smallest programming processes can, and should, use multiple methods. This may mean interviewing 20 people rather than 200, or distributing a single-page machine-readable survey rather than a detailed 10-page questionnaire, running 2 focus groups rather than 10, or conducting observations over 1 rather than 10 days. The overall picture of the organization's and its departments' and employees' requirements is likely to be more accurate and better understood using a range of these techniques, on any scale, rather than concentrating on a single technique used exclusively and extensively.

Dynamic Process

Organizations' goals, priorities, and constraints will change over the course of the project, and so mechanisms to anticipate such changes must be embedded in the process. The tendency is to blame the occupant groups for any changes or just to ignore them. A plan for accommodating information technology (IT) is made obsolete by new developments in the technology and the decision of the user department to purchase it to improve its operation. The project team must regularly communicate with informants and structure the planning process to reflect the totally predictable fact that there will be changes during a long-term building project. The old model of the programming process as a discrete (and terminal) step before design development is simply obsolete.

Types of Information Needed

Good programming requires a broad range of different types of information about the organization, the operating environment, job functions and work styles, and the physical environment. In all cases, information should be collected about the current situation and about the best guesses of various players in the process as to how things will look in the future.

About the Organization

Goals.   What are the organization's strategic goals? This kind of information can be learned in part from annual reports, but corporate strategic business plans and interviews with influential senior managers are likely to be more accurate and realistic.

Managers will not divulge this kind of information unless they know exactly why it is needed and are also confident that in doing so their trust will not be abused. There are no shortcuts to building trust. It takes time and is based on a track record. A good starting point is to take the time to explain why the facilities group needs to know about confidential business plans. Explanations and illustrations of the way facilities decisions can impact business plans, or the way facilities can be a critical factor in helping to achieve business goals, are the keys to helping managers understand this need. If the project is important enough, directives from the highest levels can help. Often, this strategic direction can be achieved by putting a building committee of senior executives in charge of the project. A workshop of senior executives to outline future strategic directions that have an impact on and can be supported by facility designs is a powerful way of discovering this information and building commitment at the highest level for the project. It can also serve as a useful organizational development tool.

Plans.   What initiatives are already in place or about to start? Why waste time designing better facilities for a group that top management has decided to spin off in 6 months? The project team needs to be kept appraised of plans regarding such business objectives so they can see how their facility plans mesh with them and also determine whether they need to collect different or additional information to answer questions that arise as these plans are implemented.

Constraints and Givens.   What are the fixed and controllable constraints and are they expected to change (or could they be changed)? Will all of the old furniture be reused? Will there be open planning? How many people will be allowed to occupy the building? Is the project time frame cast in stone? Can a height limit or parking requirement be changed? Answers to these questions can help in allocating scarce programming resources (the time and money to collect information) by targeting data collection efforts at those areas that are most open to influence and are under the control of the organization itself.

Corporate Culture.   Understanding the corporate culture is critical to effective programming. Goals and philosophy proclaimed in annual reports, posters, and other corporate communication programs should be viewed skeptically. Instead, the corporate culture is better reflected in the daily actions and activities of management and staff: how people interact with one another and management, how they dress, how they use time and space, the ways in which decisions are made, who makes what kinds of decisions, and the kind of information or evidence considered in decision making.

Asking different staff and managers to describe what the company considers a "good" manager or staff person, persuasive evidence or an effective presentation is a simple way to elicit lots of information about "the right way to do things around here." Observing how and where people interact, how presentations are made, how people are introduced to one another, how space and time are used, and so on provide invaluable clues to the validity of the archival and interview data. It also can stimulate questions to ask, "Why is it done that way?" "Why couldn't it be done this way?" "What would happen if . . . ?"

Understanding the corporate culture also helps structure the type of information to collect (interview data and anecdotes may be useful in one company, whereas quantitative survey data may be considered good evidence in another), as well as helping to identify the best people to whom to present the findings and the best way to do it. If a group makes decisions, why waste time presenting the information to a single individual (unless the objective is to test the idea with a "friend of the court" before making the official presentation)?

Many organizations are pursuing purposeful cultural change processes to improve effectiveness, agility, and their ability to attract and retain the best employees. Often, a new building or a major renovation project is seen as an integral part of this process. The process of planning and designing the new facility and the new facility itself are seen as major elements in this process.

Organizational Structure.   Organizational structure influences adjacencies and determines, at least formally, reporting relationships and decision hierarchies. The project team needs to know how decisions are made formally and whether the existing organizational structure is likely to remain stable or change. This kind of information can determine from whom to collect information and about what. If, for example, two departments are to be merged, it will be useful to know how the new manager views the situation, because the views of the new manager will drive the facility decisions.

Staff and Space.   Head-count projections for a 5-year period and their relation to space requirements are probably the most basic information collected in any programming process. But the issue is not just how many people, but also what kind of people: What kind of work do they do? What are their expectations, their work styles, and their communication patterns? What tools and equipment do they use? Are they professionals for whom demand exceeds supply (and, therefore, have a strong bargaining position)? Are they the same kinds of people who have been hired in the past, or are they different, better educated, more professional, or from different racial or ethnic backgrounds, that may affect their work styles and expectations or their ergonomic requirements?

About the Operating Environment

Business Conditions.   What are the market forces? Where is the competition coming from now and where is it likely to come from in the future? Is money tight or easily available? Answers to these kinds of questions affect the organization's attitudes about the nature, amount, and quality of space that it needs. If more international firms will be competing in the same city, for example, then information about these firms' facilities should be obtained. Some U.S. companies consciously design their facilities in Asia to a higher standard than is customary there to differentiate themselves from their competitors and attract more qualified staff. Office standards in London have improved dramatically in large part because of the influx of large American financial companies that imported higher American office standards. All of them have raised expectations among the workforce as to what constitutes an acceptable working environment.

Laws and Regulations. Codes, standards, laws, and regulations can be seen as the "given" program. These include zoning ordinances, building codes, subdivision regulations, special districts, covenants and deed restrictions, and easements. What uses does the zoning permit? How much parking is required? What are the height limit and setback requirements? It is important to do a search for applicable regulations because they can vary with location. What are the tax laws, and how will they affect building form, the decision to renovate an old building, or to build a new one? In the United States, one of the driving forces behind the success of systems furniture and open planning was the tax code, which treated panels as furniture, thereby allowing them to be depreciated at a much faster rate than conventional drywall construction. The provision of plazas (in name if not in ambience) in New York City is directly related to variances in zoning-dictated height restriction if such amenities are provided at street level.

New Technologies. From computers to automated building systems, information must be collected about the nature, extent, and use patterns of existing IT, as well as expectations about the way it may change in the future. Union Carbide's headquarters building in Danbury, Connecticut, for example, was wired with fiber-optic cable long before it could actually be used, based on the much lower cost of installing it during the original construction. Everything from workstation size to the size and location of electrical closets, risers, and wire management systems will be affected by expectations regarding technology. For many organizations today, the best approach is hedging the bet by putting in more than less. At the least, the decisions should be deliberate, based on the best information available and without believing that it comes with a guarantee.

Labor Force Patterns. Labor force patterns should be predicted with some accuracy. Will there be more women, dual-wage earners, older workers, or Spanish-speaking employees? Workers' expectations are rising with respect to pay and also to air quality, comfort and safety, time and spatial freedom, and the extent to which the environment supports their sense of personal and professional identity, their human dignity, and their sense of competence. These are not trivial issues in areas and industries in which the demand for well-educated, professional employees exceeds the supply.

Given demographic trends that show that there will be fewer qualified workers over the next 2 to 3 decades, understanding how shifts in the workforce affect facilities decisions is essential. Information about these workforce patterns has direct and immediate effects on both design requirements and design solutions. For example, families that have two wage earners put increased pressures on transportation systems, day care, and access to shops and services; more women employees work at all times of the day and night and increase demands for safety and security both inside and surrounding the workplace; age shifts in the workforce lead to more concerns about health, fitness, lighting, air temperature, and air quality.

Competitor Actions, Plans, and Experience. We all measure our own situations by comparing them with others' situations, and so competitors' facilities can become a benchmark. The project team can also learn from its competitors' experience: How did they handle space planning or churn, and how did it work? What have been the benefits and drawbacks of a new automated building system?

About Job Function and Work Style

Task Analysis. What exactly do different employees do? How do they work? Both written job descriptions and interviews are useful for understanding what different people actually do and how they relate to different jobs and parts of the organization.

Environmental Satisfaction.   How satisfied are staff with their current environment, and how do they believe it affects their ability to work effectively and productively? Just because they have survived in their current surroundings should not imply that their current situation is satisfactory or should be continued. Interviews are effective for discovering how employees see that their work environment affects their performance. Surveys are an easy, cost-effective method for quickly obtaining large amounts of quantitative information. These data can be analyzed by department, discipline, age, previous job experience, and sex to find out how widespread opinions are and to target areas that have special problems.

Communication Patterns.   Who communicates, where, when, and how often now? Is that considered acceptable? Desirable? What are some of the problems within or among departments? Before changing the environment to support a new communication pattern, it is useful to see whether the problem is perceived or real and to distinguish among different types of communication.

Adjacency Requirements.   Adjacency requirements are related to communication and range from seating location and even orientation of workers within a team or group area to the relationship of departments, buildings, and whole sites to one another. Most of the computerized adjacency software packages are based on the premise that people who have strong organizational relationships should be physically close to one another. It is more important, however, to know who should be communicating and who is likely to communicate. Then, adjacencies can be designed to support groups that should communicate but are unlikely to do so without close proximity. In other words, spatial bonds can be used to overcome organizational barriers.

Space, Furniture, and Equipment Requirements.   How do people work, what equipment do they use, and how do they use it (in what way, sequence, combination with other equipment, other materials and resources, alone, together, etc.)? What kinds of furniture and equipment do they believe reflect their personal and professional identity or job status? Functional analysis rooted in ergonomics and human factors are important, but function goes beyond lumbar support, glare, and keyboard height, and it should include desired images of a functional and effective workstation.

Information Needed for the Internal Physical Environment.   Information must be collected that describes clearly both what is and what is anticipated (to whatever degree of precision that makes sense). Many organizations have no accurate inventory of the type or amount of equipment they have or its special environmental requirements. Many have not thought through what kind of electronic networking they want or expect and in what time frame. Although the accuracy of such information is often questionable, simply seeking it forces organizational players to share their (possibly conflicting) visions of the future. It is impossible to plan realistically or effectively without such information.

An inventory should be made of existing furniture, equipment, and support spaces like cafeterias and break areas, libraries, conference rooms, computer center, and project rooms that determines their condition, location, availability, ability to support technology, and flexibility and adaptability. The principal questions are what is available (or will be by the time it is needed) and whether what is available is suitable for its intended purposes (now or in the foreseeable future). This information is invaluable for determining whether new furniture or equipment will be needed, whether what exists can be used as is, and what might be usable if it were refurbished or renovated.

Information Needed for the External Physical Environment.   Information about the external physical environment will determine what services and amenities are provided on-site and at what level. Again, it is important to understand how a site or neighborhood will be changing, and it also is important to use caution in basing amenities, such as dining or parking, on comparable services available off-site. Knowing that something will be available can, however, justify a smaller investment and modest plans during the interim period.

Collect information that will be used, and use the information that is collected. Collecting and making sense of all these data can be a frightening prospect. But remember: The objective is not to collect every conceivable bit of information but to collect that needed to make an informed decision. Often, so much effort, time, and resources are devoted to collecting information that no resources are left for data analysis or use in the decision process.

SECTION 4.2
PROGRAMMING ACTIVITIES AND TOOLS

William Sims, Ph.D., C.F.M., IMFA Fellow, Professor, and Franklin D. Becker, Professor
Cornell University, Ithaca, New York

Regardless of the exact description of the programming product you will have to make, there are activities that are common to all. In this section, we suggest tools that can help you to accomplish these activities. Some will be quite familiar; others may not be even thought of as programming tools. The less familiar will probably be those that delve more deeply into the strategies of the organization and attempt to link programming to the organizational mission and those that dig into the way facilities are used and into the preferences and knowledge of the users. The order in which we describe the tools is not necessarily the order in which they should be employed in every case. For example, we indicate that evaluative tools can be used as programming tools, yet we list them next to last, because customarily they are employed after a project is completed and often before a new one is underway. The first, structuring the process, is often omitted from programming and, if so, it is much to the disadvantage of the organization. Without this step, programming may run off in an incorrect direction or receive too little commitment from its top management or from the other stakeholders.

In this section we describe the following:

- Structuring the process
- Forming the project team
- Linking programming to organizational strategies
- Gathering information about the context
- Gathering information about the organizational subunit
- Analyzing the individual user
- Determining building requirements
- Using evaluative methods as programming tools
- Developing concepts for the future

In each programming project, not all of these activities must be carried out in the depth that may be implied by our descriptions. Nevertheless, they do comprise a range that every facilities project requires. The tools can be selected and modified as appropriate and combined into a programming process to meet the particular programming task. They should be taken only as suggestions because each programming task is unique, and the specific techniques must be custom designed to meet the particular situation based on the resources available, the complexity of the problem, and so on.

STRUCTURING THE PROCESS

Project Mission

What is this project about? What is it intended to accomplish? What is the overall purpose? The *project mission* is a statement that clearly and concisely expresses the reason that the client is undertaking the project. It defines the special purpose of the building and the needs that it must satisfy. Top management must establish the project mission. It is an opportunity for them to clarify their purpose in undertaking the project. A clear mission is helpful for management and others to communicate what the project is to achieve and to build support for it.

Top Management Support

The project team should make sure that top management understands, agrees with, and is willing to visibly support the basic programming approach and the project requirements that it produces. Top management should establish and maintain strategic direction for the project but should not get involved in day-to-day operational decisions. From a senior management perspective, the main concern should be with the way the decisions made help the organization to achieve its strategic and operational goals and plans. Management needs to know the implications of alternatives in terms of time, cost, personnel, flexibility, image, and so forth. They should review and sign off on the decisions periodically.

Clear Roles, Authority, Responsibility, and Accountability

To avoid wasted time, energy, and frustration, the project team responsible for organizing the programming process needs to identify the classes of decisions to be made by different individuals and groups. The team should not second-guess delegated authority or tolerate second-guessing of its own decisions. It should have a good sense of what it needs to do a good job and then request the resources (time, money, expertise) needed to fulfill the team's responsibilities. If these are denied, it is the team's responsibility to inform management of the likely consequences. It should also be held accountable, that is, to measure the extent to which it has achieved its goals in light of the resources and authority at its disposal.

Design Participation Matrix

Create a design participation matrix to clarify who are the key stakeholders in the process and what role each will play. This matrix identifies the different players in the process from CEO to janitor and indicates the nature of the participation of each stakeholder or group in decisions at each stage in the process. In short, it outlines who decides what. This does not mean that all employees should be involved in all decisions, but it does mean that from the outset a clear conceptual structure should identify which decisions different groups in the organizations should share. It suggests what kinds of decisions different levels should be involved in, ranging from workstations to the site as a whole. It is a good idea that all employees be given the opportunity to influence some aspects of their physical surroundings, but not that every employee be involved in every kind of decision. Some are best left to technical experts, and some are the responsibility of management.

A decision matrix serves several purposes. First, it separates tactical from strategic or operational decisions. Clearly, senior management should be responsible for and more involved in the broader, strategic decisions than in details about workstation design, furniture, and finish selection. Should we have a new building? What form should it take? Where should it be located? Where does it fit into the overall corporate strategy? Second, the decision matrix helps to set realistic expectations that give employees a sense of what kinds of decisions they will and will not be able to influence. It also helps remind senior management what kinds of decisions they should not be making.

Clear Philosophy

The project team should have a *big-picture* view of the overall project and process. The team needs to think about the implications of its philosophy; for example, "doing it right the first time" may delay the project to make it better, it may require unexpected resources, or it may suggest the need to restructure old relationships or typical work patterns. Senior management must understand these kinds of organizational implications if their support of the big-picture approach is not to be empty rhetoric.

Calculated and Controlled Risk Taking

Good projects often break the mold. The project team needs to be willing to take risks, but it should help others to understand the benefits of risk relative to cost. Top management must buy into risk with the project team. Regular reviews and sign-offs, in which the organizational implications of different choices are presented (e.g., the effect of a cutback of amenities on the ability to attract and retain staff, the impact of fewer common areas on informal communication processes and subsequent innovation, the improved response time and lowered cost and disruption of relocations if raised-access floors are installed), is one way of doing this.

Tracking and Monitoring

The team should keep track of decisions made and stress the importance of these decisions on other actions, players, and outcomes. Team members should also identify the kinds of information they need to make good decisions in a timely manner and to sell the decisions to top management and users. Finally, they should anticipate the consequences by team brainstorming, contacting others with similar experience and special expertise, and continuing contact with a wide range of users and others involved in the process.

Procedures

Checklists and procedural guides help users and others provide programming information and review program materials, proposed designs, and working drawings. Too often, the project team assumes that the representatives of occupancy groups or their managers know how to effectively solicit feedback from their staff or to explain proposed plans; often, however, they do not. Some simple tools, which may be no more sophisticated than a checklist of questions that should be asked of staff, can ensure better feedback and better-informed decisions. Procedural guides and training, to ensure that this critical input is sought and received in a credible and thorough manner, are a good idea. Also, a member of the project team should provide technical support to the project team representative, in the form of asking and answering questions and probing responses, to ensure that all needed information is obtained.

Recording Forms

The information obtained in each of the tasks must be recorded in a standard way so that it can be analyzed and communicated to designers and clients. Use cards or sheets of card-stock paper to record the information. These cards can be arranged in categories on a wall so that the programming team can interact with the client/user representatives. These cards are used graphically and with simple text to indicate the space needs that have been derived from the project goals, the strategic visioning, the unit, and user analyses. These can be taped or pinned to the wall, rearranged by category or priority, discarded, or corrected in the course of developing understanding of the program needs.

Involvement of Employees

Employees at all levels should be involved in the project in some way by completing surveys and participating in interviews, helping to identify design requirements, and reviewing preliminary design proposals and policies for allocating and using space.

The appropriateness of particular techniques for involving users and collecting data depends very much on the availability and accessibility of the users or clients. Lynch[6] developed a typology that summarizes these client conditions very nicely.

The client or user is

Present

Absent but reachable		Homogeneous		Vocal
or is	**and is**	or	**and**	or
Not reachable but known		Diverse		Silent

Unknown

In the most direct and simple programming situation the client is the user, is present on the site, and is homogeneous and vocal. In this case, if there are multiple users, then they are homogeneous in terms of their roles, functions, values, and so forth. They can be interviewed, observed, respond to simulations, or be involved in the programming process. The users/clients have concrete experience with the site, and their assessment of its adequacy for their present and future purposes is as direct and straightforward as it possibly can be. In this situation, data collection or a participatory programming process can be quite effective.

The task becomes more complicated if the clients/users are complex in terms of their roles, values, uses, and intentions. Then, each client type must be identified, and some way must be established to ensure participation either by every client of each type or representatives of each type. The process becomes much more political, and the programmer's role becomes one of mediator and catalyst.

The process is still further complicated when certain users have no choice in the process. This can arise out of the political or institutional structure of the situation or out of the cultural tradition of the users themselves. In many institutions, certain actors traditionally have no voice in deciding their surroundings. Students, for example, typically have little say in the design of classrooms in a university. Upper-level employees of a company involved in a major renovation effort are experienced and effective advocates for their points of view in a situation that involves attending meetings, lobbying, discussing, voting, and the like. Lower-level employees, on the other hand, often have no experience in such situations, and their attitudes regarding such participation may be negative. Consequently, they will not participate at all or will do so passively with the expectation that it is futile in any case. Such attitudes are very difficult to overcome, especially when they often reflect the political realities of the situation.

The problem becomes still more complex when the users/clients are removed from the site, as in the case of a move to a new facility, because they do not have immediate concrete experience to guide their assessments. This removal can be further exacerbated if there is a major organizational redesign or change process underway. But in any case, they can be assembled, interviewed, observed, and so on, and can directly participate in the process. Techniques such as envisioning or simulations can help in this situation.

The most complex situation is when the clients/users are neither assembled nor reachable, even though the type of client/user may be well known. Unfortunately, this is a very typical programming situation. It is a speculative office building, where the occupants will not be assembled until after the building is completed and the tenant organizations have made their leasing decisions. In this case, the programmer is forced to fall back on surrogates, similar to the prospective buyers/clients/users in terms of organization type, basic work processes, and for individual user needs on job type, stage in life cycle, or other relevant dimensions. Programming is made especially difficult because one is never quite sure what the relevant "dimensions" are, and it is difficult to assemble the surrogate users. Finally, given the unreality of the situation, the seriousness and reliability of the surrogate clients regarding the process is questionable.

Time and Resources

The organization must make available the time and resources that employees need to make informed judgments and decisions. It is unfair and a waste of time and goodwill to create committees and focus groups and then not give the participants access to the resources necessary to do a good job. Such resources may include access to informed experts such as consultants, site visits to broaden one's experience and image bank of alternatives, and time to do one's homework.

Timely Involvement

Employee, consultant, and staff involvement should be timely. The greatest gains are likely to be realized when people are involved at the beginning rather than at the end of any stage in the process. Also, nothing is more irritating than to find that the information that one has worked hard to provide is too late to influence the decision.

FORMING THE PROJECT TEAM

The project team composition and the way it is organized depends, to a large extent, on the size of the organization and the importance of the building in its overall operations. In large organizations, it is unlikely that the entire senior management group will be available regularly to participate in workshops and to review and approve project team proposals. A good solution is to appoint a building committee that comprises knowledgeable senior executives who are charged with overall responsibility for the project. It is important that they and the rest of the senior management understand the time commitment and importance of the responsibility—and that it cannot be delegated to their administrative assistants. The project team reports to this management committee.

The project team leader should probably be a member of the facilities department of the organization. There should also be user representatives from each of the affected organizational units. These should be members who are knowledgeable about the current and future needs of the organizational unit and who have good communication and people skills. Their role will be both to represent the needs of the unit to the project team and also to communicate the results of the project team's work to their unit. There should be a group of internal and external advisers who are experts or especially knowledgeable about some aspect of the project. Other stakeholders (e.g., customers, visitors, maintenance and operations staff, service providers, mail, delivery, food service, emergency, and security personnel) should be identified and their roles in the process determined.

Variety of Expertise

It is helpful to have a core project team that spans the kinds of expertise most central to the project. Almost all building projects today are sufficiently complex to make it impossible for a generalist programmer to have the knowledge and experience needed to make good decisions about all aspects of facility planning and design. To get the right information and to make good choices, many types of experts will be needed at different stages in the design process. Such experts include architects; interior designers; landscape architects; mechanical, structural, and electrical engineers; lighting and acoustical designers; ergonomics experts; organizational ecologists; facility managers, operations and maintenance staff; value engineers; and construction managers. When hiring or selecting consultants, it is advisable to consider not only their expertise and experience but also their approach to the process and its compatibility with that of the project team.

The *design participation matrix* (Fig. 4.2-1) is a technique that uses a simple matrix to structure a group discussion about the role of different players in a project. Its purpose is to identify the stakeholders, the kinds of decisions that need to be made, and the nature of the participation of each stakeholder group in each type of decision. This process helps to identify different views within the organization about which individuals or groups should make which kinds of decisions. The design participation matrix can help reduce the conflict and wasted effort when confusion arises over which individual or group should be making which decisions or participating in a particular way. It also forces the organization to think of all of the relevant stakeholders and to consciously consider their roles in the process.

The core project team should develop the design participation matrix. Senior management should review both the stakeholders identified and the type of participation that the core project team has assigned to different stakeholders. The objective is to increase the likelihood that no significant group is overlooked and to ensure that top management understands the role of each participant, including its own. This minimizes later second-guessing and end-runs in the project. Other key functional areas such as human resources (HR) and management information system (MIS) should also review the design participation matrix to ensure that no key parties to the project have been overlooked and that there is agreement across functional lines on the roles and the nature of the involvement of all participants.

Logistics

The time required to complete the design participation matrix may vary. At least 1 or 2 hours should be planned to allow initially for full discussion; at least 1 hour is required for top management and other groups to review the matrix. It requires group facilitation skills, knowledge of the company, and the ability to synthesize information. It also requires a clear sense of who ought to be involved and the different kinds of involvement that may be appropriate for different stakeholders.

LINKING PROGRAMMING TO ORGANIZATIONAL STRATEGIES

Increasingly, new buildings and renovations are being used to facilitate cultural change in organizations. These changes typically involve reengineering the work processes and organi-

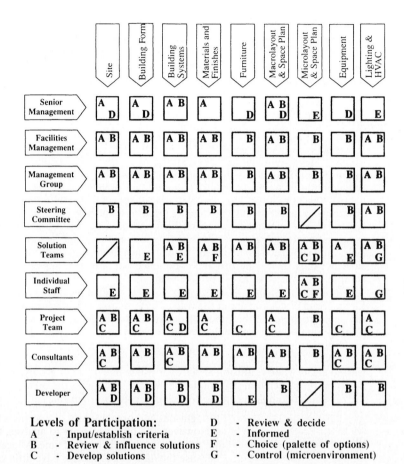

Levels of Participation:

A	- Input/establish criteria	D	- Review & decide
B	- Review & influence solutions	E	- Informed
C	- Develop solutions	F	- Choice (palette of options)
		G	- Control (microenvironment)

FIGURE 4.2-1 Design participation matrix. (*Courtesy of William Sims and Frank Becker.*)

zational structure and culture. Often, this leads to creating new ways of working and much more dynamic use of space.

Strategic organizational plans can and should have important links to facility planning. In fact, many building projects grow out of an organization's plans for transforming the way they operate or for acquiring or developing a new line of business. These are two central questions:

1. What are the implications of the strategic plans of the organization for the facilities?

2. How can the facilities help to achieve the organization's strategic vision?

Plans to start a new line of business, to grow or shrink one dramatically, or to acquire or merge with another organization all have substantial and obvious implications for facilities that must be considered as a part of the organizational strategic planning process. Where should the new plant be located? How big should it be? How long will it take to bring it on line? How much will it cost? What is our exit strategy for the facility in the event that the new product does not work out as planned?

Visions of where the organization is going, how it is trying to get there, and the best paths for success all have a strategic vision. If the organization is planning a cultural and work process change, the question of how the facility is planned, designed, managed, and used can play a central role in that process. Thus, programming establishes facility goals and performance requirements from the strategic vision and, in turn, uses both the process and the design of the new building to help achieve that strategic vision.

Management Interviews

Management interviews include a set of questions for conducting a focused interview with key management people.[7] These questions are constructed to uncover aspects of the organizational unit, such as the mission, the major activities, the nature of organizational challenges, the different aspects of organizational culture, the attitudes and aspirations around innovation and change, the nature of the business and marketplace, and how all of them are changing, that will affect the nature of the design and the overall success of the project.

Management interviews are common and their value obvious, but we provide a sample of them because they are critical to the success of any design. Furthermore, at least some of the questions one should ask may not be obvious. Understanding some of the key aspects of the culture, for instance, may influence not just which solutions are put in place, but how they are implemented and sold to the organization.

Management interviews are helpful at the beginning of a project in trying to understand the context for the new design and the factors that may both drive the change and also impede it. They are especially important in identifying business challenges that the organization is facing, so that those developing the design can frame their solutions around helping the organization meet these fundamental business objectives.

Depending on the type of information sought, interviews may involve key management in all areas of an organization. Of particular relevance are managers in information systems, HR, and the business units themselves. Executive leaders should also be interviewed to get an understanding of the vision of the corporation at the highest levels. Any managers involved in reengineering, strategic planning, technology, or TQM projects are also good candidates for interviews.

Individual management interviews usually do not require more than from 1 to 1.5 hours per manager. Having a well-thought-out and organized set of questions ready for each interview will pay huge dividends. However, the time necessary to analyze, tabulate, and/or synthesize the data collected may eat up a much larger chunk of time. Allocating time for analysis and interpretation is critical because the data by themselves have little value.

Sample: Management Interview Questions   In the following list are examples of the types of questions that should be included in management interviews directed at setting the goals of the project:

- Describe your company's business.
- Do you have a formal mission statement? What is it? Can you give some examples of the way it is reflected in what your organization (and the people in it) does/has done?
- Do you have a formal statement of organizational values and philosophy (e.g., corporate culture)? What are its major tenets? Can you give some examples of the way this philosophy and these values reflect what the organization (and the people in it) does/has done?
- How do you characterize your firm's management style or philosophy?
- Is this an unusual management style or philosophy in your field?
- Can you describe the incentive system? What are people rewarded for doing? Are there any differences between the way the formal and informal incentive systems work?
- What are the key organizational challenges that your organization is facing? Are there internal and external challenges (e.g., reducing costs, attracting and retaining staff, building teamwork, increasing market share)?

- What has your organization been doing (or considering doing) to meet these challenges?
- How do you characterize your firm's philosophy about change and innovation in the way it conducts its business? Can you give an example that reflects this character?
- Can you describe the decision-making process? Who makes what kinds of decisions? Can you give an example of a major decision that was made in your firm and how it came about?
- What is the role of rank-and-file staff in the decision process (e.g., What input do they provide in making decisions? How do they provide it? What sorts of decisions?)? Can you give some examples of decisions made by different people/levels in the organization?
- On what basis are decisions typically made (e.g., data, tradition, authority)?
- How are decisions typically made (e.g., by consensus or fiat)?
- Generally, what degree of certainty about the expected outcome is required before decisions are made?
- How do you characterize your firm's attitude about risk taking in its core business? Can you give an example that reflects this character?
- Describe the type of people who work for (or that you might want to work for) your company.
- How difficult is it for you to attract the kind of employees you want (supply versus demand of desired workforce, competitive pressures)? Is it more difficult for you to attract employees at different levels within the organization?
- What helps you to attract the kind of staff you want? Are there differences between job levels?
- How are differences in status and rank conveyed in your firm, if at all? Can you give some examples to illustrate (e.g., offices for upper management, parking spaces, executive lounges, airport club membership, etc.)?
- How important is the communication of differences in status and rank? Is it changing or has it changed recently?
- Is turnover viewed by management as healthy and natural, something to be avoided, or just not a major issue?
- What is your current client base? Where are these firms located in relation to your office (distance and time)?
- Is cost containment a significant concern in your organization?
- Is real estate cost a significant concern in your organization?
- How much is spent on real estate annually now (e.g., rental costs annually per square foot, number of square feet being paid for currently)?

Workshop to Develop the Facility Implications of the Strategic Vision

What is the process for uncovering and then translating strategic ideas for changing the culture and work processes of an organization into a program to guide the design of a facility to support that change? One approach is to use a series of workshops with the project team and the senior corporate executives involved in planning and implementing the new direction. The starting point is not the building design, cost reduction, detailed work processes, or detailed descriptions of particular spaces, their sizes, and related equipment. Rather, it is an analysis of the challenges that face the organization and the strategic ideas the organization has for meeting those challenges. Once these are identified and agreed upon by the team, they become the driving forces for the project. The next step is to derive the implications for the design of the facility. This task focuses on developing a few good rules or principles to guide the overall design of the building, so that it will build commitment to the values and new ways of working that senior management is trying to promote.

Very often, it is difficult for senior managers of the organization to comprehend the implications of some of the design principles that come out of this process. It is important that they do understand them and that they are committed to them if they involve significant changes from current practice. Nothing is worse than a sudden realization that the proposal is not what management had envisioned and that it is too radical for the culture. Second, senior managers often reject concepts out of hand that are in fact appropriate for their strategic vision because they have a negative image of the idea. Nonterritorial offices, open plans, or universal workstations all have strong negative stereotypes in the minds of many. It becomes important to enrich the image bank, so that decisions are made with a good and realistic understanding of the possibilities and drawbacks of the choices. One good way of doing this is to show real examples of the way other organizations have translated similar issues and concerns into new facilities. These presentations may include actual site visits and discussions with relevant counterpart executives where possible. When this is not possible, a photographic slide show or video of other places, combined with data illustrating how well the solutions have worked to meet the organizational challenges previously identified, can be employed. The goal is not to reach decisions about the proposed building. It is rather to stimulate a lively discussion about what the organizational challenges are and how these and the organization's strategic vision, values, and philosophy might be expressed in the building project. Following this workshop (or series of workshops if necessary), the project team would produce a set of recommendations for discussion and approval by the senior executives.

For example, a common strategic goal for many organizations these days is a flatter and more integrated organizational structure. Flatter organizations imply reductions in hierarchy to achieve quicker decisions through fewer levels of management. Commonly associated with this is a move to a more egalitarian culture, empowering employees, and getting them more engaged in and connected to the values and goals of the organization. Increased integration implies more communication and interaction and a shift from individually oriented to a more collaborative workstyle—teamwork. All of these values have significant implications for facilities:

- A horizontal building with few gate functions to indicate status.
- Workspaces sized, designed, and furnished for function rather than as status indicators.
- Work consists of movement between multiple work settings designed to support particular tasks at a high level of performance rather than at a single setting designed to support all tasks.
- Workspaces designed to support spontaneous collaborative teamwork and concentrated individual work.
- A redistribution of space—reduced individual workspaces designed to support only concentrated tasks to provide for dedicated collaborative team spaces.
- Provision for individual employees and groups to personalize their workspaces.
- Workspaces designed for quick and easy adaptation by individuals and groups to improve work processes.
- Circulation systems designed to concentrate, rather than disperse, pedestrian traffic to foster increased chance encounters.
- Clustering communication magnets, break areas, and settings for informal meetings along the major circulation path.
- Settings and places designed to accommodate displayed thinking where projects and accomplishments can be displayed for comment and information sharing.

These are just a few examples of the types of design implications that can be derived from a strategic vision for transforming the organization.

GATHERING INFORMATION ABOUT THE CONTEXT

One of the main concerns in programming is determining the existing and projected internal and external conditions that might affect the design of the building. Peña[2] describes this step as establishing the facts about the project. It is also about uncovering the list of potential goals and objectives that should be considered for the project. Use a checklist to ensure that the important conditions are considered.

Existing Conditions

Internal or endogenous conditions include the type and conditions of the existing building and its furnishings and equipment, personnel and job functions, and such things as ownership versus lease tenure in the building. It is also important to consider the external or exogenous conditions in the area surrounding the building such as the availability of parking, building maintenance, access to amenities and services for employees, and access for customers and suppliers. The quality of the surrounding buildings and uses and their appropriateness for the intended building or renovation may have a bearing on the nature of the building and its continued viability. Collect information from a variety of sources, and confirm it with site visits and interviews. Some of the information can be obtained from a simple walk-through. Others may require on-site measurements and archival searches.

Future Conditions

Because the design will be in effect for a number of years, it is important to anticipate as many as possible of the changes in the building that are likely to occur during that period. An already approved but not yet constructed addition to the existing facility, organizational changes that have been approved but not yet implemented, the effects of an emerging technology, the effects of wear and tear, the life expectancy of systems, and lease expirations are just some examples.

Identify what will be happening in the surrounding area during the planning period. Whether the area is projected to decline or improve may influence the decision of the company to stay or relocate and consequently will affect willingness to invest in improvements. A projected public action such as a freeway requiring your building site for right-of-way may nullify or make imprudent any planned improvements. Not knowing about an adjacent property owner's plans to construct a building that will block a desirable view could make a design that capitalized on that view and the project team that produced it look very foolish. Even the potential that such a building could be constructed should be evaluated. The decision of an adjacent property owner to develop a center containing shops, dining, and other services would affect the desirability or need for such services on-site.

Determine the real status of such plans before hard decisions are based on them. Often, such projects are only desires rather than real projects with a strong probability of being implemented. The status of such projects and plans and the probabilities of their being implemented should be explored and nailed down with the relevant planning officials and with knowledgeable real estate professionals.

Use the following checklists as guides to developing the information on the existing and future conditions or the context within which the design must be developed.

Local Area Checklist

- The character of surrounding environs (the status and image of the area, the reputation of organizations in the area, etc.)
- The character and quality of adjacent structures

- Civic services (fire, police, health care, libraries, schools)
- Access (air, rail, bus, etc.)
- Traffic volumes and patterns
- Climate
- Sewerage
- Electric power
- Water
- Storm drainage
- Energy sources
- Telecommunications
- Solid-waste disposal
- Land costs
- Taxes
- Development costs
- Incentives

Site Checklist
- Acreage—adequate for buildings, parking, expansion, and the like
- Shape
- Topography
- Soil and subsurface
- Vegetation
- Views
- Drainage
- Microclimate (precipitation, prevailing winds, extremes, solar orientation, etc.)
- Accessibility by cars and trucks (relation to and capacity of roads and streets; parking capacity and quality)
- Accessibility by pedestrians (routes, crossings, and distances)
- Accessibility to public transportation (bus, trains, distance to stop or station, frequency of service, etc.)
- Accessibility to airport (frequency of service, etc.)
- Accessibility for emergency vehicles (distance to fire and police stations)
- Safety and security (law enforcement, social climate)
- Amenities (shops, stores, restaurants, banks, post office)

Existing Building Checklist.   Also see the Serviceability Scales and Orbit 2.1 systems described later.
- Functionality of layout
- Building efficiency (assignable-to-rentable ratio or assignable-to-gross)
- Entrance (attractive, ADA-accessible)
- Access to common facilities, entrance, and service
- Structural and building modules
- Floor loading

- Flexibility
- Ease of subdivision
- Expansion capability
- Way-finding aids (signing, clarity of circulation system)
- Vertical transportation (elevators, stairs, service, etc.)
- Electrical capacity
- Telecommunications
- Restrooms
- Indoor air quality (temperature, smells, pollutants, air movement, humidity, control, etc.)
- Light (artificial, daylighting)
- Finishes
- Energy efficiency
- Views (from inside)
- Acoustics
- Safety and security (access control, parking, lighting, etc.)
- Maintenance and maintainability (roof, exterior walls, interior finishes, etc.)
- Waste disposal (separate room, access to loading dock, etc.)

Constraints or Givens

What are the constraints within which the project must operate? Budget and public regulations such as building codes, zoning ordinances, and the host of other laws and regulations significantly determine what can and cannot be done as do certain company policies or decisions. These conditions are not necessarily "cast in stone," but they must either be accepted, or active steps must be taken to alter them. Such changes require time and resources and often have uncertain outcomes. It is important to have an explicit agreement on these givens.

The following is a list of codes, regulations, and standards that may apply to a project. Check to see if they are required and which edition is in effect. It is important to remember that the applicability and specific provisions vary by political jurisdiction. For example, does one of the model codes apply or has the state or local government adopted a combination or written their own?

Checklist of Codes, Regulations, and Other Constraints

- Building code (BOCA National Building Code, Standard Building Code, Uniform Building Code)
- Plumbing code (BOCA National Plumbing Code, Standard Plumbing Code, Uniform Plumbing Code)
- Mechanical code (BOCA National Mechanical Code, Standard Mechanical Code, Uniform Mechanical Code)
- Electrical code (National Electrical Code)
- Life safety code (National Fire Protection Association NFPA 101)
- Americans with Disabilities Act (ADA) Guidelines
- One- and two-family dwelling code
- Model energy code
- Zoning ordinances

- Special ordinances
- Subdivision regulations
- Health codes
- Historic preservation
- Covenants and deed restrictions
- Easements
- Environmental Protection Agency (EPA) Act
- Fair Housing Act (FHA)
- Occupational Safety and Health Administration (OSHA) Act
- National Fire Protection Association (NFPA) Standards
- American Society for Testing and Materials (ASTM)
- Underwriters Laboratories (UL)
- American Society of Heating, Refrigeration and Air Conditioning Engineers (ASHRAE)

Steps in a Code Search Process. A search process would involve the following types of steps:

- First, determine which code applies.
- Then, determine the occupancy classifications, calculate the occupancy loads, and review any specific occupancy requirements.
- Determine the construction type along with the ratings for structural elements and the maximum floor area and building height allowed.
- Next, determine the requirements for egress, in terms of quantity and type, travel distances, minimum widths, and signage.
- Next, determine the requirements for fire and smoke barriers, fire and smoke detection, and fire suppression systems. Determine the types and numbers of plumbing fixtures required.
- Similar reviews for the electrical, mechanical, and other codes are needed along with a review of the special requirements for types, tests, and ratings for finishes and furniture.

GATHERING INFORMATION ABOUT THE ORGANIZATIONAL SUBUNIT

Once the strategic overall directions of the project and the context have been established, the next task is to begin the detailed programming for each of the organizational subunits, divisions, departments, and work groups. The first step in this part of the programming process is to have the unit representatives define the missions of their organizational units followed by the functions and operations that are used to accomplish it. Critical to this step is to ensure that the question of how can we best accomplish the mission, as opposed to how we currently do it, is addressed in the light of the overall strategic directions established earlier.

The next step is to create a statement that describes the mission of each of the organizational units. A mission statement explains why an organizational unit exists and what its main purpose is within the overall organization. It should explain what the organizational unit is about. This mission statement needs to be as clearly and simply stated as possible. Next, it is necessary to describe the functions that are used to accomplish the mission, and then, finally, to describe the activities that are carried out to accomplish a function.

Management Interviews[7]

As in the case of management interviews at the organizational level, management interviews at the unit level (division, department, work group) conduct a focused interview with key management people of each of the units. The questions should be similar and constructed to uncover aspects of the organizational unit, such as the mission, the major functions and activities, the nature of organizational challenges, the different aspects of organizational culture, the attitudes and aspirations around innovation and change, the nature of the business and marketplace, how all of them are changing, and how that will affect the nature of the design and the overall success of the project.

Work Process Analysis[7]

Work process analysis is a technique used to understand the nature of the work process. Patterns of work are analyzed to determine where and when different activities occur; how predictable these patterns are; and the human, technological, and printed resources involved in these activities.

Data to be collected from individuals include job title, description, and responsibilities, all work activities, and their sequence and regularity. This also includes where, how, and with whom the work is accomplished. Additionally, people are asked qualitative questions to determine what they like or dislike about their jobs, what they would preserve or change, and how could they do their jobs better. Suggestions for improvement might include changes in policies and procedures as well as physical and environmental changes.

Work process analysis should be used whenever a new workplace design is being considered. It is a major way of identifying whether modifications in an accepted workplace design need to be considered to account for variations in local conditions. For example, a field sales function that serves mostly large firms in a major metropolitan region may require a kind of office and technology support different from field sales that serves a large number of smaller clients over a wide geographical region.

Work process analysis can be used early in the programming process to find out how work is done presently and how people feel about their jobs. Any rumor of change will get employees' attention. Involving employees in a process connected with researching and planning for prospective changes will result in more employee cooperation and involvement when the actual changes are implemented.

Who to involve in a work process analysis depends on the scale of the project. However, at least one person, and preferably two to three people, from each functional group in the workplace should be involved. The analysis should also include someone from each management level, professional function, and support staff within a functional group.

Each interview will last about 45 to 60 minutes. The time needed to organize and compile the data will depend on the number of interviews conducted; it can be fairly substantial. Work process analysis is somewhat difficult. It requires a clear understanding of the issues, excellent communication, listening and interviewing skills, and the ability to synthesize information. The process will be somewhat easier if the interviewer is familiar with the activities of the work area.

The Five Why's: A Method for Analyzing the Workplace and Work Patterns

The Five *Why*'s is an exercise based on a Japanese quality technique that forces people to search for the root of a recurring problem, instead of accepting the first explanation they find. People using this technique ask *why* five times in either a group setting or individually. The premise is that the actual problem will not be uncovered until answering the fifth *why*. The first four times the question is asked, the user is simply uncovering symptoms of the problem without finding the cause. The answers to the questions should not try to lay blame on any individual or department but should attempt to find events or policies in the organization that can be changed to ensure that the problem does not recur.

The time to ask and answer the Five *Why*'s varies. Sometimes, the answers to the questions may be quite obvious. At other times, it may take several days to actually trace the root of the problem.

The Five *Why*'s method is very easy to use. Users need to be able to sift through the many random answers they may receive to trace the original problem.

***Sample: The Five* Why's.**   In conducting the work process analysis at Burdell, Inc., facility managers found that it took quite a long time for customers to receive their orders for Whickywacks, a very valuable business tool in today's global market. Facility managers hoped that the design they were proposing would improve the work processes within the organization, thus making it easier for them to push the design through. After talking to several people, however, they were not exactly sure why it was taking so long for the orders to be delivered, so they could not determine how the new design could help.

Janet, who had been working for facility management for only a year, had heard of the Five *Why*'s technique in college. She decided to try the technique to trace inefficiencies in the order process. She started with the customers' first contact with the company: field sales representatives. Her day went something like this:

As soon as Jim, one of the field sales reps who had been working for the company for 10 years, arrived at the office, Janet grabbed him and said, "I've heard that it takes several months for orders to be filled for Whickywacks. That sure seems like a long time."

Jim: "Yes, it is. I've heard from customer relations that we receive a lot of complaints about it and that we sometimes lose return business."

Janet: "*Why* do you think it takes so long?"

Jim: "I think it has to do with the people we use to deliver the Whickywacks. They are constantly losing the orders or filling them inaccurately."

Janet: "*Why* do they lose the orders?"

Jim: "The orders are sent to them in hard copy form by the finance department. I guess maybe the orders sometimes get hung up in the finance department or don't get entered correctly into the system."

Janet: "*Why* would they get hung up or entered incorrectly?"

Jim: "Well, I guess it's because the fulfillment people are supposed to verify the order with the sales administration before they enter the order into their system. That sometimes takes a while to do because sales administration then has to track down either the customer or the sales rep to confirm."

Janet: "*Why* do they have to go through sales administration? Why don't they go directly to the sales rep?"

Jim: "The orders don't have the sales rep information on them. Once we call in an order to sales administration, it becomes their order. The orders will have one of their names on it."

Janet: "*Why* does the order have to go through sales administration? Why can't you send the order directly to finance?"

Jim: "We don't actually enter the orders into the system. We just call the order in on the telephone or bring in a completed order form. Sales administration then enters the order into the computer for us."

From this brief conversation, Janet was able to uncover the following: What was one of the problems? The sales reps did not have responsibility for the orders. They did not have the means to enter the orders themselves, so they therefore had to turn over the orders to sales administration, at which point they lost control. Giving the sales reps the technology they need to do this while they are sitting with the customers would give them direct control over the orders, in addition to reducing the chance that the orders would be entered incorrectly.

Further Analysis of the User Group

Following are some of the different data collection techniques that may be useful in probing groups that comprise organizational units and discussions about their advantages and disadvantages. Detailed descriptions of these are beyond the scope of this chapter but can be found in various sources.[9-12] In general, a wide range of users and experts should be involved, but in different ways.

When the project team wants a representative viewpoint (generally a good idea) across user groups or types of employees, some form of random selection is appropriate. If it wants to gain insight into a particular problem or to explore future developments, then it makes sense to use people who are especially knowledgeable or experienced in the area. The project team can ask managers to select representatives for committees and focus groups to act as liaisons between the department or group and the project team. If the group is sufficiently small, then the whole group may be consulted.

The type of data to look for depends on the situation. If the program is for a new or existing facility for an existing staff, then most of the techniques can be used to study that staff in their current setting. If the program is for a new facility for employees yet to be hired, then surrogates must be used who are similar to the new employees. It is essential to sample a full range of users.

Communication with Users. The most commonly used data source involves communication with users. The questionnaire or survey is the most widely used technique. Questionnaires require knowing in advance the dimensions or attributes about which information is desired. Interviewing in a more informal style is useful in developing design requirements. Who to interview is at least as important as how it is done. Informal interviews with a representative range of users is often more useful than elaborate questionnaires. A good technique is to "walk people through a typical day in their work," asking them what causes problems, what they would like to keep, and what they would like to have. Direct communication allows access to feelings, plans, desires, and so on, as well as memories of past behaviors. It is reactive, however, and people do not always remember correctly or tell the truth. Consequently, interviews should be done in combination with behavioral observations. It is necessary to go through a process of deriving requirements from this data.

Interviews and Focus Groups: Pros and Cons. These are good for:

- Probing responses
- Following new directions
- Creating goodwill
- Understanding complex relationships

Interviews and focus groups are not good for:

- Developing quantitative data
- Tapping a broad sample of employees quickly

Interviews and the focus group provide a gut feeling about what people are thinking, feeling, and concerned, excited, or skeptical about. They are a good way of collecting anecdotes, which can be a persuasive source of evidence in some organizations. Because of their hands-on, face-to-face character, interviews also tend to be highly visible and liked by employees. Especially when employees are skeptical about whether management or the project team really cares about their views, interviews are a better medium than surveys, which are remote and impersonal. Pulse-taking interviews, 3- to 10-minute random discussions with staff conducted in their own workspace, are especially effective in getting a sense of a place and generating goodwill.

The interview's greatest drawback is that it is time-consuming (in terms of the number of people who can be reached in a given period of time) and expensive (in terms of the time needed to conduct a large number of interviews and to analyze the resulting data). And although it is possible to obtain quantitative data from interviews, it is more difficult than from surveys. The focus group, in which 5 to 10 people are brought together to respond to a specific proposal or charge, and group interviews, which may be more wide ranging, are quick ways of eliciting responses from many people. In efficiency, both fall somewhere between individual interviews and the survey. The group interview's main drawback is that one or two people can dominate the discussion, and if there are strong status differences among the participants, lower-level staff may choose to remain silent or mute their responses. The main benefit is that a range of opinions and views becomes apparent, and group members may stimulate one another, leading to new insights.

In conducting an interview, remember to

- Explain the purpose of the interview and who will see the results.
- Allow time for someone to answer. Do not rush the respondent.
- Find a comfortable location with reasonable privacy.
- Explain how long the interview will take.
- Watch for confusion and repeat explanations if necessary.
- Write down key words and phrases; elaborate right after the interview ends.
- Probe answers; ask for more explanation.
- Do not suggest the right answer or become defensive.

Done properly, the interview is enormously versatile and becomes even more powerful when combined with survey methods.

Surveys. Surveys are good for:
- Generating quantitative data
- Tapping a broad cross section of employees quickly
- Enabling a statistical analysis of subgroups
- Obtaining data cost-effectively

Surveys are not good for:

- Probing responses
- Understanding complex nonstatistical relationships
- Generating goodwill and confidence in the process

The survey's strength is the interview's weakness, which is why a combination of the two is so useful. In organizations that value quantitative analysis or whose top management is skeptical of interviews as a technique, surveys, with their charts, tables, and graphs (numbers!) can be a more persuasive form of data analysis and presentation than interviews. The interview's contribution, used in conjunction with a survey, is the anecdotes that bring to life the numbers and charts, giving them a human face that helps fix in memory a particular point that may slip away if presented only in a chart or table. "A good story is worth a thousand charts."

Structured Observation. These are good for:
- Raising questions to include in surveys and interviews
- Checking information given in surveys and interviews
- Generating visual evidence to support interviews and surveys
- Generating quantitative data, if systematic
- Getting at issues that employees may have difficulty verbalizing

Structured observations are not good for:

- Understanding why something is occurring
- Understanding how employees feel about a situation
- Getting at "what could be"
- Generating goodwill (unless coupled with interviews)
- Getting a representative view of the situation, if only short-term

One of the best and most objective types of data is observing what the users of a facility actually do. It is important because often people are not consciously aware of what they do each day and consequently will not mention it in an interview.

Structured observation can take many forms. One systematic and quantitative technique is called *behavioral mapping,* which uses predetermined codes to indicate where and when certain behaviors occur in a setting. This can involve observing selected individuals through a typical day or task or can focus on an area and record all the behaviors that occur in it. Mapped over several times of day and several days or months, such behavior records build a picture of which areas in a building are being used by what kinds of people, in what ways, and at what times. This information can then be analyzed to identify the physical characteristics associated with different use patterns. For example, lounges at the ends of corridors tend to be used less than those near major circulation paths and activity nodes like mailrooms and coffee areas. Or it may reveal that conference rooms designed to accommodate meetings of 30 people host meetings of 3 to 4 participants almost exclusively.

Less structured and quantitative, but quicker and still useful, are focused observations in which the programmer and another team member, who can then share their observations and check their interpretations, walk around the facility and observe certain behaviors and their residue (e.g., jury-rigged storage or lights brought in from home). Most behaviors leave traces behind which can be used to understand behavioral patterns. Wear and tear is an *erosion trace* that can also be used to infer past behaviors. Objects left behind are *accretion traces,* which may allow one to identify the behavior. Like direct observation, erosion or accretion traces usually do not tell much about the inner feelings or desires of the user. There are also problems of selective survival and deposit. Things to observe, each of which offers clues to environmental dysfunctions and actual (as opposed to expected or desired) behavior, include attempts to solve problems, typical and atypical uses, and inventive ways of using standard furniture or equipment. Heavy wear patterns, or the opposite, indicate use patterns.

In addition to noting observations on a recording sheet in predetermined categories, a photographic record of these findings is helpful for representation in which this hard evidence can be shared with people who were not present on the observation tour. Again, the value of such data can be enhanced by combining them with information from interviews and surveys to explain why certain things are occurring, how these affect the way people work, and so on. Observational data has limits—always consider before collecting data if it will offer useful insight. Often, it is a good idea to ask the users to interpret observations.

Simulation.   This is good for:

- Exploring "what if" possibilities
- Enriching the image bank of possibilities
- Eliciting responses to new designs or plans
- Removing skepticism that something will happen
- Avoiding costly mistakes
- Stimulating the imagination
- Generating enthusiasm and excitement

Simulation is not good for:

• Getting a completely realistic response

Simulations allow people to see and experience new possibilities. It is important that the search for a new design not be totally limited to the past experiences of the users. Simulations allow the programmer to obtain people's response to solutions that have been successful elsewhere or that have been envisioned but not yet built.

Simulations take many forms, including photographs, models, drawings, full-scale mock-ups, games, computer and video animation, and even site visits. They are good for getting a sense of the way people who are unfamiliar with a particular design or furniture arrangement might actually respond to it. Although the state of the technology is rapidly improving in its ability to make simulations more realistic, simulations are always just representations of reality and not the reality itself. The amount of time and the circumstances under which the simulated environment is experienced inevitably fall short of the real thing.

Yet, by having intact work groups visit and carry out typical tasks in a full-scale mock-up of a work area equipped with different furniture systems, lighting, seating, and a working computer system, quite realistic feedback can be obtained. Despite the limitations inherent in any form of simulation, the feedback is far superior to blind guessing about the way people will respond. When coupled with systematic evaluations (observations, interviews, focus groups, and surveys), they can also be used for incremental improvement of prototypes, and when the results of the user assessments and suggestions are communicated back to users, they can be very effective in generating goodwill and acceptance of designs.

Simulations do not have to be expensive. Simple working models with few details can be used for exploring different layout issues, and the cost of very realistic computer animations is reducing rapidly. The cost of a simulation should be considered in relation to the cost of making a major mistake. Buying a million dollars' worth of new workstations that do not accommodate technology, that have inadequate storage, that do not meet employees' expectations for acoustic privacy, or that are much harder to assemble or disassemble than expected can make a 50 thousand dollar mock-up study (which identifies the dysfunction before the design is implemented) look like a very good value for the money. In this context, even the costs of full-scale mock-ups are by no means unrealistic.

Archival Records.   These are good for:

• Nonreactive information
• Inexpensive data collection
• A check on other sources of information

Archival records are not good for:

• Understanding the dynamics of a situation
• Understanding why something is happening
• Looking at an issue in a detailed manner
• Interpreting data accurately
• Predicting new needs

The final technique of data collection is archival records. This information, collected by the organization as a part of its ongoing record keeping, is often a good source of readily available information. Grievance reports or annual employee satisfaction surveys may offer insights into facility-related problems that need to be addressed in the program. Records of exit interviews of people who have decided to leave the organization (did the facility contribute in any way to their decision to leave?), of job recruitment interviews (did the facility influence their decision in any way to accept or decline the job?), medical and insurance records (complaints, visits to the doctor), of absenteeism, and of turnover are all sources of data that can be used to gain insights into the design of a new facility or the redesign of an existing one. Such data can be related to parts of the building (does an area with old, large-zone HVAC have more medical complaints or higher levels of absenteeism?) and to planning processes (do projects in which employees are more involved have fewer change orders and higher levels of satisfaction with the workplace?).

Analyzing the Data.   A common difficulty in programming is that there is never enough information in exactly the right form to answer all the questions rigorously. There is never enough time or resources. Yet, to paraphrase Abraham Kaplan, we are in a taxi and the meter is running—we must decide. That is the difference between the art of programming and science. The task is to collect sufficient information to answer the main questions.

Real skill comes in collecting as few data as are needed and in analyzing them as much as possible. Unfortunately, the reverse is often the practice. Collecting data is easy (although collecting accurate and valid information is more difficult). Analyzing information and interpreting its implications for facility decisions often feels more like art than science. But they are skills that can be learned.

Simply counting instances of a given response can be very enlightening. How many staff reported (in interviews or a survey) poor air quality or lighting? Relating it to environmental characteristics is equally as important as accuracy in tabulating the responses. Did conference rooms with windows get more usage? Looking for patterns is also useful. Surveys revealed that people complained about the lack of conference room space. Observations revealed that all the conference rooms were designed to accommodate meetings of 30 people (of which there were few) but were used almost exclusively for meetings of 3 or 4 people (of which there were many). Deriving conclusions that lead to design requirements is the difference between research and programming. There is adequate space devoted to conferencing, but it is not distributed so that it matches the number and size of meetings. There is a need for more small four- to six-person conference rooms with only one or two sized for large meetings—even those should be divisible into smaller rooms if needed.

If the responses in samples of up to a few hundred are very simple (a checked box on a survey, for example), this kind of analysis can be done very quickly by hand. Many simple computer statistical or spreadsheet packages are available to summarize, display, and analyze the data quickly. These are useful for larger samples, particularly if the project team wants to do more sophisticated statistical analyses. In either case, quantitative analyses can be done quickly, accurately, and cheaply—often with impressive results in the form of computer-generated graphs and charts.

For qualitative data such as interviews or focus groups, do a thematic analysis. Read the transcripts and look for themes (poor air quality, way-finding problems, distracting noise, lack of privacy or a beautiful view, etc.).

When interpreting results from interviews, surveys, observations, or archival data, try to do the following:

Look for patterns of agreement:

- Across data sources
- By mediating variables
- With literature
- With experience

Look for contradictions:

- Across data sources
- By mediating variables

- With literature
- With experience

Try to understand contradictions:

- Through alternative plausible explanations for a finding
- By reexamining the data
- By collecting specific data to test alternative hypotheses

Identify the most important findings:

- By organizing and prioritizing them

Present the findings simply:

- Charts, tables, and diagrams
- Selected photos or videos to illustrate an important point

ANALYZING THE INDIVIDUAL USER

This section describes five approaches to the problem of developing design requirements that are oriented to the needs of individual users: the user characteristics, social/psychological functions, behavior circuits, behavior settings, and user participation approaches. These approaches complement and can be combined with those described in the section that focuses on analyzing the users in an organization by standard social science tools of observation, interview, survey, and the like. They are also complementary with each other and can be used in combination. For example, the behavior circuit approach can be used to enhance a participatory approach by having the user participants describe their typical circuits, even act them out, and then suggest design requirements. It is also possible to envision typical behavior circuits and their related design requirements as part of a user profile in the user characteristics approach. A specific methodology suited to the problem or resource conditions needs to be developed. Each of the approaches is described along with an abbreviated example to illustrate the approach and the types of information each requires and generates.

The User Profiles Approach

Most environments are used by different types of people who are likely to differ in their basic needs. Environmental needs are basically homogeneous within the user types, however. Typical types of groupings are stage in the life cycle, stage in the development process, ethnicity, social class, role, job function, and so on.

Build a detailed picture or profile that catalogs the special behavioral patterns, preferences, environmental needs, and vulnerabilities, such as mobility status, agility, dexterity, sensory acuity, and the like, that are associated with each of the particular types of users of the site. From this profile derive design requirements.[9]

Basic Steps
- Identify user types.
- Construct user profiles of each group.
- Derive design requirements for each.

For example, one of the types of users of a public pedestrian system is the elderly. Among the characteristics of this group identified in the user profile is an increasing opacity of the eye lens that makes it difficult to locate and isolate critical edge conditions in the visual field. This condition is exacerbated by declines in balance, response time, general agility, and by much greater risks of serious injury in falls, and the like. From these characteristics, it is possible to derive a number of design requirements, for example, for a pedestrian system:

- Public signs should have a contrasting border to facilitate isolating the sign from its background.
- Steps should have their leading edges marked by a 2-inch stripe that contrasts sharply with the background tread and riser color.

The major strength of this approach is that it deals holistically with all of the characteristics of a particular type of user (e.g., behavior patterns, physiological, psychological, social, economic, etc.), which might have an effect on environmental needs. Beyond simply serving as a source of programmatic material or design requirements, the material contained in the profile should also enhance the designer's understanding of the users for whom the design is intended. This should enhance the ability to go beyond simply satisfying the requirements of the program in searching for and evaluating solutions.

The major weakness is that it is dependent on a good archival system with relevant information organized so that it is accessible to a programmer/designer. Constructing a profile can be time-consuming, but this is counterbalanced by the fact that once constructed, the profile can be used for other problems and by other programmers.

The Social Psychological Functions Approach

This approach to developing design requirements involves focusing on particular processes or functions that are affected by the environment. The procedure is to identify the particular process (e.g., social interaction, way finding, crime, stress, etc.), that the design should support (or in certain instances prevent). Then, based upon knowledge of the way the environment affects that process, derive the design requirements.

Basic Steps

- Identify the function (e.g., social, psychological, physiological, etc.) to be supported or suppressed (e.g., image formation, burglary, stress, etc.).
- Identify the characteristics in the environment that affect this process.
- Derive performance statements or design requirements.

For example, the goal is to facilitate rapid, accurate, and comprehensive mental map formation for newcomers to a college campus. Recall of elements such as landmarks is directly affected by their visual exposure. From this relationship, it is possible to derive design requirements for making a college campus imageable. For example,

- Organize the pedestrian path system so that new students would be visually exposed to the major facilities, settings, and so on, of the campus during their normal daily routine.

Similarly, if one wanted to facilitate friendship formation in campus housing, a number of design requirements could be derived from the extensive work done on the relationship between physical setting and interpersonal interaction.[10] These requirements might deal with the organization of doorways and pedestrian systems, the clustering of units around courtyards, and shared facilities such as laundries, mailboxes, car wash areas, tot lots, and the like. The work of Oscar Newman[11] has resulted in a wealth of design requirements for a variety of environmental settings and population types that are aimed at reducing criminal behaviors and vulnerability to crimes of various sorts.

One of the strengths of this method is that it enables the programmer to deal with environmental needs of which the user may be unaware. On the other hand, it relies on having theory that explains how the environment affects a particular social or psychological process.

The Behavior Circuit Approach

Design requirements can be developed by focusing on the behavior of typical individual users of the environment. The stream of behavior of an individual can be broken into coherent segments or sequences that are related to particular environmental settings.[12] Because of the habitual nature of much of behavior, it is possible to describe a very significant portion of a person's stream of behavior by a relatively few recurring behavior circuits. The routines of going to work, going shopping, going to school, and so forth are relatively stable in both their spatial and temporal patterning. They tend to be similar for different individuals.[13]

In the behavior circuit approach, one identifies the major sequences of behavior of a particular user type. These sequences can be broken down into discrete behavioral units and the environmental characteristics can be specified that are required to support, prevent, or discourage them at a particular level of adequacy.[14]

Basic Steps

- Identify predominant behavior sequences for each user type (going to work, going shopping, etc.).
- Break each of the circuits into discrete behavioral units (walk to bus stop, wait for bus, etc.)
- Specify the level(s) at which the behavior is to be supported.

    Safety/survival

    Task efficiency

    Comfort

    Pleasure/enjoyment

- Using a checklist of environmental attributes, derive the design requirements for the setting(s) in which that behavior is to occur.

    Spatial form. Such things as the size, shape, and character of bounding surfaces; internal organization of objects; connection to other settings, and so forth

    Communications. Both implicit and explicit signs, symbols, and such communicating information needed to carry out the behavior in question

    Activities. The type, pattern, intensity, and scheduling of the other behaviors potentially occurring in the same space at the same time, which might act to support or constrain the behavior in question

    Ambience. Light, sound, smell, microclimate, and so forth

- Repeat for each behavior in the circuit and for each user type.

This approach is systematic and concrete. It focuses on behavior so that design implications become obvious. It provides comprehensive and detailed programmatic information. It tends to focus on overt physical behavior, not on internal processes such as environment perception/cognition or user satisfaction.

The Behavior Setting Approach

This approach focuses on areas or settings within which there are relatively stable patterns of recurring behaviors. So, instead of building up a program by looking at the sequences of behavior of typical individuals, this approach identifies the patterns of behavior within a particular setting or area. Most action settings or behavior settings have a particularly enduring form in space and schedule in time. Usually, a particular physical and social form supports a particular set of behaviors and coerces or constrains the behavior of people within it. The people, objects, and events inside the settings are ordered in a bounded pattern that takes its form in response to the recognized needs of the activity going on in it. Most such settings exist as a part of a larger structure of settings (e.g., a classroom within a university), they exist independently of particular actors and their actions, and the pattern endures over time.

To generate design requirements by using this approach, one must identify the predominant recurring patterns of behavior in the setting. Then, by breaking the patterns of behavior into discrete units, it becomes possible to derive the setting characteristics that are required to support each behavioral unit at a particular level of adequacy. The process is basically the same as that for the behavior circuit approach. As in the previous case, this approach is highly systematic and is focused on overt, concrete behaviors. This approach focuses the programmer on relatively stationary patterns of behavior and on groups, as opposed to the typical circuits of an individual.

The User Participation Approach

The approaches previously described for generating design requirements rely for the most part on experts to develop the information. User participation in the process is usually relatively passive. It takes the form of responding to questions posed by the researcher/programmer. The user's needs are developed through the programmer expert. User participation enables the user to participate much more directly in the process of developing design requirements. Professional expert programmers and designers are still a part of the process, but the professional skills that they use most are those dealing with the facilitation of group processes.

A number of specific user participation programming methods have been developed. Each method involves variants of the following approach. First, the users, their representatives, or surrogates must be assembled and their role in the process decided. Next, they begin an intelligence phase. As individuals or small groups, they systematically examine their environment and try to identify assets or strengths—those aspects that currently function well and warrant preservation in the probable future—and problems or weaknesses—those aspects that do not function well or are not anticipated to do so in the future and consequently need modification. Often, this process is done using simple disposable cameras. Each user participant is issued a camera and asked to photograph the elements they see as problems and assets. These photographs are then used to communicate with the group and with the programming team and designers. This phase of individual analysis is usually followed by a group process in which the whole group agrees upon and prioritizes a list of problems and assets. This phase is followed by an exercise wherein individuals or small groups are assigned the task of envisioning possibilities and thinking about what qualities an ideal setting might have. The group then comes together to formulate group ideals. Then follows a period wherein all three sets of programmatic conceptions, problems, assets, and ideals are combined into an agreed-upon program.

Basic Steps

- Client assemblage
- Individual assessment of problems and assets
- Group agreement on a prioritization of problems and assets
- Individual identification of ideal qualities
- Group agreement on ideals
- Group agreement on problems, assets, and ideals, or "the program"

The primary factor that most distinguishes this approach from the others is the involvement of the users in an active role. The characteristics and availability of the clients/users determine the type of participatory approach, or even whether one can be used at all.

Visual Programming

Most users have difficulty articulating their preferences and dislikes for the visual character of places, even though this character is of considerable importance to them. One way of overcoming this limitation is by using a technique called visual programming. The approach can be done in several ways, but the following is typical. First, assemble a set of 100 photographs that represent the full range of appearance types for the type of building or facility being programmed. Select a sample of the different types of users. Ask each to sort the photographs into 10 equal piles, ranging from the ones they like best to the ones they like least. Then, have them sort the least and most preferred piles in order from those least liked to those most liked. Then, have them describe what it is about the environments shown in each that causes them to like or dislike it. Keep track of the preferences for all the participants and look for patterns of likes and dislikes. Use these images to convey directly to the designers and others the preference patterns that emerge.

Basic Steps

- Assemble photographs.
- Individuals sort according to preferences.
- Individuals describe features or characteristics they like and dislike.
- Programmer analyzes all responses for patterns of likes and dislikes.
- Photographs are used to communicate preferences directly to designers and decision makers.

DETERMINING BUILDING REQUIREMENTS

Once the mission, functions, and activities of each of the units are described (and redesigned if needed), it is necessary to describe the space, equipment, and infrastructure needed to carry out the activities.

Space Requirements

There are several approaches to forecasting the amounts of space required. Because any method of forecasting the future is potentially fraught with error, it is a good idea to use multiple methods, especially if they go about the task in very different ways and use different information as the basis for the projection. Following are the descriptions of three basic methods, ideally used in combination with one another. The trend projection and the gross area method are used as checks on the total area estimate produced by the method of detailed requirements. None is very precise, and each relies heavily on informed judgment.

Detailed Requirements Approaches.   Because this approach is based on detailed descriptions of the space required for each activity or employee type, it is often seen as the most accurate. However, we must remember that small errors multiplied many times can yield large mistakes. There are several variations of this approach. The description of one version follows, and then variations are suggested. The basic process involves determining the footprint or area required for each of the different types of shared or unshared spaces such as workstations, laboratories, and conference rooms.

Project any changes in area or footprint that might be likely or planned for these space types. These may be driven by technology or work practice changes or by policies regarding space use. For example, when distributed computing using personal computers was first used, there was a net increase in the amount of space required for office workstations to accommodate the desktop personal computers and additional peripheral equipment such as printers, scanners, and modems. Policy changes may reduce the area required. This was common in the cost-driven 1990s, during which companies had consciously reduced real estate costs by reducing the sizes of workstations. Once the future area requirements for individual space types are determined, then the required numbers of such spaces that will be required in the future must be determined. For spaces assigned to individuals, this requires projecting the future number of employees of that type, which is usually done by the managers or heads of each functional unit. Shared spaces such as conference rooms also need to be projected. Often, these are based on a ratio of support spaces per number of employees or individual workstations.

Use patterns can also affect the amount of space required. The practice of *hoteling*, or unassigned nonterritorial workstations where there are more employees than workstations, is an example. In this case, it would be necessary to determine the ratio of employees per workstation to project the total. A change in work practice such as a shift to teamwork could involve a shifting of space by reducing the size of individual workstations to allocate space for team rooms for collaborative group work.

Determining the Footprint.   There are several ways to go about this process.

Space Standards.   As noted before, many organizations have adopted standards for space types. These often include the footprint enclosure type such as panels or partitions, furniture, and equipment. These standards prescribe the space, furniture, and equipment provided for each type of employee. They also often spell out the sizes, furnishings, and equipment of shared facilities such as conference rooms and a formula for determining how many should be provided. These standards are usually published as a policy guide and are available from the facilities manager.

Standard Practice.   There are also many standard architectural references that describe the typical size, layout, equipment, and other special requirements for a wide range of spaces for many different building types.[15] The standards are good guides. They are based on conventions of practice gained from wide experience over many years. They need to be examined by the user representative for their appropriateness to the particular case. They may also not reflect the latest developments in technology or practice.

Estimating the Size of Spaces.   Estimating the size of spaces for which there is no standard is one of the most difficult and important tasks in programming. The project team should work on this task with a knowledgeable representative from each of the organizational units. That person may in turn have to involve particular employees who have special knowledge about a particular space, operation, or piece of equipment. The project team should also recognize the need to assist in the task of determining the space requirements. This assistance may take the form of having a member of the programming staff simply ask good questions. Often, a checklist is a good idea to ensure that the user representatives think through all of the possible requirements and visualize and estimate the space required for equipment, access, and so forth. There is a general tendency for everyone to overestimate the amount of space they need. One good way of getting users to think more conservatively about this is to make them aware of the costs of space. For example, knowing that each additional square foot of space will cost $175 to construct with an added 10 percent of that every year thereafter for maintenance and operating costs will get most users to think much more carefully about their space needs.

GRAPHIC METHOD: Work with the user representative to lay out the space at a scale of ¼ or ⅛ inch to the foot on standard graph paper. On a separate sheet, lay out to scale the footprints of all equipment and furniture needed for the space, and cut out these pieces. These dimensions may have to be obtained from the manufacturer's specifications or product brochures. Have the user representative consider any necessary clearances, such as door swings for cabinets or refrigerators, and the like. It may also be necessary to do a cutout for typical circulation spaces. Then, work out a basic layout of the space, taking into account the circulation and access to other spaces or corridors, and their work process. This first layout may be overly generous or too small. The project team representative should work with the user representative to ensure that it is a workable and efficient layout. Discuss and explore options that might improve the layout. It is possible to do this in a group workshop with user representatives. This can shorten the time required by the project team representative. The interaction among the user representatives can stimulate useful insights and ideas for improving the layout and prevent oversights. Finally, confirm the layout with the user representative.

EXISTING EXAMPLES AND PROTOTYPES: Many spaces are simply relocations to a new facility or a new location. In these instances, the amount of space, the layout, and special requirements can be derived from an existing example. The user representative and the project representative (and the actual user if the use is highly specialized) should meet in the space. If there are several variations, then they should all be examined. Discuss the pros and cons of the existing arrangement and ways it might be improved. It is useful to work this out on graph paper, as described earlier. Visiting other facilities that have innovative and successful examples of the types of spaces being proposed is also useful.

Determining the Total Net Building Area.   After all of the individual areas or footprints for each space type are determined, the next step is to calculate the net building area. This involves totaling up the areas for the different types of spaces identified previously:

- For assigned spaces where each space, such as an office or workstation, is dedicated to one employee's sole use, it is necessary to project the number of employees who will be assigned to that type of workstation (e.g., managers, technical, clerical, etc.).

- Multiply the projected number of employees for each assigned space type by the footprint to obtain the net area required. Repeat this calculation for each assigned space type.

- Many spaces such as conference rooms and mailrooms are shared by users. Determine the number of employees per shared space type (e.g., one conference room for each 100 employees).

- Divide the total number of employees by this ratio to determine the number of shared spaces required. Multiply this figure by the footprint area required for shared spaces. This will yield the net area required for shared spaces.

Adding the assigned and shared space totals still yields only the net assignable area.[16] It does not include the area required for the building walls, circulation, core, and other building services. (See Fig. 4.2-2.)

Calculating Total (Gross) Building Area Required.   A final factor is needed to convert the net assignable area to the total or gross building area required. *Failure to include this factor can lead to a serious underestimate of the space needed for the building.* The conversion factor varies considerably with the type of building. This factor may be obtained in two ways. Derive it for current facilities by subtracting the net workstation and shared spaces identified earlier and determining the ratio. Or standard conversion factors can be used. See Table 4.2-1 for some examples.

Checking the Gross Area Estimate.   The total or gross area derived from the detailed requirements method is subject to the error produced by adding together many small errors or miscalculations. It is useful to check this against other benchmarks. There are several ways of doing this.

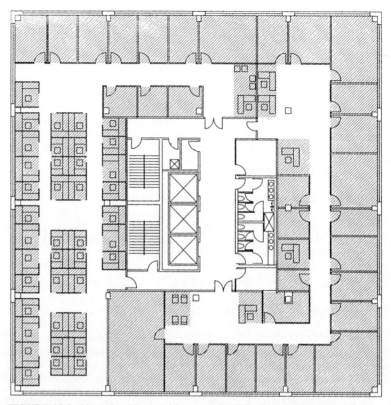

FIGURE 4.2-2 Net assignable area. (*Courtesy of William Sims and Frank Becker.*)

Trend Projection. The simplest method (and because of that simplicity, often the best method) is to project a historical trend. In certain industries, such as the high-technology area where growth has been so explosive and the actual specifics of the requirements are so unpredictable because of the rapid creation of entirely new technologies, this has been the only way that the industry could ensure that they had the needed space on hand when it was needed. Uncertainties or probabilities can be added to the trend projections to produce a range of possible outcomes.

TABLE 4.2-1 Calculation of Gross Square Footage

Building type	Net-to-gross conversion factor	
Office	1.35	1,000 nsf = 1,350 gsf
Laboratory	1.71	1,000 nsf = 1,710 gsf
Bank	1.40	1,000 nsf = 1,400 gsf
Warehouse	1.08	1,000 nsf = 1,008 gsf
Apartment	1.56	1,000 nsf = 1,560 gsf
Hospital	1.83	1,000 nsf = 1,830 gsf

Source: Means Assemblies Cost Data, 1989.

Gross Area Factor Approach.   This is also a very simple approach to space projection. The first step is to determine the current space usage for some factor that can be projected, typically employees (e.g., divide the gross area by the total number of employees), but it could be sales, product output, and so on. This current area factor can then be modified for future changes based on changing technology, work practices, or policy changes (e.g., a decision to decrease the amount of space per person by reducing the average size of workstations to some benchmark derived from industry best practices, the decision to use hoteling, or to move to a paperless office). The next step is to project the future number of employees. This may be done by asking department heads to project their future employment needs over the planning period and adding this to a total for the organization. The number of future employees may also be estimated by using a trend projection. Finally, the modified area factor is multiplied by the projected number of future employees to obtain the required gross building area.

Benchmark Data.   Finally, a number of organizations such as BOMA and IFMA maintain benchmark databases that contain data on the average space, per person, per unit of sales, and so forth, for various industries and building types. Many companies also benchmark their space use against their competitors'.

Utilization Studies[7]

Utilization studies are used to build a picture of the way the spaces are being used. The purpose is to build a picture showing who, when, and how people are using a particular workplace and associated equipment. Several times a day over a period of weeks, someone trained to work from precoded floor plans records the use pattern for the workplace. This provides data on the way spaces are being used, where people are working at different times of the day, and depending on the detail collected, what they are doing, and their pattern of movement in and out of their workstations.

Understanding how the workspace is currently being used is the indispensable starting point for determining whether a new workplace is needed. These studies provide decision makers with the basic data needed to develop recommendations for new designs and for strategies to make more efficient use of existing facilities. Utilization studies can also be helpful to support information collected through other means such as surveys and interviews. Typically, observed use patterns provide a much more accurate and valid view of work patterns than surveys or interviews. Conversely, information collected through utilization studies can be used to suggest questions to include in surveys and interviews.

For most time/space observations, only a recorder is necessary. This person can be anyone who is in the office on a regular basis and can spare a few minutes each hour to make a brief tour of the floor or area being observed. In many cases, a secretary or administrative staff person who has been given some training is a good candidate to conduct the observations.

The time required depends on the size of the operation. Typically, from less than 5 to 15 minutes is required for each data collection cycle with approximately 50 people or fewer per general location observed. For larger areas with large numbers of people, a single data collection could take an hour or more. Over a period of 3 weeks, this technique may require a total of several person hours. It is also necessary to train the recorder and to test that the forms are being filled out as expected. The only equipment needed is the observer recording form.

- Record over enough days of the week and over several weeks, months, or seasons, so that a representative sample of periods is included.
- Do not record during atypical periods (e.g., vacation season, right before Christmas).
- If there are predictable seasonable variations (e.g., the busy season for auditors), try to conduct observations that reflect these different seasonal patterns.
- Missing one or two observations periods is not a fatal flaw; the goal is to build up a credible picture of use patterns that most reasonable people within the firm will believe.

Basic Steps.   Obtain a floor plan of the office where the group is currently working. If work-stations and offices are not represented, manually add these components to the floor plan. Make a copy of the floor plan for each observation period. The person recording the information should indicate on the floor plan if

- Each workstation, office, conference room is permanently vacant [the workstation shows no signs of having been occupied (no code)], temporarily vacant [the workstation shows signs of being occupied—briefcase, papers, etc.—but is unoccupied at the moment (code = V)] or occupied (see the following).
- If occupied, the number of people using the workstation at the time.
- If occupied, the task being accomplished:

Working on the computer   (code = C)

Reading materials   (code = R)

Talking on the telephone   (code = T)

Paperwork   (code = P)

Meeting   (code = M)

Other   (code = O)

Leave appropriate space on the floor plan for the recorder to fill in information, or develop a coding system for the recorder.

Adjacency Analysis

Another important tool in the programmer's kit is *adjacency analysis* (or affinity analysis as it is sometimes called). This is a procedure that analyzes the communication patterns within an organization to determine desired spatial relationships or adjacencies among the individual spaces. Most computer-aided facility management and design programs use automated versions of this tool to carry out stack and block planning leading ultimately to the floor plan layouts of the facility. Typically, locational requirements are based on communications, joint tasks or projects, shared files, shared equipment, paper flows, task sequences, and on avoidances. In large organizations the analysis usually occurs at two levels—among units and within units. When determining desired adjacencies between organizational units, the analysis looks at group-to-group connections and group-to-ancillary facilities. Within-unit analyses focus on individual user-to-user connections and on such relationships as individual users to shared equipment and spaces.

Relationships
Within unit:

- User to user
- User to shared equipment spaces

Among units:

- Group to group
- Group to ancillary facilities

The basic process usually consists of a *communications survey* to determine the desired relationship patterns. This is summarized in an *adjacency matrix* that displays the adjacency requirements between units on a scale ranging from absolutely required, important, ordinary, and unimportant to undesired. This matrix is then transformed into a *bubble diagram* in which bubbles represent units and the lines that connect them represent the relationships. By manipulating the bubble diagram, it is possible to optimize the relationships into patterns. If the building is multistoried, then the bubble diagram can be used to produce a *stack diagram.* This is a diagram in which the vertical scale represents the floors of the building and the horizontal scale represents the area on the floor. This diagram is a very fast way to allocate units to floors on the basis of their adjacency requirements. Once the units are allocated to floors, the next step in the process is the *block diagram.* In this diagram, the bubbles are replaced by rectangles sized to represent the areas of the units. These rectangles are arranged within the basic outline of the floor plate of the building if the task is space planning an existing building. If a new building is being designed, then the basic block diagram or block plan can be used to inform the overall layout of the building.

Once the blocks are established, then the within-unit adjacency analysis can be used to guide the layout of the unit in much the same way as the overall process. (See Fig. 4.2-3.)

The Communications Survey.   The communications survey may be based on direct employee ratings or on expert ratings based on a formal communications analysis. Typically considered are the means of communication such as face-to-face, by mail, or electronically by phone, e-mail, videoconferencing, and the like. Other factors such as the frequency, urgency, and necessity of the communication may also be considered. This survey should also identify other factors that affect the need for adjacency between organizational units or between individuals such as joint tasks or projects, shared files or equipment, or work process sequences. Equally important is the need to identify avoidances such as reception or a high-security R&D area.

The Adjacency Matrix.   Analysis of the results of the communications survey will yield the information needed to construct the adjacency matrix. This is a simple two-way matrix that identifies the adjacency requirements between any two organizational units (or the spaces they occupy). The adjacency requirement is a judgment rating based on the information from the communication survey. It is a simple rating usually on a four- or five-point scale:

1. Absolutely required (immediately adjacent)
2. Important (on the same floor)
3. Ordinary (in the same building)
4. Unimportant (unimportant)
5. Not desired (located as far away as possible)

The Bubble Diagram.   The bubble diagram is a simple graphic tool for analyzing and communicating the desired spatial relationships among units. It can be simple, showing only the importance of the relationships among units, or it can contain a range of information about the relative size of units and other information about the nature of the spatial connections among them. The required adjacency relationship can be shown as a number or letter or can be represented by the thickness of the line connecting the units.

The bubble diagram can also be used to determine groupings of units. The relative spatial area of the units can be represented by the size of the bubble. The next step is to arrange the bubbles so that the lengths of the heaviest lines representing the most important connections are minimized. This may take several iterations to achieve an easily comprehended diagram that shows the relationships and identifies important groupings. In a new building design, these groupings and the relationships among them can be used to determine which units need to be immediately adjacent, near each other, or just on the same floor or whether a multifloor building is even desirable. When allocating units to an existing building, the diagram can be used to determine which units should be grouped together on the same floor and which should be on adjacent floors. Many CAD/CAFM programs have automated this process.

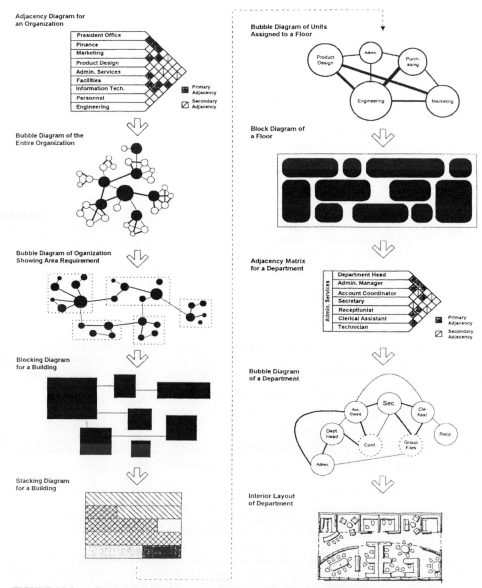

FIGURE 4.2-3   Adjacency analysis. (*Courtesy of William Sims and Frank Becker.*)

Stacking and Blocking Diagrams.   These two tools verge more on the planning and initial design of the building, but it is useful to show how they relate to the previous two phases.

If the building is multistoried, stacking diagrams are the first step in allocating units to floors in the building. The stacking diagram is an efficient way of determining this pattern because it focuses solely on fitting units with high adjacencies on the same or adjacent floors. It is a simple graphic tool resembling a bar graph. The vertical units of the stacking diagram represent the floor levels of the building. The horizontal dimension represents the area of the floor to some scale. The process of stack planning is to allocate units that have high affinities to the same floor. A bar representing each unit is then placed on the desired floor with other units to scale until all of the floor area is consumed. If floors are too small to accommodate all units that have high affinities, then some units may have to be placed on adjacent floors or in the case of a new building, the floor plate may be increased in size. Often, several trials may be necessary to arrive at the best grouping that most closely uses the areas available on the floors and achieves the best adjacency pattern.

Once the units have been allocated to floors in the building, then the process of block planning can occur. In block planning, rectangles representing the areas of each unit are arranged within a diagram of the floor to which they were allocated. This diagram also shows the main core(s) and major circulation of the floor. As in stack planning, it may be necessary to rearrange the location and configuration of the areas representing each unit several times to reach an optimal block plan.

Within-Unit Adjacency Analysis.   Once the overall block planning for the building has been completed, the next step is to repeat the process within each of the units. The basic procedure is the same, except that now the focus shifts from unit-to-unit adjacency to individual-to-individual users. Thus, the communication survey focuses on interactions between individuals within the unit and with shared equipment and spaces. It may also be necessary to include connections with individuals from other adjacent units. Similarly, a within-unit adjacency matrix is constructed, followed by a bubble diagram and a block diagram for the unit.

Equipment and Support

Factors often left out of programming exercises or given short shrift include many of those necessary to guarantee a well-functioning facility. Many of these are listed here:

Shipping and Receiving

- Loading and dock space
- Temporary storage
- Secure storage
- Berthing space
- Access to dumpster
- Centrality
- Proximity to freight elevators
- Proximity to primary users

Security

- Operations center location
- Guard posts
- Personnel access system
- Vehicular access system
- Executive access system
- Access by visitors and nonsecure personnel

Mail and Distribution

- Access to national mail system
- Distribution/collection schema
- Secure storage
- Access by external messengers

Motor Vehicle Pool

- Overnight storage
- Daytime parking
- Pickup/drop-off points

Shops

- Access to materials
- Sizing
- Access to freight elevators
- Locker room
- Security

Food Service

- Layout
- Access to staff
- Centralized or decentralized coffee bar
- Access for foodstuffs
- Egress for garbage
- Locker room
- Vending locations

Conference Services

- Stage
- Audiovisual requirements
- Acoustics
- Lighting and lighting controls
- Recording capability
- Seating—fixed or movable

Vertical and Horizontal Transportation

- Personnel elevators
- Freight or service elevators
- Escalators
- People movers
- Robots
- Location
- Access by people and goods
- Access to garages and roofs
- Security

Miscellaneous

- On-site furniture storage
- Custodial closets

Communications

- Closets
- Duct systems
- Location of file servers and network command elements

USING EVALUATIVE METHODS AS PROGRAMMING TOOLS

An evaluation of the current facility is a useful source of programming information. It is also a good idea to review other projects of similar type. Postoccupancy evaluations are especially useful because they will identify what has worked and not worked on those projects. Visits to projects that are considered especially good examples are also a good way of learning what to do and what not to do. Walk-throughs followed by meetings with informed users are an excellent way of learning about building types.

In recent years, building-evaluative tools have emerged that have important uses in programming. They were originally conceived as tools for evaluating buildings, but it soon became evident that they were equally useful for programming. One such tool, the Serviceability Tools process, provides a useful conceptual model that illustrates especially clearly how a performance appraisal system can be used for programming. The Serviceability Tools process is an ASTM standard process.

Postoccupancy Evaluation

The basic purpose of postimplementation or postoccupancy evaluations is to provide programmatic information to designers. This form of evaluative research involves systematic examination of a design once it has been completed and occupied for a reasonable length of time. This evaluation might range from a simple return visit by the designer to formal research, but its primary purpose is to identify what might be termed "the problems and assets of the design." Problems are those features of the design that through oversight or error do not meet the needs of its users and, consequently, should be modified both in the example evaluated and in future examples. Assets are those features of the design that work well and are valued by users. Assets should be preserved in renovation projects and incorporated into new designs. "Do what works, change what doesn't."

Designers can use the results of POEs as programmatic inputs. This will enable them to systematically improve the design of the overwhelming majority of projects that are repeated as slight variations on a theme: schools, prisons, parking lots, office buildings, and apartment condominiums. As Marcus has noted, this provides the designer with a way to link isolated building projects into a single design activity, aimed at discovering the optimal solutions to typical problems.[17] Robert Sommer described this process of incremental improvement of prototypes in his "Volkswagen Model."[18] He contrasts the activities of the producers of the Volkswagen Beetle, which consisted at that time of annual refinements of the basic Beetle design, with that of the Detroit designers who pursued a course aimed at producing a novel design each year with little reference to the functional successes or failures of preceding models. His analogy very nicely serves to illustrate the power of POE research in giving the designer the possibility of progressive design improvement, as opposed to purposeless variation.[19]

Using POE as programmatic input obviously requires that the designers have access to relevant evaluations or be able to conduct their own.

Basic Steps

Do a literature search for POEs that deal with:

- Similar settings (e.g., elementary school) or elements that make up the setting (classrooms, playground)
- Similar populations

Translate the findings to the specific problem.

Conduct a POE of the current setting.

Collect data that describe employee satisfaction with various elements of the current environment, the process used to plan and design it, and various aspects of behavior and performance. This may take the form of a survey, but a POE can also include other data-gathering methods, such as the use of focus groups, individual interviews, and workplace observations. A POE should be conducted of the existing environment, and then again after the initial completion of a pilot project (if possible) to provide benchmark data. Evaluation should continue at planned intervals, such as 6 months and 1 year, to allow for adjustment and to incorporate changes that are deemed appropriate.

In a pilot project, it is important to include as many participants as possible in all POE activities. The use of a survey makes it possible to collect information efficiently from a very large number of people. If users are sampled, be sure to include at least a few people from each functional area. Interview people responsible for managing, supporting, and maintaining the new workplace strategy, not just those working in it.

Some Tips for POE

- Consider adapting an existing survey to fit your specific situation rather than trying to develop a new survey from scratch.

- Closed-end questions (rating something or checking an item) are much easier to code and analyze than open-ended questions (calling for comments or opinions). A survey can be composed mainly of closed-end questions, but also provide the opportunity for comments. These help explain why people have answered as they have.

- Do not rely on a single method (e.g., a survey). Use a combination of methods that include observations of the way the space is used (which areas are not used and which are used differently than anticipated or have been modified by the users to make them more suitable), focused interviews, focus groups, and (if there are 50 or more people) surveys.

- Comparisons of the new workplace with the old are very helpful.

- Many aspects of the organization typically change at the same time there are changes in the workplace (e.g., management policies and practices, information technology, etc.) Identify all of these changes on a chart, and structure a discussion within the firm about the way these factors may influence what you are hearing and seeing about the workplace itself. Note that some changes affecting the organization may be external, relating to the economy or political situation.

Logistics. Postoccupancy evaluations can vary enormously in complexity and scope. They may involve a few focus groups and interviews with key informants during a few days, surveys sent to all participants, as well as observations and walk-throughs, and interviews and focus groups. Analysis of the information collected is likely to require several days, depending on the size of the project and the scope of the POE techniques used. Open electronic forums or databases can also be used for continuous feedback. Survey forms, focus group questions, and the like are used, as appropriate. Statistical software, including spreadsheet analyses and graphing (e.g., Excel) will also be needed to analyze quantitative data. Postoccupancy evaluations require knowledge of conducting simple statistical analyses (if surveys are done) and developing charts and graphs that accurately summarize data and the ability to synthesize information.

Sample: POE Survey. Excerpts from a survey (shown in Figure 4.2-4) designed to look at an office workplace are examples of how POE surveys can provide feedback on issues such as work effectiveness and employee satisfaction with the office design and planning process.

The major strength of a POE as a source of programmatic information is that it is based on an assessment of the performance of a real setting rather than on predictions derived from theory. Even in those rare instances where we have good theory, it is very limited in its relevance to the entire range of concerns that the designer of an environmental setting must consider. Looking at the overall performance of a real setting, on the other hand, provides one with a comprehensive and concrete list of attributes that work and do not work for particular client groups in particular situations. Given the limited state of environment/behavior theory, this holistic nontheoretical stance is probably the major strength of this approach. It does not work well for nonprototypical environments.

The Serviceability Tools.[20] The Serviceability Tools were developed for the American Society for Testing and Materials (ASTM) Subcommittee E06.25, part of Committee E6 on

International Facility Management Program
CORNELL UNIVERSITY
NYS College of Human Ecology

Cornell Workspace Survey

PART 1: Background Questions

PART 2: Workspace Ratings

PART 3: Additional Questions

This survey typically takes 15 to 20 minutes to complete.

PURPOSE:
The purpose of this survey is to identify aspects of the workplace which work well or could be improved from the employees' perspective. No data will be associated with any specific individual.

COMPANY NAME _____

Building Name or Address _____

FIGURE 4.2-4 Sample postoccupancy evaluation (POE) survey. (*Courtesy of William Sims and Frank Becker.*)

PART 1. Workspace Environmental Checklist

Please rate your workspace on each of the characteristics below. First indicate your SATISFACTION with each item by circling the number that reflects your level of satisfaction. Then rate how IMPORTANT each of the characteristics is to you in the same way (circle the correct number).

If some factor is particularly excellent or needs major improvement, please record your thoughts in the last column. Otherwise you may leave the area blank. Space is provided at the end of the questionnaire for you to elaborate upon these points.

WORKSPACE ISSUES (Please rate SATISFACTION and IMPORTANCE)	Very satisfied	Satisfied	Neutral	Dissatisfied	Very dissatisfied	Does Not Apply
1. Size of your own assigned work area	1	2	3	4	5	N/A
2. Amount of surface area for work at desk or workstation	1	2	3	4	5	N/A
3. Space for individual file storage at desk or workstation	1	2	3	4	5	N/A
4. Space for storing binders and books at desk or workstation	1	2	3	4	5	N/A
5. Space to accommodate computer/typewriter/telephone at the desk or workstation	1	2	3	4	5	N/A
6. Comfort of desk chair	1	2	3	4	5	N/A
7. Task lighting at workstation or desk (e.g. amount, color, glare, etc.)	1	2	3	4	5	N/A
8. Adjustability of task lighting at workstation or desk	1	2	3	4	5	N/A
9. Auditory privacy (e.g. not being distracted by office noise, other conversations)	1	2	3	4	5	N/A
10. Conversational privacy (e.g. not being overheard by others)	1	2	3	4	5	N/A
11. Visual privacy (e.g. not being distracted by others working or passing by)	1	2	3	4	5	N/A
12. Number of electrical outlets	1	2	3	4	5	N/A
13. Location of electrical outlets	1	2	3	4	5	N/A
14. Adjustability of temperature at workstation or desk	1	2	3	4	5	N/A
15. Opportunity to rearrange furniture and equipment at workstation or desk	1	2	3	4	5	N/A
16. Amount of individual display space for work-related items at workstation or desk	1	2	3	4	5	N/A
17. Amount of storage space for personal items at workstation or desk (e.g. coat, purse, etc.)	1	2	3	4	5	N/A
18. Accommodations for small (2-3 person) meetings at workstation or desk	1	2	3	4	5	N/A
19. Opportunity to display personal photos, art, mementos, etc. at workstation or desk	1	2	3	4	5	N/A

GROUP AREA ISSUES (Please rate SATISFACTION and IMPORTANCE)	Very satisfied	Satisfied	Neutral	Dissatisfied	Very dissatisfied	Does Not Apply
20. Access to other people with whom you need to work	1	2	3	4	5	N/A
21. Amount of group display space for work-related items within group area	1	2	3	4	5	N/A
22. Space for group file storage (near or within department area)	1	2	3	4	5	N/A
23. Artificial lighting in group area or floor (e.g. amount, color, glare, etc.)	1	2	3	4	5	N/A
24. Accommodations for small (2-5 person) meetings in group or department area	1	2	3	4	5	N/A

FIGURE 4.2-4 (*Continued*)

Very important	Somewhat important	Slightly important	Not important	Does Not Apply	COMMENTS
1	2	3	4	N/A	
1	2	3	4	N/A	
1	2	3	4	N/A	
1	2	3	4	N/A	
1	2	3	4	N/A	
1	2	3	4	N/A	
1	2	3	4	N/A	
1	2	3	4	N/A	
1	2	3	4	N/A	
1	2	3	4	N/A	
1	2	3	4	N/A	
1	2	3	4	N/A	
1	2	3	4	N/A	
1	2	3	4	N/A	
1	2	3	4	N/A	
1	2	3	4	N/A	
1	2	3	4	N/A	
1	2	3	4	N/A	
1	2	3	4	N/A	

Very important	Somewhat important	Slightly important	Not important	Does Not Apply	COMMENTS
1	2	3	4	N/A	
1	2	3	4	N/A	
1	2	3	4	N/A	
1	2	3	4	N/A	
1	2	3	4	N/A	

FIGURE 4.2-4 (*Continued*)

25. Vending and beverage services in department or group area	1	2	3	4	5	N/A
26. Location of department conference rooms	1	2	3	4	5	N/A
27. Number of department conference rooms	1	2	3	4	5	N/A
28. Reference/resource/information centers in department or group area	1	2	3	4	5	N/A
29. Dedicated project or team rooms for department use	1	2	3	4	5	N/A
30. Access to shared equipment (copy machines, printers, facsimile machines, etc.)	1	2	3	4	5	N/A
31. Informal break areas (small lounges, seating areas, etc.) in department or group area	1	2	3	4	5	N/A
32. Overall attractiveness of department or group area	1	2	3	4	5	N/A

BUILDING ISSUES (Please rate SATISFACTION and IMPORTANCE)	Very satisfied	Satisfied	Neutral	Dissatisfied	Very dissatisfied	Does Not Apply
33. Space for remote file storage (centralized storage from which files can be retrieved)	1	2	3	4	5	N/A
34. Heating and cooling/air conditioning	1	2	3	4	5	N/A
35. Air quality (e.g. stuffiness, odors, smoke, humidity, etc.)	1	2	3	4	5	N/A
36. Central, building-wide cafeteria	1	2	3	4	5	N/A
37. Location of conference rooms for entire firm	1	2	3	4	5	N/A
38. Number of conference rooms for entire firm	1	2	3	4	5	N/A
39. Central, building-wide reference/resource/ information center or library	1	2	3	4	5	N/A
40. Dedicated project or team rooms or areas for entire firm	1	2	3	4	5	N/A
41. Informal break areas (small lounges, seating areas, etc.)	1	2	3	4	5	N/A
42. Security (safety from personal harm and theft	1	2	3	4	5	N/A
43. Protection of confidential business-related information	1	2	3	4	5	N/A
44. Ability for visitors to find their way around the building	1	2	3	4	5	N/A
45. Visitor waiting/reception area	1	2	3	4	5	N/A
46. Overall attractiveness of building	1	2	3	4	5	N/A

LOCATION/SITE ISSUES (Please rate SATISFACTION and IMPORTANCE)	Very satisfied	Satisfied	Neutral	Dissatisfied	Very dissatisfied	Does Not Apply
47. Health and fitness services	1	2	3	4	5	N/A
48. Medical services	1	2	3	4	5	N/A
49. Access to public transportation	1	2	3	4	5	N/A
50. Parking (availability and affordability)	1	2	3	4	5	N/A
51. Access to child care	1	2	3	4	5	N/A
52. Location of building (access to shopping, bank, restaurants, etc.)	1	2	3	4	5	N/A

FIGURE 4.2-4 (*Continued*)

1	2	3	4	N/A	
1	2	3	4	N/A	
1	2	3	4	N/A	
1	2	3	4	N/A	
1	2	3	4	N/A	
1	2	3	4	N/A	
1	2	3	4	N/A	
1	2	3	4	N/A	

Very important	Somewhat important	Slightly important	Not important	Does Not Apply	COMMENTS
1	2	3	4	N/A	
1	2	3	4	N/A	
1	2	3	4	N/A	
1	2	3	4	N/A	
1	2	3	4	N/A	
1	2	3	4	N/A	
1	2	3	4	N/A	
1	2	3	4	N/A	
1	2	3	4	N/A	
1	2	3	4	N/A	
1	2	3	4	N/A	
1	2	3	4	N/A	
1	2	3	4	N/A	

Very important	Somewhat important	Slightly important	Not important	Does Not Apply	COMMENTS
1	2	3	4	N/A	
1	2	3	4	N/A	
1	2	3	4	N/A	
1	2	3	4	N/A	
1	2	3	4	N/A	
1	2	3	4	N/A	

FIGURE 4.2-4 (*Continued*)

PART 2. Background Information

Please circle the letter next to the appropriate response, except where blanks are provided for responses.

1. What is your sex?
    a. male
    b. female

2. What is your age?
    a. under 21   f. 41-45
    b. 21-25   g. 46-50
    c. 26-30   h. 51-55
    d. 31-35   i. 56-60
    e. 36-40   j. over 60

3. What is your position title?   _____

4. What is your department or division?   _____

5. On the average, how much time do you spend AT YOUR DESK per week?
    a. less than 8 hours
    b. 8-16 hours
    c. 17-24 hours
    d. 25-32 hours
    e. more than 32 hours

6. On the average, how much time do you spend IN OTHER AREAS WITHIN YOUR DEPARTMENT (e.g. working in a conference area, with a coworker at his or her desk, etc.) per week?
    a. less than 8 hours
    b. 8-16 hours
    c. 17-24 hours
    d. 25-32 hours
    e. more than 32 hours

7. On the average, how much time do you spend in parts of the building OUTSIDE YOUR DEPARTMENT per week?
    a. less than 8 hours
    b. 8-16 hours
    c. 17-24 hours
    d. 25-32 hours
    e. more than 32 hours

8. Please circle the letter which best describes your current individual workstation:
    a. Individual enclosed office (floor to ceiling walls)
    b. Individually assigned desk or workstation surrounded by high panels on at least two sides (i.e. cannot see over the panel when seated)
    c. Individually assigned desk or workstation surrounded by low panels on at least two sides (i.e. can see over the panel when seated) separating each desk or workstation in open plan office
    d. Individually assigned desk in a large shared group space (no panels around each desk or workstation)
    e. Shared enclosed (floor to ceiling walls) office for 2-4 persons in which each person has an assigned desk
    f. Shared group office (there are no individually assigned desks)
    g. Other (please describe briefly)
        _____
        _____

Part 3: Additional Questions
Please give your comments regarding the questions below:
1. What do you like most about your office workspace?
    _____
    _____

2. What do you like least about your office workspace?
    _____
    _____

FIGURE 4.2-4 (*Continued*)

Performance of Buildings. These tools were designed to enable organizations to rate the functionality of a building for office work. The scales are also useful as programming tools.

There are more than 100 scales in three groups (see the list in the appendix that is also useful as a checklist for programming topics). The scales in Group A, which cover the primary concerns of occupants, focus on requirements for group and individual effectiveness. The scales in Group B are about the property and its management. The scales in Group C provide a scan for issues involving laws, codes, and regulations. For example, under the category "Change and Churn by Occupants," potential occupants would rate the amount of disruption due to physical changes such as minor changes to layout and partition wall relocations that are tolerable.

"Over the years, buildings have been evaluated in many ways," says Gerald Davis, chairman of ASTM Subcommittee E06.25 on Whole Buildings. "Often, the evaluators simply reinvented the wheel, treating each project as a one-of-a-kind event and getting lost in the technical details. For building evaluations to be useful they must relate directly to the needs of the organization on issues that managers and users see as important for their organization and its employees to succeed." These same issues pertain equally to programming. Many programming issues are common among projects of similar building type, making it possible for statements of program requirements to be standardized, thus eliminating the necessity for time-consuming one-off drafting of the statements, as well as the resulting variability in quality. The task simply becomes one of picking the relevant requirements from the standard list and adapting them to the particular context, if necessary.

USING THE SERVICEABILITY TOOLS FOR PROGRAMMING

One group of Serviceability Tools, the *Occupant Requirement Scales,* are used to ascertain what functionality and quality are required. A questionnaire is used to identify the needs of the group that will occupy the proposed office building, as well as those of visitors, facility owners, and managers. The Occupant Requirement Scales, written in the language of office users, helps them to describe how their facilities should support operations. For each topic, such as meeting and conference rooms, shipping and receiving, or electrical power at the workplace, several options are provided, ranging from a very low level of requirements to the most demanding.

Each of these statements has a matching statement in the Facility Rating Scales, written in the technical language of facility managers, engineers, and architects. These more technical statements describe the generic features that are found in a typical office building in North America that meet the requirements of an occupant group with those operations.

The Serviceability Scales: Both for Setting Requirements and for Rating Buildings

Figure 4.2-5 is an example of one of the Serviceability Scales. Both the Occupant Requirement Scale (left column) and the Facility Rating Scale (right column) are printed on the same page. The left column gives occupant requirements, or the "demand side." The right column gives the "supply side"—a description of the building design strategy that will provide the required level of performance on the issue.

The scales function as a multiple-choice questionnaire. When setting the required level of serviceability, the client/user representative has only to choose which of the five paragraphs in the left column best describes what is needed. Each defines one level of serviceability, either 9, 7, 5, 3, or 1. If the actual requirement is partly described by one paragraph and partly described by the paragraph above or below it, then the requirement level would be an in-between number (i.e., an 8, 6, 4, or 2).

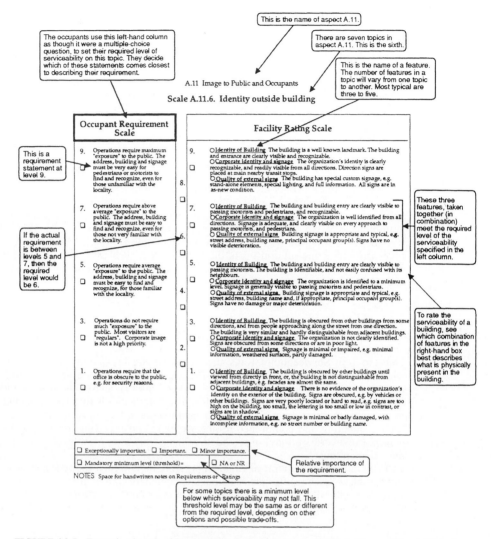

This is the name of aspect A.11.

There are seven topics in aspect A.11. This is the sixth.

This is the name of a feature. The number of features in a topic will vary from one topic to another. Most typical are three to five.

The occupants use this left-hand column as though it were a multiple-choice question, to set their required level of serviceability on this topic. They decide which of these statements comes closest to describing their requirement.

A.11 Image to Public and Occupants

Scale A.11.6. Identity outside building

Occupant Requirement Scale	Facility Rating Scale

This is a requirement statement at level 9.

If the actual requirement is between levels 5 and 7, then the required level would be 6.

9. Operations require maximum "exposure" to the public. The address, building and signage must be very easy for pedestrians or motorists to find and recognize, even for those unfamiliar with the locality.

7. Operations require above average "exposure" to the public. The building and signage must be easy to find and recognize, even for those not very familiar with the locality.

5. Operations require average "exposure" to the public. The address, building and signage must be easy to find and recognize, for those familiar with the locality.

3. Operations do not require much "exposure" to the public. Most visitors are "regulars". Corporate image is not a high priority.

1. Operations require that the office is obscure to the public, e.g. for security reasons.

9. ○ Identity of Building The building is a well known landmark. The building and entrance are clearly visible and recognizable.
○ Corporate identity and signage The organization's identity is clearly recognizable, and readily visible from all directions. Direction signs are placed at main nearby transit stops.
○ Quality of external signs The building has special custom signage, e.g. stand-alone elements, special lighting, and full information. All signs are in as-new condition.

7. ○ Identity of Building The building and building entry are clearly visible to passing motorists and pedestrians, and recognizable.
○ Corporate identity and signage The organization is well identified from all directions. Signage is adequate, and clearly visible on every approach to passing motorists, and pedestrians.
○ Quality of external signs Building signage is appropriate and typical, e.g. street address, building name, principal occupant group(s). Signs have no visible deterioration.

5. ○ Identity of Building The building and building entry are clearly visible to passing motorists. The building is identifiable, and not easily confused with its neighbours.
○ Corporate identity and signage The organization is identified to a minimum level. Signage is generally visible to passing motorists and pedestrians.
○ Quality of external signs Building signage is appropriate and typical, e.g. street address, building name and, if appropriate, principal occupant group(s). Signs have no damage or major deterioration.

3. ○ Identity of Building The building is obscured from other buildings from some directions, and from people approaching along the street from one direction. The building is very similar and hardly distinguishable from adjacent buildings.
○ Corporate identity and signage The organization is not clearly identified. Signs are obscured from some directions or are in poor light.
○ Quality of external signs Signage is minimal or impaired, e.g. minimal information, weathered surfaces, partly damaged.

1. ○ Identity of Building The building is obscured by other buildings until viewed from directly in front, or, the building is not distinguishable from adjacent buildings, e.g. facades are almost the same.
○ Corporate identity and signage There is no evidence of the organization's identity on the exterior of the building. Signs are obscured, e.g. by vehicles or other buildings. Signs are very poorly located or hard to read, e.g. signs are too high on the building, too small, the lettering is too small or low in contrast, or signs are in shadow.
○ Quality of external signs Signage is minimal or badly damaged, with incomplete information, e.g. no street number or building name.

These three features, taken together (in combination) meet the required level of the serviceability specified in the left column.

To rate the serviceability of a building, see which combination of features in the right-hand box best describes what is physically present in the building.

☐ Exceptionally important. ☐ Important. ☐ Minor importance.

☐ Mandatory minimum level (threshold)=   ☐ NA or NR

NOTES Space for handwritten notes on Requirements or Ratings

Relative importance of the requirement.

For some topics there is a minimum level below which serviceability may not fall. This threshold level may be the same as or different from the required level, depending on other options and possible trade-offs.

FIGURE 4.2-5   Example of a pair of Serviceability Scales. (*Courtesy of William Sims and Frank Becker.*)

Some requirements are more important than others. This relative importance differs from one organization unit to another, depending on their work and functions. At the bottom of the Occupant Requirement Scale is a place to indicate this relative importance. If trade-offs have to be made when setting a budget or selecting a place to lease, this can be vital information.

Many people who set requirements do not know whether a particular requirement would be more or less costly to meet. An additional Serviceability Tool deals with cost sensitivity.

In the Serviceability Rating Scale on the right side of the figure, the title of each feature is underlined. This listing at each level of serviceability makes explicit the range of features that are needed to deliver the required functionality and quality on that topic and how those features should change when higher or lower levels are required.

Using the Serviceability Scales to Program a New Facility or a Renovation

The Serviceability Tools can be used as programming instruments to set the requirements for a new building or to identify the changes needed in an existing one to meet occupant requirements.

New Facility.   In the case of a new facility, all that is necessary is for the client/occupants to identify the performance levels required on the requirement scales. The designers can then compare those performance levels with the Building Scales to identify the design strategies that will provide that performance. They will then need to adapt those strategies to the special context and other specific circumstances of the project.

Renovation.   For a renovation project, the task is similar. The Occupant Requirement Scales are used to identify the required performance levels for each feature. The Building Scales are then used to evaluate the performance of the building by comparing the descriptions of each of the design strategies with those in the building and selecting the closest match. This determines the performance level of the building for that feature. When there is a difference between the closest description, there is a process for modifying the performance score. These building performance profiles are then matched against the demand profile obtained from the Occupant Requirement Scales. When the two match, no renovation is required. When there is a mismatch, the building design strategy that matches the occupant requirement level identifies the renovation that is needed.

Using the Serviceability Scales to Evaluate a Proposed Design

When rating the serviceability of a proposed design, one looks on the Serviceability Scale for the description of the combination of features that best matches what is described in the proposal. As in renovation, a mismatch between the requirement scales and the building rating the requirement scales will identify the changes needed in the proposal. (See Fig. 4.2-6.)

Using the Serviceability Tools in Master Planning

For corporations and governments, for occupants and for building owners, the bottom line is better value for money. For building owners and custodians of real property who can more easily compare buildings in inventory and identify which are most likely to offer high functionality to the occupants or be easier to lease in future years, the bottom line is less vacancy, better fit, and higher returns.

The serviceability ratings can be used to identify those facilities that most closely match the particular requirements of an occupant organization. This would indicate which occupants could move in with the lowest fit-up cost and the highest expectation that their requirements have been met. The scales also provide an efficient and economical way of comparing properties offered for lease or purchase.

The manager or custodian of a portfolio of real property can use them to get an objective scan of the quality and functionality of properties in the portfolio. It will also identify any facilities in inventory that should be investigated in depth or that require special remedial attention, and it will call attention to any functional anomalies where the cost to cure might be significant.

Leasing agents can use this approach to scan available properties and short-list the best options for review by a prospective occupant or owner. Any professional and any occupant can use this procedure to learn how building subsystems and materials interact to provide an overall level of serviceability and how various serviceability requirements interact to determine building or facility serviceability. The process provides tangible benefits in enhanced understanding between people with different roles and particular mind-sets.

Version 2.1C

Tn=Threshold minimum level; NA=Not Applicable; NR=Not Required.
⇓ RI=Relative importance: E=Exceptionally important; I=Important; M=Minor importance.
⇓ ⇓ Level of Requirement ⇓ ⇓ Level of Rating
⇓ ⇓ ⇓ Requirement ⇓ Rating <=Less than Generic; SS=Hig
⇓ ⇓ ⇓ For A Generic ⇓ For a Notes Regarding
⇓ ⇓ ⇓ Basic General Office ⇓ Historic Building Required Level for Ba:

Code	A.1 Requirements for Group and Individual Effectiveness	Tr RI Rc	Requirement (1–9)	Rt	Rating (1–9)	Notes Regarding Required Level for Ba:
A.1	Support for Office Work.					
A.1.1	Photocopying.	T6 I	I6 (6)	5	5	Large copiers in separate rooms.
A.1.2	Training rooms, general.	T5 I	I5 (5)	3		Training is important as organize
A.1.3	Training rooms for computer skills.	T5 I	I5 (5)	3		Most staff have or use a PC, and
A.1.4	Interview rooms.	T5 I	I5 (5)	4		Normal conversation must not be
A.1.5	Storage and floor loading.	I	I5 (5)	5		Move pallets of paper for compu
A.1.6	Shipping and receiving.	I	(5)	4		Only rarely acceptable that truck
A.2	Meetings and Group Effectiveness.					
A.2.1	Meeting and conference rooms.	T6 E	I6 (6)	5		Speech privacy required for some
A.2.2	Informal meetings and interaction.	I	(6)	5		Collaboration and innovation are
A.2.3	Group layout and territory.	T5 I	I5 (5)	5		Must be able to accommodate a f
A.2.4	Group workrooms.	I	I5 (5)	5		
A.3	Sound and Visual Environment.					
A.3.1	Privacy and speech intelligibility.	T5 E	I5 (5)	3		Require speech privacy in enclos
A.3.2	Distraction and disturbance.	T5 E	I5 (5)	3		Can tolerate only limited distract
A.3.3	Vibration.	T5 E	I5 (5)	7 SS SS		Can tolerate only minor, occasion
A.3.4	Lighting and glare.	T6 E	I6 (7)	5		All or most staff use a computer.
A.3.5	Adjustment of lighting by occupants.	I	(5)	5		
A.3.6	Distant and outside views.	I	I5 (5)	1		
A.4	Thermal Environment and Indoor Air.					
A.4.1	Temperature and humidity.	T5 E	I5 (5)	3		Can tolerate moderate discomfor
A.4.2	Indoor air quality.	T6 E	I6 (7)	4		Staff must be alert at all times, a
A.4.3	Ventilation air (supply).	T5 E	I5 (5)	6 SS		Require target quality of air in bu
A.4.4	Local adjustment by occupants.	I	(5)	5		Require system to respond to spe
A.4.5	System capability and controls.	T5 I	I5 (5)	5		
A.5	Typical Office Information Technology.					
A.5.1	Office computers and related equipment.	T6 I	I6 (6)	5		Require capability to locate dens
A.5.2	Power at workplace.	T5 E	I6 (6)	5		6 plug-in points at each workplac
A.5.3	Building power.	I	I5 (5)	3		Can tolerate only minor limitatio
A.5.4	Data and telephone systems.	T5 E	I5 (5)	2		

FIGURE 4.2-6 Example of comparing a generic serviceability requirements profile to the serviceability rating of a specific building. (*Courtesy of William Sims and Frank Becker.*)

Each requirement is ranked for its relative importance and applicability, such as its relative importance for the functions or operations of office occupants, for a building's technical operation, or for its value as an investment. The scales also provide for identification of a minimum acceptable threshold level and what that level is.

Generic Requirement Packages (GRP).   The Serviceability Tools also contain packages of typical requirements for the three most common types of offices in North America: (1) basic general office, (2) basic secure office, and (3) much public contact office. Using these GRPs, occupants and office owners can focus immediately on the ways their requirements or facilities vary from what is typical.

Review the Conventions of Practice

A good source of generic information is the conventions of practice obtained over many years, and many projects can be found in reference books such as *Architectural Graphic Standards,* 10th edition, by Charles George Ramsey (John Wiley & Sons, New York, 2000), *The Architects Studio Companion: Rules of Thumb for Preliminary Design,* second edition, by Edward Allen (John Wiley & Sons, New York, 1995), *Time-Saver Standards for Architectural Design Data,* seventh edition, by Donald Watson (editor-in-chief) (McGraw-Hill, New York, 1997), *Interior Graphic and Design Standards,* by S. C. Reznikoff (Whitney Library of Design, New York, 1986). These volumes are largely devoted to descriptions of solutions and to construction practices, but they are excellent sources of information about generic and technical requirements such as parking spaces and equipment sizes and clearances. These references show the standard way of solving generic problems.

DEVELOPING CONCEPTS FOR THE FUTURE

Facilities must accommodate the user organization's current and future needs. If programming and design were focused solely on current needs, the building would be outdated before it was built. Organizations change, external conditions change, technology changes. All of them affect building requirements. A central task of programming is to anticipate as many of these changes as possible. There are several approaches to this task. The first is in the strategic visioning described earlier. It is here that changes that will be driven by the strategic plans coming from the highest levels of the organization are identified and integrated into the programming and design process for the operational units. Some aspects of strategic plans cannot be divulged even to those at lower levels of the organization. Second are the techniques for projecting changes in area requirements discussed in "Space Requirements," found in Sec. 4.2, "Programming Activities and Tools."

In addition to these steps, each of the program requirements needs to be assessed for the impact of anticipated future changes. These may range from known to probable to possible changes. Although those in leadership positions may be the only ones who are knowledgeable about the broad or even specific changes within the organization, often they are not able to describe the implications of those changes for program requirements for facilities. Thus, to the extent possible within the security requirements of the situation, each of the user representatives needs to identify the known, probable, and possible changes that might occur in the future and to identify the effects that these changes will have on the program requirements. These may include changes in the mission, functions, or activities. They may also include changes in staffing, equipment, or other factors that will affect program requirements.

It is important to have the entire programming team participate in this exercise because the changes suggested by one user representative may have implications for the area of responsibility of another. Similarly, the technical experts, the designer, and programmer representative may be able to ask questions or offer insights that trigger realizations of impacts that would have gone undiscovered if each representative were to work in isolation. Checklists of typical impacts of changes are also useful. Will this change affect the space needed, the electrical requirements, the HVAC loads, lighting, access, personnel, adjacencies, and so on?

PROJECTION TECHNIQUES

Delphi Technique[21]

The Delphi Technique can be used to arrive at consensus decisions, solve problems, or forecast future developments. It is helpful in forecasting the future needs of a group by bringing together the knowledge of all parties. After defining the issue, problem, or outcome, concerned parties are asked to provide written input on probable future outcomes based on their expertise or knowledge area. This input is made available to the entire group in full or summarized form. The group then provides written input on the results of the first round. These rounds continue until consensus is reached. The process allows participants to understand the positions of the group throughout the steps, resulting in a buyin to the final future projection.

The Delphi Technique allows people whose input is valued but are reluctant to participate in normal meetings to express their views anonymously. People from many locations may be involved because input is written rather than verbal. Experts who have specific knowledge relevant to the project should be involved. Their input may be enhanced as they synthesize the information from the rest of the group.

The Delphi Technique may be fairly time-consuming. This depends on how quickly rounds of input can be summarized and returned to the group for the next step. Also, some participants may take longer than others to complete each round in the process. The technique can be done more quickly by using automated communications systems such as e-mail or groupware software.

In its simplest form, no equipment is required to use the Delphi Technique. However, for the best results, particularly with remote participants, a computer system with e-mail capabilities or groupware can reduce turnaround time. Some organizations have even equipped conference rooms with networked personal computers especially for this type of idea sharing. The Delphi Technique can be somewhat difficult to use. It requires analytical skills and the ability to synthesize, edit, and summarize ideas succinctly. Without an automated communication system, using this tool with remote participants can be time-consuming.

Basic Steps
- Define the issue or problem.
- Have the team provide individual, written input.
- Summarize the team's input, and redistribute it.
- Have the team provide a second round of individual, written input.
- Summarize the second round, and redistribute it.
- If necessary, have the team provide a third round of individual, written input.
- Summarize the third round.

Brainstorming[21]

Brainstorming is a planning and consensus-building tool used to encourage suggestions from team members. One or two participants are specifically assigned the task of recording all comments, but brainstorming essentially involves a group of people suggesting imaginable solutions to a particular problem, issue, or situation. It can be used in small groups to seek solutions to problems such as anticipating future changes, as well as to draw out nontraditional approaches to situations. Brainstorming is an effective way of generating new ideas. The premise behind brainstorming is that no idea is too outlandish or without merit. People are encouraged to suggest ideas as they come to them.

The entire planning team should be involved in any brainstorming sessions that focus on projecting the future context of the project. Teams of more than 12 are likely to be less effective than smaller teams. Because the goal is to generate innovative solutions, people should also be invited to participate who are known to have a creative way of defining problems, who come from other fields or disciplines, and who may be from inside or outside the company. If a brainstorming session is held to tackle a specific problem, make sure that all persons who have a stake in the problem are encouraged to participate.

Brainstorming sessions varying in duration from one-half hour to several hours can be quite productive. The session should be focused and have a clear end point. A congenial setting and food and drink may facilitate the free flow of ideas. Brainstorming sessions require minimal special equipment. A flip chart or overhead projector and markers are necessary for recording ideas and suggestions so that they may be seen easily by all participants. It is also helpful to provide a closed room for any brainstorming session to encourage participation and enthusiasm without disturbing others.

A successful brainstorming session requires a leader who has the ability to persuade others to participate in the process. One of the keys to productive brainstorming is an atmosphere in which no idea is considered too outrageous and all suggestions are recorded for later evaluation. Therefore, first, it is important to appoint a person to act as the session recorder who can capture many ideas quickly in a concise, legible manner. Next, set a time limit, and stick to it. If your participants discover that a productive 20-minute brainstorming session really takes only 20 minutes, they will be much more likely to participate again, enthusiastically.

Basic Steps

- Brainstorming begins with the first idea offered. It is useful to have a few thoughts before the session starts to "prime the well."
- Monitor the flow of information as others start to voice ideas and suggestions.
- Record ideas on a flip chart, and tape the sheets to a wall.
- If only some people are participating or if the ideas are flowing much faster than the recorder can capture them, switch to a round-robin method in which each person offers one idea at a time.

The following ground rules are helpful to encourage a productive brainstorming session:

- *Do not* edit someone else's contribution.
- *Do* build on the ideas that spark other suggestions.
- *Do not* criticize each other.
- *Do* contribute wild and crazy ideas—creativity is the goal.

GENERATING DESIGN REQUIREMENTS FROM THE DATA

In addition to planning the building to meet those needs that can be reasonably anticipated, programming also needs to address the issue of uncertainty. In spite of our best efforts and our predictive tools, it is still difficult to anticipate with any precision what the future may bring. All organizations are operating in an environment of change, but many fast-growth companies are growing as fast as basic facilities and infrastructure can be provided. Lack of the right kind of space can inhibit the growth that otherwise would be possible. However, anticipating the right kind of space and the quantity needed is a virtually impossible task. The pace of change driven by global competition, changing markets, and technological innovation adds another requirement to this task of planning buildings that should last for 40 or 50 years. That is planning for unanticipated change. This is sometimes referred to as an exit strategy, but the concept is far broader than that. It involves thinking through how the building might be designed to accommodate a broader range of uses than that being planned for. There are many strategies that can be used to make buildings more receptive to changes:

- Planning the building to match needs of the broader market in the immediate area
- Providing for expansion
- Providing for internal subdivision of the building so that additional tenants can be accommodated
- Selecting structural modules, bays, and floor heights that can accommodate several different types of uses
- Providing multiple service entrances and risers that are generous in size and well distributed to accommodate a variety of service providers and increased levels of service

Several basic types of requirements become part of the program. First are those that derive from the overarching strategic vision or goals of the organization. These provide an overall direction and also help to provide for the overall organization of the project. Second are the requirements that derive from the mission, functions, and activities of each of the organizational subunits (divisions, departments, workgroups, etc.). These are followed by individual user requirements. There are also technical requirements relating to such things as the electrical power required to operate an elevator, the turning radius necessary for a semitrailer to maneuver into a loading dock, or the operational requirements of equipment. Another set relates to legal and regulatory requirements such as codes, ordinances, and standards designed to protect the public health, safety, and welfare or to implement governmental policy.

Now the task is to use the information that has been collected and analyzed to generate design requirements. Generating these requirements is an art that uses some of the tools of science. These requirement statements will be used to guide the search for solutions and to evaluate design proposals. The task is to make them

- Specific enough that it is possible to tell when a design satisfies the requirements
- Specific enough that they enable discrimination among alternative proposals
- General enough that they do not specify a particular solution

What does a good design requirement look like? Here are some examples.

Privacy

Management offices and conference rooms should enable full conversational privacy. Conversations at a normal speech level should not be discernible by someone in an adjacent space or passing by in a corridor.

Lighting

General ambient lighting should not create veiling reflections on computer screens or create direct-source glare for occupants.

Informal Communication

Selection and placement of circulation, stairs, activity generators (e.g., coffee, copy, mail) should maximize the potential for unexpected face-to-face contact in which spontaneous conversations can occur without visually or auditorily disturbing people in their workstations or offices.

These requirements do not specify how privacy, lighting, or informal communication are to be achieved. They do not say to use a 6-foot-high panel or uplighting or create small lounge areas. The design solution can reflect the designer's creativity and inventiveness. Second, they are not technical (although STC or NIC sound ratings could be specified). Rather, the intent is to state what is to be achieved, in terms that designers, the project team, and users can understand. Third, the requirements are specific enough to enable either the designer or the client to ascertain whether a design proposal satisfies them. Fourth, they are specific enough to be used to compare and choose between two design alternatives or perhaps to reject both and hence to continue the search.

Requirements can take several different forms that derive from different types of measurement and the state of the underpinning science related to a phenomenon.

Nominal. This is a simple categorical or binary measurement. Gender and job type are examples of nominal variables. "Provide a conference room" is an example of a design requirement using a nominal measure. Assessing the design is a simple yes or no question. Is there a conference room or not?

Ordinal. This is a ranking measure. Qualities can be ranked as in relative heights of people, but there is no way to determine the amount of difference. "Maximize the natural light in the classrooms" is an example of a design requirement using ordinal measurement. It does not specify by how much. In evaluating alternatives for this requirement, one would select the one that provided the most natural light.

Interval. In this case, there is a scale comprising equal intervals. But the zero point is arbitrary and does not indicate an absence of the quality being measured. The Fahrenheit temperature scale is an example. The scales used in the Serviceability Tools are examples. The requirement would specify that privacy at a level of 5 should be provided using the scales supplied.

Ratio or Metric. A scale of equal intervals exists, and the zero point is nonarbitrary. The metric system of measurement is an example. In this case, the requirement could specify that the main hallway should be a minimum of 8 feet wide or that lighting levels on the floor surface should be 30 foot-candles.

In certain highly critical areas such as acoustics or lighting, it may be useful to seek advice from human factors or acoustical experts who have a deeper understanding of the available research and more experience in assessing the adequacy of the requirements and proposed solutions.

ORGANIZING PROJECT GOALS, REQUIREMENTS, AND ASSUMPTIONS

Requirements need to be organized and displayed. A good method for this is the one developed by Peña in the "Problem Seeking Process."[2] Each statement is written on an 5-in by 8-in index card. These cards are purposely kept small so that only one idea or concept can be expressed on them. The cards are then pinned or taped to a wall where they can then be easily organized and reorganized. A card can be easily replaced as the idea it expresses is clarified or reformulated. This format allows easy reordering for editing and prioritizing.

Prioritizing the Requirements

The next step in the process is to prioritize or rank the goals and requirements that will guide the search for a solution. This is obviously a phase that must confront the fact that the project probably cannot accomplish all of the needs identified. The card system just described works well for this. The project team should rank-order the cards by using the data obtained. In the process of doing this, it is often possible to reformulate the concepts so that they can combine or consolidate requirements. Two other techniques for prioritizing are described later.

Instant Priorities[7, 22]

Instant priorities is a technique for reaching group consensus. It assists teams in establishing which requirements the group feels are most important. The process involves listing all requirements and having each group member complete an individual grid of preferences in which each requirement is paired with one other. Then individual totals are combined to produce a group total for each requirement. The group totals indicate which requirement received the greatest number of votes and how wide the gaps are between requirements, an additional piece of information that may help the group decide which requirements to consider. Instant priorities can also be used at the beginning of the project to help prioritize the goals and objectives of the project and simplify the number. Any employees who are instrumental in developing the design such as facility managers, MIS, HR managers, user representatives, senior managers, and others should participate in the instant priorities exercise. (See Fig. 4.2-7 for an example of an instant priorities chart.)

Logistics.   Depending on the size of the group involved, instant priorities can be completed in a period of between 10 min to about 1 h.

Basic Steps

1. Agree on 10 or fewer requirements, and write them on the requirements form (on the next page) so that each one is numbered.
2. Give every member of the team a photocopy of the form with the statements written on it.
3. As individuals, work your way through the chart. In every gray square, circle the number of the requirement you prefer. For instance, the top box gives you a choice between requirements No. 1 and No. 2.
4. When you are finished choosing, enter the number of times you voted for each requirement in tally col. A.
5. Then, in col. B, identify your priorities by ranking the requirements—1 for the requirement for which you voted most (the highest number in col. A) and 10 for the statement for which you voted least (the lowest number in col. A). If there are tie votes in col. A, pick one or the other as the higher ranked item in col. B.
6. When all team members are finished, add their tallies (from col. A) together to get a group total. Column C is provided here to contain the group tally.
7. Once again, identify priorities by ranking the requirements, but this time as a group. Put the group rankings (1 for the group's favorite, 10 for its least favorite) in col. D.

Note: Column D gives you the group's decision. Column C, however, tells you how wide the gaps were between the top-ranked and bottom-ranked requirements—a vital bit of information. It may show you, for instance, that the top three requirements are the only ones really worth considering over the long run.

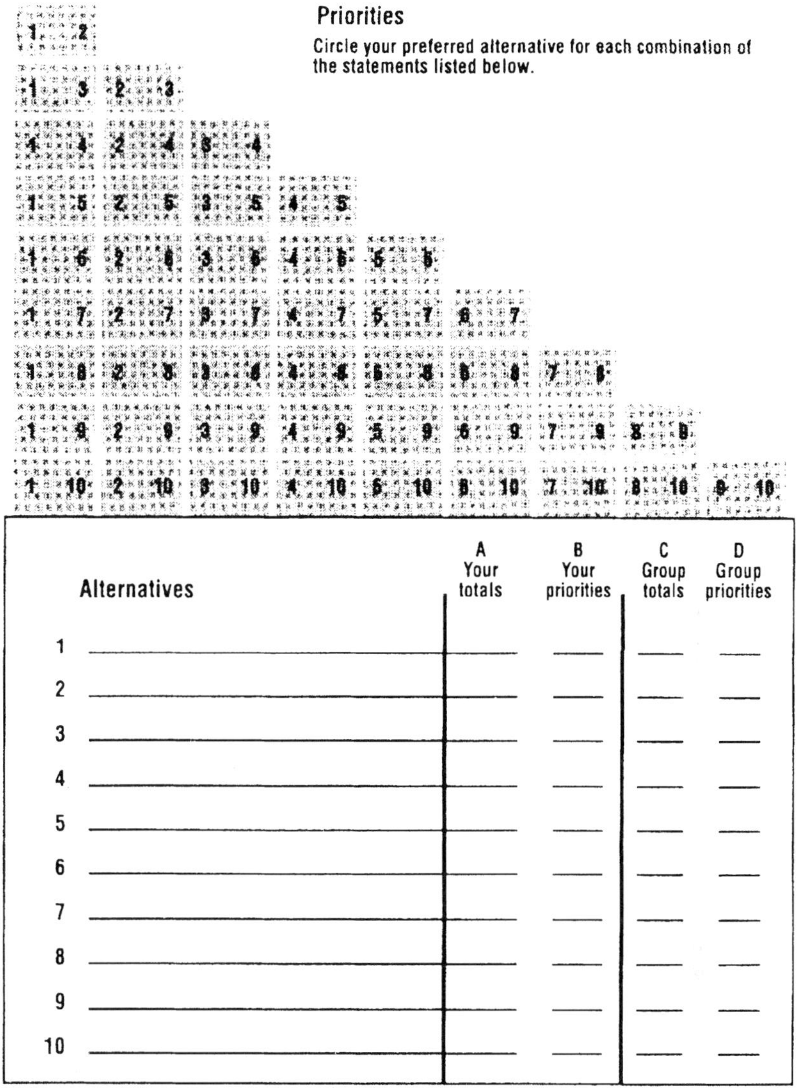

Priorities

Circle your preferred alternative for each combination of the statements listed below.

Alternatives	A Your totals	B Your priorities	C Group totals	D Group priorities
1				
2				
3				
4				
5				
6				
7				
8				
9				
10				

FIGURE 4.2-7   Sample instant priorities chart. (*Courtesy of William Sims and Frank Becker.*)

Nominal Group Technique[21]

The nominal group technique (NGT) helps teams select and prioritize requirements. Through silent voting and limited discussion, an NGT provides a way to give everyone in the group an equal voice in requirement selection. This is particularly helpful when the problem is sensitive or controversial. An NGT allows the whole team to participate in the prioritization process. An NGT produces team decisions with a minimum of conflict, yet allows discussion of the pros and cons of the alternatives. It can also be very useful in helping a group explore innovative solutions to the inevitable problems associated with a new design. All members of a team, task force, or group involved in the decision-making process should take part. This may include planning team members, management, end users, MIS, HR planners, and so on.

Logistics. This depends on the number and complexity of the problems. Two to three hours should probably be scheduled. No special equipment is necessary.

Getting a Sign-Off on the Project Program

Once the prioritizing has been completed, the project team should arrange a meeting with the decision-making body. The purpose is to present the information and seek their approval for the project program.

SECTION 4.3
USING THE PROGRAM

William Sims, Ph.D., C.F.M., IMFA Fellow, Professor, and Franklin D. Becker, Professor
Cornell University, Ithaca, New York

Whether using a structured tool such as the Serviceability Scales or a more customized programming effort, it is critical that the information actually be used in the design process. Research has shown that even in cases where a good program document exists, it was not used by the designers to generate the design and, even more surprising, it was not used by the clients to evaluate the proposals offered.[23] In part, this can be attributed to the inadequacies of the relay model process in which changes at later stages rule earlier decisions and information irrelevant. At the same time, there is a powerful inclination among designers to follow their own ideas about what is needed and what constitutes good design. These may not relate to the client's interests. What can be done to change this?

Using the Program to Guide the Search for Solutions

In the rugby model of programming, in which programming and design are seen as a continuous and interwoven process, the program is always a part of the decision-making process. It is important to have an explicit process of referencing the program during the design process. The program requirements can be reordered in priority and changed or refined as the team learns more about the problem during the design process.

Something as simple as a periodic review in which the program is used systematically as a guide can be critical to keeping the design and the program related. This simple process will go a long way toward ensuring a design that complies with the requirements of the program.

More formal procedures could involve developing a checklist with weights for priorities. Then, the design could be rated subjectively by members of the project team on a scale and the scores displayed in a matrix. It is also possible to have an organizational ecologist or environmental psychologist assess the effects of the design on things like way finding or informal communication. In other cases, there may be formal models that can be used to calculate or estimate the effects of the design. Acoustic and lighting calculations are examples. These typically occur in areas where the science is relatively well developed and, as in the case of lighting and acoustics, there are computer programs that predict the performance of a design with a high degree of accuracy. It is important to not allow these tools, as convincing as they may be, to overshadow the process and compromise a holistic and balanced assessment of the design on the other requirements that may be more important but less easily or precisely assessed.

Using the Program to Evaluate Proposals

When two or three alternative design proposals have been generated that informally seem to "satisfice,"[24] then a more careful and formal process of evaluation and refinement, is in order. Two procedures are useful for this process. They are (1) the performance profile and (2) the weighted matrix.

The Performance Profile.[25]   The performance profile is a way of evaluating proposed design alternatives using a systems perspective. It stresses the overall pattern of outcomes. Performance profiles can be used to show the level at which the proposal performs in achieving the design requirements versus what was targeted. Employees and managers who will work in the proposed design and other people who are instrumental in the implementation should agree on a core set of measures to be depicted in the performance profile.

The time required to develop the performance profile varies significantly. If the measures are already in place and will be applied before and after the implementation, the profile requires only that the results be graphically displayed. The Serviceability Tools are an example of a performance profile tool that is available for use with no preparation required. The development of new performance measures and methods of collecting information will add a considerable amount of time to the task. Many software packages are available to graph the performance profile. Performance profiles are fairly easy to use. Again, the most resource-intensive task is developing the measures to display. (See Fig. 4.3-1 for a comparison of two performance profiles.)

Weighted Matrix.   The weighted matrix is another way to compare alternatives for the program by using a systems perspective. First, construct a matrix for each solution, and in a careful way using a simple rating scale (+3 to −3), rate the degree to which it satisfies the program requirements. Also note whether any assumptions or givens were violated. Then, determine whether there are major impacts or consequences of the design that were not included as program requirements and need to be considered. Remember that the same issue may impact or affect different users in different ways. Also note that the impacts may differ in different regions or areas of the design. (See Fig. 4.3-2.)

Once the impacts are identified, it is necessary to place values on them. At the simplest level, are they good or bad? As in the case of the requirements, use a simple rating scale (−3 to +3). It is important to remember that different groups may value the same result differently. It may be necessary to recognize that certain groups weigh more heavily in the decision than others, and therefore it may be necessary to assign weights in the matrix. Similarly, different requirements may be more important than others and can be weighted. It is a good idea to leave room in the matrix for notes describing particular issues or factors about each of the evaluations.

Selecting a Preliminary Design

Now, all of the information is in the matrix that is needed to rank the alternatives. Sum the rows and columns, and determine the total score for each alternative. Remember that this is not a precise process, and it must be viewed in that way. Alternatives with relatively close scores are probably not really very different. Also, be sure to look at the pattern of scores

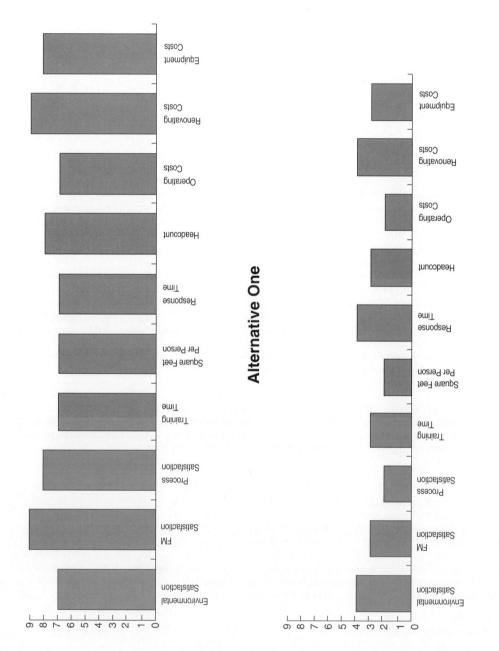

FIGURE 4.3-1 Comparing performance profiles. *(Courtesy of William Sims and Frank Becker.)*

	By User					By Goal			
	Secretary	Engineers (×3)	Managers	Total		Secretary	Engineers	Managers	Total
Privacy	1	2×3 / 6	1	8	Privacy	1	2	1	4
Storage	3	3×3 / 6	2	11	Storage	3	3	3	9
Task Support	3	2×3 / 6	2	11	Task Support (×2)	3×2 / 6	2×2 / 4	2×2 / 4	14
	7	18	6	34		10	9	8	27

FIGURE 4.3-2 Weighting. (*Courtesy of William Sims and Frank Becker.*)

because a solution with a slightly lower overall score may simply be weak in one area that can easily be modified to raise its score substantially. This information should be presented to management for a decision. (See Fig. 4.3-3.)

Design Development

The matrix can be used to improve and refine the selected design. Look at the patterns of scores. Is one user group treated poorly by the design (alternative 3 in the Fig. 4.3-3)? How can the design be changed to improve the score for that group? In the case of a single low score (alternative 2 in Fig. 4.3-3), think of ways to change the design to improve that score. Be sure that the changes do not negatively affect other scores, however. Finally, when no further improvements can be identified, a decision is needed to proceed to the development of contract documents.

CONCLUSION

The program gives the organization, architects, designers, and project managers a constant reference point for keeping the project's goals in focus. It also helps guide the development, evaluation, and eventually the selection of a design that will meet the organization's needs, not only at the moment, but for the next few years.

Programming is a process involving many different techniques and tools for collecting, analyzing, prioritizing, and communicating the requirements of a project. It requires that pro-

	Alternative 1				Alternative 2				Alternative 3			
	Secretary	Engineers	Managers	Total	Secretary	Engineers	Managers	Total	Secretary	Engineers	Managers	Total
Privacy	3	3	1	7	2	1	2	5	1	3	3	7
Storage	2	1	2	5	3	3	3	9	1	3	3	7
Task Support	3	2	1	6	3	3	3	9	1	3	3	7
	8	6	4	18	8	7	8	23	3	9	9	21

FIGURE 4.3-3 Evaluation of alternatives. (*Courtesy of William Sims and Frank Becker.*)

grammers have access to and the ability to use a range of tools, and that they are able to select the right tool for the job.

NOTES

1. Many of the ideas in this paper have been developed or alluded to by others. When we describe the approaches, we will attempt to give credit to the sources of ideas and techniques, but many have occurred in informal discussions over a number of years and may inadvertently have become "ours" through faulty recollection.

2. W. Peña, *Problem Seeking: An Architectural Programming Prime,* AIA Press, Washington, DC, 1987.

3. D. Duerk, *Architectural Programming: Information Management for Design,* Van Nostrand Reinhold, New York, 1993.

4. F. Duffy, *The New Office,* Conron Octopus, London, 1997, p. 249.

5. More detailed guidance is given in the Construction Industry Board document, *Briefing the Team,* available in the United Kingdom from Thomas Telford Publishing.

6. K. Lynch, *Site Planning,* 2nd ed., MIT Press, Cambridge, MA, 1971.

7. F. Becker, M. Joroff, and K. Quinn, *Toolkit: Reinventing the Workplace.* IDRC, Atlanta, 1995; F. Becker and K. L. Quinn, *WorkSmart: Organizational Workplace Analysis,* Unpublished working paper, Cornell University International Workplace Studies Program, Ithaca, NY, 1994.

8. S. Harmon, *The Codes Guidebook for Interiors,* Wiley, New York, 1994.

9. Edward Ostrander is the originator of this idea.

10. L. Festinger, *Social Pressures in Informal Groups,* Harper, New York, 1950.

11. O. Newman, *Defensible Space,* Collier Books, New York, 1973.

12. R. Barker, *The Stream of Behavior,* Appleton-Century-Crofts, New York, 1963.

13. Constance Perin is one of the originators of this idea.

14. The particular degree of disaggregation depends upon the type of environment being designed and is a judgmental act. For example, behavior described as recreation or work is clearly much too aggregated to be of use for most environmental design problems because work or recreation comprised many discrete behaviors. Similarly, one could go to the other extreme and describe every muscle twitch involved in walking to the bus stop. Therefore, behavior described at the level of muscle twitches is irrelevant to an urban designer designing a downtown mall but might be highly relevant to an industrial designer who is designing a new typewriter.

15. C. Ramsey and H. Sleeper, *Architectural Graphic Standards,* 9th ed., Wiley, New York, 1994; DeChiara and Calendar, *Time Saver Standards for Building Types,* McGraw-Hill, New York, 1990; Reznikoff, *Interior Graphic and Design Standards,* Whitney Library of Design, New York, 1986.

16. It may be useful to refer to standard building area measures definitions to avoid confusion. See D. Cotts and M. Lee, pp. 365–369 for the IFMA/ASTM standards, *The Facilities Management Handbook,* AMACOM, New York, American Management Association, c1992.

17. T. Marcus, "The role of building performance measurements and appraisal in the design method," in *Design Methods in Architecture,* G. Broadbent (ed.), Wittenborn, New York, 1969.

18. R. Sommer, *Personal Space: A Behavioral Basis for Design,* Prentice-Hall, Englewood Cliffs, NJ, 1969.

19. It should also be noted that a process that focuses on eliminating problems and preserving assets is not likely to produce major new breakthroughs, new forms, and so on. To use Sommer's analogy, the Volkswagen model would not produce a Rabbit.

20. ASTM, E1679-95 Standard Practice for Setting the Requirements for the Serviceability of a Building or Building-Related Facilities, E1334-95 Standard Practice for Rating the Serviceability of a Building or Building-Related Facility, American Society for Testing and Materials, Philadelphia, 1995.

21. F. Becker, M. Joroff, and K. Quinn, *Toolkit: Reinventing the Workplace,* op. cit.; P. K. Kelly, *Team Decision-Making Techniques,* Richard Chang Associates, Irvine, CA, 1994, pp. 45–57; W. F. Cascio,

Managing Human Resources: Productivity, Quality of Work Life, Profits, McGraw-Hill, New York, 1989; M. Brassard, *The Memory Jogger: A Pocket Guide of Tools for Continuous Improvement,* Goal/QPC, Lawrence, MA, 1988.

22. R. Ross, C. Roberts, and B. Smith, "Designing a learning organization: First steps," in *The Fifth Discipline Fieldbook: Strategies and Tools for Building a Learning Organization,* P. M. Senge, C. Roberts, R. B. Ross, B. J. Smith, and A. Kleiner (eds.), Doubleday, New York, 1994, pp. 53–57.

23. D. McKay, *Use of Programming in Design,* Master's Thesis, University of Manitoba, Winnipeg, Manitoba, 1994.

24. Herbert Simon's concept of limited rationality. It is impossible to find the best solution to any problem. There is insufficient time and resources to examine all of the alternatives. Typically, we "satisfice"—take the first alternative that is good enough.

25. F. D. Becker, *The Total Workplace: Facilities Management and the Elastic Organization,* Van Nostrand Reinhold, New York, 1990.

CHAPTER 5
ENGINEERING AND DESIGN PROCESS

Paul R. Smith, P.M.P., P.E., M.B.A., Chapter Editor
Peak Leadership Group, Boston, Massachusetts

Anand K. Seth, P.E., C.E.M., C.P.E., Chapter Editor
Partners HealthCare System, Inc., Boston, Massachusetts

David L. Stymiest, P.E., SASHE, C.E.M., Chapter Editor
Smith Seckman Reid, Inc., New Orleans, Louisiana

Roger P. Wessel, P.E., Chapter Editor
RPW Technologies, Inc., West Newton, Massachusetts

The organization of this chapter was a source of numerous discussions. Many books serving the facilities engineering industries are written on a *component* basis, such as boilers for providing heating and chillers for providing air conditioning. We believe the new way of thinking in facilities engineering is to look at a *systems approach* instead. We have divided the major engineered systems into one of five engineering and design categories. These five categories are as follows:

1. Architectural and structural systems
2. Electrical systems
3. Lighting systems
4. Mechanical systems
5. Instrumentation and control systems

Each of the five major engineering and design categories is further divided into the major engineered *systems* within that category. By providing information on all systems, which taken together form the facility infrastructure, we believe that the reader will have an opportunity to understand the relative merits and importance of every system. One can not have human comfort without an HVAC system, which is not possible without a robust electrical system and a well-lit, controlled, structurally capable, and architecturally pleasing environment. This is the basis of our first edition.

Our goals in this chapter are to provide sufficient information for each of the engineered systems categories so that facility managers can effectively manage new construction, additions, and renovation projects at their facilities while operating and maintaining their facilities reliably, economically, and safely. We also want facilities managers to understand the challenges at hand and to be able to communicate effectively with their engineering and construction counterparts in all phases of projects. Finally, we aim to provide a document that those facility managers will find user-friendly.

SECTION 5.1
ARCHITECTURAL AND STRUCTURAL SYSTEMS

Roger Wessel, P.E., Principal, Section Editor
RPW Technologies, Inc., West Newton, Massachusetts

INTRODUCTION

In the editor's opinion, architectural and structural systems establish the foundation, framework, form, and appearance of a building and facility for the life of the structure. Before the civil engineering work for a building or facility can begin, a geotechnical evaluation of the site is required to support the design of the building's foundations. Article 5.1.1 covers the geotechnical considerations needed to support the civil design and engineering described in Art. 5.1.2. The structural aspects, skins and facades, and roofs are addressed in Arts. 5.1.3, 5.1.4, and 5.1.5, respectively. An integral part of the architectural and structural aspects of any contemporary facility includes the design of elevators, escalators, and moving walks, which is addressed in Art. 5.1.6.

ARTICLE 5.1.1
GEOTECHNICAL CONSIDERATIONS

J. Lyndon Rosenblad, Ph.D., P.E.
Consulting Engineer, Westwood, Massachusetts

INTRODUCTION AND OBJECTIVES

This article provides information on the geotechnical issues that must be considered and addressed when siting a facility. Whether the location is in an urban or in a rural environment, the facility is totally new, or an existing facility is being expanded, many issues are the same. Some issues, however, take on a greater degree of importance in an urban area or require information in more detail than in an undeveloped rural area. The same is true for the development of a totally new facility versus adding to an existing one. Information about a totally new facility has to be developed from scratch, whereas information is already available for an existing facility, in most cases, and must be updated, enhanced, or verified. The type of facility to be built, the weight of the structures, the overall plan area, the access required, the utilities required, and the discharges emitted from the facility, however, will be unique to the facility. State and local regulations and laws can also cause variations in the number of issues that need to be addressed or the degree to which each must be addressed. For major projects, significant consideration in the planning, design, and construction must be granted to communities and citizens through community participation.

The objectives of this article are to present and discuss critical geotechnical issues to assist an owner, an engineer, a nonengineer agency inspector, or a concerned citizen in the planning process for a facility.

The approach presented focuses on the issues that must be considered in all cases. For each topic, a brief discussion is presented highlighting the most important items in that area. The first topic covers the type and number of investigations and testing that are required to obtain information on a regional basis, as well as at the site. A discussion follows that identifies hazards that could be an issue and those that need to be factored into the design with proper contingencies. Interferences can become costly, time-consuming, and a safety issue, especially in an urban area. Examples of such interferences are presented. Development of the design criteria from the exploration and testing that have been done is the ultimate purpose of the program so that the designer has the necessary parameters. The selection of foundation types for the structures is an important item in determining the cost of the facility and its performance. Some excavation is required for most facilities. Putting a hole in the ground requires planning, engineering, and good design to ensure that the desired goals are obtained, as well as to ensure that the safety of the workers and the integrity of surrounding structures are not jeopardized. The last discussion centers on construction liaison. All of the best planning, engineering, design, and specification preparations are for nothing if the construction is not done properly. Liaison during construction is extremely important and should not be ignored or eliminated.

INVESTIGATIONS AND TESTING

Regional Investigations and Testing

Investigations and testing are a necessary part of the geotechnical work that must be done in obtaining information for a new facility or in expanding an existing facility. On a regional scale, the investigations usually consist of research into work that has been done by others in the geographic region where the facility is to be constructed. The purpose is to gain information on the regional geologic picture using U.S. Geological Survey geologic and topographic maps, U.S. Soil Conservation Service soil maps, aerial photos, satellite photos, and state and county soil maps. From these sources and others, one can develop the general subsurface profile beneath the area. Unusual conditions are often highlighted or evident from these maps and photos. Topographic maps often show landslide areas or drainage patterns that may need to be addressed. Topographic highs might indicate resistant rock outcrops or deposits of unconsolidated material such as a terminal moraine left by receding glaciers. Topographic lows might indicate areas that are in a flood plain and would be subject to flooding. Local sources such as county or town maps yield important information about land usage. County maps might show land that is in agricultural production or heavily forested. City maps show developed areas and current utilization. Aerial photos may provide the most comprehensive readily available information. If the proposed site region has been flown over recently and photos are available in stereographic pairs, then one can learn about current land usage and look at the region in three dimensions.

Other sources of information that are available and should be used are rainfall records, flood records, and earthquake records in certain parts of the United States. From these records, one can ascertain whether there is a tendency for flooding or if the area has been subjected to earthquake activity in the past. The potential for flooding or earthquake activity can sometimes render an area unsuitable for a facility or can add significantly to the cost of the facility.

Depending on the development and the types of facilities in the area, another valuable source of information is from the private sector. Engineering and geologic data and design information that were collected, analyzed, and used by owners in constructing their facilities can be most helpful. Boring logs, test pit logs, soils and geologic maps, geophysical records, construction records, and data from instrumentation all have value.

If rainfall data and flood records are not available from the literature or governmental agencies, this information must also be obtained. Usually at least a year is required to obtain these records, so planning and executing this program should be done early in the process.

In addition to learning as much as possible from records and information in the public and private domain, one should also visit the area to confirm what has been written and to see what changes have occurred since the written material was prepared. The visit to the area will also point out the type of facilities that are in the area now and how each one is performing. The facilities that should be looked at are buildings and also other structures such as roadways and pavements, bridges, culverts, embankments, and geologic features such as road cuts and outcrops of natural material. Conversations with people who live in the area can sometimes be very revealing. Information that does not enter into the written record can be obtained by talking with those who experience the area every day.

Site-Specific Investigations and Testing

Once a specific site has been tentatively selected for the facility, additional specific information is needed about the site. At this time the geotechnical engineer has general knowledge of the area based upon his or her regional study and also has obtained information from the structural engineer about the type of facility, loadings, plan area, and land required. Now the geotechnical engineer is ready to determine whether the site is technically feasible, is unacceptable, or will have significant cost penalties to solve some deficiencies.

Soil Site

If the site is believed to be a soil site based upon the understanding of the regional work, the program will focus on obtaining soil foundation design information. For lightly loaded structures, the program should consist of a series of shallow borings and/or test pits that will yield information on the properties of the soils in the upper few feet beneath the structure. These could be auger borings, rotary wash borings, or borings using a sonic or air-driven rig. The depth of at least one or two borings should be a minimum of 2 times the least dimension of the building or structure that will be built above it. The number of borings should be spaced to cover the building area and extend outside of it. As the borings are being made, the geotechnical engineer or geologist must be present to log the materials as they are removed from the boring and will also determine the sampling and testing to be done in the boring. If the type of rig allows for split-spoon samples, a sample should be taken every 5 ft. At a minimum, blow counts should be recorded as split-spoon samples are taken. Blow counts are a key indicator of the integrity and physical properties of the material. Correlation from the literature is used to indicate the strength and other properties of the material. As the material is removed from the split-spoon sampler, logged, and placed into sealed jars, the engineer may also obtain more information by doing pocket penetrometer tests or vane shear tests on the samples.

For facilities that will produce heavy loads, that cannot tolerate very little or differential settlement, and will rest on thick soil deposits, a more extensive exploration program is required. The program should be planned to yield information about the soils, so that deep foundations can be designed. As such, more of the borings will be deep, and some will probably extend to or into the top of rock. These borings will require casing or the use of drilling mud and are usually rotary wash borings. If in-the-hole permeability tests are planned, drilling mud should not be used. As before, split-spoon samples should be taken every 5 ft in the boring. If soft cohesive material is encountered, undisturbed tube samples should be taken of that material. The tubes used to obtain these samples are thin-walled (to minimize disturbance) and are pushed into the soft cohesive material. Each tube should be sealed to prevent moisture loss; care should be taken in handling, storing, and transporting the tubes to the laboratory. After samples have been taken, logged, and put in sealed containers, they should be moved to an environment that does not subject them to extreme cold, extreme heat, or vibrations.

The depth to the water table is an important piece of information that must be logged by the field engineer. This reading should be made when the water level has stabilized after the hole has been completed. A 24-h period is usually long enough.

Additional in-hole testing is sometimes warranted. These tests may consist of vane shear tests, cone penetrometer tests, pressure meter tests, and standing-head or falling-head flow tests. Strength, modulus, and permeability values of the in situ material are obtained in this way.

Sometimes it is advantageous to incorporate geophysical exploration, in conjunction with borings, into the program. Geophysical techniques such as crosshole or downhole testing, along with seismic refraction or seismic reflection, can provide information on the materials between borings. This usually requires fewer borings and may result in some efficiencies in the exploration program. These geophysical techniques require a small energy source to produce waves that pass through the subsurface material and a geophone or geophones to receive the waves. The time that it takes these waves to pass through the subsurface materials is an indicator of the properties of the materials. The energy source will produce more than one type of wave, for example, primary or shear waves. Geophysicists interpret the results of the tests such as the time for travel, the shape of the curve, and the relative position of peaks and valleys to yield information on boundaries of the layers, thickness, relative strength, moisture content, and relative density of the subsurface materials.

The laboratory tests performed on the soil samples obtained from the field exploration program should yield pertinent information about the site and the type of foundation that is anticipated. These tests at a minimum will consist of Atterberg limits, moisture contents, grain size analyses, and a definitive soil classification. If undisturbed tube samples of soft cohesive material are obtained, consolidation and strength tests should be performed. The strength tests may be triaxial compression, uniaxial compression, or direct shear tests. For samples of noncohesive material, tests may consist of triaxial compression tests, permeability tests, or dynamic tests. Laboratory tests should be performed in accordance with standards and procedures established by the American Society for Testing and Materials (ASTM). The number of tests should be based on the size of the site, the variability of the materials, the thickness and number of soil deposits, and a statistically satisfactory number of values.

Rock Site

If the depth to rock is believed to be shallow, the foundation exploration program should focus on obtaining the necessary information about the rock. Core borings are the best means of obtaining these data. Drill rigs with the capacity and capabilities to core rock are required. These borings should continue at least 10 ft into sound rock. The amount of rock recovered and the amount of rock that exceeds 4 in in length are important indicators of the rock quality. The Rock Quality Designation (RQD) is the percentage number obtained by adding up the total length of the number of pieces of cored rock that are 4 in or longer divided by the total length of cored hole and multiplying by 100. The higher the percentage of core recovery and the higher the percent RQD, the better the in-situ rock quality. The recovered rock core should be placed into wooden or cardboard boxes after it has been logged and kept indoors until it is transported to the laboratory.

Additional in-hole testing may also be a part of the program. Information about the in-situ permeability of the rock formations can be obtained by performing pressure tests, falling-head tests, or other similar tests to determine the flow of water into or out of the formation. Borehole cameras are also used to get a look at the fractures in the wall of the boring and the locations where water is entering the hole. The borehole camera can be a video camera or one that provides still pictures.

The borings should be spaced so that the plan area of the entire facility is investigated and a top-of-rock contour map can ultimately be prepared. In areas where the bedrock is limestone or dolomite, other precautions need to be taken. Under certain conditions in the geologic past, a soluble limestone or dolomite formation may have been eroded by running water that left voids in the bedrock. These voids or *solution channels* are commonly filled with air, water, soft clays, or saturated silts. This is known as a *karst* condition. These caverns can be just beneath the top of rock surface or at depth as shown in Fig. 5.1.1-1.[1] If heavily loaded structures or facilities are built over these caverns, they might collapse into the void. For facilities overlying anticipated karst conditions, more borings might be required. Sufficient borings into sound rock should be made to assure the engineers and owners that voids do not exist at that location or that their presence will be incorporated into the design.

Empirical relationships have been developed that show relationships between the RQD and the in situ rock mass strength and the in situ rock mass modulus. Geophysical exploration can also be an effective tool in evaluating rock sites. Empirical relationships have also been developed that correlate the ratio of the primary velocity obtained by sonic or seismic techniques in the field to the velocity measured in the laboratory squared (called the *velocity index*) and the in-situ rock mass strength and in-situ rock mass modulus. The ratio of the field and laboratory velocities is a measure of the degree of fracturing and weathering of the rock mass. These relationships can be very helpful to the designer in gaining a better understanding of the rock mass properties, not just the intact rock core properties. Table 5.1.1-1 shows an engineering classification of in-situ rock developed by Merritt.[2]

The rock obtained from the field boring program should be sent to a laboratory with rock-testing capabilities. The laboratory tests on the rock core should be done in accordance with ASTM or International Society of Rock Mechanics (ISRM) standards. These tests are usually velocity tests and uniaxial and triaxial compression tests that yield strength and modulus values for the intact rock. Other index-type tests, such as Schmidt hammer readings, Brazilian tests, and tensile tests, are also used as indicators of strength values.

HAZARDS EVALUATION

Natural hazards must be considered in planning and designing a facility or expanding an existing facility. The designer must examine each part of the proposed facility and decide on the appropriate criteria for different hazards, the likelihood of combined loadings, and the consequences if an extreme loading situation does occur. A systematic examination of these potential loadings will identify the most critical situation. It will also identify the situations where it is not rational or economically justifiable to provide for the combined effects. Examples of items to be examined are the foundation type; structural parameters governing behavior such as stress, strain, and deformation; the materials in the facility and their resistance; and the quality of construction. Some of the natural hazards that must be considered are floods, earthquakes, faults, liquefaction, collapsible soils, shrinking and swelling soils, and slope failures.

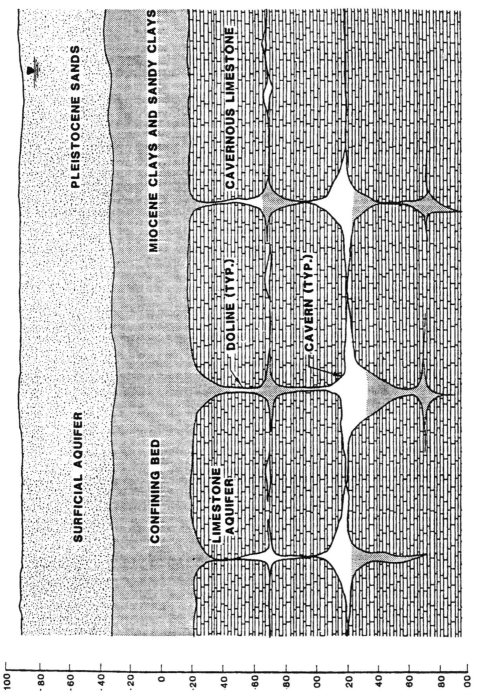

FIGURE 5.1.1-1 Conceptual subsurface profile of buried karst. (*Courtesy of J. E. Garlanger*).

5.7

TABLE 5.1.1-1 Engineering Classification of in-Situ Rock

RQD, %	Velocity index	Description
0–25	0–0.20	Very poor
25–50	0.20–0.40	Poor
50–75	0.40–0.60	Fair
75–90	0.60–0.80	Good
90–100	0.80–1.00	Excellent

Floods

The hydrologist must establish whether the site is susceptible to flooding from a nearby river or stream, or just from inadequate drainage and runoff. The problems associated with floods are those associated with unexpected buoyant effects on the facility, foundation damage and subsequent loss of support, high-velocity water loads on equipment and buildings, and submergence of the facility. The hydrologist should perform analyses using historical flood records.

Earthquakes

It is necessary that the designer establish the probability that an earthquake will occur in the area of the proposed facility. The basis for making such a determination involves judgments about degrees of acceptable risk, consideration of the life and safety of the surrounding population, the effects on the environment, and the consequences of interruption of service in the event of severe damage to the facility. The selection of the earthquake intensity by the designer is one of the important elements in the decision process. Because aboveground and belowground structures move and respond differently to earthquakes, the design criteria for the two situations may be quite different.

Faults

Surface faulting is a dangerous source of differential movement in a facility. Surface faulting, however, is a relatively rare occurrence. Care should be taken not to place a facility over a known fault where there has been movement. The engineer and geologist must prove without a doubt that the fault is not active and that there will be no movement along the fault during the life of the facility and for years beyond. If it is absolutely imperative that the facility rest on the fault or cross the fault and movement is anticipated, the design should take this into account and allow for such a contingency. Piping, conduits, utilities, and even tunnels that cross an active fault must be able to accommodate the displacement.

Liquefaction

One of the major seismic hazards for facilities built on saturated cohesionless soil deposits is liquefaction and the potential for ground movements and foundation failures. For major structures such as dams, nuclear power plants, oil storage tanks, and other structures susceptible to movement, it is imperative that the subsurface be thoroughly investigated to ascertain whether there is the potential for liquefaction. Liquefaction may result from vibrations from natural sources such as earthquakes or from construction activities such as blasting or pile driving.

One of the basic soil parameters used in liquefaction analyses is *relative density.* Various subsurface investigative techniques, such as standard penetration tests, Dutch cone soundings, and undisturbed sampling with subsequent laboratory testing, may be utilized to yield relative density data.

Where a site is believed to be prone to liquefaction, it may have to be abandoned and not used. In many cases, however, the subsurface properties can be improved or strengthened by using one or more ground improvement techniques, such as dynamic compaction, vibrocompaction, compaction grouting, blast densification, or drainage by installation of wick drains.

Collapsible Soil

Soil displacements caused by collapsible soils are a more common occurrence than surface faulting and are comparable in magnitude. Many types of deformation may result, including flow failures, lateral spreads, subsidence, loss of bearing, and buoyancy effects. Collapsible soils are soils placed in an environment that results in a loose configuration and a highly compressible material. These can be materials placed in a natural environment or in a human-made environment, such as reclaimed mined land. Collapsible soils are often cohesionless materials that have a uniform particle size. When subjected to shaking such as in an earthquake (liquefaction), to submergence, or when a load is applied, these soils may collapse. An experienced geotechnical engineer can detect soils that are susceptible to collapse. Further laboratory testing will confirm if there is such a problem. Many techniques or construction methods are used today to improve such soils. The nature, severity, and real extent of the problem, as well as the limitations imposed by the surroundings, will help determine the most suitable method. Methods commonly in use today fall into three categories: soil improvement, soil reinforcement, and soil treatment. *Soil improvement* methods include deep dynamic compaction, drainage and surcharge, electro-osmosis, compaction grouting, blasting, and surface compaction. Examples of *soil reinforcement* are stone columns, soil nails, micropiles, jet grouting, soil mixing, fiber reinforcement, mechanically stabilized earth, and lime columns. *Soil treatment* examples are soil cement, lime admixtures, fly-ash admixtures, dewatering, and freezing.

Shrinking and Swelling

Certain fine-grained cohesive deposits such as clays, silts, and organic material are other potential sources of differential settlement beneath a structure. Claystones and mudstones may contain bentonitic clay minerals that swell upon exposure to water, causing heave and subsequent shrinkage during dry periods. They may also exhibit characteristics of dispersive, erodible clays when immersed in running water, resulting in piping and erosion. These conditions are most pronounced in semiarid environments.

Slope Failures

Slope instability has the potential for damaging ground movement under either static or seismic loading conditions. This instability can occur in natural slopes or in cut slopes. The greatest seismic risks are associated with those slopes that show signs of instability under static conditions. If instability involves slumps and relatively shallow slides, slope stabilization or installation of drains may be effective in correcting difficulties. Landslides, however, involve downslope movements of large masses of soil and rock that may occur under static or seismic conditions. The principal forms of movement include rock falls, sliding of soil, and deep translation and rotation of soil and rock. Field and analytical studies should be done to assess the potential and risk for a slope failure or landslide if a facility is to be built on or at the toe of a potentially unstable slope. The costs for stabilizing and maintaining a slope can be sizable and should be factored into siting a facility.

INTERFERENCES

Siting a facility in an urban environment has interferences that are not as prevalent as in an undeveloped or relatively unpopulated area. Interferences can be existing utilities, previous structures, buried structures, and historical artifacts. The locations of existing utilities such as water lines, gas lines, sewer lines, or buried electrical lines need to be identified before any exploration or excavation is begun. Utility companies and communities have formal programs, such as *Digsafe,* which provide this service. As-built records of the existing utilities should also be utilized. In areas where there has been filling, buried foundations of previous structures may also be present. Buried structures such as seawalls and revetments can also cause major problems during exploration and excavation and can add significantly to the cost of a facility. When historical artifacts are located at a site, additional issues are involved. Not only is there an effect on the time schedule and cost because of the care that must be taken in preserving these artifacts, it may be that the site cannot be used at all.

ENVIRONMENTAL AND OTHER CONSIDERATIONS

In addition to sound technical and procedural practice, other major factors must be considered in siting and developing a facility. These include environmental and socially related issues, public health and safety, regulatory requirements, political considerations, reliability, and economic factors, to name a few.

As an example, the Environmental Impact Statement is one of the vehicles used to evaluate and present the environmental issues for a facility. Potential environmental damage or impact might include loss of an amenity as a result of noise, vibration, visual intrusion, dust, dirt, odor, traffic, or site drainage. Refer to Art. 3.2.9, "Environmental Compliance and Issues Management," for a detailed discussion of environmental issues. Each of these issues and others that might affect neighbors or landowners need to be studied. Impacts must be assessed. Where possible, the facility design should try to preclude or minimize the impact. Where that is not feasible, measures need to be taken to mitigate the impact.

Similar studies are needed for the remaining issues as well. These issues are addressed in more detail in other sections of this book.

DEVELOPMENT OF DESIGN CRITERIA

Using the data that have been obtained for the site, the geotechnical engineer should now prepare the design criteria for inclusion in the contract documents and for use by the structural engineer. These data include the information obtained from the literature, from construction of nearby structures, from road cuts, from the exploration program, from the field and laboratory testing program, and from the analyses that have been done on the data. These design criteria include values for all of the soil and rock parameters that the structural engineer will need in the design, such as soil unit weight, soil modulus of subgrade reaction, soil compressive strength, rock unit weight, orientation and strength along rock discontinuities, and rock mass modulus, among many others.

SELECTION OF FOUNDATION TYPE

Selection of the foundation type for a facility is based on several factors. These factors include the sensitivity of the structure to total and differential settlements, the properties and depth of soil, the depth to bedrock, the orientation and properties of discontinuities, the loads to be supported, the presence or absence of water and/or hazardous materials, the type of foundation and performance of nearby structures, the proximity of other structures, and the relative cost. Shallow spread footings are generally acceptable for supporting lightly loaded structures with good soil and water conditions. Foundation design considerations include bearing capacity, seismic response and liquefaction potential, settlement, and swelling. For poorer soil conditions and heavy structures, deep foundations such as piles or caissons are required. The subsurface conditions at the site are an important part of the selection process for caissons or piles. Geotechnical design considerations include the presence of a good bearing layer at a reasonable depth, skin friction, potential negative skin friction (downdrag), bearing capacity for end-bearing designs, obstructions, the location of the groundwater table, the ability of a hole to stand open, the drivability of piles, swelling soils, collapsing soils, the availability of foundation materials and installation equipment, the effect of vibrations on the subsurface and adjacent structures, and cost. If piles are selected, the right pile for the site must be selected. There are several types of piles to choose from, including timber, concrete, and steel piles. For each of these materials, there are several types from which to choose. Issues such as weathering, corrosion, drivability, displacement, noise, vibration, length, availability, and cost must all be considered.

For sites where the rock is shallow, high strength, unweathered, and massive, the foundation design is straightforward. Footings constructed on the rock can support both light and heavy loads of the proposed facility. For rock sites with weathered rock, weak rock, swelling rock, soluble rock, or limestone sites with karst conditions, additional precautions may be required in construction.

EXCAVATIONS

The construction of most facilities, whether large or small, requires some excavation of the site. For small, lightly loaded facilities, the excavation might consist of site grading and shallow excavation for footings. Larger excavations may be required for other structures where a portion of the facility is to be built underground. Geologic and groundwater conditions are a primary consideration for excavation stability, as well as for temporary and permanent ground-stabilization systems. The more complex these conditions, the greater are their influences on the various elements of construction. Some of the excavation conditions that must be addressed are dewatering, blast monitoring, support instrumentation and monitoring, and materials handling and disposal.

Dewatering

Water encountered in an excavation is almost always considered a nuisance, if not a real threat to the progress of the project. Water in this sense includes either surface water or groundwater. Predicting the location and volume of water is one of the harder tasks of the geotechnical exploration team. The prediction should come from exploration observations and measurements and evaluation of general geotechnical conditions. Methods of handling the water vary and can depend upon the contractor's preference. At other times, the method of handling the water is dictated by the need to preserve the strength of the surrounding material or to reduce or eliminate the water load on excavation support systems. If the surrounding material is not eroded by the water, as the excavation progresses and the contractor's excavation techniques can accommodate it, the least expensive and usual method is to "sump and pump." In this method, a hole is made in the lowest part of the excavation, and a pump or pumps are placed in the hole. Then water is lifted to the ground surface and carried away from the excavation. For deeper excavations where water under high pressure may be confined beneath the excavation bottom, care must be taken that the bottom does not blow. This water pressure must be relieved before the lifting force of the water exceeds the weight of the overlying material. For excavations that must be dry during the excavation process, a more elaborate dewatering system is required. This could consist of dewatering wells or well points that are placed outside of the proposed excavation and drilled to a depth well below the proposed excavation bottom. Using a system of piping and headers, water is pumped from the wells or well points and collected and carried away from the proposed excavation. This system is installed and put into operation before excavating begins. A monitoring well should be drilled at the center of the proposed excavation and readings taken before and throughout the excavation process to ensure that the water level remains well below the excavation bottom. Depending upon the quality of the water being removed, it may be necessary to collect and treat it before it can be discharged.

It is important to note that in all dewatering operations, the dewatering system must operate continuously 24 h per day, 7 days per week, while the excavation remains open. The cost for this continuous operation as well as the need for a backup power supply should be factored into the project planning process.

Blast Monitoring

For sites where the top of rock is near the ground surface and excavations are required, it is usually necessary to use explosives to break up the rock so that it can be removed. Using explosives to break rock generates air- and groundborne vibrations that could have detrimental effects on nearby structures and cause irritation to or complaints by nearby residents. A variety of complaints attributable to vibrations have always been received when blasting is done. Human response levels to ground vibrations are considerably below those levels necessary to induce damage to residential structures. To ensure that structures are not damaged, that complaints are minimized, and that actual blast data are recorded, it is imperative that each blast be monitored using a seismograph. The monitoring should be done by the contractor and also by the owner or owner's engineer.

A safe blasting limit of 2.0 in/s peak particle velocity measured from any of three mutually perpendicular directions on the ground adjacent to a structure should not be exceeded if the probability of damage to the structure is to be small. Complaints can be further reduced if a lower vibration level is imposed. Air blast does not usually contribute to the damage problem in most blasting operations. Millisecond-delay blasting should be used to decrease the vibration level from blasting because the maximum charge weight per delay interval rather than the total charge determines the resultant amplitude. Data in the literature or from other blasting jobs should be used initially to determine a conservative and safe charge weight per delay, or a test program should be done. These data are presented showing a relationship between peak particle velocity and scaled distance. *Scaled distance* is obtained by dividing the distance from the blast to the seismograph by the square root or cube root of the charge weight per delay interval. Knowing the distance to the structure in question and the safe blasting limit, the charge weight per delay can be calculated. After each blast, the measured peak particle velocity should be plotted versus the scaled distance, and appropriate changes should be made to the blast envelope, if necessary, for the design of the next blast. A safe blasting job requires good communications and cooperation between the blasting contractor and the geotechnical engineer.

Support

Excavations in soil and rock are usually made with the steepest faces possible to minimize the amount of material to be removed, the backfilling required, and the land area disrupted. These steep excavated faces are usually unstable and require support. The type of support depends upon the type of materials to be excavated, the height of the excavation, the presence or absence of water, and the size of the excavation, among many other factors. For excavations in soil, support is provided by soldier piles and lagging, sheet piles, or slurry walls. These vertical or near-vertical walls hold back the soil material as the excavation proceeds downward. High-strength tension rods anchored deep in the soil behind the walls help carry some of the load imposed by the soil and water. More than one row of anchors is required for deeper excavations. Other newer techniques may also be used, such as soil nailing and soil mixed walls (SMWs), as shown in Fig. 5.1.1-2.[3] For narrower excavations, cross excavation bracing may be used instead of soil anchors. The design of these support systems is complex and must be done by a geotechnical engineer experienced in this area.

If the excavation is to be in rock and blasting is to be used to break the rock, care must be taken to ensure that the rock surrounding the excavation is left in the best possible condition and that overbreak is minimized or eliminated. Controlled blasting techniques such as line drilling, presplitting, cushion blasting, smooth-wall blasting, or preshearing should be used. Sometimes vertical rock bolts are installed just outside the excavation line before any blasting begins to assist in holding together the rock that is to be left in place. The production blasts should be planned to minimize the amount of energy that is sent back into the remaining rock. These blasts should be directed as much as possible to a free face. Controlled blasting techniques are not a cure-all, however. Overbreak control is still primarily a function of the geology, and slopes should be designed to accommodate the geology. As the excavation proceeds downward, rock bolts or other support should be installed to help support the exposed rock face. Rock bolts should be designed with the geologic discontinuities in mind. The orientation, length, spacing, and type of bolts and anchorage should be designed to accommodate the site conditions. Timing to provide the support is important. Support should be installed as soon as possible after the face is exposed. Depending on the site conditions and availability of materials, the rock bolts used are usually solid steel bars anchored and encapsulated by either cement or resin grout. Mechanical anchored bolts also have their applications. The design of these bolts should also include the need and benefit of prestressing them. Ancillary materials such as steel straps or wire mesh may be used to minimize and collect any small pieces of rock that could fall from the slope. Shotcrete is another form of support that is sometimes used. This is cement and small aggregate that is sprayed onto the rock face as it is exposed. The shotcrete, in conjunction with wire mesh, provides support to the rock face. The shotcrete is usually 3 to 4 in thick. It is important that drainage be provided through the shotcrete so that water pressure does not build up behind the shotcrete and dislodge it.

Instrumentation and Monitoring

In an urban environment or at an existing facility where structures may be close to a proposed excavation, it is imperative that protection of adjacent structures be included in the design and planning process. The potential impacts of the proposed excavation on the existing nearby structures include lowering the groundwater table during excavation, movements of the proposed earth or rock support system, damaging or removing the foundations of the structures, or shaking the structures during blasting operations. To reduce the possibility of damage to the existing structures, an instrumentation and monitoring program should be designed and installed before the excavation begins. In addition to the instrumentation that is to be selected and installed, a preconstruction survey of the structures including extensive use of photographs should be made to document the conditions before the nearby excavation

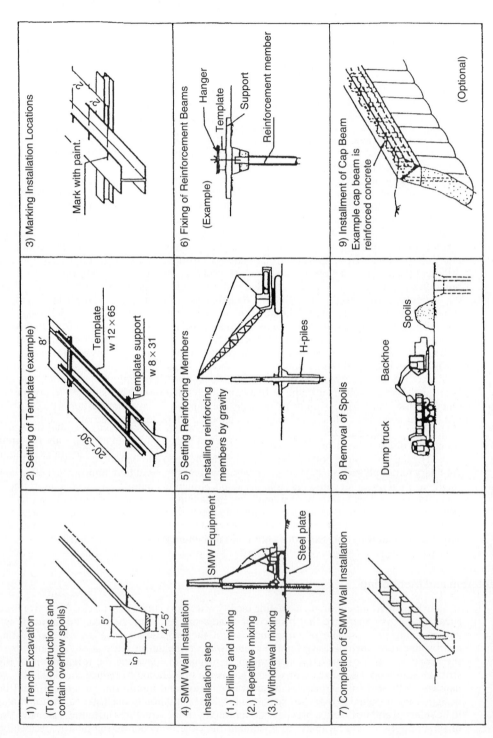

FIGURE 5.1.1-2 Soil mixed wall (SMW) installation procedure. (*Courtesy of S. L. Pearlman and D. E. Himick*).

starts. This program allows the geotechnical engineer to monitor the behavior of the excavation and the adjacent structure as the excavation proceeds and to adjust the excavation program as needed. It also provides a record for any future questions that may be raised. The geotechnical engineer should establish acceptable limits of movement of the excavation walls and the adjacent buildings. Contingency plans must be in place for implementation if the movements reach these levels. The owners and residents of the nearby structures should be included in the planning process. Their comments and concerns should be addressed before beginning the excavation. Good communication and cooperation are advantageous in this process. After completion of the excavation and construction of the new facility, a postconstruction survey of the adjacent structures should be made and documented. The monitoring process should also continue until well after the new facility is completed and all concerns have been addressed.

Materials Handling and Disposal

The nonhazardous material excavated at the site should be used as fill material at the site if possible. Unless it is processed, it is not likely that it could be used as structural fill, but it may be quite suitable for raising the grade of portions of the site that will not bear heavy structures. After obtaining the necessary permits, the contractor should transport any hazardous materials and materials that cannot be used at the site to an off-site area that has been permitted to receive such materials.

LIAISON DURING CONSTRUCTION

It is imperative that a geotechnical engineer be on-site during foundation construction to ensure that the foundation is constructed as it was designed. The geotechnical engineer acting as on-site liaison during construction represents the owner at the site, ensures that the specifications are followed, makes on-site decisions and analyses where adjustments are required, provides communication with the engineer and designer in the office, and facilitates communication with the contractor.

CONTRIBUTOR

Peter K. Taylor, P.E., Geotechnical Consultant, Weston, Massachusetts

NOTES

References

1. J. E. Garlanger, "Foundation Engineering in Deeply Buried Karst," in *The Art and Science of Geotechnical Treatment Engineering at the Dawn of the Twenty-First Century,* Prentice-Hall, Englewood Cliffs, NJ, 1989.
2. A. H. Merritt, "The Engineering Classification of In-situ Rock," Ph.D. thesis, University of Illinois, Urbana, 1968.
3. S. L. Pearlman and D. E. Himick, *Anchored Excavation Support Using SMW (Soil Mixed Wall),* Deep Foundations Institute, Pittsburgh, PA, 1993.

Bibliography

Albritton, J. A., A. W. Hatheway, and L. B. Underwood: "Improvements in Geotechnical Aspects of Contract Documents for Underground Construction," *Rapid Excavation and Tunneling Conference,* Chicago, IL, 1983, Vol. 2, Chap. 72.

American Society of Civil Engineers: *Ground Improvement, Ground Reinforcement, Ground Developments, 1987–1997, Logan, Utah,* Geotechnical Special Publication No. 69, American Society of Civil Engineers, New York, 1997.

Bell, R. A., and J. P. Singh: "Comparison of Relative Densities Estimated Using Different Approaches," in *Evaluation of Relative Density and Its Role in Geotechnical Projects Involving Cohesionless Soils,* STP 523, American Society for Testing and Materials, Philadelphia, 1972.

Coon, R. F.: "Correlation of Engineering Behavior with the Classification of in Situ Rock," Ph.D. thesis, University of Illinois, Urbana, 1968.

Deere, D. U., A. J. Hendron, F. D. Patton, and E. J. Cording: "Design of Surface and Near Surface Construction in Rock." In *Failure and Breakage of Rock,* American Institute of Mining, Metallurgy, and Petroleum Engineering, New York, 1968, pp. 237–302.

du Pont: *The Blasters' Handbook,* 15th ed., E. I. du Pont de Nemours & Co., Wilmington, DE, 1971.

Johnson, E. G., and D. A. Schoenwolf: "Foundation Considerations for the Expansion and Renovation of the Hynes Auditorium," *Civil Engineering Practice, Journal of the Boston Society of Civil Engineers Section/ASCE* 2, no. 2 (Fall 1987).

Keville, F. M., and C. D. Pizzo: "Community Participation in Public Works Projects," *Civil Engineering Practice, Journal of the Boston Society of Civil Engineers Section/ASCE* 1, no. 1 (Spring 1986).

O'Rourke, T. D., and W. J. Hall: "Engineering Planning and Practice for Pipeline Systems," in *The Art and Science of Geotechnical Engineering at the Dawn of the Twenty-First Century,* Prentice-Hall, Englewood Cliffs, NJ, 1989.

Peck, R. B.: "Advantages and Limitations of the Observational Method in Applied Soil Mechanics, Ninth Rankine Lecture, *Geotechnique,* 19, no. 2 (June 1969), Institution of Civil Engineers, London, England.

Tarquinio, F., and S. Pearlman: "Pin Piles for Building Foundations," *Seventh Annual Great Lakes Geotechnical and Geoenvironmental Conference,* Kent State University, Kent, OH, 1999.

Terzaghi, K: "Consultants, Clients and Contractors with Discussions," in *Contributions to Soil Mechanics, 1954–1962,* Boston Society of Civil Engineers, Boston, MA, 1965.

Terzaghi, K., and R. B. Peck: *Soil Mechanics in Engineering Practice,* Wiley, New York, 1962.

U.S. Bureau of Mines: *Blasting Vibrations and Their Effects on Structures,* Bulletin 656, U.S. Department of the Interior, Bureau of Mines, 1971.

ARTICLE 5.1.2
CIVIL ENGINEERING

Ronald P. O'Brien, P.E., Steven L. Bernstein, P.E., and Robert F. Daylor, P.E.
Daylor Consulting Group, Inc., Braintree, Massachusetts

This article discusses strategies for managing the civil engineering aspects of facility planning, design, and construction. Civil engineering encompasses disciplines such as geotechnical, structural, environmental, wastewater, water, fire protection, traffic, and site development engineering. Although all of these disciplines are important facets of facilities engineering, this article will guide the reader through the most critical civil engineering site development components.

The civil engineering aspects of a design define the physical constraints of a project. Civil engineers focus on the planning, design, and construction of works in or on the ground and are called upon in design teams to provide primary expertise in regulations dealing with the land, such as zoning, land development codes, wetland protection, and water quality management. Project constraints often consist of one or more of the following: zoning, access, water supply, sanitary sewage and sewerage availability, storm water discharges, existing utility infrastructural capacities, environmental resources, and subsurface conditions.

ZONING

Zoning requirements are regulated locally and are not uniform from state to state or within a specific state. Performing a thorough zoning review is a critical first step when planning a new site development or facility upgrade. The zoning district dictates allowed uses, dimensional requirements, and some of the permitting processes for all site developments. Although it is possible to have a zoning district changed or to apply for variances from certain zoning requirements, this process can be lengthy; its outcome is uncertain; and it is expensive. A facility manager should meet with the local zoning enforcement officer to discuss the proposed project, current zoning, and specific permitting requirements for the development. Selecting sites or designing new facilities in locations that are zoned to accommodate the proposed project pose the least risk to the schedule.

At a minimum, the dimensional requirements in most zoning regulations dictate the requirements for lot area, frontage, property line setbacks, building coverage, building height, number of structures, and impervious lot coverage. Other regulated items that may be in the zoning regulations include but are not limited to the following: the number of parking spaces, parking area dimensions, off-street loading, landscaping, floor-to-area ratio, outdoor lighting, and exterior signs. Other constraints found in zoning regulations or other local bylaws may relate to sensitive environmental or municipal resources such as wetland districts or aquifer protection districts. These resource-based overlay districts are drawn on top of other zoning categories. Overlay districts are more restrictive than their respective underlying zoning districts.

It may be necessary to create a lot or subdivide an existing lot for the proposed development. If the parcel to be created abuts an existing road or right-of-way (ROW), then it can be created in accordance with the local zoning requirements of the district. If the parcel does not have adequate frontage, then it may be necessary to create a new ROW to provide frontage for the lot. If the project requires creating a ROW, then additional local permits are required. The new road design must conform to local standards, as must the utilities within it.

PERMITTING

This article discusses potential time constraints, permitting issues, and costs that may be encountered during the permitting process. Permitting costs vary greatly from region to region. In some areas of the country where infrastructure is at or near capacity, regulatory agencies typically require significant contributions from potential developments. In other areas of the country, the local or state government may make the infrastructural improvements or provide tax benefits to encourage development.

There are many permits in addition to those required by local zoning. They may be more restrictive to the development than the local zoning requirements. For example, the site may be constrained by wetlands, may be designated as an environmentally sensitive habitat, or may have historic significance. The site may be upstream of an environmentally sensitive area or near a historic landmark. It is impossible to list every permit that may be required for a specific project because every site is unique, and every state, county, and municipality has unique permitting requirements. The intent of this article is to make the reader aware that many permitting agencies at many levels administer regulations that can significantly constrain or prevent a proposed development. A facility manager should retain a land development engineering company and/or land-use real estate attorney to perform a feasibility study to determine required permits and estimated permit schedules before designing a new facility or upgrade. Also, refer to Art. 3.2.9, "Environmental Compliance and Issues Management," for a detailed discussion of these and other related issues.

Permits can be administered at one or more of the following levels: federal, state, regional, county, and municipal/local. The following list of frequently encountered permits is typical but not exhaustive: wetland protection, flood-plain protection, connection to municipal sewers, subsurface sewage disposal, discharge to surface waters, connection to municipal water, groundwater discharge, groundwater withdrawal, coastal development, railroad right-of-way, and access to federal, state, or county highways.

EXTERNAL AND INTERNAL TRAFFIC AND ACCESSIBILITY

When siting a new facility, the capacities (maximum vehicle trips per day and peak-hour vehicle trips) of existing roads and intersections should be considered. The distances from major highways, airports, or rail facilities and the availability of public transportation should also be considered. The new facility will generate additional traffic that will impact existing streets and intersections. The permitting process may require a comprehensive traffic analysis and may result in some form of traffic mitigation if the proposed development generates significant traffic. Traffic mitigation can be significant and must be included in the project's budget and schedule. In high-traffic corridors, some traffic demand management, such as staggered work hours, ride sharing, shuttle buses, or subsidies to public transportation, may be required. Permit thresholds may be based upon specific traffic estimates. The standard for traffic generation estimates is the Institute of Transportation Engineers (ITE) *Trip Generation* manual. A traffic engineer should be consulted to determine if the proposed facility will generate a significant number of vehicle trips or if existing roads or intersections are at or near their respective capacities.

Internal site circulation should accommodate the following traffic types: personal vehicles, emergency response vehicles, maintenance vehicles, delivery trucks, and pedestrians. Other traffic types, such as public transit or bicycles, may be encountered. Personal vehicle and pedestrian traffic are closely tied to the parking configuration for internal site circulation. Parking areas should be located within a convenient distance to the buildings that they serve. The maximum distance from building entrances to the furthest parking space may be regulated by zoning. Parking may not be allowed directly adjacent to a building because of fire lane requirements. A facility manager should check with the zoning and/or the local fire department.

Parking areas with perpendicular parking and two-way travel aisles usually provide more parking spaces than angled parking with one-way aisles. However, angled parking is more efficient than perpendicular alignments in some instances. Locating parking spaces on the perimeter of a parking area maximizes the total number of parking spaces. However, this causes significant traffic conflicts if the perimeter travel way is a primary access for the facility. Primary access routes should be designed to be similar to public streets and should minimize traffic conflicts. Angle parking should be used where one-way traffic is desirable or where width limitations preclude right-angle parking layouts. Figure 5.2.1-1 shows a recommended 90° layout. Figure 5.2.1-2 compares 90° parking to angle parking. Figure 5.2.1-3 lists typical dimensions for parking spaces and aisles. Zoning generally dictates a minimum number of parking spaces for specific uses, as well as minimum space and aisle dimensions.

After satisfying the parking requirements in zoning and any additional owner needs, handicapped requirements conforming to Americans with Disabilities Act (ADA) Accessibility Guidelines for Buildings and Facilities must be incorporated into the design. Handicapped accessible routes must be provided to facilities with sufficient handicapped parking as close as possible to the facility that they serve. Handicapped ramps that comply with ADA must be provided whenever an accessible route crosses a curb. Figure 5.1.2-4 presents ADA handicapped parking requirements and parking space dimensional requirements. The ADA guidelines are minimum federal requirements, and the facility must also adhere to all state and local handicapped-accessibility regulations.

There are many options for pedestrian circulation. In general, pedestrian traffic in the parking area is routed along the edges of the aisles. If pedestrian traffic travels along a road or high-traffic aisle, then a sidewalk must be provided. It is important to provide well-marked pedestrian crosswalks at crossings that have significant volumes of pedestrian traffic. Visual obstructions should be avoided at intersections or other areas with vehicle and pedestrian conflicts. The facility manager should determine what level of pedestrian accessibility is necessary for internal operations and access to the site. Pedestrian routes should be vertically separated from vehicle traffic by a 6-in curb. Sidewalks should be a minimum of 4 ft wide, well lit, and handicap accessible. Sidewalks should not be used for bicycle traffic. If desired, separate lanes can be located along the shoulders of travel lanes and delineated for bicycle traffic. Many types of curbing can be used at the edges of travel ways and within parking areas. Zoning may dictate the type of curbing required for specific applications. Bituminous concrete (asphalt) curbing is the least expensive and least durable material. Vertical bituminous curbing is vulnerable to snowplow damage. Concrete curbing can be approximately 3 times more expensive than bituminous concrete curbing but is much more durable. Concrete curbing can also be damaged in cold climates by snowplows and road salt. In warmer climates, concrete curbing and gutter is the standard road edge. Granite curbing is the most durable but is not readily available in all regions and is approximately twice as expensive as concrete curbing.

Local ordinances and local fire departments dictate the portion of a building or facility to be accessible to emergency vehicles. Their decision will be based upon the intended use of a building, the level of internal fire protection, and the proximity to other structures. The design must provide room for parking fire trucks near the building while allowing other vehicles to pass. Landscape areas can accommodate heavy-duty emergency vehicles such as fire trucks if

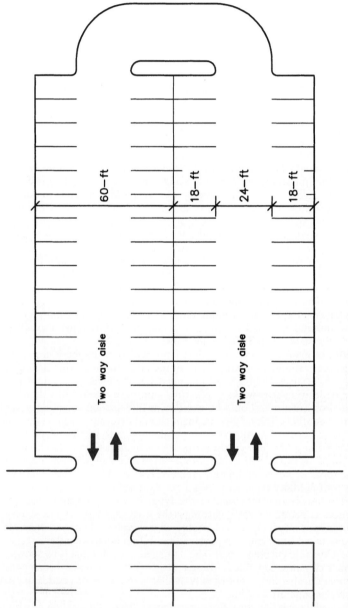

FIGURE 5.1.2-1 Typical 90° parking layout of 9- × 18-ft stalls with 24-ft aisles. *(Courtesy of Daylor Consulting Group, Inc.)*

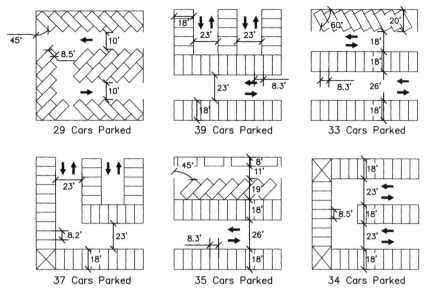

FIGURE 5.1.2-2   Six examples of 100- × 100-ft parking fields. (*Courtesy of Daylor Consulting Group, Inc.*)

A	B	C	D	E	F	G
0°	9.0'	9.0'	12.0'	23.0'	30.0'	---
	9.5'	9.5'	12.0'	23.0'	31.0'	---
	10.0'	10.0'	12.0'	23.0'	32.0'	---
30°	9.0'	17.3'	11.0'	18.0'	45.6'	37.8'
	9.5'	17.8'	11.0'	19.0'	46.6'	38.4'
	10.0'	18.2'	11.0'	20.0'	47.4'	38.7'
45°	9.0'	19.8'	13.0'	12.7'	52.6'	46.2'
	9.5'	20.1'	13.0'	13.4'	53.2'	46.5'
	10.0'	20.5'	13.0'	14.1'	54.0'	46.9'
60°	9.0'	21.0'	18.0'	10.4'	60.0'	55.5'
	9.5'	21.2'	18.0'	11.0'	60.4'	55.6'
	10.0'	21.5'	18.0'	11.5'	61.0'	56.0'
70°	9.0'	21.0'	19.0'	9.6'	61.0'	57.9'
	9.5'	21.2'	18.5'	10.1'	60.9'	57.7'
	10.0'	21.2'	18.0'	10.6'	60.4'	57.0'
90°	9.0'	18.0'	24.0'	9.0'	60.0'	---
	9.0'	19.0'	24.0'	9.0'	62.0'	---
	9.5'	19.0'	24.0'	9.5'	62.0'	---
	10.0'	19.0'	24.0'	10.0'	62.0'	---

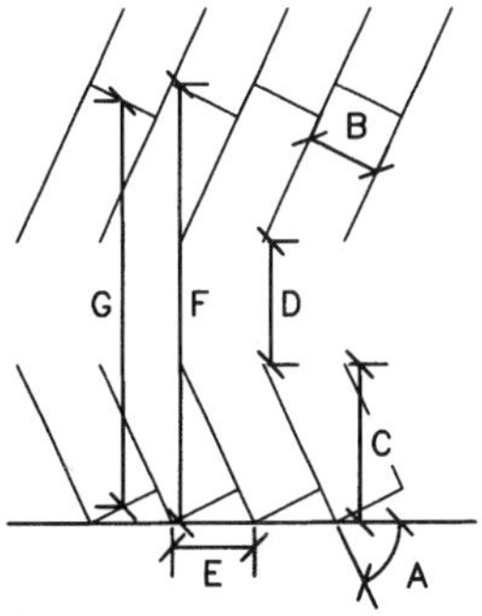

A   PARKING ANGLE

B   STALL WIDTH

C   STALL TO CURB

D   AISLE WIDTH

E   CURB LENGTH PER CAR

F   MINIMUM OVERALL DOUBLE ROW WITH AISLE BETWEEN

G   STALL CENTER (DOES NOT INCLUDE OVERHANG)

FIGURE 5.1.2-3   Typical parking area dimensions. (*Courtesy of Daylor Consulting Group, Inc.*)

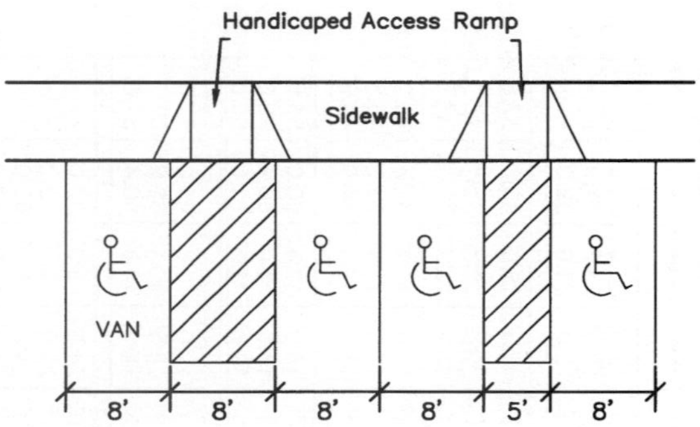

Total Parking in Lot	Required Minimum Number of Accessible Spaces
1 to 25	1
26 to 50	2
51 to 75	3
76 to 100	4
101 to 150	5
151 to 200	6
201 to 300	7
301 to 400	8
401 to 500	9
501 to 1000	2 percent of total
1001 and over	20 plus 1 for each 100 over 1000

Except as provided for van accessible spaces, access aisles adjacent to accessible spaces shall be 60 in (5 ft) wide minimum.

One in every eight accessible spaces, but not less than one, shall be served by an access aisle 96 inches (8 ft) wide minimum and shall be designated van accessible as required.

FIGURE 5.1.2-4 ADA handicapped parking requirements. (*Courtesy of Daylor Consulting Group, Inc.*)

they are reinforced with lawn pavers and are at a suitable grade and alignment. Common lawn pavers are made from concrete or high-strength plastic and allow grass to grow through, rendering them virtually invisible. Fire trucks must have access to the fire hydrants serving the parking areas. Well-marked fire lanes must be provided and maintained by the facility. Access routes for fire trucks should be designed on the basis of the criteria of the local municipality. Access should be designed for a standard ITE WB-60 truck at a minimum, unless local regulations differ. Other emergency vehicles, such as ambulances, must have unobstructed access to buildings and be able to park without blocking traffic flow.

Delivery trucks need adequate access routes. In general, it is good practice to segregate employee automobile traffic from truck routes. A route designed to accommodate a WB-60 truck should be adequate for truck access. However, a bus design vehicle has the most restrictive inner turning radius. Figure 5.2.1-5 presents scale drawings of the turning requirements for a large vehicle.

Loading areas or loading bays for delivery trucks can be accommodated in many ways. The most common loading-bay layout is perpendicular to a building face. In this scenario, a truck can approach the bay parallel to the building and back into the loading dock. It is important not to block travel ways when a truck is parked in a loading bay. Angled loading bays require less maneuvering space beyond the building face. Parallel loading bays require the least maneuvering space beyond a building face. Parallel loading bays would not be a good choice for high-volume, multiple-bay loading or transfer facilities, but would be an excellent choice for a facility with limited space beyond the building and minimal loading activity.

Loading bays can slope toward or away from the dock at the building face. If the bay slopes down toward the building, then care must be taken to prevent the top of the truck from striking the door mechanism or the building above the doorway. This can be accomplished by extending the loading bay, providing bumpers, or increasing the height of the door. High-traffic loading bays should have dock shelters to provide weathertight seals between the truck and dock. Dock shelters should accommodate the loading-bay slope and be perpendicular to the back of the truck. They should also have adjustable skirts on top to accommodate different truck heights. The slope of the loading bay should be as flat as possible (0.5 to 1 percent is recommended). Although delivery trucks can negotiate a slope of 5 percent or more, it is important to consider the safety of the fork truck operators during loading or emptying of the trailer. Dock levelers may be installed at loading docks to accommodate trucks with different bed heights.

It is advisable to use trench drains to capture water from loading bays that slope toward the building. The use of catch basins in those areas would require cross slopes that may misalign a truck at the dock. Ideally, the loading-bay slope, toward or away from the building, will be a continuous plane, perpendicular to the building without cross slopes parallel to the building.

WATER

An adequate water supply is critical to facilities for many reasons. It is used for drinking, sanitation, fire protection, process applications, cooling, irrigation, and conveying sanitary and process wastes. Water may be obtained from a municipal service, surface water, or groundwater source. If water is supplied by a municipal system, it is critical to determine how much is available, at what pressure, and at what cost. Generally, the cost of a municipal supply includes an initial hookup fee and construction costs for the connection and meter, followed by subsequent water bills based upon use. The initial connection fee could be significant depending upon the flow requirements of the new facility and the municipality's total available capacity.

Water systems from different sources cannot be cross-connected. Most municipalities have control ordinances that describe connection types for specific uses. Municipal water cannot be directly connected to fire sprinkler systems, process systems, boilers, wastewater treatment

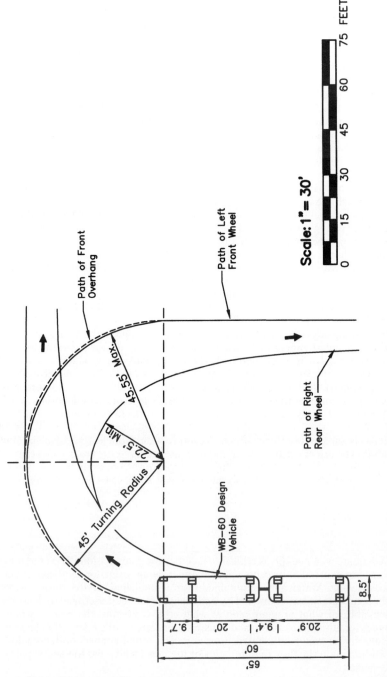

Scale: 1" = 30'

| 0 | 15 | 30 | 45 | 60 | 75 FEET |

Path of Front Overhang

Path of Left Front Wheel

Path of Right Rear Wheel

45.55' Max.

22.5' Min.

45' Turning Radius

WB-60 Design Vehicle

9.7'

20'

9.4'

20.9'

8.5'

60'

65'

FIGURE 5.2.1-5 Minimum turning path for WB-60 design vehicle. (*Courtesy of Daylor Consulting Group, Inc.*)

facilities, or other nonpotable uses without a backflow preventer. These valves require at least annual inspection and testing to assure that no backflow will occur.

Before constructing a new facility, a facility manager should check with the water supplier to determine the closest adequate water service. If a water main with adequate flow and pressure is distant, then a cost analysis should be performed to decide between extending the water main versus developing an on-site source and constructing a storage reservoir. This cost analysis should be based upon long-term planning for the facility. It should consider fire protection, fire insurance costs, water quality, water treatment costs, maintenance, and staff requirements for an on-site system. If the pressure in the existing water main is not adequate but the supply is adequate, then a booster pump may be required for domestic demand and/or fire protection. If the existing water system does not have an adequate supply for fire protection, an independent fire protection supply and distribution system must be designed and constructed.

If municipal water service is not available or the cost to connect to it is prohibitive, an alternative source is required. In most areas of the country, the alternative source is groundwater. However, in riverfront or lakefront areas, it may be surface water. Federal regulations require treatment of water taken from a surface reservoir for potable use.

Groundwater may also require treatment if it is used as a potable source. Raw water for use as a potable source must conform to federal, state, and local standards. A treatment plant permit may even require a licensed operator to operate and maintain it. Care should also be taken to properly analyze and/or pretreat raw water used for process, cooling, or even landscape irrigation. Raw water can inhibit chemical processes, cause mineral deposits in piping, corrode pipes, and stain sidewalks and buildings.

The water distribution grid should be well planned because it delivers the most critical resource to a facility. A registered professional engineer should perform the distribution design. Water mains should be looped and furnished with gate valves at each leg of major water main intersections to allow isolating segments without impacting other sections if a break occurs. Looping water mains increase the overall pressure of the system, increase the total capacity of the system, and provide needed flexibility. Dead-end pipes should be avoided because they cause water-quality problems. Gate valves should also be installed at 500-ft intervals to isolate water mains. The municipality or fire department will most likely specify the location and type of valves.

A fire protection engineer should perform the fire protection design. The associated water main should be sized to accommodate future facility expansion. Fire flow demands are determined by zoning, fire insurance underwriters, and/or by building codes and vary with the type of occupancy and the value of stored materials. Municipal water systems normally have available hydrant test data that include the static pressures, residual pressures, and fire flow in gallons per minute. If this information is not readily available, the municipality will usually conduct these tests for a fee. Pressures required in water mains for fire fighting depend upon the type of fire apparatus.

Newer water mains are typically made from cement lined ductile iron (CLDI), polyvinyl chloride (PVC)–coated steel, or concrete. Ductile iron is a type of cast iron that has ductile properties. References to cast iron and ductile iron relate primarily to their ductility. Other materials that were used for distribution mains in the past included cast-iron, asbestos-cement, and wood pipelines. Water services are typically copper or PVC, but lead, zinc, galvanized steel, brass, and bronze piping have been used in the past. CLDI piping is the most common material for city water mains. Standard sizes range from 4 to 24 in in diameter. Iron tubercles may form in unlined cast-iron pipes and seriously affect flow capacity. Lining with cement or epoxy can prevent tuberculation. Asbestos-cement pipe, although no longer available, can be easily broken by unevenly distributed external loads, such as pipe crossings, or by nearby excavation activities.

Local regulations specify the maximum allowable distance between fire hydrants. Fire hydrants should be located at high points of the water main alignment to allow the relief of air from the line. Hydrants should be the same model and from the same manufacturer as the municipal system even if they are on a separate system. If possible, hydrants should not be located too close to buildings or structures in order to allow firefighters safe access to hydrants when the facility is on fire. A fire hydrant should be located near every outside sprinkler connection on buildings in order to allow a pumper truck to pump water from the fire hydrant to the sprinkler system. The placement of the hydrants and external connections should be coordinated with the municipal fire department and the facility fire insurance underwriter.

Another issue regarding water mains is that facilities frequently upgrade and/or add new structures. Excavating to construct new utilities is inevitable; therefore, water-main fittings and elbows must be constructed with mechanical joint restraints and/or concrete thrust blocks. Constructing new buried utilities invariably disturbs soils and bedding material surrounding existing utilities. Sometimes the existing water mains fail during the new construction, sometimes they fail during the next winter or thaw, and sometimes they fail when they are stressed under extreme conditions such as during a fire. Water-main failures are never convenient, never budgeted for, and are never inexpensive. Functioning water mains are critical during fire emergencies. Water is the lifeblood of a facility and will cost the facility if it is interrupted.

WASTEWATER

Wastewater collection and disposal are also critical planning components for a new or expanding facility. Wastewater is generally classified into three categories: domestic, industrial, and infiltration and inflow. If a municipal sewer with adequate capacity is available within the immediate area, then the facility should connect to it. If a municipal sewer is not available, wastewater must be treated on-site and disposed into subsurface soils or nearby surface water. Treatment standards and allowable discharge methods vary from state to state.

The level and complexity of the required permits and treatment generally depend upon the estimated daily flows, wastewater category, and discharge location. Domestic wastewater and industrial wastewater require different levels of treatment. Generally, larger flows also require more extensive treatment if the wastewater is to be discharged on-site.

Estimates of daily wastewater flow rates for specific uses are usually established by state permitting agencies. These numbers enable a designer to determine if existing municipal sewer facilities are adequate. If the new flows exceed the capacity of the existing system and/or wastewater treatment plant, it may be possible to detain the wastewater in on-site tanks and to discharge it to the municipal system during off-peak hours. The daily flow estimates will also allow an engineer to properly design an on-site treatment and disposal system, if it is required.

The size of a sewer should be based upon peak flows and not daily average flows, because sewer flow rates vary daily and possibly seasonally depending upon the use. The ratio of peak flows to average daily flows for residential use ranges from 2 to 6.0. Peaking factors for industrial wastewater should be estimated on the basis of water use, the number of shifts worked, and pertinent details of plant operations (Metcalf & Eddy, 1989).

In some communities, industrial and commercial users must purchase capacity because many wastewater treatment plants are limited by flow. Therefore, once the wastewater volume is known, the facility should check its agreement with the municipality to determine if additional capacity is available or can be purchased. Other factors that may limit a facility's daily wastewater flows are the size of the existing sewers and pump station capacities. Purchasing capacity may be expensive and should be budgeted into the project cost.

Connecting to a municipal sewer system requires a facility to adhere to the local or state sewer use ordinance. This ordinance may restrict certain pollutants to minute quantities. If a facility discharges process water, the ordinance could require pretreatment. Common sewer ordinance restrictions apply to mercury, copper, zinc, high temperatures, high and low pH, high total suspended solids (TSS), and high biological oxygen demand (BOD). Domestic sewage does not usually require pretreatment before discharge to a municipal sewer.

Normally, a flow meter such as a parshall flume is installed in a manhole at the facility outfall to measure the sewage volume discharged to the municipal sewer system. The metering manhole is used to collect samples for analysis to determine conformance with the sewer permit or ordinance. Transmittal of the recorded data to a central facility may be required.

The sanitary sewer should be designed by a registered professional engineer and should conform to state and local standards. The following design guidelines were excerpted from New England Interstate Water Pollution Control Commission (1998):

1. Minimum slope of an 8-in gravity sewer is 0.004 ft/ft.
2. Minimum full flow velocity should not be less than 2 ft/s.
3. Velocity of 12 ft/s should not be exceeded under any flow conditions.
4. Sewers should be no closer than 10 ft to water mains.
5. Sewers closer than 10 ft may be allowed on a case-by-case basis if the crown of the sewer is at least 18 in below the invert of the water main.
6. Manholes should be installed at the end of each line; at all changes in grade, size, or alignment; and at all intersections.
7. Manholes should not be spaced greater than 300 ft for sewers with diameters of 15 in or less and spaced no greater than 400 ft for sewers 18 to 30 in in diameter.
8. Provide a drop pipe for sewers entering a manhole more than 24 in above the invert.

A facility manager should check with local regulations for sewer design criteria that may be more stringent.

The minimum size for building sewer connections is 6 in. The minimum size for sewer mains is 8 in. Service connections should be no larger than the sewer main to which they are connected. Downstream sewers should be larger than all of the upstream sewers to enable them to pass the solids contributed by the upstream sewers. When joining different sized small-diameter sewers, it is standard practice to match the inverts of the pipe to enhance free flow of sewage.

Smaller-diameter pressure or vacuum sewers (1½ to 2 in) can be used in lieu of gravity sewers but only if solids are properly handled. Solids can be handled by grinder pumps, settled out, and/or screened out in a comminutor before discharge to small-diameter pipes. Small pressure or vacuum sewers can save significant construction costs when difficult subsurface conditions are encountered because the trench depth can be kept to a minimum. Small-diameter sewers are more expensive to operate because they require solids control and sewage pump stations that need power, maintenance, and backup power.

Larger sewage pump stations or ejector stations are required when the receiving sewer is at a higher elevation than the facility's sewer system. There are several types of sewage pump stations. The sanitary district, municipal engineering department, or state regulators may specify the type of station. Common types of sewage pump stations include wet well/dry pit, submersible pumps, suction lift pumps, and ejector pots.

Generally, the preferred type of pump station depends upon the region of the country in which the system is located. Pump stations have stringent design criteria for determining the capacity of the pump, the size and shape of the wet well, alarm protocols, flow meters, control methodologies, and ventilation. At a minimum, two alternating pumps should be provided in the pump station. As with the smaller pressure and vacuum sewers, solids handling is critical. Larger pumps should be able to pass 3-in solids, at a minimum, but depending upon the specific application, it may be necessary to provide a comminutor upstream of the pumps to capture larger solids. The sewage pump station should be designed by a mechanical and/or a civil engineer assisted by an electrical engineer. The cost of sewage pump stations for individual facilities depends on the capacity, subsurface conditions, type of drives, emergency standby power, gas detection equipment, screens, comminutor, flow meters, chemical feeders for odor control, and controls.

Velocity criteria for force mains are derived from observations that solids do not settle out at a velocity of 2.0 ft/s or greater. Solids do settle at lower velocities or when the pump is stopped, and a velocity of 3.5 ft/s or greater is required to resuspend the deposited solids. For small-sized or medium-sized pumping stations where flow may be pumped intermittently at any rate up to the maximum, the desirable force main velocities range from 3.5 to 5 ft/s. Blow-offs are normally not required on force mains. However, when the force main contains a long depressed section between high points, a blow-off is desirable to allow it to be drained and pumped out. A suitable blow-off might consist of a valved connection in a manhole at the low points, discharging to a manhole or vault that would serve as a wet well for a portable pump. The size of the blow-off should not be less than 6 in. It should, if possible, be large enough to provide flushing velocities in the force main (Metcalf & Eddy, 1989).

If possible, force mains should be designed without high points, and the top of the force main should be below the hydraulic grade line at the minimum pumping rate, so that air-relief valves are not needed. If the elimination of high points is not feasible, an automatic air-release valve should be installed at each significant high point where air could become trapped. A high point maybe considered significant if it is 2 ft or more above the minimum hydraulic grade line or, when pumping is intermittent, above the static headline. An air-release valve may discharge to a sewer manhole, a vented dry well, or other suitable place (Metcalf & Eddy, 1989).

On-site wastewater treatment is regulated by the local health department, county agencies, and/or state environmental agency. Once the average daily flow is determined, site suitability for the disposal of effluent is determined. Discharge to subsurface soils is the most common method for disposing of effluent from small on-site systems. Discharge to groundwater and surface waters is permitted in some areas if the on-site system is capable of meeting proper effluent discharge standards.

For most areas, the Natural Resources Conservation Service has delineated soil types for the top 60 in of the soil. The soil maps provide a general overview of the site and are, of course, not exact. The only accurate method for determining the soil type, depth to groundwater, and soil permeability is to perform in-situ test pits and permeability tests. Most states require trained, qualified engineers to perform these tests and to prepare soil reports. Leaching systems generally require large tracts of land for disposal of treated effluent. Leaching areas must be well ventilated to allow oxygen into the subsurface biomat so that continued aerobic biological processes occur.

Septic tanks remove about 40 percent of the BOD and TSS. They also have tees or baffles to prevent scum (floatables) from leaving the tank. Septic tanks require annual inspection of the amounts of sludge and scum. If the sludge is more than one-third the depth of the tank, then the sludge should be pumped out by a licensed septic hauler and disposed of properly. It is critical to prevent the septic tank solids from discharging into the disposal field. Therefore, it is important to have a licensed septic hauler on a regular pumping schedule. A leaching field can cost tens of thousands of dollars, or in larger systems, hundreds of thousands of dollars to replace.

If the disposal site is within a sensitive environmental area, aquifer protection district, and/or nitrogen sensitive area, then additional treatment may be warranted. Several devices on the market today can be added after the septic tank and before discharge that enhance treatment and remove additional quantities of BOD and TSS, as well as nitrogen.

Large on-site systems may require comprehensive hydrogeological studies to determine the impact of treated effluent on the local receptors. These studies include a comprehensive soil-boring program that includes soil sampling and water testing.

Flows that exceed 50,000 gal per day require more conventional wastewater treatment, similar to municipal types. If the project site has inadequate receiving soil and/or is limited in area, it is not uncommon for the facility to purchase suitable off-site land for disposal purposes. Process wastewater discharged to the ground normally requires pretreatment, similar to the requirements for discharging sewage into municipal sewers. These discharges usually require groundwater discharge permits that set effluent standards and require periodic confirmatory sampling and reporting.

To reduce the amount of treated wastewater to be discharged into the ground, graywater reuse is possible. Greywater is discharge from sinks, showers, and bathtubs. In some instances, it can be used for irrigation, toilet flushing, process water, and/or cooling water. At a minimum, the treated wastewater must be disinfected before reuse.

STORM DRAINAGE

Managing stormwater involves many issues. From the standpoint of site development, it is desirable to limit the amount of stormwater flowing from off-site sources onto the developed areas of a facility. It is also important to design positive drainage patterns to direct stormwater away from facilities and to convey it efficiently from paved and landscaped areas into the storm-drain network via catch basins or other stormwater collection structures. Once in the storm-drain piping, the stormwater should be efficiently routed to discharge locations.

Increases in impervious area, altered drainage patterns, channeling of flow, and reduction of vegetation shorten the time of stormwater concentration on the site, decrease the infiltration of rainfall, and increase the flow rate and volume of runoff leaving the site. In addition, sediment, silt, and urban pollutants from paved surfaces are carried in the runoff. Stormwater is recognized now as a non–point source of pollution that can degrade downstream water quality. As such, many states regulate stormwater and require some form of treatment before discharge to the environment. Treatment may consist of one or more of the following: vegetated swales; constructed wetlands; stormwater detention facilities; stormwater retention facilities, such as swirl concentrators; or engineered facilities.

Most local regulations mandate that site improvements cannot increase the rate of stormwater runoff from a facility. This can be achieved by detaining stormwater in a detention pond or retention pond and allowing it to discharge at a low flow rate off-site (detention) or infiltrate into the ground (retention). If regulations dictate that the volume of stormwater runoff cannot be increased by a development, then stormwater infiltration into the ground is necessary. Mitigating stormwater runoff requires as much as 5 to 10 percent of the total buildable area, so it is important to account for it in the early planning stages.

Storm drainage systems are designed to convey specific storm events (10-, 25-, 50-, or 100-year). The storm designation describes the statistical likelihood that a particular design storm will occur in any given year. For example, a 10-year storm has a 10 percent chance of occurring in any given year or once every 10 years. A 100-year storm has a 1 percent chance of occurring in any given year or once every 100 years. This does not mean that once a 100-year event has occurred that it will not occur again for 100 years. Storm data have been statistically analyzed to forecast the likelihood of specific rainfall events.

Storm-drainage systems are usually designed for 10-year storm events. Culverts under roads are generally designed to convey flows from a 100-year event. Local and state regulations dictate the required design storm for specific drainage structures. Storm-drain systems are made up of a series of components that convey stormwater to downstream structures. Storm-drain systems are only as effective as the lowest capacity structure in the series. For example, if drainage piping is large enough to convey a 25-year storm but the catch basins can pass only a 10-year event, then the catch basins are the controlling factor in the system. Local and state regulations generally require that the runoff from a developed site be no larger than it was in its undeveloped condition for storms up to a 100-year event. Detention and retention ponds are designed to mitigate the maximum required storm. However, storm sewers and catch basins are not usually designed to convey a 100-year event efficiently. Catch basins will puddle if they are at a low spot in a road or parking area or will be bypassed by 100-year storm flows if they are on slopes. Storm piping will generally overcharge during large storms. Therefore, some means must be provided for a large storm to discharge into detention or retention facilities even when the drainage system is beyond its capacity. It is critical to provide surface routes for stormwater to flow safely without flooding buildings, roads, or parking lots. For example, if a low spot with a catch basin is next to a building, the rim of the low area must be lower than the finish floor of the building. The surrounding topography must be designed to allow stormwater overflows to drain away from the building to protect it in extreme storms.

The storm drainage system should be designed by a registered professional engineer and should conform to state and local standards. If no local criteria exist, the following design parameters should be adhered to:

1. To size the storm sewer mains, assume that the facility is at the maximum allowable build-out (impervious surfaces) for areas that will be developed in the future.
2. Minimum pipe size should be 12 in, and minimum full flow velocity should be 2.5 to 3 ft/s.
3. Manhole spacing and design should conform to the criteria in items 6 and 7 in the subsection on wastewater.
4. Catch basin spacing should be no greater than 300 ft. Spacing depends upon gutter flow widths for highways and other critical roads and may result in closer spacing intervals for catch basins.
5. Use a minimum of 2 ft of cover over pipes.
6. Catch basins should be located upstream of intersections and crosswalks, if possible.

SUBSURFACE CONDITIONS

It is important to investigate the subsurface conditions under a proposed facility to determine the soil characteristics (bearing capacity, permeability, soil classification, and other engineering properties). It is also prudent to determine if contaminated soils or groundwater underlie the site. In addition to contaminated material, there may be construction debris or other landfill materials beneath the site. The depth to bedrock and groundwater should also be determined. These issues impact foundation costs for the new facility, as well as construction costs for utilities and any other subsurface structure.

Groundwater may limit the depth of foundations and utilities or may require underdrain systems. High groundwater may also cause pavement frost heaves in cold regions if pavement is not properly underdrained and constructed on frost-susceptible fill. Lowering the groundwater may impact adjacent existing facilities. For example, if the site is underlain by saturated clay, lowering groundwater will cause existing facilities to settle. If the adjacent facilities are on wooden piles, then lowering the groundwater may degrade the piles.

Underdrains should be used to lower groundwater in the vicinity of utilities, roads, or subsurface building structures. For underdrains to lower groundwater effectively below utilities and/or pavement, they must discharge by gravity to open channels or drainage ditches. To lower groundwater under subsurface buildings, a system of perimeter drains and underdrains should be used to convey groundwater to either a gravity discharge or sump pump. In addition, waterproofing methods such as impermeable membranes should be installed between the drainage system and the structure. Waterproofing is critical at construction joints and seams. Waterproofing alone will not prevent groundwater from seeping into a building. Waterproofing measures decrease the rate of infiltration, but the water level inside a building tends to equilibrate to the level of the groundwater outside of the building. If the subsurface soils are very permeable and/or a structure is to be constructed deep below the groundwater (subsurface parking garage), then it may be best to construct a waterproof system outside of the underdrains in order to minimize the size of the sump pumps and pumping costs. In this scenario, the underdrain system would discharge only water that leaks through the impermeable membrane or other impervious barriers, such as clay. Any buildings that use sump pumps to lower groundwater should have a backup power supply for the pumps.

Excavation dewatering during construction requires a permit in most areas of the country. Dewatering varies dramatically depending upon the following factors: depth of excavation, depth to groundwater, soil permeability, proximity of surface water, and length of an open trench or area of an excavation. All excavations must be performed in accordance with Occupational Safety and Health Administration (OSHA) standards and regulations contained in Title 29: Subpart P—Excavations, Trenching, and Shoring. In general, all excavations deeper than 5 ft require specified side slopes or trench protection, as mandated by OSHA to protect the safety of personnel. Excavations must have proper side slopes and/or trench boxes, sheet piles, or other excavation support. Precautions must be taken to protect existing utilities and aboveground structures when excavating.

Underground utilities for a proposed or existing development can consist of any or all of the following: natural gas, electrical, cable, telephone, potable water, storm sewer, sanitary sewer, force mains, fire protection, steam, or other industry-specific utilities. Subsurface utility networks can be complicated and require careful planning to minimize conflicts (horizontal and vertical). The sanitary sewer and storm sewer are generally the most complicated to design and construct because they flow by gravity. Most of the other utilities can be redirected during construction to avoid existing utilities or other obstructions. It is recommended that the storm sewer and sanitary sewer systems be designed first and other utilities be designed to accommodate them.

As-built drawings are invaluable when constructing close to existing utilities. Construction as-built drawings are sometimes omitted or not requested by an owner. As-built drawings may also be forgotten because they are prepared at the end of the construction project when the deadline is most pressing and are therefore skipped to save some time. Buried utilities are difficult to locate in the field, and facility managers cannot rely upon their memories or even the memory of the most skilled construction manager. The need to know where a utility is buried in order to avoid it or connect to it may not occur for years or even decades. If reliable as-built drawings are not available, it will likely be necessary to dig test pits at key locations before designing utilities.

CONTRIBUTORS

David N. Hayes, P.E., BSC Group, Boston, Massachusetts

BIBLIOGRAPHY

American Association of State Highway and Transportation Officials: *A Policy on Design of Urban Highways and Arterial Streets,* U.S. Government Printing Office, Washington, DC, 1990.

Americans with Disabilities Act Accessibility Guidelines for Buildings and Facilities: "Rules and Regulations," *Federal Register* 56, no. 144 (1991).

Metcalf & Eddy: *Wastewater Engineering: Collection and Pumping of Wastewater,* McGraw-Hill, New York, 1989.

New England Interstate Water Pollution Control Commission: *Guides for the Design of Wastewater Treatment Works,* Technical Report 16 (TR-16), 1998.

Seelye, E. E.: *Design Data Book for Civil Engineers,* Wiley, New York, 1960.

ARTICLE 5.1.3
STRUCTURAL SYSTEMS

Michael L. Brainerd, P.E., and James C. Parker, P.E.
Simpson Gumpertz & Heger, Inc., Arlington, Massachusetts

UNDERSTANDING THE STRUCTURAL SYSTEM FOR FACILITY MANAGEMENT

Many engineers and facility managers believe that the structural system of a facility is built to last and that, if properly designed and constructed, the structural system will not need much of their attention throughout the facility's life. Relative to other building technologies, such as HVAC systems, electrical systems, and elevators, this perception is quite true. There are situations, however, when the facility manager must understand the behavior and limitations of the structural system. The structural system may be an important factor in managing significant changes to the other building systems and changes in the facility's use. With older buildings, the facility manager should be aware of any deficiencies in the original structural design criteria with respect to current practices and code requirements. And although the structural system usually requires less maintenance than other systems in the building, there are certain structures, such as parking garages, where diligent and continual maintenance of the structure is vital to the facility's economic life.

This article is intended to help the facility engineer and/or manager understand the behavior and limitations of the structural system. The approach to this task is twofold: first, we provide a brief primer on structural materials, design loads, structural elements and behavior, and structural system types; second, we explore some of the important structural aspects of new building design, building renovation, and building maintenance.

DESIGN LOADS

The aspects of the structural system most frequently evaluated over the facility's life are its strength and ability to safely support the intended use and to stand up to natural hazards such as snow, wind, and earthquakes. The weight that the structure is to support and the potential pressures and forces from windstorms and earthquakes influence the design loads. Structures must be safely designed with due consideration given to their service and imposed loads. Loads imposed on structures fall into two broad categories: gravity loads and lateral loads. *Gravity loads* result from gravity acting on a mass. Gravity loads, which act vertically downward on a structure, include dead loads, live loads, and snow loads. *Lateral loads,* which act horizontally on the structure, are forces generated by wind, earthquakes, and earth or fluid pressure.

Dead Loads

Dead loads include the structure's own weight and the weight of any permanent stationary components, including but not limited to roofing; floor finishes; ceilings; mechanical, electrical, and plumbing (MEP) components, such as pipes, ducts, conduits, and equipment; heavy partitions; and exterior cladding. The dead loads used for design are computed on the basis of the actual weights of the building materials and components.

Live Loads

Live loads on floors include the weights of building occupants and temporary nonstationary components such as light partitions, furnishings, light movable equipment, and stored items. Live loads on roofs include minor accumulations of water and building occupants to varying degrees, depending on intended roof access. Live loads are typically prescribed by building codes. Codes prescribe loads both in terms of a uniform load in pounds per square foot (lb/ft^2) to be applied over the entire floor or roof area and a concentrated load in pounds (lb) to be applied separately over a small area anywhere on the floor or roof. Typical code-prescribed live loads for floors range from 50 lb/ft^2 for office use to 250 lb/ft^2 for heavy manufacturing and heavy storage use.

The 50-lb/ft^2 live load for office use assumes that items included in the live load are reasonably well distributed. In office layouts involving heavy concentrations of tall file cabinets, the floor loading of the cabinets alone often exceeds 50 lb/ft^2. Judicious positioning of banks of file cabinets relative to the structural framing (near columns and over girders) may be necessary to accommodate certain layouts on office floors designed for the code minimum live load of 50 lb/ft^2. Office floors are often designed for a live load of 100 lb/ft^2, including an allowance of 20 lb/ft^2 for partitions. This approach provides flexibility in space planning by not constraining the location of localized heavy loads, lobbies, and corridors.

Typical code-prescribed live loads for roofs range from about 20 lb/ft^2 for limited access roofs to well over 100 lb/ft^2 for roof-level plazas accessible to large numbers of occupants.

Snow Loads

Roofs in cold climates are exposed to accumulations of snow. Building codes prescribe design snow loads as uniform loads in pounds per square foot to be applied over the entire roof. Where roofs of differing elevation either adjoin or are in close proximity, snow will accumulate due either to drifting or sliding from a nearby sloped roof. Drifting also occurs at projections of large mechanical units, screen walls, and parapets. Current building codes include provisions for computing localized additional snow load to account for drifting and sliding snow. Note that codes before the 1970s did not prescribe drifting snow loads, and many older roofs do not have safe capacity for drifting snow.

Rain Loads

Rain can accumulate on roofs due to deflection of the roof structure and blockage of the drainage system. Building codes address rain loading to differing degrees, but generally they require designing roofs for the maximum amount of water that will accumulate if the primary drainage system becomes blocked. The depth of water is based on the deflection of the roof framing and the height of overflow at scuppers or roof edges. Roof framing should be designed with strength adequate to carry rain loads and stiffness adequate to preclude unabated progressive deflection.

Lateral Loads to Account for Potential Windstorms

Wind blowing against and around a structure exerts lateral loads as pressure on the windward face and suction on the leeward face. The magnitude of the wind pressure is proportional to the square of the wind speed. Globally, wind speed and, therefore, wind pressure varies from region to region. The design wind speed in most of the central United States is typically about 70 mph. The design wind speed can be 100 mph and higher near the coastlines. Wind speeds throughout the life of the structure are highly variable. Therefore, U.S. codes generally require designing structures for a wind speed that has a .02 annual probability of being exceeded—in other words, for a windstorm that, on average, occurs every 50 years.

The engineer must estimate the wind pressures on the building on the basis of the wind speed. Wind speed increases with the height above ground and is influenced by the shape of the structure, the number and shape of nearby structures and obstructions, and the surrounding topography. Localized wind effects, such as vortex shedding, increase wind pressure near building corners. Building codes prescribe very detailed methods for computing wind loads on structures as a whole and on components of structures, considering all of the effects described.

Lateral Loads to Account for Potential Earthquakes

Ground motions due to earthquakes are sudden, usually caused by sudden ruptures along faults in the earth's crust. The center of the earthquake is called the *focus,* and the spot on the earth's surface directly above the focus is the *epicenter.* Complex vibrations in the form of ground waves emanate from the rupture. When the ground waves reach a structure, the base of the structure is subjected to a complex motion. As the base of a structure moves, the inertia of the upper portions of the structure resists this movement, and forces are generated within the structure. The magnitudes of these seismic forces are highly variable with time during the earthquake. Given a recording of actual earthquake records, engineers can estimate resulting forces in a particular building if the building's construction is relatively well documented. Because the exact location and magnitude of, and ground motions resulting from, future earthquakes are unpredictable, it is very difficult to predict the actual forces that a structure may experience. Therefore, current practice incorporates a probabilistic approach in designing buildings for earthquake loads. Most current codes prescribe designing for ground motions that have about .002 annual probability of being exceeded. Codes often express this ground motion as that which has a 10 percent chance of being exceeded in 50 years or that has an average recurrence interval of 475 years. (This is about 10 times less likely than the typical design wind speed.) If a building is near a known fault (a known location of weakness in the earth's crust where earthquakes have occurred in the past), then some codes require a deterministic approach to design. In this case, the design ground motions are based on the maximum expected earthquake on the given fault, the type of geology between the fault and the building, and the distance of the building from the fault.

For most common buildings, current codes do not require designing for actual ground motions from past earthquakes or ground motions predicted to occur in the future. Instead, codes prescribe a highly simplified approach of designing for horizontal static forces on the building by using prescriptive structural details. This approach does, however, account for the building's use, the building's fundamental frequency, and the site's soil conditions. The current codes, which are minimum requirements, aim to protect life, not necessarily to prevent damage to the building and its contents. Codes recognize levels of protections. Fire stations, hospitals, and communications centers require a higher degree of design. Because of the simplified and empirical nature of the code approach, the goals are only implicit and are not explicitly defined. Generally, damage from the design level ground motion is expected, but the damage should have a low probability of causing hazard to life. For ground motion 50 percent greater than the design level, damage is expected to result in hazard to life, but the probability of total collapse is still small. Future earthquake codes are likely to adopt more elaborate procedures that allow designing for a given level of performance, such as immediate safe occupancy after the earthquake or various degrees of protection from damage to the building and/or contents.

The first earthquake design rules were adopted for portions of the West Coast after the 1933 Long Beach, California, earthquake. Adoption of earthquake design rules spread along the West Coast and have been evolving as new data are collected from the responses of buildings to earthquakes and from earthquake research. Adoption of earthquake design regulations for other parts of the United States lagged far behind those adopted on the West Coast despite knowledge of the potential for earthquakes elsewhere in the country. For example, a significant historical earthquake occurred in New England in 1755, and the largest known earthquake in the continental United States occurred in Missouri in 1811. The first earthquake regulations east of the Rocky Mountains were adopted in Massachusetts in 1975. Therefore, the original designs of many buildings currently in use in the United States did not account for seismic ground motion, and these buildings present a real risk should an earthquake occur. Although many of these buildings are "grandfathered" into existing codes, exempting them from retroactive requirements for strengthening, many new codes now include triggers, such as change of use, additions, or restorations, that require seismic strengthening of older buildings.

Vibratory Loads on Structures

Human rhythmic activity, such as marching, dancing, and aerobics, can cause vibration of floor systems. Long-span structures supporting open spaces that are free of damping elements such as full-height partitions are most susceptible to vibratory loads. Vibrations can be perceptible to the point where they alarm occupants and can, in the extreme, damage the structure. Occupied spaces over convention center exhibition halls and cantilevered seating areas of stadiums and auditoriums have been known to vibrate perceptibly due to human rhythmic activity. Moderate-span light-steel structures may vibrate perceptibly if used for dancing or aerobics. Building codes are often silent on the issue of vibratory loads, although some require considering such loads in the design but do not prescribe how to do so.

Reciprocating machinery and mechanical equipment can also cause perceptible and possibly damaging vibrations. The supports for these components should be designed to minimize transmission of vibrations to the structure.

Load Combinations

Structures should be designed for the credible combination of loads that produces the most unfavorable magnitude of the structural effect being examined. In some cases this may occur when one or more loads are not acting. The least credible load combination, dead load alone or some fraction thereof, may produce the largest magnitude of some structural effects. Some loads are unlikely to occur at the same time. The design need not consider that wind and earthquake loads will act simultaneously. The chance of simultaneous occurrence of dead load, full live load, and lateral load (seismic or wind) is small. Therefore, the combined effects of these loads are typically reduced. Building codes prescribe the load combinations that must be considered in most usual situations.

BASIC STRUCTURAL ELEMENTS AND BEHAVIOR

Foundation Elements

Building foundations can be broadly classified as either shallow or deep foundation systems. The most common *shallow foundation* element is the spread footing. *Spread footings* are individual rectangular concrete elements that support one or more building columns. The spread footing transfers the load from the column to the soil. The size of the footing in plan is a function of the column load and the allowable soil-bearing pressure. When the allowable soil-bearing pressure is very low or the soils are prone to settlement, buildings are founded on mat foundations that are essentially a single spread footing covering the entire footprint of the building. The bottoms of shallow foundations that extend beyond the perimeter of the building must be located below the frost line, which can be several feet or more below grade.

When suitable bearing soils are not reasonably close to the surface, *deep foundations* are used. Deep foundations include *piles* and *drilled piers*. Unlike spread footings, which are constructed in a general excavation extended to the bearing strata, piles and drilled piers are constructed by driving or drilling through unsuitable material into the load-bearing soils. Piles and drilled piers transfer the column load to the soil by friction and/or end bearing. Many types of piles are commercially available, including steel H piles, round steel-pipe piles, and cast-in-place and precast concrete piles. Groups of closely spaced piles are installed at each column location and are covered by a single concrete cap, upon which the building columns bear. Drilled piers are relatively large diameter, round, concrete foundation elements that are constructed by casting concrete in predrilled holes. Drilled piers, also known as *caissons,* have straight shafts or are widened (belled) at the bottom. Typically, an individual drilled pier is installed at each column location and is covered with a rectangular concrete cap, upon which the column bears.

Superstructure Elements

Superstructure elements include vertical elements such as columns and load-bearing walls, and horizontal elements such as slabs, beams, and girders. The basic horizontal superstructure element is the slab. All superimposed vertical dead and live loads are carried by the slab, which spans either one way between beams or bearing walls or two ways between columns in the case of two-way slab systems. In structures with one-way slab systems and beams, the slabs are supported on the beams, which are in turn supported either directly by the building columns or by girders that span between building columns. Columns and bearing walls carry the loads from the various floors down through the building to the foundations.

Lateral Load-Resisting Systems

The lateral (wind and seismic) loads on a building are carried by the lateral load-resisting system. There are two basic types of lateral load-resisting systems: braced-frame/shear-wall systems and moment-resisting frame systems. *Shear walls* are vertical walls of concrete, masonry, or wood, depending on the structural system type. *Braced frames* comprise one or more building columns interconnected by diagonal members to create a stiff wall-like lateral load-resisting element. Braced frames are usually steel but can also be constructed of concrete. In braced-frame/shear-wall buildings, the lateral loads on the floors and roof are resisted by a series of braced frames or shear walls, two or more in each direction, that cantilever from the foundations.

Moment-resisting frames are comprised of columns and beams or girders in a plane rigidly connected to one another. Because the beams and columns are rigidly connected at the joints, the lateral loads are resisted by the bending strength of the columns and beams or girders.

COMMON STRUCTURAL SYSTEM TYPES

Concrete Systems

Concrete is a composite of cement, stone, sand, water, and chemical admixtures. Concrete is fluid when first mixed, and chemical reactions between the cement and water cause the mass to harden into a rocklike solid that is very strong in compression, but weak in tension. As a result, tensile loads are carried by reinforcing bars embedded in the concrete. Because concrete is transformed from a fluid to a solid, it can be formed into virtually any shape either on-site or in a precasting plant. Because it is versatile, concrete can be used to varying degrees in all building structures.

Concrete structural systems fall into two general categories: cast-in-place and precast systems. *Cast-in-place* concrete systems are constructed by erecting forms and casting the concrete on the project site. The components for *precast* concrete structures are cast at a precast plant, then shipped to the site, and set in place with cranes. Common cast-in-place concrete systems include one-way slab, joist, and beam systems (Fig. 5.1.3-1); one-way slab, beam, and girder systems (Fig. 5.1.3-2); and two-way flat slab and waffle slab systems (Figs. 5.1.3-3 and Fig. 5.1.3-4). Common precast systems include hollow-core plank and inverted tee-beam systems; precast, prestressed hollow-core plank and inverted tee-beam systems (Fig. 5.1.3-5); and precast, prestressed, double-tee and inverted tee-beam systems (Fig. 5.1.3-6). Precast systems usually include a cast-in-place concrete topping to provide a level floor and increase load-carrying capacity.

Concrete systems typically weigh more than other structural systems, such as steel. Because of the higher gravity loads and seismic loads, concrete superstructures require stronger and more costly foundations and seismic load-resisting systems than comparable steel superstructures. The higher mass of concrete structures makes them less susceptible to transmission of vibrations due to human activity and machinery.

Concrete is generally considered fire resistant as long as sufficient concrete cover is provided over the reinforcement. For usual fire ratings, no special fire protection is required on concrete elements.

Concrete is generally considered a very durable building material. Under certain exposures, however, concrete must be specially formulated to withstand premature deterioration. The most common exposures that deteriorate concrete include exposure to weather, particularly freeze/thaw cycling and deicing salts, and exposure to salt and sulfate in seawater and certain soils. Exposure to weather in cold climates can result in disintegration (scaling) of the exposed concrete surface due to freezing while saturated with moisture. Resistance to freeze/thaw damage is achieved by incorporating a system of small bubbles (entrained air) in

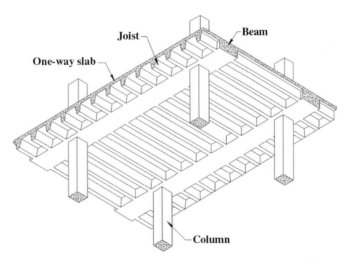

FIGURE 5.1.3-1   One-way slab, joist, and beam system. (*Courtesy of Simpson Gumpertz & Heger, Inc.*)

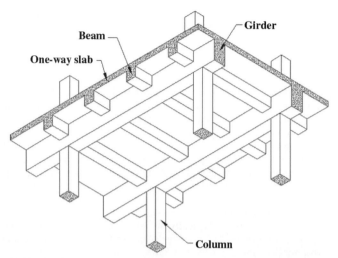

FIGURE 5.1.3-2 One-way slab, beam, and girder system. (*Courtesy of Simpson Gumpertz & Heger, Inc.*)

the concrete by adding air-entraining admixtures. Exposure to deicing salts and salt in seawater can cause corrosion of the steel reinforcement. Resistance to corrosion damage is achieved by using low permeability concrete and by increasing the distance (cover) between the exposed concrete surface and the reinforcement. Corrosion-inhibiting admixtures and coated reinforcement are also available to combat corrosion damage. Exposure to sulfate in seawater and soil can cause general disintegration of the concrete. Resistance to sulfate attack is achieved by using low-permeability concrete and special cements.

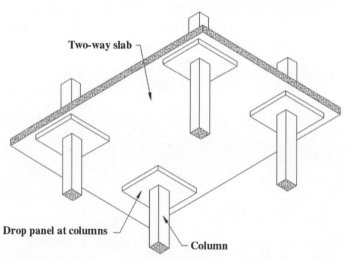

FIGURE 5.1.3-3 Two-way flat slab system. (*Courtesy of Simpson Gumpertz & Heger, Inc.*)

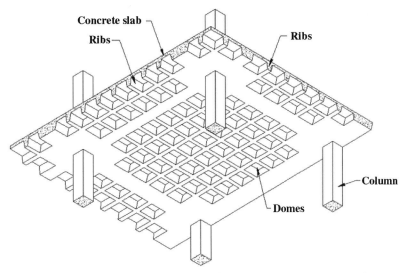

FIGURE 5.1.3-4   Two-way waffle slab system. (*Courtesy of Simpson Gumpertz & Heger, Inc.*)

Steel Systems

Steel is an alloy predominantly composed of iron and carbon. Modern structural steel for buildings falls into two broad categories: carbon steel (ASTM A36) and high-strength, low-alloy steel (ASTM A572), with minimum yield strengths of 36,000 and 50,000 lb/in², respectively.

Structural steel systems fall into two general categories: steel beam systems (Fig. 5.1.3-7) and steel joist systems (Fig. 5.1.3-8). Steel beam systems are comprised of concrete slabs sup-

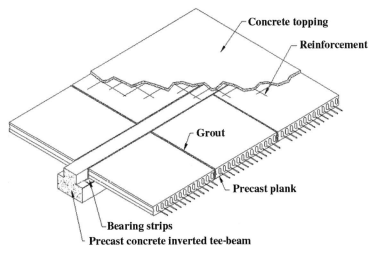

FIGURE 5.1.3-5   Precast, prestressed hollow-core plank and inverted tee-beam system. (*Courtesy of Simpson Gumpertz & Heger, Inc.*)

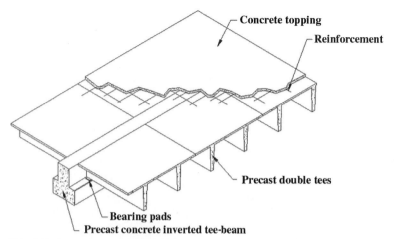

FIGURE 5.1.3-6   Precast, prestressed double-tee and inverted tee-beam system. (*Courtesy of Simpson Gumpertz & Heger, Inc.*)

ported by steel beams, girders, and columns. Common slab systems used with steel beam framing include cast-in-place concrete on metal deck, formed cast-in-place concrete, and precast hollow-core plank with a cast-in-place topping. Slabs on steel joists are usually cast-in-place concrete, either formed or on a metal deck.

Many prefabricated structural steel shapes are readily available. The most widely used are wide-flange sections (W shapes), American standard channels (C shapes), and angles (L shapes). Except for angles, structural shapes are generally designated by a letter that indicates the shape and is followed by two numbers. The first number indicates the nominal depth, and the second indicates the weight of the section in pounds per foot. For example, W14 × 22 identifies a 14-in-deep wide-flange section that weighs 22 lb/ft. It is important to realize that the depth designation is nominal only and that the actual depth may vary significantly within the nominal depth category, particularly for the heavier sections within a grouping. The designations and properties for design and detailing of commonly produced structural shapes are tabulated in the *Manual of Steel Construction* published by the American Institute of Steel Construction (AISC).

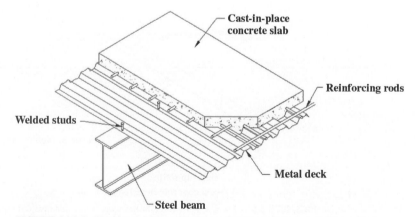

FIGURE 5.1.3-7   Steel beam system. (*Courtesy of Simpson Gumpertz & Heger, Inc.*)

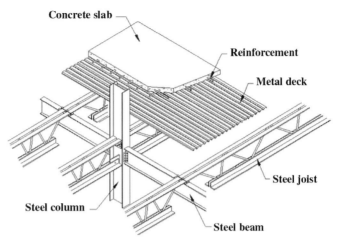

FIGURE 5.1.3-8 Steel joist system. (*Courtesy of Simpson Gumpertz & Heger, Inc.*)

Prefabricated truss-type steel components, typically referred to as open-web steel joists, are also readily available commercially. These members are often fabricated from angle shapes and round bars. Open-web steel joists, which often are comprised of double-angle top and bottom chords and a continuous round bar bent to form the web members, are produced in standard depths from 8 to 30 in. Each depth class contains many joists of varying capacity and weight. Safe load capacities of joists are contained in load tables published by the Steel Joist Institute.

Unlike concrete, steel is susceptible to fire damage and often requires additional fire protection, such as spray-on fireproofing or fire-resistant enclosures.

Masonry Systems

Modern masonry structures typically are comprised of reinforced masonry-bearing walls supporting slab systems of cast-in-place concrete or precast hollow-core plank with a cast-in-place topping. Older masonry structures may have wood floor framing systems. The masonry-bearing walls typically carry both the gravity and lateral loads. A masonry wall, by itself, does not have significant lateral load-carrying ability.

Masonry components most commonly used in modern building construction include stone, clay brick, and concrete brick and block. Clay and concrete units may be either hollow or solid. Masonry is used for exterior cladding (veneer), non-load-bearing walls, load-bearing walls, columns, and beams. Masonry walls may be either reinforced or unreinforced, although most current building codes require constructing exterior walls, walls around stairwells, and load-bearing walls of reinforced masonry. Masonry components are constructed by placing masonry units by hand with mortar in the joints between adjoining units. Horizontal reinforcement consists of truss or ladder-type constructions made from small-diameter wire and laid in the horizontal mortar joints. Vertical reinforcement usually consists of deformed reinforcing bars grouted solid in the collar (vertical) joints between multiple widths of masonry or in the hollow cores of individual widths.

The strength of masonry assemblages is a function of the compressive strength of the individual units and of the type of mortar. Clay brick is classified as NW (no exposure), MW (moderate weathering), or SW (severe weathering), and the minimum compressive strength for each grade is 1500, 2500, and 3000 lb/in², respectively. Concrete masonry (brick or block) is classified as either N (high strength and resistance to moisture penetration and severe frost) or S (moderate strength and resistance to moisture penetration and frost). The minimum strengths of Grade N and S concrete brick are 2500 and 3500 lb/in², respectively. The minimum strengths of Grade N and S hollow load-bearing block are 700 and 1000 psi, respectively. Mortars are classified as Type M, S, N, or O and have minimum compressive strengths of 2500, 1800, 750, and 350 lb/in², respectively. Structural masonry is generally constructed with Type M or S mortar.

Wood Systems

Wood structures typically are comprised of plywood flooring on wood joists supported by wood beams and bearing walls. The wood joists are either sawn lumber or any of a variety of proprietary prefabricated joist elements. Beams are either ganged pieces of sawn lumber or any of a variety of proprietary prefabricated beam members, including glulam beams and microlam beams.

The strength of wood is highly variable because of many factors, including moisture content and defects such as knots, splits, and checks. The strength of wood also varies with the load duration and the direction of the load relative to the grain of the wood. All of these factors are considered in designing wood components by using load factors for loading duration and allowable stresses based on the type, quality, and moisture content of the wood and on the type and orientation of the stress.

Common modern wood members include sawn lumber, glulam beams, and microlam beams. Glulam beams are members comprised of smaller wood members laid flat and affixed together with adhesive. Microlam beams are members comprised of wood veneers, oriented vertically and laminated together. The flexural strength and modulus values of common lumber, such as spruce, pine, and fir, are 1000 and 1,200,000 lb/in², respectively. By comparison, the flexural strength and modulus values of microlam members are 2,800 and 2,000,000 lb/in², respectively. There are also numerous proprietary wood joists and truss type components available comprised of combinations of sawn lumber chords and web members of sawn lumber, plywood, particleboard, and light-gauge metal.

NEW BUILDING DESIGN

Design Criteria

One of the first tasks of the engineer-of-record in the structural design of a new building is preparing the structural design criteria. This document lays out all of the criteria that will be the basis for the structural design. As a minimum, the design criteria should contain the following information:

- Identification of the governing building codes. Some jurisdictions publish complete building codes based on one of the national model codes, and others adopt one of the national model codes and publish supplements with jurisdiction-specific modifications.

- Identification of the report of the geotechnical investigations for the project and a summary of the foundation design recommendations contained therein.

- Design loads assumed in the design, including computations and a summary of the dead loads, a summary of the live loads, and computations and a summary of the wind and seismic lateral loads.

- Limits on deflection for floor and roof framing.

- Limits on drift for the lateral load-resisting system.
- Special loading, deflection, and vibration requirements and limitations.
- Description of the structural systems employed in the project.
- Description of the lateral load-resisting system and load paths.

Although preparing the design criteria is one of the first tasks, the criteria will evolve through the schematic design phase and into design development.

Structural Configuration

The configuration of the structure is determined by the available space, the functional requirements, and local codes. In the schematic design phase, the structural engineer and architect work closely to identify bay sizes (column spacing) that fit within the space planning constraints and are both feasible and economical. If unconstrained, story heights are selected to accommodate the desired ceiling height and structural system depth and to provide adequate space between the ceiling and structure to accommodate MEP components. However, story heights are often constrained either to match nearby or adjoining buildings or to maximize the number of stories within code-imposed building-height limitations.

Structural System Type

Once bay sizes and story heights are established, the structural engineer identifies one or more structural system types that can accommodate the desired configuration and develops schematic-level designs for cost comparison. The structural engineer identifies the advantages and disadvantages of structural system alternatives in terms of important characteristics such as cost, constructability, availability of materials and qualified contractors, future maintenance, and flexibility of structural modifications associated with future space changes. This information provides a rational basis for the owner, with the assistance of the structural engineer and architect, to select the most suitable structural system type.

STRUCTURAL IMPLICATIONS OF BUILDING RENOVATION

Interior Remodeling

Interior remodeling projects can impact the building structure to varying degrees, depending on the nature and scope of the project. Changes to layouts of interior walls often have little impact, if any, on the structure, as long as the walls are non-load-bearing and as long as the weight and/or concentration of walls are not significantly increased. Whether a wall is load bearing is not always obvious. If there is any doubt about a wall's function, it is prudent to review the original structural drawings and records of subsequent building modifications to identify load-bearing walls. In the absence of such information, inspection by a structural engineer may be necessary.

Interior remodeling projects often involve closing and/or adding door openings in load-bearing walls. Adding and closing openings affects the path of loads down through the wall and also affects the wall's ability to carry lateral (wind and seismic) loads. The structural effects of such modifications must be carefully evaluated to confirm that they do not impair the wall's ability to carry design loads safely. The level of effort required to complete such an evaluation can be substantial, depending on the number, size, and locations of openings. Record drawings of bearing-wall modifications should be maintained for use in evaluating future wall modifications. The lack of such records can lead to costly field investigations of existing conditions and errors in evaluating future wall modifications. Figure 5.1.3-9 shows an elevation of a story-high, posttensioned concrete beam taken from the structural design drawings for a building. This wall is actually a transfer girder that carries several stories of the building above it while creating a large column-free space below. An uneducated eye could easily view this wall in place and mistake it for an ordinary bearing wall which could readily accommodate a new door opening without significantly affecting the support for the floor immediately above. Depending on its location, however, such an opening could completely destroy the wall's ability to act as a transfer girder.

The addition of stairways and elevators requires large floor openings. The impact of such openings on the structure and the need for strengthening and additional structural supports depends on the size and location of the opening and the type of structural system. Generally, one-way slab systems with steel beam and girder framing can often accommodate large openings with little structural modification, particularly when the opening fits between adjacent beams or girders. Two-way flat slab systems, on the other hand, often require significant strengthening to accommodate large openings.

MEP Modifications

Modification of MEP systems often involves removing, replacing, and adding ducts, pipes, conduits, and equipment. Such changes can impact the structure by adding weight and requiring openings in slabs, beams, and walls. Openings for ducts, pipes, and conduits are usually small and easily accommodated unless they are numerous and closely spaced. As a general rule of thumb, small horizontal openings through beams can be accommodated (without impairing the load-carrying capacity of the member) if they are not numerous and closely spaced and are located within the middle third of both the depth and span of the beam. Openings through slabs are usually readily accommodated, particularly in one-way slab systems, but may require some minor supplemental framing around large duct openings or banks of pipe and conduit openings. In two-way cast-in-place concrete slabs, openings near columns must be avoided or at least limited, because they can greatly diminish the transfer of the floor load from the slab to the column. Large duct openings and numerous closely spaced pipe and conduit openings impact walls in much the same way as door openings, as discussed in the previous subsection on interior modifications.

MEP modifications that involve adding duct work and small-diameter pipes and conduits usually do not significantly affect the loads on structures that have been designed with a reasonable allowance for suspended MEP components, usually about 10 to 15 lb/ft^2 for most commercial and institutional structures. Manufactured metal buildings, on the other hand, are usually designed for ancillary loads of only several pounds per square foot and therefore require careful evaluation for even minor additional loads.

Mechanical rooms are typically designed for large live loads in order to account for miscellaneous light- to medium-weight equipment and servicing activities. In both new design and design for MEP modifications, the impact of heavy MEP components, such as cooling towers, generators, large heating units, and large-diameter liquid-filled pipes, is evaluated on the basis of the specific size, weight, and location of each component. Replacing existing units with larger units and adding units and large-diameter pipes, particularly outside of mechanical rooms, usually requires strengthening the structure.

Changes in Use

Changes in use from one of lower live loading to one of higher live loading may require structural strengthening. As an example, changing from office use to heavy storage would require that the structure be capable of carrying an additional 200 lb/in^2 of live load. Changes in use

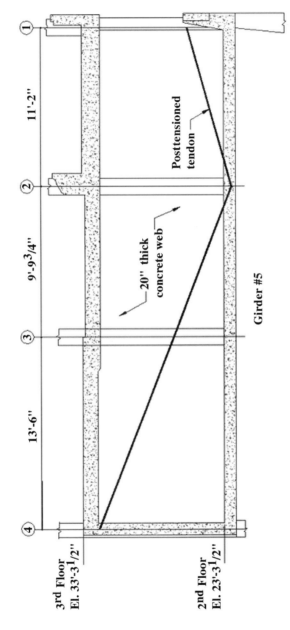

FIGURE 5.1.3-9 Posttensioned story-high concrete beam. (*Courtesy of Simpson Gumpertz & Heger, Inc.*)

from one involving minor human rhythmic activity, such as normal office use, to one involving significant rhythmic activity, such as dancing, aerobics, or reciprocating machinery, will usually require modification of the dynamic characteristics of the floor structure in order to minimize annoying vibrations and preclude damage to the structure.

Seismic Hazard Mitigation

As previously discussed, many older buildings in many jurisdictions were constructed before code-prescribed seismic design requirements were introduced. More recent buildings in many jurisdictions were designed and constructed in accordance with seismic design requirements that were less stringent than today's requirements. Most current building codes contain provisions for mitigating seismic hazard that require that existing buildings undergoing significant repair or modification be evaluated and upgraded to ensure at least some minimum level of seismic resistance; under certain circumstances, full compliance with the seismic requirements for new buildings must be met. Project teams must plan for seismic hazard mitigation early in the conceptual development of the project because it can have a wide-ranging impact on scope and cost.

Seismic hazard mitigation has two components: ensuring that the risk of life-threatening structural damage is acceptable; ensuring that the risk of life-threatening damage from nonstructural components such as walls, MEP equipment, ceilings, lights, and tall, heavy furniture is acceptable. Sophisticated owners of large numbers of buildings that are vital to their business operations have seismic hazard mitigation programs that include identifying and upgrading seismically deficient buildings. In general terms, seismic upgrading includes strengthening or augmenting the existing lateral-load-resisting system of the building and anchoring nonstructural components.

STRUCTURAL MAINTENANCE

Most building structures are enclosed and are thereby shielded from the deteriorating effects of weather. The interior conditioning of most buildings is benign. Under these circumstances, building structures require little maintenance. There are, however, some notable exceptions. Unenclosed structures, such as parking structures and outdoor sports facilities, and exposed building components, such as balconies and uncovered exterior slab edges, beams, and columns of otherwise enclosed buildings, are all exposed to the damaging effects of weather and the atmosphere. Buildings with special interior conditioning or exposure, such as natatoriums and certain industrial and food-processing facilities, are exposed to the damaging effects of humidity, high temperature, and chemicals.

Structures with deleterious exposures must be designed to include durability features above and beyond those of usual building structures. Concrete components may require specially formulated concrete, reinforcement with protective coatings, or topical treatments. Steel components may need to be fabricated from special steel or galvanized steel, or be provided with protective coatings. Most deteriorative mechanisms involve an incubation period during which the conditions necessary for initiating deterioration develop. Once deterioration begins, in most cases, it will continue at an ever-increasing rate. In many situations, maintenance and repair will generally slow but will not stop ongoing deterioration. The choice of durability measures depends on the type and severity of exposure, the cost, and the tolerance of the structure for maintenance.

The condition of the building structure should be audited periodically to identify structural conditions requiring repair or maintenance. Inspect exposed structural elements and building finishes to identify readily visible signs of deflection, cracking, movement, and deterioration (crumbling concrete, rotted wood, corroded steel, and debonding and wear of protective coatings). Also inspect structural components that are exposed in unfinished spaces or are readily accessible behind finishes which can be easily removed and reinstalled, such as acoustic tile ceilings. Pay particular attention to locations of potential water infiltration and associated deterioration, such as below roofs and plazas and where structural components adjoin exterior walls. Conditions of deflection, cracking, movement, and deterioration will require evaluation by a qualified engineer to determine the cause and seriousness of the damage and to design repairs. Facility managers can commission a general, overall assessment of the condition of a facility, or of a campus, by a qualified engineer. Effective condition assessments require blending diligent inspection, engineering judgment, and experience.

CONTRIBUTORS

David C. Foster, Ph.D., P.E., Consultant, Bolton, Massachusetts

Albert Y. Wong, Ph.D., P.E., Stone & Webster Engineering Corporation, Boston, Massachusetts

BIBLIOGRAPHY

American Institute of Steel Construction: *Manual of Steel Construction,* American Institute of Steel Construction, New York, 1994.

American Institute of Steel Construction: *Structural Steel Detailing,* American Institute of Steel Construction, New York, 1971.

American Institute of Timber Construction: *Timber Construction Manual,* Wiley, New York, 1994.

American Iron and Steel Institute: *Cold-Formed Steel Design Manual,* American Iron and Steel Institute, Washington, DC, 1996.

American Society of Civil Engineers: *Guideline for Structural Condition Assessment of Existing Buildings,* American Society of Civil Engineers, New York, 1991.

American Society of Civil Engineers: *Minimum Design Loads for Buildings and Other Structures,* ASCE 7-95, American Society of Civil Engineers, New York.

Beyer, D. E.: *Design of Wood Structures,* McGraw-Hill, New York, 1993.

Federal Emergency Management Agency: *Handbook for the Seismic Evaluation of Buildings,* U.S. Government Printing Office, Washington, DC, 1998.

Federal Emergency Management Agency: *A Non-technical Explanation of the 1994 NEHRP Recommended Provisions,* U.S. Government Printing Office, Washington, DC, 1995.

Fintel, M.: *Handbook of Concrete Engineering,* Van Nostrand Reinhold, New York, 1974.

Gaylord, E. H., and C. N. Gaylord: *Structural Engineering Handbook,* McGraw-Hill, New York, 1996.

Schneider, R. R., and W. L. Dickey: *Reinforced Masonry Design,* Prentice-Hall, Englewood Cliffs, NJ, 1993.

Wiegel, R. L.: *Earthquake Engineering,* Prentice-Hall, Englewood Cliffs, NJ, 1976.

ARTICLE 5.1.4
SKINS AND FACADES

Michael J. Kearns, C.C.S., R.C.I.
CID Associates Inc., Boston, Massachusetts

This article addresses the general architectural and engineering design criteria of external wall systems that are typically used in today's industrial and institutional facilities. This discussion will enable facility engineers and building owners to better understand the intended purpose of various design elements and to anticipate the intricacies of the design process as they relate to the facades of their facilities. Our scope is limited to the exterior walls and does not include the roofing system, which is discussed in Art. 5.1.5, "Roofs."

The building facade provides the facility with a controlled environment. The primary function, therefore, is to keep the interior of the facility free of the elements. We typically consider rain and snow when referring to "the elements." However, the building facade controls additional weather-related anomalies from intruding upon the controlled interior environment. The facade keeps wind from blowing through the workspace and provides a barrier to the flow of temperature and humidity. The facade must provide this protective environment while working with the facility structural systems and, in some cases, also acting concurrently as a significant structural element. Here are the typical performance requirements of the building facade as we have identified them thus far:

- A shield from precipitation
- A wind and airflow barrier
- A vapor retarder
- An insulating element
- A structural element
- A defining element of the building aesthetics

These are some of the basic design requirements that engineers and architects expect from the building facade. In addition to performing the basic system requirements that provide the facility with the proper controlled working environment, the facade is also the most obvious and immediately noticeable architectural feature of any building. Today the choices of facade building materials are quite extensive. The choices of materials to be used are often driven by the desire to make an architectural statement and by budgetary constraints. Facility engineers should be acquainted with the design elements of their particular facilities. This article examines the basic construction techniques and maintenance requirements for some of the typical exterior wall systems and materials.

The oldest and most common type of building facade is some form of masonry construction. This category can include natural stone, concrete block, brick, structural clay tile, and precast concrete, as well as combinations of each. Typical masonry construction from the turn of the century through the 1960s and into the present entailed a design of low-rise load-bearing exterior walls. Earlier designs consisted of 3- or, in some cases, 4-wythe-thick solid brick walls. The mass of brick and mortar was normally sufficient to keep out moisture and wind. In-wall flashing when detailed was limited to wall openings such as small windows and doors. The solid walls were typically tied together using soldier courses of brick at regular intervals and patterns. A brick placed in a soldier position is one that is turned so that the length of the brick spans 2-wythe wall sections.

Over time, masonry construction developed from the solid wall configuration to the cavity wall, shown in Fig. 5.1.4-1, that is more common today. Masonry cavity walls are still typically load bearing but use less material to achieve the same or better performance. Cavity walls also typically combine both brick and concrete block in their matrix. The use of concrete block provides construction cost savings over a solid brick wall. Thus, greater structural capacity can be obtained at lower cost by employing this construction method. However, cavity wall design and construction requires that additional techniques be employed to ensure that the facade will provide the five basic requirements listed before.

Structural integrity is maintained by the placement of wall ties that join the outer face brick to the concrete backup wall. The placement of wall ties is critical to the stability of the wall system. If an insufficient number of ties is used along with either improper placement or type of ties, the outer facade can be left in an improperly supported condition. Because the backup wall typically assumes the load-bearing components of the roof and floor framing systems, the use of inadequate ties can leave the facade in a cantilevered state. Again, depending on the wall height and slenderness ratio, an inadequately supported wall can be very unstable.

Moisture is deterred from penetrating through the wall system by placing dampproofing materials and systems within the wall cavity. Typical materials used for through-wall flashing are lead-coated copper, asphalt membranes with copper fabric backing, and other synthetic membranes. Weep holes placed at the base of the through-wall flashing materials within the cavity are necessary to convey the moisture from within the wall. These accessory items must be incorporated into the cavity wall system at specific locations in order to control and manage the flow of water through the wall without enabling penetration into the building interior.

The outer brick face provides a barrier from wind, and the structure of the wall and building system is designed to accommodate extreme wind loading. Insulating value can be added to the facade by placing insulation within the core of the concrete block or on the back of the wall system. Vapor barrier requirements are determined by geographic location and use of the facility. Vapor barriers are placed on the *warm* side of the facade, which is normally the building interior. Exceptions to this standard are controlled, chilled environments, such as freezer storage or humidity-controlled environments in hot, humid climates, in which case vapor barriers are located closer to the facade exterior. To be effective, particularly in buildings with tight humidity-control requirements such as museums, it is critical that vapor barriers be continuous. This requires close attention to detailing, especially at lap joints and penetrations. Retrofitting vapor barriers to existing masonry facades (a popular wish in museums and galleries) involves issues that are very difficult to achieve. The continuity of the vapor barrier and risks associated with a failing vapor barrier present problems. Generally, altering the vapor profile through any existing wall is risky. A dew-point analysis should be carried out to determine the risk of interstitial condensation for any humidity-controlled environment, especially if it is being added to an existing facility.

Current masonry construction includes the use of brick facades on high-rise structural steel frame buildings in which the masonry skin is primarily an architectural design element. In this use, the masonry does not contribute to the structural load-bearing capacity of the wall and must be carried by the structural framing elements. However, in this type of construction, the outer brick still serves as the primary protective barrier to the penetration of wind and precipitation into the facility, and the same cavity wall design elements must be included in the system for the desired environmental protection. This construction type is used for speed of construction (the steel contractor doesn't wait for the masonry contractor) and interior finishes can be added later.

Another current masonry construction technique that has been developed is the use of single-wythe, load-bearing masonry facades. These facades typically use a single concrete-block wall with split-face exterior block, hollow-block or cast-masonry units (CMUs), and bond beams. The interior finish wall is mounted directly to the exterior wall with light-gauge studs. Insulation is placed on the interior concrete-block surface because the cavities of the block need to be grouted and reinforced with steel bars for structural capacity. This type of construction is economical but also more difficult to make watertight. In-wall flash-

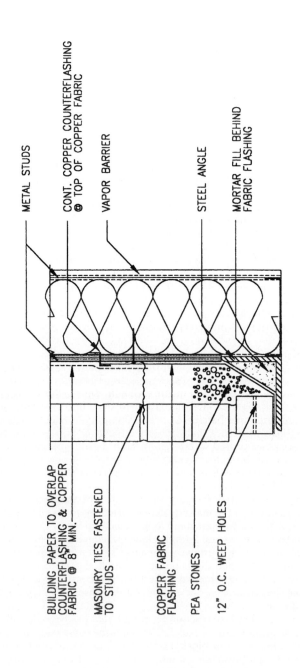

METAL STUDS

CONT. COPPER COUNTERFLASHING @ TOP OF COPPER FABRIC

VAPOR BARRIER

STEEL ANGLE

MORTAR FILL BEHIND FABRIC FLASHING

BUILDING PAPER TO OVERLAP COUNTERFLASHING & COPPER FABRIC @ 8" MIN.

MASONRY TIES FASTENED TO STUDS

COPPER FABRIC FLASHING

PEA STONES

12" O.C. WEEP HOLES

NOT TO SCALE

FIGURE 5.1.4-1 Typical cavity wall. (*Courtesy CID Associates, Inc. Boston.*)

ing detailing must include the use of split block and placement of flashing at all wall penetrations (Fig. 5.4.1-2).

Sealant is another element that is used in all current facade types. Sealant is used for a variety of purposes but primarily to form a weatherproofing seal for some aspect of the facade. In masonry wall construction, sealant may be used at locations of wall-penetrating elements, such as windows, louvers, and doors, and at expansion and control joints. Sealant in this case is in the form of what is generically referred to as *caulk* and is applied at joints. Facade expansion joints need to be properly aligned with building structural expansion joints. In addition to true structural expansion joints, the materials may require additional control joints to accommodate thermal expansion and contraction. Control-joint placement must be determined and installed to prevent splitting of the wall facade.

Liquid-applied masonry sealers are used to enhance the water repellency of outer masonry or concrete surfaces. These materials form a protective barrier either at the surface of the materials or within the matrix of the block, brick, and mortar. Sealant of this type is sometime viewed as a way to repair wall leaks. When using this type of sealant in conjunction with masonry, it is important to investigate the vapor transmission properties of the product. As mentioned earlier, vapor barriers should be placed on the warm side of the facade surface. Some liquid-applied sealants form a vapor barrier on the outer face of the masonry that restricts transmission of vapor out of the building. In cool weather, the trapped vapor condenses, and in freezing conditions, this liquid moisture changes to ice. Water expands in freezing, which places the surface of the masonry in tension. Masonry is a weak material in tension, and the end result is often spalling of the face of the affected unit. In most cases, it is advisable to use a penetrating sealant that allows vapor transmission while restricting the flow of liquid moisture. These materials are also very helpful when used with the previously mentioned single-wythe block wall systems. The architectural feature of a split face on the exterior is typically more absorbent and therefore more prone to allow water into the wall system. As noted, single-wythe construction does not allow a proper cavity to conduct drainage, which makes the system much more susceptible to moisture penetration.

When properly designed and constructed, masonry walls are extremely durable and have exceedingly long life spans with low maintenance requirements. Masonry pointing is one common maintenance need that is eventually required. The facility engineer needs to monitor the depth of the mortar joints and note the development of cracks between the mortar and brick or block. If cracks develop that are $\frac{1}{16}$ in or greater in width and the depth of the mortar joint erodes by $\frac{1}{4}$ in or more, it is time to consider pointing. This repair method requires that the joints be prepared by saw-cutting the existing mortar to a depth of $\frac{1}{2}$ to $\frac{3}{4}$ in beyond the outer face of the brick or block. The surface must also be surface saturated as with new work before applying and tooling new mortar. This process is an extensive undertaking and should include replacement of soft sealant joints. Monitoring of sealant joints is also necessary because these materials require replacement at shorter intervals than masonry pointing. The design of sealant joints is also critical to the long-term performance of the sealant material. The facility engineer should check the design limitations with the manufacturer of the particular sealant being used. In most instances, however, the width of the joint should not exceed twice the depth, and a bond-breaker tape or backer rod should be employed to prevent three-sided bonding of the sealant.

The metal wall panel is another common type of facade material construction that is prevalent today. Wall panels may be manufactured with insulated cores and light-gauge skins on either side. In common industrial buildings, this type of manufactured wall system may serve alone, or the panels may be combined with glass units to form a more complex curtain-wall system. In either instance, the metal wall panels are not structural and are mounted to the framing with a series of cleats and fasteners. These systems are optimized for weathertight integrity when the fasteners are concealed and not exposed to the weather directly. Metal-panel and curtain-wall systems depend on facade materials to provide protection from precipitation, airflow and vapor and thermal transmission. Curtain-wall systems typically employ the design of internal gutter and drainage channels to conduct water from the facade. Lock

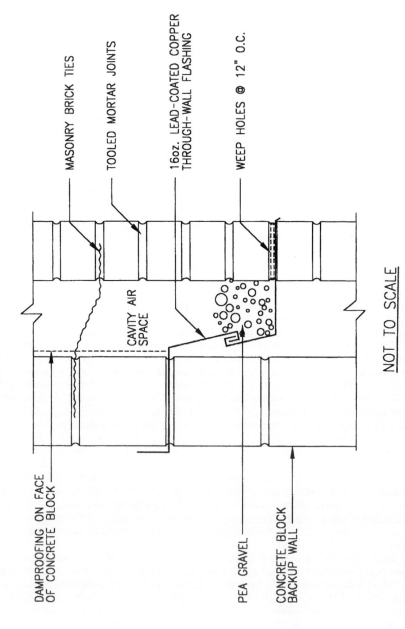

MASONRY BRICK TIES

TOOLED MORTAR JOINTS

16oz. LEAD-COATED COPPER THROUGH-WALL FLASHING

WEEP HOLES @ 12" O.C.

CAVITY AIR SPACE

DAMPROOFING ON FACE OF CONCRETE BLOCK

PEA GRAVEL

CONCRETE BLOCK BACKUP WALL

NOT TO SCALE

FIGURE 5.4.1-2 Typical through-wall flashing. (*Courtesy CID Associates, Inc., Boston.*)

joints and seals use nonskinning sealant that is placed within the joints in a permanently compressed state, rather than the exterior sealant joint design mentioned in the discussion of masonry facades. External sealant joints should not be used as the primary protection for a metal-panel or curtain-wall facade.

Metal wall panels come prefinished with anodized color or in mill-finish aluminum. Maintenance is typically limited to cleaning and the replacement of the sealant in exterior joints. Oxidized aluminum can be cleaned with specialty products that are compliant with volatile organic compound (VOC) regulations and are nonhazardous to the environment. Anodized panel, cleaning, if required, should be carefully considered and tested on an area of low visibility to ensure acceptable results before initiating a large-scale project.

These systems must also be designed to accommodate structural building expansion joints, movements due to wind loading, and thermal expansion and contraction. High-rise buildings experience significant deflection when subjected to wind loads, and curtain-wall systems have been designed to accept these conditions.

The exterior insulation finish system (EIFS) is another type of facade that is becoming more prevalent in construction. These systems are relatively new and do not have the proven track record of the systems discussed previously. EIFS systems are being used in retrofitting or refurbishing older existing buildings and also in new construction. When used in a refurbishment project, an EIFS facade system consists of placing expanded or extruded polystyrene insulation board over an existing exterior facade. The insulation boards are either mechanically attached to the existing facade substrate or bonded to the substrate with adhesives. As with the outer brick facade component of a masonry cavity-wall system, the EIFS face is not structural. Unlike masonry cavity walls and even metal and glass curtain-wall systems, the coatings and sealants placed on the outer surface of the expanded or extruded polystyrene insulation board comprises the primary means of weatherproofing. No internal backup drainage systems are included in the design of these facade walls.

After the new insulation substrate has been placed, it is finished with exterior stucco-like material and embedded reinforcement mesh. The finish material may be latex modified and applied in several coats, depending on the particular manufacturer's requirements. The reinforcing mesh can generally be specified to provide varying impact protection. A stronger reinforcing layer used in unison with a high-density extruded polystyrene insulation board will prove to be more durable than the standard EIFS system. Carefully evaluate the local environment to which your facade will be exposed when considering an EIFS facade refurbishment project.

The new EIFS facade must be made to accommodate the existing expansion and control joints, as well as all fenestration design elements (see Fig. 5.1.4-3). Failure to mirror the existing facade design elements will result in a breach and in ultimate failure of the EIFS facade. When considering this type of system to refurbish an existing facility, an in-depth evaluation of the performance implications of the system should be undertaken. Particular attention should be directed to the impact of the external insulation in determining the location of the dew-point temperature for a particular facility during winter conditions. Improperly applied EIFS systems that were not given full design consideration have been cited as the cause of structural wall failures. Relocation of the dew point has resulted in condensation within the existing wall systems that the EIFS facades were intended to refurbish. This condition is often undetected until failure of the wall is imminent. In addition, this condition has resulted in fungal and bacterial growth within wall systems, which has caused unhealthy interior environments. Design detail considerations should ensure proper backwrapping of the reinforcement materials at all perimeter and exposed locations. If not properly protected, the insulation substrate is also subject to attack by common insects.

These systems have been popular with facility owners and their engineers in recent years due to the relatively low cost and the initially dramatic face-lift effect upon completion. However, these systems do not have a long and proven history and are being viewed critically as they enter their second decade of popularity. If your facility currently has an EIFS facade, close visual inspection of the entire facade is encouraged twice a year. It is also advisable to

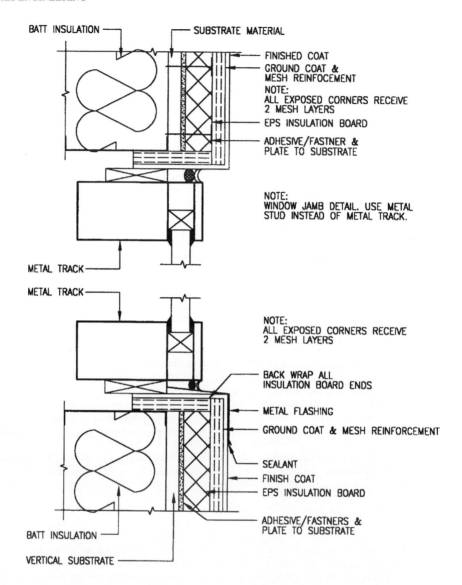

BATT INSULATION — | — SUBSTRATE MATERIAL

— FINISHED COAT
— GROUND COAT & MESH REINFOCEMENT
NOTE:
ALL EXPOSED CORNERS RECEIVE 2 MESH LAYERS
— EPS INSULATION BOARD
— ADHESIVE/FASTNER & PLATE TO SUBSTRATE

NOTE:
WINDOW JAMB DETAIL. USE METAL STUD INSTEAD OF METAL TRACK.

METAL TRACK

METAL TRACK —

NOTE:
ALL EXPOSED CORNERS RECEIVE 2 MESH LAYERS

— BACK WRAP ALL INSULATION BOARD ENDS
— METAL FLASHING
— GROUND COAT & MESH REINFORCEMENT
— SEALANT
— FINISH COAT
— EPS INSULATION BOARD
— ADHESIVE/FASTNERS & PLATE TO SUBSTRATE

BATT INSULATION —
VERTICAL SUBSTRATE —

NOT TO SCALE

FIGURE 5.1.4-3   Typical exterior insulation finish system (EIFS) window sill and head. (*Courtesy CID Associates, Inc., Boston.*)

conduct wall cuts through the interior to check the condition of the wall structure and insulation. If considering an EIFS facade as a retrofit of your facility or for new construction, be sure to have all design considerations reviewed before bidding and awarding a contract.

CONTRIBUTOR

David P. Richards, C.Eng., Ove Arup & Partners, New York, New York

ARTICLE 5.1.5
ROOFS

Harrison McCampbell, AIA
McCampbell & Associates, Brentwood, Tennessee

This article provides a mind-set to assist facility managers in the proper care and maintenance of their roofing systems, whether aging roofs or ones recently installed on new facilities. A cognitive perception and definition of the building's roofing needs is necessary when attempting to control the long-term costs related not only to the installation of the roofing system, but to its upkeep as the primary source of protection for the building's operations and occupants. Initially, analytical information about the facility should be gathered and discussed with as many of the involved parties as possible (operations, maintenance, accounting, etc.), followed by a review of the upkeep procedures currently in place and the necessary revisions to the structure of the same.

For the next phase, this article presents a general primer on roofing systems and associated terminology. This is provided as a foundation of a methodology for best applying particular roofing knowledge to a particular facility's requirements. As with any other type of operation, the degree of consistency, the quality of intent and knowledge, and the depth of persistence and tenacity determine the long-term value of participation in such a program. Once a structured line of command is created and implemented, subsequent managers will be able to more clearly understand their place on the team. This clearly defines areas of responsibility that reiterate what successful team efforts can accomplish. Roofing issues are usually one of the least desirable items to deal with when taking care of buildings. Most people do not know one roofing system from another, nor one roofing contractor from another. Too many building maintenance agendas are fixated on issues that demand and receive far more attention than roofing, yet cost much less. Roofing dollars, if spent on catastrophic conditions instead of their prevention, only increase with time. Proper analysis of problems and application of remedies provide dependable costing cycles.

ATTITUDE OF MANAGEMENT AND PLANNING

There are two primary factors that affect the efficiency of the roofing envelope, one of which begets the other. The definition of management's focus is paramount. Is this facility being cared for with profit-oriented, short-term goals in mind, or is the owner's primary outlook more toward a long-term continuity of maintenance? Certainly there is room for both views, but the primary intent of management has to be defined—is management's concern for time periods measured in 3-month quarters or in 5-year periods? Another consideration is the length of time the owner plans to keep the building.

Second is the issue of money. Once the pace and quality of care for the roofing system is defined, the financial decisions about how to sustain that particular course of action are much easier. Most costs related to roofing are derived from the direct costs of labor and materials needed to install and maintain a roofing system. But there are other costs brought about by the failure of the roofing system to keep water out. They are the indirect costs for ceiling tile, carpet cleaning or replacement, paint, computers, telephone switching equipment, and the like that may cost someone something, but are not listed under "roofing" at the end of the year.

There are a few other issues that could have an effect on the long-term care of a roof. One is whether the building is owned or leased. Find out who is responsible for its maintenance and also for its eventual replacement. Make a list of the scope of work performed and the types of materials used as an extra source of information tracking. If possible, ask that a water test be performed, following a contractor's repair attempt and prior to payment, to save further finger pointing if the work is not effective.

A roofing file should be kept to record all items of the project related to the maintenance of the roof. This file should be available to all through one source for control and monitoring. Warranties should be included in order to dictate how to handle leak calls. With the assistance of computers, automatic periodic notices should also be provided so that regular preventive maintenance can be implemented. This way the roof can function properly and the warranties can be kept intact. Any work done on any roof should be recorded on the roof plan, listed in some kind of report, and filed. Whether the entity responsible for roofing issues is in-house or not, it is imperative that files be updated on both existing conditions and available products used for repair. Knowledge of the installed system's maintenance requirements is necessary to keep the warranty intact. Not unlike insurance policies, many owners do not find out what their particular warranties cover until trouble arises and they attempt to enforce them.

EXISTING FACILITY CONDITIONS

It is important that you know what kind of roofing system you have, what you require of it, what your building's insurance carrier may require, and how best to take care of your system when considering the long-term aspects of what you want. While gathering the proper knowledge for the care of the roof itself, remember that the roof is part of a system, as are the insulation and structural deck. All are elements of the exterior building envelope that demand and deserve consideration.

One of the first courses of action is to document the existing roof conditions on all of the buildings within the company's operation. In general, you need to create a drawing or plan of each building that shows each separate roof area. This should include all fan curbs, vent pipes, roof drains, pitch pockets, dimensions, and slopes, with their locations on a key plan, if necessary. Any and all penetrations through the roof should be documented and located on the plan as close as possible to their actual positions on the roof. Along with the drawing, there should be a list of pertinent information, including the type and number of roofing systems, insulation, vapor retarder, structural deck, date of installation, information on both contractor and materials manufacturer, and any warranty information.

If a building owner has multiple buildings in different locations, is it imperative that care of the roofs be consistent. Having the technical ability to handle roofing problems is only part of the requirements for handling roofing issues of long-term concern. Education is necessary to understand and appreciate what is required for proper care of roofing. Knowing more than just roofing can only help provide the best care for a roof. Outside forces can act on a roofing system and, unless the source of the problem is addressed, it may be a waste of time and money to pursue a cure. For instance, many roof leaks have been attributed to negative air pressure in roof-mounted mechanical units—rain is literally sucked into the unit housing. The poor roofing contractor usually tires of coming out on calls to repair the roof, loses patience with the owner, and feels that compensation is in order. The owner, on the other hand, has paid for a roofing system, still has leaks, and now feels abandoned by the roofing contractor. Without proper analysis, this scenario can lead to ill will at best and lawsuits at worst.

A condition that is rarely thought about, but probably occurs on a daily basis, is that when roof insulation becomes wet from previous leaks, it slowly changes from an insulator (something that provides dead air spaces that slow down the rate of heat transfer) to a conductor of heat. This condition diminishes the heating and cooling efficiency of the building's mechanical systems.

HISTORY OF ROOFING

Roofing came about as shelters were built by early humans from overlapping leaves, bark, palms, and various natural objects. The shingle effect was found most effective and led to steeper slopes as a method of getting rid of water more efficiently. But as the need for bigger buildings grew, so did concern with building up height just to provide the shingle effect. As early as 1607, larger buildings requiring flatter roofs started using a combination of parchment paper and pine tar as an early version of what is known today as *built-up roofing.* In essence, the built-up roof is accomplished by multiple layering of a waterproofing substance (asphalt, coal tar, pine tar, etc.) with some type of reinforcing material. These are, or have been, materials made from asbestos, wood pulp, cotton rags, polyester, and fiberglass, among others. The waterproofing agent in built-up roofing is known generically as *bitumen,* and it comes in two similar forms from two different sources.

Asphalt, the most widely used type of bitumen, is composed of high-molecular-weight hydrocarbons, found in nature or obtained through petroleum processing. It is produced as a byproduct somewhat near the end of the process of manufacturing gasoline and heating oils. Sir Walter Raleigh discovered natural asphalt lakes in Trinidad and brought the substance back to England because it was effective for sealing boats against leakage.

Coal-tar pitch is obtained by the destructive distillation of coal. In nineteenth-century London, Samuel Warren began to use coal-tar pitch in built-up roofing and started roof contracting as a viable business. Pitch is a noncrystalline material, whose flow properties can "self-heal" small fissures that form during the service life of a roof. Coal-tar pitch is designated for slopes that do not exceed 2 percent. Coal tar can be bothersome, even caustic, to anyone in contact with it or its fumes, but coal tar tends to age better than asphalt because there are no oils to escape upon exposure to the elements.

Built-up roofing was the most widely used type of roofing for decades, but outside forces brought about changes that led to the multiple choices available today. By the early to mid-1970s, roofing selection was a matter of choosing from among 10 to 15 systems, ranging from gravel surfacing to a granular-surfaced cap sheet to a coated smooth-surface finish. The *1998 Roofing Materials Guide* published by the National Roofing Contractors Association listed around 170 manufacturers and more than 500 systems. So the choices of roofing systems and accessories have exploded during the last two decades, confusing even the most seasoned specifier.

The Middle East crisis in the 1970s led to still other changes in the roofing industry. One factor was the increase in the price of crude oil, which trickled down to higher asphalt prices, making asphalt-based built-up roofing more costly. The second factor was the higher cost of fuel oil and gasoline as energy concerns led to the creation of more thermally efficient buildings, which in turn led to a more competitive market in roof insulation. Politics also joined the fray, demanding that new buildings and remodeled old buildings meet certain standards of thermal efficiency. A new concern for the amounts and rates of air flow in buildings created new awareness of wall, window, and ceiling construction. Traditional insulating board, usually made from cane fiber, cork, or perlitic ore, was originally used more for a substrate under roofing material than for its insulating qualities. But after the oil crisis, roof insulation became more integral in the selection and eventual fate of most roofing systems.

As a general observation, it appears that a number of roofing problems began to rise with the increased use of roof insulation. The more the roof could work as an integral part of the building, the better it seemed to perform. But as the roofing membrane began to move further and further away from the roof deck, disparate factors began working against, instead of with, one another. Ancillary items such as fasteners, along with the use and positioning of vapor retarders, brought subtle complex concerns with their contributions to roofing systems.

For years, asbestos was used to manufacture roofing felts that provided both coal-tar pitch and asphalt-based built-up roofing with qualities of dimensional stability and resistance to ultraviolet rays, fire, and rot. However, health concerns regarding the mining and manufacturing of asbestos products overwhelmed the economic and performance advantages and virtually eliminated asbestos-related materials from the construction industry.

Changing trends in the textile industry also brought about changes in the roofing industry. As cotton and wool fibers were replaced in the 1970s with double-knit polyester, the resultant shrinking sales of cotton fabrics led to reduced availability of cotton rags, and the use of rag felts diminished, requiring an alternative for reinforcement plies. Until that time, cotton had been blended with wood fibers and waste paper—it absorbed asphalt well, provided good tensile and tear strength, and was a relatively stable material. Consequently, wood and waste were substituted for rag felts. Problems with moisture, stability, and fire and rot resistance forced the market to keep looking for a better product. Presently, fiberglass is the most widely used built-up roofing felt, and organic felt (asphalt-saturated No. 15) continues to be used as a shingle underlayment.

As the costs of built-up roofing rose in the 1970s, marketers of a relatively new generation of roofing systems began to concentrate on the American markets. Single-ply roofing had been used throughout Europe for decades, but because of higher costs of manufacturing and shipping to the United States, it could not compete with the U.S. built-up roofing market. The increasing cost of petroleum-dependent products, coupled with the demand for more thermally efficient roof systems, made built-up roofing a prime target for the rubber and plastic industries.

Single-ply manufacturers found an ally in expanded bead polystyrene (EPS) roof insulation. On the basis of dollars-per-R-value costing, EPS is probably still the most cost-effective roof insulation. Its use in a loose-laid, single-ply system allowed an attractive cost saving if the R value exceeded 12. A synthetic rubber of thermoelastic origin, ethylene propylenediene monomer (EPDM), was promoted to represent single-ply roofing, and a 0.045-in membrane was the thickness of choice.

Originally, a number of manufacturers made their own EPDM sheet goods, but because of economies of scale and mergers, now there are only two, Firestone Building Products and Carlisle SynTec Systems. The remaining suppliers rebrand the EPDM but may have their own accessories, such as adhesives, seam tapes, fasteners, roof vents, boots, and rubber flashing. Do not assume, nor fall prey to seemingly knowledgeable salespeople in believing, that accessories and components from one EPDM system are successfully interchangeable with another.

Thermoplastic roofing membranes, such as polyvinyl chloride (PVC), were introduced in the United States in the mid-1970s as competitors to EPDM. Neither sheet at that time generally used any type of reinforcement. The general idea of loose-laid, single-ply roofing was to provide a more flexible cover that could take incremental building movement considerably more easily than the typical built-up roof.

Ultimately, this proved to be a major factor in the demise of many loose-laid, single-ply roofing systems. Shrinkage at the perimeter of a considerable number of PVC roofs was the first sign that the unreinforced sheets were not as represented. The phenomenon was blamed on the migration of the plasticizers within the chemical structure of PVC that allow it to be flexible. Plasticizer migration also resulted from direct contact with EPS roof insulation.

Later on, slip sheets were introduced to minimize this problem. Currently, only reinforced PVC sheets are offered, and their plasticizer stability is greatly improved.

Built-up roofing sales began a decline in the early 1980s that persisted until the mid-1990s. This was mitigated somewhat when the desirable properties of both built-up roofing and single-ply roofing were provided in a type of roofing called *modified bitumen*. Modified bitumen also traces its early roots to Europe, where it originated in the late 1960s. Essentially, modified bitumen is the same type of waterproofing medium (recently made available with both asphalt and coal-tar pitch) that offers more durability and flexibility by the addition of polymers. Presently, modified bitumen comes in various premanufactured thicknesses, with a choice of reinforcements (usually polyester and/or fiberglass mats and scrim sheets) and surfacings.

So, as the latest market trends have oscillated back and forth between asphalt and non-asphalt roofing during the last two decades, the market is now generally divided into the three sectors of built-up, single-ply, and modified bitumen, which together comprise approximately 90 percent of the roofing market. Metal roofing and sprayed-on polyurethane foam comprise the remaining 10 percent. Metal roofing is probably the fastest-growing segment of roofing at this time, due in part to a small market share and an increase in the number of manufacturers, systems, and available applications, as well as its tie-in to the metal building industry. Depending upon the area of the country, each type of roofing may have a greater market share than another, but market shares are always changing.

ROOFING SYSTEMS

Built-up Asphalt Roofing

As discussed, built-up roofing is mainly an asphalt-oriented market, and coal-tar pitch is used as a secondary material The asphalt-based built-up roofing market can also be categorized into hot and cold applications. Hot, of course, is the most widely known and used. Asphalt comes from the processor in a solid form known as *kegs* or in heated tankers. Propane-fired kettles heat the bitumen to temperatures from 400 to 500°F. The asphalt is applied by hand mopping or mechanical spreaders onto the roof deck and/or roof insulation and between the roofing felts. The covering layer is a flood coat of asphalt and small gravel evenly distributed across the roof, with a mopped-on granular-surfaced cap sheet or a coating of some type that is usually finished in aluminum or a light color to reflect the sun's rays. All of them act in the same capacity, to protect the asphalt from the effects of the elements. Heat and the direct rays of the sun tend to bake the oils out of asphalt and render it brittle and hard.

Another form of asphalt-based built-up roofing is called *cold-process* roofing. An application of an emulsion (a dispersion of asphalt in water) is reinforced with polyester fabric or a solvent-based carrier. The bitumen is usually modified with fibers and fillers, resulting in a somewhat flexible asphalt covering that is easy to monitor for maintenance and repair.

Built-up Coal-Tar Pitch Roofing

Coal-tar pitch provides excellent water-repellent elements that are used for low-slope roofs. The self-healing characteristics of pitch make it unique and desirable, but many people object to the noxious fumes when applying it hot. Because it flows so readily, many issues must be addressed when using coal-tar pitch, such as pitch guards around roof drains and raised gravel guards (to keep pitch out of the drainage system and off building facades), installing a vapor barrier (to keep pitch from dripping down onto building occupants through the roof deck), and making proper repairs and tie-ins (pitch and asphalt are not compatible when mixed). Coal-tar pitch roof systems are surfaced with gravel or slag.

Modified Bitumen Roofing

Modified bitumen is a type of roofing that provides some of the desired elements of both single-ply roofing (premanufactured thicknesses) and built-up roofing (layers of asphalt and reinforcement). It is, essentially, a redundant system, and some of the flaws attributable to labor are diminished, if not removed. The roofing industry's goal is to finally provide a foolproof system that can be installed without problems or that never needs maintenance, or both. Some of the concerns with modified bitumen roofs, as with any other new product on the market, are such issues as who makes the modifier, and how much is enough; how one knows which products are equivalent when going out for bids; and what standards, if any, for determining product integrity are available.

Some designers and contractors bring two worlds together and call for *hybrid* systems that use layers of built-up roofing in which the top surface is a modified cap sheet with granular surfacing instead of the usual flood coat and gravel. Modified caps have a more uniform surface, have an extra layer in the reinforcement, and offer more flexibility with polymer-modified versus straight-run asphalt.

The type of reinforcement used in the modified sheet depends on factors such as whether the system is heat-applied or mopped on (sheet stability), whether there is heavy pedestrian rooftop traffic (puncture resistance), or whether certain insurance requirements have to be met (fire ratings) (see Fig. 5.1.5-1).

Styrene-Butadiene-Styrene.   Styrene-butadiene-styrene (SBS) is an elastomeric polymer additive. This roofing membrane is usually installed in an application of solid-mopped asphalt. Some cities are starting to refuse to allow open kettles of asphalt or their residual fumes, so that cold-adhesive applications are becoming more popular in the larger metropoli-

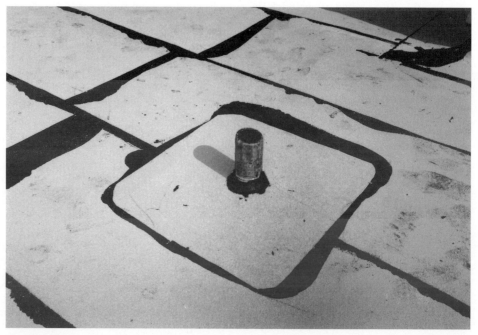

FIGURE 5.1.5-1   Vent pipe properly flashed with modified bitumen roofing. (*Courtesy of McCampbell & Associates, AIA.*)

tan areas. Heat-applied methods using open-flame torches have been used in Europe for years, but torching SBS requires a trained and certified torch operator. The melt point of SBS is considerably lower than that of atactic polypropylene, and one can burn through to the reinforcement quite easily if not diligent in using the open-flame method.

Atactic Polypropylene. Atactic polypropylene (APP) is a thermoplastic polymer that is used to modify asphalt. APP is somewhat stiffer than SBS and can tolerate higher temperatures. Hot mopping with asphalt is not possible because of the high melt point required by APP. Open-flame torching is the most popular method of application, although APP systems are now used in cold-adhesive applications where seaming is done with hot-air guns for safety.

Cold-Process Roofing

Cold-process roofing can mean different types of roofing in different parts of the country. In the middle and western United States, it may refer to multiple layers of emulsion and polyester fabric, finished with a coating that provides a smooth surface. In the southern and eastern regions, *cold process* may simply refer to the contrast with hot or asphalt-based roofing. Cold process here probably means using cold adhesives to secure the roof instead of mopping with hot asphalt or coal-tar pitch. More and more insurance companies are offering lower rates to contractors who do not use propane-fired kettles and open-flame torching methods. Workers' lost-day claims and the resultant cost to the construction market are certainly an important concern nationwide. Insurance companies are constantly looking for ways to diminish payments for injuries and property loss.

Sprayed-On Foam Roofing

Known generically as sprayed-on polyurethane foam (SPF), this type of roofing is mixed on the job site right at the point of application and is applied directly over either an existing roofing system or a new deck. Its sales niche is primarily in the reroofing market. SPF has been used for years in industrial, commercial, and institutional installations, and residential sales are generally centered in the Sun Belt states. It offers ease of application with little preparation, high insulating values, light weight, and various color finishes. There are, however, critical issues that have to be attended to when analyzing the construction conditions and application methods for SPF roofing.

As an example, improper analysis by an overly energetic SPF salesperson, as with any other system, could lead an owner to believe that an SPF system can be successfully installed directly over a water-saturated built-up roof by installing roof vents to release any trapped moisture. The deleterious effects of this scenario would probably go unnoticed for some time but eventually could cost the owner many times the amount "saved" by not removing the original roof. And the generally warm, fuzzy feeling that comes with the warranty would probably vanish as soon as a close reading reveals that the terms and conditions disclaim any incorrect technical information—even when given by an authorized representative or applicator.

Metal Roofing

For decades, metal roofing has enjoyed a small share of the American roofing market. Improved seaming techniques and coating technologies have driven some companies to offer long-term warranties on what were previously thought to be inadequate slopes. Nowadays, one can obtain a 20-year warranty on a metal building installation, either a new building or a reroofing project, that provides slopes as shallow as ¼ in/ft. This, again, calls for increased knowledge and concern when making decisions based on economic parameters. Short-term installation costs may be overshadowed by bothersome long-term maintenance and performance results. The same warning has to be applied for an installation over an existing low-slope roofing system that contains moisture. Gambling that the moisture can escape before any resultant damage is done can be costly to the owner who is not fully aware of the downside of such a decision. And the failure to design for conditions such as the possible change in the dew-point temperature and provide proper ventilation between the metal roof and the existing roof can lead to untold problems in a few years.

Metal roofing certainly has its place in building construction. Low maintenance, various styles and colors, light-weight construction, and excellent fire and wind ratings are all positive factors to consider when selecting a roofing system for a new building or providing a solution for the leaks of an older one.

ROOFING COMPONENTS

The many integral parts of a roofing system must work in concert with one another if a roof is expected to work as a full system. No one component is more important than another, although one component's failure could possibly allow more water entry than that of another of its counterparts. The vast majority of leaks in a building are found at the edges and penetrations. The edge conditions usually include gravel stops and parapet walls around a building's perimeter. Penetrations may be considered base flashings on curbs and penthouse walls, vent pipes, roof drains, pitch pockets, lightning rods, and/or window-washing stanchions.

If a contractor is not directed as to how an edge or penetration is to be flashed (sealed properly), the success of the flashing installation may rest in the hands of an unskilled laborer who is not knowledgeable in the specific requirements of a particular roofing system. Good-quality roofing documents (both specifications and drawings) are a critical ingredient in the success of any building's first line of defense against the entry of water.

Metal Edging

The two most popular elements for terminating perimeter metals are a gravel stop and a drip edge. The difference between the two is simple. The gravel-stop metal has a slight rise (usually ¼ to ½ in in height) that restricts the gravel from falling over the edge of the building. It also provides a clean line along the top edge of the building when viewed from the ground. Drip-edge metal is usually found with single-ply, modified, or built-up smooth-surfaced systems.

A drip, or hem, should be provided along the bottom face of the metal edging no matter how deep the face. Keeping the exposed metal edge fascia's profile down will lessen the chance of "oil-canning" or undulations in the planar surface. Cleats should be used on almost any metal edging. Intermittent clips, preferably made of a heavier gauge, can be installed to provide some uplift protection. Fascias on buildings located in areas of high winds should have continuous cleats to afford maximum protection. Continuous caulking along the interface of the drip edge and the exterior wall may help prevent sudden gusts from getting under the face of the metal and causing bowing or blow-offs.

Parapet Walls

Walls that encompass a building's roof areas are called *parapet walls*. There are usually two ways to seal a parapet wall. One is to extend the base flashing up the wall to a counter flashing or to take the material up and over the wall, finishing with a coping. Parapet walls must remain open on one side so that they can "breathe." Sealing in both sides of a parapet wall could seal in moisture if water somehow gets in along the top and causes hidden moisture problems.

Coping

A cap that sits atop a parapet wall is called a *coping* and is usually made of metal or stone. Some of the older buildings had terra cotta tile, but its use is practically obsolete. Coping is all too often installed with poor workmanship and little planning. The most common conditions for failure are installing the coping on a dead-level wall fastening through the top. This results in ponding water conditions around the lowest level of the coping (where the hammer impacts the metal), and bowing of the metal prevents any positive flow over either of the edges. Laps are rarely more than 1 in, and sealing is paltry at best. Most caulking is pulled apart as the metal expands and contracts.

Ideally, coping should be installed with the wall flashing continued from below and turned down at least 1 in over the outside face of the wall (Fig. 5.1.5-2). All coping should be sloped at least ½ in toward the inside to drain back onto the roof and diminish staining the outside of the building. Cleats, either intermittent or continuous depending upon wind requirements, should secure the outer face, and the inside face should be fastened through slotted holes on approximately 2-ft centers using hex-head fasteners with rubber washers. This allows for future removal for ease of wall flashing maintenance and replacement. Seam or lap techniques in the coping are almost cosmetic if the wall flashing is continuous below the coping. Otherwise, the laps are the weak link in the chain and must be designed and installed properly. Various methods are lap and caulk, flat-lock seam, and standing seam, among others.

FIGURE 5.1.5-2 Sheet-metal coping being installed atop a parapet wall with a single-ply rubber roofing membrane extending up and over the top of the wall for added protection. (*Courtesy of McCampbell & Associates, AIA.*)

Flashing

Any time a penetration is made through the roofing membrane or system, the opening in the roof must be able to afford protection against water entry, and the seal also must be able to withstand movement between the rooftop element to which it is attached and the roofing system itself. Movement is caused by the differential rates of expansion and contraction between various materials and also by the cyclical movement from hourly, daily, and seasonal changes brought about by exposure to the elements.

When considering what kind of roofing system belongs on a particular building, one must take into account the factors both below and above the roof. Manufacturing, for instance, may involve heavy vibration from machinery in the building as well as from some of the equipment installed on the roof. Air conditioning units are renowned for their cyclical vibratory demands on roofing systems. Compressors emit oils occasionally. Kitchen exhaust fans are known as one of the biggest enemies of roofing because they emit various types of animal fats and greases. When coupled with a lack of proper drainage and poor maintenance, the useful average life cycle of kitchen roofs is probably one of the shortest in the roofing industry.

Mixing flashing between systems is probably a sizable gamble unless considerable attention is paid to all of the appropriate elements of the particular building's demands. The issue is how the flashing is held in place—how the flashing is adhered to the curb and how the top of the flashing is mechanically fastened to secure it against sliding. Proper overlapping of the flashing onto the roof and onto the adjoining sheet is imperative.

Flashing generally must be somewhat heavier than the type of membrane it is sealing. This requires installation techniques specific to a particular manufacturer's sheet flashing membrane. The proper installation and maintenance of all flashing has to be done in strict accordance with the individual manufacturer's instructions. Unskilled laborers with buckets of roofing cement may do more long-term damage to the roof than the original problem.

Counterflashing

Counterflashing is as its name implies. You seal a rooftop element with flashing, and then you seal the top of the flashing with counterflashing. Most likely, the top of the flashing cannot remain sealed day in and day out. Galvanized sheet metal is probably the most widely used, but other materials, such as copper, lead-coated copper, and aluminum, are used for various reasons, including cost, compatibility, ease of installation, warranty requirements, longevity, and availability, among others. Counterflashing is usually surface-mounted, saw-cut into masonry or brick, or installed with the original masonry in a through-wall condition.

Some may try to save money by making all of their counterflashing materials in one piece. If the flashing is bent up to work on it or to remove the existing flashing for replacement, it almost never bends back down to its original position. And installing a fastener through the face is not recommended because it increases chance of leakage. Along with that, the metal is most likely now seriously bowed and looks as if it had been beaten with a hammer. (Actually, it probably was!)

The ideal remedy is to provide and install a two-piece counterflashing that allows removing the main counterflashing and reinstalling it. The initial incremental cost is far outweighed by the cost savings for fully replacing or jeopardizing the integrity of a one-piece counterflashing by turning it up and down again.

One of the worst mistakes that can occur in reroofing a building is terminating the new roofing system's edge flashing somewhere over the line of the original through-wall counterflashing. Many times extra insulation is added with tapered board to provide a more positive slope. But if this is done and the existing through-wall flashing in a higher wall, especially one with weep holes, tries to exit water encased in the wall by absorption, it will send the water out of the wall underneath the flashing of the new system.

Crickets and Saddles

Providing positive slope for rooftop drainage is one of the primary rules of good roofing practice. Many times this is ignored behind certain roofing penetrations. Mechanical units, fan curbs, skylights, and other elements wider than 18 in too often block proper water flow. This allows water to stand on the base flashing and can cause premature aging of some types of roofing materials. A *cricket* is a triangular raised area of the roof that provides positive slope in a reverse direction for the flow going toward the drain. Roof drains should be installed away from walls, and crickets are used between them and the walls to kick the water back from the walls toward the drain. A *saddle* is simply two crickets installed side by side as one unit that is centered in a valley between drains.

Expansion Joints

These are required in most roofing conditions, but many factors must be considered in proper design, placement, and installation. One of the most important places for an expansion joint is at a change in the direction of metal decking. The flutes of a metal deck expand and contract in a single axial direction. When different roof areas come together, there is most likely a need for a break in the decking system. This break has to be continued up through a relatively fixed-in-place roofing system unless it is a loose-laid or a mechanically fastened single-ply roof. Only these systems provide ample movement capabilities in their composition.

Most expansion joints are built up of vertical wood members with a bellows-type expansion joint sitting atop the nailers. Frequently, the rubber membrane that surfaces the bellows comes only in 10-ft lengths, and end laps may require periodic repair. Often the failure of expansion joints is the result of either poor installation or maintenance of the laps in this rubber membrane. Unreinforced asphalt-based roofing cement will not last more than a few weeks. And most of the caulking found on the market, especially the inexpensive types, will not properly adhere to the finished rubber bellows material.

Drains and Scuppers

Positive drainage is certainly the most critical aspect of good roofing (Fig. 5.1.5-3). At a minimum, drains must meet certain sizing requirements matched to the area of the country, the size of the immediate roof area being drained, and the design intensity of the rainfall, for anywhere from 10- to 100-year rains. Overflow requirements must also be met to prevent overloading the structure. If 4 or more inches of water are allowed to collect, the added weight could exceed the live load capacity of the structural system.

One of the most popular methods of auxiliary drainage is to provide through-wall scuppers near the primary drains. These must almost always be installed no more than 2 in above the roof drainage level so that no more than 10 lb/ft^2 of water ever builds up in a worst-case scenario. The corners of the scupper, especially on the low side, are suspect to possible water entry due to the *x-y-z* direction of the pulling forces that compose a boxlike element. Unreinforced asphalt-based roofing materials are probably the most liable to fail. Some of the modified bitumen systems, along with most of the single-ply systems, may allow more differential movement, but all of these still rely on proper installation techniques and maintenance.

FIGURE 5.1.5-3   Roof drain shown with adjacent overflow drain on separate piping system. Both have cast-iron strainer baskets for long-term performance. (*Courtesy of McCampbell & Associates, AIA*).

CONTRIBUTORS

Richard A. Fricklas, Englewood, Colorado

CODES AND STANDARDS

Building Officials and Code Administrators International (BOCA): (708) 799-2300; www.bocai.org.
Factory Mutual Research Corporation (FMRC or FM): (617) 762-4300; www.factorymutual.com.
Southern Standard Building Code (SBCCI): (800) 264-0769; www.sbcci.org.
Underwriters Laboratories (UL): (847) 272-8800; www.ul.com.
Uniform Building Code (UBC): (800) 284-4406; www.icbo.org.

ORGANIZATIONS

American Society of Civil Engineers (ASCE): (800) 528-2723; www.asce.org.
Asphalt Roofing Manufacturers Association (ARMA): (301) 348-2002; www.aisc.org.
Construction Specifications Institute (CSI): (703) 684-0300; www.csinet.org.
National Institute of Standards and Technology (NBS): (301) 975-6719; www.nbs.com.
National Research Council of Canada (NRCC) Institute for Research and Construction: (613) 993-9714; www.cistilnrc.ca/irc/roofing.

National Roofing Contractors Association (NRCA): (800) 323-9545; www.nrca.net.

Oak Ridge National Laboratory (ORNL): (423) 576-5454; www.ornl.com.

Roof Consultants Institute (RCI): (800) 828-1902; www.rci-online.org.

Roofing Industry Educational Institute (RIEI): (303) 790-7200; www.riei.org.

Sheet Metal and Air Conditioning Contractors Association (SMACNA): (703) 803-2980; www.smacna .org.

Single Ply Roofing Institute (SPRI): (617) 237-7879; www.spri.org.

ARTICLE 5.1.6
ELEVATORS, ESCALATORS, AND MOVING WALKS

James W. Fortune, M.B.A.
Lerch, Bates & Associates, Inc., Littleton, Colorado

Elevators are commonly classified by their functions and the types of people and goods that they transport. Passenger elevators, the most commonly encountered, are used to transport ambulatory and handicapped (wheelchair-bound) people (passengers) vertically between floors. Service elevators and freight elevators are generally used to transport specialized support people, goods deliveries, and bulk freight items and their handlers (Fig. 5.1.6-1). Most elevator codes do not differentiate between passenger elevators and service elevators because they are commonly equipped with sliding side doors (generally center-opening on passenger elevators and side-opening on service elevators), and the same net inside platform area requirements are used to determine their capacities (about 90 to 100 lb/ft^2). Passenger and service elevators in the United States have suspension components and platforms that are rated for 125 percent of load (capacity), so it is virtually impossible to overload them with people (Fig. 5.1.6-2). Passenger elevator platforms are typically shaped as wide, shallow units (a three-person depth is ideal) in order to move passengers rapidly in and out. Service elevator platforms are typically provided as narrow, deep units in order to accommodate carts, dollies, construction materials, furniture, ambulance stretchers, patient gurneys, and patient beds. Service elevators are often called *hospital shaped* because their inside platform depths and the maximum length items they can accommodate are often their critical design components.

Freight elevators are usually equipped with biparting doors and lift-up car gates and can have platform sizes that are designed to accommodate everything from palletized freight on hand trucks to items transported on fork lifts, and even full-size semitrailer trucks. The inside areas, capacities, and loading types of freight elevators are typically set by the type of goods or freight loadings that are to be moved on the platform. Typical freight capacities start at 4000 lb and can go up to 10,000-, 15,000-, 20,000-, or even 60,000-lb capacities. The use of freight elevators is restricted by code to an operator and a freight handler in most cases. Under certain conditions, local jurisdictions may waive this rule and permit freight elevators to transport employees.

Since the adoption of the Americans with Disabilities Act (ADA) in 1988, the most commonly used passenger elevator platform sizes that permit the mandated three-point turn for a wheelchair, when equipped with center-opening doors, are 3000, 3500, and 4000 lb. A 2100 to 2500 lb passenger elevator platform meets the ADA requirements with side-opening doors. However, in most cases where groups of passenger elevators are provided, rapid passenger

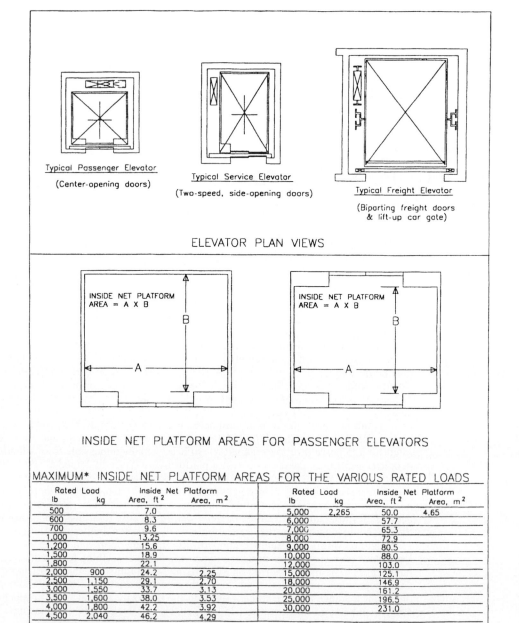

FIGURE 5.1.6-1   Elevator loading areas. (*Courtesy of Lerch, Bates & Associates, Inc.*)

5000 lbs. @ 1600 f.p.m., 48 ft² NET INSIDE AREA

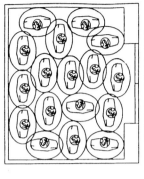

NOMINAL, MIXED SEX
CAB DENSITY = 2.5 FT²/PERSON
TOTAL CAR LOAD = 17 PERSONS / CAB

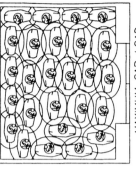

MAXIMUM CAR LOAD
CAB DENSITY = 1.6 FT²/PERSON
TOTAL CAR LOAD = 26 PERSONS / CAB

NOMINAL, NO TOUCH LOADING
CAB DENSITY = 3.0 FT²/PERSON
TOTAL CAR LOAD = 14 PERSONS / CAB

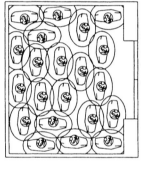

NOMINAL, SINGLE SEX
CAB DENSITY = 2.0 FT²/PERSON
TOTAL CAR LOAD = 21 PERSONS / CAB

TYPICAL BODY ELLIPSE

24" SHOULDER BREADTH
18" BODY DEPTH
= 3ft² / PERSON

TYPICAL MALE BODY

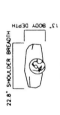

22.8" SHOULDER BREADTH
13" BODY DEPTH
= 2ft² / PERSON

7'-6" PLATFORM

47.8 ft²
INSIDE NET
AREA

7'-2" CAR INSIDE

6'-8" CAR
INSIDE

4'-0" C.O.

7'-6" PLATFORM

FIGURE 5.1.6-2 Typical people loading. (*Courtesy of Lerch, Bates & Associates, Inc.*)

movement in and out of the elevator cars is crucial, so they are almost always equipped with center-opening doors.

A high-rise building (typically 4 or more stories with heights above 75 ft) usually must be equipped with one or more elevators that can accommodate an ambulance stretcher in the horizontal position (across the car threshold). The stretcher sizes vary by the building code requirements but are generally about 24 in wide by 76 in deep. Some jurisdictions require an "ambulance stretcher" elevator in buildings of three or more floors.

ELEVATOR MOTIVE MEANS

Two motive means are used to move elevators that run vertically in their own guide rails within a hoistway: traction and hydraulics (Fig. 5.1.6-3). *Traction elevators* use hoist ropes (wire cables) that go over a reversible electric drive motor sheave (pulley) and are connected to the elevator car and a counterweight that runs alongside or behind the car in its own set of guide rails. The elevator's direction of travel, speed, and acceleration are controlled by the drive motor.

Hydraulic elevators use pressurized oil obtained from a self-contained pump unit as a force medium to move a hydraulic cylinder (ram) in a linear motion. The elevator car is directly or indirectly coupled to the hydraulic cylinder plunger, which moves the elevator up and down in the hoistway.

TRACTION ELEVATORS

When the electric hoist motor turns the drive sheave (pulley), it converts the rotary motion of the drive motor to linear motion through the hoist ropes to move the elevator and counterweight vertically in the hoistway. When a traction elevator car is at its lowest terminal floor, the counterweight is positioned near the very top of the hoistway. The counterweight is the same weight as the empty car plus 40 percent of the capacity load. The counterweights move in opposite directions as they travel through the hoistway. The use of a counterweight makes traction elevators very efficient (above 90 percent) because the hoist motor need be only large enough to overcome system friction and move (lift) the maximum rated car load.

Modern traction elevators are available with two types of hoist machines: geared traction units and gearless traction units. *Geared traction hoist machines* use an electric drive motor coupled to a worm and gearbox (can also be helical or planetary gears) to impart rotary motion to the drive sheave. Geared elevators are used in medium-height buildings with up to 15 stories and typically have speeds that range from 200 to 500 ft/min. The drive sheave for a *gearless traction hoist machine* is mounted directly to the machine motor armature. Thus the term *gearless* is used. Gearless elevators are generally used in high-rise buildings (15 to 125 stories) and employ speeds of 500 ft/min. The world's fastest units travel at 3280 ft/min. Gearless traction elevators require relatively slow-speed, high-torque drive motors that can be very precisely controlled for rapid runs between floors, express travel at up to the elevator contract speed, and accurate floor stops. Elevator speed is used to partially overcome travel distance established and express run requirements. For instance, a building of 12 to 15 stories might require elevator speeds of 350 to 500 ft/min; a 30-story building might use 500 to 800 ft/min; a 45-story building might have elevators at 1000 to 1200 ft/min; and a 60-story building might employ speeds of 1400 to 1800 ft/min. Very high speed elevators from 1800 to more than 3000 ft/min might be used as sky lobby shuttles and to service upper-level observation decks and restaurants. Virtually all geared and gearless elevators accelerate and decelerate at the same rate (about 3.0 to 4.0 ft/s^2), so it simply takes longer for the faster elevators to get up

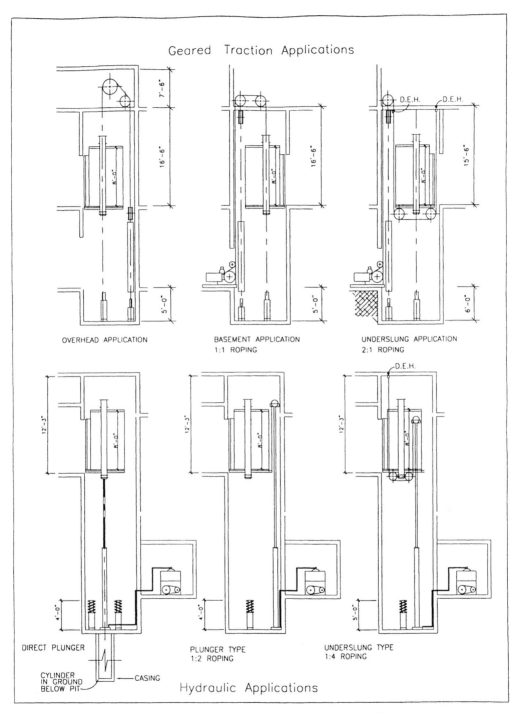

FIGURE 5.1.6-3 Elevator motive means. (*Courtesy of Lerch, Bates & Associates, Inc.*)

to and slow down from their contract speeds. On a typical one-floor run, traction elevators might achieve a speed of only about 350 ft/min, regardless of their contract speeds. Passenger elevator stops in high-rise office buildings are typically divided into multiple elevator groups and zoned so that there are no more than 6 to 8 elevators per group and no more than 12 to 15 stops per zone.

HYDRAULIC ELEVATORS

Hydraulic elevators employ ac induction motors to drive positive displacement pumps. The pressurized oil is used to actuate the hydraulic cylinder that is connected to the elevator car. The elevator acceleration/deceleration values and travel direction are controlled by a series of solenoid-operated mechanical valves. Moving the elevator in the up direction requires activating the pump motor to drive the elevator cylinder plunger. The elevator is lowered by the controlled application of gravity when the lowering valve opens to permit the elevator to travel down while the oil returns to the pump unit reservoir. Hydraulic cylinders can be located directly under the elevator (direct plunger), alongside the elevator and directly coupled to the elevator (direct offset application, also called *holeless*), or alongside the elevator and coupled to it by a series of ropes (1:2 or 2:4 roped hydraulic). Hydraulic elevators typically employ speeds of 100, 125, 150, 175, or 200 ft/min and take about one-third more time than a traction elevator to travel between floors. Hydraulic direct offset applications (single or double cylinders) are generally limited in capacity and speeds and to travels of about 25 to 30 ft and 3 stops. Direct plunger applications have travel limits up to 50 to 60 ft in 5 to 6 stops and are generally limited to 150 ft/min. Roped hydraulics can have up to 10 stops and about 100 ft of travel and can travel at speeds up to 200 ft/min. Because hydraulic elevators normally do not have counterweights, they are not nearly as efficient as comparable traction units and generally require 45- to 75-hp motors to move a given load. However, hydraulic units are relatively inexpensive to purchase, install, and maintain. They become a logical choice for specialized applications such as in suburban office buildings of 3 to 5 stories, many automobile parking applications, retail mall environments, low-rise hotels, low-rise scenic glass-cab designs, and for very large, heavy-duty freight elevators.

Because of environmental concerns about ground pollution that can be caused by leaking direct-plunger cylinders, many states have now passed very stringent buried jack cylinder protection requirements that require auxiliary casings and leakage monitoring requirements. In the future, direct-plunger buried jack cylinders might be severely restricted, and the use of roped hydraulic applications will probably increase in the United States because of their flexibility and extended duty capabilities.

ELEVATOR DOOR TYPES

Virtually all passenger and service elevators are equipped with door panels that open to the side (Fig. 5.1.6-4). Because their platforms are wide and shallow, most passenger elevators are equipped with center-opening (CO) doors, each of whose panels has to travel only one-half of the clear opening width. Center-opening doors speed up the elevator trip because they open and close very rapidly and allow for efficient passenger loading. Service elevators are typically equipped with side-opening (SO) doors because the elevator platforms are usually narrow and deep to accommodate long objects and wheeled carts and dollies. In this case, the major concern is to equip them with wide openings to facilitate the transport of large objects, palletized materials, and carts. For wider openings, both center-opening and side-opening doors can be provided as two-speed units (2SCO or 2SSO).

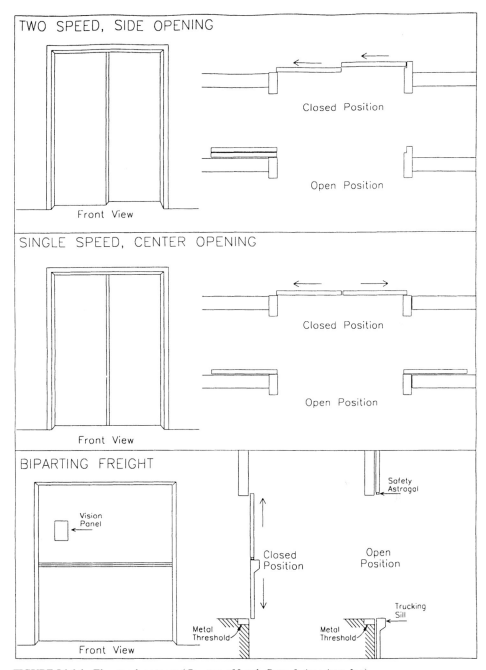

FIGURE 5.1.6-4 Elevator door types. (*Courtesy of Lerch, Bates & Associates, Inc.*)

Passenger and service elevators are equipped with master door operators that are located on top of the elevator cab. When the elevator is approaching a landing, the motion of the master door operator opens the car doors and unlocks the hoistway door interlock (an electromechanical locking device). At the landing, the car doors are physically coupled to the hoistway doors through a mechanical vane and drive-block system. This master door operator arrangement permits very smooth, rapid car and hoistway door openings with quick reversals and quiet checking.

Master elevator door operators are rated by their application and opening speed:

Operator designation	Opening speed, ft/s	Typical applications
Slow speed, light duty	1.0–1.5	Hydraulic elevators
Medium speed, medium duty	2.0–2.5	Geared elevators
High speed, heavy duty	2.5–3.0	Gearless elevators

Automatic elevator door closing speed is limited by the ASME A17.1 Elevator Code to the kinetic energy of the total door system (7 ft · lbf with a reopening device), which typically limits the door closing speed to no more than 1.0 ft/s. Table 5.1.6-1 lists some of the standard-width door times with 7-ft-high panels.

Freight elevators are typically equipped with manually operated (usually no more than 6 ft wide) or power operated vertically sliding, biparting hoistway doors and lift-up car gates. The upper and lower hoistway door widths and heights are determined by the sizes of the openings required, and the freight elevator car side panels are connected with chains or cables, so that they counterbalance one another when opening and closing. When open, the hoistway doors provide unobstructed access to the complete car width and height and can easily accommodate forklifts for moving palletized materials on and off the elevator. When the door is open, the bottom of the lower door panel acts as a trucking sill, filling the gap between the floor threshold and the elevator car sill. For power operation, each biparting door unit is equipped with an individual door operator located inside the elevator hoistway. Similarly, each vertically sliding car gate is equipped with a car-top-mounted gate operator. The ASME A17.1 Elevator Code limits the average hoistway door closing speed to 1.0 ft/s and the gate closing speed to 2.0 ft/s.

Power-operated biparting doors and lift-up car gate closings are normally controlled by constant-pressure pushbuttons located in the car and at each hoistway entrance. Automatic power closing of the doors and gates is permitted only if preceded by a 5-s audible warning signal and sequential closing of the car gate and hoistway doors. With automatic power-operated sequential operation, the hoistway doors must be at least two-thirds open before the car gate starts to open. In closing, the car gate must be two-thirds closed before the hoistway doors can start to close.

TABLE 5.1.6-1   Standard-Width Door Times with 7-ft Panels

Door width	Door type	Door times, s Opening speed, ft/s 1.0	2.0	3.0	Closing speed 1.0 ft/s
3.5 ft	CO or 2SCO	2.4	1.6	1.4	2.4
	SO or 2SSO	4.0	2.4	1.9	4.0
4 ft	CO or 2SCO	2.7	1.8	1.5	2.7
	SO or 2SSO	4.5	2.7	2.1	4.6

ELEVATOR APPLICATIONS—THE PLANNING AND PROGRAMMING PROCESS

The suggested elevator design criteria and required vertical transportation equipment applications vary by the type of facility, the expected building population, and the on-site passenger arrival and departure process. The primary elevator design terms are calculated for peak periods and are expressed as: (1) the frequency of departures of elevators from a main loading terminal, called *average interval;* and (2) the number of persons transported on the elevators during the same peak period, termed *group handling capacity.* To be meaningful, the average interval, expressed in seconds, and the group handling capacity, expressed as the number of persons or as a percentage of the zone population moved, must be measured over the same time period (usually 5 min of peak traffic). Residential, hotel, office building, and medical building traffic differ greatly from each other and must be evaluated accordingly. Vertical and horizontal transportation devices are installed in many different types of facilities. These are some of the major applications, along with the types of vertical and horizontal transportation equipment typically found in them:

Vertical and Horizontal Transportation Equipment

1. Passenger elevators
2. Service (firefighting) elevators
3. Freight elevators
4. Escalators
5. Moving walks
6. Dumbwaiters or inject/eject cart lifts

Facility Type

- Office building: 1, 2, 3, 4
- Hotel: 1, 2, 3, 4
- Hospital: 1, 2, 3, 4, 6
- Residential apartment/condominium: 1, 2
- Judicial/courts/jail: 1, 2, 3, 4
- Retail/shopping mall: 1, 3, 4
- School/university: 1, 3
- Concert auditorium/stadium/sports arena: 1, 2, 4
- Parking garage: 1, 4
- Airport: 1, 2, 3, 4, 5

Each of the facility types listed has different elevator/escalator design criteria that depend on the number of people in the facility and their arrival and departure process. For instance, in a sports facility, it may take about an hour or longer for all of the fans to arrive and be seated. At the conclusion of the game, there is usually a mass exiting queue when everyone wants to depart at once, generally within a 20- to 30-min period. The peak vertical transportation equipment design criteria are therefore selected for the exiting peak. The number of units required to satisfy the selected design criteria and their duty is determined through a statistical mathematical process called *traffic analysis.*

Office building passenger elevators are designed for a 5-min morning up-peak traffic period, when the bulk of the building tenants arrive at the main floor lobby and require vertical transport up to their office floors. If the theoretical morning up-peak elevator calculations can meet or exceed the selected design criteria, then the number of elevators provided should give an acceptable level of service at other traffic peaks (such as the lunchtime two-way-peak and the evening down-peak) during different periods of the day.

Average waiting time is the time in seconds it takes for an elevator to answer a hall call after registration. Average waiting time standards are not generally used to design elevator systems. However, they can be derived from average interval calculations (about a 65 to 70 percent ratio for office buildings) or from a computer traffic simulation. Actual average waiting times and the number of hall calls answered within certain time parameters can be physically counted and tabulated for existing elevator groups. These measurements assist in evaluating the quality level of the elevator dispatching process. They can also be used to prove the specified performance parameters for a new elevator system, to evaluate the performance of an existing elevator system, or to document the performance characteristics before and after elevator modernization.

Escalators and moving-walk waiting times are normally zero because a passenger can board immediately without a wait, unless a passenger queue has formed at the boarding area. Escalator and moving-walk designs are selected on the basis of the number of persons they can transport during a peak 5-min or hourly period. Nominal escalator people loads are determined from a maximum load (1.0, 1.5, or 2.0 persons/step) by using a normal occupancy ratio (about 70 percent of the maximum). Escalators and moving walks are rated by the width of the tread and the moving (contract) speed selected.

SERVICE ELEVATORS

Service elevators are generally designed for specific facility applications and for the amount of goods, materials, and carts that they must handle. Their motive means can be hydraulic, geared traction, or gearless traction, depending upon the building height, application, and frequency of service desired. In high-rise office buildings, the service elevators often serve separate lobbies and enclosed vestibules, so that they would be used for building emergencies such as medical assistance, firefighting, and handicapped evacuations. Because the ASME A17.1 Elevator Code generally classifies them as *passenger* units, they must also meet the ADA handicapped accessibility requirements for a three-point turn in a wheel chair. The minimum ADA-mandated service elevator platform inside width is 5 ft 8 in (a 6-ft-wide platform), and the minimum inside car depth must accommodate at least a 6-ft 4-in ambulance-type stretcher. So, typical capacity ratings are 3500, 4000, 4500, and 5000 lb. Hospital applications for moving patients in wheelchairs or gurneys might require a capacity of 5000 to 6500 lb. Some hospital trauma service (inpatient) elevator applications can be 8000 to 10,000 lb. to accommodate full-size motorized beds and the accompanying patient life support systems. A jail or courthouse prisoner transfer service elevator is typically provided with inmate handcuff restraints or even very elaborate prisoner transport cages.

The number of service elevators required in an office building is typically selected by the amount of net rentable square feet planned for the facility. The general standards shown in Table 5.1.6-2 apply. Hotel and hospital service elevator applications can be selected from a ratio of one unit for so many guest (patient) beds or as a percentage of the number of passenger (ambulatory) elevators provided. The selections vary by the type of hotel and hospital and the requirements of the facility operator. For instance, a convention facility with lots of room service might require more service elevators than a conventional businesspersons' hotel.

FREIGHT ELEVATORS

Freight elevators are classified by the loads that they are expected to carry and the type of equipment that is used to load and unload the car (Fig. 5.1.6-5). Because the elevator plat-

TABLE 5.1.6-2   General Standards for Number of Elevators

Service elevators		
Number	Type	Net rentable area, ft^2
1	Combination passenger/service	≤250,000
1	Service	≥300,000–375,000
1–2	Service	≥375,000–600,000
2	Service	≥600,000–800,000
2–3	Service	≥800,000–1,200,000
4	Service	≥1,200,000–1,500,000

forms are typically narrow and deep and are often equipped with front and rear powered, biparting hoistway doors and car gates, they lend themselves more to large bulky items, palletized cargo, and automobiles.

Freight elevators can be manually loaded and unloaded using wheeled dollies, hand trucks, forklifts, or powered "mules." Freight elevators are found in manufacturing plants, warehouses, materials distribution areas, parking garages, and at restricted building loading docks.

The use of freight elevators is typically restricted to handling freight and the freight handlers. Passengers are normally not permitted to ride on freight elevators because the platforms are not constructed to meet the code-mandated 125 percent of load passenger-loading requirements. Freight elevators are exempt from ADA compliance requirements.

Freight elevator selections are based upon the amount and type of materials to be moved and the frequency of turnover. The amount of material that can be moved is paramount in the selection process, and little thought is given to the "waiting times" for service. Freight elevator motive means can be either hydraulic or traction (normally geared units), and the type of equipment is selected by the number of stops, the height of the facility, and the car speeds required.

MATERIAL-HANDLING EQUIPMENT

Automated and manual material-handling equipment is typically found in banking, data processing, mail sorting, and medical facilities. Dumbwaiters are the most common type of material-handling equipment employed. They can be manual or automatic, floor or counter loading, and can use hydraulic or geared traction motive means. Dumbwaiter car platforms are limited by code to a maximum 500-lb capacity, 9 ft^2 of inside platform area, and a 4-ft clear car height. These restrictions are to minimize the possibility of a person riding in the dumbwaiter car. Counter-loading dumbwaiters are typically equipped with vertical, biparting doors and car gates, whereas floor-loading units usually have lift-up doors and car gates. Dumbwaiters can be equipped with automated inject/eject devices that can automatically transport plastic totes (counter loading) or specially equipped carts (floor loading).

Material lifts are special floor-loading units that can have capacities of up to 1500 lb and maximum inside heights of about 6 ft. They must also be equipped with automatic cart-loading and -unloading devices as an integral part of the platform in order to minimize the possibility of persons riding on them.

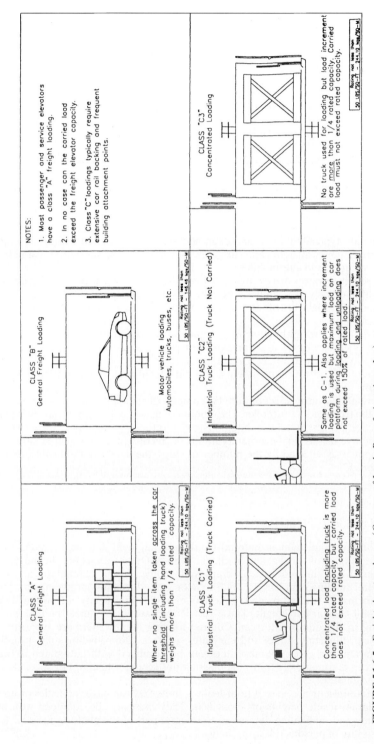

FIGURE 5.1.6-5 Freight elevator classes. (*Courtesy of Lerch, Bates & Associates, Inc.*)

ESCALATORS

Escalators are designed to facilitate the rapid movement of a high number of people, based on the escalator speed and the number of people carried on each escalator step. Maximum escalator capacities range from 1 to 2 persons per step, depending on the step width and the escalator speed (90, 100, or 120 ft/min).

The mechanics of design for escalator usage are quite well developed. Once the requirements for moving a desired number of people from one floor to the next are known, it is a relatively simple matter to select the number, speed, and width of the units necessary to handle this load.

Table 5.1.6-3 shows escalator capacities expressed in terms of maximum and nominal capacity loads.

We never utilize the maximum passenger movement capacities for escalator selection because the calculations are based upon each step being fully loaded. This is an almost impos-

TABLE 5.1.6-3   Escalator Capacities

Width	Speed, ft/min	Maximum capacity, persons	Nominal capacity, 75%
32 in (28-in step tread; 1 person/step)	90	338/5 min 4056/h	254/5 min 3042/h
	100	376/5 min 4512/h	282/5 min 3384/h
	120	451/5 min 5412/h	338/5 min 4059/h
32 in (28-in step tread; 1.25 persons/step)*	90	423/5 min 5076/h	317/5 min 3807/h
	100	470/5 min 5640/h	353/5 min 4230/h
	120	564/5 min 6768/h	423/5 min 5076/h
40 in (32-in step tread; 1.5 persons/step)	90	507/5 min 6084/h	380/5 min 4563/h
	100	564/5 min 6768/h	423/5 min 5076/h
	120	677/5 min 8124/h	508/5 min 6093/h
40 in (32-in step tread; 1.75 persons/step)*	90	592/5 min 7104/h	444/5 min 5328/h
	100	658/5 min 7896/h	494/5 min 5922/h
	120	789/5 min 9468/h	592/5 min 7101/h
48 in (40-in step tread; 2 persons/step)	90	676/5 min 8112/h	507/5 min 6084/h
	100	752/5 min 9024/h	564/5 min 6768/h
	120	902/5 min 10,824/hour	677/5 min 8118/h

Note: The ADA guidelines for transit facilities [10.3.1(16), 10.3.1(17)] require that all escalators have clear widths of at least 32 in, at least two flat steps at the top and bottom entry/egress points, and 2-in-wide, colored, nonslip demarcation strips on the nose of each step.

* Most elevator companies use 1.25 or 1.75 persons per step on these widths to allow for an adult with a child.

sible occurrence, even in the heaviest traffic situations when people depart from a rapid-transit station or a sports complex. Numerous surveys of escalator use have indicated that escalator peak nominal loads are about 75 percent of the maximum theoretical capacity, or about 1 to 2 persons on every other step.

MOVING-WALK APPLICATIONS

Moving walks were developed to provide pedestrians with an assist when they have to traverse long horizontal distances. People who board a moving walk usually have a choice of either standing in place on the unit until they arrive at their destination and then stepping off, or walking on the moving-walk treads to speed up their trip even further. It is an almost universal convention that people stand on the right and walk on the left. A traffic conflict and an impediment to passenger flow occurs when the walk treads (pallets) are not wide enough to permit some persons to walk and pass while others ride. The latest airport moving-walk applications employ double-lane pallet treads that are up to 5 ft wide. These wide units easily permit walkers to pass riders even if they are carrying bags or packages or pushing baggage carts, trolleys, or strollers.

Each moving walk is an independent, self-contained device that has its own individual drive unit, support pallets or moving belt, truss, balustrades, handrails, landing plates, and safety circuits. Moving walks are often installed in pairs so that one unit is running in the departure (outbound) direction and a parallel, adjacent unit is running in the arrival (inbound) direction. Moving walks are classified by length, tread or pallet width (single or double lane), inclination, speed (0 to 4 ft/min for a flat moving walk; 4 to 12 ft/min for a moving ramp), and reversibility.

There are two basic types of moving walks; the metal (die-cast aluminum) pallet type, which is similar to a flattened out escalator, and the rubber-belt type, which is similar to a conveyor belt. Most elevator and escalator manufacturers make only the pallet-type moving walk.

Pallet-type moving walks are provided with full support trusses (similar to escalators) that must be supported on the drive and turnaround ends and at about every 35 ft of lineal length. The truss-type units require about a 4-ft-deep truss depth (pit) that runs the entire width and length of the unit. A second pallet type of moving walk uses stanchion-mounted supports located on about 3-ft centers along the full length of the center walk. These types of units require continuous poured-concrete beds about 18 in deep between the drive and turnaround truss sections.

Moving-walk applications in airport terminals and concourses are primarily used to speed up transit passenger trips between concourses. They are also used to feed departing and arriving passengers to and from the outlying airplane loading gates. Individual moving-walk airport applications are normally installed between airplane loading gates, usually with a maximum length of about 250 to 300 ft, and are spaced along a transit axis adjacent to walking corridors, with 25- to 30-ft loading/unloading breaks between walks spaced along the travel path.

Much like escalators, moving walks are rated by the number of people that they transport within certain time segments. The smallest time period that gives a reliable average is 5 min. Moving-walk capacities also vary in whether they can accommodate single-lane (28-, 32-, or 40-in-wide treads) or double-lane (48-, 54-, 56-, or 60-in-wide treads) passenger movement. Contrary to popular belief, moving-walk capacities do not increase when passengers walk on the moving treads. The area (headway) between the passengers increases to accommodate an average adult's faster walking speed of about 250 ft/min, which is about twice the moving-walk speed. Even though passengers walking on the moving-walk treads do not increase the unit's capacity, passengers should still be encouraged to walk on the units because their relative speed is about 3 times as fast as the moving-walk tread speed, which reduces their own trip times to their destinations. Table 5.1.6-4 indicates moving-walk capacities and nominal loads.

TABLE 5.1.6-4   Moving Walk Capacities

Moving-walk speed, ft/min	Nominal loads, persons moved*					
	Single lane[†]			Double lane[†]		
	1 min	5 min	1 h	1 min	5 min	1 h
90	45	225	2700	90	450	5,400
100	50	250	3000	100	500	6,000
120	60	300	3600	120	600	7,200
125	62.5	313	3756	125	626	7,512
140	70	350	4200	140	700	8,400
150	75	375	4500	150[‡]	750[‡]	9,000[‡]
180	90	450	5400	180[‡]	900[‡]	10,800[‡]

* Capacities are figured at a 250-ft/min average pedestrian horizontal walking speed and a 24-in headway between persons with a 0.5 s/person boarding rate (see J. J. Fruin, *Pedestrian Planning and Design*, pp. 40, 108, and 110). Nominal loads are the 5-min rates.

[†] Per ASME A17.1, Rule 902.7: single lane—28-, 32-, or 40-in tread width (maximum speed 180 ft/min); double lane—48-, 54-, 56-, or 60-in tread width (maximum speed 140 ft/min). *Note:* 0 to 4 ft/min flat walk unrestricted tread width; 4 to 12 ft/min inclined walk—maximum tread width 40 in; minimum tread width is 22 in.

[‡] Not permitted by ASME A17.1 code.

CONTRIBUTORS

Lawrence Lammers, P.E., Lammers + Associates, Reston, Virginia

BIBLIOGRAPHY

American Society of Mechanical Engineers: *Elevator Code,* ASME A17.1, American Society of Mechanical Engineers, New York, 1996.

Fruin, J.J.: *Pedestrian Planning and Design,* Elevator World, Inc., Educational Services Division, Mobile, AL, 1987.

SECTION 5.2
ELECTRICAL SYSTEMS

David L. Stymiest, P.E., SASHE, C.E.M., Senior Consultant, Section Editor
Smith Seckman Reid, Inc., New Orleans, Louisiana

This section of Chap. 5 presents nine detailed treatments of the major aspects of electrical systems for all types of buildings and facilities. The first two articles (Arts. 5.2.1 and 5.2.2) on service entrances and electrical distribution systems cover the gamut of building electrical power distribution issues from the electrical utility to the basic panelboard. Then, in Art. 5.2.3, we supplement this comprehensive treatment of electrical distribution issues with a detailed presentation on emergency power supply systems, including emergency electrical generators. Following that, we focus on specific electrical components of buildings and facilities; raceway and cable systems (Art. 5.2.4), wiring devices (Art. 5.2.5), and grounding systems (Art. 5.2.6). The next two articles (Arts. 5.2.7 and 5.2.8) present engineered systems that are often not given enough attention in many buildings and facilities—cathodic protection systems for corrosion control and heat tracing systems for both freeze protection and process heating. Finally, in Art. 5.2.9, our treatment of electrical systems concludes with an important article on power quality.

ARTICLE 5.2.1
SERVICE ENTRANCES

Jack W. Dean, P.E., Electrical Department Head
Novare Engineers, Inc., Providence, Rhode Island

RESTRUCTURED ELECTRIC UTILITY INDUSTRY

Under restructuring, the American electric utility industry is being split into three major components: (1) generation, (2) transmission, and (3) distribution. Electricity is generated in hydroelectric plants, other renewable energy plants, nuclear power plants, coal and oil-fired plants, and is transmitted through the electric utility grid to local distribution companies (LDCs), also sometimes called *wires companies*. Locally, utility-supplied electric power is sometimes supplemented with steam or gas-fired cogeneration systems or other locally generated renewable sources. In the United States, transmission systems are regulated by the Federal Energy Regulatory Commission, and the distribution of electricity within states is regulated by state public utility commissions or departments. Refer to Art. 7.1.6, "Fuel and Energy Procurement," for a more detailed discussion of issues related to the restructured electric utility industry, also sometimes referred to by the misnomer the *deregulated* electric utility industry.

Because most of the electric power is still generated in large power plants and must be transported over large distances, electricity is generated and transmitted at voltages that are much higher than utilization voltage to minimize distribution losses. Therefore, it must be transformed to a lower voltage when it reaches its point of use. Discussion of regional distribution networks, the effects of deregulation on purchasing decisions, and the reliability of delivering electric power to the site is beyond the scope of this section. We will concern ourselves solely with the point at which the LDC interfaces with the facility—the service entrance and the associated electrical equipment.

THE SERVICE ENTRANCE

The electric power furnished by the LDC is connected to a facility at the service entrance. The service entrance installation consists of various types of equipment (called *switchgear*, or simply *gear*), depending on the source of power and the reliability of service required. That gear is arranged in various configurations for distributing power. The types of loads and the required continuity of service, the system capacity, the voltage required, the voltage drop, the installation conditions and restrictions, and the maintenance requirements are major considerations. The following six systems cover most basic arrangements of service entrance equipment that, in different combinations, address all reasonable service entrance schemes:

1. Simple radial system
2. Loop primary-radial secondary system
3. Primary selective-radial secondary system
4. Two-source primary, secondary selective system
5. Spot network system
6. Medium-voltage distribution system

Common Elements

After a general discussion of the common elements of these schemes, each scheme is presented in detail, along with the elements of each service entrance scheme that will be a major concern of facility managers. This section centers on the elements of ownership, cost, space considerations, reliability, and power quality and efficiency.

Service entrance equipment consists of overcurrent/disconnecting means (switches, circuit breakers, and relays); transformers (indoor, outdoor, dry, and liquid); and switchgear, switchboards, and substations—this equipment is arranged to distribute power effectively to the overall facility. Because equipment required for service entrances and that required for distribution systems overlap, this article limits detailed discussion to equipment rated for medium voltage (600 to 65,000 V). Refer to Art. 5.2.2, "Electrical Distribution Systems," which describes components such as switchgear, switchboards, and step-down transformers for low-voltage operation (600 V and below).

Metering

Metering is the means by which a supplier of electrical energy measures the amount used by a customer, and this is the amount upon which billing is based. Metering is performed at the user's site and may be measured on the primary (high-voltage or medium-voltage) side of the service entrance equipment or on the secondary (low-voltage) side. Primary and secondary metering reflect different utility rates based on the generalization that the owner provides (and owns) all equipment downstream of the meter. Although primary metering may reflect a lower rate, the owner must incur the cost of purchasing and maintaining the primary disconnects and transformers, as well as the risk of their failure. In secondary metered installations, this equipment is normally furnished and maintained by the serving utility.

Obtaining the metering information may be by direct reading (i.e., visually recording the data) or remotely via the Internet or via a modem. Remote reading may become required rather than optional due to the industry restructuring. Also of consideration is the use of analog or digital (solid-state) read-out. From a utility's standpoint, the time of peak demand of power usage is very important relative to how much generating capacity must be provided. Because of this, cogeneration and/or load peak shaving has gained increasing prominence in recent years as both owners and utilities strive to keep costs in check.

Cogeneration

Cogeneration is the production of electrical power concurrently with the production of steam, hot water, and similar energy uses. Electric power can be the main product, and steam or hot water is the by-product, or the steam or hot water can be the primary product, and electric power a by-product. Refer to Art. 5.4.5, "Cogeneration," for a detailed discussion of this strategy.

Where a cogeneration system is being considered, the electrical distribution system becomes more complex. The interface with the utility company is critical and requires careful relaying to protect both the utility and the cogeneration system. Many utilities have stringent requirements that must be incorporated into the system. Proper generator control and protection is necessary, as well. Utilities require that when the protective device at their substation opens, the device also opens that connects a cogenerator to the utility. One reason is that most cogenerators are connected to feeders that serve other customers. Utilities desire to reclose the feeder after a transient fault is cleared. In most cases, reclosing will damage the cogenerator if it remains connected to their system.

Islanding is another reason why the utility insists on disconnecting the cogenerator. Islanding is the event that occurs after a fault in the utility's system is cleared by the operation of the protective devices, when a part of the system may continue to be supplied by cogeneration. Such a condition can be dangerous to the utility's operation during restoration work. Major cogenerators are connected to the subtransmission or the transmission system of a utility. Major cogenerators have buy-sell agreements, in which case, utilities use a trip transfer scheme to trip the cogenerator's tie breaker.

Peak Shaving

Many installations now have emergency or standby generator sets. In the past, they were required for hospitals and similar facilities but were not common in office buildings or shopping centers. However, many costly and unfortunate experiences during utility blackouts in recent years have led to more frequent installation of generator sets in commercial and institutional systems for life safety and to supply other important loads. Industrial plants, especially in process industries, usually have some alternate power source to prevent extremely costly shutdowns. These standby generating systems are critical when needed, but they are needed only infrequently. They represent a large capital investment. To be sure that their power is available when required, they should be tested periodically under load.

Because of the rise in the cost of electricity in recent years and because utilities will bill on the basis of power consumed and also on the basis of peak demand over a small interval, a new use for in-house generating capacity has developed. Utilities typically measure demand charges on the basis of the maximum demand for electricity in any specified period (typically 15 or 30 min.) during a month. Some utilities have a demand "ratchet clause" that continues demand charges on a given peak demand for a full year, unless a higher peak results in even higher charges. In these cases, one large load coming on at a peak time can create higher electric demand charges for a year.

Reducing the peak demand can result in considerable savings in the cost of electrical energy. For those installations that have generator sets for emergency use, modern control systems (computers or programmable controllers) can monitor the peak demand and start the generator set to supply part of the demand as it approaches a preset peak value. The generator set must be selected to withstand the required duty cycle. The simplest of these schemes transfers specific loads to the generator. More complex schemes operate the generator in parallel with the normal utility supply. The savings in demand charges can reduce the cost of owning emergency generator equipment. In some instances, utilities with little reserve capacity have helped finance the cost of some larger customer-owned generating equipment. In return, the customers agree to take some or all of their load off the utility system and onto their own generators at the request of the utility (with varying limitations) when the utility's load approaches capacity. In some cases, the customer's generator is paralleled with the utility's to help supply the peak utility loads, and the utility buys the supplied power. Some utilities have been able to delay large capital expenditures for additional generating capacity by such arrangements. Air quality regulations and site-restricted emissions permits in certain regions may prohibit or curtail this strategy. Refer to Art. 3.2.9, "Environmental Compliance and Issues Management," for a discussion of these types of restrictions.

Utility Service Voltage

Whether the utility service is at primary or secondary voltage depends upon many factors such as type of building, total load, class of user, and the utility rate structure and standard practice. In most downtown metropolitan areas, the utility serves a single commercial or institutional building only at secondary voltage. In more open areas, especially for large buildings or multiple-building installations such as shopping centers, educational institutions, and hospitals, the utility may offer a choice of primary or secondary service.

Where a choice is available, the decision is essentially economic. Utility rate structures provide higher cost for a given load served at secondary voltage than for the same load served at primary voltage because the utility must provide and maintain the substations and pay for substation losses on the secondary service. The lower cost of primary service to the customer must be weighed against the cost of the primary distribution equipment and substations required, the space they occupy, the cost and availability of qualified maintenance for the primary distribution equipment and substations, the reliability of service, the cost of substation (mostly transformer) losses, and similar factors. It is common for industrial plants that have large loads, available room for electrical equipment, and well-qualified maintenance to take advantage of primary service. It is also usual for commercial buildings to use secondary service. Institutional services vary, depending upon the size of the institution, the number and arrangement of buildings, the continuity of service required, and the quality of maintenance available.

Designing for Maintenance

A major consideration in designing or upgrading a facility's service entrance equipment and connection to the service entrance conductors should be how the equipment will be maintained. Designs that allow isolating faults (electrical short circuits) without interrupting power to major portions of the facility are desirable. Regular maintenance to prevent faults from occurring requires isolating portions of the service entrance equipment. Designs that enable such activities easily and without undue hardship on the facility's operation should be part of the design philosophy.

Ease of maintenance can be achieved by using the following strategies:

- Using main breakers in lieu of multiple disconnecting means; up to six are allowed [per paragraph 230-71 of the National Electrical Code (NEC)][1]
- Using transformer primary and secondary tie breakers
- Using backup tie feeders between main centers of load distribution
- Including a separate bypass/isolation switch in an automatic transfer switch enclosure
- Using draw-out circuit breakers rather than bolt-in circuit breakers in main switchgear
- Using hinged covers and three-point locking systems on electrical switchgear enclosures and distribution panels
- Using double-ended unit substations rather than larger single bus substations

Primary Disconnecting Means

If the transformer is located outdoors and furnished by the utility company, the transformer will have an internal primary fused or a nonload break-type switch to offer transformer overcurrent protection and/or means of disconnecting the service. For primary metering, however, the primary switch(es) and transformer(s) will be the property of the owner and must be owner-furnished. The means of disconnecting service is located before the service transformer and must be suitable for interrupting the electric company's distribution voltage and available short-circuit current. This equipment is generally fusible or circuit-breaker type.

Fusible-Type Primary Switchgear. Fusible-type switches are less expensive than circuit breakers. They are normally furnished in what is known as metal-enclosed (ME) load-interrupter switchgear. This gear includes interrupter switches, power fuses (current-limiting or non-current-limiting), bare bus and connections, instrument transformers, and control wiring and accessory devices. This gear generally houses one or more fused or nonfused switches in a metal lineup without barriers (Fig. 5.2.1-1).

Circuit-Breaker-Type Primary Switchgear. Circuit breakers are normally provided in what is known as metal-clad (MC) switchgear and can be vacuum type (Fig. 5.2.1-2) or sulfur hexafluoride (SF_6) insulated type. This switchgear has the following four features:

1. Removable- (draw-out) type switching and interrupting devices move between a connected and disconnected position with self-alignment and self-coupling
2. Major parts of the switchgear are completely enclosed by grounded metal barriers for compartmentalization
3. Automatic shutters cover primary stabs or studs when the removable element is in the disconnected, test, or removed position
4. All primary bus conductors and connections are covered with insulation throughout

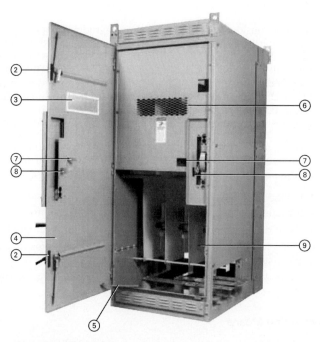

FIGURE 5.2.1-1 Example of load-interrupter fusible switchgear. (*Courtesy of Cutler-Hammer.*)

FIGURE 5.2.1-2   Example of MC vacuum breaker switchgear. (*Courtesy of Cutler-Hammer.*)

From Table 5.2.1-1, it is quite clear there are significant differences (first cost included) to consider when selecting ME or MC switchgear. The following paragraphs explain some of these differences in terms of owner and engineering value.

Construction.   The first selection factor, construction, is related to the continuity of electric service and the negative financial impact of an extended shutdown. For lineups of switchgear, MC construction offers superior short-circuit (fault) isolation from compartment to compartment because the grounded sheet-metal barriers surround all major components and devices. In ME construction, however, because of its more open and uncompartmentalized construction, an internal fault, while rare, could propagate throughout the switchgear. The probability and total cost of extended downtime, as well as the additional first cost for MC construction, need to be addressed in the selection process. (For single vertical-section application, barriering offers much less value.)

Metal-clad switchgear incorporates only insulated buses with sufficient air clearances in between-phase bussing. However, insulated bus joints are more difficult to check and maintain. Although some ME load-interrupter equipment may have insulated bus, standards permit open uncovered bussing. Insulation of the bus provides additional safety against faults due to unattached equipment or tools left on bus bars.

The draw-out feature in MC switchgear is very important, especially for lineups, for continuity of service because the entire lineup does not have to be shut down to remove a single element for maintenance and /or repair. Obviously, the draw-out feature does not offer as significant a benefit for a single load-interrupter switch on a secondary unit substation. In radial systems that have one primary upstream load interrupter feeding one or more unit substations with primary ME switches, unnecessary power outages, although unlikely, may occur if the feeder switch has to be removed from service. If the secondaries of these unit substations are fed from more than one source, the importance of this feature is considerably reduced in terms of continuity of service.

TABLE 5.2.1-1   Summary of Some Major Differences Between ME and MC Types of Equipment

Parameter*	ME-Type WLI			MC—Type VacClad-W		
Voltage Range	4.76–38 kV			4.76–38 kV		
BIL (Basic Impulse Level)	5 kV–60 kV BIL 15 kV–95 kV BIL 25.8 kV–125 kV BIL 38 kV–150 kV BIL			5 kV–60 kV BIL 8.25 kV–95 kV BIL 15 kV–95 kV BIL 27 kV–125 kV BIL 38 kV–170 kV BIL[†]		
Standards—switchgear Switch or breaker—device	C37.20.3—1987 C37.20.4 and C37.22			C37.20.2—1987 C37.04—.11		
Size/structure[†]	Width	Depth	Height	Width	Depth	Height
inches	36.0	55.3–70.0	90.4	36.0	96.3	95.0
mm	914.4	1404.6–1778.0	2296.2	914.4	2446.0	2413.0
Weight	1800 lb—816.5 kg			2650 lb—1202.0 kg		
Cost/switch/breaker	$12,000–14,000			$22,000–25,000		
ANSI minimum[§] electrical duty cycle	30			1,000		
Protective device	Power fuse			Protective relays/breaker		
Operation	Manual operation (electrical operation is optional)			Electrically operated only		
Compartmentalization	Open			Barriered		

* Some data is typical and represents close estimates.
[†] ANSI standards at 38 kV require 150 kV BIL.
[‡] Data based on 5–15 kV class equipment. Feeder breakers may be stacked two high.
[§] ME data is for 15 kV, 600-A switches. MC data is for 1200-A–2000-A breakers.
Source: Courtesy of Cutler-Hammer.

One additional advantage occurs when a spare vacuum breaker is available to minimize a particular feeder's downtime. In lineups of load-interrupter equipment, this would be impractical.

Again, estimates and analysis of the importance of continuity of service and prospective downtime need to be reviewed to make these evaluations.

Another construction feature concerns resettability, reenergizing primary circuits and/or transformers after an overload trip. In the case of ME fusible devices, fuses obviously must be replaced, whereas with protective relays, the protective relay may be reset and service can be resumed in a significantly shorter time. In addition, the protective sensing devices that use MC switchgear are typically protective relays such as induction disk or microprocessor-based protective relays, which can be tested and removed while the circuit remains energized.

Size.   Table 5.2.1-1 shows size differences per structure. For lineups of primary protective devices, size advantage in length may reside with MC vacuum circuit breakers because feeder breakers can be stacked two high. In essence, this allows two 15-kV feeder devices in a single section. In terms of depth, load-interrupter switchgear provides significant size reduction for lineups.

For single devices in a unit substation configuration, the ME sizing offers a clear advantage. The nominal 8-ft minimum depth of MC draw-out vacuum circuit-breaker switchgear is more dimensionally constraining than the smaller load-interrupter switchgear. In addition, MC switchgear requires minimum aisles in front for circuit-breaker withdrawal and to enable lift trucks to remove these breaker elements from the lineup, along with rear accessibility and working space. Metal-enclosed switchgear is accessible from the front or rear, depending upon options and cable entry and exit requirements.

For primary selective applications (two sources of primary power), ME types offer two options: (1) a duplex configuration (see Fig. 5.2.1-3) and (2) a selector configuration (see Fig. 5.2.1-4). Costs are dramatically higher for MC duplex equivalent arrangements because two vacuum breakers are required. No equivalent exists in an MC type for the selector configuration.

Method of Operation.   Most load-interrupter switchgear is manually operated, whereas all MC, vacuum circuit-breaker switchgear is inherently electrically operated either from a dc or an ac source. For single-section ME equipment, feeding a single transformer such as a unit substation, this presents no adverse problems; in fact, the simplicity of a manually operated switch may lend itself to many of these installations.

For lineups with automatic transfer (also called *throw-over*) using load-interrupter equipment, the mains and ties would have to be equipped with motor operators, which typically require auxiliary structures and are add-on devices at increased cost. (Some ME equipment is available with integral motor operators that require no additional space.) In MC construction, with automatic transfer schemes, the inherent motor operation of vacuum breakers can be more meaningful to the designer in addition to the obvious benefits of convenient operation. The speed required in a transfer also affects selection (breakers operate in several cycles; one cycle is ⅟₆₀ s, or 16.7 ms).

Duty Cycle.   Table 5.2.1-2 depicts duty cycle differences, which in essence define their useful life. In this regard, there is a stark difference between ME load-interrupter switches and MC vacuum circuit breakers. Obviously, the application must take duty cycle into account when selecting different types of equipment such as automatic transfer schemes and especially those that are exercised on a scheduled basis.

Protection with Fuses versus Relaying.   Protection is one of the major differences and reasons for selection between ME (fuse) and MC (relay) switchgear. Table 5.2.1-3 summarizes some of them. For further discussion, see Sec. 5.4 in the IEEE Gray Book.[2]

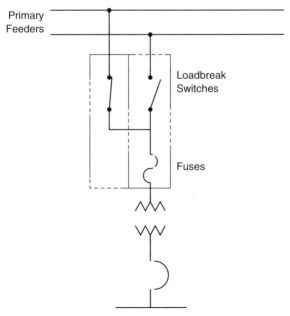

FIGURE 5.2.1-3 Duplex fused switch in two structures. (*Courtesy of Cutler-Hammer.*)

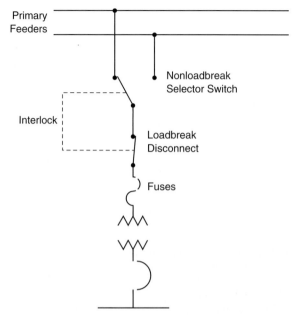

FIGURE 5.2.1-4   Fused selector switch in one structure. (*Courtesy of Cutler-Hammer.*)

Metal-enclosed switchgear typically relies on fuses for protection, which in general provide safe and cost-effective means for medium- and high-fault clearing. Several application comments are given here.

In some cases, switches have been specified on 5-kV low-resistance grounded systems where ground faults are limited to values below the switch rating, typically 600 A. Basically, external relays are called upon to signal the load interrupter to open. Because this occurs while a fault is in progress, it is possible for the fault to escalate into a higher phase-to-phase fault just as the switch is called on to open. Because of this, a vacuum circuit breaker should be used; it can interrupt up to 1000 MVA. A switch is not recommended for a ground-fault

TABLE 5.2.1-2   Duty Cycle Differences

Device*	Number of Load Current Switching Operations	Number of Mechanical Operations
ME interrupter switches (per ANSI)	30	500
MC breakers (per ANSI)	1000	10,000

* Data is for 600-A, 15-kV switches and 1200-A, 15-kV breakers.
Source: Courtesy of Cutler-Hammer.

TABLE 5.2.1-3   Protection Summary

Features	Fuses	Relays
Overload protection	No	Yes
Medium and high short circuits	Yes	Yes
May cause single-phasing of circuit	Yes	No
Adjustable time current curves	No	Yes
Maintainable	No	Yes
Additional protective functions	No	Yes
Replaced when operated	Yes	No
Ground fault protection	No	Yes
Coordination	Good	Excellent

Source: Courtesy of Cutler-Hammer.

application. Medium-voltage fuses provide poor ground-fault protection because the magnitude of the ground fault may be well below the current rating of the fuse.

Care should also be taken to prevent single-phasing of fusible load-interrupter ME equipment. (See Sec. 13.3.15 of the IEEE Buff Book.[3]) It can be shown that as little as a 3 percent of voltage unbalance can produce upward of 20 percent of overheating in induction motors. Single-phase and/or blown-fuse protection is optionally available for ME switchgear.

Possible voltage spikes/transients produced by fuses during significant current limitation should also be factored into the selection process and should be coordinated with surge protection.

References for medium-voltage fuses and load-interrupter equipment and their recommended usage can be found in Secs. 5.3, 9.2, and 9.3 of the IEEE Red Book,[4] and also in Secs. 5.12 through 5.17 of the IEEE Buff Book. These references indicate that load-interrupter fused switches are commonly used in many applications, especially for transformer protection.

Protective relays and associated current transformers, on the other hand, are for fault sensing in MC switchgear. Several application comments are given here.

Relays can provide very sensitive ground-fault protection under any condition because the vacuum breaker is a fully rated interrupting device and can be called upon to interrupt under those conditions. References for further discussion of relays include Secs. 5.5, 9.2, and 9.3 in the IEEE Red Book, 5.5 in the IEEE Gray Book, and Chap. 4 in the IEEE Buff Book.

Relays offer close overload protection. Standard power fuses do not provide overload protection and can rarely provide good backup protection for secondary main devices during overload conditions, but electronic fuses may overcome many of these imitations (see later).

Other types of protection such as differential protection can be provided by using protective relays.

Another important difference is curve shaping. Facilities likely to be expanded and /or to have changing loads may benefit from relays and MC switchgear.

Standard fuses cannot be tested, whereas protective relays can be calibrated and tested. For hospitals, waste treatment plants, and industrial plants, this may provide extended life and reduced downtime. For other types of buildings that emphasize maintenance less, the maintenance requirements of protective relays and MC switchgear may actually detract from selecting them.

Fair to good coordination is available with standard fuses, depending on the shape of the upstream and downstream overcurrent protection devices. Relays generally can be set to nest better and provide shorter wait time in tripping. Because fault damage is proportional to I^2t, significantly more damage occurs for each additional cycle during which the fault flows. (See Chap. 7 on ground fault protection in the IEEE Buff Book. Also see Chaps. 9 and 10 on equipment selection in the IEEE Buff Book.)

The previous discussion compares relay capabilities with standard-type power fuses. However, a broad selection of time–current characteristic curves, plus high continuous current and fault-interrupting ratings, makes electronic power fuses a good selection for service entrance protection, medium-voltage transformer protection, underground subloop protection, and backup protection for overdutied devices such as power circuit breakers. The fast response provides a degree of protection superior to that provided by power circuit breakers and at continuous-current levels far exceeding those available with current-limiting fuses. Moreover, the special response characteristics available with certain electronic power fuses make selecting them a good option for application where traditional types of protective equipment have not been completely satisfactory.

Maintenance.   Recommended maintenance information can be found in NFPA 70B,[5] as well as in the IEEE Buff Book maintenance section. Westinghouse Electric's *Westinghouse Maintenance Hints,* Chaps. 6, 7, and 40, provides good reference information.[6] Another reference source is Chap. 11 of the *Facility Manager's Operation and Maintenance Handbook.*[7] Also refer to Art. 7.2.3, "Electrical Testing and Maintenance," for a brief discussion of maintenance issues.

Summary.   The selection process for medium-voltage switchgear should include evaluations of building life and reliability needs, maintenance, qualifications, of operating and maintenance personnel, first cost, size and room configurations, protection of circuits, continuity of service, available ratings, and single-unit or lineup arrangements. Refer to Art. 5.2.2, "Electrical Distribution Systems," for a discussion of the various types and operation of medium-voltage fuses, circuit breakers, and relays referred to previously.

Transformers.   Most electric company–furnished transformers are oil filled for long life, lowest first cost, and quiet operation and are pad-mounted outdoors. If they are to be located indoors, a transformer vault built to utility company and building code requirements is required. If transformers are to be furnished by the facility and will be located indoors, nonflammable silicone or dry-type transformers are recommended to avoid the cost of vault construction required to minimize the potential damage from fire. Refer to Art. 450-B of the NEC for specific requirements for the different transformer categories. Although the NEC discusses the location requirements of Askarel™-filed transformers, they are no longer acceptable environmentally because of the Environmental Protection Administration and state environmental protection agency rulings against fluids containing PCBs. Refer to Art. 3.2.9, "Environmental Compliance and Issues Management," for a further discussion of related issues. It should also be noted that some states have adopted into law the efficiency levels of the National Electrical Manufacturers Association (NEMA) Standard TPI-1996[8] for distribution transformers (those designed for operation on electrical distribution systems at primary voltages of 34.5 kV and below and secondary voltages of 600 V and below).

Pad-Mounted (Compartmentalized) Transformers.   A *pad-mounted* (*compartmentalized*) *substation transformer* usually denotes a power transformer with direct cable or overhead line termination facilities that distinguish it from a unit substation transformer designed for integral connection to primary or secondary switchgear, or both, through enclosed bus connections. The substation classification is further defined by the terms *primary* and *secondary.* The primary substation transformer has a secondary or load-side voltage rating of 1000 V or more, whereas the secondary substation transformer has a load-side voltage rating of less than 1000 V.

Secondary Unit Substation Transformers.   Unit substation transformers, as distinguished from substation transformers, are designed for integral connection to primary and/or secondary switchgear via enclosed bus connections. Secondary unit substation transformers have load-side voltage ratings of less than 1000 V and typical ratings of 300 to 2500 kVA. Primary ratings do not exceed 34.5 kV.

Liquid-Filled Transformers.   Although NEMA TRI[9] and ANSI C57.12.00[10] differ to some extent on definitions of insulation class, the following is typical in the industry for unit substation transformers. Typical liquid transformers (Fig. 5.2.1-5) are designed with insulation systems to operate at a maximum temperature rise of 65°C, plus 40°C maximum ambient (average temperature of 30°C over a 24-hour period), plus a 15°C hot spot allowance for a total insulation rating of 120°C. Thus, 65°C is the standard specified for temperature rise. However, liquid-filled transformers can be specified with a 55°C rise, suitable for continuous operation up to a 65°C rise with an increase of 12 percent of the base kVA rating. For example, a 1500-kVA liquid-filled transformer designed for a 55°C temperature rise can operate continuously at 1500 kVA × 112%, or 1680 kVA. This optional 55°C temperature rise rating usually costs approximately 5 percent additional over the base price of the transformer. This extra 12 percent rating may be used for

- Unknown future load growth
- Obtaining kVA ratings slightly above standard ratings
- Double-ended unit substations to obtain increased kVA capacity when operating on only one transformer
- Operation at temperatures higher than the expected standard 40°C ambient conditions, but not exceeding 50°C

Essentially, the insulation systems of standard mineral oil, high-flash-point silicone or hydrocarbon, and nonflammable liquid-filled transformers are similar. However, some of the actual insulation materials, gaskets, and bushings may vary to ensure chemical compatibility with the fluid used.

Trade-offs between types of transformer insulation systems dramatically affect the safety performances and life cycle cost of the equipment. Both the NEC and the National Electrical Safety Code (which applies to electric utility distribution systems) allows transformers that are rated as *less flammable* with overcurrent and fault protection to be installed indoors without a 3-hour fire-resistant vault.

FIGURE 5.2.1-5 Typical liquid-filled transformer. (*Courtesy of Square D Company.*)

This can mean a lower installed cost for owners without additional protections, such as extinguishing systems, large separation distances, and barriers, both indoors and outdoors. In recent years, new dielectric liquids have emerged that meet less-flammable criteria. Although dry-type transformers are used quite widely, the insulating properties of "wet" designs are often superior.

The least expensive of the less flammable materials (those that have a fire point greater than 300°C and are self-extinguishing) are hydrocarbons such as Cooper Power System's R-Temp™ fluid. Most expensive are synthetic esters. In between in cost are silicones and natural esters (vegetable oils). ABB's Bio-Temp™ and Cooper Power Systems' Envirotemp™ (FR3) are two natural esters.

Voltage Regulation.   Voltage regulation is the percent of voltage drop of a line relative to the receiving end voltage.

$$\text{Percent regulation} = \frac{100(|E_s| - |E_r|)}{|E_r|}$$

Where E_s is the sending end voltage and E_r is the receiving end voltage.

When a system load is relatively small compared with the capacity of the source (i.e., the on-site transformer kilovolt-amperes), the difference between the sending end voltage and the receiving end voltage is relatively small, and the percent of regulation is high. As load increases relative to the available transformation, the percentage regulation tends to fall off.

Oversizing a transformer beyond a reasonable limit, however, results in transformer losses that lead to inefficiencies. Sizing transformers too close to peak demand loads reduces transformer life and negatively impacts voltage regulation.

Other Considerations.   Ordinary outdoor oil-filled pad-mounted transformers that serve an office building or factory are long-lived, have the lowest first cost, and operate quietly. In the rare instance of failure, they are easiest to obtain and replace. No hatchways in floors need to be removed, no walls chopped down; simply hire a crane. They are relatively mass-produced and are shipped most quickly.

An advantage of liquid-filled transformers is that oil (or silicone) keeps the insulating material soft, pliable, and in its original state. The liquid also dampens lamination noise, so that liquid transformers are quieter than the dry type.

Network transformers are often used on spot networks (see later). Network transformers are much more rugged, have a longer life, and a considerably higher ANSI damage point. On a true network system, they are quite desirable because they are reliable, can withstand overloads for greater periods of time, and are relatively maintenance-free.

Open Dry-Type Transformers.   Typical types of nonliquid transformers are ventilated dry-type transformers or cast transformers that use Class B 80°C rise, Class F 115°C rise, or Class H 150°C rise insulation systems. Most dry-type, noncast transformers that are constructed today for unit substations are made of Class H materials suitable for a total temperature of 220°C. This is composed of 150°C rise, plus a 40°C maximum ambient (average temperature of 30°C over a 24-hour period), plus a 30°C hot spot allowance. The 30°C hot spot allowance allows for differences in temperature at various locations inside the windings, because of their uneven heating and cooling. Transformers can be specified with all Class H 220°C materials, but with an optional winding temperature rise of either 115°C or 80°C. This optional 115°C or 80°C rise will result in a larger core and coil assembly that has typically 115 and 135 percent continuous overload capability, respectively. These transformers with an optional 115°C or 80°C winding temperature rise typically have lower total losses and slightly higher no-load losses than standard 150°C rise units. Their X/R ratio (ratio of a transformer's reactance, X, reflected by its no-load losses to its resistance, R, reflected by its load losses) is slightly higher than standard units.

Previously, conventional ventilated dry-type Class H transformers utilized NOMEX™ paper and a dip-and-bake process. Due to the increasing demand for dry-type transformers in moisture-laden or harsh service conditions and/or chemically polluted environments, vacuum pressure–impregnated (VPI) or vacuum pressure–encapsulated (VPE) transformers have become the standard of many manufacturers (Fig. 5.2.1-6).

VPI transformers are constructed with Class H 220°C materials and then are given additional environmental protection by vacuum pressure impregnation of polyester varnish suitable for a maximum of a 150°C rise. These VPI transformers have become the standard for normal applications indoors or outdoors.

VPE transformers are also constructed with Class H 220°C materials and then are given additional environmental protection by a vacuum pressure encapsulation of silicone resin suitable for a maximum of a 150°C rise. This premium VPE silicone insulation system is often specified for use in heavy moisture or outdoor locations and indoors in industrial locations, especially where harsh chemical atmospheres exist. In addition, the VPE silicone resin process is approved by the Navy for shipboard use and meets MIL-1-24092.

Both the VPI and the VPE insulation systems meet al ANSI, NEMA, and IEEE tests, including basic impulse level (BIL) tests before the environmental protection vacuum impregnation process. The VPI and VPE processes are not used for mechanical support. The glass polyester and silicone are compatible with the thermal expansion of copper and aluminum windings.

Dry-Type Design (Non-Liquid-Filled).   Cast transformers can be supplied in full cast designs or partial cast designs (Fig. 5.2.1-7). The epoxy insulation systems are available for a 155°C total temperature rise and an optional maximum total temperature rise of 115°C, plus 40°C ambient, plus 30°C hot spot, which equals 185°C. Standard cast transformers would be specified with an 80°C rise and a 155°C total temperature insulation system. As an option, transformers can be specified with an 80°C rise and with the optional 185°C total temperature system, which would have a continuous overload capability of 17 percent. Another option would be to specify transformers with a 100°C rise and with the 185° total temperature system, which would have a continuous overload capacity of 6 percent. A transformer could also be specified with a 115°C rise and a system total temperature of 185°C that would not have

FIGURE 5.2.1-6   Typical dry-type transformer. (*Courtesy of Square D Company.*)

FIGURE 5.2.1-7   Typical cast-coil transformer. (*Courtesy of Square D Company.*)

any continuous overload capability. When fully loaded, it would be operating at the maximum allowed temperature rise.

Another alternative epoxy cast design dry-type transformer, known as Resibloc™, is available with a high-voltage winding reinforced with multidirectional glass fiber. This type of transformer offers lower first cost than full cast-coil transformers.

Because all cast transformer designs are of a solid epoxy cast material, they have greater inherent short-circuit strength than any other type of transformer. The cast must be uniform and without voids to prevent a corona from occurring. The dielectric quality of a solid cast transformer is related to the quality of the cast. When cast transformers are properly manufactured, they can be subjected to severe load cycles (cold start to maximum load) and exposed to harsh climates (freezing, heat, chemicals, and moisture).

Cast-coil designs originated in Europe where sheet-aluminum windings that withstand greater short-circuit stresses than copper were initially used. Others have used copper and a different low-voltage epoxy-casting method. Cast-coil transformers offer the high basic insulation level of liquid-filled transformers, but are free from liquid containment problems.

If transformers are idle for a period of time (e.g., medium-voltage transformers serving large chillers that are shut down during the winter months), the cast-coil type is often recommended to avoid the problem of moisture buildup in coils (from condensation or humid air) that can cause serious damage to the windings when they are reenergized. The alternate to turning off the chiller transformers is to run them during the noncooling seasons at no load, which would waste energy.

General Discussion.   The VPI, VPE, and cast transformers are highly resistant to humidity and are suitable for both indoor and outdoor service when located in an appropriate NEMA Type 3R enclosure. All information discussed previously is applicable from 0 to 3300 ft in altitude. Above 3300 ft, typically, a transformer's continuous kilovolt-ampere rating and, in some cases, its short-term overload rating must be derated. All three transformer types are resistant to most normal chemicals; in addition, VPE silicone and epoxy-cast transformers are resistant to caustic and very acidic chemicals. Of the three, VPI polyester is the lowest-cost transformer, VPE silicone is in the middle range, and cast transformers generally have a much higher first cost.

Other Considerations.   This discussion of transformers divides secondary substation transformers (300 through 2500 kVA with 34.5 kV maximum primary voltage and 600 V and below secondary voltages) into two general categories:

1. *Liquid transformers.* Flammable mineral oil, less flammable liquid with 300°C flash point (silicone and hydrocarbon fluids), nonflammable liquids.

2. *Dry-type transformers.* Standard VPI polyester, premium VPE silicone, epoxy cast transformers

Transformer Initial Material Cost.   The first cost of each type of transformer will vary significantly for a specific rating if nonstandard electrical ratings are specified, nonstandard physical dimensions are required, or special options are selected. Table 5.2.1-4 provides an approximate first-cost comparison of standard units built to ANSI standards, with a 15-kV class primary and 480Y/277-V secondary. Table 5.2.1-4 also indicates the standard temperature rise for each of these units. Besides the first cost for each unit, the table also indicates the relative costs of other types of transformers, compared with mineral oil transformers of the

TABLE 5.2.1-4   Approximate First-Cost Comparison

Description	kVA	500	750	1000	1500	2000	2500	Average $/kva
Mineral oil flammable	$—65°C Rise	$12,500	$13,500	$15,000	$18,000	$22,000	$25,000	
	Ratio to Oil	1	1	1	1	1	1	
	$/kVA	25	18	15	12	11	10	15.2
Hydrocarbon—300°C less flammable	$—65°C Rise	15,000	15,750	18,000	21,000	24,000	27,500	
	Ratio to Oil	1.20	1.17	1.20	1.17	1.09	1.10	
	$/kVA	30	21	18	14	12	11	17.7
Silicone—300°C less flammable	$—65°C Rise	17,500	18,750	21,000	24,000	30,000	35,000	
	Ratio to Oil	1.40	1.39	1.40	1.33	1.36	1.40	
	$/kVA	35	25	21	16	15	14	21.0
Dry-type—VPI polyester	$—150°C Rise	15,500	17,250	20,000	23,950	30,000	35,000	
	Ratio to Oil	1.24	1.28	1.33	1.33	1.36	1.40	
	$/kVA	31	23	20	16	15	14	19.8
Dry-type—VPE silicone	$—150°C Rise	18,000	19,500	23,000	30,000	36,000	40,000	
	Ratio to Oil	1.44	1.44	1.53	1.67	1.64	1.60	
	$/kVA	36	26	23	20	18	16	23.2
Dry-Type—Epoxy full cast	$—80°C Rise	37,000	42,000	46,000	54,000	64,000	70,000	
	Ratio to Oil	2.96	3.11	3.07	3.00	2.91	2.80	
	$/kVA	74	56	46	36	32	28	45.3

Source: Courtesy of Cutler-Hammer.

same kVA, and the approximate cost per kVA for each unit. *Note:* The tables are based on standard typical BIL, sound levels, and 5.75 percent impedance (±7 percent tolerance). For these 15-kV primary transformers, the VPI and VPE transformers have a 60-kV BIL, and the cast and liquid transformers have a 95-kV BIL.

Direct and Indirect Transformer Installation Costs.   Direct transformer installation costs vary, depending on the NEC and local code special requirements previously discussed. For example, typically, a flammable mineral oil transformer installed indoors requires a code-approved transformer vault. A vault can add 40 to 80 percent or more of the first cost of the transformer to the installation costs of mineral oil transformers. Less flammable fluid transformers, containing silicone and hydrocarbons, would require sprinkler systems and liquid confinement (curbing) adding approximately 20 to 40 percent or more of the first cost to the installation cost of these transformers. Dry-type transformers, on the other hand, do not have any of these extra installation costs (Table 5.2.1-4)

Indirect transformer installation costs include the labor to handle and install the transformers that vary with the weight and size of the transformer. Table 5.2.1-5 shows approximate dimensions, floor space in square feet, and the weights of the various types of standard designs. *Note:* The dimensions and weights can vary significantly from those shown if special electrical characteristics are specified or certain options are selected. For example, a standard 2000-kVA dry-type VPI or VPE transformer, designed with a 5-kV primary, a Class H insulation system, and a 150°C temperature rise, is 90 in high by 102 in wide by 66 in deep and weighs 9400 lb. The same transformer designed for an 80°C or 115°C rise is 102 in high by 112 in wide by 66 in deep and has an increased weight of 12,000 lb.

The area in square feet shown in Table 5.2.1-5, along with the addition of code-required clearances, has to be considered in relationship to building cost per square foot. In some cases, especially in modifications to existing facilities, existing openings, and elevator capacity could be evaluative criteria in transformer selection.

Operating Cost Considerations.   Many specialized computer programs are available for transformer loss evaluations that can consider different utility rate structures, first costs as losses are varied, customer loading criteria, and economic evaluation of the customer's cost of money. This section considers only an abbreviated approximate method of loss valuation. Table 5.2.1-6 gives a relative comparison of standard approximate losses for the various types of transformers at different kilovolt-ampere ratings. The table is based on transformers that have 15-kV primaries and 480Y/277-V secondaries built to ANSI standards. Losses vary significantly, depending on types, grades, and quantities of materials utilized in the core and coil, and on manufacturing techniques for constructing the core and coil.

In addition, if special lower-than-standard temperature rises are specified (which normally result in higher first cost), no-load losses increase slightly and total losses decrease signifi-

TABLE 5.2.1-5   Approximate Dimensions and Weight Comparison

Transformer Type	500 kVA				1500 kVA				2500 kVA			
	Width (in.)	Depth (in.)	Sq ft.	Weight (lb.)	Width (in.)	Depth (in.)	Sq ft.	Weight (lb.)	Width (in.)	Depth (in.)	Sq ft.	Weight (lb.)
Liquid 65°C	61	75	31.8	6000	65	130	58.7	9800	68	135	63.8	13,000
Dry-type VPI/VPE-150°C	90	60	37.5	4600	112	66	51.3	9300	124	66	56.8	13,000
Dry-type CAST-80°C	90	54	33.8	7100	108	54	40.5	11,400	120	60	50	16,600

Source: Courtesy of Cutler-Hammer.

TABLE 5.2.1-6   Standard Design Transformer Approximate Losses (W)

Transformer Type	Temp. Rise	500 kVA NL	500 kVA TL	750 kVA NL	750 kVA TL	1000 kVA NL	1000 kVA TL	1500 kVA NL	1500 kVA TL	2000 kVA NL	2000 kVA TL	2500 kVA NL	2500 kVA TL
Liquid, all types	65°C	1200	8700	1600	12,160	1800	15,100	3000	19,800	4000	22,600	4500	26,000
Dry-type VPI/VPE	150°C	1900	15,200	2700	21,200	3400	25,000	4500	32,600	5700	44,200	7300	50,800
Dry-type VPI/VPE	80°C	2300	9500	3400	13,000	4200	13,500	5900	19,000	6900	20,000	7200	21,200
Dry-type cast	80°C	3200	3650	4000	5650	4150	7250	5700	10,550	6700	11,150	8100	16,350
Dry-type cast	100°C	2600	5300	3450	6650	4100	8600	5150	14,100	6650	14,300	7550	18,800

Source: Courtesy of Cutler-Hammer.

cantly. Dry-type VPI and VPE transformers, designed with special higher-than-standard BIL to make them equivalent to liquid transformer BIL, have slightly higher no-load losses and slightly higher total losses. In most cases, it is more cost-effective to apply a special low spark-over distribution class surge arrester at the primary connections to the VPI or VPE transformer to obtain a BIL equivalent to that of liquid transformers.

Using some abbreviated formulas for cost analysis, the differences among the various types of transformers can be evaluated approximately. Following are the key formulas and a typical example:[11]

$$I^2R \text{ losses} = \text{total losses (TL)} - \text{no-load losses (NL)}$$

I^2R losses are proportional to the load factor squared (L^2)

$$\text{load factor } (L) = \frac{\text{actual load (kVA)}}{\text{total rated load (kVA)}}$$

$$\text{operating losses} = \text{no-load losses} + I^2R \text{ loss at the appropriate } L$$

Thus, consider an example of a 1000- kVA transformer of the following three types: (1) liquid 65°C rise, (2) VPI dry-type transformer 150°C rise, and (3) cast dry-type 80°C rise, all under the following plant operating conditions:

- 5 days per week with one 8-hour shift
- Cost of electricity: $0.04 /kWH
- 800-kVA load during working hours and 100-kVA load at al other times

Thus, for working hours, $L = 800$ kVA/1000 kVA $= 0.8$. For nonworking hours, $L = 100$ kVA/1000 kVA $= 0.1$. Then, Table 5-2.1-7 can be derived by using the preceding formulas and the data of Table 5.2.1-6. Table 5.2.1-7 indicates the approximate yearly operating cost for the various units under the given conditions. Changing the loading factors, the amount of time the unit is loaded at the various load levels, and the cost of energy will change the relative analysis of savings. For this example, if we analyze first cost and operating loss expense between 65°C rise silicone liquid, 150°C rise dry-type VPI, and 80°C rise dry-type cast, we obtain the data shown in Table 5.2.1-8.

TABLE 5.2.1-7   Simplified Calculation for Loss Evaluation

Line Number	1000 kVA Transformer Description	Liquid 65°C Rise	Dry-Type VPI—150°C	Dry-Type Cast—80°C
1	Full load losses (W)	15,100	25,000	7,250
2	No-load losses (W)	1800	3400	4150
3	For $L = 1.0$ I^2R losses = (Line 1) − (Line 2) (W)	13,300	21,600	3100
4	For $L = 0.8$ I^2R losses = (Line 3) × 0.64	8512	13,824	1984
5	For $L = 0.8$ Total losses = (Line 2) + (Line 4)	10,312	17,224	6134
6	For $L = 0.1$ I^2R losses = (Line 3) × 0.01	133	216	31
7	For $L = 0.1$ Total losses = (Line 2) + (Line 6)	1933	3616	4181
8	Cost of losses for working hours = (Line 5)/1000 × 52 wk × 5 days/wk × 8 h/day × $0.04/kWh	$858	$1433	$510
9	Cost of losses for nonworking hours = (Line 7)/1000 × 52 wk × 128 h/wk × $0.04/kwhr	$515	$963	$1113
10	Total yearly operating costs = (Line 8) + (Line 9)	$1373	$2396	$1623

Source: Courtesy of Cutler-Hammer.

In this example, with the given first cost, load factor, and losses, if the liquid-filed transformer is located outdoors as part of a unit substation, it is the least expensive. Liquid confinement and spill control procedures would have to be considered in the evaluation. For indoor applications, the difference in the yearly operating costs between the VPI and the cast unit is $773.00/year, and the first-cost difference is $26,000.00. Based on this, it would take more than 33 years to pay back the additional cost of the cast unit.

Besides the specific facility requirements, a financial analysis (refer to Chap. 2) should be performed. The heat (BTUs) generated by the transformer, which varies directly in relationship to the total losses at a given load, must be considered. This heat and the possible cost of removing it must be considered for indoor applications (number of watt-hours of losses × 3.143 = number of BTUs).

Maintenance.   The availability and cost of experienced qualified maintenance personnel within the owner's organization or the cost of purchasing outside maintenance services must be considered. Most dry-type transformers normally require very little maintenance other than cleaning accumulated dust and dirt from the core-and-coil assembly and any exposed bushings or bus work. In addition, bus bar joints should be checked for tightness. Most liquid-type transformers can be furnished with liquid level gauges, winding temperature gauges, and pressure vacuum gauges, which should be inspected periodically (normally once a month during the first year and once a year thereafter) to ensure that proper liquid level, temperature,

TABLE 5.2.1-8   First Cost and Operating Cost for 1000-kVA Example

Cost	Liquid Oil—65°C	Dry-Type VPI—150°C	Dry-Type Cast—80°C
First cost	$15,000	$20,000	$46,000
Yearly operating cost (losses)	$1373	$2396	$1623

Source: Courtesy of Cutler-Hammer.

and pressure are being maintained. Readings should be recorded at each inspection and saved for future comparison. Refer to Art. 7.2.3, "Electrical Testing and Maintenance," for a brief discussion of these issues and to Chap. 11 of the *Facility Manager's Operation and Maintenance Handbook* for a more comprehensive discussion.

Lower-Than-Standard Temperature Rise. Transformers can be specified with lower-than-standard temperature rises for a first-cost price increase. The lower temperature rise generally results in higher-than-standard no-load losses but significantly lower total losses. A lower temperature rise increases the size of the core and coil and therefore generally increases the outside dimensions and weight.

For example, a standard dry-type transformer using an insulation system composed of all Class H materials suitable for a 150°C temperature rise, 40°C ambient, and a 30°C hot spot for a 220°C total temperature system, could be specified with an optional 80°C rise. For a 1000-kVA transformer (VPI), the price would increase (see Table 5.2.1-4) approximately 20 percent from $20,000.00 to $24,000.00. The losses could be expected to decrease from the 150°C rise value of 3400 W no-load and 25,000 W total load to the 80°C rise losses of 4200 W no-load and 13,500 W total load (see Table 5.2.1-6). The size and weight would probably increase to the dimensions and weight of a standard 1500-kVA transformer.

In summary, consideration of the operating loss cost savings for a particular application (based on expected loading conditions versus the increased first cost and size), along with cost of capital, most be considered when deciding which type of unit to use. Refer to Sec. 2.7, "Financial Analysis," for further guidance.

Basic Impulse Level (BIL). If system requirements dictate the need to increase standard indoor dry-type VPI and VPE transformers to equivalent liquid or cast transformer BILs, this normally results in a price increase of approximately 10 to 15 percent over the standard unit base price for primaries rated 15 kV and below. In most cases, the addition of distribution or intermediate-class surge arresters to achieve a comparable overall BIL is more economical. When dry-type transformers are part of outdoor unit substations, station class arresters should be considered because of their higher surge current rating and lower spark-over values. When the transformer BIL is increased, normally the size and weight also increase, and both no-load and total losses increase slightly.

Sound Level. Per the NEMA standards, liquid-type transformers are about 5 to 6 dB quieter than comparable dry-type transformers. Both liquid and dry-type transformers can be built to operate more quietly than these NEMA standards. The approximate cost of reducing the sound level is 2 to 4 percent of the transformer price for every decibel below the NEMA standard in the range of 3 to 4 dB below standard. A sound level lower than 5 dB below NEMA standard requires much special design and increases costs. Adding fan cooling generally raises the sound level 3 to 7 dB above the self-cooled 60 to 68 dB rating of dry-type transformers and up to approximately 67 dB on all liquid ratings from a standard level of 56 to 63 dB for liquid-filed transformers.

These criteria should include consideration of NEC requirements regarding location and flammability, first cost, losses in relationship to operating load profile, and maintenance. Special transformer design requirements, temperature rise, BIL, sound level, impedance, and options such as fan cooling, must also be evaluated in terms of the cost and capital expenditure budget. This article discusses some of these considerations and points out that most of these factors are interrelated. Once specific project criteria are established, the manufacturer should be consulted for specific transformer technical data, sizes, and cost estimates.

Secondary Unit Substations. A secondary unit substation is a close-coupled assembly that consists of enclosed primary high-voltage equipment, three-phase power transformers, and enclosed secondary low-voltage equipment. By locating power transformers and their close-coupled secondary switchboards as close as possible to the areas of load concentration, the secondary distribution cables or busways are kept to minimum lengths. This concept has obvious advantages such as reduced power loss, improved voltage regulation, improved service continuity, reduced exposure to low-voltage faults, increased flexibility, minimum installed cost, and efficient space utilization (Fig. 5.2.1-8).

FIGURE 5.2.1-8   Secondary unit substation with primary switch, transformer, and switchgear. (*Courtesy of Cutler-Hammer.*)

The incoming line section provides for connection of the incoming primary circuit or circuits and may have line selector switches, disconnect switches, fuses, or (rarely) circuit breakers. The switching equipment required depends on whether the primary distribution system is radial, primary selective, or primary loop. Circuit breakers can be relayed more exactly than fuses to provide better transformer protection and more coordination with both upstream primary protective devices and secondary overcurrent protection, but fused switches are considerably less expensive. Occasionally, oil switches or oil-fused cutouts are used for primary disconnects. If the overcurrent protection and disconnecting means are at the supply side of the feeder, the substation incoming section may be simply an air chamber for cable connection.

The substation transformer can be almost any type of dry, liquid-cooled, or gas-filled unit. Sizes usually range from about 300 to 2500 kVA, although larger and smaller substations are available. The ventilated dry-type transformer is cabled or bussed to the primary disconnect and usually bussed to the secondary distribution equipment. Liquid-cooled transformers must have a throat connection between the transformer tank and the primary and secondary gear. Gas-filled transformers have primary and secondary bushings brought out of the sealed enclosure.

Transition sections are sometimes required between the transformer and the primary and secondary equipment to provide room to make the cable or bus connection between the transformer buses or bushings and the primary section output or the secondary section buses. It is important to isolate transformer vibration from the primary and secondary structures to keep noise levels down. This requires cable or flexible braids in primary and secondary connections and vibration-absorbing gaskets between transformer enclosures or tank throats and incoming and outgoing section structures. In addition, dry transformer core and coils must be vibration-isolated from their housings.

The secondary distribution section of a unit substation is essentially a low-voltage switchgear or switchboard, close-coupled to the transformer. This section can consist of fixed or draw-out air or molded-case circuit breakers, fusible switches, or a combination of several types of equipment.

Unitized Dry-Type Power Centers.   Unitized dry-type power centers are self-contained, metal-enclosed unit substations especially designed to supply and distribute low-voltage power from medium-voltage lines in modern commercial and industrial systems (Fig. 5.2.1-9). They are ideal when equipment size, accessibility, maintainability, ease of installation, and overall economy are uppermost. Because unitized power centers are inherently compact, they are easily and conveniently applied in multiple throughout a distribution system at locations close to centers of load concentration. Thus, the distribution voltage is stepped down to

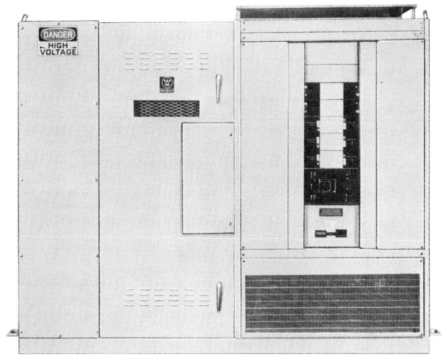

FIGURE 5.2.1-9   Unitized dry-type power center. (*Courtesy of Cutler-Hammer.*)

the utilization voltage only at or near the areas of demand, and kilovolt-amperes are allocated as required for new construction or renovation in existing buildings. Applying unitized power centers in this manner results in several advantages not available with conventional secondary unit substations. Future load growth is easily accommodated by adding unitized power centers to the system without affecting the units that serve the original load areas.

Each unitized power center is completely assembled and tested as a unit at the factory and arrives at the job site in one shipping unit. Rigging (moving) into place is facilitated by the rugged channel base construction and lifting eyes included with each assembly. Field installation consists merely of connecting the incoming and outgoing cables after rolling or lifting the unitized power center into place. All standard unitized power centers are especially designed to minimum dimensions consistent with safety and reliability. Standard unitized power centers that are accessible from the front make against-the-wall installations possible.

Lower losses in the medium-voltage portion of the distribution system result in continuous operating cost savings. Secondary output voltage may be adjusted at each unitized power center to compensate for unusual load conditions without affecting the voltage setting of other apparatuses in the system. Trouble is more quickly isolated with individual units located at or near their served loads. Primary power purchased from the utility at the lower primary power rates results in operational cost savings throughout the life of the equipment. Overall installed cost is lower because of the cost benefits of medium-voltage distribution cable compared with low-voltage cable. The apparent negative cost effect of multiple small unit substations versus a large unit substation is more than offset by the relatively low cost of secondary overcurrent devices that may be applied to each unitized power center.

Simple Radial System

The most basic system for providing service to a facility is called *simple radial*. Service from the utility company enters underground (or overhead) at the utility's distribution voltage and meets utility- (or customer-) owned switch(es) and step-down transformer(s) to achieve the facility's utilization voltage (Fig. 5.2.1-10*a*).

Components.   When the customer receives its supply of power from the primary system and owns the primary switch(es) and transformer(s) along with the secondary low-voltage switchboard or switchgear, the equipment may be a separate primary switch, a separate transformer, and a separate low-voltage switchgear or switchboard. Alternatively, this equipment may be combined as an outdoor pad-mounted transformer with an internal primary fused switch and secondary main breaker feeding an indoor switchboard. Another alternative is a secondary unit substation where the primary fused switch, transformer, and secondary switchgear and switchboard are designed and installed as a close-coupled secondary assembly.

When the utility owns the primary equipment and transformer, the power is supplied to the customer at the utilization voltage, and then the service equipment becomes low-voltage main distribution switchgear or a switchboard.

A modern, improved form of the conventional simple radial system distributes power at a primary voltage. The voltage is stepped down to utilization level in the several load areas within the building, typically through secondary unit substation transformers. The transformers are usually connected to their associated load bus through a circuit breaker, as shown in Fig. 5.2.1-10*b*. Each secondary unit substation is an assembled unit that consists of a three-phase, liquid-filled or air-cooled transformer, an integrally connected primary fused switch, and low-voltage switchgear or a switchboard with circuit breakers or fused switches. Circuits are run to the loads from these low-voltage protective devices.

Discussion of Merits.   The simple radial system is the simplest and least costly system, it is easy to coordinate, and it has no idle parts.

Ownership.   If the utility company is willing to provide the transformer for the step-down to utilization voltage, the primary switch and transformer(s) usually are outdoors, pole-mounted or pad-mounted, and utility company–owned. The customer must furnish the primary underground duct bank, as well as the transformer pad, constructed in accordance with utility company standards. Primary conductors to the transformer are generally furnished and installed by the utility. Secondary conductors (and duct bank) are supplied by the customer. Metering, which requires current transformers (CTs), is provided by the utility in a CT-and-meter enclosure provided by the customer, or remotely at the transformer pad or building exterior. The CTs, as well as the meter itself, are furnished and installed by the utility company. These divisions of responsibility are the more common, but the local utility's rules must be reviewed closely for every facility.

If space is available for an indoor primary switch and transformer or a secondary unit substation, this equipment can be owner-furnished, with provisions for metering in the secondary switchgear or switchboard.

Cost.   The initial investment in service equipment is lowest when the utility company provides the primary switch and transformation to utilization voltage. This would be the preferred arrangement if in-house maintenance capability were limited to utilization voltage experience. At a somewhat higher initial cost, but with no utility company leasing costs for the equipment they otherwise would provide, would be a secondary unit substation. Purchased as a closely coupled assembly, a secondary unit substation reduces the cost of interconnecting cabling, bus duct runs, and so forth.

Because the entire load is served from a single source, full advantage can be taken of the diversity among the loads. This makes it possible to minimize the installed transformer capacity and, hence, the overall cost.

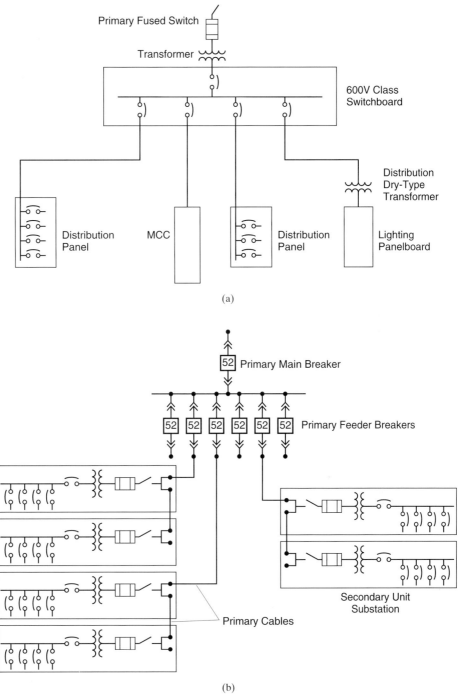

FIGURE 5.2.1-10   (*a*) Simple radial system. (*b*) Primary and secondary simple radial system. (*Courtesy of Cutler-Hammer.*)

This modern form of the simple radial system is usually lower in initial investment than most other types of primary distribution systems for buildings in which the peak load is above 1000 kVA.

Depending on the load kilovolt-amperes connected to each primary circuit and if no ground-fault protection is desired for either the primary feeder conductors and transformers or for the main bus, the primary main and/or feeder breakers may be changed to primary fused switches. This significantly reduces the first cost but also decreases the level of conductor and equipment protection. Thus, if a fault or overload condition occurs, downtime could increase significantly, and a higher cost of increased damage levels and the need for use replacement are typically encountered. In addition, if only one primary fuse on a circuit blows, the secondary loads could be single-phased, causing damage to low-voltage motors.

Another approach to reducing costs is to eliminate the primary feeder breakers completely and use just a single primary main breaker or fused switch to protect a single primary feeder circuit and supply all secondary unit substations from this circuit. Although this system would result in less initial equipment cost, system reliability would be reduced drastically because a single fault in any part of the primary conductor would cause an outage of all loads within the facility.

The costs of the low-voltage feeder circuits and their associated circuit breakers are high when the feeders are long and the peak demand is above 1000 kVA.

Reliability.   One limitation of the simple radial system is that there is no redundancy. Continuity of service depends upon the integrity of the one utility feeder, upon the operability of the primary switch, and upon the durability of a single service transformer. Fortunately, these components are usually quite dependable, provided that they are properly serviced and maintained. A fault on the secondary low-voltage bus or in the source transformer will interrupt service to all loads. Service cannot be restored until all of the necessary repairs have been made. A low-voltage feeder circuit fault will interrupt service to all loads supplied over that feeder.

The improved form of the conventional simple radial system described earlier (Fig. 5.2.1-10b) distributes power at a primary voltage, and voltage is stepped down to the utilization level in each of the building's load areas. In this improved form, a fault on a primary feeder circuit or in one transformer will cause an outage to only those secondary loads served by that feeder or transformer. When a primary main bus fault or a utility service outage occurs, service is interrupted to all loads until the trouble is eliminated. Reducing the number of transformers per primary feeder by adding more primary feeder circuits will improve the flexibility and service continuity of this system; the ultimate is one secondary unit substation per primary feeder circuit. This, of course, increases the investment in the system but minimizes the extent of an outage resulting from a transformer or primary feeder fault.

Voltage Regulation.   The voltage regulation and efficiency of the system portrayed in Fig. 5.2.1-10a may be poor because of the low-voltage feeders and single source. Because power is distributed to the load areas at a primary voltage in the Fig. 5.2.1-10b scheme, losses are reduced, voltage regulation is improved, feeder circuit costs are reduced substantially, and large low-voltage feeder circuit breakers are eliminated. In many cases, the interrupting duty imposed on the load circuit breakers is reduced.

Loop Primary System—Radial Secondary System

Description.   This system consists of one or more primary loops with two or more transformers connected on the loop. This system is typically most effective when two services are available from the utility, as shown in Fig. 5.2.1-11a. Each primary loop is operated so that one of the loop-sectionalizing switches is kept open to prevent parallel operation of the sources.

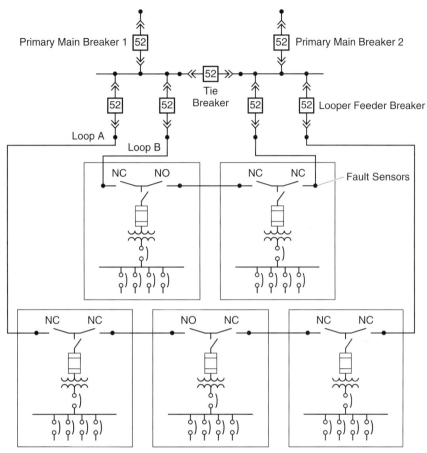

Primary Main Breaker 1 [52]

[52] Primary Main Breaker 2

[52] Tie Breaker

[52] [52] [52] [52] Looper Feeder Breaker

Loop A
Loop B

NC   NO   NC   NC   — Fault Sensors

NC   NC   NO   NC   NC   NC

Secondary Unit Substations Consisting of:
Duplex Primary Switches/Fused Primary Switches/
Transformer and Secondary Main Feeder Breakers

(a)

FIGURE 5.2.1-11   (*a*) Loop primary-radial secondary system. (*Courtesy of Cutler-Hammer.*)

When secondary unit substations are used, each transformer has its own duplex (two load-break switches with load-side bus connection) sectionalizing switches and primary load-break fused switch, as shown in Fig. 5.2.1-11*b*. When pad-mounted compartmentalized transformers are used, they are furnished with loop-feed, oil-immersed, gang-operated, load-break sectionalizing switches and draw-out current-limiting fuses in dry wells, as shown in Fig. 5.2.1-11*c*. By operating the appropriate sectionalizing switches, it is possible to disconnect any section of the loop conductors from the rest of the system. In addition, by opening the transformer primary switch (or removing the load-break draw-out fuses in the pad-mounted transformer), it is possible to disconnect any transformer from the loop.

A key-interlocking scheme is normally recommended to prevent closing all sectionalizing devices in the loop. Each primary loop-sectionalizing switch and the feeder breakers to the loop are interlocked so that they require a key to be closed (which is held captive until the

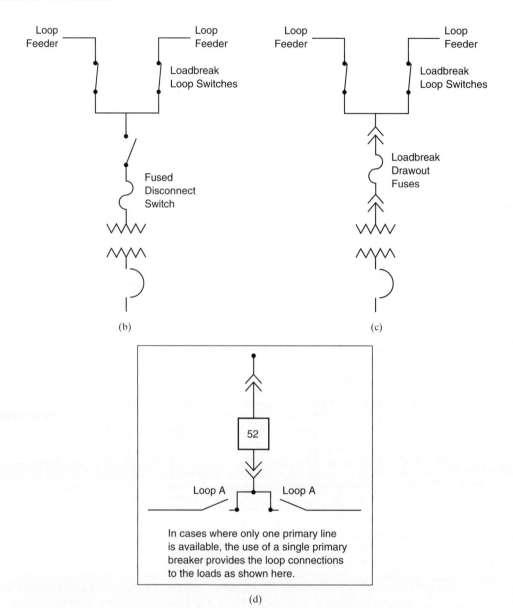

FIGURE 5.2.1-11 *(Continued)* (*b*) Secondary unit substation loop switching. (*c*) Pad-mounted transformer loop switching. (*d*) Single primary feeder loop system. (*Courtesy of Cutler-Hammer.*)

switch or breaker is opened), and one less key than the number of key interlock cylinders is furnished. An extra key is provided to defeat the interlock under qualified supervision. In addition, the two primary main breakers, which are normally closed, and the primary tie breaker, which is normally open, are either mechanically or electrically interlocked to prevent paralleling the incoming source lines. For slightly added cost, an automatic transfer (throw-over) scheme can be added between the two main breakers and the tie breaker. During the more common event of a single utility line outage, the automatic transfer scheme provides significantly reduced power outage time.

The system shown in Fig. 5.2.1-11a costs more than that shown in Fig. 5.2.1-10a but offers increased reliability and quick restoration of service when (1) a utility outage occurs, (2) a primary feeder conductor fault occurs, or (3) a transformer fault or overload occurs. If a utility outage occurs on one of the incoming lines, the associated primary main breaker can be opened, and then the tie breaker is closed manually or through an automatic transfer scheme.

When a primary feeder conductor fault occurs, the associated loop feeder breaker opens and interrupts service to all loads up to the normally open primary loop load switch (typically half of the loads). Once the faulted section of primary cable has been found, then the loop-sectionalizing switches on each side of the faulted conductor can be opened. Then, the loop-sectionalizing switch, which had been previously left open, is closed, and service is restored to all secondary unit substations while the faulted conductor is replaced. If the fault occurs in the conductor directly on the load side of the one of the loop feeder breakers, the loop feeder breaker is kept open after tripping, and the next load-side loop-sectionalizing switch is manually opened, so that the faulted conductor can be sectionalized and replaced. Under this condition, all secondary unit substations are supplied through the other loop feeder circuit breaker, and thus all conductors around the loop should be sized to carry the entire load connected to the loop. Increasing the number of primary loops (two loops shown in Fig. 5.2.1-11a) reduces the extent of the outage from a conductor fault but also increases the system investment.

When a transformer fault or overload occurs, the transformer primary fuses blow (open), and then the transformer primary switch has to be manually opened, disconnecting the transformer from the loop and leaving all other secondary unit substation loads unaffected. A basic primary loop system that uses a single primary feeder breaker connected directly to two loop feeder switches that in turn feed the loop is shown in Fig. 5.2.1-11d. In this basic system, the loop may be normally operated with one of the loop-sectionalizing switches open, as described before, or with all loop-sectionalizing switches closed. If a fault occurs in the basic primary loop system, the single loop feeder breaker trips, and secondary loads are lost until the faulted conductor is found and eliminated from the loop by opening the appropriate loop-sectionalizing switches and then reclosing the breaker.

Primary Selective-Secondary Radial System

Description.   The primary selective-secondary radial system, shown in Fig. 5.2.1-12a, differs from those described previously in that it employs at least two primary feeder circuits in each load area. It is designed so that when one primary circuit is out of service, the remaining feeder or feeders have sufficient capacity to carry the total load. Half of the transformers are normally connected to each of the two feeders. When a fault occurs on one of the primary feeders, only half of the load in the building is dropped.

Duplex fused switches, shown in Fig. 5.2.1-12a and detailed in Fig. 5.2.1-12b, are normally chosen for this type of system. Each duplex fused switch consists of two, load-break, three-pole switches, each in its own separate structure, connected together by bus bars on the load side. Typically, the load-break switch closest to the transformer includes a fuse assembly with fuses. Mechanical and/or key interlocking is furnished so that both switches cannot be closed at the same time (to prevent parallel operation) and are interlocked so that access to either switch or fuse assembly cannot be obtained unless both switches are opened.

As an alternative to the duplex switch arrangement, a non-load-break selector switch mechanically interlocked with a load-break fused switch can be used, as shown in Fig. 5.2.1-12c. The non-load-break selector switch is located at the rear of the load-break fused switch, thus requiring only one structure, and it reduces the cost and floor space compared with the duplex arrangement. The non-load-break switch is mechanically interlocked to prevent it

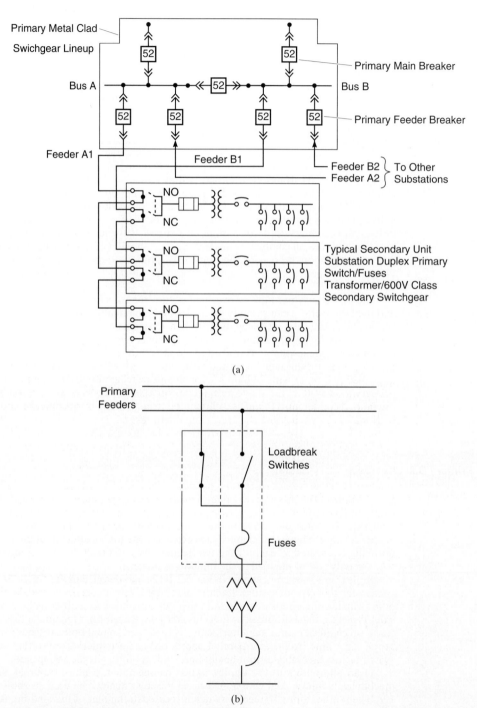

FIGURE 5.2.1-12   (*a*) Primary selective-secondary radial system. (*b*) Duplex fused switch in two structures. (*Courtesy of Cutler-Hammer.*)

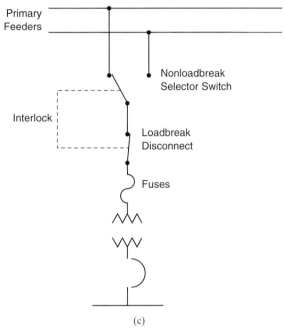

Primary Feeders

Nonloadbreak Selector Switch

Interlock

Loadbreak Disconnect

Fuses

(c)

FIGURE 5.2.1-12   (*Continued*) (*c*) Fused selector switch in one structure. (*Courtesy of Cutler-Hammer.*)

from operating unless the load-break switch is opened. The main disadvantage of the selector switch is that conductors from both circuits are terminated in the same structure. This limits cable space, especially if double lugs are furnished for each line, as shown in Fig. 5.2.1-12*a* and if a faulted primary conductor has to be changed, both lines would have to be deenergized to change the faulted conductors safely.

In Fig. 5.2.1-12*a*, when a primary feeder fault occurs, the associated feeder breaker opens, and the transformers normally supplied from the faulted feeder are out of service. Each primary switch connected to the faulted line must be opened manually, and then the alternate-line primary switch can be closed, connecting the transformer to the live feeder and thus restoring service to all loads. Each of the primary circuit conductors for Feeder A1 and B1 must be sized to handle the sum of the loads normally connected to both A1 and B1. Similar sizing of Feeders A2 and B2, and so on, is required.

If a fault occurs in one transformer, the primary fuses blow and interrupt the service only to the load served by that transformer. Service cannot be restored to the loads normally served by the faulted transformer until the transformer is repaired or replaced.

The cost of the primary selective-secondary radial system is greater than that of the simple primary radial system, shown in Fig. 5.2.1-10*a*, because of the additional primary main breakers, the tie breaker, the two sources, the increased number of feeder breakers, the primary-duplex or selector switches, and the greater amount of primary feeder cable required. The benefits derived from reducing the amount of load dropped when a primary feeder is faulted, plus the quick restoration of service to all or most of the loads, may more than offset the greater cost. Having two sources allows for either manual or automatic transfer of the two primary main breakers and the tie breaker, if one of the sources becomes unavailable.

The primary selective-secondary radial system, however, may be less costly or more costly than the primary loop-secondary radial system of Fig. 5.2.1-11*a*, depending on the location of the transformers, although it offers comparable downtime and reliability. The cost of conductors for the two types of systems may vary greatly, depending on the location of the transformers and loads within the facility, and may greatly override the differences between the two systems in the costs of the primary switching equipment.

Two-Source Primary, Secondary Selective System

Description. This system uses the same principle of duplicate sources from the power supply point using two, primary main breakers and a primary tie breaker. The two primary main breakers and the primary tie breaker are either manually or electrically interlocked to prevent closing all three at the same time and paralleling the sources. Upon loss of voltage from one source, a manual or automatic transfer to the alternate source line may be used to restore power to all primary loads.

Each transformer secondary is arranged in a typical double-ended unit substation, as shown in Fig. 5.2.1-13. Again, the two secondary main breakers and the secondary tie breaker of each unit substation are either mechanically or electrically interlocked to prevent parallel operation. Upon loss of the secondary source voltage on one side, manual or automatic transfer may be used to transfer the loads to the other side, thus restoring power to all secondary loads.

This arrangement permits quick restoration of service to all loads when a primary feeder or transformer fault occurs by opening the associated secondary main and closing the secondary tie breaker. If the secondary voltage is lost because of primary feeder fault and the associated

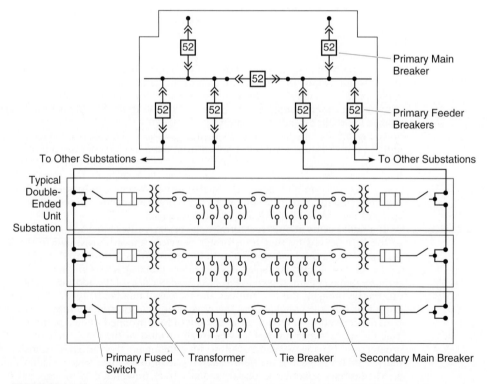

FIGURE 5.2.1-13 Two-source primary, secondary selective system. (*Courtesy of Cutler-Hammer.*)

primary feeder breaker has opened, then all secondary loads normally served by the faulted feeder have to be transferred to the opposite primary feeder. Therefore, each primary feeder conductor must be sized to carry the load on both sides of all of the secondary buses that it serves under secondary emergency transfer. If the voltage is lost because one of the transformers in the double-ended unit substation has failed, then the associated primary fuses blow and take only the failed transformer out of service. Then, only the secondary loads normally served by the faulted transformer have to be transferred to the opposite transformer.

In either of these emergency conditions, the in-service transformer of a double-ended unit substation must be able to serve the loads on both sides of the tie breaker. For this reason, transformers used in this application have equal kilovolt-ampere ratings on each side of the double-ended unit substation, and the normal operating maximum load on each transformer is typically about two-thirds of the base nameplate kilovolt-ampere rating. Typically, these transformers are furnished with fan cooling and/or a lower-than-normal temperature rise, so that under emergency conditions they can carry the maximum load continuously on both sides of the secondary tie breaker. Because of this spare transformer capacity, the voltage regulation that is provided by the double-ended unit substation system under normal conditions is better than that of the systems previously discussed.

The double-ended unit substation arrangement can be used in conjunction with any of the systems discussed that involve two primary sources. Although not recommended, if allowed by the utility, momentary retransfer of loads to the restored source may be made closed transition (antiparallel interlock schemes would have to be defeated) for either the primary or secondary systems. Under this condition, all interrupting equipment and momentary ratings should be suitable for the fault current available from both sources. Special consideration should be given to transformer neutral grounding and equipment operation for double-ended unit substations equipped with ground-fault systems. Where two single-ended unit substations are connected together by external tie conductors, a tie breaker is recommended at each end of the tie conductors.

Simple Spot Network Systems

Description. The secondary network system usually in the form of utility grids has been used for many years to distribute electric power in the high-density downtown area of cities. Modifications of this type of system make it applicable to loads within buildings.

The major advantage of the secondary network system is continuity of service. No single fault anywhere on the primary system interrupts service to any of the system's loads. Most faults are cleared without interrupting service to any load. Another outstanding advantage of the network system is its flexibility in meeting changing and growing load conditions at minimum cost and minimum interruption in service to other loads on the network. In addition to flexibility and service reliability, the secondary network system provides exceptionally uniform and good voltage regulation, and its high efficiency materially reduces the costs of system losses. The major disadvantages of the secondary network system are the relatively high levels of fault current that can be generated and the higher cost of distribution equipment necessitated by this system.

Three major differences between the network system and the simple radial system account for the outstanding advantages of the network. First, a network protector is connected to the secondary leads of each network transformer in place of, or in addition to, the secondary main breaker, as shown in Fig. 5.2.1-14. Second, the secondary windings of the transformers in a given location (spot) are connected together by a switchgear or ring bus from which the loads are served over short radial feeder circuits. Third, when any one primary feeder is out of service, the primary supply has sufficient capacity to carry the entire building load without overloading.

A network protector is a specially designed heavy-duty air-powered breaker, spring-closed with an electrical motor-charged mechanism or motor-operated mechanism, and has a net-

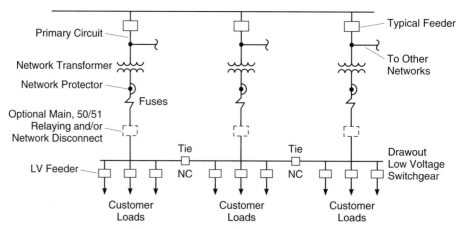

FIGURE 5.2.1-14 Three-source spot network. (*Courtesy of Cutler-Hammer.*)

work relay to control the status of the protector (tripped or closed). The network relay is usu-ally a solid-state microprocessor-based component that is integrated into the protector enclo-sure. It closes the protector automatically only when the voltage-conditions are such that its associated transformer will supply power to the secondary network loads and opens the pro-tector automatically when the power flows from the secondary to the network transformer. The purpose of the network protector is to protect the integrity of the network bus voltage and the loads served from it against transformer and primary feeder faults by quickly discon-necting the defective feeder-transformer pair from the network when backfeed occurs.

The simple spot network system resembles the secondary-selective radial system in that each load area is supplied over two or more primary feeders through two or more transform-ers. In network systems, the transformers are connected through network protectors to a com-mon bus from which loads are served, as shown in Fig. 5.2.1-14. Because the transformers are connected in parallel, a primary feeder or transformer fault does not cause any service inter-ruption to the loads. The paralleled transformers that supply each load bus normally carry equal load currents, whereas it is difficult to obtain equal loading of the two separate trans-formers that supply a substation in the secondary-selective radial system. The interrupting duty imposed on the outgoing feeder breakers in the network is greater with the spot network system because of fault contributions through more than one transformer.

The optimum size and number of primary feeders can be used in the spot network system because the loss of any primary feeder and its associated transformer does not result in the loss of any load, even for an instant. Despite the spare capacity usually supplied in network systems, the costs of primary switchgear and secondary switchgear are often lower compared with a radial system with similar spare capacity. This occurs in many radial systems because more and smaller feeders are often used to minimize the extent of any outage when a primary fault occurs.

When a fault occurs on a primary feeder or in a transformer in a spot network, the fault is isolated from the system through automatic tripping of the primary feeder circuit breaker and all of the network protectors associated with that feeder circuit. This operation does not inter-rupt service to any loads. After the necessary repairs have been made, the system can be restored to normal by closing the primary feeder breaker. All network protectors associated with that feeder close automatically.

The chief purpose of the normally closed ties of the network bus is to provide for sharing of loads and balancing of load currents for each primary service and transformer, regardless of the condition of the primary services. The ties also provide a means of isolating and sectionalizing ground faults within the switchgear network bus, thereby saving a portion of the loads from service interruptions, yet isolating the faulted portion for corrective action.

Spot network systems provide users with several important advantages. First, they save transformer capacity. Spot networks permit equal loading of transformers under all conditions. Networks also have lower system losses and greatly improve voltage conditions. The voltage regulation of a network system is such that both lights and power can be fed from the same load bus. Much larger motors can be started across the line than in a simple radial system. This can result in simplified motor control and permits using relatively large low-voltage motors that have less expensive controls. Finally, network systems provide a greater degree of flexibility in adding future loads; they can be connected to the closest spot network bus.

Spot network systems are economical for buildings that have heavy concentrations of loads covering small areas, considerable distance between areas, and light loads within the distances that separate the concentrated loads. They are commonly used in hospitals, high-rise office buildings, and institutional buildings where a high degree of service reliability is required from the utility sources. Cogeneration equipment is not recommended for use in networks unless the protectors are manually opened and the utility source is completely disconnected and isolated from the temporary generator sources. Spot network systems are especially economical where three or more primary feeders are available, principally, because they supply each load bus through three or more transformers and less spare cable and transformer capacity are required. They are also economical compared with two-transformer double-ended substations with normally opened tie breakers.

Medium-Voltage Distribution System

Single-Bus System.   Figure 5.2.1-15a illustrates a single-bus system. The sources [utility and/or generators(s)] are connected to a single bus. All feeders are connected to the same bus. Generators are used where cogeneration is employed. This configuration is the simplest system; however, outage of the utility results in total outage.

Normally, the generator does not have adequate capacity for the entire load. A properly relayed system equipped with load shedding and automatic voltage and frequency control may be able to maintain partial system operation. Note that the addition of breakers to the bus requires shutting down the bus.

Single Bus with Two Sources from the Utility.   Figure 5.2.1-15b illustrates a single-bus system that has two utility sources. It is the same as the single bus, except that two utility sources are available. This system is operated normally with the main breaker to one source open. Upon loss of the normal service, transfer to the standby normally open (NO) breaker can be automatic or manual. Automatic transfer is preferred to restore service rapidly, especially in unattended stations.

Retransfer to the normal can be closed transition, subject to the approval of the utility. Closed transition momentarily (5–10 cycles) parallels both utility sources. *Caution:* When both sources are paralleled, the fault current available on the load side of the main device is the sum of the available fault current from each source, plus the motor fault contribution. It is recommended that the short-circuit ratings of the bus, feeder breakers, and all load-side equipment are rated for the increased available fault current. If the utility requires open transfer, the disconnection of motors from the bus must be ensured by suitable time delay on reclosing, as well as supervision of the bus voltage and its phase with respect to the incoming source voltage.

This busing scheme does not preclude the use of cogeneration but requires sophisticated automatic synchronizing and synchronism-checking controls, in addition to the previously mentioned load shedding, automatic frequency, and voltage controls. When paralleling sources, reverse current, reverse power, and other appropriate relaying protection should be added as requested by the utility.

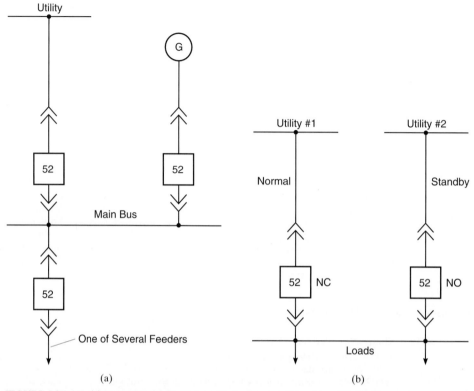

FIGURE 5.2.1-15   (*a*) Single bus. (*b*) Single bus with two sources. (*Courtesy of Cutler-Hammer.*)

This scheme is more expensive than the scheme shown in Fig. 5.2.1-15*a*, but service is restored more quickly. Again, a utility outage results in total outage to the load until transfer occurs. Extending the bus or adding breakers requires shutting down the bus.

Multiple Sources with Tie Breaker(s).   In Fig. 5.2.1-15*c*, the scheme shown is basically the same as in *b*, but the tie breaker is added. The outage to a system load for a utility outage is limited to half the system. Again, the tie breaker can be closed manually or automatically. The statements made for the retransfer of Fig. 5.2.1-15*b* apply also to this scheme. This system is more expensive than the one in Fig. 5.2.1-15*b*. The system is not limited only to two buses (see Fig. 5.2.1-15*d*). Another advantage is that if the buses are momentarily in parallel, no increase in the interrupting capacity of the circuit breakers is required as other buses are added, provided that only two buses are paralleled momentarily for switching. If a looped or primary selective distribution system for the loads is used, the buses can be extended without a shutdown by closing the tie breaker and transferring the loads to the other bus. The tie breaker can be closed manually or automatically after a main breaker is opened. However, because a bus can be fed through two tie breakers, the control scheme should be designed to make the selection.

Continuous Paralleling of Sources.   If continuous paralleling of multiple sources is planned, reverse current, reverse power, and other appropriate relaying protection should be added. When both sources are paralleled, the fault current available on the load side of the main device is the sum of the available fault current from each source, plus the motor fault contri-

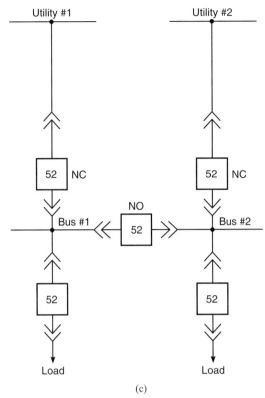

FIGURE 5.2.1-15 (*Continued*) (*c*) Two-source utility with tie breaker. (*Courtesy of Cutler-Hammer.*)

bution. Bus bracing, feeder breakers, and all load-side equipment are required to be rated for the increased available fault current.

Summary. The medium-voltage schemes described are based on using MC, medium-voltage, draw-out switchgear. The service continuity required from electrical systems makes single-source systems impractical.

In designing modern medium-voltage systems, the system should

- Be as simple as possible
- Limit an outage to as small a portion of the system as possible
- Provide means for expanding the system without major shutdowns
- Operate relays so that only the faulted part is removed from service and damage to it is minimized consistent with selectivity
- Apply all of its equipment within published ratings and meet national standards pertaining to the equipment and its installation.

System Selection

Where secondary service is delivered, most buildings will use simple radial distribution from the service. The utility will supply the load in various ways, ranging from a single pad-

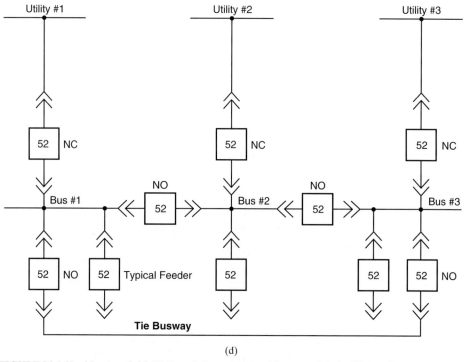

(d)

FIGURE 5.2.1-15 *(Continued)* (*d*) Triple-ended arrangement. (*Courtesy of Cutler-Hammer.*)

mounted transformer, several transformers for a multibuilding installation, or spot networks for a high-rise office building.

Where the service is at primary voltage, the distribution can be from a single substation for smaller installations or from primary distribution to multiple load-center unit substations for larger systems. Primary distribution can be radial or have multiple feeders or one or more loops to single-ended or double-ended substations. Secondary distribution can be radial, loop, secondary-selective, or even secondary network. Any of the primary and secondary distribution methods described previously may be used. The choice will depend on the continuity of service required and the cost of the system. Generally, those systems that provide higher service reliability also cost more, and the initial and operating costs must be weighed against the cost of downtime. In industrial installations, especially in the process industries, the cost of an outage can be tremendous, and distribution systems with maximum reliability flexibility and redundant equipment can easily be justified.

There has been a large increase in high-rise buildings in recent years. This type of building has very large loads and often high available short-circuit fault currents. In most cases, the electric utility company serves these buildings at a secondary voltage of 480Y/277 V from one or more spot networks.

One example would be a typical block-square 60-story office building in New York City. The utility would have one spot network in a utility vault under the sidewalk that supplies services in the basement and another in a specially constructed fireproof utility vault on the 40th floor of the building that supplies additional services to reduce the length of secondary feeder runs. Each vault might have six 2500-kVA network transformers that supply four 4000-A, 480/277-V service takeoffs. The fault current at each service would be nearly 200,000 A. Many lower high-rise office buildings are served by a utility network system at 480/277 V and have a secondary radial distribution system within the building.

Large Vertical Buildings.   A spot network system is particularly applicable to large vertical buildings such as office buildings, stores, apartments, hospitals, and hotels because it is often the most economical method of providing service reliability and good voltage characteristics (i.e., voltage regulation).

Large Horizontal Buildings.   Large horizontal buildings such as shopping centers can use a number of different systems. Size and building configuration are the determining factors. Although somewhat dated, Table 5.2.1-9, obtained from Westinghouse Electric Corporation's *Power Distribution System Design*, dated December 1966, compares the relative merits of systems for several types of shopping centers.[11]

The ratings take into account initial investment, service continuity, flexibility to handle new or changing loads, operation and maintenance costs, voltage regulation, and efficiency. Each of these factors has been weighted in proportion to its relative importance for each load magnitude and type of shopping center. The "A" ratings indicate that analysis of the system rates it as the top of the 12 systems compared in each column. The other quality ratings are given in terms of a percentage of value of the "A" system: B: 81 to 90 percent; C: 71 to 80 percent; D: 61 to 70 percent; E: 51 to 60 percent; and F: 41 to 50 percent.

Groups of Small Buildings.   The form of distribution system for a group of small buildings and the type of equipment in the distribution system depend on the number and sizes of the buildings, the spacing between them, the power requirements of the individual buildings, and the degree of continuity of service required.

The simple radial system is often used to serve groups of relatively small buildings such as the buildings on a university campus.

The likelihood of a primary feeder fault should be considered in designing the distribution system. A fault in an underground cable circuit usually requires several hours to locate and repair. Such a fault puts the feeder circuit out of service for a considerable period of time. Therefore, it becomes important to provide means of restoring service to the buildings within a reasonably short time after a primary feeder fault. This possibility is provided by the primary selective-radial secondary system.

The small additional cost of the extra primary feeder circuit and the selector switches is usually well justified for groups of smaller buildings. Although this system does not eliminate extended outages due to transformer failures, such failures are relatively rare compared with feeder failures, especially where primary feeder circuits are long.

The duration of an outage caused by a primary feeder fault is also reduced by using a secondary selective system. In addition, this type of system prevents a prolonged outage when a transformer failure occurs.

Industrial Plants.   The variation in magnitude and characteristics of loads among industrial plants of the same size is greater than the variation among other types of buildings of the same size. Within a given industrial plant, the addition of new loads and the relocation of existing loads are more frequent than in other types of buildings. These factors indicate that estimating both initial and probable future load is more critical in determining the best system for an industrial plant than in system design for other types of buildings.

A careful comparison of the characteristics of the distribution systems described and identified by designations 1 through 7A in Tables 5.2.1-10 and 5.2.1-11 was made before the publication of these tables in Westinghouse's Specification Data 470 in December 1966 for industrial plant applications. The results are given in Tables 5.2.1-10 and 5.2.1-11 (formerly Tables IV and V in the original publication). Letters in the tables indicate the relative standing or grading as in the preceding discussion of large horizontal buildings.

Based on the general comparison of isolated characteristics given in Table 5.2.1-11, it is possible to assign weighting factors to six major system characteristics and determine the best system for any particular plant. To find the relative merit of the various distribution systems for the usual, or normal, plant, the following weighting factors were assigned to the six system

TABLE 5.2.1-9 Comparison of Distribution Systems for Several Types of Shopping Centers

	Strip or "L"			Mall			Court			Ring			Group or Cluster		
Type of Shopping Center; Loads in kVA	100–1000	1000–2500	2500 and up	100–1000	1000–2500	2500 and up	100–1000	1000–2500	2500 and up	100–1000	1000–2500	2500 and up	1000–2500	2500–5000	5000 and up
Conv. simple-radial	D	D	E	E	E	F+	D	D	E	E	E	F+	E	F+	F
Modern simple-radial (a)	A	A	B	A-	B-	C	A	A	B	A-	B-	C	B-	C	D
Modern simple-radial (b)	C	C	C	C	D	D-	C	C	C	C	D	D-	D	D-	E
Loop-primary radial	A	A	B+	B+	B	C	A	A	B+	B+	B	C	B	C	D
Banked-secondary radial	C	C	B	B	B-	C	B	B	B	B	B-	C	B-	C	C
Pri-selective radial	B	B	B	B	C	D+	B	B	B	B	C	D+	C	D+	D
Sec.-selective radial (a)	C-	C	C	C-	D	D-	C-	C	C	D	D	D-	D	D-	E
Sec.-selective radial (b)	C-	C	C	C-	D	D-	C-	C	C	D	D	D-	D	D-	E
Simple network	D	D	A	B+	A-	A	D	D	A	B+	A-	A	A-	A	A
Simple spot-network	C	B	A-	B	B	B	B	B	A-	B	B	B	B	B	B
Pri.-selective network	D	C-	A	A	A	A	D	C-	A	A	A	A	A	A	A
Pri.-sel. spot-network	C	B	B	B	B	B	C	B	B	B	B	B	B	B	B

Source: Courtesy of Westinghouse Electric Corporation.

TABLE 5.2.1-10   Relative Merit of Systems, Where Characteristics Indicated Are of Primary Importance

Types of Systems	"Normal" Plant	Flexibility	Service Continuity	Regulation	Efficiency	Operation and Maint. Cost	Initial Investment
1 Conventional simple-radial	D	D	D	E	D	C	D
1A Modern simple-radial	C	D	D	C	C	C	B
1B Modified modern simple-radial	C	D	D	D	C	C	C
2 Loop-primary radial	C+	C	C	C	B	B	B
3 Banked-secondary radial	C	B	C	C+	C+	C+	C
4 Primary-selective radial	C	C	C	C	C	B	C+
5 Secondary-selective radial	D	D	C	C	D	D	D
5A Modified secondary-sel. radial	D	D	C	C	D	D	D
6 Simple network	C	C+	B	B	C	C	D
6A Simple spot-network	D	D	C	C	D	D	D
7 Primary-selective network	C+	B	B	B	B	C+	C
7A Primary-selective spot-network	C	C	C+	C	C	C	C

C+ = 79 or 80 percent
Source: Courtesy of Westinghouse Electric Corporation.

characteristics: flexibility, 3; service continuity, 2; regulation, 1; efficiency, 1; operation and maintenance, 1; and initial investment, 6.

The results of this calculation are given in the first column of Table 5.2.1-11. It is recognized that there may be no such thing as a *normal* industrial plant. The term is used here to mean a plant whose load characteristics and requirements are closest to those of the majority of industrial plants. Eight of the 12 systems are about equally applicable in a normal plant. Actually, the loop primary-radial secondary system and the primary selective network system are somewhat better than the other systems that have a C grade. This is indicated by the grade C+, which equals 79 or 80 percent.

The load characteristics and requirements of most plants will depart somewhat from the so-called normal, and the weighting factors can be changed to fit the individual plant. The weighting factors used will depend upon the characteristics and requirements of the loads in the plant; however, they will seldom differ widely from those given for a normal plant, except possibly to increase the importance of one of the characteristics.

TABLE 5.2.1-11   Relative Merit of Characteristics, Where Characteristics Are Rated as Isolated Elements

Types of Systems	Flexibility	Service Continuity	Regulation	Efficiency	Operation and Maintenance Cost	Initial Investment
1 Conventional simple-radial	C	E	F	E	A	D
1A Modern simple-radial	F	F	D	C	A	A
1B Modified modern simple-radial	D	E	D	C	A	C
2 Loop-primary radial	D	E	D	B	A	A
3 Banked-secondary radial	A	C	B	A	B	D
4 Primary-selective radial	D	D	D	B	A	B
5 Secondary-selective radial	D	B	B	D	D	D
5A Modified secondary-sel. radial	D	B	B	D	D	D
6 Simple-network	A	A	A	C	B	E
6A Simple spot-network	C	A	A	D	D	E
7 Primary-selective network	A	A	B	A	C	D
7A Primary-selective spot-network	C	A	B	A	D	D

Source: Courtesy of Westinghouse Electric Corporation.

From Table 5.2.1-11, each characteristic was taken in turn and considered of primary importance. Its weighting was increased until two of the systems emerged that are better than the others. Thus, the relative merits of the various systems applied to other than normal plants are given in Table 5.2.1-10.

For example, if flexibility is of greater importance in the plant than in an average plant, under *flexibility* in Table 5.2.1-11 it will be seen that either the banked secondary radial system or the primary selective network system should be used. The third best system for the plant would be the simple network system. Table 5.2.1-10 makes it possible to narrow the number of systems under consideration from 12 to 2 or 3, at most. Consideration of the second, and possibly third, most important characteristics usually makes it possible to eliminate all but the best system for the plant.

This comparison of systems on the basis of their major characteristics, together with the proper weighting of the characteristics, is helpful in selecting the best distribution for a plant without making a number of different system design and cost comparisons. This is possible because the comparisons given are based on a large number of such designs and studies.

CONTRIBUTORS

Joseph R. Flocco, Jr., P.E., Consulting Engineer, Account Manager, Siemens Energy and Automation, Alpharetta, Georgia

Arjun Rao, P.E., Vice President, BR+A Consulting Engineers, Inc., Boston, Massachusetts

Cornelius Regan, P.E., CLEP, Principal, C. Regan Associates, Rockland, Massachusetts

William J. Schlageter, P.E., Principal, Novare Engineers, Inc., Providence, Rhode Island

David L. Stymiest, P.E., SASHE, C.E.M., Senior Consultant, Smith Seckman Reid, Inc., New Orleans, Louisiana

William Washville, Cutler Hammer, Boston, Massachusetts

William P. Wohlfarth, P.E., Sr. Electrical Engineer, Massachusetts Institute of Technology, Cambridge, Massachusetts

REFERENCES

1. NFPA 70, *National Electrical Code,* National Fire Protection Association, Quincy, MA, 1999.

2. IEEE Gray Book, IEEE Standard 241-1990 (R1997), *IEEE Recommended Practice for Electric Power Systems in Commercial Buildings,* Institute of Electrical and Electronics Engineers, Inc., New York, 1997.

3. IEEE Buff Book, IEEE Standard 242-1986 (R1991), *IEEE Recommended Practice for Protection and Coordination of Industrial and Commercial Power Systems,* Institute of Electrical and Electronics Engineers, Inc., New York, 1991.

4. IEEE Red Book, IEEE Standard 141-1993 (R1999), *IEEE Recommended Practice for Electric Power Distribution for Industrial Plants,* Institute of Electrical and Electronics Engineers, Inc., New York, 1999.

5. NFPA 70B, *Recommended Practice for Electrical Equipment Maintenance,* National Fire Protection Association, Quincy, MA, 1998.

6. *Westinghouse Maintenance Hints, HB-6001-N,* Westinghouse Electric Corporation, Apparatus Service Divisions (Eaton Corporation combined Cutler-Hammer, purchased in the 1970s, and a portion of Westinghouse Electric Corporation, purchased in the 1990s, to form today's firm Cutler-Hammer), Pittsburgh, PA, undated.

7. B. T. Lewis, *Facility Manager's Operation and Maintenance Handbook,* McGraw-Hill, New York, 1999.

8. NEMA TP1-1996, *Guide for Determining Energy Efficiency for Distribution Transformers,* National Electrical Manufacturers Association, Rosslyn, VA, 1993.

9. NEMA TR1-1993, *Transformers, Regulators, and Reactors,* National Electrical Manufacturers Association, Rosslyn, VA, 1993.

10. ANSI C57.12.00, *American National Standard for Transformers,* American National Standards Institute, New York, 1993.

11. *Power Distribution System Design,* Westinghouse Electric Corporation, Pittsburgh, PA, 1966.

BIBLIOGRAPHY

ANSI publications C37, American National Standards Institute, New York, 1998.

S+C Electric Company 600 Series Descriptive Bulletins, Chicago, 1999.

Cooper Power Systems 210 Series Electrical Apparatus Data Sheets, Waukesha, WI, 1999.

Square D Company; October 1997 Digest, Palatine, IL, 1997.

ARTICLE 5.2.2
ELECTRICAL DISTRIBUTION SYSTEMS

Jack W. Dean, P.E., Electrical Department Head
Novare Engineers, Inc., Providence, Rhode Island

The type of electrical power distribution chosen for a facility can have a major impact on the overall usage of energy, on the ease or difficulty in expanding or modifying the existing plant, and on the general reliability of power throughout the facility. Article 5.2.1, "Service Entrances," discusses the elements required to get electricity to the service entrance and to the main distribution equipment. It also discusses the options for distribution system configurations that best meet the needs of the facility. This article discusses the elements involved in the distribution of that electricity and the relative merits of each element.

Certain common elements are required for distributing electricity once it is available at the service entrance. Among these elements are voltage, main service equipment, means of distribution, supplemental equipment, maintenance considerations, system reliability, and space requirements.

The rated voltage of an electrical distribution system varies as one moves downstream from the service entrance (the point at which electricity entering a facility is first controlled). Voltage changes are a function of several items, among them the size of the facility, the type of facility, and the availability of electricity from the serving utility. For any facility, a discussion of voltage must consider service voltage, distribution voltage, and utilization voltage.

Service voltage is the voltage of the electricity delivered to the facility by the serving utility or the generating source. It is the voltage fed to the service entrance equipment or equipment that constitutes the main control and means of cutting off the supply of electricity. For primary metered power, the voltage may be several thousand volts (i.e., 13,800 or 4160 V, etc.) because utility companies transmit power from their power plants at elevated voltages to minimize line energy losses and other expenses. The voltage is transformed at the user site (by a transformer) to a lower voltage by the owner in primary metered services or by the utility company for secondary metered services. In the case of primary metered power, the service voltage is transformed by the owner to distribution voltage, usually in the range of 277/480-V, three-phase, four-wire or, for small facilities, possibly 120/208-V, three-phase, four-wire service. In the case of secondary metered service, the voltage is transformed by the serving utility to the distribution voltage previously mentioned. For secondary metered (referring to the secondary or low-voltage side of the transformer) services, the service voltage and distribution voltage may be the same.

Distribution voltage is the voltage that a facility deems suitable for distributing electricity throughout the plant or building(s). For long distances, it is more efficient and cost-effective to distribute at a higher voltage and then transform to utilization voltage, if necessary, where the energy is needed. For example, in an office or on the factory floor, convenience receptacles require 120-V power for most plug-in equipment. Distribution at 277/480 V to the vicinity and transformation via small transformer(s) near the point of use to the utilization voltage of 120/208 V is often desirable.

Utilization voltage is the operating voltage that equipment in a plant requires. There often is more than one utilization voltage, depending on the equipment to be powered, and for that reason, the distribution voltage and the voltage required for a particular piece of equipment (utilization voltage) may be the same (i.e., 480 V).

The most prevalent secondary distribution voltage in American commercial and institutional buildings today is 480Y/277 V with a solidly grounded neutral. It is also a very common secondary voltage in industrial plants and even in some high-rise, centrally air-conditioned and electrically heated residential buildings because of the large electrical loads. Until the early 1950s, most commercial buildings, such as offices and stores, used 208Y/120-V distribution. About 1950, several simultaneous developments brought about changes. First, central air conditioning became standard practice, more than doubling the previous loads for similar non-air-conditioned buildings. Second, lighting levels were increased, and fluorescent lighting replaced most of the incandescent lighting. Third, the development of 277-V ballasts and 277-V wall switches made it possible to serve this fluorescent lighting load from a 480Y/277-V system. Finally, economical, mass-produced dry-type 480- to 208Y/120-V transformers became readily available to step down the voltage for 120-V incandescent lighting and receptacle loads.

With the increase in building electrical loads and their ability to serve air-conditioning and other motor loads at 480 V and increased lighting loads at 277 V, 480Y/277-V systems have become the most economical for distribution. This voltage allows smaller feeders or larger loads on each feeder and fewer branch circuits because powering loads at a higher voltage allows a corresponding reduction in the load current. In addition, the problem of excessive voltage drops from large loads on 208-V systems is greatly reduced with 480-V distribution. In very tall high-rise office buildings, it is nearly impossible and can be prohibitively expensive to use 208-V distribution and keep voltage drops within acceptable limits.

The choice between 208Y/120- and 480Y/277-V secondary distribution for commercial and institutional buildings depends on several factors. The most important of these are the sizes and types of loads (motors, fluorescent, or high-intensity discharge lighting, incandescent lighting, receptacles) and the lengths of feeders. In general, large motor and fluorescent or high-intensity discharge lighting loads and long feeders tend to make the higher voltages, such as 480Y/277-V, more economical. Very large loads and long runs tend to use medium-voltage distribution (4160 or 13,800 V) and make load center unit substations close to the loads more economical. Conversely, small loads, short runs, and a high percentage of incandescent lighting and receptacle loads tend to make lower utilization voltages such as 208Y/120 V more economical.

Industrial installations that have large motor loads are usually rated for 480- or 4160-V systems. These systems are often ungrounded delta, resistance-grounded delta, or wye systems. Because most commercial low-voltage distribution equipment is rated for up to 600 V and conductors are insulated for 600 V, the installation of 480-V systems uses the same techniques and is essentially no more difficult, costly, or hazardous than for 208-V systems. The major difference is that an arc (also called a *fault* or *short circuit*) of 120 V to ground tends to be self-extinguishing, whereas an arc of 277 V to ground (also called a *ground fault*) is self-sustaining and is likely to cause severe damage. For this reason, the National Electrical Code (NEC) requires ground-fault protection of equipment on grounded wye services of more than 150 V to ground but not exceeding 600 V phase-to-phase (for practical purposes, 480Y/277-V services) for any service-disconnecting means rated at 1000 A or more. The NEC permits voltage up to 300 V to ground on circuits that supply permanently installed electric discharge lamp fixtures, provided that the luminaires (fixtures plus lamps) do not have integral manual switches and are mounted at least 8 ft above the floor. This permits a three-phase, four-wire, solidly grounded 480Y/277-V system to supply directly all of the fluorescent and high-intensity discharge (HID) lighting at 277 V as well as motors at 480 V. Refer to Section 5.3, "Lighting Systems," for a further discussion of these and other issues concerning light source selection.

The principal advantage of using higher secondary voltages in buildings is that for a given load, they use less current and thus require smaller conductors and have lower voltage drops. Also, when compared with a 208-V system, a given conductor size on a 480-V system can supply a large load at the same voltage drop in volts, but a lower percentage voltage drop because of the higher supply voltage. Fewer or smaller circuits can be used to transmit the power from the service entrance point to the final distribution points. Smaller conductors can be used in many branch circuits that supply power loads, and a reduction in the number of lighting branch circuits is usually possible.

It is easier to keep voltage drops within acceptable limits on 480-V circuits than on 208-V circuits. When a 480-V system supplies 120-V loads through step-down transformers, using primary tap adjustments on the transformer can compensate for a voltage drop in the 480-V supply conductors, resulting in full 120-V output. Because these transformers are usually located close to the 120-V loads, secondary voltage drop should not be a problem. If it is a problem, primary taps may be used to compensate by raising the secondary voltage at the transformer, as previously stated.

The principal loads in most buildings can use equipment rated for either 480- or 208-V systems. Three-phase motors and their controls can be obtained for either voltage and are less costly for a given horsepower at 480 V. Fluorescent and HID lamps can be used with either 277- or 120-V ballasts (refer to Art. 5.3.3, "Light Sources, Luminaires"). However, in almost all cases, the installed equipment will have a lower total cost at the higher voltage. The breakeven point for these decisions can be determined, if necessary, although many facility designers do not go through detailed cost analyses for each facility.[1]

Incandescent lighting, small fractional-horsepower motors, wall receptacles, and plug-and-cord-connected appliances for receptacle loads require a 120-V supply. With a 480Y/277-V service, it is necessary to supply these loads through step-down transformers. If the 120-V load to be served is high, this can influence the choice of supply voltage or the relative cost of 480- and 208-V systems. Therefore, in addition to the obvious benefits of using more efficient light sources (refer to Art. 5.3.5, "Design Practices, Lighting Calculations & Economics"), it is also often economically advantageous to minimize the amount of 120-V load by using as little incandescent lighting as possible.

The higher secondary voltage system is usually more economical in office buildings, shopping centers, schools, hospitals, and similar commercial and institutional installations, as well as in industrial plants.

Grounding

Equipment grounding is essential to the safety of personnel. Its function is to ensure that all exposed non-current-carrying metallic parts of all structures and equipment in or near the electrical distribution system are at the same electric potential, and that this is the zero reference potential of the earth. Refer to Art. 5.2.6, "Grounding Systems," for a detailed discussion of this topic. Both the NEC (NEC Article 250)[2] and the National Electrical Safety Code[3] require grounding. Equipment grounding also provides a return path for ground-fault currents, permitting protective devices to operate. Accidental contact of an energized conductor of the system with an improperly grounded non-current-carrying metallic part of the system (such as a motor frame or panelboard enclosure) would raise the potential of the metal object above ground potential. Any person who comes in contact with such an object while grounded could be seriously injured or killed. In addition, current flow from the accidental grounding of an energized part of the system could generate sufficient heat (often with arcing) to start a fire. Two things must occur to prevent such an unsafe potential difference. First, the equipment-grounding conductor must provide a return path for ground-fault currents of sufficiently low impedance to prevent an unsafe voltage drop. Second, the equipment-grounding conductor must be large enough to carry the maximum ground-fault current, without burning off, for sufficient time to permit protective devices (ground-fault relays, circuit breakers, fuses) to operate and clear the fault. The grounded conductor of the system (usually the neutral conductor), although grounded at the source, must not be used for equipment grounding.

The equipment-grounding conductor may be the metallic conduit or raceway of the wiring system or a separate equipment-grounding conductor, run with the circuit conductors, as permitted by the NEC. If a separate equipment-grounding conductor is used, it may be bare or insulated; if insulated, the insulation must be green. Conductors with green insulation may not be used for any purpose other than for equipment grounding.

The equipment-grounding system must be bonded to the grounding electrode at the source or service; however, it may also be connected to ground at many other points. This will not cause problems with the safe operation of the electrical distribution system. Where computers, data processing, or microprocessor-based industrial process control systems are installed, the equipment-grounding system must be designed to minimize interference with their proper operation. Often, isolated grounding of this equipment or completely isolated electrical supply systems are required to protect microprocessors from power system noise that does not in any way affect motors or other electrical equipment.

System grounding connects the electrical supply from the utility, from transformer secondary windings, or from a generator to ground.

Ground-Fault Protection

A ground fault normally occurs by accidental contact of an energized conductor with normally grounded metal or as a result of an insulation failure of an energized conductor. When an insulation failure occurs, the energized conductor contacts normally non-current-carrying grounded metal, which is bonded to or part of the equipment-grounding conductor. In a solidly grounded system, the fault current returns to the source primarily along the equipment-grounding conductors, and a small part uses parallel paths such as building steel or piping. If the ground return impedance were as low as that of the circuit conductors, ground-fault currents would be high, and the normal-phase overcurrent protection would clear them with little damage.

Unfortunately, the impedance of the ground return path is usually higher, the fault itself is usually arcing, and the impedance of the arc further reduces the fault current. In a 480Y/277-V system, the voltage drop across the arc can be from 70 to 140 V. The resulting ground-fault current is rarely enough to cause the phase overcurrent protection device to open instantaneously and prevent damage. Sometimes, the ground fault is below the trip setting of the protective device, and it does not trip at all until the fault escalates and extensive damage occurs. For these reasons, low-level, ground-fault protection devices with minimum time delay settings are required to clear ground faults rapidly.

The NEC requires ground-fault protection only on the service-disconnecting means. This protection works so fast that for ground faults on feeders, or even branch circuits, it will often open the service disconnect before the feeder or branch circuit overcurrent device can operate. This is highly undesirable. A fine-print note (FPN) in the NEC (230-95) states that additional ground-fault protective equipment is needed on feeders and branch circuits where maximum continuity of electric service is necessary. Unless it is acceptable to disconnect the entire service on a ground fault almost anywhere in the system, such additional stages of ground-fault protection must be provided. As an example, at least two stages of protection are mandatory in health care facilities (NEC, Sec. 517-14).

Lightning and Surge Protection. Physical protection of buildings from direct damage by lightning is beyond the scope of this section. Requirements vary with geographic location, building type and environment, and many other factors (see the IEEE Green Book).[4]

Any lightning protection system must be grounded, and the lightning protection ground must be bonded to the electrical equipment-grounding system.[5] Where required, lightning protection systems should also be listed.[6]

Designing for Maintenance. Usually the simpler the electrical system design and the simpler the electrical equipment, the fewer the maintenance costs and operator errors. As electrical systems and equipment become more complicated (to provide greater service continuity or flexibility), the maintenance costs and chance for operator error increase. The systems should be designed with an alternate power circuit to take electrical equipment (requiring periodic maintenance) out of service without dropping essential loads. Ease of maintenance can be achieved by:

- Using draw-out-type protective devices such as breakers and combination starters to help minimize maintenance cost and out-of-service time (it is much easier, quicker, and safer to draw out a large device on a mechanism designed for the purpose than to unbolt it)
- Using more and smaller switchgear sections, served by main breakers and interconnected with tie breakers, rather than one large switchgear section
- Including transformer primary and secondary tie breakers in the service entrance switchgear
- Having backup tie feeders between main centers of load distribution
- Including a separate bypass/isolation switch in an automatic transfer switch enclosure
- Using hinged covers and three-point locking systems on electrical switchgear enclosures and distribution panels

Recommended maintenance information can be found in NFPA 70B,[7] as well as in the IEEE Buff Book[8] maintenance section. Westinghouse Electric's *Electrical Maintenance Hints,* Chaps. 6, 7, and 40, provide good reference information.[9] Another reference source is Chap. 11 of the *Facility Manager's Operation and Maintenance Handbook.*[10] Also refer to Art. 7.2.3, "Electrical Testing and Maintenance," for a brief discussion of maintenance.

Distribution Reliability. The degree of service continuity and reliability needed will vary, depending on the type and use of the facility, as well as on the loads or processes being supplied by the electrical distribution system. For example, a power outage of several hours may be acceptable for a smaller commercial office building, whereas only a few minutes may be acceptable in a larger commercial building or industrial plant. In other facilities such as hospitals, many critical loads permit a maximum of 10 s outage and certain loads, such as real-time computers, cannot tolerate a loss of power for even a few cycles. Refer to Sec. 9.6, "Electrical Systems (Health Care Facilities)," for a further discussion of design considerations that affect distribution reliability. Typically, service continuity and reliability can be increased by:

1. Utilizing multiple utility power sources or services
2. Providing multiple connection paths to the loads served

3. Providing alternate customer-owned power sources such as generators or batteries supplying uninterruptible power supplies

4. Selecting the highest quality of electrical equipment and conductors

5. Using the best installation methods

MAIN SERVICE EQUIPMENT

Service-Disconnecting Means

A service-disconnecting means is required in all buildings so that the electrical power serving that building or, if under one ownership, a cluster of buildings may be disconnected from one location, if necessary, in the event of a fire or other unusual circumstance. That service-disconnecting means, by code, must occur within 10 ft of the electric service entrance, where it penetrates the exterior wall of the basement slab if it enters from underground. The service-disconnecting means can be a fused switch, a circuit breaker, or a lineup of up to six such devices that may be grouped together. The switch(es) or circuit breaker(s) may be individually mounted or may be part of an assembly (a switchboard or distribution panel). Refer to Art. 5.2.1, "Service Entrances," for a discussion of the primary disconnecting means and other equipment for handling power at the point of service to the building or facility.

Overcurrent Protection and Switching Equipment

In specifying overcurrent protection and switching equipment, we must spell out both the protection and the switching requirements, whether performed by a single device or separate devices. Protection requirements include continuous current ratings, operating characteristics (time–current curves) of overcurrent, time delays and pickup values, instantaneous trip settings, ground-fault settings, special relaying such as undervoltage or reverse-current tripping, and for fuses, maximum fault-clearing and current-limiting ability. Switching requirements include voltage ratings; manual or electrical operation; ability to switch under no-load, full-load, or other load conditions; and the ability to interrupt and clear overcurrents and short-circuit currents. Each individual piece of equipment must be able to switch and protect at the available voltages and currents and in the desired manner, and all of the protective devices in the system must also be chosen to provide the desired selectivity and system coordination.

Medium-Voltage Switches and Fuses

Medium-voltage switches include many varieties, such as steel-structure-mounted air switches, vacuum breakers, oil-immersed fused or unfused cutouts, and single-pole, hook-stick cutouts. Switches can be mounted individually or as part of a switchgear lineup or unit substation.

Fuses for medium voltage can be oil-immersed types, but these are growing less common because oil-filled switches and fuses have limited fault-interrupting capabilities and present certain hazards. The fuses used most frequently are expulsion-type power fuses, usually boric acid–filled and current-limiting types. Expulsion-type fuses are available in a wide range of continuous ratings and interrupting capacities. They do not provide energy limiting and are usually used with mufflers to reduce the flame and force of the expelled gases. Because they do not interrupt until the first current-zero point, they are slow enough to provide fair coordination with secondary voltage protective equipment. Current-limiting fuses expel no hot gases, are quiet, limit the energy of the fault, and are faster than the expulsion type. They are available in a smaller range of continuous ratings, which has expanded in recent years with increased demand. Their fast operation makes them more difficult to coordinate with secondary protective devices, however.

Medium-Voltage Circuit Breakers

Medium-voltage breakers include magnetic-air types, vacuum breakers, SF_6-insulated breakers, and oil-filled units. In the class of 15 kV and below, magnetic air equipment was formerly almost universal, but now vacuum breakers are becoming quite prevalent. Oil circuit breakers are rare, even outdoors. Above 15 kV and up to 34.5 kV, outdoor oil circuit breakers are usual, and some minimum-oil types, vacuum circuit breakers, and SF_6-insulated circuit breakers are also available for this application (Fig. 5.2.2-1).

All medium-voltage circuit breakers are actually switches. They open or close the circuit but cannot sense overloads, faults, or other abnormal conditions. Sensors (current and potential transformers, for example) that actuate relays external to the circuit breaker do such fault detection. The relay then signals the breaker to open or close as required. Many different types of relays are available for protection against almost any type of deviation from normal operation, as discussed later. Proper relay selection and setting can provide excellent protection (using suitable circuit breakers) and selective coordination for any distribution system.

The selection of appropriate medium-voltage equipment depends on the available voltage and short-circuit fault currents, the flexibility and features required, and the costs. New technology and higher voltages require careful specification for effective design.

Low-Voltage Air Circuit Breakers

This category includes both draw-out power air circuit breakers and insulated-case breakers. The air circuit breakers are described in the "Low-Voltage Switchgear" paragraph of this sec-

VCP-W Breaker Element

Cut-away View of Vacuum Interrupter (Enlarged to Show Detail)

FIGURE 5.2.2-1   Vacuum-type, medium-voltage circuit breaker. (*Courtesy of Cutler-Hammer.*)

tion. The insulated-case breaker is a draw-out, molded-case, solid-state trip (static) breaker with all of the major features of a draw-out switchgear power breaker except one. These breakers do *not* have a 30-cycle momentary rating equal to their instantaneous interrupting rating; therefore, they are all provided with instantaneous overrides for faults that exceed their short-time (momentary) ratings. True power air breakers can be fully selective because backup breakers can have a short time delay and no instantaneous trip (Fig. 5.2.2-2). Insulated-case breakers can be fully selective with a short time delay only up to the setting of the instantaneous override. For faults exceeding this level, the backup breaker, as well as the downstream breaker, will trip. The advantages of these breakers are that they can be installed in switchboards similar to draw-out switchgear and cost considerably less. The disadvantage of course is the lack of protective coordination above those fault levels.

Molded-Case Circuit Breakers

Lighting-panel breakers and some larger units are available for plug-in mounting, as well as for bolting to the panel bus. Molded-case breakers are available up to about 3000-A frame sizes, and the insulated-case type described earlier is available to 4000 A (Fig. 5.2.2-3). Thermal-magnetic trips are usual in the smaller frames, 400 A and less. Both thermal-magnetic and solid-state trips are common in the 400- to 1200-A range. Above 1200 A, thermal-magnetic trips have been almost completely phased out in favor of solid-state trips.

Molded-case breakers are available with a wide range of accessories (Fig. 5.2.2-4). Most require derating to 80 percent of their *trip,* not frame, ratings for continuous duty. Some of the larger frames, however, are now 100 percent rated for continuous duty, and so labeled, which can often be a distinct economic advantage if properly specified.

The fault-interrupting abilities and ratings of molded-case circuit breakers are substantially higher than in earlier years. In addition to standard ratings, high-interrupting capacity designs are available, and if these capacities are not adequate, molded-case breakers with integral and coordinated fusible current limiters extend the ratings to faults up to 200,000 A. Several designs of true current-limiting molded-case breakers without fuses are also available that not only interrupt faults of up to 200,000 A, but do so within the first half-cycle, limiting

FIGURE 5.2.2-2 Air magnetic draw-out power breaker. (*Courtesy of Cutler-Hammer.*)

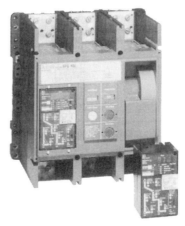

FIGURE 5.2.2-3 Insulated-case power circuit breaker. (*Courtesy of Cutler-Hammer.*)

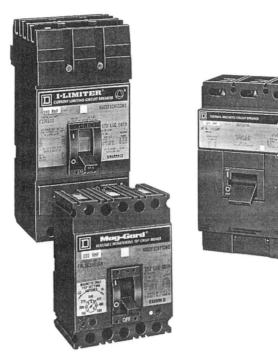

FIGURE 5.2.2-4 Molded-case circuit breakers. (*Courtesy of Square D Company, with permission.*)

the current. This current limiting is sufficient, in many cases, to protect standard lighting-panel breakers downstream. It is important to follow all NEC rules for the use of current-limiting equipment to protect other equipment.

Low-Voltage Switches

Low-voltage switches include safety switches; panelboard and switchboard devices; and large, high-current power switches. They can be unfused, for switching duty only, or fused to provide both switching and overcurrent protection. Some can be electrically opened and closed, and some can be shunt-tripped for use with ground-fault protection or other relaying.

Safety switches range from 30 to 1200 A, rated for 250 or 600 V (some also have dc ratings). Switches can be horsepower-rated for motor use, and the UL label provides the motor sizes for which the switch has been tested to open locked-rotor current safely (which may vary with time-delay or non-time-delay fuses). All switches up to 600 A can safely open at least their full-load current at full voltage; switches rated 800 or 1200 A can also do so, unless labeled "For Isolating Use Only," in which case they must not be used to break current. Switches can be equipped with fuse rejection devices that prevent insertion of any but current-limiting fuses (Class R, J, or L), and such switches are labeled with the maximum short-circuit capacity of the system in which they may be applied—usually 200,000 A at rated voltage. The National Electrical Manufacturers Association (NEMA) classifies switches as heavy-duty [30 to 1200 A, 240 or 600 V ac (and sometimes dc), or general-duty (30 to 600 A, 240 V ac only). UL does not make this distinction but labels the switches with horsepower, short-circuit, and load-interrupting data, as well as voltage and current ratings. Safety switches are also available in double-throw designs and have NEMA-rated enclosures for various environments. Some NEMA enclosure standards are NEMA 1 (general purpose), 1A (semi-dust-tight), 3R (rainproof), 4 and 5 (watertight and dust-tight), 7 and 9 (various hazardous atmospheres), and 12 (industrial use).

Switches for panelboards and switchboards are similar to heavy-duty safety switches and have similar ratings. They are usually rated for horsepower and short-circuit application. Many manufacturers use the identical switch mechanisms, and the enclosure is designed for panelboard or switchboard mounting. At least one manufacturer can provide switches from 400 to 1200 A that can be shunt-tripped for ground-fault protection or other applications. For small fusible panels, such as the residential type, switches are sometimes simply molded pull-out heads with fuses mounted in the head.

Large switches, those from 1200 up to 6000 A, fall into two basic types. One is the bolted-pressure contact switch, the other the butt-contact switch. Originally, bolted-pressure switches (BPS) were manual, slow acting, and capable of breaking full-load current. In the closed position, the bolted-blade contact is the equivalent of a low-heat, low-resistance bus joint. These switches have been adapted for modern high-power systems, including the need for closing into possible faults and tripping open on ground faults. Stored-energy closing mechanisms enable the modern BPS, with proper current-limiting fuses, to close into faults of up to 200,000 A. Shunt-tripped opening mechanisms permit the switch to break up to 12 times the full-load current, as is required for ground-fault application.

Butt-contact switches, called *power protectors* (or fused power circuit devices by UL) are essentially circuit-breaker mechanisms without automatic tripping operation, used as switches in series with integrally mounted fuses. Most of these switches have power air breaker mechanisms, although one manufacturer uses the lighter-duty, lower-cost, molded-case breaker mechanism. Because the breaker mechanisms are high-speed, stored-energy devices originally designed for remote operation and shunt tripping, they can close in (with fuses) on faults of 200,000 A, and, more easily than the BPS, open currents of 12 times rating and often much more. On opening, their speed is somewhat greater than a BPS, and they have a maximum clearing time of about three cycles for currents within their ratings. This makes them especially useful for ground-fault protection. Both the BPS and the power protectors are available with many variations and accessories, such as electrical operation (both close and trip), electrical trip, blown-fuse anti-single-phase prevention, and blown-fuse indication.

Low-Voltage Switchgear

Low-voltage switchgear is a specific class of equipment that consists of an assembly of draw-out power air circuit breakers (Fig. 5.2.2-5). All other types of low-voltage distribution equipment using molded-case circuit breakers, insulated-case circuit breakers, or fusible breakers, either fixed or draw-out-mounted (especially as a main disconnect), are *switchboards*.

Low-voltage switchgear is the top-of-the-line distribution equipment. It comes at a premium price and provides top-grade features, such as best-system selectivity and flexibility, easiest maintenance (and repair) with minimum downtime, and circuit breaker interchangeability. Switchgear consists of a structure with integrated bus work, circuit breakers in compartments or cells, and accessories. The structure is simply a housing and support for the other components, designed for indoor or outdoor application, and arranged for convenient accessibility, operation, and maintenance. Structures are often compartmentalized to isolate line-side bus from load-side bus and load cables.

Buses are usually aluminum, and copper is available as an extra-cost option, rarely justified except under atmospheric conditions harmful to aluminum. Bus joints are usually bolted.

FIGURE 5.2.2-5 Low-voltage, metal-enclosed, draw-out switchgear. (*Courtesy of Cutler-Hammer.*)

Bus plating method and thickness is important, especially on aluminum, where tin plating is preferable to silver because of the moderate joint pressures.

Circuit breakers are available in a wide range of continuous-current and fault-interrupting ratings. Interrupting ratings vary. Both the manually operated circuit breaker and the old solenoid-type electrically operated breaker have been replaced by the stored-energy, spring-operated breaker. The spring is either manually or electrically charged, but the breaker is always spring-operated. Stored-energy mechanisms are either *dependent* (the handle or motor compresses the spring for most of its travel, and then the spring is discharged to operate the breaker), or *independent* (the handle or motor compresses the spring and latches it, and in a separate action then or later the breaker is operated by releasing the spring with a pushbutton or solenoid).

All modern breakers have solid-state tripping, and an extremely accurate electronic relay gets its signal from current-transformer sensors and opens the breaker by a shunt trip coil. Solid-state relay tripping provides very close trip curve tolerances, excellent repeatable accuracy over the breaker's lifetime, and great flexibility in setting long-time and short-time delay pickup values and curves and instantaneous pickup points.

Circuit breaker compartments, called *cubicles* or *cells,* consist of the enclosure in which the breaker itself is mounted. Usually, the cell is ventilated and has a hinged front door. The movable breaker has two sets of contacts that engage fixed contacts in the cell. The main contacts are assemblies of heavy, spring-loaded contact fingers, two for each phase, line, and load, to slide onto the fixed power bus in the cell. The secondary contacts provide connections to the control power source for electrically charged breakers and to the shunt trip, auxiliary switch contacts, and other devices for interlocking, alarm, and control. The breaker can be racked or cranked into three positions—fully inserted (with all main and secondary contacts engaged), test position (main contacts disengaged, most auxiliary contacts engaged), and fully withdrawn (all contacts disengaged). The breaker is interlocked so that it cannot be moved from one position to another unless it is in the open position, and it is safely grounded in test and fully inserted positions. In most equipment, the cell doors can be closed with the breaker in any position, and some permit the breaker to be racked in and out with the door closed. Accessories are available on the breaker (various trip options, alarm contacts, auxiliary switches) or in the switchgear (metering transformers, ammeters, voltmeters, breaker hoist or lifting device, trip unit tester, and calibrator).

When the available fault current exceeds the interrupting rating of the breaker, current-limiting fuses are used with the breaker in a safe, coordinated combination. The NEC requires that such combinations be listed. In breakers in up to 1600- or 2000-A frames, the fuses are mounted integrally on the breaker. In larger 3000- or 4000-A frames, some manufacturers mount the fuses on the breaker and others mount them in a separate draw-out compartment, interlocked with the breaker compartment. As standard or as optional accessories, these fuses have indicators to locate a blown fuse and anti-single-phase protection to trip the three-pole breaker when any fuse has blown. The fuse/circuit-breaker combinations are usually good for available fault current up to 200,000 A (Fig. 5.2.2-6).

Low-Voltage Switchboards.   Low-voltage switchboards are similar to low-voltage switchgear, but they are less standardized, have many more variations available, and usually cost less. In top-of-the-line switchboards that use power-type draw-out insulated-case circuit breakers, the switchboard is almost identical to the switchgear. At the other extreme, a switchboard can be merely a series of panelboards in floor-mounted enclosures, bolted, and bussed together—or even one section containing a floor-mounted panelboard. Because of the wide variations available in equipment and construction and an equally wide range in final cost, it is important to specify carefully exactly what is required.

Switchboard structures depend on the type of distribution equipment used and the location and accessibility of the switchboard (Fig. 5.2.2-7). A switchboard can be classified as *front-connected* or *rear-connected*. A front-connected board has all incoming and outgoing connections made from the front. It is often placed with its back against a wall, but this is not

FIGURE 5.2.2-6   Low-voltage insulated-case circuit breaker switchboard. (*Courtesy of Cutler-Hammer.*)

advisable unless the equipment was specified, designed, and built for *complete* front accessibility, including all bus joints, connections, and devices that might require service or maintenance. A rear-connected board has incoming and outgoing connections in the rear. This class of construction cannot be placed against a wall because rear access is necessary, but it permits the maximum bus sizes and bracing, and there is no limitation on the size and type of feeder equipment and accessories.

Switchboards are also classified by the method of mounting the feeder-disconnecting devices. These can be individually mounted and connected and have separate line-side bus connections to each device, or they can be group-mounted. Individually mounted switches or circuit breakers can often be compartmentalized (have metal barriers) to any desired extent. This construction permits maximum flexibility, size, service, and maintenance. It also requires rear access and is more expensive than group-mounted construction. A group-mounted device consists of a distribution power panelboard chassis, mounted in a switchboard enclosure, and connected to the main switchboard bus. Group-mounted switchboards can often be designed to be completely front-accessible, if necessary. They permit somewhat less variation in layout and accessory devices. Switchboards are often a combination of both types, the main disconnect, and sometimes some large feeder devices are individually mounted and the smaller feeder devices are group-mounted. Many manufacturers can provide almost any equipment and layout desired in a switchboard. A switchboard bus is similar to a switchgear

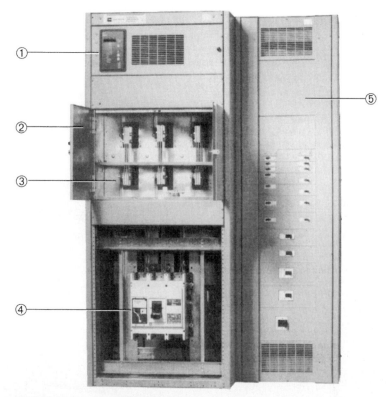

FIGURE 5.2.2-7   Low-voltage distribution switchboard, front-accessible, with group-mounted feeder devices. (1) Customer metering, (2) NEMA utility metering, (3) ground-fault panel, (4) main, (5) group-mounted distribution. (*Courtesy of Square D Company, with permission.*)

bus. Although more bus arrangements are necessary to accommodate the diversity in switchboard designs, the fundamental construction is the same.

Switchboard main and feeder devices can be any of the many switching and protective devices available, depending on the ratings required, the application, the cost, and personal preference. These include air power circuit breakers, fused or unfused, power insulated-case breakers, fused power protectors, bolted-pressure switches, molded-case breakers, fusible switches, and sometimes even motor starters. Larger equipment can be fixed or draw-out-mounted. Combinations of different units are frequent. It is not uncommon to see a large air power breaker as a main disconnect and molded-case breakers for the feeders or a bolted-pressure switch main and standard fusible switch feeders.

Accessories include devices mounted on the circuit breakers or switches; metering, relaying, ground-fault, and similar apparatus mounted in the switchboard.

Transformers

Purpose of Transformers.   Whether located outdoors at the serving utility's generating station, in a vault at the owner's premises, in a lineup of switchgear within a building, in an electric closet on a floor, or in a small piece of equipment (i.e., a control panel), a transformer's role is to change voltage from one level or value to another. Although normally used to transform voltage from a higher to a lower level (step-down), a transformer can also boost the voltage from a lower to a higher level. This occurs typically via electromagnetic induction and by using coils of wire or windings. A detailed discussion of how this occurs is beyond the scope of this book and is not necessary for the facility manager to know in detail, but it is, in simple terms, what actually occurs.

It is important for an informed facility manager to know the types of transformers and the relative merits of each. Type of transformer in this instance refers to the insulating means that is employed to keep the windings separated from each other so that voltage can be induced and so that short circuiting does not occur. The types most commonly encountered are oil-filled, non-flammable fluid, cast-resin, and dry-type.

Fluid-filled and *cast-resin transformers* are normally classified as power-type transformers, those with ratings above 500 kVA. Those classified as distribution type (from 3 to 500 kVA) are normally dry-type. Refer to Art. 5.2.1, "Service Entrances," for a detailed discussion of power-type transformers.

Dry-Type Transformers.   Generally, for economy and distribution of power, it is desirable to locate dry-type distribution transformers as near as practicable to the load (Fig. 5.2.2-8). Typical loads for dry-type distribution transformers include lighting, heating, air conditioners, fans, and machine tools. Such loads are found in commercial, institutional, industrial, and residential structures. If frequencies other than 60 cycles/s are required, such transformers must be specifically designed. The discussion of this section will be limited to standard 60-Hz dry-type units. Industry standards classify insulation systems and rise as shown in Table 5.2.2-1. The rated lives of transformers that have different insulation systems are the same—the lower-temperature systems are designed for the same life as the higher-temperature systems.

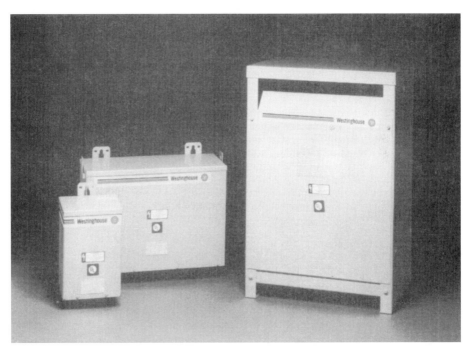

FIGURE 5.2.2-8   General purpose dry-type distribution transformers rated 1000 kVA and less. (*Courtesy of Cutler-Hammer.*)

TABLE 5.2.2-1   Dry-Type Transformer Insulation System Classification

Temperature Class	=	Ambient Temperature	+	Winding Temperature Rise	+	Hot Spot Allowance
150°C		40°C		80°C		30°C
185°C		40°C		115°C		30°C
220°C		40°C		150°C		30°C

Transformer Overload Capability.   Short-term overload capacity is designed into transformers as required by ANSI and defined by IEEE Standard C57.12.01-1989.[11] Dry-type distribution transformers deliver 200 percent of nameplate load for ½ hour, 150 percent for 1 hour; and 125 percent load for 4 hours without being damaged, provided that a constant 50 percent load precedes and follows the overload. Continuous overload capacity is not deliberately designed into a transformer because the design objective is to be within the allowed winding temperature rise for the nameplate loading.

Transformer Taps.   Primary taps are available in most ratings to allow compensation for source voltage variations. Common taps in most transformers are two 2.5 percent taps below normal and two 2.5 percent taps above normal. A primary voltage 2.5 percent below rated voltage can be compensated for by setting the primary taps at the first 2.5 percent below normal setting. This will cause the transformer secondary voltage to be at rated voltage for this condition.

Series-Multiple Windings.   Series-multiple windings consist of two similar coils in each winding which can be connected in series or parallel (multiple). Transformers with series-multiple windings are designated with a × or / between the voltage ratings, such as primary voltage of 120/240 or 240×480. If the series-multiple winding is designated by a ×, the winding can be connected only for a series or parallel connection. With the / designation, a midpoint also becomes available in addition to the series or parallel connection. As an example, a 120×240 winding can be connected for either 120 (parallel) or 240 (series), but a 120/240 winding can be connected for 120 (parallel), or 240 (series), or 240 with a 120 midpoint.

Buck-Boost Autotransformers.   An autotransformer has only one winding and therefore is smaller and more economical than the conventional two-winding transformer. In an autotransformer, the primary and secondary are electrically and mechanically connected together. The required secondary voltage is obtained by "tapping off" from the single winding.

Buck-boost autotransformers are insulated units with 120×240- or 240×480-V primaries and 12/24-, 16/32-, or 24/48-V secondaries and provide a very economical method for minor voltage adjustments where circuit isolation is not needed. Autotransformers can be used only where local electrical codes permit and where isolating the two circuits is not required.

In general, autotransformers are used when the transformation ratio is three or less and the electrical isolation of the two windings is not required. The usual application of autotransformers includes the following:

- Power distribution (lowering and raising voltage level)—buck-boost operation
- Induction motor starters on a selected basis
- Small variable-voltage power supply units

The advantages of autotransformers are that for the same input and output, the weight of the conductor, core material, and insulation of an autotransformer is less than that of a two-winding transformer. Refer to the IEEE Red Book,[12] pp. 504 and 505, for further discussion.

On-Site Power

Although virtually all facilities obtain power from a public utility and that power is distributed throughout the facility, consideration of alternate sources for code-mandated life safety needs, for the economics of cogeneration, or for peak shaving during heavy-usage periods frequently enters the picture. Depending on its intended use, complete separation of the alternate sources and their distribution may be required. Refer Art. to 5.2.3, "Emergency Power Supply Systems," for a discussion of these alternative sources.

Motor-Generator Sets.   Power from motor-generator sets (also called M-G sets) can be primary power or standby power. In most facilities, standby generators are installed that run, outside of normally scheduled testing, only when the utility source of power is interrupted. These sources are moderately sized to power only those loads that are considered critical to operation of the facility and/or the safety of its occupants.

Standby units are usually fired with diesel oil or natural gas. Gas-fired units generally cost the same or less than diesel-fired units up to about 100 kW. Beyond this size, the cost of gas units comparable in capacity with diesel units becomes less and less favorable. Refer to the section on metering in Art. 5.2.1, "Service Entrances," for a discussion of cogeneration and peak load shaving using M-G sets. Also Art. 5.2.3, "Emergency Power Supply Systems," for a more detailed discussion of standby power systems.

Protecting the Distribution System

As noted in Art. 5.2.1, "Service Entrances," the main service equipment located at the service entrance is a disconnecting means and a point of separation between the serving utility and the use of electrical power by the owner. In addition, the service equipment protects the distribution system by limiting the amount of current that can flow from the service entrance to each use location. Achieving this protection requires using fuses, breakers, and relays that are sized and coordinated to protect the premises from distribution system overloads or utility voltage anomalies.

Protection of the distribution system takes place in many locations and takes various forms. There are several manufacturers' publications, including the IEEE Buff Book, that provide guidance on protective coordination and industry standards that deal with recommended practices. At the service entrance equipment, power system protective devices provide the intelligence and initiate the action that enables circuit-switching equipment to respond to abnormal or dangerous system conditions. Normally, relays control power circuit breakers rated above 600 V and current-responsive, self-contained elements operate multiple-pole, low-voltage circuit breakers to isolate circuits that experience overcurrents in any phase. Similarly, fuses function alone or in combination with other suitable means to isolate faulted or overloaded circuits. In other cases, special types of relays that respond to abnormal electric system conditions may activate circuit breakers or other switching devices to disconnect defective equipment from the remainder of the system.

In systems that employ circuit breakers, other than those with direct-acting devices that use fault current to power relaying and trip functions, there is always a risk that the system voltage can drop suddenly during a fault to a value too low for the protective devices to function. For this reason, station battery sets or capacitor trip devices are often employed to provide tripping energy. It is important to be able to test the circuit breaker and relay systems during planned or accidental power outages. The stored-energy system should be designed to provide these functions. Large battery sets are usually provided in large systems that have centralized switchgear. Capacitor trip devices are often used in small, low-voltage systems where manual charging of springs can supply the stored energy during prolonged power interruptions. Capacitor trips are often applied in small, remote, medium-voltage systems. In systems where direct-acting devices are used for overcurrent protection, functions such as undervoltage protection, sensitive ground-fault protection, or other similar protection may require stored-energy tripping sources.

Relays.   The overcurrent relay is the most common relay for short-circuit protection of an industrial power system. Overcurrent relays are typically of the electromagnetic attraction, induction, solid-state, or bimetallic element types. The solenoid type is the simplest overcurrent relay that uses the electromagnetic attraction principle. The basic elements of this relay are a solenoid wound around an iron core and a steel plunger or armature that moves inside the solenoid and supports the moving contacts. Other electromagnetic-attraction-type relays have hinged armatures or clappers of different shapes. These relays operate without any intentional time delay, usually within one-half cycle, and are called *instantaneous overcurrent relays*. The pickup or operating current for all overcurrent relays is adjustable. When the current through the relay coil exceeds the setting, the relay contacts close and initiate the circuit breaker–tripping operation. The relay operates on current from the secondary of a current transformer.

When the overcurrent is transient, such as that caused by starting a motor or some sudden overload of brief duration, the circuit breaker should not open. For this reason, induction disk overcurrent relays are used because they have an inherent time delay that permits a current several times in excess of the relay setting to persist for a limited period of time without closing the contacts. If a relay operates faster as current increases, it is said to have an inverse-time characteristic. Overcurrent relays are available with inverse, very inverse, and extremely inverse time characteristics to fit the requirements of the particular application. There are also definite minimum-time overcurrent relays that have an operating time that is practically independent of the magnitude of current after a certain current value is reached. Induction disk overcurrent relays provide for variation of the time adjustment and permit change of operating time for a given current. This adjustment is called the *time lever* or *time dial setting* of the relay. The previous brief discussion centered on overcurrent relays. There are a number of other types of relays, directional relays, differential relays, and ground-fault relays, each designed to provide protection from various circuit anomalies. Distribution equipment often has several of these different types. More in-depth discussion of these and other relays is beyond the scope of this section. For further information, refer to the bibliography.

Electromechanical Trip Devices.   For many years, low-voltage power circuit breakers were equipped with electromechanical series trip devices as the basic form of protection. State-of-the-art technology that uses solid-state devices has replaced electromechanical trip devices on low-voltage breakers; however, these older devices may still be available in service and in replacement breakers. The moving armature-type electromechanical series trip uses heavy copper coils that carry the full-load current to provide the magnetizing force. A dashpot that restrains the movement of the armature provides overload protection. Short-circuit protection is provided when the magnetic force suddenly overcomes a separate restraint spring. A separate adjustable unit is required for each trip rating.

Several combinations of adjustable long-time, short-time, and instantaneous overcurrent trip characteristics are available (see Fig. 5.2.2-9). These units do have some inherent disadvantages. The trip point will vary depending upon age and severity of duty, the devices have a limited calibration shape with a broad tolerance band, and selective coordination of tripping with other devices is difficult.

Solid-State Trip Devices.   In contrast to electromechanical devices, solid-state trip devices operate from a low-current signal generated by current sensors or current transformers in each phase. Signals from the sensors are fed into the solid-state trip unit that evaluates the magnitude of the incoming signal with respect to its calibration set points and trips the circuit breaker if preset values are exceeded. As with electromechanical trip devices, several overcurrent trip characteristics are available. In most instances, switch settings or ratings plugs determine the trip ratings of solid-state devices; thus, separate trip units are not required for

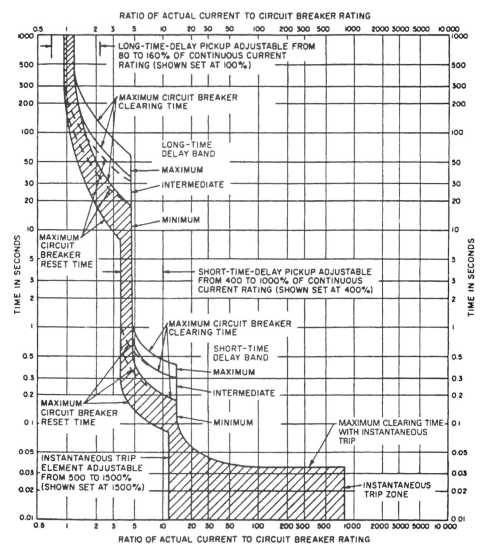

FIGURE 5.2.2-9 Typical time–current plot for electromechanical trip devices. (*Ref. 12, with permission.*)

each trip rating. In addition to phase protection, the solid-state trip device is available with integral ground-fault trip protection.

Solid-state trip devices are more accessible in the circuit breaker than trip devices and are much easier to calibrate because low-current values can be fed through the device to simulate the effect of an actual fault-current signal. Special care or provisions are sometimes necessary to guarantee predictable operation when applying solid-state trip devices to loads that have other than pure sinusoidal current wave shapes. Vibration, temperature, altitude, and duty cycle have virtually no effect on the calibration of solid-state trip devices; thus, excellent reliability is generally possible. The most important advantage of solid-state trip devices is the shape of the trip characteristic curve, which is essentially a straight line throughout its working portion (see Fig. 5.2.2-10). These devices have a very narrow and predictable tolerance, which enables several such devices to be selectively coordinated.

Continuous Current.   Standard molded-case circuit breakers are calibrated to carry 100 percent of their current rating in open air at a given ambient temperature (usually 25° or 40°C). In accordance with the NEC, these breakers, as installed in their enclosures, should not be continuously loaded over 80 percent of their current rating.

Low-voltage power circuit breakers and insulated-case circuit breakers are specifically approved for 100 percent continuous duty. These breakers can be continuously loaded to 100 percent of their current rating at 40°C ambient, when installed in proper enclosures.

Interrupting Rating.   The interrupting rating (or short-circuit current rating, as it is referred to for a low-voltage power circuit breaker) is commonly expressed in root-mean-square (rms) symmetrical amperes. It may vary with the applied voltage and is established by testing per UL or ANSI standards.

Short-Time Rating.   The short-time current rating specifies the maximum capability of a circuit breaker to withstand the effects of short-circuit current flow for a stated period, typically 30 cycles or less, without opening. This provides time for downstream protective devices closer to the fault to operate and isolate the circuit.

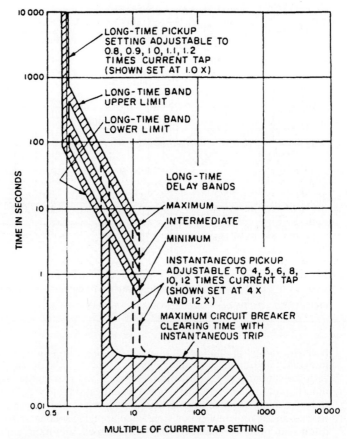

FIGURE 5.2.2-10   Typical time–current plot for solid-state trip devices. (*From Ref. 12, with permission.*)

The short-time current rating of a low-voltage power circuit breaker without instantaneous trip characteristics is equal to the breaker's short-circuit interrupting rating. Most molded-case circuit breakers are not provided with a short-time current rating; however, some higher-ampere-rated molded-case circuit breakers are provided with a short-time current rating in addition to the short-circuit interrupting rating.

A circuit breaker that has an instantaneous trip function should not be used where continuity of service requires only long-time and short-time delay functions.

Control Voltage. This is the ac or dc voltage to be applied to control devices that are intended to open or close a circuit breaker. These devices are normally supplied with a voltage rating needed for a particular control system.

Where the interrupting ratings of conventional molded-case circuit breakers or low-voltage power circuit breakers are not sufficient for a particular system application, other options are available, including current-limiting circuit breakers, integrally fused circuit breakers, and fuse/breaker coordination combinations.

Trip Unit. The trip unit is an integral part of the circuit breaker. It may be electromechanical (thermal-magnetic, mechanical dashpot) or solid-state electronic. It senses abnormal current conditions by continually monitoring the current flowing through the circuit breaker. Depending on the magnitude of the current, the trip unit will initiate an inverse-time response or an instantaneous response. This action will cause a direct-acting operating mechanism to open the circuit breaker contacts and interrupt current flow. The following discussion covers those trip unit characteristics commonly available in low-voltage circuit breakers:

- Continuous-current rating
- Long-time current
- Long-time time delay
- Short-time current
- Short-time time delay
- Instantaneous current response

The continuous-current rating may be fixed or adjustable. Some constructions may require replacing all or part of the trip unit to change the continuous rating. Overcurrent trip characteristics are a multiple or percentage function or the continuous-current rating.

Most molded-case circuit breakers are not provided with a long-time adjustment or short-time function. The inverse-time response characteristic is strictly related to how long a particular level of current has been flowing. Molded-case circuit breakers that have electronic trip units and low-voltage circuit breakers usually have long-time and short-time current adjustments. These adjustments are used to set the current level (pickup point) at which the circuit breaker trip unit begins timing to initiate tripping action. A low-voltage power circuit breaker may also be provided with or without an instantaneous trip function.

Power Fuses (over 600 V). There are four types of power fuses rated over 600 V: (1) the distribution fuse-cutout type, (2) the current-limiting type, (3) the sold-material type, and (4) the electronic type. Most fuses are available in designs that comply with ANSI E rating, R rating, or K rating requirements. Electronic fuses are very versatile power fuses that operate at a specific minimum pickup fault current, which is selectable based on coordination requirements, rather than within E or K rating requirements. Expulsion fuses can also be E-, K- and T-rated, and also meet ANSI standards. The K and T ratings refer, respectively, to relatively fast and slow melting expulsion fuses. The following is a guide to the ANSI designations:

- *E designation.* Fuses rated 100E or less melt in 300 s at a current value between 2.0 and 2.4 times the E number. Fuses rated above 100E melt in 600 s at a current value between 2.2 and 2.64 times the E number. If the current is higher than 2.0 to 2.4, or 2.64 times the E number, the user must consult the time–current curves for that particular fuse.

- *R designation.*   The fuse melts in 15 to 35 s when the current equals 100 times the R number. If the current is higher than 100 times the R number, the user must consult the time–current curves for that fuse.

- *C designation.*   The fuse melts in 1000 s at a current value between 1.7 and 2.4 times the C number. If the current is higher than 2.4 times the C number, the user must consult the time–current curves for that particular fuse.

Distribution Fuse Cutouts.   This type of fuse is generally used in distribution system cutouts or disconnect switches. An arc-confining tube with a deionizing fiber liner and fusible element is used to interrupt a fault current. The arc is interrupted by the rapid production of pressurized gases within the fuse tube, which extinguish the arc by expulsion from the open end or ends of the fuse.

Enclosed, open, and open-link types of expulsion fuses are available for use as cutouts. Enclosed cutouts have terminals, fuse clips, and fuse holders mounted completely within an insulating enclosure. Open cutouts have these parts completely exposed. Open-link cutouts have no integral fuse holders; the arc-confining tube is incorporated as part of the fuse link (Fig. 5.2.2-11). Because gases are released rapidly during the interruption process, the operation of expulsion-type fuses is comparatively noisy. They are rarely, if ever, applied in an enclosure because of the special care required to vent any ionized gases that might be released and would cause a flashover between internal live parts. Fuse cutouts and disconnect switches are used indoors to protect industrial plant distribution systems and to provide fault and overload protection for distribution feeder circuits, transformers, and capacitor-bank fault protection. They have an inverse time–current characteristic that is compatible with that of standard overcurrent relays.

Current-Limiting Power Fuses.   This type of fuse is designed so that melting of the fuse element introduces a high arc resistance into the circuit in advance of the prospective peak current of the first half-cycle. If the fault-current magnitude is sufficiently high, the arc voltage that rapidly escalates forcibly limits the current to a peak value that is lower than the prospective peak. This reduced peak value, referred to as the peak let-through current, may be a small

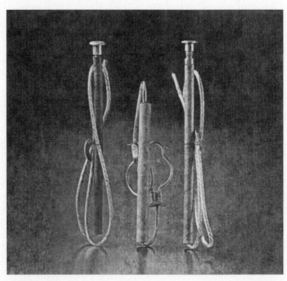

FIGURE 5.2.2-11 Medium-voltage fuse links. (*Courtesy of Bussmann Company, with permission.*)

fraction of the peak current that would flow without the current-limiting action of the fuse. If the fault-current magnitude is not sufficiently high, current limitation is not achieved.

A general-purpose current-limiting fuse is defined as a fuse that can interrupt all currents that range from the rated maximum interrupting current to the current that melts the fusible element in 1 hour. This type of fuse is not intended to provide protection against low-magnitude overload currents, because it can reliably interrupt only currents above approximately twice its continuous rating (for E-rated fuses), and usually above approximately three times its continuous rating (for non-E-rated fuses). Typical applications are for the protection of power transformers, potential transformers, and feeder circuits in primary switches. A typical time–current characteristic for this type of fuse makes selective coordination with other fuses easy but may be more difficult to coordinate with overcurrent relays.

Current-limiting fuses of the R-rated type are most commonly applied in motor starters using contactors that cannot interrupt high fault currents. The R designation is not related to the continuous-current rating, although each fuse does have a permissible continuous current that is published by the manufacturer. The R number is 1/100 of the number of amperes required to open the fuse in about 20 s. The fuse provides the necessary short-circuit protection but must be combined with an overload protective device to sense lower values of overcurrent that are within the capability of the contactor. Fuses of this type are generally designed to interrupt currents that melt the fuse elements in less than 100 s, but the fuse is not self-protecting on lower overcurrents (Fig. 5.2.2-12).

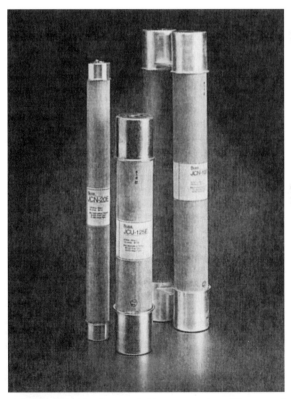

FIGURE 5.2.2-12 Current-limiting power fuse. (*Courtesy of Bussmann Company, with permission.*)

Solid-Material Power Fuses. This type of fuse uses densely molded, solid boric acid powder as the lining for the interrupting chamber. This solid-material lining liberates incombustible, highly deionized steam when subjected to the arc established by melting the fusible element. Solid-material power fuses have higher interrupting capacities than fiber-lined power fuses of identical physical dimensions, produce less noise, need less clearance in the path of the exhaust gases, and, importantly, can be applied with normal electrical clearance indoors or in enclosures when equipped with exhaust control devices. Exhaust control devices provide silent operation and contain all arc interruption products. Indoor mountings with solid-material power fuses can be furnished with an integral hook-stick-operated, load-current interrupting device, thus providing single-pole live switching in addition to fault-interrupting functions provided by the fuse. Many of these fuses also include an indicator that shows when the fuse has operated. These advantages, plus their availability in a wide range of current and interrupting ratings and time–current characteristics that conform to all applicable standards, have led to the wide use of solid-material power fuses in utility, industrial, and commercial power distribution systems (Fig. 5.2.2-13).

Electronic Power Fuses. This technological development in power fuses combines many of the features and benefits of power fuses and relays. Electronic fuses generally consist of two separate components: an electronic control module that provides the time–current characteristics

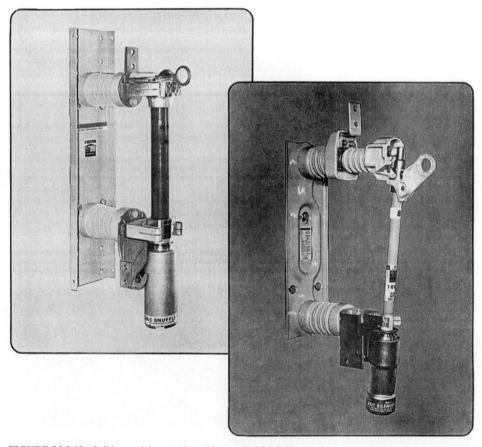

FIGURE 5.2.2-13 Solid-material power fuse. (*Courtesy of S&C Electric Company, with permission.*)

of the energy to initiate tripping, and an interrupting module that interrupts current when an overcurrent occurs. When joined together, these two modules are held in a suitable holder that fits in a mounting. A current transformer located within the control module powers the logic circuits employed in the control module that may have instantaneous tripping characteristics, time-delay tripping characteristics, or both. These two circuits may be used alone or in combination to provide a variety of time–current characteristics. When an overcurrent occurs, the control module triggers a high-speed gas generator that separates the main current path in the interrupting module and transfers the current into the current-interrupting ribbon elements, which then melts and burns back. Only the interrupting module is replaced following fuse operation.

Electronic power fuses are suitable for service entrance protection and coordination of industrial and commercial distribution circuits because these fuses have high current-carrying capability and incorporate unique time–current characteristics designed for superior coordination with source-side overcurrent relays and load-side feeder fuses. They are ideally suited for load-feeder protection and coordination in industrial, commercial, and utility substations because of their high continuous and interrupting ratings. Specific time–current characteristics are available for primary-side protection of transformers and for application at the heart of an underground loop system to provide backup protection for pad-mounted transformers that contain fuses with a limited interrupting capability. Indoor mountings with electronic fuses can be furnished with an integral hook-stick-operated, load-current interrupting device, thus providing for single-pole live switching in addition to the fault-interrupting functions provided by the fuse (Fig. 5.2.2-14).

Some electronic fuses also include indicators that make it easy to determine which fuse has operated. Because this fuse is available with many unique, time–current characteristics, it can easily coordinate with both line-side overcurrent relays and load-side power fuses.

FIGURE 5.2.2-14   Electronic power fuse. (*Courtesy of S&C Electric Company, with permission.*)

Low-Voltage Fuses (600 V and Less). Plug fuses are of three basic types, all rated 125 V or less to ground and up to a maximum of 30 A with an interrupting rating of 10,000 A (Fig. 5.2.2-15). Following are the three types:

1. Edison base with no time delay in which all ratings are interchangeable
2. Edison base with a time delay and interchangeable ratings
3. Type-S base, available in three noninterchangeable current ranges: 0 to 15, 16 to 20, and 21 to 30 A.

Cartridge fuses may be either renewable or nonrenewable. Nonrenewable fuses are factory-assembled and must be replaced after operating. Renewable fuses can be disassembled and the fusible element replaced. Renewable elements are usually designed to give a greater time delay than ordinary nonrenewable fuses, and in some designs the delay on moderate overcurrents is considerable. The renewable-type fuse is not available in ratings above 10,000 A.

Non-Current-Limiting Fuses (Class H). These fuses interrupt overcurrents up to 10,000 A but do not limit the current that flows in the circuit to the same extent as recognized current-limiting fuses. Generally, cartridge fuses should be used only in circuits in which the maximum available fault current is 10,000 A and the protected equipment is fully rated to withstand the peak available fault current associated with this fault duty, unless such fuses are specifically applied as part of an equipment combination that has been type-tested and designed for use at higher available fault-current levels.

Current-Limiting Fuses. Current-limiting fuses are intended for circuits where the available short-circuit current is beyond the withstand capability of downstream equipment or the interrupting rating of ordinary fuses or standard circuit breakers. An alternating current-limiting fuse safely interrupts all available currents within its interrupting rating, limits the clearing time at rated voltage within its current-limiting range to an interval equal to or less than the first major or symmetrical current loop duration, and limits peak let-through current to a value less than the peak current that would be possible if the fuses were replaced with a solid conductor of the same impedance. Therefore, a current-limiting fuse places a definite ceiling on the peak let-through current and thermal energy and provides equipment protection against damage from excessive magnetic stresses and thermal energy (see the ANSI/UL 198 series standards for maximum let-through limits).

These fuses are widely used in motor starters, fused circuit breakers, fused switches of motors and feeder circuits (for protection of busway and cable), and many other applications.

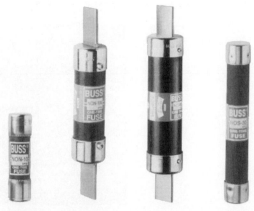

FIGURE 5.2.2-15 One-time general-purpose fuses. (*Courtesy of Bussmann Company, with permission.*)

Dual-Element or Time-Delay Fuses.   A dual-element fuse has current-responsive elements of two different fusing characteristics in series in a single cartridge. The fuse has a one-time use in operation, and the fast-acting element responds to overcurrents that are in the short-circuit range. The time-delay element permits short-duration overloads but melts if these overloads are sustained. The most important application for these fuses is motor and transformer protection. they do not open on motor starting or transformer magnetizing inrush currents but still protect the motor and branch circuits from damage by sustained overloads. Table 5.2.2-2 summarizes fuse characteristics and ratings.

Protective Coordination

As noted earlier, protective overcurrent devices are required at various locations to protect the distribution system. Besides serving as a manual means of interrupting power at a particular location, for example, to do maintenance on or replace a device or piece of equipment, these devices automatically interrupt power to protect that equipment. Necessarily, as one moves out in the system, these protective devices must be installed in series with one another, and their interrupting settings must be coordinated to prevent a small problem on one branch of the system from causing the entire system to lose power. Refer to Sec. 10.6, "Electrical Systems in Laboratories," for a detailed discussion of this important activity.

Methods of Electrical Power Distribution.   Once the service entrance and main distribution equipment for a building or a site are established, the means of distributing power throughout the facility must be determined. Refer to Art. 5.2.4, "Raceway and Cable Systems," for a detailed description of raceways, bus ducts, and so on.

Battery Sources

Where immediate power is required upon loss of utility power or where standby M-G set or cogeneration power sources are not available, battery-powered standby lighting units and/or uninterruptible power systems (UPSs) can be utilized. Refer to the Art. 5.2.3, "Emergency Power Supply Systems," for details about this equipment.

TABLE 5.2.2-2   Fuse Characteristics and Typical Ratings

Fuse Type	Voltage Range	Amperage Range	Time Delay	Interrupting Rating	Current Limiting	Comments
Misc.*	125, 250, 300, 500, 600 V		No		No	Different dimensions from UL class G,H,J,K, CC,T, or L.
Class H	250 or 600 V	Up to 600 A	Yes, if so labeled	Up to 10,000 A	No	Renewable or nonrenewable of copper or zinc link construction.
Class K	250 or 600 V	Up to 600 A		50,000 A 100,000 A or 200,000 A	No	High interrupting rating same physical sizes as type H.
Class R	250 or 600 V	Up to 600 A		200,000 A	Yes	Class H dimensions but with built-in rejection notches.
Class J	600 V or less	Up to 600 A	Yes, if so specified	200,000 A	Yes	Do not match Class H dimensions.
Class L	600 V or less	601 to 6000 A		200,000 A		Do not match Class H dimensions.

* Miscellaneous fuses are those manufactured for application in control circuits, special electronic or automotive equipment, and such.

Downstream Distribution Equipment

Electric power is distributed from the main service equipment described earlier via various types of raceway systems to remote distribution locations nearer the point of use. At these locations, downstream distribution equipment (i.e., motor control centers, distribution panels, lighting panels, and power panels) may be used to provide local distribution to equipment such as motors, lighting fixtures, receptacles, and the like.

Motor Control Centers.   In lieu of individually mounting motor starters and doing the related control interface wiring in the field, starters can be located in tailor-made assemblies of conveniently grouped control equipment primarily used for power distribution and control of motors (Fig. 5.2.2-16). They contain all necessary buses, incoming line facilities, and safety features to afford the most convenience by saving space and labor and by providing adaptability to ever-changing conditions with a minimum of effort and a maximum of safety. NEMA ICS 2 governs the type of enclosure and wiring.[13] NEMA Type 1, 2, and 12 enclosures are generally available. Wiring of motor-control centers conforms to two NEMA classes and three types. Class I construction provides for no wiring by the manufacturer. Class II requires prewiring by the manufacturer with interlocking and other control wiring being completed between compartments of the center. With Type A, no terminal blocks are provided; with Type B, all connections within individual compartments are made to terminal blocks; and with Type C, all connections are made to a master terminal block located in the horizontal wiring trough at the top or bottom of the center. The ideal wiring specification for minimum field installation time and labor is NEMA Class II, Type C wiring. The wiring specification most frequently used by industrial contractors is Class I, Type B wiring.

FIGURE 5.2.2-16   Motor control center. (*Courtesy of Square D Company, with permission.*)

NEMA ICS 2 specifies that a control center shall carry a short-circuit rating defined as the maximum available rms symmetrical current in amperes permissible at line terminals. The available short-circuit current at the line terminals of the motor control center is computed as the sum of maximum available current of the system at the point of connection and the short-circuit current contribution of the motors connected to the control center. Many manufacturers commonly show only the short-circuit rating of the bussing on the nameplate. Therefore, it is very important to establish the actual rating of the entire unit and, in particular, the plug-in units (i.e., circuit breakers, fusible disconnects, starters, etc.), especially for applications where available fault currents exceed 10,000 A.

Panelboards.   The panelboard is usually the last piece of distribution equipment between the source and the load. Although some large loads are fed directly from the switchgear or switchboard, most loads are fed from a panelboard, which takes a large block of power right from a small service or through the distribution equipment of a large service and divides it to feed multiple loads (Fig. 5.2.2-17).

There are two basic types of panelboards: (1) for power, and (2) for lighting and appliances. The code specifically recognizes only lighting and appliance branch-circuit panelboards, which are clearly defined in Sec. 384-14; all others fall outside this definition. A lighting and appliance branch-circuit panelboard has more than 10 percent of its overcurrent devices rated 30 A or less, for which neutral connections are provided. Thus, a *power* panelboard that has all three-pole circuit breakers, more than 10 percent of the poles 30 A or less that feed only motor, and a neutral bar installed (provided) even though unused, is a *lighting and appliance panelboard,* as defined by the NEC. Conversely, a panelboard that feeds only outdoor 480-V HID lighting with two-pole circuit breakers and has no neutral is *not* a lighting and appliance panelboard according to the NEC.

Lighting panelboards range from the residential load-center type to the standard-duty type. The residential load center is a smaller panel that often uses half-size or twin circuit

FIGURE 5.2.2-17   Typical panelboard. (*Courtesy of Square D Company, with permission.*)

breakers. Fusible load centers are also available. Load centers generally have plug-in circuit breakers, smaller wiring gutters, simpler hardware, and are intended for lighter duty than standard panels. Standard-duty lighting panelboards generally have a choice of plug-in or bolt-on circuit breakers, heavy-duty hinges, latches and locks, 4-in minimum wiring gutters, and can accommodate many variations and accessories. Lighting and appliance panels are limited to a maximum of 42 overcurrent devices (circuit breaker poles or fuses) in any one enclosure, not counting those in the mains.

Power panelboards have no limitations except those set by the manufacturers resulting from physical limitations and the need to dissipate heat to meet UL requirements. Main bus ratings are usually limited to 1600 A, although larger units have been built. The number of poles is limited only by the sizes of the devices used and the space available in the enclosure. Some power panels are available with built-in motor starters combined with motor disconnect switches or circuit breakers. This motor-starter panelboard is a form of "poor man's motor control center," that provides far more convenience and lower cost than trough-mounted starters, but less flexibility than higher-cost motor control centers

Branch-circuit switches or circuit breakers are usually equipped with mechanical set-screw lug terminals approved for either copper or aluminum conductors. However, many specifiers do not accept mechanical terminations for aluminum conductors; they insist on compression connectors. Two manufacturers offer compression connectors or branch circuits as a UL-labeled alternate. Otherwise, some means of aluminum-to-copper adaptation is required, and it is important to provide adequate gutter space for these splices or adapter units. All manufacturers offer compression connectors for the main buses of both power and lighting panelboards when specified.

Both lighting and power panels can be UL-labeled for use as service entrance equipment, either with a main disconnect or in accordance with the six-disconnect rule, NEC Section 230-71. For direct application in systems with high available short-circuit currents, circuit-breaker lighting panels are available with fusible main-switch disconnects. With these panels, the current-limiting fuses will limit the fault currents to values that the branch circuit breakers can withstand. The main disconnects in these applications might also be integrally fused circuit breakers or current-limiting circuit breakers.

Panelboards are available with many modifications and special features or accessories. Among these are remote-control switches or lighting contactors, time clocks, split buses, sub-feed or feed-through lugs, grounding bars for ground wires, and built-in ground-fault relays. Lighting panels can be supplied with ground-fault circuit interrupter-type breakers, and where used for switching 120-V fluorescent lighting, breakers must be switching-duty type (SWD).

Type SWD breakers are also available in 277-V ratings. Enclosure variations include raintight, dusttight, weatherproof or explosion-proof construction, increased gutter spaces, door-in-door trims (inner door over circuit breaker handles and outer door over wiring spaces), special locks, hinges and hardware, stainless-steel trim, and the like. In today's electronic environments, panels shielded and filtered against electromagnetic interference (electronic grade panels) are available at higher cost, where necessary.

Motor Starter Panelboards.   Rather than using an individual mounting, a small number of motor starters can be grouped into a panelboard. Motor starter panelboards consist of combination units using either molded-case or motor circuit protector fusible disconnects. The combination starters are factory-wired and assembled. Class A provides no wiring external to the combination starter, and Class B provides control wiring to terminal blocks furnished near the side of each unit. When a large number of motors are to be controlled from one location or additional wiring between starters and to master terminal blocks is required, conventional motor control centers (MCCs) are most commonly used.

Multiple-Section Panelboards.   Both lighting and appliance panelboards and power distribution panelboards that require more than one box are called *multiple-section panelboards*. Unless a main overcurrent device is provided in each section, each section should be furnished with a main bus and terminals of the same rating for connection to one feeder. The three methods commonly used for interconnecting multiple-section panelboards are as follows:

1. *Gutter tapping.* Increased gutter width may be required. Tap devices are not furnished with the panelboard.

2. *Subfeeding.* A second set of main lugs (subfeeds) is provided directly beside the main lugs of each panelboard section, except the last in the lineup.

3. *Through-feeding.* A second set of main lugs (through-feed) is provided on the main bus at the end opposite from the main lugs of each section, except the last in the lineup. This method has the undesirable feature of allowing the current of the second panelboard section to flow through the main bus of the first section.

CONTRIBUTORS

Joseph R. Flocco, Jr., P.E., Consulting Engineer, Account Manager, Siemens Energy and Automation, Alpharetta, Georgia

Arjun Rao, P.E., Vice President, BR+A Consulting Engineers, Inc., Boston, Massachusetts

Cornelius Regan, P.E., CLEP, Principal, C. Regan Associates, Rockland, Massachusetts

William J. Schlageter, P.E., Principal, Novare Engineers, Inc., Providence, Rhode Island

David L. Stymiest, P.E., SASHE, CEM, Senior Consultant, Smith Seckman Reid, Inc., New Orleans, Louisiana

William Washville, Cutler-Hammer Corporation, Boston, Massachusetts

William P. Wohlfarth, P.E., Sr. Electrical Engineer, Massachusetts Institute of Technology, Cambridge, Massachusetts

REFERENCES

1. Electrification Council, *Industrial and Commercial Power Distribution Course,* 8th Edition, Edison Electric Institute, New York, 1997.

2. NFPA 70-1999, National Electrical Code, National Fire Protection Association, Quincy, MA, 1999.

3. NFPA C2-1997, National Electrical Safety Code, National Fire Protection Association, Quincy, MA, 1997.

4. IEEE Green Book, IEEE Standard 142-1991, *IEEE Recommended Practice for Grounding of Industrial and Commercial Power Systems,* Institute of Electrical and Electronics Engineers, Inc., New York, 1991.

5. ANSI/NFPA 780-1997, *Standard for the Installation of Lightning Protection Systems,* National Fire Protection Association, Quincy, MA, 1997.

6. UL 96, *Lightning Protection Components,* Underwriters Laboratories, Inc. Northbrook, IL, 1994.

7. NFPA 70B-1998, *Recommended Practice for Electrical Equipment Maintenance,* National Fire Protection Association, Quincy, MA, 1998.

BIBLIOGRAPHY

ANSI publications C37 Series, *Circuit Breakers, Switchgear, Substations and Fuses,* American National Standards Institute, New York, 1998.

IEEE Gray Book, IEEE Standard 241-1990 (R1997), *IEEE Recommended Practice for Electric Power Systems in Commercial Buildings,* Institute of Electrical and Electronics Engineers, Inc., New York, 1997.

Cutler-Hammer, *Consulting Application Guide,* 11th Edition (excerpts of which are used by permission), Cutler-Hammer Corporation, Cleveland, OH, 1999.

S&C Electric Company 600 Series Descriptive Bulletins, Chicago, IL, 1999.

Cooper Power Systems 210 Series Electrical Apparatus Data Sheets, Waukesha, WI, 1999.

Square D Company Digest, (excerpts of which are used by permission), Chicago, IL, 1997.

Guidelines for Specifying Electrical Systems, *Electrical Construction and Maintenance,* Intertec Publishing Corporation, Overland Park, KS, 1980.

ARTICLE 5.2.3
EMERGENCY POWER SUPPLY SYSTEMS

Cornelius Regan, P.E., CLEP, Principal
C. Regan Associates, Rockland, Massachusetts

INTRODUCTION

The requirement for an alternative source of emergency or standby electric power is becoming more urgent as the need for electricity keeps growing, but at the same time the utility companies are cutting back on expanding their generation capacities. This results in power outages and low and unstable voltages during peak use, especially during the summer months when the use of electricity for air conditioning is at its highest. Other utility company outages result from cable failure, transformer failure, overloaded lines, and flooded manholes. Coupled with these outages are non–utility company and user-planned and unplanned outages, such as those caused by the weather and environmental conditions, inadequate electrical systems, floods, tornadoes, hurricanes, cyclones, flooded vaults, and downed trees, poles, and overhead wires. All of these events justify the installation of an on-site alternate source of electrical power, although it is required by code for most buildings.

Figure 5.2.3-1, taken from IEEE STD 446 1987, shows the average number of thunderstorm days per year for the United States in every geographic location. Figure 5.2.3-2, taken from the same IEEE standard, shows the tornado density in the southwestern United States. Both types of information are useful in determining the probability of power failures resulting from these natural events. Most utility companies that have proper shielding and surge arresters can provide service that is of equal reliability in any area.[1]

THE NEED FOR EMERGENCY POWER

The primary need for an alternate source of emergency or standby power, other than that required by code for life safety and life support, is to maintain continuous power to (1) security systems; (2) 24-hour production operations; (3) critical business computers that require 24-hour operation; (4) emergency air, sea, rail, and road transport services; (5) property protection systems such as fire alarms, fire pumps, and sprinkler systems; and (6) critical lighting

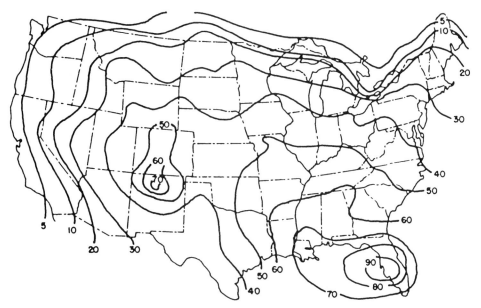

FIGURE 5.2.3-1   Average number of thunderstorm days per year from ANSI/IEEE Std 446-1987. (*Courtesy of Institute of Electrical and Electronics Engineers, Inc.*)

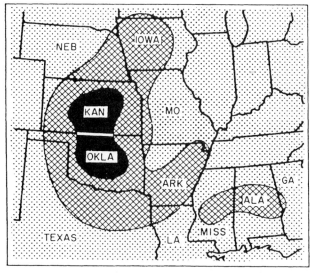

FIGURE 5.2.3-2   Approximate density of tornadoes from ANSI/IEEE Std 446-1987. Cross-hatched areas—100 to 200 in 40 years; black area—more than 200 in 40 years. (*Courtesy of Institute of Electrical and Electronics Engineers, Inc.*)

systems such as airport landing lights, warning lights, and transportation lighting. The alternate source can also be used for load sharing, peak shaving in conjunction with the utility company, and as an economic advantage to the user. The following codes and standards should be reviewed for emergency power requirements: NFPA 101, NFPA 110, NFPA 70, NFPA 111, OSHA, elevator codes, local and state codes, and IEEE-446.

Feasibility Study

Once the need has been established for an alternate emergency and standby power system, it is necessary to study the facility completely to assess its emergency power requirements. The study describes the type of occupancy (commercial, industrial, health care, etc.) (see NFPA 101 for building classifications), the physical characteristics of the building such as its floor area, the number of floors above and below grade, the location and quantity of elevators, the fire pump capacity, the smoke evacuation requirements, the emergency ventilation system requirements, and all other equipment and devices that must be operated in an emergency or fire (overhead doors, fire doors, vertical and horizontal transportation systems, fire dampers, etc.).

Information from this study, together with a conceptual design and equipment load list, forms the basis for selecting an emergency power supply for the facility. The source of power for the emergency power supply system may be a battery system, an on-site generator set, or a source of supply separate from the existing utility company or a source from another utility company.

An emergency power supply system (EPSS) consists of an emergency power source (EPS) that supplies power to an electrical distribution system consisting of overcurrent protective devices, conductors, transfer switches, disconnecting means of control, and supervisory devices up to the point of power use.

The EPS is the source of electric power of the required capacity and quality for an EPSS, including all of the related electrical and mechanical components of the proper size and/or capacity for generating the electric power required at the EPS output terminals. For a rotary energy converter, components of the EPS include a prime mover, a cooling system, a generator, an excitation system, a starting system, a control system, a fuel system, and a lube system (if used).[2] These components are discussed later.

The next tasks are to consider the design issues and equipment selection for the EPSS selected. A preliminary, one-line diagram is prepared that shows the existing utility service, the normal power distribution system, the EPS (this is the generator set complete with fuel, lube, and control system), and the EPSS (this includes the EPS and all associated cable and equipment up to the load side of the transfer switch necessary for a completely safe, reliable, and functional system).

Design Issues

EPS.   The feasibility study that provides the basis for designing an EPSS is also the basis for an EPS. Because a diesel or gas generator, if approved by the authority having jurisdiction (AHJ), is the preferred source of emergency power for the life safety and critical branches of the emergency system, it does not necessarily mean that the design of an alternate power source should be limited to a single fuel generator. A generator set supplies power to the essential loads, as illustrated in Fig. 5.2.3-3; a generator set using fuels other than diesel oil could supply the nonemergency but economically critical loads.

Reliability.   An on-site diesel generator is considered a more reliable source of power compared with a second utility feeder or other fuel-driven generator in case of a catastrophic disaster such as a hurricane, tornado or earthquake. The diesel generator has minimum piping. Readily available fuel is stored on-site, is stable, and can be stored for long periods without deterioration. There are also disadvantages to a diesel-driven generator such as noise, expense,

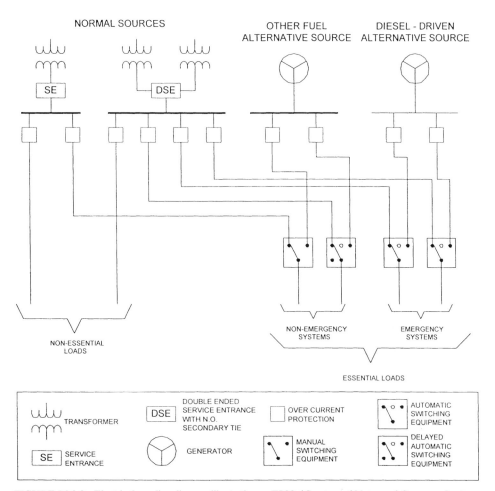

FIGURE 5.2.3-3 Electrical one-line diagram illustrating an EPSS. (*Courtesy of Motor and Generator Institute, Healthcare Circuit News.*)

and fuel odor. The pros and cons of each type of prime mover should be evaluated by an experienced engineering professional before selecting a generator set for each application.

Generator Capacity and Rating.   An EPS must have enough capacity to adequately start and carry all loads simultaneously. This is different from legally required standby systems, which must have enough capacity to start and carry the loads of all equipment that is intended to be operated at one time. Optional standby systems have to be sized only to carry loads that are selected by the user.[3]

Peak Shaving.   Emergency generating capacity is allowed for peak shaving or load sharing where automatic means is provided to assure adequate power to the emergency loads. After that, the legally required standby loads can be energized. There are no restrictions on the use of optional standby units for load sharing or peak shaving.[3] Peak shaving/load sharing operation may be acceptable for satisfying the test requirements described in the National Electric Code (NEC), Article 700.[4]

Generator Set Location.   One of the most important issues in EPSS selection is where to locate the generator set. This decision is very often made for aesthetic reasons. The generator should be seismically located on-grade and centrally positioned in relation to the emergency loads and normal switchgear. Avoid locating generators in basements where they can be subjected to flooding or on a roof structure where they may be inaccessible in case of a fire. The generator room should be well ventilated; approximately 50 to 200 cfm of air are required per kilowatt of generator set capacity. This requirement should be considered before locating a generator in a space not easily ventilated such as a basement.

Generator Enclosure.   The generator should be located in its own soundproofed room.

Fire Separation.   The generator will have the minimum fire separation from the existing facility's normal electrical system, as dictated by state and local codes. Some states require 1-hour fire separation, and other states require 2-hour fire separation.

Emergency Feeders.   The emergency feeders should not enter or traverse the normal electrical switchboard room, electrical vault, or manhole. The wiring from the EPSS to emergency loads shall be independent of all other wiring. All of the feeders emanating from the generator and generator distribution panel must have the minimum fire-rated protection stipulated by NEC Article 700. State and local codes or authorities having jurisdiction may invoke more restrictive requirements.

There are several methods of addressing the 1- or 2-hour fire separation criteria, but not all methods are approved or listed. Some authorities having jurisdiction, for example, do not accept conduit fire wraps, whereas others may. The use of fire-rated mineral-insulated (MI) cable is a widely accepted method of meeting fire separation criteria for emergency feeders. Another approach is to use architectural treatments such as listed fire-rated enclosures. Mineral-insulated cable may require less room to install than new fire-rated enclosures in existing facilities, but the required fire-rated treatment of its splice boxes and the necessity of closely following the manufacturer's recommended installation methods can make it a challenging installation as well. The engineer should carefully consider all such issues, including the cost of alternatives, before selecting the fire-rated treatment for each project.

Mechanical Issues

Engine Selection.   The generator set, its starting battery, and their maintenance will always be the most critical parts of the EPS, but in some cases the application of the generator set makes it difficult for it to perform at its rated capacity.[5] These are the five requirements for a satisfactory installation:

1. A generator set should be sized so that it always operates at more than 30 percent of its capacity to avoid wet-stacking. NFPA 110 addresses the need to avoid wet stacking during regular load testing and also discusses the correlation between generator set loading and engine exhaust temperature. Load banks can also be rented or permanently installed to provide sufficient loading, provided that the installation can connect the load banks safely into the power system. If a load bank is normally used for load-testing a generator set, then provisions are necessary for switching off the load bank and allowing the essential power system loads to be powered from the generator, if an actual power outage occurs during the test period.

2. The generator should be located and isolated on a solid foundation.

3. The generator and its ancillary equipment must be accessible for maintenance and have the proper working clearance.

4. The generator must have a proper air supply.

5. The generator must have sufficient capacity to assume the steady-state emergency load without overheating, as well as the starting kilovolt amperes (SkVA) for motors and other high-inrush current loads, while maintaining voltage and frequency variations within the maximum and minimum parameters recommended by codes, manufacturers' requirements, NEMA standards for driven equipment, NFPA standards, and so forth.

Motor Loads.   Special consideration must be given to motor loads because they can draw from 4 to 12 times their full-load running current when starting. The generator must have sufficient kilovolt amperes to start the motor loads on the emergency system. The total capacity of the generator is determined by adding the skVA of all the motors to be started simultaneously and the other loads that will be connected to the generator and using an acceptable voltage dip of between 10 and 30 percent. Single-phase loads should not exceed 50 percent of the generator rating and should be evenly balanced among the three phases of the generator.

The decision to purchase a four-stroke engine or a two-stroke engine requires a comparison of the user's needs with the advantages and disadvantages of each type of engine. Technical papers that discuss the pros and cons of each engine are available from the engine manufacturers, and it requires a detailed engineering analysis to sort out the salient points for each facility.

Derating.   The generator should be derated for altitude and temperature; manufacturers provide this information.

Seismic Design Considerations.   An EPS must be seismically supported in relation to the seismic zone in which it is located. Seismic zone maps are available to the engineers to assist them in this effort. Consideration must be given to the location of the EPS equipment both as it relates to the building structure and to the effects of an earthquake.[6]

Bolts, anchors, hangers, braces, and other restraining devices should be provided to limit earthquake-generated differential moments between the EPS nonstructural equipment and the building structure. However, the degree of isolation required for vibration and noise control of the EPS equipment should be maintained.[6]

Spring isolators could amplify the relatively small forces generated by an earthquake to a level that could damage the EPS. These isolators should incorporate seismic restraining devices whose requirements are determined by both the engineer and the supplier of the equipment. They must be secured to the equipment base and to the floor or foundation, and stops should be added to limit the motion of the isolators in all directions.[7]

Noise and Vibration.[7]   Mechanical systems with mass and elasticity are capable of relative motion. When the motion is repetitive, it is vibration. Engines produce vibrations due to combustion forces, torque reaction, structural mass and stiffness combinations, and manufacturing tolerance on rotating components. These unbalanced forces create a range of undesirable conditions ranging from unwanted noise and high stress levels to ultimate failure of the engine or generator components.[7] To reduce or eliminate vibration and consequently noise, it is necessary to analyze the engine and generator for critical linear and torsional vibration.[7]

Noisy or shaky machines exhibit linear vibration. Torsional vibration is exhibited by high-frequency noise. However, machinery vibration is very complex and has many different frequencies. Measurements of velocity, acceleration, and displacement are used to diagnose problems.[7] Vibrations carried throughout an enclosure cause early failure of auxiliary equipment such as relays, switches, gauges, and piping.[7]

There are several methods for isolating vibration in an EPS such as the use of commercially available fabricated isolators or bulk isolators. These may be spring isolators, rubber isolators, or isolation of the block foundation.

Generator sets can withstand all self-induced vibrations, and no further isolation should be required merely to prolong their service life. However, severe vibrations from surrounding equipment can harm a generator set that is inoperative for long periods of time. Bearings and shafts can beat out and ultimately fail if these vibrations are not isolated. Exterior vibrations rarely harm a running generator set. The method of isolating the unit is the same for exterior vibrations as it is for self-induced vibrations.[7]

Some rule-of-thumb guidelines are provided for generator set mounting. These general statements normally apply to all generator set mountings, but equipment manufacturer's recommendations must always be followed, and a registered structural engineer's design is required. If no isolation is required, the generator set may be placed directly on the mounting surface. This surface must be capable of supporting at least 25 percent more than the static vibratory loads. Unless the engine is driven equipment that imposes a side load, no anchor bolting should be required. Thin rubber or composition pads are suggested to minimize the unit's tendency to creep or fret the surface of the foundation.[8]

Noise levels must comply strictly with EPA and OSHA standards. Federal, state, and local noise control codes must be strictly adhered to for the installation and operation of an EPS. For example, New York City has its own code on noise control (New York City Noise Control Code).[9] The noise level permitted varies, depending on the zone in which the EPS is located (low-density residential, high-density residential, commercial/manufacturing). It also depends on whether the EPS is operating in the daytime or nighttime and the operating hours of the EPS.[9] These noise levels vary from a low of 50 dBA for a low-density residential area to a high of 150 dBA for 15 min of operation of an EPS.[9] These numbers are only representative of noise control levels and are presented here to provide a base of reference to typical noise levels.[9]

Methods of Reducing Noise.[9]   There are several significant components in the total noise emanating from an engine-driven generator set.

Exhaust System.   The muffler recommended by the muffler manufacturer cannot eliminate excessive exhaust system noise unless the entire system is solid, is without leaks, and is properly maintained. If the noise level is too high and the muffler is properly installed, it may help to install a packed stack, but replacement of the muffler is likely to be the best solution.

Cooling System.   Noise created by intake and discharge cooling air can be reduced by silencers, and it can be less objectionable by directing it away from observers. A remote radiator removes both fan and air noise from the generator set location but may introduce new noise if coolant pumps are located with the generator set.

Fan Noise.   Noise generated by the fan is increased by faults such as bent blades or damaged shrouds.

Engine Noise.   Only a sound control housing can reduce noise emanating directly from the engine and its accessories. As a rule of thumb, any noise level requirement less than 70 dBA adds appreciably to the installation costs.

Noise Reduction Recommendations.   Be sure that the muffler is recommended by the muffler manufacturer for the generator set rating and application. Direct the muffler discharge away from buildings or occupied areas. Use straight rather than curved stacks wherever possible. Minimize the flexible stack length. Changing the stack length may eliminate a sharp rapping sound.[9]

Cooling System.   Most EPSs are cooled by radiators, whether they are engine-mounted or located remote from the EPS. The radiators are sized between 15 and 20 percent larger than the engine maximum full-load heat rejection to allow for overload conditions and EPS aging. Furthermore, if the EPS is installed indoors or in an enclosure, it will require additional ventilation and/or cooling to maintain the EPS at its operating temperatures.

Remote radiators are sometimes used when the EPS is installed indoors, mainly because of space limitations in the EPS room. If an EPS is located in a basement, it is usually difficult to bring in the required volume of air to cool the EPS, thus the need for a remote radiator. This entails extra piping, an increase in coolant pump horsepower, and if the radiator is not located in accordance with the manufacturer's recommendations, it may result in pump and seal leakage. Long pipe runs also result in clamped-rubber hose leaks. The overall reliability of an EPS is reduced by a remote radiator installation. Also, the fan for the remote radiator must be supplied from the generator bus. This could result in a large parasitic load on the generator, depending on the generator capacity and the fan location.

Hot-well systems are used when the static head exceeds the manufacturer's recommendations or if the booster pump imposes an excessive dynamic head.[7] This occurs whenever the horizontal or vertical distance between the remote radiator exceeds the recommended limits. Hot-well systems are installed between the remote radiator and the EPS and must be installed in accordance with the manufacturer's recommendations.

Heat exchangers are also used instead of radiators to cool an EPS. Raw water flows through the exchanger counter to the coolant flow. This provides maximum heat transfer. Because heat exchangers use an outside utility water supply source, they are subject to disruption in the event of an earthquake. Consequently, an outside water supply is not considered a reliable source for EPS cooling. Also, unless the cooling water is obtained from an on-site well, this system cannot be used as a legally required emergency and standby system.

Fuel System.   Refer to Art. 3.2.9, "Environmental Compliance and Issues Management," for a licensing-related discussion. Fuel storage tanks are usually located as close as possible to the EPS; however, if gasoline or LP fuel is used, it cannot be stored in the same room as the EPS. Storage tanks can be located in the base of the EPS if the fuel is diesel. Storage tanks can also be located above ground, indoors, outdoors, or buried in the ground. Ideally, the storage tank is located adjacent to the EPS where it requires the minimum piping, control valves, and pumps for operation.

It can be located in the same room as the generator if the fuel used is diesel. However, local building and fire insurance regulation codes must be checked to determine if the fuel storage tank can be located in the same room as the generator or in an adjacent room, outside the building, above ground or under ground.

Type of Fuel.   Based on a comparative analysis, most emergency generators that supply power to life-safety circuits use diesel fuel. Other types that may be acceptable, depending upon local requirements, use gas or LP fuel.

Storage Tanks.   The storage tank should be located as near to the generator as practicable. The fuel capacity of the tank should be established by evaluating the historical data on power outages for that area in conjunction with the availability of fuel. Careful analysis of these data will ensure that the generator has sufficient fuel to supply power to the EPSS based on the longest historical power outage.

Day Tank.[7]   Auxiliary or day tanks are desirable if the main fuel tanks are located more than 50 ft (15.25 m) from the engine, are located above the engine, or are more than 12 ft (3.65 m) below the engine. Total suction head should not exceed 12 ft (3.65 m). Although a day tank does not aid the engine in fast starting, it offers a convenient and ready storage of fuel. Day tanks also provide a settling reservoir so that water and sediment can separate from the fuel. The auxiliary tank is located so that the level of the fuel is no higher than the fuel injection valves on the engine. The tank should be close enough to the engine to maintain a total suction lift of less than 12 ft (3.65 m).

Exhaust System.   The exhaust pipe will extend past the height of the tallest adjacent building. State and local authorities strictly regulate the height of the exhaust pipe, as well as the emissions from the generator. Refer to Art. 3.2.9, "Environmental Compliance and Issues Management," for additional licensing-related discussion. Wet stacking is a field term to describe the existence of unburned fuel, carbon, or both in the exhaust system. It is caused by operating the EPS under a light load, as defined by the manufacturer. Its presence is often indicated by the presence of black smoke during engine operation.[6] This, of course, is a very subjective approach for determining the presence of wet stacking and consequently has created much controversy and confusion among owners, engineers, and contractors. NFPA 110 includes testing recommendations that are intended to assist users in avoiding wet stacking. These recommendations now include measuring the exhaust manifold temperature at the

exhaust ports and comparing those measurements with the manufacturer's recommendations.[10] If the engine shows signs of wet stacking, an engine *load run* test is mandatory for a
period, as described by NFPA 110, 6-4.2.

Electrical Issues

Common-Mode Failure.   When designing alternate or EPSSs, the most common error is
installing emergency primary and secondary circuits in the same duct banks, manholes, pull
boxes, cable trays, and so forth. Even though the cables for the normal power and the alter-
nate/emergency power may be separated mechanically and may, in some cases, have the
required fire rating, this alone may not be sufficient to prevent emergency power cable damage during a normal power cable fault. Whether emergency circuits are primary or secondary,
they should be routed in separate duct banks, manholes, cable trays, and pull boxes, wherever
possible, to prevent a fault in one system from affecting the other.

Battery and Battery Chargers.   The battery is the heart of an EPS, and the battery charger
maintains the battery's health. Therefore, it is very important to have the battery load tested
and the specific gravity checked regularly. Because code requirements are only minimal
requirements, more frequent checks on the heart of the EPS will improve the reliability of the
system. The battery charger should be located as close as possible to the battery. The battery
charger must be derated for altitude and ambient temperature. This information is readily
available from manufacturers. The battery charger and battery pack should be seismically
supported for the seismic zone in which they are located.

Grounding.   Grounding is probably the least understood and sometimes the most confusing
part of the electrical code. The proper grounding of a generator depends on the way it is connected to the normal building power source.
    If the switch between the standby generator and utility sources (such as an automatic
transfer switch, double-throw safety switch, or interlocked breakers) does not switch the neutral conductors, then the generator source would not be considered a separately derived system (refer to NEC Sec. 250-5), and the generator neutral should not be bonded to ground at
the generator (see Fig. 5.2.3-4). A grounding electrode connection at the generator would not
be required but could be provided as a supplemental grounding electrode system by the
required normal equipment-grounding conductors routed with the phase conductors between
the generator and the building utility service.
    On the other hand, if a four-pole transfer switch is used and the neutral is switched with the
phase conductors, no circuit conductors of the generator source (including the neutral) are
solidly connected to the utility source conductors. Thus, the generator would be considered a
separately derived system and should be grounded according to NEC Sec. 250-26 (the generator neutral is bonded to ground and a local grounding electrode connection is required). The
grounding electrode connection at the generator would be connected to the utility service
grounding electrode system by way of the normal equipment-grounding conductors required
between the two systems (see Fig. 5.2.3-5).[3]

Emergency Distribution Equipment

Location.   The generator should be located so that it is central to the emergency loads to
which it is supplying power. The foundation will be seismically designed for the specific seismic zone in which the equipment will be located. In most cases, it is necessary to isolate the
generator from the foundation. This will be determined after a detailed evaluation of the site
and surrounding facilities. Refer to Sec. 5.1.3, "Structural," for further discussion of seismic
design considerations.

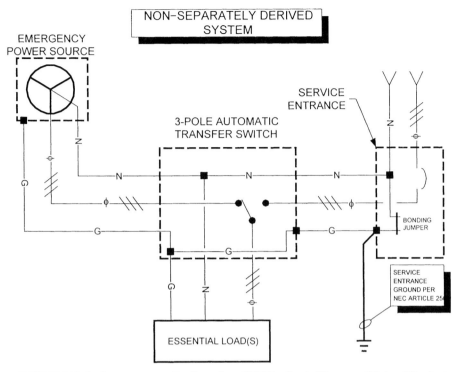

FIGURE 5.2.3-4   Generator grounding figure from NEC Handbook. (*Courtesy of National Fire Protection Association.*)

Fire Separation.   A minimum fire separation must be installed between the EPSS and the normal power supply system, as stipulated by NEC Article 700. More restrictive requirements may be invoked by state and local codes or authorities having jurisdiction (see Fig. 5.2.3-6).

Protective Coordination.   As with normal power distribution systems, coordination of overcurrent protective devices is extremely important in the EPSS. The overcurrent device nearest to the fault must open and clear the fault. A well-coordinated system isolates the fault at the source and prevents cascading operation of overcurrent devices in a short-circuit condition. The design engineer develops the coordinated system based on the one-line drawing of the EPSS and vendor information about the overcurrent device. This information should be checked, adjusted, and finalized during installation and an as-built coordinated system must be submitted to the design engineer for approval.

Voltage and Frequency Requirements.   Emergency power supply systems are used in different locations and for different purposes. The requirement for one application may not be appropriate for other applications. NFPA 110 recommends two levels of equipment installation, performance, maintenance, and testing:

1. Level 1 defines the most stringent equipment performance requirements for applications where failure of the equipment to perform could result in loss of human life or serious injuries. All Level 1 equipment shall be permanently installed.[6]

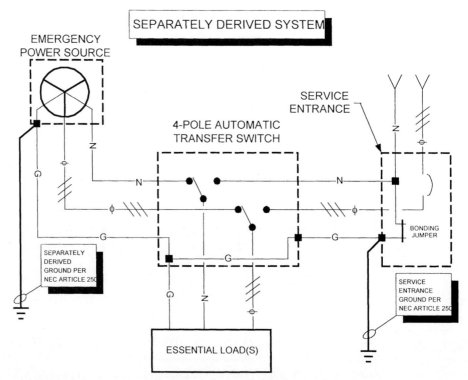

FIGURE 5.2.3-5   Generator grounding figure from NEC Handbook. (*Courtesy of National Fire Protection Association.*)

2. Level 2 defines equipment performance requirements for applications where failure of the EPSS to perform is less critical to human life and safety and where it is expected that the authority having jurisdiction will exercise its option to allow a higher degree of flexibility than provided by Level 1. All Level 2 equipment shall be permanently installed.[6]

The intent of Level 1 and Level 2 systems is to ensure that loads provided with EPSS are supplied with alternate power of a quality essentially equal to commercial power or acceptable for the load within the time specified for the type and for a duration specified for the class.[11] (See ANSI C84.1, *Standard for Voltage Ratings for Electrical Power Systems and Equipment.*)

Provisions for Expansion.   The fact that the EPS might be used for peak shaving, load sharing, and/or for supplying nonessential loads during loss of the normal power source should be considered. This would justify allowing for expansion in generator capacity, keeping in mind the disadvantages of operating a generator that is oversized for the EPSS it is serving, such as wet stacking, carbon buildup, and possible water contamination.

The design engineer will not undersize a generator; therefore, the user has the ultimate responsibility for the loads put on the EPSS. It is recommended that regular load measurements be made during generator testing and that the loads be adjusted for maximum generator usage, including the SkVA for motors.

One cannot simply sample EPS loads to obtain an estimate of the actual emergency system demand. This sampling commonly occurs with a hand-held ammeter during a field walkdown.

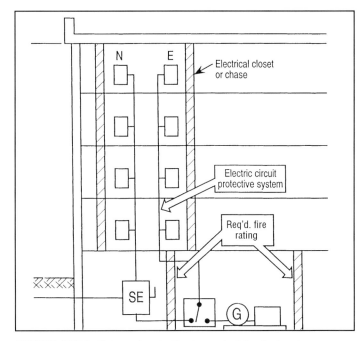

FIGURE 5.2.3-6 Enhanced protection is required for feeders to emergency equipment. One of the options is shown here for cases where a fire suppression system is not employed. Depending upon the actual configuration of the emergency system, a combination of fire suppression system, fire-resistance ratings, and electrical circuit protective systems may have to be used. (*Courtesy of* Electrical Construction and Maintenance Magazine: Practical Guide to Emergency, Standby and Other Auxiliary Systems.)

The time of day when EPS loads are sampled has an impact on the accuracy of the sampling. For example, if the maximum system demand load occurs for a short period in the afternoon, then just sampling the individual branch loading at other times of day would result in inaccurate results. Times of use load measurements of the essential system loads are very useful in predicting total system loading. These measurements can be made on the load sides of individual transfer switches or at other appropriate locations. One convenient approach using the ability of spreadsheets to add ranges graphically with stacked area graphs has been discussed with respect to health care facilities, but the approach can be applied to other types of facilities as well.[12]

Managing Emergency Power Load Growth.   It is important to avoid the possibility of overloading the EPSS by adding loads indiscriminately. The facility manager should be aware of the spare margin in an EPSS, taking into account that unplanned load additions can occur during an actual emergency.

Control Monitoring and Alarms

Location of Alarm Equipment.   A control panel for the EPS shall be provided as specified in NFPA 110 3-5.5.1.

Lighting Systems

Lighting. Sufficient light must be provided to illuminate the means of egress for safe and orderly evacuation of the building in the event of a loss of normal power, a fire in the building, or both. NFPA-101 and state and local codes set the requirements for illumination of means of egress.

The emergency illumination of a facility may be accomplished by either battery lighting units or by the part of the normal lighting system that is supplied by a prime mover-operated electric generator (EPS). It could also be supplied from both of these systems. The battery units should be so arranged so that failure of any unit does not leave the area in darkness. The emergency lighting fixtures supplied from the EPS shall be circuited so that failure of any light or circuit does not leave the area in darkness. Once you have satisfied the criteria for egress lighting, other areas should also be considered.

Lighting for 24-Hour Production Operation. Supplemental lighting for switchboard rooms and associated repair areas—a good design tip in switchboard and switchgear rooms and EPSS locations is to use lighted handle switches for emergency lighting if the emergency lighting is normally switched off when the space is unoccupied.

Perimeter Security Lighting. This includes transportation lights, traffic lights, warning lights, and so forth. High-voltage discharge lighting fixtures should not be used on emergency lighting circuits unless they are of the instant restrike type or are fitted with an auxiliary lamp that remains on during the restrike time of the high-voltage discharge-type fixture.

Table 5.2.3-1 from IEEE Std. 446-1987 provides typical areas to be illuminated for three separate criteria:

TABLE 5.2.3-1   Typical Emergency and Standby Lighting Recommendations

Standby*	Immediate Short-Term[†]	Immediate Long-Term[‡]
Security lighting	Evacuation lighting	Hazardous areas
Outdoor perimeters	Exit signs	Laboratories
Closed-circuit TV	Exit lights	Warning lights
Night-lights	Stairwells	Storage areas
Guard stations	Open areas	Process areas
Entrance gates	Tunnels	Warning lights
Production lighting	Halls	Beacons
Machine areas	Miscellaneous	Hazardous areas
Raw materials storage	Standby generator areas	Traffic signals
Packaging	Hazardous machines	Health care facilities
Inspection		Operating rooms
Warehousing		Delivery rooms
Offices		Intensive care areas
Commercial lighting		Emergency treatment areas
Displays		Miscellaneous
Product shelves		Switchgear rooms
Sales counters		Elevators
Offices		Boiler rooms
Miscellaneous		Control rooms
Switchgear rooms		
Landscape lighting		
Boiler rooms		
Computer rooms		

* An example of a standby lighting system is an engine-driven generator.
[†] An example of an immediate short-term lighting system is the common unit battery equipment.
[‡] An example of an immediate long-term lighting system is a central battery bank rated to handle the required lighting load only until a standby engine-driven generator is placed on-line.

Codes, Rules, and Regulations

Many states and municipalities have adopted their own specific codes for emergency lighting, in addition to those set down by the following four organizations[13]:

1. *Occupational Safety and Health Administration (OSHA).*   OSHA is charged with enforcing compliance and makes reference to the NEC and other NFPA codes.
2. *ANSI/NFPA 70, the National Electrical Code (NEC), Art. 700.*   Sets forth the standard of practice for installation, operation, and maintenance of emergency lighting equipment.
3. *ANSI/NFPA 101, the Life Safety Code.*   Concerns itself with the specification of locations where emergency lighting is considered essential to life safety and specifics on exit marking.
4. *Underwriter's Laboratories, Inc.*   Tests and approves equipment to uniform performance standards as established by ANSI/UL 924 (IEEE, p. 54).[13]

Power Distribution

Power.   Other than life safety equipment, the selection of equipment and systems to be installed in the EPSS depends on the type of occupancy of the building. Figure 5.2.3-7 could serve as a typical guide for an EPSS, but it is not intended as a recommendation.

Other Legally Required Standby Systems.[13]   Other legally required standby systems are subject to similar special laws and regulations for emergency power. Some examples include the following:

- Controls for pressure vessels, such as boilers or ovens, where a failure may result in a life-endangering explosion or fire
- Air-supply systems for persons in a closed area
- Fire pumps, alarms, and systems
- Communications systems in hazardous areas
- Industrial processes in which the interruption of power would create a hazard to life

To meet these or other specialized emergency power requirements, the assistance of a registered professional engineer who is experienced in the particular problem area should be obtained (ANSI/IEEE Standard 446-1987, Chap. 3).[13]

The circuit for the EPSS shall be independent from all other circuits and shall not share the same raceway, junction box, or cabinet with other circuits.

Optional Standby Systems.   Optional standby systems are intended to protect public or private facilities or property where life safety does not depend on the system's performance. Optional standby systems are intended to supply on-site power to selected loads either automatically or manually.[3]

These systems are installed in industrial and manufacturing facilities where a loss of power would result in a very expensive start-up procedure for computerized processing machinery. In some plants such as sugar refineries, it may take up to 24 hours to get the facility back on-line. In the food industry, large walk-in refrigerators and freezers would also be adversely affected. In commercial facilities such as office buildings, banks, hotels, utility companies, and stock brokerage firms, where day-to-day business depends on the computer, any extended outage would severely affect the operation and, in some cases, cause economic chaos.

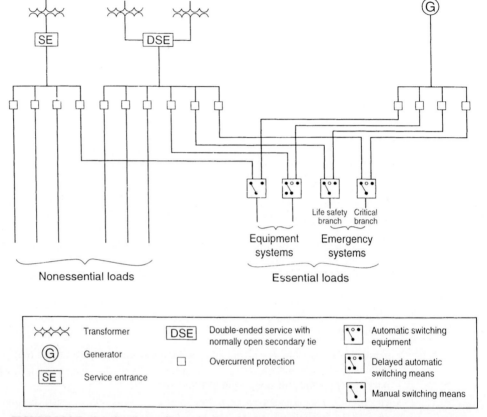

FIGURE 5.2.3-7   One-line diagram illustrating life safety system, equipment for delayed automatic connection, and other equipment for delayed manual connection. (*Courtesy of Motor and Generator Institute,* Healthcare Circuit News.)

Each facility is unique in its requirements for optional standby power. A study of the existing, normal electrical services, along with other available emergency sources, should be completed before any design is attempted. It is also necessary to have a very good understanding of the day-to-day operations of that facility. These form the basis for a design.

SYSTEM SELECTION

The selection of an EPSS is based on the type of occupancy of the building, as well as on its physical characteristics. However, once the emergency system requirements are satisfied, there are many designs to choose from. Refer to Arts. 5.2.1, "Service Entrances," and 5.2.2, "Electrical Distribution Systems," for a detailed treatment of the following three options:

1. Double-ended service with one generator
2. Single-ended service with one or more generators
3. Double-ended service with two levels of emergency service

Selection and Scope

Engine Generator.   The selection of the EPS, as well as the scope of the EPSS, is based on the type of occupancy and the physical characteristics of the building.

Generator Set Sizing.   Other than in hospitals, where the NEC requires sizing the generator set for the full capacity of the connected emergency system load, EPS capacity is based on the total connected life safety load, plus other designated emergency loads with appropriate demand factors. These designated loads may be automatic, delayed automatic, or manual transfer. The duration of the load is also an important factor in sizing the generator. This information can be obtained from utility bills, as well as from daily, weekly, and monthly load monitoring. This information is used to select the generator capacity and quantity required for best operating economy. An analysis of motor loads on the EPS is required to ensure that the generator has sufficient starting capacity for the motors. Table 5.2.3-2 from the Caterpillar *Application and Installation Manual* shows typical voltage dips for the applications listed.

Advantages and Disadvantages of Diesel-Driven Generators.   Each engine generator vendor has its own criteria for the advantages versus disadvantages of diesel versus gas engine generators, but the following is an objective comparison from IEEE Std 446-1987, Chap. 4.[13] In evaluating the merits of diesel engine versus gas turbine prime movers, the following advantages and disadvantages of each should be considered:

- *Fuel supply.*   Gas turbines and diesel engines can generally burn the same fuels (kerosene through No. 2 diesel).
- *Starting.*   Where the application requires acceptance of 100 percent load in 10 s, diesel generator sets can be provided that can meet this requirement. Most gas turbine generators sets require more than 30 s.
- *Noise.*   The gas turbine operates more quietly and has less vibration than the diesel engine.
- *Ratings.*   The gas turbine is not readily available in sizes less than 500 kW, whereas diesel engine units range from 15 kW or lower.
- *Cooling.*   Diesel engines in the larger sizes normally require water cooling, whereas gas turbines are normally air-cooled.
- *Installation.*   Gas turbines are considerably lighter and smaller. Turbines also require less total cooling and combustion air and produce minimal vibration. Installation costs are normally less and rooftop applications are more feasible.

TABLE 5.2.3-2   Typical Voltage Dip Limitations

Application	Condition	Permissible Voltage Dip
Hospital, hotel, motel, apartments, libraries, schools, and stores	Lighting load, large Power load, large Flickering highly objectionable	2% Infrequent
Movie theaters	Lighting load, large Flickering objectionable Neon flashers erratic	3% (sound tone requires constant frequency) Infrequent
Bars and resorts	Power load, large Some flicker objectionable	5–10% Infrequent
Shops, factories, mills, laundries	Power load, large Some flicker objectionable	3–5% Frequent
Mines, oil fields, quarries, asphalt plants	Power load, large Flicker acceptable	25–30% Frequent

- *Cost.*   First cost for diesel engines is lower than gas turbines, but overall installed cost sometimes becomes comparable due to the lower installation costs of the gas turbines.
- *Exercising.*   The cyclic operating requirements under load are more rigid for diesel units than for gas turbines.
- *Maintenance.*   The gas turbine is a simpler machine than a diesel engine. However, repair service for a diesel engine is generally more readily available than for a gas turbine.
- *Efficiency.*   Under full load, the diesel engine operates more efficiently than a gas turbine. However, the reduced exercising requirements for the gas turbine normally make the turbine the lower fuel consumer in standby applications.
- *Frequency response.*   The gas turbine generator is superior in full-load transient-frequency response. (ANSI/IEEE Std 446-1987, Chap. 4).[13]

Nonlinear Loading Issues.   Harmonics are current or voltage waveforms generated at frequencies other than the fundamental frequency that, in most cases, is 60 cycles/s in the United States. Because all of our equipment is designed for 60-Hz operation, any other current or voltage at another frequency is not used by the load. Consequently, harmonic current circulates between the source and the load and is manifested by overheating of motors, transformers, generators, relays, coils, contacts, and conductors.

In the past, it was customary to oversize conductors, transformers, motors, and such to compensate for the heat generated by these unwanted currents and voltages. The better solution is to design the equipment to handle the harmonic content of the loads and/or install filters at strategic locations through the EPSS sized to filter out the unwanted frequencies. When sizing a generator, it is imperative that you know the harmonic content of all equipment that is to be connected to the EPS.

Various types of loads consisting of silicon-controlled rectifiers (SCR) (thyristors) present special problems for generator sets. The problems are generator stator heating and interference with the generator voltage system.[14]

The most commonly encountered SCR loads are battery chargers used with uninterruptible power supplies (UPS systems), solid-state, reduced-voltage, motor starters, and variable-frequency drive motor speed controls.[14] In general, generators with permanent magnet exciters provide the best performance for this type of load.[11] Other excitation systems typically require voltage regulator filtering, generator derating, or both, when used with this type of load.[14] Specifications for generator sets to be used with a load should include detailed information about the equipment to be operated so that the generator supplier can advise the correct size and options to ensure satisfactory performance.[14]

Switchgear.   After the size of the generator set is established, the next task is selecting the emergency distribution equipment, switchboards, transfer switches, and generator control panels. If the EPS is used for other than pure emergency loads, such as for load sharing or peak sharing, then paralleling switchgear is also needed. Paralleling switchgear can also be used to parallel multiple generator sets with priority load shedding where desired.

The generator set output control panel can be programmed by using the paralleling switchgear to provide power to either one of these services and still maintain the emergency system on standby mode when normal power is lost. The switchgear lineup could be made up of the main circuit-breaker cubicle with associated ammeter, voltmeter, wattmeter, wall-hour meter, kVA, kW, and kVAR meters, distribution circuit-breaker cubicle, transfer switches, and synchronizing equipment cubicle, if needed. The transfer switches could also be located near the load that they serve (decentralized approach). This depends on the location of the normal power source in relation to the EPS and the loads that it serves. However, a central location is desirable for operating, maintaining, and testing these units.

Transfer Switches.   Whereas the battery is the heart of the EPS, the automatic transfer switch is the brain of the EPSS. It receives input signals from various parts of the EPSS and responds by transferring the electrical load from the normal source of power to the emergency source of power, instantaneously or within a set time delay. The manual transfer switch is used for planned load transfer and is controlled by the operating personnel.

There are two modes of operation for automatic transfer switches, the open transition transfer (OTT) and the closed transition transfer (CTT).

Open Transition Transfer.[3]   In an OTT transfer, the power output is broken or interrupted before the transfer is made to the new source. There is a definite break in power as the load is taken off one source and connected to another. Although this type of transfer is quite simple, the resulting time delay between break-and-make creates a definite interruption of power to critical loads. Refer to the ASHE 1998 conference paper by Stymiest, Dean, and Seth for a detailed discussion of such issues.[15]

Closed Transition Transfer.[3]   The most important advantage of the CTT transfer is that there is no interruption in power to a critical load, if both sources of power are energized. When using the CTT, both power sources are connected to the load before the break occurs. It is vital that the two sources have the same voltage and frequency and that they are synchronous. Therefore, a protective system is required to provide synchronization, making these transfer switches more complex than OTT types. A comparatively simple synchronizing system can be used, however, because the two sources are paralleled for only about 100 ms.

Closed Versus Open Transition Transfer Switches.[3]   Depending on the situation, OTT and CTT switch operation may be identical or totally different. For example, if a normal power source fails, OTT and CTT switches will transfer the load to an EPS in the same manner, as can be seen by comparing (A), (B), and (C) in Figs. 5.2.3-8 and 5.2.3-9. They both operate in a break-before-make manner in such a circumstance.

Because the EPS must attain a prescribed voltage and frequency before transfer, there are typically 3 to 6 s before this source can be connected and the load powered. For equipment that cannot tolerate this small period of power loss, special backup systems are required, such as a UPS. Therefore, it should be understood that neither an OTT nor CTT is a substitute for a UPS.

When the normal power source is restored, however, OTT and CTT switch operations differ greatly. In the retransfer mode, the OTT switch [Fig. 5.2.3-10 (D) through (F)] operates again in a break-before-make manner, and the load again sees a momentary power outage varying from 50 to 500 ms (3 to 30 Hz), depending on the manufacturer.

On the other hand, the CTT switch (Fig. 5.2.3-10) retransfers the load in a make-before-break sequence because it senses that both power sources are now energized. In other words, the normal and emergency sources are momentarily paralleled during retransfer, typically for about 100 ms or less.

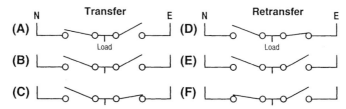

FIGURE 5.2.3-8   Operation of an open-transition transfer switch. During transfer, in (A) the load is connected to the normal source (N); in (B) the load is disconnected from the normal source; and in (C) the load is connected to the emergency source (E). When normal power is reestablished, retransfer takes place. In (D) the load is still connected to the emergency power; in (E) the load is disconnected from the emergency source; and in (F) the load is connected to the normal source. (*Courtesy of* Electrical Construction and Maintenance Magazine: Practical Guide to Emergency, Standby and Other Auxiliary Systems.)

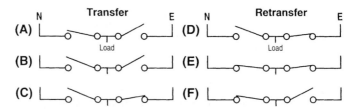

FIGURE 5.2.3-9 Operating sequence of a closed-transition automatic transfer switch. When transfer is required because of failure of the normal source, there is no paralleling of sources. On retransferring, both sources are available and there is a brief period (E) during which the sources are paralleled. (*Courtesy of* Electrical Construction and Maintenance Magazine: Practical Guide to Emergency, Standby and Other Auxiliary Systems.)

Testing an emergency source such as a generator set also results in different operational sequences between both types of switches. In this case, the main power source is still available. An OTT operates in the same fashion as if a normal power source failure had occurred. The switch functions in a break-before-make manner in both the transfer and retransfer modes. Again, the load is without power for a very small period of time during transfer.

For the CTT switch with provisions for testing (Fig. 5.2.3-10), a make-before-break operational sequence occurs because both power sources are energized. In this instance, the load does not see any power outage upon transfer to the generator for testing nor upon retransfer to the normal source

Batteries. Because the battery is the heart of the EPS, considerable care must be taken in selecting the right size of unit for the generator starting. It should be installed in a dry, well-ventilated area and must be accessible and have sufficient workspace for weekly, monthly, and yearly testing and maintenance.

Batteries are also used for tripping circuit breakers during an emergency load-shedding sequence, providing power for control systems and other emergency lighting needs. Metalwork, conduit, and wiring adjacent to the battery must withstand corrosive action.

The NEC has strict requirements for batteries that are used as a source of supply for emergency systems and legally required standby systems. IEEE Std-450 provides recommended monthly, quarterly, and yearly maintenance inspections. The battery charger maintains a float charge on the battery system. The level of the float charge must be obtained from the manufacturer and is based on the battery plate alloy and the specific gravity of the electrolyte. The

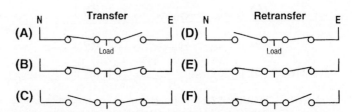

FIGURE 5.2.3-10 Closed-transition switch with provisions for testing is used when emergency source must be tested without power interruption to assure that it can carry the load. In the operating sequence, paralleling takes place both during the transfer sequence (B) and during retransfer (E). (*Courtesy of* Electrical Construction and Maintenance Magazine: Practical Guide to Emergency, Standby and Other Auxiliary Systems.)

float charge is critical to both the performance and life expectancy of the battery. The trickle charge, which is different from the float charge, is a charge given to a battery to maintain its fully charged condition when it is not supplying power to the load.

Battery chargers must be derated for both temperature and altitude (IEEE Std 446-1987, Fig. 34, p. 144).[1]

Uninterruptible Power Supply (UPS).   A UPS by its description is a source of constant power. The NEC approves the use of UPS for supplying power to emergency and legally required standby systems. It is an integral part of the essential electrical system, and receives power from both the normal and alternate source. There are many types of UPS systems available on the market that can be classified under two categories: (1) static UPS and (2) rotary UPS. A UPS should be used only for loads that cannot tolerate a 10-s outage or for critical electronic equipment sensitive to voltage and frequency variation. There can be many configurations of a UPS. For example, a nonredundant or basic UPS or a redundant UPS provides a standby unit if the other UPS fails; a nonredundant UPS with static transfer switch increases the system's overall reliability, if a rectifier fails, by providing a static transfer around the faulted UPS; finally, a parallel redundant UPS is used where reliability is of paramount importance. Other system configurations can be designed on the basis of the owner's needs. Typical units from 30 to 600 kVA at 0.8 PF have an efficiency range between 77 and 90 percent. The efficiency improves as the kilovolt amperes increase. Because a UPS is an inherent generator of harmonics, the EPS must be designed to compensate for the extra heat load or by adding filters. (See discussion under harmonics.)

The static UPS is normally used as part of the emergency or legally required standby systems to provide power to critical loads during the 10-s period between which normal power is lost and the alternate power supply comes on-line. It also maintains a steady-state voltage and frequency during internal or external voltage variations. Typically, the standard unit has sufficient capacity to supply the designated load for 15 to 30 min. Larger units are available with up to 8 hours of capacity.

The UPS is an inherent generator of harmonic current due to the firing circuit of the SCR. Therefore, filtering is required in both the input and output of the UPS. The advantage of the rotary UPS is that it can provide clean power regardless of the quality of the input power.

The batteries for a UPS should be located on a well-designed, accessible battery rack in a well-ventilated room adjacent to the UPS. It is important to keep the power wiring between the UPS and the battery as short as possible to prevent power losses.

The layout of the battery rack should allow easy inspection of all of the battery terminals. The batteries should be mounted on a stair rack so that no battery is mounted directly over another. This allows for easy replacement, inspection, and maintenance, and common-mode failure where a battery fails (leaking acid, explosion, etc.) will have a minimum effect on the other batteries. It also provides better air circulation and heat dissipation. Based on the manufacturer's recommendations, some batteries can be mounted on vertical racks, if space requirements are an issue.

To size a battery for a UPS, you must know the kilovolt-amps (kVA) of the UPS and the power factor and efficiency of the inverter. Then, the battery kilowatts are equal to the UPS kVA multiplied by the power factor and divided by the inverter efficiency. There are other factors to consider in the design of a battery system for a UPS such as load growth, the voltage window of the UPS inverter, up-time required, battery room temperature, aging requirements, and end-of-life capacity. Before you can establish the battery capacity, the kVA of the UPS must be determined.

It is best to have a measured load as opposed to a calculated load in establishing this. However, this may not be possible for a new installation. In that case, it is necessary to identify all motor loads, resistive loads, and nonlinear loads. These loads are totaled, and the appropriate demand and diversity factors are applied to obtain the actual load. Then, it is necessary to specify the percentage of the load represented by the nonlinear load and the crest factor. It is incumbent on the manufacturer to supply any derating factors for its equipment.

Many UPS systems have an adjustable input-voltage-sensitivity window. Where such UPS systems are powered by generator sets, which are commonly less stable than electric utilities, the UPS sensitivity should be broadened to account for allowable input voltage variations without transferring from the generator power to the UPS internal batteries.

Designing for Maintenance

Preventive Maintenance.[1]   Once it has been determined through reliability analysis that an emergency or standby power system is justified, available systems should be evaluated to select one that will economically satisfy the application requirements. Selection of the proper system requires that installation practices and maintenance procedures be sufficiently emphasized. After the system has been selected and the responsibility for it has been established, the plans for preventive maintenance should be completed. Preventive maintenance for electrical equipment consists of a system of planned inspections, testing, cleaning, drying, monitoring, adjusting, corrective modification, and minor repair to maintain equipment in optimum operating condition and with maximum reliability.[1] Without proper emphasis on maintenance from the design stage through to operation, the system can rapidly become unreliable and fail to meet the intended design goal.[1] Some of the maintenance problems experienced during start-up testing and operation result directly from inadequate input during the design process and design oversight. The most important of these is generator capacity—too small or too large. It is very important that the generator match the load it is serving, as well as the type of load. (See the section on generator sizing.)

Location of Generator.   The gold standard for the generator location is on-grade, indoors or outdoors. Because this is premium space for most facilities and the generator installation does not add to the aesthetics of the building, it is usually relegated to a less adaptable space such as the basement or roof.

Locating the generators in the basement is a maintenance headache and also an economic and operational issue. The generator will require a complete ventilation system or air-conditioning system that will be supplied from the EPSS. The remote radiator fan is also supplied from the EPSS. These are parasitic loads that reduce the available generator capacity. These systems—extra fans, motors, pumps, valves, and piping—add to maintenance requirements. Special isolation systems must be designed to isolate the generator, muffler, and exhaust stack from the building structure. Provisions must be made for safe installation and removal of the generator. This will involve a structural design for a removable slab on a ceiling of the generator room and a crane boom for generator installation and removal.

Provision for Load Testing.   Identifying the emergency loads or, if necessary, load banks that will be used for testing the generator should be considered. A control scheme should also be incorporated as part of this design that returns the generator to the emergency mode when an actual power outage occurs. This will reduce the maintenance effort during the weekly, monthly, and yearly testing periods.

Accessibility.   The code requires a minimum of 3 ft of working clearance around the generator. Most designers follow the minimum requirements in order to meet the space restriction criteria of the project. When the generator is undergoing a major overhaul, it may be necessary to remove the engine and shaft. This requires a minimum of 6 to 8 ft of working clearance. Good design practices and past experience should dictate the working clearance around generators.

Maintenance Downtime.   In designing an EPS, the fact that the generator and engine need to be off-line for regular yearly maintenance, emergency repairs, and even load sharing or peak sharing must be considered. In these instances, sufficient flexibility shall be built into the design to ensure that emergency power is available within the time required. Having more than one generator on a common bus is one way of meeting this criterion. Other ways of meeting this requirement would be provisions for a temporary generator tie-in. If the generator is being used for load sharing or peak shaving, a control scheme should be installed that would safely transfer the generator from load sharing/peak shaving to the EPS.

Normally, the design of an EPS system does not include provisions for installing a remote generator. Whenever the generator needs to be replaced or repaired, provisions for connecting a temporary generator should be available to the maintenance department.

Flexibility for Testing. Sufficient flexibility must be built into the EPS to allow for regular weekly, monthly, and yearly testing. This would include all of the features listed under the generator downtime and provide for testing when normal power is available, such as simulating a loss of normal power. Load testing has its own set of possible costly consequences, such as broken crankshafts, destroyed valve trains, bent rods, and failed breakers. However, the other side of the coin is that testing may bring to light incipient failures that would have occurred anyway later, perhaps during an actual normal power outage. It is good practice to switch loads in steps rather than in full-load applications, unless such full-load stepping is the intent of the test itself. In high-reliability applications such as hospitals, all transfer switches in the emergency system have bypass switches integral with the transfer switch. These bypass switches allow withdrawing the automatic transfer switch–operating mechanism for maintenance, troubleshooting, or repair without requiring that the load itself be taken out of service.[16]

Load Banks. Load bank connections to the EPS should be both automatic and manual. Extreme safety precautions must be exercised when using load banks.

Maintenance and Testing. The continuing reliability and integrity of the EPSS depend on an established program of routine maintenance and operational testing. The routine maintenance and operational testing program shall be based on the manufacturer's recommendations, instruction manuals, and the minimum requirements of NFPA 110, Chap. 6.[5]

The goal of preventive maintenance plans is to ensure that equipment is in optimum operating condition. Because the equipment is part of an emergency and standby system, it becomes more important and a greater challenge. Preventive maintenance now becomes a science of anticipation and prediction of failures. In this case, the following vital precautions should be taken:[1]

1. Make sure that the installation will not be subject to ventilation problems or obstructed by stored materials.[1]
2. Place regular test responsibilities on trained personnel, and schedule tests frequently to ensure operation when required.[1]
3. Gasoline and, to a lesser extent, diesel fuels deteriorate when stored for extended periods. Inhibitors can be used to reduce the rate of deterioration, but it is sound practice to operate a system using these fuels so that total operating time results in a complete fuel change every few months.[1]

Typical Maintenance Schedule. The following maintenance schedule from IEEE-446 is presented as a typical service guide for an 1800-rpm unit. It is not intended as a recommendation.

Every 25 h of operation (or 4 months):

- Adjust fan and alternator belt.
- Add oil to cup for distributor housing.
- Change oil in oil-type air filter.

Every 50 h of operation (or 6 months):

- Drain and refill crankcase.
- Clean crankcase ventilation air cleaner.
- Clean dry-type air cleaners.
- Check transmission oil.
- Check battery.
- Clean external engine surface.
- Perform 25-h service (preceding).

Every 100 h of operation (or 8 months):

- Replace oil filter element.
- Check crankcase ventilator valve.
- Clean crankcase inlet air cleaner.
- Clean fuel filter.
- Replace dry-type air cleaner.
- Perform 25- and 50-h service (preceding).

Every 200 h of operation (or 12 months):

- Adjust distributor contact points.
- Check spark plugs for fouling and proper gap.
- Check timing.
- Check carburetor adjustments.
- Perform 25-, 50-, and 100-h service (preceding).

Every 500 h of operation (or 24 months):

- Drain and refill transmission.
- Replace crankcase ventilator valve.
- Replace one-piece-type fuel filter.
- Check valve-tappet clearance.
- Check crankcase vacuum.
- Check compression.
- Perform 25-, 50-, and 100-, and 200-h service (preceding).

CONTRIBUTORS

Dan Chisholm, Publisher, *Healthcare Circuit News,* Motor and Generator Institute, Winter Park, Florida

Jack W. Dean, PE, Electrical Department Head, Novare Engineers, Inc., Providence, Rhode Island

Arjun Rao, PE, Vice President, BR+A Consulting Engineers, Inc., Boston, Massachusetts

David L. Stymiest, PE, SASHE, CEM, Senior Consultant, Smith Seckman Reid, Inc., New Orleans, Louisiana

William P. Wohlfarth PE, Sr. Electrical Engineer, Massachusetts Institute of Technology, Cambridge, Massachusetts

NOTES

The texts and documents listed here were also used as a source of information for this article. In some cases, the information was quoted verbatim, and in other cases, the author simplified the language to suit the presentation.

1. ANSI/IEEE Std 446-1987, *IEEE Recommended Practice for Emergency and Standby Power Systems for Industrial and Commercial Applications,* Chap. 3, Institute of Electrical and Electronics Engineers, Inc., New York, 1987. Figures 5.2.3-1 and 5.2.3-2 reprinted with permission from IEEE Standard 446-1987, "Recommended Practice for Emergency and Standby Power Systems for Industrial and Commercial Applications." Copyright © 1987 by IEEE. The IEEE disclaims any responsibility or liability resulting from the placement and use in the described manner. All rights reserved.

2. D. Chisholm, "Inspecting & Testing of Generators for Health Care Facilities," *Healthcare Circuit News,* 1999.

3. *Electrical Construction and Maintenance Practical Guide to Emergency, Standby, and Other Auxiliary Power Systems,* Intertec Publishing, Overland Park, KS, 1996.

4. NFPA 70-1999, National Electrical Code, National Fire Protection Association, Quincy, MA, 1999.

5. G. Johnson, "Integrity of Emergency Systems," *IEEE Transactions on Industry Applications,* 34, no. 4 (July/August 1998).

6. National Fire Protection Association, *Standard for Emergency and Standby Power Systems* (NFPA 110), National Fire Protection Association, Quincy, MA, 1999.

7. Caterpillar, *Application and Installation,* Generator Set, Caterpillar, Inc., Peoria, IL, 1983.

8. Caterpillar, *Selection and Installation,* Generator Set, Caterpillar, Inc., Peoria, IL, 1983.

9. Onan Technical Release, Vol. III, No. 4, Onan Corporation, Minneapolis, MN, 1988.

10. D. Chisholm, "Heading Towards Change Again," *Healthcare Circuit News,* 3, no. 6 (February/March 1998).

11. ANSI C84.1, *Standard for Voltage Ratings for Electrical Power Systems and Equipment,* American National Standards Institute, New York, 1989.

12. D. L. Stymiest, "Determining the Actual Emergency Demand Load" (4-part series of articles), *Healthcare Circuit News,* 2, no. 4 (October 1996) through 3, no. 1 (April 1997).

13. ANSI/IEEE Std 446-1987, Chaps. 3 and 4, Institute of Electrical and Electronics Engineers, Inc., New York, 1987.

14. D. Chisholm, "Inspecting & Testing of Generators for Health Care Facilities," *Healthcare Circuit News,* 1999.

15. David L. Stymiest, Jack W. Dean, and Anand K. Seth, "Managing the Impact and Cost of Emergency Power Testing on Hospital Operations," 35th Annual ASHE Conference, Denver, CO, American Society for Healthcare Engineering, Chicago, IL, 1998.

16. *Engineer's Guidebook to Power Systems,* 2nd ed., Kohler Power Systems, Kohler, WI, 1986.

ARTICLE 5.2.4
RACEWAY AND CABLE SYSTEMS

Joseph R. Flocco, Jr., P.E., Account Manager and Consulting Engineer
Siemens Energy and Automation, Inc., Alpharetta, Georgia

RACEWAY SYSTEMS

This article provides an overview of materials and methods used to distribute electrical power for lighting, heating, cooling, and power distribution. These systems are recognized and described in the National Electrical Code (NEC). The NEC is not a design guide but is used by local authorities to determine if a minimum level of reliability and safety has been achieved. Engineers, designers, and installers are cautioned to refer to the NEC for a complete treatment of permissible applications.

Nonmetallic Conduit

Nonmetallic conduit is available in polyvinyl chloride (PVC) and fiberglass varieties. These products are generally allowed in all installations except theaters. They are either directly buried or encased in concrete for added strength when passing under roadways or parking areas; these nonmetallic conduits are also useful for burial in soils that are corrosive, provided that the conduit is specifically listed for use when subjected to such chemicals. Concrete encasement is also used where these conduits enter a building to protect against shearing of the conduit by settling of the building above it. These products are commonly used for service entrance cable from the electric utility company, for large feeders under or between buildings, and for roadway and parking lot lighting. Schedule 80 PVC is a thick-walled PVC conduit that is directly buried when there is little danger of crushing or damage from traffic over the top of the ducts or excavations. Schedule 40 PVC, a thinner-walled plastic conduit, is encased in a concrete envelope and used underground to give added protection from heavy weights passing over the ducts and from excavation operations that might accidentally come in contact with it. When concrete encasement is used, conduits are generally spaced so that their connecting fittings are easily tightened and so that the spacing allows for heat dissipation from power flow in the conductors and for the concrete to flow adequately in and around each pipe during concrete pouring.

Metallic Conduit

Metallic conduit is available in several designations related to its thickness, strength, and ultimately its intended and allowed uses. Generally, the smallest trade size is ½ in.

Electrical metallic tubing (EMT).   The least expensive of the metallic conduits and commonly referred to as *tubing*. It is limited to areas where it is not subject to physical damage, and if it is exposed, these are areas generally at least 8 ft above the floor. It is not generally used to support lighting fixtures.

Intermediate metal conduit (IMC).   A thick, rigid conduit that uses threaded connections and preformed couplings and connectors. An IMC may be field-bent within NEC-allowed radii, depending on the type of conductors that are going to be installed. Intermediate metal conduit is available in aluminum or steel in diameters up to 4 in. It can be used in all locations when listed for the intended location or environment.

Rigid metal conduit.   The thickest wall conduit and, like the intermediate type, uses pre-formed couplings and connectors but can be field-bent within NEC requirements. Its maximum trade size is 6 in. Generally, rigid metal conduit is made of steel but may also be aluminum. A trade size of ⅜ in is allowed under certain conditions for motor installations. Aluminum conduit has the benefit of lower weight, which translates into easier installation and lower costs.

Cable Trays and Wireways

Cable trays and wireways are raceway systems that are capable of carrying the largest quantities of cables for power, instrumentation/control, and telephone/data. Cable trays include ladder, center-spline, or channel types. These systems consist of standard fittings and straight lengths in open and closed varieties. Large numbers of power-carrying conductors can be installed in these raceways without derating (lowering the ampacity rating of a conductor) when conductors are installed as required by the NEC. Cables for use in cable trays must be listed for such use.

Wireways.   Enclosed, square, or rectangularly shaped raceways. Standard building wire may be installed, and they are often used in laboratories for installing multiple feeds to laboratory benches or when mounting multiple motor starters on a backboard and using and tapping a single feeder as their supply.

Flexible conduit.   Used for connections to equipment. It allows for slight adjustments and limits transmission of vibrational noise from equipment to the building structure to which it is attached. Typically, jacketed liquid-type construction is used, especially in mechanical spaces where moisture may be present.

Conductors and Conductor Systems

These systems are the means by which power or signals are transmitted throughout the building or campus facility. Conductors are installed in raceways or include their own raceway system, as those described previously and listed as assemblies to carry power, control, or other electrical signals.

Building wire.   Available in aluminum or copper with insulation made of various natural and man-made materials, including rubber and plastics. The materials used in constructing insulation determine (1) whether the cable can be used in wet or dry installations, for direct burial, or for installation above hung ceilings when not installed in raceway, and (2) its environmental temperature limitations and its ampacity within the material's temperature rise.

Mineral-insulated cables.   Special fire-rated cables that are often used as emergency feeders and branch circuitry. The NEC also allows other means of establishing emergency circuits using standard conduit and cable, which are usually considered because they can offer some cost savings.

Armored cable.   An assembly of insulated conductors enclosed in a flexible metallic covering. It is allowed in concealed and exposed installations in most types of locations and buildings. Because conductors and conduit are one package, it can be easier and less expensive to install compared with conduit systems and building wire. However, because of its flexibility, it is likely that without well-written specifications requiring that the routing be "rectilinear," (i.e., parallel or perpendicular to the building structure), the cables may crisscross each other, particularly when installed above a hung ceiling.

Metal-clad cable.   Used for circuits of 600 V and higher, as well as for special remote control and signaling cables. It also has a metallic sheath. It is suitable for all locations when it has a covering over the metal jacket that is UL-listed for the environment.

Busway (also incorrectly but commonly referred to as *bus duct*).   An assembly of bare or insulated copper or aluminum conductor bars, rods, or tubes that are enclosed in a grounded-metal housing. Busway is available in ratings from 100 to 5000 A and has short-circuit ratings that allow using it as the main service entrance feeder.

Feeder busway.   Used to distribute large amounts of power and more economical than conduit and wire in many cases. A little more effort and coordination is required when planning a job using busway. However, for feeders of 800 A and larger, busway offers significant savings in installed cost because of reduced labor, and it offers increased efficiency due to its lower impedance. It is more space-efficient and weighs one-third less than the equivalent cable-and-conduit installation.

Plug-in busway.   Differs from feeder types only in its ability to attach, or plug in, devices that further distribute or consume the power being distributed. In a high-rise building, vertical bus risers are a common means of distributing power from floor to floor. The busway housing often serves as the equipment-grounding conductor (called *integral*), but some manufacturers also offer an insulated (called *internal*) ground bus, which may be an isolated ground type, if that is desired. Panelboards can be connected directly to the bus, but more often, enclosed fused switches or circuit breakers are installed. Standard conduit and wire feeders are then installed to panelboards from these devices. Direct connection of meters, motor starters, and cable drop boxes are other plug-in options often used to service large horizontal facilities, especially manufacturing types. It offers the same cost and energy advantages compared with conduit and wire as the feeder type. Recent listing rule changes have taken away a commonly used benefit of the plug-in busway—the ability to add new bus plugs without turning off the busway. Bus plugs now being shipped by manufacturers contain labels warning that they may be installed only on de-energized busways. These changes have occurred because of some isolated cases of bus plug failures during installation into energized busway that resulted in personnel injury and equipment failure.

CONTRIBUTORS

Arjun Rao, P.E., Vice President, BR+A Consulting Engineers, Inc., Boston, Massachusetts

William P. Wohlfarth, P.E., Senior Electrical Engineer, Massachusetts Institute of Technology, Cambridge, Massachusetts

BIBLIOGRAPHY

Standards and Codes

National Electrical Code, NFPA 70-1999, National Fire Protection Association, Quincy, MA, 1999.

Books

D. Fink and H. W. Beaty, *Standard Handbook for Electrical Engineers,* 13th ed., McGraw-Hill, New York, 1993.

E. C. Lister and M. R. Golding, *Electric Circuits and Machines,* McGraw-Hill Ryerson Limited, Whitby, ON, Canada, 1996.

Articles

J. Ahlstrom, "Wiring Methods Promise Savings in the Long Run," *Electrical Contractor Magazine* (October 1993).

E. Palko, "Taking Advantage of Low Voltage Plug in Busway," *Plant Engineering Magazine,* (9 January 1995).

ARTICLE 5.2.5
WIRING DEVICES

A. Todd Rocco, P.E., Principal
Diversified Consulting Engineers, Quincy, Massachusetts

INTRODUCTION

Electrical wiring devices are usually considered receptacles and switches but can also include safety switches and other devices used to connect equipment to the point of use. Wiring devices are engineered and designed to meet industry standards to provide years of safe and reliable service for the specified applications. This article provides a summary of industry standards and ratings, application ratings, and a description of the various types of wiring devices. Information about telephone and data wiring devices is provided in Art. 5.5.6, "Telecommunications and Data Distribution Systems." Occupancy sensors are described in Art. 5.3.1, "Lamps, Fixtures, and Humans—How They Relate."

Voltage and Current Ratings

American and Canadian wiring devices are designed for use at nominal voltages and currents. Codes require that switches and receptacles be labeled for the rated voltage and current at which they are applied. Devices that are not applied within their ratings violate the National Electrical Code (NEC) because they pose a potential personnel and physical safety hazard. For ac applications, the switch must have the appropriate ac voltage and current rating; for dc applications, a switch must be appropriately rated for dc voltage, current, and load type (e.g., a horsepower rating at the applicable voltage for a switch controlling a motor, or a fluorescent or incandescent rating at the applicable voltage for lighting applications). Receptacle ratings and configurations are presented in the next section.

Some devices are side-wired with screw terminals, and some have combination side and back terminals. Some devices use an internal spring mechanism to hold the conductors; others use a side wiring screw that also adjusts an internal clamp. Typically, two back wire spaces are provided per side, but four spaces per side are also available for split switching of a device, such as a lamp. The design of modern wiring devices has evolved to facilitate safe and rapid installation by electricians.

NEMA Standards

Several industry standard organizations establish guidelines for safe construction, installation, and application of wiring devices. The National Electrical Manufacturers Association (NEMA) has established ratings and configurations to prevent the misapplication of plugs into the wrong receptacles and to ensure interchangeability among devices from different manufacturers. The means for preventing misapplication include blade (or pin) shape, configuration, and size. Figure 5.2.5-1 summarizes the configuration information established by NEMA. Each receptacle and plug rating has a unique NEMA configuration number that indicates whether it is locking or nonlocking and gives its voltage rating, number of wires, current rating, and type, as shown in the chart legend.

Other agencies that publish standards or recommended practices for wiring device performance or application include the American Boat & Yacht Council (ABYC), the Canadian Standards Association (CSA), the International Electrotechnical Commission (IEC), the National Electrical Code (NEC), Underwriters Laboratories, Inc. (UL), the Defense Electronic Supply Center (for federal specifications), and the Occupational Safety and Health Administration (OSHA).

TYPES OF WIRING DEVICES

There are several types of safe and reliable wiring devices for every application. In addition to electrical ratings, application considerations include environmental exposure, safety requirements, physical requirements, and service requirements.

The number of poles is the quantity of current-carrying connections. The number of wires indicates the total number of connections including the ground connection, if it exists. For example, two-pole, three-wire grounding indicates two current-carrying conductors and a ground connection, and two-pole, two-wire indicates two current-carrying conductors without a ground connection.

The number of wiring devices mounted in an outlet box is referred to as the *number of devices ganged together*. For example, one device is called a *single-gang outlet* and requires a single-gang device plate to cover the box and the energized parts of the device. Two devices are called a *double-gang,* or *two-gang, arrangement,* and so on. Cover plates are available in various device combinations to accommodate multiple ganged installations.

Straight-Blade Devices

Typical 15-A and 20-A, 125-V receptacles and plugs are part of a subset of wiring devices called *straight-blade devices*. Straight-blade devices are used where the plug is unlikely to be accidentally disconnected or does not pose a personnel or equipment safety hazard if it does become inadvertently disconnected. The left side of Fig. 5.2.5-1 presents the range of straight-blade wiring devices. Pin-and-sleeve devices are used for higher current applications.

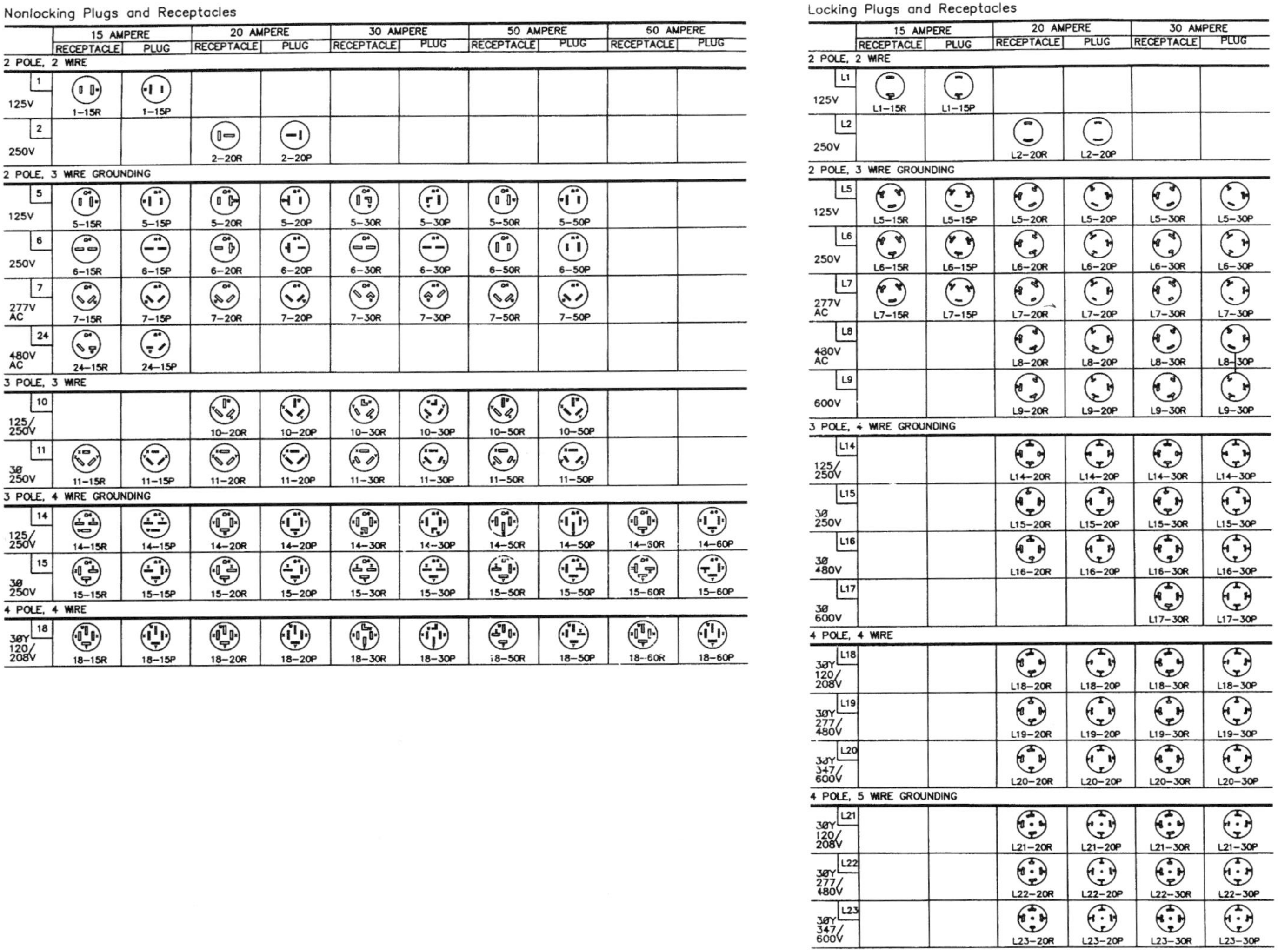

Nonlocking Plugs and Receptacles

	15 AMPERE		20 AMPERE		30 AMPERE		50 AMPERE		60 AMPERE	
	RECEPTACLE	PLUG	RECEPTACLE	PLUG	RECEPTACLE	PLUG	RECEPTACLE	PLUG	RECEPTACLE	PLUG
2 POLE, 2 WIRE										
1 125V	1-15R	1-15P								
2 250V			2-20R	2-20P						
2 POLE, 3 WIRE GROUNDING										
5 125V	5-15R	5-15P	5-20R	5-20P	5-30R	5-30P	5-50R	5-50P		
6 250V	6-15R	6-15P	6-20R	6-20P	6-30R	6-30P	6-50R	6-50P		
7 277V AC	7-15R	7-15P	7-20R	7-20P	7-30R	7-30P	7-50R	7-50P		
3 POLE, 3 WIRE										
10 125/250V			10-20R	10-20P	10-30R	10-30P	10-50R	10-50P		
11 3Ø 250V	11-15R	11-15P	11-20R	11-20P	11-30R	11-30P	11-50R	11-50P		
3 POLE, 4 WIRE GROUNDING										
14 125/250V	14-15R	14-15P	14-20R	14-20P	14-30R	14-30P	14-50R	14-50P	14-60R	14-60P
15 3Ø 250V	15-15R	15-15P	15-20R	15-20P	15-30R	15-30P	15-50R	15-50P	15-60R	15-60P
4 POLE, 4 WIRE										
18 3ØY 120/208V	18-15R	18-15P	18-20R	18-20P	18-30R	18-30P	18-50R	18-50P	18-60R	18-60P

Locking Plugs and Receptacles

	15 AMPERE		20 AMPERE		30 AMPERE	
	RECEPTACLE	PLUG	RECEPTACLE	PLUG	RECEPTACLE	PLUG
2 POLE, 2 WIRE						
L1 125V	L1-15R	L1-15P				
L2 250V			L2-20R	L2-20P		
2 POLE, 3 WIRE GROUNDING						
L5 125V	L5-15R	L5-15P	L5-20R	L5-20P	L5-30R	L5-30P
L6 250V	L6-15R	L6-15P	L6-20R	L6-20P	L6-30R	L6-30P
L7 277V AC	L7-15R	L7-15P	L7-20R	L7-20P	L7-30R	L7-30P
L8 480V AC			L8-20R	L8-20P	L8-30R	L8-30P
L9 600V			L9-20R	L9-20P	L9-30R	L9-30P
3 POLE, 4 WIRE GROUNDING						
L14 125/250V			L14-20R	L14-20P	L14-30R	L14-30P
L15 3Ø 250V			L15-20R	L15-20P	L15-30R	L15-30P
L16 3Ø 480V			L16-20R	L16-20P	L16-30R	L16-30P
L17 3Ø 600V					L17-30R	L17-30P
4 POLE, 4 WIRE						
L18 3ØY 120/208V			L18-20R	L18-20P	L18-30R	L18-30P
L19 3ØY 277/480V			L19-20R	L19-20P	L19-30R	L19-30P
L20 3ØY 347/600V			L20-20R	L20-20P	L20-30R	L20-30P
4 POLE, 5 WIRE GROUNDING						
L21 3ØY 120/208V			L21-20R	L21-20P	L21-30R	L21-30P
L22 3ØY 277/480V			L22-20R	L22-20P	L22-30R	L22-30P
L23 3ØY 347/600V			L23-20R	L23-20P	L23-30R	L23-30P

FIGURE 5.2.5-1 Wiring device NEMA configuration numbers. (*Courtesy of Hubbell Wiring Devices.*)

Single Receptacles

Single receptacles are typically used for dedicated applications to prevent other equipment from being connected to the circuit. The reasons for dedicated applications include circuit ampacity, remote switching of equipment, or eliminating the requirement for a ground-fault circuit interrupter (GFCI) where, for example, a fixed piece of equipment is not required to be GFCI-protected, but a portable piece of equipment would be required to be GFCI-protected. A common example is an automatic garage door opener, which is not required to have ground-fault protection if the receptacle is dedicated to that piece of equipment or is not accessible from the floor. However, receptacles for general use in a residential garage are required to have ground-fault protection. Another example is an overhead light fixture that is cord-connected and powered by a receptacle that is accessible from the floor in an area where receptacles are required to be ground-fault-protected. Therefore, the NEC requires that receptacles be ground-fault-protected where a person could use that receptacle, such as in commercial garages, rooftops, and kitchens.

Duplex Receptacles

Duplex receptacles are the most common receptacles. Connecting a duplex receptacle in a split-wiring configuration allows controlling one contact device in the outlet by a switch or other source and supplying the other contact device continuously (unswitched.) An example of a use for split wiring is controlling a lamp by a switch, while power is uninterrupted to an appliance connected to the other half of the same duplex receptacle.

Combination Receptacles

Combination receptacles are devices that have two receptacles each of different voltage ratings. A common application for a combination-type receptacle was for portable x-ray equipment that required 208 V, whereas the other receptacle was available for 120-V equipment. However, combination devices are no longer popular for safety reasons, and because x-ray equipment design no longer requires 208 V for portable x-ray equipment.

Ground-Fault Circuit-Interrupting (GFCI) Receptacles

Ground-fault circuit-interrupting receptacles are designed to disconnect the load when an imbalance is sensed above a certain threshold between the two current-carrying conductors. The imbalance occurs when a path, other than the intended path, of sufficiently low impedance exists and causes some of the current to flow through the other path. This other path could include equipment or personnel. The NEC has established a threshold current of 5 milliamperes (mA) for GCFI receptacles to protect personnel. This level produces a current below the threshold that causes ventricular fibrillation in a person who may be shocked by electrical current flowing from the person's hand to other extremities without involving any open wounds. A GFCI device that meets the 5-mA current threshold and operates within 25 ms is designated as a Class A GFCI, indicating that it is intended for personnel protection.

When Class A protection is desired locally for equipment, a GFCI without a receptacle can be installed similarly to a GFCI receptacle. These devices can be used for personnel protection from a single-phase motor or other device. When installing a GFCI for a motor, one should verify that the connected-motor horsepower rating is less than the device rating.

Tamper-Resistant Receptacles

Tamper-resistant receptacles are similar to standard receptacles except that they contain an internal switch that is intended to keep the receptacle deenergized until a plug is inserted. There are two switches per receptacle. Each switch is controlled by inserting an object into the other contact. This prevents a single object (like a screwdriver or knife) from being accidentally energized. For an object to become energized on one side, an object must be inserted into the contact to operate the switch on the other side.

Applications for tamper-resistant receptacles include child care areas, psychiatric patient areas, and prisons. Because this device is not tamperproof, additional precautions should be taken where personnel could defeat the tamper-resistant feature.

Transient Voltage Surge Suppression Receptacles

Transient voltage surge suppression (TVSS) receptacles (see Fig. 5.2.5-2) accept standard plugs and provide protection to the load from transient surges within a specified energy rating. Solid-state devices within the TVSS receptacle maintain the voltage at a specified limit during a transient surge. If the energy rating of the device is exceeded in either magnitude or duration, then the device no longer provides protection against similar transient modes. Many TVSS receptacles provide a visual or audible indication (LED or buzzer) that the protection is functional or is no longer functional due to a transient surge. It is important for the user to understand the meaning of the signal.

Isolated Ground Receptacles

Isolated ground receptacles provide electrical isolation from the equipment-grounding conductor contact and the device's mounting strap. To maintain isolation, a dedicated insulated

FIGURE 5.2.5-2 Isolated ground and TVSS receptacles. (*Courtesy of Hubbell Wiring Devices.*)

grounding conductor is required back to the power source at the panelboard. The mounting method for the receptacle and the device box equipment-grounding system are the same as for a standard receptacle (e.g., conduit). The purpose of the isolated ground is to minimize disturbances in the grounding system not caused by the equipment supplied. The isolated grounding circuit minimizes disturbances from another branch circuit, while still providing a ground path for personnel safety. These devices are marked with an orange triangle as required by the NEC.

Hazardous Locations

When a receptacle is required in an area that is classified as hazardous due to either a present hazard or potential hazard, special wiring devices and methods are required. These devices are designed to prevent explosions and fires. The devices typically provide a high-integrity seal when a plug is not installed, and automatically disconnect power to the load before the blades are removed to prevent any arcing that might ignite flammable or explosive materials.

Locking Devices

Locking devices consist of blades in circular configurations that accept plugs of a corresponding rating. The plugs are locked in place by a quarter turn clockwise after insertion. The locking feature provides a positive connection when subject to physical strain. Locking devices are available as either single or duplex receptacles.

A locking device is used where the plug might be accidentally disconnected or where inadvertently disconnecting it might pose a personnel or equipment safety hazard. The right side of Fig. 5.2.5-1 presents the range of twist-lock wiring devices.

Pin-and-Sleeve Devices

Pin-and-sleeve devices provide a positive locking connection that is watertight, depending on the style. Other features of pin-and-sleeve connectors are that they are available as three-, four-, and five-wire devices for all voltage configurations from 20 to 100 A, and the pins (contacts) are protected by the design of the connector body. Connectors for higher-current ratings are also available. This device could be used for a power connection to a temporary generator or for ship-to-shore connections in large vessels. Receptacles are available as an integral part of a fused or unfused safety switch.

Switches

Switches are available for variety of applications. Switch ratings must be matched to the load served (i.e., voltage rating, current rating, horsepower rating, or maximum lamp load based on lamp type). In addition, these ratings vary for ac and dc applications and must be specifically listed for the application.

Switches are available in three different styles: toggle, rocker, and push-button. This article focuses primarily on toggle switches, which are the most common style. However, rocker and push-button switches are available in many of the same types as the toggle switch.

Single-Pole Switches.   Single-pole switches are the most common type of switch used. The single-pole switch is named for its contact configuration, one set of contacts that is closed to permit current flow in the on position, and open to stop current flow in the off position.

Double-Pole Switches. A double-pole switch is similar to a single-pole switch, but it has two sets of contacts. This type of switch is used to disconnect both conductors of a two-wire ungrounded circuit. A double-pole switch can also be used to control two circuits simultaneously.

Three-Way Switches. Three-way switches are used to control devices, typically lights or groups of lights, from two locations. It is not possible to tell if three-way switches are on or off from their physical positions. Three-way switches can also be used to control a load from two different sources or to control two loads from a single source.

Four-Way Switches. Four-way switches are used in conjunction with two three-way switches to control circuits from multiple locations. It is not possible to tell if three-way and four-way switches are on or off from their physical positions. Four-way switches can also be used to reverse polarity to a load (e.g., to reverse a dc motor's direction).

Pilot and Illuminated Switches. A pilot switch is used to indicate when a load is energized. Typically, the switch handle is lit, but some have an indicator lamp light located external to the handle. Illuminated switches have a lighted handle when the load is off, and that aids in finding the switch in a dark location. Illuminated switches can also be used to indicate when the load is off. Either switch can be used to indicate if the load in a remote area is on or off when controlled by three-way switches. Users of illuminated switches need to be aware that although the lamps, such as neon, have long lives, they can burn out. Therefore, independent verification that equipment is turned off should always be done for the safety of personnel and property.

Low-Voltage Switches. Low-voltage switches are used in conjunction with lighting relays to control large groups of lights. Using low-voltage wiring eliminates running line voltage wiring to each switch point, which can be advantageous when the distance for the switch circuits is relatively long. Low-voltage wiring also allows controlling the same lights from multiple locations with minimal wiring.

Combination Switches. Combination devices are used to operate two devices in a single-gang location. This is desirable when adding a switch or receptacle to an existing single-gang device box. Some combinations include a receptacle with a single-pole or three-way switch, a single-pole switch with a three-way switch, two single-pole switches, or two three-way switches.

Key-Operated Switches. Key-operated switches can be installed in locations to prevent unauthorized or unintentional operation of a switch.

Momentary-Contact Switches. Momentary-contact switches are designed so that the handle returns to its rest position when pressure is released. This type of switch can be used as a start-stop station or where momentary control of a load (e.g., a motor-operated projection screen) is desired. The alternative to a momentary-contact switch is the maintained-contact switch that is found in most installations.

Dimmers and Motor Speed Controls. Dimmers and speed controls are devices that can be adjusted by using a slide handle or a rotary knob to proportionally control a lighting or motor load. Older devices used rheostats (variable resistors) that dissipated energy in the switch while reducing the lighting load or speed of the controlled equipment. Modern devices use solid-state components to modify the waveform and thereby reduce energy utilization while reducing the lighting output or motor speed. These devices are designed for specified load types and ratings and should be applied to control only the designed load types. Dimmers are solid-state devices with typical ranges (for incandescent lamps) from 600 to 2000 W. Dimming systems for fluorescent lighting are available but typically require special dimmable ballasts that are coordinated with fluorescent dimmers. Fluorescent dimmers for individual compact fluorescent lamps that do not require additional wiring are also available. Caution is required to make sure that stroboscopic effects are not noticeable with dimming.

Manual Motor Controllers.   Manual motor controllers are used to control motors by using a toggle-type switch when automatic operation is not required. Manual motor controllers are also available with automatically resetting thermal protection to provide overload protection to small-horsepower motors.

Device Plates.   Device plates are available in a variety of combinations for different applications. For example, toggle switch plates are available from single-gang to eight-gang. An example of a combination device plate includes one receptacle with two toggle switches.

Device plate materials include stainless steel, brass, aluminum, metallic, nylon, wood, and plastic. The application should be considered when specifying a device plate. Stainless steel is suitable for corrosive and high-use (abusive) applications. Brass and metallic plates are typically selected for appearance. Aluminum plates can be used where nonmagnetic and corrosion-resistant plates are required. Nylon plates are used primarily for safety considerations when a durable plate is required. Plates are available in many colors and with engraved lettering.

For weatherproof-when-not-in-use applications, plates with a snap cover protect the device when closed. For weatherproof-when-in-use applications, switch plates are available that allow operation of the switch without exposing the actual handle. For receptacles, the plate includes a hinged housing that allows protection while the receptacle is in use.

Temporary Wiring.   NEC and OSHA regulations usually require that temporary wiring be GFCI-protected. Refer to the earlier section on GFCIs for more detail. Cord sets with integral GFCI devices are available for temporary power applications. These units feature high-visibility housings that are durable, sunlight-resistant, and water-resistant or watertight. The GFCI devices are available in either *manual reset* or *automatic reset,* which refers to operation when power is available or restored. Both types of devices require manually resetting the device after a ground-fault trip. Manual reset requires resetting the device after a loss of power. This is an additional safety feature where reenergizing the load after power failure could lead to personnel injury or property damage. The automatic type reenergizes the load once power is reapplied; this may be useful for a pump where it is desirable to restart the load once power is applied.

Power Distribution.   Prefabricated power distribution equipment is available in a variety of configurations for temporary power applications. This equipment provides features similar to portable GFCI cord sets and can serve multiple loads. The supply ranges from 20 A, 125 V, to 50 A, 125 V/250 V, allowing simultaneous powering of several loads of different voltages.

Application Rating.   Wiring devices are manufactured in different grades to match application requirements.

Residential Grade.   Residential-grade receptacles are available in different subgrades, mainly home-center grade and higher contractor grade. Home-center grade should be used only in residential applications where minimal use is expected.

Commercial Grade, Specification Grade, and Heavy-Duty Grade.   Mil specification grades (federal specifications) are some of the more durable grades. When specifying devices, verify that the manufacturer's designation will meet the job requirements.

Hospital-grade designation (also see the right side of Fig. 5.2.5-2) refers to receptacles that are designed with higher pullout tensions and other safety features. In addition, these receptacles must meet the performance standards of Underwriters Laboratory Standard UL 498. A green dot on the front indicates that the receptacle is hospital grade.

Color and Style. Wiring devices are available in several colors to match the architecture of the installed area. Typical colors include black, brown, gray, ivory, and white. Other colors are reserved for specific representations, including blue for TVSS receptacles and orange receptacles for isolated ground installations. It is also common practice in many industries to specify red receptacles when they are supplied by emergency power.

CONTRIBUTORS

Paul Konz, P.E., Electrical Department Head, The Ritchie Organization, Newton, Massachusetts

Arjun B. Rao, P.E., Vice President, BR+A Consulting Engineers, Inc., Boston, Massachusetts

Lawrence F. Richmond, Vice President, Bennett Electrical, Inc., Quincy, Massachusetts

William P. Wohlfarth, P.E., Senior Electrical Engineer, Massachusetts Institute of Technology, Cambridge, Massachusetts

ARTICLE 5.2.6
GROUNDING SYSTEMS

Paul Konz, P.E., Electrical Department Head
The Ritchie Organization, Newton, Massachusetts

Volumes of material have been written about grounding, and yet it is still very often misunderstood. The requirement to ground an electrical power system is based primarily upon personnel safety or the process that takes place in the facility. Although there are exceptions, grounding of electrical power systems and grounding of equipment for safety is required by the National Electrical Code (NEC). The most common exceptions to system grounding involve some manufacturing processes and isolated power systems in hospitals. Equipment grounding is just as important as the system ground. Proper grounding protects facility personnel from electric shock, ensures the operation of overcurrent protective devices, and helps prevent malfunction of sensitive electronic equipment.

GROUNDING ELECTRICAL SYSTEMS

The purpose of *system grounding,* or intentionally connecting a phase or neutral conductor to earth, is to control the voltage to earth, or ground, within predictable limits. It also provides for a flow of current that allows detecting an unwanted connection between system conductors and ground that may instigate operation of automatic devices."[1] Most electrical distribution systems employ a method of grounding the system neutral at one or more points. The two methods of grounding are solid grounding and impedance grounding. The type of grounding system selected is based upon levels of available fault current and the type of protective devices used in the electrical distribution system. The most common systems in commercial buildings are solidly grounded systems. Figure 5.2.6-1 represents the diagrammatic grounding of a typical electrical distribution system. The system includes ground connections at the utility source, the main switchboard, downstream distribution panelboards, and the separately derived systems of step-down 480-208/120-V transformers.

Ungrounded Systems

Ungrounded systems are permitted under certain circumstances. Two of the most common, as stated earlier, are for manufacturing processes and hospital isolated power systems. In a manufacturing plant, where system outages can cause severe economic impact, the NEC permits the use of an ungrounded system. The facility operator risks personnel injury to ensure that the system is not interrupted from minor ground faults. To provide a level of personnel safety, the systems are equipped with ground-fault sensing equipment. The ground-fault indication allows operators to detect ground faults and prevent catastrophic failure. In hospitals where patient care areas are defined as wet locations, the NEC requires ground-fault protection. Under certain conditions, service disruption due to a ground fault can be a greater risk than the ground fault. Therefore, an exception to this requirement is the use of an isolated power system. In the ungrounded isolated power system, circuits are connected to a line isolation monitor that measures leakage current. When physiologically dangerous levels of leakage current occur, an alarm is sounded alerting the staff of the potential risk. Even when ungrounded systems are permitted by the NEC, levels of personnel protection and safety must be incorporated into the design.

Depending upon the service arrangement, system-grounding locations vary. Typically, the neutral conductor is grounded at the service entrance and is then referred to as the *grounded conductor*. Depending on the configuration of the service, the system-grounding location could be at the secondary of the service transformer or at the service-disconnecting device. NEC Art. 250 extensively describes system ground locations and requirements, the sizing of grounding electrode conductors, and bonding system and equipment grounds. Both design engineers and electricians use the requirements set forth in the NEC to ensure properly grounded systems.

Separately Derived System

The NEC defines a separately derived system as the following: a premises wiring system whose power is derived from a battery, solar photovoltaic system, or from a generator, transformer, or converter windings that have no direct electrical connection, including a solidly connected, grounded circuit conductor, to supply conductors originating in another system. In a commercial facility, the most common separately derived systems consist of delta-wye dry-type transformers that step the voltage down from 480 to 208/120 for receptacle power, generator systems installed with four-pole transfer switches, and uninterruptible power supplies (UPS). In each of these listed installations, the supply system ground is isolated from the downstream electrical distribution. A new system ground must be obtained at the location of the separately derived system usually by bonding the neutral of the separately derived system to the building grounding electrode system. It is important to bond the system ground of the separately derived system as near the installation as possible to ensure a low-impedance ground.

Grounding Electrode.   A grounding electrode is defined as a conductor used to establish a ground and to connect electrical systems and or equipment to earth.[2] Ground electrodes consist of two types. The first group consists of underground metallic piping systems, building

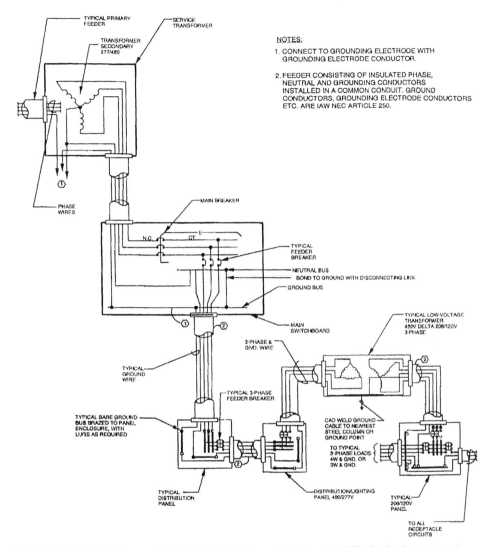

FIGURE 5.2.6-1   Typical grounded electrical distribution system. (*Courtesy of The Ritchie Organization.*)

steel systems, and other underground metal structures installed for purposes other than grounding. The second group comprises electrodes specifically designed for grounding purposes.[1] These electrodes consist of driven ground rods, Ufer grounds, ground counterpoises, and ground plates. NEC Art. 250-81 requires bonding all grounding electrodes to form a grounding-electrode system. The grounding electrodes for any given facility must be bonded together. The size of the bonding jumper is determined from NEC Art. 250-94. The objective of the grounding-electrode system is to minimize the resistance of the grounding electrode. Figure 5.2.6-2 illustrates the grounding-electrode system described by the National Electrical Code.

Due to varying soil resistivities and conditions, it is difficult to define a maximum ground resistance value. The NEC accepts a grounding-electrode value of 25 Ω or less. The objective

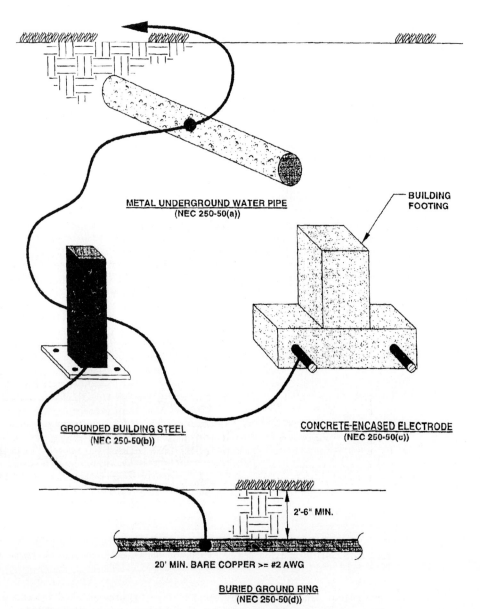

METAL UNDERGROUND WATER PIPE
(NEC 250-50(a))

BUILDING
FOOTING

GROUNDED BUILDING STEEL
(NEC 250-50(b))

CONCRETE-ENCASED ELECTRODE
(NEC 250-50(c))

2'-6" MIN.

20' MIN. BARE COPPER >= #2 AWG

BURIED GROUND RING
(NEC 250-50(d))

FIGURE 5.2.6-2   Grounding-electrode system. (*Courtesy of The Ritchie Organization.*)

is to minimize the resistance so as to limit the voltage to ground during a ground-fault condition. To complicate the process further, it is extremely difficult to measure the actual resistance of the grounding electrode system. The Institute of Electrical and Electronics Engineers (IEEE) has published a standard for measuring ground resistance: IEEE Standard 81-1983, *The Recommended Guide for Measuring Earth Resistivity, Ground Impedance, and Earth Surface Potentials of a Ground System.* The standard discusses the accepted methods of measurement and their advantages and disadvantages. Regardless of which test method is used, "the system should be tested within a few months of completion of the system and after the soils have had a chance to consolidate around the electrodes. The test should be performed again at the time of building completion and then at periodic intervals (change of seasons) to provide baseline data for future testing comparisons.[3]

Equipment Grounding. Equipment grounding is defined as the "interconnection and grounding of the non-electrical metallic elements of a system."[1] Equipment ground systems reduce electric shock to personnel, provide adequate current-carrying capability to accept the ground-fault current permitted by the overcurrent device, and provide a low-impedance return path for ground-fault current that is necessary for timely operation of overcurrent devices. "Electric shock injuries result from contact with metallic components that are unintentionally energized."[4] Possible sources of potential shock include failed electrical insulation systems, damaged components, or improperly installed equipment. Such hazards are minimized by providing a low-impedance path to ground with an equipment-grounding system. Equipment-grounding systems must be sized and suitable to carry the available ground-fault current in the circuit. "All parts of the fault circuit including the terminations and other parts, must be capable of carrying the fault current without distress."[1] To ensure a low-impedance path to ground, it is essential that the equipment-grounding system be installed properly and maintained. Periodic inspections are recommended. Supervision of contractors is also important. Breakdowns in equipment-grounding system integrity often result from modifications to existing electrical distribution systems where ground connections are loosened or damaged when a new component or device is installed. Grounding systems should be tested periodically and when the system is modified.

Equipment-grounding methods are detailed in Art. 250 of the NEC. Additional requirements are often established for special applications. For example, branch circuits supplying patient areas in hospitals must have two separate equipment ground paths. It is important to note that the NEC requirements set a minimum level of installation to ensure personnel safety. It is good practice to provide equipment-grounding systems that exceed the NEC minimum requirements. The NEC permits the use of metal raceway systems as an equipment ground. Installing an additional ground conductor in the cabling provides an inexpensive safeguard against loose conduit connections or improperly installed equipment. "The intended overall purpose of the grounding rules is to achieve as nearly practical, a zero potential difference condition between equipment electrical grounding conductor, the frames of electrical equipment, metal raceways that enclose electrical conductors, and various items of exposed metal building frames and metal piping within the building."[1]

CONTRIBUTORS

Arjun Rao, P.E., Vice President, BR+A Consulting Engineers, Inc., Boston, Massachusetts

Cornelius Regan, P.E., CLEP, Principal, C. Regan Associates, Rockland, Massachusetts

William P. Wohlfarth, P.E., Sr. Electrical Engineer, Massachusetts Institute of Technology, Cambridge, Massachusetts

NOTES

1. IEEE Green Book, Standard 142-1991, *Recommended Practice for Grounding of Industrial and Commercial Power Systems,* Institute of Electrical and Electronics Engineers, Inc., New York, 1991.

2. Simmons, *IAEI Soares Book on Grounding,* 6th ed. International Association of Electrical Inspectors, Richardson, TX, 1996.

3. Power Plant Electrical Reference Series, Vol. 5, *Grounding and Lighting Protection,* Electrical Power Research Institute, Palo Alto, CA, 1987.

4. J. A. Gienger and R. L. Lloyd, *Bibliography on Electrical Safety 1930 through 1953,* Sec. C, AIEE Publications S-69 (now Institute of Electrical and Electronics Engineers, Inc.), New York, 1969.

ARTICLE 5.2.7

CORROSION CONTROL BY CATHODIC PROTECTION

Cornelius Regan, P.E., CLEP (retired), and Harold L. Gordon, P.E.
Stone & Webster Engineering Corporation, Boston, Massachusetts

Cathodic protection is an electrochemical method to reduce or eliminate many forms of corrosion. It is commonly used to mitigate external corrosion on ship hulls, buried fuel oil piping and tanks, gas pipelines, carbon steel water lines, and other miscellaneous equipment such as heat exchangers. The most common form of corrosion occurs when two dissimilar metals are in contact in the presence of an electrolyte.[1] A battery represents a common example of dissimilar metal corrosion. A battery consists of a case (typically zinc for a flashlight battery) that serves as the anode and a rod (typically carbon) that serves as the cathode, both in an electrolyte (mild acid). When the anode and cathode are connected externally (as through a flashlight bulb), current flows in the external circuit from the rod (+) to the bulb to the case (−) through the electrolyte back to the cathode, completing the electrical circuit. Corrosion occurs where current leaves the anode and enters the electrolyte.

Cathodic protection is considered a specialized technical area within the electrical discipline. Systems are typically designed by either specialty cathodic-protection–consulting firms and by equipment vendors who provide turnkey services that include design, furnishing of materials, installation, and start-up services. This article provides an overview of the problems associated with corrosion and presents an introduction to cathodic protection technology. The major technical society dealing in cathodic protection is the National Association of Corrosion Engineers based in Houston, Texas. It is a good source for industry standards and recommended vendors.

CONDITIONS FOR CORROSION

Three conditions must exist for most forms of common corrosion to occur. First, there must be a difference in potential between two different metallic structures or between one part of a metallic structure and another. The area of higher potential (more positive) is the cathode; the area of lower potential (more negative) is the anode. This potential difference can be caused by many factors but most often results from the use of dissimilar metals. The second condition for corrosion is that there must be a metallic path that electrically connects the anode and the cathode. Finally, the anode and cathode must be in an electrically conducting electrolyte. This can be soil, water, or any other chemically suitable material. Thus, when these three conditions exist, current flows from the anodic material into the electrolyte to the cathode and back to the anode via a direct electrical path. Corrosion occurs where the current leaves the anode and enters the electrolyte; the corrosion rate of steel is 20 lb A^{-1} year^{-1}. Corrosion rates are increased when the potential difference [electromagnetic field, (EMF)] between the anode and cathode is large (e.g., interconnecting buried aluminum pipe to copper pipe (0.5- to 0.8-V difference), or when the resistivity of the electrolyte is low. For example, clean dry soils have resistivity values that can range from 100,000 Ω/cm to well over 1 million Ω/cm. These soils are generally classified as relatively noncorrosive, and corrosion tends to be minimal because of the overall high circuit resistance. Corrosive soils and water, however, have resistivity values as low as 100 Ω/cm; the resistivity of seawater is 25 Ω/cm. Structures and equipment located in these environments can be subject to rapid corrosion. Additionally, facilities personnel can have a direct impact on corrosion rates. For example, the widespread use of deicing salts in northern environs can significantly decrease soil resistivities and cause increased corrosion to structures that otherwise would not be adversely affected.

Finally, local, state, and federal regulations may dictate the type of corrosion protection that is required. Title 49 of the Department of Transportation Regulations for hazardous materials pipelines mandates that new piping systems that transport hazardous materials such as natural gas (1) be provided with protective coatings, (2) be electrically isolated from any other structure, and (3) have a functioning cathodic protection system.

CORROSION CONTROL METHODS

Three principal methods are used to reduce or eliminate corrosion: (1) protective coatings, (2) insulating flanges, and (3) cathodic protection. Protective coatings reduce corrosion in two ways. First, metallic structures whose surfaces are completely coated do not corrode because there are no exposed locations on the structure for corrosion current to leave (the structure) and enter the electrolyte. Second, coatings significantly reduce the amount of cathodic protection current that the structure requires by reducing the total exposed surface area. For example, the ratio of required cathodic protection current for a coated structure (effective coating resistance of 100,000 Ω/ft^2) compared with an uncoated structure can be 3×10^{-4} less for the coated structure than the uncoated structure. In practice, there is no such thing as a perfect coating; thus, all coated structures are susceptible to some form of corrosion.

Insulating flanges operate by breaking the electrical path between anodic and cathodic structures. They are widely used in piping systems to isolate buried piping from aboveground piping. Aboveground piping is normally electrically grounded and creates dissimilar metal cells.

The third method for reducing corrosion is cathodic protection. Cathodic protection systems work by forcing dc current to flow from an external source (called the *anode*) through the electrolyte to the structure to be protected (called the *cathode*). When all exposed cathodic surfaces receive sufficient current, then corrosion is eliminated. The design of cathodic protection systems takes into account total current requirements, anode location, anode life, and other conditions, including operations and maintenance.

In practice, it has been found that all usual forms of corrosion are prevented when the cathodic protection makes a pipe or other metallic structure 0.25 to 0.30 V more negative than the soil or liquid surrounding the pipe.[1] There are two basic methods for applying cathodic protection: impressed current and sacrificial anode.

Impressed Current Systems

Impressed current systems use rectifiers to energize long-lived anodes that are positioned at critical locations to prevent corrosion. These systems are designed for long life and can produce high protective currents to eliminate even the worst problems. Anodes for impressed current applications are constructed from high-silicon cast iron, graphite, or from more exotic materials such as platinized titanium, niobium, or mixed-metal oxides. Rectifiers generally are the power source for these systems. Rectifiers can be specified in dc output voltage, and current ranges from 4 V to 120 V and from 4 A to well over 400 A. Rectifier units are manufactured as air-cooled, oil-cooled, or can be constructed as explosion-proof for hazardous locations.

Sacrificial Anodes

Sacrificial anodes include magnesium and zinc for underground applications and aluminum alloys for seawater uses. They can be directly connected to the structure to be protected, or they can be grouped together and connected to the structure via cable connections and function because of their inherent anodic voltage with respect to other materials. For example, magnesium is the most anodic material followed by both zinc and aluminum. Sacrificial anodes are used for smaller surface areas or areas that are electrically shielded. Typically, less maintenance is required for sacrificial anodes than impressed current systems.

Impact of Electrical Grounding.   Electrical grounding is required for the safety of personnel and for the proper and safe operation of equipment. However, electrical grounding can accelerate the corrosion of metals due to an increase in dissimilar metals. Grounding systems that typically include the bare-ground grid cable system and ground rods are generally constructed of copper, whereas buried metallic pipelines and tanks are generally constructed of carbon steel. Because most pipelines are installed with a protective coating, corrosion is concentrated at locations where the coating is damaged. These locations are called *coating holidays.*

Corrosion is aggravated when a carbon steel pipe is electrically connected to the grounding system because the corrosion is concentrated at the coating holiday. It is important, therefore, that precautions be taken to ensure that when the pipes are grounded in a corrosive environment, cathodic protection is installed to mitigate dissimilar metal corrosion.

Stray Currents.   Stray currents cause one of the most severe forms of corrosion. Typically, stray currents are generated by two major sources: dc transit systems and interference currents caused by adjacent cathodic protection systems.

Dc transit stray currents are the most severe cause of rapid and potentially dangerous corrosion. Affected structures can range from underground pipelines, tunnels, foundations, reinforcing steel, underground conduits, and hydraulic elevator shafts. Corrosion currents in the thousands of amperes have been known to cause rapid failure of equipment. Because carbon steel corrodes at a rate of 20 lb A^{-1} year^{-1}, these large stray currents have caused failure of underground pipelines within days. It is essential, therefore, that this potentially diverse problem be quickly identified and that mitigative measures be implemented in a technically correct and thorough manner.

Stray currents from electric traction equipment may be reduced by welding sections of track together and by welding flexible connections across the track, joints which cannot otherwise be made electrically continuous. This reduces the electrical resistance of the track. Also, maintaining good, dry ballast below and between the ties helps to increase the resistance of the ground circuit, thus reducing stray currents.

The second major source of stray currents is due to the nearby presence of a functioning cathodic protection system. These systems can cause an interference condition on underground structures and equipment not connected to the cathodic protection system. Cathodic protection current always takes the path of least resistance. If a structure is in that path and is not part of the cathodic protection system, severe corrosion can occur where the current leaves the structure and enters the earth or water. Solutions to both of these forms of corrosion usually require extensive testing programs to identify and resolve.

Due to a decrease in the rights-of-way for new underground pipelines, many pipelines are being installed in a joint right-of-way with high-voltage transmission lines. This proximity to the high-voltage line, however, can create several problems including adverse electrical conduction and induction effects. If pipelines are installed under high-voltage transmission lines, then a consultant who is familiar with this application should be contacted to provide the required expertise.

Maintenance

Proper maintenance of a cathodic protection system is essential to its proper performance. As part of any installation, provisions should be made for periodic testing. In the case of impressed current systems, rectifier outputs should be read and recorded often, possibly once per month. Structure-to-electrolyte measurements every 3 or 6 months may be adequate to ascertain whether the cathodic structure is receiving sufficient current. Sacrificial anode systems typically require less testing; however, anode current measurements should be made periodically to make certain that the anode is still functioning. Structure-to-electrolyte measurements to ascertain whether the cathodic structure is receiving sufficient current should be made every 3 or 6 months or the same as the impressed current system.

CONTRIBUTORS

Paul Konz, P.E., Electrical Department Head, The Ritchie Organization, Boston, Massachusetts

Arjun B. Rao, P.E., Vice President, BR+A Consulting Engineers, Boston, Massachusetts

William P. Wohlfarth, P.E., Senior Electrical Engineer, Massachusetts Institute of Technology, Cambridge, Massachusetts

NOTE

1. D. Fink and H. W. Beaty, *Standard Handbook for Electrical Engineers,* 13th ed., McGraw-Hill, New York, 1993.

ARTICLE 5.2.8

ELECTRIC HEAT TRACING SYSTEMS

Cornelius Regan, P.E., CLEP
Stone and Webster Engineering Corporation (ret.),
Boston, Massachusetts

Richard G. Brunner, P.E.
Stone and Webster Engineering Corporation, Boston, Massachusetts

INTRODUCTION

The terms *electric-resistance heat tracing, heat tracing, electric pipe heating,* and *trace heating* have all been used to describe the practice of following, or tracing, a pipeline with a heat source. Originally, a steam line fastened to the pipe provided pipe heating; today, however, most pipe heating is done with electric heating wire or cable. Generally, *heat tracing* means directly applying heat to pipes or vessels, as compared with space heating.

An electric freeze protection or process heating system consists of electric heaters (elements), controllers, sensors, a dedicated power system, transformers, panelboards, cables, and alarm devices. The electric heating elements generate heat, and when strapped to surfaces, they transfer heat by conduction and radiation.

The two major reasons for heat tracing are to prevent freezing and to maintain a fluid at a constant temperature. Electric heating is sometimes also used to "heat up" a pipeline to operating temperature.

For freeze protection, the electric heating cable is usually the variable-wattage parallel-circuit type. This cable has a composite heating element that regulates or varies its heat output with respect to its temperature and decreases its output as the temperature increases. Thus, it is self-limiting and ideal for freeze protection.

For process heat tracing or heat-up heat tracing at temperatures above freezing (32°F, 0°C), series heater heating elements are used that have a specific resistance at a given temperature. They are not self-limiting and may require more controls.

Electrical heat tracing can also be used for ice and snow melting. The same information that is required for pipe or tank heat tracing systems is also needed for its proper installation.

DESIGN

Design Requirements

All heat tracing design and installation must comply with the following codes and standards, where applicable:

- IEEE 515, *IEEE Standard for the Testing, Design, Installation, and Maintenance of Electrical-Resistance Heat Tracing for Industrial Applications,* Institute of Electrical and Electronics Engineers

- IEEE 622, *IEEE Recommended Practice for the Design and Installation of Heat Tracing Systems for Nuclear Power Generating Stations*

- National Electrical Code (NEC)
- National Electrical Manufacturers' Association (NEMA)
- National Electrical Safety Code (NESC)
- Occupational Safety and Health Administration (OSHA) Standards
- American National Standards Institute (ANSI)

As with all electrical equipment and systems, electric heat tracing systems must be listed if they are of a type that can be listed. Two sources for listing electric heat tracing systems are

1. Underwriters' Laboratories, Inc. (UL 746B)
2. Factory Mutual Research Corp. (FM)

The following nine design requirements must be known to design an efficient heat tracing system:

1. Temperatures—highest and lowest ambient process temperature to be maintained, and minimum and maximum operating temperatures
2. Size, length, and material of the pipes and vessels
3. Thickness of the thermal insulation
4. Type of fluid or gas being traced
5. Type of heat tracing—freeze protection, process-maintained temperature, heat-up, or prevention of condensation
6. High- and low-temperature alarm set points
7. Classification of the area (hazardous or nonhazardous) as defined in the NEC, Art. 500
8. Isometric pipe drawing showing locations of valves, hangers, and so on
9. Wind speed

Mechanical Requirements

The piping system must be accessible for installation and maintenance of the heat tracing system. The heat tracing system must be continuous through floors and walls (to obviate cold spots), should be segregated from other piping where possible, and should be supported on insulated clamps.

Electrical Requirements

The power supply for heat tracing circuits should come from dedicated transformers and distribution panels that provide independence. Connecting additional heating circuits to lighting panels for convenience should be avoided.

Grounding.   An effective ground must be provided from the outer metallic covering of the heating cable to the power distribution system. If the sheath or braid covering the heating cable does not provide a ground path, ground-fault currents must be investigated.

Heating Elements.   All heating elements should be selected to utilize a standard voltage: 120, 208, 240, 277, or 480 V, as applicable. The heating element protective sheath must be capable of withstanding the maximum pipe operating temperature, and the sheath temperature must not exceed the ignition temperature of any gas or vapor that could be present. The type and size of the heating element are selected on the basis of the pipe size, type, and temperatures.

Design Calculations

Rather than make fundamental thermal calculations each time that a heat loss determination is required, it is entirely practicable to use the universal design guides prepared by the major heat trace vendors. Heat tracing software available to users can give tabulated results for virtually every heat tracing condition, including various pipe and tank sizes, types of thermal insulation, wind and safety factors, and operating conditions.

Many heat loss formulas and sample calculations appear in IEEE Standard 515 and can be applied to parallel constant- and variable-wattage and series cables. Heat tracing installation should be designed under the supervision of a competent electrical engineer.

Isometric drawings should be prepared to facilitate heat trace application, showing pipe and tank size and insulation and length of runs. The amount of heat needed equals the heat loss of the system; thus, the thermal conductivity of the insulation, the dimensions of the pipe and insulation, and the temperature difference to be maintained must be known.

Control, Alarm, and Monitoring Systems

The control system provides the means to electrically energize and deenergize the heating elements in response to changing temperature conditions, and the alarm and monitoring systems provide information about the system's condition. Temperature alarms indicate operating conditions, and current and voltage detection monitor the condition of the heaters.

Freeze Protection Application

Most freeze protection is accomplished by using ambient temperature sensing to actuate the heat-traced circuits through a contactor. Some energy savings are possible by using variable-wattage cable. Another method is to sense pipe temperature.

Process Control Application

A critical process control system is redundantly traced and has two independent controllers for each pipe. These controllers should be physically and electrically isolated from each other.

Heat-Up Control Application

The heat required for an initial heat-up is greater than that for maintaining a normal operating level. Other requirements are identical.

Snow Melting Application

Heat is supplied to keep sidewalks clear of ice and snow.

EQUIPMENT SELECTION

Selection of Heater Element

The three basic types of heating cables are constant-wattage series-circuit, constant-wattage parallel-circuit, and variable-wattage parallel-circuit, as well as blankets and strip heaters. Each type has distinct advantages for different applications.

In operation, an electric heating cable develops a sheath temperature higher than the pipe or vessel and thereby heats the pipe. If the pipe temperature exceeds the temperature rating of the heater material, it is a major problem. Typically, thermoplastic elastomers have a maximum temperature rating of 121°C; fluorinated ethylenepropylene (FEP) Teflon, 204°C; Tefzel, 150°C; Kapton, 371°C; copper-sheathed mineral-insulated (MI) cable, 148°C; and Inconel and 304 stainless-steel MI cable, 537°C. Heater characteristics should be obtained from the manufacturer at the time of heater selection to avoid inappropriate application.

Constant-Wattage Series-Circuit Heating Cable.   Mineral-insulated cable is recommended when it must operate under high-temperature conditions, is subject to high levels of radiation, or requires mechanical protection. For less rigorous applications, cables insulated with Teflon, Tefzel, or silicone rubber are available for various temperature limits.

Constant-Wattage Parallel-Circuit Heating Cable.   Constant-wattage parallel-circuit cable is available in various insulating materials for different temperature limitations. Constant-wattage cable is advantageous for long cable lengths and can be cut to length in the field. Constant-wattage parallel-circuit heaters also come in blanket form. They are available in standard and custom sizes and are well suited for tanks, valves, flanges, and ducts.

Variable-Wattage Parallel-Circuit Heating Cable.   Variable-wattage parallel-circuit cable is ideal for freeze protection and can also be cut to length in the field. It is also a good application for plastic and fiberglass-reinforced epoxy pipe because of the self-limiting temperature feature.

Strip Heaters.   Strip heaters provide constant-wattage heat output and are available in a wide range of voltage, heat output, and size. They are used for convection air heating and clamp-on heat transfer applications.

Bulb and Capillary Thermostats.   Bulb and capillary thermostats are the most commonly used mechanical controllers. They have limited load-switching capability, and the freezing temperature of the fluid may limit low-temperature operation. Copper, chrome-copper, and stainless-steel bulbs and capillaries are available and should be selected to suit the environment.

Bimetallic Thermostats.   Bimetallic thermostats are factory-preset to temperature. They may be paralleled to sense temperature at several places along the pipe. Because they are factory-preset, they are not suitable for variable set-point applications.

Electronic Controllers.   Electronic controllers may be located a significant distance from the heated pipes. They use resistance temperature detectors (RTDs), thermistors, or thermocouples to sense temperature and control the power circuits to the pipe heaters. They may be designed for either on/off or proportional control. On/off control turns power on at the low set point and off at the high set point. This results in some temperature overshoot or undershoot from the set-point temperatures because the temperature lags the heat input. Proportional control provides just enough power to maintain the set-point temperature and minimizes temperature cycling. Electronic controllers show the actual pipe temperature and the desired set-point temperature, usually as a digital display.

Microprocessor.   Microprocessors have expanded the scope of control and alarm functions and permit management as well as control of heat tracing systems. The microprocessor is an independent system connected via communication lines to a central computer. The microprocessor provides data for temperature sensing and alarm, for the operator's display, and for power switching in the heat trace system. The computer periodically scans all sensors and provides an alarm and readout in the event of trouble.

Maintenance

No heat tracing system is complete until it is checked and tested in accordance with the codes and standards listed in this section and also with the manufacturer's recommendation. Records must be kept of the original readings for comparison with the periodic test readings. These readings will be helpful in spotting moisture penetration into the electrical system (gradual decline in isolation resistance) or physical damage to the heating cable (sharp decline in the insulation resistance). It is of paramount importance that the heater cable be completely tested before installation. This will prevent the installation of damaged heater cable whose cost is minimal when compared with the costs of cable installation and thermal insulation. The following insulation resistance testing is recommended and should be performed:

- Upon receipt of the heating cables
- Before the thermal insulation is installed
- After the thermal insulation is installed
- During periodic maintenance testing (as mentioned previously)

Temperature maintenance circuits should be checked at least twice a year. Freeze protection circuits should be checked before the season approaches that requires using them.

The following checks should be done periodically:

- Visually inspect the inside of connection boxes for corrosion, moisture, and loose connections.
- Check heating cable connections, grounding, and ensure that the heating cable connections are insulated from the connection box.
- Megger test at power connections to make sure that both bus wires are disconnected from the power source.
- Check circuit amperage and voltage.
- Check thermostats for moisture, corrosion, switch operation, and capillary damage.

Check for damage to insulation seals at valves and pumps, and so on. Also, check for leaks throughout the system.

All metal parts of temperature controls should be sprayed with a moisture repellent or corrosion repellent yearly. If a cable is damaged, it must be replaced in its entirety. Moisture may have migrated into the good section of the cable during the damage, and it may short out after repair of the damaged section.

CONTRIBUTORS

Arjun B. Rao, P.E., Vice President, BR+A Consulting Engineers, Boston, Massachusetts

William P. Wohlfarth, P.E., Senior Electrical Engineer, Massachusetts Institute of Technology, Cambridge, Massachusetts

BIBLIOGRAPHY

NFPA 70-1999, National Electrical Code, National Fire Protection Association, Quincy, MA, 1999.
Master Catalog, Nelson Heat Tracing Systems, Tulsa, OK, 1999.

General Catalog, Thermon Heat Tracing Systems, San Marcos, TX, 1999.
Design Guide, Raychem Heat Tracing Systems, Menlo Park, CA, 1999.

ARTICLE 5.2.9
POWER QUALITY

A. Todd Rocco, P.E., Principal
Diversified Consulting Engineers, Quincy, Massachusetts

Power quality has become a prime concern of facilities engineers and managers because of the increasing use of voltage-sensitive electrical and electronic equipment. Compounding the need for awareness of this issue is that this same modern electrical and electronic equipment can be both the source and that which is affected by many of the power quality phenomena. This article presents a basic explanation of power quality and definitions, along with the reasons that power quality is a concern and a description of the power quality phenomena that cause equipment malfunction or misoperation. Voltage considerations for equipment include the susceptibility profile for computer equipment and the causes of sags, swells, and transients external or internal in a facility, such as lightning, power factor capacitor switching, improper wiring, or a motor starting. Design considerations and solutions are presented. The information presented is based on and refers to publications of the Institute of Electrical and Electronics Engineers, Inc. (IEEE) and other recognized industry standards and guidelines, in addition to information gathered from experience.

DEFINITIONS

Power quality can be considered an electrical system abnormality that causes an undesirable result in system loads or components. The IEEE Standard 100, *Standard Dictionary of Electrical and Electronic Terms,*[1] provides definitions for various types of electrical disturbances, as shown in Table 5.2.9-1. The names, waveforms, and descriptions of common disturbances are also presented in Table 5.2.9-1.

These and other terms are commonly used in inappropriate contexts. The reader should be aware that other interpretations of these terms exist and that terms should be questioned to reduce ambiguity. More detailed descriptions are presented in IEEE Standard 100,[2] IEEE Standard 1159,[3] and NFPA 70B.[4] Two articles that describe waveforms well are "Storing Power for Critical Loads" [*IEEE Spectrum* (June 1993)][5] and "Protecting Computers Against Transients" [*IEEE Spectrum* (April 1990)].[6] The IEEE Standard 1159-1995, *Recommended Practice for Power Quality Monitoring,* contains additional terms taken from other standards, which are being adopted more widely to be consistent with other international standards. This standard also includes a list of terms to avoid.

Work has also been done on the electric utility side of the meter. The Electric Power Research Institute (EPRI), a research and development organization funded by electric utilities, has developed indexes to describe the power quality level for electric utility distribution circuit areas. Facilities that are considering procuring *custom power* (a term that describes advanced

TABLE 5.2.9-1   Summary of Common Power Quality Disturbances and Associated Waveforms

Disturbance Type	Description	Waveform
Harmonic distortion	Nonlinear distortion that appears as harmonics (i.e., harmonic components) of a single-frequency input.	
Harmonic components	The components of the harmonic content as expressed in terms of the order and root-mean-square (rms) values of the Fourier series terms describing the periodic function.	
Interruption	The complete loss of voltage for a time period. The time-base of the interruption is characterized as follows: —Instantaneous: 0.5–30 cycles —Momentary: 30 cycles–2 s —Temporary: 2s–2 min —Sustained: >2 min	
Notch	A switching (or other) disturbance of the normal power voltage waveform, lasting less than a half-cycle; which is initially of opposite polarity than the waveform, and is thus subtractive from the normal waveform in terms of the peak value of the disturbance voltage. This includes complete loss of voltage for up to a half-cycle.	
Overvoltage	An rms increase in the ac voltage, at the power frequency, for durations greater than a few seconds.	
Sag	An rms reduction in the ac voltage, at the power frequency, for durations from a half-cycle to a few seconds. Note: The IEC terminology is *dip*.	
Swell	An rms increase in the ac voltage, at the power frequency, for durations from a half-cycle to a few seconds.	

TABLE 5.2.9-1   Summary of Common Power Quality Disturbances and Associated Waveforms (Continued)

Disturbance Type	Description	Waveform
Transient	A subcycle disturbance in the ac waveform that is evidenced by a sharp brief discontinuity of the waveform. May be of either polarity and may be additive to or subtractive from the nominal waveform.	
Unbalanced load regulation	A specification that defines the maximum voltage difference between the three output phases that will occur when the loads on the three are of different levels.	Va, Vb, Vc, Vmin
Unbalanced three-phase system	A three-phase system in which the rms value of at least one phase voltage (or current) or line-to-line voltage is significantly different from the others, or in which the phase angle displacement between any pair of phases significantly differs from 120°. Note: In an unbalanced three-phase system, negative or zero-sequence components exist.	Va, Vb, Vc, Vmin
Undervoltage	An rms decrease in the ac voltage, at the power frequency, for durations greater than a few seconds.	

Source: Definitions courtesy of Institute of Electrical and Electronics Engineers, Inc. Charts courtesy of Diversified Consulting Engineers.

technologies that can be applied to the utility system to improve the power quality delivered to utility customers) are advised to research the costs and benefits of alternative technologies.[7]

EQUIPMENT MALFUNCTION AND DAMAGE

Any of the electrical disturbances defined in the previous section can cause equipment malfunction or damage, but the most severe damage to equipment typically results from overvoltages or high-energy transients (impulses). Damage to electrical equipment is usually from long-term exposure to such conditions that can cause equipment (e.g., semiconductor devices, fluorescent ballasts, motors, and transformers) to operate above its designed temperature range, or electrical or mechanical stress on the system beyond the intended design. Interruptions to processes and damage to products during manufacturing are industrial examples of the result of sags (the most common power quality phenomena), swells, or interruptions. For example, medical imaging equipment or industrial process controllers can be damaged by an interruption or sag if the equipment requires a sequenced shutdown procedure. Restoring the equipment to operation can take days if components are damaged, software is corrupted, or the product must be cleaned out of the machines, and the process slowly restarted. In addition to the effects of overvoltage and undervoltage conditions described for motors and other equipment, these conditions can also have a considerable financial impact on operating and maintenance costs of a facility's electrical system or cause a loss of revenue. Swells and overvoltage situations can cause premature equipment failure (e.g., lamps and motors). Sags and undervoltage situations, which are the most common types of power quality disturbances, can cause premature equipment failures, operational problems such as adjustable-speed drive tripping,[8] and higher electrical energy costs resulting from increased system losses.

The presence of significant levels of harmonic components may require derating motors, transformers, and other parts of the electrical system, including neutral conductors and panel ampacities. Other problems can include premature failures and misoperation of equipment.

Unbalanced load regulation (phase imbalance) along with transients, overvoltages, undervoltages, sags, and swells can cause the equipment's power supply to perform unsatisfactorily or be damaged, resulting in erratic or undesirable equipment operation. Table 5.2.9-2 summarizes the potential effects on equipment as fundamental as a motor and the failure that can result from various electrical disturbances.

Voltage Considerations

Standards and Guidelines. Standards and guidelines establish performance criteria for electrical power systems and equipment. The American National Standards Institute (ANSI) has standardized nominal voltages and voltage ranges for electrical power systems and equipment. Table 5.2.9-3 presents excerpts from Table 1 of the ANSI Standard C84.1a-1980 for

TABLE 5.2.9-2 Potential Effects on Motors and Resultant Failures From Electrical Disturbances

Disturbance Type	Potential Effects and Problem	Resultant Failure
Interruption	Reenergizing a motor while coasting causes severe torque and current transients.	Motor winding insulation failure due to heat or physical damage from high magnetic fields. Shaft, drive, or driven equipment damage from large torques.
	Frequent stops and starts can exceed motor duty cycle and temperature specifications and stress shafts.	Motor winding insulation failure from heat.
Phase imbalance	Excessive heating due to negative sequence (180° out of phase) currents.	Motor winding insulation failure from heat.
Sustained undervoltage and overvoltage	Excessive heating due to I^2R losses, and reduced fan cooling (undervoltage).	Motor winding failure from heat.
Sag and swell	Increased heating due to torque fluctuations and produces stresses on shafts.	Motor winding fails prematurely.
Transients	Excessive winding voltage causes insulation stress.	Motor winding insulation develops localized deterioration and failure.
Harmonics	Eddy current losses increase as square of harmonic number, and negative sequencing currents result in heating.	Motor winding fails prematurely; fluting of bearings.

TABLE 5.2.9-3   Electrical System Voltage Ranges

Nominal System Voltage	Minimum Utilization Voltage	Minimum Utilization Voltage Without Lighting	Minimum Service Voltage	Maximum Utilization Voltage	Maximum Service Voltage
208Y/120	191Y/110*	187Y/108*	197Y/114*	218Y/126*	218Y/126*
	184Y/106[†]	180Y/104[†]	191Y/110[†]	220Y/127[†]	220Y/127[†]
480Y/277	440Y/254*	432Y/249*	456Y/263*	504Y/291*	504Y/291*
	424Y/245[†]	416Y/240[†]	440Y/254[†]	508Y/293[†]	508Y/293[†]

* Range "*" is for satisfactory design of equipment.
[†] Range "[†]" allows for slight deviations from range "*" but should be restored to range "*" in a reasonable amount of time.
Source: Courtesy of Institute of Electrical and Electronics Engineers, Inc.

three-phase, four-wire systems.[9] Refer to IEEE Standard 241 for a thorough presentation of standard voltages.[10]

The Computer Business and Equipment Manufacturing Association (CBEMA) has established guidelines for voltage disturbances as a function of time for electronic equipment (Fig. 5.2.9-1). The CBEMA Curve, as it is commonly known, has been widely used but has been replaced by the Information Technology Industry Council (ITIC) curve. However, a much more useful set of curves can be found in IEEE Standard 1346.[11] This standard shows how to plot the power quality phenomena from a survey versus equipment susceptibility to determine how many problems to expect in a typical year.

Sources of Voltage Disturbances.   There are many causes of voltage disturbances in facilities that can be caused by external or internal events. Recent surveys have found that most power quality phenomena result from disturbances within the facility itself. This brief summary is representative of common disturbances and causes of the disturbances.

Interruptions.   Interruptions can be caused externally by weather-related events (ice, high wind, or lightning), traffic accidents that affect utility equipment such as power poles, distribution or transmission equipment malfunctions, or overloads that result in substation circuit-breaker operation or fuses blowing. Internal causes include equipment failures, improper wiring, or system overloads that result in circuit-breaker or fuse operations. Planned outages, also called shutdowns, are scheduled interruptions of selected portions of the power system.

Undervoltages and Overvoltages.   The causes of undervoltages and overvoltages (externally or internally) include poor system voltage regulation (i.e., the inability to provide a voltage within a specified range for the anticipated load fluctuations), equipment malfunction, or system overload. The cyclical loading of an electrical system can cause overvoltage conditions during periods of minimum load or undervoltage during maximum load. Many facilities experience cyclical loading [e.g., steel mills, schools, office buildings, and commercial and health care facilities (treatment and support areas)]. Other external causes include planned utility voltage reduction (brownout).

Externally Generated Sags or Swells.   Externally generated sags or swells can result from disturbances in the electric utility system. The causes are similar to those that cause interruptions but are buffered from the disturbance only by remoteness and the location of the fault relative to the protective equipment of the electrical system source(s) (breakers) and the facility. A swell can result when system load is abruptly reduced until the utility's voltage regulation system responds by lowering the system voltage. An unbalanced fault (in a three-phase system) can also cause a voltage increase on the unfaulted phase.

Internally Generated Sags or Swells.   Internally generated sags or swells can be caused by motors that start and stop, the operation of other high-power loads, elevators, improper wiring and grounding, or system overloads and faults. Internally generated transient voltages

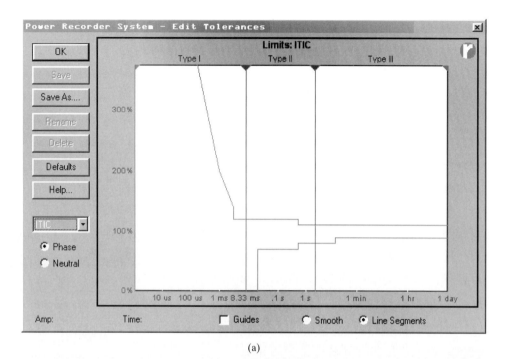

(a)

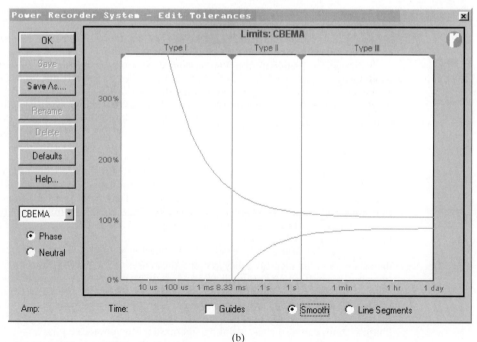

(b)

FIGURE 5.2.9-1   CBEMA and ITIC voltage tolerance curves. (*Courtesy of Reliable Power Meters, Inc.*)

can be caused by arcing contacts and poor connections (e.g., loose bus bar connections), loads such as copy machines, medical imaging equipment, adjustable-speed drives, and other equipment that draws high current for a very short duration.

Externally Generated Transient Voltages.   Causes of externally generated transient voltages include utility system switching from capacitor bank operation, abnormalities, and lightning. The utility system abnormalities may include system faults, circuit reclosers (a practice utility companies use when a faulted line is reenergized with expectations that the fault has cleared), or equipment failures (depending on the duration, these events can also cause sags or interruptions). Lightning is another source of external transient voltages. Certain geographical areas are highly prone to lightning strikes. For example, some health care facilities in Florida have had power system disturbances caused by summertime lightning storms so often that their emergency generators operate more frequently than the required monthly testing.

Electromagnetically Conducted Disturbances.   Electromagnetically conducted disturbances can be caused by a variety of equipment that produces an electromagnetic field that is conducted and disturbs other equipment. When a radiated electromagnetic field causes a disturbance, typically, it is not considered a power quality issue.[12]

Design Considerations and Solutions.   Using a dedicated feeder is an effective practice for minimizing electrical disturbances from high-power devices. This provides a connection point where the system impedance is lower, thus improving voltage regulation.

Transient voltage surge suppression (TVSS) devices are designed to limit the system voltage to a specified level when the energy of the surge is less than the rating of the TVSS device. Therefore, the maximum anticipated surge energy should be calculated for the TVSS device to ensure that it is adequately sized. The location of the TVSS device connection depends on the source of the anticipated voltage transients. If protection is desired from external causes, then, typically, the equipment is installed at the electrical service. If protection is desired from internal causes, then the location of the TVSS device should be analyzed with respect to the cause of the disturbances and the equipment to be protected. Proper installation and grounding is necessary for such devices to be effective.

Power cables can emit electromagnetic fields that interfere with the operation of adjacent equipment. Conversely, power cables can also be subject to electromagnetic fields produced by adjacent equipment. To minimize these effects, cable routing (to reduce proximity) and cable configuration (to reduce magnetic fields) should be studied.

Electrical infrastructure, including the transformers, circuit protection, and wiring, may need to be derated if harmonic currents are high enough to cause significant losses (see "Harmonic Components," later in this chapter).

Power Quality and Grounding Audit

When facilities are experiencing problems caused by suspected power quality, the surest way to determine the causes and obtain the most appropriate solutions is to retain a qualified firm to conduct a power quality and grounding audit. Professionals in the field of power quality have known of the importance of power quality and grounding audits for many years, and this knowledge is becoming more widely disseminated to facilities managers as well. Faulty wiring and grounding cause many of the power quality problems in facility power systems, and an audit is often necessary to isolate the causes.[13] Performing a power quality and grounding audit a priori provides a benchmark to help identify any system changes.[14]

Harmonic Components.   Harmonic components are emphasized in this section because the cause of some power quality problems can be traced to harmonic components. This section focuses on an explanation of both harmonic components and mitigation techniques on a fundamental engineering level that is technical enough to explain the issues. This depth is usually required to adequately understand and solve the actual source of the problem rather than the apparent problem. Sources of harmonic components are presented with solutions. The text includes background information and elementary concepts, including basic Fourier analysis (illustrating representations in both the frequency and time domains); causes of harmonic components; problems associated with harmonic components, including resonance; and solutions, including system design and mitigating techniques. Design techniques include equipment sizing (e.g., feeders, transformers, and panels) and applying K-rated transformers.

Understanding Harmonic Components.   A harmonic is a frequency that is an integral multiple of a fundamental frequency. For example, perturbing (plucking) a guitar string makes the string oscillate at a predominant frequency and integral multiple frequencies based on the length of the guitar string. Other frequencies produced by the initial perturbation quickly decay because the string has an appropriate length for a certain note.

The harmonic number represents a frequency that is an integral multiple of the fundamental frequency. Therefore, the fundamental frequency can also be considered the first harmonic, contrary to the way to which harmonic components are typically referred (i.e., the first harmonic has sometimes been incorrectly called the "double of the fundamental frequency"). A mathematical technique to represent a periodic waveform using a linear combination of sine and cosine waveforms is called a *Fourier representation.* For example, Fig. 5.2.9-2 demonstrates that the summation of a fundamental frequency and the first seven odd harmonic components resembles a square wave. The Fourier representation provides an intuitive basis of the way harmonic components can distort the fundamental waveform.

Figure 5.2.9-2 illustrates the magnitudes of the fundamental frequency and the harmonic frequencies (note that the harmonic components are represented as integral multiples of the fundamental frequency). This representation provides straightforward data about the harmonic components. The Fourier transform converts the waveform from the time representation to the frequency representation.

Causes of Harmonic Components.   Much of the equipment in the modern electrical environment, including computers, printers, copiers, fluorescent and high-intensity discharge

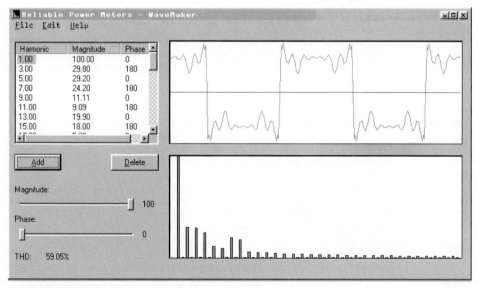

FIGURE 5.2.9-2   Example of square-wave voltage waveform and frequency spectrum. (*Courtesy of Reliable Power Meters, Inc.*)

lighting, electronic instrumentation, uninterruptible power supplies, welders, adjustable-speed drives, solid-state controls, radiology equipment, and so on is classified as a nonlinear load. Nonlinear loads distort current and voltage waveforms by drawing current mainly at the peak of the voltage waveform. This flow of current is called *nonlinear* because the relationship between the voltage and the current is not constant and therefore must include other frequencies, in addition to the fundamental, to comprise the waveform. Figure 5.2.9-3 illustrates the current and voltage waveform distortion due to a nonlinear load. The current waveform shown is not a pure sine wave and comprises many harmonic components. This nonlinear current can also cause voltage distortion, as shown in Fig. 5.2.9-3.

Resonance.   Resonance is a condition in which an ideal system (i.e., no losses) oscillates. Electrical, as well as mechanical, structural, and other systems can have natural resonant frequencies. A classic example of resonance in a structural system is the suspension bridge in Washington State that was perturbed by the wind (which provided the energy required to start the bridge oscillating), and the amplitude of the oscillations increased until the bridge self-destructed.

Electrical power systems have natural resonant frequencies based on the system inductance and capacitance. Like the bridge example, an electrical system needs to be excited near a resonant frequency before resonance starts.

Problems Associated with Harmonic Components and Concerns.   Resonance occurs when a harmonic produced by a nonlinear load excites the system at (or near) the resonant point. Although the system has losses that make the resonance decay, it continues to provide harmonic components while the nonlinear load is connected, and the system remains in reso-

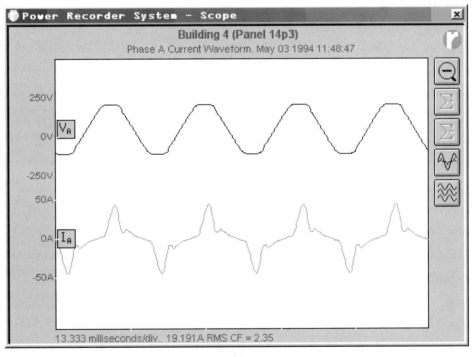

(a)

FIGURE 5.2.9-3   Example of (*a*) nonlinear load voltage and current waveforms, (*b*) frequency spectrum, and (*c*) contents. (*Courtesy of Reliable Power Meters, Inc.*)

Harmonic	RMS Value	Phase	Percent
0	112.6mA	180°	0.679%
1	16.56A	124.8°	100%
2	56.45mA	126.6°	0.340%
3	7.744A	41.21°	46.74%
4	52.72mA	24.67°	0.318%
5	5.211A	318.1°	31.45%
6	33.18mA	290.5°	0.200%
7	2.608A	206.7°	15.74%
8	37.70mA	254.0°	0.227%
9	967.8mA	47.60°	5.841%
10	27.31mA	211.4°	0.164%
11	820.1mA	266.1°	4.950%
12	5.790mA	206.5°	0.035%
13	338.2mA	142.7°	2.041%
14	14.47mA	206.5°	0.087%
15	108.3mA	294.7°	0.653%
16	2.589mA	270°	0.015%

Building 4 (Panel 14p3)
Phase A Current Harmonics, May 03 1994 11:48:47

Odd Harmonics: 59.060%.   Even Harmonics: 0.618%.   Total: 59.063%.

(b)

FIGURE 5.2.9-3   *(Continued) (Courtesy of Reliable Power Meters, Inc.)*

nance. Resonance decays when the nonlinear load is disconnected or the system resonant frequency changes sufficiently to a frequency that is not excited by the harmonic components. The system resonance frequency can change by applying inductive loads like relatively large motors. Capacitance is usually introduced into the system to improve the power factor. The capacitance can also change (e.g., from automatic power factor correction equipment) which also changes the system's resonant frequency. This resonance can result in overvoltage conditions and other stresses on the system.

Effects on Electrical System.   The presence of harmonic components in an electrical system reduces the system's ability to carry the design load. For example, cable ampacity, transformer capacity, and circuit-breaker capacity are all reduced by harmonic components. A cause of the capacity reduction can be demonstrated by the total increase in current magnitude illustrated in Fig. 5.2.9-3. Another cause is that the increase in frequency increases cable and transformer losses. This phenomenon is known as the *skin effect,* which describes the way electrons move to the outer surface of a conductor as a function of frequency. The operating temperature of electrical equipment increases as equipment losses increase. Eddy current losses in transformers increase as the square of the harmonic number. Hence, higher-order harmonic components cause much larger losses, which results in increased heating of the transformer.

Potential Problems of Harmonic Components.   Harmonic components can cause serious problems in electrical equipment, especially if resonance occurs. Harmonic components alone, but more so during resonance, can overload equipment (e.g., cables, transformers, and panels). The overload can go undetected until serious damage or failure occurs. For example,

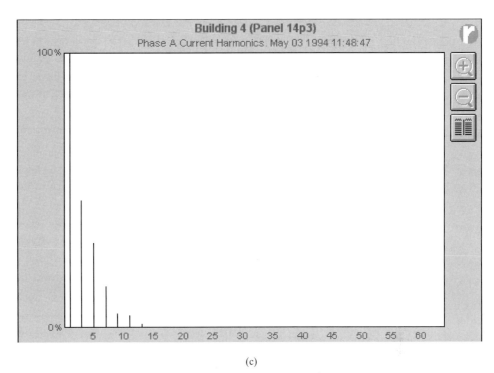

(c)

FIGURE 5.2.9-3   (*Continued*) (*Courtesy of Reliable Power Meters, Inc.*)

a capacitor connected to a motor for power factor correction could be part of the resonant system. This condition can cause a high current to flow in the capacitor and in its connection leads and cause excessively high current until either an overcurrent device opens or something else burns sufficiently to open the circuit (see the following examples). Examples of other high-current situations that can go undetected are: transformers that are overloaded due to the harmonic components, but the overload is not apparent to the overcurrent device; and panelboard neutrals and shared neutrals (e.g., in a furniture partition system), because only the phase currents are monitored for overcurrent. When single-phase loads are supplied from different phases of the same three-phase source, harmonic components whose number is divisible by three [referred to as *triplen harmonic currents* (e.g., 3, 6, 9, 12, 15 . . .)] will add together in the neutral conductor. The even triplens cancel each other out, whereas the odd triplens mutually contribute to increased current. The result can be a neutral current that is higher than that of the phase conductors. This is why the NEC recommends neutral conductors larger than the associated phase conductors when significant harmonic levels are present.

Nuisance Tripping.   Some electronic circuit breakers (dating back to the time when the technology was introduced) use a current-sensing method based on the waveform peak rather than on the root-mean-square (rms) value. When the current waveform is linear, this method provides satisfactory protection. However, when the current waveform is nonlinear, the ratio between the peak value and the rms value is no longer valid. The result can be a nuisance trip—the circuit breaker opens from what was improperly detected as an overload, despite the fact that an overload never occurred.

Effects on Conductors.   The increased heating due to the skin effect, the proximity effect, and the conductor impedance, all of which increase with frequency, reduces the effective ampacity of conductors.

Effects on Transformers.   Transformer losses increase due to increased eddy currents (which increase by the square of the load and frequency) and hysteresis losses (which increase with frequency). In addition, transformers can also be part of a resonant circuit that increases the load on the transformer.

Effects on Capacitors.   Capacitors can experience load increases due to the harmonic voltages, additional heating from an increase in dielectric losses, overvoltage conditions, shortened life expectancy, and possible failure from a resonant condition.

Effects on Motors.   The effect of harmonic components on motors should be considered because of the high probability that they could be exposed to harmonic components, the capital investment, and the potential downtime due to a failure. Large facilities typically have three-phase motors that are supplied by a 480-V system. These motors are typically exposed to harmonic components produced by an adjustable-speed drive (ASD) for this motor or for adjacent motors. The harmonic components produced by an ASD with rectifier input is a function of how continuous the current draw is. This current waveform depends on the number of paths of conduction (or poles). A six-pulse converter produces harmonic currents of the 5th, 7th, 11th, 13th, 17th, 19th, and so on. A 12-pulse converter produces harmonic currents of the 11th, 13th, 23rd, 25th, and so on, and the total harmonic distortion is lower than the six-pulse converter because the current waveform is a closer representation of a sinusoid. Intuitively, based on the Fourier representation of both current waveforms, the 12-pulse converter will have less harmonic content (because it is closer to a sinusoid), and the fundamental frequency will be predominant. A motor will operate at a higher temperature for the same reasons as described for a transformer, but, in addition, other heating issues involve the rotating magnetic field of the motor.

The rotation of a motor's magnetic field is due to the alternating current supplied to the motor. When the motor is supplied by a linear current waveform, the resultant magnetic field is also linear; that is, one magnetic field that causes the mechanical rotation is rotating. When the motor is supplied by a nonlinear current waveform, several fields are rotating, one corresponding to each harmonic, as demonstrated by the Fourier representation. The negative-sequence harmonic components cause the magnetic field to rotate in the direction opposite to the positive-sequence harmonic components. Therefore, the sum of the harmonic components causes rotating magnetic fields in both directions. This results in additional heat in the motor, increasing insulation and bearing temperatures and thereby reducing motor life. In addition, losses are greater due to harmonic components, including higher losses due to the skin effect and eddy currents.

Harmonic currents can also be induced in grounded motor shafts. These harmonic currents result in damage to the bearings, called *fluting*. This can result in premature failure of the motor bearings.[1]

Examples.   Practical examples are presented from actual troubleshooting and analysis of power quality problems, demonstrating problems associated with power quality and also data to indicate that power quality should not have been the cause of equipment malfunction.

High-Current THD with Low-Circuit Current.   The total harmonic distortion (THD) quantifies the harmonic content with respect to the total or fundamental voltage or current. However, the current THD can be misleading by itself. Figure 5.2.9-4 illustrates a circuit where the current THD is nearly 500 percent. Based on the current THD, it may appear that this is a problem. However, the total current is less than the circuit ampacity, and Fig. 5.2.9-4 illustrates that the voltage waveform distortion is negligible.

Power Factor and Resonance.   When power factor correction capacitors are connected to a power system, a resonant frequency can exist at which the capacitive and inductive reactances are equal. This frequency can cause high impedance on the system bus, and voltage distortion can result because the high impedance will not shunt this resonant current. In addition, high current will flow between the inductive and capacitive resonating components. Figure 5.2.9-5 illustrates a system that had a resonance condition around the 17th harmonic. The facility had power factor–correcting capacitor banks for each chiller motor. When ASDs were installed in

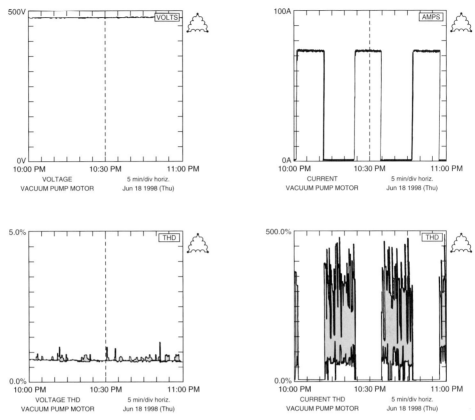

FIGURE 5.2.9-4   Example of nonlinear load current waveform, frequency spectrum, and total harmonic distortion. (*Courtesy of Diversified Consulting Engineers.*)

the building, the capacitors started to burn out. The current THD graph demonstrates that the circuit resonance was caused by the ASD exciting the circuit near its resonant frequency (in this case, the 17th harmonic).

Elevated Neutral Voltage.   The odd triplen harmonic components produced in a three-phase system when supplying rectified-input, single-phase loads are additive and make the neutral conductor possibly carry more load than the phase conductors. The harmonic components can cause an excessive neutral voltage drop. The voltage drop will elevate the neutral with respect to ground (because the ground conductor is intended as a non-current-carrying conductor). This potential between the neutral and ground can cause misoperation of sensitive electronic equipment, even though the voltage supplied to the equipment may be within its specified voltage range.

Solutions

There are several ways to reduce harmonic components and improve power quality. This section presents some practical methods.

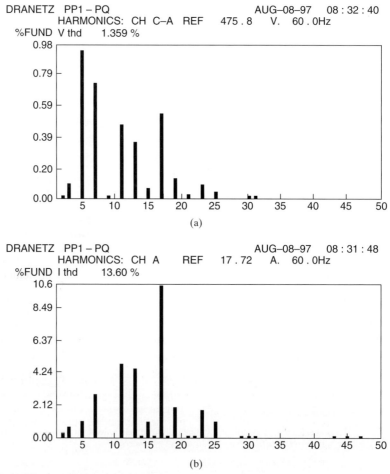

DRANETZ PP1 – PQ   AUG–08–97   08 : 32 : 40
            HARMONICS: CH C–A REF   475 . 8   V.   60 . 0Hz
  %FUND V thd   1.359 %

(a)

DRANETZ PP1 – PQ   AUG–08–97   08 : 31 : 48
            HARMONICS: CH A   REF   17 . 72   A.   60 . 0Hz
  %FUND I thd   13.60 %

(b)

FIGURE 5.2.9-5 Example of current frequency spectrum with resonance condition at the 17th harmonic. (*Courtesy of Diversified Consulting Engineers.*)

System Design.   There are many considerations for electrical systems with nonlinear loads. The basic premise is to understand the load and the way it will affect the electrical system and equipment that may be susceptible to harmonic components. The fundamental cause of harmonic components in an electrical system is seen as an increase in load due to the harmonic currents. The concern for other equipment lies in what way the harmonic components could cause an undesirable result (e.g., computer data loss, misoperation of process equipment). This section introduces some techniques for use when a circuit supplies nonlinear loads.

K-Factor and Transformer Sizing.   The K-factor-rated transformer is a class of transformer designed specifically to satisfactorily supply nonlinear loads. This transformer is designed to handle a neutral current higher than the phase current, as specified by its K-factor. Standardized K-factors have been established in ANSI/IEEE 57.110, which specifies the harmonic content for a transformer load.[15] For example, a K-4-rated transformer should be capable of supplying loads of 100 percent at the fundamental frequency (60 Hz), 16 percent for the 3rd harmonic current, 10 percent for the 5th harmonic, 7 percent for the 9th harmonic, and specified values continue to the 25th harmonic.

A standard transformer may be used to supply a nonlinear load but should be properly derated, a process that is comprehensively covered in ANSI/IEEE 57.110. The transformer derating is based on calculating the rms current of the load (considering the fundamental current plus the harmonic current values) and establishing the transformer rating to achieve a normal service life.

Electrical Equipment Derating. Harmonic components can cause a load current higher than the fundamental frequency, as calculated for the transformer capability. A method similar to ANSI/IEEE 57.110 for calculating the rms current of the load can be applied to circuit breakers, feeders, panelboards, branch circuits, and outlets. Based on the rms current, the equipment rating can be reduced so that the heating effects from the I^2R losses (and other factors) do not make the equipment operate above its designed temperature.

Circuit breakers should be rms sensing and derated if recommended by the manufacturer. Feeder and panelboard maximum-phase current may have to be limited to less than their ratings to keep the neutral current within the equipment's ampacity. A common practice is to increase the size (some say double) of the neutral conductor and panelboard neutral bus. To provide a well-engineered and cost-effective system, the load characteristics should be reviewed to calculate the neutral current and determine the appropriate rating(s). If a K-factor is calculated for the transformer, it can also be used to size the neutral conductor and the panelboard neutral bus for wye-connected (three-phase, four-wire) loads. The K-factor can be used similarly to determine the phase currents for delta-connected (three-phase, three-wire) loads.

Mitigation Techniques. ANSI/IEEE Standard 519 establishes two criteria: "a limitation on the harmonic current that a user can transmit into the utility system," and "specifies the quality of the voltage that the utility must furnish the user."[16] This standard does not apply to the electrical system within a facility, even though many use it as a guideline for establishing limits within a facility.

Branch Circuit Configuration. A common problem (particularly in an office environment) is the effect of harmonic components on branch circuit wiring with a common (shared) neutral. The problem is that the triplen harmonic components cause a higher-than-expected neutral current based on balancing the load on the phase conductors. The effects can be as serious as an overheated neutral or cause erroneous operation of equipment (refer to "Elevated Neutral Voltage" in the previous section). A simple way to eliminate these problems is to run the branch circuits individually back to the panel, rather than together using a common neutral. This limits the effects of the triplen harmonic components to the panelboard and upstream, thereby eliminating the hazards of overheated neutral branch circuits.

Filters. A filter can be an effective way to reduce harmonic components. However, passive filters (i.e., inductors or capacitors) are typically matched to the system. When applied to loads that are not operated continuously, all load scenarios for the system should be studied. There are two basic types of filters: (1) shunt filters that act as a sink (i.e., a short circuit) for the harmonic components, and (2) series filters that allow current of desired frequency to flow, but block the undesired harmonic components. Active filters are electronic devices that can sense the load harmonic components and produce and inject a current that is 180° out of phase, thereby canceling the undesired harmonic components.

Telephone Interference. Telephones or other devices may be affected by interference from the electric and magnetic fields produced by harmonic components. Mitigation of the harmonic components, grounding, filtering for the affected equipment, or physical separation are effective means of reducing interference. The TIF weighting factor quantifies the frequencies that can cause interference in a telephone.

Isolation Transformers. Isolation transformers were primarily used for personnel safety or to provide electrical isolation before the problems associated with nonlinear loads became well known. An isolation transformer provides a separately derived system that limits the ground and neutral circulating currents and makes the system less susceptible to power quality issues. When an isolation transformer includes an electrostatic shield, it attenuates the common-mode noise (i.e., it reduces the amount of common-mode noise that transfers from one winding to the other). Transformers are more susceptible to passing high-frequency noise due to their improved coupling characteristics as frequency increases. The ability of a device to attenuate the common-mode noise is quantified as the common-mode rejection ratio.

Drive Isolation Transformers. Drive isolation transformers are used to introduce impedance between the facility power system and the load. As the power circuit of an ASD switches, it may short two phases together for a short duration, an action known as *commutation*. The impedance of the isolation transformer reduces notching on the facility's electrical system caused by commutation, thereby reducing the harmonic content of the supply. If the transformer has an electrostatic shield, it will attenuate the high-frequency, common-mode noise, as previously described for the isolation transformer.

Voltage-regulating transformers are also available that keep the output voltage envelope (the difference between minimum and maximum voltage) within a smaller range than the broader input voltage envelope.

CONTRIBUTORS

Richard P. Bingham, Manager, Products/Technology, Dranetz-BMI, Edison, New Jersey

Arjun Rao, P.E., Vice President, BR+A Consulting Engineers, Inc., Boston, Massachusetts

Cornelius Regan, P.E., CLEP, Principal, C. Regan Associates, Rockland, Massachusetts

David L. Stymiest, P.E., SASHE, C.E.M., Sr Consultant, Smith Seckman Reid, Inc., New Orleans, Louisiana

William P. Wohlfarth, P.E., Sr Electrical Engineer, Massachusetts Institute of Technology, Cambridge, Massachusetts

NOTES

1. Bill Howe, *Protecting Motor Bearings from Electrical Damage in Adjustable-SpeedDrive Applications,* Publication No. TU-95-11, E-Source, Inc., Boulder, CO, 1995.

2. IEEE Standard 100-1996, *The IEEE Standards Dictionary of Electrical and Electronic Terms,* Institute of Electrical and Electronics Engineers, Inc., New York, 1996.

3. IEEE Standard 1159-1995, *Recommended Practice for Power Quality Monitoring,* The Institute of Electrical and Electronics Engineers, Inc., New York, 1995.

4. NFPA 70B-1998, *Electrical Equipment Maintenance,* National Fire Protection Association, Quincy, MA, 1998.

5. The Institute of Electrical and Electronics Engineers, Inc., "Storing Power for Critical Loads," *IEEE Spectrum* (June 1993).

6. The Institute of Electrical and Electronics Engineers, Inc., "Protecting Computers Against Transients," *IEEE Spectrum* (April 1990).

7. M. McGranaghan, "Custom Power—When Does It Make Sense?" *Power Quality Assurance Magazine* (January/February 1999), 24.

8. R. Langley and A. Mansoor, "What Causes ASD's to Trip During Voltage Sags?—Part 1," *Power Quality Assurance Magazine* (September 1999), 12.

9. ANSI Standard C84.1a-1980, *American National Standard for Electric Power System and Equipment Voltage Ratings (60 Hz),* American National Standards Institute, New York, 1989.

10. IEEE Standard 241-1990 (R1997), *IEEE Recommended Practice for Electric Power Systems in Commercial Buildings,* Institute of Electrical and Electronics Engineers, Inc., New York, 1997.

11. IEEE Standard 1346-1998, *System Compatibility.* Institute of Electrical and Electronics Engineers, Inc., New York, 1998.

12. M. H. J. Bollen, *Understanding Power Quality Problems: Voltage Sags and Interruptions,* IEEE Press Series on Power Engineering, Institute of Electrical and Electronics Engineers, Inc., New York, 1991.

13. R. Natarajan, J. Oravsky, and R. W. Gelhard, "Conducting a Power and Grounding Audit," *Power Quality Assurance Magazine* (September 1999), 16.

14. "Description of Scenario Software," in *News You Can Use,* Reliable Power Meters, website www.Reliablemeters.com, 1999.

15. ANSI/IEEE C57.110-1986, *IEEE Recommended Practice for Establishing Transformer Capability When Supplying Nonsinusoidal Load Currents,* Institute of Electrical and Electronics Engineers, Inc., New York, 1986.

16. ANSI/IEEE Std 519-1992, Revision of IEEE Std 519-1981, *IEEE Recommended Practices and Requirements for Harmonics in Electric Power Systems* (formerly titled *IEEE Guide for Harmonic Control Compensation of Static Power Converters*), Institute of Electrical and Electronics Engineers, Inc., New York, 1992.

Other Reference Publications and Web Sites

Power Quality Assurance Magazine, published by Adams/Intertec International, Inc., Ventura, CA (available on the Internet at www.powerquality.com).

Power Quality Internet Search Engine, sponsored by *Power Quality Assurance Magazine,* www.powerquality.com/ise-search.html.

D. W. Carroll, *The Power Quality Tutorial Series: Quantifying, Measuring and Protecting The Modern Facility,* Current Technology, Inc., Irving, TX, 1997.

M. A. Freeborn, *Diagnosing and Solving Power Disturbance Problems,* Square D Company, Groupe-Schneider—North America, Smyrna, TN, 1998.

J. V. Leger and L. Scott, *Power Quality for Medical Electronics—Taking Charge of the Electrical Environment,* Paper 31.4.47, On Power Systems, Inc., Brossard, Canada, 1995.

T. P. Moss, *Power Quality Issues and Interactions in Modern Electrical Distribution Systems,* American Society for Healthcare Engineering, Chicago, 1997.

W. D. Paperman, Y. David, and K. A. McKee, *Electromagnetic Interference: Causes and Concerns in the Hospital Environment,* American Society for Healthcare Engineering, Chicago, 1994.

Power Quality: New Opportunities to Save Money, Increase Uptime. http://hpcc923.external.hp.com/mpgcsd/probe/spr97/feature.html, Hewlett-Packard Company, Dallas, TX, 1997.

Power Quality Terms and Definitions, PQNetwork, www.pqnet.electrotek.com/pqnet/main/backgrnd/terms/paper/paper.htm, Electrotek Concepts, Inc., Knoxville, TN, 1994–1997.

Power Quality for Healthcare, Electric Power Research Institute, EPRI Healthcare Initiative, BR-109172, White Plains, NY, 1997.

IEEE Standard 1100, *IEEE Recommended Practice for Powering and Grounding Sensitive Electronic Equipment,* Institute of Electrical and Electronics Engineers, Inc., 1999.

The Power Quality Survey; Do It Right the First Time, Richard P. Bingham, NETA World, InterNational Electrical Testing Association, Inc., Morrison, CO, 1999.

SECTION 5.3
LIGHTING SYSTEMS

Paul R. Smith, P.M.P., P.E., M.B.A., Section Editor
Peak Leadership Group, Boston, Massachusetts

David L. Stymiest, P.E., SASHE, C.E.M., Section Editor
Smith Seckman Reid, Inc., New Orleans, Louisiana

Good lighting is a fundamental component of all successful buildings. The purpose of this section is to emphasize the positive role that effective illumination can play and to comment on some of the techniques and tools for achieving it initially and for maintaining it over time. Furthermore, as Harvey Bryan states in Art. 5.3.2, inspired by the demand for more energy-efficient products, the lighting industry has been at the forefront of innovation during the last two decades. During this period, lamp manufacturers have introduced an array of new products, such as T-8 and T-5 fluorescents, compact fluorescents, MR16s, color-improved low-wattage HIDs, and electronic ballasts. One would expect that with all these new products, the luminous environment would be significantly enriched. Unfortunately, this has not been the case. It is becoming increasingly clear that the balance that once existed between technical requirements and design has been eroding. Today, good lighting is all too often defined as meeting some quantitative standard or energy target, and visual accuracy, required esthetics, and comfort considerations are neglected. This section will show that good lighting depends on all of these variables and will provide readers with the background necessary to make informed lighting decisions for their facilities.

ARTICLE 5.3.1
LAMPS, FIXTURES, AND HUMANS— HOW THEY RELATE

Bonnie Seaberg
Wolfers Lighting, Allston, Massachusetts

Peter Coxe, A.I.A.
Peter Coxe Associates, Marblehead, Massachusetts

Penny Henderson-Maher
Lightolier, Div. of Genlyte Thomas, Garland, Texas

Good lighting is a fundamental component of all successful buildings. The purpose of this article is to emphasize the positive role that effective illumination can play and to comment on some of the techniques and tools for achieving it initially and for maintaining it over time.

The widespread introduction of electric light as we know it began with Thomas Edison. We take his achievements for granted today and tend to shift our focus to detailed specifics of lamp, ballast, and fixture efficiencies. We rely heavily on computerized computations and turn to their increasingly useful visual simulations to fine-tune the design. Individual enthusiasm and a very few published recommendations still occasionally promote higher light levels. But a host of regulations initiated at the federal level and filtering down through the states has effectively and appropriately combined with new lighting tools to turn much of that former tide in the direction of sensible conservation. Industrial, commercial, and institutional facilities have been in the forefront of this trend. They have come to appreciate that intelligent lighting design and systems management can reduce costs and simultaneously contribute to the well-being and effectiveness of employees. Technological improvements in lamps, fixtures, and controls are fundamental, but old-fashioned common sense that Edison would appreciate remains useful for evaluating any aspect of lighting. And the fundamental metric—the needs of the human user—remains unchanged.

LIGHTING A SPACE

A space can be as small as a one-room office or as large as an airport terminal. What really defines the space is its function—often, its numerous functions. These are the primary factors that determine the type of illumination, the appropriate intensities, and the distribution of illumination in the space. A corridor or circulation area seldom ought to be as brightly or uniformly lit as a task-oriented space such as a manufacturing area, an operating room, a laboratory area, or an office. Outside spaces will seldom be as intensely illuminated as interior spaces, even for security surveillance. So the first step is to identify the users, their tasks, and their needs. Then the many available lighting tools and techniques can be evaluated for effectiveness and efficiency.

Contrary to much common practice and some guidelines, the best lighting and the best resulting environments adopt a broad definition of *tasks* and *needs*. The lighting should enable the staff members to see their work tasks, and it should also address subtler needs, such as permitting them to do so comfortably and, perhaps, even with occasional pleasure. One of the lighting decision maker's primary challenges is to strike an appropriate balance between the often competing trade-offs that various lamps offer. For example, high-pressure sodium delivers high efficacy and very long life, but its color rendering is so poor that it is almost never found in offices. The decision maker must remember that the *quality* of the illumination, above modest threshold levels of intensity, is almost always more important than the quantity. It is impossible to overemphasize this.

The following are a few of the most useful general lighting design guidelines:

- *Use light and energy effectively.*   Highlight that which is a hazard, is important, is relevant, is interesting, or is attractive. Provide less light on the remainder.

- *Use light selectively, more as a rifle bullet than as the blast of a shotgun.*   Recognize that critical visual tasks usually occur in a small zone and that the bright illumination needed there is seldom required uniformly over a large area or it is needed only for limited periods. Important savings in power consumption and maintenance costs result from the selective use of higher light levels. Furthermore, studies show that spaces where there is visible variation in lighting tend to be preferred.

- *Provide positive highlighting of walls and spatial edges.*   More time is spent looking at or toward walls than at the floor. A dollar spent on strong illumination of vertical surfaces is especially effective in creating the appearance of a well-illuminated space. This is as true in a warehouse as in an office.

- *Favor light-colored room finishes.*   Surface reflectances play an essential role in the perceived brightness of a space. Low-reflectance surfaces look dark and also absorb incident light, reducing overall room illumination. Part-height cubical partitions introduce extensive areas of light-absorbing surface. Consider here an indirect lighting system that forcefully illuminates a white ceiling and upper wall. When a preponderance of low reflectances is unavoidable, try to incorporate a few light finish accents to provide some visual relief. Recognize that large quantities of electrical energy are required to compensate for very dark finishes. A coat of white paint is a highly cost-effective lighting tool.

- *Be aware that glare reflected directly from bright fixtures and indirectly off other surfaces can be a source of discomfort and a significant interference with vision.* Figure 5.3.1-1 illustrates this problem; the solution might be as simple as changing a fixture lens or shifting the aim of an adjustable fixture. Avoiding source glare is one of the most important concerns in fixture selection. Minimizing glare is essential for indoor lighting, and outdoors at night it is critical.

- *Use the expanding tools available for affording localized control of the user's lighting system.* Save money by increasing the ability of individuals to direct and control lighting within their individual areas.

- *Understand that lighting energy conservation can save money, and it can also easily be an opportunity to provide superior working environments.* Much of the luminous environment in North America today is far superior to that of the 1950s and 1960s, when designers indiscriminately shoveled much larger quantities of electric light into facilities. The difference is due more to a change in attitude and approach than it is to technological progress.

WHO PARTICIPATES IN THE DESIGN?

The scope of the project will determine the participating parties, usually some combination of owner, architect, interior designer, lighting consultant, electrical distributor lighting sales rep-

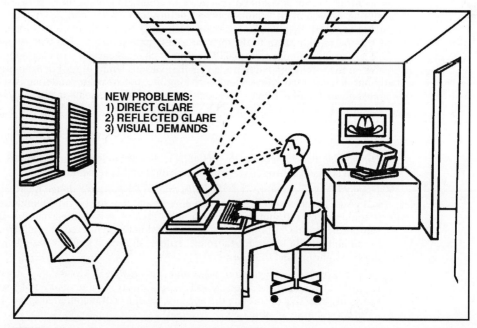

FIGURE 5.3.1-1   New problems of glare have arisen with the introduction of video display terminals (VDTs) into the work environment. (*Courtesy of National Lighting Bureau.*)

resentative, and electrical engineer. The owner or representative usually initiates the design process by outlining the goals, user needs, budget, and schedule for the project. Awareness of the language of lighting is useful, but the owner's primary responsibility is to communicate project objectives and evaluate the recommendations being made. The architect, interior designer, lighting consultant, or electrical engineer will help formulate the overall lighting schematic. They will work with the client to identify, recommend, evaluate, and select appropriate illumination techniques, levels, lamps, and equipment. Electrical distributors and lighting sales representatives may assist in obtaining lighting calculations and by offering equipment pricing and additional fixture suggestions. But as they may have a financial interest in their own product lines, others must exercise independent judgment on the appropriateness of their recommendations.

Lamp manufacturers identify bulb (lamp) types, their physical size, power input, lumen output, average rated life, color temperature (color appearance—warm or cool), color rendering, and photometric distribution. Fixture manufacturers provide information on the light output and illumination efficiencies of their equipment with various lamps. Ballast manufacturers provide detail on fluorescent and high-intensity discharge ballast performance. Specifics drawn from these data may be used in computer programs to compute illumination intensities and verify compliance with the lighting power budget restrictions of applicable energy codes.

The overall arrangement and configuration of the fixtures remains in the hands of the designer. The lighting professional is concerned with the following factors:

- *Design compatibility—selecting lamps, fixture types, and equipment layouts that support users' needs and the overall architectural design.* Possibilities for enhanced luminous and spatial enrichment will often be in the back of the designer's mind.

- *Light levels and illumination quality—identifying and incorporating the proper quality and electrical and natural light quantities to meet user needs and preferences.* This will include considering where the various intensities are wanted and the most effective techniques for delivering and controlling them. It will address the merits of light directed downward versus indirect uplighting systems. Often field measurement with project staff or users in an existing facility is an especially effective method of establishing confidence that the recommended intensities and illumination characteristics (color, glare, distribution, variation, and uniformity) are appropriate.

- *Controls and energy conservation—establishing the equipment needed to minimize power consumption and to provide appropriate levels of control.* The costs and merits of energy management systems, timing devices, occupancy sensors, dimmers, and individual user controls are likely to be considered.

- *Specifications and procurement procedures—developing fixture schedules and/or performance-oriented specifications for the products desired.* Thoughtful formulation of these may ease construction procurement, encourage competitive bidding, and smooth the process of evaluating alternatives. A checklist under development by the International Association of Lighting Designers/Lighting Industry Resource Council (IALD/LIRC) identifies a five-stage process. It consists of (1) foundation elements (establish policies, specifications, performance criteria, and owners' requirements), (2) design development phase (establish product, budget, and design), (3) construction document phase (establish fixture schedule, cost estimates, and integration of architectural and lighting drawings), (4) bidding phase, and (5) construction phase (establish procedures for alternatives, samples, and drawings).

- *Maintenance—ensuring that the owner and facilities personnel understand the operating life and maintenance characteristics of the equipment under consideration.* There should be clear understanding of how fixtures will be accessed for maintenance. The project record should include complete, detailed lamp identifications and the replacement information necessary to maintain the lighting system.

EVALUATING THE HUMAN ELEMENT IN LIGHTING

The lighting design for a space ought to be determined by the intended functions and user needs. We view *function* as the data basis for determining the lighting needs. A bus terminal must have many travel areas from outside to inside and might be used 24 hours a day. Light levels might be brought up to higher levels at night for security but dimmed down during daylight hours (sometimes it is the exact reverse). A grocery store must enhance the appearance of the food in its deli areas while providing sufficient light for reading labels when selecting items. The intent of the space provides valuable information to bring to the equation of lighting the space properly.

What might constitute the best illumination is subject to many interpretations and to many constraints. All must be valued and weighed against what is most important in lighting the space. Initial costs and ongoing operating expenses are always a consideration.

The end users place the greatest restrictions on lighting. They deserve visual comfort and quality lighting. Good lighting can enhance worker performance, create a better working environment, and aid in reducing stress. The by-product may be increased productivity, customer satisfaction, and profits for management. The following are examples of companies that have altered or changed their lighting for the benefit of their customers and employees and have achieved better than average results.

At Hyde Tools of Massachusetts, a small manufacturing plant upgraded its lighting from fluorescent fixtures to new high-pressure sodium and metal halide fixtures. The employees were given the option of restoring the original fixtures after 6 mo if they were not satisfied. At first, the employees complained about the color of the lamps, "an orange hue," but when they converted back to the original fluorescent fixtures, there was an outpouring of disapproval. They could not see the specks of dirt on their equipment; this, in turn, resulted in production of an inferior and defective product, and productivity fell back to its original levels. When they returned to the new lamps, productivity improved and led to increased sales and a payback for the lighting in 1 year.[1]

Pennsylvania Power and Light upgraded its engineering drafting room. Veiling reflection, a form of indirect glare in which the light washes out the contrast between detail and background on a task surface, was causing eyestrain, headaches, and other ailments. The company upgraded its lighting to a task-oriented system with 8-cell parabolic louvers and local controls with multiple circuits. The new systems reduced employee complaints so that the payback for the new systems went from the anticipated 4.1 years to just 69 days.[1]

In 1993, Wal-Mart built a new store in Lawrence, Kansas. The company usually allows a 3- to 5-yr payback for new construction, but costs were running 20 percent higher than anticipated. As a cost-cutting measure, skylights were installed in half the roof, a passive lighting solution. Tracking results showed that sales per square foot were higher where the sunlight became a source of light in this store, and a bonus was that the employees were more satisfied. Wal-Mart is considering adding skylights to more stores.[1]

If we accept the conclusion that better lighting is a key element in the degree of customer satisfaction, in productivity, and in profits, then we accept the fact that the end user can have an impact on the function and value of lighting. We should take into consideration the fact that the national average cost of operation per square foot is $130 for workers (Fig. 5.3.1-2) and only $1.81 for total energy consumption. Costs of operation show that total electric usage in an office averages about $1.53 per square foot. This is a tiny fraction of the total operating costs, in which employee salaries are more than 80 times as much ($130). Designers and owners must continue to conserve energy but when doing so, must also realize that worker productivity can directly be related to specific changes in the quality of lighting within the working environment.

TYPES OF ILLUMINATION

Standards and regulations are increasingly forcing reduced use of incandescent lamps. They are promoting the highest-efficacy lamps and ballasts for reduced energy consumption. More stringent energy budgets are increasing the reliance on lighting selectivity rather than carpet-

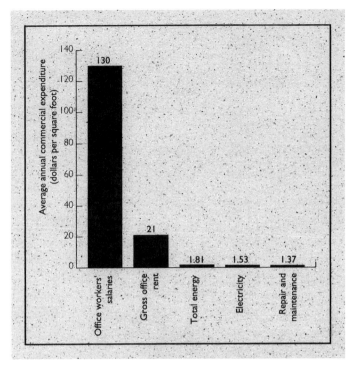

FIGURE 5.3.1-2 National average annual costs of operation per square foot. (*Courtesy of Building Owners and Managers Association; Electric Power Research Institute; Statistical Abstract of the United States, 1991.*)

ing an area with uniform illumination intensity. Most states have their own codes, and some states, such as California, Massachusetts, and Arizona, have more stringent regulations.

Another influence and opportunity in the design of the space is natural light or a daylight lighting fixture. Natural light can add charm as well as brightness. It falls into two categories, top lighting with skylights or sidelighting with windows or clerestories. Both must take into account the angle of the sun as it progresses through the seasons and the geographical location of the facility. If this light plays a significant role in the illumination, calculations are made and factored into the overall equation for the light levels. The addition of other added items such as blinds or screens should carefully be evaluated; glare or too much brightness can easily accompany a large glass aperture.

Artificial light, or lighting with luminaires, can be general downlighting, general, indirect uplighting, or a combination of direct and indirect systems. Wall-washing perimeter illumination and selective task lighting constitute the major artificial lighting techniques.

As ongoing lamp technology offers smaller wattages and more compact sources, traditional distinctions in lamp application continue to blur. The specifier is presented with more choices and is required to make more decisions than in past decades. The following generalizations usually apply to commercial, institutional, and industrial applications.

Linear fluorescent lamps are favored when priorities are for introducing broad washes of general illumination, both in downlighting and uplighting; for linear fixtures and arrangements; for moderate intensities; for very long average life; and for consistent, very good to excellent color. As a rule, the physical size of linear lamps limits the ability to control the beam. Nevertheless, the smaller-diameter T-5 lamps are challenging this limitation, at least on the short axis. Dimming of fluorescent lights for energy management or functional reasons continues to improve, although few systems currently can dim to the 1 to 2 percent intensity required by the most demanding applications, such as projector presentations.

Compact fluorescent lamps are favored when priorities are for general light from point sources and compact fixtures; for moderate intensities delivered to modest sized spaces from average height ceilings; for a combination of moderately high efficacy and long average life; and for consistent, very good to excellent color. As a rule, the size of the tube prevents achieving very tight beams and the dramatic effects characteristic of many incandescent lamps. Compact fluorescent lamps are most frequently employed in ceiling recessed and some surface-mounted fixtures. Dimming lamps and systems are available, but current technology dims to only about 5 percent, and a distinctive color shift toward the blue suggests that dimming be approached with caution.

Metal halide lamps are favored when priorities are for point sources and compact fixtures that have very good color, long life, and high efficacy. Compared to high-pressure sodium lamps, metal halide lamps offer distinctly superior color, slightly inferior efficacy, and usually a markedly short life. The compact arc of the clear lamp can deliver a highly controlled beam. In recent years, an increasingly wide range of sizes has become available, significantly expanding the extent of common use. Concurrent improvements in lamp color stability and appearance have further contributed to the lamps' acceptance. These trends are projected to continue. The high-bay fixtures of industrial floors and high-intensity office uplighting are most common, but the smaller lamps are increasingly being employed in the ceiling recessed fixtures of office and institutional areas. Track fixtures are also available. The lamps can be expected to continue challenging indoor incandescent lamps and high-pressure sodium lamps in wide area exterior installations.

Induction lamp technology is a recent and interesting addition to the options available. This new family of lamps offers extremely long life, often promising an average of up to 100,000 hours. Current color characteristics and operating efficiency are generally comparable to other discharge sources. Some of the lamps are quite large, which makes precise control difficult but renders them most effective in large fixtures or in applications for wide, diffuse illumination. Prices today are very expensive, but may prove particularly cost-effective where access for relamping is unusually difficult or costly. Induction lamps warrant continued scrutiny as the product line expands and matures.

High-pressure sodium lamps are especially favored when very long average life is the priority. The lamps offer high efficacy and are available in a wide range of sizes in compact, point-source fixtures. A color appearance and color rendering visibly inferior to that of most other sources has relegated high-pressure sodium use in North America primarily to outdoor area and street installations. A compact arc lends itself to controlled beams. Developments in sodium lamp technology have been increasingly few, and many project its slow eclipse by ongoing improvements in metal halide lamps.

Low-pressure sodium lamps are especially favored when color is not an issue and very long life and very high efficacy are the priorities. Low-pressure sodium, the most efficient of the lamps regularly employed, has a large arc tube that significantly limits the ability to control light distribution and the operating efficiency of the installation. Illumination is a monochromatic orange hue, which ranks as 0 on the color rendering scale of 0 to 100; most objects under it are perceived as being shades of gray. More popular in Europe, the lamp has never achieved wide acceptance in North America. It is most frequently encountered in infrequently accessed storage areas, all-night security lighting, and roadway illumination.

Incandescent lamps are favored when priorities are for very localized illumination; for narrow, very precise, long- and short-throw beams; for superior color rendering and natural appearance; and for straightforward, full-range dimming. Incandescent lamps can be employed very efficiently. However, incandescent lamp efficacy is typically only 15 to 25 percent that of competing fluorescent and discharge sources. With few exceptions, nominal average incandescent life is only about 10 percent that of its competition. The use of this lamp family is limited and decreasing in commercial and institutional applications.

Most lamps strive to render all colors faithfully, but their success at this varies widely. Some lamps strongly impact their intended areas. The color rendering of vegetation and exterior materials is a matter of choice. Lamping can enhance green foliage or turn it brown. It can enhance the appearance of the exterior of the building or change it significantly. A marble facade might be strikingly strengthened by metal halide lamps or muted by high-pressure sodium lamps.

LIGHTING CONTROL FOR EFFECT AND ENERGY MANAGEMENT

Controls are the only part of the lighting system that the user touches; therefore, they should be easy to adjust to user preferences and easy to use. Lighting controls can aid in managing and limiting the energy consumption in a space, while still providing the individual user with flexible light levels in multiple-use spaces. There are six basic strategies to save energy, enhance performance, and offer convenient light level changes: (1) occupancy sensing, (2) scheduling, (3) daylight dimming, (4) lumen maintenance, (5) tuning, and (6) preset dimming. *Occupancy sensing* and *scheduling* are a means of limiting the hours of lighting to the hours of occupancy. *Daylight dimming* and *lumen maintenance* aim to dim electrical light to achieve a constant light level in a space during the day and over time. *Tuning* and *preset dimming* allow the user to change light levels for specific tasks or events. In employing any combination of these strategies, the designer should keep in mind that the employee is the most expensive and sensitive piece of equipment in the space. Thus, the lighting controls should be easy to adjust to the user's needs and work habits. Further, the designer must take care to specify mutually compatible components (fixture, ballast, lamps, and controls) and must ensure that they are properly installed and commissioned.

Occupancy Sensing

Simply turning lights off when a space is vacant can save tremendous amounts of electricity. Estimates of savings range from 30 to 70 percent in rooms used intermittently. Sensors are reasonably priced. A switch sensor has a cost comparable to that of a specification grade dimmer, so the payback period is short where electricity rates are at least $0.10 per kilowatt hour. Private offices, conference rooms, restrooms, and storage areas favor using occupancy sensors, which automatically turn the lights off a few minutes after the space becomes vacant. The best sensors work like switches; they preserve personal on/off control, are noninvasive, and are sensitive to small movements by the occupant to keep the lights on.

Occupancy sensors may be wall mounted in a standard switch location, ceiling mounted, or mounted high on a wall in a corner viewing position. The choice of mounting location depends on sight lines to the occupant and concerns for equipment security. In a private office, a switch-mounted sensor works well and gives the user easy control of the lighting. In a classroom, ceiling-mounted sensors see the space well and are out of reach of curious fingers. Wherever the sensor is located, care should be taken to block the sensor from seeing out doorways, into circulation areas, or other spaces where detection of movement is not desired. A common placement error with ceiling sensors is locating the sensor in the center of the room, giving it a clear path to look out an open door into the adjoining corridor. A better location would be close to the doorway wall and in the center of the wall, so that the viewer sees the room but has no viewing angle into the corridor.

Some sensors are manually activated, and some turn lights on automatically. In private offices and conference rooms, manual activation provides a sense of personal ownership and control that allows the occupant to turn lights on only when they are needed and to use the space without electric light, if desired. In shared office spaces, areas with multiple entrances, and areas with special security requirements, automatic activation may be desirable. Manual activation results in greater energy savings because natural ambient light often provides sufficient illumination for noncritical visual tasks and the user may choose not to turn the lights on. Manual-deactivation (off) control is desirable in some applications and allows the occupant to turn lights out as required. *Time-out* is the length of time the sensor waits to turn the lights off after detecting the last occupancy movement. It should be convenient to adjust the time-out response of the sensing device to the occupancy pattern of the user, allowing the device to ride through brief periods of inactivity by the occupant, as well as brief absences from the room, without turning the lights off. Turning the lights off too quickly and too often will result in frequent restarts, which can increase energy consumption and shorten lamp life. Conversely, rooms with infrequent occupancy, such as storage areas, should be set for short time-outs to maximize energy savings. The designer must take care in placing the device and in specifying a time-out period long enough to accommodate the user's work habits and must ensure that the installer complies with these instructions to avoid user rejection of the device.

Currently, occupancy sensors use two technologies to detect movement in a monitored space, passive infrared (PIR) and ultrasonic. Ultrasonic sensors emit a high-frequency sound wave into the space and, with their built-in receivers, look at the returning waveform as it is reflected by the adjoining surfaces. Occupancy is detected by sensing changes in the frequency of the returning waves deflected by the human moving through the space. Ultrasonic sensors for occupancy can occasionally read nonhuman moving objects, such as fans and draperies moved by HVAC currents. Ultrasonic sensors are highly effective in public restrooms, where modesty partitions interfere with the line of sight required by PIR sensors. The major drawback of ultrasonic sensing is that it cannot be stopped, or masked, from looking into an area where sensing is not desired. In addition, ultrasonic sensors may interfere with some security alarm systems and may be heard by some individuals with hearing aids.

PIR is a noninvasive approach that senses occupancy by detecting changes in the thermal image in the sensor's field of view. Usually, the PIR sensor regards the fixed heat pattern in its field of view as the unoccupied state, and a human that enters the field of view will be seen as a moving, warm image against a cooler background, signaling occupancy to the device. A well-designed PIR device has logic in its processor that allows it to identify and ignore nonoccupant heat-change signals, such as HVAC units and windows. PIR sensors are easily masked to block their viewing lenses from seeing into circulation areas or other spaces outside the desired area of coverage. Because PIR devices work on a line-of-sight principle, to function they must be placed where they can see the occupant or occupants. Thus, PIR devices may require multiple sensor viewers where spaces are large, obstructed, or contain multiple high partitions. In addition, if the ambient temperature is high, a PIR sensor may not be able to distinguish between a human (98.6°F) and the background.

Scheduling

In large shared office spaces, circulation areas, and lobbies, turning lighting circuits on and off on a timed schedule is an efficient strategy for energy conservation. Typically, low-voltage switching systems, or controllable circuit breakers, are combined with a time clock or building management system to turn lighting on and off at times to accommodate the average occupancy of the space. Individual workers who arrive early or stay late normally activate their personal zone of light by sending a code to the controlling system via telephone or networked computer. It is easier to design this type of zone control than it is to plan for correct coverage using occupancy sensors. However, less energy is saved due to the averaging nature of the larger areas controlled and the longer hours of lamp operation. To be completely energy-efficient, a low-voltage switching system should have zones that are small enough to respond to a few individuals working outside normal business hours. Very little energy is saved if one individual working after hours keeps lights on for an entire floor. A combination of low-voltage switching for large shared work areas and circulation areas, plus occupancy sensors for private offices, is an energy-efficient solution.

Daylight Dimming and Lumen Maintenance

These energy-saving strategies employ photocells combined with electronic fluorescent dimming ballasts in conjunction with a controller to maintain a constant illumination level that is appropriate for the visual task being accomplished in the space. Lighting designers and engineers must compensate for the effects of lamp aging and dirt accumulation by providing higher-than-needed light levels when a space is new, usually 20 to 30 percent more. Lumen maintenance dimming is achieved by using a device that dims the new lamps to the design criteria level and gradually gives the system more voltage, as the lamps age, to maintain a constant light output. Because electronic dimming ballasts save just about as much electrical energy as the amount of light output removed, an average of 10 to 15 percent is saved over an average 3-year lamp life.

The next concern is the issue of photocell placement, or where the photocell should look. If the controlled space is an office or workplace, the photocell needs to monitor the work surface. A photocell that looks at the floor and sees an area of brightness there from a nearby window can cause the electric lights to dim when no useful daylight is striking the work plane. It would be most useful to monitor only the work plane, the top of a desk or table, but changing the materials resting on the desktop could send false information to the photocell. Some daylight dimming systems use a neutral surface, such as a wall that does not receive direct sunlight, and calibrate the system's minimum and maximum by "teaching" the device the illumination levels on the wall when useful daylight is present and when only electric light is available. In some daylight systems, the photocell sends a signal that turns lights off or prevents them from being turned on when certain levels of daylight are sensed. This strategy may work well if the controlled area is used for circulation, but it is not so appealing to employees who must work under these fixtures. The perception is often that there is insufficient light if fixtures are not illuminated. If the same fixtures are on, but dimmed to a very low level, user perception is that light levels are acceptable. Payback periods are typically long in daylight dimming designs unless the architectural design carefully incorporates daylighting from the inception.

Tuning and Preset Dimming

Tuning is a strategy that may be combined with lumen maintenance or daylight dimming or may be incorporated into an occupancy sensor. It simply gives the individual a manual dimmer to adjust light settings below the design criteria. This strategy works well in private offices or small shared spaces where all occupants are performing similar visual tasks and have similar visual abilities. It works less well in large shared spaces where the light output of all fixtures is required to provide average luminance throughout the space. In this type of space, giving individuals the ability to dim fixtures near their spaces may result in depriving a nearby worker of needed light.

Preset dimming is an important control strategy in commercial spaces for increasing the flexibility of multiuse rooms, such as boardrooms, conference rooms, training rooms, and auditoriums. In these spaces, the ability to touch a button and have the lighting change from meeting level to video-viewing level without fussing with multiple switches is a welcome addition. It is no longer an expensive installation because microprocessors have become so small and inexpensive that there are whole dimming systems small enough to be installed in a small multigang wall box. Dimming systems can be integrated into audiovisual control systems so that the touch of one button closes room-darkening shades, brings down the projection screen, starts the video and sound presentation, and dims the lights. Dimming systems may be combined with occupancy sensors or may be signaled by a building management system that combines the convenience of light-level selection with energy efficiency. Employee cafeterias, reception areas, and lobbies are also candidates for preset dimming that allows different light levels for various times of day, creates different moods, takes advantage of daylight, and saves energy by reducing lighting during off-peak hours.

EVALUATING LIGHTING NEEDS

In the ever-changing workplace, lighting needs that blend the aesthetic with the functional are constantly being reexamined. Functionality and attractiveness need not conflict with one another. The lighting design can be fine-tuned by replacing fixtures, by reselecting lamping, or by adhering to the specification guide provided by the lighting specifier.

Lamping is the key to evaluating lighting needs for the end user. As the life of a lamp varies and decreases with time, the lumen output will also decrease. Thus, the color rendering and illumination of the space may change in a direct ratio to the lamp life. When a few lamps burn out it is often most economical to group relamp all of the fixtures at the same time. This ensures that the lumen levels and color will be constant as intended. The continuity and the quality of light ought to be maintained to ensure the satisfaction of the end user and in turn provide savings. In the business world, the smarter companies know that heightening employee satisfaction increases productivity, customer satisfaction, and profits.

A more in-depth discussion of lighting and its functions is found in the Arts. 5.3.2 through 5.3.6.

CONTRIBUTOR

David L. Stymiest, P.E., SASHE, C.E.M., Senior Consultant, Smith Seckman Reid, Inc., New Orleans, Louisiana

NOTE

1. J. U. Romm, and W. D. Browning, *Greening the Building and the Bottom Line,* Rocky Mountain Institute, Snowmass, CO, 1994, rev. 1998.

ARTICLE 5.3.2
LIGHT, VISION, AND COLOR

Harvey J. Bryan, Ph.D. F.A.I.A.
Arizona State University, Tempe, Arizona

Inspired by the demand for more energy-efficient products, the lighting industry has been at the forefront of innovation during the last two decades. During this period, lamp manufacturers have introduced an array of new products, such as T-8 and T-5 fluorescents, compact fluorescents, MR16s, color-improved low-wattage HIDs, and electronic ballasts. One would expect that with all these new products, the luminous environment would be significantly enriched. Unfortunately, this has not been the case. It is becoming increasingly clear that the balance that once existed between technical requirements and design has been eroding. Today, good lighting is all too often defined as meeting some quantitative standard or energy target, and visual accuracy, required esthetics, and comfort considerations are neglected. This article will show that good lighting depends on all of these variables and will provide readers with the background necessary to make informed lighting decisions. However, this article is not intended to transform the reader into a lighting designer. It is best to rely on a qualified lighting design professional to design and specify a lighting system in new construction or modernization.

A good understanding of lighting systems requires basic knowledge of physics, physiology, and psychology. A brief discussion of the relevance of these subjects to lighting follows.

THE NATURE OF LIGHT

There are several competing theories as to the exact nature of light, but the lighting profession most frequently uses the electromagnetic theory. Light occupies a very narrow band within the electromagnetic spectrum, which defines radiation in terms of its wavelength (the distance traveled in one harmonic cycle). Light, also called the visible spectrum, is defined as that part of the electromagnetic spectrum which excites the retina of the eye and produces a visual sensation. For the average human eye, the visible spectrum ranges from wavelengths of 380 nanometers (nm) for violet through blue, green, yellow, orange, and into red before finally ending around 780 nm (Fig. 5.3.2-1).

Before continuing with a further discussion of the nature of light, it is important to define several terms that will be used in this article. The *candela* (cd), also referred to as *candlepower*

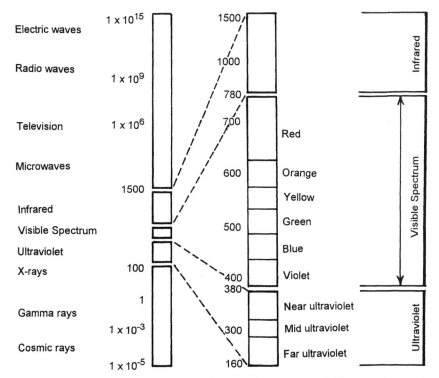

FIGURE 5.3.2-1   Electromagnetic spectrum (wavelength in nanometers). (*Courtesy of Harvey J. Bryan.*)

(cp), is the unit used to measure luminous intensity and is the base unit in lighting. It derives its name and the measure of its intensity from the flame of a standard candle. A good analogy for defining related terms and units is to take a sphere and place at its center a 1-cd source radiating equally in all directions. Because radiation expands spherically from the source, as the sphere expands, the radiant energy, although constant, is being spread over a much larger area. To take this phenomenon into consideration, the sphere is divided into solid angles called *steradians* (sr) whose origin is at the source. Assuming a 1-cd source, the quantity of light in 1 sr is defined as 1 *lumen* (lm). A sphere contains 4π sr; thus, 1 cd = 12.57 lm. Now, if we assume that the sphere is 1 ft in radius, the amount of light falling, or illuminance, on 1 ft^2 of the sphere is 1 lm/ft^2 or 1 *foot-candle* (fc). If metric units are needed, we need only to increase the sphere to 1 m in radius; the amount of illuminance on 1 m^2 of the sphere is 1 lm/m^2 or 1 *lux* (lx). The conversion from foot-candles to lux is equal to the number of square feet in a square meter, which is 10.76; thus 1 fc = 10.76 lux; or, conversely, 1 lx = 0.0929 fc (Fig. 5.3.2-2).

VISION

For vision to occur, there must be light, an object, a receptor (the eye), and a decoder (the brain). Light can be transmitted through or reflected from an object whose brightness stimulates electrochemical receptors in the eye (color receptors are called cones; monochromatic receptors are called rods), which then transmit signals to the brain, where they are interpreted. To a large extent, this interpretation is based on experience and is the least understood aspect of the visual process. It is for this reason that lighting professionals cannot approach problem

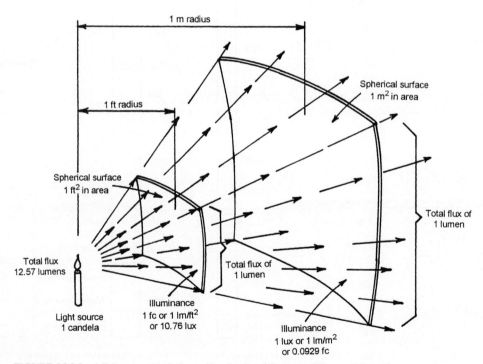

FIGURE 5.3.2-2   A light source's relation to illumination. (*Courtesy of Harvey J. Bryan.*)

solving within their field strictly analytically. Their response has been to focus on what is known, which is understandable. However, the field must not lose sight of the fact that certain lighting conditions evoke various *feelings,* or subjective responses which are poorly understood, and that an ongoing effort is needed to define these factors or responses clearly.

What is known is that these four conditions must be present for the eye to distinguish detail, which is also referred to as *visual acuity:*

1. *Size.* Visual acuity depends greatly on the size of the object viewed. Vision research has found that the most important aspect in determining size is not the physical size of the object but the visual angle that the image projects onto the retina. Thus, the nearer an object, the easier it is to see.

2. *Luminance.* Luminance is often confused with brightness; the difference is that brightness is the subjective evaluation of the amount of light that leaves a surface and reaches the eye, whereas luminance involves an objective evaluation. Surfaces that have high reflective values (light in color) reflect more light than low reflective values (dark in color). Hence, more illumination is needed on a dark surface to produce equal luminance than on a light surface. For diffuse surfaces, luminance is defined as

$$L = \rho \times E$$

where L = luminance
ρ = reflectance
E = illumination

The eye sees luminance (the exiting energy from surfaces), not illumination (the incident energy on surfaces). Therefore, prudence should be maintained when using many of the calculation procedures that are based on illumination because they can lead to inaccuracy unless surface reflectances are properly considered. The unit of luminance is either the candela per square foot [also referred to as *footlambert* (fL)], or the candela per square meter.

3. *Contrast.* Contrast is the difference between the luminance of an object and that of its immediate background. Black type on white paper is quite legible because the contrast approaches 100 percent. Contrast is defined as

$$C = \frac{L_o - L_b}{L_b} = \frac{\Delta L}{L_b}$$

where L_o = luminance of the object
L_b = luminance of the background
ΔL = difference between L_o and L_b

No matter how much luminance is available, contrast is the single most important factor in visual acuity.

4. *Time.* Seeing is not instantaneous; a lag exists between the time it takes to move visual information from the object to the eye and then into the brain. The impact of time on seeing is not unlike that on the camera, which requires a longer time exposure in dim light than in bright light. Thus, as luminance decreases, the time it takes to interpret detail increases.

COLOR

Color is a major factor in lighting design because light influences the way things should appear. The grass should be green, the sky blue, and the tomato red. When things do not appear as they should, we often become troubled. The two most important aspects of color are the spectral distributional characteristics of the light incident on an object and the spectral reflective characteristics of the object itself. What we see as color is light that is selectively reflected from an object. For example, if a red tomato and a green leaf were lit with only red wavelengths, the leaf would appear black (without color); all other light energy is absorbed. If green light were used, the opposite would occur. Thus, if the color is not present in the light source, it will not be seen in the object. In the design of any space, careful attention must be paid to the spectral characteristics of both the light source and the object (i.e., surfaces). Many facility managers do not well understand this interaction between the source and surfaces. For example, if relamping is not done carefully and incorrect lamps are used, the original interior colors of a space may appear dull or drab.

The color characteristics of a light source can be plotted on what is called the *spectral energy distribution* (SED) curve, which displays the amount of energy emitted at each wavelength of the visible spectrum. The SED curve is a good indicator of the color-rendering properties of a light source. Incandescent sources enhance the "warm" colors such as red and orange and dull "cool" colors such as blue and green. The SED curves for fluorescent and other discharge lamps are much more varied due to the variety of phosphor coatings that are available; they range from the warm to the cool whites in color. SED curves can also be called *spectral power distribution* (SPD) curves (Fig. 5.3.2-3).

Another method of expressing the color characteristic of a light source is by using color temperature or chromaticity, which indicates the warmth or coolness of a source. A blackbody radiator is used to reference color temperature. As the temperature of a blackbody radiator increases, the color it gives off changes from red to orange, yellow, white, and bluish white. The colors from these temperature changes are plotted on what is called the *International Commission on Illumination* (Commission Internationale de l'Eclairage) (*CIE*) *chromaticity chart*. Then, color temperature (CT) is determined by matching the source's color exactly to the blackbody chromaticity curve (also called the *planckian locus*). However, because many light sources do not lie exactly on this curve, an approximate color matching or corrected color temperature (CCT) is used. Incandescent lamps, for example, have a CCT of about 2800 K, whereas fluorescent lamps vary widely, from the warmer, or reddish yellow, at 3000 K to the cooler, or bluish white, at 5000 K. Color temperature does not indicate the color-rendering ability of a light source, but it is important in creating mood within a space and can greatly influence employee performance and consumer behavior (see Table 5.3.2-1).

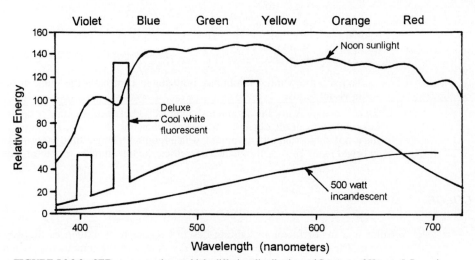

FIGURE 5.3.2-3   SED curves can have widely differing distributions. (*Courtesy of Harvey J. Bryan.*)

TABLE 5.3.2-1   Influence of Color Temperature on Mood and Lighting Applications

Color temperature	Warm	Neutral	Cool	Daylight
Kelvin range	3000 K	3500 K	4100 K	5000 K
Associated effects and mood	Friendly Intimate Personal Exclusive	Friendly Inviting Nonthreatening	Neat Clean Efficient	Bright Alert Exacting coloration
Appropriate applications	Restaurants Hotel lobbies Boutiques Libraries Office areas Retail stores	Public reception areas Bookstores Office areas Showrooms	Office areas Conference rooms Classrooms Mass merchandisers Hospitals	Galleries Museums Jewelry stores Medical examination areas Printing companies

Source: Philips Lighting.

The most popular but often the most misunderstood method of expressing the color characteristic of a light source is by using *color rendering,* which is the ability of a light source to represent colors in objects. The relative index that is used to determine this ability is called the *color rendering index* (CRI) which rates light sources on a scale of 0 to 100. This is a method of measuring the change in color of an object lit by a test source and then comparing it to a reference source at the same color temperature. When changes are small or have a close color match, the CRI is high; when changes are great, the CRI is low. Generally, the higher the CRI, the better the spectrum is distributed, and the better vibrant colors will look under a particular light source. However, a number of concerns are associated with the use of the CRI. The first relates to the fact that the index averages across 8 color samples, which means that the pattern of difference among colors can be lost. Thus, two light sources that have identical CRIs can have varying color-rendering abilities. The second relates to the fact that to compare the CRIs from different light sources directly, they must have nearly the same color temperature. For example, a metal halide lamp (at 4200 K) that has a CRI of 65 can be compared to a triphosphor fluorescent lamp (at 4100 K) that has a CRI of 85 but not to a warm white fluorescent lamp (at 3100 K) that has a CRI of 55. Although nowhere near perfect, the CRI is the only method presently available that indicates how colors will look under a given light source. Good color rendering (CRI of 70 to 80) is critical in settings where people desire to appear natural, in retail environments where merchandise must look attractive, and in offices where pleasant and productive work environments are required.

CONTRIBUTOR

David L. Stymiest, P.E., SASHE, C.E.M., Smith Seckman Reid, Inc., New Orleans, Louisiana

ARTICLE 5.3.3
LIGHT SOURCES

Harvey J. Bryan, Ph.D., F.A.I.A.
Arizona State University, Tempe, Arizona

Light sources can be divided into two general classifications: light that comes from the sun and sky, natural light, and light generated by chemical or electrical processes, or some combination of the two, artificial light. This article focuses on electrical processes that generate light; natural light is the focus of Art. 5.3.6.

Nearly 6000 different electric lamps are currently produced. Probably one of the hardest tasks in lighting design is choosing the lamp that is best suited for the application. In choosing a lamp, issues of energy efficiency, color, life, lumen depreciation, and cost need to be considered.

Electric lamps differ significantly in their efficiency. However, in lighting, because the output and input units are mixed, it cannot properly be called efficiency, but rather *efficacy*. Thus, lamp efficacy is defined as the rate at which a lamp can convert electrical power (watts) into light (lumens) and is measured in terms of lumens per watt (lm/W or LPW). Lamp efficacy can range from 10 to 20 lm/W for an incandescent to nearly 160 lm/W for low-pressure sodium. Although lamp performance is extremely important, it must be remembered that to operate a number of lamp types, particularly discharge lamps, auxiliary equipment such as ballasts are required. In such cases, system efficacy needs to be determined, which includes charging both lamp and auxiliary equipment energy to the input side of the efficacy equation. System efficacy is crucial in evaluating lighting system performance because lighting typically represents 30 to 50 percent of the operating costs of a building and also has a major secondary impact on such things as air conditioning demand. In addition, lighting accounts for approximately 25 percent of the electricity used in the United States; thus conservation efforts in this area can greatly benefit the environment, particularly by reducing greenhouse gas emissions. For this reason, the U.S. Congress passed the Energy Policy Act (EPACT) in 1992 that mandates lighting performance standards for both lamp efficacy (measured by LPW) and color rendering [measured by color rendering index (CRI)]. Lamps that do not meet these standards are being phased out, and their manufacture or importation will be prohibited.

The policymakers responsible for drafting EPACT included color rendering because of the poor reception in the marketplace for many of the first-generation energy-efficient lighting products. Many in the industry argued that this poor reception was due in a large part to the unacceptable color rendering properties of these early products. Thus, EPACT was a clear recognition that any attempt at introducing energy-efficient lighting must operate in conjunction with qualitative improvement. As was previously discussed, the CRI is an important, but not the only, indicator of the color characteristics of a lamp; the spectral energy distribution (SED) curve and color temperature may also need to be considered. Most lamp manufacturers include CRI and color temperature information in their product catalogs; SED curves can often be requested from manufacturers, and generic SED curves can be found in lighting handbooks.

Lamp life can range from a low of 750 to 1000 h for an incandescent to more than 24,000 h for high-pressure sodium. The measure of lamp life is the number of hours for 50 percent of a large group of lamps of the same type to fail. Selecting lamps based on lamp life makes the most sense when labor cost for replacement is high, such as in high-bay or hard-to-reach locations. Mortality curves published by lamp manufacturers show the rate at which a group of lamps survive as a percentage of rated life.

As all lamps age, their light (lumen) output deteriorates. This deterioration is caused by numerous factors and can typically result in a lamp losing 70 to 80 percent of its initial lumens when it reaches its rated life. Thus, it is important to consider *lamp lumen depreciation* (LLD) in lighting design. For example, if a task requires 500 lx, a designer might design to provide more than 500 lx initially, which will deteriorate to less than 500 lx at the lamp's rated life. Depreciation curves published by lamp manufacturers show the rate at which lumens depreciate as a percentage of rated life.

It is extremely difficult in any field to make generalizations about the inevitable trade-offs between first cost and operating cost. However, as far as lighting is concerned, if the comparison can be kept simple, a few generalizations can be attempted. This is the most interesting: as the cost of a lamp increases, operating cost decreases because the more expensive lamps generally have better lamp efficacy (LPW rating). The dilemma that emerges from this generalization is that, as the LPW increases, color rendering (CRI rating) does not necessarily increase. For example, the best CRI lamps (incandescent and tungsten-halogen) have a low LPW, and the highest LPW lamps (high-pressure and low-pressure sodium) have a poor CRI. Fortunately, there are a number of light sources that provide a good balance between LPW and CRI.

Electric light sources (lamps) generally can be divided into two main categories: incandescent and gaseous discharge. The gaseous discharge lamp can be either high pressure or low pressure. Fluorescent and low-pressure sodium lamps are considered low-pressure gaseous discharge sources. High-pressure gaseous discharge sources are the mercury, metal halide, and high-pressure sodium lamps.

INCANDESCENT SOURCES

Incandescence is the visible radiation released from a heated object. An incandescent lamp consists of a wire or filament that incandesces (glows) when heated by a passing electric current. The filament material is usually a tungsten alloy, and the bulb (glass housing) is usually filled with an inert gas, such as nitrogen, argon, or a halogen, to prolong filament life. Generally, standard incandescent lamps are limited in luminous efficacy to about 10 to 20 lm/W; however, higher LPW ratings can be achieved by increasing voltage and/or operating temperature, but this shortens lamp life. Such low luminous efficacy makes incandescent lamps much better sources of heat than of light and is the prime reason why energy conservation legislation (i.e., EPACT) has targeted many incandescent lamps for phaseout. Although incandescent lamps rank poorly in energy performance, in most other categories—simplicity of use, low unit cost, color rendering quality, flexibility of shape and size, and controllability—they have a good to excellent ranking. Such rankings will ensure this lamp's continued popularity, particularly in residential, merchandising, and display applications.

There are two general classifications of incandescent lamps, those that use line voltage and low voltage. Line-voltage incandescent lamps operate directly off 110 or 220 V lines and do not require any transformers or ballasts. However, if dimming is required, a dimmer can be easily placed within the circuit. On the other hand, low-voltage incandescent lamps require step-down transformers to reduce voltage from 110 or 220 V to usually 12 or 24 V. Low-voltage lamps are very popular in merchandising and display applications where precise control of beam pattern and flexibility is required. Flexibility is further enhanced by the fact that low-voltage applications do not require rated wiring from the transformer to the lamps. Thus, the lamp or luminaire can be moved and repositioned by nonelectricians.

For each of the two incandescent classifications, there are three technologies, which are, in increasing order of luminous efficacy: tungsten filament (tungsten), tungsten-halogen (halogen), and infrared dichroic (IR).

The tungsten lamp is the classic filament-and-bulb technology previously described. A major limitation of this lamp has been poor lumen maintenance and short lamp life. As the filament is heated, the tungsten slowly evaporates and deposits itself on the inner wall of the bulb, thus reducing light transmission through the bulb and decreasing lamp life by structurally weakening the filament. In the 1950s, it was discovered that by introducing a halogen gas (such as iodine) and operating at higher temperatures, the evaporation of the tungsten filament could be controlled. The halogen bulb is made of quartz (also commonly referred to as a *quartz lamp*) to withstand the higher operating temperature. The halogen process or cycle works when the evaporated tungsten from the filament unites with the iodine to form tungsten iodide. However, instead of depositing itself on the bulb wall, the tungsten iodide recombines with the filament when the lamp cools, resulting in higher lumen maintenance and an extended lamp life. Because incandescent lamps produce so much heat, they have been known to cause thermal discomfort to people or damage the objects they light. To resolve this problem, *dichroic coatings* are used to produce a light beam of lower temperature. The dichroic process works by acting as a selective reflector of certain wavelengths. By placing a dichroic coating on the reflector surface, light (visible energy) can be reflected to the front of the lamp, while heat (infrared energy) is transmitted through the back. Care must be used in specifying these lamps, and they should never be used in luminaires not properly designed to dissipate the exiting heat. Incandescent lamps are labeled to denote wattage, shape, and size. The first designation is usually a number that indicates wattage; the letter that follows represents the shape of the bulb (see Fig. 5.3.3-1); and the final number indicates the maximum diameter of the bulb in eighths of an inch. For example, a 75A19 lamp is a 75-W A-shape bulb that is ¹⁹⁄₈ or 2⅜ in at its widest point. Typical incandescent bulb shapes are defined as follows:

- *A and PS.* *A* is for arbitrary shape and *PS* is for pear-shape. Both are considered general-service lamps and are the most widely used. Their wattage usually ranges from 15 to 1500

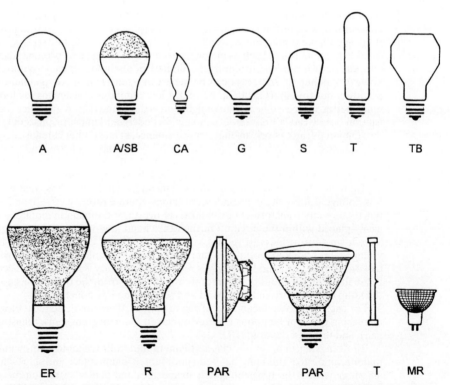

FIGURE 5.3.3-1   Typical incandescent lamp shapes (not to scale). (*Courtesy of Philips Lighting.*)

W, and their average life is 750 to 1000 h. Coatings can be either clear, inside frost, or silica powder "soft white," which is used extensively in residential applications.

- *PAR.* Parabolic aluminized reflector lamps provide excellent beam control, are available in a variety of beam patterns, usually are heavy and robust, and can be used in unprotected outdoor applications.
- *R.* Reflector lamps are less accurate in beam control, lighter in weight, and less expensive than PARs.
- *ER.* Elliptical reflector lamps cross light a few inches ahead of the lens and produce a narrow beam and less light loss when used in luminaires with small openings or deep baffles.
- *SB.* Silvered bowl lamps are A bulbs in which the inside of the upper portion of the bulb is silvered. They are used in open luminaires to create an indirect lighting effect.
- *S and T.* S is for straight-sided shape and T is for tubular shape. Both of these lamps are usually reserved for either appliance or specialty applications.
- *B, C, G, and F.* Decorative lamps are left exposed and take the form of candle (B or C), spherical (G), or flame (F) shapes.
- *H, HAL, and MR.* Tungsten-halogen lamps are compact incandescent sources that usually take a capsule or tubular shape. The bulb may be as small as 0.5-in in diameter and less than 1 in long. These bulbs are small enough to be incorporated or *encapsulated* into large lamp housings such as a PAR or R. These lamps are indistinguishable from ordinary PAR or R lamps except that an H or HAL is denoted in the label after the lamp diameter (e.g., 75PAR38/HAL). Tungsten-halogen bulbs have also been integrated into small dichroic-coated glass reflectors, called *multifaceted reflector* (MR) lamps; the MR-16 is the most common. This lamp is available in a variety of beam patterns from very narrow spot to wide flood and in the low-voltage version ranges from 20 to 75 W. The MR lamp has become very popular in recent years, and because of its small size and excellent beam control, it is widely used in merchandising and display applications.

Lamp bases are designed to connect the lamp to the electrical system. There is a wide range of base types depending on such factors as position of the filament, current requirements, bulb size, and weight. However, these are the most common base types:

- *Screw.* The most commonly used base which is available in several sizes: candelabra (mostly for low-wattage and decorative lamps), intermediate (for appliance and decorative lamps), medium (the standard base for general service lamps of 300 W or less), and Mogul (usually reserved for lamps of 300 W or more). Within the screw base are several variations, such as the three-contact base for three-way switching lamps and the skirted base for thick-necked lamps.
- *Bayonet.* Has either a single or double contact designed to be pushed into a socket and twisted. These bases are most commonly used in transportation applications or where vibrations could loosen other types of bases.
- *Prefocus.* Similar to the bayonet, except that a flange is present to locate the bulb depth precisely. Used where precise optics is required.
- *Prong or post.* Has either a two-contact prong or post on the end or on the side. The prong base is common in PAR lamps, and the post base is used in many high-wattage lamps.

LOW-PRESSURE GASEOUS DISCHARGE SOURCES

The fluorescent lamp is the most commonly used lamp in this category. It operates by generating a low-pressure mercury arc across two electrodes that emits ultraviolet (UV) radiation, which in turn excites the phosphor coating on the inside of the bulb, causing it to fluoresce or produce visible light. This process is about 4 to 5 times more efficient in producing visible light than incandescence and can result in luminous efficacy of 100 lm/W, although 50 to 90 lm/W is more typical. In addition to improved efficacy, fluorescent lamps have lamp lives 7 to 20 times greater than incandescent sources. The concern for energy efficiency has directed considerable attention to fluorescent technology in the belief that it could someday replace the poorer-performing incandescent lamp. This attention has paid off with the successful development of improved color rendering and compact lamp designs that have made the fluorescent lamps look and act more like incandescent sources. Considerable advances have also been made in traditional fluorescent tubular lamps. In the years since the 1973 oil embargo, the luminous efficacy of the tubular lamp/ballast combination has nearly doubled.

Fluorescent lamps vary in performance and color as a function of gas fill, phosphor type, and tube length and diameter. Fluorescent lamps are also available in a wide range of specifications. The smallest is a 4-W, 6-in-long T-5 lamp, and the largest is a 215-W, 8-ft-long T-12 VHO, a very high output (1500 mA) lamp. In addition, because fluorescent lamps are arc sources, they require current-limiting devices, called *ballasts*. Ballasts provide a high initial voltage to start the lamp, then reduce the lamp current to sustain normal operation. Generally, ballasts are designed to operate a particular lamp at specified conditions. Interchanging lamps and ballasts is not recommended.

Fluorescent lamps are labeled to denote wattage, shape, size, and color. In tubular lamps, the first designation is F (fluorescent), the number that follows indicates the wattage, followed by T (tubular) and a number that represents the diameter of the tube in eighths of an inch, and the final designation is for color. For older lamp colors, designations such as cool white (CW) and warm white (WW), have been standardized; however, the newer lamp colors have not (e.g., a lamp with a CCT of 3000 K is designated as SP30, D30, and SPEC30 by GE, Osram/Sylvania, and Philips, respectively). Thus, a F32T8/D30 lamp is a 32-W fluorescent tube, 4 ft long, 1 in in diameter, and has a CCT of 3000 K. Care should be taken in specifying fluorescent lamps, and particular attention should be given to their color-rendering properties. Most lamp manufacturers include cross-referencing guides in their catalogs for determining lamp labeling among manufacturers. Typical fluorescent lamp shapes can be found in Fig. 5.3.3-2.

Historically, fluorescent technology evolved to produce light for the lowest possible cost rather than to create a high-performance product. This trend resulted in a 40-W (F40T12) lamp that was argon gas filled and coated with a halophosphor to produce a cool white light. This became the lamp of choice for lighting nonresidential buildings during the 1950s and 1960s. After the 1973 oil embargo, lamp manufacturers began to introduce a number of energy-saving lamps to the market. Early attempts such as the F40T12/ES (energy-savings) did save energy—they draw 34 W rather than 40 W—but did it at the expense of lumen output and color rendering. Subsequent efforts focused on making fundamental changes to both lamp and ballast design and resulted in improved lighting quality. However, as in the case of incandescent lamps, energy conservation legislation (i.e., EPACT) has targeted a number of older fluorescent lamps for phaseout. The three most important developments in fluorescent technology in recent years have been the introduction of the triphosphor coatings, the development of compact lamp shapes, and the emergence of the electronic ballast.

Triphosphor coatings can produce visible light in three of the most important wavelengths for color rendering—red, green, and blue. Compared to halophosphor lamps, such as cool white (with a CRI of 62 and 50 to 60 lm/W), triphosphor lamps produce better color rendering (CRI of 70 to 89) and higher luminous efficacy (70 to 95 lm/W). Although triphosphor technology has improved T-12 lamp performance, it does a much better job of improving performance in smaller-diameter lamps, such as T-8, T-5, and compact fluorescents. Smaller-diameter lamps also provide more optical control. Thus, with accurately designed reflectors, luminaires that use these lamps are more efficient in delivering light to where it is needed. In addition, reducing the lamp diameter from 1½ in (T-12) to 1 in (T-8)

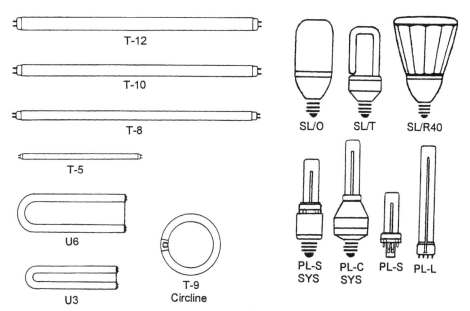

FIGURE 5.3.3-2   Typical fluorescent lamp shapes (not to scale). (*Courtesy of Philips Lighting.*)

and to ⅝ in (T-5) reduces the lamp surface area significantly (55 percent reduction for the T-8 and 83 percent reduction for the T-5); this is important in the case of triphosphor, which is very costly compared to halophosphor. Because T-12, T-8, and T-5 have similar luminous output (2700 to 2900 lm per 4-ft lamp), the reduced surface area means that the T-8 is twice as bright and the T-5 is 5 times as bright as a T-12 lamp. Such high brightness suggests that these new lamps have a greater potential for glare and must be used with caution, particularly in direct downlighting applications. Triphosphor lamps are available in CCTs of 3000, 3500, and 4100 K.

Compact fluorescent lamps are named for their relatively small size, usually about 6 to 7 in high. These lamps have good luminous efficacy (60 to 80 lm/W), very good color rendering properties (CRI of 75 to 85), and long life (7500 to 10,000 h), and are available in various wattages, sizes, shapes, and base types. Originally designed to replace incandescent lamps in residential applications, they have been so popular that they are now available for a wide range of commercial applications. Compact fluorescent bulbs can either be housed in a single assembly incorporating both lamp and ballast, or work as a replacement bulb which can be inserted into a ballasted holder. The single-assembly lamps usually have a medium screw base to facilitate direct substitution for an incandescent lamp. Compact fluorescent bulbs have also been incorporated into R lamps and are indistinguishable from their incandescent counterparts. Similar to T-8 and T-5 lamps, compact fluorescent lamps have high surface brightness, and care should be given when using them in direct downlighting applications. In addition, like other fluorescent lamps whose life is proportional to hours per start, compact fluorescent lamps are not appropriate for storage areas, closets, or other spaces where on/off cycles are frequent.

All gas discharge lamps require ballasts to start and maintain operation. Historically, magnetic or electromagnetic ballasts were the industry standard. These ballasts were relatively inexpensive and had a useful life of 12 to 15 years in typical applications. However, they consumed nearly 20 percent of the lamp wattage in parasitic losses. Because of this, they can no longer be sold in the United States, although they still make up the majority of the ballasts in buildings. Today, ballast replacement has to be done with either energy-saving (ES) magnetic or more typically with electronic ballasts. ES magnetic ballasts are about 8 to 10 percent more efficient than previous ballasts and work well with T-12 lamps. However, this ballast still operates lamps at 60 Hz, which is not the optimum operating frequency to reduce internal losses within many of the newer lamp types (T-8 and T-5). Conversely, electronic ballasts possess the appropriate solid-state components to operate these newer lamps at operating frequencies in the range of 20 to 40 kHz. In addition to their high efficiency, electronic ballasts are noise-free and flicker-free, extremely light in weight, and continuously dimmable. Although electronic ballasts have been available since the mid-1980s, high cost and reliability problems plagued the early years. Today, several highly reliable electronic ballasts are on the market, and costs have fallen so that electronic ballasts in combination with T-8 and T-5 lamps are cost-effective in many lighting applications.

HIGH-PRESSURE GASEOUS DISCHARGE SOURCES

This group of lamps, also known as *high-intensity discharge* (HID), includes mercury, metal halide, and high-pressure sodium. A related source, low-pressure sodium, technically a low-pressure gaseous discharge source, is discussed here because it is used in many applications similar to those where HID lamps are used.

As in fluorescent technology, HID lamps produce light by passing electric current through a conducting gas; however, unlike fluorescent lamps, the gaseous arc, not the phosphor coating, is the primary source of light. Typically, these lamps consist of an inner tubular envelope (usually made of quartz or ceramic) which is designed to withstand high pressure and temperature. In addition, there is an outer jacket or bulb, which protects the arc tube, blocks ultraviolet radiation from the arc, and regulates the operating temperature inside the lamp. The outer bulb is usually made from a borosilicate glass and often has a phosphor coating, which modifies the color characteristics of the light being emitted from the arc. The HID process of producing light is about 2 times more efficient than the fluorescent process and can result in luminous efficacy of 140 lm/W, although 60 to 120 lm/W is more typical. HID lamp life is the longest of all light sources. Some can operate up to 24,000 h, although 10,000 to 20,000 h is more typical. Historically, the color-rendering properties of HID lamps have not been very good, which is the reason that these lamps have been used in applications where color is of little concern, such as in industrial and outdoor uses. However, the concern for energy efficiency has directed considerable attention to HID technology in the belief that its color-rendering properties could be improved. This attention has paid off with the successful development of improved-color-rendering HID lamps.

HID lamps usually take several minutes to ignite or strike because time is required for the metallic gases inside the arc tube to evaporate and ionize (i.e., create the arc). It can take upwards of 5 min for HID lamps to reach proper operating temperature, and during this process there is a significant increase in brightness and often a noticeable shift in color. If power is interrupted, even briefly, an HID lamp must cool down before the arc can restrike, a period that can last several minutes. Thus, in a space lit by HID where emergency lighting is required, auxiliary lighting must be provided to start operating immediately after a power interruption. This is often accomplished with a specially designed HID luminaire that is fitted with an incandescent lamp that remains on until power is restored and the HID lamp approaches normal lighting output. As in fluorescent lamps, HID lamps require ballasts to start and maintain operation. The shift from magnetic to electronic ballasts has also been occurring for HIDs, but this changeover has not been occurring as rapidly. One possible reason for this is that a major advantage of electronic ballasts is their ability to reduce a lamp's lighting output (i.e., dimming ability). Unfortunately, the starting and operating characteristics of HID lamps do not easily lend themselves to dimming. It seems that changes in voltage in many HID lamps can seriously alter lamp operations and have a negative effect on lamp life, color rendering, and color temperature. If dimming is required, the lamp manufacturer's specifications should receive careful attention.

HID lamp designations are not as standardized as those used for other lamp types, but most manufacturers use the ANSI designation: *H* for mercury, *M* for metal halide, and *S* for high-pressure sodium. A number usually follows that indicates the electrical or ballasting characteristic of the lamp. The two letters that follow are an arbitrary manufacturer's code for the lamp's physical characteristics. The remaining item is a number that represents the lamp wattage; often, several letters that indicate the lamp's color characteristics or allowable operating position follow this number. Hence, an M57PE-175/HOR is a 175-W metal halide lamp that operates in a horizontal position and uses a ballast type 57. Many HID lamps are designed to operate in a particular orientation, hence the need for that designation. Using lamps in the wrong position shortens lamp life and may result in rupturing the arc tube. Another important consideration in HID operations is that ballasts and lamps are not interchangeable. For example, placing a mercury lamp in a high-pressure sodium luminaire or vice versa may result in a lamp explosion. To ensure safe and proper operation, manufacturer's catalogs should be consulted for any HID relamping (particularly the cross-reference guide if switching between lamp manufacturers). As with other lamp types, letters are used to represent bulb shape, and the number following the letter indicates the bulb diameter in eighths of an inch. Typically, HID lamps are available in six bulb shapes: BT, ED, E, R, PAR, T-D and T, as shown in Fig. 5.3.3-3.

The mercury lamp, also called a mercury vapor lamp, is the oldest of the HID family of lamps. It is available in wattages from 40 to 1000, has a relatively low lamp efficacy in the range of 35 to 65 lm/W, and has poor color characteristics; however, it has an excellent longevity of more than 24,000 h. Typically, mercury lamps produce a very bright bluish light and exhibit poor color-rendering properties. Adding phosphors to the inner surface of the outer bulb has improved color rendering. Mercury lamps have CRIs that range from 20 (for a clear lamp) to 50 (for a phosphor-coated lamp) and color temperatures that range from 3000 to 6000 K. The major advantage of mercury lamps has been their long life; in fact, these lamps are known to many as "the lamps that never burn out." It is not uncommon to find mercury lamps operating for 80,000 to 100,000 h; although still operating, they may be producing only 25 percent of their rated light output. This can be a particular problem in exterior lighting, where maintenance is not pursued as diligently as in interiors; here light levels can easily deteriorate below design conditions. This problem is usually resolved by group relamping scheduled at the rated life of the lamp. Over the years, a number of improvements have been attempted, such as the development of a self-ballasted mercury lamp that was to

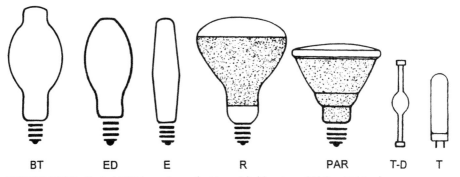

FIGURE 5.3.3-3   Typical HID lamp shapes (not to scale). (*Courtesy of Philips Lighting.*)

be used as a direct replacement for high-wattage incandescent lamps. However, the high cost of these lamps usually made it less costly to replace the incandescent luminaire with a standard HID luminaire. Because of mercury lamps' low efficacy, poor color, and low lumen maintenance, other HID sources usually perform the job better. Today, most lighting professionals consider the mercury lamp an obsolete light source, except possibly in some exterior and low-wattage applications.

Metal halide lamps are essentially mercury lamps with metallic halides, typically iodine, added to the arc tube. The addition of these halides broadens the spectrum from the typical blue mercury colors and doubles lamp efficacy to approximately 110 lm/W; however, they also reduce lamp life to 10,000 to 15,000 h. Wattages for metal halide lamps range from 35 to 1500 W; lamps of less than 100 W are the most recent addition. Metal halide lamps emit light across the entire visible spectrum. They produce good color rendering without the need for a phosphor coating, although these coatings are often applied. Metal halide lamps have CRIs that range from 65 to 85 and color temperatures that range from 3000 to 4300 K. These color properties make this lamp well suited for most interior applications. Most lighting designers have found that metal halide lamps, particularly those that have color temperatures of 3000 K, are the most appropriate HID solution for interior environments and mix rather well with 3000-K and 3500-K triphosphor fluorescent lamps, as well as with most incandescent lamps. The most problematic aspect of metal halide technology is that there can be a noticeable color shift from lamp to lamp and during the life of the lamp. This is primarily due to the manufacturing process, as well as variations in the operating conditions maintained by the ballast. This problem can be partially addressed by specifying or group relamping with lamps purchased from the same production batch (the batch number is usually included within the lamp's serial number). Another important concern is that metal halide lamps are particularly prone to violent rupturing. To protect building occupants and furnishings from hot glass, metal halide lamps should be used only in enclosed luminaires. If open luminaires are required, specially designed plastic-coated lamps are available for these applications.

The high-pressure sodium (HPS) lamp is the newest addition to the HID family (first introduced in 1965). Since its introduction it has rapidly replaced mercury lamps for street lighting; today most street lighting is HPS. As its name implies, it contains sodium, whose corrosive effect on quartz requires that a ceramic material be used for the arc tube. HPS lamps are available in wattages from 35 to 1000 W, have excellent lamp efficacy in the range of 40 to 140 lm/W, acceptable color characteristics, and lamp life in the range of 10,000 to 24,000 h. Typically, HPS lamps produce a bright yellowish orange color with CRIs that range from 22 (for standard HPS) to 65 (for color-improved HPS) and a color temperature that ranges from 1900 to 2200. In 1990, several lamp manufacturers introduced a new HPS lamp, called white sodium, which has very good color rendering properties (CRI of 85 at 2600 to 2800 K). However, color performance was achieved at the expense of lamp efficacy, which was reduced to the range of 35 to 45 lm/W. Lamp manufacturers were optimistic that white sodium lamps would become an ideal candidate for interior applications, but their poor lamp efficacy has hindered acceptance; triphosphor fluorescent and metal halide still prove better choices. Undoubtedly, improvements to HPS technology will continue, but poor color performance still makes it a poor choice for nonindustrial interior applications.

Low-pressure sodium (SOX) lamps have the highest lamp efficacy known. It ranges from 100 to 180 lm/W but produces a hideous monochromatic yellow light with 95 percent of its energy at the 589-nm wavelength. Thus, anything seen under it will be viewed as either black or yellow. SOX lamps are used only where color rendering is of absolutely no importance, such as for some security lighting.

Table 5.3.3-1 compares several of the light sources and performance characteristics described in this article. Although this table presents a meaningful overview of lamp performance, matching the correct light source to the task at hand is only one of many considerations that need to be made in any lighting design. Proper lamp selection can still lead to a poor design if the total lighting system (lamp + ballast + luminaire + surface + task) is not completely analyzed.

TABLE 5.3.3-1 Performance Characteristics of Standard Light Sources

				Light source			
Characteristic	Incandescent	Tungsten	Fluorescent	Compact fluorescent	Mercury	Metal halide	High-pressure sodium
Wattage	≦1500	≦1500	4–215	8–50	40–1000	35–1500	35–1000
Lamp life, h	750–2000	2000	20,000	7500–10,000	12,000–24,000+	10,000–15,000	10,000–24,000
Efficacy, lm/W	10–20	15–30	50–90	60–80	35–65	65–110	40–140
Color	2300–	2300–	3000–	2800–	3000–	3000–	1900–
K	3500	3500	6500	5000	6000	4300	2800
CRI	100	100	40–90	75–85	20–50	65–85	22–85
Optical control	Good/excellent	Excellent	Fair	Good	Fair/good	Good	Very good
Lumen maintenance	Good	Excellent	Good	Good	Fair	Good	Very good
Ballast required	No	No	Yes	Yes	Yes	Yes	Yes
Ignition, time, min							
Strike	—	—	—	—	5–7	4–7	2–4
Restrike	—	—	—	—	4–7	5–8	1–2
Cost							
Initial	Low	Moderate	Moderate	Moderate	High	High	Very high
Operating	Very high	High	Moderate	Moderate	Moderate	Low	Very low

Source: Courtesy of Harvey J. Bryan.

CONTRIBUTOR

David L. Stymiest, P.E., SASHE, C.E.M., Smith Seckman Reid, Inc., New Orleans, Louisiana

ARTICLE 5.3.4
LUMINAIRES

Harvey J. Bryan, Ph.D., F.A.I.A.
Arizona State University, Tempe, Arizona

Luminaires range from a simple bare bulb in a lamp holder to a crystal chandelier. Generally, the luminaire's function is to control and distribute the light from the lamp, to position and protect the lamp, to house the control gear, and add to the appearance of the space. These functions may conflict at times, but it is important at the onset of the lighting-design process to establish a clear order of priorities for luminaire selection. Luminaires can be either standard products, modifications of off-the-shelf items, or custom designs. Often they can be integrated into architectural details, such as coves and shelves, and the surrounding surfaces become in effect a luminaire. In many contemporary designs, off-the-shelf products are also being used in innovative ways, such as augmenting a standard industrial luminaire with a metal screen or turning it upside down to mimic an indirect luminaire. A good understanding of lighting equipment is crucial in using the range of commercially available luminaires properly.

Of the luminaire properties, classification of the light distribution patterns is the most important. The manner in which light is distributed by the luminaire has a major impact on such important factors as total lighting system efficiency, glare, and the overall appearance of a space. Luminaires are classified by the International Commission on Illumination (CIE) according to the percentage of total output emitted above and below the horizontal centerline of the luminaire (see Fig. 5.3.4-1).

Light distribution in a luminaire begins with the surface that is closest to the lamp. That surface can be either specular (usually polished aluminum) or matte (usually white enamel paint) and may be contoured to further control light distribution. Contours (either parabolic or elliptical) used in conjunction with specular surfaces are very effective in directing light. For example, by simply adjusting the distance between the lamp and the focal point of a parabolic reflector, light distribution can be changed from a parallel beam (lamp at the focal point) to a converging beam (lamp outside the focal point) or a diverging beam (lamp inside the focal point).

Another means of light distribution is achieved by using lenses such as diffusers or prismatic refractors. The *diffuser* distributes light in all directions; it controls the brightness of the lamp by obscuring the lamp. Such lenses are usually made of white opal glass, frosted glass, or white plastic. Diffusers are generally good at distributing lamp output over a large area, but they produce high levels of direct glare and have a low overall system efficiency. Their use is most appropriate in corridors, high-bay areas, stairwells, and applications that do not have visually demanding tasks. *Prismatic refractor* lenses, on the other hand, are very good at controlling light and come in varying distribution patterns. These lenses use nonparallel surfaces to change the direction of the light (refract) permanently. Thus, the angles of the nonparallel surface can be specifically designed for the appropriate viewing angles. Prismatic refractor

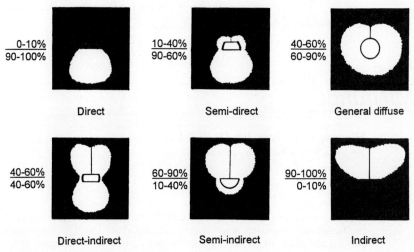

FIGURE 5.3.4-1 CIE luminaire classification by light distribution pattern. (*Courtesy of Illuminating Engineering Society of North America,* IES Lighting Handbook.)

lenses are usually made of clear plastic, have good control, high overall system efficiency, and low direct glare. Their use is appropriate for most visually demanding applications.

Shielding or blocking of the bright lamp surface is done to eliminate direct glare. This can also be accomplished by positioning the lamp in relationship to the luminaire's edges or by using louvers or baffles. In general, louvers provide shielding in two directions, whereas baffles provide shielding only in a single viewing direction. In either case, shielding is effective within what is referred to as the *shielding zone,* which is defined as the angle above the horizontal that the eye can rise before seeing through the louver or baffle. Louvers or baffles can be either aluminum or plastic and can have specular or matte finishes. Louver designs have become complex in recent years with the advent of wedge-shaped systems composed of specular parabolic sections. Luminaires that have such louvers have excellent system efficiency and extremely low surface brightness when viewed from the shielding zone (i.e., little direct glare), and are very good at distributing light straight down. This latter attribute suggests that parabolic luminaires should be used only in applications where high surface reflectances are present (especially for the floor); otherwise, the space can appear extremely dim.

The combination of lamp, reflective housing, and lens or louver/baffle results in a final distribution of light which can be plotted on what is called a *candlepower distribution curve* or *photometric curve* (which is illustrated in the upper right hand corner of Fig. 5.3.4-2). Because every luminaire has a particular distribution profile, understanding these curves is crucial in luminaire selection. The photometric curve is usually plotted on a polar graph, which represents the intensity of luminous energy (i.e., candlepower) from a luminaire in a particular direction. The luminaire is at the center of the diagram, and lines that radiate from the center represent the number of degrees from the vertical (or nadir). Candlepower values are on the vertical scale of the graph. For the illustration shown in Fig. 5.3.4-2, one can find the candlepower for 30° from the vertical by following the 30° line radiating from the center until it hits the curve and following that point over to the vertical scale, which should read 1000 cd. For symmetrical luminaires, only half the polar graph needs to be shown. For more complex distributions, several lines (e.g., solid, dashed, dotted) which represent different angles through the luminaire (e.g., normal, parallel, 45°) would also appear on the polar graph. Photometric analyses are usually generated by an independent testing laboratory, thus guaranteeing objectivity, and are available from the luminaire manufacturer on request (Fig. 5.4.3-2 is an actual photometric report from a testing laboratory). However, most manufacturers provide some type of abbreviated photometric information in their product catalogs.

Photometric data can be used in a number of ways. The most common is to use the data to calculate illuminance levels within a space (this is discussed in detail in Art. 5.3.5, "Lighting Design"). Another important use is determining whether the luminaire's luminous distribution minimizes excessive brightness or glare. Glare is a complex subject to address; it is generally divided into two major types, direct and indirect. *Direct glare* is due to excessive brightness from either a luminaire or a window that is in the direct field of view, resulting in discomfort and/or the loss of visibility. Direct glare is most often associated with heads-up tasks, which are visual activities that occur as someone looks around the environment. Luminaires that produce high brightness in the 45° to 90° zone are the major cause of direct glare. Direct glare can be evaluated by a method called *visual comfort probability* (VCP). VCP values represent the percentage of people in a space that would not complain of the presence of direct glare. Luminaire properties, room size, and viewing direction are the important variables when determining VCP. Often, luminaire manufacturers generate precalculated VCP tables for a standard room and present these data along with their photometric reports. Several computer programs are available for calculating VCP.

Indirect glare, on the other hand, is much subtler and consists of the loss in visibility due to light that is reflected from the task into the eye. Indirect glare is most often associated with heads-down tasks, which are visual activities that occur as someone looks down at the work surface. Indirect glare is often divided into two forms, reflected glare and veiling reflection. *Reflected glare* is caused when bright light is reflected into the eyes from specular or glossy surfaces. *Veiling reflection* occurs when reflected light reduces the contrast between detail and

SKETCH

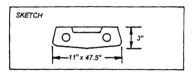

3"
←—11" x 47.5"—→

Description

1 x 4 surface mounted
wraparound prismatic lens
w/clear acrylic
Test No. 48109   Cat. No. CH 409
Lamps 2F40CW   Volts 120
Lumens/Lamp 3110   Test Dist. 20'
Shielding Angle N 90°   L 90°
Bare Lamp 2360 fL   Date

LDD Category V

Distribution Data

Angle	C.P.	Angle	C.P.
0°	1298	95°	169
5°	1307	105°	165
10°		115°	162
15°	1253	125°	160
20°		135°	158
25°	1159	145°	155
30°		155°	150
35°	937	165°	141
40°		175°	132
45°	757	180°	125
50°			
55°	515		
60°			
65°	365		
70°			
75°	189		
80°			
85°	180		
90°	175		

Brightness Data

	Angle	Maximum	Average	Ratio Max./Min.
Along (90°)	45°	2483	1148	2.2
	55°	1939	909	2.1
	65°	1089	678	1.6
	75°	850	533	1.6
	85°	510	384	1.3
Across (90°)	45°	2721	1076	2.5
	55°	2211	827	2.7
	65°	1701	628	2.7
	75°	1429	560	2.6
	85°	1463	597	2.5

OUTPUT DATA

Zone Degrees	Lumens	Percent Total Lamp Lumens	Percent Total Distribution
0-40°	1838	29.5	41.0
40-90°	2019	32.5	45.1
90-180°	625	10.1	13.9
0-180°	4482	72.1	100.0

COEFFICIENT OF UTILIZATION

Floor		Pfc = 20%								
Ceiling		80%			50%			10%		
Walls		50%	30%	10%	50%	30%	10%	50%	30%	10%
	1	73	71	68	66	64	62	57	56	55
	2	65	60	56	58	55	52	51	49	47
	3	57	52	48	52	48	45	46	43	41
	4	51	45	41	47	42	39	41	39	35
Room Cavity Ratio	5	46	39	35	42	37	33	37	33	31
	6	41	35	31	37	33	29	33	30	27
	7	37	31	27	34	29	25	30	26	24
	8	33	27	23	30	25	22	27	23	21
	9	30	24	20	27	22	19	24	21	28
	10	27	21	18	25	20	17	22	18	16

Maximum recommended spacing-to-mounting height ratio above work plane is : 1.4

FIGURE 5.3.4-2   Typical photometric test report. (*Courtesy of Harvey J. Bryan.*)

background and thus reduces visibility. In both cases, altering the position of the task or moving the light source from the reflected zone usually corrects the problem. Veiling reflection can be evaluated by a method called *equivalent sphere illumination* (ESI). ESI defines a reference lighting situation that is free of veiling reflection (i.e., a photometric sphere) and then calibrates the actual visual task to it. Thus, the ESI for an actual visual task is the equivalent illuminance for making that task as visible in the sphere as it is in the actual setting. In using the ESI system, it is possible to have higher ESI foot-candles than classical foot-candles. For example, two identical luminaires at the same distance from a task would provide the same amount of foot-candles. However, if one luminaire was placed outside of the veiling reflection zone and the other was not, the luminaire outside would provide higher ESI foot-candles than the other. The ESI method requires considerable data about the luminaire, the room, the viewing position, and the task; thus, ESI can be calculated only via computer analysis, for which there are several programs available.

Photometric reports usually contain other valuable information, such as overall luminaire efficiency, spacing–to–mounting height ratio, and coefficient of utilization tables. *Luminaire efficiency* is the ratio of total lumens leaving a luminaire to the total initial lamp lumens. It usually appears on the photometric report, but if not, it can be easily calculated by dividing the total lumen output by the total initial lamp lumens. For example, the luminaire presented in Fig. 5.3.4-2 would have an efficiency of (4482 lm output)/(2 lamps × 3110 lm/lamp) = 72 percent. The *spacing–to–mounting height* (S/MH) *ratio* is the ratio of the distance between luminaires to the height above the work plane. The S/MH is used to maintain a uniform light across the work plane, which is defined as ±⅙ the average level of illuminance. For example, in a room with a ceiling height of 9.0 ft, a work plane height of 2.5 ft, and the luminaire that appears in Fig. 5.4.3-2 (which has an S/MH of 1.4), the luminaires should not be placed more than (9.0 ft − 2.5 ft) × 1.4 = 9.1 ft apart. The *coefficient of utilization* (CU) refers to the percentage of the total lamp lumens that actually reaches the work plane. In a number of ways, the CU is similar to luminaire efficiency, except that it also considers room size and shape, as well as the reflectance of ceiling, walls, and floor. The CU tables are used in conjunction with the zonal cavity method calculation (which is discussed in detail in the next article) to provide an easy method for determining average uniform illuminance within a space.

CONTRIBUTOR

David L. Stymiest, P.E., SASHE, C.E.M., Smith Seckman Reid, Inc., New Orleans, Louisiana

ARTICLE 5.3.5

LIGHTING DESIGN

Harvey J. Bryan, Ph.D., F.A.I.A.
Arizona State University, Tempe, Arizona

In addition to the qualitative issues that must always be addressed in any lighting design, the most crucial task is effective lighting of the workspace. Getting the quantity of light right for the variety of luminous tasks that exist is no simple matter. The amount of light required for good visibility varies by the size and contrast of the task and the brightness and time needed to view the visual information. Up to a point, visibility improves as illuminance is increased. However, there is some disagreement as to what that point is. For many years, European illuminance recommendations were half those in the United States. In 1979, the Illuminating Engineering Society of North America (IESNA) changed its illuminance recommendations and stopped recommending single absolute values; this resulted in recommendations that conformed more to European standards.

The procedure introduced in 1979 involves selecting one of nine letter categories (A through I) that are based on a specific task activity (these tables can be found in the 1993 *IES Lighting Handbook*)[1] or on a general description of the activity (see Table 5.3.5-1). Each letter category corresponds to a three-number illuminance range. For example, Category B is 5–7.5–10 fc and E is 50–75–100 fc. To determine if one is on the high or low end of the range, three weighting factors are introduced. A weighting of −1 to +1 is assigned to the age of the worker (<40 = −1; >55 = +1), the importance of speed and/or accuracy (not important = −1; critical = +1), and the reflectance of the task background on which the visual information is displayed (>70 percent = −1; <30 percent = +1). If the three weightings sum to −2 or −3, the lower end of the illuminance range should be used; conversely, if it is +2 or +3, the higher end should be used. For weightings of −1 to +1, the middle value should be used. The resulting illuminance values are considered for the task only for categories D through I and for the entire room (or ambient condition) for categories A through C.

TABLE 5.3.5-1   General Illuminance Categories

		Ranges of illuminances		
	Illuminance			
Type of activity	category	lx	fc	Reference workplane
Public spaces with dark surroundings	A	20–30–50	2–3–5	
Simple orientation for short temporary visits	B	50–75–100	5–7.5–10	General lighting throughout spaces
Working spaces where visual tasks are only occasionally performed	C	100–150–200	10–15–20	
Performance of visual tasks of high contrast or large size	D	200–300–500	20–30–50	
Performance of visual tasks of medium contrast or small size	E	500–750–1000	50–75–100	Illuminance on task
Performance of visual tasks of low contrast or very small size	F	1000–1500–2000	100–150–200	
Performance of visual tasks of low contrast and very small size over a prolonged period	G	2000–3000–5000	200–300–500	
Performance of very prolonged and exacting visual task	H	5,000–7,500–10,000	500–750–1000	Illuminance on task, obtained by a combination of general and local (supplementary lighting)
Performance of very special visual tasks of extremely low contrast and small size	I	10,000–15,000–20,000	1000–1500–2000	

Illuminance categories and illuminance values for generic types of activities in interiors

Source: Illuminating Engineering Society of North America, *IES Lighting Handbook,* 8th ed., 1993.

POINT-BY-POINT METHOD

The point-by-point method is used only for calculating the direct component from a light source and assumes that the source is a point source, as opposed to an area or line source. This method utilizes the fundamental law of illumination, the inverse square law:

$$E = \frac{I}{D^2}$$

where E = illuminance at the receiving surface, fc
I = the luminous intensity, cd, at the source viewed in the direction of the receiving surface, ft
D = the distance from the source to the surface

The law states that the illuminance at a point on a surface normal (perpendicular) to the beam is equal to the luminous intensity divided by the square of the distance between the source and the surface. This law is appropriate only for surfaces normal to the beam; because most surfaces are at an angle of incidence to the beam, this law needs to be modified by the cosine law of illumination. This combined formula is given as

$$E = \frac{I \cos \beta}{D^2}$$

where β = the angle of incidence (the angle between the source to the surface and a line normal to the receiving surface).

When this method is used, the luminous intensity I for the angular direction of the receiving surface is determined from the photometric curve (available from the luminaire manufacturer). Figure 5.3.5-1 illustrates how the relevant variables used in the point-by-point method relate to both horizontal and vertical receiving surfaces.

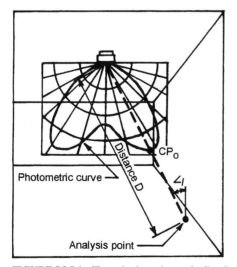 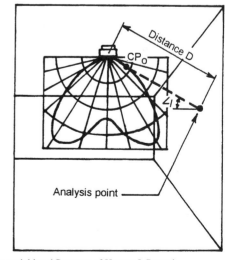

FIGURE 5.3.5-1 The point-by-point method's relevant variables. (*Courtesy of Harvey J. Bryan.*)

ZONAL CAVITY METHOD

It would be tedious and expensive to use the point-by-point method for a room that has a large number of points. In addition, the interreflected component should be considered for an accurate analysis. Thus, a simplified alternative was developed to determine average uniform illuminance on the work plane. This method is called the *zonal cavity method* or *lumen method* and is based on precalculated tables of the coefficient of utilization (CU). The CU refers to the percentage of the total lamp lumens that actually reach the workplane, and it considers the room size and shape, as well as the reflectance of the ceiling, walls, and floor. Most luminaire manufacturers provide CU tables with their products' photometric reports (see Fig. 5.3.5-2). The zonal cavity method uses a simple formula for determining average uniform illuminance on the workplane:

$$E = \frac{N \times n \times LL \times LLF \times CU}{A}$$

where
E = average uniform illuminance
N = number of luminaires
n = number of lamps per luminaire
LLF = combined light loss factor
LL = lumens per lamp
CU = coefficient of utilization for the luminaire
A = area of the work plane

This formula can also be manipulated to solve for the number of luminaires needed to maintain the desired average uniform illuminance:

$$N = \frac{E \times A}{n \times LL \times LLF \times CU}$$

Before all of the variables can be introduced into these formulas, the determination of the CU requires the *room cavity ratio* (RCR), and the components of the *light loss factor* (LLF) have to be found.

The zonal cavity method divides a room into three cavities: a *ceiling cavity* (CC), that area composed of the ceiling and wall surfaces above the luminaire; a *room cavity* (RC), that area composed of the wall surfaces from the luminaire to the workplane; and the *floor cavity* (FC), that area composed of the floor and wall surfaces below the workplane (see Fig. 5.3.5-2). In

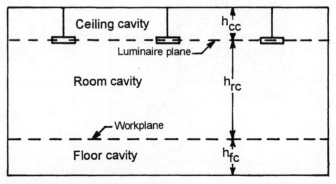

FIGURE 5.3.5-2 The three cavities used in the zonal cavity method. (*Courtesy of Harvey J. Bryan.*)

rooms that have surface-mounted or recessed-mounted luminaires, the ceiling cavity is eliminated. Once the cavities have been defined, the cavity ratios can be determined for each:

$$\text{Cavity ratio} = \frac{5h(L + W)}{L \times W}$$

where h = height of cavity
          L = length of cavity
          W = width of cavity

The final step before the CU can be determined is to find the effective reflectances for both the ceiling cavity ρ_{cc} and floor cavity ρ_{fc}. The effective reflectances are the weighted average reflectances of the wall and ceiling for ρ_{cc} and the walls and floor for ρ_{fc}. Given the respective cavity ratios, ρ_{cc} and ρ_{fc} are found by using Table 5.3.5-2. Knowing the effective reflectances, the room cavity ratio, and the luminaire, the CU can be determined from the CU table. Often, interpolation is needed between the various reflectance columns, as well as between room cavity ratios.

Finally, the LLFs need to be determined. A number of factors can be included in this category, but two require important consideration, the *lamp lumen depreciation* (LLD) and the *lamp dirt depreciation* (LDD). LLD accounts for the effects of lamp aging on output. The impact of this factor varies greatly depending on lamp type. It is for this reason that lamp manufacturers usually distinguish between initial lumen output and maintained lumen output in their catalogs. If the initial lumen output is used for the lumen per lamp (LL) value, then the LLD must be used. If the maintained lumen output is used, then no LLD is necessary. LDD considers the amount of dirt that the luminaire gathers, either on the luminaire's reflective surfaces or on the surface of the lamp. LDD is a function of the design of the luminaire (open, closed, etc.), the maintenance schedule, and the cleanliness of the surroundings. Six basic luminaire categories have been defined by the IESNA, along with an LDD chart for each. The luminaire's LDD category is usually presented in the photometric report. Once the LDD category is known, the LDD can be determined by using Fig. 5.3.5-3 and by knowing the cleanliness of the surroundings (five levels from very clean to very dirty) and the maintenance schedule between major luminaire cleanings (1 to 36 mo). If both the LLD and LDD are known, then the product of these becomes the LLF; if only one is known, then that becomes the LLF.

Now all the variables can be entered into either of the two zonal cavity methods' formulas. If the number of luminaires is the desired answer, the result should be rounded off to the next even number to aid in symmetrical luminaire spacing. The spacing between luminaires should not exceed the spacing–to–mounting height ratio that appears in the photometric report.

The zonal cavity method cannot deal with nonuniform illuminance within a space and therefore has fallen into disfavor. As recommended illuminance levels decreased in response to energy concerns, lighting designers increasingly began to design task/ambient solutions that created greater illuminance ranges than the zonal cavity method could accurately address. Fortunately, personal computers have allowed designers to overcome the tediousness of the point-by-point method and the limits of the zonal cavity method by computer simulation.

COMPUTER SIMULATION

During the last decade, a number of computer programs were developed that have proved to be an asset in lighting design. Today, there are more than two dozen lighting design programs available that range from simple automated versions of the zonal cavity method to sophisticated luminous radiosity programs. Selecting the appropriate program depends greatly on the

TABLE 5.3.5-2   Effective Ceiling and Floor Reflectances

Percent base reflectance*	90										80										70										60										50									
Percent wall reflectance	90	80	70	60	50	40	30	20	10	0	90	80	70	60	50	40	30	20	10	0	90	80	70	60	50	40	30	20	10	0	90	80	70	60	50	40	30	20	10	0	90	80	70	60	50	40	30	20	10	0
Cavity ratio																																																		
0.2	89	88	88	87	86	85	85	84	84	82	79	78	78	77	77	76	76	75	74	72	70	69	68	68	67	67	66	66	65	64	60	59	59	59	58	57	56	56	55	53	50	50	49	49	48	48	47	47	46	44
0.4	88	87	86	85	84	83	81	80	79	76	79	77	76	75	74	73	72	71	70	68	69	68	67	66	65	64	63	62	61	58	60	59	59	58	57	55	54	53	52	50	50	49	48	48	47	46	45	45	44	42
0.6	87	86	84	82	80	79	77	76	74	73	78	76	75	73	71	70	68	68	65	63	69	67	65	64	63	61	59	58	57	54	60	58	57	56	56	53	51	51	50	46	50	48	47	46	45	44	43	42	41	38
0.8	87	85	82	80	77	75	73	71	69	67	78	75	73	71	69	67	65	63	61	57	68	66	64	62	60	58	56	55	53	50	59	57	56	54	51	51	48	47	46	43	50	48	47	45	44	42	40	39	38	36
1.0	86	83	80	77	75	72	69	66	64	62	77	74	72	69	67	65	62	60	57	55	68	65	62	60	58	55	53	52	50	47	59	57	55	53	51	48	45	44	43	41	50	48	46	44	43	41	38	37	36	34
1.2	85	82	78	75	72	69	68	63	60	57	76	73	70	67	64	61	58	55	53	51	67	64	61	59	57	54	50	48	46	44	59	56	54	51	49	48	44	42	40	38	50	47	45	43	41	39	36	35	34	29
1.4	85	80	77	73	69	65	62	59	57	52	76	72	68	65	62	59	55	53	50	48	67	63	60	58	55	51	47	45	44	41	59	56	53	49	47	44	41	39	38	36	50	47	45	42	40	38	35	34	32	27
1.6	84	79	75	71	67	63	59	56	53	50	75	71	67	63	60	57	53	50	47	44	67	62	59	56	53	47	45	43	41	38	59	55	52	48	45	42	39	37	35	33	50	47	44	41	39	36	33	32	30	26
1.8	83	78	73	69	64	60	58	53	50	48	75	70	66	62	58	54	50	47	44	41	66	61	58	54	51	48	42	40	38	35	58	55	51	47	44	40	37	35	33	31	50	48	43	40	38	35	31	30	28	25
2.0	83	77	72	67	62	56	53	50	47	43	74	69	64	60	56	52	48	45	41	38	66	60	56	52	49	45	40	38	36	33	58	54	50	48	43	39	35	33	31	29	50	46	43	40	37	34	30	28	26	24
2.2	82	76	70	65	59	54	50	47	44	40	74	68	63	58	54	49	45	42	38	35	66	60	55	51	46	43	38	36	34	32	58	54	49	45	41	37	34	31	29	28	50	46	42	38	36	33	29	27	24	22
2.4	82	75	69	64	58	53	48	45	41	37	73	67	61	58	52	47	43	39	34	31	65	60	54	50	46	41	37	35	32	30	58	53	48	44	41	36	32	30	27	26	50	46	42	37	35	32	27	25	23	21
2.6	81	74	67	62	58	51	46	42	38	35	73	66	60	55	50	45	41	38	34	31	65	59	54	49	45	40	35	33	30	28	58	53	48	43	39	35	31	28	26	24	50	46	41	37	34	30	26	23	21	20
2.8	81	73	66	60	54	49	44	40	36	34	73	65	59	53	48	43	39	36	32	29	65	59	53	48	43	38	33	30	28	26	58	53	47	43	38	34	29	27	24	22	50	46	41	36	33	29	25	22	20	19
3.0	80	72	64	58	52	47	42	38	34	30	72	65	58	52	47	42	37	34	30	27	64	58	52	47	42	37	32	29	27	24	57	52	46	42	37	32	28	25	23	20	50	45	40	36	32	28	24	21	19	17
3.2	79	71	63	56	50	45	40	36	32	28	72	65	57	51	45	40	35	33	28	25	64	58	51	46	40	36	31	28	25	23	57	51	45	41	36	31	27	23	22	18	50	44	39	35	31	27	23	20	18	16
3.4	79	70	62	54	48	43	38	34	30	27	71	64	58	49	44	39	34	32	27	24	64	57	50	45	39	35	29	27	24	22	57	51	45	40	35	30	26	23	20	17	50	44	39	35	30	26	22	19	17	15
3.6	78	69	61	53	47	42	36	32	28	25	71	63	54	48	43	38	32	30	25	23	63	58	49	44	38	33	28	25	22	20	57	50	44	39	34	29	25	22	19	16	50	44	39	34	29	25	21	18	16	14
3.8	78	69	60	51	45	40	35	31	27	23	70	62	53	47	41	36	31	28	24	22	63	58	49	43	37	32	27	24	21	19	57	50	43	38	33	29	24	21	19	15	50	44	38	34	29	25	21	17	15	13
4.0	77	69	58	51	44	39	33	29	25	22	70	61	53	46	40	35	30	26	22	20	63	55	48	42	36	31	26	23	20	17	57	49	42	37	32	28	23	20	18	14	50	44	38	33	28	24	20	17	15	12
4.2	77	62	57	50	43	37	32	28	24	21	69	60	52	45	39	34	29	25	21	18	62	55	47	41	35	30	25	22	19	16	56	49	42	37	32	27	22	19	17	14	50	43	37	32	28	24	20	17	14	12
4.4	78	61	58	49	42	36	31	27	23	20	69	60	51	44	38	33	28	24	20	17	62	54	46	40	34	29	24	21	18	15	56	49	42	36	31	27	22	19	16	13	50	43	37	32	27	23	19	16	13	11
4.6	76	60	55	47	40	35	30	26	22	19	69	59	50	43	37	32	27	23	19	15	62	53	45	39	33	28	24	21	17	14	56	49	41	35	30	26	21	18	16	13	50	43	36	31	26	22	18	15	13	10
4.8	75	59	54	46	39	34	28	23	21	18	68	58	49	42	38	32	27	23	18	15	62	53	45	38	32	28	23	20	16	13	56	48	41	34	29	25	21	18	15	12	50	43	36	31	26	22	18	15	12	09
5.0	75	59	53	45	38	33	28	24	20	16	68	58	48	41	35	30	25	21	18	14	61	52	44	36	31	26	22	19	16	12	56	48	40	34	28	24	20	17	14	11	50	42	35	30	25	21	17	14	12	09
6.0	73	61	49	41	34	29	24	20	16	11	68	55	44	38	31	27	22	19	15	10	60	51	41	35	28	24	19	16	13	09	55	45	37	31	25	21	17	14	11	07	50	42	34	29	23	19	15	13	10	06
7.0	70	58	45	38	30	27	21	18	14	08	64	53	41	35	28	24	19	16	12	07	58	48	38	32	26	22	17	14	11	06	54	43	35	30	24	20	15	12	09	05	49	41	32	27	21	18	14	11	08	05
8.0	68	55	42	35	27	23	18	15	12	06	62	50	38	32	25	21	17	14	11	05	57	46	35	29	23	19	15	13	10	05	53	42	33	28	22	18	14	11	08	04	49	40	30	25	19	16	12	10	07	03
9.0	66	52	38	31	25	21	16	14	11	05	61	49	36	30	23	19	15	13	10	04	56	45	33	27	21	18	14	12	09	04	52	40	31	26	20	16	12	10	07	03	48	39	29	24	18	15	11	09	07	03
10.0	65	51	38	29	22	19	15	11	09	04	59	48	33	27	21	18	14	11	08	03	55	43	31	25	19	16	12	10	08	03	51	39	29	24	18	15	11	09	07	02	47	37	27	22	17	14	10	08	06	02

Percent base reflectance*

Percent wall reflectance →	40										30										20										10										0									
Cavity ratio \ wall %	90	80	70	60	50	40	30	20	10	0	90	80	70	60	50	40	30	20	10	0	90	80	70	60	50	40	30	20	10	0	90	80	70	60	50	40	30	20	10	0	90	80	70	60	50	40	30	20	10	0
0.2	40	40	39	39	38	38	37	36	36	36	31	31	30	30	29	29	29	28	28	27	21	20	20	20	20	20	19	19	19	17	11	11	11	10	10	10	10	09	09	09	02	02	01	01	01	01	00	00	00	0
0.4	41	40	39	38	37	36	35	34	34	34	31	31	30	30	29	28	28	27	26	25	22	21	21	20	19	19	19	18	18	16	12	11	11	11	11	10	10	09	09	08	04	03	03	02	02	02	01	01	00	0
0.6	41	40	39	38	37	36	34	33	32	31	32	31	30	29	28	27	26	26	25	23	23	21	21	20	20	19	18	18	17	15	13	13	12	11	11	10	10	09	08	08	05	05	04	03	03	02	02	01	01	0
0.8	41	40	38	37	36	35	33	32	31	29	32	31	30	29	28	26	25	25	23	22	24	22	21	20	19	19	18	17	16	14	15	14	13	12	11	10	10	09	08	07	07	06	05	04	04	03	02	02	01	0
1.0	42	40	38	37	35	33	32	31	29	27	33	32	30	29	27	25	24	23	22	20	25	23	22	20	19	18	17	16	15	13	16	14	13	12	12	11	10	09	08	07	08	07	06	05	04	03	02	02	01	0
1.2	42	40	38	36	34	32	30	29	27	25	33	32	30	28	27	25	23	22	21	19	25	23	22	20	19	17	16	15	14	12	17	15	14	13	12	11	10	09	07	06	10	08	07	06	05	04	03	02	01	0
1.4	42	39	37	35	33	31	29	27	25	23	34	32	30	28	26	24	22	21	19	18	26	24	22	20	18	17	16	15	14	12	18	16	14	13	12	11	10	09	07	06	11	09	08	07	06	04	03	02	01	0
1.6	42	39	37	35	32	30	27	25	23	22	34	33	29	27	25	23	22	20	18	17	26	24	22	20	18	17	16	15	13	11	19	17	15	14	12	11	09	08	07	06	12	10	09	07	06	05	03	02	01	0
1.8	42	39	36	34	31	29	26	24	22	21	35	33	29	27	25	23	21	19	17	16	27	25	23	20	18	17	15	14	12	10	19	17	15	14	13	11	09	08	06	05	13	11	09	08	07	05	04	03	01	0
2.0	42	39	36	34	31	28	25	23	21	19	35	33	29	26	24	22	20	18	16	14	28	25	23	20	18	16	15	13	12	09	20	18	16	14	13	11	09	08	06	05	14	12	10	09	07	05	04	03	01	0
2.2	42	39	36	33	30	27	24	22	19	18	36	32	29	26	24	22	19	17	15	13	28	25	23	20	18	16	14	12	10	09	21	19	16	14	13	11	09	07	06	05	15	13	11	09	07	06	04	03	01	0
2.4	43	39	35	33	29	27	24	21	18	17	36	32	29	25	23	21	18	16	14	12	29	26	23	20	18	16	14	12	10	08	22	19	17	15	13	11	09	07	06	05	16	13	11	09	08	06	04	03	01	0
2.6	43	39	35	32	29	26	23	19	17	15	36	32	29	25	23	21	18	16	14	11	29	26	23	20	17	15	14	11	08	08	23	20	17	15	13	11	09	07	06	04	17	14	12	10	08	06	05	03	02	0
2.8	43	39	35	32	28	25	22	19	16	14	37	33	29	25	23	21	17	15	13	11	30	27	23	20	17	15	13	11	09	07	23	20	18	16	13	11	09	07	05	03	17	15	13	10	08	07	05	03	02	0
3.0	43	39	35	31	27	24	21	18	16	13	37	33	29	25	22	20	17	15	12	10	30	27	23	20	17	15	13	11	09	07	24	21	18	16	13	11	09	07	05	03	18	16	13	11	09	07	05	03	02	0
3.2	43	39	35	31	27	23	20	17	15	13	37	33	29	25	22	19	16	14	12	10	31	27	23	20	17	15	12	11	09	06	25	21	18	16	13	11	09	07	05	03	19	16	14	11	09	07	05	03	02	0
3.4	43	39	34	30	26	23	20	17	14	12	37	33	29	25	22	19	16	14	11	09	31	27	23	20	17	15	12	10	08	06	26	22	18	16	13	11	09	07	05	03	20	17	14	12	09	07	05	03	02	0
3.6	43	39	34	30	26	22	19	16	14	11	38	33	29	24	21	18	15	13	10	09	32	27	23	20	17	15	12	10	08	05	26	22	19	16	13	11	09	06	04	03	20	17	15	12	10	08	06	04	02	0
3.8	44	38	33	29	25	22	18	16	13	10	38	33	28	24	21	18	15	13	10	08	32	28	23	20	17	15	12	10	08	05	27	23	19	17	14	11	09	06	04	02	21	18	15	12	10	08	05	04	02	0
4.0	44	38	33	29	25	21	18	15	12	10	38	33	28	24	21	18	14	12	09	07	33	28	23	20	17	14	11	09	07	05	27	23	20	17	14	11	09	06	04	02	22	18	15	13	10	08	05	04	02	0
4.2	44	38	33	29	24	21	17	15	12	10	38	33	28	24	20	17	14	12	09	07	33	28	23	20	17	14	11	09	07	04	28	24	20	17	14	11	09	06	04	02	22	19	16	13	10	08	06	04	02	0
4.4	44	38	33	28	24	20	17	14	11	09	39	33	28	24	20	17	14	11	09	06	34	28	23	20	17	14	11	09	07	04	28	24	20	17	14	11	08	06	04	02	23	19	16	13	10	08	06	04	02	0
4.6	44	38	32	28	23	19	16	14	11	08	39	33	28	24	20	17	13	10	08	06	34	29	24	20	17	14	11	09	07	04	29	25	20	17	14	11	08	06	04	02	23	20	17	13	11	08	06	04	02	0
4.8	44	38	32	27	22	19	15	13	10	08	39	33	28	24	19	16	13	10	08	05	35	29	24	20	17	13	10	08	06	04	29	25	20	17	14	11	08	06	04	02	24	20	17	14	11	08	06	04	02	0
5.0	45	38	31	27	22	19	15	13	10	07	39	33	28	24	19	16	13	10	08	05	35	29	24	20	16	13	10	08	06	04	30	25	21	17	14	11	08	06	04	02	25	21	17	14	11	08	06	04	02	0
6.0	44	37	30	25	20	17	13	11	08	05	39	33	27	23	18	15	11	09	06	04	36	30	24	20	16	13	10	08	05	02	31	26	21	18	14	11	08	06	03	01	27	23	18	15	12	09	06	04	02	0
7.0	44	36	29	24	19	16	12	10	07	04	40	33	26	22	17	14	10	08	05	03	36	30	24	20	15	12	09	07	05	02	32	27	21	17	13	11	08	06	03	01	28	24	19	15	12	09	06	04	02	0
8.0	44	35	28	23	18	15	11	09	06	03	40	33	26	21	16	13	09	07	04	02	37	30	23	19	15	12	08	06	04	01	33	27	21	17	13	10	07	05	03	01	30	25	20	15	12	09	06	04	02	0
9.0	44	35	26	21	16	13	10	08	05	02	40	33	25	20	15	12	09	07	04	02	37	29	23	19	14	11	08	06	04	01	34	28	21	17	13	10	07	05	02	01	31	25	20	15	12	09	06	04	02	0
10.0	43	34	25	20	15	12	08	07	05	02	40	32	24	19	14	11	08	06	03	01	37	29	22	18	13	10	07	05	03	01	34	28	21	17	12	10	07	05	02	01	31	25	20	15	12	09	06	04	02	0

Note: Values in this table are based on a length-to-width ratio of 1.6.

* Ceiling, floor, or cavity.

Source: Illuminating Engineering Society, *IES Lighting Handbook*, 8th ed., 1993.

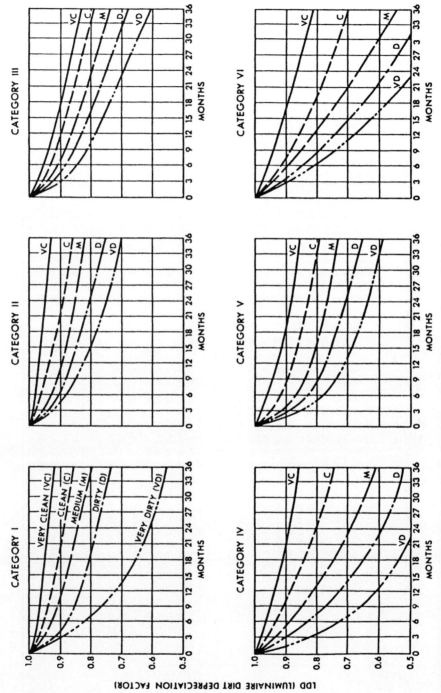

FIGURE 5.3.5-3 Luminaire dirt depreciation factors. (*From Illuminating Engineering Society of North America, IES Lighting Handbook, 8th ed., 1993.*)

intended use. Programs are highly recommended that perform qualitative analysis [such as visual comfort probability (VCP) and equivalent sphere illumination (ESI)], as well as traditional quantitative analysis. One does not have to be an expert in lighting to use many of these programs. Several are designed to allow importing data from computer-aided design (CAD) programs; luminaire and photometric data are available from many manufacturers on CD-ROMs that are readable by most of these programs. The output from these programs can vary from tabular printouts to iso-illuminance plots to photorealistic renderings.

CONTRIBUTOR

David L. Stymiest, P.E., SASHE, C.E.M., Smith Seckman Reid, Inc., New Orleans, Louisiana

NOTE

1. Illuminating Engineering Society of North America, *IES Lighting Handbook,* 8th ed., IESNA, New York, 1993.

ARTICLE 5.3.6
DAYLIGHTING

Harvey J. Bryan, Ph.D., F.A.I.A.
Arizona State University, Tempe, Arizona

Daylight has traditionally been an integral part of building design for the vast majority of buildings. Candles, kerosene lamps, and gas lamps were used for lighting only at night, and their quality was very poor. Building forms tended to be narrow and had high ceilings, large windows, and highly reflective surfaces, so that daylight could be used to the best advantage. However, with the invention of the fluorescent lamp and advances in HVAC technology in the late 1940s, deep-bay building forms, which relied exclusively on electric light, became the dominant building form. Thus, by the early 1960s, daylighting—the once-significant building-design strategy—was all but forgotten by the vast majority of building designers. This lack of interest in daylighting continued through the 1960s and for the first few years following the oil embargo of 1973 to 1974. It was not until after the design community first focused on energy in terms of thermal load reduction that the realization surfaced that in many large buildings, the single largest energy load was electric lighting and not heating or cooling. Thus, daylighting did not begin to emerge as a viable energy strategy until the late 1970s.

Good daylighting design requires sensitivity beyond that provided by simple rules of thumb. For example, a simple rule of thumb would suggest that the larger a window opening, the better the daylight. However, if the thermal considerations for such a window are not considered, overheating and thermal discomfort may result. Thus, the heat/illumination trade-off should not be overly simplified. For a large project that is considering the use of daylight, one of the large energy simulation programs that include integrated daylighting calculations (such as the DOE-2 program) should be used.

The strategies for daylighting design generally can be divided into two categories, depending on building form. Daylighting design strategies for offices, apartment buildings, and multistory buildings rely on what is known as *sidelighting,* whereas warehouses, factories or other single-story buildings generally fall in the category known as *toplighting.*

SIDELIGHTING

The most familiar approach to daylight design is sidelighting. Unfortunately, daylight from a window wall rapidly falls off as it travels deeper into a space. A typical practical limit for daylight penetration into a space is on the order of 15 to 20 ft from the window wall. The basic approaches to sidelighting were well developed in the 1940s and 1950s; strategies such as unilateral and bilateral daylighting design have remained unchanged. The limited cross-sectional width that sidelighting allowed (usually a hallway with a room on each side) evolved into floor plans that had high perimeter areas and typically took such forms as E-, H-, U-, L-, and O-shaped floor plans.

Recently several techniques have emerged to increase the penetration of daylight under sidelighting conditions, thus allowing deeper spaces to be effectively daylit. The most often discussed strategy is to use spectrally coated (mirror-like) light shelves to bounce reflected direct-beam sunlight deep into a space. Although the light quality achieved by such a device is satisfactory, the control of the reflected direct beam as sun angles change is not a trivial problem. Traditionally, light shelves using a highly reflective diffuse coating (a white surface) have been a very effective solution for bringing a more even distribution of daylight into a space. Light shelves can also work as shading devices and in imaginative roof designs as clerestories.

TOPLIGHTING

Toplighting is the other daylight design strategy. It provides excellent uniform general illumination but is less appropriate for task lighting. Historically, toplighting has proved most successful in large one-story spaces, such as factories and warehouses, where sawtooth roofs and roof monitors have been commonplace. Today, most skylights lie parallel to the roof plane and receive both direct and diffuse radiation. Although they provide good daylight below, such skylights can generate considerable heat. Thus, some type of solar controls should be considered when using such skylights.

To overcome the limitations of toplighting with respect to daylighting more than one story, several concepts have been proposed to capture direct sunlight at the roof and distribute it through the building by using mirrors, lenses, and other optical devices. These schemes are feasible, but their optical performance requirements and complexity limit their practicality for many building applications.

Calculation Methods

Procedures for calculating the effects of daylight can be divided into two approaches, those using the lumen method and those using the daylight factor method. The *lumen method,* which is used primarily in the United States, is based on a *coefficient of utilization* and is very similar to the zonal cavity method used for electric-lighting calculations. This method is easy to use and is recommended by the Illuminating Engineering Society of North America (IESNA), but it makes several assumptions that make it applicable only to a limited range of window configurations and reference points within a room. The *daylight factor method,* which is recommended by the *Commission Internationale de l'Eclairage* (CIE), is used in Europe—particularly in Britain, where design aids such as protractors, nomographs, tables, and graphs have been developed for easily determining daylight factors at various stages of the design process. Unlike the lumen method, the daylight factor method is derived from first principles that allow the calculation of daylight at any point within a room and can be used for a wide range of window configurations. In recent years, the daylight factor method has been expanded to include clear-sky conditions.

Manual calculations for either the lumen method or the daylight factor method are too tedious to be done for anything other than a few reference points under a specific design condition. Thus, in recent years, numerous automated procedures for calculating illumination from daylight have been developed. Today, there are a number of very reliable PC-based daylighting calculating programs, several of which are in the public domain and are available at little or no cost.

Daylighting Controls

Daylight can provide good visibility and amenity to indoor space, but it cannot save energy unless the electric lights can be turned off. This may sound obvious, but many buildings cited for good daylight design never save energy, partially because of the lack of user awareness and poor control-system design.

Most daylight control systems employ some type of photosensor that can either switch (having an on/off function) or dim the electric light. On/off switching has a low to moderate cost but may not realize the full daylight savings potential and may result in uncertain user response from abrupt changes in light levels. By selecting a larger number of control levels, the designer can exercise finer control over the system (i.e., fewer abrupt changes in the light levels); for example, a designer could choose to tandem-wire or split-wire the luminaires within a space. This would result in a two-lamp luminaire that has three levels of control (0, 50, or 100 percent) and in a three-lamp luminaire, four levels of control (0, 33, 66, or 100 percent). If more control is preferred, a dimming ballast should be used. Dimming systems are typically more complex and costlier, but provide significantly more daylight savings. As previously discussed, electronic ballasts, which are continuously dimmable, are increasingly being used. It makes sense to make them work in a daylighting setting by integrating photosensors into their control systems. Several manufacturers have developed off-the-shelf systems that do just this.

SECTION 5.4
MECHANICAL SYSTEMS

Roger Wessel, P.E., Section Editor
Principal, RPW Technologies, West Newton, Massachusetts

Anand K. Seth, P.E., C.E.M., C.P.E., Section Editor
Director Utilities and Engineering, Partners HealthCare System, Inc., Boston, Massachusetts

INTRODUCTION

In the editors' opinion, mechanical systems are the most vital components of building systems. They can be compared to the heart, lungs, and arteries of the human body. This section is divided into three main themes, and the subject matter is presented in a manner that is considered useful to the reader. The three themes are environmental control, facility sanitation and water supply, and fire suppression.

The modern building requires environmental controls. This subject is discussed in some detail in several articles. Art. 5.4.1 discusses issues in human comfort; Art. 5.4.2 describes heating ventilating and air conditioning (HVAC) systems; Art. 5.4.3 discusses boilers for buildings and process heating; Art. 5.4.4 addresses chilled water plants; and Art. 5.4.10 discusses noise and vibration controls. Together, these articles provide a comprehensive view of the environmental control requirements in modern facilities. The need to make these facilities more efficient is discussed in Art. 5.4.5, on cogeneration, and Art. 5.4.9, on thermal storage. Balancing the need for facility efficiency and its effect on environmental quality is discussed in Art. 5.4.8, on energy efficiency and indoor air quality.

Water is the stuff of life, and the level of sanitary facilities in buildings is an indicator of the level of civilization. We marvel at the ancient societies where running water was provided by various ingenious means. The need for sanitation and plumbing systems, as well as special systems, in modern facilities is discussed in Art. 5.4.6.

Fire has been a source of danger to facilities since ancient times. Modern facilities are designed with fire suppression systems that are described in Art. 5.4.7.

ARTICLE 5.4.1
ISSUES IN HUMAN COMFORT

Walter E. Henry, P.E.
Syska & Hennessy New England, Inc., Cambridge, Massachusetts

Human comfort in a facility is affected by many factors, including noise, color, air quality, and, most important, thermal comfort. The majority of complaints received by facility managers center around thermal comfort. What is thermal comfort? ISO 7730 defines it as "That condition of mind which expresses satisfaction with the thermal environment." The HVAC system in a building is solely responsible for creating and maintaining thermal comfort and air quality. This article discusses the issues that affect perception of thermal comfort. A description of air quality as it relates to human comfort is introduced in Art. 5.4.8. Other contributors to human comfort in facilities such as noise, light and color are dealt with in other articles.

HUMAN PHYSIOLOGICAL FACTORS

Heat Generation

Like any machine, the body requires energy for its mental and physical activities. The food that we eat provides this energy, and the process that turns food into energy is called the *metabolic process.* Similarly to the machine, the body generates waste heat that it must be able to dissipate to prevent overheating. When the body suffers from too much heat loss, lowering the core temperature, it is said to suffer from *hypothermia,* and when the waste heat cannot be dissipated quickly enough, it is said to be suffering from *hyperthermia.* Much like a variable volume pumping system, the body dissipates this heat by increasing or decreasing the flow of blood to the skin, the body's radiator. This function is regulated by the central nervous system.

Body core and skin temperatures vary with activity and are inversely related. Basal (at rest) core temperatures normally range from 97.6 to 98.2°F and rise with increased activity level. The skeletal muscles largely generate the increased heat that must be dissipated during periods of physical activity. The brain is also a large heat generator whose temperature is related to activity level. Internal core temperatures greater than 115°F can cause brain damage, and temperatures less than 82°F can cause death. On the other hand, skin temperatures at rest range from 91.5 to 93°F and actually decrease with increased activity. In warm surroundings, skin temperature is higher than it is in cold surroundings. This is due to increased warm blood flow in warm surroundings and decreased flow in cooler ones. When the air around the skin is very cold there is little blood flow, allowing the skin layer to provide some insulating qualities.

The rate of metabolic heat generation by the body varies somewhat with the surrounding temperature but is mostly affected by the level of activity. The unit of measure for heat generation is the British thermal unit (Btu). We normally speak in terms of Btus per hour, or Btu/h. Figure 5.4.1-1 shows the range of such heat generation. The heat produced is converted to watts for a better understanding. Note that a person engaging in substantial activity such as jogging releases as much heat as a small space heater. The upper limit of heat generation is defined by the amount of oxygen that the body can take in and use in the metabolic process. Although other "fuels" can be stored, oxygen cannot. Some of the heat generated is sensible, and some is latent. The percentage changes with the level of activity and ranges from two-thirds sensible heat at low levels of activity to one-third sensible heat at high activity levels. These levels of heat generation are net of the actual work that the body does. That work is very little in most cases, because many human tasks do not require that high levels of work be done, and because the body is not a very efficient machine. At best, only 20 percent of metabolic activity is used in work done, and this number is usually closer to 5 to 10 percent.

Heat Loss

The body loses heat in four ways: convection, radiation, evaporation, and conduction. *Convection* accounts for the largest amount of body heat loss. Some researchers estimate that the heat loss by convection is approximately 40 percent of the total body heat loss. In this heat-loss mechanism, air that is cooler than the skin circulates around the body and removes heat

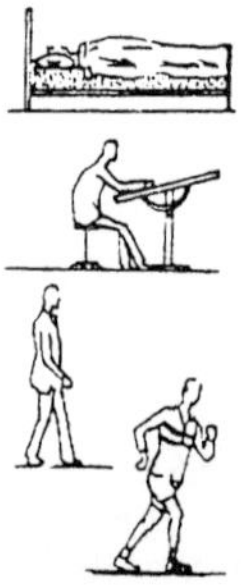

Activity	Heat Produced per Hour (btu/hour)		Heat Equivalent to (W)
Sleeping	300		100
Light work	600		200
Walking	900		300
Jogging	2400		800

FIGURE 5.4.1-1   Body heat production as a function of activity. (*From N. Lechner,* Heating, Cooling, Lighting Design Methods for Architects, *Copyright © 1991, reprinted by permission of John Wiley & Sons, Inc.*)

from it, just as an automobile radiator works. Increased air velocity results in greater heat loss. This is often described as the *wind chill factor,* which is discussed in more detail later in this article. *Radiation,* the second most important heat-loss mechanism, accounts for an additional 40 percent of total heat loss. A body whose surface temperature is higher than that of its surroundings will lose heat to those surroundings by radiation. The rate of heat loss is independent of the surrounding air temperature. This heat loss is discussed more fully when the concept of mean radiant temperature (MRT) is introduced. *Evaporation* accounts for most of the remaining heat loss. In this mechanism, the body is exhaling warm and moist gases with the breath. Additional evaporation occurs from the surface of the skin. This heat-loss mechanism is relatively small until a person exercises. Then the body increases the skin's ability to reject heat by increasing the rate of evaporational cooling. *Conduction* is heat loss that occurs when the skin touches something cold, for example, when it loses heat to a cold floor through uninsulated shoe soles. This mechanism usually results in very little of the total heat loss. Clearly, the proportions of heat loss attributed to the four mechanisms described vary with the amount and type of clothing, the velocity of the surrounding air, and the temperatures of the surrounding surfaces.

ENVIRONMENTAL FACTORS

Air Temperature and Humidity

The dry-bulb air temperature determines the rate of convective heat transfer from the skin to the air. The lower the dry-bulb temperature, the higher the rate of heat transfer. The comfort range, as defined by the American Society of Heating, Refrigerating, and Air-Conditioning Engineers (ASHRAE), ranges from a low of 68°F in the winter to a high of 81°F in the summer. A more generally used and narrower range is 73 to 77°F. During heating the temperature difference from the ankle to the neck should not be greater than 4°F. When cooling, that difference should not be greater than 2°F.

Both relative humidity (RH) and wet-bulb temperatures are measures of the amount of moisture in the air. Relative humidity levels below 25 percent allow static electricity to be generated, and those below approximately 20 percent result in a dry, uncomfortable feeling. These lower levels of RH allow the air to absorb moisture easily from the skin, creating evaporational cooling of the body. However, these lower RH levels usually occur in the winter when greater body cooling is not required. High relative humidity levels, on the other hand, reduce the body's ability to be cooled by evaporation. For best comfort conditions, the RH should be below 60 percent and above 30 percent. The outside limits of RH are 20 to 80 percent.

Air Velocity

Air velocity affects the rate of convective and evaporative heat transfer. In both cases, a higher velocity increases the rate of heat transfer. This is usually a benefit in the summer, but is a liability in the winter. The maximum air velocity in the occupied zone (6 in to 6 ft vertically and within 2 ft of walls) varies from 50 ft/min for cooling to 30 ft/min for heating. ASHRAE Standard 55-92 states that no air motion is required for human comfort as long as the temperature is in the recommended zone. An air velocity change of 15 ft/min produces approximately the same effect as a 1°F temperature change. Therefore, people generally feel cooler when the air velocity increases. The phenomenon known as wind chill results from the fact that heat transfer and evaporation both increase as the wind speed increases. This affects plants and animals only and has no effect on buildings, vehicles, or other inanimate objects.

Radiative Heat Loss

Radiative heat loss is proportional to the difference between the skin and clothing temperature and the temperatures of the surrounding surfaces. The greater the temperature difference, the greater the rate of heat transfer. The weighted average of the surrounding surface temperatures is called the *mean radiant temperature* (MRT). This weighted average is calculated by multiplying the temperature of a surface by the angle of exposure of the body to that surface. The total of all of the angles of exposure must add up to 360°. Then, the divide the sum of these products by 360° to get the mean radiant temperature. An example of this calculation is shown in Fig. 5.4.1-2. This mechanism accounts for the fact that people may be sit-

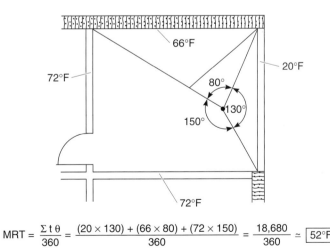

$$MRT = \frac{\Sigma t\,\theta}{360} = \frac{(20 \times 130) + (66 \times 80) + (72 \times 150)}{360} = \frac{18,680}{360} \approx \boxed{52°F}$$

FIGURE 5.4.1-2 Determining mean radiant temperature. (*From* Concepts in Thermal Comfort, *M. David Egan copyright 1975, Prentice Hall.*)

ting in 70°F surroundings and may still feel cold. They feel cold because they are losing body heat to a cold window or wall that is nearby.

Clothing

Clothing reduces the body's heat loss and is classified according to its insulation value. The unit of clothing classification is the clo. The scale is based on a clo value of 0.0 for a naked person, and someone wearing a suit has a clo value of 1.0. The moisture permeability of clothing is also a factor. Table 5.4.1-1 contains a list of clo units. These are additive and can be used to calculate the total insulating value of the clothing worn by a person. Note that the total clo value is the sum of the individual pieces of clothing multiplied by 0.82. By insulating the body and therefore reducing the heat flow from it, clothing effectively alters the air temperature at which a person is comfortable. A person wearing minimal clothing (a bathing suit, for example) would be comfortable at an air temperature of 81°F. A person wearing light slacks and a short-sleeved shirt would require an ambient temperature of 76°F for comfort because of the increased insulation of the additional clothes.

THE COMFORT ZONE

ASHRAE Standard 55-92 sets standards for human comfort. The foreword states that "The Standard specifies conditions in which 80% or more of the occupants will find the environ-

TABLE 5.4.1-1   Garment Insulation Values

Garment Description*	$I_{clu,i}$, clo[†]	Garment Description*	$I_{clu,i}$, clo[†]	Garment Description*	$I_{clu,i}$, clo[†]
Underwear		Long-sleeve, flannel shirt	0.34	Long-sleeve (thin)	0.25
Men's briefs	0.04	Short-sleeve, knit sport shirt	0.17	Long-sleeve (thick)	0.36
Panties	0.03	Long-sleeve, sweat shirt	0.34	**Dresses and skirts[‡]**	
Bra	0.01	**Trousers and coveralls**		Skirt (thin)	0.14
T-shirt	0.08	Short shorts	0.06	Skirt (thick)	0.23
Full slip	0.16	Walking shorts	0.08	Long-sleeve shirtdress (thin)	0.33
Half slip	0.14	Straight trousers (thin)	0.15	Long-sleeve shirtdress (thick)	0.47
Long underwear top	0.20	Straight trousers (thick)	0.24	Short-sleeve shirtdress (thin)	0.29
Long underwear bottoms	0.15	Sweatpants	0.28	Sleeveless, scoop neck (thin)	0.23
Footwear		Overalls	0.30	Sleeveless, scoop neck (thick),	
Ankle-length athletic socks	0.02	Coveralls	0.49	i.e., jumper	0.27
Calf-length socks	0.03	**Suit jackets and vests (lined)**		**Sleepwear and robes**	
Knee socks (thick)	0.06	Single-breasted (thin)	0.36	Sleeveless, short gown (thin)	0.18
Panty hose	0.02	Single-breasted (thick)	0.44	Sleeveless, long gown (thin)	0.20
Sandals/thongs	0.02	Double-breasted (thin)	0.42	Short-sleeve hospital gown	0.31
Slippers (quilted, pile-lined)	0.03	Double-breasted (thick)	0.48	Long-sleeve, long gown (thick)	0.46
Boots	0.10	Sleeveless vest (thin)	0.10	Long-sleeve pajamas (thick)	0.57
Shirts and blouses		Sleeveless vest (thick)	0.17	Short-sleeve pajamas (thin)	0.42
Sleeveless, scoop-neck blouse	0.12	**Sweaters**		Long-sleeve, long wrap robe (thick)	0.69
Short-sleeve, dress shirt	0.19	Sleeveless vest (thin)	0.13	Long-sleeve, short wrap robe (thick)	0.48
Long-sleeve, dress shirt	0.25	Sleeveless vest (thick)	0.22	Short-sleeve, short robe (thin)	0.34

* "Thin" garments are made of light, thin fabrics worn in summer; "thick" garments are made of heavy, thick fabrics worn in winter.
[†] 1 clo = 0.880°F · ft^2 · h/Btu.
[‡] Knee-length.
Source: Reprinted with permission of the American Society of Heating, Refrigerating, and Air-Conditioning Engineers, Inc., from the 1997 *ASHRAE Handbook.*

ment thermally acceptable" (ASHRAE, 1997). The local thermal environment will not be uniform. Differences in mean radiant temperature, vertical temperature differences, drafts, or cold floors may cause local differences in the thermal environment that result in additional dissatisfaction.

Acceptable ranges for dry-bulb temperatures and for relative humidity are shown in Fig. 5.4.1-3. The ranges vary slightly depending on whether the occupant is in a heated (winter) or a cooled (summer) environment. They are also intended for a person clothed in typical summer or winter clothes who is performing light, mainly sedentary, work. Assumed in the ranges is air motion that is within the limits mentioned previously, no greater than 30 ft/min in winter and 50 ft/min in summer. A higher summer limit can be tolerated with increased air motion up to 160 ft/min, when paper and other light objects may be blown about. As noted earlier, the Standard states that there is no required minimum air velocity if the temperatures are within the stated ranges. Figure 5.4.1-4 shows that the comfort zone shifts up and to the right in the diagram with increased air velocity.

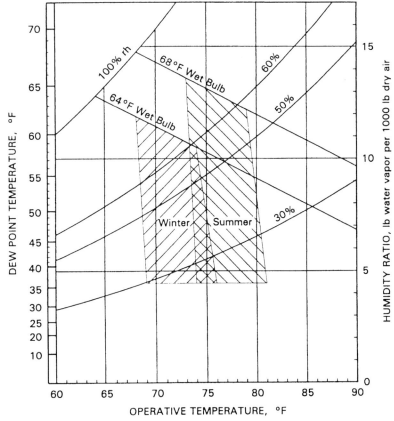

FIGURE 5.4.1-3 Standard effective temperature and ASHRAE comfort zones. (*Reprinted with permission of the American Society of Heating, Refrigerating, and Air-Conditioning Engineers, Inc. from the 1997 ASHRAE Handbook*).

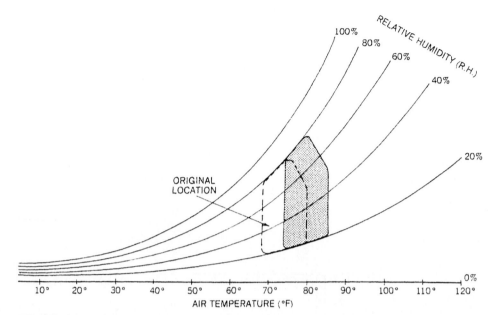

FIGURE 5.4.1-4 With high air velocity, the comfort zone shifts to the right. (*From N. Lechner,* Heating, Cooling, Lighting Design Methods for Architects, *Copyright © 1991, reprinted by permission of John Wiley & Sons, Inc.*)

A high mean radiant temperature increases heat gain and allows a person to be comfortable in colder ambient air temperatures. Figure 5.4.1-5 shows that the comfort zone shifts down and to the right with increased mean radiant temperature. The comfort zone would be shifted similarly if the space occupants were involved in increased physical activity.

Additional factors affect human comfort. Space temperatures cycle due to control dead bands or the limitations of the HVAC systems. To maintain comfort levels, the rate of cyclical temperature change should not exceed 4°F per hour. Noncyclic temperature changes may also occur. These changes should not exceed 1°F per hour and should not remain outside of the temperature range for more than an hour. Conductivity represents a small amount of the total heat loss. Generally, the feet are the largest part of the body in constant contact with a warm or cold surface and therefore able to lose or gain heat. To minimize discomfort to the feet, floor surface temperatures should range between 65 and 84°F for people wearing normal footwear. Drafts may also cause local discomfort. The probability that a draft will be a problem is a function of the speed, turbulence, and temperature of the air. The head and ankles are the body areas most susceptible to drafts.

THE PROBABILITY OF THERMAL COMFORT

There is a method of predicting thermal comfort called the *predicted mean vote* (PMV) *index.* This index predicts the mean value of the satisfaction ratings of a group of people exposed to a given thermal environment. The scale has 7 points that range from –3 (cold) to +3 (hot). A neutral or satisfied rating is 0. Remember that even at 0 as many as 20 percent of the people may be dissatisfied. ISO 7730 suggests that for thermal comfort the PMV should be within ±0.5. The equation by which the PMV is calculated takes into account the metabolic rate at which body heat is generated and the rate of heat loss. Although useful for academic pur-

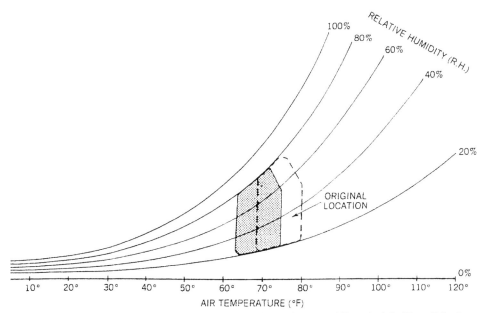

FIGURE 5.4.1-5   With a high mean radiant temperature, the comfort zone shifts to the left. (*From N. Lechner, Heating, Cooling, Lighting Design Methods for Architects, Copyright © 1991, reprinted by permission of John Wiley & Sons, Inc.*)

poses, the calculation is too theoretical and complex for use by engineers, architects, and building operators. For this purpose, the American Society for Testing and Materials (ASTM) has developed Standard E1334-95, *Standard Practice for Rating the Serviceability of a Building or Building-Related Facility.* This document provides two scales, one called the Facility Rating Scale and the other the Occupant Rating Scale. The former allows one to rate a building on its ability to meet ASHRAE human comfort standards, and the latter deals with the importance of human comfort conditions in a particular situation.

DESIGN ISSUES

Determination of design temperatures should take into account the activity level and clothing customs of the building occupants. Those design temperatures should recognize that the ASHRAE human comfort envelope has different parameters for winter and summer. These parameters suggest different temperature set points for each season. This is much more easily implemented with energy management systems than with traditional controls. Those controls can be programmed to vary temperatures and humidity levels with changes in outdoor conditions.

Air Quality

Although thermal comfort is the largest source of complaint about the indoor environment, air quality is often the second largest complaint. One estimate states that 20 to 30 percent of all nonresidential buildings in the United States is defined as a problem building from the standpoint of indoor air quality. Typical complaints are headaches, stuffiness, eye and throat irritations, and tiredness. A facility manager could expend a significant amount of money and time addressing these problems.

Contaminants can be brought in from outside the building through the ventilation system or through infiltration, or they can be generated inside the building in the occupied space, or in the HVAC system itself. To prevent contaminants from entering the building from outside most effectively, those issues must be addressed during the design of the building. The location of fresh-air intakes, type and amount of air filtration, and other basic HVAC system parameters are difficult for the facility manager to change or improve after the building is built. The generation of contaminants inside the building is easier for the facility manager to control. Occupants may create some of the problems. These might include the use of chemicals, such as cleaning agents; or office equipment, such as blueprinting machines that generate contaminants. In a large facility, such activities can easily escape the notice of the management staff. Odors are part of the air quality measure. Air of good quality has an undetectable concentration of odors. Though manufacturers are becoming more aware of the issue, the use of solvents, glues, and other chemicals in new carpeting or furniture is often a significant source of contaminants. A significant source of contaminants is the HVAC system itself. Dirty filters and ducts can generate large quantities of dust, and plugged drain pans, dirty cooling coils, and malfunctioning humidifiers can be significant sources of bacterial growth. Even the water treatment of cooling towers, if not properly monitored, can generate airborne contaminants that can enter the system through the fresh-air intakes. A more detailed discussion of these issues is contained in Art. 5.4.8.

CONTRIBUTORS

David M. Elovitz, P.E., Energy Economics, Inc., Natick, Massachusetts

BIBLIOGRAPHY

American Society for Testing and Materials: *Standard Practice for Rating the Serviceability of a Building or Building-Related Facility,* Standard E1334-95, ASTM, West Conshohocken, PA, 1995.

American Society of Heating, Refrigerating, and Air-Conditioning Engineers: *ASHRAE Handbook of Fundamentals,* ASHRAE, Atlanta, GA, 1997.

ASHRAE Standard 55-92, The American Society of Heating, Refrigerating & Air Conditioning Engineers, Inc., Atlanta, GA, 1997.

International Organization for Standardization (ISO): Standard ISO 7730, ISO, Geneva, Switzerland, 1994.

Lechner, N.: *Heating, Cooling, Lighting Design Methods for Architects,* Wiley, New York, 1991.

ARTICLE 5.4.2
HVAC SYSTEMS

David M. Elovitz, P.E.
President, Energy Economics, Inc., Natick, Massachusetts

GENERAL CONCEPTS

The primary function of a heating, ventilating, and air conditioning (HVAC) system, in its simplest terms, is to add heat to a space at the same rate that heat is leaking out of that space or to remove heat from a space at the same rate heat is coming into the space, as well as to provide appropriate ventilation. To be sure, a well-designed HVAC system performs many other important functions in connection with the heat balance task: it recirculates air, controls humidity levels, cleans the circulating air, controls drafts off cold surfaces, and maintains an acceptable level of acoustical background noise.

Load Calculations

How do we know how fast we must add heat into or remove it from the space? We make heating and cooling load calculations.

The heat that leaks into a space and is generated within that space is called *heat gain*. The heat that leaks out of that space is called *heat loss*. Neither heat loss nor heat gain is constant over time. The simplest example is the outdoor temperature changes during each day and during the year that directly affect heat gains and losses. The sun also moves around a building during a day and changes its position in the sky in the course of the year, changing the amount of solar energy that strikes the various outside surfaces of the building. People, lights, and equipment generate heat inside the space. The amount of heat generated varies, depending on how many people are in the space, what they are doing (people give off less heat when sitting or standing quietly than when they are active), how many light fixtures are on and for how long, which of the appliances are on at the same time inside the space, and at what portion of full load they are running. All of this information helps in determining the heat loss and the heat gain of the space.

What we need to know to determine the design capacity for that space is what the peak simultaneous heat gain and the peak simultaneous heat loss will be. Once the heat loss and the heat gain are determined, the equipment can be sized.

When we size equipment that serves a number of spaces, we have another consideration. Not all of those spaces will usually need their peak capacities at the same time, so we might be able to do the job with equipment capacity that is less than the sum of the individual space peak capacities. The calculation of the simultaneous peak load for a number of different spaces is called a *block load*. As we will see later on, the block load that the equipment will see may, or may not, be less than the sum of the individual peak loads, depending on the system type, even though the simultaneous peak capacity required by the spaces may be less.

The details and specific techniques of calculating heating and cooling loads are beyond the scope of this handbook. The methods and techniques are continually being refined and made more accurate as a result of ongoing research. The most up-to-date information is delineated in great detail in ASHRAE's *Handbook of Fundamentals,* published every 4 years in updated form with the newest technology.[1] ASHRAE also periodically publishes a *Load Calculation*

Manual that contains the most up-to-date calculation methodology.[2] A simpler load calculation methodology, suitable for less critical residential and light commercial applications, is published by the Air Conditioning Contractors Association as the *Load Calculation Manual N*.[3] Also on the market are several proprietary computer load calculation programs that automate the calculation process.

Ventilation

In addition to heat that is generated in the space and heat that is transferred into or out of the space through the space enclosure, the heating and cooling loads must consider the amount of outside air that enters the space directly by infiltration and is introduced into the space through the HVAC system as mechanical ventilation. The amount of outside air ventilation required is established by the larger of two considerations:

1. The amount of outside air required to replace air removed from the space by exhaust systems. Many designers also include an additional allowance over and above the amount exhausted to pressurize the occupied space relative to the outdoors.

2. The amount of outside air required to dilute odors and other contaminants generated within the space in order to maintain an acceptable indoor environment.

Ventilation rates are discussed in Art. 5.4.8.

Heating Loads

Heating loads consist of transmission through opaque walls and roof, transmission through fenestration, transmission to the ground (if applicable), and the heat required to bring outside air up to room temperature. The major factors that determine heat loss are the outdoor and indoor design temperatures. The indoor design temperature is established by the designer in consultation with the owner and is typically of the order of 70°F. The desired space conditions are described in Art. 5.4.1.

The outdoor design temperature is normally based on historical weather data for the location. Extensive historical weather data are available from the National Weather Service of the National Oceanic and Atmospheric Administration (NOAA). The ASHRAE *Handbook* and *Load Calculation Manual* include tables for thousands of locations of the outdoor temperatures that have been exceeded historically 99 percent of the hours in the year (only 88 hours in the typical year are colder than that temperature) and 99.6 percent of the year (only 35 hours in a typical year are colder than that temperature).[2,3] Sometimes, local practice uses a design temperature even lower than the 99.6 percent value.

The difference between indoor design temperature and outdoor design temperature affects all of the calculated heat losses except heat loss to the ground (which is usually based on a seasonal average temperature), so selecting the design temperatures is a critical decision. Once that design temperature difference (TD) has been established, the heat loss Q is calculated by two general formulas:

$$Q = U \times A \times \text{TD}$$

and

$$Q = 1.08 \times \text{cfm} \times \text{TD}$$

The first formula is for heat transmission through walls, roof, and floor, where U is the overall coefficient of heat transfer (which is calculated from the thermal resistance properties of the various layers through which the heat is transmitted) and A is the area in square feet through which the heat is transmitted. The thermal properties are calculated for all of the different kinds of surfaces that enclose the space, the heat flow Q in Btu per hour is calculated for each surface, and then all the heat flows are added up for the transmission heat loss. Insulated walls or roofs have U factors of the order of 0.03 to 0.12. Regular insulating glass in a thermal break metal sash might have a U factor of the order of 0.60, so it is apparent that glass area generally is the predominant determiner of transmission heat loss. Heat loss to the ground is a smaller factor, and it is generally determined by reference to special tables and charts prepared for that purpose.

The second formula is for the heat loss due to outside air that may be either by mechanical ventilation or infiltration. The factor 1.08 includes the density and specific heat of air and corrects for minutes and hours; cfm is the quantity of air (cubic feet per minute); and TD is the temperature difference in degrees Fahrenheit. It is not unusual for the outside air heat loss to be as large as, or even larger than, the transmission heat loss.

Heating load calculations are generally done without taking credit for any internal heat gains because the peak heating requirement may occur when lights and equipment are off. Heat loss calculations should consider three different sets of conditions: occupied hours at normal temperature with mechanical outside air ventilation; unoccupied hours at night setback temperature with no mechanical ventilation, but with infiltration; and morning warm-up, bringing the space back up to daytime temperature from the night setback temperature. Most often, the ventilation load will remain off during morning warm-up, but there will still be infiltration. The maximum capacity requirement could occur under any one of those three conditions, depending on the parameters for any specific project, which is why all three must be evaluated to determine the required capacity. When sizing heating systems, it is usual to include both a pickup factor (to account for warming up an idle system at the start of a cycle) and a safety factor to allow for less than ideal envelope construction. Article 5.4.3 discusses boilers that supply heat energy in the facility.

Cooling Loads

The cooling load calculation includes the same two components, heat transmission and outside air heat, but it is complicated by additional factors related to the effect of solar radiation from the sun and to the moisture content of the air. Although the same two general formulas apply in principle, they have to be adapted to reflect these additional considerations.

The basic variables for peak load calculations include weather conditions, building envelope, internal heat gain, ventilation, and to a lesser extent, infiltration. Less obvious but nonetheless important are the diversity among the various elements and the effects of thermal mass. The diversity of loads is a measure of the simultaneous occurrence of varying peak loads. In other words, it is a measure of the likelihood that the occupancy, lighting, and plug loads will each peak at the same time that the envelope load peaks.

In typical summer weather in most areas, cooling the outside ventilation air also entails removing moisture from the air. Lowering the temperature of the air means removing *sensible* heat from the air. As the air temperature is reduced to the point where it can no longer absorb all of the moisture it contains (called *reaching saturation*), water vapor begins to condense out of the air as liquid water and gives up *latent* heat as it condenses. That latent heat must be removed along with the sensible heat, so the change in the moisture content of the air must be considered in the cooling load calculations along with the change in temperature.

Solar radiation affects the cooling load calculation in two ways. The first is through direct solar radiation that enters the space through transparent fenestration, strikes opaque surfaces within the space, and is absorbed by the opaque surface, raising its temperature. The warmed opaque surface transfers heat to the air in the room, and it also radiates heat to other opaque surfaces within the room. Some of the surfaces warmed by direct solar radiation are lightweight and give up heat to the room air very quickly, and other surfaces are heavy and may hold the heat for several hours before it is released into the room air. Some of the solar radiation is also absorbed by the window glass itself on its way through the glass, raising the glass temperature so that the glass transfers heat into the room air.

The second way by which solar radiation affects the cooling load calculation is that it reacts with opaque surfaces. When solar radiation strikes the outer surface of a wall or roof, it raises the temperature of that surface. Some energy is absorbed by the outdoor air, and some of it is reradiated from the outdoor surface to other opaque surfaces outdoors, but much of it is transmitted through the wall or roof to the inside surface. The inside surface transfers some of that energy directly to the room air and radiates some to other opaque surfaces within the room. A significant time lag exists between the time the solar energy strikes the wall and the time it reaches the inside surface of the wall, and a significant additional time lag also exists between the time the energy reaches the inside surface and the time it has all been transferred into the room air. Many complex calculations are involved in accurately accounting for the time lags from fenestration, opaque walls, and roofs; as well as for the changes in the incident solar radiation due to the different positions of the sun relative to the receiving surfaces at different times and due to the different properties of the many varying construction materials. Continuing ASHRAE research is developing improved calculation and approximation methods for cooling load calculations to develop the optimal balance between accurate results and manageable procedures. The most up-to-date specific information is made available by ASHRAE in its publications each year.

Finally, cooling load calculations have to consider internal heat gains from occupants, lights, and equipment or appliances, none of which are typically factors in heat loss calculations. In a modern commercial, institutional, or industrial building, the amount of heat generated within the space by internal heat sources is often as large as, or even larger than, the heat gain through the building shell by transmission and solar radiation. Thus, a careful and accurate determination of the heat generated within the space can have a major effect on the selected system capacity.

All cooling load calculation methodologies necessarily include approximations and simplifications to make them practical. It may seem that anyone would be able to enter the wall, glass, and window areas and the lighting and appliance information into one of the computer load programs without any extensive background in HVAC design. However, except for the simplest, noncritical applications, the programs give the best results only when used by someone who has the thorough training and years of experience that provide the basis for understanding the program limitations and for making sound judgments.

The cooling load due to outside air ventilation has two components: (1) sensible heat to change the dry-bulb temperature of the air from the outdoor temperature to the indoor temperature and (2) latent heat to change the moisture content of the outside air from the outdoor specific humidity level to the indoor specific humidity level. No credit for the change in latent heat is taken, however, in dry climates where the moisture content of the outdoor air is lower than that of the indoor air. There is a more complete discussion of the heating and cooling loads associated with outside air in the *ASHRAE Handbook of Fundamentals.*[1]

HVAC SYSTEM CONTROL CONCEPTS

As discussed earlier, maintaining a relatively constant temperature within a space requires that the system removes heat from, or adds heat to, the space at exactly the same rate at which heat is being gained or lost by the space. To simplify the discussion, we are going to talk only in terms of cooling loads; the same principles apply, in reverse, for heating loads. You have determined the peak rate at which heat will leak into your space, and you know what you must install to have that much capacity. But heat does not always leak into your space at the peak rate, does it? No, on cloudy days, very little solar radiation reaches the building, and on only a handful of days in the year is the outdoor temperature as high as the design outdoor temperature. Sometimes you need less cooling than the peak capacity, so you need a system that can vary its capacity. There are basically four different approaches.

Cycling or On-Off Control

Sometimes heat leaks into the space more rapidly than at other times, so you need a way of controlling the cooling system's capacity. The simplest way is simply to turn the cooling on and off. That is the way package unitary air conditioners are controlled. We use a thermostat to measure room temperature. When the room temperature rises above the thermostat set point, the cooling turns on. When the room temperature drops below the thermostat set point, the cooling turns off. This simple type of control is inexpensive, reliable, and widely used in household refrigerators, window air conditioners and package terminal air conditioners, and for large single-zone package air conditioners like rooftop units. An inherent characteristic of this type of control is a difference in temperature between the time when the cooling turns on (*cut-in temperature*) and the time when the cooling turns off (*cut-out temperature*). That characteristic means that the temperature in the space must rise and fall over time, and the difference in space temperature is often a few degrees. That may be quite noticeable to the occupants, particularly when the load is small relative to the cooling capacity. Room temperature drops rapidly, then drifts up slowly, then drops rapidly again, and so on. The larger the difference between the cut-in and cut-out temperatures, the more noticeable it is. On the other hand, making the difference very small can result in short-cycling the unit, which can shorten its life. So there is a practical limit as to how closely an on-off thermostat can control space temperature.

Modulating Control

How can we get more even control of the temperature as the rate at which heat leaks into the space changes? Instead of turning the cooling on and off as the temperature became too high or too low, we could somehow vary the rate at which the system removed heat from the space.

This can be accomplished by linking the thermostat that senses room temperature to the control mechanism. Then, we can automatically make the system cool at the exact same rate that heat is leaking in, and the temperature would stay constant.

As the space temperature rises, we increase cooling capacity; as space temperature falls, we decrease cooling capacity. When capacity and load are exactly equal, the temperature stays constant. Modulation of water flow via a control valve into a water-cooling or -heating coil or modulation of the quantity of supply air to a room, as in a variable–air volume (VAV) system, are examples of such modulation.

Reheat Control

Another method for controlling our system at a constant temperature is to supply enough cooling to the space to meet the peak load but then add heat to the space to make up for difference in the rate at which heat leaks in. No matter what the rate of leakage, the total of the heat that leaks in and the added extra heat stays constant, so the temperature stays constant. If your first reaction is that this is a very wasteful way to control the temperature, you are correct, but that is the way the once-popular terminal reheat systems controlled space temperature. The amount of cooling capacity delivered to each space is constant, always enough to absorb the peak cooling load in the space. When the actual cooling load is less than the peak, a reheat coil adds enough heat to the supply air before it enters the space, so that the heat gain in the space is just enough to absorb the amount of cooling capacity that is left in the supply air.

Terminal reheat systems are generally not designed and installed today in commercial office buildings, but there are some special applications, where dehumidification is required even when the space does not need cooling, or where reheat energy from a heat recovery application is free. But it is not that uncommon to find VAV/reheat systems today, where the cooling capacity is reduced to a selected minimum level by modulating control of a VAV box, and cooling capacity below that level is controlled by adding heat to that minimum airflow by a reheat coil.

Hybrid Control

Actually, multizone systems and most dual-duct systems are hybrid forms of reheat systems. These types of systems control the cooling capacity of the constant volume of air delivered to each zone by mixing warm and cool airstreams, so that the supply air temperature to each zone just balances the heat that is generated in that zone. In multizone systems, the mixing occurs inside a central unit that has a dedicated duct for each zone. With dual-duct systems, the mixing takes place in a mixing box near the zone, and there are two ducts from the unit to all zones, one that contains warm air and one cool air. In both systems, a constant volume of air goes through the unit, some air goes over a cooling coil (the *cold deck*), and the rest goes over a heating coil (the *hot deck*). The amount of air that flows over each coil depends on how much air is being drawn from each deck by the zones as they mix air to get the required zone supply air temperatures. Mixing systems are hybrids between modulating and reheat control. Although less cooling air is used when less cooling is required, warm air is added to the cooling air to keep the total supply air volume constant, while matching the capacity to the load.

There is a special type of multizone unit called the *three-deck,* or *Texas,* multizone that avoids using reheat energy. Return air that has not been over either the hot deck or the cold deck is mixed with the cool supply air from the cold deck to reduce the cooling capacity to a zone, except when actual heating is required.

Bypass-VAV Control

There is a fourth control concept. The system is equipped with a bypass path that dumps part of the cooling flow right back into the unit. The cooling unit always cools at a constant rate, but the amount of cooling actually delivered into the space is varied, depending on the temperature in the space. This is called the *bypass-VAV system* and is generally used only with small package units to provide individual zone control when the system is so small that a VAV package unit is not available. The zone VAV box has a damper that is modulated to maintain space temperature just like a regular VAV box, but rather than choking off the airflow, the box diverts part of the airflow into the return plenum or into a return duct where it returns to the unit. The unit and the supply duct handle a constant volume of air, but only part of that air is delivered into the occupied spaces—whatever amount is needed to meet the cooling load—and the rest is intended to return directly to the cooling unit without going through the space. This control concept allows the relatively inexpensive provision of multiple zones on small systems, and there is some cooling energy saving with this system. The bypassed (*dump*) air reduces the mixed-air temperature entering the cooling coil, but caution is required in design and installation so that bypassed air does not find its way into the occupied space via return grilles.

HVAC SYSTEM TYPES

As a practical matter, there is no such thing as an HVAC system type. Every HVAC system is made up of a combination of system building blocks, blocks that generate heating or cooling from an energy source or sources, blocks that transport heating and cooling from that generator to where it will be used, and blocks that deliver heating and cooling into an airstream and into the space where it is needed. The first five chapters of the *ASHRAE Systems and Equipment Volume* are a good source for detailed descriptions of practical system configurations that implement these system concepts.[4] The first chapter describes the process of selecting an HVAC system type for a project. Chapter 4 describes cooling and heating production equipment, which is discussed elsewhere in this section. Chapter 2 describes building air distribution systems; Chap. 3 describes in-room terminal systems, and Chap. 5 describes unitary refrigerant-based systems, all three of which we review briefly here. The reader will find more detailed explanation of each system type in the *ASHRAE Systems and Equipment Volume.*[4]

Refrigeration Cycle

Heat is ultimately removed from the airstream by a refrigeration process or cycle. In very simple terms, heat is transferred to a liquid refrigerant that is at a low pressure and changes to a vapor as it absorbs the heat. Then, the refrigerant vapor is raised to a higher pressure at which the heat can be rejected at a higher temperature, and the heat is rejected to the atmosphere either directly (by a refrigerant to air heat exchanger coil) or indirectly through a refrigerant to a water-cooled heat exchanger. The water rejects its heat to the atmosphere most commonly through a cooling tower. The cooled refrigerant vapor condenses back into a liquid and is returned to the low-pressure heat exchanger (where it absorbs more heat) through a metering device. Mechanical cooling systems use a compressor to move the refrigerant vapor from the low-pressure evaporator to a higher-pressure condenser. Absorption cooling systems use very small vapor pressure differences between an absorbent at different solution concentrations to provide the pressure difference, using heat to create the different concentrations. See the *ASHRAE Refrigeration Volume* for a complete description of the various refrigeration cycles and equipment types.[5]

The cooling effect may be achieved indirectly by a circulated heat transfer medium like chilled water or directly by heat transferred from the airstream directly to an evaporating refrigerant. The latter is called a *direct-expansion system.* The various types of refrigeration machinery and processes for chilled-water systems are described in Art. 5.4.4, "Chilled-Water Plants." Further information on chilled-water machines and on direct-expansion refrigeration is available in the *ASHRAE Refrigeration Volume.*[5]

Building Air Distribution—Air-Handling Units

Air-handling units can have a great variety of configurations, depending on the application. The simplest air-handling unit consists of a filter bank, heating/cooling coil, and fan. A more elaborate air handler may include a mixing box (with outside air and return air dampers,) prefilters, main filters, preheat coil (perhaps with face and bypass dampers), cooling coil, heating coil, fan, humidifier, and final filters or any combination in between. More complete information on air-handling units will be found in Chap. 2 and the air-handling equipment chapters of the *ASHRAE Systems and Equipment Volume.*[4] The air-handling unit configuration depends in part on the system type as described later, and in part on the special requirements of the application. Air-handling units may be large central units that serve large areas, perhaps many floors in an office building, or they may be small local units that are located in or adjacent to a single area served by that unit. The air handler may be either field-erected or a factory-assembled package unit. The factory-assembled package unit offers a significant first-cost advantage, but the upper size limit for a factory-assembled package unit is usually of the order of 35,000 to 40,000 cfm (Fig. 5.4.2-1).

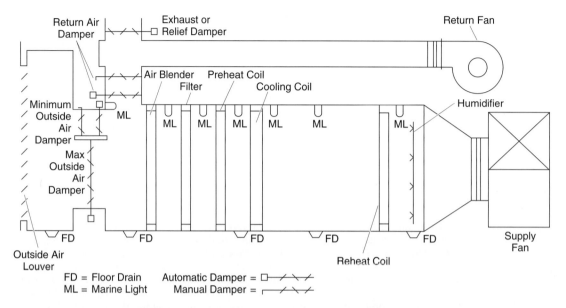

FIGURE 5.4.2-1   Equipment arrangement for central system draw-through unit. (*Courtesy American Society of Heating, Refrigeration, and Air-Conditioning Systems from 1996* ASHRAE Handbook.)

Building Air Distribution—Single-Duct System.   Constant-volume systems vary the capacity delivered by varying the supply air temperature of a constant volume of air. Variable-volume systems vary the capacity delivered by varying the amount of air supplied at a constant temperature.

Single-zone, constant-volume systems are the simplest system concept. Air temperature is varied at the air handler in response to space temperature.

Multiple-zone reheat systems are also called *terminal reheat systems* because the reheat is added at the end of the duct near the zone being served. One air-handling unit serves multiple spaces, and the unit produces enough cooling to meet the sum of the peak loads in every space. A reheat coil is provided in the duct that supplies air to each space. When a space needs less than peak cooling, the reheat coil for that space adds heat to the supply air stream to raise the supply air temperature to that space (Fig. 5.4.2-2).

Some would consider that a bypass system is a type of VAV system, but the system actually handles a constant volume of air at all times. The volume of air delivered to a room varies to meet the actual cooling load, but the volume of air through the air-handling unit is constant. When the room needs less cooling, some of the supply air is bypassed directly to the return, reducing the airflow through the room. The bypass can be ducted directly to the return duct or dumped into a return plenum. If the bypass air is dumped into a return plenum, extra care is needed to ensure that it does not find its way into occupied space through return grilles (Fig. 5.4.2-3).

A true VAV system reduces both the amount of air delivered into the room and the amount of air flowing through the air handler as the load decreases. Only enough cooling is needed to meet the maximum simultaneous (*block peak*) load imposed on the air handler, and there are substantial savings in fan energy due to the reduction in the amount of air circulated. The air-handling unit has to include special provision to control both the amount of cooling taken from the airstream and the static pressure in the supply duct as the load varies.

The variable-diffuser system is a special type of VAV system where the supply outlet itself throttles the supply airflow to match the cooling load and also varies its geometry to keep a

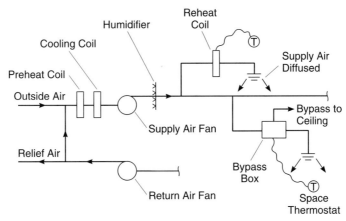

FIGURE 5.4.2-2   Constant volume with reheat and bypass terminal devices. (*Courtesy American Society of Heating, Refrigeration, and Air-Conditioning Systems from 1996* ASHRAE *Handbook.*)

more or less constant discharge velocity as the airflow varies. The air-handling unit requirements are the same as those of a conventional VAV system, but the variable diffuser system inherently provides more, smaller control zones.

Building Air Distribution—Dual-Duct Systems.   There are three general types of dual-duct systems. In all three types, two parallel main ducts generally deliver air from a central air-handling unit to multiple spaces. One duct carries cold air, the other warm, and there is a mixing box for each space that takes the appropriate proportions of air from each duct to deliver the appropriate supply air to each space.

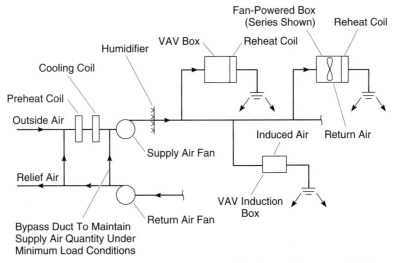

FIGURE 5.4.2-3   Variable air volume with reheat, induction, and fan-powered devices. (*Courtesy American Society of Heating, Refrigeration, and Air-Conditioning Systems from 1996* ASHRAE *Handbook.*)

Constant-volume, single-fan duct systems with no reheat actually have only one duct and have a face and bypass damper at the coil to deliver a mixture of cooled and recirculated air to the space through that duct. They might be considered a special case of a single-duct, constant-volume system but are classified as a dual-duct system because they control capacity by mixing two airstreams.

Constant-volume, single-fan duct systems with reheat use blow-through air-handling units where the air discharged from the fan can flow through either a cooling coil and into the cold duct or through a heating coil and into the hot duct. The respective coils and the sections of the air-handling unit downstream of the coils are often referred to as the *cold deck* and the *hot deck*. Each room is supplied with a constant volume of supply air from a dual-duct mixing box connected to the hot and cold ducts. Appropriate proportions of air are taken from each duct and mixed in response to a space temperature signal to provide the supply air temperature to match the load in the space (Fig. 5.4.2-4).

Variable air volume dual-duct systems can be single-fan or dual-fan (separate fans for the hot and cold ducts.) Some means of fan capacity control is required as the air volume from each duct varies. Interior zone terminal units may be connected only to the cold duct, as in a single-duct VAV system, and perimeter zone units can function like a single-zone VAV cooling terminal and a single-zone VAV heating terminal (Fig. 5.4.2-5).

A dual-conduit system has one heating/cooling supply duct for the perimeter zones that is used to offset the solar and transmission gains or losses. That duct may carry cold air or warm air, depending on outdoor conditions. A separate cooling-only duct provides cold air to cooling-only terminals in interior zones and also to a second (cooling-only) terminal in each perimeter zone. The perimeter duct terminal handles the gain or loss through the building shell, and the second terminal handles the internal heat gain in the perimeter zone (Fig. 5.4.2-6).

Building Air Distribution—Multizone Systems. Multizone systems are constant-volume systems that control capacity the same way as dual-duct systems, by mixing warm and cold air, but multizone systems do the mixing inside the air-handling unit, not near the space in a terminal unit. Instead of a hot duct and a cold duct going to all zones, the multizone unit has a dedicated duct from the air-handling unit to each zone. The multizone unit has a hot deck and a cold deck like a dual-duct unit but has mixing dampers inside the unit to mix air from each deck to provide the required supply air temperature into each zone duct that leaves the unit (Fig. 5.4.2-7).

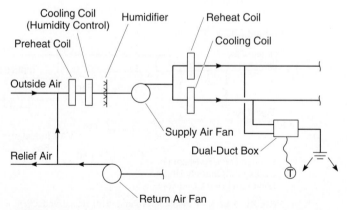

FIGURE 5.4.2-4 Dual duct-single fan. (*Courtesy American Society of Heating, Refrigeration, and Air-Conditioning Systems from 1996* ASHRAE Handbook.)

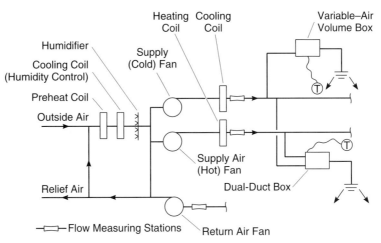

FIGURE 5.4.2-5   Variable air volume, dual duct, dual fan. (*Courtesy American Society of Heating, Refrigeration, and Air-Conditioning Systems from 1996* ASHRAE Handbook.)

The three-deck multizone system saves reheat energy by having three decks: (1) hot, (2) recirculated, and (3) cold. Zone dampers mix cold deck air and recirculated air to reduce cooling capacity. When the space needs heat, zone dampers mix recirculated air and hot deck air to supply the required heating supply air temperature.

The Texas multizone is a special variation on the three-deck multizone concept. There is no heating coil in the unit, which has only two decks: cooling and recirculated air. Heating coils are placed in each perimeter zone duct. Cooling capacity is varied by mixing cold-deck air and recirculated air, and heat is provided, when required, to the perimeter zone reheat

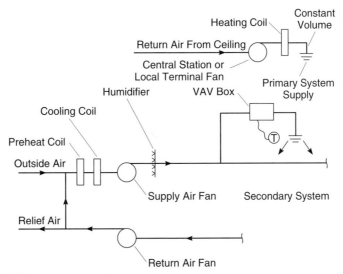

FIGURE 5.4.2-6   Dual-conduit system. (*Courtesy American Society of Heating, Refrigeration, and Air-Conditioning Systems from 1996* ASHRAE Handbook.)

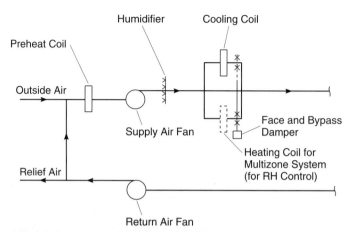

FIGURE 5.4.2-7   Single-fan no reheat system. (*Courtesy American Society of Heating, Refrigeration, and Air-Conditioning Systems from 1996* ASHRAE Handbook.)

coils. The arrangement saves quite a bit of reheat energy, especially in humid climates where low supply air temperatures are selected for dehumidification capacity.

Building Air Distribution—Special Systems.   There are a number of special systems that do not fit into the general categories mentioned. The primary/secondary system uses two interconnected air distribution systems to serve spaces like clean rooms where circulated air volumes must be very high and the space-sensible heat gain would require only a very small supply air temperature difference. A smaller amount of air is conditioned to a low enough temperature to provide necessary dehumidification and sensible cooling in a smaller air-handling unit, and then that air is mixed with a much larger recirculating air stream (Fig. 5.4.2-8).

The up-air system is sometimes called the displacement ventilation system. More common in Europe, the system supplies air from the floor rather than from overhead and has special terminal devices in the floor or built into the furniture.

Wetted-duct/supersaturated system air-handling units spray enough water into the airstream to create a controlled carryover. This concept is used where high humidity is desirable for process reasons, as in the textile or tobacco industries. The compressed air and water spray system is a variation on the supersaturated system except that the water is atomized in nozzles using compressed air (Fig. 5.4.2-9). Low-temperature air systems (as low as 40°F) can absorb large amounts of heat with much smaller air quantities. They require special terminal units that mix the supply air with more than the usual amount of room air. Low-temperature air systems are most commonly used with ice storage systems.

Standard air-handling systems are sometimes adapted for special uses. A good example is the use of the building air-conditioning system to assist in managing smoke during a fire by the addition of special dampers and controls. Smoke management principles are discussed more completely in the Smoke Management Chapter in *ASHRAE HVAC Applications Volume*[6] and Art. 5.5.5, "Smoke Control Systems."

Building Air Distribution—Terminal Units.   A terminal unit is a device that deals with an individual zone. The most common terminal units are used to control the amount of heating or cooling delivered to a zone. Reheat terminals vary supply air temperature; VAV terminals vary supply air quantity, and sometimes supply air temperature as well; and dual-duct terminals may vary supply air temperature, supply air volume, or both. Hot-water perimeter base-

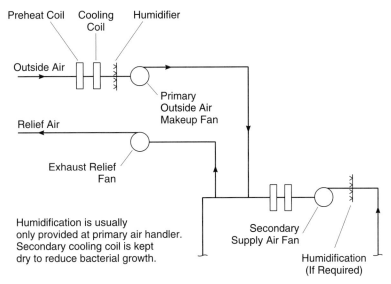

FIGURE 5.4.2-8   Primary/secondary system. (*Courtesy American Society of Heating, Refrigeration, and Air-Conditioning Systems from 1996* ASHRAE Handbook.)

board and fan-coil units are also terminal units; they are discussed later under "In-Room Terminal Systems" because they are not associated with building air distribution systems.

A wide variety of strategies are used to control terminal units, depending on type, but the subject is too extensive to cover in this overview. The section "The Zone Control Systems" of the chapter entitled "Automatic Control" in the *ASHRAE HVAC Applications Volume* provides a complete discussion.[6]

Terminal units, normally called *boxes,* may be *pressure-dependent* (the temperature signal controls the damper position and the airflow varies if duct pressure changes) or *pressure-*

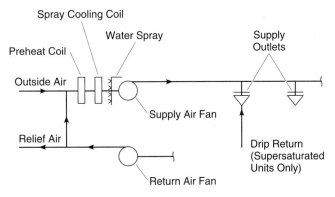

FIGURE 5.4.2-9   Supersaturated/wetted coil. (*Courtesy American Society of Heating, Refrigeration, and Air-Conditioning Systems from 1996* ASHRAE Handbook.)

independent (the temperature signal resets an airflow controller that repositions the damper, as necessary, to maintain the supply airflow when the duct pressure changes). The two types are described in the following text.

Constant-Volume Box.   A constant-volume box provides a constant volume of supply air to the space. Constant-volume reheat terminals include a reheat coil and may or may not include a volume regulator to maintain a constant supply air volume as the duct static pressure changes. A space temperature signal varies the supply air temperature by varying the amount of heat from the reheat coil.

Variable Air Volume (VAV) Box.   There are a number of different VAV terminals. All of them vary the amount of cooling air taken from the central system and delivered to the space, but some perform other related functions. The simplest VAV terminal is the throttling, or pinch-off, box without reheat. The box is essentially a damper that modulates to provide more supply air as the space temperature rises and less as the space temperature falls. The box must close fully (no supply airflow) when the space temperature falls to the lower limit of the control range, or the space will overcool.

A throttling box with reheat is created by adding a reheat coil (and controls) to a throttling VAV box. It is used where it is necessary to provide some minimum amount of airflow to the space, regardless of space temperature, or where the space will need heating at some times of the year. When the space needs heat, the reheat coil must have enough heating capacity to heat the minimum supply air from the cooling supply temperature to room temperature in addition to the capacity required to heat the space. Variable air volume boxes with reheat can use a lot of extra energy for reheat if the minimum supply air setting is not very low.

The induction VAV box is a special form of throttling box with reheat. When the induction VAV box throttles the cooling airflow in response to the falling space temperature, it also draws warm return plenum air into the supply airstream by induction. This induction is possible only if the supply air duct pressures are high, approximately 1.5 in watergauge (wg). This eliminates the need for reheat to maintain a minimum supply volume, and also it tends to keep the supply air volume into the space more or less constant, as the load changes.

Throttling VAV boxes can be combined with local fans in what are then called *fan-powered boxes*. There are two types: (1) series boxes, in which the box fan operates whenever the system is on and the total supply air to the space is constant; and (2) parallel boxes, in which the box fan operates intermittently, usually only when the VAV portion of the box has throttled back to some selected flow. Both types of boxes draw air from the return plenum. Series boxes deliver a constant volume to the space that is the total of primary air from the cooling duct, plus air drawn in from the plenum. Parallel boxes deliver a decreasing airflow to the space, all from the cooling duct, until the airflow drops to a selected minimum amount. At that point, the parallel fan comes on and adds air from the plenum to the primary air from the cooling duct. Both types of fan-powered boxes usually have reheat coils when they are used in perimeter zones. The reheat coils can provide heat to the space using the box fan, even when the main air-handling system is shut down during unoccupied hours.

Dual-duct terminals have dampers that control the amount of cold air from the cold deck and hot air from the hot deck and usually include some form of volume-regulating mechanism to maintain constant supply airflow to the space. Some VAV dual-duct terminal units function as two separate VAV boxes, one heating and one cooling, and space supply air volume varies, depending on the amount of cooling or heating required.

Terminal humidifiers are sometimes used when close control of space humidity is required or when humidity requirements differ in different zones. The main supply air is humidified at the air-handling unit somewhat below the lowest level desired, and a terminal humidifier in the supply duct to each zone brings the supply air to that zone to the level required to satisfy that zone.

Terminal filters can also be located in the supply duct just before supply air is delivered to the space where particle control is of particular concern. The terminal location allows the filters to intercept any contamination that might have been introduced within the air-handling unit or supply duct system. Such an arrangement is most commonly found in specialized applications like hospitals or clean rooms.

In-Room Terminal Systems.   A number of system types are based on a terminal unit within, or immediately adjacent to, the conditioned space, and the terminals may not rely on any central air-handling unit. Such systems, sometimes classified as *air-and-water* systems, may be two-pipe, three-pipe, or four-pipe, and the in-room terminal units may be fan-coil units or induction units. These system types are discussed more fully in Chap. 3 of the *ASHRAE Systems and Equipment Volume.*[4]

In climates where both heating and cooling are required at different times of the year, two-pipe systems require a seasonal changeover. Either hot water or chilled water is supplied to the terminal units. Some two-pipe systems have multiple two-pipe zones where units on some faces of the building can get hot water and units on other faces get chilled water, but generally heating and cooling are not available at the same time with two-pipe systems. This may result in significant comfort problems in the "shoulder" seasons (spring and fall.) The four-pipe system makes heating or cooling available at every terminal unit all the time (providing the chiller and boiler are not shut down seasonally) and provides more flexible control. The three-pipe system supplies hot water to the terminal unit in one pipe and chilled water in another pipe, but uses a single common return pipe to return hot or cold water that leaves the terminal unit to the central plant. The three-pipe system has been infrequently used in recent decades because of the resulting extra energy use.

Induction Units.   Induction units offer the advantage of having no fans or moving parts in the space and also offer constant outside air ventilation with no risk of local coil freeze-up. The induction unit uses jets of high-velocity primary air to induce air from the space and carry it over a heating or cooling coil inside the induction unit (Fig. 5.4.2-10). The primary air is often at a temperature lower than usual supply air temperatures during the cooling season.

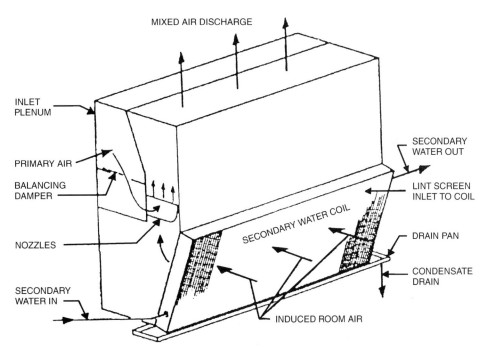

FIGURE 5.4.2-10   Air-and-water induction unit. (*Courtesy American Society of Heating, Refrigeration, and Air-Conditioning Systems from 1996* ASHRAE Handbook.)

This has two advantages: (1) The primary air does all the dehumidifying required in the space, so that the coil in the unit does not engender any condensation; this does away with the need for a condensate disposal system and reduces the likelihood of mold growth inside the unit; and (2) the larger supply air temperature difference, coupled with the high duct velocities usual in induction systems, allows using very small supply ducts that may fit inside the walls. Induction unit systems can sometimes be arranged to make heating and cooling simultaneously available in shoulder seasons. Cold primary air is supplied while hot water is circulated to the in-unit coil until the outdoor temperatures are so high that the primary air alone cannot provide all of the cooling that spaces typically require. At that point, cool water is circulated to the coil for additional cooling, but that cool water must not be so cool as to remove moisture from the air unless the induction unit is equipped to drain away condensate. The high-pressure requirement for the supply air in these systems was already discussed in this article.

Fan-Coil Units.   A fan-coil unit is a small, individual, air-handling unit that is usually dedicated to a single space. It can perform all of the following functions for one space that a central air-handling unit performs for many spaces: filtration (the level of filtration is limited), cooling, dehumidification, heating, and air circulation. The fan-coil unit consists primarily of a fan and a coil (or coils) in a cabinet. The cabinet may be a vertical console with a finished exterior mounted on the floor under a window. It may be a console concealed inside the wall construction. Or it may be a horizontal unit mounted in or above the ceiling with either direct grilles between the unit and the space or with a small duct system (Fig. 5.4.2-11).

Unitary Refrigerant-Based Systems.   Unitary systems based on more or less self-contained package units are probably the most common type of HVAC system we see in use today. Unitary systems range from simple window units that cool (or heat and cool) a single room (Fig. 5.4.2-12) to large factory-built rooftop penthouse units that heat and cool an entire building. Some package unit or unitary systems are complete systems that provide heating and cooling in the same way as the in-room terminal units discussed earlier. Other package units are applied in conjunction with building air distribution systems for single zones or multiple zones. They have been used in all sizes and all classes of buildings [e.g., for buildings without central plants and also for special areas of buildings with central plants either to provide cool-

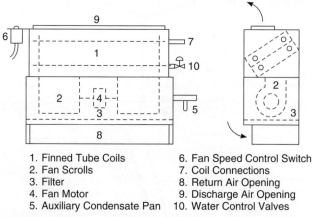

1. Finned Tube Coils
2. Fan Scrolls
3. Filter
4. Fan Motor
5. Auxiliary Condensate Pan

6. Fan Speed Control Switch
7. Coil Connections
8. Return Air Opening
9. Discharge Air Opening
10. Water Control Valves

FIGURE 5.4.2-11   Typical fan-coil unit. (*Courtesy American Society of Heating, Refrigeration, and Air-Conditioning Systems from 1996* ASHRAE Handbook.)

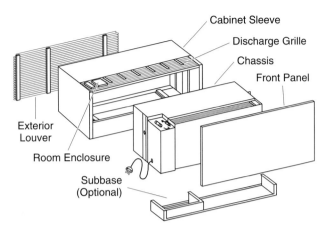

FIGURE 5.4.2-12   Packaged terminal air conditioner with combination heating and cooling chassis. (*Courtesy American Society of Heating, Refrigeration, and Air-Conditioning Systems from 1996* ASHRAE Handbook.)

ing to an area with specialized needs (like a computer room) or to provide cooling to small areas that need cooling even when the central plant is shut down].

Some unitary systems, like rooftop package units, are complete and self-contained and need only a duct system to connect them to the conditioned spaces. Others, like the floor-by-floor VAV systems used in high-rise buildings, rely on a central condenser water system and separate cooling tower. Sometimes, as in residential whole-house air-conditioning systems, the packaged equipment is in two parts connected with piping in the field. Sometimes heating may be from a furnace built into the package unit, and other installations will use a heating coil in the package unit that is connected to a separate boiler. But what unitary systems have in common is that the cooling machinery—the refrigeration cycle—is included in the packaged unit as it comes from the factory. Chapter 5 of the *ASHRAE Systems and Equipment Volume* provides more complete information about unitary refrigerant-based systems (Fig. 5.4.2-13).[4]

Ductwork.   Heat gain in the space is generally absorbed by cool, dry supply air that flows through the space and returns to the air handler as return air after absorbing heat and moisture in the space. The air is transported to and from the space by ductwork. The size of the ductwork is determined by the air volume and the velocity in the duct. The air volume is determined by the amount of heat to be absorbed and the supply air temperature. Space for ductwork is a major consideration in planning the system. Therefore, if the ductwork is installed above suspended ceilings, the clear space becomes a very critical issue. For example, a typical cooling system with supply air volume of the order of 1 cfm/ft^2, and using the typical supply temperatures and velocities described later, might use the equivalent of 3½ ft^2 of main-duct cross-sectional area to serve 5000 ft^2 of space; or 1¼ ft^2 for 1000 ft^2 of space.

The designer selects the supply air temperature based on the amount of moisture that must be removed from the space. Higher supply air temperatures can absorb less humidity. Supply air temperatures may be as high as the low 60s in some cases, or well below 50° in low-temperature air systems, but are commonly of the order of 55° to 58°F.

The designer selects duct velocity based on the following factors: noise (higher velocities tend to be noisier); the balance of higher operating cost due to higher pressure drop and higher fan horsepower with high velocities, compared with the saving in first cost with the

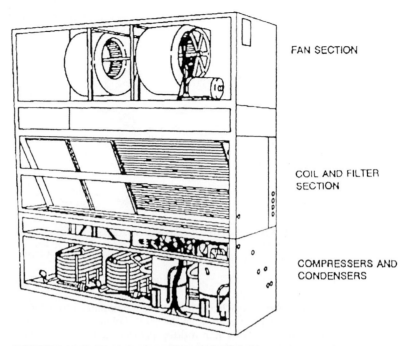

FAN SECTION

COIL AND FILTER SECTION

COMPRESSERS AND CONDENSERS

FIGURE 5.4.2-13   Vertical self-contained unit. (*Courtesy American Society of Heating, Refrigeration, and Air-Conditioning Systems from 1996* ASHRAE Handbook.)

smaller ductwork that can be used for high velocities; and the available space. A velocity of 1500 to 2000 ft/min is the commonly used range of velocities for main ducts, with lower velocities in branch ductwork. Higher velocities are often used in primary ductwork supplying terminal boxes, especially where space for ductwork is at a premium. Construction requirements and, thus, the cost of ductwork to handle velocities over 2000 ft/min and pressures over 2 in wg are more stringent than for low-velocity ductwork.

Duct construction is defined by two classifications: pressure and seal classes. Ductwork with less than (both) 2000 ft/min velocity and 2 in wg static pressure is considered low-pressure ductwork and has the least stringent, standard construction requirements. If either the velocity is 2000 ft/min or faster or the pressure is more than 2 in wg, the ductwork is considered high-pressure and requires heavier-gauge material and greater reinforcing. Construction requirements and ductwork costs vary, depending on how much the static pressure exceeds 2 in wg. Ductwork is also classified according to how leak-tight it must be and the extent to which construction joints must be sealed. Seal classes vary from the tightest, Seal Class A, where sealant must be applied to every joint and opening, to Seal Class C, where only the transverse joints get sealant. The leakage rate from a Seal Class A rectangular duct is only 6 ft^3 min^{-1} ft^{-2} of duct surface with 1 in wg static pressure inside the duct, compared with 24 ft^3 min^{-1} ft^{-2} of duct surface for Seal Class C, or 48 ft^3 min^{-1} ft^{-2} of duct surface for an unsealed duct. Although it would seem intuitively that the minimum duct leakage, and thus the tightest seal class, would be desirable, sealing ductwork represents a significant cost. The degree of sealing specified should represent an economic balance of the additional sealing cost against the operating cost with the higher leakage rate. Refer to *HVAC Duct Construction Standards* for more complete information on duct construction.[7] Exhaust air ducts are generally similar to supply air ducts.

Duct Material and Construction. HVAC ductwork is most commonly fabricated of galvanized steel, but ducts may also be made of aluminum or stainless steel where high levels of moisture are involved. Laboratory ducts that convey corrosive vapors may be made of plastic or plastic-coated steel, and ducts for exhaust from commercial cooking equipment are generally fabricated of welded, heavy-gauge black iron. On some projects the ductwork is fabricated of a glass fiberboard that provides inherent thermal insulation and sound absorption. Galvanized ductwork is often wrapped with a vapor barrier–jacketed insulating blanket or lined internally with insulating material. When the temperature difference between the air inside the duct and the air surrounding the duct is small, insulation may be omitted. However, if the temperature of the air inside the duct is in the range of the dew point of the air around the duct, both thermal insulation and a vapor barrier are necessary to prevent condensation on the duct surface.

Duct construction for supply ducts generally differs from that for return and exhaust ducts for two reasons: (1) the pressure and velocity in the ductwork and (2) the temperature of the air in the ductwork. Supply duct pressures and velocities are normally significantly higher than velocities and pressures in either return or exhaust ducts, and that pressure difference requires different duct construction both for considerations of strength and for the effects of leakage. Return duct velocities are usually lower than supply duct velocities to limit the duct pressure drop, especially if there is no return fan. Exhaust duct velocities, on the other hand, are often even higher than supply duct velocities, so that the airflow conveys the contaminants effectively. The effect of duct leakage may also be different in supply or exhaust ducts, where leakage results in additional capacity and operating cost requirements. In many cases, return duct leakage has no such impact, especially if the return duct is in a conditioned space. *HVAC Duct Construction Standards* discusses the requirements for duct construction in detail for various pressure classifications and for various leakage goals.[7]

Because the air in cooling supply ducts is often at a temperature below the dew point in the space, insulation and a complete vapor barrier are necessary to prevent condensation on the ductwork. That is usually not a consideration for return or exhaust ducts, where the air in the duct is at room humidity or higher. Exceptions, when insulation should be considered in return or exhaust ducts, include a return duct running through an unheated space (to prevent condensation inside the duct) and high-temperature exhausts like kitchen exhausts, where insulation may be used to separate the surrounding construction from the risk of fire when there is a grease fire inside the duct.

Filtration. Circulating contaminated air to the occupied space would obviously decrease the quality of the indoor air. Further, contaminated air is harmful to the performance of coils and air-moving devices. Contaminants can be brought in from outdoors with outdoor ventilation air, and contaminants generated in the occupied space can also be brought back to the air-handling unit in the return air. Air filters are normally provided in the air-handling unit immediately after the outdoor air and return air are mixed and before the air goes through any coils or fans. Sometimes it is more cost-effective in very dirty areas to filter outdoor air separately from return air. It is also not unusual to provide afterfilters at the point where the air leaves the air-handling unit when higher levels of cleanliness are desired. A wide variety of filter types offers different levels of effectiveness and are aimed at controlling different types of contaminants.

The most common filter types are for removing particulates. Filters for removing particulates are rated for efficiency or effectiveness, pressure drop, and dust-holding capacity. When selecting a filter type, it is important to recognize that the ability of each type of filter to remove particulates differs for different particle sizes. The filter selection should consider the range of particle sizes that are of concern. From the standpoint of operating cost, it is important to remember that the cost of the filter itself is generally small compared with the labor cost to remove a dirty filter, dispose of it, and install a new filter. Furthermore, the costs of both the filter and the labor are relatively small compared with the cost of the fan energy required to offset the pressure drop through the filters. Thus, in addition to providing less effective particle removal, the least expensive filter may not have the lowest overall annual operating cost. For more complete information on filters generally, see *ASHRAE Systems and Equipment Volume.*[4] For guidance in selecting filters for specific applications, see the *ASHRAE Applications Volume.*[6] Specific maintenance and installation issues are discussed later in this article.

Special filter types are also used to remove vapors and gases and are applied particularly when circulation of odors through the HVAC system is a potential problem. The most common of these types of filters is the charcoal, or activated carbon, filter that removes gases and vapors from the airstream by adsorption onto the surface of the carbon granules in the filter. Other types of media for gas cleaning work by chemically, rather than mechanically, bonding the gas molecules to the media. Selection of media for gas filtration is a highly specialized field, and expert assistance is strongly recommended. The February 1996 *ASHRAE Journal* article, "Chemical Filtration of Indoor Air: An Application Primer," is an excellent source of general background information on this topic.[8]

Piping. HVAC systems can involve several different types of piping systems:

- Chilled water
- Hot (heating) water
- Steam
- Steam condensate returns
- Coil condensate drains
- Refrigerant
- Fuel (gas or oil)

The number of different piping systems for a specific project depends on the type of system selected. On some projects, certain of the piping systems will be designed and installed as part of the plumbing system, rather than as part of the HVAC system, depending on local trade practices and regulations. The materials of which the piping will be made and how it will be assembled also often depend on local practices and code requirements, as well as on cost and technical considerations.

The most common practice is to use copper for water piping up to 2 in and steel for larger sizes and for all steam and steam condensate piping. Steel piping is most often used with threaded fittings up through about 3 or 4 in. Welded fittings or mechanical joint fittings are used in sizes above that; the choice depends on cost, and to some extent, individual preference. High-performance plastic piping is sometimes used for chilled water or condenser water, although it is less widely accepted. Fuel gas piping is normally black iron, although gas utility companies are increasingly using plastic piping for underground lines. Oil piping is most often steel, but soft drawn copper tubing is common in small sizes.

Steam and steam condensate piping must be laid out and installed with consideration for pitch, a requirement that does not generally apply to piping for other uses. Heat losses from steam supply lines result in the formation of liquid condensate that must be drained from the supply line at intervals through steam traps at drip points. Supply lines are pitched toward those drip points to enable the flow of condensate. Although it is possible to drain condensate against the direction of flow, that greatly reduces the steam-carrying capacity of the steam line, so the line should be pitched to drain condensate in the same direction as steam flow wherever possible. Steam condensate lines must carry condensed steam and also vented air and gases back to the condensate receiver. If they are not pitched to drain freely all the way back to the receiver, additional separate means of venting gases must be included. Because of the presence of air and particularly carbon dioxide, steam condensate lines often corrode more than other piping systems, and many designers specify heavier wall pipe for steam condensate systems to extend piping life.

Piping layout and installation must also consider thermal expansion. The rate of thermal expansion differs for different piping materials, but the growth in the length of pipe with an increase in temperature can result in tremendous forces if provision is not included to accommodate that growth. In most system layouts, thermal expansion can be taken up by the inherent changes in direction along the pipe route. It may be necessary, however, to add extra expansion loops or mechanical expansion joints to keep piping stresses due to expansion within acceptable limits, especially in long runs of steam or high-temperature, hot-water piping.

Pumps and Pump Types

Single-Suction.   Base-mounted single-suction pumps can be either close-coupled or flexibly coupled. Close-coupled pumps use a special motor that has an extended shaft to which the pump impeller is directly connected. Flexibly coupled pumps allow removing either the motor or pump without disturbing the other. The flexible coupling requires very careful alignment and a special guard. The flexibly coupled pump is usually less expensive than the close-coupled pump. Single-suction pumps are usually preferred for volumes up to 1000 gal/min but are available up to 4000 gal/min.

In-line pumps have the suction and discharge connections arranged so that they can be inserted directly into a pipe or they can be mounted on a base like other pumps. In the past, these pumps were used almost exclusively for small loads with low heads, but now they are available in the full range of sizes. Because of the restricted inlet condition, these pumps are not as efficient as single-suction pumps. These pumps can save considerable space, but extra care must be taken to assure that pipe stresses are not transferred to the pump casing.

Double-Suction.   In the double-suction pump, the fluid is introduced on each side of the impeller and the pump is flexibly connected to the motor. These pumps are preferred for larger-flow systems (typically, greater than 1000 gal/min) because they are very efficient and can be opened, inspected, and serviced without disturbing the motor, impeller, or piping connections. Typically, the pumps are mounted horizontally but can be mounted vertically. The pump case can be split axially (parallel to the shaft) or vertically for servicing. This pump takes more floor space than end suction pumps and is more expensive.

Vertical Turbine.   Vertical turbine pumps are axial-type pumps that are used almost exclusively for cooling tower sump applications. These pumps can be purchased with enclosures, or "cans," around the bowls when not sump-mounted.

Pump Limitations

Inlet Conditions.   The boiling temperature of water is a function of the absolute pressure surrounding the water. In a pump, the pressure at the eye of the impeller can be the lowest in the system, and depending on the temperature, the water can boil (vaporize). As the liquid moves through the impeller and gains pressure, the water vapor collapses back into liquid. This process is called *cavitation* and can be very harmful to the impeller. The pressure at the inlet of the pump must be high enough to prevent the water from boiling. The higher the velocity of water into the eye of the impeller, the lower will be the pressure, and the more likely that cavitation will occur.

Manufacturers publish *net positive suction head required* (NPSHR) curves with the pump curve. The designer must ensure that the system will have enough *net positive suction head available* (NPSHA) to prevent cavitation. In closed systems with minimum inlet pressure control (usually not less than 12 lb/in^2 gauge), cavitation is rarely a problem. In open systems (e.g., cooling towers), cavitation is a very real concern, and all efforts must be made to ensure that the NPSHA is greater than the NPSHR.

Another pump inlet problem to be avoided is *vortexing,* or air entrainment. Any time water is drawn from an open tank or sump, there is a potential that a vortex will occur. Vortexing will cause air to enter the pump suction line and will decrease the effectiveness of the pump. Vortexing will occur even with very deep sumps. Any time water is drawn from a sump or open tank, antivortexing devices should be installed. When using axial pumps (i.e., vertical turbines), care must be taken to ensure that the manufacturer's recommendations are followed to maintain a minimum submergence distance above the inlet bell. Adequate clearance must also be maintained from the tank's bottom to the pump's inlet.

Radial Thrust.   When pumps operate at points on the pump curve other than *best efficiency point* (BEP), nonuniform pressures can develop on the impeller. This is called *radial thrust* and can cause severe shaft deflection, excessive wear on pump bearings, and even shaft failure. Radial thrust occurs when pumps are operated at or near the shut-off pressures or near the end of the curve.

Controls.   Control systems used for HVAC systems range from simple on-off electric thermostats to complex computer-based, solid-state systems that calculate optimum control strategy. The simplest are electric control systems that use on-off control (with one or more stages), floating control, or modulating control. On-off, or cycling, control simply turns the system on when output is needed, and then turns it off when the set point is reached. Modulating electric controls adjust the system output in proportion to the difference between the set point and the measured condition. The further the room temperature is above the set point, for example, the greater the amount of cooling that is delivered. Floating control has two set points and a dead band. If the room temperature is above the upper set point, for example, the rate at which cooling is delivered to the space is increased very slowly until the room temperature falls below the upper set point. Then, the rate of cooling stays constant. If the room temperature then falls below the lower set point, the rate at which cooling is delivered is very slowly decreased until the room temperature rises to the lower set point. The rate of cooling stays constant at whatever it was when the temperature entered the dead band between the two set points. Electric control systems have been widely used because they are relatively simple and inexpensive, but they did not offer control as close as other systems. The modern solid-state electric controls available today, such as programmable thermostats, offer much closer control.

Traditionally, when an electric control system did not meet the control needs for a project, a pneumatic control system would be installed. Pneumatic control systems offered more flexible control strategies and provided modulating control of valves and dampers with less expensive operators, but pneumatic control devices need more frequent calibration. As the cost of solid-state, direct digital control (DDC) systems has fallen, these computer-based control systems have largely supplanted pneumatic controls in new HVAC systems for buildings.

Energy Management Systems

Energy management systems (EMS) [also called *building management systems* (BMS) or *facility management systems* (FMS)], are increasingly common in modern buildings, and more and more often, they perform most or all of the functions of the conventional HVAC control system. Energy management systems are discussed in detail in Art. 5.5.1. Article 3.1.3 also discusses these and other facilities management systems.

SPECIAL SYSTEMS

Special types of HVAC systems are used for special process facilities like clean rooms and computer rooms and for special applications like health care facilities and laboratories. Chapter 9 describes systems for health care facilities, and Chap. 10 addresses laboratories.

Commissioning

Commissioning for HVAC systems is highly recommended. For details of the process, refer to Art. 6.2.10, "Commissioning Programs for HVAC Systems."

Maintenance Issues

HVAC system maintainability starts with access. Equipment that is inaccessible is ignored. Access doors or panels should be installed wherever a control, a valve, or a damper is located behind a wall or above a ceiling. Access panels are also required to gain access to ductwork interiors for cleaning.

The amount of dirt that may accumulate in ductwork or inside HVAC units is a function of how effective the air filters are and how carefully they are installed. Filter manufacturers provide performance data to help select proper filters. The most important of those data is the filter efficiency, an indicator of how much dirt a given type of filter will capture. Many manufacturers also publish dirt-holding capacity, which determines how frequently that type of filter will need to be changed. A filter with high dirt-holding capacity and good efficiency may cost more, but it can be used longer with less frequent changes, so it will often have a lower total annual cost. If filters are to be effective, they must be installed properly with no bypass air leakage between or around filters. To this end, filters must have the dimensions for which the filter rack was designed, and they must be installed so that they are firmly held in place, pressing tightly against each other or against a filter frame. Air-handling unit filter racks are often equipped with a springy accordion-fold filler piece that presses the filters together in the rack.

Energy Conservation

Interest and effort in conserving energy was widespread in the late 1970s and 1980s, but attention has waned as energy prices stabilized and even fell. Yet, energy-using systems are much more efficient today than they were before the energy crises of the early 1970s. To a considerable extent, that is the legacy of mandatory measures requiring that manufacturers produce equipment to meet certain energy efficiency standards.

Attractive opportunities for investment in energy-conserving technologies still abound, in operating existing buildings and in designing and constructing new projects. Chapter 34 of the 1999 *ASHRAE Applications Volume* outlines in detail how to organize an energy management program, the alternatives for financing a program, and the steps in implementing a program.[6] The chapter also provides a convenient summary of the large database of building energy use characteristics compiled by the U.S. Department of Energy/Energy Information Administration. Some information is also provided in Art. 5.4.8, "Energy Efficiency and Indoor Environmental Quality."

CONTRIBUTORS

William J. Goode, P.E., C.P.M., President, W.J. Goode Corporation, Walpole, Massachusetts

Gregory O. Doyle, Director Buildings & Grounds Department, Massachusetts General Hospital, Boston, Massachusetts

Anand K. Seth, P.E., C.E.M., C.P.E., Director Utilities & Engineering Department, Partners HealthCare System, Inc., Boston, Massachusetts

Paul DuPont, P.E., Associate, DuPont Dobbs & Kearns Engineers L.L.C., Portland, Oregon

Olga Leon, P.E., President, Diversified Concepts, Sharon, Massachusetts

BIBLIOGRAPHY

1. *ASHRAE Fundamentals Volume,* American Society of Heating, Refrigeration, and Air Conditioning Engineers, Atlanta, GA, 1997.

2. *ASHRAE Load Calculation Manual,* 2nd ed, American Society of Heating, Refrigeration, and Air Conditioning Engineers, Atlanta, GA, 1992.

3. *Load Calculation Manual N,* Air Conditioning Contractors of America, Washington, DC, 1988.

4. *ASHRAE Systems and Equipment Volume,* American Society of Heating, Refrigeration, and Air Conditioning Engineers, Atlanta, GA, 2000.

5. *ASHRAE Refrigeration Volume,* American Society of Heating, Refrigeration, and Air Conditioning Engineers, Atlanta, GA, 1988.

6. *ASHRAE Applications Volume,* American Society of Heating, Refrigeration, and Air Conditioning Engineers, Atlanta, GA, 1999.

7. *HVAC Duct Construction Standards,* 2nd ed., Sheet Metal and Air Conditioning Contractors National Association, Chantilly, VA, 1995.

8. M. A. Joffe, "Chemical Filtration of Indoor Air: An Application Primer," *ASHRAE Journal* (February 1996); *ASHRAE Handbook of Fundamentals,* American Society of Heating, Refrigeration, and Air Conditioning Engineers, Atlanta, GA, 1996.

ARTICLE 5.4.3
BOILERS

Howard McKew, P.E., C.P.E.
Sebesta Blomberg and McKew, Topsfield, Massachusetts

William J. Goode, P.E., C.P.E.
W. J. Goode Corporation, Walpole, Massachusetts

Boilers used for building and process heating are basically ASME-certified heat exchangers that generate hot water or steam from the combustion of fossil fuels. The term *boiler* also applies to an electrically heated pressure vessel that generates steam and hot water. Before the turn of the twentieth century, boilers were extremely dangerous to operate and caused many fatalities and injuries. As a result, the American Society of Mechanical Engineers (ASME) was formed to standardize the design, construction, control, testing, and safe operation of boilers. Since the early 1900s, boiler design and performance have stabilized and achieved an extraordinary safety record. Boilers are constructed primarily of steel or cast iron. Copper vessels are sometimes specified for residential hot water heating, but are seldom used in larger industrial or commercial applications. Today's heating boilers are smaller and, at the same time, have greater energy output than older units.

CLASSIFICATION OF BOILERS

Boilers are classified by working pressure and temperature, size, steam or water, shape or configuration, material of construction, fire tube or water tube design, and the fuel used. The greater safety of steel water tube boilers enabled the development of large high-pressure and critical-pressure power boilers used in central station electric generating plants. High-pressure power boilers are not generally used in facilities and are not discussed further. With few exceptions, all boilers used in facilities are constructed to meet ASME Boiler and Pressure Vessel Code, Section IV, Rules for Construction of Heating Boilers (low pressure) and Sec. I, Rules for Construction of Power Boilers (medium and high pressure). A typical heating boiler for a facility is depicted in Fig. 5.4.3-1. The working pressure and temperature of ASME-designed and -constructed boilers are classified as low-, medium-, or high-pressure boilers. Low-pressure boilers are designed and constructed for a maximum working steam pressure of 15 psig and up to a maximum of 160 psig for hot water. A low-temperature hot-water boiler is lim-

Cleaver Brooks®

4
design
standards...

*your guarantee
of peak
performance!*

1 FOUR PASS
DESIGN

*maintains high flue gas
velocity for high heat transfer...
greater operating economy*

Advantage of the four pass principle used by Cleaver-Brooks is expressed in a simple basic formula:

$$\text{Gas Velocity} = \frac{\text{Gas Volume}}{\text{Area}}$$

CB models maintain a continuously high flue gas velocity essential to good heat transfer. As hot gases travel through the four passes, they transfer heat to the boiler water, then cool and thus occupy less volume. The cross section of gas flow area (number of tubes) is reduced proportionately to maintain high gas velocity, and produce constant and complete heat transfer.

Rapidly moving gases through the four passes of the CB scrub the tube surfaces clean and increase the heat transfer by improving the gas film coefficient. Every square foot of tube surface becomes more effective in transferring heat to the water. The ultimate result: excellent overall efficiency proven with low temperature combustion exhaust gases.

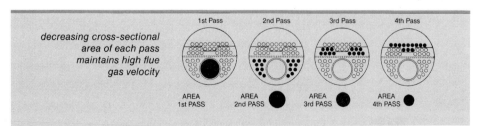

*decreasing cross-sectional
area of each pass
maintains high flue
gas velocity*

FIGURE 5.4.3-1 Typical facility heating boiler. (*Courtesy of Cleaver-Brooks, Inc.*)

ited to an operating temperature of 250°F. The controls and relief valves that limit temperature and pressure must be installed to protect the boiler according to ASME rules. ASME-classified medium- and high-pressure boilers are designed and constructed to operate above 15 psig steam pressure and/or above 160 psig pressure or a 250°F water temperature. Steam boilers are available in standard sizes up to (and above) capacities of 100,000 lb of steam per hour with heat rates of 60,000 to more than 100,000,000 Btu/h. Water boilers are available in standard sizes from 50,000 to more than 100 million Btu/h heat rates. Many low-pressure–class boilers are used for space heating and are referred to as heating boilers.

Most cast-iron and copper tube boilers are low-pressure, residential boilers that are configured as in Fig. 5.4.3-2. Cast-iron boilers are assembled at the factory or on the job site from

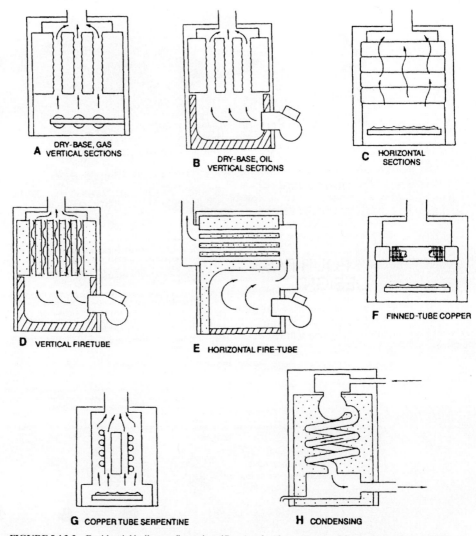

FIGURE 5.4.3-2   Residential boiler configuration. (*Reprinted with permission of the American Society of Heating, Refrigerating, and Air-Conditioning Engineers, Inc., from the* 2000 ASHRAE Systems and Equipment Handbook.)

individual, modular segments. Gasket joints ensure pressure tightness for water, steam, and products of combustion that pass through the sections. Each added section increases the boiler's heat output, energy input, and volumetric capacity. Three firebox configurations, shown in Fig. 5.4.3-3, that have similar efficiencies and test requirements, are used in cast-iron boilers. In dry-base boilers, the burner chamber is beneath the fluid section. In a wet-leg configuration, the burner chamber is enclosed on the top and sides by the fluid sections. In wet-based units, the fluid sections surround the burner chamber, except for necessary openings.

Fire-tube boilers used in medium and large boiler applications incorporate a central, dry-back or wet-back cylindrical firebox surrounded by multiple-pass fire tubes, as shown in Fig. 5.4.3-4. A steel shell contains the entire apparatus. Fire-tube boilers are typically rated in

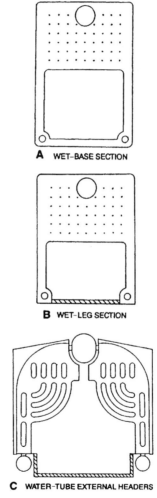

A WET–BASE SECTION

B WET–LEG SECTION

C WATER–TUBE EXTERNAL HEADERS

FIGURE 5.4.3-3 Cast-iron boiler configurations. (*Reprinted with permission of the American Society of Heating, Refrigerating, and Air-Conditioning Engineers, Inc., from the 2000 ASHRAE Systems and Equipment Handbook.*)

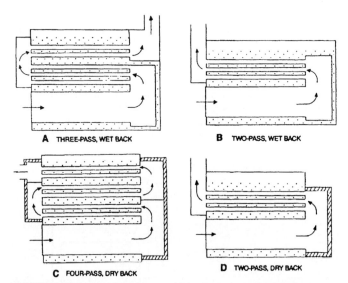

A   THREE-PASS, WET BACK   B   TWO-PASS, WET BACK

C   FOUR-PASS, DRY BACK   D   TWO-PASS, DRY BACK

FIGURE 5.4.3-4   Scotch marine-type fire-tube boilers. (*Reprinted with permission of the American Society of Heating, Refrigerating, and Air-Conditioning Engineers, Inc.,* 2000 ASHRAE Systems and Equipment Handbook.)

boiler horsepower. One boiler horsepower is equivalent to 33,476 Btu of output per hour. Fire-tube boiler capacities typically increase in increments of 50 or 100 horsepower up to a maximum of 1000 boiler horsepower. Water-tube boilers for low-, medium-, and high-pressure applications are usually made of welded construction and are rated from 10,000 Btu/h to the largest units made. Larger commercial units employ water tubes and are configured as shown in Fig. 5.4.3-5. The heat-exchanger surface consists of tubes located around the burner chamber in a vertical configuration. Water-tube units have fluid inside the tubes and the combustion chamber on the outside.

Fuels and Efficiency

Fired boilers may be designed to burn coal, wood, various grades of fuel oil, and/or various fuel gases. The use of coal and wood fuel in facility boilers has diminished during the last few decades. Most heating boilers operate on propane, methane, natural gas, or fuel oils (No. 2, 4, or 6). Dual-fuel burners enable using gas or oil, depending on availability, EPA stack discharge limits, or cost. Considering these factors can help in choosing a fossil fuel: Is natural gas available? Does the owner prefer two sources of fossil fuel? If oil is used, where will it be stored and how much storage is needed? If the storage tank is located below grade, will vehicular traffic be passing over it? If so, a concrete pad will be needed to protect the tank from heavy trucks. State and federal regulations may also dictate the type and grade of fuel that may be used under certain circumstances and at specific locations. Boiler efficiency is expressed as a percentage and defined either as overall efficiency or combustion efficiency. Engineers most often specify overall efficiency in boiler selection. To determine overall efficiency, the unit's gross energy output is divided by the total energy input. Because the output measurement requires evaluating steam and water leaving the boiler, it can be determined precisely only at the factory under fixed test conditions. The overall efficiency of electric boilers is in the range of 92 to 96 percent. Fossil-fueled boiler performance depends on several

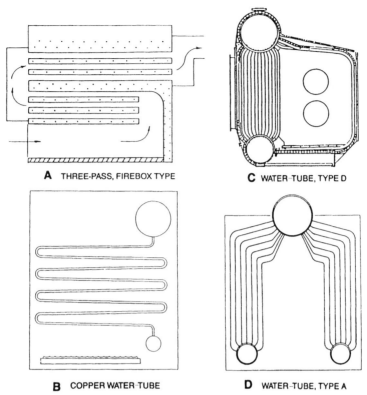

FIGURE 5.4.3-5 Commercial fire-tube and water-tube boiler configurations. (*Reprinted with permission of the American Society of Heating, Refrigerating, and Air-Conditioning Engineers, Inc., from the* 2000 ASHRAE Systems and Equipment Handbook.)

factors and varies more widely than electric boiler performance, but 85 percent is a good rule of thumb for estimating natural gas-fired or oil-fired boiler efficiency.

If a new boiler is a replacement for an existing unit, changing fuels or adding dual-fuel burners may lower operating costs. Any economic evaluation to determine the feasibility of changing fuels should consider rebate programs available from local utilities that may make switching a more attractive option. For example, some utilities offer low-cost rates during the air-conditioning season, which reduces the cost of boiler operation to provide steam or high-temperature water to absorption chillers.

Boiler Selection

Simpler is better! Generally, the lowest technology that satisfies the project or site needs should determine the type of boiler to be installed. When one speaks of a type of boiler, it generally infers a type of system (e.g., usually a low-pressure steam boiler will deliver steam through a low-pressure steam distribution system). Following is the hierarchy of boiler and system types in order of increased complexity:

- Low-temperature, hot-water
- Low-pressure steam

- Medium-temperature hot-water
- Medium-pressure steam
- High-pressure steam
- High-temperature, hot-water

The higher the operating temperature and pressure, the more complex and expensive is the operation of the system. Boiler selection is a response to system needs (i.e., the system requirements determine the boiler type). Several examples follow:

System Requirements	Boiler Type
Building heating	Low-temperature, hot-water
Humidification	Low-pressure steam
Institutional kitchen	Medium-pressure steam
Laundry	High-pressure steam
Large campus heating	Medium-temperature hot-water or high-pressure steam
District heating of a city	High-temperature, hot-water or high-pressure steam
Utility power plant	High-pressure steam

Generally speaking, steam boilers and systems are more complicated than hot-water systems, because steam requires more complicated equipment and maintenance (e.g., steam systems require condensate return pumps, steam traps, continuous water treatment, daily boiler blow-down, constant makeup water, preheating, and deaerating of feed water). In addition, steam inherently imparts a higher life safety risk if the system ruptures.

For relatively small steam requirements, consider a small steam boiler located adjacent to the steam load or attempt to buy process equipment that has built-in steam generators. When the choice is available, buy equipment designed to operate on low-pressure steam (e.g., cooking kettles or laboratory cage washers).

Process for Selecting Steam and Hot-Water Operating Pressures and Temperatures

1. Requirement for steam system. For example, is steam needed for humidification, to drive a turbine, sterilizers, steam absorption chillers, or a manufacturing process? If steam is not needed, a hot-water boiler may be more appropriate for the facility.
2. Steam temperature and correct steam pressure. For example, if steam is needed only for humidification, a low-pressure steam system is adequate. On the other hand, for sterilizing or laundries, higher temperatures are needed that can be supplied only by high-pressure steam.
3. Miscellaneous steam flow requirements.
4. Steam quality and need for superheat (e.g., turbines or high-temperature process).
5. Pressure drop from a steam generator to the point of use is especially important.

If steam is not needed, then select a hot-water system at a temperature that will satisfy heating or process requirements. Higher temperatures will require higher pressure hot-water systems to prevent flashing water into steam. This adds a higher level of hazard and complexity to the system. Consider the temperature drop from a hot-water generator to the point of use. Distribution systems, especially long ones greater than 1000 ft, influence the selection of steam pressure or hot-water temperature, pipe size, and insulation thickness on transmission piping. Low-pressure steam and low-temperature, hot-water systems have practical limits for transmission (typically 1000 ft for low-pressure steam systems and 2000 ft for low-temperature, hot-water systems). The long transmission for district heating of cities and very large industrial, educational, or hospital campuses generally requires a steam pressure greater than 15 psig and a hot-water temperature higher than 220°F.

Matching Boiler Types to Steam and Hot-Water Systems

- Low-pressure steam and low-temperature hot-water systems use sectional cast-iron water-tube, vertical water drum, and fire-tube boilers.
- Medium-pressure steam and medium-temperature hot-water systems use water-tube, vertical water drum, and fire-tube boilers.
- High-pressure steam and high-temperature hot-water systems use fire-tube boilers and water-tube boilers up to 350 psig.

State Licensing and Staffing Requirements

The devastating and horrific losses caused by boiler explosions during the nineteenth century resulted in the passage of state laws that required minimum levels of competence for boiler operators and required that high-pressure steam boilers be under constant visual observation by operators licensed by each state.

As boiler fueling and operating controls became automated and safety controls became more reliable, the need for continuous attendance by licensed operators decreased. Most of the states in the United States relaxed their requirements for continuous attendance. A few states maintain their restrictive regulations. The result is a significant labor cost penalty, which can be as high as $200,000 to $300,000 per year. Therefore, the decision to install a high-pressure steam system may have a significant operating cost impact. It is recommended that the engineer and/or owner research their local and state laws before deciding on a type of heating system.

Boiler Sizing.   Boilers are sized to provide for peak demand, plus a factor of safety, plus piping heat losses, plus any boiler plant auxiliary equipment that requires steam or hot water to perform its function (e.g., a fuel oil heater, a feed-water heater, a deaerator, soot blowers, etc.). The resulting total load is divided among one or more boilers. The number of boilers is determined by the need for reliability and the capital available to finance the boiler plant. Multiple boilers cost more than one boiler! A residential house typically has one boiler to minimize construction cost. A hospital has a spare boiler equal in size to the largest boiler in the plant, because it must provide for the continuous needs of inpatients. Large industrial or research facilities may have fully redundant equipment because they require a high level of reliability. Speculative commercial and office complexes are likely to have very little or no backup capacity, because they are sensitive to construction costs.

Boiler Plant Auxiliary Equipment.   The auxiliary equipment that helps the boiler to do its job should also be subjected to the same analysis regarding equipment redundancy. Typically, the redundancy of the auxiliary equipment matches that of the boilers. The auxiliary equipment includes condensate transfer and boiler feed-water pumps, fuel oil pumps, fuel oil heaters, steam to hot-water heat exchangers, and system hot-water pumps. Physically large and very expensive items, like a deaerator, are usually made redundant by using piped bypasses around the equipment, which are only used in an emergency basis.

A steam or hot-water system requires numerous pieces of auxiliary equipment to support the operation of the boilers and to deliver the steam or hot water to the point of use. Steam systems require much more equipment than hot-water systems for the following three reasons:

1. Steam boilers require continuous makeup of water, which must be treated to remove contaminants.

2. Steam is condensed to water before it is returned to the boiler. This condensing process requires special equipment to collect and pump the condensed water back to the boiler room.

3. High-pressure steam distribution systems typically require pressure reductions to control the process at the various points of use. This is achieved with elaborate pressure-reducing valve stations.

Steam Boiler Plant Auxiliary Equipment.   Figure 5.4.3-6 shows the relationships between the various components of a steam delivery and a condensate-return system. This 200-psig steam system is typical of a medium-pressure system that supplies heating or process steam to a manufacturing building or to a district heating system for a campus or city. Steam is delivered at full boiler pressure to each load point, where the pressure is reduced to accommodate the pressure limitation of the process or the building heating system. One or more pressure reducing stations operate in a series arrangement to step the steam pressure down so that it does not damage the equipment and is easily controlled for the process or heating requirements. As the steam gives up its heat, it condenses to water and flows through a steam trap that separates the steam and condensate piping systems. The steam traps ensure that only condensed steam (water) is returned to the boiler plant.

Steam traps come in many varieties (e.g., float, thermostatic, bucket, inverted-bucket, disc, etc.) They are the single most problematic maintenance issue in a steam system because of their relatively short operating lives. It is recommended that all steam traps be cataloged, inspected, and repaired annually. Typically, condensate drains by gravity from the steam trap to a low point, where a condensate receiver collects it. The condensate receiver is constructed of cast iron or welded steel plate and has one or two condensate transfer pumps mounted on its side. The transfer pumps lift and force the condensate back to the boiler plant.

When the condensate arrives back at the plant it is collected in a large steel tank, which is a surge tank, a feed-water heater, or a deaerator. Sometimes a surge tank is used in combination with a feed-water heater or a deaerator. A surge tank will hold a sudden surge of steam condensate that cannot be used immediately by the boiler. Such a surge might occur during the warm-up of buildings during a cold winter morning or the start-up of a process in a manufacturing plant. Condensate transfer pumps move the steam condensate to the feed-water heater or deaerator to maintain a constant water level in the latter.

When the condensate is inadequate to maintain the water level in the feed-water heater or deaerator, "makeup water" is introduced to make up the difference. This makeup water contains dissolved and undissolved solids and gases that will clog and corrode the boiler if not removed.

Feed-water heaters and deaerators contain perforated tubular lances that introduce steam into the condensate. The steam heats the condensate to a point, where oxygen and carbon dioxide are driven out of the condensate. This process reduces these gases to a level of 0.03 cc/L of condensate. These gases can be further removed by using a pressurized deaerator that allows a higher operating temperature in conjunction with mechanical agitation of the condensate. This type of deaerator achieves an oxygen and carbon dioxide residual level of 0.005 cc/L of condensate. Pressurized deaerators operate with a minimum steam pressure of 3 psig or higher depending on the plant's operating pressure.

The amount of cleanup required for the makeup water is determined by the amount of contaminants contained in this water and the operating temperature of the boiler. Dirt is removed by cartridge filters, hydrocarbons by carbon filters, and iron and carbonates by reactions with other chemicals and deionizers. A water treatment consultant should determine the amount of cleanup required. Once the condensate and makeup water have been treated and heated, it is called *boiler feed water*. The boiler feed water is introduced into the boiler by feed-water pumps. These pumps are controlled to maintain a constant water level in the

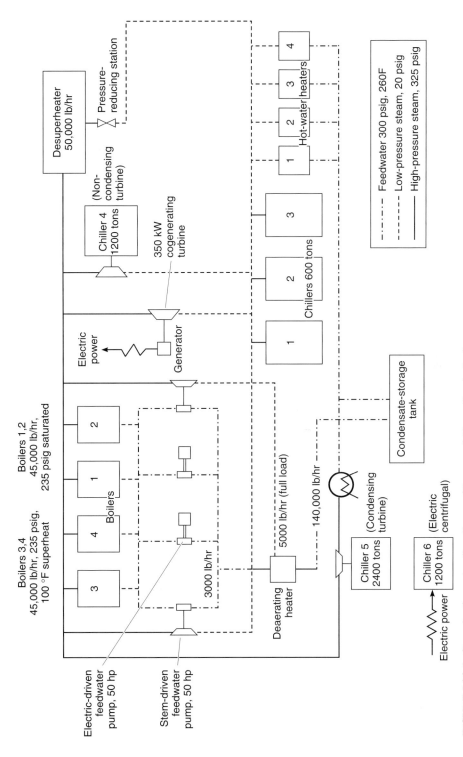

FIGURE 5.4.3-6 Steam heating system flow diagram. (*Courtesy of Cleaver-Brooks, Inc.*)

steam drum of the boiler. It is a general practice to introduce chemicals into the deaerator or feed-water piping to further reduce the oxygen concentration of the feed water, to keep any solids from precipitating onto the boiler's heating surfaces, and to protect the condensate return piping from corrosion. These chemicals are introduced into the feed water by a metering pump. Generally, the pump's output is set by testing daily samples of the boiler water and adjusting the pump's displacement per revolution.

Other equipment that is unique to a steam boiler plant includes:

- Sample cooler to cool a boiler water sample before testing.
- Blowdown separators to allow blowdown, or contaminated boiler water, to flash into steam and condensate. The condensate is discharged to a sewer after further cooling.
- Feed-water preheaters that use waste heat streams to increase the feed-water temperature and thus the boiler's overall efficiency.
- Blowdown heat recovery units that recover heat from continuous "surface blow" of the boiler to preheat makeup water.

Hot-Water Boiler Auxiliary Equipment. Figure 5.4.3-7 shows a simple hot-water heating system. It has fewer components than a steam heating system because a hot-water system is sealed. It does not normally require continuous makeup water and boiler blowdown. Therefore, a hot-water boiler plant does not require special chemical treatment, preheating, and deaerating of the makeup water; nor does it require blowdown with its attendant separators, heat recovery, and coolers. However, a hot-water heating system does require an air separator to remove the air bubbles when the system is initially filled with cold water. Hot water holds less dissolved air than cold water, and this air should be removed for optimum heat transfer.

Hot-water heating systems also require expansion tanks to absorb the increase in the volume of the water in the system as it is heated to its operating temperature. Today's expansion tanks usually have neoprene bladders that separate the system's water from a charge of compressed air or nitrogen. The initial fill pressure of the compressed gas is set so that it offsets the static head required to fill the system, the expansion of the heated water, and a few extra pounds of pressure to prevent boiling the water at the top of the system.

Chemical treatment of a hot-water system is usually limited to an initial treatment, when the heating system is filled with cold water. The chemical is added through a device called a *shot feeder,* which gradually mixes the chemical with the hot water in the system. Subsequent draining and refilling of the system should be accompanied by retreatment. Hot-water systems that contain propylene or ethylene glycol for protection against freezing need to have a sample of the glycol and water mixture tested annually or biannually to ensure that there is an adequate percentage of inhibitor to prevent corrosion of the system.

Whereas steam systems use the pressure generated by the boiler's conversion of water to steam to move the steam through the heating system, hot-water systems must use pumps to push the water through the system and back to the boiler. These pumps are usually centrifugal types with mechanical seals to prevent leakage of water from the system. Typically, there is a spare pump to maintain heating in the event of a pump failure during the winter.

PERFORMANCE CODES AND STANDARDS

For rating and performance standards, the heating industry traditionally has looked to the Hydronics Institute [formerly the Institute of Boiler and Radiator Manufacturers (IBR)], the Steel Boiler Institute, the American Gas Association, and the American Boiler Manufacturers Association. Following is a sample of performance codes and standards required for burner units:

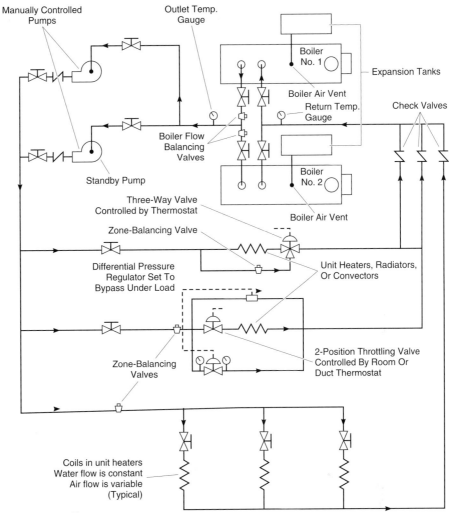

FIGURE 5.4.3-7 Hot-water system flow diagram. (*Courtesy of Cleaver-Brooks, Inc.*)

- Burner and all components will comply with Underwriter's Laboratory (UL), Factory Mutual (FM), or Industrial Risk Insurer (IRI) standards.
- Relief valves are required and must be ASME stamped and rated.

Boiler Room Layout

Design and facility engineers should coordinate efficient equipment layout within the boiler room. Access must be provided on both sides of boilers. Tube removal space also is required at one end of fire-tube boilers. The room must provide adequate headroom for safety. Usually 16 ft to the underside of the structure is a minimum elevation to allow pipes and flue gas breeching to pass over boilers and to allow for the deaerator tank in steam systems. Hot-water pumps, fuel oil pumps, water meters, backflow preventers, heat exchangers, boiler feed pumps, and deaerators require access; a 3-ft minimum space should be provided on as many sides as possible to permit servicing.

Ventilation also is critical for a successful installation. The airflow rate in cubic feet per minute (cfm) required to replace combustion air drawn into the boiler burner can be calculated on the basis of its size. Ideally, more supply air will be introduced into the room than the air exhausted from the chimney flue. The resulting positive-pressure design enhances equipment performance. Summer ventilation must also be considered. If the equipment is to operate in warm weather, heat rejection from the boiler surface, associated pipes, and boiler breeching combined with hot outdoor air may make the space temperature conditions intolerable. Although air conditioning the space may be economically impractical, increased ventilation may improve the environment sufficiently. Spot cooling, an often-overlooked option, can be a cost-effective means of introducing conditioned air at strategic locations.

Although electric boilers have no such requirement, fossil-fueled units require a chimney to discharge flue gases using one of the three following methods:

1. A boiler may have a forced-draft fan that pushes the combustion air and products of combustion through the boiler, its breeching, and out the chimney.
2. A boiler may operate on a natural draft. A natural draft is caused by heating the flue gases, which reduces their density and causes them to rise up the chimney.
3. A third method utilizes a forced-draft fan at the burner and an induced-draft fan at the exit of the boiler to push and pull the flue gases through the boiler to the outside air.

The chimney should be engineered to ensure removal of the flue gases under all operative conditions. Important engineering considerations for the chimney include satisfying state and federal clean-air requirements; foundation and earthquake requirements; materials of construction (impacts the cost and expected life of the chimney); accurate draft calculations; aircraft warning lights; and accessibility for clean-air compliance monitoring.

Automatic Controls and Metering. Boilers are usually furnished with self-contained automatic controls. These controls comprise the basic components required for efficient and safe operation. In addition, there are local, state, and federal laws that must be complied with. The state compliance will in part refer to other standards (i.e., ASME, NFPA, etc.). Federal requirements may come from standards established by OSHA, the Clean Air Act of 1989, and other applicable regulations. Following are some of the boiler and burner controls:

- Fully modulating flame-retention natural gas burner with a minimum 10-to-1 turndown while maintaining a stable and efficient flame
- Gas flow control valve that maintains a constant pressure throughout its modulating range to maintain a constant fuel-to-air ratio

- Safety gas shutoff valve and the auxiliary safety gas shutoff valve
- Variable-input flow control valve

Most automatic boiler controls are direct digital (DDC) and are packaged in the boiler's self-contained panels. The DDC controls enhance boiler efficiency, emissions control, safety, turndown capability, and troubleshooting. The boiler controls can be interlocked with the central building automation system to simplify normal operating and emergency response protocols.

The ability to determine a boiler plant's output, the amounts of fuel and water consumed, and the plant's efficiency require metering. Microprocessors now make it possible to record and calculate instantaneous measures of fuel consumption and efficiency. The classic displacement meters and their totalizing registers are giving way to mass, ultrasonic, Doppler, vortex shedding, magnetic, and other forms of flowmeters that integrate well with microprocessors to provide instantaneous data, histories, and data trending.

Equipment Service Life and Maintenance

The American Society of Heating, Refrigerating, and Air-Conditioning Engineers (ASHRAE) has estimated the median equipment service life for steel water-tube boilers at 24 years, steel fire-tube boilers at 25 years, and cast-iron boilers at 35 years for hot-water applications. Steam boilers vary somewhat from these numbers. Service life is directly proportional to the level of maintenance. Programmed maintenance consists of planned inspections, adjustments, and repairs designed to minimize equipment downtime and extend service life. These programmed operations occur in daily, weekly, monthly, and annual cycles, each of which is critical. Documentation (e.g., system flow diagrams, valve charts, pipe system identification, operation manuals, and equipment shop drawings) plays a large role in implementing programmed maintenance.

CONTRIBUTORS

Anand K. Seth, P.E., C.E.M., C.P.M., Partners HealthCare System, Boston, Massachusetts

Roger Wessel, P.E., RPW Technologies, Inc., West Newton, Massachusetts

RESOURCES

ASHRAE Systems and Equipment Volume, American Society of Heating, Refrigeration, and Air Conditioning Engineers, Atlanta, GA, 2000.

The Boiler Book, Cleaver-Brooks Co., Division of Aqua-Chem-Inc., Milwaukee, WI, 1997.

V. Ganapathy, *Steam Plant Calculations Manual,* Second Edition, Marcel Dekker, New York, 1994.

Erwin G. Hansen, *Hydronic System Design and Operation, A Guide to Heating and Cooling with Water,* McGraw-Hill Book Company, New York, 1985.

Jason Makansi, *Managing Steam, An Engineering Guide to Commercial, Industrial, and Utility Systems,* Hemisphere Publishing Corporation, New York, 1985.

1998 ASME Boiler & Pressure Vessel Code, An International Code, 1998.

S. C. Stultz and J. B. Kitto, *Steam: Its Generation and Use,* 40th Edition, Babcock & Wilcox, McDermott Company, 1992.

Everett B. Woodruff and Herbert B. Lammers, *Steam Plant Operation,* Third Edition, McGraw-Hill Book Company, New York, 1965.

ARTICLE 5.4.4
CHILLED WATER PLANTS

Paul J. DuPont, P.E., Principal
DuPont Dobbs & Kearns Engineers L.L.C., Portland, Oregon

A central chilled water plant represents a major investment in a facility, and it also can be one of the primary consumers of energy. Many facilities depend on reliable air conditioning to manufacture their products or provide their services. Central chilled water plants can be highly reliable when designed and maintained properly.

This article describes the components that make up a central chilled water plant. A wide variety of equipment is available on the market, and selecting the correct types is the first step in owning and operating a central chilled water plant. Chilled water plants are systems that meld many different components—including chillers, heat rejection devices, pumps, piping, and controls—into a comprehensive entity to provide mechanical cooling to a facility. Also discussed in this article are the design considerations for formulating an effective plant. We briefly look at the instrumentation and controls that are so important to a successful plant. Finally we provide an overview of the start-up and commissioning process that is necessary for a well-designed and well-constructed plant to ensure delivery of efficiently produced chilled water to air condition the facility.

CHILLER PLANT EQUIPMENT

Refrigerants

To be useful, refrigerants must have low toxicity, low flammability, and a long atmospheric life. Recently, refrigerants have come under increased scrutiny by scientific, environmental, and regulatory communities because of the environmental impacts attributed to their use. Some refrigerants—particularly chlorofluorocarbons (CFCs)—destroy stratospheric ozone. The relative ability of a refrigerant to destroy stratospheric ozone is called its ozone depletion potential (ODP). CFCs and halogenated chlorofluorocarbons (HCFC) are being phased out according to the 1987 Montreal Protocol (see Table 5.4.4-1).

TABLE 5.4.4-1 Provisions of 1987 Montreal Protocol for Common Refrigerants Used in Chilled Water Plants

Refrigerant	Year	Restrictions
CFC-11	1996	Ban on production
CFC-12	1996	Ban on production
HCFC-22	2010	Production freeze and ban on use in new equipment
	2020	Ban on production
HCFC-123	2015	Production freeze
	2020	Ban on use in new equipment
	2030	Ban on production
HFC-134a	—	No restrictions

The global warming potential (GWP) of refrigerants is another significant environmental issue. Gases that absorb infrared energy enhance the "greenhouse" effect in the atmosphere, leading to warming of the earth. Refrigerants have been identified as greenhouse gases. Table 5.4.4-2 lists the ODP versus GWP of various refrigerants.

Theoretically, the best refrigerants would have zero ODP and zero GWP, like R-717 (ammonia). Although some refrigerants used in a particular system may have a direct effect on global warming, there will also be an *indirect* effect on global warming as a result of that system's energy consumption. The burning of fossil fuels and subsequent release of carbon dioxide are the indirect effect. To reduce greenhouse gases to the greatest extent possible, it is critical to focus on the system's overall energy efficiency, not just to consider the refrigerant's GWP.

When comparing the theoretical and practical efficiencies of different refrigerants, it becomes apparent that there are only slight differences among various refrigerants (R-123 is somewhat better than the refrigerants it is designed to replace). Calm et al. note "that efficiency is not an inherent property of the refrigerant, but rather achieving the highest efficiencies depends on optimization of the system and individual components for the refrigerant."[1]

Water Chiller Types

A number of different types of water chillers and different energy sources are available. To select properly among types of chillers and energy sources, an understanding of the different options available and their differences is important. The following is a review of different types of chillers:

Reciprocating Chillers.   Reciprocating chillers are widely used in capacities ranging from 50 to 230 tons, although they are available in capacities up to 400 tons (see Fig. 5.4.4-1). Water-cooled machines are usually selected at a *coefficient of performance* (COP) of 4.2 to 5.5, and

TABLE 5.4.4-2   ODP versus GWP for Common Refrigerants

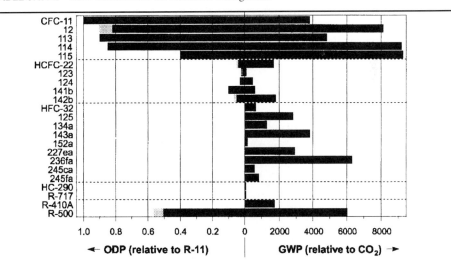

Source: Courtesy Calm and Didion, "Trade-Offs in Refrigerant Selections: Past, Present, and Future," in *ASHRAE/NIST Refrigerants Conference, Refrigerants for the 21st Century,* copyright 1997 by the American Society of Heating, Refrigerating and Air-Conditioning Engineers, Inc.

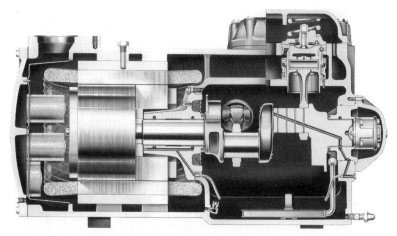

FIGURE 5.4.4-1   Reciprocating compressor.

air-cooled machines are selected at a COP of 2.8 or higher. Typically, these chillers use R-22 but have also been designed to use R-134a and R-717 (ammonia). The larger machines have multiple compressors that are grouped (piped in parallel) in several circuits. This creates some redundancy if a compressor should fail. As positive displacement machines, they retain nearly full cooling capacity even when operated at conditions above the design conditions, and they are very suitable for air-cooled applications. For the same reason they are also suitable for use as heat recovery machines. Control is achieved by stepping unloaders and cycling compressor on/off, which creates a choppy part-load performance curve. Reciprocating chillers tend to be low first-cost machines. Manufacturers are moving away from producing this type of machine in favor of the screw (rotary) chiller.

Single- and Twin-Screw Rotary Chillers.   Both single- and twin-screw chillers are positive displacement machines. The refrigerants typically used include R-22, R-134a, R-410a, and R-717 (ammonia on open machines only). The sizes available range from 30 tons to 1250 tons, but the most common sizes are from 70 to 400 tons. They are particularly suitable as air-cooled chillers. Water-cooled machines are selected for a COP range of 4.9 to 5.8 (0.72 to 0.61 kW/ton), and air-cooled machines are selected for a COP range of 2.8 and higher.

There is no practical design advantage of single- versus twin-screw except that the single screw may be slightly quieter. The machines have excellent turndown capability. Some chillers incorporate two compressors, which provides additional efficiency advantages during part load and allows unloading below 10 percent. Screw chillers are inherently more efficient than reciprocating types because they incorporate refrigerant economizers, have very few moving parts, and have balanced forces on the main bearings. Consequently, these machines are very reliable. Screw machines are controlled with a slide valve and are fully modulating. Screw machines are very cost-effective in the ranges cited previously (see Fig. 5.4.4-2).

Centrifugal Chillers.   Centrifugal chillers have the highest full-load efficiency ratings of all of the chillers discussed. They are available in sizes from 80 tons to 10,000 tons, but the most common sizes are from 200 to 2000 tons. Centrifugal chillers use high-pressure refrigerants R-22 and R-134a and low-pressure refrigerant R-123. They are available in both air-cooled and water-cooled versions, but because of very low COPs, air-cooled centrifugal chillers are very seldom used. Water-cooled centrifugal chillers have COPs that range from 5.5 to 7.1 (0.64 to 0.49 kW/ton). Centrifugal chillers are usually controlled with inlet guide vanes that allow full modulation to as little as 10 to 15 percent of capacity (with condenser water relief). Variable

FIGURE 5.4.4-2   Screw compressor.

speed drives can be added to enhance the part-load operation characteristics, but because of the high cost of the drives the value of this option must be carefully evaluated. Hot gas bypass can also be used but should be considered only if very low turndowns (at elevated condenser temperatures) are required. Centrifugal chillers are the flagship products of the major manufacturers and as such generally have the highest quality and reliability (see Fig. 5.4.4-3).

Absorption Chillers.   Absorption chillers can be either single- or double-effect. Single-effect chillers have COPs of 0.60 to 0.70, and double-effect chillers have COPs of 0.92 to 1.20. The 50 to 100 percent improvement in efficiency between single- and double-effect machines

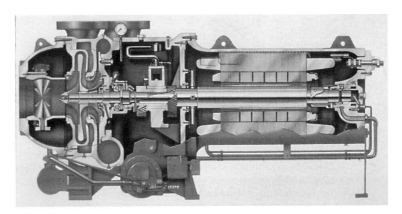

FIGURE 5.4.4-3   Centrifugal chiller.

leaves little doubt about which to choose if absorption is being considered. Triple-effect and quadruple-effect machines are being developed but are not yet on the market.

Although the principles of operation may be difficult to understand, absorption machines are quite simple and require just a few moving parts, including pumps and burner. Modulating the steam valve or the burner controls the capacity, and the part-load characteristic is very linear over the range of operation to a minimum of about 30 percent of peak capacity.

Absorption machines can be direct- or indirect-fired (see Fig. 5.4.4-4). Direct-fired machines have the advantage that they can also be used to heat the building and/or domestic hot water. If a direct-fired absorption machine is also used as a heater, the avoided cost of a separate boiler and boiler room (space) may help offset some of the added cost of the machine. Sizes for absorption chillers range from 100 to 1700 tons. Absorption machines typically cost two or more times the cost of electric-driven chillers. Because a double-effect absorption machine requires as much as 4.5 gpm/ton of condenser water, cooling towers are larger than for an electric chiller plant. Sometimes the economics of using this type of chiller make it the best choice. These are some of the reasons for using absorption chillers:

- Low natural gas cost combined with high electrical cost including demand
- Hybrid of absorption and electric chillers to reduce demand charges
- Electrical service not available or too costly to upgrade
- Gas available from landfill, solar, or biomass
- Waste steam or low-cost steam readily available
- Need for chiller during prolonged periods of normal utility power outage with only emergency power available
- No CFCs, no ODP, and GWP comparable with other alternatives

Engine-Driven Chiller.   Although engine-driven chillers do not comprise a large segment of the chiller market, this type is sometimes economically viable. The engine-driven chiller uses the same vapor compression cycle as an electric machine, except that it uses a reciprocating engine or a gas- or steam-driven turbine as the prime mover. For larger applications, the refrigeration component is usually an open screw or centrifugal chiller. A range of refrigerants may be used, including R-22, R-123, R-134a, and R-717. COPs for engine-driven machines are somewhat higher than for absorption machines and run from 1.5 to 1.9. Because they use variable speed technology, the part-load characteristics are very good, with integrated part-load values (IPLVs) up to COP 2.2. The engines use natural gas or diesel fuel. Some are hybrid units that have both an engine and an electric motor so that the fuel may be switched, depending on the most favorable utility rates at the time. Engines require heat rejection from the jacket water. Heat can be rejected through a cooling tower (through a heat exchanger) or by air cooling for smaller units. The jacket water heat is available for heating domestic water or other loads that occur when the engine runs.

Engines need additional maintenance; top-end overhauls are required every 12,000 hours, and complete overhauls at 35,000 hours. Reciprocating engines are much louder than electric-driven or absorption machines and may require special enclosures or acoustical abatement.

Chiller Performance and Energy Efficiency Ratings

A number of variables determine the operational characteristics and energy performance of water chillers. A chiller is selected to meet a specific maximum capacity requirement at certain design conditions, to have maximum energy consumption at these conditions, and to have specific part-load operation characteristics.

At peak design conditions, the efficiency of water chillers is rated by the *coefficient of performance* (COP). The COP is the ratio of the rate of heat removal to the rate of energy input

TWO STAGE STEAM-FIRED ABSORPTION UNIT

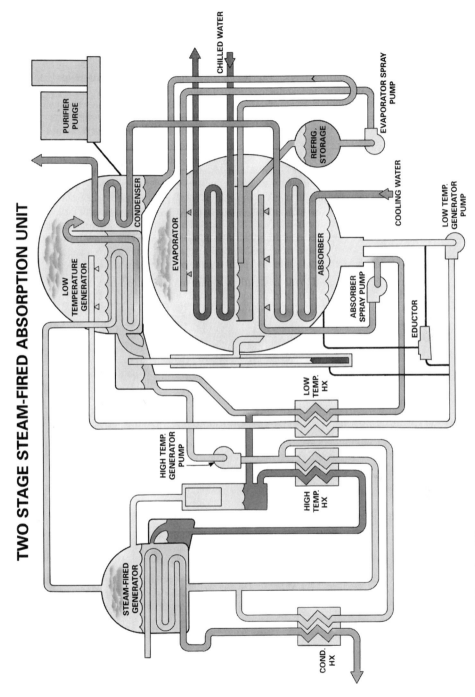

FIGURE 5.4.4-4 Absorption chiller.

in consistent units for a complete refrigerating system or some specific portion of that system under designated operating conditions. The higher the number, the more energy efficient the machine. ASHRAE Standard 90.1-1999 provides minimum energy efficiency standards for water chillers. Many local jurisdictions have adopted the ASHRAE standard as code minimum performance.

Another useful energy efficiency rating is the *integrated part-load value* (IPLV). The IPLV is a single-number figure of merit based on part-load COP or kW/ton. The part-load efficiency for equipment is based on the weighted operation at various load capacities. The equipment COP is derived for 100, 75, 50, and 25 percent loads and is based on a weighted number of operational hours (assumed) at each condition. A *weighted average* is determined to express a single part-load efficiency number.

The *nonstandard part-load value* (NPLV) is another useful energy efficiency rating. This is used to customize the IPLV when some value in the standardized IPLV calculation is changed. Electrically driven chiller efficiencies are also discussed in terms of "kW/ton" for peak ratings, IPLV, and NPLV. This is another way of describing the COP (COP = 3.516/[kW/ton]). The lower the kW/ton, the more energy efficient the machine.

Table 5.4.4-3 is a comparison of energy efficiency ratings of various types of water chillers.

Heat Rejection

One of the prime purposes of a chilled water plant is to reject unwanted heat to the outdoors. This is accomplished in a number of different ways. Although a number of methods exist to reject heat, including cooling tower ponds, lakes, rivers, groundwater, and city water, the primary means of heat rejection in the HVAC industry are the cooling tower, the air-cooled refrigerant condenser, and the evaporative refrigerant condenser.

Cooling Towers.   Simply put, evaporation is a cooling process. More specifically, the conversion of liquid water to a gaseous phase requires the latent heat of vaporization. Cooling towers use the heat from water to vaporize in an adiabatic saturation process. A cooling tower's purpose is to expose as much as possible the water surface area to air to promote the evaporation of water. The performance of a cooling tower is almost entirely a function of the ambient wet bulb temperature. The dry bulb temperature has an insignificant effect on the performance of a cooling tower.

Types of Cooling Towers.   Cooling towers come in a variety of shapes and configurations. A *direct* tower is one in which the fluid being cooled is in direct contact with the air. This is also known as an *open* tower. An *indirect* tower is one in which the fluid being cooled is contained within a heat exchanger or coil and the evaporating water cascades over the outside of the tubes. This is also known as a *closed-circuit fluid cooler.*

TABLE 5.4.4-3   Energy Efficiency Ratings of Typical Water Chillers

Chiller type	Capacity range (tons)*	COP range[†]	IPLV range (COP)[†]
Reciprocating	50–230 (400)	4.2–5.5	4.6–5.8
Screw	70–400 (1250)	4.9–5.8	5.4–6.1
Centrifugal	200–2000 (10,000)	5.8–7.1	6.5–7.9
Single-effect absorption	100–1700	0.60–0.70	0.63–0.77
Double-effect absorption	100–1700	0.92–1.2	1.04–1.30
Gas engine driven	100–3000 (10,000)	1.5–1.9	1.8–2.3

* Capacities in parentheses are maximum sizes available.
[†] COP units are in Btu/h output ÷ Btu/h input

The tower airflow can be driven by a fan (mechanical draft) or can be induced by a high-pressure water spray. Mechanical draft units can blow the air through the tower (forced draft) or can pull the air through the tower (induced draft). The water invariably flows vertically from the top down, but the air can be moved horizontally through the water (cross-flow) or can be drawn vertically upward against the flow (counterflow).

Spray Towers.   Spray towers distribute high-pressure water through nozzles into a chamber where air is induced to flow with the water spray. There are no fans. The air exits out the side of the tower after going through mist eliminators. Spray towers are seldom used. One problem is that the nozzles are easily plugged by the precipitation of mineral deposits and by airborne particulates that foul the water. Varying the water flow through the tower controls capacity. Because air velocities are very low, spray towers are susceptible to adverse effects from wind. On the positive side, spray towers are quiet and can have low first cost.

Forced-Draft Cooling Towers.   Forced-draft towers can be of the cross-flow or counterflow type and can have axial or centrifugal fans. The forward curved centrifugal fan is commonly used in forced-draft cooling towers. The primary advantage of a centrifugal fan is that it can overcome high static pressures that might be encountered if the tower were located within a building or if sound traps were located on the inlet and/or outlet of the tower. Cross-flow towers with centrifugal fans are also used where low-profile towers are needed. These towers are relatively quieter than other types of towers. Forced-draft towers equipped with centrifugal fans are not energy-efficient. This type of tower requires more than twice as much energy as a tower with an axial fan. Another disadvantage of forced-draft towers is that they are more susceptible to recirculation than induced-draft towers because of low discharge air velocities.

Induced-Draft Cooling Towers.   The induced-draft tower is by far the most widely used and energy-efficient cooling tower available in the HVAC industry. These towers can be cross-flow or counterflow and use axial fans. Because the air discharges at a high velocity, they are not as susceptible to recirculation. The large blades of the axial fan can create noise at low frequencies that is difficult to attenuate and could cause problems, depending on the location on the property. The axial fans are either belt-driven or direct (shaft) driven.

Closed-Circuit Fluid Coolers.   One advantage of a closed-circuit fluid cooler is that the fluid is located within a coil (rows of tubes) rather than being open to the environment. A pump draws water from a sump and delivers it to a header where the water is sprayed over the coil. With proper initial chemical treatment, the fluid (usually some form of glycol solution) does not foul the condenser tubes, so chiller maintenance is reduced, and energy efficiency is always at peak. Closed-circuit fluid coolers are much larger and significantly more expensive than conventional open towers of the same capacity because of the additional heat exchange process.

Cooling Tower Application Issues

Siting and Recirculation.   When the saturated air leaving the cooling tower is drawn back into the intake of the tower, the recirculation that occurs degrades the performance of the tower. Wind forces create a low-pressure zone on the downwind (lee) side of the tower that causes this phenomenon. Recirculation can also occur when cooling towers are located so that the discharge from one tower is directed into the intake of an adjacent tower. Recirculation is a greater problem when cooling towers are confined within pits or have screen walls surrounding them. If the tower is sited in a pit or well, it is essential that the tower manufacturer be consulted to determine the proper location of the outlet and minimum clearances for the air intake.

Capacity Control.   Like most air conditioning equipment, cooling towers are selected for peak capacity at design weather conditions. Of course, most of the time they operate at less than peak capacity. A number of methods are used to control the temperature of the water leaving the cooling tower:

- *On/off.* Cycling fans is a viable method but leads to an increase in wear on belts and drives (if used) and can lead to premature motor failure. This is the least-favorable method of controlling temperature.

- *Two-speed motors.*   Multiple wound motors or reduced voltage starters can be used to change the speed of the fan for capacity control. Because of basic fan laws, significant energy savings are achieved when the fans are run at low speed.

- *Pony motors.*   This is another version of the two-speed approach. A second, smaller motor is belted to the fan shaft. For low-speed operation, the larger motor is de-energized, and the smaller motor is energized for a lower speed. This is a cost-effective and energy-efficient approach.

- *Variable speed drive (VSD).*   Adjustable frequency drives can be added to the motors for speed control. This method provides the best temperature control performance and is the most energy-efficient method of control. It is also the most expensive and may not be the most efficient life-cycle cost application compared to two-speed.

- *Modulating discharge dampers.*   Used exclusively with centrifugal fans, discharge dampers built into the fan scroll can be modulated for capacity control. This is a cost-effective way to accomplish close temperature control. Although it does save energy by "riding the fan curve," other methods of capacity control may provide better energy savings.

Chemical Treatment and Cleaning.   Cooling towers are notorious for high-maintenance requirements. The use of cooling towers has been linked with the outbreak of legionellosis (Legionnaires' disease). Cooling towers are very good air scrubbers and can accumulate substantial quantities of dirt and debris. Because they are open to the atmosphere, the water is oxygen-saturated, which can cause corrosion in the tower and associated piping. Towers evaporate water, leaving behind calcium carbonate (hardness) which can precipitate out on the tubes of the condensers and decrease heat transfer and energy efficiency.

Towers must be cleaned and inspected regularly. Well-maintained and regularly cleaned cooling towers have generally not been associated with outbreaks of legionellosis. It is best to contract with a cooling tower chemical treatment specialist.

Air-Cooled Refrigerant Condensers.   Another method of heat rejection commonly used in chiller plants is the air-cooled refrigerant condenser. This can be coupled with the compressor and evaporator in a packaged air-cooled chiller or can be located remotely. Remote air-cooled condensers are usually located outdoors and have propeller fans and finned refrigerant coils housed in a weatherproof casing. Some remote air-cooled condensers have centrifugal fans and finned refrigerant coils, and are installed indoors. The maximum size for remote air-cooled refrigerant condensers is about 500 tons, but 250 tons maximum is more common. Air-cooled chillers are available with capacities up to 400 tons. Air-cooled chillers are used for a number of reasons, including the following:

- Water shortages or water quality problems
- Lower cost than water-cooled equipment
- No need for machine rooms with safety monitoring, venting, etc. with packaged air-cooled chillers
- Less maintenance required than cooling towers

Air-cooled chillers are not as energy efficient as water-cooled chillers. When comparing the energy efficiency of air-cooled to water-cooled chillers, care must be taken to include the energy consumed in the water-cooled chiller by the condenser water pump and cooling tower. Air-cooled chillers have very good part-load performance; the COP improves significantly as the air temperature drops. Remote air-cooled condensers in chilled water plants are very seldom used because of the excessive size of the larger-tonnage machines.

Application and Maintenance of Remote Air-Cooled Condensers

Designing the piping for remote air-cooled condenser application is somewhat tricky. Care must be taken to size the piping properly to limit the pressure drop and also to ensure that oil is carried by the refrigerant and does not accumulate in the piping. The minimum oil-carrying capacity of the hot gas piping needs to account for the lowest load (considering compressor unloading). Double hot gas risers may be necessary. If the condenser is located below the evaporator, the liquid line must be carefully sized to prevent flashing caused by the pressure drop due to pipe friction and also the change in the elevation of the fluid. Additional sub-cooling may be required in this instance. When the condenser is located above the compressor, care must be taken to prevent the liquid refrigerant and oil from flowing backward by gravity into the compressor. Using remote refrigerant condensers greatly increases the likelihood of refrigeration leaks from the piping.

Controls.   When an air-cooled condenser operates, typically the fan runs continually in conjunction with the compressor. When the outside temperature falls, it is possible to decrease the liquid refrigerant pressure too much to overcome the thermal expansion valve (TXV) pressure drop adequately. In this case, controls including the following are required to limit heat rejection:

- *Flooded coil.*   Control valves back up liquid refrigerant into the condenser to limit the heat transfer surface. This requires a receiver and a large refrigerant charge.
- *Fan cycling.*   Usually need multiple fans with one or more cycling on and off to maintain minimum head pressure.
- *Dampers.*   Discharge dampers on condenser fan restrict airflow.
- *Variable speed fans.*   Fan speed modulates airflow.

For systems not intended to run at cold temperatures (less than 40°F), fan cycling is usually the most appropriate choice for control.

Evaporative Condensers.   A pump draws water from a sump and sprays it on the outside of a coil. Air is blown (drawn) across the coil and some of the water evaporates, causing heat transfer. These devices are primarily used in the industrial refrigeration business and have little application in the HVAC industry. Some manufacturers produce small packaged water chillers with evaporative condensers as an integral component.

The effectiveness of evaporating water and the refrigerant in the heat transfer process means that for a given load, evaporative condensers can have the smallest footprint of any heat rejection method. The evaporative condenser results in lower condensing temperatures and consequently is far more efficient than air-cooled condensing. Maintenance and control of evaporative condensers are similar to those for the closed-circuit fluid cooler.

Pumps.   In a chilled water plant, centrifugal pumps are the prime movers that create the differential pressure necessary to circulate water through the chilled and condenser water distribution system. A motor in the centrifugal pump rotates an impeller that adds energy to the water after it enters the center (eye). The centrifugal force coupled with rotational (tip speed) force imparts velocity to the water. The pump casing is designed to maximize the conversion of the velocity into pressure.

Pump Limitations

Inlet Conditions.   It is critical that the pump design and location address the issues of cavitation, net positive suction head (NPSH), and vortexing or air entrainment. For a more detailed discussion, see Art. 5.4.2, "HVAC Systems."

CHILLER PLANT DESIGN

Fuel Choice

The primary fuel choices for central chilled water plants include electric power, fossil fuels, or a combination of the two. Fossil-fuel-driven equipment can utilize direct-fired or steam absorption chillers or engine-driven chillers. The choice of fuel depends on many factors but is primarily based on *life-cycle cost analysis*. A life-cycle cost analysis compares different alternatives and takes into account the first cost, annual operation and maintenance cost, future operations and maintenance cost inflation, and the time value of money.

One of the most important elements of the selection process is a very accurate estimate of the energy usage of each option. The appropriate accuracy and detail necessary for energy calculations depends on the size of the project and the engineering budget. Even for very small projects, chiller plant modeling tools accurately estimate chiller plant performance. For existing projects, measured performance data may be used.

The utility rates used in the analysis are very critical because they will vary over time and are difficult to predict. Given this uncertainty, it is often necessary to assume that current rates, or something similar, will be in effect during the chiller plant's life cycle. Virtually all utilities charge both for energy consumption and for demand. Because chillers are one of the largest energy users in typical buildings, it is essential that demand charges be properly taken into account. This is particularly true when demand charges are *ratcheted,* which means that the owner pays some percentage of the maximum peak demand over the year regardless of the actual monthly demand. Other factors that enter into the fuel choice include the need to operate the chiller plant under prolonged power outages, or the availability of waste heat from a cogeneration solar or biomass source. In some cases, the availability of the fuel or the cost of bringing the fuel onto the site may determine the best alternative.

Sizing the Chiller Plant

Before a chilled water plant is designed, it is essential to understand how the plant will be operated and what loads it will be expected to handle throughout its service life.

Peak Load Estimation.   The process for estimating peak cooling loads in new construction is well understood from *ASHRAE Handbook—Fundamentals*[2] and is also discussed in Art. 5.4.2, "HVAC Systems."

When faced with an expansion or remodeling, the opportunity exists to monitor the existing plant for peak load. The plant may have an automated system that has trend logs for monitoring peak loads. Many times, a good operator can accurately report the percentage of full load that the plant experiences during peak weather conditions.

For most designers, the perceived risks of understating the peak load condition (undersizing the cooling plant) are much greater than overstating the peak load. An undersized cooling plant may not meet the owner's expectations for comfort and may impact the ability to manufacture products or provide essential services. Conversely, oversizing the cooling plant carries an incremental first-cost penalty that is not always easy to identify. An oversized plant may not be as energy-efficient as a smaller plant. The tendency is for designers to maximize assumptions for peak load and to add safety factors at several levels in the calculation process. Conversely, diversity of loads is not always well understood. It is this author's opinion that load-estimating techniques used today tend to overestimate peak cooling loads, so that most central cooling plants are oversized.

Annual Cooling Loads

A *cooling load profile* is a time series of cooling plant loads and correlated weather data. The primary role of the cooling load profile is to facilitate the correct relative evaluation of alternative design options. Having an accurate understanding of the cooling load profile will affect the plant configuration. For example, a plant that serves a hotel complex that has long periods of very low loads would be designed differently from a plant that serves widely varying loads only in mild and warm weather during the daytime (as for an office building).

DESIGN APPROACH

Optimizing Energy Efficiency

Normally, chilled water plants run at peak load for only a few hours a year. During the remainder of the time the plant will operate at part load. The following factors are key in designing a chilled water plant for optimum efficiency:

- Number and size of chillers
- Type and size of heat-rejection device
- Peak and part-load efficiency of chillers
- Evaporator and condenser water temperatures
- Temperature difference across evaporator and condenser
- Type of chilled water distribution system
- Method of control.

Number and Size of Chillers.   The number and size of chillers has significant impact on part-load operating performance. The load profile of the building plays a very important role in the selection of the number and size of chillers. For example, buildings that operate for long hours at low loads may run more efficiently with multiple chillers, one of which is sized to handle the low load. In this example, the use of a variable-speed drive on the small chiller may also be cost effective. A single chiller may be most appropriate for small plants.

Providing a life-cycle cost analysis based on a customized load profile is a time-tested way of determining the optimum number and size of chillers to purchase. Understanding the first-cost implications and the energy benefits for various size chillers is beneficial. One way to secure the optimum selection is to provide a procurement process that allows the vendors to mix and match their products over a wide range that meets the peak load requirement. This allows pricing to take advantage of a particular "sweet spot" in the vendor's selections. Based on the equipment selected, the part-load operating characteristics can be evaluated on a computer simulation model to determine the annual energy impact of the selections. This can be put into a life-cycle cost analysis to determine the lowest life-cycle costs for the project.

Type and Size of Heat-Rejection Devices.   Heat-rejection devices are not as readily adaptable to the chiller procurement method mentioned previously. Water-cooled units are invariably more energy efficient than air-cooled units, but air-cooled units may have a first-cost advantage. Again, the first costs should be analyzed along with the annual energy costs to determine the optimum life-cycle costs. When selecting a cooling tower, many times the incremental first cost for increasing the size of a cooling tower (oversizing the towers) can be justified by the increased energy efficiency of lower condenser water temperatures or increasing the number of hours during which the fans are at low speed.

Optimizing Evaporator and Condenser Water Temperatures.   The energy efficiency of a water chiller is a direct function of the temperature of the entering condenser water and the leaving evaporator temperature. Raising the evaporator temperature increases the efficiency of the cooling process but also has an impact on the amount of water that needs to be pumped to achieve a given load. The greater amount of water pumped may have an impact on the sizing of the piping or pump head and hence the first cost of the project. Conversely, lowering the evaporator temperature may have the opposite effect. Likewise, lowering the condenser water temperature will increase the chiller efficiency but may require more cooling tower fan energy.

Similarly, the operating temperature difference between inlet and outlet water temperatures for the evaporator and condenser of the chiller may have a significant impact on the energy efficiency of the plant. With higher differential temperatures, less water is pumped, which reduces piping first cost and pumping energy cost. Yet, higher temperature differences may mean decreased energy efficiency at the chiller. This interplay among the variables is not always easy to understand, and consequently few chiller plants are truly optimized. Yet, using modern computer simulation programs, models can be made to test the options available to the designer, so that optimum designs are more likely.

CHILLED WATER DISTRIBUTION SYSTEMS

Constant-Flow versus Variable-Flow Systems

The chilled water distribution system melds the chillers, pumps, piping, cooling coils, and controls into a dynamic system that provides mechanical cooling for a large number of air conditioning applications. Because this comprises one of the most energy-intensive systems utilized in buildings, understanding how the systems react with varying loads and interactions among the components is essential for designing the system with the most effective life-cycle cost.

Chilled water distribution systems should not be more complex than needed for the application, but they should not be less complex than needed to satisfy performance and energy efficiency. Originally, chilled water plants were constant-volume designs that supplied a constant stream of chilled water to three-way valves at cooling coils or to entire buildings. When applied with multiple chillers during off-peak periods, problems with water distribution and low return temperatures made shutting down unneeded machines for energy conservation very problematic. The application of variable-flow systems was the next logical step, but a whole new series of issues surfaced.

Constant-Flow Systems

The simplicity of a constant-flow, chilled water system is one of the primary attractions of this approach. In constant-flow systems, the flow through the chiller(s), as well as the flow in the distribution piping and at the cooling coil, is constant. Most constant-flow systems use three-way valves at the cooling coils. The following are examples of constant-flow systems.

Single Chiller Serving a Single Cooling Load.   When a single chiller serves a single cooling coil, the simplest approach is to use a constant-volume pump to circulate water between the evaporator and the coil and to delete the traditional three-way control valve. One caution when applying this approach is that manufacturers will insist that there be a sufficient volume of water in the piping system to prevent unstable temperature swings at the chiller. Often, small storage tanks are required when the chillers are closely coupled to the coils.

Single Chiller with Multiple Cooling Loads.   When applying a single chiller with multiple cooling coils, using a constant-flow chiller with three-way valves at the cooling coils is a simple time-tested way to achieve a life-cycle cost-effective system. An energy-saving control strategy for this approach is to reset the leaving water temperature of the chiller based on the position of the coil valves that require the coldest temperature.

Multiple Parallel Chillers with Multiple Cooling Loads.   On the surface this approach seems simple, but problems arise during periods of part-load operation. When both (or all) of the chillers and pumps operate with a nearly full load, the system will work well, but there is little or no opportunity for pumping or cooling tower energy savings. At some point, the load is reduced enough so that one chiller and pump could theoretically handle the load. By turning off one chiller and pump, the reduction in flow from the central plant basically starves all of the coils in the system. This design can still work for many applications, provided that all the loads in the building tend to change together; for example, no one unit demands full flow while others require very little flow.

Multiple Series Chillers with Multiple Loads.   One solution to providing multiple chillers on a constant-flow system is to arrange the chillers in series. Then, all of the flow goes through each machine. This method is effective for systems designed with a very high temperature difference. During off-peak periods, the lag machine is turned off, and the lead machine continues to deliver chilled water at the correct temperature. This system works well, although it does not provide chilled water pump energy savings during periods of low load.

Variable-Flow Systems

As can be seen from the discussion of constant-flow systems, the idea of varying the flow in the system has appeal in larger systems that have multiple chillers and multiple loads. The basic advantage is that the plant can effectively be turned down during periods of low load and that an opportunity for significant energy savings is available. One of the most popular design concepts for multiple machine chiller plants is primary/secondary pumping. Systems that use a primary-only, variable-volume design approach are being applied more because of greater simplicity and cost-effectiveness.

Primary-Only Variable-Flow Design.   Primary-only, variable-flow systems consist of single or multiple chillers with system pumps that move the water through the chillers and distribution system to the cooling load (see Fig. 5.4.4-5). The cooling loads are controlled with two-way valves. Typically, a bypass line with a control valve diverts flow from the supply piping to the return piping to maintain either a constant flow through the chiller(s) or to maintain a minimum flow through the chiller(s). This approach has a simplicity that makes it very attractive. However, there are several issues of concern. The bypass valve acts against a relatively high pressure differential, so it is susceptible to wear, cavitation, and unstable operation at low loads. In some cases, the bypass valves have been located at the ends of the distribution loops. This ensures circulation in the main loops and reduces the pressure differential across the control valve. If the bypass valve maintains a constant flow through the chiller(s), there will be no pump energy saved as the loads vary, but pumps can be shut off as chillers are disabled. Variable speed drives can be added to the primary pumps so that as the demand falls from maximum to minimum, the speed can be adjusted downward, thus saving pump energy.

Primary/Secondary Variable-Flow Design.   Primary/secondary variable flow design has become the standard approach for designing large central chilled water plants using multiple chillers with multiple cooling loads (see Fig. 5.4.4-6). The beauty of the primary/secondary approach is that the piping loop for chillers (primary) is hydraulically independent (decoupled) from the piping loop for the system (secondary). The key to the design is that two independent piping loops share a small section of piping called the *common pipe*. A review of the flow patterns in the common pipe reveals that when the two pipe loops have the same flow

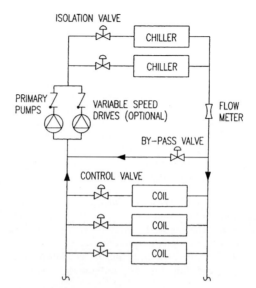

FIGURE 5.4.4-5 Diagram of primary-only, variable-flow system.

rate, there is no flow in the common pipe. Depending on which loop has the greater flow rate, the flow direction in the common pipe is subject to change. The primary pumps are typically constant-volume low-head pumps intended to provide constant flow through the chiller's evaporator. The secondary pumps deliver the chilled water from the common pipe to coils with two-way valves and then return it to the common pipe. These pumps are variable speed pumps controlled from differential pressure sensors located remotely in the system.

Chilled Water Plant Machine Room Considerations

Refrigerants are classified into groups according to toxicity and flammability. Refrigeration systems are also classified according to the degree of probability that a leakage of refrigerant could enter a normally occupied area. In "high-probability systems," leakage from a failed connection, seal, or component could enter the occupied space. Such is the case with refrigeration components located within an occupied space. "Low-probability" systems include systems whose joints and connections are effectively isolated from the occupied space. This is the case with chillers, condensers, and other equipment located in refrigeration machine rooms that are isolated from the normally occupied space.

ASHRAE Standard 15 *Safety Code for Mechanical Refrigeration* requires that when the quantity of refrigerant in the system or compressor size exceeds established limits, the equipment must be located outdoors or within a "machinery room."[3] A machinery room must be designed within strict guidelines, some of which include

- Continuous and emergency ventilation separate from other building systems
- Continuous monitoring, equipment shut-downs, and remote alarms
- No open flames
- Tight fire-rated room and door construction
- Exiting to outdoors

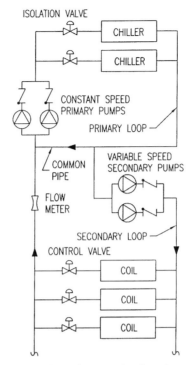

FIGURE 5.4.4-6 Diagram of primary/secondary variable-flow system.

- Special signage
- Restrictions on the location of relief and exhaust outlets and intakes

The chiller plant should not be located within the same space as other mechanical or electrical equipment except that equipment directly needed for operating the chillers. Under no circumstances should be the chiller plant be located within the boiler room, unless the boiler combustion air is ducted directly to the boiler intake. When retrofitting or upgrading an existing chiller plant, care should be taken to include the upgrade of the machinery room when needed. This typically means that the chillers must be separated from the boilers and other equipment in accordance with the requirements of the standard.

Chiller Plant Controls and Instrumentation

Controls. The chilled water plant is one of the most energy-intensive spaces within a facility, and therefore special care must be taken to ensure that the plant is operated to conserve energy and provide long-term, reliable service. At the heart of this effort is the automatic control system. The control system must be selected to operate in the chiller plant environment. Chillers and other plant equipment generate heat and vibration that may adversely affect controls and instrumentation.

Microcomputer-based direct digital control (DDC) systems and networks are normally provided in a modern chiller plant. Chiller plant controls incorporate factory-installed instrumentation that can be accessed through network connections. This enables control and monitoring functions to occur over a network. The instrumentation points must be carefully chosen to ensure the proper level of data so that the systems can be optimally controlled. Too much instrumentation can be confusing to the operations staff and difficult for them to maintain. The selection of control and monitor points should be based on a careful analysis of the chiller plant's control and operating requirements. To justify including a particular control or monitoring point in a chilled water plant, that point must meet at least one of the following criteria:

- It must be necessary for effective control of the chiller plant as required by the sequence of operations established for the plant.
- It must be required to gather necessary accounting or administrative information such as energy use, efficiency, or run time.
- It must be needed by the operations staff to ensure that the plant is operating properly or to notify the staff that a potentially serious problem has occurred (or may soon occur).

Controllers used in the system must have a powerful and flexible program language and the ability to be interfaced with chiller networks, variable frequency drives, and power monitoring networks. The most economical method of integrating the instrumentation into the DDC system varies by manufacturer. It is advisable to specify a BACnet gateway between the chiller(s) and the DDC system. BACnet is a set of rules for designing software protocol that allows different vendors the ability to share and communicate data between different programs. If for some reason BACnet is not used as the system protocol, the chiller and controls manufacturers must have an interoperable system so that there is full compatibility without any special gateways or connections.

Performance Monitoring

Performance monitoring can help identify opportunities for energy efficiency. Integrating chiller plant monitoring with the control system helps the plant operating staff to determine the most efficient equipment configuration and settings for various load conditions. It also helps the staff to schedule maintenance activities at proper intervals, so that maintenance is frequent enough to ensure the highest levels of efficiency, but not so frequent that it incurs unnecessary expense. Most DDC systems that can operate chiller plants effectively are well-suited to providing monitoring capabilities. Because chiller plant efficiency is calculated by comparing the chilled water energy output to the energy (electric, gas, or other) required to produce the chilled water, efficiency monitoring requires only the following three items:

- *Chilled Water Output.*   Because the control instrumentation already includes chilled water supply and return temperatures, only a flow sensor must be added to normal chilled water plant instrumentation.
- *Energy Input.*   To obtain the total energy input, it is necessary to install kW sensors on the tower fans, condenser pumps, chillers, and chilled water pumps. It may also be possible to use only one or two kW sensors to measure the total energy used by the plant. Finally, it is often acceptable to use a predetermined kW draw for constant-speed fans and pumps whenever they are operating.
- *DDC Math and Trend Capabilities.*   In addition to the instrumentation requirements, efficiency monitoring requires that the DDC system chosen have good mathematical functions, so that the instrumentation readings can be easily scaled, converted, calculated, displayed, and stored in trend logs for future reference.

Start-Up and Commissioning

Most chilled water plants are custom-designed for a particular facility. The process of design and construction involves many skilled professionals, who must perform their tasks well if the project is to be a success. Because we are all human, errors can occur. Commissioning is intended to achieve the following objectives:

- Ensure that equipment and systems are properly installed and receive adequate operational checkout by the installation contractors
- Verify and document the proper operation and performance of equipment and systems
- Ensure that the design intent for the project is met
- Ensure that the project is thoroughly documented
- Ensure that the facility operating staff is adequately trained

Commissioning Activities

Depending on the size, complexity, and budget of the project, the tasks involved in commissioning can vary widely. Consequently, there can be a number of phases and levels in the commissioning process.

Design Phase. In the design phase of the project, commissioning activities can include assisting in selecting the system configuration, providing design review, and preparing the commissioning specification.

Construction Phase. Construction-phase activities prepare for the acceptance-phase testing procedures. The goal during the construction phase is to complete the installation efficiently with minimal problems to hamper acceptance-phase activities. Commissioning tasks during construction include submittal review, construction observations, preparation of test procedures, and coordination with the team.

Acceptance Phase. The acceptance phase is where the commissioning process pays great dividends. The acceptance tests are singularly the most important part of the commissioning process. These activities include verification of control devices, functional performance testing, coordination of operator training, and development of a systems manual.

CONTRIBUTORS

Anand K. Seth, P.E., C.E.M., C.P.E., Director of Utilities and Engineering, Partners HealthCare System, Boston, Massachusetts

REFERENCES

1. J. M. Calm and D. A. Didion, *Trade-Offs in Refrigerant Selections: Past, Present, and Future,* American Society of Heating, Refrigerating, and Air Conditioning Engineers, Inc., Atlanta, GA, 1997.
2. ASHRAE, *1997 ASHRAE Handbook—Fundamentals.* American Society of Heating, Refrigerating, and Air Conditioning Engineers, Inc., Atlanta, GA, 1997.

3. ASHRAE, *Standard 15 Safety Code for Mechanical Refrigeration,* American Society of Heating, Refrigerating, and Air Conditioning Engineers, Inc., Atlanta, GA, 1994.

General References

AGCC, *Natural Gas Cooling Equipment Guide,* American Gas Cooling Center, Arlington, VA, 1996.

ASHRAE, *1998 ASHRAE Handbook—Refrigeration.* American Society of Heating, Refrigerating, and Air Conditioning Engineers, Inc., Atlanta, GA, 1998.

ASHRAE, *1996 ASHRAE Handbook—HVAC Systems and Equipment,* American Society of Heating, Refrigerating, and Air Conditioning Engineers, Inc., Atlanta, GA, 1996.

Carrier, *Centrifugal Refrigeration Equipment,* TDP Manual 791-329, Carrier Corporation, Syracuse, NY, 1983.

Carrier, *Reciprocating Liquid Chilling Equipment Technical Development Manual,* TDP Manual 791-332, Carrier Corporation, Syracuse, NY, 1983.

Cler, G. L., *TechUpdate: GasChiller Buyer's Guide,* TU-97-11, E-Source, Boulder, CO, 1997.

Fryer, L., *Tech Update: Electric Chiller Buyers Guide,* TU-95-1, E-Source, Boulder, CO, 1995.

Marley, *Cooling Tower Fundamentals,* The Marley Cooling Tower Company, Mission, KS, 1982.

Pacific Gas and Electric Company, *Chilled Water Plant Design and Performance Specification Guide,* Pacific Gas and Electric Company, San Francisco, CA, 1999.

Smit, K., et al., *Electric Chiller Handbook,* EPRI TR-105951, Electric Power Research Institute, Palo Alto, CA, 1996.

Trane, *Trane Air Conditioning Manual,* The Trane Company, LaCrosse, WI, 1974.

ARTICLE 5.4.5
COGENERATION

Friedrich W. Finger III, P.E.
The Finger Company, Inc., Duxbury, Massachusetts

DEFINITION

Cogeneration is sequentially converting a primary fuel energy source such as oil, gas, coal, or biomass into power, typically electricity and thermal energy. By capturing and converting the useful energy (from an engine, turbine, or other mechanical or electrical generating device) that would otherwise be rejected to the environment, the cogenerator operates at significantly higher efficiencies than those achieved by a basic generating device (Fig. 5.4.5-1).

INTRODUCTION

The engineering principles behind cogeneration have long been understood, and the technology has been refined and developed over the years. Modern cogeneration systems can achieve efficiencies of up to 90 percent. Cogeneration offers a great amount of flexibility, and usually a combination of plant and fuels can match most individual requirements.

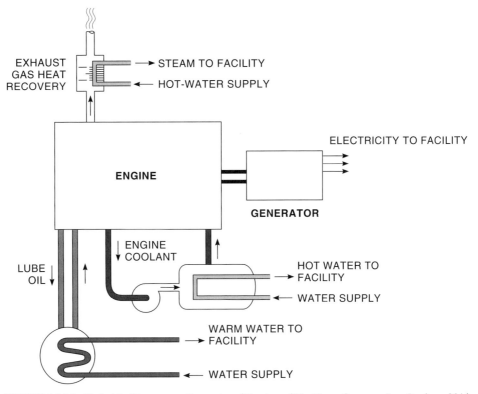

FIGURE 5.4.5-1 Typical facility cogeneration system. (*Courtesy of The Finger Company, Inc., Duxbury, MA.*)

Cogeneration recovers the waste heat that is always produced in power generation and puts it to good use rather than letting it escape into the atmosphere. In conventional power generation, about two-thirds of the energy input is wasted in this way. Cogeneration can recover the majority of this waste heat, uses resources economically, and often results in energy savings of 20 to 40 percent.

Cogeneration contributes to a more sustainable energy future by minimizing the environmental damage caused by commercial activities. The most obvious and well-known benefits are the carbon dioxide savings. Cogeneration routinely results in savings of up to 50 percent of CO_2 emissions compared with conventional sources of heat and power. Reduced emissions of sulfur dioxide and particulates are also added benefits.

A well-designed and well-operated cogeneration plant will always improve energy efficiency and significantly reduce carbon emissions. With a typical supply efficiency of between 70 and 95 percent, cogeneration is the best all-around solution for combined power and heat production.

Cogeneration is a field of renewed interest in the energy-conscious climate that is prevalent in industry today. It can satisfy both the electrical and process heat needs of a facility. Because natural gas is plentiful and the cost of small gas-powered electric generators is decreasing, small industries can produce power at a cost that is comparable with that of local utilities. In addition, because the thermal energy normally disposed of in producing power is captured and used, the energy utilization efficiencies previously mentioned are attainable.

Cogeneration can be accomplished in a number of ways. The most common is by using a gas turbine generator or reciprocating engine generator coupled with a waste heat recovery boiler or a steam boiler coupled with a steam turbine generator. The main difference between the two types of systems is the order in which the electricity is obtained. The gas turbine or reciprocating engine produces electricity first, then the hot exhaust gases are sent to the waste heat boiler to generate steam, a process known as a *topping cycle*. When a boiler produces steam that is first sent to a steam turbine to generate electricity, and the extraction and/or exhaust steam is used for a thermal process, this is also a topping cycle. The less common bottoming cycle is when the steam produced is first sent to a thermal process and the exhaust is used to produce power.

Depending upon the requirements and the nature of an installation using cogeneration systems, each system has advantages. The gas turbine and reciprocating engine systems are much better for new installations. The amount of power produced by a gas turbine or reciprocating engine system, for a given heat demand, is superior to that of the boiler and steam turbine system. For retrofit applications, where a boiler is already installed and operating, the steam turbine may be ideally suited. Many plants generate steam at a higher pressure than necessary and then throttle the steam to a lower pressure before sending it to the intended process. Replacing the pressure-reducing valve with a steam turbine recovers the energy wasted in the throttling process and converts it to electricity. Because small steam turbines are relatively inexpensive, the initial cost is relatively minor.

COGENERATION HISTORY

Many thermal power plants built in the early part of the twentieth century were designed as cogenerating units. Power plants that were used to supply electric street railways and electric streetlighting used simple back-pressure steam turbines in many cases, and their discharge provided district heating steam or hot water. Steam plants at mills and factories generated electric or mechanical power to drive machinery, and the discharge supplied steam and hot water for processes such as cooking, drying, and heating. During the midcentury, more efficient central power generators and electric drives were emphasized without considering cogeneration. During the latter part of the century, cogenerators and combined cycle units gained prominence as reducing discharges to the environment and more efficient utilization of fuel supplies became paramount.

Following the 1972 fuel crisis, a number of regulations and laws were enacted in the United States that encouraged the use of cogeneration. The Public Utilities and Regulatory Policy Act (PURPA) provides specific benefits to cogenerators. Other provisions by the Federal Energy Regulatory Commission (FERC) and the Securities and Exchange Commission (SEC) favor the development of cogeneration. Consequently, many cogenerating plants of widely differing configurations have been developed, ranging from small (kilowatt output) units for facilities to medium-size (megawatt output) plants for campuses, industrial parks, and manufacturing facilities, and to large central power stations (100+-MW output) located adjacent to a thermal host. For the purposes of this handbook, the smaller kilowatt output range cogenerating units for facilities and the medium megawatt output range plants for facility complexes are addressed (Fig. 5.4.5-2).

The application of cogeneration depends upon the ability to match the electrical and thermal output of the cogenerating unit to the demands of the facility, and backup for both output energy streams is available when the cogenerator is out of service.

ASSESSMENT OF COGENERATION OPTIONS AND NEEDS

Power companies have established a vast number of rate structures for determining the cost of electrical usage. Deregulation has increased the number and the complexity of the rates available to electrical consumers. Most rates for industrial and commercial users include an

FIGURE 5.4.5-2 Blodgett Pool cogeneration module at Harvard University. Tecogen CM-60 cogeneration module maintains the 750,000-gal, 75-m Blodgett Pool at Harvard University at 80°F. The cogeneration module was installed July 1984. Key data from installation to February 1993 is as follows: total run hours—46,000 h; thermal output—202,400 therms; and electric output—2,760,000 kWh. The cogeneration module produces over 300,000 kW hours of electricity per year for building lighting, reducing electric utility charges by more than $20,000 annually. (*Courtesy of Tecogen, Inc., Waltham, MA.*)

energy charge and a demand charge. The energy charge is the charge for the actual electrical energy usage. Power companies often vary this rate depending upon the time of day during which the electricity is used. The demand charge is a fixed cost that is related to the highest electrical usage in a predetermined period (such as a billing month) and added to each month's bill. The newest electrical rates in the deregulated environment vary on an hourly basis, and on a given day they can spike up in excess of $0.80/kWh.

A decision to cogenerate electricity must be made only after a detailed analysis to determine the effect on utility charges, for which an in-depth understanding of the electric rate structure is essential. Most utilities now employ customer representatives who have detailed knowledge of the new rate structures. These representatives can suggest which rate structure will yield the lowest energy cost for a given electrical profile. However, they are representatives of the energy supplier. It is in the best interests of the suppliers to maximize profits from each of their customers. Because cogeneration causes the energy supplier to lose electrical sales, the advice from a utility representative should be considered along with the capable advice of a selected consulting partner.

Cogeneration represents a unique opportunity to control energy costs with a high-efficiency, on-site power plant. To get the best return on the cogeneration investment, however, four conditions should be met.

Energy Efficiency

Before cogeneration can be cost-effective, the facility itself must be energy-efficient. It makes no sense to install a high-efficiency cogeneration plant to supply steam to an inefficient distribution system or to generate electricity with a high-efficiency cogeneration plant when the facility itself uses electricity inefficiently. Improve efficiency first, and then consider cogeneration.

Seasonal Load Requirements

Efficient use of cogeneration systems requires that there be a year-round demand for both electricity and thermal energy as steam or hot water. Most facilities use steam or hot water during the heating season, but demand drops off significantly during the summer months. Examine the load profiles for your facility to determine if there is sufficient demand for heat energy during the summer to make economical use of steam or hot water from the cogeneration system. If not, consider installing absorption chillers to increase steam load requirements. By teaming or partnering with a consultant or a consulting firm that knows your industry and process and that has experience in cogeneration applications, you can determine the right mix of electrical and heat energy requirements for the process, or economical and efficient facility changes that take advantage of a cogeneration system to reduce total energy consumption.

Daily Load Profile

Just as loads change with seasons, electrical and heat loads change with the time of day. To be most effective, the right mix of loads must be present 24 h/day. If not, the system will sit idle for long periods of time each day, or will have to dump excess heat or reduce the quantity of electricity generated. All of these factors decrease the efficiency of the system and extend the payback period.

There are ways to modify the energy load profile to make cogeneration more cost-effective. For example, if excess heat energy is generated at night during the cooling season, consider installing an ice storage system driven by absorption cooling. Again, your consulting partner experienced with cogeneration systems can examine energy use in your facility to determine if a suitable mixture of loads exists for cogeneration.

Utility Company Support

In a deregulated market, the facility manager will have to work closely with the energy supplier or local utility. The best approach is to involve the utility and your knowledgeable consulting partner early. The utility can help evaluate the feasibility of the project. In addition to being a potential customer for excess electricity generated by the system, the utility can address issues such as backup power and the requirements for interconnection.

COGENERATION CRITERIA

A general discussion of the most important parameters that affect the choice of a cogeneration system should be undertaken. Once the most important factors have been identified and reviewed, a system that best satisfies the requirement can be selected.

The first selection criterion is the magnitude of each type of load, both thermal and electrical. If either of these is relatively low, or even nonexistent, then a cogeneration system is obviously not an option. In many cases, no thermal load exists. In such cases, a combined cycle generating system may be an alternative to cogeneration. If it does appear that pure power generation is an economic possibility, a detailed study of the power company rate structure that serves the facility should be performed. It is likely that changing to another rate structure would lower electrical costs enough to make the pure power generation option economically undesirable.

A second selection criterion is the size of the electrical and thermal loads relative to each other. This should not be confused with the first criterion. We assume that the magnitude of each type of load is sufficient to consider a cogeneration system. For high electrical usage versus thermal usage, a system with a higher electrical efficiency is desirable (e.g., a reciprocating engine generator). If the opposite is true and the thermal load outpaces the electrical load, then a steam turbine would better suit the application. Finally, if both are relatively equal, then a gas turbine system might be the initial system to analyze.

Not only is the relative magnitude of the thermal and electrical loads a criterion, but also the time dependence of each load must be considered. Loads that vary considerably with time can cause undesirable effects on certain systems, much more so than on others when a load-following operational strategy is used. This holds for both electrical and thermal loads, although if properly sized, transient thermal loads can often be handled by adjusting the firing of a heat-recovery steam generator (HRSG). A reciprocating engine generator responds much better to changing loads than a gas turbine, in terms of efficiency and reliability. Steam turbines can match loads well by simply throttling the steam flow through the turbine.

The type of fuel most readily available is an important consideration when choosing a cogeneration system. For almost every fuel, there is a system capable of using it. Gaseous fuel, such as natural gas, is most commonly used in gas turbines, but it is also used in natural gas-fired reciprocating engines. Fuels such as No. 2 and No. 6 oils are burned in reciprocating engines, and No. 2 oil is a backup fuel for a gas turbine. Solid fuels, such as coal and biomass, are almost exclusively used in the large base-loaded boiler cycles. Except for solid fuels, almost any fuel can be used a cogeneration system, so a certain amount of flexibility exists. However, using a fuel other than the ideal increases operating and maintenance costs, increases forced and scheduled outage periods, and decreases equipment life.

The type of industry that chooses to cogenerate often determines the fuel, and thereby the cogeneration system to be used. The paper industry, which generates a great deal of biomass and chemical by-product fuel, generally opts for a conventional boiler cycle system to use the readily available fuel source. The boilers burn bark removed from the incoming logs and a chemical liquid (black liquor) generated in the pulping process. Similarly, the petroleum industry most often relies on fuel oil as a heat source. The available supply and low cost of using their own fuel oil make it economical to do so. However, for those industries that do not generate a fuel source in their production process, natural gas is often the best choice due to low cost, high efficiency, ease of transport, and low capital cost of the storage and distribution equipment.

Pollution concerns have become particularly important in recent years, especially in heavily populated areas. Gaseous fuels tend to have the lowest emissions, followed by fuel oils, and finally solid fuels. Large industrial locations that use solid fuels are equipped with scrubbers and/or precipitators to remove contaminants, particulate matter, and other pollutants from the exhaust stream. These pollution abatement and control systems add considerable capital cost to a cogeneration installation. An economic analysis should include the additional capital outlay. High-sulfur fuel oils may also require expensive pollution control equipment. Low-sulfur fuel oils are available but also at a higher cost. Finally, the efficient operation of each of the systems will minimize the pollutants generated in the combustion process. If the combustion process for any of the systems is poorly managed (through combustion, air, etc.), or maintenance is not performed at required intervals, pollutants can dramatically increase.

The physical space available for a cogeneration system often affects the type of equipment used. Gas turbines and reciprocating engine generators are compact, packaged units that are simply dropped into place, attached to the fuel, steam, and electrical systems, and started.

Steam turbine systems usually require more on-site preparation, but only because drop-in packaged units do not exist. For completely new systems, steam turbine cogeneration systems are the most expensive due to the high cost of the boilers, condensers, and other associated operating equipment.

Some general guidelines have been developed through experience for selecting prime movers. Specifically, reciprocating engines are well suited to smaller systems up to 3000 kW or to systems that operate only during peak-use energy periods (because of the relatively short operational time). Gas turbines perform best in moderately larger applications from approximately 5000 kW up to several hundred megawatts. Steam turbines are ideal for the largest applications or applications where solid fuel is used, because the large boilers that use this type of fuel produce enough steam to allow for huge extraction turbines to produce sizable amounts of electricity. Steam turbines also perform well when steam is required at different pressures.

COGENERATION ALTERNATIVES AND OPTIONS

Depending on the results of a study of a facility's specific needs integrated with a detailed analysis of the available utility rate structures and the facility load profile, the cogeneration option can be addressed. Specific cogeneration systems should be evaluated. These systems may include the following:

- A gas-fired boiler to produce steam and a back-pressure steam turbine to generate electricity and provide low-pressure process steam
- A gas turbine generator system (GTG) to produce electricity and an HRSG to produce low-pressure process steam
- A gas turbine generator system (GTG) to produce electricity and an HRSG to produce high-pressure steam that supplies a steam turbine generator (STG) and low-pressure process steam
- A reciprocating engine generator to produce electricity and a heat recovery steam generator to produce steam and hot water for process requirements (Fig. 5.4.5-3)
- A reciprocating engine generator to produce electricity and an absorption cycle chill water cooling system for air conditioning or process cooling

A combination of these systems or a noncogeneration system (that is incorporated into an existing plant facility or used to develop a new energy facility) that meets the process requirements and reduces the total energy costs, is considered before selecting the final system.

LIFE CYCLE CONSIDERATIONS

Operational cost is a key factor in choosing a cogeneration system. Systems that have high fuel, high maintenance, or high supervisory costs will undermine any savings gained from cogeneration. Systems that have low availability, a low capacity factor, or are unreliable in any way will also undermine any savings gained from cogeneration.

Generally, reciprocating engine generators have the highest operating costs in terms of downtime and preventive maintenance because of the high number of moving parts in the system. Steam turbine systems have lower maintenance costs than reciprocating engines, and gas turbines have the lowest of all.

Life cycle considerations should address replacement power and thermal energy costs and systems, electric supplier connection charges, and electricity capacity demand charges that may be a condition of maintaining backup power.

FIGURE 5.4.5-3 Fairbanks Morse six-cylinder Model 38ETDD8-1/8 reciprocating engine generator set installed at a midwestern hospital. (*Courtesy of Cobb Industries, Fairbanks Morse Engine Division, Beloit, WI.*)

Fuel costs, replacement or standby fuel sources, contract maintenance costs, and replacement or lease engine charges for the cogeneration system should be considered when evaluating the project. Your consulting partner, who is experienced in operating and maintaining cogeneration system options, and the original equipment manufacturer (OEM) can assist in developing and evaluating the life cycle considerations.

CONCLUSION

A simple payback can be calculated for each operating strategy, each type of cogeneration system, the location of the cogeneration plant, the generator type and size, a grassroots project, or a modification of an existing facility. The economic analysis can be carried to the degree of detail required to demonstrate the economic viability and to meet facility capital planning and approval requirements. Ultimately, the relative attractiveness of each cogeneration option strategy can be demonstrated by using the payback method.

Reasons for considering cogeneration at a facility include, but are not limited to, the following:

- Balancing the facility energy profile by changing the electrical and thermal energy mix and using a cogeneration cycle that reduces the cost of energy
- Modifying an existing energy facility by using a cogeneration cycle that improves efficiency and reduces the total cost of energy

- Replacing an obsolete facility energy plant to improve operational reliability, to increase process production, and to reduce energy cost
- Reducing environmental emissions to comply with regulatory requirements or to improve local environmental quality and lower total energy costs

The opportunities for cogeneration are limited by the physical plant arrangement, the primary mission and process of the facility, the ability to deal with the local electric energy supplier, the availability of a reliable fuel source, the imagination and leadership of the facility manager, and most important, the economic viability of the proposed cogeneration project. There are many excellent references and available expertise on all aspects of cogeneration that can assist in developing and implementing a cogeneration project.

For many facilities, cogeneration will be an attractive alternative to conserve precious resources and minimize environmental effects. Cogeneration can be a win-win solution: an efficient, low-cost producer for the owner and a consistent good neighbor in the community.

CONTRIBUTORS

Robert Foley, P.E., Consultant, Duxbury, Massachusetts

William P. Bahnfleth, Ph.D., P.E., The Pennsylvania State University, State College, Pennsylvania

ARTICLE 5.4.6

PLUMBING: PROCESS, GAS, AND WASTE SYSTEMS

**Robert V. DeBonis, CIPE, CET, Senior Associate,
Eugene B. Kingman, CIPE, Principal,
and John A. Rose, Jr., Master Plumber**
Robert W. Sullivan, Inc., Boston, Massachusetts

INTRODUCTION

The life cycle of a typical building plumbing project may be thought of in four distinct ways: (1) conception, (2) design, (3) construction, and (4) completion.

Conception, Design, and Construction

Once a project is conceived, the site utility and public utilities plans should be obtained and reviewed. A site survey will be necessary to help ensure that utility services indicated on the site drawings are true, "as-built" current site conditions. It is essential during the site survey to determine the availability, size, and location of domestic water, natural gas, sanitary, and storm systems. The site plan should be prepared for use by each engineering discipline to include pertinent aboveground geographical features of the project, and that site utility plan will display the water, sewer, gas, and electric utilities for use by the plumbing engineering design firm. The project team should develop a preliminary budget. Using the project's intended square footage, and multiplying by current square-foot costs of the type of construction project being built, can achieve early-stage budgeting. There are various engineering guides to obtain the square-footage costs, such as those published by RS Means Company, Inc.,[1] or elsewhere contained in various building trade journals. Use of *square-footage cost estimation factors* should be applied carefully, as many unforeseen factors can come into consideration during subsequent design phases. Therefore, a liberal cost contingency percentage should be applied to these square-foot cost budgets. Chapter 6 describes various construction delivery processes.

Design Drawings and Specifications.   The site drawing is usually the first in the set showing the building location and how it is situated on the building lot. The site drawing should indicate the location of the services entering the building. The streets or various accesses to the building should also be identifiable. Geographic contours and landscaping, which are important in determining entrances, exits, and grade levels as related to windows and doors, will be clearly identified as well.

The plumbing engineer will be tasked with preparation of a complete set of design drawings to include sanitary drainage systems, potable-water systems, storm water drainage systems, venting systems, fuel gas systems, special waste piping systems, and steam piping as appropriate for the facility's intended use. The plumbing engineer's drawings should be based on architectural building features and proposed occupant use. The architectural drawings will indicate the location of restrooms, laboratories, kitchens, cafeterias, drinking fountains, and so on, which are building functions for which the plumbing engineer will be required to design mechanical services. Besides the drawings, the engineer will prepare a set of specifications relevant to the materials [e.g., copper, brass, iron, steel, polyvinyl chloride (PVC), pipe, valves, and fittings], equipment (e.g., water heaters, water meters, etc.), and installation techniques (e.g., welding, soldering, solvent joining, etc.) required on the project.

Plumbing Code

The plumbing code provides a set of rules, regulations, and standards adopted by local or state governments to insure minimum requirements for the design, installation, and maintenance of the plumbing systems and to help protect the health and safety of the public. The plumbing code is the most important tool in starting a good design. The engineer must adhere to design, materials, and installation requirements that are regulated by the code. The Massachusetts Plumbing Code[2] is an example of these codes.

The number of fixtures required is the best place to start. Various types of buildings require different sanitary requirements. Restaurants' sanitary requirements are based on seating capacity and kitchen requirements. Hand wash sinks, water temperature, antiback-flow requirements, and many Board of Health requirements come into play. Theaters and entertainment facilities require fixtures and pipe sizing related to seating capacities.

Water Supply

If the city water pressure is unable to provide adequate pressure to operate plumbing fixtures on upper floors in the building, water booster pumps should be provided. This equipment should be duplex, and possibly a triplex, depending on the peak flow demands. Pump sets are for the most part instantaneous with pressure regulator controls. Alternate types of water pressure booster equipment such as hydropneumatic tanks and roof gravity tanks are no longer practical. Booster pump sets shall be arranged to assure that the instantaneous peak flow demand can be met in the building. If the building requires a dual pressure zone system (i.e., a city pressure zone and a boosted pressure zone), separate hot-water heaters for each zone are also required.

The municipal water supply quality should be reviewed to determine if any water treatment is warranted. Chemical and bacteriological analysis reports are readily available from the municipality. In some instances, treatment such as softening and filtration, if implemented, can prolong the life of interior water distribution systems dramatically, and reduce maintenance requirements.

Sizing Water Supply. Once the pressure analysis is completed and all the criteria proved acceptable, such as available city pressure in relation to pressure loss due to elevation and friction loss, or if a booster pump or storage tank is required, the water supply demand can be established.

Most codes establish fixture and demand factor values for each type of fixture (see Tables 5.4.6-1 and 5.4.6-2, respectively). Adding the number of fixtures and their respective fixture unit values, the sum of which should be multiplied by the demand factor, the result is water demand that can be used for pipe sizing.

Hot-water requirements are based upon the demand of the building that is served, and in accordance with the minimum standards of the applicable sanitary or plumbing code. The maximum temperature of the domestic hot water in residential buildings, or for use where human contact is possible, is usually 120°. Plumbing fixtures requiring higher temperatures for their proper use and function, such as dishwashers or other process applications, will normally be exempt from this limitation. The hot-water pipe sizing is similar to cold water requirements.

Potable Water Systems. The *A.S.S.E. Plumbing Dictionary*[3] defines potable water as follows: *1. Water that is suitable for drinking, culinary, and personal purposes. 2. Water free from impurities present in amounts sufficient to cause disease or harmful physiological effects.* Water quality is normally controlled by public health regulations. The U.S. Environmental Protection Agency sets forth parameters for quantities of organic and inorganic contaminants that are allowed in drinking water. Table 5.4.6-3 shows the standards. These primary and secondary contaminant levels, if adhered to in drinking water supplies, are acceptable for human consumption.

TABLE 5.4.6-1 Minimum Sizes of Fixture Water Supply Lines and Factor Values

Type of Fixture or Device	Nominal Pipe Size (in.)	Factor Value
Bathtubs	½	2
Combination sink and tray	½	2
Drinking fountain	⅜	1
Kitchen sink, commercial	¾	6
Lavatory	⅜	1
Shower (single-head)	½	2
Sinks (service, slop)	½	2
Sinks (flushing rim)	¾	6
Urinal (flush tank)	½	2
Urinal (direct flush valve)	¾	6
Water closet (flush valve-type)	1	12
Hose bibbs	½	2

Source: Abstracted from Massachusetts Plumbing Code (Ref. 2).

TABLE 5.4.6-2 Demand Factor Values

Occupancy Use	Demand Factors
School	
General	0.75
Shower room	1.00
Institutional	
General	0.45
Assembly	
General	0.25
Restaurant, café	0.70
Clubhouse	0.60
Business and Mercantile	
General	0.25
Laundry	1.00
Industrial	
General, exclusive of process piping	0.90

Source: Abstracted from Massachusetts Plumbing Code (Ref. 2).

Potable water is typically used for flushing toilets and urinals, hand washing, showers, food preparation and cleanup, lawn watering, and building cleaning. Safety fixtures, such as eyewashes and emergency showers, also use potable water.

Nonpotable Water Systems.   Water that is used for glass/cage washing, equipment cooling, cooling tower, boiler, chiller, and other HVAC or process makeup, solution preparation, creating suction through aspiration, diluting chemicals; as well as any other use where water comes in contact with a substance, or a piece of equipment that renders it, or could render it, unsuitable for human or animal consumption or contact, is defined as nonpotable.

Cross-Connection Control.   By definition, water from the supplier is considered to be potable. It is the responsibility of the supplier to protect their distribution system and the public from what is happening inside a facility. It is the responsibility of the user of the facility to protect their personnel from these same activities within the facility.

The following excerpt from a Commonwealth of Massachusetts regulation is an example of a typical regulation:

> Cross connections between potable water systems and other systems or equipment containing water or other substances of unknown or questionable safety are prohibited except when and where, as approved by the Massachusetts Department of Environmental Protection or its designee, suitable protective devices such as the reduced pressure zone backflow preventer or equal are installed, tested and maintained to insure proper operation on a continuing basis.[3]

One primary way both parties have of fulfilling their obligations is to prevent the intermingling of potable and nonpotable water. This is called *cross connection control,* which simply means preventing the possibility that nonpotable water can cross into potable water.

Cross-connection control is accomplished by different means, depending on the degree of hazard presented by the conditions that prevail. The most positive methods of cross-connection control are the air gap and the barometric loop. The air gap is exactly what the names suggests—a gap that separates potable from nonpotable uses. A barometric loop is a pipe rising 34 ft above the highest use to discharge into a tank or container. The air gap, though quite effective, reduces pressure in the system to atmospheric, thus requiring a means of putting that pressure back into the system; this option, for all practical purposes, requires the use of pumps to provide the pressure necessary to operate the system properly. This is

TABLE 5.4.6-3 Environmental Protection Agency Primary Maximum Containment Levels (MCLs)*

Inorganic contaminants			
Parameter	MCL	Parameter	MCL
Arsenic	0.05	Fluoride	4.0
Asbestos (fibers/L)	7 million	Lead	0.015[†]
Barium	1	Mercury	0.002
Cadmium	0.005	Nitrate (as N)	10
Chromium	0.1	Nitrate + Nitrite (as N)	1
Copper	1.3[†]	Selenium	0.01
Fluoride	4.0		

Organic contaminants			
Alachlor	0.002	Ethylene dibromide	0.00005
Atrazine	0.003	Heptachlor	0.0004
Benzene	0.005	Heptachlor epoxide	0.0002
Carbofuran	0.04	Lindane	0.0002
Carbon tetrachloride	0.005	Methoxychlor	0.04
Chlordane	0.002	Monochlorobenzene	0.1
2,4-Dichlorophenoxyacetic	0.07	Polychlorinated biphenyls	0.0005
Dibromochloropropane	0.0002	Styrene	0.1
o-Dichlorobenzene	0.6	Tetrachloroethylene	0.005
para-Dichlorobenzene	0.075	Toluene	1
1,2-Dichloroethane	0.005	Toxaphene	0.003
1,1-Dichloroethylene	0.007	Total trihalomethanes	0.10
cis-1,2-Dichloroethylene	0.07	2,4,5-TP (Silvex)	0.05
trans-1,2-Dichloroethylene	0.1	Trichloroethane	0.005
1,2-Dichloropropane	0.005	1,1,1-Trichloroethane	0.20
Endrin	0.0002	Vinyl chloride	0.002
Ethylbenzene	0.7	Xylenes	10

Radiological contaminants	
Parameter	MCL
Total radium (radium-226 + radium-228) (pCi/L)	5
Gross alpha activity (pCi/L)	15
Beta particle and photon activity (millirem/year)	4

Secondary maximum contaminant levels (SMCLs)*			
Parameter	SMCL	Parameter	MCL
Aluminum	0.05 to 0.2	Manganese	0.05
Chloride	250	Odor (TON)	3
Color (PtCo units)	15	pH	6.5 to 8.5
Copper	1.0	Silver	0.1
Corrosivity	Noncorrosive	Sulfate	250
Fluoride	2.0	Total dissolved solids	500
Foaming agents	0.5	Zinc	5.0
Iron	0.3		

* Units in mg/L unless otherwise noted.
[†] USEPA has set "action levels" for copper and lead in public drinking water.

obviously very expensive, both in initial equipment costs and ongoing operating costs, and is not usually done. The barometric loop, which is useful only when there is no possibility of back-pressure, works because back-siphonage, at sea level or higher, cannot develop sufficient pressure to lift water to 34 ft (14.7 psia) in height.

Given the limitations and expense of these most-positive methods, lesser-positive (but nonetheless satisfactory) methods have been developed, accomplished by the use of backflow preventers. There are numerous types of backflow preventers, ranging from reduced-pressure backflow preventer devices (RPBFP), to double-check valves, pressure vacuum breakers, and atmospheric vacuum breakers (AVBs). The type of backflow preventer selected depends upon the potential hazard presented. RPBFP devices are suitable for use in systems constantly subjected to pressure, with the potential present for both back-pressure and back-siphonage to occur, and are considered to be the best mechanical protection against severe hazards.

Double-check valves are suitable for use in systems that are subjected to constant pressure, and where back-pressure and back-siphonage are the concern. They are suitable where the hazard is less severe, such as from water-based fire protection systems into which fire departments can pump water by means of the "siamese connection." Double-check valves are frequently used in fire protection systems, but not too often, if at all, in plumbing and process systems in facilities. However, when a fire department can use a pond, river, stream, or the ocean as a source of water for fire fighting, and this water can be introduced into the siamese connection, then an RPBFP is recommended. In some jurisdictions, this design is mandated by the cross-connection regulations.

Pressure vacuum breakers are suitable for use in systems constantly subjected to pressure, with the potential present for both back-pressure and back-siphonage. They are suitable where the hazard is not severe at all, such as in lawn irrigation systems. Atmospheric vacuum breakers are suitable for systems in which the outlet of the device is not subjected to constant pressure and the hazard is not severe, such as on hose end faucets on outside spigots, on service sinks, or on laboratory faucets with hose barb tips.

Cross-connection control usually is implemented on two levels: (1) protecting the water main and the public from activities within a facility, and (2) protecting personnel within the facility from these same activities. This protection of the public is often referred to as *containment,* and the protection of personnel within the facility is often called *separation.* Containment is accomplished by placing backflow preventers (which, because many municipalities usually require testable devices, RPBFP are used) on the water service immediately after the utility water meter. Separation is accomplished by placing backflow preventers in the water supply line to any potential or actual source of contamination. Sometimes point-of-use backflow preventers accomplish this, which results in a rather large number of devices in a facility of any size. To help defray the cost of the required testing, the authorities enforcing these codes and regulations may impose a fee for the testing. They may also require reporting of the testing results. A facility with any significant number of backflow preventers would be required to spend a large portion of their time testing devices and filing reports, not to mention the costs for testing that would be incurred over a period of time. As a result, when testable devices are required for the separation function, a central backflow preventer installation is commonly used. This central backflow preventer installation is accompanied of necessity by a piping system (labeled in many instances as nonpotable, but also as *nondomestic, protected,* or *process*) that is completely independent from the potable water system. This separation extends to all components of the water system, including water heaters, circulation pumps, and other necessary appliances. To economize on floor space and maintenance costs, containment and separation backflow preventers are typically placed adjacent to each other, at the water service entrance location. Because of the relatively large volume of water used for nonpotable purposes, the potable water is often a branch of the main, taken off the main before the separation backflow preventers.

In many jurisdictions, completely separate water systems are not mandated by the regulations. In these instances, each fixture, device, and point of connection must be assessed for its potential to present a hazard to the potable water system and to the health and welfare of the personnel of the building. Sinks might be equipped with AVBs. Cooling connections to equipment in which the water outlet is submerged in a tank or container of water, such as in certain types of humidifiers, might be equipped with RPBFP. Boiler and cooling tower makeup connections might be equipped with RPBFP. Water service connection to low-hazard facilities might be equipped with double-check valves, in lieu of the more expensive RPBP. In each instance a decision has to be made about what the hazard potential is, and the appropriate type of device to be installed.

When choosing between point-of-use and central backflow preventers, the following should be considered:

- The central scheme will require duplicate but separate potable and nonpotable water piping systems, including hot-water heaters and all of their accessories and accoutrements. This added piping and equipment must be planned for in chases, shafts, and ceilings. Additionally, duplicate steam/gas/oil/electric connections to supply the necessary energy will also be required.

- Most authorities have developed standard for placement of RPBPs, including clearances, height of installation, access requirements, and drainage needs, as well as lighting and heating of space requirements. These standards tend to involve a relatively significant amount of floor space and headroom. These space and volume needs must be considered when planning mechanical spaces.

- Reduced-pressure backflow preventer devices, when operating properly, discharge water from the atmospheric pressure chamber (the "vent") to the floor. If the back-pressure or back-siphonage that caused this discharge lasts for more than a few seconds, the amount of water could be hundreds of gallons. Floor drainage, designed to deal with this amount of water, is required to prevent the creation of tripping hazards or of flooding problems for other equipment in the area around the RPBP.

- When an RPBFP vent opening fails in the open position, it is possible that this discharge can be driven by the municipal water pressure, resulting in hundreds of gallons discharging for an extended period of time. The various backflow preventer manufacturers have tables for the amount of water to be expected to be discharged from these vent openings. The manufacturer should be contacted about the magnitude of this volume of water, which is based upon the water pressure at the inlet to the device. Few plumbing drainage systems can deal properly with hundreds of gallons of water, especially over an extended period of time.

- If these devices are located so that the waste from floor drains has to be pumped up to reach a sewer, this potential volume of water must be considered both when selecting the pumps and when sizing the sump.

- Point-of-use devices, if required to be testable, need space for testing to be performed, and a means of draining the wastewater from the test. If these devices are RPBFPs, then a means of dealing with the large volume of water described above needs to be provided.

The separate potable and nonpotable water systems should be clearly labeled.

Building Sanitary Drainage Systems

The soil, waste, and vent system is, for the most part, dictated by code. Key features in these systems are the location and accessibility of soil and vent stacks and of cleanouts for maintaining the system and adequate slopes on the horizontal collection systems to maintain self-cleansing velocities. The venting should be arranged to provide a 1-in w.c. pressure differential within the system to adequately protect trap seals at plumbing fixtures. All fixtures should be trapped and vented. Uncontaminated cooling water and clear water drains from areaways, HVAC equipment, sprinkler drains, and the like should not connect to the sanitary drainage system. Minimum slope of a building drain or branch is dictated by codes and is commonly set at ⅛ in/ft. Some codes allow 1/16 in/ft slope on larger-size piping. Some codes offer the option to hydraulic-design the piping system to attain self-cleansing velocity.

All fixtures should discharge from the building by gravity. The sole possible exception is when the basement floor elevation is below the street. The reader will note this says *street,* not sewer. This is because anytime the basement is below the street, there is a potential for a surcharge being experienced first by flooding of floor drains and fixtures in the basement, before the surcharge is evidenced in the street. In some instances, especially in older municipalities, especially those with combined storm and sanitary sewers, facilities have had to resort to pumping the grade-level floor also. It is highly recommended that wastes from fixtures in a basement be pumped to the sewer.

Pumping by the use of duplex sewage pumps is the minimum recommended. Vertical wet pit pumps or ejectors are recommended, as opposed to submersible pumps. Submersible pumps tend to expose maintenance and repair personnel to the hazards associated with raw sewage. Another option, when the lift is not very high (10–15 ft), is pneumatic ejectors. Pumping should be by automatic means, based upon level in the sump, with automatic alternation of lead and lag positions. Emergency power should be considered whenever the flow to the pumps cannot be stopped during power outages. Most sewer authorities prohibit discharge of certain metals or chemicals. Figure 5.4.6-1 provides a typical example.

Building Roof Drainage. Code requirements, materials, and installation concerning interior roof conductor piping systems is similar to the soil and waste system. Roof drain horizontal piping above the basement floor not buried should be equipped with a vapor barrier, to avoid condensation that can occur when warm, moist air in the building comes in contact with the surface of piping that is carrying colder rainwater. The metallic piping materials are particularly of concern in this issue. The plastic materials may have some advantages in this issue over the metals. Plumbing and building codes should be consulted to determine if overflow roof drains are required. Some codes offer the alternative of designing the structure to withstand the load imposed by water or snow standing on the roof to the depth of the point of roof overflow (parapet).

Roof drains on lower-building roof areas, where windows, intakes, or other building openings may exist, should have running traps installed, as a precaution against odors being drawn into the building. Most municipalities make a distinction between sanitary waste and storm drainage. Usually, separate storm and sanitary piping systems are required. The following paragraph, abstracted from the Boston Water & Sewer Commission in Boston, Massachusetts, sewer use regulations, offers an example of specific language that prohibits introducing clean water into a sewer or combined sewer.

> No person shall discharge or cause or allow to be discharged directly or indirectly into a Commission combined sewer or into a combined sewer tributary thereto any of the following:
> Groundwater, dewatering drainage, subsurface drainage, tidewater, accumulated surface water, non-contact cooling water, non-contact industrial process waters, uncontaminated contact cooling water and uncontaminated industrial process water, or waters associated with the excavation of a foundation or trench, hydrological testing, groundwater treatment/remediation, removal or installation of an underground storage tank or dewatering of a manhole without expressed approval.

Building codes sometimes mandate that roofs be provided with a certain minimum pitch. Some roofing systems manufacturers have minimum pitch requirements if their warranties are to be respected. The means of pitching roofs varies, but the two most common are to slope the structure or to use tapered installation.

Tapered insulation can be problematic when large roof areas are involved. The drop in elevation that results from minimum pitches over long distances can be significant. At times, this

Reduced Pressure Zone
Backflow Preventer

Double Check Valve
Backflow Preventer

Dual Check Valve
Backflow Preventer

Atmospheric Vacuum
Breakers

Pressure Vacuum
Breakers

FIGURE 5.4.6-1   Basic types of backflow preventers. (*Courtesy of Watts Regulator Co.*)

drop can be excessive. One of the concerns encountered is that the depth of insulation can exceed that allowed by roof system manufacturers. Another concern is that the insulation becomes subject to crushing, especially if there is the possibility that heavy equipment will be moved across the roof, such as when rooftop chillers or cooling towers are added.

Roof drains are typically fabricated of cast iron and equipped with accessories to allow installation in all types of roof systems and structures. Some of these accessories include flashing clamps, underdeck clamps, strainer screens, extensions, and domes. A typical roof drain is shown in Figure 5.4.6-2.

Building Process Waste Drainage Systems.   Wastes requiring traps, separators, or treatment prior to connecting to the general sanitary system include wastes from plaster sinks, laboratory sink and equipment wastes, hazardous wastes, and kitchen grease wastes. See Sec. 10.5 for additional details about special concerns in laboratories.

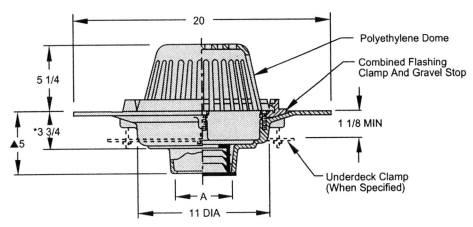

FIGURE 5.4.6-2   Typical roof drain.

Grease Waste Systems.   Commercial kitchens, office buildings, schools, colleges, hospitals, and other such facilities have kitchens where meals are prepared for customers, patients, visitors, and staff. Large quantities of grease can, and usually are, generated in these facilities. Types of equipment of concern are pot sinks, dishwashers, and floor drains in close proximity to griddles, ranges, soup kettles, and "fryolaters." Also of concern are cooking line exhaust hoods equipped with water washdown systems. Drains from these fixtures should not be introduced into the sanitary drainage systems without first passing the wastewater through a grease interceptor. When grease cools down, it congeals and adheres to the walls of drainage piping and causes blockages. Point-of-use grease interceptors should be located as close to the grease-producing fixtures as possible. Grease is lighter than water and gets trapped in the interceptor. At a minimum, point-of-use traps should be installed.

A facility-wide grease interceptor should also be considered to protect the municipal sewer. This can sometimes be a tank of many thousands of gallons' capacity. The probability of a facility having the space to locate such a tank within the building is low. As a result, facility-wide grease traps are usually located outside, underground. Issues to be considered when placing these devices are accessibility for pumping, which is required to be performed on some regular basis; the possible need to vent the trap separately from the inlet piping; and the possibility that the odors associated with congealed grease will be found objectionable by all who are exposed to them.

In most jurisdictions, the sewer use regulations will set a limit on the oils and grease concentrations allowed into the municipal system. This limit and the facility's ability to keep below it will determine if a facility grease trap will be required. In some jurisdictions, an automatic assumption is made that the facility will not be able to meet the limits and a facility trap is mandated.

In most jurisdictions, codes and authorities require grease-laden wastes to have separators or grease traps installed. Maximum allowable concentration of fats, oils, and grease (FOG) is from 100 to 300 mg/L by most local sewer authority standards. It is recommended that the engineer review all applicable local regulations. For example, the Massachusetts Assembly Code[3] describes where such units should be installed. Traps should be accessible for cleaning, emptying, and maintenance. Grease separators can vary from local cast-iron traps at sinks and dishwashers to dedicated grease waste collection systems piped separately to exterior, buried concrete tanks. Grease emptied from such traps should be disposed of via a disposal firm that is licensed to handle such waste.

There are several different types of point-of-use interceptors that may be used at fixtures. Undercounter interceptors are installed adjacent to the grease-producing fixtures. Below-slab interceptors do not require any floor space and therefore may be more desirable than the on-the-floor type. Garbage disposers should not be piped through grease traps; they clog the unit.

Sizing of a grease interceptor is a function of the volume of waste produced by the grease-producing fixture and the time required to drain the fixture. Most manufacturers provide instruction on how to properly size a local trap. Most local and regional sewer authorities have sizing criteria for facility traps. These criteria include volume of flow, retention time, baffling, venting, materials of construction, and placement of the trap for ease of pumping.

Problems facing the facilities manager include buildup of solids in the interceptor and infrequent or too long a period between grease removal. Point-of-use interceptors and facility interceptors should be provided and maintained properly in order that large slugs of grease do not close cast-iron sewers and drainage lines within the facility.

Pure-Water Systems.   This type of water can be known as distilled water, reverse-osmosis (RO) water, deionized (DI) water, or a combination reverse-osmosis deionized (RODI) water. The terms *distilled water, reverse-osmosis (RO) water, deionized (DI) water,* or *RODI water* refer to the processes by which potable water is cleaned to the level desired. Each of these types of water has different properties. A number of associations and societies, including the federal government, have developed and published standards for the characteristics that define "pure" water. It is to be remembered that the various interests represented by these associations and societies have different goals and concerns in mind when describing these characteristics. It is important that facilities are aware of the differences between these goals, and that they apply to the facility, when considering pure water.

The College of American Pathologists (CAP), The American Society for the Testing of Materials (ASTM), and the National Committee for Clinical Laboratory Standards (NCCLS) (see Tables 5.4.6-4, 5.4.6-5, and 5.4.6-6) provide standards for water to be used for research,

TABLE 5.4.6-4   Reagent-Grade Water Specifications[a]

	Type		
College of American Pathologists (CAP)	I	II	III
Bacterial content [colony-forming units per milliliter (cfu/mL) (maximum)]	10^b	10^3	NS
pH	[c]	[c]	5.0–8.0
Resistivity [megohm centimeter (MΩ/cm)][d]	10 (in line)	1.0	0.1
Silicate mg/L SiO_2 (maximum)	0.05	0.1	1.0
Particulate matter[e]	0.22 μm filter[e]	NS	NS
Organics[f]	Activated carbon[e]	NS	NS

[a] Specifications are subject to change.
[b] Preferably, Type I water should be bacteria free.
[c] Type I and II water contain so few ions that it is unnecessary to obtain pH measurements.
[d] At 25°C.
[e] This is a process specification and is not measured by the user.
[f] At a minimum, some form of activated carbon should be included in a reagent water system as pretreatment.

TABLE 5.4.6-5   Reagent-Grade Water Specifications[a]

National Committee For Clinical Laboratory Standards (NCCLS)	Type		
	I	II	III
Bacterial content [cfu/mL (maximum)]	10[b]	10[3]	NS
pH	[c]	[c]	5.0–8.0
Resistivity (MΩ/cm)[d]	10 (in line)	1.0	0.1
Silicate [mg/L SiO_2 (maximum)]	0.05	0.1	1.0
Particulate matter[e]	0.22 µm filter[e]	NS	NS
Organics[f]	Activated carbon[e]	NS	NS

[a] Specifications are subject to change.
[b] Preferably, Type I water should be bacteria free.
[c] Type I and II water contain so few ions that it is unnecessary to obtain pH measurements.
[d] At 25°C.
[e] This is a process specification and is not measured by the user.
[f] At a minimum some form of activated carbon should be included in a reagent water system as pretreatment.

laboratory, and other departments within a facility. The federal government, in the form of the Food & Drug Administration, also publishes guidelines for water that is to be used in the production of food, medicine, or equipment used for medical care.

Pure-water systems consist of many and varied components, depending upon the quantity and quality of water desired. In most instances, a system of any size has a treatment component, a storage component, and a distribution component. The treatment component is divided into three segments: (1) pretreatment, (2) primary treatment, and (3) tertiary treatment (or polishing). The pretreatment segment consists of equipment designed to remove particulate and organic contaminants and usually consists of particle filters, water softeners, multimedia filters, and then carbon filters. The primary treatment segment consists of equipment designed to remove dissolved and ionic contaminants and usually consists of reverse osmosis, followed many times by deionization. In some instances, deionization without reverse osmosis is considered sufficient. The polishing segment of the system removes trace contaminants prior to distribution to the facility. Polishing may consist of a second DI treat-

TABLE 5.4.6-6   Reagent-Grade Water Specifications[a]

American Society For Testing & Materials (ASTM)	Type			
	I	II	III	IV
Total matter [maximum (mg/L)]	0.1	0.1	1.0	2.0
Specification, conductance (µΩ/cm)[b]	0.06	1.0	1.0	5.0
Specification, resistance (MΩ/cm)[b]	16.67	1.0	1.0	0.2
pH[b]	[c]	[c]	6.2–7.5	5.0–8.0
Potassium permanganate reduction (min)	60	60	10	10
Soluble silica, max.	[d]	[d]	[d]	[d]
Microbiological classification				

[a] Specifications are subject to change.
[b] At 25°C.
[c] Type I and II water contain so few ions that it is unnecessary to obtain pH measurements.
[d] At a minimum some form of activated carbon should be included in a reagent water system as pretreatment.

ment, ultraviolet sterilization for bacteria, other organics control, and submicron filtration. The presence or absence of any of these unit purification processes is dependent upon the incoming water profile coupled with the desired product water quality.

Storage of volumes of pure water is a common factor in almost all systems of any size. This storage is usually accomplished in storage tanks of suitable materials. These materials will be determined based upon the quality of the water involved, the size and weight of the empty tank, and the possibility that sanitizing may be required or desired. The various plastics may not be suitable for sanitizing with steam. The system control panel is integral to the RO unit at most pure-water systems. Control panels can be the electromechanical type or PLC processor type. Controls allow the various components of a system to operate as an integrated whole. Backwash signals from carbon and multimedia filters signal the RO system to stop. Water levels in the storage tank activate the RO system. Controls also provide alarms, which can be used to alert operators to conditions which require intervention or action. Controls should activate contact closures for remote-alarming capabilities. System controls can be designed to function in almost any way the user/owner wants and can be monitored by the building automation system.

A factor common to all water treatment processes is that significant pressure loss is entailed in all segments. This means that pumping is a major portion of these systems. If city pressure or the building-provided pressure is insufficient, pumps may be needed to provide pressure for the pretreatment segment. Reverse-osmosis units invariably require very high pressures to work properly, so pumps are built into the units. Pumps are needed to distribute the water to the facility. In some instances, especially when the facility is a multifloor building or the distribution involves significant losses due to the length of piping, or when equipment pressure requirements demand it, pumps have to be capable of generating significant pressures. In these instances, it may be necessary to have steps of pumping, to avoid exceeding the pressure ratings of the various components.

In many cases, the water being fed to a pure-water system is tempered when an RO system is utilized. A thermostatic mixing valve set at approximately 77°F should be installed along with duplex reduced-pressure backflow preventers for in-plant protection prior to any pretreatment components. This 77°F temperature ensures that the RO unit will function at maximum efficiency, although systems are commonly run on cold water as well.

Multimedia filters consist of a mineral filled with graded layers of sand and gravel. Tanks are usually constructed of fiberglass or lined carbon steel. Some applications may require FDA-approved tank linings, which should be specified when required. Multimedia filters are used to remove particulate, organic and suspended solids removal.

Carbon filters consist of a mineral tank filled with carbon and gravel and are constructed of the same materials as the multimedia filter. The sizing of carbon filters is important and should be based on local water conditions to allow sufficient contact time for proper disinfectant removal from the water supply. Carbon filters may be installed in parallel to reduce their size, or in series to ensure effective disinfectant removal.

Water softeners consist of a mineral tank filled with water-softening resin and gravel and are constructed of the same materials as the multimedia filter and carbon filter. Media consists of a cation resin with a quartz or gravel underbedding. Water softeners may or may not be required as part of a pure-water system. Softeners are critical in boiler feed applications and may be required prior to reverse osmosis, depending upon the incoming water hardness, the degree of concentration of salts by the RO system, or the requirement of a particular RO membrane.

Reverse osmosis is the pressurization of feedwater on one side of a semi-permeable membrane. The pressure should be high enough to exceed the osmotic pressure to cause osmotic flow of water. If the membrane is highly permeable to water but essentially impermeable to dissolved solutes, pure water crosses the membrane and is known as product water. Simply put, high-pressure water is forced across a membrane where the majority of dissolved solids and impurities are removed. Different types of RO membranes include cellulosic, fully aromatic polyamide, and thin-film composite. Thin-film and polyamide membranes have replaced cellulosic membranes as the mainstays in RO systems. The RO process removes between 90 and 99 percent of dissolved ions, greater than 99 percent of bacteria and pyrogens, 99 percent of organics over 200 molecular weight, and 99 percent of colloids and particles. As part of the process, RO units today can recover as much as 75 percent of the feedwater, depending upon feedwater chemistry. Care should be taken with these units, as some laboratory units recover as little as 10 percent of the feedwater. Feedwater parameters need to fall into a certain range in order to maximize efficiency and minimize fouling of the membranes. The silt density index (SDI), which measures the level of suspended solids, cannot be greater than 5, the pH range between 4 and 11, the temperature range from 36°F to 95°F, and the free chlorine level 0 for thin-film composite membranes. The pretreatment portion of the system should bring feedwater to within these parameters. A moderately sized RO unit with an 8-gpm capacity will produce 480 gph of product or 11,530 gpd of product if the unit were to run 24 h/day.

Reverse-osmosis (product) storage tanks, where steam sanitation is not a factor, should be provided of high-density polyethylene or polypropylene, FDA-approved, for product storage. Tanks should include level controls, vent filter housings, vent filters, and access ports. Storage tank controls are to be wired to the system control panel. In some instances, nitrogen blanketing of the tank, by means of an automatic nitrogen supply, is warranted. Factors to be considered in this decision are the amount of water expected to be drawn out of the storage tank over a period of time, the amount of air associated with the rising and falling water level, the effect on the filters of that amount of air movement and the frequency and cost of filter replacement. These costs then are compared with the cost of nitrogen. Also to be considered is whether filters can be effective enough in dust and vapor removal to meet the quality requirements for the stored water.

- Recirculation pumps should be stainless steel in construction and provide adequate capacity (flow and pressure) to maintain proper velocities through piping distribution loops. Recirculation pump controls are to be wired to the system control panel. Depending upon the size of the system, a spare or duplex pump may be a wise investment.

- Organics control may be necessary depending upon the organics levels required for product water quality. This may be accomplished by using ozone or by adding a total organic carbon- (TOC-) reducing UV unit to the system at this point. Controls for this unit should be wired to the central control panel.

- Portable ion exchange bottles are typically used to improve the water quality, producing water quality as high as 18 MΩ/cm. Many different types of ion exchange resins can be used. The higher the desired water quality, the more important it is to specify Type 1 resins.

- Resin trap filters may be used as a roughing filter prior to any additional treatment or filtration. These filters protect any downstream equipment from a break or rupture in the integrity of a portable ion exchange bottle.

- Ultraviolet (UV) units are typically used after portable ion exchange bottles to destroy any bacteria that may be present in the water.

- Final filters are used after UV units to remove fine particles and bacteria from the water prior to distribution to the facility. Filter elements rated at 0.2 μm absolute are typically used in this location, although water quality requirements may dictate using finer filters.

- Skid mounting of equipment may be specified to speed up the installation process. Skid mounting permits the pure water equipment manufacturer to assemble more of the system components into an integrated system before it reaches the end user site.

Installation supervision by the pure-water equipment manufacturer or supplier along with system start-up is very important. Piping and instrumentation diagrams are part of the shop drawing submittal process for the owner's use for troubleshooting purposes. Start-up and training services to ensure proper system operation should be provided for 1 to 2 weeks.

Monitoring pure-water system performance is critical for proper system operation. On-line monitoring such as pressure gauges, flowmeters, and resistivity monitors tell the owner how various system components are functioning. Off-line monitoring (i.e., testing for hardness, chloramines, bacteria, etc.) should be performed. Periodic checks of product water for compliance with pure-water quality specifications and feedwater testing should be performed several times a year. Bacterial contamination is a continual problem with water treatment systems and should be watched for. In some cases, *sanitation* of the whole system may be required. Daily logs should be kept, monitoring on-line parameters such as production rate, percent rejection of the RO units, and pressures at various system locations.

Piping.   Piping distribution design of pure-water systems is probably the most critical component of the system design, particularly if it is improperly designed. Pure water should be distributed through a fully recirculated piping system with velocities in the range of 5 ft/s. Proper velocities, organic control, and elimination of dead legs are critical to minimize bacterial growth. Pipe sizes, velocities, flow rates, and pressure drop across piping in most systems are the governing factors as to how large your system will be. Distribution pumps and polishing equipment sizing are all a function of the aforementioned characteristics. Most systems, which are very large, are that way because of the volumes of water that must be pumped through the piping loop in order to maintain velocities and flow rates. Instrumentation locations, gauges, flowmeters, and valves are essential for proper piping system operation.

Piping system design for the pure-water system depends upon the type of distribution desired, the facilities physical size and height, the number of outlets and quantity of water to be used, the equipment connected, the means of disinfecting that is desired, and the criticality of maintaining water quality. Fully recirculated systems piping design can be one of two general schemes. One scheme is a single line feeding the entire facility then returning to the storage tanks. All outlets and equipment connections are supplied, in a series configuration, through this one pipe. This can require a significant amount of pressure to overcome friction loss and elevation changes and still provide adequate pressure and volume at the outlets and equipment connections. This type of system can also mean that a mishap to the main disrupts service to the entire facility. This type of scheme can have short dead legs to serve outlets and equipment connections, or it can have the main coming within inches of the outlets and equipment connections by dropping from the ceiling, or rising up from the floor below. The other general scheme is to have a common supply line running through the facility, including up through the floors on multistory facilities, in a given location. Branches run across the floor to a return line. This return line can be adjacent to the supply line, or it can be on the opposite end of the facility. On multistory facilities, the topmost branch acts as the supply main to keep water flowing at all times. The pressures in all branches must be regulated as they rejoin the return line, to allow water to flow evenly from each branch as it connects in to the return. It may be necessary to also control flow volume in each of the branch lines, to assure that flow reaches all branches, instead of short-circuiting to the easiest path. In this scheme, the system can be shut down a branch line at a time and still maintain service to the other branches.

Unpigmented polypropylene, PVDF, stainless steel, and CVPC are some of the pipe materials being used in systems today, and PVC is still commonly used as well. Butt-welded joints, socket-welded joints, glued joints, and mechanical joints are types of joining systems commonly used. For pressurized systems, the user should minimize the number of mechanical joints present (wherever possible) to minimize leaks. Unpigmented polypropylene is a popular choice used at many institutions. Allowable internal pressure, at a given temperature, is also a factor with the material selection. PVC and polypropylene have limitations on the pressure ratings at elevated temperatures. CPVC has a somewhat higher rating. Stainless steel and PVDF have the highest of these materials. In instances where steam sanitation is being considered, stainless steel is the more common choice, due to its ability to withstand high temperature and pressure.

Replacement of portable deionization bottles is periodically required based on the water quality being delivered to the facility. Installing multiple bottles in parallel will allow for bottle replacement without any downtime experienced by the system. Media replacement in the carbon filters will be required about once every year or so. Water-softening media will require replacement about every 5 to 10 years. Multimedia filters, if properly backwashed, may require media replacement as infrequently as every 10 to 15 years. Many facility managers find that proper system maintenance should be accomplished through a service agreement with the pure-water system manufacturer or other qualified vendor.

High water and sewer costs make recovery of reject water from water purification systems a cost-effective option. New RO units are now being designed to permit a higher conversion rate and less frequent backwashing cycle, thereby reducing the amount of water being sent to drain. In some jurisdictions, volume of this wastewater is monitored as well.

Special Waste

There are many special wastes that require special treatments. A few examples and their unique requirements follow.

Laboratory Wastes.   Laboratory wastes, which can be at acidic or be at alkalinity levels beyond parameters allowed by code, should have automatic pH adjustment tanks installed to neutralize the chemistry of the wash water prior to connecting to the building sanitary system. Tank capacities (determined by flow rates and detention time) are usually equipped with automatic chemical pumps, flow recorders, and sampling ports. For additional details for laboratories, consult Sec. 10.5.

pH Neutralization Systems.   pH neutralization systems are an integral part of a facility's building drainage system. Neutralization systems insure that the effluent exiting the facility into the municipal sewer system is of the correct pH value for safe discharge. Testing laboratories, pathology departments, radiological departments, research labs, and production facilities produce wastes, which contain acids and bases that need to be neutralized before exiting to the city sewer. For additional details for laboratories, consult Sec. 10.5.

Photographic Process Waste (Fixer, Developer) and Silver Recovery Systems.   Fixer and developer are chemicals used in the film-developing process that can require pH adjustment before being introduced into the municipal sewer system. Silver recovery is the removal of silver from the waste effluent of photographic processing machines commonly found in many hospitals and medical care facilities, as well as many commercial facilities. Federal, state, and local agencies regulate the quantity of silver that are allowed to be introduced into municipal sewerage systems.

Used fixers from these devices contain from 1 to 3000 parts per million (ppm) of silver. Developer and wastewater contain from 1 to 2 ppm each. Most jurisdictions limit the discharge of silver to 5 ppm or less. It is a common experience of many facility managers that when there is a large number of these processors in their facilities, it is economically advantageous to collect spent fixer at a central point for recycling and recovering of the silver for its salvage value. One option sometimes employed is to pipe from the processors to a central recovery system. Another option is to use local recovery systems.

Developer and wastewater, on the other hand, because they both are chemicals that can need pH adjustment, should be piped to the pH neutralization system described in Sec. 10.5.

Hazardous Waste Collection Systems.   Certain process waste should not be dumped down the sanitary drainage system or into the pH neutralization system. Chemicals such as xylene, which is used in the tissue typing process in a pathology laboratory, is highly flammable and cannot be neutralized, and should not be introduced into any building drainage system or the municipal drainage system. Formalyn or formaldehyde, a toxic organic used in pathology departments in hospitals, will pass through the pH neutralization system without any effect because acids and caustics do not have an effect on formalyn. Such chemicals should be collected and dealt with as hazardous waste.

Hazardous waste collection systems may consist of an independent piping system, resistant to the chemical or waste being transported. In some instances, double-containment piping may be required. The piping system should be piped to a collection tank or tanks at a location where a hazardous waste collection agent can remove these chemicals. Collection tanks need to be vented independently to the outside. As part of some systems designs, it may be necessary to provide a pumping system in order to pump to a location where waste can be collected and transported by a removal-type company. All piping, valves, tanks, pumps, and so forth should be compatible with the waste being handled. Controls, including tank storage levels, pumps running, and so on should be incorporated into the system design and reviewed with the owner. Alarm connections to the building automation system may be warranted.

Systems should be designed to handle the number of devices or continuous flows being served with a sufficient safety factor to allow the facility flexibility in arranging materials pickup schedules. A concern to keep in mind is that limiting the amount of waste hazardous materials on-site, and the time they are kept there, are both highly recommended by safety regulators. With some substances, codes may have special construction and operation requirements when quantities exceed mandated limits.

Water Conservation

The need to conserve water is being addressed more and more in certain areas of the plumbing code and by water utilities. The use of gray-water systems, especially in the process water areas of industry, is being researched. There is a possibility to reuse wastewater, other than sanitary, in some commercial processes such as cooling water. This undoubtedly will be a necessity of future considerations. Water recycling is becoming more accepted, and indeed even mandated, in areas of the country with limited water supplies.

Plumbing Fixtures and Equipment

The visual part of any plumbing system is the fixtures. The fixture is both the delivery point for water and the beginning point of the sanitary and vent system. Fixtures should chosen for their availability, aesthetics, dependability, and practicality. Are they easily maintained and cleaned? Will they function properly over the long term? In the case of water-conserving water closets and urinals, do they do the proper job of cleaning the fixture or are numerous flushes required? Are replacement fixtures or parts thereof readily and economically available?

Plumbing equipment such as water heaters, booster pumps, hot-water circulation pumps, sewage ejectors, compressors, and cross-connection protection devices will need to be selected. Purchasing these items and all equipment from respected, well-established manufacturers is recommended to ensure that warranties are serviced when there is a problem, that spare parts and technical advice and service are all readily available, and to minimize the possibility of encountering discontinued items or suppliers going out of business, with no notice, leaving the facility with no recourse when there is a problem.

Handicap Accessibility

Title III of the Americans with Disabilities Act (ADA) of 1990 (42U.S.C. 12181), also known as 28 CFR US[4] handicap coding prohibits discrimination on the basis of disability by public accommodations and requires places of public accommodation and commercial facilities. Residential, places of assembly, schools, detention facilities, day care, and office areas are all unique in their design. The local codes will provide minimum requirements for most all services. The number of handicap fixtures required is always a main consideration. A common misconception is the assumption that handicapped access only means wheelchair access. It could also benefit those with arthritis or many of the muscular debilitating diseases (e.g., wrist lever handles on faucets). Fixture spacing and heights should be decided upon. These recommendations can be researched through the various manufacturer of the particular piece of equipment and national standards. The facilities manager therefore must ensure that facilities are accessible by all.

Compressed-Air Systems

There are many compressed-air systems in a facility. They may range all the way from piped breathing air in hazardous industrial situations (such as spray paint chambers in auto body finishing facilities), to process air to control the operation of pneumatic doors, to controls for HVAC or other systems, to medical air for health care facilities, and so on. In each case, the specific application will determine the required quality and pressure of the air.

Among other criteria that come under the heading of *quality* are particulate matter, moisture content, hydrocarbon content, and bacteriological purity. These should all be considered when selecting compressors and accessories.

The issue of duplex versus simplex, of all of the components of the system, should also be addressed. The desired or required system pressures will also be a factor in the selection of the type and size of the equipment selected.

The use of water versus oil versus Teflon© for sealing elements is also to be considered, as is the matter of possibly having to deal with the sealing fluid or substance as a hazardous waste. The matter of the temperature to which the compressor can operate properly is a consideration also.

Natural Gas/Liquefied Petroleum (Commercial Propane)

Various agencies and regulatory bodies in various regions govern natural gas and liquefied petroleum (LP) installations. In some instances, uniform codes have been developed and enacted by local or state agencies. In other instances, NFPA Standard 54[5] for natural gas and for LP[6] have been adopted as the codes to be enforced.

The gas utility, to its specifications and space and location requirements, usually provides natural gas services and meter installations. Often power, lighting, and a telephone connection, or combination of these, is required. If the project's gas demands are great enough to generate sufficient revenues so that the gas company can expect a reasonable payback, it will perform the installation work for no cost.

Interior gas distribution system pressures vary, from as low as 4-in water column (w.c.) to as high as 125 psig. The pressure provided depends upon:

- The pressure requirements of the equipment involved (i.e., large boilers or chillers can require 5 psig or higher)
- The pressure available from the utility (i.e., pressures can be as low as 12- to 14-in w.c. in the utility mains and up to as high as 50 to 60 psig)
- The economics of pipe sizing to meet the demands (pipe sized for 14-in w.c. could be 12 in in size, when piping sized for 50 psig could need to be only 4 in in size
- The regulatory restrictions upon pressures within a facility

Natural gas physical characteristics vary somewhat from region to region and from supplier to supplier, but generally the heating value is approximately 1000 Btu/ft^3 and the specific gravity is 0.65. Manufactured-gas characteristics may be different. The facility should be aware of changes in the characteristics of the gas supply over time.

In some jurisdictions, the use of elevated-pressure gas (and the definition of *elevated pressure* varies) inside a building requires a waiver or a special permit from the code authorities.

Natural gas piping is commonly black steel pipe. Valves are commonly iron body valves of the plug cock type requiring a tool to operate the valves. Ball valves are used in many areas.

Special valving is recommended in educational occupancies where classroom use of gas by students is planned, and mandated in kitchens where gas cooking is utilized. In school laboratories, where students can have access to gas outlets for purposes of using open flames, strategically located master gas valves are recommended (and mandated in some jurisdictions). These valves provide a single shutoff point for gas in an individual room, area, or on a floor. These valves can be located at the teaching station of an individual room, adjacent to the doorway at the exit from that room, in the corridor outside an individual room or group of rooms, or at strategically placed locations that lend themselves to quick and easy control of the gas flow to the protected areas.

In kitchens, where exhaust hoods are present, automatic gas shutoffs are mandated, by NFPA 96,[7] to shut down the gas flow to that hood when the automatic fire suppression system operates.

In labs when gas and compressed air enter rooms, check valves should be installed on gas lines to prevent pressurizing gas systems with air.

The installation of propane gas in buildings where natural gas is present is prohibited in some jurisdictions without special permission and a waiver by the regulatory agency.

Propane gas is regulated by the same agencies that regulate natural gas, and the systems are installed in the same manner and with the same materials. The source is usually a tank or tanks located on-site, at locations that are severely proscribed, based upon the tank volume and the surrounding hazards (both real and potential).

The physical characteristics of propane are somewhat different from natural gas, as are the normal operating pressures allowed without special permissions. Propane is allowed by NFPA to be used normally at up to 20 psig in a building or structure. Specific gravity of LP (commercial propane vapor, at 60°F) is approximately 1.5, and heat value after vaporization is approximately 2500 Btu/ft^3.

CONTRIBUTORS

A. Eugene Sullivan, M.S.P.E., FASC, FACEC, Chairman, Robert W. Sullivan, Inc., Boston, Massachusetts

Mark J. Sullivan, B.S., C.E.T., CIPE, CDA Treasurer, Robert W. Sullivan, Inc., Boston, Massachusetts

Christopher Russo, P.E., Partner, Newcomb & Boyd, Atlanta, Georgia

Robert S. Capaccio, P.E., President, Capaccio Environmental Co., Inc., Sudbury, Massachusetts

Arlene Parquette, Fluid Solutions, Inc., Boston, Massachusetts

Brian Hagopian, Fluid Solutions, Inc., Boston, Massachusetts

REFERENCES

1. Mechanical Cost Data, R. S. Means Co. Inc., Kingston, MA, 2000.
2. Massachusetts State Plumbing Code (248 CMR 2.00), Board of State Examiners Plumbers and Gas Fitters, Commonwealth of Massachusetts, Boston, MA.

3. I. D. Jacobson, *A.S.S.E. Plumbing Dictionary,* American Society of Sanitary Engineers, Westlake, OH, 1996.

4. 28 CFR U.S. Handicap Code, 28 CMR is formally titled: *28 CFR Part 36 Nondiscrimination on the Basis of Disability by Public Accommodations and in Commercial Facilities.*

5. NFPA 54, National Fuel Code, National Fire Protection Association, Quincy, MA, 1999.

6. NFPA 57, Liquefied Natural Gas (LNG) Fuel Systems Code, National Fire Protection Association, Quincy, MA, 1999.

7. NFPA 96, *Ventilation Control and Fire Protection of Commercial Cooking Operations,* 1999.

ARTICLE 5.4.7
SPRINKLERS AND FIRE PROTECTION

Edwin A. Kotak, Jr., P.E., C.E.T., CIPE, Senior Associate
Robert W. Sullivan, Inc., Boston, Massachusetts

HAZARD ASSESSMENT AND SELECTION OF EXTINGUISHING AGENTS

The engineering and design process for any facility should begin with an analysis to determine the types of hazards that will, or do, exist. Hazard assessment must consider a variety of issues such as: the building construction type, size, and area; the proximity to other areas within the facility; and the proximity to other areas that may be outside the building (known as *exposures*). What is the function of the area you intend to protect? Requirements for patient care areas in a hospital differ greatly from mechanical equipment areas, warehouses, offices, laundry areas, and other hazards. Each area should be evaluated separately and in conjunction with all of the other parts. Review individual operations that could inadvertently be combined to result in a very hazardous occurrence. For example, the storage of incompatible laboratory chemicals and cleaning agents in the same place can cause a very hazardous condition. Once a thorough assessment has been completed, selection of extinguishing agents can be discussed with a greater level of confidence.

Water is almost always the most practical and inexpensive extinguishing agent. A reliable, adequate source of water is usually readily available. In some cases, even if a reliable, adequate source of water is not readily available, it is very cost-effective to create a stored water supply rather than be burdened with the long-term costs of maintaining, testing, and servicing an alternative.

However, water-based suppression systems may not be suitable for hazards such as high-voltage electrical equipment, computer equipment, flammable liquids, sensitive process and/or storage areas, and similar hazards. Although water-based systems could be specially designed to protect against these hazards, consideration of all the aspects involved during the hazard assessment phase of the project may determine that extinguishing agents other than water are a more feasible approach to protection. Extinguishing agents that are alternatives to water are clean agents (commonly referred to as Halon replacements), carbon dioxide (CO_2), a variety of foam mixtures, dry-chemical combinations, and wet-chemical combinations (Table 5.4.7-1).[1]

Several variations of the Halon family of gases were widely used as extinguishing agents in computer rooms and other areas containing electrical equipment during the 1970s. However, in the early 1980s, researchers found that the Halons were contributing to the depletion of Earth's ozone layer, and the production of Halon was banned. Consequently, production of the

TABLE 5.4.7-1   Guide for Suppression System Selection

Hazard	Water sprinklers or spray	Foam	Carbon dioxide or clean agent	Dry chemical	Wet chemical
Ordinary combustible materials (wood, cloth, paper, rubber and many plastics)	X	X	X	X	X
Flammable or combustible liquids, flammable gas, greases, and similar materials	X	X	X	X	X
Energized electrical equipment	X		X		

Source: *Massachusetts State Building Code,* 5th ed., Commonwealth of Massachusetts, Boston.

Halon derivatives used in fire protection was also banned. The elimination of Halon as a fire-extinguishing agent has spurred the development of several generations of *clean agents.* Almost all of the clean agents suppress fire in the same manner as the Halons. They reduce the amount of oxygen in the protected area to a level that sustains life but does not allow fire to propagate. Each of the clean agents has its idiosyncrasies that must be considered. However, the main consideration must be whether the area to be protected is considered *occupied* or *unoccupied.* This determines the quantity of clean agent that can be used and impacts the overall design of the system as well. Systems for unoccupied areas could be less expensive. However, the restrictions to have the area considered unoccupied may not be worth the savings.

The selection of a clean-agent system will impact other aspects of the project as well. Special consideration must be given to the way the protected area will be enclosed. The design of the system must take into account walls, penetrations, and other passive enclosures; doors and other openings must close automatically before the clean agent is discharged; and climate control systems (HVAC systems) that are within or that breach enclosures must cease operation before clean-agent discharge. When the clean-agent system discharges, the pressure within the protected area is slightly elevated. If the protected area is too tightly enclosed, physical damage may result. Also, many clean agents must be exhausted to the atmosphere from the protected area after suppression has been achieved. This normally requires separate supply and exhaust capabilities or the design of a climate control system serving the protected area that takes these functions into consideration.

Carbon dioxide (CO_2) systems are usually used in industrial and other special applications (Fig. 5.4.7-1).[2] These systems have both the same considerations as clean-agent systems and additional considerations. The use of carbon dioxide may cause a condition known as *thermal shock.* To contain carbon dioxide gas in cylinders, the gas is cooled to a liquid state and then placed in the cylinders. When the carbon dioxide is discharged, it exits the cylinder still in the liquid state and then converts back to the gaseous state. This phenomenon is a complicated physical reaction. However, one end result is the immediate lowering of the temperature within the enclosure. Although this change in temperature may be only slight, the impact must be considered during your hazard assessment.

Another significant limitation of carbon dioxide systems is their risks to personnel. Carbon dioxide systems can reduce the oxygen in the area of application to levels not suitable for breathing. Thus, they are not usually used in occupied areas. When carbon dioxide systems are used in occupied areas, special considerations are required to ensure the safety of personnel who may be present for maintenance or other purposes.

A variety of *foam mixtures* are suitable agents for extinguishing fires that involve alcohols and flammable and/or combustible liquids. Foam concentrates either are mixed with water before discharge or are entrained in the discharge stream upon activation of the system. There are several different types of foam systems. High-expansion foam systems work by creating a foam that totally fills an enclosure and smothers the fire while cooling the environ-

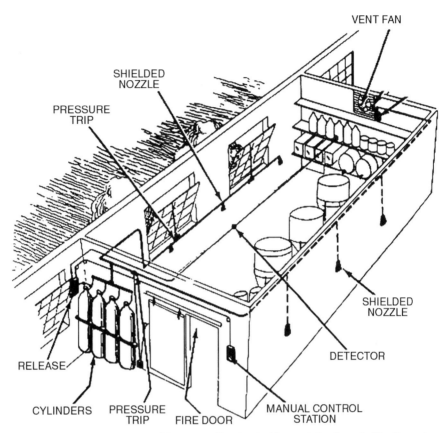

FIGURE 5.4.7-1 High-pressure CO_2 system. (*Reprinted with permission from the* Fire Protection Handbook, *18th ed., © 1998 National Fire Protection Association, Quincy, MA.*)

ment. High-expansion foams are commonly used in aircraft hangars and storage applications and would most likely not be considered for use in your facility. Low-expansion foam systems work by creating a foam blanket that totally covers an entire area and smothers the fire while cooling the surrounding area (Fig. 5.4.7-2). Low-expansion foams are commonly used on dip tanks, plating baths, and other similar applications.

Foam systems can also be of the foam-water type. These are very effective in fighting certain flammable liquid fires. The foam-water system extinguishes much in the same manner as the low-expansion foam system. The combination of the foam concentrate with water creates a blanket upon discharge that coats the surrounding area with foam, and the water helps to provide a cooling effect.

Dry-chemical systems discharge various types of agents in powder form. The powder coats the protected area and extinguishes the fire by reducing and/or eliminating the air supply, much in the same manner as a foam system. Dry-chemical systems are normally used to protect areas where water alone may not be feasible as an extinguishing agent. These could include areas where water-reactive metals, such as sodium and potassium, are stored. This type of application may not be necessary in your facility.

Dry-chemical systems are usually used to protect kitchen equipment such as deep fat fryers and griddles and the exhaust ductwork that serves this equipment (Fig. 5.4.7-3).[3] However,

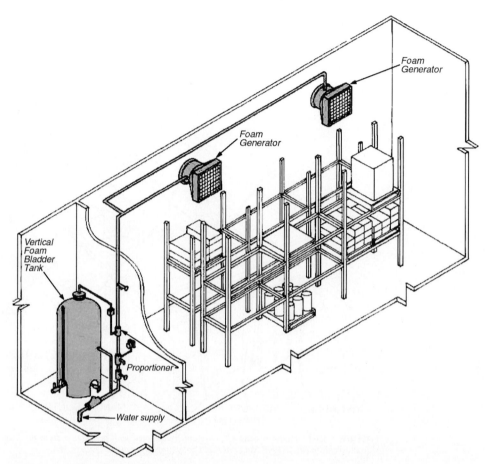

FIGURE 5.4.7-2   Low-expansion foam system. (*Courtesy of Automatic Suppression Systems, Inc.*)

due to sanitary conditions that must be maintained in food service and preparation areas, dry chemicals are being replaced by systems that use wet-chemical extinguishing agents. Dry-chemical discharge tends to permeate everything within the enclosed area, even items that are not being protected.

Wet-chemical systems are mainly used to protect kitchen equipment and the exhaust duct-work that serves the equipment (Fig. 5.4.7-4). Unlike dry-chemical systems, when wet-chemical systems discharge, the area affected is limited to that being protected and the area immediately adjacent to it. This helps to minimize the impact on sanitary conditions that must be maintained in food service and food preparation areas and therefore reduces the amount of food product that must be discarded, the amount of cleanup involved, and the overall impact on food service operations. Wet-chemical system discharge covers the protected item and/or area with a blanket of chemical extinguishing agent that smothers the fire and cools the area. This blanket of most wet-chemical agents also forms a crust upon cooling that prevents fire from recurring and aids in cleanup.

One factor that all systems other than automatic fire sprinkler systems usually share is that they require separate detection systems to initiate discharge. In automatic fire sprinkler systems, the fusible element in the head acts as its own detection system. It reacts to heat and

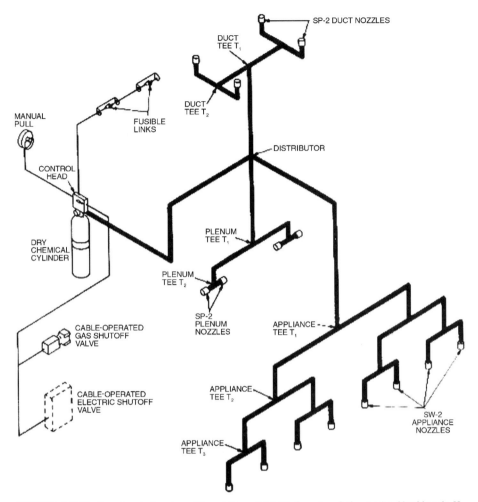

FIGURE 5.4.7-3 Dry-chemical system. (*From UL ex. 2-53, Underwriters Laboratories, Northbrook, IL, November 1998.*)

releases a seating mechanism that covers the discharge orifice, thus discharging water onto the fire. However, preaction, most foam-water, and other types of special systems usually incorporate a separate detection system to ensure that water or other agents are not discharged unless the presence of fire is confirmed by a combination of responses from both fusible elements and detectors. The need to activate these systems by separate detection systems adds to the initial cost. Long-term maintenance and testing issues must be considered, as well.

Clean-agent, carbon dioxide, foam, and chemical extinguishing-agent systems cannot be expected to continue to control a fire if the initial discharge does not produce the desired results. Many of these systems are designed with attached backup (reserve) supplies that can discharge additional agent into the protected area. Many are also designed to consider special issues such as deep-seated fires, openings that cannot be closed, and HVAC fan system run-down times. Once the additional agent has been expelled, however, the capability of that system to extinguish or control a fire has been lost. For this reason, areas that require protection

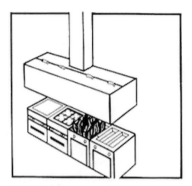 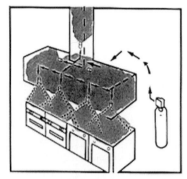

FIGURE 5.4.7-4 Wet-chemical system. (*Courtesy of Walter Kidde Fire Systems.*)

systems other than automatic fire sprinkler systems supplied by adequate sources of water must be carefully designed, and all considerations regarding the limits of these systems must be addressed as a part of the overall hazard assessment. Suitable protection against any fire hazard can be developed from a combination of water and alternative extinguishing agents.

DEVELOPING DESIGN CRITERIA

Once a hazard assessment has been completed and extinguishing agents have been selected, the design criteria for the protection required are established by reviewing applicable codes and standards. The local building codes are usually in the form of a state-wide building code that is enforced by the local building official. In turn, most state building codes are usually based on one of the national model building codes (e.g., BOCA-NBC, ICBO, and SBCCI).[4]

Contact the local building official to confirm the existence of any local bylaws, ordinances, and/or amendments that could change certain requirements of the applicable codes. Counties, cities, towns, and fire districts can legislate changes to applicable codes for specific reasons. Local issues such as special requirements for impaired persons or special requirements of the local fire department, historical commission, access barriers board, or zoning board may need to be incorporated into the design. The owner's insurance underwriter should also be contacted at this stage of the project because they may require variations in certain design criteria that must be addressed. For example, insurers such as IRI, Kemper, and Factory Mutual have design requirements more stringent than those of the National Fire Protection Association (NFPA) standards and many state codes for similar occupancy classifications.

Developing a Suitable Water Supply

Water will most commonly be used as the extinguishing agent because it is inexpensive compared to other extinguishing agents, and it is usually readily available. The design of the water supply for the facility should consider many issues. In new construction, future considerations should be addressed. Will more buildings eventually be added at this site? Will hazards or processes be added at a later date that could increase the demand that has been established? Can the buildings that are planned for development at this time be expanded, or could they eventually house operations that will place them in a higher hazard category that requires more water for suppression systems?

Considering future issues at the inception of developing a water supply could help to avoid costly changes in the future. The current placement and location of buildings or other site features could severely impact the future location of municipal connections or stored water supplies that might be needed in the future. The initial expense of installing larger pipe sizes, additional valved and capped connections, and even larger stored water supplies may be easier to absorb at the inception of a facility if future needs are considered in the overall planning.

If a site plan is available, consideration must also be given to the location of any developed water supply. The location of the supply could become very important if it becomes unfeasible to pump fire protection water from one end of the complex to the other. Other considerations should include the actual site itself. Could a ledge be encountered? Will it be more feasible to locate any stored water supplies and pumping equipment at the highest elevation on the site? All issues should be considered.

If the water supply is to be provided through an existing fire pump, the most current results of a fire-pump flow test can be used to determine the available water supply. Bear in mind that NFPA 20, *Installation of Centrifugal Fire Pumps,* requires flow testing a fire pump at least annually.[5] Therefore, test results more than a year old should not be considered current, and a new fire-pump test should be performed. Previous records can then be used as a baseline to help determine the condition of both the fire pump and its water supply.

High-rise buildings (usually those over 70 ft above grade) almost always require a fire pump. One reason for this is that NFPA 14, *Installation of Standpipe Systems,* requires a minimum flow of 500 gal/min at a pressure of 100 lb/in^2 at the top of the most remote standpipe.[6]

Even if the local water authority can produce records of fire-hydrant flow testing, additional testing by competent, qualified persons is strongly recommended. Comparison of the results provided by the water authority with these additional test results will either confirm that the supply is consistent or they could reveal inconsistencies that should be investigated. See Figs. 5.4.7-5 and 5.4.7-6.

Local water authorities establish requirements for items such as materials and methods to be used in conjunction with connection to the public system, fees for initial connections, user fees, metering requirements, cross-connection control requirements, valve types, and hydrant types. In addition, regulations could also impact valve and hydrant locations, as well as the direction of operation for valves and hydrants.

Water Supply from Sources Other Than, or in Addition to, Connections to the Public Distribution System

Ponds, lakes, rivers, and/or wells used as water supply sources must be proven to be reliable. Various nationally recognized fire protection codes and standards, as well as other reference

Location: Adams St. between Cox St. and Baker St.						
Hydrants: 2½-in. square shape assumed $C = 0.80$						
Pressure	Pitot pressure				gpm	Total gpm
Hyd 1	Hyd 2	Hyd 3	Hyd 4			
72	–	–	–		0	0
62	18	–	–		633	633
50	10 10		–		472 472	944

Gage 103 at hydrant No. 1 — gage 79 for pitot readings.

FIGURE 5.4.7-5 Public water flow-test data. (*Reprinted with permission from the* Fire Protection Handbook, *18th ed., Copyright © 1998 National Fire Protection Association, Quincy, MA.*)

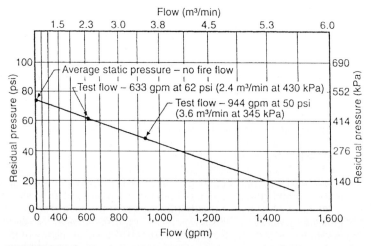

FIGURE 5.4.7-6   Curve plotted from flow-test data given in Fig. 5.4.7-5. (*Reprinted with permission from the* Fire Protection Handbook, *18th ed., Copyright © 1998 National Fire Protection Association, Quincy, MA.*)

sources, contain data that can be used to confirm the reliability of a naturally occurring body of water. Historical variations in water volume during each season for several years or decades, evidence of man-made or natural contamination, cycles of wildlife in the area that use or could use the body of water, and occurrences during short-term dry spells and long-term drought are some items that must be considered. Wetlands restrictions, Environmental Protection Agency (EPA) regulations, and flood-plain restrictions must also be considered.

Water obtained from naturally occurring sources can be used to supply a fire-pump system directly through a wet pit, or it could be used to fill a stored water supply in the form of a cistern. If pumping directly from the source is not feasible, smaller pumps could be used to fill stored supplies (tanks), which can in turn be used as a source of supply for fire pumps.

In areas where both supply from the public distribution system and naturally occurring bodies of water are readily available, it may be desirable to consider a water-supply scenario where the primary water supply will come from one source and the other can be used as an alternate supply.

The Maximum Reliability Benefit

A *maximum reliability* design is recommended and may be required for critical facilities. It should be considered for other applications where implementing it may not be excessively costly. Connections to the public distribution system, the on-site water-supply system, the interior water-distribution system, and all water-based suppression systems within the facility should be arranged so that each system is supplied from at least two sources. Multiple connections to the public distribution system, a stored water supply in addition to the public water supply, main and back-up fire pumps, looped distribution systems within the building, and fire sprinkler systems using cross-connected combination risers that can provide the required demand solely from the most remote riser are all examples of maximum reliability. In some cases, this may be a code requirement.[4]

Developing the On-Site Water Supply

Once the source of water has been established, the next consideration should be the development of the water supply available on the site for fire-protection purposes. For a facility that will consist of only one building located within the limits of a city block, water-supply considerations may be obvious, straightforward, and easily provided. However, items such as the availability of and access to fire hydrants, fire department connections, hose houses, and fire pumps must be considered at large sites where many buildings exist or may be constructed. As with the considerations particular to connection to the public distribution system, planning for any on-site distribution system (exterior and/or interior) should give serious attention to all applicable codes, standards, and other concerns regarding the needs of current projects as well as future considerations.

Particular attention should be directed to future needs. If a master plan is developed, the on-site water supply should be part of that planning. If no such plan exists, then an on-site water supply should be developed that accounts for as many future needs as feasible. Prudent planning at this stage of development will make future changes less of a problem.

Once plans for the on-site water supply have been developed, a specification should be added to address the materials to be used, installation methods, testing procedures to be followed, and any specific requirements of local authorities and the facility itself. Specifications should reference all applicable codes and standards and should contain language that makes contractors responsible for executing the work in accordance with all applicable codes and standards, for performing testing in the presence of representatives of the facility, for providing proper documentation of all testing, and for providing as-built drawings that show the system as it was when accepted by the facility. Of particular importance is the need to verify that the installing contractor has submitted a flush and test certificate as required by NFPA 24, *Standard for the Installation of Private Service Fire Mains.*[7] Flushing the supply piping at the volumes outlined in NFPA 24 will ensure that no foreign objects (gravel, sand, nuts and bolts, rags, etc.) remain. If such items are drawn into the supply piping, they can damage fire pumps and obstruct piping by lodging in offsets, valves, and even in individual fire sprinkler heads.

To make the on-site distribution system as reliable as possible, more than one water-supply connection should be investigated. Certain codes and standards require more than one connection to the water supply so that if a water supply is impaired for any reason, the alternate supply will keep all water-based suppression systems in service. Care should be taken at the time of design to assure that water-supply connections are arranged to be as independent as possible. Multiple connections on the same side of a building, or entire facility, to the same public main serve no purpose if the public main is taken out of service. Connections on opposite sides of the building, or entire facility, will help to avoid this situation. Most public distribution systems (especially in municipal areas) are arranged in a grid system that allows supplying many piping segments in the grid from adjacent segments, even when one segment must be taken out of service. This arrangement is not possible on "dead-ended" mains. In dead-ended areas, the possibility of connecting dead-ended mains to form a loop by using at least two independent mains should be considered. In areas where looping dead-ended mains is not possible, and depending on the need to keep water-based suppression systems in service at all costs, the possibility of using a stored water supply with a fire pump could be considered.

If multiple supplies are considered, hydraulic calculations for the water-based suppression systems should be based on the worst-case water-supply scenario. Also, many highly protected risk (HPR) insurers (e.g., FM, IRI, and Kemper) require calculations to prove that hazards can be partially protected using the worst-case water-supply scenario, even if fire pumps are a part of the overall water-supply arrangement.

Installation of the on-site water-supply piping should be observed at all stages of the process. If the system is installed below grade, proper excavation, bedding and backfilling of pipe trenches, proper depth of cover to keep pipe from freezing, proper joining of full lengths of piping to each other, proper joining of pipe to fittings (retainer glands and rodding), proper size and location of thrust blocks at pipe offsets, installation of flexible connections, proper penetration and proper restraint of piping at foundation walls or floor slabs, and seismic considerations must all be monitored. The cost of rectifying problems encountered after piping trenches have been backfilled will be far greater than that of holding construction schedules until the installation has been properly inspected. It should be noted that several types of pipes and fittings are available for underground installation; the best type will depend on conditions at the installation site. If the piping is to be installed within buildings, many of the same considerations must be addressed. However, without the earth to aid in restraining forces produced by water pressure within the piping system, issues concerning piping restraint must be reviewed in much more depth.

All valves on the system that will be controlled above grade (nonrising stem-gate post-indicator valves [PIVs]) should be electrically supervised to indicate operation of the valve. If electrical supervision of these valves is not feasible, these valves should be locked and/or sealed in the full-open position (Fig. 5.4.7-7).

(a)

(b)

FIGURE 5.4.7-7 Underground piping valves: (*a*) adjustable-type post-indicator valve; (*b*) NRS gate valve. (*Courtesy of Mueller Company.*)

As-built drawings should be reviewed to ensure that all valves have been accounted for and marked, and it should be verified that they are in the position intended by the design (full open, normally closed, etc.).

Developing the Interior Distribution System

Some large facilities consist of separate large buildings or properly fire-separated areas (which can be considered as separate buildings[8]) that could be joined at these separations to form one, or several, contiguous "buildings." The goal in designing the interior distribution system should be to provide a stand-alone system that supplies all water-based suppression systems within individual and/or contiguous buildings. The best way to approach the initial design is to think of this system in much the same manner as the on-site water-supply system. To start, the piping should be supplied from at least two independent sources, even if not required. This will allow the interior distribution system to remain in service if one source of supply is impaired. Hydraulic calculations for all connected water-based suppression systems should be based on the worst-case water-supply scenario.

Piping for the interior distribution system should be installed so that all portions of this system are accessible. Valves should be visible and marked so that they can be easily located and operated. Ideally, piping for the interior distribution system should be located in areas where it will be totally exposed and accessible. Normally, this will be where main lines for other building systems (steam, domestic water, storm and sanitary drains, etc.) are present. However, if this is not practical, piping and valves can be located above ceilings, as long as access to valves is provided. Once a location for the distribution system main has been established, feed mains to the standpipes in stairwells, to fire department connections, and to other connected systems can be located.

Consideration must also be given to drain piping. The floor control stations (discussed later in this article) and any pressure-regulating devices (2½-in fire department valves and pressure-reducing valves at floor control stations[8]) must be flow tested periodically to assure that they are operating properly. Although no sizes are mandated for this drainage-system piping, it must be able to accommodate a minimum 250-gal/min flow from a 2½-in fire department valve. Drainage-system piping sized at 3 in is recommended. The most practical location for drain piping is adjacent to standpipes in the stairwells or out on floors adjacent to intermediate standpipes, if they are required. Vertical drain-riser piping should terminate above grade and discharge at locations where the water will not pose a safety threat. Direct discharge onto sidewalks and public ways may be prohibited in many municipalities. In many cases, it may be practical to provide drain inlets in the storm-drainage system outside the building into which drain piping can discharge directly. Plumbing codes and standards, as well as local ordinances, may place restrictions on such an arrangement. Therefore, the arrangement should be reviewed against local requirements before review by local authorities. In any case, an air gap must be maintained between the two systems.

If it is not feasible to discharge to the outside at the base of vertical drain risers, then all risers should connect to a main drain line that can discharge to a single, central location. The main drain line should be above grade. This will allow the vast majority of drainage to flow by gravity. The dead legs of vertical drain risers and any below-grade ancillary drain piping can be drained to floor drains, sinks, etc.

Supply connections to the main piping should be valved using supervised indicating valves, so that each supply can be individually isolated from the main piping without taking the main piping out of service. In systems using a loop design, each leg of the loop should be isolated as well, so that as many segments of the loop as possible can remain in service if a portion of the loop must be removed from service. Standpipes, combination risers, and supplies to any other water-based suppression systems that will connect to the main interior distribution system should be valved, using supervised indicating valves, so that the they can be isolated from the main piping without taking the main piping out of service.

Standpipe System. Standpipe systems are an important part of the infrastructure that support firefighting operations and supply water to fire sprinklers and other water-based fire-suppression systems. The standpipes and their associated drain and test risers are normally located in stairwells. This location places them in a 2-h rated enclosure that could serve as a staging area for firefighting operations and as egress for occupants.[7] For this reason, care should be taken so that no standpipe, drain riser, 2½-in fire department valve, first-aid hose rack, sprinkler system control-valve, and/or sprinkler system test and drain valve is inadvertently placed in any part of the egress pathway.

Care should be taken to assure that the proper types of hose valves are installed. If pressures in all parts of the standpipe system will not exceed 100 lb/in^2, only a standard hose valve is required. It should be noted that systems supplied through a fire pump are likely to generate churn (static) pressures in excess of 100 lb/in^2. For pressures between 100 and 175 lb/in^2, pressure-restricting valves are required. Pressure-reducing valves are required for pressures of more than 175 lb/in^2. NFPA requires that pressure-restricting and/or -reducing valves must be tested to ensure that they are performing as required. For this reason, a 3-in drain riser must be installed parallel to the standpipe. The drain must be equipped with 2½-in connections to allow connecting a flowmeter between the hose valve and drain riser. Standpipe and drain risers, as well as hose valves themselves, should be located in the stairwells so as not to restrict the egress pathway.

In buildings that have large floor areas, 2½-in hose valves may have to be located out on the floor to allow the fire department accessibility to all areas. If the hose valves are located in the same area on each of multiple floors, then it may be feasible to supply them with an additional standpipe. This intermediate standpipe should also be provided with a 3-in drain riser. If additional hose valves are required only in certain areas, then it may make more sense to pipe these individual valves from the standpipe in the nearest stairwell. Proper supply to these individual hose valves (minimum 250 gal/min) should be verified by hydraulic calculations if the equivalent length of piping exceeds 25 ft. Piping larger than 2½ in may be required because of the pipe and fittings required to reach individual valves out on the floors.

True standpipe systems supply only 2½-in fire department valves and 1½-in first-aid fire-hose racks. The 2½-in valves are used by the fire department to supply heavy hose lines during a fire. Most municipalities allow the installation of 2½ × 1½-in reducers at the fire department valves. Consult your local building code to verify any requirements for 1½-in first-aid hose racks. Many municipalities do not want hose racks installed unless the facility has a fire brigade or emergency response program that provides training in the proper use of this equipment.

If the facility has two or more standpipes, chances are that one, both, or all will be used as *combination risers*. Combination risers supply both hose valves and water-based fire-suppression systems from the same piping system. Combination risers are very common and should be encouraged in order to reduce the amount of piping in stairwells, which will, in turn, reduce overall initial and long-term maintenance costs.

FIRE SPRINKLER SYSTEMS

Fire sprinkler systems are what you will most commonly see installed in all types of facilities. In low-rise individual buildings, the common piping arrangement is a water supply entering the building to provide water to a single or multiple wet-alarm valves. The alarm valve is a sophisticated check valve. The alarm valve operates similarly to a check valve. However, the casting of the alarm valve is designed to contain threaded ports for connecting main drain and other drain valves, system and supply pressure gauges, and alarm lines that operate both mechanical (water-motor gong) and electrical (pressure switch) alarms. Although an alarm valve is not required by codes and standards, all require a check valve, main drain valve, and system supply pressure gauges. For many individual systems supplied through fire pumps, there should be a sufficient number of check valves, drain valves, and pressure gauges in the system to consider eliminating an additional alarm valve. However, multiple systems supplied from fire pumps should be provided with at least a check valve, main drain, and system-pressure gauge for each individual system. Although a check valve downstream from the main water supply may not be required for each system, it will allow the system to remain fully charged when all control valves are in the full-open position, even when the water supply on the supply side of the individual check valve is out of service. These check valves, if installed as a part of the original installation, will pay for themselves many times over by eliminating the need to shut down other system control valves on the same supply to avoid draining all connected piping. One of the main causes of destruction by fire in fully sprinklered buildings is failure to reopen control valves after servicing piping on the connected system.

In facilities where combination risers are used to supply fire sprinkler systems, if two combination risers are available for each area to be protected, then the combination risers should be connected through the sprinkler system mains to provide maximum reliability. If combination risers are cross-connected, then a check valve must be installed at the floor control station at each connection to the riser. The installation of a check valve at each of the floor control stations will allow the system control valve for each floor to remain in the full-open position, even when the water supply on the supply side of the individual check valve is out of service. However, if work is required on a system that is supplied by two cross-connected combination risers, then both control valves for that area must be shut down to avoid water flow at that location. Control valves on other floors, or in other areas, not directly connected to the piping that is to be worked on may be left in the full-open position.

By cross-connecting combination risers, the area being protected can remain fully protected by one riser if the other is out of service. Hydraulic calculations of pipe sizes for the area to be protected should be based solely on the supply from the most remote individual riser. At the present time, NFPA standards allow using the available supply from both risers to size piping. The use of flow split through both combination risers will almost always allow a decrease in pipe sizing. However, the use of both risers will defeat the intent to create maximum reliability.

Fire sprinkler piping that will protect individual areas should be designed in accordance with the current issue of NFPA 13, *Standard for the Installation of Sprinklers,* as a minimum requirement.[8]

The approach to the installation of fire sprinkler systems will differ greatly depending on where they are to be installed within your facility. The exposed piping system that may protect basement storage and mechanical areas will not be the same as the system that will protect the elegant atrium entry that may be the showplace of your facility. Although the fire sprinkler systems in these two areas may differ drastically in appearance, they have many similarities in their basic design and operation.

The installation of fire sprinkler systems must start in much the same manner as an overall fire-protection design. First, a hazard assessment must be performed to develop the design criteria for protecting the defined hazard. For the most part, the current issue of NFPA 13 should suffice as a guide in determining which occupancy classification to use to determine the design criteria for a fire sprinkler system to protect a designated area. Bear in mind that the term *occupancy* as used in NFPA 13 is specific to NFPA 13. It differs from the same term as used in NFPA 101, *The Life Safety Code,*[9] and in state and local building codes. As an example, *health-care occupancy* in NFPA 101 could contain light hazard, ordinary hazard group I, and ordinary hazard group II occupancies as defined in NFPA 13. All three occupancy groups, as noted from NFPA 13, have significantly differing design requirements that impact water supply requirements, hose stream demands, fire sprinkler head location and spacing, and even the types of fire sprinklers that can be used in these areas. Fire sprinkler systems may be required to interface with other fire-protection systems, such as the smoke-control fire-alarm zones or other zoning arrangements. Designing the piping system to interface with these other systems will provide a more efficient and effective fire-protection system overall.

The careful review of all aspects of new construction and renovation work is a must before trying to establish fire sprinkler design criteria. Selecting inappropriate design criteria and incorporating them into bid documents can have a major impact on the overall cost of a project when it is found that the criteria are not those required to protect the hazard, or occupancy classification, that will exist. This reason alone should make it clear that your project should have a registered architect and/or licensed professional engineer on board who is bound by statute to be responsible for designing the project in accordance with applicable codes and standards.

Although most new facility construction will require the involvement of a registered professional, many renovation projects may not. In renovation, the primary information to consider at the inception of the design process is the water supply (volume and pressure) at the point of connection to the renovated area and the NFPA 13 occupancy classification of the area. The point of connection for the water supply will usually be at an existing standpipe or fire sprinkler main. Once a point of connection has been established, it is also imperative that all piping from the point of connection back to the point of supply (fire pump, city connection, etc.) be documented and calculated to ensure that an adequate water supply will be available at the point of connection. By determining the available water supply during the inception of the project, measures can be taken to compensate for a marginal water supply within the scope of the renovation. Or, as is usually the case, once it has been determined that an adequate water supply is available, the peace of mind in knowing that a major consideration has been confirmed will help the renovation process proceed with a greater degree of confidence.

For environments that cannot be heated to 40°F, standard wet-pipe sprinkler systems are not practical because the water may freeze. For these locations, a dry-pipe system should be used. The system piping is filled with pressurized air instead of water, and a dry-pipe valve is used to hold back the water supply. The dry-pipe valve is constructed and arranged to allow the differential between the air pressure above the dry-pipe valve clapper and the incoming water pressure to equalize in the valve chamber that contains the clapper. Once a sprinkler is activated or a pipe breaks, the air escapes, and the valve opens because the pressure difference has been upset. Dry-pipe valves are more complex than wet-pipe alarm check valves. They also have special design considerations, such as piping pitch requirements, the use of dry pendant sprinklers, and an increase in the required hydraulically remote design area. Some locations where dry-pipe systems are required include cold-storage warehouses, air-intake plenums, large freezers, crawl space and attic areas, and some outdoor applications.

The least desirable alternative to a dry-pipe system is an antifreeze system filled with chemicals such as ethylene glycol. These systems are effective, but the chemicals used are of special environmental concern. The sizes and uses of these systems are restricted by most codes and standards because of special backflow devices, expansion tanks, drainage requirements, and other considerations.

Fire Sprinklers

Fire sprinklers themselves present a myriad of considerations. Upright, pendant, exposed, concealed, quick-response, extended-coverage, extended-coverage quick-response, dry type, intermediate temperature rating, and many more sprinkler types present many possibilities for cost-effective designs—as well as for inappropriate designs that will add significant cost to your renovation project for changes to bring the fire sprinkler system into compliance with applicable requirements (Fig. 5.4.7-8). Some of the different sprinkler types are for specific

FIGURE 5.4.7-8 Some different sprinkler types. (*Courtesy of Central Sprinkler Company.*)

purposes; others are designed to be more aesthetically pleasing when used in certain construction types.

OTHER TYPES OF WATER-BASED FIRE-SUPPRESSION SYSTEMS

In addition to the standard wet-pipe and dry-pipe fire sprinkler systems that you may encounter at your facility, there are other types of water-based fire-suppression systems that may be considered. The most common of these is some variation of the preaction system. To a lesser extent, you may also need to consider installing a foam-water system or even a water mist system.

Preaction systems are dry systems that are activated by adding a detection system (Fig. 5.4.7-9). Preaction systems are usually used in areas where the discharge of water is undesirable. These areas could include computer rooms, record storage areas, and the like. Bear in mind that preaction systems are more expensive than standard-wet-pipe systems. The added cost is mainly due to the addition of a detection system. There are several types of preaction systems that operate at varying levels of complexity. All levels take the concern of an unwanted discharge of water into consideration. The standard preaction system operates as follows. The piping system is a closed system; the preaction valve is actually a deluge valve (usually electrically actuated). The valve is held closed by an electrically actuated solenoid connected to a release mechanism. The solenoid is actuated through a release panel by the detection system. Two conditions must be met for water to discharge onto the fire: (1) the detection system must activate and release the solenoid—this action will cause the piping system to fill with water; and (2) heat from the fire must fuse a fire sprinkler head to release

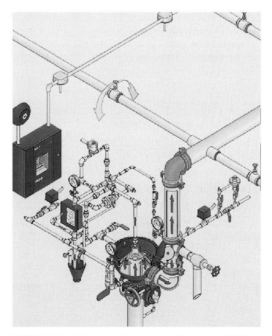

FIGURE 5.4.7-9 Preaction system. (*Courtesy of Viking Corporation.*)

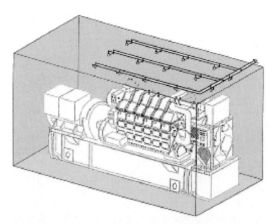

FIGURE 5.4.7-10   Water-mist system (*Courtesy of Fireline Corporation.*)

water. The other variations of the preaction system operate in basically the same manner. However, the combinations of conditions that must be present for water discharge become more complex as levels of redundancy are added.

Water-mist systems are new to the market (Fig. 5.4.7-10). These systems extinguish fire by dispersing certain-sized droplets throughout the protected area. These droplets cool the fire and reduce the volume of air available to the fire for combustion. Although water-mist systems hold a promising future, they are limited in use at this time to certain-sized enclosures, local area applications, and shipboard protection. Water-mist systems are most useful where the volume of water is a concern. This is one reason why they have been so popular in marine applications. To consider a water-mist system for a particular application, one should contact only knowledgeable, qualified persons who have experience with this equipment and the design of the systems that must be installed.

CONTRIBUTORS

Michael Hickey, P.E., Partner, Thomson Consultants, Inc., Marion, Massachusetts

Anand K. Seth, P.E., C.E.M., C.P.E., Director of Utilities and Engineering, Partners HealthCare System, Inc., Boston, Massachusetts

Paul D. Sullivan, P.E., M.S., President, Robert W. Sullivan, Inc., Boston, Massachusetts

Sean Duffy, B.S., Robert W. Sullivan, Inc., Boston, Massachusetts

Mark J. Sullivan, B.S., C.E.T., CIPE, CDA, Treasurer, Robert W. Sullivan, Inc., Boston, Massachusetts

REFERENCES

1. *The Massachusetts State Building Code,* 5th ed., Commonwealth of Massachusetts, Boston,
2. *Fire Protection Handbook,* 18th ed., National Fire Protection Association, Quincy, MA, 1998.
3. UL ex. 2-53, Underwriters Laboratories, Northbrook, IL, November 1998.

4. *National Building Code* (BOCA-NBC), Building Officials and Code Administrators International, Country Club Hills, IL, 2000; International Conference of Building Officials (ICBO), Whittier, CA; Southern Building Code Congress International, Inc. (SBCCI), Birmingham, AL.

5. *Installation of Centrifugal Fire Pumps,* NFPA 20, National Fire Protection Association, Quincy, MA, 1996.

6. *Installation of Standpipe Systems,* NFPA 14, National Fire Protection Association, Quincy, MA, 2000.

7. *Standard for the Installation of Private Service Fire Mains,* NFPA 24, National Fire Protection Association, Quincy, MA, 1995.

8. *Standard for the Installation of Sprinklers,* NFPA 13, National Fire Protection Association, Quincy, MA, 1999.

9. *The Life Safety Code,* National Fire Protection Association, NFPA 101, Quincy, MA, 2000.

ARTICLE 5.4.8

ENERGY EFFICIENCY AND INDOOR ENVIRONMENTAL QUALITY

John F. McCarthy, Sc.D., C.I.H.
Environmental Health & Engineering, Inc., Newton, Massachusetts

The primary function of buildings is to provide a healthy, comfortable, indoor environment where people can work and live. The elements of a productive indoor environment to be optimized include ventilation, temperature, humidity, noise, and light levels. Maintaining these elements is an expense and, as with any other business expense, must be justified. By detailed analysis, Fisk found that the potential financial benefits of improving indoor environments in the United States exceed costs by factors of 9 to 14.[1] It is ironic that the means to an end—controlling energy use to create a productive environment—can often become the end in itself simply because its cost impact is more readily quantifiable.

This article focuses on a way of simultaneously maintaining good indoor environmental quality (IEQ) and energy efficiency in new and existing buildings. In this context, we propose an operational definition in which *energy conservation* is defined as the more efficient or effective use of energy, rather than simply reducing energy consumption. The ASHRAE definition implies that the purpose of energy is to help achieve an objective.[2]

Energy conservation projects can take several forms, depending on the targeted end point. The different, generally exclusive goals for energy efficiency in new or retrofit projects include the following[3]:

- Lowest energy consumption over the operating life of a building while meeting the users' needs. Here, first costs are secondary to energy savings.

- Lowest life cycle costs over the operating life of the building while meeting users' needs by designing a system geared toward minimized energy consumption.

- Minimizing energy consumption at lowest first costs while meeting the users' needs. In this design approach, energy savings are secondary to first costs and often result from value engineering. Generally, operating, maintenance, and energy costs are not included in the cost evaluation.

- Rationalizing the use of energy to optimize user performance by incorporating productivity metrics in the life-cycle cost analysis.

Using the last goal as the preferable end point in optimizing occupant productivity and using energy efficiently, the objectives of this article are (1) to focus on reducing energy use in buildings while simultaneously ensuring indoor environmental quality, and (2) to encourage the practice of assessing the impacts of energy strategies on the indoor environment in the early stages of design and programming. Taking a more holistic view of design through *sustainability* review will offer several alternative procedures to the more common *dilution* approach.

ENERGY CONSERVATION AND INDOOR AIR QUALITY

Air exchange in a building occurs by (1) uncontrolled infiltration and exfiltration of air through the building's envelope; (2) natural ventilation through windows, doors, or specially designed openings; and (3) mechanical ventilation. The impact of these elements is described in other articles.

When people think about energy conservation, reducing ventilation rates is often considered a first step. Much has been written to indicate that reducing ventilation rates degrades indoor air quality (IAQ), but it is important to note that indoor air quality is a balance between pollutant production and dilution by the fresh air supply.

In 1993, Levin, writing about the myth of energy conservation and IAQ, correctly stated that two types of conservation measures directly impact indoor air quality: (1) inappropriately reducing ventilation (either intentionally or unintentionally), and (2) using sealants and caulks that emit pollutants.[4] Although dilution ventilation can be effective in diluting pollutants, it is far more efficient to control the sources and reduce the pollutant load, as shown in Fig. 5.4.8-1.[5] To these measures, a third can be added: relaxing controls on temperature and relative humidity to save energy, which often has poorly understood ramifications on indoor air and environmental quality.

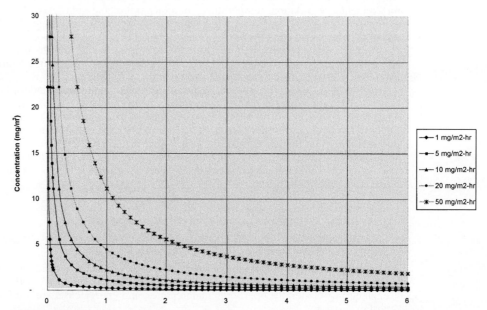

FIGURE 5.4.8-1   Dilution ventilation as a means to dilute pollutants. (*Reprinted with permission from H. Levin, "VOCs: Sources, Emissions, Concentrations, and Design Calculations," Indoor Air Bulletin 3, no. 5 (1996), p. 5.*)

TABLE 5.4.8-1   Potential Reduction in Infiltration Impacts

Modification	Reduction in heating load from infiltration	Reduction in cooling load from infiltration
Reduce envelope leakage (25–50%)	26%	15%
Better ventilation system control	19%	58%

Source: Ref. 8.

ENERGY USE IN BUILDINGS

The energy used for heating, cooling, lighting, and other basic services represents nearly 40 percent of the total energy used in the United States and European countries.[6] Of this, approximately 32 percent is used in the service sector, and 62 percent is used in the residential sector. In service buildings, approximately 68 percent is used for space conditioning (heating and cooling) and the rest for lighting and other services, such as hot water. Therefore, the contribution from the thermal loads of conditioning the air from either mechanical ventilation or from infiltration can be a significant proportion of the total energy budgets of buildings.

Ventilation represents approximately 33 percent of the space-conditioning component. Therefore, energy consumption due to ventilation corresponds to approximately 22 percent of the energy delivered to the building, and is generally seen as an area for significant savings.

It is essential that ventilation sufficient to satisfy occupant requirements for health and comfort be provided in a controlled manner. Currently, ASHRAE Standard 62-1999, *Ventilation for Acceptable Indoor Air Quality,*[7] and its precursor, ASHRAE 62-1989, are the most generally recognized guidance documents in this area. Because prescriptive requirements for delivering ventilation air are available, other methods of controlling the amount of conditioning are required. This could include improving ventilation efficiency to the occupied space and/or reducing uncontrolled infiltration. Tightening the building envelope can also reduce energy losses from uncontrolled infiltration. Table 5.4.8-1 shows that the potential savings by reducing infiltration can be significant.[8]

TABLE 5.4.8-2   Steps to Implementing an Energy Conservation Program

1. Develop a detailed understanding of how energy is used in the building.
2. Systematically evaluate all potential opportunities for energy conservation.
3. Clearly specify the energy and cost savings analysis methods to be used.
4. Identify, allocate, and prioritize the resources required to implement energy conservation opportunities.
5. Implement the energy conservation measures in a rational order. These are often a series of independent activities that take place over a period of years. Typical sequences may be:
a. Evaluate lighting conservation program.
b. Implement a building tune-up program to meet original design.
c. Evaluate possible load reduction, e.g., window insulation and roof upgrade.
d. Upgrade for systems, e.g., variable-speed drives, heat-recovery systems.
e. Upgrade heating and cooling plant, e.g., replace chiller equipment.
6. Commission all energy conservation measures to ensure optimal operation.
7. Periodically reevaluate performance of the conservation measures to ensure their continued effectiveness. This is especially important if the building functions change over time.

Source: Ref. 2.

TABLE 5.4.8-3   Energy Measures That Are Compatible with IAQ

Improve building shell.
Reduce internal loads (e.g., lights, office equipment).
Install fan/motor/drives.
Install chiller/boiler.
Implement energy recovery.
Install air-side economizer.
Implement night precooling.
Perform preventive maintenance (PM) of HVAC.
Install demand-controlled ventilation.
Reduce demand (kW) charges.
Reset supply air temperature.
Downsize equipment.

Source: Ref. 9.

METHODS FOR IMPROVING IEQ AND CONSERVING ENERGY

Energy conservation often focuses on reducing ventilation rates in buildings even though there may be other energy conservation strategies that are equally beneficial. The strategy of reducing ventilation rates has garnered the greatest attention for its potentially detrimental impact on occupant health, comfort, and productivity. A structured method that can be used to improve energy efficiency and indoor air quality is presented in Chap. 32 of the 1995 *ASHRAE Handbook, "Energy Management,"*[2] and is summarized in Table 5.4.8-2.

Tables 5.4.8-3 and 5.4.8-4 present information adapted from Hall et al.[9] All of the IAQ-compatible energy strategies noted in Table 5.4.8-3, except demand-controlled ventilation, have little to do with modifying ventilation rates. The energy activities that could potentially degrade IAQ (Table 5.4.8-4) generally revolve around inappropriate operation of the mechanical systems.

Buildings are complex systems that have many nodes of interaction. These interactions occur simultaneously and dynamically among building systems and subsystems, the outdoor environment, and building occupants. The ways that we use buildings, our expectations of building performance, and the evolving demands caused by heating, cooling, lighting, and equipment usage indoors have brought dramatic changes in the ways structures are designed and built. Reducing energy consumption has become a major driving force in the design of many buildings that are to be renovated or retrofit. Figure 5.4.8-2 provides a guide to the basic elements to consider in maintaining the balance between improvements and energy efficiency and acceptable IAQ.

TABLE 5.4.8-4   Energy Activities That May Degrade IAQ

Reducing outdoor air ventilation
Installing variable-air-volume (VAV) systems with fixed percentage outdoor air
Reducing HVAC operating hours
Relaxing thermal controls

Source: Ref. 9.

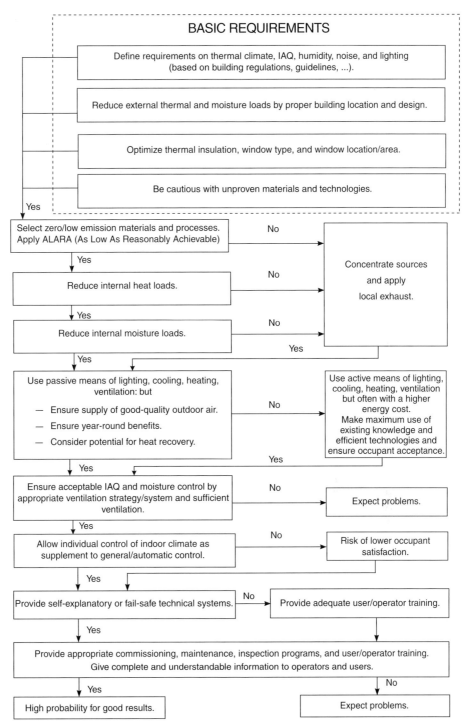

FIGURE 5.4.8-2 Strategy for achieving good IAQ and efficient use of energy. (*From ECA 1996 [Ref. 5].*)

CONCLUSIONS

Both conserving energy and providing acceptable IAQ are important aspects of building design. Unfortunately, an inherent conflict between these requirements often surfaces. Because energy costs are more readily calculable, they often become the focus of control. Often this focus is so limited that the broader goals and objectives of productively using energy are overlooked or minimized. It is essential that the impact of energy conservation measures be fully assessed before implementation; when an unacceptable outcome is expected, the measures should be reevaluated or avoided.

Architects and engineers can promote acceptable IAQ and maintain energy-efficient operation by accounting for expected loads and identifying effective source-control strategies and thoughtful operational procedures.

The following is a recommended approach for building design and retrofit:

- Control sources of pollution by eliminating, substituting, or localizing ventilation, wherever possible.
- Minimize energy loss by improved sealing, adding insulation, adding weather stripping, or replacing windows with insulated glass.
- Design ventilation systems using energy-efficient technologies, where appropriate.
- Design energy systems to meet the required ventilation rate, as determined by relevant guidance documents.
- Use ventilation strategies, where appropriate, to improve ventilation efficiency in occupied spaces.
- Design systems for ease of operation and maintenance.
- Commission all systems after construction or refurbishment.

In this manner, the original goal of having buildings serve the intended purpose of providing a healthy, comfortable environment where people can work and live will be met at an energy cost that is appropriate.

CONTRIBUTOR

James O. Turner, P.E., C.E.M., Utilities Manager, Partners HealthCare System, Inc., Boston, Massachusetts

REFERENCES

1. W. J. Fisk, "Estimates of Potential Nationwide Productivity and Health Benefits from Better Indoor Environments: An Update," in J. Spengler, J. M. Samet, and J. F. McCarthy, (eds.), *Indoor Air Quality Handbook,* McGraw-Hill, New York, in press.
2. ASHRAE, *The ASHRAE Handbook CD* [CD-ROM], American Society of Heating, Refrigerating, and Air-Conditioning Engineers, Atlanta, GA, 1998; *ASHRAE Handbook-HVAC Applications,* 1995. "Energy Management," in Chap. 32.
3. N. R. Grimm and R. C. Rosaler, *HVAC Systems and Components Handbook,* 2d ed., McGraw-Hill, New York, 1998, Chap. 8.4.1–8.4.57.
4. H. Levin, "The Myth of Energy Conservation and IAQ," *Indoor Air Bulletin* 2 no. 8 (1993), 1–6.

5. ECA, "Indoor Air Quality and the Use of Energy in Buildings," Report No. 17. EUR 16367 EN, *European Collaborative Action, Indoor Air Quality and its Impact on Man,* Office for Official Publications of the European Community, Luxembourg, 1996.

6. IEA/AIVC, "Energy Impact of Ventilation," Technical Note AIVC 49, Air Infiltration and Ventilation Centre, International Energy Agency, 1998.

7. ASHRAE, *Ventilation for Acceptable Indoor Air Quality,* Standard 62-1999, American Society of Heating, Refrigerating, and Air-Conditioning Engineers, Atlanta, GA, 1999.

8. S. J. Emmerich and A. K. Persily, "Energy Impacts of Infiltration and Ventilation in U.S. Office Buildings Using Multizone Airflow Simulation," in *IAQ and Energy 98: Using ASHRAE Standards 62 and 90.1,* conference proceedings, New Orleans, LA, October 24–27, 1998, American Society of Heating, Refrigerating, and Air-Conditioning Engineers, Atlanta, GA, 1999, pp. 191–205.

9. J. D. Hall, D. H. Mudarri, and E. Werling, "Energy Impacts of Indoor Environmental Quality Modifications to Energy Efficiency Projects, in *IAQ and Energy 98: Using ASHRAE Standards 62 and 90.1,* conference proceedings, New Orleans, LA, October 24–27, 1998, American Society of Heating, Refrigerating, and Air-Conditioning Engineers, Atlanta, GA, 1999, pp. 171–179.

ARTICLE 5.4.9

THERMAL ENERGY STORAGE

Douglas T. Reindl, Ph.D., P.E.
HVAC&R Center and Thermal Storage Applications Research Center,
University of Wisconsin, Madison, Wisconsin

William P. Bahnfleth, Ph.D., P.E.
Pennsylvania State University, State College, Pennsylvania

Facility energy management has always been an area of concern and resource expenditure for facility managers. Early efforts in facility energy management grew as a result of the energy crises of the early 1970s. During this era, energy cost indexes were relatively high, and uninterrupted supplies of gas and electricity became increasingly scarce.

During the mid- to late 1970s, U.S. electric utilities were facing significant increases in peak electrical demands (especially during summer periods). Much of the growth in the electricity market resulted from applying direct mechanical refrigeration technologies for facility air conditioning (in the United States, more than 95 percent of air conditioning is delivered from electrically powered equipment). Electrically driven mechanical refrigeration presented two problems for electric utilities. The first was the absolute growth in the number of refrigeration systems installed to meet the craving of building occupants for comfort cooling. The second problem presented by direct mechanical refrigeration was that the aggregate demand profiles from the many mechanical refrigeration systems operating to meet building cooling loads coincided and nearly always peaked in midafternoon. The combination of these two factors greatly exacerbated a shrinking electricity-supply market. Electric utilities had no choice but to explore alternatives to meet their customers' demands.

One obvious solution was to build more power plants to meet these growing demands. Utilities pursued this route on a limited basis, but it was not a viable alternative by itself because the projected electrical demand growth rate was significantly outpacing the rate at which new generation could be added. Another potential solution was conservation. From the public's perspective, conservation seemed honorable, but many found utility-promoted conservation hard to comprehend because it is the utility itself that sells the electricity. Many utilities had rate structures that discouraged the use of electricity during high-demand periods; the most common was a time-of-use (TOU) rate structure. Application of TOU rate structures grew to encompass virtually all facilities that had peak electrical demands exceeding 250 kW.

During the early 1980s, many electric utilities throughout the United States had undertaken initiatives to develop demand-side management (DSM) programs. Thermal energy storage (TES) was identified as a key technology for managing a strategic electrical load—space cooling. TES allowed end-use customers to shape their electrical demand profiles by operating high-demand refrigeration equipment during low-utility-aggregate-demand periods to produce a cool storage medium, and subsequently use to that medium to provide space cooling during high-utility-demand periods. But utilities could not easily reap the benefits of this technology because decisions to implement it were made by the end users. One of the biggest challenges that utilities faced was identifying strategies that encouraged end-use customers to adopt and utilize thermal storage technologies for their mutual benefit. Many utilities throughout the United States were so convinced of the value of this load-shaping technology that they attempted to "drive the market" by extending financial incentives to end-use customers who installed the technology.

One of the most controversial and market-damaging types of financial incentives was the *capital cost incentive*. Many utilities provided capital cost incentives in the form of cash rebates to thermal storage customers. In this scenario, utilities paid approximately $200 to $400 per kW of electrical demand that was shifted to off-peak periods. Compared to the cost of new generation, ranging from $390 to $1100 per kW, rebates seemed like a bargain to utilities for deferring power plant construction and improving load factors. From the perspective of a facility manager, rebates provided a means of implementing a technology that would deliver considerable operating cost savings, maintenance cost savings (in some cases), and facility operating flexibility. Unfortunately, capital cost incentives allowed TES systems to be installed in otherwise uneconomical situations.

Today, only a small number of U.S. electric utilities still offer financial incentives to customers who install thermal storage technologies. Such a radical change in the market transpired over the last 3 to 4 years. Although many utilities eliminated their thermal storage rebate programs because they achieved their DSM objectives, the vast majority suspended rebate programs because of the advent of electric market restructuring.

Today, facility managers are still responsible for managing energy use and energy costs that represent a significant component of their total operating costs which range from $1.50 to $4.00 per ft^2/year. Furthermore, some market experts expect that future electric prices will increase substantially for high-demand periods (daytime) and decrease moderately for lower-demand periods (nighttime). Some industry analysts expect that facilities with operating flexibility in terms of their electrical demand will be uniquely positioned to negotiate electric-supply contracts at more favorable rates by establishing electric usage patterns that benefit the supplier. Although some suggest that technologies such as on-site generators and gas-driven chillers can accomplish end-use electrical demand shaping, these alternatives have not provided operating and maintenance cost benefits comparable to that of TES. Facility managers should view TES as one technology that can allow them to achieve effective energy and energy cost management—a powerful weapon in their arsenals.

The remainder of this subsection presents an overview of two classes of TES technologies that have been proven in the marketplace during the last quarter of a century: *latent energy change* and *sensible energy change* technologies. Specific system types within each of these categories are discussed, as well as fundamentals of their operation.

LATENT ENERGY CHANGE TECHNOLOGIES

In 1994, Potter conducted a survey of TES across the United States.[1] At that time, Potter estimated that 1300 to 1722 ice TES systems were installed and operational. More recent estimates of ice storage system installations place the total at 3000 to 4000 systems. Of 116 ice storage facilities that Potter directly surveyed, 45 percent are commercial, 35 percent are institutional, and 11 percent are industrial. The average system capacity is 10,938 kWh_T (3110 ton · h) somewhat larger than the 7000 kWh_T (2000 ton · h) system capacity cited by Dorgan and Elleson in 1993.[2]

Ice is the most frequently used latent energy change medium in the United States. Three types of ice storage technologies can be considered mature: *static ice, internal melt; static ice, external melt;* and *encapsulated ice. Ice slurry* is one emerging ice storage technology that shows future promise. Each of these ice storage technologies is presented and discussed in the following subsections.

Static Ice, Internal Melt

The static-ice, internal-melt system (sometimes referred to as an *ice-on-coil* system) is a technology that is configured by immersing a series of coiled tubes (piped in parallel to a common header) within a water-filled container. Operationally, a secondary fluid (e.g., an ethylene glycol and water mixture) is circulated through the water-filled tank (via the tube side) to store and discharge energy. If desired, a plate-and-frame heat exchanger can be used to isolate the storage/chilling system from the building's cooling hydronic system (Fig. 5.4.9-1).

FIGURE 5.4.9-1 CALMAC static-ice internal-melt system. (*Courtesy of CALMAC Manufacturing.*)

During charging, cold [~–4°C (25°F)] secondary fluid is supplied to the coils immersed in the storage tank. As the secondary fluid sensibly absorbs energy, ice forms on the outside surfaces of the immersed coil, and the secondary fluid exits the tank at a temperature near the freezing point of pure water [~–0.6°C (31°F)]. Charging continues until the return fluid temperature from the tank begins to decrease rapidly—a sign that nearly all of the water in the tank is frozen. A design feature that has improved the performance of this technology is the arrangement of adjacent coil passes in a counterflow fashion (Fig. 5.4.9-2).

During discharge, warm secondary fluid returning from the load is circulated through the coils in the tank and melts the ice from the tube wall outward. The annulus of water formed between the outside surface of the coil and the inside surface of the ice front can represent a significant source of thermal resistance. To improve the discharging rate characteristic of this technology, some manufacturers agitate this relatively stagnant annular water layer by pumping small bubbles of air into the annulus (Fig. 5.4.9-3).

Figures 5.4.9-1, 5.4.9-2, and 5.4.9-3 show three static-ice, internal-melt systems that are commercially available today. Static-ice, internal-melt systems are a proven technology that offers facility managers a number of benefits. First, the systems in today's marketplace are extremely reliable (as one might infer from the designation *static*). Second, the technology offers ease of maintenance. Finally, static-ice, internal-melt systems deliver excellent performance. Capital costs for this technology alternative range from \$23 to \$28 per kW_T (\$80 to \$100 per ton · h) of nominal capacity.

FIGURE 5.4.9-2   FAFCO static-ice internal-melt system. (*Courtesy of FAFCO, Inc., Redwood City, CA.*)

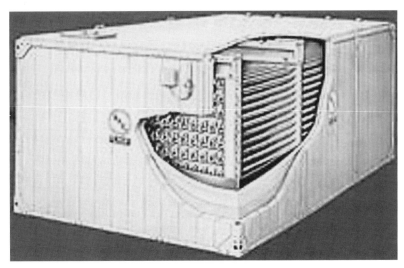

FIGURE 5.4.9-3   BAC static-ice internal-melt system. (*Courtesy of Baltimore Air Coil, Inc.*)

Static Ice, External Melt

The static-ice, external-melt system is a class of latent energy change technology that was once very popular. Early system designs (refrigerant-based ice-on-pipe) used direct refrigerant supplied to the tube side of a heat exchanger immersed in a water-filled tank. High costs of refrigerants coupled with the trend toward reducing refrigerant inventories in refrigeration plants have significantly reduced the application of refrigerant-based external-melt systems (less than 3 percent of systems sold today use the refrigerant-based approach). Today, external-melt systems use a secondary coolant, ethylene glycol and water, circulated on the tube side of the tank heat exchanger (brine-based ice-on-pipe).

The primary difference between the internal-melt and external-melt systems is the manner in which the storage device is discharged. In internal-melt systems, the secondary fluid reverses its flow through the iced heat exchanger, melting the ice from the inside outward. In external-melt systems, free water within the storage tank is circulated to melt ice from the outside inward.

During a charge cycle, cold [~–4°C (24°F)] secondary fluid is pumped through a heat exchanger (typically composed of hot-dipped galvanized steel) immersed in a water-filled tank. Ice forms on the outside surface of the heat exchanger, and it continues to grow thicker as the build cycle progresses. The build cycle is terminated when the ice reaches a predetermined thickness of ~3.8 cm (1.5 in). During discharge, warm return water from the building or load is circulated through the tank (rather than circulating glycol through the heat exchanger) and melts the ice from the outside surface inward. Activation of optional air pumps helps promote good heat transfer during the ice melting process.

Operationally, it is desirable to melt out as much of the ice from the heat exchanger as possible because subsequent ice builds are less efficient if ice insulates the surface (the charging efficiency decreases as the ice thickness increases). Capital costs for this technology alternative range from \$23 to \$28 per kW_T (\$80 to \$100 per ton · h) of nominal capacity.

Encapsulated Ice

Encapsulated-ice systems are configured with a multiplicity of storage modules placed in containers large enough to provide the thermal capacity to shift a given cooling load. Individual encapsulant containers are filled with water (usually containing compounds to minimize supercooling and promote ice nucleation). There are a number of geometric encapsulant configurations, including spherical, parallelepiped, and toroidal. The typical encapsulating container is made of a polymer such as polyethylene.

During a charging cycle, cold secondary fluid is circulated through the tank containing the multiplicity of storage modules. Ice forms inside each individual storage module, building from the outside inward. During discharge, warm secondary fluid is pumped through the tank containing the storage modules, and melting the ice from the outside inward.

Encapsulated ice systems have a number of disadvantages that have limited their penetration into the U.S. marketplace. First, most of the systems based on this technology have used field-erected storage tanks that have resulted in variable charge and discharge performance. (A European manufacturer which uses this technology provides packaged storage modules and tanks, leading to higher-quality installations). In some vessel configurations, the circulating secondary fluid may bypass a significant fraction of the storage modules (due to "dead zones" of fluid circulation); the result is that usable storage capacity is significantly less than the theoretical capacity. Another disadvantage of this system type results from the characteristic expansion of water as a result of solidification. The cyclic expansion and contraction of water in the encapsulating module results in fatigue failure over time. A telltale sign of this phenomenon in field-installed systems is the apparent dilution of the secondary fluid over time (indicating that the encapsulant has ruptured and has allowed water to dilute the secondary fluid). Another disadvantage is the cost of shipping storage modules over long distances. Typically, the individual storage modules are injection molded and filled with water in a factory. The significant weight increase of adding water to the modules in the factory creates a shipping-cost penalty compared to systems filled on-site.

Encapsulated-ice systems are the most modular storage systems available today. Storage can be added in extremely small increments due to the nature of this technology. Another advantage of the encapsulated-ice technology is the potential (albeit with risk) for using existing on-site tanks for storage.

Ice Slurry

Worldwide development of ice storage technologies is continuing. Although many of the "new" technologies do not depart significantly from current ice storage technologies, one development stands apart—ice slurry systems. Simply defined, *ice slurry* is a mixture that consists of a suspension of ice crystals in a liquid. The working fluid in its liquid state is typically a mixture of a solvent (water) and a solute (e.g., glycol, ethanol, calcium carbonate, etc.). Depending on the specific ice slurry technology, the initial solute concentration varies from 2 percent to more than 10 percent by weight. A primary purpose of the solute is to depress the freezing point of the solvent and buffer the production of ice crystals in solution.

The primary motivation for pursuing the development and application of ice slurry is its high energy transport density. Most direct chilling and thermal storage technologies available today rely on sensible changes to deliver energy for cooling loads. Significantly higher rates of heat absorption can be achieved by transporting an ice slurry compared with chilled water or a chilled brine. The successful development of a technology that eliminates energy transport as the limiting factor in space conditioning represents the next breakthrough in the HVAC and process-cooling markets. Another benefit of ice slurry is the ability to separate the production of the cooling medium from its storage. Separating the production from storage provides flexibility in system layout and field installation. Because the slurry contains a high fraction of liquid, it readily takes the shape of and fully fills a storage tank of any size or geometry (unlike the large ice fragments produced by ice harvesters that leave a significant portion of the storage tank unusable).[3,4] This characteristic maximizes the use of a significant system component—the storage tank. Although several organizations are pursuing research, development, and demonstration of ice slurry systems, significant advances are required to make this technology viable.

SENSIBLE ENERGY CHANGE TECHNOLOGIES

Sensible energy is the energy that results from a change in the temperature of a storage medium. The predominant medium for sensible cool thermal energy storage is water. The use of water with additives to increase the density of storage by extending the practical range of operating temperatures has also received limited application in sensible storage systems. Both approaches are reviewed in this subsection. Properties and representative sensible storage densities for water are summarized in Table 5.4.9-1.

Even with large temperature differentials, sensible media typically have lower storage densities than latent media. However, sensible storage has other advantages that may make it a better choice in certain circumstances. One of these is sheer size. Tanks for sensible storage drop rapidly in unit cost as they increase in size up to roughly 7500 m³ (4 million gal). Other advantages include compatibility of sensible storage with existing system temperatures in retrofit situations and lower charging-energy consumption for the refrigeration plant.

The average size of chilled-water storage systems surveyed by Potter in 1994 was 50,261 kWh$_T$ (14,291 ton · h)—nearly 5 times the average size of ice storage systems surveyed.[1] At the time of the survey, Potter estimated that 10 percent of all cool storage systems used stratified chilled water. Because of their size, however, this amounts to 34 percent of the total cooling capacity of all cool storage systems.

Applications of sensible storage can be found in virtually every type of facility in which latent storage is also used; however, the most common applications are district cooling systems, industrial facilities, university campuses, and other large loads. Applications of sensible storage in these settings have been very successful. Two exemplary installations described in detailed case studies are the systems at Texas Instruments, Dallas, Texas,[5] and Cornell University, Ithaca, New York.[6]

Because sensible storage depends on temperature, it is essential that the cooler, charged medium be thermally separated from the warmer, discharged medium. Thermal blending between the two destroys available capacity. Sensible storage devices are distinguished primarily by their method of thermal separation. Thermal separation technologies used at various times include empty-tank systems, series-tank systems, membrane tanks, and naturally stratified tanks. However, two decades of operating experience have shown that naturally stratified tanks are the superior technology, and most new systems employ them. For this reason, only naturally stratified systems will be described in some detail. For information regarding the other technologies, the reader may consult Dorgan and Elleson.[2]

Stratified Chilled-Water Storage

Stratified chilled-water storage uses the variation of water density with temperature to maintain thermal separation between warmer and cooler storage media. The density of pure water is at its maximum at 4.3°C (39.8°F). If a volume of water at a temperature greater than or equal to

TABLE 5.4.9-1 Typical Chilled-Water Storage Properties

Property	Temperature range, °C (°F)	
	6 (10)	12 (20)
Density, kg/m³ (lb/ft)	1000 (62.4)	
Specific heat, kJ/kg · °C⁻¹ (Btu/lb · °F⁻¹)	4.2 (1.0)	
Storage density		
kJ/m³ (Btu/ft³)	23,250 (624)	46,500 (1248)
L/kWh$_T$ (gal/ton · h⁻¹)	155 (144)	77 (72)

4.3°C (39.8°F) is stored beneath a volume of warmer water, this buoyancy difference suppresses mixing between the two volumes even though there is no physical barrier between them. Mixing is confined to a relatively thin transition region called a *thermocline*. Thermocline thickness in full-scale tanks is of the order of 1 to 2 m (3 to 6 ft). To move water through a stratified tank without disturbing the thermocline, cooler water must always be introduced and withdrawn at the bottom of the tank, and warmer water must always be introduced and withdrawn at the top of the tank. This is done through a pair of diffusers designed to minimize mixing. The temperature profile in a representative stratified water tank is shown in Fig. 5.4.9-4.

The thermal performance of large stratified tanks is typically excellent. It can usually be expected that less than 10 percent of the theoretical capacity of a fully charged tank will be lost to mixing and ambient heat gains. Because chilled-water storage is charged at a higher temperature than ice storage, the charging coefficient of performance (COP) may be equal to or larger than the COP for direct cooling. Net energy savings, in addition to substantial operating cost savings, have been documented for both the Cornell and Texas Instruments systems mentioned previously.[7,8]

Stratified storage tanks are commonly cylindrical welded-steel or wire-wound prestressed concrete vessels built to typical standards such those of the American Water Works Association or other organizations.[9,10] Steel tanks are less expensive in first cost for aboveground installations but may require more maintenance than concrete tanks. Concrete tanks may be bermed, partially buried, or completely buried. Berming or burial can be advantageous both from the perspective of minimizing the visual impact of a large tank and also because it reduces or eliminates the need for insulation, which is always required for steel tanks. Rectangular tanks have been built, often when the tank is located beneath a building or parking structure. These tanks are generally of reinforced concrete. They are more prone to leakage than cylindrical tanks and must be carefully designed and built to minimize this risk. In particular, an interior sealant should be considered.

Thermal storage tanks of typical capacity are relatively tall. No statistics are available on the distribution of water depth for full-scale systems. However, many are 14 m (45 ft) deep, and some tanks with water depth of 27 m (90 ft) have been built. Tall tanks are preferred for several reasons. Thermocline thickness is determined mainly by diffuser performance and does not depend strongly upon tank dimensions. Consequently, simply making a tank taller reduces capacity lost to mixing. Another reason to prefer taller tanks has to do with system hydraulics, a subject that cannot be covered in detail within the scope of this handbook. Because chilled-water storage tanks are vented atmospheric pressure vessels, it is necessary to provide booster pumping to move water from the storage tank into the chilled-water system during both charge and discharge if the tank water level is below the top of the chilled-water

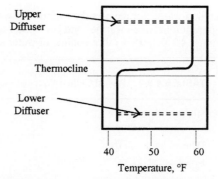

FIGURE 5.4.9-4 Stratified tank temperature profile. (*Courtesy of William P. Bahnfleth, Pennsylvania State University, State College, PA.*)

system. This adds control complexity as well as increased pumping energy that increases in proportion to the difference in tank and system height. If the tank can be made the high point of the system (often impossible more for political than technical reasons), first cost, operating cost, and maintenance are reduced. When transfer pumping is required, the pumping system should be designed to minimize energy consumption. A full discussion of these issues is given by Bahnfleth and Kirchner.[11]

The first-cost economics of chilled water storage for large systems are good if site-specific factors do not add excessively to tank cost. These may include soil conditions that increase foundation requirements, seismic conditions, and aesthetic treatment of the tank exterior. Costs as reported by Potter varied from \$25 to \$33 per kWh_T (\$88 to \$117 per ton · h).[1] At the low end of installed costs, chilled-water storage systems have actually resulted in capital cost savings (without utility rebates) relative to chilled-water plant capacity additions.

Chilled-water storage systems are among the most successful applications of thermal storage and continue to be built despite the removal of most electric utility incentives. In general, chilled-water storage systems have met and exceeded owner expectations. Potter reports that only 12 percent of systems surveyed exceeded budget; actual peak demand shift was 91.2 percent of the design estimate; and actual annual savings were 111 percent of design (possibly affected by rising rates, of course). In addition to providing operating cost savings, system owners have noted that chilled-water storage makes system management easier once it is integrated into normal operations, provides capacity to cover planned maintenance outages, and can also function as emergency capacity.[12]

Density-Depressed Chilled-Water Storage

As noted previously, attempting to store water in stratified tanks at temperatures below 4.3°C (39.8°F) will induce mixing within the tank, thereby reducing the effectiveness of stratification. A practical upper limit in return water temperature for building space conditioning applications is 21°C (70°F). As a result, a practical upper limit on the overall water-side temperature rise for building space conditioning is 17°C (30°F).

Stratified chilled-water storage systems have faced two major barriers because of these restrictions on supply and return water temperatures. First, the energy density of a chilled-water storage system that serves a building is limited by the lower bound on storage supply water temperature. Second, concepts of cold air distribution to minimize total system power have limited potential in stratified chilled-water storage systems with a 4°C (39°F) temperature limit. These limitations motivated investigation of the use of additives to depress both the freezing point and the point of maximum water density. Additives that have been considered include sodium chloride and calcium chloride (corrosive), glycols (expensive), and a mixture of sodium nitrate and sodium nitrate. The latter is the basis of a commercially available additive.[13]

As an example of the effect of additives on water properties, information furnished by Thermal Technologies regarding their additive indicates that a 7-percent-by-weight solution, costing approximately \$0.024 per L (\$0.09 per gal) plus a one-time \$0.57 per rated kWh_T (\$2 per ton · h) technology license fee would permit stratified storage at 1.1°C (30°F).[13] The additive also contains corrosion-inhibiting and antimicrobial treatments, so it would essentially replace conventional water treatment. Taking a conventional chilled-water storage system with a charging temperature of 3.9°C (39°F) and a system return temperature of 12.2°C (54°F) as a basis of comparison with a 7 percent solution of this additive, the volume of the system would be reduced through use of the additive from roughly 103 to 66 L per kWh_T (96 to 61 gal/ton · h) for an initial cost of approximately \$2.13 per kWh_T (\$7.50 per ton · h) and \$1.56 per kWh_T (\$5.50 per ton · h) for replacement additive. The initial cost penalty relative to water would trade off with savings in tank cost due to increased storage density of nearly \$3.98 per kWh_T (\$14 per ton · h) and any other size reductions that might result from the larger operating temperature differential.

The advantages and disadvantages of density-depressed stratified sensible storage technologies are still being evaluated. One advantage of a density-depressed system is the ability to increase the energy density of the thermal storage tank by increasing the temperature differential across the storage apparatus. Increased energy density translates into lower storage tank cost per kilojoule of energy stored. The increase in storage tank temperature differential also allows reductions in pumping power to meet the same cooling loads. Even with the increase in energy density of density-depressed chilled water storage, this system still has about one-quarter the energy density of ice storage systems. Table 5.4.9-2 provides a summary of advantages and disadvantages of density-depressed stratified sensible energy storage systems.

The concept of a density-depressed storage system is still new. Two large installations in the United States currently use density-depressed sensible storage. One is McCormick Place Convention Complex in Chicago. This system can store as much as 42,573 kWh$_T$ (123,000 ton · h) and delivers fluid as cold as 0.6°C (33°F).[14,15] The technology has potential merit in some applications; however, the disadvantages listed in Table 5.4.9-2 are significant. The viability of this technology for new installations is yet to be proved. A more likely application of density depressants may be in the area of retrofitting existing chilled-water storage systems. Adding density depressants to an existing system offers a means of increasing the design storage capacity of the water storage tank.

APPLICATION CONSIDERATIONS

Utility Rate Structures

As described earlier, interest in thermal storage in the 1980s and 1990s was driven to a great extent by the desire of electric utilities to reduce their peak demand as a less costly alternative to building new generating capacity. Electric rates of this period were structured to provide an incentive for load shifting by penalizing low load factors. Typical rates had high peak-demand charges during the middle of the day as well as time-of-use energy charge differentials. The use of such rate structures remains widespread.

In general, the greater the difference between on-peak- and off-peak-demand charges and energy rates, the greater the potential operating cost advantage of thermal storage. However, thermal storage may be cost-effective even when flat rates are applied for cases when a first-cost advantage exists.[16] This is most likely to occur either in retrofit situations with chilled-water storage in which a large storage addition eliminates the need for a large plant-capacity increase. Another example is a fully integrated new building design in which ice storage and low-temperature air distribution are exploited to reduce distribution system cost and floor-to-floor height.

TABLE 5.4.9-2   Advantages and Disadvantages of Density-Depressed Systems

Advantages	Disadvantages
Increased storage energy density/decreased storage volume (relative to chilled water)	Significantly lower energy density than ice
Decreased pumping power	Refrigeration system penalty as a result of lower evaporator temperatures
Ability to incorporate concepts of cold-air distribution	Increased cost for working fluid additives (density depressant)
Lower space humidity with cooler supply air temperatures	Increased storage tank insulation requirements to minimize ambient gains
Potential for improved comfort when coupled with cold-air distribution	

As the deregulation of the U.S. electric industry has proceeded, the use, or contemplated use, of *real-time pricing* has increased. A real-time rate is intended to reflect the actual cost of a unit of electric power at the time it is used.[17] This will vary with load on the electric grid, weather conditions, the fuel mix of generating stations, and other factors. Instead of separate demand and energy charges, electricity is billed on the basis of usage, but rates may vary from a few cents to several dollars per kilowatt-hour. In a typical real-time pricing rate structure, a price signal indicating the rates for each hour of a day is sent to utility customers on the preceding day. Electric demand-shifting technologies such as cool thermal energy storage may become even more valuable to the owner faced with real-time pricing because of the enormous hour-to-hour energy cost differences that may occur.

Facility Cooling Loads

The role of utility load factors in creating the initial impetus for the thermal storage boom of the late 1980s has been discussed. For a facility to take advantage of electric rates intended to reduce the utility electric-load factor, it must itself have electric- and cooling-load factors that permit load shifting. A low cooling-load factor creates the ability to achieve substantial reduction both in the size of the refrigeration plant required by a facility and the possibility of a large reduction in its electric demand peak. For smaller systems, for example, typical commercial office buildings and schools, it is practically essential that the load factor be low because of the added cost of thermal storage. For large loads, in which economies of scale shift the relative cost of refrigeration equipment and thermal storage in favor of storage, load factor is less critical. Some of the best applications for chilled-water storage are university campuses, which often have a minimum load under design conditions that is 60 percent or more of the peak load.

Operating Strategies

To meet any given design cooling load, an infinite number of combinations of cooling-plant capacity and storage capacity is available. For every combination of storage and plant capacity, an infinite number of potential operating strategies is available, constrained only by the size of storage. Sizing is usually based on a 24-h charge-discharge cycle, although longer cycles are used in some cases.

At one extreme, storage can be sized to minimize the refrigeration capacity required. This is typically called *level-load sizing* because the chiller capacity required is approximately the average cooling load over the design day. This approach to sizing results in minimum chiller size, moderate storage capacity, and moderate load shift. It is often the most economical approach because it results in the least first-cost storage system, and payback is rapid even though potential savings from load shifting are not maximized. At the other extreme, storage can be sized to completely displace chiller operation during a specified period of the design day. This is commonly called *full storage*. It creates a large load shift, but both the refrigeration capacity and storage capacity required will generally exceed the level-load values. It is feasible when peak electric cost is sufficiently high to offset the added first cost that must be borne.

Between the extremes of level load and full storage, any combination of partial storage may be selected. In some cases, partial storage is sized to maintain a maximum electric demand on a facility, not to limit peak cooling demand. This practice can lead to as much as a doubling of operating cost savings relative to a strategy that maintains a refrigeration plant load limit without regard to its impact on electric demand.[18]

It is important to distinguish between operation of a system in terms of the design day and operating it under part-load conditions. Part-load operating strategies range from extremely simple *chiller-priority* controls, in which storage is used only when the cooling load exceeds the capacity of the plant, to *storage-priority* demand-limiting controls, in which the objective of the control sequence is to use as much storage as possible every day and to maximize the peak-electric-demand shift. At the chiller-priority extreme, no load prediction is required, but operating cost savings may be much smaller than they could be. At the storage-priority demand-limiting extreme, some form of load prediction is required, and controls are more complex, but the potential operating-cost savings are substantial. Sizing and operating strategies are discussed in a number of design guides, for example, Dorgan and Elleson and ASHRAE.[2,19]

Maintenance Issues

Thermal storage systems do not require substantially more maintenance than nonstorage cooling plants. However, deferred maintenance for storage is very ill-advised because of its consequences for both operating cost savings and system life. Each type of thermal storage technology has unique maintenance issues. Taking chilled-water storage as an example, preventing corrosion of steel tanks and treatment of the large volumes of water involved to prevent biological contamination are important concerns. Many ice storage systems use water-glycol mixtures as a secondary coolant. For these systems, glycol maintenance is a critical issue. Guidelines for equipment maintenance can be obtained directly from the manufacturer. Water-treatment professionals should be consulted regarding appropriate programs for start-up and maintenance.

For all systems, calibration of controls is critical because implementation of the selected operating strategy depends upon it. Faulty controls will certainly have an adverse impact on operating cost savings. All temperature and flow instrumentation associated with a thermal storage system should be specified for acceptable accuracy and calibrated at least once per year.

There is emerging awareness of the potential of thermal storage systems to provide maintenance benefits. As noted previously, the ability of storage to displace refrigeration equipment operation provides a capability to do emergency or routine maintenance on these system components that is not available otherwise.

Characteristics of Successful Applications

The following are several common characteristics of successful thermal storage systems:

1. A knowledgeable design professional and proactive owner collaborated to develop a system concept based on the specific opportunities available.

2. Systems were designed primarily to stand on their own merits (i.e., they would have been successful systems with or without first-cost incentives from a utility)

3. In some sense, a comprehensive approach to project delivery was employed. Design intent was clearly defined and tracked through project development; measures were taken to ensure that the system constructed was the system designed; performance of the system was verified after construction and start-up; training was provided for system managers and operators; and maintenance and performance review continued on a regular basis. A framework for delivery of successful thermal storage projects developed through ASHRAE research is essential reading for any facility manager contemplating a thermal storage project and for any engineer interested in designing storage systems.[20]

4. Management and operating staff were committed to achieving the full benefits of thermal storage. This depends, to a great extent, on the success of the orientation and training that should be part of the start-up of any new system.

CONTRIBUTOR

Roger P. Wessel, P.E., RPW Technologies, Inc., West Newton, Massachusetts

REFERENCES

1. R. Potter, *Study of Operational Experience with Thermal Storage Systems,* Final Report 766-RP, American Society of Heating, Refrigerating, and Air-Conditioning Engineers, Atlanta, GA, 1994.

2. C. E. Dorgan and J. S. Elleson, *Design Guide for Cool Thermal Storage,* American Society of Heating, Refrigerating, and Air-Conditioning Engineers, Atlanta, GA, 1993.

3. G. Gute, W. Stewart, and J. Chandrasekharan, "Modelling the Ice-Filling Process of Rectangular Thermal Energy Storage Tanks with Multiple Ice Makers," *ASHRAE Transactions* 101 no. 1 (1995), 56–65.

4. W. Stewart, G. Gute, and J. Chandrasekharan, "Modelling the Melting Process of Ice Stores in Rectangular Thermal Energy Storage Tanks with Multiple Ice Openings," *ASHRAE Transactions* 101 no. 1 (1995), 66–78.

5. D. Fiorino, "Case Study of a Large, Naturally Stratified, Chilled-Water Thermal Energy Storage System," *ASHRAE Transactions* 97 no. 2 (1991), 1161–1169.

6. W. Joyce and W. Bahnfleth, "Cornell Thermal Storage System Saves Money and Electricity," *District Heating and Cooling* 77 no. 4 (1992), 22–29.

7. D. Fiorino, "Energy Conservation with Thermally Stratified Chilled-Water Storage," *ASHRAE Transactions* 100 no. 1 (1994), 1754–1766.

8. W. P. Bahnfleth and W. S. Joyce, "Energy Use in a District Cooling System with Stratified Chilled Water Storage," *ASHRAE Transactions* 100 no. 1 (1994), 1767–1778.

9. AWWA, *Wire & Strand Wound, Circular, Prestressed Concrete Water Tank,* ANSI/AWWA Standard D110-95, American Water Works Association, Denver, CO, 1995.

10. AWWA, *Welded Steel Tanks for Water Storage,* ANSI/AWWA Standard D100-96, American Water Works Association, Denver, CO, 1996.

11. W. Bahnfleth and C. Kirchner, "Analysis of Transfer Pumping Interfaces for Stratified Chilled Water Thermal Storage Systems—Part 1: Model Development," *ASHRAE Transactions* 105 no. 1 (1999), 3–18; "Analysis of Transfer Pumping Interfaces for Stratified Chilled Water Thermal Storage Systems—Part 2: Parametric Study," *ASHRAE Transactions* 105 no. 1 (1999), 19–40.

12. W. Bahnfleth, A. Musser, and C. Peretti, "Design Considerations for the Use of Chilled Water Thermal Storage as Emergency Cooling Capacity," *Proceedings of the International District Energy Association 11th Annual College/University Conference,* 1998, pp. XVI-1–XVI-13.

13. Thermal Technologies product literature, 1999.

14. P. Winters and J. Andrepont, "TES for Today and Tomorrow: Capital Cost Savings Without Utility Cash Rebates," *International Sustainable Thermal Energy Storage Conference Proceedings,* Thermal Storage Applications Research Center, University of Wisconsin, Madison, 1996, pp. 79–81.

15. J. S. Andrepont, "An Alternative Medium for Thermally Stratified Thermal Energy Storage (TES)," *Proceedings of the International District Energy Association 11th Annual College/University Conference,* 1998, pp. VIII-1–VIII-12.

16. J. Caldwell and W. Bahnfleth, "Chilled Water Storage Feasibility Without Electric Rate Incentives or Rebates," *ASCE Journal of Architectural Engineering* 3 no. 3 (1997), 133–140.

17. W. P. Bahnfleth and D. T. Reindl, "Prospects for Cool Thermal Storage in a Competitive Electric Power Industry," *ASCE Journal of Architectural Engineering* 4 no. 1 (1998), 18–25.

18. R. T. Tamblyn, "Optimizing Storage Savings," *Heating/Piping/Air-Conditioning* 62 no. 8 (1990), 43–46.

19. ASHRAE, "Thermal Storage," *ASHRAE Handbook-1999 HVAC Applications,* American Society of Heating, Refrigerating, and Air-Conditioning Engineers, Atlanta, GA, 1999, Chap. 33.

20. J. Elleson, *Successful Cool Storage Projects: From Planning to Operation,* American Society of Heating, Refrigerating, and Air-Conditioning Engineers, Atlanta, GA, 1997.

ARTICLE 5.4.10

NOISE AND VIBRATION CONTROL

Kurt N. Milligan, INCE

*Acen*tech *Incorporated, Cambridge, Massachusetts*

Noise and vibration control is an increasingly important aspect of facilities design. Whether protecting employees' hearing in loud factory environments, affording reasonable privacy between condominiums or offices, limiting noise emissions from industrial equipment to nearby residential neighbors, providing appropriate acoustics for teleconferencing spaces, or designing manufacturing facilities for vibration-sensitive production equipment, there are few facilities where appropriate consideration of noise and vibration will not be necessary in some form.

The goal of this article is to provide the reader with an overview of how sound works, the terminology that is commonly used to describe and quantify it, typical noise and vibration issues to be aware of, and commonly-used noise and vibration control techniques. As many modern facilities also include sound reinforcement systems of some type, we have also included a short section on electro-acoustic systems. This article is not meant to be a detailed engineering guide. Rather, it is hoped that the reader will come away with a general understanding of sound and vibration that will help to avoid the common pitfalls (many of which that could have been avoided with proper planning) that we so often see as acoustical consultants.

Several industry-standard terms introduced along the way; they will be briefly defined and some included in the glossary. More in-depth definitions may generally be found in the chapter entitled "Sound and Vibration Control" of the *1991 ASHRAE Applications Handbook,* in *Noise and Vibration Control* (Beranek, 1988), or in *Architectural Acoustics* (Egan, 1988). Several other general references that provide detailed engineering information about most of these topics are listed in the bibliography.

PRELIMINARY CONCEPTS

Sound can be defined as fluctuations in the ambient pressure; small perturbations in the air move the eardrum back and forth, causing a signal to be sent to the brain that we can "hear." *Noise* is generally considered to be any unwanted sound. The components of sound important for our consideration here are *amplitude* (level) and *frequency* (pitch).

Amplitude and Frequency

The amplitude or loudness of a sound depends on the magnitude of the pressure fluctuations; the sound level is a measure of the average amplitude of these pressure fluctuations. The range of pressures to which the human ear is commonly exposed is enormous. A barely audible whisper produces about 0.0002 *pascals* (Pa; the standard mks units for pressure), while a loud jet engine can generate about 200 Pa. To accommodate this huge range and to make the numbers more reasonable to work with, sound level is commonly expressed in units called *decibels* (dB), which are simply logarithmic ratios of the measured pressure fluctuations to a standard reference pressure. The louder the sound, the higher the decibels; the previously-mentioned whisper would correspond to 20 dB, while the jet would be 140 dB. To clarify which standard reference is used in the decibel, a suffix is typically added to the number; for example 140 dB (re: 20 μPa) indicates the use of 0.00002 Pa as the reference in the logarithmic ratio.

The human ear will typically judge a 10 dB increase (or decrease) in sound pressure level as twice (or half) as loud, while a 3-dB change will be just noticeable (although in some situations smaller changes may be noticeable), and a 5 to 6-dB change will be considered a substantial difference. It is important to note that these metrics do not match the numerical changes in pressure; a ten-fold change in average pressure is required for a 20-dB difference, and halving the average pressure will reduce the level by only 6 dB. The reason to mention this is that some noise control product manufacturers have claimed that their products will cut the sound in half based on a 6-dB loss. Though this may be numerically correct, it is misleading, as the human ear will not consider the new to be half of what it was.

It is important to note that decibels do not add linearly. For example, one might measure a sound level of 70 dB, 50 ft from a given cooling tower. If a second cooling tower (located next to the first cooling tower, and of the same type and size), is switched on, the new sound level at the same measurement location will not be 140 dB. The new sound level will be closer to 73 dB. The rule of thumb is that if the difference in sound level between two sources is 0 to 1 dB, add 3 dB to the level of the louder source. If the difference is 2 to 3 dB, add 2 dB to the level of louder source, and for a difference of 4 to 9 dB, add 1 dB to the level of louder source. So, at a given location, if one source is has a measured level of 65 dB, and a second source has a measured level of 70-dB, the combined sound level will be about 71 dB.

Frequency is a measure of the rate at which the sound pressure fluctuations occur, and it is given by the number of oscillations (cycles) per second in units of *hertz* (Hz). Sounds consisting of a relatively small number of cycles per second and a smaller number of hertz, such as the notes produced by a tuba, are considered lower in frequency than the notes produced by a flute, which can be considered high frequency. Humans are typically able to hear sound with frequencies ranging from about 20 Hz to 20,000 Hz, although the upper limit tends to decrease with age. The frequencies produced by human speech range from about 250 Hz to about 4000 Hz. Figure 5.4.10-1 shows the range of frequencies typically produced by several common sources.

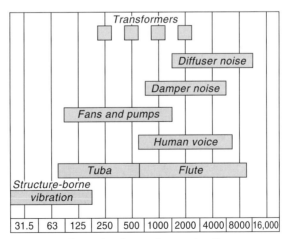

Octave band center frequencies, Hz

FIGURE 5.4.10-1 Approximate range of frequencies generated by common sources. (*Courtesy of Acentech, Inc., Cambridge, MA.*)

It is often useful to evaluate how much sound is present at various frequencies of the audible spectrum. Sound measurements made over a large frequency range are often separated into *octave bands* that break the audible spectrum into 10 smaller ranges, allowing one to assess the relative amounts of low-, middle-, and high-frequency energy present in the sound. For more detailed analysis, one-third octave bands are used; as the term implies, there are 3 one-third octave bands per octave band. For investigation of pure tones or sound with very small bandwidths, *narrowband* analysis is used.

Sound Propagation

In general, sound level decreases with distance from the source. This is primarily due to *geometrical spreading.* Dropping a rock in a pond shows strong ripples close to where the rock enters the water, but as the ripples propagate out, the same amount of energy is spread over a larger circumference, and consequently the ripples become smaller in amplitude until they eventually become insignificantly small. Sound behaves in a similar manner—close to the source the sound level is generally highest, and the level drops as one moves further away. However, unlike the rock in the water, where the waves are essentially two-dimensional, sound propagates spherically in three dimensions, spreading the energy over larger areas as distance increases. This is commonly referred to as *spherical spreading.*

There are other mechanisms by which sound is attenuated over distance in addition to geometrical spreading, such as atmospheric effects at the molecular level, but they do not typically become significant until one is at least 300 ft from the source (in air). Atmospheric losses are dependent on temperature, humidity, and frequency; more detail on how to estimate these losses can be found in ANSI 1.26—1995.

Source, Path, and Receiver

The most important concepts relating to the analysis of sound as applied to noise and vibration control are those of *source, path,* and *receiver.* All sound must have a *source,* be it a loudspeaker, a personal computer's fan, a violin, a cooling tower, a truck, or a person speaking. The *path,* as one might infer, is any route that the sound can travel to arrive at the receiver. There are typically several potential paths, including through walls, inside air ducts, or through cracks in doors. Although here we will mainly be concerned with sound propagation in air, it is important to remember that sound can also travel in liquids and solids (however, when the disturbances travel in solids, they are typically termed *structure-borne noise* or *vibration*). A *receiver* could be a microphone, a student listening to a lecture, or a piece of equipment that is sensitive to noise and vibration, such as a microprocessor fabricator or electron microscope.

Understanding the source, path, and receiver methodology is essential to effective noise and vibration control. For example, let us consider a noisy HVAC mechanical room (source) and an executive boardroom (receiver) that should be relatively quiet. Ideally, the receiver will be located as far from the source as possible when sound isolation between the two is desired, as sound generally decays with distance. If the two spaces must be close together, increasing the attenuation provided by the path by locating a closet, corridor or other acoustically nonsensitive buffer zone between the spaces can significantly increase the sound isolation between the source and receiver. Another option addressing the path may be to make the walls between the spaces more massive, as heavier materials tend to block sound better than lightweight materials. Alternatively, the source might be modified to make it quieter; assuming the major noise source is a fan, one might operate the fan at a lower speed, or select an inherently quieter fan. Changing the source may eliminate the need to alter the path and vice versa. In some cases modifications to both may be necessary. For this example it is probably not possible (or at least not practical) to modify the receiver; it makes little sense to provide hearing protection for the people in the boardroom. If the receivers were instead work-ers on a factory production floor, this might be a necessary approach after exhausting the pos-sibilities for modifying the sources and paths.

It is generally considered ideal to modify the source to provide the most effective noise and vibration control, as this will reduce the perceived level at the receiver, regardless of the number and types of paths. The overwhelming majority of options available for noise and vibration control, however, deal with altering the path. Often, the closer to the source the path can be modified, the more cost-effective the noise control will be. Identification of all of the source-to-receiver paths can be quite difficult. And, unfortunately, sound transmission is very dependent on construction details; an 8-in concrete wall may only stop as much sound as a ½-in piece of plywood if there is a gap left open around a pipe or ductwork penetration in the concrete.

Sound Power versus Sound Pressure

Consider a room with one 100-W lightbulb and one 40-W lightbulb. The 100-W lightbulb emits more light than the 40-W lightbulb. Similarly with sound, a source with a high *sound power level* will emit more sound than a source with a lower sound power level, regardless of its physical setting.

If only one of the lightbulbs is on, and you hold a piece of paper close to it, there will be more light on the paper than if you hold it some distance away from the lightbulb. Similarly with sound, the *sound pressure level* will be decrease (it will sound quieter) the further you are to the source. This is primarily due to geometrical spreading, which we discussed earlier. Another consequence of geometrical spreading is that there will be some distance from the 100-W lightbulb where the level of light on the paper will be equivalent to the level of light on the paper at some significantly shorter distance from the 40-W lightbulb.

Now, consider two rooms of equal size and shape. All of the surfaces in one room are painted flat black, and all of the surfaces in the other room are painted white. Given that a 100-W lightbulb turned on in both rooms, the white room will appear significantly brighter than the dark room. There will be more light falling on a piece of paper at a specified distance from the bulb in the white room than on a piece of paper at the same distance from the bulb in the dark room. The perceived amount of light in each room is different, but the lightbulbs are both rated at 100 W—they are both emitting the same amount of light. Power ratings can also be determined for noise sources; the sound emitted by a source will remain the same regardless of where it is located. However, the sound pressure level perceived at the receiver will depend on the distance from the source and the environment in which the sound is heard. Sound power and sound pressure are both expressed in decibels, so it is important to note when a manufacturer provides sound data for their equipment whether it is sound power or sound pressure. In addition, if sound pressure is given, the distance from the source and the type of environment in which the measurement was made must also be provided or the data is of no value.

Sound Absorption

Sound absorption occurs when acoustical energy is dissipated as heat and/or transmitted through a surface. When a sound wave strikes a surface some of the energy is reflected back and some is absorbed. The previous example of the lightbulb in the room can also serve to illustrate the concept of sound absorption. If the walls in the white room are gradually painted black, the light level reflected from a surface decreases as the reflectivity changes, and the room appears dimmer. Conceptually, sound behaves in a similar fashion; there is an obvious difference between the way a conversation sounds in a tiled bathroom versus the same con-versation in a living room with plush furnishings, draperies, and thick carpet. Absorption is generally described in terms of *absorption coefficient* (α), ranging from 0 to 1, with 0 representing no absorption (total reflection) and 1 being total absorption (an open window, which reflects no sound, has an absorption coefficient of 1.0). Real-world materials are neither perfect reflectors nor perfect absorbers, and numbers between 0 and 1 essentially indicate the fraction of acoustical energy absorbed every time sound strikes the surface. The absorption coefficient usually varies with frequency, and most building materials are more absorptive at higher frequencies. Porous or fibrous materials (such as glass fiber, mineral wool, or open-cell foam) tend to make good absorbers, and their absorptive properties at low frequencies tend to increase with thickness. The efficiency of absorptive materials is greatest when evenly distributed throughout a given space.

It is important to note that the absorptive properties of a material not only depend on the material itself, but also on how the material is mounted. Many relatively lightweight materials exhibit significant low frequency absorption when mounted over an air space. For example, if one directly applied a single layer of gypsum board to dense concrete wall, the low frequency absorption (around 125 Hz) would be rather low, probably less than 0.1. However, if this gypsum board was mounted to 3.5-in studs on 24-in centers (creating an air space between the gypsum board and the concrete), the absorption coefficient around 125 Hz might be greater then 0.3.

To simplify the presentation of the absorption coefficients, a single-number rating of absorption called the *noise reduction coefficient* (NRC) is commonly used. This is an average of the midfrequency absorption coefficients, rounded to the nearest 0.05. The absorption coefficients and NRC ratings of some common building materials is given in Table 5.4.10-1.

Reverberation is directly related to how much absorption there is in a room. Reverberation is the residual sound that remains in a room after the source has been switched off. If you clap your hands in a cathedral or concert hall, the reverberation is obvious, and the residual sound might last from 2 to 5 s before becoming inaudible. However, in a plush office with furnishings and acoustical ceiling tile, the reverberation may last less than a ½ s, practically unnoticeable. The length of time it takes the residual sound to decay by 60 dB is called the *reverberation time* (RT_{60}). This value is directly proportional to the volume of the space, and inversely proportional to the amount of absorption in the space. Long reverberation times are desirable in spaces used for nonamplified music, such as concert halls and cathedrals, as it gives the music warmth and provides a sense of envelopment for the listener (the qualities of warmth and envelopment can actually be further quantified by other acoustical metrics that we will not discuss here, but they are still related to the reverberant characteristics of the room). Short reverberation times, however, are desirable in spaces used for speech and amplified programs; long reverberation times make speech (or any sounds of short duration) more difficult to understand. Short reverberation times are also desirable in factory and manufacturing spaces, simply

TABLE 5.4.10-1   Absorption Coefficients and NRC Ratings of Common Building Materials

Material	Octave band center frequencies, Hz						NRC
	125	250	500	1000	2000	4000	
Painted concrete block	0.10	0.05	0.06	0.07	0.09	0.08	0.05
1-in non-foil-backed glass-fiber acoustical tile, suspended	0.78	0.98	0.76	0.98	0.99	0.99	0.95
Single gypsum layer on studs	0.29	0.10	0.05	0.04	0.07	0.09	0.05
1-in fabric-wrapped glass-fiber panel, mounted on wall	0.05	0.30	0.70	0.99	0.99	0.99	0.80
Ordinary window glass	0.35	0.25	0.18	0.12	0.07	0.04	0.05
Carpet on concrete	0.02	0.06	0.14	0.37	0.60	0.65	0.30

to reduce the build-up of noise generated by the room's reverberation. Here it is important to note that the use of absorptive materials will not have a significant effect on the sound level close to a piece of noisy equipment. In close proximity to sound sources (the *direct field*), the sound coming only from the source (not from reverberation) dominates the sound pressure level. As one moves away from the source (remaining in the same room, of course), into what is called the *reverberant field,* the reverberant sound will dominate. It is in this distant area where the effects of general absorption will be most effective.

Sound Isolation

Sound isolation is a measure of the reduction of sound as it propagates from a source to a receiver. Often, the natural distance attenuation between a source and receiver is not sufficient, and the sound isolation needs to be enhanced. This is often accomplished by the use of some sort of barrier or other impediment to the propagation of sound. In general, massive barriers stop more noise than lighter ones: visualize how much more difficult it is to shake a thick concrete wall than it is to shake a lightweight plywood partition. In addition, a given barrier will typically block high-frequency sound better than it will low frequencies: it would be comparatively easier to shake a wall slowly (low frequency) rather than quickly (high frequency).

The amount of sound isolation provided by a building partition is measured by its *transmission loss* (TL). TL is a measure of the reduction of acoustical energy expressed in decibels, and is independent of the partition's environment. TL is measured in laboratories over a wide range of frequencies. The single-number *sound transmission class* (STC) rating is derived from the TL data using a special frequency-weighted technique that is useful for identifying how well a construction blocks transmission of sound in the speech-frequency region. The TL and STC ratings for several common building partitions are listed in Table 5.4.10-2.

It is important to note that TL is measured under optimal conditions that ensure that the path of least resistance from the source to the receiver is through the partition under test. This

TABLE 5.4.10-2   Transmission Loss and STC Ratings of Common Building Constructions

Construction	Octave band center frequencies, Hz						STC
	125	250	500	1000	2000	4000	
2 layers ½-in gypsum board, laminated together	21	26	30	30	29	36	31
1 layer ½-in gypsum board on both sides of 3⅝-in metal studs, 24-in o.c.	25	29	39	49	41	38	39
Add 3-in glass-fiber batt in stud cavities	30	38	51	55	48	44	46
Add one layer ½-in gypsum board on one side + 3-in cavity insulation	35	42	51	57	55	47	49
Two layers of ½-in gypsum board on both sides of stud, cavity insulation	33	48	56	60	56	56	55
Double wall, two layers of ½-in gypsum outside of both stud rows, cavity insulation	38	47	57	60	60	58	58
8-in lightweight cast masonry units (CMU), loose fill	36	41	46	52	59	63	51

is not always the case in the real world, as there are often a number of paths by which sound travels from source to receiver. There may be piping penetrations, doors, and ducts punching through perfectly good floors or walls. Any path other than the direct path between a source and receiver is considered a *flanking path*. Flanking paths are important considerations in sound isolation, as even a relatively small flanking path can drastically reduce the isolation effectiveness of an otherwise good partition. The sound isolation between two spaces is also affected to some degree by the amount of sound-absorbing materials within the receiving space. To take into account all of these considerations, a field-measured adaptation of TL is used, called *noise reduction* (NR). NR has a single-number metric similar to an STC rating called *noise isolation class* (NIC). There are no tables for NR and NIC because they depend on the installation conditions as well as the partition itself. However, even for a well-built construction the NR is usually at least a few decibels lower than the TL, primarily due to flanking paths.

A common technique used to improve the transmission loss of a given partition is the use of sandwich-type constructions involving an air space. A stud wall is a good example: the TL of a partition consisting of one layer of gypsum board on each side of the studs is significantly better than if both layers of gypsum board were simply attached to the same side of the studs and nothing was attached to the other side. Further improvements may be obtained by placing glass-fiber insulation in the stud cavities, adding more layers of gypsum board, and increasing the depth of the air space. In addition, decoupling the layers of gypsum board by using two separate nonconnected rows of studs or resilient channels will significantly improve the isolation.

Sound isolation can also be obtained outdoors using freestanding barriers. Barriers are much less effective than full-height partitions in a building because sound *diffracts* over and around barriers. For a given barrier to provide any significant isolation, it must at least block line-of-sight between the source and receiver. Barriers become more effective as they get taller, and as the source and/or receiver get closer to the barrier.

Sound can also be transmitted effectively through structures when the source is directly coupled to the structure. Structure-borne noise is a common complaint in condominiums, for example, when the people in the upstairs condominium are walking around in hard-soled shoes on a tile floor. The *impact insulation class* (IIC) is used to rate the amount of isolation a given construction provides from impacts. It is a single-number rating scheme very similar to STC. In the condominium case, the solution may be as simple as using a carpet with a thick backing; however, it is important to note that while this may significantly increase the IIC, it will have almost no effect on the STC. The carpet treatment may not be an option, for example, where there is a gymnasium above an auditorium; more drastic measures are required. To reduce the impact transmission in this scenario, a secondary partition that is not rigidly connected to the first may be required. This might involve suspending several layers of gypsum board on vibration isolators from the structural deck (a *resiliently suspended ceiling*), or placing a concrete slab on vibration isolators on top of the primary floor (a *floated floor*). The use of a separate secondary structure will increase both the STC and the IIC; however, it should be noted that these types of constructions are typically very expensive and difficult to install properly. It is almost always better to space plan properly in advance; the gymnasium should not be directly adjacent to the auditorium (vertically or horizontally).

Note that *sound isolation* is often confused with *sound absorption,* and most materials that are good sound absorbers are not good sound isolators. For example, consider a condominium directly adjacent to a mechanical room with a large, noisy pump. Wrapping the pump with sound absorbing material (glass-fiber insulation, for example) will not appreciably change the sound level in the mechanical room or the condominium. Adding mass (layers of gypsum board, concrete block, etc.) to the demising wall (hence improving the sound isolation), will not change the noise levels in the mechanical room, but it will likely reduce the sound level from the mechanical room into the condominium. Adding absorption in the mechanical room may reduce the reverberant build-up of pump noise, which can reduce the noise levels in the mechanical room, and hence reduce the total amount of noise available to be transmitted through the wall. Of these three scenarios, improving the sound isolation provided by the demising wall is generally the most effective. The goal of sound isolators is to stop sound transmission, while sound absorbers reduce reverberation. When used together appropriately, they can help achieve the acoustical goals of the space.

ACOUSTICAL GOALS AND RATING SCHEMES

Generally, the most important acoustical goals for a facility are to achieve appropriate background noise levels inside a given space and/or to control the level of noise emitted from the facility to the surrounding area. The desired or acceptable noise levels vary significantly depending on the use of the space or its location; what may be considered quiet in a typical office may be much too loud for a music recital hall, and the level of noise produced by a factory in a large city probably will not be acceptable for a similar plant in the middle of a rural neighborhood. Fortunately, there are numerical standards that assist the engineer in setting appropriate goals for noise levels.

The most commonly used metric for assessing background noise is *A-weighted decibels* (dBA). The human ear interprets sound levels differently at different frequencies. Middle-frequency sounds will seem louder than low or very high frequency sounds of the same decibel level. To compensate for this, filters have been developed that alter what a microphone measures to reflect the sensitivity of the ear. Sound levels measured using the A-weighting filter are expressed in A-weighted decibels. Examples of sound levels produced by various sources are given in Fig. 5.4.10-2.

Another important assessment standard is the *noise criteria* (NC) rating. This is typically used in interior environments where the spectral qualities of the background sound and the signal-to-noise ratio are more important. The NC rating is derived from comparing the measured octave band spectrum to a pre-defined set of curves called the noise criteria curves. The NC curves are similar to A-weighting in that they both account for the ear's non-flat frequency response. A typical HVAC noise spectrum is depicted on a NC chart in Fig. 5.4.10-3.

There are several other interior noise rating systems similar to NC in current use including *room criteria* (RC) and *noise criteria balanced* (NCB), but they generally represent refinements to the NC system that are appropriate in certain applications. Refer to ANSI 12.2—1995 for details on these rating systems.

Although the dBA value and the NC rating cannot be directly compared with one another, the dBA value for an average background noise spectrum similar to a ventilation system will be about 5 to 8 dB higher than the NC rating. Table 5.4.10-3 lists common room uses or zoning adjacencies and their appropriate (or maximum) dBA or NC goals.

Measurement Types

There are also a variety of measurement types in addition to the rating schemes. Commonly used metrics include L_{EQ}, L_{DN}, and L_X.

The equivalent sound level L_{EQ} is the average of all of the sound energy measured over a specified time period. It is important to note that because L_{EQ} uses a logarithmic energy-averaging technique (as opposed to an arithmetic average), it tends to weigh loud, impulsive sounds more heavily than constant, lower-level background noises.

The day–night sound level L_{DN} is similar to the L_{EQ} in that it is an energy average, but it is measured over a 24-h period, and the sound measured between 10 P.M. and 7 A.M. is penalized by adding 10 dB. It is often specified for land-use planning purposes.

The percentile level L_X is the level exceeded X percent of a specified measurement period. Commonly used values for X are 10, 50 and 90. For example, the L_{50} for a 1-h measurement represents the sound level that was exceeded 30 out of the 60 min. L_X measurements are not as responsive to impulsive sounds as L_{EQ}, and are therefore commonly used to assess residual

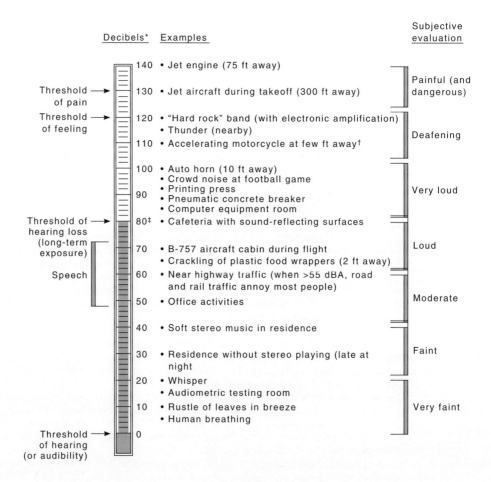

*dBA are weighted values measured by a sound level meter.

†50 ft from a motorcycle can equal the noise level at less than 2000 ft from a jet aircraft.

‡Continuous exposure to sound energy above 80 dBA can be hazardous to health and can cause hearing loss for some persons.

FIGURE 5.4.10-2 Sound levels produced by various sources. (*Courtesy of M. David Egan; from* Architectural Acoustics, *McGraw-Hill, New York, 1988*).

background levels in the presence of other random events (L_{90}), the median average level (L_{50}, similar to L_{EQ} for average time-history distributions), or peaks of noise (L_{10}).

EMPLOYEE HEARING PROTECTION

In the United States, the Occupational Safety and Health Administration (OSHA) regulates maximum permissible employee noise exposure in the workplace. The OSHA requirements are aimed at limiting employees' overall time-weighted average (TWA) exposure to noise. Essen-

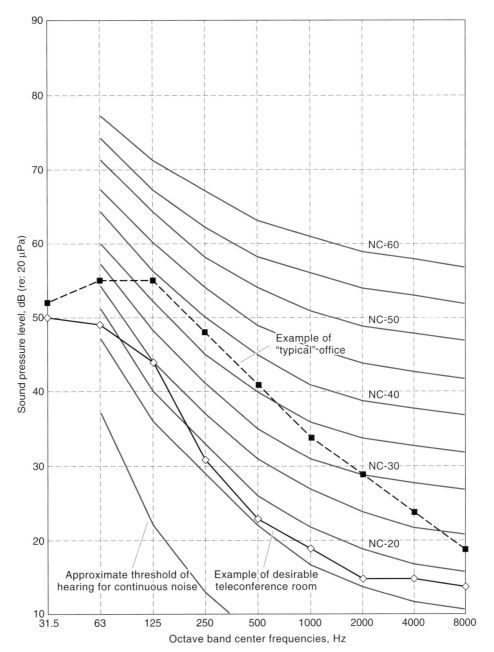

FIGURE 5.4.10-3   Noise criteria (NC) chart showing typical spectra from an HVAC system. (*Courtesy of Acentech, Inc., Cambridge, MA.*)

TABLE 5.4.10-3 Background Noise Goals and Limits for Various Occupied Spaces

Criteria	Space or location	Desired or acceptable level
Hearing protection	Factory: 8 h	Max 90 dBA (OSHA)
Freedom from annoyance	Residential property line near commercial zone	50 dBA
Clarity of communication	Typical office	NC-35–NC-45
	Teleconference room	NC-20–NC-25
	General classroom	NC-25–NC-30
	Recording studio	≦NC-15
	Concert hall	NC-15–NC-20

tially, OSHA limits exposures to less than an 8-h TWA of 90 dBA, with a 5-dB trade-off for doubling of exposure time. A 90-dBA exposure is permitted for 8 h per day; 95 dBA for 4 h per day; 100 dBA for 2 h, and so forth. Exposure to continuous noise levels higher than 115 dBA is not permitted, nor is exposure to impulsive noise levels higher than 140 dBA. A *hearing conservation program* is mandated by OSHA when employee noise exposures exceed an 8-h TWA of 85 dBA.

It is important to recognize that OSHA does not directly regulate in-plant noise levels; their concern is employees' time-weighted exposure to noise. In many industrial settings, it is difficult to differentiate between the two: where employees work in the same general location for their entire work shifts, and the noise levels in the area are fairly constant, the levels in the work area completely determine the daily noise exposures. In these cases, efforts made to reduce area noise levels will typically result in overall reductions in employee noise exposures.

However, in many work settings, employees are exposed to varying noise levels, either because noise levels in fixed work areas fluctuate, or because workers move through areas with different sound levels throughout the course of their workday. Maintenance workers are, perhaps, some of the best examples of employees who are exposed to widely varying noise levels in the performance of their normal duties. In cases like these, it is often difficult to directly tie worker noise exposures to the noise emissions of any particular noisy piece of machinery, and specific noise-control efforts targeted toward reducing equipment noise emissions (or even area noise levels) may not adequately address excessive employee noise exposures. *Administrative* noise controls may be useful in such situations; careful scheduling of employees' times in and out of noisy areas may help reduce their overall daily noise dose. Designating (and posting) of certain areas as *noise exposure areas,* where use of personal hearing protection devices is mandatory, may also be necessary.

One fairly recent trend that seems to be coming into consideration is employers' desire to maintain or enhance the overall quality of the work environment for their employees. Many companies have instituted quality assurance and other total quality management measures that are geared toward increasing employee productivity, and have recognized the benefit of providing in-plant noise levels significantly quieter than are required to meet OSHA requirements. A number of manufacturing facilities, for example, have moved historically office-bound engineering staff to main production floors to better integrate design and manufacture of products. This has necessitated many functional and operational changes, not the least of which are modifications needed to maintain reasonably low-noise work areas for face-to-face and telephone conversation and other quiet-intensive tasks.

Environmental Noise Issues

Another major force driving many noise-control efforts is a facility's noise emissions to noise-sensitive offsite receptors, most typically, nearby residential neighborhoods. In many cases, state or local noise-control regulations establish maximum permissible noise levels that can be emitted from the facility, often to be measured at the property line of the industrial facility, or, more equitably, at the lot line of the receiving property (this permits intervening non-sensitive property to be used as a buffer zone between the emitter and the receptor). However, many noise complaints arise in situations where there are no numerical regulatory limits against which compliance can be assessed, and both the complainant and the offending noise emitter are left without clear acceptability criteria. In these cases, the issue often boils down to the reasonableness of both the complainant and the emitter. A strong good-neighbor policy on the part of the facility can often prove very effective tool in mitigating the situation with the complainant, or, as it sometimes evolves, in a legal setting. Efforts made to identify and treat the source of the offending noise, whether or not legal compliance issues are involved, are generally appreciated, and can go a long way towards defusing the contentious relationships that often develop over noise complaints.

MECHANICAL SYSTEM NOISE CONTROL

The primary source of noise in most buildings is usually the mechanical equipment associated with the building's HVAC system. Standard commercial and institutional building design, without any special provisions for noise, generally yields background noise levels of NC-35 to NC-45 in average occupied office-type spaces, which is usually considered acoustically acceptable, though levels of NC-30 to NC-40 are generally found to be more desirable. As shown in Table 5.4.10-3, however, there are often spaces within these facilities that require quieter background levels, such as conference rooms and teleconferencing suites. These spaces often require special attention, and are deemed acoustically sensitive. In the commercial space, this would apply to spaces where it is desirable to have background noise levels of NC-30 or less. Another example is a laboratory utilizing local terminal units; the untreated level will be somewhat higher, NC-55 to NC-65, while levels ranging from NC-50 to NC-55 are considered more desirable.

The desired noise goals can usually be attained given proper planning and consideration.

SPACE PLANNING

For interior spaces, the major noise-control issues usually include separating a noisy mechanical room from the surrounding occupied spaces, and reducing the noise traveling the HVAC duct system to an appropriate level at the occupied space. This first can be dealt with using careful space planning; do not locate the conference room next to the mechanical room, for example; it is also not advisable to locate the chiller plant over the exhibition space. These seem like relatively obvious situations to avoid, but they manage to be built often enough. Adjacencies such as these require special, expensive construction to provide the desired sound isolation, which still is not adequate to achieve the desired background noise goals.

HVAC Noise Control

There are two main components of HVAC noise: the noise generated by the fans themselves, and noise generated by flow in the ductwork. One of the most effective methods of reducing noise from both sources is internally lining the ductwork. There has been much debate regarding the potential health hazards of glass fiber in ductwork; the primary concerns are that the fibers will shed as the air stream passes over it, and that bacteria will grow in the glass fiber. However, recent studies have found no relationship between exposure to airborne glass fibers and human health. In addition, it has been found that if the conditions are right, bacteria will grow in ductwork regardless of whether it is lined or unlined. In an attempt to reduce the potential for these problems to arise, liner manufacturers are now treating the airflow side of the glass fiber to prevent the shedding of the fibers at normal flow velocities; these coatings also contain anti-microbial agents to reduce the chance of bacterial growth.

Although internal lining reduces both fan and flow noise, other means of reducing the often excessive levels of low frequency noise produced by the fans are often required. The most commonly used device is a *duct silencer,* which is a type of dissipative muffler. They generally come in lengths of 3, 5, 7 and 10 ft, and are sized similar in cross section to the duct in which they are inserted. The insides of these devices consist of several parallel baffles of glass fiber. An example is pictured in Fig. 5.4.10-4.

Flow noise in ductwork is the result of turbulence generated at the duct surfaces, through elbows and intersections, and when the flow passes over objects such as volume dampers, grilles, and diffusers. The level of this noise is proportional to the flow velocity; therefore, the quieter the system needs to be, the lower the required flow velocity in the ductwork. This directly translates to the need for appropriate space planning to accommodate larger ductwork and larger diffusers and grilles in acoustically sensitive spaces.

The noise generated by flow over volume dampers and grilles can be significant. An excellent way to reduce noise generated at the dampers is to eliminate the need for dampers; this can be accomplished by designing a self-balancing duct layout. Utilizing a layout that branches out symmetrically, with approximately equal lengths of ductwork to each diffuser, will significantly reduce the amount of dampering required, which consequently reduces the noise. In addition, to the extent that smooth transitions and turning vanes can be used in the ductwork, the noise generated at elbows and intersections can be reduced. Another common noise control tactic is to locate the volume dampers as far away (at least 5 to 10 ft) from the grille or diffuser, placing internally lined ductwork or a length of flexible ductwork in-between. Placing dampers directly behind the grilles or diffusers is not recommended, as there is nothing between them and the occupied space to provide any attenuation. Appropriate selection of the grilles and diffusers themselves is also important; typically it is advisable to select these devices such that the manufacturer's sound rating is at least 5 dB below the noise goal of the space.

Terminal units are often used to provide control of the amount of air entering a space, may have heating or cooling capacity, and possibly a small fan. These devices can radiate a significant amount of noise, and when they serve acoustically sensitive spaces, they should be located in an adjacent corridor or other nonacoustically sensitive space. Similarly, devices such as toilet exhaust fans should not be located in ceiling space of acoustically sensitive spaces. Silencers and/or internal duct lining are used downstream of terminal boxes to reduce

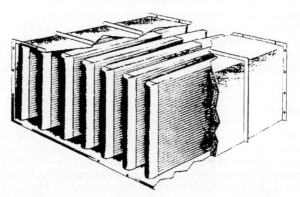

FIGURE 5.4.10-4 Cutaway view of a parallel-baffle duct silencer. (*Courtesy of Industrial Acoustics Corporation, Bronx, NY.*)

duct-borne noise. Noise generated by terminal units can be kept within reasonable limits if the pressure drop across the box is kept to a minimum; it is generally more important to design the noise control measures for the terminal units close to the beginning of a long duct run because the pressure drop across those units will be the greatest.

VIBRATION ISOLATION

Sources of noise and vibration, if directly connected to building structures, will transmit vibratory energy to the structure; this vibratory energy can travel very efficiently though the structure to occupied areas of the building, where it may be felt as vibration or heard as structure-borne noise. Generally, palpable vibration in facilities is not acceptable; the goal is to keep vibration levels below the threshold of perception. Structure-borne noise from equipment is really indistinguishable from noise that might be transmitted through a partition and then radiated from the wall surface into the receiving space. The goals for controlling structure-borne noise are the same as those for other noise-control issues.

The first rule in controlling structure-borne noise and vibration is to create as much distance as reasonably possible through the structural path between the source and receiver locations. For critical applications, such as a microelectronics fabrication facility, this is taken to the point of having separated structures so that the vibration from the source will have to travel down the structure on one side of the separation, travel through the ground to the columns of the vibration-sensitive area, and then up though the structure of the sensitive area. Similar approaches might be taken in performance facilities where very low levels of noise are desired. The further the source is from the receiver, the easier it will be to resolve the noise and vibration concern, and the construction requirements to control noise and vibration will be reduced accordingly.

Unless equipment is located far from noise and vibration sensitive areas and/or on slab-on-grade structures, it is usually wise to provide vibration isolation for the equipment. This provides control of vibration and structure-borne noise close to the source where control is most economically implemented. Isolation can be provided by a variety of devices (with varying isolation characteristics) which need to be properly applied. The Sound and Vibration Control chapter of the *1991 ASHRAE Applications Handbook* has an excellent section on vibration isolation, including specific recommendations for numerous common applications. Typically, isolation can be provided by spring hangers or floor mounts achieving static deflections under load in the range of ¾" to 3.5 in, neoprene mounts or hangers achieving deflections of about 0.3 in, and neoprene pads. Basic isolators of these types are illustrated in Fig. 5.4.10-5, but there are numerous variations on these that provide specific additional features for special purposes.

All isolators have a resonance frequency at which the vibratory energy is stored in the isolation element with little energy dissipation, and no isolation occurs. At this frequency the vibration forces transferred to the supporting structure are actually amplified (due to the stored energy) and the supported equipment can move excessively. Obviously the isolation systems must not be driven at this resonance frequency. In fact, to achieve effective isolation, the resonant frequency of the isolators needs to be carefully selected to be lower than about ⅓ of the lowest drive speed of the equipment. The greater the difference between the drive speed and the resonance frequency of the isolators, the better the isolation efficiency. The resonant frequency of the isolators would preferably be as low as 1/10 the drive speed of the equipment, although achieving this is not always necessary or practical. The resonant frequency of isolators commonly used are a function of the inverse of the square root of the deflection of the isolator under load and is given by $F_n = 188 \sqrt{1/d}$, where d is the deflection in inches and F_n is in cycles per minute or rpm. If the drive speed of the piece of equipment is known, the required isolator deflection to achieve effective isolation can be computed. Where equipment is operated with a variable-speed drive, the lowest

FIGURE 5.4.10-5 Examples of vibration isolators: (*a*) single-spring mount with base plate, (*b*) neoprene mount, (*c*) spring and neoprene hanger, and (*d*) neoprene waffle pad. (*Courtesy of Mason Industries, Inc., Smithtown, NY.*)

operating speed should be no less than 1.5 to 2 times the resonant frequency of the isolation system. The lower the drive speed of the equipment, the lower the required resonant frequency of the isolators. This translates into higher spring deflection, which translates into larger and more expensive springs. It is wise to check to be sure that there is space planned for the appropriate isolators. It is also appropriate to avoid providing substantially more isolation deflection than is needed because this can become very costly without much improvement in isolation effectiveness.

There are other factors that go into the selection of recommended deflections for isolation systems which cannot be purely assessed from just the comparison of the equipment drive speed and isolator resonance frequency. In particular, the magnitude of vibration that has to be isolated is not included; this has been accounted for based on experience, and is reflected in the various recommended deflections given in the *1991 ASHRAE Applications Handbook.* In addition, to support equipment having different weights and weight distributions, manufacturers of equipment typically provide different stiffnesses of springs available in several deflection ranges.

When mounted on resilient supports, equipment with significant unbalanced forces and/or large starting torque can exhibit substantial movement. To minimize this movement and avoid excessive stress on connected components, massive bases, commonly called *inertia bases*, are used. It is worthwhile noting that the use of such bases only reduces the motion of the attached equipment; they do not provide any isolation and do not reduce the magnitude of vibratory force transmitted to the building. Only the resilient mounts provide isolation.

In many areas of the country equipment needs to have seismic restraints to restrict movement and prevent damage in the event of an earthquake. Seismic restraint is often integrated into isolators for convenience while still supporting and isolating the equipment. The seismic-restraint aspects of the isolators are not related to the isolation that the isolators provide. It is important that the seismic restraint system not defeat the isolation.

In addition to isolating individual pieces of mechanical equipment it is necessary to isolate piping that connects to vibrating equipment to minimize vibration transfer to the building. Flexible connections are often used in attaching pipes to equipment. The best type of flexible connections for vibration control purposes are those made of multiple spheres of reinforced rubber, but even these rarely provide sufficient isolation because vibration in the fluid carries across the flexible connection and couples back into the pipe. Corrugated-metal flexible connections provide very little vibration isolation. The preferred approach to preventing vibration transmission to building structures via piping is to include vibration isolation pipe supports within some defined zone near the connected equipment. Electrical connections to vibrating equipment require some flexibility, either by use of a slack section of flexible conduit or by using a special expansion and deflection coupling. Ducts are routinely connected to air handling equipment with flexible canvas connections, but where there are acoustically sensitive spaces nearby, the magnitude of flow being handled is great, or the noise level in the duct is high, ductwork is occasionally vibration isolated.

For vibration isolation systems to be effective, the structures supporting the equipment must be suitably massive and significantly stiffer than the isolation system. For example, placing a mechanical unit on spring isolators in the middle of a lightweight metal roof deck will render the isolators almost useless because the roof deck has little mass and is relatively flexible. The unit needs to be supported either on a substantial structural floor or on dunnage that transfers the load of the equipment to major beams or columns.

Structural Dynamics and Structures for Vibration Sensitive Equipment

Not only does vibration in structures need to be controlled to avoid perception by people so there will not be complaints about noise, but there is an ever-expanding array of vibration-sensitive equipment used in buildings: diagnostic equipment, metrology equipment, microelectronics manufacturing equipment, electron microscopes, surgery equipment, and the like. The vibration requirements for this equipment are often significantly more stringent than those for human occupancy. It is wise to obtain specific vibration sensitivity information from the equipment manufacturer. When such data is available, the specific vibration criteria for equipment will vary dramatically from one frequency to another because of the nature of resonances of lightly damped internal structures. The relative motion of internal components at their respective resonances determines how sensitive the equipment actually is. The general tendency for the vibration requirements of many (but not all) pieces of vibration-sensitive equipment is to follow lines of constant vibration velocity throughout most of the typical sensitivity range from 4 to about 100 Hz. In the extreme low frequency range, somewhat more vibration can typically be allowed because there are few resonances that create relative motion between internal elements. At these very low frequencies the

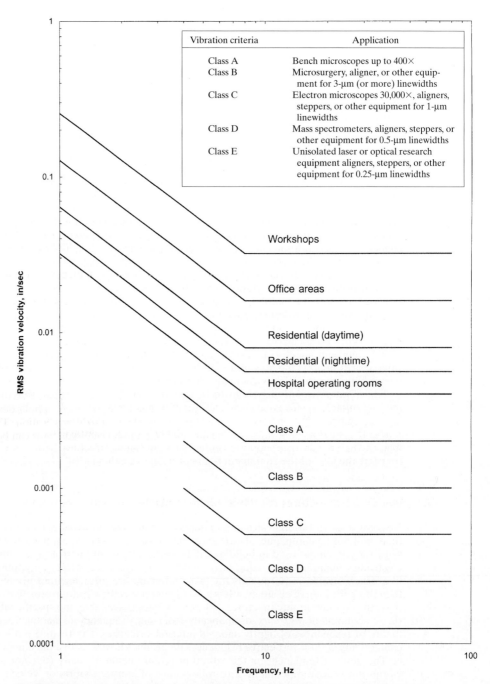

FIGURE 5.4.10-6   Generalized vibration criteria. (*Courtesy of Acentech, Inc., Cambridge, MA.*)

internal components tend to move together, which does not create large relative motions. In absence of specific sensitivity criteria the chart presented as Fig. 5.4.10-6 gives preliminary guidance as to the acceptable level of vibration often associated with the use of various types of equipment.

Clearly one would like to locate vibration-sensitive facilities far from sources of ground-borne vibration. Distance and the attenuation that typically comes with distance is one of the most cost effective ways to control ground-borne vibration. There are few other practical methods by which to control such vibration in the very low frequency range.

Assuming that vibration in the ground is suitably low, the most inexpensive location for vibration-sensitive equipment is on a slab on-grade with well-packed soil beneath. In this case, there is little extra installation cost attributable to vibration control. When vibration-sensitive equipment is located on a structure above grade, substantial extra structural cost can be incurred; the extra depth of structure that is required can impact the amount of space available in the ceiling space below. Plan for relatively short column bays in areas supporting vibration-sensitive equipment because the level of vibration that is generated is related to the span to the third power; therefore, shorter spans will provide the required stiffness with less structure. To the extent that vibration-sensitive equipment can be planned to be located in limited areas of the floor, such as near columns, where the floor is inherently stiffer than the middle of a bay, this will help economize on structure. Similarly, to the extent that the area in which vibration-sensitive equipment is located can be limited, this will limit the area where extra structural cost is incurred.

People walking in buildings are often the most significant source of vibration with which you need to be concerned. The faster people walk, the greater the vibration they produce. Where vibration-sensitive equipment can be kept out of structural bays with major corridors where people will walk rapidly, a substantial structural penalty can be avoided.

Concrete structures that are designed based on strength to support a given load typically are stiffer than steel structures designed for the same strength. So, a concrete building designed for a given strength will typically have lower vibration levels than a steel building designed to the same strength level. This is not to say, however, that concrete is the best way to build structures housing vibration-sensitive equipment. The cost-effectiveness of various design solutions needs to be evaluated for each case and will vary from location to location. There are also other considerations that need to be addressed. Certainly, steel structures can be designed to achieve the same low level of vibration that is achieved with concrete structures.

Mechanical-equipment–induced vibrations need to be carefully considered in buildings housing vibration-sensitive equipment, but with good space planning at the outset, it is not usually necessary to use exotic and expensive isolation systems. Usually it is sufficient to have well conceived, well specified, and well executed isolation systems of the conventional variety.

Sometimes, to resolve a particular problem with a single piece of vibration-sensitive equipment, or if in a new facility there will only be a few pieces of vibration-sensitive equipment, it might be more cost effective to simply provide a special local isolation table or mounting assembly to provide the required isolation in a situation where higher than otherwise acceptable vibrations exist. For these applications, pneumatic isolation systems are usually employed, which are available from specialty isolation system manufacturers. The application of such devices should be thoroughly investigated in advance, because not all pieces of equipment are well suited for mounting on such devices (for example, if the elevation of the operating console were changed by placement on an isolation system).

ELECTROACOUSTIC SYSTEMS

There are several types of electroacoustic systems to be aware of, including active noise control, sound masking, sound reinforcement, and audiovisual teleconferencing systems.

Active Noise Control

Noise cancellation and *antinoise* are popular buzzwords in engineering circles recently. Active noise control can be quite effective in particular applications, but it is important to understand what it actually is and when it can be beneficial.

If one imagines two sine waves having the same frequency that start at the same time, adding the two together will result in a wave that has twice the amplitude of either original, maintaining the same frequency. If one of the waves is delayed by one-half of the wave period, however, the two waves will sum to zero. In one type of active noise-control system, a primary microphone detects the noise, which is then processed to become out of phase when produced by a loudspeaker. A second error-detection microphone is placed after the loudspeaker to verify the change.

Active noise control can work well in confined spaces at low frequencies, where the wavelengths are large with respect to distance between the microphone and loudspeaker and with respect to the physical dimensions of the locale. Where noise problems are more broadband in nature, conventional passive noise-control techniques are generally preferred.

Sound Masking

The process by which the audibility of one sound is reduced by covering it with another sound is called *sound masking*. The open office is a good example of a space where electronically added sound masking is often appropriate. Typical sound-masking systems provide broadband noise (similar to the static heard between FM radio stations) that is most prominent in the speech frequency range. This noise acts to mask speech, which reduces the level of distractions due to sources outside the cubicle. These systems work best when the ceiling of the space is treated with a highly absorbent acoustical finish (NRC 0.80 or greater), and if the cubicles have tall, acoustically absorbent walls as well. Sound-masking systems consist of several distributed loudspeakers located within the ceiling plenum (usually about one per 150 to 250 ft², depending on the ceiling type and plenum conditions), and a rack of equipment that can be located anywhere in the office.

Sound Reinforcement

Sound reinforcement systems are generally required in all venues seating more than about 100 people to ensure that the speaker can be heard everywhere in the room. Many styles of music also require amplification, especially in large venues. These systems range from small ceiling mounted loudspeakers in a boardroom to large centralized clusters of loudspeakers in arenas, capable of generating sound levels greater than 110 dB. Open outdoor venues have become quite popular recently, and the high levels of sound produced by these systems can be annoying to nearby neighbors if the systems are not properly planned. At a smaller scale, the increasing popularity of multimedia systems in classrooms has made sound isolation between classrooms even more critical; students that are taking an exam in one classroom do not want to be disturbed by the adjacent classroom observing a multimedia presentation.

Sound systems generally consist of four main components: input devices (microphones, CD players), control systems (mixers, signal processors), amplifiers, and loudspeakers. Generally, the control systems and amplifiers are mounted in equipment racks. In smaller installations, the control systems and amplifiers may fit in a closet, whereas in a concert facility, there is typically an entire room close to the stage for the racks of amplifiers, and a separate rack for the control equipment is located in the control booth. It should be noted that the control rooms and amplifier rooms for larger sound-reinforcement systems have above-average power and cooling requirements. These spaces require separately controlled HVAC independent from other building systems. In addition, sound systems typically require conduit, power, and grounding systems that are separate from the other building electrical systems. This minimizes the potential for radio frequency (RF), 60-cycle, and other electrical noise to be introduced into the wiring for the sound system.

The loudspeakers used for larger venues usually have middle- and high- frequency horns, which direct sound to a certain area. This *coverage pattern* (typically given in terms of horizontal and vertical angles, for example, 90° by 40°), allows one to aim the loudspeakers so that most of the sound only goes to the listeners, instead of radiating sound in all directions. Directional loudspeakers increase speech intelligibility in a reverberant environment. However, appropriate acoustical treatments still need to be considered. For example, the rear walls of large, closed venues should typically be covered with either sound absorbent materials or constructions that *diffuse* sound (scatter in all directions rather than reflect in one direction like a mirror). It is also important to note that it is very difficult to effectively direct low frequency sound.

Audiovideo Teleconferencing

Audio teleconferencing systems are becoming increasingly important in the business world, as they allow several people to communicate without having to be at the same location. These systems can be coupled to live video to make the meeting more realistic.

Audio teleconferencing systems are essentially sound-reinforcement systems with some specialized digital signal-processing equipment that mix and route the signals to and from the various locations. These systems are extremely sensitive to background noise and reverberation, even though they have some echo-canceling capability. Rooms used for audio teleconferencing work best if the ceiling and at least two adjacent walls are completely covered with highly absorptive acoustic treatment. The background noise in these spaces should be no more than about NC-25 (though NC-30 is sometimes acceptable). If video is involved, the patterns of the finish materials should be plain (as opposed to busy), and there will be special lighting requirements to consider. It is generally not appropriate to locate video teleconference rooms on a window wall, as outside light is very disruptive, and the glass (unless specially specified) generally does not provide sufficient sound isolation from outside noise. To reduce the amount of noise that enters these rooms, the partitions are generally more substantial than normal, and the doors require special consideration. Avoid having doors leading from noisy or busy public areas, but if this has to occur, consider providing a vestibule into the space. When the door to the room leads from a quiet area of the building it may be acceptable to have a single gasketed door.

OTHER FACILITIES—NOISE AND VIBRATION CONCERNS

We have discussed general noise- and vibration-control principles. The following subsections outline some of the common acoustical concerns that are encountered in common buildings that were not previously addressed in the preceding text. In many cases, the assistance of a qualified acoustical consultant may be required, and the following information is simply an initial listing of items to be wary of.

Performance Spaces

Concert and recital halls, opera houses, theaters, and other similar spaces have a wide range of acoustical concerns. Critical spaces need to be shielded from any loud outdoor areas, including underground sources such as subways. Noise levels typically need to be extremely low. The implications of this on the mechanical system need to be addressed early in the design process due to the large amount of space required for ducts (to move the air at low velocities) and high amount of sound isolation required at the mechanical room. Achieving desirable reverberation (and other appropriate room-acoustics parameters) in the performance spaces requires a large space. The sound-reinforcement systems for these spaces tend to be very complex, often having multiple clusters of loudspeakers, significant amounts of wiring and conduit, and extensive control systems. Room shapes will need to be carefully considered, along with room finish materials. Appropriate sound isolation around critical rooms will mean substantial construction, including vestibules or special acoustical doors. Good space planning (keeping noisy spaces far from acoustically sensitive spaces) can help avoid unnecessarily expensive sound-isolation construction.

Hospitals

Many hospitals now contain vibration-sensitive diagnostic equipment that requires that vibration levels be kept low to function properly. The structural design needs to be carefully handled to minimize the cost to accommodate this. Narrower-than-normal structural bays may be considered to help minimize the structural impact. Beams that are stiffer than normal may be required to support this equipment, which may squeeze the ceiling space for other systems. Recognition of this requirement in the early design stages can help avoid excessively tight or unworkable conditions in the ceilings.

Educational Facilities

There have been a plethora of studies recently that directly link students' abilities to learn with the acoustical conditions of the classroom. Low background noise levels and low reverberation times are important for good hearing and hence good learning. Although there are a lot of educational facilities with unit ventilators or fan-coil units to provide heating and cooling, their use is rarely consistent with achieving the desired low noise levels. Centralized, ducted HVAC systems are generally quieter. More stringent noise standards (perhaps national standards) are in the works to help improve acoustical conditions in educational spaces. Reverberation generally needs to be well controlled within desirable limits to promote clarity of speech. Sound isolation between spaces also requires careful consideration.

Correctional Facilities

The key issue in correctional facilities is to incorporate sound-absorptive materials to help control activity noise and make the spaces a bit more comfortable. The lack of sound-absorptive materials in these facilities makes living in them like living in a reverberation chamber (one might compare this with living in a racquetball court with several people). It is important that the absorptive materials be especially tamper-resistant.

Sports and Convention Centers

These facilities usually include large-volume spaces where the generous use of acoustically-absorptive materials is important to control activity noise and to allow the sound system to be intelligible. Limiting noise from the mechanical system is also important to maintain modest noise levels for comfort and intelligibility of the sound system. This can often be a challenge in large-volume spaces where there are limitations on duct placement and there is the desire to throw air long distances. Convention centers often pack, adjacent to one another, spaces with potentially high-sound-level activities, and sound isolation is key. Operable walls are often considered in such facilities for flexibility. It is important to note that the sound-isolation performance of operable walls depends greatly on their proper installation, operation, and maintenance. The perimeter seals especially are weak points that require careful handling and occasional replacement to allow the wall to achieve its rated acoustical performance. The sound and paging systems are often very complex and require a great deal of planning.

Office Buildings

Although the noise goals in office buildings are generally not that critical, it is appropriate to identify high-sound-isolation areas such as conference rooms and executive office areas where full-height partitions will be required and lower than typical background noise levels may be desired. Avoid placing these rooms adjacent to (next to, under, or over) mechanical rooms. For open-plan offices, good space planning, care in the selection of cubicle panels (use of tall panels is significantly better than use of half-height panels), ceiling finishes, and a sound-masking system are key to achieve acceptable speech privacy. If glass-fiber tile ceilings are used for high sound absorption, remember that these types of tile typically have low sound-transmission loss—noise will be easily transmitted from equipment in the ceilings to occupied spaces below. Rooftop HVAC equipment can be of concern for occupied spaces below if placed on a lightweight metal deck roof. Special consideration is required for vibration isolation and control of break-out noise from ducts which drop directly out of equipment into the ceiling plenum above occupied spaces. The pressure within the air-distribution systems should be kept as low as possible to avoid unnecessarily high noise generation at terminal boxes.

Cafeterias, Dining Halls, and Restaurants

In any large dining space, the most important acoustical problem is excess noise due to reverberation. It is rather impressive how much more comfortable a dining space becomes when appropriate acoustical treatment is applied. Provide a substantial area of highly sound-absorptive material in the space to reduce the reverberant noise levels in the space; generally, the most cost-effective approach is to cover the entire ceiling. There are treatments available to match just about any decor.

Retail Stores

These spaces are not usually too critical from the acoustical standpoint, but they require basic reverberation control for comfort, such as with an acoustical tile ceiling. Moderate attention to noise control is required to avoid high levels of mechanical-system noise. If there are residential neighbors nearby, outdoor noise emission (especially from rooftop equipment) may be a concern.

Industrial Spaces

Although typical industrial spaces may not have particularly stringent noise requirements, this does not mean that noise control should be ignored or any-old piece of equipment should be used. Although not always possible, attempt to select reasonably quiet equipment to start. Be mindful of the noise emissions from special equipment and high-pressure devices. They may be significant sources of vibration, and this may be in conflict with vibration-sensitive equipment that is also part of the process. Incorporate sound-absorptive materials into the ceiling to control reverberant noise levels. Create barriers around particularly noisy activities to minimize noise intrusion into adjacent spaces.

Exterior noise problems generally fall into two broad categories: continuously operating devices and intermittent sources that may be addressable with administrative controls. Many situations fall into both categories. Noisy cooling towers, ventilation equipment, and so forth, often must be treated to reduce emissions to acceptable levels. Complaints about noisy trucking operations for delivery and shipping activities, for example, may be able to be addressed through changes in schedule, offsite queuing, and other nonhardware noise-control measures.

High-Tech Buildings

In high-tech buildings, vibration control is often a key element of the project that needs to be taken into account in the early space-planning phases to avoid undue costs. Mechanical equipment is best located on grade to avoid a transmission path through the structure. Where there are high-velocity airflows, carefully assess noise-control requirements. Noise emissions to the outdoors from fume exhausts and air intakes may be a concern if there are sensitive receivers nearby. Avoid siting the building near a major source of ground-borne vibration, such as a highway or busy commerce road.

Hotels and Function Rooms

High sound isolation is usually required between guestrooms. Avoid having guestrooms close to or above function rooms, as the activities in the function rooms usually generate a significant amount of noise. Guestroom HVAC units are often quite noisy. The noise can be good from the standpoint of providing masking sound to help with sound isolation between rooms, but be careful that the units are not too loud, or the guests will be annoyed. It is best if the fans in the units run all of the time in a low operating mode to ensure that a modest level of background noise is maintained.

Laboratories

Although laboratories are generally not overly acoustically sensitive, basic noise control for is appropriate to avoid excessive noise levels. The more airflow-intensive the laboratory, the greater the noise level is likely to be. Be careful of terminal-box noise, which is often the single greatest source of acoustical concern. Mechanical equipment often needs to be well vibration-isolated to avoid creating excessive vibration for a wide variety of even moderately vibration-sensitive pieces of laboratory equipment. Noise emissions to the outdoors from fume exhausts and air intakes may be a concern if there are sensitive receivers nearby. Locate extremely vibration-sensitive equipment on grade to avoid the need for extra stiffening of above grade structures.

CONCLUSION

As you may have surmised, there is a significant range of potential acoustical issues in almost every type of facility. The goal of this article is to provide a general sense of what some of these issues are and some of the methods used to address them. There are many situations, however, where an expert is required. The National Council of Acoustical Consultants (NCAC), Springfield, New Jersey, and the Institute of Noise Control Engineers (INCE), Saddle River, New Jersey, are two organizations that maintain a list of qualified acoustical consultants and engineers.

CONTRIBUTORS

Robert S. Berens, INCE, Carl J. Rosenberg, A.I.A., and Douglas H. Sturz, INCE Bd. Cert., Acen*tech* Inc., Cambridge, Massachusetts

Robert D Celmer, Ph.D., P.E., University of Hartford, West Hartford, Connecticut

Marshall Long, Ph.D., P.E., Marshall Long Acoustics, Sherman Oaks, California

BIBLIOGRAPHY

ANSI 1.26-1995, American National Standards Institute, New York, 1995.

ANSI 12.2-1995, American National Standards Institute, New York, 1995.

Beranek, L. E., *Noise and Vibration Control*, Institute of Noise Control Engineering, Saddle River, NJ, 1988.

Egan, M. David, *Architectural Acoustics*, McGraw-Hill, New York, 1988.

Bies, D. A., and C. H. Hansen, *Engineering Noise Control Theory and Practice*, 2nd ed., E & FN Spon, London, 1996.

Harris, Cyril M. (ed.) *Handbook of Acoustical Measurements and Noise Control*, McGraw-Hill, New York, 1991.

Harris, Cyril M. (ed.), *Noise Control in Buildings*, McGraw-Hill, New York, 1994.

1991 ASHRAE Handbook, Heating, Ventilating, and Air Conditioning Applications, American Society of Heating, Refrigeration and Air-Conditioning Engineers, Inc., Atlanta, GA, 1991, Chap. 42.

1997 ASHRAE Handbook, Heating, Ventilating, and Air Conditioning Applications, American Society of Heating, Refrigeration and Air-Conditioning Engineers, Inc., Atlanta, GA, 1997, Chap. 7.

Wilson, C. E., *Noise Control: Measurement, Analysis, and Control of Sound and Vibration*, Krieger Publishing Co., Malabar, FL, 1989.

GLOSSARY

A-weighted decibels (dBA)   A frequency-weighted measurement designed to mimic the response of the human ear.

absorption coefficient (α)   The fraction of sound that will not be reflected when striking a surface.

decibel (dB)   See *Level*.

diffraction   Upon striking the edge of a barrier, sound will reradiate from the edge, giving the appearance of bending around the barrier (although it does not actually bend).

frequency   The number of vibrations per second, which is heard as a pitch or tone.

hertz (Hz)   A unit measurement of frequency.

impact isolation class (IIC)   A laboratory-measured single-number rating of impact isolation of a floor-ceiling assembly.

level   Typically, 10 times the logarithm of the ratio of two quantities, expressed in decibels.

noise criteria (NC)   A single-number method for rating a continuous sound spectrum, such as that produced by an HVAC system.

noise criteria balanced (NCB)   A single-number method for rating a sound spectrum that is more detailed than NC ratings, but useful in certain applications.

noise isolation class (NIC)   A field-measured single-number rating of noise reduction of a wall or floor system.

noise reduction (NR)   The difference in sound level between two spaces (in a given frequency band) when a source is activated in one of the spaces. This will include the contributions of flanking paths and room conditions.

noise reduction coefficient (NRC)   A single-number average measure of sound absorption.

octave band   A limited range of frequencies like an octave on a piano. One octave is a doubling in frequency.

room criteria (RC)   A single-number method for rating a sound spectrum that is more detailed than NC ratings and is useful in certain applications.

sound transmission class (STC)   A laboratory-measured single-number rating of transmission loss.

transmission loss (TL)   The amount of sound a partition can stop in absence of any flanking paths and independent of the room conditions in a given frequency band.

SECTION 5.5

INSTRUMENTATION AND CONTROL SYSTEMS

Anand K. Seth, P.E., C.E.M., C.P.E., Director Utilities and Engineering, Section Editor
Partners HealthCare System, Inc., Boston, Massachusetts

David L. Stymiest, P.E., SASHE, C.E.M., Senior Consultant, Section Editor
Smith Seckman Reid, Inc., New Orleans, Louisiana

This portion of Chap. 5 presents detailed treatments of special equipment and systems that are sometimes classified as belonging to either the electrical systems or the mechanical systems in facilities. At other times, however, they are considered as independent or stand-alone equipment and systems. Regardless of how they are considered, they are important enough to be given separate treatment in a facilities engineering handbook. The systems discussed here could be considered as information systems for the facility infrastructure.

We start off with an overview of energy management systems. This overview is then followed by an authoritative presentation on motor controls equipment, a topic that is often given short shrift in facility design but can also have a lasting impact on facility operation. The next three systems presented—security systems, fire alarm systems, and smoke control systems—are sometimes installed as separate systems but are also often installed as a portion of a comprehensive building management and control system. Finally, our treatment of instrumentation and controls systems concludes with an article on telecommunications and data distribution systems.

ARTICLE 5.5.1

ENERGY MANAGEMENT SYSTEMS

Charles Kalasinsky, P.E.

R. G. Vanderweil Engineers, Inc., Boston, Massachusetts

This article outlines in detail the components and functionality that define an energy management system (EMS) or an energy management control system (EMCS). The fundamentals of control systems are defined in the unit of this article on control fundamentals. Four types of control systems are presently employed in HVAC applications: pneumatic, electric, electronic, and microprocessor-based direct digital control (DDC). These are described in the unit of this article on types of control systems. The basic operating elements of a control system are the controlled devices, the controllers, and the sensors, as well as other auxiliary elements. These are detailed in the unit on system devices. The unit on energy management control systems details the explicit architecture, elements, and software characteristics of the EMCS. System evaluation considerations are summarized in the last unit of this article.

CONTROL FUNDAMENTALS

Control system components are combined together to control equipment in the form of control loops. A typical simplified control loop consists of a sensor, controller, and controlled device functioning together to maintain a present condition as shown diagrammatically in Fig. 5.5.1-1.

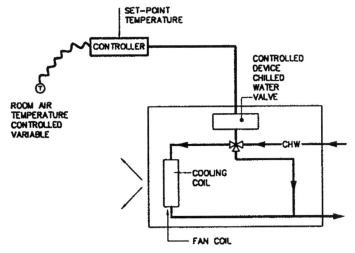

FIGURE 5.5.1-1   Simple control loop. (*Courtesy of R. G. Vanderweil Engineers, Inc.*)

The sensor measures a defined controlled variable and sends information about its condition to a controller. The controller compares this information to a previously defined set point and computes an error signal (the difference between the predefined set point and the controller-inputted signal); the controller sends this error signal modified with gain constants to a controlled device. Gain is simply amplification of the signal that governs the change in the controller output per unit change in the sensor input. The controlled device receives this signal and causes a change in the controlled variable.

This summary describes a *closed control loop* with feedback from the sensor to the controlled device (in this case, a proportional control loop). Closed-loop control systems respond to internally measured disturbances in system variables and provide corrective action continuously. Another type of control loop is an *open control loop* that responds to external disturbances in system variables with corrective action via a fixed relationship. An outdoor air reset control is an example of an open control loop. Here, an outdoor thermostat could control the operation of perimeter radiation solely based upon anticipated requirements, regardless of the need for internal heating or temperature setting. Control systems that control HVAC equipment can operate in a multitude of operating modes; however, the two most common are *two-position* (on/off) and *modulating*. These modes refer to the position that the controlled device takes upon receipt of an error signal. To avoid rapid cycling (on/off, open/closed) and control instability, a differential is used. Thermal lags in the physical response of the IIVAC system dictate the magnitude selection of the control differential. In a two-position control mode, the differential is the difference between the value that causes the control to turn on and the value that causes the control to turn off. This differential is defined as the controlled offset. Pure two-position control is crude at best when used for steady-state operation. Its functionality can be improved greatly by adding multistage control. This variation segments the operating range into a multitude of two-position stages, effectively creating a step-like process control. The process operation becomes smoother with more stages. Modulating controllers operate in one of four control modes: proportional, proportional-integral (PI), proportional-integral-derivative (PID), or adaptive (see Fig. 5.5.1-2).

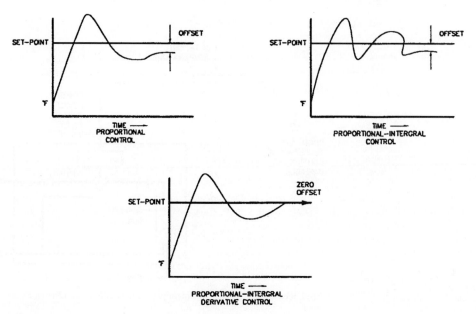

FIGURE 5.5.1-2 Control modes. (*Courtesy of R. G. Vanderweil Engineers, Inc.*)

Proportional control generates a continuous linear relationship between the steady-state displacement of the controlled variable and the corrective action of the controller to maintain steady-state conditions. An offset (the difference between the set point and the actual measured condition) always exists with proportional control.

Under ideal average conditions, this offset will be minimized; however, at lighter loads the offset can cause instability in the system control. To eliminate proportional offset, resetting the control point in response to changing load conditions is required. Proportional-integral control continuously integrates a measured error signal with respect to time. Adding an integral term to the process equation that introduces time as a dependent variable reduces the average-cycle offset that is created by proportional control. The process response time either speeds up or slows down as a function of the error signal. Control becomes more gradual. A further refinement adds derivative control to create PID control. This mode adds a time derivative to the process along with a derivative constant to fine-tune the process control. The derivative function opposes any change and is proportional to the rate of change. Direct digital control (DDC) utilizes PID algorithms as its software basis. Adaptive control is available in some DDC controllers. Adaptive control algorithms enable a controller to adjust its response under all load conditions. A control loop that has been tuned with certain gains under one set of conditions cannot always respond well when the conditions change. An adaptive control algorithm monitors the performance of a system and attempts to improve its performance by adjusting control gains or parameters.

TYPES OF CONTROL SYSTEMS

The four basic types of control system employed in HVAC systems are *pneumatic, electric, electronic,* and *microprocessor-based DDC.* Hybrid systems are combinations of these four.

Pneumatic systems generally use compressed air at 15 to 30 psig for large valves. Higher pressures such as 60 to 80 psig are used to operate valves, relays, and other pneumatic control equipment. System components consist of:

- Compressor(s) to supply clean, dry air
- Air lines to distribute air to control devices
- Sensing devices to measure changes in the controlled variable
- Controllers such as thermostats, humidistats, and pressure controllers
- Controlled devices such as actuators
- Positioned devices such as valves or dampers

Pneumatic systems use varying analog air pressure signals to transmit sensor inputs and controller outputs. Typically, pneumatic systems utilize two-position, multistage, and proportional control. The principal installation advantage of this system type is lower first cost. However, accuracy is sacrificed and higher maintenance costs will apply.

As previously noted, proportional control inherently results in offsets from control set points. Depending upon measured conditions and system equipment status, controlled device "cycling/hunting" is frequently prevalent. System components require frequent recalibration to remain effective.

Electric systems utilize either low or line voltage to operate the system devices. System components consist of sensing devices, controllers such as electric thermostats, and controlled devices such as motors, relays, contactors, and electromechanical and electromagnetic regulating devices. Modulation typically occurs by using floating control or proportional potentiometer control.

Electronic systems use low-voltage electrical energy. System components consist of sensing devices, solid-state controllers, and controlled devices such as actuators. Electronic controller circuits use solid-state components such as transistors, diodes, and integrated circuits. Typically, electronic systems use two-position, proportional, or PI control.

Microprocessor-based DDC systems are similar to electronic systems in the sensing devices and controlled devices used. The difference is in the processing of input signals. In an electronic control system, the analog signal is amplified and compared to a set point through solid-state control circuits. In a DDC system, the input signal is converted to a digital value, where software algorithms provide the control function. The direct digital controller puts out electric, electronic, or pneumatic signals, and the controlled devices are electric, electronic, or pneumatic. System control accuracy is enhanced with PID and adaptive control modes. Although the first costs of DDC systems are significantly more than those of typical pneumatic control systems, annual building operating costs are lower. Response times to HVAC control failures are accelerated due to the centralizing monitoring function of the system. Identification of HVAC system performance issues is provided and can be analyzed. System expansion and changes in sequences are readily accommodated. Unique control strategies can be digitally implemented with complex algorithms without requiring additional hardware.

SYSTEM DEVICES

Controlled Devices

Controlled devices are usually *valves, dampers,* or *motors.* As previously noted, these devices are controlled in two-position or modulating modes.

Valves can be classified as *single-seated* valves (tight shutoff under normal pressures), *double-seated* valves (high-pressure shutoff, not tight), *three-way mixing* valve, *three-way diverting* valve (proportional flow to two outlets), and *butterfly* (mixing/diverting). Flow characteristics impact performance across a valve's full operating range. Two-way valves are characterized as quick opening (two-position or a large change in flow as the valve opens), linear (percentage valve stem travel = percentage flow), or equal percentage (similar changes in stem travel at any point change the existing flow by an equal percentage). Valve operators are typically solenoids (magnetic coil), electric motors, pneumatic operators (diaphragm or bellows), or electric-hydraulic actuator (incompressible fluid used). The valve size is determined by a valve capacity constant C_v

$$C_v = Q/H^{.5}$$

where

Q = flow rate

H = head loss

Valves are viable control devices at 30 to 50 percent pressure drop of the subsystem.

Dampers can be classified as *modulating dampers* and *two-position dampers.* Two damper arrangements are used: parallel blade (all blades rotate in the same direction) and opposed blade (contiguous blades rotate in opposite directions). Modulation accuracy differs significantly between the two. Damper operators are typically electric actuators (reversible, spring return, or unidirectional) or pneumatic operators.

Controllers

Controllers are classified on the basis of the control action and the type of energy used to send an error signal to a control device. Pneumatic controllers are typically combined with sensing elements and are classified as *nonrelay* type (small volumes of air for long response times) or *relay* type (amplification for quick response times). Electronic/electric controllers are typically *two-position, floating control* (control is activated outside of high and low limits), and *proportional control.* These controllers can be *indicating, nonindicating,* or *recording.* Direct digital controllers employ a digital computer and control software logic. Control algorithms are typically preprogrammed in read-only memory (PROM) or user-programmable controllers.

Sensors

A sensing element is a device that measures the value of the controlled variable. Some sensors, such as thermistors, are nonlinear. Typical sensors include temperature sensors (bimetal type-dissimilar metals; bulb and capillary; vapor-filled, sealed bellows; solid-state thermistor; resistance temperature-detector-RTD; and thermocouple-electromotive force), humidity sensors (hygroscopic, thin-film absorbent sensors), dew point sensors, pressure sensors (piezometer-crystalline, bellows-corrugated cylinder), flow sensors (sail switch, hot wire anemometer, pitot tube total pressure and static pressure, orifice plate, and turbine flow meter), and current sensors.

Other system devices for electric/electronic control systems include:

- Transformers that provide current at the required voltage
- Electric relays when the electric load is too large to be handled by a controller
- Potentiometers that allow remote set point adjustment of electronic controllers
- Switches for manual operation

Other system devices for pneumatic control systems include:

- Air compressors and accessories
- Electric pneumatic relays (EP switches) that control pneumatic equipment with electric inputs
- Pneumatic-electric relays (PE switches) that are pressure actuated to make/break an electrical circuit
- Electric pneumatic transducers that convert electronic signals to pneumatic outputs to control pneumatic actuators
- Pneumatic relays (two-position and proportional relays)
- Pneumatic manual switches
- Gradual switches

ENERGY MANAGEMENT CONTROL SYSTEMS (EMCS)

An EMCS typically consists of a host computer (central console), direct digital controllers, communication links to the controllers, field interface devices, sensor devices, and controlled devices. System control within an EMCS is distributed by implementing direct digital controllers. Therein, remote field digital controllers each programmed to its unique required strategy independently perform complete control functions. This architecture is outlined in Fig. 5.5.1-3. The system comprised of *digital controllers* (DC) or *direct digital control field panels* (DDCFP) can share information in the network with other devices without a front-end device such as a central computer, gateway, or network managing device. Each DC monitors and controls its connected mechanical equipment through its own internal microprocessor. DCs perform automatic self-test diagnostics and report malfunctions automatically to central I/O devices.

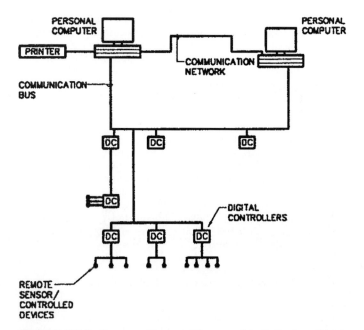

FIGURE 5.5.1-3   System architecture. (*Courtesy of R. G. Vanderweil Engineers, Inc.*)

Two types of communication protocols are used in EMCS for data transfer: *poll/response* and *peer to peer.* Poll/response protocols use a central unit or DC master that is responsible for polling all controllers for information such as alarms, status, and device data. The process can be extremely slow due to the resultant response time inherent in a large system. Terminal controllers will often employ poll/response protocols because communication densities are low. With peer to peer (token passing) protocol, a time slot is passed automatically from one DC device to another. Each DC device sets as a master when it is in possession of the time slot. Not only is a front-end device unnecessary but also communication does not depend upon any one device. Communication transfer is quicker and more effective.

DDC controllers are electronic and have no moving parts. Minimal maintenance and high reliability is obtained. Two types of controllers are prevalent in the industry today: *zone controllers* and *system controllers.* The former provide many less external points than system controllers and are typically used to control HVAC terminal units such as VAV terminal units and fan coil units. DDC controllers may have fixed programs with parameter changes available only to the user or may be fully programmable.

System controllers contain a higher point and memory capacity than zone controllers and typically control major HVAC equipment such as chillers, boilers, and air handlers. System controllers are even configured to control multiple zone controllers.

The sensors and controlled devices that are connected to controllers are called *points.* These points can be external to the control device or internal, built-in to provide specific functions. Four types of points can be classified:

- Digital input (DI) that sends a two-state (on/off) or two-position signal to a controller
- Digital output (DO) that controls a two-position (on/off) output
- Analog input (AI) that sends a variable signal to a controller; some signals are nonlinear

- Analog output (AO) that controls a variable output; typical ranges are 0 to 10 Vdc or 4 to 20 mA

  Points are summarized in point lists that contain:

- A list of components to be sensed or controlled
- A list of hardware associated with each point, such as solenoids, contactors, and transducers; transducers change a modulating pneumatic signal into a modulating electric signal, or vice versa
- A list of analog inputs to be sensed at each point such as temperature and relative humidity
- Alarms associated with each point
- Control functions associated with each point, such as start/stop and set point control

As shown in Fig. 5.5.1-3, the communication bus ties the DDC system together. The media are typically twisted shielded copper (16 to 24 gauge), fiber-optic cable, or direct-dial telephone, with various Ethernet media available at the highest communications level.

Fiber-optic cable is the most costly transmission medium but is also the most reliable. The permanent operator interface is maintained at the personal computer (central console). Herein are contained the functions of the keyboard, printer, display, computer, and software. The latter is also resident at DCs as previously noted. EMCS use software to perform two basic functions: control the operation of equipment and minimize the use of energy (electrical/thermal) through intricate sequences (digital algorithms). Energy reductions are realized:

- When equipment is operated only as needed; fans and pumps are turned off when they are not needed, for example, during unoccupied periods; and lighting is used as needed
- When simultaneous heating and cooling are minimized; centralized heating coils and cooling coils should not be energized together; supply air temperatures should be reset with loads
- When energy is recovered

Energy Reduction Algorithms

Some of the following software algorithms are used to effect these energy reductions.

Optimum Start-Stop.   Software monitors building and outdoor temperatures before occupied hours begin to determine when building mechanical equipment needs to be started. Projected histories based upon day type and weather conditions are built over time to determine start times. The reverse condition applies to stop times. Two examples follow:

1. *Optimal start and warm-up control.*   The EMCS will determine an optimal start time (adjustable). The optimal start time is calculated on the basis of the outdoor air temperature, the space temperature, the warm-up capacity of the air handling unit (AHU) for the space, and the time of day. If the space temperature at this time is below 68°F (adjustable) and the outside air temperature is less than 55°F (adjustable), the supply fan will start and the electric heating coil will stage to provide 90°F discharge air (adjustable). The AHU will stay in the warm-up mode until the space temperature rises above 68°F (even if occupancy time arrives and the space has not reached 68°F). At this time, control will shift to the occupied mode of operation for temperature control. Before occupancy, there is no requirement for minimum outside air from the AHU.
2. *Optimal start and cool-down control.*   The EMCS will determine an optimal start time (adjustable). If the space temperature at this time is above 78°F (adjustable), the supply fan will start, and the mixed air dampers and direct expansion cooling will modulate in sequence to provide 78°F (even if occupancy time arrives and the space has not fallen to

78°F). At this time, control will shift to the occupied mode of operation for temperature control. Before occupancy, there is no requirement for minimum outside air from the AHU (see Fig. 5.5.1-4).

Unoccupied Night Purge. Building spaces can be subcooled at night with cooler evening outdoor air by initiating 100 percent ventilation cycles. This process reduces the amount of work that mechanical cooling systems need to perform during occupied periods.

Enthalpy Control. Typical outdoor air economizer control cycles provide varying amounts of outside air greater than the established minimum ventilation setting to provide free building cooling. This control modulates outdoor air and return air dampers off the system enthalpy. Controlling the ventilation amount off the respective air stream enthalpy greatly increases the number of hours of free cooling that is available, compared to dry-bulb temperature economizers.

Cooling Coil/Heating Coil Reset. HVAC systems are designed and sized to maintain building comfort under peak load conditions. These peak conditions occur only during a minimal number of hours per year. Supply air temperatures can be reset either to satisfy conditions in the most highly loaded zone (discriminator control) or to modulate on the basis of outdoor weather conditions (linear).

Unoccupied Period Setback. A night cycle is initiated during periods of low or no occupancy. During this period, set points are changed and ventilation strategies are modified. Fans may even cycle to save energy.

Load Shedding. Because electricity is priced on a usage ($/kWh) and demand basis ($/kW), limiting the demand during the occupied period saves operating costs. Demand charges are often based on a *ratchet clause.* Under this costing structure, the peak demand that occurs during a given month establishes the demand charge for the succeeding 12-month period. Electrical consumption is monitored during each and every demand interval (typically 15 minutes). Equipment is turned off as appropriate and required to reduce demand. Equipment is prioritized per the building utility, and low-priority equipment is shed first and restored last. High-priority equipment is controlled in reverse. Multispeed and variable speed motors are shed proportionally, and set points are raised or lowered to reduce electrical demand.

Duty Cycling. Fans are turned on/off when the temperatures in the building areas they serve are above the midpoint of their preprogrammed comfort range. Unnecessary electrical demand peaks caused by synchronized equipment "on" times are avoided. This control strategy is most appropriate for HVAC equipment that is oversized, and electrical energy is saved through this process. Duty cycling accelerates wear and increases maintenance costs.

Lighting Control. Three types of lighting control are appropriate:

- Occupied/unoccupied control based on the time of day
- Multilevel lighting control based on the daylight level
- Modulated lighting control based on the daylight level

Cooling Plant Chilled Water Reset. The chilled water supply temperature is raised whenever the facility's cooling load decreases. The control tracks either the outdoor wet-bulb temperature or the chilled water return temperature, so that the chillers can be operated at higher suction pressure for low loads. Compressor power is decreased, effectively increasing chiller efficiency.

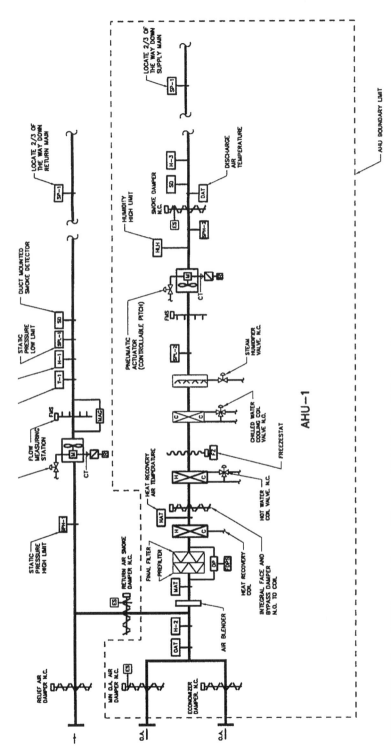

FIGURE 5.5.1-4 Typical AHU control diagram. (*Courtesy of R. G. Vanderweil Engineers, Inc.*)

Cooling Plant Condenser Water Reset.   The condenser water supply temperature is lowered whenever the cooling tower capacity and compressor-head temperature allow it. The controls track the condenser water return temperature, so that the chillers operate at lower head pressures but increased cooling tower capacities. Chiller compressor power is reduced at the expense of operating the cooling tower fans at full speed without cycling until the lower condenser water temperature is reached. The compressor energy savings outweigh the additional energy consumption of the cooling tower fan.

Cooling Plant Chiller Sequence.   The facility's chillers are sequenced on line and off line to maximize total system efficiency. Chilled-water pumps and condenser water pumps are also interlocked to operate as needed.

Performance Algorithms

Other software algorithms address the optimum performance of specific equipment, such as:

Variable air volume boxes

- Parallel fan boxes
- Series fan boxes

Boiler plants

- Burner optimization
- Oxygen trim control
- Blowdown recovery
- Hot water reset

SYSTEM EVALUATION

Implementations of all energy management systems should include economic and functional evaluations. Included should be an estimate of hardware point costs and estimated energy cost savings. This evaluation can prioritize point inclusion based upon the economic payback of points. The installed cost of a control point can range from $300 per point to $3000 per point, depending upon the functionality of the point and the installation requirements.

Given the wide diversity of functionality among various system suppliers, detailed efforts should be made to identify the following required system characteristics:

- The system architecture, including file servers, workstations, DDC field panel types, and network communications levels
- The input/output point capacity and types for each DC (point/unit)
- The network speed at each communication level (baud rate)
- Advanced PID loop tuning
- The ability to view the status of all points from a CRT or laptop connected to any DDC field panel
- The protocol for each communication level: poll/response, peer-to-peer, other
- The open protocols supported and the implementation method
- Trend data storage capacity in workstations and each DC type (memory limited)
- An optional built-in display and/or keyboard for each DC type
- The memory capacity for each DC type

- The point history supported for each DC type and the system as a whole (point quantity and maximum samples)
- Run concurrent third-party software
- Transfer data to third-party software such as a spreadsheet or database
- Full PID control at each DC type
- Preprogrammed applications
- DC types that have fixed programs or are fully programmable by a user
- The number of application-specific controllers supported
- Password level control
- The costs of adding points after the system is installed
- Backward compatibility to prior products
- Access to future software revisions
- Stocking of repair parts and skill required

CONCLUSION

An EMS or EMCS is used regularly in most modern buildings. These systems provide excellent control, and when combined with other maintenance management systems described in Chap. 7, they have enormous capability for efficient physical plant operation.

CONTRIBUTOR

Robert Clark, Vice President, BCM Control Systems, Woburn, Massachusetts

BIBLIOGRAPHY

ASHRAE Handbook of Fundamentals, "Controls," Chap. 37, American Society of Heating, Refrigeration, and Air Conditioning Engineers, Atlanta, GA, 1997.

ASHRAE Practical Guide to Building Control (Supplement to *ASHRAE Journal*), American Society of Heating, Refrigeration, and Air Conditioning Engineers, Atlanta, GA, September 1997.

N. E. "Bill" Battilkha, *The Management of Control Systems,* Instrument Society of America, Research Triangle Park, NC, 1992.

Control Systems for Heating, Ventilating, and Air Conditioning, Van Nostrand Reinhold, New York, 1987.

Engineering Manual of Automatic Control, Honeywell, Minneapolis, MN, 1988.

Paul G. Friedmann, *Economics of Control Improvement,* Instrument Society of America, Research Triangle Park, NC, 1995.

Handbook of Automatic Refrigerant Controls, ALCO Valve Co., St. Louis, MO, 1954.

Béla G. Lipták, *Optimization of Industrial Unit Processes,* Second Ed., CRC Press LLC, New York, 1999.

Mulley, Raymond, *Control System Documentation, Applying Symbols and Identification,* Instrument Society of America, Research Triangle Park, NC, 1994.

Bill Swan, *The Language of BACnet,* Engineered Systems, July 1996.

ARTICLE 5.5.2
MOTOR CONTROLS

Syed M. Peeran, Ph.D., President
Peeran Engineers, Inc., Burlington, Massachusetts

Joseph R. Flocco, P.E., Sen. Mem. IEEE
Siemens Energy and Automation Inc., Atlanta, Georgia

ELECTRIC MOTORS IN COMMERCIAL AND INDUSTRIAL FACILITIES

The electric motor has been aptly called the workhorse of the industry. Since it was invented approximately 110 years ago, it has found numerous uses as a drive for a variety of household and industrial machinery because no other form of motive power matches its versatility, reliability, and economy. The electric motor can give years of trouble-free service when it is specified correctly and controlled and operated within its ratings. In this article, we discuss starting, speed control, and protection of the electric motors most commonly used in commercial and industrial facilities.

Motors are classified on the basis of the source of input electric power such as dc, ac single-phase, and ac three-phase, and based on types such as dc shunt-wound, dc series-wound, dc compound-wound, ac squirrel-cage induction, ac slip-ring induction, and ac synchronous motors.

DC Motors

Historically, the dc motor preceded the ac motor. It is generally more expensive than the ac motor of the same rating. It also requires more maintenance because of the commutator and the brushes. However, it has several distinct advantages over the ac motor, which is the reason that it is still widely used in industry. Continuous operation over a wider speed range is possible with the dc motor. The ac motor tends to stall, but the dc motor can produce a high starting torque, typically five to six times its full-load torque, which makes it ideal for starting loads of high inertia such as electric trains and elevators. Reversal of the direction of rotation is possible without power switching. The torque-speed characteristic of the motor can be easily altered to match that of the load.

AC Motors

The ac motor has almost completely replaced the dc motor in most general applications because of its low cost and ruggedness. Even for variable speed operation, which was so far considered as a forte of the dc motor, the ac motor has now become a preferred drive because of the advent of and great advancement in electronic variable-frequency controllers.

Single-phase ac motors operate from the commonly available 120V/240V 60Hz supply. Ceiling and portable fans, window air conditioners, electric washers and dryers, blowers, and small pumps use single-phase motors.

Three-phase ac motors are widely used for drives ranging from 1 hp to several thousand hp. The two types of three-phase ac motors commonly used are the *induction motor* and the *synchronous motor*.

The induction motor is probably the most common type of electric motor in industry. More than 70 percent of all motors in industry are induction motors. Three-phase induction motors are usually rated at voltages of 208, 480, 565, 4,160, and 13,800 V. Most blowers in air-handling units use 10 to 50 hp motors. Chiller motors are rated at 2000 to 5000 hp. Boiler induced-draft and forced-draft fans and boiler feed pumps are rated from 300 to 5000 hp.

The synchronous motor differs from the induction motor in the construction of the rotor. Instead of a squirrel-cage of copper or aluminum bars or a three-phase winding, the rotor of a synchronous motor has poles that carry windings. A direct current is passed through the pole windings. This current is obtained either from an ac/dc converter (a rectifier) or an independent dc source.

There are two important differences between the operation of induction motors and synchronous motors:

1. The speed of the induction motor decreases slightly with load, typically 1 to 10 percent depending upon the motor design and type. The speed of the synchronous motor is, however, constant at all loads, determined only by the supply frequency.

2. The power factor of the input current is typically 80 to 90% lagging for the induction motor. The power factor of the input current to the synchronous motor can be adjusted to any value, typically between 80 percent lagging and 80 percent leading, by adjusting the dc field current.

In small and medium sizes the synchronous motor is generally more expensive than the induction motor. The synchronous motor is a preferred drive in ratings of thousands of hp and low speeds such as 300, 450, 600 and rpm.

DC Motor Starting

The current drawn by a dc motor, when it starts, is limited only by the low resistance of the armature winding. Consequently, a large current would flow if not limited by the starter. The starter limits the starting current to a value low enough to prevent damage to the armature but high enough to develop the required starting torque. As the motor starts and attains the desired speed, the armature winding develops a voltage (known as the back emf) because it rotates in a magnetic field produced by the stator poles. The direction of this back emf is opposite to that of the source voltage. Thus, the motor current reduces it to a value required to drive the load.

DC Motor Speed Control

Conventional methods of speed control are series-parallel controls and Ward-Leonard (motor-generator set) controls. Modern static methods of speed control are the *thyristor phase control* and the *transistor chopper control.* Series-parallel controllers are used for the dc series motors in subway cars. Ward-Leonard controls (motor-generator control) are used for elevator motors in facilities, large dc motors in the steel industry, and for dc motors driving various shipboard machinery such as winches and pumps.

DC Thyristor Drives.   A dc *thyristor drive* is a static electronic controller that provides starting, speed control, and braking of a dc motor. It gets its name from the semiconductor device known as the *thyristor* (also known as the *silicon-controlled rectifier* or the *SCR*). The controller has many thyristors connected to form an ac/dc converter. When supplied with an ac voltage, the thyristor conducts only during the positive half-cycle of the voltage, thus passing current in only one direction. This is the basis of ac/dc conversion. DC thyristor drives are either single-phase (for fractional hp motors) or three-phase.

Reversal of the motor direction is required in many applications such as elevators and cranes. Because the single-phase and three-phase drive thyristor permits current in the motor in only one direction, the direction of the motor rotation cannot be reversed unless the voltage applied to the field winding is reversed. The field winding is a highly inductive circuit. Reversing the direction of current in the field winding would produce considerable arcing at the reversing switch contacts and would wear out the contacts when frequent reversing is required. This is avoided by providing dual converters that reverse the direction of the motor armature voltage. The dual converter is the static equivalent of the Ward-Leonard motor-generator control.

Induction Motor Starting

Starting Current and Voltage Dip.   When the induction motor starts, it draws a large current, typically 6 to 10 times the full load current. The magnitude of the starting current is independent of the load, but the duration depends upon the motor and the inertia of the load.

The motor is designed to withstand the large starting current without damage. However, if the rotor is blocked, for example, by a clogged pump, the starting current flows continuously, undiminished in magnitude. Then, the motor would overheat and fail if it is not tripped out. Therefore, the protective device in the motor controller should permit the starting current to flow, so as to let the motor start, but should detect the blocked-rotor condition and trip out the motor before it is damaged. (The starting current is frequently referred to as the *blocked-rotor current* or the *locked-rotor current.*)

Another problem in starting motors is the system voltage dip caused by the starting current. The starting current (or the locked-rotor current) has a low power factor, typically 20 percent lagging and, in addition, is several times the full load current. This causes a significant dip in the system voltage. Every time a large motor starts, a brief flicker in the lights results from this voltage dip. If the voltage dip is less than 20 percent, then service is not normally disrupted. If the dip is more than 20 percent, some of the other motor controllers in the system may trip out due to undervoltage. The voltage dip is the primary reason for using reduced-voltage starting for large motors.

The starting current of an induction motor depends upon the design of the motor. A code letter that specifies the starting kVA or the locked-rotor kVA range designates each motor. Code letters range from A to V. The corresponding locked-rotor kVA ranges from 3.14 for code letter A to 22.4 or more for code letter V. See Table 430-7(b) of the National Electrical Code for the locked-rotor kVA range of each code letter. The code letter is generally stamped on the nameplate of the motor together with the other ratings. Most industrial motors have a code letter G or H.

Selection of a Starter.   A starter's primary requirements are that the motor starting torque is adequate to start the motor under the worst-case line voltage (minimum line voltage) and load condition (maximum shaft load) and that the line current does not exceed the limits set by the utility or the plant voltage dip. Secondary requirements in starter selection include smoothness of acceleration, maintenance, power factor, reliability, and efficiency. A comparison of the performance of a typical 60-hp, three-phase, 60Hz, 460V, code letter H motor with various starters is shown in Table 5.5.2-1.

Full-Voltage Starter.   This type of starter is also known as the *direct-on-line* or the *across-the-line* starter. This is the simplest type of starter, the least expensive one, and easiest to install and maintain. It is most commonly used for motors up to about 150 hp. The starter connects the motor directly to the line through a contactor. Each contactor is designated by a contactor size that corresponds to its continuous current rating. Table 5.5.2-2 shows typical ratings of ac contractors used in full voltage starters.

TABLE 5.5.2-1 Performance of a Typical 60-hp Three-Phase 60-Hz 900-rpm Motor
with Different Types of Starters

Type of starter	Starting current % of full-load current	Starting torque % of full-load torque
Full voltage starter	470	160
Autotransformer, 80% tap	335	105
Part-winding starter	235	70
Autotransformer, 65% tap	225	67
Solid-state reduced voltage	300	65
Wye-delta	158	54
Autotransformer, 50%tap	140	43

Autotransformer Starter. Most large low-voltage and medium-voltage motors require reduced-voltage starting to limit the starting current and the voltage dip. The autotransformer starter applies a low voltage (typically 50, 65, or 80 percent) to the motor when it is starting. Then, full voltage is applied when the motor reaches full speed.

Part-Winding Starter. Dual-winding, dual-voltage motors use this type of starter. The motor winding is in two parts, and at least six terminal leads are provided. The starter connects one section of the winding to the supply through one contactor when the start button is depressed. After a preset time delay provided by a timing relay, a second contactor connects the second section of the winding to the supply, in parallel with the first section. The starting current is approximately one-half what would be required when both sections of the winding are connected at the same time.

Wye-Delta Starter. The wye-delta starter is a reduced-voltage starter that applies 57 percent of the voltage to the motor when it is starting and 100 percent of the voltage after a preset time delay. This starter is used for three-phase motors in which all six terminals of the windings are brought out. When the motor is starting, the three windings are connected in wye, thus impressing only 57 percent of the voltage on each winding. After a preset time delay or by a manual action, the windings are connected in delta, thus impressing 100 percent of the voltage on each winding.

TABLE 5.5.2-2 Typical Ratings of AC Contactors Used as Full-Voltage Magnetic Starters
with Three-Phase Motors

Contactor size	Rating, amperes	Horsepower at 110V	Horsepower at 220V	Horsepower at 440–550V
00	9	¾	1½	2
0	18	2	3	5
1	27	3	7½	10
2	45	—	15	25
3	90	—	30	50
4	135	—	50	100
5	270	—	100	200
6	540	—	200	400
7	810	—	300	600
8	1215	—	450	900
9	2250	—	800	1600

Solid-State Reduced-Voltage Starter.   This type of starter uses back-to-back connected thyristors in two or three lines to the motor. The thyristors are phase-controlled during the starting period to maintain a current of about 300 percent of the full load current by gradually increasing the voltage. The starting current and torque, as well as the acceleration of the motor, can be adjusted to any desired value. The solid-state starter is generally more expensive than other types of starter. However, there are no moving parts or contactors to wear out.

Synchronous Motor Starting

Unlike the ac induction motor, the synchronous motor is not self-starting. However, all synchronous motors are provided with damper windings in the rotor pole faces. Because of these damper windings, the synchronous motor starts like an induction motor when the stator is energized. When the speed of the motor is near the rated speed, the field supply is switched on. The motor then "pulls up" to the rated speed. Full-voltage as well as reduced-voltage starting can be provided just as for the induction motor.

AC Motor Speed Control

Voltage Control.   AC induction and synchronous motors are inherently constant-speed motors because they are normally operated at constant voltage and frequency. However, a limited variable speed operation can result from varying the applied voltage by inserting a resistor or a reactor in series with the motor or by using a static voltage controller of the same type as the solid-state reduced-voltage starter.

Typical torque-speed characteristics of an induction motor are shown in Fig. 5.5.2-1 for 100, 50, and 30 percent voltage. Also shown in Fig. 5.5.2-1 is the speed torque characteristic of the load (a pump in this case). The speed at which the motor operates is S1 for 100 percent voltage, S2 and S3 for 50 and 30 percent voltage. Therefore a speed variation of S3 to S1 is possible. This type of speed control is used in small fractional-horsepower motors such as domestic fans and blowers. As the voltage is reduced, the efficiency of the motor is also reduced because the motor operates at higher slip and consequently there are greater power losses in the rotor bars. Frequency control is a better method of speed control for larger three-phase motors.

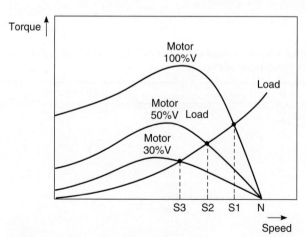

FIGURE 5.5.2-1   Results of voltage control of induction motor. (*Courtesy of Stone & Webster Engineering Corp.*)

A synchronous motor does not respond to voltage control because its speed is strictly proportional to the frequency of the current.

Frequency Control.   Frequency control is used for single-phase and three-phase induction and synchronous motors. Typical torque-speed characteristics of an induction motor for 100, 50, and 30 percent frequency are shown in Fig. 5.5.2-2. A wide range of speed control is possible, theoretically 0 to 100 percent, or even higher if permitted by the bearings. Frequency control is a highly efficient method of speed control and is used for small, medium, and large motors. For this reason variable-frequency drives (VFDs) are widely used in industry for all types of applications, and the VFD technology is well developed. Modern VFDs are sophisticated controllers that incorporate microprocessor control, advanced digital and analog metering, and local/remote manual and automatic control and monitoring. Several types of VFDs are used. Some of the most common types are described in the following paragraphs. Because of the increasing use of the VFDs, some problems related to power quality and waveform distortion have surfaced. These are also discussed briefly in the following paragraphs.

Types of Variable-Frequency Drives.   Three basic types of VFDs have evolved:

1. Pulse-width-modulated (PWM) drive

2. Current-source-inverter (CSI) drive

3. Load-cumulated inverter (LCI) drive

The PWM and the CSI drives are used for induction motors, and the LCI drive is used for synchronous motors. These drives differ in their component configuration and in the method of control and operation. Refer to Art. 5.2.9, "Power Quality," for a discussion of the line harmonics caused by VFDs and their relation to facility power quality.

PWM Drives.   The rectifier in the PWM drive is a simple bridge rectifier that uses silicon diodes. The dc link consists of an inductor and a large capacitor so as to obtain an almost ripple-free dc voltage at the input to the inverter. The inverter uses power transistors or IGBTs (integrated base bipolar transistor). The inverter controls both the frequency and the voltage to maintain a constant volts/Hertz ratio over the operating speed range.

The PWM drive provides soft motor starting. Because of the nearly sinusoidal waveshape of the output current, the additional heating of the motor due to harmonic currents is only

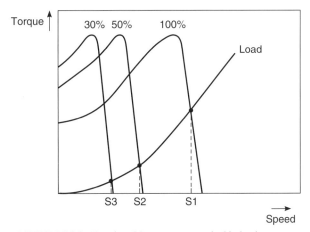

FIGURE 5.5.2-2   Results of frequency control of induction motors. (*Courtesy of Stone & Webster Engineering Corp.*)

minimal. Although the currents in the input are generally nonsinusoidal, the power factor is high in the range of 90 to 95 percent.

Because of the diode rectifier at the input, the PWM drive is incapable of regenerative operation. Therefore, no regenerative braking can be provided. In addition, the high-frequency pulses in the output of the drive cause high-voltage spikes at the motor. The problem due to the spikes is more severe when there are long leads between the drive and the motor. The usual solution is to use either an output filter called the *dv/dt filter* or simply an output reactor.

PWM drives are most commonly used for induction motors rated from 1 to 1000 hp. These drives are less expensive than the other types of drives and can be used to retrofit existing motors.

CSI Drives. The CSI drive is a rugged, simple drive for large motors. The inverter controls the motor speed. Because the switching of the devices in the inverter is at the fundamental frequency or at a lower frequency, rectifier-grade thyristors can be used. They are less expensive than high-frequency, inverter-grade thyristors.

It is necessary to derate the motor when it is driven by a CSI drive. A rule-of-thumb derating is 15 percent. The output voltage is somewhat sinusoidal but contains spikes due to the switching of the inverter thyristors or **G**ate **T**urn **O**ff/**O**n semiconductor device (GTO). However, the spikes are less severe than those from the PWM drives. Therefore, generally no output filtering is required. The distance between the VFD and the motor is not critical.

The CSI drive can work as a regenerative drive. Therefore regenerative braking of the motor is possible. The input current to the drive has a nonsinusoidal waveshape that consists of the fundamental frequency and higher harmonic components. Some harmonic filtering or harmonic mitigation is normally required to prevent excessive distortion of the system voltage. The power factor of the input current is not constant, as it is for PWM drives, but typically varies in the range of 30 to 85 percent. In addition, the thyristors in the rectifier of the drive create *notches* in the waveshape of the input voltage, which can result in power quality problems. An isolation transformer or an ac line reactor is required to reduce the notching.

Because of the large dc link reactor and commutating capacitors, the CSI drive is bulkier and has a larger footprint than an equivalent PWM drive. It is also more expensive because of the thyristors in both the rectifier and the inverter. For this reason and despite its many advantages, the CSI drive is justifiable only in large sizes, typically 1000 hp or larger.

Load-Commutated Inverter Drive. The load-commutated inverter (LCI) drive is used for synchronous motors. The drive has the same basic components as the CSI drive except that the inverter does not require commutating capacitors because the synchronous motor is a source of voltage just as a generator is because of the field excitation. All the advantages of and problems associated with the CSI drive also hold in this case. Consequently, the drive is justifiable only in large sizes.

Selection and Specification of VFDs

The decision whether to use a VFD is based primarily on economic considerations and payback. VFDs are expensive equipment. The first cost varies between $200 and $600 per hp of the rating. The primary reason for using a VFD is the savings in energy. If the motor is expected to run close to its rated speed most of the time, and only occasionally at a reduced speed, then the VFD may not be justifiable. For air-handling units, water pumps, boiler feed pumps, and primary pumps in wastewater treatment plants, there is significant energy saving when a VFD is used. The payback period is normally three to five years, a strong incentive to use VFDs.

Specifying, installing, and testing of VFDs must be done carefully. A preliminary harmonic analysis must be performed to determine the expected voltage and current distortion level. This study would guide the specifier in selecting the type of drive. Unless otherwise indicated by the harmonic study, a PWM-type VFD can usually be specified for most general applications to 1000 hp.

When specifying PWM VFDs, the following must be addressed as a minimum:

1. Installation compliance with IEEE Std. 519
2. Input power factor
3. AC line reactors or isolation transformers
4. AC harmonic filters
5. Output filters or reactors
6. Microprocessor control, instrumentation, and protection
7. Ability to ride through brief voltage sags or to restart automatically
8. EMI and RFI limits
9. Shop testing at full load for a period of about 4 hours in an enclosed space that can maintain the specified ambient temperature.
10. Field-testing, including harmonic measurements

When specifying CSI and LCI drives, the issue of compatibility between the drive and the motor must be addressed in addition to the issues listed before. It is best to specify the drive and the motor as a package so that the supplier guarantees compatibility.

Motor Protection

Motor protective devices are expected to detect an abnormal operating condition and quickly disconnect the motor. Abnormal operating conditions include single-phasing due to one blown fuse, sustained overload, locked rotor due to mechanical jam, excessive unbalance in the line currents, stator overheating, winding fault, undervoltage and undercurrent. The motor protective device is part of the motor starter, speed controller, or the VFD. For small motors, overload protection is provided by thermal overload relays as explained in the unit on motor starting. For larger motors, electronic motor protection relays are now available that provide all of the protective functions. Motor performance curves can be programmed into these relays so that departure from the normal operation can be detected.

Motor Heaters.   Overload protection is provided on a motor starter to prevent damage to a motor's windings when it is required to produce horsepower above its service factor for too long a time. It will also operate quickly when the motor stalls and is drawing locked-rotor current. These *heaters,* as they are called, are designed for a specific current range and can be changed for different motor loads within the NEMA starter horsepower range.

Bimetallic Relays.   Bimetal relays employ a mechanism that is actuated by the movement of two dissimilar metals sandwiched together that expand at different rates when heated by the load current. The bimetal relays allow adjustment of full-load current and therefore fewer sizes are required within a NEMA starter horsepower range, compared to melting alloy types. They can be arranged for manual or automatic reset and are available with ambient compensation to allow accurate motor protection at high ambient temperatures.

Electronic Relays.   Electronic relays are self-powered devices that have current-sensing ability and electronics that offer advanced protection and availability of functions in addition to overloads. They generally provide a wide range of adjustment, and therefore the least amount of sizes covers the NEMA horsepower range. Electronic trip units should be specified that have repeatability over the operating range to ensure that the unit will trip in the same time each time under the same conditions. The relay is unaffected by ambient conditions because it senses motor current. Relays should also have a thermal memory that "remembers" the heat buildup in a motor and trips faster when an overload occurs. An additional feature is single-phase protection or phase loss protection that trips a motor in a few seconds when a phase is lost.

Contributor

David L. Stymiest, P.E., SASHE, C.E.M., Senior Consultant, Smith Seckman Reid, Inc., New Orleans, Louisiana

ARTICLE 5.5.3
SECURITY SYSTEMS

Darryl E. Galletti, P.E., Senior Associate
R. G. Vanderweil Engineers, Inc., Boston, Massachusetts

In today's "high tech" business environment, facility managers are faced with the complex challenge of developing a comprehensive security program to handle the risks of today and the future. The buzzword for today's security professional is *integrated security systems*.

The physical building, contents, and people are the assets that the facility manager must protect. The competitive and litigious nature of business no longer allows the facility manager to wait for an event to occur before taking action. In fact, the facility manager must demonstrate a clear and concise plan for all possible exposures of a facility's assets. Risks and safety must be identified and evaluated on an ongoing basis.

The facility manager must then adopt standards and enact procedures that clearly and unequivocally demonstrate a protection plan that is most appropriate for the facility. Documentation of all decisions (with special attention paid to the way and the reason that these decisions were made) and the exceptions, if any, is of utmost importance. The increased threat of workplace violence, theft, assault, fraud, drug abuse, piracy, and terrorism make this task increasingly more complicated. It is important to note that the integrated security system program must fit into the company's corporate culture.

This article describes concepts and technologies from which the facility manager can choose to install an integrated security system that meets the challenges in today's high-tech business environment. This article also describes how intrusion detection systems (IDS), closed-circuit television (CCTV), and access control systems can be used independently or can be combined in an integrated security system.

CHOOSING AND IMPLEMENTING AN INTEGRATED SECURITY SYSTEM

Installing a security system is similar to installing other complex building systems. The following seven steps will help a facility manager to choose and implement an integrated security system:

1. *Analysis of requirements.*   The first step in implementing an exhaustive well-structured risk analysis that thoroughly examines all the possible exposures to a facility's assets is the most critical phase in the process. Evaluating a facility's layout and surrounding environment enables the security professional to determine potential trouble spots, architecture and the quality of construction of the facility, personnel and vehicle entry and exit points, communications systems, type of parking, and lighting.

The goal of an integrated security system is to unite people, electronics, and procedures so that the result is a reliable security system that operates under normal and emergency conditions and supports organizational goals.

2. *System design and approval.*   Preliminary design criteria must be established, plans prepared, and presented. A review with the facility manager, owner, and/or occupants must be conducted that includes an explanation of the rationale, concept, procedure, and discussion of the harm that could result if the security system is not implemented. The final design, cost estimate, and impact studies are presented to the management and owner for their approval.

3. *Plans and specifications.*   This step is a key to the project's success. Developing site drawings and preparing a detailed outline of the system components are critical. These documents will specifically detail types of equipment and wire, location of equipment and wiring, installation methods, standards and codes, and operational protocol. Important parameters for selecting a contractor such as insurance, bonding, service capability, project management capability, financial stability, and experience are also to be included in these documents. Professionals at an architectural or engineering firm, security consultants, or management and owners can develop these criteria.

4. *Pricing and project award.*   The owner or owner's representative may contact prospective qualified vendors to submit proposals on the project, may negotiate with a selected vendor, or may choose to have the installation implemented by in-house personnel. Upon receiving the price, the owner reviews the documents, chooses the qualified company, and awards the project.

5. *Submittal and approval.*   The drawings and specifications together form the contract documents for pricing, purchasing, and installing the system. Before construction, it is recommended and common practice to require shop drawing submittal from the installer and equipment supplier. The design engineer should verify system and component compliance with the contract documents and provide specific manufacturer's requirements to the installer for wiring and installation.

6. *Installation.*   The installer performs this step in the process. The installer's work is governed by adherence to the plans and specification and planned periodic meetings. These meetings provide a forum for discussing installation issues, questions, status, and problem resolution

7. *Testing and sign-off.*   The installation is complete, and the company performs a system acceptance test. An acceptance test document should be prepared in advance. This test should verify and document proper operation of all components of the system as outlined in the plans and specifications. Training for the system operators is generally conducted at this time. Once the system has passed the acceptance test with 100 percent compliance, the final documents are compiled, and the owner acknowledges that the project is completed.

COMPONENTS OF A SECURITY SYSTEM

These are the three components of a security system:

1. Intrusion detection system (IDS)
2. Closed-circuit television system (CCTV)
3. Access control system

Intrusion Detection System

The intrusion detection system (IDS) is generally comprised of detection devices, wiring, power supplies, and a head-end controller or control panel.

Detection Devices.   Many types of intrusion detection devices are available. Each has its advantages, disadvantages, and appropriate applications.

Magnetic Contact.   A magnetic contact is comprised of a magnet, which is usually installed on the movable surface, and a reed switch. The switch is held closed by the magnet. When the magnet is moved away from the switch, the switch opens and reports an alarm.

Contacts can be used to protect movable openings such as doors and windows. The movable member must be fairly rigid without play or vibration. An important item to remember is that these devices can protect against openings, but they do not protect against cutting through a door, or breaking the glass in a window. They should be mounted as close as possible to the leading edge of the door or window.

Magnetic contacts are low in cost and easy to install and maintain. However, they are considered a fairly low security device because they can be easily defeated with an external magnet and can be fragile.

Balanced Magnetic Contact.   A balanced magnetic contact is similar in operation to a magnetic contact; however, the switch assembly also contains a balancing magnet for gap and sensitivity adjustment, a tamper switch, and a resistor to protect the unit from voltage spikes. Higher-security variations of this device are available that incorporate three or more sealed reed switches, two of which are polarized in one direction and the third with reversed polarity. The magnet housing contains multiple magnets polarized with the switch. The purpose of the multiple polarities is to prevent defeat by moving another magnet close to the switch.

Balanced magnetic contacts are available for both surface mounting and flush mounting. Caution should be taken when mounting to steel doors because the steel may dampen the magnetic field. When mounting to a steel surface, nonmetallic spacers should be installed beneath the magnet. Weatherproof, explosion-proof, and intrinsically safe versions are available. This is a relatively high-security device, is relatively false/nuisance-alarm–free, and is flexible. It is somewhat large and bulky and requires careful installation and adjustment.

Glass Break Detector.   As the name indicates, these devices sense glass breaking either through stress, frequency, or position changes.

Shock Sensor.   A shock sensor uses a small weight that is balanced on two or more sets of electrical contacts and completes a circuit. The impact of an object that strikes the mounting surface is transmitted to the sensor. This causes the weight to slide or jump off the contacts and creates a momentary open circuit. Pulse count circuitry included in the device minimizes nuisance alarms. An accumulator circuit is included to counter slow penetrations and can be set to hold accumulated pulses for some minutes.

Shock sensors are used to protect glass, metal, and similar surfaces against forcible entry. There are appropriate models for specific materials, and proper selection is necessary to provide protection. Shock sensors are small, easy to conceal, and easy to install. They are very sensitive to the mounting position; sensors on different materials cannot be mixed in a common zone; and they may require a separate signal processor.

Mercury Switches.   Mercury switches, also known as *window bugs,* are made up of two contact wires suspended above a reservoir of mercury. Any impact splashes the mercury onto the wires, closing the circuit, and reporting an alarm. The contact made by the mercury is very brief; therefore, a pulse extender is usually necessary to increase the alarm signal duration so that the control unit can recognize the signal.

These devices are quite inexpensive but are highly sensitive to mounting position. A detector is required on each pane of glass; heavy traffic or high wind may cause nuisance alarms; and mercury is very temperature-sensitive.

Mechanical/Inertia Sensors.   These sensors have either spring-suspended disks, each of which is connected to a side of the circuit, or a weight resting near a set of contacts. In both cases, impact causes movement within the sensor, opening or closing the circuit, and reporting an alarm. Cost is moderate; however, contacts can fail or corrode, the moving parts can become sticky, and they are very sensitive to mounting position. They are prone to nuisance alarms when installed in areas subject to heavy traffic or high winds.

Piezoelectric Frequency Sensor.   This device uses a piezoelectric crystal in contact with a pane of glass to detect the frequency associated with breaking glass. It is more reliable than mechanical units, and is not sensitive to position, but is fairly expensive because it requires a separate signal processor, and because being a powered device, it requires additional wiring.

Stress-Sensing Piezoelectric Sensor.   This device, installed in contact with a pane of glass, detects the flexing of the glass as it breaks. Generally the most reliable of glass break detectors, it is not position sensitive, requires no external power, and is very resistant to nuisance alarms. It is quite expensive and may require a separate signal processor.

Passive Audio Discriminator.   This device listens for the sounds of breaking glass and attempts to differentiate these sounds from other background noise. It is usually mounted on a wall or ceiling near the windows to be protected, not on the glass. It can protect several windows. Cost is moderate, and it is relatively nuisance-alarm-resistant. Curtains or blinds can reduce coverage.

Photoelectric Beam.   The photoelectric sensor is made up of a transmitter (or light source) and a receiver. Light is transmitted to the receiver, and if the light beam is broken, an alarm is reported. The transmitter and receiver may be separate or may be in the same enclosure. If the latter, a mirror is used to reflect the beam back to the receiver. The detection area is confined to the diameter of the beam and is fairly narrow. However, the range can exceed 1000 feet. Most models use a pulsed light source to make them more resistant to defeat or nuisance alarms caused by light.

Photoelectric beam detectors are appropriate for protecting long rows of windows, warehouse and office corridors, and interior ceiling spaces. They are often deployed in stacks to broaden the detection area. If mirrors are used to increase pattern complexity, the range is reduced by about 30 percent for each mirror used.

Photoelectric beams are very appropriate for long, open aisles and hallways and are effective as a secondary protection barrier.

Foil Tape.   Foil is a thin strip of lead taped or glued to the surface to be protected. The foil carries the protective loop onto the surface to be protected. Foil is most frequently applied to glass surfaces; however, it can also be applied to walls. Operation is such that when the surface on which the foil is applied is broken, the circuit is interrupted, and an alarm is reported.

Foil can be installed on various types of glass; however the installation patterns are specific to the type of glass on which the foil is to be installed:

- Plate glass: A continuous strip around the perimeter of the glass is required, three to seven inches from the edge. When plate glass is broken, cracks usually will extend to the edges.
- Wired glass: Strips three inches apart are required across the entire surface.
- Tempered glass: A single strip four to seven inches long at the top of the glass is required. Tempered glass usually disintegrates when broken.
- Safety glass, Lexan™, or other glass substitutes cannot be protected with foil.

Foil is a low-cost material, but it is labor-intensive to install. Sometimes its visibility can act as a deterrent.

Tape Switch.   The tape switch, or mat, is made of two conductive layers separated by a thin layer of compressible material encased in plastic or rubber. This is a normally open device, and application of pressure will close a circuit, and report an alarm. This device is often installed at entrances to give warning of movement across the entry. It is a low-cost, low-nuisance alarm device. Maintenance is low; its detection area is narrow.

Grid Wire.   Grid wire is fine, 24-gauge or smaller enamel-insulated wire. It is stapled to the surface to be protected with four-inch spacing between runs. Two circuits are usually installed, one horizontal and one vertical, that overlap each other. A break in either circuit or a short between them causes an alarm.

Grid wire may be installed directly to surfaces that can be stapled. Alternatively, it can be applied to brick or block by stapling the grid to plywood or paneling and attaching the plywood or paneling to the concrete surface, and the grid wire is sandwiched between the two surfaces.

Grid wire can also be used to protect skylights, ducts, windows, and so forth by inserting the wire into slotted dowels, which have been built into a frame that will fit over the opening; it can also be woven through window screen fabric.

Grid wire can be installed in many areas; maintenance is low, as is material cost. It is expensive to install because it is labor-intensive.

Vibration Detector.   Similar to glass break detectors, vibration detectors respond to the low frequencies generated by attacks on rigid material, such as block, brick, or drywall. Devices are made for specific types of materials to be protected. Coverage is variable, depending upon the surface to which the device is mounted.

Volumetric and Space Protection Devices.   Volumetric and space protection devices comprise ultrasonic, microwave, passive infrared (PIR) technologies, and combinations of these technologies, such as PIR and ultrasonic. Ultrasonic and microwave devices operate on the Doppler shift principle.

A pattern of inaudible (to humans) sound is transmitted into the space to be protected. The receiver monitors the frequency of the sound. Motion within the monitored space causes a change in the frequency of the received signal. Motion toward the device causes an increase in frequency, and motion away from the device causes a decrease. Up or down shifts are measured against a preset threshold. When this threshold is exceeded, an alarm is reported. Most units are referred to as "balanced," that is, a time window is factored in to balance up and down shifts, and an alarm is reported only on a net up or down shift to minimize nuisance alarms.

Ultrasonic devices are available as separate transmitters and receivers, transceivers that require separate signal processors, and transceivers that incorporate signal processors. The detection pattern depends on the device used and its position with respect to the area that it monitors. Detection patterns can be adjusted by the addition of baffles and deflectors. Separate or split devices produce a dog-bone pattern between heads, wall-mounted transceivers produce a teardrop pattern, and ceiling-mounted devices produce a circular pattern.

Care must be taken when using this device because it is susceptible to nuisance alarms from air turbulence, fans, heaters, telephones, bells, jet aircraft, and other sources of ultrasonic noise, and is not suited for use in poorly constructed facilities or for exterior applications. Ultrasonics is a widely used technology and is effective when properly installed and aimed. The device should be tested in place before final installation.

Microwave devices emit an energy field into the area to be protected and monitor that energy field for changes caused by motion. Microwave devices use a precisely shaped antenna to determine the shape of the detection pattern, and most units come with interchangeable deflectors to accommodate on-site adjustments.

Microwave energy penetrates glass, wood, and plasterboard or gypsum wallboard. Concrete permits slight penetration, and metal will contain the emitted pattern. Metal-faced insulation and metallic or foil wallpaper also are effective in containing the energy. This ability to penetrate many materials can be useful because the device can be concealed above dropped ceilings to protect the space beneath. It can be used to protect several spaces separated with drywall; it can be directed at air-conditioning ducts, and the energy will travel along the duct to protect an adjacent space. Microwave energy will reflect off shiny metal surfaces and has been known to detect liquid movement in PVC piping.

Because microwave energy does not use air as a transmission medium, many of the nuisance alarm sources associated with ultrasonics do not affect it. Nuisance alarms can be caused by EMI/RFI, fluorescent lighting, two-way radios, police radar, and nearby airports. Specially constructed microwave units are installed in exterior environments, such as along fence lines and parking areas. The range is limited to about 500 feet. If the device is not aimed directly at or parallel to fluorescent lights, there generally will be no problems associated with these lights.

Passive infrared devices detect movement by monitoring the background infrared (IR) energy level. An object entering the covered area that both moves and creates a change in the IR level will initiate an alarm. This is achieved by an optical system that focuses the IR energy onto a pyroelectric element. The element transforms the IR energy into electric current. If the IR energy level changes, a corresponding change takes place in the electrical output. The sensor considers this change an alarm condition.

PIRs incorporate two types of lenses: *mirror optics* and *Fresnel lens optics*. Mirror optics reflect IR energy onto the pyroelectric element. Mirror optics have longer detection ranges and are more accurate than Fresnel systems. Fresnel lenses detect over a shorter range and have more detection patterns because they have interchangeable lenses, but they are less precise.

PIRs are suitable for most interior applications. They are not sensitive to the common nuisance alarm sources of other motion detectors, but they are affected by sources of rapid temperature change. Detection patterns are stable and easily contained by the usual construction materials. PIRs should always be aimed at a background rather than into open space because a stable reference must be provided for the IR energy level.

Capacitance Device. This sensor consists of a coaxial cable detection loop and a signal processor. Objects that are to be protected must be metal and insulated from ground. The sensor radiates an electrical field around the object, and disturbance of this field will report an alarm. The device is commonly used to protect safes, cabinets, and desks, but tool cages and artwork can also be protected with this device by using insulating blocks and metallic screen. Insulating blocks made of wood are not acceptable because wood absorbs moisture and eventually creates a path to ground.

Heat Detector. Heat detectors are generally related to fire detection and are often installed in safes, vaults, and similar spaces to protect against attack by cutting torches, thermal lances, and the like. Heat detectors operate in one of two ways. One type uses a bimetallic strip to generate an alarm as the temperature increases. The other uses a metallic fusible link that has a low melting point. Most heat detectors operate at a predetermined temperature of 135 or 190°F. However, units that incorporate fixed temperature with rate of temperature increase are readily available.

Wiring. The device and the manufacturer dictate the wiring for IDS system devices. Typically the wiring is two-conductor, 18 AWG, or four-conductor, 18 AWG, but this can vary depending on building conditions and wire lengths. Careful consideration must paid to the type of wire selected, which is generally plenum-rated or PVC.

Power Supplies. IDS generally operate on low-voltage power supplies, either 12 Vdc or 24 Vdc. The field devices are wired and connected to low-voltage power supplies, which in turn are powered by standard 110 V outlets.

Head-End Controller. The choices for both the manufacturer of and type of head-end controller are important decisions. There are many manufacturers, and each one has a variety of system types. The facility manager or security professional should conduct a careful study, using literature, industry articles, Web sites, and references to choose the manufacturer that can provide a system to fulfill the facility's particular security requirements.

As with any system that is installed in a facility, the facility manager must be aware of state and federal laws regulating security and life safety, as well as applicable municipal rules, codes, and laws.

Closed-Circuit Television System

The advent of digital video technology as an alternative to conventional analog storage media holds great promise for CCTV security applications. New digital compression technologies and reduced mass storage costs are rapidly reducing the cost difference between conventional

CCTV and digital video systems. Advantages include substantially improved search capabilities and the transmission of video over local area networks (LANs). The latter can potentially offer significant reductions in installation costs for select applications because separate runs from individual remote camera locations are typically not required. When considering digital video, it would be prudent to ascertain that a vendor's approach to tamperproofing the recorded video meets the judicial system's requirements for admissibility.

CCTV system technology is constantly evolving. In the 1970s, the typical CCTV system was a camera and a monitor. Advances in technology have brought us digital recorders, multiplexers, and matrix switchers. However, the CCTV system starts with the camera that creates a video image.

CCTC is not strictly a detection device. It does not report alarms, although it can be made to do so. It will not prevent an intrusion. Rather, it should be considered an enhancement to or an extension of security. It expands the capabilities of an existing security force by enabling personnel to monitor a number of areas from a single location. To discuss all of the pieces and parts, technologies, and alternatives would require a book, and there are such comprehensive books available. I will discuss components that will assist the facility manager in the design of a CCTV system that is appropriate for the facility.

The first issue to consider is the purpose of the system. If the intent is to view your parking area, do you want to see the entire area, observe movement, identify vehicles, identify people, or read a license plate? Do you want to view this area only during working hours or 24 hours a day? What kind of light is available to light the area? Are there any obstacles to clear view, such as trees or small buildings? If the lighting is insufficient to support nighttime operation, is the cost of upgrading the lighting worth the expense?

If the area under consideration houses valuable merchandise, integrating the CCTV system with an intrusion detection system may be a practical approach. If evidentiary material is required, adding video tape recording media would be appropriate.

Another consideration is whether the camera should be visible or hidden. Visible cameras can provide a level of deterrence. Hidden or covert cameras provide surveillance without giving people the feeling that they are being watched.

Cameras.   Determine the areas that the cameras are to view. Cameras differ in size, weight, sensitivity, resolution, and power. Light is the greatest external factor in determining the effectiveness of CCTV. In addition to the light issues mentioned before, other issues impact lighting. Is the light constant, as in a hallway, or variable, as in an exterior environment? Does the scene have light, such as from windows or vehicles, which could impact the camera's capability? Will the scene have a background brighter than the objects you wish to see? If so, consider another camera location or adding light from the same direction as the camera.

If nighttime operation is planned, and additional lighting is necessary, it may be more cost-effective to use infrared (IR) lighting. IR light is invisible to the human eye and therefore would be unobtrusive. IR light sources can also be added to the camera mount, which would generally make it much less expensive to install.

The area in which the camera is to be installed must be considered. Is vandal protection needed? Does it have to match existing decor? Is surveillance to be covert? Is the environment dusty, dirty, very hot, or very cold? Many types of housings are available to protect the camera from harsh or unfriendly environments. Most cameras can operate satisfactorily below –30°F; however, the lubricant used in lens iris, focus, and zoom rings tends to solidify or turn to a gel in subzero weather, and this causes the lens to freeze up.

The camera mounting location has to be considered. If it will be mounted high on a pole, a lowering device or a lift truck will be needed to perform maintenance. If the viewing angle is too steep, only the tops of heads will be seen.

Lens.   Quite possibly the most difficult decision in the surveillance process is lens selection. What do you want to see? There are four items to consider in lens selection: (1) the camera format, (2) the distance from the camera to the scene, (3) the field of view, and (4) color versus black and white.

Camera format is determined by the size of usable image area on which the lens focuses light. Common formats for charge-coupled device (CCD) cameras are $\frac{1}{4}$, $\frac{1}{3}$, and $\frac{1}{2}$ inch. For tube-type cameras, the formats are $\frac{2}{3}$ and 1 inch. The significance of format is that if a smaller-format lens is used on a larger-format camera, the result will be tunnel vision or poor focus. On the other hand, a larger-format lens can be used on a smaller-format camera without picture degradation.

The distance from the camera to the scene determines the focal length of the lens. Lenses are available in four varieties: (1) telephoto, (2) standard, (3) wide-angle, and (4) zoom. Zoom is a combination of wide-angle, standard, and telephoto. Standard lenses produce pictures similar to those of the human eye. Wide-angle lenses produce pictures that are generally wider than the human eye's peripheral vision.

Cameras that are to be installed in areas where the light is fairly constant can use a manual iris lens. This simply means that the lens opening is set to render the best possible picture, and is left that way. In scenes that have variable light conditions, an automatic iris lens should be selected. This type of lens samples the video levels. Decreasing light produces a corresponding decrease in video level. This causes the electronic circuitry sampling the video level to open the iris electronically. If color cameras are selected, ensure that the lenses are color corrected. A monochrome lens mounted on a color camera produces poor definition and color.

Power Requirements. Cameras installed in the field can be powered locally by standard 110-V outlets or by low-voltage power supplies that originate centrally or strategically. The preference is mostly for convenience and/or cost.

Wire. Coaxial cable is the most commonly used connecting cable, although fiber optics are rapidly gaining in popularity. Probably the single largest cause of system failure is cheap coaxial cable. Consideration must be given to cable paths, cable lengths, cable environment, and connector installation. RG-59 is used for runs up to about 1000 feet. RG-6 is used for runs up to 1500 feet. RG-11 is used for cable runs up to about 2500 feet. All coaxial cable used in CCTV installations must be rated at 75 ohms, regardless of jacket, center core, or shield, because all CCTV equipment is designed to emit a video signal balanced to a 75-ohm impedance.

Monitor. Monitors are available in a variety of sizes and resolutions. The distance between the monitor and the viewer has a direct effect on the clarity of picture. The smaller the space, the smaller the screen size that is needed. A rule of thumb that may be followed is to subtract 4 from the size of the monitor (measured diagonally in inches) to arrive at the most comfortable viewing distance in feet.

A monitor will produce a picture in direct proportion to the resolution of the picture transmitted to it. A monitor that can produce 700 lines of resolution will produce 700 lines of resolution whether the camera inputs 300 or 1000 lines. Also, if a 500-line camera is fed through a 450-line VCR and a 700-line monitor produces the picture, the input to the monitor is 450 lines.

Switchers. Switchers are used to view multiple cameras on a single monitor, to view single cameras on multiple monitors, and to view multiple cameras on multiple monitors. The determining factor is who is to view what. The output of a switcher can be manually selected or can automatically sequence through all cameras connected to it at a preset time interval.

Multiplexers. A multiplexer allows the CCTV system to view and control up 16 video images from one monitor. The use of multiplexers in a CCTV system is one of choice based on need versus costs. Multiplexers give the facility manager increased features and benefits for the CCTV system. However, this piece of equipment is costly and therefore optional.

Recorders.   Video recorders are a crucial element of the CCTV system. When an event (theft, assault, vandalism, etc.) occurs, it is critical to have evidence, and video recorded images can be powerful evidence. As with all head-end type of equipment, careful review and consideration is required when choosing a manufacturer and model.

As with any system that is installed in a facility, the facility manager must be aware of state and federal laws regulating security and life-safety issues, as well as municipal rules, codes, permit requirements, and laws.

Access Control

Access control is a term that describes the way a facility manager controls who goes where and when in the facility. One school of thought is to stop crime at the front door. This measure deters common problems such as workplace violence and theft. Another goal is to secure the perimeter entrances and exits during nonbusiness hours without jeopardizing life safety for maintenance personnel and the ever-increasing number of late workers, and at the same time provide access to authorized people. The need for customized equipment, which takes into account the operating style of the facility, corporate culture, and public perception, has driven this industry to consider new solutions. These new technological advances have allowed the facility manager to create a highly secure facility and at the same time to reduce manpower costs.

Card Readers/ID Devices.   There are many different types of card readers and reading devices to choose from today. This is the first and one of the most important decisions the facility manager must make. Again the preference here will be based on the level of security desired, ascetics, operability, and costs:

1. A typical solution is the magnetic stripe swipe reader and card. This offers a low-cost, flexible system.
2. Bar-code swipe readers and cards are another flexible, low-cost solution.
3. Wiegand swipe style readers and cards are the next level; these offer a higher level of security at a higher cost.
4. Proximity style readers and cards are currently the most widely used equipment. These offer a high level of security combined with ease of operation and a slightly higher cost.
5. Biometric readers employ technology that includes voice recognition, fingerprint recognition, and iris or retinal scans. These are mostly used in high-security applications and are the most costly.

All of these styles can be used independently or in tandem, again depending on the level of security one desires to obtain and the impact of costs on the budget.

Electronic Locking Devices.   The choice of electronic locking devices is generally dictated by door and door frame construction and cost. Other considerations in the choice of electronic locking devices are the impact this will have on life safety issues. Almost all systems are required to release the locks in the event of a fire alarm, so careful attention must be paid to interfacing the access control to the existing or proposed life safety systems in the facility:

1. Electromagnetic locks are a commonly used solution. These devices are simple and efficient, easily interfaced to fire alarm systems, and provide a high degree of security.
2. Electric strikes are another commonly used device. They are available in numerous configurations and have many options.
3. For customized doors, there are other devices such as electrified mortise lock sets. If this type of device is required, it would be advisable to seek the advice of a qualified locksmith.

Wire.   Wiring for the access control system will vary according to the manufacturer chosen. Each manufacturer's system will require a cabling layout that will be basic with minor differences. In general, the readers need one five-conductor, 22 AWG, shielded cable, although some systems use a two-pair, 22 AWG, shielded cable. The electronic locking device will need one two-conductor, 18 AWG cable, and other similar type wires will be needed for door status switches or egress devices. As with all systems, careful attention should be paid to the need for installing plenum-rated cable.

Power Requirements.   Most access control systems are low-voltage (24 Vdc). The electronic locking devices are usually installed with 24 Vdc or 24 Vac power requirements. The readers are normally passive and do not require power.

Controllers/Head End/Software.   This choice requires the facility manager to select a vendor, and again this is an important decision and needs to be done after careful consideration, fact-finding, consulting with consultants and colleagues, system demonstrations, and references. The vendor and the type of system chosen must be suitable for the facility and meet the goals of the user.

The controllers are generally configured to control anywhere from a single door up to eight doors per enclosure. These controllers can be installed remotely in the facility and cabled to the head end, or they can all be installed in the same area as the head end. It is advisable that the controllers are of the type that has distributed processing. This simply means that the panels will operate the doors in the event the head end has a failure or communication of data is interrupted.

The head end and software are the brains of the system. Most systems sold today are PC-based with Windows- or UNIX-based operating platforms, as well as a number of networking and communications options and features. The choice of software and PC is one of the most important steps in selecting a vendor.

CONTRIBUTORS

Michael Colburn, SES Company, Inc. (SESCO), Hingham, Massachusetts

Charles Dever, SES Company, Inc. (SESCO), Hingham, Massachusetts

Roderick A. Dike, Honeywell, Boston, Massachusetts

John P. White, Police and Security Department, Massachusetts General Hospital, Boston, Massachusetts

David L. Stymiest, P.E., SASHE, C.E.M., Senior Consultant, Smith Seckman Reid, Inc., New Orleans, Louisiana

RESOURCES

Access control & security systems: www.securitysolutions.com

American Society for Industrial Security (ASIS): www.asisonline.org/

Security technology & design: www.securitysales.com/educ.cfm

Security sales: www.securitysales.com/educ.cfm

ARTICLE 5.5.4
FIRE ALARM SYSTEMS

Steve Collins
R. G. Vanderweil Engineers, Inc., Boston, Massachusetts

Fire alarm systems are required in most facilities by applicable codes and standards. A fire alarm system plays a critical role in alerting the facility's occupants that a dangerous situation exists. It can also notify the local fire department and in-house or contract personnel who are designated to respond to a fire condition. Any discussion of a fire alarm system must necessarily include the codes, standards, and other regulations that apply; perhaps even more than for other types of systems. These include local requirements that will usually take precedence over federal and state requirements. Facility engineers, managers, and designers must develop close working relationships with the local fire authorities as recommended in Chapter 8, *Codes and Standards.* Close working relationships will benefit all parties involved in managing buildings and facilities. In most cases the "authority having jurisdiction" for building fire alarm and signaling system design is the local fire department. It is always best to keep the authority having jurisdiction fully informed of the proposed fire alarm system design and installation parameters and to include all of the authority's recommendations and requirements.

Because one cannot consider fire alarm systems independently from the code requirements, this article includes numerous references to NFPA 72, the National Fire Alarm Code.[1] NFPA 72 includes requirements for fire alarm system design, installation, inspection, maintenance, testing, and use. All facility and building owners, designers, contractors, and operation/maintenance personnel should obtain and comply with the latest version of NFPA 72. Information from this code is not copied in this article because of the potential for misleading by omission.

The references to NFPA 72 in this article are not intended to minimize the importance of knowing and complying with other local codes and regulations. However, it is impractical to attempt to cover all of the different local codes and regulations here.

Scope of National Fire Alarm Code

NFPA 72 covers the following:

1. The fundamentals of fire alarm systems; including qualifications for fire alarm system designers and installers, power supplies, compatibility, system functions, performance and limitations, zoning and annunciation, monitoring integrity of installation conductors and other signaling channels, and documentation.

2. Performance, selection, use, and location of initiating devices (e.g., automatic fire detection devices such as smoke and heat detectors, flame detectors, sprinkler water-flow detectors, manually activated fire alarm stations, and supervisory signal initiating devices). Fire alarm system inputs include those that initiate alarm signals (detection devices, water-flow alarms, and signals from types of fire suppression systems other than water flow). Inputs also include those that initiate supervisory signals indicating that sprinkler and other fire suppression systems are in an off-normal condition that could adversely affect the performance of the system. Such supervisory signals include fire pump supervision and automatic fire suppression system panel supervision. Trouble signals from fire suppression systems are also classified as fire alarm system inputs.

3. Performance, location, and mounting of notification appliances. Notification appliances initiate evacuation or relocation or provide information to occupants and staff. Pertinent requirements include listings, interconnections, physical construction, mechanical protection, mounting, audible characteristics, visible characteristics, spacing criteria, and other detailed performance criteria. Fire alarm system outputs include occupant notification, signal annunciation, suppression system annunciation, and off-premises signals.

4. Requirements of fire alarm systems for protected premises, including signal annunciation, software and firmware control, system performance and integrity, initiating device circuits, signaling line circuits, notification appliance circuits, fire alarm control units, fire alarm system inputs and outputs, guard's tour, fire safety functions for protected premises (HVAC system fan control, door control, elevator recall, and elevator shutdown).

5. Supervising station fire alarm systems, including those for central station service, proprietary supervising station systems, and remote supervising station systems. Requirements include third-party verification, facilities, equipment, personnel, operations, testing, maintenance, and communications methods.

6. Requirements of public fire-alarm reporting systems, including management and maintenance, equipment and installation, street boxes, receiving equipment and facilities, and coded reporting systems.

7. Requirements for inspecting, testing, and maintaining both new and existing systems including test methods, inspection and testing frequency, and record-keeping requirements. Requirements for test methods include detailed instructions for testing all of these portions of the fire alarm system: control equipment, generators and secondary (standby) power supplies, uninterruptible power supplies, batteries, public reporting systems, transient suppressors, control-unit trouble signals, remote annunciators, conductors, initiating devices, alarm notification appliances, special hazard equipment, transmission and receiving equipment for supervising station fire alarm systems, emergency communications equipment, interface equipment, special procedures, and wireless systems. Frequencies for visual inspection include initial and monthly, quarterly, semiannual, and annual reacceptance testing. NFPA 72 also provides examples of inspection and testing forms.

8. Fire warning equipment for dwelling units as a performance-based code, including limitations, performance criteria, verification methods, and acceptable solutions of the primary and optional functions for this equipment. The standard also provides general reliability requirements and general performance criteria.

9. Explanatory material and engineering guidance on types of equipment, applications, requirements, wiring methods, detector mounting, spacing, placement, and layout, and a fire alarm system input/output matrix.

NFPA 72 defines the means of signal initiation, transmission, notification, and annunciation and the features associated with these systems. It also provides the information necessary to modify or upgrade an existing system to meet the requirements of a particular system classification.

CATEGORIES OF FIRE ALARM SYSTEMS

NFPA 72 classifies fire alarm systems in three major categories:

1. *Household fire warning system.*   This system only warns the occupants.

2. *Protected premises fire alarm system.*   This system is also called a local fire alarm system. It is the building block upon which connections to the other types of systems listed following are built.

3. *Supervising station fire alarm system* (off-premises system).   This includes the following:

- *Auxiliary fire alarm system.*   This system is connected to a municipal fire alarm system to transmit an alarm to the public fire-service communications center.
- *Remote supervising station fire alarm system.*   This system transmits the fire alarm, supervisory, and trouble signals from the protected premises to a remote location.
- *Proprietary supervising station system.*   This system serves single-owner, contiguous and noncontiguous properties from a proprietary supervising station at the protected property. This type of system is often used in multibuilding complexes such as large educational, medical, and industrial facilities.
- *Central station fire alarm system.*   This system requires a listed central station that has on-duty personnel.
- *Municipal fire alarm system.*   This system covers from street locations (municipal fire alarm boxes) to a public fire-service communications center.

NFPA 72 generally defines the required levels of performance, extent of redundancy, and quality of installation but does not define the methods to meet these requirements.

Drawings and Specifications.   Engineering documents for designing and installing a fire alarm and signaling system are generally made up of two major parts, the contract drawings and the contract specification.

The contract drawings generally show the layout and locations of the main system equipment central panels, annunciators, and so forth, and the locations and coverage requirements for the systems initiating and signaling devices (manual-pull stations, detectors, audiovisual devices, water flow, and tamper switches, etc.). Also generally included with the contract drawings is a fire alarm riser diagram that represents the overall system configuration and requirements. The fire alarm riser diagram does not necessarily show every device used but is information supplemental to the fire alarm plan drawings to define the overall system and requirements.

The fire alarm system specifications are generally divided into three sections. The first section (or Part 1) generally contains the general system requirements; the level of quality required for the specified system; shop drawing submittal, delivery, shipping and handling; and other specific project requirements. The second section (or Part 2) generally describes the specific system, equipment, operation, capacities, and materials, and includes the technical system requirements. The third section (or Part 3) of the fire alarm system specification generally includes installation requirements for the fire alarm system components, requirements for wiring and conduit methods, and workmanship and testing.

The drawings and specifications together form the contract documents for the bidding, purchasing, and installation of the fire alarm system. Once the contract is bid on and awarded, it is recommended (and is common practice) to require that the installing contractor and equipment supplier submit shop drawings for review by the design engineer (to verify system and component compliance with the contract documents) and provide specific manufacturer's requirements to the installing contractor for wiring and installation. The authority having jurisdiction may also have to be involved in this preinstallation approval of shop drawings.

Annunciation.   The primary purpose of a fire alarm system annunciator is to enable responding fire service personnel to quickly and accurately identify the location of the alarm or fire condition and to indicate the status of life safety emergency equipment functions that may affect the safety of the building occupants. The annunciator must to be readily accessible to the fire service responding personnel and shall be located as coordinated with and required by the local authority having jurisdiction to a facilitate an efficient and effective response to the alarm/fire situation.

Annunciator panels are generally of three basic types:

1. *Lamp-type annunciators* use a series of different-colored lamps to indicate the alarm or trouble conditions. Specific lamps associated with specific characteristics (zones, devices, trouble and alarm indications, etc.) are illuminated on a panel annunciator to define the alarm condition.

2. *Graphic-type annunciators* generally use a plan of the building and backlit or outline-type lamps to visually locate the area in alarm and also identify with additional backlights the type of device, alarm, or trouble indication, and so forth.

3. *LCD panels* with system controls or graphics only.

Zoning. A fire alarm zone is "a defined area within the protected promises." A zone may define an area from which a signal can be received, an area to which a signal can be sent, or an area in which a form of control can be executed.

In older, hardwired fire alarm systems, a zone would typically be defined as an area of the building on a per floor basis. The floor area is not to exceed 20,000 square feet. Therefore, on projects whose large floor plates exceed 20,000 square feet, multiple fire alarm "area" zones would be provided or established for each floor. Within each building area zone, the annunciator would typically provide status information for each type of fire alarm system device used (i.e., manual-pull station, smoke detector, water-flow switch, etc.) and whether the condition is an alarm or trouble condition.

In newer addressable fire alarm systems, each connected fire alarm system device has its own associated "address" (identity and location), and therefore each device is treated as a separate zone within a defined building area.

Documentation. Proper documentation should be prepared and maintained for the fire alarm system, including but not limited to the following:

1. The original engineering design documents that are part of the contract documents and include floor plan layouts, riser and wiring diagrams, and system specifications

2. Any change, alteration, or addendum to the documentation

3. The complete record system shop drawings as submitted by the equipment manufacturer/supplier to the contractor for system purchase and installation requirements, including manufacturers' catalog cut sheets for all system components and devices showing model numbers and all operating characteristics

4. A detailed written sequence of operation for system functions

5. A complete system riser diagram showing interconnections, including conduit and wiring requirements

6. All system calculations required to size the system equipment components (i.e., batteries, amplifiers, etc.) properly

7. Reproducible as-built installation drawings

8. Manufacturer's operating and maintenance manuals, including installation instruments, operating instructions, routine maintenance requirements and instructions, and detailed troubleshooting procedures

9. A certificate showing that the system has been successfully and completely tested by the installing contractor or testing agency

10. A certificate verifying that the system is complete

11. A permanent record of all inspections, testing, and maintenance per Chap. 7 of NFPA 72[1]

12. All approval, testing, and acceptance documents received from the authority having jurisdiction

13. Any other documentation required by the authority having jurisdiction

14. Records of training of owner's personnel

CONTRIBUTORS

Arjun Rao, P.E., Vice President, BR+A Consulting Engineers, Inc., Boston, Massachusetts

Cornelius Regan, P.E., CLEP, Principal, C. Regan Associates, Rockland, Massachusetts

David Stymiest, P.E., SASHE, CEM, Senior Consultant, Smith Seckman Reid Inc., New Orleans, Louisiana

REFERENCES

1. NFPA 72. *National Fire Alarm Code,* National Fire Protection Association, Quincy, Massachusetts, 1999.

BIBLIOGRAPHY

R. W. Bukowski and R. J. O'Laughlin, *Fire Alarm Signaling Systems* (2nd ed.), National Fire Protection Association, Quincy, Massachusetts, 1994.

ARTICLE 5.5.5
SMOKE CONTROL SYSTEMS

Charles Stein
R. G. Vanderweil Engineers, Inc., Boston, Massachusetts

Smoke is an airborne mixture of solid and liquid particulates and gases that evolves when a material undergoes pyrolysis or combustion, together with a quantity of air that is entrained or otherwise mixed into the mass. This toxic mass causes more than 90 percent of the deaths in fires from smoke inhalation and significant asset damage in adjacencies. In his report, O'Hagan succinctly states that "smoke causes over 90% of deaths in a fire."[1]

This article discusses fire protection and smoke control systems in buildings related to HVAC systems. Smoke control in modern buildings is a very complex process. This article presents the most important issues, which are as follows:

- Approaches to fire protection
- Reason for providing smoke control systems
- Reasonable expectations for a smoke control system
- General smoke control system issues
- Stairwell pressurization and overpressure relief
- Elevator smoke control
- High-rise building and zoned smoke control
- Atrium smoke control
- Smoke control system commissioning and acceptance testing

The article concludes with a reference section listing codes, publications, standards, and handbooks that provide additional information and resources.

BASIC APPROACHES TO FIRE PROTECTION

The two basic approaches to fire protection are (1) to prevent fire ignition and (2) to minimize fire impact. Building occupants and managers have the primary role in preventing fire ignition. The building design team should incorporate features into the building to assist the occupants and managers in this effort. Because it is impossible to prevent fire ignition completely, minimizing fire impact has become significant in fire protection design. Examples include compartmentation, fire alarm systems, suppression systems, use of noncombustible construction materials, exit systems, and smoke management.

Prevention of Fire Ignition

- Eliminate ignition sources including open flames, arson, smoking materials, and spontaneous combustion.
- Keep fuel away from ignition sources, including materials in storage, flammable liquids, trash, clutter, and combustible building materials.

Minimize Fire Impact

- Minimize threat by using standpipes, sprinklers, fire walls, fire doors, fire dampers, smoke dampers, and, most important, a fire and smoke detection system.
- Minimize exposure, including exit systems, smoke management, smoke control, smoke venting, and smoke barriers. The need to seal floor and wall penetrations properly cannot be overly emphasized. Without such efforts, even an excellent smoke management system is rendered useless.

SMOKE CONTROL SYSTEMS GENERAL

There are some general basic concepts that must be understood. Smoke rises and can travel very fast. Smoke control methods may differ in different types of buildings (e.g., in a high-rise office building compared with a high-rise hospital), and may differ in atria, malls, and in buildings that have mixed occupancy. However, the basic concept remains to contain the smoke, to prevent its spread in other areas, and in the event of fire to provide safe means of egress for the building's occupants.

For life-safety considerations, the smoke control system should be designed to inhibit the flow of smoke into the means of egress, exit passageways, or other similar areas of a building. Smoke control systems cannot be expected to maintain an area free of smoke, but can be expected to provide a tenable means of egress during the initial evacuation of occupants from the zone of incidence. Therefore, the primary mission of a smoke control system is to control the migration of smoke, to maintain a tenable environment in the means of egress from the area of incidence, during the time required for evacuation.

The systems should be engineered for the specific occupancy and building design and should be coordinated with other life safety systems, such as the HVAC system (see Art. 5.4.2), fire alarm system (see Art. 5.5.4), sprinkler system (see Art. 5.4.7), and emergency power system (see Art. 5.2.3), so that they complement, rather that counteract, each other.

Frequently, a building's ventilation system is used for smoke control. If this approach is taken, the systems are called *engineered smoke control* systems.

Brief Code Review and Historical Perspective

The Building Officials and Code Administrators (BOCA) International, Inc., published the *BOCA National Building Code* in 1996, which is one of the standards adopted by many states and jurisdictions in the United States.[2] This code may not apply in specific jurisdictions. Local codes and regulations may be different, but they always govern.

In the late 1960s, the idea of using pressurization to prevent smoke infiltration of stairwells began to attract attention. This concept was followed by the idea of the "pressure sandwich," venting or exhausting the fire floor and pressurizing the surrounding floors.

Before 1993, the BOCA code required smoke control systems in covered malls, high-rise buildings, atria, and underground structures. In 1993, the BOCA code deleted the requirement for smoke control systems in covered malls and high-rise buildings because both occupancies require an automatic-sprinkler fire protection system.[2] Most of the new covered malls provide smoke management systems—even single-story—to qualify for the less stringent (and more realistic) exit distance requirements required by NFPA 101 24-4.1 exception (e).[3]

BOCA considers that automatic-sprinkler fire protection systems provide the best smoke control within the zone of incidence in a high-rise occupancy because, typically, the zone is not more than 40,000 ft^2 and a life safety area of refuge is easily accessed via an exit stair. Smoke control in a single-level covered mall is not required because life safety egress to the outside is readily available. A mall that has two or more levels typically requires an atrium smoke control system.

BOCA recognizes, however, that local code officials may require, and/or owners may choose to include, smoke control systems in buildings that are not specifically required by the code. These are considered *voluntary smoke control systems,* but, when implemented, they should conform to the provisions of BOCA 1999 Section 922.0, "Smoke Control Systems," and should be consistent with systems required by the code.

Neither NFPA nor BOCA have any general maximum area requirement for a smoke compartment. There are maximum area requirements in NFPA 101[3] and BOCA for smoke compartments in two specialized institutional occupancies (Type I-2 and I-3), hospitals and jails. The maximum for Type I-2 occupancy is 22,500 ft^2. The Type I-3 maximum is in terms of the number of beds. Some local codes may have such area limits. A smoke barrier is different than a 2-h-rated fire barrier. Any ventilating air duct that passes through that barrier must be controlled by an automatic smoke damper, and all penetrations through that barrier must be sealed. The general recommended practice on smoke zone size is that smoke zones should coincide exactly with sprinkler zones and fire alarm zones.

NFPA 90A contains three conditions that affect duct-installed smoke detector requirements: None is required for an air-handling system whose capacity is less than 2000 ft^3/min.[4] Smoke detectors are required only in supply air ducts whose capacities range from 2000 to 14,999 ft^3/min. Detectors are required in supply and at return air connections to each floor for systems whose capacities are 15,000 ft^3/min and larger. When these detectors sense smoke, the air handler is required to shut down. If the building has a fire alarm system, the duct detector must also signal the fire alarm system. Some jurisdictions may require sending an alarm to the fire department on a duct detector trip; others prefer that the signal be just a supervisory signal to be checked by the facility manager because of false alarm problems with smoke detectors. NFPA 92A and NFPA 92B are also excellent guides.[5, 6]

Smoke control is a very difficult challenge in the design and operation of buildings. Continuous research is ongoing to improve the design concepts, processes, and tools. The United States Fire Administration, National Fire Academy in Emmitsburg, Maryland, has developed a scale Plexiglas high-rise building model for interactive testing and observation of smoke flow under various conditions and smoke control scenarios. This facility provides valuable information for code officials, firefighters, landlords, building operators, architects, and engineers. For additional information, contact The National Fire Academy at (301) 447-1209.

Recommendation

The owner's insurance carrier may provide incentives for a voluntary smoke control system for asset protection, when a smoke control system is not yet required by building code. Such a system should be designed to control and reduce migration of smoke between the area of fire incidence and adjacent spaces. Smoke management systems are designed primarily for life safety. There is controversy as to their functional effectiveness and cost-effectiveness in protecting property in general occupancies. There are definite advantages for certain high-value occupancies that are especially vulnerable to smoke damage, but smoke control for property protection is an area where insurance companies and risk managers hold differing views.

Design Process

Before initiating the design of a smoke control system, the facilities manager and design engineer should meet with the authority having jurisdiction (AHJ), the owner's insurance underwriters, and the safety officer to resolve conflicting codes and to develop criteria for life safety and asset protection

Each member of this group has different primary concerns. Typically, building codes address the life safety requirements. Owners and their insurance carriers address property and asset protection to mitigate fire and smoke damage. Working together, a comprehensive approach can be developed that is useful throughout the design process and can save considerable time.

Stairwell Pressurization Systems.   Properly designed and built pressurized stairwells provide smoke-free escape routes in the event of a building fire. They also provide smoke-free staging areas for firefighters. To operate according to design, a pressurized stairwell on the fire floor must maintain a positive pressure difference across a closed stairwell door so that smoke infiltration is prevented.

Section 1015.0 of the 1999 BOCA Code requires that

> All exit stairways serving occupants of a floor level more than 75 feet above the level of exit discharge, or located more than 30 feet below the level of exit discharge serving such floors shall be protected by a smokeproof enclosure.

However, high-rise buildings are typically required to have fire protection sprinkler systems. Therefore, the design team or facilities manager may choose to provide stair pressurization systems for all interior exit stairways in lieu of a smokeproof enclosure in accordance with the provisions of BOCA paragraph 1015.7, "Stair Pressurization Alternative." The BOCA code requires that all stairs be enclosures but accepts a pressurized stair as an acceptable alternative to a smokeproof enclosure. Some states, however, may not accept that substitution as an equal and require that at least one stairwell be a smokeproof enclosure.

During building fires, some stairwell doors are opened intermittently during evacuation and fire fighting, and some doors may even be blocked open. Ideally, when the stairwell door is opened on the fire floor, airflow through the door should be sufficient to prevent smoke from entering the stair. Designing such a system is difficult because of the manner that combinations of open stairwell doors and weather conditions affect airflow.

Stairwell pressurization systems may be single- or multiple-injection systems. A single-injection system will supply pressurized air to a stairwell only at one location.

Single-injection systems in tall stairwells can fail when a few doors are open near the air supply injection point. Such a failure is especially likely when a ground-level stairwell door is open in bottom-injection systems.

To prevent such failures in tall stairwells, air should be supplied at multiple locations over the height of the stairwell. The supply fan may be located at ground level or on the roof, but ground level is preferred because smoke rises and a roof fan could reintroduce smoke into the stairwell. Supply duct risers should be located in a separate shaft.

Stairwell Overpressure Relief.   The total airflow rate is selected to provide the minimum air velocity when a specific number of doors are open. When all of the doors are closed, part of this air is relieved through a vent to prevent excessive pressure buildup, which could cause excessive door-opening forces. This excess air should be vented to the outside. Because exterior vents can be subject to adverse effects from wind, windshields are recommended.

Barometric dampers that close when the pressure drops below a specified value can minimize the air losses through the vent when doors are open. In systems with vents between the stairwell and the building, the vents typically have a special fire damper in series with the barometric damper. As an energy conservation feature, these special fire dampers are normally closed, unlike traditional fire dampers, but they open when the pressurization system is activated. This arrangement also reduces the possibility of the annoying damper chatter that frequently occurs with barometric dampers. This design is not recommended, however, because it is likely to fail over time.

An exhaust duct can provide overpressure relief in a pressurized stairwell. This system is designed so that the normal resistance of the exhaust duct maintains pressure differences within the design limits. No exhaust fan is provided. This is a simple design but provides little control.

Exhaust fans can also relieve excessive pressures when all stairwell doors are closed. A differential pressure sensor should control the exhaust fan, so that it does not operate when the pressure difference between the stairwell and the building falls below a specific level. This control should prevent the fan from pulling smoke into the stairwell when a number of open doors have reduced stairwell pressurization. Such an exhaust fan should be specifically sized so that the pressurization system performs within design limits, which are 0.15 to 0.35 in of water gauge. Because an exhaust fan can be adversely affected by wind, a windshield is recommended.

An alternate method of venting a stairwell is through an automatically opening stairwell door to the outside at ground level. Under normal conditions, this door would be closed and, in most cases, locked for security reasons. Provisions need to be made so that this lock does not conflict with the automatic operation of the system. Possible adverse wind effects are also a concern with a system that uses an open, outside door as a vent. Occasionally, high local wind velocities develop near the exterior stairwell door, and such winds are difficult to estimate without expensive modeling. Nearby obstructions can act as windbreaks or windshields. This approach is not recommended, because it degrades the pressurization control.

Another method of pressure control is a supply fan bypass system. In this system, the supply fan is sized to provide at least the minimum air velocity when the design number of doors is open. Modulating bypass dampers that are controlled by one or more static pressure sensors that sense the pressure difference between the stairwell and the building vary the flow rate of air into the stairwell. When all of the stairwell doors are closed, the pressure difference increases, and the bypass damper opens to increase the bypass air and to decrease the flow of supply air to the stairwell. In this manner, excessive pressure differences between the stairwell and the building may be prevented. Care must be taken if this approach is adopted. Without careful design, proper maintenance, and operation, smoke can be reintroduced into the building.

ELEVATOR SMOKE CONTROL

Elevator smoke control systems intended for use by firefighters should keep elevator cars, elevator shafts, and elevator machinery rooms smoke-free. Small amounts of smoke in these spaces are acceptable, provided that the smoke is nontoxic and that the operation of the elevator equipment is not affected. Elevator smoke control systems intended for fire evacuation of the handicapped (or other building occupants) should also keep elevator lobbies smoke-free.

Conceptual studies of elevator smoke control systems for handicapped evacuation indicate that the major problem was maintaining pressurization with open doors, especially doors on the ground floor. Of the systems evaluated, only the one that has a supply fan bypass with feedback control maintains adequate pressurization with any combination of open or closed doors. There may be other systems capable of providing adequate smoke control. The study procedure used by Klote and Tamura is an example of a method for evaluating the performance of a system to determine whether it meets the particular characteristics of a building under construction.[7] Transient pressures caused by the piston effect of an elevator car moving in a shaft have been a concern in elevator smoke control. The piston effect is not a significant concern for slow-moving cars in multiple car shafts. However, for fast cars in single-car shafts, the piston effect can be considerable.

HIGH-RISE BUILDING AND ZONE SMOKE CONTROL

In a high-rise building that has only code stairwell pressurization, smoke can flow through cracks in floors and partitions and through shafts to damage property and threaten life at locations remote from the fire. Although not required by the BOCA Code, a voluntary-zone smoke control system for limiting such smoke movement may be desirable. In some areas, this feature is part of code.

In a zoned smoke control system, the building is divided into a number of smoke control zones, each zone separated from the others by partitions, floors, and doors that can be closed to inhibit smoke movement. In the event of a fire, pressure differences and airflow produced by mechanical system fans limit the smoke spread from the zone in which the fire initiated. The concentration of smoke in the zone of incidence is exhausted at the rate of six air changes per hour. In buildings that have zone smoke control systems, the occupants should evacuate the zone of incidence immediately after detecting the fire. Testing of full-scale fires has demonstrated that zone smoke control can restrict smoke movement to the zone where the fire started.

A smoke control zone can consist of one floor, more than one floor, or a floor can be divided into more than one smoke control zone. Sprinkler zones and smoke control zones should be coordinated so that the sprinkler water flow will activate the zone's smoke control system.

Zoned smoke control is intended to limit smoke to the smoke zone by maintaining an air pressure difference between the smoke zone and adjacent zones. Pressure differences across the barriers of a smoke zone can be achieved by supplying outside (fresh) air to nonsmoke zones, by venting the smoke zone, or by a combination of these methods.

Venting smoke from a smoke zone by exterior roof exhaust, wall vents, smoke shafts, or mechanical exhaust reduces overpressure due to the thermal expansion of gases caused by the fire. However, this venting may result in creating a pressure difference, and the smoke concentration in the smoke zone may be only slightly reduced.

The information provided is only a short introduction to a very important subject. More research is being conducted in this field. For more information, the reader is referred to the resource material listed at the end of the article. ASHRAE is also a valuable resource.[8]

ATRIUM SMOKE CONTROL SYSTEMS

An atrium is an occupied space that includes a floor opening or a series of floor openings, connecting two or more stories.

The BOCA Code paragraph 404.4 requires installing a smoke control system complying with Section 922.0 in all atria that connect more than two stories. However, some AHJs require a smoke control system in a two-story atrium. Accordingly, the local authority should be consulted. Per BOCA, atrium smoke control systems must be designed to keep the smoke interface layer 6 ft above the highest floor level of exit access open to the atrium for a period of 20 min. An active smoke control system is not required by the BOCA Code where calculations demonstrate that the smoke interface level requirement will be met without operating a smoke exhaust.

Sections 922.0 and 922.2.1.2 of the 1999 BOCA Code provide formulas for calculating the required smoke interface layer volume (smoke reservoir) 6 ft above the highest floor level of atrium exit access for a period of 20 min. When the available volume does not meet the calculated smoke-filling criteria, a mechanical smoke control system should be provided to maintain the required interface layer. This section also provides the formula for determining the volumetric rate of smoke production and the minimum rate of exhaust to maintain the interface layer. The mechanical smoke control system should be designed to provide a ventilation rate of at least two air changes per hour from the space where smoke control is required.

BOCA paragraph 922.2.4 allows alternatives to achieve the same level of smoke control required by section 922.0; for example, the atrium in the Copley Place Retail Mall and Office mixed-use complex, constructed in 1982 in Boston, Massachusetts, was designed to provide four air changes in the volume of the areas that have a common atrium. This system consists of a series of propeller exhaust fans at the top of the atrium that provide approximately 600,000 ft^3/min of exhaust. Makeup ventilation air is introduced from gravity intakes designed for 200,000 ft^3/min at each end of the mall to provide a flow of fresh air along the egress route that flows toward the atrium exhaust. The top of the Copley Place atrium is 150 ft above the main floor. Therefore, two 100,000-ft^3/min smoke entrainment air jets were integrated in the water sculpture at the base of the atrium to direct smoke up into the reservoir and exhaust above and to provide a tenable egress path. Testing and operating the Copley Place atrium smoke control system demonstrated the success of this system.

Other large-volume spaces besides atria—covered malls, arenas, exhibition halls, and auditoriums—where compartmentation cannot be used to separate occupants from smoke for egress have other challenges.

Commissioning and Acceptance Testing

Regardless of the care, skill, and attention to detail with which a smoke control system is designed and built, an acceptance test is needed to ensure that the system, as built, operates as intended. An acceptance test should be composed of two levels of testing. The first is functional to determine if everything in the system works as designed. The functional checkout is intended to discover misoperation of the equipment (e.g., fans that operate backward, fans to which no electrical power was supplied, and controls that do not work properly).

The second level is testing to determine if the system performs adequately under the required modes of operation. This can consist of measuring pressure differences across barriers under various modes of smoke control system operation. When airflows through open doors are important, they should be measured. Chemical smoke from smoke candles (sometimes called smoke bombs) is not recommended for performance testing because it normally lacks the buoyancy of hot smoke from a real building fire. Smoke near a flaming fire has a temperature in the range of 1000°F to 2000° F. Heating chemical smoke to such temperatures to emulate smoke from a real fire is not recommended unless precautions are taken to protect life and property. The same comments about buoyancy apply to tracer gases. Thus, pressure difference testing is the most practical performance test.

Guidelines for commissioning smoke management systems are available in the ASHRAE *1999 Applications Handbook*[8] and in NFPA 92A[5] and NFPA 92B[6] for large-volume spaces.

CONTRIBUTORS

Arjun Rao, P.E., Vice President, BR+A, Boston, Massachusetts

William J. Goode, President, W. J. Goode Corporation, Walpole, Massachusetts

David M. Elovitz, President, Energy Economics, Natick, Massachusetts

Frederick M. Gibson, Associate, Taylor & Partners, Boston, Massachusetts

REFERENCES

1. J. T. O'Hagan, *High Rise/Fire and Life Safety,* R. L. Donnelly and Sons Company, Chicago, IL; 3 NFPA, *Life Safety Code Handbook,* 7th ed., National Fire Protection Association, Quincy, MA, 2000.
2. *The BOCA National Building Code 1999,* The Building Officials and Code Administrators, International, Inc., Country Club Hills, IL, 1999.
3. NFPA 101, *Life Safety Code,* National Fire Protection Association, Quincy, MA, 2000.
4. NFPA 90A, *Installation of Air-Conditioning and Ventilating Systems,* National Fire Protection Association, Quincy, MA, 1994.
5. NFPA 92A, *Recommended Practice for Smoke Control Systems,* National Fire Protection Association, Quincy, MA, 1996.
6. NFPA 92B, *Guide for Smoke Management Systems in Malls, Atria, and Large Areas,* National Fire Protection Association, Quincy, MA, 1995.
7. J. H. Klote and G. T. Tamura, "Smoke Control and Fire Evacuation by Elevators," *ASHRAE Transactions* (1986).
8. ASHRAE, "Smoke Management" (Chap. 51), in *1999 Applications Handbook,* American Society of Heating, Refrigeration, and Air Conditioning Engineers, Inc., Atlanta, GA, 1999.

BIBLIOGRAPHY

R. W. Bukowski, "Fire Models, the Future Is Now!" *NFPA Journal* 85, no. 2 (1991).

J. H. Klote, A computer program for analysis of smoke control systems. *NBSIR 82-2512.* U.S. National Bureau of Standards, Washington, D.C., 1990.

J. H. Klote, "Fire Experiments of Zoned Smoke Control at the Plaza Hotel in Washington, DC," *ASHRAE Transactions* 96, no. 2 (1990), 399–416.

J. H. Klote and J. A. Milke, *Design of Smoke Management Systems,* American Society of Heating, Refrigeration, and Air Conditioning Engineers, Inc., Atlanta, GA, 1992.

J. A. Milke and F. W. Mowrer, "Computer Aided Design for Smoke Management in Atriums and Covered Malls," *ASHRAE Transactions* 100, no. 2 (1994).

NFPA 204M-98, *Guide for Smoke and Heat Venting,* National Fire Protection Association, Quincy, MA, 1988.

G. T. Tamura, "Field Tests of Stair Pressurization Systems with Overpressure Relief," *ASHRAE Transactions* 96, no. 1 (1990), 955–958.

G. T. Tamura and A. G. Wilson, "Pressure Differences for a 9-Story Building as a Result of Chimney Effect and Ventilation System Operation," *ASHRAE Transactions* 72, no. 1 (1996), 180–189.

UL, *UL Standard for Safety Fire Dampers,* 4th ed., UL 555-90, Underwriters Laboratories, Northbrook, IL, 1990.

UL, *UL Standard for Safety Leakage Rated Dampers for Use in Smoke Control Systems,* 2nd ed., UL 555S.93. Underwriters Laboratories, Northbrook, IL, 1993.

ARTICLE 5.5.6
TELECOMMUNICATIONS AND DATA DISTRIBUTION SYSTEMS

Dennis R. Julian, P.E., RCDD
Carter & Burgess, Inc., Boston, Massachusetts

TELECOMMUNICATIONS SYSTEMS, THE FOURTH UTILITY

Telecommunications, often called *tel/data* by the architectural community, is a global term that encompasses voice, data, and video systems, as well as structured cabling to support these and other systems. Adequate access and availability of telecommunications systems are critical to performing the missions of almost all organizations. This inherent need requires that telecommunications systems become part of initial project planning to ensure adequate physical space, pathways, and building support functions. These systems are subject to obsolescence as technology progresses. The goal is to plan today for tomorrow's technology by providing adequate infrastructure and pathways to accommodate advancing technology and changing business needs. Telecommunications is the fourth utility, critical to proper facility operation, and is as important as electricity, HVAC, and water.

In addition to cabled telecommunications systems, wireless systems can provide some distinct advantages. Where cabling is difficult or impossible to install, as in historic buildings, in wide open spaces between buildings, or where mobility or temporary installations are required, wireless systems may be the optimum solution. The development of industry standards allows designing and installing infrastructures that enables the investment made to be beneficially used into the foreseeable future by the variety of vendors and equipment that comply with these standards. Facility designs must be based on standards. This article provides the information and resources needed to verify and require that the consultant's proposed design at least meets these standards.

The design of the pathways and grounding must comply with the ANSI/TIA/EIA standards for telecommunications infrastructures, which have distinct differences from standard electrical installations. If cabling systems are not installed properly, the specification and purchase of high-performance materials are invalidated. Improper installations can result in inferior performance (e.g., Category 3 performance from Category 5 components). When the installation is complete, the cabling infrastructure must be tested and documented. To help ensure that installers are qualified and well-trained, the specifications should require the manufacturer to provide a system performance extended warranty that covers material and labor.

The telecommunications infrastructure comprises only a few percent of the total cost of a project but is responsible for many of the operational problems. Therefore, it makes sense to develop a system that is reliable, is installed well, has plenty of operational headroom (performance), and is well-documented, so as to simplify troubleshooting and future moves, adds, and changes (MAC).

General Issues

The telecommunications industry has its own terminology that, when understood, demystifies the work and the requirements. The typical industry definition, as it is used in this book, is that *telecommunications* is an all-encompassing term that includes all forms of communications, including but not limited to telephone, data, and video. The word *technology* refers to the equipment such as telephone switches, PBXs, computers, and network electronics, including routers, data switches, and hubs, that enables communications to occur.

The designer should be licensed and experienced in designing cabling system infrastructures and knowledgeable in the applicable codes, standards, and building design. Telecommunications pathways and cabling must follow electrical and building codes and therefore require a professional engineer's stamp on permit drawings. Qualifications should be reviewed for past performance on projects, references, and membership in applicable industry organizations such as Building Industry Consulting Service International (BICSI), Society of Cable Telecom Engineers (SCTE), Association of Cabling Professionals (ACP), and industry certifications such as BICSI's Registered Communications Distribution Designer (RCDD) or equivalent education and training.

Telecommunications cabling infrastructure must follow electrical and building codes, as well as industry standards and manufacturers' requirements. Industry standards allow designing cabling system infrastructures that support the systems to be connected without knowing all of the particulars of the technology to be deployed. By designing systems and specifying components that comply with industry standards, we can install systems that are cost-effective and usable in the foreseeable future. Specified systems should comply with standards, and a third-party tester should verify manufacturers' claims. It is possible to specify systems that exceed the minimums set by the standards, but for our protection they must be verified after installation. When exceeding the standards, we must be careful not to pay excessively for features that are available but may have no future value. Decisions to provide advanced component and cable designs should be reviewed and justified.

Telecommunications systems require dedicated physical spaces in buildings for equipment, cabling pathways, wire management, and cabling terminations. These spaces have specific requirements, including environmental needs, grounding, security, and location of EMI sources. The standards have strict overall cable length restrictions governing the location of telecom rooms with respect to the locations of the outlets being served. Therefore, telecommunications rooms must be located early in the design process (schematics), so that they are properly located and integrated into the architectural layout and program.

Installing the telecommunications cabling infrastructure according to regulations and standards is as important as quality of the materials specified. Installers must be licensed for the work they perform. Installers should be trained and certified by the manufacturer whose products they install. It is also beneficial to have third-party certification of their knowledge as evidenced by BICSI installer certification or other equivalent certification.

Physical Topology.   There is a variety of network protocols that use different types of wiring configurations. The goal of a structured cabling infrastructure system is to provide cabling that can be configured as required for any network topology. The token ring topology uses a ring bus configuration, and an Ethernet topology uses a bus configuration. By making the appropriate connections to the horizontal cabling in the telecom room, either configuration can be accomodated.

The TIA/EIA 568 Telecommunications Wiring Standard indicates that a facility should be designed with a *hierarchical star topology*. This is a design in which the workstations are cabled (i.e., horizontal wiring) to a telecom room *horizontal cross connect* (HC), sometimes called an *intermediate distribution frame* (IDF). Each telecom room is cabled to the telecom room *main cross connect* (MC), sometimes called the *main distribution frame* (MDF), using what are called *backbone cables*. The standard includes strict length limitations for horizontal and backbone cabling.

A *local area network* (LAN) is a network that is within the facility. A *wide area network* (WAN) is a network that interconnects multiple facilities in various geographical areas, typically over circuits provided by various service providers. A *metropolitan area network* (MAN) is a network based within a city.

Internet/Intranet.   Connections to the Internet or to a company intranet server (a private Internet-like system dedicated to the company) are typically accessed via the LAN. High-speed connections outside the company are typically provided by the telecommunications service provider through switched or leased lines, or through a dedicated wide area network using T-1 or other high-capacity lines.

Convergence of voice, data, video, CATV, and the Internet is beginning. This will mean that all systems will become server-based and will require data-grade wiring with sufficient bandwidth to transmit all of these types of signals simultaneously. Design should organize all cabling and equipment into one multipurpose room to allow convenient interconnection as these services migrate together.

Multiple System Use (Intelligent or Smart Building)

One of the advantages of a well-planned cabling infrastructure for a facility is the ability to use it for multiple building systems such as fire alarm, security, building automation and management, and lighting controls. These systems can be monitored from a computer on the LAN or Internet, providing centralized management or remote monitoring. By using the installed cabling infrastructure, there are economies in cost and added flexibility in adding or changing the system, including the head end. An additional benefit is in knowing where the wiring for these various systems is located and that it is protected in the telecom rooms. The common cabling systems can be used as dedicated cabling for each system, or if designed appropriately, could operate over a common network system.

Reliability/Redundancy.   Telecommunications systems are the lifeblood of most organizations. Therefore, providing the appropriate level of reliability and redundancy must be considered during planning and design.

Redundancy can be provided by supplying multiple separate services to the facility, starting at the incoming telecommunications services. Each service could be diversely routed to the facility from the same or different central office(s). Service could be provided from different service providers. When reviewing different service providers, find out if they use the same pathways (utility poles and duct banks) or are colocated in the same central office. If fiber optic services are required, is the fiber system a SONET-type ring system that will route signals in the other direction if one side of the ring is interrupted?

Once inside the facility, separate backbone cabling systems diversely routed to separate telecom rooms should be considered. This would allow a loop design for the backbone system and permit signals to travel in either direction.

Issues to consider during design include protecting the installed cabling. Cabling installed in enclosed raceways is less likely to be accidentally damaged. Termination locations and equipment located in dedicated locked rooms minimizes the possibility of accidental interruptions. Proper support and location of cabling will minimize failure of cabling systems due to stress or accidental damage.

The need for reliable power with surge protection and battery or generator backup for the electronic equipment should be determined. Surge protection should always be provided to protect the electronic equipment. Power spikes and surges damage electronic equipment, lessen its useful life, and increase maintenance and data corruption. An uninterruptible power supply (UPS) system that has continuous filtering and surge protection is the best method of providing good quality power to electronic equipment.

Reliability and redundancy for the cooling system should be considered and provided if justified by the critical nature of the systems served.

Backup power should be provided for the telephone systems and systems that affect life safety or critical processes. Emergency lighting should be provided in telecom rooms where backup power is required.

Telecommunications Service Choices

Telecommunications is an integral part of most businesses, and therefore it is important that the connection to the outside world (i.e., telecommunications service) is appropriate for the business and its use. Arrangement for service is an owner's responsibility. There may be more than one telecommunications service provider available, in addition to the local telephone company or local exchange company (LEC), or the building landlord may have made agreements with alternate telecommunications providers to bring enhanced offerings into the building. Services may also be available via satellite, microwave, or other wireless technology, or through the cable TV system.

The availability of service from multiple vendors and the need for redundancy (whether multiple incoming conduits or diverse routing to multiple telephone central offices) are decisions that should be made early in the project, so that the same level of design can take place in the building. The need for multiple vendors and/or redundancy is a business decision. It may be required for the proposed tenants for the building, for the proposed services to be provided, or to allow using the most cost-efficient telecommunications provider. In designing the incoming service, the need for multiple service points, conduit paths, and termination points in the building and the need for satellite dishes or cable TV service as supplemental or redundant services should be considered.

Contact telecommunications service providers early in the design process. The time needed to bring high-speed or redundant services or fiber to a project could be considerable and may affect the occupancy date. For projects in new locations that require wireless access, verify before committing to the new location that a suitable location also exists for satellite dishes, microwave dishes, or radio towers.

Electromagnetic Interference (EMI).   The flow of electricity creates electrical and magnetic fields. The strength of these fields varies directly with the amount of electrical power available and the distance from the source. Typical sources are electrical power lines, motors, transformers, lighting ballasts, copy machines, radio transmitters, radar facilities, and microwave antennas. These electrical fields impose electrical currents in conductors that are within their field of influence. Telecommunications and electronic systems operate at voltages under 5 V, and newer systems that are more compact operate at 3.3 V. The low voltage level of these signals means that even small amounts of interference can disrupt transmissions. The interference can show up as false signals on cables or as interference with the operation of equipment. The problems can occur intermittently, and they can sometimes be difficult to troubleshoot. The problem may appear only when certain motors, lighting circuits, or dimmers operate. It could show up as a computer monitor operating poorly, or with a shrunken screen, or with wavy lines because of a magnetically induced field from electrical panels, cables, or transformers located in adjacent spaces. Magnetic fields of 8 milligauss can affect monitors, and electric fields of 1 to 3 volts per meter can affect electronic equipment.

There are many ways to counter these problems, but the best way is to avoid them in the first place. EMI fields drop off rapidly with distance. Therefore, locate sensitive equipment and wiring away from sources of interference. Practically, this is only partially possible because of the proliferation of electrical sources and sensitive equipment and wiring. Therefore, during the design process, it is necessary to be aware of these concerns and locate and install wiring and equipment appropriately. Shielding can be effective in reducing EMI effects. Shielding for magnetic low frequency fields such as from transformers is difficult, requires special materials and designs, and, other than increasing separation, typically requires expert advice. The use of totally enclosed raceways that are properly grounded can reduce some kinds of EMI.

Telecommunications equipment rooms should be located away from large electrical sources such as transformers and panels in adjacent electrical rooms, copy machines, or large motors. A wall of masonry or plasterboard does not provide much protection other than distance. Locations adjacent to elevator shafts or mechanical rooms are also a concern. Locating telecommunications equipment on the far side of a room, with an aisle along the wall that is common with the potential sources, puts additional distance between the equipment and may be enough to avoid problems.

Cabling must be installed properly to avoid problems and interference. The construction of unshielded twisted pair (UTP) cabling provides some protection. Additional protection can be attained from installing cable in a properly grounded ferrous metal raceway that will act as a shield. Shielded twisted pair (STP) or fiber-optic cabling can also be used in areas of concern. Typical sources of interference result from poor installation techniques such as installing cables over or too close to lighting ballasts or adjacent to or parallel to electrical power lines. Installing cables a minimum of 2 ft from electrical equipment or raceways and 5 ft from large or high-voltage power sources or transformers will minimize potential interference problems.

The references listed at the end of this article contain additional sources of information on EMI issues and solutions.

Voice System Cabling

Telephone systems can be analog, digital, or a combination of the two. The majority of newer systems use digital technology. The cabling systems required are the same, except that digital systems normally require higher-performance cabling. It is common to install Category 3 cabling for a telephone system, which provides better quality and more pairs of conductors than may typically be required. This is done to help combat interference on the telephone line and to provide capabilities for future technologies or to allow multiple telephone extensions. Higher-performance cabling equal to data cabling is sometimes provided to allow future flexibility or to allow for the convergence of voice and data systems.

Analog modems are typically provided for external communications at computers, building management systems, and equipment that notifies vendors of equipment status or for troubleshooting. An analog telephone line or electronics are required to allow the modem to work over a digital line.

Telephone service cabling to the central office (CO) of the telephone service provider can be copper or fiber-based, analog or digital. Selection of the appropriate service connection depends on the bandwidth capacity (amount and type of traffic expected) and an economic evaluation of the installation and maintenance costs involved. Installation costs may vary by the distance from the service provider's central office, the installed capacity available in the area, and the revenue that the provider estimates it will get. There may be several service providers in addition to the major utility in the area. Alternate providers can be used to keep costs competitive and to provide an alternate redundant path for communications with the outside world.

If the telecommunications service required warrants it, the telecommunications service provider may provide fiber for voice and data services. The voice backbone cabling in a building is typically copper because of the high cost of converting light signals on fiber-optic cables to the electrical signals carried on copper cabling and used by equipment.

Telephone Systems

Telephone systems come in many configurations but break down into three types. Centrex systems (Central Office Exchange) have dedicated lines from the local telephone service provider central office for each phone line. PBX systems have a limited number of dedicated lines from the central office and then switch them within the PBX to the appropriate phone extension. Key switches (KSU) typically have up to 16 telephone lines that appear on each telephone in a small office.

In a Centrex system, the telephone service provider owns the switching equipment. This can result in lower initial costs. Centrex-type systems also allow facilities with multiple locations to call different extensions in other locations by entering only a 3- or 4-digit number instead of a 7- or 10-digit phone number. The operating costs and maintenance costs of a Centrex system have to be evaluated to determine if it will be a cost-effective approach. The operating costs can be high because in essence the necessary switching equipment is being rented from the telephone service provider and each line is subject to monthly charges. Modifications required by changing or moving personnel incur additional costs. The owner also does not necessarily have ready access to enhanced management and billing features. The advantages are that a Centrex system may not require additional space, power, or other infrastructure on the premises and is in fact managed by the service provider. Centrex services may require special cabling terminations, depending on the type of instrument installed at the desk.

A PBX is a telephone switching system. It is located on the premises and is maintained and operated by the owner or technicians contracted from the manufacturer or from a third-party vendor. A PBX can be part of a stand-alone system in which an appropriate number of outside phone lines are brought into the facility from the telephone service provider and these phone lines are switched to the phone extension. This capability allows renting a smaller number of phone lines from the telephone service provider. It is also possible to train in-house personnel to manage the system features and perform the routine changes required by changes and moves of employees, thereby reducing monthly maintenance costs. Independent telephone systems or key switch units networked to PBX systems also have many other features that may be beneficial and must be evaluated. A PBX can be used in conjunction with a Centrex system to allow a cost-effective method of connecting to many small remote facilities such as doctors' offices or to provide a central operator to answer general phone calls.

Voice mail does not affect the type of cabling infrastructure installed because it is integrated with or connected to the main telephone system. Voice mail is digital recording of phone messages. Multiple features are available to customize the system. One of the most critical features of the system is the size of the storage device for messages. The more message minutes stored, the larger and more expensive the device. Therefore, when specifying the design criteria for the system, the number and length of the messages to be stored are important. It is desirable to get a system that uses variable-length messages so that a message will take up only the space it actually requires rather than a set length of time. Enhanced features, such as an automated attendant to answer and process incoming calls, interactive voice response (IVR) for callers to access a database with account information (such as a bank), and unified messaging that offers the ability to download the voice mail onto a PC, are available in many systems. Multiple options exist in phone systems, including the ability to have portable phones or a local wireless system. A local wireless system allows the flexibility, range, and features of cellular phones to be used within the facility. It is sometimes possible to have a system that interfaces with the local cellular provider, thereby also allowing the same phone to be used off-site as a standard cellular phone. Range limitations and security are concerns. High-frequency transmission signals such as 900 MHz or 2.4 GHz extend the usable range. Spread-spectrum or frequency-hopping protocols enhance security. Range is affected by the type of building structure and construction materials. A distributed antenna system throughout the facility can extend the usable signal range.

The most appropriate way to purchase a new telephone system is to use the request for proposal (RFP) method. The RFP method is appropriate because the systems and features offered by each vendor are not exactly alike and may vary in ways that may be critical to an appropriate selection. When an RFP is issued, it should outline the capabilities and services that the owner requires. The RFP should require interested and qualified vendors to make proposals and presentations on their offerings. The RFP should specify a specific response format so that features, capabilities, and pricing can be compared for the initial purchase and the life of the system. Upon reviewing the proposals, the costs and features that a vendor offers can be selected, or a modified request for proposal can be issued, to allow the vendors to bid on similar system capabilities and features. The owner should investigate potential affiliations with government or industry organizations that have bulk purchasing arrangements with system manufacturers. The specification and purchase of the telephone system should be done when the facility is almost ready for installation (with consideration of the required time to manufacture and install the system). Waiting allows purchasing the latest technology at the best price.

Data Systems

Data system networks operate on a variety of protocols and topologies (e.g., Ethernet, fast Ethernet, token ring, 1000 BaseT [gigabit], CDDI, FDDI, ATM, etc.). The advent of industry standards published by the Institute of Electrical and Electronic Engineers (IEEE), the Telecommunications Industry Association (TIA), and the Electronics Industries Alliance (EIA) have simplified the installation of cabling infrastructures. Most vendors have agreed to provide equipment that will operate correctly on the basis of parameters given in these standards, although some proprietary configurations still exist. This allows installing a cabling infrastructure during the design phase of a project before determining the exact type of data network system. Implementing a standards-based infrastructure design allows installing any standard compliant equipment with the assurance that it will operate correctly. This allows the design of the project to take into account the needs of the telecommunications systems, including the physical space, cooling, and power, and to incorporate them seamlessly into the design cost-effectively and with minimal disruption.

Video Systems

Video systems can be operated over a variety of media types, including 75-ohm coax cable (750-MHz rated), fiber, or twisted-pair wiring. Video systems used in teleconferencing can operate over Integrated Services Digital Network (ISDN) or T1 connections from the local telephone service provider or through the PBX. However, coax is still viable for cable or broadcast television, in-house recorded programs, or closed-circuit security cameras (CCTV). Videoconferencing over the telephone system can be marginally done on normal phone wiring, but a minimum bandwidth of three ISDN lines is required for full-motion video of good quality.

In the future, more video will be carried over the LAN and will require a lot of bandwidth. It is useful for videoconferencing at the workstation but will require each workstation to have the required pipeline (cabling and electronics) to handle the large amount of traffic it will generate. If an adequate pipeline can be provided, the resulting video can be satisfactory. High-performance copper cabling operating above 100 MHz should be adequate, but fiber-optic cabling could also be considered.

Systems can be designed as *baseband* or *broadband* systems. Baseband systems carry one signal or channel at a time. A multichannel system requires a home run from each outlet to the head end, and the channel selection occurs at the head end. In a broadband system, multiple signals are sent over the cabling at the same time, and the tuner at the outlet selects the channel or signal desired. Broadband cabling does not require a dedicated cable from each outlet to the head end, and therefore it is typically more economical to install.

Most broadcast television signals are analog, but there is a movement to digital television in normal broadcast TV. The main advantage of this is better picture quality. Another advantage is the ability to store programs on digital storage devices (e.g., hard disks) that would allow multiple viewers at one time, each with different starting, stopping, and control of the program. This is known as *video on demand.* Interactive video systems that allow users to request information interact with an information source such as a Web site or collaborate as in a typical videoconference will become increasingly popular.

Video systems can be designed for remote media control. These are systems that allow locating media sources such as modulators for TV channels, VCRs, CD-Is, laser discs, and

DVD at a central location and controlling them remotely at the outlet. These systems can be controlled through the telephone keypad, the computer, and by using infrared remote control.

Cabling System Categories/Levels of Performance

The specification of the level of performance of telecommunications system components comprises many different requirements and values that vary by manufacturer and type of design. To simplify system specification and review, a classification system was developed. The TIA/EIA standards define categories of performance with specific requirements. Because the performance requirements are specified, third-party verification of manufacturer's claims is possible by nationally recognized testing laboratories (NRTL) such as UL or ETL. Refer to Table 5.5.6-1, "Telecom Cabling Systems Performance Categories."

The development of standards is a process of consensus; therefore, standards always lag behind the advances in the state of the art. The telecommunications industry has pushed beyond the performance parameters of the latest standards, especially with the available types of cabling systems that can be installed in the facility infrastructure. The concern with the various types of cabling systems is to determine what is mostly marketing hyperbole and what will be needed in the future. To assist in selecting cables, some common benchmark is required for the many parameters that must be determined and verified. Manufacturers have foreseen this concern and have expanded upon the available levels of performance. These levels, or proposed enhanced categories, have specific performance requirements of the cable and also of the jacks, patch panels, patch cords, and equipment cords that connect a piece of equipment to the backbone connection. These performance levels correspond to the antici-

TABLE 5.5.6-1 Telecom Cabling Systems Performance Categories*

TIA/EIA 568 Category	CENELEC EN 50173 (Equivalent) ISO/IEC-IS 11801 Class	Canadian CSA T529-95	Frequency	Remarks
1	A	1	Up to 100 kHz	Plain old telephone service (POTS) voice and low speed application—modem
2	B	2	1 MHz	Low-speed data, ISDN
3	C	3	16 MHz	Minimum for horizontal cabling Typical voice-grade cable
4	—	4	20 MHz	Token ring—cable is rarely used
5	D	5	100 MHz	100 megabits, superseded by Category 5e
5e		5e	100 MHz	Higher performance than category 5 (gigabit level)
6	E	6	250 MHz	Proposed—approval expected in 2000–2001
7	F	7	600 MHz	Proposed—approval expected in 2001; fully shielded pairs and an overall shield may allow multiple signals in one cable

* Some vendors have programs to certify cabling systems to specific levels of performance that do not directly correspond to the categories identified in the standards.

Note: Mexico's standard is a combination of TIA/EIA 568 and ISO/IEC 11801 and is called NMX-I-248-1998-NYCE.

pated requirements of the various proposed uses of the networking cabling infrastructure, including gigabit speeds at frequencies beyond 100 MHz. Vendor's claims must be third-party verified and tested. In addition, it must be determined whether the performance requirements apply to the system (i.e., channel compliance) or to the individual components (i.e., component compliance).

Copper Cabling. Copper cabling comes in a number of constructions to provide the performance demanded by the high-speed, high-frequency applications it serves. To control performance parameters, cables are constructed with tightly controlled twists and orientation of the pairs of conductors. If the twists or orientations of the individual conductors are disturbed, the performance of the cabling is reduced. That is the reason that the installation methods used to install these systems are important and why it is necessary to test the final installation. Testing will verify that the performance specified and bought through the bid documents is actually provided. Unshielded twisted-pair (UTP) cabling is the most common cabling construction in the United States. Shielded twisted-pair (STP) and screened twisted-pair (ScTP) cabling use a metallic covering over the conductors that minimizes the electrical interference given off by the cable and also the electromagnetic interference (EMI) that affects the cable from outside sources. Shielded cabling has a shield over the individual pairs of conductors in a multipair cable, and screened cables have shielding only over the outside of the multipair cable. European facilities in particular use mostly shielded cable because of stricter requirements to minimize EMI radiation. The proposed Category 7 system is a fully shielded cable that provides high performance and immunity from EMI. Category 7 systems may be cost-competitive with fiber-optic systems in particular when considering the cost of electronics and the ability for Category 7 systems to carry multiple signals under one sheath without mutual interference.

Fiber-Optic Cabling. Fiber-optic cabling is available in various configurations, depending on the application requirements. Fiber-optic cabling uses a light source to transmit a signal. This process requires converting the normal electrical signal to light and than back again at the receiving end. This necessary conversion increases the cost of the total fiber-optic system. The advantages of fiber are in its electrical isolation properties (no electrical connection from end to end eliminates the concern over ground differentials), its immunity to electromagnetic interference, and the extreme difficulty in tapping or eavesdropping without being detected. Typically, fiber is made of glass. However, plastic fiber is available and is acceptable for certain technologies over short distances.

Fiber is also available with *tight* or *loose* buffer construction. Tight buffer construction is typically used indoors and consists of fiber in a plastic covering with strength members added (preferably nonmetallic). Loose buffer construction is typically used outdoors and consists of fiber in an oversized plastic covering that is then filled with gel. The gel prevents moisture penetration and cushions the fiber. Fiber is available in either *multimode* or *single-mode* construction. Multimode fiber is typically used within buildings for shorter distances, and uses lower-cost LED light sources. Single-mode fiber is typically used for long distances or extreme bandwidth applications and uses a more expensive laser light source. Multimode fiber comes in various sizes. The present TIA/EIA size standards call for multimode fiber of 62.5/125-μm. A 50/125-μm fiber that uses a laser light source to increase transmission speeds is under review by the standards committee.

Coax Cabling. Coax cable is no longer the preferred cable type for data systems but is still widely used for video applications even though video over twisted pair or fiber is becoming more feasible and popular. Coax should be installed for video applications to specific areas such as conference centers, auditoriums, patient rooms, or other areas that require virtual full-time use of CCTV or CATV. However, coax to UTP baluns are available now that allow UTP some flexibility in distributing these video signals without installing coax to every workstation faceplate. Coax is still used by some vendors in proprietary applications.

Cabling Insulation.   Cabling is available with a variety of insulation types, depending on the usage and characteristics required. The most important criteria are the requirements of the electrical and building codes. The codes define requirements based on the areas in which the cable will be used. In air-handling plenum spaces (e.g., above ceilings used for supply or return air), a low smoke and fire propagation type of insulation is required. Similar but less stringent criteria are required for cables rising through a building. In addition to the code requirements, insulation also affects the ease of installation and the electrical characteristics of the cable. One type of insulation that provides enhanced performance of the cable electrically and has low smoke and low fire propagation properties is a fluoropolymer resin, more commonly known as Teflon® FEP, manufactured by Dupont. FEP is used as conductor insulation because of its superior electrical characteristics, and as the cable jacket for its smoke and fire properties.

For enhanced performance and for the cabling to properly handle future networking protocols, it is important that all four pairs of the cable have the same insulation. Different types of insulation on individual pairs of wires in the same cable create differences in the performance characteristics of each pair (i.e., skew). When networking protocols start using all four pairs simultaneously (e.g., gigabit Ethernet), these differences affect the performance of the cable.

Wireless Transmission

Wireless (radio frequency, microwave, infrared, cellular, satellite) communications are also possible and sometimes the best solution in some circumstances. In most cases, these technologies supplement a good telecommunications cabling infrastructure. In general, wireless transmission technologies are slower and more expensive than hardwired systems. For particular situations they may be the most advantageous solution to overcome the inability to install cabling because of the need for mobility, because spaces are wide open, or because installations are temporary. Some potential applications include historic buildings, trade shows, temporary installations to support ad hoc teams, or for connecting remote buildings. The building structure, environmental conditions (weather), and other signal sources can affect wireless transmissions, thereby reducing transmission distance, speed, and disrupting or blocking transmission completely. Security of wireless transmissions must also be considered. Because any airborne signal can be intercepted, appropriate precautions such as encrypting transmissions should be employed. It may not be desirable to broadcast sensitive company information over the airways.

Microwave transmission is used outdoors in line-of-sight applications as an alternative to dedicated landlines. Radio frequency can be used indoors or outdoors over short distances and is also an alternative to dedicated landlines. Radio frequency and cellular transmission are similar technologies and suffer from some of the same concerns for the way they interact with the human body and other electronic equipment. There is a perceived concern by the public that these technologies may cause medical problems. There is a realistic concern that certain frequencies used by these technologies and/or the intensity of the transmitted signal will interfere with the proper operation of electronic equipment, and therefore, they have been banned or restricted in some facilities. Examples of this are cellular phones or walkie-talkies that can cause false alarms on fire alarm systems, affect building management systems, and interfere with the proper operation of electronic equipment. As these systems begin to enhance productivity, customer service, and patient care, they will become more prominent. All wireless systems must be engineered for use and frequency allocation within the building so that they function properly and do not interfere with any other systems.

Infrared systems require a line of sight between the transmitter and receiver. The signal transmitter locations must be coordinated with architectural features such as wall coverings and lighting fixtures to ensure proper coverage. Presently, infrared systems have transmission speeds up to 10 Mbps. Technology is improving, and the transmission speeds are also expected to improve.

Satellite transmission is an alternative for long-distance communications. The main issue with a satellite dish is proper location and structural support to provide a direct line of sight to the communicating satellite(s). Many types of physical obstructions, including trees and structures, will interfere with the transmitted signal. A connection between the satellite dish and the telecommunications system backbone is required. Power is required at the dish for operation and for deicing. Ice buildup may interfere with the signal and will increase the weight and stress on the dish and associated structural support. When a dish is installed, it is in essence a large sail. Even if a screened type of dish is used, it becomes a solid dish when it is covered with snow and ice.

PLANNING/PROJECT MANAGEMENT

Telecommunications equipment and network requirements must be considered during the initial project planning. These are some of the issues that need to be reviewed and determined:

- Planned usage and user requirements
- Requirement for space and centralization or decentralization of network
- Multitenant or multiuser considerations and security requirements
- Flexibility and move-in (phasing) requirements
- Twenty-four-hour cooling and power backup requirements
- Architectural features and furniture coordination
- Maintenance and moves, adds, and changes (MAC)
- Future plans

By including telecommunications in initial planning, it can be worked into the project with minimal disruption and can be accounted for in the budgeting phase of the project. The installed cost of the telecommunications systems is substantial. The costs can be equivalent to the cost of the mechanical or electrical systems, especially when the cost of network electronics is considered. Planning provides for an infrastructure design that will create maximum affordable performance and service life and the optimum design for the project.

Design

The design of a cabling infrastructure should be based on the TIA/EIA standards and the recommendations in the BICSI telecommunications distribution methods manuals (TDMM). The standards indicate that at least two cables must be installed to each workstation outlet. One cable shall be Category 5 UTP (or higher performance) and the other cable should be one of the other acceptable media (fiber, STP, etc.) or an additional Category 5 (or higher performance) cable. Common designs include one Category 3 cable for a voice circuit and one Category 5e or higher performance cable for data, although this differentiation is disappearing. In some designs, the voice cable is the same as the data cable to allow for future flexibility. Because a voice circuit is usually required, installing a Category 3 cable and multiple cables of Category 5e or higher performance is recommended for economy. The decision as to the type of horizontal cabling should be based on the anticipated usage and bandwidth requirements. The total costs of the cabling and the associated electronics should be considered when comparing fiber-optic and copper cabling systems.

Fiber to the desktop can be provided where it is advantageous for noise immunity, security, distance, or bandwidth. The installed cost of fiber-optic cable is similar to high-performance copper cable, but the connecting electronics are six to eight times more expensive. The cost of electronics is coming down; therefore, it is an option that may be worth considering if the requirements justify its use. Copper cabling is providing gigabit capabilities, and therefore the bandwidth advantage of fiber to the desktop is being narrowed.

To take advantage of a standards-based cabling design, the standards suggest making any modifications to the cabling system to change its electrical characteristics or wiring configurations outside of the cable and jack (e.g., baluns, line splitters, etc.). For example, if two voice lines are desired from a Category 3 cable, the cable would be terminated on an eight-pin modular jack, and then a splitter would be plugged into this jack to provide two voice jacks for telephones. Each jack could be connected to two separate pairs of conductors, which would allow multiple telephone extensions.

During design, the need or advantage of wireless systems should be evaluated and integrated into the project.

The spaces required for telecommunications cabling infrastructures consist of the *building entrance facility* (BEF); the *main telecom room,* and the *main cross-connect* (MC), sometimes called the *main distribution frame* (MDF); the *equipment room* (ER) that typically houses the head-end equipment; and the *telecom room* (TR) that houses the *horizontal cross-connect* (HC), sometimes called the *intermediate distribution frames* (IDF). These typical terms actually refer to usage and in practice may be contained within the same spaces. The space requirements given in the standards are based on one workstation per 100 sq ft and capacity for three cables to each workstation. Smaller workstation spaces or more cables per workstation may increase the size of the space required. Depending on the layout of the racks and the types of electronic equipment (switches, hubs, routers, etc.) used, the typical 10 ft × 11 ft telecom room will support 200 to 250 data jacks terminated on patch panels in the racks. These spaces do not include space for user equipment but only for equipment needed for the cabling infrastructure, network electronics, and telephone system (see Fig. 5.5.6-1).

The equipment room size can be estimated at 0.75 sq ft per workstation in the facility but no smaller than 150 sq ft. This size is for one service provider and if multiple providers or redundant services are required, this space requirement will increase. In addition to the building equipment room, each tenant may need an equipment room for its equipment.

A telecom room (TR) is required on each floor to terminate cabling and to house the network electronics. The space required per the standards is one 10 ft × 11 ft room for each 10,000 sq ft of floor area. The location of these rooms is determined by the requirement that the total cable length from the room to the telecommunications outlet (TO) can not exceed 90 m (295 ft). After subtracting cabling up and down the wall or rack and spare cable, the allowable horizontal distance for cable is 250 ft (this assumes 10 ft up, 10 ft down, 10 ft slack in the TR, 1 ft slack at the TO, and 14 ft of cable in the TR to reach the patch panel). One method of reviewing a floor plan for adequate telecom room coverage is to draw parallel and perpendicular lines starting from the center of the telecom room, 250 ft in each direction. Then draw a line connecting the end points to create a series of isosceles triangles which when combined form a square 500 ft on a side, centered in the telecom room. Any area within these triangles can be reached within the 250-ft limitation when the cables are installed parallel and perpendicular to the building structure. This method indicates that a telecom room location could possibly service approximately 25,000 sq ft of floor space. This is possible, but the size of the telecom room would have to be approximately 12 ft × 20 ft to accommodate the number of cables to be terminated (assuming three cables per 100 sq ft).

If the cable to be installed must double back on itself, for example, when modular furniture is fed from an exterior wall, then the maximum horizontal distance available will be reduced. If the modular furniture to be served consists of three 10 ft × 10 ft cubicles, then an additional 35 ft of cable may be necessary to connect to an outlet in the middle of the wall of the last cubicle in line (see Fig. 5.5.6-2).

The cabling infrastructure includes various components, including telecommunications outlets (TO) with faceplate jacks and work area equipment line cords, horizontal cross-connect (HC) areas with patch panels, patch cables, cross-connecting blocks, and equipment patching cables. These components and the cable must be matched for the level of perfor-

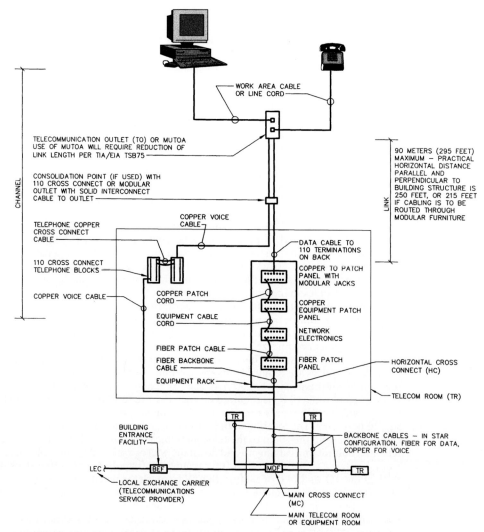

FIGURE 5.5.6-1 Typical building cabling system.

mance required. The typical outlet consists of voice and data jacks and cabling. The jacks are eight-conductor modular rated to match the performance level required. The wiring configuration of the jacks per the standard should be 568A. Also available is 568B that may be used in existing facilities to match existing wiring configurations. The wiring configuration at equipment may depend on the equipment manufacturer.

Adequate pathways must be provided for telecommunications cabling between the incoming service (BEF), the main telecom room (MC), equipment rooms (ER), telecommunications rooms (HC), and workstation areas (TO). The type of pathway (conduit, cable tray, J hooks, inner duct, etc.) should be selected on the basis of the areas in which it will be installed and the need for future addition of cabling. The raceways selected should be appropriate for high-performance communications cabling. The support methods must ensure that

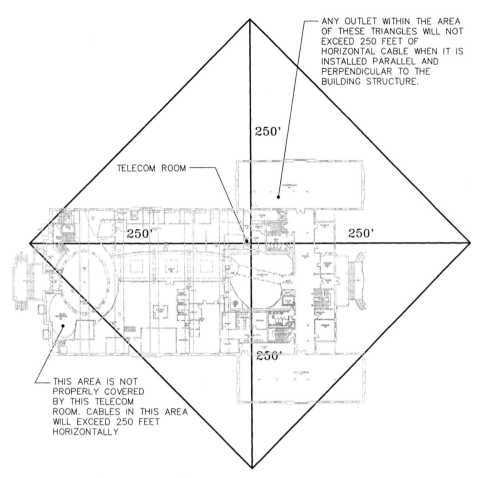

FIGURE 5.5.6-2   Telecom room area of coverage.

the cabling will not be overstressed and that its configuration will not be altered during or after installation. If the cables are kinked, overbent, or subjected to undue pressure, thereby changing the relative positions of the conductors or the twists of the conductors, or if the insulation is deformed, the cable performance will be degraded. Cables are affected by environmental conditions; hot areas will degrade the performance of the cable due to changes in its electrical characteristics. Heat mainly causes the cable's attenuation to increase, thereby reducing the performance of the cable.

Instead of raceway, cabling (copper or fiber) preassembled in a metal armor similar to electrical type MC cable can be furnished. This is a flexible metal raceway that protects the cabling and could be installed instead of conduit or wireway or where segregation in a cable tray or support system is desired. The main drawbacks to this assembly are that it needs closely spaced supports and the cable installed inside the armor cannot be added to or changed. This assembly could be used in a cable tray to provide protection for fiber-optic cabling in lieu of an inner duct.

Pathway design criteria for telecommunications cabling have some distinct differences from typical electrical system requirements, even though they both fall under the electrical code. The bending radii, pulling tensions, and raceway fill all affect the size and type of raceways required. The National Electrical Code (NEC) allows filling a conduit to a maximum of 40 percent of the area. The telecommunications cable pulling tension limited to a maximum of 25 lbf limits the number of cables installed to less than the 40 percent fill rate. Per the standards, three Category 5 cables fit in a ¾-in conduit, six Category 5 cables fit in a 1-in conduit, and 10 Category 5 cables fit in a 1¼-in conduit. Additional fill recommendations are provided in the Building Industry Consulting Services International (BICSI) *Telecommunication Distribution Methods Manual (TDMM)*.

Surface raceways have similar capacity limitations. The largest conduit connection to a 4¹¹⁄₁₆ in-square electrical box or surface raceway entrance fitting is 1¼ in. Therefore, a conduit feed is required for every 10 cables or approximately every 20 ft of surface raceway (one jack per 30 in). The maximum number of cables that can be installed in surface raceways must be decreased by the amount of space an outlet takes up in the channel and the space required to properly terminate and bend the cable at the outlet without stressing the terminations.

Based on a Category 5 cable, the minimum radius for a cable is 4 times 0.25 or 1 in and 1.2 in for fiber-optic cables. This bend radius must not be reduced. The typical surface raceway or modular furniture system raceway has 90° bends and must be specified with inserts to assure that the minimum bend radius is provided. Raceways should be installed per TIA/EIA standard 569. This standard specifies the following bend radii that are greater than those required in the electrical codes:

Conduit less than 2-in diameter: 6 × radius

Conduit for fiber optics of 2-in diameter or more: 10 × radius

The NEC allows as much as 360° of bends between pull boxes. The TIA/EIA standards limit this to 180° or a maximum of 100 ft between pull boxes. The pull box size indicated in the TIA/EIA standards is larger than that required by the NEC.

Grounding is an important consideration for telecommunications. There should be a ground bar (TMGB) in the main equipment room that is connected to the main electrical service grounding point and to building steel if it exists. There also should be a grounding bar (TGB) in each telecom closet that is connected back to the main equipment room, building steel, and to the local power source equipment grounding bar. All communications equipment and raceways in the room must be bonded to the grounding bar and labeled. This grounding system is in addition to the safety personnel grounding provided with the electrical power systems and is provided to minimize the ground currents flowing over the telecommunications systems and to provide a path for high-frequency grounding. Grounding should comply with TIA/EIA standard 607.

Lightning and surge protection must be provided on incoming telecommunications circuits with outdoor exposure and on the electrical circuits that provide power to the electronic equipment in the system. Proper grounding and direct pathways are required to minimize voltage gradients. Power conditioning is recommended for electronic equipment that makes up the cabling infrastructure. Uninterruptible power supply (UPS) systems designed to provide full-time power conditioning and filtering protect sensitive electronic equipment from power sags, surges, and spikes, and in addition provide short-term battery backup to get through short power outages or brownouts.

The design of telecommunication rooms must consider the requirement for cooling and ventilation. The main equipment room should be treated as a computer room due to the large amount of expensive electronic equipment that is vital to the facility's operation. Typically air conditioning is required 24 hours per day, 7 days per week (24 × 7) for equipment rooms. Equipment rooms may also require humidity control, depending on equipment requirements. Equipment vendors should provide power and environmental requirements for their equipment. Humidity control has the added benefit of helping to control static electricity, which can damage electronic equipment. Telecom rooms for horizontal cross connects may require only positive ventilation, depending on the amount of heat-generating electronics in the room. Emergency lighting may be required, depending on the critical nature of the facility. The room should be secured with lock or card-key access and should be monitored for temperature or moisture with remote alarms.

Telecommunications systems are critical to the efficient cost-effective operation of a facility, and therefore the location of rooms and equipment should be given appropriate consideration. The locations should be in dry areas, not subject to accidental flooding from groundwater, roof leaks, or aboveground piping sources (e.g., do not locate under toilets, roof drains, locker rooms, and kitchens). Locate away from sources of electrical interference (e.g., power lines, transformers, lighting ballasts, elevators, large motors, and copy machines).

Modular furniture systems pose potential problems for high-performance data cabling systems. The typical furniture system with built-in raceways has telecommunications cabling pathways that are located close to the power wiring, are typically plastic (which provides no shielding from the power wiring), and may not have the appropriate wire bending radii required at corners and at outlets. Furniture systems can be specified to provide compliant telecommunications raceways, but these are typically the more expensive models. Typically, no more than 12 to 16 cables can be installed in the pathways provided. It is common to provide twice the number of feed points for telecommunications compared to electrical feed points.

In open office layouts or in areas subject to regular changes, more flexibility is required than is typically attainable with home-run cables from each outlet to the telecom room. TIA/EIA has issued a standard designated TSB 75 that allows for what is called a *consolidation point* (CP) between the telecom room and the telecom outlet. This CP is similar to a local patch panel. The local outlets are connected to the CP, which is then cabled to the telecom room. When modifications are required, only the cabling from the telecom outlet to the CP must be revised. Without a CP, if the cable to the workstation outlet is not easily relocated to the new location, a new cable installed from the telecom room to the outlet would be required because cable splicing is not allowed under the standards. Another option under TSB 75 is to use a *multiuser telecommunications outlet assembly* (MUTOA). This allows a work area equipment line cord (that connects the outlet to the equipment) longer than 10 ft. When it exceeds 10 ft, the allowable length of the link cable back to the telecom room must be decreased, so that the overall cabling does not exceed the standard's requirements for performance.

Wire management is a concern in designing and installing telecommunications cabling. The large number of cables required necessitates specifying adequate wire management at termination points (e.g., racks and cross-connect blocks) and for the cables installed between telecom rooms and telecommunication outlets (e.g., cable trays, hooks, conduits). Adequate supports are required, and the cables must also be properly supported at regular intervals. Methods that compress the cables and distort their electrical characteristics result in performance below that specified and expected. Hook-and loop-type (Velcro®) support straps are one method of attachment that meets these requirements and easily allows adding or removing cables. The use of ty-wraps is not preferred but if done correctly (after installation the ty-wrap is loose enough to rotate around the cables) could be used to support cables on J hooks but should not be used in telecom rooms due to the need to add or remove cables. Plenum-rated hook and loop straps are available.

To reduce costs, provide flexibility, and simplify installation, it is typical to install cables exposed above accessible ceilings or under accessible floors. Two factors must be addressed for these types of installations. If the area is used as an HVAC air plenum, then the cables and the supports must be rated for plenum use if not installed in enclosed raceways (except for electronic data processing rooms that follow the requirements of NEC NFPA 70). If the area is not used as an air plenum and the area does not have sprinkler or smoke detection for protection, the large amount of combustible material located in an unprotected space may be a concern. The authority having jurisdiction (AHJ) may require smoke detectors, sprinklers, or the use of plenum-rated materials and methods.

Construction/Installation

To achieve the performance expected, the installation of the cabling infrastructure is as important as specifying the correct components. The following are some of the items that should be addressed in the installation specification and observed during construction:

- Wire management and labeling
- Cabling support spacing and methods (minimal compression, ty-wraps rotate easily, hook and loop straps)
- Bend radius of cables (exposed, conduit, furniture, cable trays, wireways, and surface raceways)
- Fire and smoke seals and plenum-rated materials
- Proper terminations (Category 5 cables untwisted no more than 2 in, Category 3 not more than 1 in)
- Distance from EMI sources
- Only qualified licensed installers trained by the manufacturer of the material to be installed
- Extended system performance warranty by the manufacturer and installed by manufacturer-approved installers

Testing/Documentation

The only way to know if the cabling infrastructure system installed meets the desired performance specified is to test the completed installation. Every cable installed should be tested to verify specified performance and to provide a record of the cable to be used as a reference point for future comparison testing if a problem occurs. The testing should be performed per TIA/EIA standards, and for enhanced cable installations the parameters used to specify the cabling infrastructure should be tested and verified (e.g., attenuation, delay skew, propagation delay, NEXT, powersum NEXT, ACR, power sum ACR, return loss, ELFEXT, and powersum ELFEXT). The entire channel from the equipment patch cord in the telecom room to the work area equipment line cord at the workstation outlet should be tested, not just link testing from the patch panel to the workstation outlet. The tester used shall be a standard-compliant Level 2 or 3 tester, as required by the performance specified for the cabling system.

The typical average for a telecommunications systems move, add, or change (MAC) is 125 percent annually. Because of this, complete accurate record documents (true as-builts) are required at the end of construction. In addition to drawings, a complete set of administration documents should be assembled as recommended in TIA/EIA Standard 606. These documents can be on paper or computerized with the ability to generate work orders for moves, adds, or changes (MACs).

Record Keeping.   Proper records are vital to quick, efficient troubleshooting when problems occur and to facilitate moves, adds, and changes. Therefore, records obtained at the end of construction must be kept up to date as MACs are made. Many network problems typically turn out to be related to the physical infrastructure. Because downtime affects the productivity and operations of the firm, quickly troubleshooting a problem and repairing it are important. Nothing helps to accomplish this faster than accurate, up-to-date records.

Future Proofing

Future proofing is designing a cabling infrastructure that will be acceptable for use by future technologies. This is difficult to do without knowing what will be required in the future. The electronics portion of the systems will need to be replaced as technologies and requirements change. This is expensive but is relatively easy to do and can be done with minimum downtime. The need to replace the cabling infrastructure will be minimized by installing systems (cabling and equipment) that at least meet TIA/EIA standards and provide high-quality, high-performance cabling, follow good installation practices, have proper grounding, and avoid EMI sources. Standards-based systems are supported by vendors and have a large installed base of users. As technologies advance, it is a distinct advantage to the manufacturers and vendors to design their systems to use the standard designs and infrastructure as much as possible. One method to provide for future changes is to design with adequate spaces and accessible pathways to allow installing additional or new cabling or equipment, including the supporting power and cooling infrastructure.

Additional methods for future proofing the cabling infrastructure are to provide additional spare cabling, to minimize cabling distances, and to use state-of-the-art high-performance cables, equipment, and installation techniques. Care must be exercised in specifying high-performance cabling systems. Manufacturers make many claims of improved performance. These claims often rely on using components from a single manufacturer to function properly, which may be not be a realistic or cost-effective limitation. The key is to know what parameters will be important to future technologies. Will it be simply an increase in frequency (200, 300, 600 MHz, or higher?), or will it be a different encoding technique, number of pairs used, or better electronics? Present network protocols use only two of the four pairs of conductors in a cable simultaneously. Gigabit network protocols use all four pairs of conductors simultaneously. This will require cables with better characteristics and will require testing additional parameters not normally tested in standard Category 5 systems. These systems use high-performance cables, outlets, and patch panels, and these components must work together as a system. The only way to verify a manufacturer's claims is third-party verification. Until IEEE and TIA/EIA have formally adopted higher performance standards, it is not known exactly what will be required.

Designing systems to facilitate moves, adds, and changes (MACs) will allow future modifications with minimal disruption to current operations. Incorporating oversized raceways will allow adding additional cables or higher-performance cables. Category 6 cables are larger than Category 5, and Category 7 cables will be larger than category 6. Provide easy access to cable trays and installed cabling. Do not install above ductwork or piping or high up in the ceiling so as to limit easy access. Provide 12 in around the top and sides of supports to facilitate cabling installation by laying in or for pulleys to allow pulling cables. Telecom rooms should be sized to allow adding cable terminations and electronic equipment.

Labeling all cables, jacks, terminations, and components is critical for quick troubleshooting and to facilitate changes. Proper system documentation begins at the time of installation with comprehensive labeling following a logical labeling scheme that identifies end points of cables. Require contractors to provide accurate as-built documents including test reports at the completion of any work, and update them for any MACs after the initial installation.

Provide space or spare raceway in the backbone cabling system to allow installing additional or upgraded fiber-optic cables. This can be accomplished by installing multiple inner ducts in a 4-in conduit or tubing to allow fibers to be blown in.

INFORMATION SOURCES

Codes and Standards

American National Standards Institute (ANSI): www.ansi.org

Telecommunications Industry Association (TIA): www.tiaonline.org

Electronics Industries Alliance (EIA): www.eia.org/

ANSI/TIA/EIA—568:   Commercial Building Telecommunications Cabling Standard

ANSI/TIA/EIA—569:   Commercial Building Standard for Telecommunications Pathways and Spaces

ANSI/TIA/EIA—606:   Administration Standard for the Telecommunications Infrastructure of Commercial Buildings

ANSI/TIA/EIA—607:   Commercial Building Grounding and Bonding Requirements for Telecommunications

TIA/EIA—TSB67:   Transmission Performance Specifications for Field Testing of Unshielded Twisted-Pair Cabling Systems

TIA/EIA—TSB72:   Centralized Optical Fiber Cabling Guidelines

TIA/EIA—TSB75:   Additional Horizontal Cabling Practices for Open Offices

TIA/EIA—TSB95:   Additional Category 5 requirements (for qualifying existing cabling)

TIA/EIA—455 Series: Fiber-Optic Test Procedures

TIA/EIA—492 Series: Specifications for Optical Waveguide Fibers

ANSI/TIA/EIA—570:   Residential and Light Commercial Telecommunications Wiring Standard Building Industry Consulting Service International (BICSI): www.bicsi.com

BICSI Telecommunications Distribution Methods Manual (TDMM)

BICSI Telecommunications Cabling Installation Manual

Federal Communications Commission Documents (FCC):   www.fcc.gov

Federal Information Processing Standards (FIPS):   www.itl.nist.gov/fipepubs/

Institute of Electrical and Electronics Engineers, Inc. (IEEE):   www.IEEE.org

Rural Utilities Services USDA/RUS (RUS):   www.usda.gov/rus/

International Organization for Standardization (ISO):   www.iso.ch/

International Electro-Technical Commission (IEC):   www.iec.ch/

ISO/IEC 11801—Generic Wiring Standard

UL Telecommunications Page: www.UL.com/Telecom/index.htm

World Standards Services Network: www.wssn.net/WSSN/

Informational Web Sites

Association of Cabling Professionals (ACP):   www.wireville.com

Society of Cable Telecom Engineers (SCTE):   www.SCTE.org/

Anixter Corporation levels program: www.anixter.com

EMC Information Center: www.emc-journal.co.uk/

IEEE Electromagnetic Compatibility Society: http://emclab.umr.edu/ieee.emc

UL EMC Page: www.ul.com/emc/emc1.html

Magnetic Shield Corporation: www.magnetic-shield.com

Cable—telecom Web links: http://cable.doit.wisc.edu/

Telecom virtual library: www.analysys.com/vlib/default.htm

On-line searchable and readable books: www.netlibrary.com

Technical definitions and reference material: www.whatis.com

Premises network virtual community: www.premisesnetwork.com

Power Quality Magazine: www.powerquality.com/
Fiber Optics Online: www.fiberopticsonline.com/content/homepage/
Cabling Installation & Maintenance Magazine: www.cable-install.com/
Cabling design information: www.cabling-design.com
Computer Telephony: www.telecomlibrary.com/
Wireless Networks Online: www.wirelessnetworksonline.com/content/homepage
TechWeb IT network information: www.teledotcom.com
Search and purchase technical books: www.telecombooks.com/
Wiring information: www.wiring.com

CONTRIBUTORS

Thomas F. McNamara, RCDD, Manager of Telecommunication/Data Engineering Services, BR+A Consulting Engineers, Inc., Boston, Massachusetts

Diane L. Morgan, M.P.A., Senior Telecommunications Analyst, BR+A Consulting Engineers, Inc., Boston, Massachusetts

Dennis L. Wagner, P.M.P., (formerly) Vice President of Operations, Stone & Webster Communications Services Group, Boston, Massachusetts

CHAPTER 6

CONSTRUCTION, MODIFICATIONS/ RENOVATION, AND DEMOLITION/ SITE RESTORATION

Paul R. Smith, P.M.P., P.E., M.B.A., Chapter Editor
Peak Leadership Group, Boston, Massachusetts

Bart Bauer, P.E.
Edwards and Kelcey, Inc., Boston, Massachusetts

Kenneth H. Stowe, P.E.
George B. H. Macomber Co., Boston, Massachusetts

The organization of this chapter was a source of numerous discussions. Many contributors within the construction industry traditionally organize tasks and processes around the Construction Specification Institute's (CSI) 16 Divisions for specifying materials and thus recommended that we use this breakdown (into divisions) as the organizing system. Both design and construction professionals use the CSI document for preparing specifications, grouping work type, and cataloging cost information. The project management contributors believe that the Project Management Institute's (PMI) knowledge areas or process steps should define the chapter format (see Subsec. 3.2.1, "The Project Management Process"). Project managers (PMs) believe that the stages, or steps, of the construction process, provide just-in-time information for the reader, especially one seeking an integrated approach. Because the remaining sections of this handbook follow the process management format, we decided to continue to use this as the principal format. However, we support both points of view by providing a quick cross-reference table between the process categories and the CSI Division items.

This chapter identifies the physical and mental processes involved in constructing a facility. Our goals in this chapter are to provide sufficient knowledge in the following areas so that facility managers (FMs) can oversee the construction projects going on at their facilities and can communicate with their construction peers in both the design and construction professions.

- To help the FM become part of the senior construction project management team
- To help the FM understand the major knowledge areas of the construction process
- To identify and describe the phases of the facility construction process and the risks at each phase
- To describe construction process variations due to such factors as climate, geography, working within existing buildings, and preservation of building fabric
- To combine theory and applications

And, finally, we aimed to provide a document that FMs find *user-friendly.*

SECTION 6.1
CONSTRUCTION MANAGEMENT

ARTICLE 6.1.1
CONSTRUCTION MANAGEMENT

Kenneth H. Stowe, P.E.
George B. H. Macomber Co., Boston, Massachusetts

This chapter complements the discussion of facility design in Chap. 5. Facility construction and renovation requires building experience with the material, equipment, labor, the market conditions, the site, the climate, and other variables that challenge project success. The builder directs the procurement, schedules and supervises the construction forces, and coordinates the turnover to the owner. The builder is charged with handling vast amounts of data, often changing during the project, and as such is often an information manager as well as a construction manager.

Many projects benefit by employing a builder's construction expertise during design. In these cases, the preconstruction services provided by the builder are designed to yield an affordable, buildable design accompanied by procurement and contracting strategies to meet the project goals.

The articles in this chapter are in a chronological order that defines the construction process. They include discussions of construction management, the facility construction process, and facility construction variations. They describe the builder's responsibilities and help the reader to understand the factors that enter into the decisions associated with contracting methodology and how to select the builder. The term *builder* is used interchangeably to describe a general contractor (GC) and a construction manager (CM). Some discussion of the different roles follows.

ARTICLE 6.1.2
CONSTRUCTION DELIVERY METHODS

Christopher L. Noble, Esq.
Hill and Barlow, Boston, Massachusetts

Bart D. Bauer, P.E.
Edwards and Kelcey, Inc., Boston, Massachusetts

At some point in a facility's life, the owner determines in the roughest sense that the facility no longer meets the needs of the occupant. Therefore, the owner needs to engage the services of an architect or engineer (referred to here as an *A/E*) to evaluate the facility's and the occupant's needs. At this early point, as part of this decision to retain professional services, the FM must evaluate the "delivery methodology" for the new facility. Design and construction services can be obtained on the basis of the traditional arrangement or on the basis of increasingly common variations, such as construction management and design/build.

THE TRADITIONAL SYSTEM

In the traditional system for project delivery, the architect, supported by the design team engineers, prepares detailed plans and specifications (see Fig. 6.1.2-1) under an owner/architect agreement. The intricacies of the actual contract and its language are discussed in Chap. 3 Art. 3.2.3, "Different Types of Agreements: Which Is Best for You?"). Then, that design package is given to a general contractor who estimates and builds the facility under a separate owner/contractor agreement. When the construction contract is signed, the GC becomes responsible for building the facility within the contractual constraints of time, quality, and price, as described in the scope of work, the A/E's drawings and specifications.

THE CONSTRUCTION MANAGEMENT SYSTEM

The construction management system involves adding the construction manager (CM) to the traditional process described previously and often involves eliminating the general contractor, as shown in Fig. 6.1.2-2. Unlike the traditional general contractor, the CM ideally comes on the job during the design phases or even earlier. This role continues through the construction phase. The CM's role varies from project to project, but the following generalizations can be made about the construction management process. During the design phases, the CM furnishes such services as a constructability design review, cost estimating, project scheduling, and early purchasing of long-lead-time items. The CM administers an open-book bidding or negotiation process during which trade contractors are selected to perform most or all of the

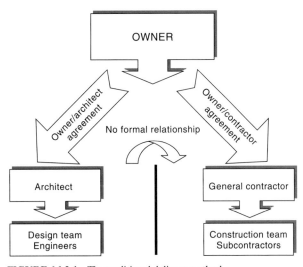

FIGURE 6.1.2-1   The traditional delivery method.

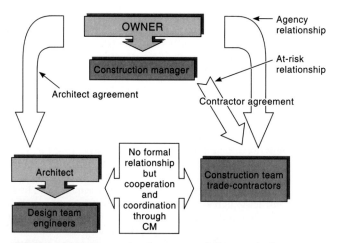

FIGURE 6.1.2-2   The construction manager delivery method.

actual construction work. In an agency construction management project, the trade contractors have direct contractual relationships with the owner or with the CM as agent for the owner. In an at-risk, or constructor, CM project, the trade contractors are subcontractors to the CM, who is contractually responsible for their work to the owner.

Construction managers have developed sophisticated cost estimating capabilities using historical data and unit costs that give greater certainty of costs and outcome earlier in the facility design process. By developing a detailed project schedule during the design phases and enforcing it during the construction phase, CMs give owners the analytical and administrative capacity required to overlap and shorten various elements of the development process. Construction managers offer themselves as nonadversarial members of the owner's construction team and have a role more akin to that of the A/E than that of the conventional general contractor.

The continuity provided by the presence of the CM during both the design and construction phases permits the phases to overlap, if necessary, to meet the owner's schedule. In general, the CM concentrates on cost estimating, scheduling, early purchasing, and other nondesign activities during the preconstruction period. It is sometimes difficult to distinguish the CM from the conventional general contractor after the actual process of construction begins.

Construction of some portions of the project can begin while the A/E is completing the design of other portions of the same project. This process is known as *fast-track* construction, and its management and control are among the CM's most important advantages.

Construction managers are generally believed to be closer to the construction and, therefore, can predict more accurately the cost of constructing the facility shown in drawings and specifications. This closeness also allows analyzing practical alternatives that could increase construction efficiency and/or reduce construction costs. Finally, construction management is one methodology for avoiding a potentially adversarial relationship between the owner and the general contractor, in which the A/E may have the ambiguous role of both owner's representative and impartial mediator.

THE DESIGN/BUILD SYSTEM

The design/build system is one in which the design and construction functions are integrated with one another, as shown in Fig. 6.1.2-3. This is unlike the traditional system where there is

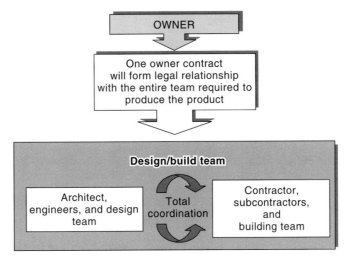

FIGURE 6.1.2-3   The design/build method.

no direct contractual relationship between the A/E and the contractor. By electing to use the design/build system, the owner usually hopes that the close relationship between the designer and the constructor will result in efficiencies, time savings, and single-point responsibility for construction activities and also for design. To achieve these objectives, however, the owner is normally required to give up the benefit of having an independent A/E who owes duties directly to the owner and who can be depended upon to protect the owner's interest.

Design/build services can be procured in several forms: An owner hires a general contractor who, in turn, hires an A/E as an independent subcontractor; or an owner engages a design/build entity that subcontracts the design services to an A/E and the construction services to a contractor. Under a less common model, a general contractor and an A/E form a contractual joint venture with which the owner will contract for design and construction services. The fourth design/build model is that of the integrated design/build firm, which provides both design and construction services through one or a number of affiliated entities. Instead of putting together a new team for each project, the integrated design/build firm develops internal working relationships, and the owner receives the benefit of efficiencies generated by the firm's experience and familiarity with the design/build process.

The owner's objective is that the design/build firm can deal with the problems arising out of this fast-track process in a more centralized and coordinated manner than an A/E and CM who are engaged separately. As the design progresses, the design/builder consults with potential subcontractors and develops increasingly detailed cost estimates. When documents are sufficiently complete, the design/builder offers the owner a guaranteed maximum price (GMP) for the remainder of the design and construction services.

In theory, the owner has the option of rejecting the design/builder's proposed GMP and getting someone else to build the project. In reality, however, this option will almost certainly result in the loss of the time that the owner hoped to save by embarking on the design/build process in the first place.

There are many other variations of project delivery; some are regionally quite common, and others are uniquely strong in a particular market discipline. Some of the variations are expressed in the following matrix. It is often possible to use the items in the matrix as a collection of parts, and the client can select the delivery of the products (e.g., a facility construction design and contract and assembly of a particularly unique delivery that suits the clients needs).

Which design and construction system is best for any particular project? The selection of the best project delivery methodology is best determined by an integrated team approach. This team generally makes the decision based upon the need to control risk. The book, *Client Advisor,* published by American Associated General Contractors of Massachusetts (AGC) and the Boston Society of Architects (1997), presents a risk pyramid that defines three mutually exclusive areas of risk. The three sides of the pyramid represent the three common constraints of scope, time, and budget. As the client team determines that one or more of these constraints is critical, it suggests that one type of delivery method is better suited for that project (see Fig. 6.1.2-4).

Often, there is no clear answer. Every project has different needs, and as such there is not one method that serves all projects. In general, construction management is selected when the owner senses that there will be valuable construction input during design and that teamwork will be required during construction that can best be fostered during a long-term relationship. The choice of construction management contract is most often chosen, for example, when the project is a renovation of occupied space that requires complex (and often daily) interactions to execute the work. The GC solution, or traditional delivery method, is most often picked when the scope development does not require a builder's input, the scope is expected to be relatively simple and stable, and competitive pricing is deemed more important than early occupancy.

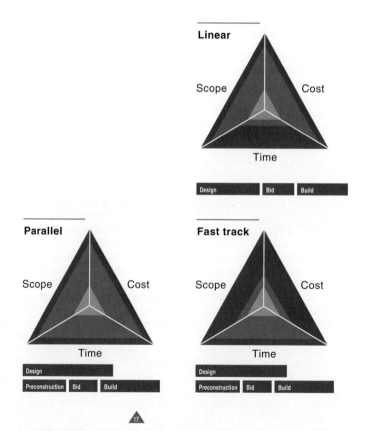

FIGURE 6.1.2-4 Pyramids representing different elements risks that will help determine the type of construction delivery that is best for your particular project. (*Courtesy of the Associated General Contractors of Massachusetts and the Boston Massachusetts Society of Architects.*)

The traditional sequential system, the construction management system, and the design/build system all have inherent advantages and disadvantages. Some of the relevant factors will be clear, but the success or failure of a project will also turn on the identities, the experience, and the capabilities of the members of the design and construction team, as much as on the system of contracting and procurement. For this reason, it is extremely important for owners to develop procedures to prequalify and evaluate A/Es, CMs, contractors, and/or design/build firms, so that each member of the design and construction team is a known quantity before the process is started. The team, including architects, engineers, contractors, and subcontractors, should have had an opportunity to get to know one another and to establish working relationships on prior projects. Each firm should have financial, management, and technical capabilities commensurate with the project's requirements as described in the scope of work documents, the plans, and specifications. Owners should always keep in mind the fact that they expect their buildings to perform satisfactorily for a long period of time and that a primary factor in ensuring such performance is the quality of the members of the design and construction team.

SUBARTICLE 6.1.2.1
Construction Contract Decisions

Refer to Art. 3.2.3, "Different Types of Agreements," for further information.

ARTICLE 6.1.3
BEYOND CONTRACTING TO CUSTOMER SERVICE

Kenneth H. Stowe, P.E.
George B. H. Macomber Co., Boston, Massachusetts

The builder often delivers more than a facility and the associated documents and as-built files. The owner requires planning, procurement, and resources for moving and other services for successful occupancy. The builder also is an advisor to design, and has an eye for achieving cost and schedule goals.

HOW DO YOU SELECT A METHOD AND THEN THE RIGHT CM OR GC?

The best builders create strong, dependable companies by adhering to the following principles:

- Finding, selecting, and promoting ongoing relationships with the best-qualified subcontractors
- Selecting and training the best-qualified PMs and field supervisors

- Providing a rigorous process to adhere to a best-practices approach
- Providing customers with a high-quality construction experience that creates value

This means that the builder delivers professional services that include competent reporting, comprehensive cost estimating and control, meeting scheduled milestones, and supplying the quality of materials and workmanship that meets or exceeds the specified standards.

The builder should adhere to a quality assurance program. The major attributes of a good quality assurance program include the following:

- The builder's personnel should be highly professional and proud of their work.
- The builder should foster a corporate culture where the needs of customers are a high priority. For example, the builder should have safety/awards programs throughout the year that highlight good procedures and decisions made by field people to protect the workers and the customers.
- The builder should work in a partnering manner, seeking win-win solutions.

The builder should be selected on the basis of appropriate experience, individual skills, personal chemistry (to maximize teamwork), references, and price. The builder is often selected before the scope of the work is well defined and months before a firm price can be estimated for the hard construction costs. The builder candidates provide a fee and often propose general conditions, with the understanding that, after some scope definition and budgeting exercises, the owner and builder will often agree on a contract price, schedule, and levels of quality.

PARTNERING

Partnering is a term for an approach to project management that calls on all members of the team to honor a set of project goals and think of the other members as partners rather than adversaries. The most common method of partnering is to employ a skilled facilitator to conduct one or more team building sessions at which goals, roles, skills, and challenges are agreed upon. Often, a formal document is created and signed by the entire project team. Project communication strategy is a critical component of the partnering.

RESOURCES

The builder should have the following six principal resources that form the foundation of a quality program:

1. *Project executive.* The project executive fosters a close relationship with the owner, understands the goals of the project and the contributions of every team member, and is responsible for owner satisfaction at every level.
2. *Superintendent.* The builder's superintendents generally have total responsibility for implementing the scope of work relative to schedule, safety, means, and methods. Additionally, they are responsible for obtaining performance and quality from subcontractors. The quality of the product is the superintendent's responsibility.
3. *Project managers (PMs).* The role of the PM is defined in detail in Chap. 3, Art. 3.2.1, "The Project Management Process." Generally, however, a PM's responsibility includes all of the business aspects of a project, including cost and schedule models, the procurement

process, the overall quality assurance program, and keeping project information current and accessible.

4. *Systems.*   The builder should have dependable project control systems, conduct training on current means and methods, and provide rewards for those superintendents and PMs who stay current with new practices.

5. *Philosophy.*   The builder should earn profits and retain clients by good-quality work, team effort, and fair treatment of the changes that inevitably occur.

6. *Lessons learned.*   The builder should have a lessons-learned program that continually feeds the organization with current best practices and practical insights from previous projects.

Smaller projects often employ one person to perform more than one of these staff roles.

ARTICLE 6.1.4
SCHEDULE

Kenneth H. Stowe, PE
George B. H. Macomber Co., Boston, Massachusetts

The builder is charged with creating a comprehensive critical path method (CPM) schedule (see Chap. 3, Art. 3.2.1, "The Project Management Process"). For a builder, that responsibility often begins in preconstruction, sometimes before design. In that role, the builder's planner will consult with, advise, assist, and make recommendations to the owner and A/E on all aspects of planning for project construction. The responsibility of creating a comprehensive schedule can take shape as a facilitating role as well as a creating role. Input and commitment from many stakeholders are most often vital because varying expertise and commitment by many parties are needed. The stakeholders include architect, engineers, PM, superintendent, subcontractors, owner's PM, users, permitting and other governmental agencies, and so on. A good builder's contribution here often consists of practical construction ideas that stem from experience on-site. Examples include anticipating muddy conditions, knowing the production rate of crews in specific weather conditions, suggesting an opening in a wall to allow installation of large equipment, and so forth.

The plan emerges in stages—first, a starting point for discussion, then increasing detail and more focused presentations until a version is accepted as a baseline and is documented. An example of a summary schedule is shown in Fig. 6.1.4-1. Now progress can be measured against that baseline version. The CM closely monitors the schedule during project construction and is responsible for providing all parties with periodic reports on work status with respect to the project schedule. If there is variance from the schedule on important activities, the project team can take appropriate actions.

Schedules can grow in detail up to hundreds or thousands of activities. Traditionally, both project management staff, as well as field staff, use the builder's schedules. An integrated project team's use of schedule-based knowledge is discussed in Chap. 3, Art. 3.2.1, "The Project Management Process." Properly managed, the vendor deliveries and subcontractor work performance will be forecast and measured according to the commitments in the project schedule.

Vital resources should be identified and if any resource demand exceeds supply or if selected subcontractors must be screened for manpower or equipment limitations, the scheduling process must identify this and facilitate a solution.

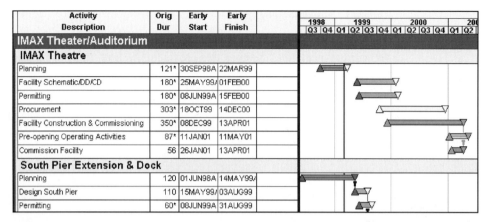

Activity Description	Orig Dur	Early Start	Early Finish		1998 Q3 | Q4	1999 Q1 | Q2 | Q3 | Q4	2000 Q1 | Q2 | Q3 | Q4	200 Q1 | Q2
IMAX Theater/Auditorium								
IMAX Theatre								
Planning	121*	30SEP98A	22MAR99					
Facility Schematic/DD/CD	180*	25MAY99A	01FEB00					
Permitting	180*	08JUN99A	15FEB00					
Procurement	303*	18OCT99	14DEC00					
Facility Construction & Commissioning	350*	08DEC99	13APR01					
Pre-opening Operating Activities	87*	11JAN01	11MAY01					
Commission Facility	56	26JAN01	13APR01					
South Pier Extension & Dock								
Planning	120	01JUN98A	14MAY99A					
Design South Pier	110	15MAY99A	03AUG99					
Permitting	60*	08JUN99A	31AUG99					

FIGURE 6.1.4-1   A typical summary construction schedule. (*Courtesy of the George B. H. Macomber Company.*)

ARTICLE 6.1.5

CONSTRUCTION COST CONTROL

Kenneth H. Stowe, P.E.

George B. H. Macomber Co., Boston, Massachusetts

The builder relies on historical cost data, experienced and skilled personnel, and good procedures to conceptualize, forecast, and control the project cost.

CONCEPTUAL ESTIMATING

As a key part of the cost management process, the construction manager (CM) provides a wide range of advisory services during the planning and design phases. Specific assignments include conceptual estimating, means/methods/materials options, detailed estimating, and time–cost trade-offs.

The CM has the responsibility of preparing a conceptual budget estimate at the preliminary stage of development. At this point, the estimator is charged with filling in the gaps because the scope of the project will not have been completely defined, but assumptions, contingencies, and allowances must be developed to anticipate details that will emerge as the design matures. Continuous review and refinement of this estimate is vital as the development of the plans and specifications proceeds. The CM will advise the owner and the architect/engineer (A/E) if the budgeted targets for project cost and/or completion may be threatened and will recommend corrective action. The CM prepares a final cost estimate when plans and specifications are completed to a firm scope and, if the CM is contracting at-risk, executes a guaranteed maximum price (GMP) contract based upon these estimates. Fig-

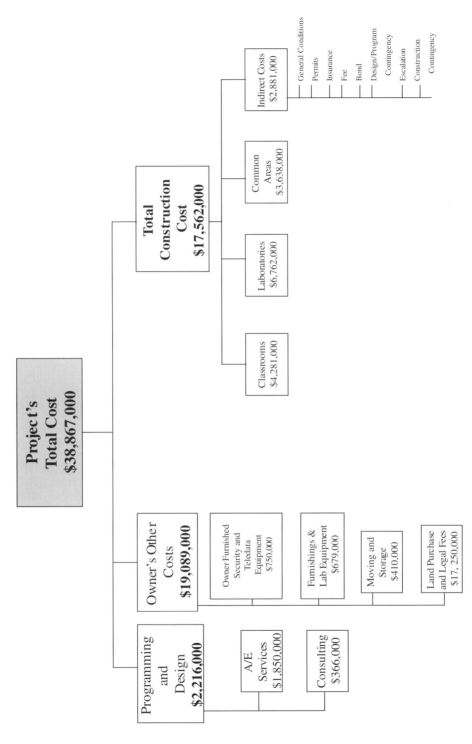

FIGURE 6.1.5-1 A summary of project costs.

6.11

ure 6.1.5-1 shows a typical organizational chart of the relationship of the costs and sources of the costs.

COST ESTIMATING

Good estimating services provide good decision support. The estimator's products must have these characteristics:

- Be accurate in quantities and prices that reflect current market conditions
- Be flexible in format to be understood by the team
- Be responsive to changes and suggestions
- Be easily understood, even by untrained audiences
- Be able to fills gaps with clear and logical assumptions
- Be able to reflect good version control with references to schedule and scope documents
- Be auditable so that future versions track migration

VALUE ANALYSIS/CONSTRUCTION FEASIBILITY

As they are developed, the CM reviews architectural, civil, mechanical, electrical, and structural plans and specifications at each phase and advises and makes recommendations with respect to such factors as construction feasibility, possible economies, availability and dependability of materials, equipment and labor, time requirements for procurement and construction, and projected costs. This is often called *value engineering* and is defined as the process of identifying the performance of elements of the proposed facility design and evaluating the costs and benefits of various means of achieving the same or similar performance. Figure 6.1.5-2 shows a tool that can be used to evaluate the true value received from the different options evaluated.

The estimator should create a flexible database that can break down the costs into departments for the owner's accounting, into systems for value engineering, and finally into the divisions established by the Construction Specifications Institute (CSI). The estimator should

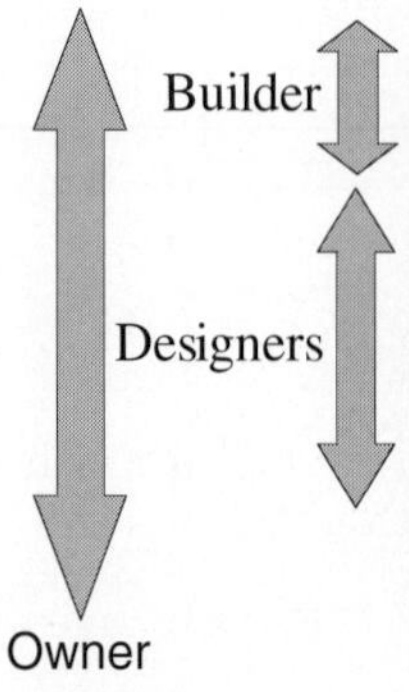

	Option 1	Option 2	Option 3
Capital cost			
Schedule			
Safety			
User satisfaction			
Life-cycle performance			
Maintainance			
Operations			
Aesthetic			
Codes			
Environment			
Labor relations			
Political			

Builder

Designers

Owner

FIGURE 6.1.5-2   Options matrix.

create a cost breakdown structure to integrate with the schedule, to maintain continuity, and to illuminate what the budget includes and, equally important, what it does not. Too often, a number is taken out of context, such as a construction total, as opposed to a project total. The builder's carefully structured and consistent budgets avoid this unnecessary confusion. Work breakdown is considered in detail in Chap. 3, Art. 3.2.1, "The Project Management Process."

CONTRACT DOCUMENT PACKAGING

The CM makes recommendations to the owner and A/E regarding the division of work in the plans and specifications to facilitate bidding and awarding of trade contracts, taking into consideration such factors as time of performance, availability of labor, overlapping trade jurisdictions, and provisions for temporary facilities.

Finding Savings in Time-Based Costs

The project can often save a significant amount of money by condensing the schedule. By using an integrated approach to cost and schedule (discussed further in Chap. 3, Art. 3.2.1, "The Project Management Process"), the cash flow can be established and time–cost trade-offs can be considered. Time-sensitive costs can be analyzed against the baseline schedule to find every efficiency possible within the project plan. The builder will look at the durations for rentals, staffing, and temporary construction to forecast their effects accurately and tightly on the budget. These are important ingredients to identify early so they do not come as a surprise later. Creative efforts to reduce these temporary expenditures can result in reduced project cost.

Budgeting General Conditions

General conditions are those elements of the project budget that are provided by the CM to facilitate construction but are not elements of the final product. They include, for example, construction management staffing, construction trucks, construction office and storage trailers, temporary fencing, temporary wiring and power, construction phones, and computers. Many of these costs vary with time, and so they emphasize the need for an efficient schedule because time slippage can result in cost overruns. Owners should take care when requests for proposals solicit a "fee and costs for general conditions" from each CM. "General conditions" is a budgeting category first and a price comparison tool second. Any price comparison requires a line-item analysis to guarantee that the prices reflect the same services.

Cost Accounting/Reporting—Earned Value—Deliverables

The CM is charged with cost accounting, knowing precisely how much has been budgeted, committed, spent, and how much remains in each account. There is a vital difference between earned value and actual costs. Owners may require earned value reporting, which is more revealing of real productivity than a simple updating schedule. Earned value is the value of the work in-place. For example, a mason may have consumed half of the time allowed and 60 percent of the budgeted cost for erecting a brick façade, but only 25 percent of the brick is actually in place. If the brick activity has a value of $100,000, the earned value is $25,000. Stored materials, formally accepted by the owner's representative, are often considered in the pay requisition to be fair to the subcontractor and vendors for their cash flow requirements. Figure 6.1.5-3 shows a typical cash flow throughout the project life cycle.

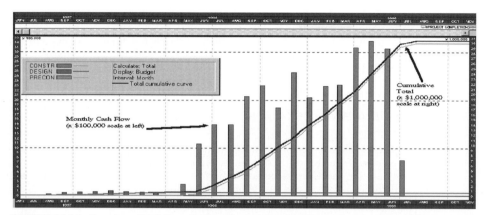

FIGURE 6.1.5-3   Summary of cash flow relative to project duration. (*Courtesy of the George B. H. Macomber Company.*)

Change Management

Changes in scope occur in virtually every project and are considered in detail in Chap. 3, Art. 3.2.2, "Controlling Change." Every effort should be made to minimize the number and the size of the changes because they can be expensive—in time, morale, disruption, and in money. Changes can be avoided by doing thorough research, enlisting the right people to develop and review the project scope, and by producing quality drawings, specifications, and bidding instructions.

Pay Requisitions, Retainage, and Deliverables

The wise owner will employ a well-documented procedure for paying the builder who, in turn, will do the same thing for the subcontractors and vendors. Key to the procedure is being sure that the product is in place per the contract documents and is free of all liens. A schedule of values is often agreed upon early in the contract arrangement so that there is little disagreement. When a discrete portion of the work is clearly finished, the builder is paid. A sum of 5 to 10 percent is often held until the owner is satisfied with the punch list, warranties, and operations and maintenance manuals. (See Art. 6.2.10, "Construction Closeout.") A good pay requisition procedure follows the guidelines established when using AIA forms G702 and G703.

FIELD ACTIVITIES, PLANNING, SUPERVISION, AND SUPPORT

Construction Logistics/Site Utilization

The CM prepares a site logistics plan indicating the locations of field offices, site access and egress, staging areas, delivery and storage areas, hoisting, and trash collection areas. The CM also prepares a traffic plan in accordance with local requirements. The CM may also prepare a detailed Affirmative Action Plan to implement Equal Employment Opportunity on the project in close coordination with the owner and local community and employer groups. The CM may also be charged with acquiring permits—soil conservation, wetlands, historical, archeological, and so on.

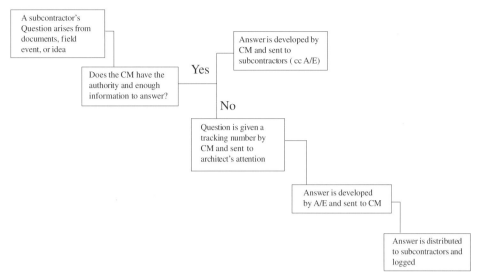

FIGURE 6.1.5-4   The request-for-information process.

Requests for Information (RFIs)

During the execution of a design and construction project, there is a continuous need for correct and current information to support good decisions. Requests for information are the result of findings at the site, such as unsuitable soil conditions. Also, there is often unclear, conflicting, or erroneous information on the contract documents. Construction progress can be held up until an answer is received. A well-managed RFI process gets questions to the decision makers quickly, prioritizes them for urgency, tracks them for response, keeps all parties informed of the decision status, and identifies and approves impacts on cost and schedule.

If an answer is given immediately by the owner's representative(s), it is important that the field personnel document any directives given in the field by the design team. Any deviations from the contract documents must then be documented in the form of a confirming RFI. The builder's RFI process helps to keep the schedule on track and at the same time maintains the integrity of the contract documents. (See Fig. 6.1.5-4.)

ARTICLE 6.1.6
CHANGE ORDER MANAGEMENT

Kenneth H. Stowe, P.E.
George B. H. Macomber Co., Boston, Massachusetts

Changes to the scope of the work as described in the contract documents are a fact of project life. They stem from various dynamic changes of the project environment (e.g., the market changes; the owner employs new personnel; there are unanticipated site conditions; ideas are received from the subcontractors' personnel; etc.). Successful project teams manage changes carefully so that the cost and schedule implications do not have undesirable impact. Proposed change orders are numbered, submitted, logged, and tracked by the builder until they are approved or finally rejected.

ARTICLE 6.1.7
PAYMENT CONTROL

Kenneth H. Stowe, P.E.
George B. H. Macomber Co., Boston, Massachusetts

The CM supervises and controls payments to all subcontractors and vendors for the work performed on the contracts. The builder additionally reviews and processes all applications for payments by involved trade contractors and material suppliers, in accordance with the terms of the contract.

PAYING THE VENDORS FOR DELIVERED MATERIALS, PROTOCOL, INSURANCE, LIEN WAIVER

Payment for materials is negotiated with each subcontract. Some vendors require a portion of the payment when the shop drawings begin, another portion when the owner takes delivery, payment when the material is installed, and final retainage of 5 or 10 percent of the payment is released when the work is substantially complete. Retainage is the owner's leverage to ensure that the work is completed to specifications and that any and all warrantees, waivers, and as-built documents are in hand.

ARTICLE 6.1.8
PROCUREMENT

Kenneth H. Stowe, P.E.
George B. H. Macomber Co., Boston, Massachusetts

Equipment and material procurement can often take months or even years. Procurement can involve many steps and many parties from specification to bidding, to submission, sometimes mock-ups, approvals, fabrication, and delivery. An effective process also includes identifying potential long-lead equipment and materials and a procurement strategy that ensures that the materials arrive on the site when they are needed and not before or after—often called *just-in-time* deliveries.

DELIVERIES AND INVENTORY MANAGEMENT

Deliveries must be managed to ensure a safe, secure, controlled environment for all delivered material and equipment and to allow construction to proceed without undue crowding and rehandling of materials.

The CM breaks the contract work into work packages and controls the bidding process. Care must be taken that the scope is covered in exactly one subcontract. Gaps and overlap can lead to confusion, waste, legal problems, and cost overruns. The speed of some projects may demand that unfinished drawings must be ultimately distributed as bidding documents. The builder can use two means to reduce the exposure to change orders in this situation. The CM can build a strategy with the architect regarding which elements of design must be detailed on the documents, and which can be communicated easily in the bid process by the builder's staff. The CM will look for every opportunity to prioritize drawing submissions, reduce unnecessary duplication of design effort, and look for opportunities to provide performance specifications as a substitute for full specifications.

The building team will take responsibility for communicating both the details and the intent of design packages to subcontractors, and the builder ensures that the bidder understands the schedule requirements and site logistics. This will ensure that the builder closes all of the gaps during the bidding process—not after final negotiations with the subcontractors.

ARTICLE 6.1.9
OWNER-FURNISHED ITEMS

Kenneth H. Stowe, P.E.
George B. H. Macomber Co., Boston, Massachusetts

The owner is often the best member of the team to purchase furniture, fixtures, and movable equipment (e.g., large computers, networking equipment, manufacturing equipment, and medical devices) because he or she may have special expertise and/or buying power. However, the project schedule should be comprehensive and integrated and should include the procurement and installation of these items. The builder is most often charged with coordinating the timing and the physical condition of the space when the equipment is to be installed. Equipment may be bought by the owner and installed by the builder. In this case, careful enumeration of roles and responsibilities is key to preventing gaps and overlaps that result in unnecessary expense and confusion.

ARTICLE 6.1.10

SUBCONTRACT-QUALIFIED SUBCONTRACTORS, BID COMPARISON, CHAIN OF COMMAND, REFERENCES

Kenneth H. Stowe, P.E.
George B. H. Macomber Co., Boston, Massachusetts

The builder must contract with multiple subcontractors, and the subcontract strategy must include a process that identifies the most competitive and qualified subcontractors. The process should be fair and orderly, allowing for sufficient time for the subcontractors to develop an understanding of the work and the environment and to research the best, most competitive vendors. The CM makes every effort to ensure that the prices cover all of the intended scope and must compare bids. When the best price–performance subcontractors are selected, there should be a final information/award meeting. The owner often requires that a representative be present at the subcontractor information sessions to ensure complete and common understanding of the requirements. In addition, the builder checks a subcontractor's insurance, bonding, permits, certifications, and ability to perform jobs of this nature. The builder also checks a subcontractor's workload, the qualifications of the proposed supervisors, and the individuals' willingness to agree to the builder's and the owner's requirements. The builder ensures conformance to safety, sequence, and quality expectations through strict on-site supervision, inspections by the superintendent and field engineers, and ongoing communication with the subcontractors' representatives.

ARTICLE 6.1.11

FILE-SUB-BID

Kenneth H. Stowe, P.E.
George B. H. Macomber Co., Boston, Massachusetts

There is a contracting methodology called *file-sub-bid* that has been promoted by some and mandated for some state-funded work. With this method, the GC submits a bid for coordinating and managing the work but must use the lowest responsive bid filed by the subcontractors in each trade. Proponents of this system claim that it prevents bid shopping by the GC. Others oppose the system, claiming that it disallows teamwork and promotes adversarial behavior during construction.

ARTICLE 6.1.12
SELF-PERFORMED WORK

Kenneth H. Stowe, P.E.
George B. H. Macomber Co., Boston, Massachusetts

Most GCs and CMs subcontract the majority of their work, but some work is not readily purchased from the subcontract community. Examples include selective demolition of elements not easily accessed, unknown elements of construction that must integrate with operating requirements of the owner, and construction activities that resist ready quantification and definition. In these cases, the CM may elect to *self-perform* the work and submit a lump-sum cost, a unit price, or labor and markup rates to accomplish the work on a time-and-material basis.

ARTICLE 6.1.13
CONSTRUCTION WORK PACKAGING

Kenneth H. Stowe, P.E.
George B. H. Macomber Co., Boston, Massachusetts

Work to be accomplished must be broken down into packages that can be priced and awarded in a competitive environment. The work-packaging strategy should invite competition (four bidders per package is a good minimum target), align with the schedule requirements, and attract subcontractors with a long-term interest in serving the owner.

ARTICLE 6.1.14
SUBMITTAL MANAGEMENT PROCESS

Kenneth H. Stowe, P.E.
George B. H. Macomber Co., Boston, Massachusetts

Submittals are formal documents, mock-ups, and demonstrations to prove that a material, method, or piece of equipment meets the specifications. In an effort to be proactive in managing the submittal process, the builder manages a process to ensure that the subcontractors fully understand their submittal obligations as soon as they are awarded the project. After work-package approval, the builder reviews the specifications for all submittal requirements for the scope of work awarded. After compiling a list of these requirements, the builder then consults the project schedule to assign due dates to each submittal item. Submittal requirement lists and due dates are transmitted to the subcontractors and are reviewed with them and updated on an ongoing basis to ensure that the subcontractors meet or exceed these requirements (Fig. 6.1.14-1).

Another important step in managing the submittal process is the initial submittal review performed by the builder's personnel. This is important to the success of the project because it is the CM's responsibility to streamline the review process and ensure that only conforming submittals are forwarded for review. Forwarding nonconforming submittals may hold up the project schedule because then it would be possible that unnecessary review cycles would be necessary for final submittal approval.

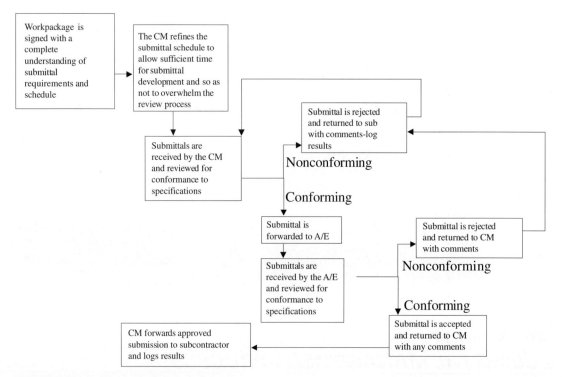

FIGURE 6.1.14-1 The steps in the submittal process.

ARTICLE 6.1.15
QUALITY ASSURANCE

Kenneth H. Stowe, P.E.
George B. H. Macomber Co., Boston, Massachusetts

The builder is charged with quality assurance, which consists of establishing expectations, managing the submittal, approval, and fabrication, and inspecting the work, as it is being performed, until final completion and acceptance by the owner, to ensure that the materials furnished and work performed are in accordance with working drawings and specifications.

ARTICLE 6.1.16
COORDINATION AND TRADE INTERFERENCE DETECTION

Kenneth H. Stowe, P.E.
George B. H. Macomber Co., Boston, Massachusetts

In cooperation with the A/E, the builder establishes and implements procedures for expediting and processing all shop drawings, catalogs, and other project documents. Trade coordination drawings involve a formal process to allow multiple trades to route their piping, ductwork, cable-tray, lighting, wiring, and so on to avoid interference, to facilitate construction, and to allow access for maintenance and future expansion. This process is managed by the CM/GC and results in a signed-off document that serves during fabrication and installation and, finally, as as-built documentation.

SUBARTICLE 6.1.16.1
Mechanical and Electrical Coordination

Bart D. Bauer, P.E.
Edwards and Kelcey, Inc., Boston, Massachusetts

During the construction of a facility, no trade or process requires more coordinated effort by every member of the project team than constructing and installing the mechanical and electrical components of a building. These efforts are expended in the areas of coordination, layout, mobilization and demobilization, commissioning, and decommissioning. For the purposes of this article, I use *mechanical* in the broad sense to include the trades of plumbing, HVAC, and fire suppression systems.

The installation of the electrical and mechanical systems is the most complex because of the large number of steps involved. Each and every step requires intense coordination between the electrical or mechanical subcontractor with the remaining trades and the other building components. Every step also includes gearing up the subcontractor's on-site crew to complete a specific scope of work. Then, the subcontractor must demobilize a large portion of its production staff to wait for other trades to advance the project to the next step.

Early in the construction process as part of the site work, the site work contractor begins to lay sewer pipes, water pipes, and electrical conduit. These pipes and conduits often run perpendicular to each other because the point of connection at the utility and the building are very diverse points. While the site work contractor is laying pipes, the foundation is also being placed. Both the electrical and mechanical contractors need to be involved at this stage. There are holes, or bond-outs, that need to be placed in the concrete formwork to allow the utilities to enter the building at a later date after the concrete foundation has been placed. Then, the electrical and mechanical contractors leave the project until the foundation is placed and backfilled.

The next time that the electrical and mechanical contractors will be on-site is to install underslab conduit and especially the main sanitary drains. The installation of these components relies upon heavy coordination with the GC's layout team. Both electrical and mechanical contractors are digging trenches to route major distribution pipes and conduits into walls with little or no allowance for layout discrepancies. Because the walls are not yet there, the layout of the future location is critical. Once the slab is placed, these components cannot be relocated. After the underslab work is completed, the contractors leave the site again.

After the major structure is complete and the wall framing has begun, the subcontractors return again to the site in force. This time they are installing and connecting the major supply lines to the underslab distribution system. This includes the rough-in work, installing items like electrical outlet boxes and conduits, plumbing pipes for the sinks and bathrooms, and HVAC ductwork for the mechanical contractor. Once this rough-in phase is completed, the contractors again demobilize a large crew or leave the site.

After the wall sheathing and the finishes such as paint and countertops are installed, the contractors return, this time to install the finish items. For a mechanical contractor, this includes sinks, toilets, grilles, and louvers. For the electrical contractor, the finishes include plugs, switches, and light fixtures. Along with the finishes, the contractors are beginning the commissioning phase. This involves turning on and testing the equipment and distribution systems that have been constructed in the project. This commissioning process is discussed in Art. 6.2.11, "Commissioning Programs for HVAC Systems."

For a large project that is constructed in phases, it is possible to keep the subcontractors busy on the different steps in different parts or floors of the project. Even in that case, however, there is a transition in processes as the contractors travel between the different phases of building construction.

Whether it is a large project or a small renovation, the coordination and sequencing among the electrical and mechanical subcontractors and the different trades and their effects on the project schedule are a crucial and critical effort that cannot be overlooked on a successful project.

ARTICLE 6.1.17
SCOPE CONTROL

Kenneth H. Stowe, P.E.
George B. H. Macomber Co., Boston, Massachusetts

What is the builder to build, how is it described, and how does it change? The word *scope* is used here to describe what the builder is charged with building, to what level of quality, and to some extent, how it will be built. Managing the information required to direct the design, procurement, and construction forces requires a multifaceted process that includes meeting minutes, design review, bidding instructions, submittal management, fabrication checks, on-site inspections, and commissioning.

ARTICLE 6.1.18
QUALITY ON-SITE

Kenneth H. Stowe, P.E.
George B. H. Macomber Co., Boston, Massachusetts

There are several ways for the builder's on-site management team to ensure quality control:

- Proper time and sequence for quality craftsmanship
- Daily on-site inspections
- Skilled supervision
- Horizontal and vertical layout and control
- Coordinating drawings where necessary

MAINTAINING DIMENSIONAL CONTROL

The builder provides qualified field engineering support. This staff is primarily in charge of setting control lines and making sure that subcontractors understand and adhere to them. The field engineer position is often considered an important step in the career path to the positions of superintendent and project manager.

ARTICLE 6.1.19
CHANGE PROCESS

Refer to Art. 3.2.2, "Controlling Change."

ARTICLE 6.1.20
LINKING PROJECT ISSUES WITH COSTS

Kenneth H. Stowe, P.E.
George B. H. Macomber Co., Boston, Massachusetts

The builder matches its document-tracking process with the internal financial control system. The builder should be able to track costs seamlessly directly from the initial commitment (purchase order) with the subcontractor through to final payment and invoicing. The link allows the project team to instantly capture the project cost(s) of any events reported during the document-tracking process. The builder's control systems are designed to allow for the maximum flexibility in the midst of dynamic project schedules while still maintaining rigorous control over every project event.

ARTICLE 6.1.21
FIELD REPORTING—FIELD SUPERVISION

Kenneth H. Stowe, P.E.
George B. H. Macomber Co., Boston, Massachusetts

The builder establishes an on-site organization and lines of authority to carry out the overall plans of the construction team. The builder also must maintain a complete and competent full-time supervisory staff at the job site to coordinate and provide general direction of the work and to ensure progress of the trade contractors on the project.

The builder also is responsible for establishing effective programs for safety, job site records, labor relations, and progress reports.

ARTICLE 6.1.22
WARRANTY AND O&M MANUALS

Kenneth H. Stowe, P.E.
George B. H. Macomber Co., Boston, Massachusetts

The owner's forces will be taking responsibility to operate, maintain, and later perhaps to renovate and/or expand the facility. There can be a large volume of vital information that should be organized and presented to the owner as part of the builder's service. As-built drawings, warranties, and operating and maintenance (O&M) manuals for all equipment and installations should be described in the specifications and followed up. The completed as-built drawings shall be submitted to the owner for its records upon final completion of the project.

SECTION 6.2
FACILITY CONSTRUCTION PROCESS

ARTICLE 6.2.1
CONSTRUCTION DOCUMENTS

Robert Morin, P.E.
Pizzagalli Construction Co., South Burlington, Vermont

The documents for a construction project include drawings, specifications, and the contract. The drawings are the graphic presentation of the construction project. They are usually broken down into site, structural, architectural, mechanical, and electrical sections. Floor plans, elevations, sections, and details of what is to be constructed are included in the drawings. At times, written specifications may be included on the drawings.

The specifications are used in conjunction with the drawings. They provide the information about the materials to be used, when they can be used, and how they are installed. The breakdown of the specifications is usually by the format of the Construction Specifications Institute (CSI). The CSI has developed a standardized specification format broken down into divisions that denote a trade or basic unit of work. Table 6.2.1-1 is a table of the 16 major divisions in the CSI format with a brief description of the most commonly used items for each division.

TABLE 6.2.1-1 Sixteen Major Divisions of the CSI Format

No.	Division	Major items	Handbook reference
01	General conditions	Construction expenses (temporary utilities and personnel) required to support the project for the duration of construction	3.2
02	Site work	Earthwork, foundation support systems, outside utilities, paving, and landscaping	5.1
03	Concrete	Foundation	5.1
04	Masonry	Structural walls, divider walls, and skin	5.1
05	Metals	Steel structural framing systems (columns, beams, girders, angles, joists, metal deck), cold-rolled framing (metal studs and joists), and miscellaneous metals (metal stairs, handrail, ladders, grating, etc.)	5.1
06	Carpentry and Plastics	Rough carpentry (wood framing, blocking, and roof nailers), finish carpentry (wood siding, moldings, paneling, cabinets, and casework), and structural plastics (structural shapes, grating, handrail and ladders)	5.1
07	Thermal/ moisture protection	Waterproofing, dampproofing, roofing, preformed siding and roofing, and caulking	5.1
08	Doors and windows	Doors, door frames, finish hardware, windows, store fronts, and curtain walls	5.1
09	Finishes	Fireproofing, plaster, drywall, flooring, ceiling, acoustical treatment, painting, and wall covering	5.1
10	Specialties	Toilet compartments, louvers and vents, access flooring, identifying devices (signs), lockers, fire extinguishers, and toilet and bath accessories	5.1
11	Equipment	Security and vault equipment, loading-dock equipment, food-service equipment, and laboratory equipment	5.1
12	Furnishings	Manufactured cabinets and casework, window treatment, furniture and accessories, rugs and mats, and multiple seating (auditorium seats)	5.1
13	Special construction	Cleanrooms, insulated rooms, pre-engineered structures, vaults, special instrumentation, utility control systems, and industrial control systems	5.1 to 5.5
14	Conveying systems	Elevators, escalators, hoists and cranes, and material handling systems	5.1
15	Mechanical	Process piping, heating ventilating and air conditioning (HVAC), plumbing, fire-protection systems, and process equipment	5.4 & 5.5
16	Electrical	Service and power distribution, lighting, and equipment connections	5.2 & 5.3

A construction project may not use all 16 divisions, and at other times one or more divisions may be added to the list. This breakdown is to be used as a guide to find specification information.

ARTICLE 6.2.2
SITE LOGISTICS

Robert Morin, P.E.
Pizzagalli Construction Co., South Burlington, Vermont

Contractors use vertical and horizontal controls for construction that are established by field engineers. The horizontal control (baseline) and the vertical control (benchmark, usually a USGS survey marker) need to be set before the contractor mobilizes (Fig. 6.2.2-1). Once the contractor is on site, the contractor maintains these controls. It is critical to layout the baseline and benchmark correctly because all construction is related back to the baseline and benchmark. Any errors in control layout will result in increased project costs and possible delay.

Segregating construction access and activities by using fencing and gates is recommended. This will assist with the flow of people, equipment, and deliveries, and it will help in avoiding possible labor issues (see Art. 6.3.4, "Renovation and Restoration"). Proper behavior of construction personnel, usually defined by the facility and the contractor, should be required during construction. Any construction personnel who exhibit improper behavior should be disciplined by removal from the site.

All construction should be done safely. Check to see that the construction specifications meet government guidelines. If the facility has safety guidelines, they should also be incorporated into the contract documents, and it should be stipulated that the contractor follow these guidelines.

FIGURE 6.2.2-1   A confined construction site needs control of people, equipment, and deliveries. (*Courtesy of Pizzagalli Construction Company.*)

Construction materials will be delivered to the construction site throughout the duration of the construction project. Certain materials will need to be placed in a lay-down area (a location that is used to store materials before they are used in the construction). The lay-down area location should be as close as possible to the construction for increased efficiency and to reduce material handling costs. Due to site constraints, an off-site lay-down area might be needed. If an off-site lay-down area is used, either the contractor or owner can provide this area. Another possibility is using just-in-time delivery of materials, but this can be very difficult, frustrating, and unsuccessful. Flexibility needs to be built into the construction process, and that is why a lay-down area is critical to the construction project (Fig. 6.2.2-2).

A pictorial record of construction activities will be needed. This record can be made up of photos, slides, and/or videos. A digital format is the most flexible type of record. Before any construction starts, a pictorial record of the construction site and a survey of surrounding buildings need to be made. Key items to focus on during construction are areas that will be hidden once the construction is completed (for example, underground piping). Each record should have the date, time, and location of where it was taken. If there are claims on the construction project, these pictorial records will be valuable documentation.

FIGURE 6.2.2-2   Lay-down areas are a major factor in keeping the project on schedule and within costs. (*Courtesy of Pizzagalli Construction Company.*)

ARTICLE 6.2.3
SITE MANAGEMENT TEAM AND ROLES

Robert Morin, P.E.
Pizzagalli Construction Co., South Burlington, Vermont

Someone in the organization must be responsible for overseeing the construction process. This person should understand the basics of construction and also know which people in the organization to contact for specific answers and information. If the project is large enough, a person (clerk of the works) or a group of people may need to monitor the project. Key factors to monitor are the cost of the project (is it escalating?), progress payment requests (are they spending as anticipated?), project schedule (will it be on time?), scope increases and changes, and quality assurance. See Art. 3.2.3, "Different Types of Agreements," for more information on the structural interface between owner, designers, contractor, and subcontractors.

A construction person should be designated as the contact for any construction/existing plant issues to ensure that facility operations are not adversely disturbed or disrupted. There should also be a backup person if the primary person cannot be contacted. A beeper, cell phone, e-mail, voice mail, or Web site can all be very effective tools for reaching this person. Each project should be evaluated if 24-h contact is needed. The designers should also have a designated contact person who monitors construction and answers questions. The designers will also use inspectors for monitoring and testing. During the majority of the time, the owner will work through the designers for inspection services. Depending on the project size, the designers might also have personnel on-site. The contractor will have a project manager (project leader) for the construction. The PM will be the main contact person, and all other contractor and subcontractor personnel will answer to this person. The contractor's field staff is broken down into three parts: (1) the supervisory personnel (superintendent and supervisors), (2) engineering staff (project engineer, field engineer, and office engineer), and (3) the clerical staff. See Fig. 6.2.3-1 for a typical field organization.

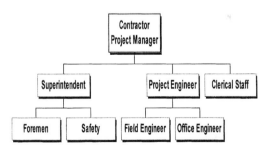

FIGURE 6.2.3-1   A typical project organization.

ARTICLE 6.2.4
CONSTRUCTION START-UP

Robert Morin, P.E.
Pizzagalli Construction Co., South Burlington, Vermont

Before there is activity on the site, establish a checklist for permit and fee requirements for the project. These permits and fees should have been set early in the project design, and responsibility for execution should have been assigned (owner, designers, or contractor); see Sec. 6.1, "Construction Management," for more information. The permit/fee checklist should include the following: name of the permit/fee, what it covers, costs, who pays the cost, procurement, when it expires, and lessons learned. The checklist should be maintained throughout the project. Any permit delays will increase the project duration and cost.

Electrical power is the most critical utility needed for construction. Water supply is practically a necessity to have on the construction site. Temporary utilities can be the responsibility of the owner or the contractor. If the contractor is required to pay for owner-furnished utilities, then a metering system will need to be installed to monitor consumption. If outside power supply is required, then the power company should be alerted early in the design stage to schedule their work to coincide with the construction project.

A preconstruction meeting should be held before any construction work starts. The people who attend are the owner, designers, contractor, and subcontractors. This meeting is for the major parties to meet each other, talk through the project, and state major issues that affect the owner and that the contractor needs to know. Some of the items that should be discussed are scope of work, logistics, schedules, methods, and the interface between the owner, contractor, and designers.

Contractor personnel who are assigned to the project will set up a field office for the construction period. Office trailers are the most commonly used structure if the owner does not provide office space. The office trailers should be located as close as possible to the construction. It is also desirable to locate the trailers at the construction entrance; this will assist in controlling the movement of personnel, equipment, and material onto the site.

A plan mapping out the steps for shutting down the existing plant operations affected by construction will need to be completed. Careful timing is required to coordinate shutdowns with construction. This information should be available to the contractor. Utilities may require temporary rerouting for items that are removed from service.

A hazardous material survey needs to be done as early as possible in the design phase for those materials that are to be demolished and removed. If hazardous materials exist (most common are asbestos and lead paint), then abatement of these items will need to occur before any demolition can begin. Because of liability issues, it is common for the owner to employ an abatement contractor who is independent of the construction contractor to remove the hazardous materials. It is best to have the abatement done before the main construction work starts. When abatement work is going on, no other work should be happening in the same area.

Demolition can occur at the beginning of the construction, during construction, at the end of construction, or a combination thereof. A plan for the demolition will need to be developed. Hazardous material abatement, demolition responsibility, and salvage of demolition materials should be addressed in the demolition plan.

At times it might be advantageous for the owner to salvage certain equipment. If the owner plans to reuse the equipment or if the equipment is complex enough that damage might occur in removal, it is in the owner's best interest to perform the removal. If the owner plans to have the equipment salvaged by the contractor, then a storage location should be established. The location should be accessible to both the contractor and the owner, and any weather-related issues for equipment storage should be addressed at that time. Materials that will not be salvaged will need to be disposed of or recycled properly.

During demolition, weather conditions will need to be addressed. If the demolition penetrates existing building skin, the contractor should erect temporary barriers that will keep out adverse weather conditions. These temporary barriers can be used to segregate the construction from existing activities and should also meet security requirements.

ARTICLE 6.2.5
FOUNDATIONS

Robert Morin, P.E.
Pizzagalli Construction Co., South Burlington, Vermont

Erosion control should be in place before any excavation begins. Erosion control consists of silt fences, hay bales, and inlet protection designed to control surface water erosion. A stone construction entrance is designed to keep earth materials from being tracked onto other areas of the facility or local roadways. The requirements for erosion control will vary by site and by state. Critical areas are installing a stone construction entrance before the earthwork starts and maintaining the silt fence system throughout the project.

Before excavation starts, ensure that all underground utilities are identified. Using local utility identification services (Dig Safe in the United States) or as-built drawings are a common practice in accomplishing this, but neither method is always successful. There should be a backup plan if utilities are damaged during excavations. Two key issues to address in the plan are who should be contacted and by whom, and what has to be done to resume operations. If an excavator digs through a power, communication, or water line, it could shut down the facility, so be prepared to use your backup plan (Fig. 6.2.5-1).

Now, excavation for the building foundation can start (Fig. 6.2.5-2). Any required building supports (piling, caissons) and excavation supports (steel sheeting, shoring) will also be

FIGURE 6.2.5-1 Site utility installation.

FIGURE 6.2.5-2 Formwork for concrete foundation wall placement. (*Courtesy of Pizzagalli Construction Company.*)

FIGURE 6.2.5-3 Formwork and reinforcing steel for building footings. (*Courtesy of Pizzagalli Construction Company.*)

installed at this time. If a high groundwater table exists, then dewatering will be needed to lower the groundwater table below the excavation. Ensure that the dewatering system does not adversely affect surrounding structures (building settlement).

Once the building pad is prepared, foundation structural work starts (Fig. 6.2.5-3). Most foundation designs use concrete for footings, slabs, and walls. Depending on the project schedule, the concrete for slabs-on-grade may be placed after the building superstructural frame is erected. When the concrete foundation has cured, then backfill can be placed around the foundation. If waterproofing and/or insulation of the foundation walls are required, this will take place before the backfill operation. See Art. 6.2.7, "Building Envelope," for more detail.

ARTICLE 6.2.6
SUPERSTRUCTURE

Robert Morin, P.E.
Pizzagalli Construction Co., South Burlington, Vermont

Once the foundation work has progressed, the superstructural framing starts. Structural steel is the most common type of framing system (Fig. 6.2.6-1). Other types of systems are precast concrete, poured-in-place concrete, masonry, or a combination thereof.

There will be a steady stream of materials delivered to the site as the frame is erected. Continuous access to the construction area is needed. If continuous access cannot be provided,

FIGURE 6.2.6-1   Delivery of structural steel is critical in a limited access site. (*Courtesy of Pizzagalli Construction Company.*)

then using the lay-down area next to the site for stockpiling framing materials will keep the erection operation running smoothly.

Major equipment is required for this phase of work. Trucks will be moving materials onto the site. Cranes will be used to erect structural members. The cranes will be placed as close as possible to the structure and at times might be inside the structure. Three major types of cranes are truck, crawler, and tower cranes. Truck cranes are usually hydraulic, and crawler and tower cranes are cable types. Truck and crawler cranes are mobile; they can move around the site. Tower cranes have a fixed base and cannot move around. Tower cranes are used when the site has very limited access.

A few items must be considered when a crane is moved onto a construction site. The crane should have been inspected and certified that it is in good running order. The area under the crane swing should not have any other work going on when the crane is in use. There should not be any overhead obstructions like power lines in the way of the crane swing.

After the framing is in place, flooring will be installed. The most common type of subfloor is concrete on metal deck. The metal deck is used as a form to hold the concrete, and once the concrete has cured (hardened), it becomes the subfloor system. Other types of subfloor systems are precast concrete and cast-in-place concrete. Concrete is usually conveyed from the concrete transport truck to the structure by a concrete pump or bucket and crane. The method chosen will depend on where the concrete is placed, the quantity of concrete needed, and equipment availability.

ARTICLE 6.2.7

DEMOLITION AND CONSTRUCTION OF VERTICAL TRANSPORTATION

James Fortune

Lerch Bates and Associates, Inc., Littleton, Colorado

Most elevators, escalators, moving walks, and dumbwaiters in the United States are installed by International Union of Elevator Constructors (IUEC) personnel. These IUEC constructors (mechanics) are employed by the elevator manufacturers to install their specific brands of vertical transportation equipment.

New equipment installations start after the layout (shop) drawings for the equipment installation are approved by the elevator consultant, architect, and general contractor and after the equipment is delivered to the job site.

Elevator components are installed in four different sequences:

1. *Pit equipment.*   Includes pit channels, buffer supports, car and counterweight buffers, compensation sheaves, and hydraulic jacks

2. *Hoistway equipment.*   Includes the vertical car and counterweight steel guide rails, supports and brackets, the counterweight and frame, the hoistway entrances, supports and fascia, hoist cables and wiring, the car sling, platform door operator, and the elevator cab.

3. *Machine room equipment.*   Includes the hoist machine or hydraulic pump unit, the power converter/inverter drive and motor control and car controllers, and the electrical components

4. *Signal fixtures.*   Includes the hall push-button stations, hall lanterns, car position indicators, car operating panels and buttons, communication devices (intercoms or self-dialing telephone), remote monitoring stations, and life/safety monitors

Escalators and moving walks are now almost always shipped as self-contained, preassembled components and are simply hoisted into the finished wellways. They typically require little in the way of construction labor. Elevator modernization is usually accomplished by upgrading and replacing various machine room components such as providing new power converter drives, new microprocessor controls, and sometimes, new door operators. During this program, the signal fixtures and sometimes the cabs are also replaced. Most times, the hoist machines and much of the existing mechanical hoistway equipment is reused in the modernization program.

There is little demolition of existing elevator or escalator equipment that is required. Existing buildings that contain elevators are either torn down or replaced with larger, more modern facilities with new elevators. The existing elevators are usually modernized/upgraded, or the old escalators are simply removed and replaced by new, self-contained units. If an existing elevator is to be completely removed and replaced by a completely new unit in the same hoistway, IUEC construction personnel do this demolition work.

ARTICLE 6.2.8
VERTICAL TRANSPORTATION—
ONGOING? MAINTENANCE

James Fortune
Lerch Bates and Associates, Inc., Littleton, Colorado

Like all electromechanical devices, elevators, escalators, and moving walks require ongoing maintenance to keep the units operating properly. All major installing elevator companies also provide continuing, contractual maintenance programs to the vertical transportation (VT) equipment owner. After the VT equipment is installed or modernized and during the 1-year equipment warranty period, a representative from the contractor will usually contact the facility owner/manager and try to sell him or her a continuing contract for VT equipment maintenance. The best time to negotiate the ongoing maintenance price, number of work hours, and the contract terms is during the installation bidding process before the equipment is actually installed in the facility. During this period, the facility owner is negotiating from a position of strength because it is almost a given that the installing VT equipment manufacturer/contractor will also become the continuing maintenance contractor. Today's microprocessor-based logic, motor, motion control, and proprietary software make it almost impossible for any maintenance contractor, other than the installing manufacturer, to provide this continuing maintenance service. This is the reason that we often recommend that the prudent facility owner include the continuing maintenance contract form and have the potential VT equipment supplier submit the completed contract with their equipment installation bids. The most profitable part of the elevator market is in maintaining the equipment rather than manufacturing or installing it.

A proper VT equipment maintenance program includes equipment cleaning, lubrication, replacement of worn parts and components, software upgrades, system testing, and adjustments. It should also include annual and 5-year, A.S.M.E. A17.1, code-mandated governor, safety, and buffer tests (system pressure tests on hydraulic elevators).

Virtually all of the major OEM elevator/escalator companies also maintain their own equipment after installation. They usually hire IUEC personnel for maintenance and typically train them in the proper maintenance procedures and operating standards for the particular brand of equipment being maintained. Maintenance mechanics are assigned to maintenance routes, and it is their job to see that the maintenance schedule and tasks are performed in a timely manner. A typical maintenance route can include up to from 30 to 40 units in various locations. In larger facilities, it is not uncommon for one or two maintenance mechanics to be assigned to that job or building full-time for 40 h/wk.

Most elevator maintenance contractors employ IUEC maintenance mechanics and pay virtually the same wages and benefits. Because maintenance labor is about 80 percent of the cost of a typical maintenance contract, if one contractor is substantially cheaper in his quote than another, then obviously the number of labor hours allocated to the contract is being cut. Typically, on-the-job labor content for a maintenance contract (exclusive of repairs and tests) should run about 1.5 to 2.0 h/wk unit for gearless elevators, 1.0 h/wk for a geared elevator or escalator, and about 0.5 to 0.75 h/wk for a hydraulic elevator or dumbwaiter.

Newer elevator microprocessor-based dispatch controls employ sophisticated, self-diagnostic strategies and fault-reporting capabilities to assist the maintenance personnel in diagnosing and correcting problems, sometimes before they actually occur! The latest technical innovations include remote elevator monitoring (REM) through a modem, the Internet, or radio frequency (RF) links from the elevator controller to the maintaining company's central monitoring stations. This interactive communications is used to schedule maintenance tasks and personnel assignments based upon actual equipment usage (much like changing the oil in your car after every 3000 mi). The REM system is also used to log faults and dispatch a mechanic in response to a trouble call and to assist trapped elevator riders by keeping them appraised of the status of the ongoing rescue efforts (a priority trouble call).

ARTICLE 6.2.9
EXTERIOR SKIN DAMP-PROOFING/ WATERPROOFING

Bart D. Bauer, P.E., and John F. King, Jr., L.C.S.
Edwards and Kelcey, Inc., Boston, Massachusetts

This article complements Arts. 6.2.10 and 6.2.11 of the exterior skin section and covers the below-grade areas of the building exterior.

There is a difference between below-grade damp-proofing and waterproofing in all phases of a building's life. *Damp-proofing* is the effort and materials that are required to stop water from entering the foundation through the areas of the foundation that are above the water table (Fig. 6.2.9-1). *Waterproofing* is defined as the efforts and materials used to stop the flow of water through the foundation wall when the wall in question is below the water table. The water table is the level (height or depth) of the groundwater in the soil. The reason that this is important is that below the water table, the water against the foundation is under pressure and, therefore, a different amount of effort is required to keep the water from penetrating the foundation wall. The terms *waterproofing* and *damp-proofing* are also used to describe

FIGURE 6.2.9-1   The potential of failed below-grade damp-proofing after several years of neglect.

above-grade installations, but in this article we will only discuss their application below grade. Although the difference between the two is critical at the design stage to the designer, product specifier, and the product manufacturer, it is a less critical difference for the reader of this section, which limits the topic to construction-related issues. For the purposes of this article, the difference is a moot point, because the concerns that are raised during the construction process are similar in either application.

Materials range from natural materials (e.g., bentonite clays) to highly engineered plastics and related materials [e.g., ethylene propylene diene monomer (EDPM) asphalt derivatives and copolymer rubbers such as butyl] to cementitious materials. Each material has good qualities and drawbacks in permeability, toughness, and application technology and therefore, should be selected by the designer. In addition, the selection of warranties available is very limited. The design process is reviewed in Art. 5.1.4, "Skins and Facades," and is not discussed further.

There are two basic approaches to damp-proofing and waterproofing. These are described in the trade as *positive-side* and *negative-side applications*. The positive side is the exterior side of the foundation wall, and the negative side is the interior of the foundation wall. No matter which side of the wall the designer has selected for the application of material, the major issues that should be monitored during construction to ensure a watertight product include substrate conditions, edges or terminations of the membrane application, dealing with cracks, sequencing the construction efforts, application restrictions, the means and methods of application, and finally, the kind of warranty received.

The substrate is the surface of the foundation on which the waterproofing is going to be applied. Generally, this is a cast-in-place concrete surface but could be concrete masonry units (CMU), commonly referred to as concrete block, wood, or any number of modern hybrid composite materials. There are three major concerns when designing for the substrate. These include the smoothness and condition of the finish, the moisture that is absorbed in the material, and cracks, either active or static.

Concrete by nature may not provide a smooth surface. During the placement of the concrete, these are areas where the concrete was not properly consolidated, called *honeycombs* or *rat holes* in the trade, and where the concrete finish is rough because of used-and-abused or misaligned forms. No matter what the cause, concrete is not a material that can be forced or persuaded into proper alignment once it is set. The defects are discovered only after the concrete has set and the forms are removed, and at that point there is no easy remedy. Waterproofing or damp-proofing materials are generally designed for a smooth substrate, so these substrate irregularities must be filled or ground down. The second substrate-related issue is moisture that is in the substrate due to a recent rain or to incomplete cure of the concrete or mortar. Most of the materials that would be used to waterproof or damp-proof a building will apply only to a dry substrate. Each manufacturer defines *dry* differently. No matter what the material or the application, the individual manufacturer has the ultimate authority to approve the substrate. On large projects, the manufacturer's representative should be on-site to approve the substrate and monitor the first day of installation. All parties who monitor construction need to understand that once the foundation walls are backfilled, the opportunity to correct any major or minor error inexpensively is lost.

The second general area of concern during the installation is the terminations of the damp-proofing or waterproofing material. These terminations occur at both the top and bottom of the foundation and at the sides and corners on each side. In theory, the goal is to create a water-tight "reverse bathtub" for the building to sit in (one that keeps all the water out, not keeps all the water in), so it may be hard to envision a side-to-side termination. However, most applications cannot be completed in a single day. So there are construction-related terminations to consider from day to day, called *night seals,* as well as the more permanent terminations.

Most waterproofing and damp-proofing materials are designed to perform in the dark regions below ground and are not resistant to ultraviolet rays from the sun. Therefore, it is very critical to deal appropriately with the top at grade level. There is still a need for damp-proofing in this area, and in fact some codes require extending it above the grade lines. But it becomes an aesthetic concern as well as an ultraviolet issue because the waterproofing or damp-proofing material may be exposed. Care must be taken to ensure that the application selected for this location addresses both of these criteria. Corners are another area of concern when certain types of sheet membranes are used. The membranes are not designed to withstand tremendous forces acting to puncture the material. However, if the material is stretched around the corner, the actual foundation wall can be seen as an object trying to puncture the material. The last termination area to be concerned with is penetrations, where various drains and other services go through the foundation walls. These can be treated like cracks that must remain able to absorb small amounts of movements.

Cracks can be categorized into four groups on the basis of their causes and characteristics (Fig. 6.2.9-2). Shrinkage cracks occur as the concrete cures and are generally dormant, or static, cracks. Settlement cracks that develop after the building is built are generally caused by larger-scale differential movements among different parts of the building. Depending upon the action that caused the settlement, these are considered active, or dynamic, cracks. The third group may not be considered a crack or an independent group. These are control joints, so named because they are joints that are created in the concrete to control the location of the shrinkage cracks. Therefore, these are also static cracks. The lines in a sidewalk are good examples of control joints. The crack that is formed between different concrete placements is a "cold" joint and often acts as a weak link in the chain, allowing movement in either a dynamic or static condition.

Due to the causes of shrinkage cracks either in control joints, if they were placed well, or wherever they occur in the wall, it can be assumed that the crack developed before the installation of the waterproofing or damp-proofing materials. Most waterproofing or damp-proofing materials have very limited crack-bridging capabilities; therefore, additional materials and application details must be used. All reputable manufacturers have a complete "kit of parts," compatible materials for all typical application conditions, which contain a specific material for crack bridging. It is critically important to use compatible materials for all special conditions to ensure that the materials work as one cohesive unit as they repel water. Before the application of the waterproofing, all visible static and active cracks must be sealed. Cracks that develop thereafter or remain active depend upon the bridging capabilities of the waterproofing materials.

Settlement cracks are much harder to address for several reasons. First, the crack does not show up until years later when pristinely clean construction conditions do not exist (if it occurred during construction, there are some other even more serious problems). The water-

FIGURE 6.2.9-2   A crack in a foundation wall and evidence of a recurring leak.

proofing/damp-proofing effort is usually a *fix* and includes three parts, each equally important. First, the structure must be investigated to ascertain if the crack is still active. Second, there is an attempt at the structural level to "glue" the building components back together (Fig. 6.2.9-3). If successful, the glue will also seal the crack to water penetration, but there is generally also a waterproofing or damp-proofing application as well. Although these are gen-

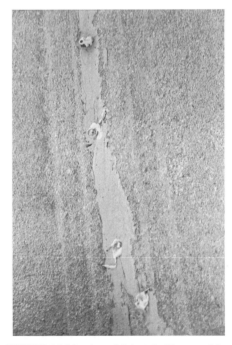

FIGURE 6.2.9-3   A crack injected with epoxy. Injection ports are still evident.

erally negative-side (inside) applications, applications can be made to the exterior as a positive side by excavating down to the affected area.

The proper sequencing of construction activities during this phase of the project also needs to be monitored to ensure proper applications. After the concrete has been placed and the forms have been removed, the substrate should be inspected. This will identify corrective work needed to address any shortcomings. After the corrective work has been completed, the waterproofing/damp-proofing application can commence, depending upon the ambient temperature and moisture and the substrate temperature and moisture. These conditions are very specific and vary tremendously from manufacturer to manufacturer based upon the material selected. Once the material has been installed and the proper cure time has elapsed, it is time for backfill. Perhaps the most overlooked aspect of the entire waterproofing/damp-proofing application is the failure to complete the next step properly. That is the installation of the protection board. This is a board as simple as a sheet of masonite, or as complex as a highly engineered sheet of composite plastic with drainage channels. No matter what the composition is, the purpose is the same, to keep the rocks and dirt off the newly applied membrane so as not to puncture it as the excavation is backfilled and the backfill material is compacted. Quality installation of the protection board requires coordinating the site work contractor, the heavy equipment operator, and the waterproofing contractor. Once the backfill is complete, all of the material is covered, and the only thing to do is to wait for rain.

ARTICLE 6.2.10
EXTERIOR SKIN/FACADE

Bart D. Bauer, P.E.
Edwards and Kelcey, Inc., Boston, Massachusetts

This article complements Arts. 6.2.9 and 6.2.11 of the exterior skin section and covers above-grade vertical areas of the building exterior.

The exterior skin or facade creates the exterior image, and most important, it keeps out the natural elements such as wind and rain (Fig. 6.2.10-1). There are facade materials that have been used since the dawn of man, such as stone and clay, and there are new modern materials,

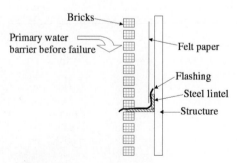

FIGURE 6.2.10-1 Primary water barriers in a brick veneer wall.

such as exterior insulation and finish systems (EIFS) and different plastic and composite spandrel panel materials. All are designed to perform the same function, to keep out the elements.

This article is restricted to the construction processes that are involved in the actual installation and to the items to watch for to ensure that the materials are installed functionally. This article also assumes that the drawings and specifications were properly prepared detailing the specific exterior materials and methodology of assembly, which is discussed in Art. 5.1.4, "Skins and Facades." There are a few designers who try to reinvent a proven system with every new building, at times with disastrous results. It is the responsibility of the construction team to request guidance and further instructions when details are insufficient.

As with any component in any portion of the building, if the appropriate details are not followed, parts of the facade will not perform as designed. However, unlike most material failures, there is no easily visible inspection point. The problem will be noticed only at the point of facade failure. Even a minor deviation from the drawings and specifications could result in a compromised exterior skin. When the exterior is compromised, the failure of the skin system or consequential damage may not be discovered for years, leading to extensive, expensive, and perhaps catastrophic results. These negative results include falling building elements and water penetration into the structural system.

The exterior skin is a composite system made up of several components that work together as an impervious barrier to protect the interior of the building from the natural elements. Often, a lack of understanding of the complete system and how the components work together leads to failures during the construction phase. Often, different individuals or different trades assemble the components, each trade working in its own vacuum on a finite task. There are many critical items related to all types of facade systems that can go wrong in the construction process. These items can be generally grouped into five categories: allowances for (1) expansion and contraction, (2) sealant, (3) internal drainage, (4) adequate substrate, and (5) flashing.

All materials, buildings, and building components experience natural forces resulting from expansion and contraction, structural flexure, and variation in temperatures and pressures. The mechanism that a building facade uses to absorb movement and resulting horizontal and vertical forces is an *expansion,* or *control,* joint. The size, location, and frequency of these joints are a well-planned engineering exercise. Despite the importance of these joints, they often appear as single lines on the drawing and therefore are easy to deemphasize during construction. Critical components of constructing expansion joints that should be part of any oversight or inspection include the width and depth of the joints, the locations, and the quantity of joints. Then, as the building facade goes through its natural expansion and contraction cycles, the anticipated movement of materials can be absorbed by the building as calculated and designed for rather than creating extreme unplanned stresses on the materials, resulting in cracking, warping, and failure of the integrity of the barrier. Additionally, the movement of the materials must be fully coordinated among the different building systems. A second very important issue related to control joint construction is that the control joint in the foundation must align with the expansion joint in the wall and must continue in alignment with an expansion joint in the roof. As an example, let us examine a brick facade wall where it meets the foundation. If the control joint (installed by the foundation contractor) in the foundation is not aligned with the expansion joint in the facade (installed by the mason), the brick will crack as the building moves, generally in a vertical or diagonal line above the foundation joint. The building's reaction to the movement will connect the expansion joints even if the joints are not in alignment. This creates a failure in the exterior skin.

The expansion joints in the materials mentioned earlier and the joints between dissimilar materials that occur at windows, doors, and other similar intersections of facade elements are sealed with sealants. Although commonly called *caulking,* caulking is a particular type of sealant of a particular chemical composition and is generally based on an older technology. These sealants are highly engineered materials designed to work extremely well within finite limits.

The shape of the joint, the preparation of the joint, the size of the bead, the actual piece or string of sealant in the joint, and the adjoining materials are absolutely critical to the success or failure of the joint. No one sealant is the panacea for all joints. If the contractor is not properly educated and does not understand the use of different sealants for different conditions, one material will be used in all situations.

A well-designed facade system will have both a primary and secondary method for removing water from the facade because of the variety of materials used in any one facade, different life spans, and different maintenance needs (Fig. 6.2.10-2). Often, one material will fail unnoticed by reaching the end of its useful life.

A well-detailed facade is designed for this inevitable fact. This failure is planned for or accommodated by planning a secondary or internal method for the water to exit from the facade. This secondary means relies even more upon coordinating the installation of the various materials and trades than the primary protection. However, this secondary system of protection is even less understood than the primary system. This secondary system can be found in the weep hole system in a brick cavity wall, in internal gutters and drains in a curtain wall system, or in a simple application of asphalt-impregnated felt paper in a traditional stucco application. It is generally misunderstood, so it is difficult for an ill-informed inspector to inspect the integrity of the secondary barrier. Often, weep holes are filled in or forgotten, interior channels are filled with sealant, and felt paper is lapped in the wrong direction, creating a failure in the secondary system. Because the secondary barrier is covered by the construction of the primary barrier, the discovery of errors in the secondary barrier is often associated with significant damage to the interior finishes. The fix for the failures generally requires deconstruction of the primary barrier.

When considering the elements of the facade that make up the substrate, these are the components that are never seen after the installation is completed. Therefore, it is critical that a diligent effort is made to install these materials in complete accordance with the designer's plans and specifications. The substrate materials in question can generally be categorized as sheathing materials and flashings (Fig. 6.2.10-3).

Sheathing materials as a group have their faults, but the greater percentage of failures comes from an inappropriate method of attachment, for example, installing sheathing board in an incorrect horizontal or vertical direction (thus preventing planned movement or creating a weakened structural diaphragm), or installing the board upside down. The fasteners may not fully engage the supporting structure because they are too short or the fastener is incorrect, either creating a galvanic reaction between dissimilar metals or inappropriately treated for exterior use, causing premature rusting and subsequent failure. No primary facade material can survive a failure of the material to which it is connected. Second, because the substrate is out of sight and out of mind, the failure is undetected until it is widespread.

Flashing is the final typical area of concern. Every window, door, expansion joint, penetration, and change of plane or elevation requires some sort of flashing. This is as simple as understanding that water runs downhill. Equally important in the detailing is the consideration of differential pressures and wind-driven elements that may force water in an uphill direction. In

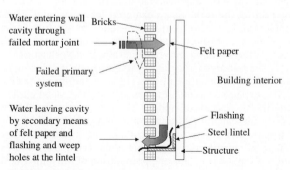

FIGURE 6.2.10-2 Secondary water barrier.

FIGURE 6.2.10-3 Failure of a steel beam supporting the facade material resulted from long-term water penetration.

the secondary or internal drainage system, flashing is critical to allow the water that is caught by the secondary system to exit from the building. Although understood as a concept, the application of this simple technology is the most overlooked or misapplied item in the composite facade. Whether it is called positive drainage, positive pitch, or just a simple overlapping of materials, any individual, no matter what the experience level, should be able to look at the wall/flashing system and see that the water will be directed away from the building.

Although this discussion has been limited to the extremely general concepts of constructing a composite facade, these basic concepts can be more specifically investigated case by case. Evaluating these general topics will provide the FM with a groundwork or foundation on which to build a thorough evaluative plan to review the construction of the individual facade elements on any particular building.

ARTICLE 6.2.11
EXTERIOR SKIN/ROOFING

Bart D. Bauer, P.E.
Edwards and Kelcey, Inc., Boston, Massachusetts

This article complements Arts. 6.2.9 and 6.2.10, which deal with the exterior skin section and covers the top, flat, or roof areas of the building exterior.

Roofing projects in general are extremely critical to a successful building project. A large majority of roofing projects are actually reroofing projects, for the roof is perhaps the largest and most critical portion of an exterior skin that appears in a reasonable capital renewal projection. Most other major facade materials, except sealants, have longer life spans than roof membranes. The FM also has greater knowledge and a vested interest in maintaining a water-tight roof because of the potential and magnitude of consequential damage from a roof failure.

There are many different types of roofing membranes, such as ethylene propylene diene monomer (EPDM), built-up-roofing (BUR), modified bitumen, and polyvinyl chloride (PVC), to name a few, but all roofing and reroofing can benefit from aggressive management throughout the roofing or reroofing project. Selecting and designing a roofing system is discussed in Art. 5.1.5, "Roofs," and is not discussed further in this article. Management of the project should begin at the stages of the prebid meeting and carry through the warranty inspection by the manufacturer.

The FM should attend the prebid meeting. Attendance at this meeting will show the contractors that there is an active owner, and it will acquaint the FM with the details of the project and enable him or her to view the project from the designer's and contractor's perspective. This is an important image or impression to leave in the contractor's mind. The contractor is less likely to try to cut corners or plan on cutting corners if a strong owner's presence is anticipated.

A fully engineered roof is a composite system made up of three basic materials: (1) a roof deck, (2) insulation and protection boards, and (3) the membrane. There are many different types of each of these materials; however, the majority of components for any roof project fall into these categories. The nature of the fully engineered roof system is very similar to a wrapped package. Once the wrapping is on the package, it is extremely difficult to see what is inside. The roof membrane acts as that package wrapping, and can hide a plethora of cut corners, inappropriate product substitutions, and poor workmanship. So many issues that are critical to a successful product can be covered up by the membrane that it is imperative for the FM to ensure that the entire job is inspected by someone other than the contractor doing the work. One common method of inspection is for the owner to employ a full-time, third-party observer to oversee the installation. This individual would be working in the owner's interest throughout the roofing and reroofing project.

All major roofing manufacturers have Approved Applicators programs in which the contractor's superintendents go to a school run by the manufacturers to learn to install the product correctly. That is all well and good and can often transfer to good information and activities on a particular site. There are two common disconnects that happen in this manufacturer-driven training installation system. First, in the construction industry, the available labor pool does not look upon roofing work as a highly technical or challenging employment opportunity. Therefore, the actual manpower staff to complete a roof project is built out of a very transient, poorly skilled workforce. The second issue is that roofing work is extremely weather dependent; therefore, seasonally dependent. This creates a large workload bubble in the summer months across most of the country. This bubble of work, coupled with a transient labor pool, creates opportunities for prime roofing contractors to subcontract the work. This process of having your roofing contractor award the work to a perhaps lesser-skilled roofing contractor allows the work to be transferred from a manufacturer-approved applicator to a "jobber"-type contractor who is not an approved applicator. At that point, oversight or understanding the manufacturer's standard details and installation procedures is limited.

These situations are the primary cause for needing a full-time observer. The particular type of roof system was selected at the design stage for its abilities to deal with specific material and design idiosyncrasies of the project and its performance standards or requirements. During the construction phase is not the time to have low-skilled, semi-experienced roofing laborers creating new details to suit their limited understanding of the engineered roof system or the project's specific requirements. The Roofing Consultants Institute (RCI) has created a program with Registered Roof Observers (RRO) and Registered Roof Consultants (RRC). This program is designed to raise the level of the full-time observers and the professional designers that select and design the roofing systems. The RRO receives individual registrations in each of the four major categories: (1) EPDM, (2) BUR, (3) modified bitumen, and (4) thermoplastics. Therefore, the RRO is fully knowledgeable about the roof system that is being installed. An FM who retains the services of an RRO for full-time observation buys cheap insurance against the catastrophic consequential damage that can occur.

During construction, each type of roof membrane—EPDM, BUR, modified bitumen, and thermoplastics—has its own individual installation constraints that need to be monitored. However, there are some general topics that can act as a set of guidelines for good installation practices, and actual material specific standards can be researched through the product submittals. These include deck preparation, material storage, clean work habits, and contractor coordination.

A properly prepared deck, ready for the installation of a new roof system either in a new project or a reroofing project must be firmly attached to a clean, dry substrate. The individual panels of the roof deck should not be able to flex to the extent that it will dislodge the insulation that is installed on top. Nor should there be gaps in the decking. All previous, unused penetrations must be filled in. Insulation is not designed to bridge gaps in the decking. The greatest enemy that a roof membrane has is a sharp object. Especially in reroofing projects, total and complete removal of all old material is essential. This includes all old rusted fasteners and broken bits of flashing. Finally, there should be no standing water on the roof. If water is left on the roof and the membrane is installed on top of the water, when the watertight roof membrane heats up in the sun, the heat will turn that water into steam. If the steam cannot escape, then it will either create an escape path through the membrane and cause failure, or will rain back down into the facility as it condenses at night when the roof cools off.

Storage of materials is another area of concern. One concern is the temperature of the storage location, and the second is the concentration of the storage of materials on the roof. Roofing materials are very sensitive especially to temperature, which affects the viscosity of the sealants and adhesives and also the volatile materials and the flexibility of the sheets of roofing membrane. Temperatures that are too cold or too hot will adversely affect the materials. Some of the effect will show up immediately; other effects will appear in a failure several years down the line as the coat or adhesive was too thin, or the sheet cracked during application and created a weakened section that failed under normal use and temperature cycling. In terms of the location of storage, the issue is structural. The easiest place to put all the material is on the edge of the roof by the crane. However, this pile of roofing material will soon exceed the capacity of the roof structure and deck. Roofing materials should be evenly distributed throughout the roofing area to spread out the load.

The final issue of common concern is that other contractors must also work on a roof. These include masons and carpenters working on the parapet walls, HVAC and other mechanical contractors working on rooftop mechanical units, exhaust fans, and roofing drains, as well as the sheet-metal contractor who may be installing flashings. All of this work must be coordinated to minimize the foot traffic on the roof after the installation. Depending upon the type of membrane that has been selected, the small cuts and punctures that are caused by the excessive and uncaring foot traffic are invisible to the untrained, and often trained, eye. Therefore, no warranty inspection will pick it up.

One final word about warranties. A good roofing system will carry a manufacturer's extended warranty for from 5 to 15, and even 20, years in some situations. This time is split between the contractor who generally carries the first year or two, and the manufacturer who carries the remainder. The thought is that the workmanship issues will show up in the near future and the material defects in the long run. This leaves the fox evaluating the quality of the security of the hen house in the short term, and the manufacturer who governs the approved applicator program is not concerned about the workmanship of the installation. Neither is working in the FM's interest. Again, this is good rationale for a full-time RRO to enforce quality as the membrane is installed.

As part of a warranty inspection on a large project, the manufacturer's representative will actually walk the roof with the contractor before issuing the warranty. For a small job, this site inspection will be performed across the table in a coffee shop down the road from the project. By requesting to be part of the walk, the FM forces the parties to come together and actually get up on the roof. This is the last opportunity that the FM has to bring any concern to the authorities attention. The people who attend the meeting are, in fact, working for the FM and have the responsibility of providing a satisfactory watertight roof for the project.

ARTICLE 6.2.12
INTERIOR FINISHES

Robert Morin, P.E.
Pizzagalli Construction Co., South Burlington, Vermont

Interior work starts after the structural frame and building skin are constructed (Fig. 6.2.12-1). This work involves interior doors, partitions, floor and ceiling finishes, specialties, and furnishings. It is divided into rough and finish work. *Rough work* involves interior framing (metal studs), blocking, piping, ductwork, and wiring (items you do not normally see once the project is complete). *Finish work* involves drywall, flooring and ceiling finishes, painting, plumbing fixtures, and light fixtures (items you normally do not see once the project is completed). See Subart. 6.1.16.1, "Mechanical and Electrical Coordination," for more information about mechanical and electrical systems.

Owner-furnished equipment needs to be coordinated in this phase for delivery and installation. Those items that cannot fit through doors must be installed before or during the rough-work phase. The contractor, designers, and owners should all be clear as to who delivers the owner-furnished equipment, who carries the insurance for the equipment, when it

FIGURE 6.2.12-1 The building has been enclosed, and interior finish work has started. (*Courtesy of Pizzagalli Construction Company.*)

FIGURE 6.2.12-2   Installation of finish work and mechanical/electrical systems can be complex. (*Courtesy of Pizzagalli Construction Company.*)

will be delivered, who moves the equipment to its location, and who installs and connects the equipment.

Interior work involves multiple trades, and sequence and spatial coordination are key (Fig. 6.2.12-2). The rough work must first be completed before the majority of finish work can start. When finish work starts, the contractor should divide the project into areas to limit access and prevent damage to the finish work. The contractor must assure that the construction does not damage already-completed finish work by coordinating the order in which the finish work activities occur. Once the finish work is completed, the punch-list operation starts. See Art. 6.2.13, "Construction Completion," for more information.

The major part of a building's complexity with which a FM must deal in maintenance, repair, and renovation is in this phase of work and in the mechanical/electrical phase. The greater the attention FMs direct to this phase of construction, the more effective they will be in maintaining the operations of the facility after construction.

End users will want to tour the construction site in this phase of work. This is the time they can see their space three dimensionally. This is also the time when they will most likely request changes in the design. Control of changes will be necessary to keep the project within the construction budget and completion date. See Art. 3.2.2, "Controlling Change," for more information on facility maintenance, repair, and renovations.

ARTICLE 6.2.13
CONSTRUCTION COMPLETION

Bart D. Bauer, P.E.
Edwards and Kelcey, Inc., Boston, Massachusetts

The construction completion process once again allows the FM to play a leading role in the construction process. Acting as the representative for the end user in conjunction with the design team of the architects and engineers, the FM has the opportunity to assist in establishing acceptable standards of workmanship and a date to accept the facility. This process begins with a pre–punch list, and carries through the punch-list process, operational introductions to the physical plant and its components, substantial completion, certificates of occupancy, final certificates of payment, and eventual occupancy of the completed project.

The first step in the construction closeout process is to establish the acceptable level of workmanship for finish quality of work performed. This standard-setting exercise should be accomplished preemptively and is considered the beginning of the end of the project.

A punch list is a list of items, determined by the design team, that are not complete or have not been installed to an appropriate standard and need to be corrected. The process of generating a punch list can be very politically charged for several reasons. The architect's role previously was restricted to answering questions and providing additional information but will now take a commanding role when the project is complete. The FM's role changes from a casual bystander to a full-time reviewer of material installed on the project.

After working with subcontractors to perform throughout the project, contractors see the punch-list process as a constant and personal criticism of their efforts and the company's product.

From the owner's perspective, this pre–punch list is the beginning of the process of accepting the building. Therefore, it is an appropriate venue to bring all parties from the facilities side back into the project. This inserts a totally new group of personalities into the project team, which disrupts the working status quo. Finally, it is a politically charged process because all designers and contractors look upon the punch-list process as a necessary evil. Although it is the last part of the construction process, it is lengthy, difficult, tedious, and often monotonous.

To initiate this punch-list process, one typical room should be completely finished before the rest of the building. This room will be used to set the standard of appropriate workmanship through a pre-punch-list process. Every single item in this room, except furnishings (if not built in), should be complete. The review of this typical room allows all parties, political or otherwise, to be involved in setting the workmanship standard. Once that room has been thoroughly reviewed and all the remaining or remedial work has been cataloged, the acceptable standard has been set. Then, the task of the large politically motivated team is complete. The contractor must be challenged to make the rest of the project meet that clearly defined standard or template. If this first pre-punch-list step is accomplished successfully, the pomp and circumstance will be out of the way and "punching out" the rest of the building becomes a cooperative venture among the seasoned members of the project team.

If the full punch list is started without the standard-setting process described before, the punch-list process becomes much more tedious. This will be obvious by the number of remedial items found in each room. If there are more than two or three major items per room, the contractor did not do its job and must be rechallenged to meet the established standard. If the contractor is not rechallenged, the procedure of generating the punch list becomes bogged down in an exhaustive list-making exercise of collecting 20 to 30 items per room, with tempers flaring on both sides.

The next major issue in the process of project closeout is the mutually exclusive but often misrepresented and intertwined concepts of certificate of occupancy versus certificate of substantial completion. Both certificates are major milestones in the project and signify that the project is nearly complete.

The local code enforcement official, usually a building inspector, issues a *certificate of occupancy*. It signifies that the work required by the building codes and permit has been completed. Normally, it is concerned with functionality and life safety issues. For instance, is the fire suppression (sprinkler) system installed and operational? Are all fire doors in place? Is there a functional restroom? Depending upon the type of occupancy, finishes such as final coats of paint and finished flooring are not a concern of the building official.

The architect issues a less formal but equally important *Certificate of Substantial Completion*. This is a point when the work described in the contract documents (drawings and specifications) is sufficiently complete that the owner can occupy the space for its intended use. This certificate also establishes a date on which the project is complete. This will determine if the contractor has met the contracted completion schedule or not.

As an example of the difference between the certificate of substantial completion and the certificate of occupancy, let us look at a typical office nearing completion. This hypothetical office is finish painted and carpeted, ceilings, lights, and phones are in place, and it is complete with furniture. The only thing lacking is one sprinkler head, which is back-ordered. Because the office is 99.9 percent finished, the office is substantially complete, and the architect would issue a certificate of substantial completion. However, the building official will not issue the certificate of occupancy because the life safety system (sprinkler) is not complete and operational.

Once it has been determined that the project is substantially complete, the FM should receive a project operation manual. This manual should contain a full list of all the equipment that has been installed in the project complete with all operational instructions, parts lists, and warranty information. A tour should be conducted matching the appropriate maintenance personnel and physical plant operators with the appropriate subcontractors. The tour should be a full operational introduction with training on the new equipment, not just a visual overview.

The FM needs to be even more acutely aware than ever before of the financial status of the project at the construction closeout stage. At this point in the project, the monthly requisition will begin to draw down the retainage. Retainage is a withholding from the monthly requisition that creates a pool of funds for the explicit purpose of correcting remedial work. Reducing the amount of retainage is a critical item of discussion because it is often the FM's only leverage to encourage the contractor to complete the punch-list items. Often, it is very difficult to motivate a contractor to complete the minor detailed items on a typical punch list. The FM should evaluate the contractor's effort and commitment concerning the punch list before agreeing to reduce the retainage.

The other part of the financial equation is a release of liens, or lien waivers. These are certificates from the GC and subcontractors that, in a general sense, state that the contractors have been paid for their work already in place. This is a release of liability and works toward obtaining a free and clear title. Before the final certificate of payment is processed, all of these lien releases need to be in place.

The last pieces of the project closeout puzzle are designer affidavits and a transfer of insurance. Some local jurisdictions require affidavits from the various design disciplines that the project was built according to plans and specifications. These affidavits should be obtained while the contractor is still available to answer questions that may arise. The formal transfer of insurance from the contractor to the owner may be the last piece that needs to be in place before occupancy.

If these items are given the critical attention they require, the FM can play an important and decisive role in the final phases of construction and occupancy of the new facility.

ARTICLE 6.2.14

COMMISSIONING PROGRAMS FOR HVAC SYSTEMS

John F. McCarthy, Sc.D., C.I.H., and Michael Dykens
Environmental Health & Engineering, Inc., Newton, Massachusetts

The American Society of Heating, Refrigerating, and Air-Conditioning Engineers, Inc. (ASHRAE) published its original guidance document on commissioning heating, ventilating, and air-conditioning (HVAC) systems in 1989.[1] The document presents a framework for viewing the commissioning process and general requirements that can be adapted to various projects. Although this is still the most widely cited guideline on commissioning, many commissioning projects involve building systems beyond heating, ventilating, and air-conditioning. In this manner, a well-conceived commissioning program can serve as an overall quality assurance measure for integrating complex building systems. These additional building systems could include the building envelope, mechanical and electrical systems, power and communication systems, occupant transport systems, fire and life safety systems, water systems, specialized control areas, and building management systems.

Commissioning has been defined as a systematic process that begins in the design phase and lasts at least for a year after the completion of construction. Properly executed, the process includes the preparation of operating staff and ensuring through documented verification that all building systems perform interactively according to the documented design intent and the owner's operational needs.[2] Commissioning is most valuable when the system performance is evaluated under the full range of load and climatic conditions. The ability to assess system performance under part-load or extreme conditions is often the best way to discover problems in buildings and correct them before occupancy.

BENEFITS

The value of commissioning projects is most often cited with respect to new building construction projects. Part of HVAC design intent is formally documenting performance objectives; basing the criteria of acceptability on these performance objectives, from construction through to system operation, results in a better-functioning building. As useful as these procedures are, they can be equally valuable in renovation projects and energy conservation programs. Many times, the start-up, control, and operational problems that occur due to minor changes in localized areas can often compromise the performance or efficiency of entire buildings. Furthermore, when one considers the impact that HVAC systems that are deficient in operation can have on indoor air quality, the benefits of incorporating proper building commissioning activities in all HVAC-related projects is obvious.

The benefits that can be realized from a complete commissioning program include

• Higher-quality building systems and the knowledge that the facility operates consistently with the original design intent and meets occupant needs.

- Identification of system faults and discrepancies early in the construction process so that they can be resolved in a timely manner while appropriate contractors are still on the job. This will reduce the number of contractor callbacks.
- Improved documentation, training, and education for operators and FMs to ensure longer equipment life and improved performance.
- Increased equipment reliability by discovering system problems during construction. In this way, commissioning prevents costly downtime due to premature equipment failure and reduces the wear and tear on equipment by ensuring that it operates properly.
- Reduced operating and maintenance costs.
- Improved occupant comfort and indoor air quality. Managing these factors effectively can reduce employee absenteeism and improve productivity and morale. Furthermore, the reduction in occupant complaints of discomfort minimizes service calls to building operators during the life of the building.
- Reduced potential for liability and litigation. This is true for minimizing liability of owners due to occupant personal injury cases and minimizing litigation of engineers and contractors due to claims from owners.

SELECTING A COMMISSIONING AUTHORITY

Because we recognize that commissioning does not clearly come under the purview of any single discipline, it becomes obvious that a separate commissioning agent, or as ASHRAE now prefers to designate the role, *commissioning authority* (CA), is required or available to carry out these functions. For this function to be effective, it is essential that the CA be independent and report directly to the building owner.

The following section describes the various options that have been proposed to fill the role of CA and the pros and cons for using each one.

The Owner

Owners are often the most obvious choice for the CA because they have a vested interest in ensuring that the work is held to the highest-quality standards. By using in-house staff, the owner can take control of the commissioning process to ensure that the contractor delivers the building properly. Disadvantages are that permanent staff must be assigned to deal with these ongoing projects and this may result in delays in other areas of the project or that in-house staff may lack the appropriate expertise to effectively serve as a CA.

An Outside Expert

The owner can still act as a CA by hiring an outside expert to serve in the role of the CA. The expert would report directly to the owner on the contractor's performance and provide effective monitoring of the commissioning progress. This also requires that the outside consultant be given appropriate authority to coordinate outside subcontractors for the many commissioning activities that are required. For this program to be effective, the line of authority from the owner to the commissioning authority must be clearly defined.

General Contractor

It is logical that the contractor be held accountable for quality control, which takes into account many of the activities required for effective commissioning of a building. Furthermore, it is the GC who is responsible for construction sequencing and who can effectively police the quality of workmanship on the job. The GC has a stake in the successful completion and timely delivery of the entire project. There is also direct financial benefit if the GC can reduce warranty and service calls in the future. Of course, the major drawback is the possibility of a conflict of interest because the contractor would be responsible for replacing any items found deficient. To try to avoid this conflict of interest, as commissioning agent, owners should retain the prerogative of approving of the GC's work by performing spot checks and providing quality assurance of the contractor's commissioning efforts. However, by doing this, the owner may generate animosity and ill will because exercising this authority will directly undermine the activities of the GC as CA.

The Engineer of Record

The advantages of using the design engineer as a CA include the fact that the engineer has full knowledge of the system design and is intimately familiar with its sequence of operation. This could achieve significant economies. However, there is a potential conflict of interest because design engineers may not acknowledge problems that are, in fact, design errors for which they are responsible. In addition, a major benefit of commissioning, that of outside peer review, is lost in this approach. Areas of design that could be deficient may not be captured because the engineer may not see them as deficient.

THE COMMISSIONING PROCESS

Phase 1: Programming Review

The building/HVAC commissioning process begins by

- Designating a commissioning authority (CA)
- Establishing the parameters for design and acceptance
- Designating the responsibilities of the various parties
- Delineating the documentation requirements for the entire project

The CA, the design team, and the owner review the building program and identify the information required for effective design and performance criteria for building acceptance.

Phase 2: Design Review

The CA is often thought of as a quality control element of the design team. As such, the CA is responsible for reviewing and documenting discrepancies between architectural/HVAC design and specifications and the owner's building system performance criteria. As the design review proceeds, the CA is also responsible for reviewing proposed value-engineering options for conformance to codes, occupant needs, and general sensibility and for providing an opinion regarding the resulting limitations. Early in the design phase, the CA is responsible for preparing and distributing a commissioning plan that identifies the responsibilities of each of the key members of the team and scheduling commissioning activities and deliverables. This should be in sufficient detail so that the required submittals designate parties and instrumentation that need to be present for each test. In addition, the master construction schedule should include the schedule of commissioning activities and link commissioning activities with other construction activities.

The CA must ensure that the design team takes explicit responsibility for documenting the following items:

1. *Design criteria and underlying assumptions.*   The design criteria should include all of the following environmental considerations:

Thermal conditions	Special loads
Humidity	Air quality design criteria
Occupancy (hours and levels of activity)	Pressurization and infiltration requirements
Lighting	Fire safety
Noise	Life safety
Vibration	Energy efficiency
Total and outside air requirements	Maintainability
Code requirements and impact on design	

2. *Functional performance test specifications.*   These specifications are developed during the design phase and allow the design team to better anticipate the commissioning process requirements. These specifications are required at a minimum to

   - *Describe the equipment or systems to be tested.* The description of the HVAC system includes type, components, intended operation, capacity, temperature control, and sequences of operation.
   - *Identify the functions to be tested.* Operation and performance data should include each seasonal mode, seasonal changeover, and part-load operational strategies, as well as the design set points of the control systems and the range of permissible adjustments. Other items to be considered include the life safety modes of operation and any applicable energy conservation procedures.
   - *Define the conditions under which the test is to be performed.* It is important to consider all possible operating modes, for example, full and partial loads and the extremes of operating temperatures and pressures. The documentation that is provided to support this is critical because it will clearly show the completeness of the engineer's design.
   - *Specify acceptance criteria.* It is essential that the acceptance criteria be presented in clear, unambiguous terms. Where possible, the acceptance criteria should be quantitative, and the accuracy and precision required should be consistent with the limitations of the equipment and system design.

Phase III: Construction Oversight

During the construction phase, the CA is responsible for on-site inspection of materials, workmanship, and installation of HVAC system and components, including pressure tests of piping and duct systems. The CA should also observe and/or independently use audit testing, adjusting and balancing, and calibrating system components.

Other activities that an effective CA performs during the construction phase include (1) review of warranty and retaining policies before construction, (2) obtaining copies of contractor's approved equipment submittal for review, (3) ensuring that effective construction containment techniques are used, and (4) documenting and reporting discrepancies for the owner.

Finally, it is essential that personnel who will be responsible for operating the completed system receive adequate training before system acceptance. This is best done during the construction phase. The CA should take responsibility to ensure that appropriate personnel (often equipment manufacturers through the design engineer) provide this training for the numerous components of the HVAC system.

Phase IV: Acceptance Phase

The acceptance phase should follow the commissioning plan established during the design phase. The functional performance test specification forms the basis for documenting the performance tests. The CA either conducts or observes the appropriate parties testing the functional performance of each system. This testing should start at the lowest reasonable level (i.e., system components, then on to subsystems, then finally systems) until every piece of equipment has been tested. The CA must also ensure that all essential activities have valid performance tests (e.g., hydrostatic testing, testing and air balancing (TAB) work, and calibration of automatic controls), and that the tests have been completed to a satisfactory level before starting the acceptance verification procedures. As stated earlier, it is critical that testing be done in all modes of system operation, including full-load and emergency conditions.

All required documentation should be compiled to form the basis of the system operations manual. Furthermore, as-built documents should be revised to ensure that accurate plans are available showing all relevant control points and values.

The CA will produce and distribute to the appropriate parties a document detailing all discovered deficiencies in the form of an action list. After the required work has been completed, the CA will revisit the site and perform follow-up performance testing, where required, to verify that all action list items have been successfully resolved.

Phase V: Postacceptance Phase

The postacceptance phase can best be thought of as an ongoing audit function of the building systems and the building's occupancies. Periodic retesting is often advisable, especially during the first year. This can be particularly important during extreme seasonal variations from the original commissioning period or during design extremes. The CA should also document the building operator's adjusted set points to ensure that they are consistent with the original design. Where differences exist, the CA should evaluate the impact and reconcile them in a written report.

COST AND OFFSETS

There is currently no standard approach to costing commissioning services. The following are some of the more common methods

- Budget a percentage of the total mechanical/electrical cost of a project. A range of between 2.0 and 6.0 percent is generally considered reasonable. The higher percentages are generally used for those projects that are smaller in scope or those that are more complex.

- Set up a separate commissioning budget that is independent of the project budget. This is often useful when the owner has an ongoing construction program, such as that found in many hospitals. Setting aside a commissioning budget that represents between $0.10 to $0.28 per square foot allows carrying out the work over a number of projects during a year's time. Most owners use an operations budget, although some do capitalize this work.

- Use a payment schedule based on time estimates provided by the CA to the owner. In these types of projects, it is important that all parties agree in writing as to what constitutes a completed plan, as well as an appropriate payment schedule.

No matter what budgeting approach is selected, contracts with the general and specialized contractors must clearly state that although the CA is initially paid by the owner, additional charges incurred by the CA will be paid by the contractors if systems fail or cause delays to the schedules established for the commissioning.

Although commissioning is often seen as an added cost to a project, owners experienced with commissioning do not find an overall cost increase in constructing buildings. The costs provided before compare favorably with several cost parameters normally associated with building construction. For example, the 2 to 6 percent of M/E project cost compares favorably with the 9 to 18 percent range of change orders and claims generally encountered on capital projects.

In addition to the benefits cited earlier, experience shows that an effective commissioning program can also

- Reduce change orders and claims by 50 to 90 percent.
- Provide energy savings in the first year of operation that generally exceed the cost of commissioning.
- Reduce overall system maintenance costs during the first year by an amount that is comparable with or often exceeds the cost of the commissioning program.

Levin lists the benefits reported by Portland Energy Conservation, Inc.'s survey, including 146 case studies.[3] The results are reproduced in Tables 6.2.14-1, 6.2.14-2, and 6.2.14-3. The data reported in these tables clearly indicate that commissioning programs provide important economic and operational benefits to the construction process.

Although readily available data may not exist to demonstrate improvements in worker productivity or illness rates due to commissioning activities, many groups have done extensive evaluation on the impact of commissioning and on implementing energy conservation or energy efficient measures. Piette studied the commissioning of energy conservation programs in 16 buildings in the Pacific Northwest and found that the investment in commissioning was cost-effective based on energy savings alone.[4]

CONCLUSION

Commissioning is a systematic, detailed process that requires the mutual commitment of owners and the CA to ensure its success. The goal of commissioning is to turn over to the

TABLE 6.2.14-1   Benefits of Commissioning

Benefits of commissioning	Percentage reporting the benefits
Energy savings*	82
Thermal comfort	46
Improved operation and maintenance	42
Indoor air quality	25
Improved occupant morale	8
Improved productivity	8
Reduced change orders	8
Timely project completion	7
Liability avoidance	6

Source: Reprinted with permission from Ref. 3.
* More than 70 percent documented energy savings by metering or monitoring.

TABLE 6.2.14-2 Thermal Comfort Benefits of Commissioning

Benefit	Percentage reporting the benefit
Improved thermal control	90
Reduced humidity control requirements	52
Benefited from improved air balances	30
Reduced occupant complaints	30

Source: Reprinted with permission from Ref. 3.

TABLE 6.2.14-3 Indoor Air Quality Benefits of Commissioning

Benefit	Percentage reporting the benefit
Improved ventilation	70
Better contaminant control	22
Improved carbon dioxide levels	19
Improved moisture control	11
Improved containment: clean rooms or laboratories	8

Source: Reprinted with permission from Ref. 3.

owner a building that meets the design intent with the appropriate safeguards (such as operator training and required documentation) to ensure that it will continue to function properly. As owners, contractors, architects, and engineers have begun to see the benefits of commissioning, they are incorporating it into their building projects. Although there are many definitions of commissioning, it is important to bear in mind that this is the ultimate quality assurance program in the life of a building. As such, it must clearly and unequivocally set the standards of acceptability. The CA has a responsibility to the owner and to the community of professionals involved in the building process to ensure that the highest standards are met and to ensure that a building performs according to its design intent and its occupant needs.

REFERENCES

1. ASHRAE, *ASHRAE Guideline 1-1989, Commissioning of HVAC Systems.* American Society of Heating, Refrigerating, and Air-Conditioning Engineers, Inc., Atlanta, GA, 1989.
2. PECI, "Summary Report," in *Proceedings of the Fourth National Conference on Building Commissioning,* St. Petersburg Beach, FL, 29 April–1 May 1996, Portland, Oregon, Portland Energy Conservation, Inc., Portland, OR, 1996.
3. H. Levin, "Commissioning: Life cycle design perspective," in *Proceedings of the Fifth National Conference on Building Commissioning.* Huntington Beach, CA, 28–30 April 1997, Portland Energy Conservation, Inc., Portland, OR, 1997.
4. M. Piette and B. Nordman, "Cost and benefits from utility-funded commissioning of energy efficient measures in 16 buildings," *ASHRAE Transactions: Symposia* 102, no. 1 (1996), 482–491.

SECTION 6.3
CONSTRUCTION PROCESS VARIATIONS

ARTICLE 6.3.1
CLIMATIC, GEOGRAPHIC, AND LOCAL INFLUENCES

Joseph J. Albanese
Shawmut Design and Construction, Boston, Massachusetts

Mark Hasso, PhD, P.E.
Wentworth Institute of Technology, Boston, Massachusetts

Climatic, geographic, and other local influences must be considered in planning, designing, and executing any construction project. Projects of similar scope that are planned and executed in different parts of the world or even in different parts of the same country will have different considerations that will influence decisions about means, methods, and materials. Generally positive weather conditions and virtually minimal disruption of activities such as concrete placement or excavation influence a construction schedule in a warm climate. Planners of a similar project in a cold climate will have to account for adverse weather conditions from late November to early March, adding premiums for heat and winter protection to place concrete, excavate, or perform masonry work. Geographic and local influences such as terrain and site condition of the project, labor costs versus material costs and availability, the quality and experience of tradespeople, and local economic issues will shape many of the design and construction decisions that impact the project.

CLIMATE

The atmosphere under which the project is executed influences the working conditions, cost, productivity of work, and the choice of materials for construction. Consideration should be given to climatic conditions outdoors and indoors. The climatic conditions to be seriously considered include weather. Local weather conditions must be considered during the conceptual stage of a project and continuing through the planning, scheduling, and execution of the work. In colder climates, there are quantifiable costs for temporary heat and other provisions that must be made for winter conditions during the construction project. In extremely hot climates, safety and productivity concerns could result in scheduling certain trades early in the morning or on an off-shift to avoid the strong midday sun and radiant heat. Temperature plays a major role in the productivity of the labor force. In snow and freezing conditions, it is important to use heating equipment indoors to provide reasonable working conditions that will maintain labor productivity levels. Heat can also be applied within tented enclosures surrounding concrete pours or building facades to allow mortar and concrete to cure without freezing. Special insulated forms and blankets, along with concrete admixtures, are also used to place concrete in cold weather. In hot weather conditions, it is important to provide an environment that will reduce its impact on the progress of work and the productivity of labor. Air quality must be maintained by using ventilation and by exhausting fumes and dust. It is important to keep workers hydrated, with plenty of potable water and work breaks to avoid heat stress. Special construction techniques may need to be used in hot weather, such as cooling the concrete mix and hydrating the setting concrete to avoid cracking from the rapid vaporization of liquid.

Wind velocity must also be considered, particularly in urban locations and in high-rise construction. In windy conditions, temporary construction, stockpiled materials, storage containers, and gear must be secured to ensure that the wind does not displace them. Such displacement could create severe safety hazards to the workers and the public. The effects of wind on partially installed work or hoisted materials should not be underestimated. Gale forces can cause deflections that lead to failures. Temporary shoring and bracing must be used until the structural integrity of the building system is completed to design capacity. Hoisted material can catch the wind and sail like a kite, creating an unsafe condition for workers and the public, and potentially causing damage to the work in progress. Tethers, guy lines, and additional manpower need to be employed to control hoisting and installation of materials when wind is a consideration.

Rain and the water table must also be considered. Erosion controls must be employed during excavation and foundation construction. Well points, cofferdams, and sumps can be used to control the water table and to evacuate water collected during periods of precipitation. Different regions have historical precipitation conditions that vary throughout the year. Construction costs associated with potential precipitation should be considered for the project's location.

Humidity is also an important consideration, particularly when contemplating the installation of finishes. Plaster, millwork, wood doors and frames, wood floors, fabrics, and acoustical tiles are particularly sensitive to the effects of humidity. The building environment must be stabilized before installing these elements. In addition, the moisture content of the finished elements must be stabilized before installing them. The building environment can be controlled by using temporary HVAC systems or by accelerating the completion of the permanent systems.

Expected weather conditions should be considered in pricing and scheduling the project. Ordinary construction contracts dictate that costs associated with the risk of normally expected local weather conditions are the responsibility of the contractor. This may address harsh weather conditions such as freezing, rain, and snow. The contractor may be relieved from responsibilities for damages and delays due to extreme natural phenomena such as tornadoes, hurricanes, mudslides, and floods, which could be considered acts of God.

Weather conditions may also dictate using specific materials in construction. For example, stucco is much more widely used in areas of moderate to hot climate, partially because of its durability and resistance to hot weather conditions and to the adverse affect that the freeze/thaw cycle has on it.

SEISMIC CONDITIONS

In the past 15 to 20 years, project owners have been alerted to the fact that earthquakes may happen in any region and at any time. In renovating or restoring existing facilities, most building codes require that the structure be modified to comply with seismic requirements. It is important to recognize seismic risk during construction. Even minor tremors can have disastrous effects on the work in progress.

GEOGRAPHY

The term *geography* in this context refers to site conditions such as terrain and accessibility. A site visit is an essential part of the planning and cost-estimating process. Terrain and accessibility to the site are major determining factors that impact the project design, schedule of delivery, and cost. Issues such as lay-down of materials and site use, discussed in detail in Art. 6.1.5, "Construction Cost Control," are influenced by the geography of the site. The construction of a ski lodge at the top of a mountain is fraught with cost-impacting challenges that limit the material and equipment that can be delivered and used in the work. Site accessibility and local terrain are important logistical issues that must be considered.

LOCAL INFLUENCES

A region's economic conditions will impact the execution of a construction project. Engineering and construction firms enjoy good economic conditions but also experience downturns in the economy that make the construction industry cyclical. During a downturn in the economy, profit margins are lower, despite the high risks of construction. Bankruptcies or downsizing of firms often result from the impacts endured during economic plight. Before embarking on a project and engaging a construction firm, an FM must consider the sustainability of the activity level of the construction firm. In dynamic economic times, bankruptcies, acquisitions, mergers, dynamic growth and rapid downsizing can have riddling effects on the stability of the firm. The associated risks are the availability of resources to complete the work. A review of the new construction put in place from 1970 to 1993 in 1987 dollars indicates that cyclical activity affects the construction by about $65 billion a year every 5 to 6 years, as shown in Fig. 6.3.1-1.

Labor Supply Information and Practices

Sources of Labor.   The supply of construction labor becomes an important factor at the construction development phase and continues until the closeout of the project. Close attention to the availability of construction labor should be given, starting with the planning and conceptual phase. Cyclical economies will naturally impact the availability of skilled labor. The supply of unskilled and semiskilled labor is not as problematic as that of skilled labor, especially in times of high economic activities. Unskilled and semiskilled labor may be trained on the job and will generally satisfy the needs of the construction project. There are constraints in union contracts as specified in their collective-bargaining agreements. These constraints limit the number of semiskilled and unskilled workers (apprentices) as a ratio against trained journeymen. Con-

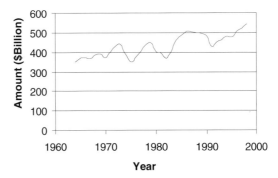

FIGURE 6.3.1-1 Construction put in place in the United States from 1960 to the present, in 1992 dollars. (*U.S. Bureau of Census.*)

versely, in nonunion environments during boom times, firms are constrained in growth and technical capacity by the availability of trained skilled labor. To achieve growth and greater technical capacity, these firms must initiate focused training programs for their craftspeople.

The supply of construction workers is provided from the following sources:

- Workers already employed in the industry (both union and nonunion)
- Unemployed construction workers
- Workers employed at other jobs, who are waiting for construction employment
- Travelers
- Persons enrolled in apprentice-training programs
- Persons enrolled in vocational-training programs
- New entrants to the construction industry

Local Labor Practices. Labor in different parts of the world comes with varying degrees of skill and cost. In construction operations, the cost of labor is often compared to the cost and availability of material and equipment. Depending on the local cost comparisons, choices in different regions of the world could lead to approaching common activities in very different ways. For example, in an area where a specialty piece of equipment has a premium cost, low-cost labor might be employed as a cost-saving measure.

Cost comparisons and decisions based on the most economic solution also bring up the concept of efficiencies in labor practices. It is important to understand and it is strategic to address the levels of efficiency of labor operations. The high cost of labor inefficiencies can be attributed in part to these factors:

- Weak project management, technology, and project controls
- Lack of continuous training and upgrading of labor skills
- Improper tools and unavailability of equipment
- Safety programs and proper personal protective equipment

These impacts can be regarded as geographical in the sense that technology, skills, tools, equipment, and safety programs are part of an economy, sophistication, and culture that influences labor practices.

Materials and Equipment Availability

Local practice and expertise often dictate methods and materials. Cast-in-place concrete is used more than structural steel in Europe and in the southwest region of the United States (Fig. 6.3.1-2). Tilt-up construction is widely used in the western part of the United States due to local practices and mild climatic conditions (Fig. 6.3.1-3). The same product used in the Northeast would require additional thermal resistance, making a cavity wall assembly more appropriate.

Given these examples, in planning a project in different regions, it is important to understand the region's predominant practices, materials, and equipment availability. Equipment and material availability depends on these factors:

- Local, regional, national, and global economies
- Project location (e.g., difficulty of access to the project site)
- Normal shortages of materials and equipment due to manufacturing environments

FIGURE 6.3.1-2   Cast-in-place concrete in place and formwork in the foreground.

Code Utilization

Numerous codes are involved in delivering any construction project. These codes change from region to region throughout the world. These three building codes are used in the United States:

- Building Code Officials Administrators (BOCA)
- Southern Building Code Congress (SBCC)
- Uniform Building Code (UBC)

It is important to understand the codes that govern construction requirements in the region for the project under consideration. In many regions, specialty code consultants can and should be engaged to help with interpretations and special requirements specific to the project location.

Infrastructural facilities such as highways, tunnels, and bridges have other codes of practice that are different from the previously listed building codes. National and state organizations and agencies publish these codes, such as the American Association of State Highways and Transportation Officials (AASHTO). Facilities managers should be aware of these codes and their requirements for compliance.

FIGURE 6.3.1-3   Tilt-up construction.

Licensing Requirements

Professional practice laws and regulations govern architects and engineers. Individual states issue licenses to architects and engineers. Thus, for an architect or engineer to practice in any state, it is necessary to apply for a license and to obtain approvals from the relevant board. Most states require an examination, especially California, due to the extraordinary seismic design requirements. Construction, plumbing, and electrical trades must be licensed to work in their respective trades. Again these licenses vary from state to state.

In the last 2 decades, environmental licenses have become more commonly required in the United States for handling radon, asbestos, lead paint, and other hazardous waste. Check with your local authorities on these requirements because there is no national uniformity. For example, to practice in the field of hazardous waste management and abatement in Massachusetts, one must be a licensed site professional (LSP).

ARTICLE 6.3.2
GOVERNMENT VERSUS PRIVATE WORK

Mark Hasso, Ph.D. P.E.
Wentworth Institute of Technology, Boston, Massachusetts

Joseph J. Albanese
Shawmut Design and Construction, Boston, Massachusetts

National, state, or local regulations govern public-sector construction projects. Each government agency determines the needs for construction projects that encompass all kinds of facilities, including public works projects, municipal buildings, defense projects, and other services. Private-sector construction is driven by private-sector needs or may be driven by creative forms of project development. These common issues are encountered by both sectors:

- Construction project development
- Funding
- Bidding and bidding practices
- Progress payments
- Procurement of labor, materials, and equipment

The main difference in a publicly funded contract is the accountability that the contracting officer has to the public. The aforementioned regulations dictate rigid rules of engagement between the contracting authority and the architect/engineer and general contractor/construction manager. Recent developments have created newer and more sophisticated forms of contracting for project development, funding, and execution, but there are still stark differences between private- and public-sector construction. An FM should thoroughly understand the major differences before choosing to fund a project with public funds and potentially exposing the process to a myriad of regulations. For example, funding a housing project with Housing and Urban Development (HUD) funds rather than totally financing the project through private means will require executing the project in accordance with specific regulations and statutes. These differences are outlined here.

Construction Project Development

Public-sector project development varies with the type and size of project and with the public agency involved. Public works and municipal projects go through a long process of approval that explores and addresses the cost, benefits, and value of the project to the community that it will serve. Typically public works and municipal projects originate at the local or state level. Public hearings at these levels discuss the project benefits, costs, values, and impact on the community. The project proponents push the positive impacts of the project, and its opponents stress the costs and negative impacts of the project, such as traffic impacts, environmental impacts, and construction impacts, as well as other potential issues imposed on property values. A school project could have positive social and property-value impacts while a prison project may have adverse impacts on both counts. This process could take several months or years. For example, highway projects can take 10 to 15 years from concept to completion. This lengthy period is exacerbated by the regulatory influence of the public process.

Private-sector construction projects are driven by the demand of an industry to serve a specific goal. These projects are ordinarily time-sensitive, though public statutes that require permitting and code compliance may impact them. The severity of these impacts depends on the type of project, the zoning of the property, and the project size. The myriad of delivery method choices is discussed in greater detail in Art. 6.1.2, "Construction Delivery Methods." Some of these methods are creative and respond to the time and cost sensitivity of many of these private-sector projects.

Public-sector contracting regulations have also evolved to more progressive delivery methods, amending regulations to meet schedule, cost, and quality demands that the traditional delivery method would not allow. Prequalified select bid lists, competitive negotiations, quality-based selection, design/build, turnkey, and build-own-operate-transfer agreements have been used in federal, state, and municipal procurements when full and open competitive procurements are not appropriate.

Project Funding

Public projects are funded from available capital funds at the agency, or bonds may be floated to raise money for these projects. Private-sector projects are financed from corporate funds, private venture capital, stock issues, bank financing, or other forms of creative financing. For example, real estate construction funds may become available from real estate investment trusts (REITs). Creative-financing methods are becoming popular in projects that may include a combination of private and public funding or an outlay of investments from investment bankers. When even a portion of the funding is public, statutes that oversee the procurement process regulate the project as if the project were public.

Bidding and Bidding Practices

In the United States, the Freedom of Information Act makes all information about bids and contractor selection available to the public. Formal procedures are available to dispute an award if a citizen has reason to believe that the process was not conducted equitably. The agency is obliged to award the project to the most responsive bidder in accordance with the criteria set by the formal invitation for bidders (IFB) or request for proposals (RFP). As mentioned earlier, public-sector projects are procured through more progressive methods when appropriate. The traditional public procurement demands full and open competition, formal public bid openings, and the award to the lowest responsible and eligible bidder. In this case, *responsible bidder* is defined as one who possesses the skill, ability, and integrity necessary to faithfully perform the work called for by the contract, based upon a determination of competent workmanship and financial resources. The details of the bidding process and alternative delivery methods are described in Art. 6.1.2, "Construction Delivery Methods."

Private-sector bidding practices are more flexible in the selection and negotiation process. Regulations mandate that the process be free of collusion. Although competitive bids are used in private procurements, the results are not required to be public, and the selection process is more subjective and is not bound by the lowest price. The dynamics of disputes in this arena are much more influenced by long-term relationships and the sense of fair play by owners and contractors in the marketplace.

Progress Payments

Progress payments are regulated by statute in public-sector construction. In federal procurements, the Prompt Payment Act provides for a stipulated period between requisition, approval, and payment and sets provisions for interest on late payments. In private-sector construction, progress payments are made in accordance with the contractual agreement between the parties involved. The payment period is agreed upon before signing the contract. Industry-standard contracts, such as AIA and AGC forms, outline the norms of the progress payments process, including retainage, payment for stored material, and other payment rules. These are private agreements not regulated by statute. They may be modified to meet the demands of both parties and usually include provisions for interest in the case of late payments.

Procurement of Labor, Materials, Equipment, and Professional Services

In public-sector construction projects, the procurement of labor, materials, equipment, and professional services such as design services and contract administration services are mandated by statute. It is incumbent upon the agency to enforce the statutes. The Department of Labor regulates contractor and subcontractor labor hours, wages and benefits under the Davis–Bacon Act. Specifications for materials or equipment must be performance based and must foster competition at the bidding level among at least three vendors that can meet the specifications. In some cases, by going through an exhaustive approval process, proprietary specifications may be used. In addition, in federally funded projects, the specification and procurement must be in accordance with the Buy American Act, mandating that the purchase be made from a domestic source. Procurement in the private sector is much more subjective in the selection of materials and equipment and in the procurement of professional services. Some regulations mandate minimum payments for labor, but these are usually below the rates that the marketplace commands.

ARTICLE 6.3.3
RENOVATION AND RESTORATION

Bart D. Bauer, P.E.
Edwards and Kelcey, Inc., Boston, Massachusetts

Joseph J. Albanese
Shawmut Design and Construction, Boston, Massachusetts

The scarcity of vacant real estate in most urban centers makes reuse and adaptive reuse of existing buildings a serious option for facility managers in their search to find new space to meet the expanding space needs of their corporations. In addition, there are many benefits from reuse and adaptive reuse of existing buildings that may make this option even more attractive. These benefits include the following:

- *Economic benefits.*   Reuse is often less expensive then building new.
- *Geographic benefits.*   Older existing buildings are often found in choice locations that support premium levels of investment in the property and are also often more plentiful than vacant development sites.
- *Social benefits.*   Many older buildings are culturally significant to a community, and the reuse of the building will gain community support, rather than destroying the open space of a development site.

This article addresses the unique concerns that should be taken into account by the facility manager and the design team when considering the use or reuse of an existing structure.

RENOVATION VERSUS RESTORATION

A *renovation* project is generally considered to be any construction project that rejuvenates an existing structure. This project can either maintain the original use of the building or change the use of the building. For example, an old school building is changed and becomes a residential apartment building, a mill building is changed to office space or laboratories, or a classroom is changed to administrative space. This second case is considered an *adaptive reuse* renovation project. A *restoration* project is a project that is within the context of a historic building of some recognized significance. In a restoration project, special attention is paid to preserving the historic character, fabric, and materials that are found in the original building. This work is generally more specialized, but carries some additional economic considerations in most communities. Any FM who is considering undertaking a renovation or restoration project should understand that there may be certain compromises that must be made to fit the ideal space program into the existing structure because the designer is not starting with a clean slate. However, the benefits of a renovation or restoration project far outweigh these minor space-planning compromises.

The Renovation or Restoration Project

Any renovation or restoration project begins with selecting and analyzing a new site and building. Among the issues to consider during this site-selection process are unique issues such as the following: What is the physical condition of the building infrastructure, and is it sufficient to support the new use? How much of the existing building fabric can be used in the new use? Are there existing environmental issues that need to be remedied? Is the new use compatible with the current zoning? And, considering the answers to all of these questions and the risk of additional concealed conditions, is the project financially feasible?

The analysis begins with a detailed property condition assessment. During this assessment period, the FM needs to determine what exists and also what investment the building owner must make to upgrade the building to meet current codes, address environmental issues, satisfy infrastructural deficiencies, and bring the building to a satisfactory base building condition.

A complete property condition assessment (PCA) addresses the status and condition of the building, its individual components and infrastructural systems, and issues pertaining to code compliance, and it provides budget estimates for any necessary remedial work and anticipated capital replacements during the next stipulated number of years. This PCA should also be available in an electronic format so that the current building deficiencies can be used by the estimator, the scheduler, and the rest of the design team as the project moves forward (Fig. 6.3.3-1).

The firm that prepares the PCA should provide the following services:

- Conduct a noninvasive visual review of the property and its major components and systems, including site environmental issues; site drainage; site utilities; pavements; traffic issues; foundations; structural framing; roof; exterior skin, including doors and windows; interior finishes; appliances; elevators; plumbing systems; HVAC systems; fire-suppression systems; electrical systems; controls; and fire-alarm systems.

- Conduct an ADA compliance review (not a thorough audit of all components, but a general evaluation of accessibility and major deficiencies).

- Interview individuals familiar with the property, such as property managers, property maintenance engineers, and local code officials, to obtain a recent history of the major issues that the building has faced and how they were addressed.

The building roof is a fully adhered rubber EPDM membrane over polyisocyanurate insulation surrounded by a false mansard with a standing seam metal roof. The roof is frequently penetrated by the HVAC units and appears to be well maintained. Only one active leak was observed. The roof is 20 years old and is reaching the end of its useful life.	

Immediate deficiency correction cost	*Repair leak over corner office, $1000*
Total replacement value	*New EPDM roof for entire building, $75,000*
Year in which replacement is required	*Year 5*

FIGURE 6.3.3-1 Stylized excerpt from a property condition assessment report.

- Review available documents, such as original construction documents, warranties, prior engineering reports, permits, variances, and citations.

Additional Analysis

When a PCA report determines that a building is in good condition, additional analysis is still required. At this point, the proposed new use is only a conceptual plan with very few specific details drawn. However, general design, proposed plan guidelines, and current industry standards may dictate some of the following upgrades. Present-day building code requirements usually far exceed the code requirements of the original construction. This may require modifying the structures to meet new loading criteria; additional egress options may need to be added, and fire separation walls may need to be built. Telecommunications and data networking require high-technology premise wiring and telephone and data-support rooms to distribute cables for connectivity. This presents challenges in locating distribution routes within the building. Data centers and equipment, computer rooms, and high-density layouts of present-day cubicle landscapes require additional cooling and ventilation. Along with additional cooling and ventilation comes larger ductwork. These distribution challenges affect ceiling heights and the coordination of other infrastructural elements. Power supplies in older buildings are likely to be insufficient due to the huge load requirements of present-day office equipment and computers, as well as the increased voltage requirements of heavier HVAC equipment. Uninterrupted power supply and redundant services backed up by multiple generators have become standard requirements for buildings uses that support communications, security, and data-processing uses. Designing to meet current seismic requirements may impose radical modifications on older buildings. Although most older buildings are sturdy and substantial, they are often prone to crumble in reaction to the lateral loads imposed by seismic activity.

At the completion of this report and additional analysis, the owner can, for the first time, clearly determine if there is a favorable cost-benefit ratio to renovating, demolishing, and reconstructing, or seeking a new site to analyze. This additional analysis of the way the proposed new use fits into the confines of the existing structure must be constantly revisited as the design process moves forward.

As the design process moves forward, the facility and design team must address unique existing conditions and renovation considerations to reuse an existing building successfully.

Today's access requirements are dictated by the need to provide access for all segments of the physically challenged population. This is often the greatest challenge in renovating older buildings. Locating appropriate numbers of accessible means of egress creates challenges in coordinating elevations, landings, handicapped lift locations, and ramps. Elevator cab dimension requirements could cause the replacement and reconfiguration of elevators and existing shafts. Toilet rooms must be adequately sized to meet requirements for the increased numbers of disabled employees in the workforce.

THE RESTORATION PROJECT

All of the aforementioned renovation challenges need to be considered in restoring and preserving historic buildings. In addition to these significant renovation and reuse challenges, the true restoration project has an additional challenge placed upon it. This is the challenge of the careful and sympathetic treatment of the historic fabric and significant features. Meeting this particular challenge is paramount in designing and implementing a restoration project. All buildings suffer from years of exposure to water, weather, ultraviolet rays, and temperature and humidity shifts that adversely impact the shell, structure, utilities, and finishes. On a restoration project, the design team must consider how the building and its significant and sometimes sensitive fabric can best withstand the stress of the restoration as the implementation plan is mapped out. How aggressive can the abatement, demolition, and new construction work be if the goal is to minimize the adverse effect on the fabric? What is the right balance between restoration and renovation? Individual pieces and entire components can be repaired or replaced. In historic restorations, complex climate-control systems are often integrated into the infrastructure for the comfort of building occupants and also to stabilize the environment and its effect on the sensitive building fabric. Adding to the complexity of the restoration process, the painstaking work that is involved in these projects is performed to the highest standards of workmanship under the careful scrutiny of preservation groups and civic organizations. Their concerns need to be addressed while accomplishing the goals of the client in performing the work as cost-effectively as possible.

With the many challenges of renovation and restoration come many advantages. Reusing an existing building recycles an existing product and creates obvious ecological benefits; it can also net real cost savings. Traditional new-construction costs of site preparation, excavation, and foundation work are eliminated. Reusing existing building elements, such as the structure and exterior skin, reduces the project's hard costs in these areas. The floor slabs or decks and roof deck are already in place and are generally reused. Although selective demolition costs and potential abatement costs may exceed similar costs in a new building project, they will rarely exceed the cost savings from eliminating the work items listed before. Less construction also creates a shorter schedule. This shorter project duration results in savings in the contractor's general conditions, tremendous savings in the owner's construction financing costs, and a shorter stay in the current outgrown facility.

As a company expands and outgrows its facility, the FM is faced with the challenge of finding new space. Often the choice is to look close to the current location for new space opportunities. In most areas, there is a shortage of vacant developable parcels, but many existing buildings are available. In many cases, these existing buildings can be modified to meet the facility needs of the organization, regardless of the specific needs of the new facility. A restoration or renovation project may offer the FM the right opportunity to provide a needed solution that will be less expensive than new construction, be in the location that is desired, and receive significant support from the community.

CHAPTER 7
FACILITY OPERATIONS, MAINTENANCE, AND ASSESSMENT

Anand K. Seth, P.E., C.E.M., C.P.E.,
Director of Utilities and Engineering, Chapter Editor
Partners HealthCare System, Inc., Boston, Massachusetts

INTRODUCTION

In Part I of this handbook we discuss facilities management. The part we are in now is devoted to facilities engineering. Chapter 5 presents the engineering and design process, and Chapter 6 is devoted to construction issues. Once the facility is built or commissioned, it must be properly operated and maintained; otherwise, in the long run, it would cease to function. Chapter 7 deals with the operation and maintenance of the facility, and assessment of its needs.

Operations and maintenance expenses are part of the facility's operating cost. Whenever any business comes under pressure to reduce operating costs, it becomes very tempting to achieve this by reducing maintenance quality and maintenance frequency in the facility. Facility managers should be cautious about succumbing to that temptation. The results of reducing the operations and maintenance level should be carefully analyzed. Does it mean that the facility would not be adequately staffed on the second or third shift? Does it mean that important infrastructure equipment will not receive adequate preventive maintenance to prevent breakdown? It is very possible that without proper maintenance many infrastructural systems could be significantly harmed and would require significant reinvestment in later years, at the risk of harming systems reliability. It is very easy for a facility manager to become a hero and show savings for a few years, whereas the systems in the facility may be declining severely and only a very large infusion of funds in later years will return the facility to an appropriate level.

In the editor's experience, the concept of facility operations has been transformed during the past several years. Even 20 years ago, "operating engineers" in a large facility made rounds, manually recorded various parameters of the operating equipment, and brought them back to the office for a supervisor's review. Today, in a modern facility, computerized building management systems perform this function. Now, building systems can be operated remotely in a safe and efficient manner. In fact, in most facilities, the ratio of operating and maintenance personnel per square foot of facility space has been declining. At the same time, the facility today is much more complex and user-friendly. The facility management approaches discussed in Part I are applied increasingly in facility operations.

Chapter 7 is divided into three sections. Section 7.1 discusses facility operations, Sec. 7.2 discusses facility maintenance, and Sec. 7.3 is devoted to facilities assessment. A systematic, periodic facility assessment is suggested so that facility owners, managers, and engineers can understand the condition of the facility and its ability to serve the core mission of the business in a changing marketplace. The facility assessment points out the nature and scope of the deferred maintenance for the facility. The facility may also be assessed as part of "due diligence" before a facility changes hands.

SECTION 7.1
FACILITIES OPERATIONS

Anand K. Seth, P.E., C.E.M., C.P.E.,
Director of Utilities and Engineering
Partners HealthCare System, Inc., Boston, Massachusetts

This section discusses most of the critical issues involved in setting up a viable operations program in a modern, cost-effective manner. Facilities operations cover maintenance planning and budgeting, systems records keeping, and document control. Following these subjects, we present specific facility operation issues such as physical security, environmental services, groundskeeping, and fuel and energy procurement.

ARTICLE 7.1.1
MAINTENANCE PLANNING
AND BUDGETING

Brian T. Herteen, Blake Stewart, and William Hughes
Johnson Controls, Inc., Milwaukee, Wisconsin

WHY FACILITY PERFORMANCE IS IMPORTANT

According to ASHRAE, constructing a building represents only 11 percent of total building costs over 40 years. Operations, on the other hand, make up 50 percent. Ignoring maintenance means ignoring the largest single component of building costs. It means wasting limited budgets on equipment replacement and higher energy costs. Worse yet, it can mean random disruptions or even production delays due to equipment failure.

This article is designed to help facilities managers make the most of maintenance budgets. It discusses the approaches to maintenance that facilities managers have used—and the resulting problems. It also outlines how to design an effective maintenance strategy for a facility.

MAINTENANCE APPROACHES AND EVOLUTION

There are four common approaches to maintenance: *reactive, preventive, predictive,* and *proactive.* Maintenance practices have evolved since the 1940s, when reactive maintenance was the only approach. The evolution mirrors advances in technology. Reactive maintenance relies only slightly on technology. Proactive maintenance, the most recent approach, relies heavily on the latest maintenance technologies. A discussion of each approach and its pros and cons follows.

Reactive Maintenance

Reactive maintenance is also known as *run-to-fail* maintenance. It means fixing or replacing equipment only when it breaks. Reactive maintenance is acceptable for noncritical equipment if the cost to replace or repair the equipment is less than the cost of monitoring it and preventing problems. For example, this may be the case with a small motor that would cost only $400 to replace. It may also be the right choice for inexpensive items, such as lightbulbs.

But reactive maintenance is typically the most expensive and least effective approach. Although it has its place in an effective maintenance strategy, using only reactive maintenance is comparable to having no strategy at all. Facilities managers who rely on reactive maintenance for important equipment can expect:

1. *Costly downtime.* Machinery often fails with little or no warning, so equipment is out of service until replacement parts arrive. If the equipment is critical to the area, business is disrupted or stopped entirely. If parts are hard to find, a long out-of-service period can result. Even inexpensive equipment can cause downtime and a significant negative business impact.

2. *Higher overall maintenance costs.* Unexpected failures mean costly overtime to make emergency repairs. Parts costs increase because delivery may need to be expedited and there may be insufficient time for competitive bidding. In addition, failures are more likely to be severe when failure is unexpected, possibly damaging or destroying other parts. Just as a failed timing belt on a car can cause valve damage, a failed bearing in a facility can lead to damage in shafts, couplings, impellers, fan cages and blades, gearing, and housings.

3. *Safety hazards.* The failure of equipment, especially vane axial fans, can injure persons nearby. For example, parts of fan blades can cut through ductwork.

Preventive Maintenance

Preventive maintenance means scheduling maintenance or tasks at specific times. For example, it means changing the oil in a car every 3000 miles or changing the timing belt every 60,000 miles. In an HVAC system, it includes such tasks as changing the oil and filter and cleaning equipment.

By offering a first line of defense, preventive maintenance avoids many of the problems of a reactive approach. Preventive maintenance has a part in an effective maintenance strategy. For example, changing oil filters and lubricating bearings are inexpensive and important ways to reduce the chances of problems. Unfortunately, preventive maintenance also has several disadvantages:

1. *It is often wasteful.* Preventive maintenance replaces equipment that may still have a long useful life ahead. A car's timing belt may last 100,000 miles, so replacing it at 60,000 to

avoid failure may be wasteful. Similarly, a chiller teardown that is unnecessarily scheduled wastes $15,000 and may replace good bearings.

2. *It does not prevent all failures.*   For example, if oil is leaking onto a new belt, the new belt will immediately begin to break down. Similarly, if unbalance or misalignment is causing bearing wear, bearings could fail before the next scheduled maintenance.

3. *It can introduce problems.*   Preventive maintenance fails to catch some problems, and it can actually cause new ones. Every disassembly creates the potential for mistakes during reassembly or the early failure of a new component. Both events can lead to failure sooner than if the machine had been allowed to run with its original components.

4. *It requires large inventories.*   Preventive maintenance requires a larger parts inventory to address all of the potential problems that could arise with a piece of equipment or could be required during a scheduled teardown.

Predictive Maintenance

Predictive maintenance checks the condition of equipment as it operates. Equipment condition, rather than time intervals, determines the need for service. If an analysis shows problems, facilities managers can schedule repairs before total failure occurs. Knowing problems early helps to avoid unscheduled downtime and the costs of secondary damage.

Predictive maintenance squeezes the greatest possible life out of parts without letting them fail. By doing so, it reduces maintenance costs and downtime. For a car, knowing that a timing belt would not fail until 110,000 miles would allow the owner to forego the scheduled replacement at 60,000 miles. In a facility, predictive maintenance allows managers to eliminate scheduled overhauls when predictive techniques show that equipment is in good condition.

Numerous predictive technologies are used today, including vibrational analysis, oil analysis, eddy current tube tests, motor current analysis, and megger tests. For information on these and other predictive techniques, see Art. 7.2.1, "Mechanical Testing and Maintenance."

Role of Vibrational Analysis.   Vibrational analysis is the cornerstone of a predictive maintenance program because it reveals so much about the condition of a machine. It removes guesswork by alerting facilities managers early to deteriorating parts. For equipment operating within specification, it allows them to extend service intervals.

Proactive Maintenance

Proactive maintenance is also known as *reliability-centered maintenance.* It can reduce equipment downtime and maintenance costs below predictive levels. Predictive maintenance diagnoses problems but does not correct them; proactive maintenance moves beyond diagnosing problems by isolating and correcting the sources of the problems. Facilities managers must correct the root causes of equipment problems to have the greatest impact on their goals.

Root-Cause Approach.   Proactive maintenance relies on predictive methods (such as vibrational analysis) to point out which parts are deteriorating. However, rather than being satisfied with knowing when parts will fail, proactive maintenance takes a root-cause approach that eliminates the sources of failure altogether. For example, rather than simply replacing worn bearings, proactive maintenance seeks to eliminate the causes of wear. By getting at the root causes (unbalance and misalignment in the case of fan and pump failure, for example), the proactive approach:

- Reduces downtime costs
- Eliminates recurring problems

- Extends machine life
- Reduces energy costs
- Reveals evidence of ineffective approaches

Maintenance was once considered a largely uncontrollable cost of doing business. Today, despite all evidence to the contrary, 47 percent of organizations still follow a reactive maintenance strategy.[1] In the book *Reliability-Centered Maintenance,* Anthony Smith discusses the most common maintenance problems in facilities today[2]:

- Maintenance is performed only when equipment fails (reactive maintenance).
- Repeated and frequent equipment failures occur, without focus on solving the root cause.
- Preventive maintenance is sketchy, inconsistent, or unnecessary.
- No predictive maintenance, or excessive maintenance is performed on equipment.
- No standardized best maintenance techniques exist.
- Blind acceptance of OEM schedules (e.g., chiller teardowns are based on time intervals).

Overspending is another sign of ineffective maintenance. The Edison Electric Institute found that without preventive maintenance, a facility's distribution system wastes between 1 and 4 percent of the facility's total electric utility cost.[3] An *Industry Week* article stated that the maintenance approach selected has a large impact on equipment maintenance costs.[4] The article listed the following maintenance costs per horsepower:

- $18 for a reactive approach
- $13 for a preventive approach
- $8 for a predictive approach

To see how widely costs vary by approach, consider equipment totaling 400 horsepower. The average reactive maintenance cost per year is $7200, versus $3200 for predictive maintenance.

DESIGNING AN EFFECTIVE MAINTENANCE STRATEGY

Designing an effective maintenance strategy makes the best use of a facility manager's three key assets—operational budgets, capital dollars, and in-house resources. It does so by tying equipment to an organization's goals. Equipment that has the greatest business impact receives the most attention.

To design a strategy, the facilities manager needs to:

- Determine business goals, challenges, and risks
- Relate business goals to critical areas and functions
- Relate equipment to critical areas
- Determine the appropriate service approach
- Define roles and responsibilities

A discussion of each step follows.

Determine Business Goals, Challenges, and Risks

Goals for the organization can be stated broadly. For example, a company may want to launch a new product. Goals may also address some of the following areas common to most organizations: competition, financial health, legal/regulatory compliance, staffing, risk management, image, reputation, and productivity. An important distinction in this step is to think of the organization's goals—not simply the facility's.

Next, record the challenges related to the goals the facility faces. For a company launching a new product, one challenge might be an infrastructure that would not support expanded operations. If the existing production environment needed to be reconfigured, another challenge might be the expertise of the in-house staff. Some other common challenges include:

- Cost containment
- Deferred maintenance
- Limited maintenance budgets

Risks are the consequences of not being able to overcome the challenges identified. For a new product launch, loss of investor or board support might be critical risks.

Relate Business Goals to Critical Areas and Functions

Once goals are identified, connect them to areas in the facility. Critical areas and functions are those that enable the achievement of the most important goals. For example, operating rooms, production areas, and tenant spaces may be critical if they bring in revenues that help a company achieve its profit goals. Downtime or problems in those areas hurt revenue. In addition, think of areas whose problems act as barriers to achieving goals.

Relate Equipment to Critical Areas

In many facilities, all building systems and components receive a blanket approach. For example, a 3-year-old motor receives the same maintenance as a 20-year-old one. Similarly, a pump serving a space that is critical to the organization receives the same treatment as one that serves a low-priority area of the building. Such an approach ignores the link between equipment and business goals. Relating equipment to critical areas begins to highlight which systems and components have the most impact.

Determine the Appropriate Service Approach

To determine the best service approach (reactive, preventive, predictive, or proactive) for each piece of equipment, first determine criticality. Criticality (high, medium, or low) is based on three factors:

1. *Downtime risk.* To determine downtime risk, consider whether the equipment is critical to the mission. This means that if equipment fails, the organization cannot do its primary business. In other words, there is a critical link between equipment and business processes. This holds true for any organization where a space must maintain a specific temperature, humidity level, or pressure level.

2. *Redundancy.* Think of total redundancy as having a duplicate set of equipment for a backup. If one piece of equipment fails, the other takes over with little or no change in service. Partial (or prorated) redundancy means having some equipment for backup, so services can continue at a limited level. No redundancy poses the greatest risk because there is no alternate equipment if a machine fails. For example, if a nonredundant air handler for a set of offices fails, the offices are without ventilation until the equipment is repaired.

3. *Repair risk.* If a customer has older machinery or equipment for which parts are not readily available, the risk of failure is higher. The need for specialized labor to repair equipment adds to repair risk. Repair risk also includes the cost of secondary damage. Such damage on large centrifugal chillers can easily top $50,000, whereas it may be only $2000 for a small pump.

Based on criticality, there may be many machines with a high rating. Use these factors to prioritize the equipment further:

- *Repeat offenders.* Equipment that has a history of failure most likely has problems whose root cause has not been addressed. These pieces of equipment waste a large portion of the repair budget. Review computer maintenance management systems (CMMSs), overtime labor, and incomplete work orders to see the level and types of failures that are occurring.
- *Condition and machine age.* Consider the machine's age and also its expected life. A piece of equipment that is 10 years old and has a useful life of 10 years requires a different approach from one the same age that has a useful life of 30 years.
- *Equipment environment and its application.* Equipment in a dirty environment is likely to be in poor condition. For example, dirt on motor insulation causes the motor to run hotter. The insulation can also deteriorate and lead to burnout. Equipment in a harsh environment should have a higher priority than equipment in a clean one. Using these criteria, prioritize the high-, medium-, and low-criticality equipment. For example, the highs may now be broken down into 1s, 2s, and 3s. In this final listing, the most critical equipment should receive predictive and proactive maintenance. The service frequency should also increase for the most critical machinery. Machines near the bottom of the list are better candidates for preventive and reactive maintenance.

Once the facilities manager begins to implement the maintenance program, the strategic approach needs to be updated as new information becomes available. For example, predictive maintenance may show that a critical piece of equipment should be replaced or retrofitted. Another piece may need less attention than anticipated.

Define Roles and Responsibilities

Once each maintenance task is identified, consider who will perform the services. Where practical, identify in-house resources to perform the maintenance work. For tasks that the in-house staff cannot handle, consider contractors or consultants. Ensure that the contractor uses best maintenance practices and has specific training. To save time and avoid communication problems, make sure that the contractor can provide both diagnostic and corrective services. Finally, to ensure that the root causes of problems are addressed, find a contractor who takes a holistic view of the facility.

REFERENCES

1. Thomas Marketing Information Center, *Predictive Maintenance User Study,* Thomas, New York, 1997.
2. A. Smith, *Reliability-Centered Maintenance,* McGraw-Hill, New York, 1992.
3. Edison Electric Institute, Washington, DC, 1995.
4. "The Power of Prediction," *Industry Week,* July 4, 1994, pp. 45–47.

ARTICLE 7.1.2
SYSTEMS RECORDS MANAGEMENT AND DOCUMENT CONTROL

Brian T. Herteen, Blake Stewart, and William Hughes
Johnson Controls, Inc., Milwaukee, Wisconsin

Systems records management and document control were previously handled manually on paper. Many companies used 3×5 cards to record maintenance dates, and invoicing was paper-based. Today computerized maintenance management systems (CMMSs) handle many of these tasks automatically. A CMMS is a computer software program designed to help plan, manage, and administer procedures and documentation required for effective maintenance. The CMMS consists of a series of modules or functional areas, each of which is responsible for a specific maintenance operation. The CMMS is often referred to simply as *maintenance tracking software*. However, it can also provide the following benefits:

- Reduced unscheduled downtime
- Reduced inventory costs
- Faster access to needed parts
- Improved equipment performance
- Better use of maintenance time, resources, and budget dollars
- Improved access to diagnostic and repair data, history, and other information

For all the things a CMMS can do, it does not exist in a vacuum. A CMMS can be a pivotal tool for improving maintenance operations. However, it only *supports* an effective maintenance system. The system must be in place first. The tools, such as a CMMS, come later. Just as a builder does not start a project without a blueprint, it is an invitation to failure to expect a CMMS to streamline a maintenance operation if there are not good processes in place.

How a CMMS Works

The following is the general process a CMMS follows for equipment maintenance. The CMMS may also perform more advanced functions that are discussed later in this article.

1. Equipment information—combined with periodic or unscheduled tasks—triggers work orders. Work orders can be triggered on dates, meter readings, and process or equipment variables.

2. Work orders are generated with work plans and resource requirements. The work order steps a mechanic through a particular task.

3. When the work is completed, the electronic work order is modified to reflect the actual work performed and the resources used.

4. The work order is closed. The information becomes part of the permanent service history for the equipment.

5. Analysis and reports provide critical feedback for equipment monitoring and improvements to the maintenance process.

Selecting a CMMS

When selecting a CMMS, one important consideration is its intended use. Although some companies use a CMMS primarily for tracking maintenance work, it can include various other modules. For example, some CMMSs can generate work orders automatically and deduct the parts used from inventory. They can also be used for purchasing and receiving, vendor management, and customized reporting. The CMMS can even be integrated with a facilities management system (FMS), which allows it to use alarm and run-time data that the FMS collects.

When purchasing a CMMS, some companies focus exclusively on a problem at hand. Features that seem unimportant when buying the system may become critical later. Therefore, be sure that the CMMS can grow with the company and that it can accommodate new uses. For a comprehensive list of questions to ask vendors when considering a CMMS, see Joel Levitt's *Managing Factory Maintenance.*[1]

Implementing the CMMS

Implementing the selected CMMS can take three to five months. Gathering data and procedures for equipment requires the most effort. These are the general implementation steps:

1. *Develop the equipment asset list.*   Think carefully about organizing equipment information because it relates directly to the flexibility and completeness of the system. Take advantage of existing numbering schemes with which employees are familiar. Also consider which items to include individually and which to group.

2. *Create a trade/building code/employee list.*   Determine which trade groups and trade codes to track. For building codes, use the commonly used name or number of the facility. For employees, use an employee number or some combination of employees' first and last names.

3. *Determine which assets are to be included on the PM list.*   Decide which items on the equipment list are going to have preventive maintenance records. Enter the maintenance requirements text for these items and establish the frequencies. Prepare a preliminary schedule that is based on the last preventive maintenance date or the beginning of the next quarter.

4. *Identify/write PM procedures (task lists).*   Adding and creating preventive maintenance schedules for all equipment is a major undertaking. For example, many procedural tasks for maintaining a chiller may exist only in binders. To be effective in the CMMS, procedures from the binders must be added to the system manually. To save work, some companies add equipment only as trouble calls come in. This can be useful because those are the pieces of equipment that need the most attention and whose costs and failure rates managers are most interested in tracking.

5. *Establish employee and contractor records.*   There are two possibilities for employee and subcontractor records. Each employee can have a record, or generic records can be created for groups of employees. The advantage of individual records is that time reports for each employee can be stored. The advantage of group records is that an average labor rate can be established and the overhead of tracking individual employees can be avoided.

6. *Determine the work flow and work processes.*   Decide when to run management reports, issue inventory, and distribute work orders for the following week.

After these implementation steps are complete, the CMMS can be used to produce preventive maintenance work orders. Over time, reporting requirements can be reviewed. New reports can be added as needed.

REFERENCES

1. J. Leavitt, *Managing Factory Maintenance,* Industrial Press, New York, 1997.

ARTICLE 7.1.3

PHYSICAL SECURITY

Bonnie Michelman, CPP, CHPA, Director of Police and Security
John Driscoll, CHPA, Assistant Director of Police and Security
Massachusetts General Hospital, Boston, Massachusetts

INTRODUCTION

Providing a safe environment is an inherent responsibility of every group within an organization. Every employer should aim to minimize the potential for injury, harm, or threat to anyone who has business at a facility and to minimize theft, destruction, and the misuse of equipment or property. The facilities and security functions are often the most critical contributors to this goal. Providing an effective physical security function is a confluence of effective facilities management with the security function. Appropriate physical security is a merger of security systems, personnel, and protocols.

The goal of a security department is to provide a safe, secure environment that offers everyone in the facility an opportunity to deliver its mission without fear of personal safety. Security departments protect tangible assets (people, property, etc.) as well as intangible assets (reputations, goodwill). Progressive security functions proactively assess vulnerabilities inherent in the type of business, are actively aware of the types of incidents and issues that are occurring in their organizations, and take appropriate steps by using safeguards and countermeasures to reduce these vulnerabilities or risks.

A balanced approach to security operations involves excellent, well-trained staff, good procedures, and state-of-the-art equipment and technology, mixed with heightened staff awareness and consciousness. A good physical security program has identified the potential risks and has developed countermeasures to address and combat these risks.

Security Personnel

There are several options in providing security services. An internal (proprietary) department can perform the security function; it can be outsourced to a (contract) outside agency; off-duty police officers can be used; or, in some instances, a combination can perform the function. There are advantages and disadvantages to each option. Most experts agree that the advantages of having an in-house or proprietary staff are that you can attract better-educated, higher-quality personnel, provide site-specific training, have a lower turnover of employees, and the employees are integrated more easily into the culture of the organization. There is also greater control over policy and protocol with an in-house staff. The disadvantage of a proprietary staff is cost, because salaries, fringe benefits, and training costs are higher.

Contracting for security services also has benefits. There are usually lower payroll costs, a limited administrative burden on the organization (because those functions are the duty of the contracting agency), and replacement of staff is expeditious. The disadvantages of employing a contract security company are that the experience, training, and education of the staff are generally less than that of a proprietary staff. Contract security personnel are not employees of the organization to which they are assigned and may have less commitment and loyalty to the institution.

The advantage of using off-duty police officers is that they have extensive training in law enforcement, which usually enhances cooperation with the local police department. The disadvantage is that the off-duty police officers may not understand the overall mission of the organization.

Shared services or using a combination of contract and proprietary staff can also provide appropriate security services. This method provides several benefits of the proprietary and contract options. Shared services can provide experienced, specially trained security personnel who can supply those functions deemed critical and have ancillary functions provided by others. The disadvantage of shared services is that friction may occur between the two sections and create difficulties in communication and collaboration.

It is important that the security function (proprietary, contract, police officers, or shared services) is assessed periodically to ensure that it is appropriately addressing the needs and concerns of the organization and employees.

Procedural Expectations

Organizational policies and procedures are important components of a well-run organization. The rules of an organization should be clearly defined and available to every employee. Policies that have been developed to ensure safety standards and to protect employees should contain the reasoning and explanation for the policy. For instance, a policy that requires all employees to show their identification badges or sign in after hours may seem restrictive. It is, however, an appropriate mechanism to ensure that only appropriate personnel are allowed into the facility.

Security Needs

The security needs of an organization are based on the assessment of risk and vulnerability. The best approach is an integration of people and systems. This may include constant security staffing, a variety of security equipment (locks, fences, safes) and technologies (CCTV, alarms, metal detectors, card readers), as well as crime prevention and training programs in managing aggressive behavior to educate the staff in controlling/deescalating upset people or protecting themselves from physical confrontation.

The role of security may differ slightly from one organization to another, but there are generally accepted functions that are common to most physical security programs. The role of the security program is to protect the people, property and assets of an organization. This includes basic security functions such as preventive patrols, providing escorts, responding to unruly or disruptive visitors, providing visitor control, investigating thefts, and responding to a chemical spill, bomb threat, or serious injury of an employee or visitor.

Three Lines of Defense

Protection planning usually includes looking at a physical site and dividing it into three protection zones, called the *three lines of defense*. The three lines are *perimeter barriers, exteriors of buildings,* and *interior controls.* Perimeter barriers define the outside line or perimeter of a site. These barriers can be natural (trees, shrubs, water) or structural (fences, walls), and are designed to deter anyone from coming onto the site uninvited. Barriers help to channel personnel and vehicles through a designated entry area that can be controlled. Barriers deter or delay intruders, act as a psychological deterrent, and supplement security personnel. Perimeter barriers are an important security device for certain types of businesses; for example, a manufacturing plant where highly combustible materials are stored, or a business that provides top-secret work for the government. The second line of defense is the building exterior and includes doors, windows, walls, skylights, and exterior lighting. The third line of defense is interior controls. Interior controls are defined as the internal areas and policies within a building. Some businesses, such as health care facilities, use various physical security controls within the facilities to protect the environment. Security countermeasures are focused primarily on the interior controls or the third line of defense. Alarm systems, locks, closed-circuit television cameras, panic buttons, policies, and procedures are all part of interior protection.

Communications Center

One of the most essential components of good security operations is a communications center. The communications center is where access control systems, CCTV, radio communications systems, intrusion detection systems, and panic/urgent response alarms are usually integrated. It is extremely important that the design and engineering of this area is compatible with the functions and priorities of the security department. It is also important to consider the support systems needs (HVAC, electrical, telecommunications). An effective physical plan integrates the personnel and procedures integrate into a comprehensive security design. An important consideration in developing or updating a communications center is to analyze the needs and the type of vulnerabilities and risks that the organization faces. For instance, in locations where the risk of terrorist attack or the probability of violence is higher, enhanced technology and systems would be appropriate. In other locations where the risks are more moderate, these systems may be unsuitable. The ultimate goal is to provide the appropriate level of systems and protection.

Security Systems

Security systems have become one of the fastest-growing technologies in the world today. Improved computer technology has had an immediate impact on access control technology and other security products. There has also been substantial progress in research and development in security system technologies. As a result of the downsizing of the defense industry, many large companies have focused their expertise on security systems. Many products that were used exclusively to protect high-security areas are now being used differently. Access control systems are now commonplace, as are closed-circuit television monitoring, working synergistically to provide integrated security technology. In many organizations, the use of this new technology means that facilities personnel are getting more involved with the security function. The need for security technology is certain to grow and become even more sophisticated and expansive, given the growing concerns and prevalence of violence, substance abuse, computer security, research competition, terrorism, and fraud. Refer to Sec. 5.5., "Security Systems," for a detailed discussion of these technologies.

Vulnerable Areas. Different facilities may have unique vulnerabilities that need to be identified and controlled. The following are some examples:

- *Cashier's areas.* These areas are security-sensitive where large quantities of cash are kept. They should be secured with appropriate security systems (locks, cameras, and safes), and protocols and standards should be implemented.
- *Parking lots and garages.* The security of parking areas is critical because many assaults

occur there. If the facility operates on a 24-hour basis, the parking facilities are used continuously, and many people use the lots during odd hours, when they are alone and need protection. Well-lit, well-designed parking areas with access to emergency help (intercoms, CCTV, voice-activated alarm) can minimize incidents and increase peace of mind. Security escorts and frequent patrols of any parking facility should be part of any security plan.

- *Areas of confidential information.*   Access to confidential information (e.g., personnel and medical records, trade secrets, copyrights, patents) must be restricted to those people who need the information. Very specific security systems and protocols are important components to protect this information. With the advent of the computer age, protecting information is a multifaceted problem because information can be retrieved remotely, sabotaged, and destroyed.

- *Research Areas.*   If the facility is engaged in complex research, some of which involves animal experimentation, it is critical that a good security plan exist, given the nature of terrorist antivivisectionist groups, which are strongly against the use of animal research and other medical research. Infiltration, sabotage of experiments, vandalism, serious assaults, and bombings have occurred at research facilities whose security departments do not have adequate intelligence gathering, investigative abilities, or protective methods. This planning needs to be done proactively.

Risk Assessment

The process of risk assessment/threat analysis is very important. People need protection for a variety of reasons. The risks to an organization must be thoroughly analyzed, and measures should be adopted to deal with those risks that are specific, balanced, and comprehensive for that person or organization's needs. A safe and secure environment is necessary to maintain good public and employee relations. Good security adds to the financial bottom line when incidents are reduced, turnover is lowered, worker's compensation is diminished, litigation is avoided, and there is less abuse in the workplace.

CONTRIBUTORS

Roland Ouellette, President, REB Training International, Stoddard, New Hampshire

BIBLIOGRAPHY

Broder, J., *Risk Analysis and Security Survey,* Butterworth, Boston, MA, 1984.

Burstein, H., *Security: A Management Perspective,* Prentice-Hall, Englewood Cliffs, NJ, 1996.

Ouellette, R., *Management of Aggressive Behavior,* Performance Dimensions, Avon, CT, 1993.

Colling, R., *Hospital Security,* Butterworth, Boston, 1992

Green, G., *Introduction to Security,* Butterworth, Boston, 1998.

Meadows, R., *Fundamentals of Protection and Safety for Private Protection Officer,* Prentice-Hall, Englewood Cliffs, NJ, 1995.

Wheeler, E., *Violence in Our Schools, Hospitals and Public Places,* Pathfinder, New York, 1994.

ARTICLE 7.1.4
ENVIRONMENTAL SERVICES

David Blanchard, B.S., B.A., Consultant
Blanchard Management Services, Inc., Hingham, Massachusetts

Industry has recognized the importance of appropriate environmental services to protect, preserve, and extend the life of capital assets. Perhaps a more important reason for good environmental services is their effect on indoor air quality and employee health, as well as their impact on the balance of the overall environment.

DEFINITION OF TASK

The environmental services in a facility are also called *housekeeping services*. The first step in developing a successful environmental services program is to develop a very clear definition of the task, or expected output. Task/output is a combination of a set of responsibilities performed to a predetermined level of quality. A successful environmental services program depends on the correct balance between effective department management, trained production staff, correct equipment, appropriate cleaning agents, reasonable production standards, and employee training to meet the expectations for quality and service within a company or institution. Because expectations and needs vary, a well-structured environmental services department will be a unique mix of resources and specific requirements.

The use of space—heavy industry, light manufacturing, general office, health care, or education—is the first factor that determines what type of cleaning is appropriate or necessary. The facilities manager also has to consider which external factors such as regulatory agencies, licensing authorities, and competitive pressures establish specific cleaning functions and set a minimally acceptable quality standard. In addition, within each type of space, these other factors affect the required quantity and mix of resources:

- Age of the building
- Types of surface materials
- Condition and age of cleanable surfaces
- Internal and external air quality (surrounding area)
- Atmospheric conditions
- Climate (region of the world)
- Intensity of use (hours per week)
- Type of use (light office, light industry, dirty, health care)
- Special requirements (clean rooms, high security)

Basic Tasks

Regardless of specific or unique responsibilities, there are basic duties commonly associated with housekeeping departments.

Removal of Waste Material

- General trash
- Hazardous material
- Infectious waste
- Radioactive waste

Removal of Dirt, Dust, and Litter from All Surfaces

- Exterior vents and returns
- High and low ledges
- Furniture
- Walls (various materials)
- Floors—hard (wood, vinyl) or carpet
- Protection and enhancement of the appearance of floor materials
- Sanitation of rest rooms (toilets, sinks, tubs/showers)
- Cleaning of interior and exterior glass
- Response to unscheduled demands (spills, snowstorms, emergencies)
- Moves and/or setups of furniture and equipment
- Response to other calls for service
- Specific health care issues
- Operating rooms, delivery suites
- Discharge cleaning of beds and patient rooms
- Precaution and reverse precaution

Provision of Service—Options. A professional management team is always considering options and seeking alternatives to provide the best product or service most cost-effectively. Environmental Services are no exception.

Alternatives. Management will often consider alternatives when personnel problems develop. Is there difficulty in maintaining the workforce during times of high employment? Does the employer have the commitment and supervisory talent to train new employees and measure performance if there is significant employee turnover? Are wage rates in the department higher than the market wages for comparable skills? Can management change the mix of employees and reduce the labor cost by using part-time employees who have a lower wage rate and a smaller benefit package?

There are several ways that companies can provide necessary cleaning services. In a *traditional* department, all personnel including managers, supervisors, and production workers are company employees, and the company provides training, chemicals, and equipment. Department management has responsibility for financial and quality performance.

In support of a traditional department, some companies are adding the services of a competent *consultant* to evaluate the operation, coach the management team, monitor cost and quality, evaluate new processes and systems, and generally keep the program on track.

The company can use a *management service* to support its own production team. The contracted management team provides employee training, equipment and supplies, a quality assurance program, and financial and performance guarantees. The service company is paid a fixed fee or an incentive fee based on some combination of cost and quality.

In a *full-service* program, a service company provides all management and production personnel, tools, equipment, and supplies. As in a management service, financial options can include a fixed cost or incentive-based fees.

In *combination* programs, the company uses its employees to perform part of the cleaning responsibilities and contracts with a service company for the remainder. This option can meet the needs of specialized cleaning that requires specific equipment and training. In health care, hospital employees may work in the patient care areas and the contracted employees may work in offices, labs, or production sites where there is less public interaction and it may be possible to use high-production equipment.

Although each buyer has unique needs or goals in using a service company, several general questions apply to each decision.

Why is the buyer considering a service? Some factors to consider are the need for knowledge, skills, and equipment that require specific training.

How much of the cost and quality can be identified and guaranteed? Agreement can guarantee service for a price—no questions or accountability to the buyer for the mix of resources—the output is all that counts.

What are the unique needs of the buyer? Is a special skill a full-time requirement, or needed only occasionally? Are the environmental services responsibilities predictable and are the labor requirements reasonably consistent? If the labor requirements fluctuate due to peaks and valleys in demand, how can the department meet its routine functions and also the occasional peak demand?

The terms of an agreement will reflect the specific conditions and are limited only by the imagination of the buyer and provider. Agreements can identify all cost components, including fees, and have verification as part of the deal. The agreement can pay the equivalent of an hourly rate that includes only payroll-related costs or also include costs for equipment, supplies, management, and company expenses.

Technical Issues

For many years, environmental services departments have been labor-intensive. Much of the actual effort to remove dirt and dust has been through manual effort using a variety of absorbent materials such as rags, sponges, dust mops, and mop heads to apply cleaning solutions and chemicals on furniture, equipment, production lines, walls, floors, and ceilings. As new cleaning agents or application techniques (e.g., aerosol cans) became available, the results may have been better, and the productivity may have improved, but the work was still manual.

Despite the development of effective vacuum cleaners, floor machines (buffers), and battery-powered floor equipment to decrease the physical effort to maintain floor surfaces, environmental services has remained labor-intensive. For this reason, equipment and chemical manufacturers are constantly challenging present techniques to provide products that will increase the productivity of the worker.

High-speed equipment, chemicals that clean more quickly and easily, and chemicals and floor finishes that last longer and require less frequent attention are some areas of major improvement. New technology in vacuum equipment has produced clean-air vacuums, portable backpacks, and battery-powered sweepers for floor care. New equipment has encouraged the development of new methods and systems. For example, team cleaning is not a new idea, but recent equipment developments have finally made it a viable option for specific applications. When considering of any new or replacement piece of equipment, the buyer has to separate the manufacturer's claims from reality and also has to determine whether the product is appropriate for the intended use. The department has to balance the total cost of the equipment (acquisition plus repair and maintenance for its life) against the value of increased quality and any changes in labor costs.

The equipment supplier or distributor can be a source for references, performance standards, maintenance requirements, and projected costs, and costs of expendable supplies. However, the environmental services department evaluation should address the following factors:

- Required skill level of operator
- Available training for personnel

- Time to set up and clean up
- Ease of filling/emptying solutions or changing filters and attachments
- Any claims of faster production speeds
- Useful hours of battery-powered equipment before recharging and time to fully recharge
- Performance of new technology
- Flexibility of equipment (does it perform multiple tasks?)
- Determination of appropriate size considering congestion, use, and noise level

Equipment is only part of a successful system for cleaning surfaces and preserving indoor air quality. Without the proper family of protective finishes, cleaning agents, and chemicals, any effort will be wasted and also may damage the surface you are trying to protect and maintain.

Each manufacturer of a product (flooring, paint, paneling, ceilings) should have its recommendations for the combination of equipment, chemicals, cleaning agents, protective finishes, frequencies, and techniques that will preserve the product for its optimal economic life.

In addition, the use and location of the surface or product will affect the maintenance requirements:

- Traffic—heavy to light frequency.
- Atmospheric variations—temperature, humidity.
- Airborne particles—dust, dirt, chemicals, industrial byproducts, volatile organic compounds
- Geographic—sand, dirt, snow, harmful light rays

As builders and designers select surfaces and products, they should determine the expected life of the asset and consider the maintenance expense as part of the life-cycle costs. At this time of review, the type and quality of surface materials should match the expected function and useful life.

After obtaining and considering the manufacturer's recommendations for products and procedures, an evaluation of the products should consider the following:

- Desired level of shine (floor finishes)
- Frequency of maintenance (floors—buffing, minor repair, stripping, refinishing)
- Ability to withstand black marks, scuffing, scratches
- Wearability
- Color retention
- Slip resistance
- Ability of paint and wall coverings to withstand spot cleaning
- Ease of replacing or repairing soft surfaces (e.g., carpet)

Because more than 80 percent of an environmental services budget is labor, the largest potential for cost reduction and improved quality lies in those issues that directly affect labor costs. Management should consider anything that improves productivity (e.g., laborsaving devices, training techniques, and systems). Continual evaluation of the tasks, techniques, frequencies, and equipment is the best way to deliver the best service for the lowest cost.

Personnel Issues

A strong, effective department management team is the best support that senior management can provide for any labor-intensive department. The most effective environmental services management team assumes the role of a coordinator of labor and equipment resources to deliver a service through a team of well-trained, properly supervised, highly motivated employees.

As the department has attained professional status, industry has recognized the benefits that good management skills contribute to an effective department. Good management teams concentrate on the following factors:

- Proper selection of personnel
- Initial skills training
- Establishing quality standards
- Establishing productivity rates
- Establishing open, honest relationships with all department personnel
- Communicating with users of the service
- Using budget preparation as a management tool
- Challenging historical practices in search of improvements
- Using a quality assurance program
- Providing state-of-the-art equipment and cleaning agents

Training.   Training is the single most important factor in a successful environmental services department. The manager or supervisor who does the initial and follow-up training establishes in the mind of the employee that ongoing training and high quality are a condition of employment.

Employees learn how to deliver an established level of quality within the established rates of productivity. When the management team can demonstrate the technique as well as the productivity rate, employees become believers.

A good training program contains the following elements:

- Qualified trainers—managers or supervisors
- Established training procedures
- Scheduled training periods
- Follow-up sessions
- Semiannual review

Productivity.   The productivity (output per unit of time) of environmental services departments has increased as department management seeks better and faster ways to produce a clean and safe environment. Manufacturers of equipment and cleaning agents have responded to the competitive pressures for new and better equipment and systems. As a result, increased productivity is possible if the department management recognizes the components that affect final productivity and make a conscious effort to address each factor:

- Pace—the speed at which someone performs the task (slow, average, fast)
- Equipment—state-of-the-art, powered, performs multiple functions
- Chemicals and cleaning agents—remove soil, preserve surface, protect existing finish
- Supplies—appropriate hand tools and applicators for cleaning solutions
- Technique—most labor-efficient and follows prescribed sequence
- Presence on site—on-location-prepared to perform the task

Standards.   Industry-wide standards for productivity are so broad that their application should be used only as an indicator of functions that deserve detailed study. For example, if an industry standard for cleaning office space is 4000 square feet per hour and the present staff is performing at 2500 square feet per hour, management cannot automatically conclude that the staff is not efficient.

For the same reason, management should not automatically assume that the staff is efficient if it meets or exceeds industry standards. If the staff is performing at 4000 square feet per hour, it may be possible to increase that rate to 4500 square feet per hour with a different cleaning system.

Management should determine if the present output is, in fact, appropriate for the conditions. The wide range of some of the accepted industry standards shows the importance of an individual assessment to determine efficiency.

General cleaning rates	Sq. ft. per hour
General office buildings	4000
Computers, high tech, laboratories, clean rooms	1300–1500
Heavy industry–dirty (steel)	6000–8000
Medium manufacturing	5000–7000
Hospitals	1000–1500
Nursing homes	1000–1500
Research laboratories	1000–1500
Cafeterias, lunchroom, floors only	3500–4000
Conference rooms	3000–4000
Washrooms/restrooms, complete	2.5 min/fixture
Washrooms/restrooms, check and clean	1.0–5.0 min each

Floor care—tile, concrete, carpet	Sq. ft per hour
Daily dust removal, vacuuming, spot mopping	
Obstructed, manual	3000–4000
Unobstructed, battery-powered	20,000–30,000
Detailed, obstructions, battery-powered	10,000–13,500
Minor repair, burnishing, and major refinishing	Depends on equipment, size, speed, automatic power

As a management team examines its performance and makes positive changes, the team will also be developing its own standards, applicable to its particular company and conditions. If these standards are well-developed and accepted by senior management, they become the basis for professional budgets.

Budgets. Budgets, in turn, become the basis for planning the work of the department and assuring that there are adequate resources to deliver the expected level of quality and service. If asked to reduce or increase the budget, management has the information, data, and credibility to show where and how the changes will affect quality and service.

CONTRIBUTORS

Raymond J. Taylor, Director, Housekeeping Management Specialists, Inc., South Hamilton, Massachusetts

Guillermo J. Banchiere, Director of Environmental Services, Massachusetts General Hospital, Boston, Massachusetts

Arthur Mombourquette, Director of Environmental Services, Brigham & Women's Hospital, Boston, Massachusetts

Anand K. Seth, P.E., C.E.M., C.P.E., Director of Utilities & Engineering, Partners Health-Care System, Inc., Boston, Massachusetts

ARTICLE 7.1.5

LANDSCAPING SERVICES

Bernard T. Lewis, P.E., C.P.E., Consultant
Potomac, Maryland

GENERAL RESPONSIBILITIES OF THE LANDSCAPE SERVICE PROVIDER

The landscape service provider can be either a landscape contractor or in-house personnel of the facilities maintenance department. This article is written as if a landscape contractor were providing the service; however, the same principles apply if in-house personnel provide the service.

The landscape contractor should be responsible for furnishing all labor, equipment, and materials necessary to perform all specified landscape maintenance tasks. All equipment should be of the type needed to perform the required task effectively and to avoid any unreasonable hazards or dangers to the properties, occupants, and pedestrians. The equipment should be well maintained and should not produce excessive noise or noxious fumes when operated under normal conditions. Some communities have ordinances restricting noise. These restrictions may apply to mowers, leaf blowers, and other landscape equipment. Check with your local jurisdiction to determine if any restrictions apply.

SELECTING A LANDSCAPE SERVICE PROVIDER

The service provider chosen should be a trained professional and have the appropriate equipment for the job. Look for a contractor who listens to your ideas and problems. Ask for references, visit the company's offices, and visit properties that the service provider is currently maintaining.

Lawn Maintenance

Lawn areas require a consistent maintenance program to stay healthy. If lawn areas are neglected for only a couple of seasons, they decline. A maintenance program should consist of proper mowing, pest and weed control, fertilization, watering, aerating, and overseeding. This program encourages the growth of healthy turf, which in turn crowds out unwanted weeds.

Mowing and Trimming. Mowing should begin in March or April at intervals of five to ten days (maximum) between mowings. Mowing should be done frequently enough so that no more than one-third of the leaf area is removed at one time. This will help the turf to develop a more extensive root system and withstand environmental stresses. The type of grass determines the optimum height for turf.

Horizontally spreading grasses are typically cut shorter than vertically growing grasses. Frequent mowing tends to produce a finer-textured turf because cutting frequently stimulates new growth. If the turf is neglected and becomes too tall, the growth becomes coarse and may produce seeds. Mowing the turf too short causes the grass to dry and burn, which allows weed seeds to germinate.

Mowing should be done in alternate directions at least every four mowings. This eliminates ruts and a striped or streaked look. Mowers should be well maintained and cutting blades kept sharpened at all times to prevent tearing the leaf blade.

Litter and debris should be removed from all lawn areas before mowing. Clippings can be left in the turf area. This is more cost-effective because the clippings put nutrients back into the soil. If the turf becomes excessively long due to long periods or wet weather, the clippings should be removed because they can form a mat on the turf that shades and kills the grass. In areas where there is concern that clippings are unsightly, such as entrances, clippings can be bagged, or raked, and removed. Mowing should be done so that clippings are not blown into the shrub beds and tree rings because this can be unsightly.

Areas around posts, signs, buildings, and trees should be trimmed at the same height as the lawn. Lawn mowers and string trimmers should not be used at the bases of trees and shrubs because they can damage the base of the plants.

Controlling Weeds and Pests.   Weeds are simply plants that grow in the wrong place. There is no such thing as a weed-free lawn, but weeds can be minimized with proper control. To minimize weeds, the following recommendations are generally applicable:

In the early spring, when daytime air temperatures reach 55 to 60°F, a broad-spectrum preemergent (applied before weeds emerge) herbicide that controls both noxious grasses and broadleaf weeds should be applied to all lawn areas in accordance with the manufacturer's recommendations. Additional applications of preemergent weed control may be necessary to effectively control all weeds.

In the late spring and again in the early fall, when daytime air temperatures are not above 80°F, the contractor should apply a broad-spectrum, postemergent (applied after weeds emerge) herbicide to control all weeds. The presence of certain weed species that are difficult to control may require additional applications of herbicides.

The contractor should regularly monitor all turf areas for insect, disease, and weed infestations, and treat as needed. The contractor should be responsible to replace with sod all turf areas damaged as a result of pest and disease problems. The sod should match the surrounding healthy turf.

Fertilization.   Soil fertility is one of the major considerations in any lawn management program. A healthy lawn requires a soil that is fertile from year to year. Because grass can quickly deplete soil of essential nutrients, the nutrients should be added into the soil regularly. The essential nutrients for turf areas are nitrogen, phosphorus, and potassium. Nitrogen is critical because it stimulates leaf growth and keeps turf green. Phosphorus is needed to produce flowers, fruits, and seeds, and to induce strong root growth. Potassium is valuable in promoting general vigor and increases resistance to certain diseases. Potassium also plays an important role in sturdy root formation.

Soil pH is critical to growing a healthy stand of turf (or any plants). Soil pH is the acid-alkali balance of the soil. The pH scale divides the range of alkaline and acidic materials into 14.0 points. The middle value of 7.0 is neutral and marks a balance between acidic and alkaline soil values. Some plants thrive in neutral conditions. Others prefer a more acidic or alkaline soil. Lawns grow best at a pH of 6.0 to 7.0; consequently, for a healthy stand of turf, it is important to make sure that the pH is correct. The pH can be changed by adding lime if the soil is too acid or by adding sulfur if the soil is alkaline.

A soil test gives the pH, as well as the level of nutrients available in the soil. From the soil test results, a contractor can determine the amount of fertilizer to be applied and if an application of lime or sulfur is necessary.

Maintenance of Trees, Shrubs, and Other Plantings

Mulch and Weed Control. Mulch regulates soil temperature, insulates plant roots from temperature extremes, reduces water loss from the soil surface, and minimizes the time and labor that is required to maintain the garden by minimizing the germination of weed seeds. The most important function of mulch is moisture retention. Mulch allows water to percolate through and protects the soil from the drying effects of the sun.

Mulch is available in a number of organic and inorganic forms. Organic mulch is typically recommended because it eventually decomposes and adds humus to the soil. This in turn improves the soil composition and texture. In addition, nutrients are released during decomposition, which increases the fertility of the soil. There are a number of factors to consider in selecting mulches, including the availability of the material, the cost, and the appearance. Local nurseries or an extension service should be able to advise you on the type of mulch most suitable for your needs.

All beds and tree rings should be defined and edged before mulching. The edge should be maintained throughout the season to give the landscape a clean, crisp appearance. Edging debris should not be placed in the beds or rings but should be removed from the site because excess soil at the base of plants can be detrimental to their health. All tree rings should be evenly concentric around the tree, and all bed edges should be maintained as smooth, continuous lines.

Pruning. There are many reasons to prune: to keep plants healthy, to restrict or promote growth, to encourage bloom, to repair damage, to remove structurally weak or otherwise undesirable branches, to clear a building, or to allow light to penetrate to the ground.

Shrubs. It is important to remember that different types of shrubs have different growth habits and characteristics. Plants are selected for a particular area based on form, color, and texture. If all the plants on a site are sheared into hedges, individual balls, or squares, the characteristics of the plants are lost. Therefore, it is important to follow proper pruning techniques so that the natural beauty of a plant is recognized.

The best time to prune shrubs depends on their flowering habits. Shrubs that flower on new growth should be pruned in early spring before the new growth emerges or during the last weeks of winter. Shrubs that flower on old growth should be pruned directly after flowering. If these shrubs are pruned through the growing season, the flower buds will be removed, and there will be no floral display. As a general rule, spring-flowering shrubs should be pruned immediately after blooming. Broadleaf evergreen trees and shrubs should be pruned after new growth hardens (except for hollies, which should be tip-pruned in early spring). Conifers should be pruned by pruning new growth (candling) and again, only if necessary, after the new growth hardens. Shrubs that flower in summer should be pruned in late fall to early winter or early spring. Hedges should be pruned by hand as necessary to maintain a neat and trim appearance.

Trees. Trees may require pruning, particularly if they have been neglected for many years. All dead, diseased, weak, and cross branches should be removed to improve the structural integrity of the tree. To avoid having to prune large trees extensively, the trees should be pruned and trained while young. Properly pruned trees will grow into structurally sound trees as they mature. Cross branches should be removed; permanent branches should be carefully selected; and a strong branch structure should be developed.

Vines and Ground Covers. Vines and ground covers should be pruned regularly to maintain a neat and manicured appearance (but should not be sheared). String trimmers are never to be used to prune ivy or other ground covers. Ground covers should be pruned at the nodes, with the cut hidden. Depending on aesthetic preferences, ground covers should be maintained within or partially overhanging all planters and off all paved surfaces. Ground covers should be kept 4 to 6 in away from the trunks of trees and shrubs.

Fertilization. Plants need different amounts and proportions of nutrients to stay healthy. Supplemental fertilization is necessary in most areas, even if the soil has ample amounts of organic matter. Many soils have an insufficient amount of one essential nutrient and an over-abundance of another. As discussed earlier, the amount and proportions of fertilizer required should be determined by a soil test.

Cleanup.   All areas, including planting areas, plant materials, lawns, and paved areas, should be kept clean at all times. During each visit the contractor should remove and dispose of any and all trash (including cigarette butts), sticks, natural debris (including soil, sand, rocks and gravel, withered flower buds, seed pods, and leaves) from all landscape areas, including all raised planters, turf, and ground cover beds.

In autumn, leaves should be raked and removed regularly. All leaves should be removed from all lawn and bed areas before mowing, including leaves and branches that drop throughout the spring and summer months.

Pest Management.   The contractor should be responsible for detecting, monitoring, and controlling all pests. The contractor should be aware of the pests that might be encountered and should make regular and thorough inspections of all plant material. Treatments should be applied as necessary using products and methods that target the insect pest with minimal residual effects.

If there is a need for chemical application, the contractor must adhere to the Department of Agriculture regulations for commercial application of pesticides.

Watering.   The key to watering is to water deeply and infrequently. This helps the vegetation to develop an extensive, deep root system. Frequent, light watering encourages roots to stay near the surface. This encourages the plants to be more and more shallow-rooted. Shallow-rooted soils tend to be less porous, and plants suffer during drought. It is always preferable to water early in the morning, because the sun will then burn off the excess moisture. This will decrease the potential for fungus and disease.

If plants are too dry, their leaves and flowers wilt, and eventually the plant will wither and die. Plants can also die from too much water, especially if the water accumulates around the roots of the plant. With too much water, the leaves and flowers turn black and fall off, and the roots rot. When the roots are exposed, they will be black, and the plant will have no healthy white roots. The exposed roots will often smell foul. The correct amount of water will vary according to soil type, plant, and turf species, climate, and weather. Once the correct amount of water is determined, plants will be healthy and vigorous.

Seasonal Color.   Seasonal color is essential for distinguishing one property from the next. For very little money, annuals, perennials, and bulbs can enhance the overall appearance of a property.

Annuals are defined as plants that complete their life cycle in one season. In most areas of the country, they flower throughout the growing season but need to be removed at the end of the season. With careful planning and design, the display will perform for the entire season.

Perennials are plants that come back each year but typically have a shorter blooming time. Mixing perennials that bloom at different times will give a constant display of color. Because perennials come back year after year, it is recommended that a landscape architect or garden designer be involved in the layout and design to ensure a successful planting.

Bulbs have a short blooming period but give a beautiful display and are always welcome in the early spring. Many bulbs, such as tulips, are removed after they bloom. Others, such as daffodils, can be left in place for many years, as long as the bed is not disturbed.

The seasonal color displays should be unified rather than disjointed groupings throughout the property. To achieve the greatest visual impact, the displays should be planted at key points of a building such as entrance drives and the front door. Keep the designs simple, using a few colors and a limited number of plant types. A single color has more impact than five or six colors mixed together.

Annual planting beds require high maintenance, and this should be considered in determining the size of the bed and the location. Annual plantings may need daily watering and regular fertilization. Scattered beds throughout a property that do not have access to water will decline quickly and show poorly. On the other hand, a small bed near the front entrance will be noticed by everyone entering a building and will be easier to maintain.

CONTRIBUTOR

Anand K. Seth, P.E., C.E.M., C.P.E., Director of Utilities and Engineering, Partners HealthCare System, Inc., Boston, Massachusetts

GENERAL REFERENCE

Lewis, B. T., *Facility Manager's Operation and Maintenance Handbook,* McGraw-Hill, New York, 1999.

ARTICLE 7.1.6
FUEL AND ENERGY PROCUREMENT

Sheridan A. Glen, Principal
Thermetric Inc., Madison, Wisconsin

The procurement of fuel and energy is fast becoming one of the most important economic responsibilities for facility engineers. Now, in a rapidly emerging move away from regulated to unregulated markets, facility managers can manage their energy-related financial obligations to lower costs, based on three key factors: size, location, and load factor. Traditional gas and electric utilities ran their business as integrated companies. Now they are separating (that is *unbundling*) their commodities, and delivery services and markets have opened up to non-franchised energy marketers. This article provides managers with background about the energy markets and offers a comprehensive approach to managing fuel and energy procurement so that these emerging deregulated markets are understood and the facility managers can maximize cost savings without sacrificing security of supply.

During the next few years, the $300 billion electricity market will undergo a wave of innovation, and the companies involved will consolidate in many ways—including through financial mergers, combining of traditional franchise territories, and electric and gas convergence. Interestingly, although more than 50 percent of the electric power in the United States is generated by coal, the growth of future electricity generation will be fueled almost exclusively by relatively inexpensive, very clean, domestically produced natural gas. The electric-gas convergence will create accessibility to the fuel assets required for load growth and for the new plants that will replace the aging fleet of generators in this country. Facility managers must keep up to date with this evolution to understand the motivations of the marketers bringing them new ways to save on their energy bills.

THREE PRIMARY FUELS TO MANAGE

Most commercial buildings use one or more of the following fuels: *electricity, oil,* and *natural gas.* Generally speaking, electricity provides light, gas provides heat, and oil provides refined fuels for transportation. However, many permutations of these basic applications are found in the commercial marketplace. Many buildings use oil for space heat, for a boiler that provides steam, or for a boiler that generates electricity. Gas is used for the same reasons, but it is the most popular and certainly the cleanest fuel for space heat. Electricity is used for lights and appliances but rarely for its least efficient use, heat.

ENERGY MARKETS, OLD AND NEW

Deregulation of electricity is already resulting in open markets in some states, but gas has been deregulated in many states for about seven years. Oil, of course, is not a regulated energy source at the customer level because no particular retail oil vendor operates in a regulated franchised territory. Interestingly, however, in the Northeast, where retail oil for commercial and residential space heating remains a major business, oil jobbers have staked out territories based on the most economic areas to serve with their 2500 to 8000 gallon trucks and the location of their storage tanks. Here the market has effectively created virtual service territories based on economic efficiency and localized branding of the most prominent jobber. Although facility managers will be solicited by each of the oil jobbers in a chosen area of service, their biggest source of leverage is represented by the size of their facility's oil storage tanks. The less frequently that the trucks must make a delivery, the better the price is likely to be. Facility managers who must purchase oil weekly and have less than four local jobbers from whom to choose are better off signing longer-term contracts with one company because their ability to leverage the three principal factors (size, location, and usage pattern) is limited. The manager who has the benefit of large storage capacity can purchase a base amount of fuel on periodic bids and play the spot market as well, topping off tanks when his price points are met, either by polling other facility managers or by an informal bid process among the local jobbers. Adroit oil vendors will anticipate this kind of customer behavior and solicit the facility manager's business with market-sensitive and aggressive bids. The retail oil business, particularly in the Northeast, is very mature, and competition is measured by cents per gallon. Whatever strategies are employed by the facility manager are likely to bring success by a modest sum because the market behavior of oil will dictate the principal offers by jobbers, and the deviations will be small. As a result, the most successful jobbers have become those who offer the best service: 24-hour on-demand delivery, guaranteed discount pricing, tank and boiler maintenance, customized billing, and contract billing. The commodity price may be distinguished by cents, but success is defined by volume and customer satisfaction.

NEW MARKET STRATEGIES FOR GAS AND ELECTRICITY

The oil purchasing strategy serves as a useful template for understanding the other two principal energy markets. Until about five years ago, contracting for natural gas supplies was a matter of signing up or connecting with the local distribution company (LDC). Like electricity, gas was a regulated business, and the customer was a ratepayer, just a number in a franchised service territory. The "upstream" side of that was changed first, when shippers were allowed open access to all interstate pipelines, opening up the business for sales to the "city gate," which is defined as the intersection between the interstate pipeline and the local distribution pipe leading to the customer facility. "Downstream" of the city gate, marketers have now become allowed to solicit the LDC customers for sale of the gas commodity. The LDC was formerly responsible for purchasing and creating the gas path upstream and had the downstream relationship all to itself. This unbundling process has created problems for LDCs because they make their earnings only on the throughput of gas from their city gate and do not realize earnings from the sale of the gas commodity. The cost to them is passed directly to the customer on a basis that is adjusted after seasonal costs are incurred. LDC earnings come solely from the local transportation of the gas they purchase.

Because the market downstream of the city gate was opened to marketers earlier last decade in virtually every state, marketers now have the opportunity to transport their own gas over the open access interstate pipelines, and the LDC must then transport the marketers' gas to the customer. A marketer's ability to serve the customer still depends wholly on the very system built over decades through regulation. Although more costly, the system has certainly provided security of supply, even on the coldest winter day. The facility manager may have heightened expectations that the deregulated marketplace will provide savings of 25 to 30 percent or even higher. This range of savings is not likely to materialize.

The marketer can succeed only in capturing the gas customer by selling the commodity more cheaply than utilities to a range of customers. Marketers accomplish this by buying and managing their interstate pipeline commodity and capacity more efficiently by using, for example, interruptible and released capacity and avoiding some of the fixed charges that can elevate the long-term costs incurred by LDCs. A large marketer may be able to spread fixed costs over a larger set of aggregated customers whose load profiles are less sensitive to seasonal usage than those customers behind the LDC because LDC customers pay an "average" price for their gas commodity and transportation. Still, the marketer's margins must be wide enough to cover marketing expenses, transport, and costs associated with risk management and still provide a suitable incentive for the customer to switch away from the familiar utility.

Marketers have significant risks as start-ups in the gas franchise territory. They must plan properly for the amount of gas for their customer base, figuring in weather and customer growth, and must grapple with the issue of *balancing,* the difference between the amount of gas they have asked the LDC to transport on their behalf and the actual amount used by their customer base. If that customer base is a small group, there may be customers whose gas demand far exceeded or was less than the historical record available to the marketer and the marketer will be "long." If marketers finds that they have provided too much gas to the LDC, the LDC can absorb the overage, in accord with tariffs, and return a portion of the value to the marketer—often at less than their cost. Commensurately, if the marketer fails to designate a sufficient amount to cover demand, that is, his requests are "short," then the LDC will purchase the amount of gas from the daily market, again at a preprescribed tariff cost and bill the marketer for the actual cost difference. Only when the marketer's customer base becomes more mature (like the LDCs' already is), that is, some of the customers use more than anticipated and others use less, can the marketer keep his customer pools in relative balance. The beleaguered gas marketer must deal with a myriad of change factors, new relationships, and margins squeezed by the deregulation process and the need to provide the customer with a compelling reason to switch from the LDC. It is safe to say that few marketers became profitable in any retail markets for the first three years.

Facility managers can exploit the marketplace most effectively if they have size (peak-day requirements exceed 300mmBtu/day), location (inside a cooperative LDC), and load factor (relatively flat and/or predictable usage). The better the historic gas usage record and the ability to analyze meter-specific usage, the better in line the marketer will be to achieve maximum savings from the marketplace. Most LDC meters provide monthly readings, which do not show daily fluctuations in usage. Historic usage and monthly readings are frequently upgraded with meters installed by the marketer that can provide daily reports.

A NEW PARADIGM FOR THE FACILITY MANAGER

The facility manager must weigh the benefits of this process against the likely savings and judge the capability and service fortitude of the new supplier. The electricity market is more complex, certainly driven by the same competitive urges, but usually implemented through state-level legislation that has been drawn through the political process. There are dramatic differences in electricity prices among states and regions. Generally, those states that have indigenous fuels for steam-electric generation fare the best; states where fuel must be transported long distances before it is converted to electricity pay more. The reason is simple: fuel represents about 70 percent of the total busbar (at the plant) cost of electricity. People in Laramie, Wyoming, the state that produces more than one-third of the nation's steam coal, pay about 5 cents/kWh. Residents of Long Island, New York, pay almost three times as much, but they are saddled with the long-term costs of an expensive nuclear power program and costly non-utility-generator (NUG) contracts. The facility manager should view electric deregulation as the beginning of an era in which unique, customized products and services support operational and financial objectives. Although facility managers can expect overall electricity prices to decline, along with that will come increased risk and retail price volatility. It is not inconceivable that price variability will trend similarly for wholesale and retail markets simultaneously. That means the facility manager must not just worry about base price or contract term, but also about volatility and risk management.

THE EMERGING MARKET FUNDAMENTALS

The natural gas industry has become a major source of expertise for the electricity market. Having gone through the deregulation process, there are many analogous operations, financial exposures, and customer issues. For the present, deregulating utilities have an urgent issue—stranded assets—those plants and purchased power contracts that are not supportable by today's electric prices, unless part of a regulated rate base. Nuclear power plants, once thought to be able to produce power "too cheap to meter," are the bane of the utility balance sheet. Non-utility-generator (NUG) contracts, a result of misguided 1978 legislation that forced utility companies to purchase power developed by private cogeneration companies at an avoided cost formula based on a rapidly escalating oil price that did not materialize, represent another source of stranded assets.

The Deregulation Process Begins

In the deregulation process, these financial burdens must be dealt with as the quid pro quo of deregulation: stranded costs get to be recovered in the rate mechanism, and the customers have the freedom to choose their power producers. In Massachusetts, the first state to effectively monetize the stranded costs of the investor-owned utilities whose assets are located in that state (municipal utilities are thus far exempt from the process), this cost was added to the access charge to transmission. Now, any company can move power over transmission lines, much like the gas in interstate pipelines. In return, the utility companies can sell their generation assets to outside, unaffiliated companies and exit the power production business. Now, the new generation of asset owners can penetrate any markets using open access to move, or wheel, power over transmission lines into the distribution network. This forms the basis of the first unregulated market for electricity in this century. The utility companies have been further required to offer the electric energy (representing about one-third of the total delivered cost of generated, transmitted, and distributed power) as a "standard offer" available to everyone, from the largest industrial titan to the smallest homeowner at the same discounted price. Further, the first year of deregulation in this state would mandate an across-the-board 10 percent discount off the total electric bill.

Although only about 20 states have a fully open marketplace for retail electricity the time of this writing, this arrangement appears to be one useful model: Allow the utility companies to recover their above-market costs, force them to exit the generation business, and assure some guaranteed savings to all classes of ratepayers. Like the retail gas business, new electric industry marketers face the same vexing operational issues such as transmission access (like interstate pipelines), relationships with the distribution utility (who, like the gas LDC, is having to relinquish the commodity sales relationship with the customer), and load following, the service of supplying power to follow the customer's instantaneous demand (like the issue of balancing the gas customers' differential usage between the designated amount and the actual consumption).

Although utility commissions still regulate the delivery and service aspects of gas and will set the parameters for transmission and distribution of electricity, the logistics of power pool dispatch—making sure power is available on demand—will fall to regional Independent System Operators (ISO). These agencies will create an entirely new tier of products that may be purchased and traded. They include charges for transmission levied during demand periods that create congestion, backup supplies to ensure uninterrupted service, and other services that can be used by marketers to make sure that power not produced indigenously (that is, out of the pooled region) is deemed reliable by the market.

The Endless Possibilities

As important to everyday life as the electric market is, it is easy to envision the business becoming very innovative and product oriented, much more so than retail oil, where the companies making the delivery distinguish themselves not on the quality of the commodity, but on price and service. These opportunities will be huge for the more ubiquitous electric industry. The future for the electric industry to create new value for the customer is a vision of endless technology-based possibilities.

To manage these many customers, each with a different size, location, and load factor, aggregations will be formed based on the common denominators of the customer group. Rather than just being a member of a rate group, as in "industrial," "commercial," or "residential," customers will be treated as part of a commercial alliance. Consider a group of restaurants or hospitals with similar load factors, an association of building owners or developers, or a large industrial park—aggregations like these have something important related to size, location, and/or load factor that makes the purchase and delivery of electric power more predictable and less risky, and therefore more desirable, to the marketer. Rather than waiting to see if an aggregation can be formed by the marketer around a facility manager's domain, many smaller companies can band together to form their own aggregations and, in so doing, make their own factors more attractive to the power marketers and generation owners in the region. The diminution of risk represented by an aggregation is making this concept one of the sure drivers in the rapidly emerging retail electric market.

The overall convergence of the electric and gas markets at the wholesale level supports the logical concept at the retail level. Companies that serve a large regional electric load will logically try to become the one-stop energy shop. Many are already having some degree of success applying this concept to the customer who wants to deal with energy as efficiently as possible. In a more mature marketplace, think of the exciting and very plausible services that will extend from the core electric product. One only need muse for a moment about where we were 15 years ago, before the advent of telecommunications revolution spawned by deregulation. Would one have asked for cellular phone service? How about a pager? Could anyone have imagined the technology and economic revolution called the Internet? E-mail? There is one indisputable fact at work here: Deregulated markets are breeding grounds for competitive offerings from companies who want your business and are willing and able to create value from the new commercial relationship.

A BRAVE NEW WORLD

During the next few years, electric and gas service networks that optimize energy consumption at multiple sites will be inevitable. Internet sites may allow the customer or aggregation to select a power supplier, then review a menu of goods and services, including financial products to protect the customer from price volatility. Perhaps the customer can select a green power option, which, for a premium, will secure specific nonpolluting sources of electric generation. Energy services can also be selected from a menu of unbundled Btu modifiers, the cost of which can be financed over term on the electric commodity bill. The bill itself may be downloaded at any time, complete with time of day and a comparative analysis of any historical usage patterns. What about a distributed generation option? A complete analysis will be made available to compare bundled service with installing a gas-fired turbine to meet generation requirements for a group of buildings or facilities electronically lashed together.

Power quality versus an interruptible power contract will surely be offered as a way to select your power quality requirements, from absolutely fail-safe service for the computer chip manufacturer to an interruptible rate for the flexible manufacturing facility that has installed a full requirements backup diesel or gas-fired generator. Real time pricing, monitored over the Internet, will allow the facility manager to decide how manpower and shipping schedules can be modified to take advantage of inexpensive off-peak power. Meter monitors will be strategically placed at various economic points in the facility to understand where usage, cost, and time intersect to reflect real savings and provide a point of analysis. Financial products will quickly trickle down to the customer or aggregation in the form of forward price protection mechanisms, vehicles to hedge the fuels and market volatility that is the by-product of a free and unfettered marketplace.

Creditworthiness will be a big issue for the marketer. A default rate of 2 percent of revenue will cancel probable profit margins early in the deregulated sales relationship. Look for the marketer to strongly suggest a prepayment of electric bills and service invoices and for that to reflect further incentives in commodity and service pricing that perhaps even exceed the net present value for the period. Electric prices may be offered at long-run prices indexed to other commodities, even indexed to the products made by the customer. Price collars will be a popular method to cap upside price swings and to participate in price movements below a predetermined floor.

How the Evolution Will Work

Now that facility managers are armed with this information, how are they to proceed? In many states, solicitations from energy marketers are already commonplace. Even when deregulation has been legislated with a long enough lead time for the host utilities to deal comfortably with their issues of stranded costs and divestiture of generating assets, facility managers may receive transitional pricing offers, bonus pricing for early sign-up, or "free" energy audits and other services for establishing an early allegiance to a particular marketer. Experience suggests that these teasers are best left on the table because no one can foresee how the market is going to develop itself in any given location, particularly in view of the early economic covenants mandated by legislatures that are eager to appear as populist protectors and not safety nets for the utilities. There are several ways to prepare to take advantage of innovations in energy supply and in turn create competitive advantages for the facility manager's own company. Oil has been discussed in some detail, and it is not a recently deregulated market. The following ideas are to be used in analyzing unregulated supplies of natural gas and electricity. These are concepts to be treated as the prime components of the facility manager's own request for proposal (RFP).

Quality of Service.   Most customers will want firm service for gas and electric supplies, but it does not hurt to understand the parameters and cost savings inherent in having an interruptible supply. Not applicable to gas unless there is dual fuel capability for heat or steam, the alternative cost for electricity may be compared to installed generation, the cost of which may be quoted from the commodity supplier as a benchmark for your analysis. As deregulation spawns new distributed generation opportunities, this analysis will be even more important because it will provide a ready market comparison.

Reliability of Supply.   One of the most vexing issues in a deregulated market is who to trust. In the electric market, deregulation takes on a level of importance greater than other nascent markets of the past. It is imperative that the facility manager understands the supply path for natural gas. What capacity contracts are in place to move the gas in the interstate pipeline system? What is the contractual source of the commodity? What is the anticipated customer load attuned to weather demand, and what contracts are in place to handle these obligations? What arrangements have been made for LDC coordination and registering? What level of pipeline capacity for moving gas from regional delivery points to LDC city gates does the marketer have? Are some of these services bundled together? Are physical contracts in place or financial resources committed to pay balancing penalties? For electricity, has the marketer registered with the ISO and/or the state? Is there a comprehensive understanding of the resources available and the requirements to conduct business behind each electric distribution company? What is the source of the marketer's electric energy? Has he contracted with particular generation owners? Is backup service arranged? Are his trading operations underpinned by asset or contract? What future supply considerations are there?

The facility manager should never be bashful in getting to the bottom of the marketer's supply capability and its portfolio components. The early era of the deregulated gas market was rife with marketers who ambitiously oversold their supply portfolio and had to be bailed out by the LDC. The customer should demand and get seamless service. Make sure that your supplier is equipped to handle the hottest day in August or the coldest day in January. Although there should be a primary focus for each RFP and separate RFPs for each fuel, the manager should always inquire about fuel oil and natural gas in an electric supply RFP. The adroit marketer will find ways to add value if the opportunity exists to provide other services.

Administrative Capability.   The quality of administrative services that is provided by the prospective energy supplier is a necessary component of analysis. What are the capabilities in coordinating data acquisition, nominating, processing, reconciling accounts, and billing and customer care functions? Is the back office manned 24 hours per day, seven days per week? What communications links and protocols have been established for emergencies? Where are primary functions performed in relation to the market served? Does the company have a local office with remote links to a home office or operations center? What systems are employed to monitor wholesale accounts and path energy at the trading floor? Are risk management procedures in place so that the marketer's supply position relative to outstanding obligations can be tracked? How are customer inquires handled? Collections? Dispute resolution? Bundled services? Internet capability tied to customer account information? A marketing company that has a thinly staffed back office and little home office capability should be regarded as a marketer with short-term objectives. Ask for references and a way to judge past account performance. Even visit the operations center.

Price.   Of course price is important, but it is not the most important element. If the marketer fails to perform core duties, then price could pale in consideration to operational difficulties during the contract. The price should be requested for varying terms. The longer-term customer should be rewarded with a lower price, but in energy commodity markets, sometimes short-term prices are lower than longer term. At a minimum, the facility manager should request prices for six, twelve, eighteen, and twenty four months, plus ask the marketers if longer-term options, even up to several years, will provide greater guaranteed savings. Some companies are set up to provide fixed price guarantees for as long as 10 years. Will the company provide competitive price protection after the contract is signed? How about index pricing to a commodity or manufactured product? For electric customers, price has always assumed load-following capability; the facility usage is served irrespective of load shape and demand criteria. Make sure that all price offers for term are based on the same level of service. Then analyze the other considerations based on tolerance for risk and your ability to assume responsibility for pricing that has variable provisions.

Contract Terms. Some facility managers will want to have prepackaged contracts for negotiation. The manager is urged to show some flexibility because a contract can be reasonably simple and the marketer may even have an improved version. The degree to which contract negotiations become a pleasant or dreadful experience will reflect much on the future course of the commercial relationship. A contract negotiation that finds the supply party anguishing needlessly over every detail should cause the facility manager to reconsider the entire relationship. Do not sign a contract with a company with which you or your colleagues are uncomfortable, no matter how good the price. It is a relationship that will be a source of regret.

Consulting

The facility manager should not assume that all of this information and strategic processing can be administered and managed without outside assistance. Generally speaking, if the manager has a peak-day gas requirement in excess of 300 mmBtu/day and/or an electric load of greater than 200 kV, then unregulated supplies of gas and electricity should be obtained through the RFP solicitation process. Smaller facilities may do best as part of an aggregation. This procurement process does not have to be an expensive deployment of resources or time; most energy consultants will negotiate a contract that is remunerative on a performance basis, not just on an hourly basis. Having a consultant in the room to evaluate offers and particularly to assist in the negotiations, will assure you that the marketer has been properly evaluated and is being forthright with representations. Make sure that any retained consultant is free of commercial conflicts, particularly regarding supplier relationships.

Data. Facility managers should have two years of historic data for each gas or electric meter under consideration. They will be asked to adhere to a utilization schedule that is plus or minus 10 percent of historic usage unless a 60-day notice is given the marketer; that is a reasonable requirement. The manager should be guided generally by the notion that more knowledge about facility energy usage will pay big dividends in negotiating and contract terms.

CONCLUSION

Deregulation is quickly shifting the role of energy consumers from passive ratepayers to active, knowledgeable, price-sensitive, and savvy customers, whose business must be earned. Significant new benefits will quickly accrue to emboldened facility managers who arm themselves with the knowledge and the will to insert their companies into the energy value chain and demand their share of the savings. New products that may redefine the nature of energy value will become available in this huge market. Regional and national suppliers will develop menus of products and services and vie in this exciting and very competitive new marketplace. Having assessed their risks and size, location, and load factors in the marketplace, facility managers are now positioned to reduce energy costs meaningfully during the next few years.

CONTRIBUTORS

Anand K. Seth, P.E., C.E.M., C.P.E., Director of Utilities and Engineering, Partners HealthCare System, Inc., Boston, Massachusetts

David L. Stymiest, P.E., SASHE, C.E.M., Senior Consultant, Smith Seckman Reid, Inc., New Orleans, Louisiana

ARTICLE 7.1.7

OPERATIONS AND MAINTENANCE PLANS

Bernard T. Lewis, P.E., C.P.E., Consultant
Potomac, Maryland

MANAGEMENT OPERATIONAL PLANS

Work Control Method and Procedure

The facility manager's primary goal is to manage resources wisely by providing responsive, high-quality maintenance and repair services to all entities being supported. To accomplish this mission, the facility manager must establish well-defined procedures and the organizational structure to fulfill this work.

A typical work control cycle is shown in Fig. 7.1.7-1. It is important to point out that there is no standard model to follow. One method would be to establish a work control center capable of controlling all work received in the facility department. Work control centers can range in size from a one-person operation, as an added duty, to an entire department that supports a large municipality or a billion-dollar corporation.

Work Control Center.   The work control center, as shown in Fig. 7.1.7-2, is the "heartbeat" of any facilities organization. This is the central point where all work requirements are funneled, then coordinated, planned, cost accounted, scheduled, and measured. As the primary interface between the customer and the organization, the work control center can significantly influence the facility management image. Continuous coordination with workers and "closing the loop" or providing feedback to customers is essential in developing the professional reputation of the department internally within the facility management department and externally with customers.

Organizational Elements

Work Reception.   Work requirements are generated from various means: customers, either verbally or through the completion and submission of a work request form; and operations and maintenance personnel as a result of preventive and predictive maintenance tasks, facility tours, and facility inspection programs. These requirements are received by telephone, from walk-in customers, in writing, and in some cases through electronic mail.

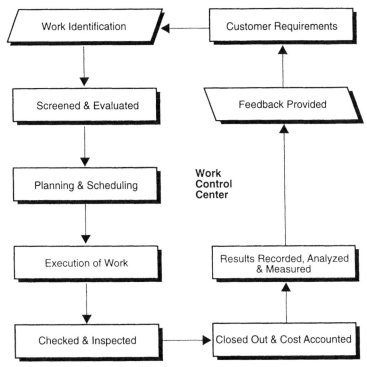

FIGURE 7.1.7-1 Work control cycle. (*From* Facility Manager's Portable Handbook *by Bernard T. Lewis, McGraw-Hill, 1999.*)

Upon receipt, work requests are evaluated for validity, reviewed for compliance with regulatory requirements and feasibility, checked for funding source, prioritized, assigned a project number, and entered into the computer maintenance management system (CMMS). If additional information is required, the caller or initiator of the request is contacted and requested to provide additional detail. Service work requests are then generated and forwarded (electronically or manually) to the appropriate shop or zone for accomplishment. In the event of emergencies, calls are made via radio or other expeditious means to field mechanics.

Planner-Estimator. The normal shortage of facility management resources dictates that work be carefully planned and estimated to ensure that it is efficiently accomplished. The planner-estimator must be an experienced, highly competent individual who can generate work that is easily understood by facility workers. This individual is responsible for planning and estimating in-house work, which could involve only one trade or, in more complex projects, multiple trades and requirements.

Materials Control. The materials control section stores materials, parts, and supplies to support planned operating and maintenance requirements, preventive maintenance tasks, and other repair and new work activities. Its main purpose is to provide timely parts, supplies, and materials while concurrently controlling the availability of the inventory which ultimately impacts the organization's budget; in simplest terms, providing materials, parts, and supplies at the lowest possible cost to the right place and at the right time.

Work Control Center

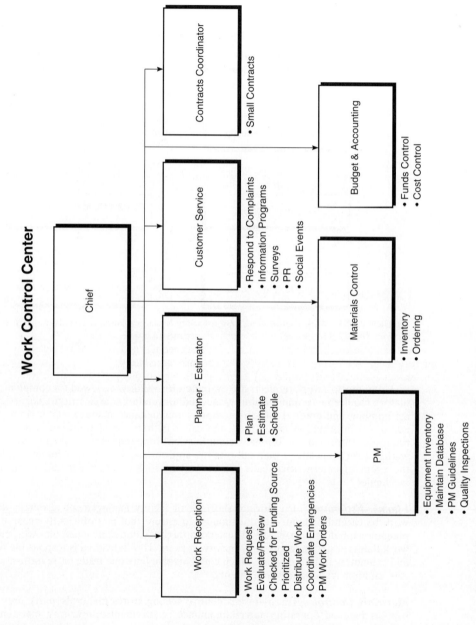

FIGURE 7.1.7-2 Work control center. (*From* Facility Manager's Portable Handbook *by Bernard T. Lewis, McGraw-Hill, 1999.*)

Budget and Accounting.   The work control center should be responsible for developing the budget and performing cost accounting and fund control for the facility department:

- *Funds control.*   The status of various facility management accounts must be closely monitored to ensure that the budget is not overobligated.
- *Cost control.*   This requires good tracking and record keeping; without it, the facility manager does not have good job control. A good cost control system provides for tracking by budget item, work, classification, and job function. Although not totally exact, it gives the facility manager a good approximation of cost.

Preventive Maintenance (PM).   The preventive maintenance manager operates the PM program for the department. This can be done manually or with a computerized maintenance management system. The latter is necessary to manage large electrical and mechanical systems consisting of thousands of items.

The PM section should also maintain records, publications, and specialized tools to help in diagnosing and predicting potential equipment and systems defects. Records include PM inspection lists and schedules, access to "as built" drawings, either paper copies or through the computer-assisted digital data (CADD) system, and repair histories of equipment and components (organizations that are computerized can have this information located in their CMMS). Publications include parts, service, and operations manuals, and other engineering data. Specialized tools common to predictive maintenance include infrared imaging cameras, vibrational analysis collectors, bore scopes, temperature measuring devices, ultrasonic testing devices, listening devices such as a stethoscope, and other high-cost equipment. All of this equipment can be used by various entities throughout the facility department when it is necessary to determine the repair history of a system or component, repair parts reference, asset value, and for planning and scheduling information.

Customer Service.   Everyone in the facility department must be customer service–oriented, but it is beneficial to have a section dedicated to this function that focuses on improving the image of the department and marketing its positive contributions. Excellent customer service is a prerequisite for success. It exists when the facility department meets and exceeds customer expectations for service. For this to happen, the department must have a program that allows it first to provide excellent service and, second, to find ways of reminding the customer of the great service it does give. What customers perceive is reality to them. This work control section, then, has the goal of shaping perceptions.

Work Identification.   Work for the facility department is generated when a customer identifies a requirement and asks that it be satisfied. Requests can come from externally supported customers or facility staff personnel through the normal discharge of their duties. Facility personnel generate routine work requests, for example, in the performance of preventive maintenance, through scheduled inspections such as monthly fire extinguisher inspections, and by simply walking through and looking at their facilities on a daily basis. Normally, changes occur in the physical condition of the facility from age, environment, and use; by changes in regulatory requirements; and by requirements for construction, including alteration, due to expansion and changes in mission and operational needs.

The work receptionist, located in the work control center, is the primary point of contact for all customers. The work receptionist in a facility department is responsible for receiving all requests for work and entering them into the work management and control system. A formal method must exist for receiving these work requests. The work request form is the management tool established for just this reason. It is an official document that describes the work to be done, who requested it (and how to contact them), and when it must be completed, plus a budget number if the work is reimbursable, and, finally, the signature of an approving individual who is authorized to obligate funds for that particular department. Work requests can be generated in several different ways: by fax, telephone, mail, and walk-up. Figure 7.1.7-3 depicts a sample written work request form, and Fig. 7.1.7-4 illustrates an electronic work request form.

Facility Work Request

Section - 1

Control Number	Time:	Date: ____ / ____ / ____
Title		

Section - 2

Facility Name:	Location:

Description of Work:

Justification:

Request for Cost Estimate Only ? Yes ☐ No ☐ | Sketch Attached? Yes ☐ No ☐

Requested Completion Date: ____ / ____ / ____

Requesting Organization:	Cost Center:
Requestor Contact:	Telephone:

Requestor Authorized Signature:

Name: | Date: ____ / ____ / ____

Section - 3

Type of Work:	Classification:
Cost Estimate: $	Date: ____ / ____ / ____

Funding	Cost Center	Amount	Percentage
Source 1	_____	$	_____ %
Source 2	_____	$	_____ %

Division of Facilities Authorization	Date: ____ / ____ / ____
Division of Facilities Contact:	Telephone:
Division of Facilities Comments:	

FIGURE 7.1.7-3 Sample written work request. (*From* Facility Manager's Portable Handbook *by Bernard T. Lewis, McGraw-Hill, 1999.*)

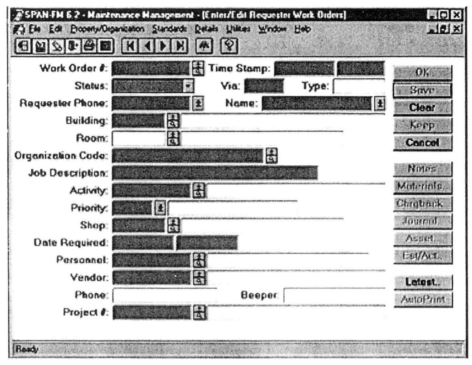

FIGURE 7.1.7-4   Electronic work request form. (*Copyright © 1998, Innovative Tech Systems, Inc.; from* Facility Manager's Portable Handbook *by Bernard T. Lewis, McGraw-Hill, 1999.*)

Once the work request is received and entered into the work management and control cycle, it is a generic work order to which a work order number is assigned. The work control center then tracks each work order as to status: *waiting scheduling, waiting materials, ongoing,* or *completed.* A typical work order flow is shown in Fig. 7.1.7-5.

Work Screening and Evaluation.   Then, the work receptionist classifies each work request as *maintenance, repair,* or *new work* (see Fig. 7.1.7-5). Maintenance work is defined as work performed to keep the facility operating and prevent equipment and systems breakdowns. Repair work is necessary to fix something that has already failed. Finally, new work is work that is used to expand, enhance, or reconfigure a facility.

Once work has been classified, it is categorized into one of four categories: (1) *service orders,* (2) *work orders,* (3) *standing operating orders,* and (4) *preventive maintenance work orders.*

Service Orders (SO).   Service orders are small, service-type maintenance jobs that require immediate attention and cannot be deferred. Normally, time and cost limitations are established for this type of work. The number of service orders can be reduced by strictly enforcing work priorities, forcing customers to pay for damages that result from vandalism or for repairs other than fair wear and tear, adhering to established performance standards, and applying strong preventive and predictive maintenance programs.

Work Orders (WO).   Work orders can include maintenance, repair, and new work. They are formally planned, estimated, scheduled, and cost accounted. This type of work can be accomplished by the in-house work force or by contract, depending on limitations that have been established. Work performed in-house is planned, estimated, and scheduled by the

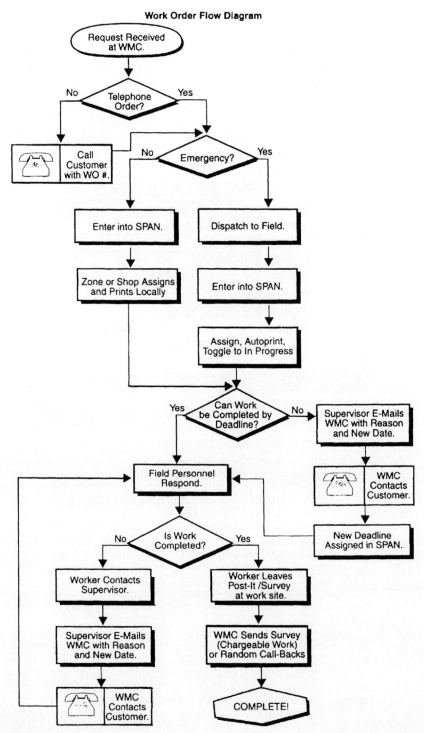

Work Order Flow Diagram

FIGURE 7.1.7-5 Work order flow diagram. (*From* Facility Manager's Portable Handbook *by Bernard T. Lewis, McGraw-Hill, 1999.*)

planner-estimator. The contract coordinator schedules work performed by contract. In either case, upon completion of the job, completion documents showing costs and usage data are returned to the budget and accounts section for appropriate action.

Standing Operating Orders (SOO).   Standing operating orders are special work orders for which the specific work and the labor hourly requirements are relatively constant and predictable. They include operations, services, and routine maintenance whose requirements are repetitive and can be planned annually. Examples include fire protection inspections, custodial services, utility plant operations, refuse collection and disposal, and grass cutting.

Work is planned, estimated, and scheduled before the start of the fiscal year. Careful analysis should be made of all standing operations. Scheduling must be complete and detailed. There should be periodic inspections and ongoing work sampling for each standing operation to determine the quality of the provided service and quantity.

Preventive Maintenance (PM).   Preventive maintenance is routine, recurring work such as regular servicing of HVAC systems and equipment. PM reduces emergency responses, allows work to be efficiently scheduled, and ensures that replacement parts are available when needed. PM work orders are generated in a manner similar to normal work orders.

Work Priority System.   After the work is classified and categorized, it should be prioritized for work performance scheduling. Each facility department should establish its own work priority system. Normally, there are three standard levels of work priorities: (1) *emergency,* (2) *urgent,* and (3) *routine.* These priorities should be carefully and clearly defined and distributed to all customers for their understanding. By doing this, the job of the work receptionist should be made easier and become a little more pleasant.

Emergency Priority.   This work request designation takes priority over all other work. It requires immediate action. Work continues until the emergency is corrected. Normal response time is within one hour. Examples include lock-ins, gas leaks, broken water and steam lines, utility failures, hazardous and toxic spills, elevator entrapment calls, and stopped-up commodes (when only one is available).

Urgent Priority.   This priority involves correcting a condition that could become an emergency. Normal response times range from immediate up to 72 hours, depending on the availability of the workforce. Once started, the work should continue until it is completed, depending on the availability of materials and parts. Examples include garbage disposal problems, lights that are out, pest control problems, lock changes, and appliance (refrigerator, range) problems.

Routine Priority.   This is work that is an inconvenience to the customer. The completion of this work is normally accomplished within seven calendar days. Examples include dripping faucets, screen door replacement, countertop repair, and floor replacement. All of these orders must be completed. To perform this function, planning and scheduling are required.

Execution of Work.   This is the actual conduct of maintenance and repair work in response to a work request. It can be accomplished by an in-house work force or by contract. It is the point at which all the planning, estimating, and scheduling efforts pay off. Many activities (e.g. ordering materials and equipment; requesting and supplying support services such as scaffolding, land surveys, and utility marking and identification; safety inspections; and developing trade and individual work schedules) are synchronized to complete the work within the estimated cost and time frame.

The in-house work force should record labor hours, material, supplies, parts, and equipment usage on a daily employee activity sheet. This document is a record of worker activities for that day. It should be completed daily by the worker and turned in to the supervisor, who then checks the time for each work order. Supplies and materials usage can be noted on this form or included on the work request.

For contracted (outsourced) work, an inspector should be assigned to periodically check and inspect as the work progresses. Nothing should be placed or approved for payment unless the department's quality control inspector has inspected it.

Emergency work should be handled on a case-by-case basis. This work requires special attention. Emergencies called into the work receptionist should be verified by requesting someone (either a supervisor or craftsperson) to verify the emergency. Once verified, requests to shops or zones should be made via radio or pager for quick response. Emergency work takes priority over all other work.

Checked and Inspected.   Good facility managers check! check! check! Even with continuous quality improvement programs, wherein quality is built into everything that is done, a need for inspection still exists. Everyone in a facility department, whether he or she believes it or not, is part of the inspection program. Inspection is a continuous undertaking. It never ceases! Craftspersons should be trained in their trades and pride instilled in them to produce the best-quality work they can. They should not be satisfied with mediocre results. Supervisors and managers should be trained the same way—to strive to improve continuously upon everything they do.

Inspections are also a major component of the PM program. It is important to note here that as PM inspections are properly accomplished, there will be a decrease in reactive maintenance and a corresponding increase in scheduled maintenance. A mechanic conducts PM on a specific piece of equipment and annotates any unusual information on the PM work ticket so that it can be entered into the CMMS PM module database. In most cases, the PM inspector will verify the information in the field. If severe enough, repair can be planned and scheduled without the need for an emergency outage. As equipment ages and deteriorates, the importance of PM inspections is obvious. Serious problems resulting from breakdowns can be reduced. Downtime is reduced through planning and scheduling.

Closed and Cost Accounted.   Work is properly closed out when excess materials, parts, and supplies are returned to materials control; punch-list items are completed; required tests, demonstrations, or verifications are completed; operating and maintenance manuals have been collected, reviewed, and accepted; "as-built" drawings have been submitted; lessons learned are documented; labor, material, and equipment used have been assembled and added to the work order; and if reimbursable work, the customer has been formally charged for the work through the accounts payable system.

Results Recorded, Analyzed, and Measured.   Once a work order has been completed and closed out, it has to be analyzed and measured. This procedure is to compare actual performance against predetermined standards. At a minimum, the quantities of materials used, as well as the charges that may have been credited, should be spot-checked. Actual labor hour charges should also be compared to those estimated by the planner-estimator to improve the value and validity of the planner-estimator's database. All of these actions are critical in the continuous quality improvement loop. This step is important to determine if the trends, the lessons learned, and potential improvements can be synthesized from the work that took place. Tools that can be used to assess the effectiveness of the maintenance and PM programs can be incorporated into weekly and monthly management reports.

Feedback.   A facility department has to receive timely and accurate feedback to improve its service and the way it conducts business. This feedback in the form of review and analysis provides the information to ensure that actions are being performed as planned. It also allows the facility manger to focus efforts on organizational improvement.

Customer Feedback.   Customer feedback can be solicited in various ways. Mechanics can leave "while you were out" cards when they respond to a service order and the initiator is not available. They key here is that customers are satisfied if they know what's happening. They only become disgruntled when they're kept in the dark. A second means of feedback is for the work reception center to call back a sampling of customers daily to follow up on completed work orders and solicit feedback. In this case it would be worthwhile to standardize questions.

These questions will provide immediate feedback and serve as a good measure of customer perceptions of the services they receive. These questions can be as simple as:

- Was the work completed to your satisfaction?
- Were facility department personnel courteous?
- Was a "while you were out" card left?

Another, more formal, means of obtaining customer feedback is to use a formal survey. This technique is covered in detail later in this article.

BUILDING OPERATIONAL PLAN

Facility managers are responsible for overseeing the safe, efficient operation and maintenance of each facility under their control. Their decisions impact every member of their organization every day. These common issues lie within the numerous processes for which they are answerable: how to schedule and coordinate work efficiently, how to control startups and shutdowns of equipment and operating systems, how to handle emergency situations, how to coordinate diagnosis and repair of trouble, and how to use benchmarking and performance indicators to determine the operational status of equipment and systems.

Shutdowns

Scheduled shutdowns are not considered emergencies. Customers should be notified in writing at least 72 hours before the scheduled outage. Alternative support may be required if the shutdown will last more than several hours. This support could consist of lighting, heat, potable water, or cooling and can be planned for and connected in advance. Typical planned shutdowns include emergency generator testing; utility repairs; equipment repair, replacement, or overhaul; sprinkler and fire protection system testing and repair; and building heating, ventilating, and air-conditioning (HVAC) system repair or adjustment.

Start-Ups

The objective is to get equipment operating efficiently, safely, cost-effectively, and as quickly as possible. The following points should be considered:

- Expect equipment problems. Depending on how long the equipment has been down, fluids and lubricants will have drained from metal surfaces, systems will become air-bound, and equipment will lose calibration.
- Be cautious of pressurized systems. Because of no demand, systems that are normally pressurized may have higher than normal pressures.
- Ensure that all necessary materials and supplies are on hand. This will reduce workforce idle time and provide a more balanced workload.
- Before start-up, ensure that all maintenance and utility crews are on hand early and available to minimize and respond to problems.

Emergencies

Emergencies require immediate action. Compared with scheduled maintenance, emergency maintenance represents a very small percentage of total maintenance requirements (usually

less than 10 percent). The detailed planning and scheduling that precede scheduled maintenance are impractical in emergency situations because of the immediacy of the situation. This
is the reason that emergency situations must be planned for ahead of time and written procedures put in place. (This will be covered in more detail.) As a minimum, the facility manager
should plan for three areas of immediate concern.

Fire Protection

- The first step is to develop a fire protection protocol unique for each building. The protocol
  should include the following: a definition of the "authority having jurisdiction" (prior to the
  fire department arrival) who could be the building security officer, the property manager,
  or the facility manager; building evacuation procedures; a description of the alarm system
  and the locations of annunciator panels, sprinkler valves, and the fire pump; shutdown procedures; a description of the emergency generator and its location; and a description of the
  actions to take if the systems fail.

- Training is extremely important. Through periodic fire drills, occupants of buildings should
  become familiar with evacuation and also know the locations and activation procedures for
  alarm pull stations. Maintenance and operations personnel should be trained semiannually
  on the location of specific monitoring panels, shutoff valves, and disconnects, and should
  have the protocol etched in their minds.

- Designate a spokesperson to coordinate public relations and news media issues and statements.

Power Outage

- Control devices should always be set for "fail-safe."

- All electrical start-stop switches should be shut off. This will prevent damage to certain
  types of electrical equipment should partial electrical power be restored. Also, circuit
  breakers may switch off if they sense an overload when power is restored.

- Verify that all emergency systems are operational. If the outage lasts for several hours,
  check the fuel tanks of the emergency generators.

- In the winter, check throughout the building for possible pipe freeze-ups. Crack open water
  faucets at various locations throughout the building.

Flooding

- If the building is located in a flood plain, contact the state flood control agency or appropriate federal agencies (e.g., U.S. Army Corps of Engineers) to obtain information on the
  expected time of arrival and water levels.

- Provide property protection. Time permitting, all doors and windows should be sealed and
  sandbagged. Emergency generators should be obtained to operate pumps and provide
  emergency lighting. Electrical equipment should be deenergized. Where feasible, delicate
  equipment should be removed to safe locations.

- Implement procedures that define who makes decisions to evaluate and shut down equipment, details evacuation procedures, and determines emergency crews.

Trouble Diagnosis and Coordination

As mentioned before, coordination is extremely important. You can have the most competent
mechanics and the best maintenance plan, but if there is no coordination, the facility department is looked on as incompetent.

Diagnosing a Maintenance Problem.   Diagnosis is the identification of maintenance problems, causes, and effects. This initial step eventually leads to correcting and eliminating such

causes as misapplication and operator abuse. The "diagnostician" plays an important role here. This individual must have the technical background and personality to find a problem and teach mechanics and operators the proper maintenance techniques without coming across as a "know-it all" or trying to find fault. It is not easy to find an individual who has these characteristics. Before any diagnosis of a maintenance problem can begin, the technician must have the proper training and experience, understand the steps involved in diagnosing a problem, and be knowledgeable and skilled in using diagnostic methods and instruments. In some cases, it may be more cost effective to contract for this service.

Coordination with Third Parties.   If customers have been impacted by an equipment outage or if they are going to pay for the repair, they must be told what happened, why it happened, and the corrective action taken to prevent it from happening again. Unless this is done, the facility department loses credibility. Even the best operating facility department that has an excellent PM program to minimize equipment downtime and maximize scheduled maintenance will experience occasional emergencies wherein customers cannot be warned of unscheduled outages. This is where good customer relations come into play.

Benchmarking.   The purpose of benchmarking is to improve an organization's performance by comparing its practices to those of another. To conduct a benchmarking study, it is important that facility managers know their own processes. This is necessary to get the maximum value from study comparisons with other facility organizations.

Statistical Process Control.   Once thorough understanding of a procedure is attained, major advances in service quality, productivity, and cost can be achieved. To develop this knowledge, a tool such as *statistical process control* (SPC) can be used. SPC is used in industrial and manufacturing environments as a process for improving quality and minimizing waste. It is a statistical technique suited for use whenever something can be measured or counted to show trends that will guide future actions. It is typically used in continuous-improvement processes and can be creatively adapted to various building operating systems. Facility managers can use SPC to help better understand their processes, measure and analyze quality, and ultimately improve maintenance.

COMPREHENSIVE FACILITY OPERATING PLANS

All facility managers are responsible for effectively planning for and responding to an emergency and bringing that situation to a suitable conclusion. To do this, preparation is vital! Establishing a formal written plan that defines how to handle emergency procedures is essential. This plan should be disseminated to everyone who is responsible for putting emergency procedures into practice. Additionally, emergency drills and problem exercises should be periodically scheduled to ensure that facilities personnel are properly trained and conditioned to react quickly and skillfully when an emergency occurs.

Emergency Response Plan

Emergencies occur! A good emergency response plan defines duties and responsibilities; identifies the various emergencies that may occur; explains what to do before, during, and after an emergency; and details the procedures for resolving most conceivable emergencies.

Organization.   Top management should appoint an emergency coordinator. This individual should be assigned to a central location [emergency operations center (EOC)] and should coordinate actions during both the emergency and the restoration period. The EOC is the primary control point for coordinating and handling responses to emergencies. Staffing of the EOC depends on the type, magnitude, and location of the emergency.

Concept of Operation.   The *concept of operation* is a statement of the emergency coordinator's visualization of the way an emergency should be handled from start to finish. It should be stated in sufficient detail to ensure appropriate action. After normal work hours, the organization's security office should be the first office to be notified of an emergency, because security officers patrol buildings and can physically detect a problem, alarm systems (fire and environmental) terminate at the security office, and an individual who detects an emergency calls the security office to initiate the emergency response plan.

Command and Control.   During major emergencies, a positive chain of command must exist and be functional. There must be a clearly identified chain of organization as shown following:

1. *Command.*   The responsibility for resolving any emergency situation rests with top management. However, the emergency coordinator should be the executor of the emergency response plan and should assist top management in planning, training, responding to, and mitigating emergencies.
2. *Responsibility succession.*   In the absence of the ultimate top manager responsible, a successive line of authority and responsibility for decision making must be defined.
3. *Control.*   Initial control responsibility for emergency situations rests with the security manager. As a minimum, this responsibility should include the following activities:

   • Immediate evacuation of affected buildings as necessary for the life safety of occupants.
   • Initial response to emergencies in the facility or building. This response determines the type and size of the emergency.
   • Coordination with external support agencies such as police and fire departments.
   • Exterior crowd control at all emergency scenes.
   • Assisting the emergency coordinator with initial activation of the EOC.

Communications.   Adequate, effective, redundant communications systems must be available to effectively respond to and control emergencies. These systems must be an integral part of the EOC. They must be reliable and capable of functioning during periods of power loss.

Types of Emergencies

Depending on the type of facility or property, the following emergencies should be addressed. For each of these emergencies, the plan should address preparations before an emergency occurs, training requirements, response procedures (which include description of personnel actions for each credible emergency requiring specialized response), and restoration procedures.

• Hazardous material
• Fire emergencies
• Natural disaster
• Bomb threat
• Utility outage
• Labor unrest

Support services.   These are services that can be provided to measure, control, or mitigate emergencies. Information that should be incorporated here includes:

• Resources available
• Temporary housing support
• Updated list of telephone numbers and points of contact

- Local building codes and regulatory requirements
- Insurance information
- Flowchart of the chain of command
- Reporting procedures and required documentation
- Federal, state, and city disaster assistance procedures
- Information on how to track costs

Guidelines for Facility-Specific Plans

Specific emergency procedures for each individual facility should be prepared. At a minimum, the plans should contain the following information:

1. A brief description of the facility or type of operation
2. Procedures for evacuating and accounting for any visitors and for all personnel who normally work in the facility, including:
   - Determining when evacuation is necessary
   - Designating and using assembly areas to account for personnel
   - Selecting evacuation routes out of the facility
   - Establishing a personnel accountability system
   - Procedures for evacuating physically impaired personnel
3. Description of the means of communicating the emergency to all personnel
4. Identification of specialized equipment needed by units
5. Responding to the emergency
6. Procedures for shutting down all utilities to the affected area and securing the facility from unauthorized entry
7. Floor plans and blueprints
8. Building equipment and systems information

Hazardous Materials Plan

Hazardous and toxic materials are all materials (gaseous, liquid, or solid) that can cause physical or chemical changes in the environment, affect property, or affect the well-being of living organisms. A release or spill of hazardous material requires that the controlling organization immediately notify people in the surrounding vicinity that a release or spill has occurred. If deemed necessary, the area (room, floor, or building) is evacuated. A specific hazardous material concern to all facility managers is the use of refrigerants for cooling and refrigeration equipment.

Training. Facility managers are responsible for the facilities they maintain, and they also have to ensure that their personnel are appropriately trained in hazardous materials usage in the workplace. The following training exercises should be considered:

- Hazards communication (right-to-know)
- First responder awareness
- First responder operations

Assessment. According to the Code of Federal Regulations, 29 CFR 1910.132, employers are required to perform a workplace condition assessment to determine whether hazards are present, or likely to be present, that necessitate using personal protective equipment (PPE). If such hazards exist or are likely to, the employer must select appropriate PPE and certify that a workplace hazard condition assessment was performed.

Safety Plans

All organizations want to reduce the loss of downtime, lost production, and workers' compensation claims related to accidents. To improve safety within an organization and, hopefully, decrease the number of accidents and injuries on the job, a formal written safety plan is needed. The safety plan provides a system to constantly monitor the work environment to minimize potential safety and health threats.

Purpose.   The safety plan's purpose is to establish standard procedures for normally occurring activities. These procedures, when followed, save the organization money by keeping safety continuously in everyone's minds.

Involvement.   For employees to buy into the safety plan, they have to see the policy put into practice. This requires that safety be included in the organization's mission statement; that various safety topics germane to the functions of the organization should be included in the plan; and that safety training, consistent enforcement of safety policies and procedures, and positive reinforcement of safe working behavior should be included. Listed here are a variety of methods used by organizations to promote safe working areas and prevent accidents:

1. Foster area involvement in the process. It is very important to have input from all areas affected by the safety topic. Office areas, the loading dock, production areas, distribution areas, mailroom, supply and storage areas, laboratory areas, mechanical rooms, and others should be included.
2. Organize safety and health committees. These committees are formed to serve a number of basic functions:

   - Create and establish safety guidelines.
   - Plan and implement facility safety programs.
   - Establish and review safety-training programs
   - Investigate accidents.
   - Conduct and evaluate safety inspections.
   - Initiate ideas and plans to correct safety problems.
   - Hold weekly and monthly safety meetings.
   - Develop a safety suggestion program.
   - Design safety-training sessions with employee input.
   - Stress the organization's long-term safety goals in the light of employees' involvement.
   - Institute and communicate an organizational philosophy that accidents are not acceptable and will not be tolerated.
   - Use statistical techniques to categorize accidents into system shortcomings or human error.

Accident Reporting.   To facilitate a uniform response to accident investigation and provide for the safety and health of employees, a standardized accident investigation report form should be implemented. The appropriate supervisor should complete this form within a specified time period after the accident for any accident or on-the-job injury. The facility manager should review each form for completeness, determine what happened, and take corrective action to prevent similar accidents in the future.

Categories of Fire Protection Responsibility

The facility manager is normally responsible for fire protection and prevention within a facility. This responsibility can be divided into two categories: (1) regular prevention and protection inspections, and (2) controlling fires that occur. A planned fire prevention and protection program must provide the maximum protection against fire hazards to life and property consistent with the mission, engineering, and economic principles. Additionally, facility managers should have a plan to combat fire emergencies.

Prevention and Protection.   Inspections should be conducted monthly and are performed to reduce the potential fire hazard:

- *Prevention.*   Environmental situations known to be causes of fires are inspected first. These situations include storage and handling of flammable materials and liquids; operation of electrical equipment and associated wiring; high-temperature generating equipment, such as hot work involving welding and soldering equipment; smoking noncompliance; and finally, proper housekeeping procedures.
- *Protection.*   Inspecting and testing fire protection equipment are intended to ascertain the operability of that equipment. The inspection should include these activities: identifying each fire protection control valve and verifying that it functions correctly; verifying the availability of the water supply; completing a fire pump checklist for each pump; examining the critical components of special extinguishing systems; testing all sprinkler systems and alarms; checking all fire extinguishers; identifying each hydrant; determining that all fire doors work properly; verifying that smoke and heat detectors operate properly; and finally, ensuring that protective signaling devices function correctly.

Fire Control.   Upon the discovery of a fire, the fire alarm should be sounded. If the fire is small and an extinguisher is nearby, the fire should be extinguished with the extinguisher. If the fire is larger and spreading, clear the building. If time permits:

- Close all doors and windows
- Turn off all oxygen and gas outlets
- Disconnect electrical equipment
- Turn off all blowers and ventilators

Fire Protection Planning

Fire emergencies can occur at any time. Having a fire plan is good insurance against total chaos. As a minimum, the fire plan should cover the following:

- Authority having jurisdiction. Define the overall organization that has ultimate jurisdiction until the fire department arrives.
- Define what measures should be taken before, during, and after the fire emergency.
- Training requirements.
- Formation of an organizational fire brigade.
- Ensure that plans and drawings for each building are available. These plans should identify electrical, chilled water, steam, telephone, and data cable locations, and other utility distribution systems.
- Specific requirements should be spelled out for the operations and maintenance staff, contract section, purchasing staff, and public relations staff.
- Requirements for fire watch.

Facility Occupant Support Plan

One of the biggest challenges facing facility managers today is to provide high-quality customer service. To do this, facility managers must ensure that work is scheduled routinely and also that it is done in a caring manner. High-quality service exists only when the expectations of customers are exceeded.

Many facility managers believe that their departments provide excellent service, because all of the mechanical and electrical systems in their buildings operate and meet required life safety and building codes, burned-out lights are replaced immediately, the maintenance and repair backlog work has been reduced, and the service technicians respond quickly when a service request is submitted. All of this may be true—and still customers may not be satisfied because their expectations are not exceeded. A possible cause for this may be the way the customers perceive the service that is provided.

Improving Perceptions.   There must be caring and consistency in the way service is provided before customers will perceive facility support as efficient, effective, and customer-oriented. This change will come each time they have a positive experience as a result of the service provided. Expectations must be exceeded every time. Every service opportunity must be seized to influence customer perceptions positively. Some ideas that will help include:

- Minimize interference with customer activities.
- Curtail unplanned outages. Ensure that customers have, as a minimum, 72 hours' notice.
- Coordinate and communicate with customers to provide them information. They will accept almost anything as long as they know the impacts and why a situation has to exist.
- Provide proper and timely service for customer work requests. As discussed earlier, work requests should be screened, evaluated, and then prioritized. Time frames should be established for each level of work. Assigned priorities must be adhered to. A system should be established to "close the loop" with customers if work cannot be accomplished in a timely manner. Electronic mail lends itself nicely to doing this.
- Do not let problems go unanswered. If a facility manager learns of a problem wherein a customer is dissatisfied, investigate the reasons on both sides to "close the loop" with the customer. If the facilities department erred, acknowledge that, apologize, and let the customer know that you will implement corrective action. . . . but respond to the customer! Again, customers will accept almost anything as long as they know that the department is trying to improve and that it does care. In his book *How to Win Customers and Keep Them for Life,*[1] Michael LeBoeuf states, "A rapidly settled complaint can actually create more customer loyalty than would have been created if it had never occurred. Customers are much more likely to remember the 'extra touch,' fast action, and genuine concern that you exhibited when they felt dissatisfied."

Determining Wants.   To exceed customer expectations, facility managers must determine (1) what customers need and (2) what they expect:

- *Need.*   One effective method of determining need is to go and see the customer face-to-face. Personal contact has a way of cementing relationships and developing rapport.
- *Expectations.*   To determine customer expectations, a formal written survey is a good technique to use. Once you have it developed, plan on distributing it annually. Start a mailing list of persons to mail the survey to and keep adding to it. Include people who have experienced problems with service in the past; budget and financial people; and key department, floor, or building coordinators.

Provide Feedback.   Most customer complaints and dissatisfaction are generated because of a lack of communication. Implementing a simple customer communications program will eliminate many routine complaints before they mushroom into major problems of customer relations. These are some ideas that can be used:

- *Callbacks.* On a daily basis, have the work receptionists make callbacks to customers whose work has been completed. A good percentage to use would be to call back 20 percent of those whose work orders have been completed. This technique lets 20 percent of the customers who called in a problem know that the facility department measures its performance, identifies follow-up problems which can then be proactively corrected, and finally provides a daily gauge to measure customer satisfaction.

- *Customer-contact employees.* Train employees who have daily contact with customers to "close the loop" and provide information that they should have. These employees are the first line of contact with these customers and know when something is not right. They can be an early warning system to alleviate potential problems. Additionally, a periodic group meeting with all customer-contact employees can be instructional for them through the sharing of ideas and can also alleviate customer concerns.

- *Reports.* Distribute an annual report to customers. This lets them know what and how well the facility department is doing to meet their expectations.

Inspections by Facility Department (FD) Personnel. The facility manager, chief engineer, operating engineers, and maintenance mechanics should inspect completed work, including service calls and maintenance and repair work orders. The chief engineer, operating engineers, and maintenance mechanics should tour the facility each day, conducting quality control (QC) inspections informally. When their daily QC inspections are completed, their reports should summarize all deficiencies noted during the inspections and recommend corrective actions. The facility manager should review the QC reports and forward them to the FD administrative assistant for processing. All daily QC reports should be maintained in a hard-copy file for review by corporate managers.

REFERENCES

1. M. LeBoeuf, *How to Win Customers and Keep Them for Life,* G. P. Putnam's Sons, New York, 1987.

GENERAL REFERENCE

Lewis, B. T., *Facility Manager's Operation and Maintenance Handbook,* McGraw-Hill, New York, 1999.

SECTION 7.2
FACILITIES MAINTENANCE

Anand K. Seth, P.E., C.E.M., C.P.E.,
Director of Utilities and Engineering, Section Editor
Partners HealthCare System, Inc., Boston, Massachusetts

This section on facilities maintenance discusses the most critical issues involved in setting up a viable maintenance program in a modern, cost-effective manner. It addresses three common types of maintenance: *breakdown, preventive,* and *predictive.* It also emphasizes that maintenance cannot be provided in a vacuum. Designing for maintenance is critical.

The specific and unique maintenance requirements for health care facilities are described in Sec. 9.8, "Maintenance for Health-Care Facilities," which contains an excellent discussion on preventing Legionnaires' disease in building systems. Some of that information can apply to other types of facilities as well. Section 9.8 also discusses renovation and construction plan review by the maintenance and operations department. That discussion is included in Sec. 9.8 rather than here because of the critical nature of the activity in health care facilities where change is more frequent. However, much of the information can also apply to other facilities.

This section does not address in detail architectural systems maintenance such as painting, pointing, roof maintenance, window maintenance, and waterproofing. Some of these issues are discussed in Arts. 6.2.8 ("Exterior Skin Dampproofing and Waterproofing"), 6.2.9 ("Exterior Skin Façade"), and 6.2.10 ("Exterior Skin Roofing"). Some of the predictive maintenance techniques mentioned in Art. 7.2.1, "Mechanical Testing and Maintenance," may be very helpful in early prevention of serious mechanical problems in large systems. This early knowledge sometimes makes possible inexpensive remedies that prevent a major failure in the future.

ARTICLE 7.2.1

MECHANICAL TESTING AND MAINTENANCE

Brian T. Herteen, Blake Stewart, and William Hughes
Johnson Controls, Inc., Milwaukee, Wisconsin

PREDICTIVE MAINTENANCE

The appeal of predictive maintenance is threefold. First, it uncovers problems before they cause failures. Second, it extends service intervals for machines in good condition. Finally, it determines the condition of a machine as it operates—without taking the machine apart. Predictive maintenance techniques reduce expenses by revealing the optimal time for maintenance. This article discusses the following predictive techniques:

- Vibrational analysis
- Infrared thermographic inspection
- Motor current analysis
- Oil analysis
- Refrigerant analysis

This article also discusses two proactive maintenance tasks—shaft alignment and dynamic balancing—that can increase equipment life and reduce maintenance costs. These proactive steps address the root causes of many machine failures.

Vibrational Analysis

Vibrational analysis is one of the most effective techniques for analyzing the condition of rotating equipment. It is the cornerstone of a predictive maintenance program because it detects a wide range of equipment problems before they can cause failure:

- Misalignment and imbalance (these account for 60 to 80 percent of fan and pump problems)
- Resonance and bearing defects
- Gear and belt problems
- Sheave and impeller problems
- Looseness and bent shafts
- Flow-related problems (cavitation and recirculation)
- Electrical problems (rotor bar problems)

In environments that are critical to an organization's mission, the greatest benefit of vibrational analysis is that it forecasts the most appropriate time to correct machine problems. This eliminates unscheduled downtime.

Performing Vibrational Analysis.   Vibrational analysis involves attaching small sensors to predetermined locations on equipment. Then, a technician connects these sensors to an accelerometer. The accelerometer collects data and converts the mechanical motion (vibration) into electrical signals. Plotting these signals produces a graph called a *vibrational signature,* which tells technicians which components are vibrating and how much.

Figure 7.2.1-1 shows a typical vibrational signature for a chiller. The *amplitude* and *frequency* are the two characteristics of vibration used to diagnose equipment problems.

Amplitude is the *amount* of vibration. It indicates the severity of a problem. The greater the amplitude, the greater the problem. Amplitude is measured in inches per second (ips), mils of displacement, or g's of acceleration.

Frequency is the *type* of vibration. It identifies the source of the vibration. For example, a motor shaft may vibrate at 50 Hz, and a compressor may vibrate at 120 Hz. In addition, different mechanical problems cause vibration at different frequencies.

Frequency is measured in revolutions per minute (rpm), cycles per minute (cpm), and cycles per second (cps, Hertz, or Hz). The rpm of machinery is a measure of frequency. In

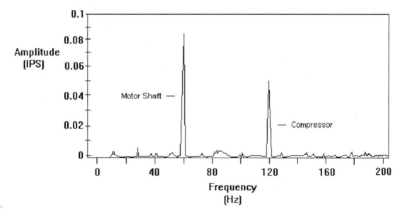

FIGURE 7.2.1-1   Typical vibrational signature for a chiller. (*Courtesy of Johnson Controls, Inc.*)

imbalance, one cycle occurs during each revolution. Therefore, the frequency for imbalance is 1 × rpm. Different machines run at different rpms. A motor that operates at 1800 rpm has a frequency of imbalance of 1 × rpm or 1800 cpm.

Trending Vibrational Levels. The big picture offered by *trending*—measuring vibrational levels over time—helps determine more precisely when a machine will fail. A single vibrational measurement gives a snapshot of a machine's condition, but trending shows a full view of the equipment's performance.

In Fig. 7.2.1-2, a machine's vibration in August 1997 was 0.13 ips, which is within specifications. Given that measurement alone, the machine does not appear to have a problem. The trend chart, however, shows that vibrational levels have been rising at an increasing rate, a sign of upcoming problems.

Trending forecasts the future condition of equipment, and it provides time to prepare for necessary maintenance. Rather than make emergency fixes, managers can schedule repairs for planned outages on off-peak hours. Every trend measurement gathered reduces the risk of unscheduled downtime. The importance of the space served by the equipment dictates how often to take measurements. If a space is critical to the business, perform a vibrational analysis at least four times per year. Article 7.1.1, "Maintenance Planning and Budgeting," discusses how to identify critical spaces in a facility.

Overall Vibration. Overall vibration (measured in ips) is the total vibration within a piece of equipment, the vibration caused by all of the equipment's problems. Measuring the overall vibration of a machine quickly reveals whether it is in good condition. However, it does not tell you what the specific problem is. High overall vibration points to a need to analyze the vibrational signature further.

Infrared Thermographic Inspection

Thermography involves analyzing heat transfer by electromagnetic radiation. All animate and inanimate objects (including electrical control panels, motors, and boiler doors) emit electromagnetic radiation in the infrared spectrum. An infrared camera can see this radiation. Thermographic inspection is an accurate, quick, and effective technique to avoid equipment breakdowns by gathering and presenting information about a system. However, it does not ensure proper equipment operation; other tests and proper maintenance are necessary to ensure reliable performance.

An infrared scanner resembles a video camera, and it records site information on diskettes

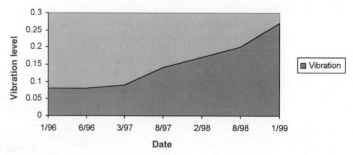

Vibration Trending (twice annually)

FIGURE 7.2.1-2 Vibration trending. (*Courtesy of Johnson Controls, Inc.*)

or on standard VHS videotapes for later review and investigation. A display screen helps to identify potential problem areas immediately.

Conducting a Thermographic Inspection.   In a thermographic inspection, plant equipment is systematically scanned for temperature profiles to find and correct developing problems before equipment failure occurs. Analysis can isolate the source of overheating or other problem areas. Temperature anomalies in equipment—both hot spots and cold spots—can be investigated. The relative severity of a hot spot can be determined, and its root cause can be isolated and identified.

Other Uses of Thermographic Inspection.   Electrical inspections are one of many applications of thermographic technology. Think of an electrical system as a chain. Stress causes the chain to break at its weakest link. In an electrical system, hot spots caused by a small temperature rise weaken the chain. When the component deteriorates, the temperature rises, and eventually it burns up or short-circuits. The job of the thermographer is to find these weak links in the electrical system before they fail. An infrared camera can easily detect defective components, such as poor connections, overloads, and phase imbalances. Refer to Art. 7.2.5, "Electrical Testing and Maintenance," for a more detailed discussion of these issues.

Motor Current Analysis

Motor current analysis diagnoses rotor problems, including:

- Broken or cracked rotor bars or shorting rings
- Bad high-resistance joints between rotor bars and shorting rings
- Shorted rotor lamination
- Loose or open rotor bars not making good contact with end rings

Motor current analysis eliminates the need for Variac or Growler tests that diagnose the same problems but require that equipment be turned off and disassembled. It can generally be performed with the equipment running. (One exception is high-voltage machines; shut these down to avoid the risk of electrocution.)

How Is a Motor Current Analysis Performed?   A motor current analysis is performed with a vibrational analyzer and a motor current clamp that measure the current drawn by the motor. Motor current can be measured on either the main phase circuit or on the secondary control circuit. The secondary circuit is safer; always use it on equipment rated at above 600 volts.

When performing a motor current analysis, an analyst measures the three-phase power line leads one at a time. Then the analyst compares the amplitudes of current at the electrical line frequency in each phase lead. The current in each phase should be within approximately 3 percent of the others. Variations higher than 3 percent point to stator problems such as those listed earlier.

Oil Analysis

Oil analysis is one of the oldest, most common, and most useful predictive technologies in use. It helps to prevent failure and unscheduled downtime by displaying the wear metal count and types of contaminants in oil. The wear metal count indicates whether equipment is experiencing unusual wear. The types of contaminants in the oil, as well as its physical characteristics, determine whether the time interval between oil changes can be extended.

What Are the Most Common Types of Oil Analyses? Common methods for determining oil quality include *spectrochemical analysis, physical tests,* and *ferrography:*

1. Spectrochemical analysis identifies wear particles (metals such as zinc, aluminum, copper, nickel, and chromium) in the oil. Friction between bearings and gears causes these metals to wear from the surfaces and circulate in the lubricant. A high level of metals indicates that components are wearing.

2. Physical tests show how well a lubricant is doing its job. Then, contaminated lubricants can be changed before they accelerate component wear. The most common physical tests include:

   • *Viscosity* is a lubricant's internal resistance to flow. It is the single most important physical property of oil. Changes in viscosity indicate lubrication breakdown, contamination, or improper servicing. Each of these changes leads to premature component failure.
   • *Water* in the lubricant promotes oxidation and rust in components. It also prevents a lubricant from doing its job.
   • *Total acid number* (TAN) is the level of acidic material in the lubricant. It indicates acidic contamination of the oil or increased oil oxidation. Both increase the potential for corrosive wear.

3. Ferrography is a useful technique for analyzing centrifugal equipment that has transmissions and screw compressors. It determines the condition of a component by directly examining wear metal particles. Wear metals and contaminant particles are magnetically separated from the oil and arranged according to size and composition. Direct reading (DR) ferrography monitors and measures trends in the concentration of ferrous wear particles. DR trends indicate abnormal or critical wear that can be used to trigger analytical ferrography. Direct reading ferrography is usually unnecessary if vibrational analysis is being used, because vibrational analysis assesses the condition of gears more accurately.

When Should Tests Be Performed? For equipment with less than 50 horsepower, perform a TAN test once per year. For larger equipment or smaller equipment that is critical to the mission, perform physical tests and a spectrochemical analysis twice per year. Perform more frequent testing if a system is critical or has a history of problems. Use ferrography when a vibration or spectrographic analysis detects an abnormal condition.

Regular sampling is important. It determines the suitability of the oil for continued service. It can also provide crucial information about the presence of wear metals, acids, moisture, and other contaminants.

Refrigerant Analysis

Refrigerant analysis checks physical properties, vapor phase contaminants, and liquid phase contaminants to determine the condition of the refrigerant. *Moisture* and *acidity* are the two most important levels to monitor. High moisture levels lead to increases in acid levels. This, in turn, causes motor insulation to deteriorate and tube metal to erode. Once acid is in the system, it migrates into the oil. In the oil, the acid accelerates the wear of rotating components such as bearings and gears. This leads to premature component failures.

A refrigerant analysis can also verify whether the refrigerant you buy meets acceptable standards. ARI Standard 700–95[1] is typically used to assess refrigerant condition. Finally, perform an analysis after repairing leaks, adding refrigerant, or performing major repairs that have a high potential for moisture contamination.

The accuracy of a refrigerant test depends on the sampling technique. Be careful not to contaminate the sample with outside moisture, because the moisture level is an important indicator of condition.

Shaft Alignment

Improper alignment may be the largest cause of high vibration and premature failures in equipment. High vibrational levels lead to excessive wear on bearings, bushings, couplings, shaft seals, and gears. Proper alignment can slow the deterioration of equipment. Alignment means adjusting a piece of equipment so that its shaft is in line with the machine to which it is coupled. When the driver and driven machines are connected through a common coupling and rotate together at operating equilibrium, the unit rotates along a common axis of rotation as one continuous unit without excessive vibration.

The three most common types of alignment are:

- *Parallel,* in which the coupling hub faces are parallel, but the two shaft centerlines are off-set. In short, there is a distance between the two shaft centerlines
- *Angular,* in which the coupling hub faces are not parallel and the shaft centerlines are not concentric
- *Perfect,* in which the coupling hub faces are parallel and shaft centerlines are concentric

Correcting Shaft Misalignment.   When alignment and shimming procedures are performed, the adjustments are made only to one machine. This machine is the driver unit. The machine not adjusted is usually the stationary machine or driven unit because of size or other physical constraints.

Newer alignment methods and tools make alignment relatively fast and easy. Alignment methods include:

- Reverse indicator
- Laser
- Optical
- Straightedge

Any alignment, even a straightedge alignment, is better than no alignment at all.

Before aligning equipment, perform a phase analysis to check for and eliminate *soft foot* and *bent shaft* conditions. Soft foot means that the machine's feet do not support the weight of the machine properly, much like a chair that has one short leg.

Make the initial alignment at ambient temperatures without pipe strain on the equipment. Recheck the alignment for pipe stress and thermal growth. Investigate and correct any significant changes in alignment.

Dynamic Balancing

Unbalance occurs when the center of the mass of a rotating system does not coincide with the center of rotation. An excessive mass on one side of the rotor results in an unbalance. The centrifugal force that acts on the heavy side exceeds the centrifugal force exerted on the heavy side by the unequal forces. The magnitude of the rotating speed vibration due to unbalance is directly proportional to the amount of unbalance. It can be caused by a number of things, including incorrect assembly, material buildup, or rotor sag.

An unbalanced rotor causes elevated vibrational levels and increased stress in the rotating element. Elevated vibrational levels in the rotor of an assembly affect the entire machine and cause excessive wear on the supporting structure, bearings, bushings, shafts, and gears.

An unbalanced condition can be in a *single plane* (*static unbalance*) or *multiple planes* (*couple unbalance*). The combination is called *dynamic unbalance* and results in a vector that rotates with the shaft and produces a once-per-revolution vibrational signature.

Dynamically balancing the unit:

- Extends the life of the bearings, bushings, shafts, and gears
- Reduces vibration to an acceptable level that will not accelerate equipment deterioration
- Reduces stress that causes equipment fatigue
- Minimizes audible noise, operator fatigue, and dissatisfaction
- Reduces energy losses

Identifying Unbalance. Unbalance needs to be distinguished from other sources of vibration before beginning any balancing procedure. A vibrational peak at or near the rotating speed of the rotor can have several causes, such as misalignment, a bent or cracked shaft, eccentricity, open rotor bars, or unbalance. Therefore, verify the presence of unbalance before proceeding with the balancing procedure. Analytical techniques—such as spectrum, waveform, or phase analysis—can isolate unbalance as the cause of vibration. Unbalance is characterized by:

- Dominant vibrational magnitude at the rotating speed of the rotor
- Highest vibration in the radial and vertical planes and lower vibrational levels in the axial plane
- An amplitude and phase angle of vibration that is repeatable and steady
- Radial versus vertical phase angle vibrational measurements of about 90°

REFERENCE

1. *ARI 700-95 Specification for Fluorocarbon and Other Refrigerants,* Air-Conditioning and Refrigeration Institute, 4301 North Fairfax Drive, Arlington, VA 22031.

ARTICLE 7.2.2
INSTRUMENTATION AND CONTROL SYSTEMS MAINTENANCE

Brian T. Herteen, Blake Stewart, Gregory Kurpiel, and William Hughes
Johnson Controls, Inc., Milwaukee, Wisconsin

THE STATE OF INSTRUMENTATION AND CONTROL SYSTEMS MAINTENANCE

In specialized industries, especially those that need FDA approval, maintaining control systems and instrumentation is a high priority. Poorly maintained controls in the pharmaceutical industry can mean losing FDA approval. Similarly, an inaccurate controls system in the food industry can cause problems with the FDA if product batches are contaminated. In the paper and pulp industry, as well as in the petroleum industry, control and instrumentation problems can lead to poor product quality. In each of these industries, the cost of problems due to improper control system or instrumentation maintenance is enormous.

This article discusses maintenance of controls systems and instrumentation in facilities in general. Highly regulated industries may have more stringent maintenance requirements than those presented here.

Today's building controls systems are quite reliable. Nevertheless, failing to maintain the systems and instrumentation can lead to wasted energy and reduced comfort. In the case of fire and security systems, poor maintenance can also lead to life safety issues.

MAINTAINING CONTROLS IN A FACILITIES MANAGEMENT SYSTEM (FMS)

Being able to sit at a single computer and control all interactions within a building is a great blessing. For building occupants, an FMS increases comfort and improves air quality. For building owners, it cuts energy and maintenance costs, reduces training, improves safety and security, and integrates reporting.

The Golden Rule

The facilities management database contains the most complete picture of a facility available. If there is a golden rule for maintaining a facilities management system, it is to back up the facilities management database. Any time that changes or upgrades to the system are planned, back up the database *before* proceeding with the work. Having an accurate backup on hand may save hundreds of hours of recreating work from scratch. The backup copy should be kept in a separate location from the primary database.

Replacing Pneumatics

The beauty of a modern FMS is its simplicity. With few exceptions, a control is working or it is not working. If it is not working, the board or the control itself can be replaced inexpensively. Pneumatic controllers are more complex. A change in temperature or air pressure can cause old pneumatic controls to go off calibration, which quickly erases energy savings. The maintenance costs are high, as well. Therefore, replace pneumatic controllers with direct digital controllers to reduce future maintenance headaches. Note that it is not necessary to replace the pneumatics in the mechanical system, such as those that open and close valves.

Maintaining the Central Workstation

The operator's console, head unit, or central workstation is at the heart of the FMS. Therefore, do not overlook maintenance for it. Check the monitor regularly for clarity, focus, and color. Make sure that there is no unusual bearing or motor noise. Clean the read/write heads of any removable drives, and clean all exterior surfaces. Finally, ensure that the system restarts properly and that it shows the correct system date, time, and hardware status.

In addition to maintaining the physical unit, follow these general rules to avoid problems with the workstation:

- *Avoid "playing."* Having power over an entire facility from a single graphic interface can be tempting. Unfortunately, changing building settings may have unintentional side effects. Change settings only when required and when the full effects of a change are known. Always back up programs before modifying them.
- *Avoid games.* Games or unnecessary programs on the central workstation pose several problems. *Storing* games limits the computer's storage capacity and can slow disk access.

Playing games reduces available memory and can severely affect the operation of the facilities management system.

- *Avoid Internet downloads.*   As with games, downloading information to the central workstation wastes hard disk space. It also puts the machine at risk of computer viruses. The best policy is to use the machine exclusively for monitoring the building system, which means having no unessential software on it.

- *Defragment the hard drive.*   A computer that is used daily will slow down without regular maintenance. Disk fragmentation creeps up slowly and causes longer file access times. To improve hard disk performance, run a defragment utility once per month. In the Windows® environment, the disk defragmenter utility is available.

- *Use a UPS.*   An uninterruptible power supply can protect the computer during a power failure, brownout, or power surge. During a short power failure, it eliminates the need to reboot the workstation. For longer power failures, the UPS can act as a parachute, allowing adequate time to shut down the computer properly.

Verifying What the System Says It Is Doing

Despite the accuracy of today's control systems, it pays to verify periodically that the system is doing what it says it is doing. For example, the system can signal that it has opened a particular damper. Maintenance means verifying that the damper actually opens when the control system tells it to.

Investigate reported abnormalities, even if the control system shows no problems. In many cases, the control system is operating correctly. For example, the control system may tell a damper to open, but a mechanical problem is keeping the damper closed. Verifying in person can provide a definitive answer.

Enlisting Trend Logs

Trend logs are useful features of a facilities management system. Trend logs record and display data over a specified time interval. They are especially helpful for maintenance because they reduce the need for dedicated personnel to monitor and record data. Setting trends on equipment allows you to see whether controls work as they should. Trend logs also provide a quick check against other control system indicators.

Maintaining Other Controllers

Application-specific controllers for air-handling units (AHUs), variable air volume boxes (VAVs), and unitary/heat pumps should be maintained twice per year. The first step is to ensure that each controller is controlled at the appropriate value. To check operation, change a set-point value and verify that there is a smooth transition and stable control at the new set point. Also, verify that controlled valves and dampers stroke fully in both directions and seal tightly where appropriate. If problems are present, other points associated with these units may need to be verified or calibrated.

Maintaining Fire System Controls

If a building control system is UL-approved for fire management, the fire control systems can be integrated with the FMS. This can improve building safety by permitting a total building response to a fire. For example, in response to a fire signal, the FMS can:

- Turn on lights on the floor to improve evacuation
- Turn on air-handling units on the floors above and below the floor with the fire to contain the egress of smoke
- Send audible evacuation messages based on the occupancy of the zone
- Send textual and graphical evacuation messages to public displays and information kiosks
- Unlock selected security doors to help with the evacuation
- Automatically print reports of occupied zones to manage evacuation

If the control system is UL-approved for fire management, maintenance tasks become even more critical. The fire control system should be checked at least monthly or more often if required by local codes or the NFPA.

Monthly Maintenance Tasks.   The following steps are general and may not provide complete coverage of all services:

- Notify the facility owner and the fire department or central station of the fire alarm test.
- Deactivate extinguishment and control interlock circuits.
- Initiate a lamp test, verify the proper test sequence, and return to normal indicators.
- Activate the fire alarm.
- Confirm the receipt of the alarm at all appropriate local and remote reporting stations.
- Reactivate extinguishment and control interlock circuits.
- Notify the owner of any problems uncovered during testing, as well as recommended corrective actions.

Semiannual Maintenance Tasks.   The following steps are general and may not provide complete coverage of all services:

- Deactivate the alarm signaling and control circuits.
- Inspect the panel interior for loose components and connections.
- Clean and vacuum the interior of the panel enclosure, and clean all exterior surfaces.
- Check the battery voltage, clean the terminals, and test it under load.
- Check for proper LED and LCD indications.
- Initiate a system test of all intelligent detectors, and record the results.
- Simulate alarms on each zone in conjunction with field device tests.
- Reactivate the alarm signaling and control circuits.

Ongoing Maintenance for Changing Buildings.   Building spaces are seldom static, so be sure that maintenance plans address modifications. For example, a conference room may be converted to office space. Doing so changes room sizes, adds walls, and may change occupancy hours. New room sizes and walls will change the airflow within the new offices. Changing occupancy patterns may affect the occupied/not occupied mode for the zone in which the offices are located.

ARTICLE 7.2.3
ELECTRICAL TESTING AND MAINTENANCE

Dana L. Green
Substation Test Company, Forestville, Maryland

A detailed listing of maintenance criteria and guidelines is beyond the scope of this article. An excellent overall reference for equipment descriptions, operational issues, and specific maintenance guidelines is Chap. 11, "Electrical Systems Operations and Maintenance," from Dr. Bernard Lewis' handbook.[1] Also refer to NFPA 70B for industry-recommended practices for electrical systems and equipment maintenance.[2]

A complete electrical testing and maintenance program is often developed by the joint efforts of in-house personnel and supporting contractors, and it begins with identifying the electrical apparatus to be maintained. To be effective, the program must identify the types of equipment to be serviced, their characteristics, the conditions that affect their use and environment, and must address all needs identified.

Electrical Safety

Electrical safety is an integral part of overall facility safety, and general facility safety concerns and rules apply to all electrical workers. However, all work actions include safety concerns peculiar to that particular type of work. Electrical maintenance work is certainly no exception to this rule. Although this is not a complete listing, these are a few important safety reminders.

Ladders should be made of nonconductive materials such as wood or fiberglass, although these materials contribute to weight. Wooden ladders should be sealed with a clear sealer. Temporary extension cords and drop lights should always include ground-fault protection. Lockout-tagout is more than an OSHA requirement; it may be a matter of life and death. It is necessary to adhere to sound lockout and equipment tagging procedures developed and performed by knowledgeable, well-trained individuals. All maintenance and operating personnel must understand and comply with the procedures implemented. Lockout-tagout is discussed in OSHA 1926.269(d).

Electrical switching to de-energize equipment and make it safe for hands-on maintenance or repair work includes recognizing potential hazards and including safety measures that are not ordinarily necessary during normal equipment operation. All potential sources of energy of any type must be recognized and considered.

Testing for dead is a procedure to determine if electrical equipment is de-energized and is performed before hands-on maintenance. An appropriate voltage detector, rated for the system voltage to be detected and used in accordance with the manufacturer's instructions, is necessary. Voltage detectors usually provide audible, visual, and/or a vibrating indication of energization, either by contact with or coming into proximity to an energized conductor. When using a voltage detector, first it is necessary to check the detector by intentionally testing a known energized line to prove that the detector is working. Then it is necessary to recheck the detector immediately after testing the intended de-energized equipment to ensure that the detector did not itself fail during the test of the equipment thought to be de-energized.

Temporary grounding of electrical equipment or conductors is often required after testing for dead but before hands-on maintenance of normally energized conductors. Temporary grounds must be placed between the source of electrical energy and the work site. Procedures must include the removal of any temporary grounds before locks and tags are removed.

Confined spaces are found throughout most plants and facilities. All workers must be trained to recognize and properly deal with confined spaces and permit-required spaces. Permitted confined spaces require completing an approved permit before entry. OSHA 1910.269(e) further describes confined space concerns and requirements.

Personal protective equipment might include eye and face protection, respiratory protection, head protection, work gloves, foot protection, hearing protection, body belts, and so forth, that are appropriate for the conditions present. Clothing should be of a less flammable material, such as wool or cotton, or materials specially made and maintained as fire resistant. Many popular synthetic materials that serve well under other conditions are extremely dangerous when exposed to fire or electrical arcing and may substantially contribute to the wearer's injury or death. Fire-retardant suits or coveralls are available for use by electrical personnel during switching operations.

Insulating rubber gloves, sleeves, blankets, mats, footwear, and disconnect (hot) sticks are available at most facilities. Proper care and use of the needed protective goods are vital to safety. Unfortunately, too often the user is uninformed, and the protective equipment itself is unsafe to use. Many facilities are not aware that all these items, including aerial lifts, require scheduled testing. Periodic electrical testing of protective rubber goods and insulating tools and apparatus should comply with the appropriate OSHA and ASTM test standards. Testing these protective goods requires special training and test equipment.

Insulation Resistance Measurements

Insulation resistance measurements are one of the most useful, commonly performed field tests and are applicable to most but not all low-, medium-, and high-voltage apparatus. These tests are reasonably easy to perform and are a very valuable indication of the condition of insulation systems or components. *Caution:* Do not attempt to perform this test with equipment energized under normal power or electrically charged. Insulation resistance is tested by applying a known direct-current (dc) voltage across the insulated system under test, by measuring the minute current flow, and by converting the readout to resistance. The test instrument has become known as a *megohmmeter* or *megger*.

Winding-Turns Ratio Measurements

These tests measure the ratio between windings of current or voltage transformers (i.e., the number of turns in one winding to that of the other winding that shares the same core). Therefore, the winding-turns ratio of a transformer with high-voltage windings rated 13,800 V and low-voltage windings rated 115 V is 120:1. A current transformer that has primary rating of 150 A and a secondary rating of 5 A has a ratio of 30:1. Winding-turns ratios are tested to ensure that intended transformer output with respect to input complies with applicable accuracy standards. These tests are performed when the transformer is installed and as a routine maintenance procedure. Record and compare results to previous measurements to determine if turn-to-turn winding short circuits (shorts) have developed.

Power Factor and Dielectric-Loss Measurements

The ac dielectric-loss and power-factor test is recognized as one of the single most effective methods for locating defective insulation. In the jargon of the electrical utility maintenance engineer, the dielectric-loss and power-factor test is often called the *Doble test* because of the extensive use and unique capabilities of Doble field test equipment and the orderly test methods that have been developed by the Doble Engineering Company in cooperation with its clients.

The basic principle of nondestructive tests such as dielectric loss and power factor is the detection of a change in the measurable characteristics of an insulation that can be related to the effects of moisture, heat, ionization (corona), lightning, and other destructive agents that reduce the breakdown strength of insulation.

Ohmmeter Readings of Closed Contacts

Ohmmeter readings can be performed on closed circuit breaker contacts. In the jargon of the electrical maintenance engineer, a common ohmmeter used for this purpose is the *ductor.* These readings are valuable in determining the condition of the contacts.

Thermographic Inspection

Nondestructive and noncontact thermographic inspection, also commonly known as *infrared scanning,* is another excellent tool. Modern thermographic cameras can detect temperature differentials within 0.2°C. A thermographic camera cannot "see" (read temperatures with accuracy) through metal covers, but it can detect hot joints, even through taped connections.

TYPES OF EQUIPMENT THAT REQUIRE PREVENTIVE MAINTENANCE AND TESTING

The following types of electrical equipment require regular preventive maintenance and testing by qualified professionals: switchboards, switchgear, busways, protective relays, ground-fault protection, ground-fault circuit interrupters (GFCI), low-voltage molded-case (insulated) circuit breakers, low-voltage air circuit breakers, medium-voltage breakers, air circuit breakers, vacuum breakers, oil breakers, switches, fuses, motor control centers and starters, batteries and battery chargers, transformers, voltage-regulating apparatus, wire and cable, rotating machines, surge arresters, grounding systems, and equipment grounding.

Cautions

Important cautionary rules of thumb are included here. By no means do they constitute a complete list, but they are included to raise the managers' awareness of issues that could cause hazardous conditions, and could decrease the reliability of the electrical power system, if not followed.

- Opening energized current transformer secondary circuits must be avoided to prevent possible introduction of hazardous high voltage to those circuits that will endanger personnel and equipment.
- The test technician must be fully knowledgeable of all control and protection circuits, devices, and schemes to be tested.
- Removing the cover of an in-service relay can result in inadvertent relay operation.
- All air circuit breakers operate at high speed and very forcefully. Additionally, most air circuit breakers include stored energy for opening and/or closing. It is necessary to understand the design of each particular mechanism and to block or discharge all stored energy safely before working on the breaker.
- Facility managers should carefully consider maintaining their own stock of all fuses, especially large power fuses and all medium-voltage fuses. In one instance, we worked around the clock to rebuild and make ready for service a faulted 15-kV switchgear assembly, only

to wait the next four days with the equipment de-energized while locating and obtaining three specific 15-kV fuses. Some large fuses may cost $1000 to $2000 each, and they may never be needed. But the cost of one otherwise unnecessary plant shutdown of 8 to 24 hours may make a once-in-30-years fuse inventory purchase a relatively small expense.

- Batteries emit highly explosive hydrogen gas when charging. Open flames and sparks must not be allowed in battery areas or rooms. Adequate ventilation is required. When working around electric storage batteries do not cause or allow short circuits or arcs to occur. Protect against battery acids by wearing protective gloves, goggles or face shields, and aprons.
- It is not safe to interrupt ground connections or neutral connections when the associated electrical system is energized.

Frequency of Performance.   How frequently various maintenance activities are required depends on the type of equipment, its characteristics, operating demands, environment, and any significant occurrences. Excess dirt, surface contaminants, heavy demand on moving parts, and unusual mechanical or electrical stresses might be causes for more frequent service. Industrial plants often place harsher demands on equipment than university or government usage imposes. Some contactors, breakers, and switches operate much less frequently than others, so wear patterns will differ.

All electrical apparatus should be visually inspected while in service at least weekly, if not daily. A worker with a trained ear should also listen for variations in the normal operating sounds.

Batteries require weekly inspection, monthly service, and annual complete service and testing.

All low-voltage (600-V class) draw-out style breakers and molded-case breakers larger than 400 A should be serviced and tested at least every three years, or even more frequently if required by the authority having jurisdiction or the insurer. Under harsher environments, yearly service may be necessary. Longer intervals may allow breakers with failing trip units to remain on-line unnecessarily, leaving the appearance of safety where none exists. Critical or life safety breakers should be tested annually.

Most low-voltage apparatus requires inspection, service, and testing at least every two to three years.

Regardless of voltage rating, contactors, breakers, and switches that are operated frequently to energize and de-energize loads may require annual or even more frequent service to monitor contact wear and other deterioration.

Medium-voltage insulation can deteriorate to flashover in much less than two to three years. Medium-voltage apparatus operates under greater electrical stress and can intentionally or undesirably deliver much more electrical energy. All medium-voltage apparatus, including protection and control devices, should be inspected, serviced, and tested annually. Insulating liquids should be drawn and tested annually.

In general, motor inspection and service should not always include significant disassembly, unless tests indicate a need to do so.

Personnel Qualifications.   Performance usually centers on the facility's electrical staff, whose knowledge of the electrical system and equipment at that facility is invaluable. These are the people who maintain daily operation, perform ongoing maintenance and repair, and carry out or oversee system and control modifications and additions. Depending upon their availability and technical training, much electrical maintenance work can be completed by in-house personnel. At the time of a major shutdown, however, the facility staff is often concerned with other downtime tasks, which limits available time for completing a thorough maintenance program. Some facilities employ electrical personnel who have in-depth knowledge of the mechanical measurements and clearance adjustments necessary and the ability to do good testing; however, many do not.

A reliable electrical contractor may well provide electricians who are trained and experienced in industrial apparatus and controls and who may provide strong support in many areas of shutdown maintenance, especially with regard to cleaning and tightening work.

Inspection, testing, calibration, and evaluation is best performed by a fully qualified electrical testing service company that provides in-house electrical engineering staff and certified test technicians. Such test companies routinely work on all makes and models of electrical apparatus to provide evaluating opinions and recommend actions that are completely unbiased by the manufacturer, supplier, or installing contractor. The testing company's in-house engineering staff often specializes in performing acceptance and maintenance inspections and tests, helps shape and technically define the program, reviews test procedures and data collected, and prepares the test report.

Test technicians who perform this work can be certified by the InterNational Electrical Testing Association (NETA) or by the National Institute for Certification in Engineering Technologies (NICET) in the field of electrical testing engineering technology. NETA and NICET are both nonprofit certification sources, independent of training institutions, equipment manufacturers, and installers.

CONTRIBUTORS

Daniel J. Caufield, Chief Electrician, Massachusetts General Hospital, Boston, Massachusetts

Edwin J. Fortini, Senior Electrical Engineer, Siemens-Westinghouse Technical Services, Inc., Needham, Massachusetts

David L. Stymiest, P.E., SASHE, C.E.M., Senior Consultant, Smith Seckman Reid, Inc., New Orleans, Louisiana

REFERENCES

1. B. T. Lewis, *Facility Manager's Operation and Maintenance Handbook,* McGraw-Hill, New York, 1999.
2. NFPA 70B. *Recommended Practice for Electrical Equipment Maintenance,* National Fire Protection Association, Quincy, MA, 1998.

ARTICLE 7.2.4

ELEVATOR/ESCALATOR/MOVING WALK MAINTENANCE

James W. Fortune, Chairman
Lerch, Bates & Associates, Inc., Littleton, Colorado

Like all electromechanical devices, elevators, escalators, and moving walks require an ongoing maintenance program to keep the units operating properly.

All major installing elevator companies also provide continuing maintenance programs that they offer to the vertical transportation (VT) equipment owner on contract. After the VT equipment is installed or modernized and during the one-year equipment warranty period, representatives of contractors usually contact the facility owners or facility managers and try to sell them continuing contracts for VT equipment maintenance. The best time to negotiate the ongoing maintenance price, number of work hours, and the contract terms is during the installation bidding process and *before* the equipment is actually installed in the facility. During this period, the facility owner is negotiating from a position of strength because it is almost a given that the installing VT equipment manufacturer/contractor will also become the continuing maintenance contractor. Today's microprocessor-based logic, motor, motion control, and proprietary software make it almost impossible for any maintenance contractor other than the installing manufacturer to provide this continuing maintenance service. This is the reason that we often recommend that prudent facility owners include the continuing maintenance contract form and have the potential VT equipment suppliers submit completed maintenance contracts with their equipment installation bids. The most profitable part of the elevator market is in maintaining the equipment, rather than manufacturing or installing it.

A proper VT equipment maintenance program includes equipment cleaning, lubrication, replacement of worn parts and components, software upgrades, system testing, and adjustments. It should also include annual and five-year ASME A17.1, code-mandated governor, safety, and buffer tests (system pressure tests on hydraulic elevators).

Virtually all of the major original equipment manufacturer (OEM) elevator/escalator companies also provide services to maintain their own equipment after installation. They usually hire International Union of Elevator Constructors (IUEC) personnel for this maintenance and typically train them in the proper maintenance procedures and operating standards for the particular brand of equipment being maintained. Then, maintenance mechanics are assigned to maintenance routes, and their responsibility is to see that the maintenance schedule and tasks are performed in a timely manner. A typical maintenance route can include up to 30 to 40 units in various locations. In larger facilities, it is not uncommon to have one or two maintenance mechanics assigned full time, 40 hours per week to that job or building.

Most elevator maintenance contractors employ IUEC maintenance mechanics and pay virtually the same wages and benefits. Maintenance labor represents about 80 percent of the cost of a typical maintenance contract. Therefore, if one contractor offers a substantially cheaper quote than another, obviously the number of labor hours allocated to the contract is being cut. Typical on the job labor content for a maintenance contract (exclusive of repairs and tests) should run about 1.5 to 2.0 hours/week/unit for gearless elevators, 1.0 hour/week for a geared elevator or escalator/moving walk, and about 0.5 to 0.75 hours/week for a hydraulic elevator or dumbwaiter.

Newer elevator microprocessor-based dispatch controls employ sophisticated, self-diagnostic strategies and fault reporting capabilities to assist maintenance personnel in diagnosing and correcting problems, sometimes before they actually occur! The latest technical innovations include remote elevator monitoring (REM) through a modem, the Internet, or radio-frequency (rf) links from the elevator controller to the maintaining company's central monitoring stations. This interactive communication is used to schedule maintenance tasks and personnel assignments based on actual equipment usage (much like changing the oil in your car after every 3000 miles). The REM system is also used to log faults and dispatch a mechanic in response to a trouble call and to assist trapped elevator riders by keeping them apprised of the status of the ongoing rescue efforts (a priority trouble call).

In certain facilities (health care facilities, for example) the VT equipment generally receives more use because of the critical nature of its mission. Special care should be taken in designing maintenance programs for these facilities.

CONTRIBUTORS

John W. Basch, Cini-Little Schachinger, LLC, Dagsboro, Delaware

Anand K. Seth, P.E., C.E.M., C.P.E. Director of Utilities and Engineering, Partners HealthCare System, Inc., Boston, Massachusetts

SECTION 7.3
FACILITIES CONDITION ASSESSMENTS

Harvey H. Kaiser, M. Architecture, R.A., Ph.D., President
HKK and Associates, Inc., Syracuse, New York

PURPOSES AND GOALS

The facilities audit is an essential tool for determining the condition of an organization's capital assets, including buildings, infrastructure, and fixed and movable major equipment. Organizations perform facilities audits for strategic planning associated with financial planning of capital assets, for cost allocations and analyses of the productivity of capital assets, and for strategic decisions on retaining or disposing of assets.

The audit has multiple purposes that include use by consultants and the facilities management staff of an organization to evaluate its buildings and infrastructure. Consultants retained to evaluate conditions for an owner can provide analyses that guide decision making for acquisition, renovations, or disposal. The audit provides a database that can determine the relative value of a capital asset by the effect of deterioration and other deficiencies that will require capital renewal for major systems and components, and corrections necessary to comply with life safety, environmental, and accessibility regulations.

An organization's facilities management staff responsible for maintenance, capital renewal, and capital budgeting will find the audit useful for establishing a baseline of conditions and for identifying specific deficiencies that need corrective action. Circumstances may differ between organizations that undertake a comprehensive survey of all facilities for the first time and those that have a limited set of goals for determining existing conditions. The basic principles and methodologies presented here can be used for all types and sizes of organizations, from a single structure to a facility consisting of multiple building complexes in dispersed locations. A continuous process of facilities audits beyond a one-time program provides up-to-date capital renewal priorities and can generate a significant portion of routine maintenance workloads. A related benefit is the development of a culture in which maintenance staff personnel do routine inspections and report facilities conditions. An effective audit program will extend the useful life of facilities, reduce disruptions in the use of space or equipment downtime, and improve facilities management relationships with facilities users.

The facilities condition assessment is a process of:

- Developing a database of existing building and infrastructural conditions by conducting a facilities audit
- Assessing plant conditions by analyzing the results of the audit
- Reporting and presenting the findings

The facility audit systematically and routinely identifies building and infrastructural deficiencies and functional performance of facilities by an inspection program and reporting of observations. The audit process assists maintenance management and the institution's decision makers by recommending actions for major maintenance, capital renewal, and deferred maintenance planning.

A well-designed audit has the following goals:

- Assessing the value of a property by determining the cost of corrective actions necessary because of the deterioration of systems and components and required to comply with life safety, environmental, and accessibility requirements
- Developing a baseline of the current condition of capital assets
- Identifying capital renewal and replacement projects to eliminate deferred maintenance
- Providing guidance for developing and prioritizing capital projects to correct observed deficiencies
- Forecasting future capital renewal needs for capital budgeting and scheduling
- Providing for a routine inspection process for all facilities
- Restoring functionally obsolete facilities to a usable condition
- Eliminating conditions that are either potentially damaging to property or present life safety hazards

Audit Process

A successful facilities audit program requires the support of senior financial and facilities management. Planning for an audit program should incorporate their review of the process and form of results to ensure that requirements are met for capital asset management planning and allocation of resources. Senior management's involvement in facilities audit planning can result in reliable sources of information on plant conditions and for determining funding needs.

Conducting a facilities audit requires a clear set of objectives before committing staff and financial resources. Whether an organization has previously conducted facilities audits or is beginning one for the first time, there should be thorough preparation to ensure understanding and support of all staff involved in the process. Management and inspectors must make a commitment to collect accurate data and to identify deficiencies as objectively as possible. Cost estimates to correct deficiencies should be based on current local data and priorities set by established criteria. A key ingredient to success is the assignment of an audit manager to oversee the process, whether it is conducted by in-house staff, by consultants, or by a combination of the two.

An audit can be comprehensive, collecting information on building and infrastructure components and functional performance of all facilities, or it can be limited to a condition inspection program, eliminating the functional performance evaluation. It can also be selective for specific components of building types such as roofs, or a unique collection of information for safety or new regulatory requirements. Coordinating an assessment of facilities conditions with functional performance is encouraged to gain a complete picture of funding requirements necessary to correct all deficiencies in a comprehensive capital improvement program. Integration of condition and functional needs provides greater credibility for defined improvement projects.

Audit results are used to plan routine and major maintenance, urgent and long-term measures to correct facilities conditions through a deferred maintenance reduction program, and capital renewal and replacement budgeting and planning. For example, audit data collection forms can be used to provide a description of conditions for a special facility type. The audit methodology can also be applied to a survey of one or more systems and/or components that demand information required to comply with new regulations or codes.

Condition Inspections.   Condition inspections are designed to provide a record of deficiencies and estimated costs to eliminate the deficiencies. The deficiency-cost methodology requires training inspectors for a "self-audit" (or giving clear instructions to consultants) to produce an objective and consistent database for future reference. Incorporated into the training is the development of a process of continually observing and reporting deficiencies, flowing results into maintenance work, and capital budgeting and planning. A successful audit program will introduce a culture of observing and reporting conditions as a regular part of the work of supervisory and tradespeople, not just on a one-time basis.

The condition inspection part of a facilities audit is a visual inspection of buildings and infrastructure systems and components. Figure 7.3-1 outlines the systems and components for building and infrastructure inspections. Primary systems include foundations, the structural system, the exterior wall system, and the roofing system. Secondary systems include interior work that make the facility usable, including ceilings, floors, interior walls and partitions, and specialties. Service systems include all operating systems, such as HVAC, plumbing, and electrical systems. Safety standards, including life safety and code compliance, are grouped together. Decisions on where to record data for a unique system or alternatives to the component definitions should be flexible and decided in the initial organization of the audit and instructions to inspectors. Major infrastructure components are listed in groups that can be inspected by appropriate specialized staff or consultants. These lists should also be reviewed and flexibility retained until a final list is adopted.

The design of the inspection forms and methodology is based on the way a building or infrastructure is constructed and how inspectors would logically proceed to make observations and collect data for deficiencies and costs of corrective measures. A building inspection begins with the way a structure is placed in the ground, then travels upward to structural framing, exterior wall enclosures, and roof, and then moves to the interiors. Each service system—heating, ventilating, air-conditioning, plumbing, electrical—is inspected separately. A comprehensive audit provides a complete inspection of architectural, civil and structural, mechanical, electrical, and safety components of each facility. Infrastructure inspections are conducted in a similar, methodical manner.

Infrastructure Evaluations.   The *infrastructure* is nonbuilding improvements that directly support a facility's operation. Included are central utility plans, utility distribution systems, roads, walks, parking surfaces, culverts, drainage and watering systems, and other structural site improvements. These often represent a substantial investment, and malfunction of infrastructure components can result in interruptions in normal operations, catastrophic failures that result in loss of life, and energy inefficiencies. Unlike buildings, infrastructure is generally not observable, that is, until a disruption occurs and immediate corrective action is necessary.

An infrastructure assessment methodology should provide an organization with a comprehensive overview of the current condition and code liabilities present, as well as a predictive model that projects the costs that result from the continuing degradation of these facilities. The desirable characteristics of a methodology are to maximize value by providing significant management capabilities and utilities information without the cost of extensive forensic analyses or extensive direct observation of inaccessible utility components. In a campus setting, a survey based on photogrammetric mapping provide by aerial photography and confirmed by field observations can supplement information available from construction documents and maintenance staff knowledge and experience.

A proposed infrastructure assessment methodology is based on a series of actions for implementation that can progress simultaneously:

- Initial data gathering on existing conditions in a standardized format of mapping and infrastructure categories
- Calculating the condition of nonobservable infrastructure components and their remaining life cycle based on industry standards for life cycles of each type of component, refined through collection and analysis of the organization's experience with regard to life expectancy of various infrastructure components
- Projecting the "expired" life cycle to determine capital renewal needs

The challenge of infrastructure condition assessment is that a significant portion of the infrastructure generally is not directly observable. Extensive forensic work is expensive and time-consuming and may not yield information more readily available from campus maintenance staff. Alternatively, a great deal of insight can be gained by analyzing readily accessible

Primary systems

1. Foundation and substructure
   Footings
   Grade beams
   Foundation walls
   Waterproofing and under-drain
   Insulation
   Slab on grade
2. Structural system
   Floor system
   Roof system
   Structural framing system
   Preengineered buildings
   Platforms and walkways
   Stairs
3. Exterior wall system
   Exterior walls
   Exterior windows
   Exterior doors and frames
   Entrances
   Chimneys and exhaust stacks
4. Roof system
   Roofing
   Insulation
   Flashings, expansion joints; roof hatches, smoke hatches, and skylights
   Gravel stops
   Gutters and downspouts

Secondary systems

5. Ceiling system
   Exposed structural systems
   Directly applied
   Suspended systems
6. Floor covering system
   Floor finishes
7. Interior wall and partition systems
   Interior walls
   Interior windows
   Interior doors and frames
   Hardware
   Toilet partitions
   Special openings: access panels, shutters, etc.
8. Specialties (Examples)
   Bathroom accessories

Kitchen equipment
Laboratory equipment
Projection screens
Signage
Telephone enclosures
Waste handling
Window coverings

Service systems

9. Heating, ventilating, and cooling
   Boilers
   Radiation
   Solar heating
   Ductwork and piping
   Fans
   Heat pumps
   Fan-coil units
   Air-handling units
   Packaged rooftop A/C units
   Packaged water chillers
   Cooling towers
   Computer room cooling
10. Plumbing system
    Piping, valves, and traps
    Controls
    Pumps
    Water storage
    Plumbing fixtures
    Drinking fountains
    Sprinkler systems
11. Electrical service
    Underground and overhead service
    Duct bank
    Conduits
    Cable trays
    Underfloor raceways
    Cables and bus ducts
    Switchgear
    Switchboards
    Substations
    Panelboards
    Transformers
12. Electrical lighting and power
    Lighting fixtures
    Wiring

(Continued)

FIGURE 7.3-1 Building component descriptions.

Motor controls
Motors
Safety switches
Telecommunications and data
Emergency/standby power
Baseboard electric heat
Lightning protection
13. Conveying systems
Dumbwaiters
Elevators
Escalators
Material handling systems
Moving stairs and walks
Pneumatic tube systems
Vertical conveyors
14. Other systems
Clock systems
Communications networks
Energy control systems[a]
Public address system
Satellite dishes, antennas
Security systems
Sound systems
TV systems

Safety standards
15. Safety standards
Asbestos
Code compliance
Egress: travel distance, exits, etc.
Fire ratings
Extinguishing and suppression
Detection and alarm systems
Disabled accessibility[b]
Emergency lighting
Hazardous/toxic material storage

Infrastructure component descriptions
1. Site work
Curbing and wheel stops
Fencing
Parking lots
Roads and drives
Walks, plazas, and malls
Water retention
2. Site improvements
Landscaping
Lighting
Furniture: benches, bike racks, bus stops and shelters, waste receptacles, kiosks, signage
Emergency telephones
Flag poles and stanchions
Sculpture, memorials, and fountains
3. Structures
Bridges
Culverts
Retaining walls
4. Utility Delivery Systems
Central energy plants
Chilled water distribution
Compressed air
Distilled water
Domestic water
Electrical distribution
Energy monitoring and control
Fire protection
Heating hot water
Irrigation water
Natural gas
Steam and condensate
Storm drainage
Sanitary sewage
Utility tunnel structures
Wastewater treatment and collection
Water treatment and distribution

[a] Energy audit.
[b] Disabled accessibility audit.

FIGURE 7.3-1   (Continued)

existing utility information. A preferred approach for both observable and nonobservable infrastructure components is based on available documentation, interviews with maintenance staff most familiar with the components, and field inspections. Maintenance staff information and campus records can be supplemented by forensic techniques for unique conditions, as necessary.

Functional Performance Evaluations.   Providing a comprehensive indication of a facility's capital needs with analyses of its functional characteristics and deficiencies are recommended. Analyses of the facility's condition and functional requirements should be integrated. Information on the condition deficiencies of systems and components that require corrective action should be complemented by an analysis of the functional performance aspects of a facility. Before summarizing the results of the condition inspection, the functional performance of a facility in terms of suitability and adequacy in satisfying the requirements of existing or planned activities should be integrated into a complete summary of the facility's capital needs.

For example, the condition inspections may reveal that a building has significant deficiencies, and the costs of corrective measures may exceed the replacement value. Nevertheless, for historic, aesthetic, or other reasons, the building may be retained for remodeling and extended use. Major renovations resulting from functional performance evaluation can include all identified priorities in a single improvement program. On the other hand, a facility may be considered for remodeling but demolition may be recommended because of conflicts with plans for future land use or sale as a source of revenues.

The functional performance evaluation, integrated with the results of a condition inspection, requires standards to guide the assessment. The specific nature of a facility will determine the content of condition standards for finishes, equipment, mechanical, electrical, lighting, and communications requirements. A complementary set of standards is necessary for functional performance evaluation.

Audit Phases

There are four phases in the facilities audit process: (1) designing the audit, (2) collecting the data, (3) summarizing the results, and (4) presenting the findings (see Fig. 7.3-2). In this systematic approach, the scope of the audit is first determined, the audit team selected, and the

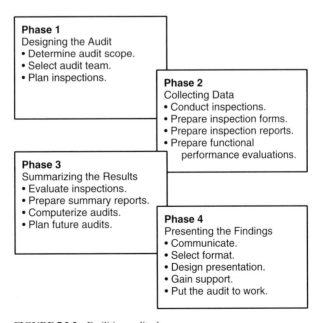

FIGURE 7.3-2   Facilities audit phases.

inspection planned. In the second phase of the audit, data is collected through inspection of buildings and infrastructure and functional performance evaluations. In phase three, the data collected are evaluated and summarized, priorities set, and future audits planned. In the final phase, results are presented to various audiences.

The audit is a method of collecting information on the current maintenance conditions and functional performance of a facility. Included in the audit are

- An inventory of facilities that provides descriptions of characteristics
- Inspections of existing buildings and infrastructure conditions
- Evaluations of functional performance
- Recommendations for correcting observed deficiencies

The inspection process is accomplished by a systematic inspection of buildings, infrastructure, and major fixed and movable equipment by systems and components following the sequence of construction. Functional performance is evaluated for future planning purposes, capital renewal, and replacements. The methodology can be implemented in a comprehensive review of all facilities and infrastructure systems and components or can be adapted to meet special conditions and needs of an organization.

Phase 1—Designing the Audit.   The first phase of the audit process—designing the audit—includes:

- Determining the audit scope
- Selecting the audit team
- Designating the audit manager
- Planning the inspection

Building and infrastructure inspections can represent a significant commitment of resources. The audit design should be prepared to adjust staff schedules for in-house inspectors and allow adequate time for managing consultants when they are used. Direct costs must be budgeted for reproducing plans, preparing building histories, laboratory testing, and producing other information. A cost-benefit analysis by in-house staff, consultants, or both should include a schedule for completion, costs (including staff downtime or replacement personnel), and intangibles of retention of inspection knowledge and introduction of a culture of regular inspections by maintenance staff. Thoroughness of preparation, including staff training, will ensure the usefulness of facility audit results.

An audit's scope is determined by goals and objectives for conducting condition inspections, methodology for inspections and assessment, the use of functional performance evaluations, and the intended use of audit results. The goals and objectives for conducting building and infrastructure inspections can include a broad overview of plant conditions and deficiencies, determining projects for major and minor maintenance, developing deferred maintenance reduction programs, and providing forecasts of future capital renewal and replacement.

The choice of specific facilities to be inspected depends on the audit's goals and is often influenced by deadlines, available resources (staff and budgets), building access, and seasonal conditions. In determining the scope of the audit, keep in mind that the inspection methodology produces quantitative information on building and infrastructure deficiencies and the costs of correcting them. Information obtained from inspections should be tailored to staff ability and the time to analyze results and implement corrections of observed deficiencies.

An inspection program should be designed to assess all facilities on a three-year cycle. Facilities with special concerns (for example, public assembly and unique structural design) should be inspected more frequently and should be coordinated with safety inspections and other mandatory inspection requirements (sprinklers, elevators, load testing, etc.).

The inspection is based on an exterior and/or surface observation of conditions. Invasive or other destructive methods, including disassembly of materials for testing, is warranted only where conditions indicate the need for further examination. Audit design should include the following:

- Define audit goals and objectives.
- Determine inspection methodology (in-house staff and/or consultants).
- Define the intended use of the results and report formats.
- Review the available information on the facilities to be inspected.
- Prepare a preliminary outline of facilities and components to be inspected.
- Establish deadlines, availability of staff and/or consultants, and access to facilities.

The methodology for inspection and assessment includes choosing the forms to be used and deciding whether to use staff or consultants or both. Defining the use of information and planning how to computerize the inspection results at an early stage of the process can avoid the need to redo work. The intended results should be defined and the final report formats designed to assure that adequate inspection data are collected and analyses can be prepared.

Thorough planning of the building and infrastructure component inspections is essential in producing accurate, timely, and useful results. There are several critical factors to be considered in planning inspections. These include:

- Scope of the audit
- Scheduling of inspections
- Responsibility for results
- Information requirements for inspection assignments
- Training
- Tools and equipment
- Notification to building occupants
- Emergency work

Phase 2—Collecting the Data.   The second phase of the audit is collecting data by building and infrastructure system and components and preparing the inspection reports. Collecting data for the facilities audit in the deficiency-cost methodology should use a standard inspection form that is completed for each of the suggested 15 building and 4 infrastructure components (see Fig. 7.3-1). The design of a standard form provides consistency for the inspection process, enabling modifications to the number or terminology of components.

Building and infrastructural inspections provide observations of conditions to be summarized on inspection report forms for individual components. A sample condition inspection form (see Fig. 7.3-3) contains three sections:

- Facility inspection data
- Component deficiency description
- Component deficiency evaluation

Use of the form provides a reference source for capital renewal planning and project management. Section 1 of the form identifies the background on the name of the facility and component, the inspector's name, and the date of inspection. Section 2 describes the deficiency and location and suggests a corrective action and the need for specialized testing. Section 3 provides additional information (including priority rating, suggested year of implementation, and cost estimates) that can be incorporated into the summary of projects for a capital improvement program.

The inspection thoroughly examines the component and documents all deficiencies that require corrective measures. In a deficiency-cost audit, the inspector has the responsibility for recording deficiencies, assigning a priority, and estimating the cost of corrective measures.

1. FACILITY INSPECTION DATA

Facility #: _____   Facility Name: _____

Component #: _____   Component Name: _____

Inspector: _____   Date: _____

2. COMPONENT DEFICIENCY DESCRIPTION

#1. _____

_____

#2 _____

_____

#3 _____

_____

#4 _____

_____

#5 _____

_____

#6 _____

_____

FIGURE 7.3-3 Sample condition inspection form.

3. COMPONENT EVALUATION

Project Title	Proj. #	Priority	Year	Project Estimate		
				Const $	"Soft" $	Total $

FIGURE 7.3-3   (Continued)

Inspectors use the component form as a guide in making field notes of observations that describe the deficiency and define corrective measures. Photographs or video recordings can supplement the inspector's field notes when preparing the report. Drawings of building floor plans or infrastructural installations should be noted to locate the deficiency. Each deficiency is assigned a sequential number corresponding to the written description, designating it on the drawing, and providing a reference for job planning and future audits. Any inaccurate information on the component description should be noted and corrected in the information base. Conditions observed by inspectors that require testing or specialized inspection skills, such as removal of material or inspecting hidden conditions, should be identified for follow-up work.

The inspection report form is prepared after fieldwork is completed. The inspection form is the basic information-collecting instrument that requires a simple process of organization before issuance to inspectors. The form is designed so that it can generate data for a variety of summaries and is easily adapted to simple word-processing or spreadsheet formats, or software database applications. Report preparation following fieldwork enables the inspector to compare notes with other members of the audit, request any specialized testing, and consult resources available for estimating the costs of corrective measures.

A project prioritization rating system is necessary for standardized and consistent capital planning and budget management. The reality of limited funding for capital projects requires that priority rating systems sort out the relative importance of each project, that is, when a project should occur. Projects given a high priority are those that entail a recognizable risk or significant benefits, for example, enhancing life safety; preventing loss of a component, a system, or an entire facility; meeting legal or code requirements; conserving energy; or saving other costs. Low-priority projects are those that are not time-sensitive.

A preferable approach is a rating system that has multiple uses for field inspections of facility conditions for both buildings and infrastructure and budget preparation for capital improvement programs. A suggested priority rating system is shown in Fig. 7.3-4.

A comprehensive facilities audit includes an evaluation of building functional performance that can be conducted simultaneously with a facility condition inspection. A *functional performance audit* of a facility is a survey of the programmatic capability and adequacy of interior building spaces or an entire facility. The goal of this audit component is to provide information that enables an evaluation of the adequacy and suitability of existing space to meet current and possible future uses. In combination with the facilities condition inspection, the functional performance audit can guide decisions on the appropriateness of an investment in a facility.

The functional performance evaluation form (see Fig. 7.3-5) provides a record of the characteristics of a facility with a choice of five ratings for programmatic adequacy (optimum, adequate, fair, poor, or unsatisfactory). These ratings can be assigned a numerical value for comparative analyses. Individuals in an organization who are knowledgeable about functional requirements should provide information to assist the inspectors in the evaluation. The completed form is used in the project prioritization process and can help decide whether a deferred maintenance or renovation project should go forward as a high priority or be delayed until program decisions are finalized. This sequence ensures that major maintenance and deferred maintenance backlog work will be appropriate for the projected use of a facility.

Priority 1: Immediate concerns (under 1 year).   Urgent needs to be completed within one year, such as correcting a safety problem, eliminating damaging deterioration, or complying with life safety and building electrical and environmental codes.

Priority 2: Short-term concerns (1–2 years).   Potentially urgent deficiencies that should be corrected in the near future to maintain the integrity of the building, including systems that are functioning improperly or not at all and problems which, if not addressed, will cause additional deterioration.

Priority 3: Long-term concerns (3–4 years).   Deficiencies that are not potentially urgent but which if deferred longer than from 3 to 4 years will affect the use of the facility or cause significant damage. Included are building and infrastructure systems and components that have exceeded their expected useful life but are still functioning.

Priority 4: Delayed projects.   Projects required or desirable to bring the facility to perform as it should, including systems upgrades and aesthetic issues. Included are projects that are not time-sensitive and can be delayed indefinitely (e.g., work to conform with codes instituted since the construction of the building and therefore grandfathered in their existing condition, and can be addressed in any major renovation effort; unresolved program use of a facility; a facility under consideration for removal from the building inventory, etc.).

FIGURE 7.3-4   Suggested priority rating system.

1. SPACE INVENTORY

Building #:   Building Name:

Gross Square Feet _____   Building Use: _____

Assignable Square Feet _____   Date: _____

Year Occupied _____

2. SYSTEM EVALUATION

	Yes	No	Comments
Finishes			
Equipment			
Mechanical, Electrical, Lighting Systems			
Communications Systems			
Safety Requirements			

3. SUITABILITY EVALUATION

	Suitable			Unsuitable		Comments
	A	B	C	D	E	
Suitability for Current Use						
Circulation						
Functional Relationships						
Conflicting Uses						
Code Conformance						
Disabled Accessibility						
Deferred Maintenance						
Other						

(Continued)

FIGURE 7.3-5 Sample functional performance evaluation form.

4. COMMENTS

5. OVERALL BUILDING /PROGRAM RATING

_____ (A)   Programmatically Optimum Space   _____ (D)   Programmatically Poor Space

_____ (B)   Programmatically Adequate Space   _____ (E)   Programmatically Unsatisfactory

                                                                         Space

_____ (C)   Programmatically Fair Space   _____

Prepared by: _____   Date: _____

FIGURE 7.3-5   (Continued)

Phase 3—Summarizing the Result.   The third phase of the audit assesses the results of the condition inspections, reviews their effectiveness, and develops summary reports of the inspection data. The audit manager reviews the completed inspection forms and plans or maps the indicated deficiencies and discusses the process with the inspectors. Any testing or laboratory work results are evaluated to confirm the recommended corrective measures and the accuracy of the cost estimates.

Thoroughness and consistency of inspections are important considerations for the audit manager in reviewing the results. A random selection of inspections and facility visits by the audit manager can confirm results or suggest reinspections. The subjectivity of the inspectors is a factor that should be considered in evaluating recommended corrective measures. The manager should suggest any needed improvements in inspector training programs.

The information base provided by the inspection reports can be the source for a variety of summary reports. Using data collected in an audit of buildings and infrastructural conditions, various report formats offer information on the overall condition of facilities, facilities types, individual facilities, systems, or components. Further sorting of data can identify projects by priority, cost ranges, use of in-house crafts or trades, or contractors. Reports can be designed for presentations for different purposes and are limited only by requirements determined by senior facilities and financial administrators.

Summary reports should be more than facts and figures. An executive summary should provide an overview and the highlights of the findings. A narrative should describe the audit process, objectives, and priority selection criteria. Conclusions should include a summary of overall facility conditions, an assessment of needs for capital renewal and deferred maintenance projects, and an assessment of the adequacy of operating budgets for maintenance. Audit summaries can be organized in many ways: by all facilities audited or individual facilities, by all building systems or individual building systems, by all building components or individual building components, by project costs, and by project schedules.

Comments obtained from the audit are also helpful in producing feasibility studies of changes in building use that can result in renovations or replacements. The inspections and reports of building and infrastructure conditions and the functional performance evaluation allow addressing several critical questions: Is the facility suitable for its current uses, or will it require remodeling? What is the actual cost compared to a new building, and is relocation to another building feasible and desirable? Is disposal (sale, lease, or demolition) a preferred alternative?

A comprehensive facilities audit provides data on deficiencies that can be used in several types of analysis. A method of measuring the relative condition of a single facility or group of facilities is useful in setting annual funding targets and the duration of deferred maintenance reduction. The facilities quality/condition index (FQCI) serves this purpose. The FQCI is the ratio of the cost of remedying facilities deficiencies to current replacement value (CRV):

$$FQCI = Deficiencies/\ Current\ Replacement\ Value$$

Deficiencies are the total dollar amount of existing maintenance repairs and replacements and functional deficiencies identified by a comprehensive facilities audit of buildings, fixed equipment, infrastructure, and a facilities functional performance. *Current replacement value* is the estimated cost of constructing a new facility containing an equal amount of space which is designed and equipped for the same use as the original building, meets the current commonly accepted standards of construction, and also complies with environmental and regulatory requirements.

The FQCI provides a readily available and valid indication of the relative condition of a single facility or group of facilities. It also enables comparing conditions with other facilities or groups of facilities. The higher the FQCI, the worse the conditions. Suggested ratings for comparative purposes are assigned FQCI ranges as follows:

FQCI range	Condition rating
Under .05 (5%)	Good
From .05 to .10 (5 to 10%)	Fair
More than .10 (10%)	Poor

In addition to data on current conditions and functional performance, the audit provides data to enable projecting capital renewal for ongoing facilities funding. Often absent from facilities management annual budgeting practices, capital renewal for facility components that deteriorate on varying life cycles require a pool of funds available annually to supplement annual maintenance budgets. A guideline for an annual capital renewal allowance, *in addition to operations and maintenance funding,* is 1.5 to 3.0 percent of a facility's current replacement value. Data available from the facilities audit and the calculated FQCI enables a scenario of various alternatives to address the reduction of capital renewal backlogs—deferred maintenance—by testing alternative funding levels and goals for an acceptable FQCI over differing time periods to reduce backlogs.

Phase 4—Presenting the Findings.   The final step in the facilities audit process is presenting the findings. A well-developed and well-presented facilities audit is one of the most valuable tools available for facilities management in performing its responsibilities. But even a flawlessly performed assessment of audit findings is useless unless the information can be communicated to audiences in a readily understandable format. Careful consideration should be given to the audience, its interests, knowledge of the subject, and the issues it faces as a decision-making group. Conclusions and recommendations stand on their own merits.

The documentation that a facilities audit provides will be of more than passing interest because of its effect on an organization's financial conditions. Thorough preparation is necessary in designing presentations to establish and maintain credibility for facilities management and the reliability of its information. Before beginning the audit process, consider an overall communications strategy that includes presentation formats and the potential audiences. The wide array of summaries available from the inspections should focus on priorities and costs, and supporting material should be keyed to concise presentations. If the report of facilities conditions is to be submitted in print form only, without oral presentation, consider what graphic material would be helpful. The report itself may be presented as a brief statement of facts with graphics or as an extensive narrative that includes background, description of methodology, findings, and conclusions.

In all cases, the chief facilities officer should provide material that is concise, easily understandable, and attractively presented. It should be free of jargon, confusing terms, or acronyms that are not self-explanatory. It should not be oversimplified for readability but should be designed so that the documents are readily cross-referenced.

Material should be developed in anticipation that the sharpest minds in the organization will receive the information. The documentation must be meticulous in detail and accuracy. Simple arithmetic errors and broad generalizations should be avoided by thoroughly checking financial data, priority selections, and cost-benefit analysis. Expect the unexpected. Be prepared to answer the question, "What will happen if we postpone or don't do the work at all?" Organize the presentation so that the train of thought can be followed. Above all, keep the presentation simple and to the point.

Adapting the Audit

Although the maintenance staff may collect and analyze some information to ensure the performance of buildings and infrastructure components, they do not typically collect planning data to assess long-term needs. The audit goes beyond a single snapshot of current conditions for planning capital renewal and eliminating deferred maintenance. Using the audit procedures systematically, the in-house staff or consultants produce a database of information on facilities that can be used as a baseline for future condition inspections. Facilities managers and/or consultants should incorporate the special characteristics of an organization and its facilities into an individualized facilities audit. The design of forms for recording inspections is adaptable to unique building and infrastructural systems and components not included in the illustrative forms (see Figs. 7.3-3 and 7.3-5).

By careful design and preparation of a facility audit, the facilities manager can predetermine the level of information to be obtained and ensure that the information gathered is appropriate for the projected application of the findings. However, selecting the approach must be driven by the nature of an organization's facilities, budgeting methods, and organizational structure. Formats for reporting the audit findings should be tailored to match the input requirements for the maintenance work order system and for the capital budgeting and planning process.

As the facilities manager and staff gain experience with the audit, they will recognize its potential for familiarization with the building and infrastructure condition, creation of an information source for maintenance planning, and restoration and maintenance of the physical and functional adequacy of facilities. The audit contributes to overall effectiveness of the facilities management organization by developing condition inspections as a routine part of operating activities.

Automated Applications

A desirable goal for effective facilities management is an information system that integrates space management with maintenance work planning, major maintenance, capital renewal, and deferred maintenance reduction projects. An interrelational database built on space characteristics has broad applications for facilities and financial management and other applications within an organization (see Fig. 7.3-6). The power of personal computers, commercially available services, and software products can integrate a space database with the facilities audit process of inspecting and reporting building and infrastructure conditions.

The facilities audit process and methodology are readily adaptable to automated applications through computer-aided facilities management (CAFM) techniques. The use of inter-

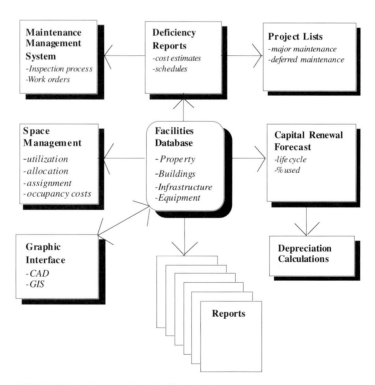

FIGURE 7.3-6   Interrelational facilities database.

relational database programs allows integrating the audit into a facilities management information system. Organizations have used internal resources to develop automated facilities audit programs with varying degrees of success. Consulting firms provide various options, including digitizing drawings, performing facilities audits, preparing condition assessments, and training staff in software and assessment techniques. Services and software products offer automated data collection and estimates of cost deficiencies, cross-referencing data to drawings, and graphic presentation of results. Evaluation of appropriate methodologies, techniques, and applicability of results is recommended in selecting CAFM assistance from consultants.

Automated applications of the facilities audit introduce a powerful tool for the facilities manager. Important benefits are readily retrievable information and capability for updating. Computerizing makes it easier to change the perspective of facilities management from collecting data for a one-time or occasional evaluation of conditions to updating facilities conditions continually. Thus, the assessment can become a useful operational tool for predicting needs and a valuable component in decision making for maintenance management and prioritizing capital improvement projects.

SUMMARY

Although some information may be collected and analyzed by the maintenance staff to ensure the working order of a facility, planning data needed to assess long-term needs are not typically included. The facilities audit process goes beyond maintenance planning. By following the audit procedures and using the suggested forms, an information database is created on facilities and infrastructure deficiencies and functional performance of a facility that is a baseline for future condition surveys and capital renewal planning and budgeting.

An interrelational database makes it easier to conduct special audits from the database. For example, surveys of hazardous materials, conditions of components (roofs, roads, etc.), or building or space types (housing, classrooms, etc.) are possible with less start-up effort and will contribute to expanding a database. Feasibility studies for acquisition, remodeling, or consideration of disposal of a facility can be based on the information provided by a facilities audit. The audit process provides a powerful tool for budget planning and allocation of resources and offers a method for regularly incorporating information obtained with a high level of confidence into facilities management decision making.

CHAPTER 8
CODES AND STANDARDS

David L. Stymiest, P.E., SASHE, C.E.M., Chapter Editor
Senior Consultant, Smith Seckman Reid, Inc., New Orleans, Louisiana

Any authoritative treatment of facilities engineering and management requires a discussion of the rules and regulations that apply to designing, constructing, operating, and maintaining commercial, institutional, and industrial buildings and facilities. These rules and regulations are most often promulgated as codes and standards. We present Chap. 8, "Codes and Standards," as the final chapter in Part 2, "Facilities Engineering." The content of this chapter applies, of course, to all types of buildings and facilities. Chapter 8 commences with an introduction to building codes, including a discussion of the history of codes. The treatment then presents access codes and consensus standards. Finally, the chapter concludes with a presentation on performance-based codes.

SECTION 8.1
OVERVIEW

**A. Vernon Woodworth, A.I.A., C.B.O.,
and Paul D. Sullivan, P.E., President**
Sullivan Code Group, Robert W. Sullivan, Inc., Boston, Massachusetts

When does repair work require a permit? Which codes apply? When does an existing building need to be accessible? These are just a few of the questions facilities managers must be prepared to answer regarding the impact of code requirements on their facilities. This chapter will serve as a reference for a general overview of codes and standards, providing as well some historical background and a glimpse into the future evolution of the world of codes and standards. First, some practical advice: all building owners and managers should familiarize themselves with the workings of their local building and fire departments. This includes determining which codes apply, what inspections are required and when, and who are the best people to contact when questions arise.

Although not covered in this chapter, facilities managers should also be aware of the zoning requirements that apply to their buildings. Zoning ordinances are adopted by local governments and vary widely in scope and nature. Typically, zoning ordinances regulate uses within a building, as well as any proposed additions. So, it is wise to verify that a use is allowed before proceeding with any lease arrangements.

Although the specifics of obtaining a permit also vary by jurisdiction, typically any new construction or replacement of existing construction (other than ordinary repairs) requires a building permit. Usually a licensed builder must be in charge of the work, and the services of a registered architect or engineer may be required, depending upon the scope of the work. Failure to secure a permit or to adhere to the terms of a permit's issuance can result in sanctions against the property owners, who are ultimately responsible for the code compliance of their properties.

Facilities managers cannot be expected to know all of the applicable code requirements for the properties that they manage. It is hoped that this chapter will provide a brief overview of the world of codes and standards to put each manager's individual experience into perspective.

FROM HAMMURABI TO THE ICC: A BRIEF HISTORY OF PRESCRIPTIVE CODES

The first known code of laws was compiled by King Hammurabi (1792–1750 B.C.), the founder of the Babylonian Empire. Hammurabi's code included specific provisions for buildings and set standards for construction and punishments for any failure to meet these standards. An example of the strictness of the Code of Hammurabi is found in the following provision: "In the case of collapse of a defective building, the architect is to be put to death if the owner is killed by accident; and the architect's son if the son of the owner loses his life."

Early Greek and Roman laws also contained provisions regulating construction. In general these laws sought to limit loss of life and property from building failure by requiring construction inspections, specifying the use of materials, and setting limits on building size.[1] In the New World, building regulations were enacted as early as the Plimouth Plantation in Massachusetts. In 1788, Wachovia, North Carolina, enacted regulations "which shall be basic for all construction in our community so that no one suffers damage or loss because of careless construction by his neighbor . . . ," a precedent for both zoning and building codes in the newly independent United States.[2]

The National Fire Protection Association's *Fire Protection Handbook* contains the following definition of a building code: "A building code is a law that sets forth minimum requirements for design and construction of buildings and structures. These minimum requirements, established to protect the health and safety of society, generally represent a compromise between optimum safety and economic feasibility."[3]

Building code requirements are often enacted in response to particular catastrophic events. This was the case in London after the great fires of 1664 and 1666, when thatched roofs were banned and detailed provisions and restrictions regarding other building materials were enacted. Restrictions on building materials were among the first laws enacted in colonial Boston in response to the threat of fire, and the requirement that exit doors swing out was introduced after 492 people died in 1942 in a fire at Boston's Cocoanut Grove nightclub. Code requirements that mandate building materials, direction of door swings, and other design features are called *prescriptive* codes, because the nature of construction is prescribed by the code. Typically, a prescriptive code includes requirements for fire-resistant construction, the number and location of exits, and fire-suppression and fire-detection systems, among others.

Although every jurisdiction (town, city, or state) could potentially adopt its own building code, the trend nationwide has been for adoption of statewide codes based on one of the *model codes*. The three model codes currently in widespread use in the United States are the *Uniform Building Code* (UBC), the *National Building Code* (NBC), and the *Standard Build-*

ing Code (SBC). The International Conference of Building Officials (ICBO) has been publishing the Uniform Building Code since 1927. The UBC has been adopted in many western states as the basis for statewide codes. The Building Officials and Code Administrators (BOCA) have been publishing a model code since 1950, now referred to as the *BOCA National Building Code*. Adoption of BOCA is common in the Midwest and Northeast. The Southern Building Code Congress International (SBCCI) first published the Standard Building Code in 1945. The SBC has been adopted in many southern states.[4] A complete list of current code adoptions and a graphic map are shown in Table 8.1-1 and Fig. 8.1-1, respectively. It is important to determine the specific code, edition, and amendments applicable to your buildings. This information is available from your local building department.

A dramatic step in the evolution of model codes is anticipated in the year 2000. The Council of American Building Officials (CABO), which was formed by ICBO, BOCA, and SBCCI in 1972, has formed the International Code Commission (ICC) to produce one model code to replace the three existing codes. To be known as the *International Building Code* (IBC), the first edition will be available for adoption in the year 2000 and is expected to gradually replace the existing model code adoptions nationwide.

THE GOALS OF A BUILDING CODE

A building code has two principal goals: structural integrity and life safety. *Structural integrity* is the outcome of design and construction that accommodates all anticipated loads within the properties of the building materials selected. Structural integrity during a fire is also a goal of the fire-resistance requirements of a building code. The structural elements of buildings must be able to support both anticipated daily loads and the occasional anomalous loading of high winds, fire damage, and earthquakes.

Life safety includes structural reliability, as well as safe egress. A building subjected to the anomalous conditions described before must have the structural ability to resist collapse for a minimum period of time to allow safe evacuation. In addition, fire-resistant construction contributes to the safety of an egress path or an area of refuge during a fire.

The two principal goals of structural reliability and life safety are based on a standard of expectation that is constantly evolving through an ongoing modification of code requirements. For instance, before the Cocoanut Grove fire, the direction of exit-door swing was not subject to code restrictions. Advances in engineering and in the understanding of earthquake forces are resulting in constant modifications to the structural sections of building codes in many areas. Technology is having an important impact on fire protection and notification, and the increasing adoption of *performance-based* code provisions is also contributing to an evolving code environment.

Because all codes are being amended on an ongoing basis, it can be difficult and confusing to keep track of the many requirements that affect your building or facility. The services of a registered professional (either an architect or engineer in the relevant discipline) are essential when construction is contemplated. There are also registered professionals who specialize in code consulting. This chapter will serve as an introduction to codes and their uses; however, confidence regarding a code's intention and application can be achieved only as the result of constant patient use, with professional guidance as needed.

Use and Occupancy

All building codes classify structures according to their occupancy. Because an office building is used differently from a nightclub or a hotel, for instance, different requirements are prescribed. Because many buildings accommodate more than one use, provisions are made for multiple occupancies, a condition often referred to as *mixed-use*. Under certain circum-

TABLE 8.1-1   Statewide Model Code Adoptions

State	Code	Year	Building	Mechanical	Plumbing	Electrical	Other	Comments
Alabama	SBC	1994	SBC	SMC	SPC	1996 NEC	State amendments	Voluntary
Alaska	UBC	1994	UBC	1991 UMC	UPC	1993 NEC	State amendments	Mandatory
Arizona	None				1994 UPC	1998 NEC	None	None
Arkansas	SBC	1991	SBC	SMC	SPC	1996 NEC	State amendments	Mandatory
California	UBC	1994	UBC	1991 UMC	UPC	1993 NEC	State amendments	Mandatory
Colorado	UBC	1991	UBC	UMC	UPC	1993 NEC	Hotels, multi-family	Mandatory
Connecticut	BOCA	1990/92	BOCA	BNMC	BNPC	1993 NEC	State amendments	Mandatory
Delaware	None					1996 NEC	NFPA 101—1997	Mandatory
District of Columbia	BOCA	1990	BOCA	BNMC	BNPC	1990 NEC	City amendments	Mandatory
Florida	SBC	1997	SBC	SMC	SPC	1993 NEC	Minimum standards	Mandatory
Georgia	SBC	1994	SBC	SMC	SPC	1996 NEC	State amendments	Mandatory
Hawaii	None						None	UBC voluntary
Idaho	UBC	1994	UBC	UMC	UPC	1996 NEC	State amendments	Mandatory
Illinois	None				State		State accessibility	BOCA voluntary
Indiana	UBC	1991	UBC	UMC	1990 BNPC	1993 NEC	State amendments	Mandatory
Iowa	UBC	1994	UBC	UMC	UPC	1996 NEC	State amendments	Mandatory
Kansas	UBC	1994	UBC	UMC	UPC	1993 NEC	State bldgs. only	Mandatory
Kentucky	BOCA	1996	BOCA	1993 BOCA	State	1996 NEC	State amendments	Mandatory
Louisiana	SBC	1991	SBC	SMC	SPC	1993 NEC	State amendments	Mandatory
Maine	None				State	1996 NEC	Minimum standards	Plumbing only
Maryland	BOCA	1996	BOCA	1996 IMC	1993 NSPC	1996 NEC	State amendments	Mandatory
Massachusetts	BOCA	1993	BOCA	BNMC	State	1996 NEC	State amendments	Mandatory
Michigan	BOCA	1993	BOCA	BNMC	BNPC	1993 NEC	State amendments	Mandatory
Minnesota	UBC	1994	UBC	UMC	State	1996 NEC	State amendments	Mandatory
Mississippi	SBC	1994	SBC	SMC	SPC	1993 NEC	State bldgs. only	Mandatory
Missouri	BOCA	1996	BOCA	1996 IMC	1995 IPC	1996 NEC	State bldgs. only	Mandatory

(Cont.)

TABLE 8.1-1 (*Continued*)

State	Code	Year	Building	Mechanical	Plumbing	Electrical	Other	Comments
Montana	UBC	1994	UBC	UMC	UPC	1996 NEC	State amendments	Mandatory
Nebraska	UBC	1991	UBC	None	None	1996 NEC	State bldgs. only	Mandatory
Nevada	UBC	1991	UBC	UMC	UPC	1993 NEC	State amendments	Mandatory
New Hampshire	BOCA	1996	BOCA	None	1993 BNPC	1996 NEC	State amendments	Mandatory
New Jersey	BOCA	1993	BOCA	BNMC	NSPC	1993 NEC	State amendments	Mandatory
New Mexico	UBC	1991	UBC	UMC	UPC	1996 NEC	State amendments	Mandatory
New York	State written					1993 NEC	State written	Mandatory
North Carolina	SBC	1994	SBC	SMC	SPC	1996 NEC	State amendments	Mandatory
North Dakota	UBC	1994	UBC	UMC	State	1996 NEC	State amendments	Mandatory
Ohio	BOCA	1993	BOCA	BNMC	State	1996 NEC	State amendments	Mandatory
Oklahoma	BOCA	1993	BOCA	BNMC	BNPC	1993 NEC	State bldgs. only	Mandatory
Oregon	UBC	1994	UBC	UMC	UPC	1996 NEC	State amendments	Mandatory
Pennsylvania	None						Fire & panic regs.	BOCA voluntary
Puerto Rico	Territory written							
Rhode Island	BOCA	1996	BOCA	1996 IMC	1995 IPC	1996 NEC	State amendments	Mandatory
South Carolina	SBC	1994	SBC	SMC	SPC	1993 NEC	State bldgs. only	Mandatory
South Dakota	UBC	1994	UBC	UMC	1996 NSPC	1996 NEC	State amendments	Mandatory
Tennessee	SBC	1994	SBC	SMC	SPC	1993 NEC	State amendments	Mandatory
Texas	None							UBC/SBC local
Utah	UBC	1994	UBC	UMC	1991 UPC	1996 NEC	State amendments	Mandatory
Vermont	BOCA	1987/88	BOCA	BNMC	1990 BNPC	1996 NEC	State amendments	Mandatory
Virginia	BOCA	1996	BOCA	1996 IMC	1995 IPC	1996 NEC	State amendments	Mandatory
Washington	UBC	1994	UBC	UMC	1991 UPC	1993 NEC	State amendments	Mandatory
West Virginia	BOCA	1996	BOCA	1996 IMC	1995 IPC	1003 NEC	State amendments	Mandatory
Wisconsin	State written					1996 NEC	State written	Mandatory
Wyoming	UBC	1994	UBC	UMC	None	1996 NEC	State amendments	Mandatory

Source: Kelly P. Reynolds & Associates, Inc.

ICBO Uniform

BOCA Basic/National

SBCCI Standard

state written

Information should be verified with state and local building departments. Many areas use different codes for other construction areas.

FIGURE 8.1-1   Statewide model code adoptions: ICBO Uniform, BOCA Basic/National, SBCCI Standard, and state written. Information should be verified with state or local building departments. Many areas use different codes for other construction areas. (*Courtesy of Kelly P. Reynolds & Associates, Inc.*)

stances, the different occupancies existing in one building are required by code to be separated by fire-rated construction. This condition is called *separated mixed-use.*

Types of Construction

All building codes also describe typical construction types and require that all structures be classified according to one of these types. In general, a construction type describes the fire-resistance qualities of the building's components, such as columns or bearing walls, floor and ceiling assemblies, and exit enclosures. Construction types range in fire resistance from standard wood-frame construction (least fire resistance) up to noncombustible materials such as concrete and masonry (greatest fire resistance). It is essential to know the construction type for a given building because alterations or additions to a building can unintentionally lower the building's assigned type if the wrong materials are used.

Fire-Rresistant Materials and Construction

Another important reason to know the occupancy and construction type of a given building relates to maintaining the fire-resistance ratings of the building's components. For instance, if the construction type or occupancy of a building requires that a corridor wall maintain a fire-resistance rating, it is important that alterations or renovations maintain this rating. Something as simple as installing an unrated door, removing self-closing hardware, or even just propping a door open can negate the intended protection of a fire rating.

When a partition is required to have a fire-resistance rating, any penetration through that partition is subject to code requirements. For instance, if ductwork is being installed, the code may require that a fire-rated damper be installed within the duct at the wall penetration. Fire stopping—that is, a material tested and rated for its fire-resistant qualities—is required around all penetrations of fire-rated walls or floor and ceiling assemblies. Fire stopping is particularly important in concealed cavities, such as within exterior walls, where fire can potentially spread from floor to floor.

Interior Finishes. Although construction type dictates the fire resistance of structural elements, another important factor in fire safety involves the nature of interior finish materials. Most fires typically involve furnishings and interior finishes for their initial fuel, and the smoke and gas released from the rapid combustion of these elements can cause loss of life without involving the structural building materials. For this reason, careful control of decorations and other flammable materials introduced into a building is an important aspect of all codes and a principal responsibility of a facilities manager.

Means of Egress. Because building construction cannot, of itself, guarantee occupant safety, an important aspect of any building code is the requirement for adequate and safe egress. Adequate egress is typically prescribed in terms of both *number* (separate means available for exiting the building) and *capacity* (such as the aggregate width of exit doors, corridors, and stairways). Safe egress involves adequate signage, lighting, hardware, and maintenance. Travel distance and occupant familiarity are two additional important factors in the safe evacuation of a structure under emergency conditions. Egress as a phenomenon combines construction, design, and human behavior to provide maximum safety for the building's occupants. Regular inspection to ensure that exits are not obstructed, exit enclosure doors are not being improperly blocked or held open, and emergency signs and lighting are installed and functioning properly are key to maintaining safe egress.

Alteration, Addition, and Change of Use of an Existing Structure

The provisions found in a prescriptive model building code apply principally to new construction. All codes recognize that existing buildings cannot be upgraded to the latest standards each time a new code provision is enacted. Although any new construction is typically required to meet the code for new construction to the fullest extent possible, model codes and local amendments attempt in various ways to set minimum standards for existing buildings without imposing hardships on property owners. For facilities managers, the provisions for existing buildings that apply in their areas could be the most important section of the building code to read and refer to on an ongoing basis.

Loss Prevention and Building Codes

As previously mentioned, the standards established by a building code apply principally to new construction. With certain exceptions, a property owner is generally not liable for conditions in an existing building that may not meet the standards of the code for new construction.

These exceptions generally include maintaining egress and other building systems. Because egress is not considered complete until a public way is reached, maintenance of exterior steps and walks, including the systematic removal of snow and ice, is often a specific building code requirement. Although a lease may specify tenant responsibilities, usually the authority that has jurisdiction (the building official) is required to cite the property owner when a code violation is identified.

SUMMARY

Building codes attempt to provide minimum standards for construction and life safety. Many building code provisions have been enacted in response to specific, often catastrophic, events. Codes are constantly evolving and often require professional experience to interpret properly. It is important for facilities managers to understand that any alterations or additions must comply with current codes, that egress must be maintained, and that the services of a registered professional within the appropriate specialization are necessary when construction or alteration is anticipated.

CONTRIBUTORS

Thomas J. McNicholas, Consultant, Retired Commissioner of Inspectional Services for the City of Boston, Massachusetts, Charlestown, Massachusetts

Jim King, Senior Fire Protection Specialist, Office of State Fire Marshal, Delaware Fire Service Center, Dover, Delaware

NOTES

1. D. F. Boring, J. C. Spence, and W. G. Wells, *Fire Protection Through Modern Building Codes,* American Iron and Steel Institute, Washington, DC, 1981, p. 1.
2. Ibid., p. 2.
3. NFPA, "Building and Fire Code Standards," in A. E. Cote (ed.), *Fire Protection Handbook,* 17th ed., National Fire Protection Association, Quincy, MA, 1991, pp. 6–141.
4. Ibid., pp. 6–142.

SECTION 8.2
ACCESS CODES

A. Vernon Woodworth, A.I.A., and Paul D. Sullivan, P.E., President
Sullivan Code Group, Robert W. Sullivan, Inc, Boston, Massachusetts

INTRODUCTION TO THE HISTORY, ENACTMENT, SCOPE, AND ENFORCEMENT OF ACCESS CODES

It is a fact that the built environments of buildings, transportation systems, even roads and walkways, have been principally constructed based upon the physical abilities of able-bodied individuals. Where does this leave people who have physical disabilities, such as the inability to walk, visual or hearing impairment, or lack of stamina or coordination?

Building features that facilitate use by people who have disabilities are described as providing accessibility. Although buildings have been part of human society since the Stone Age, the widespread application of accessible design is a very new idea that originated during the 1950s from research at the University of Illinois under a grant from the Easter Seals Research Foundation. This research that resulted in the first standard for accessibility was published by the American National Standards Institute (ANSI) in 1961 and revised in 1971. Early pioneers in accessibility research include Ed Steinfeld, of the State University of New York at Syracuse, and Tim Nugent, at the University of Illinois at Urbana. Nugent's work related specifically to physical barriers faced by returning disabled Vietnam veterans. These parallel researchers helped plant the seed for a revolution in designing and constructing buildings that resulted in a greatly expanded understanding of the interaction between human behavior, biomechanics, disabilities, and the built environment.

Early research looked for answers to questions that no one had asked before. What is the maximum slope of a ramp that a person in a wheelchair can negotiate comfortably and safely? How many pounds of force can a physically impaired person readily expend to open a door? What are the optimum heights for forward and side reach from a wheelchair? Applying this research data in the design of buildings and facilities challenged architects, builders, and owners to rethink their assumptions about the way people use buildings and to gradually recognize that the disabled have been excluded from participation in many aspects of society due to the barriers imposed by inaccessible building design.

The first Architectural Barriers Act, enacted in 1968, established general requirements for accessibility to federal buildings. In 1973, the federal government enacted legislation requiring that all new construction with government funds be accessible, referencing ANSI-71 as the standard for accessibility. This act also established the *Access Board,* an independent federal agency whose primary mission is ensuring accessibility for people with disabilities. Among the Access Board's initial mandates was the development of minimum accessibility standards for government buildings.

The Southern Building Code referenced the original ANSI standard in the 1970s, and the *BOCA Basic Building Code/1975* included a new section, 316.0, "Physically Handicapped and Aged." By the late 1970s, the Department of Housing and Urban Development (HUD) became the secretariat of the ANSI committee responsible for developing accessibility standards. In 1974, HUD contracted with Ed Steinfeld to prepare a revised standard that was issued in 1980 as ANSI-80.

Simultaneous with the efforts of HUD and the Access Board, the Department of Defense, the General Services Administration, and the U.S. Postal Service were also developing their own accessibility guidelines. With different agencies and organizations developing independent standards, conflict and confusion in accessible design was inevitable. The development of Uniform Accessibility Standards was an attempt to correct this situation. Under the acronym UFAS, these uniform standards were developed in coordination with the ANSI-sponsored standards and released in 1984. UFAS-84, along with subsequent amendments, is still enforced under the Architectural Barriers Act (ABA). Section 504, the Rehabilitation Act, requires that all programs conducted under federal guidelines must adhere to the UFAS standards.

The adoption of accessibility standards by individual states that had begun in the 1970s picked up momentum during the 1980s. In most cases, the ANSI standard was used as the basis of state accessibility codes. Simultaneous with this increased activity at the state level, a national movement to enact a "bill of rights" for mobility-impaired individuals was growing. This movement culminated in the 1990 enactment of the Americans with Disabilities Act (ADA). The ADA took many of its concepts from Sec. 504 of the Rehabilitation Act. Title III of the ADA covers public accommodations and services and incorporates accessibility guidelines for building and site design. These Americans with Disabilities Act Accessibility Guidelines (ADAAG), prepared by the U.S. Access Board, are usually referred to by their acronym.

Both the Department of Justice and the Department of Transportation adopted the Access Board's minimum guidelines contained in the ADAAG as standards. Because public services include transportation facilities, van shuttles operated by hotels, as well as the hotels themselves, were now subject to accessibility guidelines. The adoption of the Americans with Disabilities Act brought awareness of accessibility issues to a national level.

Currently a new level of coordination between the Architectural Barriers Act and the Americans with Disabilities Act is being developed. Although the scope and application of each act differ, an effort is being made to develop generic technical guidelines for both. These guidelines would then function like model codes with periodic updates. These revised guidelines require a thorough review and public-comment period, and it is expected that they will be enacted by 2001.

With the gradual coordination of standards and several decades of experience integrating accessibility into building and site design, the seed planted 50 years ago has taken root and created a new dimension in the landscape of the built environment. Accessibility is now an aspect of virtually all facilities and construction projects. The following subsections explore aspects of accessibility and their relevance for the facilities manager.

ANSI/ICC A117.1-98: ACCESSIBLE AND USABLE BUILDINGS AND FACILITIES

ANSI/ICC A117 is a standard, and, as such, does not constitute a code in and of itself. The A117 standard, however, forms the basis of most accessibility codes adopted by jurisdictions within the United States. Because it is a standard and not a code, the administrative authority that adopts it establishes the *scoping* or application of the standard. A117 provides technical criteria for making sites, facilities, buildings, and elements accessible, but does not specify how these criteria are to be applied.

A117 analyzes accessibility features in a rational sequence beginning with the "building blocks" of accessibility (changes in level, openings in ground surfaces, maneuverability and knee clearance for wheelchair users, etc.) through site features, building elements, transportation and communication elements, to built-in furnishings and dwelling units. Following are certain specific examples that are of general interest to facilities managers.

Doors

When a door or doorway is part of an accessible route, numerous features are required to maintain accessibility. These include minimum maneuvering clearances, threshold dimensions, door hardware, opening force and closing speed, and floor surface. Before a door is replaced, compliance with any applicable accessibility requirements must be ensured.

Stairs

Accessible stairs are required to have uniform treads and risers, a specific radius of curvature at the leading edge of the tread (nosing), continuous handrails at specific heights and clearances from adjacent walls, and continuous gripping surfaces uninterrupted by newel posts or other obstructions.

Plumbing Elements

Accessibility to plumbing elements incorporates specific dimensional criteria for drinking fountains, mirrors, coat hooks, water closets, toilet compartments, grab bars, toilet paper dispensers, urinals, lavatories, bathtubs, showers, and laundry equipment. Operable parts such as faucet controls must be operable with one hand without "tight grasping, pinching, or twisting of the wrist."[1]

Signage

For managers of existing buildings, accessible signage is a relatively inexpensive way to make a facility user-friendly to disabled individuals. The signage specifications of A117 encompass tactile characters (raised lettering), Braille, and pictograms (visual symbols such as the international symbol of accessibility).

Alarms

Life safety is a critical feature of accessibility. What good is an audible-only fire alarm to a deaf person? For this reason, the ANSI standard incorporates pulsing-strobe visual alarms in all accessible fire-alarm systems.

ADA ACCESSIBILITY GUIDELINES FOR BUILDINGS AND FACILITIES

Title III of the Americans with Disabilities Act prohibited "discrimination on the basis of disability by public accommodations." 28 CFR Part 36 was enacted by the federal government to implement the intent of Title III by establishing standards for the design, construction, and alteration of "places of public accommodation and commercial facilities."[2] 28 CFR Part 36 is made up of scoping and legislative provisions, general requirements, specific requirements, requirements for new construction and alterations, enforcement provisions, and certification requirements for local jurisdictions. The ADA Accessibility Guidelines for Buildings and Facilities, incorporated as App. A to Part 36, is at the heart of this legislation. Because the specific technical data contained in the guidelines is very similar to the ANSI/ICC A117.1-98 standard, this subsection focuses on the unique aspects of 28 CFR Part 36.

Application

The requirements of Part 36 apply to any public accommodation, commercial facility, or private entity "that offers examinations or courses related to applications, licensing, certification or credentialing for secondary or postsecondary education, professional, or trade purposes."[3] A public accommodation is defined as "a private entity that owns leases or operates"[4] a place that accommodates public use. Both the landlord who owns the building that houses a place of public accommodation and the tenant who owns or operates the place of public accommodation are subject to the requirements of Part 36. A commercial facility is defined as any facility "whose operations will affect commerce"[5] with certain exemptions, such as aircraft and certain railroad cars.

Alterations to a place of public accommodation or a commercial facility must be made "to ensure that, to the maximum extent feasible, the altered portions of the facility are readily accessible to and usable by individuals with disabilities, including individuals who use wheelchairs."[6] Because of the difficulties often encountered in existing buildings, Part 36 provides alternatives to barrier removal. To invoke this provision, a public accommodation must demonstrate that barrier removal is not readily achievable and must provide alternative methods to deliver goods or services (this provision is often entitled *appropriate accommodation*). Structural impracticability is specifically mentioned as an exception to the requirement for full compliance but "only in those rare circumstances when the unique characteristics of terrain prevent the incorporation of accessibility features."[7]

CONTRIBUTORS

Thomas J. McNicholas, Consultant, Retired Commissioner of Inspectional Services for the City of Boston, Massachusetts, Charlestown, Massachusetts

Jim King, Senior Fire Protection Specialist, Office of State Fire Marshal, Delaware Fire Service Center, Dover, Delaware

NOTES

1. ANSI/ICC A117.1-98, *Accessible and Usable Buildings and Facilities,* International Code Council, Falls Church, VA, 1998, Sec. 309.4.
2. 28 CFR Part 36, revised July 1, 1994, paragraph 36.101.
3. Ibid., paragraph 36.102.
4. Ibid., paragraph 36.104.
5. Ibid.
6. Ibid., paragraph 36.402.
7. Ibid., paragraph 36.401.

SECTION 8.3
STANDARDS

Kevin S. Hastings and Paul D. Sullivan, P.E., President
Sullivan Code Group, Robert W. Sullivan, Inc., Boston, Massachusetts

INTRODUCTION TO THE DEVELOPMENT, JURISDICTION, AND APPLICATION OF STANDARDS

Building, access, fire prevention, and other codes contain the primary criteria that regulate building design, but they often reference other standards to provide information that is more detailed for a particular building element. In general, standards either contain specific design requirements for a particular building system (e.g., sprinkler system, elevators, electrical equipment, etc.) or provide test procedures for a building system that is required by the code to achieve a minimum performance level (e.g., fire rating, structural capacity, etc.). Although standards provide a valuable reference source, they become legally adopted codes only when referenced by a building, fire-prevention, or other code within a particular region. For example, although ICC/ANSI A117.1, *Accessible and Usable Buildings and Facilities,* is a well-known standard containing requirements for designing buildings and facilities accessible to individuals with disabilities, it is not applicable within a particular region unless referenced by the building code or otherwise adopted within the region.

Some of the primary organizations involved in developing standards include the following:

American Concrete Institute (ACI)

American Institute of Steel Construction (AISC)

American Society of Heating, Refrigeration, and Air-Conditioning Engineers (ASHRAE)

American Society of Mechanical Engineers (ASME)

American Society for Testing and Materials (ASTM)

Factory Mutual Research Corporation (FMRC)

National Fire Protection Association (NFPA)

Underwriters Laboratories, Inc. (UL)

In addition to these organizations, the American National Standards Institute (ANSI) also plays a major role in developing standards. Although ANSI does not develop standards, it does act to regulate the development of standards by other organizations. Almost all standards are revised periodically to keep up to date with current technology and design practices. In fact, all standards certified by ANSI must be reviewed and reaffirmed, modified, or withdrawn every 5 years. Although standards are generally kept current, the particular editions referenced by building codes and other regulations may not be the most current because the revision schedule for the referenced standards may not coincide with that of the building code. Therefore, it is important to determine the correct edition of a standard for your particular region, because in some cases the latest edition may not have been adopted, and the earlier edition may have significantly different requirements.

The Development and Application of NFPA Standards

The mission of the National Fire Protection Association (NFPA) is to "reduce the burden of fire on the quality of life by advocating scientifically based consensus codes and standards, research, and education for fire and related safety issues." The NFPA's primary means of accomplishing this objective is by developing and maintaining more than 300 standards relevant to fire safety. The NFPA is the leading source of information for all building design and maintenance issues pertaining to fire, hazard, or life-safety protection. NFPA standards can be a valuable tool for a facilities manager to ensure that systems within the building are properly maintained and to provide a general understanding of the way a particular building system is designed to function.

NFPA standards cover a broad range of fire-safety topics, including all types of fire-protection systems design, hazardous materials storage and use, fire-test methods, and fire-fighting equipment design. In addition to providing design and testing criteria, a number of the NFPA standards also include maintenance schedules and information for fire-protection systems, means of egress, hazardous material storage, and other building elements. The following are some of the most common NFPA standards with respect to building design and maintenance:

NFPA 1: *Fire Prevention Code*

NFPA 13: *Installation of Sprinkler Systems*

NFPA 70: *National Electrical Code*

NFPA 72: *National Fire Alarm Code*

NFPA 101: *Life Safety Code*

Most NFPA standards are typically adopted by reference in a building or fire code, but a number of the standards such as NFPA 1, *Fire Prevention Code,* and NFPA 101, *Life Safety Code,* are commonly adopted as full codes within a particular region. The *Life Safety Code* is a unique document that addresses occupant safety by specifying design criteria for means of egress, fire-protection systems, and fire rating of building elements. Although the *Life Safety Code* is not a complete building code, it is often adopted and enforced in conjunction with the local building code. The NFPA also publishes fire-safety, maintenance, and procedural standards for a variety of specific types of buildings and facilities ranging from airports to recreational campgrounds. Even if they are not adopted by a local code, these standards can be a valuable resource for operating and maintaining a facility.

Fire-protection and life-safety systems should be tested and maintained in all types of facilities in accordance with the applicable NFPA standard, and in many areas the local codes require that these systems be maintained in accordance with a particular NFPA standard. Procedures and schedules for testing and maintaining sprinkler systems, fire-alarm systems, emergency lighting, and all other types of emergency systems and equipment are included in NFPA standards. Whether facility maintenance staff or an outside company tests emergency systems, the facility manager should maintain an accurate record of all testing and maintenance.

The Development and Application of ANSI Standards

The American National Standards Institute was founded in 1918 by a group of engineering societies and governmental agencies. ANSI was formed to serve as an administrator and coordinator of standards development to support the development of standards based on consensus, due process, and openness. The majority of the nationally recognized standards currently in use within the United States are accredited by ANSI. In addition to involvement with the development of standards, ANSI also acts to promote the use of standards developed in the United States throughout the world. As part of this effort,

ANSI was a founding member of the International Organization for Standardization (ISO) and continues to play a major role in the adoption of U.S. standards by the ISO as international standards.

All standards accredited by ANSI must follow developmental guidelines to ensure that the standard represents a consensus and that due process was followed in its development. To achieve consensus, ANSI requires that a standard represent substantial agreement among all interested and affected parties. The approval of the standard must represent more than a simple majority vote, and all objections to the standard must be considered, with a significant effort made to resolve them. Due process requires that all interested parties (individuals, organizations, companies, government agencies, etc.) be given the opportunity to participate in developing the standard. Parties involved in developing the standard must represent a balance of the interested parties and cannot be dominated by a single interest group. There can be no fee or required membership in an organization for parties involved in developing the standard, and adequate notice must be provided of any actions that affect the development of the standard.

Facilities managers should ensure that systems within their buildings are designed and maintained in accordance with nationally recognized standards. When encountering an unfamiliar standard and/or standard organization, facilities managers should look for ANSI accreditation to validate the document.

The Development and Application of ASTM Standards

The American Society for Testing and Materials was formed in the late 1800s and early 1900s. The society was primarily formed by representatives of the railroad and steel industries seeking uniformity in the quality of steel production within the United States. Most of its early work was focused on developing specifications and testing standards for steel. Eventually the ASTM expanded to include technical committees and standards in many other areas. In addition to testing standards for construction materials such as steel and concrete, the ASTM currently maintains standards covering a wide range of areas such as consumer products, environmental issues, and occupational safety.

In general, the standards developed by ASTM can be classified into one of the following six types of standards:

1. *Standard test method.*   A definitive procedure for identifying, measuring, and evaluating one or more qualities, characteristics, or properties of a material, product, system, or service that produces a test result.

2. *Standard specification.*   A precise statement of a set of requirements to be satisfied by a material, product, system, or service that also indicates the procedures for determining whether each of the requirements is satisfied.

3. *Standard practice.*   A definitive procedure for performing one or more specific operations or functions that does not produce a test result.

4. *Standard terminology.*   A document comprised of terms, definitions, descriptions of terms, explanations of symbols, abbreviations, or acronyms.

5. *Standard guide.*   A series of options or instructions that do not recommend a specific course of action.

6. *Standard classification.*   A systematic arrangement or division of materials, products, systems, or services into groups on the basis of similar characteristics such as origin, composition, properties, or use.

As is the case with all standards, compliance with ASTM standards is required only when they are legally adopted or referenced by a legally adopted code in a particular area.

CONTRIBUTORS

Thomas J. McNicholas, Consultant, Retired Commissioner of Inspectional Services for the City of Boston, Massachusetts, Charlestown, Massachusetts

Jim King, Senior Fire Protection Specialist, Office of State Fire Marshal, Delaware Fire Service Center, Dover, Delaware

SECTION 8.4
PERFORMANCE-BASED CODES

Paul D. Sullivan, P.E., President
Sullivan Code Group, Robert W. Sullivan, Inc., Boston, Massachusetts

INTRODUCTION TO PERFORMANCE-BASED DESIGN

In the past, the nature of code development has always been to require certain materials or specific design features, often in response to a major loss. Such *prescriptive* methods dictate significant aspects of a building's design and serve as inflexible mandates with which design professionals are required to comply. Because of technological advances in building design and evaluation, the performance of buildings and building systems subjected to adverse conditions can be predicted more accurately. These advances have provided design professionals with the ability to create building designs that provide an equivalent level of safety to the prescriptive code requirements, but do not comply with the letter of the code. In response to this, recent code developments have included increasing amounts of *performance-based* design criteria that establish certain minimum standards for the performance of building systems, along with consensual evaluation criteria. This performance-based code language allows professionals the opportunity to design building systems free of prescriptive mandates.

At present, the model codes within the United States contain primarily prescriptive code requirements. Other countries, however, such as Australia and New Zealand, have already developed and implemented performance-based building and fire-safety codes. Both the International Code Council and the National Fire Protection Association are expected to release performance-based model codes in 2000. The development of these codes represents a major advancement toward the adoption of performance-based codes within the United States.

In addition to providing design professionals with greater design flexibility and the ability to safely design structures that are not adequately addressed by prescriptive codes, in some cases performance-based codes will also require that facility managers basically understand the principles and design objectives involved in a performance-based design within their facilities. This basic knowledge about a particular performance-based design may be required to ensure that future renovations do not compromise the design or that special maintenance procedures associated with the design are followed.

CURRENT APPLICATIONS OF PERFORMANCE-BASED DESIGN

An example of performance criteria establishing code compliance can be found in the BOCA-1996 model code, describing smoke exhaust design that is applicable to all atriums more than two stories high: "The smoke control system shall be designed to keep the smoke layer interface above the highest of either the highest unprotected opening to adjoining spaces or six feet above the highest floor level of exit access open to the atrium for a period of 20 minutes"[1] (see Fig. 8.4-1). The BOCA commentary notes that, "The intent of this section . . . is that if the smoke layer is not lower than these points, smoke control has been achieved passively (i.e., without reliance on mechanical exhaust). If the smoke layer is lower than the required level, then mechanical exhaust must be introduced in an amount sufficient to keep the smoke layer above the required level." Before the adoption of this performance-based language, smoke control required by the BOCA code could be achieved only by mechanical means. (*Note:* The Standard Building Code, SBCII/97, also allows designers to provide their own solutions for controlling the migration of combustion products in an atrium.)

This example is a good illustration of two aspects of performance-based design: a greater involvement of the designer in life-safety issues and the opportunity to avoid expensive equipment or construction when adequate life safety can be otherwise provided. Through the

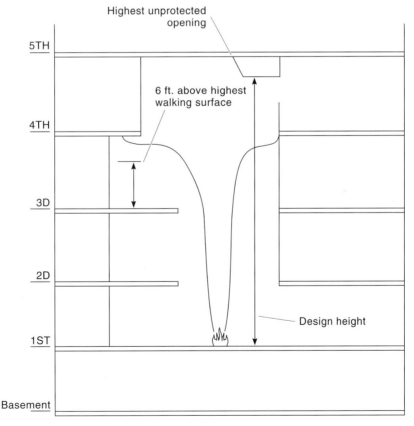

FIGURE 8.4-1   Atrium smoke control.

design of the atrium space, the architect can create conditions that eliminate the need for mechanical smoke exhaust. The increased involvement of the designer includes performing the calculations to demonstrate that the compliance criteria for the smoke layer have been met. A willingness to take responsibility for the technical dimension of performance-based design is the required price for regaining creative control over life safety in buildings. When the tools for designing performance-based solutions are understood and confidence in their results is established, this price will willingly be paid, for it will buy back a key role of the architectural profession.

Another example of performance-based design involves using computer modeling in supporting an alternative egress configuration for 4-story, multifamily, residential buildings. The BOCA National Building Code recognizes enclosed interior corridors and open exterior balconies, but it does not contain specific prescriptive code requirements for a *breezeway,* an interior corridor that is open to the exterior at both ends (see Fig. 8.4-2).

To evaluate the safety of various buildings incorporating breezeways in the means of egress, computer programs developed by the Building and Fire Research Laboratory at the United States National Institute of Standards and Technology (NIST) were used. Approximately 50 fire simulations were performed for each breezeway configuration at six different apartment complexes. This multiplicity and diversity of simulations yielded a sensitivity analysis indicating which variables had significant effects on the fire-modeling results. The simulations for each complex included four different fire types, four different fire locations, two different corridor configurations, and several different configurations of openings among the compartments.

Simulations focused on one scenario of a fire originating in a dwelling unit and another scenario of a fire originating in the breezeway. Three fire types were evaluated in the dwelling-unit fire scenario: a mattress fire in the bedroom, a couch fire in the living room, and a fast-growing arson fire in the living room. For the fire scenario with a fire in the breezeway,

FIGURE 8.4-2   Open breezeway.

an arson fire set with wood and a gallon of gasoline was evaluated. Of these two fire scenarios, the arson fire in the breezeway had the more adverse impact because of its location in the path of egress.

Smoke production and location for these breezeway scenarios were compared with those for enclosed corridor designs. The computer simulations demonstrated that the upper smoke layer descends to the breathing height of an average standing adult (approximately 5.7 ft) in an enclosed corridor before it reaches this point in a breezeway design.

This detailed analysis demonstrated that open breezeways are safer than enclosed corridors due to the opportunity for smoke dissipation. As a result, builders could provide a design with proven market appeal and avoid the significant additional expenses associated with enclosed, conditioned corridors.

PERFORMANCE-BASED CODES: PAST, PRESENT, AND FUTURE

Prescriptive codes have often been called a cookbook approach to building design: prescribed ingredients in predetermined amounts are required to achieve the desired outcome. Following this analogy, a performance-based code recognizes the creative abilities and technical expertise of the chefs (architect and/or engineer) and allows them to throw away the cookbook and create an original recipe.

Chefs can experiment with ingredients and cooking methods until the result meets their expectations. An architect does not have the same luxury, which is a major reason that rated assemblies and prescribed materials have become central to the building industry. However, the advent of computer modeling as a building design tool now allows architects to preview life-safety features and predict building performance under various hypothetical scenarios. Therefore, computer models can revolutionize the code-compliance process by providing a scientific basis for developing and reviewing performance-based designs.

Computer modeling is still new to the design and construction industry, yet the potential for this design tool to revolutionize the way life safety is incorporated into the built environment is already evident. By understanding the applications of computer fire modeling in performance-based design, architects can regain a decision-making role in the life-safety dimension of building design while offering important new cost-saving opportunities to their clients. To avoid mechanical ventilation for smoke evacuation in an atrium, for instance, a computer-modeling program can predict smoke production for various fire scenarios and determine the potential impact on evacuation times for the atrium occupants. Variables, such as the volume and configuration of the atrium space, the location and response time of sprinkler heads, and the ceiling heights of occupied spaces, can be adjusted until the *smoke layer interface,* that is, the lowest point of smoke accumulation, remains above the required level for the time necessary for occupant safety.

Who decides which set of variables is valid for a computer-based fire-hazard analysis of a complex scenario such as this atrium example? Some variables, such as fire size, are prescribed in the model codes, whereas others must be supported by referenced literature and engineering judgment (preferably an *authoritative consensus document*) and approved by the authority having jurisdiction. Conservative variables should always be used. For instance, although most fires are suppressed by sprinkler activation, the modeler of fire events generally assumes a steady-state fire, with an ongoing constant rate of heat release, even after the sprinklers have activated. The travel time for evacuating occupants is input by the modeler at a rate of speed less than the average walking pace of an able-bodied adult. Because the actual conditions of every different life-threatening scenario cannot be predicted, conservative assumptions and criteria provide a safety factor analogous to the equivalent factor used in determining allowable structural loads.

CONTRIBUTORS

Thomas J. McNicholas, Consultant, Retired Commissioner of Inspectional Services for the City of Boston, Massachusetts, Charlestown, Massachusetts

Jim King, Senior Fire Protection Specialist, Office of State Fire Marshal, Delaware Fire Service Center, Dover, Delaware

NOTE

1. BOCA, *National Building Code,* Building Officials and Code Administrators International, Country Club Hills, IL, 1996, Sec. 922.2.

P · A · R · T · 3

FACILITIES: BUILDINGS AND COMPLEXES

Paul R. Smith, P.M.P., P.E., M.B.A., Chapter Editor
Peak Leadership Group, Boston, Massachusetts

Anand K. Seth, P.E., C.E.M., Chapter Editor
Partners HealthCare System, Inc., Boston, Massachusetts

Roger P. Wessel, P.E., Chapter Editor
RPW Technologies, Inc., West Newton, Massachusetts

David L. Stymiest, P.E., SASHE, C.E.M., Chapter Editor
Smith Seckman Reid, Inc., New Orleans, Louisiana

William L. Porter, Ph.D., F.A.I.A., Chapter Editor
Massachusetts Institute of Technology, Cambridge, Massachusetts

Mark W. Neitlich, M.B.A., Owner, CEO, and Chief Engineer of Chemical Manufacturing (Retired), Chapter Editor
New Haven, Connecticut

Whereas Part 2 addresses facilities planning, programming, and engineering for common building types, Part 3 addresses the specific planning, programming and engineering features for many of the types of buildings and facilities that have unique needs. We believe that Part 3 addresses most of the building and facility types in use today: health care facilities in Chap. 9; laboratory facilities in Chap. 10; industrial and manufacturing facilities in Chap. 11; educational institutions and universities in Chap. 12; airports, government installations, and prisons in Chap. 13; and data center facilities in Chap. 14. Parts 2 and 3 should be taken together, with the reader referring to Part 2 for common facilities planning, programming and engineering issues, and then to Part 3 for issues unique to each type of usage.

CHAPTER 9
HEALTH CARE FACILITIES

Anand K. Seth, P.E., C.E.M., C.P.E., Chapter Editor
Partners HealthCare System, Inc., Boston, Massachusetts

David L. Stymiest, P.E., SASHE, C.E.M., Chapter Editor
Smith Seckman Reid, Inc., New Orleans, Louisiana

Health care facilities are among the most complex types of facilities. People who work within their walls provide care to the sick and preventive medicine to the healthy. Today there is a virtual explosion in health care technology that places unprecedented demands on these facilities. This chapter covers the full gamut, including planning and programming, process design, special systems and needs (e.g., medical waste transportation systems), and traditional facilities issues such as structural, mechanical, and electrical systems.

Important aspects of ongoing operations are also discussed in the maintenance, environmental management, and utilities management articles. The electrical utility management program, a vital aspect of ongoing operations, is discussed in detail.

SECTION 9.1
MANAGING THE PLANNING AND DESIGN PROCESS

David Hanitchak, R.A.
*Director of Planning and Construction, Massachusetts General Hospital,
Partners HealthCare System, Inc., Boston, Massachusetts*

DEMANDS ON HEALTH-CARE FACILITY MANAGERS HAVE BECOME INCREASINGLY COMPLEX

Traditionally, health care facility management is part of the service operations of an organization, managed along with dietary, security, and information systems services, and other service providers. Responsibilities vary widely, but managers generally direct facility and infrastructural operations and projects of varying sizes and ensure that environmental safety, code, and regulatory requirements are met. They may have planning, design, and project or construction management staff, with responsibility for data collection, analysis, and reporting, and may be responsible for capital financial management as well. Larger institutions may divide roles into separate real estate, planning, construction, engineering, and buildings and grounds departments to tap specific professional expertise and focus on specific organizational needs. Changes in the way health care facilities are managed, financed, and controlled, as well as how health care is delivered, are changing the mix and emphasis of the facility manager's responsibilities.

REORGANIZATION OF HEALTH-CARE DELIVERY CONTINUES

Continued turbulence in the way health care is organized, controlled, financed, and regulated places a premium on facility flexibility and adaptability. Market changes in penetration of managed care and capitation, in operations control and organization, and in business practice and management techniques all contribute to new complexities in funding and operating facilities. Functional demands resulting from changes in the way medicine is practiced—with new therapies, translational research, and new technologies in equipment, imaging, information systems, and pharmaceuticals—affect facility requirements and flexibility. Regulatory, code, and legislative issues such as the 1997 Balanced Budget Act, Health Insurance Portability and Accountability Act of 1996 (HIPPA), state determinations of need, departments of public health, and the Joint Commission on Accreditation of Healthcare Organizations contribute to the complexity of requirements. Even patients are changing, with increased age, increased acuity, and increases in outpatient care and the need for rehabilitation services. The doctors' central controlling role is challenged by increased media interest, direct marketing of drugs and therapies to patients, Web access to medical information, and knowledge and mechanisms of accountability. All of these changes make the provision and management of facilities more complex.

HEALTH CARE FACILITY MANAGER'S ROLE MUST BROADEN

Health care facilities are increasingly obsolete: wrong type, wrong place, and older stock of buildings at the same time that competition for capital by information systems, medical technologies, and programs has increased. Deferred maintenance has increased because of the reduced availability of capital. The metabolic rate of buildings is slower than changes in health care—building decisions have great inertia; in the time it takes to design and construct a building, the new facility can easily become obsolete.

One way to approach better integration of real estate and facilities with the mission of the organization is by tighter integration of facility management with strategic vision and operational and program planning. Master planning and maintaining flexibility of facilities to meet changing demand can help the institution to prioritize resource allocation, ensure the highest and best use of facilities, and provide decision support through strategic facility planning. As hospitals are increasingly valued for their efficient delivery of quality care rather than for the availability of beds, real estate must become a strategic asset to be managed, rather than the corpus of the institution.

KNOWLEDGE AND EXPERTISE FROM OTHER FIELDS IS COMING TO HEALTH CARE

Managers may be well served to understand how other industries approach real estate. Knowledge and application of benchmarking, use of metrics and their analysis, development of functional and performance standards, and development of mechanisms for project- and project-management accountability become important expectations of management, and fiduciaries become more sophisticated. Real estate expertise is no longer enough to manage health care facilities.

The following articles outline approaches to the planning and design of health care facilities. Context is important. Without a federal health care policy, states and providers define issues and priorities that result in wide regional differences. A facility manager must know the core business of health care, understand and help interpret the national and local regulations and market forces that affect the facility, and become a member of the strategic planning team. Needs, goals, and missions of different types of health care facilities differ widely. One size does not fit all for the clinic, rehab facility, community hospital, or quaternary academic medical center.

ARTICLE 9.1.1
PLANNING AND PROGRAMMING

Rick Gibson, A.I.A., Partner
Taylor & Partners, Boston, Massachusetts

INTRODUCTION

Advances in health care that improve the precision of care and minimize its impact on the patient, and increasing pressures on health care organizations to improve their performance, are powerful forces that shape the way health care is provided. These forces move diagnostic and treatment procedures from the inpatient setting to ambulatory settings, and they keep challenging decision makers who are faced with positioning their organizations and facilities for the future. Although reimbursement, regulation, and competition drive the process, planning for the delivery of health care is about human needs and the prudent use of resources to meet those needs.

As health care organizations strive to improve performance, comprehensive plans for the array of services offered and their deployment are necessary today to position these organizations for the future. The planning process is the consideration of opportunities (and the ensuing rigorous analysis) that form the framework for decision making to implement its initiatives.

Preparing for the future is a continuous process of anticipating needs, forming strategies, allocating resources, implementing solutions, and monitoring the results, whether on a broadband enterprise-wide basis or targeted at specific services where opportunities for substantial improvements in services and operations are identified.

This article presents an overview of the key issues in the planning process for health care organizations, briefly highlighting the significant factors considered in strategic planning, program planning, space programming, and master facility planning (Fig. 9.1.1-1).

STRATEGIC PLAN

The strategic plan is the formulation of the organization's mission and tactical initiatives to strengthen its performance. The strategic planning process determines the array of health

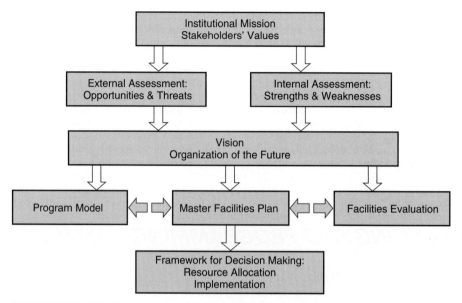

FIGURE 9.1.1-1   The planning process. (*Courtesy of Taylor & Partners.*)

care services necessary to meet the needs of the population served and specific initiatives for configuring them to ensure high quality and low cost.

The first step is determining the regional market for the organization's mission. The organization's constituencies create its mission and influence the scope of its services. The mission becomes broad and complex when there is a research or teaching mission in addition to regional patient care.

Strategic Issues

Regional demographics are analyzed to establish the population profile and to understand trends that impact its size, concentration, and makeup. Using detailed U.S. census and state health department data, planners can identify regional trends in the age and distribution of the population to establish the basis for the total volume of health care services that the regional population will generate.

The services provided by the organization and its competitors within the region are analyzed to determine how well the population's overall health care needs are met and whether there are unmet needs for certain services. This review is strongly influenced by complex access issues. The region's geography and the locations of other providers are simple determinants in understanding how far the population will travel to access urgent care services. The structures of health insurance plans and reimbursement are a more complex determinant of access.

The organization's role in providing its specific services within the overall continuum of services needed by the region varies if the organization is part of a coordinated network of health care providers or is a stand-alone provider. Whether in a stand-alone or networked context, the organization's planning process identifies overall regional needs for health care services and establishes the segments of the needs that the organization serves.

The external assessment of the organization should consider both quantitative and qualitative measures of the organization's performance in fulfilling its mission. The measures include market share, competition, and the community's perception of the organization's performance. The organization's strengths will become apparent immediately, but its weaknesses will become troublesome threats that result either in continuing diminished performance or in an opportunity for a competitor.

The internal assessment establishes a baseline from present operations to measure the quality and cost of care provided. Measurement of the cost of operations is quantitative, and its largest component is staffing. Measurement of quality is a qualitative process that assesses the perspectives of the organization's constituencies and consumers.

Vision—Organization of the Future

The strategic-planning team establishes the organization's objectives for providing services to meet the region's needs and sets goals for the organization's performance. The organization of the future is part of an integrated delivery system that involves many entities that together provide the entire continuum of health care services needed in the region. Strategic planners rationalize the delivery of services to eliminate overlaps and to fill gaps with health care services provided in the most appropriate setting, at the best quality, and the lowest cost. The goal of the strategic plan is a clear and meaningful vision of the organization at a specific interval in the future and a set of guidelines and tactics for reaching that vision (Fig. 9.1.1-2).

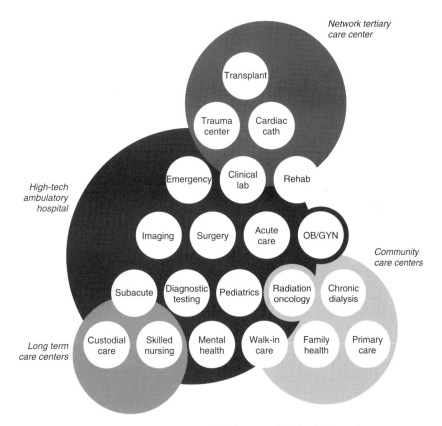

FIGURE 9.1.1-2   The network program model. (*Courtesy of Taylor & Partners.*)

Framework for Decision Making

The strategic plan provides short-range and long-range initiatives for improvement within the context of continuous improvement of services and operations. Although the strategic plan's view of the future is usually a 5- to 10-year period, the strategic plan is a framework for decision making and should be updated continuously to address new opportunities and threats.

HEALTH CARE PROGRAM PLANNING

Program Model

Determining the program model is the first step in translating the broad goals of the strategic plan into specific initiatives. Having assessed overall regional health issues, planners formulate the array of health care services needed and the context for each service as part of an integrated delivery system. Documentation of the organization's present services forms the baseline for considering alternatives for improvement and expansion.

Alternative program models are considered to establish the optimal array and deployment of services for the organization's regional market and its position either as a stand-alone provider or as a member of a networked delivery system. The organization's mission for primary, acute, and specialty care and its role in the continuum from primary care to tertiary care influence the selection of its program model (Fig. 9.1.1-3).

The analysis of alternative program models includes adding new services to address unmet needs identified by the strategic plan and reconfiguring present services to improve quality and access or to reduce cost.

Patient Flow Modeling

Patient flow models track all aspects of the patient's encounter with the organization's services and processes. For each clinical service, planners construct flow charts that show each step from the initial access point to discharge. Analysis of existing processes provides opportunities to simplify and streamline processes and to eliminate redundancies. Coordination between clinical and support services is necessary to achieve improvements in operational systems that are necessary for efficiency and satisfaction. This process entails redesigning work and has spatial implications for clinical and support functions (Fig. 9.1.1-4).

Patient Registration.   Many health care organizations have implemented point-of-service registration to streamline the patient's visit by decentralizing this function to ambulatory departments. This model reduces the central registration department by deploying its activities to patient care centers and allows arriving patients to proceed directly to their destinations without an additional visit to another department. Bedside registration moves this process even farther, enabling registrars to interact with patients or family members in patient care settings.

Alternative Service Delivery Methods

Traditionally, clinical services have been organized in departmental units that reflect the needs of individual clinical disciplines and their organizational administrative structure. Planning alternatives include grouping diagnostically related services, alternative care models, and off-site deployment.

Planners can consider functional groupings of services related by patient diagnosis as an

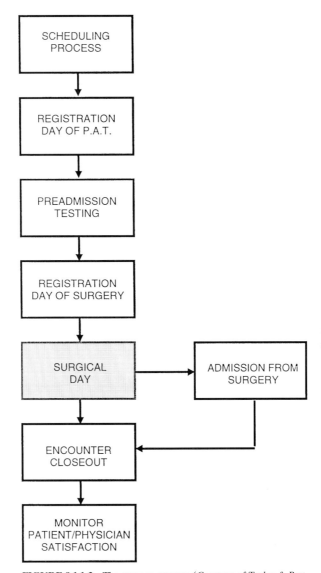

FIGURE 9.1.1-3 The surgery process. (*Courtesy of Taylor & Partners.*)

alternative to freestanding units and thereby achieve benefits in patient satisfaction and operational efficiency. The analysis must consider whether the volume of service generated by a new grouping of services can offset duplication of some support functions and whether it will generate new demand for services or simply divert services from an existing department and render that service inviable.

Women's Health.   Women's health centers are an example of market-driven synergy among clinical services that are traditionally separated by their own organizational structures. Comprehensive centers provide gynecology, imaging, oncology, and alternative therapies in one

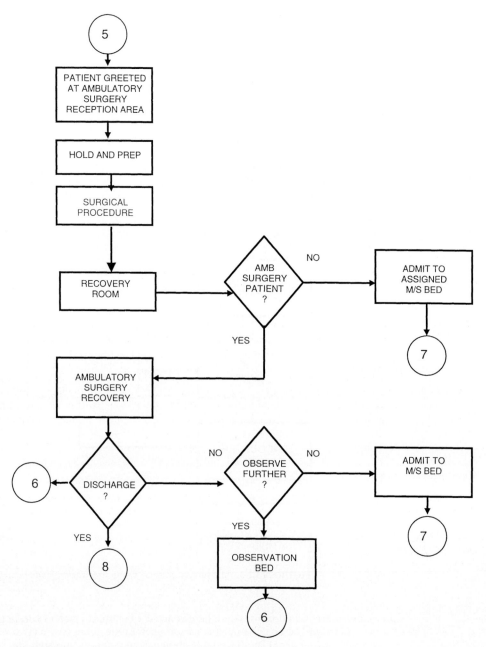

FIGURE 9.1.1-4 The surgical day process. (*Courtesy of Taylor & Partners.*)

setting and bring the services to the patient rather than sending the patient to several departments for the same care.

Alternative care models are program-driven reorganizations of clinical services that have profound impacts on space requirements and support systems. Planners construct program models for each alternative scenario to compare requirements for space, equipment, and staffing.

Birthing Center. The choice between a traditional model; a labor, delivery, and recovery (LDR) model; and a labor, delivery, recovery, and postpartum (LDRP) model for obstetrical services is an example of the options for realigning functions within a single clinical unit. Although the advocates for the LDRP model may argue that it offers the best in patient comfort, it does come at a cost of additional space and equipment for a high-volume service. Critical factors in deciding among these models are the volume of service to be provided, medical staff awareness, and external market factors. Generally, smaller-volume services can implement the LDRP model efficiently because fewer delivery rooms are duplicated.

Off-site deployment of services is an alternative service model that places clinical services where the patients are and relieves congestion at the organization's main campus. These models improve access and visibility for the organization's ambulatory services, frequently at a lower cost than on the main campus. Planners consider the volume of visits to determine whether there is sufficient demand to offset costs of duplication of space, equipment, and staff for an off-site location and to determine whether sufficient volume remains for services provided to inpatients on the main campus.

Freestanding Ambulatory Surgery. Freestanding ambulatory surgery centers capitalize on the growing trend away from inpatient surgery: faster recovery from less invasive procedures using significantly reduced anesthesia. The strategic advantage of an off-site center must be weighed against impacts on the surgical services at the inpatient and emergency care site. Where ambulatory cases account for over 80 percent of surgical cases, there may not be sufficient volume to operate an efficient inpatient surgical center alone and maintain its quality, if all ambulatory cases are moved off-site. Physician practice patterns require careful consideration where many surgeons regularly perform both inpatient and outpatient surgery (Fig. 9.1.1-5).

Investment in Technology

Advances in clinical technology, information systems, and automation are opportunities to streamline diagnostic and clinical support services throughout the organization.

Direct Digital Radiography. Technological advances have profound implications for space planning and workflow for clinical units. The implementation of direct digital radiography transforms imaging departments from film-based work flow and its space requirements. The new imaging systems improve turnaround time in radiographic rooms by eliminating waiting time for film quality-control checks This allows rooms to handle a larger volume of patient examinations and potentially reduces the number of rooms needed to handle the projected volume. Electronic transmission of digital images allows interpretation to occur anywhere on-site or off-site and facilitates specialists' consultations. Images are stored in far less space, and fewer images are lost in patient care spaces.

Program Space Standards

Room sizes are based on functional uses and the furnishings and equipment needed. Many are prescribed by regulatory standards, such as the American Institute of Architects

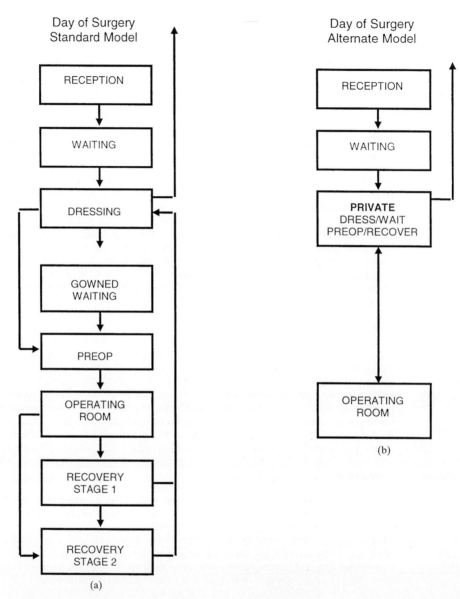

Day of Surgery
Standard Model

Day of Surgery
Alternate Model

(a)

(b)

FIGURE 9.1.1-5   Ambulatory surgery patient destination models: (*a*) day of surgery, standard model, and (*b*) day of surgery, alternate model. (*Courtesy of Taylor & Partners.*)

Academy of Architecture for Health Care with Assistance from U.S. Department of Health and Human Services (AIA/DHHS) Guidelines for Design and Construction of Hospital and Health Care Facilities, state regulations, and facility accessibility codes. It is important to understand that these are minimum standards for regulatory compliance and not necessarily sufficient for an organization's present or future needs. The organization should adopt a philosophy for sizing spaces that considers present and future uses of functional spaces and establishes a consistent allocation of space and resources throughout the facility.

Facility space standards are useful tools for maintaining control and consistency of space allocation. These guidelines establish the size, furnishings, and equipment for typical spaces throughout the medical center and provide a basis for enforcing a uniform approach to space allocation and standards compliance. They are also helpful in ensuring that the space needs of operational support systems are provided consistently throughout the facility.

Space standards should be established for offices, meeting rooms, and waiting rooms, as well as for repetitive support spaces such as toilet rooms, utility rooms, food-service galleys, medication rooms, and housekeeping rooms, so that logistical support functions can operate consistently.

Space standards should also be established for clinical spaces, so that flexibility in specialty assignment and future needs for more advanced equipment can be accommodated. Basic departmental building blocks such as examination and treatment rooms, imaging rooms, and operating rooms are frequently reassigned between specialty uses and are frequently upgraded with new medical equipment. There are benefits in sizing these rooms to accommodate multiple specialty uses, not making them skintight to the present use or equipment.

Universal Patient Rooms.   As the population ages and as the length of inpatient stays decreases, many organizations are finding that private patient rooms are advantageous in managing the demands of an older and sicker population. Traditional advantages of private rooms in managing admissions, infection control, and patient preference issues are reinforced by the need for more intensive nursing care and more bedside equipment. Anticipating the need for more intensive inpatient care, many organizations are developing private patient rooms sized for intensive care that accommodate the needs of acute care patients or can be converted with minor adaptations to serve as intensive care rooms.

Document Program Space Requirements

The facility space program is a document that enumerates the functional spaces required to provide the organization's services, establishing the size of each room based on a functional and operational narrative that describes the uses and operational systems that will support the services. The space program lists the number and net area requirements for each functional unit, as well as the furnishings and equipment requirements for each space. The functional and operational narrative describes the clinical, administrative, and operational support systems on which the program spaces are based and identifies any important adjacencies between spaces within and between related functional units.

Planners and designers use the space program to develop facility design alternatives, and it is periodically updated during the facility design process, as additional information becomes available either from changes in clinical service needs or from operational and financial planning.

Size Matters.   The space program describes the area needed for facilities using three terms:

• *Net area* is the clear area provided in each functional room for its equipment, furnishings, and casework, excluding wall thickness.

• *Departmental gross area* is the area that each functional unit occupies in the facility, including wall thicknesses and internal corridors serving only that unit, excluding the area of building support functions such as lobbies, major circulation corridors, stairs, elevators, mechanical spaces, and the thickness of the exterior wall. Planners estimate the departmental gross area required for each net area program space list using departmental gross area factors, which are mathematical multipliers that approximate the overall size of the facility and aid designers in laying out preliminary block plans. Multipliers vary depending on the type of functional unit, the number and size of its rooms, and its internal circulation system. Departmental gross factors range from 1.15 for large units with a few large rooms and little internal corridor space to 1.45 for emergency departments that have many small rooms and a lot of wide stretcher-access corridors.

- *Building gross area* is the overall construction area of the facility, measured from the outside face of the exterior wall and including all interior space. Building gross area includes lobbies, major circulation corridors, stairs, elevators, the thickness of the exterior wall, mechanical equipment rooms, and systems distribution spaces.

Equipment Planning

Determining the organization's immediate and long-term needs for medical equipment within each functional unit is an important factor in setting the facility space program. Accommodating the special requirements of medical equipment for access, shielding, proximity to other systems or devices, and physical environment, influences space planning and establishes special criteria for developing structural, mechanical, and electrical building systems. In the early phases of program and plan development, equipment planners should compile a comprehensive list of medical equipment to be accommodated, together with budgets and acquisition strategies, that will form a valuable ongoing data base for facility planning, as well as for subsequent design, construction, and commissioning phases.

Operational Planning

Operational planning considers the feasibility of various initiatives to improve the management structure and the work-flow processes necessary to support the planned array of patient care services. Planners and managers review requirements for staffing to meet the clinical unit's protocols and consider work design alternatives to optimize operational efficiency. The level and type of staffing are determined by each operational unit's patient flow model, its intended hours of operation, and its scheduling practices. Potential improvements can be achieved by using automation and information technology, where available, and by cross-training support staff members.

A work plan is prepared for each operational unit to document the selected work-flow alternative, hours of operation, anticipated needs for automation technology, training programs for quality improvement, performance goals, and staffing requirements, that financial planners will use to analyze business plans.

Financial Planning

Financial planners determine the viability of the level of resource allocation required to achieve the goals of the selected clinical services program model, and they consider the costs of implementation and operations and return on the investment necessary. Business plans for each operational unit document costs for facilities, equipment, staffing, and operational support and anticipated revenues for services provided. The business plan is used to evaluate opportunities for enhancing clinical service and improving operational performance and to select programming and facility-planning initiatives for implementation.

Financial planners determine affordability through an analysis of the organization's reserves, its debt capacity, and opportunities for additional sources of funding with the business plan's cost and revenue streams. The resulting financial plan model for the proposed program initiative determines the value to the organization of an improved volume of service and operational efficiency that is achieved by augmenting a core service or by entering a new initiative. This is compared with the cost of doing nothing through loss of opportunity and obsolescence.

FACILITIES EVALUATION

Parallel with the determination of the program model, architects and engineers conduct a thorough analysis of the organization's existing facilities and infrastructure to document their present condition, their capacity to support current programs, their adaptability for other uses, and any capacity for expansion. The facility evaluation includes an assessment of the age and condition of the buildings, their mechanical and electrical systems, and their suitability for present and future uses.

Many core hospital buildings were constructed 35 years ago following the historic Hill-Burton Act. These buildings were designed to provide health care to inpatients and were built before modern codes for life safety and accessibility were instituted. They are frequently considered candidates for modernization and conversion to ambulatory services. The facilities evaluation documents the weathering of the exterior envelope's walls, windows, and roofs; the capacity and condition of the superstructure and foundations; the capacity, condition, and useful service life expectations for mechanical and electrical systems; and compliance with life-safety and accessibility codes. The facility evaluation team identifies corrective work necessary to bring the existing buildings up to date and prioritizes the work into scope packages for implementation.

The facility evaluation provides a basis for the master facility plan by allowing planners to determine the suitability of existing buildings to house new programs and services based on the fitness of their location and configuration, as well as the additional costs necessary to bring them up to date.

MASTER FACILITY PLAN

The master facility plan is the site-specific embodiment of the space program for the organization's selected services model. Facility planners, architects, and engineers analyze alternative organizational concepts to define and evaluate facility design alternatives.

Site Access and Circulation Options

The master facility design team reviews the site for its capacity and its limitations in the light of the facility's development goals and regulatory and zoning restrictions. The master site plan establishes access points; on-site routes for patients, staff, emergency, and service circulation; parking areas; site utility infrastructure; and building locations. The master site plan creates clear site-access points for each type of circulation and direct routes to each building entry, and it emphasizes simplicity and visibility for patient and emergency services. The master site plan creates zones for the present need of each functional zone on the site, and it anticipates the direction of future expansion of each zone.

Facility Organizational Concepts

The master facility planners organize the functional units of the space program into alternative facility organizational concepts that embody the patient flow and required interdepartmental adjacency relationships that the organization's functional and operational program establishes. Using departmental gross-area block diagrams, the planners create facility concept plans showing access points, major circulation routes, and functional service support networks. The planners integrate building systems requirements for structural, mechanical, and electrical systems with the facility space program to optimize the placement of these services and their distribution routes.

Planners consider a variety of facility organizational concepts interactively with the master site plan based on alternative models for patient access to the organization's services. These alternatives vary due to the selection of plan models for single or multiple entry points for patient services.

Facility Circulation Models

The design of internal circulation promotes user-friendly facilities by separating various types of incompatible traffic and by differentiating them from each other. Arrival and destination points for the public and ambulatory patients should be clear, recognizable, and distinct from internal routes for inpatients, staff, and logistical services. Clear circulation networks support ease of finding one's way within the facility, and they are a foundation for facility design.

Transportation Systems

Most health care facilities rely on elevators and conveyors to move people and materials. The configuration and location of elevators support the facility circulation model by separating incompatible traffic. Either they physically separate the elevators used for public and ambulatory patients from other uses, or they schedule the travel of incompatible functions outside the hours of patient use. Inpatient and logistical support traffic can be separated in larger facilities where the volume of traffic justifies the additional elevator cars. Dedicated service elevators and conveyors can be specialized for scheduled and unscheduled transport of materials where the volume of service or the immediacy of need warrants their use.

Building Systems

The integration of infrastructural building systems is a central component of the master facility plan. The facility organizational concept addresses the size and location of central plant equipment and its distribution networks to the various functions served as an organizing principle similar to the facility circulation model. Space allocation for infrastructure considers present and future needs based on the understanding that intensification of clinical services will place increased demands on ventilation, plumbing, and electrical services. Planners accommodate these additional future needs by creating utility distribution networks that allow an orderly layout of main distribution trunks vertically and horizontally throughout the facility.

Flexibility

Facility planners have several methods for providing flexibility in future development. The methods range from planned building expansion zones to deployment of "soft space" within the facility. The master site plan identifies locations for future building additions that are coordinated with the facility, circulation, and infrastructural plans. *Soft space* is created within the building by placing unfinished shell space or functions that can be moved easily and inexpensively adjacent to clinical programs that are expected to grow. Within functional units, internal support and administrative spaces that are not mission-critical can be placed adjacent to clinical facilities to allow for internal zones for expansion.

Image

The facility's planning and design project the organization's culture and values to its customers and staff. Its planning and implementation should project prudent use of resources to create an orderly, functional environment that is pleasant for those who interact with it.

Project Scope Definition

Each master plan alternative is documented to quantify the proposed allocation of space for each functional area, as well as the construction and building systems necessary for the function. Scope definition should focus particular attention on areas of implied scope in addition to the cost of program space and general construction. Items of implied scope are special conditions specific to the organization's buildings and site that materially affect the cost of construction. Unusual site conditions can create the need for unusual building foundations or for site utilities. Unusual facility designs may require unusual superstructures. The scope-definition process identifies these special conditions as a premium to the cost of constructing the functional program spaces.

In master facility plans that involve renovation of existing buildings, each space allocation is categorized to differentiate the anticipated renovation work into various levels of intensity, ranging from *move-in,* the reuse of existing facilities without alteration, to *extensive,* the total reconfiguration of existing facilities to define scope more finely. Again, planners identify implied cost factors for building-specific conditions that affect overall renovation cost, including remedial building systems improvements to support the programs, corrections for code compliance, management or remediation of hazardous materials, and the like.

Project Scheduling

Planners develop schedules to implement the master facility plan alternatives based on estimates of construction duration and any special sequencing of construction or interim steps necessary to maintain operations on an active site. The project schedule should be designed to maximize opportunities for the earliest implementation of high-priority programs and improvements. The master project schedule includes all of the regulatory activities, as well as design, construction, and commissioning phases.

Project Cost Estimates

Comprehensive statements of anticipated project cost are compiled to establish the project cost model. These statements include all hard- and soft-cost items for planning, design, construction, furnishings, equipment, financing, and other items necessary to complete the project. In these cost studies, particular attention should focus on special requirements that are unique to each site or program, that will significantly impact its cost, and that may not be present on another site—for example, the need for an unusual building foundation because of local soil conditions. Attention should also focus on the costs of medical equipment and information systems that will be needed to implement the project. Because these cost studies are based on preliminary conceptual planning documents rather than fully detailed design plans and specifications, planners incorporate allowances for design and construction contingencies that are appropriate for the level of detail of the information available.

In master facility plans that envision altering existing facilities, planners should establish two additional project cost models for comparison with various alteration alternatives. The cost of total replacement for new facilities on a new site and the cost of doing nothing other than the remedial and corrective items identified in the facility evaluation are two scenarios that are useful in analyzing the cost and value of new construction and renovation alternatives.

MASTER FACILITY PLAN SELECTION

The preferred master facility plan alternative is selected by analyzing each of the alternatives and ranking each for its capacity to promote the strategic goals set forth in the organization's strategic plan, as well as its cost and affordability. Planners rate the alternatives and compare their ability to promote the institutional goals in important areas, such as the following:

- Patient issues
- Physician and provider issues
- Cost and business plan
- Schedule
- Future flexibility

Because the rating process compares value against cost, each organization will give varying weights to each factor in arriving at the selection of a master facility plan alternative.

The selection of the preferred master facility plan alternative provides the organization with a framework for decision making about programs and facilities, and it provides a basis for implementing interim and long-range improvements in an action plan that is coordinated with the organization's strategic initiatives. Although the master facility plan's view to the future is usually a 5- to 10-year period, it is a framework for decision making within the context of continuously improving services and operations. It must be revisited periodically and updated continuously to address new opportunities and threats.

CONTRIBUTORS

Mary Cancian, Taylor & Partners, Boston, Massachusetts

David Hanitchak, R.A., Massachusetts General Hospital, Partners Health Care System, Inc., Boston, Massachusetts

David Beard, RTKL Associates, Baltimore, Maryland

William Porter, Ph.D., F.A.I.A., Massachusetts Institute of Technology, Cambridge, Massachusetts

ARTICLE 9.1.2
DESIGN PROCESS

Harvey Kirk, A.I.A.
Associate, Stubbins Associates, Inc., Cambridge, Massachusetts

The health care facilities of today are undergoing significant change. The financial structure has become more unpredictable, as federal reimbursement shrinks and the full impact of managed care evolves. More than ever, projects require rationalization based on the expectation of return on investment. The members of the project team must understand that projects that generate revenue will have a much better chance of success in terms of both administrative approval and return on investment. This means that the architect's and engineer's expertise have to include an understanding of the financing of the health care facility and the long-term strategy for providing cost-effective care.

As the financial landscape is changing, so is technology. In existing facilities, much of the equipment installed in the 1970s has reached the end of its useful life and needs upgrading or replacement. In new and existing facilities, the issues of indoor air quality, energy efficiency, life-cycle costs, and even the deregulation of utilities are major influences on design.

Everyone involved with the process should understand that they all play a part in the advancement of health care and can positively affect the lives of the patients and create a healthy environment for the caregivers.

REVIEW OF BUILDING TYPES

Health care building types are divided into five main categories on the basis of the health of the patient and the type of procedure provided:

1. *Inpatient facilities.* These buildings contain the most acute level of care possible. This includes procedures requiring that the patient remain under the care of a physician for one or more days. As a new building project, this type is almost extinct because hospitals throughout the country have found that they have too many inpatient beds and are converting space to other uses.

2. *Outpatient facilities.* These contain day surgery, clinics, and offices. These services have expanded faster than any other, placing new demands on space resulting both from advancements in medicine that make operations less stressful on the patient and from insistence on shorter stays by third-party payers.

3. *Primary care/medical office buildings.* Buildings of this type contain physicians' offices and can be licensed to perform procedures that do not require sophisticated anesthesia. Often, physician group practices own the facility and contract with an adjacent hospital.

4. *Long-term care.* Facilities are sometimes integrated with a hospital, and the emphasis is shifting toward rehabilitation.

5. *Rehabilitation hospitals and clinics.* Space for wheelchair access and unimpeded paths of travel for patients are important planning issues in this project type. The inpatient population often has additional health problems and needs more comprehensive care. The clinics provide many treatment modalities, such as therapy pools and physical and occupational therapy equipment.

DESCRIPTION OF THE DESIGN PROCESS

The design of health care facilities begins by understanding the financial and medical parameters involved in creating the structural, architectural, and mechanical infrastructure for patient care. The road to a successful project has many steps. The first is master planning. Programming details are covered in Art. 9.1.1, "Planning and Programming."

Master Planning

The participants must understand that design is a continuing process of planning for the future. The master plan is the initial phase that focuses on physical and organizational elements of the hospital in a broad perspective, the so-called big picture. The master plan should be reviewed and updated regularly. This is an investigation that explores the effects of various ideas on the physical working of the hospital.

Group interviews form the core of the master planning process. Representatives of all groups are invited to respond in question-and-answer sessions. Ideas are judged for their financial feasibility and impact on existing functions. The needs of groups are assessed and a logical order is proposed before implementation. Master planning explores the big questions, and programming creates the recipe and structure for the design options.

The results of the master plan are a conceptual plan and a model of the physical change to the institution. A narrative that describes each phase will acquaint the reader with the changes in the structure. This document will be used as the road map for future planning.

Feasibility Study

The feasibility study is a cost-effective way of creating budgets and assessing the impact of a project on hospital space and operations. Design architects and engineers are called on to produce sketches of various options to establish a cost estimate. This gives the hospital a more focused approach and identifies the areas of major investment. A common problem with this approach is that if the cost is incomplete, the budget will be too small to achieve the results. If the budget remains inflexible at this point, the project will suffer.

To avoid a shortfall, the hospital should create a budgeting form that includes all anticipated costs and have all affected parties participate in its review. The form should be easy to understand, easy to adjust, and as complete as possible.

The most important aspect of the design phase is building consensus. This is the time when the components of a project are examined in detail. Interviews with client groups, including physicians, nurses, administrators, and sometimes patients, are essential.

FACILITY DESIGN

Successful design requires careful attention to the following key areas:

- *Site.*   In beginning the construction process, it is essential to document all existing conditions, including site utilities, soil conditions, subsurface water levels, bearing conditions, and topography.

- *Plan footprint.*   Hospitals come in all shapes and sizes. These forms are created as a result of specific functions, usually inpatient floors and intensive care units. There is a risk that specific shapes will limit the flexibility of future change. Because the rate of change in health care is accelerating, the facility needs to be designed for flexibility, and any new design needs to be tested to determine if change can be reasonably accomplished.

- *Circulation.*   The critical design element in any health care project is clear circulation for people, equipment, and materials. Patient and visitor traffic should be as simple and direct as possible. 8 ft is the standard width of corridors that accommodate beds and stretchers, but traffic flow may require even more space. All elevators should be large enough to accommodate patients transported in beds, plus space for equipment and personnel accompanying the patient. Many hospitals have separate elevator banks for the public and for staff and equipment. Materials management and waste-disposal routes need special attention to reduce contamination.

- *Departmental adjacencies.*   Health care designs require an emphasis on the efficient zoning of functions to allow the critical departments to relate properly. Patients who enter the hospital may encounter many different departmental areas. Departmental access and adjacencies affect patients' health outcomes.

- *Codes and standards.*   Numerous codes and standards set design requirements for health care. A major reference is the *Guidelines for Design and Construction of Hospital and Healthcare Facilities,* published by the American Institute of Architects (1996). Most public health authorities use these guidelines to judge the acceptability of designs. This text outlines the basic components of hospital, nursing, outpatient, rehabilitation, and psychiatric facilities. NFPA 99, *Standard for Health Care Facilities* (1996), and NFPA 101, *Life Safety Code* (1997), published by the National Fire Protection Association, are other important references. State and local codes should always be followed, but these texts reference additional pertinent standards.

Special Design Features

- *Operating rooms.*   The operating room is evolving into a multipurpose space for many different procedures. As technology improves, the room needs to be as flexible as possible. Imaging and monitoring are becoming increasingly important as minimally invasive surgery advances. Equipment is usually ceiling mounted to provide clearance for personnel and to eliminate wiring and tubing on the floor. The equipment can be positioned in various locations to maximize the surgeon's viewing for different cases. The room should have ceilings of at least 9 ft, 6 in to allow for the swing arms of the lighting and equipment supports. Certain specialty services such as neurosurgery, orthopedic surgery, and cardiac surgery require rooms of at least 600 ft^2 to provide space for equipment and personnel. Adequate storage for equipment and supplies is critical to the efficient functioning of operating room suites. There is increasing concern about performing procedures on patients who have airborne infectious diseases such as tuberculosis. The possibility of creating a system of infection isolation that has local exhaust ventilation should be considered. This is counter to the requirement for positive air pressure in the operating room and needs careful design consideration and review from the hospital staff and local health regulators.

- *Endoscopy suites.*   As less-invasive procedures become commonplace, many patients are treated and released in the hours shortly after surgery. Endoscopy rooms can be smaller than typical operating rooms and include less support space. The typical suite includes the procedure room, instrument decontamination and sterilizing facilities, and patient preparation and recovery areas.

- *Infectious isolation.*   Infectious isolation is achieved by controlling airflow and providing gowning and hand-washing vestibules. There are two types of infectious isolation: protective environment rooms that require positive air pressure to protect an immunocompromised patient from outside infection and airborne-infection isolation rooms that require negative pressure to isolate an infected patient. Protective environment rooms have high-level HEPA filters for patients who are receiving organ or bone-marrow transplants. Airborne infection isolation rooms need to be provided for each patient unit and incorporate hand-washing and gowning vestibules with storage for supplies.

- *Emergency services.*   These services are grouped in levels defined by the Joint Commission on Accreditation of Healthcare Organizations. Level I is the most comprehensive and has services and in-house specialists available 24 h a day. Levels II, III, and IV provide lesser availability of physician coverage. A clear ambulance route to the emergency room (ER) without crossing other traffic and a clear path to a triage station for determining the patient's status are important. The path should provide access to a trauma area to quickly treat critical patients. This path should be separate from walk-in traffic that will receive different types of treatment. Proximity to a security station is important to control patients who have emotional problems. Storage is necessary for wheelchairs, emergency medical service (EMS) equipment, decontamination supplies, and other medical equipment. Many patients require imaging of some type and therefore proximity to radiology is a very important adjacency.

- *Intensive care unit.*   The most critically ill patients require special care and immediate access to emergency equipment and highly trained staff. Most units are separate specialized treatment areas such as coronary, neonatal, pediatric, medical/surgical, and respiratory care. Access to emergency services, radiology, laboratory, and surgery is essential. The nurse station should be located so that the patient is under constant visual surveillance. Many units are experimenting with satellite nurse stations by integrating the computer as an information resource. This allows the staff to monitor the patients directly and still be in communication with a central station. Daylight is important to both patients and staff. This reduces stress and provides a way for the patients to orient themselves after surgery. Infection control is of primary importance, and hand-washing facilities should be located near every bed area. Noise from monitors, respirators, and other equipment should be controlled, and most units provide separate rooms for each patient with sliding doors designed for emergency access.

- *Radiology/imaging.* Patient imaging is undergoing a revolution. Digital radiography is enabling the physician to access data without film. The data can be transmitted on-line to many departments and sent to the treatment and operating rooms. The growth in the use of angiography, computerized tomography, magnetic resonance imaging (MRI), mammography, ultrasound, and the like, has increased the physician's access to information about the patient's condition. In the future, the equipment will become smaller and more accurate, and more uses will the found for imaging. For example, MRI equipment can now be used interoperatively during surgery. The units require magnetic shielding and special instruments, as well as radio-frequency shielding, because the images are constructed by using radio frequencies and can be degraded by interference. Smaller, portable MRI units do not need elaborate magnetic shielding but still need radio-frequency protection. X-ray diagnostic equipment requires lead radiation protection. The equipment is often ceiling mounted, and weight requirements should be carefully considered.

- *Commissioning.* Many projects in health care require that all systems function properly and that the staff receives the proper documentation and training. Commissioning the systems is usually undertaken by a separate engineering team that can review specifications, verify that the systems have been installed properly, and test for proper operation through all modes of operation. Staff training verifies that the people who work with the systems understand their operations and proper maintenance.

BIBLIOGRAPHY

American Hospital Association/American Society for Hospital Engineering: *International Conference and Exhibition on Health Facility Planning, Design and Construction,* 1993, Proceedings Manual, Vols. 1 and 2.

American Institute of Architects Academy of Architecture for Health, with assistance from the U.S. Department of Health and Human Services: *Guidelines for Design and Construction of Hospital and Health Care Facilities,* American Institute of Architects Press, Washington, DC, 1996.

Blyth, P. (ed.): *Health Care Interior Finishes: Problems and Solution—An Environmental Services Perspective,* American Society for Healthcare Environmental Services of the American Hospital Association, Chicago, 1993.

Bush-Brown, A., and D. Davis: *Hospitable Design for Healthcare and Senior Communities,* Van Nostrand Reinhold, New York, 1992.

Carpman, J., and M. Grant: *Design That Cares: Planning Health Facilities for Patients and Visitors,* American Hospital Publishing, Chicago, 1993.

Committee on Architecture for Health: *The Health-Care Architect of Tomorrow,* American Institute of Architects, Washington, DC, 1991.

Franta, G.: *Environmentally Sustainable Architecture in a Health Care Facility,* American Institute of Architects, Washington, DC, 1992.

Hardy, O., and L. Lamers: *Hospitals: The Planning and Design Process,* Aspen, Rockville, MD, 1986.

Hemmes, M.: *Managing Health Care Construction Projects: A Practical Guide,* American Hospital Publishing, Chicago, 1993.

Lebovich, W.: *Design for Dignity: Accessible Environments for People with Disabilities,* Wiley, New York, 1993.

NFPA: *Standard for Healthcare Facilities,* NFPA 99, National Fire Protection Association, Quincy, MA, 1996.

NFPA: *Life Safety Code,* NFPA 101, National Fire Protection Association, Quincy, MA, 1997.

SECTION 9.2
ENGINEERING AND DESIGN PROCESS

Eugene Bard, P.E., President, and Kevin Sheehan, Senior Associate
BR+A Consulting Engineers Inc., Boston, Massachusetts

OVERVIEW

As part of health care facility planning and programming, facility managers and their teams of engineering consultants must work in conjunction and in parallel with the project's planner and architect during the early phases of the planning process to ensure the success of a health care project.

One of the key planning factors for any health care facility is identifying and selecting the most appropriate mechanical and electrical systems for the mechanical, electrical, and plumbing (MEP) infrastructure that is required to support the various functions of the building type. The major system selection should be by life cycle analysis.

The location and placement of the MEP rooms (and equipment) must be correctly inserted into the building's program in terms of diagram adjacencies and functional relationships. The decision as to central or decentralized HVAC systems must be evaluated as early as possible in planning because this will have a major spatial impact.

Integrating the MEP spatial demands into the design process while the project is in the conceptual or preschematic phase is an important step in obtaining a facility that can be properly maintained and serviced, as health care buildings demand. Locating and sizing vertical shafts for pipes and ducts must be coordinated with the architectural planning at the earliest stages.

RECOGNIZING THE HEALTH CARE FACILITY TYPE

Health care facilities are rapidly becoming increasingly diversified in the types of medical services that may be provided within the various facilities, and this results in customization of the MEP infrastructural systems to suit the medical services program. AIA *Guidelines for Design and Construction of Hospital and Healthcare Facilities* is an excellent resource for more detailed information.[1]

Facility managers and their design teams must thoroughly understand the anticipated medical services to be provided within the facility before developing the energy utility requirements, spatial requirements, and architectural envelope models.

1. A general (community) hospital has a core of critical care spaces that include operating rooms, intensive care units, labor rooms, delivery rooms, and nurseries. In addition, the following functions should be anticipated by the building infrastructural requirements:

   - Radiology
   - Laboratory
   - Central sterile and support
   - Pharmacy
   - Inpatient nursing units
   - Intensive care and critical care units
   - Emergency room
   - Kitchen
   - Dining and food service
   - Morgue
   - Central housekeeping support
   - Docking facility for mobile MRI, etc.
   - Laundry
   - Isolation patient room
   - Isolation ER rooms
   - Isolation operating rooms

2. Teaching hospitals include all of the normal services of the general hospital and may also include one (or several) of the following specialized services:

   - Cancer treatment center
   - On-site MRI or lithotripter
   - Linear accelerator
   - Burn unit
   - Bone marrow transplant unit
   - Organ transplant unit
   - Orthopedic surgical unit

3. Outpatient health care facilities are also becoming a building norm for health care providers. *Hospital without beds* is a common term utilized within the present organizational model of medical care delivery.

4. Nursing homes fall into the following classifications:

   - Extended care
   - Skilled nursing homes
   - Residential care homes

5. Doctor's office building

LICENSING AGENCIES FOR HEALTH CARE FACILITIES

National, state, local, or regional agencies are involved in accrediting the various health care facilities, and they have established the minimum standard of services that are required and that influence the design team's perspective of the project and facility. Among the national agencies are the following:

- U.S. Department of Health and Human Services
- U.S. Public Health Services
- Medicare and Medicaid
- U.S. Department of Veterans Affairs
- Joint Commission on Accreditation of Healthcare Organizations (JCAHO)

In general, it is important that the design team be familiar with and experienced in the requirements of all of these agencies. In addition, the design team must be familiar with the state's public health services requirements because the conceptual integration of the building's life-safety systems (i.e., smoke management, fire alarms, smoke barriers, and areas of refuge) varies substantially from state to state. Thus, the overall MEP strategy for system layout and design is a critical issue, and these guidelines and references directly influence layout and design.

APPLICATION OF ENGINEERING CRITERIA (SELECTION OF MEP ENGINEERING INFRASTRUCTURE)

The selection and logical application of the MEP system plays an extremely important role in the building's overall ability to provide the proper environment for health care services and its ability to maintain that level of care and service for the life span of the facility. This concept may be highlighted by the following conceptual viewpoints.

Methodology for Equipment Selection

Due to the nature of this building type, the majority of the MEP services are generally 24-h continuous-duty applications. In contrast to an office building application that may function only 8 to 10 h per day, the hospital system must provide continuous duty 24 h per day.

The medical design community must recognize that conventional commercial-grade HVAC, plumbing, and electrical equipment may not be constructed to provide continuous operation. It is an important practice to specify *institution*-grade equipment as an initial step in the process of establishing the MEP infrastructure. This equipment should be capable of being maintained and serviced for its projected life span, and this concept should be incorporated during the early stages of the design process because it will affect the construction budget. In renovation projects, new equipment should be coordinated with existing equipment. In some facilities, standardization of equipment is highly desirable.

Methodology for Equipment Location

In addition, conceptual location and placement of the MEP infrastructure must be addressed in the initial planning phase, as shown by the following example.

It is common practice to locate the air-handling components of the HVAC system on the roofs of many building types. However, exposing any critical component to the weather makes it impractical to service it during the winter, and exposure to the elements limits its expected life. Thus, the configuration of the mechanical rooms and/or penthouse must be addressed in conjunction with the architectural massing study and models and must be factored into the overall gross-square-foot construction figures and project budget.

Specific Design Criteria

The American Institute of Architects Academy of Architecture, assisted by the U.S. Department of Health and Human Services, has defined the most common guidelines for a hospital facility.

Life Cycle Cost Analysis

Recognizing that MEP systems represent approximately 40 to 45 percent of the construction budget for a new hospital facility, the facility manager must be provided with the critical construction cost and system evaluation data to make the proper decisions for implementation. The features and details of the HVAC system and its subsystems are commonly evaluated by a life cycle cost analysis to assist in the overall decision and the cost-control aspects of project management.

Strategic HVAC Decisions

Consult Sec. 9.6, "Mechanical Systems," regarding special issues. It is important to address reliability issues.

Strategic Electrical Systems Decisions

Consult Sec. 9.5, "Electrical Systems," regarding special issues. It is important to provide reliable, normal and emergency power sources for the heath care facility.

Strategic Plumbing, Medical Gases, and Fire-Protection Systems Decisions

Consult Sec. 9.7, "Medical-Gas, Plumbing, and Fire-Protection Systems," regarding special issues. It is important to provide a reliable source of water for the heath care facility.

Strategic Decisions on Other Systems

Consult Sec. 9.3, "Special Systems and Needs," regarding special systems used in health care facilities. The MEP requirements of these systems need to be carefully evaluated.

CONTRIBUTORS

Richard T. Battles, P.E., President, Thompson Consultants, Inc., Marion, Massachusetts

Teerachai Srisirikul, P.E., C.E.M., D.S.M., Senior Facilities Engineer Mechanical, Partners HealthCare System, Inc., Boston, Massachusetts

John Saad, P.E., Senior Vice President, R.G. Vanderweil Engineering, Inc., Boston, Massachusetts

Anand K. Seth, P.E., C.E.M., C.P.E., Director of Utilities and Engineering, Partners HealthCare Systems, Inc., Boston, Massachusetts

REFERENCE

1. American Institute of Architects Academy of Architecture for Health Care, with assistance from the U.S. Department of Health and Human Services, *Guidelines for Design and Construction of Hospital and Healthcare Facilities,* American Institute of Architects Press, Washington, DC, 1996.

SECTION 9.3
SPECIAL SYSTEMS AND NEEDS

Anand K. Seth, P.E., C.E.M., C.P.E.
Director of Utilities and Engineering, Partners HealthCare System, Inc.,
Boston, Massachusetts

INTRODUCTION

There are many special needs in health care facilities, especially in modern hospitals. This section does not even attempt to address all the special systems and needs for health-care facilities. Only those systems are mentioned that impact the facilities directly.

All health care facilities generate medical waste. These wastes are by-products of patient care, and they are contaminated with patients' bodily fluids. These wastes are regulated and must be disposed in an appropriate manner. Article 9.3.1, "Medical Waste Management," is an authoritative summary of the various issues and options.

Article 9.3.2, "Pharmacy Department and Functions," describes some of the special drug preparatory requirements and their impact on the facility infrastructure. There is a need to ensure that patients, visitors, and staff members can go where they need to go and that patient test samples and records can be moved quickly. Article 9.3.3, "Transport Systems," provides an excellent discussion of transportation issues and their solutions.

We were unsure whether to include a discussion of patient-monitoring systems in this section, but we decided against it because those systems are clinical in nature. Similarly, many hospitals have large food-service cafeterias with special needs. We decided not to include any discussion of this matter because the cafeterias are similar to large restaurants, and we have excluded restaurants and other food-service areas from this handbook. If any readers feel that any particular system was omitted that should be included, they should feel free to contact the editor. Perhaps that omission can be remedied in a future edition.

ARTICLE 9.3.1
MEDICAL WASTE MANAGEMENT

Jorge Emmanuel, Ph.D., P.E., CHMM
Consultant, EPRI Healthcare Initiative, Palo Alto, California

An estimated 3.3 million tons of waste are generated annually from health care facilities in the United States.[1] Hospitals generate 2.4 million tons of this amount, and physicians' and dentists' offices, long-term care facilities, outpatient clinics, laboratories, and blood banks account for another 880,000 tons. About 15 percent of this waste is considered infectious. A study by the Agency for Toxic Substances and Disease Registry concluded that medical waste contributes to overall environmental problems and that occupational health concerns exist for workers who are involved with medical waste.[2] This article deals with the medical waste management issues that health care facility managers face.

The Medical Waste Stream

Medical waste is waste generated by the diagnosis, treatment, and immunization of humans or animals. Because there is no one common definition of medical waste, each facility must determine this based on applicable federal, state, and local regulations. In general, wastes from a health-care facility fall into four major categories: (1) general trash or municipal solid waste; (2) infectious waste, also called regulated medical waste, biohazardous waste, or red bag waste; (3) hazardous waste; and (4) low-level radioactive waste. In practice, some of these waste streams may be intermingled.

Many regulatory definitions of regulated medical waste are based on the following 10 general categories defined by the 1988 Medical Waste Tracking Act: (1) cultures and stocks; (2) pathological wastes, for example, tissues, organs, and body parts; (3) blood, blood products, and other body fluids; (4) contaminated sharps; (5) animal wastes; (6) isolation wastes; (7) contaminated medical equipment; (8) surgery wastes; (9) laboratory wastes; and (10) dialytic wastes.[3] Regulatory agencies differ as to which of these categories are infectious.[4,5]

Infectious waste varies considerably in the rate of waste generation and in composition, heat content, moisture, and bulk density, as shown in Table 9.3.1-1.[6,7] In one study, the generation of hospital waste ranged from 8 to 45 lb per bed per day wih an average of 23 lb per bed per day.[8] A nationwide survey of infectious waste generation in hospitals gave a national average of 1.38 lb per bed per day (0.627 kg per bed per day) or 2.29 lb per patient per day (1.04 kg per patient per day).[9]

Waste Minimization

Waste minimization, an important aspect of medical waste management, is the reduction, to the extent feasible, of waste generated at a facility, as well as of waste subsequently treated, stored, or disposed of. Source reduction and recycling are the two major techniques of waste minimization. *Source reduction* means reducing or eliminating the waste at the source. This can be achieved by product substitution, changes in technology, and good operating practices. *Recycling* is the use, reuse, or reclamation of materials from the waste stream.

The development of a waste minimization program involves four basic stages: (1) planning and organization, (2) assessment, (3) feasibility analysis, and (4) implementation. The commitment of top management is essential. During the planning and organization phase, overall goals are set and a task force that involves key departments is organized. The assessment phase includes data collection, prioritization of waste streams, and an assessment by the staff and/or consultants to select waste minimization options. A wide range of options exists for general waste[10–13] and hazardous waste.[12,14–16] A product management approach can identify

TABLE 9.3.1-1   Typical Composition and Characteristics of Infectious Waste

Composition	
Cellulosic material (paper & cloth)	50–70%
Plastics	20–60%
Glassware	10–20%
Fluids	1–10%
Typical characteristics	
Range of heating values (Average)	1,500–20,000 Btu/lb (8,500 Btu/lb; 3,400–45,000 kJ/kg [19,800 kJ/kg])
Moisture	8.5% by weight
Incombustibles	5.0% by weight
Bulk density	4.0 lb/ft³ (64 kg/m³)

opportunities for reducing waste through purchasing, inventory control, changes in packaging, and working with suppliers.[17] An effective program entails education and periodic evaluations.

A medical-waste audit is a tool for getting data on sources of waste, compositions, generation rates, and flow patterns within a facility. Sample self-audit forms and questionnaires are found in various references.[10,11,14,18] The need for representative sampling determines the time period for data collection: 1 day provides a quick snapshot of waste generation, whereas longer periods could reveal weekly, monthly, or seasonal variations. Long-term monitoring can be accomplished by using a computerized waste-tracking system. Data on waste generation rates and composition can be used to evaluate classification and segregation practices. Overclassification (treating noninfectious waste as infectious waste) and commingling trash with infectious or hazardous waste add significantly to disposal costs. A waste audit can uncover inefficiencies and establish a baseline for waste minimization.

Segregation and Collection

Segregation means separating certain types of waste from the general waste stream and placing them in appropriate containers at the point of generation for separate disposal. It is key to minimizing infectious waste, thereby reducing disposal costs and protecting public health and the environment. Each facility must first establish its definition of potentially infectious and hazardous wastes in keeping with applicable regulations. To improve the efficiency of segregation, containers must be properly located to minimize incorrect use. General trash containers placed beside infectious waste containers could result in better segregation. Too many infectious waste containers tend to inflate waste volume, but too few containers may lead to noncompliance.

Medical waste collection practices should be designed to move waste efficiently from points of generation to storage or disposal and to minimize the risk to personnel. Generally, carts are used to transport waste within a facility. Carts used for infectious waste should not be used for other purposes. They should be closed during transport to prevent spillage and avoid offensive sights and smells. A program of regular cleaning and disinfection of carts should be followed.[5]

Containment, Labeling, and Storage

Infectious waste should be segregated in clearly marked containers appropriate for the type and weight of the waste. Except for sharps and fluids, infectious wastes are generally put in plastic bags, plastic-lined cardboard boxes, or other leakproof containers that meet specific performance standards, such as the ASTM Standard Test Methods for Impact Resistance of Plastic Film.[19] In the United States, red or orange bags are commonly used to designate infectious waste, and general waste is placed in black, white, or clear bags. Labels affixed to containers must include the international biohazard symbol. For sharps, the primary containers must be rigid, leakproof, break-resistant, and puncture-resistant. If the primary container could leak during transport, a secondary leakproof container is used. Liquid infectious waste should be collected in leakproof containers that are then placed in pails, cartons, drums, or bins for transport. Reusable containers should be washed thoroughly and disinfected.

If infectious waste has to be stored, the storage site should have good drainage, easy-to-clean surfaces, good lighting and ventilation, and should be safe from weather, animals, and unauthorized entry. The space should be sized according to generation rates and collection frequency. Some states require refrigeration of regulated medical waste if storage times exceed a specified limit. To prevent putrefaction, the following maximum storage times are suggested: in temperate climates, 72 h in winter and 48 h in summer; in warm climates, 48 h in the cool season and 24 h in the hot season.[20] Infectious waste should not be compacted before treatment. Containment, labeling, and storage specifications for medical waste containers should comply with applicable regulations such as OSHA's Bloodborne Pathogen Regulation (29 CFR Part 1910.1030).

Transportation

Off-site transportation of regulated medical waste must meet regulations covering interstate and intrastate transportation promulgated by the Department of Transportation (DOT; 49 CFR Parts 171 to 180), the Public Health Service (42 CFR Part 72), the U.S. Postal Service (39 CFR Part 111.1), the state, and localities. DOT regulates packaging, labeling, and shipping of infectious substances and regulated medical waste. Packages for regulated medical waste must be rigid, leak-resistant, and impervious to moisture; must not tear or burst under normal use; must be sealed to prevent leakage during transport; and must be puncture-resistant. Labels must meet general labeling requirements for hazardous waste and have the universal biohazard symbol with instructions to notify the Centers for Disease Control in case of damage or leakage. The Public Health Service regulates packaging, labeling, and mailing of etiological agents (microorganisms or their toxins that may cause human disease). The U.S. Postal Service (USPS) has its own regulations for packaging and mailing of sharps and medical devices.

MEDICAL WASTE TREATMENT AND DISPOSAL

Health-care facilities have three basic options for treating and disposing of medical waste: on-site incineration, hauling and off-site treatment, or on-site treatment using an alternative technology.

Medical Waste Incineration

Incineration is the process wherein waste is burned to produce combustion product gases, ash, and incombustible residues. A typical incinerator has a waste-loading system, a refractory-lined primary chamber (hearth or furnace), a secondary chamber, an ash removal system, and a flue gas stack. Some incinerators also have a heat recovery boiler, air pollution abatement equipment, and controls. There are three basic incinerator designs: (1) controlled-air incinerators, also called starved-air, two-stage, or modular incinerators; (2) multiple-chamber incinerators, also called retort or excess-air incinerators; and (3) rotary kiln incinerators.[6]

In 1997, the U.S. EPA promulgated "Standards of Performance for New Stationary Sources and Emission Guidelines for Existing Sources: Hospital/Medical/Infectious Waste Incinerators." Under the rule, new and existing medical waste incinerators are subject to emission limits depending on the incinerator's burning capacity (see Table 9.3.1-2). Although the EPA does not specify a pollution control device, incinerators may need wet scrubbers, dry scrubbers, or lime- and carbon-injected baghouse filters to meet emission limits. Secondary chambers may have to be retrofitted to achieve proper temperatures and residence time for good combustion. Periodic stack tests must be performed to show compliance with the rules, and facilities must continuously monitor operating parameters such as secondary chamber temperature, as well as parameters related to the pollution control device. The EPA regulations also require operator training and qualification, inspection, waste management plans, reporting, and record keeping.

TABLE 9.3.1-2   Emission Limits for Hospital/Medical/Infectious Waste Incinerators

	Emission limits					
	Small*		Medium*		Large*	
Pollutant	New[†]	Existing	New[†]	Existing	New[†]	Existing
Particulate matter, mg/dscm	69	115	34	69	34	34
Carbon monoxide, ppmv	40	40	40	40	40	40
Dioxins/furans, ng/dscm total or ng/dscm TEQ	125 or 2.3	125 or 2.3	25 or 0.6	125 or 2.3	25 or 0.6	125 or 2.3
Hydrogen chloride, ppmv or % reduction	15 or 99%	100 or 93%	15 or 99%	100 or 93%	15 or 99%	100 or 93%
Sulfur dioxide, ppmv	55	55	55	55	55	55
Nitrogen oxides, ppmv	250	250	250	250	250	250
Lead, mg/dscm or % reduction	1.2 or 70%	1.2 or 70%	0.07 or 98%	1.2 or 70%	0.07 or 98%	1.2 or 70%
Cadmium, mg/dscm or % reduction	0.16 or 65%	0.16 or 65%	0.04 or 90%	0.16 or 65%	0.04 or 90%	0.16 or 65%
Mercury, mg/dscm or % reduction	0.55 or 85%	0.55 or 85%	0.55 or 85%	0.55 or 85%	0.55 or 85%	0.55 or 85%

* Capacities: small ($\leqq$200 lb/h), medium (200 to 500 lb/h); and large (>500 lb/h).
[†] New incinerators are those whose construction commenced after June 20, 1996.

Hauling

Waste management firms offer transport, storage, treatment, and disposal; waste brokers may offer additional services such as testing, repackaging, temporary storage, and compliance assistance. Off-site transportation should meet Department of Transportation, Public Health Service, state, and local regulations for interstate and intrastate transport of regulated medical waste. The waste should be transported by approved haulers (as required in many states) and sent to a facility that meets all local, state, and federal regulations for storage, treatment, and disposal. To determine the full costs of hauling, facilities should take into account the cost of labor, boxes and other supplies, insurance, penalties for containers that do not meet specifications, and other costs. Potential liabilities associated with improper disposal, occupational injuries, or transportation accidents should be considered because waste generators ultimately bear responsibility for what happens to their waste.[18]

On-Site Treatment Technologies

On-site treatment technologies use five basic processes to treat different kinds of medical waste: (1) thermal (low-heat and high-heat) processes, (2) irradiation, (3) chemical treatment, (4) biological processes, and (5) mechanical processes.[21]

Low-heat processes are those that use thermal energy (heat) to decontaminate the waste at temperatures insufficient to cause combustion or pyrolysis. In general, low-heat thermal or "non burn" technologies operate between 200°F (93°C) to around 350°F (177°C). The two basic categories of low-heat thermal processes are steam disinfection and dry heat treatment. Steam disinfection can be done in an autoclave or retort and has a minimum time–temperature requirement for proper disinfection.[22] Steam sterilizers may involve prevacuuming, internal shredding, agitation, and compaction. Microwave treatment is in effect a steam disinfection process. In dry heat processes, no steam is added. Instead, the waste is heated by convection and/or thermal radiation using infrared or resistance heaters. One technology uses high-velocity heated air.

High-heat processes generally operate at temperatures ranging from around 1000°F (540°C) to more than 15,000°F (8300°C). Electrical resistance, induction heating, natural gas combustion, and/or plasma energy provide the heat. High-heat processes involve chemical and physical changes that result in total destruction of the waste and a significant reduction in mass and volume. Some technologies involve molten metal slag or molten glass systems. Depending on the level of oxygen in the treatment chamber, combustion or pyrolysis dominates the process. If the waste is heated in the absence of air or in the presence of an inert gas, pyrolysis dominates, and the gaseous reaction products are offgases with low to medium heat content. In plasma-based pyrolysis, the solid residues can be metals, carbon, and/or an unleachable vitrified slag. Some processes involve superheated steam and steam-reforming reactions, and others use pyrolysis and controlled oxidation in tandem.

Irradiation-based technologies use electron beams, cobalt-60, or UV irradiation. These technologies require shielding to prevent occupational exposures. The pathogen destruction efficacy of electron-beam irradiation depends on the dose absorbed by the waste, which in turn is related to waste density and electron energy. Cobalt-60, a gamma-emitting radioactive isotope, is used mainly for sterilizing medical supplies. Germicidal ultraviolet radiation (UV-C) has been used to disinfect gas or liquid streams. Irradiation does not alter the waste physically.

Chemical processes employ disinfectants such as dissolved chlorine dioxide, sodium hypochlorite (bleach), peracetic acid, or dry inorganic chemicals. Chemical processes involve shredding or grinding to enhance exposure to the chemical agent. One new technology uses ozone to treat medical waste, and another uses alkali to hydrolyze tissue waste in heated stainless-steel tanks. Besides chemical disinfectants, there are also encapsulating compounds that can solidify sharps waste, blood, or other fluids into a solid matrix before disposal. *Biological processes* employ enzymes to destroy organic matter.

Mechanical processes such as shredding, grinding, hammer milling, mixing, and compaction supplement other treatment processes. Mechanical destruction renders the waste unrecognizable and prevents the reuse of needles and syringes. It can also improve the rate of heat transfer or expose more surfaces to disinfectants. A mechanical process is itself not a treatment process and should not be applied before the waste is decontaminated, unless it is part of a closed treatment system.

The factors to consider in selecting an on-site technology are microbiological inactivation efficacy,[23] environmental emissions, regulatory acceptance, capacity, types of waste treated, space requirements, volume and mass reduction, the nature of the solid residue, occupational safety, automation and ease of use, reliability, track record, maintenance requirements, public acceptance, the level of commercialization, the availability of technical support, and cost-effectiveness.

Land Disposal

Regulated medical waste must first be treated before being disposed of in a solid-waste landfill. Moreover, various states require that treated waste be rendered unrecognizable. Some states specify interment as an acceptable method for disposing of anatomical parts. Federal guidelines for land disposal facilities that deal with infectious waste are found in 40 CFR 241.101(h), but many states have developed their own guidelines. In practice, some landfill operators may not accept treated but recognizable medical waste. The number of solid-waste landfills has declined in the last two decades, and some states have limited disposal capacities. Hence, waste minimization programs are important.

Contingency Planning

Health care facilities should be prepared to respond to contingencies such as spills, exposures to infectious waste, or failure of waste-treatment systems. Spill containment and cleanup kits that consist of containment supplies, absorbent material, disinfectant, and safety equipment can be used to clean up most spills in health care facilities. (See Ref. 5 for containment and cleanup procedures.) Strategies should also be developed in response to exposure incidents. Follow-up procedures after an exposure are required under OSHA's bloodborne pathogen rule. Sample contingency plans are provided in Ref. 18.

Hazardous Waste

Health care facilities generate a variety of hazardous waste. Lists of common chemical wastes in hospitals are provided in Refs. 16 and 18. The storage, transportation, and disposal of hazardous waste are regulated under the Resource Conservation and Recovery Act (RCRA). The basic framework is a cradle-to-grave approach. Generally, a health care facility must identify hazardous waste on the basis of RCRA definitions, obtain an EPA identification number, reduce and properly manage hazardous waste at the site, label the waste with information about the chemicals and date when the waste accumulation started, track and document the transport and disposal by using a uniform hazardous waste manifest, and provide employee training. Only licensed transporters can haul hazardous waste to permitted treatment, storage, and disposal facilities (TSDFs). Many health care facilities follow standard lab packing procedures to prepare hazardous waste for bulk transport to TSDFs. A hazardous waste management plan should include education and procedures for dealing with hazardous waste and spills.

Unplanned releases of hazardous substances fall under the Comprehensive Environmental Response, Compensation, and Liability Act (CERCLA), which requires reporting of releases depending on the type of substance and amount released. The Emergency Planning and Community Right-to-Know Act (EPCRA) regulates hazardous release emergency response and hazard communication. OSHA's hazard communication standards (29 CFR 1910.1200) and laboratory safety standard (29 CFR 1910.1450) involve, among others, hazard identification, evaluation, and control; material safety data sheets (MSDS) that are readily available to employees; labeling; training programs; reporting and record keeping; and a hazard communication plan. See Refs. 16 and 18 for more information about hazardous waste management.

Low-Level Radioactive Waste

Radionuclides are used in diagnosis and treatment, as well as in clinical and research studies. The U.S. Nuclear Regulatory Commission and the state where the facility is located must approve the use of radioactive materials; disposal of radioactive material should comply with regulations and conditions in the facility's license. The common disposal methods are storage for decay (or "decay in storage"), shipment to an authorized radioactive waste disposal site, incineration, and disposal in the sewer. Radionuclides with short half-lives are generally stored for a period (typically 10 times the half-life of the longest-lived radionuclide in a container) to allow decay to background levels as confirmed by a radiation survey, and are then disposed of as regular waste. Beta- and gamma-emitting radionuclides require special handling and shielding. Facilities can segregate radioactive waste according to the time needed for storage: short-term (half-life less than 30 days) and long-term (half-life from 30 to 65 days). Storage facilities must be secured and designed to prevent human exposure. Waste may be minimized by limiting the quantity of radioactivity purchased, by using nonradioactive materials or shorter-lived radionuclides where possible, and by designing laboratory procedures to reduce the volume of mixed waste.[14,16,18]

Regulatory Compliance

At least five federal agencies have regulations dealing with medical waste: OSHA, EPA, DOT, PHS, and USPS. Relative to medical waste, OSHA has enforceable standards on hazard communication; occupational exposure to tuberculosis; and standards related to ethylene oxide, formaldehyde, toluene, and other chemicals commonly found in a health care setting; and a bloodborne pathogens standard. EPA was directed to regulate medical waste under the Medical Waste Tracking Act of 1988, but the statute expired in 1991 and was not renewed. The EPA regulates hazardous waste under RCRA, CERCLA, and EPCRA. Under the 1990 Clean Air Act amendments, the EPA enforces standards and emission guidelines for medical-waste incinerators. The Department of Transportation, the Public Health Service, and the U.S. Postal Service regulate packaging, labeling, transportation, and shipping of medical waste. Other agencies such as the Centers for Disease Control and the National Institute of Occupational Safety and Health develop guidelines. Within state governments, specific areas may be regulated by agencies that deal with waste and air quality (under environmental or natural resources departments), transportation, public health, and licensing. Local governments issue ordinances that may affect health practices, recycling, landfills, incinerators, hauling, etc. The Joint Commission on Accreditation of Healthcare Organizations has standards that deal with waste management under the Management of the Environment of Care Standards. Other nongovernmental agencies such as the American Biological Safety Association, the American Hospital Association organizations, the Association of periOperative Registered Nurses, the College of American Pathologists, and the National Committee for Clinical Laboratory Standards develop their own standards, guidelines, or recommended practices relevant to medical waste.

MEDICAL WASTE MANAGEMENT PLAN

The medical waste management plan describes a facility's program for managing waste from generation to disposal. The plan should address the following issues: (1) compliance with regulations, (2) responsibilities of staff members, (3) definitions and classification of medical waste, (4) procedures for handling medical waste, and (5) training plans. The procedures should cover identification, segregation, containment, labeling, storage, treatment, transport, disposal, monitoring, record keeping, and contingency planning.[5] The plan should be designed to protect the health and safety of the staff, patients, and visitors; to protect the environment; and to comply with applicable regulations. It should be read by all staff members, reviewed periodically, and kept at easily accessible locations in the facility. The waste management plan can incorporate or be linked to the facility's waste minimization plan. Also related are a chemical safety plan, a hazard communication plan required under EPCRA, and an exposure control plan regarding exposure to blood or other potentially infectious materials, as required by OSHA.

REFERENCES

1. U.S. Environmental Protection Agency, *Medical Waste Incinerators—Background Information for Proposed Guidelines: Industry Profile Report for New and Existing Facilities,* EPA-453/R-94-042a, Washington, DC, 1994.

2. U.S. Department of Health and Human Services, Agency for Toxic Substances and Disease Registry, Public Health Service, *The Public Health Implication of Medical Waste: A Report to Congress,* Washington, DC, 1990.

3. U.S. Environmental Protection Agency, *EPA Guide for Infectious Waste Management,* EPA/530-SW-86-04, National Technical Information Service, Springfield, VA, 1986.

4. W. A. Rutala and C. G. Mayhall, "Medical Waste," *Infection Control and Hospital Epidemiology* 13 no. 1 (1992), 38–48.

5. W. L. Turnberg, *Biohazardous Waste: Risk Assessment, Policy and Management,* Wiley, New York, 1996.

6. C. R. Brunner, *Medical Waste Disposal.* Incinerator Consultants Incorporated, Reston, VA, 1996.

7. U.S. Environmental Protection Agency, *Operation and Maintenance of Hospital Medical Waste Incinerators,* EPA-450/3-89-002, Washington, DC, 1989.

8. U.S. Environmental Protection Agency, *Hospital Waste Combustion Study: Data Gathering Phase,* EPA-450/3-88-017, National Technical Information Service, Springfield, VA, 1988.

9. W. A. Rutala, R. L. Odette, and G. P. Samsa, "Management of Infectious Waste by U.S. Hospitals," *JAMA* 262 no. 12 (1989), 1635–1640.

10. C. L. Bisson, G. McRae, and H. G. Shaner, *An Ounce of Prevention: Waste Reduction Strategies for Health Care Facilities,* American Society for Healthcare Environmental Services (American Hospital Association), Chicago, 1993.

11. G. McRae and H. G. Shaner, *Guidebook for Hospital Waste Reduction Planning and Program Implementation,* American Society of Healthcare Environmental Services (American Hospital Association), Chicago, 1996.

12. J. Emmanuel, *Waste Minimization in the Healthcare Industry: A Resource Guide,* EPRI Healthcare Initiative Report CR-106627. EPRI, Palo Alto, CA, 1999.

13. Ohio Hospital Association, *Recycling Revisited: A Comprehensive Recycling Manual for Hospitals,* Columbus, Ohio, 1995; California Integrated Waste Management Board, *Waste Reduction Activities for Hospitals,* Publication 500-94-04, Sacramento, CA, 1994; Minnesota Hospital Association, *The Waste Not Book,* Public Affairs Division, Minneapolis, MN, 1993.

14. U.S. Environmental Protection Agency, *Guides to Pollution Prevention: Selected Hospital Waste Streams,* EPA/625/7-90/009, Washington, DC, 1990.

15. H. G. Shaner, *Becoming a Mercury Free Facility: A Priority to Be Achieved by the Year 2000,* American Society for Healthcare Environmental Services (American Hospital Association), Chicago, 1997; National Wildlife Federation, *Mercury Pollution Prevention in Healthcare: A Prescription for Success,* Ann Arbor, MI, 1997.

16. K. D. Wagner, C. D. Rounds, and R. A. Spurgin (eds.), *Environmental Management in Healthcare Facilities,* W. B. Saunders, Philadelphia, PA, 1998.

17. C. D. Rounds, "A Product-Focused Waste Management Model: Materials Management and Waste Minimization," in K. D. Wagner, C. D. Rounds, and R. A. Spurgin (eds.), *Environmental Management in Healthcare Facilities,* W. B. Saunders, Philadelphia, PA, 1998.

18. P. A. Reinhardt and J. G. Gordon, *Infectious and Medical Waste Management,* Lewis, Chelsea, MI, 1991.

19. American Society for Testing and Materials (ASTM), "Standard Test Methods for Impact Resistance of Plastic Film by the Free-Falling Dart Method, Designation: D 1709-91," *1994 Annual Book of ASTM Standards,* ASTM, Philadelphia, PA, 1994, Sec. 8, Vol. 8.01:393-400, PCN: 01-080194-19.

20. A. Pruss and W. K. Townend, *Teacher's Guide: Management of Wastes from Health-Care Activities,* WHO/EOS/98.6. World Health Organization, Geneva, Switzerland, 1998.

21. J. Emmanuel, *New and Emerging Technologies for Medical Waste Treatment,* EPRI Healthcare Initiative Report CR-107836-R1, EPRI, Palo Alto, CA, 1999.

22. J. L. Lauer, D. R. Battles, and D. Vesley, "Decontaminating Infectious Laboratory Waste by Autoclaving," *Appl. Environ. Microbiol.* 44 no. 3 (1982) 690–694; W. A. Rutala, M. M. Stiegeland, and F. A. Sarubbi, Jr., "Decontamination of Laboratory Microbiological Waste by Steam Sterilization," *Appl. Environ. Microbiol.* 43 (1982), 1311–1316.

23. EPRI Healthcare Initiative, *Technical Assistance Manual: State Regulatory Oversight of Medical Waste Treatment Technologies: A Report of the State and Territorial Association on Alternative Treatment*

Technologies, EPRI Report TR-112222, EPRI, Palo Alto, CA, 1999; State and Territorial Association on Alternative Treatment Technologies, *Technical Assistance Manual: State Regulatory Oversight of Medical Waste Treatment Technologies,* www.epa.gov/epaoswer/other/medical/index.htm, April 1994.

ARTICLE 9.3.2
PHARMACY DEPARTMENT AND FUNCTIONS

Harold J. De Monaco, M.S.
Director of Drug Therapy Management, Massachusetts General Hospital, Massachusetts General Physicians Organization, Boston, Massachusetts

Today's institutional pharmacy represents a crossroads of the majority of functional design elements in health care facilities. Warehousing, distribution, asset management, security, distribution, specialized manufacturing facility design requirements with governmental and organization oversight, and information systems are but a few of the considerations in planning the average pharmacy. As drug therapy becomes a more predominant feature in the care of patients, the need to use these assets efficiently will become paramount. Increasingly, information systems play a key role in providing efficient drug therapy.

In years past, institutional pharmacy facilities required little more than warehousing space with adequate ventilation, heating, and cooling systems. Electrical requirements were within the norm for office space. All of this has changed dramatically in the past decade, as reduced warehousing requirements and the need for more sophisticated air-handling and electrical service have become key elements. A return to a reliance on pharmacy-based manufacturing along with an increasing reliance on computers for both data storage and processing has significantly increased facilities requirements.

The pharmacy department has two basic functions:

- Providing drug products
- Providing drug information

Adequate facility design addresses the needs of both of these critical functions.

FUNCTIONAL FACILITY DESIGN REQUIREMENTS

To design a facility properly, its present and future functions need to be well understood. The practice of institutional pharmacy is evolving, and demands are increasing for efficient warehousing and distribution coincident with specialized drug preparation and asset management.

Warehousing

Inventory warehousing requirements have been gradually reduced in recent years. The rationale for the decrease is directly related to the large increase in inventory value and the fiscal need for increasing turnover rates. Whereas the average institutional pharmacy achieved an inventory turnover rate of 11 to 12 turns per year, well-run pharmacies strive for turnover rates in the range of 20 to 25. Some very specialized institutions have achieved turnover rates as high as the 40s. This large increase in inventory turns has been accomplished by establishing *prime vendor* arrangements with local and national drug wholesalers and by installing relatively sophisticated electronic ordering systems that feed directly off pharmacy drug distribution. With increasing turnover rates, consideration should be given to receiving-dock requirements because dock requirements are likely to increase consistently with inventory turnover rates. The greater the turnover rate, the more attention should be directed to docking facilities. In addition, many institutions place a greater reliance on frozen, ready-to-use drugs. Facilities design should allow for delivery of frozen drugs from the supplier directly to the pharmacy.

The majority of line items warehoused within a pharmacy may be stored under routine ambient conditions. Temperature and ventilation requirements depend on both the physical plant and the storage system to be used. Consideration should be given to such key variables as shelving (fixed or rolling), the use of robotics, and the number of employees.

As drug therapy becomes more sophisticated and relies more on bioengineered proteins such as monoclonal antibodies, specialized distribution and storage requirements increase. Adequate attention should be paid to both the electrical and cooling requirements for the increased use of refrigerators and specialized freezers. Provisions should be made for adequate alarm systems for refrigerator or freezer failure. The growing trend toward manufacturing ready-to-use drug containers (so-called compounding in advance) sold by local and national vendors increases the need for additional refrigerator and freezer storage.

The extent of clinical research conducted within the institution may also play a key role in the requirement for specialized storage space. The need to segregate research drugs from routine inventory and to provide unique environmental requirements (storage in liquid nitrogen, for example) should be considered during the initial design discussions.

Regardless of the environmental requirements, adequate security measures must be ensured. The potential for employee and outsider theft of pharmacy supplies is high, and drug loss has significant financial and social implications. Narcotics and other controlled substances should be housed in limited-access areas based on federal requirements for such storage. Distribution to and from narcotic storage areas should be under controlled circumstances that require using large, bulky storage carts. Computerized card access or other methods of employee identification are essential, along with the provision for multitiered security zoning. Consideration should be given to video surveillance equipment, either on a routine or on an ad hoc basis.

Production

Increasingly, pharmacies are called upon to produce both sterile and unsterile specialized drug products for local consumption. These products may be used in routine clinical care or in clinical research. The most demanding area of the pharmacy from the facilities perspective is likely to be in the area of sterile manufacturing of injectable and ophthalmic drugs.

Sterile products may be prepared either from sterile pyrogen-free ingredients or from unsterile ingredients that are terminally sterilized by autoclaving or microfiltration. Each has different facility requirements for air handling, temperature control, electric power, fume hoods, and piped or bottled gases. Key to developing an adequate facility are air handling and the ability to quarantine critical areas of production to ensure the sterility of products and also to prevent environmental contamination. The majority of sterile product preparation is conducted under laminar airflow conditions, either in laminar flow hoods or laminar airflow rooms in larger facilities. Attention should be paid to the routine cleaning needs of such facilities. Requirements for seamless walls, floors, and ceilings should be considered, as well as an adequate segregation process that has limited access, ensures adequate airflow, and prevents environmental contamination.

Facilities that prepare sterile and pyrogen-free products routinely are likely to need both steam and ethylene oxide sterilization facilities. In addition, specialized equipment such as lyophilizers may be needed. Larger facilities may require high-volume bottle washers as well. In all circumstances, attention should be paid to preparation areas for both material and personnel. Ideally, each area within the drug preparation facility will have access to eyewash fountains and sinks.

Although federal good manufacturing practice (GMP) requirements are presently waived for institutional pharmacies, it is likely that these facilities may come under the same strict facilities requirements as the pharmaceutical industry. Refer to Chap. 11, "Industrial Facilities," for a detailed discussion of GMP.

Open processing of nonsterile products such as ointments, tablets, and capsules presents much less of a burden on facility design than the production of sterile products. Attention should be directed, however, to airflow as a means of segregating and maintaining cleanliness. Regardless of the products produced, adequate facilities must be provided for storing raw materials and finished products. In addition, facilities designers should consider the need for quarantine facilities and laboratory facilities for raw-material and finished-product analysis.

Drug Distribution

Institutional pharmacies routinely deliver multiple medications to individual patients on an ongoing basis. Most pharmacies warehouse hundreds, if not thousands, of different products. In contrast to dietary delivery requirements, delivery from the pharmacy is routinely ad hoc. As a result, thousands of transactions are likely to occur daily in larger institutions. Because the delivery process usually involves a comparatively small number of items to patients at one time, the use of computer-based pneumatic-tube systems has become commonplace. The costs of acquiring and maintaining these systems are high; therefore, expectations for reliability are high. Great care should be taken in both the structural and operational design of such systems. They must be managed to provide maximum use by both senders and receivers. Because they are easy to use and are inherently reliable, they are prone to excessive use.

Hospitals are increasingly relying on automated drug-dispensing machines to meet the drug storage needs of patient care units. In addition to the space requirements, electrical and telephone service need to be considered. As noted earlier, the security of controlled substances should always be ensured. This is especially true at the nursing unit level. Regardless of where they are stored, medications should be under constant lock and key. Drugs that fall under the jurisdiction of the Drug Enforcement Agency of the U.S. Justice Department must be doubly locked. Conventional keyed systems are plagued by misplaced and lost keys. Newer locking systems with key card access should be considered for both pharmacy and nonpharmacy storage

Information Requirements

The goal of drug therapy is achieving a defined therapeutic outcome. Unfortunately, no therapeutically active drug is without clinical liability or risk. This risk can be reduced by properly applied information and information technology.

The reasons for adverse drug events are complex, but a key element is often the lack of vital information at critical times during the drug-ordering process. Sophisticated computer systems are being deployed in unprecedented numbers. The roles of the computer systems are multiple and range from collecting and aggregating medication orders to internal routing of drugs through the distribution process, identifying patients, and verifying as well as documenting drug orders. On-line real-time patient demographic data along with pertinent laboratory findings and concomitant drugs provide prescribers and other caregivers with information essential for the safe and efficient use of drugs. As drug therapies such as those applied in treating cancer become more specific and powerful, the provider's need for information is essential to a safe and efficient process. A physician order-entry system will play a larger and larger role in the future. Facilities design will need to take this increasing need for computers into consideration.

CONTRIBUTORS

Margaret Dempsey Clapp, M.S., Director of Pharmacy Operations, Massachusetts General Hospital, Boston, Massachusetts

Frank Massaro, Director of Pharmacy, North Shore Medical Center, Salem, Massachusetts

ARTICLE 9.3.3
TRANSPORT SYSTEMS

John W. Basch
Vice President, Cini-Little Schachinger, LLC, Dagsboro, Delaware

Today's health care facilities cover a wide range of building designs, from small outlying clinics to large medical center complexes. Each of these of these facilities has its own unique set of transport and access requirements that must be considered during planning and design. This article concentrates on the larger types of facilities, ranging from ambulatory surgery and special care centers to the large medical center complexes.

Most large health care facilities are complexes that consist of a variety of functional building types, including the inpatient hospital, medical office buildings, a variety of special care buildings (cancer centers, ambulatory surgery centers, imaging buildings, etc.), educational buildings, and various support-service facilities. Most of these buildings have requirements for transporting people and materials. Separation of traffic types and clean and soiled materials is also a major consideration and is desirable.

The balance of this article is a discussion of the unique transport requirements of health care facilities and the unique functional requirements of the various systems found in today's health care facilities.

UNIQUENESS OF HEALTH CARE TRAFFIC AND TRANSPORT REQUIREMENTS

Health care facilities are different from many other large buildings or building complexes because they are populated by a variety of types of people; they also have substantial vehicular transport requirements, transport time constraints, and demands that are not found in other facilities. The vehicular transport requirements include the movement of patients in wheelchairs and stretchers, various types of supply carts, patient meal carts, and the like, many of which have time constraints. An inefficient transport system can be the cause of delays that in turn result in lost time and money.

People-Transport Requirements

People-transport requirements can be divided into two very broad categories: *ambulatory pedestrians,* which includes patients, staff, and visitors, and *patient transport.*

Ambulatory pedestrian traffic consists of inpatient and outpatient traffic, staff, and all types of visitors. Both types of patient traffic require special consideration. The ambulatory inpatient moves about the hospital for a number of reasons that range from a therapeutic walk to trips to the gift shop or cafeteria. Some of the patients may be using walkers, crutches, or canes. Outpatients are generally more mobile; however, they too can require walkers and crutches. Patients tend to be slower and less steady than other types of passengers and require additional time moving in and out of elevators.

Staff traffic includes physicians, other types of patient care personnel, and a variety of other support personnel. Staff traffic creates other requirements and challenges for transport-system planners. The staff in a larger health care facility tends to generate a great deal of interfloor traffic during the day. It also generates substantial peak requirements during shift changes and the lunch period.

Visitors in a health care facility generally fall into two categories: *patient visitors* and *business visitors.* Both types of visitors create a fairly light demand throughout the day and early evening.

Patient-transport requirements are unique to health care facilities. In smaller facilities, these requirements include moving patients in wheelchairs or allowing room for those who have walkers and crutches. In hospitals or larger medical center complexes, the patient-transport requirements must also include movement of patients on stretchers and in beds. The transport requirements can be routine movement of patients to and from treatment and/or diagnostic areas or the urgent transport of a trauma patient accompanied by a trauma team. Examples of urgent transport requirements include trauma situations, movement of patients from surgery to recovery, moving cardiac patients from the emergency department to the cardiac care unit, and so on. In many cases, an escort and/or family member accompanies the patient during these movements.

Material-Transport Requirements

Material-transport requirements themselves are not unique to health care facilities; however, the volume of material-transport activities in a health care facility is unique. Health care facilities also demand separating clean and soiled materials during transport, a consideration that must be incorporated into all planning and design.

There are two major categories of material-transport requirements in a health-care facility: *bulk-material-transport* items and *unit-material-transport* items. The bulk-material-transport requirements include medical and surgical supply replenishment, clean linen distribution and soiled linen collection, delivery of administrative and laboratory supplies, prompt delivery of patient meals, waste removal, delivery of surgical case carts, and a wide variety of other bulk-transport requirements.

Unit-materials-transport requirements include stat delivery of critical medications and laboratory specimens, first dose or unanticipated medications, and blood products, and non-stat delivery of laboratory specimens. The main issue for all of the pharmacy and laboratory transport requirements is reducing the turnaround time from the time a medication or test is ordered until the medication or laboratory result is delivered.

No One System Answers All Needs

Obviously, when one reviews the basic transport requirements discussed before, it becomes apparent that no single system satisfies the health care facility's transport needs. Every facility will have a combination of systems to satisfy its transport requirements. The challenge is to get the right combination of systems (elevators, escalators, materials handling systems, etc.) to satisfy the facility's transport needs without leaving gaps or being overly redundant. One approach to determining those needs and developing appropriate systems is a master transport study. Details of the master transport study are discussed next.

CONCEPT OF THE MASTER TRANSPORT STUDY

A master transport study addresses a facility's transport requirements and a wide range of potential alternatives for meeting those needs. This technique can be used for improving the transport systems in an existing facility, as well as in developing the concepts and systems for new facilities. It is a comprehensive look at a facility's total transport requirements, systems, and equipment. It also examines how the facility's operational philosophies interface with the planned transport systems.

These are the primary goals of a master transport study:

- Define and analyze the current and foreseeable transport needs of the facility.
- Evaluate current and future circulation patterns along with the vertical and horizontal transport systems to determine their ability to meet the established transport requirements.
- Develop alternative transport solutions that respond to the established transport needs (current and future) of the facility.
- Review and develop recommendations pertaining to operational areas or functions (i.e., materials management, materials handling, waste management, patient transport, etc.) impacting or being impacted by the proposed systems.
- Provide a recommended course of action that responds to the facility's transport and operational needs.

Vertical Transport Systems

Virtually every health care facility has a vertical transport system. Large facilities have a combination of a number of systems. Coordinated planning, designing, and implementation are required to provide efficient, cost-effective vertical transport systems for larger facilities. The balance of this article addresses the various vertical transport systems normally found in a health care facility.

Elevators.   Elevators are by far the most common transport system in health care facilities. The planning and design of elevators for larger health care facilities can be a complicated process. It is extremely important to get the appropriate type, speed, and number of elevators for the facility. A number of different types of elevators are available to answer the facility's vertical transport requirements, including passenger, service, freight, and special-purpose elevators. Each type of elevator answers a specific transport requirement. An elevator study should be performed to determine the appropriate mix, number, speed and configuration of elevators in a health care facility. The persons performing the study should be thoroughly familiar with the operations of health care facilities and the particular facility under study.

Passenger elevators should be designed to transport ambulatory pedestrians efficiently and safely. The elevator cabs should be wide and shallow to expedite the transfer of people in and out of the elevator. Ambulatory patients, visitors, and staff (depending on the facility's policies) normally impact these elevators. Because of the types of passengers using these elevators, there is a need for an excellent elevator door and passenger protection system.

Service elevators, also known as hospital elevators, should be designed to transport vehicular traffic (stretchers, wheelchairs, beds, a variety of carts and equipment, furniture, etc.) effi-ciently and safely. These elevators are sometimes mistakenly called "freight" elevators. Ser-vice elevator cabs should be narrow and deep to expedite the transfer of the vehicular traffic in and out of the elevator. Because these elevators normally handle patient transport, special attention must be given to the types of patient-transport activities, the size of the transport team and vehicles, and the urgency of the demand. All of these factors affect the final config-uration of the elevators. Areas that require thought include the size and type of door opening, the platform size, access to clean the cab interior, the type of control system, and the eleva-tor's ability to handle emergency transport.

Freight elevators have limited application in health care facilities. Most often, they are found connecting a warehouse facility with the medical center. The true freight elevator has vertical biparting doors and is designed to handle very concentrated loads—for example, a forklift with pallet loads of materials.

Special-purpose elevators are found in many health care facilities. Special-purpose eleva-tors are provided for applications such as trauma team and open-heart patient movement and dedicated material transport (automated guided vehicle systems, cart lifts, etc.)

Consideration must be given to the proper design of the elevator lobbies. Elevators should have dedicated lobbies whenever possible. Under no circumstances should elevators open directly into heavily traveled corridors.

Escalators. Escalators are found in large health care facilities—normally in ambulatory care or medical educational institutions. They are designed to move large numbers of people in a short time. Care must be taken in using escalators in health care facilities. They should not be the only means of vertical transportation. Convenient elevators should be close by for peo-ple who do not wish to use an escalator, for handicapped access, and for passengers who require a walking assistance device. Patients who use crutches or walkers or are elderly or not steady should not use escalators. Escalators should not be used where there is a large pedi-atric population. Thought must be given to specifying the width of an escalator unit. Narrower-tread units enable holding onto both handrails at the same time.

Inclined Ramps. The most common use of inclined ramps in health care facilities is in mak-ing a transition from one building to another where there is a difference in floor heights. Some of the factors that should be considered in designing ramps include the type of traffic that will be using the ramp, the type of floor surface, and the grade of the ramp. The traffic type and mix should be defined. The type of anticipated traffic has a bearing on other considerations. For example, if a ramp will be handling vehicular traffic, the type of floor surface, the width, and the grade of the ramp are extremely important. Will the vehicular traffic be manual (stretchers, wheelchairs, carts, etc.) or powered (forklifts, robotic devices, and/or vehicles, etc.)? All of these questions must be answered before finalizing the design of the ramp or ramp system.

Cart Lifts. Cart lifts were very common in health care facilities at one time; however, due to poor design and operational problems, they have waned in popularity. There are still applica-tions that are appropriate for the health care setting—namely, as dedicated cart lifts connect-ing vertically stacked areas such as the central sterile-processing department with the operating room and/or labor and delivery suites, where there is controlled access to the system.

Dumbwaiters. Most dumbwaiters in health care facilities are dedicated units. They are fre-quently used to move laboratory specimens, small supply items, food-service items, and records.

Material-Transport Systems

Material-transport requirements in health care facilities fall into two very broad categories: *bulk-material transport* and *unit-material handling.* In addition to this division, material transport is further divided into clean and soiled materials transport. The presence of or absence of an automated material-handling system will have a significant impact on a facility's elevator system. For example, a well-functioning pneumatic-tube system can take as many as 6 trips per occupied bed per day off the facility's elevator system.

Bulk-Material-Transport Systems.   By far the most common type of bulk-material-transport system found in health care facilities is the manual system. Manual systems are chosen because of the low capital investment and because automated systems must be planned for early in the design process. Manual systems have a significant impact on elevator systems. As previously indicated, bulk-material transport includes medical and surgical supply replenishment, clean linen distribution and soiled linen collection, delivery of administrative and laboratory supplies, prompt delivery of patient meals, waste removal, delivery of surgical case carts, and a wide variety of other bulk-transport activities. Each of these activities has a set of transport criteria that must be considered when planning the transport systems. These issues must be considered:

- Separation of clean and soiled traffic
- Potential conflict with patient-transport activities
- Coordination of transport requirements with those of other departments
- Separation of pedestrian and vehicular traffic
- Separation of public traffic from service-related traffic.

Automated bulk-material-transport systems can be found in larger health care facilities. Examples of automated bulk-material-transport systems include automated guided vehicle systems (AGVSs), overhead rail conveyor systems, and large-diameter pneumatic trash and linen systems. During the last 20 to 30 years, about 30 bulk distribution systems have been designed for health care facilities. Within the last few years, a number of these systems have been upgraded and/or replaced by systems with new technology. Several of the systems have been decommissioned and removed from service. Any facility that considers these systems should analyze its transport needs in detail and carefully examine the cost justification for the proposed systems.

The AGVS is a driverless vehicle system that transports bulk materials throughout the health care complex. These systems are generally used in larger health care facilities (500 beds and more). They should be designed to move in dedicated rights-of-way and areas where there are not any patients. They also normally have dedicated elevators for vertical movement. The AGVS can be mixed with staff traffic. The proper design and implementation of an AGVS requires very detailed analysis and planning to maximize the system's cost-effectiveness. The primary advantage of the AGVS is its long-term flexibility and its economic impact on a facility's bulk-material-handling costs. In addition to the actual bulk-material-transport requirements, consideration must be given to patient and employee safety, the interface with other systems, and the codes that regulate these types of systems. Other systems that will potentially interface with the AGVS include material management information systems, automated cart-wash systems, waste-dumping systems, facility maintenance systems, and security systems.

Overhead-rail conveyor systems were fairly common at one time in health care facilities. Today, there are very few still running in health care facilities. These systems also require dedicated rights-of-way and vertical transport systems. Because they are diminishing in number, this article does not address them in any further detail.

Pneumatic trash and linen systems are still found in health care facilities. These systems should always be designed so that trash and soiled linen are transported in separate systems. Factors that need to be considered when planning this type of system include the volumes of waste and soiled linen that will be transported, the types of waste, and the layout of the systems. Because of the sizes of the blowers required for these systems, they are very noisy, and care must be taken in locating the blowers. Gravity trash and linen chutes are still used in some facilities.

Two very important considerations in reviewing any automated system include the ability of the facility to maintain the system and the ability of the employees to use the system. The systems can be maintained by in-house staff or by contract services. Employee operational training is an extremely important consideration for all of the automated systems. Facilities must ensure that their employees receive initial as well as ongoing in-service training on these systems.

As indicated at the beginning of this article, there is no one solution to all of a facility's transport requirements. Virtually all facilities have a combination of systems. Care must be taken when planning the overall bulk-transport system to ensure that all requirements are satisfied and that the facility has some redundancy in its systems if one of the systems is out of service for a time. All of these systems are mechanical, so they will be out of service at times, at the very minimum for maintenance.

Unit-Material-Transport Systems

Similarly to bulk-material-transport systems, unit-material-transport systems also can be classified into manual, automated, and combination systems. The commonest unit-material-transport systems are manual courier services, pneumatic-tube systems, and automated box-conveyor systems. Pneumatic-tube systems are by far the most used automated unit-materials-transport systems in health care facilities and are found in three sizes: 4-in-diameter, 6-in-diameter, and 4- × 7-in-oval systems. Automated box-conveyor systems are used to transport larger payloads, such as medical records, x-ray films, and small supply items. Both of these systems routinely move medications, laboratory specimens, small supply items and records. Today's systems are computer controlled and extremely reliable. As electronic technology becomes more prevalent in moving data (records, images, etc.), these systems truly are unit-material-handling systems.

Planning and design of unit-material-transport systems requires a thorough analysis and understanding of the facility's unit-material-transport needs. Planners and designers must take the following into consideration when developing unit-material-transport systems:

- Which items require transport?
- What is the expected volume for each item that requires transport?
- What is the timing or frequency of the transport requirements?
- What is the time sensitivity of the transports (stat requirements, priorities, etc.)?
- Can the items be safely and successfully transported by the systems under consideration?
- What security issues are there for items such as medications and blood products?
- What are the payload requirements (how big are the items being transported)?
- What areas of the facility must be served by the proposed unit-material-transport systems?

Once the preliminary data have been collected, the analysis of potential alternatives can begin. During this process, the planners must also determine which systems provide viable solutions to the facility's unit material transport requirements. After the potential alternative systems have been determined, the planners must determine if the proposed system will fit into the existing facility, if the system is economically justified, and how the effectively the system will serve the needs. Preliminary layouts should be developed to review geographic considerations, the size of the system, etc. The economic justification process should use techniques such as life cycle costing analysis. There will always be a manual system in place to handle the items that cannot be accommodated by the automated material-transport systems of choice. It is extremely important that the planners develop the complete unit-material-transport system. Nothing should be forgotten. The final decision on unit-material-transport systems should be based on the life cycle costs of the various alternatives.

INTERFACE WITH OTHER SYSTEMS

Any of the automated material-transport systems can and should interface with other information- and material-handling systems in the health care facility.

Facilities Management Information Systems

Automated material-transport systems should interface with appropriate facilities management to monitor operation of the system, troubleshoot the system when problems occur, and track preventive maintenance requirements and activities.

Material Management Information Systems

From a material management perspective, the automated material-transport systems can be interfaced with accounting, patient billing, and management engineering systems. The systems can provide material management with records of deliveries (where, when, etc.), inventory levels, and employee performance data.

Other Material-Handling Systems

Automated material-transport systems can be designed to interface with automated pharmacy stock-picking systems, automated storage and retrieval systems, automated waste-dumping systems, and many other systems, if properly planned and designed.

CODE AND REGULATORY REQUIREMENTS

Following is a partial list of the various codes and regulatory agencies that must be considered when planning and designing automated material-transport systems.

- *ASME/ANSI A17.1* is the basic safety code covering elevators and other vertical transportation equipment. It is used in conjunction with the National Electrical Code, ANSI/NFPA 70, and ANSI/NFPA 80, *Fire Doors and Windows,* to govern all aspects of vertical transport design and installation. Other codes include ANSI/UL 10B, *Fire Tests of Door Assemblies.*
- *OSHA* regulates safety requirements for material-handling systems and facilities.
- *The Americans with Disabilities Act (ADA) Accessibility Requirements* apply to elevators and escalators.
- *NFPA* rules apply to vertical transport and material-handling systems.
- *JCAHO* requirements must be taken into consideration when planning and designing material-handling systems.
- *Various state and local codes* will take precedence over the previous codes whenever they are more stringent.

CONTRIBUTORS

Anand K. Seth, P.E., C.E.M., C.P.E., Director of Utilities and Engineering, Partners HealthCare System, Inc., Boston, Massachusetts

David Beard, RTKL Associates, Baltimore, Maryland

SECTION 9.4
STRUCTURAL SYSTEMS

Michael L. Brainerd, P.E., Principal
Simpson Gumpertz & Heger, Inc., Arlington, Massachusetts

LOADS

Live load requirements vary somewhat between building codes, but typical code-prescribed uniform live loads for health-care facilities are 100 lb/ft^2 for operating rooms and laboratories, 80 lb/ft^2 for corridors above the first floor, and 40 lb/ft^2 for private rooms and wards. Along with the usual dead load of the structure's self-weight (the weight of finishes and mechanical, electrical, and plumbing (MEP) components), health care facilities typically contain relatively heavy equipment that must be considered individually, such as specialty medical equipment, process equipment, operating-room lights, and the like. Many diagnostic equipment units today are mobile, and these units are brought to the patients. The wheel loads of this heavy equipment must be considered as concentrated live loads on the structure. Such concentrated loads are often considered separate from and not coincident with the full uniform live load. Radiology, radiation therapy, nuclear medicine and magnetic resonance imaging (MRI) diagnostic units, and other similar devices may contain very heavy medical equipment that requires special consideration. These units can weigh many tons, and the rooms that contain them must often be shielded with lead lining, concrete enclosures, or other material, although self-shielding units are now available. The heavy weight of the units and shielding significantly impact the gravity load and also the seismic loads. Health care facilities, especially hospitals, typically require large emergency generators, which are also very heavy, particularly if equipped with a load bank for testing. Isolating generator noise and vibration from the remainder of the hospital may require a heavier structure and base isolation. Chillers pose similar challenges. Often, significant savings in the cost of the structure can be achieved by locating such equipment on-grade.

Due to the importance of health care facilities, especially hospitals in posthazard emergency management, code-prescribed lateral loads (wind and seismic) and lateral load resistance requirements for health care facilities are usually more stringent than those for similar buildings with other occupancies, such as office use. The lateral loads are increased about 10 to 15 percent to ensure that such facilities are operational after the wind or seismic event. In seismic zones, most codes require supporting all but the lightest MEP components so that the lateral loads of the components are transferred to the structure.

Modern microsurgical equipment and laboratory equipment such as microscopes can be very sensitive to vibration; therefore, vibration of structural elements resulting from foot traffic and reciprocating equipment must be evaluated for its effect on such sensitive equipment.

NEW CONSTRUCTION

Modern hospital space layouts tend to be modular. Functional modules such as groups of examination rooms and adjoining support spaces are repeated in a pattern throughout the floor space. Structural bay sizes are selected to accommodate module dimensions. Modern hospital bay sizes are typically moderate and are in the range of 25 to 30 ft. Health-care facilities tend to have more MEP components, such as pipes, ducts, conduits, and equipment, which must fit above the ceiling and below the structure. A minimum clear space of about 3 ft between the ceiling and the structure is desirable. This requires story heights of about 14 ft for typical hospital ceiling heights of about 9 ft. Many older hospital facilities were constructed with story heights of about 11 ft. Consequently, the story heights of additions to such facilities are often constrained and thereby require a delicate balance between bay size, structural system depth, and ceiling height.

In the design of modern hospital buildings, consideration should be given to the adaptability of the structure to future changes in medical equipment, functional layout, and MEP systems. Framed systems of steel or concrete are the most flexible in accommodating future floor and roof openings. Often new buildings on nearly fully developed hospital campuses are designed for future vertical expansion. This requires identifying the number of future floors and designing the initial structure and foundations for the gravity and lateral loads of both the current and future configurations.

If a new building is expected to connect to an existing facility, there is a great temptation to match the floor heights. Obviously, this floor connection will aid greatly in the flow of patients, staff, and visitors from existing buildings to the new building. This requirement poses a considerable challenge for the structural systems to allow as much clear space above the ceiling as possible. The structural systems design of the new building may be influenced significantly by this need.

If floor heights are not similar, there may be ramping, tunnels, and bridge connections from the new building to the existing structure. It is important that these structures be capable of carrying the weight of utility (e.g., steam and chilled water) lines.

RENOVATION

Older hospital facilities are often renovated to accommodate newer functional layouts and modern equipment. Modern framed hospital buildings can fairly readily accommodate interior modifications and MEP changes, whereas older bearing-wall structures often require significant structural modifications to accommodate such changes. One-way slab-and-beam systems more readily accommodate new floor openings that can be located between beams. Steel-framed structures best accommodate the addition of supplemental framing and the strengthening of existing framing, due to the ability to field-weld connections and strengthening elements.

Hospital buildings can be expected to undergo multiple renovations during the life of the facility. The facilities manager must keep complete records of the original structural design and construction and of all subsequent structural modifications, including calculations, design drawings, as-built drawings, and shop drawings. These records are invaluable for evaluating and designing structural modifications efficiently and avoiding design errors.

CONTRIBUTOR

Fredrick M. Gibson, A.I.A., Partner, Taylor and Partners, Boston, Massachusetts

SECTION 9.5
ELECTRICAL SYSTEMS

Arjun B. Rao, P.E., Vice President
BR+A Consulting Engineers, Inc., Boston, Massachusetts

RELIABILITY

The design of every hospital electrical power distribution system should have reliability, availability, and flexibility as the primary goals, along with provisions for maintenance while maximizing uptime. Although these terms involve advanced statistical analysis, they can be explained as follows:

- *Reliability.* A method by which one evaluates equipment and systems on the basis of failure probability
- *Availability.* The probability that a unit will be operational at a randomly selected time in the future
- *Flexibility.* The ability of a system to recover from power irregularities

Health care facilities operate 24 hours a day, 7 days a week, which requires systems to be available when needed. Further, any required shutdown of electric power should be scheduled with all affected departments, and the duration of the outage should be kept to a minimum consistent with the tasks to be accomplished and employee safety. Both system design and system operation should respond to the criteria that many types of loads cannot easily be turned off in a hospital. Refer to Sec. 9.11, "Electrical Utility Management Program," for a discussion of the issues surrounding planning and conducting hospital electrical system shutdowns.

ELECTRICAL POWER SYSTEMS

Electric Services

Select electric service capacity with adequate spare capacity to provide for future load growth as health care facilities become more equipment-intensive and space utilization becomes denser. An example of this is providing a 13.8-kV rather than a 4.16-kV primary source. Many hospitals request that utility companies provide multiple power sources, such as spot networks, dual primary lines, and other features like automatic throw-over switches between power sources. (Refer to Art. 5.2.1, "Service Entrances," for definitions and examples of these concepts.) Although totalized metering and establishing annual charges for standby sources are not reliability issues, they can be pursued to lower the cost of electricity. The use of static transfer switches between utility sources (not commonly used for cost reasons) should be discussed with the owner and the utility company. Although static transfer switches that switch between utility sources can enhance service availability and provide more reliable normal power to hospital power systems, their costs and benefits should be analyzed.

Primary Power Distribution

Refer to Art. 5.2.1, "Service Entrances," for a discussion of the merits of different electric service configurations. Loop feeders should be considered for primary power to hospital power distribution systems. Radial feeders are weak links when used solely in conjunction with a single transformer. Primary services should be designed with total redundancy to allow the loss of one transformer or one utility line. Nonredundant transformation requires load shedding upon the loss of one transformer. The use of liquid-filled transformers instead of dry-type transformers may be considered in order to reduce noise and vibration. Spot network utility transformers are not commonly offered by many utility companies but are a great feature to have in hospital design because the loss of any single primary line will be transparent to hospital operations. The $N + 1$ (redundant) transformation offered in spot network systems provides a highly reliable power source. The disadvantages of a spot network system are the possibility of a common-mode failure on the collector bus, higher available fault currents, and the resulting higher cost. A main–tie–main system requires power interruptions if the tie breaker must remain open until one main breaker is opened. Designs with main–tie–main systems may or may not include load shedding.

Power Monitoring

Appropriate metering devices should be provided on the main switchboards and other large distribution panels to monitor and record power fluctuations, harmonic content, and the like. These devices will be useful for preventive maintenance scheduling and troubleshooting, and as early warning devices and operational data storage. They can also provide power-factor and peak-demand data, which can be used in turn to lower the cost of electricity.

Low-Voltage Systems (480Y/277 V and 208Y/120 V)

Power distribution systems at both 480Y/277 V and 208Y/120 V are required in almost all hospitals. Lighting and motor power loads usually require 277-V and 480-V electrical service, respectively. The main distribution voltage choice of 480Y/277 V provides larger incoming power capacity for the *same* current rating, and requires less equipment space.

Double-Ended Versus Single-Ended Substations. Hospital power sources should be designed with double-ended substations rather than single-ended substations. Failure of the main transformer, the main breaker, or the single primary-source radial feeder in a single-ended substation could result in total normal power loss to that portion of the hospital. In double-ended substation systems, consider the use of automatic main–tie–main transfer schemes for higher uptime instead of manual key interlocks. In addition, the concentration of the load on one substation bus hampers the hospital's ability to schedule planned shutdowns for maintenance.

Circuit Breaker Trip Units. Refer to Art. 5.2.2, "Electrical Distribution Systems," for a detailed discussion of protective coordination and the equipment used to provide protective coordination. The following discussion assumes a grasp of the content of that article. Most applicable codes mandate a coordinated power distribution system. It is very difficult to attain full protective coordination by using thermal magnetic breakers. Breaker trip units with electronic trip units should be considered at least to the third level of the power distribution system. Two levels of ground-fault (G) protection are required by codes on the main 480-V switchboards of hospitals. Main breakers should have long-time (L), short-time (S), and ground-fault (G) trip units. A unit with all three features is designated LSG. Instantaneous (I) trip units on main and tie breakers should be avoided if possible because of the need to avoid tripping those breakers until breakers downstream have had a chance to clear the short circuit. Feeder breakers in the main switchboard should have either LSG or LSIG trip units, depending upon the type of protection offered downstream from those breakers. Using ground-fault trip units on bus duct plug-in breakers is strongly recommended to avoid tripping the entire busway riser when a local ground fault occurs. The use of LSI trip units on 208 Y/120-V system breakers may be considered.

Fully Rated Versus Series-Rated Overcurrent Devices.   The notion that a fully rated thermal magnetic distribution system is better coordinated than a series-rated system is untrue. A thermal magnetic breaker will open or trip on a fault condition anywhere between 8 to 12 times its ampere rating, which could result in opening or tripping the upstream breakers, as well as breakers nearest to the fault, depending on the magnitude of the fault currents. Refer to Art. 5.2.2, "Electrical Distribution Systems," for a more detailed discussion of these issues.

Ground-Fault-Sensing Overcurrent Devices (480-V System Only).   Because most faults in a distribution system are phase-to-ground faults rather than phase-to-phase faults, provision of ground-fault-sensing trip units for better protective coordination should be considered in the following equipment:

- *Site lighting panel.*   At least the main breaker in the site lighting panel should have ground-fault trip sensing.
- *480-V normal-power busway plug-in units.*   Busway risers generally emanate from the main switchboard feeder breakers that are required to have ground-fault sensing in hospitals. Plug-in-units on the normal-power busway riser should be provided with ground-fault sensing to separate each particular plug-in-unit load in a ground-fault condition in lieu of tripping the busway feeder breaker in the switchboard.
- *Motor control centers.*   All (normal power) motor control starters should have motor circuit protectors with ground-fault sensing to isolate the failed motor or motor circuit rather than tripping the entire motor control center off-line.
- *Lighting fixtures (277-V ballast failures).*   One approach is to provide ground-fault trip units on branch circuit breakers that serve lighting fixtures to isolate the failed ballast. Because it is inadvisable to provide ground-fault tripping on emergency lighting circuits, another more reliable method would be to provide all 277-V ballasts with built-in fuses.

Fire Separation.   Fire separation of emergency distribution equipment from the normal power distribution equipment in electric rooms is mandated in certain states and counties and by certain NFPA standards. Regardless of the applicable codes and regulations, it is always a good idea. Designs should strongly adhere to this concept to avoid failure of both the normal and the emergency services in any single event, such as a fire, short circuit, or explosion.

Radiology Equipment—Power Distribution.   Due to the ever-changing technology of radiology equipment and its increasing use throughout health care facilities, flexible power requirements should be planned for it. Provision of the following features in the power distribution system is highly recommended:

- Oversized and adjustable trip breakers
- Oversized feeders
- Allowances in construction costs to accommodate unknown construction changes
- Close coordination with x-ray vendor documents
- Portable radiology equipment—provisions for patient rooms if portable x-ray equipment is used.

- Interfacing room light controls with radiology equipment
- Choice of the type of light source, depending upon the frequency of on/off cycles
- Special switching to shut down radiology equipment when doors are opened
- Special emergency shutdown switches
- Engineering study required when using one transformer for multiple rooms.

EMERGENCY POWER SUPPLY SYSTEMS

An on-site diesel- or gas-engine-driven generator is required in all hospital power distribution systems as a backup or a standby source to the normal power or the utility source. Engineering emphasis and evaluation is required in sizing and locating the generators, intake and exhaust air systems, and fuel supply methods and sources, and in complying with local emission regulations. Refer to Art. 5.2.3, "Emergency Power Supply Systems," for general concepts and definitions regarding emergency power in all facility types, including health care facilities.

Generator-Set Sizing

Some engineers believe that 4 to 6 W/ft^2 would be a good rule of thumb in sizing generators that are used as the standby source in hospitals. This rule of thumb does not include chiller loads, high-density radiology loads, and computer loads, if such loads are required on emergency power. Others believe that "Unless a hospital has electric heat or an inordinate amount of equipment (i.e., HVAC chillers) on the standby generator, a generator sized at 3 to 3.5 watts per square foot has virtually no chance of being overloaded."[1] The demand on an essential power system will vary, depending upon the type of facility being served. Typically, research laboratory buildings are usually more heavily loaded than patient care buildings, and critical care areas are more heavily loaded than general patient care areas. The essential power system demand will also vary, depending upon the quantity of elevators permitted to operate simultaneously on emergency power. Therefore, both methods may be correct, depending upon the type of facility being considered, the issues just mentioned and others, and whether the generators being sized are for just one building or for a multibuilding facility. Those sizing generators need to know as much as possible about the facility under consideration.

Emergency Power Distribution (480Y/277V and 208Y/120V)

The generator voltage in many hospital distribution systems is at 480Y/277 V and is stepped down to 208Y/120 V via local delta-wye step-down transformers. Regulations require that hospitals test their emergency power supply systems monthly by transferring all transfer switches and running the generator at a minimum of 30 percent of its rated capacity. These monthly load tests must be for at least 30 min at a temperature that is sufficient to prohibit exhaust wet stacking from occurring.[2] Because of these required monthly live load transfers, it is possible for step-down transformer magnetizing currents to trip the instantaneous elements of circuit breakers that are not large enough to pass the magnetizing current. Therefore, it is important to ensure that such primary overcurrent protection design for each essential power system dry-type transformer considers magnetizing current inrushes. Selection of primary overcurrent protection and associated primary feeders on emergency systems should be carefully considered.

Essential System Branches

Three separate essential distribution subsystems are required in hospitals by NFPA 99.[3] Depending upon the size of the emergency power supply system (EPSS), the separation can be as little as a panelboard or as much as independent automatic transfer switches and all associated distribution equipment.

Life-Safety Branch of Emergency System.   This branch strictly provides power to egress lighting and exit signs, fire alarms, medical gas alarms, and only the other loads that are specifically listed in NFPA 99. The most recent edition of NFPA 99 must always be consulted because some load classifications have been switched between the various branches and systems, as new editions of NFPA 99 have been approved.

Critical Branch of Emergency System.   This branch strictly allows a majority of hospital emergency loads and restricts motor loads to fractional horsepower. The required loads are listed in NFPA 99, along with statements allowing hospitals to choose additional loads to be supplied from this branch.

Equipment System.   All motor loads are allowed on this branch. The use of the phrase "other selected equipment" on this branch should be verified with the local authorities, where required. This system is typically used for HVAC systems that serve critical areas.

Elevator Equipment Subsystem.   Separate automatic transfer switches and distribution systems for elevator motor loads are suggested in larger hospitals. This approach avoids the undesirable situation where the inrush current and typically nonlinear characteristics of the elevator controller loads adversely affect other equipment. It also allows the hospital to control the transfer switching times of the elevators for testing independently of the other equipment. The automatic transfer switch should be designed to carry all of the elevator loads in the normal power mode of operation and a predetermined (usually lesser) number of elevators when on the emergency power source. It is common practice in many jurisdictions to allow the operation of just one elevator per bank on emergency power. In some hospitals, however, this level of service may not provide adequate vertical transportation capacity, which in turn can negatively affect hospital operations. Also refer to Art. 5.1.6, "Elevators, Escalators, and Moving Walks," for a further discussion of vertical transportation issues generally (including codes) and Art. 9.3.3, "Transport Systems," for a further discussion of health care facility vertical transportation issues.

Automatic Transfer Switches Integral with Manual Bypass Switch.

At minimum, the life-safety and critical-branch automatic transfer switches should be provided with manual bypass switches to facilitate maintenance and repairs without interrupting power to the critical loads. This feature is recommended, however, for all transfer switch classifications. The greatest flexibility will occur when the bypass feature allows the load to be bypassed either to normal power or to emergency power.

Closed-Transition Transfer Switches

Closed-transition automatic transfer switches are becoming increasingly popular in hospital distribution systems because they prevent a load-break transition of power from the emergency to the normal power source and vice versa. Some electric utility companies do not recommend using closed-transition transfer switches, so it is wise to consult with the utility regarding this issue. The hospital engineer should understand the differences between the closed-transition bypass switch (where the bypass feature only is without interruption of power) and the closed-transition transfer switch (where the transfer feature between two live sources is without interruption of power.) Care should be exercised in determining the fault currents to the distribution system because the momentary fault currents are cumulative—that is, from both the emergency and normal power sources. There are also other factors to be considered, such as block loading of generators and voltage fluctuations when specifying closed-transition systems. In addition, it is important that users do not become complacent about the possibility of power outages with a closed-transition system because a short outage will be caused by an actual utility power outage before the standby generator set starts.

NEC-Mandated Minimum Circuits and Receptacles at Each Patient Bed Location

Even though NEC specifies the number of receptacles and circuits at each patient bed location (headwall), the users and caregivers may require many more outlets at a particular type of bed location based on the medical procedures to be performed. Some consider it good engineering practice to provide 6 to 8 receptacles at each general patient care bed location, and 8 to 16 outlets at a critical care bed location. The following outlets should be on the emergency circuits:

- Physiological system monitoring outlets
- Patient bedroom night light
- Patient reading light
- One bathroom light

Code-Mandated and Recommended Loads on Emergency Power in Hospitals

- *Air-handling units serving operating rooms.*   This is good engineering practice.
- *Selected cooling loads for operating rooms, elevator machine rooms, and procedure rooms.* This is not code mandated in most areas but is highly recommended in all hospitals.
- *Air-temperature control (ATC) system.*   All ATC systems should be on emergency power to allow the proper operation of mechanical equipment on emergency power. It is generally good practice for ATC systems on emergency power to have uninterruptible power supplies (UPSs) or battery backup so that the ATC systems do not lock up, reset, or shut down mechanical systems upon transfer of the power source from normal (utility) power to emergency (generator) power.
- *Compressed-air and vacuum system pumps.*   This is code mandated in most areas.

Operating Rooms

The operating rooms (ORs) are classified as critical care areas. The decision as to whether specific ORs are to be considered wet locations or dry locations should be made with input from the hospital's clinical administration staff. If an OR is classified as a wet location, an ungrounded system (isolated power system [IPS]) should be used. Two isolation panels are required in a wet OR as mandated by the NEC. One isolation panel will be on normal power and the second isolation panel will be on critical power, or the two isolation panels will be fed from the two seperate critical branch automatic transfer switches and related distribution systems. Using breakers with ground-fault sensing and tripping is not allowed in a hospital emergency distribution system. If the hospital elects to classify its OR as a dry location, a grounded system may be specified. Some hospitals require an IPS in all ORs because their clinical staffs have determined that the design allows the surgeon greater control of the entire OR environment. Although single-phase isolation panel ratings typically vary from 3 to 10 kVA, typical isolation panel kVA ratings for an OR range from 5 to 10 kVA, depending upon the size of the OR and the procedures done therein. Locate the isolation panel so that its face is in the clean corridor, not in the OR. This allows panel monitoring and troubleshooting without non-OR personnel in the OR. Large operating rooms (larger than 400 ft^2) require 20 to 25 branch circuits, and dual isolation panels are highly recommended. Separate isolation panels, centrally located, can be used for outlets for YAG and argon lasers. These are usually single-phase or three-phase units rated at 25 kVA.

Battery-powered lighting units should be provided in the general OR lighting fixtures to span the 10-s delay allowed in the transfer of power sources.

Designers should provide multiple distribution panels, feeders, and automatic transfer switches for redundant paths of power to critical care areas such as ORs and recovery rooms.

Other Systems

Nurse-Call Systems. Visual and audiovisual nurse call systems are required in all patient care areas. The medical staff assigned to each patient care unit should be consulted in determining the features of the nurse-call system for each unit. The AIA booklet *Guidelines for Design and Construction of Hospital and Healthcare Facilities* is a good resource for system design.[4] The following features should be considered in developing nurse-call systems:

- Microprocessor central control units
- Master control station dedicated to multiple patient care units
- Nurse-follower feature
- Nurse-locator feature
- Wireless pocket pager reply stations
- Code-blue systems
- Interphase outlets in patient headwall units
- Zone lights and zone annunciation
- Patient privacy rights

Clock Systems and Elapsed-Time Indicators. Synchronized clock systems are highly recommended and required in clinical care areas such as operating rooms, nurse's stations, and recovery rooms. Synchronized clock systems are available that are either hard-wired or that use transmitter technology. Clock systems use either analog or digital display units. Elapsed-time indicators are used mostly in operating rooms and procedure rooms. Another issue to be considered is using frequency generators for resetting the time versus using electronic or local controls.

Central Paging System. The use of the central paging system has declined with the ever-growing use of pocket pagers and cell phones, although many hospitals severely restrict cell-phone use on the premises. Overhead paging continues to provide a widely used form of public announcement in hospitals during emergencies and for staff notifications. The following points should be considered in designing paging systems:

- Speakers strategically located in patient care and non–patient care areas
- Controls—all-page and selected-area paging.
- Access to speakers from nurse-call systems, designated telephone sets, etc.
- Backup amplifiers to provide redundancy
- Interfacing and prioritizing between the central paging system and other systems, such as fire alarms, code calls, etc.

Lighting.   The Illuminating Engineering Society of North America (IESNA) has issued a recommended practice for lighting in health care facilities.[5] It provides guidelines for lighting in areas that are unique to health care facilities, and it is intended to be used by health care professionals, as well as by lighting designers. Also refer to Sec. 5.3, "Lighting Systems," for a further discussion of lighting for all facility types.

Telecommunications (Telephone and Data), Cable Television, and Patient Monitoring.
All of these systems are very specialized in health care facilities, and professional engineering and design assistance is required. Designers should consider the following in the design of telecommunication systems:

- Fiber-optic and copper cable plant design
- Main distribution frame (MDF) and intermediate distribution frame (IDF) room layouts
- Interface with diagnostic systems, such as filmless radiology systems and medical records
- Early input from system suppliers who are specialists in the field

Testing

Regular testing of electrical systems in hospitals is of paramount importance, all the way from the main service entrance equipment down to the patient care area receptacles. Refer to Sec. 9.11, "Electrical Utility Management Program," for a further discussion of some of these issues in health care facilities, and to Art. 7.2.3, "Electrical Testing and Maintenance," for a discussion of electrical testing in general.

Hospitals are required to perform receptacle testing and to maintain test records, including documentation of historical performance data in accordance with NFPA 99 and the requirements of the Joint Commission on Accreditation of Healthcare Organizations (JCAHO).

CONTRIBUTORS

Dan Chisholm, Principal, Motor and Generator Institute; Editor and Publisher, *Healthcare Circuit News,* Winter Park, Florida

Cornelius Regan, P.E., C.L.E.P., Principal, C. Regan Associates, Rockland, Massachusetts

A. Todd Rocco, P.E., Principal, Diversified Consulting Engineers, Quincy, Massachusetts

David L. Stymiest, P.E., SASHE, C.E.M., Senior Consultant, Smith Seckman Reid, Inc., New Orleans, Louisiana

REFERENCES

1. H. O. Nash, "The Science of Sizing Generators," *Healthcare Circuit News* 4 no. 5 (1999), www.mgi-hcn .com/resources/Articles/V4N5/science_of_sizing_generators.htm, Motor and Generator Institute, Winter Park, FL.

2. NFPA, *Standard for Emergency and Standby Power Systems,* NFPA 110, National Fire Protection Association, Quincy, MA, 1999.

3. NFPA, *Standard for Health Care Facilities,* NFPA 99, National Fire Protection Association, Quincy, MA, 1999.

4. AIA, *Guidelines for Design and Construction of Hospital and Healthcare Facilities,* American Institute of Architects Press, Washington, DC, 1996–1997.

5. IESNA, *Lighting for Hospitals and Health Care Facilities—ANSI Approved,* IESNA RP-29-95, Illuminating Engineering Society of North America, New York, 1995.

SECTION 9.6
MECHANICAL SYSTEMS

Eugene Bard, P.E., President, and Kevin Sheehan,
Senior Vice President
BR+A Consulting Engineers, Inc., Boston, Massachusetts

HVAC SYSTEMS WITHIN A HEALTH CARE FACILITY

The health care facility manager must be aware that a properly designed heating, ventilation, and air conditioning (HVAC) system is an effective and critical tool for preventing cross-contamination of patients, medical staff, and facilities and for providing comfort. The HVAC systems within a health care facility must have the following attributes that are not commonly found in other building types:

- Restrict air movement in and between the various departments or zones within the facility
- Maintain specific air movement and filtration to dilute and remove contamination in the forms of airborne microorganisms, viruses, and odors
- Control temperature and humidity within different areas of the facility for accurate control of environmental conditions that suit the health care treatment program

HVAC System as an Integrated Tool of Medical Treatment

The HVAC system becomes an effective tool in providing quality health care by controlling the following environmental conditions:

- Temperature
- Humidity
- Air-pressure relationships
- Filtration
- Air changes
- Air movement
- Air quality

Temperature.   The control of space temperature is an important factor in patient treatment. Various patient conditions and treatments may demand a swing in the local environment (i.e., the patient bedroom) from a cool and dry environment to one that is warm, hot, and humid to prevent or minimize dehydration, depending on the diagnosis and treatment.

Humidity.   The control of space humidity may also be critical in patient treatment. The HVAC system should be able to control the humidity level within numerous microclimates in a hospital facility, as described in Ref. 1.

Air-Pressure Relationships.   The HVAC system must be able to move air (i.e., control air-pressure relationships) from clean areas to less-clean areas within the hospital facility. The advent and use of direct digital controls (DDCs) has made it common practice for the HVAC system to account for all airflow movement, which can be identified and alarmed as needed. Table 9.6-1 lists a small sample of air pressure relationships that need to be identified, controlled, and alarmed. For further information concerning air-pressure relationships between hospital areas, refer to Table 2, Sec. 7.31, "Mechanical Standards," of the *Guidelines for Design and Construction of Hospital and Healthcare Facilities.*[1] This guideline is now being revised, but the reader should consult the most recent edition.

Filtration Strategies.   Specific health care design manuals clearly designate the number and placement of the air-filtration beds required for an air-handling system that serves various areas within a hospital facility. However, one of the more common points of miscommunication between the project team and the surgery department arises from filter efficiency. Virtually all surgeons assume that every operating room (OR) is equipped with *HEPA* filtration (99.97 percent efficient), whereas the design manuals call for filter efficiencies of 90 percent.[2] Proper selection of filters must be agreed upon before the final configuration of the air-handling system or unit.

It is important to install filters in such a manner that filter replacement is not a problem. Improper filter installation can cause air to bypass the filter, thereby obviating any advantage. In some cases, especially for HEPA filter installations, some method of certification is suggested.

Air Changes per Hour.   The design community commonly uses the term *air changes per hour* to define the amount of air to be exchanged within any designated area.

A key factor in the design process revolves around the interpretation of the published minimum air changes per hour guidelines, as identified within the design standards. One *cannot* assume that the minimum air changes per hour are *sufficient* to offset the internal heat gains from specific spaces or functions, as illustrated from practical experience in Table 9.6-2. Table 9.6-2 illustrates commonly used levels of air changes per hour in a typical health care facility. The figure in the comment column in the table should be reviewed. Note that arbitrarily increasing the number of air changes per hour does not, by itself, improve the environmental conditions within the space. Proper air distribution by careful placement and selection of air inlet and outlet devices is needed.

Air Movement.   The method of introducing air into a critical space must also be factored into the process because the bacterial count within a surgery room may be affected or controlled (or minimized) by the location of air supply, return or exhaust devices. The effectiveness of laminar flow systems is still a hotly debated subject among medical authorities. However, laminar flow ceiling diffusers are commonly used effectively to deliver some 25 air changes per hour into a designated space (e.g., ORs) without generating noise and air drafts.

Air Quality.   The design of the HVAC system must provide, to the best of its ability, conditioned air to the health care environment that is virtually free of dust, spores, dirt, odors, and

TABLE 9.6-1   Air-Pressure Relationships

Room	Air-pressure relationship
Operating room or suite	Positive (out)
Isolation operating room	Negative (in)
Autopsy	Negative (in)
Contagious isolation	Negative (in)
Immunocompromised isolation	Positive (out)
Intensive care units	Positive (out)

TABLE 9.6-2   Typical Levels of Air Change (AC) per Hour for a Health Care Facility

Space	Guideline minimum, AC/h	Comments
Operating room	15	It is not uncommon for a fully equipped OR theater to require 20–25 AC/h to offset the heat gain and to provide comfort cooling. This high rate of air change could be offset by a lower discharge air temperature, say 50–52°F, versus the traditional, say 55°F.
Radiology and other imaging systems	6	As the trend of digitized radiology becomes more common, the introduction of electronic computer hardware will mandate 10–12 AC/h for comfort cooling.
Patient corridor	4	The designer must recognize that the patient-unit corridor functions as a multitask room for a host of health care tasks and functions and is commonly underventilated.
Infectious isolation	10–15	Infectious isolation rooms are provided for use with contaminated patients and are maintained under negative air pressure with respect to the adjacent corridor.
Protective isolation	10–15	Protective isolation rooms are used for immunosuppressed patients, including bone marrow or organ transplant, leukemia, burn, and AIDS patients. They are designed to be under a positive air-pressure condition to protect the patients. Typically, an isolation anteroom may be provided on the corridor side. HEPA filtration may be provided for supply air serving the room.
Transplant room	10–15	Transplant rooms are used as protective isolation rooms. Use of HEPA filter in supply air is required.
Intensive care unit	6	A great deal of medical equipment is used for patient care, which introduces a large heat load in the space. The minimum total air change rate per hour may not be sufficient to meet the heat load.

airborne pollutants. The location of the outside air-intake louvers must be strategically and logically located as far as practical from the following list of air pollution sources:

- Boiler stacks
- Emergency generator stacks
- Ventilation exhaust outlets
- Cooling towers
- Plumbing or medical gas vent stacks
- Loading dock facilities
- Vehicular exhaust
- Any other sources of noxious fumes

SPECIFIC DESIGN CRITERIA

The American Institute of Architects Academy of Architecture, assisted by the U.S. Department of Health and Human Services, has defined the most common guidelines for a hospital facility.[1]

STRATEGIC HVAC DECISIONS

In addition to the issues identified earlier, the hospital planners may also elect to develop strategic positions on these issues.

Outside Air Versus Recirculation Versus Economizer Cycle

A very common technique for controlling the energy consumption of an air-conditioned facility is using an air-side *economizer control* strategy for the central air-handling systems, which can reduce the amount of mechanical cooling required to provide comfort cooling throughout the facility. However, the implementation of this control strategy introduces numerous mechanical dampers and actuators, which must be maintained in proper working order. The use of this control must also take into consideration the geographic location of the facility and its influence on the air-pressurization issues identified earlier.

HEPA Filtration for the Entire Facility

The concept of a totally HEPA-filtered facility may be introduced into the facility planning process for selective high-end health care facilities that will be treating more difficult immunocompromised patients. It is virtually impossible to prevent the growth of some molds and spores (*Aspergillus*) in the entire facility (e.g., elevator shafts, ceiling plenums, mechanical rooms, loading docks, and wall cavities). Thus, the concept of HEPA filtration for the entire facility is a positive approach to this issue.

Smoke-Control Systems.   The configuration of the ventilation system should also conform to the designated smoke-control divisions and areas of refuge. The present engineering trend is to implement *active* smoke-control systems that use the central ventilation system in conjunction with an integrated system of fire and smoke dampers and the building's central fire alarm. All parties should recognize that using the central ventilation ductwork system might offer an effective method for removing smoke and pressurizing nonfire and smoke zones during a fire or smoke incident. However, introducing any smoke may leave a smoke residue on the inner lining of the sheet-metal ductwork, which must be thoroughly cleaned before reusing it as a component of the return air system.

Patient Room HVAC Systems.   During the 1960s and 1970s, many patient bedrooms were heated and cooled by a local room device, either a through-the-wall unit or a two-pipe or four-pipe fan-coil unit. However, several clinical tests have indicated that the wet drain pans of the subject units, unless properly maintained, may support the growth of bacteria and be sources of viral infection. The present trend in controlling the patient environment is to use all-air HVAC systems.

Additional Capacity.   The HVAC system must be able to change with the demands of the health care facility today or in the future. In addition, the system must have standby or redundant capacity.

Ventilation Requirements. The ventilation required of HVAC systems for health care facilities can be sizable. This results in a large heating and air conditioning load. Therefore, energy recovery systems should be part of the design.

Energy Conservation. Energy conservation approaches such as variable air-volume (VAV) systems and variable-speed drives (VSDs) should be used to optimize the system's operation and reduce operating cost.

Duct Systems. Duct systems should be designed for easy duct cleaning, and they must conform to NFPA 90A.[3]

Steam Injection for Humidification. Direct steam injection is the most favored method of humidification in health care facilities. The quality of steam and the boiler treatment compounds used at the boiler plant should be carefully evaluated. In some cases, steam generators may be needed.

CONTRIBUTORS

Richard T. Battles, President, Thompson Consultants, Inc., Marion, Massachusetts

Teerachai Srisirikul, P.E., C.E.M., D.S.M., Senior Facilities Engineer Mechanical, Partners HealthCare System, Inc., Boston, Massachusetts

John Saad, P.E., Senior Vice President, R.G. Vanderweil Engineering, Inc., Boston, Massachusetts

Anand K. Seth, P.E., C.E.M., C.P.E., Director of Utilities and Engineering, Partners HealthCare System, Inc., Boston, Massachusetts

REFERENCES

1. American Institute of Architects Academy of Architecture for Health Care, with assistance from the U.S. Department of Health and Human Services, *Guidelines for Design and Construction of Hospital and Healthcare Facilities,* American Institute of Architects Press, Washington, DC, 1996

2. ASHRAE, *Gravimetric and Dust Spot Procedures for Testing Air Cleaning Devices,* ASHRAE/ANSI Standard 52.1, American Society of Heating of Heating, Refrigerating, and Air-Conditioning Engineers, Atlanta, GA, 1992.

3. NFPA, *Standard for Installation of Air Conditioning and Ventilating Systems,* NFPA 90A, National Fire Protection Association, Quincy, MA, 1999.

SECTION 9.7
MEDICAL-GAS, PLUMBING, AND FIRE-PROTECTION SYSTEMS

Kenneth S. Charest, C.I.P.E., C.E.T., Associate
Robert W. Sullivan, Inc., Boston, Massachusetts

OVERVIEW

Health care facilities require special medical-gas, plumbing, and fire-protection systems. Medical-gas systems are unique to health care facilities and, although mandated by codes and subject to accreditation inspection, are primarily maintained and inspected by the facility manager. A review of Art. 5.4.6, "Plumbing, Gas, Process, and Waste Systems," will provide the facility manager with a basic knowledge of plumbing infrastructure and better prepare the facility manager to understand the special requirements of health care plumbing systems. Fire-protection systems described in Art. 5.4.7, "Sprinkler and Fire Protection," are common to health care facilities, and selection of the proper system is the concern of the facility manager. Most health care facilities have diagnostic and/or research laboratories with systems described in Art. 10.5, "Plumbing and Fire Protection Systems in Laboratories."

For additional details on maintenance need of these systems, check Sec. 9.8, "Maintenance for Health Care Facilities." For utilities management of these systems, consult Sec. 9.10, "Utilities Management Program."

MEDICAL-GAS SYSTEMS

Medical-gas systems fall into three broad categories: pressurized gases, vacuum systems, and environmental systems. These systems are centrally piped and have unique design, installation, testing, certification and maintenance requirements. Health care facilities must adhere to the requirements of their accrediting bodies and also must comply with local and federal requirements. The Joint Commission on Accreditation of Healthcare Organizations (JCAHO)[1] accredits about 80 percent of the nation's hospitals. The JCAHO has adopted NFPA 101, *Life Safety Code*[2] as its standard and by the way of reference requires that medical-gas and vacuum systems comply with NFPA 99, *Standard for Healthcare Facilities.*[3] The American Institute of Architects Academy of Architecture for Health with the U.S. Department of Health and Human Services (AIA/DHHS) *Guidelines for Design and Construction of Hospital and Healthcare Facilities*[4] requires compliance with NFPA 99. Plumbing codes[5,6] also reference NFPA 99 as the standard for the installation of nonflammable medical-gas systems. Uniform Plumbing Code—2000[7] does not directly reference NFPA 99, yet its requirements are very similar to those of NFPA 99. NFPA 99 has become the rule because it works and it is the single best source for the facility manager who is responsible for medical-gas systems.

Pressurized Gases

Centrally piped gas systems include oxygen, nitrous oxide, nitrogen, carbon dioxide, and medical air-distribution systems. The distribution piping shall be silver-soldered copper piping purged with nitrogen during installation.

Oxygen. Used for respiratory treatment and anesthetic mixtures, oxygen systems are installed at remote tank-farm locations. Such central supply systems are required to be installed per NFPA 50, *Standard for Bulk Oxygen Systems at Consumer Sites,*[8] and additional sizing, alarm, and distribution requirements are given in NFPA 99.

Nitrous Oxide. Used for anesthetic mixtures, nitrous oxide systems are commonly cylinder and manifold assemblies located at storage locations within the facility (see Fig. 9.7-1). In accordance with NFPA 99, systems whose volumes are 3200 lb or more shall be installed per CGA Pamphlet G-8.1, *Standard for the Installation of Nitrous Oxide Systems at Consumer Sites.*[9] Because of the potential that nitrous oxide can be abused, the Compressed Gas Association (CGA) and the National Welding Supply Association (NWSA)[10] developed a set of sales and security guidelines for nitrous oxide.

Nitrogen. Used to drive orthopedic surgical tools and inflate tourniquets, nitrogen systems are commonly cylinder and manifold assemblies similar to nitrous oxide systems. Nitrogen is a nonoxidizer, and therefore there are no quantity limitations for interior storage locations. For economy, pressurized cylinders are often replaced by liquid-nitrogen dewars that normally offgas within the room. These configurations require vaporizers and an oxygen-depletion alarm system.[11]

Carbon Dioxide. Used for abdominal insufflation and laparoscopic techniques, carbon dioxide systems are commonly cylinder and manifold assemblies similar to nitrous oxide systems. Carbon dioxide is a nonoxidizing gas contained in cylinders; therefore, carbon dioxide poses no special concerns for fire or oxygen depletion.

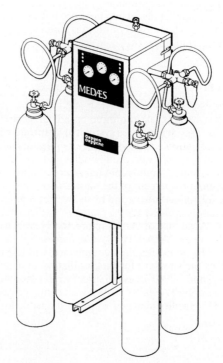

FIGURE 9.7-1   Cylinder and manifold. (*Courtesy of Medaes.*)

Medical Air.   Used for respiratory treatment and anesthetic mixtures, a medical air system presents the facility manager with a special set of concerns. Unlike other compressed gases, medical air is produced at the hospital. Equipment selection, sizing, and location are all dictated by NFPA 99. By law, the air produced must meet the requirements of the U.S. Pharmacopoeia (USP).[12] NFPA 99 has additional requirements. The Emergency Care Research Institute (ECRI)[13] has compiled a recommended standard that includes the USP and NFPA 99 requirements (see Table 9.7-1).

Vacuum Systems

Centrally piped vacuum systems serve both patient and surgical needs for suction. Equipment selection, sizing, and location are dictated by NFPA 99. The distribution piping must be metallic but need not be copper piping purged with nitrogen during installation. Although one vacuum system could be sized to meet all suction needs, dedicated vacuum systems are recommended.

Patient Vacuum.   This system should serve the general care and critical care areas, except for anesthetizing locations, as defined in NFPA 99. Although the majority of vacuum inlets within the institution would be on this system, due to minimal inlet flow requirements and diversity demand, this system could be kept relatively small. In addition, these areas would not normally see anesthetizing agents, thus permitting various types of vacuum producers to be considered. A separate system serving these areas could also act as a backup for the surgical vacuum system.

Surgical Vacuum.   This system should serve the anesthetizing locations defined in NFPA 99. Because operating rooms have the highest suction demands, these systems could become relatively large. As noted in the NFPA 99 handbook commentary, a joint study between the CGA and the American Society for Healthcare Engineering (ASHE) found that the number

TABLE 9.7-1 Comparison of Medical Air Standards

	Standard		
Component	NFPA 99	USP	ECRI
Oxygen, %	19.5–23.5	19.5–23.5	19–23
Carbon monoxide, ppm	10	10	5
Carbon dioxide, ppm	500	500	500
Methane, ppm	25	—	25
Nonmethane hydrocarbons	—	—	½ TLV
Total halogenated hydrocarbons, refrigerants, and solvents, ppm	2	—	5
Anesthetic agents, ppm	—	—	1
Nitrous oxide, ppm	—	2.5	5
Particulate matter, mg/m^3	1	—	1
Liquid water or hydrocarbons	None	None	None
Sulfur dioxide, ppm	—	5	1
Dew point @ STP	–15°C @ 14.7 lb/in^2	—	–15°C @ 14.7 lb/in^2
Other components	—	—	½ TLV
Odor	—	None	None

Note: — indicates no standard has been set; ppm = parts per million; STP = standard temperature and pressure (0°C and 14.7 lb/in^2 [760 mmHg]); TLV = threshold limit value put forth by the American Conference of Governmental Industrial Hygienists.

of operating rooms was the single most important parameter affecting suction demand. Given the potential for anesthetizing agents to enter this system, the surgical vacuum pumps should be water liquid-ring units. NPPA 99—1993 required water seal or inert materials in the compression chamber. Subsequent editions have removed the word *water.*

Environmental Systems

Waste anesthetic gas disposal (WAGD) and surgical smoke evacuation (SSE) systems are centrally piped nonmedical vacuum systems.[14] Equipment selection, sizing, and location of WAGD systems are dictated by NFPA 99. Equipment selection and sizing of SSE systems are based on good engineering practices and manufacturers' recommendations.

WAGD.   The purpose of this system is to remove residual waste anesthetic gases from anesthetizing locations. NFPA 99 handbook commentary (the second, third, fourth, and fifth editions of the handbook make this argument) suggests that this system should be independent of the vacuum system, although the standard does allow the vacuum system to remove waste anesthetic gases. The vacuum producer for this system should be a water liquid-ring unit.

SSE.   The purpose of this system is to evacuate electrosurgical and laser smoke from the procedure room. Because this is a centrally piped system that uses piping methods and vacuum producers similar to those used in vacuum and WAGD systems, this system becomes the responsibility of the engineering department. These proprietary systems are often provided by the owner and installed by the contractor, and the manufacturer establishes many of the design requirements.

Design

As noted, codes and standards dictate medical-gas design requirements. In addition, the facility manager should review designs as they pertain to facility operations. Particular attention should be paid to gas-cylinder delivery routes (loading dock to storage room) and medical-gas shutdowns required to construct the design. The facility manager should establish the best location for required independent master medical-gas alarm panels and should determine what information these alarms should relay to a building automation system. Where manufactured assemblies (ceiling units or rail systems) are included in the design, the facility manager should request documentation that these units comply with NFPA 99.

Installation

Once construction begins, the facility manager gets an immediate sense of the quality of installation by making a few simple observations:

- Are the installers certified per NFPA 99?
- Is the delivered pipe capped and properly marked for oxygen service?
- Are valves and fittings bagged?
- Does the purge nitrogen comply with NFPA 99?
- Does the installer check piping for obstructions?
- Are cutting tools sharp and free of oil and grease?
- Does the installer label the pipe during installation?
- Does the installer leave newly installed pipe capped and filled with nitrogen?

Testing

The installing contractor tests during installation. Test result documentation required by NFPA 99 includes reports for the following tests:

- Blowdown test
- Initial high-pressure test
- Cross-connection test
- Purge test
- Operational pressure test

Certification

An independent contractor does certification (verification, to use the NFPA 99 term). It is best that the owner hire the certification contractor directly. The facility manager should ensure that the installation contractor and the certification contractor meet early in the project to discuss the division of responsibilities, shutdown protocol, and tie-in strategies, and scheduling. Certification test result documentation, as required by NFPA 99, includes reports on the following points:

- Cross-connection test
- Valve test
- Flow test
- Alarm test
- Purge test
- Purity test
- Gas-source test
- Outlet test
- Final tie-in test
- Operational pressure test
- Concentration test

Maintenance

Charest points out that the facility manager is responsible for maintaining medical-gas quality.[15] At a minimum, maintenance should consist of periodic flow and pressure testing of each medical-gas outlet. NFPA 99 does not mandate a periodic test protocol, and the JCAHO simply wants to know if the facility manager is familiar with the system. Documentation for JCAHO should consist of: (1) an inventory of medical-gas outlets, (2) periodic flow and pressure test results for each medical-gas outlet, (3) established in-house design standards, (4) medical-gas shutdown protocols, and (5) inspection and certification test results. The ASHE has published sample documentation formats.[16]

PLUMBING SYSTEMS

Health care plumbing systems have distinctive requirements within two broad categories: water and plumbing fixtures. The clinical laboratory is a unique occupancy in health care facilities. In accordance with AIA/DHHS guidelines, plumbing systems within hospitals shall be designed and installed in accordance with the National Standard Plumbing Code (NSPC).[17] Clinical laboratories shall be designed in accordance with NFPA 99. Additional requirements may be found in local plumbing codes and health regulations.

Water

The local public health department monitors the water delivered to a health care facility. How the water arrives, how the water is protected, at what temperature the water should be distributed, and what additional treatment the water receives are all distinctive health care requirements.

Service.   According to the NSPC, health care facilities shall have dual water services to maintain a water supply if a water main fails.[18] The American Water Works Association (AWWA) recommends installing water services with reduced-pressure-principle backflow preventers to protect the public from the health care facility.[19] If water pressure is inadequate for plumbing fixtures on upper floors of the building, then water booster pumps must be provided. This booster system should be on emergency power, as described in Sec. 9.5, "Electrical Systems."

Protection.   To prevent contamination of the water distribution system, point-of-use backflow preventers should be installed. These point-of-use devices include atmospheric vacuum breakers on bedpan washers, indirect waste on ice machines, reduced-pressure-principle backflow preventers on autoclaves and sterilizers, and atmospheric vacuum breakers on hose connections. For a discussion of separate nonpotable water distribution systems see Art. 5.4.6, "Plumbing, Gas, Process, and Waste Systems." Note that some plumbing codes permit intermediate atmospheric vacuum breakers on sterilizers.

Temperature.   Safety issues concerning hot water must be carefully evaluated. Hot-water generators must be duplex, and water must be stored at a minimum temperature of 140°F. The general consensus is that the growth of legionella bacteria and opportunistic waterborne pathogens will be prevented at 140°F. Proposed changes to the AIA/DHHS guidelines note this concern. Thermostatic mixing valves should be installed to ensure maximum delivery temperature of 110°F to plumbing fixtures. A separate 140°F distribution system should supply water to dishwashers and laundry equipment where point-of-use booster heaters can provide 160°F and 180°F water, respectively.

Pure Water.   Contaminants found in potable drinking water can be toxic to hemodialysis patients, who are exposed to more water in 3 years than most people are exposed to in a lifetime. (Additional information is available in Ref. 20.) The Association for the Advancement of Medical Instrumentation (AAMI)[21] provides hemodialysis water standards. (see Table 9.7-2). Comparison of this table with potable water standards in Table 5.4.6-1 shows that the AAMI standard for fluoride is 20 times less than the level accepted by the EPA. The EPA standard allows levels of aluminum that are 5 to 20 times greater than that allowed by the AAMI. Other areas within the health care facility that may require pure water are bone-marrow transplant areas and central processing departments where instruments are sterilized. For a discussion of pure water system design, see Art. 5.4.6, "Plumbing, Gas, Process, and Waste Systems."

Plumbing Fixtures

Facility user needs determine plumbing fixture counts, types, locations, and elevations. The AIA/DHHS guidelines stipulate minimum requirements for various departments and recommends wall-hung vitreous china fixtures for water closets, urinals, and lavatories. Shower types vary widely from built-in-place showers to prefabricated fiberglass or acrylic units. Counter sinks are available in stainless steel, vitreous china, enameled cast iron, or integral molded units. A primary concern is the ability to clean the fixtures.

Typically, handwash sinks with wrist blades are adequate for infection-control purposes. Some occupancies—such as operating rooms, intensive care units, and postanesthesia care units—are required to have hands-free handwash sinks (Fig. 9.7-2). Here, foot, knee, or ultrasonic controls are required. Users tend to have a preference for foot controls, with knee controls being the standard on freestanding operating-room scrub sinks. Electronic controls that require additional maintenance are slowly being accepted. Proposed changes to the AIA/DHHS guidelines require supplying scrub sinks with emergency power. Special health care fixtures include those discussed next.

Bedpan Washer. These units are either integral to a patient water-closet flush valve or a spray hose assembly adjacent to the patient water closet. Regardless of the type, water closets meant to receive the waste from bedpans should be equipped with bedpan lugs.

Clinical Sinks. These flushing-rim sinks are located in soiled utility rooms and are used to discard bedpan fecal matter.

Bedpan Steamer or Boiler. These units combine the function of a human waste-elimination and sterilization. Because of the development of separate soiled utility and clean utility rooms these fixtures are becoming less common. Some models require separate independent vent systems.

Sterilizers and Autoclaves. These units are located in central-processing departments and are used to sterilize instruments. Point-of-use sterilizers are located in substerile rooms adjacent to operating rooms. Some models require separate independent vent systems.

Emergency Showers and Eyewashes. These units are located where they are accessible to users who must leave the location of an accident. The type of device shall comply with the

TABLE 9.7-2 AAMI Water-Quality Standards for Dialysis

Contaminant	Suggested maximum level, mg/L
Calcium	2 (0.1 mEq/L)
Magnesium	4 (0.3 mEq/L)
Sodium	70 (3 mEq/L)
Potassium	8 (0.2 mEq/L)
Fluoride	0.2
Chlorine	0.5
Chloramines	0.1
Nitrate (N)	2
Sulfate	100
Copper, barium, zinc	0.1 each
Arsenic, lead, silver	0.005 each
Chromium	0.014
Cadmium	0.001
Selenium	0.09
Aluminum	0.01
Mercury	0.0002
Bacteria	200 (cfu/mL)

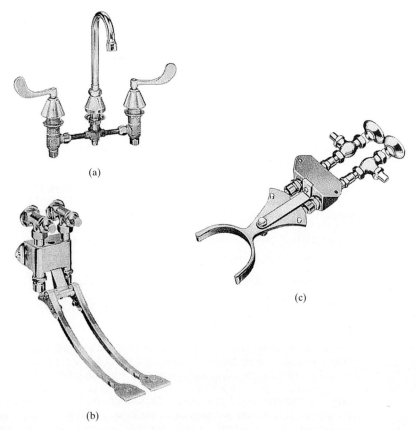

FIGURE 9.7-2   Handwash sink controls: (*a*) wrist blade, (*b*) foot control, and (*c*) knee control. (*Courtesy of Chicago Faucet.*)

American National Standard for emergency eyewash and shower equipment, ANSI Z358.1-1998, as well as OSHA guidelines.[22,23]

Decontamination Showers.   Increasingly, these units are becoming self-contained systems and the only plumbing connection is a water supply to the shower. Local codes dictate what can drain into the sanitary system. Some propose that drainage into the sanitary system should not be a concern.[24]

Cystoscopy Room Floor Drains.   The AIA/DHHS Guidelines permit floor drains provided that they contain a non-splash, horizontal-flow flushing bowl beneath the drain plate. Some propose that these drains should be eliminated and provide alternative drainage designs for cystoscopy rooms.[25]

Ice Machines.   The NSPC prohibits locating ice makers or ice storage chests in soiled utility rooms or similar areas where they may be subject to possible contamination.

Clinical Laboratories

As noted, most health care facilities have diagnostic and/or research laboratories. For a discussion of laboratory design, see Arts. 5.4.6, "Plumbing, Gas, Process, and Waste Systems," and 10.5, "Plumbing and Fire Protection Systems." The laboratory air and vacuum systems must comply with NFPA 99.

Laboratory Compressed Air.   A separate laboratory compressed-air system should be designed for laboratories. NFPA 99 goes to great lengths to keep medical and laboratory air separate. Typically, the quality of laboratory air should be equivalent to medical air, but the redundant medical air appurtenances (dryers, filters, and regulators) are not required on laboratory air systems.

Laboratory Vacuum.   A separate laboratory vacuum system is recommended by NFPA 99. Laboratory vacuum inlets can be piped to the medical vacuum system, providing that the collection piping connects directly to the receiver tank through its own isolation valve and fluid trap located at the receiver.

FIRE-PROTECTION SYSTEM

Local codes and insurance requirements mandate protecting health care facilities with fire-suppression systems. NFPA 101, the JCAHO standard, mandates that health care facilities shall be fully protected with an automatic wet sprinkler system.[26] The AIA/DHHS guidelines further mandate that clinical laboratories be designed in accordance with NFPA 99.[27] The facility manager should be cognizant of three general areas of concern: building design, obstructions to sprinkler spray, and occupancy hazards in regard to the health care facility sprinkler system.

Building Design

Building design issues fall under the category of passive fire-protection design.

Stairwells.   In renovation work there is a tendency to use stairwells as supplementary mechanical pipe chases. NFPA 101 prohibits this, and emphases the safe egress functions of stairwells.[28]

Cylinder Storage.   Like centrally piped medical gas cylinder and manifold assemblies, cylinder storage area design shall comply with NFPA 99.[29]

Smoke Compartments.   Per NFPA 101, smoke compartments that contain patient sleeping areas shall be protected with listed quick-response sprinkler heads.[30]

Obstruction to Sprinkler Spray

NFPA 13 requires an 18-in clearance between the sprinkler head deflector and any obstruction to the spray pattern.[31]

Privacy Curtains.   Curtains may act as an obstruction to sprinkler spray. "Privacy curtains with a mesh panel 70 percent open to a distance or 18″ beneath the deflector are acceptable."[32]

Way-Finding Signs.   The proximity of the sprinkler head deflector to way-finding signs must be considered during the design review.

Filing Systems.   Movable filing systems must not obstruct the sprinkler head spray pattern, which requires 18 in between the deflector and the top of the filing system.

Occupancy Hazards

Fire-suppression systems other than automatic wet-sprinkler systems should be considered for various occupancies.

Computer Rooms.   Non-water-based and clean agents can be used in these rooms. Special architectural features to ensure that the rooms are gas tight are required for these systems.

Radiology.   Preaction wet-sprinkler systems can be used in these occupancies where there is concern about water damage to the equipment. In magnetic resonance imaging (MRI) rooms, sprinkler components shall be nonferrous.

Loading Dock.   In exposed cold occupancies such as loading docks, antifreeze or dry-pipe wet-sprinkler systems can be used. Where loading docks are enclosed, sprinkler protection shall be provided below garage doors in the open position.

Hyperbaric Chambers.   NFPA 99 has special requirements for these oxygen-enriched environments.

CONTRIBUTORS

Anthony T. DiStefano, P.E., C.I.P.E., Principal, Robert W. Sullivan, Inc., Boston, Massachusetts

Robert V. DeBonis, Associate, Robert W. Sullivan, Inc., Boston, Massachusetts

REFERENCES

1. R. Bergman, "Is it worth it? Hospitals Question the Value of JCAHO Accreditation," *Health Facilities Management,* December 1994.
2. NFPA, *Life Safety Code,* NFPA 101, National Fire Protection Association, Quincy, MA, 1997.
3. NFPA, *Standard for Healthcare Facilities,* NFPA 99, National Fire Protection Association, Quincy, MA, 1999.
4. American Institute of Architects Academy of Architecture for Health, with the U.S. Department of Health and Human Services, *Guidelines for Design and Construction of Hospital and Healthcare Facilities,* American Institute of Architects Press, Washington, DC, 1996–1997.
5. ICC, *International Plumbing Code—1995,* International Code Council, Falls Church, VA, 1995.
6. ICC, *National Standard Plumbing Code—1996,* International Code Council, Falls Church, VA, 1996.
7. ICC, *Uniform Plumbing Code—2000,* International Association of Plumbing and Mechanical Officials, Walnut, CA, 2000.
8. NFPA, *Standard for Bulk Oxygen Systems at Consumer Sites,* NFPA 50, National Fire Protection Association, Quincy, MA, 1996.
9. CGA, *Standard for the Installation of Nitrous Oxide Systems at Consumer Sites,* Pamphlet G-8.1, Compressed Gas Association, Arlington, VA, 1990.
10. NWSA, *Safety in the Public Interest, Nitrous Oxide Sales and Security—Recommended Guidelines,* Compressed Gas Association and National Welding Supply Association, Philadelphia, PA.
11. CGA, *Oxygen-Deficient Atmospheres,* Safety Bulletin SB-2-1992, Compressed Gas Association, Arlington, VA, 1992.
12. United States Pharmacopeia and National Formulary, USP 23 / NF 18, United States Pharmacopeial Convention, Rockville, MD, 1995.

13. Emergency Care Research Institute, "Medical Gas and Vacuum Systems," *Health Devices Source-book* 23, no. 1–2 (1994).

14. K. Charest, "Environmental Plumbing Systems for the Operating Room," American Society of Plumbing Engineers, 1999 Technical Symposium.

15. K. Charest, "Medical Gas Quality," *Health Facilities Conference Proceedings,* American Society for Healthcare Engineering, Annual Conference, 1995.

16. ASHE, *Medical Gas and Vacuum Systems,* Management and Compliance Series, vol. 3, American Society for Hospital Engineering, Chicago, 1994.

17. ICC, "Medical Gas and Vacuum Piping Systems," *National Standard Plumbing Code—1996,* International Code Council, Falls Church, VA, 1966, Sec. 14.3.

18. ICC, "Water Service," *National Standard Plumbing Code—1996,* International Code Council, Falls Church, VA, 1996, Sec. 14.2.

19. AWWA, *Recommended Practice for Backflow Prevention and Cross-Connection Control,* AWWA M14, 2d ed., American Water Works Association, Denver, CO, 1990.

20. U.S. Department of Health and Human Services, *Manual on Water Treatment for Hemodialysis,* Washington, DC, July 1989.

21. AAMI, Association for Advancement of Medical Instruments, Arlington, VA, Dialysis, 1998.

22. ANSI Z358.1-1998, American National Standards Institute, New York, 1998.

23. OSHA, "Requirements for Emergency Eyewashes and Showers," Occupational Safety and Health Administration Standards Interpretation and Compliance Letter, Washington, DC, September 30, 1994. The requirements reference the 1990 ANSI standard.

24. C. Kampmier, "Decon Design," *NFPA Journal,* 94 no. 2 (March/April 2000); 59–61.

25. H. Laufman, "Surgeons' Requirements for Operating Room Air Environment," *Healthcare Facilities Management,* June/July 1999.

26. NFPA, *Life Safety Code,* NFPA 101, National Fire Protection Association, Quincy, MA, 1997, Sec. 12-3.5.1, 13-3.5.1.

27. AIA/DHHS, *Guidelines for Design and Construction of Healthcare Facilities,* American Institute of Architects Press, Washington, DC, 1996–1997, Sec. 7.12

28. NFPA, *Life Safety Code,* NFPA 101, National Fire Protection Association, Quincy, MA, 1997, Sec. 5.2.2.5.3.

29. NFPA, "Gas Equipment," *Standards for Healthcare Facilities,* NFPA 99, National Fire Protection Association, Quincy, MA, 1999, Chap. 8.

30. NFPA, *Life Safety Code,* NFPA 101, National Fire Protection Association, Quincy, MA, 1997, Secs. 12-3.5.2, 13-3.5.3(d).

31. NFPA, *Standard for the Installation of Sprinkler Systems,* NFPA 13, National Fire Protection Association, Quincy, MA, 1999, Sec. 5-6.5.2.1.

32. W. Koffel, "In Compliance," *NFPA Journal,* 94 no. 2 (March / April 2000), 30.

SECTION 9.8
MAINTENANCE FOR HEALTH CARE FACILITIES

Kimball Ferguson, P.E., Engineering Manager
Duke Medical Center, Durham, North Carolina

INTRODUCTION

This section highlights the unique feature and the importance of the maintenance function in healthcare facilities. For a general discussion of maintenance and operations issues, consult Chap. 7.

The maintenance function in health care facilities can be provided in different ways. Many facilities have in-house maintenance personnel who provide minimum to extensive maintenance, often on very sophisticated and technically complex systems. Some perform the minimum, usually the life-safety–related functions, and outsource all other functions. The other model would be a completely outsourced maintenance department. Some departments are unionized, with specialized work practices; some are nonunionized, and have more diverse functionality. It is not the intent of this article to discuss merits of various maintenance delivery systems.

PACE OF MAINTENANCE

The pace of operation in health care facilities is in general more intense than in other types of facilities. The demand to provide quality patient care puts enormous pressure on the maintenance department. All systems must operate and must be reliable.

To meet these demands, many maintenance departments provide some support to the second shift and some even to the third shift. This is not an easy decision, as staffing for the second and third shifts can be very operating-cost–intensive. It is suggested that facilities managers perform a careful needs analysis to determine the correct staffing levels and shift staffing.

NAME OF DEPARTMENT

The maintenance department in health care facilities is called by many names. In the writer's experience it has been called the maintenance department, engineering department, engineering and maintenance, buildings and grounds department, facilities maintenance, and engineering and operations. Whatever name is selected, it is important that the facility's occupants or the department's clients understand the department's mission: to ensure that the health care facility's buildings and systems operate in a safe and reliable manner in order to provide the core mission, which is patient care, and to ensure that the department can be contacted promptly when the need arises. Many departments include qualified engineers as a part of their staff; some health care facilities have separate engineering departments, and some outsource this function.

CONSTRUCTION

Many maintenance departments provide in-house construction services for renovation or new construction projects. If well managed, they can usually provide construction at a lower cost than outside contractors. They also have an advantage in conducting shutdowns and if needed can move from location to location in a short time. Construction and maintenance functions must be clearly defined and separate. There is always a danger that too much construction can divert departmental resources from maintenance functions.

CONTINUITY OF SERVICES

Hospitals never close. Hospitals operate 24-h, 7-days-per-week. Facilities must be designed in a manner to allow maintenance shutdowns and to add new features to systems. Two words can easily sum up your design philosophy: *isolation* and *redundancy.*

Similarly, *outpatient facilities* or *clinics* are becoming increasingly more common and more complex in the procedures that are performed. Many clinics house hospital-grade systems that are operated less than 24 h per day. Although classified as a business occupancy rather than a hospital, the facilities manager should look carefully at the requirements for each function within the building. Be familiar with the equipment and procedures. Above all, ask questions. The health care field is changing rapidly in all areas.

Clinics differ from hospitals in that they are operated less than 24 h per day. Some are operated 23 h and 59 min per day, meaning there are not patients housed in the facility around the clock; others are operated 8 A.M. to 5 P.M. The dividing line between hospital and clinic can be confusing. The same basic requirements are present. What redundancy should be in place to protect patient care? What systems can be replaced overnight rather than in hours or minutes, as in a hospital? How much redundancy should be provided? Will emergency power be needed? These are all good questions that the facilities managers and owners should face together.

Nursing homes are similar to hospitals in that many of the patients are not typically mobile. Some patients may require observation 24 h per day. Moving them is impractical. Again, we are faced with the same scenario as in a hospital environment: nonambulatory or marginally ambulatory patients that require facilities system support.

In a hospital, 100 percent redundancy is advised and at times required. Multiple electrical services, HVAC systems, medical-gas systems, water supplies, and redundant communications are all recommended. At the same time, the need for space for future growth and for technological advances which require additional service is inevitable.

One example of how redundancy in mechanical systems can be achieved is by looping piping systems. Two paths of distribution within the plant allow more options in case of emergency. Looping can be used effectively for gases and liquids. Piping systems also benefit from loop configurations. Examine your requirements carefully, and look at costs as well.

There are never enough valves. Let's qualify that statement by saying that valves are seldom in the right locations to properly isolate equipment for maintenance. Valves are inexpensive in comparison to the alternative. Emergency stop valves (or emergency line plugs) are typically 100 times more expensive than a valve in the same location. It is typical to provide a valve between each piece of equipment on a loop or header, but almost never on the header.

Some of the preceding items may not appear to be cost effective during the design phase, but facilities managers must think ahead and consider whether the systems in question can be shut down for repairs or maintenance when patients are being served. In a hospital, the answer is typically no. In many cases, the owner will have to add isolation valves or bypass loops after the building is operational. This is not cost effective. The valves should be included in the original building design.

MAINTENANCE

Everything that has moving parts breaks or fails eventually. A hospital, like any other facility, requires maintenance. Chapter 7 contains detailed information on maintenance needs and functions. This subsection discusses only items unique to health care facilities.

To ensure that the health care facility's buildings, facilities and systems operate in a reliable manner to support the core mission, which is patient care, a good maintenance and operation program must be implemented. It is also critical that all maintenance activities be documented. Some documentation is required by agencies such as JCAHO, while other documentation is good practice and may provide the paper trail that is needed in the event that a major failure or disaster occurs.

Close collaboration and teamwork is also required for the maintenance department to work with other entities of the hospital. These entities may include the infection control, respiratory therapy, or biomedical engineering departments, police and security, or environmental services. Some of the collaboration issues are described here.

Infection Control

The maintenance department works in collaboration with other departments to ensure that buildings systems operate in a proper manner to reduce infection. For example, improper use of a negative-pressure isolation room may allow infectious agents from inside the room to come into the corridor and infect the hospital workers, other patients, or visitors; improper temperature of hot-water heating systems may allow microbial growth; or improper filter application or maintenance may create infection-control problems.

Construction Plan Review

Maintenance departments should be involved in reviewing new construction and renovation projects. The following list can be very helpful and is adapted from work done by Hardin.[1]

General Mechanical and Electrical Equipment Rooms. Mechanical rooms for major equipment like air-handling equipment and chillers ideally could have direct access from the outside of the building for replacement. This feature though ideal may not be always practical. At a minimum, the location of equipment should minimize maintenance personnel intrusion onto the medical floors. If possible, direct vehicle transport for maintenance items and equipment is desirable. For mechanical spaces in upper floors, direct elevator access is most helpful.

Rooftop-Mounted Equipment. In general, rooftop mounting should be avoided for critical applications due to the usual difficulty of maintenance access and safety considerations. However, rooftop HVAC equipment is a very cost-effective option for clinics. Exhaust fans, cooling towers, or other heat-rejection equipment must be on the roof. It is highly recommended that whenever any rooftop equipment is used, pavers or other access pathways which will not damage the roof should be provided. For any equipment requiring maintenance access (including valves) that is not readily accessible from a 6-ft-high portable ladder, a fixed ladder and/or catwalk should be considered.

Mechanical Room Layout. Layouts should include sufficient space to enable access to equipment for operation, maintenance, and replacement, including permanent catwalks or ladders for access to equipment not reachable from the floor. It should be verified that practical means are provided for the removal and replacement of the largest and/or heaviest equipment items located in the facility and that pull space for all coils, heat exchangers, chillers, boiler tubes, and filters is provided.

Chillers and Boilers.   Provisions should be made to get the unit in and out of the building. For large chillers, consider installing a beam attached to the structure in order to move or replace large compressors or motors.

General Personnel Access.   Safe and practical personnel access should be provided. A minimum of 2 ft of clearance is generally required at all service points for mechanical equipment to allow personnel access and working space. Greater space may be required for particular equipment and maintenance applications.

Electrical Equipment Clearances.   Minimum clearances for electrical equipment must be at least the minimum clearances required by NFPA 70.[2]

Separate Energy Plants.   When chilled water, heating water, or steam generators are located in a separately located energy plant exterior to the primary facility, connecting utility line installation in a tunnel or other accessible enclosure, providing maintenance and inspection access and protection from the elements, is highly desirable. Accessibility to entire utility main runs is desirable in order to facilitate inspection and repairs of insulation, fittings, thermal expansion compensation, air vents, etc., as well as to facilitate future replacement or expansion. Safe and convenient accessibility is essential for those elements requiring periodic inspection or service, including isolation valves, condensate drainage traps (both manual and automatic), sump pumps, ventilation fans, etc.

Cooling Towers.   Their location and placement should be reviewed. The spray or plume could be a source of legionella.

Cooling Coils.   There should be no more than 6 rows, to facilitate cleaning. When additional rows are required for dehumidification, the coil could be separated into two 4- or 6-row units with access to both upstream and downstream coil faces.

Stainless Steel Drain Pans.   Drain pans should be provided to optimize cleaning and reduce microbial growth. To ensure pan drainage, facilities managers should require that designers provide dimensioned trap details, which compensate for the effects of fan static pressure. Ongoing drain-pan treatment is also suggested where needed to reduce microbial growth.

Freeze-Protection Features.   This is a very important feature. Freezestats are designed to protect the air-handling equipment and coils from freezing. If this system is not correctly designed, the air-handling equipment will frequently trip off and shut down the air-handling equipment. This nuisance tripping not only creates a loss of airflow pressure control but can also be a safety hazard. Many maintenance personnel attempt to compensate for this situation by increasing the supply air temperature. This high temperature causes difficulty in providing cooling.

Balancing Features.   To facilitate future troubleshooting or system balancing, check for measurement devices and balancing dampers in all HVAC equipment. This may include temperature and pressure measurement ports or devices on inlet and outlet connections of all coils, as well as balancing valves and flow measurement apparatus, and temperature measuring ports or devices at various locations in the air-handling unit (AHU). Pressure ports or gauges upstream and downstream of the fan and ports for pitot traverse and airflow should also be provided. To facilitate periodic rebalancing or future modification, manual balancing dampers should be provided on all branch duct runouts. These should be located as far upstream from the terminal fixture (diffuser, register) as practicable to reduce air-generated noise.

Ductwork Design Considerations. Access panels for inspection or servicing of duct-mounted equipment, including fire dampers, smoke dampers, and controls, and to facilitate periodic cleaning or disinfecting must be located and sized for ease of use. New NFPA requirements for fire damper testing may require two doors per damper on each side of the damper.

Fire-Protection Systems. The trend has been that these systems are oversized, resulting in large and expensive overpressure relief piping. Examine the submittals carefully and be certain that street pressure has been properly accounted for in the calculations. A bypass line with a flow meter is a good option to add. This saves tremendous amounts of water since hospitals systems must be checked weekly.

Emergency Generators. The facility manager must determine if cogeneration, load-sharing programs, or peak shaving will be part of the emergency power system. Generator placement is extremely important but is generally dictated by architectural issues, rather than performance or ventilation issues. Be very cautious about generator specifications. It is common for specifications to name NFPA 70, *National Electric Code,* as part of the generator specification, but NFPA 101, *Life Safety Code,* is referred to by NFPA 70, and has some very specific requirements about emergency generator construction and operation.[3] Review these documents carefully and adjust the specifications if needed.

Emergency Recovery Plan. Check to determine if the health care facility can run its *entire* critical operation on emergency power. It is also common to find that there is simply not enough fuel storage to run the buildings for an extended period. The facility manager should carefully review the facility's requirements, and codes for required run times.

Outside Air Intakes. Improper location of outside air louvers near a contamination source can cause IAQ problems. Do not allow the architect to put the fresh-air louvers near the loading dock. Similarly, do not allow the architect to put the diesel generator near fresh-air louvers. Diesel exhaust is detectable by humans in concentrations as low as 6 parts per billion! It is unclear why, but too many buildings are built this way.

Domestic Water Pumps. Application of variable-speed drives for domestic water pumps must be carefully evaluated. Otherwise the system may not respond properly to rapid changes in building water demand. Multiple pumps are also recommended, as loss of domestic water is one of the most disruptive utility failures that a hospital can experience.

Emergency sources of domestic water should be considered during design. A pool, for example, is a good source of emergency water.

Water Heaters. Quality units are stainless-steel or glass lined and are thermally efficient. Inspection ports must be accessible. Temperature and pressure gauges on the outlet and inlet of each unit should be provided. Water flows must be properly balanced. Venting is a major concern for gas-fired units. Gas-fired units may not share space with refrigeration equipment.

Construction Project Acceptance

Generally, hospital projects are phased (not done all at once) in order to keep critical health care functions in operation. Maintenance department acceptance of projects after construction can be very complex depending on the contract with the builders. Fixed boundaries and scope for each area under construction must be defined. Project scope should be clear and well documented. All additions to scope, impact on schedule, and most important, health care operations, should be clearly delineated to the caregivers as well as to the maintenance department.

Caregivers usually have an idea what they want but do not understand the construction process and the cost and time consequences. In most instances, there is a panicked dash near completion to move into the new project and generate revenue. Maintenance personnel are then called upon to solve problems, which by right should have been the contractor's responsibility. If this describes your projects of late, consider a *commissioning* process for the construction projects. The commissioning agent observes the work, points out problem areas, helps define and resolve schedule problems, and generally tries to ensure that the owners get what they are paying for. The commissioning agent approves payment based on percent completion. The agent is also responsible for final drawings, operation and maintenance manuals, and holding back funds from the constructors for work not performed or not in accordance with the contract. This reduces the burden on the maintenance department personnel.

SPECIAL MAINTENANCE CONSIDERATIONS

HVAC Systems and Equipment

Fan-Coil Units.   Each fan-coil unit with a cooling coil has a drain pan which could become a reservoir for microbial growth. Periodic inspection of the condensate pan and sometimes even drain-pan treatment may be necessary to prevent blockages from causing overflow and wetting surrounding materials, thereby creating additional microbial amplification sites. As these units are typically located within spaces served, maintenance personnel need access into occupied areas.

Fin-Tube Radiation and Convection Units.   These units may also require frequent cleaning to minimize the collection of dirt and debris. This equipment also requires frequent maintenance personnel access into the spaces served.

Fan-Powered Terminal Units.   Fan-powered terminal units represent additional components that require inspection and maintenance.

Secondary Air Systems.   An example of this type of unit may be a laminar flow system in an orthopedic OR or a HEPA-filtered recirculating unit for a bone-marrow transplant unit. The unit may be provided with filters that require replacement, and a motor that may require periodic service or replacement.

Central Station Air-Handling Units.   To reduce the possibility of microbial growth in unit insulation, air-handling units used in medical facilities should be of the internally insulated, double-wall type, with corrosion-resistant inner-wall material. Perforated inner-wall surfaces are in general not recommended. When final filtration is provided in an air-handling unit in a location downstream from cooling coils, some provision must be provided to avoid wetting the filters. Carefully evaluate a draw-through versus a blow-through design.

Plumbing Systems and Equipment

Domestic Water Pump.   The control systems must be periodically checked for proper performance and overpressure.

Hot-Water System.   Medical facilities generally require tempered water for domestic hot water. This is universally done with blending or tempering valves. These valves should be checked on a monthly preventive maintenance schedule for proper operation and minimize risk of scalding patients and visitors. In many cases, two separate systems are provided because hot water is generally needed at two distinct temperatures (i.e., kitchen equipment and sterilizers may require >140°F water versus 110°F for patient care areas).

Fire-Protection Systems

Sprinkler heads do occasionally need replacing. Be sure to have a few of each type available. Patient rooms generally require the fast-action-type heads.

Remote Alarm Annunciation.   It is critical that malfunctions of certain critical equipment and systems are annunciated on a remote basis.

Fire-Alarm Systems.   The fire-alarm and fire-suppression systems require special attention, because many times the patients are too sick to be able to move from their beds to other locations. Evacuation is typically the last resort and requires advanced planning and coordination. In general, hospital personnel tend to fight a fire at its location. The maintenance department needs to have an established maintenance and testing program to ensure reliability of the fire-alarm systems.

Electrical Systems and Equipment

Electrical power systems are one of the most critical systems in health care facilities. The facility's maintenance department is responsible for maintaining the high degree of reliability required for both the normal and emergency power systems. This responsibility requires focused attention on preventive maintenance and management of operational decisions. Article 7.2.3, "Electrical Testing and Maintenance," describes electrical system maintenance and testing requirements in detail. Section 9.11, "Electrical Utility Management Program," describes utility management requirements for the electrical power systems. As described in Sec. 9.11, preventive maintenance will necessitate periodic major shutdowns of portions of the electrical power systems. Protective coordination of electrical power systems is also very important, and this issue is discussed in Sec. 9.5, "Electrical Systems," and Art. 5.2.2, "Electrical Distribution Systems." In modern health care facilities there is a large amount of electrical nonlinear load, which may result in electrical harmonic issues. Maintenance personnel must be trained to be aware of these issues. Experienced electrical engineers should periodically review electrical power systems in health care facilities and make recommendations to the maintenance department to implement prudent preventive and corrective actions.

Transportation Systems

The maintenance department is responsible for maintaining various transportation systems in the health care facility. These systems could be vertical transportation systems like elevators or escalators (see Art. 7.2.4, "Elevator/Escalator/Moving Walk Maintenance") or special transport systems for patient records or specimens. These systems are described in detail in Art. 9.3.3, "Transport Systems." Even if the department chooses to outsource the maintenance function, some of the staff members need to be trained in dealing with elevator entrapments and stoppages in the specimen transport systems.

COMPLYING WITH JCAHO REQUIREMENTS

The maintenance department works very closely with the safety committee and other departments in the health care facility for Joint Commission on Accreditation of Healthcare Organizations (JCAHO) compliance. See Secs. 9.9 ("Environmental Health and Safety Management Program"), 9.10 ("Utility Management Program"), and 9.11 ("Electrical Utility Management Program") for additional details.

Statement of Conditions (SOC)

JCAHO requires that all health care facilities keep up-to-date information on the condition of the facility. This document is called the *statement of conditions* and is generally the responsibility of the facility manager. It lists all plan for correction (PFC) corrective measures, which must have a schedule of completion and a known funding source. The SOC is a living document. It should be continuously updated as the facility is changed, renovated, and improved. The maintenance department plays a central role in preparing SOC documents and correcting various PFC items.

Hospital Disaster Preparedness

State licensure and JCAHO mandates usually require that a health care facility has a disaster plan. The maintenance department plays a very important part in formulating and implementing this plan. Maintenance departments are also required to have written contingency plans for the failure of all critical utility systems. This is another example of the close collaboration required in working with other departments. The disaster planning committee usually includes representatives from the following departments:

- Medical staff (ER physician or trauma surgeon)
- Administration (includes risk manager)
- OR manager
- Nursing staff
- Emergency department
- Security
- Communications
- Public relations
- Medical records
- Admissions
- Laboratory
- Radiology
- Respiratory therapy

Interim Life Safety

Creating a safe building environment is the goal of life-safety codes and standards, which study egress, stairs, fire-detection devices, and general occupancy. As long as the building design remains unchanged, the integrity of life safety remains. However, health care facilities are always changing. As buildings are subjected to renovation (both planned and unplanned), the integrity of life safety may diminish during the construction. This decrease in life safety has resulted in the creation of interim life safety measures. Even though the maintenance department plays a vital role, this process again must be a team effort.

Interim life safety measures (ISLMs) are generally overlooked during the design process and are not dealt with until construction actually begins. It is never too late to make necessary adjustments to the design and construction process. The ILSM also requires that construction activities be physically separate from patient care areas by at least a 1-h barrier. Otherwise, the patients and visitors may be exposed to grave dangers. The maintenance department may be called upon to support additional fire and evacuation drills and exercise control over cutting, soldering, or the use of flame in the construction process. Some maintenance departments issue an internal flame permit to the outside contractor, and secure fire-alarm zones as needed to allow construction processes in existing buildings, but put the alarm system back in normal operating condition after the construction is complete.

Utilities Management

Utility management has become a complex function in today's health care facility. Quality improvement, along with trends occurring in utility equipment, will help the facilities manager reduce the amount of maintenance service calls due to recurring problems. Steam, chilled water, medical-air, HVAC, plumbing, medical-gas, security, and nurse-call systems need to be tracked for failures and recurring problems. Some of the critical issues associated with utilities management are described in Sec. 9.10, "Utility Management Program."

OTHER IMPORTANT ISSUES

Self-Assessment and Resolution of Indoor Air-Quality Problems

Indoor air-quality (IAQ) problems have their origins in many different generating sources within facilities. These sources may include building systems, processes and procedures, management, employees, and sometimes even outside influences. Maintenance departments usually get the first call informing them of these issues and they need to follow a systematic process. Communication with the customer is also critical in containing the problem and possibly preventing unwarranted concern within a building.

Legionnaire's Disease Prevention

Health care facilities could be prone to outbreaks of legionellosis. Maintenance departments again are the first line of defense against this problem. Butkus et al.[4] and ASHRAE[5] provide excellent discussions of the problem and solutions. The name of the disease comes from a well-known outbreak at an American Legion convention in Philadelphia in 1976. This outbreak was attributed to a cooling tower. Since then, legionella have been found to be present in untreated stagnant water. Construction or renovation activities should always consider this fact when capping lines for an extended period of time.

Legionella are bacteria. They occur in natural water sources and municipal water systems in low or undetectable concentrations. However, under certain conditions, the concentration may increase dramatically, a process called *amplification*. Conditions favorable for amplification are as follows:

- Water temperature of 77 to 108°F
- Stagnation
- Scale and sediment
- Biofilms
- Presence of amoebas
- Certain materials—natural rubber, wood, and some plastics

Transmission to humans occurs when water containing the organism is aerosolized in restorable droplets (1 to 5 μm) and inhaled by a susceptible host. Infections initially occur in the upper or lower respiratory tract. The risk is greater for older people, those who smoke, those who have chronic lung disease, and those who are immune suppressed.

In health care facilities, nursing homes, and other high-risk situations, cold water should be stored and distributed below 68°F and hot water above 140°F. It should be noted, however, that in certain jurisdictions this high-temperature storage may not be allowed. Recirculation of hot water should be above 124°F. These facilities should have appropriate antiscald devices. If this is not possible, periodic chlorinating or heating water to 150°F followed by flushing should be considered. When practical, hot water should be stored above 120°F, with appropriate safeguards.

In high-risk situations (such as for immune-suppressed patients), monthly showerhead and aerator removal for cleaning and disinfection is recommended. For systems that have stagnant pipelines, or were recently built or repaired, flushing and possibly chlorinating or high temperature flushing is recommended.

Flush safety showers and eyewash stations at least monthly; some facilities flush them weekly (check with your local cross-contamination official). Fire sprinkler system protection recommendations include protective respiratory gear for firefighters and evacuation of non-firefighting personnel.

Cooling towers have long been known as a legionella risk. The key recommendations for cooling towers are clean surfaces and a biocide program. Professional help with chemical treatment is recommended. Mechanical filtration should be considered to minimize fouling. Drift eliminators should be regularly inspected and cleaned and repaired as needed. It is a sound practice to alternate biocides used for cooling-water treatment to avoid developing resistant strains of microbes. Weekly changes in dose and frequency are recommended.

Shutting down and starting a tower system needs specific attention. When a system is shut down for more than 3 days, it is recommended that the entire system be drained to waste. When not practical to do so, stagnant water must be pretreated with a biocide regimen before tower start-up. After adding biocide, circulation of water for up to 6 hours is suggested before tower fans are operated.

Steam Plants

Many health care facilities, especially hospitals, have a need for high-pressure steam (60 lb/in^2) for sterilizer or laundry operation. These boilers, as their operating pressure is in excess of 15 lb/in^2, may require dedicated licensed operators. This requirement causes staffing challenges for the facilities manager. As 24-h operation is required, 5 to 7 dedicated personnel are usually needed. Some facilities managers outsource this function. In the case of facilities supplied from a central plant or district heating system, the steam is supplied from another source, and on-site personnel are not needed.

FIRST RESPONSE

Maintenance departments are the first responders whenever there are any problems in health care facilities. In case of floods, heat or cold complaints, or whatever, the staff is first to respond. The personnel must therefore be trained in their respective trades as well as be knowledgeable in first-response techniques. For example, they should know where the shut-off valves and switches are.

Capital Investment Planning

In general, maintenance departments are responsible for budgeting for facility infrastructure upgrades. Careful attention must be given to assessing the facility's future growth needs when presenting a capital budget. Maintenance departments should also request funding for repairing or replacing mechanical electrical items which need replacing on a regular basis.

CONTRIBUTORS

Gregory O. Doyle, Director, Buildings and Grounds, Massachusetts General Hospital, Boston, Massachusetts

George Player, Director of Engineering, Brigham and Women's Hospital, Boston, Massachusetts

Joseph Gibson, Manager, Duke Hospitals, Durham, North Carolina

REFERENCES

1. Personal communication from Jeffery Hardin, PE, 1999, in preparation for the ASHRAE design manual for HVAC systems for hospitals and clinics.
2. NFPA, *National Electric Code,* NFPA 70, National Fire Protection Association, Quincy, MA, 1999.
3. NFPA, *Life Safety Code,* NFPA 101, National Fire Protection Association, Quincy, MA, 1997.
4. Alexander S. Butkus, Daniel L. Doyle, Jean O'Brien Gibbons, and Laurence V. Wilson, *Grumman/Butkus Bulletin,* Grumman/Butkus Associates, Evanston, IL, September 1999.
5. ASHRAE, *Minimize the Risk of Legionellosis Associated with Building Water Systems,* American Society of Heating, Refrigerating, and Air-Conditioning Engineers, Atlanta, GA, 1999.

SECTION 9.9
ENVIRONMENTAL HEALTH AND SAFETY MANAGEMENT PROGRAM

John F. McCarthy, Sc.D., C.I.H., and Nanette E. Moss, M.S., C.I.H.
Environmental Health and Engineering, Inc., Newton, Massachusetts

Health care facilities present numerous potential hazards to employees because of the variety of work processes conducted and the nature of services provided within them. These may range from chemical exposures or infectious diseases to ergonomic injuries, such as neck and back strains. Developing and implementing a comprehensive safety program is essential to afford employees, patients, professional staff, and visitors the highest levels of personal safety and protection in a facility, as directed by federal, state, and the Joint Commission on the Accreditation of Healthcare Organizations (JCAHO)[1] standards and guidelines. The program should be proactive and should consistently seek to correct conditions that can contribute to accidents or injury, as well as to develop new and better ways to avoid those situations that may cause accident and injury in the future.

Under its Management of the Environment of Care (EOC) standards EC 1 to 5, JCAHO provides specific guidance in designing, implementing, and continuously improving safety programs. The JCAHO EOC plan focuses on four main processes for developing safe, functional, and effective environments: (1) planning space, equipment, and resources to support the services provided effectively; (2) educating the staff about the role of the environment in effectively and safely supporting patient care; (3) developing standards to measure the performance of the staff and the hospital in managing and improving the EOC; and (4) implementing plans to create and manage the hospital's EOC, including an information collection and evaluation system (ICES) to continuously measure, assess, and improve the status of the EOC.

The JCAHO Environment of Care consists of seven integrated programs for managing safety, life safety, emergency preparedness, hazardous materials, utilities, security, and medical equipment. The first steps toward establishing a safe and effective EOC should be to develop management plans for all seven of the programs cited, in accordance with the specific JCAHO design requirements (EC 1.3 to 1.9).[1]

FIRE AND LIFE SAFETY

Fire and life safety are elements of the health care facility's safety program that require constant vigilance. Fire safety and life safety are integrated programs designed to ensure that the environment is free from known fire hazards and that the facility is properly maintained to provide all occupants with access to an area of safety in the event of a fire.

Code Compliance

Local and national standards have been developed to guide architects, builders, and property owners in their compliance efforts. The codes applicable in the health care environment include state and local building and fire codes and the National Fire Protection Association (NFPA 99)[2] and JCAHO fire-safety requirements (EC 1.7, 2.6, 2.10, and 2.12).[1]

Inspections

Self-inspection is a key element of a health care facility's safety program. Fire-protection systems require periodic testing. Table 9.9-1 outlines the primary testing and inspection criteria contained in the code documents listed before.

Drills and Training

Initial training in fire safety during orientation is critically important for all newly hired employees. When the staff member is introduced to the assigned work area, specific job-related fire-safety training is required. Fire drills serve as the primary instrument for maintaining staff readiness for proper response to fire emergencies. Under the standards, fire drills are to be conducted quarterly in all areas and on all shifts (NFPA 101 12-7.1.2 through 12-7.2.3, and 13-7.1.2 through 13-7.2.3[3]; JCAHO EC 2.10[1]).

In the inpatient care areas, fire drills normally do not involve evacuation. The staff members in these areas are drilled in the proper response to the discovery of a fire. The steps to be taken are generally associated with an acronym that reminds personnel of the key issues to be addressed. Many facilities use the word *RACE,* which stands for *rescue, alarm, confine,* and *extinguish.* In most cases, evacuation of the fire area is not necessary unless the fire develops

TABLE 9.9-1 Testing Requirements for Life-Safety Equipment

Fire-alarm and -detection equipment	
Device	Testing frequency
All supervisory signal devices (except valve tamper switches)	Quarterly
Valve tamper switches and waterflow devices	Semiannually
Smoke detectors (including duct detectors)	Annually
Electromechanical releasing devices	Annually
Heat detectors	Annually
Manual fire-alarm boxes	Annually
Occupant alarm-notification devices (audible devices, speakers, and visual devices)	Annually
Transmission equipment for fire department notification	Quarterly

Automatic extinguishing equipment	
Device	Testing frequency
Fire pump—churn (no flow)	Weekly
Fire pump—full flow	Annually
Main drains (at all system risers)	Annually
Fire department connections	Quarterly
Package suppression systems over cooking equipment	Semiannually
Package gaseous suppression systems (i.e., CO_2, FM200)	Annually

Portable fire extinguishers and standpipe systems	
Device	Testing frequency
Portable fire extinguishers	Monthly (visual)
Portable fire extinguishers	Annual maintenance
Occupant hose (1½ in) where installed	5 years after installation, 3 years thereafter
Standpipe systems	5 years

Building fire-protection equipment	
Device	Testing frequency
Fire dampers	4 years
Smoke dampers	4 years
HVAC shutdown	Annually
Sliding and rolling fire doors	Annually

beyond the ability of building systems to control smoke and fire conditions. Staff knowledge about areas of refuge, evacuation routes, and any special duties related to fire response should be evaluated at the time of the drill by conducting either staff interviews or written tests.

Outpatient and business portions of health care facilities participate in evacuation drills that involve the actual evacuation of all personnel from the fire-alarm area. The standards allow an alternative to total evacuation in some cases. Where applicable, occupants are trained to move to the exit route assigned and, at the direction of a drill coordinator, remain in place while a designated staff member follows the evacuation route to the outside. The designated staff person checks to see that the route is clear of all obstructions and that it is well lit. Training and review of the fire plan should be held at the time of the drill and documented as part of the annual training requirements listed in the standards.

The frequency and coverage of fire drills may be altered as building conditions change. Large facilities are not required to conduct drills in all occupied areas 4 times per year but may reduce the areas covered by using criteria outlined in the standards (JCAHO EC 2.10).[1]

Evacuation Plans

The *Life Safety Code* (NFPA 101 12-7.1.1 and 13-7.1.1) requires that all health care facilities have written plans for protecting all persons in the event of a fire.[3] These plans must indicate the evacuation routes and areas of refuge that occupants will use. The standard does not require floor plans or any special drawings, but it does require that written plans be available and that the information be presented to the occupants periodically. Maintaining the written plan and documentation of training is essential for compliance with this element of the fire-safety program. Consideration must be given to special building features and occupancies such as locked facilities, operating rooms, and clinical laboratories. The *Life Safety Code* provides guidance in this area.[3]

INTERIM LIFE-SAFETY MEASURES (ILSM)

Health care facilities need to be continuously aware of any factors that may inhibit response to fire emergencies and establish appropriate plans to deal with them. Construction projects and renovation and repair work can impact the effectiveness of life-safety measures in buildings. Fire- and life-safety programs need to be proactive in their oversight of these activities. Whenever fire-signaling systems, protective systems, or egress routes will be altered or impaired by establishing a work site, plans must be in place and communicated to all affected staff. These plans must also provide alternate evacuation routes, supplemental fire protection, temporary emergency notification systems, access to the work site for emergency personnel, separation of the work site from the occupied areas by suitable smoke and fire barriers, and monitoring of the work site for safe handling of hazardous materials. In addition, where egress routes may be affected, supplemental fire drills must be conducted in the construction zone.

Local emergency response personnel, building inspectors, hospital insurance carriers, and contractors should be consulted during the organization of ILSM plans. In large-scale projects, the fire- and life-safety function will be essential in pulling together all applicable agencies to ensure the safety of the facility and its occupants. Changes to the plan, added drills, daily inspection reports, and communications with all parties involved must be documented. Follow-up actions on deficiency items are critical to ensure safety and to validate compliance.

HAZARD COMMUNICATION

The Occupational Safety and Health Administration (OSHA) introduced the Hazard Communication standard (HAZCOM), 29CFR1910.1200, also known as the Employee Right-to-Know Law, to ensure that the hazards of all materials encountered in the workplace are effectively communicated to employees.[4] This may be accomplished by implementing a comprehensive hazard communication program that includes requirements for container labeling and other forms of warning, maintenance of material safety data sheets (MSDSs), and employee training.

Container Labeling

All containers of hazardous materials received from manufacturers, distributors, or importers must be clearly labeled to indicate the identity of the hazardous contents, hazard warnings, routes of entry into the body and target organs, and the name and address of the manufacturer, importer, or responsible party.

Material Safety Data Sheets

Material safety data sheets (MSDSs) are detailed informational documents prepared by manufacturers or importers of chemical products. An MSDS describes the chemical and physical hazards of the product or substance, when and how it may be hazardous, the effects of exposure, precautions for safe handling, emergency and first-aid procedures, and applicable control measures such as localized ventilation or personal protective equipment. Employers are required to maintain or make readily available to employees a complete and accurate MSDS for each hazardous material that is used in the workplace. This can be provided in the form of paper copies; however, the information is also available on CD-ROM and on Internet Web sites where, for an annual subscription fee, MSDSs can be accessed.

Contractors who work at a hospital site should also submit MSDSs for the materials such as paints, adhesives, and waterproofing agents that they bring to the work site and for the process by which they will apply or use these materials. The hospital safety department may require review and approval of the MSDS as a condition for using the materials at the facility or to determine appropriate control measures to minimize any impact that the materials may have on occupants. Providing a general form specifically for contractors to complete is the best way of ensuring that they acknowledge and understand the elements of the institution's HAZCOM program.

Training

Under the HAZCOM standard, employees who work in an area where hazardous materials are present must be trained by their employers when they are initially assigned and whenever a new hazard is introduced. The training curriculum should address the OSHA Hazard Communication Standard generally and should deal with the following specific issues: (1) activities in the employee's work area where hazardous materials, including pharmaceuticals, are present and their possible health effects; (2) safe work practices; (3) ways to minimize exposures; (4) emergency procedures to follow in case of accidents; (5) ways to read labels and MSDSs; and (6) the locations of the MSDSs and written hazard communication programs.

Contractors should also be informed of hazards they may encounter on-site under normal working conditions or during an emergency situation before beginning their contract work. This should include a review of what hazardous materials are located near the job site, procedures for safe handling, and the availability and location of appropriate MSDS.

BLOODBORNE PATHOGENS

Bloodborne pathogens are infectious agents present in blood and other body fluids that can pose a serious health hazard to hospital employees. The pathogens of greatest concern include hepatitis B virus (HBV) and the human immunodeficiency virus (HIV). Exposures to bloodborne pathogens may occur in a number of ways, including needle sticks, lacerations, and body fluid splashes, and may occur among various job functions, including housekeeping, laundry services, nursing, laboratories, and engineering and facilities.

In 1991, OSHA released the Occupational Exposure to Bloodborne Pathogens standard (OSHA 29CFR1910.1030)[5] in an effort to prevent transmission of disease to health care workers from exposure to bloodborne pathogens and to provide appropriate treatment and counseling to employees who do have such exposures. The first step toward compliance with this standard is the development of an exposure-control plan.

Exposure-Control Plan

All of the required elements for complying with the bloodborne pathogens standard should be addressed in the exposure-control plan. The plan should minimize the exposure of employees by using control equipment, such as leak-proof and puncture-resistant sharps disposal and specimen containers, needle safety devices, and personal protective equipment, as well as through work-practice controls such as handwashing, cleaning, disinfecting, and segregating food from infectious materials. Special labeling with the orange biohazard symbol is also required for specimen and waste containers to identify bloodborne pathogen hazards. The standard defines procedures that must be implemented following an exposure incident, including immediate medical evaluation, laboratory testing, and treatment. All incidents involving potential bloodborne pathogen exposures must be investigated and documented. Medical records for employees with exposures must be kept for the duration of their employment plus 30 years. The institution must provide hepatitis B vaccinations at no cost to employees who may be assigned to duties with exposure to blood and body fluids.

Training

Training must be conducted upon initial hiring and annually thereafter for employees who may be exposed to bloodborne pathogens. Training is also required when there are specific changes in products or procedures that may have an impact on the employee's exposure risk. The standard specifically identifies the elements that must be covered in the training curriculum and requires the opportunity for employees to have questions answered by a knowledgeable individual.

INCIDENT REPORTING AND INVESTIGATION

The principal purpose of accident and incident analysis is to determine the root causes of an accident or incident so that similar incidents may be prevented in the future by implementing process changes, product or equipment substitution, better supervision, or employee education.

Incident Reports

Written documentation should be maintained in a standardized format on all incidents. This should include information about the employee such as job title, name of department, and name of supervisor, and the location, nature, and severity of the incident. Documentation should also include information about treatment, corrective action, and follow-up activities to prevent a recurrence of similar incidents. Supervisors should review and sign off on all incident reports. Employees should also be encouraged to report hazardous conditions that do not result in injury.

Coordination of Activities

Corrective actions and preventive measures may require allocating resources and coordinating among several departments. These activities may often be successfully identified, prioritized, and delegated through the safety committee. This should include a requirement that completed actions are reported back to the safety committee.

Interaction with Risk Management

Many hospitals handle employee incidents separately from patient and visitor incidents (including reporting under the Safe Medical Device Act). It is important that there be communication between these two groups to identify potential hazards that may impact both populations.

Trend Analysis

Collecting incident information in a standardized format allows analyzing trends. This may include evaluations of departments or job titles with the highest incident rates overall or those with specific types of injuries. This will allow for targeted efforts such as implementing an ergonomic training program among environmental services workers to reduce back-injury rates. Trend analysis is also useful for monitoring any positive or negative impact that product changes may have on employees' safety, such as needle-stick injuries related to changes in brands of syringes or sharps containers.

Hospitals are also required to complete and maintain certain records under the Occupational Safety and Health Act of 1970. This requires hospitals under the Standard Industrial Classification (SIC) code 8062, General Medical and Surgical Hospitals, to prepare OSHA Form 200 (Log and Summary of Occupational Injuries and Illnesses), as well as OSHA Form 101 (Supplementary Record of Occupational Injuries and Illnesses). The OSHA 200 log or annual summary must be posted every year from February through March, and the OSHA Form 101, a document of recordable injuries, must be kept for a minimum of 5 years.

CHEMICAL SPILLS

Hazardous chemicals and medications are used in many departments throughout the hospital. They may be used in clinical laboratories, the morgue, respiratory therapy, and the maintenance and engineering departments, as well as in patient care areas. Despite well-defined safety practices for handling and storing chemicals, accidents and spills are always a possibility. It is important to be well prepared for such events.

A hazardous material spill procedure should be developed and communicated to all personnel. Generally, small chemical spills of low-toxicity materials or those unlikely to produce a harmful concentration in the air may be cleaned up by the users, provided they have received appropriate training. Individuals specifically trained to respond to hazardous material incidents should handle a large spill or a spill of highly toxic material. These trained individuals should be prepared in the following areas:

- Must be medically cleared and fit-tested for respirators
- Should be knowledgeable in selecting and using personal protective equipment and in proper waste disposal procedures
- Must understand chemical terminology and chemical-exposure limits and be proficient in using chemical-monitoring equipment.

A spill-response cart or kit should be available for emergency responders. The contents of the kit should include the following items:

- Reference guides such as the National Institute of Occupational Safety and Health (NIOSH) pocket guide to chemical hazards
- Acid/base neutralizer and pH paper
- Solvent or charcoal absorbent and vermiculite
- Absorbent pillows and blanket
- Mop, bucket, towel, sponge, dust pan, tongs, and broom
- Signs and barricade tape
- Respirator and cartridges, suits, goggles, and gloves
- Hazardous waste disposal bags and labels
- Emergency telephone numbers

The basic steps in spill response should include the following procedures:

- Attending to people in the immediate area and alerting others to evacuate
- Consulting the MSDS for the material spilled
- Donning appropriate personal protective equipment (gloves, respirator, goggles, lab coat, smock, or jumpsuit)
- Confining the spill to a small area
- Using appropriate materials from the spill kit to neutralize acidic or basic chemicals and absorb other chemicals
- Collecting the residue, placing it in a container, labeling it, and disposing of it appropriately
- Cleaning the spill area with water
- Completing a spill report or other required documentation

In the event of an incident involving large quantities of toxic materials or unknown materials or mixtures, it may be prudent to call in an outside hazardous material emergency response contractor. Numerous companies offer 24-h emergency response for these types of situations.

REFERENCES

1. JCAHO, "Fire Safety Requirements," EC 1.7, 2.6, 2.10, and 2.12, *Comprehensive Accreditation Manual for Hospitals: The Official Handbook (CAMH)*, Joint Commission on Accreditation of Healthcare Organizations, Chicago, 2000.
2. NFPA, *Standard for Health Care Facilities,* NFPA 99, National Fire Protection Association, Quincy, MA, 1999.
3. NFPA, *Life Safety Code,* NFPA 101, National Fire Protection Association, Quincy, MA, 1997.
4. OSHA, *Hazard Communication* (HAZCOM), 29CFR1910.1200, Occupational Health and Safety Administration, Washington, DC., 1999.
5. OSHA, *Occupational Exposure to Bloodborne Pathogens,* 29CFR1910.1030, Occupational Health and Safety Administration, Washington, DC., 1999.

SECTION 9.10
UTILITY MANAGEMENT PROGRAM

Kenneth S. Weinberg, M.Sc., Ph.D.
Director of Safety, Massachusetts General Hospital, Boston,
Massachusetts

Anand K. Seth, P.E., C.E.M.
Director of Utilities and Engineering, Partners HealthCare Systems, Inc.,
Boston, Massachusetts

RELATIONSHIP OF HEALTH AND SAFETY MANAGEMENT AND UTILITIES

The Environment of Care Model of the Joint Commission on Accreditation of Healthcare Organizations (JCAHO) has made clear that close relationships must exist among various operational components of hospital infrastructure, including safety and utilities management.[1] The idea of close working relationships comes as no surprise to those who have worked in health care in recent years. The concept has evolved as a result of a number of lessons learned, as well as practical requirements that have come to the fore. Not the least of them has been the necessary attention to regulatory mandates and expectations. In addition, hospitals have finally come to the realization that all activities that support patient care must be safe, a concept that has existed for a long time, and also at the same time all activities must be environmentally friendly, a somewhat newer idea. As systems have evolved and become more complex, testing and maintenance requirements have become more stringent. Peak operation of all equipment has become the expected, rather than the exception. There is a need for several types of expertise that must interface and work together constructively to ensure that all of the goals of operating and maintaining patient care support systems are met.

The interaction between safety management and utility management is ongoing and occurs virtually daily. The relationship may also be heightened and intensified during routine and nonroutine operations, such as generator testing and shutdowns or equipment replacement and modification. The role of the safety professional is to ensure that when systems are shut down or tested, alternate and backup systems are available and adequately tested to provide patient safety and continuity of the environment of care. During system upgrades or changeovers or during renovation and construction, when utility systems are modified or added, the safety and utility functions must interact and integrate to ensure that maximum patient, staff, employee, and environmental safety are maintained.

The utilities management professionals are responsible for specifying and purchasing only that equipment that meets all appropriate regulatory and compliance criteria. The safety professional's role is to act as the guide to ensure that all of the safety and environmental criteria are, indeed, met and adhered to. Safety must work closely with those who manage the utility systems to ensure that routine checks and inspections are performed. The safety function also must ensure that all of the permitting and discharge requirements for planned and new systems can be adhered to before the systems are installed. Of course, after the systems are installed, the safety professional must provide the expertise to help the utilities professional to ensure that the systems installed or modified comply with regulations.

Finally, the safety professional must be able to help the utilities professional plan for the future. The safety professional must be knowledgeable about current practices and future changes in hospital operations and techniques, as well as about current regulatory requirements and planned changes in those requirements. With this knowledge in hand, future budgets and long-range plans can be developed. Such planning and preparation will play a role in ensuring that the institution remains up to date and in compliance.

JCAHO recognizes that utility systems are critical. As a result, JCAHO requires that a formal management plan for these systems be created and reviewed by the appropriate professional. The safety committee should also approve such plans. The management plan should include the following components:

- How the organization will promote the safe and effective use of utility systems
- How those who use or are responsible for maintaining the system will receive ongoing training and education
- Identification of equipment which should be included in the program
- Standards for equipment maintenance and repairs
- A process for selecting and acquiring medical equipment that interfaces appropriately with the institution's existing utility system

UTILITY MANAGEMENT AS A TEAM EFFORT

Utility systems in health care facilities are never under the purview of any one department. Rather, utility systems are a shared commodity. One department may be responsible for producing or procuring a particular utility, whereas another department actually uses it. Therefore, teamwork among various departments is critical. This situation is illustrated in the following example, which deals with medical-gas use in a hospital: Usually, the engineering or maintenance department is responsible for procuring or producing and maintaining centrally piped medical-gas systems. Medical-gas systems are the responsibility of engineering because the piping system is viewed as a plumbing system, whereas the operating equipment, such as compressors and vacuum pumps, are mechanical. The respiratory care department is the biggest user of medical gases, particularly of air and oxygen for patient care. This department often becomes involved in managing of these two systems. As a result, respiratory care may closely monitor the quality of the air delivered to the terminals and leave maintenance of the air compressors to engineering. (Respiratory care may also manage any shutdowns of the air or oxygen systems.) The anesthesiology department also plays a support role in managing medical-gas systems. As the primary user of nitrous oxide, the anesthesiology department may oversee this system. In recent years anesthesiologists have also been encouraged to take an interest in centrally piped medical-gas systems. Biomedical engineering usually maintains the equipment that provides the nitrous oxide to the patient and is an additional interested party in the centrally piped medical-gas system.

This example illustrates why teamwork and cooperation among a variety of seemingly unconnected departments are important to ensure proper utilities management in a hospital setting.

Inventory Criteria

All utility systems are not equal! The level of impact of any given utility system on the hospital's mission can vary. Therefore, to determine the impact of a utility system, an inventory criteria weight list should be developed to identify and evaluate equipment for inclusion in the program. Inventory criteria should consider the impact of the following systems:

- Life-support systems
- Infection-control systems
- Environmental-support systems
- Equipment-support systems
- Communication systems

To make the inventory procedure more effective and meaningful, numerical weighting should be developed based on critical impact and the impact of the respective systems on patient care, so that systems can be compared to one another in terms of importance. The evaluation procedure and the criteria should be specified in the utility management plan (see Fig. 9.10-1). The potential advantages of such an inventory procedure include a reduction of items included in the plan, modification of testing intervals, and a multilevel tier system of equipment that is included in the preventive maintenance program.

Using this criteria-based inventory procedure, utility systems should be categorized according to the testing priority, degree of risk, type of risk, risk of failure, and risk of harm. Some examples of risk include harm to the patient if critical power branches fail or if systems are unable to support critical life-safety systems in an emergency, or the breakdown of barriers that prevent the spread of infection by negative-pressure isolation of patient care locations.

When selection criteria are used, the following utilities should be included in the inventory:

- Emergency power supply system and components
- Medical gas, including vacuum system and components

Figure 9.10-1 shows a typical format used to create numerical values. Each question can be answered on a scale of 1 to 5, where 5 is the highest risk and 1 is the lowest.

The following utilities should be included in the weighting system due to their importance in patient care, life safety, or maintaining the environment:

- Fire-alarm system
- Normal power
- Steam
- Air-handling system
- Chilled water
- Communication
- Data
- Overhead paging
- Nurse call
- Emergency power

MANAGEMENT PLAN

A management plan should include the following items:

- An operational plan to ensure the reliability of the systems
- A method for risk minimization and system-failure reduction
- Inspection procedures
- Testing guidelines

FIGURE 9.10-1   Utilities classification. Answer each question on a scale of 1 to 5, where 5 is the highest risk and 1 is the lowest. (*Courtesy of MedSafe, Inc.*)

Utility System	Name	Discipline

System disruption of the following utilities, categorized by the Roman numeral heading, may result in a number of consequences, categorized by capital letters.

I.   Life support

   A. Death

   B. Injury

   C. Inability to provide diagnosis, treatment, therapy

   D. Inconvenience or insignificant risk

   Score:____

II. Infection control

   A. Infection-control incident

   B. No infection-control incident

   Score:____

III. Environmental support

   A. Inability to provide critical patient care services

   B. Inability to provide diagnostic patient care services

   C. Inability to provide therapeutic patient care services

   D. No effect on the delivery of patient care services

   Score:____

IV. Equipment support

   A. Inability to provide critical patient care services

   B. Inability to provide diagnostic patient care services

   C. Inability to provide therapeutic patient care services

   D. No effect on the delivery of patient care services

   Score:____

V.   Communication systems

   A. Inability to communicate with unattended patients or high-risk areas

   B. Inability to communicate via paging or beeper

   C. Data systems used in diagnosis and treatment monitoring

   D. No effect on patient care services

   Score:____

Total score ____ included in the utilities management program

Using a scale of 1 to 5, where 1 is the least problematic and 5 is the most problematic, select a cutoff point to determine which utility systems are more critical.

a Courtesy of MedSafe Inc.

- Maintenance protocols
- A layout of utility systems shown in a one-line diagram
- Locations of labeled control valves in the event of an emergency
- A description of utility system management problems, failures, and user errors noted and documented by incident reports
- An annual evaluation of the program's effectiveness

Documentation

It is critical to document every utility system. A one-line drawing is very useful for routine management as well as disaster response. A detailed description of the process of documenting electrical systems is given in Sec. 9.11, "Electrical Utility Management Program." Similar steps could be taken to document other utility systems.

Orientation and Training

Orientation and training are needed to use utility system staff effectively. JCAHO requires that a formal program be developed. This program should include the following items:

- Limitations, capabilities, and special applications of utility systems
- Preparation and implementation of emergency procedures in the event of system failure
- Information and skills necessary for assigned maintenance responsibilities
- A method of reporting failures, problems, and user errors
- Instructions for using emergency shutoff controls
- Locations of emergency shutoff controls

Users of the utility system must be trained in the capabilities, limitations, and special applications of equipment, as well as in operating and safety procedures for proper use of the system.

Emergency Procedures

Any utility system failure in a health care facility can have a significant impact on the safety and welfare of patients. Therefore, it is important to establish procedures to address the actions that must be taken if a utility system fails. These procedures must include a plan for alternatives when a utility system fails, including how to obtain repair services, the location and instructions for using emergency shutoff controls, methods for staff notification in affected areas, and training of clinical staff to perform emergency clinical interventions if a system fails.

Maintenance, Testing, and Inspection of Utilities

Adopting a regular program in maintenance testing and inspection of utility systems is critical. Good documentation is necessary for a successful maintenance and inspection program. The records should contain a current, accurate, separate inventory of all utility systems and components, as well as records of performance and safety testing of all critical components and equipment before initial use. Regular equipment testing of the utility systems is also needed. Testing at 12-month intervals is very typical, but a longer interval between tests may be justified on the basis of historical experience. The testing frequency should be determined after a review of system needs. In addition, professional judgment that takes into account the known risks and hazards of utility equipment needs to be applied when establishing testing intervals. Testing all systems is important, but emergency generator testing stands out as one of the most crucial elements of a utility management program (see Sec. 9.11).

Effective utility management requires a strong preventive management (PM) program. See Chap. 7 for more information on maintenance. Facility managers need to review the completion of the various components of the PM program. The completion of PM elements on time can also be used as an indicator of appropriate staffing, an issue that may come up in leadership meetings.

The utility systems defined in the utility management plan should be cross-referenced with the PM program to ensure completeness. *All* systems should be included, regardless of their simplicity or complexity, for example, nurse-call systems, negative-pressure rooms, panic alarms, and paging systems.

Other components covered under utility management systems that affect patient, staff, and visitor comfort and safety also need to be considered. Included in this category are grounds and parking-lot lighting, safe location of aboveground or belowground storage tanks for gases or fuels, and the location and accessibility of connectors and shutoffs.

Critical Switch and Valve Labeling

To avoid confusion, critical switches and valves need to be labeled. A surveyor may question this judgment and ask, for example, whether a particular bypass valve on the front of a heat exchanger is critical for shutting down the system or unit in an emergency. When in doubt, *label it.*

A simple criterion can be established to test the criticality of a valve. If the valve or switch is used to bring a unit or system to an operational level, whether in an emergency mode or not, it should be labeled. *All* systems listed in the management plan should be so labeled.

SPECIFIC UTILITY MANAGEMENT PROGRAMS

Many of the steps that must be taken to maintain a utility management program are site-specific. The following specific utility programs are described in general terms and are only meant to give readers a guide to implementing their own utility management programs.

Performance Indicators

The facility manager should use numerical indicators to understand how well a particular utility system is operating. These indicators aid in evaluating deficiencies in any system and generate either a process or practice correction. JCAHO leaves the formulation of these indicators largely to the facility manager. The following examples under each utility described might help readers formulate their own performance indicators.

Electrical.   Section 9.11, "Electrical Utility Management Program," describes this program in detail.

Steam.   Steam is one of the critical utilities because it is used for sterilization and for heating, humidification, and in some instances cooking. Some of the highlights of a successful utility management program for steam include regular maintenance, as described in Chap. 7, proper record keeping, careful attention to the boiler-treatment program, and testing in place to ensure that the backup system will be operational when needed. A good steam-trap maintenance program reduces steam leakage in the condensate system and ensures that steam systems operate at optimal efficiency. An example of a performance indicator for this program is the frequency of loss of steam pressure.

HVAC.   The HVAC system is also one of the critical utilities because it provides code-required ventilation and pressure balance, as described in Sec. 9.6, "Mechanical Systems." The HVAC system is crucial for all patient care spaces, including operating rooms, recovery rooms, intensive care rooms, isolation rooms, treatment rooms, and patient bedrooms.

Some of the highlights of a successful utility management program for an HVAC system are regular maintenance, as described in Chap. 7, sound record keeping, and a regular testing program to ensure that an appropriate air-exchange rate exists in operating rooms and isolation rooms.

Chillers may be part of a health care facility's HVAC system and therefore should be part of the utilities management program. A regular PM program, including some preventive maintenance, as described in Art. 7.2.1, "Mechanical Testing and Maintenance," may be very beneficial to understanding long-term maintenance needs.

NFPA 99 requires regular inspection and testing of *fire and smoke dampers.*[2] This is a very difficult requirement to meet in an existing facility. In an existing facility, a detailed survey must be conducted to determine the location of all duct penetrations where such dampers are required. In locations where such dampers are located and not required, the facility manager needs to assess the condition and determine if the excess dampers can be removed. Then a careful testing program can be implemented.

A variety of obstacles to regular inspection and testing of fire and smoke dampers exist. These obstacles include lack of access panels, access panels that are not appropriately sized, or panels that are located too far away to allow actual access to the fire and smoke dampers themselves. Even when access to fire and smoke dampers is possible, testing is difficult. Of the two types of dampers, correct operation of a smoke damper is easier to confirm because the operating signal can be simulated. The only way to close an existing fire damper is to melt the fusible link or remove it and verify that the fusible link would indeed melt at the required temperature. If a fire damper is closed when an air-handling unit (AHU) is operating, the sudden loss of airflow will change the air-pressure balance in the space. Such a sudden change of air pressure may not be desirable and may not be in the best interest of the patient.

In new facilities, a combination smoke and fire damper correlated with an energy management system (EMS) could be considered. These combination dampers can be checked by remote access and tested by the EMS to ensure that the damper closes in the event of smoke and fire. An example of a performance indicator for such a system is the frequency of failures among dampers during testing.

Sprinkler and Other Fire-Suppression Systems. A sprinkler system is a very important part of a health care facility's fire-prevention program. A program that monitors valve conditions to ensure that valves are tamper resistant and open is critical. Regular testing of the fire pump and maintaining records is also important. Other fire-suppression systems—for example, a powder system in kitchen hoods—also require regular testing and record keeping. An example of a performance indicator for fire-suppression systems is the frequency of failures in the system during testing.

Fire-Alarm and Smoke-Control Systems. Every health care facility requires an approved fire-alarm system. In certain high-rise buildings, a smoke-control system may also be required. Some jurisdictions may even impose specific requirements for the frequency and duration of these tests. Fire-alarm testing can be part of other JCAHO-required practice fire drills. In the absence of any other requirement, it is recommended that smoke-control systems be checked at least annually. An example of a performance indicator is the frequency of system failures during a practice drill.

Water. Water is one of most critical utilities in a health care facility. Failure of this system may be due to a line break either inside or outside the facility. Such a break may be the result of a flood or a hurricane. More and more in today's environment, there is concern that water supplies may be interrupted by terrorist activity. Water loss could also be due to a failure of the municipal water-treatment system or the municipal water-distribution system.

A loss of potable water affects a number of routine hospital activities, including food preparation, dishwashing, sanitary facilities, linen washing, and handwashing. Firefighting may also be affected by losing sprinkler capability and lack of fire-hydrant water pressure. Loss of water may be catastrophic and can have far-reaching effects on patient care due to failure of medical vacuum systems, sanitation and hazard-control systems, and safety systems.

Therefore, it is critical for the facility manager to prepare for loss of water and the effect such a loss might have on the operation of a health care facility.

A water-loss plan must be addressed at several levels. First, if at all possible, a health care facility should be connected to at least two different water-supply lines, fed from different municipal lines. Thus, a break in one line would still allow uninterrupted water supply from the other line. Second, the facility manager can provide an alternate source of potable water by contracting with a local vendor for bottled water, by an agreement with the local fire department or local civil defense authorities to provide tanker truck supply outside the hospital, or by using a trailer-mounted water-treatment or -purification system that takes water from a local pond or river and purifies it. These supplies could also be used for some limited manual toilet flushing, as well as handwashing. However, to conserve water supplies, a hand-disinfectant solution that does not depend on the need for water should be available to caregivers. Third, a firefighting plan must be arranged in cooperation with the local fire department. Fourth, the loss of makeup water to the HVAC systems should be analyzed and remedial action considered. This might include, if needed, a separate air-cooled direct-expansion system for the most critical applications and an analysis to determine how long the HVAC system would operate without makeup water. Fifth, installing modifications to critical systems in advance to eliminate or reduce the amount of water required for operation can mitigate the effects that the loss of water-cooled systems such as medical air or vacuum might have. An example of a performance indicator for water-system operation is the frequency of system failure.

Medical Gases. When the piped medical-gas system has been compromised, testing of all downstream fittings is needed to ensure that no cross-connections or contamination of the system has resulted. There are usually no requirements for frequency of routine testing (e.g., annually or triennially).

The batch purity certification data for bulk liquid oxygen and bottled oxygen gas should be readily available. Even though the facility manager might not be directly responsible for keeping these records, the facility manager should know who is responsible and even have copies of the most recent certifications.

NFPA 99 defines the testing requirements and requires that installation and certification test results be kept on file. NFPA 99 also recommends periodic checks of various systems and components. These periodic checks include the following items:

- Pressure check of pressurized gas systems
- Air compressor intake, receiver, and quality air checks
- Testing area and master alarms
- Testing medical-gas valves for leaks
- Testing medical-gas outlets for leakage and flow

An example of a performance indicator for a medical gas system is the frequency of system failure.

Process Waste Systems. Many health care facilities may also have clinical laboratories whose wastestream effluent is regulated by state and local authorities as it enters the sanitary sewer system. The effluent stream may also require pH neutralization and pretreatment to remove metals or other restricted materials that should not enter the municipal wastewater-treatment system. Even the grease discharged from the kitchen must be accounted for as the wastewater effluent enters the sanitary sewer line. These systems need to be maintained in the same manner as outlined in Chap. 7. Systems failures must also be recorded and corrected in a similar manner.

An example of a performance indicator is the number of incidents when the effluent stream pH or metal content exceed limits as determined by routine analysis of samples collected as the wastewater leaves the facility.

Vertical Transportation Systems.   Vertical transportation systems are items such as elevators or escalators. These systems should be maintained in the manner described in Art. 7.2.4, "Elevator/Escalator/Moving Walk Maintenance." Systems failures must also be recorded and corrected.

An example of a performance indicator for vertical transportation systems is the frequency of system failure or the length of waiting time before an elevator is available.

Other Patient Records and Specimen Transportation Systems.   Many health care facilities have pneumatic-tube or other systems described in Art. 9.3.3, "Transport Systems," to transport patient records or specimens. These systems should be maintained in the manner described in Chap. 7. Systems failures must also be recorded and corrected. Examples of a performance indicator for such systems are the frequency of system failure, the specimen delivery time to its destination under routine and nonroutine operations, or the availability of the system to users.

Communications.   In modern health care facilities, both voice- and data-systems communications are becoming increasingly important. Some of these systems are described in Art. 5.5.6, "Telecommunications and Data Distribution Systems." These systems should also be maintained in the manner described in Chap. 7. Systems failures must also be recorded and corrected. An example of a performance indicator for communication systems is the frequency of system failure.

It is critical that the utilities management plan address backup systems. For example, loss of normal phone service may activate a plan where the radios or cell phones in the facility are called into service. Sometimes the pay phones, if connected to different circuits, can be used as a backup to the facility's dedicated phone system. If this option is used, it is not practical to have to use change to initiate a call. Some other way to activate these phones must be arranged. Another potential backup in the event of failure of a phone system is the overhead paging system, if it has been installed to cover the affected locations in the hospital. Some facility managers have arranged to use the dedicated fire phones in high-rise patient care buildings as the emergency means of communication. Such use of fire phones is not desirable because if a fire condition and communication failure were to occur simultaneously, the fire phones would be used by fire department personnel and would not be available for patient caregivers. An example of a performance indicator for communication systems is the frequency of system failure.

Utility systems are crucial for patients' safety and well-being. A well-defined utility management program sets in motion the steps necessary to maintain the routine operation of utility systems and provides a framework for dealing with emergencies. Such programming and planning cannot be executed in a vacuum. Rather, utility systems, like other hospital programs, must be planned and conducted in collaboration with other departments that either use the utility or physically maintain the systems and their components. In today's fast-paced, competitive, and highly regulated world of health care, each hospital department occupies its own unique niche in providing patient services, and each department must be ready to meet the demands placed upon it by patients, administrators, staff, and regulators. This section provides what the authors believe are useful guidelines, based on practical experience and the application of prudent practice, to assist utilities managers, facilities managers, and others to prepare for and meet these demands.

CONTRIBUTORS

Douglas Erickson, FASHE, President, DSE Consulting LLC, Christiansted, Virgin Islands

Dawn LeBaron, C.H.P.A., Director, Facilities Services, Newtown Wellesley Hospital, Newton, Massachusetts

REFERENCES

1. JCHMO, *Comprehensive Accreditation Manual for Hospitals: The Official Handbook* (*CAMH*); Comprehensive Accreditation Manual for Ambulatory Care (*CAMAC*); *Comprehensive Accreditation Manual for Long Term Care* (*CAMLTC*), Joint Commission on Accreditation of Healthcare Organizations, Oakbrook Terrace, IL, 1994.
2. NFPA, *Standard for Healthcare Facilities,* NFPA 99, National Fire Protection Association, Quincy, MA, 1996.

SECTION 9.11
ELECTRICAL UTILITY MANAGEMENT PROGRAM

David L. Stymiest, P.E., SASHE, C.E.M., Senior Consultant
Smith Seckman Reid, Inc., New Orleans, Louisiana

To manage a hospital electrical utility system effectively, you must:

1. Know the system, including its utility service and details of internal distribution.
2. Design it for flexibility, so that any portion may be shut down with minimal or no disruption.
3. Manage and control modifications, additions, and alterations by having a master plan and ensuring that all changes to the system conform with this master plan.
4. Maintain a monitoring system at each main switchboard and at each major normal and emergency feeder.
5. Plan for growth at the transformers, switchboards, feeders, transfer switches, switchboard rooms, and electric closets.
6. Plan for power outages with contingency plans for both scheduled and unscheduled power outages.

MANAGING ELECTRICAL SYSTEM CHANGES IN HEALTH CARE FACILITIES

The Need to Manage Changes

Ongoing improvements in clinical and research technologies require periodic modifications in the electrical and mechanical infrastructure.[1] Space reconfiguration to meet the changing needs of the health care industry can increase, realign, or occasionally decrease the electrical load. Research density is likely to increase over time and increase the demand for electrical power.

Management System Components

The management system's required components are as-built field verification walkdowns, electrical load monitoring systems, infrastructure databases, engineering and design standards, and master electrical one-line and riser diagrams. This information is then used in computer-based electrical system demand analyses and load-flow, short-circuit, and protective coordination studies.

Each component of the process complements the others and reduces duplication of effort. Field walkdowns allow one to create accurate, as-built, one-line and riser diagrams, as well as wiring and loading information. Future drawing updates are easier and less costly if the drawings are prepared with computer-aided design and drafting (CADD) software. The monitoring systems provide regular loading and load-growth information. System study results (protective device settings) are then factored back into CADD drawings. The engineering and construction standards help ensure that future changes do not adversely affect protective coordination.

Field Walkdowns and As-Built Verification. Most facilities can use field walkdowns as the mechanism for obtaining a set of accurate as-built documents. Using bound record books ensures that all of the recorded details are available later. One recording method is to draw the equipment and interconnecting conduits in standard riser-diagram format. Another, often more useful, recording method with engineering assistance is to draw a one-line diagram of the equipment and wiring observed in every space, and record all equipment ratings, feeder sizes, and loads where available, along with panel and load names and ID numbers. The one-line diagram provides an accurate representation of the flow of electricity but should be supplemented by sketches or photos of the equipment's physical arrangement.

Load Monitoring and Equipment Databases. The loading of switchboards, distribution feeders, and automatic transfer switches (ATSs) should be measured annually. Unanticipated load changes can then be probed further, if needed. Additional monitoring should occur approximately 3 months after a renovation because not all of the new equipment may be installed and running normally until after such a stabilization period. Recording instrumentation can provide strip-chart or floppy diskette recordings. The information can then be translated into a spreadsheet (Fig. 9.11-1) for further graphical analysis of the load profiles.

Monitoring of x-ray, cardiac catheterization, and other types of radiology equipment needs instrumentation that can respond to and record the pulse-type load profile. The engineer (who should be qualified to do this level of work) should assess the peak current drawn by the radiology equipment and also the average loading when the radiology equipment is not firing.

The electrical load measurements are then entered into equipment databases or spreadsheets. Equipment databases or spreadsheets are very useful in providing accurate information on the loading of existing equipment. These tools also allow the facilities engineer to sort for many different conditions and can provide easily updated single-source documentation for analyses of utility failures.

Electrical Service Demand Load Profiles. Because of electrical industry restructuring, which is discussed in more detail in Art. 7.1.6 many health care facilities will generate electrical load profiles at their main service entrances as a necessary part of the power procurement process.[2] Such documentation can be used for several other important purposes as well, including utility management. Figure 9.11-2 illustrates sample 15-min demand load profiles for a large, multibuilding, academic medical research center. The data are representative samples of the actual metered 15-min campus electrical load profiles at points during the same year. A few simple comparisons of the data in Fig. 9.11-2 illustrate the value of load profiles. The sample winter Sunday load profile illustrates the base load of the facility, including lighting, ventilation, and the medical processes that occur 24 h per day all year. The variation from the winter Sunday to the average winter weekday illustrates the variable *process* load of the

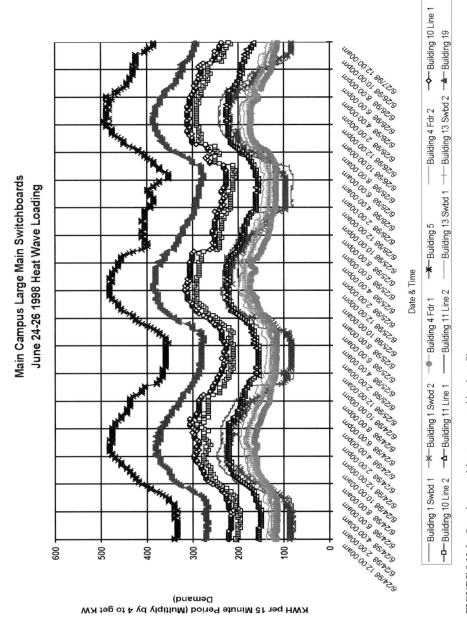

FIGURE 9.11-1 Sample spreadsheet-generated load profiles.

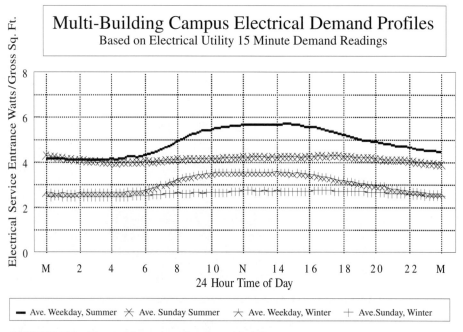

FIGURE 9.11-2   Electrical demand profiles of a multibuilding campus.

hospital, which can also be seen from the midnight to midday variation. The difference between the sample winter Sunday and the sample summer Sunday load profiles shows the additional cooling load in the summer. Daily energy usage can easily be learned from demand load profiles by simply calculating the area under the load profile curves.

Building Load Profiles.   These sampled load profiles illustrate the value of real-time switchboard metering. One truism often stated by energy managers is that "one cannot manage what one does not meter." Time-of-use metering installed for energy management also provides important demand load information when the metered period includes a *design day,* or series of days when the ambient conditions cause peak facility loading.

Essential Power System Load Profiles.   It is important for hospital facility engineers to have load profile data for the hospital's essential power systems for several reasons. A recent series of articles discussed the importance of emergency power system load profiles and presented an approach to documenting generator loading from graphical analysis of individual transfer-switch load profiles.[3]

Engineering and Design Standards.   Any modifications to the electrical systems, whether due to a renovation, addition, or infrastructure upgrade, should complement the master plan whenever possible. Each hospital should have its own set of engineering and design standards that should be issued to the consulting engineers before they start work on renovation or expansion projects. Engineering and design standards should establish clear criteria for equipment, materials, methods, and nomenclature. In multibuilding facilities, equipment numbering is often a source of confusion. It is common in older facilities to see many devices that have the same identification number, such as "DPI" or "ATS-1." A complete numbering convention that identifies the building and floor when it is important to accumulate data on equipment located in several buildings (such as switchboards, generators, and ATSs) will help facility operating personnel to manage their environment properly.

CADD Master One-Line and Riser Diagrams.   Master one-line diagrams can show, on one drawing, the flow of electrical power from the normal and emergency power sources (utility and emergency generators) to the major loads. In multibuilding campuses, the master, or main, one-line diagram cannot usually show all panelboards because of limitations caused by the maximum allowable drawing size and minimum text size. The drawings should show measured demand information at selected locations such as ATS, and switchboards, and can also show system study node numbers for consistency among documents.

The electrical riser diagrams are used to supplement the master one-line diagram. Each riser diagram is generally limited to one building and sometimes to only one (normal or emergency) power subsystem if the total power system is extensive. The riser diagram shows various parts of the electrical distribution system in their relative locations, along with interconnecting conduits between the pieces of equipment. The purpose of this drawing is to show the physical as well as electrical relationships among pieces of equipment.

Riser diagrams may be less useful than master one-line diagrams as a disaster management tool when electrical distribution systems interconnect different buildings. This situation is common in larger facilities. However, within discrete buildings, riser diagrams prepared with CADD software can be very useful in understanding the coverage of the emergency power branches in different building areas.

Electrical Demand Analysis.   After the field verification has been done, electrical feeder information is entered into the computerized analysis system database, using common commercially available software. The connected loads and demand factors for each piece of distribution equipment are also entered into the database. The facilities engineer determines demand factors for the various types of electrical loads on the basis of the historic information at the specific facility. The facilities engineer can then use the software to calculate the anticipated demand at each distribution point in the facility. In major multibuilding facilities, the analysis can be broken into the following discrete segments:

- *Each building's emergency power system analysis* (*assuming a full utility power outage*) includes all elements from the emergency generators through the ATSs to the normal and emergency distribution panelboards.

- *Each building's normal power system analysis* (*assuming no utility power outage*) builds on the previous analysis by considering each ATS as an end-use device. Because both normal and emergency power feeders supply an ATS, it is unnecessary to model for both. The remainder of the normal power system is then modeled, and the degree of detail is decided by the size of the system and the facilities engineer's needs.

- *Multibuilding services and distribution analysis* takes the calculated (and verified) main switchboard loading at each building switchboard element as end-use loads. Again, this avoids duplication of effort. The modeling then extends beyond the individual buildings and includes all upstream elements, including medium-voltage transformers, up to the medium-voltage utility service.

Load-Flow, Short-Circuit, and Coordination Studies.   All other variations of the electrical system studies use the common database previously modeled. Then, after the demands are calculated, various operating parameters of the system can be analyzed. These operating conditions might include (1) utility voltage reductions, (2) loss of portions of the power system at any level, and (3) the open versus closed conditions of primary and secondary tie breakers. Then, it is easy to analyze all operating scenarios independently with little additional effort. The short-circuit studies also use the common database previously created. Protective coordination for both phase overcurrent and ground-fault conditions is necessary in health care facilities. Without protective coordination, the negative impact of any one event is magnified needlessly, possibly increasing the adverse impact on patients, employees, and visitors. Another issue that deserves special attention is the protective coordination between the utility and the hospital's main electric service, especially in cases involving cogeneration systems.

Reducing the Cost of Engineering and Design

Owners often require that consulting engineers field-verify any owner-provided as-built drawings during the predesign or design phase of a renovation project. If the facilities engineering staff has accurate documentation of existing conditions, the drawings can then be provided to the consulting engineers as accurate documents. Avoiding costly duplication of effort in repeated field verification allows a reduction in fee proposals for engineering and design services. A word of warning is appropriate if this approach is being considered—the existing drawings must be accurate and up to date. The responsibility of accuracy would then be the owner's.

Crisis Management of Internal Disasters

Internal disasters are those that occur inside the facility. The accurate main one-line diagram is critical for properly managing internal electrical disasters. With this document, it is not necessary to rely only on human memory during stress-filled crises. Because of this important need, the facility electricians should participate in deciding the types of data shown on the one-line diagram. Up-to-date load information is also necessary for properly analyzing utility failure scenarios. This is especially important when tie breakers and secondary-spot network systems are in use and where partial failures can propagate into much larger catastrophes if the resulting loads on remaining portions of the system are not accurately predicted. Mimic bus diagrams on switchboard and switchgear front panels are also excellent visual aids for crisis management.

Electrical System Master Planning

The master one-line diagrams are an important resource in electrical system master planning. When they are field-verified and regularly updated, they can be relied upon as rapid factual sources of information when space changes or load rearrangements are contemplated.

Improved Management of Infrastructure Upgrades and Shutdowns

Electrical systems must be regularly shut off for preventive maintenance and testing because the equipment cannot be serviced while it is energized. Thus, it is necessary to schedule shutdowns of the electrical distribution equipment that will affect entire buildings or portions of buildings. The clinical and research staffs will have more confidence in the management of the shutdown process when the facilities staff displays detailed knowledge of system design, system operating parameters, and system flexibility. Accurate, as-built, one-line and riser diagrams help the engineering and maintenance staffs satisfy this requirement.

Optimum Use of Available Funding

Cost-containment pressures on the health care facility limit available capital funding for equipment replacement and infrastructure upgrades. The optimum use of limited funding requires that the health care facilities engineer prepare an infrastructure master plan that complements the facility's strategic plan. With a master plan in place, the engineering consultants and facilities engineer can consider global needs when reviewing local projects. Sometimes, local projects can be guided into cost-effective solutions that also address the needs of both local projects and the global master plan.

MANAGING HOSPITAL ELECTRICAL SYSTEM SHUTDOWNS

This section introduces a comprehensive program for analyzing the need for, planning, and conducting electrical shutdowns in operating hospital buildings.[4] The complete program is described in detail in Ref. 4. An electrical shutdown as described here is a preplanned and scheduled partial or full electrical distribution system outage. Shutdowns are necessary for preventive maintenance, equipment repairs, support staff training, and unit training. Shutdowns meet needs in all of these areas that cannot be met any other way. Electrical shutdowns must be conducted safely and effectively. Communication and comprehensive analysis are most important in building the credibility necessary to accomplish this end.

Need For Electrical System Shutdowns

Hospitals must train their personnel to respond effectively to internal and external disasters and to assess the operating capabilities of the clinical and support services during utility power outages. Electrical distribution equipment requires periodic preventive maintenance or corrective action necessitating that it be turned off. Meeting these needs requires partial, building-wide, or even multibuilding normal power (NP) shutdowns, depending upon the electrical system configuration. Operating experience shows us that if electrical equipment is not shut down when the facilities staff wants to, it will shut itself down when the staff does not want it to.

Adequate advance planning, communication, pretesting, and scrutiny can ensure adequate protection for patients, staff, workers, and visitors during a shutdown. The shutdown planning process, described in detail in Ref. 4, involves detailed planning at several levels, comprehensive multilevel communication, brief scheduled power outages for assessment, assessment of the shutdown's impact on each space and activity, action plans, shutdown contingency plans, a hold-point list to permit clinical control of the process, feedback forms, an overall shutdown white paper, critiques of timely lessons learned, and test reports.

Limitations of Monthly Emergency Power System Testing. Hospitals must test their emergency power (EP) systems monthly by switching their distribution systems from normal utility power to emergency generator power for at least 30 min. These monthly generator and EP system tests demonstrate the capabilities of the electrical equipment to perform its automatic functions and to serve the emergency load. These tests also provide important training for plant electricians. However, these tests do not demonstrate the capabilities of the clinical personnel and the personnel of other supporting services to perform their functions during a full or partial utility blackout because the normal power is not usually turned off for the monthly test. These tests also do not demonstrate the shortcomings of EP coverage. They simply test the existing system. A controlled outage of normal electrical power can adequately train medical and support personnel in what to expect if the power goes out and how to respond to that scenario.

Power outages can occur at any time. They can be caused by external influences on the utility system, such as underground cable faults (short circuits), lightning strikes, or vehicle accidents, or by internal influences on the in-plant distribution system, such as equipment failures, cable faults, or motor faults in uncoordinated systems. The hospital clinical and support staff must be able to respond effectively to the NP outages caused by all of these situations.

Shutdown Scope of Work. Many complementary tasks can be addressed simultaneously during shutdowns, including preventive maintenance (PM), testing, and upgrades. The PM includes cleaning and tightening loose connections. The testing scope of work includes circuit breakers, switchboards, switchgear, transformers, and meters. Refer to Art. 7.2.3, "Electrical Testing and Maintenance," for further discussion and references. The upgrade scope of work may include new circuit breakers or just overcurrent trip-unit upgrades, switchboard upgrades, and transformer replacements.

Clinical Unit and Support Service Training. This training is achieved through the numerous meetings, communications, and personal contacts necessary to conduct the shutdown. Each affected area and service should prepare written action plans that then enable them to manage their environment of care more proactively in the event of a complete or partial utility NP outage.

Short, Selected, Local, Planned NP Outages. Short planned outages of the NP system allow assessing the adequacy of existing EP coverage. This approach is necessary before the first major shutdown in each building or area and is a good idea after any space renovations. There are several reasons for this, including confusion about EP outlets, changes in the use of space, changes in personnel or tasks even though the space usage itself does not change, and mistakes in emergency circuit installation or marking.

It is common practice to identify outlets powered by EP branch circuits with special nameplates or symbols, panel and circuit numbering, or special receptacle colors, such as red. However, this approach is not foolproof. Installation wiring mistakes can occur that will be discovered only by 100 percent tracing with a high-frequency signal.

The experience of a short, planned outage provides a gauge, but not complete information, about the impact of a much longer power outage. As a result, a well-informed assessment must be made of all outages of short duration when assessing the impact of longer outages.

Shutdown Contingency Planning

Once the upgrade or maintenance function of a shutdown has started, it can be impossible to return to NP instantly in the event of an electrical malfunction. This occurs because shutdowns often involve withdrawing circuit breakers from their normal locations for primary injection testing, perhaps withdrawing some cabling from its normal terminations for modification, and because personnel are actually inside larger switchgear for better access when cleaning and tightening components.

If a generator failure occurs during a shutdown, it is critical for patient care and life safety that power be restored to those branches of the EP system immediately. This contingency planning is a necessary part of the preshutdown planning. One approach could be to bring in one or more backup lines with rapid means of connection. Using quick-connects along with clear instructions for isolating the back-fed equipment by opening switches or circuit breakers can permit the hospital electricians to restore power to life-safety and critical branches very quickly. Contingency walkthroughs with the electricians are important to review the procedures to follow if a failure occurs.

It is very important to have battery-backed lighting installed before a shutdown, so that the generator failure contingency does not throw the electrical equipment rooms into total darkness. The battery-backed lighting should be tied into the emergency ac lighting circuit so that the batteries are not drained during the first 90 min of the NP shutdown.

Shutdown contingency planning is somewhat simplified if the only work of the shutdown is circuit-breaker testing. In this event, the specific circuit breakers to be tested at any time can be the only ones withdrawn from their cubicles. In case of a generator failure, power can be returned to the switchboard quickly.

Effect of Electrical System Design on Shutdown Cost and Duration

The main electrical switchgear in buildings provides power to all spaces in the building. The spaces often involve separate uses with different needs. If distinct uses are powered by separate sections of switchboard, it may be possible to limit the scope of an individual shutdown to less than the entire building. This approach can reduce shutdown cost and disruption but may not provide adequate training for unit clinical and support services.

Reporting to Other Groups.   The shutdown task force should report to several groups, including the safety committee, safety department, medical advisory committee, and department heads. The purpose of the medical advisory committee is to ensure that the medical community participates fully in the basic decisions and management of the shutdown. This approach alleviates clinical concerns about a shutdown that may develop a momentum of its own.

All affected department heads should receive shutdown task force meeting notes and notices. Meeting notices should always identify the date, time, and scope of the shutdown, as well as the planned dates, times, and scope for interim EP testing and NP planned outages.

Shutdown Task Force.   The initial membership of the shutdown task force should include those departments or services located in the affected buildings, along with departments that have facility-wide responsibilities and would be affected by shutdowns in *any* building. As planning progresses, the initial representatives can identify others that either (1) could be affected by the shutdown and should be part of its planning, or (2) could help them or others cope with the challenges presented by the shutdown.

The shutdown task force makes many important collaborative decisions. Among these decisions are the dates and times for local outages, the recommendation for the shutdown day and times, and the level of communications. The shutdown task force is also the mechanism for communicating information about special electrical system testing or modifications that may be required. It is important that all affected groups and services be represented in this process. It is also important that different services have the opportunity to hear the issues of other services. This stimulates broader consideration of the impact of an unexpected power outage, as well as of the shutdown itself.

Working Groups.   It is impossible to cover all of the communication and analysis required for a building power shutdown in one meeting per week. Therefore, it is important that the shutdown task force limit its deliberations to testing issues, reports of common interest, areas where cross-fertilization of ideas is important, important decisions affecting broad areas, and interdepartmental issues.

It is necessary to establish and empower working task forces to coordinate the multiple inputs to a shutdown. These working groups will perform much of the background work necessary for a successful shutdown. The following subcommittees or working groups, as a minimum, are necessary: building systems, patient care, vertical transportation, shutdown information communication, laboratories, physician and tenant offices, code-call team, radiology, operating rooms, information resources, and network services. Other departments, such as communications, dietary services, research, and cardiac services, also form their own specific interest areas to deal with the shutdown's impact on their missions. The chairs of each group, along with representatives from each individual department not otherwise covered by other working groups, will belong to a white paper subcommittee. The chiefs of service or department heads should be included on standard distribution lists for all meeting notes and notices, and those individuals should also be asked to provide contact persons who can attend the task force meetings.

Building Systems Working Group.   A critical working group is the building systems working group. This group will review the impact of the building systems on others and will assess the best way to respond to requests for backup power to make the shutdown as tolerable as possible for the remainder of the hospital. The building systems group should meet for several weeks before the rest of the shutdown task force starts meeting and should continue during the remainder of the planning period. The following are among the issues determined by this group:

- Identify and track the EP system's load-monitoring program in the shutdown buildings.
- Identify NP equipment and lighting to be shifted to one of the EP branches or to receive a backup NP source during the shutdown. Identify the need for new NP or EP backup feeders, and track their progress.
- Identify and evaluate the scope, schedule, duration, cost, action assignments, and overall impact of the required preventive maintenance before, during, and after the shutdown.
- Identify and evaluate the need, scope, schedule, cost, action assignments; track the progress of all necessary infrastructure upgrades before, during, and after the shutdown.
- Review the results of NP system planned outages and other events that affect shutdown planning or operations. Use the results of each planned outage to plan equipment, system, and/or set-point modifications required because of the lessons learned.
- Review and respond to requests for additional EP and emergency lighting that come from the departments and areas affected by the shutdown.

Choosing the Day and Times—Building a Consensus.   The hardest part of the shutdown in some buildings is building a consensus about the day of the week and the time for a shutdown. This process requires that all parties collaborate to determine the date and time that will have the least overall impact on the institution's patients, visitors, other customers, and staff.

Command Station.   The shutdown command station should be a location typically used for internal disaster control. The shutdown task force or safety committee may elect to use the commencement of the shutdown as a disaster drill, although this approach is not a necessary part of a shutdown. A radio unit and contact person should be stationed in every patient unit and major service department in the building.

Communicating Shutdown Information.   Communication about a shutdown is critical. People need to know that the shutdown is about to happen, what its impact on their activities will be, and what they should be doing to prepare for it. A special shutdown logo can build a consistent thread and raise staff awareness of all electrical system management activities. Over time, this approach helps occupants to understand what will be required of them. All documents (such as agendas, meeting notes, and other communications) should carry the same logo.

Target Audiences and Discrete Communications Vehicles for Each Audience.   The shutdown task force determines who else in the hospital needs to know about the shutdown. These are typical target audiences in a patient care building:

- *Patients and their families.*   A letter should be prepared by the patient care unit subcommittee and approved by the task force. Patient unit nursing representatives will normally hand out this notice.
- *Other members of the public, including visitors.*   The shutdown task force may decide that this target group should receive its own type of communication. As an alternative, the shutdown task force may elect to post notices in the elevator lobbies and stairwell doors to reach this target group and any building occupants who might not have heard about the shutdown. Posters in many areas should be multilingual.
- *Hospital Community and Administration.*   This group should receive a letter or e-mail from a high-ranking task force spokesperson, such as the director of facilities engineering.
- *Individuals responsible for patient care and supporting services.*   This group normally includes chiefs of service, nurse managers, operations coordinators, and department heads. This group should receive the white paper and feedback forms.

Multimedia Communications Approach. It is important to reach the hospital community through as many media as possible because an individual may selectively ignore certain types of media. Among the options commonly available to the shutdown task force are e-mail, posters, individual letters, official memoranda, network computer sign-on screens, the shutdown summary (white paper), and voice-mail systems.

Generator Loading—Planned Shutdowns Versus Unanticipated Utility Outages

Hospital generators power several types of loads. This discussion focuses on the electrical or usage characteristics of these loads rather than the branch classification. Among these loads are the base load that rarely changes, even for a shutdown; the occupancy-related load that has a predictable profile during a typical weekday or weekend day; task-related loads that will also be severely reduced during a shutdown; and extra or optional loads that may be connected to the EP system specifically for the shutdown.

A facility's engineers should identify available EP capacity for new loads before commencing shutdown task force planning sessions. This limits unrealistic expectations. However, when shutdown planning commences, several classes of loads may not have EP. Although this is an existing condition and therefore is in most cases grandfathered because code changes are not retroactive, it is not necessarily acceptable during a shutdown. This issue is more problematic with older buildings where consistent design philosophy was not applied as building electrical systems evolved over time.

As an example, code requirements do not dictate supplying EP to research laboratory freezers or incubators. However, long shutdowns of power to incubators in tissue-culture rooms can cause the loss of valuable tissue cultures that may represent many years of research. Backup feeders may be necessary in this event, even if there is inadequate generator capacity to provide EP. Freezers, on the other hand, may be able to get through a shutdown safely by using dry-ice slices.

Nonemergency loads can be powered from generators during a shutdown only if facility engineers verify that temporary loads will not overload the generators.

Elevators—A Special Concern

Elevators should have battery-backed lighting. The shutdown starts with a complete normal and EP outage while the generators start. Because of the possibility of patient or public anxiety from a complete blackout in the elevator cars, it is important to determine carefully exactly how the elevator cars will be controlled during the shutdown switching operations themselves. For the initial switching, the shutdown task force may elect to station a task force representative with a flashlight in each operating elevator car.

Alternatively, the shutdown task force may elect to recall all elevator cars just before the switching occurs so as to avoid short entrapments while the elevator control system goes through its normal EP start-up cycle. Even when the system works properly, personnel can feel trapped briefly while each car is brought to its designated floor.

SHUTDOWN PLANNING SCHEDULE

However critical the electrical shutdown is to the reliability of the electrical distribution system, it is nevertheless an intrusion into the normal work environment of the hospital's staff. It is important that the critical shutdown targets be communicated clearly to all necessary parties. The critical shutdown targets include special EP tests required for system verification, scheduled NP outages for training and verification, scheduled normal or EP outages for system upgrades in anticipation of the shutdown, target dates for submittal of draft and final action plans, and target dates for preparing the white paper.

Action Plans

The predetermined script (action plan) should be as carefully considered and detailed as possible. This is necessary because it is very difficult to communicate changes to all parties who need to know of them during or immediately before the event.

A predetermined written hold-point list would indicate every affected patient care unit and the name of the clinical contact and shutdown personnel radio contact. The clinical contact (usually a nurse manager) is responsible for verifying and relaying the responsible physician's concurrence that the shutdown can proceed. The list should be polled by radio (with telephone backup) by the operations manager on duty from the shutdown communications center.

White Paper and Feedback Forms

The action plans provide training for the individual units and services, but the white paper is the means to train the hospital as an entity. The white paper also is the vehicle for communicating to others who were not part of the detailed planning process. Because the white paper summarizes important information common to many groups, it also becomes a convenient reference for later disaster management. Surprises are anathema to a shutdown. The intent of the feedback form is to record all surprises for future improvement of the entire process. If the shutdown task force anticipated an occurrence that someone records as a surprise, then the communication itself fell short.

Lessons Learned

The final shutdown task force meeting, a week after the shutdown, is billed as a *lessons learned* meeting. This is an opportunity for people to report their perceptions and their recommendations for improving future shutdowns.

Costs of Building-Wide Electrical Shutdowns

The costs of shutdowns, which can vary widely, consist of these elements, listed in descending order of importance: (1) the scope of corrective work before the shutdown, (2) how long it has been since the last shutdown (with training as well as infrastructure field verification), (3) the scope of system modifications that are programmed during the shutdown; (4) the scope of preventive maintenance during the shutdown; and (5) the amount of time required to plan the shutdown.

Shutdown Test Report and Cost Analysis

The shutdown test report should be prepared within a month of the shutdown, and the first draft should be distributed within 2 weeks if possible. This should be done while issues are still fresh in participants' memories. The shutdown test report is the official record of the training, upgrades, findings, and lessons learned. It should also include a cost analysis for future budgeting purposes.

MANAGING HOSPITAL EMERGENCY POWER TESTING PROGRAMS

Hospitals must have emergency power testing programs.[5] These programs include requirements for generator load testing and maintenance of emergency power supply systems. Both topics have been well represented in an excellent JCAHO monograph.[6] Several other comprehensive analyses deal with designing the testing program[7] and the code requirements that apply to hospital emergency power system testing.[8] More recent publications address code changes and other pertinent issues.[9] This section, however, introduces a management program that uses the lessons learned from the emergency power testing program to improve the hospital's facilities and training. Reference 5 provides more information.[5]

It is important to analyze test results and trends, not just record test results. Several references address recording and analyzing the test results for engine operating parameters. In the following analysis, we address test results that describe the kinds of interactions among the various electrical distribution system components and their emergency power loads.

The emergency power testing program should comply with regulatory requirements but should not adversely affect the mission and patient focus of the hospital. It is important to be as comprehensive as possible and leave little to chance in dealing with these increasingly complex systems.

Description of Emergency Power Testing Program

The emergency power testing program should involve more than just monthly testing. It should also involve annual measurement of emergency load profiles, determination of total emergency power system loading under emergency conditions, monthly testing of the emergency power system using actual installed loads, monthly review and analysis of test results, and trend analysis of problems. The results of the trend analysis can help the hospital's engineers to identify training and/or systemic issues that require further investigation or resolution.

The emergency power testing program is not only the vehicle for maintaining the emergency power system in a reliable operating state, but it is also a vehicle for maintaining a high state of disaster readiness among the hospital staff. This includes the hospital's clinical staff, not just its facilities staff. If equipment is affected by the 10-s outage that normally precedes power transfer to the emergency generator in an actual power outage, then the hospital's clinical staff should know how to deal with these effects. Testing without simulating these effects may cause the clinical staff to be unaware of the real impact of a power outage on their clinical equipment and critical processes. When the monthly emergency power system testing program is combined with regular electrical system shutdowns, then the hospital's entire staff is better trained for disaster management.

Emergency Power System Test Procedures

These are the benefits of written test procedures:

- They provide control by the hospital's facility managers of the test process itself.
- They empower the testing personnel to act when required.
- They require that the testing personnel take responsibility for performing all required tasks.
- They reduce the chances of incorrect actions by testing personnel that may increase the risk to the hospital's patients, visitors, or staff.
- They provide written documentation of the actions taken during the test in the event that something goes wrong.
- They provide a mechanism for exploring potential trends.
- They provide the source documentation for later trend analyses.

The test procedures for the generator personnel will be specific to the needs of the generator documentation. Those procedures are beyond the scope of this paper but have been covered in the referenced articles and in NFPA 110.[10]

Avoiding Elevator Entrapments Due to Testing.   The interactions between hospital emergency power systems and automatic elevator-recall controls are very complex. Careful attention can help ensure that entrapments do not occur during emergency power tests. Not all elevator failures or entrapments result from testing, particularly in early-morning testing. Sometimes elevator door problems such as dirty tracks or dirty motion-sensor screens may be masked by the emergency power test and mistaken as test-related failures.

Analyzing Monthly Test Results

Monthly test results should be reviewed shortly after each test. One effective method is for the testing personnel to return the signed test procedure to the supervisor immediately after the test, along with a short verbal report of any important events or surprises that occurred during the test. All surprises, failures, and other unexplained occurrences should be entered into a testing database. Keywords should be used to aid in later trend analysis. The monthly test results should be recorded in an emergency power testing database that allows for later revisiting and analysis.

Analyzing Equipment Failures

All equipment failures should be analyzed to find out if they were caused by human error, problem system interactions, test procedure inadequacies, equipment malfunction, or another cause. Corrective action should be planned for the failed equipment, of course. Each failure should be considered for its generic relevance as well, allowing for the circumstances of the failure and its potential for occurrence elsewhere in the hospital or again under the same set of conditions.

Effects of Monthly Testing on Emergency Power Equipment

The normally-off emergency power equipment is loaded very sporadically. Most months will find 30 days (43,000 min) of standby conditions followed by 30 to 45 min at more than 50 percent rated load. This will often cause electrical terminations to work loose due to the expansion and contraction of the sporadic loading. Hospital staff should consider this fact and regularly use infrared scanners on the normally-off portions of the emergency power system during the tests.

The monthly testing will cause emergency power system failures to occur. The benefit of this situation is that the failures will be much more likely to occur during the test itself, when plant electricians are on duty and focused on the generators, the transfer switches, the systems, and the buildings being tested. The other important benefit is that normal power will be available during this test. Many hospitals that report emergency power equipment failures during tests also report that the failures would have occurred anyway.

DEVELOPING AN ELECTRIC UTILITY FAILURE RESPONSE PROGRAM

Reliable electrical power systems are absolutely critical to the successful operation of modern health-care facilities.[11] However, like all mechanical equipment, portions of electrical power systems can fail and cause electrical system outages. These failures can be internal or external to the health care facility. They can also cause partial or full outages. The most effective response depends upon the nature of the failure.

A recent article discusses one decision matrix for deciding the scope and criteria for a backup, or emergency, power system.[12] This subject has been discussed comprehensively there and elsewhere. This section focuses on needs that go beyond the usual emergency power system. Whereas emergency power systems are provided to meet code and regulatory requirements, much of the day-to-day business of the hospital requires that normal power systems also be available. Something as simple as adequate and timely vertical transportation, for example, is essential to hospital operation. Many elevator banks have six or eight cars, but most high-rise codes require that only one car at a time operate on emergency power. This limitation may be adequate for getting people out of a high-rise building, but it may be inadequate for the operational needs of the hospital in a disaster-response mode.

When an electrical outage occurs, there is a great deal of pressure on hospital facilities personnel to do something. Actual outage experience shows that important system design constraints or unusual configurations may be forgotten in the heat of the moment. Because of this, it is important to have well-thought-out and documented disaster plans in which personnel have confidence. The utility outage procedures meet this need.

Code Requirements

Code requirements are typically satisfied by a permanently installed emergency power system. The only exception to this rule should be where code requirements have changed and the infrastructure has not yet caught up with the code changes. Electrical infrastructure upgrades should then be programmed and funded.

OTHER NEEDS FOR EFFECTIVE FACILITY OPERATION DURING ELECTRICAL OUTAGES

Many emergency power systems were designed for code requirements in effect at the time of the design, along with predicted additional operational needs. The dependence on electrical power has increased with time and with the concomitant load growth. Because current failure analyses often identify operational needs that exceed previous code requirements or existing emergency power system capacities, the challenge is to develop a response program that meets the new needs within existing constraints.

Effective hospital operation may demand other backup power needs that go beyond code requirements. Examples are activities that require temporary backup power during planned building shutdowns, cafeterias, patient floor nourishment stations, elevators in addition to the code-required single elevator for increased vertical transportation capacity, lighting in addition to the code-required lighting in patient care areas, pneumatic tube systems, patient television systems, environmental rooms, freezers, incubators, ice machines, and selected rooms for radiological procedures that do not have emergency power.

Existing Emergency Power System Limitations

A review of utility-failure scenarios may suggest areas where additional emergency power is warranted. However, existing loading of the presently installed emergency generators may preclude adding indicated loads, despite their newly determined importance. In this event, the hospital may find it necessary to triage the total emergency power system when funding is obtained and emergency power upgrades are designed, licensed and installed.

Electrical System Designs That Inhibit Effective Responses to Utility Failures

Because the actual location of a failure cannot be predicted, the easiest designs to work with are those that give the hospital operations team the most flexibility. Main breakers, transformer primary and secondary tie breakers, and even backup tie feeders between centers of load concentration can provide the desired flexibility.

Designs that inhibit effective emergency responses are those that do not include main breakers, tie breakers, and the like. If the design does not allow plant electricians to isolate the internal switchboard from the external utility service, such as designs that use multiple service-entrance disconnects rather than main breakers or switches, the plant electricians' ability to isolate and backfeed switchboards under normal power outage conditions is hampered. Although this design may be allowed by the applicable electrical codes, such as in paragraph 230-71 of the National Electric Code, in which up to six service-entrance disconnects are allowable, it is not good design practice for health care facilities.[13]

Other designs that inhibit effective emergency responses are those that keep large switchboard ampere ratings in one lineup without bus sectioning. A switchboard fault or other event that disables one switchboard will also disable all loads fed from that switchboard section. If it is the only section, then the entire building will be without normal power. A more flexible design is one that splits the switchboard into more than one bus section and uses separate main and tie breakers for isolation. This approach allows hospital electricians to isolate the problem area and then return normal power to the rest of the building by closing tie breakers to unaffected sections.

Another design that does little for hospital operational flexibility is one that uses more than one main transformer (usually a good first step) but locates the secondary tie breaker on the line side of (ahead of) the main breakers. The loss of one utility line or one transformer can be accommodated by this design, but it does not allow the flexibility to isolate the main service entrance cables on the line side of the main breaker.

Utility Failure Procedures

It is important to assess the current situation before taking any action, unless such actions have been repeatedly practiced before the failure and personnel are comfortable that they are the most appropriate actions. Such assessments need not take much time, but they can help the plant electricians avoid operational errors. Reference 11 includes several sample procedures.

Backup Feeders.   The carefully considered installation of backup feeders between areas of heavy distribution can be very useful for disaster response. Any kind of backfeed can be dangerous if it is not done properly. For more than any other reason, this would be the reason to engineer, design, and install permanent backup feeders and then equip them with sufficient administrative controls for personnel protection. Properly installed backup feeders *must have* sophisticated engineering oversight to ensure that dangerous scenarios with neutral conductors do not occur during the backfeed condition. Temporary or interim backfeeds hastily installed in response to an outage and the driving need to do something can be much more dangerous than well-engineered backup feeders that are normally locked and tagged open. Even well-designed backup feeders can cause problems or injury if their use is not controlled by well-thought-out procedures.

Interbuilding Backup Tie Feeders in a Multibuilding Campus.   Building switchboard shutdowns for preventive maintenance, testing, and upgrades require bringing 480-V power for primary injection testing of circuit breakers into the vicinity of the switchboard room.

Because shutdowns must occur with some regularity to maintain the switchboards in reliable operating condition, the cost of rental generators or temporary cabling from other buildings can be avoided by installing backup feeders between buildings or switchboards. These same backup feeders then provide the means to meet the challenge of an emergency power equipment failure. A simple backup feeder, and one that provides flexibility, is a 480/277-V, 400-A, 3-phase, 4-wire feeder that has fused 400-A disconnects on each end. This feeder can move 333 kVA from one place to another within a multibuilding campus. These feeder termination points are then left disconnected on their outboard ends because the nature of the need cannot be predicted until it occurs. Several control wires should also be pulled into the conduits with the power cables, thus allowing the use of the backup feeder as a temporary emergency feeder with its own engine-start leads. A well-thought-out plan for using these backup feeders can also have the benefit of lower insurance costs, when the hospital's insurance company sees the easy business-loss limitations that the backup feeders represent.

Designs for Temporary Connections of Rented Generators.   Whenever new generators are installed in hospitals, a spare backup feeder should be installed from the generator location to a convenient location at the street. The feeder should be rated for full generator output. Then, bringing in a rental generator and connecting it to the outside end of the backup feeder can accommodate a generator failure. In addition, the backup feeder can also accommodate generator load-bank tests, if necessary, to meet regulatory requirements or to help in troubleshooting emergency generator malfunctions. The backup feeder eliminates the need for rigging the heavy load bank through the hospital for troubleshooting.

CONTRIBUTOR

Cornelius Regan, P.E., CLEP, Principal, C. Regan Associates, Rockland, Massachusetts

REFERENCES

1. D. Stymiest, "Managing Hospital Electrical System Change," *Proceedings of the ASHE 31st Annual Conference,* American Society for Healthcare Engineering, Chicago, 1994.

2. D. Stymiest, "Energy Benchmarking with Hospital Electrical Load Profiles," *Proceedings of the ASHE 34th Annual Conference,* American Society for Healthcare Engineering, Chicago, 1997.

3. D. Stymiest, "Determining the Actual Emergency Demand Load" (four-part series), *Healthcare Circuit News* 2 no. 4 (1996) through 3 no. 1 (1997).

4. D. Stymiest, *Managing Hospital Electrical System Shutdowns,* Healthcare Facilities Management Series, No. 055126, American Society for Healthcare Engineering, Chicago, 1996.

5. D. Stymiest, *Managing Hospital Emergency Power Testing Programs,* Healthcare Facilities Management Series, No. 055142, American Society for Healthcare Engineering, Chicago, 1997.

6. JCAHO, *Emergency Power: Testing and Maintenance,* PTSM Series No. 1, Joint Commission on Accreditation of Healthcare Organizations, Oakbrook Terrace, IL, 1994.

7. B. Thurston, *How to Test Out Your Emergency Generators,* Healthcare Facilities Management Series, American Society for Healthcare Engineering, Chicago, 1992.

8. H. O. Nash, Jr., *Hospital Generator Sizing, Testing, and Exercising,* Healthcare Facilities Maintenance Series, No. 055851, American Society for Healthcare Engineering, Chicago, 1994.

9. *Healthcare Circuit News,* 1 no. 5, and other articles; www.mgi-hcn.com, Motor & Generator Institute, Winter Park, FL, December 1995.

10. NFPA, *Standard for Emergency and Standby Power Systems,* NFPA 110, National Fire Protection Association, Quincy, MA, 1999.

11. D. Stymiest, "Developing an Electrical Utility Failure Response Program," *Proceedings of the ASHE 33rd Annual Conference,* American Society for Healthcare Engineering, Chicago, 1996.

12. M. Anastasio, "Backup Power: Finding the Right System," in *Building Operating Management,* Trade Press, Milwaukee, WI, 1996, p. 24.

13. NFPA, *National Electrical Code,* NFPA 70, National Fire Protection Association, Quincy, MA, 1999, para. 230–71.

CHAPTER 10
LABORATORIES

David L. Stymiest, P.E., SASHE, C.E.M., Senior Consultant, Chapter Editor
Smith Seckman Reid, Inc., New Orleans, Louisiana

Anand K. Seth, P.E., C.E.M., C.P.E., Director of Utilities and Engineering, Chapter Editor
Partners HealthCare System, Inc., Boston, Massachusetts

Laboratories are among the most intensive and demanding types of buildings and have complex engineering systems requirements that must be integrated with the building program and structure. Laboratories are found in most major types of facilities, including research, clinical, and industrial buildings. We are pleased to present comprehensive treatments of laboratory programming and laboratory facility layout, followed by a discussion of the engineering and design process for laboratories. This chapter then presents the special requirements of the major engineering fields as they pertain to laboratory design. Finally, the critically important procedures for decontaminating and decommissioning laboratory buildings prepared by a team of experts round out our treatment.

SECTION 10.1
LABORATORY PROGRAMMING

Janet Baum, A.I.A., Principal
HERA, Inc., St. Louis, Missouri

OVERVIEW

Facility programming is a process for identifying and assessing needs; a program of requirements (POR) documents the owners' and occupants' needs for a facility, whether the facility is a new laboratory or a laboratory renovation. The programmer investigates the goals and objectives of the organization that will own and operate the proposed laboratory facility. The programmer analyzes prospective occupants' stated space needs and performance criteria and reconciles them with the owner's goals and financial objectives. The size, quality, and estimated cost of the proposed project must meet the owner's budget. If the owner does not have an established project budget, the owner may also use the facility program to inform the budget formulation process. The owner can also use a facility program as a tool to raise funds for the proposed project or even to recruit new scientists to the organization.

A laboratory facility program also guides and informs the architects and engineers about the size, occupants, functions, performance, and spatial relationships that are required in designing the new facility. In-house science or facility staff, laboratory program and planning specialists, and architectural design consultants can develop laboratory facility programs. Normally, the program team interviews the prospective occupants, including scientists, technical and administrative staff, and students, or their designated representatives, and other managers and personnel who will be involved in the health, safety, operation, and maintenance of the facility. It is very helpful for good laboratory engineers to participate in the interview process and contribute to the program document. Refer to task #11 later.

If the owners can identify only a few or no future occupants, the programmers can analyze examples of built laboratory facilities that the owners recognize as similar in function and performance to their goals and objectives. As the owner recruits future occupants, the program should be updated with more detailed information. Laboratory building planners can use conceptual programs to complete basic, initial planning, but conceptual programs do not achieve sufficient detail needed for actually designing laboratories.

PROGRAM DEVELOPMENT

Detailed program development that is adequate for a complete facility design is a 12-step process. Programmers can do tasks #1 to #3 to produce a simple conceptual program, or tasks #1 to #8 for outline programs that do not go into much detail about laboratory spaces. Programs should include task #12, the initial construction cost estimate or budget confirmation. These are the 12 tasks that this chapter explains:

1. Analyze the existing facility.
2. Interview future occupants and administrative personnel.
3. Establish space standards.
4. Develop a list of room types, and estimate room areas.
5. Prioritize users' needs, especially if the proposed project is a renovation.
6. Diagram room types, and confirm room areas.
7. Determine the numbers of each room type.
8. Calculate the building's net and gross areas.
9. List room performance specifications.
10. Diagram zoning, adjacencies, and functional relationships.
11. Describe the mechanical, electrical, and plumbing systems.
12. Estimate the construction cost, or reconcile the program with the budget and project costs.

Analyze the Existing Facility

The purpose of analyzing an existing facility is for the programmer to gain an understanding of the occupants' current working conditions. The programmer collects the following information:

- Total building net assignable and gross areas. Area used by those occupants moving to the new facility.
- Net assignable area totals for each primary function in the laboratory building used by occupants moving to the new facility:

  Assigned laboratories, shared and common laboratory support, pilot plant

  Scientists' and lab workers' office area, administrative and managerial office area

  Conference areas, such as auditorium, conference and seminar rooms, library

  Building support functions, such as maintenance shops, loading dock

  Personnel support, such as first aid station, locker rooms, cafeteria/lounge

- Totals of net assignable area for each principal investigator and organizational unit, such as a department. Also identify the total area of shared facilities that are not assigned, but are shared by occupants moving to the new facility.
- Building occupant population totals, using the most current and accurate data available. List by name, position, primary location in the building, and full-time or part-time status. Calculate full-time equivalent (FTE) population for the following categories:

  Scientific program managers/directors

  Principal investigators

  Scientific staff (including paid graduate and postdoctoral students)

  Technical staff (including animal caretakers, glass wash, shop technicians)

  Administrative and clerical staff

  Facilities maintenance staff

  Other support staff

- Building layout with a map of the primary functions listed before.
- Building layout with a map of the territory, or space, assigned to each principal investigator and organizational unit.
- Takeoff average linear measure of several typical laboratory modules with average population density in these typical conditions:

  Linear feet (meters) of fixed benches and chemical fume hoods

  Linear feet (meters) of scientific equipment, lab sinks, and biosafety cabinets

Using the previously mentioned data gathered for an existing facility, programmers can derive several critical ratios to characterize the occupancy patterns. Total net assignable area (NASF) per FTE lab occupant indicates the laboratory population density. Programmers can benchmark this ratio with other laboratory buildings in the organization and nationally to identify laboratories that are overcrowded and those that are inefficiently used. Total gross area per FTE lab occupant is less helpful because of wide variations in the net to gross area ratio from building to building.

Ratios of net area per FTE in separate functional categories (second bullet point in preceding list) give further insight into the deficiencies, efficiencies, or excesses in existing building use. Programmers assess scientists' complaints quantitatively by using these statistics. Programmers identify functional ratios that need to be adjusted in the new facility to improve scientific productivity and building efficiency. For example, research dedicated to molecular biology research works well with ratios of 1:0.75 to 1:1 in general lab area to specialized lab equipment and support area, whereas synthetic chemistry research ratios of 1:0.1 to 1:0.2 perform well.

Interview Future Occupants and Administrative Personnel

Programmers conduct in-depth interviews with future occupants who are available and identified: lead scientists, principal investigators, faculty, technicians and research associates, students and postdoctoral students, research executives, and administrative staff. In addition, programmers should interview operations and maintenance managers, environmental and laboratory health and safety directors, chemical hygiene officers, the security director, and, if possible, the organization's financial manager and scientific director. Interviews with future occupants focus on the following topics:

- Interrelationships of work flow and processes
- Flexibility parameters
- Level(s) of hazards anticipated, overall quantities, and distribution of hazardous materials
- Level of safety and security required
- Current functional and operational deficits
- Current area deficits and excesses for specific functions
- Room types that individuals will require for their use in the proposed facility
- Room types that individuals will use and can share in the proposed facility

Programmers document performance and environmental considerations, as well as technical requirements. Data sheets provide a standard format for recording the detailed requirements for each lab. Goals, productivity, and quality-of-life issues are important to discuss with research leaders and staff. Intangibles are very difficult to document, other than in completely detailed meeting notes. Design team members need to understand important aspects of the organization's culture before entering the design process. The interview process is to reveal and document consensus and conflict in goals for the proposed facility that must be resolved in programming. Discovery of critical long-term problems and disputes among occupants or between proposed occupants and administrators critically informs recommendations that programmers offer.

Establish Space Standards

Programmers review occupancy statistics and ratios developed in task #1 in relation to observations and complaints documented in the interviews. Then, programmers evaluate recommended adjustments to linear feet or area per FTE laboratory occupant in the following factors:

- Wet and dry bench
- Freestanding equipment
- Lab waste collection
- Chemical fume hood and/or biosafety cabinet
- Safety station

Percentages of the total net assignable area occupied by general assigned laboratories, laboratory support functions, offices, and conferencing, and of building support functions are also important to review and adjust, if existing operations are deficient or inefficient. Programmers recommend new standards for the following:

- Population density and optimal net usable area per researcher (including internal circulation area)

- Percentage of each function, including general laboratories, lab support, building support, personnel support, and administrative functions
- Percentages of specialty facilities such as pilot plant, core instrumentation, and animal facilities

Generate the List of Room Types

Programmers develop the list of rooms for the proposed facility from interview meeting notes and area standards. Room type names should express the explicit functions of rooms in simple terms. Programmers can use numbering systems to identify each generic room type. These room identification numbers are used throughout the design process to refer to specific program spaces. Room type categories include the following:

- Research module types (assigned to principal investigator or department)—organic chemistry lab, molecular biology lab, high energy physics lab, and so on
- Teaching laboratory types—introductory biology lab, advanced mechanics lab, analytical chemistry lab, and so on
- Specialized laboratory types (assigned and/or shared)—molecular beam epitaxy, electrophysiology lab, radiophysics lab, and so on
- Laboratory support—glass wash, darkroom, electrophoresis, chemical storage, equipment storage, and so on
- Animal facility—cage room, quarantine room, cage wash room, feed storage, and so on
- Administrative areas and support—principal investigator or faculty office, conference room, mail and copy room, and so on
- Personnel support facilities—locker room, cafeteria, break room, first aid station
- Building support facilities—loading and shipping docks, data closet, housekeeping supply storage, and so on

Estimate the area for the typical laboratory module (see Sec. 10.2, "Laboratory Facility Layout," for the definition and use of modules). From the space standards developed in task #3, the areas for typical modules can be derived. The area of one laboratory module is the sum of the NASF per FTE laboratory occupant for general laboratory functions (bench, circulation in the lab, equipment, fume hood, and safety equipment) multiplied by the number of FTEs who will occupy a single module. Laboratory rooms are multiples of the module.

$$\text{Module area} = (\text{bench} + \text{equipment} + \text{fume hood} + \text{safety equipment area}) \times \text{FTE count/module}$$

Programmers develop possible layouts of typical research modules to reconfirm the module area derived from the space standard. Programmers should diagram laboratory egress and entry in conjunction with hazard zoning strategies within the module. Hazard zoning strategy development starts with investigating the best possible location(s) for the most hazardous processes to be conducted in the laboratory and the safety equipment for these processes, such as chemical fume hood(s). Strategies for other safety equipment, electric power, and piped utility distribution can also be explored. Programmers diagram these strategies. Owner representatives and users should review and discuss these diagrams because a laboratory module is the planning unit that laboratory designers use to organize the facility.

After the owner and users confirm the proposed module area, programmers estimate the area of each specific space on the room list. For room types on typical laboratory floors, areas are multiples (double, triple, etc.) or simple fractions of the typical module area (half, third, quarter). This method of estimating area supports modular planning and enhances long-term flexibility.

If the programmer tries to estimate the area of each room by summing the estimated net areas for each and every component of furniture, fixture, and equipment, the areas turn out to be erratic and nonmodular. Program area inconsistency makes laboratory planning far more difficult, and the facility is then less flexible. Internal circulation (within the room) is often left out or grossly underestimated. If this happens, there will not be enough area to accommodate all of the room requirements when the laboratory designer proceeds to lay out each room. The functionality of the room will already have been sacrificed in the program phase.

Determine Users' Priorities, Particularly for a Renovation

Programmers investigate the functional priorities for the new facility before finalizing the list of room types. It is easier for future occupants to objectively discuss functions that are most important to them and those that are least important, although desirable, before rooms are finally assigned to a specific department, research group, or individual. This discussion should take place with all proposed occupants or their representatives before budgets are set. After the building committee or occupants' representatives review the conceptual budget in task #12, their priorities may well change. An in-process benchmark of their priorities helps the programmer gauge the political and budgetary pressures versus the functionality and performance expected of the new facility.

Diagram Room Types and Confirm Area

Programmers should diagram all typical room types and special functions, if there is any doubt about the adequacy or appropriateness of the areas. Room diagrams should show furnishings, equipment, casework, and circulation required. These scale diagrams reveal most major difficulties in the area estimates. Program areas are adjusted up or down, as required.

Determine the Numbers of Each Room Type

Interview meeting notes again provide information to the programmer for estimating the numbers of each room type. If particular room types must be provided on each laboratory floor or wing (darkrooms, equipment rooms, glass wash rooms, etc.), the program must indicate this. Programmers investigate the range of laboratory floors that may be required to accommodate the program area so as to estimate the total number of redundant room types. During the concept design phase, designers may further adjust the program area to include multiples.

Calculate the Building's Net and Gross Areas

The first part of this task is simply computing the sum of the net areas of all of the room types.

$$\text{Net area} = \text{room area} \times \text{number of rooms}$$

Estimating the gross area is more difficult. Typical net to gross efficiency falls between 50 and 67 percent for entire buildings that are primarily for laboratory use, especially if chemicals are stored and used. In predesign phases of the planning process, there is often a conflict between the need for a conservative or realistic estimate and the optimistic desire to make the numbers look good. Programmers need to evaluate these factors carefully and not depend on casual relabeling of functional areas to resolve these issues. Owners often push for the highest efficiency numbers but do not understand either the definitions or the effects on the project, such as loss of design time and reduction of scope and value, that this may cause.

Efficiency may drop, for example, if there is an animal facility in the building, if there is a central utility plant, or if there is significant area in classified clean rooms or biosafety level 3 or 4 laboratories. Efficiency may rise above the normal range under several conditions. If the building houses less area than normal for mechanical and electrical equipment and building utility requirements are supplied from another building or mounted on the roof, efficiency rises. If the circulation area between laboratories is defined as net assignable, is used for assigned or shared program functions, and meets egress and fire protection requirements specified in building codes and industry standards, efficiency can rise. Just labeling an area as net and assignable is not enough; the design must support code and functional requirements.

List Room Performance Specifications

This task is complex and time consuming. Programmers do this for detailed programs. Designers may also complete this task at the beginning of the design development phase, instead of during the normal programming phase. Room performance specifications cover, in detail, the following categories of requirements that the laboratory design and engineer need to know to design each space:

- Room identification, assignment, program number
- Hazard and/or clean room category and specific room type
- Functional adjacencies desired, immediate, and on the same floor
- List of major freestanding and bench equipment (or processes) with utility requirements
- Chemical inventory, organized by class
- Architectural finishes for floors, walls, and ceilings
- Door width, height, and special characteristics or hardware
- Laboratory casework, types, and linear feet
- Light fixture types, switching, and light levels required
- Laboratory safety equipment
- Fire protection and alarm system characteristics
- Laboratory ventilation devices, type, size, and quantity
- Heating, ventilating, and air-conditioning design standards and tolerances
- Piped utilities types, quality levels, quantities, pressures, and flow rates
- Gas cylinders, type of gases, and quantities

Model building codes in the United States limit the quantities of chemicals in buildings in the following hazard categories: flammable, combustible, oxidizer, unstable reactive, water reactive, organic peroxide, radioactive, corrosive, highly toxic, toxic, irritant, asphyxiate, cryogenic/flammable, and explosive. Programmers should obtain quantities of all of these materials during the programming phase. Too many laboratory and building code design decisions are based on chemical use and storage in science buildings to delay gathering this information in the late stages of the design process.

Scientific equipment occupies a lot of area in laboratory buildings. Scientific equipment drives electric power and piped utility loads, as well as cooling loads in laboratory ventilation systems. Ideally, programmers gather detailed information on all major equipment during the programming phase, including either catalog cut sheets or installation manuals. Design engineers need this very specific information to proceed with load calculations and to size mechanical and electrical equipment. The design process can be delayed if owners and users do not provide this information early.

Diagram Zoning, Adjacencies, and Functional Relationships

Programmers document adjacencies and functional relationships by using a number of methods such as simple line and bubble diagrams or by more formal "blocking and stacking" diagrams that look similar to floor plans. The purpose is to use information gathered in the interviews (task #3) to develop the building organization strategies, functional unit by functional unit. Each functional unit also should be diagrammed to show desirable adjacencies and the preferred work flows within the unit.

Describe Mechanical, Electrical, and Plumbing Systems

Engineers need to participate actively in the programming process, as stated before. Their role in this task is to gather and analyze information on mechanical, electrical, emergency, and data communications systems in the interviews, in the room data sheets, and in scientific equipment lists provided by future occupants. This information comes from prospective occupants and from the facilities engineers, planners, and health and safety professionals on the owner's project team. Program team engineers describe the scope of utility services and equipment and outline their performance characteristics. Cost estimators use this information to develop conceptual-level construction budgets.

Estimate Construction and Project Costs

One purpose of the program process is to align the proposed scope of the project with the owner's construction and project budgets. The construction budget can be estimated in several ways: by a professional construction cost consultant, by a construction management firm, by an architect and engineers, or by the owner's representatives who have extensive and current laboratory construction experience. To achieve a higher level of confidence in the conceptual estimate of the program phase, the owner is wise to seek estimates from two of the four sources listed. Each source will have a different perspective and varying levels of understanding of the project that lend valuable input to the owner at this critical stage, especially if the owner has not fixed the budget. These estimates then can be reconciled to arrive at a well-considered figure, or if the budget is already fixed, the project scope can be adjusted to meet that budget. In the subsequent design phases, the design team will use the budget as a primary goal.

Construction cost is simply what a general contractor or construction manager would charge the owner to erect the building or complete the renovation. Project cost is the sum of all of the owner's costs related to the project, including construction. This is a partial list of project cost categories that must be considered:

- Nonbuilding construction: demolition, land acquisition, site utilities, special foundations, site work and landscaping, permits, owner supervision and institutional surcharges, and others
- Mock-up construction
- Fees for services: architectural and engineering design, including basic services; additional services and reimbursable expenses; specialty consultants; economic feasibility; construction manager; construction supervision; legal
- Site and materials testing
- Zoning amendments and hearings

- Environmental impact assessment or study
- Surveys: land, soils, geotechnical, traffic, other
- Furnishings, fixtures, and equipment
- Telephone and data installation
- Finance costs: interim financing and bonds, including bid, performance, and payment bonds
- Cost escalation
- Insurance costs: public liability, vehicle liability, property damage, fire and extended coverage, vandalism, workers' compensation, employees' liability, and other
- Moving costs: packing and moving, commissioning, equipment calibration and installation, temporary facilities to accommodate phasing

One of the most frequently overlooked cost items is funding for operating and maintenance (O&M) expenses over the life span of the facility. Without recognizing the magnitude of and budgeting for ongoing O&M costs, owners may find that the laboratory performance deteriorates rapidly. Pro forma statements for laboratory project financing should include an adequate amount to fund these expenses. Facilities or operations managers can estimate reasonable budgets by using historical data from other laboratory buildings on the campus, whose size, program scope, and performance level are comparable. If historical data is not available, managers can survey managers of other similar laboratories in the region to gain some perspective of critical issues and costs. Key cost factors are energy and piped utilities such as water, sewage, and gas. During design, engineers can work with the facility and operations managers to provide the level of conservation that the proposed budget requires by careful equipment selection and systems strategies. Architects can also contribute to durability and ease of maintenance of laboratory buildings by specifying suitable materials and building systems and planning adequate areas for mechanical and electrical maintenance. The area for mechanical/electrical plumbing (MEP) equipment and maintenance is directly affected by the estimate for net to gross ratio that the program provides (task #9). If the proposed net to gross ratio is ambitious, for example, higher than 60 percent, maintenance expenses are likely to be higher. The programming process should open this complex issue to discussion and review and continue to evaluate it through design and construction.

CONCLUSION

This 12-step process can lead to program documents that accurately define the owner's goals and expectations for new or renovated facilities. However, other tasks can be added to the basic 12. For instance, proposing construction phasing for a complex facility or renovation that cannot be constructed or funded at one time is a valuable task that programmers can provide. Code and zoning ordinance investigations may also be needed during the programming phase for proposed functions that are hazardous or particularly difficult

Programming lays the foundation for the planning and design phases. If design teams are impatient or unduly eager to start design concepts before substantially completing the programming process, owners must be wary that the design will co-opt an objective determination of needs. Owners use program documents to check the design to ensure that rooms listed are in the plans and that functional areas are in the same proportion. For example, when the laboratory area shrinks and the office area expands to accommodate the architects' design ideas, the owner must carefully reconsider whether these modifications of the program area meet the occupants' needs or whether they are unacceptable and do not comply with the program.

Programming is the most direct and effective path for future users to communicate with the design team; it is the most inclusive phase of the laboratory design process. It is the basis of building consensus among owners and prospective user groups. Organizational and, sometimes, personal disagreements and conflicts surface during this process. Programming begins their resolution. Optimal solutions are usually not found by deferring programmatic and scope difficulties until design starts.

CONTRIBUTOR

William L. Porter, Ph.D., F.A.I.A., Massachusetts Institute of Technology, Cambridge, Massachusetts

SECTION 10.2
LABORATORY FACILITY LAYOUT

Janet Baum, A.I.A., Principal
HERA, Inc., St. Louis, Missouri

OVERVIEW

Many innovative designs in laboratory construction have been produced during the past 20 years. One helpful way for the owner to evaluate different planning options and layouts is for a group that represents the owner and future occupants to tour several laboratory buildings that have some of the desired characteristics, similar scientific disciplines, and approximate size of the proposed laboratory building. Visiting groups should request that the tour be conducted by building facilities and management staff who were involved in the design or who currently operate the building. These staff members can provide more complete answers to visitors' questions about the strengths and weaknesses of the scope, design concept, performance, and design/construction process.

Owners and building committee members can review actual laboratory building plans of many projects that are published in architectural and laboratory trade journals. Other "ideal" layouts are illustrated in facility books, such as this one and American Institute of Architects' *Guidelines for Design and Construction of Hospital and Healthcare Facilities.*[1] Owners can find projects in these books and articles that they can arrange to tour. The National Institutes of Health, National Academy of Science, Veterans Administration, Project Kaleidoscope (funded by the National Science Foundation), and the American Chemical Society have published excellent research and teaching laboratory planning guidelines.

Some critical studies must be completed before planning renovations to existing buildings can start. These include feasibility studies and a cost-benefit analysis for continued use of existing buildings for laboratory use or converting nonlab buildings to laboratory use. Planning should not proceed without basic information on building code and zoning compliance, structural integrity and capacity, the life spans and capacities of utility systems, and a thorough environmental audit. Without understanding the costs for bringing the building up to current codes and environmental regulations, owners cannot even guess if it will be financially feasible to renovate proposed buildings for any function.

Planning a new facility layout differs from building design, though it is closely related. Laboratory planners and architects work together to develop concept options for owners' reviews early in the design process. The difference between planners and architects lies in their approaches to organizing buildings. Planners do this from the inside out; building designers typically from the outside in. Owners benefit from well-coordinated team efforts on fundamental decisions that form the basis of the design and the function of the proposed facilities. Facility layouts have several categories of elements that must be organized for good building performance and functionality and are normally identified in a facility program:

- General (generic) laboratories
- Teaching laboratories
- Special laboratory types such as magnetic resonance imaging (MRI) suite, pilot plant, biosafety level 3 suite, Class 100 clean room
- Laboratory support such as darkroom, glass wash, controlled environment room, electronics shop
- Animal facilities such as animal holding, cage wash, food and bedding storage
- Offices such as administration, clerical, reception, principal investigator, technician
- Meeting and conference such as seminar room, auditorium, conference room, breakroom
- Personnel support such as locker/shower, first aid or health center, recreation, day-care center
- Building support such as receiving/loading dock, maintenance shop, housekeeping storage

FACILITY RELATIONSHIPS

Planners explore two sets of facility relationships: horizontal and vertical. *Horizontal relationships* determine room adjacencies on each floor of proposed facilities. *Vertical relationships* determine functional proximity from floor to floor. Planners use the following steps:

1. Develop priorities for the functions that should be located on each floor
2. Design a typical general (generic) laboratory module (if this is not in the program)
3. Develop options for the sizes of floors (floor plate)
4. Develop floor circulation and exit strategies and connections to adjacent buildings
5. Plan the distribution of utility and ventilation systems
6. Test the options for vertical circulation
7. Arrange the program room types on each level
8. Explore the options and constraints of structural systems

The following paragraphs describe each step and the alternative or special considerations needed in planning renovations.

Step 1. Develop Priorities for the Functions That Should Be Located on Each Floor

In this first step, the planner considers the functions that must be on the first or ground level, some that should be located in the basement, and others that are better on the top floor. The program document may provide this information. Investigate the program room list and sort functions and room types by desirable floor level. Demand is often greatest on the ground level because direct access to the outside is required to a driveway or terrace for egress or materials transport into the building.

Laboratory buildings may contain significant quantities of chemicals and other hazardous materials. If the detailed program did not include a comprehensive chemical inventory by hazard class, existing location, and corresponding proposed location, the owner should provide an up-to-date inventory for the planner at this step (see Table 10.2-1). National model building codes set limits on the quantities of a wide range of hazardous chemical categories. Codes limit total quantities in buildings as well as on each floor level. Allowable quantities decrease dramatically on each level above grade level. Above the third floor, allowable quantities are minimal.

Building code regulations are based on the experience of firefighters and emergency responders in effectively dealing with emergencies above or below ground level that may involve or be caused by hazardous chemicals. The first objective of emergency personnel is to save lives and second to reduce property damage. The objective of building codes is to prevent loss of life and property.

Many institutions and corporations that own and operate laboratory buildings have traditionally located the most intensive chemical use and chemical fume hoods on upper floors to reduce the length of risers to exhaust fans on the top of the building. This strategy certainly offers some economy. However, because building codes severely restrict the amounts of flammable and combustible liquids on upper floors, this tradition may wither away. Chemical usage and storage should be one of the major considerations for determining the preferred floor level for program functions.

Step 2. Design the Typical General (Generic) Laboratory Module

The module should be planned in detail, if this was not already done in the program phase. Using an understanding of the floor plate size and proportions, estimate the module length and width (see Fig. 10.2-1.)

Investigate the egress pathway and options for the internal circulation of the lab module. Refer to the program document for recommended space standards, including the linear measure of bench, freestanding equipment, and chemical fume hood per person in the module. Investigate the best layout to provide hazard zoning in the module and clear and safe egress from the laboratory. As stated before, there are many different ways to design the module to achieve the performance level required in the program. The planner should offer several for the owner's and building committee's review and comment.

In renovation projects, there may be fewer options for module size because of the constraints of the existing building's dimensions and structural grid. In this case, the planner should bring a similar thoughtful process to the module design to optimize the existing building's functionality. This may even mean moving the existing corridor location to develop a laboratory that is long enough for the proposed functions. For renovations, utility and ventilation systems distribution strategy becomes of utmost importance because older buildings have lower floor-to-floor heights than new laboratory buildings. This often forces more vertical distribution of utilities and directly impacts the module design. The planner should work on task #5 concurrent with the design of the module.

Step 3. Develop Options for the Sizes of Floors

The floor size and configuration are based on actual site constraints and opportunities. The planner and designer need to investigate the zoning regulations for easements and setback requirements. Building codes also have many conditions that must be met, particularly when the site has existing buildings and other impediments nearby that affect fire protection and emergency access to all parts of the site. The topography of the site and subsurface conditions may also influence the floor plate size and configuration.

For laboratory buildings in particular, fire protection systems and the fire-rated construction category selected for the new building also influence the floor plate size. For Class A sprinkler systems, allowable building areas for both new construction and renovation are much more liberal. In a new building, noncombustible and fire-rated construction assemblies can be designed to increase area allowances. In renovations, the area limits may be more

TABLE 10.2-1 Chemical Inventory Form

EXISTING RESEARCH LAB CHEMICAL INVENTORY *Provide form for each room type*

Client Name _____ Date _____

Room Numbers _____

Project Name _____ Biosafety Level

Lab Assignment _____ BSL-1 _____

College _____ BSL-2 _____

Lab Name or Type _____ BSL-3 _____

Department _____ BSL-LS _____

Funding Source(s) for Lab _____ Clean Room Class _____

1. GAS AND CHEMICALS

Please complete the following tables. Where volume amounts are requested, estimates are acceptable.

GAS CYLINDERS

Arsine	Qty ___	Cyanide	Qty ___	Methane or Propane	Qty ___	P-10 or P-5	Qty ___
Carbon dioxide	Qty ___	Helium	Qty ___	Nitrogen	Qty ___	Phosphine or Silene	Qty ___
Carbon monoxide	Qty ___	Hydrogen	Qty ___	Oxygen	Qty ___		Qty ___

CHEMICALS

	Estimated Volume			Estimated Volume		Check if you have any	
Flammable Liquids							Check if you have any
		Unstable Reactive				Acrylamides	___	Monomers ___
Class IA	Qty ___	4		Alcohols/glycols	___	Nickel compounds ___
Class IB	Qty ___	3		Aldehydes	___	Ninhydrins ___
Class IC	Qty ___	2		Amine/alkanolamine	___	Nitriles ___
Combustible Liquids		*Water Reactive*		Benzene/benzotriazoles	___	Perchloric acid ___
Class II	Qty ___	3	Qty ___	Carcinogens	___	Peroxides ___
Class IIIA	Qty ___	2	Qty ___	Cationic surfacants	___	Phenol ___
Class IIIB	Qty ___	*Organic Peroxide*		Cyanohydrins	___	Phosphorus ___
Oxidizer		1	Qty ___	Ethylene glycol ethers	___	Petroleum oils ___
4	Qty ___	II	Qty ___	Epoxides	___	Silver compounds ___
3	Qty ___	III	Qty ___	Formaldehyde/formalin	___	Stilbene ___
2	Qty ___	*Radioactive*		Fluoride compounds	___	Tolulene, Xylene ___
1	Qty ___	*Ionizing Radiation*	Qty ___	Halogenated comp	___	Xylene ___
Corrosive	Qty ___	*Highly Toxic*	Qty ___	Other heavy metals	___	
Irritant	Qty ___	*Toxic*	Qty ___	Mercury	___	

THANK YOU FOR THE TIME YOU AND YOUR STAFF HAVE SPENT TO PROVIDE THIS INFORMATION

Width in feet

Minimum	Maximum
10′ 4″	11′ 6″

Length in feet

Minimum	Maximum
20′ 8″	35′ 0″

FIGURE 10.2-1 Recommended range of module dimensions, centerline to centerline.

restrictive, unless the owner decides to make considerable improvements in or to replace combustible or non-fire-rated construction.

Based on the site for a new facility, estimate the size for the ground level, basement floors, and typical floors above. The planner will do this from a functional standpoint; the building designer will do it from a formal aesthetic standpoint. The planner's objective is to provide a floor plate size that allows a critical mass of research or science activity on each floor, yet is not so large as to discourage any sense of neighborhood or community among the occupants of each floor. Communication and materials handling both benefit from horizontal, not vertical, connections. Fewer floor levels and floor plates between 20,000 to 30,000 gross square feet (1800 to 2800 square meters) are desirable planning goals for research laboratory buildings. Small floor plate sizes are less efficient; a higher proportion of floor area must be used for elevators, egress stairs, and vertical chases for utilities. Very large floor plates can be zoned into ideal size sectors that give the desired proportions of people, space, and interconnects.

In a renovation, the existing building perimeter determines the size of each floor. Determine if any limited addition is required to meet program requirements.

Step 4. Select or Develop a Floor Circulation Strategy

There are seven common circulation strategies (see Fig. 10.2-2):

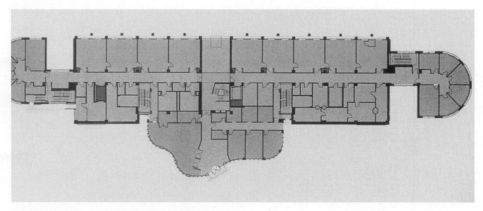

FIGURE 10.2-2 Lab floor circulation patterns.

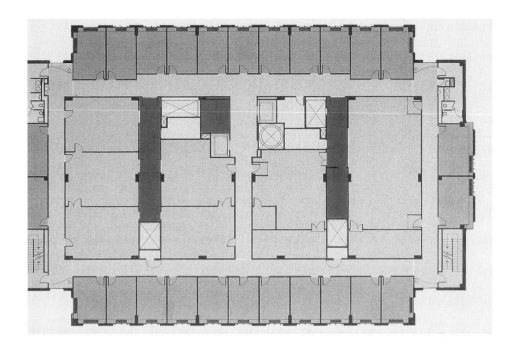

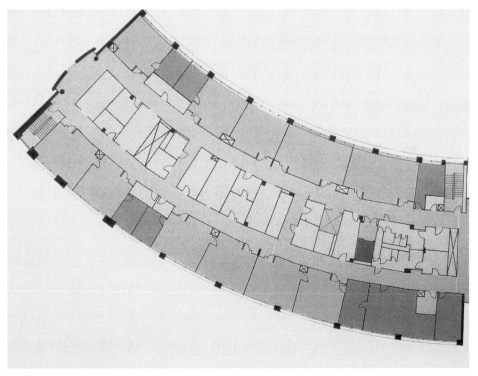

FIGURE 10.2-2 (*Continued*)

1. Single corridor that has functions on both sides
2. Racetrack corridor that has functions on the perimeter and in the center
3. Perimeter (or exterior) corridor that has all functions in the center
4. Service corridor combined with racetrack or perimeter corridor
5. Grid corridor system, corridors intersect
6. Central corridor that has only vertical circulation and building core elements in the center and functions arrayed around the perimeter
7. No corridors, just a hierarchy of laboratory aisles with direct exterior egress

A single corridor, double-loaded, provides simple and relatively efficient horizontal circulation. Laboratories and lab support functions can be arrayed on both sides, or offices can be on one side and laboratories on the other. A racetrack corridor functions well when the center area has shared support functions, accessed from both outer sides, arrayed with laboratories. A perimeter corridor surrounds all of the laboratory functions. This is effective especially when direct light is not desired in the laboratories—for example in physics. This strategy also allows the maximum contiguous area for reconfiguration. In mild climates, the perimeter corridor can simply be an out-of-doors passage. This technically improves the net-to-gross ratio, because outdoor area does not count in building gross area.

A very popular and effective variation on the perimeter and racetrack corridor strategies is adding a service corridor between laboratories. Laboratories back up to the service corridor that can actually function in a number of ways. A service corridor can be used strictly for maintenance and operations staff to access utility and control distribution systems. The service corridor function can expand to laboratory materials deliveries, pick-up, and some selected nonhazardous, noncombustible storage, as well as O&M staff access. This service corridor can be used safely and effectively as a second egress from laboratories.

Some service corridors have even been converted to actual laboratory support use by being designed with adequate width for large pieces of scientific equipment, such as refrigerators, freezers, and centrifuges, as well as storage. Potentially hazardous materials storage and processes in the service corridor severely compromise its performance as a second fire egress from laboratories. Fire marshals do not allow equipment and any combustible material storage in egress corridors, so other means of secondary egress must be provided. Laboratory occupants like service corridors that accommodate some lab support functions because they function as secure "back streets" that promote good informal interaction from one side of the building to the other.

Complex multiple service corridor layouts can be found in the microelectronics industry, where each corridor forms a barrier between varied clean room classifications. This circulation system is sometimes called the *Texas tunnel* layout.

A grid corridor system in which single corridors intersect provides circulation in very dense, large laboratory and mixed-use facilities. Single corridors are normally double-loaded and have functions arrayed on both sides. Hospital planners frequently use this circulation strategy and research, and clinical laboratories on hospital campuses sometimes adopt it. Finding your way in grid buildings can be a challenge. Unless many exterior courtyards or interior atriums are introduced in the grid, the majority of laboratories and offices will be without light or views.

A central corridor that has only vertical circulation and building core elements in the center and functions arrayed around the perimeter is a good strategy for small floor plates in tall buildings. Because of this minimal vertical circulation and virtually no "public" circulation, one to three research groups would share an "open" laboratory floor that has no locking doors between them because unimpeded egress pathways must be maintained from every corner of the floor. The original Salk Institute in La Jolla, California uses a variation on this strategy with a long, large, interstitial building to maximize efficiency and communication (Fig. 10.2-3).

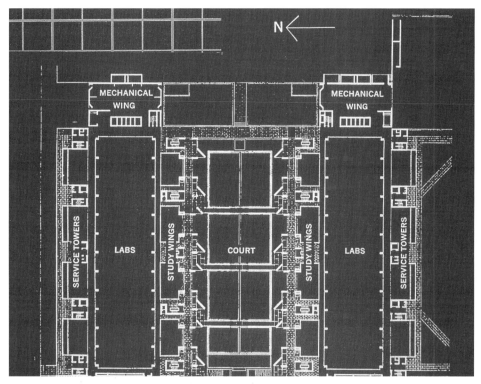

FIGURE 10.2-3   Salk Institute layout.

This strategy eliminates fire-rated egress corridors entirely and simply plans a hierarchy of wide laboratory aisles or "ghost corridors" that egress directly outside. Production and industrial laboratories in single-story, on-grade structures are common examples of this layout. Adequate egress, by code, usually demands multiple exits. Multiple exits can be of concern in high-security facilities. Informal corridors, most without any walls, are a great advantage for flexibility. The disadvantages, however, are considerable. Chemical spills or other hazardous or obnoxious air contaminants cannot be contained easily—or at all. There are no fire barriers to contain smoke and limit damage from water and fire. This laboratory layout calls for state-of-the-art fire detection and suppression systems.

Step 5. Plan the Distribution of Utility and Ventilation Systems

Design engineers explore ways to distribute the piped utility systems and ventilation systems through the building. Utilities must be supplied to each module or must pass from module to module to service the fixtures and equipment in laboratories. If the project is renovation of an existing laboratory building, utility distribution may already be in place and too costly to move. However, consideration should still be given to options for redistributing ventilation systems because these often undergo the greatest modifications in laboratory renovations.

As in personnel circulation strategies, there are several distinct strategies for distributing utilities and ventilation. They vary from predominantly horizontal to predominantly vertical, and there are many variations in between. In general, horizontal distribution options take more clear height and volume beneath the floor structure above than vertical options (see Fig. 10.2-4). On the other hand, vertical distribution often takes up more floor area and adds more fixed obstructions in the floor layout than horizontal options (Fig. 10.2-5). Many horizontal and vertical strategies parallel personnel and service corridors. So the circulation pattern can strongly influence utility and ventilation distribution. The exception to this rule of thumb is interstitial floor distribution. Utilities and ventilation air systems drop down or up into each occupied floor (Fig. 10.2-6).

Operations and maintenance managers and laboratory health and safety directors should participate in evaluating the distribution options for utility and ventilation systems. Safety and ease of maintaining utility and air systems are the primary goals for selecting a distribution strategy. Refer to Articles 10.4 and 10.6 in this chapter on mechanical and electrical systems for further information about the engineering design process for both new and renovated laboratory buildings.

Step 6. Test the Options for Vertical Circulation

Vertical circulation elements include code-mandated egress stairs, monumental and convenience stairs, escalators, and passenger and freight elevators. Dumbwaiters and pneumatic tube systems are other means strictly for materials transport. Because vertical circulation is an essential component of emergency egress and fire and smoke control in multistory buildings, building codes set many parameters for the number, location, design, and fire protection of vertical circulation. In multistory laboratory facilities, vertical circulation is an important path that can support easy communication between floors. Productive research science and educa-

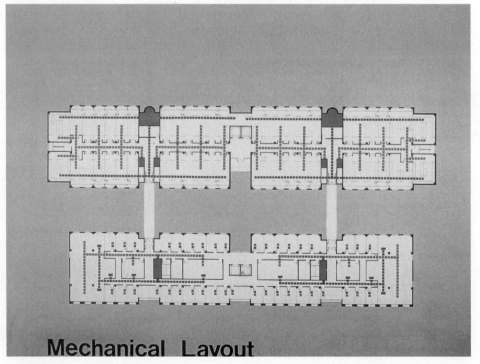

FIGURE 10.2-4 Horizontal distribution.

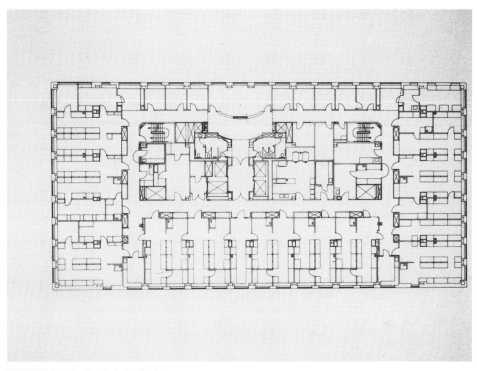

FIGURE 10.2-5   Vertical distribution.

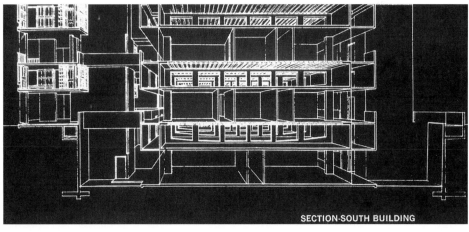

SECTION-SOUTH BUILDING

FIGURE 10.2-6   Salk Institute section.

tion depend upon good communication among laboratory workers and students. Stairs need to be truly convenient for building occupants to use them, instead of waiting for an elevator.

Step 7. Arrange the Program Room Types on Each Floor

After estimating the floor size and determining the circulation pattern, arrange the program rooms on each level. Refer to the program document section on room adjacencies to understand which functions need to be close to each other and which functions should be separated. A common but difficult decision in laboratory layout involves the relationship among laboratories and offices, offices for principal investigators and their staff members. Some researchers insist that their own private offices be located in the lab (Fig. 10.2-7, *a* and *b*). Even more frequently, researchers insist that their staff, particularly students and technicians, occupy desks within their laboratories. Owners should examine this practice carefully. Owners face liability for possible chronic exposure to chemicals for staff who spend 8 hours a day or more in the laboratory environment. The counterargument to these health concerns is concerns for scientific productivity and learning for students.

A second important decision in laboratory layout is the relationship of laboratory support to laboratories. There are several good options. Support functions can be distributed in the same band and modules as the generic laboratories. In the racetrack circulation layout option, support works well between the corridors. A layout that has area adjacent and parallel to modular laboratories gives immediate access to support functions. Other layouts cluster support functions along the bands of laboratories and/or at the ends of the bands (Fig. 10.2-7*c*).

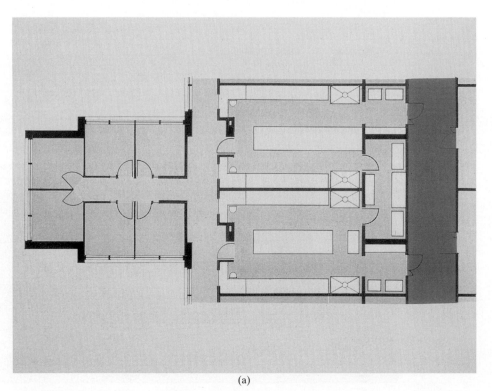

(a)

FIGURE 10.2-7   Office/laboratory support relationship options.

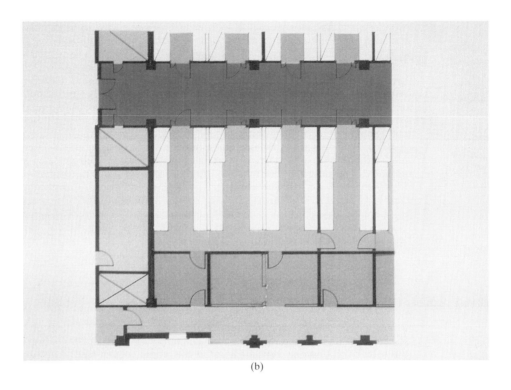

(b)

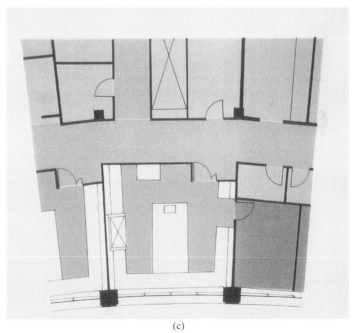

(c)

FIGURE 10.2-7   (*Continued*) Office/laboratory support relationship options.

Another planning challenge in arranging room types on each floor is to understand functions that need to be accessed sequentially, such as clean room or biosafety level 3 suites.

The overarching objective of arranging program rooms is long-term flexibility. This objective applies in the vertical axis as well. Arrays of laboratories above laboratories allow some economy for consolidated utility and ventilation air risers. Concerns for access to under-slab utilities and controls force careful stacking of functions. Biosafety level 3 suites and other specialized laboratory environments are sometimes located on the top floors of laboratory buildings because there can be direct access to supporting mechanical equipment above and no intervening occupied space. Nuclear magnetic resonance imaging (MRI) instruments are often located on the bottom floor or basement because the 5-gauss line from these instruments cannot intersect any other occupied space on any side, including above and below. Vibrational considerations also affect functional vertical relationships. Elevators, for instance, cause local vibration and also disturb magnetic fields along their vertical paths. Basements are often desirable for equipment that is sensitive to vibration because the supporting slab can be isolated from the building structure. Some equipment needs greater vertical separation from fan and pump rooms than the top floor beneath a mechanical penthouse.

Pay attention to functional relationships during the planning phase.

Step 8. Explore the Options and Constraints of Structural Systems

This activity should be deferred in laying out new laboratory buildings until the module is designed and the building is substantially laid out. Premature determination of the structural grid before the planning steps are substantially complete can lead to unfortunate coordination problems.

CONCLUSION

The steps described show how rational laboratory building planning can be. There are many considerations, but each step can build toward truly innovative, creative design and planning solutions. Inside-out planning need not diminish the architect's role in overarching human issues such as communication and a delightful and safe environment in which building occupants and users can do great science.

CONTRIBUTOR

William L. Porter, Ph.D., A.I.A., Massachusetts Institute of Technology, Cambridge, Massachusetts

REFERENCES

1. *Guidelines for Design and Construction of Hospital and Healthcare Facilities,* American Institute of Architects, Washington, DC 1996–1997.

SECTION 10.3
ENGINEERING AND DESIGN PROCESS

Kevin T. Sheehan, Principal
BR+A Consulting Engineers, Inc., Boston, Massachusetts

OVERVIEW

As a component of the laboratory planning and programming process, the design engineer must work in conjunction with and in parallel with the laboratory planning process. One of the most important factors in the development of any laboratory project is integrating the mechanical and electrical space requirements into the laboratory design process. In the various specific building types reviewed within this handbook, the laboratory building application must be recognized as the most intensive and demanding of engineering system requirements in terms of building square footage, floor-to-floor heights, and locations and sizes of vertical shafts. It is common that the mechanical/electrical/plumbing (MEP) systems require 10 to 15% of the building's gross square footage to house the equipment and systems for the infrastructure needed by the laboratory functions.

Additionally, the location and placement of the MEP rooms must be correctly inserted logically into the building's program in terms of diagrammatic adjacencies and functional relationships.

As the project moves into various design phases, integrating the MEP spatial demands into the design process is a key factor in obtaining a good, functional laboratory building.

The following sections identify the required steps, which must be anticipated within the process, to establish the "most appropriate" selection of engineering systems and components and to integrate them into the laboratory architecture.

Refer to Sec. 10.4, "Mechanical Systems in Laboratories;" Sec. 10.5, "Plumbing and Fire Protection Systems in Laboratories;" and Sec. 10.6, "Electrical Systems in Laboratories," for more detailed discussions of the issues surrounding each of these disciplines.

RECOGNIZE THE LABORATORY TYPE

Laboratories may be classified in these generic types.

Biological laboratories include any kind of laboratory in which biologically active materials are found or chemical manipulation of these materials takes place. This includes laboratories in scientific disciplines such as biochemistry, microbiology, cell biology, biotechnology, immunology, pharmacology, and toxicology. Both chemical fume hoods and biological safety cabinets are commonly installed in these areas. This type of facility is most common in institutions, colleges, and medical schools; pharmaceutical companies; and biotechnology industries.

Chemical laboratories include both organic and inorganic synthesis and analytical functions. Substantial numbers of fume hoods are commonly installed in chemical laboratories.

Animal housing suites and laboratories include areas for manipulation, surgical modification, and analytic observation of laboratory animals. It is common for an animal housing suite to be a large component of either of the previously mentioned types of laboratory buildings.

Dry laboratories are spaces in which no chemical or biologically active materials are present. These spaces may be used for research in physics, lasers, optics, material testing, or electronics, or psychological testing of humans.

After thoroughly understanding the laboratory type and assessing the MEP requirements, as detailed within the laboratory planning document, the design team is ready to begin the process of selecting the most appropriate engineering systems to meet the laboratory needs.

Identify Hazards

All laboratories generally involve some level of hazards by their implied definitions and contain some type or form of hazardous materials. It is common practice for a knowledgeable representative of the owner to create a comprehensive hazardous assessment of the laboratory's specific procedures and protocols to assist the design team's understanding of the projects needs and requirements. The hazard assessment may include the following components:

Chemical hygiene plan
Radiation safety plan
Biological safety protocols

The nature of the hazards should dictate the levels of containment and the various types of engineering support systems and devices that will be required to create a safe laboratory environment.

Example.   The understanding of the various biological safety protocols and the resultant implementation of the engineering support systems are illustrated in the following matrix.

Service	BL-2 Lab	BL-3 Lab
Supply air flow	Required	Required
Redundant air source	Desired	Required
Exhaust air flow	Required	Required
Redundant exhaust source	Desired	Required
HEPA filtration at exhaust	—	Required
Lab pressure control	Required	Required
Lab pressure alarms/monitor	Desired	Required
Dedicated exhaust system	—	Required
Entrance vestibule	—	Required

This generic interpretation of the BL-3 laboratory guidelines shows that the building infrastructure demands for BL-3 suites are greater than those for BL-2 suites, and thus the engineering systems and equipment will require more building space and result in higher first costs.

This matrix represents a simplistic interpretation of the guidelines. The project team must work in conjunction with the client's biological safety protocols, as established by the hazardous assessment program, and must review the details of all engineering systems to obtain clear direction and understanding among all parties.

Define Laboratory Unit Size

One of the initial steps for the design and client team is to define the laboratory unit size in accordance with NFPA 45 (Standard on Fire Protection for Laboratories Using Chemicals). Although most building codes do *not* recognize or reference this NFPA Standard, it is commonly referenced as a benchmark of design standards within the laboratory design community and has influenced building liability claims.

Working with the defined laboratory unit size, the design team is then enabled to create various zoning compartments in terms of HVAC supply and exhaust air system configurations, so that the design engineer can begin with the most appropriate system configuration.

Specific NFPA 45 references include:

6-5.9 Manifolding of Laboratory Hood and Ducts

6-5.9.1 Exhaust ducts from each laboratory unit shall be separately ducted to a point outside the building, to a mechanical space, or to a shaft

6-5.9.2 Connection to a common laboratory hood exhaust duct system shall be permitted to occur at these points

6-5.9.3 Exhaust ducts from laboratory hoods and other exhaust systems within the same laboratory unit shall be permitted to be combined within that laboratory unit

Safety Protocols

The goal of the entire designer/owner project team must be to provide a safe, reliable, environmentally comfortable laboratory space. It is common practice for any institutional or industrial client that uses the laboratory to designate a manager (or department) who is responsible for industrial hygiene and safety within its facilities. Thus, the design team must work in conjunction with this office (or department) to satisfy and exceed all goals toward creating a safe working environment.

Level of Equipment Redundancy

The design team must also recognize that most engineering systems are in operation continuously. Thus the selection of the individual equipment must be viewed accordingly. The individual components (fans, pumps, and terminals) must be institutional grade rather than commercial grade.

Additionally, the designer must recognize that, even with a complete and thorough preventive maintenance program, equipment components will need to be taken out of service periodically. Thus, for continuous service, most engineering systems and services must be equipped with either standby equipment or preengineered system cross-connections to minimize interruptions to the laboratory process.

CONTRIBUTORS

Eugene M. Bard, P.E., President, BR+A Consulting Engineers, Inc., Boston, Massachusetts

Anand K. Seth, P.E., C.E.M., C.P.E., Director of Utilities and Engineering, Partners HealthCare System, Inc., Boston, Massachusetts

SECTION 10.4
MECHANICAL SYSTEMS IN LABORATORIES

Eugene M. Bard, P.E., President, and Kevin T. Sheehan, Principal
BR+A Consulting Engineers, Inc., Boston, Massachusetts

GENERAL

Refer to Sec. 10.3, "Engineering and Design Process," for a general discussion of issues applicable to all trades. This article also contains examples of biological safety protocols that affect air systems, as well as a discussion of the application of NFPA 45, Standard on Fire Protection for Laboratories Using Chemicals.

ESTABLISH HVAC SYSTEM DESIGN CRITERIA AND FEATURES

From the viewpoint of an HVAC system, a laboratory building is the most demanding of the various building types discussed in this handbook.

An initial step in identifying the system requirements is to establish design criteria sheets for the subject spaces as benchmarks and targets within the engineering design process. Table 10.4-1 shows sample HVAC design criteria.

Air Filtration Levels.   Working with the building users, the level of air filtration—that is, the quality of the air filter components within the HVAC air handling units—should be defined to enable configuring the correct air-handling unit. Based on several laboratory projects, the standard for laboratory air filters levels detailed in Table 10.4-2 is suggested.

A common industry misapplication is derived from the confusion between 95% filters (ASHRAE) and 99% HEPA filters (DOP). There is a substantial difference between the two grades of filters in size of parts, methods of testing, weight, configuration, pressure drop, and first cost. It is important that all parties agree on the most suitable filter for the laboratory's use, function, and requirements.

Minimum and Maximum Air Changes per Hour.   Our recent reviews of various engineering guidelines and industry standards have identified the following benchmarks:

Federal Register—OSHA	4 to 12 air changes
ASHRAE Handbook (Application 13.8)[2]	6 to 10 air changes

As one reviews this information, it is obvious that one individual standard cannot be nor should be used to fit all laboratory types and classifications. Thus, it is important that the project team establish the maximum and minimum air changes per hour to suit the particular installation and laboratory program.

Maximum Air Changes per Hour.   The HVAC system should have sufficient air-handling capability to satisfy the upper limits of the following ranges and demands:

TABLE 10.4-1   Sample HVAC Design Criteria

Outside Air Design Criteria	Example	Footnote
Summer	95° db 75° wb	
Winter	0° db	(*a*)

Interior Design Condition (Temperature and Humidity)	Winter (°F)	Summer (°F)	Footnote
1. Laboratory	72/20%	75/50%	(*a*)
Lab support room	72/20%	75/50%	(*a*)
Equipment room	72/20%	75/50%	(*a*)
Office conference room	72/20%	75/50%	(*a*)
Tissue culture room	72/20%	75/50%	(*a*)
Computer room	72/40%	72/50%	(*b*)
2. Animal house suite			
Mouse, rat, hamster	64/30%	79/70%	(*b*)
Gerbil, guinea pig	64/30%	79/70%	(*b*)
Rabbit	61/30%	72/70%	(*b*)
Cat, dog, primate	64/30%	84/70%	(*b*)
Farm animal, poultry	61/30%	81/70%	(*b*)
3. Support space			
Electric rooms	65°	Ventilation only	(*a*)
Mechanical rooms	65°	Ventilation only	(*a*)
Teledata rooms	75°	80°	(*a*)
Stairways	70°	—	(*a*)

[a] These values are given to illustrate the concept. The project design team must customize the actual design criteria for each project's geographical location, the applicable codes and standards, and the owner's standards and requirements.
[b] Refer to *Guide for the Care and Use of Laboratory Animals* and additional information presented in the following text for further discussion.[1]

Sufficient air (changes) to provide heat removal and comfort cooling to the laboratory

Sufficient air (changes) to satisfy the makeup air required to offset the air exhausted by the laboratory's fume hoods and local exhausts

Clean room standards—selected laboratory areas may need a local environment that satisfies a clean room standard (e.g., Class 1,000, Class 10,000). For further assistance, refer to Federal Standard 209E, Institute of Environmental Sciences.[3]

Minimum Air Changes for Energy Conservation.   The HVAC system should also be able to reset the volume of air that must be supplied to the laboratory based on the following variables:

1. Internal heat gain diversity

2. Day/night mode of operation

TABLE 10.4-2   Standard for Laboratory Air Filters

Type of Filter	Common Application
30% particulate filters (ASHRAE testing)	Prefilter
65% particulate filters (ASHRAE testing)	Laboratories
85% or 95% particulate filters (ASHRAE testing)	Research laboratories
99% HEPA filters (DOP testing)	Animal holding suite

3. Variable volume fume hood operation and functions

4. Minimal volume of air required to dilute room air and potentially airborne hazardous material to maintain the contamination level at parts per million (ppm) less than the various health standards for chemical vapor

5. Minimal volume to maintain the program's current internal air quality (IAQ) standard from the viewpoint of nuisance odor

Conceptual Solutions for Maximum and Minimum Air Changes

Recognizing that the laboratory usage and program are dynamic and subject to change because of shifts in research programs or classroom usage, it is most logical to design an HVAC system that provides these features (within reasonable limits):

1. Selectable maximum air flow by room and/or program

2. Selectable minimum air flow by room and/or program

3. Reversible pressure control feature (the ability to balance the individual rooms at a positive or negative pressure)

In view of the number of variables to which an individual laboratory room/area may be subjected across a span of 30 years, changing technology, and new service demands, the HVAC system must be sufficiently flexible to satisfy a wide range of conditions and variables.

Internal Heat Gains.   As the nature of laboratory functions has evolved and computer technology has advanced, the laboratory unit has become exposed to greater internal heat gains, which the HVAC system must have the air conditioning capacity to offset. HVAC engineers have discovered that during the early stages of the design process, it is virtually impossible to correctly forecast the exact amount of heat-generating equipment that will be located within a laboratory suite. Thus, it is common practice to establish cfm ranges for individual room types based upon location and function.

Southern exposure research lab	1.5 to 3.0 cfm/sq ft ±
Northern exposure research lab	1.2 to 2.0 cfm/sq ft ±
Equipment support rooms	3.5 to 5.0 cfm/sq ft ±

Future Growth of Fume Hood Density and Program Changes.   As the project design team begins to establish the infrastructure capacity of the various mechanical, electrical, and plumbing services, all parties must be cognizant that the laboratory demands will change throughout the MEP projected lifetime. The basic mechanical changes will need to support the research activities for some 30 to 40 years and must be viewed and selected accordingly. The need for additional electric circuits and power should be anticipated above and beyond the initial program. Likewise, the installation of future fume hoods must also be allowed for in terms of supply air fan capacity, exhaust air fan capacity, exhaust ductwork pathways, and so on. The project design team is responsible for understanding and agreeing on the potential growth factor in terms of first cost and spare capacity.

HVAC SYSTEM ALTERNATIVES (PRIMARY)

Overview

As the project design team begins to consider the most appropriate HVAC system, the following key directions must be reviewed.

Primary Equipment

The laboratory building is generally the most energy intensive of the various building types included within this handbook and creates the largest demands for utilities represented by electric demand and consumption, air conditioning tonnage, and boiler horsepower.

Most Common Energy Sources

Electric power:	From local utility
Heating source:	On-site boiler plant
	Remote boiler plant (satellite)
	Steam purchase from local utility
Chilled water:	On-site chiller plant
	Remote chiller plant
	Purchased chilled water

This article does not intend to explore the various methods of creating these utility demands and supplies, but rather focuses on the delivery of the services to the laboratory space. However, within the planning process, the project design team must recognize the following key points concerning the primary energy sources:

1. If on-site boiler and chillers are required, the requirement for building square footage is substantially increased.
2. The selection of the building's energy source must be reviewed in context with the individual project's location, the regional cost of energy, and available sources.

HVAC SYSTEM ALTERNATIVES (SECONDARY)

Table 10.4-3 shows a short list of HVAC system types that have been used by the building industry with various levels of success, both good and bad.

The project team (engineer and owner) should have a thorough discussion about the best HVAC system to satisfy the demands of the individual laboratory functions and to control the annual energy and operating cost. Each of the system advantages and disadvantages listed in Table 10.4-3 should be understood before a consensus decision. The majority of laboratory

TABLE 10.4-3 HVAC System Types

System	Footnote
All air systems	
Constant volume	(a)
Two-position constant volume	(b)
Variable air volume (VAV)	(c)
Dual duct	(d)
Hybrid hydronic and air systems	
Induction units	(d)
Fan coil units with minimum air	(e)

[a] Simplistic, but may be energy intensive/wasteful.
[b] Simplistic with some level of energy conservation.
[c] Currently, the most common lab application.
[d] A favorite in the design community during the 1960s and 1970s.
[e] Widely utilized in renovation projects that have limited floor to floor heights.

spaces require a 100 percent outside air/100 percent exhaust air type system, which results in high operating costs.

Heat Recovery Options and Payback Analysis

As part of HVAC system selection, most institutions have an obligation to review the practical implementation of heat recovery within their facilities. In this age of accountability, every laboratory facility manager and project manager must balance the construction cost against future building operating and energy charges. A heat recovery analysis should be based on the following variables:

1. Type of heat recovery system
2. Cost of heat energy
3. Cost of cooling energy
4. Climatic/weather data

The systems listed in Table 10.4-4 are the most common in the HVAC industry.

Additional Energy Conserving Strategies

1. VAV air system (in lieu of constant volume)
2. VAV (constant sash velocity) fume hoods
3. Low-temperature hot water reheat from condenser water
4. Intelligent automatic temperature controls (reset capability)

Stack Heights and Air Intake Locations.   As part of the schematic design phase, the project design team must analyze the building's configuration in terms of air intake locations and exhaust air stack locations and heights, in conjunction with prevailing wind patterns, to ensure all parties that air reentrainment does not occur. It is common practice for the project design team to engage the services of a special (wind) consultant to assist in this analysis. The ASHRAE 1999 Handbook, Chap. 43 provides an excellent description of modeling techniques.[4]

Obviously, separating the stack discharge and air intake locations allows the atmosphere to dilute the effluent. Computer modeling (or wind tunnel testing) may be required to verify that adequate atmospheric dilution of the effluent occurs, taking into the building's air intakes and other surrounding buildings.

Stack Height and Air Exit Velocity.   As part of the computer modeling, the stack height and air exit velocity become critical factors within the outcome of the wind study. Working with the architectural team, the details for the individual exhaust air system must be reviewed to

TABLE 10.4-4   The Most Common HVAC Systems

System	Remarks
Glycol run-around coil system	Most common
Heat recovery (enthalpy) wheel	—
Twin-tower enthalpy recovery loops	—
Heat pipe	—
Fixed-plate (air) exchangers	—

assist in preventing reentrainment. For additional information, refer to ANSI/AIMA Standard Z9.5.[5]

Fume Hoods and Laboratory Exhaust Containment Devices. The laboratory building industry has manufactured a series of devices to contain, capture, and exhaust any chemical fumes, vapors, and airborne particulate matter within a designated enclosure.

Working with the HVAC system features, the most appropriate fume hood should be selected from the types listed in Table 10.4-5.

The following additional special hoods and devices also affect the HVAC system application:

1. Radioisotope hoods
2. Perchloric acid hoods
3. Walk-in hoods
4. Distillation hoods
5. Canopy hoods (these are not fume hoods; they are used only to capture heat!)
6. Local spot exhausts (snorkels)
7. Laminar flow clean benches

Fume Hood Containment. The most commonly used benchmark in the laboratory design community for fume containment is the term "100 feet per minute (fpm) across the open sash area." The 100-fpm rate successfully contains the interior airborne particulate matter within a fume hood in a variety of applications.

However, the HVAC/research community continues to challenge this applied benchmark because various acceptable fume hood test methods (ASHRAE Standard 110) indicate that safe containment may occur at lower velocities. The 100-fpm benchmark may also be insufficient for containment in selected fume hood arrangements or configurations.

It is interesting to note that since 1982, NFPA 45 has strongly recommended installing fume hood monitors (flow indication) on all new or modified fume hoods. This safety feature has been ignored until recently by many within the laboratory industry.

Student Workstations. Facility managers should also be cognizant of a recent trend in the methods of teaching chemistry. Several recent graduate and undergraduate-level chemistry teaching laboratories and classrooms have implemented a student workstation concept, in which individual students have their own designated fume containment hoods and workplaces, thus eliminating the need for chemical experiments on open lab benches. Two laboratory fume hood manufacturers now offer this product line.

Exhaust Air Strategies

The HVAC designer has the following methods available for removing laboratory exhaust air and fume hood exhaust:

TABLE 10.4-5 Fume Hood Types

Fume hood type	Remarks
1. Standard (bypass)	Constant volume with variable face velocity (within ranges)
2. Variable volume	Constant face velocity
3. Auxiliary air	Constant volume with variable face velocity (within ranges)

1. An individual exhaust fan per fume hood
2. An individual exhaust fan for a (given) set of fume hoods (mini-ganged system)
3. A centralized ganged or manifold system

These applications may also be connected to the following types of ductwork classifications:

1. Low-pressure exhaust ductwork
2. Medium- to high-pressure exhaust ductwork
3. Constant or variable volume applications

The critical decision to be made by the project design team in conjunction with the hazard assessment manager/program revolves around the most appropriate exhaust fan configurations on a project-by-project basis. Exhaust systems for higher hazard applications such as radioisotopes, perchloric acid, and BL-3 exhaust tend to require the individual fan arrangement.

The facility manager should recognize that the centralized, ganged (manifolded) exhaust concept has these key advantages:

1. Exhaust air is diluted by multiple sources.
2. There is less exhaust fan equipment to maintain.
3. It enables the project to maintain standby fan(s) in all/most applications, thus providing a redundant system.

Biological Safety Cabinets (BSC).   A biological safety cabinet is a very common containment device and is normally used in biohazardous laboratory applications. Some sections of the laboratory community suggest that all BSCs are the same. From the viewpoints of containment and the HVAC system application, nothing can be further from the truth.

It is critical that the individual applications of the various BSC types be identified so that proper HVAC services may be designated (see Table 10.4-6).

For additional information, refer to CDC NIH guidelines for Biosafety in Microbiological and Biomedical Laboratories, U.S. Dept. of Health and Human Services.[6]

Guidelines for the Care and Use of Laboratory Animals (HVAC Viewpoint).   Of the various organizations involved with the housing and care of laboratory animals (ILAR, IACUC, ILARC, AALAS, NIH, and AAALAC), the majority of committees agree upon the guide published by AAALAC[1] as the basis for designing an animal housing suite and laboratory.

The interpretation of this guideline is also a critical part of the design process. It should be a mandatory practice for the design engineer and the director of the animal suite to openly discuss all aspects and expectations of the HVAC system features and capacities required to deliver the level of service demanded by the site requirements.

TABLE 10.4-6   Applications of Various BSC Types

Classifications[a]	Exhaust requirements	Remarks
Class I	Vent into room or outside	—
Class II Type A or B3	Vent into room or outside	—
Class II Type B1	Vented to outside	Special fan required
Class II Type B2	Vented to outside	Special fan and external HEPA fixtures required
Class III	Vented to outside	Custom application

[a] As manufactured by the Baker Company.

Example of Interpretation Problem

	Temperature	Humidity
Room criteria for mouse room:	64–79°	30–70%

Broad interpretation. The HVAC system must provide continuous climatic conditions between a low of 64°/30% (winter condition) and a high of 79°/70% (summer condition).

Narrow interpretation. The HVAC system must provide continuous climatic conditions between 64° (and 30 to 70% RH) and 79° (and 30 to 70% RH).

A substantial amount of additional HVAC equipment will be required to deliver the system capability for the narrow interpretation, if it is at all possible to obtain the extremes of the ranges.

The facility manager must be aware that the animal house suite and laboratory is most likely to be the most expensive portion of the laboratory building. A cost of $300 to $400 per square foot is not uncommon. Thus, it is imperative that all parties be "on the same page" in understanding the system's features and capabilities.

The current trend in designing HVAC systems for animal laboratories is to include these features:

1. A sufficient number of air changes
2. A reversible pressure feature for individual rooms
3. Individual room temperature and humidity controls
4. Temperature and humidity monitoring (and alarms)
5. Lighting level control
6. Proof of lighting control cycle
7. Proof of space pressurization
8. High room exhaust (new concept)
9. Ability to add microisolator type cages
10. Multiple supply air fans/units
11. Multiple exhaust air fans/units
12. Odor control devices
13. Dedicated automatic temperature control (ATC) automation system and terminal

Automatic Temperature Controls. With the development of direct digital control (DDC) technology and its wide range of applications, the majority of new laboratories are beginning to implement total DDC technology to all HVAC system components to facilitate the intelligent laboratory (smart building) application.

Although many designers have utilized DDC technology for the main HVAC components, CFM tracking air terminals allow direct monitoring (and alarming) of individual room temperatures, humidity levels, room pressurization modes, and occupancy modes. The use of CFM tracking air terminals is a control strategy that accurately controls the air volumes being supplied and exhausted from individual laboratory spaces. With this concept, the facility managers have finally been given the operating tools that enable quick response to various laboratory problems (from an HVAC and equipment viewpoint).

In addition, when this technology is employed throughout the laboratory floor plans, it is a relatively minimal expense to extend the DDC system to include monitoring points for BSC alarms, fume hood alarms, environmental room alarms, equipment failure alarms, and so on, for delivering the information gathered to a central command center.

The subject alarms can now be annunciated to all interested parties via automatic telephone dialers, pagers, and direct e-mail messages for critical alarms and conditions.

TABLE 10.4-7 Methods of Monitoring Space Pressure Relative to Adjacent Spaces

Type/Application	Remarks
Direct-pressure measurement	Utilized for high-end applications
Volumetric flow tracking	Most common applications

Control of Laboratory Pressure (Negative to Surrounding Areas). The designer should also discuss the various methods of monitoring the appropriate space pressure relative to adjacent spaces (see Table 10.4-7).

CONTRIBUTOR

Anand K. Seth, P.E., C.E.M., C.P.E., Director of Utilities and Engineering Partners HealthCare System, Inc., Boston, Massachusetts

REFERENCES

1. *Guide for the Care and Use of Laboratory Animals,* available on the World Wide Web at http://www.nap.edu/readingroom/books/labrats/index.html (National Research Council of the National Academy of Sciences, Washington, DC, 1996), used and referenced by *Association For Assessment and Accreditation of Laboratory Animal Care International (AAALAC International),* Rockville, MD.

2. *1999 ASHRAE Handbook—HVAC Applications,* American Society of Heating, Refrigeration and Air Conditioning Engineers, Inc., Atlanta, GA, 1999.

3. Federal Standard 209E, Airborne Particulate Cleanliness Classes in Cleanrooms and Clean Zones, written 1992 by Institute of Environmental Services, Mount Prospect, IL, approved by Commissioner, Federal Supply Service, General Services Administration, Washington, DC, September 11, 1992.

4. *1999 ASHRAE Handbook, HVAC Applications,* American Society of Heating, Refrigeration, and Air Conditioning Engineers, Inc., Atlanta, GA, 1999.

5. American Industrial Hygiene Association (AIHA). *ANSI/AIHA Z9.5 American National Standard for Laboratory Ventilation,* AIHA, Fairfax, VA, 1992; American Industrial Hygiene Association (AIHA) Z9.5 Committee. *Clarification of ANSI/AIHA Z9.5 Standard "Laboratory Ventilation."* AIHA, Fairfax, VA, 1993.

6. Biosafety in Microbiological and Biomedical Laboratories. HHS Publication No. (CDC) 93-8395, U.S. Department of Health and Human Services, Public Health Service, Centers for Disease Control and Prevention and National Institutes of Health, 3rd Edition, U.S. Government Printing Office, Washington, DC, March 1993.

BIBLIOGRAPHY

NIH Ventilation Design Handbook on Animal Research Facilities Using Static Microisolators, Volumes I and II, Division of Engineering Services, Office of Research Services, National Institutes of Health, Bethesda, MD, September, 1998.

Occupational Safety and Health Administration (OSHA), *29 CFR Part 1910.1450,* US Department of Labor, OSHA, Washington, DC, 1996.

SECTION 10.5
PLUMBING AND FIRE PROTECTION SYSTEMS IN LABORATORIES

Eugene B. Kingman, C.I.P.E., Principal, and Robert V. DeBonis
Robert W. Sullivan, Inc., Boston, Massachusetts

PLUMBING SYSTEMS IN LABORATORIES

Plumbing in laboratory facilities uses many of the same systems as all other buildings, namely potable water, sanitary waste, storm drainage, and natural gas. In laboratories, plumbing includes other systems, such as nonpotable water, special waste, lab gases, lab-grade water, and vacuum.

Water Supply

In addition to the uses water shares in all other types of buildings, water in laboratories is used for glass/cage washing, equipment cooling, solution preparation, ice making, creating suction through aspiration, diluting chemicals, and protecting personnel from spills, splashes, and burns. Separate potable and nonpotable water systems may be required.

Potable water, defined as water suitable for human and animal consumption/contact, is typically used for toilets, hand washing, showers, food preparation and cleanup, lawn watering, and building cleaning. Safety fixtures, such as eyewashes and emergency showers also use potable water. *Nonpotable water,* which is generally defined as water from a nonpotable source, is used for glass/cage washing, equipment cooling, solution preparation, creating suction through aspiration, and diluting chemicals, as well as any other use where water comes in contact with a substance or a piece of equipment that renders (or could render) it unsuitable for human or animal consumption or contact. By definition, water from the supplier is considered to be potable. It is the responsibility of the supplier to protect its distribution system and the public from what is happening inside a facility. It is the responsibility of facility users to protect their personnel. Consult Article 5.4.6, "Plumbing, Process, Gases, and Waste Systems," for additional details on cross-connection.

Wastewater

In addition to the sanitary wastes generated by all types of facilities, wastewater in labs is generated from cage/glass washing, sterilizing processes, aspirators, equipment cooling, experiment by-products, spent chemical solutions, and water-sealed rotary equipment. This wastewater may be suitable for disposal in the sanitary system. Characteristics that could make the wastewater unsuitable for the sanitary system might include the presence of fats, oils, or grease; a high solids content; or a high or low pH. This wastewater could contain heavy metals, solvents, flammable liquids, human or animal blood and other body fluids, viable organisms, and recombinant DNA. Federal, state, regional and local codes/regulations, as well as sewer use regulations from the entity that operates and maintains the sewer system, will dictate the prohibited substances, as well as the maximum concentrations of the allowed substances. Violations of these regulations can result in the levy of significant fines on the offending facility. See Figs. 10.5-1 and 10.5-2 for standard sampling ports that are used to procure samples for testing. It is recommended that such sampling ports be provided in the wastewater lines.

Wastewater that contains some of the prohibited substances, that has unacceptable concentrations of otherwise acceptable substances, or that has a low or high pH is allowed into the sewer system after pretreatment. Solids and grease/oil/fats can be removed from the wastewater by properly sized separators. Proper sizing includes quiescent flow. Adequate retention time is the key to proper separator design.

The pH of wastewater can be adjusted up or down as needed with proper equipment. If the pH varies between acidic and basic levels or is constantly higher than the regulations allow, then a chemical addition system can be used. If the pH is constantly lower than the regulations allow, then limestone chips can be used, which contribute hydroxyl ions, raise the pH, and negate the need for adding chemicals and using mixers, tanks, and so on. *pH systems* typically consist of tanks, mixers, chemical feed pumps, flow monitors, pH monitors, and controls. pH systems can be batch-type or continuous. In the continuously operating type, one or more tanks could be required, depending upon the variations in the wastewater pH. All tanks would be equipped with mixers that would operate continuously to promote thorough mixing in the tanks. Chemicals, usually hydrochloric acid and sodium hydroxide, would be added automatically from storage drums to raise or lower the pH as determined by the control system. The regulations typically require equipping all systems with pH and flow recorders. In some instances, bypasses may be required or warranted, depending upon the consequences of a shutdown and stoppage of flow if repairs are needed.

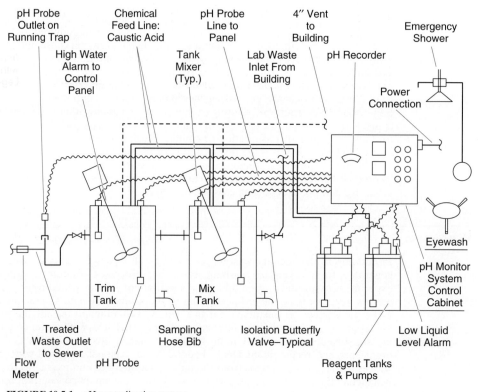

FIGURE 10.5-1 pH neutralization system.

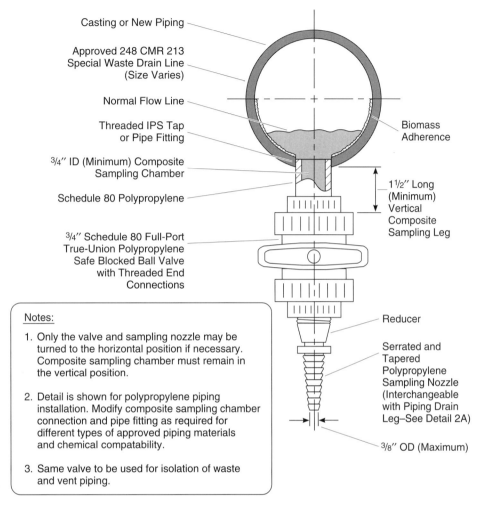

Casting or New Piping

Approved 248 CMR 213
Special Waste Drain Line
(Size Varies)

Normal Flow Line

Threaded IPS Tap
or Pipe Fitting

3/4" ID (Minimum) Composite
Sampling Chamber

Schedule 80 Polypropylene

3/4" Schedule 80 Full-Port
True-Union Polypropylene
Safe Blocked Ball Valve
with Threaded End
Connections

Biomass
Adherence

1 1/2" Long
(Minimum)
Vertical
Composite
Sampling Leg

Reducer

Serrated and
Tapered
Polypropylene
Sampling Nozzle
(Interchangeable
with Piping Drain
Leg–See Detail 2A)

3/8" OD (Maximum)

Notes:

1. Only the valve and sampling nozzle may be
turned to the horizontal position if necessary.
Composite sampling chamber must remain in
the vertical position.

2. Detail is shown for polypropylene piping
installation. Modify composite sampling chamber
connection and pipe fitting as required for
different types of approved piping materials
and chemical compatability.

3. Same valve to be used for isolation of waste
and vent piping.

FIGURE 10.5-2 Cross-sectional view of standard MWRA sampling port.

Systems for pH adjustment require maintenance and incur operating costs for pumps, mixers, and controls. Resupplying feed pumps with chemicals can involve transfer of chemicals from drums to holding tanks. This is a hazardous operation that involves risk to the personnel performing it. Proper training and safety equipment (face shields, aprons, gloves, wash-up sinks, spill control provisions, and drum handling equipment) are required for this procedure. Locating the pH system to allow gravity flow is often desirable. Pumping raw wastewater into the treatment tanks could require an increase in the tank sizing for adequate retention and contact time in the tanks. The location of this system in relation to the loading docks used to receive the chemicals should be considered. See Fig. 10.5-3 for a typical pH adjustment system.

Heavy metals can also be removed by pretreatment, but these removal processes are relatively expensive and maintenance intensive, and it is typically considered easier and simpler to prevent heavy metals from entering the waste stream in the first place. Finding acceptable substitutes for the substances that contain heavy metals can do this. This is called *source reduction*. Solvents and flammable liquids are best not introduced into the waste stream but should be containerized and disposed of as hazardous waste.

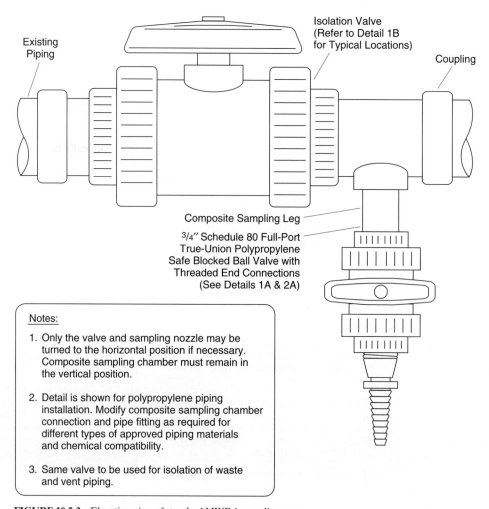

FIGURE 10.5-3   Elevation view of standard MWRA sampling port.

Blood and other body fluids, viable organisms, and recombinant DNA are usually used in specially designed labs within a facility. Special care must be taken to prevent contamination of areas outside the specially designed labs. A means of deactivating these substances must be employed to protect the waste stream and the personnel who maintain the systems. This could be steam sterilizing by autoclave or by chemical sterilization. Once these fluids or the organisms are deactivated, it is sometimes possible to discharge them into the sanitary sewer. In some jurisdictions, however, it is still necessary to dispose of these substances as hazardous wastes.

Lab-Grade Water

This term is commonly used for many different types of treated water systems for laboratories. See Art. 5.4.6, "Plumbing, Process, Gases, and Waste Systems," for a general overview of lab-grade water.

Gases

Gases commonly used in laboratories include natural gas, propane, compressed air, nitrogen, carbon dioxide, helium, and argon. Other gases that are sometimes used are carbon monoxide, hydrogen, acetylene, and other toxic, flammable, corrosive, or explosive gases. (*Vacuum,* although not really a gas, is considered one for installation, maintenance, and control purposes.) A utility company supplies natural gas by a piped connection. Propane, used where natural gas is not available, is supplied from a tank or tanks. Codes (NFPA Fuel Gas Code,[1] local, or state level codes) or the supplying utilities usually regulate the installation and use of natural gas and propane. Compressed air is usually supplied from a compressor system. When quality or pressure requirements are higher than can economically be obtained from a compressor, gas cylinders are used. The compressor can be a local bench-top device or a central system complete with dryers, filters, and reservoir. Depending upon the size of the facility and the demand, nitrogen, helium, and argon can be supplied from central tank locations or from local cylinders located at the point of use. Carbon monoxide, hydrogen, acetylene, and the other toxic, flammable, corrosive, or explosive gases are usually stored and used locally at locations that are dictated by building codes. Storage (defined as when not in use) of full and empty cylinders is regulated by building codes and by entities such as Factory Mutual, Kemper, IRI, and other property loss prevention organizations. Use of cylinders is also regulated by building codes. Cylinders in use are usually placed in gas control cabinets, located as close as possible to the intended point of use to minimize the length of piping needed to carry the gas. This helps reduce the probability of leaks or ruptures. The gas control cabinets are sometimes equipped with gaseous fire suppression systems or wet sprinkler systems, depending upon the gas used.

Vacuum can be provided locally by aspiration using water. Larger-scale demands are supplied from a central system complete with pumps, separators, and a receiver. When using vacuum, care should be taken to reduce or eliminate the entrapment of liquids in the vacuum stream, so as to keep the amount of liquid entering the piping system as low as possible. This is needed to help reduce blockages in the piping and reduce the risk to maintenance personnel when working on the piping. Vacuum bottle traps can do this. These are devices that separate liquid from the vacuum stream and collect it before it can enter the piping stream. Bottle traps should be provided at any outlet at which liquid could be drawn into the piping system. The contents of these bottles, which must be emptied periodically, could be considered hazardous waste, depending upon the processes and procedures performed in the lab. For more details, see Art. 5.4.6, "Plumbing, Process, Gases, and Waste Systems."

Perchloric Acid Hood Washdown Systems

Fume hoods in which perchloric acid is used pose a special problem. The accumulation of deposits of perchloric on the hood surfaces, on the walls of the exhaust duct, and on the exhaust fan can create fire and explosion hazards. To help counteract the natural tendency of perchloric to precipitate out of the airflow, a water washdown system is usually installed. This system consists of a series of small-orifice nozzles located within the hood, within the duct along its entire length, and at the fan housing. These nozzles apply water in sufficient quantity to wash the deposits away. The exact amount of water and pressure required are based on engineering judgment. The facility is ultimately responsible for determining what is sufficient. The flow of water must completely wet all interior surfaces of the hood, duct, and fan. Nozzles must be carefully located so as to expose all interior surfaces to the water flow. Hoods equipped with nozzle systems can be purchased from manufacturers. Nozzles and piping for ducts and fans are installed when the hood is installed. Access to the nozzles for inspection and/or maintenance must be considered when determining locations and attachment methods. Various nozzle manufacturers make attachment devices for the different types of duct materials.

Ducts and fans must be arranged to promote complete drainage of the water from this washdown. Hoods are provided with means to drain the nozzles within them, but this cannot handle the flows from the duct/fan portion of the system. A means of collecting the wash water before it can drain into the hood is required, or the hood could become flooded. If the hood exhaust system involves a stack higher than a floor or two, the amount of water generated could be significant. Connections from the duct and the fan housing must be equipped to overcome the static pressure of the exhaust system. Direct connection into a waste piping system can be equipped with a deep seal trap. A 4-inch water column trap seal is usually sufficient to prevent the air flow in the duct from siphoning the trap seal.

The wastewater flow from this process must be treated by pH adjustment. Perchloric acid is not an explosion hazard in the concentrations encountered in a properly designed washdown system. The volume of wash water must be considered, especially its effect on the pH adjustment system. It may be necessary to stage the washdown or to increase the size of the pH adjustment system to accommodate this flow. This is especially of concern if more than one hood is so equipped. Because the volume that the nozzles pass depends on the water pressure at the inlet, it may be necessary to carefully regulate water pressure for this system.

The use of the hood during the washdown cycle also must be considered carefully. Apparatus within the hood may be hot. Applying cold water to hot apparatus may be another source of problems. The use of the hood must be considered when determining the proper controls on the washdown system.

Safety Fixtures

Safety fixtures in labs include eyewashes and emergency showers. Body drench hoses are sometimes supplied, but these should supplement eyewashes and showers. The need for these devices is dictated by what is being done in the particular lab. In some jurisdictions, codes require emergency showers and/or eyewashes. Some codes require tepid or tempered water. OSHA requirements for worker safety also impose the need for these devices. If it is determined that such items are necessary, then ANSI Standard Z358.1 advises how, where, and what is to be used.[2] Travel distances, mounting heights, and pressures are defined in this standard. The standard also requires "tepid" water, but does not define the term. *All devices must be suitable for hands-free use.* See Figs. 10.5-4 and 10.5-5 for typical units.

FIRE PROTECTION SYSTEMS IN LABORATORIES

Fire protection systems are extremely important in laboratories because of the number and variety of chemicals in use and because of the presence of open flame. All personnel must be trained in safe work practices and in the correct use of protection systems. Fire protection systems in laboratory facilities use the same standpipes and automatic wet sprinklers systems that are used in other types of buildings, but laboratories can present specific problems that call for specialized fire suppression systems, such as preaction sprinklers, foam, dry chemicals, carbon dioxide, and Halon substitutes such as Inergen or FM200.

Sprinklers

Sprinklers save lives. A properly designed, installed, and maintained sprinkler system will reduce the loss of life and property. Sprinkler systems in laboratories are typically designed to meet the criteria set forth in NFPA 13 for protecting Ordinary Hazard Group II uses.[3]

FIGURE 10.5-4 Emergency shower station (*Courtesy of Haws.*)

This designation controls the design and installation of the system by setting spacing requirements for the sprinkler heads and the design density requirements. Design density is the amount of water delivered per square foot per second to the design area. The greater the density, the more water on the fire. A typical design density for a laboratory would require 0.20 gpm per sq ft for a design area of 1500 sq ft. NFPA 13 sets minimum requirements for systems.[3] Often, the facilities insurance carrier will have more stringent requirements. Additional details about sprinkler systems are described in Art. 5.4.7, "Sprinkler and Fire Protection."

Hazards presented by airflow through ducts, especially exhausts from fume hoods and other special exhaust systems, may warrant installing sprinklers inside ductwork. Access to the heads for inspection and or replacement must be considered. Special coatings on the heads may be warranted, depending on the vapors carried by the ducts.

Standpipes

Standpipes can be effective for suppressing fires in laboratories, but personnel expected to use standpipes and hoses should be properly trained, so as to minimize the possibility of injury. Additional details about standpipe systems are described in Art. 5.4.7, "Sprinkler and Fire Protection."

FIGURE 10.5-5 Eyewash station (*Courtesy of Haws.*)

Special Fire Suppression Systems

Sometimes special fire suppression systems are required in laboratories because of the chemicals present or other considerations that make the discharge of water inappropriate. Preaction systems require an electrical signal to open the water supply after a sprinkler head has activated. This signal may come from a smoke, heat, or flame detector or from a manual pull station. These systems help prevent the accidental discharge of water while still providing the safety of a water-based sprinkler system.

Many extinguishing agents other than water can be used when water poses a greater threat to the situation, such as for sodium or other water-reactive substances, or when water would simply be ineffective, such as for flammable liquids. Carbon dioxide, FM 200, Intergen, dry chemicals, and foam water systems are examples of special extinguishing systems. They all have very specific advantages and disadvantages, and the facility manager must work closely with the design team, safety professionals, and sometimes even the local fire department to select the correct system for each individual situation. NFPA regulates the design, installation, and testing of these various special suppression systems under various standards.[4]

Fire Extinguishers

All laboratories must have fire extinguishers available for personnel to use. Building codes and NFPA 101 regulate when fire extinguishers are required.[5] NFPA 10 addresses the proper location and spacing of extinguishers.[6] The facility is responsible for supplying the proper extinguishing agents for the types of fires that are apt to be encountered.

Fire Blankets

Fire blankets are required for the safety of personnel. They must be accessible and clearly marked. Personnel must be trained in their proper use and made aware of their location.

CONTRIBUTORS

Anand K. Seth, P.E., C.E.M., C.P.E., Director Utilities and Engineering Department, Partners HealthCare System, Inc., Boston, Massachusetts

REFERENCES

1. *NFPA Standard 54, National Fuel Gas Code,* National Fire Protection Association, Quincy, MA, 1999.
2. *ANSI Standard Z358.1-1998. American National Standard for Emergency Eyewash and Shower Equipment,* American National Standards Institute, Washington, DC, 1998.
3. *NFPA Standard 13, Installation of Sprinkler Systems,* National Fire Protection Association, Quincy, MA, 1999.
4. *NFPA Standard 11, Low Expansion Foam,* National Fire Protection Association, Quincy, MA, 1998.
5. *NFPA Standard 11A, Medium and High Expansion Foam Systems,* National Fire Protection Association, Quincy, MA, 1999; *NFPA Standard 12, Carbon Dioxide Extinguishing Systems,* National Fire Protection Association, Quincy, MA, 2000; *NFPA Standard 12A, Halon 1301 Fire Extinguishing Systems,* National Fire Protection Association, Quincy, MA, 1997; *NFPA Standard 16, Deluge Foam-Water Sprinkler Systems,* National Fire Protection Association, Quincy, MA, 1999; *NFPA Standard 17, Dry Chemical Extinguishing Systems,* National Fire Protection Association, Quincy, MA, 1998; *NFPA Standard 17 A, Wet Chemical Extinguishing Systems,* National Fire Protection Association, Quincy, MA, 1998; *NFPA Standard 30, Flammable and Combustible Liquids Code,* National Fire Protection Association, Quincy, MA, 1996; *NFPA Standard 45, Fire Protection for Laboratories Using Chemicals,* National Fire Protection Association, Quincy, MA, 1996.
6. *NFPA Standard 101, Safety to Life from Fire in Buildings and Structures,* National Fire Protection Association, Quincy, MA, 2000; *NFPA Standard 10, Portable Fire Extinguishers,* National Fire Protection Association, Quincy, MA, 1998.

Additional Recommended Reading

P. C. Ashbrook and M. M. Renfrew (Eds.), *Safe Laboratories: Principles and Practices for Design and Remodeling.* CRC Press, Boca Raton, FL, 1991.

National Research Council, *Prudent Practices for Handling Hazardous Chemicals in Laboratories,* National Academy Press, Washington, DC, 1981.

National Research Council, *Prudent Practices for Disposal of Chemicals from Laboratories,* National Academy Press, Washington, DC, 1983.

SECTION 10.6
ELECTRICAL SYSTEMS IN LABORATORIES

Arjun B. Rao, P.E., Vice President
BR+A Consulting Engineers, Boston, Massachusetts

Electrical systems for laboratories need to be designed as high-reliability systems, similar in many respects to the level of reliability designed into health care facilities. Therefore, many references will be to Sec. 9.5, "Electrical Systems," in the Health Care Facilities chapter. Those interested in laboratory electrical systems should read that article and Sec. 9.11, "Electrical Utility Management Program." The guidelines that follow should be carefully reviewed before any laboratory facility renovation or construction project design commences and will serve as a good first step in assessing existing laboratory electrical distribution systems.

LOW-VOLTAGE ELECTRICAL DISTRIBUTION

Laboratory buildings should have low-voltage electrical distribution systems (i.e., 277/480 and 120/208 V) similar to health care facility electrical distribution systems. Features to be included are redundant transformation, draw-out breakers in the main switchgear to allow ease of maintenance, and two levels of ground fault sensing in the main switchgear.

Busway Distribution Versus Dedicated Feeder for Each Laboratory Floor

Refer to Sec. 9.5, "Electrical Systems," in the health care facilities chapter for the distribution concepts of busway versus dedicated feeder application in low-voltage systems. The use of ground fault sensing/trip units is strongly recommended in the bus plug-in overcurrent device to provide maximum system reliability and breaker trip coordination.

K-Type Transformers to Compensate for Harmonic Distortion

The use of K-type transformers to serve laboratory loads is highly recommended to compensate for the harmonic currents generated by laboratory equipment. Refer to Art. 5.2.9, "Power Quality," for a discussion of harmonics and their effects on electrical distribution equipment. K-type transformers require larger spaces in electric closets and require higher air flow to cool the closet spaces. The use of K-type transformers also requires 200 percent neutral conductors in both the main feeders to the distribution panel and the branch circuits in the laboratory panel.

LABORATORY ELECTRICAL POWER DISTRIBUTION CONCEPTS

Laboratory electrical power distribution designs should minimize the potential for failures in one laboratory area (principal researcher, for example) to propagate into another laboratory area. To attain a high level of reliability in each laboratory module (600 to 800 ft^2), each laboratory module should be provided with a dedicated panel that serves only the loads of that laboratory. The panel should be on a dedicated feeder from the distribution panel. A typical laboratory panel is rated from 100 to 225 A, has a single or double tub, is located within the laboratory module, and has a shunt-trip main breaker with a remote trip device.

UPS System—Centralized Versus Bench-Top Units

Designers and facility managers should discuss the need for an uninterruptible power system (UPS) with the laboratory users before designing either a centralized UPS unit for multiple laboratory loads or a laboratory bench-top UPS unit for a dedicated load. UPS systems may not be cost effective and should be used only after a detailed analysis of the electrical loads and the costs and benefits.

Bench-Top Power Raceways—Factory-Wired Versus Field-Wired Units

Almost all biomedical research, chemistry research, and teaching laboratories use factory-wired or field-wired bench-top raceways to house wiring devices for plug-in, bench-top laboratory equipment. Some of the raceways are used to house both power wiring devices and communication devices such as data outlets and low-voltage wiring for remote monitoring of equipment and have metal barriers in the raceway to provide code-mandated segregation between power and communications systems. Most popular raceways used in research laboratories are made of anodized aluminum with duplex receptacles 18 to 24 in. on center, sometimes as few as two outlets on a 20-A, single-phase circuit and with an insulated grounding conductor. Raceways can also house outlets from other than the normal power systems, such as from a UPS or an optional standby power source. These other outlets are often identified on the raceway by using different colored outlets, such as red for emergency power, and so on. All raceways should have cross-sectional areas to comply with the 20 percent NEC required raceway fill and for future expansion capabilities.

Ground Fault Wiring Devices on Bench Tops in Proximity to Sinks.   The NEC requires ground fault circuit interrupter protection for outlets near sinks in residential and some other types of occupancies. In a laboratory, however, the user should be consulted before providing such a feature to avoid affecting the outcome of an experiment if power is interrupted to the equipment that is performing the experiment. Other design approaches, such as keeping these outlets away from sinks, may provide higher reliability along with sufficient personnel safety.

Standby or Emergency Power

Refer to Sec. 9.5, "Electrical Systems," and Art. 5.2.3, "Emergency Power Supply Systems," for discussions of the equipment required to supply standby power to critical equipment in laboratory buildings. In this and the following discussion, it is important to realize that we are not discussing the local or state building code, state or National Electrical Code, or Life Safety Code definitions of required emergency power; rather, we are discussing what the NEC would classify as optional standby power. Since that term, although correct, is not well known, the following discussion uses the much more commonly recognized term, *emergency power.*

Before laboratory facilities are designed or renovated, the engineers should discuss equipment that requires emergency power, and the reason for such a requirement, with all end users in each laboratory. Following those discussions, the engineers should prepare a menu of all emergency loads, that are code-mandated and are requested by users, to arrive at a recommended generator rating. All parties should jointly review the impact of all requests. Listed in the following text are the loads that generally require standby power in laboratory buildings. This list does not include any code-required loads that would also be included without discussion, because those loads are determined by applicable codes and by authorities having jurisdiction.

Outlets on Laboratory Bench Tops.   Generally, these are requested by laboratory users. The locations should be clearly agreed upon by all and are often on the bench top or on a wall inside the lab module. Loss of power for a short period of time—for example, during a power failure—could affect an experiment's outcome. Users need to be aware that a backup or standby power source will indeed involve a short loss of power to the equipment for at least 10 sec during transfer from the normal power source to emergency power. Alternatives such as a UPS are available for equipment that cannot tolerate such short-term power outages.

Laboratory Exhaust Hoods.   This requirement could be requested by the user; it could even be code-mandated based on the type of laboratory; or it could be required to allow proper operation of the mechanical system during a power failure.

Animal Rooms on Standby Power.   This requirement for animal room lighting and cooling/heating systems on emergency power is often a user requirement and is also strongly recommended as good engineering practice to provide the proper environment for the animals.

Walk-in Boxes, Incubators, and Freezers.   It is strongly recommended that walk-in boxes, incubators, and freezers be on emergency power. Some walk-in boxes can be classified as "warm rooms," and others are classified as "cold rooms." Warm rooms have the same needs for emergency power as incubators, and cold rooms have the same needs as freezers or refrigerators, depending upon their temperature ratings. Other types of equipment, such as some smaller refrigerators, may also need to be on emergency power. User input is usually necessary to identify the equipment that should be on emergency power. Because emergency power capacity is often only a fraction of the maximum normal power demand, it is often used to prevent irreparable harm to science through the loss of critical or irreplaceable cells and materials, not to keep the lab operating as usual.

Chillers on Emergency Power.   Refer to the standards established by the Association for Assessment and Accreditation of Laboratory Animal Care International (AAALAC)[1] for requirements and recommendations on providing emergency power to selected building cooling systems. These requirements often increase the size of the standby power plant and the resulting cost. The design engineer should provide and identify the standby power ratings (kW) both with and without the cooling system on emergency power, so that the facility manager and laboratory manager can make informed decisions.

Lighting.   The IESNA Lighting Handbook[2] provides recommended lighting levels (now called illuminance) appropriate for the facility and visual task. Refer to Sec. 5.3, "Lighting Systems," for a detailed discussion of lighting design criteria and practices. One of the important features to be considered before the illuminances are calculated is the reflectance or the finish of the laboratory bench top. Most laboratory bench tops are black and provide very low reflectance. Of similar importance in lighting illuminance calculations are reflectances from reagent shelves.

Fixture Types.   Most laboratory lighting uses fluorescent fixtures, typically two lamps, either recessed in the ceiling cavity or ceiling-hung to provide indirect lighting. In many laboratories, the fixtures are oriented parallel to the laboratory bench tops, aligned with the edge of the bench top. A large variety of high-efficiency lighting fixtures have glare-free reflectors to provide proper lighting in laboratories, and these are available from many fixture manufactures. Making a mock-up of the proposed laboratory lighting fixture is strongly recommended before selection of a particular type of fixture.

Task Lighting.   The use of task lighting to supplement overhead ambient lighting is recommended when the illuminance provided from the ceiling-hung light fixtures cannot meet user requirements. The users should determine where task lighting is most appropriate. Improperly located task lighting fixtures could hinder or interfere with daily laboratory activity.

Lighting Switch Control.   It is common practice to provide multiple switching in laboratory areas. An analysis of the usage patterns in the area will assist in determining the most appropriate degree of switching.

Emergency Lighting.   Some laboratories require emergency lighting only for safe egress of personnel from the laboratory. However, in other laboratories, the experiments performed are so critical that both emergency power and lighting are required for continued safe operation of the equipment.

Animal Room Lighting.   Many research laboratories have areas assigned for housing animals, popularly called *vivariums.* These areas can have stringent electrical and environmental control requirements that need careful planning of the systems. Some of the features generally provided are water-resistant lighting fixtures, automatic lighting controls from the building's automated system to create day/night ambient illusion, and local timer control switching of lighting fixtures to allow observation of animals by researchers.

CONTRIBUTORS

Joseph R. Flocco Jr., P.E., Consulting Engineer, Account Manager, Siemens Energy and Automation, Alpharetta, Georgia

A. Todd Rocco P.E., Principal, Diversified Consulting Engineers, Quincy, Massachusetts

Dennis G. Stafford, Account Manager, Drives; Siemens Energy and Automation, Alpharetta, Georgia

David L. Stymiest, P.E., SASHE, C.E.M., Senior Consultant, Smith Seckman Reid, Inc., New Orleans, Louisiana

REFERENCES

1. *Guide for the Care and Use of Laboratory Animals,* available on the World Wide Web at http://www.nap.edu/readingroom/books/labrats/index.html (National Research Council of the National Academy of Sciences, Washington, DC, 1996), used and referenced by *Association For Assessment and Accreditation of Laboratory Animal Care International,* Rockville, MD.
2. *The IESNA Lighting Handbook* (9th ed.), Illuminating Engineering Society of North America, New York, 1999.

SECTION 10.7
PROCEDURES FOR DECONTAMINATING AND DECOMMISSIONING LABORATORY BUILDINGS

Robert Edwards, Louis DiBerardinis, Richard Fink, Mitchell Galanek, and William VanSchalkwyk
Massachusetts Institute of Technology, Cambridge, Massachusetts

Kurt Samuelson
Beacon-Skanska USA, Boston, Massachusetts

Decontaminating and decommissioning laboratory buildings must be a cooperative effort between various disciplines that may involve architecture, engineering, safety, industrial hygiene, health physics, biosafety, facilities, construction management, and the occupants affected. Spaces scheduled for reoccupation, renovation, and/or demolition projects should be free from all physical, chemical, radioactive, and biological hazards. These hazards may arise from construction materials or from materials and equipment used inside the space. When cleared of these hazards, the laboratory should be posted, and work may begin.

The principal investigator (PI) or area supervisor and the department are responsible for the laboratory, including its hazardous materials. Before moving, the researcher should be required to leave behind a laboratory that is free from any equipment or surface contamination and hazardous materials. This insures that the new occupants and renovation crew have a safe working environment free from unknown hazards. This is one of the more difficult tasks. It requires a clear statement of responsibilities and continued vigilance as groups begin to move, particularly in the early stages. There needs to be a clear policy on who bears the cost of disposal. For some buildings, disposal costs are carried as part of the demolition budget, whereas for others the department or specific laboratory group must bear the cost. These costs could easily run into the tens of thousands of dollars.

Facilities should establish policies regarding which departments are responsible and ultimately are required to clean the area if laboratories are vacated without following proper waste disposal and decontamination procedures.

PLANNING

An architect, engineer, and/or coordinator should be chosen to plan and manage the decontamination process. These people's authority should be outlined before the start of decontamination or demolition. Their responsibilities should include:

• Researching the building records to identify any hazardous building materials such as asbestos and lead paint. Along with this, they should interview current and past occupants and researchers to identify any potential chemical, biological, radiation, or physical hazards associated with the laboratory spaces.

- Coordinating and scheduling all work associated with the project, such as working with regulatory agencies, permitting agencies, building owners, and contractors. Lockout and tagout issues should be addressed in the construction contract.

- Coordinating a group of environmental health and safety (EH&S) professionals to assist in planning the demolition. By knowing how the contractors perform their work, the EH&S group may be able to identify potential hazards that contractors may create.

- Coordinating surveys by the EH&S group to locate and identify hazardous materials that may impact worker health and safety, as well as the disposal of construction waste.

- Planning for the relocation and transportation of the occupants, equipment, and hazardous material.

Regular progress meetings and reports should be conducted and reported to the building owner, occupants, and EH&S team.

Part of the planning stage should include meeting with the occupants and contractors to outline requirements and responsibilities. During this phase, hazards are identified and mitigation techniques are specified. This will minimize any confusion as to who is accountable for each phase of the project. A representative of each group should review the scope of the work and scheduling. Procedures should also be developed for cleaning out laboratory and office spaces. Schedules and responsibilities should be made clear, including the decontamination, disposal, and transportation of hazardous material; a definition of when a space is sufficiently clean; and a protocol to address complaints.

SPACE CLEANOUT

Responsibilities should be clearly spelled out to all owners, occupants, and contractors involved. Their responsibilities include but should not be limited to the following:

Who supplies and removes rubbish containers?

Who removes building property?

Who removes occupant property?

Who decontaminates equipment?

Who checks the cost effectiveness of salvaging equipment?

Who packages and removes hazardous waste?

Who packages and transports hazardous material and sensitive equipment?

Who decommissions spaces, including a walkthrough with the EH&S team?

Who shuts down telephone, water, and electric utilities?

Survey of Building Materials

The building survey conducted by a knowledgeable and in some cases licensed person(s) should include chemical, radiological, biological, and safety hazards. Table 10.7-1 lists some of the relevant federal and state regulations that may apply to a laboratory renovation before decontamination and decommissioning.

Asbestos.   All work associated with asbestos must be done by state licensed or certified workers following approved methods.[1,2] All suspect materials should be sampled and analyzed for type and amount of asbestos present.[3] All state and federal laws must be followed during the removal and disposal of any asbestos material. Table 10.7-2 shows a partial list of possible asbestos-containing materials.

TABLE 10.7-1   Chemical Hazards[a]

Hazard	Governmental agency	Regulation	Comments
Asbestos	OSHA	29 CFR1926.1001	Worker protection
		29 CFR1926.1101	Work practices
	EPA	40 CFR part 61 NESHAPS	Asbestos in schools
		40 CFR part 763 AHERA	Waste disposal
			Accidental releases
	M. DLWD	453 CMR 6.00	Worker protection
			Work practices
Lead	OSHA	29 CFR1926.62	Worker protection
		29 CFR1910.1025	Work practices
	EPA	40 CFR part 141&142	Lead in drinking water
		40 CFR part 261	Waste disposal
		40 CFR part 50.12	Lead in ambient air
		40 CFR part 745	Training and disclosure
			Identification of dangerous levels of lead
	M. DLWD	454 CMR 22	Work practices
			Worker protection
	M. DPH	105 CMR 460	Lead poisoning prevention and control
Hazardous waste	EPA	40 CFR part 261,262	Accumulation, disposal
	M. DEP	310 CMR 30.00	Heavy metals
Unknown chemical	EPA	40 CFR part 261	Identify unknowns
Chemical containers	M. DEP	310 CMR 30.00	May present unforeseen hazards
Unknown gas cylinders	EPA	40 CFR part 261	Identify unknowns, disposal limited
	M. DEP	310 CMR 30.00	May present unforeseen hazard
Perchlorates	EPA	40 CFR PART 261	Reactive waste
	M. DEP	310 CMR 30.00	Reactive waste
PCBs	EPA	40 CFR parts 302.4&761	Disposal and reportable quantities
	M. DEP	310 CMR 30.00	Accumulation and disposal
Hazardous waste	EPA	40 CFR parts 261,262	Accumulation and disposal
	M. DEP	310 CMR 30.00	Accumulation and disposal
Drain traps	EPA	40 CFR 261	Mercury contamination
	M. DEP	310 CMR 30.00	Mercury contamination
	MWRA	360 CMR 10.0	Plumbing codes
Radiation	NRC	Title 10 CFR	Radiation standards
	DOT	Title 44 CFR	Transportation
	EPA	Title 40 CFR	Mixed waste
	M. DPH	105 CMR 120	Radiation standards
Biohazards	M. DPH	105 CMR 480	Waste disposal
	HHS	71 CFR part 71	Transportation of materials
	OSHA	29 CFR part 19910.1030	Bloodborne pathogens
	USPS	39 CFR part 124	Transportation of materials

[a] U.S. and Commonwealth of Massachusetts regulations are shown only for illustrative purposes. For other localities, local, state, and federal regulations must be researched and complied with.

TABLE 10.7-2 Sample List of Suspect Asbestos-Containing Materials

• Cement pipes	• Elevator brake shoes
• Cement wallboard	• HVAC duct insulation
• Cement siding	• Boiler insulation
• Asphalt floor tile	• Breeching insulation
• Vinyl floor tile	• Ductwork flexible fabric connections
• Vinyl sheet flooring	• Cooling towers
• Flooring backing	• Pipe insulation
• Construction mastics (floor tile, carpet, ceiling tile, etc.)	• Heating and electric ducts
• Acoustic plaster	• Electrical panel partitions
• Decorative plaster	• Electrical cloth
• Textured paints and coatings	• Electrical wiring insulation
• Ceiling tiles and lay-in panels	• Chalkboards
• Spray applied insulation	• Roofing shingles
• Blown-in insulation	• Roofing felt
• Fireproofing material	• Base flashing
• Taping compounds (thermal)	• Thermal paper
• Packing materials (for wall/floor penetrations)	• Fire doors
• High-temperature gaskets	• Caulking and putties
• Laboratory hoods/table tops	• Adhesives (especially flooring)
• Laboratory gloves	• Wallboard/vinyl wall coverings
• Fire blankets/curtains	• Joint compound
• Elevator equipment panels	• Spackling compounds

This list does not include every product or material that may contain asbestos. It is intended as a general guide to show which types of materials may contain asbestos.

Lead.   A state licensed inspector should survey the building to determine if there is any lead paint on building surfaces. If present, appropriate work practices need to be followed for worker protection during demolition.[4] The presence of lead in construction waste may affect the disposal of the material. To help in this determination, a toxic characteristic leaching procedure (TCLP)[5] may need be done.

Mercury.   Laboratory plumbing systems contaminated with mercury are not uncommon. Over the years, there have been many uses of mercury in laboratories. Because mercury is a liquid at room temperature, it can find its way into sink drains and collect in traps. It may also roll behind benches and under floors. Mercury has a very low vapor pressure, and once it becomes covered with dust or water, it does not evaporate until disturbed. To avoid mercury spills, care should be taken when disconnecting and removing laboratory sink traps and adjoining pipes. For demolition work, remove and inspect all traps before initiating demolition. Mercury can also be found in fluorescent lights, manometers, thermometers, heating thermostats, and electrical switches.[6] Care should be taken to avoid breaking any mercury-containing device when cleaning spaces.

PCBs.[7]   Reviewing old building records can help determine the presence of any electrical transformers that may contain PCBs. Fluorescent light ballasts used to contain PCBs. It may be possible to identify PCB-containing ballasts by using the date of manufacture and serial numbers. If they are not identifiable, it should be assumed that they contain PCBs. Smaller electrical equipment that contained oil for cooling or used oil as a lubricant may also need to be tested for PCBs.

Perchlorates.   Perchloric acid, a very strong oxidizer, may react with organic matter to form explosive products.[8] If perchloric acid was used in a laboratory hood that was not equipped with a washdown system, explosive perchlorates can collect within the hood and ductwork. To avoid mishaps, all hoods scheduled for dismantling should be tested for perchlorates. One test is the methylene blue qualitative test.[9]

Miscellaneous Items. A number of the following items should also be addressed when cleaning out a space before renovation or demolition:

1. The freon in window air conditioners must be reclaimed.
2. Incinerator ash typically contains heavy metals. A toxicity characteristic leaching procedure (TCLP) test will help determine the waste stream for the ash.
3. Many consumer products, including cleaning solvents, paints, paint thinners, oils, and pesticides, may need to be disposed of as hazardous wastes. Look to reuse these in other laboratories before disposal.
4. Sharp items need to be properly disposed of to avoid any cuts or puncture wounds.

DECONTAMINATION AND DECOMMISSIONING

If necessary, all items in the space should be decontaminated before moving or disposal.

Laboratory Chemicals

All chemical containers should be properly labeled and closed. Laboratory glassware should be emptied and cleaned before moving or disposal.

Refrigerators, autoclaves, and any contaminated equipment should be decontaminated following recommended guidelines before disposal or reuse. A label, as shown in Fig. 10.7-1, should be placed on the equipment. Chemicals stored in the cold need to be stabilized at room temperature before removal.

Laboratory bench tops and hood work surfaces should be washed with a solution of soap and water. For information on specific decontamination procedures, contact the health and safety office or the chemical manufacturer. Gas cylinders should be closed and capped before being returned or moved. Moving gas cylinders any distance should be done by experienced personnel. Try to identify the contents of gas cylinders, because disposal of unknown gas cylinders can cost thousands of dollars.

All laboratory drain traps should be removed and inspected for mercury and other hazards before any demolition work begins.

Considerations for Moving Regulated Biological Materials

Regulated biological materials include all genetically engineered microorganisms, recombinant plants and seeds, human and/or animal tissue, blood or body fluids,[10] tissue culture cells, and organisms of class 2 and higher. General move considerations are outlined here. Contact the biosafety office for additional information or assistance.

1. Plan to minimize the liquid volume and weight of transported materials. In addition, reduction of active materials should be planned the week before the move.
2. Provide new location information to the biosafety office (recombinant DNA regulatory requirements). In the United States, notify the USDA if you hold veterinary service or plant service permits. Permits are location specific. The move can trigger inspections by those agencies. Other rules apply elsewhere.

I certify that the cold/warm rooms and laboratory rooms at the following location have been emptied of biological materials.

_____

_____

The surfaces in these rooms have been decontaminated with: _____

_____

All sink traps have been bleached and flushed with water: use 1 cup of concentrated bleach, wait 20 minutes, and then flush thoroughly with water.

4. Please note location(s) of new lab area(s): _____

_____

_____   _____
Date   Signature of Principal Investigator

                                    _____
                                    Print Name

                                    _____
                                    New Lab Phone Number

FIGURE 10.7-1   Laboratory decontamination certification.

3. All biological materials must be packaged before the move. Proper packaging consists of a primary sealed container placed within a secondary, sealed, unbreakable container with enough absorbent material in between to contain and absorb any spill. Some examples are: petri dishes in a plastic sleeve within a plastic-lined box using paper towel spacers, sharps (items that can puncture or stab, such as syringe needles) in a sealed plastic container with a snap-on cover and with paper towels to cushion vials, sealed tubes in a rack placed into a plastic sealable container with enough paper towels to absorb any spilled contents, and tissue culture dishes placed into a plastic-lined dishpan or a sealable cardboard box with an absorbent. Freezers can be moved intact, provided all contents are in sealed, unbreakable containers and the freezer remains closed. Because contents may shift, enclose loose items in boxes or secure in some other way to avoid breakage and spills when the freezer is reopened. Other equipment, such as fermenters, refrigerators, incubators, and biosafety cabinets must be empty and decontaminated before the move. If there are any questions concerning packaging or movement of equipment or materials, contact the biosafety office.

4. Once packaged, all biological materials must be properly labeled. Labels must include name, principal investigator (PI), new location, identity of agent, biosafety level, telephone number for assistance in the event of any breakage, and a FRAGILE notice if applicable. Also use the universal biohazard label whenever packaging a biosafety agent of level 2 or higher.

5. A spill kit consisting of disinfectant, paper towels, plastic bags, tongs, dustpan and brush, laboratory coat, gloves, and safety glasses must be available during the move. Report all spills to the PI, the supervisor, and the biosafety office. Spill cleanup procedures and emergency plans should be established before working with or moving biological materials. For personal injury or direct contact involving biological materials, wash the affected area with soap and water and report to the appropriate medical facility.

6. Collect all packaging items needed before the move date. Carts, plastic bags, toweling and other cushioning absorbent materials, dishpans, boxes, labels, sturdy tape, and a spill kit should be readily available. Do not cut corners when packaging and labeling biological materials. Plan ahead.

7. Move only during the designated hours.

Biological Decontamination of Surfaces

When decontaminating against biological agents used in the laboratory, make sure that the decontaminant is effective against the specific agent. For example, if spore-forming bacteria were spilled, a phenolic-based disinfectant would be ineffective. One would have to use a 1:10 dilution of chlorine bleach with pH about 7 and a contact time of about 20 min. For vegetative bacteria, fungi, and many viruses, most of the common disinfectants such as 1:100 chlorine bleach (1:10 for spills), 1 percent organic iodine, 700 to 1000 ppm phenol, 70 percent ethanol or propanol, a quaternary ammonium compound, or 3 percent hydrogen peroxide are satisfactory. The key is correct concentration and contact time. Any commercially available product must be used in accordance with the manufacturer's instructions. All disinfectants take time to react.

To decontaminate a hard surface with no obvious spill, wear appropriate personal protective equipment (PPE), spray the surface with the disinfectant, and wipe with paper towels to coat the surface evenly and remove dirt. Let the disinfectant remain on the surface for 5 to 20 min (depending upon the organism present and the disinfectant), and then dry. Reapply the disinfectant if it evaporates too soon. Dispose of the used towels in an autoclave bag. Any surface or equipment that has been decontaminated should be labeled (see Fig. 10.7.1).

If a spill has occurred, again while wearing appropriate PPE, contain the spill with paper towels or an absorbent, being careful to avoid any broken glass or other sharp items. Cover the towels/absorbent with disinfectant, and let it sit for 5 to 20 min (see earlier). Pick up the towels/absorbent being careful to avoid any sharp materials, and place the towels/absorbent into a biohazard bag for autoclaving.

Unless high-level disinfection or sterilization is required, the use of formalin, glutaraldehyde, or peracetic acid is not recommended because of their toxicity and/or irritating vapors. If chemical sterilization is needed, >6 percent hydrogen peroxide or chlorine dioxide should be considered.

For a list of laboratory equipment and fixed structures that may need to be decontaminated, see Table 10.7-3.

TABLE 10.7-3   Laboratory Equipment That May Need to Be Decontaminated

Centrifuges
Fermenters
Microscopes
Gamma and beta counters
Hoods and biological safety cabinets
Incubators: water-jacketed incubators need to be drained and decontaminated by laboratory personnel.
   Remember to avoid algae growth and have water conditioners on hand upon refilling.
Vacuum pumps: oil must be drained.
Freezers: pack tightly to avoid internal breakage, decontaminate outside, and lock.
Refrigerators: empty and decontaminate inside and outside.
Radioactive waste containers
Radiation shielding
Radioactivity handling tools

Decommissioning and/or Moving Radiation Laboratories

Anyone using radioactive materials in the United States requires either federal U.S. Nuclear Regulatory Commission (NRC) or individual state licensing. Moving or decommissioning a laboratory may require prior approval of the regulators via an amendment to the license. This process is time consuming and, if not carefully planned, will delay the actual move schedule. In addition, if the laboratory is to be moved to an entirely new facility, the new facility will need appropriate licensing and permitting. Never ship radioactive materials to another licensee before obtaining a copy of the license to possess these materials. A record of the license and material shipped must be maintained in accordance with the NRC or sate regulations.

Note that some radioactive materials do not require the user to have a license or permit. Two examples of unlicensed sources are natural radioactive uranium and thorium. Many chemistry, biology, and materials science laboratories use these materials routinely without the knowledge of the radiation safety office. To avoid inadvertently mixing radioactive waste with chemical waste, these laboratories should be identified.

Decontamination/Move Preparation

All equipment apparatus and fixed structures must be decontaminated if necessary. Whatever equipment and materials are to be moved must be properly packaged. See Table 10.7-3 for a list of apparatus that may require attention. An outline of decommissioning guidelines is presented later.

When any apparatus and equipment is contaminated with radioactive material, it is identified and labeled. If radioactive contamination is not removable, the laboratory may choose to move the material in an as-is condition and not decontaminate to release-level criteria. See the following instruction regarding surveying and preparing facilities and equipment for unrestricted use.

> If radioactive contamination is not removable, the laboratory may choose to move the material in an as-is condition and not decontaminate to release level criteria. Contact your health and safety office regarding surveying and preparing facilities and equipment for unrestricted use.

Moves within facilities may be expedited if radioactive materials are packaged in refrigerators/freezers. The same basic principles described earlier for biological materials apply to moving radioactive material in freezers/refrigerators. Before actually moving a refrigerator/freezer containing radioactive materials, the radiation safety office should make a final inspection of the packaging and secure the individual packages or freezers as a whole to move them to their new location. Once secured by the radiation safety office, the freezers/refrigerators should not be opened until they arrive at their new location. Moving should take place only during designated hours and should be coordinated between the movers and safety personnel. If the support of your radiation safety office is available, it is advisable to let the radiation safety office move radioactive sources within a facility.

It is advisable to have Equipment Decontamination Record stickers available (see Fig. 10.7-2). Affix this sticker to all equipment before moving. Also, place stickers on any equipment that will be discarded or placed in storage. Moving personnel should not move equipment that lacks the decontamination sticker.

An inspection should be done by the EH&S group after the move to determine if the laboratory has been properly decontaminated. If it has, a notice should be completed and posted at the entrance to the space (see Fig. 10.7-3).

Moving Phase

This phase involves moving radioactive materials and equipment to a new laboratory facility or to a holding area until the laboratory renovation is completed. Other components of the

```
Principal Investigator _____

name _____ phone _____ date _____

This piece of equipment was used with the following:

[ ] No hazardous materials   [ ] Biologicals

[ ] Chemical          [ ] PCBs surveyed by IHO
_____

              initials _____     date _____

[ ] Radiation         [ ] Surveyed by RPO _____

              initials _____     date _____

[ ] Other hazards (specify)
_____

Decontaminated with _____

By (name) _____     Date _____

Equipment OK for removal or reuse:     _____ YES

                                       _____ NO

REMOVE THIS LABEL BEFORE REUSING EQUIPMENT
```

FIGURE 10.7-2   Equipment decontamination record.

move not covered in the move preparation phase may include transporting materials and unpacking.

For radioactive materials, it is important that all handlers be properly trained. Training includes potential health hazards and handling techniques. It is also important to establish emergency procedures. Emergencies may include material spills, fires, slips and falls, and cuts. Protective clothing and spill-absorbing material should be readily available during packing, moving, and unpacking. Ample fire extinguishers should be available, and personnel should be trained appropriately. Labeling is of prime importance. Each container or piece of equipment must have adequate labels identifying the agent, potential hazards, and precautions.

Radioactive materials require appropriate record-keeping and packaging in accordance with applicable NRC, state, and DOT regulations. Within your own facility or institution, this packaging and moving should be coordinated with the radiation safety professionals. A complete inventory of radionuclides, amounts, and storage requirements (i.e., room temp, 0 to 70°C) must be provided. Materials to be sent to another facility require proper packaging and permission of the receiving facility along with a copy of the appropriate license or permit.

Construction/Renovation Phase

The EH&S group should conduct a historical review of the space by looking at records and interviewing researchers to gather information on radioactivity used, where it was used, ventilation requirements, and contamination incidents.

Once this is done, a thorough radiological survey should be done by the radiation safety office to identify any areas that need to be decontaminated. Radiation safety or laboratory personnel should do the decontaminating.

A radiation survey should be done after the decontamination process to confirm that the equipment or surface qualifies for unrestricted use. A record of this should be maintained indefinitely.

Room #(s) _____ have been surveyed and visually
inspected for the following contaminants:

Hazardous materials	[ ] NA	[ ] cleared	[ ] decon. required (see comments)

Surveyor's name _____   Department _____   Date _____

Chemical	[ ] NA	[ ] cleared	[ ] decon. required (see comments)

Surveyor's name _____   Department _____   Date _____

Radiation	[ ] NA	[ ] cleared	[ ] decon. required (see comments)

Surveyor's name _____   Department _____   Date _____

Biological	[ ] NA	[ ] cleared	[ ] decon. required (see comments)

Surveyor's name _____   Department _____   Date _____

Fume Hoods	[ ] NA	[ ] cleared	[ ] decon. required (see comments)

Surveyor's name _____   Department _____   Date _____

Drain traps	[ ] NA	[ ] cleared	[ ] decon. required (see comments)

Surveyor's name _____   Department _____   Date _____

Sharps	[ ] NA	[ ] cleared	[ ] decon. required (see comments)

Surveyor's name _____   Department _____   Date _____

PCBs (with the exception of ballasts)	[ ] NA	[ ] cleared	[ ] decon. required (see comments)

Surveyor's name _____   Department _____   Date _____

Comments _____

_____

_____

[ ] Ready for clean-out, abatement, and demolition

FIGURE 10.7-3   Room decommissioning survey.

Decommissioning Radiation Laboratories

The goal of the decommissioning process is to ensure that intermediate (movers, construction workers) and future users of the laboratory are not exposed to unacceptable levels of radiation or radioactive contamination.

The following guidelines for laboratory surveys, recommendations on radiation survey instruments, and Nuclear Regulatory Commission (NRC) Acceptable Surface Contamination Levels may be used to prepare radiation laboratories for release for unrestricted use.

Facility Characterization: Document and History Review.

1. Obtain a description of the types and locations of experiments performed using radioactive material.
2. Obtain a list of the specific radionuclides used and the amounts stored and used in the laboratory.

3. Obtain information about the use of the ventilation system for volatile radioactive materials and the sanitary sewage systems for the disposal of low-level radioactive waste.

4. Collect information about the history of radiation surveys and any incidents or spills that resulted in facility contamination.

This information should be readily available from the facility radiation safety officer. Interviews with the laboratory occupants may also provide valuable information.

Radiation Surveys. Radiation surveys must be performed to determine the extent of radiological characteristics. They include the exposure rate or ambient radiation levels and the surface activity whether fixed radioactive contamination or removable radioactivity contamination levels.
Exposure Rate.

1. Federal or state licensing agencies determine guideline values for radiation exposure.

2. Determine background radiation levels in μR/hr. Choose an on-site building that has similar construction but no history of using radioactive materials.

3. Measure exposure rate levels at 1 m from the floor/lower wall surfaces. Measurements may be averaged over a 10-m^2 area (the approximate size of a small office).

4. The maximum exposure rate should not exceed two times the guideline values above background.

Surface Activity. Fixed contamination:

1. An effort should be made to survey all suspect and contaminated surfaces and equipment for fixed contamination, including bench tops, drawers, floors, hood bases, hood walls, ductwork, sink basins, and sink drain traps.

2. Contaminated areas should be cleaned to the fullest extent possible.

3. Results should be averaged over a 100-cm^2 area and should not exceed guideline values.

4. Do not cover contaminated areas with paint, cement, or other forms of covering to reduce contamination levels to guideline levels.

Removable contamination:

1. Grid the laboratory into squares whose sides are approximately 1 m.

2. Perform swipe tests for removable contamination by rubbing a 1 to 2-in. filter disc over approximately 100 cm^2 of area to be tested. A 47-mm glass fiber filter works well in a liquid scintillation counter.

3. Analyze swipe tests in an appropriate analyzer for the radionuclides of interest.

4. All surfaces and equipment should be decontaminated of removable contamination that is above guideline values. With reasonable effort, background levels may be achieved.

Note that it is important to survey the insides of hoods and ductwork thoroughly if volatile radioactive materials were handled. Likewise, it is important to thoroughly survey sink basin and sink drain traps for contamination. Some laboratories contain individual waste neutralization "chip" tanks at each sink drain. The insides of these tanks or the calcium carbonate chips may be a source of radioactive contamination and should be tested.
A detailed map showing the test locations and a description of the way the tests were performed should be kept with the results of the radiation surveys.

Impact on Adjoining Spaces

The health and comfort of occupants of neighboring offices and buildings can be greatly impacted during any construction or demolition project.[11] Meetings should be held to inform neighbors of the project and steps that will be taken to ensure their health and comfort.

1. Monitoring should be done to document exposure levels and monitor work practices during asbestos or lead abatement.
2. Provisions should be made to reduce the amount of construction dust and odors from being brought into the surrounding building through supply ventilation systems. Monitoring can be done to document this.
3. Noise monitoring may need to be done to ensure that noise levels will not cause any permanent hearing loss.
4. Set up a reporting system where complaints can be made, investigated, and acted on.

REFERENCES

1. OSHA, Occupational Health and Safety Standards (29 CFR1915.1001), "General Industry Standard," revised July 1998.
2. OSHA (29 CFR1926.1101), "Safety and Health Regulations for Construction," revised July 1998.
3. EPA (40 CFR763), "Asbestos Hazard Emergency Response Act," revised July 1998.
4. OSHA (29 CFR1926.62), "Safety and Health Regulations for Construction," revised July 1998; (29 CFR1910.1025), Safety and Health Standards, revised July 1998.
5. EPA publication SW-846, "Test Method for Evaluating Solid Waste, Physical/Chemical Methods," updated December 1996.
6. National Association of Demolition Contractors, *Recommended Management Practices for the Removal of Hazardous Materials from Buildings Prior to Demolition*, Doylestown, PA, 1998.
7. EPA (40 CFR part 302), "Designation, Reportable Quantities and Notification," revised July 1998; (40 CFR part 761), "Polychlorinated Biphenyls (PCBs) Manufacturing, Processing, Distribution in Commerce and Use Prohibitions," revised July 1998.
8. EPA (40 CFR part 261), "Identification and Listing of Hazardous Waste," revised July 1998; (40 CFR part 30), "Grants and Agreements with Institutions of Higher Education, Hospitals and other Non-Profit Organizations," revised July 1998.
9. F. D. Snell and C. T. Snell, *Colorimetric Methods of Analysis*. Van Nostrand, Princeton, NJ, 1949, pp. 718–721.
10. OSHA (29 CFR part 1910.1030), "Blood-Borne Pathogens," revised July 1998.
11. Sheet Metal and Air Conditioning Contractors' National Association, Inc., *IAQ Guidelines For Occupied Buildings Under Construction*, Chantilly, VA, 1995.

CHAPTER 11
INDUSTRIAL AND MANUFACTURING FACILITIES

Gregory A. Schmellick
Symmes Maini & McKee Associates, Cambridge, Massachusetts

**Michael K. Powers, P.E.; Leslie A. Glynn, A.I.A.;
Cathleen M. Ronan, A.I.A.; Heather C. McCormack;
Brian W. Lawlor, P.E.; John A. Stevermer, A.I.A.;
Michael J. Reilly; Patrick E. Halm, P.E.;
Michael J. Sweeney, P.E.; Kevin Tunsley;
Peter S. Glick; and George D. Combes, P.E.**
Symmes Maini & McKee Associates, Cambridge, Massachusetts

SECTION 11.1
INTRODUCTION

WHY ARE INDUSTRIAL AND MANUFACTURING FACILITIES UNIQUE?

In this chapter, we examine the unique aspects of planning, programming, design, and construction of advanced technology facilities, including facilities for the semiconductor, microelectronics, biopharmaceutical, biomedical, and pharmaceutical industries. Advanced technology facilities are unique because they are designed from the "inside out" and they require intensive environmental systems to support the manufacturing processes housed within (see Fig. 11.1-1).

A "time-line" approach is used to assist owners and facility managers in understanding when concurrent information, decisions, activities, and tasks are ongoing among the architectural and engineering (A/E) disciplines. Milestone phases of the time line are concept or schematic design, design development, construction documents, bidding, construction administration, and commissioning. Subsequent sections discuss architectural and structural design, mechanical systems design, chemical process design, and bidding and construction approaches for advanced technology facilities. Business needs often drive a "fast-track" or

FIGURE 11.1-1   Exterior view of Intersil Semiconductor, Mountaintop, Pennsylvania. (*Courtesy of Symmes Maini & McKee Associates.*)

"time-to-market" design and construction approach. Such approaches require specific protocols for understanding the evolving facility's program, while addressing critical design and construction needs, such as very early foundation, structure, facade, and roof construction documents packages.

Often, specialized physical environments such as clean or sterile suites, exhaust/toxics discharge systems, or liquid/solid waste containment and handling systems are required. These specialized requirements elevate the cost per square foot far above normal building cost models. Such programs often require a high degree of interstitial or support spaces.

Specialized approaches are necessary for the site, architectural, structural, and mechanical designs to achieve efficient and cost-effective advanced technology facilities. The site and architectural disciplines normally initiate the design process through a master planning phase with the owner to bracket the project's scope and establish criteria that become guidelines for the other design decisions. This up-front planning includes input from all design disciplines, and the outcome is a plan for a building on a site that evaluates and incorporates the owner's immediate and future needs and corporate goals.

Advanced technology facilities focus on larger structural bay spacing or long clear spans to maximize the density of production equipment. Often, the manufacturing need is supported by ancillary spaces for environmental and process support equipment above and below the actual manufacturing floor levels. Vibration mitigation is often a key concern, particularly in the semiconductor and flat panel display market sector. The need to manufacture materials of less than one micron in width, while producing a high volume of quality yields, means that structures require sophisticated isolation and stiffening/mass designs.

Recent studies show that the use of light and interior design elements, sometimes referred to as *humanization,* can result in higher productivity and employee retention. In times of low unemployment and a reduction in the skilled labor force, this has become a growing program theme in advanced technology environments that can easily become dehumanizing amidst the highly controlled equipment environments.

Lastly, advanced technology facilities are beginning to promote themselves as marketing tools. Customers can watch manufacturing from carefully isolated corridors, introducing them to the quality precision and efficiency of the manufacturing process.

TEAM APPROACH AMONG OWNER, ARCHITECT/ENGINEER, AND CONSTRUCTOR

The complexities and the pace of the programming, design, and construction process make it imperative that a team partnership be developed to plan and commit to the entire project delivery approach. This involves planning that maximizes overlap of design and construction by procuring individual elements as the design progresses. The architect/engineer (A/E) and constructor become team players, sharing scheduling and cost estimating responsibilities very early in the schematic design phase, while carefully planning a well-conceived delivery system with proper review and approvals by the owner.

The importance of team compatibility cannot be overstated. The project pace requires that the owner, A/E, and constructor work in harmony and have an effective approach to project and crisis management. The inside out programming and design approach inherent in advanced technology projects may dictate early decisions on space and form that create spatial risk if the problem cannot be identified and resolved in a timely manner. Project manager interface is imperative in defining and controlling the design process as it unfolds. Frequent progress scheduling and team meetings document the process to allow for recorded and timely communication among all team members.

THE OWNER IS AN IMPORTANT MEMBER OF THE DESIGN TEAM

An owner who understands the building process can provide unique efficiencies. Recognizing the A/E as a professional services provider and the constructor as a knowledgeable procurement/construction specialist is the first step. Owners can seriously restrict a potential return on the investment of each construction dollar spent by being too conservative in identifying scope and fees to support the execution of the appropriate design and construction approach. Knowledge of the value management process can optimize value.

The owner's role is critical in providing assistance and participating in the programming and data collection phases. Early definition of criteria leads to efficiency and ultimately results in higher-quality services from the A/E team. An understanding of the construction market with respect to delivery times and vendor/subcontractor quality is also beneficial.

UNDERSTANDING THE IMPORTANCE OF DECISION MAKING IN A TIME-TO-MARKET FRAMEWORK

The speed of delivering the manufactured product drives the ability of firms to optimize their businesses' profit margins before similar products from competitors reach the market. Significant pressure is often present to speed up the design and construction process. Decisiveness is critical and risks must be understood. Scheduling and cost modeling become the primary management and control tools of the owner, A/E, and constructor team. Well-conceived work plans for the management process are essential to the successful execution of work.

GUIDE TO OBTAINING A/E DESIGN SERVICES

It is critical to recognize the A/E design component as a professional services provider and as such, a partner in the process with the owner. The degree of attention paid to the execution of the design concept through construction document phases will provide an added value benefit to the owner. Section 11.5, "Bidding and Construction Approach," provides insight into the way the members of the project team can integrate their respective skills and experiences into one team focused on a successful project. Successful projects are created by quality design and construction teams. With this in mind, owners should select design teams based on the following criteria:

- Team member expertise
- Firm qualifications
- References of both the firm and the individuals
- Project management approach
- Design approach
- Quality control process
- Chemistry
- Construction knowledge

SECTION 11.2
ARCHITECTURAL AND STRUCTURAL DESIGN

PLANNING AND PROGRAMMING

Advanced technology design and construction are participatory processes. As each issue is examined in this section, a brief generic synopsis is provided, followed by a more in-depth discussion of aspects or applications unique to biotechnology/pharmaceutical and semiconductor/microelectronics facilities. Discussions are generally applicable to both new construction and renovation work, and the differences are highlighted.

Successful facility design and construction require a large, diverse team that can meet the technical challenges unique to the industry in a time-to-market framework. Examining the impact of the following issues is fundamental:

- Effect that time-to-market framework has on the design and construction process
- Importance of the team

- Need to set goals collectively
- Specialized environments required

Historically, design and construction have consisted of a linear approach in which the entire project is first permitted and designed, then procured, and finally constructed. The owner's input and decisions are made in a similar linear fashion before the start of each successive phase.

In the advanced technology industries, the design and construction of new and renovated facilities are typically done by a "fast-track" method to minimize the time to market. The fast-track approach requires that portions of the project be designed, procured, and often built before the whole design is complete. To be effective, the owner's input and decisions must be made before all information is known. The owner must be able to devote the company's resources (financial, personnel, time, and effort) at the outset of the project and throughout the design and construction process. The owner must understand the additional risks and their associated costs and be prepared to incur premium design and construction costs to expedite procurement and accelerate construction.

Setting Goals

In the advanced technology industry, the complexity of the design, the diversity and size of the team, and the time-to-market delivery requirements make it imperative to identify project, operational, and technical goals at the outset of the project. Project goals address the *process*—"how will we achieve our goals?" Operational and technical goals address the *resulting product*—"what will our goals achieve?"

Project Goals

- Who is on the team? How can the team best achieve decision making, cost-effective design, and construction in a time-to-market framework?
- How much is an appropriate budget? The operational and technical requirements critical to production of the project drive the budget.
- When is the Ready for Equipment (RFE) date? Is there one RFE date for the entire project or several for different portions of the facility?
- How does the RFE date impact phasing for design and construction? Is there a cost premium for achieving this date?

Operational Goals

- What is the primary purpose of the facility (R&D, pilot production, full manufacturing, etc.)?
- How clean do the operations in the facility need to be?
- What types of chemicals are required for the process?
- What types of by-products will the process produce?
- What level of system redundancy is required, desired, and/or affordable?
- What operational safety needs must be achieved?

Technical Goals

- How tight is the tolerance for temperature and humidity?
- What range of vibration is acceptable?
- What level of purity is required for each process system?

The Importance of the TEAM

The most important decision an owner/facility manager makes on an advanced technology project is selecting the team. Who makes up the "typical" team? The team consists of the owner's team, the A/E's team, the constructor's team, and the government agency's team.

The Owner's Team.   The owner's team consists of key individuals within the organization who represent the various operational and maintenance aspects of the facility (e.g., development, manufacturing, tooling, process, safety, HVAC, and electrical). The owner's team may also include the insurance underwriter and representatives of testing/inspection agencies.

Owners often find that the best people for the team are also the key individuals who operate and/or maintain the facilities, but these individuals often find that the time required for the project conflicts with their day-to-day operational and managerial responsibilities. An organization may have a separate facility design and construction department that can assist the key individuals in fulfilling both commitments. In organizations that do not have separate facility design and construction departments, the owner may assist the team by using the following techniques:

- Designating a dedicated project manager, and/or leveraging the key individuals. A dedicated project manager would have no other operational or management responsibilities within the organization. The individual may come from within the organization or be hired as a consultant. The project manager acts as a facilitator, assisting the team in gathering critical information and expediting time-sensitive management decisions.

- Leveraging key individuals by creating two- or three-person teams that are familiar with the status and direction of the project throughout the process. This serves to distribute the time commitment among several individuals. These subteams serve as conduits for information and decision making, but to be successful this requires effective communication among the subteams.

- Investing in team-building strategies. People working toward common goals are usually more effective and work more efficiently.

The Architect/Engineer's (A/E's) Team.   The A/E's team includes the disciplines necessary to design the project and may include individuals who specialize in site/civil, architectural, structural, plumbing, HVAC, fire protection, process, electrical, geotechnical, and vibration design.

Retaining the A/E early in the process can assist the owner in assessing potential sites, establishing baseline height and areas, evaluating the expansion potential of systems in existing facilities, and preparing early cost models. Many early project decisions require completing feasibility studies before the project can progress.

Once the owner's and A/E's teams are selected, the key individuals from the owner's team are typically "partnered" with the corresponding A/E counterparts. This allows tailoring systems design to meet the unique requirements of the facility. Because of the concentration of engineering systems required in advanced technology manufacturing, the assignment of architectural and site design counterparts may be overlooked initially. However, these disciplines are essential in driving the overall layout, design, and permitting process.

The Constructor's Team.   The constructor's team consists of the construction manager (CM), one or more general contractors, subcontractors, and major equipment suppliers. The constructor's team will typically consist of a project manager, and based on the scope of work involved, one or more assistant project managers, site superintendents, expeditors, clerks, and so forth, as needed to keep both the procurement and construction work on schedule, within budget, and in conformance with the drawings and specifications.

The owner should retain the services of the constructor as soon as possible, so that the owner, A/E, and constructor can begin to work together as a team. The constructor's team should be retained at or before the start of the A/E's design development phase. Retaining the constructor early benefits the owner in several ways:

- The team dynamics and overall design and procurement scheme are established earlier in the project, reducing the potential for crisis management by anticipating project needs before they become critical. The faster the project pace, the more important the project team planning and synergy become. Significant increases in project costs can result when project planning deteriorates into crisis management.

- The amount and accuracy of the information available to the owner is increased, thus potentially reducing the risks to the owner when timely decisions are required.

- Value management can be incorporated during the design so that the owner may realize cost savings without reducing design benefits. Value engineering conducted during construction to reduce cost often results in reducing quality and design benefits and increasing design coordination issues.

The Government Agency Team.   The government agency team includes environmental and zoning review boards, building code officials, and other regulatory agencies. The government agency team should be considered an essential component of the overall team. This is especially true with semiconductor, biotechnology, and pharmaceutical projects where government agency review time and approval can seriously impact the overall project schedule. A relationship with the various review agencies needs to be established early and continued throughout the design and construction process.

THE TEAM—TEAM-BUILDING STRATEGIES

Owners understand how challenging it may be to foster team synergy within their own organizations. Design and construction teams are naturally large and diverse, and for a successful project, the individual teams must form a synergy to become the team. Therefore, team building is one of the most important investments an owner makes.

One method of team building that has been used successfully in many applications is *partnering*. Partnering is a process that examines group dynamics and techniques to improve communications in large, diverse teams to assist the team in establishing and achieving common project goals. It is recommended that this process occur each time a new entity is added to the team, for example, initially with the owner's team; then with the owner's and the A/E's team; then with the owner's, A/E's, and the constructor's team; and then with the owner's, the A/E's, the constructor's, and the subcontractors' teams.

How well the team functions determines to a large extent how successful a project will be. All team members need to focus on common goals and objectives and be sensitive to each other's needs and expectations. Although this concept is not unique to the advanced technology industry, it is more crucial in a time-to-market scenario of a semiconductor or biotechnology project where the tasks of the individual team members are highly interdependent.

Programming—General

Programming is both (1) a conversation between the owner and the A/E to identify and understand the owner's needs and goals and (2) a systematic approach with which to document the project's needs and goals. Programming is especially critical in the design of complex facilities typical of the advanced technology industries. This process can be done at the building level, as well as at a site master planning level. Programming gives the A/E a better understanding of the owner's operations/processes; determines the type and size of functional areas needed; establishes adjacency requirements; and identifies the engineering design parameters, such as temperature, humidity, lighting levels, vibration characteristics, process systems, special construction, cleanliness classifications, and safety protocols that are needed to support the intended processes and define "program drivers." When the owner already has program or facility standards established, these serve as the basis from which to begin or verify programmatic requirements.

In advanced technology design, it is important first to identify potential code and regulatory issues and understand their implications. In addition to the zoning and environmental regulations, state and local codes, and insurance underwriter requirements, funding source specifications and other regulatory agency guidelines have been established to protect personnel, minimize shutdowns and property loss, and set guidelines for the processes carried out in these facilities. Understanding their benefits to and impacts on the facility early in the design process will result in a more safely designed facility that successfully supports the owner's manufacturing processes.

The three model building codes used in the United States are the Uniform Building Code (UBC), most often used in the western part of the United States; the Southern Building Code (SBCCI), primarily used in the Southeast; and the Building Officials & Code Administrators International Code (BOCA), which is typical in the Northeast. States may adopt one or more of the model building codes, base their code on a modified version of a model code, or draft a unique code. Refer to Chapter 8, "Codes and Standards," for a detailed treatment of these issues. See Table 11.2-1 for representative advanced technology codes and standards.

The implications and impacts of regulations and/or guidelines such as the National Fire Protection Association (NFPA), Factory Mutual (FM), or Industrial Risk Insurers (IRI) underwriting agencies, as well as the owner's individual standard operating procedures, must be identified, understood, and considered early in the design process. Biotechnology and pharmaceutical projects must also consider Current Good Manufacturing Practices (CGMP), Food and Drug Administration (FDA) guidelines, Good Lab Practices (GLP), and Good Clinical Practice (GCP). See Table 11.2-2 for FDA applicable codes and standards.

Whether the project is a new facility, a facility addition, a renovation, or a combination, performing a building code analysis is essential to understanding how the applicable codes and regulations will shape overall design and operation. A building code analysis will establish appropriate use group classification(s), set height and area limitations, determine types

TABLE 11.2-1   Representative Advanced Technology Codes and Standards

1. Building Code, latest edition in the state that has jurisdiction
2. Fire Prevention Codes, latest edition in the state that has jurisdiction, if applicable
3. Other Codes and Regulations referenced by the Building Code
4. Factory Mutual (FM) Loss Prevention Data Sheet 1-56, Clean Rooms
5. Factory Mutual (FM) Loss Prevention Data Sheet 7-7, Semiconductor Fabrication Facilities
6. Industrial Risk Insurer's (IRI) IM 17.1.1, Guiding Principles for the Protection of Semiconductor Manufacturing Facilities
7. Industrial Risk Insurer's (IRI) IM 17.11, Clean Rooms
8. National Fire Protection Association NFPA 30, Flammable and Combustible Liquids Code, 1996 Edition
9. National Fire Protection Association NFPA 318, Standard Protection for Clean Rooms, 1995 Edition
10. Food and Drug Administration (FDA)
11. Good Manufacturing Practice (GMP)
12. Good Lab Practice (GLP)
13. Good Clinical Practice (GCP)

TABLE 11.2-2 Food and Drug Administration Applicable Codes and Standards

• Parts 1–99:	General
• Parts 100–169:	Food for Human Consumption
• Parts 170–199:	Food for Human Consumption
• Parts 200–299:	Drugs: General
• Parts 300–499:	Drugs for Human Use
• Parts 500–599:	Animal Drugs, Feeds, and Related Products
• Parts 600–799:	Biologics and Cosmetics
• Parts 800–1299:	Medical Devices, Radiological Health, and Other Acts
• Parts 1300–End:	Drug Enforcement Administration

and characteristics of construction required, influence internal building organization, and identify necessary safety features.

Because advanced technology facilities tend to use and/or produce hazardous materials, identifying and analyzing the nature, extent, and quantity of the hazardous material(s) is of prime importance in determining the appropriate use group classification(s). The owner's understanding of the nature, extent, and quantities of the materials used in the manufacturing process is essential. The A/E will work with the owner's team to determine the aspects of the handling, storage, dispensing, and transportation of hazardous materials that have critical impacts on facility design. The results will indicate whether all or parts of the facility need to be classified as hazardous, including hazardous production materials (HMP) and biohazardous material, factory/industrial, business, or other use group. The use group classification in combination with the proposed type of construction and local zoning requirements (see site evaluation for a discussion of these requirements) will set the maximum height and area allowed for the facility.

The greater the hazard involved, the more the height and area of the facility will be restricted. However, this can be offset in part by the type of construction and the safety features utilized. The model codes define five "type of construction" classifications, designated as Type 1 (1A and 1B), Type 2 (2A, 2B, and 2C), Type 3 (3A and 3B), Type 4, and Type 5 (5A and 5B). The type of construction defines the rating, if any, required for the structural components of the building; the exterior wall construction/rating; the size of exterior openings; ratings for internal building components such as stair enclosures, shafts, corridor walls, floor and roof ratings; and the frequency, extent, and rating of internal subdivisions. Only Types 1 and 2 are typically used in manufacturing and industrial facilities. In older, academic, and/or in some laboratory facilities, Types 3 and 4 construction may be encountered but are not typical.

Manufacturing process requirements typically drive maximization of the building's height and area. Increases in the facility height and area are generally allowed when safety features such as automatic fire suppression and alarm systems and increased fire department access around the perimeter are incorporated into the design. Internal subdivisions such as firewalls can also increase heights and/or areas by in effect creating "separate buildings," each having its own height and area limitations. However, because this creates potential operational issues, the use of the "separate building" approach should be considered carefully.

Facility additions can often be separated from existing facilities with firewalls to minimize requirements for upgrading the entire facility up to current code or when an addition would cause the building to exceed its permitted height and area. In new construction, this approach can be effective when used to segregate mixed uses, such as business and HMP, to allow each use to have its maximum height and area.

Once the appropriate use group classification, height and area, and type of construction have been determined, the implications and potential issues due to the nature, extent, and quantities of the hazardous materials must be identified, understood, and incorporated into the design. The three model codes allow using or storing certain exempt amounts of hazardous material within the facility before the building or portion of the building containing the materials must be classified as a hazardous or hazardous production materials (HPM) use group. Research and development (R&D) facilities, academic laboratories, pilot operations, or other operations that may use quantities of hazardous materials that are exempt by code are often classified in one of the nonhazardous use group classifications.

Pharmaceutical manufacturing is typically included in the factory/industrial use group. This use group has low and moderate hazard classifications, depending on the quantities of combustible materials. The amount of hazardous materials is determined by the same tables as for hazardous and HPM facilities. The type, quantity/density, and delivery methods for each hazardous material establish this. The quantity and density that are exempted for each type of hazardous material are based on the nature of the hazard posed, which is typically categorized by NFPA 704 with a health, flammability, and/or reactivity rating.

The quantity of hazardous materials in other types of advanced technology facilities such as semiconductor facilities routinely exceeds the code exempt amounts. The following is a brief code synopsis, using a generic semiconductor facility to highlight the potential code issues encountered in the design. Note that code analysis is unique to each facility. The following illustrates the issues encountered but is not a comprehensive checklist.

Because of the nature of the materials used in the microelectronics and semiconductor industries, these facilities typically fall under a subsection of the hazardous use group, hazardous production materials (HPM). HPM by definition is "a solid, liquid, or gas that has a degree of hazard rating in health, flammability or reactivity of Class 3 or Class 4 as defined by NFPA 704 . . . which is used directly in research, laboratory, or production processes which have as their end product materials which are not hazardous." An HPM facility consists of fabrication areas (which can include ancillary rooms, dressing areas, and offices), HPM storage rooms, offices outside fabrication areas, and mechanical, electrical, and process support spaces.

Programmatic Cautions

1. Plan flexibility for expansion or change. Don't landlock spaces that are most likely to expand. Does this include the support spaces like staging areas/wipe in place (WIP), staff locker rooms, gowning rooms, parking, etc.?
2. Plan for future expansions. Future growth, horizontal or vertical, should not impede the building or site circulation. Is there space for utilities to expand?

THE OWNER AND A/E ESTABLISH FDA REQUIREMENTS, CLEANLINESS, BIOHAZARD LEVELS, TOXICITY, AND NFPA HAZARD CLASSIFICATIONS

Based on the manufacturing processes and regulatory criteria, the owner establishes the standards that the facility must meet with respect to FDA regulations, clean room classifications, biohazard levels, toxic levels, and so forth. The owner and A/E determine the most appropriate way to meet the standards. Decisions are usually based on safety, cost, flexibility, redundancy, labor, and maintenance.

The A/E will develop traffic patterns for:

- People (staff, manufacturing and administrative, visitors)
- Material (raw, waste, rejected, quarantined, hazardous, dispensing, WIP)
- Product (finished, rejected, awaiting quality control)

A/E Issues a Program Document That Incorporates the Decisions Made to Date

The *program document* is the foundation of subsequent design phases. It records decisions made to date and organizes the data into a format that the A/E can use to make design decisions. In pharmaceutical facilities, in addition to standard program functions, the program document must record GMP compliance; diagrammatically establish people, product, and material flows; and establish the clean and hazard ratings of spaces.

Develop Containment Strategies.   Many containment or separation methods that the process may require (such as clean or aseptic environments) are extremely expensive to engineer and install. Clustering of like environments together in a unified zone allows for better operation and maintenance of the facility and more reasonable construction and operating costs.

Resolve Building Code Issues and Requirements.   The design must meet all building code and regulatory requirements necessitated by the current process and should forecast future needs. For instance, if a process currently does not require dispensing flammable chemicals, but there are plans to dispense on-site, then such spaces should be planned and located accordingly. Dispensing areas may require explosion relief panels that are more difficult to locate subsequently.

Strategies to Humanize the Environment.   Pharmaceutical facilities revolve around process needs. OSHA and other worker safety agencies enforce minimum safety standards. Adding windows to the outside or to an adjacent space is an inexpensive design option that can make a safer and more pleasant work environment. Windows can also enhance the "facility tour," enabling clients to view operations without disturbing the work or requiring guests to gown. Outside views can break the tedium of the job and allow employees to feel connected with the rest of the plant. From a safety viewpoint, windows can enable rescue workers to gauge the severity of a situation before putting anyone else at risk (see Fig. 11.2-1).

Select Appropriate Materials for Construction and Finishes.   Depending on the requirements of the space, interior finishes may need to be chemically resistant, antimicrobial, nonconductive, nonstatic, heat/steam-resistant, and so forth. The A/E should compile a set of standard finishes for each different type of environmental condition for review and approval by the owner's process engineers.

Normally, the most critical spaces in a pharmaceutical production or biopharmaceutical facility are the clean rooms. Depending on the classification of the clean room, Class 100,000 through Class 1, there are increasing degrees of HVAC and process filtration. The architectural challenge is to provide an envelope that does not harbor contaminants (i.e., one that can be completely cleaned and sterilized). The protocol specifies the cleaning agents and processes that set the performance specifications for the finishes required. Ideally, the construction of the finishes should be seamless (floor to wall, wall to ceiling, wall to door, light fixture to ceiling, air grille to wall/ceiling, etc.) and impervious. Both of these attributes are to ensure that there are no spaces, voids, pinholes, or ledges where contaminants can accumulate/grow or from which cleaning agents cannot naturally drain. Depending on the chemical cleaning solutions, water and steam temperatures, and so forth, common finishes are:

- *Floors:* Epoxy resin, welded PVC, epoxy terrazzo
- *Walls:* Epoxy coating, polyester coating, seamless, spray-on PVC coating, welded sheet PVC
- *Ceilings:* Epoxy coating, polyester coating, seamless, spray-on PVC coating, gasketed clean room panels and grid

The closer one must get to the ideal, the more costly the finishes of a space become. Therefore, it is important to work with the protocol to determine the appropriate level of finishes.

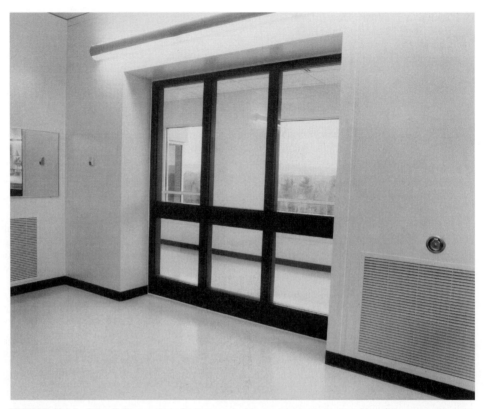

FIGURE 11.2-1   Intersil Corporation clean room, looking out into viewing corridor. (*Courtesy of Symmes Maini & McKee Associates.*)

Facility Siting.   The maximization of the building's height and area are typically process driven. In the semiconductor industry, this is particularly true of the increased heights needed for process equipment and the mechanical equipment needed to supply the increasingly cleaner environments.

The minimum distance of the building from lot lines and other structures depends on the rating and, therefore, the construction materials used for exterior walls. Siting also depends on the requirements for explosion venting and increased separation for chemical and gas storage areas. The structural framing system used will be determined on the basis of the use and type of stresses imposed by the manufacturing process.

Exiting requirements such as travel distance, number of exits serving a room, and door swing directions have specific requirements in this use group:

- Travel distance, which is limited to 100 ft to an approved exit, poses significant limitations to the configuration of the facility. Additional travel distance restrictions may also apply once the occupant reaches the approved exit.

- Two means of egress are required from each fabrication area that exceeds 200 GSF. Doors must swing in the direction of travel.

- Fabrication areas must be separated from means of egress corridors, from each other, and from other parts of the building by a minimum of one-hour-rated construction.

- The quantity and/or density of the hazardous materials present determine the maximum size of a single fabrication area. Ventilation in fabrication areas must be a minimum of an average of 4 cfm per square foot of floor area, and no location must have less than a minimum of 3 cfm per square foot of floor area to preclude having to use electrical devices classified for use in hazardous locations.

- Storage of hazardous materials in excess of the exempt amounts and methods of transportation are regulated. Storage rooms have significant limitations regarding size, location, enclosure ratings, minimum ventilation, egress requirements, containment, leak detection and monitoring, and explosion relief.

- Service corridors are used for transportation of hazardous materials, not enclosed piping systems. Travel distance in service passages is limited to 75 ft, and a minimum of two exits is required. Hazardous materials may not be transported by use of egress corridors, except under very limited conditions in existing buildings.

When the letter of the code cannot possibly be met, which often occurs in existing facilities, compliance alternatives are sometimes used to afford a similar level of protection. Owners should give particular attention to the use of compliance alternatives because these may increase the owner's liability in the event of a disaster.

Due to the critical nature of code compliance in the safety, design, and operation of advanced technology facilities, it is important to identify the *code team* at the outset of the process. In large organizations, this begins with the owner identifying the appropriate process engineers, the safety department personnel, and emergency response team representatives, and establishing a relationship with the local code officials, if one has not been previously established. The typical code team consists of the owner's team identified previously, the local building and fire officials and their consultants, the A/E team and their consultants, and the equipment vendors.

Before a specific HPM code existed, buildings were often classified as factory- or high-hazard use. As it became apparent that neither classification was appropriate, the semiconductor industry and building officials in California arrived at a compromise. The HMP classification allows the building owner to take into consideration additional safety features such as fully automatic sprinklers, air changes, and reduced travel distance that are incorporated into this type of facility. In return, other code requirements were revised to be more in line with a business use classification.

Current Good Manufacturing Practices (CGMP) basically state that ventilation, air filtration, heating, and cooling be adequate to allow for the manufacture of drugs. Because this is a wide-open requirement, the experience of the engineer is required. The A/E needs to indicate those areas that are within the scope of CGMP because all drug production areas must be high-efficiency particulate air (HEPA) filtered, a requirement that affects the design and operating cost of the mechanical system.

In pharmaceutical facilities, the key to a successful project and cost control is early awareness of the different classes of spaces and their regulated interconnections, separations, and circulation paths. FDA regulations and clean and sterile protocols further complicate the customary product and material flow of manufacturing. Biological and chemical safety may also be factors.

Of these regulations, 99 percent address protocol, procedures, and record keeping. Only 1 percent refers to the physical facilities and process systems. However, close familiarity with the required protocols allows the A/E to use the physical plant to facilitate meeting these exacting regulations by planning for adequate space and clearly defining the flow of separated materials (e.g., incoming, finished product by batch, rejected raw materials, quarantined materials).

During the schematic design phase, the plans and concepts for the facility should be introduced to the building and regulatory agencies before it becomes costly to incorporate change in the design.

The manufacturing process and flow must be established because they generate the program:

- They establish the product, material, equipment, and people flow.
- They govern the types of chemicals (quantities, container sizes or centrally piped waste, etc.).
- They mandate any specialized systems, clean rooms, robotics, environmental rooms, and so forth.

Validation.   It is the responsibility of the owner to provide the validation requirements so that the A/E may ensure that the physical plant supports the protocol and process changes. Process engineers and quality control representatives must be intimately involved in the initial design of the facility because they are responsible for the success of the process, the viability of the final product, and the validation.

Validation is the term the FDA uses to describe the documentation of a repeatable process and protocol in the manufacture of drugs. "There shall be written procedures for production and process control designed to assure that the drug products have the identity, strength, quality, and purity they purport or are represented to possess. . . . Establishing documented evidence which provides a high degree of assurance that a specific process will consistently produce a product meeting its predetermined specifications and quality attributes" ("FDA Guideline" General Principles of Validation, May 1987). Historically, a validation package for the FDA consists of:

- Installation qualification (IQ) documentation proving that the equipment and instrumentation are installed correctly within the appropriate and proper functioning environment.
- Operational qualification (OQ) documentation that the equipment and instrumentation performs as projected in all facets of their functions.
- Process qualification (PQ) documentation that the production process is repeatable and consistent in quality.
- Protocol is the set of specifications that will ensure that the product/process can be validated. It normally consists of a sequence of tests and acceptable test result parameters.

Pharmaceutical manufacturing is classified under BOCA-based codes as a factory and industrial use group. An analysis of the applicable standards and regulatory requirements will further define building organization and required safety features. Within this group are low and moderate hazard ratings (high-hazard use is more common in semiconductor manufacturing). The hazard rating is established by the hazardous material usage, type, quantities, delivery methods, waste, storage, and so forth. The owner must provide this information to the A/E. Ideally, the A/E will supply the owner with a matrix to fill in, so that the A/E receives the data in a usable design input format. With this information, the A/E can classify the building and determine the type, materials, and capacity of any chemical distribution or dispensing systems. Such information is an important factor in developing an accurate project cost model.

Depending upon the hazard involved, special precautions may be required. A description of the biosafety levels follows.

OWNER ESTABLISHES INSURER AND OUTSIDE FUNDING REQUIREMENTS

Many insurers or outside funding agencies such as NASA, NIH, and NCH have additional requirements for the physical plant (i.e., fire separations, alarms, system redundancies, etc.), especially if these facilities are not duplicated elsewhere in the company's holdings. This information should be given to the A/E early in the design process.

The A/E Identifies Imminent Changes in the Codes and in Regulations

Often an experienced A/E will be aware of pending changes in Code, CGMP, and so forth that might affect the project. The owner may benefit either by being included under the new regulation or by staying with the old requirements. The owner and A/E should determine the pros and cons of each alternative and decide the appropriate course of action.

At this point in the project, process engineers and environmental specialists become involved. The process engineer develops the facility design to a point where air and water emissions from effluents can be estimated. Once these are known, the environmental specialist (often an outside consultant) studies the state and federal regulations to determine the acceptable emissions limits. Working together, the process engineer and the environmental specialist determine the best control technology. These recommendations are communicated to the owner throughout the design process.

Program square footage should be projected on the basis of similar operations and adjusted to meet the individual process needs. Before finalization, the program needs to be tested. Broad scope checks may include the following:

- Does the program meet the strategic plan of the company?
- How does the program break down into "billable" and "nonbillable" space?
- How does a layout of a prototypical space work?
- Is there flexibility for future growth or changes in the process?
- Will future growth, horizontal or vertical, impede the building or site circulation? Is there also space for utilities to expand?
- Is there any way that currently dedicated support space could be shared to be more efficient?

Tool Sets

Early identification of tool sets or consensus on the assumptions to be made in lieu of definitive information is essential in the programming phase. From a manufacturing perspective, tool sets drive production, and therefore, profits. From a building design perspective, tool sets drive the spatial layout and organization of the facility, including the types, locations, and quantities of process services; electrical, HVAC, and exhaust requirements; structural systems; vibration specifications; and cleanliness classifications. These same factors also drive the structural system that can be economically used. During the programming phase, the A/E's chemical process and industrial engineers work with the owner and the owner's industrial consultant to establish preliminary tool sets based on the owner's operational and output needs. When tool manufacturers can be identified, tool vendors should be contacted and initiated into the team. Once the tool vendor is identified, the A/E can work with the vendor and the owner to keep current with proposed changes and keep the owner aware of potential issues that might affect the facility design. Their safe use and storage can become a reason for special containment. One such approach is to contain any potential blast; therefore, areas involved must be reinforced to resist the force of the blast, and the design must provide a safe venting path to the outside. Use of an integrated database and three-dimensional modeling of the layouts by the A/E will enable the owner to visualize the space and process flow before the tool sets are finalized.

SCHEMATIC DESIGN

The program can be graphically depicted in *bubble* or *block* diagrams. These diagrams, which are precursors to the resulting schematic plan, graphically illustrate major functional areas, their adjacency requirements, and people and product flows. Examples of functional areas include clean and nonclean manufacturing, clean and nonclean support areas, offices, office support areas, and building support areas. Circulation in advanced technology facilities is more complex than in other buildings because of separate clean and nonclean, horizontal and vertical circulation/egress for people, materials, products, and chemicals. Identification and diagramming of the required circulation is necessary to establish appropriate organization and hierarchy of the functional areas. The code requirements for egress travel distance, which is limited typically to 100 ft and 75 ft in service passageways, may suggest maximum widths for the building.

Product requirements often dictate that a "clean" environment be provided for manufacturing. In the semiconductor industry, particles are the problem because they can alter the electronic circuits on a semiconductor chip by establishing alternate paths for current flow. In the biotechnology arena, sterility is important because a few bacteria or microbes introduced into a fermentation batch can rapidly grow to a major contamination of the broth.

Cleanliness protocols determine the types, sizes, quantities, and classifications of ancillary functions and areas such as preentry, gowning, and air showers needed to enter and exit clean manufacturing areas (see Fig. 11.2-2). Separation of "clean" and "cleaner" processes within the manufacturing (such as photo and GMP areas) areas may also be required. This in turn drives HVAC supply, exhaust and zone requirements, egress concerns, and the need for additional walls and doors and other dedicated equipment.

To minimize the effects of contamination, the owner and the A/E should establish protocols for movement of equipment, material, and personnel. These should target eliminating

FIGURE 11.2-2   Clean room gowning area. (*Courtesy of Symmes Maini & McKee Associates.*)

opportunities for introducing contaminants into the clean environment. Equipment should be thoroughly washed and wiped down on the exterior before introduction into a clean space. Depending on the cleanliness levels desired, personnel should "gown-up." This can vary from simple lab coats to full body suits with face coverings. In many cases, the personnel in a clean space are the dirtiest things there.

A plan should be developed to identify the location of one part of the facility in relation to the next. To minimize traffic congestion or reduce the possibility of product contamination, a diagram should be developed to show people, product, raw material, and equipment flows. These should specify entering and exiting activities. The diagrams will visually point out areas where large flows need to be redirected to avoid congestion. Product purity issues need to be addressed by carefully avoiding any areas where incoming "clean" material passes near any outgoing "dirty" material. This is especially true in the biotechnology industry, where sterility of the product is of utmost concern.

A structural system needs to be devised to complement the function that is performed in the space. The majority of advanced technology facilities will realize specific operational and manufacturing benefits if an economical structural framework is devised. For example, in the semiconductor industry, the floor plan or "ballroom" has inherent long-term benefits from the operational perspective because of the ease of implementing future tool layout modifications when the next-generation chip is in demand. The challenge is to devise a structural system that provides both long-term operational as well as up-front cost savings when the facility is built or renovated. The bay size, along with the way the physical plant and manufacturing equipment are supported, located, and maintained to achieve the operational and functional requirements, influences the facility cost.

The operational, maintenance, and expansion requirements for the unique manufacturing and physical plant equipment must be considered before determining the structural system. Concerns for operational access, functions isolation, equipment-generated forces, and maintenance access must be factored into the structural system analysis. For example, in the semiconductor industry, it is often a requirement to minimize or eliminate vibration. If the structural system chosen does not factor this component into its solution, the manufactured product could be impacted. Seismic requirements for the facility also must be considered in the structural system analysis. Seismic conditions impact new ceiling, piping, electrical, and HVAC systems that are supported and anchored to the structure.

Lastly, depending on the type of manufacturing undertaken in the facility, the storage and handling of the various explosive, corrosive, flammable, and/or radioactive materials will be reviewed as to what protective measures must be taken to minimize or eliminate the operational risk and the risk of injury. Blast containment, blowout panels, containment basins, and so forth might be warranted by the raw materials used.

Equipment Selection and Layout

The first step in selecting equipment is to determine how it is to be utilized. Will the process be set up for batch or continuous operation? Will the production requirements be met all at once or campaigned over a much longer time frame?

The answers to these questions require completing a comprehensive study and cost estimate. Is it better to use one or a few large tanks or multiple small tanks? The answers may depend upon many factors. Is the value of the product such that a lost batch results in significant costs to the firm? Some pharmaceutical or biotech products, when completed, ultimately may be worth hundreds of thousands of dollars and be contained in only a 1000-liter vessel. The final dosage form may be a few milliliters or less diluted with water-for-injection (WFI)–grade water and ready for injection into a patient at $10 to $1000 per dose.

Campaigning can be considered when the processing time becomes very long. Some fermentation steps in the biotechnology industry take many weeks to complete. For these processes, equipment costs versus time lost must be weighed carefully when deciding on the ultimate solution. Once this thorough analysis is completed, an equipment list can be prepared that details the size and number of each piece of equipment required in the process plant. Layouts can follow that consider the operational and maintenance aspects of each.

During the equipment layout, attention should be given to the areas needed for special maintenance, such as pulling tube bundles from heat exchangers or lifting agitators out of tanks. Space should be allocated for these efforts and marked as required areas on flow plan drawings so that nothing gets placed into these zones to prevent proper access and maintenance. Often, the height needed to pull an agitator out of the top of a tall tank determines the height of the room. Allow room for rigging and adequate clearance.

Utilities

Now that the equipment list has been created, the utility needs of the process can be determined. Are the vessels heated or cooled? How much power do the motors consume? Does the chemistry produce or require energy? Is there a need for a gas (e.g., oxygen, nitrogen) feed or purge? The answers to these questions can be tabulated and summed to produce a matrix of utility usage by each piece of equipment.

For continuous operation, these utilities must be available throughout the process. However, for batch operation, a rule of thumb is that at any one time only about one-third of the equipment is calling for a utility. This rule can become the initial basis for utility equipment sizing.

At this point, unless a totally new facility is being built, the existence of the rest of the plant must enter into the design. Are existing utilities adequate? Are the temperatures and pressures compatible with the new additions, or must adjustments be made? This information determines the basis for the new (if any) utility needs.

As is the case with most capacity upgrades or modifications, some tie-ins to existing systems may be required. Care should be given to develop designs that minimize the impacts of these tie-ins to ongoing production operations. As much as possible, all tie-in pieces should be prefabricated and available ahead of time. The utility outage should be coordinated with the owner and planned well in advance to maximize coordination and minimize downtime.

To control utility costs, companies look for ways to reduce energy consumption by looking into steps such as waste heat boilers that use low-pressure steam as a preheat material for other fluids. Companies that embrace strategies of energy conservation will achieve lower production costs.

Process and Instrumentation Diagram

The focal point for the engineer during the design phase is the *process and instrumentation diagram* (P&ID). This flow drawing details the equipment, piping components, and instrumentation needs for all of the systems that go into the final process design.

One last step to utility and equipment programming is to review the redundancy requirements with the owner. *Redundancy* is defined as installed backup systems ready to take over in an instant for any failed piece of equipment or system. A 100 percent on-stream factor is the goal of any production facility. When the equipment is off-line, products (and profits) are not being made. The solution oftentimes is to build in oversize utilities that are multiples of the same size. For instance, install three units, each capable of carrying 50 percent of the design load. Then, two units are on-line all the time, and one unit is ready to come on if needed. A failed unit can be repaired off-line while 100 percent of the utility needs are maintained.

Deliverables and Approvals Needed to Progress to the Next Phase

The owner must review and approve the program as the basis for the next phase. This review is especially important in pharmaceutical and semiconductor facilities because of their complex nature and high costs. This is also an excellent time to bring the insurance agent into the project team.

Semiconductor Facility Design Drivers

The following are the primary design drivers in semiconductor facilities:

- Height and area needed
- Clean and nonclean circulation
- Identifying tool sets
- Building code requirements

Height and Area Needed.   The production process, the owner's standard methods of operation, the project budget, the space available, and the height allowed determine the height and area needed for the facility. In renovations, space available and height allowed tend to have an increased impact on design decisions. The horizontal and vertical spatial organizations are also of key importance in semiconductor facilities. The efficiency of this organization is a major driver of both the initial and long-term operation, maintenance, and cost-effectiveness of the facility.

The four basic clean room plan types of a semiconductor facility are the *clean bay with service chase, clean aisle with tool core,* the *ballroom,* and a *hybrid clean room.* These layouts may occur in one-, two-, and three-story configurations (see Figs. 11.2-3 and 11.2-4).

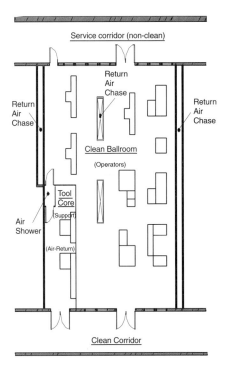

- Large Open Clean Areas Offering Unlimited Tool Layout Flexibility
- Tool Cores Can be Incorporated Into the Layout as Necessary for Bulkheaded Tools
- Tools are Exposed Inside the Clean Areas
- Return Air Through Perforated Raised Access Floor Plenums
- Return Air Chases are Required Sporadically Throughout the Layout
- Expensive to Operate Due to Large Areas Which Must be Maintained Clean
- Common to Use Lower Cleanliness Classifications with Mini-Environments on Tools to Offset Operating Cost Issue

FIGURE 11.2-3   Intersil Corporation ballroom clean room. (*Courtesy of Symmes Maini & McKee Associates.*)

FIGURE 11.2-4   Intersil Semiconductor ballroom clean room. (*Courtesy of Symmes Maini & McKee Associates.*)

Clean Bay with Service Chase.   Early clean rooms used the clean bay with service chase arrangement. Clean bays were limited to widths of 12 to 14 ft with sidewall returns and narrow service aisles. The limited width of the clean bay minimized the costs by limiting the area under filtration. Main tools were located in the filtered area, and small ancillary equipment and services were located in the service aisle.

Clean air circulation was typically supplied from ceiling-mounted filters and returned through the service aisle via the sidewall return grilles. In two-story buildings, mechanical units feeding the filters may have been housed on a mechanical level directly above the clean room. In existing one-story buildings, where penthouses are not permitted, the mechanical units may have been hung from the roof structure above the clean room where there was sufficient height, or floor-mounted in an adjacent area.

Bay and chase clean rooms are still in use today. However, renovated and newly designed nonballroom clean rooms use the modified version of clean aisle (bay) and tool core (service chase) described here, in lieu of this older-version layout.

Clean Aisle with Tool Core.   As the wafer sizes and therefore tool sets became larger, the need for larger areas to house the tools arose. The aisle with tool core responds to these needs by reversing the proportions of the clean bay with the service aisle. In this arrangement, the tool core expands to allow bulkheading the main tools through a smaller clean aisle wall. The tool core is sized to accommodate both the main tools as well as large and small ancillary pieces of equipment. Services are typically run overhead in the tool core.

The clean air supply and return may be the same as for the bay and service chase in one-

and two-story buildings. As an alternative, return air may be brought through an access floor located in both the clean aisle and tool core. The advantages of using an access floor are increased laminar airflow through the clean aisle and the potential for running services to tools and between tool components under the floor, as well as overhead. The disadvantages of using an access floor are the additional costs and difficulty of coordinating floor heights between adjacent areas. A depressed slab on grade is sometimes used to resolve this issue.

The Ballroom.   Clean-room ballrooms are large, continuous, and relatively unobstructed areas under filtration. Although ballrooms are potentially more costly because of the larger areas under filtration, the advantages can often outweigh this disadvantage.

This type of clean room layout is typically found in a three-story building configuration consisting of a process level or *fab,* a mechanical level, and a subfab level. The main tools reside in the ballroom at the process level; the ancillary equipment is located in a subfab; and the mechanical equipment is located on the mechanical level.

In two-story buildings with a mechanical level, the ancillary equipment may be located partially in the ballroom and/or in adjacent tool cores. In the absence of a full mechanical level, the ancillary tools and portions of the mechanical equipment may be located in the lower subfab level. This type of arrangement requires the use of fan filter modules versus ducted or plenum filters to supply clean air to the clean room.

Clean air is supplied through ceiling-mounted filters and returned via raised-access flooring to the subfab level or up through perimeter and interior return air chases. Ballrooms require larger areas under filtration but generally permit the use of lower overall cleanliness classifications with the use of minienvironments. Minienvironments provide the higher cleanliness classifications required at the operating portion of the equipment where the product is exposed, whereas the remainder of the room can remain at a lower classification, often with mixed airflow.

Hybrid Clean Room.   As the name implies, a hybrid clean room is a combination of the three basic types of clean rooms described previously. Examples of this are:

- Facilities utilizing more than one type of clean room
- Bay and chase arrangements in a three-story or a two-story building with a penthouse that exhibit some advantages of ballroom design

The pros and cons of the design of these facilities must be discussed on an individual basis. The following factors need to be considered in selecting and designing clean rooms:

- Suitability for production process and building configuration.
- Preliminary tool set arrangement should take into consideration the need for periodic return air chase. Required egress and maintenance clearances must also be considered.
- Which equipment will be designed to reside in the clean room and which will be required to reside in the subfab?
- When equipment is bulkheaded, the spacing of equipment should consider the space required for a wall support system and adequate return air grilles.
- Horizontal and vertical organizations of the services in the subfab and below access flooring and/or above clean room ceiling.
- Degree of access required above the clean room ceiling.
- Clean room ceiling heights.
- Floor-to-floor heights.

Clean and Nonclean Circulation.   Identification of the required circulation types is essential for establishing the appropriate organization and hierarchy of functional areas. Circulation in semiconductor facilities is more complex than in other building types due to:

- Separate clean and nonclean, horizontal and vertical circulation

  People, materials, and tools

  Owner's cleanliness protocols

- Separate routing (supply and waste) required for hazardous chemicals
- More stringent egress requirements that may apply

Semiconductor facilities require accessible horizontal and vertical circulation for people and small equipment/materials. The clean and nonclean circulation must be separated to preclude product contamination from microscopic dust and dirt particles, yet be joined to the nonclean circulation to allow continuous transportation of materials, equipment, and personnel from the clean rooms.

To minimize the effects of contamination, protocols for movement of equipment, material, and personnel should be established between the owner and the A/E during the programming and schematic design phase. Diagrams showing people, product, raw material, and equipment flows can be developed with the owner. The diagrams should show proposed incoming and exiting activities by type and cleanliness classification required to test the proposed circulation.

Cleanliness protocols determine the types, sizes, quantities, and classifications of ancillary functions/areas needed to enter and exit clean manufacturing areas.

- For personnel circulation, this includes functional areas such as preentry, gowning, air showers, and separation of "clean" and "cleaner" processes within the manufacturing [such as photo and good manufacturing practice (GMP) areas]. Egress requirements for semiconductor facilities are governed by the HPM subsection of most codes and are typically more stringent than other building types.
- Tool movement requires similar consideration. The width and height of corridors and doors must accommodate the tool size, including equipment used to transport the tools and necessary turning radii. Wipe-down areas and oversized freight elevators may be necessary.
- Nonhazardous material circulation will require a pass-through or use of tool wipe-down areas.
- Product circulation requires either clean circulation until it reaches a noncritical production stage (i.e., is no longer required to be under filtration) or containment to allow it to pass between clean and nonclean areas during production.
- Hazardous material (supply and wastes) handling, transportation, and dispensing are governed by building codes. Its "circulation" must comply with the manufacturing, clean/nonclean, and applicable code requirements.

The added circulation requirements have several implications for the building design:

- Additional square footage is required to provide partially redundant clean/nonclean circulation.
- Additional square footage is required for ancillary facilities such as gowning and wipe-down areas.
- HVAC supply, exhaust, and zone requirements are driven by the need for additional separations and cleanliness levels required.

Identifying Tool Sets.   Early identification of the tool sets or a team consensus on the assumptions to be made in lieu of definitive information is essential in the programming phase. Tool sets drive both the spatial needs and the applicable building code requirements (refer to code analysis section for a discussion of code requirements) for the facility.

From a manufacturing perspective, tool sets drive production, therefore profits. From a building design perspective, tool sets drive:

- Spatial layout and organization of the facility
- Types, locations, and quantities of process services
- Electrical, HVAC, and exhaust requirements
- Vibration specifications
- Cleanliness classifications
- Structural systems
- Applicable code requirements due to the types and quantities of hazardous materials

DESIGN DEVELOPMENT

The primary purpose of the design development phase is to further refine and integrate the results of the programming and schematic design phase into a well-articulated, coordinated, constructable, cost-efficient, and code-compliant facility that is the basis for the construction documents phase. Architecturally, the design development phase involves the following:

- Integration of code-required construction
- Insurance underwriter's requirements
- Integration of engineering system requirements
- Mechanical (HVAC), process, plumbing, fire protection and electrical systems
- Future flexibility considerations
- General engineering system coordination and layout considerations

    Horizontal and vertical zones

    Clean and nonclean circulation

    Specialty systems

- Structural system considerations
- Selection and integration of clean room, wall, and ceiling systems

The building and the equipment must fit together to function smoothly and operate properly. The A/E should watch for interference problems and ensure that adequate floor space and clear height have been provided to maintain the equipment. Plan drawings and piping/equipment elevation drawings should be prepared in sufficient detail to show that all of the parts that make up the "whole" facility fit and work well together.

Integration of Code-Required Construction

During design development, the A/E begins to coordinate all disciplines (architecture, site/civil, structural, mechanical, electrical, plumbing, process piping, and fire protection) to create an efficient, functional facility. This process will test the success of the sectional relationships (i.e., the three-dimensional stacking of the process). In the pharmaceutical or biotechnology facility, this relates to the product flow (tanks, mixers, piping, etc.). When planning for future expansion or flexibility, it is crucial to check this third dimension.

Added challenges in the design of a semiconductor facility are the two main code requirement drivers—*the tool layout* and *the hazardous materials list*—which are usually evolving throughout the design process and sometimes into the construction phase. The further into the design and construction process that the finalization of tool sets and hazardous materials lists slip in time, the more costly it becomes to incorporate the code requirements into the design.

The delay of definitive tool information may also impact the construction schedule. Therefore, code analysis becomes an ongoing and evolutionary process in most semiconductor facilities.

Early in the design development phase, the overall height and area of the building is determined. The number of exits and travel distances are estimated, and strategies for raw material storage and distribution are identified. A code analysis plan, indicating rated construction, is developed both as an internal communication tool for the A/E and as a communication tool for the constructor.

Manufacturing processes that tend to push the design envelope may develop code issues that do not have straightforward solutions. This is especially true for facilities designed under previous editions of the codes. When developing potential solutions, the "cheapest" solution can become the most expensive one in the final analysis. The design team needs to evaluate alternatives and invest in the most appropriate and cost-effective solution.

Insurance Underwriter's Requirements

Factory Mutual (FM) or Industrial Risk Insurers (IRI) companies have published requirements for clean room facilities. The insurance underwriter is generally looking to minimize exposure to physical loss and to mitigate production downtime. Ideally, the owner's insurance underwriter representative is already part of the team. Although owners often prefer to minimize the underwriter's involvement to limit their opportunity for comment, some dialogue with the underwriter is required. Underwriter's requirements may exceed code requirements, and early identification can reduce the cost of incorporating these requirements.

Integration of Engineering System Requirements

Relative to code requirements, integration of engineering systems can be divided into (1) hazardous materials management and (2) coordination issues and requirements between engineering systems and code-required construction. The following considerations need to be addressed:

Hazardous Material Storage and Dispensing Rooms

- The storage and handling of various explosive, corrosive, and/or flammable materials need to be reviewed to ensure that protective measures are incorporated so as to minimize or eliminate the operational risk and the risk of injury resulting from inadequate evaluation of the structural system(s) used.
- Rated blast-resistant and/or blast-relieving wall and ceiling construction. Blast-relieving walls are often hinged at one end to permit the force of explosion to be released without damaging the building structure. Engineering calculations are necessary to determine the height-to-depth ratio of rooms and the area of blast relief required to direct the explosion safely along a predetermined path.
- Appropriate containment strategies must be developed for both chemicals and fire suppression system discharges. Areas involved must be reinforced to resist the force of the blast and designed to provide safe venting to the outside.
- Chemical- and acid-resistant surfaces, as required.
- Additional ventilation, as required.

Transportation of Hazardous Materials

- Dedicated horizontal and vertical rated circulation paths.
- Elevators and dumbwaiters may require specific material finishes and may need modifications that require local government review and approval.
- Additional ventilation, as required.

HVAC, Process, and Electrical System

Ventilation, special electrical, spill containment, monitoring systems, and other life/safety issues will be further defined and incorporated into the design.

Coordination issues and requirements between engineering systems and code-required construction include the following:

- Ductwork may require dampers to penetrate rated construction or internal sprinklers or be constructed of rated materials (refer to Sec. 11.3, "Mechanical Systems Design," of this chapter for a further discussion of these requirements).
- Rated walls are required to extend full height—ductwork, piping, and so forth cannot run parallel with walls.
- Flammable materials that have low flash points can be easily ignited by sparks. Facilities that process these materials need special electrical equipment.
- The use of explosion-proof or intrinsically safe design can minimize any potential sparking hazard. Brass tools are often recommended because they are nonsparking. Large quantities of flammables contained in a room may necessitate the use of blast walls.

Future Flexibility Considerations

Future flexibility has both short- and long-term implications for code-required construction. The short-term future flexibility arises out of the typical inability to identify all of the hazardous materials and/or the systems necessary to transport, handle, dispense, and/or contain hazardous materials. Long-term future flexibility takes into consideration processes that may not be required today but can be identified as likely to be required in the future. For instance, if a process currently does not require dispensing of flammable chemicals, but there are future plans to dispense these on-site, then such spaces should be planned and located accordingly. Dispensing areas may require explosion-relief panels that are more difficult to locate after the initial design concepts have been finalized.

General Engineering System Coordination and Layout Considerations

In developing equipment layout drawings, the A/E will identify interference problems and ensure that enough floor space and clear height has been provided to maintain the equipment. Plan drawings and piping and equipment elevation drawings should be prepared in sufficient detail to show that all the parts that make up the whole facility fit and work well together.

Horizontal and Vertical Zones.   The size and quantity of mechanical process and electrical services make it imperative that horizontal and vertical zones for utilizing distribution are established early in the design and are maintained throughout the building. In facilities where only part of the building is initially fit up, a master plan for future expansion should be developed.

Clean and Nonclean Circulation.   The circulation strategies for clean/nonclean people, product, and chemical routes that were developed in the concept/schematic design phase should be reviewed and revised as required to include new information or to respond to refined design criteria. The circulation paths need to be finalized during this stage of the project.

Specialty Systems.   During design development, specialty systems, such as conveying systems, minienvironments, waste recovery and recycling, high-purity water, chemical dispensing units, and so forth, need to be incorporated into the overall design (refer to the mechanical and chemical process sections in this chapter for further discussion).

Structural System Considerations

Semiconductors, which use a ballroom approach, need the structure designed to allow clear and open spaces. This will provide a major benefit because of future rearrangement of tool sets and their associated process changes and will ensure that they can be installed unencumbered.

Selection and Integration of Clean Room Floor, Wall, and Ceiling Systems

In three-story semiconductor facilities, modularity begins at the subfab level and extends up through the building. Centerline modularity begins at the process-level waffle slab, which sets the layout of the subfab, and access floor systems and is maintained through the wall and ceiling systems up to the layout of the mechanical equipment on the fan deck above.

Economically designed waffle slabs typically have a 30×30 in or 36×36 in centerline modularity. Taking vibration into consideration, this translates to a 15×15 ft or 18×18 ft structural bay spacing on the subfab level. Typical access floor systems have a 24×24 in centerline modularity. In installations that do not require excessive penetration of the waffle slab, the difference in modularity is often resolved by careful planning of the penetrations, so that these do not coincide with the access floor pedestals. In installations where waffle pan openings are frequent (for service routing) or throughout (where waffle slab is used for return air), the use of newer systems with a 72×72 in pedestal centerline spacing may be considered. The 6×6 ft access flooring systems are relatively new and not fully tested, compared to the traditional 2×2 ft systems required.

Then walls are set on the access flooring 2×2 ft modularity and extended to the ceiling grid that has a 2×4 ft modularity. Alignment of the access floor and ceiling systems is essential. The ceiling is supported from the structure of the fan deck (mechanical level) above. This requires that both the structural system and the mechanical unit layout fall within the module to maximize the cost-effectiveness of ceiling and structural support systems and to allow duct openings and ductwork runs between the support points of both systems.

Advanced technology facility design requires specialized materials for interior construction, including wall, floor, and ceiling finishes; process equipment; and piping materials. Material selection considerations include longevity desired, suitability for the environment in which it will be used, whether it supports bacteria, whether it sheds particulates, if it has appropriate static electrical properties, how it will be cleaned, and what loads will be superimposed upon it.

In interior construction, the first criteria to ascertain are the types of environments in which the material will be used. Is the environment corrosive, clean—low particulate, clean—antimicrobial, heavy use/wear, exposed to chemicals, and so forth? The next question is: what is the desired length of service? Consideration may be given to materials that have a shorter in-service use in pilot or R&D facilities that are intended only for short-term operation. Which material characteristics are the most important: chemical resistance, vibration absorption, static dissipation, conductivity, microbial resistance, nonparticulating, subjected to aggressive cleaning, and so forth? Nonparticulating surfaces are of prime importance in the semiconductor and microelectronics industries. Smooth, non-outgassing, low-particulating surfaces that form an airtight clean room envelope are the norm. Silicones are the most frequently used sealant materials.

Floor Systems. Floor surfaces are often exposed to chemical spills, heavy rolling, and point loads. They may need to dissipate or conduct static electricity away from personnel, product, or equipment, or they may be required to resist or minimize transmission of vibrations. Epoxy floor systems, solid vinyl flooring, stainless steel plates, and high-grade epoxy paints are materials that have high chemical resistance. For the A/E to recommend the most appropriate material(s), the A/E must know the type, frequency, and concentration of chemicals likely to come in contact with the surface. Although some chemicals are typical of the particular manufacturing process and are easily identified, others, which are unique to the owner's process and not readily identifiable, may require further investigation.

The owner needs to verify the vibration criteria used for structural design. Usually, these are determined by the recommended vibration tolerances set by the tool vendors. Vibration mitigation is primarily accomplished through structural design of the floor systems and vibration isolation of mechanical and manufacturing equipment. Where access flooring is used, highly vibration-sensitive equipment is typically isolated from the surrounding floor by a vibration base. Isolation detailing between walls and ceilings may also need to be considered. For special situations, a vibration engineer may need to be consulted.

From the weights and vibration requirements, strategies for tool support can be discussed. How the various strategies interface with the design should be reviewed to develop a list of potential approaches. Actual tool support design will typically occur during the later stages of the construction document phase or in tool fit-up phases which occur after the design of the building has been completed and the building is in advanced construction stages. Costs of tool support, as well as other tool fit-up requirements, should be identified in the overall project cost model.

Equipment weights, especially in the semiconductor industry, have been increasing in recent years, and this trend appears to be continuing. In addition to designing sufficient load capacity in the structural floor systems, the resistance to point and rolling loads by the floor covering or access flooring should also be reviewed. Often, larger or additional feet or base plates that distribute the load over more of the surface are recommended to accommodate equipment point loads. Larger-diameter wheels or special rigging are recommended to reduce the impact of rolling loads.

Conductive flooring is produced by impregnating materials and adhesives with metal chips, strips, or wires and grounding the floor to the building's grounding grid. Solid vinyl flooring and epoxy systems are two materials that are standardly manufactured with dissipative or conductive properties. Conductive surfaces are not typically recommended around high-voltage equipment such as ion implanters, due to their ability to discharge quickly, potentially through attending personnel. Manufacturers of conductive materials also require that owners have a program that includes special footwear grounding straps and special maintenance procedures to achieve the full effectiveness of the flooring's conductive properties.

Wall Systems.   Where wall surfaces will come in contact with chemicals, chemical resistance should be considered. PVC or chemically resistant epoxy finished surfaces are often used. When vibration transmission is of concern, isolation detailing at the floor and/or ceiling is employed. Impact loads should be considered and impact-resistant surfaces or treatment of surfaces with impact-resistant materials such as bumper systems may be required. Wall surfaces may be required to be antimicrobial or nonshedding. The need for static dissipative or conductive wall surfaces is often determined by the owner's internal standards. Conductive finishes such as conductive epoxy paint are often utilized with ionization systems.

Ceiling Systems.   Superimposed structural loads are a significant concern in the semiconductor industry. In addition to the dead load from the grid, lights, and clean air filters, the system typically is required to support hanging loads from piping, minienvironments, and conveying systems. Seismic considerations and uplift from impacts to the minienvironments must be incorporated into the design. Choosing ceiling grids with a greater section modulus, increasing the number or size of hangers, decreasing the distance from the ceiling to the structure above, and the use of welded construction are several means of increasing the loading capacity of a ceiling system. Identification of the desired types of loads by the owner early in the process will facilitate the ceiling design.

Whether integrated into the tool or exterior to the tool, minienvironments may create additional HVAC requirements. When they are exterior to the tool, a supporting method must be considered. For ceiling-mounted minienvironments, the ceiling or its support system must withstand the gravity load imposed by the weight of the minienvironment. The effects of an overturning moment or uplift caused by seismic forces, people and/or cart impact on the integrity of the ceiling envelope, and potential deformation of grid members must also be considered.

Cleaning Methods.   Cleaning methods should be considered when selecting the appropriate material. If required, the material should be able to withstand aggressive cleaning methods such as high-pressure water or steam. This type of cleaning is more prevalent in the specialty chemical, pharmaceutical, and biotechnology industries.

SECTION 11.3
MECHANICAL SYSTEMS DESIGN

The physical plant includes mechanical systems that support an advanced technology facility. The systems are separate from those used for the manufacturing operation. Mechanical systems include heating, ventilation, air conditioning (HVAC), air supply, and exhaust for clean room systems; dehumidification; heating and cooling for environmental rooms; exhaust; dust collection; fume scrubbing; fire and life safety ventilation; humidification; and piping distribution for condenser water, chilled water, fuel oil, natural gas, steam and condensate, heating hot water, engine exhaust, primary and secondary cooling, primary and secondary heating, and temperature and humidity control. See Table 11.3-1 for representative mechanical codes and standards (also see Figs. 11.3-1 and 11.3-2).

TABLE 11.3-1   Representative Mechanical Codes and Standards

All codes are considered including, where applicable,
- Local Codes and Ordinances
- City, County, and/or State Building Codes
- National Electrical Code
- Occupational Safety and Health Administration Regulations
- Energy Codes

The design professional relies upon industry standards that include the following:
- ARI—Air Conditioning and Refrigeration Institute
- ADC—Air Diffusion Council
- AMCA—Air Movement and Control Association
- ANSI—American National Standards Institute
- ASHRAE—American Society of Heating, Refrigeration and Air Conditioning Engineers
- ASME—American Society of Mechanical Engineers
- ASTM—American Society for Testing and Materials
- AWWA—American Water Works Association
- CGA—Compressed Gas Association
- IBR—Institute of Boiler and Radiator Manufacturers
- IES—Institute of Environmental Sciences
- NBS—National Bureau of Standards
- National Council on Radiation Protection
- NFPA—National Fire Protection Association
- SMACNA—Sheet Metal and Air Conditioning Contractors National Association, Inc.
- UL—Underwriters Laboratories

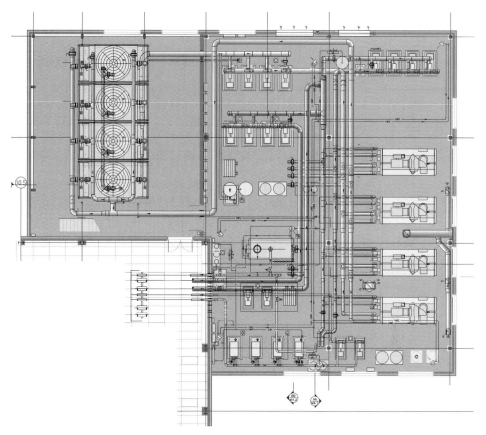

FIGURE 11.3-1 Typical central utility plant (CUP) plan for advanced technology facilities. (*Courtesy of Symmes Maini & McKee Associates.*)

Each pharmaceutical, microelectronics, or biotech facility is unique. Often, the product to be produced is unique and at the leading edge of technology. There are no precedents except those established in the pilot plant or in the scale-up installation. With this in mind, it is critical that the A/E work with the owner early in the design process to establish design criteria defining the characteristics of a specific project.

Design criteria set conditions upon which the design will be based—climatological and environmental conditions, exterior and interior conditions, and requirements for the product and for people. Mechanical equipment selection and system considerations applicable to advanced technology facilities are enumerated here.

EQUIPMENT SELECTION

In equipment selection, consideration is given to equipment operating conditions and capacities that are included to allow flexibility and variability in the manufacturing process. Selection will fix the minimum overall capacities of the systems. Properly selected equipment will ensure the conditions of temperature and humidity that the facility will be willing to accept in

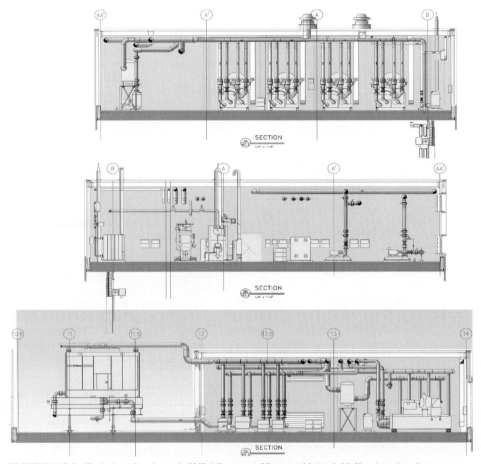

FIGURE 11.3-2   Typical section through CUP. (*Courtesy of Symmes Maini & McKee Associates.*)

each room or space. What conditions must be met so that the manufactured product meets the standards upon which it has been developed? What provisions may be necessary to meet CGMP for the consumables? Are these standards set by governmental bodies? What was determined during pilot production? Can the process be scaled up? How will the minimum conditions for indoor air quality be met? Are there limitations to effluent?

If excess capacity or redundancy is required, this will be considered in describing and selecting the primary system components.

SYSTEM CONSIDERATIONS

Air Conditioning Systems

How a space is conditioned will be determined as much by the source of primary heating and cooling as by the need for precise control and the variability of loads in the space. In people spaces, this variability is the primary driver. In production spaces, the driver may be the cleanliness level or the maximum allowable humidity level. The acceptable range of temperature and humidity are determining factors in the design of cooling systems. Residential spaces at one extreme (although not part of this discussion) will operate with an acceptable temperature range of 6°F from the time when the thermostat is satisfied to the time when it again calls for heating or cooling. A Class 10 clean room, at the other extreme, would be unacceptable if the temperature were to vary much more than 1°F.

The following discussion will deal with clean room classification and characteristics; air supply and conditioning in microelectronic facilities; air supply and conditioning in pharmaceutical facilities; humidification systems; dehumidification systems; dust collectors; fume scrubbers; exhaust systems for microelectronic facilities; and ventilation for life-safety.

Clean Room Classification and Characteristics

Clean rooms are used in R&D where new products are developed; in service, where products that are created and sealed in clean manufacturing areas are returned for service or upgrade; in assembly areas, where components are assembled and where assembled equipment is packaged for shipment; and (usually the most stringent) in the clean manufacturing areas.

These areas range from Class 100,000 to Class 1, or in some cases, to sub-Class 1. These classifications are described as the numbers of 0.5-micron-diameter particles present per cubic foot of air volume. If upon measurement, 630 particles were found, the space would be classified as Class 1,000. The basic unit of measurement is a micron—one millionth of a meter, or 39 millionths of an inch (0.000039 in). The human eye can see particles approximately as small as 25 microns—equivalent to the finest speck of dust seen through a window on a sunny day.

Tests are prescribed at three phases, depending upon their cleanliness levels, during the *as-built,* the *at-rest,* and the *operating* phases. When a space is tested in an as-built phase, it is often one or two classes better than when the space is active as in an at-rest phase, and when people are present, as in an operating phase. A space without people or equipment that is certified as Class 1 would, most likely, be Class 100 when people and equipment are in place (see Tables 11.3-2 and 11.3-3).

Air Supply and Conditioning In Microelectronic Facilities

Means for Makeup Air.   Systems that sense extremes of outdoor air temperature and humidity will be dedicated to providing conditioned makeup air. These units are sized to handle all air from outside the facility although, in some cases, provisions are included to accept air returned from the clean rooms. Makeup air systems are designed to deal with the sensible and latent characteristics of air. Makeup air units are often constant-speed units because they offset exhaust air

TABLE 11.3-2   Classifications and Characteristics of Clean Room Levels

Classification	Particle count	Velocity	Range	ACH (10′ Clg)
1	No more than 1	>100	>110	>600
10	No more than 10	100	90–110	600
100	No more than 100	90	50–90	540
1 000	No more than 1000	20	13–30	120
10 000	No more than 10000	10	5–10	60
100 000	No more than 100000	5		30

TABLE 11.3-3   ES-RP-006, Recommended Tests by Clean Room Type

Section	Test	Laminar airflow	Turbulent airflow	Mixed airflow
4	Velocity/uniformity	1		0
5	Filter leak	1	1	1
6	Parallelism	1, 2		0 (1, 2 only)
7	Recovery	1, 2	1, 2, 3	1, 2
8	Particle count	1, 2, 3	1, 2, 3	1, 2, 3
9	Particle fallout	0	0	0
10	Induction	1, 2	1, 2	1, 2
11	Pressurization	1, 2, 3	1, 2, 3	1, 2, 3
12	Air supply capacity	1	1	1
13	Lighting level	1	1	1
14	Noise level	1, 2	1, 2	1, 2
15.2	Temperature	1, 2, 3	1, 2, 3	1, 2, 3
15.3	Humidity	0	0	0
16	Vibration	0	0	0

Key:
0 Test optional, depending upon process requirements
1 Test suited to as-built phase
2 Test suited to at-rest phase
3 Test suited to operating phase

As shown in the table, although a space can be classified strictly by the particle count, this is not usually the case when this space is certified as a clean room. To be compliant with the IES standard, the certification must consider temperature, humidity, laminarity, air velocity, air changes per hour (ACH), lighting levels, and possibly ambient noise levels and vibration. In addition to meeting the standard, the certification has the added benefit of being repeatable because the standard sets the criteria that the testing protocol must follow.

quantities that are generally constant. Cooling coils in makeup air units are either direct exchange (D/X) with remote air cooled condensers, or chilled water. Heating coils in makeup air units use electrical energy, hot water, or steam as a source of heat. (See Figs. 11.3-3 and 11.3-4.)

Means for Recirculating Air.   Depending on the size and classification of clean rooms, there will be systems dedicated to providing the large quantities of recirculated air necessary to sustain the substantial air change rate. These units do not usually have heating coils. If required, heating is provided by heating coils within the duct system. Recirculating air units have filters, sometimes at both the inlet and at the discharge. They have comparatively small cooling coils because their role is usually one of sensible cooling. Each unit is arranged to recycle all of the air it delivers and to accept small quantities of outside air from the makeup air unit to maintain pressurization and to offset exhaust quantities. The cooling coils are either D/X with remote air-cooled condensers or chilled water.

Air Supply and Conditioning in the Pharmaceutical Industry

Means for Makeup Air.   Outdoor air is brought into the production area of a pharmaceutical facility by makeup air units. Each unit provides a constant quantity of supply air to the controlled air environment by automatically proportioning the proper ratio of outdoor air and recirculated air. In summer, makeup air is dehumidified by chilled water cooling coils to maintain a constant maximum dew point of approximately 58°F. In winter, the makeup air is heated and humidified by low pressure steam to maintain a constant minimum condition of approximately 58°F dry bulb and approximately 50 percent relative humidity (i.e., a dew point of approximately 40°F).

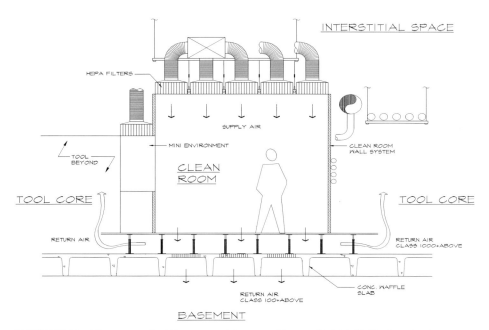

FIGURE 11.3-3   Clean room raised floor return. (*Courtesy of Symmes Maini & McKee Associates.*)

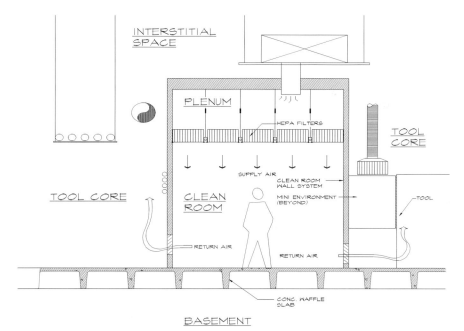

FIGURE 11.3-4   Clean room low wall return. (*Courtesy of Symmes Maini & McKee Associates.*)

Makeup units will include insulated drain pans to remove condensate. Makeup air is filtered with roughing filters and 60 percent ASHRAE filters.

Controlling Facility Makeup Air Units.   The controlled air environmental space pressure is read as a velocity reading in a small duct between the space and a neutral pressure room. Space pressure controls the ratio of outdoor air intake and recirculated air from the controlled-air environmental space (see Fig. 11.3-3).

In summer, the controlled-air environmental space is controlled to a maximum dew point of 58°F by a dew point controller and a three-way control valve in the secondary chilled water circuit serving the cooling coil. As the dew point of the outdoor air drops below 58°F, more outdoor air needs to be brought in through the makeup unit so that less air is required from the dehumidified air main.

In winter, the controlled-air environmental space is typically maintained at 62°F dry bulb and 50 percent relative humidity by controlling the face and bypass dampers of the low-pressure steam heating coil and the steam grid humidifier.

Means for Recirculated Air.   Numerous air supply units are provided to maintain the prescribed environmental conditions in occupied spaces. Return air from the occupied spaces and makeup air from the controlled-air environmental space are filtered in accordance with accepted standards. The return air is withdrawn from the room by horizontal linear registers near the floor or through the floor system for spaces containing clean manufacturing processes. Purge air, in accordance with GMP requirements, is exhausted to the outside from the return air discharge plenum for spaces where hazardous materials are present (see Fig. 11.3-4).

Clean room supply air is filtered in accordance with CGMP. If an area served by an air supply unit has one or more dust collectors, the dust collector discharge air is returned to the air supply unit after filtering.

Supply and exhaust fans are provided with dual, variable-speed motors, directly connected to the fan wheel shaft ends by magnetic clutches. The second motor starts automatically to drive the fan if the operating motor fails. These fans have automatically controlled variable speed motors operated in conjunction with air monitoring devices to provide the proper air quantity and static pressure, efficiently, accurately and at minimum energy consumption through a full range of filter life and heating, cooling, and dehumidification requirements.

Branch ductwork is generally provided with constant-volume regulators (with or without remote automatic reset, as required) to compensate for pressure drop variation in terminal filters, variations in room environmental conditions, or cyclical operation of intermittent process exhaust (see Fig. 11.3-4).

Terminal filters are 90 percent ASHRAE and are located in the controlled-air environmental space immediately above the area being served. All ductwork, grilles, and so forth after terminal filters are stainless steel or aluminum, in accordance with CGMP.

Required dehumidification is provided by automatically proportioning the proper ratio of makeup air from the dehumidified air main and makeup air from the controlled-air environmental space. Additional sensible cooling is provided, as required, either by a central cooling coil in the air supply unit or by cooling coils in branch ductwork serving individual rooms. In a typical arrangement, all supply units contain a hot water heating coil located in the opening for makeup air from the controlled-air environmental space. It may be necessary for these coils to operate at the changeover point from cooling/dehumidifying to heating/humidifying during marginal seasons.

Controlling Facility Air Supply Units.   The makeup air damper and the supply and return fans are modulated to provide the proper quantity of airflow by signals received from air monitoring devices, from process exhaust pick-ups, and from process discharges to the outside (such as can wash, truck wash, etc.)

For air supply systems requiring dehumidified supply air in summer, makeup air is a controlled mixture of air from the controlled-air environmental space and from the dehumidified main, as determined by a dew point sensor in the supply duct. Means are provided to permit a designated unit to supply dehumidified air if the space requirement changes. In winter, makeup air is taken entirely from the controlled-air environmental space.

For air supply systems in which only a portion of the rooms served require dehumidified supply air in summer, supply air to each of those rooms should be a controlled mixture of air from the air supply unit and air from the dehumidified main, as determined by a dew point sensor in the supply branch duct. The space temperature in each individual room is typically provided by a space temperature sensor that first regulates the flow of 62°F chilled water through the sensible cooling coil and then the volume of air supplied to the room.

Humidification

The humidification process adds moisture to the air and crosses constant moisture lines. If the dry bulb remains constant, humidification involves adding only latent heat. Air is humidified by various means including steam injection, air washers, and vaporizing.

The most popular humidifier is the steam-jet type. It consists of a pipe with nozzles partially surrounded by a steam jacket. The steam-filled jacket is fed through the nozzles, and moisture is sprayed into the air stream. The jacket minimizes condensation when the steam enters the pipe with the nozzles and ensures dry steam for humidification. Steam is typically sprayed into the air at a temperature of 212°F or higher.

With water injection, a system that mixes compressed air and water is used to introduce water into the air stream through nozzles at high pressure. This method atomizes the water into micro-droplets allowing easy absorption by the passing air stream. The water can be basic potable (suitable for human consumption), of a higher quality such as highly filtered (reverse osmosis), or deionized. In fab facilities, the object is to introduce humidity without introducing particulates into the air stream.

Air washers are also used as humidifiers, particularly for applications requiring added moisture without introducing significant heat as in warm southwestern climates. Air washers can be recirculating or heated. In recirculating washers, the heat necessary to vaporize the water is sensible heat changed to latent heat that reduces the dry-bulb temperature. Because the water is recirculating, the water temperature becomes the same as the wet-bulb temperature of the air.

Air must be in a state that can accept additional moisture. Air at 55°F is likely to be very near saturation and will not accept enough moisture to be, for example, at 60 percent relative humidity when it gets to the desired room condition of 72°F. In fact, if an attempt is made to add significant moisture by either method, it is very likely that the added moisture will condense on the downstream ductwork.

To prevent downstream condensation, the temperature of the air is raised high enough to accept the necessary moisture. The resulting air may have the right relative humidity but be too warm to provide the desired cooling effect—a classic conundrum. Another solution is to heat and humidify a small portion of the total air and to then mix this air with the remainder of the air to achieve the desired room conditions.

Dehumidification

Humidity is described as *relative* because a given volume of air at a given temperature holds a certain amount of moisture. When the air is saturated, the RH is 100 percent. Because a given volume of air cannot be at more than 100 percent relative humidity and warmer air can hold more moisture than cooler air, it follows that a given volume of air at a given temperature and relative humidity will be dehumidified by cooling. As air is driven across a cooling coil, it tends to approach the temperature of the coil surfaces and become "saturated."

In the microelectronics industry, when microcircuits are manufactured, hygroscopic polymers called *photoresists* are used to mask circuit lines in the etching processes. If these polymers absorb moisture, microscopic circuit lines are cut or bridged, resulting in circuit failure. In pharmaceutical manufacturing, many powders are highly hygroscopic. When moist, they are difficult to process and have limited shelf life.

For these and other reasons, clean rooms are equipped to remove moisture from the air (dehumidify). Various means include cooling the air to condense the water vapor; increasing the total pressure of the air; or presenting a desiccant that pulls moisture from the air through differences in vapor pressure. Cooling dehumidification systems include *direct expansion cooling, chilled liquid cooling,* and *dehumidification-reheat systems.* These remove moisture from the air by cooling to condense the water vapor. Increasing the total pressure of the air reduces the relative humidity but also causes condensation. It is impractical for the types of facilities being considered here. Presenting a desiccant to the air, which pulls moisture from the air through differences in vapor pressure, however, is often used in pharmaceutical facilities.

Dehumidification by Cooling.   In cooling dehumidification, as the air continues through the conditioning process and whenever the air temperature rises, the relative humidity falls. For example, in a draw-through system, air passing across the supply fan (and the supply fan motor, if it is in the air stream) causes the air temperature to rise and the relative humidity to fall. This cooling and subsequent reheating is *dehumidification.* It is limited by the conditions required within the space served by that particular unit. A desired room condition of 75°F and 50 percent RH can be achieved by a conventional setup, where the air is cooled to 53°F by passing it across a cooling coil with a 52°F surface temperature and then allowing it to warm to room conditions after it is delivered to the space. Similarly, conditions of 72°F and 50 percent RH or possibly even 70°F and 45 percent RH can be achieved, if the cooling coil is cold enough and if there are sufficient rows of cooling coil to minimize bypass air. If desired temperatures are much lower or if the desired RH is much lower, say 65°F and 40 percent RH, the coil temperature approaches a level that requires water, if that is the cooling medium, at a temperature below freezing. This is still possible (with appropriate additives to prevent the water from freezing), but operation can become problematic.

Desiccant Dehumidification.   Desiccant dehumidifiers are available in five configurations: *liquid spray-towers, solid packed towers, rotating horizontal beds, multiple vertical beds,* and *rotating honeycombs.* Each chemical dehumidifier consists of a rotating wheel containing a desiccant and a means of regenerating the desiccant. The humid air passes through a portion of the rotating wheel that contains a dehumidified desiccant. The desiccant absorbs moisture from the process air, and then is regenerated by passing a stream of high-temperature air (250 to 300°F) through the wheel.

The regenerating air is taken in from the outside and heated with high-pressure (approximately 150 psig) steam. It is filtered with roughing filters and 55 percent ASHRAE filters to keep the wheel clean, so as not to contaminate the process air. After the regenerating air passes through the wheel, it is exhausted to the outside. Regenerating air intake ductwork from the outside to the unit inlet is generally insulated standard galvanized sheet metal. Regenerating air outlet ductwork from the unit discharge to the outside is usually insulated aluminum ductwork brazed or welded watertight.

Process air is taken in from the controlled-air environmental space and blown through the wheel by a centrifugal fan controlled to provide a constant static discharge pressure. An automatically controlled bypass duct is provided to ensure that the flow of process air through the wheel does not drop below the recommended minimum flow.

Variable quantities of air are withdrawn from the dehumidified air main, as determined by the specific process or space requirement. Whenever outdoor conditions permit reducing the operation of the chemical dehumidifier, automatic controls modulate the capacity of the dehumidifier or take it out of service completely.

The dehumidified air main and branch ducts are round, medium-pressure galvanized ducts, generally uninsulated because the ducts will be at a surface temperature of approximately 95 to 115°F.

Dust Collectors in Pharmaceutical Facilities

Exhaust air pickup stations are generally provided at large process workstations. These pickup stations contain only dust (i.e., no consequential fume level). The exhaust air is conveyed at medium to high velocity to a dust collector. One dust collector may serve many pickup stations. Because the operation of each pickup station is generally intermittent, the fan serving the dust collector is controlled to maintain a constant negative pressure in the dust collector. The amount of return air from a room served by one or more exhaust air pickup stations automatically adjusts for reduced flow to maintain proper space pressurization relationships when one or more exhaust air pickup stations are in operation.

Generally, air discharged from a dust collector is recirculated. Dust collector ductwork is round, medium-pressure galvanized sheet metal, with smooth full-radius elbows. All ductwork, hood pickups, and so forth, located in a process room, are stainless steel.

Dust collectors are typically shaker type, with self-contained fabric filters, a hopper-type dust storage base, and a manual-dump slide gate. Dust collectors contain one set of high-efficiency (98.4 to 99.75 percent by weight), fire-resistant, fabric-envelope-type filters for dusts to one micron.

Fume Scrubbers in Pharmaceutical Facilities

Scrubbers are provided to remove alcohol fumes from process exhaust air. Scrubbers are fabricated of chemically inert laminated plastic sheets, chemically welded to form a low-pressure vessel. Effluent from the scrubber is piped to underground tanks to be held approximately 24 hours before discharge to the sanitary sewer.

Ductwork from the process to the scrubber is round, medium-pressure galvanized spiral-seam uninsulated sheet metal. Ductwork from the scrubber to the atmosphere is round, galvanized spiral-seam sheet metal, brazed or soldered watertight, thermally insulated to minimize condensation within the duct.

Fume Scrubbers in Microelectronics Facilities

For facilities designed for manufacturing microelectronics, scrubbers are provided to remove acidic fumes from the process exhaust air stream. A typical scrubber installation includes a packaged, factory-assembled and tested acid fume scrubber with water recirculation pumps and exhaust fans. Scrubber units include complete stand-alone programmable logic controller (PLC)–based control panels to control the unit and fans completely and to communicate fully with the facility-wide distributed digital control system for monitoring and control.

Ventilation Exhaust Systems

General Ventilation.   Ventilation systems are provided for mechanical rooms, transformer rooms, switchgear rooms, telephone equipment rooms, power generation rooms, and so forth, as required, to maintain space temperature and meet code requirements.

Fire and Life Safety Ventilating Systems.   Facility ventilation must be arranged to function under various scenarios including those based on preserving viable conditions for worker safety. These provisions may include changing operating conditions to pressurize a space if an alarm should sound, purging an area after an incident to remove the hazardous gases inadvertently released, purging an area that was secured using an agent suppression system, and smoke control and removal.

Venting systems with fire-rated enclosures are provided for duct and pipe shafts, as required by code. Systems typically include motorized smoke dampers with venting ducts or openings to atmosphere sized at 10 percent of the shaft area.

Passenger elevator lobbies and service elevator vestibule pressurization systems with fire-rated enclosures may be required.

Garage Ventilation. Garage ventilating systems (with heating in cold climates) are provided for garages by using supply fans, electrical heating coils with multiple-step controls, motorized air intake dampers, CO monitoring systems with one sensor per 25 ft radius area, and corresponding exhaust fan(s) with motorized exhaust dampers. Controls are arranged so that they are off when there is no vehicular activity, on at low speed when moderate activity occurs, and on at full speed during high traffic periods.

SECTION 11.4
CHEMICAL PROCESS DESIGN

This section focuses on process design parameters for the semiconductor, biotechnology, microelectronic, biomedical, and pharmaceutical industries. Common components for these industries include clean room space; use of ultrapure water; use of solvents and gases; utilities such as boilers, chillers, and cooling water systems; air abatement systems; and waste treatment systems. Because of clean room space, these facilities are often constructed as two distinct building areas, one where the product is manufactured and one to house the associated utility systems. Although these industry facilities have much in common, there are significant differences, especially in the ways system designs may be integrated.

SETTING PROJECT, OPERATIONAL, AND TECHNICAL GOALS

Project, operational, and technical goals need to be identified, understood, and mutually agreed upon by the team at the project outset. Goals established before the A/E joins the team need to be reviewed, tested, and mutually agreed upon. Project goals address the process and include a budget establishing how the various systems are integrated to complement operational and technical objectives, determining what design/construction phasing is required to achieve the operation/start-up schedule, and developing a framework for decision making.

Operational and technical goals typically address the resulting product, and operational goals answer questions like these:

1. Will the facility be used for R&D, pilot production, or full manufacturing?
2. Will the operations facility be primarily electronic, clean, wet/dry lab, biohazardous, toxic, or mixed?
3. What are the current and projected populations and hours of operation?
4. What are the present and future production outputs?
5. How much required/desired redundancy can be afforded?

6. What is the desired life span for materials and major equipment/systems?

7. How can the facility's design be cost-effectively humanized?

Operational goals examine future flexibility versus initial expenditures and the initial cost of major systems versus the cost of effective operations during the facility's life.

Technical goals set performance requirements for individual systems or components. These performance goals can include system requirements for temperature, humidity, vibration, acoustics, process purity, quantities, and flows. Goals for the performance of building components to meet vibration, acoustical, and cleanliness specifications are also set.

Another objective is to determine if it is more cost effective to use in-house maintenance staff or to try to locate an agency to perform maintenance. The owner in concert with the A/E should develop a list of the tasks and skills required by the manufacturing staff. In general, the higher the skill level, the greater the need for an in-house staff because special (and usually ongoing) training is required.

PROJECT COSTS AND SCHEDULES

In most projects, the "nice-to-have" items and systems drive costs higher. Value management is used area by area and system by system to review and determine what is essential to meet the process/manufacturing goals. The purpose of value management is twofold—to develop priorities ranging from bare bone essentials to levels of enhancements that would be "nice to have" and/or to find a more economical solution that will permit the owner to afford the project.

The findings of this review identify both cost reductions now and opportunities to incorporate provisions now in the building and system design that will permit additions in the future, all at minimal costs. If these provisions are not incorporated early but are identified after the plant is being constructed or operational, they could have major building design and cost impacts and/or disrupt the manufacturing operation.

In addition to cost, schedule is also a major factor driving most projects. The owner's plant must be on-line and producing a saleable product in advance of its competition. This criterion forces innovative approaches to designing, procuring, and constructing the facility. One approach is *fast-track,* whereby many design, procurement, and construction tasks are performed in parallel instead of sequentially. Shortening the overall project schedule requires purchasing some equipment, system, and/or material items much earlier to permit completing design and engineering tasks in time for the required construction sequence.

Long-lead items need to be identified during the schematic design and design development phases. The time for a long-lead item is the time between its purchase, review of submittals, and the start of the manufacturing process, plus the time it takes to manufacture the item and deliver it to the site. For example, structural steel currently takes 20 weeks after the purchase order is awarded until the first steel arrives on-site. Other examples are electrical switchgear, refrigeration compressors, cooling towers, fired steam boilers, and special manufacturing tools. Each provides input into the design process so that the plans and specifications can be finalized. Early procurement is required to ensure that equipment and materials arrive on-site in time to be inserted into the building and connected to the system early enough to coincide with the owner's commissioning and start-up schedule.

It may be worthwhile for the owner to hire an expediter for these long-lead items, who is responsible for staying in regular contact with the vendors to ensure that the document submittal and manufacturing schedule of critical items are in accordance with the contracted scope and schedule. This equipment should be thoroughly checked out at the factory by the owner's agent and/or A/E to ensure that it performs in complete conformance with the purchase specifications before it is released for shipment to the site. The cost to ensure the equipment's operability up front could prevent future major additional cost and schedule delays if the equipment is inoperable at the site.

Early in the schematic design phase of a project, the A/E works to develop a realistic estimate of the cost to build the facility to the owner's specifications. Based on prior projects of similar square footage, manufacturing process conditions, and the size and type of supporting physical plant equipment and systems, the A/E and/or constructor develops a cost model with an accuracy of ±25 percent. This cost model promotes design interaction with the owner and enables the owner to put in perspective the alternative design approaches presented by the A/E with respect to potential costs.

Once the design development phase starts, the constructor must be on board if it is a fast track or design/build project. The Constructor verifies and adjusts the cost modeling. Because of the A/E's greater definition of the project scope and the cost of tools used by the constructor, the accuracy of the project estimate can be increased to ±15 to 20 percent.

By the time the A/E issues construction documents, the constructor can provide a lump sum or *guaranteed maximum price* (GMP) to the owner. This enables confirmation of the project's budget near the start of the construction phase. If costs are higher than expected, additional "value management" might be proposed by the constructor to stay within the budget. Usually, the A/E advises the owner of the alternatives proposed by the constructor and the potential effect of actions the owner may take.

RISK MANAGEMENT ASSESSMENT

Early in the design development phase, the owner is advised to perform or have his insurer perform a risk management assessment of the design. This assessment should take into account the A/E's code review, the facility's layout, the type of facility construction, the physical plant systems, and the systems required for the manufacturing process.

By doing this exercise early, the owner achieves these benefits: an outside and independent review of the planned facility from an informed operating risk view; identification of any system risks inherent in the owner's plant; elimination or minimization of any life safety and operational risks; and ultimately a buyin by the insurer or the identification of any additional insurance/risk costs inherent in the planned design. This is accomplished early in the design development phase, with adequate time to permit resolution by the insurer, owner, and A/E during the construction document development phase.

The first step in process design is to use the target production numbers and calculate the required equipment trains to produce the required volume. Detailed heat and material balances are the basis for this determination. Factoring in such criteria as yields, safety concerns, economic batch size, and redundancy results in a selected set of equipment. Usually this means multiple tanks, multiple reactors, and so forth. The owner and A/E jointly decide whether issues of redundancy or economic batch size are realistic for the project. The owner's insurance company reviews the plan at this stage as well, so that all parties agree with the equipment decisions and costly changes later are avoided.

Continuous processes change very little over time. Any change is usually slow and gradual, and drastic swings in operating conditions are minimized or eliminated.

On the other hand, batch processes involve filling, starting, running, stopping, and emptying, then repeating the cycle or recipe. Both products and operating conditions can change. Run time lengths often vary by product. One batch may be run with the vessel almost empty, whereas in the very next one the vessel is completely filled. These changes put a tremendous demand on the flexibility of the utility system. One method used to handle the large swings in demand is to provide modular utility units, arranged with some installed excess capacity and variable frequency drives (VFD). For example, suppose that two circulating pumps are running at 60 percent of capacity. As demand rises, the VFD increases motor output until 100 percent capacity is reached. At that time, a third pump is brought on line that allows capacity to grow to even higher levels.

Materials handling in these specialized piping systems requires great care in design. Fluid velocities should be kept in the low range (2 to 7 ft/s) to minimize erosion and to help preserve the pipe longer. "Saltlike" solutions are abrasive to piping, and a study should be undertaken at the start of the design process to understand these project-specific effects.

MATERIALS HANDLING

The owner and A/E need to develop appropriate containment strategies and review methods for handling difficult materials. Radiation hazards require special lead shields. Isolated "hot" boxes are used to contain these substances, and the operators often use remote manipulator arms to handle the radioactive hazard properly. The goal is to achieve a facility design with the proper precautions built in. Operator safety is always paramount.

Handling hazardous materials presents different problems for the piping designer. The integrity of the piping components is of utmost concern. When laying out a run of pipe, continuous welding/joining of the pipe is desirable, and flange connections should be minimized. Splash guards around flange joints are used to contain any leak, spray, or drip from exposure because the material within the pipe may cause bodily harm.

SEMICONDUCTOR AND MICROELECTRONICS INDUSTRIES

The semiconductor and microelectronics sector has some of the most demanding and dynamic industries. The useful life of a generation of a particular semiconductor product may be only 18 months. The company that gets its product to market first may capture 70 to 90 percent of the market. But to get there, a company needs to plan properly. A state-of-the-art semiconductor facility can cost over $1 billion. The company usually needs to commit the funds to start construction before it is absolutely sure that the next-generation product will work and with only a vague idea of needed utilities. The A/E needs to be experienced in designing these facilities and able to lead the owner in deciding the system requirements.

Process systems required in semiconductor facilities include:

- Deionized (DI) water, sometimes called *ultrapure water* (UPW)
- Process cooling water
- Process vacuum
- House vacuum, sometimes called *janitorial vacuum*
- Bulk gases
- Specialty gases
- Chemical distribution systems
- Wastewater collection and treatment
- Reclaim water collection and treatment
- Solvent collection
- Acid gas abatement
- Alkali gas abatement
- Solvent vapor abatement
- Compressed air systems

Deionized (DI) Water, Sometimes Called Ultrapure Water (UPW)

Deionized water is used throughout a microelectronics facility. In fabricating a given wafer, it will usually be rinsed several hundred times. Any particles or ions that are present in the DI water can stick to the wafer and alter the electrical properties of one or more chips on it. This results in reduced yields. Generally, these plants consume a tremendous amount of water; a good starting assumption would be about 0.01 gpm of makeup water for every 100 square feet of fab floor space.

The DI system can be configured in a number of ways. The water that is used in a plant can come from a variety of sources: lakes, rivers, oceans, or underground streams. In the industrial arena, there is considerable need to clean up the raw incoming water to a higher level of purity. This step is called *pretreatment*. It can be as simple as coarse filtering or as fine as microfiltration. The A/E needs to evaluate the end use requirements of the water to determine the required degree of pretreatment. If the source is surface water, pretreatment is often provided to remove organics and suspended solids that may occur at various times of the year. In that case, the process usually has filters, followed by granular activated carbon filters to remove organics and chlorine. After this, the pretreatment systems use some combination of ion exchange and reverse osmosis to remove most of the dissolved ions. A typical target level for resistivity at this point is about 17.5 to 18 MΩ. The water is then usually sanitized using ultraviolet light that partially ionizes the water again. Mixed beds, vessels that contain both anion- and cation-exchange resins, are used to remove the last vestiges of particles from the water and discharge what is called *DI water*.

Although the dissolved and suspended solids have been removed, the water still has bacteria that will grow in this ultrapure environment, and these stick to any surface in the piping system that is not experiencing turbulent flow. This means that DI water is distributed in the fab in a continuously recirculating loop of water that is recleaned after each pass through the facility.

The tools in the facility are usually configured to operate at one common delivery pressure. This requirement also needs to be met under all operating flow conditions that could range as high as ±30 percent around the design average flow rate. Usually, this requirement is met by providing a reverse return loop or by providing standard return loops and a sophisticated control loop between the recirculation pumps and a control valve located at the end of the supply main.

Process Cooling Water

Process cooling water is a cooling water system separate from the HVAC systems. It is a closed-loop system that is usually charged and made up from DI water. This water is delivered to tools that require cooling water. The reason for the separate system is that many tools operate at very high voltages and need cooling water that has a moderately high resistivity. If normal water were used, the high voltage would generate a current in the cooling water and reduce tool efficiency. Also, care must be taken not to use only DI water because it may be too pure and corrosive for this application. Many tools use copper, brass, and other materials that may be corroded by ultrapure water. Many systems operate at a resistivity of about 50,000 Ω-cm.

Process cooling water systems usually consist of recirculation pumps, a heat exchanger, cartridge filters, and a blowdown and makeup system. A slipstream ion-exchange vessel may be included to help maintain the proper resistivity of the water.

Process Vacuum

Process vacuum systems are the high vacuum systems whose normal operating pressures are around 27 to 28 inches of mercury vacuum. These systems normally use stainless steel piping in their distribution systems.

House Vacuum, Sometimes Called Janitorial Vacuum

House vacuum systems are low-pressure vacuum systems whose normal operating pressures are around 6 to 18 in of water vacuum. These systems typically use galvanized sheet-metal pipe joined with heat-sensitive shrink wrap.

Bulk Gases

Bulk gases are those that are delivered to the facility in bulk by container truck or tube trailer. Many larger plants have bulk nitrogen and oxygen generating facilities. Bulk gases include nitrogen, oxygen, helium, hydrogen, argon, and forming gas (a combination of hydrogen and carbon dioxide). Hydrogen and forming gas are considered flammable and their distribution systems usually are more consistent with the approaches used for specialty gases than for the other bulk gases.

Specialty Gases

Specialty gases are those that are supplied to the facility in bottles. These gases have various properties and hazards. The hazards are grouped as corrosives, oxidizers, flammables, pyrophorics, and inerts. Significant building code requirements regulate the allowable quantities of specialty gases, how they are stored, how they are distributed, and how the rooms that store them need to be designed.

Gases that are classified as corrosives, oxidizers, or flammables usually are stored inside a ventilated cabinet. This cabinet provides secondary containment for the gases and protects workers if a leak occurs in one of the many fittings of bottles and manifold valve assemblies. Gases classified as inert are relatively harmless, other than as potential asphyxiants, and storage in vented cabinets is not usually required. Pyrophoric gases ignite immediately upon contact with air. Usually, these gases must be stored in an outdoor area at least 50 feet from any buildings.

At least one gas, chlorine trifluoride (ClF_3), poses unique risks to the manufacturing environment. This gas is a highly corrosive oxidizer. There is no known personnel protection gear that will stand up to this gas. If a leak occurs, personnel cannot suit up and try to isolate the bottle manually. The only course of action is to allow the bottle to release gas to the local environment until all of it is exhausted. If this gas is located in a room filled with other gases, none of the other gas bottles can be changed until the ClF_3 bottle has exhausted itself and the room has been decontaminated by the room exhaust system. This type of failure event could impact many aspects of the fab operation. Because of this scenario, many fabs that use ClF_3 locate this gas inside a dedicated room to prevent other systems from being impacted.

Chemical Distribution Systems

The semiconductor industry uses a large variety of ultrapure chemicals during the manufacture of silicon chips. Virgin as well as waste chemicals must be considered. The owner provides the A/E with this information, much of which is available from the owner's Industrial Safety Sheets (ISS), Material Safety Data Sheets (MSDS), and vendor purchase orders. This information, collected in matrix form, can be used to plan the capacity and design of the chemical distribution and dispensing systems. This information is important in developing the overall project cost model.

Chemicals are usually brought into the fab in tote bins that are hooked to specially designed chemical distribution units. These units are supplied by a number of vendors and are usually designed to maintain chemical consistency and purity. The chemicals are usually located in groups as acids (except hydrofluoric acid), hydrofluoric acid, alkalis, solvents, and slurries. This grouping facilitates the control of any spills or leaks. Control areas specifically designed to store hazardous materials may be used to increase the amount of hazardous materials allowed in an area without the need to classify the entire facility as a hazardous use group. Building code requirements for separating incompatible materials such as acids, alkalis, and oxidizers from flammable solvents must be considered. Hydrofluoric acid is generally kept separate from all the other materials because most facilities have a wastewater discharge limit for the amount of fluoride that can be released. This material and any wastes containing it are usually collected in a separate system and pretreated on-site before combining with the wastewater flow from the rest of the facility.

Chemical process and environmental engineers must become involved in the project design in the schematic design phase. The process engineer develops the facility design to a point where air and water emissions from effluents can be estimated. Once these are known, the environmental engineer (often an outside consultant) studies the state and federal regulations to determine the acceptable limits for the emissions. Working together, the process engineer and the environmental engineer determine the best control technology to provide the required emission reduction. These recommendations are communicated to the design team throughout the design process.

Wastewater Collection and Treatment

Program standards need to be developed in the schematic design phase to define what to do with process wastes. Clear guidelines must be established because governmental agencies closely regulate the treatment and disposal of industrial wastes. A matrix should be established that shows the flows and expected chemical composition of all waste streams. This matrix can be used as a management tool to segregate wastes into similar categories and make treatment easier and less costly. If segregated carefully, valuable products in waste streams can often be recovered and reused.

Wastewater is generated in many places within a semiconductor facility. The largest source is the tools used in the fab to etch, rinse, and polish the wafer. The next-largest source is usually the *reverse osmosis* (RO) reject stream from the DI water system. Other significant sources are the cooling tower blowdown and the scrubber blowdowns. Cooling tower and scrubber wastes all generally have high flow rates but contain relatively low concentrations of contaminants. The reject from ion exchange in the DI water system is usually not significant in volume but is a significant source of the facility's ionic discharge. Some tools in the fab will also have multiple baths. Some of these baths contain concentrated acids, and others contain rinsing baths. Some manufacturers collect the concentrated acids in segregated systems without combining them with the rinse waters. Other manufacturers allow all of the waste from the tools to drain to a common drain system.

Wastewater is discharged either directly into a stream or river or indirectly into a sewer system, where it will be combined with sanitary wastewater and ultimately be treated in a publicly owned treatment works (POTW). Regardless of the way the wastewater is discharged, it must be treated to meet certain discharge limits imposed by state and local authorities. At a minimum, treatment usually requires pH adjustment to maintain an effluent pH between 6.0 and 9.0. Depending on the facility and its location, other constituents such as fluoride, phosphorus, copper, cyanide, and total biological oxygen demand (BOD) may also need to be controlled. Cost-effective treatment for these other constituents usually requires that wastes containing these materials must be collected in separate collection systems. Treatment for fluoride and phosphorus normally requires precipitation using lime. Treatment systems such as ion-exchange, reverse osmosis, or chemical oxidation may be required to remove other contaminants.

An important logistical issue occurs in larger facilities—the slope of the drain piping places the final main pipe leaving the building 10 to 20 feet below ground level. This requires either a very large hole that is almost 40 feet deep to hold the facility wastewater treatment system or a pit this deep that is large enough to hold pumping stations to allow the location of treatment systems at grade level. Minimizing the excavation will usually minimize the cost of the system and save a significant amount of time during construction.

Reclaim Water Collection and Treatment

To minimize the consumption of off-site water, many facilities install reclaim wastewater collection and treatment systems. These systems collect wastewater from the last one or two rinses from each tool. This water contains only trace amounts of contaminants that make it too dirty to use for rinsing a computer chip but very clean by normal industry standards. This water is treated to control pH and any significant inorganic contamination and then is usually treated with activated carbon to remove any biodegradable organics. The reclaimed water is used to provide makeup to the cooling towers and scrubbers and to supply dilution water to any sulfuric acid waste discharges and any concentrated slurry waste discharges. Some facilities reclaim as much as 70 percent of the initial DI water.

Solvent Collection

A variety of solvents are used in a typical semiconductor facility, including alcohols and ketones, assorted glycols, and a solvent called *N*-methylpyrolidone. These chemicals become contaminated with various substances in the course of their use. Because the material specifications for these solvents are very stringent, it is rarely cost-effective to try to clean them for reuse. The industry standard is to collect these wastes and haul them off-site for disposal as hazardous wastes. Because many of these solvents are flammable, a number of building code issues need to be addressed during the design of these systems. In addition, because these wastes are designated as hazardous wastes, a number of federal regulations must be met during the design and construction of the collection and storage system.

Acid Gas Abatement

Acid gases emanate from many sources in a fab, including wet benches, vapor deposition devices, and point-of-use abatement systems. These gases are collected by acid gas vent systems and are treated using scrubbers. Scrubbers are generally packed towers using either water or a caustic solution to remove the acid gases from the vent air before it is discharged to the atmosphere. The blowdown from the scrubbers is usually sent to the wastewater treatment facility. Past experience has favored the use of water only for scrubbing the acid gases. This medium is not as efficient as caustic and thus requires larger scrubbers, more water, and more blowdown. However, using caustic as a scrubber medium usually results in the absorption of carbon dioxide from the air which does not require removal. The concentration of carbon dioxide in the acid gas is usually much higher than the total amount of acid. If caustic is used in the scrubber, significantly more caustic is required than the stoichiometrics of the acid gases. Then, the resulting blowdown puts even more of a chemical load on the wastewater treatment system. The cost of the caustic supply system and the operating cost for supplying caustic essentially to remove carbon dioxide can become very high. Because these plants have lots of gray water available for scrubbing system makeup, it usually is more cost-effective to use only water in the scrubbers.

If a facility is using chlorine gas and needs to abate this source, then scrubbers using caustic need to be used. Water alone does not remove chlorine gas emissions.

Alkali Gas Abatement

Some of the more common gases used in the semiconductor industry include ammonia and trimethylammonium hydroxide. Removal of these gases is achieved by using packed towers similarly to the acid gases. Sulfuric acid is typically used in the tower to help remove the alkali gases. Water alone is not sufficient to remove these gases.

Solvent Vapor Abatement

Volatile solvents used in the semiconductor industry include isopropyl alcohol, acetone, hexanol, and N-methylpyrolidone. Some of these gases need to be controlled in accordance with the Clean Air Act, which requires a minimum removal efficiency of 90 percent. Several technologies exist to remove these gases, including:

• Activated carbon with off-site regeneration
• Activated carbon with on-site steam regeneration
• Activated carbon with on-site hot gas regeneration and on-site oxidation of the gas
• Zeolite adsorption with on-site hot gas regeneration and on-site oxidation of the gas
• Direct oxidation of the gas using several technologies such as catalytic oxidation and regenerative thermal oxidizers

Selecting the technology to use is typically determined by a study that evaluates issues pertinent to the particular facility, such as removal efficiency required, actual chemicals to be used and estimated concentrations, capital and operating cost, system size and weight, and prior plant experience. One compound that needs special attention is dimethyl-silazane. This compound produces silica when burned and if not handled correctly, will quickly foul an abatement system.

Waste Management

An additional step in determining program standards is defining what to do with wastes generated by the process. A clear set of guidelines must be established because governmental agencies closely regulate the treatment and disposal of wastes.

A matrix should be established that shows the flows and expected chemical composition of all waste streams. This matrix can be used as a management tool to segregate wastes into similar categories and make their treatment easier and less costly. If segregated carefully, waste streams can often be recovered as valuable products and reused. The owner and the A/E should look for ways to recycle them because substantial cost savings can often result.

The semiconductor and biopharmaceutical industries have many unique material handling devices to solve their toughest requirements. In the semiconductor field, computer chips must be processed through many steps to become completed devices. Each step along the way is usually done in a dedicated machine, and then the chip is automatically moved to the next one. To minimize contact with humans (and thus risk of contamination), robotics is used extensively. The operator loads a tray of material at the start, and the robotics passes it from machine to machine automatically. When individual chips must be handled often, vacuum wands (a small tube with a large surface at one end) are used to pick up items without touching them by hand.

Compressed Air Systems

Compressed air is generated to drive instrumentation pneumatic systems: air-operated diaphragm pumps, pneumatics associated with the fabrication tools, and mechanical aspirators to humidify the air during winter operations.

Compressed air systems usually consist of one or more compressors followed by a receiver tank, coalescing filters, dryers, and particle filters. The compressors are generally oil-free rotary-screw-type compressors, but some facilities use centrifugal-type compressors. For relatively small amounts of compressed air, rotary-screw compressors usually are more cost-effective than centrifugal compressors. A receiver tank is generally used to help the system respond to fluctuating conditions in the plant. Variable frequency drives on rotary-screw compressors or inlet vane dampers on centrifugal-type compressors are also used to control system demand. The receiver tank is usually sized based on one gallon of volume for every cubic foot per minute (cfm) of compressor capacity, and as a pressure vessel, the tank is required to meet ASME pressure vessel requirements. Receiver tanks can be made of carbon steel, but because of the very high humidity in the tank, it is recommended that the tank interior be coated with a corrosion inhibitor. Air under high pressure cannot hold as much water vapor as air at atmospheric pressure. Because of this, coalescing filters are used to help remove the water that exists as liquid water particles.

Semiconductor plants normally use compressed air with a dew point of at least −40°F, and many use a dew point of −100°F. Many facilities use desiccant dryers to achieve these dew points. "Heatless" dryers use about 15 percent of the inlet air flow rate to regenerate the desiccant. This demand needs to be accounted for when sizing the compressed air system. Because the desiccant in the dryer will shed dustlike particles, particle filters are generally installed downstream of the dryers. After filtration, this air is then split into streams such as instrument air and plant air. Air to be used within the tools is often filtered again with a very-small-pore filter to remove as many particles as possible. Cleanliness of tool air is important because when a tool in the fab uses air, the air is vented into the air space of the fab and can contribute particles that may affect chip yields and performance.

BIOTECH, BIOMEDICAL, AND PHARMACEUTICAL INDUSTRIES

Water used in an advanced technology facility requires some form of pretreatment. Purification follows pretreatment and involves ion exchange, reverse osmosis, and other extensive process steps.

Pharmaceutical and biotechnology companies look at purified water as a substance that contains no microorganisms. These industries would like pure water (no or low chemicals) but focus on the sterility content as their most important concern.

Process

Specialty process utilities are used in the biopharmaceutical industries. These include, but are not limited to, deionized (pure) water, clean stream, and water for injection (WFI). These utilities require special expertise from both the engineer and the mechanical contractor because exotic materials might be needed to achieve high purity. The biopharmaceutical industries often have to deal with materials that are difficult to move (in solid form, very viscous, or toxic). Here conveyors are often the solution because materials can be sucked through a pneumatic conveying line using a vacuum source. Screw conveyors work well with viscous materials because the materials can be loaded in bulk into a hopper and screw-conveyed (much like a meat grinder) to their destination. The drawback to these systems is distance. Oftentimes they can be operated only at 35- to 40-ft distances, so equipment must be placed close together.

It makes good economic sense to control emissions because it is becoming more expensive to dispose of wastes. Landfills are scarce, and the costs to use them are rising rapidly. In keeping with this theme, recycling has gained popularity. If wastes can be treated on-site and returned to a point where they are acceptable for reuse, then raw material costs decline, and waste disposal costs are drastically reduced. A side benefit to the recycling/reusing approach is that the overall yields of products rise and the cost to produce declines. The companies that are successful in dealing with their wastes through either recycling or sale of by-products will realize competitive advantages in the marketplace.

Pharmaceutical facilities must record GMP compliance by diagrammatically establishing personnel, product, and material flows, as well as establishing the clean and hazard ratings of all spaces. Based on manufacturing processes and all regulatory criteria, the owner must establish those standards that the facility must meet (e.g., FDA regulations, clean room classifications, biohazard levels, toxic levels). The owner and A/E should work together to determine the most appropriate way to meet the standards. Such decisions are usually based on safety, cost, flexibility, redundancy, labor, and maintenance.

Negative pressurization is required for a containment area. Negative pressure control is intended to protect surrounding areas from biological and process hazards. The National Institutes of Health (NIH) guide *Bio-safety in Microbiological and Biomedical Laboratories* describes various facility design considerations in detail. These range from relatively safe environmental controls (BL-1) to dealing with lethal agents (BL-4), as shown in the following listing (lethal agents are those for which there is no antidote):

- Biosafety Level 1: Work involving well-characterized agents not known to cause disease in healthy human adults.

  Minimal potential hazards to lab personnel and the environment.

  Lab is not necessarily separated from the general traffic patterns of the building.

- Biosafety Level 2: Similar to Level 1 but contains moderate potential hazards to personnel and to the environment.

  It differs from Level 1 in that:

  1. Lab personnel have specific training in handling pathogenic agents.
  2. Access to the lab is limited when work is being conducted.
  3. Extreme precautions are taken with contaminated sharp objects.
  4. Infectious aerosols are created if the work is done in a biological safety cabinet.

- Biosafety Level 3: Work is done with indigenous or exotic agents that may cause serious or potentially lethal disease as a result of exposure by inhalation.

  Lab personnel have specific training in handling pathogens and potentially lethal agents and are supervised by competent scientists.

  Materials involving the manipulation of infectious materials are conducted within biological safety cabinets.

  The lab has special engineering and design features such as a difference in pressure between it and the next room or hall.

- Biosafety Level 4: Required for work with dangerous and exotic agents that pose a high individual risk of aerosol transmittal infections or life-threatening disease.

  Members of the staff have specific and thorough training in handling extremely hazardous infectious agents.

  Work is done in biosafety cabinets.

  Personnel are trained in primary and secondary containment protocols.

  Access to the lab is strictly controlled.

  The facility is in a separate building or in a controlled area within a building that is completely isolated from all other areas of the building.

SECTION 11.5
BIDDING AND CONSTRUCTION APPROACH

OVERVIEW

In an advanced technology project, the bidding and construction phase doesn't begin with the completion of the construction documents (plans and specifications, "contract documents"). It begins during the programming and schematic design phase, is refined during the design development phase, and finalized before the construction document phase. By agreeing on issues at these times, the owner, A/E, and constructor have an execution strategy for the schedule, budget, and tasks/approaches that complement the owner's goals.

The owner's product release date usually sets the schedule goals for portions of the facility that must be up and running significantly earlier than other systems, permitting its integration with existing production or systems. These dates help focus the owner and A/E on the strategies available when planning the timing and order of tasks and which construction approach is best-suited to the project objectives (i.e., general contractor, construction management at risk, design-bid-build, design-build, guaranteed maximum price, etc).

To achieve success from the schedule and budget perspective, the owner's team must be established during the programming and schematic design phases and must participate in the day-to-day evolution of the design strategies and decisions. Involving the owner's key personnel at this time, and subordinate personnel as the project progresses into the design development phase, ensures that decisions made and actions taken will not be changed later at great expense by new decision makers.

The bidding process evolves from the agreed-upon execution strategies. It will be influenced by the schedule objectives, as well as by the type of construction contract planned, and should be formalized in a contracting plan. The long-lead equipment and construction materials usually are defined in the design development phase and placed on order at the start of the construction document phase. This permits vendor information, and ultimately the equipment or material, to arrive in time to complement the finalization of the contract documents and its timely installation. The rest of the procurement tasks are usually accomplished during the construction phase. The exceptions to this are fast-track projects, which require a phased approach to bidding and construction. For example, site preparation work; foundations and building structural framework; the building envelope; mechanical, electrical, plumbing, and fire protection rough-in work; and internal process equipment/systems all might be initiated while the construction documents are in various stages of development.

During construction, the owner and A/E must provide timely support to the constructor. The owner coordinates the project and operation/maintenance teams to plan, schedule, and integrate the existing plant operation with the tie-in needs of the project. The owner's team members should be assigned day-to-day inspection duties to ensure that the operation/maintenance, safety, and quality are consistent with project requirements. By using future plant operation and maintenance personnel to perform these inspection tasks, the owner will enhance their knowledge of the new plant and their readiness to commission, operate, and maintain it. The owner's duties parallel the inspection, quality, and safety tasks "owned" by the constructor. Concurrently, the A/E reviews the in-progress work for compliance with contract documents. The A/E reviews trade submittals, shop drawings, and system mock-ups for compliance with plans and specifications.

The constructor develops a construction schedule and an approach that complements the owner's project goals and manages so that the trades perform with quality workmanship in a timely and safe manner. The owner, A/E, and constructor must function as a team, working to interact and provide timely support and direction.

The owner's commissioning phase is planned by the team (owner, A/E, and constructor) usually during the design development phase. This is required to ensure that all parties are in step with each other, all equipment and materials are available when required, and all trades have a clear understanding of work priorities. No owner wants to find that key components of a process or system can't be completed because something is missing or that tasks take significantly more time to install and bring on-line than anticipated. Therefore, all parties must fully understand what is going to happen and when. This will ensure that all contributors have adequate time to review and discuss their roles in the construction and commissioning process; to agree on the activities, materials, and equipment required to permit the work to proceed; and to achieve overall team buyin to the construction and commissioning plan.

BIDDING PROCESS

The A/E and owner work with the constructor to establish the realistic lead times necessary to ensure adequate turnaround time for bidding and procurement. This might mean pursuing alternate sources or using payment incentives to achieve the required delivery or finding an innovative adjustment to the schedule that won't impact the commissioning date(s). The strategies for bidding or the contracting plans are "set in stone" to ensure that all parties work toward the same objectives.

The A/E assists the owner and constructor in developing approved vendors lists for critical materials and equipment. These lists contain the names of prescreened vendors for unique scope of work; a minimum of three vendors, usually a maximum of five, per bid inquiry forms the basis for the constructor to "let" bid inquiries. Each bid inquiry identified in the contracting plan usually stipulates an allotted budget; delivery schedule; release date; and the necessary plans and specifications that define the scope of material, equipment, and/or service.

The design schedule is updated by the A/E and constructor to reflect the strategy identified in the contracting plan. For example, on fast-track projects it is common to issue in-progress A/E documents for bid before the documents are formally released for construction. In this approach, design work can be checked, coordinated, and finalized while proposals are being received, providing a potential overlap of two to six weeks in each team's work schedule. Then the "Issued for Construction" documents are released and used by the constructor to confirm the two lowest bidders, just before awarding the contract. The A/E assists the constructor and is a loyal ally to the owner—putting issues in the proper perspective, identifying trade-offs, and answering Requests for Information (RFIs).

The A/E normally performs a technical review of the bids received by the constructor and makes recommendations to the owner regarding alternatives proposed by bidders. The A/E participates in the value engineering function to identify alternative solutions within the domain of the agreed-upon quality and standard for an item. Occasionally, this might require a redesign of specific portions of the work (as an additional service) to implement an innovative strategy.

The owner is responsible for timely responses to A/E and/or constructor queries. For example, the owner receives a Letter of Agreement from the constructor with his recommendation for the purchase of a specific material, piece of equipment, and/or service. It is important that this be reviewed and a timely decision made. Delays by any party beyond the period allocated can cause schedule delays.

Using the project schedule, contracting plan, and approved vendors list, the constructor lets the bid inquiries, coordinates bid evaluations with the A/E, evaluates bids received against the project budget and schedule requirements, and makes recommendations to the owner for contract awards.

The contracting plan for long-lead items includes all materials and equipment that must be purchased before the awarding of the applicable construction trade contract. Some of this long-lead procurement involves obtaining specific vendor equipment and material information that is required by the A/E to adequately complete construction documents so that the time it takes to manufacture the unique material and/or equipment and deliver it on-site will dovetail with the construction schedule date for its installation.

CONSTRUCTION PROCESS

The project construction phase is the most critical one of all in that this is when team actions are most likely to determine whether quality, budget, and scheduling goals are to be realized. All of the up-front work in planning and strategizing is compromised unless each party is diligent in performing competently during this phase. Initially, the A/E team reviews numerous submittals from each vendor and trade. The A/E reviews each submittal for compliance with plans and specifications and visits the site periodically to ensure that approved materials and equipment are being used and installed correctly. This is a double-check system; the constructor. is ultimately responsible for ensuring that the subcontractors comply with plans and specifications.

The A/E has found that by requiring the trade to provide a unique mock-up of a specific exterior wall system, pump piping manifolds, typical process operator station, special architecture features, and so forth, a higher quality and more consistent result can be achieved. Now the A/E, owner, and constructor have an approved standard (before the specific trade work is allowed to proceed) by which the trade's future work can be fairly judged for uniformity, functionality, color, and quality.

The submittal review evolves into the construction shop drawing review from each trade. During this process, it is common to receive Requests for Information (RFIs) from the various trades to clarify and reconcile issues or details. The trades might suggest alternate details for the A/E's review and approval that reconcile or modify details of field conditions found during renovation and/or construction.

The A/E is usually requested by the owner to review the constructor's change requests for accuracy and completeness, advising the owner of alternatives. The A/E normally reviews the constructor's monthly requests for payment and evaluates the percentage of the work completed.

Ultimately, the A/E completes "punch lists" and final inspection and develops a list of outstanding items or issues that must be addressed before the constructor receives the Certification of Substantial Completion. This "punch list" activity is initiated by each trade/contractor, indicating known incomplete or unfinished work, and added to by constructor's and owner's teams before it is given to A/E for its input.

The owner's team must work with their manufacturing, production, and maintenance personnel to plan and coordinate tasks that interact with the construction program. This might include site access and traffic changes; tie-ins to existing power supply and other public utilities, process, and physical plant services; keeping pedestrian traffic safely away from the construction activities; and informing employees of emergency egress during different phases of the construction work, tie-ins to operating systems/services, or phases of any renovation work.

The constructor monitors his subcontractors for quality, productivity, and safety issues on a day-to-day basis. The constructor's job is to manage the subcontractors' work sequence and to ensure that the subcontractors' means and methods are providing the timely and quality product purchased by the owner. The constructor coordinates the RFIs from each of his trades to achieve a timely resolution. On a fast-track project, the subcontractors' productivity is of major concern; therefore, the constructor will be watching the arrival of materials and manning levels provided to ensure that the work sequence and schedule objectives are achieved. The constructor's major challenge is to keep each of the subcontractors focused on project priorities, working in concert with other trades in adjacent areas, and working smoothly without disruption to each other. This is usually accomplished via a weekly coordination meeting with the trades, A/E, and owner teams. By using a three-week look-ahead work scheduling approach, the constructor identifies design, construction, and owner issues that require support and coordination by all parties to achieve their expectations.

COMMISSIONING

Most equipment with moving parts such as motors, compressors, and pumps is bought with a warranty or certificate that guarantees a level of performance and quality of workmanship. During the commissioning phase of the project, the A/E's job is to monitor this equipment and verify that (1) the equipment provided is correct, and (2) the equipment works as specified. This process can take many forms, from in-factory to in-plant tests, or from bench top to full production loading. The goal is to verify that in each instance the right piece of equipment is installed to do the job that it is designed to perform.

In the biotechnology and pharmaceutical industries, the certification step described previously is carried one step further. Here, a three-step program is undertaken for commissioning: *installation qualifications* (IQ), *operational qualifications* (OQ), and *performance qualifications* (PQ). This is formally documented, signed, and dated by the participants.

Installation qualifications (IQ) is a step-by-step method of review to ensure that the proper way to install a piece of equipment is implemented. How to move, how to bolt down, alignment, any vibration isolators required, and other such tasks are part of this review.

Operational qualifications (OQ) covers both operator training and something that other industries call a systems checkout. Proper training results in fewer mistakes by the operator and less maintenance or unscheduled downtime. During this phase, switches are activated and motors are "rolled over" to verify that all parts of the equipment are fully functional and in accordance with engineered function. Usually this phase involves the constructor coordinating with equipment vendor representatives on-site to provide instruction to the owner's team in the operating and maintaining of any unique equipment and systems.

Performance qualifications (PQ) involves running the equipment under design load conditions. When the Food & Drug Administration (FDA) is licensing a new facility, the "PQ" step is the primary way of determining that products can be made in the same exact way each time, time after time. This testing becomes the base case against which product yield and purity levels are to be monitored and constantly measured.

The last phase of commissioning is start-up. Here the A/E and constructor's subcontractors work to bring all of the equipment on-line and to begin manufacturing the product. Measured planning and organizing from project inception are the keys to a successful start-up. Everyone must be familiar with the proper operation of equipment and know its proper sequence within the overall scheme. Computer simulations or virtual reality modeling are invaluable. The operators can be trained in a short time to face many simulated crises and be taught the most effective ways to correctly deal with them—all without affecting production or costs.

The constructor's and A/E's primary roles now are to run down the "punch list" items and be available to assist the owner's team in making modifications and adjustments during the start-up and initial operation of equipment and systems. Additionally, skeleton staffs of electricians, millwrights, mechanics, and instrumentation/communication trades are present to assist the owner's team, as required.

CONTRIBUTOR

Donald G. Munson, P.E., Raytheon Engineers & Constructors, Inc., Cambridge, Massachusetts.

CHAPTER 12
COLLEGE AND UNIVERSITY FACILITIES

William L. Porter, Ph.D., F.A.I.A., Chapter Editor
Massachusetts Institute of Technology, Cambridge, Massachusetts

INTRODUCTION

College and university facilities represent a huge proportion of annual construction activity. According to the authors, who draw upon extensive experience with academic facilities, what distinguishes these from most other types of facilities is the combination of three factors. The history and character of the context within which the facility will be built require that the design "reflect the past and anticipate the future." The long-term interest in each facility on the part of the client organization requires that it be adaptable and cost-effective over a long life span. And the complex and participatory nature of the design process requires that "architects and planners must possess process and team management skills, as well as design and technical skills, to navigate effectively." The authors then go on to discuss types of university facilities, the nature of clientship, and how the processes of master planning, programming, design, and construction should be handled.

SECTION 12.1
COLLEGE AND UNIVERSITY FACILITIES

Scott Simpson, F.A.I.A.
The Stubbins Associates, Cambridge, Massachusetts

Fred Clark, AIA
Shepley Bulfinch Richardson and Abbott, Boston, Massachusetts

Except for health care projects, the college and university market represents the largest annual construction volume in the United States, primarily because of the wide variety of projects in new work and renovations that are undertaken every year, including instructional, research, residential, athletic, administrative, and support space. When the impact of landscape architecture is factored in, the college and university market offers more challenge and more opportunity for architects, engineers, and planners than the market for any other building type. The college and university market is complex, and each different project type has its own special requirements.

Many factors make this building type unique. First and foremost are the issues of context, precedent, and identity. Every institution, no matter what its size, location, or special focus, projects a sense of place. No new project is ever the first, nor will it be the last, to be built on a given campus. Each building has its predecessors, and each helps set the context for future campus development in some important way. Thus, respect for the history and traditions of the institution and an understanding of its mission and master plan are extremely important factors in the design process. Each and every project must make its place as an architectural citizen of the campus and must reflect the past and anticipate the future.

Another distinguishing characteristic of college and university buildings is that they are likely to be in service for many years—even generations—and during this long period of useful life they will probably undergo multiple additions, renovations, or even conversions to new uses. This need for built-in flexibility is reflected in the way the buildings are programmed, constructed, and maintained, and even in the choice of materials and engineered systems. The fact that such projects are often partially or fully funded by donors can also have an impact on the way the design and construction are handled.

Finally, college and university projects differ from others in the nature of the decision-making process that is used to create them. Unlike private sector enterprises (for example, commercial or industrial clients), colleges and universities are not traditionally driven by bottom-line economics (though of course economics does play a large role). Decision making in the academic world often reflects a blend of administrative and academic perspective, and when the politics of town-gown relations are factored into the mix, it is often true that decision making is more consensus based, more complex, and generally slower than for other building types. This means that architects and planners must possess process and team management skills, as well as design and technical skills, to navigate effectively.

All of these factors combine to make design for colleges and universities a special challenge, but one that is also ripe with opportunity. In this chapter, we touch on the range of projects and outline an approach to planning that takes into account the special conditions that must be handled. It should also be noted that, as with many other project types, the advent of new technologies, funding mechanisms, and changing demographics will have a profound effect on the way environments for learning are conceived of in the future.

SECTION 12.2
PROJECT TYPES

Scott Simpson, F.A.I.A.
The Stubbins Associates, Cambridge, Massachusetts

Fred Clark, A.I.A.
Shepley Bulfinch Richardson and Abbott, Boston, Massachusetts

College and university campuses function like villages or small towns. They comprise a collection of special-purpose buildings that must accommodate a very wide range of activities. Obviously, teaching and learning spaces are at the top of this list and include large and small classroom spaces, lecture halls, seminar rooms, auditoriums, laboratories, and the like. To support teaching and learning space, there are administrative and faculty offices. The students (both graduate and undergraduate) require residence halls, dorms, or apartments, and of course dining halls as well. In addition, indoor and outdoor athletic facilities for both league and intramural sports, such as exercise and weight training rooms, practice courts and fields, natatoriums, hockey rinks, football and track stadiums, tennis courts, golf courses, and lockers and showers are necessary. Science buildings—laboratories for chemistry, biology, physics geology, and the related fields of engineering and mathematics—are frequently built in their own special precinct on campus. Many colleges and universities also provide art studios, galleries, and special facilities for practicing and performing music. One building type common to all institutions of higher learning is the library. However, because of the advent of new computer technology for storing and retrieving information, a "library" can exist in a variety of forms and in multiple locations. Many institutions have developed specialty libraries for specific disciplines, and some have resorted to creating off-site storage for books to handle the explosion of new materials that are published every year.

CLASSROOMS

The basics of pedagogy have not changed for thousands of years. There is a teacher, and there are students. Class size may vary, but the essential act of teaching depends upon one individual who disseminates special knowledge. Traditional classroom settings are the lecture hall for large groups, the seminar room for small groups, and the laboratory for the sciences. The basic tools of teaching, even in today's high-tech environment, are the blackboard and chalk. However, the explosion in new information and communications technology is fundamentally changing the concept of what constitutes a learning environment. In the past three decades, xerography has replaced the mimeograph machine, computers have replaced typewriters, and the Internet has emerged as an incredibly powerful informational tool. Cell phones, voice mail, cable TV, VCRs, fax machines, compact discs, and video games have changed our notions of the way information is gathered, stored, retrieved, and transferred. Many colleges and universities provide broadcast services for classes, and some have produced videocassettes of the most popular courses. *Long-distance learning* or *teleconferencing* means that a teacher, who could normally reach a few dozen people at a time, can now teach thousands. Books are bought on the Internet, select portions are downloaded, and on some campuses the "library" is as close at the nearest computer terminal. How does all this affect the design of classroom space?

More and more colleges and universities are constructing high-tech classrooms that are based on auditorium or parliamentary seating. These rooms are tiered to maximize sight lines, and seating is generally fixed so that each student "station" can be equipped with a computer connection. The rooms are also equipped with high-tech audiovisual systems (slides, video projection, and touch-screen controls) that can record the lecture on video or broadcast it live. The sizes and shapes of the rooms are designed to accommodate a variety of class sizes, and some are arranged for both lecture and "break-out" spaces so that students can learn in small groups (Fig. 12.2-1).

The newer-style classrooms are much more complicated to design and engineer and much more expensive to build. Specialty consultants are often engaged to assist with technical issues such as acoustics, lighting, and audiovisual systems. In tiered classrooms, special attention should be paid to ADA guidelines to ensure that provisions are included for the disabled.

When planning classroom space, it is especially useful to coordinate the program requirements with the expectations of the registrar's office to make sure that the size, number, configuration, and classroom equipment are as flexible and adaptable as possible.

FACULTY AND ADMINISTRATIVE OFFICES

As is the case with teaching spaces, the requirements of technology have also affected offices. The traditional faculty office consists of a desk, a chair, a bookcase, and some files, and not much more. However, faculty offices must adapt to technology, and this means including sufficient space for computers, printers, fax machines, and so on. Most contemporary offices have several phone lines and a campuswide Ethernet connection. The additional equipment will affect office size somewhat. In general, it is good policy to standardize office size for flexibility; a normal range is 150 to 170 ft^2. Depending upon departmental layouts, conference space can be shared. The most common complaint about general administrative areas is lack of storage space.

RESIDENTIAL HALLS

In the past several years, there has been a clear trend away from the traditional dorm layout (double rooms arrayed along a central corridor, with a shared bath) to apartment-style living. The new dorms consist of suites (usually two to four bedrooms each) that have living rooms, private baths, and some kitchen space as well. Storage is always an issue, as is the requirement for plenty of connections for phones, faxes, and computers. Two current examples of this trend are the Armory dorms at Boston University and the "Swing dorm" at Yale. Colleges and universities are finding that students are willing to pay a premium for living quarters that are more spacious and comfortable. Some of the newer dorm projects are designed to be convertible to market-rate housing, and many contain substantial amenities such as food service, laundries, computer labs, recreation rooms, seminar or study rooms, and activity spaces for clubs and campus organizations. The college dormitory is one project type that has evolved rapidly with market conditions—no longer content to live in a barracks-like place, students are seeking space that is more akin to a home environment.

FOOD SERVICES

It has been said that time spent in the dining hall can be just as educational as that spent in the classroom. Dining halls are traditional social places for meals and also for dialogue and debate. Food plans vary from institution to institution, but today's students demand more

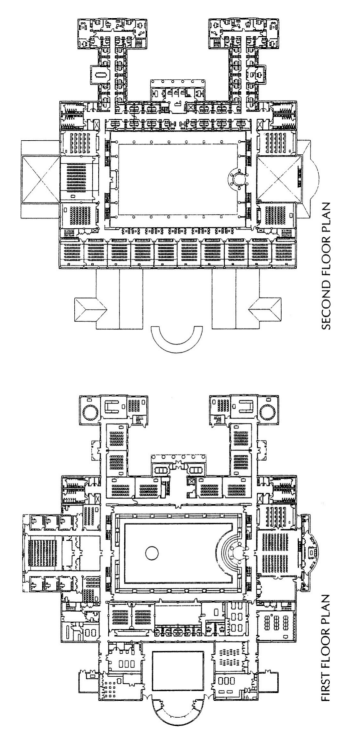

SECOND FLOOR PLAN

FIRST FLOOR PLAN

FIGURE 12.2-1 Example of multipurpose classroom and instructional project at Coastal Carolina University.

variety and choice and a higher level of quality. Many institutions have set up food courts to augment or replace traditional dining halls, and some have turned to contract management, outsourcing all food services to control cost and quality. New menus and new cooking techniques have migrated from restaurants into dining halls, and it is not unusual to see special menus for ethnic food or vegetarian dishes as standard menu items. When planning such projects, it is usually good practice to seek the assistance of a qualified food service consultant.

ATHLETIC FACILITIES

Many colleges and universities are focusing on upgrading their athletic facilities for two reasons. First, they need to respond to the requirements of Title IX, a federal mandate for equality in athletic programs for both men and women. Second, institutions are discovering that updated athletic facilities are extremely important as a marketing device because they attract potential students. In addition, media exposure has made college athletics a big business that can produce millions of dollars in revenue and that promotes loyalty among the alumni. Most institutions understand and support the importance of vigorous athletic exercise as part of the overall educational program—producing both healthy minds and healthy bodies. For obvious reasons, athletic facilities tend to be clustered in their own precinct, usually at the edge of campus, because of the space required. New facilities are usually designed to be multiple-purpose spaces. For example, it is not uncommon for one building to house a hockey rink, basketball courts, an indoor track, a swimming pool, and general exercise facilities or a weight room in the same structure. These buildings use long-span structural members and high-volume HVAC systems, so engineering is of particular concern. This article will not attempt to deal in detail with the specific requirements for the many sports that need accommodations. It is sufficient to note that, like food service, it is usually helpful to engage appropriate consulting expertise for each specific project.

SCIENCE AND TECHNOLOGY BUILDINGS

One of the most significant nationwide trends in colleges and universities is the renewed interest in investing in buildings for science and technology. Curricula are being revamped, existing labs are being renovated, and construction for new laboratories is at an all-time high. New technologies, new equipment, and new building code requirements need to be accommodated. A trend toward student-teacher collaboration on research projects and an emphasis on team teaching have influenced the way science is taught. All of these factors influence the planning, design, and construction of science facilities.

The traditionally separated disciplines of physics, chemistry, and biology are evolving into overlapping studies (biophysics, biochemistry), and this requires a different approach to lab planning. New research techniques such as microchemistry, which requires using much smaller amounts of chemicals for experiments, and the introduction of computer simulation into the lab environment have created a trend away from wet bench research to more dry bench research. Labs and classroom spaces are now being combined into single rooms, so that lectures and experiments can be conducted side by side. The lab benches themselves are arranged in new configurations that permit teams of students to work together.

The pace of change in scientific research and teaching is very likely to accelerate in the coming years, and this places an extra premium on designing for flexibility. Whereas in the past separate buildings might be built for each discipline, now it makes more sense to develop a universal lab/classroom/office cluster that can be used by any member of the faculty. Shared support areas for functions such as glass washing or equipment storage are now common, as is the emphasis on better daylighting and brighter colors in the interiors. Dark, dingy labs are fast becoming relics on campuses. The recent and rapid growth of the computer and biotechnology industries and the availability of funding to support them bode well for the future of science.

SPACES FOR THE ARTS

Not every college or university offers studies in art, music, or theater, but even for those that do not, it is highly likely that there will be an auditorium or performance space of some kind on campus. As with athletic facilities, designing for the arts requires special expertise, and such projects are often shaped by the demand for long-span structures, special lighting, and carefully designed acoustics. In recent years, studio spaces for painting, graphics, and sculpture have gotten significantly larger to accommodate the trend toward bigger works of art. Schools of architecture are discovering that they must retrofit their curricula and their studio space to accommodate computer technology. To create new theater spaces and also avoid the cost of new construction, some institutions have purchased and renovated churches. In many cases, performance spaces such as auditoriums are being designed for multiple uses as convocation halls, lecture halls, and the like. As one innovative example, Boston University has shared the use of a movie theater complex that shows films in the afternoons and evenings but doubles as classroom space in the mornings. Arts programs have traditionally been poor cousins in terms of raising money from alumni and foundations, but the ability of a well-developed arts program to contribute to the overall quality and brand image of an institution should not be overlooked. It is predicted that investment in projects for fine arts will continue to accelerate in the years to come.

LIBRARIES

The library is probably the most common building on any campus of higher education—every college or university has one. Access, storage, and retrieval of information is one of the fundamental activities of learning. However, computer technology has changed the very notion of what a library or even a book actually is. Like automatic teller machines, which changed the way banking is conducted and in fact made banks relatively obsolete as a building type, the Internet, campus Ethernets, widespread use of personal computers, and the ability to access books in the library electronically or even purchase them online is profoundly affecting library organization and management. As computer power is increasing exponentially, it is now possible to put an entire encyclopedia (and more) on a single compact disc. Most academic libraries have already converted from hard copy card catalogues to computers. Some have even instituted bar coding for books, which increases both efficiency and security. The phenomenon of storing more information in less space and accessing it faster is likely to accelerate substantially in the years to come. This means that access to books may no longer be the primary reason to build a library. It is not much of an exaggeration to suggest that in the future, access to a personal computer will constitute a library. Recent library designs include more spaces for group study, seminar rooms, computer labs, and even lounges equipped with food service. The library is becoming a gathering space that has as a social as well as an academic dimension. As a result of these factors, the traditional library as a building type could very well disappear.

SECTION 12.3
GROWTH FACTORS IN COLLEGE AND UNIVERSITY FACILITIES

Scott Simpson, F.A.I.A.
The Stubbins Associates, Cambridge, Massachusetts

Fred Clark, A.I.A.
Shepley Bulfinch Richardson and Abbott, Boston, Massachusetts

The pressure for growth can come from many directions. Most institutions of higher learning have substantial investments in their buildings. In fact, real estate represents more than half of the net assets of most colleges and universities. These buildings also represent significant depreciation and operational cost. Upkeep costs approximately $7 to $10 per square foot annually. Older facilities are often cramped, are out of date, need substantial maintenance, do not comply with codes [especially the Americans with Disabilities Act (ADA)], or do not lend themselves to new teaching technologies or methods. As noted earlier, in recent years there has been a marked trend toward upgrading or replacing science buildings in campuses across the country, a movement that is supported by government initiatives such as Project Kaleidoscope (PKAL). There is also a strong trend toward upgrading, renovating, or replacing student housing and food service. Classrooms are now being designed for auditorium or parliamentary-style seating, and are equipped with state-of-the-art audiovisual equipment, as well as computer jacks, at every seat. A new building is a tangible and visible symbol of progress—a way for a college or university to declare to its faculty, staff, alumni, and students (and the parents who pay tuition) that the academic mission is being successfully carried out.

Much of the pressure to upgrade facilities is a simple function of demographics. Today's generation of students is the offspring of the baby boomers (the largest cohort ever to attend college). Another factor is the availability of capital from generous alumni and government grants and subsidies. For example, a portion of funding from agencies such as the National Science Foundation or the National Institutes of Health goes to support the construction and maintenance of the facilities that are needed to conduct research.

SECTION 12.4
THE NATURE OF CLIENTSHIP

Scott Simpson, F.A.I.A.
The Stubbins Associates, Cambridge, Massachusetts

Fred Clark, A.I.A.
Shepley Bulfinch Richardson and Abbott, Boston, Massachusetts

One of the most important first steps in assessing a new project for a college or university is to understand the nature of the clientship involved (Fig. 12.4-1). The many and varied stakeholders in the process include the administration (which must authorize, approve, and pay for the new project), faculty members (who are often the primary, though not always the exclusive, users), the buildings and grounds staff (which must maintain the structure against very significant wear and tear), the students (who are vocal but transient users), the parents (who actually

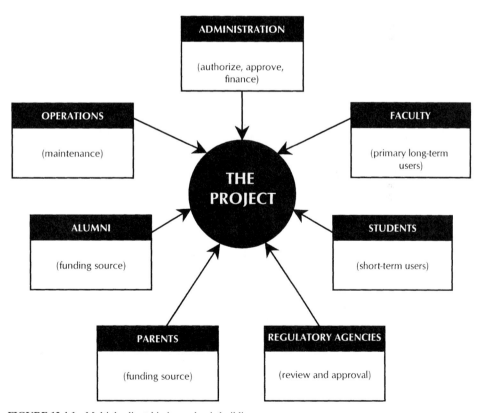

FIGURE 12.4-1   Multiple clientship in academic buildings.

pay the bills), and even the alumni (whose donations make many projects possible). Each and every stakeholder will have a slightly different influence on decision making, and it is very important at the outset to understand how this multiple clientship will affect the design process. A design team should never assume that the stated goals and objectives for the project are the only ones that matter. Very often, the hidden agendas ultimately govern the outcome.

Because colleges and universities are frequent buyers of professional design and construction services, they are likely to be sophisticated consumers. Most, if not all, major institutions maintain highly trained professional project management staffs that often include licensed architects and engineers. It is not uncommon to encounter strong preferences or even published standards for everything ranging from mechanical, electrical, and plumbing (MEP) systems to what kind of faucet to use in a janitor's closet. Such standards can be extremely beneficial in helping to communicate the client's expectations. Most likely, the project management staff has developed relationships with the contracting community by virtue of work on multiple projects. Such relationships are especially useful because many projects must be completed on a rigorous schedule to be placed in service before the student body arrives in the fall. Universities and colleges are desirable, high-profile clients. Architects, engineers, and contractors who establish good working relationships can often look forward to repeat business for many years.

By virtue of their size and economic impact on the community, plus their nonprofit tax status, institutions often develop a unique relationship with local regulatory authorities. This relationship, traditionally termed *town-gown,* can be either contentious or cooperative. It is not uncommon for institutions of higher learning to be subject to special rules and regulations, such as the requirement to file institutional master plans or the negotiation of payments in lieu of taxes. Many are in special use districts that require variances or permits for each new project. The relationship between the institution and the local political authority often plays out during the design process. It should be thoroughly assessed at the beginning of each project and managed throughout.

All of these clientship factors must be taken into account when establishing the basic parameters for a project (program, schedule, and budget). How will information be gathered and processed? What are the political influences that may govern? What are the standing policies and procedures that must be respected? Who are the key decision-makers? How long will the process of review and approval take? Successfully navigating this process requires a cooperative effort from all hands. Planners, architects, and engineers can certainly assist in this process, but should it be made clear that the primary responsibility to manage it rests with the client representatives because they are the only people who have legal standing.

SECTION 12.5
PROJECT INITIATION
AND PREDESIGN

Scott Simpson, F.A.I.A.
The Stubbins Associates, Cambridge, Massachusetts

Fred Clark, A.I.A.
Shepley Bulfinch Richardson and Abbott, Boston, Massachusetts

To be successful, every new project, no matter what size, no matter whether new construction or a renovation, no matter where it is located, needs three essential things: a clear, concise, and comprehensive program, a realistic schedule, and a sufficient budget. The program spells out the goals and the ingredients of the project (the size, location, and relationship of each proposed space in the building). The schedule must account for the traditional design and construction phases and should also recognize the time required for the client's review and approval process. The budget must address the cost of construction itself (hard cost) and also the soft costs for fees, furnishings, equipment, and financing, including a reasonable contingency (5 to 7 percent is recommended), plus an inflation factor, because most projects take two to three years to complete. When the program, schedule, and budget have been carefully thought out and communicated to all the interested stakeholders in the process, then the formal design process can begin with confidence. If any of the three prerequisites is missing, insufficiently developed, or misunderstood, then it is recommended that a feasibility study be conducted first. Ultimately, this saves considerable time, money, and frustration.

PREDESIGN

The desire to add new space or improve existing facilities can be based on many things—increasing enrollments, addition of sophisticated new equipment, the necessity to address maintenance of building code issues (especially the ADA), or pressures that come from new consumer expectations such as those for better dorms and athletic facilities. Whatever the project, whatever the scale, and whatever the cost, things will run better and more predictably if the planning process includes a defined phase for research and predesign. To put it simply, before you drive off to Shangri-La, make sure that you have gasoline, maps, and a spare tire. Knowing what you are up to and why makes all the difference between a successfully executed project and one that simply happens on its own accord (Fig. 12.5-1).

A good analogy for predesign is painting. When a surface is properly scraped, sanded, and primed, then the new coat of paint makes everything look better. However, when the proper preparation is not done, the new coat of paint only exposes and exaggerates the flaws in the wood. In other words, it is worth doing predesign. The purpose of this initial phase is to determine the parameters that will govern the project. Predesign should focus on asking the right questions—it is a discovery phase, not a problem-solving phase, and it is important to maintain objectivity so that premature conclusions and hidden agendas can be avoided.

Predesign has several facets, which can be characterized by the following questions: What? Where? Why? When? and How much? The *what* refers to the scope of work and the func-

Predesign	Programming	Schematic design/ design development construction documents	Construction/ closeout	Occupancy
Principal activities				
• Establishes project goals and parameters • Tests feasibility • Identifies stake-holders and resources • Outlines overall process	• Establishes size, number, and function of spaces • Sets preliminary budget • Specifies special conditions	• Creates design alternatives • Gives project physical and architectural form • Confirms budget • Documents projects specifics	• Construction control • Budget control • Schedule control • Quality control	• Move-in • "Shakedown" • Postoccupancy evaluation
Primary responsibility				
Owner	Owner/architect	Design team (architect/engineers/ consultants)	Contractor	Owner
Typical timeframes				
1–3 months	2–6 months	4–12 months	12–36 months	1–6 months

FIGURE 12.5-1 Typical phases in college and university projects.

tional programs. It is very important to relate this question to the institution's overall strategic plan (and if such a plan does not exist, make one!). What is proposed for construction? What are the functional requirements in general? How does the proposed scope of the project relate to similar facilities on campus and at peer institutions? What degree of loose fit should the program provide to allow for growth beyond the immediate need? How will demographic or pedagogic trends affect the projected space needs? When the project is completed, what, exactly, are the needs to be made clear to all concerned? It is important to ask a generous cross section of those who are likely to be affected by the outcome: students, faculty, administration, and support staff. Sometimes even the alumni come into play, so don't forget to contact the development office while you are at it. The question of what the project is about is the single most important point of reference.

Once there is a *what,* the next step is to determine *where.* It is less important that a specific site be chosen in advance and more important that a process for choosing a site be established as part of predesign. There is a world of difference between the two. Choosing a site involves many variables, including size, location, the impact of zoning restrictions, access to utilities, and pedestrian and vehicular circulation. Determining a site is an exercise in itself. In predesign, it is sufficient to determine the parameters that the site must ultimately satisfy, or at the very least, to determine that at least one site exists that can accommodate the new project. In the case of renovations only, of course, this choice is predetermined, but it still pays to investigate in advance if there are any new code requirements, zoning or code restrictions, or accessibility issues that must be solved to ensure that the project is still viable.

One of the most fundamental questions in predesign is the question of *why* this particular project is being proposed at this particular time. Sometimes this is answered by the demands of faculty recruitment (a new lab space must be made ready by September) or perhaps it is justified by alumni largesse (a generous donor is ready to endow a new building). Sometimes code requirements, especially the ADA, demand quick remedial action, and sometimes it is just the accumulated effect of long-deferred maintenance. Whatever the reason, it is very important to express the justification for the project clearly, concisely, and convincingly, in terms of program space and financial commitment, so that key decision-makers in the university community can understand and support the project. The simple reason for this is that resources are always scarce and carefully guarded, even for the most well-endowed institutions. Given the money that can be spent now, is this particular project the best and highest use of available resources? Are there other, more productive ways to use the space or the funds? It is surprising how often this simple but very important question is overlooked to the peril of the project.

When the *what,* the *where,* and the *why* are understood, the schedule is the next critical issue. *When* should all this happen? What is the impact of the project on the school schedule? In areas of the country where the climate affects construction, what is the best time to start the project? What influence will the regulatory and approvals process have? What about fundraising—how will this affect the timing? How about inflation? Even at a modest rate of 2 to 3 percent per year, the overall project budget should contain a contingency to cover this unpredictable cost. Schedule is also critically important because it may impact the logic of the program. This is especially true of projects in the sciences and in teaching spaces requiring high-tech audiovisual systems—they may be obsolete even before they are completed. This should be factored into the planning process. All projects have a dimension of time, and understanding how time affects the project at hand is an extremely important point of reference.

Finally, there is the question of *how much* all of this should cost. Note that this is the last, not the first, question that is posed during predesign. It is certainly possible to conduct a project that is entirely cost-driven, in which all decisions are made on the basis of how much money is available. But it is often not the best policy because it is just as easy (and dangerous) to spend *too little* on a project as too much. It is generally better to determine need and feasibility first and then deal with the question of how to finance the project. Money is just another ingredient in the design and planning process, and if it becomes the only lens through which the future is seen, distorted vision is sure to result. This is not an argument for unlimited spending—quite the contrary. The point is to decide how to spend the right amount of money to accomplish the goals of the project—not too little and not too much. During predesign, general parameters should be established, but detailed budgeting for construction and soft costs should be done at a later date when more data are available. It is true that form follows function, but it is also true that form follows finance. Understanding this fundamental management principle is extremely important in producing a successful project.

The questions of what, where, why, when, and how much will set the strategy for the project. When this is done, there is still more homework in predesign. Make sure that campus maps, site surveys, zoning reviews, code analyses, topographic surveys, and existing conditions drawings (including technical drawings for mechanical, electrical, and plumbing systems) are available and accurate. If these do not exist, conduct surveys and create them. If there are standards for equipment, hardware, finishes, or suppliers, also make sure that this information is part of the predesign process. This effort will save countless hours of speculation and wrong turns during planning and design.

A thorough and comprehensive predesign process is the best way to ensure that the proper foundation has been laid for all future phases. This determines who the stakeholders and key decision-makers should be, what the strategic goals are, and what kinds of resources (both time and money) are available. Any new project involves questions of predictability and uncertainty. The purpose of predesign is not to answer the questions themselves, but rather to set up a process by which they are answered in due time. Effective predesign is like a good map—the document will not take you there by itself, but it will illustrate the proper path.

When the research and predesign phase has been completed, the actual project work can begin, starting with general issues and working to the specific.

SECTION 12.6
A WORD ABOUT MASTER PLANNING

Scott Simpson, F.A.I.A.
The Stubbins Associates, Cambridge, Massachusetts

Fred Clark, A.I.A.
Shepley Bulfinch Richardson and Abbott, Boston, Massachusetts

Colleges and universities have the advantage of long-term planning that maps the goals and growth of the campus in organized stages. The best plans grow out of a vision for the academic shape and future of the college that covers the physical element of buildings and also academics, personnel, and the community. In essence, the best plans result from a clear sense of mission coupled with a plan for achieving that vision. A college master plan must be written (both verbal and graphic), dynamic (responds to events), relevant, practical, and subscribed to by the college community.

Not to be overlooked is the stewardship of the existing building inventory. The facilities department should establish a maintenance and renovation plan for all buildings on campus. Data should be collected and shared about problem areas or systems. Criteria for building performance across campus should be established, and items that fall below the threshold should be addressed. Lessons learned can be incorporated into the standards for construction issued to architects and contractors working on campus. All buildings and their use are dynamic, and all will need renovation as finishes and systems approach the ends of their useful lifetimes. In the long run, it makes sense to plan for staggered renovations on a rotating schedule, no matter how old the building is. The projects most likely to come on line in the next 10 years should be quantified (in general terms) for scope and cost of renovations. Then, this plan can be incorporated into the university's long-term financial plans.

One of the unique aspects of planning college and university facilities is the issue of determining how each project is expected to respond to the established context. Each new building, whether new construction or a renovation, is part of a larger whole. Some campuses have already established a strong and cohesive architectural style, but others are more eclectic. Either way, it is important to consider how the location, function, material, and style of the building will impact the entire institution. Most campuses have evolved over time into distinct and recognizable precincts—areas devoted primarily to teaching, residential halls, student activities, athletics, and administration. For example, it is not uncommon for the science buildings to be clustered together in their own special neighborhood. Many campuses (the University of Virginia is an obvious example) have a strong architectural image that should be respected.

Thus, when planning a new project, one of the first and most important things to do is to review the existing master plan to understand the history of the development of the campus and the anticipated patterns of future growth. This will immediately involve questions of architecture and also landscape design, vehicular and pedestrian circulation, and access to utilities (Fig. 12.6-1). A new project is also very likely to result in a town-gown dialogue about zoning regulations and the costs associated with providing civic services to tax-exempt organizations.

A thorough discussion of master planning for college and university campuses could fill its own chapter, but the essence can be addressed by the following principles:

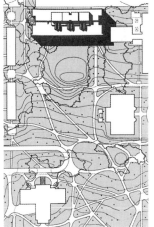

FIGURE 12.6-1   New construction integrated with landscape design at Vanderbilt University.

1. The proposed project should be sited in an appropriate functional zone of the campus.
2. The design should anticipate how existing vehicular and pedestrian circulation patterns will be affected.
3. Services and utilities should be adequate for both short-term and long-term demand.
4. The massing and materials should be compatible with the existing context, unless it is a conscious design goal to create a signature building (such as chapel or a library).
5. The design should anticipate subsequent projects in the future.
6. Particular attention should be paid to the landscape design, including pathways, lighting, planting, and hardscape.

This last point—landscaping—deserves special attention for a very simple reason. It is true that a college or university comprises a collection of buildings, but it can only be a campus by virtue of the spaces among those buildings. The character, beauty, and function of a campus are very much functions of the landscape design as opposed to the architecture. For this reason, every project team should include expertise in landscape design.

SECTION 12.7
PROGRAMMING

Scott Simpson, F.A.I.A.
The Stubbins Associates, Cambridge, Massachusetts

Fred Clark, A.I.A.
Shepley Bulfinch Richardson and Abbott, Boston, Massachusetts

Programming is the act of creating a comprehensive yet concise description of all of the required functional elements in a given project. (For detailed information on programming, see Chapter 4.) This includes all of the rooms and also other support spaces such as mechanical space, corridors, and toilets. Most often programming is done by conducting interviews with a cross section of the stakeholders, who are encouraged to create a wish list of desired elements. One problem with this approach is that it is very difficult for many people to anticipate and articulate desired changes in their physical environment, and this is particularly true in high-tech projects such as laboratories. (Also refer to Secs. 10.1, "Laboratory Programming," and 10.2, "Laboratory Facility Layout.") Thus, although interviews are an effective way to take the pulse of an institution, outside references should also be used. For example, when planning a new dormitory, it is a relatively simple matter to survey dorms at peer institutions, conduct research on emerging trends, and review construction costs of similar projects over a certain time span. This external focus helps validate the anecdotal information provided by interviews.

Generally speaking, a program is a tabulation of all rooms by size, number, function, and adjacency, with special notations for special equipment and requirements. Programs are usually expressed in terms of net square feet, with a grossing factor added to account for structure, wall thickness, and vertical penetration such as plumbing and HVAC shafts. A good program will describe what is in the building, but it will not attempt to prescribe how those elements might be arranged—that is properly the responsibility of the design team. Most architects prefer to program a building themselves because the process helps create an intimate relationship between what is intended and how the design evolves, but there are many good non-architecture firms that specialize in programming academic buildings. A properly written program (what) can be used to derive parameters of cost (how much) and schedule (when) with reasonable accuracy.

SECTION 12.8
DESIGN

Scott Simpson, F.A.I.A.
The Stubbins Associates, Cambridge, Massachusetts

Fred Clark, A.I.A.
Shepley Bulfinch Richardson and Abbott, Boston, Massachusetts

During design, the opportunities and constraints prescribed by the program, budget, and schedule are manifested. This is generally an iterative process, which means that various alternatives are proposed, tested, and then refined into the single option that will be built. It is also during design that aspiration is tempered by compromise.

The architect, assisted by engineers and consultants, leads the design phase, and the outcome can be greatly affected by how active or passive the client chooses to be in the decision-making process. It is not uncommon for college and university buildings to last for many generations of use, so it is important to anticipate this during design. The project must be specific for its intended function, but it should also be adaptable over time. This is one of the distinguishing characteristics of college and university architecture. Another factor that greatly influences the design process is that educational institutions, unlike private sector businesses, are less sensitive to bottom-line economics. Longer life cycles mean that higher-quality materials and systems can and should be used. Designing with maintenance in mind, especially for building types such as dormitories and classrooms that must withstand very heavy student use, is important and also cost effective. For this reason, it is not uncommon for buildings and grounds staff to participate in design review.

In addition to the architecture itself, several other elements of design should receive particular attention. Graphics and signage are frequently overlooked, but they are extremely important in providing an identity that is consistent with those of other campus buildings. Interior design, which includes both furnishings and materials, deserves special attention for two reasons. First, most projects of this type must withstand heavy wear and tear, and second, deferred maintenance cycles are generally long, which means that it can be many years between renovations. Finally, as mentioned elsewhere, the quality of the landscape design is extraordinarily important. Walkways, lighting, benches, bicycle racks, and plantings should be chosen with great care because they must last for many years.

The design phase of a project is not just about creating appearances—it also includes construction documents and specifications. The complexity and longevity of college and university projects usually demand that a team of professionals be engaged. Architects, structural engineers, and mechanical, electrical, and plumbing engineers are just the beginning. Other consultants that might be needed include those who specialize in security, telecommunications, specialty lighting, food service, acoustics, vertical transportation (elevators), equipment (especially for science-related projects), and building codes. This means that communication and coordination are especially important during design, and that the nature of the documentation must be tailored depending upon which delivery method is chosen. Delivery methods can include the traditional design-bid-build, construction management, fast track, design-build, or aspects of all four. Because the construction documents are in a very real sense the road map for the entire project, it is extremely important that they be reviewed with great care. No set of documents will be perfect, and change orders during construction are inevitable.

SECTION 12.9
CONSTRUCTION

Scott Simpson, F.A.I.A.
The Stubbins Associates, Cambridge, Massachusetts

Fred Clark, A.I.A.
Shepley Bulfinch Richardson and Abbott, Boston, Massachusetts

University and campus construction, whether renovation or new, is driven by the academic year calendar. Most renovation projects take advantage of the 10- to 15-week summer term break. With careful planning, an enormous amount of work can be accomplished during the summer, including projects of significant size. Larger renovation projects will be phased over two summers, requiring careful coordination. More substantial projects, such as new buildings, are always coordinated with the academic calendar. Often major construction in the middle of campus that involves heavy equipment, numerous trucks, or major site work will be scheduled for the summer when construction noise will impact fewer classes. The delivery date of the new building will be scheduled for the beginning of a semester.

Most universities have increased their use of buildings to the point where it is not possible to shift classroom space during the year. Buildings and classrooms are also being used in the summer more than ever before. Many schools now generate additional revenue by running executive seminars for businesses and adult education programs. It is increasingly common and cost effective to use overtime and weekend labor to accelerate the completion of projects. For this reason, construction management has become a favored form of project delivery.

The construction manager must be aware of the significant dates of the university calendar. The first day of the summer break will generally mark the first day of construction. The date of substantial completion except for punch-list items remaining will generally be 1 or 2 weeks before the beginning of the fall semester, so that the university can move furniture in and to learn to operate new equipment. No extensions are possible when the final delivery date in the fall schedule has been published a year in advance. Other significant dates are the start of each term and the term breaks. These are the times when the building may be unoccupied and available for clearly defined short-term construction operations or investigations. It is not uncommon to schedule installation of a new air handler or other new equipment to coincide with the spring break. Projects such as carpeting and new wall covering are possible, but the extent of the work should be limited by what can realistically be accomplished. During these breaks, dormitories and other support facilities such as student centers and dining commons may continue to be occupied.

The breaks around Thanksgiving, Christmas, and spring are also times when buildings are available for preconstruction investigations into structure, building components, mechanical system, and other utilities that are a vital part of planning for summer construction. The construction manager will look at the project from a different perspective, reviewing issues of constructibility and phasing.

Planning for summer construction is a special challenge. The completion date is usually a given determined by the university schedule. Significant dates in the project calendar are established by counting back from the final delivery date. The ideal schedule for a project would allow sufficient time in late summer for final punch list, testing and balancing, and move-in. Once a construction start date is determined, the bid and contract award dates are set by subtracting the number of weeks required and allowing the contractor several weeks after the award to mobilize. A March 1 bid date seems to work well for a June 1 construction start when dealing with a major construction project. However, significant long-lead items may require an earlier date to be on site at the appropriate time. The exact time required for the bid/award cycle will be determined by the size and complexity of the project; public bidding laws, if applicable; the bidding climate; and the timelines of the university decision-making process. This last can very greatly depending on whether decision authority is vested in a small group of facilities people or in a larger committee that represents a variety of users.

College and university projects are more schedule sensitive than most. Therefore, it is very important to choose a qualified contractor and the proper delivery method. In selecting a contractor, experience with the particular building type should be balanced with experience working on a particular campus. Projects are certainly price sensitive, but choosing the lowest bidder can actually be expensive. The specific personnel assigned to the job, especially the project manager and the supervisor, are extremely important, as is the ability to work in a team manner. The bidding process can be significantly improved by prequalifying bidders. The university is in a unique position in the construction marketplace because it has many construction projects continually going on and has had experience with local contractors. The university has a record of contractors with whom it has successfully worked and those with whom it has had problems. University work, despite its schedule constraints, is desirable for most contractors. The projects are prestigious, high in quality, and profitable, and often result in repeat business. In the best of circumstances, a long-term symbiotic relationship can develop where the university gets a reliable contractor with whom it builds several projects.

LONG-LEAD ITEMS

Because of a shortened summer construction schedule, in university construction long-lead items are even more critical than usual. The project manager must review the construction documents and note any items whose schedule from order to delivery may affect the project. Typical long-lead items include steel, HVAC air handlers and other components, electrical switchgear, and elevators. The experienced construction manager will read through the documents and spot the items of concern. However, even the most experienced project manager can be surprised on occasion, so it pays to check on the availability and lead time of every item required. It is common practice to pull together an early bid package with long-lead items. Often the university or owner will contract directly with the supplier for a major piece of equipment such as an air handler before the contractor is on board. If the university makes such an early purchase, it may unwittingly take on added responsibility normally vested in the contractor. Care must be taken to precisely coordinate all of the options and variables involved in such a purchase. Early construction drawing packages carry inherent risk because they are released before the design is fully detailed and coordinated. Some changes may be required when the full construction documents are complete, but on balance, the benefits outweigh the liability of eventual changes.

During construction, complications can occur if it is discovered that a custom component with a long lead time must be reordered, in which case it may be more practical to modify the item on hand even if it may be more costly. When a component must be reordered, all favors owed should be called in, and the owner should authorize rush freight charges.

The summer takes on its own pace because one eye is always on the schedule. Cooperation among contractors is vital for smooth progress during the limited construction period. Recently, it has become popular to involve everyone in a team-building event to foster a sense of teamwork. The key players are invited to a lunch off-site where they meet each other and university representatives. The afternoon concludes with everyone signing a written pledge to work together.

A weekly job meeting led by the construction manager and attended by each trade, the owner's representative, and the architect can cover current issues and coordinate work among contractors. The construction manager records the minutes, gives each issue its own bullet, and lists under whose responsibility the item falls. The minutes provide a written record of the progress of the job and define expectations. The construction manager also prepares a schedule in the form of a bar chart that is distributed at the job meeting. Construction is a sequential process, and the work of one trade depends on that of another. (Refer to Chap. 6, "Construction, Modification/Renovation, and Demolition/Site," for detailed discussions of the construction process.) Electricians cannot run wiring until the drywall contractor builds the wall. The schedule provides the framework for the necessary interfaces among contractors. It can help to show problem areas and define specific dates when a contractor must be done. Because of the compressed summer construction period and careful scheduling, there is no room for time extensions.

Construction is a complex activity, involving multiple variables, and it can be affected by unforeseen circumstances (such as weather, strikes, delays in delivery, and so on). In addition, each project will require a period of time for "shakedown" (testing, balancing, and so on) and move-in, and so the schedule should allow 4 to 6 weeks before the actual opening date to ensure that everything is in order. This means that a project expected to be in service by September 1 should actually be finished by July. This leads directly to the issue of how to actually finish the project.

PROJECT CLOSEOUT

When the construction phase of a project is nearing completion, closeout can begin. In many ways, this is the most difficult and telling phase because it is the culmination and the test of everything that has gone before—the programming, planning, financing, design, scheduling, project management, and construction. During project closeout, the successes and all of the flaws in the project are laid bare, the actual budget and actual schedule are finally made manifest, and how the project will work becomes unavoidably obvious. Project closeout is simply a confirmation of a job well done for a well-planned, well-executed project. For a project that has experienced bumps in the road, the effects of the bumps are magnified. It is for these reasons that it is important not to begin project closeout too soon, because early closeout does not allow sufficient time for the contractor to test and install systems properly. Closeout done too early is simply a waste of time. Why bother to generate a punch list of 100 items if 1 week later the list is reduced by normal processes to only 25 items? At the same time, scheduling closeout is necessary because all projects of whatever size and complexity contain surprises. The wrong equipment may be ordered, or equipment may not arrive on time. Sprinkler systems and fire dampers may not work properly. A certificate of occupancy is delayed or withheld because of code violations that have previously gone unnoticed. During project closeout, these and dozens of other glitches rear their heads, and all of them must be attended to quickly if the building is to open in time for the students and faculty.

As with all the previous phases of the project, closeout depends on an organized and predictable process, in which the various stakeholders have defined roles and responsibilities. One of the biggest mistakes that can be made during closeout is to cross lines of authority or communication in a noble effort to get something done quickly. The irony is that in many cases, doing something quickly can actually delay overall completion; it is better to do it once and do it right than to have to do it over. Those charged with specific responsibilities to review, test, and certify systems or equipment should be permitted—and required—to fulfill their individual duties. Expediency always has a hidden cost.

What are the specifics of closeout and how should this important phase be approached? The roles and responsibilities that are maintained during planning, design, and construction do not change in any fundamental way. The architect and engineers are there to see that the construction has been accomplished according to specifications and in most cases to document deviations from the plan by producing record drawings. The construction manager and the subcontractors are there to complete their specific portions of the work. The owner has an especially difficult job, because now is the time to take the keys to the new car, get in, and drive off. It is one thing to design a building, another to build it, and yet another to own and operate it properly. Experience has shown that most problems that occur during early occupancy of a new project are due to lack of knowledge or experience on the part of the owner's staff in operating and maintaining the new space. This is especially true because many systems such as HVAC, alarms, security, audiovisual, and telecommunications are much more sophisticated than even a few years ago. A savvy owner will anticipate these growing pains and take special care to provide sufficient training for all staff members who will be using the building or are responsible for taking care of it.

Each party in the closeout process has specific responsibilities that must be addressed if the other parties are to complete their work. These are outlined in the following text.

Owner

1. Provide training for maintenance and operations staff
2. Provide existing equipment to be installed in the new project
3. Provide sufficient information to tie into centralized systems (phones, security, and so on)
4. Coordinate move-in
5. Provide final project accounting and accept legal responsibility for the project

Contractor

1. Finalize construction, including all punch-list items
2. Provide manuals and warranties for all equipment and systems
3. Provide information to the architect/engineer team for processing record drawings
4. Arrange all final inspections, approvals, and certifications
5. Prepare a final certificate for payment

Architect/Engineer

1. Generate a final punch list and certify its completion
2. Prepare record drawings based on information supplied by the contractor

3. Certify that all specified manuals and warranties are transmitted to the owner

4. Address any last-minute corrections required by final inspections

5. Review and approve the contractor's final application for payment

Within the broad categories listed, there are dozens of details to take care of. For example, keying systems, automatic clock systems, telecommunications systems, alarm systems, and the campus computer network have to be hooked up (the contractor's job), and they have to be tested to ensure that they actually work as intended. In some cases, it might take months for problems to surface. For this reason, it might make sense for the owner to contract for post-occupancy services to ensure that the relevant expertise is at hand when needed.

Because colleges and university campuses operate on a predetermined schedule and because the inherent uncertainties in the construction (and especially renovation) process are not always predictable, there is almost always tremendous schedule pressure during closeout. It is important to realize that closeout is not the same as move-in. Once the building is legally turned over to the owner for occupancy, the time has come to move in tables, chairs, desks, filing cabinets, and even the wastebaskets. Thus, when planning for project closeout, allow sufficient time afterward for the owner to handle the actual move-in. Generally speaking, because of the large number of books, files, and research materials (especially for laboratories), a month for the initial move, followed by 6 to 8 weeks for a shakedown period, is not unusual. As a general rule of thumb, a building has to undergo at least one complete heating and cooling cycle before it can be reliably balanced.

One of the most frustrating aspects of project closeout is that no matter how carefully planned and executed a project might be, it will not be perfect. There are simply too many variables for this to happen. It takes patience, flexibility, and wisdom to determine which details are mission-critical and which can be deferred or overlooked entirely. To make the cleanest possible transition, it is practical and sensible to delay owner occupancy as long as possible so that as many punch-list items as possible can be completed without interference. Even a brand-new building begins to wear out the day of move-in. It is not always obvious which scuff marks, dings, and broken hardware are untended punch-list items and which are due to normal wear and tear of the move-in and occupancy. Undone punch-list items are the bane of every project. The only way to solve this problem is to make a clear and concise list and attack each item one at a time. If the schedule expires before the punch-list items are completed, then sometimes it is best for the owners to take on minor items using their own forces and negotiate a credit. Under no circumstances, however, should an owner accept a project that cannot function for the purpose intended.

The best way to manage project closeout is to demystify it. Include closeout on the overall project schedule at the very beginning. Allow sufficient time for completing construction and the shortest possible punch list before formal closeout begins. Similarly, make sure that there is time remaining after closeout for project move-in. Develop a process by which members of the project team are responsible for their own closeout activities, and do not overlap these responsibilities. Pay special attention to tying in campus-wide systems such as telecommunications, computers, and alarms. As a safety valve, make advance arrangements for postoccupancy services by the architect/engineer team and the contractor to cover problems that are sure to emerge after move-in. Project closeout is often a frustrating and time-consuming activity, but if it is properly planned for, it need not be. The best approach to closeout is to do it once and do it right to avoid the need to do it over.

SUMMARY AND CONCLUSION

College and university buildings represent a unique challenge and opportunity for planners, architects, and engineers. They comprise an unusually large spectrum of specialized building types that range from instructional spaces to residential halls to laboratories to athletic facilities. Each campus has its own special heritage and sense of place, which must be respected. College and university projects are shaped by a consensus-based decision-making process that is unique to academia. Other important influences include funding (a large proportion of funding is donor-driven) and politics (the effect of town-gown relations). Projects include new and renovated work, and given issues of deferred maintenance and requirements to adapt to new technologies, there is a continuous need to upgrade facilities to comply with ever evolving building codes and regulations such as ADA. All of these factors combine to make this one of the largest market segments in the United States.

The need to adapt the design and construction process to the special requirements of the academic calendar means that college and university projects need to be carefully programmed, budgeted, and scheduled. Project closeout is of particular interest, because each project needs to be brought online on time to be considered successful.

Though the range of project types is large, the need for specialists is great. Many campuses run central utilities systems or even co-generation plants, and engineering for mechanical, electrical, and plumbing systems is often centrally controlled and requires sophisticated expertise. Athletic structures and buildings for the performing arts frequently demand long-span structures, and there is a great demand for consultants experienced in dealing with such areas as food service, acoustics, specialty lighting, signage, and graphics. This requires a team approach and means that the planners and designers must possess process management skills as well as design and technical skills.

College and university buildings are usually designed to last for generations and are often expected to undergo multiple additions, renovations, or even conversions to new uses. Another critical factor is the importance of landscape design in strengthening the sense of place that makes each campus unique. When undertaking such projects, it should come as no surprise that a period of research and program verification is a prerequisite.

Successful design for college and university projects demands a special range of skills—the ability to reach consensus, the ability to work within schedules and budgets, and vision that can create projects of long-term value. No other building type offers this kind of challenge or the rewards that go with it. A well-designed campus is not just good architecture—it can inspire an entire generation of students.

CONTRIBUTOR

Paul J. Rinaldi, Director of Space Management Department, Boston University, Boston, Massachusetts

CHAPTER 13
AIRPORTS, GOVERNMENT INSTALLATIONS, AND PRISONS

Roger Wessel, P.E., Principal, Chapter Editor
RPW Technologies, West Newton, Massachusetts

INTRODUCTION

A significant portion of the design and construction capacity in the United States, as well as worldwide, is devoted to expanding, upgrading, and building new facilities for airports, government installations, and prisons. The annual increase in world travel has necessitated a more or less continuous improvement and expansion of airport facilities in all countries. Critical issues facing airport facility managers who are dealing with the need to expand, renovate, or rebuild airport facilities are presented in Sec. 13.1, "Airports."

The pervasive expansion of government throughout society in the United States and the ever-changing focus of government functions lead to major construction of new government facilities and constant change at many existing government facilities. The government owns more than 25 percent of the land area, as well as a huge portion of facilities within the United States. A partial discussion of the design, engineering, contracting, construction, and management aspects of government facilities is presented in Sec. 13.2, "Government Installations."

In recent years, the prison system and associated facilities within the United States have been expanded significantly. Many of the older prison facilities have been condemned, others have been phased out in favor of new modern facilities, and still others have been expanded and modernized to keep up with current demand. Prison facilities design, specification, and construction are covered in Sec. 13.3, "Prisons."

SECTION 13.1
AIRPORTS

Stephen M. Sessler, P.E.
Newcomb & Boyd, Atlanta, Georgia

James L. Drinkard, P.E.
HNTB Corporation, Atlanta, Georgia

Recent and projected growth in the airline transportation industry has placed unprecedented demands on airports. Air-carrier airports ranging from small commuter stops to major airline hub and gateway cities are faced with the need to expand, renovate, or rebuild. Critical issues facing facility managers dealing with the needs of aviation facilities include environmental regulations, land use, security, continuity of operations, and funding sources and associated requirements. These issues affect planning, design, construction, and operation in ways that are detailed in this section.

MASTER PLANNING

Airports are similar to other major facilities in that any construction, maintenance, and expansion in accordance with a well-thought-out plan of action can and should avoid inefficiency and waste in development or renovation projects. The creation and periodic updating of an airport master plan is a fundamental requirement of an airport operator, and the document is a tremendous resource to a facility manager. A comprehensive airport master plan (which is required if a sponsor is to be eligible to receive federal aviation trust funds) answers the basic questions of (1) how the airport is accommodating its current demand, (2) how the airport is expected to grow and what facility adjustments are needed to meet the growth, and (3) how the current and future facility requirements can be met financially.

Master plans created for today's airports have to be more flexible than at any previous time and must illustrate how the airport would adapt to changes which can and do arise quickly in the aviation and airline industries. The ability to adapt to new airline growth initiatives and to individual airline demise is a significant change in the process and theories used in master planning 20 or 30 years ago. Airport master plans must be prepared using the best data and resources available, and those participating in the plan's development should represent a broad and diverse cross section of affected parties, usually those representing technical, business, and political interests. As controversial as airport development can be for a community, consensus building among those representing these interests, as well as active public participation, give any plan a much greater chance for acceptance and implementation.

Today's airport plans are also becoming more strategic and business oriented, in that the basic questions to be answered go beyond how the facility will grow. Indeed, answers are needed to address how a community feels about its airport, what the community's vision is for the facility, and what type of business and financial plan must be implemented to achieve the overall goals. Furthermore, participants must agree on the basic precepts used in formulating the forecasts for the facility's growth.

Within the Airport Boundaries

Once actual physical planning begins, the major issues to be addressed for those areas within the airport boundaries include ways to expand or renovate the basic airfield components (aprons, taxiways, and runways), the terminal complex, the air-cargo complex, and the general-aviation complex to meet the projected demand. Key issues to be considered are the efficient interplay among all of these facilities and the multiple interfaces with Federal Aviation Administration (FAA) facilities and associated regulations. These facilities generally involve on-ground and imaginary surfaces that must be protected for the safety of air and ground navigation. Successful master plans also address the long-term infrastructural needs to support the major facility expansion or renovation. Separate planning efforts should be developed within the overall master plan for the long-term needs for water, sewer, gas, power, communications, and roadway systems, and parking.

External to the Airport Boundaries

Because an airport frequently has a significant impact on areas external to its boundaries, the master plan should consider these external areas in relation to any proposed development. Airport master plans must be coordinated closely with the plans of surrounding cities, counties, and municipalities. These interfaces are most important in relation to current and planned ground transportation links such as roadway and/or transit networks and to overall surrounding land-use development strategies, which may recommend land uses incompatible with airport noise or land uses that can adversely affect airport operations.

Airport ground-transportation planning must be developed in conjunction with the long-range regional transportation plans prepared by metropolitan planning organizations (MPOs). Traditionally a state or local department of transportation (DOT) function, the Intermodal Surface Transportation Efficiency Act (ISTEA) transportation legislation enacted in 1991 empowered MPOs to take a stronger policymaking role in developing regional and metropolitan transportation plans. Inasmuch as an airport's operational efficiency can be handicapped on the land side, as well as on the air side, airport planners must look closely at the regional multimodal issues and incorporate them into the overall plan.

As to off-airport land uses and environmental matters, noise and water-runoff quality have historically been the major concerns; however, there are new land-use issues that also require attention. For example, recent restrictions imposed by the FAA concerning the adjacency of solid-waste facilities and detention ponds to airports have had a significant impact on community and airport planning. Birds near and attracted to these facilities have proven to be a significant hazard to air navigation as the number of aircraft-bird collisions has steadily increased. Because most people do not consider a solid-waste facility or an airport "good neighbors," the politics involved in deciding how a community can accommodate both of these necessary facilities can be complex and extremely divisive.

Business Planning

In many cases, the business and financial planning for an airport facility precedes the physical planning and often drives the programming and design aspects of the facility. A wide variety of revenue sources is available to the airport owner, including traditional federal trust funds for capital improvements, landing fees, lease revenues, passenger facility charges, concessions, and parking. The business plan should assess the opportunities for maximizing the revenues from these sources commensurate with the facility's overall needs.

AIRFIELD ENGINEERING AND DESIGN

Civil Engineering

It is mandatory that the civil engineering and design of airfield components comply with the recommendations of the Federal Aviation Administration, published in its numerous advisory circulars relating to airport development, when federal funds are used. Many of these published guidelines are more than "advisory" and are in fact mandatory for all airport development when relating to distances separating runways, taxiways, buildings, fixed objects, restricted zones around navigational aids, gradient criteria, and design life. Beyond these issues, however, the federal guidelines allow and encourage an airfield designer to apply local practices and materials.

The major airfield civil-design issues that affect today's facility manager involve selecting pavement and stormwater-runoff systems and controls. Federal guidelines allow the use of either flexible (bituminous) or rigid (portland cement concrete) pavements on airfields, but the design, construction, and maintenance of these pavement systems vary greatly. The factors that should be considered in selecting the pavement type include locally available materials; native subgrade conditions; intended pavement usage; local contractor expertise; life cycle engineering as it relates to long-term maintenance and rehabilitation schemes; and the effect on airport operations of constructing, maintaining, and rehabilitating the pavement, now and in the future. The life cycle analysis will generally show that a flexible pavement design has the lowest initial costs, but as long-term maintenance and rehabilitation costs are included, the gap between the two pavement types can narrow greatly or reverse. Beyond the initial and long-term costs involved, experience indicates that rigid pavements are better suited to areas subject to fuel spills, such as aircraft aprons, and to areas subjected to constant slow aircraft turning movements.

An airport facility manager should be involved with pavement selection discussions inasmuch as a significant effort is required to monitor and maintain pavements for safe aircraft operations. For a rigid pavement system, maintenance of the jointing system and joint seals will constitute the major ongoing effort (Fig. 13.1-1.). For this reason, facility managers should thoroughly familiarize themselves, during design, with the processes and materials used in initially sealing the joints and the corresponding process which would be used for repairing damaged joint seals. For flexible pavement systems, facility managers should familiarize themselves with the composition of the total pavement section so that the repair approach in distressed areas is consistent with the structural and drainage characteristics of the original design.

The pavement condition index (PCI) should be used to evaluate pavement condition. In many cases, the success of any pavement system can be traced to the components of design, which facilitate the removal of water from the pavement base, subbase, and subgrade. Free-draining base layers or longitudinal and transverse underdrains can dramatically increase pavement performance and pavement life. This aspect of the design is particularly important for airports, where heavy aircraft tire loads can induce the pumping of fines from a saturated subsurface pavement section through small pavement cracks or pavement joints, eventually causing voids and lack of structural support of the section.

Stormwater and pollutant runoff are major factors in today's airfield design. On aircraft aprons where fueling takes place, FAA gradient regulations contained in Advisory Circular 150/5300-13 must be applied in concert with National Fire Protection Association (NFPA) codes. NFPA 407 specifically addresses the drainage and runoff design of aircraft fueling ramps. These regulations were written to safeguard life and minimize property damage in the event of a major fuel spill. Even without major fuel spills, routine fueling and the use of motorized ground service equipment on aircraft aprons create a constant level of contaminated runoff. The disposition of this runoff is generally governed by state or local guidelines that meet or exceed guidelines established by the federal Environmental Protection

FIGURE 13.1-1   New concrete runway at Kansas City International Airport. (*Courtesy of HNTB Corporation, Atlanta, GA.*)

Agency (EPA). Procedures to manage and control this runoff vary from state to state. Examples of airport methodologies to meet these regulations include incorporating holding tanks in the primary drainage system, inlets that include baffles, and downstream retention ponds that can be skimmed for contaminants. In locales where freezing precipitation is common, deicing fluids also contaminate the stormwater runoff. Because certain deicing fluids can be captured for reuse applications, many airports and airlines have opted to create a separate area for deicing operations, so that the fluid is contained apart from all other runoff. Where deicing fluid is not recycled, it must generally be filtered and diluted to have a maximum 1.5 percent glycol before it can be discharged to the sanitary sewer. The significant impervious area on an airport due to large pavement and building masses creates significant runoff flows. For this reason, many airports design and incorporate a series of detention ponds that are used to maintain the rate of runoff into adjacent waters at predevelopment levels.

Electrical and Lighting

FAA Advisory Circulars and Orders govern the design of airfield lighting and electronic navigation systems. The extent of airfield lighting and electronic navigation systems varies with the visibility conditions under which the airfield operates. Airports that operate with line-of-sight aircraft approaches may have no ground-based navigation systems and only low-intensity runway approach and edge lights. Airports that operate under all visibility conditions have complex lighting and navigation systems. The FAA defines visibility conditions as follows:

Instrument category (visual)	Runway visual range, ft	Decision height, ft
Nonprecision	N/A	N/A
I	2400	200
II	1200	100
IIIA	600	—
IIIB	150	—
IIIC	0	—

The FAA usually owns and operates navigation and approach lighting systems. A nonprecision navigation system can consist of a nondirectional beacon (NDB) and very high frequency omni range (VOR) and localizer antennas to assist pilots in locating the airport and runway. Airports supporting precision approaches for aircraft with low-visibility instrument approaches need glide slope, localizer, and marker antennas and the like to further define the path for incoming aircraft. A runway visual range meter is used to determine the local visibility conditions. The FAA also owns and operates approach lighting systems, which are an array of ground-mounted lights in front of the runway medium-intensity approach lighting system (MALS), medium-intensity approach lighting system with runway alignment indicator lights (MALSR), or high-intensity approach lighting system with sequenced flashers for Category II operations (ALSF2), and lights on the side of the runway indicating the glide slope visual approach slope indicator (VASI) or precision approach path indicator (PAPI).

An extensive ground-based lighting system is required for an airport to operate under low-visibility conditions. The runway needs high-intensity edge lights, centerline lights, and touchdown-zone lights. The taxiways need centerline lights and edge lights along the curves. Controllable stop bars are needed at runway taxiway intersections. Illuminated signs are also used to guide aircraft.

The lighting systems at large airports require a substantial infrastructure to support them. A duct bank and manhole system should permit routing cable to any point on the airfield that requires lighting. The airfield lights should be served by a series of lighting circuits to ensure that every light on a circuit has the same intensity. Constant current regulators should be used to convert utility electrical power into series circuits. The utility power source should be backed up by an emergency power system capable of carrying the entire airfield. Primary power from the utility should also include alternative or dedicated feeders to achieve the highest reliability possible.

Fire Protection and Firefighting

Airports are required by the FAA to locate airport rescue and firefighting (ARFF) facilities so that response time to airport landing and takeoff areas is within defined limits; consequently, a large airport may have several ARFF facilities. An ARFF is similar to a municipal fire station that houses the firefighting equipment and normally has quarters and locker rooms for firefighters. In addition to the ARFF facility, some existing major airports should have areas known as *burn pits* or *live-fire training facilities* that are set aside for firefighting practice. These facilities allow firefighters to simulate crash conditions and to practice emergency response techniques for both firefighting and victim rescue. Such areas are discouraged in newly constructed airports.

Security

Federal Regulations and FAA Advisory Circulars largely dictate airport security requirements. For security purposes, the airfield is referred to as the air operations area (AOA), which FAR 107.1 defines as the portion of an airport designed and used for landing, takeoff, or surface maneuvering of airplanes. The focus of security in the AOA is to prevent access by persons other than those employed by the airport or one of its entities; these security measures include prompt cancellation of access privileges for persons terminated or who no longer need access. The most common method of achieving this is via a computer-based access-control system that grants or denies access through selected portals (doors, gates, etc.) based on the user's preprogrammed access rights, the specific portal, time of day, and day of week. One significant problem occurs where the screened public must exit to the AOA in the event of a fire or other emergency. A common approach used to provide security, while still complying with the life-safety code, is to use delay-egress devices that prevent the release of the door for 15 or 30 s while allowing a security response before release. These doors should also be under closed-circuit television (CCTV) camera surveillance to allow remote observation of the door from the security console. A second significant problem is providing access to the AOA by authorized vehicles while preventing tailgating. Tailgating can be prevented with a sally port or simply by having authorized users stop just clear of the gate and wait for it to close completely before proceeding. CCTV cameras should be deployed at these locations to enforce the stop-and-wait policy.

Fueling

Aircraft fueling at airports varies with the size and operations of the facility. Most major airports have opted exclusively for in-pavement hydrant systems, whereas other airports use truck refueling on the ramp. An airport with a hydrant system will normally have a fuel storage area, called a *fuel farm,* where the fuel can be stored, filtered, and pumped through transmission lines to the aircraft aprons and the individual hydrants at each gated parking position. Fuel farms should be provided with redundant primary power sources. For airports that use truck refueling, a fuel storage area is also needed where the refuelers can access the filtered fuel. Aviation fuel is transported to the storage areas by over-the-road tankers and/or commercial, underground, fuel transmission lines. The decision to utilize refuelers versus in-pavement hydrants is an economic and time issue that requires airline participation. In the brief turnaround time typical of major hub airports, relying on a refueling fleet for the bank of aircraft is impractical. The hazard of the increased major tanker traffic on the ramp itself is a factor, as is the rate of refueling, which is significantly faster using in-pavement hydrants. Conversely, at less busy airports, the investment required for a hydrant system, in many cases, cannot be justified.

The design of an apron hydrant system requires that final aircraft parking positions be firmly established. Hydrants can be located to serve several different aircraft types, but in these cases require detailed studies to determine the final locations. Modern hydrant system design also incorporates isolation valve pits and looped pipe networks, so that portions of the fuel system can be shut down for maintenance while other portions remain operational. Cathodic protection and leak detection of fuel lines are also major components of the system design.

Airfield Traffic Control and Communication Systems

The modern-day airport will have a complex network of traffic-control and communication systems. Air-traffic control is under the primary control of the Federal Aviation Administration, but it has interfaces with the airport's airfield lighting system. This interface will normally be in the form of control wiring between the airfield lighting vault and the airfield lighting-control system located in the air-traffic-control tower. Tower controllers determine which components of the airfield lighting system should be operational consistent with the operation of the airspace and airfield at any given time. Air traffic controllers select the airfield lighting levels through switches or touch-screen computers located in the cab of the tower.

Components of communication systems located on the airfield will be routings of the systems required for the airport's various building systems, its security systems, and the airfield's lighting and navigational aids. Because of the number of different communications systems required and the massive volume of information carried, the modern airport complex should be provided with a fiber-optic backbone that facilitates all types of data and systems, provides excellent reliability, and is immune to lightning-induced transients.

PASSENGER TERMINALS (BUILDINGS)

The passenger terminal is a dynamic, complex building type whose operation requires a comprehensive knowledge of systems, people, and material (baggage) flow. Many air carriers have specific requirements for dedicated areas such as ticket counters and gate hold rooms. But the global issues that affect long-term owning and operating costs must also be addressed in the planning and design stages. All systems should be designed to be flexible to accommodate the constant modifications needed by airlines and tenants. Likewise, future expansion should be anticipated in both the building structure and in all systems. All of the equipment used in a terminal should be supported by local parts and service organizations.

Architecture

The architect's role in designing the terminal has at least two distinct functions. The architect charged with terminal planning and layout must have a true understanding of all of the functional elements of an airport. For example, the terminal planner is charged with creating efficient relationships of ticketing to baggage, terminal to concourses, and parking to the terminal. Aesthetic considerations are also important to establish local identity and to create a sense of place (Fig. 13.1-2). The terminal planner must thoroughly understand passenger circulation and plan spaces accordingly for aisles, people movers, elevators, and queuing at ticket counters and at bag-claim devices. Changes in the airline business and the desire to generate more revenue from passengers during the past 10 years have forced terminal planners and architects to create more flexibility in their space planning. Airports have become much more mall-like, and the location of retail spaces must be planned to allow maximum visibility and accessibility to potential customers.

Of greater interest to the facility manager is the role the architect plays in specifying the materials used in the building spaces. Material selection, including flooring, ceilings, doors, walls and wall coverings, and door hardware, should be made for extreme durability and longevity. Material selection should also be standardized as much as possible so that the airport facility manager does not have to stock and maintain an inventory of numerous parts and supplies. Attention must be paid to the accessibility of systems in the building for ease of maintenance.

Electrical

Reliability is the single most important attribute of electrical systems in passenger terminals. Primary switchgear should be arranged to provide each main electrical distribution area with two primary feeder circuits. Half of the distribution transformers should be normally connected to each of the feeders. Main 480-V distribution equipment should be arranged in a main–tie–main configuration for redundancy. An emergency power system should serve the essential life-safety functions and also maintain critical airport operations and should be designed for a full capacity redundancy ($N+1$) configuration. An uninterruptible power supply (UPS) should be provided for critical functions such as security, voice and data communi-

FIGURE 13.1-2   T. F. Greene Airport, Providence, Rhode Island. (*Courtesy of HNTB Corporation, Atlanta, GA.*)

cations, and airport computer systems. Each airline will generally provide a UPS for its ticketing system. To the extent practical, lighting should have central controls, and the different types and wattages of lamps should be minimized. Provisions should be made for easy access for relamping all lighting fixtures.

Fire Protection

Generally, airport terminal facilities, including exterior canopies, must have sprinklers. The sprinkler system design should particularly address baggage-handling areas, spaces above and beneath conveyors, and conveyor penetrations of fire walls. Where glass windows face an aircraft ramp, waterspray protection of the glass may be required if fuel spills can occur within 100 ft horizontally from the glass. Class I fire department fire hose connections should be provided in accordance with NFPA 14 and local codes. Fire hose connections may be required within automated transit system tunnels. Fire hydrants are required around the terminal perimeter, including ramp areas. Pipe bollards should be provided to protect all hydrants on the ramps.

Speakers are recommended as the fire-alarm audible notification device. They offer a multitude of tones and can also broadcast live or recorded verbal instructions. The paging system should not be used for this purpose. The fire-alarm system should have a sufficient number of speakers so that voice instructions are clearly intelligible (in multiple languages) in spaces with hard surfaces. Because of long line-of-sight areas, the flash rate of fire-alarm visual signals should be synchronized. Distribution of fire-alarm power, initiating circuits, and notification circuits may be required because of long wire runs.

Mechanical Systems

HVAC system selection is critical in a passenger terminal. The HVAC system must provide some backup capability (but not necessarily redundant capacity), so that an unplanned shutdown of one piece of major equipment (chiller, cooling tower, pump, boiler, or air-handling unit) will not jeopardize the operation of the entire facility. Cooling towers located on or near a terminal may need defogging coils so that pluming does not obstruct traffic-control sight lines. Air-handling systems should be zoned to allow shutting down areas that are occupied only periodically. Variable-volume systems should be used to accommodate the wide variations in load that occur. Outside air must be introduced to meet the latest ANSI/ASHRAE 69 standard and also to minimize infiltration of unconditioned air through passenger doors, jetway openings, and baggage-system openings. Jet fuel and exhaust odors pose a significant threat to indoor air quality and can best be mitigated by judiciously locating air-intake louvers. Activated charcoal filters can be used where necessary but, because of their high life cycle cost, are recommended only where no other solution exists.

Communication Systems

Generally, telephone system needs are met with a dedicated telephone switch or private branch exchange (PBX). Basic PBX systems will provide for the multitude of administrative telephones, house telephones, and, perhaps on a lease-back basis, concessionaires' telephones. Metallic or fiber-optic lines should connect the PBX system to the local telephone company via voice lines, T1 circuits, and/or integrated services digital network (ISDN) services. Radio systems play a critical role in operating the airport. A VHF, UHF, 800-MHz, or 900-MHz trunked radio system should be installed to facilitate wireless conversations between emergency response teams, operations and maintenance personnel, and air- and ground-traffic control. Special attention should be paid to grounding and transient-surge protection.

Information display systems are located at strategic points to advise passengers of flight schedules and locations (flight information display systems [FIDS]) and baggage-claim locations (baggage information display systems [BIDS]). Most FIDS and BIDS are dedicated, single-airline systems. Common-use systems are increasingly being installed by airports to gain flexibility in using gates. These systems can be either baseband (local-area networks that use computer monitors to display the information) or broadband (cable television systems with character generators that use television monitors to display the information). Cable television systems for public viewing should be provided with sets typically located in restaurants, lounges, and concourse waiting areas. Data-communications systems are required to support the many business entities and various operational functions ranging from parking to ticketing. Although the infrastructure or supporting systems such as cable trays, conduits, duct banks, structured cabling systems, power, and HVAC provisions will be provided in the base building contract, the actual local-area networks and high-speed data links will be generally designed and implemented by each different entity.

Security Systems

Security systems in terminals include a wide variety of subsystems, some of which are CCTV surveillance, emergency call stations, metal detectors and baggage x-ray systems, transition portals, and access control. More elaborate provisions are required in international airports with Federal Inspection Service (FIS) facilities. In addition to the CCTV surveillance of AOA access points and portals that provide transition from non-AOA areas to AOA areas, FIS facilities should have CCTV surveillance of points that provide transition from the sterile area to the screened area, a security control console with sight lines over the various inspection stations, emergency call systems for summoning police or medical assistance, CCTV surveillance of the sterile corridors, and CCTV surveillance of the airplanes' baggage compartments and the entire route to the inspection station.

The security process must be designed to ensure that all persons on the AOA or in sterile areas (those areas past the preboarding screening checkpoints) have been screened for possession of any explosive, incendiary, or deadly or dangerous weapon. Airport tenants or fixed-base operators are required to conduct screening at any access points to the AOA or sterile areas under their control. These tenants or operators are required to screen according to an FAA-approved security-screening program (as defined in paragraphs 108.5 and 108.7 of 14 CFR Part 108) which typically includes metal detection and x-ray equipment.

Aircraft-Support Services

At small airports, aircraft use their auxiliary power units (APUs) to provide power and air conditioning. But at large facilities, aircraft-support services originate from the systems in the passenger terminal, thus providing greater flexibility and efficiency. Cooling for aircraft parked at gate positions is provided via preconditioned air (PCA) that is delivered to the aircraft by special fan-coil units usually built into the loading bridges. These units typically require 20°F glycol and thus cannot be served by the terminal's chilled-water system. A separate, low-temperature glycol system (chillers, pumps, and piping) should be provided for that service, but the terminal's condenser water system can be used for heat rejection by the glycol chillers. These PCA fan-coil units can also provide heat, where needed.

In colder climates, aircraft deicing is usually done at a designated ramp location where the deicing fluid can be collected via surface grates and recycled. The fluid is sprayed on the aircraft by tanker trucks with booms. The deicing fluid is mixed and heated to 180°F at a central location, usually a maintenance hangar.

The aircraft's potable-water supply is replenished at the gate from a connection to a potable-water cabinet. This cabinet at each gate is served from the building domestic cold-water system. Where freezing temperatures can occur, exterior domestic cold-water piping serving these cabinets must be heat traced.

400-Hz power supports aircraft power and lighting needs while the aircraft is sitting on the ground without internal power. This power may be provided by portable or fixed-frequency converters. Fixed equipment within the terminal requires wire in aluminum conduit to disconnects and flexible power cords and connectors at each aircraft parking position on the ramp. This service should be supplied by 480-V service in the terminal.

An emergency fuel-shutoff system should be provided with at least one annunciator and emergency shutoff station located at each aircraft parking position to sectionalize the pressurized fuel lines, so that a fuel spill at any aircraft location can be immediately halted by local system activation. Remote annunciation and notification of fuel and firefighting authorities should also be possible.

Baggage-Handling Systems

The baggage-handling system necessarily varies greatly depending on the specific passenger traffic types and volumes for that airport. The system design should consider such things as originating, terminating, and connecting (transfer) passengers; average peak periods, daily and hourly; type of aircraft; load factors; and passenger-processing time (usually specific to the carrier).

Baggage-handling systems at modern airports are grouped into three categories: outbound, transfer, and inbound. Typically, the outbound system takes baggage from the ticket counter or curbside to the baggage-processing area(s). These outbound systems can be as simple as a take-away belt behind the ticket counter that takes the baggage to a simple sorting belt. It could also be a system in which an agent assigns the flight number, and then the baggage-handling system delivers that bag to the appropriate point for that flight. This could be a common baggage-sorting system that places the baggage in the appropriate spur or delivers the baggage to an aircraft gate delivery point adjacent to the aircraft parking area. One of the frequently overlooked aspects of the baggage-handling system is the transfer system.

The facility manager should be aware of the various requirements for each system. For example, the gate delivery system necessitates baggage corridors through the operating level of the terminal and concourse. These corridors would have many points of interface with other critical airline-support functions. It is crucial that these interfaces be identified early and addressed. On the inbound side, the baggage delivery to the passenger will also vary. Typically, domestic baggage is delivered via a baggage cart to the secured area of the terminal, where the baggage is placed on a conveyor system. The delivery point is usually located in the terminal area, requiring baggage cart drives. Interface with other critical terminal support functions must be addressed early in the planning process. The facility manager should take into account operational procedures, equipment preferences, equipment cost and life cycle costs, and flexibility and expandability.

At major airports, some airlines may have bulk-mail-sorting operations. These facilities are owned and operated by the sponsoring air carrier, but the U.S. Postal Service provides inspectors to oversee the operations.

People-Mover Systems

It has become more common in today's airport layouts to include people movers to transport passengers between distant functional areas in the airport, such as the terminal building and the concourses. Many large U.S. airports, including Atlanta, Cincinnati, Dallas–Ft. Worth, Denver, Houston, Newark, Orlando, and Tampa have automated people-mover systems. People movers can be located either outdoors or indoors and can be elevated, at grade, or underground (Fig. 13.1-3).

In planning for people-mover systems, a study must be made to examine the passenger circulation peaks with respect to the physical layout of the airport. From such a study, the general type, size, and frequency of a people-mover operation can be determined. Most people movers have trains serving two-way travel to or from a concourse. Of the several people-mover manufacturers, each has a slightly different technological approach to its individual system. A *pinched-loop* system allows individual trains to switch tracks, whereas a *shuttle* system will ride only on one set of tracks. The pinched-loop system requires a large maintenance area at the end of the rail system, whereas the shuttle system may require maintenance from underneath the system while on-line.

Other decisions to be made concurrently with selecting people-mover technology are the layout of the station platforms and any provisions for redundant means of travel should the people mover be shut down or become too crowded. When people movers are located indoors or in underground tunnels, a number of emergency and life-safety systems and provisions must be in place.

Tunnel ventilation must provide adequate conditioning, as well as smoke control for occupied areas during emergency conditions. Air conditioning for passenger stations and walkways is primarily based on lighting and intermittent people loads. Air conditioning for passenger cars is typically handled through systems installed by the train manufacturer. Smoke control can be complicated in long tunnels with multiple stations. In the event of a fire, the smoke should be exhausted away from occupied areas and fresh air supplied to occupants during their egress. Exhaust and supply systems need to be controllable to accommodate various occupancy configurations because train positions are constantly changing. Reversible fans with emergency power service are typically used.

FIGURE 13.1-3   Automated people mover provides access from terminal to air-side building at Orlando International Airport. (*Courtesy of HTNB Corporation, Atlanta, GA.*)

AIR CARGO

All airports have air-cargo facilities. These facilities can range from small buildings with a contract management vendor serving all carriers to large individual structures for each airline at major airports. Lease rates are very important, and creative strategies such as the use of third-party developers are allowing air carriers and airport authorities to use capital funds better. Air-cargo buildings are basically warehouse/distribution centers that do not require many of the special features that are found in other airport facilities.

Architecture

Architectural design of air-cargo facilities is primarily a space-planning issue. The function of an air-cargo building is to facilitate the temporary storage and movement of goods and packages as efficiently as possible. Accordingly, the building should be designed by considering the personnel, equipment, and space for that purpose.

At a major airport, there are likely to be several types of air-cargo facilities. Major companies such as DHL, Emery, Federal Express, and United Parcel Service, which generally handle all aspects of the shipping process, will desire to customize their own facilities, based on their proprietary processes. An airport sponsor will generally provide the infrastructure, such as trunk utilities and perhaps even the aircraft ramp, as part of a leasing agreement with the company, and the building will be designed and constructed by the individual company. Major airlines may also have their own facilities for handling cargo similar to those described.

For generalized cargo handling, an airport may build a facility for multiple tenants. A prototype cargo building will be designed to accept air cargo from several different types of aircraft on a ramp; to allow transporting the cargo easily from the aircraft inside the building for handling and sorting; and finally, to allow for expeditious loading onto trucks or for pickup or transfer. The architectural design and space planning will need to examine the types of equipment used, such as fork lifts, automated conveyors, in-floor makeup lifts, sorting areas, transfer accommodations, and storage areas. The entire cargo movement process is enhanced by large open spaces free of intermediate support columns. As such, special framed roof systems are often found in cargo buildings.

Office space inside the building is also necessary. Because of the large open space required for efficient movement inside the building and because of high ceilings, it is common to find office space located on a mezzanine level above the land-side delivery area. The office space in this location takes advantage of the smaller door openings required on the land side versus the air side and also provides supervisors with a bird's-eye view of the operations below.

On the land side of the building, traditional truck docks are normally used with automated doors. Careful planning is required to anticipate the circulation and parking needs on the land side. In addition to employee and visitor parking, it may be necessary to consider and incorporate overnight parking areas for trucks and trailers that unload and then must wait for the next scheduled cargo shipment to load.

Aesthetic issues are usually not a primary consideration for air-cargo buildings. Frequently, these buildings are located in an industrial area of the airport away from the more public functions. However, many airports opt to set aesthetic guidelines for all development on their property, including cargo and industrial buildings. These guidelines may include the color schemes, general architectural shape, and any height issues relative to the adjacent airfield.

Lighting

Lighting should provide even, energy-efficient illumination and should generally be metal halide. Refer to Sec. 5.3, "Lighting," for a detailed discussion of this topic.

Fire Protection

Air-cargo facilities, including all exterior covered loading docks, should be fully sprinklered. The selection of sprinkler discharge rates and operating areas must be determined on the basis of storage heights and storage methods. Water demand can be high and must be coordinated with available water supplies.

Mechanical

Air-cargo facilities are heated and ventilated but are not air conditioned. Winter design temperatures of 50 to 60°F can be maintained by low-intensity infrared heaters, except in extreme southern climates where no heating is required. Ceiling fans are effective in minimizing the stratification that occurs in high-bay structures. Ventilation is typically designed for 6 or 8 air changes per hour (in extreme southern climates, 10 air changes per hour is generally used as the criterion). Facilities that handle perishable products and other temperature-sensitive items need cool- or cold-storage capability that can generally be provided by pre-engineered package coolers. The condensing units on smaller coolers and freezers can be mounted on top of the cooler, but for larger units, the condensing units should be mounted on the roof or outside on grade. The battery-charging area will require an eyewash station and a ventilation system for fumes.

Security

Typical warehouse security provisions are adequate for all areas of air-cargo facilities except two. Public access to the front of the facility must be separated from the AOA in the rear. The other concern is secure storage for in-bond freight that is awaiting clearance by customs. Generally, a lockable fenced area with CCTV surveillance is adequate. Where loading door openings are used for ventilation air makeup, rolling security grilles are required.

Fueling

Aircraft fueling at air-cargo facilities is similar to fueling at the passenger terminal complex. It can be handled through a dedicated underground hydrant system, by over-the-road tanker trucks, or by on-airport tanker trucks that are filled at an on-site airport storage facility. Fueling of support vehicles at an air-cargo facility is best accomplished through a dedicated automobile and truck fuel storage and distribution system.

Aircraft-Support Services

Where needed to support the activity and the turnaround time, a 400-Hz frequency converter can be provided for aircraft. Optical lead-in indicators should also be provided where pilots will power the aircraft into position.

Material-Handling Systems

For storage, many larger cargo facilities now are incorporating stacker or rack systems. These systems take advantage of the ceiling heights required in cargo facilities and can be automated and computerized for ease of tracking and movement.

SUPPORT FACILITIES

Civil Engineering

The civil engineering design for airport support facilities such as fire stations, post offices, ground-service equipment maintenance areas, and rental-car support facilities will conform to the overall design standards that apply to the airfield and terminal areas. For each of these facilities, choice of pavement types should be based on the type and frequency of traffic. Environmental factors should also be considered; for example, areas routinely subjected to oils and other fluids are best served by concrete. The design of drainage systems where maintenance activities occur may require incorporating oil/water separator holding areas to contain contaminants. Roads and parking areas for service vehicles and public parking should be designed in accordance with local practices and codes. The geometric and structural design should take into account the operating characteristics of vehicle types encountered at an airport; these can include tugs, refuelers, baggage carts, and other nonlicensed vehicles.

Electrical

Electrical service provided to emergency support services such as fire stations and communications facilities should be provided with emergency standby power service. Other services that should be considered for emergency power backup include storm and sanitary lift stations, water-pumping and -treatment facilities, and fuel-distribution pumping systems.

Communications

Communication paths must be provided between all facilities and the airport fire and police stations. These communication paths can be radio, dedicated telephone lines, local-area computer networks, or a combination of them. Fiber-optic systems often link these critical facilities with the terminal, concourses, and support facilities.

Security

The fire station and security or airport police station may be housed in the same building and thereby share some resources. If so, some police or security forces should be stationed in the terminal near the security checkpoint to provide rapid response in case of an incident.

Fueling

Fueling facilities for the vehicles related to support buildings can be located in areas with the aviation fueling facilities but must be clearly separate because the functions and fuels are totally different. One reason to have these facilities in the same area of the airport adjacent to aircraft-fueling facilities is that there will normally be a roadway network to support the delivery of fuels, as well as a roadway network accessing the airfield, that can serve both demands. However, each support facility must examine its own needs in this respect. Rental car companies, for example, will normally incorporate their own exclusive fueling facilities at the facility where the cars are prepared for return to the ready lot. Individual airline ground-equipment-support facilities may also benefit from having separate fueling operations at the building site.

Central Plants

Airport complexes with multiple land-side and air-side buildings (concourses) will generally benefit from central refrigeration and heating plants. Although the initial costs of a central plant may be higher than that for separate local systems, the benefits of central plants include higher-efficiency equipment, the ability to take advantage of load diversity, centralized maintenance, and equipment redundancy. Piping distribution systems should be designed to be easily accessible, and means should be provided to allow isolating major runs and all branches. Central emergency power systems offer similar benefits.

INFRASTRUCTURE

Ground Transportation

The ground-transportation system at an airport can be a complex design issue, particularly at the major hubs. As an initial consideration, the design of the on-airport system must begin with a thorough examination of the transition from the state or local roadway system to that of the airport. There must obviously be consistency in data used to determine traffic demand and corresponding roadway capacity for the on-airport and off-airport systems, and off-airport roadway signing must be coordinated between the two agencies. The general design of the structural pavement, drainage, marking and striping, sight distances, laneage, horizontal alignment, and vertical alignment for an airport roadway system should follow state DOT and American Association of State Highway and Transportation Officials (AASHTO) standards and guidelines.

As the layout and design of an airport roadway system evolve, the primary consideration is the numerous functions which must be accommodated, such as access to ticketing, baggage claim, parking, rental car returns, courtesy and commercial vehicle access corridors, service and delivery vehicle routes, terminal return points, and airport exits. Because these numerous functions must be accommodated, the single most important issue besides roadway capacity is horizontal (and to some extent, vertical) design which allows signing for adequate decision time for lane changes and merges for new users. The frequent user can and will ignore signs, but with airline fare wars and the increasing desire for expedited travel, airports are attracting new users each year and the roadway and signing systems must anticipate the unfamiliar driver.

Unfamiliar drivers need continual reinforcement that they are on the correct road and in the correct lane. Major decision points for the driver should be signed well in advance, and these signs ideally should be repeated as many as three times (Fig. 13.1-4). The operating speeds on the road govern the ability to digest sign content and make necessary maneuvers. The geometric design of the road can assist in limiting operating speeds, thereby facilitating decision making and lane-changing times.

The design of the terminal curbside is a specialized effort, but traditional DOT and AASHTO design manuals do not address this area. Curbside design must be closely coordinated with the design of the terminal itself; basic decisions as to how many levels will be used and where the various functions (e.g., ticketing, baggage-claim, rental car, and commercial

FIGURE 13.1-4 Variable message signs provide current additional information on traffic and parking at Raleigh-Durham International Airport. (*Courtesy of HNTB Corporation, Atlanta, GA.*)

vehicle areas) will be located are fundamental to both designs. Experience has shown that terminal curbsides must be a minimum of 44 ft wide to operate efficiently for the entire length of the curbside area. This width allows routine double parking, one maneuvering and merging lane, and one through lane. This area can occur on several levels if necessary to accommodate the demand for vehicles needing ticketing, vehicles meeting arriving passengers, and for commercial vehicles (taxis, hotel shuttles, buses, and off-airport parking shuttles). At busier airports, it has proven effective to separate the commercial vehicles from private automobiles. Besides the safety aspects of keeping large buses and shuttles away from the automobile traffic, the commercial driver's perspective and familiarity with the airport roadway system versus that of the private automobile driver can be quite different.

The length of the curbside can be determined from studying peak vehicular and passenger flows in concert with an estimate of the dwell times of each vehicle. The *dwell time* is that time during which a vehicle actually stops at the curb and either drops off or waits to pick up a passenger. Generally, dwell times will be greater where passenger pickup is accompanied by loading bags and the possibility that the passenger is not waiting at curbside when the vehicle approaches. Accordingly, curbside dwelling times at baggage claims or at areas designated for arrivals will be longer than at the departure curb.

Because of the concern that a vehicle laden with explosives might be parked adjacent to a busy terminal and left for detonation, the FAA frequently restricts the parking of unattended vehicles in the passenger pickup and drop-off zones during periods of heightened security. It may well be just a matter of time before this becomes standard operating procedure. FAA regulations should be consulted for current required clearances between the terminal and unattended parked nonairport vehicles before providing such spaces near the terminal.

The facility manager should be involved with the airport roadway and curbside design process and should clearly understand the issues discussed here. Traffic congestion on the roadways and at the curb can be a functional and public-relations dilemma which, if unsolvable without a major design and reconstruction effort, must be solved with increased enforcement. Curbside dwell times must be minimized for a curb to function properly. In reconstruction or renovation projects, meeting the basic criteria discussed before can be a significant challenge and may require constructing a number of "throwaway" roads, curbs, and canopies to meet the basic operating parameters during construction. The interface of pedestrian traffic and the curbside and roadway network on renovation projects is a major concern. It may be necessary to highlight crosswalks with flashing lights and bold striping during these projects. In some cases, a true stop condition for vehicles may also be warranted. The Transportation Research Board (TRB) maintains an excellent library of technical papers written about the design of airport ground-transportation components, including the roadway network and the curbside.

Utilities

Airports are often called minicities, and in the case of utilities this is a particularly accurate description. Airport utilities normally include water (both supply and fire protection), wastewater, irrigation, power, communications (e.g., telephone, local-area networks, and cable TV), security, and natural gas. In addition to these utilities, the FAA will have its own network of utilities in an airport to serve the air-traffic-control tower and all of the navigational aids. With such a plethora of utilities, airport owners are best served by having a master utility plan that is prepared in coordination with the overall master plan of the airport. A thorough understanding of the utility network in an airport is necessary from the standpoint of facility management because it represents the very backbone of the entire facility.

A major consideration in designing any utility at an airport is redundancy or looped systems. A utility failure at an airport can have devastating financial and operational effects on the airport and all of its tenants. Of primary concern in this respect are power and communications. Airports should coordinate with these providers to have at least two distinct points of service; this creates a near 100 percent reliability factor for continuous service.

Engineering and design standards and criteria for utilities should conform to local practices because there are no applicable federal guidelines. Lightning protection and surge arrestors should be included in all power, alarm, signaling, and communications systems at each facility serving the airport. Trunk utilities are frequently located along the major transportation corridors for ease of access and maintenance. These locations often create competition between the aesthetic desires of most airport owners and the practical needs of the utility companies and the airport facility manager. Innovative landscaping can be developed to accommodate both interests.

An airport sponsor may find itself with the choice of owning and operating its utilities. This decision on any utility must be based on an economic assessment of the airport's cost to provide and maintain staff for that purpose, compared to the cost of traditionally provided service and maintenance; the potential economic benefits of selling utility service to its tenants; and an assessment of the reliability gained by having the utility under its direct control. In many locales, this issue can be politically volatile, but in this age of deregulation, more and more airports have found benefits in taking control.

Parking

The business of parking at airports constitutes a major benefit to the user and the owner. Airport users demand and appreciate the convenience of close-in parking, as well as less expensive long-term parking, and airport owners derive a huge amount of revenue from this concession. The overall parking layout at an airport must be developed in concert with the terminal complex and roadway network.

CONSTRUCTION

Green-field airport construction poses very few problems that are not typical in other complex building projects. During construction, observation of critical components and systems is essential—for example, reinforcing steel in aprons and flatwork. The personnel who will be responsible for maintenance in the completed facility should be hired at the beginning of construction and should use the construction and commissioning process to become thoroughly familiar with the buildings, utilities, and systems.

Most airport construction, though, is renovation and addition work. These projects require carefully planned protocols to prevent unplanned utility interruptions, security breaches, and interference with ground and air operations, including emergency response. Full-time site representation by the design team and a close working relationship among the airport authority, contractor, design team, and construction manager are absolutely essential.

OWNERSHIP AND MANAGEMENT

Ownership

Commercial-service airports in the United States are owned by public or quasi-public agencies. The owning entity can be a city, a county, a joint city-county agency, a state agency, a joint state-city or -county agency, or a separate legislatively enacted authority whose members are appointed by a governor or by a city or counties represented by the authority. This public form of ownership is facilitated and encouraged by the fact that the federal government has traditionally provided airport improvement funds to publicly owned and operated airports.

Management

Airports in the United States are managed by trained and experienced airport professionals who have business and/or technical backgrounds. Only a few years ago, many airport managers tended to be World War II or Korean War aviators, or individuals who had some other military aviation background, but a transition has occurred as these managers have reached retirement age. The newer generation of airport managers often comprises individuals who started as assistants and have primarily on-the-job training or who were educated at the growing number of colleges and universities that offer airport management curricula. The American Association of Airport Executives (AAAE), an industry organization, has also implemented a program for certifying airport executives. To qualify for this status, one must have a set number of years of hands-on experience and pass both written and oral examinations about all aspects of airport management.

A typical airport management team will consist of a director (who may be called a manager or an executive director) and a team of direct-report assistant managers heading up areas such as planning and development, operations and maintenance, and business and finance. The remainder of the staff composition depends on the size of the airport. For the planning and development functions, some airports maintain professional engineering and planning staffs that are capable of planning and designing capital improvement projects; however, airport consultants are most often used because of professional liability requirements and the peaks and valleys of airport development programs.

For the operations and maintenance side, an airport will normally have a staff responsible for ensuring the daily operation of the airport, primarily in terms of safety and security. FAA requires that runways and taxiways be inspected routinely to check for any debris or other airfield maintenance matters that could affect aircraft operational safety. Airport maintenance staffs are capable of keeping airport lighting and other communication systems on the airport operational. Police and firefighters may report through this management team also.

The business and finance team will be involved with all the various lease and property arrangements with individual concessionaires, airlines, and other tenants, as well as being responsible for the overall fiscal health of the airport.

Privatization

As cities and other public entities are looking at privatizing traditional services provided in-house, such as water and sewer departments, there is also increasing interest in privatizing aspects of airport operations. A few U.S. airports have already started the process beyond services that are traditionally outsourced, such as engineering and construction inspection. Many aspects of airport operation, such as airfield and terminal maintenance and the overall management function itself, can be candidates for privatization.

Local, State, and Federal Funding

Commercial-service airports have many sources of revenue generated by their tenants and customers. Airline landing fees, lease payments, fuel surcharges, parking fees, and food and beverage concession fees are but of a few of these revenue sources. Although these funds can be used for airport capital improvements, for many years the federal government has maintained a trust fund specifically for use at commercial-service airports. This fund is generated from a tax imposed on tickets and is redistributed to the airports through the Federal Aviation Administration and its Airport Improvement Program (AIP). The total funds distributed in a given year are determined by legislative action by the U.S. Congress, with the president's signature.

An individual airport's share of these funds depends on a variety of factors, including its activity level and a project priority system that has safety-enhancing improvements as its cornerstone. The funds can be used only for areas of the airport that benefit the public as a whole and for areas that do not otherwise generate revenue themselves. For example, through this program the public areas of a terminal and common-use airfield pavements such as runways are eligible for funding, whereas a public parking lot and a terminal gift shop are not. Another popular use of these funds is for community noise abatement, such as soundproofing or land and home acquisition. AIP funding requires that an airport sponsor provide matching funds at some level. For larger airports, the formula calls for a 25 percent match by the sponsor. For smaller airports, the match can be as low as 10 percent. The sponsor's matching share is frequently a combination of state and local funding. Many states create their public airport-funding programs by providing a portion of the federal program's matching funds. States with more aggressive programs will assist airport sponsors with improvement projects where no federal assistance is available.

A newer program for funding airport improvements is through passenger facility charges (PFCs). An airport can elect to charge a ticket fee of up to $3 that can be used to fund airport improvements. Before the fee can be collected, the airport must have FAA approval and must notify the carriers, give them an opportunity to comment, and consider their comments. To gain approval, the airport sponsor must clearly demonstrate need for the projects that it intends to fund. Approval to collect and use PFCs also affects the share of AIP funds an airport will receive. For the large airports, the total PFC fees gained from passengers using the airport far exceeds the grants historically received through the Airport Improvement Program.

CONTRIBUTORS

Richard Marchi, President Technical Affairs, Airports Council International, Washington, DC

Richard de Neufville, Ph.D., MIT Technical and Policy Program, Cambridge, Massachusetts

SECTION 13.2
GOVERNMENT INSTALLATIONS

Brian Brenner, P.E.
Parsons-Brinckerhoff, Boston, Massachusetts

Nicholas N. Timpko, AIA, CSI
Anderson-Nichols, Boston, Massachusetts

GOVERNMENT FACILITIES IN GENERAL

Government facilities such as post offices, federal buildings, and court houses (Fig. 13.2-1) located within the United States are designed, specified, and constructed to comply with local, state, and generally applicable federal requirements and ordinances. Special federal

FIGURE 13.2-1   The new U.S. Courthouse, Boston, Massachusetts. (*Courtesy of Parsons-Brinkerhof, Boston, MA.*)

requirements are added only as required. Generally applicable federal requirements such as security, handicapped access, and no smoking on the premises are equally applicable to many facilities but must be especially invoked at government facilities.

Government facilities such as embassies, military bases, or U.S. Federal Buildings located in foreign countries are designed and constructed to meet additional federal requirements that generally center on security. All types of security measures, including physical, operational, electronic, light, and sound, are imposed on foreign-located government facilities. In many cases, foreign-located facilities must be designed to survive for protracted periods independent of all outside services and utilities, including power, water, sewer, fuel, food, and communications. The specialized federal guidelines and specifications that apply to such foreign-located government facilities go well beyond the scope of this handbook.

Government installations comprised of many facilities can include, but are not limited to, military bases, stations, campuses, national laboratories, testing stations, proving grounds, reservations, and space, research, and other special-purpose centers. Most government installations are located within the United States on large and sometimes huge tracts of land. Each government installation has a general function in line with the federal agency or department, such as NASA, DOD, NHA, GSA, and the Department of the Interior, that is responsible for its operation. Some government installations can be rather small, such as the Anthrax Experimental Station on Plum Island, off Long Island's Orient Point. Others, such as Edwards Air Force Base in California and the Idaho National Engineering and Environmental Laboratory, cover hundreds of square miles. Figure 13.2-2 depicts the Test Reactor Area, one of several sites at the 889-mi^2 Idaho National Engineering and Environmental Laboratory. The primary mission of this site is operation of the Advanced Test Reactor to study the effects of radiation on materials. Several isotopes used in medicine and industry are produced here and distributed nationally.

Special-purpose facilities such as weapons testing or disposal facilities, highly contagious disease centers, and military reactor stations are largely specified and constructed in accordance with federal or military specifications and standards supported by conventional codes and standards. The operation and management of government installations, whether executed by government employees, contractors to the government, or military personnel, generally follow federal and/or military procedures unique to the responsible government agency or department. In many cases the management, operation, and maintenance procedures employed at a given government facility or installation vary significantly, depending upon the direction of the responsible agency or department of the government. These methods and procedures differ greatly in most cases from the operations and management material presented in Part 1 of this handbook and should be addressed on a case-by-case basis. Figure 13.2-3 depicts the Idaho Nuclear Technology and Engineering Center at the Idaho National Engineering and Environmental Laboratory. This center currently receives and stores spent nuclear fuel; converts liquid high-level radioactive waste into a dry, granular material called

FIGURE 13.2-2   The Test Reactor Area at the Idaho National Engineering and Environmental Laboratory, Idaho. (*Courtesy of Bechtel BWXT Idaho, LLC, Idaho Falls, ID.*)

FIGURE 13.2-3   The Idaho Nuclear Technology and Engineering Center at the Idaho National Engineering and Environmental Laboratory, Idaho. (*Courtesy of Bechtel BWXT Idaho, LLC, Idaho Falls, ID.*)

calcine and stores it; and carries out environmental restoration and technology development activities.

Security measures for government installations are equally as diverse as those imposed upon any given facility. Installations must adhere to federal and military requirements that are more extensive than the conventional security measures addressed in Part 2 and again should be addressed on a case-by-case basis.

DIFFERENCES THAT NEED TO BE UNDERSTOOD

The range of differences in the codes, standards, specifications, and procurement, operating, and maintenance procedures that apply to a government facility or installation can, and do, vary enormously. This diverse range of differences and the mix between conventional standards and practices and the particular federal and military specifications, procedures, and standards that apply to the facility's engineering, operation, and management extend beyond the scope of this handbook. The design, engineering, construction, management, operation, and maintenance of government facilities and installations should be addressed on a case-by-case basis.

Government Idiosyncrasies

Many unique aspects of government contracts, such as proposal efforts, contracting, conceptualizing, designing, engineering, soliciting, constructing, operating, maintenance, and the overall life cycle management of a facility or installation, differ depending upon the agency or branch in control. Every branch of the government, every set of requirements, every competitor, and every contracting environment is different. The government recognizes these idiosyncrasies and is attempting to unify, release, or standardize the records, specifications, contracting, operation, and maintenance practices of its facilities through programs such as the Freedom of Information Act (FOIA).

From a facilities manager's point of view, the operation and maintenance (O&M) and management activities of a government facility or installation are discretely different from the design, engineering, and construction related issues discussed herein. This is addressed in the conclusion.

CASE STUDY: RECENT CONSTRUCTION OF THE FEDERAL COURTHOUSE ON FAN PIER, BOSTON, MASSACHUSETTS

One successful example of an executed government contract is the design and construction of the recently built Federal Courthouse on Fan Pier in South Boston (Fig. 13.2-1). The work for this unique facility included constructing the new courthouse and a new harbor park on the waterfront site. The building includes a court library, a secure prisoner circulation and detention system, a cafeteria/café, and a one-level basement parking garage. Its facade is extensively detailed in brick veneer and precast backup with granite trim, thermal window panels, and a full-height conoid-shaped glass curtain wall that encloses a six-story atrium. The building interior is also extensively detailed with architectural millwork in all courtrooms, judges' chambers, and executive offices. The GSA awarded a construction quality management contract (CQM) to the G. A. Fuller/Parsons-Brinkerhof team to act as GSA's agent in care of all managerial, supervisory, and administrative activities related to the design, engineering, and construction of the courthouse project. This novel approach indicates a change in government agency control and practices in the design, engineering, and construction of government facilities.

GOVERNMENT RESOURCES

To facilitate the execution of government contracts, the Freedom of Information Act (FOIA) requires the release of records held by government agencies. These are offered as part and parcel of the contract but still must be requested in writing. Before entering contract negotiations with a government agency or office, the designer is advised of this act and is encouraged to procure the necessary documentation. It is necessary for the designer to keep the document library complete and current. Each branch of the government has its own unique set of rules and guidelines. However, the intended spirit of government guidelines is to provide consistent policies, guidance, information, and direction. The guidelines facilitate accuracy in estimating proposals and assist in executing contracts accurately and uniformly.

The underlying philosophies of government guidelines are responsiveness, responsibility for the construction, and a commitment to executing a contract using quality design principles. The practices used are based on logical and conservative design approaches. Some are more rigid than others, leaving organizations that work for different branches of the government with an unclear vision of the way to get things done.

Without collaboration among the many branches of government, a burden is imposed on all those who plan, design, procure, construct, use, operate, maintain, and retire government facilities. This burden has resulted from failure to take full advantage of the cross section of knowledge within the federal government, as well as in the private construction arena.

Congress has recognized this lack of collaboration and has endeavored to improve the experience of conducting business with different branches of government. In the 1970s, Congress empowered a nonprofit organization, the National Institute of Building Sciences (NIBS), located in Washington, DC, to become a collective center for information. The objective of NIBS is to be the one authoritative voice in matters related to different government documents and branch management policies.

Information regarding advice on building-science technology and the government construction arena can be found at NIBS. This organization brings together representatives of the different branches of government, private industry, labor unions, and consumer interests to identify and resolve conflicts and potential problems that hamper facility construction throughout the United States. The NIBS is referred to by most branches of government, and its products, such as the Construction Criteria Base (CCB) system, must be used in executing many government contracts. CCB is currently distributed on CD-ROM and can also be found on the Internet. It includes guidelines, specifications, manuals, a handbook, regulations, reference standards, and documents covering other essential design and construction criteria. These documents come directly from more than 22 federal agencies and from more than 100 private industry organizations. They are complete, current, updated quarterly, and are governmentally approved resources. Working in concert with NIBS, NASA has provided the CCB system with an advanced document-processing system called Specsintact. This system greatly enhances the abilities of government authorities, designers, contractors, and facility managers to conduct quality control reviews during the design phases and through construction. The marriage of political influence, information, and technology has facilitated contractors' ability to execute contracts by a set of rules that can be measured against by governing authorities throughout the life cycle of the project.

In another proactive approach, government planners are working to augment the standardization and automation of government construction contracting. "Smart buildings" are not absent from the government's vision for future facilities. The standardization and integration of electronic systems will provide full automation of an entire building and its life cycle. Examples of these concepts and approaches can be found in the efforts of the Facility Information Council, which provides industry-wide forums on standardizing and integrating electronic systems. This council aims to provide integrated quality performance from the conception of a project through to the retirement of the facility or installation, including its design, engineering, construction, operations, and maintenance. The desired processes can be accomplished by using the facility's life cycle as its frame of reference.

The government has progressed a great deal during the past decade. It acknowledges many of its own shortcomings and is receptive to change. The use of partnering and its desire to embrace the new advancements in technology to create an improved result is refreshing. In recent actions, Congress has addressed the need for one authoritative voice and created a platform for one focal source of commentary. This translates into constant improvement in the processes and products created. The responsible use of technology has also made it possible to undertake much larger and far more complex projects than ever before. The commitment to a project's life cycle and the general overall effect on public opinion has opened the once-closed eyes of the government, allowing much more freedom in doing business with it.

CONCLUSION

Government agency practices in the design, engineering, and construction of new facilities are in flux and subject to evolving technology. The same is true for operations, maintenance, and facilities management of government facilities and installations. We have not addressed this more significant aspect (O&M and facilities management) of government installations. A complete treatment of the range of operations, management, and maintenance of government installations from a facilities manager's point of view will be covered in the next edition of this handbook.

CONTRIBUTOR

Vaughn L. Lotspeich, P.E., INEEL-Bechtel BWXT Idaho, LLC, Idaho Falls, Idaho

SECTION 13.3
PRISONS

Robert A. Broder, A.I.A.
Monacelli Associates, Inc., Cambridge, Massachusetts

Most of us have experienced the environments of airports, hotels, hospitals, office complexes, and other spaces illustrated earlier in this handbook and therefore have first-hand knowledge of these spaces, but fortunately, few of us have experienced the environment of a prison, jail, or lockup. Yet such places involve more than 1½ million persons in the United States alone, awaiting trial for a variety of offenses or serving penalties imposed by the criminal justice system. At the same time, these facilities are staffed by correctional officers, counselors, social workers, medical personnel, and other civilians who, it is sometimes said, also serve lengthy sentences out 8 at a time. This section is an introduction to an environment that has evolved over time and reflects various influences of the society. The prison environment is addressed, including some of the specific requirements for the inmates (incarcerated persons) and for protecting and providing for the prison staff. The mechanical, electrical, and security systems employed in today's prison facilities play a critical role in their safety. Basic engineering for the HVAC, plumbing, security, and electrical systems are identified by state building codes for institutional (restrained) occupancies. However, no other type of construction involves a clientele that, having the luxury of idle time, coupled with boredom, loss of freedom, and possible violence, develops ways, often ingenious, of defeating the system, and therefore making life more interesting, albeit for short periods. No other occupants of any other building type have this noted proclivity to trashing the environment that the designers have created. Therefore, the mechanical and electrical systems of modern prisons must address this phenomenon and provide as vandal-resistant an environment as can be had in any building type.

HISTORICAL CONTEXT

Prisons have been around for most of recorded history, differing mainly in the type and level of punishment imposed as a societal imperative. Humans have always required the capability of incarcerating others who violate the laws of the land, sovereign, or religion. For many years, the imprisoned offenders have been something less than second-class citizens, and few people have been overly concerned with their well-being or comfort. The images that motion pictures and television have given of historical prisons, such as the subterranean prisons of ancient Rome, the Bastille, and the Hulks on the Thames, reflect the society of the period. In fact, one of the earliest prisons in U.S. colonial history was an underground copper mine in Newgate, Connecticut, that is currently a tourist attraction. The West abounds with one-cell prisons (or jails) of the frontier days of the nation. The Charles Street Jail in Boston, Massachusetts (Fig. 13.3-1), was originally built in 1849 and was replaced in 1990 by the new Suffolk County Jail facility, shown in Fig. 13.3-2. In the Charles Street Jail, due to overcrowding, it was not uncommon to find three inmates confined in a 6- by 9-ft cell, served by a portable camp toilet, and having the bar-grate cell front and door as its sole method of ventilation.

CURRENT TRENDS IN INCARCERATION

The United States has long been an innovator in planning penal, correctional, or prison facilities dating from precolonial times. The strong moral background of the colonists escaping persecution in Europe, coupled with their religious beliefs, gave rise to a series of different prison plans. The term *penitentiary* was created by these early Americans based on an early concept that incarceration should be a time for penitence and contemplation of sins. It has only been relatively recently that convicted criminals have been granted rights; before that, they were considered "slaves of the State." The evolution of the modern prison directly reflects a growing concern for prisoners' civil rights and the safety of the persons assigned to operate the facil-

FIGURE 13.3-1   Charles Street Jail, built in 1849 in Boston, Massachusetts. (*Courtesy of The Stubbins Associates, Inc., Cambridge, MA*).

FIGURE 13.3-2 Suffolk County Jail, Boston, Massachusetts. (*Courtesy of The Stubbins Associates and Voinovich-Monacelli Associate Architects, Cambridge, MA; photo by Edward Jacoby, Boston.*)

ities. Each improvement in the physical plant of prison facilities is geared toward these concerns. Improvements in mechanical systems have addressed stress levels within the prisons caused at least in part by environmental conditions: excessive heat, cold, filth and unsanitary conditions, cold food, and the dangers inherent in confining a large number of potentially violent people in relatively close quarters. Along with specifically developed management techniques, such as direct supervision, which removes physical barriers between inmate and officer, the nation's extraordinary technology has been applied to prison systems, providing new methods and techniques to assist staff in surveillance, communications, and control. Current prison methods, management techniques, systems, and equipment are technologically driven and are evolving rapidly. The forum for keeping current in the rapid development is vested in the related societies, such as the American Correctional Association, the American Jail Association, and the American Institute of Architects Committee on Architecture for Justice.

OVERVIEW OF PRISON SYSTEMS

HVAC and Plumbing

The typical prison HVAC and plumbing systems are rarely proprietary, nor do they present significant engineering breakthroughs that have not been covered in Sec. 5.4, "Mechanical Systems," of this book. The overarching criteria employed in the design of prison systems are (1) ease of operation, (2) simplicity of maintenance, and (3) vandalism resistance.

Ease of operation is necessary due to the relatively rapid turnover of operating personnel in this stressful public occupation. Generally, there is little time to train new staff. The consulting design engineer should investigate system-wide usages within the jurisdiction to determine the possibility of designing the installation to provide continuity between one facility and another.

Maintenance of equipment within a facility varies in difficulty with accessibility and the security level of the facility. Maintenance of HVAC elements inside a secure space, such as a dayroom, requires a lockdown of inmates to allow the mechanic to work in safety of person and tools. Where possible, the design architect should design a secure environment for the equipment. The most common method is by creating a *chase,* usually serving two cells, that houses the entire plumbing tree for the cells, as well as the supply and exhaust ductwork and the cell's electrical system of conduits, junction boxes, and the like. This chase is provided with a high-security door or access panel that protects the mechanical system and gives the mechanic easy access during the lockdown. It does, however, also require some level of design ingenuity to locate the systems in a very close and confined environment, because space is always at a premium.

Vandalism resistance is always foremost in the minds of the vendors of prison-related products. Heavy-gauge sheet metal, concealed and vandal-resistant fasteners, and basic design to avoid the concealment of drugs, razor blades, and other contraband are called for. For this reason, ducted air is the dominant method for heating and cooling inmate areas of prisons. Fin-tube or cast-iron radiators should be limited to areas of staff use because they are vulnerable and could easily be damaged and/or converted to weapons. Inmates have been known to dismantle air grilles and fashion their vanes into weapons. Mechanical components should be concealed where possible, to avoid both vandalism and possible suicide venues. Exposed pipes provide convenient attachment points for hanging injuries and/or death.

Central Controls

Continuing the concept of security and safety in the institution, control systems for the secure areas of the facility should be centralized, with duct-mounted sensors, and no inmate-accessible thermostats or controls other than security light switches. With increasing frequency, lighting and temperature controls are integrated into the programmable logic controller/graphical user interface described later, so that the duty officer has full control of the cell-house environment. By the same token, it is possible and frequently desirable to integrate the shower water-temperature control into the system to avoid the possibility of using a scalding or cold shower as a weapon.

During the past decade, building codes have mandated installing energy management systems in facilities of this size. This concept is entirely consistent with the requirements outlined earlier. Energy management systems have proven to be the rule in prisons. Initial cost is offset by efficiency and operating-cost savings. Properly designed, the system can reduce staff requirements to the minimum required to operate the various HVAC, plumbing, and electrical systems of the facility.

Electrical and Security Systems

Prison electrical systems are generally reliable and break down similarly and to the same degree as lighting, power, and control systems in commercial or institutional buildings. This includes fully integrated security systems and facilities management systems installed in any modern prison. Included in a fully integrated security system are CCTV, access-control, door-control, interior- and exterior-communications, personal-alarm, and intrusion-detection features. As an example of the unique aspects of electrical and security systems in a modern prison, the following describes a fully integrated security system as outlined by the Sousa-Baranowski maximum security facility in Shirley, Massachusetts.[1] The basic elements of the integrated security system for this facility may include but are not limited to the following:

1. *Perimeter security system*

   - *Physical barriers.*   Fencing and vehicle and pedestrian gates, fence protection systems, and deterrents such as barbed wire and razor ribbon.
   - *Electronic detection systems.*   Microwave intrusion detection and electronic taut-wire array.
   - *Perimeter surveillance system.*   CCTV, patrol road and vehicles, and illumination.

2. *Access-control system.*   An access-control system allows selective access to certain areas by staff or visitors. Generally, these consist of routine traffic or areas of a security level such that control-room staff need not visually monitor and inspect each and every person who gains access. It also includes doors where only specific persons may gain access. The system automatically logs all transactions on a computer in inner control.

   - *Door controls.*   The door-control system is part of the fully integrated electronic security system. It consists of programmable logic controllers (PLCs) in outer control, inner control, and all major control centers such as housing-pod officer's stations and special management control rooms. All PLCs will be linked by an LAN and programmed to control specific doors by graphical user interfaces (GUI)—computers with touch-screen and/or mouse-controlled video monitors. Security and maintenance staff using portable computers, controllers, or other specialized equipment, program PLCs, and the graphic programming software of the GUIs is an integral part of the security system. Doors to be controlled will be equipped with electric security locks or strikes, according to their security level, and are generally outlined section by section in this security program. All doors are controlled on a selective, hierarchical basis, as determined by the DOC, to avoid unnecessary duplication, allow the staff greater operating efficiency, and enhance the security of the facility. Doors are programmed to suit the required function: individually, group release, group release with exclusions, all release, exterior only, interlocks, etc.
   - *Inner control* is the primary center of facility control; the system allows for redundancy of control from outer control, if required during disturbances, when inner control is compromised. At least three GUI stations are installed in inner control.
   - *Outer control* is the secondary control point and has primary responsibility for facility access screening and perimeter security. The nature of the PLC and LAN will allow for redundancy of operation of all security systems. At least three GUI stations are installed in outer control.

3. *Communications systems*

   - *Radio communication system.*   A radio communication network is installed in the facility to allow backup communication between officers and control. The network will have as many channels as required to allow for redundant functioning, including transportation vehicles, maintenance, and coverage of areas not served by telephone or intercom. The system will include a staff-down feature at each transceiver.
   - *Intercom system.*   This system will provide full duplex communication to all locations where it is installed. With a head-end cabinet in inner control, armored remote push-button stations and smaller intercom handset consoles, as appropriate, will allow call-up from doors, sally ports, offices, workrooms, rooms, and other such locations as required. Control of the system in inner and outer control will be by switches integrated into the GUI, and from an IC console and handset when required for more private conversation.
   - *Public address system.*   This will allow selective, group, and all call to areas throughout the facility. A head-end cabinet will be located in inner control.
   - *Personal-alarm system.*   A personal-alarm system will be installed in the facility to provide staff protection. The system will be installed facility-wide (indoor) only within the security perimeter, and both duress-alarm transmitters will send out coded signals. The design of the personal-alarm system will include receivers located in the ceiling that have a receiving range of approximately 50 ft (100-ft diameter). No receiver installation will be required in utility spaces such as closets, toilets, and mechanical spaces, and outdoors except at the perimeters of the building. Large spaces may require fewer receivers for

coverage. The receivers will be tied into the control LAN and integrated into the general security system as alarm points to the inner control PLC, and therefore can be displayed on any of the facility GUI control points by appropriate software programming. It will be the responsibility of the proposer to submit a layout design for the personal-alarm system and ensure adequate coverage.

4. *Surveillance system*

- *Closed-circuit TV (CCTV).*   A CCTV system will be installed in the facility to facilitate the control of traffic, safeguard staff, and maintain security. The system will be full-color video, and cameras will be located at strategic points to assist staff in maintaining visual coverage. Cameras will be installed in a variety of housings as noted herein, in keeping with the desired view, the vulnerability of the equipment, and the size and shape of the space to be monitored.
- *Primary monitoring* of the interior video cameras will generally be in *inner control,* which will be the location of the main matrix switcher, the multiplexers, and the primary video controls. Monitoring will be on three or more 20-in quad bracket-mounted color monitors. Each monitor will be capable of showing up to 16 camera images at one time, although 4 would be more common. A multiplexer and time-lapse videocassette recorder will be associated with each monitor, therefore affording the capability of recording the 16 images and playing any of them back individually. Each monitor will be programmable to display automatically or manually a single image called up during an alarm, as required by the duty officer. Tilt, pan, and zoom functions may be done by joystick control or on the door-control GUI, manually, or by mouse (Fig. 13.3-3).
- *Secondary monitoring* will be in *outer control,* which will have similar equipment. Outer control will also perform the primary monitoring of the perimeter security systems and video, thereby relieving inner control of this task and ensuring efficient and effective staff use. In an emergency, outer control will have the capability of assuming the function of inner control for total control of all doors and cameras in the facility.
- *Monitoring of other areas* such as inmate receiving and processing, housing, recreation decks, special management housing, and inpatient inmate rooms in the health services unit will be accomplished locally, but conduit and cable will be run to the inner-control matrix for selective monitoring from inner or outer control during disturbances.
- *Other monitoring locations* may include internal perimeter security (IPS; specifically for monitoring and recording contact visitation), the superintendent's office area (the command post during periods of unrest), and the office of the chief of security or other locations noted in the room data sheets.
- The system will incorporate time-lapse video recorders (TLVCR) in all major control areas that can record all images shown on the monitors. TLVCRs will be provided to record at the rate of 1 per 16 cameras installed. Playback will have the capability to playback individual cameras as identified by the titling.

5. *Systems integration.*   The systems described will be fully integrated into a control, communication, and surveillance system. Main control functions will be 20-in color computer monitor graphical user interfaces (GUIs). These will have inputs and outputs (I/Os) to the intercom system, the access-control system, the CCTV system, the fire-alarm system, the main door-control PLC, and, of course, to the LAN onto which other control stations will be connected. A future inmate identification system (wrist bracelet with bar code) may also be integrated into the system to allow for a total electronic data trail of activity. This integrated system will allow for extensive programmability and for all of its separate subsystems to work together. In this way, for example, pushing an intercom button at any door in the facility may instantly call up a video image of that caller at the control center, establish an audio path, and upon identification (by future bar-code reader), will allow the monitoring officer to push the appropriate door-lock-release button or icon. Alarm conditions from all sources will be monitored and logged in a highly efficient manner. For the purposes of this security program, it should be assumed that although a primary-control locus is indicated, other locations on the LAN

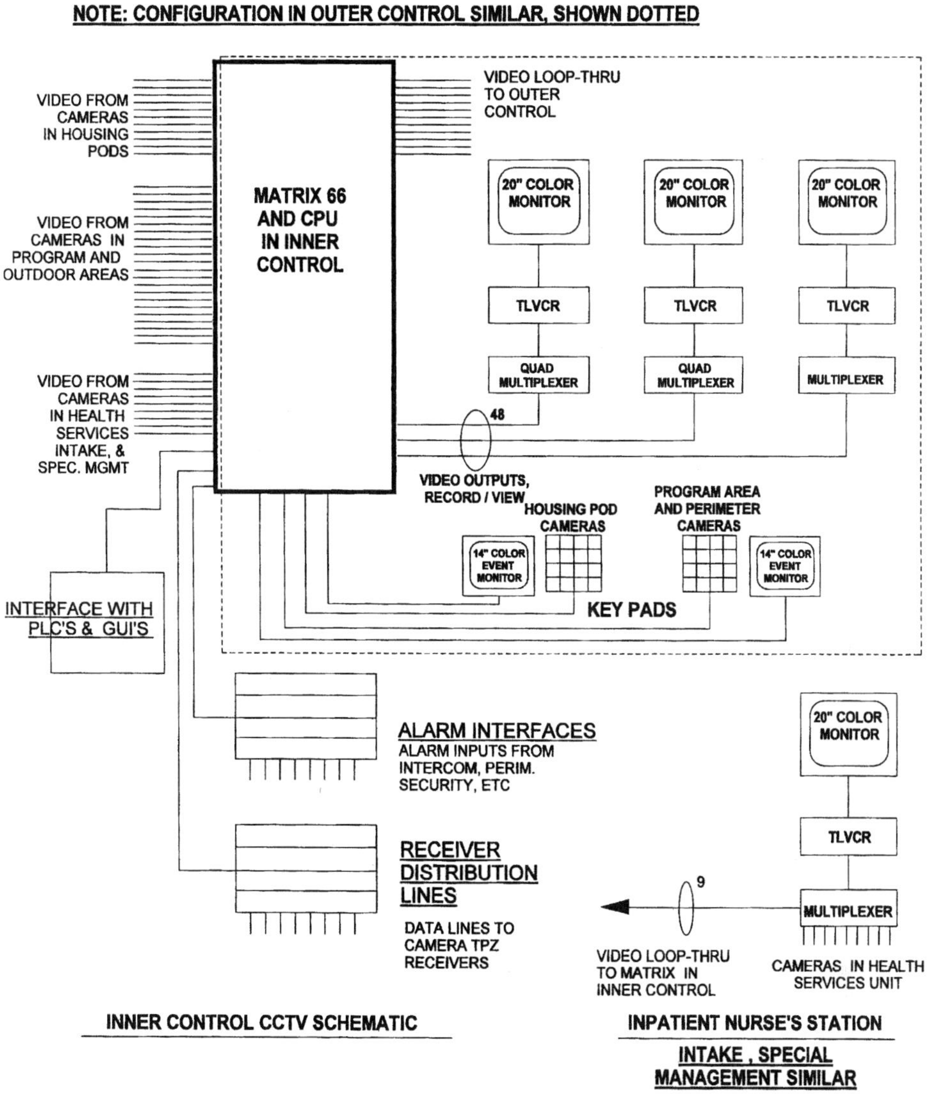

NOTE: CONFIGURATION IN OUTER CONTROL SIMILAR, SHOWN DOTTED

FIGURE 13.3-3 Central control room CCTV schematic. (*Courtesy of Monacelli Associates Incorporated, Cambridge, MA.*)

can be so configured by appropriate software programming to allow for alternative control locations.

The security program outlines the basic security concepts specific to the program area. Included are concepts pertaining to access control, door control, CCTV, communication, physical hardening, and special equipment or systems, all of the basic elements of security, surveillance, and detention. The program is not intended to identify each and every camera, assault-resistive glass panel, security door, etc. Rather, it establishes certain minimum standards for these components while allowing the designer the flexibility to integrate these concepts into the design proposal. The overall security program for the facility is outlined in an integrated security systems overview, and the individual space program areas integrate with this outline.

The project under consideration is a maximum-security prison, to be designed and built to house, feed, and provide a safe, stable, and otherwise suitable living environment for the most violent offenders in the Commonwealth. Security concepts outlined herein reflect this. The design/build (contractor) team is required to include in its design and construction components personnel who have a high level of expertise in designing and constructing high-security facilities and in the technology that is available today and possibly in the near future. These persons should have a working knowledge of the way a prison functions and the dynamics of prison management. The team should be well versed in the appropriate application of available technology, as well as the techniques of constructing these most secure institutions.

CONCLUSION

This section outlines some very basic concepts in prison design. Many have strong parallels to other building types, but few buildings have the comprehensive integration of these systems. The section illustrates a maximum-security facility. Prisons vary in classification. In what are frequently called *super-max* institutions, inmates are confined up to 23 h per day and are always escorted in full ankle and wrist restraints by a minimum of two, and frequently three, officers. On the other side of the system, lesser security facilities may be built whose security components are greatly reduced, making the facility less costly to build and to operate. Each has its own dynamic and special requirements, and each has its role in the overall state criminal justice system. The engineer who contemplates working with the prison design team must understand and be willing to accept the following conditions:

- Recognition of the teamwork required by the criminal justice planner, prison architects and engineers, the detention-equipment consultant, and the other design team members
- The integration of the various prison systems, some of which will be familiar, others with which those will be only peripherally associated, yet all must work together effectively and efficiently
- The ability to work with the operating staff of the prison, empathizing with them and remembering that they must work in a dangerous environment being designed and will be in it long after the facility is built and the design team has completed its work
- An understanding that the technology of incarceration is still evolving and will require superior understanding of advanced construction methodology and electronic systems integration
- Finally, willingness to understand the dynamic of an extraordinary environment that frequently requires extraordinary design solutions and teamwork

CONTRIBUTOR

Roy A. Pedersen, R.A., The Stubbins Associates, Inc., Cambridge, Massachusetts

REFERENCE

1. Sverdrup, Gilbane & Monacelli Associates, *Request for Proposals for Design/Build of Maximum Security Facility in Shirley, Massachusetts,* Commonwealth of Massachusetts Division of Capital Planning and Operation, Boston, April 1996.

CHAPTER 14
DATA CENTERS

Paul R. Smith, P.M.P., P.E., M.B.A., Chapter Editor
Peak Leadership Group, Boston, Massachusetts

David L. Stymiest, P.E., SASHE, C.E.M., Chapter Editor
Smith Seckman Reid, Inc., New Orleans, Louisiana

Glen J. Goss
GJ Associates, Stow, Massachusetts

Thomas Stratford
McClair, Chicago, Illinois

INTRODUCTION

Not very many years ago, the common wisdom was that data centers would not be required in the future. Networked PCs could perform all of the functions of large mainframe computers. They were inexpensive and did not require the controlled environment recommended for mainframes. It did not take very long to discover that PCs are made up of the same components used in larger computers and that they, too, run much more reliably in a protected environment. The present trend is toward server and mass storage consolidation, which look to a facilities manager like minicomputers, mainframes, and disk farms of old. There has been an unprecedented growth of data center construction embodying some new twists that are covered in this chapter.

Modern telecommunications switching equipment is electronic and has nearly all of the same environmental requirements as computers. One result of the PC revolution has been to put computers on desktops and to educate millions of people in directly creating and using computer data. This data is useless unless it is shared. Company wide area networks (WAN) and the Internet are the paths of data transfer over telecommunications lines. Telecommunications companies have been hard-pressed to keep up. Internet service provider (ISP) and telecom facility construction has exploded as the result of this demand and the deregulation of the communications industry. The principles outlined in this chapter can be applied to communications sites and data centers, except for power, which is usually supplied by 48-V dc rectifier plants and batteries, rather than uninterruptible power supply (UPS) systems.

Entrepreneurs and real estate companies have recognized the demand for data center and telecommunications facilities. The latest developments are computer hosting sites and communications hotels. These are speculative facilities built to house client computer or telecommunication equipment. They range from web-hosting facilities that have highly sophisticated infrastructures to communications hotels, where clients are usually responsible for their own environmental equipment. Presently, these facilities are being filled as soon as they are built. These are exciting times for those of us who are involved with data centers and telecommunications facilities.

SECTION 14.1
INTRODUCTION TO DATA CENTERS

Glen J. Goss
GJ Associates, Stow, Massachusetts

Data centers are specifically built to house electronic computing equipment and the support for that equipment. The computer equipment can be any combination of large mainframe computers and personal computers. There are several physical configurations for these computers, including individual large cabinets, two-drawer filing cabinet-sized boxes, or standard tower and horizontal desktop PC enclosures. The most popular computer configuration today is a 19-in rack-enclosed cabinet. Shelves make it possible to install equipment that is not specifically designed to be rack-mounted. Racks allow data center managers to mix and match various pieces of equipment in one cabinet. Racks also allow them to stack many pieces of equipment, one above the other, in a small footprint. Another popular arrangement is an open shelving system for PCs and other small computers. These systems often include work surfaces at desk height, plus shelf space for video monitors.

Data centers must house more than just computers. They contain mass storage equipment, network equipment, telecommunications equipment, control consoles, and printers, plus storage for documentation, media, and data. Data centers differ in many ways from any other area of a facility. Of foremost importance is that data centers are nearly always critical to the mission of the business. Even momentary disruptions in data center operation can cause severe consequences to the business. Second, the environment must be maintained within tighter limits than other spaces. Electronic equipment is a concentrated source of dry heat that can present a challenge to air-conditioning systems. Most electronic loads draw nonlinear electrical current, yet demand undistorted voltage. Finally, the data center must be protected from physical threats such as fire, heat, water, and unauthorized access. It is important for the business to define the cost of the data center's downtime in terms of cost per minute, hour, and day. These costs should include the total cost of employees who cannot perform their jobs, the cost of expensive computer equipment that is idle, and the loss of client goodwill. The cost of downtime can then be used to evaluate the viability of alternative infrastructure designs that provide various levels of computer uptime. Figure 14.1-1 contains a check list for evaluating potential data center sites.

Project # _____ Project Name _____ Location _____.

Date _____

A.	**General**	Yes	No	**Comments**
1.	Are there up-to-date as-builts? Are they accurate?	☐	☐	
2.	In what year was the data center constructed?			
3.	Has the data center been renovated?	☐	☐	
4.	If so, in what year?	☐	☐	
5.	Is the building owned by the data center's corporate entity?	☐	☐	
6.	Are maintenance records kept for all equipment?	☐	☐	
7.	Describe location of data center.			
8.	Indicate square footage of raised floor.			
9.	Are there other tenants in this building?	☐	☐	
10.	Is there an inventory of data center equipment?	☐	☐	
11.	List types of systems.			
12.	Indicate location of support equipment.	☐	☐	
13.	Has the data center experienced outages?	☐	☐	
14.	Is there space for expansion?	☐	☐	

B.	**Architectural**	Yes	No	**Comments**
1.	Are there up-to-date equipment floor plans?	☐	☐	
2.	Is there evidence of roof leaks?	☐	☐	
3.	Is there evidence of other leaks?	☐	☐	
4.	Have all penetrations been fire/smoke sealed?	☐	☐	
5.	Fire rating of data center perimeter walls	☐	☐	
6.	Are there any exterior windows in the center?	☐	☐	
7.	Is the data center ADA-compliant?	☐	☐	
8.	Has the data center had a pressurization test?	☐	☐	
9.	Is there an access floor?	☐	☐	
10.	What height?	☐	☐	
11.	Is the slab depressed (i.e., no ramp or stairs)?	☐	☐	
12.	Is the floor system bolted stringer?	☐	☐	
13.	Indicate type of floor tile.			☐ Hollow metal ☐ Wood core ☐ Concrete
14.	Is the underfloor sealed?	☐	☐	
15.	Is the floor system grounded?	☐	☐	
16.	Indicate condition of floor.			☐ Poor ☐ Acceptable ☐ Excellent
17.	Is the floor clean/free of excess wiring?	☐	☐	
18.	Are the doors fire-rated and labeled?	☐	☐	
19.	Indicate type of ceiling tile.	☐	☐	

FIGURE 14.1-1   Data center checklist.

B.	Architectural	Yes	No	Comments
20.	Indicate condition of ceiling system.			☐ Poor ☐ Acceptable ☐ Excellent
21.	Does the data center have adequate working clearances?	☐	☐	
22.	Indicate floor rating.	☐	☐	
23.	Is a freight elevator available?	☐	☐	
24.	Are there drains or any other piping located above the data center?	☐	☐	
25.	Is there a vapor barrier in the perimeter wall?	☐	☐	
26.	Is there a master console area?	☐	☐	
27.	Provide description of console area.	☐	☐	
28.	Is there a print operation in the data center?	☐	☐	
29.	Is there tape storage?	☐	☐	
30.	Are there any other auxiliary spaces?	☐	☐	

C.	Mechanical	Yes	No	Comments
1.	Does the data center have a dedicated mechanical system?	☐	☐	
2.	Does the data center use computer-room-rated equipment?	☐	☐	
3.	Does an outside vendor maintain the equipment?	☐	☐	
4.	What is the current capacity of the A/C system?	☐	☐	
5.	Indicate type of system.			☐ Chilled Water ☐ Glycol ☐ DX and tower
6.	Is the system designed for redundancy?	☐	☐	
7.	Is there a fresh air system?	☐	☐	
8.	Indicate location of heat-rejection equipment.	☐	☐	
9.	Are CRAC units individually piped?	☐	☐	
10.	If the system is looped, describe the loop.	☐	☐	
11.	Has the data center experienced hot spots?	☐	☐	
12.	Is there a humidification system?	☐	☐	
13.	Is there a leak detection system?	☐	☐	
14.				
15.				
16.				

D.	Generator	Yes	No	Comments
1.	What is the size of the generator set?	☐	☐	
2.	Indicate generator set make and model number.	☐	☐	
3.	Indicate type of fuel.			☐ Diesel ☐ Gas ☐ Propane
4.	Indicate location of generator.	☐	☐	
5.	Is the generator dedicated to the data center?	☐	☐	

FIGURE 14.1-1   *(Continued)*

D.	Generator	Yes	No	Comments
6.	How much demand load is on the generator?	☐	☐	
7.	What type of load is on the generator?			☐ UPS ☐ CRAC ☐ Lighting ☐ Other (Describe):
8.	Indicate location of fuel tank.			☐ Above ☐ Below ground
9.	Indicate size of fuel tank (in gallons).	☐	☐	
10.	If the generator is gas-fueled, is there a propane backup system?	☐	☐	
11.	Do the generator, tanks, pumping or piping have redundant components?	☐	☐	
12.	Where is the generator located?			☐ Outside ☐ Indoors
13.	Is there remote annunciation for the generator?	☐	☐	
14.	How often is the generator run?	☐	☐	
15.	How often is the generator tested either with an actual load or a load bank?	☐	☐	
16.	Indicate the rating, make, and model of the transfer switch.	☐	☐	
17.	Does the transfer switch have a bypass feature to both power sources?	☐	☐	
E.	**Electrical**	**Yes**	**No**	**Comments**
1.	Indicate the size and voltage of the data center power source.			
2.	UPS			Size:
3.	UPS			Make: Model: Year:
4.	UPS			Input voltage: Output voltage:
5.	Does the UPS have an external maintenance bypass?	☐	☐	
6.	Indicate the present demand loading on the UPS.			
7.	Has the UPS recently been load bank tested?	☐	☐	
8.	Describe any redundant components in the UPS system.			
9.	Indicate type of batteries.			☐ Wet ☐ Sealed
10.	Indicate UPS location.			☐ Remote room ☐ Data center
11.	Are PDUs used to distribute UPS power to the equipment?	☐	☐	
12.	Indicate ratings of the PDUs.			
13.	Give a general description of the electrical distribution system.			
14.	Are full-sized grounds used in all conduits?	☐	☐	

FIGURE 14.1-1   *(Continued)*

E.	**Electrical**	Yes	No	Comments
15.	Is the raised floor grounded?	☐	☐	
16.	Is there an emergency power-off button?	☐	☐	
17.	Has the EPO system been tested?	☐	☐	
F.	**Plumbing**	**Yes**	**No**	**Comments**
1.	Potable water.	☐	☐	
2.	Sanitary sewer.	☐	☐	
3.	Storm sewer.	☐	☐	
4.	Natural gas.	☐	☐	Interruptible
5.	Hot water.	☐	☐	
6.	Toilets.	☐	☐	
7.				
8.				
9.				
F.	**Fire Protection**	**Yes**	**No**	**Comments**
1.	Water supply.	☐	☐	
2.	Fire pump.	☐	☐	
3.	Fire hydrants.	☐	☐	
4.	Gaseous fire suppression.	☐	☐	
5.	Standpipes.	☐	☐	
6.	Does the data center have a detection system?	☐	☐	
7.	Is the system cross-zoned?	☐	☐	
8.	Has system been tested?	☐	☐	
G.	**Security/Monitoring**	**Yes**	**No**	**Comments**
1.	What type of entry system is used?			
2.	Is there a card-out system?	☐	☐	
3.	Is there a building security system?	☐	☐	
4.	Is there a CCTV system?	☐	☐	
5.	Are support rooms monitored?	☐	☐	
6.	Indicate locations of all cameras.			
7.	Are there perimeter alarms?	☐	☐	
8.	Provide a description of the perimeter alarms.			
9.	Is there an alarm for remote support equipment?			
10.	Is there an environmental monitoring system?	☐	☐	
11.	Provide a list of monitoring points.			
12.	Is there a building energy management system?	☐	☐	

FIGURE 14.1-1   *(Continued)*

H.	Telephone Company	Yes	No	Comments
1.	Smart ring.	☐	☐	
2.				
3.				

FIGURE 14.1-1   *(Continued)*

SECTION 14.2
RELIABILITY

Glen J. Goss
GJ Associates, Stow, Massachusetts

Calculations of reliability are the basis for determining uptime or computer system availability. Reliability is stated in hours of mean time between failures (MTBF). Great caution should be exercised in accepting MTBF figures. Two consultants can calculate very different figures using different industry standards for component reliabilities, calculation procedures, and assumptions. Comparison of the reliability of various systems can be made if the system MTBF is calculated by using consistent methods and assumptions. Manufacturer's claims of product reliability should be generally disregarded without first gaining complete understanding of the data used. When equipment is located in a data center, system reliability is critical. For example, the failure of a single desktop computer may put one employee out of work for half a day once every three years when his or her PC fails. If the failure of one of 100 personal computers in a data center causes the entire system to crash, it could prevent hundreds or even thousands of people from doing their work. Furthermore, if each of the 100 computers in the data center failed randomly once every three years, the system would crash an average of once every 11 days. For these reasons, it becomes necessary to provide the best possible ambient and electrical environment to ensure the maximum reliability of computer equipment. Theoretical calculations of reliability for the electrical power and air-conditioning systems are performed to determine the viability of redundancy for any device that is determined to be a single point of failure (SPOF) for the data center.

SECTION 14.3
GENERAL CONSTRUCTION

Glen J. Goss
GJ Associates, Stow, Massachusetts

Tom Stratford
McClair, Chicago, Illinois

DISASTER AVOIDANCE

A disaster that has a major negative impact on an enterprise could range from a major database software failure to complete destruction of the facility. Traditionally, data have been held sacred, and computer equipment has been considered expendable. Data were stored on-site in fire-rated rooms or in vaults, and backup data were stored off-site. Disaster recovery plans were put in place so that computer hardware and software suppliers were ready to deliver replacements quickly for destroyed systems. Backup data and applications would then be reloaded and the business would be up and running.

The treatment of disasters has changed from disaster recovery to disaster avoidance because of the increasingly on-line transactional nature of business. A number of steps can be taken by data center managers to avoid such disasters. They can run mirrored systems where data is duplicated and processes are run in parallel, always comparing results. These parallel operations may be at different locations to protect against location-specific natural disasters. Facilities that do not have the luxury of duplicate data centers must be designed with redundant infrastructure systems to prevent the data center from shutting down periodically for maintenance. Data centers must be carefully designed to ensure the maximum protection possible against disasters, including fire, heat, acrid gases, dust, and firefighting water. They need to be protected against unauthorized access, including employee vandalism or terrorist attacks. Data centers also need to be protected against industrial or international espionage by electronic means. The walls, ceiling, subfloor, doors and all penetrations of those barriers can be designed to complement the security and fire protection systems.

Fire-Rated Walls

Fire-rated walls are required by code for wet-cell battery rooms. A two-hour rating is recommended. A two-hour fire rating has also been used as a standard for data storage rooms. Special construction detailing by an experienced professional engineer is required to properly maintain fire ratings and room isolation required by the fire suppression system. Most data center managers mistakenly believe that their data would be safe for two hours from flame impinging on the fire-rated wall. Although the wall may not be breached by flames for two hours, the inside temperature of the wall would rise in a few minutes above the 40°C limit for storing magnetic media and cause destruction of data. There could also be damage from smoke, water, and humidity caused by fire-fighting water unless the room was carefully sealed. Better protection requires special building materials. Prefabricated insulated wall, ceiling, and floor systems are available that maintain the internal temperature below 40°C for up to 90 minutes when an 1100°C flame impinges directly on the outside surface. These systems also protect against smoke, water, humidity, and gas when used with specially designed doors, automatic air dampers, and self-sealing cable and piping ducts.

Vapor Barrier

The data center is usually the only portion of the facility that has humidity control. The space adjacent to the data center may be temperature-controlled, but the relative humidity may vary widely with outdoor air conditions. In cold climates, heated winter air is exceedingly dry, as is hot summer air in arid climates. Hot, humid days present the opposite condition. A vapor barrier is required on all walls, ceilings, and floors to prevent humidity from escaping to dry adjacent spaces or excess vapor from entering the data center in humid conditions. Vapor loss causes unnecessary operation of humidifiers to maintain room relative humidity at the design level. Excess vapor forces the air conditioners into a dehumidification mode and calls the compressors on, along with reheat coils to maintain the temperature. Both of these conditions cause an unnecessary increase in the annual consumption of energy.

Raised Flooring

Raised flooring serves multiple purposes in a data center. The most obvious purpose is to provide a space for power distribution and data cables to interconnect the various pieces of data equipment. Power conductors can be a source of electrical noise for data cables. Noise transmission is greatest when the data and power cables run in parallel. It is best to separate the data cables from the power cables and have them cross at right angles wherever possible. One way to encourage this is to install separate cable trays at two different heights. The larger, more stable power cables are usually installed at the lower level, and data cables are at a higher level to pass over the power cable trays. Cable trays make changes to power distribution cables in the lower tray somewhat easier because the data cables are kept separated in their own trays. Power distribution cables should be installed in flexible or solid steel conduit to increase shielding from electrical noise. Larger permanently installed power feeds to air conditioners and power distribution units are usually installed in hard conduit at the lowest level possible. Pipes for humidifier makeup water, condensate drains, and insulated pipes for coolant also need room under the floor. Water detection cables need to be attached directly to the subfloor under the raised floor. A high-frequency grounding signal reference grid (SRG), made of copper conductors on two-foot centers, is attached and bonded to the raised floor pedestals. The grounding grid wires are attached to each other at each crossing. Then, the grid is bonded to the building ground system.

One of the greatest advantages of a raised floor is that it can be used as a supply air plenum for air-conditioning. One of the greatest challenges in designing a data center is not to overly restrict the flow of cooling air with cables and pipes. Whenever possible, raised floors should be 18 in or higher to ensure that everything fits under the floor and that air gets to all of the loads. A 12-in raised floor is the minimum for a small room if it is to be used for air distribution.

Raised flooring is available with floor tiles constructed of steel, wood encased in steel, concrete, or concrete-filled steel. The most economical type is wood encased in steel. It is quieter to walk on than steel, lighter in weight than concrete, and easy to cut for cable access. The floor type should be specified for the duty that it must withstand. Particular attention should be paid to the floor specification if the area is to be subjected to continual heavy rolling loads such as paper carts in printing areas. Pedestals are attached to the subfloor with mastic, and stringers are attached between the pedestals for the tiles to lie on. Stringers may be snap-in or bolted for maximum rigidity. Tiles may also be screwed to the pedestals where maximum strength is required and frequent underfloor access is not required.

Room EMI/RFI Shielding

External sources of electromagnetic energy such as radar, radio transmitters, and electrical arcing can interfere with the operation of high-speed electronic devices or computer equipment. Nearby strong sources of interference will require that the walls, floor, and ceiling incorporate a faraday shield to eliminate directly radiated energy. Filter capacitors supplied with some surge suppression devices or shielded isolation transformers in the critical power distribution path will reduce EMI/RFI that is conducted on the power feeds. Other sources of EMI are high-power conductors or transformers in close proximity to the perimeter surfaces of the data center. The most common complaint about this type of EMI is bending of the image on video monitors. Thus, a portion of the space is unusable for video monitors. Although these types of fields are not likely to erase magnetic media, it is prudent to store them elsewhere. Shielding against this type of problem can be accomplished with lead sheets installed in partitions to block its effect on the data center.

SECTION 14.4
MECHANICAL

Glen J. Goss
GJ Associates, Stow, Massachusetts

Tom Stratford
McClair, Chicago, Illinois

CONDITIONS TO BE MAINTAINED

Temperature

The optimum environmental conditions for computer equipment have not changed substantially with advances in technology. The optimum temperature of the air entering the inlet air grille of each piece of computer equipment is 75°F. It should not be higher for extended periods, nor should it be below 52°F. Higher temperatures increase the stress on the computer components, shortening their life. Lower temperatures promote thermal shock and condensation. A computer will stabilize at the entering air temperature when it is powered off with cold air forced through it. When power is applied to the computer, the internal temperature of the larger chips changes almost instantaneously. The thermal expansion inside the chip flexes the tiny internal connections and stresses them. Although a chip is unlikely to fail from a single thermal shock (unless it is turned on in subzero temperatures), each flex takes its toll. The temperature of the air that enters the computer equipment should also be as constant as possible. The internal computer components must never be cooled below the dew point of any possible entering air. The resulting condensation can make dust, dirt, or mineral deposits on the surfaces of the printed circuits conductive.

Humidity

The optimum ambient humidity for computer equipment is 45 to 50 percent RH. Low humidity results in the buildup of static electricity on surfaces in the data center. Electrostatic discharges are very damaging to semiconductor devices in the computer equipment. On the other hand, humidity that is too high promotes gold scavenging and silver migration at the cable and circuit board contacts. Over an extended period of time, the connection reliability is reduced. As mentioned earlier, condensation must be completely avoided. Humidifiers are arguably the most maintenance-intensive part of the computer room infrastructure. Required maintenance is directly proportional to humidifier operating time. Excessive humidifier operation caused by the lack of a vapor barrier, a low air-conditioning sensible heat ratio, or excessive ventilation will result in higher humidifier maintenance and failure rates.

Dust Control

Dust buildup on surfaces inside the computer produces a thermally insulating layer that prevents heat from escaping. The resulting temperature rise increases the stress on the semiconductors and reduces their life span. Conductive mineral dust combined with water such as condensation is extremely damaging. Four sources of dust deserve special attention; these include dry-wall sanding, concrete subfloors, humidifiers, and paper handling (such as printers).

- Drywall dust is usually produced and removed before the computer equipment is moved into the room. The use of joint compound containing tin oxide should be avoided, and wet sanding is recommended to reduce conductive dust. Infrastructure equipment that is pre-installed—such as uninterruptible power supplies, power distribution units, fire protection systems, monitoring systems and computer room air conditioners—should be turned off and wrapped in plastic during construction. The filters of any air conditioning equipment that is operated during construction should be replaced, and the internal surfaces should be inspected and cleaned as necessary before the computer equipment is moved in.

- All concrete surfaces and especially the concrete subfloor under the raised floor must be sealed with at least two coats of sealer. Wet concrete dust is conductive.

- Humidifiers can also produce mineral dust. The only type of humidifier that completely eliminates mineral dust is an ultrasonic humidifier supplied with deionized water. Ultrasonic humidifiers that are not designed for use with deionized water should be avoided. Infrared and other pan-type humidifiers deposit minerals in the drain pan that can be carried into the air stream, particularly after being cleaned by scraping.

- High-speed printers may produce two types of dust—paper dust and ink or toner dust. Paper dust has a high insulating affect. The effects of toner and ink dust are more of a health problem than a computer issue. Some printers may require installing local electrostatic filter systems.

Air-Conditioner Sizing

Most of the air-conditioning load in a data center consists of sensible heat given off by the electronic equipment. UPS equipment, electrical distribution transformers, and lights produce additional sensible heat. People and ventilation air are the only latent heat sources in the room. Ventilation air should be kept to the minimum required by code for the safety and comfort of people in the data center. Excess ventilation will cause difficulty in maintaining humidity control. It also causes the humidifiers to use more energy in cold weather and results in additional refrigeration operation in the summer.

The sensible heat ratio of air-conditioning units should be specified to nearly match the high sensible heat ratio of the load. Air conditioners that have a lower sensible heat ratio will waste total cooling capacity and energy by unnecessarily removing too much moisture from the air. Excess dehumidification wastes cooling energy and also wastes the energy required to return the humidity to the room to maintain the proper level. Condensate carryover into the supply air stream from excess dehumidification can cause water to drip into the raised floor area. Packaged computer room air conditioners are usually designed with the required high sensible heat ratio.

Airflow Designs

Heat loads in a data center vary widely from area to area and are always in flux. Data and media storage areas produce no internal heat gain other than lights. Mass storage devices and hardware stacked in vertical cabinets can produce local loads of up to 2000 Btu per hour per square foot of floor area. Equipment in a data center is continually being updated, removed, and added. Most electronic equipment that has the highest heat losses has forced ventilation with small internal fans. Most take the air in the front and discharge it from the back or top. Heat then rises naturally after being discharged from the equipment. The natural air flow should be encouraged by introducing cooling air into the data center at floor level and removing the heat with return air at the ceiling. The requirement for bottom-to-top airflow and the ability to redistribute cooling air easily makes a raised floor supply plenum desirable for any data center and the only acceptable solution for centers of more than a few hundred square feet. The ability to easily redistribute cooling air by moving perforated floor tiles is crucial to maintaining acceptable temperatures throughout the data center (see Fig. 14.4-1).

Multiple packaged computer room air-conditioning units are the most common source of cool air supplied into a raised-floor plenum. These units sit on the raised floor, take warm return air in at the top, and discharge cool conditioned air down into the raised floor plenum. Cables and refrigeration or water pipes restrict underfloor air distribution. The pressure losses due to these restrictions cannot be calculated as easily as they can in ducted supply air systems. Careful planning of pipe and cable routing can minimize these restrictions.

Redundancy.   Of all of the infrastructure components, air-conditioning systems require the most service and are the units most prone to failure. They are mechanical devices, and the heat-rejection sections of the system are exposed to outside weather conditions. It is important to design the systems so that any component of the system can fail or be taken off-line for maintenance without losing acceptable conditions in any part of the data center. Redundancy of air conditioners must be maintained throughout the data center. Redundancy must be maintained in air distribution, as well as in total room cooling capacity. Redundant and available cooling capacity is of no value if the cool supply air from the redundant unit cannot reach the area served by a failed unit. Single points of failure in the piping, heat rejection, and controls must be identified and eliminated.

Water Detection

Water and electronic equipment do not mix. Plumbing systems not directly required by critical operations should be routed around these facilities. If pressurized piping must be located over any critical support or operational equipment, a drain pan system should be installed. Unfortunately, there are always one or more sources of water in a data center:

- Humidity control requires a water supply for humidifiers, and condensate is formed during the dehumidification process.
- Heat-rejection piping carries water in chilled water, water-cooled, and glycol-cooled systems.
- The local code or the facility insurance company may require sprinkler systems.

Water detection systems consist of continuous water-sensing cable and readout of the distance along the cable where the water is detected. The cable is run close to all of the possible sources of water, both under the floor and overhead. The location of a leak is indicated by a map of the data center that shows the cable location and is marked at various points along its length (see Fig. 14.4-2). The readout of the cable distance to the leak is compared to the length on the map. The location of the water should be automatically reported to the data center operations manager and the facilities maintenance department by a computer room environ-

FIGURE 14.4-1   Air-conditioning system. (*Courtesy of Stulz of North America.*)

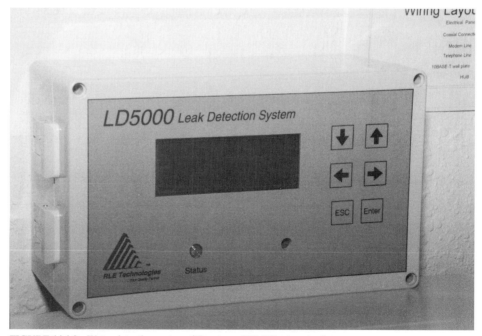

FIGURE 14.4-2   Water detection system. (*Courtesy of RLE Engineering.*)

mental monitoring system. Drain piping should be installed under the raised floor to quickly remove any water that appears from any of these sources.

SECTION 14.5
ELECTRICAL

Glen J. Goss
GJ Associates, Stow, Massachusetts

POWER QUALITY REQUIREMENTS

Electrical power quality is defined by the requirements of the electronic equipment to be protected. Only the voltage presented to the input of the electronic equipment determines the power quality and is defined by its frequency, waveshape, and root-mean-square (rms) value. Transient voltage events and rms voltage events are described by the percentage variations from nominal voltage and their durations. The Computer Business Equipment Manufacturers Association (CBEMA) and Information Technology Industry Council (ITIC) curves describe allowable events. These curves show the acceptable voltage variance that can be tolerated for various durations.

Frequency

The frequency of utility power generated in the United States is very stable at 60 Hz, and most modern electronic equipment is insensitive to frequency. Older computer equipment and some specialized equipment require constant frequency within 0.5 Hz of 60 Hz nominal. Frequency only becomes an issue when the facility is operated on a local backup generator.

Harmonic Distortion

Computer power supplies are designed to expect the input voltage as a perfect sine wave. Any distortion of that shape that repeats on every cycle can be described as the sum of multiple sine waves at various frequencies that are exact multiples of the fundamental 60 Hz, called *harmonics*. The third harmonic, for example, is 180 Hz for a 60 Hz fundamental frequency. The amplitude of each harmonic is expressed as a percentage of the value of the fundamental component's amplitude. Then, voltage waveshape power quality is described by total harmonic distortion (THD), which is the sum of the percentages of the individual harmonics. The maximum value acceptable to most computer manufacturers is 5 percent THD with 3 percent in any single harmonic. Excessive voltage harmonic distortion causes additional power losses and resultant heating in computer power supplies. The expected life of the computer power supply and any other computer components that are heated in turn will be reduced.

Transient Voltages

A second type of waveshape fault is called a *transient voltage*. It consists of a single, fast rising, high-energy, high-voltage event. According to the CBEMA curve, a 1200-V impulse that lasts one millisecond can be easily tolerated. High-speed, high-energy transients above the CBEMA limits impressed on the supply voltage by a nearby lightning strike can cause immediate and extensive damage. Energizing and deenergizing of reactive devices inside the facility such as motors, transformers, and capacitors generate lower energy transients. Multiple low-energy transients eat away at the semiconductors in the electronic equipment and increase their failure rate.

RMS Voltage Sags, Surges, and Outages

The rms value for voltages such as 120, 208, or 480 V, with which we are most familiar, is the root-mean-square calculated over one complete cycle. Reductions in the rms voltage below nominal are called *sags,* and increases are called *surges.* Continuous rms voltage must be maintained within +8 to −13 percent of nominal. Extended low voltage causes regulating computer power supplies to draw additional current to keep their dc output voltage constant. The extra current results in additional losses and heating in the electronic equipment power supplies, stressing them and other nearby circuits. Rms voltage sags that last a few cycles are the second-most-common voltage aberration after transients. Sags can be caused by operation of utility power protective devices or by switching on the grid. They can also be caused by operation of devices within the facility such as large motors that have high starting currents. Sags can cause computers to reset if the voltage drops low enough and long enough to exceed the CBEMA curve limits. According to the CBEMA curve, a complete outage for eight milliseconds should not cause a reset (see Fig. 14.5-1).

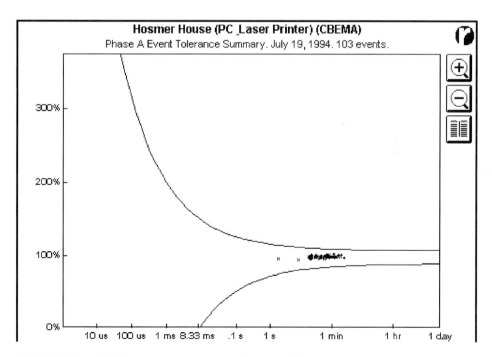

FIGURE 14.5-1   CBEMA voltage tolerance curve. (*Courtesy of GJ Associates.*)

TRANSIENT VOLTAGE SURGE SUPPRESSION

Surge-protection devices (SPD) protect against high-energy, high-voltage, short-duration transients. SPDs should be located between the source of the transient and the equipment to be protected. That means locating them on the service entrance for transients from the utility that are induced by lightning, and between any reactive device that might be switched and the critical load for transients produced inside the building. Placing SPDs on the secondary side of the isolation transformer in addition to the service entrance is a good practice. SPDs are designed to shunt energy to ground when the voltage exceeds the device threshold. The most common component of these systems is the metal oxide varistor (MOV). MOVs are connected between the power conductors and ground and between conductors. MOVs become depleted as they pass transient energy, and finally fail. Multiple MOVs connected in parallel provide high-energy capability and longer life. Silicon avalanche diodes are sometimes used with MOVs to increase the reaction speed and further reduce the energy allowed to pass through to the protected equipment. Because transients have a fast rise and fall time, much of the energy is contained in high frequencies. Capacitors are also sometimes used to provide a low impedance path for high frequencies to ground, thus improving the transient performance and adding some high-frequency noise protection. All SPDs must carry the UL 1449 label. Devices are rated in their ability to limit the maximum pass-through voltage for standard ANSI defined transients. This rating is is assigned by UL. Let-through voltages of 1600 V for 480-V circuits, 800 V for 208-V circuits, and 400 V for 120-V circuits are all within the tolerances of electronic equipment shown in the CBEMA curve. SPDs are also rated in amps of total diverting capacity. Unfortunately, there is no standard for determining that rating (see Fig. 14.5-2). It is usually stated as the sum of the parallel MOV single-impulse ratings. Manufacturer's overall ratings for devices should be carefully analyzed for comparison. It is necessary to ask for the MOV ratings for each individual mode of protection, line to ground, line to neutral, neutral to ground, and line to line. SPDs should also be compared for the maximum single-impulse capacity. Some SPDs have current-limiting fuses to meet UL 1449 that severely limit the single- and multiple-impulse current capacity of the device. The capacity of the device to withstand a number of repetitive standard ANSI defined impulses should also be evaluated. A building lightning protection system around the perimeter of the facility and on all roof-mounted equipment is recommended in addition to SPDs.

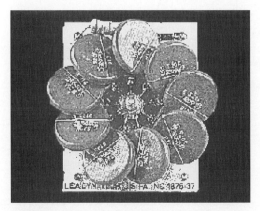

FIGURE 14.5-2 MOV surge-protection device array. (*Courtesy of LEA International.*)

LOAD DETERMINATION

Electronic loads have a high power factor and can be highly nonlinear. Load calculations for determining the size of power equipment such as UPS and backup generators should be done in kilowatts (kW). The load for sizing the data center electrical systems is usually determined by the data center manager, who uses the best data available to add up the requirements of all of the various pieces of equipment. It usually comes from a combination of computer installation planning manuals, power supply nameplate ratings, circuit breaker sizes, and just plain guesses when necessary. A value of two to three times the actual load can result from using this method, which raises construction costs higher than necessary for critical support systems. An existing load can be very accurately measured with a recording meter to better determine the current and future load projections. Caution must be used to ensure that the monitoring equipment accurately measures nonlinear current; that the monitoring period includes a full business cycle; and that it therefore captures the true peak load. Finally, the sizing can be based on rules of thumb. This last method should be used as a comparison with the other methods to keep the sizing in the right range. There is some logic to using the rule of thumb as the minimum design load if the calculated load is less. Computer rooms are continually changing, because the life expectancy of computer equipment is only three to five years. The computer room is part of the facility and should be designed so that it is flexible enough to accommodate unforeseen requirements. The power density of medium to large data centers has been constant over the years at 40 to 50 watts per square foot. Estimates by other means that are substantially greater should be questioned. Estimates that are substantially less should be ignored to allow for the unknown future.

DESIGN CRITERIA

All data centers must meet minimum power quality requirements. Beyond that, the design of the data center electrical distribution system depends on the cost of downtime and the requirements for reliability, capacity, and availability. Establishing these design levels as early as possible in the project firmly establishes the goals of the project.

Reliability and Availability

If you ask data center managers what their requirement is for availability of the computer operation, they will tell you that it must be kept up and running 7 days per week, 24 hours per day. Having said that, most expect that a UPS and backup generator will suffice (see Fig. 14.5-3). In fact a single UPS and generator will reduce unplanned power interruptions to about

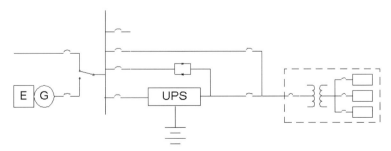

FIGURE 14.5-3   Typical UPS and generator. (*Courtesy of GJ Associates.*)

one every 8 to 10 years, but this does not reduce the requirement for planned downtime to maintain the systems. True around-the-clock availability requires redundant power paths from the power generating plant to the computer equipment power cord. Designs to ensure these levels of uptime can be complex. A national user organization called 7X24 Exchange has grown up to share ideas and experiences with facilities that have that requirement.

UTILITY POWER SOURCES

The most reliable electrical power system possible would consist of power from two utility generating stations fed by independent distribution all the way to two redundant dc power supplies in each piece of computer equipment. Obviously this is impossible. All utility generating stations and distribution systems are interconnected in one way or another in a power grid. It does serve to illustrate the concept of redundant power sources. Any one component in either source could fail, and the load would be unaffected. Any one component or all of the components of one distribution system could be removed from service for maintenance without affecting the load. Most small facilities have one utility connection fed from one utility distribution system as the primary power source and backup generators dedicated to the data center as an alternate source of power. Larger facilities have multiple connections and primary transformers but still are connected to one utility distribution system. A few facilities generate their own primary power on-site with a cogeneration plant. Even fewer sites have two feeds from independent utility systems that do not have a history of simultaneous failure or are not affected by a single recloser. Two utility sources are highly desirable for the data center that absolutely must be in operation without interruption for maintenance.

FACILITY PRIMARY ELECTRICAL DISTRIBUTION

A second primary transformer or double-ended substation should be considered so that the critical load can be transferred to the remaining feed when the primary transformer fails. As a minimum, the backup generator should be considered the second power source. The utility and the generator should feed opposite sides of a double-ended switchboard, so that the critical load can be automatically or manually switched from one source to the other. UPS battery backup will keep the critical load powered during the transfer. Either source, including one-half of the double-ended switchboard, can be taken off-line for maintenance without interrupting power to the critical load that is supported by the UPS.

BACKUP GENERATORS

Generators can be used as the primary source of power, a cogenerator in parallel with the utility, or as a standby system. Cooperation with the local utility for cogeneration or peak shaving can result in a significantly reduced electrical cost for some facilities. Systems that are designed to run more often or for longer periods of time will have more stringent environmental permitting issues and higher maintenance costs. Peak shaving or negotiated interruptible electric rates means that one of the two alternate power sources may be unavailable at certain times. Data center managers should resist any scheme that increases the risk of power loss to the data center. Potential savings must be balanced against any increased risk of data center downtime. Independent generators should be dedicated to support only the critical operations. In the most critical facilities, redundant generators are installed to back up each of two utility feeds. In less critical installations, redundant or even nonredundant generators become the primary alternate source.

Generators must be sized appropriately for nonlinear loads typical of data centers. Severely nonlinear loads can fool generator governors because of multiple zero crossing of the current waveform. A permanent magnet generator with an isochronous governor will ensure the most reliable generator operation. Harmonic currents also cause heating in the generator windings. Capacitive loads that result from lightly loaded computer power supplies or UPS input filters can cause overexcitation and resulting loss of generator voltage control. Six-pulse rectifier loads should represent no more than 60 percent of the rated kW capacity of the generators. Twelve-pulse rectifiers should represent no more than 70 percent of the generator capacity. Single-phase electronic loads fed from a three-phase delta-wye transformer can be considered a six-pulse rectifier load. Techniques for reducing the harmonics presented to the power supply including the generators are discussed later under secondary electrical distribution.

UNINTERRUPTIBLE POWER SUPPLIES

Uninterruptible power supply (UPS) systems provide power conditioning and short-term power outage protection for the data center. In spite of its name, a UPS is a power conditioner more than 99.9 percent of the time and is occasionally a backup system. The UPS minimizes stress on computer equipment by supplying constant high-quality voltage and frequency with little harmonic distortion. Of course the UPS does supply uninterrupted power when the power does go out. It is important that the UPS supply quality power under all conditions and that it does not itself become the source of a power failure.

UPS Sizes and General Usage

UPS systems are available in sizes from 100 watts or less to large paralleled systems of several megawatts. UPSs defined by the International Electrotechnical Commission Standard IEC 62040-3 are *passive-standby, line-interactive,* or *double-conversion* systems. Double-conversions systems continuously provide 100 percent of the load power from the inverter. Line-interactive inverter systems are called by many names including *off-line, line-interactive, tri-port,* or *delta conversion* (see Fig. 14.5-4). Line-interactive and passive-standby systems supply 100 percent of the output power from the inverter only during battery discharge. Line-interactive systems often have a one- or two-step buck voltage regulation that introduces voltage transients into the critical load. The line-interactive UPS must discharge the battery to maintain output power quality when the input power becomes unsuitable for the critical load. If the problem persists, the battery will be depleted, and the critical load is dropped, even though input power is available to the UPS. Traditionally, line-interactive UPS designs have

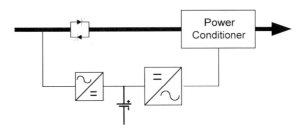

FIGURE 14.5-4   Line-interactive UPS. (*Courtesy of GJ Associates.*)

been used for small single-phase UPS due to the lower cost to manufacture. Recently, three-phase line-interactive systems up to 500 kVA have been introduced.

On-line double-conversion systems continuously supply constant voltage and frequency with low distortion, no matter what happens to the UPS input power. A double-conversion system corrects power quality without battery operation (see Fig. 14.5-5). An on-line UPS system must have an automatic uninterrupted static bypass to keep supplying power to the load should the UPS fail, because the inverter always supplies power to the load. A solid-state, static bypass switch is also used temporarily to supply current directly from the utility that would otherwise overload the inverter, such as load fault or transformer saturation currents. Double-conversion UPS systems are available from 700 VA and to 4500 kVA.

UPS Maintenance Bypass

UPS systems also require a mechanical maintenance bypass switch arrangement to allow servicing the UPS without interrupting the load. The maintenance bypass is in parallel with the static bypass and therefore must come from the same power source to ensure that they are always synchronized. It is best to supply the UPS input from a circuit different from the one that feeds the bypasses. Dual-input circuits eliminate a single point of failure in the breaker that feeds the UPS. Without dual inputs, failure of the upstream breaker will cause the UPS to go to the battery until the battery is depleted. Then, the UPS would attempt to switch the load to the static bypass. Bypass power would not be available if the bypass is fed from the same failed breaker, and the load would be shut down. Power would not be restored to the critical load until the breaker is repaired or replaced. In redundant UPS arrangements discussed later, a second UPS system may become the maintenance bypass.

UPS Batteries

Batteries are a major part of the purchase price and the cost of owning a UPS. UPS systems up to 750 kVA are available with sealed valve-regulated batteries in cabinets that match the UPS. Large systems often use open rack lead calcium wet-cell batteries. Valve-regulated sealed batteries are less expensive and need no ventilation but need to be maintained twice per year and replaced every five to seven years. Wet-cell batteries can be expected to last 15 to 20 years but require more extensive maintenance, ventilation, and acid spill containment.

UPS Backup Time

It is important to select the battery backup time carefully to provide the necessary protection without spending more than is necessary. Battery backup times are stated and selected based

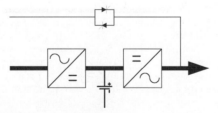

FIGURE 14.5-5 On-line double-conversion UPS. (*Courtesy of GJ Associates.*)

on fully loading the UPS to its maximum kW capacity. The actual backup time provided depends on the load on the UPS. Even though UPS systems are rarely loaded to full capacity, it is good practice to size batteries to allow using the full capacity of the UPS system. If the system has a backup generator, the UPS will see a utility outage that lasts for the length of time required to start and transfer to the generator. This is commonly about 10 seconds but can be extended to avoid unnecessary generator starts. The minimum battery capacity for a UPS system is about 5 minutes. A battery with less capacity may not be able to handle the step load of transfer to the battery and still maintain an acceptable dc voltage for the UPS inverter. Many MIS managers want the UPS to have enough backup time to be able to gracefully shut down the computer systems should the generator fail to start. UPS systems today can be supplied with optional Simplified Network Management Protocol (SNMP) software and hardware that automatically shuts down computer systems gracefully before the battery time expires. The battery must allow enough time for the automatic shutdown. Battery backup times of more than 20 to 30 minutes are often impractical for data centers because the air conditioning systems are off during a power failure if generator power is not available. Without cooling, the computer equipment will overheat and incur damage, or the UPS will overheat at 90 to 104°F ambient and shut down. Small telephone systems, on the other hand, give off very little heat and are often supplied with UPSs that have two, four, or eight hours of backup time. A valve-regulated sealed battery loses capacity very rapidly at the end of its life. Some UPS owners who have experienced battery failures insist on having at least two parallel strings of batteries. Then, if one battery fails, it incapacitates only one string and leaves the other to support the load during a generator start. This approach is no longer necessary with the latest battery monitors. Specifying a UPS that performs an automatic periodic battery test avoids the delay in discovering a bad battery until the power fails and the load is lost. There are stand-alone battery monitors that continuously track battery health and also automatically equalize the charge on the individual batteries (see Fig. 14.5-6). These systems extend the lives of valve-regulated batteries substantially beyond their otherwise expected lives. 5 to 15 minutes of battery time is usually sufficient for data centers. A good battery monitoring system guarantees the integrity of any battery and extends its useful life.

Centralized Versus Distributed UPS

The next consideration is whether to use individual small distributed UPSs or a larger central UPS system. The best application for distributed UPSs is multiple single-user loads such as workstations (see Fig. 14.5-7). Individual UPS systems avoid the cost of running dedicated UPS circuits to each work area. If a UPS should fail, the user can either move to another workstation or disconnect the UPS until it is replaced. In any case, only one user is affected. On the other hand, a centralized UPS should always be used to supply a data center. Multiple small UPS systems become unmanageable when failure of any one of several UPS systems can disrupt the network. There is also no way for the data center manager to know which part of the system is going to run out of battery first when utility power fails. Multiple UPS systems result in substantially more UPS installed capacity since the capacity of each UPS must be larger than the sum of the individual loads connected to it. Central UPS systems naturally take advantage of any load diversity. On-site preventive and emergency repair service are available for larger UPSs that ensure the proper operation of backup power for the entire system. Service contracts with the UPS manufacturer relieve the IT manager of responsibility for UPS management. Reliability is the best reason to use a centralized UPS system (see Fig. 14.5-8). If two UPS systems are required to operate properly to keep the computer network running, the reliability of the network will be exactly half that of a system supplied by one UPS of the same type. Line-interactive UPS systems most often fail when they are required to accept the load during a power failure, just when it is needed the most. Redundant, centralized UPS systems can be used to increase the reliability of on-line double-conversion systems.

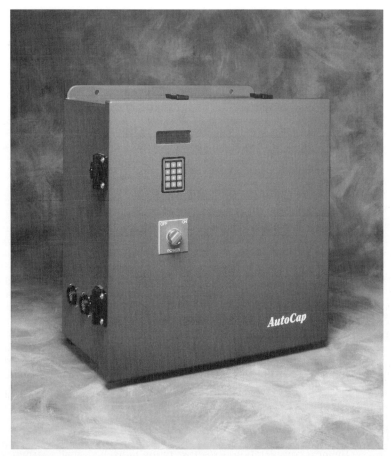

FIGURE 14.5-6   Battery management system. (*Courtesy of AutoCap Inc.*)

Redundant UPS Options

Traditional, central, redundant UPS systems consist of multiple paralleled UPS modules sized so that any one UPS module can fail and the remaining modules can support the load. This is known as an $N + 1$ system (see Fig. 14.5-9). The reliability of a single-module UPS is about 100,000 hours mean time between failures (MTBF). Two modules paralleled for redundancy have a reliability of about 250,000 hours MTBF. If the load on two paralleled modules increases to more than the capacity of one module, the system will become parallel for capacity, that is, both UPS modules will be required to support the load. Failure of either will result in loss of the load. Then, UPS reliability would immediately drop to one-half that of a single module or 50,000 hours MTBF. Paralleling more modules for capacity would reduce the reliability even further. Parallel redundant UPS systems have several drawbacks, including single points of failure in the parallel output bus, output distribution, and single static bypass switch. A variation of the parallel system that has a maximum of two modules can be selected in which each UPS module retains its own static bypass, thus avoiding the static bypass as a single point of failure.

FIGURE 14.5-7 Desktop UPS system. (*Courtesy of MGE UPS Systems.*)

Other redundant UPS schemes have become more popular. Isolated redundant systems are simply two or more standard UPS systems that have the bypass of the primary UPS systems fed from the output of the redundant UPS (see Fig. 14.5-10). If a primary system fails, the primary system static bypass automatically transfers the load to the redundant system. This system increases the output power reliability to about 450,000 hours MTBF. The most reliable system is one in which there are two completely independent power distribution systems, each with its own UPS and backup generator, that automatically switch the load between the two systems by load static transfer switches. See "Secondary Electrical Distribution."

Secondary Electrical Distribution

The burden of supplying reliable, high-quality power lies on the secondary distribution system that actually connects power to the sensitive electronic equipment. At this point in the system, particular attention must be paid to grounding, high-frequency electrical noise, and voltage harmonics.

Computer Room Grounding.   The computer room should be supplied by a solidly grounded ac supply system. All metallic equipment parts and metallic systems should be solidly interconnected to form a continuous electrically conductive system, including all enclosures, raceways/conduits, equipment grounding conductors, earth ground rods, main and interior cold water piping, building structural steel, and neutral-to-ground bonds of each separately

FIGURE 14.5-8 Central UPS system. (*Courtesy of MGE UPS Systems.*)

derived ac system. In addition, a high-frequency signal reference grid (SRG) must be established in the data center with a two-foot-square copper grid at or below the floor level. This keeps the ground connector length to a minimum when connecting the equipment ground to the ground grid. At high frequencies, even a short ground conductor has significant impedance. The SRG must be connected to all metal objects, including the transformer ground, the mechanical systems, and local building steel within six feet. Single-point grounding of the SRG is not recommended.

Isolated grounds, once recommended by some computer equipment manufacturers, are no longer recommended.

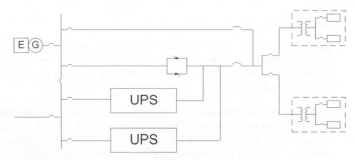

FIGURE 14.5-9 Parallel redundant UPS. (*Courtesy of GJ Associates.*)

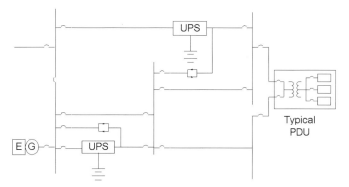

FIGURE 14.5-10   Isolated redundant UPS. (*Courtesy of GJ Associates.*)

Each separately derived source such as an isolation transformer or UPS inverter output must have its neutral-to-ground bond solidly grounded to building steel if it is in turn bonded to the service entrance ground. If building steel bonded to the service entrance is not accessible, the neutral-to-ground bond must be connected to the building service ground. High-frequency *common mode noise* is one of the most serious concerns in the secondary power distribution system. Common mode noise is electrical noise that appears between the neutral and ground conductors. Common mode noise is always zero at the separately derived source such as the secondary of a delta-wye transformer because the neutral and ground are bonded at that point and are therefore at the same potential. It is important to keep the separately derived source as close as possible to the load to minimize common mode noise that can build up over the impedance of long branch circuits. In general there should be no more than 100 feet of conductor between the load and its separately derived source.

Voltage Harmonic Distortion

Voltage harmonic distortion is caused by the nonlinear nature of the current demanded by electronic power supplies and appears across the source impedance of the power source. The lower the source impedance, the less voltage distortion at the critical load. For example, assume a single PC that draws one amp from a 100-kVA transformer that has 5 percent impedance and is connected to a 4000-amp, 480 V service. This load would distort the voltage very little at the transformer or at the service, even though the current harmonic distortion of the PC is very high. On the other hand, 250 PCs connected to the same transformer would distort the voltage at the transformer beyond that allowable by the computer manufacturer. The effect on the 4000-amp service would still be insignificant, but the power quality is unacceptable; also, the third harmonic and its multiples are circulating in the delta winding of the transformer, producing additional heating in the transformer. The transformer heating issue can be handled by specifying a K-20 transformer that is built to stand the additional heat. K-20 transformers do nothing for the power quality problem of distorted voltage, however. Specially built harmonic cancellation transformers are available that eliminate harmonic currents by canceling one phase against another or a three-phase load against another three-phase load by using phase-shifting techniques. Reduction of harmonic currents reduces the heat losses of the transformer and the voltage harmonic distortion seen by the load. Filters or harmonic cancellation transformers can also be selected to greatly reduce the harmonic currents presented to the utility or backup generator power supply. Transformers can also be provided with galvanic shields to reduce high-frequency noise.

Computer Power Distribution Units

Power distribution units (PDUs) provide a convenient way to supply a single-point ground, a separately derived source close to the load, specially shielded transformers for nonlinear loads, load management metering, and a shunt trip main breaker for emergency power off (EPO) function. The PDU is a cabinet that contains a transformer, distribution panelboards, transient voltage surge suppression (TVSS), and electrical metering. Metering provides a means to manage the electrical loads to ensure that they are evenly distributed among the phases and do not exceed the capacity of the power system.

Panelboards in the data center put control of electrical circuits into the hands of the data center manager, who can manage the load on individual circuits and switch individual circuits on or off if panelboard circuit schedules are kept up to date. Panelboard management systems are available that keep a computerized database of the loads on each circuit and print out revised panelboard schedules to post at the panelboards.

Computer manufacturers are furnishing more and more computer equipment with redundant power supplies, and each has its own input power cord (see Fig. 14.5-11). Dual-cord computer equipment can be plugged into independent power sources, thus retaining power redundancy from the power source to the dc power bus inside the computer. Computer equipment that is not dual-cord can be connected to automatic switches that switch the power for the single cord from one source to another. These switches must be open transition and fast enough to prevent a problem for the computer equipment. Fast electromechanical or solid-state transfer switches are available from 15 amp single-phase to 4000 amp three-phase. Small rack-mounted single-phase switches can be used to switch individual loads or circuits between power sources such as two independently fed PDUs. PDUs with 4-ms solid-state transfer switches incorporated into them can switch the entire PDU load from one source to another (see Fig. 14.5-12). Static switch PDUs cost substantially less than individual circuit switches, unless most of the computer equipment is dual-cord that does not require switches for redundancy. Keep in mind, however, that power source redundancy is maintained only to the point of the switch output. The load must be shut down in case of failures or maintenance required on the load side of the switch. Static switch PDUs can be furnished with static switches located on the primary or secondary side of the transformer (see Fig. 14.5-13). Static switches on the secondary side allow the PDU to have redundant transformers and are sized for the higher low-voltage current. The additional cost and floor space required by this arrangement are often not justified by the minimal additional reliability of adding a redundant transformer.

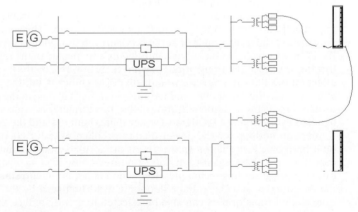

FIGURE 14.5-11   Dual distribution from substation to outlet. (*Courtesy of GJ Associates.*)

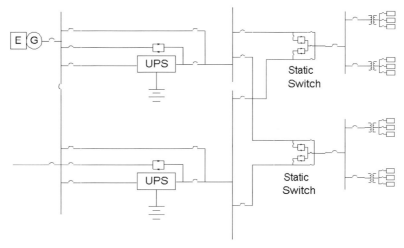

FIGURE 14.5-12   Static switch redundant UPS. (*Courtesy of GJ Associates.*)

FIGURE 14.5-13   Power distribution unit. (*Courtesy of MGE UPS Systems.*)

SECTION 14.6
FIRE PROTECTION

Tom Stratford
McClair, Chicago, Illinois

The fire-extinguishing medium for a suppression system can be either water or a gas such as FM-200. The choice between the two will be partly determined by the building codes and the company's insurance carrier. In either case, disaster has already struck once the extinguishing agent is released. The fire protection system is normally tied into the data center emergency power off (EPO) circuit so that the power to the room is turned off when f the extinguishing agent discharges.

Water will cause damage to electronic equipment, paper records, and documentation, as well as some magnetic data storage media. Recovery to normal operation will be faster with the gas agent, but evacuation and replacement of the gas still takes several hours before recovery operations can begin. Very sensitive combustion detection devices can give early warning of smoldering conditions common to overheated wires and circuit boards. Early detection can prevent the disaster by giving enough warning to locate and treat a developing situation before flame erupts. Sensors should be located at the return air inlets of air conditioners and in any dead air spaces such as above a raised ceiling.

Most data centers are located within larger office buildings that require, by the building codes, protecting all rooms by an inexhaustible supply of fire suppression agent. A gas suppression system does not comply with this code requirement. Therefore, a sprinkler system connected to a municipal water source or to a water-storage tank is required to protect all rooms within a facility. A preaction sprinkler system using semirecessed or fully recessed sprinkler heads is recommended to minimize exposure to damage from unwanted water discharged into the data center from a leaking or accidentally damaged sprinkler head or leaking pipe joint.

A preaction sprinkler system is similar to a conventional sprinkler system, except that water within the piping system is not allowed into the protected area until a preevent occurs, such as activation of one sensor. The piping system is filled with monitored compressed air, and a preacting valve assembly keeps the water out of the system. There are two types of preaction valves, single- or dual-interlock system. Upon a signal from a fire detection system, the valve is released, and water fills the piping system. Should a dual interlocked preaction system be used, water is further delayed from entering the piping system until a sprinkler head's fusible link is melted by heat from the fire. Then, water is released and discharges onto the fire. Water is discharged into the affected room only at the location of an activated sprinkler head.

A gas suppression system is considered a supplemental suppression system by most building codes and insurance carriers. A gas suppression system for a data center can be based on FM-200, Inergen, or carbon dioxide (CO_2). Carbon dioxide systems are not recommended for occupied spaces or for use on electronic equipment. CO_2 is toxic to humans and requires a breathing apparatus on-site within the protected areas. Residue and condensation created by CO_2 gas can damage electronic equipment.

Halon was the gas suppression agent of choice for the past 10 years. This gas is similar in chemical makeup to Freon. Both are ozone-depleting agents. Now, production and use of Halon are restricted except for special conditions. The industry has developed two approved new extinguishing agents, FM-200 and Inergen.

The Inergen system is a mixture of argon, nitrogen, and oxygen, elements that occur within the air we breathe. Therefore, people can remain within a room that contains the agent with only minor effects or discomfort. When Inergen is discharged, the gas lowers the oxygen levels within the protected room below the amount needed to maintain combustion. Because Inergen is made up of non-ozone-depleting gases, it has the least effect on the environment. The major drawback to using Inergen within a data center is that the required volume of gas requires considerable floor space for the gas storage tanks. Barometric dampers are also required to prevent overpressurizing the room. Additional space is generally the reason that this gas system is not used exclusively within the industry. An Inergen system costs roughly the same as an FM-200 system.

FM-200 is a gas system similar to Halon in chemical makeup but which has far less detrimental affects on the ozone layer. Because FM-200 requires less floor space for tanks, a second or reserve gas system can be provided to allow additional protection in the same amount of space required by a basic Inergen system. Use of any gas suppression system requires that special attention be given to the envelope of the protected room. Partitions, doors, pipe, and ductwork penetrations into the protected space must be properly sealed to contain the gas at the effective concentration and for the required duration. Air-conditioning and outside ventilation systems in the protected area must be shut off quickly to avoid interfering with the operation of the gas suppression system.

It is recommended (and in some municipalities required) that a gas suppression exhaust system be installed within a data center. This system removes the gas from within a room and removes any smoke or steam that results from an event. The sooner a space can be cleared and put back on-line, the lower the cost and the less impact there will be on the critical business function of the data center.

Hand-held portable fire extinguishers are also required within a data center. These units are available in two forms, a standard ABC general-purpose unit, which works well on paper, wood, or plastic fires, and the newer "clean agent" extinguishers. The clean agent units use a gas similar to Halon or FM-200 and work best on electronic equipment. Carbon dioxide extinguishers raise the same concerns within a data center as a CO_2 fire suppression system and are not recommended.

FIRE DETECTION

An activation system must be provided to operate a preaction or gas suppression system. This activation system is a fire detection system that detects particles of combustion before an open flame becomes apparent to an operator. Upon detecting of a possible event, the fire detection system can annunciate to the occupants of the room and other facilities personnel that a problem has occurred and may be escalating. Because the staff can react to an event before major flame or smoke is produced, the operators may have a chance to shut down the computers, identify the event, and extinguish the fire or vacate the facility in a timely manner.

Fire detectors come in two forms: (1) traditional heat detectors that alarm at a fixed temperature, and (2) new rate-of-rise detectors. The rate-of-rise-type sensor sounds the alarm when the temperature rises at a rate that increases more rapidly than normal or activates at a fixed temperature setting. These are generally used to activate a preaction sprinkler system. A second system uses a combination of photoelectric, ionization, or electronic smoke detectors to annunciate an event at the first sign of smoke. A third type of detection is an early warning and detection system that uses an air sampling system to detect the particles of combustion earlier than smoke detectors, thus giving operators more time to react to the alarm before the event escalates.

Other features that should be considered within the fire detection and suppressions system include graphic annunciator panels, abort switches, visual and audible annunciation devices, and signage. These accessories assist the occupants and facilities personnel in locating a device that is in alarm or trouble. An item as simple as adequate signage to assist in extinguishing a fire can minimize the effects of a major event in a facility.

SECTION 14.7
SECURITY

Tom Stratford
McClair, Chicago, Illinois

Data security (other than the EMI/RFI shielding mentioned earlier) is the responsibility of the data center manager. Physical security, on the other hand, is mostly the responsibility of the facility manager. Card access to the data center itself should be required. Proper management of card activation and deactivation functions is essential to successful implementation, as is employee training. Employees must be aware that it is their responsibility to report lost cards and control the access of guests without cards. Inside and outside security camera systems are useful for discovering violations of procedures. Building entrance and lobby security must be set up so that access to the data center entrance door is limited to authorized personnel. Infrastructure equipment that supports the data center and is located outside the building—such as backup generators, fuel storage, and cooling towers—also need as much protection as the data center itself.

Active security systems can take on many forms and levels of protection and include entry control systems, visual surveillance, special wall construction, mantraps, and tracking of material and personnel movement. Entry control systems are available as simple keyed doors, mechanical combination locks, card access, biometric type, and beyond. Visual surveillance can consist of on-site security guards, closed-circuit camera systems, motion sensors, and ultrasensitive temperature and audio detection systems. A properly engineered security system provides various levels of protection and tracking of personnel and material to address the needs of the project.

The levels of security required for various spaces, including physical limitations, should be established during the early stages of a project's design. The levels of security required, along with the required movement of personnel and material to and from the critical rooms and support areas of a data center, are used to establish the physical security guidelines for wall construction, the need for mantraps, material and personnel tracking, and monitoring of areas outside of the data center. The security design should also include a plan and budget to implement security upgrades in the future.

SECTION 14.8
MONITORING

Glen J. Goss
GJ Associates, Stow, Massachusetts

INFORMATION FOR THE DATA CENTER MANAGER

The data center manager will have many unpredicted issues to handle every day (e.g., hardware failures, software configuration, and user issues). The infrastructure is comparatively stable and is generally outside of his or her area of expertise, but can nevertheless have a major and sudden impact on the data center. A monitoring system dedicated to the data center infrastructure can help to give the data center manager confidence that the infrastructure is sound and being kept in reliable condition. A monitoring system that presents only streams of data about the environment and power conditions is useless. The monitoring system must recognize the difference between an equipment maintenance alarm and a critical system failure alarm, and provide a message to the operating manager with a required action completely described. The message should also describe the consequences of not taking the action. The monitoring system can also identify trends that could lead to more serious problems. Another useful function of the monitoring system is to record data logs in sufficient detail to reconstruct undesirable events, so that they can be analyzed, corrected, and avoided in the future. Integration of the building management system (BMS) with the critical monitoring systems or security devices must be done carefully. It is possible for the critical monitoring system to send so much information that the BMS system becomes slow and unresponsive in its primary function. Proper separation of the monitoring system from the building management and controls is imperative to avoid making the BMS system a weak link within the alarming, management, and controls capabilities of a system.

Each piece of infrastructure equipment that is critical to the operation of the data center should be monitored in as much detail as possible. In general, that means gathering data through a digital communications connection directly to the microprocessor controller of the equipment being monitored. Then, analog values such as temperature, humidity, voltage, and electrical current can be recorded and trends analyzed. Equipment to be monitored includes computer room air conditioning, outdoor heat-rejection devices, pumps, UPS/battery room air conditioning, water leak-detection systems, UPS systems, battery monitoring equipment, power distribution units, solid-state transfer switches, engine generators, automatic transfer switches, primary and secondary electrical distribution boards, power quality meters, security systems, and fire detection and suppression equipment.

COMPUTER NETWORK INTERFACE

A data center monitoring system must be fully capable of informing all who must react to an event, no matter where they are located. The system should be capable of generating e-mail and pager alarm messages in addition to those presented on the system screen. Conditions that mean the imminent interruption of operation should be communicated directly to the computer on-line users and the network operating system software. The most common example is that the UPS is approaching the end of battery operation. When the battery is depleted, the computer will be shut off. If users are warned early enough, they may be able to shut down operations gracefully and close files for easier and faster recovery when power is restored. Many data centers with networks have a simplified network management protocol system (SNMP) in place that monitors the network components. A standard information database is established for each type of network equipment. UPS systems are one of the devices monitored by SNMP. This is another message system that the data center monitor should be able to accommodate. It should be capable of generating standard SNMP traps such as "UPS low battery" that can be interpreted by client software running on the network servers and by SNMP network management software. These systems will in turn send messages to users that the system is about to be shut down and to complete work in process. Then, the network servers automatically initiate a programmed shutdown of the network before power is lost.

INTEGRATION WITH BUILDING MANAGEMENT SYSTEMS

The facilities manager is responsible for the routine and emergency service of the data center infrastructure equipment. Monitoring information interface with facilities can take place in a number of ways. The data center monitoring system can be set up to e-mail or page facilities personnel for selected alarm conditions. A second data center client computer can be set up in the facilities department to provide full access to the information available in the system. A serial link can be set up between the data center monitoring system and a building management system (BMS). This link can be used to send messages generated by the data center monitor to the BMS. In this case the data center monitor gathers raw data, analyzes it, and sends formatted messages to the BMS. The link can also be used for direct access to the data center data points so that the BMS can be used for analysis and alarm generation. The second method requires close coordination between the data center manager and the facilities manager. Every time data center monitoring points are modified, a corresponding change has to be made in the BMS.

INDEX

ABOUT THE EDITORS

Paul R. Smith, editor-in-chief, is a principal with Peak Leadership Group in Boston, Massachusetts, and has more than 25 years of diverse experience focused on defining and improving business operations, including development of high-performance cross-functional teams. While program manager of configuration management, he directed the process of determining the design basis, documentation, and physical plant for a 20-square-mile facilities complex and support facilities with approximately 8000 employees. As program manager of business process improvement for EG&G Idaho and Lockheed/Martin, his teams saved millions of dollars, improved customer satisfaction and quality, and reduced cost and cycle time. Improved functional areas included all administration department processes of health care, procurement, human resources, materials management, asset management, information systems, budgeting, and so on. He writes and presents workshops nationwide for the American Management Association (AMA), the Project Management Institute (PMI), the Northeastern University MBA Program, and Boston University (BU). He is a Registered Professional Engineer (P.E.) and Licensed Project Management Professional (PMP). He has written a book published by McGraw-Hill entitled *Piping Systems and Their Supports* and has presented more than 15 papers at national conferences, 2 of which were reprinted in international magazines. He has a Master's Degree in Business Administration from Suffolk University and Master's Degree in Mechanical Engineering from Worcester Polytechnic Institute. Mr. Smith may be reached at psmith31 @hotmail.com.

Mark W. Neitlich, the retired owner, Chief Executive Officer, and Chief Engineer of a chemical manufacturing company, has a Bachelor of Engineering degree in Chemical Engineering from Yale University and a Master of Business Administration in Management from New York University. He has copyedited a wide range of scientific books and journals for the past five years.

William L. Porter is Professor of Architecture and Planning at the Massachusetts Institute of Technology School of Architecture and Planning, and he is codirector of the Space Planning and Organization Research Group (SPORG) at MIT. SPORG was created in 1990 to explore the interdependencies among space, organization, finance, and technology as they contribute to shaping the workplace. He has taught courses on programming and workplace design and conducted research with his students on the relationships between physical and informational settings for work. He has coauthored *Excellence by Design: Transforming Workplace and Work Practice* (John Wiley & Sons, New York, 1999). Dr. Porter was formerly the Dean of the MIT School of Architecture and Planning. He is past President of the National Architectural Accreditation Board and was Chairman of the Designer Selection Board of the Commonwealth of Massachusetts. He is a Registered Architect and a Fellow of the American Institute of Architects. He has Bachelor of Arts and Master of Architecture degrees from Yale University and a Ph.D. from MIT.

Anand K. Seth is the Director of Utilities and Engineering for Partners HealthCare System, Inc. (PHS) in Boston, Massachusetts. PHS was created in 1995 when Massachusetts General Hospital and the Brigham and Women's Hospital, both teaching hospitals for the Harvard Medical School, merged. The system now has grown to include 12 hospitals. Before joining PHS, he was Director of Facilities Engineering at Massachusetts General Hospital. He has a Bachelor of Science degree from Gorakhpur University, a Bachelor of Engineering degree from Allahabad University in India, a Master of Science in Mechanical Engineering from the University of Maine, and has taken several postgraduate courses at other universities. He has published numerous papers on HVAC, energy conservation, and other facilities issues. He coauthored *Laboratory Design Health and Safety Considerations* (John Wiley & Sons, New York, 1993). Mr. Seth has more than 33 years of experience in all aspects of facilities engineering. He has taught extension courses at Franklin Institute of Boston and Cambridge College and regularly teaches extension courses at the Harvard School of Public Health. He is a very active member of the American Society of Heating, Refrigeration, and Air Conditioning Engineers (ASHRAE). He is a member of TC 9.8; Large Building Air Conditioning Systems and is chair of a special project committee SP91 in charge of writing a design manual for HVAC systems for hospitals and clinics. He is former president of the Metropolitan Chapter of the Massachusetts Society of Professional Engineers. He is a Registered Professional Engineer in several states, a Certified Plant Engineer, and a Certified Energy Manager. Mr. Seth may be reached at aseth@partners.org.

David L. Stymiest, Senior Consultant for Smith Seckman Reid, Inc., in New Orleans, Louisiana has more than 27 years of intensive professional experience in all phases of electrical systems analysis, facilities engineering, and project electrical design, engineering, and construction for health care, institutional, industrial, commercial, civil works, transportation, alternative energy, and utility projects. He was Senior Electrical Engineer for Partners HealthCare System and Massachusetts General Hospital during much of the development of this handbook. He is a Registered Professional Engineer and a Certified Energy Manager. He is a Senior Member of the American Society for Health-care Engineering. He was a member of the Board of Advisors of *Healthcare Circuit News*. He is a member of the NFPA 110/111 Technical Committee on Emergency and Standby Power Systems. He was a member of the Illuminating Engineering Society of North America (IESNA) Energy Management Committee. He is a member of the IESNA Emergency Lighting Committee and the IESNA Health Care Facilities Lighting Committee. He has a Bachelor of Science degree in Electric Power Engineering and a Master of Engineering degree in Electric Power Engineering from Rensselaer Polytechnic Institute and a Certificate of Special Studies in Administration and Management from Harvard University Extension. He has written and presented numerous papers on electrical engineering and facilities engineering and management topics. A paper he coauthored was published by *Consulting-Specifying Engineer Magazine.* Two of his papers were published by the American Society for Healthcare Engineering as Technical Documents in their Healthcare Facilities Management Series. Mr. Stymiest may be reached at DStymiest@ssr-inc.com.

Roger P. Wessel is the Principal of RPW Technologies, Inc., consultants, involved largely with the engineering, management, and operations of commercial and industrial facilities. He is a Registered Professional Engineer with 40 years of experience involving power plants and facilities design, engineering, construction, and operations. Formerly, as a Senior Program Manager at Raytheon Engineers and Constructors, Inc., he was responsible for various energy conservation, waste-to-energy, and thermal energy storage projects at several Raytheon facilities, as well as other industrial and commercial facilities. In addition, he was responsible for providing engineering and design services at operating nuclear power plants. While at Stone & Webster Engineering Corporation for more than 20 years, he held many positions including Senior Consulting Engineer, Head (and founder) of the Facilities Engineering Group, Chief Engineer of the Engineering Mechanics Division, and Project Engineer/Manager of various nuclear plant projects. He started a West Coast operation with a project for designing the production facilities for the AVLIS program at the Lawrence Livermore Laboratory. He graduated from SUNY Maritime College with a Bachelor of Marine Engineering degree and worked for 25 years on the design, engineering and construction, and testing of nuclear power plants and various support facilities for naval submarines and commercial central stations throughout the United States and Europe.

ABOUT THE ARCHIBUS/FM EXPRESS CD-ROM
MCGRAW-HILL EXECUTIVE EDUCATION SERIES:
ARCHIBUS/FM EXPRESS PROFESSIONAL
SELF-STUDY PROGRAM

This self-study course has been designed for executives to get up to speed with facility management automation and infrastructure management concepts. We want you to be able to immediately deliver professional results. Not just to produce a CAD plan, spreadsheet or Access-style database—but true, professional, relevant, rich, integrated results from an advanced system. To do so, this evaluation is going to use a highly focused subset of the ARCHIBUS/FM product: ARCHIBUS/FM Express Professional. We hope that through this evaluation process, you will see that **ARCHIBUS/FM Express is:**

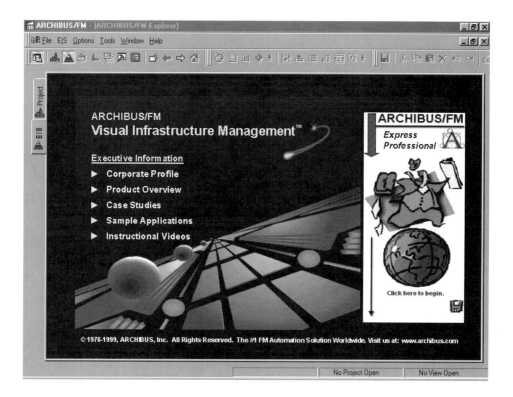

- **Easy.** There is no magic to getting automated if you put the ARCHIBUS/FM methods to work for you.
- **Maintainable.** The data you develop is a by-product of your day-to-day activities. You do not have to update your occupancy plan: It happens as a by-product of moving people.
- **An FM power tool.** ARCHIBUS/FM Express works with your other programs, like the Microsoft Office suite, but it delivers clear advantages over the common functionality of a

spreadsheet, or Access database, or CAD plan. Your spreadsheet does not understand your CAD plan; ARCHIBUS/FM does. Your CAD plan does not understand moves; ARCHIBUS/FM does. ARCHIBUS/FM Express gives you the features you need.

- **Designed to get you results.** ARCHIBUS/FM Express is based on the ARCHIBUS/FM product structure and FM Best Practices™ procedures. Express is designed to get you directed results in ways that other packages cannot. Express was built around the methodologies that will give you the highest return for every minute you spend in automation activities.

- **Designed so that you can learn ARCHIBUS/FM Express by yourself.** We hope that in following the evaluation guide and course materials, you will be confident that you can succeed in facility automation. You will be confident that for a limited, reasonable, predictable effort, you can become up-to-speed and get results.

To supplement your understanding of the benefits and features of this software system, you can read Arts. 3.3.4, 3.3.5, and 3.3.7 in this *Handbook.*

Estimated Time Required to Complete This Course: 16 Hours

This self-study course is modular in design with eight sections. Each section takes approximately 2 hours to complete. In each section, there are *required* activities and *optional* activities. When time permits, we recommend that you elect to undertake some of the optional activities.

The goal from your standpoint is to use this 90-Day Evaluation Version of ARCHIBUS/FM Express to accelerate your FM automation and infrastructure management efforts, and accelerate your career.

Who Is ARCHIBUS/FM Express For?

The ARCHIBUS/FM Express Edition is a focused bundling of the ARCHIBUS/FM product. It uses the same core program as the ARCHIBUS/FM Enterprise Edition, but is smaller in scale and concentrates on those activities that will speed computer-aided facilities management (CAFM) implementation at new sites. It is intended to be a solution for:

- Small to midsized facilities.
- Midsized to large organizations that are starting to automate facilities management and want to focus on a limited set of procedures. These users can start with the Express Edition, and later step up to the Enterprise Edition, if necessary.
- Remote or branch offices of large companies.
- Businesses offering facilities-related services, such as engineering, interior design, real estate advisement, and consulting firms.
- Architects and other CAD users who want to get started with CAFM. These users may wish to purchase the ARCHIBUS/FM overlay for AutoCAD as part of their Express package.

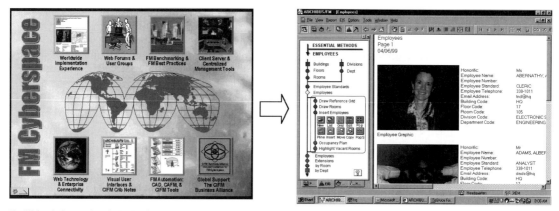

Facilities information in FM cyberspace is only a mouse click away.

Best-of-Class Applications for Facilities Management Automation and Infrastructure Management

The ARCHIBUS/FM solution consists of a series of application modules and activities focusing on specific areas of facilities management.

- Real Property and Lease Management
- Strategic Master Planning
- Space Management
- Overlay with Design Management
- Furniture and Equipment Management
- Telecommunications and Cable Management
- Building Operations Management
- Fleet Management
- Room Reservations
- Hoteling
- Move Wizards
- Work Wizards

The modules and activities are supported by a variety of ARCHIBUS/FM Visual User Interfaces™: ARCHIBUS/FM core programs, the ARCHIBUS/FM Overlay for AutoCAD™, ARCHIBUS/FM Web Central™, and numerous authoring tools. As a COTS (commercial-off-the-shelf) product, ARCHIBUS/FM can readily reflect your own business processes and optimize your corporate facilities environments.

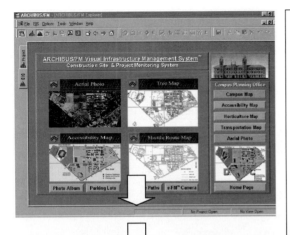

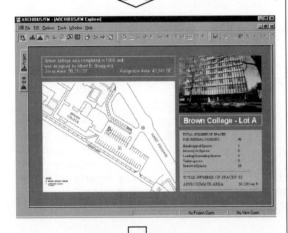

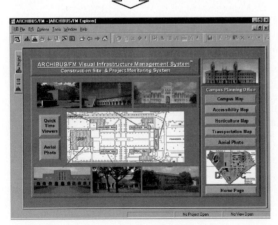

Executive drill-downs make learning easy.

ORIENTATION TO ARCHIBUS/FM

By using ARCHIBUS/FM Express, you join a global community of ARCHIBUS/FM users: infrastructure automation professionals dedicated to enhancing the bottom line of their organizations through the efficient and productive management of their facilities.

Since 1987, ARCHIBUS/FM has been designing FM automation tools that enable organizations to coordinate their spatial and nonspatial assets with their people and strategic goals. By doing this, ARCHIBUS, Inc. helps companies optimize business results for greater productivity and profitability through more efficient cost controls and cooperative work processes.

The ARCHIBUS/FM family of products enables users to create, integrate, manage, and communicate asset data and information that can be rapidly distributed at the local, regional, and global levels. ARCHIBUS/FM enables organizations of any size to benefit from powerful capabilities by allowing them to choose the facilities automation tools best suited to their workplace style and needs. ARCHIBUS/FM, the foundation of FM Cyberspace™ and Visual Infrastructure Management™, supports timely, results-driven, strategic and operational decision making.

ARCHIBUS/FM is an FM Cyberspace™ product combining the power of Computer-Aided Facilities Management (CAFM) and Computer-Integrated Facilities Management (CIFM™) solutions with new millennial technologies into a globally accessible depository of FM data, FM information, FM knowledge, and FM wisdom.

What Does ARCHIBUS/FM Express Do?

ARCHIBUS/FM Express contains everything you need to get results. ARCHIBUS/FM Express can take you from start to finish, whether you are working with CAD diagrams, spreadsheet-style grids, database information, data-entry dialogs, or reports. It can consolidate large volumes of facility data into presentation-quality reports. For example, you can easily perform the following typical FM tasks:

- Benchmark different rental properties using a consistent basis.
- Make sure you haven't missed an option exercise date.
- Know how much space each department occupies.
- Highlight vacant stations and identify underutilized rooms to accommodate a few extra people.
- Know where everyone sits and their current phone extensions.
- Manage groups of employee moves.
- Find jack numbers quickly, without crawling under desks or moving furniture.
- Generate a complete list of data and voice equipment.
- Verify that your audit of software licenses is complete.
- Prioritize and schedule a constant stream of maintenance requests.

The ARCHIBUS/FM Express Methodologies

ARCHIBUS/FM Express can cover this much material because it is the result of an intensive analysis of industry best practices that have identified the most critical skills for getting results. It includes not just data and drawings, but the methodologies you need to get facility jobs done. ARCHIBUS/FM Express represents these fast-track methodologies with unique Process toolbars, which take you from ground zero directly to results. For example, to generate a report that presents a floor-plan drawing that highlights rooms by department assignment, you work through the steps outlined in the Rooms Process toolbar.

Because the job of facility management is broad and demanding, ARCHIBUS/FM Express covers a full range of applications. With the ARCHIBUS/FM Express bundling, you receive the complete set of Essential Process toolbars, which cover the most common aspects of the Real Property & Lease Management, Space Management, Furniture & Equipment Management, Telecommunications & Cable Management, and Building Operations Management modules. In addition to the Essential Process toolbars, the Express bundling includes the complete ARCHIBUS/FM core program and access to the required portions of the ARCHIBUS/FM database, but it limits the number of database records you can create. The ARCHIBUS/FM Express Professional is a 90-day evaluation version of ARCHIBUS/FM Express, which includes additional presentation and learning materials.

The Integrated ARCHIBUS/FM Product Family

For many corporations, ARCHIBUS/FM Express provides a complete solution. But your company may grow or merge. Or your automation effort may expand in scope beyond the record limits of ARCHIBUS/FM Express, or may need features not provided with Express' selected subset. If so, you can move to the other ARCHIBUS/FM products with no conversion. One hundred percent of your data, your drawings, and your skills are preserved. For example, you might move to the ARCHIBUS/FM Enterprise bundling or FM Web Central, because of:

- Networking features, the Expanded Process toolbars, the ARCHIBUS/FM Navigator, security, and customization features.
- You may find that you want to broadly share—both internally and externally—your FM data by making it available through the Internet or a corporate intranet.
- You may need live enterprise access to FM data, need to remotely collect FM data, and remotely manage move orders and work orders.

Because a company's facilities management needs vary greatly, based on the amount of data they must manage and their current stage in the FM automation process, they may seek other options. ARCHIBUS, Inc. offers three product bundlings: (1) ARCHIBUS/FM Express Edition—for small to midsized companies or those just getting started with computer-aided facilities management (CAFM); (2) ARCHIBUS/FM Enterprise Edition—for larger companies seeking an enterprise-wide solution; and (3) ARCHIBUS/FM Web Central Edition—for those who are looking to automate via the World Wide Web.

ARCHIBUS/FM is a top provider of FM automation and infrastructure management solutions in the world. For more information, please contact: ARCHIBUS, Inc., 100 Franklin Street, Boston, MA 02110; Tel.: 617-338-1011; Fax: 617-338-1012; or visit us at *www.archibus.com.* Also, please refer to this *Handbook.*

Executive Education Series: Self-Study Program Curriculum

16-Hour Self-Study Program A/FM-EPO1	*ARCHIBUS/FM Express Professional:* **Unleashing the Power of FM Automation** © 1976–2000 ARCHIBUS, Inc. All rights reserved • ARCHIBUS, Inc. Boston, MA • Tel: 617-338-1011 • Fax: 617-338-1012 • www.archibus.com
Module 1 (total time: approximately 2 hours)	**Topics:** **General Overview to Facilities Automation, Infrastructure Management, and ARCHIBUS/FM** **Requirements: ARCHIBUS/FM license diskette (3½″) and ARCHIBUS/FM Express Professional—McGraw-Hill Executive Education Series CD**
	1. Insert ARCHIBUS/FM Express Professional CD into CD drive (A self-start help program will automatically begin).
	2. Select **ARCHIBUS/FM Express Professional Study Course** (for reading and/or printing documents).
5 minutes	☐ Document 0: *ARCHIBUS/FM Express Self-Study Program Curricula* and *ARCHIBUS/FM Express Professional Notes & Additions* (print and read).
5 minutes	☐ Document 1: *ARCHIBUS/FM Product Specification Sheets* (print and read).
2 minutes	☐ Document 2: *Venture into FM Cyberspace Brochure* (print and read).
10 minutes	☐ Document 3: *An Evaluation Guide for ARCHIBUS/FM Express* [print entire document for use in subsequent activities (49 pages)].
	☐ Return to ARCHIBUS/FM Express Professional's Main Help Menu.
	3. Select **ARCHIBUS/FM Express Professional Self-Study Course** (for viewing videos).
2 minutes	☐ Instructional Video 1: *Introduction to Self-Study Program.*
2 minutes	☐ Instructional Video 2: *Introduction to ARCHIBUS/FM Express—Part 1.*
2 minutes	☐ Instructional Video 3: *Introduction to ARCHIBUS/FM Express—Part 2.*
2 minutes	☐ Instructional Video 4: *Example: Executive Information System—Part 1.*
2 minutes	☐ Instructional Video 5: *Example: Executive Information System—Part 2.*

3 minutes	☐ Instructional Video 6: *Example: Executive Information System—Part 3.*
3 minutes	☐ Instructional Video 7: *Example: Executive Information System—Part 4.*
3 minutes	☐ Instructional Video 8: *Example: Real Property and Lease Management.*
3 minutes	☐ Instructional Video 9: *Example: Creating a Work Order at a Help Desk.*
3 minutes	☐ Instructional Video 10: *Example: Creating a Room in the Database.*
3 minutes	☐ Instructional Video 11: *Example: Creating a Room Using CAD and Databases.*
	☐ Return to ARCHIBUS/FM Express Professional's Main Help Menu.

60 minutes

4. Select **ARCHIBUS/FM Essentials CBT Program** (computer-based training).
☐ CBT Module 1-1: ARCHIBUS/FM Essential Skills CBT.
☐ Return to ARCHIBUS/FM Express Professional's Main Help Menu.

5. Install ARCHIBUS/FM Express Professional Program.
☐ Read instructions for installing ARCHIBUS/FM Express Professional (pages 7 and 8 of *An Evaluation Guide for ARCHIBUS/FM Express Professional*).
☐ Go to www.archibus.com to register and download the ARCHIBUS/FM License File, which is required for installation (an e-mail address is required) or call ARCHIBUS, Inc., at 1-617-338-1011 to have your license file mailed to you. You should also refer to this *Handbook.*

10 minutes
☐ Select **ARCHIBUS/FM Express Professional Program** and then select **Install ARCHIBUS/FM Express Professional.** The Installation Program will self-start.
☐ Return to ARCHIBUS/FM Express Professional's Main Help Menu.

Optional activities:
Select **ARCHIBUS/FM Express Professional Self-Study Course** (for reading and/or printing documents).
☐ Document 4: *ARCHIBUS/FM Start Here Manual* (view and/or print).
☐ Document 5: *Benefits of Utilizing FM Automation and Infrastructure Management Solutions* (view and/or print).
Independent Exploration of ARCHIBUS/FM Express Professional.
☐ Select **ARCHIBUS/FM Express Professional Program.**
☐ Return to ARCHIBUS/FM Express Professional's Main Help Menu.

Module 2 (total time: approximately 2 hours)	**Topics:** **Executive Overview to the Essentials of Working with ARCHIBUS/FM** **Requirements: ARCHIBUS/FM Express Professional—McGraw-Hill Executive Education Series CD** **Printout of *An Evaluation Guide for ARCHIBUS/FM Express***

1. Place ARCHIBUS/FM Express Professional CD in disk drive (ARCHIBUS/FM Express Professional's Main Help Menu will appear).

10 minutes
☐ Independent review of previous activities.

2. Select **ARCHIBUS/FM Express Professional Self-Study Course** (review videos—launch from desktop).

5 minutes
☐ Instructional Video 12: *e-FM Real Property and Lease Management— Commercial Use.*

3. ARCHIBUS/FM Basic Exercises (requires printout of *An Evaluation Guide for ARCHIBUS/FM Express.* If you do not have a printout: select **ARCHIBUS/FM Express Professional Self-Study Course** on the ARCHIBUS/FM Express Professional's Main Help Menu; click on **Document 3: An Evaluation Guide for ARCHIBUS/FM Express;** and print).

5 minutes
☐ Read Chap. 1.

100 minutes
☐ Work through exercises in Chap. 2, starting on pg. 9.

Optional activities:

Select **ARCHIBUS/FM Express Professional Self-Study Course** (for reading and/or printing documents).

 ☐ Document 23: *FM Automation and Infrastructure Management Solutions* (view and/or print).

Independent Exploration of ARCHIBUS/FM Express Professional.

 ☐ Select **ARCHIBUS/FM Express Professional Program.**

 ☐ Return to ARCHIBUS/FM Express Professional's Main Help Menu.

Module 3 (total time: approximately 2 hours)	**Topics:** **Requirements:**	**FM Automation and Infrastructure Management Applications** **ARCHIBUS/FM Web Central** **ARCHIBUS/FM Express Professional—McGraw-Hill** **Executive Education Series CD**

1. Place ARCHIBUS/FM Express Professional CD in disk drive (ARCHIBUS/FM Express Professional's Main Help Menu will appear).

10 minutes ☐ Independent review of previous activities.

2. Select **ARCHIBUS/FM Express Professional Self-Study Course** (review videos).

2 minutes ☐ Instructional Video 49: *ARCHIBUS/FM Web Central—Lease Management.*

2 minutes ☐ Instructional Video 50: *ARCHIBUS/FM Web Central—Space Management.*

2 minutes ☐ Instructional Video 51: *ARCHIBUS/FM Web Central—Furniture & Equipment Management.*

2 minutes ☐ Instructional Video 52: *ARCHIBUS/FM Web Central—Telecom. & Cable Management.*

2 minutes ☐ Instructional Video 53: *ARCHIBUS/FM Web Central—Building Operations Management.*

2 minutes ☐ Instructional Video 54: *ARCHIBUS/FM Web Central—New Employee Management* (integrated ERP, CAD, IT infrastructure, and Help Desk offerings).

3. Select **ARCHIBUS/FM Express Professional Self-Study Course** (print document).

 ☐ Document 29: *User's Manual—ARCHIBUS/FM Web Central* (pp. 1–59).

 ☐ Document 30: *Developing the Winning Web Strategy.*

4. ARCHIBUS/FM Basic Exercises (requires printout of the preceding documents).

5 minutes ☐ Go to www.archibus.com and select "e-FM Forums."

5 minutes ☐ Go to www.archibus.com and select "Benchmarking and Best Practices."

15 minutes ☐ Go to www.archibus.com and select "The Test Drive—ARCHIBUS/FM Web Central."

8 minutes ☐ Go to www.archibus.com and select "Case Studies."

10 minutes ☐ Go to www.archibus.com and select "Products."

50 minutes ☐ Read documents and work through exercises.

Optional activities:

Select **ARCHIBUS/FM Express Professional Self-Study Course** (for reading and/or printing documents).

 ☐ Document 29: *User's Manual—ARCHIBUS/FM Web Central* (pp. 60–135).

Independent Exploration of ARCHIBUS/FM Express Professional.
  ☐ Select **ARCHIBUS/FM Express Professional Program.**
  ☐ Return to ARCHIBUS/FM Express Professional's Main Help Menu.

Module 4
(total time:
approximately
2 hours)
	Topics: **FM Automation and Infrastructure Management Applications**
	               Real Property and Lease Management—Space Management
	Requirements: ARCHIBUS/FM Express Professional—McGraw-Hill
	               Executive Education Series CD

1. Place ARCHIBUS/FM Express Professional CD in disk drive
   (ARCHIBUS/FM Express Professional's Main Help Menu will appear).

10 minutes   ☐ Independent review of previous activities.

2. Select **ARCHIBUS/FM Express Professional Self-Study Course** (review videos).

2 minutes   ☐ Instructional Video 13: *Real Property & Lease Management—Introduction.*
2 minutes   ☐ Instructional Video 14: *Real Property & Lease Management—Properties.*
2 minutes   ☐ Instructional Video 15: *Real Property & Lease Management—Leases.*
2 minutes   ☐ Instructional Video 16: *Real Property & Lease Management—Buildings and Floors.*
2 minutes   ☐ Instructional Video 17: *Real Property & Lease Management—Suites.*
2 minutes   ☐ Instructional Video 18: *Real Property & Lease Management—Summary.*
2 minutes   ☐ Instructional Video 19: *Space Management—Introduction.*
2 minutes   ☐ Instructional Video 20: *Space Management—Gross Areas.*
2 minutes   ☐ Instructional Video 21: *Space Management—Rooms.*
2 minutes   ☐ Instructional Video 22: *Space Management—Employees.*
2 minutes   ☐ Instructional Video 23: *Space Management—Summary.*

3. Select **ARCHIBUS/FM Express Professional Self-Study Course** (print documents).
   ☐ Document 6: Cover—*ARCHIBUS/FM Essential Methods Training Exercises.*
   ☐ Document 7: *ARCHIBUS/FM Essential Methods Training Exercises—Fundamentals.*
   ☐ Document 8: *ARCHIBUS/FM Essential Methods Training Exercises—Real Property and Lease Management.*
   ☐ Document 9: *ARCHIBUS/FM Essential Methods Training Exercises—Space Management.*

4. ARCHIBUS/FM Basic Exercises (require printout of the preceding documents).

88 minutes   ☐ Read documents and work through exercises.

Optional activities:
Select **ARCHIBUS/FM Express Professional Self-Study Course** (for reading and/or printing documents).
   ☐ Document 13: *User's Manual—ARCHIBUS/FM Real Property and Lease Management.*
   ☐ Document 14: *User's Manual—ARCHIBUS/FM Space Management.*
   ☐ Document 24: *Facilities Management Automation and Infrastructure Management Audits.*
Independent Exploration of ARCHIBUS/FM Express Professional.
   ☐ Select **ARCHIBUS/FM Express Professional Program.**
   ☐ Return to ARCHIBUS/FM Express Professional's Main Help Menu.

Module 5
(total time:
approximately
2 hours)
	Topics: **FM Automation and Infrastructure Management Applications**
	        Furniture and Equipment Management
	Requirements: ARCHIBUS/FM Express Professional—McGraw-Hill
	        Executive Education Series CD

1. Place ARCHIBUS/FM Express Professional CD in disk drive
   (ARCHIBUS/FM Express Professional's Main Help Menu will appear).

10 minutes
   ☐ Independent review of previous activities.

2. Select **ARCHIBUS/FM Express Professional Self-Study Course** (review videos).

2 minutes
   ☐ Instructional Video 24: *Furniture & Equipment Management—Introduction.*

2 minutes
   ☐ Instructional Video 25: *Furniture & Equipment Management—Furniture.*

2 minutes
   ☐ Instructional Video 26: *Furniture & Equipment Management—Furniture Survey.*

2 minutes
   ☐ Instructional Video 27: *Furniture & Equipment Management—Equipment.*

2 minutes
   ☐ Instructional Video 28: *Furniture & Equipment Management—Moves.*

2 minutes
   ☐ Instructional Video 29: *Furniture & Equipment Management—Summary.*

3. Select **ARCHIBUS/FM Express Professional Self-Study Course** (print document).
   ☐ Document 10: *ARCHIBUS/FM Essential Methods Training Exercises—Furniture and Equipment Management.*

4. ARCHIBUS/FM Basic Exercises (requires printout of the preceding documents).

98 minutes
   ☐ Read documents and work through exercises.

Optional activities:
Select **ARCHIBUS/FM Express Professional Self-Study Course** (for reading and/or printing documents).
   ☐ Document 17: *User's Manual—ARCHIBUS/FM Furniture & Equipment Management.*
   ☐ Document 26: *Managing the Total Costs of Ownership (TCO) of Facilities.*
Independent Exploration of ARCHIBUS/FM Express Professional.
   ☐ Select **ARCHIBUS/FM Express Professional Program.**
   ☐ Return to ARCHIBUS/FM Express Professional's Main Help Menu.

Module 6
(total time:
approximately
2 hours)
	Topics: **FM Automation and Infrastructure Management Applications**
	        Telecommunications & Cable Management
	Requirements: ARCHIBUS/FM Express Professional—McGraw-Hill
	        Executive Education Series CD

1. Place ARCHIBUS/FM Express Professional CD in disk drive
   (ARCHIBUS/FM Express Professional's Main Help Menu will appear).

10 minutes
   ☐ Independent review of previous activities.

2. Select **ARCHIBUS/FM Express Professional Self-Study Course** (review videos).

2 minutes
   ☐ Instructional Video 30: *Telecommunications & Cable Management—Introduction.*

2 minutes
   ☐ Instructional Video 31: *Telecommunications & Cable Management—Data.*

2 minutes	☐ Instructional Video 32: *Telecommunications & Cable Management—Voice.*
2 minutes	☐ Instructional Video 33: *Telecommunications & Cable Management—Faceplates.*
2 minutes	☐ Instructional Video 34: *Telecommunications & Cable Management—New Faceplates.*
2 minutes	☐ Instructional Video 35: *Telecommunications & Cable Management—Faceplate Connections.*
2 minutes	☐ Instructional Video 36: *Telecommunications & Cable Management—Jacks.*
2 minutes	☐ Instructional Video 37: *Telecommunications & Cable Management—Patch Panels.*
2 minutes	☐ Instructional Video 38: *Telecommunications & Cable Management—Software.*
2 minutes	☐ Instructional Video 39: *Telecommunications & Cable Management—Summary.*

3. Select **ARCHIBUS/FM Express Professional Self-Study Course** (print document).
   ☐ Document 11: *ARCHIBUS/FM Essential Methods Training Exercises—Telecommunications & Cable Management.*

4. ARCHIBUS/FM Basic Exercises (require printout of the preceding documents).

90 minutes   ☐ Read documents and work through exercises.

Optional activities:
Select **ARCHIBUS/FM Express Professional Self-Study Course** (for reading and/or printing documents).
   ☐ Document 18: *User's Manual—ARCHIBUS/FM Telecommunications & Cable Management.*
   ☐ Document 27: *Life Cycle Costs of Facilities.*
Independent Exploration of ARCHIBUS/FM Express Professional.
   ☐ Select **ARCHIBUS/FM Express Professional Program.**
   ☐ Return to ARCHIBUS/FM Express Professional's Main Help Menu.

Module 7 (total time: approximately 2 hours)	**Topics: FM Automation and Infrastructure Management Applications** **Building Operations Management** **Requirements: ARCHIBUS/FM Express Professional—McGraw-Hill Executive Education Series CD**

1. Place ARCHIBUS/FM Express Professional CD in disk drive (ARCHIBUS/FM Express Professional's Main Help Menu will appear).

10 minutes   ☐ Independent review of previous activities.

2. Select **ARCHIBUS/FM Express Professional Self-Study Course** (review videos).

2 minutes	☐ Instructional Video 40: *Building Operations Management—Introduction.*
3 minutes	☐ Instructional Video 41: *Building Operations Management—Work Order.*
2 minutes	☐ Instructional Video 42: *Building Operations Management—Work Request.*
2 minutes	☐ Instructional Video 43: *Building Operations Management—Estimate Work.*
2 minutes	☐ Instructional Video 44: *Building Operations Management—Schedule Work.*
3 minutes	☐ Instructional Video 45: *Building Operations Management—Update Work Orders.*
2 minutes	☐ Instructional Video 46: *Building Operations Management—PM Scheduling.*
2 minutes	☐ Instructional Video 47: *Building Operations Management—PM Analysis.*
2 minutes	☐ Instructional Video 48: *Building Operations Management—Summary.*

3. Select **ARCHIBUS/FM Express Professional Self-Study Course** (print document).
   ☐ Document 12: *ARCHIBUS/FM Essential Methods Training Exercises— Building Operations Management.*

4. ARCHIBUS/FM Basic Exercises (requires printout of the preceding documents).

90 minutes
   ☐ Read documents and work through exercises.

Optional activities:
Select **ARCHIBUS/FM Express Professional Self-Study Course** (for reading and/or printing documents).
   ☐ Document 19: *User's Manual—ARCHIBUS/FM Building Operations Management.*
   ☐ Document 28: *FM Best Practices and Benchmarking for FM Automation.*
Independent Exploration of ARCHIBUS/FM Express Professional.
   ☐ Select **ARCHIBUS/FM Express Professional Program.**
   ☐ Return to ARCHIBUS/FM Express Professional's Main Help Menu.

Module 8 (total time: approximately 2 hours)	**Topics:**	**FM Automation and Infrastructure Management Applications Strategic Master Planning—Room Reservations—Hoteling— Personalized-Off-the-Shelf (POTS) Solutions (Example: Hotlists)**
	Requirements:	**ARCHIBUS/FM Express Professional—McGraw-Hill Executive Education Series CD**

1. Place ARCHIBUS/FM Express Professional CD in disk drive (ARCHIBUS/FM Express Professional's Main Help Menu will appear).

10 minutes
   ☐ Independent review of previous activities.

2. Select **ARCHIBUS/FM Express Professional Self-Study Course** (review videos).

2 minutes
   ☐ Instructional Video 55: *ARCHIBUS/FM Strategic Master Planning— Programming.*

2 minutes
   ☐ Instructional Video 56: *ARCHIBUS/FM Strategic Master Planning— Forecasting.*

2 minutes
   ☐ Instructional Video 57: *ARCHIBUS/FM Strategic Master Planning— Allocation.*

2 minutes ☐ Instructional Video 58: *ARCHIBUS/FM Strategic Master Planning—Stacking.*

2 minutes ☐ Instructional Video 59: *ARCHIBUS/FM Activity—Hoteling.*

2 minutes ☐ Instructional Video 60: *ARCHIBUS/FM Activity—Room Reservations.*

2 minutes ☐ Instructional Video 61: *ARCHIBUS/FM POTS Example—Hotlists: Strategic.*

2 minutes ☐ Instructional Video 62: *ARCHIBUS/FM POTS Example—Hotlists: Business.*

2 minutes ☐ Instructional Video 63: *ARCHIBUS/FM POTS Example—Hotlists: Operations.*

2 minutes
   ☐ Instructional Video 64: *ARCHIBUS/FM POTS Example—Hotlists: Administration.*

3. Select **ARCHIBUS/FM Express Professional Self-Study Course** (read and/or print document).

15 minutes
   ☐ Document 15: *User's Manual—ARCHIBUS/FM Strategic Master Planning.*

4. ARCHIBUS/FM Basic Exercises.
Independent Exploration of ARCHIBUS/FM Web Central.

15 minutes
   ☐ Go to *www.archibus.com* and select "The Test Drive—ARCHIBUS/FM Web Central" (Independent Exploration).
Independent Exploration of ARCHIBUS/FM Express Professional.

60 minutes
   ☐ Select ARCHIBUS/FM Express Professional Program.
   ☐ Return to ARCHIBUS/FM Express Professional's Main Help Menu.

Optional activities:

Select **ARCHIBUS/FM Express Professional Self-Study Course** (for reading and/or printing documents).

    ☐ Document 22: *User's Manual—ARCHIBUS/FM for MS SQL Server.*
    ☐ Document 21: *User's Manual—ARCHIBUS/FM for Oracle.*
    ☐ Document 20: *User's Manual—ARCHIBUS/FM for Sybase.*
    ☐ Document 16: *User's Manual—ARCHIBUS/FM Overlay with Design Management.*

Independent Exploration of ARCHIBUS/FM Express Professional.

    ☐ Select **ARCHIBUS/FM Express Professional Program.**
    ☐ Return to ARCHIBUS/FM Express Professional's Main Help Menu.

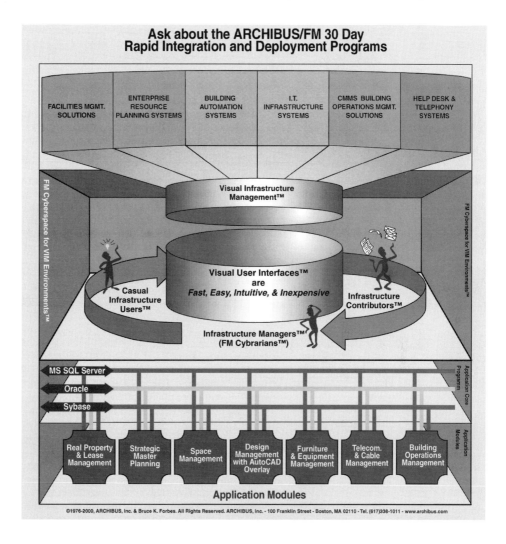

Ask about the ARCHIBUS/FM 30 Day Rapid Integration and Deployment Programs

If you need help selecting your Information Systems, please contact:

Paul R. Smith, Peak Leadership Group, Two Hawthorne Place, Suite 2J, Boston, Massachusetts 02114; 617-367-2772; psmith31@hotmail.com.

ABOUT CURRENT ACTIVITIES THAT SUPPORT THE *FACILITIES ENGINEERING AND MANAGEMENT HANDBOOK*

Paul R. Smith has recently joined **PSMJ Resources, Inc.,** in Newton, Massachusetts, as Editor-in-Chief of a new newsletter for facilities professionals, and will be the principal presenter for PSMJ's Facility Management seminars.

A leading provider of management publications and training to the A/E/C industry for nearly 30 years, PSMJ Resources, Inc. is producing a 2-day intensive training course for facility managers.

This seminar, part of PSMJ's signature **"BOOTCAMP"** series, covers all phases of the facility management life cycle:

- Strategy and finance
- Planning and budgeting
- Design and construction
- Information and knowledge management
- Operations management

PSMJ FACILITY PROFESSIONAL newsletter serves as an adjunct "correspondence course" to both the *Handbook* and the training. Each month, skilled writers and communicators working within the FM profession deliver helpful tips for the PSMJ *Facility Professional* about topics such as:

- Facility financial modeling
- Risk management
- Contracts
- Automation
- Environmental compliance

For further information, contact:

PSMJ Resources, Inc.
10 Midland Avenue
Newton, Massachusetts 02458
Telephone: 617-965-0055
E-mail: editor@psmj.com

Paul Smith is also Principal of **Peak Leadership Group** (**PLG**), management and technical consultants, trainers and coaches. Located in Boston, Massachusetts, PLG has helped numerous firms to design and implement business solutions for their toughest challenges.

Today, the facilities management profession is going through *revolutionary* and *evolutionary* processes. The facility manager must be computer literate, Internet comfortable, a planner, a scheduler, a motivator of people, an integrator, a problem solver, and, most important of all, a great manager. Paul Smith and the PLG help the FM professional to attain knowledge and achieve success through:

- Assessments (needs, wants, desired results, gap analysis, and project implementation)
- Benchmarking
- Program management

For further information, contact:

Paul R. Smith, Principal
Peak Leadership Group
Two Hawthorne Place
Suite 2J
Boston, Massachusetts 02114
Telephone: 617-367-2772
E-mail:
psmith31@hotmail.com

SOFTWARE AND INFORMATION LICENSE

The software and information on this CD ROM (collectively referred to as the "Product") are the property of McGraw-Hill Companies, Inc. ("McGraw-Hill") and are protected by both United States copyright law and international copyright law and international copyright treaty provision. You must treat this Product just like a book, except that you may copy it into a computer to be used and you may make archival copies of the Products for the sole purpose of backing up our software and protecting your investment from loss.

By saying "just like a book," McGraw-Hill means, for example, that the Product may be used by any number of people and may be freely moved from one computer location to another, so long as there is no possibility of the Product (or any part of the Product) being used at one location or on one computer while it is being used at another. Just as a book cannot be read by two different people in two different places at the same time (unless, of course, McGraw-Hill's rights are being violated).

McGraw-Hill reserves the right to alter or modify the contents of the Product at any time.

This agreement is effective until terminated. The Agreement will terminate automatically without notice if you fail to comply with any provision of this Agreement. In the event of termination by reason of your breach, you will destroy or erase all copies of the Product installed on any computer system or made for backup purposes and shall expunge the Product from your data storage facilities.

LIMITED WARRANTY

McGraw-Hill warrants the physical CD ROM(s) enclosed herein to be free of defects in materials and workmanship for a period of sixty days from the purchase date. If McGraw-Hill receives written notification within the warranty period of defects in materials or workmanship, and such notification is determined by McGraw-Hill to be correct, McGraw-Hill will replace the defective CD ROM(s). Send request to:

Customer Service
McGraw-Hill
Gahanna Industrial Park
860 Taylor Station Road
Blacklick, OH 43004-9615

The entire and exclusive liability and remedy for breach of this Limited Warranty shall be limited to replacement of defective CD ROM(s) and shall not include or extend in any claim for or right to cover any other damages, including but not limited to, loss of profit, data, or use of the software, or special, incidental, or consequential damages or other similar claims, even if McGraw-Hill's liability for any damages to you or any other person exceed the lower of suggested list price or actual price paid for the license to use the Product, regardless of any form of the claim.

THE McGRAW-HILL COMPANIES, INC. SPECIFICALLY DISCLAIMS ALL OTHER WARRANTIES, EXPRESS OR IMPLIED, INCLUDING BUT NOT LIMITED TO, ANY IMPLIED WARRANTY OR MERCHANTABILITY OR FITNESS FOR A PARTICULAR PURPOSE. Specifically, McGraw-Hill makes no representation or warranty that the Product is fit for any particular purpose and any implied warranty of merchantability is limited to the sixty day duration of the Limited Warranty covering the physical CD ROM(s) only (and not the software or information) and is otherwise expressly and specifically disclaimed.

This Limited Warranty gives you specific legal rights, you may have others which may vary from state to state. Some states do not allow the exclusion of incidental or consequential damages, or the limitation on how long an implied warranty lasts, so some of the above may not apply to you.

This Agreement constitutes the entire agreement between the parties related to use of the Product. The terms of any purchase order shall have no effect on the terms of this Agreement. Failure of McGraw-Hill to insist at any time on strict compliance with this Agreement shall not constitute a waiver of any rights under this Agreement. This Agreement shall be construed and governed in accordance with the laws of New York. If any provision of this Agreement is held to be contrary to law, that provision will be enforced to the maximum extent permissible and the remaining provisions will remain in force and effect.